ADVANCES IN BUSINESS, MANAGEMENT AND ENTREPRENEURSHIP

Advances in Business, Management and Entrepreneurship

ISSN 2639-8249
(Online) ISSN 2639-8257

Book series editor

Ratih Hurriyati
Universitas Pendidikan Indonesia, Bandung, Indonesia

PROCEEDINGS OF 3RD GLOBAL CONFERENCE ON BUSINESS MANAGEMENT AND ENTREPRENEURSHIP (GC-BME 3), BANDUNG, INDONESIA, 8 AUGUST 2018

Advances in Business, Management and Entrepreneurship

Editors

Ratih Hurriyati
Universitas Pendidikan Indonesia, Indonesia

Benny Tjahjono
Coventry University, Indonesia

Ikuro Yamamoto
Kinjo Gakuin University, Japan

Agus Rahayu, Ade Gafar Abdullah & Ari Arifin Danuwijaya
Universitas Pendidikan Indonesia, Indonesia

CRC Press
Taylor & Francis Group
Boca Raton   London   New York

CRC Press is an imprint of the
Taylor & Francis Group, an **informa** business

A BALKEMA BOOK

Published by:
CRC Press/Balkema
P.O. Box 447, 2300 AK Leiden, The Netherlands
e-mail: Pub.NL@taylorandfrancis.com
www.crcpress.com – www.taylorandfrancis.com

First issued in paperback 2022

**Visit the Taylor & Francis Web site at
http://www.taylorandfrancis.com**

**and the CRC Press Web site at
http://www.crcpress.com**

Typeset by Integra Software Services Pvt. Ltd., Pondicherry, India

Library of Congress Cataloging-in-Publication Data

ISBN: 978-1-03-240085-3 (pbk)
ISBN: 978-0-367-27176-3 (hbk)
ISBN: 978-0-429-29534-8 (ebk)

DOI: https://doi.org/10.1201/10.1201/9780429293348

Table of Contents

ix

Section 6: Strategic management, entrepreneurship and contemporary issues

Preface/foreword

The GCBME Book Series aims to promote the quality and methodical reach of the Global Conference on Business Management & Entrepreneurship which is intended as a high-quality scientific contribution to the science of business management and entrepreneurship.

The Contributions are expected to be the main reference articles on the topic of each book. It will be strictly peer-reviewed by experts in the fields. This book provides opportunities for the delegates to exchange new ideas and implement experiences to establish business or research connections and to find global partners for future collaboration. This book is expected to be published annually and the 2019 theme is: **"Creating Innovative and Sustainable Value-added Businesses in the Disruption Era".** GCBME ultimately intends to provide a medium forum for educators, researchers, scholars, managers, graduate students and professional business persons from the diverse cultural backgrounds, presenting and discussing their research, knowledge and innovation within the fields of business, management and entrepreneurship. I hope the readers experience the usefulness of this book, in developing business analysis and implementation.

The GCBME conferences cover the major thematic subject areas, yet opens to other relevant topics: Organizational Behavior, Innovation, Marketing Management, Financial Management and Accounting, Strategic Management, Entrepreneurship and Green Business

I hope the readers will find the usefulness of this book in understanding the development of the analysis and application of business management and entrepreneurship for decision-making support.

With warmest regards,
Prof. Dr. Ratih Hurriyati, MP

Editorial board

Advances in Business, Management and Entrepreneurship – Hurriyati et al (eds)
© 2020 Taylor & Francis Group, London, ISBN 978-0-367-27176-3

Scientific committee

Prof. John Paul (*Kedge Business School, France*)
Prof. Varakorn Samakoses (*President of Dhurakij Pundit University, Thailand*)
Prof. Dr. Ikuro Yamamoto (*Kinjo Gakuin University, Nagoya, Japan*)
Prof Dr Taehee Kim PhD (*Youngsan University, Busan, South Korea*)
Prof. Dr. Mohamed Dahlan Ibrahim (*Universiti Malaysia Kelantan, Malaysia*)
Prof. Dr. Nanang Fattah, MPd (*UPI, Indonesia*)
Prof. Dr.Agus Rahayu MP (*UPI, Indonesia*)
Prof. Dr. Tjutju Yuniarsih SE, MPd (*UPI, Indonesia*)
Prof. Dr. Disman MS (*UPI, Indonesia*)
Prof. Dr. Suryana MS (*UPI, Indonesia*)
Prof. Dr. Eeng Ahman, MS (*UPI, Indonesia*)
Prof. DR. Ratih Hurriyati, MP (*UPI, Indonesia*)
Prof. Ina Primiana SE, MT (*UNPAD, Indonesia*)
Prof Lincoln Arsyad, MEc, PhD (*UGM, Indonesia*)
Prof. Gunawan Sumodiningrat MEc, PhD (*UGM, Indonesia*)
Prof Dr Badri Munir Sukoco , Msc, PhD (*UNAIR, Indonesia*)
Dr. Phil Dadang Kurnia MSc. (*GIZ Germany*)
Assoc.Prof. Arry Akhmad Arman, MT, Dr (*ITB, Indonesia*)
Assoc.Prof. Dwilarso , MBA, PhD (*ITB, Indonesia*)
Assoc.Prof. Hardianto Iristiadi MSME, PhD (*ITB, Indonesia*)
Assoc.Prof. Rachmawaty Wangsaputra, MSc, PhD (*ITB, Indonesia*)
Assoc.Prof. Teungku Ezni Balkiah, MSc, PhD (*UI, Indonesia*)
Assoc.Prof. Ruslan Priyadi MSc, PhD (*UI, Indonesia*)
Assoc.Prof. Sri Gunawan, MBA, DBA (*UNAIR, Indonesia*)
Assoc.Prof. Yudi Aziz, MT, PhD (*UNPAD, Indonesia*)
Assoc.Prof. Lili Adiwibowo, MM, DR (*UPI, Indonesia*)
Assoc.Prof. Vanessa Gaffar, MBA, DR (*UPI, Indonesia*)
Assoc.Prof. Chaerul Furqon, MM, DR (*UPI, Indonesia*)
Assoc.Prof. Tutin Aryanti, ST, MT, PhD (*UPI, Indonesia*)
Vina Andriany MEd, PhD (*UPI, Indonesia*)

Organizing committee

Conference Chair:
Prof.Dr. Ratih Hurriyati MP, *Universitas Pendidikan Indonesia, Indonesia*

Technical Chairperson:
Assoc. Prof. Lili Adiwibowo, *Universitas Pendidikan Indonesia, Indonesia*

Members:
Dr. Hari Mulyadi MSi
Dr. Ade Gafar Abdullah, M.Si
Dr. Eng Asep Bayu Nandiyanto MSc
Drs. Rd Dian H Utama, Msi
Drs. Girang Razati, Msi
Lisnawati SPd. MM
Sulastri SPd, Mstat., MM
MasHaryono SPd, MM

Acknowledgements

This book is primarily supported by the Universitas Pendidikan Indonesia and co-sponsored by the Airlangga University, Padjajaran University, Ngurah Rai University, Universitas Garut, Bank BJB and telkom sigma. We would also like to express our gratitude to CRC Press/Balkema, Taylor & Francis Group for their efforts and cordial cooperation in publishing this book.

Section 1: Marketing management

Advances in Business, Management and Entrepreneurship – Hurriyati et al (eds)
© 2020 Taylor & Francis Group, London, ISBN 978-0-367-27176-3

The origins and consequences of trust in online shopping

T. Handriana & D. Herawan
Universitas Airlangga, Surabaya, Indonesia

ABSTRACT: The past decade has been marked by an increase in the number of products sold through online shops. Various types of businesses that originally were only available offline have now shifted to online as well as offline methods. Some have even abandoned the offline concept. Due to the nature of online shopping, direct meetings between sellers and buyers are not required. Thus, the trust factor plays a crucial role. Questionnaires were distributed online to consumers who have purchased products from online shops. Data from a total of 271 respondents was eligible to be analyzed using Structural Equations Modeling (SEM) analysis techniques with AMOS 16.0 software. From the 10 hypotheses tested, there were nine supported hypotheses, including influence of perceived reputation, perceived risk, and perceived ease of use in online trust, perceived value and online trust toward satisfaction, as well as influence of perceived reputation, perceived ease of use, satisfaction, and online trust in online repurchase intention. Meanwhile, the influence of e-commerce knowledge on online trust was not supported. The findings contribute to the development of the online trust concept and also are beneficial to online shopping businesses.

1 INTRODUCTION

Shopping is an activity that cannot be separated from people's lives. An alternative to physically going shopping is online shopping. Online shopping is a form of electronic commerce used in business-to-business (B2B) and business-to-consumer (B2C) transactions. Online shopping provides convenience and many advantages when compared with traditional shopping.

The rapidly changing Internet environment has created a competitive business landscape that provides opportunities and challenges for various businesses (Lee et al. 2011). Types of web-based online companies are more frequently appearing in Indonesia, including online stores such as Lazada, Bhinneka, OLX, Tokopedia, Bukalapak, and so on. The main feature of online shopping is the absence of direct interaction between the buyer and seller. Hence, consumer trust in online shopping holds a central and important role. Chiu et al. (2009) explained that general trust is a set of certain beliefs, especially those related to benevolence, competence, and honesty on the part of the other party. As the study conducted by Li et al. (2007) indicated, there is a significant relationship between trust and purchase intention. Similarly, research conducted by Bulut (2015) showed that trust and satisfaction influence repurchase intention in online shopping in Turkey.

Meanwhile, several studies show that perceived reputation (Berman & Evans 2004; Muhammad et al. 2014), perceived risk (Berman & Evans 2004), e-commerce knowledge (Davision & Carol 2008), and perceived ease of use

(Li et al. 2007) all influence trust. Thus, this study aims to analyze the origins and consequences of trust in online shopping.

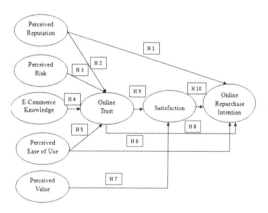

Figure 1. Research model.

2 METHODS

2.1 *Conceptual definition of research variables*

The consumer's perception of assurance about the ability of an online shopping company informs the perceived reputation of that company. Perceived risk develops from an uncertainty surrounding the circumstances that consumers take into account when deciding to undertake online transactions. E-commerce knowledge is the online shopping consumer's knowledge when searching for a product, online sale, purchase sites, and online purchasing methods. Perceived ease of use is the ease of using facilities provided by online shopping, which are in accordance with the consumer's expectation of online shopping. Perceived value is the assessment of the online shopping consumer when taking into account the overall values offered by online shopping companies. Online trust is the consumer's trust of products and services from an online shopping company. Satisfaction is the degree of joy or satisfaction that online shopping consumers feel after comparing expectations and performances of the companies from whom they buy as well as the products. Online repurchase intention is the desire of online shopping consumers to make a repurchase in the future.

2.2 Research and measurement instruments

To measure each construct, a 5-level Likert scale was used: 1 = strongly disagree, 2 = disagree, 3 = neutral, 4 = agree, 5 = strongly agree.

2.3 Sampling design

There were 271 samples (respondents) in this study. The sample units were individuals, that is, people who have made purchases of products using online shopping. In this study, analysis was performed using Structural Equations Modeling with AMOS 16.0 software.

3 RESULTS AND DISCUSSION

3.1 Testing of measurement model

Table 1 shows a convergent validity test for each construct of this research. Convergent validity test results show that each construct is valid, as well as the reliability test result, indicating that all of the construct research is reliable.

The result of the discriminant validity test shows that the AVE square root value of each construct is greater than the correlation value with the other variables. Thus, all constructs in this study fulfill the discriminant validity.

Table 1. Results of convergent validity and construct reliability test.

Construct	Indicator	Stand. Reg. Weight	Remark	CR	Remark
Perceived Reputation	Prep 1	0.800	Valid	0.792	Reliable
	Prep 2	0.774	Valid		
	Prep 3	0.666	Valid		
Perceived Risk	Prisk 1	0.500	Valid	0.814	Reliable
	Prisk 2	0.585	Valid		
	Prsik 3	0.886	Valid		
	Prisk 4	0.877	Valid		
E-Commerce Knowledge	EK 1	0.782	Valid	0.775	Reliable
	EK 2	0.779	Valid		
	EK 3	0.627	Valid		
Perceived Ease of Use	PEOU 1	0.708	Valid	0.803	Reliable
	PEOU 2	0.846	Valid		
	PEOU 3	0.841	Valid		
	PEOU 4	0.500	Valid		
Perceived Value	PV 1	0.675	Valid	0.70	Reliable
	PV 2	0.674	Valid		
	PV 3	0.629	Valid		
Online Trust	OT 1	0.824	Valid	0.796	Reliable
	OT 2	0.632	Valid		
	OT 3	0.793	Valid		
Satisfaction	S 1	0.897	Valid	0.921	Reliable
	S 2	0.906	Valid		
	S 3	0.871	Valid		
Online Repurchase Intention	ORI 1	0.914	Valid	0.930	Reliable
	ORI 2	0.917	Valid		
	ORI 3	0.879	Valid		

3.2 Testing of structural model

Hair et al. (2014) explain that for the purpose of testing the goodness of fit of the research model, it is required that at least one absolute fit index (GFI, RMSEA, or RMR) and at least one incremental it index (CFI, NFI, IFI, AGFI, or TLI). In Table 2, it can be seen that these provisions are met. Thus, it can be said that the model is fit.

The influence test between one variable and another variable is also found in the model and assessed through the critical ratio (CR) value and its probability value. The results of the causality test by using regression weight on the structural model can be seen in Table 3.

Table 2. Test results of goodness of fit at structural model.

Index Goodness of Fit	Cut of Value	Obtained Value	Conclusion
Chi square	Expected Small	536.468	Not Fulfill
Probability	≥ 0.05	0.000	Not Fulfill
CMIN/df	< 2 or < 3	1.692	Fulfill
GFI	≥ 0.9	0.854	Not Fulfill
AGFI	≥ 0.9	0.926	Fulfill
TLI	≥ 0.9	0.824	Not Fulfill
CFI	≥ 0.9	0.850	Not Fulfill
RMSEA	≤ 0.08	0.054	Fulfill

Table 3. Regression weight structural.

Hypotheses	Relationship between Variables	C.R	P Value	Remark
H1	Perceived Reputation – Online Repurchase Intention	2.291.	0.022	Accepted
H2	Perceived Reputation – Online Trust	5.536	0.000	Accepted
H3	Perceived Risk – Online Trust	-3.893	0.000	Accepted
H4	E-Commerce Knowledge – Online Trust	0.285	0.775	Not Accepted
H5	Perceived Ease of Use – Online Trust	3.983	0.000	Accepted
H6	Perceived Ease of Use – Online Repurchase Intention	2.865	0.004	Accepted
H7	Perceived Value – Satisfaction	3.154	0.002	Accepted
H8	Online Trust – Online Repurchase Intention	2.807	0.000	Accepted
H9	Online Trust – Satisfaction	6.373	0.000	Accepted
H10	Satisfaction – Online Repurchase Intention	6.235	0.000	Accepted

3.3 Analysis

Based on the result of the first hypothesis test, it can be concluded that perceived reputation had a positive influence on online repurchase intention. The result of this study supported the research conducted by Broutsou and Fitsilis (2012) and Hess (2008), which showed that perceived reputation is also one of the factors influencing repurchase because it includes the reputation of the sold product. Perceived reputation also has a huge impact on online repurchase. In their study, Wen et al. (2011) found that a reputable company will lead consumers to greater online repurchase intentions. Similarly, Lee et al. (2011) stated that perceived reputation influences repurchase intentions by online.

The result of the second hypothesis test showed that perceived reputation had a positive influence on online trust. If consumers perceive that a company has a strong online shopping reputation, then this will help to establish a sense of customer trust in online shopping. This is in line with the findings of Berman and Evans (2004) and Muhammad et al. (2014). Reputation is a guarantee of the capability and reputation of an online seller. Berman and Evans (2004) stated that perceived reputation embedded by the company to consumers reflects a position desired by the company. Buyers will feel safe when shopping at an online shopping site owned by a reputable company. Broutsou and Fitsilis (2012) stated that perceived reputation has a huge impact on customer trust.

In testing the third hypothesis, it was found that perceived risk negatively influences online trust. The customer or buyer always considers the advantages and disadvantages of making a transaction. Increasing risks can reduce a customer's trust in the online store, and this is related to perception of product quality and suitability. According to Oglethorpe (1994), risk stems from consumer perceptions of the negative situations resulting from online transactions, and therefore risk can reduce customer trust in online shopping. The results of the present study support findings from studies conducted by Berman and Evans (2004). Any activity must contain risk. The inherent difference between offline and online shopping is the absence of direct meetings between the seller and the buyer; the risk factor will therefore be greater for online rather than offline shopping. Masoud (2013) reveals that perceived risk has a huge negative impact on online trust.

The fourth hypothesis in this research was not supported, meaning that e-commerce knowledge does not influence online trust. E-commerce knowledge includes one's skills for online transactions and uses different types of technologies and methods necessary to conduct online transactions (Baek et al. 2006). Consumers with good knowledge of e-commerce do not tend to directly choose or trust an online shopping site without considering various aspects, including risks and ease of shopping. The result of the present study supports the findings of

5

Davision and Carol (2008), which explained that e-commerce knowledge does not influence consumers' confidence in online trading sites. Li et al. (2007) revealed that e-commerce knowledge positively influences online trust. The greater the level of the individual consumer's e-commerce knowledge, the more trust they will have in an online store.

Based on the results of the fifth hypothesis testing, it can be concluded that perceived ease of use positively influences online trust. Consumers who understand the use of the Internet or information technology well, especially with regard to online shopping sites, generally have a good understanding of online shopping procedures. Therefore, they can distinguish between good and poor online stores, and this has an impact on online trust. According to Li et al. (2007), the consumer will trust a website when it is easy to use. The role of online trust is more important than in conventional business, due to distance (i.e., virtual) factors as well as personal factors. In a study conducted by Wen et al. (2011), it was found that consumer trust in online sellers were influenced by the ease of accessing the website. This is in accordance with the research of Li et al. (2007), which stated that perceived ease of use has an impact on online trust by online.

The sixth hypothesis was supported, meaning that perceived ease of use has a positive influence on online repurchase intention. The ease provided by online shopping sites influences consumers' intentions to make a repurchase. This indicates that the greater the ease of use, the higher the intention of making a repurchase. The result of this research supports the findings of Lee et al. (2011), that is, that perceived ease of use has an enormous impact on repurchase intention online. This is because when the customer is comfortable with the ease of the online shopping site, a habit can be formed, and this can influence the intention to repurchase.

The seventh hypothesis, perceived value, positively influences satisfaction. This hypothesis was accepted, indicating that the benefits perceived by customers are greater than the amount that they spend to purchase the products offered by online shops. This finding is in line with those of Spais and Vasileiou (2006). Perceived value is the determining factor indelivering and communicating with the customer where it will create a pleasure and satisfaction to the customer in buying a product and it also has an impact on customer satisfaction that will elicit a repurchase intention in subsequent actions. (Hellier et al. 2003).

The result of the analysis of the eighth hypothesis was accepted, meaning that online trust positively influences online repurchase intention. According to the research of Lee et al. (2011), the party with a high level of trustworthiness tends to engage in the transaction relationship or repurchase due to the belief that the other party will fulfill all of its obligations as expected in terms of both quality and timeliness. This indicates that the higher the level of consumer trust, the higher the occurrence of repurchase. Trust cannot simply be recognized by other parties or business partners but must be established at the beginning and can be proven. Yousafzai et al. (2003) stated that trust is a catalyst in various transactions between seller and buyer, leading to the expected level of customer satisfaction.

The result of the ninth hypothesis test was supported, that is, online trust has a positive influence on satisfaction. Trust encourages customers to behave in a certain way when they believe that the transaction activity will occur as expected with respect to the quality of goods and services provided by the online store. The result of this study is in line with the findings of previous studies by Papadopoulou et al. (2001) and Urban et al. (2000). The relationship between online trust and satisfaction is that trust can have an influence on customer satisfaction because the consumer has chosen to select a trusted online shopping site to make a purchase. This leads to the customer's belief to obtain the expected item based on an advertisement or offer received following a previous transaction. Papadopoulou et al. (2001) explained that trust has a positive influence on customer satisfaction and experience.

Finally, the test on the tenth hypothesis indicates that this hypothesis was accepted, meaning that satisfaction has a positive influence on online repurchase intentions. A customer's satisfaction with the online shopping is related to their previous experiences. This aligns with the research of Bulut (2015), which stated that satisfaction can influence, as well as have an impact on, consumer intentions to make repurchases. Law et al. (2004) explained that consumer repurchase decisions are influenced by the consumer's experience. The basis of customer satisfaction then triggers consumer behavior to make a repurchase.

4 CONCLUSIONS

From the results of analysis and testing of each research hypothesis, it was found that nine out of the ten hypotheses were supported and one hypothesis was rejected. It is necessary for online businesses to appreciate the number of problems experienced by consumers related to products and services provided by online shopping, which can impact consumer trust in online shopping and ultimately affect loyalty. As this study took place in developing countries, it is suggested that further research should be done via comparative studies involving developing and developed countries.

REFERENCES

Baek, C.H., C.S. Seo, J.W. Hong; and W.J. Suh. 2006. Empirical research about the influence on customer trust and purchase intention in China internet shopping mall. *Spring Semi Annual Conferences of KMIS.*

Broutsou, A. and P. Fitsilis. 2012. Online trust: the influence of perceived company's reputation on consumers' trust and the influences of trust on intention for online transactions. *Journal of Service Science and Management*. 05: 365–372.

Bulut, Z. A. 2015. Determinants of repurchase intention in online shopping: a Turkish consumer's perspective. *International Journal of Business and Social Science*. 6 (10): 55–63.

Chiu, C.M.; C.C., Chang; Cheng, H.L.; and Y. H. Fang. 2009. Determinants of customer repurchase intention in online shopping. *Online Information Review*. 33(4): 761–784.

Hair, F.H.; W.C. Black; B.J. Babin; and R.E. Anderson. 2014. *Multivariate data analysis*. Pearson New International Edition. USA: Pearson.

Hellier, P. K., G. M. Geursen, R. A. Carr and J. A. Rickard. 2003. Customer repurchase intention: a general structural equation model. *European Journal of Marketing*. 37:1762–1800.

Hess, R.L. 2008, The impact of firm reputation and severity on customers' responses to service failures. *Journal of Services Marketing*. 22 (5): 385–398.

Law, A. K.Y., Y.V. Hui and X. Zhao. 2004. Modeling repurchase frequency and customer satisfaction for fast food outlets. *The International Journal of Quality and Reliability Management*. 21 (5): 545–563.

Lee, C. H., Eze, U. C., and N. O. Ndubisi. 2011. Analyzing key determinants of online repurchase intentions. *Asia Pacific Journal of Marketing and Logistics*. 23 (2): 200–221.

Masoud, E.Y. 2013. The influence of perceived risk on online shopping in Jordan. *European Journal of Business and Management*. 5 (6): 76–87.

Oglethorpe, J. E and K.B. Monroe. 1994. Determinant of perceived health and safety risk of selected hazardous product activities. *Journal of Consumer Affairs*. 28(2): 326–346.

Papadopoulou, P.; A. Andreous; P. Kannelies; and D. Martakos. 2001. Trust in relationship building in electronic consumer. *Internet Research: Electronic Networking Applications and Policy*. 11(4): 322–332.

Spais, G. S., and K. Vasileiou. (2006). Path modeling the antecedent factors to consumer repurchase intentions for advanced technological food products: some correlations between selected factor variables. *Journal of Business Case Studies*, 2(2), Second Quarter.

Wen, C., R.P. Victor, and Chenyan. 2011. An integrated model for customer online repurchase intention. *Journal of Computer Information Systems*. 52 (1): 14–23.

Yousafzai, S. Y.; J.G. Pallister; and G.R. Foxall. 2003. A proposed model of e-trust for electronic banking. *Technovation*. 23: 847–860.

Advances in Business, Management and Entrepreneurship – Hurriyati et al (eds)
© 2020 Taylor & Francis Group, London, ISBN 978-0-367-27176-3

How digital certificate affects e-commerce consumers trust and purchase intention

P.K. Sari & A. Prasetio
Telkom University, Bandung, Indonesia

ABSTRACT: The government of Indonesia develops e-commerce roadmap to encourage the development of e-commerce industry. One of government's initiatives is through consumer protection. Government here creates a local Certificate Authority that publishes a digital certificate for e-commerce actors. With a digital certificate, it is expected that consumer trust will increase. They will purchase goods through e-commerce websites in Indonesia. This study aims to find out the effect of consumers' awareness of digital certificate on their trust and purchase intention. Three dimensions of digital certificate become the focus of investigation. Those are: knowledge, behavior and attitude. The population were the users of top-five e-commerce site in Indonesia. The data were collected through online survey. It was then analyzed by using PLS-SEM (Partial Least Squares – Structural Equation Modelling) method. The findings showed that the dimension of knowledge had no significant effect on trust. On the contrary, the other two dimensions, attitude and behavior, had a significant effect on trust. Furthermore, the findings also revealed that trust had a significant effect on purchase in-tention. In conclusion, consumers' awareness of digital certificate affects their trust and purchase intention.

1 INTRODUCTION

The development of information and communication technology affects the quality of life Pramiyanti & Millanyani (2014), including on how people do a business. Online transaction through e-commerce site becomes a trend in Indonesia. In the e-commerce industry, companies should focus their attention on establishing consumer confidence through good management.

Indonesia government continually encourage the development of domestic e-commerce industry. One of them is by arranging national e-commerce road-map as a part of the economic policy package (Daryl 2016). There are seven focuses in the roadmap, two of them are consumer protection and cyber security (Daryl 2016). Related to consumer protection, it is expected that all local e-commerce actors utilize electronic certificates or digital signature, by which Ministry of Communication and In-formation becomes the root of CA (Certificate Au-thority). Local CA was established along with Cyber Security Regulatory Plan on June, 2017. The use of root and local CAs is governed by PP 82/2014 on the Imple-mentation of Electronic Transaction Sys-tems and Transactions, and ITE Law on articles 13-14. Cur-rently, some large e-commerce and banking still use international CA, such as DigiCert and Ver-isign.

Trust is the most important factor in buyer-seller relationships, especially on e-commerce (Grabner-Kräuter & Kaluscha 2003). It is one of the reasons to buy online (Lee & Turban, 2001). It is also con-sidered as a long-term barrier, which blocks the po-tential for online buying, more particularly when the business is in an uncertain condition. Therefore, trust is central to all economic transactions, whether con-ducted at retail outlets in the offline world or inter-net (Gefen & Straub 2004, Pavlou & Fygenson 2006). Related to consumers' trust, it is influenced by various factors such as online store reputation, online store competence, and capacity to meet con-sumer needs (Koufaris & Hampton-Sosa 2004).

To build the consumers' trust on e-commerce, the government attempts to improve the security of elec-tronic transactions by producing digital certificates issued by local CA (in this case is Ministry of Com-munication and Information) for Indonesian e-com-merce players. This digital certificate is also ex-pected to increase the number of on-line transactions and to create a better e-commerce industry climate in Indonesia.

Based on the above explanation, the research question of this study is "How is the relationship be-tween security awareness, trust, consumers' interest in e-commerce, and the existence of digital certifi-cate?"

To get the answer of the above problem, this study conducted a survey to a number of e-com-merce users in Indonesia. Based on preliminary survey on some of leading e-commerce sites, most of them have used digital certificates to protect the security of information on their sites. However, there were some sites with a famous brand that have not used it. Therefore, this study only selected e-com-merce sites with a popular brand that have ap-plied digital certificate.

1.1 Information security management

Information security is a combination of systems, operations and internal controls to ensure the integrity and confidentiality of data and operate procedures within an organization (Hong et al. 2003). The purpose of information security is to ensure business continuity and to minimize business losses by preventing and minimizing the impact of security incidents (Kruger et al. 2010).

Information security has three basic components that must be managed, namely (Mitchell et al. 1999):

a. Confidentiality of sensitive information, protecting sites from unauthorized access
b. Integrity (integrity), ensuring the accuracy and completeness of the information
c. Availability, ensuring that vital information and services are available for the users whenever needed.

One of the most important parts of information security management is information security awareness. Information security awareness is a control or rule designed to reduce the incidence of information security violations, as a result of negligence or planned action Whitman & Mattord (2011). According to Kruger & Kearney (2006), the primary goal of information security awareness is to ensure that computer users are aware of the risks associated with the use of information technology as well as an understanding of prevailing policies and procedures. This information awareness program needs to be done by system owners as part of information technology management. As Peltier (2014) states that the system owner is responsible for providing a qualified knowledge of the existence and prevailing level of control so that all users are confident that the system is secure.

According to some experts Kruger & Kearney (2006) and Sari & Candiwan (2014), there are three components that can be used to measure the level of information security awareness, namely:

a. What one knows (knowledge),
b. How do they feel about the topic (attitude),
c. What they do (behavior).

1.2 Purchase intention

In general, intention is defined as indications of how hard people are willing to try, of how much of an effort they are planning to exert, in order to perform the behavior (Ajzen 1991). Online interest buying studies on the web site have several different definitions. Ha & Janda (2014) define online purchasing interest as a consumer's willingness to purchase products or services from a particular web site. Alternative definitions are offered by Broekhuizen & Huizingh (2009). They say that online purchasing interest is a consumers' willingness to buy a given product at a specific time or in a specific situation.

This study will use the definition proposed by (Ha & Janda 2014) as it represents research conducted online on the web site. This definition also has a more general character when compared to the definition offered by (Broekhuizen & Huizingh 2009).

In accordance with the definition proposed by Ha & Janda (2014), there are three indicators that can be used to measure interest in online purchases: 1) the desire to visit the web site again, 2) the desire to buy other products/services, and 3) the desire to buy new products/services. Meanwhile, the studies conducted by Kim et al. (2012) and Ponte et al. (2015) use the following indicators: 1) possibility to consider buying, 2) considering online store if going to buy, 3) possibility of buying, and 4) buying desire. Other indicators for online purchasing interest are offered by (Lu et al. 2014). They mention five indicators, such: 1) will consider to buy the product, 2) do not have interest to buy the product, 3) possibility of buying the product, 4) buying the brand when needing a product, and 5) will buy the product if necessary. This study adapted all indicators from Ha & Janda (2014), Kim et al. (2012), Ponte et al. (2015), and (Lu et al. 2014) by combining several similar indicators.

The desire to re-visit an online store is a reflection of the interest in online purchase. Returning visits, however, will increase the chances of a purchase. The intention to buy more products either new products or other products also indicates an interest in online purchases on a sales website. In addition, if an online store visitor feels that he or she might buy a product is also an indication of an interest in online purchases. Based on the above explanation, the indicators used in this study are:

1) The desire to re-visit the online store,
2) The willingness to buy other products/services,
3) The willingness to buy new products/services,
4) The consideration to choose online store as a place to buy a product,
5) The Possibility of buying, and
6) The desire to buy.

1.3 Trust

Trust is the foundation of business. A business transaction between two or more parties will occur if each trusts each other. Trust cannot simply be acknowledged by other parties/business partners. It must be built from the beginning, and it can be proven. Trust has been considered as a catalyst in various transactions between sellers and buyers in order that customer satisfaction can materialize as expected (Yousafzai et al. 2003).

There are three indicators of trust, namely (Gefen 2002):

1.3.1 *Integrity*

It is a consumer perception that company follows acceptable principles such as keeping their promises or behaving ethically and honestly. The integrity of a company depends on the company's past consistency, credible communication or non-credible communication of a company to another group, and whether the actions of the company are consistent with its promises or words spoken.

1.3.2 *Benevolence*

It is based on the magnitude of partnership trust. It has a purpose and motivation. It provides an advantage for other organizations at a new condition that is the condition where commitment is not formed.

1.3.3 *Competence*

Competence is the ability to solve the problems faced by consumers and meet all the needs. Ability refers to the skills and characteristics that enable a group to have a dominant influence.

2 METHOD

This study used quantitative method. In analyzing the data, descriptive analysis was applied. Continuum line and statistical analysis with Partially Least Square Structural Equation Modelling (PLS-SEM) was conducted to know the relationship between variables. The operational variables of this study included:

1) Information Security Awareness adopted from (Kruger & Kearney, 2006). It had three dimensions that were measured by using four in dicators.
2) Trust (Gefen, 2002). It had four indicators.
3) Purchase Intention (Ha & Janda, 2014). It had six indicators.

Data collection was done by distributing questionnaires to respondents through online media. Questionnaire was in the form of seven-points Likert Scale. The population of this study was consumers of big-five e-commerce sites in Indonesia who have applied digital certificate, either the consumers already know or not know about digital certificate. The sampling technique used nonprobability sampling.

The research framework for statistical analysis using PLS-SEM can be seen in Figure 1.

3 RESULTS AND DISCUSSION

Data was collected through online survey using third party service. There were 178 respondents who filled the questionnaire. After being validated, there were only 158 respondents who had experience in online transactions on five major e-commerce sites in Indonesia (Tokopedia, Bukalapak, Lazada, Blibli and Shopee). Demographic description of respondents can be seen in Table 1.

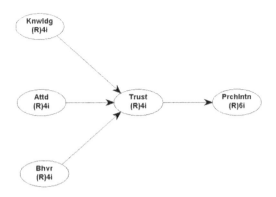

Figure 1. Research framework.

3.1 *Outer model*

In the outer model, composite reliability (CR) and average variances extracted (AVE) values were used to determine whether the latent variable was reliable or not. The expected value of the three units was higher than 0.7. If they were less than 0.7, then the latent variable was not feasible and could not be used in the formation of PLS model. From the calculation results using WarpPLS 5.0 software (Table 2), it can be seen that the value of all latent variables was higher than 0.7. This indicates that the variables is reliable to be used.

The model also tested whether each indicator of variables was feasible. Testing was done by calculating the cross-loadings value of each indicator against other

Table 1. Demographic description of respondents.

Criteria	Options	Freq.	%
Gender	Female	26	16.45%
	Male	132	83.54%
Age	< 20 years	9	5.69%
	20 - 29 years	113	71.52%
	30 - 39 years	34	21.52%
	≥ = 40 years	2	1.27%
Location	urban	105	66.45%
	rural	53	33.54%
Education	Elementary–Senior High School	78	49.36%
	Diploma	18	11.39%
	Under graduate	61	38.61%
	Post graduate	1	0.63%
	Doctoral	0	0%
Online transaction frequency in a month	Less than once	63	39.87%
	1 – 3 times	72	45.57%
	> 3 times	23	14.56%
Frequently visited e-commerce site (checkboxes type)	Lazada	117	74.05%
	Tokopedia	116	73.42%
	Shopee	94	59.49%
	Blibli	44	27.84%
	Bukalapak	100	63.29%

Table 2. Latent variable validity test.					
	Attd	Knwldg	Bhvr	Trust	PrchInt
CR	0.947	0.934	0.945	0.933	0.952
Cronbach's alpha	0.925	0.905	0.922	0.905	0.939
AVE	0.816	0.779	0.813	0.778	0.766
Reliability	Yes	Yes	Yes	Yes	Yes

indicators. An indicator should have higher cross-loading value against itself than against other indicators. From the calculation of cross-loading values, it was found that all indicators used were reliable. The cross-loading values also fulfilled convergent validity test in which all the factor loading were higher than 0.50.

3.2 Inner model

After doing the calculation of outer model, the next step in PLS-SEM analysis was to build inner model. The aims were to know the influence of independent variable to dependent variable and its effect. Figure 2 shows the construct model used in this study. The model had four structural paths that were formed from PLS model.

Table 3 illustrates the calculation of p-value for each structural path of the PLS model. The expected p-value was less than 0.05, to indicate the significant influence of independent variable on the dependent variable. From all the paths, it was only knowledge to trust path that was higher than 0.05. This means that knowledge of information security does not have a significant influence on consumer confidence in e-commerce sites. Meanwhile, the p-value for other paths were less than 0.05. Thus, it can be said that attitude and behavior have a significant effect on trust which subsequently affects the buying interest. Nevertheless, the result of p-value to indirect influence from knowledge to interest was less than 0.05. This implied significant effect. In conclusion, consumer knowledge of digital certificates as one of the information security

Table 3. P-values for direct and indirect effect.			
Structural Path	P-values for direct effects	P-values for indirect effects	P-values for total effects
Knowledge → Trust	0.072		0.072
Attitude → Trust	0.028		0.028
Behavior → Trust	< 0.001		< 0.001
Trust → Purchase Intention	< 0.001		< 0.001
Knowledge → Purchase Intention		0.038	0.038
Attitude → Purchase Intention		0.011	0.011
Behavior → Purchase Intention		< 0.001	< 0.001

controls affect consumer buying interest through established trust.

After knowing the influence of independent variables on dependent variable, the next step was calculating the path-coefficient value of each structural path. Table 4 shows that trust variables had the greatest influence on purchase intention that was equal to 86%. It was followed by variable behavior to trust (46.5%) and attitude to trust (14.8%). Thus, behavior that is in accordance with the control of information security is very important to be improved because it has a great influence on trust and buying interest.

R-square value was used to find out how much the dependent variable was determined by the independent variable. From the Table 5, it can be seen that trust variables had a value of 0.473 or in moderate category. This means that 47.3% of trust variables can be measured by independent variables in this study,

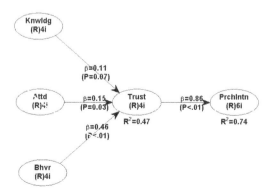

Figure 2. Construct model PLS-SEM.

Table 4. Path-coefficient.	
Structural Path	Path-coefficient
Knowledge → Trust	0.114
Attitude → Trust	0.148
Behavior → Trust	0.465
Trust → Purchase Intention	0.862

Table 5. R-squared coefficients.		
Variable	R Square	Criteria
Trust	0.473	Moderate
Purchase Intention	0.743	Strong

which is the information security awareness consisting of three dimensions (knowledge, attitude and behavior). Meanwhile, the purchase intention variable (buying interest) had a value of 0.743 or in the strong category. This indicates that 74.3% purchase intention variables can be measured by trust variables.

4 CONCLUSION

From the results of this study, it can be concluded that there is significant influence between attitude and behavior that are aware of information security on consumer trusts. This subsequently has a significant effect on buying interest. Although consumer knowledge does not directly influence, but the other three dimensions of information security indirectly have a significant effect on buying interest. Meanwhile, there are other factors encouraging consumers to do transaction on e-commerce sites that have not implemented information security protection well as digital certificates. Some of the factors are website quality (Hidayat & Hidayat 2017) and system quality (Pratomo & Hapsari 2017). Thus, the use of digital certificates still does not become a primary consideration in choosing e-commerce sites they will use to do transaction.

REFERENCES

Ajzen, I. 1991. *The Theory of Planned Behavior.* Organizational Behavior and Human Decision Processes 50(2): 179–211.

Broekhuizen, T., & Huizingh, E. K. 2009. Online Purchase Determinants: Is Their Effect Moderated by *Direct Experience? Management Research News* 32(5): 440–457.

Daryl, M, 2016. Road Map e-commerce RI 2016–2019 & Geliat Industri e-commerce. (Retrieved from: https://komite.id/2016/11/13/road-map-ecommerce-ri-2016-2019-geliat-industri-ecommerce/).

Gefen, D. 2002. Reflections on the Dimensions of Trust and Trustworthiness among Online Consumers. *ACM SIGMIS Database* 33(3):38–53.

Gefen, D., & Straub, D. W. 2004. Consumer trust in B2C e-Commerce and the importance of social presence: experiments in e-Products and e-Services. *Omega, 32* (6), 407–424.

Grabner-Kräuter, S., & Kaluscha, E. A. 2003. Empirical research in on-line trust: a review and critical assessment. *International Journal of Human-Computer Studies*, 58(6): 783–812.

Ha, H.-Y., & Janda, S. 2014. The effect of customized information on online purchase intentions. *Internet Research: Electronic networking applications and policy*, 24(4): 496–519.

Hidayat, R & Hidayat, A.M. 2017. Contribution of Environment Online Store for Purchase Decision (A Case Study Online Store Hypermart—Bandung). *Advanced Science Letters*, 23(1): 292–294.

Hong, K-S., et al. 2003. An Integrated System Theory of Information Security Management. *Information Management & Computer Security* 11(5): 243–248.

Kim, H. W., et al. 2012. Which is more important in Internet shopping, perceived price or trust?. *Electronic Commerce Research and Application*, 11(3): 241–252.

Koufaris, M., & Hampton-Sosa, W. 2004. The development of initial trust in an online company by new customers. *Information & Management*, 41(3): 377–397.

Kruger, H.A & Kearney, W.D. 2006. A Prototype for Assessing Information Security Awareness. *Elsevier Journal: Computers & Security* 25(4): 289–296.

Kruger, H. et al. 2010. A vocabulary Test to Assess Information Security Awareness. South Africa: South African Information Security Multi-conference.

Lee, M. K. & Turban, E. 2001. A trust model for consumer internet shopping. *International Journal of electronic commerce*, 6(1): 75–91.

Lu, L. C. et al. 2014. Purchase intention: The effect of sponsorship type, product type, and brand awareness. *Computers in Human Behavior*, 34: 258–266.

Mitchell, R. C. et al. 1999. *Corporate Information Security Management. New Library World*, 100 (1150) pp 213–227. MCB London: University Press.

Pavlou, P. A. & Fygenson, M. 2006. *Understanding and predicting electronic commerce adoption: An extension of the theory of planned behaviour.* MIS Quarterly, 30 (1): 115–143.

Peltier, T. R. 2014. *Information Security Fundamentals, 2nd Edition.* Boca Raton: CRC Press.

Ponte, E. B. et al. 2015. Influence of Trust and Perceived Value on the Intention to Purchase Travel Online: Integrating the Effects of Assurance on Trust Antecedents. *Tourism Management* 47: 286–302.

Pramiyanti, A. & Millanyani, H. 2014. *The Role of ICT On Quality Of Life* (Study On Indonesian Domestic Workers In Singapore). *Journal of Theoretical & Applied Information Technology* 63(3): 684–693.

Pratomo, D. & Hapsari, D. 2017. *The Factors Affecting Information System Success in Inventory Retail System. Advanced Science Letters*, 23(1): 620–622.

Sari, P.K & Candiwan. 2014. *Measuring information security awareness of Indonesian smartphone users. TELKOMNIKA* 12(2): 493–500.

Whitman, M. E., & Mattord, H. J. 2011. *Principles of Information Security, 4th Edition.* Atlanta: Cengage Learning.

Yousafzai, S. Y., et al. 2003. A proposed model of e-trust for electronic banking. *Technovation*, 23(11), 847–860.

Advances in Business, Management and Entrepreneurship – Hurriyati et al (eds)
© 2020 Taylor & Francis Group, London, ISBN 978-0-367-27176-3

Encouraging pro-environmental behavior through consumer innovativeness

R. Kuswati
Universitas Muhammadiyah Surakarta, Surakarta, Indonesia

B.M. Purwanto & B. Sutikno
Universitas Gadjah Mada, Yogyakarta

ABSTRACT: Environmental issues are important in the context of sustainability development goals. Pro-environmental is an idea with a new paradigm that is able to influence consumer lifestyles. As a relatively new concept or idea, pro-environmental becomes consumers' challenge in terms of adopting and applying it to be a part of their lifestyle. This shows that pro-environmental behavior becomes a challenge to be examined. The aim of this literature review is to describe pro-environmental behavior and the antecedents of pro-environmental behavior. Theories that explain pro-environmental behavior include Value Belief Norm Theory, Theory of Planned Behavior, Ecological Theory, Attitudinal-Behavior-Context Theory, etc. While the antecedents of pro-environmental behavior are consumer innovativeness, values, belief, attitudes to pro-environmental behavior, age, gender, income, level of education and others. This literature review presents an opportunity for researchers to conduct validation testing of pro-environmental behavior instruments, which can be either unidimensional or multidimensional. Consumer innovativeness as an alternative predictor of pro-environmental behavior becomes an opportunity for researchers to explore in further research. The gap theory of consumer innovativeness arises with the existence of several different theoretical standpoints that explain consumer innovativeness.

1 INTRODUCTION

Sustainability of economic growth and development is the goal of all countries of the United Nations (UN). It is written in the resolutions issued by the UN in September 2015, related to New Sustaina-bility Development Goals (New SDGs) that must be achieved until 2030. Why is the issue currently important in emerging market? In fact, environmental issues have been discussed for more than 25 years and researchers' concern was initially on the rela-tionship of marketing with consumer behavior and environment (Peattie 2010). It is proven by the presence of green marketing discussion in textbooks. In the next stage, many studies link environmental concerns to purchasing behavior (Modi & Patel 2013). In 1990s, business practices began to make strategic decisions related to environmental management and marketing. Environment-based segmentation became a trend and there was an increase in market demands for pro-environmental products at the time. However, pro-environmental behavior (PEB) based segmentation had not been widely studied (Modi & Patel 2013). The need for environment based segmentation is growing towards environmental care behavior; information of pro environmental behavior becomes important for marketers because it is able to bridge business needs in environmental business practice so as to meet the environmental market demands or

consumer behavior. In other words, demand side has to meet supply side (Kastadinova 2016).

The aims of this literature review are first, to serve as a theoretical review of pro-environmental behavior (PEB) and explore the types of pro environmental behavior measurement. Second, to review the theories used by the researchers in modeling pro environmental behavior. Third, to study predictor variables of pro environmental behavior. Of the many pro environmental behavior predictors or antecedents, the researchers tried to directly and indirectly explore consumer innovativeness as the pro environmental behavior antecedent. There have been many researches on pro environmental behavior that involve individual demographic and psychographic variables, particularly values and attitudes. However, researches that make use of consumer innovativeness as the pro environmental behavior predictor are still relatively limited.

2 METHOD

The current study was conducted by reviewing the secondary data obtained from double-blind peer review from academic journals, book chapters, and commercial reports. Relevant materials were identified through keywords pro-environmental, pro-environmental consumer behavior, sustainable consumer, eco-friendly consumer, environmetal behavior, environmentally sensitive behavior, environmentally responsible behavior, environmentally friendly behavior, eco-sensitive behavior, environmentally conscious consumer behavior, environmental sustainability, and conservation behavior. This study used several databases, including EBSCO host, JSTOR, ProQuest New Platform and Emerald Insight. In total, 130 journals and 82 works were collected and assessed to gain the understanding of the theoretical and methodological foundation of pro-environmental behavior. The current study was limited to the topic of pro-environmental behavior and consumer innovativeness.

3 RESULT AND DISCUSSION

The result and discussion of this study included the definition of pro-environmental behavior, measurement, theoretical background, antecedents of pro enviromental behavior, consumer innovativeness as a predictor of pro-environmental behavior, theoretical gap of innovativeness as an antecedent of pro-environmental behavior.

3.1 *Definition of pro environmental behavior*

The definition of pro-environmental behavior according to the United Nations is the behavior showing the use of products and services that make quality of life better by minimizing the use of resources and toxic materials so that they will not interfere the needs of future generations (Park & Ha 2011). Kollomus & Agyeman (2002) define pro-environmental behavior as the behavior that tries to reduce the use of energy and natural resources, toxic materials, or energy consumption. Another definition of PEB is individual efforts to reduce actions capable of damaging the natural environment. According to Steg & Vlek (2009), pro-environmental behavior is the behavior that has the least negative impact or behavior that benefits the environment.

3.2 *Measurement of pro environmental behavior (PEB)*

Pro environmental behavior measurement can be categorized into two measurements: unidimensional and multidimensional measurement. Some unidimensional measurements include: Kaida & Kaida (2016) where pro environmental behavior is measured by unidimensional measurement with water usage, electric lamp usage, and home heating efficiency indicator measured by 1-5 Likert scale. Onel (2016) measured pro environmental behavior using six unidimensional indicators including: how often do you: recycle cans and bottles, buy pesticide free fruits and vegetables, avoid purchasing product, drive less for environmental reasons, reduce fuel, and save water for environmental reasons. Pensini et al. (2012) measured pro environmental behavior using The New Ecological Behavior Scale consisting of 22 question items, with 1-4 scale from never to

always, while Stern et al. (1995), Noe & Snow (1989, 1990) used NEP (The New Ecological Paradigm) to measure pro environmental behavior. Dermody et al. (2015) measure pro environmental behavior in sustainable consumption behavior context with five unidimensional indicators with 1-5 Likert scale.

Multidimensional pro-environmental behavior measurement uses scaling in household conservation behavior context. There are four multidimensional pro environmental behavior, namely recycling, water conservation, electricity conservation and energy conservation. Bronfman et al. (2015) observed pro-environmental behavior using multidimensional measurement with six dimensions. The dimensions are power conservation, ecologically aware, biodiversity protection, water conservation, rational automobile use and ecological waste management.

3.3 *Theoretical background of pro-environmental behavior*

Background theory used in various researches of pro-environmental behavior is quite varied. The researchers used some theories such as: Theory of Planned Behavior, Value Belief Norm Theory, Eological Value Theory, Attitude-Behavioral–Context Model, Theory of Interpersonal behavior and The Role Theory to predict and analyze pro-environmental behavior.

3.4 *Antecedents of pro environmental behavior*

Some factors that influence pro-environment behavior can be categorized into socio-demographic and psychographic factor. Socio-demographic variables are commonly used by researchers in various research contexts. Socio-demographic antecedents of pro environmental behavior are age, level of education, gender, income (Lee 2008), and socio-economic status (Onel & Mukharjee 2014). Psychographic factors or psychological constructs related to pro environmental behavior are value, belief attitude, consumer knowledge, environmental concern, pro-environmental self-identity, consumer innovativeness, etc., as explained below. However, there is no conclusion of psychological factors that mostly influence pro-environmental behavior.

3.5 *Values, beliefs and attitudes*

The results of many researches indicate that these variables have an influence on pro-environmental behavior on specific context in adopting eco-friendly products. Being consistent with Ajzen's (1991) study, using specific context in attitude and belief will be stronger to predict specific context of the behavior, compared to using common context in belief and attitude on specific behavior (Lee 2008,

Stern et al. 1999, Jansson et al. 2010, Untaru 2014, Moser 2015).

3.6 Pro-environment value orientation (PVO)

Values are suspected to affect attitude; value is considered as criterion or driver variable determinant of consumer effectiveness. Schwartz stated that ecology behavior is affected by two values: self-transcendence and self-enhancement are negatively correlated with eco-centric attitude. Pro-environmental value orientation (PVO) has three approaches as explained by Dunlap & Liere (1978), Stern et al (1993) and Thompson & Barton (1994).

First, Dunlap & Liere (1978) stated that PVO is a general world view. Post-materialism hypothesis changes materialism value into post-materialism. Hence, PVO is important because individuals develop involvement and relation with ecosystem. Thus, human influences nature substantially. The new environmental paradigm (NEP) is a Pro-Environmental Value Orientation (POV) measurement with environment that emphasizes unity with nature (Soyev 2012).

The second approach is from Stern et al. (1993) in which they introduce three dimensions in PVO: socio-altruistic dimension based on norm activation theory (Schwartz 1977). This dimension is expressed through a view that environmental decline or worse environmental condition gives bad influence on human life in general. Stern also developed a theory from Schwartz, that individuals care not only about themselves but also about others, however they also considered the consequences of cost on certain behavior. Therefore, egoistic value dimension appears and reflects self-interest. The three bio-spherically oriented dimensions—individual values towards nature—are only for their self-interest. Tripartite model of Stern's value orientation covers three object value classes: other, self and non-human object (Soyev 2012).

The third approach is from Thompson & Barton (1994) who explained three values and most of them are reflected in Stern et al. (1993). The values are eco-centric value orientation which is similar to Stern's biosphere, anthropocentric value orientation related to social altruistic in Stern, and environmental apathy that represents individuals who have apathy towards environmental issues and who do not care to environment (Soyev 2012).

3.7 Consumer knowledge, environmental concern

Consumer knowledge or eco-literacy is a variable that influences eco-friendly environment. However, eco-literacy findings have weak influence. The misunderstanding of the findings shows that this variable has to be re-examined to predict pro environmental behavior. Meanwhile, environmental concern is individual awareness on environmental issues which is shown by personal supports and their involvement in solving an environmental issue (Dunlop & Jones 2000). Environmental concern is considered as the main element in encouraging an individual to or not to behave sustainably. Environmental concern is a set of perceptions, emotions, knowledge, attitudes, values, and behavior. Environmental concern has indirect relationship of pro-environmental behavior with specific situation (on sustainability-oriented behavior). Environmental concern is considered to have positive influence on sustainability consumption behavior (Dermody et al. 2015).

3.8 Environmental consciousness and materialism

Environmental Consciousness (EC) is a cognitive dimension from attitude towards environment or pro-environmental behavior. EC is proven to be able to change green product purchase (Peatti 2010, Young et al. 2010). Materialism is defined as orientation of ownership value considered as one's happiness formed by three dimensions: success (seen by the properties they have), happiness (ownership is a center of one's happiness and material success), and centrality (important materials which have to be attached to a person). Social status of materialism orientation is measured by what they own. This orientation has negative relationship with pro-environmental behavior and serves as a dominant consumer ideology in modern society, both in developed and in developing countries (Dermody et al. 2015).

3.9 Pro environmental self identity (PSI)

PSI is a self-identity owned by an individual related to pro environmental behavior (Werff 2013). PSI is the main driver of a person to socially behave (Stets & Biga 2003). A study by Dermody et al. (2015) state that PSI as a PEB predictor or sustainable consumption is better than attitude. The results of their study in UK and China showed that in both locations PSI is strongly correlated to sustainability consumption behavior, PSI variable is also an alternative PEB predictor, in addition to attitude variable which has been used as a direct or indirect pro environmental behavior predictor.

3.10 Perceived behavioral control, self-efficacy

According to TPB theory, this variable is believed to affect pro-environmental behavior. Other terms that have the same meaning with perceived behavioral control are self-perceived ability, efficacy of successfully performing pro-environmental behavior, environmental knowledge, an internal environmental locus of control, and perceived behavioral control (Jasson et al. 2010, Soyez 2012). Another study showed that trait variables considered to correlate with pro-environmental behavior are locus of control, self-efficacy and self-construal (interdependence of others). Self-efficacy is strongly correlated because it is an individual belief related to organizing capability, so that he/she is able to behave based on the situation he/she has.

3.11 Social norms and reference group influences

Based on social normalization and social practices theory, it is indicated that social factors such as environmental behavior group or belief group and government policy can realize pro-environmental attitude and make the attitude more common. Some re-searches related to pro-environmental consumer behavior are influenced by social group (Gupta & Ogden 2009, Connell 2010, Gerpott & Mahmudova 2010, Stern 2000, Kim 2017).

3.12 Consumer innovativeness

Consumer innovativeness in a study by Englis & Phillips (2013) is proven to mediate the relationship of attitude and PEB, although not in all dimensions. In this context, the researchers used global innovativeness or innate innovativeness term. Innate innovativeness is defined by Midgley & Dowling (1978) as the degree to which individuals accept new ideas and make innovative decision independently, apart from communication experience with other individuals. The relationship of consumer innovativeness and pro-environmental behavior is also supported by a study from Lao (2014) and Li et al. (2015).

3.13 Discussion: consumer innovativeness as predictor of pro environmental behavior

n the context of pro-environmental behavior antecedents, there is consumer innovativeness which has not been explored in depth in pro environmental behavior research. However, some researchers have initiated a study of PEB with consumer innovativeness as a predictor (Englis & Phillips 2013, Chen 2014, Dibrell et al. 2015). Consumer innovativeness has several terms.

The common terms of consumer innovativeness are innate innovativeness and global innovativeness (Midgley & Dowling 1978, Hirschman 1980, Tellis & Bell 2009). Both of them have similar meanings, and they review innovativeness in general. These constructs are considered as dispositional or constant personality traits in the individual.

Another specific term of innovativeness is domain specific innovativeness. The concept of this construct is that the degree of innovativeness can be affected by external factors (Steenkamp, Hofstede & Wedel 1999) except for individual factors. Specific context is attached to domain specific innovativeness construct in various researches.

Both of these groups must be observed by researchers when modeling with consumer innovativeness construct. Some researches use general terms such as innate or global innovativeness in predicting pro environmental behavior (Englis & Phillips 2013). Global innovativeness as personality trait conceptually cannot be influenced by other variables because it is innate and predisposition. The researchers have to use this general innovativeness variable as a mediating or dependent variable because it is in accordance with the concept of personality traits theory.

On the other hand, specific innovativeness term such as domain specific innovativeness construct can be used as a dependent or mediating variable which connects innate innovativeness to attitude toward eco-innovative behavior, innovative behavior, eco innovative adoption, pro-environmental behavior and so on.

Consumer innovativeness is a part of personality trait theory. Consumer innovativeness has the meaning of individual openness to new ideas. The meaning is similar to personality trait variable in big five theories, that is openness to experiences. Individuals who are open to new things tend to be innovative rather than those who are closed. In another pro-environmental behavior study used personality traits predictor in the big five personality theory. Among the five personality traits, openness to experience tends to influence pro environmental behavior. In Spain, a study indicated that open to experience (curiosity, creativity, preference novelty and seeking for variations) is related to pro-environmental activities, similar to a research in USA that openness is identical to pro-environmental behavior both in community and in individual level. The relationship of openness to pro environmental behavior is completely mediated by environmental attitude and connection to nature (Markowitz et al. 2012). The results of a study in Germany showed that concern of environment was found in those who are open (Hirsh 2010). Generally, openness strongly influences or correlates to environmental involvement.

Englis & Phillips (2013) conducted a study of pro-environmental attitude towards pro environmental behavior which was mediated by consumer innovativeness variable. The results showed that consumer innovativeness is proven to mediate the relationship of pro-environmental attitude toward pro environmental behavior although it does not mediate all dimensions. Consumer innovativeness was measured by three dimensions in a research by Englis & Phillips (2013): affinity for new ideas, early product adoption, and distrust of new product. The influence of attitude towards pro-environmental behavior is proven to be mediated by affinity for new ideas and early product adoption.

Hirschman (1980), Midgley & Dowling (1978), developed innovativeness concept from natural process of adopting new ideas, known as dominant consumer behavior. One of the consumer innovativeness terms is innate innovativeness. It is a degree to which individuals accept new ideas and make innovative decision independently, from the experiences of communicating with others (Midgley & Dowling 1978).

Some definitions show that innovativeness is almost similar to openness personalities. With the assumption, innovativeness can be submitted as a quite strong pro-environmental behavior predictor (Lee 2008, Lao 2014, Li & Wang 2015), in addition to some antecedents that have been widely discussed. This innovativeness concept is relatively not stable with various perspectives in defining and measuring the construct. Thus, it is important for researchers to test innovativeness concept in PEB context.

3.14 *Theoretical gap of consumer innovativeness*

Consumer innovativeness not only becomes a part of constant and dispositional personality but also can be influenced by external or situational factor. This view does not use personality traits but attitude theory approach. The debate of consumer innovativeness discussion has not been discussed clearly. It challenges the researchers to explore consumer innovativeness context from different perspectives, theories and contexts, such as eco-innovative behavior, eco-innovative adoption behavior in pro environmental behavior.

The view contains a meaning that innovativeness can be reviewed by some different theories (Hirschman 1980, Midgley & Dowling 1978). *First, personality traits theories*. This approach is known as trait-behavioral model. Innovativeness categorized as general variable is known as innate or global innovativeness that has personality trait theory review (Goldsmith & Hofacker 1991, Im et al. 2007, Im et al. 2003, Chao 2012), this theory assumes that traits are dispositional, innate in individuals and constant (Ewen 2003, Ellis et al. 2009). The assumption has consequence that innate or global innovativeness has a position as independent or moderating variable in the modeling.

Second, attitude theories. This approach is known as attitude-behavioral model. It assumes that individual behavior is caused by not only internal factors, but also environmental factors outside the individuals. Therefore, strong or weak degree of consumer innovativeness can be influenced by other environmental and external factors. Domain specific innovativeness represents innovativeness from attitude theory's point of view, so that domain specific innovativeness is appropriate if it is positioned as a mediating or dependent variable. Third, the combination of both approaches, known as contingency theories (Midgley & Dowling 1978, Hirschman 1980).

4 CONCLUSIONS

Pro-environmental behavior has some concepts in terms of how to define the meaning. Different concepts cause various pro-environmental behavior measurements, either unidimensional or multidimensional scales. It becomes a concern in the pro-environmental behavior to conduct various pro-environmental behavior measurement validation tests. It is expected that the results will give correct measurement tool. Enhancing environmentally friendly behavior can be done by influencing pro-environmental behavior through various antecedents, such as consumer innovativeness. Consumer innovativeness is an alternative predictor for pro environmental behavior, in addition to other antecedents including values, beliefs and norms. This literature review engages many researchers to consider consumer innovativeness as a direct or indirect predictor pro-environmental behavior. Besides, consumer innovativeness construct has a gap that is theoretically feasible to be observed, especially in pro environmental behavior context.

ACKNOWLEDGEMENT

We thank Dr. Anton Agus Setyawan for our pleasant discussion. We also thank the team of LPPI UMS for supporting this article.

REFERENCES

Ajzen, I. 1991. Theory of planned behavior. *Organizational Behavior and Human Decision Processes* 50:179-211.

Bronfman, N.C., Clsternas, P.C., Lopez-Vazquez, E., de la Maza, C. & Oyanedel, J.C. 2015. Understanding Attitudes and pro-environmental behaviors in a Chilean community. *Sustainability* 7: 14133-14152.

Chao, C. W., Reid, M. & Mavando, F. T. 2012. Consumer in-novativeness influence on really new product adoption. *Australian Marketing Journal* 20: 211-217.

Chen, K.K. 2014. Assessing the effects of customer inno-vativeness, environmental value and ecological lifestyles on residential solar power systems install intention. *Energy Policy* 67: 951-961.

Connell, K. Y. H. 2010. Internal and external barriers to eco-conscious apparel acquisition. *International Journal of Consumer Studies* 34(3): 279-286.

Csutora, M. 2012. One more awareness gap? The behaviour–impact gap problem. *Journal of Consumer Policy* 35:145–163.

Dermody, J., Hanmer-Lloyd, Stuart, Lewis Nicole Koenig & Zhao, A.L. 2015. Advancing sustainable consumption in the UK and China: the mediating effect of pro-environmental self-identity. *Journal of Marketing Management* 31 (13–14): 1472–1502.

Dibrell, C., Johnson, A.J., Craig, J.B., Kim, J. 2015. Establishing how natural environmental competency, organizational social consciousness, and innovativeness relate. *Journal of Business Ethics*, 127:591–605.

Dunlap, R.E.V.L., Kent D., Mertig, A.G., & Jones, R.E. 2000. Measuring endorsement of the new ecological paradigm: A revised NEP scale. *Journal of Social Issues*, 56: 425–442.

Dunlap, R.E. and van Liere, K.D. 1978. The new ecological paradigm: a proposed measuring instrument and preliminary results. *Journal of Environmental Education*, 9 (3): 10-19.

Ellis, A., Abrams, M. & Abrams, L.D. 2009. *Personality Theories*. California, USA: SAGE Publication Inc.

Englis, B.G. & Phillips, D. M. 2013. Does innovativeness drive environmentally conscious consumer behavior?. *Psychology and Marketing* 30(2): 160–172.

Ewen, R.B. 2003. *An introduction to theories of personality (6th ed.)*. London: Lawrence Erlbaum Associates.

Gerpott, T. J., & Mahmudova, I. 2010. Determinants of green electricity adoption among residential customers in Ger-many. *International Journal of Consumer Studies*. 34(4): 464-473.

Goldsmith, R. E., & Hofacker, C. F. 1991. Measuring consumer innovativeness. *Journal of the Academy of Marketing Science* 19: 209–221.

Gupta, S., & D. T. Ogden. 2009. To Buy or Not to Buy? A So-cial Dilemma Perspective on Green Buying. *Journal of Consumer Marketing*, 26(6): 376-391.

Hirschman, Elizabeth C. 1980. Innovativeness, Novelty Seek-ing and Congsumer Creativity. *Journal of Consumer Re-search* 7: 283-295.

Im, S., C. H. Mason & M.B. Houston. 2007. Does innate con-sumer innovativeness relate to new product/service adop-tion behavior? The intervening role of social learning via vicarious innovativeness. *Jounal of the Academic Market-ing Science*. 35:63–75.

Im, S., B. L, Bayus., & C. H, Mason. 2003. An empirical study of innate consumer innovativeness, personal characteristics, and new-product adoption behavior. *Journal of the Academy of Marketing Science*, 31, 61–73.

Jansson, J., Marell, A., & Nordlund, A. 2010. Green consumer behavior: Determinants of curtailment and ecoinnovation adoption. *Journal of Consumer Marketing*, 27 (4):358–370.

Kaida, Naoka and Kosuke Kaida. 2016. Pro-environmental be-havior correlates with present and future subjective well-being. *Environment, Development and Sustainability*. 18: 111-127.

Kim, Junyong. 2017. An empirical comparison of alternative models of consumers' environmental attitudes and eco-friendly product purchase intentions. *Seoul Journal of Business* 23 (1).

Kollmuss, A. & J. Agyeman. 2002. Mind the Kesenjangans: Why do people act environmentally and what are the barri-ers to pro-environmental behavior?. *Environmental Educa-tion Research* 8(3): 239-260.

Kostadinova, E. 2016. Sustainable consumer behavior: Literature Overview. *Economic Alternative* Issue 2:.224-234.

Lao, K. 2014. Research on mechanism of consumer innovativeness influencing green consumption behavior. *Nan-kai Business Review International* 5(2): 211-224.

Lee, K. 2008. Opportunities for green marketing: young con-sumers. *Marketing Intelligence and Planning*. 26 (6): 176-199.

Li, G., Zhang R. & Wang, C. 2015. The role of product orig-inality, usefulness and motivated consumer innovativeness in new product adoption intentions. *Journal of Product Innovation Management*, 32(2): 214–223.

Midgley, F. D., & Dowling, G. R., 1978. Innovativeness: the concept and its measurement. *Journal of Consumer Re-search* 4: 229–241.

Modi, A.G., & Patel, J.D. 2013. Classifying consumers based upon their pro environmental behaviour: an empirical investigation. *Asian Academy of Management Journal* 18 (2): 85–104.

Moser, A.K. 2015. Thinking green, buying green? Drivers of pro-environmental purchasing behavior. *Journal of Consumer Marketing*. 32(3): 167–175.

Noe, F. P., & Snow, R. 1989. Hispanic cultural influence on environmental concern. *Journal of Environmental Educa-tion*,21: 27–34.

Noe, F. P., & Snow, R. 1990. The new environmental para-digm and further scale analysis. *Journal of Environmental Education*, 21: 20–26.

Onel, Naz., Mukherjee, A. 2014. Analysis of the predictors of five eco-sensitive behaviors. *World Journal of Science, Technology and Sustainable Development* 11 (1): 16-27.

Onel, Naz., Mukherjee, A. 2016. Consumer knowledge in pro-environmental behavior an exploration of its antecedents and consequences. *World Journal of Science, Technology and Sustainable Development* 13 (4): 328-352.

Park, J. & Ha, S. 2011. Understanding pro-environmental behavior, a comparison of sustainable consumers and apathetic consumers. *International Journal of Retail & Distribution Management* 40 (5): 388-403.

Peattie, K. 2010. Green Consumption: Behavior and Norm. *Annual Review of Environmental Resources* 35: 195-228.

Pensini, P.M., Slugoski, B.R. & Caltabiano, N.J. (2012). Predictors of environmental behavior: a comparison of known groups. *Management of Environmental Quality: An International Journal* 23 (5): 536-545.

Schwartz, S.H. 1977. Normative influences on altruism in Berkowitz, L. (Ed.). *Advances in Experimental Social Psychology*. New York: Academic Press.

Soyez, Katja. 2012. How national cultural values affect pro-environmental consumer behavior. *International Marketing Review* 29 (6): 623-646.

Steenkamp, J. E. M., Hofstede, F. & Wedel, M. 1999. A cross-national investigation into the individual and national cultural antecedents of consumer innovativeness. *Journal of Marketing*, 63: 55–69.

Steg, L. & Vlek, C. 2009. Encouraging pro-environmental behavior: An integrative review and research agenda. *Journal of Environmental Psychology* 29:309–317.

Stern, P. 2000. Toward a coherent theory of environmentally significant behavior. *Journal of Social Issues*, 56 (3): 407-424.

Stern, P., Dietz, T. Abel, Guagnano, G & Kalof, L. 1999. A value belief norm theory of support for social movements: the case of environmental concern. *Human Ecology Review* 6:81-97.

Stern, P., Dietz, T., Abel, Guagnano, G. 1995. The new eco-logical paradigm in social psychological context. *Environ-ment and Behavior* 27: 723-743.

Stern, P.C., Dietz, T. & Kalof, L. 1993. Value orientations, gender, and environmental concern. *Environment and Be-havior* 25(5): 322-348.

Stets, Jan E. & Biga, Chris F. 2003. Bringing Identity Theory into Environmental Sociology. *Sociological Theory* 21(4): 398-423.

Tellis, G. J., Yin, E. & Bell, S. 2009. Global consumer innovativeness: Cross-country differences and demographic commonalities. *Journal of International Marketing*, 17: 1–22.

Thompson, S.C.G. & Barton, M.A. 1994. Eco-centric and anthropocentric attitudes toward the environment. *Journal of Environmental Psychology*, 14(2): 149-157.

Untaru, E.N., Epuran, G. & Ispas, A. 2014. A conceptual framework of consumers' pro-environmental attitudes and behaviours in the tourism context. *Bulletin of the Transilvania University of Braşov. Series V. Economic Sci-ences*. 7 (56/2).

Van Der Werff, E., Steg, L., & Keizer, K. 2013. The value of environmental self-identity: the relationship between bio-spheric values, environmental self-identity and environmental preferences, intentions and behaviour. *Journal of Environmental Psychology*, 34: 55–63.

Young, W., Hwang, K., McDonald, S. & Oates, C.J. 2010. Sustainable consumption: green consumer behaviour when purchasing products. *Sustainable Development*. 18 20–31.

Advances in Business, Management and Entrepreneurship – Hurriyati et al (eds)
© 2020 Taylor & Francis Group, London, ISBN 978-0-367-27176-3

Entrepreneurial marketing and marketing performance: The moderating role of market-sensing capability

D.A.A. Mubarok, R. Hurriyati, D. Disman & L.A. Wibowo
Universitas Pendidikan Indonesia, Bandung, Indonesia

ABSTRACT: Marketing performance in small to medium-sized enterprises is explained by the success of marketing strategies and programs implementation. Several studies have described that small to medium-sized enterprises often encounter difficulties in marketing. It is hard to conduct market-sensing capability to utilize the market opportunity that effects market profit. This study aims to determine the effect of market-sensing capabilities on entrepreneurial marketing and marketing performance. A survey was conducted on 50 small to medium-sized enterprises of food and beverage sector with a simple random sampling technique. The data was analyzed by using linear regression. Results of the study revealed that entrepreneurial marketing as a marketing activity in small to medium-sized enterprises affected marketing performance by market-sensing capabilities. In running the business, small to medium-sized enterprises manager should be able to make the dynamic change in the marketing environment to that of a profitable one.

1 INTRODUCTION

Marketing performance is a description of company's marketing capabilities Cacciolatti and Lee (2016) to achieve profitability, sales growth, market share, customer satisfaction, customer loyalty and brand equity (Clark 1999). This condition explains how well a company would be capable to execute marketing strategies and programs (Katsikeas et al. 2015). Marketing performance is inseparable from the influence of competitors' marketing activities and the changes of marketing environment (Najafi et al. 2016; Katsikeas et al. 2015).

The success of a company to achieve its best marketing performance describes company capability to manage and utilize resources efficiently and effectively (Clark 2000). The changes in marketing environment require dynamic and multidimensional parameters to measure company success (Katsikeas et al. 2015), such as return on investment, customer life value or return on customers (Ambler et al. 2004).

Marketing performance measurement is adjusted with the outcomes that each firm intends to achieve (Ambler et al. 2004). The adjustment often is influenced by firm size and type, time horizon and firm management perspectives of success (Katsikeas et al. 2015). The differences in perspective provide a clear definition that marketing performance is a form of a company's marketing ability in utilizing its resources so that it can adapt to competitive situation and be able to deliver better value than its competitors (Clark 1999a, 2000b, Morgan et al. 2002).

Small to medium-sized enterprise marketing performance is determined by the ability to respond to the marketing environment (ranging from market conditions, competitors, consumers) and to manage limited resources, knowledge, and experience as well as marketing activities (Frank 2005, Baker & Sinkula 2009, Haroon Hafeez 2012). Thus, value and efficiency is created (Putri et al. 2016).

According to Irjayanti and Mulyono (2012), the capability of small to medium-sized enterprises in West Java to achieve marketing performance faces several obstacles. One of the obstacles comes from marketing environment, such as the level of competition, access to financial resources and technology, fluctuation of fuel prices, inefficient production costs, economic factors, management skills, limited sales, and lack of raw materials. These obstacles affect the company's goals in achieving the best marketing performance. Therefore, small to medium-sized enterprises need a comprehensive approach on marketing strategy in order to be more effective in dealing with environmental turbulence that will affect the company both in productivity and profitability (Morris et al. 2002).

Based on this problem, small to medium-sized enterprises should apply entrepreneurial marketing as an option of marketing strategy to achieve growth by exploiting opportunities and creating value for the target market (Kilenthong et al. 2016). Entrepreneurial marketing is an alternative choice in carrying out a comprehensive marketing activity process to create superior value (Morrish & Morrish 2011). This is integrated with entrepreneurial philosophy, being proactive, innovative and risk-averse (Kraus et al. 2010).

In entrepreneurial marketing, there is the identification and exploitation process of opportunities. The aim is to obtain profitable consumers through innovative approaches, risk management, resource utilization, and value creation (Morrish & Morrish

2011). In this concept, marketers can proactively build consumer equity (Morrish et al. 2010) through innovation in creating product competitiveness (Dewi et al. 2016). Thus, business performance can be improved (Hamali et al. 2016).

As augmented process, entrepreneurial marketing focuses on achieving business growth by utilizing its resources through innovation activities in marketing activities (Kilenthong et al. 2016). The synergy of entrepreneurial marketing elements (such as proactive, risk taking, innovation, opportunity focus, leverage resources and value creation) can be seen from its application in the company. Here, the focus should be in consumers behavior change and activities conducted by competitors in serving the target market (Gorica & Buhaljoti 2016) and (Morrish et al. 2010).

Entrepreneurial marketing is able to manage marketing activities as a driving factor in business development (Gorica & Buhaljoti 2016, Nyuur et al. 2016) and as part of an innovation to face competition and consumers behavior change (Dwyer et al. 2009). It is also able to overcome the limitations of small to medium-sized enterprises in running marketing activities (Miles et al. 2015).

Company ability basically describes the capability of a company to learn and respond to all market changes (Merrilees et al. 2011), to create value for consumers (Cass & Viet 2012) by utilizing marketing resources, and to increase marketing advantage (Fulka et al. 2018) as part of a company's competitive advantage (Hou & Chien 2010). A company's ability to adapt the environmental changes encourage the improvement in marketing management to achieve the business growth (Ardyan 2016). The achievement, however, should be based on the market-sensing capability elements, such as environmental analysis, understanding market trends, tracking competitors' strategies, and ability to adapt with the changes that occur in the marketing environment.

Market-sensing capability is a form of organizational knowledge that helps a company to learn and respond to marketing environment changes. Several studies show that market-sensing capabilities can improve organizational adaptation toward environmental change (Mu 2015). Hence, it can increase the organization's speed in responding the markets by creating innovative products (Ardyan 2016) to achieve the business growth (Lindblom et al. 2008).

Based on this explanation, the purpose of this study is to determine the effect of entrepreneurial marketing on marketing performance with market sensing capability as a moderation variable.

2 METHOD

2.1 Sample and data analysis

The research used quantitative research design. The population was small to medium-sized enterprises of food and beverage sector in West Java. Regarding the sample, 50 small to medium-sized enterprises were selected. The data was analyzed by using linear regression.

2.2 Research variables

Entrepreneurial marketing was measured by using a scale proposed by Gorica and Buhaljoti (2016). There were seven dimensions of entrepreneurial marketing, namely: proactive, risk taking, innovation, opportunity focus, leverage resources, customer intensity, and value creation. Marketing performance was measured by using Clark's (1999) scale. There were three dimensions of marketing performance: profit, sales growth, and market share. Market-sensing capability as a moderating variable was measured by using Ardyan's (2016) scale, which used four dimensions: learning about environment, tracking competitor strategy, understanding market trend, and responsive.

2.3 Validity and reliability of research variables

The validity and reliability test was used to find out the relevance of each variables. The result of validity and reliability test is provided here.

Table 1 shows that the validity of each variable was higher than 0.3. Meanwhile, the reliability of each variable dimension was higher than 0.6. This indicated that variables can be used to measure the influence of entrepreneurial marketing and market sensing capability on marketing performance.

Table 1. Validity and reliability of variables research.

Research variables	N of item	Validity	Reliability
Marketing Performance	3		0.705
Sales growth		0.821[**]	
Profitability		0.814[**]	
Market share		774[**]	
Entrepreneurial Marketing	7		0.842
Proactive		0.528[**]	
Risk taking		0.794[**]	
Innovation		0.723[**]	
Opportunity focus		0.772[**]	
Leverage resources		0.800[**]	
Customer intensity		0.717[**]	
Value creation		0.701[**]	
Market sensing capability	4		0.903
Learning about environment		0.866[**]	
Tracking competitor strategy		0.890[**]	
Understanding market trend		0.850[**]	
Responsive		0.912[**]	

2.4 Research model

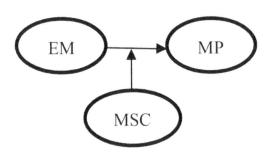

Figure 1. Research models.

MP = Marketing performance
EM = Entrepreneurial marketing
MSC = Market sensing capability

3 RESULTS AND DISCUSSIONS

Linear regression was used in analyzing the data. The aim to determine the effect of entrepreneurial marketing on marketing performance. The result can be seen in Table 2.

From Table 2, it can be seen that the value of R was 0.75. This indicated that there was a strong correlation between entrepreneurial marketing variables and marketing performance variables. The Table also shows that the value of R^2 was 0.56. This implies that the entrepreneurial marketing used in the research model contributes to the achievement of marketing performance. The contribution is 56 percent.

In conclusion, entrepreneurial marketing as marketing strategy and process can increase marketing performance. This is in line with the role of entrepreneurial marketing that prompts a company to be more proactive and innovative (Morrish & Morrish 2011, Jones et al. 2013) in improving its performance (Ahmadi & O'Cass 2015). This result is in accordance with Gorica and Buhaljoti (2016). Related to small to medium sized enterprises, its marketing activity basically can be integrated with entrepreneurial philosophy to create superior value that is focuses on target market (Kraus et al. 2010).

The effect of market-sensing capability as a moderating variable on the relationship between entrepreneurial marketing and marketing performance can be seen in Table 3.

Table 3 shows that the effect of entrepreneurial marketing on marketing performance increased 77percent. R^2 value reveals that after there was an influence of market sensing capability, the contribution of entrepreneurial marketing to marketing performance increased for about 60 percent. This means that market sensing capability increases the influence of entrepreneurial marketing on marketing performance. This result is supported by Lindblom et al. (2008) and Ardyan (2016), who say that market-sensing capability has positive influence on marketing performance. In other words, market-sensing capability can enhance the ability of small to medium-sized enterprises to learn and know the changes in marketing environment. This ability can be used to achieve good company performance (Lin & Wang 2015).

In addition, small to medium-sized enterprises also should be proactive and innovative to apply market-focused strategy and to create value (Mort et al. 2012). However, the ability of a small to medium-sized enterprises or a company in achieving marketing performance is a description of company's adaptability in managing marketing resources effectively and efficiently (Clark 2000). Small to medium-sized enterprises of food and beverage sector in West Java have to set marketing performance as a goal. The achievement of marketing performance is influenced by entrepreneurial marketing as an augmented strategy that plays a role in driving and seeking opportunity (Morris & Schindehutte 2002) for creating customers' value (Hamali et al. 2016).

Marketing performance achieved small to medium-sized enterprises of food and beverage sector will increase if the implementation of entrepreneurial marketing is influenced by the ability of small to medium-sized enterprises in environmental scanning. This is in line with the Ardyan (2016), which states that SMEs performance is influenced by the ability to observe changes in consumer behaviour, to know competitors' strategies, and to be able to respond to each change as a profitable opportunity.

Table 2. Model summary EM to MP.

Model	R	R Square	Adjusted R Square	Std. Error of the Estimate
1	0.750[a]	0.563	0.554	34.61193

a. Predictors: (Constant), Entrepreneurial Marketing

Table 3. Model summary EM*MSC to MP.

Model	R	R Square	Adjusted R Square	Std. Error of the Estimate
1	0.777[*]	0.603	0.577	33.69665

* Predictors: (Constant), Entrepreneurial Marketing
* Market Sensing Capability, Entrepreneurial Marketing, Market Sensing Capability.

4 CONCLUSION

Based on the results of this research, it can be concluded that small to medium-sized enterprises of food and beverage sector in West Java are able to achieve marketing performance through the implementation of entrepreneurial marketing strategies. Entrepreneurial marketing strategies contribute to the enhancement of marketing performance achievement by using market-sensing capabilities to respond to marketing environment changes and entrepreneurial marketing strategies. This study proposes the framework of entrepreneurial marketing as marketing strategy framework to enhance marketing performance. This framework is aligned with the other marketing elements, such as product innovation, branding, capability to access marketing channel, and integrated marketing communication.

REFERENCES

Ahmadi, H., & O'Cass, A. 2015. The role of entrepreneurial marketing in new technology ventures first product commercialisation. *Journal of Strategic Marketing*, 4488(June), 1–14.

Ambler, T., Kokkinaki, F., & Puntoni, S. 2004. Assessing Marketing Performance: Reasons for Metrics Selection. *Journal of Marketing Management*, 20(3–4), 475–498.

Ardyan, E. 2016. Market Sensing Capability and SMEs Performance: The Mediating Role of Product Innovativeness Success. *Business & Economic Review*, 2 (March), 79–97.

Baker, W. E., & Sinkula, J. M. 2009. The Complementary Effects of Market Orientation and Entrepreneurial Orientation on Profitability in Small Businesses. *Journal of Small Business Management*, 47(4),443–464.

Cacciolatti, L., & Lee, S. H. 2016. Revisiting the relationship between marketing capabilities and firm performance: The moderating role of market orientation, marketing strategy and organisational power. *Journal of Business Research*.

Cass, A. O., & Viet, L. 2012. Creating superior customer value for B2B fi rms through supplier firm capabilities. *Industrial Marketing Management*, 41, 125–127.

Clark, B. H. 1999. Marketing Performance Measures: History and Interrelationships. *Journal of Marketing Management*, 15(8),711–732.

Clark, B. H. 2000. Managerial perceptions of marketing performance: efficiency, adaptability, effectiveness and satisfaction. *Journal of Strategic Marketing*, 8(1),3–25.

Dewi, L. K. C., Wardana, I. M., Kertiyasa, N. N., & Sukaatmadja, I.P.G. 2016. Effect Of Entrepreneurial Marketing And Customer Relationship Marketing On Sme'S Competitiveness In Bali Indonesia. International Journal of Economics, Commerce and Management, IV (9), 512–525.

Dwyer, M. O., Gilmore, A., & Carson, D. 2009. Innovative marketing in SMEs. *European Journal of Marketing*, 43 (1/2), 46–61.

Frank B, G, J, M, K. 2005. Successful brand management in SMEs: a new theory and practical hints. *Journal of Product & Brand Management*, 14(4), 228–238.

Fulka, B. M., Ramli, A., & Bakar, M. S. 2018. Marketing Capabilities, Resources Acquisition Capabilities, Risk Management Capabilities, Opportunity Recognition. *Asian Journal of Multidisciplinary Studies*, 6(January), 12–22.

Gorica, K., & Buhaljoti, A. 2016. Entrepreneurial Marketing : Evidence from SMEs in Albania. *American Journal of Marketing Research*, 2(2), 46–52.

Hamali, S., Suryana, Y., Effendi, N., & Azis, Y. 2016. Influence of Entrepreneurial Marketing Toward Innovation and Its Impact on Business Performance. *International Journal of Economics, Commerce and Management*, IV(8), 101–114.

Haroon Hafeez, M. 2012. Relationship between Entrepreneurial Orientation, Firm Resources, SME Branding and Firm's Performance: Is Innovation the Missing Link? *American Journal of Industrial and Business Management*, 02(October), 153–159.

Hou, J., & Chien, Y. 2010. The Effect Of Market Knowledge Management Competence On Business Performance: A. *International Journal of Electronic Business Management*, 8(2), 96–109.

Irjayanti, M., & Mulyono, A. 2012. Barrier Factors and Potential Solutions for Indonesian SMEs. *Procedia Economics and Finance*, 4(Icsmed), 3–12.

Jones, R., Suoranta, M., & Rowley, J. 2013. Entrepreneurial marketing: A comparative study. *Service Industries Journal*, 33(7–8), 705–719.

Katsikeas, C. S., Morgan, N. A., Leonidou, L. C., & Hult, G. T. M. 2015. Assessing Performance Outcomes in marketing. *Journal of Marketing*, Online first (March), Online first.

Kilenthong, P., Hultman, C. M., & Hills, G. E. 2016. Entrepreneurial Orientation as the Determinant Of Entrepreneurial Marketing Behaviors. *Journal of Small Business Strategy*, 26(2),1–21.

Kraus, S., Harms, R., & Fink, M. 2010. Entrepreneurial marketing : Moving beyond marketing in new ventures. *International Journal of Entrepreneurship and Innovation Management*, (December).

Lin, J., & Wang, M. 2015. Complementary assets, appropriability, and patent commercialization: Market sensing capability as a moderator. *Asia Pacific Management Review*, 1–7.

Lindblom, A., Olkkonen, R., & Kajalo, S. 2008. Market-sensing capability and business performance of retail entrepreneurs. *Contemporary Management Research*, 4 (3),219–236.

Merrilees, B., Rundle-Thiele, S., & Lye, A. 2011. Marketing capabilities: Antecedents and implications for B2B SME performance. *Industrial Marketing Management*, 40(3),368–375.

Miles, M., Gilmore, A., Harrigan, P., Lewis, G., & Sethna, Z. 2015. Exploring entrepreneurial marketing. *Journal of Strategic Marketing*, 4488(May), 1–18.

Morgan, N. A., Clark, B. H., & Gooner, R. 2002. Marketing productivity, marketing audits, and systems for marketing performance assessment: Integrating multiple perspectives. *Journal of Business Research*, 55(5),363–375.

Morris, M. H., & Schindebutte, M. 2002. Entrepreneurial Marketing: A Construct for Integrating Emerging Entrepreneurship and Marketing Perspectives. *The Journal of Marketing Theory and Practice* (October), 1–21.

Morris, M. H., Schindehutte, M., & LaForge, R. W. 2002. Entrepreneurial Marketing: A Construct for Integrating Emerging Entrepreneurship and Marketing Perspectives. *Journal of Marketing Theory & Practice*, 10(4), 19.

23

Morrish, S. C., Miles, M. P., & Deacon, J. H. 2010. Entrepreneurial marketing: acknowledging the entrepreneur and customer-centric interrelationship. *Journal of Strategic Marketing*, 18(4),303–316.

Morrish, S., & Morrish, S. C. 2011. Entrepreneurial marketing: a strategy for the twenty-first century? *Journal of Research in Marketing and*, 13(2),110–119.

Mort, G. S., Weerawardena, J., & Liesch, P. 2012. Advancing entrepreneurial marketing: Evidence from born global firms. *European Journal of Marketing*, 46(3–4), 542–561.

Mu, J. (2015). Marketing capability, organizational adaptation and new product development performance. *Industrial Marketing Management*.

Najafi-Tavani, S., Sharifi, H., & Najafi-Tavani, Z. 2016. Market orientation, marketing capability, and new product performance: The moderating role of absorptive capacity. *Journal of Business Research*.

Nyuur, R. B., Brecic, R., & Simintiras, A. 2016. The Moderating Effect of Perceived Effectiveness of SMEs ' Marketing Function on the Network Ties — Strategic Adaptiveness Relationship*. *Journal of Small Business Management*, 00(00),1–19.

Putri, A. K., Suryana, Y., Tuhpawana, & Hasan, M. 2016. The effect of market orientation and competitive strategy on marketing performance. *International Journal of Economics, Commerce and Management*, IV(7), 274–288.

Advances in Business, Management and Entrepreneurship – Hurriyati et al (eds)
© *2020 Taylor & Francis Group, London, ISBN 978-0-367-27176-3*

The role of brand equity in making decisions to choose higher education for new middle-class students

A.M. Ramdan, A. Rahayu, R. Hurriyati & M.A. Sultan
Universitas Pendidikan Indonesia, Bandung, Indonesia

ABSTRACT: The aim of this research is to measure the influence of brand quality dimensions on the deci-sion of new middle-class students in selecting their higher education. The research method that is used is the quantitative method. Data collection was conducted through a questionnaire aimed at 100 new students from 20 different departments. The result of this research, through simultaneous measurements, found that brand equity has a significant influence on students' decisions in selecting their higher education, and at the same time four dimensions were analyzed; quality awareness and quality perception dimension do not significantly influence the decision of middle-class students in selecting higher education.

1   INTRODUCTION

The selection of higher education by prospective stu-dents is determined by two factors; they are brand equity and firm equity (Muafi 2002). Brand equity has an important role to company Clow & Donald (2005) similarly to private owned higher education. For this reason, it is necessary to manage brand equity well. High brand equity is an advantage and benefit for a company (Futrell & Stanton 1989), moreover it is profitable in the future (Aaker 1997). Managing brand equity is not a matter of improving the higher education image. Brand management must be able to direct the decision of prospective students to determine the place where they want to study.

Brand equity is defined as the strength of a brand. It means that brand equity is a benefit from the brand to a product (Farjam & Hongyi 2015). Before Farjam & Hongyi, brand equity has been defined by previous experts. Leuthesser (1988) stated that brand equity is a group of association and consumer behav-ior, channel member, and corporation which make a brand get more volume and a higher margin com-pared to without the brand, and which makes the brand become strong, sustainable and competitive compared to the competitors (Aaker 1997). The value consumer given is based on the consumer's association to the brand, as it is reflected in a brand awareness, a brand association, quality perception, a brand loyalty and other brand assets (Winter 1991). Brand equity gives added value to a product with consumer's brand association and perception to the brand (Keller 1993). Brand equity represents the extent of the consumers' familiarity to the brand and the capability of recalling the brand based on a good, strong and unique brand association. Vazquez et al. (2002), define it as the total utility of the consumer's association to use and consume a brand, including a functional and symbolic association. Inline with

this conceptual framework, brand equity is formed through brand awareness, brand association, quality perception and brand loyalty.

Brand awareness plays an important role in most equity conceptual models and it can produce a high level of preference for higher education; it happens because the consumer tends to buy a familiar brand and it can enhance company sale and profitability (Baldauf et al. 2003). It means that brand equity can become the reason for a consumer to choose his product (Huang & Sarigollu 2012). Brand awareness is through top of mind, brand recall, brand recogni-tion, and unaware brand dimension (Aaker 1997). The brand association is an information related to brand knot in the consumer's memory and it has meaning for the consumer (Keller 1993). The brand association reflects the bond between a consumer and a brand and it is a key of a product attribute, such as a logo, a slogan, or a famous personality (Grewal & Levy 2008). Quality perception is defined as consumer's assessment on the whole excellence of product or superiority of objective quality (Zeithaml 1988). Quality perception is an attitude which results from the comparison between con-sumer's expectation and actual performance (Para-suraman et al. 1988). The quality perception's dimension in this reasearch is based on the dimen-sions offered by Parasuraman et al. (1988) known as SERVQUALwhich consists of physic, empathy, guarantee, responsive and reliability. Brand loyalty has been defined in many ways by many experts. Brand loyalty is defined as the consumer bond to a brand (Aaker 1997), reselect higher education (Keller & Lehmann 2006), and consumer loyalty to a product (Rangkuti 2009).

Many researchers have conducted researches about brand equity, among them are Aydin, Gokhan and Ulegin (2015), Asif et al. (2015), Kim et al. (2009). The specific researches about brand equity in higher education were conducted by Mourad et al. (2011), Mupemhi (2013) and Ardyan (2015), but the unit analyzed in those researches are in general characteristics of the consumer. In the next research, the research that will be conducted in Universitas Muhammadiyah Sukabumi, the researcher will focus on the new middle-class consumers, those who have an expenditure of around US$2–US$20 per day (World Bank 2007). The problem in this study is "How is the influence of brand equity dimensions on the decision of new students from the middle-class in choosing their higher education?"

2 METHOD

The research is conducted to measure brand equity dimension influence on the decision to choose higher education for new middle-class students by using a quantitative approach, meaning data collection through a set of questionnaires distributed to 100 new students in 20 different programs in Universitas Muhammadiyah Sukabumi; the respondents have an income of around US$2–US$ 20 per day.

3 RESULT AND DISCUSSION

3.1 Simultaneous test

Based on the result of double regression on F test with the extent of significance value 0.05 (5%) it is found that brand awareness, brand association, and quality perception variables have an influence on the decision to choose higher education. It is shown by the result of $F_{statistics}$ 27.962 > F_{table} 2.46 and the value of probability significance α 0.00 < 0.05. Similarly, based on determination coefficient (R^2) it resulted in the value of 0.541 (54.1%). It means thatthe brand equity influence contribution on the decision to choose is 54.1% and the rest (46.9 %) is influenced by other variables that are not analyzed in this research. This result is in line with the ones from Aydin et al. (2015), Asif et al. (2015) in which it is concluded that brand equity has a significant influence on the decision to choose higher education. The research that was conducted at the Universitas Muhammadiyah Sukabumi finds that brand equity strength is very influencing on new students to choose their higher education. This is also in accordance with the result of the reasearch from Muafi (2002), that the prospective students will make brand equity a consideration to select higher education.

3.2 Partial test

3.2.1 The infleunce of brand awareness toward the decision to choose higher education
Based on the data analysis, the brand awareness (X1) and the decision to choose higher education (Y) show that $t_{statistics}$ is 1.810 and t_{tabel} (α = 0.05) resulted in t_{tabel} 1.983. The significant value that resulted from it is 0.074 > 0.05 and $t_{statistics}$ Value < t_{tabel} is 1.810 < 1.983, in which case brand awareness on the decision to choose higher education is not significant. Based on the analysis, it can be concluded that the brand awareness infleunce is not significant on the decision to choose higher education and it is not in line with the result of the research from Baldauf et al. (2003), which states that consumers tend to choose the brand they are familiar with.

3.2.2 The influence of brand association on the decision to choose higher education
The reasons of new middle-class students to choose the place to study are based on a unique value, strength and a distinction from others. They think that these three dimensions make higher education have a good competitive value. Based on the data analysis, the influence of the brand association on the decision to choose higher education shows $t_{statistics}$2.460 and t_{table} (α = 0.05) resulting in t_{table} 1.983. The significance value is 0.016 < 0.05 which means that brand association on the decision to choose higher education is significant. This research suppports the result of the research from Aydin et al. (2015).

3.2.3 The influence of quality perception on the decision to choose higher education
Based on the data analysis, the influence of quality perception on the decision to choose higher education shows $t_{statistics}$ -0.105 and t_{able} (α = 0.05) resulting in t_{tabel} 1.98 with the significance value 0.917 > 0.05, from which it can be concluded that quality perception does not influence the decision to choose higher education significantly. New middle-class students of Universitas Muhammadiyah Sukabumi do not have quality perception as the reason to choose higher education. This is contratry to the result of the study from Mupemhi (2013) which states that quality perception has an important role in directing the student's choice for higher education.

3.2.4 The influence of brand loyalty on the decision to choose higher education
Based on the data analysis resulted from t test between the brand loyalty (X4) with the decision to choose higher education (Y) shows $t_{statistics}$ 5.500, and t_{table} (α = 0.05) resulting in t_{tabel} 1.983. The significance value is 0.000 < 0.05 and the value of $t_{statistics}$ > t_{table} is 5.500 > 1.983 which means the brand loyalty on the decision to choose higher education is significant. The result shows that the decision to choose higher education for new middle-class students is influenced by their loyalty to the brand of higher education, the reluctance to move to other higher education, the willingness to recommend others to join the higher education to which they belong and the pride to be alumni. This proves brand loyalty. This result is in line with the theory

from Aaker (1997), that someone tends to make recommendations to others and will use the product continuously.

4 CONCLUSION

Based on the simultaneous test result, the decision of new middle-class students to choose higher education is mostly influenced bybrand equity. It means that brand equity highly contributes to influence the decision of new middle-class students to choose higher education. Similarly, based on the partial test result, brand equity and quality perception do not contribute to form the decision of new middle-class students to choose Universitas Muhammadiyah Sukabumi.

Further research can develop a research involving more than one higher education institution which can represent a wider research scope with prospective student respondents to see the brand equity concept objectively.

REFERENCES

Aaker, D. 1997. *Manajemen ekuitas merek. Memanfaatkan nilai dari susatu merek*. Jakarta: Mitra Utama.

Ardyan, E. 2015. Memahami ekuitas merek perguruan tinggi: penelitian empiris pada Stie Surakarta, Jawa Tengah, Indonesia. *Jurnal Manajemen dan Kewirausahaan* 17(2).

Asif, M., Abbas, K., Hussain, S. & Hussain, I. 2015. Impact of brand awareness and loyalty on brand equity. *Journal of Marketing and Consumer Research* 12.

Aydin, G. & Ulegin, B. (2015), Burc effect of consumer-based brand equity on purchase intention: Considering socioeconomic status and gender as moderating effects. *Journal of Euro Marketing* 24: 107-119.

Baldauf, A., Cravens, K. S. & Binder, G. 2003. Performance consequences of brand equity management: Evidence from organizations in the value chain. *Journal of Product & Brand Management* 12(4): 220–236.

Clow, K. E. & Baack, D. 2005, *Brand and brand equity. Concise Encyclopedia of Advertising*. Haworth Press, Inc.

Eun Young Kim, E. Y., Knight, D. K. & Pelton, L.E. 2009. Modeling brand equity of a U.S. apparel brand as perceived by generation y consumers in the emerging Korean market. *Clothing & Textiles Research Journal* 27(4): 247–258.

Farjam, S. & Hongyi, X. 2015. Reviewing the concept of brand equity and evaluating consumer-based brand equity (CBBE) models. *International Journal of Management Science and Business Administration* 1: 14-29.

Futrell, C. & Statnton, W. J. 1989. *Fundamental of marketing, 8th Edition*. Singapore: McGraw-Hill.

Grewal, D. & Levy, M. 2008. *Marketing*. New York: McGraw-Hill.

Huang, R. & Sarigollu, E. 2012. How brand awareness related to market outcome, brand equity, and the marketing mix. *Journal of Business Research* 65: 92–99.

Keller, K. L. & Lehmann, D. R. 2006. Brand and branding: Research finding and future priority. *Marketing Science* 25(6): 740–759.

Keller, K. L. 1993. Conceptualizing, measuring, and managing customer based brand equity. *Journal of Marketing* 57(1): 1–22.

Leuthesser, L. 1988. Defining, measuring, and managing brand equity. *Paper presented at the Conference Summary*. Cambridge.

Mourad, M., Ennew, C. & Kortam, W. 2011. Brand equity in higher education. *Marketing Intelligence & Planning* 29(4): 403-420.

Muafi, M. 2002. Mengelola ekuitas merek: Upaya memenangkan persaingan di era global, manajemen usaha Indonesia, p.44-50.

Mupemhi, S. 2013. Factors influencing choice of a university by students in Zimbabwe. *African Journal of Business manajemen*.

Parasuraman, A., Zeithaml, V. A. & Berry, L. L. 1988. SERVQUAL: A multiple-item scale for measuring customer perceptions of service quality. *Journal of Retailing* 64(1): 12–40.

Rangkuti, F. 2009. *The power of brands*. Jakarta: Gramedia Pustaka Utama.

Vazquez, R., Del Rio, A. B. & Iglesias, V. 2002. Consumer-based brand equity: Development and validation of a measurement instrument. *Journal of Marketing Management* 18: 27–48.

Winter, L. C. 1991. Brand equity measures: Some recent advances. *Marketing Research* 3: 70–73.

World Bank. 2007. *Global economic prospects 2007: Managing the next wave of globalization*. Washington, DC: World Bank.

Zeithaml, V. A. 1988. Consumer perception of price, quality, and value: A means-end model and synthesis of evidence. *Journal of Marketing* 52(3): 2–22.

Advances in Business, Management and Entrepreneurship – Hurriyati et al (eds)
© 2020 Taylor & Francis Group, London, ISBN 978-0-367-27176-3

Dynamic marketing capabilities and company performance: Marketing regression analysis on SMEs in Indonesian

A. Riswanto, R. Hurriyati, L.A. Wibowo & H. Hendrayati
Universitas Pendidikan Indonesia, Bandung, Indonesia

ABSTRACT: One of the most important issues is the role of the marketing function during the development through the concept of dynamic capabilities, even today there is a new term in marketing, it is "Dynamic Marketing Capabilities" (DMC). This is interesting, as some previous studies have conducted assessments related to building DMC. The main problem that often exists in developing a company is the poor quality or performance of the marketing department. This research will test DMC with company performance in marketing department, conducted on several small and medium-sized enterprises (SMEs) in Indonesia as one of the developing countries. The results showed that there is a positive influence between the variables studied.

1 INTRODUCTION

Dynamic capability refers to "organizational capacity to create new products and to process and respond to change market conditions" (Adner 2003, Helfat et al. 1997). The important role of marketing functions is in dynamic capability research (Fang & Zou 2009). This means that Dynamic Marketing Capabilities (DMC) as the response and efficiency of cross-functional business processes, and include product development management, supply chain management, and customer relationship management, to create and deliver customer value, market changes. Product development management is a cross-functional process of designing, developing, and launching new products to meet the needs required by customers (Srivastava, Shervani & Fahey 1999).

Successful companies not only respond to the needs their current customers desire but also anticipate future trends and develop ideas, products or services to meet the demanding future with process and fast and effective stages (Day & Schoe-maker, 2008, Dessler 2002). Therefore, with the various forms of businesses being built, a company can stay ahead of their competitors.

The perceptible on the success of a company that is measured in a specified time period shows the managerial conditions within a company (Beveridge & Nelson 1981, Dess & Robinson J.R 1984). This is the result from a collection of activities that have been planned, arranged and implemented to find out what strategies can be used appropriately to achieve company goals (Andrews 1997, Gilmore & Hart). By using the right strategy this tends to have a system within the company so that the company's performance can be maximized (Cordeiro & Sarkis 1997).

The main objective of organizational science is to find the determinants of company performance (Bower 1979, Lubatkin 1986, March & Sutton 2016,

Timilsina 2017). A number of tests are carried out to assess how a construct can be used, and various assessments of company performance tend to have conceptual problems and also empirically serious things must be done (Rangone et al. 2000, Rauch et al. 2009, Taffler 1983).

In marketing research, as a conceptualization, marketing innovation is suggested to improve company performance through: marketing-product space, marketing process space, and space marketing relationships. This improvement process, however, is thought to be moderated by the level of radical product innovation the company is currently undergoing as well as the level of the company's practice innovation process (Cascio 2011). Marketing capabilities have a stronger impact on company performance than research and development capabilities, and operations (Krasnikov & Jayachandran 2008) and customer analysis has a positive influence on company performance (Čermák 2015).

Furthermore, the marketing research process plays an important role in business organization performance, which means that there is a favorable relationship between marketing research and business organization performance. Providing appropriate and adequate facilities to improve the business environment and making it more responsive to customer needs and developing strategies must be put in place to improve staff performance and increasing their contribution to the organization (Ayuba & Kazeem 2015) and market orientation have a positive impact on financial and nonfinancial performance in the intensive knowledge industry. It is important for high-tech companies to improve their performance by implementing a market-oriented strategy, placing emphasis on effective market research and being strong in customer and competitor orientation and can influence the company's success in developing country (Charles, Joel & Samwel 2012, Protcko & Dornberger 2014). As far as entrepreneurs who have innovation can play a role in economic advancement marked by the percentage the number of entrepreneurs in developing countries and matured countries (Riswanto 2016).

Based on this background, the company's performance needs to get important attention to develop and advance the business that is carried out. If you are not able to manage, it will cause the condition of the business unit to be disrupted and in turn be bankrupt. One of the objectives of establishing a business is to be able to maintain the business life cycle so that the business can compete with other units. If it is ignored, company performance will cause the influence other business elements that exist within the company. Therefore, a marketing strategy needs to be formulated and designed to overcome the deterioration of company management which may affects the company's performance. Companies must be able to design the best solutions to overcome existing problems that might be emerged in the future.

Some of the factors that influence incorporated performance are dynamic marketing capabilities (DMC) (Bruni & Verona 2009, Cabañero, Ros, & Cruz 2015, Fang & Zou 2009, Kachouie, Mavondo & Sands 2018); develop marketing innovation by fostering marketing insights and marketing imagination (Cascio 2011); marketing capabilities (Krasnikov & Jayachandran 2008); customer analysis (Čermák 2015); marketing research process (Ayuba & Kazeem 2015); market orientation (Charles et al. 2012, Protcko & Dornberger 2014); company environmental pro activity (Cordeiro & Sarkis 1997); ownership of business units (Anderson & Reeb, 2003, Douma, George & Kabir 2006, Li, Meng, Wang & Zhou 2008, Martinez, Stohr & Quiroga 2007); political connections (Sheng, Zhou & Li 2011, Wang, Yao & Kang 2018); decision of capital structure (Akingunola, Olawale & Olaniyan 2018); fiscal incentives/incorporated income tax exemptions (Amendola et al. 2018); and ownership of human resources (Human Resource) (Jackling & Johl 2009, Shin et al. 2018, Xu et al. 2018).

Dynamic marketing capabilities (DMC) have consistency in influencing company performance (firm performance). DMC can contribute to a more detailed understanding of management practices and heterogeneity of performance in science-based settings and conclude our work by identifying its limitations and providing useful directions for future research (Bruni & Verona 2009).

2 METHOD

This study was conducted to examine the extent relationship between DMC, and the company's performance carried out on 80 Micro, Small and Medium Enterprises (MSMEs) in Sukabumi City and the company's marketing manager as a sample. This study is used descriptive quantitative studies accepting SPSS V 24 statistical tools and Amos 24 applications. The hypotheses are as follows:

H1: There is a favorable influence between Market Orientation and Dynamic Marketing Ability (DMC);

H2: There is a positive influence between the Development Process and Dynamic Marketing Ability (DMC);

H3: There is an explicit influence between Dynamic Marketing Ability (DMC) and Company Performance.

3 RESULT AND DISCUSSIONS

The extension of this study rests on two important assumptions. Number one, the ability to co-ordinate tasks and also present working dimensions (Grant 1996). In this case, marketing is very significant functional area and requires knowledge in the firm's value chain (Grant 1991). Secondly, middle managers are an immediate and influential part as a forceful managerial determinant, they have important roles as much as highest managers (Peteraf 2005). Dynamic marketing capabilities involve decisions from various parties, not just prime management companies (Marketing VP and other members of the top management team) but also middle management (managers from the marketing department) involved in manufacturing and using advertise knowledge and marketing resources. The key point is the extent to which vigorous capabilities are different from operational capabilities (Zollo 2002). Dynamic marketing abilities have differences with common or general marketing skills.

The results from this research can be described by the results of data processing and functions that show the influence of market orientation and process development factors on dynamic marketing capacity, below: $Y1$ (DMC) $= 0.221 + 0.505 X1$ (MO) $+ 0.742 X2$ (DP), with the value of $R = 0.996$ and the value of $R2 = 0.992$ and the test value of $F = 4556.329$ with a weighty level of 0.000.

While the results from the calculation and the functions obtained, related to the influence of marketing capacity on the company's performance are as follows: $Y2$ (CP) $= 7.986 + 0.619 Y1$ (MO), with a value of $R = 0.776$ and a value of $R2 = 0.602$ and the test value $F = 117.778$ with a significant level of 0.000.

Marketing capabilities help companies achieve equilibrium by satisfying the various needs of today's customers, by exploiting existing products and distributing channels and disseminating their own brands. According to Day, for example, maintaining the bond of a channel is a marketing capability that can strengthen relationships with distributors (Day 1994). Dynamic marketing capabilities (DMC), by contrast, support companies through the process of changing from their stationary processes. In fact, DMC is specifically focused on releasing and inte-grating market knowledge that helps companies evolve. Under dynamic marketing capabilities, it is intended to include what others have through the label of market sensing and conjure it with customer capabilities (Day 1994), customer-centric skills and the concept of competence customer in second order. (Slater & Narver 1994).

Furthermore, in developing countries such as Indo-nesia and others associated with DMC, because this tends to have different context/character and process than those in developed countries (Zhou et al. 2012). This needs to be accounted for the difference between matured and developing countries as the majority of customers with relatively low disposable income lead to greater attention and the importance of price by customers (Morgeson et al. 2015). Corporations need to build good deals with customer perceptions of the product/service, by bringing up new things to develop the economic market. This requires the use of marketing activities to build quality customer perceptions of quality and product knowledge to enable the creation and development of markets capable of sustaining growth (Ramaswa-my et al. 2000, Thompson & Chmura 2015). The nature of the underdeveloped marketing channels in many developing countries often requires the company to advanced DMC to build key elements of the marketing channel to enable an effective system to deliver products. By Karen, this market development process is essential to enable new entrants to build a product/ service profile to achieve good sales performance in a challenging and sometimes winning market environment in a growing economic environment (Helm & Gritsch 2013, Sumartini & Riswanto 2017). In short, the DMC rate may be related the performance within the company, especially in terms of sales because even though this capability is important than the market, whether they are in developed or developing countries, the specific conditions prevailing in later countries further strengthen cases to create and maintain the capability.

4 CONCLUSION

Based on the results from the assessment can be concluded that DMC has a favorable relationship between market orientation and development process, in addition DMC has a positive influence on company performance. The implications of this research include the fact that the marketing department should be enhanced in terms of dynamic marketing capabilities (DMC) including strategic marketing capabilities, marketing capability planning and advertise knowledge capabilities.

This study only discusses two variables affect the DMC taken from internal facilitators and external factors, recommendations for further research so that more factors that influence DMC can be assessed,

besides it would be better to use moderator variables as the factors that influence DMC and also equipped with research using control variables, such as company size, company age, type industry and soon.

REFERENCES

Adner, R. & C.H. 2003. Corporate effects and dynamic managerial capabilities. *Strategic Management Journal* 24: 1011–1025.

Akingunola, R.O., Olawale, L.S. & Olaniyan, J.D. 2018. Capital structure decision and firm performance: Evidence from non-financial firms in Nigeria. *Œconomica* 13(6): 4136–4373.

Amendola, A., Boccia, M., Mele, G. & Sensini, L. 2018. Fiscal incentives and firm performance evidence from the Dominican Republic, 8382.

Anderson, R.C. & Reeb, D.M. 2003. Founding-family ownership and firm performance: Evidence from the S&P 500. *Journal of Finance* 58(3): 1301–1328.

Andrews, K.R. 1997. 5 The concept of corporate strategy. Resources, firms, and strategies: A reader in the resource-based perspective (2). Homewood, IL: Irwin.

Ayuba, B. & Kazeem, O.A. 2015. The role of marketing research on the performance of business organizations. *European Journal of Business and Management* 7(6): 148–157.

Beveridge, S. & Nelson, C.R. 1981. A new approach to decomposition of economic time series into permanent and transitory components with particular attention to measurement of the "business cycle." *Journal of Monetary Economics* 7(2): 151–174.

Bower, J.L. & Y.L.D. 1979. *Strategic management: A new view of business policy and planning. in strategic management: A new view of business policy and planning. (D. E. S. and C. W. Hofer, Ed.)*. Boston: Little, Brown and Company.

Bruni, D. S. & Verona, G. 2009. Dynamic marketing capabilities in science-based firms: An exploratory investigation of the pharmaceutical industry. *British Journal of Management* 20(1): 101–117.

Cabañero, C.P., Ros, S.C. & Cruz, T.G. 2015. The contribution of dynamic marketing capabilities to service innovation and performance. *International Journal of Business Environment* 7(1): 61–78.

Cascio, R.P. 2011. Marketing innovation and firm performance research model, research hypotheses, and managerial implications. University of Central Florida.

Čermák, P. 2015. Customer profitability analysis and customer life time value models: Portfolio analysis. *Procedia Economics and Finance* 25 (May): 14–25.

Charles, L., Joel, C. & Samwel, K.C. 2012. Market orientation and firm performance in the manufacturing sector in Kenya. *European Journal of Business and Management* 4(10): 2222–2839.

Cordeiro, J.J. & Sarkis, J. 199. Environmental proactivist and firm performance: Evidence from security analyst earnings forecasts. *Business Strategy and the Environment* 6(2): 104–114.

Day, G.S. & Schoemaker, P.J. 2008. Are you a vigilant leader? *MIT Sloan Management Review* 49(3): 43–451.

Day, G.S. 1994. The capabilities of market-driven organizations. *Journal of Marketing* 58(4): 37–352.

Dess, G.G. & Robinson JR, R.B. 1984. Measuring organizational performance in the absence of objective measures: the case of the privately held firm and conglomerate business unit. *Strategic Management Journal* 5(1983): 265–273.

Dessler, G. 2002. *Human resource management.*

Douma, S., George, R., & Kabir, R. 2006. Foreign and domestic ownership, business groups, and firm performance: Evidence from a large emerging market. *Strategic Management Journal* 27(7): 637–657.

Fang, E. & Zou, S. 2009. Antecedents and consequences of marketing dynamic capabilities in international joint ventures. *Journal of International Business Studies* 40 (5): 742–761.

Fang, E. & Zou, S. 2009. Antecedents and consequences of marketing dynamic capabilities in international joint ventures. *Journal of International Business Studies* 40 (5): 742–761.

Gilmore, J.H., Ii, B.J.P. & Hart, S.L. (n.d.). CO PY Harvard Business Review. Harvard Business Review.

Helfat, C.E. 1997. Know-how and asset complementarity and dynamic capability accumulation: The case of R&D. *Strategic Management Journal* 18(5): 339–360.

Helfat, C.E.S., Finkelstein, W., Mitchell, M.A., Peteraf, H., Singh, D.J.T., & S.W. 2007. *Dynamic capabilities. Understanding dynamic change in organizations.* Oxford: Blackwell.

Helm, R. & Gritsch, S. 2013. Examining the influence of uncertainty on marketing mix strategy elements in emerging business to business export-markets. *International Business Review* 23(2): 418–428.

Jackling, B. & Johl, S. 2009. Board structure and firm performance: Evidence from India's top companies. *Corporate Governance: An International Review* 17(4): 492–509.

Kachouie, R., Mavondo, F. & Sands, S. 2018. Dynamic marketing capabilities view on creating market change. *European Journal of Marketing* 52(5/6): 1007–1036.

Krasnikov, A. & Jayachandran, S. 2008. The relative impact of marketing, research-and development, and operations capabilities on Firm Performance. *Journal of Marketing* 72(4): 1–11.

Krasnikov, A. & Jayachandran, S. 2008. The Relative Impact of Marketing, Operations Capabilities on Firm. 72 (July): 1–11.

Li, H., Meng, L., Wang, Q. & Zhou, L.A. 2008. Political connections, financing and firm performance: Evidence from Chinese private firms. *Journal of Development Economics* 87(2): 283–299.

Lubatkin M,S.R. 1986. Towards reconciliation of market performance measures to strategic management research. *Accad Management Rev* 11: 497–512.

March, J.G. & Sutton, R.I. 2016. Organizational Performance as a dependent. *JSTOR for this article* 8(6): 698–706.

Martinez, J.I., Stohr, B.S. & Quiroga, B.F. 2007. Family ownership and firm performance: Evidence from public companies in Chile. *Family Business Review* 20(2): 83–94.

Morgeson III, F.V., Sharma, P.N. & Hult, G.T.M. 2015. Cross-national differences in consumer satisfaction: mobile services in emerging and developed markets. *Journal of International Marketing* 23(2): 1–24.

Peteraf, M.A. 2005. *Research complementarities: A resource-based view of the resource allocation process.* Oxford: Oxford University Press.

Protcko, E. & Dornberger, U. 2014. The impact of market orientation on business performance—The case of tatars

tan knowledge-intensive companies (Russia). *Problems and Perspectives in Management* 12(4): 225–231.

Ramaswamy, B.R., Alden, D.L., Steenkamp, J.E.B.M. & R. S. 2000. Effects of Brand Local/Non-Local Origin on Consumer Attitudes in Developing Countries. *Journal of Consumer Psychology* 9(2): 83–95.

Rangone, A., Paolone, F. & Felice, I.S. 2000. Assessment Criteria and Perspectives in Italy. (286): 1–11.

Rauch, A., Wiklund, J., Lumpkin, G.T. & Frese, M. 2009. Entrepreneurial orientation and business performance: An assessment of past research and suggestions for the future. *Entrepreneur-ship: Theory and Practice* 33(3): 761–787.

Riswanto, A. 2016. The Role of the Entrepreneur in Innovation and in Economic Development. *In Proceedings of the 2016 Global Conference on Business, Management and Entrepreneurship* (15): 729–732.

Sheng, S., Zhou, K.Z. & Li, J.J. 2011. The Effects of Business and Political Ties on Firm Performance: Evidence from China. *Journal of Marketing* 75(1): 1–15.

Shin, J. Y., Hyun, J. H., Oh, S. & Yang, H. 2018. The effects of politically connected outside di-rectors on firm performance: Evidence from Korean chaebol firms. *Corporate Governance: An International Review* 26(1): 23–44.

Slater, S.F. & Narver, J.C. 1994. Does competitive environment moderate the market orientation-performance relationship? *Journal of Marketing* 58(1): 46–55.

Srivastava, R.K., Shervani, T.A. & Fahey, L. 1999. Marketing, business processes, and shareholder value: An organizationally embedded view of marketing activities and the discipline of marketing. 63(Special Issue): 168–1179.

Sumartini, S. & Riswanto, A. 2017. Indonesian Economic Growth Rate: Inflation and Unemployment Rate Analysis. *In in Proceedings of the 2nd International Conference on Economic Education and Entrepreneur-ship (ICEEE 2017)* 714–717.

Taffler, R.J. 1983. The assessment of company solvency and performance using a statistical model. *Accounting and Business Research* 13(52): 295–308.

Thompson, F.M. & Chmura, T. 2015. Loyalty programmed in emerging and developed markets: the impact of cultural values on loyalty program choice. *Journal of International Mar-keting* 23(3): 2015.

Timilsina, B. 2017. *Gaining and sustaining competitive operations in turbulent business environments what and how? Vaasa.* Finland: University of Vaasa.

Wang, Y., Yao, C. & Kang, D. 2018. Political connections and firm performance: Evidence from government officials' site visits. *Pacific-basin finance journal* 15 (7):2018

Xu, H., Guo, H., Zhang, J. & Dang, A. 2018. Facilitating dynamic marketing capabilities development for domestic and foreign firms in an emerging economy. *Journal of Business Research*: 141–152.

Zhou, L., Wu, A. & Barnes, B.R. 2012. The effects of early internationalization on performance outcomes in young international ventures: the mediating role of marketing capabilities. *Journal of International Marketing* 20(4): 25–45.

Zollo, M. & S.G.W. 2002. Deliberate learning and the evolution of dynamic capabilities. *Organization Science* 13: 339–351.

Advances in Business, Management and Entrepreneurship – Hurriyati et al (eds)
© 2020 Taylor & Francis Group, London, ISBN 978-0-367-27176-3

Is e-service quality required to develop customer satisfaction? A case study of Grab Indonesia

C.T. Sudrajat & M.A. Sultan
Universitas Pendidikan Indonesia, Bandung Indonesia

ABSTRACT: Online transportation service has become a disruption phenomenon in Indonesia. Grab Indonesia, one of the transportation companies providing online transportation, tries to maintain its service to attract loyal customers to choose online transportation as their main option. Customer satisfaction is one of the most important things in the online transportation business to survive. One of the factors that affects customer satisfaction is e-service quality, which is positive towards customer satisfaction. This study used a descriptive survey administered to Grab bike customers in Bandung. The hypothesis in this research is that there is a positive effect of e-service quality towards customer satisfaction. Data were analyzed using regression analysis to measure the e-service quality effect on customer satisfaction. The results show that there is a positive effect of e-service quality on customer satisfaction. Thus, Grab Indonesia should optimize and direct e-service quality to get customer satisfaction.

1 INTRODUCTION

Service quality delivery through electronic platforms for electronic commerce is one of the most critical issues for marketers in the service sector (Yang & Fang 2004, Zavareh et al. 2012, Quan 2010). Electronic service quality (e-SQ) becomes increasingly important in determining the success or failure of an electronic commerce application. There is an ongoing discussion in the literature about the relation and underlying difference between customer quality perceptions and satisfaction in internet-based services. Several studies found that specific website e-services can positively affect customer satisfaction with website and online purchasing in the long run (Khalifa & Shen 2005).

2 CONCEPTUAL FRAMEWORK

Based on concepts from both the service quality and retailing literature (Wolfinbarger & Gilly 2002) used in online and offline focus groups, a sorting task, and online survey of a customer panel, a scale named comma was developed consisting of four factors: website design, reliability, privacy/security and customer service. The contents of each dimension were: (1) website design involves the expected attributes associated with design, as well as items dealing with personalization; (2) reliability involves accurate representation of the product, on time delivery and accurate orders; (3) privacy/security focuses on feeling safe and trusting the site; and (4) customer service combines interest in solving problems, willingness of personnel to help and prompt answering to enquiries.

Based on the definitions in the literature (Özer et al. 2013, Herington & Weaven 2009, Bressolles et al. 2014, Ahmed 2011, Khalifa & Shen 2005), customer satisfaction is defined as the psychological reaction of the customer with respect to his or her prior experience, with a comparison of expected and perceived performance. Most marketing researchers seem to accept a theoretical framework in which quality leads to satisfaction (Özer et al. 2013), which in turn influences purchasing behavior (Kim & Jackson 2009). These arguments suggest that e-service quality is likely to affect customer satisfaction. This leads to the first research hypothesis:

H1: E-service quality is positively associated with customer satisfaction.

Furthermore, this study attempts to investigate the influence of the e-SQ to customer satisfaction.

3 VALIDITY AND RELIABILITY

Table 1 shows the results of validity and reliability testing.

4 RESULT AND ANALYSIS

The results of the SPSS, as presented in Table 2, indicated that there were four dimensions with 13 items that constitute the e-SQ for Grab services in Bandung.

The regression equation is made of Y = 1.383 + 0.206 X and it is translated so that if there is no value e-service quality, the participation value of 1.383, and the coefficient of 0.206 express a positive increase of

Table 1. Validity and reliability testing results.

Research construct and research items	Validity result	Reliability result
E-service quality		0.764
Website design		
This website has a good selection	0.349[**]	
The site doesn't waste my time	0.726[**]	
The level of personalization at this site is about right, not too much or too little	0.618[**]	
It is quick and easy to complete a transaction at this website	0.588[**]	
Reliability		
The product is delivered by the time promised by the website	0.344[**]	
The product that came was represented accurately by the website	0.417[**]	
I get what I order from this website	0.403[**]	
Security/privacy		
I feel that my privacy is protected at this site	0.449[**]	
This website has adequate security features	0.726[**]	
This website has adequate security features	0.618[**]	
Customer service		
This website offers diversiform contact channels (FAQ, email, toll-free number, etc.)	0.588[**]	
This website offers the information about customers' policies (privacy and dispute details)	0.344[**]	
I can always enquire online about the delivery of my orders	0.417[**]	
Customer satisfaction		0.508
I am satisfied with my decision to purchase from this website	0.553[**]	
If I had to purchase again, I would feel differently about buying from this website	0.842[**]	
My choice to purchase from this website was a wise one	0.719[**]	

Table 2. The results of testing dimensions.

Model	R	R Square	Adjusted R Square	Std. Error of the Estimate	R Square Change	F Change	df 1	df 2	Sig. F Change
					Change Statistics				
1	.805[a]	0.648	0.645	0.64690	0.648	180.60	1	98	0.000

a. Predictors: (Constant), E-SERVICE QUALITY

Table 3. The results of T-test.

Model	Unstandardized Coefficients B	Std. Error	Standardized Coefficients Beta	t	Sig.	Correlations Zero-order	Partial	Part
1 (Constant)	1.383	0.845		1.636	0.105			
e-service quality	0.206	0.015	0.805	13.439	0.000	0.805	0.805	0.805

a. Dependent Variable: CUSTOMER SATISFACTION

variable e-service quality to variable customer satisfaction.

From this output, a known t value of 13.439 and a significance value of 0.000 < 0.05, leads to H0 being rejected and H1 being accepted, which means there is a positive influence between variable e-service quality to the customer satisfaction variable.

5 CONCLUSIONS

This research has validated that the construct of E-SERVQUAL, with some modification, can be used to measure customer satisfaction for Grab Indonesia in Bandung. Adjustments of dimensions and items of ESERVQUAL were required to ensure validity of the instrument in the online transportation environment. In addition, dimensions such as site aesthetic and customization should be considered when determining customer satisfaction.left

REFERENCES

Ahmed, K. 2011. Online service quality and customer satisfaction: A case study of bank Islam Malaysia Berhad. *Politics* (3920): 6.

Bressolles, G., Durrieu, F. & Senecal, S. 2014. A consumer typology based on E-service quality and E-satisfaction. *Journal of Retailing and Consumer Services* 21(6): 889–896.

Herington, C. & Weaven, S. 2009. E-retailing by banks: E-service quality and its importance to customer satisfaction. *European Journal of Marketing* 43(9/10): 1220–1231.

Khalifa, M. & Shen, N. 2005. Effects of electronic customer relationship management on customer satisfaction: A temporal model Sciences. *Proceedings*: 171a–171a.

Kim, E.Y. & Jackson, V.P. 2009. The effect of E-servqual on E-loyalty for apparel online shopping. *Journal of Global Academy of Marketing Science* 19(4): 57–65.

Özer, A., Argan, M.T. & Argan, M. 2013. The effect of mobile service quality dimensions on customer satisfaction. *Procedia-Social and Behavioral Sciences* 99: 428–438.

Quan, S. 2010. Assessing the effects of e-service quality and esatisfaction on internet banking loyalty in China. *Proceedings of the International Conference on E-Business and E-Government* 93–96.

Wolfinbarger, M. & Gilly, M.C. 2002. Quality of the E-tail Experience.

Yang, Z. & Fang, X. 2004. Online service quality dimensions and their relationships with satisfaction. *International Journal of Service Industry Management* 15(3): 302–326.

Zavareh, F.B., Ariff, M.S.M., Jusoh, A., Zakuan, N., Bahari, A.Z. & Ashourian, M. 2012. E-Service quality dimensions and their effects on E-customer satisfaction in internet banking services. *Procedia-Social and Behavioral Sciences* 40: 441–445.

Advances in Business, Management and Entrepreneurship – Hurriyati et al (eds)
© 2020 Taylor & Francis Group, London, ISBN 978-0-367-27176-3

The influence of brand personality dimension on brand equity

P.D. Dirgantari, M. Permatasari, L.A. Wibowo & H. Mulyadi
Universitas Pendidikan Indonesia, Bandung, Indonesia

ABSTRACT: This study aims to investigate the influence of brand personality dimension on brand equity. The method used was explanatory survey. There were 98 respondents. The sample technique used was Tabachnick and Fidell. The results of this study indicated that companies in the cosmetic industry should increase aggressiveness in brand personality to improve brand equity.

1 INTRODUCTION

Brand is an important part of business. The company believes that the brand is one of the valuable assets that can simplify the customer decision-making process, define their expectations, and reduce risk (Nasrabadi & Zandi 2015). Companies can take advantage of certain values of brands as a method to achieve stable competitive advantage (Aghaei et al. 2014). The goal of many organizations is to build strong brands. What makes customers interested and choose a brand is known as brand equity (Far & Sadat 2015).

Brand equity was first introduced in marketing in the 1980s. It became an important concept by academics (Ahmad & Sherwani 2015). It is generally accepted as a critical success factor in differentiating the company from its competitors. For corporations, the increase of brand equity is the main goal to be achieved by acquiring more profitable associations (Farjam & Hongyi 2015). Brand equity is no longer assessed by the large amount of money the company invests (Virvilaite et al. 2015), but the result of investing in past marketing activities (Nasrabadi & Zandi 2015). Strategies to create brand equity can help companies and organizations achieve their performance goals (Majid 2016).

Brand equity research has been carried out in various industries, such as the supermarket industry (Arthur et al. 2011), luxury brands (Stefano et al. 2013), beverages (Çal & Adams 2014), and insurance (Gholami et al. 2016). These researches have produced several findings including: 1) Marketing mix can affect brand equity (Nasrabadi & Zandi 2015); 2) brand equity offers a comprehensive framework that shows how social media marketing efforts affect brand equity against five luxury brands in four countries (Godey et al. 2016); and 3) increasing industrial brand equity can be done by utilizing cash flow and improving network power, clarifying and strengthening the corporate or industrial brand identity (Beverland et al. 2007), and building brand equity through the company's brand image and company reputation (Chi-Shiun et al. 2010, Cretu &

Brodie, 2007 in Liu et al. 2016). Based on the above results, the usefulness of brand equity in the business world is considered important (Norzalita & Yasin 2010).

Indonesia provides large opportunity for cosmetics industry. It becomes a consideration for consumers to choose one brand that is considered appropriate. Each brand of cosmetic product has its own advantages. It also has distinctive features that distinguish it from similar competitors' products. Therefore, consumers are now required to be more selectively in choosing alternative brands produced by various cosmetic manufacturers (Arinita et al. 2014). Brands can create profits for businesses that match customer's demands and expectations. If there is consistency between business and consumer's expectations, then business is considered to have a good personality (Alavije 2014).

Brand personality is a vehicle of consumer self-expression. It helps consumers to express their true self, ideal self, or certain aspects of themselves (Ahluwalia 2013). Building a suitable brand personality is an important factor by which it can create a consistent image for the brand in the customer's mind. Therefore, organizations should establish the basic foundations of proper and long-term communication with customers (Ekhlassi et al. 2012).

The purpose of this research is to investigate the influence of brand personality dimension on cosmetic brand equity in Indonesia.

2 METHOD

This research was conducted to analyze the influence of brand personality dimension on brand equity of Oriflame lipstick product on Fanpage Facebook Oriflame Indonesia Community. In August 2017, this fanpage had 9,043,038 users that were used as population of this research. Regarding the sample, 98 respondents were chosen by using sampling technique from (Tabachnick & Fidell 2013).

Table 1. The effect of the brand personality dimension on brand equity.

Var	Path Coeff	Direct effect (in%)	In direct influence (through)in %					Total indirect effect	Total
			X_1	X_2	X_3	X_4	X_5		
X_1	0.190	3.61	-	4.299	2.376	3.467	2.451	12.593	16.2
X_2	0.295	8.7	4.299	-	3.287	5.330	3.281	16.197	24.9
X_3	0,,55	2.4	2.376	3.287	-	2.547	1.948	10.158	12.6
X_4	0.263	6.91	3.467	5.330	2.547	-	2.937	14.281	21.2
X_5	0.156	2.43	2.451	3.281	1.948	2.937	-	10.617	13.1
TOTAL									88.0

* Note: responsibility (X_1), activity (X_2), aggressiveness (X_3), simplicity (X_4), and emotionally (X_5).
Source: SEM PLS Processed Data, 2018.

Independent variable in this research was brand personality (X). This variable had five dimensions, which were responsibility (X_1), activity (X_2), aggressiveness (X_3), simplicity (X_4), and emotionally (X_5). Meanwhile, the dependent variable was brand equity (Y). Similar to independent variable, this variable also had five dimensions, namely: product differentiation, relevance, reward, and knowledge.

This study was conducted in less than one year i.e. in 2017. The collection of information was done only once in a period of time or one-shot or cross sectional (Maholtra 2010) with the type of explanatory survey research. Data collection techniques used were literature studies, field studies with question Meanwhile, direct and indirect influence of each dimension is presented in Table 1. naires, and literature studies. While the data analysis technique used was path analysis (path analysis).

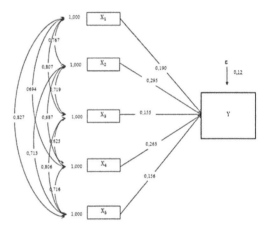

Figure 1. Chart of brand personality dimension path to brand equity (path coefficient and correlation coefficient).

3 RESULTS AND DISCUSSION

The path coefficient and the correlation coefficient of brand personality dimension to brand equity are shown in Figure 1.

Table 1 shows five result. First, the value of t_{count} responsibility (X_1) was higher than t_{table} (2.488 > 1.986). This meant that responsibility (X_1) had an effect on brand equity (Y). The direct influence of responsibility (X_1) on brand equity (ρYX_1) was 3.61%, indicating that responsibility variable (X_1) directly affected the brand equity (Y). Meanwhile, the indirect effect of activity on brand equity through responsibility (X_1) increased for about 0.689%. The effect of activity through aggressiveness (X_3) decreased 1.234%; simplicity (X_4) decreased 0.143%, and emotionality (X_5) decreased 1.159%. This result indicates that indirect influence of responsibility (X_1) through activity (X_2) on brand equity increases rather than direct influence of responsibility (X_1) on brand equity (Y). Meanwhile, the other dimensions, such as aggressiveness (X_3),

simplicity (X_4), and emotionality (X_5), reduce the influence of responsibility (X_1) on brand equity (Y). This result is in accordance with the study conducted by Wibi et al. (2016) which found the positive effect of responsibility on brand equity. In addition, Hsu (2012) argued that responsibility encourages a positive brand awareness/association on buyers, while the quality of the offered product also increases. As a result, brand loyalty to suppliers is built and brand avoidance takes place. Brand awareness and brand associations make up brand equity.

Second, the value of t_{count} activity (X_2) was higher than t_{table} (4.805 > 1.986). This meant that the activity affected the brand equity. The direct effect of activity (X_2) on brand equity (ρYX_2) was 8.7%, indicating that activity (X_2) directly affected brand equity for about 8.7%. The indirect effect of activity on brand equity through responsibility (X_1) decreased 4.401%; through aggressiveness (X_3) decreased 5.413%, through simplicity (X_4) decreased 3.37%, and through emotionality (X_5) 5.419%. This means that the direct impact of activity on brand equity is greater than the indirect effect of activity through other dimensions on brand equity. This is in accordance with previous researches (Wang et al. 2016) which found the influence of activity on brand equity. Being an active brand leads to higher brand awareness, and telling to the consumers about the active brand of company is good for company's business progress (Molinillo et al. 2017). In addition, personality activity can be a key variable that influences the formation of brand identity and loyalty (Alexandris et al. 2017).

Third, the value of that t_{count} aggressiveness (X_3) was higher than t_{table} (2.247 > 1.986). This meant that aggressiveness (X_3) had an effect on on brand equity (Y). The direct effect of aggressiveness (X_3) on brand equity (ρYX_3) was 2.4%, meaning that aggressiveness (X_3) affected brand equity directly by 2.4%. The indirect effect of aggressiveness (X_3) on brand equity through activity (X_2) and simplicity (X_4) increased 0.887% and 0.147%. Meanwhile, the indirect influence of aggressiveness (X_3) on brand equity through responsibility (X_1) and emotionality (X_5) decreased 0.024% and 0.452%. This implies that indirect influence of aggressiveness through activity and simplicity increases its influence on brand equity. In contrast, aggressiveness through responsibility and emotionality reduces its impact on brand equity. People of agreeableness are looking for brands that show aggressiveness. They love brands with aggressive personalities. This result is supported by Aaker (1997) and Ekhlassi et al. (2012) who say that customers use brands to express themselves. Thus, if a company is able to build a brand well, then it can face any competitor aggression and retains its customers.

Forth, the value of t_{count} simplicity (X_4) was higher than t_{table} (4.703 > 1.986). This meant that simplicity (X_4) had an effect on brand equity (Y). The direct effect of simplicity (X_4) on brand equity (ρYX_4) was 6.91%. This implied that simplicity (X_4) affected direct brand equity by 6.91%. The indirect effect of simplicity on brand equity through responsibility (X_1) was 3.467%; activity (X_2) was 5.330%; aggressiveness (X_3) was 2.547%; and emotionality (X_5) was 2.937%. This result indicates that direct influence of simplicity to brand equity is higher than the indirect influence of simplicity through other dimensions. This is in accordance with the research conducted by Wang et al. (2016) that simplicity has an influence on brand equity. Simplicity based on the need to provide the right amount of brand information (not exaggerated or reduced) to consumers may be able to contribute to brand equity (Keller 2013).

Last, the value of t_{count} emotionality (X_5) was higher than t_{table} (2.091 > 1.986). This emotionality (X_5) had an effect on brand equity (Y). The direct effect of emotionality (X_5) on brand equity (ρYX_5) was 2.43%. This meant that emotionality (X_5)

directly affected the brand equity of 2.43%. The indirect effect of emotionality (X_5) on brand equity through responsibility (X_1) was 2.951%; activity (X_2) was 3.281%; aggressiveness (X_3) was 1.948%; and simplicity (X_4) was 2.937%. This result implies that indirect influence of emotionality through responsibility, activity, and simplicity increases the effect of emotionality on brand equity. On the other hand, indirect influence of emotionality through aggressiveness reduces the effect of emotionality on brand equity. This is in accordance with Veloutsou et al. (2013) who convey that a brand tends to have strong brand equity if it is well known to consumers who evaluate it positively and emotionally associated with it. In addition, Malär et al. (2011) argue that brands are said to be emotional if they do not have stable emotions through the formation of a particular brand personality. The dimension of emotionality refers to characteristics such as romantic and sentimental. This dimension causes customers to form strong emotional connections with brands.

4 CONCLUSION

The overall dimensions of Oriflame brand personality have significant impacts on brand equity. Nevertheless, there are some things that need to be improved. The dimension of aggressiveness has the lowest value impact on brand equity. To solve it, company needs to take a more aggressive stance in a way that is 'introductions of new products', advertising, and promotions in order to reduce risk when there is potentially damaging competitive action. If this is done carefully by the company, it can increase its brand equity.

Brand equity can be improved on the knowledge aspect because it has the lowest influence. The knowledge given by the company is a good way to represent how brand knowledge exists in the consumer's memory. Company should pay attention to knowledge in general and brands in particular, especially related to brand awareness and brand image through advertising and promotion.

Weaknesses in this study include research conducted at one time (cross sectional) so that there is the possibility of individual behavior that quickly changes over time. In addition, the variables used in this study were not represented all the factors that affect brand equity. For further research, it is suggested to update the theory of brand personality and brand equity as well as to conduct research on more specific objects. Further research also can investigate the other factors of that can affect brand equity, such as: marketing mix, gender brand, advertising, sales promotion, product innovation, word of mouth, marketing communication quality, community commitment, and brand experience.

REFERENCES

Aaker, J. L. (1997). Dimensions of brand personality. *Journal of marketing research*, 347-356.

Aghaei, M., Vahedi, E., Safari, M., & Pirooz, M. 2014. An examination of the relationship between Services Marketing Mix and Brand Equity Dimensions. *Procedia - Social and Behavioral Sciences*, 109, 865–869.

Ahluwalia, R. 2013. When Brand Personality Matters: The Moderating Role of Attachment Styles. *Journal of Consumer Research*, 35(6), 985–1002.

Ahmad, F., & Sherwani, N. U. K. 2015. An Empirical Study on the effect of Brand Equity of Mobile Phones on Customer Satisfaction. *Nternational Journal of Marketing Studies*, 7(2), 59–69.

Alavije, K. 2014. The Effect of Perceived Business Ethic on Brand Personality Dimensions & Creation of Brand Equity in Developing Countries. *Central European Business Review*, 3(3).

Alexandris, K., Du, J., Funk, D., Alexandris, K., Du, J., & Funk, D. 2017. Managing Sport and Leisure The influence of sport activity personality on the stage-based development of attitude formation among recreational mountain skiers The influence of sport activity personality on the stage-based development of attitude formation, 472(March).

Arinita Febrianti, W., Bambang, I., & N. Ari, S. 2014. Pengaruh Citra Merek, Kualitas Produk, Dan Harga Terhadap Perpindahan Merek. *Artikel Ilmiah Mahasiswa.*

Arthur W, A., Patricia, H., Judith, W., & Alexander E E. 2011. Customer-based brand equity, equity drivers, and customer loyalty in the supermarket industry. *Journal of Product & Brand Management*, 20(3), 190–204.

Beverland, M., Napoli, J. and Yakimova, R. 2007, *"Branding the business marketing offer: Exploring brand attributes in business markets", Journal of Business and Industrial Marketing*, Vol. 22 No. 6, pp. 394-399.

Çal, B., & Adams, R. 2014. The Effect of Hedonistic and Utilitarian Consumer Behavior on Brand Equity : Turkey - UK Comparison on Coca Cola. *Procedia - Social and Behavioral Sciences*, 150, 475–484.

Chi-Shiun, L., Chih-Jen, C., Chin-Fang, Y. and Da-Chang, P. 2010, "The effects of corporate social responsibility on brand performance: The mediating effect of industrial brand equity and corporate reputation", *Journal of Business Ethics Vol. 95 No. 3, pp.* 457–469.

Cretu, A. E., and Brodie, R. J. 2007, "The influence of brand image and company reputation where manufacturers market to small firms: A customer value perspective", *Industrial Marketing Management, Vol. 36 No. 2, pp.* 230-240.

Dagustani, D. 2014. The Brand Building: Developing Brand Asset Valuator and Brand Association. *International Journal of Business, Economics, and Law*, 5(2), 58–69.

Ekhlassi, A., Nezhad, M. H., Far, S. A., & Rahmani, K. 2012. The relationship between brand personality and customer personality, gender and income: A case study of the cell phone market in Iran, 20(3–4), 158–171.

Far, E., & Sadat, L. 2015. Investigating the affective factors on building brand equity in banking Industry of Iran (Case study: Tosee Saderat Bank), 4(8), 1–5.

Farjam, S., & Hongyi, X. 2015. Reviewing the Concept of Brand Equity and Evaluating Consumer- Based Brand Equity (CBBE) Models. *International Journal of Management Science and Business Administration*, 1(8), 14–29.

Gholami,S., Roushanghias, E., & Karimiankakolaki, M. 2016. Examination of Factors Influencing On Enhancement of Brand Equity with Emphasis on Advertising and Sales Promotion. *Journal of Current Research In Science*, 320–326.

Godey, B., Manthiou, A., Pederzoli, D., Rokka, J., Aiello, G., Donvito, R., & Singh, R. 2016. Social media-marketing efforts of luxury brands: Influence on brand equity and consumer behavior. *Journal of Business Research*, 69(12), 5833–5841. http://doi.org/10.1016/j.jbusres.2016.04.181

Hsu, K. 2012. T*he Advertising Effects of Corporate Social Responsibility on Corporate Reputation and Brand Equity: Evidence from the Life Insurance Industry in Taiwan.* Journal of Business Ethics, 189–201.

Keller, K. L. 2013. *Strategic Brand Management* (Fourth). Pearson Education, Inc.

Liu, M., Loh Hui, Y., TC, M., Nguyen, B., Syed, A., & Faridah, S. 2016. Explicating industrial brand equity Integrating brand trust, brand performance and industrial brand image. *Industrial Management & Data Systems*, 116(5), 858–882.

Maholtra, K. N. 2010. *Marketing Reseach: An Applied Orientation Sixth Ed Pearson Education* (Sixth edit). Pearson Education.

Majid, E. 2016. Effect of Dimensions of Service Quality on Brand Equity in The Fast Food Industry, 11(11), 30–46.

Malär, L., Krohmer, H., Hoyer, W. D., & Nyffenegger, B. 2011. Emotional Brand Attachment and Brand Personality: The Relative Importance of the Actual and the Ideal Self. *Journal of Marketing*, 75, 35–52.

Molinillo, S., Japutra, A., Nguyen, B., Hao, C., & Chen, S. 2017. Responsible brands vs active brands? An examination of brand personality on brand awareness, brand trust, and brand loyalty. *Marketing Intelligence & Planning*, 35(2), 166–179.

Nasrabadi, E. M., & Zandi, P. 2015. Investigating the Impact of Marketing Mix on Brand Equity (The Case Study: Consumers of Samsung's Appliances in Tehran-Iran), 2015(2), 20–26.

Norzalita, A. A., & Yasin, N. M. 2010. Analyzing the Brand Equity and Resonance of Banking Services: Malaysian Consumer Perspective. *International Journal of Marketing Studies*, 2(2), 180–189.

Stefano, B., Armando, C., Domenico, C., Guendalina, C., Roberta, C., & Francesca, D. P. 2013. The Effects of Online Brand Communities on Brand Equity in the Luxury Fashion Industry Regular Paper. *International Journal of Engineering Business Management*, 1–9.

Tabachnick, & Fidell. 2013. *Using Multivariate Statisrics, Sixth* Edition. Boston: Pearson Education, Inc.

Veloutsou, C., Christodoulides, G., & Chernatory de, L. 2013. A taxonomy of measures for consumer-based brand equity: drawing on the views of managers in Europe. *Journal of Product & Brand Management*, 22(3), 238–248.

Virvilaite, R., Tumasonyte, D., & Sliburyte, L. 2015. The Influence of Word of Mouth Communication on Brand Equity: Receiver Perspectives. *Procedia - Social and Behavioral Sciences*, 213, 641–646.

Wang, J. J., Zhang, J. J., Byon, K. K., Baker, T. A., & Lu, Z. L. 2016. Promoting Brand-Event Personality Fit as a Communication Strategy to Build Sponsors ' Brand Equity. *International Journal of Sport Communication*, 294–320.

Wibi, G., Pratama, S., & Nurcaya, I. N. 2016. Pengaruh Corporate Social Responsibility Terhadap Ekuitas Merek yang Dimediasi oleh Citra Perusahaan, 5(7), 4253–4280.

Consumer's purchase intention on halal detergent in Jakarta

E. Saribanon, R. Hurriyati, A. Rahayu & M.A. Sultan
Universitas Pendidikan Indonesia, Bandung, Indonesia

ABSTRACT: The aim of this study is to analyze the purchase intention of Indonesian female consumers on the detergent with halal label and the customers' most dominant factors influencing the purchase intention. This study is based on the Theory of Planned Behavior model. It applies associative and comparative quantitative approach. The data is obtained through questionnaires with 5-point Likert scale, which are distributed to 50 respondents using purposive sampling in several areas in Jakarta. Those respondents are women who are exposed on halal detergent. This study uses Structural Equation Modeling (SEM), which is based on the variant with statistical technique of Partial Least Square (PLS) and uses SmartPLS software. This study finds that there are three factors of intention, namely attitude, subjective norm, and perceived behavioral control have positive influence on consumer's intention to buy halal detergent.

1 INTRODUCTION

Concerning the research model used by (Ajzen 1991, Alam & Sayuti 2011, Aziz & Wahab 2013, Hashim & Musa 2014, Endah 2014) this study applies the same model with the a difference on the object of research. The researches carried out by Ajzen and others focus on the consumer behaviour in general. Meanwhile, the researches carried out by (Lada et al. 2009, Aziz & Wahab 2013, Hashim & Musa 2014, Endah 2014) focus on the behaviour of halal product users. This study, similar to what has been done by Endah (2014), concerns on the consumer behaviour of halal cosmetic users. Different from (Endah 2014), this study makes analysis focusing on consumer's purchase intention on the detergent with halal label, which is included in the product category of cleaner. Some literatures focus on halal food and halal cosmetics but very few studies concern on the product of halal cleaner.

Detergent is a (or a combination of some) compound(s), which facilitates the process of cleaning. Detergent is a formulation product combining some chemicals, aiming at improving the cleaning ability. The important component of detergent is surfactant. In relation to the aforementioned facts, the connotation of halal can also be embedded on several kinds of consumables, such as cosmetics, soap, and detergent. Therefore, there have been cosmetic, soap and detergent with halal certification, one of which is Total Almeera detergent which guarantees its product to have obtained halal certificate from the Indonesian Council of Ulama or MUI.

Consuming halal goods is specifically expressed by religion as the product orientation starting from the raw materials, process, until the distribution. The aspect of consuming halal product becomes a priority in line with the importance of providing quality product (Zakaria 2008). This is stated in Islamic teachings as mentioned in the Holy Qur'an surah Al-Baqarah: 168, surah Al-Maidah: 88 and surah An-Nahl: 114 which order Muslims to consume halal and quality goods and services. The importance of halal for a Muslim finally is not only individual do-main, but it becomes the obligation of ulama (Islam-ic scholars) and umara (government) to prepare the guidance, regulation, and various policies that regu-late and bind people to obey the system.

Considerable demand for halal products will increase the supply for halal certificate and logo (Noordin et al. 2009). Having halal logo is assumed that the products are allowed by religion to be consumed. The need for establishing a standard of global halal is a requisite for halal products all over the world, because today the most urgent problem in Halal industry is that there has been no consensus on the standard of halal (Afifi et al. 2017).

The consumer expectation for halal products is to get quality, safe, healthy products and those which do not hurt animals in the production process (Rezai et al. 2012). Although the prices of halal products are relatively higher, consumers are satisfied with the quality and after sales service of halal product will keep buying (Khraim 2011). However, it is necessary to realize that not all producers of daily products are concerned with the halal guarantee for their products.

Marketers need appropriate strategies to make their product become a brand which is acceptable for consumers. In this case, in order that the chosen strategy can give effective results, marketers need to know how purchasing behavior is built in buying detergent with halal label and to know the factors driving consumers to buy the detergent with halal label. Halal has become the main attention in the community of Noreen Noe consumers.

The Theory of Planned Behaviour (Ajzen 1991) is a common model used to see the factors which influence consumer's purchase decision and behaviour toward a certain product. In accordance with this theory, an act is initiated by an intention where it is influenced by three internal factors: attitude, subjective norms, and perceived behavioral control.

Attempts to use of TPB approach in a research have been much done by, among others, (Ajzen 2011, Dharmmesta 1998, Alam & Sayuti 2011, Aziz & Wahab 2013, Aziz & Chok 2013, Ramdhani 2011, Kim & Chung 2011, Hashim & Musa 2014, Endah 2014). Nevertheless, studies using the same approach are still relevant to be carried out in order to strengthen TPB approach so that it becomes an established and proven approach for revealing consumer's intention and purchase behavior, especially if it uses the research subject and recency scope which is different from previous studies.

The importance of attitude in decision making is the illustration of an individual act based on belief. Ajzen (1991) states that attitude influential in choosing a product is the stage of positive or negative feeling and certain intention to make an act. It is also underlined by Ajzen & Fishbein (2005) that the attitude toward this behavior is determined by the belief in the consequence of a behavior or what so called behavioral beliefs.

Subjective norm is the individual perception on the people who are important for him toward an object (Ajzen & Fishbein 2005). If the attitude is driven by the result of self-evaluation, it is different from subjective norm which comes from external in-fluence (normative belief). Ajzen (2006) further adds, the social reference in behaviors, which is considered important, also includes the social reference coming from parents, couples, friends, peers, and other references related to behavior.

Previous researches concerning halal products using TPB model show that attitude and subjective norm are significant factors influencing someone's intention to consume halal product (Soesilowati 2010, Tarkiainen & Sundqvist 2005, Bonne et al. 2007, Suparno 2017).

Ajzen (2006) state that intention and perceived behavioral control are influential to the behavior performed by an individual, however those two things in general do not have significant relationship. This is because every individual has full control for the behavior he will perform (Abrams et al. 1999). Tanti et al. (2008), add that perceived behavioral control has an important

meaning when individual's self-confidence is still in the low condition.

It is underlined by Ajzen & Fishbein (2005) that intention is the cognitive and conative representation of individual's readiness to perform a behavior. Intention is the determiner and disposition of behavior, so that an individual has right chance and time to perform the behavior in real. Meanwhile, Dharmmesta (1998) asserts that intention is the mediator of motivational factors which has an impact on behavior.

This study aims to analyze Jakarta female consumer's intention to purchase the detergent with halal label, including the dominant factors influencing that behavior.

2 METHOD

The endogenous variable in this study is the consumer's purchase intention. Whereas the exogenous variables are attitude, subjective norm and perceived behavioral control. This study is a quantitative approach which applies associative and comparative techniques. The research location is East Jakarta area. The research object is the Female Lecturers of Trisakti Institute of Transportation and Logistics who are assumed to have knowledge about Halal Detergent.

The population in this study is the Female Lecturers of Trisakti Institute of Transportation and Logistics. The size of sample in this study is 50 respondents. The sampling technique used here is non-probability sampling with the purposive-sampling method (Saribanon et al. 2016). Data is collected through questionnaires distributed to respondents. Data is measured using 5-point Likert Scale with the range 1 up to 5 from very disagree up to very agree.

Inferential statistical analysis is used to examine the relationship among variables. It is used to analyze the sample data and its results can be explained and applied to the population (Sugiyono 2014). For overall associative hypothesis and model testing, statistical analysis of Partial Least Square (PLS) is used.

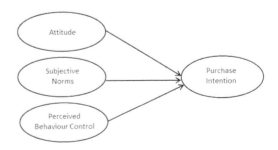

Figure 1. Research framework.

3 RESULTS AND DISCUSSION

The results of the reliability test resulted in all latent variables producing AVE values above 0.500, namely for attitude variables (0.706). Subjective norm (0.8225), Perception of Behavioral Control (0.712) and Consumer Purchase Intention (0.751) means that all convergent validity values are met because of the Composite Reliability and Cronbach Alpha values both exceed the specified minimum standard of 0.700. The results of the correlation between constructs and AVE root values of all variables indicate that the root value of AVE is for latent variables Attitude (0.881), Subjective Norm (0.902), Perception of Behavioral Control (0.844) and Consumer Purchase Intention (0.867) greater than the correlation value between constructs. Thus this result shows that discriminant validity of all latent variables is good and meets the requirements when viewed based on Fornell-Larcker criterion.

The resulted value of R^2 is 0.609 or 60.9%, indicating that simultaneously the three independent latent variables (Attitude, Subjective Norm, and Perceived Behavioral Control) give fair contribution (fairly strong) to Consumer's Purchase Intention as many as 60.9% whereas the rest of 39.1% is influenced by other variables which are not studied.

H_1: Simultaneously Attitude, Subjective Norm, and Perceived Behavioral Control significantly influence Consumer's Purchase Intention.

To test the above hypothesis, F test is used with the following formula:

$$F_{Hitung} = \frac{R^2/(k-1)}{(1-R^2)/(n-k)}$$

$$= \frac{0.609/(3-1)}{(1-0.609)/(50-3)} = 36.602 \qquad (1)$$

with $\alpha = 0.05$; $k = 3$ and $n = 50$, it is obtained $df_1 = k = 3$ and $df_2 = n - k - 1 = 50 - 3 - 1 = 46$, so that the value of F table is found 2.807. From this value, it is seen that F statistics (36.602) > F table (2.807) thus we reject H_0 and accept H_1, meaning that simultaneously Attitude, Subjective Norm and Perceived Behavioral Control significantly influence Consumer's Purchase Intention.

Table 1. Result of R square.

	R Square
Consumer's Purchase Intention	0.609

H_2: In partial, Attitude significantly influences Consumer's Purchase Intention.

From Figure 1 it is found the value of t statistics $\gamma_1 = 3.630$. This value is bigger than t table 1.960, so the conclusion is to reject H_0 and accept H_2, meaning that Attitude in partial is proven to significantly influence Consumer's Purchase Intention. The magnitude of partial influence of Attitude on Consumer's Purchase Intention 41.9%. Thus, it can be concluded that Attitude gives significant partial influence to Consumer's Purchase Intention with the contribution as many as 41.9%.

H_3: In partial, Subjective Norm significantly influences Consumer's Purchase Intention.

From Figure 1 it is found the value of t statistics $\gamma_2 = 2.167$. This value is bigger than t table 1.960, so the conclusion is to reject H_0 and accept H_3, meaning that Subjective Norm in partial is proven to significantly influence Consumer's Purchase Intention. The magnitude of partial influence of Subjective Norm on Consumer's Purchase Intention 29.2%. Thus, it can be concluded that Subjective Norm gives significant partial influence to Consumer's Purchase Intention with the contribution as many as 29.2%.

H_4: In partial, Perceived Behavioral Control significantly influences Consumer's Purchase Intention.

From Figure 1 it is found the value of t statistics $\gamma_3 = 1.484$. This value is smaller than t table 1.960 so the conclusion is to accept H_0 and reject H_1, meaning that Perceived Behavioral Control in partial is proven not significantly influence Consumer's Purchase Intention. The magnitude of partial influence of Perceived Behavioral Control on Consumer's Purchase Intention -10.2%. Thus, it can be conclud-ed that Perceived Behavioral Control in partial does not give significant influence to Consumer's Pur-chase.

A number of studies support the positive relationship between consumer's attitude and behavioral intention as performed by Chan & Lau (2001) (Tarkiainen & Sundqvist 2005, Bonne et al. 2007, Kim & Chung 2011).

The dominant construct affecting consumer buy-ing intention is attitude. It is a tendency that is learned to respond to objects or classes of objects consistently both in liking and disliking. In terms of of halal products, someone considers it useful both for him and he will give a positive response. One of the benefits that individuals perceive from using it is the feeling of comfort, healthy products and calm in using halal products in accordance with religious.

Meanwhile, this study indicates the same findings as (Ajzen 2006), although with different objects, where for halal products Perception of Behavior

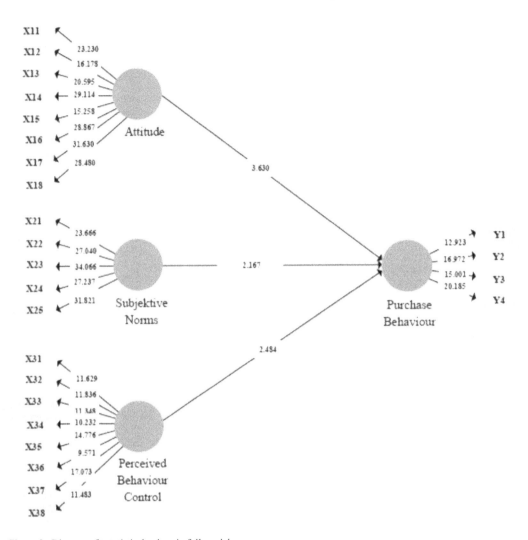

Figure 2. Diagram of t statistical values in full model.

Control does not have a significant partial effect on Consumer Purchase Intention, meaning that either there is or not a driver, the consumer's buying interest will be the same.

4 CONCLUSIONS AND SUGGESTIONS

Based on the analysis and discussion, it can be concluded that from the calculation, Attitude, Subjective Norm, and Perceived Behavioral Control are simultaneously proven to significantly influence Consumer's Purchase Intention with the total contribution as many as 60.9%, whereas the rest of 39.1% is the contribution of other factors which are not studied. The result shows that attitude is the strongest contributing aspect as (Lada et al. 2009, Endah 2014) research.

Concerning Attitude, there is a partial significant influence on Consumer's Purchase Intention with the contribution as many as 41.9% which, in the context of this study, is the biggest influence, where Subjective Norm shows a significant partial influence on Consumer's Purchase Intention with the contribution as many as 29.2% and Perceived Behavioral Control does not give significant partial influence to Consumer's Purchase Intention, where the contribution given is negative as many as -10.2%.

From the above description, the producers of halal detergent can make a more effective marketing strategy by creating a positive perception related to the consumer's belief that can influence their atti-tude. From this side, quality, security, health and after sales service are the benchmarks for consumers to evaluate the product. Therefore, halal detergent will tend to become the main choice of consumers in Jakarta.

The limitations of the study are usually halal-labeled products make prices far more expensive, in that we need to re-examine whether prices can affect the intention to purchase a product. With the limitations of this research, it is expected that it can encourage further research related to consumer buying intentions for halal products.

REFERENCES

Abrams, D., Hinkle, S. & Tomlins, M. 1999. Leaving Hong Kong?: The roles of attitude, subjective norm, perceived control, social identity and relative deprivation. *International Journal of Intercultural Relations* 23(2): 319–338.

Afifi, M., Halim, A., Mahyeddin, M., Salleh, M., Syariah, F., Sains, U. & Sembilan, N. 2017. The possibility of uniformity on halal standards in organization of Islamic Countries (OIC) country. *World Applied Sciences Journal* 17: 6–10.

Ajzen, I. & Fishbein, M. 2005. The influence of attitudes on behavior. *Journal of Social Psychology*, 44(1): 115–127.

Ajzen, I. 1991. The theory of planned behavior. *Organizational Behavior and Human Decision Processes* 50: 179–211.

Ajzen, I. 2006. *Attitude personality and behavior (second)*. New York: Open University Press.

Ajzen, I. 2011. The theory of planned behaviour: Reactions and reflections. *Psychology and Health* 26(9): 1113–1127.

Alam, S.S. & Sayuti, M.N. 2011. Applying the theory of planned behavior (TPB) in halal food purchasing. *International Journal of Commerce and Management* 21(1): 8–20.

Aziz, N.N.A. & Wahab, E. 2013. Understanding of halal cosmetics products: TPB Model. 1–6.

Aziz, Y.A. & Chok, N.V. 2013. The role of halal awareness, halal certification, and marketing components in determining halal purchase intention among non-Muslims in Malaysia: A structural equation modeling approach. *Journal of International Food and Agribusiness Marketing* 25(1): 1–23.

Bonne, K., Vermeir, I., Bergeaud-Blackler, F. & Verbeke, W. 2007. Determinants of halal meat consumption in France. *British Food Journal* 109(5): 367–386.

Dharmmesta, B.S. 1998. Theory of planned behaviour dalam penelitian sikap, niat dan perilaku konsumen. *Kelola* 7(1998).

Endah, N.H. (2014). Perilaku pembelian kosmetik berlabel halal oleh konsumen Indonesia. *Jurnal Ekonomi dan Pembangunan* 22(1): 11–25.

Hashim, A J..btC.M. & Musa, R. 2014. Factors influencing attitude towards halal cosmetic among young adult urban Muslim women: A focus group analysis. *Procedia-Social and Behavioral Sciences* 130: 129–134.

Khraim, H.S. 2011. The influence of brand loyalty on cosmetics buying behavior of UAE female consumers. *International Journal of Marketing Studies*, 3(2): 123–133.

Kim, H.Y. & Chung, J. 2011. Consumer purchase intention for organic personal care products. *Journal of Consumer Marketing* 28(1): 40–47.

Lada, S., Tanakinjal, G.H., Amin, H., Lada, S., Tanakinjal, G. H. & Amin, H. 2009. Theory of reasoned action predicting intention to choose halal products using theory of reasoned action. *International Journal of Islamic and Middle Eastern Finance and Management* 2(1): 66-76.

Noordin, N., Noor, N.L.M., Hashim, M. & Samicho, Z. 2009. Value chain of halal certification system: A case of the Malaysia halal industry. *European and Mediterranean Conference on Information Systems* (EMCIS) 2009 (2008): 1–14.

Ramdhani, N. 2011. Penyusunan alat pengukur berbasis theory of planned behavior. *Buletin Psikologi* 19(2): 55–69.

Rezai, G., Mohamed, Z. & Shamsudin, M.N. 2012. Non-Muslim consumers' understanding of halal principles in Malaysia. *Journal of Islamic Marketing* 3(1): 35-46.

Saribanon, E., Sitanggang, R. & Amrizal. 2016. Kepuasan pengguna jasa transportasi untuk meningkatkan loyalitas. *Jurnal Manajemen Transportasi & Logistik* 03 (03): 317–326.

Soesilowati, E.S. 2010. Business opportunities for halal products in the global market: Muslim consumer behaviour. *Journal of Indonesian Social Sciences and Humanities* 3(May 2007): 151–160.

Sugiyono, S. 2014. *Metode penelitian bisnis*. Bandung: Alfabet.

Suparno, C. 2017. Atenden niat untuk memilih jasa halal beauty center: Aplikasi theory of planned behavior dan peran religiusitas. *Media Ekonomi dan Manajemen* 32 (1): 85–100.

Tanti, C., Stukas, A.A., Halloran, M.J. & Foddy, M. 2008. Tripartite self-concept change: Shifts in the individual, relational, and collective self in adolescence. *Self and Identity* 7(4): 360–379.

Tarkiainen, A. & Sundqvist, S. 2005. Consumers in buying organic food subjective norms, attitudes and intentions of finish consumers in buying organic food. *British Food Journal* 107(11): 808-822.

Zakaria, Z. 2008. Tapping into the world halal market: Some discussions on Malaysian laws and standards. *Jurnal Syariah* 16(Keluaran Khas): 603-616.

Advances in Business, Management and Entrepreneurship – Hurriyati et al (eds)
© 2020 Taylor & Francis Group, London, ISBN 978-0-367-27176-3

The usage of digital marketing channels in micro, small and medium enterprises in Bandung District, Indonesia

M.E. Saputri
Telkom University, Bandung, Indonesia

N. Kurniasih
Institute Technology of Bandung, Bandung, Indonesia

ABSTRACT: Today, businesses are beginning to shift due to digital technology. Every single process of business is related to social media, internet and application based. That's why Micro Small Medium Enterprise should begin to adopt digital technology because the market is already shifted towards it. The problems in this research are the switch of marketing trends from conventional (offline) to digital (online); the disproportionate number of business players in Micro Small Medium Enterprise who utilize digital marketing compared to the number of Internet users growth and their inability to understand how to take part in digital marketing using social networking. This research combined quantitative and qualitative approaches through several phases in its process. The results of this study showed that the use of IT among Micro Small Medium Enterprise in Bandung District has been relatively common in terms of operating computer and accessing the internet, however, the use of computers and the Internet to support the management is relatively low.

1 INTRODUCTION

Owing to a wealth of new digital technologies, marketers now can interact and reach customers anywhere and any time. Technology changes the way we do marketing since today most of those activity are related to the internet. Based on data from a survey by the Indonesian Internet Service Providers Association, or APJII, Internet users in Indonesia reached 143.26 million in 2017, a 7.9 percent increase from 2016 (APJII 2017). Many of these users are on the Internet for so activities, including making payments, information exchange, communication media, social relation, media for business and so on. Digital marketing is basically promotion of brands using all available forms of digital advertising media to reach the target segment. There are so many types of media that can be used for digital marketing, such as radio, mobile, the Internet, television, social media marketing and other less popular forms of digital media like digital signage, digital bill boards, and so on. Digital marketing is a new approach to marketing, not just traditional marketing boosted by digital elements (Liu et al. 2011, Kotler et al. 2011).

Micro Small Medium Enterprise (MSME) is one of business entities that should utilize digital marketing since every single process of business is related to social media, internet and application. That is why today MSMEs should begin to adopt digital technology because the market is already shifted towards there.

According to a survey, about 44.16 percent of the respondents accessed the Internet with their mobile phones, 39.28 percent used computers and mobile phones, while 4.49 percent used computers only. The study revealed that 89.35 percent of them used the Internet for chatting, 87.13 percent to access social media, 74.84 percent to use search engines, 72.79 percent to view images, and 70.23 percent to download videos. Several features are usually accessed simultaneously (Yuniarni 2018). Indonesia is ranked as the fifth highest internet penetration in the world (Internet World Stats 2017). More than 40 percent of Internet users in Indonesia are social media users (Instagram, Facebook and Twitter). This number can be further seen as containing potential customers of MSMEs. In fact, the marketing reality of MSMEs is far from that of large corporations and, hence, digitization is a greater challenge for them (Taiminen & Karjaluoto 2015).

Social media plays a vital role in marketing and creating relationships with customers. With limited barrier to entry, micro small businesses are beginning to use social media as a means of marketing. Unfortunately, many small businesses struggle to use social media and have no strategy going into it. As a result, without a basic understanding of the advantages of social media and how to use it to engage customers, countless opportunities are missed (Adegbuyi et al. 2015).

SMEs seem not to be keeping pace with digital developments, mostly due to the lack of knowledge of digital marketing. Most of the studied SMEs do not apply the full potential of the new digital tools and hence are not benefitting fully from them.

There is no standard definition of marketing in MSME, especially digital marketing. Optimization of using smartphone to do business is a big achievement. Most of their marketing is spontaneous, informal and disorganized.

Online marketing is the fastest growing form of marketing. Widespread use of the Internet and other powerful new technologies are having a dramatic impact on both buyers and the marketers who serve them. The major online marketing domain include Business to Business (B2B), Business to Consumer (B2C), Consumer to Consumer (C2C) and Consumer to Business (C2B) (Kotler et al. 2011). MSMEs can be a player in those four types of domain. So it would be very potential to MSMEs to use digital marketing in their business process.

Social media is one of the subunits of Internet marketing which is part of digital marketing. Social media is one of internet application that has played an important role in modern marketing and retailing. Most of business activities use social media to promote and also to engage the customer. The Internet has been highly influential to combine "IT" and "the business" (Nuseir 2016).

Many companies view the use of social media as a profitable marketing tool from which they can derive several benefits such as to attract new customers and to reach existing customers more efficiently, to increase performance and efficiency and to improve growth and competitiveness (Taiminen & Karjaluoto 2015, Lindqvist 2017, Adegbuyi et al. 2015, Nuseir 2016). In addition, the internet marketing can benefit SMEs by reducing costs (Srinivasan et al. 2016).

2 METHOD

This research combined quantitative and qualitative approaches through several phases of the research process. Data were gathered from 30 in-depth semi-structured interviews conducted with micro small business owners located in Bandung District.

A case study approach was applied in this study, with the field interviews of micro small business owners. The Bandung District micro small business enterprises were presented as a sample in the context of developing countries. Interview was chosen as a data collection technique in this study as it gave opportunities to participants to deliver their views and thoughts in greater depths as compared to other data collection techniques such as questionnaires. The objective of the interview was to gather knowledge of the companies' digital marketing activities, their usage of digital channels and the difficulties experienced in using them. The survey charted digital marketing usage from a wider perspective and contextualized the results from the interviews.

The participants had to fulfill a few criteria to be included in interviews among others were being an owner with minimum employees of four; having an independent business which did not belong to any companies or subsidiaries and having already attended digital marketing workshop by Telkom University (Community Service Program by Telkom University in Cooperation with Koperasi Pengembangan Ummat Darul Tauhid-KOPMU-DT).

3 RESULT AND DISCUSSION

Respondents in this research were mostly woman, more than 20 years old, married and categorized as necessity entrepreneur. Necessity entrepreneur is entrepreneurs who are forced into starting a business out of necessity because of the lack of other options in the labor market. Most of them were uneducated, some finished elementary school. They cannot find a job or another way to make income, so the only option is to become an entrepreneur.

According to the survey findings, almost all (76 percent) of the respondents, or 23 MSMEs, said that they did not utilize digital marketing because they could not access the Internet nor did they have any intention to learn about it. Seventeen percent of the respondents (5 MSMEs) said they utilized the digital marketing poorly or extremely poorly, and only 3 percent (2 MSMEs) described their current digital marketing as very good or excellent. Most of the MSMEs still running the business in traditional way did not optimize social media such as Facebook and Instagram to sell and promote their products and services. Instead of using social media for their personal life, most of them used their personal account to run their business.

MSMEs who reported investing more in digital marketing perceived that they utilized it better. This indicated that the MSMEs having adopted digital marketing and made the necessary investments was able to take advantage of these tools.

Cader and Al Tenaiji (2013) and Goi (2009) stated that by understanding the Internet, particularly social media (Facebook, Twitter, Instagram), as marketing tools, MSME could reach out and connect with the customers. Nuseir (2016) concluded that people were influenced by marketing strategies using the Internet and digital media, especially most of people in this era who use internet and social media for many purpose. According to the survey, digital marketing channels were used most often by business owners around 20 to 35 years old. They knew how to operate the social media and sometimes used Facebook Ads or Instagram Ads to promote their business and joined e-commerce (Bukalapak, Tokopedia). However, none of these channels was utilized very actively. On the other hand, the other age groups used word of mouth to spread information about their product and services. Small businesses often relied on word of mouth recommendations to get new customers.

Word-of-mouth marketing provided small businesses with an opportunity to give customers a reason to talk about products, making it easier for word of mouth to take place (Walsh & Lipinski 2009).

Among the MSMEs whose representatives were interviewed, none of them had set a clear goal to attract new customers or maintain the existing ones through social media. Facebook page was used to facilitate communication and once or twice to facilitate their product. Although almost half of the interviewed companies used social media, none of them used social media for the purposes of dialogical communication; instead they used the channel primarily to post personal news. Social media was still seen as a largely informal and relaxed communication channel.

All the MSMEs were also well aware that it was possible to measure the influence of marketing practices through digital channels; however, only a few respondents were able to clearly state what should be measured and primarily, what marketing goals they should set. It seemed that digital marketing within the companies studied was mostly implemented in spontaneous manner rather than planned.

Based on this research, we could say that there are some barriers for using social media: First, lacking of sufficient skills for using digital marketing or lacking of need to use social media in order to run their business. It happened almost in all the MSMEs.

"it is difficult to adopt digital marketing or technology because we have no education". We want to join the BL, Tokopedia but don't know how".

"we think internet is not necessary for our business. We have facebook and instagram but for personal use".

Second barrier is lack of monetary and time resource which leads to minimum access to the internet (internet quota limit).

"we don't have budget to buy internet quota, and sometimes do not have time to use it".

Third, uncertainty about how to use new digital tools and difficulties to find the right person to teach them digital marketing. Solution to these barriers might be by giving continuous education to MSME and to improve access to the Internet (Wi-Fi).

4 CONCLUSION

The use of the Internet or social media among MSMEs in Bandung District has been relatively common in terms of operating computer and accessing the Internet; however, the use of computers and the Internet to support the management is relatively low. The results of this study also very clearly illustrate that the level of education of MSMEs is highly correlated with their ability to utilize IT as supporting facilities for management of MSMEs.

ACKNOWLEDGEMENT

The authors would like to thank the two anonymous reviewers for their valuable comments regarding this research. The authors further would like to thank to KOPMU DT as partner of this research.

REFERENCES

Adegbuyi, O.A., Akinyele, F.A. & Akinyele, S.T. 2015. Effect of social media marketing on small scale business performance in Ota-Metropolis, Nigeria. *International Journal of Social Sciences and Management* 2(3): 275-283.

APJII. 2017. Penetrasi & perilaku pengguna internet Indonesia.

Cader, Y. & Al Tenaiji, A.A. 2013. Social media marketing. *International Journal of Social Entrepreneurship and Innovation* 2(6): 546-560.

Goi, C.L. 2009. Perception of internet users and marketers on internet marketing activities: Malaysia perspective. *International Journal of Business Forecasting and Marketing Intelligence* 1(2): 181-199.

Internet World Stats. 2017. Top 20 countries with the highest number of internet users. Retrived https://www.inter networldstats. com/top20.htm. Accessed 15 March 2018.

Kotler, P., Amstrong, G., Ang, S.H., Leong, S.M., Tan, C.T. & Ming, O.Y.H. 2011. *Principle of marketing an Asian perspective*. Singapore: Pearson.

Lindqvist, M. 2017. *Social media marketing within small medium sized tourist enterprise*. Vaasa: Hanken School of Economics.

Liu, Q., Karahanna, E. & Watson, R.T. 2011. Unveiling user-generated content: Designing websites to best present customer reviews. *Business Horizons* 54(3): 231-240.

Nuseir, M.T. 2016. Exploring the use of online marketing strategies and digital media to improve the brand loyalty and customer retention. *International Journal of Business and Management* 11(4).

Srinivasan., Bajaj, R. & Bhanot, S. 2016. Impact of social media marketing strategies used by micro small and medium enterprises (MSMEs) on customer acquisition and retention. *IOSR Journal of Business and Management (IOSR-JBM)* 18(1): 91-10.

Taiminen, H.M. & Karjaluoto, H. 2015. The usage of digital marketing channels in SMEs. *Journal of Small Business and Enterprise Development* 22(4): 633-651.

Walsh, M. & Lipinski, J. 2009. The role of the marketing function in small and medium sized enterprises. *Journal of Small Business and Enterprise Development* 16(4): 569-585.

Yuniarni, S. 2018. Indonesia had 143m internet users in 2017: APJII. http://jakartaglobe.id/business/indonesia-143m-internet-users-2017-apjii. March 2018.

Advances in Business, Management and Entrepreneurship – Hurriyati et al (eds)
© *2020 Taylor & Francis Group, London, ISBN 978-0-367-27176-3*

Experiential marketing: A review of its relation to customer satisfaction in online transportation (a study of GO-JEK company)

M.E. Saputri
Telkom University, Bandung, Indonesia

N. Kurniasih
Institut Teknologi Bandung, Bandung, Indonesia

ABSTRACT: GO-JEK company provides customers with a different experience through the use of online ojek services by downloading the mobile application. GO-JEK uses Experiential Marketing as its best value as this marketing type creates an impression among customers. Providing unique experiences that match with the customers' mind is deemed to be important. Good or positive experiences will result in the customers' satisfaction, leading to the repeated use of the service, and they will eventually become loyal customers. GO-JEK stimulates the five senses of the customers through the unique experiences offered to them. It combines all the dimensions of Sense, Feel, Think, and Act that will Relate the feelings and experiences between the company, drivers and customers. This study aimed at determining whether the Experiential Marketing of GO-JEK- Sense, Feel, Think, Act and Relate- has any influences on the customers' satisfaction. This research was quantitative research employing multiple regression analysis technique. The number of respondents in this study was 400 respondents. The results obtained in this study indicated that the experiential marketing of GO-JEK Indonesia had a significant effect on customer satisfaction. Partially, however, there were only four variables that had significant effects on customers' satisfaction, such as Feel, Think, Act and Relate, while Sense had no effects on customer's satisfaction.

1   INTRODUCTION

GO-JEK company or better known as GO-JEK was established in 2011 by Nadiem Makarim as a response to the problem of traffic congestions that have occurred in the capital city of Jakarta. The headquarter of GO-JEK is on Jalan Kemang Selatan Raya, South Jakarta and it has some branch offices located in several areas, namely Bandung, Surabaya and Bali. Nadiem created GO-JEK, a shuttle service with a modern ojek-based order. Ojek or motorcycle taxi is a very effective means of transportation for mobility during congestions. GO-JEK has been running since 2011 and has already had more than 10,000 ojek fleet of drivers in Indonesia. Every day, GO-JEK can serve more than 150 personal orders, not including the company's order. Ojek is an online motorcycle taxi service that helps all people to cope with the traffic jams. The services provided by GO-JEK are not limited to passenger shuttle, but also instant courier/Go-Box, such as Go-food, Go-transport, Go-Shopping, Go-busway, Go-Massage, Go-Glamb, Go-clean.

Providing experiences of using or consuming unique products and services that match the customers' needs and lifestyles will create a long-lasting memory in the minds of the customers. The fun experience gained during the consumption process makes the customer satisfied. The satisfaction of the customers will make them reuse the products or services, and they will therefore recommend the products and services to the others.

According to Kotler (2009), customer satisfaction is "a person's feeling of pleasure or disappointment which resulted from comparing a product's perceived performance or outcome against his/her expectations". More specifically, Joby (2003) defines customer satisfaction as "the number of customers or percentage of total customers, whose reported experience with a firm, its products, or its services (ratings) exceeds specified satisfaction goals". The customer satisfaction can be reached, among others, by the sense-related experiences gained from using a product or service.

Experiential Marketing, according to Schmitt (1999) in Lupiyoadi (2013), is a way to get customers to have experiences through the five senses, generate creative thinking, and create an experience that deals with the physical body, behaviors and lifestyles. Customers also experience interactions with others and connections with social, lifestyle, and cultural circumstances, all of which can reflect the brand, which is the development of sensations, feelings, cognitions, and actions (relate).

In other words, the five aspects of Experiential Marketing: sense, feel, think, act, and relate, will touch the customer's emotion and therefore create pleasurable sensations and result in a positive experience for the customers, hence an unforgettable experience (memorable experience) will ensure the creation of satisfaction (Kartajaya 2006).

A study on the application of Experiential Marketing at a hotel in Taiwan conducted by Lin et al. (2009) showed that every guest staying or visiting the hotel felt satisfied with the experience and therefore recommended the hotel to others. This study concluded that the experience had an influence on satisfaction.

This is in line with the claim that the five aspects of Experiential Marketing have impressed the emotion of the customers as such that it has created pleasurable sensation and positive experience among the customers. It has become an unforgettable (memorable) experience that is deemed essential in creating satisfaction among the customers (Kartajaya 2006).

GO-JEK provides a different experience in using their services to the customers. Therefore, this research was conducted to 30 customers in Jakarta, Bandung, Surabaya, Makassar, and Bali about their experience of using GO-JEK services. The application of Experiential Marketing strategy is appropriate for GO-JEK as it distinguishes GO-JEK from other conventional motorcycle taxi (ojek) services. The varied innovation that customers can directly experience is the aspect that distinguishes GO-JEK from the conventional motorcycle taxis, from reservations or placing order for a GO-JEK service to the facilities provided by GO-JEK to customers.

In relation to the Experiential Marketing concept of sense-feel-think-act-relate (Lupiyoadi 2013), GO-JEK stimulates the five senses of its customers by generating their 'Sense' of unique experience through, for instance, the color of their attributes, such as the use of green jackets and green helmets. 'Ride Carefully' and 'Respond Quickly' to customers' orders are the examples of the 'feel' among the customers. GO-JEK customers will 'think' that GO-JEK provides many services such as Go-transport, Go-instant courier (Go-box), Go-food. Another example is customers can easily perform their transactions/orders for the service.

Customers can easily download GO-JEK apps from iOS or Playstore in their own smartphones and android gadgets. They can directly perform all transactions according to their needs. The GO-JEK drivers are also provided with facilities in the form of smartphones to be able to conduct transactions to customers. The facility that GO-JEK drivers are provided with helps them to make quick responses to customers' requests/orders, and this is the example of the sense of 'Act' that GO-JEK provides to gain its customer satisfaction. GO-JEK has combined all the dimensions of sense, feel, think, act that will make up a relationship (relate) between feelings and experiences, between GO-JEK and its customers, leading to customer satisfaction.

Based on the background discussed above, the aim of this study was to explore the effect of Experiential Marketing on GO-JEK's Customer Satisfaction.

2 METHODS

As stated by Kotler & Amstrong (2014), customer satisfaction is the feeling of pleasure or disappointment after they compare their expectations with the product's perceived performance or outcome. Customer satisfaction has developed into a central concept in the business and management discourse. Customers become the focus in many discussions about satisfaction and quality of services, including in this present research.

This research used quantitative method, a research method that is based on the philosophy of positivism and used to examine a particular population or sample. The data collection was conducted using research instruments, and the data analysis performed quantitatively/statistically. The objective of the research was to test the hypothesis readily set (Sugiyono 2013). The framework for this research is as depicted in Figure 1 as follows.

This study used non-probability sampling technique. It is a sampling technique that does not provide the same opportunity for each element or member of population to be selected as samples. This research used multiple linear regression analysis; a method of analysis used to find out to what extent a variable has influenced other variables. There are one dependent variable and more than one independent variable.

3 RESULTS AND DISCUSSIONS

3.1 (Simultaneous) F-test

The value of F_{table} for n = 400 (df1 = 2 and df2 = n-k-1) with α = 0.05 is 2.23. To calculate the extent of the influence between the variables, this study used SPSS 20.

In Table 1, F_{count} is 139.399 with a significance level of 0.000. Therefore, in both calculations $F_{count} > F_{table}$ (139.399 > 2.23) and the significance level is 0.000 < 0.05. This indicates that H6 is accepted, which

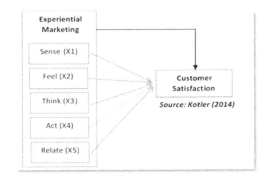

Figure 1. Theoretical framework.

Table 1. Simultaneous hypothesis test (F-test).

ANOVA$_a$

Model	Sum of squares	Df	Mean square	F	Sig.
Regression	122.880	5	24.576	139.399	0.000
Residual	69.462	394	0.176		
Total	192.342	399			

* Dependent Variable: customer satisfaction
Predictors: (Constant), relate, think, sense, act, feel

means the independent variables Sense, Feel, Think, Act, and Relate have simultaneous or joint influence on the customer satisfaction of GO-JEK Indonesia.

3.2 (Partial) t-test

The hypothesis was tested using the method of Multiple Linear Regression analysis. The t-test was performed to find out the partial influences between variables. Based on the criteria for conducting the statistical analysis using t test, Ho is rejected when the $t_{count} > t_{table}$. With the confidence level of 95% or alpha is 0.05, the result of the analysis of each hypothesis is as depicted in Table 2.

1. The variable 'Sense' (X1) has a smaller t_{count} value than the t_{table}. As t_{count} (-0.217) < t_{table} (1.966), Ho is accepted. It can therefore be concluded that partially there was a significant influences of the variable Sense (X1) on the variable of customer satisfaction (Y). Further interviews with random GO-JEK customers were also conducted and the result showed that when customers placed an order or reservation using the GO-JEK apps on their smartphone, they did not take into account the logo and color of GO-JEK, or the tidiness of the GO-JEK drivers. Some customers stated that they often came up with GO-JEK drivers who did not wear GO-

JEK's typical identities. Thus, the variable Sense did not influence customer satisfaction.
2. The variable 'Feel' (X2) has a greater tcount value than ttable. Since the value of t_{count} (8,060)> t_{table} (1,966), Ho is rejected. Therefore, it can be concluded that partially there is a significant influence of the variable 'Sense' (X2) on Customer Satisfaction (Y)
3. The variable 'Think' (X3) has a greater tcount value than ttable. As the t_{count} (2.662) > t_{table} (1.966), then Ho is accepted. Therefore, it can be concluded that partially there is no significant effect of the variable 'Think' (X3) on the customer satisfaction (Y).
4. The variable 'Act' (X4) has a greater tcount value than ttable. As the value of t_{count} (5,080) > t_{table} (1,966), Ho is accepted. Thus, it can be concluded that partially there is no significant influence from the variable 'think' (X4) on Customer Satisfaction (Y).
5. The variable 'Relate' (X5) has a greater t_{count} value than t_{table}. The value of t_{count} (5.195) > t_{table} (1.966), and therefore Ho is rejected. It can be concluded that partially there is a significant influence of the variable 'relate' (X5) on the Customer Satisfaction (Y).

Based on the data processing results shown in Table 3, a model of multiple linear regression equation can be formulated as follows.

$$Y = a + b1X1 + b2X2 + b3X3 + b4X4 + b5X5$$
$$Y = (0.524) - 0.11X1 + 0.408X2 \tag{1}$$
$$+ 0.133X3 + 0.217X4 + 0.161X5$$

The above equation can be explained as follows.
a = 0.524, there is no influence from Sense (X1), Feel (X2), Think (X3), Act (X4), Relate (X5). Therefore, the value of the Customer Satisfaction (Y) is 0.524 unit of measurement.
b = -0111, if Sense (X1) increases as much as one unit of measurement and the other variables remain

Table 2. Result of partial test coefficients.

Model	Unstandardized Coefficients		Std. Coeff Beta	T	Sig.
	B	Std. Error			
(Constant)	0.524	0.164		3.187	0.002
Sense	-0.011	0.049		-0.217	0.829
Feel	0.408	0.050		8.060	0.000
Think	0.133	0.050	0.126	2.662	0.008
Act	0.217	0.043	0.214	5.080	0.000
Relate	0.161	0.031	0.205	5.195	0.000

Table 3. Result of multiple regression analysis of coefficients.

Model	Unstandardized Coefficients		Std. Coeff Beta	T	Sig.
	B	Std. Error			
(Constant)	0.524	0.164		3.187	0.002
Sense	-0.011	0.049		-0.217	0.829
Feel	0.408	0.051		8.060	0.000
Think	0.133	0.050	0.126	2.662	0.008
Act	0.217	0.043	0.214	5.080	0.000
Relate	0.161	0.031	0.205	5.195	0.000

constant, the Customer Satisfaction (Y) decreases as much as 0.101 unit.

c = 0.408, if Feel (X2) increases as much as one unit of measurement, and the other variables remain constant, the Customer Satisfaction (Y) increases as much as 0.408.

d= 0.133, if Think (X3) increases as much as one unit of measurement, and the other variables remain constant, the Customer Satisfaction (Y) increases as much as 0.133.

e = 0.217, if Act (X4) increases as much as one unit of measurement, and the other variables remain constant, the Customer Satisfaction (Y) increases as much as 0.217.

f = 0.161, if Relate (X5) increases as much as one unit of measurement, and the other variables remain constant, the Customer Satisfaction (Y) increases as much as 0.161.

3.3 Coefficients of determination (R2)

The test for Coefficients of Determination (R2) was performed to find out the extent of the independent variable (X) influence on the dependent variable (Y). The value of R2 is expressed in percentage; the nearer to 1 or 100%, the more the influence of the independent variable on the dependent variable, and vice versa. The nearer the value to 0, the less the influence is.

Table 4 indicates that the value r = 0.634 which means that the relation between the independent variables Sense, Feel, Think, Act, and Relate and the dependent variable (Customer Satisfaction) is 63.4%, implying that they are closely related. The value of Adjusted R Square = 0.634 means that 63.4% of the factors for customer satisfaction can be explained by the independent variables (Sense, Feel, Think, Act, and Relate), while the rest 36.6% can be explained by other factors not covered in this research.

The Sense variable had a total value of 6349 or 79.36% of the ideal score of 2000. Therefore, it can be concluded that the 'Sense' variable of GO-JEK Indonesia (Jabodetabek, Bandung, Surabaya, Makassar, and Bali) was in the Good category. Among all the items of the Sense variable, there was an item with the lowest score of 69.6%, i.e. 'I think the appearance of the GO-JEK drivers is neat'. This is because many of the GO-JEK drivers were not tidy.

Table 4. Table of coefficients of determination.

Model Summary[b]

Model	R	R Square	Adjusted R Square	Sts. Error of the Estimate
1	0.799	0.649	0.634	0.419881

a. Predictors: (Constants), relate, think, sense, act, feel
b. Dependent Variable: Customer Satisfaction

Based on the results of the partial hypothesis test (t-test), the variable Sense had no effect on the Customer Satisfaction. This can be seen from the value of t_{count} (-0.217) < t_{table} (1.966), and therefore Ho was accepted. It can be concluded that partially there was no significant influence of Sense on Customer Satisfaction. A follow-up interview was conducted to random GO-JEK customers and it was revealed that when GO-JEK customers placed their orders or reservations using GO-JEK application, they overlooked the logo and the color of GO-JEK. They did not consider the drivers' neatness either. Customers often encountered drivers who did not use any identity of GO-JEK. Accordingly, Sense was deemed to have no effect on customer satisfaction.

The 'Feel' variable got a total value of 6433 or 80.41% of the ideal score 2000. Therefore, it can be concluded that the condition of the feel variable was in good category. Of all items contained in the Feel variable, there was an item with the lowest value of 77.45%, which is 'the customers felt secure when they were using GO-JEK'. Some customers tended to disagree with the statement that using GO-JEK gives the feeling of safety.

According to the results of the partial hypothesis test, the 'Feel' variable had positive and significant effects on customer satisfaction. It was evident from the t_{count} (8.060) > t_{table} (1.966). Based on the results of the partial calculations, it can be assumed that the 'Feel' variable had significant influences on the customer satisfaction of GO-JEK Indonesia (Jabodetabek, Makassar, Bandung, Surabaya and Bali).

The 'think' variable got an overall score of 6484 or 81.05% from the ideal score of 2000. It can be concluded that the 'think' variable was in the Good category. Among the items of the 'think' variables, there was an item with the lowest value of 75.3%, which is 'the affordable GO-JEK rate'.

Based on the partial hypothesis test result, the 'think' variable had a positive and significant effect on the customer satisfaction. This can be seen from the t_{count} (2.662) > t_{table} (1.966). From the results of the partial calculations, it can be stated that the "think" variable significantly affected the customer satisfaction. This indicates that GO-JEK customers' satisfaction was influenced by their thinking variables.

The 'Act' variable earned a total value of 3089 or 77.22% from the ideal score of 2000. Thus, it can be concluded that the act variable was in the Good category. Among the items of act variable, there were some items that had the lowest value of 71.35%, which is 'using all the services of GO-JEK'.

The 'act' variable showed a positive and significant influence on customer satisfaction. This can be seen from the value of t_{count} (5.080) > t_{table} (1.966). From the results of the partial calculations, it can be stated that the 'act' variable act had a significant effect on GO-JEK customer satisfaction. This indicated that customer satisfaction was influenced by the act given by GO-JEK.

The 'relate' variable reached a total value of 2848 or 71.2% of the ideal score of 2000. Therefore, it is concluded that the 'relate' variable was in the Good category. Among the items of the 'relate' variable, there are items that have the lowest value of 69.7%, which is 'the feeling of having 'attention' when using GO-JEK'.

The 'relate' variable had a positive and significant effect on the customer satisfaction. It is evident from the t_{count} (5.195) > t_{table} (1.966). Based on the results of the calculations, it can be stated that the Relate variable had a significant influence on the customer satisfaction of GO-JEK. This indicated that customer satisfaction was influenced by the 'Relate' provided by GO-JEK.

Based on the hypothesis simultaneous Test result in Table 4:11, the F_{count} is 139.399 with a significance level of 0.000. Therefore, in both calculations, the F_{count} > F_{table} (139.399 > 2.23) and the significance level is 0.000 < 0.05. This indicates that the hypotesis was accepted, implying that the independent variables Sense, Feel, Think, Act, and Relate had a simultaneous or joint influence on Customer Satisfaction of GO-JEK Indonesia. It is in line with the research conducted by Rahardja, et al (2010) which revealed that experiential marketing had influence on the customer satisfaction.

Indrawati & Fatharani (2016) concluded based on the result of their research that Experiential Marketing simultaneously had a significant effect towards Zalora Customer Satisfaction. However, partially only four sub-variables of Experiential Marketing had significant effects towards Customer Satisfaction, namely Feel, Think, Act, and Relate. This research rejected the hypothesis that Sense had a significant effect towards Customer Satisfaction.

4 CONCLUSION

The result of this study strengthens the argument that experiential marketing is a key driving factor to the customer satisfaction. The Experiential Marketing (Sense, Feel, Think, Act, and Relate) had simultaneous or altogether influences on Customer Satisfaction of GO-JEK Indonesia. Otherwise, partial effects of Experiential Marketing on the Customer Satisfaction of GO-JEK Indonesia showed that Feel, Think, Act and Relate significantly influenced the customer satisfaction of GO-JEK Indonesia, whereas the Sense variable had no positive or significant effects on the customer satisfaction.

REFERENCES

Indrawati & Fatharani, U.S. 2016. The effect of experiential marketing towards customer satisfaction on online fashion store in Indonesia. *Asia Pacific Journal of Advance Business and Social Studies. Vol 2 Issue 2.* Institute of Advanced Research (APIAR).

John, J. 2003. *Fundamentals of customer focused management: competing through service.* Westport, Conn.: Praeger.

Kartajaya, H. 2006. *Hermawan Kartajaya on marketing.* Jakarta: Gramedia.

Kotler, Philip. & Armstrong, G. 2014. *Principle of marketing, 15th edition.* New Jersey: Pearson Prentice Hall.

Lin, Kuo-Ming, et al. 2009. "Application of experiential marketing strategy to identify factors affecting guests' leisure behaviour in Taiwan hot-spring hotel." *WSEAS Transactions on Business and Economics* 6.: 229-240.

Lupiyoadi, R. 2013. *Manajemen pemasaran jasa berbasis kompetensi.* Jakarta: Salemba Empat.

Rahardja, Christina and Anandya, Dudi 2010: Experiential marketing, customer satisfaction, behavioral intention: timezone game center Surabaya. *Proceedings the first international conference business and economic* (1): 1-6.

Sugiono (2013). *Metode penelitian kuantitatif kualitatif dan R&D.* Bandung: Alfabeta.

Advances in Business, Management and Entrepreneurship – Hurriyati et al (eds)
© 2020 Taylor & Francis Group, London, ISBN 978-0-367-27176-3

Redesigning the e-commerce Banyuwangi Mall for small and medium enterprises

A.A.G.S. Utama & P.P.D. Astuti
Universitas Airlangga, Surabaya, Indonesia

ABSTRACT: Banyuwangi became the main pioneer providing facilities for the sale of its SME products through integrated digital marketplace platforms. Therefore, the purpose of this paper is to evaluate and redesign the e-commerce system at Banyuwangi Mall for selling SME products in Banyuwangi. The approach used in this research is qualitative explorative with case study method from SME seller of goods product. The theory used is (UTAUT2). The results of this study indicate that SMEs welcome a redesign of the e-commerce Banyuwangi Mall, as the system will become more efficient. Then the researchers redesigned to facilitate the operation of the system for SME by considering its costs and benefits and its internal control.

1 INTRODUCTION

Banyuwangi was the main pioneer providing facilities for the sale of Small and Medium Enterprise (SME) products through an integrated digital marketplace platform (banyuwangikab.go.id). In September 2017, based on data obtained from the administrative site of Banyuwangi Mall, the number of SMEs that had been registered reached 82 sub-districts, from various areas in Banyuwangi. Meanwhile, the number of experienced products contain as many as 552 products. In addition to 82 SMEs, there are still 30 SMEs on the waiting list, which cannot be registered to Banyuwangi Mall because there are still some incomplete requirements. However, in the digital market, this place still has problems, including the number of SMEs registered. The problem is that SME owners of products that are marketed on product details have a complete address but their contacts cannot be contacted. Only users can see the SME name of the product owner.

Based on the preliminary survey that has been done, not all SME owners understand the performance of Banyuwangi Mall. Another disadvantage of the current system is that there is no number or number of web visitors who have seen one product being sold. Testimonials also can not be shown, the number of items provided doesn't sow in every month. Sellers do product updates only by admin using the admin Banyuwangi Mall while the Banyuwangi Mall can list all products everyday. Therefore, redesigning e-commerce systems in Banyuwangi Mall as a media for selling SMEs products in Banyuwangi is necessary.

E-commerce is the part of electronic business that performs transactions, that is, buying and selling goods and services through the internet. These transactions include advertising, marketing, customer support, security, delivery, and payment (Laudon & Laudon 2016). Studies by Turban et al. (Turban et al. 2018, Turban et al. 2015) show that e-commerce is a process of transactions ranging from purchasing, selling, transferring, exchanging products in the form of goods, services, or information made through computer media, whether through internet or intranet. The definition of e-commerce has three aspects, according to Radovilsky (2005): aspects of technology, business aspects, and aspects of value.

There are seven constructs categorized into the UTAUT 2 model (Venkatesh et al. 2012): performa expectancy, effort expectancy, social influence, facilitating condition, hedonic motivation, price value, and habit.

Bodnar and Hopwood (2013) explain that accounting information system is a collection of resources, an equipment and a human designed to transform financial data and other data into an information.

Hall (2011) explains that the development of an information system is also included in an investment. Something is considered an investment if there has been a sacrifice of resources in order to gain something in the future. The process of developing information systems also requires a sacrifice of resources that aims to obtain new benefits. If the expected benefit is greater than the resources issued, then the information system is considered to have value and is feasible. So, before implemented the system development, it needs to analyze the cost and benefit.

Various studies on a system evaluation of online shopping sites have been done, including research on the quality of website design and user behavior. This study used the Unified Theory of Acceptance and Use of Technology (UTAUT) model (Al-Qeisi et al. 2014). Research conducted by Azis and Kamal (2017) using the Unified Theory of Acceptance and Use of Technology 2 model (UTAUT 2) explains that there is an influence of consumer behavior towards the desire to buy SME products in West Java. Another study was conducted in China by Li, Chung and Fiore (2017), which explains that security, social motivation and playfulness factors influence consumer behavior in e-auction transactions, a type of C2C e-commerce. The test uses the Technology Acceptance Model (TAM) model.

Past researchers focused solely on evaluating the system. In this study, researchers will provide a system design to complement the existing system. So that the design of the new system will provide more benefits for users, namely Seller (SME) and Buyer.

2 RESEARCH METHODOLOGY

The approach used in this research is qualitative explorative with a case study method. The data collection procedures used in this research are interview, observation, and documentation.

The type of data used in the study is qualitative data, expressed in the form of words, sentences, and images. Sources of data used in this study are primary and secondary data. The research locations are Rumah Kreatif (hereafter, Creative House) Banyuwangi as manager in Banyuwangi Mall, Department of Cooperative and Micro Enterprise, and SMEs Location.

3 RESULT AND DISCUSSION

3.1 Banyuwangi Mall

Banyuwangi Mall is one of the marketplaces or online stores of all SME products in Banyuwangi formed by the Government of Banyuwangi Regency in collaboration with Bank BNI'46 Jakarta. The address of this online shopping website is www.banyuwangi-mall.com. Banyuwangi Mall was opened on 20 April 2016 by Rini Soemarno (Minister of State Own Enterprise), accompanied by Achmad Baiquni (President Director of BNI'46) and Abdullah Azwar Anas (Banyuwangi Regent).

The difference between Banyuwangi Mall and other online shopping sites (private property) is in terms of product prices displayed on the website. Online shopping sites generally cost more than the original seller (the product owner). In accordance with economic principles, it is because the distributor wants to get the maximum profit by expending the least capital. However, at Banyuwangi Mall, all products are sold at normal prices, where that determines the price of products installed on the website. So that the perpetrators of SMEs will benefit according to their wishes.

The customer traffic on Banyuwangi Mall, in 2016, increased every month. In just ten days, the number of customers in Banyuwangi Mall reached 301. In August, the number of consumers reached more than 500, which is 508 buyers. By the end of 2016, the number of consumers reached 605, not only from the Banyuwangi area but also from outside Banyuwangi, including Jember, Malang, Surabaya, Jakarta Bandung, Medan, Bali, and Sulawesi.

3.2 Creative House

Creative House is a place entrusted by the Government of Banyuwangi Regency under the line of the Department of Cooperative and Micro Enterprise to form a digital economic ecosystem through coaching for the perpetrators of SMEs in Banyuwangi in order to improve capacities and capabilities of SMEs. This Creative House will be the management center of Banyuwangi Mall's market place. The research and development (R&D) of Banyuwangi Mall will be fully controlled by Creative Homes, but still in the line of supervision from Department of Cooperatives and Micro Enterprises. Creative House was inaugurated with the date of the inauguration of Banyuwangi Mall on April 20, 2016. The location of Creative House is on Achmad Yani no. 97, Banyuwangi.

3.3 Current system

Banyuwangi Mall is an online store that is included in e-commerce customer to customer (C2C). This is because Banyuwangi Mall is only a third party. Sales transactions are done by one consumer to another consumer. This consumer is the perpetrators of SMEs who sell their products through Banyuwangi Mall and other consumers are buyers who are interested in products sold through Banyuwangi Mall (Laudon & Laudon 2016, Pradana 2015, Turban et al. 2015).

All of these menus in official account of one of the sellers in Banyuwangi Mall can be edited by the seller themselves. So, it isn't necessary to wait for the admin Banyuwangi Mall to change the data.

3.4 Analysis using unified theory of acceptance and use of technology 2 (UTAUT2)

This analysis uses theory of UTAUT2 from Venkatesh et al. (2012).

3.5 System design

3.5.1 Context diagram

The sales cycle through Banyuwangi Mall is a repeat business activity that describes the provision of

Table 1. UTAUT2.

No	Construction	Result
1	*Performa expectancy*	Banyuwangi Mall can improve the performance of the sellers. The increase is in the form of product promotions in local Banyuwangi, all over Indonesia, even abroad. The number of products sold is still not a priority. The important products are better known in the community.
2	*Effort expectancy*	The seller's account at Banyuwangi Mall is easy to understand and does not take long. If there are SMEs who do not know the ease of using Banyuwangi Mall just because they never opened the account.
3	*Social influance*	All informants suggested other SMEs to join there. The goal is to help them expand their marketing beyond Banyuwangi.
4	*Faciliating condition*	Facilities owned to access Banyuwangi Mall is enough through smartphone, laptop, or other gadgets that can open the website Banyuwangi Mall. It is also supported by the technical guidance provided by the Creative House Team as the manager of Banyuwangi Mall.
5	*Hedonic motivation*	The sellers willing to join because in the beginning before the establishment of Banyuwangi Mall, there are invitations from related agencies, Department of Cooperatives and Micro Enterprises on the direction of Banyuwangi Mall system.
6	*Price value*	All sellers feel that the benefits provided when joining the Banyuwangi Mall are far more numerous. This is supported by an online shopping site managed by the Government of Banyuwangi Regency, so all financing on the registration of sellers is all borne by the Government of Banyuwangi Regency.
7	*Habbit*	All informants have never opened seller accounts again.

products from the seller to the buyer based on prior orders until the seller receives cash or cash equivalents for the sales transactions that have been made. Context diagram Banyuwangi Mall connects the four parties involved in the sales cycle. Parties in question are Seller (Seller), Customer (Buyer), Administration (Administration Section in Creative House), and Bank.

Sellers are the perpetrators of SMEs, parties who provide products that will be sold through e-commerce Banyuwangi Mall. The registered seller is not just one but is in fact many sellers with different product types.

Buyers are interested in purchasing products sold through Banyuwangi Mall e-commerce. Buyers can interact directly with the seller.

Administration is the party that manages the e-commerce system in Banyuwangi Mall. Each activity can be made and known by the Administration Section. This administrative section also serves as a subordinate of the Cooperative Office and MSMEs who must report any activities conducted to the Office, including the Sales Report from Banyuwangi Mall.

Bank is the party in charge of receiving payment every transaction made through Banyuwangi Mall from the buyer and transfer it to the account of each seller. In addition, the Bank also has the authority to see every activi-bag carried out in Banyuwangi Mall. This is because the main server system is the Bank.

3.5.2 *Business process*

Based on Business Process Banyuwangi Mall, it shows the following activities.

1. The seller receives an order notification.
2. Products ordered by customers in check availability.
3. Products that have been checked availability then checked the status of payment in the seller's account.
4. When payment has not been made, the order will not be processed but when the customer has made the payment, then the packaging of products that have been ordered and automatically reduce inventory in the website display.
5. The product that has been packaged and then will be sent to the customer through the delivery service by providing the cost of the sender himself who will be replaced by the Bank.
6. When the goods have been sent, then the seller will get the product report sent.
7. The seller updates the items sent on the system as a basis for making payment requests to the Bank.
8. The proof of the item will be processed by the Bank and then transferred the payment money in accordance with the selling price and substitute the shipping cost to the account of each seller.
9. Transfered money will directly enter the system and reduce the amount of balance listed in the system Banyuwangi Mall and sales activities finished shining.

3.5.3 *Entity relationship diagram*

Each transaction in will involve the database. The database is a product, seller, and customer. Each database is given each ID to distinguish any similar data. Each transaction is also given each ID. The ID is used to adjust the transaction data entered into the sales report in the seller account with transactions recorded by the Bank. So the risk of error will be smaller.

Table 2. Cost and benefit for a year.

Cost						
One-Time Cost						
No.	Information	Frequency	Unit	Unit Price	Total	Source
1	Training	12	Meeting	Rp. 1.000.000	Rp. 12.000.000	http://mafia.com/ (Accessed 13 April 2018)
Total cost					Rp. 12.000.000	
Benefit						
Tangible benefit						
No.	Information	Frequency	Unit	Unit Price	Total	Source
1	Reduction salary expense	12	Month	Rp. 1.500.000	Rp. 18.000.000	http://www.qerja. com/company/ salary/(Accessed 13 April 2018)
Total benefit					Rp. 18.000.000	

3.6 Cost and benefit analysis

The costs required are for training SME players or representatives who participate in assisting their business with the use of updated systems. The training is held weekly with different participants for three months in a row. However, every week is done with participants of different SMEs. So, out of 92 MSEs divided into four groups (1 month = 4 weeks). Up to each group there are 23 perpetrators of SMEs.

The benefits of the development of the system is that Creative House will no longer need an additional employee. The salaries that would be used to pay new employees can be allocated to training or other needs. Table 2 is a breakdown of costs and benefits in one year.

$$\text{Payback period} = \frac{12.000.000}{18.000.000} \times 12 \text{ months} = 8 \tag{1}$$

The total training cost of the latest system can be fulfilled in the eighth month of salary when accepting a new employee.

4 CONCLUSION

The analysis, using Unified Theory of Acceptance and Use of Technology 2 (UTAUT2), indicates that this system is good enough but still can be improved to make its usage easier and more efficient. The results of the analysis show that the cost is less than the benefits. Using the payback period indicates that within eight periods it has covered all expenses incurred.

REFERENCES

Al-Qeisi, K., Dennis, C., Alamanos, E. & Jayawardhena, C. 2014. Website design quality and usage behavior: Unified theory of acceptance and use of technology. *Journal of Business Research* 67(11): 2282–2290.

Azis, E. & Kamal, R.M. 2017. Adopsi teknologi belanja online oleh konsumen umkm dengan model unified theory of acceptance and use of technology 2. *Creative Research Journal* 2(1):19-38.

banyuwangikab.go.id. (1970). Genjot Pemasaran UMKM, Banyuwangi Luncurkan Digital Market Place. https://www.banyuwangikab.go.id/berita-daerah/genjot-pemasaran-umkm-banyuwangi-luncurkan-digital-marketplace.html. Accessed 5 October 2017.

Bodnar, G. H. & Hopwood, W. S. (2013). Accounting information systems.

Hall, J. A. 2011. *Accounting information systems seventh edition.*

Laudon, K. C. & Laudon, J. P. 2016. Information systems, organizations, and strategy. *Management Information Systems: Managing The Digital Firm.*

Li, R., Chung, T. L. (Doreen). & Fiore, A. M. 2017. Factors affecting current users' attitude towards e-auctions in China: An extended TAM study. *Journal of Retailing and Consumer Services* 34(January 2016): 19–29.

Pradana, M. 2015. Klasifikasi bisnis e-commerce di Indonesia. *Modus* 2727(22): 163–174.

Radovilsky, Z. 2015. Application models for e-commerce Zinovy Radovilsky application models for e-commerce. 36.

Turban, E., King, D., Lee, J. K., Liang, T.-P. & Turban, D. C. 2015. *Electronic commerce: A managerial and social perspective.* Springer.

Turban, E., Outland, J., King, D., Lee, J. K., Liang, T.-P. & Turban, D. C. 2018. Electronic Commerce 2018.

Venkatesh, V., Thong, J. & Xu, X. 2012. Consumer acceptance and user of information technology: Extending the unified theory of acceptance and use of technology. *MIS Quarterly* 36(1): 157–178.

Advances in Business, Management and Entrepreneurship – Hurriyati et al (eds)
© *2020 Taylor & Francis Group, London, ISBN 978-0-367-27176-3*

The effect of differentiation strategy on competitive advantage

R.A. Aisyah
Universitas Airlangga, Surabaya, Indonesia

ABSTRACT: A strategy can provide support for a company to compete. One of the competitive advantage indicators is differentiation strategy. A company can differentiate strategy targeted to the market by performing product differentiation, service differentiation and company's image differentiation strategies. This research examined the effect of product, service, and image differentiation strategies on competitive advantage. The objective was to determine the influence of product differentiation, service differentiation and image differentiation strategies on competitive advantage conducted in one of the restaurants located in Surabaya. The research model used was questionnaires. The sampling technique used was accidental sampling involving 157 respondents. The analytical method employed multiple linear regression analysis. Furthermore, correlation calculation was used to determine the degree of the influence and the significance of product differentiation, service differentiation and image differentiation strategies on competitive advantage.

1 INTRODUCTION

A strategy can provide support for a company to compete. One of the competitive advantage indicators is differentiation strategy (Porter 1980). Kotler (2011) states that a company can perform differentiation strategy by identifying the source of the existing competitive advantage, having company's major distinguishing feature, choosing an effective positioning in the market and communicating the decisive position in the market. According to Sulistiani (2014), all the efforts performed by a company to create a difference between competitors aim at providing the best value to consumers. A company can differentiate strategies given to the market from three aspects comprising product differentiation, service differentiation, and image differentiation (Kotler & Keller 2016).

Kotler & Keller (2016) state that product differentiation is a better, faster, and cheaper company's product offering that will create higher value for the customers than competitor's product. A unique and different product can be used as the characteristic of a company. In this case, the advantage can be in the form of superiority position in an industry or market, so that the company can improve its performance because the competitive advantage can be achieved from the various competencies owned and enhanced through the company's distinctive strategic assets.

In addition to performing product differentiation, company can carry out service differentiation. Service differentiation is a form of service and quality improvement containing different value in providing service offering to the customers (Kotler 2011). Furthermore, Ariyasti (2013) states that a company which is able to give better service than its competitors will win the competition and eventually have higher market share. In addition to product and service differentiation, company can perform image differentiation.

Image differentiation is an appropriate image of imaging elements in creating brand image (Keller 2013). In image differentiation, consumers can form company's image in the society, so that the company must give good image to the consumers. Good company's image will give competitive advantage in consumers' minds (Keller 2013). Ong & Sugiharto (2013) state that competitive advantage is basically originated from value and benefits created by a company for its buyers, which are higher than the cost that must be incurred to create them. If the company is then able to create excellence through one of the three generic strategies, it will get competitive advantage. Based on the background explained, the formulated title of this study is "The Effect of Differentiation Strategy on Competitive Advantage".

1.1 The effect of product differentiation on competitive advantage

Kotler (2014) states that product differentiation is a better, faster and cheaper company's product offering that will create higher value for customers than competitors' products. Kotler & Keller (2016) argue that product differentiation is an action to design some significant differences to distinguish a company's offer with its competitor's.

The unique and different products can be used as a company's characteristic. In this case, the significance is in the form of a superiority position in an industry or market, so that the company can improve its performance because competitive advantage can be achieved from a variety of competences and enhanced through the company's distinctive strategic assets (Kotler 2011).

H1: The effect of product differentiation on competitive advantage.

1.2 The effect of service differentiation on competitive advantage

Johnson & Sirikit (2002) argue that the provision of good service today is regarded as a major strategy for the company to succeed and survive. They also reveal that the service differentiation is a profitable strategy because the outcomes can bring new customers, grow company's activities with the existing customers, reduce the impact of customers' loss, and reduce errors in service. The key to company's success not only depends on products but also services offered so as to provide value added to consumers (Kotler & Keller, 2016).

H2: The effect of service differentiation on competitive advantage.

1.3 The effect of image differentiation on competitive advantage

Image has an important meaning in business. The message of an advertisement significantly affects the customers' trust and attitude to buy the products contained in the advertisement (Kotler & Armstrong 2010). Introducing a unique image through advertising and sponsorship activities is proved to be more effective to achieve the creation of brand equity. If the company has a high brand equity, then customer loyalty can be formed or in other words the company already has a competitive advantage in the minds of customers (Keller 2013). Various concepts developing today imply the importance of the image as a means or a tool to achieve competitive advantage in the marketplace. Once a company has clearly defined its image to customers, then the next step is to communicate the image since this element is a source of company's competitive advantage in the long term (Porter 1980).

H3: The effect of image differentiation on competitive advantage.

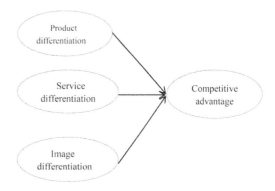

Figure 1. Research framework.

2 METHODS

2.1 Variable measurement

For measuring the different variables included in this study, a five-point Likert scale (1 = Strongly disagree; 5 = Strongly agree) was used. Indicators used to measure product differentiation were taken from research conducted by Tambunan (2010). Indicators used to measure service differentiation were taken from research conducted by Ariyasti (2013). Indicators used to measure image differentiation were taken from Kotler (2011). Indicators used to measure competitive advantage were adapted from Rangkuti (2003).

2.2 Data collection

The research site was one of the restaurants in Surabaya. Restaurant X is a restaurant that serves a variety of food menu and various types of chili sauce in which very spicy as its characteristic. As a characteristic or distinguishing feature of other restaurant, Restaurant X branch in Surabaya has 31 kinds of chili sauces with various levels of spiciness. The level of spiciness ranges from at least 2 chilies to as many chilies preferred by the customers.

Out of 160 respondents, 157 respondents were eligible to be processed into data processing. The respondents as the samples of this study were customers who visited Restaurant X Branch Surabaya characterized based on gender, age, occupation and monthly income. The majority of respondents who visited Restaurant X Branch Surabaya was men for 52 respondents (52%), the remaining 48 (48%) were women. The age of respondents ranges from younger than 18 years old for 26 respondents (26%), between 18 to 25 years old for 35 respondents (35%), between 26 to 34 years old for 22 respondents (22%), between 35 to 42 years old for 11 respondents (11%) and older than 42 years old for 6 respondents (6%).

The occupations of the respondents were students for 40 respondents (40%), private employees for 19 respondents (19%), civil servants for 12 respondents (12%), entrepreneur for 10 respondents (10%) and the rests 19 respondents (19%) were others. The monthly income of the majority respondents were less than 1 million rupiah for 41 respondents (41%), above 3 million rupiah for 24 respondents (24%), about one to two million for 18 respondents (18%) and about 2 to 3 million for 17 respondents (17%).

2.3 Method and data analysis

According to Malhotra (2010), regression analysis is a statistical procedure to generate the relationship between the dependent variable and the independent variable. If there are two or more independent variables then multiple linear regression analysis used. Thus, the degree of the influence between independent variables to the dependent variable can be identified.

3 RESULTS AND DISCUSSION

The significance values of product differentiation variables, service differentiation, image differentiation and competitive advantage were lower than 0.05 and considered as valid, or can be seen by comparing the value of r-arithmetic with r-table, rcalculate since each indicator of product differentiation, service differentiation, image differentiation and competitive advantage variable generated were greater than the value of r-table. Indicators or the questionnaires used by each variable of product differentiation, service differentiation, image differentiation and competitive advantage were stated as valid to be used as a measuring tool. The values of cronbach's alpha (α) of product differentiation variables, service differentiation, image differentiation and competitive advantage were greater than 0.60, so it was considered reliable.

3.1 Results of regression analysis

Based on Table 1, the significant value of the multiple regression of product differentiation is 0.019, service differentiation is 0.034, and image differentiation is 0.000. The significant value of each variable is < 0.05, so that product differentiation, service differentiation, and image differentiation strategy significantly affect competitive advantage.

Where XI is product differentiation, X2 is service differentiation, X3 is image differentiation, and Y is competitive advantage. Based on Table 1, hypothesis 1 stating that product differentiation strategy affects competitive advantage is accepted with significant value of .019.

Table 1. Results of multiple regression analysis.

Hypothesis	Unstandardized Coeff.		Standardized Coeff.		
	Beta	Std. Error	Beta	t	Sig.
H1	0.131	0.055	0.189	2.385	0.019
H2	0.190	0.088	.0196	2.149	0.034
H3	0.683	0.103	0.550	6.615	0.000
(Constant)	0.977	1.312		0.745	0.458
F	85.597				
df	99				
R square	0.728				
Adjusted R Square	0.719				

Based on Table 1, the regression equation model is as follows:

$$Y = \beta o + \beta\_1 X\_1 + \beta\_2 X\_2 + \beta\_3 X\_3 + e$$

$$Y = 0.977 + 0.131 X\_1 + 0.190 X\_2 + 0.683 X\_3 + e \tag{1}$$

It indicates that product differentiation strategy improvement affects the competitive advantage. It means that Restaurant X Branch in Surabaya improves the resulted product differentiation, so that its competitive advantage is getting higher. Kotler (2011) states that product differentiation is company's way to win the competition through a different power or unique attributes, so that it is perceived as a product that has higher value by consumers.

Based on Table 1, the results show that H2 is accepted with a significance value of 0.034. It means that service differentiation significantly influences competitive advantage. Teece et al. (1997) state that company which is able to provide better service than its competitor will win the competition, so that it has greater growth of market share. Johnson & Sirikit (2002) revealed that different service was a profitable strategy because it could bring new customers, company's activity could be more developed with the existing customers, the impact of customer lost was reduced and errors in service could be reduced.

Based on Table 1, H3 is accepted with a significance value of 0.000. It means that image differentiation significantly affects competitive advantage. Unique image formation through advertising activity and sponsorship is proven to be more effective in achieving the formation of brand equity. If customers perceive that the brand of the company has high brand equity then customer loyalty can be built by itself or in other words the company has a competitive advantage in the minds of customers (Sulistiani 2014). Once the company has clearly defined its image to customers, the next step is to promote that image as a source of competitive advantage owned by the company for a long term (Porter 2008).

4 CONCLUSIONS

4.1 *Conclusion*

Based on the research findings, the following conclusions can be drawn. Product differentiation partially and significantly affected competitive advantage, service differentiation significantly affected competitive advantage, image differentiation partially and significantly affected competitive advantage.

4.2 *Managerial implications*

Image differentiation was a dominant influencing variable. Therefore, it is suggested that the restaurant can retain its image, such as by carrying out advertising both through brochures and over the internet more frequently with a more unique and creative appearance, establishing more branches of Restaurant X throughout Indonesia as a way of promotion to be more publicly recognizable, as well as doing sponsorship activities at an event, especially the big ones.

Product differentiation variable in Restaurant X had a low effect on competitive advantage. In terms of its product, it is suggested that the products are further enhanced by the restaurant, for example by creating many more unique sauce or Indonesian taste to suit customers' taste, and evaluating and improving the taste quality of food/beverage. Service differentiation variable in Restaurant X in Surabaya had a low effect on competitive advantage. Thus, it is suggested that the restaurant improves its service by training its employees to provide better and improved service and facilities.

REFERENCES

Ariyasti, Y. 2013. Analisis Strategi Diferensiasi Terhadap Keunggulan Bersaing Pada Hotel Grand Elite Pekanbaru. Jurnal Universitas Islam Negeri Sultan Syarif.

Johnson, W. C., & Sirikit, A. 2002. Service quality in the Thai telecommunication industry: a tool for achieving a sustainable competitive advantage. Management Decision, 40(7), 693–701.

Keller, K. L. 2013. Strategic Brand Management: building, measuring, and managing brand equity. Brand (Vol. 58).

Kotler. 2011. Manajemen Pemasaran di Indonesia : Analisis, Perencanaan, Implementasi dan Pengendalian. Jakarta :Salemba Empat.

Kotler, K. 2014. Manajemen Pemasaran. Jakarta :Salemba Empat.

Kotler, P., & Armstrong, G. 2010. Principles of Marketing. World Wide Web Internet And Web Information Systems.

Kotler, P., & Keller, K. L. 2016. Marketing Management. Global Edition (Vol. 15E).

Malhotra, N. K. 2010. Marketing Research: An Applied Orientation. Pearson.

Ong, I. A., & Sugiharto, S. 2013. Analisa Pengaruh Strategi Diferensiasi, Citra Merek, Kualitas Produk dan Harga Terhadap Keputusan Pembelian Pelanggan Di Cincau Station Surabaya. Jurnal Manajemen Pemasaran, 1(2),1–11.

Porter, M. E. 1980. Competitive Strategy. Techniques for Analyzing Industries and Competitors, 1(2), 396.

Porter, M. E. 2008. Competitive Strategy. In Competitive Strategy (pp. 73–118).

Rangkuti, F. 2003. Business Plan. Jakarta: Gramedia.

Sulistiani, D. 2014. Mencapai keunggulan bersaing dengan strategi diferensiasi. EL MUHASABA Jurnal Akuntansi, 4(2), 1–17.

Tambunan, P. M.C. 2010. Pengaruh Diferensiasi Produk terhadap Kepuasan Konsumen Mobil Truk Ringan Isuzu. Medan: Jurnal Universitas Sumatera Utara.

Teece, D. J., Pisano, G., & Shuen, A. 1997. Dynamic capabilities and strategic management. Strategic Management Journal, 18(7), 509–533.

Advances in Business, Management and Entrepreneurship – Hurriyati et al (eds)
© 2020 Taylor & Francis Group, London, ISBN 978-0-367-27176-3

Determinants of consumer purchase intention toward luxury brands

M. Kurniawati & R.A. Ramadhan
Universitas Airlangga, Surabaya, Indonesia

ABSTRACT: The emergence of smartphones has become a trend in society. There is fierce business competition among existing smartphone brands to compete for market share. There are many types of smartphones available on the market, including luxury smartphones. One of the luxury smartphone brands in demand is the iPhone. According to IDC (International Data Corporation), the iPhone has been the top smartphone, breaking many shipment records. The purpose of this study is to determine the effect of luxury brand perception, social influence, vanity and materialism on consumer purchase intentions of the iPhone. Questionnaires were distributed to 244 respondents and data was analyzed using Partial Least Square (PLS). The findings of this study indicate that the perception of luxury brand and materialism affect the purchase intention of consumers, while social influence and vanity has no effect on purchase intention. In addition, it was found that vanity does not have a moderating effect on the relationship between luxury brand perceptions and social influences on purchase intention.

1 INTRODUCTION

The process of globalization has brought great changes to human life, particularly in the field of information and communication technology. Smartphones has become one of the most important devices in people's lives today. A smartphone is a device that functions as more than just a device to make and receive phone calls, text messages, and voice messages. Other features includes enabling access to the internet, collecting and storing digital media such as pictures, music and videos, and also running computer programs, called apps (Weinberg 2012). Smartphone sales have increased in the last half decade as many countries have transitioned into digital-based communities. Currently, the smartphone is turning into the primary way for people to connect with family, friends, colleagues and the world. The smartphone industry has become the fastest growing industry in the world (Becker et al. 2012).

The large number of smartphone users indicates that there is a great opportunity in the smartphone market so that smartphone manufacturers should be able to seize this opportunity well (Lay-Yee et al. 2013). One of the premium-class smartphone brands in demand is the iPhone. This is evident from the results of research published by International Data Corporation (IDC), which states that the iPhone has been the top smartphone manufacturer, beating other brands by successfully breaking the latest product shipment record of 7,800,000 units in the fourth quarter of 2016, an increase of about 4.7% from the same quarter in 2015 (www.idc.com). This is interesting, considering that the price of the iPhone is more expensive than other brands with similar specifications.

With its higher than average price, the iPhone can be regarded as a luxury brand. Some researchers such as Horiuchi (1984), Dubois and Laurent (1994), Pantzalis (1995), Dubois and Paternault (1997), and Wong and Ahuvia (1998) posit that high price is an important attribute for luxury brands. In addition to high price, other characteristics of luxury brands are high quality (Aaker 1991, O'Cass and Frost 2002, Vigneron and Johnson 2004, Horiuchi 1984, Dubois and Laurent 1994, Dubois and Paternault 1997), uniqueness and exclusivity (Srinivasan et al. 2014). In addition, other researchers also mention that luxury brands go beyond functional benefits (Nueno & Quelch 1998) and are used to denote prestige and status (Hur et al. 2014).

There are some factors that contribute to a consumer's motive to purchase luxury branded products: luxury brand perception (Hung et al. 2011), social influence (Kulviwat 2009), vanity (Hung et al. 2011), and materialism (Richins & Dawson 1992).

Thus, this research seeks to reveal the internal and external factors of consumers that affect consumer intention to purchase luxury goods.

Luxury brand perception refers to customer perceived value in terms of the attractiveness or benefits of a luxury brand (Hung et al. 2011) and consists of functional value, experiential value, and symbolic value (Berthon et al. 2009). Functional value emphasizes the function of a luxury brand, manifested as the actual quality of a product or service (Hung et al. 2011) because quality can be a sign of what an object can do and how well it works (Berthon et al. 2009, Sweeney & Soutar 2001). Schiffman and Kanuk (1997) suggest that consumer evaluation of product quality will help them to consider which products they will buy. Vigneron and Johnson (2004) suggest that individuals who value luxury functions (e.g. emphasizing product quality) tend to have a positive attitude toward the purchase of luxury products.

The experimental dimension of experiential value relates to a subjective perceived experience of consumers' thoughts and feelings toward a luxury brand (Hung et al. 201I, Berthon et al. 2017). These thoughts and feelings are generated by brand-related stimuli that are part of the design, identity, packaging, communication, and environment of related brands (Berthon et al. 2009). Thus, luxury brands are often regarded as rare, valuable and unique (Hung et al. 2011). Vigneron and Johnson (2004) also support this idea that consumer's experimental value has a positive impact on their attitude about luxury purchases.

According to Truong et al. (2008), consumers also consume luxury brands for symbolic meanings. In other words, luxury brand ownership provides a sign of one's social status, wealth or power to others (Belk 1988, O'Cass 2004, Dubois & Duquesne 1993, Alden, Steenkamp & Batra 1999). Thus, our proposed hypothesis is:

H1: luxury brand perception affects consumer purchase intention.

When considering external factors, social influence affect consumers' intentions to purchase luxury brand (Hung et al. 2011). The concept of social influence asserts that social groups can influence consumers to perform certain behaviors including the purchase of a luxury brand. According to Hofstede et al. (2004), individual consumption decisions are systematically influenced by their cultural and social values and norms. This is because consumers use luxury brands as a way to be compatible with their social groups (Han et al. 2010). Therefore, our proposed hypothesis is:

H2: social influence affects consumer purchase intention.

Netemeyer et al. (1995) proposed dimensions of vanity, physical vanity and achievement vanity. Physical vanity emphasizes the importance of physical appearance, while achievement vanity emphasizes the importance of personal achievement. Mamat et al. (2016) posit that as part of the dimension of consumer vanity, consumers will be embarrassed if they do not look the way they want, which lead them to use luxury brands. This reflects that consumers who have excessive attention to physical appearance tend to purchase and consume luxury brands to support their performance. In addition, luxury brands can also be a way to show off their personal achievements. This is because luxury brands symbolically indicate personal success or achievements (Durvasula et al. 2008). Thus, our proposed hypothesis is:

H3: vanity positively affects consumer purchase intention.

In addition to the perception of luxury brands associated with the product, individual trait can also strengthen the effect of perception of luxury brands on consumer purchase intention toward luxury brand, that is, vanity (Hung et al. 2011). Also, vanity can strengthen the effect of social influences on consumer purchase intentions toward luxury brands (Hung et al. 2011). Social influences will affect consumption decisions including purchases of luxury brands to suit their social groups (Hofstede et al. 2004, Han et al. 2010). Consumers with vanity will prefer luxury brands. They will continue to desire and consume new products to satisfy their excessive desires and their need for self-esteem (Sedikides et al. 2007). Thus, our proposed hypothesis is:

H4: vanity moderates the effect of perception of luxury brands on consumer purchase intention.

H5: vanity moderates the effect of social influences on consumer purchase intention.

According to Richins and Dawson (1992), materialism consists of three dimensions: centrality, happiness, and success. Materialism refers to consumers' mind who believes that wealth is important to their lives. Wealth are thought to provide a meaning and complement one's life (Csikszentmihalyi & Rochberg-Halton 1981), and luxury brands are natural choices for materialistic consumers (Wong & Ahuvia 1998, Tatzel 2002). This will affect materialist consumers in their purchase intentions toward luxury brands. This is because materialist consumers tend to place wealth at the center of their lives (centrality). They believe that purchasing material goods or possessing possessions is important in their lives and capable of providing satisfaction in their lives (happiness). Moreover, materialists see success both of themselves and others by the extent to which they can have products that project the desired image (Richins & Rudmin 1994). They believe that the number and quality of one's possessions is an indicator of one's success. For that reason, our proposed hypothesis is:

H6: materialism positively affects consumers' buying intentions.

2 METHOD

The population in this study is iPhone owners, who have the latest version of the iPhone and are familiar with information about the iPhone. Questionnaires were distributed to 244 respondents.

Perception of luxury brands refers to values felt by consumers related to the attractiveness or benefits of luxury brands. Indicators are adapted from Hung et al. (2011), Shukla et al. (2015), and Cerqueira (2015). Social influence refers to the extent to which reference group members influence consumer behavior in purchasing luxury brands. Indicators are adapted from Hung et al. (2011). Vanity refers to an excessive consumer concern or a positive outlook on physical appearance and/or personal achievement. Indicators are adapted from Hung et al. (2011). Materialism refers to major belief in the importance of wealth in the lives of consumers. Indicators are adapted from Sun et al. (2014) and Heaney (2005). Purchase intention refers to consumer desire in purchasing luxury brands. Indicators are adapted from Hung et al. (2011). All indicators were measured using a 5-point Likert scale. Data was processed using PLS.

3 RESULTS AND DISCUSSION

Table 1 shows the estimated values for the path relationships in the inner model obtained with the bootstrapping procedure. The value considered significant is if the statistical t value is greater than 1.65 for each of its path relations.

Based on the results of statistical tests, social influence does not affect consumer purchase intention (T-stats < 1.65) because consumers tend to be more concerned with aspects of technological sophistication, high quality, better design and price when they consider purchasing smartphone. This finding is similar to the results of research by Yang et al. (2007) that the influence of reference groups or social groups is a less important aspect when choosing a smartphone.

Vanity also does not affect consumer purchase intention (T-stats < 1.65). This finding is consistent with research by Park et al. (2007), whose respondents' characteristics were similar with this study. Majority of respondents are young people with average income while the price of the iPhone is quite expensive so it becomes a barrier. This is supported by the results of a follow-up survey of 20 respondents who have no intention to purchase iPhone. The majority of respondents stated that the expensive price prevents them from wanting to purchase

iPhone. So it can be said although consumers have a high level of vanity, it does not make them intend to purchase a luxury brand because they are not financially independent (Hung et al. 2011).

Furthermore, based on the results of statistical tests, vanity does not moderate the effect of luxury brand perceptions on consumer purchase intention. Despite the fact that H1 supported, even though consumers have a high perception of the iPhone and have a high level of vanity, but their purchase intention is still lacking for financial reasons. In addition, according to the results of further surveys, other reason that also contributed is that respondents feel there is still no need to change the phone at this time.

Based on the results of statistical tests, it also can be seen that vanity variables do not moderate the effect of social influence on consumer purchase intention. These findings are consistent with the results of research conducted by Hung et al. (2011). Despite the fact that H2 is supported, consumers in this study were not influenced by their social groups in conjunction with their purchase intention although their level of vanity is high because of financial reason.

4 CONCLUSIONS

This study found that luxury brand perception that is formed from functional, experiential, and symbolic value affects purchase intention toward luxury brand. So it is important for manufacturers and marketers of premium-class smartphone brands to improve functional aspects characterized by high quality or sophistication of the product, experiential aspect that can make consumers feel that the brand is valuable, engaging and unique through an exclusive design, and symbolic aspect that is able to signify social status users through price and the form of marketing stimuli such as advertising or other forms of promotion.

This study found that materialism affected consumers' intention to purchase luxury branded smartphone, so it is important for manufacturers and marketers to develop stimuli that can improve the materialistic side of their target customers. It is expected that this will make consumers pay more attention to their public rather than personal self-image, so they will choose to adapt to their public image through ownership of products offered.

Since this study has limitations on respondents with average incomes, further research especially in researching a luxury brand should include respondents with above-the-average income to better reflect true purchase intentions.

Table 1. Path coefficient dan t-statistics.

Independent Variable: Purchase Intention	Original Sample	T Statistics	Remark
Perception of Luxury Brands	0.289	4.166	H1 Supported
Social Influence	0.054	0.789	H2 Not supported
Vanity	0.067	0.809	H3 Not supported
Vanity x Perception of Luxury Brands	0.057	0.750	H4 Not supported
Vanity x Social Influence	0.060	0.778	H5 Not supported
Materialism	0.345	5.243	H6 Supported

REFERENCES

Aaker, D. 1991. Brand equity. *La gestione del valore della marca* 347–356.
Alden, D.L., Steenkamp, J.B.E. & Batra, R. 1999. Brand positioning through advertising in Asia, North America,

and Europe: The role of global consumer culture. *The Journal of Marketing* 75–87.

Belk, R.W., 1988. Possessions and the extended self. *Journal of consumer research* 15(2): 139–168.

Berthon, P., Pitt, L., Parent, M. & Berthon, J. P. 2009. Aesthetics and ephemerality: observing and preserving the luxury brand. *California management review* 52(1): 45–66.

Cerqueira, D., Matos, M.V.M., Martins, A.P.A. & Pinto, J.J., 2015. Avaliando a efetividade da Lei Maria da Penha.

Dubois, B. & Duquesne, P. 1993. The market for luxury goods: Income versus culture. *European Journal of Marketing* 27(1): 35–44.

Dubois, B. & Laurent, G. 1994. Attitudes towards the concept of luxury: An exploratory analysis. *ACR Asia-Pacific Advances*.

Dubois, B. & Paternault, C. 1997. Does luxury have a home country? An investigation of country images in Europe. *Marketing and Research Today* 25(2): 79–85.

Durvasula, S. & Lysonski, S. 2008. A double-edged sword: understanding vanity across cultures. *Journal of Consumer Marketing* 25(4): 230–244.

Han, P., Niu, C.Y., Lei, C.L., Cui, J.J. & Desneux, N. 2010. Use of an innovative T-tube maze assay and the proboscis extension response assay to assess sublethal effects of GM products and pesticides on learning capacity of the honey bee Apis mellifera L. *Ecotoxicology* 19(8): 1612–1619.

Heaney, J. G., Goldsmith, R. E. & Jusoh, W. J. W. 2005. Status consumption among Malaysian consumers: Exploring its relationships with materialism & attention-to-social-comparison-information. *Journal of International Consumer Marketing* 17(4): 83–98.

Hofstede, G., Noorderhaven, N.G., Thurik, A.R., Uhlaner, L. M., Wennekers, A.R. & Wildeman, R.E., 2004. Culture's role in entrepreneurship: self-employment out of dissatisfaction. *Innovation, entrepreneurship and culture: The interaction between technology, progress and economic growth, 162203.*

Horiuchi, Y. 1984. *A systems anomaly: consumer decision-making process for luxury goods.* Doctoral dissertation, Graduate School of Arts and Sciences, University of Pennsylvania.

Hung, K. P., Huiling Chen, A., Peng, N., Hackley, C., Amy Tiwsakul, R. & Chou, C.L. 2011. Antecedents of luxury brand purchase intention. *Journal of Product & Brand Management* 20(6): 457–467.

Hur, W.M., Kim, M. & Kim, H. 2014. The role of brand trust in male customers' relationship to luxury brands. *Psychological reports*, 114(2): 609–624.

Kulviwat, S., Bruner, G.C. & Al-Shuridah, O. 2009. The role of social influence on adoption of high tech innovations: The moderating effect of public/private consumption. *Journal of Business Research* 62(7): 706–712.

Lay-Yee, K. L., Kok-Siew, H. & Yin-Fah, B. C. 2013. Factors affecting smartphone purchase decision among Malaysian generation Y. *International Journal of Asian Social Science* 3(12): 2426–2440.

Mamat, M. N., Noor, N. M. & Noor, N. M. 2016. Purchase intentions of foreign luxury brand handbags among consumers in Kuala Lumpur, Malaysia. *Procedia Economics and Finance* 35: 206–215.

Netemeyer, R.G., Burton, S. & Lichtenstein, D.R. 1995. Trait aspects of vanity: Measurement and relevance to consumer behavior. *Journal of consumer research* 21(4): 612–626.

Nueno, J.L. & Quelch, J.A. 1998. The mass marketing of luxury. *Business Horizons* 41(6): 61–61.

O'cass, A. & Frost, H. 2002. Status brands: examining the effects of non-product-related brand associations on status and conspicuous consumption. *Journal of product & brand management* 11(2): 67–88.

O'Cass, A. 2004. Fashion clothing consumption: antecedents and consequences of fashion clothing involvement. *European Journal of Marketing* 38(7): 869–882.

Pantzalis, I. 1995. *Exclusivity strategies in pricing and brand extension.*

Park, H. J., Rabolt, N. J. & Sook Jeon, K. 2008. Purchasing global luxury brands among young Korean consumers. *Journal of Fashion Marketing and Management: An International Journal* 12(2): 244–259.

Richins, M.L. & Rudmin, F.W. 1994. Materialism and economic psychology. *Journal of Economic Psychology* 15 (2): 217–231.

Schiffman, L.G. and Kanuk, L.L., 1997. Comportamiento.

Sedikides, C., Gaertner, L. & Vevea, J.L. 2007. Evaluating the evidence for pancultural self-enhancement. *Asian Journal of Social Psychology* 10(3): 201–203.

Shukla, P., Singh, J. & Banerjee, M. 2015. They are not all same: variations in Asian consumers' value perceptions of luxury brands. *Marketing Letters* 26(3): 265–278.

Srinivasan, R., Srivastava, R.K. & Bhanot, S. 2014. A Study of the antecedents of purchase decision of luxury brands. *Journal of Business and Management* 16 (5): 99–101.

Sun, G., D'Alessandro, S. & Johnson, L. 2014. Traditional culture, political ideologies, materialism and luxury consumption in China. *International Journal of Consumer Studies* 38(6): 578–585.

Sweeney, J.C. & Soutar, G.N. 2001. Consumer perceived value: The development of a multiple item scale. *Journal of retailing* 77(2): 203–220.

Tatzel, M. 2002. "Money worlds" and well-being: An integration of money dispositions, materialism and price-related behavior. *Journal of Economic Psychology* 23(1: 103–126.

Vigneron, F. & Johnson, L.W. 2004. Measuring perceptions of brand luxury. *The Journal of Brand Management*, 11 (6), 484–506.

Wong, N.Y. & Ahuvia, A.C. 1998. Personal taste and family face: Luxury consumption in Confucian and Western societies. *Psychology & Marketing* 15(5): 423–441.

www.idc.com. 2017. Smartphoe Market Share [Online]. Retrieved https://www.idc.com/promo/smartphone-market-share/os. Accessed 29 May 2018.

Yang, J., He, X. & Lee, H. 2007. Social reference group influence on mobile phone purchasing behaviour: a cross-nation comparative study. *International Journal of Mobile Communications* 5(3): 319–338.

Advances in Business, Management and Entrepreneurship – Hurriyati et al (eds)
© *2020 Taylor & Francis Group, London, ISBN 978-0-367-27176-3*

Analysis of direct premium influence on brand attitudes and consumer purchase intention

M. Kurniawati & T. Widianto
Universitas Airlangga, Surabaya, Indonesia

ABSTRACT: Competition in automotive industries, particularly in the diesel-fueled MPV (Multi-Purpose Vehicle) category, is getting tougher. This prompts every company in the automotive industry to intensify their efforts in order to promote their sales volume. Isuzu is one of these companies. Panther is an Isuzu product that is classified in the diesel-fueled MPV category. To increase its sales, Isuzu uses sales promotion techniques in the form of direct premium. Direct premium itself consists of two categories: fit and non-fit with the core product. Direct premium is applied in order to create a positive attitude towards the brand that will ultimately affect consumers' purchase intention. This study aims to investigate whether there are the differences in consumers' attitudes toward brand and purchase intention on the application of direct premium fit and non-fit with the core product (Isuzu Panther). In addition, it also tests whether the attitude toward brand (Isuzu) affects consumers' purchase intention on product (Isuzu Panther). This research was conducted using experimental design. The hypothesis was tested by using One Way ANOVA and simple linear regression. The results of this study found that in the case of direct premium, there were the different in attitude toward brand and consumers' purchase intention. Attitude toward brand in the case of direct fit was higher than non-fit. Consumers' purchase intention in direct fit was also higher than non-fit. It is also found that attitude toward brand had positive significant influence on consumer purchase intention.

1 INTRODUCTION

Isuzu is a growing brand in Indonesia. Panther has become the flagship brand of Isuzu. Panther has two types of vehicles, the minibus and the pickup truck, and both types fall into Low Commercial Vehicle (LCV) category. In the automotive industry, the Isuzu Panther minibus belongs to the diesel-fueled Multi-Purpose Vehicle (MPV). Current market condition indicates intense competition among inter-brand and inter-products on the diesel-fueled MPV category. This intense competition in the automotive industry, especially in diesel-fueled MPV category, requires manufacturers to be more active and aggressive in fulfilling consumers' need.

It is important for manufacturers to maintain the existence of their products or services. One way to achieve it is through effective marketing communications. One component of marketing communication mix that is often used by companies is sales promotion. Sales promotion is a various sets of incentive tools that are mostly short-term designed to stimulate the purchase of certain products or services faster and larger by consumers or business (Kotler & Keller 2016). The purpose of sales promotion is to reduce the risk of consumers switching brands and accelerate the purchase decision process. One method of sales promotion is premium (gift-giving). Premium is the goods offered with

a relatively low price or free as an incentive to buy certain products (Kotler & Keller 2016).

Isuzu implements sales promotions in order to attract consumers and accelerate consumers' purchase decisions. Specifically, the method used is direct premium. With this method, Isuzu provides direct prizes to consumers for the purchase of Panther's minibus. Direct Premium provided by Isuzu to its customers is in the form of goods and services, that is, free spare parts, TVs, smartphones, GPS, free service, tablets, and cash back. All of the prizes provided by Isuzu to its customers can be grouped into fit and non-fit direct premium. Fit refers to the level of proximity or conformity between a product being promoted or awarded with the main product (Montaner et al. 2011). Direct Premium is intended to create a brand attitude (Landreville & d'Astous 2003). Brand attitude affects purchase intention. According to Wu and Lo (2009), the higher the brand attitude, the higher consumers' purchase intention.

This study aims to confirm the affect of fit and non-fit direct premium toward brand attitude and its impact on consumers' purchase intention. Specifically, this study tries to investigate the difference of brand attitude toward consumers' purchase intention on fit and non-fit direct premium of Isuzu product and also to confirm whether the attitude on the brand affects consumers' purchase intention on Isuzu products.

1.1 Premium-based sales promotion

According to Kotler and Keller (2016), premium-based sales promotion is goods offered with a relatively low price or free as an incentive to buy certain products. d'Astous and Landdreville (2003) define premium-based sales promotion as goods or services offered for free or at a relatively low price on a single or many purchases of products or services.

1.2 Brand attitude

According to Assael (2001), brand attitudes refers to a mental statement of the recipient of a message that assesses positive or negative, good or not good, like or dislike, toward a product. Howard (1994) conveys consumers' brand attitudes can emerge after they get knowing the brand or exposing to advertising message (information). The tendency of consumers to judge a brand on the basis of likes or dislikes is represented by three factors: belief, evaluation of a brand, and a tendecy to act:

1. The feeling of belief affected by brands
2. Brand attitude can affect consumers
3. Intention to purchase

When the brands belief matches the perceived benefit, the consumers' judgment of the brand will be favorable. Assessment of this positive brand often raises the intention to purchase toward the brand (Assael 2001).

1.3 Purchase intention

The purchase intention is the stage before the purchase decision in the purchasing decision process (Kotler & Keller, 2016). According to Schiffman and Kanuk (2008), buying intention is an attitude of pleasure towards the object that makes the individual trying to get the object by paying it. Howard (1994) defines purchase intentions as a mental statement from consumers that they plan to purchase a number of products with a particular brand. Consumers' intentions to purchase are influenced by attitudes toward advertising and brand (Goldsmith et al. 2000).

1.4 Relationship between direct premium and brand attitude

According to D'astous and Landreville (2003), premium can be regarded as a reward that might lead to a positive reaction to a product or brand purchased or can be considered a persuasive trick set up to make the product more attractive. The studies conducted by d'Astous and Landreville (2003) suggest that unattractive premium offerings may affect brand image and consumers' attitudes toward brands. This is in line with the study by Simonson et al. (1994), which found that consumers reacted negatively to brands which provided gimmicks like useless premiums in an attempt to attract potential customers. From these explanations, the first hypothesis that can be put forward is:

H1: Consumers' attitudes toward brands will be higher in fit compared to non-fit direct premium.

1.5 Relationship between direct premium and purchase intention

Landreville and d'Astous (2003) note that attractively packaged premiums are a significant factor in shaping consumer evaluation reactions. In their study, Simonson et al. (1994) mention that a promotion offered which includes attractive premiums will be better evaluated by consumers. Consumers show a positive response on attractive premiums, which fit with the main product. Although consumers generally value promotional offers with attractive premiums, the results show that the suitability between premiums and products has a positive and significant impact on the appreciation of promotional offers when premiums are unattractive (Landreville & d'Astous 2003). The level of fit between premium and the main product will increase purchase intention (Montaner et al. 2011). Premium that fit with the main product will have a significant impact on consumers' evaluation. The use of premium promotions on strong product brands contributes to greater purchase intentions. In addition, a high fit between the product and the premium offered will increase consumers' purchase intentions (Montaner et al. 2011). Application of sales promotion techniques such as premiums will increase customer satisfaction and purchase intentions. This is evident in the Malaysian market, where sales using these techniques increase sales (Weng & Run 2013). From this explanation, the second hypothesis that can be put forward is:

H2: Consumer purchase intentions will be higher in fit compared to non-fit direct premium.

1.6 Effect of brand attitude toward purchase intention

Attitude toward a particular brand often affects whether consumers' will buy a product or not. A positive attitude towards a particular brand will lead consumers to make purchases of that brand. Conversely, a negative attitude will prevent consumers to make a purchase (Sutisna 2002). Positive and negative feelings towards brand will eventually create purchase intention. The higher the brand attitude, the higher the purchase intention (Kurniawati 2009). From this explanation, the third hypothesis that can be put forward is:

H3: The attitude of the brand will affect the consumers' buying intention.

2 METHOD

2.1 Population and sample

Population is a composite of all elements that have a set of similar characteristics that encompass the universe for the sake of research problems (Malhotra & Birsk 2005). According to Indriantoro and Supomo (2002), if the population is relatively large or even difficult to quantify, then research on all elements of the population will be difficult to do. The population of this study was prospective customers (hot prospect) who would buy Isuzu Panther vehicles. The prospective customers included in the hot prospect category were potential customers who had asked the price in detail, credit calculations, and had a plan to buy in the near future. To facilitate researchers in conducting research, a sample was used to represent the existing population. The method used in sampling was accidental sampling, which took the respondent as a sample by chance (Sugiyono 2004). In this research, respondents were shown two different car advertisements with different direct premiums. Each advertisement was shown to 40 respondents. Therefore, the sample used for this research was 80 people. Based on the opinions expressed by Gay and Diehl (1996), the number of samples used for experimental studies should be at least 15 subjects per group.

2.2 Variable measurement

The independent variable of this study was direct premium. Direct premium is a product that is offered for free at the time of purchase of Isuzu Panther products. Direct premium is in two categories: fit and non-fit (Montaner et al. 2011). A fit direct premium is GPS (Global Positioning System). A non-fit direct premium is a 32 "LED TV.LED TV 32."

The dependent variables of this study were brand attitude and purchase intention. Brand attitude in this study is the mental revelation of message recipients who rate positive or negative, happy or unhappy, likes or dislikes, qualified or not qualified to Isuzu Panther. Variable measurements are adapted from Gubta and Dang (2009). The indicators are:

1. Interest levels of Isuzu Panther
2. Fondness levels of Isuzu Panther
3. Happiness levels of Isuzu Panther

In addition, another dependent variable was purchase intention. Purchase intention is a mental statement from consumers who reflect the purchase plan of Isuzu Panther car. Variable measurements are adapted from Gupta and Dang (2009) as follows:

1. Plan to purchase Isuzu Panther products.
2. Intent to purchase Isuzu Panther products.

2.3 Data analysis method

In this study, one-way Anova was used to analyze the influence of direct premium (fit and non-fit) on consumers' attitude toward brand and their purchase intention. Furthermore, simple linear regression analysis was conducted to predict how far the functional or causal relationship of consumer attitudes towards the brand to purchase intention.

3 RESULTS AND DISCUSSION

From the ANOVA analysis, it was found that there was a significant difference in consumers' brand attitude toward fit and non-fit direct premium ($F = 8.761$; Sig = 0.004). In other words, there was a difference in brand attitudes from both conditions (the significance level was less than 5%). The mean of consumers' attitude toward brand was higher in fit direct premium condition than in non-fit ($Mean_{fit}$: 5.634 $Mean_{non-fit}$: 4.950). This means that H1 is accepted. Results from ANOVA also shows that there was a significant difference in consumers' purchase intention on fit and non fit direct premium ($F = 6.166$; Sig = 0.015). In other words, there were the differences in purchase intentions from both conditions (signification rate less than 5%). The mean of purchase intention in fit direct premium condition was higher than in non-fit ($Mean_{fit}$: 5.312; $Mean_{non-fit}$: 4,575). This means that H2 is accepted. The result of simple linear regression is shown in Table 1.

Since the significant value for brand attitude coefficient was below 5% (Sig. 0.000), the third hypothesis is also accepted.

The result for the first hypothesis shows that there is a significant difference on consumer brand attitude toward fit and non-fit direct premium. Therefore, it can be concluded that prospective consumers prefer fit direct premium to non-fit direct premium. This is similar to Montaner et al. (2011)'s study, which found that fit direct premium produced positive consumer reaction and non-fit direct premium produced negative reaction. Simonson et al. (1994) also argue that unattractive premium deals can affect brand image and consumers' attitudes toward brands. It is also supported by D'astous and Jacob (2002) who found that consumers reacted negatively to brands that provide gimmicks such as useless premiums in an attempt to attract potential customers. Another study from D'Astous and Landreville (2003) also found that on non-fit direct premium condition, the promotional offerings may be considered inconsistent and may be considered opportunistic.

Table 1. Result of simple linear regression.

Variables	Coefficient	t-value	Sig.
Constanta	-0.049	9.857	0.000
Brand attitude (X)	0.944		
	R square = 0.555		
	R = 0.745		

The result for the second hypothesis shows that there is a significant difference on consumers' purchase intention toward fit and non-fit direct premium. Fit between products and direct premium has a significant impact on consumer evaluation. The use of premium promotions on strong brands contributes to greater purchase intentions. In addition, Montaner et al. (2011) also found that a strong fit between product and premium offered increased consumers' purchase intentions.

The tests for the third hypothesis shows that brand attitude significantly affects consumer purchase intention. This result is similar to Kurniawati's (2009) study, which found that positive and negative feelings toward brand ultimately produced purchase intentions. The higher the attitude toward the brand, the more consumer purchase intention will increase.

4 CONCLUSIONS

From the results of this study, it can be concluded that:

1. There is a significant difference on consumers' brand attitude toward fit and non-fit direct premium, where brand attitude toward fit direct premium is higher than non-fit direct premium.
2. There is a significant difference on consumers' purchase intention toward fit and non-fit direct premium, where consumers' purchase intention toward fit direct premium is higher than non-fit direct premium.
3. The brand attitudes positively affect consumers' purchase intentions.

It is recommended for marketers to apply fit direct premium as it has higher positive response on brand attitudes and consumers' purchase intention compared to non fit direct premium with the main product. Also, it is suggested that marketers should understand the direct premium that is considered 'fit' with the main product. Sometimes a premium that is considered fit with the main product does not have to have the same or strong relationship (e.g. a car with GPS), but has the benefit or an effect on the main product. When consumers purchase oatmeal, for instance, they would also get a plastic box that can be used to store the product.

For further research, researchers can use direct premium fit and non-fit with different price or value to investigate the response from consumers if the value of premium offered have different value or price. This is in line with d'Astous and Jacob (2002), who argue that consumers will prefer fit direct premium than non-fit even when non-fit premiums have a higher price or value.

REFERENCES

Assael, H. 2001. *Consumer Behavior and Marketing Action 6th Edition*. New York: South Western College Publishing.

D'Astous, A. and Jacob, I. 2002. Understanding Consumer Reactions to Premium-Based Promotional Offers. Canada: *European Journal of Marketing* Vol 36, pp. 1270–1286.

D'Astous, A. and Landreville, V. 2003. An Experimental Investigation of Factors Affecting Consumers' perceptions of Sales Promotions. Canada: *European Journal of Marketing* Vol 37, pp. 1746–1761.

Gay, L.R. and Diehl, P.L. 1996. *Research Methods for Business and Management*. Singapore: Simon & Schuster Ltd.

Goldsmith, R.E., Lafferty, B.A., and Newell, S.J. 2000. *The influence of corporate credibility on consumer Attitudes and purchase intention*. Corporate Reputation Review, 3 (4): 304–318.

Gubta, M.A. and Dang, P.J. 2009. Examining Celebrity Expertise and Advertising Effectiveness in India. *South Asian Journal of Management, 16(2):61*

Howard, J.A. 1994. *Buyer Behavior in Marketing Strategy. Englewood Cliffs*. New Jersey: Prentice Hall.

Indiantoro, N. and Supomo, B. 2002. *Metodologi Penelitian Bisnis: Untuk Akuntansi dan Manajemen. Edisi pertama. Cetakan kedua*. Yogyakarta: BPFE Yogyakarta.

Kotler, P. and Keller, K. 2016. *Marketing Management*. New Jersey: Prentice.

Kurniawati, D. 2009. *Studi Tentang Sikap Terhadap Merek dan Implikasinya Pada Minat Beli Ulang*. Tesis tidak diterbitkan. Semarang Pasca Sarjana Universitas Diponegoro.

Malhotra, N. K and Birks, D. F. 2005. *Marketing Research: An Applied Approach*. New Jersey: Pearson Education, Inc, Prentice Hall.

Montaner, T., De Chernatony, L., & Buil, I. (2011). Consumer response to gift promotions. *Journal of Product & Brand Management, 20*(2), 101–110.

Schiffman, L. G. and Kanuk, L.L. 2008. *Consumer Behavior, 9th Edition*. New Jersey. Pearson Prentice Hall.

Simonson, I., Carmon, Z., & O'curry, S. (1994). Experimental evidence on the negative effect of product features and sales promotions on brand choice. *Marketing Science*, 13(1),23–40.

Sugiyono. 2004. *Metode penelitian Bisnis*. Bandung; CV Alfabeta.

Sutisna, 2002. *Perilaku Konsumen daqn Komunikasi Pemasaran*. Bandung: PT Remaja Rosdakarya Offset.

Weng, J. T. and de Run, E. C. 2013. Consumer's Personal Values and Sales Promotion Preferences Effect on Behavioural Intention and Purchase Satisfaction for Consumer Product. Malaysia: *Asia Pasific Journal of Marketing and Logistics* Vol 25 PP 70–101.

Wu, S. I. and Lo, C. L. 2009. The Influence of Core-Brand attitude and Consumer Perception on Purchase Intention Towards Extended Product. *Taiwan: Asia Pacific Journal of Marketing and Logistics* Vol. 21, No. 1, pp. 174–194.

Stimulus of social media: The influence of e-wom towards visiting interest in the Lodge Maribaya through trust as mediation variable

A. Widodo, R. Yusiana & F.N. Aqmarina
Telkom University, Bandung, Indonesia

ABSTRACT: The presence of the Internet brings many advantages that can be seen in how technology can be enjoyed by almost all communities without any limitations and difficulties. The growth of the Internet must be accomplished by the increasing number of users of social media and social media for business. One of them is Instagram where there are great opportunities to do promotion. The result of this research shows that variable of Electronic Word of Mouth partially influences the Trust. Based on *Test T* Trust partially has positive effect on Visiting Interest. It can be concluded that the Electronic Word of Mouth has a positive and significant impact on the Trust at The Maribaya Lodge. Trust directly influences interest in visiting The Lodge Maribaya. The indirect influence between the Electronic Word of Mouth variables on Trusted Visibility has a significant positive effect and has a large role. Electronic Word of Mouth has a significant effect on Visiting Interest at The Lodge Maribaya.

1 INTRODUCTION

The advance of Internet has many advantages and benefits for a wider community. The Internet has become an important part of human life in this modern era. Various activities can be conducted by using the Internet, such as working, looking for a job, communicating, selling and many more.

According to APJII (Asosiasi Penyelenggara Jasa Internet Indonesia, an association for Internet providers in Indonesia), the number of Internet users in Indonesia in 2016 was 132.7 million users, or about 51.5% of the total population of Indonesia of 256.2 million. Most Internet users are on the Java Island with a total of 86,339,350 users or about 65% of the total users of Internet. If it is compared to the users of Internet in Indonesia in 2014 which was 88.1 million users, then there is an increase of 44.6 million within 2 years (-2014–2016). The growing number of Internet users is also accompanied by the increasing number of users of social media services. From 79 million users in 2015, that number has now risen to 129.9 million users. The users who actively use social media on mobile devices also rose from 66 million to 92 million (APJII 2017).

The results of the survey from APJII also lays out some other interesting things, for example that the use of Internet is still dominated by the use of social media and entertainment. The users of Instagram and Snapchat are said to have doubled since 2014. This is in line with the growing engagement on both platforms with such high visuals. The increase of Instagram users in Indonesia has come about because the people of Indonesia perform daily activities with more frequent use of social media such as uploading photos to share a moment with others and do activities that are considered to be 'fun'. Social media has developed into a medium for consumers to obtain company information or selling products as well as connecting with family and friends.

The conversations that appear in Instagram social networks, especially social media for sharing photos, can be treated as an integrated marketing communications process which needs planning and building all relevant marketing communications, so that, all aspects can work together in harmony to provide greater and more efficient effect in marketing (Bataineh 2015). It is not only companies that provide information about products but also consumers who have been using the product and who do not hesitate to share information about their experiences. Seeing great opportunities for promotion, the Lodge Maribaya uses Instagram, Facebook, and Twitter for publicity and promotion. However, the Lodge Maribaya, among all the available social media, they are most active using Instagram under the name @thelodgemaribaya.

The development of social media makes that the spread of Electronic Word of Mouth is not limited to face-to-face communication. The spread of electronic word of mouth can be intentional or unintentional. Intentionally means electronic word of mouth is formed consciously and planned. Unintentional means the spread of electronic word of mouth is random and unplanned so that electronic word of mouth just emerges (Lin et al. 2013). To add, the number of Electronic Word of Mouth on social media can make people curious to see.

Hasan (2010) defines electronic Word of Mouth as an important aspect of marketing programs, as well as in developing consumer expression of the brand. Since electronic word of mouth shows all the important signs, its application will become more widespread in the future as a social networking marketing tool. The three components of brand knowledge are the key areas where electronic word of mouth has a direct influence and future knowledge about brand-related purchases (brand trust, brand satisfaction, brand loyalty) that has a strong impact as a recommender in social marketing networks.

Cakim (2010) defines consumer trust as all knowledge possessed by consumers, and all conclusions made by consumers about objects, attributes, and benefits. The objects can be products, people, companies, or anything in which a person has trust and attitude (Zhang et al. 2010). Jansen et al. (2009) states that interest in visiting is the impulse of the consumer in the form of a desire to visit a place or region that attracts one's attention. The relation with tourism theory of interest is taken from the theory of buying interest to a product, so that some categories of buying interest can be applied to the interest of visiting (Paquette 2013).

1.1 Marketing communications

According to Chang, Lee and Huang (2010), marketing communication occurs between producers, intermediaries, marketing, and consumers, and is an activity to help consumers make decisions in marketing and direct exchange or transactions to be more satisfying by encouraging all parties to think, do, and behave in a better way.

According See-To and Ho (2014). marketing psychology is essentially the integration of psychology and human behavior into marketing activities to achieve marketing objectives, namely maximum sales.

Lin et al. (2013) states that EWOM is a type of communication that greatly affects decisions and changes in customer behavior. Word of mouth affects seven times more brand switching over news and journals, four times more than a salesperson's sales, and twice as much on radio advertising.

Electronic word of mouth is an important aspect of marketing programs in developing consumer expression of the brand (Lin et al. 2013). The effect of online branding shows that current purchase is due to the strong role of electronic word of mouth in building and activating the brand image, there are dimensions that can be used to measure electronic word of mouth, which are:

a. Intensity in electronic word of mouth: The number of opinions written by consumers on a social networking site.
b. Positive Valence: This is a positive consumer opinion about products, services, and brands.
c. Negative Valence: This is a negative consumer opinion about products, services, and brands.
d. Content: The content of information from social networking sites related to products and services.

According to Boztepe (2012), belief is the power that a product has certain attributes. The belief is often called an object-attribute linkage where the consumer believes there is a possible existence of a relationship between an object and its relevant attributes (Cakim 2010). These are the factors that

make up one's belief in others: ability, benevolence, and integrity.

According to Jansen et al. (2009), interest in visiting is the impulse of the consumer in the form of a desire to visit a place or region that attracts the attention of a person. The association with tourism interest theory of interest is derived from the theory of buying interest on a product, so that in some categories of buying interest, can be applied to the visiting interest (Goyette 2010). There are indicators of buying interest of a potential consumer which are: attention, interest, desire and action.

2 METHOD

This study uses quantitative methods based on the philosophy of positivism to examine a particular population or sample. The data are collected by research instruments, quantitative/statistical data analysis which aim to test the hypotheses that have been determined (Paquette 2013). This research uses descriptive research type and causality.

This research uses Electronic Word of Mouth as an independent variable, Interest of Visiting as a dependent variable and Trust as a mediation variable. In this research, the sampling technique is non-probability sampling with incidental sampling type.

To support the analysis in this study, the data are primary data in the form of distributed questionnaires and secondary data such as books, literature, journals, scientific papers or previous research.

3 FINDINGS AND DISCUSSIONS

Based on the calculation results from SPSS, researchers got the results of path analysis as illustrated in Figure 1.

Based on the results of data processing information between the variables, Electronic Word of Mouthhad a significant effect on the Trust variable at The Lodge Maribaya on Instagram users in Bandung. This can be seen from t hitung > t table (13.614 > 1.960) and significant value 0.000 < 0.05. Based on the calculation of coefficient of determination (R^2) it can be seen that the variable of electronic Word of Mouth to Trust is equal to 55.6%. The remaining amount of 44.4% by other factors

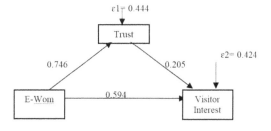

Figure 1. Two line equation regression analysis.

69

was not examined in this study. This means that the Word of Mouth Electronics on Trust at The Lodge Maribaya is 55.6%.

Based on a T test of the Trust variable, a partially significant effect on the Visiting Interest variable can be seen from t count > t table resulting in 2.544 > 1.960 and a significant value equal to 0.012 < 0.05. The magnitude of Trust's influence on Interest in The Lodge Maribaya is 4.2% and the rest of 95.8% is influenced by other variables beyond this study. It shows that Trust has a significant influence on the Visiting Interest at The Lodge Maribaya amounting to 4.2%. The encouragement of factors that form the trust of The Lodge Maribaya can increase the sense of interest from consumers in visiting the place and with positive information, it can bring the consumer confidence.

Based on the coefficient of indirect influence of the variable of Electronic Word of Mouth to the Interest of Visiting variable of Trust, it equals to 46.5% and the rest equals to 53.5% because it is influenced by other variables beyond this research. This indicates that the indirect influence between Electronic Word of Mouth variables on Trust Visibility has a significant influence and has a big role compared to direct influence. This happens because of the stimulus through Trust and it can increase the indirect effect.

Based on the results of correlation data of the variable Electronic Word of Mouth, there is a partial effect on the variable Interest of Visiting at The Lodge Maribaya on Instagram users in the city of Bandung. This can be seen from t count > t table (7.369 > 1.960) and a significance value of 0.000 < 0.05. It can be seen that the influence of Electronic Word of Mouth variables on Visiting Interest is 35.2% while the remaining 64.8% is influenced by other factors which are not examined in this research. This shows that Electronic Word of Mouth has a significant effect on the Visiting Interest at The Lodge Maribaya. The existence of positive Electronic Word of Mouth can directly affect the interest of consumer visiting. In addition to the indicator of buying interest of a potential, a consumer can influence the decision-making process.

4 CONCLUSION AND SUGGESTIONS

Based on the results of the research that has been done, conclusions that can be drawn include:

a. Effect of Electronic Word of Mouth on Trust at The Lodge Maribaya

Based on the results of data processing correlation, the Electronic Word of Mouth variable significantly influences the variable Trust at The Lodge Maribaya on Instagram users in Bandung. This shows that Electronic Word of Mouth has a positive and significant impact on the Trust at The Lodge Maribaya.

b. The Effect of Trust on the Visiting Interest in The Lodge Maribaya

Based on the T test of the Trust variable, it has

a partially significant effect on the Visiting Interest variable. This indicates that the Trust directly influences interest in visiting The Lodge Maribaya.

c. Effect of Electronic Word of Mouth on Visiting Interest mediated by the Trust at The Lodge Maribaya

Based on the coefficient of the indirect influence of the Electronic Word of Mouth variable to the Visiting Interest variable of Trust variable, is 46.5% . This indicates that the indirect influence between Electronic Word of Mouth variable to Interest Trust-mediated visits has a significant positive influence and has a large role compared to direct influence.

d. Effect of Electronic Word of Mouth on Visiting Interest.

Based on the results of correlation data processing between variables Electronic Word of Mouth has a partial effect on the variable Interest Visits at The Lodge Maribaya on Instagram users in Bandung. It is known that the influence of Electronic Word of Mouth variable on Visiting Interest is 35.2%. This shows that Electronic Word of Mouth has a significant effect on Visiting Interest at The Lodge Maribaya.

REFERENCES

APJII. 2017. Hasil survei penetrasi dan perilaku pengguna Internet Indonesia 2016 [Online]. Retrieved https://apjii. or.id/survei. Accessed 26 June 2018.

Boztepe, A. 2012. Green marketing and its impact on consumer buying behavior. *European Journal of Economic and Political Studie*. 5(1): 7.

Cakim, I. 2010. *Implementing word of mouth marketing*. New Jersey: John Wiley & Sons.

Chang, L.Y., Lee, Y.J. & Huang, C.L. 2010. The influence of e-word-of-mouth on the consumer's purchase decision: A case of body care products. *Journal of Global Business Management* 6(2): 1.

Goyette, L. R. 2010. E-WOM Scale: Word of Mouth Measurement Scale for e- service Context. *Journal of Administration Sciences* 27 (1): 5–23.

Hasan. A. 2010. *Marketing dari mulut ke mulut*. Yogyakarta: Media Pressindo.

Jansen, B.J., Zhang, M., Sobel, K. & Chowdury, A. 2009. Twitter power: Tweets as electronic word of mouth. *Journal of the American society for information science and technology*, 60(11): 2169–2188.

Lin, C., Wu, Y. S. & Chen, J.C.V. 2013. Electronic word-of-mouth: The moderating roles of product involvement and brand image. *TIIM 2013 Proceedings*: 39–47.

Paquette, H. 2013. Social media as a marketing tool: A literature review.

See-To, E.W. & Ho, K.K. 2014. Value co-creation and purchase intention in social network sites: The role of electronic Word-of-Mouth and trust–A theoretical analysis. *Computers in Human Behavior*, 31: 182–189.

Zhang, J.Q., Craciun, G. & Shin, D. 2010. When does electronic word-of-mouth matter? A study of consumer product reviews. *Journal of Business Research* 63(12): 1336–1341.

Advances in Business, Management and Entrepreneurship – Hurriyati et al (eds)
© *2020 Taylor & Francis Group, London, ISBN 978-0-367-27176-3*

Factors influencing normative community pressure in brand community: A study of young entrepreneurs community of *Mandiri* Bank

L. Lindiawati
STIE Perbanas Surabaya, Surabaya, Indonesia

I. Usman & S.W. Astuti
Universitas Airlangga, Surabaya, Indonesia

ABSTRACT: Brand community has been extensively built and developed by either the companies or the customers voluntarily. These two conditions with different perspective hold different purposes. For the companies, brand community is considered an effective marketing strategy which is expected to achieve customer brand engagement. However, often companies focus more on the purposes of the brand community, but do not realize the policies, rules, or traditions firstly created as norms in order to lead the way community members perceive and act. Often norms become pressures for the brand community members causing them to be reluctant to engage the brand. This study develops a model of normative community pressure that is affected by both brand community identification and inter-member relationship quality that is previously affected by brand relationship quality. The empirical analysis derived primary data of young entrepreneurship community created by one of the big five banks in Indonesia and was analyzed using Structural Equation Modeling. The result explains community pressure in a brand community is affected by both brand community identification and inter-member relationship quality, and these two variables are affected by brand relationship quality. This implies that within a brand community, companies must be aware the relationship and interaction of their customers and how they feel toward the brand community to minimize pressure from the brand community members.

1 INTRODUCTION

Brand community is considered an effective marketing strategy which is expected to achieve customer brand engagement. Often companies focus more on the purposes of the brand community, but less on the policies, rules, or traditions firstly created as norms in order to lead the way the community members perceive and act (Algesheimer et al. 2005). This norm can be a pressure although at the beginning of the setting is not aimed at burdening the members.

Normative community pressure naturally emerges from the community that implies more formal rules and very possibly from the individual level of the individual member of the community that falls at the more informal level (Leckie et al. 2016, Mcmillan & Chavis 1986).

If members measured from both perspective–individually or in group–feel that the brand has emotional value a good relationship will be created.

The power of brand may bring positive emotion that will lead to customers loyalty and (Palmatier et al. 2017) that is why companies of certain products, usually of high involvement ones, eagerly set, maintain and develop a brand community. A bottom-up brand community voluntarily set up by customers also causes benefits for companies. However, in dealing with a brand community, companies must also be aware of community members' emotions; whether they are happy with the community or not. Some companies focus more on the purposes of the brand community but are not aware of the internal aspect of an individual in the brand community. Should companies go to such detail level? The answer must be 'yes' since brand community is a domain in the discourse of the decision making process in purchasing a product. Being happy or pressured in a brand community reflects the needs and wants of customers of the community members.

Within a brand community, companies or the community leaders often set norms to lead the effectiveness of the brand community. Companies should realize the policies, rules, or traditions firstly created as norms in order to lead the way the community members perceive and act. Often norms become pressures for the brand community members causing them to be reluctant to engage the brand.

Bottom-up brand communities which are set up voluntarily by members as customers, are mostly those set up due to buying products not by experiencing services. Service industry is considered fragile so actually it should be leveraged and held by the existence of brand community, especially for service industry which easily encounter crisis such as banking industry. Usually, banks accommodate their roles to maintain customers by implementing CSR programs, but this is not enough to lead customers to be loyal and engaged. Banks should design a specific program leading to the forming of customer community. They can have conversation with each other that can strengthen the bank's image, recommend the bank to others, spread positive word-of-mouth, and show engaging behaviors.

One of the big banks in Indonesia dealing with brand community to leverage the image and loyalty is Mandiri Bank. Mandiri Bank through its CSR program has been educating young entrepreneurs, inviting them to join a series of competition that ends up with final stage. All of the participants are selected through a very strict selection, so they are proud of being the participants of the bank program called *Wirausaha Muda Mandiri* (WMM) or young entrepreneurs. The participants of this WMM pro-grams keep in touch using social media of WhatsApp and they are called the WMM alumnae. In this case, these alumnae are the brand community members of Mandiri bank. The interaction among members is very important to make them feel comfortable being the community and will lead to the effectiveness of developing a brand community (Algesheimer et al. 2005, Schau et al. 2009).

Brand community is considered a new way of understanding a loyalty (Mcalexander et al. 2002). Brand community is like the mediator between brand and customer. Customers do not only need utilitarian support, intellectual support, social support, and actualization, but also need positive emotion resulting from interaction with other members. When they feel pressured, all purposes of the brand community will not be achieved.

Being in a pressure is not the purpose of joining a brand community. Being pressured in a brand community may be caused by two factors first the rule and norms of the community as well as how members identify their relationship with the community, and second the attitude and behaviors of other members. If the members feel that they are part of the community and have perspectives in line with the community policy, they do not tend to be in pressures. Referring to the relationship among member, when a member feel good attitude and behaviors from other members, then he or she will not feel pressured. Those two factors might be strongly determined by the quality of their relationship with the brand. The brand relationship quality influence how an individual perceive the community policy and how they react on the attitude of the members.

This study develops a model of normative community pressure that is affected by both brand community identification and inter-member relationship quality that are previously affected by brand relationship quality.

2 METHOD

2.1 *Measurement development*

The measures for four variables were adopted from the literature of some studies, then they were adapted to suit the context of the research. Measures for member relationship quality were newly developed. These variable measurements were adopted from brand community identification variables that were more suitable to measure inter-member relationship quality instead of measuring brand community identification which it was adapted from. While some items to measure normative community pressure were developed based on some literature (Cialdini & Goldstein 2004, Mcmillan & Chavis 1986).

2.2 *Population and sample*

The population of this research was entrepreneurs that was trained by the CSR program of Mandiri Banks called *Wirausaha Muda Mandiri* (WMM) alumnae. These WMM participants are grouped under batches. Using convenience sampling techniques, the respondents were invited by the help one main member. She then appointed some sub-coordinator to help attract and achieve the minimum number of respondents (100) spreading in Java and one respondent was from East Kalimantan. Luckily, all respondents were young entrepreneurs that were, although under different groups, always connected to each other using WhatsApp.

To make sure that the questionnaire was well designed, the data were analyzed to test the validity and reliability of the research instrument.

After showing acceptable validity and reliability scores, the data were analyzed using Multi Structural Equation Estimation using WarpPLS to test the causality of exogenous variables on the endogenous variable and to confirm the model of the normative community pressure in WMM brand community.

3 RESULT AND DISCUSSION

3.1 *Measure evaluation*

The validity test shows that all items of the four variables totaling twenty two items are successfully explains the measure of the variables. This can be seen by the discriminant validity score which are all > 0.7 while the accepted score is minimum 0.7. The discriminant validity score are explained by normalized Structure Loading and Cross-Loading.

Reliability of the measure items is shown by the composite reliability which are ideally should be > 0.6.

3.2 Structural model estimation

This research includes three exogenous variables and one endogenous variable that all were tested statistically using Structural Equation Estimation supported by WarpPLS 6.0 version. The following is the research model that has been successfully tested. The causality of each construct effects show significant result showed by all p values which are < 0.01.

The effect of brand relationship quality (BRQ) to brand community identification (BCI) shows quite strong effect (β: 0.66). Brand relationship quality explains 44% of the variance in brand community identification (BCI).

The effect of brand relationship quality (BRQ) to inter-member relationship quality (I-MRQ) shows quite strong effect (β: 0.64). Brand relationship quality explains 41% of the variance in inter-member relationship quality (I-MRQ).

Normative community pressure (NCP) is explained by the antecedents which are brand community identification (BCI) and inter-member relationship quality (IMRQ) as of only 14%. Normative community pressure is affected significantly by brand community identification (BCI) with β: 0.44 while the effect of inter-member relationship quality (IMRQ) also significantly affects normative community pressure with β:24%.

3.3 Discussion

Based on the data processed, the research found support from the four variables. The impact of brand relationship quality to brand community identification showed that the closeness of young entrepreneurs who had been trained and supported by Mandiri Bank lead to their awareness on membership and existence in *Wirausaha Muda Mandiri* (WMM) community. They perceive that Mandiri Bank contributed value to their life. Based on the open questions, most young entrepreneurs agreed

that Mandiri Bank had excellent performance, professional management and high integrity. These meant that Mandiri Bank was very demanded by its customers. Brand relationship quality significantly and positively affected inter-member relationship quality. Young entrepreneur community perceived that the good traits that the bank owned made them feel that WMM community members also showed good attitudes to fellow members, causing members to have comfortable interaction with each other, and helped each other. However, based on the open question there were 21% respondents expressed their feeling of being left by other members and that they were apart emotionally and in term of jobs. After having explored more on this point, it was perhaps because Mandiri Bank did not set sustainable programs for the young entrepreneur community members anymore.

Every community must set norms, rules, or hold certain ways or practices to regulate the community members. But not all the members perceive them with the same enthusiasm. Some might feel burdened with the norms and rules. Normative community pressure did not appear as the effect of both inter-member relationship quality and brand community identification since the beta values of these two antecedents were very low. Since the relation was negative, it meant that both inter-member relation quality and brand community identification did not cause burden for the respondents. They did not feel any burden in mingling with other members nor in behaving and acting.

4 CONCLUSION

Based on the analysis on all variables, it can be highlighted that WMM community members have good emotional relationship with Mandiri Bank, and this relationship causes two aspects namely WMM Community identification and the relationship quality of inter-members of WMM. WMM community members identify themselves in the community positively, and the quality of inter-member of WMM community is also good. The good quality of inter-member relationship and good identification of WMM community members do not affect pressure caused by norms and rule in the community.

ACKNOWLEDGEMENT

The authors wish to say thank you to all parties that gave them assistance during the research.

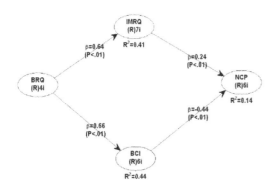

Figure 1. Research model.

REFERENCES

Algesheimer, R., Dholakia, U.M. & Herrmann, A. 2005. The social influence of brand community: Evidence from European car clubs. *Journal of Marketing* 69(3): 19–34.

Cialdini, R.B. & Goldstein, N.J. 2004. Social influence : Compliance and conformity. *Annu. Rev. Psychol* 55: 591-621.

Leckie, C., Nyadzayo, M.W. & Johnson, L.W. 2016. Antecedents of consumer brand engagement and brand loyalty. *Journal of Marketing Management* 32(5–6): 558-578.

Mcalexander, J.H., Schouten, J.W. & Koenig, H.F. 2002. Marketplace communities: A broader view of brand community.

Mcmillan, D.W. & Chavis, D.M. 1986. Sense of community: A definition and theory. *Journal of Community Psychology* 14(1): 6-23.

Palmatier, R.W., Kumar, V. & Harmeling, C.M. 2017. Customer engagement marketing. *Customer Engagement Marketing*: 1–328.

Schau, H.J., Muñiz Jr, A.M. & Arnould, E.J. 2009. How brand community practices create value. *Journal of Marketing* 73(5): 30-51.

Advances in Business, Management and Entrepreneurship – Hurriyati et al (eds)
© 2020 Taylor & Francis Group, London, ISBN 978-0-367-27176-3

Smartphone operating systems based on consumer perceptions in West Java

E. Azis & Y. Fachruddin
Telkom University, Bandung, Indonesia

M.M.A. Rohandi
Universitas Islam Bandung, Bandung, Indonesia

ABSTRACT: Today, smartphones are available in different brands and types, using operating systems such as Android, iOS, Windows Phone and BlackBerry. The purpose of this study is to determine the positioning of OS based smartphones based on the perception of consumers in West Java and to know the best judgement from consumer-based attributes such as design, durability, product features, brand, social influence and price. This research used descriptive method to compare the four brands and the number of respondents was 503 respondents using Bernoulli with 5% error margin. Multidimensional Scaling was used for mapping the position of operating system with 6 perceptual attributes. Each operating system has different target market and customer perception shoed that iOS got the first position on design, brand and social influence, while Android got the first position on durability, product features and price.

1 INTRODUCTION

Growth of technology enables consumers to access information quickly and precisely thanks to the capabilities of advanced features on the smartphone. One of the very important decisions in choosing smartphone is Operating System (OS), it can make consumers confused because many options available. However, many of them do not understand the differences between operating systems. In Indonesia, many smartphones users do not necessarily focus on the features of each operating system, they simply use it to support their lifestyle. There are four operating systems in Indonesia: Android, iOS, Windows Phone and BlackBerry.

Brand manufacturers must realize that marketing promotions play less important roles in consumer attraction, as many consumers purchase handsets based on their opinion of a particular mobile OS or mobile handset. Since highly qualified people prefer branded mobiles, manufacturers work towards this area of branding as it implicates trust, reliability and reduces psychological risk. Nowadays, it is becoming highly essential for younger generation to stay connected on the go, this requires a handset with all the essential connectivity tools like Wi-Fi network, instant messaging, social, and so on (Bagga et al. 2016).

This study was conducted on the smartphone operating system based on the attributes that have been selected. Based on data from gs.statcounter.com for the last five years. The Android operating system currently has a 91.06% market share of smartphones in Indonesia, while iOS, Windows Phoneand BlackBerry have only 3.41%, 3.27% and 0.6%, respectively.

Positioning is critical to marketing and marketing programs attempt to be relevant to a certain consumer group (Kotler & Keller 2016). Positioning is the act of designing a company's offering and image to occupy a distinctive place in the minds of the target market. The goal is to locate the brand in the minds of consumers to maximize the potential benefit to the firm (Kotler & Keller 2016). Positioning helps consumers find the unique way to get the best product based on value proposition and get information from word of mouth which is the best marketing channel (Rohandi 2016).

Multidimensional Scaling (MDS) is a class of procedures to represent spatial respondent perceptions and preferences using a visual display. The perceived relationship or the psychological relationship between the stimuli is represented by the geometric relationships between points in a multidimensional space. This geometric representation is often called a spatial map. The spatial map axes are assumed to represent the psychological basis or basic dimension that respondents use to shape the perceptions and preferences of the stimuli (Malhotra 2006).

Perception is a process that begins with consumer exposure and attention to marketing stimuli and ends with consumer interpretation (Hawkins & Mothersbaugh 2010). It is highly selective; the consumer processes just one small thing from the information given. The process by which people select, organize, and interpret information to form a meaningful picture of the world (Kotler & Armstrong 2016) and the objective is to generate high levels of awareness (Belch & Belch 2003).

This research used some attributes to ensure it can be trusted and accountable, including: design, durability, product features, brand, social influence and price. If the attribute is not in accordance with the benefits of the product, it reduces the level of customer satisfaction (Tjiptono & Chandra 2012).

Based on the phenomenon and previous data, the research problem for this study is how consumers perceive the positioning of smartphone operating systems in West Java Indonesia.

2 METHOD

This research used descriptive analysis, defined as a study that attempts to describe a phenomenon/event systematically in accordance with what it is. In such research, researchers try to determine the nature of the situation as it is at the time of the study (Dantes 2012). An ordinal scale was arranged in sequence from low to high according to a certain characteristic, with different distance between the sequences (Suharsaputra 2012). Sampling technique used was nonprobability sampling, the researcher chose purposive sampling method with certain considerations with the intention of obtain information from a specific target group.

The population in this study was the users of smartphone in West Java. The population was not known and samples should be representative of the population of respondents (Sugiyono 2010). The proportion of the number of samples was based on Bernoulli method (Zikmund et al. 2010) with the level of trust 95% and level of accuracy (α) 5%. From the distribution table, it was found that (5%)/2 = 0.025 so Z = 1.960. The error rate was set at 5% with the probability of acceptable questionnaire 0.5. From the equation, the sample was 385, but in this study, there were 503 respondents.

3 RESULT AND DISCUSSION

The previous research discovered that when Android was compared to iOS, 72% of the respondents used Android. People were satisfied with the performance of iOS as 71% respondents rated excellent for iOS performance. 56% respondents rated excellent for iOS in terms of user interface. Seventy-nine percent of respondents rated security and data protection excellent in iOS (Kaur et al. 2016).

Public perceptions in Israel revealed that iOS was more user-friendly than Android and women invested the necessary effort and quickly learned how to use it. People also selected the operating system that suited

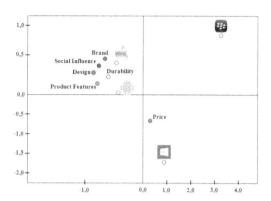

Figure 1. The positioning map of smartphone operating system based on overall attribute.

their preferences and adjusted or rationalized their buying decisions (Gafni & Geri 2013).

Users were highly satisfied with the touchscreen quality and features which iOS provided. iOS had more useful free applications than Android, but Android had a better categorization of applications and a better user interface. Some consumers chose iOS because of user-friendly criteria followed by prestige and brand name and comfort, while others chose Android because of its user-friendly interface and value for money (Jain & Shamra 2013).

To get positioning map of smartphone operating system, the data obtained from respondents was processed using multidimensional scaling method in SPSS version 21. The perceptual map provided an overview on how consumers perceived the four operating systems based on their selected attributes.

The matrix input for multidimensional scaling processing used intact conditional proximity matrix. Table 1 presents the input data matrix for MDS processing.

From the Table 1 matrix we can see that the stress value is 0.032 and RSQ (Squared Qorrelation) is 1. It means that this model is perfect for multidimensional scaling.

Figure 1 clearly shows the position of the operating system (Android, iOS, Windows Phone and BlackBerry) with their attributes.

In Table 2, it can be seen that each smartphone operating system obtained different positions depending on

Table 1. Matrix forms.

Stimulus	Stress	RSQ
1	0.002	1
2	0	1
3	0	1
4	0.049	0.998
5	0.12	1
6	0.022	1
Average Over Stimuli	0.032	1

Table 2. The positioning of smartphone operating system based on overall attribute.

Attribute	Position 1	Position 2	Position 3	Position 4
Design	iOS	Android	Windows	BlackBerry
Durability	Android	iOS	Windows	BlackBerry
Product Features	Android	iOS	Windows	BlackBerry
Brand	iOS	Android	Windows	BlackBerry
Social Influence	iOS	Android	Windows	BlackBerry
Price	Android	iOS	Windows	BlackBerry

Table 3. Euclidean distance smartphone operating system based on overall attributes.

Operating	System	Android	iOS	Windows	BlackBerry
Design	Ed	0.334	0.320	2.671	4.166
	Rank	2	1	3	4
Durability	Ed	0.174	0.220	2.524	3.997
	Rank	1	2	3	4
Product Features	Ed	0.283	0.332	2.611	4.151
	Rank	1	2	3	4
Brand	Ed	0.441	0.170	2.769	4.004
	Rank	2	1	3	4
Social influence	Ed	0.307	0.237	2.657	4.089
	Rank	2	1	3	4
Price	Ed	1.016	1.272	1343	3.475
	Rank	1	2	3	4

each attribute. According to consumer perceptions, android competed closely with iOS. Android occupied position 1 on the attributes of durability, product features and price, while the iOS occupied position 1 on design, brand and social influence. Both operating systems had three times the opportunity to occupy position 1. Windows Phone occupied position 3 on design attributes, durability, product features, brand, social influence and price and BlackBerry occupied position 4 on design attributes, durability, product features, brand, social influence and price.

Euclidean Distance (ED) was used to reinforce the previous result, and the results of these calculations can be seen in Table 3.

Results from the ED calculation revealed that Android was superior in durability, product feature and price while iOS was superior in design, brand and social influence. Windows Phone was in the third position based on all features. The most inferior operating system was BlackBerry.

4 CONCLUSION

The Android operating system is the market leader in Indonesia and dominates smartphone market share. Based on the research, it can be concluded that consumers have different assessment regarding six main attributes. In design, brand and social influence attributes, iOS was the best smartphone operating system confirmed by the best ED point, Android in second position and BlackBerry in third. In durability, product feature and price, the result shows that Android is the best operating system because it has the best ED point; iOS is number two and the worst is BlackBerry. Every operating system has a different target market; if consumers want design, brand and social influence they can choose to use iOS, but if consumers want to get something user-friendly, with value for money and durability, they can chose Android. Meanwhile, the

Windows Phone is clearly the third choice and the last choice in all categories is the BlackBerry.

REFERENCES

Bagga, T., Goyal, A. & Bansal, S. 2016. An invetigate study of the mobile operating system and handset preference. *Jurnal of Science and Technology* 9(35).

Belch, G.E. & Belch, M.A. 2003. *Advertising and promotion (6th Ed)*. England: McGraw-Hill Companies.

Dantes, N. 2012. *Metode penelitian*. Yogyakarta: Andi.

Gafni, R. & Geri, N. 2013. Do operating systems affect perception of smartphone advantages and drawbacks? *Journal Informing Science and Information Technology* 10.

Hawkins, D.I. & Mothersbaugh, D.L. 2010. *Consumer behavior: Building marketing startegy (11th Ed)*. McGraw-Hill Irwin.

Jain, V. & Sharma, A. 2013. The consumers preferred operating system: Android or iOS. *International Journal of Business Management and Research (IJBMR)* 3(4): 29-40.

Kaur, A., Kanupriya, K. & Rita, R. 2016. Consumer preference towards major mobile operating systems. *Orbit-bizdictum* 1.

Kotler, P. & Armstrong, G. 2016. *Principles of marketing (16th Ed)*. Pearson.

Kotler, P. & Keller, K.L. 2016. *Marketing management 15th Ed*. England: Pearson Education Limited.

Malhotra, N.K. 2006. *Riset pemasaran pendekatan terapan*. Jakarta: Indeks.

Rohandi, M.M.A. 2016. Effective marketing communicaton: word of mouth. *Journal Performa*: 1–14.

Sugiyono, S. 2010. *Statistika untuk penelitian*. Bandung: Alfabeta.

Suharsaputra, U. 2012. *Metode penelitian kuantitatif, kualitatif dan tindakan*. Bandung: Refika Aditama.

Tjiptono, F. & Chandra, G. 2012. *Pemasaran strategik*. Yogyakarta: Andi.

Zikmund, W.G., Babin, B.J., Carr, J.C. & Griffin, M. 2010. *Research business methods*. Boston: South Western Cengage Learning.

Statcounter Globalstas. 2018. Mobile operating system market share in Indonesia. http://gs.statcounter.com/os-market-share/mobile/indonesia/#monthly-201303-201803. Accessed 24 April 2018.

Advances in Business, Management and Entrepreneurship – Hurriyati et al (eds)
© *2020 Taylor & Francis Group, London, ISBN 978-0-367-27176-3*

The role of personal innovation in online purchasing behavior among Indonesian consumers

C.K. Dewi
University Sains Malaysia, Pulau Pinang, Malaysia
Telkom University, Bandung, Indonesia

Z. Mohaidin
University Sains Malaysia, Pulau Pinang, Malaysia

ABSTRACT: The development of digital technology has the potential to transform all aspects in life, including the behavior of consumers. Innovation is considered to be a key driving factor of technological acceptance. Therefore, the aim of this study is to examine and investigate the role of personal innovation on online purchasing behavior among Indonesian Consumers. The UTAUT's framework was extended by proposing new casual pathways between personal innovativeness and behavioral intention. The partial least squares technique was employed to evaluate the statistical significance of the proposed pathways by analyzing 200 sets of survey responses. The result indicates that personal innovativeness was one of the most significant predictors of behavioral intention. This strengthens the argument that personal innovativeness is a key driving factor of technological acceptance and plays a significant role in innovation adoption, particularly consumers' intention to adopt online shopping channels as a new technology alternative for their purchasing activities.

1 INTRODUCTION

Indonesia is a country with potential and remains one of the most attractive, developing and emerging markets. Indonesia is characterized by its huge population size, and they are going online. In 2017, there were 143.26 million internet users in Indonesia, representing about 54.68% of its total population by APJII 2017.

Internet used for business activities (e-commerce) is still rather new for Indonesian consumers but it has begun to gain trust (Nurudin 2016). The potential of e-commerce industries in Indonesia cannot be underestimated as retail e-commerce sales have increased every year. Specifically, Indonesian retail e-commerce sales reached $7.056 billion in 2017. The Indonesian government believes that the stabilization of the e-commerce will play an important role for the country in achieving its full potential (Gryseels et al. 2015). Therefore, the President of Indonesia, Joko Widodo, has a vision to make Indonesia become Southeast Asia's largest digital economy by 2020 in order to facilitate maximum benefits for local products and actors.

The development of e-commerce has changed the behavior of consumers. Consumers nowadays are more demanding, informed, smarter and innovative. Along with the changes in Indonesian consumer behavior, it is important to note the online purchasing behavior and consider the factors that will affect internet users' behavior in adopting online shopping channel as a new technological alternative for their purchasing activities. One of the theories which can help in explaining and predicting the behavior of an individual, whether to accept or reject the adoption of a new technology is the Unified Theory of Acceptance and Use of Technology (UTAUT) (Venkatesh et al. 2003).

UTAUT is an integrated model that combines alternative theories of user acceptance and innovation which are applied in studying the adoption of technologies. It has been identified as the most comprehensive theory in overcoming the limitations of any previous technology acceptance models and is thus making it the most appealing model for researching an individual's intention and use behavior of the specific system (Tan et al. 2013, Williams et al. 2015).

It generates four core determinants of intention and usage, and up to four moderators of key relationships. Direct determinants for behavioral intention consist of performance expectancy, effort expectancy, and social influence. Direct determinants for use behavior are facilitating conditions and behavioral intention. There are four moderators of key relationships including age, gender, experience and voluntariness of use. It has been widely employed in researches on technology adoption and diffusion as a theoretical lens for researchers conducting empirical studies of user intention and behavior including in e-commerce (Celik 2016, Chang et al. 2016, Escobar et al. 2014, Pahnila et al. 2011, Tan et al. 2013, Yeganegi & Elias 2016).

In information technology adoption studies, there is a gap between the integration of technology acceptance and personal issues, and thus there is a need to integrate some personal traits, such as personal innovativeness together with UTAUT (Agarwal & Prasad 1998, Thakur & Srivastava 2015, Turan et al. 2015). Personal innovativeness represents innovations to target customers in technology acceptance (Thakur & Srivastava 2014). Purchasing online can be categorized as innovative behavior compared to purchasing in traditional physical stores (Zhou et al. 2007). As more and more consumers are innovative consumers nowadays, (San Martin & Herrero 2012, Sriniva-san 2015) argued that innovativeness is considered as a key driving factor of technological acceptance, which means that online purchasing intention is positively influenced by the level of user innovation. Therefore, personal innovativeness has received considerable attention in the technology adoption studies specifically in online purchasing adoption.

This study adopts the UTAUT model and extends it by integrating construct which complements the online purchasing behavior investigated. This study added the role of personal innovativeness construct in behavioral intention of online shopping for the Indonesian context. Therefore, the aim of this study is to examine and investigate the role of personal innovativeness along with other factors on online purchasing behavior among Indonesian consumers.

2 METHOD

The initial research framework was based on the UTAUT, with an additional causal pathway by which personal innovativeness influences behavioral intention (Figure 1), for the advancement of the theory building progress in the online shopping adoption domain.

Performance expectancy has an important role in individual behavior for accepting or rejecting new technology (Ratten 2015, Ali Tarhini et al. 2016). Numerous researches have tested that the performance expectancy positively influences behavioral intention and it has proved to be the major determinant of behavioral intention (Escobar et al. 2014, Pascual-Miguel et al. 2015, Tan et al. 2013).

Effort expectancy has a significant role in the intention to use new technology to influence behavioral intention in using technology. The relationship between effort expectancy and behavioral intention has been tested by various studies whose outcome is that effort expectancy positively influences behavioral intention (Escobar et al. 2014, Ho et al. 2016, San Martin & Herrero 2012).

Previous researchers have also examined the role of social influence on the behavioral intention of technology acceptance (Escobar et al. 2014, Lian & Yen 2014). Social influence contains the presumption that individual behavior is influenced by how people believe that others will

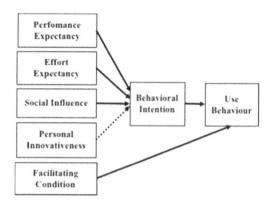

Figure 1. Research framework.

view them as a result of having used a particular type of technology and positively affects the behavioral intention (Venkatesh et al. 2003, Venkatesh et al. 2012).

Personal innovativeness plays an important role in innovation adoption and also has a strong and direct impact on consumers' decision to adopt a new technology, because individuals become aware of the existence of new technology based on personal traits such as personal innovativeness (Hyunjoo & Ha 2016). Several studies in the e-commerce literature have demonstrated the important role of personal innovativeness in purchasing intention within different contexts (Agarwal & Prasad 1998, Escobar et al. 2014, Hyunjoo & Ha 2016, San Martin & Herrero 2012).

The role of the facilitating condition on predicting use behavior has been examined by numerous researches (Tarhini et al. 2016, Venkatesh et al. 2003), and the results notified that the facilitating condition has a positive influence on use behavior.

Based on the explanation, the following related hypotheses were developed:

H1: Performance expectancy has a positive and significant relationship with behavioral intention.
H2: Effort expectancy has a positive and significant relationship with behavioral intention.
H3: Social influence has a positive and significant relationship with behavioral intention.
H4: Personal innovativeness has a positive and significant relationship with behavioral intention.
H5: Facilitating conditions has a positive and significant relationship with use behavior.
H6: Behavioral intention has a positive and significant relationship with use behavior.

In order to test the hypotheses, a sample of Indonesian consumers was surveyed with an online survey. The questionnaire was employed using five-point scale items, and modified from (Celik 2016, Venkatesh et al. 2003, Agarwal & Prasad 1998, & Pahnila et al. 2011). A professional translator translated the original items in English into Indonesian because the questionnaire was administered in Indonesia.

All of the questions were first pretested by two academics in Indonesia for readability. Then the instrument was pilot-tested using 30 respondents. For the actual data collection, the target respondents were Indonesian consumers specifically from Java, Indonesia. The initial selection of respondents is based on the study of APJII 2017 & Deloitte 2016, which stated that the majority of Internet users in Indonesia lived in the western part of Indonesia, which is on the island of Java.

This study employs G*Power 3.1 for power analysis and sample size calculations. This study is using typical significance level in business research which is 5% (0.05) and five predictors. The computed minimum acceptable sample size for this study is 138. Data were collected from 200 respondents. About 57.5% of the respondents were women, and 46.7% of them were around 18–30 years old. In terms of education, most of them held a university degree (55.9%), and worked in the private sector (41%).

To test the theories and concept and to analyze the cause-effect relationship that was used in this study, one of the techniques that can be employed is Structural Equation Modelling (SEM). A PLS-SEM approach was used as it is primarily used to develop theories in exploratory research or an extension of an existing structural theory (Ramayah et al. 2016). The statistical analysis tools were SmartPLS 3.0 (Ramayah et al. 2018).

3 RESULTS AND DISCUSSION

The two main criteria used for testing the quality of the measures are validity and reliability. Construct validity testifies to how well the results obtained from the use of the measure fit the theories around which the testwas designed (Sekaran & Bougie 2016). This is assessed through convergent and discriminant validity. As suggested by (Hair et al. 2017), to evaluate convergent validity of reflective constructs the following factors must be considered: loadings, composite reliability (CR) and the average variance extracted (AVE). The result of convergent validity can be seen in Table 1. The loadings for all items exceeded the recommended value of 0.5. CR values exceeded the best value of 0.708, and the AVE values exceeded the recommended value 0.5.

The discriminant validity was assessed by examining the correlations between the measures of potentially overlapping constructs. As shown in Table 2, the items load more strongly on their own constructs in the model.

In terms of reliability it can be concluded that the measurements are reliable, as seen from Table 1,

Table 1. Result of the convergent validity and reliability.

Constructs	Items	Loading	CR	AVE	Cronbach's Alpha
PE	PE1	0.828	0.884	0.656	0.825
	PE2	0.781			
	PE3	0.860			
	PE4	0.766			
EE	EE2	0.762	0.912	0.722	0.871
	EE3	0.868			
	EE4	0.876			
	EE5	0.888			
SI	SI1	0.826	0.916	0.731	0.878
	SI2	0.909			
	SI3	0.827			
	SI4	0.854			
PI	PI1	0.854	0.892	0.734	0.819
	PI2	0.832			
	PI4	0.883			
FC	FC1	0.855	0.900	0.751	0.835
	FC2	0.866			
	FC3	0.878			
BI	BI1	0.922	0.954	0.874	0.928
	BI2	0.943			
	BI3	0.940			
UB	UB1	0.945	0.922	0.856	0.835
	UB2	0.905			

Table 2. Result of the discriminant validity.

	BI	EE	FC	PE	PI	SI	UB
BI	**0.935**						
EE	0.464	**0.850**					
FC	0.557	0.447	**0.867**				
PE	0.533	0.508	0.432	**0.810**			
PI	0.422	0.339	0.407	0.335	**0.857**		
SI	0.351	0.247	0.177	0.390	0.195	**0.855**	
UB	0.388	0.425	0.319	0.437	0.347	0.203	**0.925**

where all Cronbach's alpha values are between 0.80 and 0.90, so that it can be regarded as a satisfactory measurement (Hair et al. 2017).

Next, we proceeded with the path analysis to test all the hypotheses generated. As we can see from Figure 2, the R^2 value for behavioral intention was 0.398, suggesting that 39.8% of the variance in extent of behavioral intention can be explained by performance expectancy, effort expectancy, social influence, and personal innovativeness. The R^2 value for use behavior was 0.170, suggesting that 17% of the variance in extent of use behavior can be explained by facilitating condition and behavioral intention.

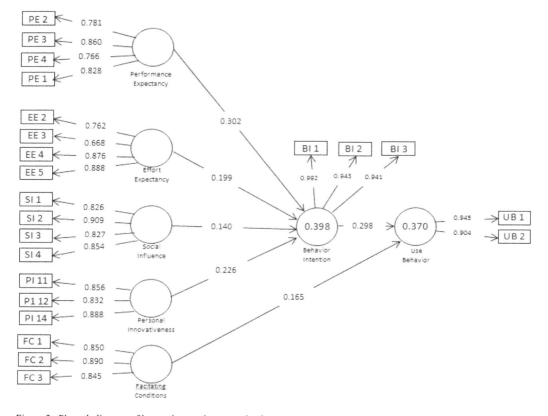

Figure 2. Pls path diagram of beta values and r-squared values.

Based on the hypothesis testing result in Table 3, it can be seen that all the hypotheses (H1–H6) were supported. In this study it was found that performance expectancy was the most significant predictor of behavioral intention followed by personal innovativeness. It was also found that the higher the extent of the facilitating condition and behavioral intention, the better the use behavior.

The result of this study reinforces the statement of (Venkatesh et al. 2003) which argued that performance expectancy is the strongest predictor of behavioral intention in using new technology, particularly in online purchasing context.

This study also supports Hyunjoo and Ha (2001), San Martin and Herrero (2012), and Srinivasan (2015), who state that personal innovativeness plays an important role in innovation adoption and also has a strong and direct impact on consumers decision to adopt a new technology, because individuals become aware of the existence of new technology based on personal traits such personal innovativeness. Personal innovativeness is most effective in determining the adoption of innovations, since it captures an individual's natural tendency to try out a new technology in multiple domains of acceptance (Lu 2014). Innovators are likely to be more knowledgeable about the internet and online shopping

Table 3. Path coefficients and hypothesis testing.

Hypothesis	Relationship	Coefficient	t value	P values	Supported
H1	PE→BI	0.302	4.564	0.000	YES
H2	EE→BI	0.199	2.819	0.005	YES
H3	SI→BI	0.140	1.986	0.048	YES
H4	PI→BI	0.226	3.638	0.000	YES
H5	FC→UB	0.165	2.167	0.031	YES
H6	BI→UB	0.298	4.399	0.000	YES

than consumers who do not purchase goods or services online. Innovators are more willing to adopt new ideas and are able to cope up with financial risk or a high degree of uncertainty arising from innovation adoption (Thakur & Srivastava 2015).

4 CONCLUSION

The result of this study strengthens the argument that personal innovativeness is a key driving factor of technological acceptance and plays a significant role in innovation adoption, particularly a consumers' intention to adopt online shopping channels as new technology alternative for their purchasing activities. For the theoretical contribution, this study is expected to have several contributions to the body of knowledge in new technology adoption and usage behavior particularly broadening the understanding of relationships among the variables of UTAUT as a basic theoretical model. From a practical standpoint, this study's findings may help online retailers to redefine their strategies so that they can increase sales, reach higher profits and increase market shares. Finally, the results will be different if the model includes the four original moderators of key relationships: age, gender, experience and voluntariness of use. Therefore, future research should add to this study by exploring the role of original moderators in the model.

ACKNOWLEDGEMENT

The authors would like to thank the two anonymous reviewers for their valuable comments regarding this research. The authors further would like to thank to qualified and experienced translators from Language Center-Telkom University who are helping in back-to-back translation process.

REFERENCES

Agarwal, R. & Prasad, J. 1998. A conceptual and operational definition of personal innovativeness in the domain of information technology. *Information Systems Research* 9(2): 204–215.

Celik, H. 2016. Customer online shopping anxiety within the Unified theory of acceptance and use technology (UTAUT) framework. *Asia Pacific Journal of Marketing and Logistics* 28(2): 278–307.

Chang, H.H., Fu, C.S., & Jain, H.T. 2016. Modifying UTAUT and innovation diffusion theory to reveal online shopping behavior. *Information Development* 32(5): 1757–1773.

Deloitte. 2016. Consumer Insights: The evolution of the Indonesian consumer.

Escobar-Rodriquez, T. & Carvajal, T.E. 2014. Online purchasing tickets for low cost carriers: an application of the unified theory of acceptance and use of technology (UTAUT) model. *Tourism Management* 43: 70–88.

Gryseels, M., Manuel, N., Salazar, L. & Wibowo, P. 2015. Ten ideas to maximize the socioeconomic impact of ICT in Indonesia.

Hair, J.F., Hult, G.T.M., & Ringle, C.M. 2017. *A primer on partial least squares structural equation modeling (PLS-SEM) (Second Ed)*. Los Angeles.

Ho, C.B., Chou, Y.D. & Fang, H.V. 2016. Technology adoption of podcast in language learning : Using Taiwan and China as examples. *International Journal of E-Education, E-Business, E-Management and E-Learning Technology* 6(1): 1–13.

Hyunjoo, I. & Ha, Y. 2016. Determinants of mobile coupon service adoption: Assessment of gender differences. *International Journal of Retail & Distribution Management* 42(5): 441–459.

Lian, J.W. & Yen, D.C. 2014. Online shopping drivers and bar-riers for older adults: Age and gender differences. *Com-puters in Human Behavior* 37(2014): 133–143.

Lu, J. 2014. Are personal innovativeness and social influence critical to continue with mobile commerce. *Internet Re-search* 24(2): 134–159.

Nurudin, A. 2016. Indonesia E-commerce: Winning the Smart Consumer.

Pahnila, S., Siponen, M. & Zheng, X. 2011. Integrating habit into UTAUT: the Chinese eBay case. *Pacific Asia Journal of the Association for Information Systems* 3(2): 1–30.

Pascual, M.F.J., Agudo, P.Á.F. & Chaparro-Peláez, J. 2015. Influences of gender and product type on online purchasing. *Journal of Business Research* 68(7): 1550–1556.

Ramayah, T., Cheah, J., Chuah, F., Ting, H. & Memon, M. A. 2016. *Partial least squares structural equation modeling (PLS-SEM) using smartPLS 3.0: An updated and practical guide to statistical analysis (First Edition)*. Kuala Lumpur: Pearson.

Ramayah, T., Cheah, J., Chuah, F., Ting, H. & Memon, M. A. 2018. *Partial least squares structural equation modeling (PLS-SEM) using smartPLS 3.0 (second edition)*. Kuala Lumpur: Pearson.

Ratten, V. 2015. Factors influencing consumer purchase intention of cloud computing in the United States and Turkey. *EuroMed Journal of Business* 10(1): 80–97.

San Martin, H. & Herrero, A. 2012. Influence of the user's psychological factors on the online purchase intention in rural tourism: Integrating innovativeness to the UTAUT framework. *Tourism Management* 33(2): 341–350.

Sekaran, U. & Bougie, R. 2016. Research methods for business. Wiley: 1–407.

Srinivasan, R. 2015. Exploring the impact of social norms and online shopping anxiety in the adoption of online apparel shopping by Indian consumers. *Journal of Internet Commerce* 14(2): 177–199.

Tan, K.S., Chong, S.C., & Lin, B. 2013. Intention to use internet marketing: A comparative study between Malaysians and South Koreans. *Kybernetes* 42(6): 888–905.

Tan, P.J.B. 2013. Applying the UTAUT to understand factors affecting the use of English e-learning websites in Taiwan. *SAGE Open* 3(4): 1–12.

Tarhini, A., El-Masri, M., Ali, M., & Serrano, A. 2016. Extending the UTAUT model to understand the customers' acceptance and use of internet banking in Lebanon. *Information Technology & People* 29(4): 830–849.

Thakur, R. & Srivastava, M. 2014. Adoption readiness, personal innovativeness, perceived risk and usage intention across customer groups for mobile payment services in India. *Internet Research* 24(3): 369–392.

Thakur, R. & Srivastava, M. 2015. A study on the impact of consumer risk perception and innovativeness on online shopping in India. *International Journal of Retail & Distribution Management* 43(2): 148–166.

Turan, A., Tunç, A.O. & Zehir, C. 2015. A theoretical model proposal: personal innovativeness and user involvement as antecedents of unified theory of acceptance and use of technology. *Procedia-Social and Behavioral Sciences* 210: 43–451.

Venkatesh, V., Morris, M. G., Davis, G. B. & Davis, F. D. 2003. User acceptance of information technology: toward a unified view. *MIS Quarterly* 27(3): 425–478.

Venkatesh, V., Thong, J. Y. & Xu, X. 2012. Consumer acceptance and use of information technology: extending the unified theory of acceptance and use of technology. *MIS Quarterly* 36(1): 157–178.

Williams, M., Rana, N. & Dwivedi, Y. 2015. The unified theory of acceptance and use of technology (UTAUT): A literature review. *Journal of Enterprise Information Management* 28(3): 443–488.

Yeganegi, R. & Elias, N.U.R.F. 2016. Measuring the user acceptance on online hypermarket shopping system based on UTAUT model. *International Journal of Management and Applied Science* 2(2): 6–9.

Zhou, L., Dai, L. & Zhang, D. 2007. Online shopping acceptance model: Critical survey of consumer factors in online shopping. *Journal of Electronic Commerce Research* 8(1): 41–62.

Whether service differentiation can add competitive advantage to enhance consumer satisfaction: Internet services at Indihome

R. Hadiantini, R. Hurriyati, V. Gaffar & M.A. Sultan
Universitas Pendidikan Indonesia, Bandung, Indonesia

ABSTRACT: In addition to product differentiation, there is also service differentiation to increase competitiveness on every company. Differentiation is one of the two competitive advantages which can be owned by a company apart from fee advantage (Porter 1994). Service differentiation is a kind of service and quality enhancement which consists of different value in offering service to the customers. Nowadays, one of the services that is mostly used is internet service. The use of internet service is increasing in this digital area. The competition in internet service is becoming tight which is seen by competitors coming in. This condition makes PT. Telkom decreased. This research will discuss service differentiation and service quality towards customer satisfaction on internet service provider of Telkom (IndiHome). The results of the research revealed that the service differentiation perception did not affect the customer satisfaction of IndiHome product. The positive and significant influence are shown by the variable of service quality perception with tangible indicator (real), reliability, responsiveness, assurance, and empathy towards the satisfaction of IndiHome product customers.

1 INTRODUCTION

A service cannot be separated from process and system. Service is a process such as people processing, mental stimuli processing, possession processing, and information processing. The fourth categorization is made up based on who and what received the service (whether people or ownership), and based on its own action (whether tangible action or intangible action). Service as a system includes three following steps: the service operation system, service delivery system, and service marketing system. (Lovelock & Wright 2002).

In addition to product differentiation, there is also service differentiation to increase competitiveness on every company. Differentiation is one of the two competitive advantages which can be owned by a company; the other is fee advantage (Porter 1994). One way to differentiate service offer from competitors is through one or more basic performance both from image and product feature (Dickson 1997). In other words, it is a process of adding some important and valuable differentiations to differentiate company offer and competitor offer (Kottler 2005). Differentiation is an activity to make a significant differentiation to distinguish product and competitor's product that have several dimensions inside such as product, service, personnel, distribution line, and image (Kottler 2005). In conclusion, differentiation is an activity that is held by a company in creating or designing something different, unique, and valuable comparing to the other competitors to achieve competitive advantage.

Nowadays, one of the services that is mostly used is the internet. The use of the internet is increasing in this digital area. It can be seen from the results of a survey undertaken by IISPA (Indonesia Internet Service Provider Association), which states that 51.8% of Indonesian citizens are active users of internet service. Mobile internet service does not necessarily eliminate business opportunity of fixed broadband internet. Currently, fixed board band internet business perpetrator is coming continuously. It signifies that business service of cable internet based on technology fiber to the home (FTTH) is potentially developing.

There are several companies that provide internet service in Indonesia, including BizNet, First media, MNC Play and IndiHome. They provide Fiber to the Home (FTTH) service. Internet service providers keep on compete by giving interesting offer with various offers both from internet service specification and speed. One of the internet providers is IndiHome which is owned by the biggest telecommunication company in Indonesia, PT. Telkom Indonesia (previously known as Speedy). IndiHome provides different choice to the customers by bundling together home telephone, cable TV, and internet service.

The competition in internet service is getting tighter, which is seen by competitors coming in. This condition makes the PT. Telkom decreased. The previous research (Bettega 2016) which is taken by the report of Intelligent marketing from October 2015 to February 2016 said that market share for fixed board band service of PT. Telkomsel was decreasing from 85.37% to 83.28%.

It is revealed from the previous research that the reason of market shared decline is caused by some problems that occurred such as inappropriate product quality, slow service and low priority to the customers. Based on the above elaboration, thus this research will discuss service differentiation and service quality towards customer satisfaction on internet service provider of Telkom (IndiHome).

2 METHOD

Survey method is used in this research, that is, a research held to obtain facts from available symptoms and find out factual evidence from ongoing practice (Sugiyono 2003). Suharyadi and Purwanto (2003) on other hand says that the survey method is a sample method that takes one population sample and uses the questionnaire as a data collector tool.

The type of the research is descriptive. It is called descriptive research because of this research held by giving a description or systematic, factual, accurate depiction about fact or object nature as well as interpreting the relationship of investigated phenomenon.

In analytic descriptive research, a population means generalization area which consists of an object or subject that has quality and certain characteristic which is applied by the researcher to be analyzed and to draw a conclusion. Furthermore, a sample is a part of the population (Sugiono 2012), that is, customer of Telkom (IndiHome) internet service are 35 people. This research was conducted at one of the subsidiary of PT Telkom under Yakes Telkom, Health Center. It refers to the researcher consideration of time, energy, and fund here are more or less 100 questionnaires spread in two weeks, and the questionnaire returned and pass the screening only 35.

A good sample should describe population furthermost. Regarding the previous statement, the researcher used non-probability sampling technique in a group of purposive sampling, that is, a sampling technique that the feature or characteristic is known earlier based on population feature needed in the research.

This elaboration reveals that the sample taken is 35 in a questionnaire form and the result of the questionnaire become primary data in this research. It refers to the researcher consideration of time, energy, and fund. There is 35 questionnaire spread, and they are the sample of this research, 35 respondents.

The variable in this research is divided into two, they are, dependent variable and independent variable. Dependent variable used in this research is customer satisfaction perception. Meanwhile, the independent variable used in this research is service differentiation perception and service quality perception. Figure 1 shows the model of variable relationship.

This model tells several hypotheses that will be tested in this research, such as:

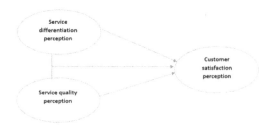

Figure 1. Model of variable relationship.

H_1: There is a relationship between service differentiation perception towards customer satisfaction.

H_2: There is a relationship between service quality perception towards customer satisfaction.

H_3: There is a relationship between service differentiation perception and service quality perception towards customer satisfaction perception.

3 RESULT AND DISCUSSION

The analyzed variable in this research is an independent variable of service differentiation perception and service quality perception, meanwhile dependent variable of this research is customer satisfaction of IndiHome product. The analyzed data is a collected data from the result of the research then is processed by software SPSS.

Based on the research, H1, the influence of service differentiation perception towards customer satisfaction, cannot be accepted. The result of analyzed data taken from the linearity test shows that X1 does not influence significantly towards Y (0.314 > 0.005). The result of hypothesis analyzed, moreover, showed that H2, the influence of service quality perception towards customer satisfaction, is accepted. The result of the analyzed data is taken from a linearity test, which shows that X2 influences significantly towards Y (0.0 < 0.05).

ANOVA[a]

Model		Sum of Squares	df	Mean Square	F	Sig.
1	Regression	1789.760	2	894.880	20.036	.000[b]
	Residual	1429.212	32	44.663		
	Total	3218.971	34			

a. Dependent Variable: y

b. Predictors: (Constant), x2, x1

Coefficients[a]

Model Summary[b]

Figure 2. Data result by SPSS.

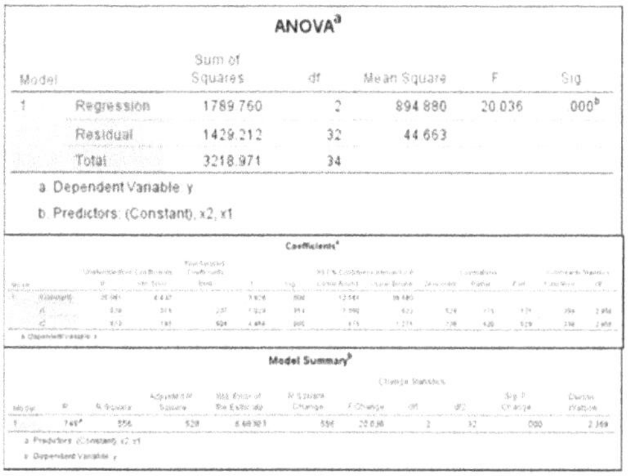

Table 1. Indicator table.

Indicator	Note
0.00 < r < 0.199	Extremely low relationship
0.2 < r < 0.3.99	Low relationship
0.4 < r < 0.5.99	Medium relationship
0.6 < r < 0.7.99	Strong relationship
0.8 < r < 1	Extremely strong relationship

The result of analyzed H3, the influence of service differentiation perception and service quality perception towards customer satisfaction, is accepted, and it can be seen from the analyzed data that shows X1 and X2 influence significantly towards Y (0.0 < 0.05). Figure 2 shows the research.

The correlation between X1 and X2 towards Y = 0.746 show a strong relationship which is told by the following indicator table.

4 CONCLUSION

Competitive advantage is an acquired ability because it requires the characteristics and resources of a company to achieve a higher level of performance than other companies in the same market (Porter 1994).

To create a competitive advantage is by doing differentiation strategies, one of them is service differentiation. In another way to differentiate is to consistently provide a better quality of service than competitors. This can be achieved by meeting or even exceeding the quality of services expected by customers (Tjiptono 2001).

Based on the result of the research and elaboration, it can be drawn a conclusion that generally service differentiation perception does not influence customer satisfaction with the IndiHome product. The service that is given to the customer that is suitable to the standard determined by the company does not give satisfaction to the customer, particularly the speedy to cope customers complain. The

positive and significant influence is shown by the variable of service quality perception with a tangible indicator (real), reliability, responsiveness, assurance, and empathy towards the satisfaction of the IndiHome product customers.

Differentiation is a way of concretizing a company's marketing strategy with all kinds of related aspects in a company that distinguishes it from competing companies (Kartajaya 1996).

The purpose of the differentiation strategy is to make something that is considered different by consumers in a particular industry. Differentiation requires a harder effort with greater costs. Consideration of the willingness of consumers to pay more for the value offered by differentiation is very necessary. Non-significant differentiation for consumers will have a negative effect on the company (Kartajaya 2004).

REFERENCES

Bettega, M. 2016. Pengaruh Brand Equity terhadap Brand Preference dan Purchase Intention Produk IndiHome. Thesis, Telkom University.

Dickson, P. R. (1997), *Marketing Management*, Second ed. Florida: The Dryden Press, Harcourt Brace College Publishers.

Kartajaya, H. 1996. *Marketing Plus 2000 Siasat Memenangkan Persaingan Global*. Jakarta: PT. Gramedia Pustaka Utama.

Kartajaya, H. 2004. *Positioning, Diferensiasi, dan Brand*. Jakarta: PT. Gramedia Pustaka Utama.

Kotler, P & Susanto, A. B. 2001. *Manajemen Pemasaran di Indonesia*. Salemba, Jakarta.

Lovelock, C & Wright, L. K. 2002. *Principles of Service Marketing and Management*. Prentice Hall.

Porter, M. E. 1994. *Developing Competitive Advantage in India*. Confederation of Indian Industry.

Sugiyono. 2003. *Metode Penelitian Bisnis*. Bandung: Alfabeta.

Sugiyono. 2012. *Metode Penelitian Kuantitatif Kualitatif dan R&D*. Bandung: Alfabeta.

Suharyadi & Purwanto. 2003. *Statistika untuk Ekonomi dan Keuangan Modern*. Jakarta: Salemba Empat.

Tjiptono, F. 2001. *Service, Quality, Satisfaction*. Yogyakarta: Andi.

Advances in Business, Management and Entrepreneurship – Hurriyati et al (eds)
© 2020 Taylor & Francis Group, London, ISBN 978-0-367-27176-3

The effect of service quality to customer satisfaction: A case study from Timor Leste, Dili

I. Indrawati & M.P. Henriques
Telkom University, Bandung, Indonesia

ABSTRACT: This research was conducted at Bank Mandiri Branch of Timor Leste, Dili. Bank Mandiri is one of foreign banks in Timor Leste. The main reason for choosing it as the object of this research is that it is still facing obstacles in its services, especially on the front liner such as teller and customer service. The data is obtained from customer complaints when researchers performed direct observations. On a normal day, a customer needs approximately 1.5 to 3 hours to make transactions. If there is a big day, customer complaints are increasing because sometimes it takes more than 1 day to do transactions. The purpose of this study is to determine respondents' assessment about the dimensions of service quality, satisfaction level, and the influence of the dimensions of service quality to customers' satisfaction either partially or simultaneously. This research applies quantitative method by using purposive sampling technique in order to obtain the samples. The data were collected from 400 respondents through questionnaires distributed both on line and off line. The result of the descriptive analysis indicates that the score for each dimension from the highest to the lowest are Tangibles, Responsiveness, Assurance, Reliability and Empathy respectively. Customer Satisfaction indicates the score of 59.6%, while the Multiple Linear Regression Analysis result shows that Service Quality simultaneously influences the Customer Satisfaction. Improving the quality of teller service should be pursued in relation to tangibles, responsiveness, assurance, reliability, and empathy variables, such as providing comfortable seating with attractive designs, providing a convenient queue for customers, increasing teller's knowledge of Bank Mandiri products, giving service time as promised, and applying ethical attention when communicating with customers.

1 INTRODUCTION

The success of the State of Timor-Leste's development is determined by the participation and cooperation of various parties such as government, society and other private institutions (such as banking and private sector of both national and international levels). In the framework of the State of Timor Leste's development, the society often face problems related to fund, such as funds for consumption needs, productive needs and other needs such as education and housing. In order to foster economic development of the State of Timor-Leste, the government has undertaken various ways, one of which is to provide opportunities for foreign parties to open business in the country in such field as property and the provision of services.

Banking services are established in the country to address issues related to the needs of the people. On 11 August 2003, Bank Mandiri initiated a branch in Timor-Leste to serve the people of Timor-Leste. Currently there are four banks operating in Timor-Leste: Banco National Ultramarino (BNU), which is originating from Portugal with a total 27,000 (22.5%) customers, ANZ Bank of Australia and New Zealand with a total 13,000 (12.5%) customers, Banco Nacional Comércio de Timor-Leste (BNCTL)

originating from Timor-Leste with a total 40,000 (25%) customers and Bank Mandiri from Indonesia with a total 48,000 (40%) customers (BNCTL 2017).

Banks act as financial intermediaries that raise funds in the form of credit (UU No 10 1998). The role of banks is to serve people who have excess money and then save the money in the form of demand deposits, deposits and savings and serve the needs of the community through the provision of credit. The bank's strategic management pattern should be oriented towards quality and service that apply the principles of modern management. The strategic management pattern needs to be supported by good planning and prioritizing services that can satisfy customers.

Kotler & Keller (2016) explain that holistic marketing of the quality of a company's services is tested in every meeting internally and externally to create service expectations in comparing between perceived services and expected services. The problem faced by customers is the quality of service provided by the company to its customers. This complaint comes from customers who publicly make complain, for example, some companies impose a tortuous and unclear service system that confuses customers, even offends their employees In addition, human resources (HR) companies did not realize the importance of customers for the success of the company. In Indonesia, Bank Mandiri is the largest bank that implements an efficient customer service system. Unlike the Bank Mandiri Dili branch, located in Timor Leste, Bank Mandiri's service in Dili is very inefficient. This results in customers' dissatisfaction, so that customer complaints in Timor-Leste increase more than customers in Indonesia, especially in terms of teller and customer service.

The main obstacles faced by Bank Mandiri in Timor-Leste are teller and customer service. Customers always complain about the slow service time in making transactions. This is known from customers' direct conversation among them during waiting time in the bank, verbal complains on the bank through teller and customer service and also through the researcher's direct observation where on a normal day per customer takes approximately 1.5 to 3 hours queue to make a transaction.

Researchers found some of the initial facts faced by Bank Mandiri customers of Dili branch: (1) Customer Service and Teller staffs are not sufficient; (2) Customer service sometimes leaves the place while serving the customers; (3) Customer services do not concentrate when they serve the customers; (4) Customer services do not have comprehensive knowledge of products of Bank Mandiri; (5) Teller is unfriendly and disrespectful while serving the customer; (6) Service is not in line with queue; (7) Service procedure in customer service is complicated. These facts can influence the customer satisfaction which is very important for the continuation of Bank Mandiri. Therefore, this study intends to find out the customers' satisfaction level of Bank Mandiri and to determine the dimensions of service quality influencing the customer satisfaction. The aims of this study are to determine (1) the score of service quality dimensions, (2) the level of customer satisfaction, and (3) the effect of service quality dimensions to customer satisfaction partially and simultaneously. Research regarding the influence of service quality to customer satisfaction can be found in several existing studies, but the studies were not done in Timor leste, hence the result may be different. The result of research done in one country can be different from others because of different the difference in social, economical and cultural backgrounds (Indrawati et al. 2015, Indrawati & Adicipta 2017).

Kotler & Keller (2016) explain that service is any action or performance that one party can offer to another, essentially intangible and does not result in ownership of something. Its production may be related or not related to physical product. Thus services can be interpreted as; any activity or benefit offered by a party to any other party and is essentially intangible, and does not result in the ownership of anything. Sunyoto & Susanti (2015) state that services are activities that have some related intangibles involving multiple interactions with consumers or with property in ownership, and do not result in ownership transfers.

Kotler & Keller (2016) distinguish four categories of service characteristics that vary from goods to services: (1) Intangibility, services have an intangible property because they cannot be seen, felt, heard, or smelled before a purchase transaction; There are several ways and strategies applied in the sale of services; (1) place in the form of interior or exterior service of a bank capable of giving a convincing impression or attracting customers; (2) people, in the form of hospitality, speed, neatness; (3) equipment, including equipment used; computers and others; (4) communication, such as printed and well laid out brochures and other forms of communication; (5) symbols, in the form of a short name or symbol, interesting, and gives the impression of glory; and (6) price, in the form of clear and competitive interest; (2) Inseparability, Service cannot be separated from the source, in the form of a person or machine, present or not, tangible physical products remain; (3) Variable, Service is easy to change because it depends on the party who presents, the time and place of presentation. (4) Perishability, the endurance of a service will not be a problem if demand always arises from the consumer.

Kotler & Keller (2016) state that quality is the totality of features and characteristics of a product or service to meet the needs and wants of the customer. Quality service is the service received relatively more satisfactory than what is expected by the customer. The concept of service quality is basically relative in nature, depending on the perspective used to determine the characteristics and specifications. There are basically three quality orientations that should be consistent with each other; (1) consumer perception; (2) Products (services); and (3) Process. For services that intangible goods these three orientations are almost always clearly distinguished, even the product is the process itself. Consistency of service quality for these three orientations can contribute to the success of an enterprise in terms of customer satisfaction, employee satisfaction, and organizational profitability (Sunyoto & Susanti 2015). When analyzed further, the relationship between quality and long-term profit is seen in two ways, namely external profit factor obtained from customer satisfaction and internal benefits obtained from the improvement of product efficiency (Sunyoto & Susanti 2015).

The external advantages in question can be implied in the production process of a good (service), i.e. where the quality of the product (service) provided by the firm can create a positive perception of the customer to the company and produce a customer satisfaction and loyalty. Internal gains appear at the same time as external gains, where the company's focus on quality can bring the company's internal positive value in the upgrading process. The success of the company in providing quality services to its customers, achieving high market share, and increasing profits of the company is largely determined by the approach used (Matos et al. 2013).

Service Quality (SERVQUAL) is built on the basis of a comparison between two main factors, namely customer perceptions of the actual service they receive (perceived service) and the actual service expected (expected service). Service quality can be defined as how far the difference between reality and customer expectations for the services they receive (Zeithaml & Bitner 2000).

Research on SERVQUAL conducted by Parasuraman (1998) can give conclusion that there are five dimension of SERVQUAL as follows: (1) Tangible that is ability of a company in showing its existence to external party; (2) Reliability namely the ability of the company to provide services in accordance with the promised accurately and reliably; (3) responsiveness, which is a policy to assist and provide prompt and responsive service to the customer, with clear information delivery; (4) Assurance, i.e. knowledge, courtesy, and the ability of company employees to cultivate trust in customers to the company, which includes several components, among others; communication, credibility, security, competence, and courtesy; and (5) Empathy, which is to give a sincere and individualized or personal attention given to the customer by attempting to understand the desire of the consumer.

Satisfaction is a person's feeling of pleasure or disappointment resulting from his computation of products perceived in relations to his or her expectations (Kotler 2000). Customer satisfaction is the feeling of happiness or disappointment of someone that arises because it compares the perceived performance of the product (outcome) to the expectation.

In the service industry, customer satisfaction is always influenced by the quality of interaction between customers and employees who make service contact (service encounter) that occurs when customers interact with the organization to obtain service they purchased. Customer satisfaction in the banking services industry by Kotler & Keller (2016) companies can systematically measure based on these indicators: perceptions of customer feelings developed from service performance dimensions, cost burden, corporate image, and the decision to use services to conduct operations and marketing services.

This research has reason that if banking company perform service with quality in accordance with the customers' expectation, it will give good appraisal between banking company Bank Mandiri Branch, Timor-Leste and customer that is intertwining harmonious relationship. To measure the satisfaction level of Bank Mandiri customers in Timor Leste, empirically customer satisfaction can be understood by the company by examining 5 (five) service quality dimensions consisting of reliability, assurance, responsiveness, empathy, and tangibles. The systematic framework of thought in this study can be seen in Figure 1.

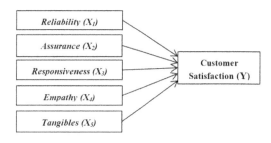

Figure 1. The influence of variable quality of service to customer satisfaction, a conceptual model.

2 METHOD

The researchers implement a quantitative research method, while the purposes of this research include descriptive verification research and causal research or Expost Facto. The population in this study is customers of Bank Mandiri of East Timor Branch in Dili City, Timor-Leste, conducting transactions at the teller more than 3 (three) times. This research use purposive sampling method and sampling determination of this research is purposive sampling technique (Indrawati 2015).

The number of selected samples is determined by the Slovin formula with 400 respondents as the result. The risk of error (α) = 0.1 (10%) and bound of error is set at 0.05 (5%). The sampling technique used is non-random sampling and the method applied is purposive sampling method. Data collection methods used are research questionnaire and observation. The secondary data in this study was obtained from Bank Mandiri Branch of Timor-Leste concerned in the history of Bank Mandiri establishment, organizational structure and the development of the number of customers. To test the validity of this research, this study uses Product Moment Correlation formula from Pearson and reliability test is measured using Alpha Cronbach formula (Indrawati 2015).

3 RESULTS AND DISCUSSION

3.1 *Respondent characteristics*

The respondents of this research is 57.5% female and 42.5% male, in term of age, most of the respondents are between 21 to 50 years old. They occupation are mostly employees and their education background is mostly diploma. It can be seen in Figure 2.

3.2 *Respondents response regarding reliability (X₁)*

In order to achieve the first and second research aims, this research conducts the descriptive calculation. The expected score for the respondent's answer to 6 (six) statements is 12,000. The calculation results show that the value obtained for Reliability is 6434 or

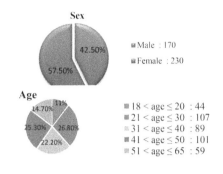

Sex

42.50%
57.50%

■ Male : 170
■ Female : 230

Age

11%
14.70%
26.80%
25.30%
22.20%

■ 18 < age ≤ 20 : 44
■ 21 < age ≤ 30 : 107
■ 31 < age ≤ 40 : 89
■ 41 < age ≤ 50 : 101
■ 51 < age ≤ 65 : 59

Occupation

25.70%
25%
28%
21.30%

■ Entrepreneur : 100
■ Student : 85
■ Government/Private Employees : 112
■ Others : 103

Education

13%
20.8%
16.50%
26%
23.80%

■ Primary School : 52
■ Junior High School : 66
■ Senior High School : 95
■ Diploma : 104
■ Bachelor, Master, Doctor : 83

Figure 2. Respondents' characteristics.

53.6% of the ideal score of 12,000. Thus Reliability (X_1) is in fairly good category based on respondents' point of view.

3.3 Respondents response regarding assurance (X_2)

Ideally, the expected score for the respondent's answer to 5 (five) statements is 10,000. The calculation results show the value obtained 5379 or 53.8% of the ideal score of 10,000. Thus Assurance (X_2) is categorized in the quite well category based on respondents' point of view.

3.4 Respondents response regarding responsiveness (X_3)

The expected score for the respondent's answer to 6 (six) statements is 12,000. The calculation results show the value obtained 6626 or 55.2% of the ideal score of 12,000. Thus Responsiveness (X_3) is categorized in fairly good category based on respondents' point of view.

3.5 Respondents response regarding empathy (X_4)

Ideally, the expected score for the respondent's answer to 7 (seven) statements is 14,000. The calculation results show the value obtained 7428 or 53.1% of the ideal score of 14,000. The calculation descriptively shows Empathy variable gives the lowest influence to Customer Satisfaction variable. Thus

Empathy (X_4) is categorized in fairly good category based on respondents' point of view.

3.6 Respondents response regarding tangibles (X_5)

Ideally, the expected score for the respondent's answer to 7 (seven) statements is 14,000. The calculation results show the value obtained 8085 or 57.8% of the ideal score of 14,000. The calculation of descriptive Tangibles variable gives the highest influence on Customer Satisfaction variables. Thus Tangibles (X_5) are in fairly good category based on respondents' point of view.

3.7 Respondents response regarding customer satisfaction (Y)

The expected score for the respondent's answer to 6 (six) statements is 12,000. The calculations in the Table shows that the value obtained is 7150 or 59.6% of the ideal score of 12,000. Thus the Customer Satisfaction (Y) is considered in the quite well category. This indicates that the respondents believe that Bank Mandiri's Customer Satisfaction (Y) of Branch of Timor-Leste is quite good.

3.8 The influence of service quality on customer satisfaction

In order to achieve the third research objective, this research uses inferential analysis. Inferential analysis method used in this research is multiple regression analysis. In doing multiple regression analysis, this study should fulfill classic assumptions, such as normality test, multicollinearity test, and heteroscedasticity. In the Normality Test, the normality analysis is conducted based on the Kolmogorov-Smirnov method which shows the normal curve when the Asymp value. Sig. is above the maximum error limit of 0.05. As for the regression analysis, the normality test is residual or random stochastic variable disorder. In the Multicollinearity Test, detecting the presence or the absence of multicollinearity is done by using Variance Inflation Factors (VIF). The VIF value in the data is less than 10, so it can be concluded that there is no multicollinearity in the data. To test whether there is absolute heteroscedasticity of residual value using Spearman Rank Correlation. The result of calculation show that there is no absolute heteroscedasticity.

Multiple correlation analysis (R) is used to know the relationship between the quality of service and customer satisfaction. Based on the calculation results, correlation coefficient (R) obtained is 0.84. This shows that there is a very strong relationship between service quality and customer satisfaction. Partially, the values obtained are as follows: X_2 value 19.1%, X_5 value 18.9%, X_3 value 18.7%, X_4 value 7.7%, and X_1 value of 6%. Simultaneously, the total influence of service quality to customer satisfaction is 70.5%.

To conceive the effect of service quality on customer satisfaction, multiple linear regression analysis with equation can be described as follows:

$$Y = a + b_1X_1 + b_2X_2 + b_3X_3 + b_4X_4 + b_5X_5 \quad (1)$$

Meanwhile, the form of multiple linear regression equation can be described as follows:

$$Y = -0.114 + 0.141X_1 + 0.266X_2 + 0.275X_3 \\ + 0.124X_4 + 0.318X_5 \quad (2)$$

From the multiple linear regressions equation above, the obtained value of constant equal to -0.114. This indicates that if Customer Satisfaction variable is not influenced by the five independent variables– Reliability, Assurance, Responsiveness, Empathy and Tangibles is zero, then the average value of Customer Satisfaction will worth -0.114.

The sign of the independent variable regression coefficient shows the relationship direction of the variables concerned with Customer Satisfaction as shown in Figure 3.

3.9 Assessment of respondents on service quality at Bank Mandiri branch of Timor-Leste

The result of descriptive analysis shows that service quality variables consisting of Reliability has a value of 53.6%, Assurance has a value of 53.8%, Responsiveness has a value of 55.2%, Empathy has a value of 53.1%, and Tangibles has a value amounting to 57.8%. The results of this service quality variable fall into the category quite well.

3.10 The level of customer satisfaction at Bank Mandiri branch of Timor-Leste

The results of descriptive analysis and data processing from 400 respondents show that the customers' satisfaction variables have satisfaction value of 59.6%. The results of customer satisfaction variables included in the category is quite good. This indicates

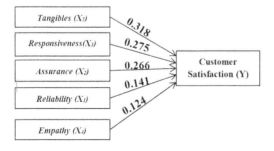

Figure 3. The influence of variable quality of service to the customers' satisfaction, a result model.

that the satisfaction of Bank Mandiri customers in East Timor Branch is categorized as good enough to the service of teller.

3.11 The partial effect of service quality on customer satisfaction

Based on the results of analysis with data processing using SPSS version 25, the influence of service quality to customer satisfaction partially is positive and significant. This can be seen from the partial hypothesis test with the provision: if t-value > t-table, then the result is significant. Each variable forming the quality of service has a significant effect on customer satisfaction.

3.12 The simultaneous effect of service quality on customer satisfaction

Based on the results of simultaneous test calculations or F test, it has F-value of 188.49. Since the value of F-value (188.490) > F-table (2.237), then Ho is rejected. Thus it can be said that simultaneously there is a significant influence of service quality on customer satisfaction.

Coefficient of Determination, or R^2 with service quality variable, influences 70.5% of customer satisfaction variable. Meanwhile, the rest of 29.5% is influenced by other variables which are not discussed in this study.

4 CONCLUSIONS

4.1 Conclusions

Service quality variables consisting of Reliability, Assurance, Responsiveness, Empathy and Tangibles are generally well categorized as X_1 = 53.6%, X_2 = 53.8%, X_3 = 55.2 %, X_4 = 53.1%, and X_5 = 57.8%. Customer satisfaction variables are categorized in the good enough category with a percentage of 59.6%. Simultaneous test results (Test F) shows that there is a significant influence between Quality of Service and Customer Satisfaction where the value shows that F-value 188.490 > F-table 2.237. Effect of Service Quality on Customer Satisfaction simultaneously is 70.5%. While 29.5% is the contribution of other variables that are not discussed in this study.

4.2 Suggestions

Service quality variables consist of five variables which give the highest influence to customer satisfaction variable: Tangibles 57.8%, and Empathy 53.1%. Improvements need to be made related to tangibles variables, i.e. the provision of comfortable seats with attractive designs, convenient queue provision for customers, increased knowledge of tellers about Bank Mandiri products, service time as promised, and ethics of communication towards the customers.

REFERENCES

Ariyanti, Fiki. 2016. Bank Mandiri targetkan 3.000 nasabah baru di Dili [Online]. Retrieved: http://bisnis.liputan6.com/read/2466014/bank-mandiri-targetkan-3000-nasabah-baru-di-dili. Accessed 2 September 2018.

BNCTL. 2017. Annual Report Banco Nacional Comércio de Timor-Leste 2017.

Indrzawati & Adicipta, S.R.M. 2017. Factors influencing internet banking acceptance: A case study of ABC internet banking in Bandung Indonesia. *Journal of Engineering and Applied Sciences* 12(7): 1705-1709.

Indrawati. 2015. *Metode penelitian manajemen dan bisnis. Konvergensi Teknologi Komunikasi dan Informasi.* Bandung: PT. Refika Aditama.

Kotler, P. & Keller, K. L. 2016. *Marketing management. 15th Global Edition.* Edinburgh Gate: Pearson Education Limited.

Kotler, P. 2000. *Management, analysis, planning, implementation, and control. Edisi Kedelapan.* Englewood Cliffs: Prentice Hall International, Inc.

Matos, C.A., Henrique, J.L. & Rosa, F. 2013. Customer reactions to service failure and recovery in the banking industry: the influence of switching costs., *Journal of Services Marketing* 27(7): 526-538.

Parasuraman, A. 1998. SERVQUAL: A multiple-item scale for measuring consumer perceptions of service quality. *Journal of Retailing* 4 (1).

Sunyoto, D. & Eka, S.F. 2015. *Manajemen pemasaran jasa, Cetakan pertama,* 2015. Penerbit PT Buku Seru.

UU No 10 Tahun 1998 tentang Perubahan Atas Undang-Undang No. 7 Tahun 1992 tentang Perbankan.

Zeithaml, V.A. & Bitner, M.J. 2000. *Services marketing: integrating customer focus across the firm, 2nd ed.* Boston: Irwin/McGraw-Hill.

Optimization of customer retention programs

E. Amelia & M.A. Sultan
Universitas Pendidikan Indonesia, Bandung, Indonesia

ABSTRACT: This study aims to obtain a description of customer retention BRI unit Paseh Sumedang. Customer retention is a strategy which goal it is is to retain corporate customers and maintain the contribution of revenue from customers. There is competitive competition in the banking industry, especially regarding savings marked by the decline in customer satisfaction. Customer satisfaction cannot be ignored in a company. The main solution that must be carried out by the company is to increase customer satisfaction and maintain the life of the company by conducting various strategies, especially customer retention. The concept of customer retention arises because many companies tend to ignore existing customers and prioritize acquisition programs more. In fact, according to a number of studies, the costs of new customers get higher than the costs of retaining old customers. If you have to choose because of resource constraints, you better prioritize customer retention programs than customer acquisition programs. The research method used is explanatory survey with systematic random sampling technique with a total sample of 110 respondents. The object studied is the customer of Simpedes Bank Rakyat Indonesia Unit Paseh Kabupaten Sumedang.

1   INTRODUCTION

Competition is considered as how the company can provide services that satisfy customers to be loyal to the company. This decrease in satisfaction is a threat to the company because customers who feel dissatisfied will move to another bank. Customer satisfaction with a bank should always be maintained and enhanced, so that customers believe in the products produced. Customer satisfaction is a state where the customer's wants, expectations and needs are met. Service is considered satisfactory if the service can meet the customer's needs and expectations. If the customer is not satisfied with a service provided, the service can be ineffective and inefficient. The company's experience shows that to attract new customers costs five times more than maintaining existing customers, while attracting customers who have already moved to a competitor bank requires 10 to 12 times the cost to be incurred (McKenzie 2008). The customer will be very satisfied and loyal to a bank but will also be quick to move to another bank that can provide better services. Customer satisfaction (CS) is where the wishes, expectations, and customer needs can be met. The nature of customer satisfaction always changes from time to time; customer expectations of a product or service will never be the same and can rise or even go down. The problem that became the basis of this study is the decline in the overall number of saving customers. From 2006 to 2009 the number of customers experienced a continuous decline. In 2009 the number of customers was 4774, which is quite lower than the 8603 in 2006. This decrease in the number of customers shows that customers can easily move from one bank to another. There are five factors that drive customer satisfaction, including product quality, price, service, emotional, and convenience.

Customers will feel satisfied if the products purchased and used have good qualities. In terms of quality, the product has six commonly used global elements: performance, reliability, features, durability, conformance, and design. For consumers considering the price, a cheap price contributes greatly, as it amounts to value. But for other consumers, low prices are not so influential, so these elements tend to be stable in the consumer's view.

Customer satisfaction is more sensitive. Another phenomenon is also the rapid changing consumer preferences. Customer satisfaction is moving in an emotional direction so that marketers should touch the heart and explore the emotional consumer. Customer retention identification and resolution of these issues can increase the customer efficiency ratios as satisfied customers normally have greater purchase volume than non-satisfied customers, and the relative cost per customer drops (Nataraj 2010). The concept of customer retention arises because many companies tended to ignore existing customers and prioritize acquisition programs more. In fact, according to a number of studies, the costs of obtaining new customers is higher than the costs of maintaining old customers. If you have to choose because of limited resources, for example, it is better to prioritize customer retention programs than customer acquisition programs. This customer retention program is expected to increase customer satisfaction because this program provides an appreciation for its customers to continue saving and increase the balance.

Customer retention strategies can be viewed as a mirror image of customer defection, where high retention rates have the same impact as low defect rates. There are five types of defectors/customers who switch to competitors. Citing Tjiptono et al. (2008) those include price defectors, customers who switch suppliers because of the pursuit of cheaper prices; product defectors, customers who switch suppliers because they find superior products; service defectors, customers who switch suppliers because they get better service elsewhere; market defectors, customers who switch suppliers due to moving to another market; and technological defectors, customers who switch suppliers because of switching from certain technologies to other technologies.

This customer retention program is about activities in optimizing its performance of products, prices, services, target markets, technology and the convenience of having a saving product for its customers. Companies, especially banks, use customer retention programs to satisfy customers and increase customer loyalty. The research problem is how to describe the customer retention program.

According to Simbolon (2011), the argument about customer retention explains that customer retention activities become more economical compared to the activities of finding new customers. Activities to acquire new customers cost a lot more than retaining customers. Existing customers usually trade more, and if they are satisfied they will recommend the bank to their colleagues, and indirectly they provide a word-of-mouth promotion which is profitable for the company.

Existing clients are usually less sensitive in case of changes in transaction costs (Healy 1999 in Cohen et al. 2007). According to Ahmad and Buttle (2001) a 5% retention rate increase could affect the increase in net present value of consumers by 25% to 85%. The results of Page et al., Payne and Frow in Ahmad and Buttle (2001) support his opinion; Dawkins and Reicheld argue that customer retention is able to generate real financial benefits for the company. Rust and Zahorik in Cohen et al. (2007) suggest that financially, the activity of attracting new consumer costs five times more when compared to preserving existing customers. Customer retention is a reflection of the low number of consumers who leave the product or service company. Customer retention can be problematic if it is not properly defined according to the company's business objectives. Companies that offer diverse products or services should understand customer retention from the consumer's point of view, and not just from the number of customers they have. In the banking industry, for example, banks should define customer retention as a percentage of total savings, credit expenses and customer purchases, and not just define it based on the number of customers owned by the bank (Ahmad & Buttle 2001).

The results of Crutchfield (2001) suggest that trust and commitment factors significantly influence customer retention, where trust has a stronger influence compared to commitment. The same is also found by Gounaris (2005), the results of whose research says that trust and commitment factors have an important role in maintaining relationships between clients and providers, and also that the trust factor has a more important role than commitment.

The results of research by Ranaweera and Prabhu (2003) found that consumer satisfaction and trust significantly influence customer retention, where consumer satisfaction has a stronger influence when compared with trust. Research conducted by Gustaffsson et al. (2005) found that customer retention can be predicted through customer satisfaction, affective commitment, and commitment commitments. Customer satisfaction is the overall evaluation of the performance of a company's product or service, affective commitment describes trust and reciprocity in a relationship, and the calculative commitment explains the switching costs and the absence of alternatives.

Research conducted by Gerpott et al. (2000) on Customer retention, loyalty, and satisfaction in the German mobile cellular telecommunications market aims to research construct differences in consumer retention, loyalty of the consumer, and consumer satisfaction on mobile cellular in Germany which turned out to have variable linkages which influence each other. The results of this study prove that customer satisfaction is significantly influence to consumer loyalty indirectly gets influence from the consumer. The results of research conducted by researchers prove that the retention program is one of CRM strategies that will be able to increase the satisfaction of consumers, so that every company will be able to achieve customer satisfaction by increasing some features in the retention program. Retention and variables that are very influential toward consumer loyalty are consumer retention. These results are similar to other studies about consumer loyalty that is influenced by the factors consumer satisfaction and switching barrier in the cases of mobile telecommunication services in Korea, giving the result that consumer satisfaction influences consumer loyalty positively and significantly (Sukmaputra 2017).

The high adoption of the low interest rate of the bank and the amount of fee compared to the competitor banks is not the main reason for the customers to switch to other banks or not. Other factors such as the existence of emotional bonds between the bank and the customer can be the decisive factor in managing the satisfaction that leads to customer loyalty. The increasing level of consumer retention will automatically increase the number of consumers owned by an organization. In addition, increasing levels of retention will increase customer loyalty (customer tenure).

The higher rate of consumer retention leads to a greater positive impact. Consumer retention and loyalty management have two benefits. First, marketing costs can be reduced. Companies do not need to spend additional funds to find a vague consumer. Second, along with the increasing customer tenure, the more comprehensible willingness of consumers, and the understanding about what the company can do for them. The deepening of the relationship and the trust and commitment of both parties will develop by itself (Kurniawan et al. 2014).

2   METHOD

This study uses a quantitative approach. This research is conducted to understand the ability of the customer retention program. Independent variables or exogenous variables found in this research are customer retention with the dimensions of welcome, reliability, responsiveness, recognition, personalization, and access. The dependent variable or endogenous variable in this research is customer satisfaction and can be seen from the gap between customer perceptions with customer expectation.

The object of analysis in this research is Simpedes Bank Rakyat Indonesia saving account of Paseh Sumedang unit. This research was conducted in less than one year, so the data collection technique used in this research is cross-sectional method. The research method used is explanatory survey with systematic random sampling technique with a total sample of 110 respondents. Data collection techniques used were literature studies, field studies with questionnaires, and literature studies. The data analysis techniques used computer software tool SPSS 17.0.

3   RESULT AND DISCUSSION

Hypothesis testing of this study was conducted to determine the magnitude of the influence of customer retention on customer satisfaction. Hypothesis testing was done simultaneously and partially using path analysis technique. The hypothesis of this study is that customer retention affects customer satisfaction simultaneously and partially. This hypothesis is tested simultaneously or partially by using computer software tool SPSS 17.0. The results of hypothesis testing as a whole (simultaneous) can be seen based on the variable factors of customer retention having a direct impact on customer satisfaction of 64.29%. Based on the result of research between customer retention program and Simpedes Bank Rakyat Indonesia saving customer satisfaction, the result of the correlation is: Welcome (0.582), Reliability (0.379), Responsiveness (0.712), Recognition (0.604), Personalization (0.644) and Access (0.722). Access obtained the largest correlation of 0.722; this happens because Bank Rakyat Indonesia consistently builds communication networks by providing convenience for customers. The lowest correlation value is Reliability (0.379); Bank Rakyat Indonesia should be able

to improve its ability in products and services to customers further.

The ability to deliver the services as promised accurately and reliably can prevent customer removals. According to Lewis and Blooms, in Tjiptono and Chandra (2005) service quality as a measure of how good the level of the service provided is, is able to match customer expectations. Simultaneously, (overall) there is a positive influence between the effect of the customer retention program on customer satisfaction of Simpedes Bank Rakyat Indonesia savings account. Based on the calculation results it can be seen that the customer retention strategy to customer satisfaction is equal to 0.838. While the coefficient of other variable paths outside the customer retention variables of Welcome, Reliability, Responsiveness, Recognition, Personalization and Access explains that welcome (X1), Reliability (X2), Responsiveness (X3), Recognition (X4), Personalization (X5) and Access (X6) jointly affect customer satisfaction (Y), with a rate of 83.8% 0.4025) 2 = 0.1620 x 100% = 16.20% influenced by other factors not included in the research.

Influence directly and indirectly know the lowest correlation between variable X and variable Y is reliability with customer satisfaction which is equal to 0.379, while the lowest correlation between variable X is reliability with access of 0152. The customer retention program that has the highest impact on customer satisfaction is responsiveness, which directly affects the satisfaction of Simpedes Bank Rakyat Indonesia's savings account with 11.09%, indirect influence through welcome is 1.64%, through reliability 0.93%, through recognition 2.46%, personalization 1.7% and through access 5.85%. It means that the program retention provided by BRI turned out to give satisfaction to consumers who become loyal customers. This is also appropriate with research conducted by Sari (Sukmaputra 2017) that the retention program (CRM) gives satisfaction to consumers because it brings many benefits and increases the profitability of the company. Responsiveness besides each unit increase the retention program will increase 0.712 units consumer satisfaction, which means that the bigger the efforts made to improve the program retention, the moreincrease in customer satisfaction. This matter is also in accordance with research conducted by Bolton and Kannan that the program retention carried out by the company will have a big influence on consumer satisfaction. The retention program will provide relationships and influence on consumer loyalty because the aim of the program is an effort to gain customer loyalty (Sukmaputra 2017).

4   CONCLUSION

To achieve customer satisfaction, the company must continue to improve customer service. The company must also retain existing customers as new customers cost more than maintaining customers. Customer retention at Simpedes BRI savings, consists of welcome, reliability, responsiveness, recognition, personalization and access. The welcome dimension gets the highest score in satisfying the customer.

Meanwhile, the responsiveness dimension gets the lowest score in satisfying the customer. But overall, this study found that it turns out that the program retention from the service company can provide benefits for the consumer and the company itself, through retention corporate programs. It can be said that if competitiveness is high, they are able to suggest their image in the minds of consumers. Regarding further explanation about why consumer satisfaction influences loyalty, more research is needed on this matter. Further research is expected to reexamine the level of customer retention program to customer satisfaction and for other variables that have not been studied in this study. In addition, continuous research on the customer retention program must be developed, continuously adjusting to customer expectations to achieve customer satisfaction. Customer satisfaction is important and should be improved to get loyal customers.

The customer retention program should be able to carry out innovative activities to retain customers who are increasingly exposed to the number of quality service options. BRI must look at customer satisfaction and the survive customer. It is a key that must be pursued seriously, and looked at in the overall state (total quality management) for facing the competition that will come.

REFERENCES

Ahmad, R. & Buttle, F. 2001. Customer retention: a potentially potent marketing management strategy. *Journal of strategic marketing*, 9(1): 29–45.

Cohen, D., Gan, C., Au Yong, H. H. & Chong, E. 2007. Customer retention by banks in New Zealand. *Bank and Bank Systems*, 2(1).

Crutchfield, T.N. 2001. The effect of trust and commitment on retention of high-risk professional service customers. *Services Marketing Quarterly*, 22(2): 17–27.

Gerpott, T.J., Rams, W. & Schindler, A. 2001. Customer retention, loyalty, and satisfaction in the German mobile cellular telecommunications market. *Telecommunications policy*, 25(4): 249–269.

Gounaris, S.P. 2005. Trust and commitment influences on customer retention: insights from business-to-business services. *Journal of Business research*, 58(2): 126–140.

Gustafsson, A., Johnson, M.D. & Roos, I. 2005. The effects of customer satisfaction, relationship commitment dimensions, and triggers on customer retention. *Journal of marketing*, 69(4): 210–218.

Kurniawan, D.F., Suroso, I. & Irawan, B. 2014. Analisis Pengaruh Customer Retention Program (CRP) terhadap Loyalitas melalui Kepuasan Nasabah tabungan PT. BNI 46, Tbk Cabang Jember. *Jurnal Ekonomi Akuntansi Dan Manajemen*, 13(1): 1–15.

McKenzie, B. 2008. Customer relationship management and customer recovery and retention: the case of the 407 express toll route. *Knowledge Management Research & Practice*, 6(2): 155–163.

Nataraj, S. 2010. Customer Retention–CRM Application. *Issues in Informations Systems*, 11(2): 44–47.

Ranaweera, C. & Prabhu, J. 2003. On the relative importance of customer satisfaction and trust as determinants of customer retention and positive word of mouth. *Journal of Targeting, Measurement and Analysis for Marketing*, 12(1): 82–90.

Simbolon, F. 2011. Pengaruh Switching Costs terhadap Customer Retention pada Industri Perbankan di Indonesia. *Binus Business Review*, 2(2): 698–707.

Sukmaputra, E.R., Esarianita, Y. & Megandini, Y. 2017. Pengaruh Program Retention dan Kepuasan Konsumen terhadap Loyalitas Konsumen di CGV Blizt Bandung. *In Prosiding Industrial Research Workshop and National Seminar*, 8(3): 362–368.

Tjiptono, F. & Chandra, G. 2005. *Service, quality & satisfaction*. Yogyakarta: Andi Offset.

Tjiptono, F., Chandra, G. & Adriana, D. 2008. *Pemasaran strategik*. Yogyakarta: Andi Offset.

Advances in Business, Management and Entrepreneurship – Hurriyati et al (eds)
© 2020 Taylor & Francis Group, London, ISBN 978-0-367-27176-3

Engaging young consumers with advergames: The effect of presence and flow experience

S. Soebandhi
Universitas Narotama, Surabaya, Indonesia

S. Hartini & S. Gunawan
Universitas Airlangga, Surabaya, Indonesia

ABSTRACT: This study aimed to analyze children's evaluation of advergame, brands integrated in a game and their intention to purchase the advertised products through presence and flow experience. An advergame with its fictional products was specially designed for this study, with 120 children aged 10-12 years old as the participants. Four-point Likert scale with smiley face icon to express respondent's opinion was used in the questionnaire. The results show that flow does not influence the attitude towards advergame and the brand, while other hypotheses are supported.

1 INTRODUCTION

Fast growing internet and technology have made children more familiar with digital media and trigger them to spend a lot of time in multimedia environments (Moore & Rideout 2007, Waiguny et al. 2011). Children are considered as potential customers since they are able to influence purchase decisions made by parents or family (Buckingham 2000, Calvert 2008), causing more and more companies to use techniques that blur the boundaries between advertising, entertainment, and information to get their attention (Raney et al. 2003).

One of the most widely used marketing media is advergame. Advergame or advertising game is a branded entertainment that displays advertising mes-sages, logos, and characters in a game format (Mallinckrodt & Mizerski 2007). In other words, advergame is a game that is designed as a media campaign with an emphasis on aspects of entertainment (Gross 2010, Terlutter & Capella 2013, Youn & Mira 2012). In advergame, the brand or product is a central of the game, thus the message conveyed integrated into gaming experience (Cauberghe & Pelsmacker 2010). In general, the use of advergames for children is intended to make them learn about the products and the companies (Bogost 2007) and the result is expected to form a positive attitude toward the brand and to enhance their intention to purchase the advertised product (Waiguny et al. 2012).

Advergame is a unique medium for marketing products to children because it can provide an immersive experience, moreover, the hidden message can affect their attitudes without them being aware of it (Nuijten et al. 2013, Rosen & Singh 1992, Staiano & Calvert 2012). However, the use of advergame as an advertising medium faces some

challenges. According to Limited Capacity Model (LCM) of mediated message processing (Lang 2000), a person's ability to process information is limited. If he/she has focused on the main task, then he/she will tend not to pay attention to the secondary tasks. In advergame, the main task of the player is to finish the game or mission. Their cognitive is used to understand the storyline, to anticipate the events, or to think of the strategies to win the game. This causes them to pay more attention to their main task (i.e. completing the game) and not paying attention to the emergence of the secondary information (i.e. advertising messages) (Herrewijn & Poels 2014, Vashisht & Royne 2016). Additionally, the existing ads or persuasive messages are integrated in an interactive and fun game that can make it difficult for players to understand the advertising messages in the advergame (Terlutter & Capella 2013). Considering that the main purpose of advertising is to influence consumer attitudes, that will lead to consumers' intention to buy, the effectiveness of advergame as a media campaign, especially for children is need to study further.

2 METHOD

2.1 Procedure

The explanation about the guideline of how to play the advergame and how to fill out the questionnaire with the "smiley face" scale as well as the meaning of each scale were delivered out first. The link to the game was also provided. The game was played for 15 minutes, continued with filling out the questionnaire assisted by the researcher. In this game, players were required to collect snacks as much as possible to win the game. The snack displays the product brand on its package and appears during the game. Players can also choose the characters they play.

One hundred and twenty respondents aged 10-12 years old from public and private schools in Surabaya participated in this study. Beforehand, consents from school and parents were required.

Four point Likert scale was designed using an appealing emoticon for children (Hall et al. 2016). The expressions on these icons ranged from happy to unhappy (Reynolds-keefer & Johnson 2011).

2.2 Measurement

Presence in entertainment media is a sense of "being there" which can make someone forget that they are actually playing the game and is in front of a TV or PC (Kim & Biocca 1997, Minsky 1980). A person who feels presence will feel as if he is in the game he is playing. The experience of the brands provided in this video game is remembered as if the experience is real, and if the perceived experience is fun, then it can produce a positive evaluation of the advergame and brand in the advergame (Debbabi et al. 2013, Debbabi et al. 2010). In this study, presence was measured using two items adapted from Hyun & O'Keefe (2012).

Game flow is a situation where a person is immersed in the game he is playing. This immersed feeling can make players ignore advertising messages embedded in advergame, thus players can accept advergame as entertainment, but not as an advertis-ing medium (Lang 2000). Furthermore, advertising messages in advergame are not displayed explicitly but integrated with the game. A positive experience when playing a game (e.g. defeating the enemy) is always associated with the brand. Hence, every time a player sees the brand, this experience will always be remembered and positive feelings towards the brand will increase (Nelson & Waiguny 2012). In this study, flow was measured using three items based on Vanwesenbeeck et al. (2016).

Attitude is a person's evaluation of an object (Mitchell & Olson 1981). In marketing communication, attitude is a strong predictors that can shape intention and behavior (Chang et al. 2013, Sicilia et. al 2006). In the advergame context, when a player feels entertained while playing, he will form a positive attitude on the advergame being played (Rifon et al. 2014). This positive experience can also be forwarded or transferred to the inherent brand in the game (Nelson & Waiguny 2012, Waiguny et al. 2011). When playing games, positive feelings about the game are associated with the brand and when players see the brand, the positive feelings will be remembered. Attitude toward advergame (Att_{ad}) was measured using four items adapted from Hernandez (2008), Soebandhi & Andriansyah (2017).

Brand attitude is an evaluation of a brand based on a belief in the attributes of the brand (Mitchell & Olson 1981). Compared to other advertising media, advergame offers an entertainment features to its customers. The experience felt when playing this advergame can form a like or dislike feeling on the brand in the game. This positive attitude is important because it can boost the formation of behavioral intentions (Ajzen 1991). This idea is in line with Baker (1999) who argued that a unified brand in an advertising format can influence brand attitude and behavioral intentions through a person's sub consciousness. Attitude toward brand (Att_b) was measured using three items from Roedder et al. (1983). Finally, the two items measuring purchase intention (PI) were adapted from Rozendaal et al. (2013).

3 RESULT AND DISCUSSION

3.1 Analysis and finding

We only processed 120 out of 125 responses because the remaining does not fit the criteria required. Most respondents were 11 years old (66.7%). 53% of the total respondents were girls and the respondents

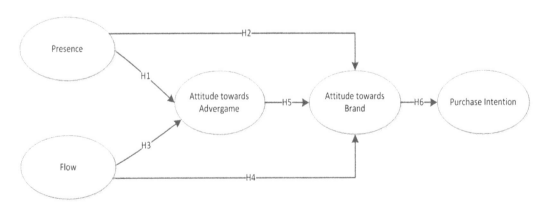

Figure 1. Conceptual framework.

Table 1. Respondents' profile.

		Frequency	Percentage
Age	10	22	18.3%
	11	80	66.7%
	12	18	15.0%
School	Private	63	52.5%
	Public	57	47.5%
Gender	Boy	56	46.7%
	Girl	64	53.3%
Video game play in a week	Almost every day	29	24.2%
	Occasionally	70	58.3%
	Only on school holidays	21	17.5%

Table 3. PLS path coefficient and hypothesis result.

Hypothesis	Path	Original Sample (O)	T Statistics (|O/STERR|)	Result
H_1	Presence -> Att_{ad}	0.3877	4.67	Supported
H_2	Presence -> Att_b	0.2496	2.4739	Supported
H_3	Flow -> Att_{ad}	-0.0533	0.5379	Not Supported
H_4	Flow -> Attb	0.1974	1.5104	Not Supported
H_5	Attad -> Attb	0.2732	2.7688	Supported
H_6	Attb -> PI	0.5463	7.6251	Supported

from private schools were greater in number at 52.5%. The respondents only occasionally play video games in a week because they have to prioritize their study. Table 1 presents the profile of respondents.

PLS-SEM (Partial Least Square-Structural Equation Modeling) was used to analyze conceptual models in this study. There were five loading indica-tors below the cut-off value recommended, as stated by Hair et al. (2011). However, some authors (i.e. Chen & Tsai 2007, Lin & Filieri 2015) have used the recom-mended cut-off of 0.5, therefore only one in-dicator is omitted from the model. The convergent and discrim-inant validity of the model was assessed using the Average Variance Extracted (AVE) and Composite Reliability (CR). The AVE values were higher than the recommended level (0.5), while CR values were above the acceptable values (0.7), which indicated a good internal consistency (Hair et al. 2011). Table 2 indicates that these conditions have been satisfied. The cross loadings of the indicators were then examined to confirm the discriminant va-lidity. The result showed that all loadings of each indicator on its construct were higher than their loadings on other constructs. To measure the model's predictive accuracy of endogen-ous constructs, coefficient of determination (R square) was as-sessed. The results demonstrated that R square value for Att_{ad}, Att_b, and PI in the structural model

Table 2. PLS estimation measurement model.

	AVE	Composite Reliability	R Square
Presence	0.660	0.795	0.000
Flow	0.622	0.825	0.000
Att_{ad}	0.541	0.778	0.150
Att_b	0.596	0.812	0.235
PI	0.767	0.867	0.298

were around or less than 0.25 which described weak level of predictive accuracy.

Hypothesis test was done by comparing t value with critical t-values, with t value should be greater than 1.96 (significance level = 5%), hence the influ-ence between variable was significant (Hair et al. 2011). Thus, from Table 3 we can conclude that hypotheses three and four are not supported.

3.2 Discussion

The findings showed that all constructs, excluding flow, had an effect on independent variables despite low R square values, which meant relatively weak impacts. This was contrary to previous studies, which suggested that flow is an important compo-nent of the game enjoyment (Csikszentmihalyi 1990, Vanwesenbeeck et al. 2016) and flow is able to influ-ence the attitude towards advergame or the inte-grated brand (Hernandez 2011, Waiguny 2013). It might happen because the game can be played repetitively during the observation. This repetitive action cause boredom which has negative impact on the stimuli (Cauberghe & Pelsmacker 2010).

4 CONCLUSION

This study aimed to analyze children attitudes and intention towards the brands in advergame by inte-grating presence and flow experience during game-play. In contrast with television commercials that is broadcasted with exact duration set due to govern-ment policy, there is no limit on advergame expo-sure, especially in children. Although the impact of advergame on children in this study was relatively weak, policies regarding this new ad format need to be prepared. It is expected that future research can use larger sample size and add other variables to get better analysis of the impact of advergame exposure on children.

ACKNOWLEDGMENT

This research was funded by Doctoral Research Grant from Directorate of Higher Education of Republic of Indonesia, year 2018.

REFERENCES

Ajzen, I. 1991. The theory of planned behavior. *Organizational Behavior and Human Decision Processes* 50(2): 179-211.

Baker, W.E. 1999. When can affective conditioning and mere exposure directly influence brand choice?. *Journal of Advertising* 28(4): 31-46.

Bogost, I. 2007. *Persuasive games: The expressive power of videogames.* Cambridge: MIT Press.

Buckingham, D. 2000. *After the death of childhood: Growing up in the age of electronic media.* Cambridge: Polity Press.

Calvert, S.L. 2008. Children as consumers: Advertising and marketing. *The Future of Children* 18(1): 205-234.

Cauberghe, V. & Pelsmacker, P.D. 2010. Advergames. *Journal of Advertising* 39(1): 5-18.

Chang, H.H., Rizal, H. & Amin, H. 2013. The determinants of consumer behavior towards email advertisement. *Internet Research* 23(3): 316-337.

Chen, C.F. & Tsai, D.C. 2007. How destination image and evaluative factors affect behavioral intentions?. *Tourism Management* 28(4): 1115-1122.

Csikszentmihalyi, M. 1990. *Flow: The psychology of optimal experience.* New York: Harper & Row.

Debbabi, S., Baile, S., Des Garets, V. & Roehrich, G. 2013. The impact of telepresence in an online ad on forming attitudes towards the product: The relevance of the traditional experiential approach. *Recherche et Applications en Marketing* 28(2): 3-24.

Debbabi, S., Daassi, M. & Baile, S. 2010. Effect of online 3D advertising on consumer responses: The mediating role of telepresence. *Journal of Marketing Management* 26(9-10): 967-992.

Gross, M.L. 2010. Advergames and the effects of game-product congruity. *Computers in Human Behavior* 26: 1259-1265.

Hair, J.F., Ringle, C.M. & Sarstedt, M. 2011. PLS-SEM: Indeed a silver bullet. *The Journal of Marketing Theory and Practice* 19(2): 139-152.

Hall, L.E., Hume, C. & Tazzyman, S. 2016. Five degrees of happiness: Effective smiley face likert scales for evaluating with children. *Paper presented at The IDC '16 Proceedings of the The 15th International Conference on Interaction Design and Children Manchester.*

Hernandez, M.D. 2008. Determinants of children's attitudes towards "Advergames": The case of Mexico. *Young Consumers* 9(2): 112-120.

Hernandez, M.D. 2011. A model of flow experience as determinant of positive attitudes toward online advergames. *Journal of Promotion Management* 17: 315-326.

Herrewijn, L. & Poels, K. 2014. *Rated A for advertising: A critical reflection on in-game advertising. In* M.C. Angelides & H. Agius *(Eds.), Handbook of Digital Games.* New Jersey: Wiley-IEEE Press.

Hyun, M.Y. & O'Keefe, R.M. 2012. Virtual destination image: Testing a telepresence model. *Journal of Business Research* 65: 29–35.

Kim, T. & Biocca, F. 1997. Telepresence via television: Two dimensions of telepresence may have different connections to memory and persuasion. *Journal of Computer-Mediated Communication* 3(2).

Lang, A. 2000. The limited capacity model of mediated message processing. *Journal of Communication* 50(1): 46–70.

Lin, Z. & Filieri, R. 2015. Airline passengers' continuance intention towards online check-in services: The role of personal innovativeness and subjective knowledge. *Transportation Research Part E: Logistics and Transportation Review* 81: 158-168.

Mallinckrodt, V. & Mizerski, D. 2007. The effects of playing an advergame on young children's perceptions, preferences and requests. *Journal of Advertising* 36(2): 87-100.

Minsky, M. 1980. Telepresence. *OMNI*: 45–51.

Mitchell, A.A. & Olson, J.C. 1981. Are product attribute beliefs the only mediator of advertising effects on brand attitude? *Journal of Marketing Research,* 18(3): 318-332.

Moore, E.S. & Rideout, V.J. 2007. The online marketing of food to children: Is it just fun and games?. *Journal of Public Policy & Marketing* 26(2): 202-220.

Nelson, M.R. & Waiguny, M.K.J. 2012. Psychological processing of in-game advertising and advergaming: Branded entertainment or entertaining persuasion?.In L. J. Shrum (Ed.). *The Psychology of Entertainment Media: Blurring the Lines Between Entertainment and Persuasion.* Mahwah, NJ: Taylor and Francis.

Nuijten, K.C.M., Regt, A.D., Calvi, L. & Peeters, A.L. 2013. Subliminal advertising in shooter games: Recognition effects of textual and pictorial brand logos. *International Journal of Arts and Technology* 6(1): 5–21.

Raney, A.A., Arpan, L.M., Pashupati, K. & Brill, D.A. 2003. At the movies, on the web: An investigation of the effects of entertaining and interactive web content on site and brand evaluations. *Journal of Interactive Marketing* 17(4).

Reynolds-keefer, L. & Johnson, R. 2011. Is a picture is worth a thousand words? Creating effective questionnaires with pictures. *Practical Assessment, Research & Evaluation* 16(8).

Rifon, N.J., Quilliam, E.T., Paek, H.J., Weatherspoon, L.J., Kim, S.K. & Smreker, K.C. 2014. Agedependent effects of food advergame brand integration and interactivity. *International Journal of Advertising* 33(3): 475-508.

Roedder, D.L., Sternthal, B. & Calder, B.J. 1983. Attitude-behavior consistency in children's responses to television advertising. *Journal of Marketing Research* 20(4): 337-349.

Rosen, D.L. & Singh, S.N. 1992. An investigation of subliminal embed effect on multiple measures of advertising effectiveness. *Psychology & Marketing* 9(2): 157-173.

Rozendaal, E., Slot, N., Reijmersdal, E.A.V. & Buijzen, M. 2013. Children's responses to advertising in social games. *Journal of Advertising* 42(2-3): 142-154.

Sicilia, M., Ruiz, S. & Reynolds, N. 2006. Attitude formation online how the consumer's need for cognition affects the website and attitude towards the brand. *International Journal of Market Research* 48(2): 139-154.

Soebandhi, S. & Andriansyah, Y. 2017. In-game advertising: Analyzing the effects of brand congruity, integration, and prominence towards IGA attitude and purchase intention. *Jurnal Manajemen Teknologi* 16(3): 258-270.

Staiano, A.E. & Calvert, S.L. 2012. Digital gaming and pediatric obesity: At the intersection of science and social policy. *Social Issues Policy Review* 6(1): 54-81.

Terlutter, R. & Capella, M.L. 2013. The gamification of advertising: Analysis and research directions of in-game advertising, advergames, and advertising in social network games. *Journal of Advertising* 42(2-3): 95-112.

Vanwesenbeeck, I., Ponnet, K. & Walrave, M. 2016. Go with the flow: How children's persuasion knowledge is associated with their state of flow and emotions during advergame play. *Journal of Consumer Behaviour* 47: 38-47.

Vashisht, D. & Royne, M.B. 2016. Advergame speed influence and brand recall: The moderating effects of brand placement strength and gamers' persuasion knowledge. *Computers in Human Behavior* 63: 162–169.

Waiguny, M.K.J. 2013. Investigating the entertainment–persuasion link: Can educational games influence attitudes toward products? In S. Rosengren, M. Dahlén & S. Okazaki (Eds.). *Advances in Advertising Research* 4: 173–186.

Waiguny, M.K.J., Terlutter, R. & Zaglia, M.E. 2011. The influence of advergames on consumers' attitudes and behaviour: An empirical study among young consumers. *International Journal of Entrepreneurial Venturing* 3(3): 231-247.

Waiguny, M.KJ., Nelson, M.R. & Terlutter, R. 2012. Entertainment matters! The relationship between challenge and persuasiveness of an advergame for children. *Journal of Marketing Communications* 18(1): 69-89.

Youn, S. & Mira, L. 2012. *In-game advertising and advergames: a review of the past decade's research*. In S. Rodgers & E. Thorson (Eds.). Advertising Theory. New York: Routledge.

Advances in Business, Management and Entrepreneurship – Hurriyati et al (eds)
© 2020 Taylor & Francis Group, London, ISBN 978-0-367-27176-3

Increasing fish consumption: A perspective theory of planned behaviour and role of confidence

M.M.L. Tambunan & T.E. Balqiah
Universitas Indonesia, Depok, Indonesia

ABSTRACT: The government is implementing fish consumption programs to overcome malnutrition in society and improve health and public welfare. This study aims to examine the antecedent factors that influence consumer intentions toward increased fish consumption, through the perspective theory of planned behaviour model (TPB) and the role of consumer confidence. An online questionnaire was distributed to 200 participants in the DKI Jakarta area, aged 25–60 years, through purposive sampling. The analysis was performed using stepwise regression analysis to test the strength of five hypotheses of the variable relationships in the model. The results showed that attitude and perceived consumer effectiveness had a direct and positive influence toward intentions to increase fish consumption. Confidence only affects attitudes toward consumer intentions, while perceived availability and subjective norms did not significantly influence intentions toward increased fish consumption.

1 INTRODUCTION

As a maritime country with a potential for sustaining marine life and fishery industries, Indonesia still has a low fish consumption level. In 2014, the realization of aquaculture production reached 14.52 million tons, capture fisheries amounted to 6.2 million tons and processed fishery products to 5.37 million tons (KKP 2015). Potential marine products and fisheries should be able to increase fish consumption in the Indonesian society. Fish contains many functional compounds such as proteins, fats with omega 3, vitamins, minerals and taurine that are beneficial to our health (Susanto & Fahmi 2012). Although the nutrient content in fish is higher and fish resources in Indonesia are abundant, it is not followed by high fish consumption levels. In 2014, the total of fish consumed by the community was 38.14 kg/cap/yr, having increased 8.32% from 2013. Despite the fact that the total fish consumption is increasing, the number is still far below the availability of fish consumption of 51.8 kg/cap/yr. The level of fish consumption in Indonesia remains low.

Low levels of fish consumption are influenced by various factors, including the availability of fish, misperceptions about fish, knowledge about the benefits of consuming fish, economic ability and lifestyles. Misperception about fish is the belief that eating fish can cause itching, child worms and mother's milk to taste fishy. Enhancing the population's eating habits remains an important public health challenge and identifying factors that promote the consumption of healthy foods constitutes the first step (Lacroix et al. 2016).

Currently, the trend toward living a healthy lifestyle is growing in the community. This lifestyle is influenced by those who began to feel there is something wrong with their health condition, i.e., suffering

from illness, and by those who have long implemented a healthy lifestyle. Fish protein can be an important component in the diet of densely populated countries with a low total protein intake. Dietary patterns in many of these countries reveal a high level of dependence on staple foods. Eating fish is important in helping to increase the ratio of calories/protein. In addition to fish representing an affordable source of animal protein, it is also cheaper than other sources of animal protein and part of local and traditional recipes (FAO 2016).

Consumer behaviour is the way of how consumers buy goods (Buerke et al. 2017). Conceptually, consumer retention and consumer switching behaviour are two different marketing constructs that have unique theoretical and managerial implications (Nimako 2012). Consumer behaviour toward food preferences makes for an interesting study because understanding consumer interest in a product can be a reference in developing programs which can increase consumer loyalty. This study aims to determine the factors that influence consumers' intention in increasing their fish consumption which is ultimately expected to change certain consumer behaviours.

Vermeir and Verbeke (2008) refer to the theory of planned of behaviour (TPB) to describe how to increase the intention to purchase dairy food. Their research included attitudes, subjective norms, personal efforts and perceived product availability as antecendents and they also used two moderating personal characteristic variables: confidence and human value. The theory of planned behaviour (TPB) indicates that a person will not form a strong intention to display a behaviour if he/she is unsure of having the source or opportunity to do so, even if he/she has a positive value or believes others will approve it.

This particular belief stands out in influencing individual behaviour (Ajzen 1991). In the theory of planned behaviour, the intention to behave is influenced by three components: attitude, subjective norm and perceived behaviour control.

In the present research, we focused on attitude, subjective norm, personal effort, and product availability in explaining how to increase the fish consumption. The last two factors are reflected by perceived behavioural control. Because there are many varieties and qualities of fish, we also exercise confidence as moderating variables–but not human value. We assume fish is a daily food that is not related to certain values to consume it.

Attitude is a psychological emotion directed through the evaluation of the consumer; if the evaluation is positive, then behavioural intentions will tend to be more positive (Chen & Tung 2014). If the results of consumer evaluation from the experience of buying and consuming fish are positive, it can be expected that the intensity to consume fish will increase.

H_1: Attitudes toward increased fish consumption have a positive effect on the intention to increase fish consumption.

Perceived consumer effectiveness (PCE) related to consumer beliefs that personal efforts can contribute to the solution of a problem is another control factor (Vermier & Verbeke 2008). Consider fish consumption: if consumers believe that their efforts to increase fish consumption are good for them, they will increase this consumption.

H_2: Perceived consumer effectiveness has a positive effect on the intention toward increased fish consumption.

The term "perception of availability" indicates that a consumer feels he or she can easily obtain or consume a particular product (Vermeir & Verbeke 2008). The availability of products will also make it easier for consumers to obtain the products they want. Product availability is among the factors influencing consumer buying interest (Kotler 2005).

H_3: Perceived availability has a positive effect on the intention toward increased fish consumption.

Subjective norms are defined by normative beliefs (i.e. social pressure to adopt behaviours) (Lacroix et al. 2016). Selection of food or habits that have been influenced by family or relatives will affect the intention of consuming certain foods.

H_4: Subjective norms have a positive effect on the intention toward increased fish consumption.

Many consumers are not confident in their ability to evaluate food quality in general (Verbeke et al. 2007). Confidence can develop via information or previous experience. If consumers are convinced of the benefit of consuming a product (i.e. fish), their responses or behaviour should be different from those of less confident consumers. Consumers who feel more confident about the true content and claim of a product

will have different determinants that guide their behaviour compared with consumers who lack confidence (Vermier & Verbeke 2008).

H_{5a-d}: Positive influence of attitude, perceived consumer effectiveness, perceived availability, and subjective norms of the intention toward increased fish consumption are moderated by confidence.

2 METHOD

The research model was adapted from Vermier and Verbeke (2008). In this research, the human values variable is eliminated (Figure 1).

2.1 Research design

Data were collected by cross-sectional survey in Jakarta, using self-administered questionnaires from 200 respondents who were selected by purposive sampling. The respondents are mothers, 25–60 years old, who are fish consumers in the Jakarta area.

The hypotheses were tested through stepwise regression analysis.

2.2 Measurement

The questionnaire was developed by conducting literature reviews and through examining previous research. The questionnaire consisted of 22 questions measured on a 6-point Likert scale for six research constructs: five items of attitude (Paul et al. 2016, Tomić et al. 2016), four items of perceived consumer effectiveness (Roberts 1996, Kabadayi et al. 2015), three items of perceived availability (Paul et al. 2016), four items of subjective norm (Paul et al. 2016, Tomić et al. 2016), two items of confidence (Verbeke et al. 2007), and four items of intention (Pawlak et al. 2008).

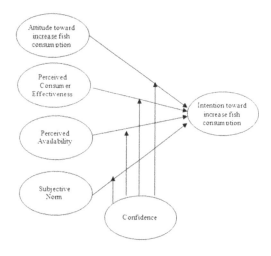

Figure 1. Research model.

Table 1. Validity and reliability.

No	Variables	Factor loading	Mean	Cronbach Alpha
Attitude				0.88
1.	Buying fish is a great idea	0.86	5.41	
2.	I like the idea of buying fish	0.90	5.33	
3.	I like to consume fresh fish	0.64	5.65	
4.	I was happy when the fish was on the menu	0.89	5.26	
5.	I feel better after consuming fish	0.80	5.03	
Perceived Consumer Effectiveness				0.84
6.	Consider its impact on the environment and other consumers	0.74	4.26	
7.	Positive impact on society	0.80	4.86	
8.	By eating fish, I can contribute to the solution of environmental problems	0.89	4.57	
9.	Solve the problem of natural resources	0.86	4.44	
Perceived Availability				0.67
10.	Fish is generally available in the market where I usually shop	0.81	5.42	
11.	I have the resources, time and willingness to buy fish	0.94	5.04	
12.	There are probably many opportunities for me to buy fish	0.92	5.01	
Subjective norms				0.78
13.	My family consumes fresh fish	0.55	5.33	
14.	My friends thought I should consume fresh fish	0.85	4.61	
15.	A positive friend opinion influenced me to buy fish	0.84	4.47	
16.	People whose opinions I appreciate would rather me buy fish	0.83	4.54	
Confidence				0.66
17.	Great opportunity I made a bad choice in buying fish	0.87	5.03	
18.	I never knew if I made the right decision when buying a fish	0.87	5.17	
Subjective norms				0.94
19.	Intent to consume fish daily	0.89	4.77	
20.	Will try to consume fish daily	0.95	4.76	
21.	Plan to consume fish daily	0.94	4.74	
22.	Probably consume fish daily	0.92	4.70	

We conducted a pretest (n = 30) to evaluate the validity and reliability of items, and all items were valid (loading factors > 0.5), and reliable (Cronbach > 0.6). In the main test, all variables were valid and reliable (Table 1).

2.3 Questionnaire design

There were three sections: screening, main variables, and demography. In the screening section, we asked after the participants' fish consumption frequency and whether they are the decision-makers in purchasing food in the family. Researchers used mothers as respondents in this study because of the role they play in the family. Therefore, a mother who liked to consume fish was considered playing an important role in increasing the fish consumption in the community because her penchants can affect those around her to consume fish.

3 RESULT AND ANALYSIS

Most respondents were at the productive age of 25–44 years, dominated by housewifes, with 53% having two to three family members, and household expenditures per month ranging from Rp 3.000.000 to Rp 5.000.000 at 29.5%, which indicated that the majority of respondents were upper middle class (Table 2).

Hypotheses were analyzed with stepwise and interaction effects of moderating. As can be seen in Figure 2, only three hypotheses are supported.

Table 2. Respondent's profiles.

Demography	Category	Frequency	
		n	%
Frequency of consumption/week	1 – 2 days	47	23.5
	3 – 4 days	95	47.5
	5 – 6 days	25	12.5
	Everyday	33	16.5
Age (Years)	25 – 34	108	54
	35 – 44	51	25.5
	45 – 54	19	9.5
	55 – 60	22	11
Number of family members	3 – Feb	106	53
	5 – Apr	74	37
	> 5	20	10
Education	Secondary school	8	4
	High school	49	24.5
	Diploma	9	4.5
	Bachelor	110	55
	Master	24	12
Occupation	Housewife	65	32.5
	Civil servant	44	22
	Private	54	27
	Entrepreneur	17	8.5
	Doctor	3	1.5
	Student	2	1
	Other	15	7.5
Expense/month	< Rp 1 million	2	1
	Rp 1 million – Rp 1.5 million	3	1.5
	Rp 1.5 million – Rp 2 million	16	8
	Rp 2 million – Rp 3 million	38	19
	Rp 3 million – Rp 5 million	59	29.5
	Rp 5 million – Rp 7.5 million	28	14
	> Rp 7.5 million	54	27

Based on data processing, not all TPB dimensions have a positive effect on the intention to increase fish consumption (Figure 2). Attitudes toward increased fish consumption has a strong and positive effect toward the intention to increase fish consumption. This result is in accordance with that of Kotchen and Reiling (2000), which explains that attitude is the most important and major predictor of intention.

Perceived consumer effectiveness has a positive effect on the intention to increase fish consumption, which can be caused by several factors. Mothers in Jakarta on average had a higher education, which affects the perception of consumers who are more concerned about the quality and benefit of fish. Usually this construct is used to explain why consumers show certain behaviours toward green products. Furthermore, this contruct is concerned about how certain human behaviours help to protect the environment. Consumers typically believe that his or her own efforts can make a difference and contribute to solving social and environmental problems (Ghvanidze et al. 2016). In this study, we refer to this concept with the assumption that consumers must also be concerned with other health conditions by increasing their fish consumption compared to other foods, for example, beef, chicken, and so on.

The term "perception of availability" indicates that the consumer feels that he or she can easily obtain or consume a particular product (Vermeir & Verbeke 2008). Availability will also make it easier for consumers to obtain certain products. Product availability is among the factors that influence consumer buying interest (Kotler 2005). In this study, perceived availability did not affect the intention toward increased fish consumption. Despite the high availability of fish products, this did not encourage the intensity of fish consumption. This can happen

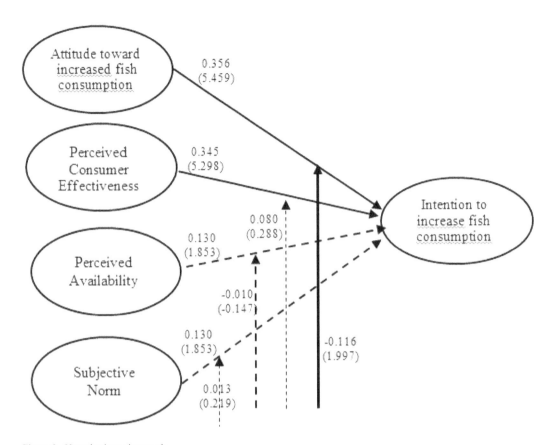

Figure 2. Hypothesis testing result.

due to the results of this research, the frequency of daily fish consumption in Jakarta is only 16.5%; furthermore, the majority of fish consumption is three to four days a week (47.5%). Because most of the waters in Jakarta are contaminated, Jakarta is not a fish-producing area. Fish supplies come from Thousand Islands, North Jakarta, and from the surrounding producer areas. This can raise fish prices in the market as well as affect the intention toward increased fish consumption.

The subjective norm did not affect the intention to increase the fish consumption. Previous researchers, who have applied the framework of theory of planned behaviour, identify subjective norms as the weakest link in the intention model (Ajzen 1991). Consumers felt approval or input from "people who were considered important," which is not a deciding factor toward making a purchase.

Confidence only affects the influence of attitude toward intention to increase fish consumption. For respondents who demonstrated a higher confidence that they made good decisions in buying fish, this will strengthen the influence of attitude toward intention compared with that of respondents with a lower confidence. This result

shows that consumers who feel more confident about the true content and claim of a product will have different determinants that guide their behaviour compared to consumers who lack this confidence (Vermier & Verbeke 2008).

4 CONCLUSION

The attitude and perceived consumer effectiveness influenced the intention to increase fish consumption. Perceived availability and subjective norms did not influence intentions. Consumers' believes about the benefit of fish consumption for themselves and society will enhance their intention to increase fish consumption.

This finding suggests that, in accordance with the effectiveness of the "Gemarikan" campaign by the Ministry of Marine Affairs and Fisheries and the Ministry of Health, the authority must be concerned with increasing positive attitudes, perceived consumer effectiveness, and confidence. The authority could deliver messages about the utilitarian and hedonic value of fish consumption in a creative format to attract the attention of society, not only by delivering information about the health benefits of consuming fish but also on how to process fish into various foods for children. The campaign is also intended for children, to help them develop a positive attitude towards fish as well as influence their mothers to cook fish for them.

Future research should include other relevant variables beyond the existing variables in this study, and provincial coverage can be extended, especially to big cities in Indonesia that still have low fish consumption levels.

REFERENCES

Ajzen, I. 1991. The theory of planned behavior. *Organizational Behavior and Human Decision Processes* 50: 179–211.

Buerke, A., Straatmann, T., Lin-Hi, N. & Müller, K. 2017. Consumer awareness and sustainability-focused value orientation as motivating factors of responsible consumer behavior. *Review of Managerial Science* 11(4): 959–991.

Chen, M.F. & Tung, P.J. 2014. Developing an extended theory of planned behavior model to predict consumers' intention to visit green hotels. *International Journal of Hospitality Management* 36: 221–230.

FAO. 2016. *The state of world fisheries and aquaculture 2016: Contributing to food security and nutrition for all.* Rome.

Ghvanidze, S., Velikova, N., Dodd, T. H. & Oldewage-Theron, W. 2016. Consumers' environmental and ethical consciousness and the use of the related food products information: The role of perceived consumer effectiveness. *Appetite* 107: 311–322.

Kabadayi, E.T., Dursun, İ., Alan, A.K. & Tuğer, A.T. 2015. Green purchase intention of young Turkish consumers: Effects of consumer's guilt, self-monitoring and perceived consumer effectiveness. *Procedia-Social and Behavioral Sciences* 207: 165–174.

KKP. 2015. Laporan kinerja kementerian kelautan dan perikanan tahun 2014. *Retrived* www.kkp.go.id.

Kotchen, M. & Reiling, S. 2000. Environmental attitudes, motivations, and contingent valuation of nonuse values: A case study involving endangered species. *Ecol. Econ* 32: 93–107.

Kotler, P. 2005. *Manajamen pemasaran.* Jakarta: PT. Indeks Kelompok Gramedia.

Lacroix, M.J., Desroches, S., Turcotte, M., Painchaud, G. G., Paquin, P., Couture, F. & Provencher, V. 2016. Salient beliefs among Canadian adults regarding milk and cheese consumption: A qualitative study based on the theory of planned behaviour. *BMC Nutrition* 2(1): 48.

Nimako, S.G. 2012. Consumer switching behaviour: A theoretical review and research agenda. *The International Journal's Research Journal of Social Science & Management* 2(3): 74–82.

Paul, J., Modi, A. & Patel, J. 2016. Predicting green product consumption using theory of planned behavior and reasoned action. *Journal of Retailing and Consumer Services* 29: 123–134.

Pawlak, R. & Malinauskas, B. 2008. The use of the theory of planned behavior to assess predictors of intention to eat fruits among 9th-grade students attending two public high schools in Eastern North Carolina. *Family and Consumer Sciences Research Journal* 37(1): 16–26.

Roberts, J.A. 1996. Green consumers in the 1990s: Profile and implications for advertising. *Journal of Business Research* 36: 217–231.

Susanto, E. & Fahmi, A.S. 2012. Senyawa fungsional dari ikan : Aplikasinya dalam pangan. *Jurnal Aplikasi Teknologi Pangan* 1(4): 95–102.

Tomić, M., Matulić, D. & Jelić, M. 2016. What determines fresh fish consumption in Croatia? *Appetite* 106: 13–22.

Verbeke, W., Vermeir, I. & Brunsø, K. 2007. Consumer evaluation of fish quality as basis for fish market segmentation. *Food Quality and Preference* 18(4): 651–661.

Vermeir, I. & Verbeke, W. 2008. Sustainable food consumption among young adults in Belgium: Theory of planned behaviour and the role of confidence and values. *Ecological Economics* 64(3): 542–553.

Advances in Business, Management and Entrepreneurship – Hurriyati et al (eds)
© 2020 Taylor & Francis Group, London, ISBN 978-0-367-27176-3

Understanding millennial decision making in buying a car: Lifestyle and personality

O. Pramoedya & T.E. Balqiah
Universitas Pendidikan Indonesia, Bandung, Indonesia

ABSTRACT: The study of millennial consumers has become a trending subject in marketing practice in order to further understand their consumption behavior. It is believed that millennials differ from one generational group to another. This study aims to analyze the behavior of millennials in Indonesia and identify the factors that they are concerned with in buying a car (i.e., Brand T or Brand H). This study refers to their perception toward the roles of the 7Ps marketing mix (i.e., product, price, place, promotion, people, process, and physical evidence), personality, and lifestyle in car buying decisions. For that purpose, qualitative and quantitative research methods were applied with in-depth interviews of eight respondents and surveys of 300 respondents based on snowball sampling via off- and online questionnaires. Brand T millennials see their personality comes first when deciding to buy a car, as they seek a car brand that has a similar personality to them, while Brand H millennials see the basic marketing mix as their first consideration, that is,, covered by product, process, physical evidence, and people.

1 INTRODUCTION

Millennials are defined as being individuals born between 1981 and 2000 and are part of a broader categorization of generations (Ali & Purwandi 2016). Based on different social and economic conditions, millennials reported having lifestyles that differ from those of previous generations (Ali & Purwandi 2016, Circella et al. 2016).

Determining the specific factors that influence millennials and their purchasing attitudes and patterns has become an important focus of consumer research, as a result of millennials' potential spending power, the ability to be trendsetters, adoption to new products, and potential for becoming lifetime customers (Martin & Bush 2000, Ordun 2015).

Generally, buying behavior is structured by a complex interaction influenced by habit, inertia, experience, advertising, peer pressure, environmental constraints, accumulated opinion, household, and family constraints (Louviere et al. 2000). In the sense of private cars, it is recognized that its selection influence is based on a built environment, which is represented by the degree of density. With the degree of density of urban areas, individuals will choose certain vehicles to connect their needs and opportunities. Studies revealed that sedans and wagons/hatchbacks are preferred in areas with a high density and are suitable for urbanized areas (Kitamura et al. 2000).

It is recognized that decision-making for the fulfillment of the need for purchasing private vehicles is complex. In vehicle purchase decision-making, consumers usually take certain attributes into consideration, for example, price, operational cost, residual value, dimension, power, turning radius, airbags, and transmission (Mannering et al. 2002). These factors cover a basic marketing mix of concepts that are enriched by adding people, physical assets, and processes that form the 7Ps (Rafiq & Ahmed 1995, Lin 2011):

1) Product – must provide value to a customer but does not have to be tangible at the same time. Basically, it involves introducing new products or improving existing products.
2) Price – competitive and entails profit, as the pricing strategy can comprise discounts, offers, and so on.
3) Place – where the customers can buy products and how the product reaches out to that place.
4) Promotion – activities that communicate the merits of a product and persuade target customers to buy it.
5) People – refers to customers, employees, management, and everybody else involved.
6) Process – methods and the process of providing a service that is essential for a thorough knowledge on whether the services are helpful to customers, if they are provided in time, and if the customers are informed first hand about the services, and so on.
7) Physical evidence – the experience of using a product or service.

However, it remains rare to study millennials' behavior in car-buying decisions and the role it plays in regard to personality and lifestyle. One can, for the factors above, take note of the study conducted by Choo and Mokhtarian (2004), which explores the role of attitudes, personality, and lifestyles for car choice behavior for commuters in the San Francisco Bay Area.

In order to create a better understanding, this study will focus on two dominating hatchback brands in GAIKINDO (Automotive Industry Association of Indonesian) and try to combine the 7Ps with personality and lifestyle to analyze millennial behaviors in Indonesia and identify the factors that make up the process of buying a car.

2 METHOD

2.1 Research design

Exploratory research with in-depth interviews was conducted before the descriptive research with surveys through questionnaires was applied. In-depth interviews were conducted with eight Brand H customers and analyzed by NVivo 11 (Malhotra 2010).

2.2 Survey and sampling method

Data was collected by using a questionnaire previously enriched with findings from in-depth interviews with a quantitative survey method. The questionnaire was administered via both self-administered survey as well as online surveys. A snowball sampling method was used to deal with the collected data from Brand H and Brand T customers.

A total of 300 questionnaires was distributed to the target audience (60 questionnaires were generated online and 240 offline). Greater Jakarta (Jakarta, Tangerang, Bekasi) was selected as the geographic location for this study, being the biggest area composition (32%) of passenger cars over the past five years, released byGAIKINDO 2016. The limitation of the targeted audience was millennials only (20–35 years old) who have or used Brand H and Brand T from 2016 forward.

2.3 Instrument development

The scales were adopted from previous studies (Lin 2011, Choo & Mokhtarian 2004). Items related to the marketing mix (7Ps), lifestyle, and personality were measured by a five-point Likert scale, where "1" denoted "strongly disagree" and "5" denoted "strongly agree" (Hair et al. 2003).

Additional items were added to the scale from the in-depth interview that represented a variable marketing mix in product (i.e.,,, baggage roominess and compact car size), promotion (i.e., source of information from social media), physical evidence (i.e., showroom cleanliness), and variable personality (i.e., quiet, cheerful).

The questionnaire was divided into different sections to assess all the variables used in this study. Section A consisted of screening questions to ensure respondent criteria matched the researcher needs. Section B comprised the marketing mix (7Ps), lifestyle, and personality variables (total 60 items) and it

ended with Section C including questions regarding personal sociodemographic information.

3 RESULTS AND DISCUSSION

3.1 Exploratory

This step was conducted with in-depth interviews of eight Brand H buyers or users. All 7Ps variables, personality, and lifestyle openly asked by the researchers were then analyzed through verbatim and NVivo. The objective is to gain insight into what respondents are concerned with when buying a car. This result, combined with previous studies, will enrich the measurements in the survey.

There are additional insights into what respondents consider in deciding to buy Brand H. Here is a list of answers for some variables: large baggage compartment and compact size car (product), social media as source of information (promotion), showroom cleanliness (physical evidence), and a quiet, cheerful, expressive, and friendly sales person (personality).

3.2 Sample profile

Demographic profiles of Brand H and Brand T users or buyers who participated in this study are summarized in Table 1. The majority of respondents were the decision making and final determinants (57% and 52%). Knowledge about the car from friends and family (31% and 35%), use of the car (66% and 71%), male (68% and 75%), between 24–27 years old (43% and 33%), and mostly still single for Brand H (41%) while married with children for Brand T (41%).

3.3 Dimension of decision-making attributes

A factor analysis with varimax rotation was employed to the 60 statements to identify the underlying dimensions of the marketing mix, personality, and lifestyle. Items with a factor loading lower than 0.5 were removed (Malhotra 2010), and the remaining items were factor-analyzed again. Different factor numbers were found, where each of the factors was given the name based on the items contained.

3.3.1 Dimensions of brand T

There are 18 dimensions (factors) that respondents of Brand T take into consideration, but the first factor had the largest share (27.201 with an eigenvalue of 21.217), whereas the 18th factor had the smallest share (1.346 with an eigenvalue of 1.050). In sum, all of the 18 factors with eigenvalues larger than 1 explained 74.171% of variance of 60 statements.

The components of each factor are introduced in the following: Research findings showed that the

Table 1. Demographic profile of respondents (n = 300).

	Brand H Frequency abs. (%)	Brand T Frequency abs. (%)
Role on decision making		
Only decision maker	27 (18)	31 (21)
One of the decision maker & final determinant	86 (57)	78 (52)
One of the decision maker & not final determinant	36 (25)	40 (27)
Source of Information		
Dealers	33 (22)	30 (20)
Website	10 (7)	8 (5)
Friends & Family	46 (31)	52 (35)
Brochure	22 (15)	11 (7)
Social Media	5 (3)	5 (3)
Magazine	1 (1)	22 (15)
TV Ads	6 (4)	12 (8)
Car Exhibition	16 (11)	6 (4)
See Car on the Street	11 (7)	4 (3)
User		
My Self	99 (66)	106 (71)
Spouse	29 (19)	21 (14)
Parents	7 (5)	13 (9)
Other Family Members	12 (10)	10 (7)
Gender		
Male	103 (68)	111 (75)
Female	46 (32)	38 (25)
Age		
20–23	14 (9)	27 (18)
24–27	64 (43)	50 (33)
28–31	43 (29)	40 (27)
32–35	29 (19)	33 (22)
Marriage status		
Married w/o child	39 (26)	34 (23)
Married w/child	48 (33)	61 (41)
Single	59 (41)	54 (36)

Table 2. Factor analysis of brand T.

	Rotation Sum of Squared Loadings		
Factor	Eigenvalue	Variance (%)	Cumulative Variance (%)
High Performance	21.217	27.201	27.201
Performance Seeker	5.277	6.765	33.966
Convenience Seeker	3.692	4.734	38.699
Price Concern	3.165	4.058	42.757
Full of Challenge	2.639	3.384	46.141
Social Status	2.467	3.162	49.303
Approach	2.349	3.012	52.315
Safe & Sound	2.228	2.857	55.172
Happiness	2.156	2.764	57.936
Friendly	1.870	2.397	60.333
Easy Process	1.769	2.268	62.601
Controlling	1.606	2.059	64.660
Aggressive	1.545	1.981	66.640
Independent	1.374	1.761	68.401
Effective	1.355	1.737	70.138
Functional	1.283	1.644	71.783
Me Time	1.184	1.518	73.300
Tolerant	1.105	1.417	74.717

Table 3. Ranking of factors considered by millennial when buying a car.

Priority	Factors	Mean of 5
1st	Independent	3.958
2nd	Easy Process	3.925
3rd	Safe & Sound	3.900
4th	Full of challenge	3.890
5th	Social Status	3.860
Intermediate Factors 6 Through 15 Omitted		
16th	High Performance	3.263
17th	Functional	3.231
18th	Convenience	3.027

following 19 factors were identified and prioritized as factors influencing millennials when buying car.

Result of the Kaiser–Meyer–Olkin (KMO = 0.730) and Bartlett's test at a significant level of 0.000 suggested that factor-analysis was suitable for these statements, where statements with a factor loading score below 0.5 (Amerioun et al. 2018) were excluded.

Therefore, based on the means result, factors categorized as the three most important for millennials of Brand T in decision making are independent, easy process, safe, and sound with a score of 3.958, 3.925, and 3.900, respectively.

Four of the five most important factors came from personality and lifestyle, while the additional factor is from a marketing mix that consists of speed in ordering and delivering the unit.

The top two personalities included independent, responsible, like to stay near home, being alone, and quiet. The third and fourth personalities seem highly different from the previous, as it includes adventurous, variety-seeking, spontaneous, bravery, seeing the car bought as social status symbol, and workaholic. This proves that millennials leaning toward Brand T are not all that similar.

These results were supported by a previous study that explained that millennial buyers select and consume products that help them define who they are, what is important to them, and what they value in life and also serves to express aspects of their own personality or image (Ordun 2015).

Especially in choosing car as their social status, Gen Y is concerned about how he or she is seen by friends and social acquaintances. This behavior will be most noticeable when they decide to buy clothes or cars, where choices will fall on products that exceed utility requirements and become status symbols (Parment 2013, Bucuta 2015).

The convenience factor that consists of showrooms being available in many locations, easy to find and easy to access is the least important for millennials when buying a car, with a means score of 3.027 out of 5.

With these results, millennial buyers of Brand T could be approached by placing advertisements in lifestyle magazines, and the content itself should contain more than simple specification information; it should contain how the car could accompany them in their daily life and support their lifestyle.

Lifestyle magazines could vary because there are millennials who like to be in a quiet place and stay near their comfort zone, while others are adventurous, spontaneous, brave, and appreciate symbols of social status.

Another approach highlights the quality delivery (Sharma & Patterson 1999, Yoo et al. 1998, Odeker-ken-Schröder 2003) as the main message in order to gain millennial trust because their concern is getting the car quickly.

Friends and dealers as their source of information and millennials who see their car as social status could be used as strategy implementation, in which Brand T could invite the millennial and their friends to the dealer (Noble et al. 2009). Millennials could show off their car to their friends and take their friend's suggestion about the car they would like to buy.

3.3.2 Dimensions of brand H

The first factor had the largest share (23.751 with an eigenvalue of 18.526) of variance, whereas the twentieth factor had the smallest share (1.485 with an eigenvalue of 1.159) of variance. In sum, all of the 18 factors with eigenvalues larger than one explained 73.826% of variance of 60 statements.

The factors and components of each factor are introduced in the following. Research findings showed that the following 19 factors were identified and prioritized as factors influencing millennial when buying a car.

Results of the Kaiser–Meyer–Olkin (KMO = 0.720) and Bartlett's test at a significant level of 0.000 suggest that factor analysis was suitable for these statements, where statements with a factor loading score below 0.5 (Amerioun et al. 2018) are excluded.

Table 4. Factor Analysis of Brand H.

Factor	Eigenvalue	Variance (%)	Cumulative Variance (%)
	Rotation Sum of Squared Loadings		
1	18.526	23.751	23.751
2	5.441	6.975	30.727
3	3.945	5.058	35.785
4	3.338	4.280	40.065
5	2.733	3.504	43.569
6	2.452	3.143	46.712
7	2.256	2.892	49.604
8	2.185	2.802	52.406
9	2.085	2.672	55.078
10	1.812	2.324	57.402
11	1.742	2.233	59.635
12	1.692	2.170	61.805
13	1.532	1.964	63.769
14	1.435	1.840	65.609
15	1.380	1.769	67.378
16	1.349	1.729	69.107
17	1.300	1.667	70.773
18	1.223	1.568	72.341
19	1.159	1.485	73.826

Therefore, based on the means result, factors categorized as the three most important for millennials of Brand H in decision making are:

Four of five of the most important factors were from the marketing mix, while the additional factor is from personality and lifestyle, which consists of effectiveness and being on time.

Marketing mix factors that were considered when millennials purchase a car included product (performance, safety, interior design, equipment, and technology, power), process (fast ordering, delivery, document readiness, and easy payment terms), physical evidence (facility, pleasant environment, clean, size, lighting, quietness, color), and people (attractive, enthusiast, knowledge, consistent, and courtesy).

These factors are supported by the previous study in which the common attributes considered were price, operational cost, residual value, dimension, power, turning radius, airbags, and manual owner (Mannering et al. 2002). Other research related to buying a car revealed that attributes of safety, speed, performance, and technology were considered important by consumers (Kate & Handa 2016).

On the other hand, personality and lifestyle factors such as loners, life to fullest, and friendly were the least important for millennials when buying a car with a means score for each being 3.477, 3,436, and 3,364 out of 5, respectively.

With these results, millennials from Brand H could be approached by putting effort in the marketing mix of product, process, physical evidence, and people.

Table 5. Ranking of Factors Considered by Millennial When Buying a Car.

Priority	Factors	Mean of 5
1st	Product Attribute	4.044
2nd	Power	3.980
3rd	Effective	3.933
4th	Easy Process	3.920
5th	Physical & People Concern	3.910
Intermediate Factors 6 Through 16 Omitted		
17th	Loner	3.477
18th	Life to fullest	3.436
19th	Friendly	3.364

Highlighting the product features through brochures is important due to brochures clearly explaining the product attributes (Odekerken-Schröder 2003). Other than product, maintaining where the car unit is shown and the activity of selling are also important.

Having a well-trained sales person is one of the strategies for millennials. Fluent in explaining product features or other benefits to the customer (i.e., millennials) or to their family as one of the decision-making factors could introduce advantages (Nayeem & Casidy 2013). Sales persons could also do direct marketing to build long-term relationships (Kotler & Armstrong 2012).

4 CONCLUSION

Millennials of Brand T agree that personality comes first when deciding to buy a car, as they seek a car brand that has a similar personality to them. Gen Y purchases a product that is perceived to fit or match a certain image (Noble et al. 2009). On the other side, millennials of Brand H are different compared to those of Brand T, even though they are from the same class-category of the hatchback medium. Millennials of Brand H see the basic marketing mix as their first consideration when buying a car (i.e., covered by products, processes, physical evidence, and people). Personality and lifestyle come later, as they count as the least important factors.

REFERENCES

Ali, H. & Purwandi, L. 2016. *Indonesia 2020: The Urban Middle-class Millenials*. Alvara Research Center.

Amerioun, A., Alidadi, A., Zaboli, R. & Sepandi, M. 2018. The data on exploratory factor analysis of factors influencing employees effectiveness for responding to crisis in Iran military hospitals. *Elsevier Inc.*, 19: 1522–1529.

Bucuta, A. 2015. A review of the specific characteristics of the Generation Y consumer. *International Conference "Marketing – from Information to Decision"*: 38–47.

Choo, S. & Mokhtarian, P. L. 2004. What type of vehicle do people drive? The role of attitude and lifestyle in influencing vehicle type choice. *Transportation Research Part A: Policy and Practice*, 38(3): 201–222.

Circella, G., Fulton, L., Alemi, F., Berliner, R.M., Tiedeman, K., Mokhtarian, P.L. & Handy, S. 2016. What Affects Millennials' Mobility? PART I: Investigating the Environmental Concerns, Lifestyles, Mobility-Related Attitudes and Adoption of Technology of Young Adults in California. *A National Center for Sustainable Transportation* (No. CA16–2825).

Hair, J.F. Black, W.C. Babin, B.J. Anderson, R.E., & Tatham, R.L. 2006. *Multivariate Data Analysis, sixth edition*. Upper Saddle River, NJ: Person Education.

Kate, N. & Handa, A. (2016). Empirical analysis of factors influencing the purchasing of luxury cars in Pune City. *Indian Journal of Applied Research*, 6(4): 624–626.

Kitamura, R., Golob, T.F., Yamamoto, T. & Wu, G. 2000. Accessibility and auto use in a motorized metropolis. *Presented at the 79th Annual Meeting of the Transportation Research Record*.

Kotler, P. & Armstrong, G. 2012. *Principles of Marketing, 14th Edition*. New Jersey: Pearson Prentice Hall.

Lin, S.M. 2011. Marketing mix (7P) and performance assessment of Western fast food industry in Taiwan: An application by associating DEMATEL (Decision Making Trial and Evaluation Laboratory) and ANP (Analytic Network Process). *African Journal of Business Management*, 5(26): 10634–10644.

Louviere, J.J. Hensher, D.A. & Swait, J.D. (2000). *Stated choice methods: analysis and applications*. Cambridge university press.

Malhotra, N.K. 2010. *Marketing research: An applied orientation*, 6th Edition. London: Pearson.

Mannering, F., Winston, C. & Starkey, W. 2002. An exploratory analysis of automobile leasing in the United States. *Journal of Urban Economics*, 52(1): 154–176.

Martin, C.A. & Bush, A.J. 2000. Do role models influence teenagers' purchase intentions and behavior? *Journal of consumer marketing*, 17(5): 441–453.

Nayeem, T. & Casidy, R. 2013 .The role of external influences in high involvement purchase behavior. *Marketing Intelligence & Planning*, 31(7): 732–745.

Noble, S.M., Haytko, D.L. & Phillips, J. 2009. What drives college-age Generation Y consumers? *Journal of Business Research*, 62(6): 617–628.

Odekerken-Schröder, G., Ouwersloot, H., Lemmink, J. & Semeijn J. 2003. Consumers' trade-off between relationship, service package and price: An empirical study in the car industry. *European Journal of Marketing*, 37(1/2): 219–242.

Ordun, G. 2015. Millennial (Gen Y) Consumer behavior their shopping preferences and perceptual maps associated with brand loyalty. *Canadian Social Science*, 11 (4): 1–16.

Parment, A. 2013. Generation Y vs. Baby Boomers: Shopping behavior, buyer involvement and implications for retailing. *Journal of retailing and consumer services*, 20 (2): 189–199.

Rafiq, M. & Ahmed, P. K. 1995. Using the 7Ps as a generic marketing mix. *Marketing Intelligence & Planning*, 13 (9): 4–15.

Sharma, N. & Patterson, P.G. 1999. The impact of communication effectiveness and service quality on relationship commitment in consumer, professional services. *Journal of Services Marketing*, 13(2): 151–170.

Yoo, C., Park, J. & MacInnis, D. J. 1998. Effects of store characteristics and in-store emotional experiences on store attitude. *Journal of Business Research*, 42(3): 253–263.

Advances in Business, Management and Entrepreneurship – Hurriyati et al (eds)
© 2020 Taylor & Francis Group, London, ISBN 978-0-367-27176-3

Sport motivation and decision to participate in the Tahura trail running race

V. Gaffar, O. Ridwanudin & D. Inassa
Universitas Pendidikan Indonesia, Bandung, Jawa Barat, Indonesia

ABSTRACT: Participant's decisions to be involved in sports tourism activities tends to be low. The Tahura Trail Running Race is a big event which is always held annually in Bandung City. The event faces the problem of low interest and participation from inside and outside of West Java. The purpose of this study is to obtain findings about the influence of participant motivation on the decision to join the Tahura Trail Running Race. The sample of the study covers 90 respondents, consisting of 32 participants from West Java and 58 participants from outside West Java. The data analysis technique used Multiple Regression. Results showed that psychological motives and achievement influenced the sport decision, while physical incentives didn't have any influence on it. Social motives influenced the decision to participate although it wasn't significant. It is important for the organizer to consider what motivates participants in joining sport events, as the base of formulating its policies.

1 INTRODUCTION

The level of enthusiasm of Indonesian people in being involved in sports activities is still very low. This can be seen in a survey conducted by Susenas in 2012. The survey explains that the population aged 10 years and over who do sports only covers about 25 percent. Sports participation increased from 2000 (22.6%) to 2003 (25.4%). Meanwhile, in the periods of 2003, 2006 and 2009, the population's participation in sports continued to decline, from 25.4% in 2003, down to 23.2% in 2006 and lastly, down to 21.8% in 2009 (Kemenpora 2014). Based on this phenomenon, it is important to know why the level of sport participants tends to be so low. What is the motivation behind it? Does motivation influence the decision to participate in a sports activity?

A previous study shows that it is important to understand the motivation of a participant for a sports activity. Van Heerden (2014) stated that reasons for participating in sports are various. Intrinsic as well as extrinsic motivation can influence the decision to participate. Another study stated that the reason people participate in sport is more to prove themselves rather than for the love of the game (Recours et al. 2004). Funk (2008) mentions that the components of motivation are need, tension, drive, desire, and purpose that determine the exercise of the consumer decisionmaking, also the desire and willingness of individuals to engage in a cognitive activity or behaviour engage. It is in line with Schwarz and Hunter (2008), who argue that motivations include accomplishment, fun, improvement of skill, health and fitness, or the desire for affiliation with or love for a player or team.

Another study by Shank and Lyberger (2015) states that people who have made a sports decision have done so almost automatically because participant decisionmaking is a complex cognitive process that brings together memory, thoughts, information processing, and evaluates judgement.

The emergence of sports activities in urban areas such as running 5 kilometers or 10 kilometers, walking on car-free days, and various mass gymnastic events are activities that invite people to participate in sport activities. In Bandung City, the Tahura Trail Running Race, with the theme "Satu Pelari Satu Pohon" (One Tree One Runner) aims to invite lovers of cross-country running to contribute to natural environment preservation by planting trees. This activity was opened not only for the people of Bandung, but also from outside the city of Bandung. In addition, this activity is divided into several categories, namely family, short courses, long courses, half marathons, and marathons. Although this activity is not just to run, it is still less likely to increase people's interest to participate. This is evidenced by the decrease in the number of participants from previous years. In addition, the participation level of local communities is smaller than the participation from outside the city.

The low level of community participation illustrates that there are still many people who have a low interest to participate in sporting events. Therefore, it is important to explore more about what motivates people to participate in sporting events. The aim of this study is to explore the influence of sports motivation on the decision to participate in sporting events, such as the Tahura Trail Running Race.

2 METHOD

This study analyzed the influence of sports motivation on to decision to participate in the 2016 Tahura Trail Running Race event. Therefore, the hypothesis of this study is that there is a positive influence of sports motivation, which consists of psychological motives, physical motives, social motives, and achievement to join, in the decision to participate in the 2016 Tahura Trail Running Race. These are the independent variables of this study. The dependent variable of the study is a sports decision that consists of the physical surroundings, the social surrounding, time, task definition, and antecedent state.

The population of the study covers 1359 participants from West Java and outside of West Java who participated in the running event Tahura Trail Running Race 2016. The sample included 90 respondents. This study used multiple regression as the technique of analysis, while the T-test was used for comparison purposes.

3 RESULTS

The data covered demography aspects and experiences of respondents. Table 1 shows that most of the respondents were male (75.6%) while 24.4% were female. The majority are 18–25 years old (46.7%), followed by 26–30 years old (28.9%), 31–35 years old (16.6%) and > 40 years old (3.3%). The highest proportion of education background fell to employees (42.2%), followed by students (34.4%), entrepreneurs (13.3%), government officers (10%). As for their monthly income, the majority of respondents have > IDR 5,000,000 (44.4%), followed by IDR 500,000– IDR 1,000,000 (20%), IDR 1,100,000-IDR 2,000,000 (12.2%), IDR 2,100,000-IDR 3,000,000 (8.8%), IDR 3,100,000- IDR 4,000,000 (7.7%) and IDR 4,100,000-IDR 5,000,000 (6.6%) respectively. Most of respondents came with friends (81.1%), family (6.7%), coworkers (5.6%), and others (6.6%). Regarding the source of information, most of them got it from social media (42.2%), followed by friends (36.7%), website (15.6%), family (2.2%) and others (3.3%).

Correlation analysis aims to find the relationship between the two variables studied. Both the correlations and regression have a very close relationship. The coefficient of determination states the size of the values of variable X to Y. Both the correlation test and the coefficient of determination aim to find the variable X to Y. The study of correlation is to detect the extent to which variations in a factor are associated with variations on one or more other factors based on the correlation coefficient. Table 1 shows the correlation and determination coefficient data to find out the influence of sports motivation on the participants' decision to join the Tahura Trail Running Race 2016.

Table 1. Profile of the respondents.

	Category	Frequency	Percent
Gender	Male	68	75.6
	Female	22	24.4
Age	18–25 years old	42	46.7
	26–30 years old	26	28.9
	31–35 years old	15	16.6
	36–40 years old	4	4.4
	> 40 years old	3	3.3
Occupation	Student	31	34.4
	Entrepreneur	12	13.3
	Employee	38	42.2
	Government Officer	9	10
Monthly income (IDR)	500,000–1,000,000	18	20
	1,100,000–2,000,000	11	12.2
	2,100,000–3,000,000	8	8.8
	3,100,000–4,000,000	7	7.7
	4,100,000–5,000,000	6	6.6
	> 5,000,000	40	44.4
Participation	Family	6	6.7
	Coworker	5	5.6
	Friends	73	81.1
	Others	6	6.6
Source of Information	Social Media	38	42.2
	Website	14	15.6
	Family	2	2.2
	Friends	33	36.7
	Others	3	3.3

Table 2 shows that the correlation value (r) of the sports motivation on the participants' decision to join Tahura Trail Running Race 2016 is 0.713. It can be interpreted that the value indicates that the correlation strength between each dimension of sports motivation (X) to the decision variable (Y) is simultaneously included in the strong category as proposed by Sugiyono (2010), which is between 0.600–0.799.

This data also shows that the coefficient of determination (R Square) is 0.509. R Square is the result of the square value correlation. If the correlation is interpreted in the correlation table, then the relationship between the sports motivation and sport decisions has a strong relationship. Based on the calculation result, the coefficient of the determination value is 0.486 or 49%. It shows that 49% lies in the participants' sport decisions while the rest of

Table 2. Model summary.

Model	R	R Square	Adjusted R Square	Std. Error of the Estimate
1	0.713[a]	0.509	0.486	5.53019

Table 3. Coeficient (T-test).

Model	Unstandardized Coefficients		Standardized Coefficients		
	B	Std. Error	Beta	T	Sig.
(Constant)	14.551	4.529		3.213	.002
1 Psychological	.494	.140	.388	3.520	.001
Physical	.224	.212	.092	1.057	.293
Social	.380	.194	.181	1.965	.053
Achievement	.448	.203	.223	2.202	030

a. Dependent Variable: Sport Decision.

51% is explained by other factors that are not examined in this study.

Table 3 shows the partial influence of the variables of sports motivation to the sport decision. To find the t_{table} on degree of freedom (df) and $\alpha = 10\%$ with two-party test becomes 5%.

By comparing the value of t_{count} and t_{table} it can be explained as follows.

1. There is a significant influence between the psychological sub-variable on the sport decision with a value of t count > t table equal to t count 3.520 > t table 1.663 with a significant value of 0.01 < 0.05, so that H_o is rejected and H_a is accepted.
2. There is no significant influence of the physical sub-variable to the sport decision with t count > t table which is t count 1.057 < t table 1.663 with a significant value 0.293 > 0.05, so that H_o is accepted and H_a is rejected.
3. There is influence of the social sub-variable to the sport decision with t count > t table equal to t count 1.965 > t table 1.663. However, the influence received is not significant because the significant value is equal to 0.053 > 0.05, so H_o is rejected and H_a is rejected.
4. There is a significant influence of the achievement sub-variable to the sport decision with t count > t table equal to t count 2.202 > t table 1.663 with a significance value equal to 0.030 < 0.05, so H_o is rejected and H_a is accepted.

4 DISCUSSION

There are four sub-variables in this study that were used to measure sport motivation: psychological, physical, social, and achievement (Ferrer 2015). The psychological motive is an effort to maintain or to improve self-esteem, to give a sense of meaning to life or for aesthetics. In other words, it is to make life more focused and to solve problems or to overcome negative emotions so that the performer can become less depressed. The physical motive is an activity of running that covers common health benefits such as improving health and weight care by helping to control weight. The social category includes the affiliation with other runners, for example to socialize with other runners and to gain recognition or approval from family or friends. This category is intended to get the respect from people in general. Achievement includes competition with other runners. It is to see how a runner can achieve in the race and the achievement of personal goals, that is, to try to run faster.

Previous study shows that motivation influences sports decisions. Schwarz and Hunter (2008) describe decision-making as a cognitive process that is carried out consciously to make tactical or strategic choices followed by carrying out an action. By contrast, Blakey (2011) mentions that sport consumption decision making influences both the internal and external factor and that the decision process brings those factors together. It consists of physical and social surroundings, time, task definition and antecedent states. Physical surroundings are location, weather, and environmental aspects. In sports participation, the physical surroundings play a very important role in decision making. Social surrounding focused on the influence of others in the participation of sports activities. Time is another issue; nowadays there is an increase in time pressure on all of us. Changes in family structure increase the number of families with multiple incomes. Then, there is also the emergence of the single-parent phenomena. They all cause time to participate in sports become rare. Another component or other elements are task definitions, referring to the reason of opportunities in the participant's need to participate in sports activities. In other words, the participants' reasons influence the decision-making process. Antecedent state refers to temporary physiology and mood brings to the mood of the participants.

Based on these results, the psychological and achievement sub-variables have a significant influence on the decision to participate in the Tahura Trail Running Race 2016. These mean that participants join the race mainly because they want to be happy and to make them feel much better. This can be done by how they compete with other runners. Reaching their personal best in their running records and finishing the race under cut-off time are things that matter. This is aligned with the study of Jordalen and Lemyre (2015) that states marathon runners find the challenge of participation stimulating and motivation is associated with feelings of deep personal awareness and positive self-perception.

Social sub-variables influence sport decisions, but are not significant. Recognition from others also seems to influence the decisions, although it is not significant. Participants usually join in the race with their own runners' community so mostly they don't to make new friends. This is related to the study by Wiid and Cant (2015), which stated that group affiliation is one of the sport consumption motivations. Havenar and Lochbaum (2007) examined that dropout runners rated social motives more important than the finishers did. Gender indicated that men are more highly motivated by performance and ego-related factors, such as challenge, strength and endurance, competition, and social recognition, compared to women, regardless of the activity type (Kilpatrick & Hebert 2005).

The physical sub-variable doesn't influence the decision to participate in the event. It can be concluded that, interestingly, improving health and weight care are not the reasons of participants for joining the race. These can be carried out personally, outside of the race event. It is contradictory with a previous study that mentioned the physical factor influencing the sport decision (Havenar & Lochbaum 2007). Participants assess that physical health or better physical fitness results from participation in sports and exercise (Ferrer 2015).

5 CONCLUSION

The results of hypothesis testing showed that sports motivation has a high level of influence on sports decisions. However, partially, only psychological and achievement motivation significantly influenced the sport decision. There is no influence of physical sub-variables to the decision to join the event. This is because participants do not focus on changes in the physical form, such as controlling weight and shaping their body by joining the Tahura Trail Running Race. On the other hand, participants did not consider that social sub-variables were a major factor in influencing the participants' goal to join the running event because they mostly came with their own community. It influenced the decision but not significantly.

It is important for the event organizer to consider participant motivation when formulating their strategy. By understanding it, management would be able to create an event that is more interesting, such as the existence of a pre- and post-event race. Management also needs to involve the running community to help motivate its members to joining this kind of event. This is surely to increase the level of interest in participating in the event.

This research has some limitations. First, there was no categorization of respondents. It was general and not depending on the running category. Future research should differentiate respondents based on their running category. Second, this research didn't explore whether participants were from Bandung or outside of the city. Future research should compare the motivation based on whether someone is from the city or from outside the city.

REFERENCES

Blakey, P. 2011. *Sport Marketing*. Newcastle: Northumbria University Press.

Ferrer, A. D. 2015. Physical Motivation Influences Race Performance Over a 24-hour Ultra-Marathon. *International Journal of Sport Studies* 5(10): 1162–1169.

Funk. 2009. *Consumer Behaviour in Sport and Events Marketing Action*. Burlington, VT: Elsevier Inc.

Havenar, J & Lochbaum, M. 2007. Differences in participation motives of first time marathon finishers and pre-race dropouts, *Journal of Sport Behaviour* 30: 270–279.

Jordalen, G & Lemyre, P.N. 2015. A Longitudinal Study of Motivation and Well-being Indices in Marathon Runners. *International Journal of Sport and Exercise Science* 7(1): 1–11.

Kemenpora. 2014. *Penyajian Data dan Informasi Statistik Keolahragaan Tahun 2014*. Jakarta: Sekretaris Kemenpora Republik Indonesia.

Kilpatrick, M Hebert E, & Bartholomew J. 2005. College students' motivation for physical activity: Differentiating men's and women's motives for sport participation and exercise. *Journal of American College Health* 54: 87–94.

Recours, R. A., Souville, N & Griffet, J. 2004. Expressed motives for informal and club association-based sports participation. *Journal of Leisure Res* 36:1-22.

Schwarz, E.C & Hunter, J. D. 2008. *Advanced Theory and Practice in Sport Marketing*. UK: Elsevier Inc.

Shank, M. D & Lyberger, M. R. 2014. *Sports Marketing: A Strategic Perspective, 5th edition*. Routledge.

Sugiyono. 2010. *Metode Penelitian Kuantitatif Kualitatif dan R&D*. Bandung: Alfabeta.

Van Heerden, C. H. 2014. The relationships between motivation type and sport participation among students in a South African context. *Journal of Physical Education and Sport Management* 5(6): 66–71.

Wiid, J. A. & Cant, M. C. 2015. Sport fan motivation: are you going to the game? *International Journal of Academic Research in Business and Social Sciences* 5 (1).

Advances in Business, Management and Entrepreneurship – Hurriyati et al (eds)
© 2020 Taylor & Francis Group, London, ISBN 978-0-367-27176-3

Visual perception in improving learning motivation: Gender and developmental study

E. Eriyansyah & H. Hendrayati
Universitas Pendidikan Indonesia, Bandung, Indonesia

ABSTRACT: Visual perception is a process in organizing and interpreting input or obtained information visually. There have been several studies revealing that there is a relationship between visual perception and academic ability, but there is no study related it to learning motivation. What individuals see provides information that is processed into something meaningful. This meaning can stimulate attitudes and motivations for what is seen, in this case is learning materials. What is visually perceived is linked to student motivation in learning. The research was conducted quantitatively and the data was collected by distributing questionnaire to the respondents. The research respondents were male and female students at elementary school, junior high school, and senior high school in Bandung City. Based on the research findings, there is no difference in learning motivation in students who have different gender. However, there are differences in motivation in students who are at different levels of education. Elementary school students have a higher visual perception and greater influence on the level of reviews their learning motivation. This can provide input and suggestions for improving students' learning motivation in general.

1 INTRODUCTION

Technology now plays an important role in pedagogy. When teachers use it in their classrooms, they actually want to attract the attention of students, so they can improve the way of effective learning. According to Ybarra & Green (2003), the process of learning a new language can be tedious and painful for students, so they need a lot of support and the teachers who teach English also know that support is beneficial to the ability to master the language. One of the most important factors for learners is the method used by teachers in their teaching and learning processes to facilitate learning (Running 2014). Ilter (2009) refers to the positive aspects of the use of technology in both parties, students and teachers. When students are active in the learning environment, their interest increases. Teachers should try to motivate their students to become eager learners. Technology makes learning more fun by putting control over learning in the hands of students (Wartinbee 2009).

Therefore the use of technology may be one way to attract the attention and motivation of students to learn. Technology is closely related to the new visual aspects encountered by students. Something visually presented either in the form of information or other content to attract attention so that augment learning motivation if it is packed with new technology that is different from the way usual (conventional).

In addition to the technological aspects, visual effects can also be associated with the use of other media to facilitate student learning and improve learning motivation. Visual effects are very closely linked to students' learning process and interaction in it. Computer-based language learning program within the scope of the visual effects allow for the integration of text, graphics, video animation and phonetic symbols in various combinations that facilitate the learning of "pronunciation" (Meskill 1996). Learners will be given a different visuals when learning to use a computer-based language learning program.

Sovorov (2008) suggests that the main shortage in the software is less precise visual education feedback. Shimizu and Taniguchi (2005) reported that feedback from visual interaction can affect the Japanese students in raising the bar in terms of intonation in English in all four types of materials/media and their viewpoints. In their experiments, only the test group who have access to feedback visual interaction, using a dynamic hand movements using computer software so that its visual interaction can appear on the computer screen as a visual feedback from the intonation of the students themselves or their teachers.

In addition to its software and computer media in terms of the ability to speak English intonation, students will interact visually in the process of learning the computer program with media presentations known as PowerPoint. They also reported, the teachers believed that the PowerPoint presentation make learning content becomes more interesting. Therefore, using a PowerPoint presentation can help them to attract the attention of students. Results Corbeil study (2007) showed that students prefer a PowerPoint presentation given rather than a textbook presentation. He believes that students learn better when their attention is captured through visual reinforcement, change colors, different fonts, and other visual effects. Stepp-Greany (2002) reported in his study several benefits for students associated with the use of technology in the classroom among which increased motivation, improved self-concept and the mastery of basic skills, learning more student-oriented and student engagement in the learning process.

The use of technology and computer-based language program can trigger learning patterns that appeal to students in school so that in turn can trigger the motivation to learn. Both of these aspects (technology and computer programs) are closely related to the visual effects posed to attract the interest and attention of students. Aside from the visual effect of media, there are also other aspects that affect student learning environment that is the interior and exterior of the classroom. It can also be one of the factors that influence students' motivation in learning.

So many previous studies have discussed the effects of visual technology on students' interest in learning, but not many have examined the influence of other visual effects that surround the student learning environment. Learning stimulation can be obtained from many things other than learning media. Visually, students are more easily affected in their motivation to learn because this aspect often interacts with them. Both men and women will get visual stimulation from their learning environment. But are there differences in visual effects obtained from the learning environment between students with different gender? This will be explained further and in depth in this study. In addition, this study will also discussed further the influence of visual effects they get around the learning environment at different levels of elementary, middle and high school. Ozaslan & Maden (2013) in his research, students study better if the courses are presented through multiple visual tools. Good learning is also related to students' motivation in learning, if learning motivation is high, it can also improve the quality of learning. A learning environment that can stimulate students visually consist of the interior and exterior of the room, the visual appearance of the teaching staff and the learning media. This study will also discuss the four factors that can affect student learning motivation.

This study aims to determine the contribution of the visual effects on students' motivation both in the classroom and outside the classroom. Therefore, to find out more about this, this study will attempt to find the right answers to the following questions:

a. Is the indoor visual media exposure units contributed positively to the motivation of learners?
b. Is the outdoor visual media exposure units contributed positively to the motivation of learners?
c. Does exposure to visual media in the form of props contribute positively to the motivation of learners?

d. Does exposure to visual media in the form of uniform appearance employee contributes positively to the motivation of learners?

The questions above have been reformulated in the form of the following hypothesis:

a. Interior visual media exposure units have a significant impact on the motivation of learners.
b. Exterior visual media exposure units have a significant impact on the motivation of learners.
c. Exposure to visual media in the form of props have a significant effect on the motivation of learners.
d. Exposure to visual media such as employee uniforms have a significant effect on the motivation of learners.

2 METHOD

2.1 Respondents

The methodological approach in this study used quantitative methods. Quantitatively data was obtained by questionnaire using 1-5 Likert scale then processed it using SPSS and AMOS. Respondents from this study were students from Tridaya Educational Institution in Bandung. Seventy four students are the subjects of this study. They consisted of 25 men and 49 women among them were 21 students from elementary school grade, 30 from junior high school and 23 from senior high school students.

2.2 Data collection instrument

The research instrument used a questionnaire survey consisted of several questions about the visual perception of the respondents to the students motivation. The data was obtained in about 2-3 weeks and then processed using SPSS and AMOS to determine acceptance or rejection of the hypothesis of research that have been made. Data collection instruments very effectively distributed through a Google form because it was very efficient in terms of time and number of respondents obtained in a short time. Students in Tridaya Education Institutions also had known how to fill the form because it was also used to obtain customer satisfaction data and other data associated with the service and quality improvement in Tridaya Education Institutions itself.

2.3 Data analysis

Data was analyzed using SPSS and AMOS with a complementary manner, the Excel spreadsheet was also used to process the raw data obtained from Google response form that has been distributed. We have implemented a descriptive analysis and Path Analysis in SPSS and AMOS accordance with the objectives of the proposed research.

The results of the data analysis was used to determine the relevant conclusions obtained from students' visual perception. It was very beneficial for the development of visual media in Tridaya Education Institutions so that the quality of service to the customer can be improved.

3 RESULTS AND DISCUSSION

3.1 Classic assumption test

Score obtained from questionnaire was calculated with Excel, SPSS and AMOS and path analysis showed that there were no differences in the learning motivation of students who had different genders. Both male and female students showed a high learning motivation (average score 4) so that gender was not significant differences related to student motivation. However, there were differences in motivation in students who were at different levels of education. Elementary school students had higher visual perception and had a greater influence on the level of motivation compared to middle and high school students, as shown in Table 1 and Table 2.

Based on the table above, student motivation would increase if they obtained visual stimuli both for male and female students. This is in line with previous research related to visual perception with students' academic abilities delivered from Corbeil (2007). Students with different gender had motivation that was not much different from the presence of visual perception stimulus in their learning environment. This meant that the influence of visual perception on students' learning motivation had nothing to do directly with gender differences.

As shown in Table 2 above, the average score of elementary school students was the highest compared to the average of middle and high school students. This meant that the visual perception of

Table 1. Level of student motivation by gender.

Gender	The average scores
Male	4:18
Female	4:03

Table 2. Visual perception level school students based study.

Ladder	The average scores
Elementary School (SD)	4:01
Junior High School (SMP)	3.73
School (SMA)	3.83

elementary school students had the strongest influence on their learning motivation compared to students in junior and senior high school. Elementary students used more visual effects to receive and process information in their learning environment. This visual stimulus made the level of learning motivation higher compared to junior and senior high school level. Students at junior high school had the lowest score of 3.73. This showed that the learning motivation of junior high school students was not strongly influenced by visual perceptions obtained from their learning environment. Visual aspects obtained by junior high school students were not used optimally to improve their learning motivation.

From the test results of the four variables related to visual perception and motivation to learn, which is interior of rooms, exterior of rooms, instructional media and staff uniforms obtained the following results:

a. Interior visual media exposure units contributed positively to the motivation of learners (the hypothesis is accepted).
b. Exterior visual media exposure unit does not contribute positively to the motivation of learners (the hypothesis is rejected).
c. Exposure to visual media in the form of props did not contribute positively to the motivation of learners (the hypothesis is rejected).
d. Visual media exposure in the form of uniform appearance employee does not contribute positively to the motivation of learners (the hypothesis is rejected).

4 CONCLUSION

The results showed that the interior of the unit made a great contribution in student motivation, while other visual effects, the exterior of the unit, a medium of learning and uniform employees did not contribute positively to students' motivation. This study supports previous studies on the influence of visual effect on student learning, and the results of the questionnaire showed that most students show a positive perception of the existing interior visual effects unit.

The implication of this study is that the interior of the classroom becomes an important factor in increasing their learning motivation. Room interior is the strongest factor in influencing students' learning motivation, especially students at the elementary level, both for male and female students. There are several recommendations for subsequent research in terms of visual perception of learning motivation. First, further research can be developed into a study of students' learning motivation only influenced by visual factors or by the audio and kinesthetic effects of the environment around their learning space. Second, visual perceptions that can affect students' learning motivation, can also affect their learning achievement so that the ultimate goal of student learning to obtain optimal performance can be achieved by the visual perception that stimulates them around their learning environment.

REFERENCES

Caballero, A. O. 2017. 7th International Conference on Intercultural Education "Education, Health and ICT for a Transcultural World. Visual perception in art education. *Gender and Intercultural Study, Social and Behavioral Sciences* 237. pp. 588-593.

Corbeil. 2007. Are You Ready for Mobile Learning: Frequent use of mobile devices does not mean that students or instructors are ready for mobile learning and teaching.

Felder, R. M. 2006. Learning and Teaching Style in Engineering Education. *Engr. Education* 78 (7). pp. 674-681.

Fotovatnia, Z. 2013. The Effect of Exposure to the Visual Medium on Learning Pronunciation and Word Stress of L2 Learners. *Theory and Practice in Language Studies*, Vol. 3, No. 5, pp. 769-775.

Ilter. 2009. Effect of Technology on Motivation in EFL Classrooms. *Turkish Online Journal of Distance Education-TOJDE*. Volume: 10 No. 4.

Meskill. 1996. Listening Skills Development Through Multimedia. *Journal of Educational Multimedia and Hypermedia*. Vol. 5 No. 2.

Ozaslan & Maden. 2013. The use of Power Point Presentations at in the Department of Foreign Language Education at Middle East Technical University. *Middle Eastern & African Journal of Educational Research*, Issue No. 2.

Running, F. S. 2014. International Conference on Current Trends in ELT. The Impact of Using PowerPoint Presentations on Students' Learning and Motivation in Secondary Schools. *Social and Behavioral Sciences* pp. 1672-1677.

Sovorov. 2008. Context Visuals in L2 Listening Tests: The Effectiveness of Photographs and Video vs Audio- only Format. *Retrospective Theses and Dissertations*.

Sugiyono. 2011. Research Methodology Quantitative. *Alfabeta*. Bandung: Indonesia.

Wartinbee. 2009. The Value of Technology in the EFL and ESL Classroom.

Advances in Business, Management and Entrepreneurship – Hurriyati et al (eds)
© *2020 Taylor & Francis Group, London, ISBN 978-0-367-27176-3*

The implication of social media marketing in modern marketing communication

J. Waluyo & H. Hendrayati
Universitas Pendidikan Indonesia, Bandung, Indonesia

ABSTRACT: In the current era of globalization, the development of the business world takes place in a very competitive, fast, and unpredictable situation and condition. Companies are compelled to make continuous improvements and innovations in all fields, including marketing. The purpose of this research is to analyze the development of social media marketing, especially the implication in Facebook marketing in Small and Medium Enterprises (SMEs). This research uses descriptive qualitative research method and provides an overview of the current digital marketing phenomenon based on the theories, the study of various literature review, and the study of other relevant research. Facebook Marketing has the following advantages: Providing Demographic Information, Focus Market Segmentation, Cross Platform Access, Simplify Communication, Zero Time Feedback, Always Connected, Low Budget High Impact, and The New Wave Marketing in modern marketing communication.

1 INTRODUCTION

In the current era of globalization, the development of the business world takes place in a very competitive, fast, and unpredictable situation and condition. Companies are compelled to make continuous improvements and innovations in all fields, including marketing. One of the innovations to solve this problem is one related to the increase of company's existence (Utami & Purnama 2012).

In facing market competition that is increasingly open and competitive, market domination is one of the main requirements to improve the competitiveness of the companies. Therefore, the role of information and communication technology is needed in encouraging the success of companies to expand market share through the use of internet-based technology (Muttaqin 2011).

The internet is a loose network of thousands of computer networks that reach millions of people around the world. The initial purpose of making the internet was to provide a means for researchers to access data from a number of computers. But now the internet has developed into a very effective means of communication, so it has deviated far from its original goal (Soemirat 2003).

Today, the internet becomes as large and powerful as an information and communication tool that cannot be ignored. The internet is a computer network that allows interactive communication to occur in it, where millions of people in the world can connect to a new community in the network, until the emergence of social network (Afriani 2011).

Social network is an online network where users can easily participate, share and create content. Blogs and wikis are the most common forms used by people around the world. Social network is a community that can collect people together on the internet (Fowdar 2013). Users use social network to connect with others in a variety of ways, including dating, meeting and gathering with other users who share common interests, and sharing information (Leung 2012).

At present, many people have used social networks to help the process of marketing, especially Small and Medium Enterprises (SMEs). SMEs have an important and strategic role in national economic development as well as their role in the development of products distribution. So far, SMEs have contributed 57.60% of the Gross Domestic Product (GDP) and have a labor absorption rate of around 97% of the entire national workforce (LPPI & Bank Indonesia 2015).

But on the contrary, according to data from Bank Indonesia (2015), around 60-70% of SMEs in Indonesia have not received bank funding, which causes limitations in funding and procurement of infrastructure, including the procurement of tools and technology to assist the marketing process. The good news is there are currently alternative marketing tools to help overcome the problem, namely Facebook.

Facebook is an easy and affordable marketing tool to save marketing expenses (Arifin 2015). Facebook was firstly introduced on 4 February 2004 by Mark Zuckerberg, an expert computer programmer from Harvard University (Gemilang 2011). Facebook is one of the social networking sites that emerged in the Web 2.0 era (Nurcahyo et al. 2009). Facebook is a social network site that connects users with various demographic characteristics such as gender, age, address, workplace, school, hobbies, likes and so on. Facebook users can interact and connect with other users, including marketing communications.

Marketing communication is a means used by companies in an effort to inform, persuade, and remind consumers, either directly or indirectly about the products and brands they sell (Kotler & Keller 2007). Facebook as a social network is expected to help marketing communication better.

Facebook is in the first position with the number of users reaching 2.1 billion. The second place is occupied by Youtube with the number of users reaching 1.5 billion, and the third place is occupied by Whatsapp and FB Messenger with the number of users reaching 1.3 billion. Furthermore, table 1 and table 2 is the data of the top 10 countries and the top 10 cities with the most number of Facebook users in the world (We Are Social & Hootsuite 2018).

According to table 1 and 2, India ranks first with the most Facebook users with 250 million users. Following, the United States ranks second with 230 million users. Then, Brazil ranks third with 130 million users, and Indonesia is the fourth with 130 million users.

Not only the countries but also the top 10 cities with the highest number of Facebook users in the world are shown in the table. The first rank is occupied by Bangkok has 22 million users, following Dhaka has 20 million users, Bekasi has 18 million users, and Jakarta has 16 million users.

Facebook as a social networking site stores a lot of information about its user demographics, opening up opportunities for business owners to develop their businesses by advertising and selling their products through this platform (Utami & Purnama 2012).

Facebook is the most widely used social media to promote products and services (Hasanah et al 2015). It has become a social media with the highest number of users and access volume (Hsu 2012).

Facebook provides a place to meet and communicate with fellow users. The more users gather, the better Facebook offers advertising facilities to market or promote their products and services. Facebook is a promotional media that is very easy and inexpensive that can reach markets throughout the world (Arifin 2015). Facebook is one of the most interesting media to be examined and studied more deeply related to modern marketing communication and its impact on the company (Triyono 2011).

In line with this, the purpose of this study is to analyze the implications of using Facebook in relation to the efforts to optimize the marketing communication of Small and Medium Enterprises (SMEs) in Indonesia.

2 METHOD

This study uses descriptive qualitative method. The research develops concepts and fact compilation with internet and literature studies, yet does not conduct hypothesis testing (Utami & Purnama 2012).

3 RESULTS AND DISCUSSION

3.1 Results

Based on studies from a variety of relevant literature, the results show that the advantages and supporting features of marketing communication through Facebook lie in the completeness of features and ease of use. Marketing communication using Facebook is equipped with a variety of diverse features. It starts only from a feature of status updates, writing notes, sharing links, images, videos, and sending messages. In addition, Facebook also provides invitations to events, groups and other features.

Facebook is unique as a social network that can make it easier to find and establish friendships, including setting the target market. Marketers can search and make friends with consumers through Facebook personally by first analyzing the costumer's Facebook profiles. The Facebook feature allows users to view profiles of other users easily and quickly. This is confirmed by previous studies which suggested that: "On Facebook, the level of visibility to see other users' profiles is quite high. Users who are part of the same network can freely view each other's profiles, unless the profile owner decides to close their profile, limiting it to only the circle of closest friends." (Ellison et al. 2007).

In addition to provide space to update user information or status updates, Facebook also provides other facilities that are rarely known by people that the status update can also contain other intentions, such as promotion of products.

According to Awl (2010) the use of Facebook features can be combined with each other so that it can have a greater effect. The following are Facebook features that can be used to further optimize marketing communications.

a. Fan Page
   Facebook provides free fan page creation services as a marketing tool. A well-managed fan page is the foundation of every successful ad promotion campaign on Facebook. Fan page is a perfect tool for communicating and building relationships with loyal fans and customers. Fan page is a valuable marketing tool to be applied to a marketing approach (Cox & Park 2013).

b. News Feed
   The key to spreading information about something on Facebook is to produce news feeds. After creating a fan page, event, or other marketing content on Facebook, share the content to a personal profile wall. If the consumer comments on the content that has been posted, be sure to answer immediately, so that the communication is established until the sale occurs.

c. Updating Status
   Updating status is an activity carried out to share any news, as well as asking questions or starting discussions with customers to encourage them to

make regular visits to fan page and make purchases.

d. Note

Note can be used to draw attention to more detailed things because they can be used to explain more information. Notes can be used to post ongoing projects and can also be used to ask questions, advice or input from consumers.

e. Event

Event posts on Facebook help marketers to invite consumers that potentially also invite their friends. Marketers can use events to schedule meetings and promote business.

f. Photo and Video

Photo and video posts on Facebook are the best way to attract consumer attention. Sharing photos and videos is the best feature in using Facebook as a marketing tool (Handayani & Lisdianingrum 2011) and (Hansson & Wrangmo 2013).

g. Link

After posting other content on the web, such as YouTube videos, blogs, or podcasts, it is important to also make sure to post the content link on Facebook fan page so that the opportunity to be seen by consumers is greater.

h. Group

Use group to facilitate communication between people who share the same interests. The key to success in managing groups is to create discussion and maintain the management and activity of the group members (Xia 2009).

i. Facebook Advertising

If marketers want to reach consumers more specifically with relatively more affordable promotional costs, Facebook advertising service facilities can be an alternative.

3.2 *Discussion*

Based on the data obtained, researchers found that Facebook has helped hundreds of companies, especially SMEs to grow and develop better. Here are some success stories of SMEs that use Facebook as a marketing tool (Facebook 2018).

First is Ertos, an SME that provides health and care products. Facebook has helped Ertos Indonesia since it first started a business with only 1 employee, now Ertos has more than 70 employees in less than 6 years. Facebook advertising facilities helped Ertos marketing process during the December-May 2017 period and generated 60% increase in gross income and 8.5 times return of advertising spend. This is also experienced by Life HAF, a clothing provider. Live HAF uses Facebook advertising during the February-April 2017 period, successfully generated 60% reduction in costs per purchase, 50% increase in revenue per quarter, and 21 times return of advertising spend.

A similar condition is experienced by IDN, media companies that make content for millennial generations. IDN used Facebook advertising during the period May-July 2016 and managed to generate 10 times increase in the number of reach in just 12 months, 5 times increase in total duration for 3 months of advertising, and 35 million unique audiences per month.

One of the most important things when using Facebook as a marketing tool is marketers must pay attention that Facebook is a place where people socialize, communicate, interact, and share. Marketers should not overdo their products or services, because people join Facebook not to buy something, but to connect with others. Marketers can attract people attention by creating fan page to build brand, loyalty, and managing the consumer base by implementing innovative, effective and efficient marketing communication strategies (Nurcahyo 2009).

4 CONCLUSION

Based on 2018 We Are Social and Hootsuite data regarding the number of users of social networks in the world, placing Facebook in the first position with the number of users reaching 2.1 billion, and the number of Facebook users in Indonesia reaches 130 million users. Limitations in terms of marketing costs experienced by 60-70% of Small and Medium Enterprises (SMEs) in Indonesia, getting solutions with the help of Facebook. Considering the huge number of Facebook users, it is not surprising that Facebook can be used as a very potential marketing tool.

Based on the results of studies from various relevant literature, several advantages of using Facebook as a marketing communication media are as follows:

a. Providing Demographic Information

Facebook provides a column to fill in various user information, such as age, location, hobbies, activities and other information that are needed to determine market segmentation in marketing activities.

b. Market Segmentation Focus

Marketers can set the target market appropriately using demographic information in Facebook's audience insight feature.

c. Cross Platform Access

Facebook can be accessed at any platform, either from a computer, tablet, or from a mobile phone.

d. Simplify Communication

Facebook provides direct communication features using Facebook messenger, so communication with the target market becomes easier.

e. Zero Time Feedback

Feedback from consumers is faster and easier to respond.

f. Always Connected

Facebook as a marketing tool can always connect business owners and consumers, so that business relationships can be established properly.

g. Low Budget, High Impact
The costs incurred for advertising on Facebook are much cheaper than advertising in other media, such as television.

h. New Wave Marketing
Hermawan Kartajaya, one of the 50 teachers who influenced the future of world marketing, in Lasmadiarta (2011) said that the marketing world has changed and has entered the New Wave Marketing era where Facebook is one of the main causes.

Research on the implications of Facebook Marketing for the optimization of corporate marketing strategies, especially Small and Medium Enterprises, has its own appeal. The results of this study are expected to be able to help business people, especially the actors of Small and Medium Enterprises (SMEs) in Indonesia. The results of this study are also expected to help other researchers who have the same interests regarding the impact of using Facebook Marketing in the company's marketing communication strategy.

REFERENCES

Afriani, U. F. 2011. Rown Division's Strategy in Utilizing the Media Je-Social Social Facebook as a Promotion Tool Online. Universitas Muhammadiyah Surakarta: Indonesia.

Arifin, R.W. 2015. The Role of Facebook as a Promotion Media in Developing Creative Industries. *STMIK Bina Insani*. Bekasi: Indonesia.

Awl, D. 2010. Facebook Me! A Guide to Socializing, Sharing, and Promoting on Facebook, 2nd Edition. *Mass: Peachpit Press*.

Cox, T. & Park, J. H. 2013. Facebook Marketing in Contemporary Orthodontic Practice: *A Consumer Report. Journal of the World Federation of Orthodontists*.

Ellison, N. B. & Steinfield, C. & Lampe, C. 2007. The Benefits of Facebook "Friends": Social Capital and College Students Use of Online Social Network Sites.

Journal of Computer-Mediated Communication 12: 1143-1168.

Fowdar, R.R. & Fowdar, S. 2013. The Implications of Facebook Marketing for Organizations. *Contemporary Management Research Pages* 73-84, Vol. 9, No. 1.

Gemilang, R. D. A. 2011. The role of Facebook as an Online Business Communication Media. *Universitas Pembangunan Nasional Veteran*. Surabaya: Indonesia.

Handayani, P.W. & Lisdianingrum, W. 2011. Impact Analysis on Free Online Marketing Using Social Network Facebook: Case Study SMEs in Indonesia. *University of Indonesia*. Jakarta: Indonesia.

Hansson, L & Wrangmo, A. 2013. Optimal Ways for Companies to Use Facebook as a Marketing Channel. *Journal of Information, Communication, and Ethics in Society Vol*. 11 No. 2, 2013 pp. 112-126.

Hasanah, N. 2015. Analysis of the Effectiveness of Social Network Advertising as Promotional Media Using EPIC Model. Yogyakarta: Indonesia.

Hsu, Y. 2012. Facebook as international eMarketing strategy of Taiwan hotels. *International Journal of Hospitality Management* 31 (2012) 972-980.

Kotler, P. & Keller, K.L. 2007. Marketing Management. *12 Edition*, Jakarta: Indonesia.

Leung, X. Y. 2012. The Marketing Effectiveness of Hotel Facebook Pages: From Perspectives of Customers and Messages. *UNLV Theses, Dissertations, Professional Papers, and Capstones*.

LPPI, Bank Indonesia. 2015. Profile of Micro, Small and Medium Enterprises. Jakarta: Indonesia.

Nurcahyo, B. 2009. Inside Facebook: A Prospective Marketing Channel. The International Conference on Economics and Administration. *Faculty of Administration and Business, University of Bucharest*, Bucharest: Romania.

Remaja R. & Triyono, A. 2011. The Influence of Facebook's Social Networking Site as an alternative Media for Promotion, Semarang: Indonesia.

Soemirat. 2003. Basic Public Relation, Bandung: Indonesia.

Utami, A. D. & Purnama, E. B. 2012. Use of Social Networks (Facebook) as an Online Business Media. *Universitas Surakarta*, Surakarta: Indonesia.

Xia, Z. D. 2009. Marketing library services through Facebook groups, Library Management, Vol. 30 Iss: 6 pp. 469-478.

Advances in Business, Management and Entrepreneurship – Hurriyati et al (eds)
© *2020 Taylor & Francis Group, London, ISBN 978-0-367-27176-3*

Loyalty of the $1 barbershop customers: Investigating the roles of service quality, satisfaction, and trust

U. Suhud, S.F. Wibowo & L. Namora
Universitas Negeri Jakarta, Jakarta, Indonesia

ABSTRACT: This study aims to examine the impact of service quality, customer satisfaction, and customer trust on the loyalty of $1 barbershop's customers. There aren't any studies focusing on consumer behaviour in barbershop services. This study involved 200 male participants who visited a barbershop. Data were analysed by using exploratory and structural equation model for hypothesis testing. This study found that service quality significantly affected customer satisfaction, loyalty, and trust. In addition, customer satisfaction and customer trust significantly changed customer loyalty, and customer satisfaction significantly affected customer trust. This result is useful as a recommendation for practitioners and future research.

1 INTRODUCTION

In Indonesia, villages can produce thousands of professional barbers who then roam the countryside and move to other cities (Herdiana 2015). Most of the barbers who work in barbershops come from Bagendit, a village in Garut, West Java Province (200 kilometres from south of Jakarta). Most of them come to barber skills naturally, handed down from their ancestors, from generation to generation (Soekirno 2017). In order to become a master at applying modern techniques of hair cutting and hairstyles, some of these barbers take courses from their seniors in their villages (Supriadin 2017).

Most of the barbers open a barbershop with cheap prices. They equip their barbershop with or without air conditioning. They also use simple razors, sometimes even without considering hygiene. Nevertheless, they have visitors, generally from low-income communities, who are loyal. These services include haircuts, shaves, and hair colouring, both for adults and children. For this study, the barbershops investigated are those that have rates ranging from IDR 10,000 to IDR 15,000 (approximately USD1) for one haircut. When many barbershops set low prices, use simple shavers, have simple furniture, some without air conditioning, and even some without a brand name, how can customers be loyal?

Scholars have investigated customer loyalty in the hospitality service industry, including hotels, restaurants, cafés, spas, and salons (Alsaqre et al. 2012, Han & Ryu 2009). However, customers who are loyal to barbershops suffer from a lack of scholarly attention. Therefore, this study aims to measure the loyalty of barbershops' customers by employing service quality, customer satisfaction, and customer trust as predictor variables.

1.1 *Theoretical background*

1.1.1 *Service quality, customer satisfaction, and customer loyalty*

Service quality has been used by prior studies to predict customer satisfaction, purchase intention, customer trust, and customer loyalty (Saleem & Raja 2014, Lai 2015). Hafeez and Muhammad (2012) found that service quality, customer satisfaction, and loyalty programs increased customer loyalty. Taking place in Malaysia in the banking industry setting, Kheng et al. (2010) tested the impact of service quality on customer satisfaction and customer loyalty, and the impact of customer satisfaction on customer loyalty. As they reported, customer satisfaction significantly influenced customer loyalty. In addition, in that study, Kheng et al. (2010) tested the impact of each dimension of service quality both on satisfaction and loyalty. As a result, tangible and responsiveness were insignificant to influence customer loyalty whereas reliability, empathy, and assurance were. Additionally, responsiveness, empathy, and assurance were significant to influence satisfaction, whereas tangible and reliability were insignificant. A similar approach with different findings has been shown by Khan and Fasih (2014b). These scholars included five dimensions of service quality to be used to measure satisfaction and customer loyalty. They found that all the dimensions were good predictors of satisfaction and loyalty.

Loyalty of restaurant customers in Hong Kong was investigated by Lai (2015), employing service quality, perceived value, satisfaction, and affective commitment as predictor variables. Two of the findings were that there was a significant impact of service quality on satisfaction and loyalty. Furthermore, a study with a bank industry setting was conducted by Hidayat et al. (2015). According to these scholars, service quality and customer trust had a significant impact on customer satisfaction. Additionally, service quality and customer trust had a significant impact on satisfaction.

Rasheed and Abadi (2014) investigated customer loyalty in the service sector in Malaysia. In this study, service quality was connected to trust and perceived valued, trust to loyalty, and perceived value to loyalty. Some of the findings demonstrated that there was a significant impact of service quality on trust, and trust on loyalty.

Chou et al. (2014) focussed on the loyalty of high-speed passengers in Taiwan by using service quality and satisfaction to measure it. According to them, passengers had concerns regarding aspects of service quality including car cleanness, neat appearance of employee, employee service attitude, comfort of air conditioning, and on-time performance. They found a significant effect of service quality on customer satisfaction and customer loyalty, as well as the effect of customer satisfaction on loyalty.

In the hotel industry, Saleem and Raja (2014) assessed the impact of service quality, customer satisfaction, and loyalty on brand image in a Pakistani setting. They found that service quality had a significant influence on customer satisfaction and loyalty, and customer satisfaction on loyalty.

Based on the discussion of this literature, here are the hypotheses to be tested in this study:

H₁–Service quality will significantly affect customer satisfaction.

H₂–Service quality will significantly affect customer loyalty.

H₃–Service quality will significantly affect customer trust.

1.1.2 *Customer satisfaction, customer trust, and customer loyalty*

Customer satisfaction is a great mantra for developing a customer loyalty. Therefore, managers work hard to satisfy customers. Particularly in service businesses, scholars have demonstrated the crucial effect of satisfaction towards customer loyalty, trust, attitude, purchase intention, and intention to revisit (Chiou & Droge, 2006, Hazra & Srivastava 2009, Lombart & Louis, 2014, Poku et al. 2013, Suhud & Wibowo 2016). In this study, customer satisfaction is linked to customer trust and customer loyalty.

In a study undertaken by Lee et al. (2015), satisfaction was linked to customer trust and loyalty. This study was addressed to measure loyalty of the mobile phone users towards brands. Participants were university students and adults in South Korea. The findings showed a significant impact of satisfaction on trust and loyalty, as well as a significant impact of trust on loyalty.

In Turkey, Orel and Kara (2014) invited customer of a supermarket customers to be involved in their study. They linked service quality on customer satisfaction and loyalty. They found a significant impact of service quality on satisfaction and satisfaction on loyalty. In contrast, they failed to show a significant impact of service quality on loyalty.

Loyalty of customers towards a cosmetic product was studied by Veloutsou (2015) involving women participants in Scotland who experienced buying lipsticks in certain period of time. This study examined brand evaluation, trust, satisfaction, and brand relationship as predictors. They showed all predictor variables significantly affected loyalty.

Furthermore, Stathopoulou and Balabanis (2016) focussed on the customers of fashion retailers in the United States. In this study, satisfaction and trust in the loyalty program were employed to predict store loyalty. They compared between customers who purchased in low and high-end stores. Although the findings indicated significant impact of satisfaction and trust on store loyalty, however, there was no significant between these two groups of samples.

A study took place in the banking industry was conducted by Khan and Fasih (2014a). In this study, service quality was represented by tangibles, reliability, assurance, and empathy. Furthermore, service quality was linked to customer satisfaction and loyalty, and customer satisfaction was linked to loyalty. They found a significant impact of all dimensions of service quality, unless tangibility on customer satisfaction. In addition, there was a significant impact of satisfaction on customer loyalty.

Therefore, the hypotheses that can be made to be tested are as follows:

H₄–Customer satisfaction significantly affect customer loyalty.

H₅–Customer satisfaction significantly affect customer trust.

1.1.3 *Customer trust and customer loyalty*

Prior studies mentioned trust affected customer satisfaction, loyalty, attitude, and purchase intention (Choi & La 2013, Martínez & del Bosque 2013, Nguyen et al. 2013 Lombart & Louis 2014). In this study, trust is linked to customer loyalty.

Akbar and Parvez (2009) put customer satisfaction as a mediating variable. They found customer trust and satisfaction had a significant impact on customer loyalty. They also found a significant impact of service quality on customer satisfaction. Customer trust was also selected by Nguyen et al. (2013) to predict customer loyalty in the financial industry. Their study took place in Canada. They found a significant impact of trust on customer loyalty.

A Spanish study was conducted by Martínez and del Bosque (2013) to investigate the role of CSR on customer loyalty in hospitality companies. These scholars employed CSR, trust, customer identification, and satisfaction to predict loyalty. Some of the

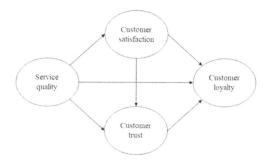

Figure 1. The theoretical framework.

findings revealed a significant impact of trust on satisfaction and loyalty, and satisfaction on loyalty. Another study employed CSR was conducted by Choi and La (2013). In their study, CSR was connected to trust and loyalty. In addition, they linked satisfaction on CSR and trust, and trust on loyalty. They found that satisfaction significantly affected trust, and trust significantly affected loyalty.

Loyalty of customers relating to online shopping has been explored by Bilgihan (2016) by involving generation Y as participants. In his study, loyalty towards an online shop was measured using brand equity, flow, and trust. Bilgihan demonstrated that loyalty was significantly influenced by trust. Furthermore, in the banking industry, Amin et al. (2013) compared loyalty of Muslim and non-Muslim customers of Islamic banks. To measure loyalty, these scholars employed satisfaction, image, and trust. They documented a significant impact of customer trust on loyalty between two groups of samples.

Based on this explanation, a hypothesis to be tested is as follows:

H–Trust will significantly affect customer loyalty.

1.2 The theoretical framework

Based on this literature review, the authors outline hypotheses and illustrate the proposed research model to be tested.

2 METHODS

2.1 Sample and data collection

Data was collected in 2015 from several barbershops in Karawaci, a developed area in Tangerang, Province of Banten, Indonesia. These barbershops were dedicated to serving lower-income consumers, charging approximately $1 for a hair cut. All of the participants were males. They were approached conveniently. They were asked to fill out a self-administered instrument. All collected instruments were usable.

2.2 Measures

For validation purpose, indicators from prior studies were adapted. To measure trust, indicators from Auh (2005) and Sari (2007) were adapted. Furthermore, service quality was measured adapting indicators from Rinanda (2013) and Chow et al. (2012). In addition, customer satisfaction was measured adapting indicators taken from (Sari 2007, Chow et al. 2012, and (Lustinayanti, 2011). Lastly, indicators from (Auh 2005, Rinanda 2013) were adapted to measure customer loyalty. The indicators were translated into *Bahasa* Indonesia and measured using a 5-point Likert's scale ranging from 1 for strongly disagree to 5 for strongly agree.

2.3 Data analysis

Data were analysed in three stages. The first stage was conducting exploratory factor analysis (EFA) to form dimensions of each variable. For the EFA calculation, the authors only considered the indicators with loadings of 0.4 and greater. Therefore, number of participants should be 200 or larger (Hair Jr et al. 2006). Furthermore, a reliability test was conducted. Constructs with a score of 0.7 and greater was included for further analysis (Hair Jr. et al. 2006). However, construct with a score less than 0.7 would also be included. The next stage was conducting structural equation model to assess the theoretical framework model. The fitted model was expected to have a probability score of 0.05 (Schermelleh-Engel, Moosbrugger & Müller 2003), CMIN/DF score of $\leq$ 2 (Tabachnick & Fidell 2007), CFI score of $\geq$ 0.97 (Hu & Bentler 1995), and RMSEA score of $\leq$ 0.05 (Hu & Bentler 1999).

3 RESULTS AND DISCUSSION

3.1 Participants

In total, 200 male participants involved in this current study. The participants were 21–30 years (120 participants; 60%), less than 20 years old (37 participants; 18.5%), between 31 and 40 years (24 participants; 12%), and 40 years old and older (19 participants; 9.5%). Participants predominantly were students (114 participants; 57%), followed by employees (36 participants; 18%), self-employed people (25 participants; 12.5%), and others.

3.2 Exploratory factor analysis and reliability test

In total, exploratory factor analysis produced seven factors, including trust with two dimensions. The first dimension consisted of seven indicators with a Cronbach's alpha score of 0.903 and the second dimension consisted of three indicators with a Cronbach's alpha score of 0.850. Furthermore, service quality formed two dimensions. The first dimension possessed nine indicators with a Cronbach's alpha score of 0.881 and the second dimension had four indicators with a Cronbach's alpha score of 0.738. Additionally, customer satisfaction had two dimensions. The first dimension sustained three indicators with a Cronbach's alpha score of 0.645 and the second dimension had six indicators with a Cronbach's alpha score of 0.931. Lastly, customer loyalty survived three indicators with a Cronbach's alpha score of 0.733. All construct had a Cronbach's alpha score was greater than 0.6, and they were considered reliable for further analysis.

3.3 Hypotheses testing

Figure 2 was the result of structural equation model. This model achieved a fitness with a C.R. score of 0.052 and CMIN/DF score of 1.228. In addition, it

Table 1. Result of exploratory factor analysis.

		Factor loadings
	Trust (1) ($\alpha = 0.903$)	
T9	The attitude of the barbershop employee I used to visit provoked my trust in the barbershop service	0.883
T8	I rely on barbershop that I can visit	0.864
T7	I believe in the barbershop I used to visit	0.856
T10	I feel safe carrying goods to the barbershop I used to visit	0.814
T2	I believe the regular barbershop employee I visit can be relied upon in terms of hair care	0.699
T4	I believe that my usual barbershop always develops and trains his employees	0.661
T6	I am willing to follow the new treatment and hairstyle recommendations from the barbershop employee I used to visit	0.589
	Service quality (1) ($\alpha = 0.881$)	
S14	The employees of the barbershop I regularly visit are patient	0.810
S1	The barbershop I regularly visit gave me a good impression of service at my first visit	0.717
S11	The employees of the barbershop I regularly visit are the polite in serving customers	0.658
S13	The employees of the barbershop I regularly visit are warm in serving customers	0.638
S6	The barbershop employee I used to pay the customer well	0.579
S10	The barbershop employee I used to visit treated all his customers the same	0.573
S2	The barbershop I used to visit prioritizes error-free service	0.564

Table 1. (Cont.)

		Factor loadings
S4	The barbershop employee I used to visit has a good skill in explaining the options of hair guards	0.541
S5	The usual barbershop clerk I visited provided quick service to the customer	0.490
	Trust (2) ($\alpha = 0.850$)	
T1	I believe that the barbershop I used to visit provides the best service to its customers	0.918
T3	I believe that the barbershop employee I used to visit is able to fulfil my request	0.881
T5	The barbershop parties I used to visit always try to understand the opinions of their customers	0.822
	Customer satisfaction (1) ($\alpha = 0.645$)	
C2	The barbershop that I used to visit provide a comfortable waiting room so that customers are satisfied with the waiting time while waiting for their turn	0.781
C5	I am satisfied with all the processes in the barbershop that I used to visit from start to finish	0.734
C7	I am satisfied with the regular barbershop I visit	0.706
	Customer satisfaction (2) ($\alpha = 0.931$)	
C8	I am happy with the barbershop I regularly visit	-0.854
C6	I have a good opinion about the barbershop I regularly visit	-0.804
C9	I have a positive thought about the barbershop I regularly visit	-0.795
C4	In general, I am satisfied with the service given by the barbershop	-0.756
C1	I am satisfied with the barbershop I regularly visit for the price and quality	-0.708
C3	The barbershop I regularly visit is easy to be accessed	-0.695
	Customer loyalty ($\alpha = 0.733$)	
L5	I will try to do more than one type of hair care in the barbershop I used to visit	-0.682
L4	I will come revisiting the barbershop I regularly visit although the tariff increases	-0.651
L8	I will be loyal using the service of the barbershop I regularly visit	-0.647
	Service quality (2) ($\alpha = 0.738$)	
S18	There is a waiting room in the barbershop I regularly visit	0.862
S19	The parking bays of the barbershop I regularly visit are adequate	0.720
S8	The barbershop cashier clerks that I regularly visit is rigorous in calculating payment and change	0.485
S12	The usual barbershop clerk I visit is always ready when it's needed by the customer	0.419

(*Continued*)

128

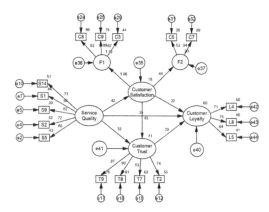

Figure 2. Structural model of hypotheses testing.

Table 2. The result summary of hypotheses testing.

				C.R.	P
H₁	SQ	→	CS	2.502	0.012
H₂	SQ	→	CL	3.668	***
H₃	SQ	→	CT	3.454	***
H₄	CS	→	CL	2.497	0.013
H₅	CS	→	CT	2.952	0.003
H₆	CT	→	CL	2.786	0.005

had a CFI score of 0.984 and RMSEA score of 0.034. Service quality survived 5 of 13 indicators, and customer satisfaction had two dimensions with three and two indicators, respectively. Furthermore, customer trust sustained 4 of 10 indicators, and customer loyalty had three indicators.

Table 2 shows the results of hypotheses testing. All path tested had a C.R. scores that was greater than 1.96. They indicated significances and, therefore, all hypotheses were accepted.

3.4 Discussion

Every business will find its market and service quality to be an important keyword even in a micro-shave business as in the case studied in the current study. Although paying very cheap for each time comes, the visitors still demand to get good service from the barbershop they visit. This study found that service quality significantly affected customer satisfaction, customer loyalty, and customer trust. The first hypothesis predicted the effect of service quality and customer satisfaction. This path had a C.R. score of 2.502. This finding was significant with existing studies (Akbar & Parvez 2009, Chou et al. 2014, Khan & Fasih 2014a, Orel & Kara 2014). Furthermore, the second hypothesis predicted the influence of service quality and customer loyalty. This study found that a C.R. score of this path was 0.628.

Therefore, H₂ was accepted. Prior studies documented the same finding (Chou et al. 2014, Orel & Kara 2014, Saleem & Raja 2014). In addition, the third hypothesis assumed the impact of service quality and customer trust. With a C.R. score of 0.503, the path indicated significance, and therefore, H₃ was accepted. The finding supports prior studies (Hidayat et al. 2015, Rasheed & Abadi 2014).

If the service quality as a keyword in the success of the shaving business, another keyword that should be considered important is customer satisfaction. As in many other business categories, in this case customer satisfaction can lead to changes in customer loyalty and customer trust. The more satisfied a customer becomes, the more loyalty he will be and trust him to the barbershop he continuously visits. The fourth hypothesis predicted the impact of customer satisfaction on customer loyalty. The path owned a C.R. score of 0.439, and it was significance. This finding is in line with existing studies (Khan & Fasih 2014a, Lee et al. 2015, Orel & Kara, 2014, Stathopoulou & Balabanis, 2016, Veloutsou 2015). Furthermore, the fifth hypothesis predicted the influence of customer satisfaction on customer trust. According to prior studies (Choi & La 2013, Lee et al. 2015), customer satisfaction would have a significant effect on customer trust. This prior study is relevant with the finding of this current study. The path had a C.R. score of 0.431 showing H₆ was accepted.

The visitor comes to a barbershop and hands his head touched; his hair is cut, his whiskers and beard are cleansed. What would happen if the customers do not trust the barber he is visiting? The barbers who work at barbershop-barbershop are mostly young and low-educated. However, their skill in cutting hair cannot be doubted. Of course, this applies only to standard pieces that do not require complex and modern cutting techniques. However, as mentioned at the beginning of this paper, some young barbers will be happy to develop their skills to master the latest cutting techniques. Consequently, customers judge that dropping trust in these barbers is not something wrong. Therefore, it is proven that customer trust can affect customer loyalty. In this case, based on the structural model calculation, this path had a C.R. score of 0.285. This finding supports prior studies (Akbar & Parvez 2009, Amin et al. 2013, Bilgihan, 2016, Choi & La 2013, Martínez & del Bosque 2013, Nguyen et al. 2013, Rasheed & Abadi 2014, Veloutsou 2015).

4 CONCLUSION

This study aims to measure the impact of service quality, customer satisfaction, and customer trust on customer loyalty. This study shows a significant impact of service quality on customer satisfaction, customer trust, and customer loyalty. In addition, there is significant impact of customer satisfaction on customer trust and customer loyalty. There also is a significant impact of customer trust on customer loyalty. Theoretically, customer loyalty towards a barbershop can be related to customer loyalty in other settings, including restaurants, hotels, banks, and shops and online shops.

Looking at frequency of visit, most participants had visited the barbershop once or twice. The author considers one of the limitations of this study is lack of participants screening. As this study was planned to measure loyalty, only participants with more than three or more visits should be included. In other words, this study should employ revisit intention as the predicted variable instead of customer loyalty.

Noticing the two limitations mentioned earlier, future study is expected to test the theoretical framework by recruiting only participants who have visited barbershops three times or more. This approach can be expected obtaining a better result in measuring customer loyalty of a $1 barbershop. Besides, globally, there is a limited study focussing on barbershop customer behaviour. By exploring this, it would contribute much to the literature of this particular area.

REFERENCES

Akbar, M. M., & Parvez, N. 2009. Impact of service quality, trust, and customer satisfaction on customers loyalty. *ABAC Journal*, 29(1).

Alsaqre, O. E. (n.d.). The impact of physical environment factors in hotels on Arab customers' loyalty.

Amin, M., Isa, Z., & Fontaine, R. 2013. Islamic banks: Contrasting the drivers of customer satisfaction on image, trust, and loyalty of Muslim and non-Muslim customers in Malaysia. *International Journal of Bank Marketing*, 31(2), 79–97.

Auh, S. 2005. The effects of soft and hard service attributes on loyalty: the mediating role of trust. *Journal of Services Marketing*, 19(2), 80–92.

Bilgihan, A. 2016. Gen Y customer loyalty in online shopping: An integrated model of trust, user experience and branding. *Computers in Human Behavior*, 61, 103–113.

Chiou, J.-S., & Droge, C. 2006. Service quality, trust, specific asset investment, and expertise: Direct and indirect effects in a satisfaction-loyalty framework. *Journal of the academy of marketing science*, 34(4), 613–627.

Choi, B., & La, S. 2013. The impact of corporate social responsibility (CSR) and customer trust on the restoration of loyalty after service failure and recovery. *Journal of Services Marketing*, 27(3), 223–233.

Chou, P.-F., Lu, C.-S., & Chang, Y.-H. 2014. Effects of service quality and customer satisfaction on customer loyalty in high-speed rail services in Taiwan. *Transportmetrica A: Transport Science*, 10(10), 917–945.

Chow, H. W., Tan, H. L., Thiam, B. N., & Wong, P. J. 2012. *The effects of service quality, relational benefits, perceived value & customer satisfaction towards customer royalty in hair salon industry*. UTAR.

Hafeez, S., & Muhammad, B. 2012. The impact of service quality, customer satisfaction and loyalty programs on customer's loyalty: Evidence from banking sector of Pakistan. *International Journal of Business and Social Science*, 3(16).

Hair Jr., J. F., Black, W. C., Babin, B. J., Anderson, R. E., & Tatham, R. L. 2006. *Multivariate data analysis* (6 ed.). New Jersey: Prentice-Hall, Inc.

Han, H., & Ryu, K. 2009. The roles of the physical environment, price perception, and customer satisfaction in determining customer loyalty in the restaurant industry. *Journal of Hospitality & Tourism Research*, 33(4), 487–510.

Hazra, S. G., & Srivastava, K. B. L. 2009. Impact of service quality on customer loyalty, commitment and trust in the Indian banking sector. *IUP Journal of Marketing Management*, 8(3/4), 74.

Herdiana, I. 2015. Looking into the village of barbers in Garut. Retrieved from Merdeka.com website: https://www.merdeka.com/peristiwa/melongok-kampung-tukang-cukur-di-garut.html

Hidayat, R., Akhmad, S., & Machmud, M. 2015. Effects of service quality, customer trust and customer religious commitment on customers satisfaction and loyalty of Islamic banks in East Java. *Al-Iqtishad: Journal of Islamic Economics*, 7(2), 151–164.

Hu, L.-t., & Bentler, P. M. 1995. Evaluating model fit. In R. H. Hoyle (Ed.), *Structural equation modeling. Concepts, issues, and applications* (pp. 76–99). London: Sage.

Hu, L.-t., & Bentler, P. M. 1999. Cutoff criteria for fit indexes in covariance structure analysis: Conventional criteria versus new alternatives. *Structural Equation Modeling: A Multidisciplinary Journal*, 6(1), 1–55.

Khan, M. M., & Fasih, M. 2014a. Impact of service quality on customer satisfaction and customer loyalty: Evidence from banking sector. *Pakistan Journal of Commerce & Social Sciences*, 8(2).

Khan, M. M., & Fasih, M. 2014b. Impact of service quality on customer satisfaction and customer loyalty: Evidence from banking sector. *Pakistan Journal of Commerce and Social Sciences*, 8(2), 331–354.

Kheng, L. L., Mahamad, O., & Ramayah, T. 2010. The impact of service quality on customer loyalty: A study of banks in Penang, Malaysia. *International Journal of Marketing Studies*, 2(2), 57.

Lai, I. K. W. 2015. The roles of value, satisfaction, and commitment in the effect of service quality on customer loyalty in Hong Kong–style tea restaurants. *Cornell Hospitality Quarterly*, 56(1), 118–138.

Lee, D., Moon, J., Kim, Y. J., & Mun, Y. Y. 2015. Antecedents and consequences of mobile phone usability: Linking simplicity and interactivity to satisfaction, trust, and brand loyalty. *Information & Management*, 52(3), 295–304.

Lombart, C., & Louis, D. 2014. A study of the impact of Corporate Social Responsibility and price image on retailer personality and consumers' reactions (satisfaction, trust and loyalty to the retailer). *Journal of Retailing and Consumer services*, 21(4), 630–642.

Lustinayanti, W. E. (2011). *The influence of service quality and service value on customer satisfaction and behavioral intention of spa customers (in Bahasa Indonesia)*. (Bachelor), Universitas Udayana Denpasar.

Martínez, P., & del Bosque, I. R. 2013. CSR and customer loyalty: The roles of trust, customer identification with the company and satisfaction. *International Journal of Hospitality Management*, 35, 89–99.

Nguyen, N., Leclerc, A., & LeBlanc, G. 2013. The mediating role of customer trust on customer loyalty. *Journal of service science and management*, 6(01), 96.

Orel, F. D., & Kara, A. 2014. Supermarket self-checkout service quality, customer satisfaction, and loyalty: Empirical evidence from an emerging market. *Journal of Retailing and Consumer services*, *21*(2), 118–129.

Poku, K., Zakari, M., & Soali, A. 2013. Impact of service quality on customer loyalty in the hotel industry: An empirical study from Ghana. *International Review of Management and Business Research*, *2*(2), 600–609.

Rasheed, F. A., & Abadi, M. F. 2014. Impact of service quality, trust and perceived value on customer loyalty in Malaysia services industries. *Procedia-Social and Behavioral Sciences*, *164*, 298–304.

Rinanda, N. (2013). *Effect of service quality on customer loyalty: Case study on V Salon Jl. Osamaliki 78 Salatiga*. (Bachelor), Satya Wacana Christian University.

Saleem, H., & Raja, N. S. 2014. The impact of service quality on customer satisfaction, customer loyalty and brand image: Evidence from hotel industry of Pakistan. *Middle-East Journal of Scientific Research*, *19*(5), 706–711.

Sari, N. M. 2007. *The effect of service attributes on customer loyalty (in Bahasa Indonesia)*. (Bachelor), Universitas Indonesia, Depok.

Schermelleh-Engel, K., Moosbrugger, H., & Müller, H. 2003. Evaluating the fit of structural equation models: Tests of significance and descriptive goodness-of-fit measures. *Methods of Psychological Research Online*, *8* (2), 23–74.

Soekirno, S. 2017. The Asgar story maintains the tradition of barber Garut (in Bahasa Indonesia). Retrieved from Kompas.com website: http://regional.kompas.com/read/2017/01/17/09454341/cerita.para.asgar.merawat.tradisi.tukang.cukur.garut

Stathopoulou, A., & Balabanis, G. 2016. The effects of loyalty programs on customer satisfaction, trust, and loyalty toward high-and low-end fashion retailers. *Journal of Business Research*, *69*(12), 5801–5808.

Suhud, U., & Wibowo, A. 2016. Predicting customers' intention to revisit a vintage-concept restaurant. *Journal of Consumer Sciences*, *1*(2), 56–59.

Supriadin, J. 2017. Banyuresmi, the hometown of the Indonesia's barber school (in Bahasa Indonesia). [online] Retrieved from Kompas.com website: http://regional.liputan6.com/read/2968462/banyuresmi-kampung-sekolah-tukang-cukur-andal-indonesia

Tabachnick, B. G., & Fidell, L. S. 2007. *Using multivariate statistics* (5 ed.). Boston: Pearson/Allyn & Bacon.

Veloutsou, C. 2015. Brand evaluation, satisfaction and trust as predictors of brand loyalty: The mediator-moderator effect of brand relationships. *Journal of Consumer Marketing*, *32*(6), 405–421.

Advances in Business, Management and Entrepreneurship – Hurriyati et al (eds)
© 2020 Taylor & Francis Group, London, ISBN 978-0-367-27176-3

The influence of birth certificate application process service toward public satisfaction in the population and civil registration agency of Garut Regency Indonesia

I. Kania, D.T. Alamanda, N. Nurbudiwati & D.H. Fauzan
Universitas Garut, Garut, Indonesia

ABSTRACT: The sustainability of national development and public services must include a special attention to the population factor as the subject and object. A birth certificate is an identity form and an integral part of citizen civil and political rights. Unfortunately, the people in Garut still consider the service quality of the local Population and Civil Registration Agency is not yet effective in serving the society during birth certificate application process. Therefore, this study aims to find out the influence of service quality provided during the process of birth certificate application toward public satisfaction in the Population and Civil Registration Agency of Garut. Using a quantitative approach, data were collected through observations, interviews, literature studies and questionnaire distributions. Questionnaires were spread to 100 respondents using non probability sampling technique which is a purposive sampling. The result of this study indicates that there is an influence of service quality during the process of birth certificate application toward Public Satisfaction in the Population and Civil Registration Agency of Garut as big as 36%. The results of this study can be treated as a reflection of the performance of the Garut government in serving a birth certificate in Garut.

1 INTRODUCTION

The quality of service will create a strong relationship between customer perception and the service provider agency (Ramseook-Munhurrun & Lukea-Bhiwajee 2010). According to Ministerial Decree no. 25/M.PAN/2/2004, public satisfaction is measured by public opinion and judgment on the performance of public service administrators (Menteri 2004). Meanwhile, refer-ring to the Decree of the State Minister on Administrative reform Number: 63/KEP/M.PAN/7/2003 that every public service must have a standard and it must be published as a guarantee of certainty for the service users (Menteri 2003).

In the profile of local Population and Civil Registration Agency of Garut Regency in 2016, it is mentioned that the population data resource is from the population registration service through the Population Registration and Record unit of Garut Regency Population and civil registration agency which has been consolidated with the Ministry of Internal Affairs by integrating the PAIS Database (Population Administration Information System) functioning to improve the service effectiveness and validity of the resulted data. The good relationship between Public and the Civil Registration Office of Garut regency is important in understanding the expectations and needs of the society.

A birth certificate is an identity form of every child which is an integral part of citizen civil and political rights. The right for identity is a state recognition form of a person existence before the law, and each child is entitled to a name as an identity and

citizenship status. Gerber et al. (2017) stated that it is important to include the right to birth certificate.

Government apparatus, especially in the Population and Civil Registration Agency of Garut Regency as the service provider should provide the best service to public in order to fulfill the need for birth certificates which are valid for a lifetime and useful for dealing with various needs in society. Therefore, service becomes one of the most important things to be provided by the Population and Civil Registration Agency of Garut regency to influence the level of public satisfaction.

This is in line with the mission of the Population and Civil Registration Agency of Garut Regency to realize a professional and trusted population and civil registration service which is elaborated more in the goal to improve public services which can be accessed easily and precisely through the availability of quality IT-based services. In 2018, Garut Regency Government through the Agency of Population and Civil Registration has issued 1,500 free birth certificates for the residents of Kadungora Sub-district (Galamedianews.com). Head of Population and Civil Registration Office of Garut regency, Rina Siti Syabariah, appreciates the Head of Sub-district and the service staff at the Sub-District Office who have provided good service in the issuance of the documents.

The measures of public satisfaction assessment as a global concept are still considered limited and still need to be developed (Vreugdenhil & Rigby 2010). Salim et al. (2017) state how important community satisfaction to trust the government. Research on the index of public satisfaction for government service was conducted in Malang using IPA method (Hakim 2017).

The purpose of this study is to find how high the service quality provided to the society during birth certificate application process is. The result of this study is expected to become a consideration and input for the Population and Civil Registration Agency of Garut regency.

2 METHODOLOGY

This research adopts descriptive quantitative method. The survey was conducted through distribution of questionnaires to 100 respondents representing 42 sub districts in Garut regency. The population used are those who applied for the Birth Certificates in the Population and Civil Registration Agency of Garut Regency until the end of 2017 with a number of 544,081 people. The sampling technique used is a simple random cluster in which the number of respondents taken in each sub-district is the proportion of sampling using slovin. Sub-districts with few applicants were represented by 1 respondent, and sub-districts with the largest number of applicants were represented by 6 respondents.

Results of the survey through questionnaires are then processed with spearman ranks analysis technique for finding the magnitude of relationship between the service quality of Garut Regency Population and Civil Registration Agency and public satisfaction. Meanwhile, KD (magnitude of determinant coefficient) is used for finding out how high the service quality level of birth certificate application process can describe public satisfaction toward the Population and Civil Registration Agency of Garut Regency.

So, the proposed Hypotheses are:

H_0: The service quality of birth certificate application process does not explain public satisfaction in the Population and Civil Registration Agency of Garut Regency.

H_1: The service quality of birth certificate application process explains public satisfaction in the Population and Civil Registration Agency of Garut Regency.

3 RESULT AND DISCUSSION

To find out and measure the influence magnitude of variable X (service quality of birth certificate application process) toward variable Y (public satisfaction in the Population and Civil Registration Agency of Garut Regency) in this research, non-parametric statistic with Rank Spearman Cor-

relation Coefficient Test is used. From the calculation result, the following score is obtained:

$$\sum x^2 = 83,132 \text{ and } \sum y^2 = 82,955.5$$

Having obtained the values of $\sum x^2$ and $\sum y^2$, then the following formula is used for the *Rank Spearman* test:

$$r_s = \frac{\sum x^2 + \sum y^2 - \sum d_i^2}{2\sqrt{\sum x^2 \cdot \sum y^2}} \tag{1}$$

$$r_s = \frac{83.132 + 82.955,5 - 66.357,5}{2\sqrt{83.132 \times 82.955,5}}$$

$$r_s = \frac{99.730}{166.087,406}$$

$$r_s = 0.6$$

So the relationship magnitude between variables of X and Y is $r_s = 0.19$ and based on the interpretation guidance of correlation coefficient, it is categorized strong. Meanwhile, to find out how big the percentage of service quality variable of birth certificate application process can explain public satisfaction in the Population and Civil Registration Agency of Garut Regency is measured using Determinant Coefficient (KD) with the percentage as follows:

$$KD = r^2 \times 100\%$$

$$KD = (0,6)^2 \times 100\% \tag{2}$$

$$KD = 36\%$$

This means that the service quality of birth certificate application process explains public satisfaction of 36% in the Population and Civil Registration Agency of Garut regency and the remaining 64% is determined by other variables. The hypothesis proposed is "The service quality of birth certificate application process explains public satisfaction in the Population and Civil registration Agency of Garut regency".

Based on the above calculation with the error level of $\alpha = 0,1$ and the degree of freedom is (df) = n - 2 (100 - 2 = 98) and consulted with the t distribution table, then the value obtained is 1.661 because the t_{count} = 7.42 and the t_{table} = 1.661, the rule is $t_{count} \geq t_{table}$, hence the significance is 7.42 ≥ 1.661 = significant, meaning the existence of variable X (service quality of birth certificate application process) explains variable Y (public satisfaction in the Population and Civil Registration Agency of Garut Regency).

Therefore, from the calculation of *Rank Spearman* correlation coefficient, the following result is obtained:

$$Correlation(r_s) = 0.6$$

$$t_{count} = 7.42$$

$$t_{table} = 1.661$$

The above data of t_{count} and t_{table} are compared under the condition of:

a. If $t_{count} \leq t_{table}$, then H_0 is accepted and H_1 is rejected which means that variable X does not explain variable Y.
b. If $t_{count} \geq t_{table}$, then H_0 is rejected and H_1 is accepted which means variable X explains variable Y.

The results of the study reveal that the proposed hypotheses of H_0 is rejected and H_1 is accepted, because the t count ≥ t table, or 7.42 ≥ 1.661 with the correlation level of r_s = 0.6, the significant level of 10% or $\alpha = 0.1$ and the confidence level of 90%, therefore the service quality variable of birth certificate application process (X) significantly explains the variable of public satisfaction (Y) in the Population and Civil Registration Agency of Garut Regency.

Based on 16 statements of the questionnaires given to respondents, it can generally be said that the service of birth certificate application process is good and it can give satisfaction to society. Indicator with the highest percentage of responses is "the officers serve and respect every applicant" which is equal to 73.8% or within a good category. While, indicator with the lowest percentage of responses is "easy service process" which is equal to 62.8% or within good enough category. It is expected in the future that the employees at the Population and Civil Registration Agency of Garut Regency can simplify and improve the service to be better, especially in the process of birth certificate making. It can be done among others through improvement of the existing infrastructure such as to enlarge the service room and provide seats for queue.

Meanwhile for the variable of public satisfaction, the indicators with the highest percentage of responses are "the process of applying for a birth certificate is carried out efficiently", and "the result of the birth certificate application process service is in accordance with the standard operating procedures" respectively of 69.2% or in a good category. Meanwhile the indicator with the lowest percentage of responses is "the applicants consider that the service is well-delivered" at 57.6% and categorized as good enough. It is expected that the apparatus in the Population and Civil Registration Agency of Garut Regency can improve the service by adding the number of service officers to become 4 (four) people. The importance of improving the quality of providers is delivered by Schneider et.al (2017) to improve the accuracy of birth certificates.

The service of making birth certificates in every region in Indonesia is still very diverse. Haselman (2015) links the issue of ownership of birth certificate with ownership of ID Card and Family card of parents in Merauke, Indonesia. The same case happens in East Nusa Tenggara, West Nusa Tenggara, and West Java (Duff et al. 2016).

The problem of government services and system differences in dealing with birth certificates is not only happening in Indonesia, but also in India (Mohanty & Gebremedhin 2018), Zimbabwe (Chereni 2016) and UK (Crawshaw et al. 2016). As well as in Nigeria (Makinde et al. 2016, Isara & Atimati 2015) the problem was the communication between the providers of birth certificate with community. In Iraq, Bah (2014) found most of birth certificate were disappeared since the war in 2003. Unlike the case in developed countries such as the UK, the importance of birth certificates due to the many cases of donor conception (Blyth et al. 2009, Crawshaw et al. 2016), and adoption (Clapton 2014). In US, issues of birth certificates occur in coordination between hospitals with 52 Jurisdictions (Kim et al. 2015).

4 CONCLUSION

Based on the results of the hypothesis test, it can be concluded that the Service of Birth Certificate application process explains Public Satisfaction in the Population and Civil Registration Agency of Garut regency. The service in the birth certificate application process explains public satisfaction at 36%.

ACKNOWLEDGMENT

Garut University for the Grant of Internal Funding

REFERENCES

Bah, S. 2014. The Iraqi civil registration system and the test of political upheaval. *Canadian Studies in Population* 41(1-2): 111-119.
Blyth, E., Frith, L., Jones, C., & Speirs, J. M. 2009. The role of birth certifi cates in relation to access to biographical and genetic history in donor conception. *International Journal of Children's Rights* 17: 207-233.
Chereni, A. 2016. Underlying dynamics of child birth registration in Zimbabwe. *The International Journal of Children's Rights* 24(4): 741-763.
Clapton, G. 2014. The birth certificate, 'father unknown' and adoption. *Adoption & Fostering*.

Crawshaw, M. A., Blyth, E. D. & Feast, J. 2016. Can the UK's birth registration system better serve the interests of those born following collaborative assisted reproduction? *Reproductive Biomedicine & Society Online*.

Duff, P., Kusumaningrum, S. & Stark, L. 2016. Barriers to birth registration in Indonesia. *The Lancet Global Health* 4 (4): 234-235.

Galamedianews.com. 2018. Disdukcapil Garut bagikan akte kelahiran secara cuma-Cuma. Retrieved on January 28, 2018 from http://www.galamedianews.com/daerah/178600/disdukcapil-garut-bagikan-akte-kelahiran-secara-cumacuma.html.

Gerber, P., Gargett, A. & Castan, M. 2017. Does the right to birth registration include a right to a birth certificate? *Netherlands Quarterly of Human Rights*.

Hakim, A. N. 2017. Assessment of community satisfaction index of population and civil registration office in Malang municipal. *3rd International Conference of Planning in the Era of Uncertainty. 70*. IOP Publishing Ltd.

Haselman, S. 2015. The Implementation of issuance service of birth and death certificates indepartment of population and civil registrar of Merauke District. *International Journal of Scientific & Technology Research* 4(1): 120-124.

Isara, A. R. & Atimati, A. O. 2015. Socio-demographic determinants of birth registration among mothers in an urban community in southern Nigeria. *Journal of Medicine in the Tropics* 17(1): 16-21.

Kim, S. Y., Ahuja, S., Stampfel, C. & Williamson, D. 2015. Are birth certificate and hospital discharge linkages performed in 52 jurisdictions in the united states?. *Matern Child Health Journal* 19(12): 2615–2620.

Makinde, O. A., Olapeju, B., Ogbuoji, O. & Babalola, S. 2016. Trends in the completeness of birth registration In Nigeria: 2002-2010. *Demographic Research* 35(12): 315-338.

Menteri. 2003. Keputusan Menteri PAN Nomor 63/KEP/M.PAN/7/2003 tentang Pedoman Umum Penyelenggaraan Pelayanan. Kementerian Pendayagunaan Aparatur Negara republik Indonesia.

Menteri. 2004. Keputusan Menteri PAN Nomor: KEP/25/M.PAN/2/2004 tentang pedoman penyusunan indeks kepuasan masyarakat unit pelayanan instansi pemerintah. Kementerian Pendayagunaan Aparatur Negara republik Indonesia.

Mohanty, I., & Gebremedhin, T. A. 2018. Maternal autonomy and birth registration in India: Who gets counted? *Plos One*, 1-19.

Ramseook-Munhurrun, P. & Lukea-Bhiwajee, S. D. 2010. Service quality in public service. *International Journal of Management and Marketing Research* 3(1): 37-50.

Schneider, P., King, P. L., Finnegan, K., Goel, S., Grobman, W., Oh, E. & Borders, A. 2017. Optimizing accuracy of birth certificates through quality improvement. *American Journal of Obstetrics & Gynecology* 216(1): 254-255.

Salim, M., Peng, X., Almaktary, S. & Karmoshi, S. 2017. The impact of citizen satisfaction with government performance on public trust in the government: Empirical evidence from urban Yemen. *Open Journal of Business and Management*, 5(2): 348-366.

Vreugdenhil, A. & Rigby, K. 2010. Assessing generalized community satisfaction. *The Journal of Social Psychology* 127(4): 367-374.

Advances in Business, Management and Entrepreneurship – Hurriyati et al (eds)
© 2020 Taylor & Francis Group, London, ISBN 978-0-367-27176-3

The influence of the restaurant atmosphere on customer loyalty through a hedonic experience

T. Handriana & A.R. Meyscha
Universitas Airlangga, Surabaya, Indonesia

ABSTRACT: With the increasing income of the Indonesian people, lifestyle changes impact the functions of a restaurant, which is not only a place to eat and drink, but also a place to gather, socialize and release fatigue. Furthermore, many restaurants now offer more and more varied food, beverages and entertainment. Similarly, the competition among restaurants is becoming more challenging. This study aims to analyze the influence of the restaurant atmosphere on customer loyalty through a hedonic experience. The restaurant atmosphere consists of the design factor, ambient factor, and social factor. The sample in this study consisted of 130 visitors. The analysis technique used is Structural Equations Modeling (SEM) with AMOS software. The research finding showed that the design factor and social factor have a significant effect on the hedonic experience. Likewise hedonic experience has a significant effect on customer loyalty. Meanwhile, the ambient factor has no effect on the hedonic experience. The findings of this study contribute to the development of the concept of hedonic experience, as well as being beneficial to culinary practitioners

1 INTRODUCTION

The restaurant business is a promising business type because today many people experience a change in habits or lifestyle. They prefer a place that can provide various benefits, so that the restaurant has additional functions, from just a place for enjoying food and beverages to a place to gather and socializing with others and releasing fatigue from daily activities.

So far, the development of the restaurant business has been increasing year by year. In East Java, Indonesia, according to APKRINDO (Association of Cafe and Restaurant Entrepreneurs), throughout 2014 the growth of restaurants increased by 20%, and in 2015 the growth was around 12% and it is predicted that by 2016 the number of restaurants of various classes would grow by 15% (http://pressreader.com/indo nesia/jawa-pos/20160217/281741268477759).

To differentiate offerings beyond making products and delivering services, a company creates and manages customer experiences with their company (Kotler & Armstrong 2013). This can be done by applying to the store/restaurant as a brand strategy, which requires retailers to integrate the store atmosphere and product imagery to form a cohesive experience for their customers. The created store atmosphere should be unique and impressive, and consider all senses in creating a longer effect to form the customer experience (Pine & Gilmore 1998; Burt & Davies 2010). The atmosphere can enhance the shopping experience through environmental changes that influence the emotional response of customers (Gilbert 2003). The store atmosphere according to Baker (1986) consists of the elements design, ambience, and social factors.

The findings of a study conducted by Ballantine et al. (2010) show that ambient factors have a positive relationship to the hedonic experience. Research conducted by Muhammad et al. (2014) shows that the design factor and social factor have a positive relationship to the hedonic experience. In addition, the hedonic experience formed on the customer has a positive influence on loyalty. Customer loyalty is the key to business success. Thus, this study aims to analyze the influence of the restaurant atmosphere on customer loyalty through a hedonic experience. The restaurant business is a promising business type, because today many people experience a change in habit or lifestyle. They prefer a place that can provide various benefits, so that the restaurant has additional functions, from just a place for enjoying food and beverages to a place to gather for socializing with others and releasing fatigue from daily activities.

2 METHOD

2.1 Operational definition of research variables

The design factor is a consumer's perception of the visual element related to restaurant aesthetics. The ambient factor is a consumer's perception of environmental factors which include non-visual elements in the restaurant room. The social factor variable is defined as the operational perception of people in the restaurant environment. The hedonic experience variable is a consumer's perception in terms of seeking pleasure and emotion as the main motivation for consumption. The last variable of customer loyalty is the probability of repurchase, long-term choice, and recommendation of a product or service restaurant.

2.2 Research and measurement instruments

The instrument used in this study is a questionnaire. The Likert scale is applied to measuring each construct. The scale contains a five-level response to the proposed statement, namely: 1 = strongly disagree; 2 = disagree; 3 = neutral; 4 = agree; and 5 = strongly agree.

2.3 Sampling design and technique of analysis

Sample units in this study are diners who are at least 17 years or older. The number of samples in this research comprise as many as 130 people. In this study, the analytical techniques used are Structural Equations Modeling (SEM), with AMOS software.

3 RESULTS AND DISCUSSION

3.1 Testing of measurement model (measurement model fit)

Convergent validity test results show that the variables in this study meet the Exploratory Factor Analysis (EFA) criteria of data having a minimum loading factor of 0.5 and a Construct Reliability (CR) of data larger than 1.65. So it can be concluded that the data meet the convergence validity test. The discriminant validity test results show that the square root of AVE is higher than the correlation value among the variables. So it can be concluded that the data meet the discriminant validity test. The next consideration is the reliability test. The results of the reliability test on the variables show that the value of construct reliability in the variable design factor = 0.771; ambient factor = 0.680; social factor = 0.760; hedonic experience = 0.807; and customer loyalty = 0.849. Thus, it can be concluded that the constructs in this research are reliable.

3.2 Testing of structural model (structural model fit)

Based on the data processing in this study, the results show that there are two parameters of the goodness of fit that meet the cut-off value, as shown in Figure 1. Based on the principle of parsimony, if two or more of the parameters of the goodness of fit already meet the cut-off value, then the model is considered fit.

A further test was run to examine the relationship between latent variables in order to verify the research hypothesis. The results of the hypothesis test can be seen in Table 1.

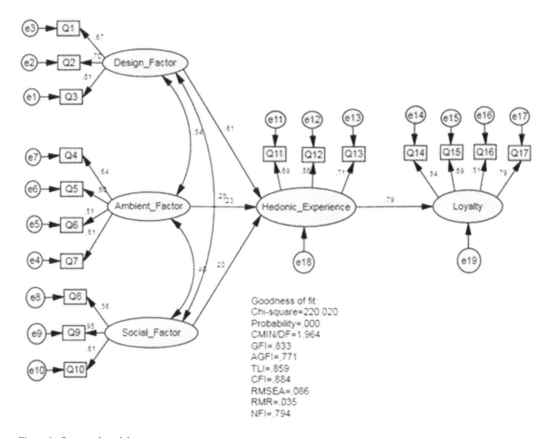

Figure 1. Structural model.

137

Table 1. The results of the hypothesis test.

Hypotheses	Path	Estimate	C.R.	P label	Remark
H1	Design Factor →Hedonic Experience	0.572	4.916	***	Significant
H2	Ambient Factor →Hedonic Experience	0.269	1.906	0.057	Not Significant
H3	Social Factor →Hedonic Experience	0.246	2.282	0.022	Significant
H4	Hedonic Experience → Customer loyalty	0.972	7.052	***	Significant

3.3 *Analysis*

3.3.1 *Design factor and hedonic experience*

The design factor becomes one of the significant variables of forming hedonic experiences in restaurant consumers. This result is in accordance with previous research by Jang and Namkung (2009), which states that the design factor becomes one of the stimuli in the customer's dining experience. Ballantine et al. (2010) also conclude similarly that the design factor is one of the most interesting cues in influencing the customer's hedonic experience. The research findings of Matthews et al. (2009), Slatten et al. (2009) and Muhammad et al. (2014) also show that design factors have a positive relationship with the hedonic experience. From the results of the questionnaire, it appears that the majority of respondents agree on the statement that the style of restaurant architecture attracts attention so that the consumer's interest in restaurant buildings can lead to a consumer's hedonic experience of restaurants.

Innovative interior design and decoration can be an important marketing tool in influencing consumer responses such as attitude, emotion, perception, value, satisfaction, and behavior (Berry & Wall 2007; Liu & Jang 2009; Ryu & Jang 2007). Color also has the power to attract consumers (Belizzi et al. 1983). Effective store design provides a useful shopping experience that encourages repeat visits and has long-term effects on building customer loyalty (Levy et al. 2012).

3.3.2 *Ambient factor and Hedonic experience*

The results of this study indicate that the ambient factor does not influence the hedonic experience. These findings support the research findings conducted by Muhammad et al. (2014). By contrast, the results of this study are in contrast to the research findings of Ballantine et.al. (2010), Milliman (1986) and Matthews et al. (2009) who state that ambient factors bind the meaning of the hedonic experience to the customer. A study conducted by Jang and Namkung (2009) finds that the ambient factor becomes an important stimulus in the customer's dining experience. The research hypothesis is not supported, probably due to demographic factors of restaurant consumers. Wysocki and Gilbert (1989) describe that women are more sensitive to certain odors and have a greater ability than men to identify aromas that can react to their emotional side.

The ambient factor is the characteristic of an intangible background (e.g. music, aroma, temperature) that tends to influence the nonvisual senses and have a subconscious effect on the customer. The played music influences the attitudes and behaviors of the consumer. Milliman (1982) tested the musical tempo and showed that background music tempo influences consumer behavior. The atmosphere factor plays an important role in dictating how customers show the meaning of their hedonic experience (Matthews et al. 2009). Kim et al. (2009) and Dutta et al. (2013) recognize that the atmosphere influences consumers' behavior and their perception of the restaurant. Research from Ballantine (2010) also finds that the ambient factor has a positive relationship to the hedonic experience.

3.3.3 *Influence of social factor on hedonic experience*

The result of the hypothesis test shows that the social factor has a positive and significant influence on the hedonic experience of restaurant customers. This result is in accordance with previous research conducted by Matthews (2009), Ballantine et al. (2010), and Muhammad et al. (2014) who stated that the social factor has a positive influence on hedonic experience. Thus it can be said that the social factor is an important factor in the hedonic experience (Nguyen et al. 2012). From the results of the questionnaires, it appears that most of the respondents agree with the statement that consumers assess the appearance of other restaurant consumers who look polite. This assessment can be a stimulus to create a hedonic experience in restaurant consumers.

The social factor refers to people (employees and customers) in a service setting. The social variable includes employee appearance, number of employees, employee gender, and clothing or physical appearance of other customers (Ryu & Han 2010). Employee appearance carries a hidden message that instills meaning through the language of the object (Ruesch & Kees 1956), thus helping customers form an assessment of the service before and after the meal is consumed.

3.3.4 Influence of the hedonic experience on customer loyalty

Based on the results of data processing, it is known that a hedonic experience has a positive and significant influence on customer loyalty of restaurant customers. This result is in accordance with previous research by Nowak and Newton (2006), Smilansky (2009), Cetin and Dincer (2014), and Muhammad et al. (2014) that the hedonic experience has a positive relationship with customer loyalty. In the questionnaire distribution results, more respondents agree on a statement about the willingness of consumers to recommend restaurant-related information to others. This means the hedonic experience experienced by consumers in restaurants can lead to positive word-of-mouth behavior.

Biedenbach and Marell (2010) explain that customer experience is directly proportional to brand loyalty. Meanwhile, Muhammad et al. (2014) state that the hedonic experience has a positive relationship with loyalty. Hollyoake (2009) concludes that a good customer experience is gained from an understanding of customer expectation, proper product acceptance at every opportunity, and any other factor that will generate loyalty. Similarly, Jones et al. (2006) argue that hedonic value is related to loyalty because customers develop a favorable attitude about experiences that provide psychological incentives such as customers who have a pleasant shopping experience.

4 CONCLUSIONS

The greatest positive influence was the design factor on the hedonic experience variable, so it is expected that restaurant managers pay more attention to matters related to the design factor, including: (1) the indoor environment can be added with books, up-to-date novels/magazines, so consumers can stay longer in the restaurant and further strengthen their hedonic experience; (2) in outdoor environments, decorations can be maximized to look more attractive. This can be achieved by increasing lighting and adding moving decorations to the interior of restaurant buildings to provide entertainment to restaurant customers.

As for future researchers, it is advisable to conduct a study with a qualitative approach to explore more detailed information related to restaurant customer loyalty. Future research is also recommended to focus on target market-based restaurants, such as target markets of families or young people. Future studies can also be based on the type of restaurant franchise/chain or independent restaurants.

REFERENCES

Baker, J. 1986. The role of the environment in marketing services: The consumer perspective. *The services challenge: Integrating for competitive advantage*, 1(1): 79–84.

Ballantine, P.W., Jack, R. & Parsons, A.G. 2010. Atmospheric cues and their effect on the hedonic retail experience. *International Journal of Retail and Distribution Management*, 38(8): 641–653.

Bellizzi, J.A., Crowley, A.E. & Hasty, R.W. 1983. The effects of color in store design. *Journal of retailing*.

Berry, L.L. & Wall, E.A. 2007. The combined effects of the physical environment and employee behavior on customer perception of restaurant service quality. *Journal Cornell Hotel and Restaurant Administration Quarterly*, 48(1): 59–69.

Biedenbach, G. & Marell, A. 2010. The impact of customer experience on brand equity in a B2B services setting. *Journal of Brand Management*, 17(6): 446–458.

Burt, S. & Davies, K. 2010. From the retail brand to the retailer as a brand: Themes and issues in retail branding research, *International Journal of Retail and Distribution Management*, 38(11/12): 865–8878.

Cetin, G. & Dincer, F. I. 2014. Influence of customer experience on loyalty and word-of-mouth in hospitality operations. *Anatolia: An International Journal of Tourism and Hospitality Research*, 25(2): 181–194.

Dutta, K., Parsa, H.G., Parsa, A.R. & Bujisic, M. 2013. Change in consumer patronage and willingness to pay at different levels of service attributes in restaurants: a study from India. *Journal of Quality Assurance in Hospitality and Tourism*, 15(2): 149–174.

Gilbert, D. 2003. *Retail Marketing Management*. 2nd ed. Pearson Education Limited.

Hollyoake, M. 2009. The four pillars: Developing a 'bonded'business-to-business customer experience. *Journal of Database Marketing & Customer Strategy Management*, 16(2): 132–158.

Jang, S. & Namkung, Y. 2009. Perceived quality, emotions, and behavioral intentions: application of an extended Mehrabian-Russell model to restaurants. *Journal of Business Research*, 62: 451–460.

Jones, M.A., Reynolds, K.E., & Arnold, M. 2006. Hedonic and utilitarian shopping value: Investigating differential effects on retail outcomes. *Journal of Business Research*, 59(9): 974–981.

Kim, W.G.K., Ng, C.Y.N. & Kim. Y. 2009. Influence of institutional DINESERV on customer satisfaction, return intention, and word-of-mouth. *International Journal of Hospitality Management*, 28(1): 10–17.

Kotler, P. & Armstrong G. 2013. *Principles of Marketing 15th ed*. Pearson Education.

Levy, M. Weitz, B. A. & Grewal, D. 2012. *Retailing Management* 8th ed. McGraw-Hill.

Liu, Y. & Jang, S. 2009. The effects of dining atmospherics: an extended Mehrabian-Russell model. *International Journal of Hospitality Management*, 28(4): 494–503.

Matthews, S.J., Mark A.B. & David S. 2009. Atmospherics and consumers' symbolic interpretations of hedonic services. *International Journal of Culture, Tourism and Hospitality Research*, 3(3): 193–210.

Milliman, R.E. 1986. The influence of background music on the behavior of restaurant patrons. *Journal of consumer research*, 13(2): 286–289.

Muhammad, N.S., Musa R. & Ali, N.S. 2014. Unleashing the effect of store atmospherics on hedonic experience and customer loyalty. *Journal of Social and Behavioral Sciences*, 130: 469–478.

Nguyen, D.T., DeWitt, T. & Russell-Bennet, R. 2012. Service convenience and social servicescape: retail vs.

hedonic setting. *Journal of Services Marketing*, 26(4): 265–277.

Nowak, L.I. & Newton, S.K. 2006. Using the tasting room experience to create loyal customers. *International Journal of Wine Marketing*, 18(3): 157–165.

Pine, B.J. & Gilmore, J.H. 1998. Welcome to the experience economy. *Harvard Business Review*, 76: 97–105.

Ruesch, J. & Kees, W. 1956. *Nonverbal communication*. Univ of California Press.

Ryu, K. & Han, H. 2010. Influence of the quality of food, service, and physical environment on customer satisfaction and behavioral intention in quick-casual restaurants: Moderating role of perceived price. *Journal of Hospitality & Tourism Research*, 34(3): 310–329.

Ryu, K. & Jang, S. 2007. The effect of environmental perceptions on behavioral intentions through emotions: the case of upscale restaurants. *Journal of Hospitality and Tourism Research*, 31(1): 56–72.

Slåtten, T., Mehmetoglu, M. Svensson, G. & Sværi, S. 2009. Atmospheric experiences that emotionally touch customers: a case study from a winter park. *Managing Service Quality: An International Journal*, 19 (6): 721–746.

Smilansky, S. 2009. *Experimental marketing: A practical guide to interactive brand experience*. Kogan Page Publisher.

Wysocki, C.J. & Gilbert, A.N. 1989. National Geographic Smell Survey: effects of age are heterogenous. *Annals of the New York Academy of Sciences*, 561(1): 12–28.

Advances in Business, Management and Entrepreneurship – Hurriyati et al (eds)
© 2020 Taylor & Francis Group, London, ISBN 978-0-367-27176-3

The dark side of life insurance in achieving sales targets

A. Nirmala & G.C. Premananto
Universitas Airlangga, Surabaya, Indonesia

ABSTRACT: In this study, the researchers wanted to know more about what unethical things marketers did, and also what factors make marketers unethical, and to what extent supervisors knows about unethical actions performed by insurance marketers. The research method used was a qualitative approach with a snowball sampling technique. Thematic analysis was used as the basis for qualitative research analysis with an in-depth interview. The findings show that there are unethical actions conducted by marketers, including misrepresentation, premium payment, churning, pooling, discrediting competitors, discount/indocument, premium deposit delivery, account opening and policy insurance without customers knowledge, falsifying customers data information, misappropriation of customers fund, "suicide," faking signature in form that forgot, and forging customer's signature. The unethical actions performed by these marketers can be categorized into several indicators, namely, the applicable special regulatory indicators, legal indicators, cultural indicators and ethical indicators. In addition, the factors affected marketers to perform such unethical actions are due to factors of need for money, obligations to meet sales targets and work environment culture.

1 INTRODUCTION

Carrigan et al. (2005) recognizes that the need and value for firms to develop strategy is based on long-term relationship orientation with stakeholders and with ethical principles as the main foundation (McDevitt et al. 2007). Such recognition has led some marketing experts to devote particular interest to factors that can influence ethical behavior, especially to salespeople (Chonko et al. 1996). Whereas salespeople are generally responsible for generating profits, and always facing short-term pressures, they may also experience ethical dilemmas (Dubinsky et al. 1980). In the face of such ethical dilemmas, the seller may engage in many unethical practices such as lying or exaggerating the merits of a product, unfairly criticizing competitors, bribing, and falsifying information (Ferrell & Gresham 1985, Dubinsky et al. 1991, Lagace et al. 1991, and Román & Munuera 2005). Obviously, it tends to seek and exploit the various weaknesses that exist in the procedures as well as the lack of supervision.

In the pharmaceutical industry, for example, medical representatives (MR) provide misleading and incomplete information, provide incentives, make disparaging remarks about competitors and their competitors' products, falsify daily reports and misuse samples (Skhandrani & Sghaier 2016). With the increasing number of ethics-related issues faced by companies, researchers want to examine ethical issues or unethical actions in one of the emerging financial services industries that also require salespeople in marketing their products, that is, the life insurance services industry. Currently the insurance is increasingly recognized by the community both local and international insurance companies. In contrast to the 1990s, insurance is still underestimated. Now insurance has become the needs of the community with the various services offered (Kuntadi 2013).

Based on the results of early in-depth interviews with marketers that work in a life insurance company, it is not as easy as imagined which ultimately resulted in them doing various ways to get consumers, such as manipulating customer data information and falsified signatures. This is certainly very detrimental to consumers. Not only in Indonesia, at the 2000 Malaysian Institute of Insurance Lecture at that time, Assistant Governor of Bank Negara Malaysia Datuk Awang Adek Husin reaffirmed that Bank Negara Malaysia has received complaints from the public against insurance companies and health-related entities that have worked with him, improper sales practices and claims settlement (Haron et al. 2011). This means that even in Malaysia unethical sales practices still occur.

If doing all of these things benefit the salesman and harm the company and the consumer, this can be fatal, because in marketing or selling its products, a salesman must meet ethical standards. If an insurance salesman is doing unethical things, then it becomes an ethical issue that needs to be noticed because it has violated rules set by the company and the Indonesian Life Insurance Association (AAJI) on Practice Standards and Code of Ethics of Life Insurance Marketers. 03/AAJI/RAT/2012.

Based on this, problems that can be formulated include: What unethical things did the salesman do in achieving the sales target set by the company? To what extent do SPV or company managers know the ways in which salespeople do to achieve sales targets? How is the unethical link in achieving sales targets conducted by salesmen with business ethics? What are the factors affecting salesmen in doing unethical in achieving sales targets?

Business ethics is a way to conduct business activities, covering all aspects related to individuals, companies and society. Velasquez (2005) says that business ethics is a study devoted to true and false morals. Here, business ethics is an area of application of general moral principles to the area of human action in the economic field, especially business. So, in essence, the target of business ethics is the moral behavior of businessmen with economic activity. According to Ernawan (2011) In general, the principles used in business will never be separated from our daily lives.

2 METHOD

The research method that will be used is qualitative approach research method. In this study, researchers used the term informants who will be interviewed in depth with regard to issues to be researched and discussed in this study. Informants in this study were selected and determined by certain considerations that have been determined by the researchers or commonly called purposive sampling. Criteria of informants used by researchers are:

- Marketers who work in life insurance companies.
- Marketers who have worked for at least 2 years.
- Marketers who market their products directly to individual customers.
- Marketers who have known unethical sales.

The use of this technique will cease if the data obtained is considered saturation point.

3 RESULT AND DISCUSSION

To obtain more in-depth information, this study examined eight marketer informants and one supervisor informant in accordance with the criteria of research informants.

The results of data analysis obtained through inter-views show some unethical actions performed by salesmen in achieving targets set by the company.

The unethical actions undertaken by the salesman not only fall into one indicator, but an unethical act can fall into two indicators.

It can be seen in Table 1, first in the applicable special regulatory indicators there are six category unethical actions by the salesman. The specific regulations that apply are the regulations of the life insurance industry, the Indonesian Life Insurance Association (AAJI) on Practice Standards and the Code of Ethics of Life Insurance Marketers. 03/AAJI/2012. Where in the regulation there are administrative sanctions violating the code of ethics that should be able to minimize unethical acts committed by the salesman?

Table 1. Unethical actions of salesmen.

Indicator	Unethical Actions
1 Indicators in the Applicable Special Rules	*Misrepresentation*
	Premium Payments
	Churning
	Pooling
	Disparaging competitors
	Discount/Rebate/Indocument
	Suspension Premium Deposit
	Falsifying Customer Data Information
	Misuse of customer funds
	Signed Form blank
2 Legal Indicators	Account Opening and Insurance Policy without Customer's Knowledge
	Falsifying Customer Data Information
	Misuse of customer funds
	Signed Form blank
3 Cultural Indicators	Immoral
4 Indicators of ethics of each businessperson	"Suicide"

Second, there is also a legal indicator that becomes an indicator of unethical action, where the salesman opened the savings account and insurance policy without the knowledge of the customer. This is very detrimental to customers and contrary to the regulations made by the government. Referring to Law no. 7 of 1992 concerning Banking as amended by Act no. 10 of 1998 Article 49 and Law no. 40 Year 2014 About Insurance Article 78.

By contrast, if the customer signs the blank form and the salesman fills the SPAJ form themselves, there are often writing mistakes or customer signature deficiencies, so the salesman falsified the customer's signature. This has become a habit that has been rooted in the salesman, so that salesmen always take it easy by falsifying the signature of customers both with the knowledge of customers and without the knowledge of customers.

Signature forgery is prohibited by the government according to Criminal Code Article 263 on Counterfeiting of Documents and Law No. 2 of 1992 on Insurance Business Article 21 paragraph (5).

When the salesman fills out his own SPAJ form, frequent changes to customer data information deviate from the actual conditions for SPAJ submission are accepted by the company. These changes can later be categorized in the form of data information forgery. Falsifying customer data information violates the provisions of Government Administration and Population Act no. 23 of 2006; KUHD Article 251; Criminal Code Article 372 and Law no. 02 Year 1992 About Insurance Business. In addition to unethical things violating the law mentioned earlier, there are also illegal acts of embezzlement of funds that violate the Criminal Code (Criminal Code) Article 372 on embezzlement that sets a maximum imprisonment of four years, more specifically the Act No. 2 of 1992 concerning Insurance Business has also regulated related premium embezzlement in Article 21 paragraph 2, which mentions the threat of imprisonment for a maximum of 15 (fifteen) years and a maximum fine of IDR. 2,500,000,000 (two billion five hundred million rupiah).

Third, to provide maximum services to customers, sometimes salesmen provide deviant services to customers such as establishing a special relationship with customers, this is very violate the cultural norms of Indonesia, in addition there will certainly be some parties who are harmed by this deviant act.

The last unethical act can be categorized into the ethical indicator of each businessperson that the sales-man's habit of "suicide" to close the sales target, in this case even the salesman ask for help from relatives to apply for the policy where the funds originated from the salesman funds for payment a one-time premium. So the insurance policy does not continue for the next month. This has become a habit of salesmen who can harm the company.

Salesman who are only tempted by short-term profits do not understand the importance of business ethics. Salesman make the target of the company as a reason to do things that are not ethical so as to justify all kinds of ways to pursue profit. In addition, unethical acts done by the salesman due to weak supervision of the leader or supervisor, where the leader or supervisor turn a blind eye to what the salesman as long as it can meet the target company.

As a result, they often ignore the values of business ethics. As found in this study, salesmen perform various unethical and even unlawful ways to gain personal gain and fulfill their obligations to the company.

This finding is also in accordance with the findings of research conducted by Haron et al. (2011) in examining the extent to which unethical intentions of insurance agents in Malaysia. The study found that the influence of supervision contributed to agency intentions in achieving sales targets, due to lack of firmness from supervisors in performing unethical actions. To achieve a higher level of performance, there is a tendency to do unethical things. Supervisor performance correlates with agency performance. When an agent can achieve the sales target desired by the company, supervisors also benefit greatly from this. From these explanations, it can be described what ethical actions are included in legal or illegal actions in Figure 1.

It can be seen from Figure 1 that there are actions that violate the code of ethics, but legally it is still said to be legitimate, suicide, churning, funding for premium payment, pooling, customer relationships and delays in depositing premiums. There is no written regulation such as the state legislation

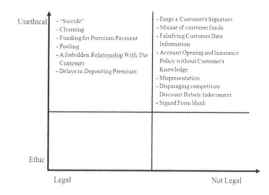

Figure 1. Category of unethical behavior.

to regulate it. Other unethical acts violate the law because it has been written in the legislation of the state.

This research can be associated with the concept of deontology, meaning duty or obligation. According to Fahmi (2014), ethics deontology emphasizes the human obligation to act well. Such an action is neither judged nor justified on the basis of the good effect or purpose of the action, but on the basis of the action itself as good in itself. In the business world if the obligations imposed on a person then concerned is worth to do, especially if he does not want to disappoint the consumer, because consumers always want satisfaction when dealing with a product.

This study found unethical actions conducted by salesmen to achieve the target company, which in the long term can harm customers and companies. These unethical acts are carried out because of the needs and obligations and pressures of the company. Therefore, according to the deontology theory, the things that are done by this salesman can be seen from two points of view. Based on the point of view of the need and the obligation of unethical things done by the salesman to meet the company's target is the obligation that must be fulfilled. So, it is considered correct to do, but if seen from the standpoint of customers and companies, unethical things done by the salesman is something bad. Because it is contrary to the code of ethics and laws and regulations applicable in this country.

It can be concluded that unethical acts committed by salesmen are contrary to business principles and are not in accordance with ethical principles in general. One example of the contradiction between unethical actions and business principles is the principle of mutual benefit, unethical acts committed by salesmen tend to harm one party (customer or company) where the profit is only obtained by the salesman in the form of fulfillment of sales target. Eventually, the salesman will get the benefit of a commission, while the other party (the company and customers) lose out in the short or long term.

The unethical actions carried out by the salesman were also found in Cooper and Frank (1991), who stated that there are 6 (six) unethical actions. In the study, there are similarities with this research that are misleading or misleading presentation of the product, misrepresentation or hide limitations in service and also underestimate competitors.

While in this study, researchers found 9 (nine) new findings about unethical actions made by salesmen in achieving the target company, namely premium payments, pooling, delayed deposit premium, ac-count opening and insurance policies without the knowledge of customers, misuse of customer funds, "suicide," blank form signatures, falsifying customer's signatures, and immoral deeds in relationships with customers.

4 CONCLUSION

After passing a series of analysis process and discussion it can be drawn conclusion as follows:

In performing the obligations as an employee, the salesman justifies any means to achieve the company's target. Unethical actions are carried out by the salesman without thinking of long-term losses for customers and companies. The unethical actions performed by the salesman can be categorized into several indicators: the applicable special regulatory indicators, legal indicators, and cultural indicators. Of the three indicators are explained that the things done by the salesman contrary to business ethics. The unethical actions carried out by the salesman also conflict with business principles that are inconsistent with ethical principles in general.

The factor of consideration of the salesman to per-form unethical actions is in addition to the obligation as an employee in achieving the target company also the desire to gain profit from the sale of insurance policies that is the commission. In addition, unethical acts done by the salesman due to weak supervision of the leader or supervisor, where the leader or supervisor turn a blind eye to what the salesman as long as it can meet the target company.

ACKNOWLEDGEMENT

Thank you to the Faculty of Economics and Business Universitas Airlangga for the research grant.

REFERENCES

Carrigan, M., Marinova, S. & Szmigin, I. 2005. Ethics and international marketing: research background and challenges. *International Marketing Review* 22(5): 481–493.

Chonko, L.B., Tanner, J.F. & Weeks, W.A. 1996. Ethics in salesperson decision making: a synthesis of research approaches and an extension of the scenario method. *Journal of Personal Selling and Sales Management* 16 (1): 35–52.

Cooper, R.W. & Frank, G.L. 1991. Ethics in the Life Insurance Industry: The issues, helps and hindrances. *Journal of the American Society of CLU & ChFC* 45: 54–66.

Dubinsky, A.J., Berkowitz, E.N. & Rudelius, W. 1980. Ethical problems of field sales personnel. *MSU Business Topics* 28(3): 11–16.

Dubinsky, A.J., Jolson, M.A., Kotabe, M. & Lim, C.V. 1991. A cross-cultural investigation of industrial salespeople's ethical perceptions. Journal of International Business Studies 22(4): 651–670.

Ernawan. E. 2010. *Etika Bisnis*. Jakarta: Alfabeta.

Ferrell, O.C. and Gresham, L.G. 1985. A contingency framework for understanding ethical decision making in marketing. *Journal of Marketing* 49 (3): 87–96.

Haron, H., Ismail, I., & Razak, S. H. A. 2011. Factors influencing unethical behavior of insurance agents. *International Journal of Business and Social Science*, 2(1).

Kuntadi. 2013. Kesadaran Berasuransi Masyarakat Kian Tinggi. Sindo News. [Online]. http://www.sindonews.com. Accessed: May 13, 2013.

Lagace, R.R. 1991. An exploratory study of reciprocal trust between sales managers and sales-persons. *Journal of Personal Selling & Sales Management* 11(2): 49–58.

McDevitt, R., Giapponi, C. & Tromley, C. 2007. A model of ethical decision making: the integration of process and content. *Journal of Business Ethics* 73(2): 219–229.

Román, S. & Munuera, J.L. 2005. An exploratory study of reciprocal trust between sales managers and sales-persons. *European Journal of Marketing* 39(5/6): 49–58.

Skhandrani, H. & Malek, S. 2016. The dark side of pharmaceutical industry. *Marketing Intelligence & Planning* 34(7): 905–926.

Velasquesz, M.G. 2005. *Etika Bisnis, Konsep dan Kasus*, *Fifth*Edition. Yogyakarta: ANDI.

Advances in Business, Management and Entrepreneurship – Hurriyati et al (eds)
© 2020 Taylor & Francis Group, London, ISBN 978-0-367-27176-3

Exploration of Ludruk as potential icon in Indonesia show business for the millennial generation

G.C. Premananto & M. Ikhwan
Airlangga University, Surabaya, Indonesia

ABSTRACT: The purpose of the research is to make Ludruk as a traditional cultural art become iconic again in Indonesia. A qualitative approach was conducted. The result of the study showed the inability of managers and artists to produce and features Ludruk products art that meet consumer expectation caused by reluctance of Ludruk artists itself to develop and improve existing art products. Paradoxes of creativity identified in this research. The support of government and CSR from private companies are needed to help them.

1 INTRODUCTION

The next era, to follow that of the era of agriculture, then the industrial era, and the information age, is the creative era. This is evidenced by the high intensity of countries such as Britain, the United States, and China to make the creative industries as one of the supporters of their economic growth (Kemenperin.go.id 2016). The development of creative industries in Indonesia is still relatively good, although it still lags behind other Asian countries such as South Korea and Japan (Kurnia 2015). The creative industry has several sub-sectors, and one of the sub-sectors of the creative industry is the performing arts.

The audience or visitors of the performing arts vary widely, especially regarding age of the audience. The percentage of spectators or visitors of performing arts by age is as follows: 12–25 years old has 42% attendance, 26–45 years old has 38%, and audiences over the age of 45 hold a 20% rate (Rejeki 2012). Millennials are therefore the main target of artistic performances.

Ironically from all types of performing arts that are present and displayed, performing arts with traditional values are even considered ancient. For example, it was found out that the Kecak dance from Bali, which required 150 personnel per performance was lower than the band performance which required only 3 to 6 personnel in each performance (Rejeki 2012).

One of the traditionally based business shows in East Java is Ludruk. Ludruk which has been declared since 1907 is currently in a state of concern. In this period ludruk already abandoned the audience, especially after the emergence of private television in the country (Hurek 2015).

Some researchers have expressed the importance of marketing management in the marketing of art products (Botti 2000, Colbert 2003, Evrard & Colbert 2000, Kerrigan et al. 2009, Schroeder 2006). But there are some things that can distinguish the

two. Butler (2000) states that the art of show is a marketing context that has its own uniqueness, often art products can not be subject to consumer desires. Marketers can not simply apply consumer orientation, it can even be the producers who push the product to consumers for consumption. Yet many other marketing academics assume that the science of marketing management can be easily transferred to any product no exception art performances. So research relating to the understanding of the art show industry becomes very interesting and important to do.

Based on these matters, it is interesting to explore the obstacles and marketing opportunities of ludruk art performances that later can make a ludruk art show become a worthy show business selling and become a mainstay art show in Surabaya, Indonesia. These interesting questions were explored via qualitative research with a case study approach conducted either by observation or by in-depth interview.

As for the characteristics of the implementation process, to some interesting discussions, namely the definition, development, and delivery of value. The process begins with an understanding of the process of determining the value to be built, searching for novelty effects, socializing and developing artists and audiences, determining access and pricing.

2 METHOD

This study is very suitable and appropriate if done using qualitative methods, based on the goals to be achieved by researchers.

a. Observation of participants. The researcher in this case becomes an "insider" who also simultaneously or at the same time as "outsider." The result of observation is field notes and other forms of documentation. Researchers are also expected to provide the context of the situation conditions to improve understanding of the data taken.

b. Interview. This research is the object of his research is the ludruk humanist in Surabaya like a humanist. In this research the object of research is the cultural Ludruk Irama Budaya, and also in the next generation ludruk group that is Ludruk Luntas.

In addition to the culture also conducted interviews to consumers and potential consumers ludruk millennial generation. In accordance with the data mentioned in the background of this study, audiences or young consumers of the 12–25 year age range are selected as one of the objects to be studied.

Informants in this research are ludruk managers, directors, ludruk players, gamelan players and people involved in ludruk artists' performances.

To the players are asked to continue to be motivated in the world ludruk, how to continue to survive and thrive. What keeps them in the industry. How they expect ludruk to persist and continue by the next generation.

For interviews and formal FGDs conducted to teenagers as potential markets as well as to domestic and foreign tourists.

3 RESULT AND DISCUSSION

Art can also be classified as a product that must be marketed in the form of ideas, as the performing arts can be regarded as a product of a show event (Botti 2000, Colbert 2003, Evrard & Colbert 2000, Kerrigan et al. 2009, Schroeder 2006). But art products are often not offered by following the tastes of the market, or not based on consumer orientation (Dholakia et al. 2013). Producers have their own idealism that is offered specifically to its customers. But nevertheless art products and performance art still have to be marketed and follow the concept of "describe, price, package, enhancement and deliver" (Kotler & Scheff 1997). Schroeder (2006), states that art and commerce are fast friends.

More specifically, Butler (2000), whose research is also supported by Lange (2009), shows the business of performing arts has its own characteristics divided into two groups, namely structural characteristics and process characteristics. Structural characteristics consist of product, organizational and market specificity. The product discusses the existence of the cultural domain, the appearance of the person, as well as the cultural location. The organization deals with artist management, the contrast between commercial value and cultural value, as well as network management. For market conditions, problems arise, consumer diversity, critics' criticism.

This research divides the analysis in two parts, namely: the internal side of ludruk groups, that is, players, directors and managers (player/supply perspective), and the external side, that is, consumers and potential consumer performances Ludruk (perspective consumer/demand).

3.1 *Perspective of player*

Based on the results of observations and interviews conducted by researchers during this study, found some things that become the most dominant cause of art performances Ludruk be unpopular or abandoned by consumers or audience.

a. Hidden shows venue
"*Well if this time the place is fargoes deep into, how people will know if the place is here*" (Cak Sapari, senior Ludruk player).

"*The location of the building although in the middle of town, but go inside*" (Deden Irawan, manager Ludruk).

Statement about inappropriate location and difficult to access supported by result of observation done by researcher, that access to go to place of performances that is THR is very difficult and very minimal direction directed to Ludruk staging location, also there is no adequate lighting along road which leads to the location of art performances Ludruk it.

b. The venue does not provide comfort
The state-of-the-art performance building that is physically building looks obsolete and does not give the impression that the building is a building that is used as a place of art performances ludruk, facilities that exist in the building such as seats spectators, toilets, and atmosphere of the room is less artistic and not maintained clean further adding to the impression that the building is not maintained. This is reinforced by informant statements:

"*What else is the building right! The young people wouldn't go here if the building is dirty and not appropriate. I several times asking my friendsto watch and finally they come out because of heat, smell and other things*"(Ari Setiawan, junior player Ludruk).

c. Paradox of creativity
Ludruk performances require high improvisation from the players, given the art of this show does not provide detailed scenarios and dialogue between players. The players are only given a description of the theme of the story. It takes creativity from the players. But on the way the artists perform on stage only rely on their experience of stories that they had memorized before. This raises a paradox in building creativity. Building creativity on every show week is not easy. If they are asked to make different and new things, senior ludruk players have difficulty running it.

"*No scripts, no scripts. The script is only written on the board, the road, ADG (scene) 1, 2, 3 so wrote. But when he performs they have the dialogues*" (CakHengky, cultural, observer and coach Ludruk East Java).

"*Nowadays it has a lot of memorization, yes, jokes are used in only particular, because it is commonly played story and the scene was so memorized*" (CakSapari Senior Player).

Senior peludruk even have experienced pessimistic conditions against the future ludruk and feel that ludruk in time will experience the faded.

"That conventional Ludruk more than it. Yes so may be fixated on some stories. Who often brought. Not complicated term. So there is less willingness to dig things new. If ludruk the senior is hard to work, because it is hard to learn, less innovation, then most of them. Say it's okay. Ludruk time to die. Will not exist, they just wait dead, not want to revive"(Ari Setiawan, junior young Ludruk artist).

d. Professionalism

Another problem found by researchers from the observations and interviews is the availability of players at the Ludruk performance will be played, often at the Ludruk art performance will be played or when the performance schedule of artists or Ludruk players that there is not in accordance with the needs of the performances or the number of players and the existing role is not satisfied.

As a result of the lack of players or artists ludruk that often happens this, then Ludruk art performances are played at that time does not work in accordance with the way the story. To overcome the problem of shortage of players then the manager and the artists ludruk choose to show or bring a story that has often been done and already memorized by most artists Ludruk that exist at that time, so the performances displayed seem less interesting and entertaining. The statement is supported by a statement from the informant at the time of interview conducted by the researcher that is as follows:

"Lha who came to play anyone not certainly mas, sometimes who came a lot but often also come a little" (CakSapari, senior player).

e. Old equipment

Another problem that became one of the causes of art performance ludruk abandoned or not interested by consumers or spectators based on observations and interviews conducted by researchers at the time of the research is supporting equipment when performances of art Ludruk done, the supporting equipment referred to here is a costume or wardrobe, good quality musical instruments, screen background or screen, sound system, lighting and others that support ludruk art performance when performed on the consumer or audience. The statement is supported by the statement submitted by the informant at the time of the interview as follows:

"Kendang and other tools have been in use a long time or from the first. But if you say according to my opinion is less appropriate. Used to have a new but even sold" (CakSapari player's kendang and also the actor/artist Ludruk old).

Beyond the existing problems, there is also high expectation of the future Ludruk. The statement about the potential possessed by ludruk art performance to become a show business that is in demand by consumers is supported by a statement from the informant as follows:

"It Possible to exist with Cak Kartolo. Play on Ismail Marzuki with two theme Sarip and Joko Sambang. Ticket costing six hundred thousand. Taman Ismail Marzuki building so much, exhausted! And later in May I will play in Taman Ismail Marzuki" (CakHengky, Cultural, Observer and Builder ludruk East Java).

Table 1. Perspective of player: The constraints.

Constraints		
Place	Location	• Less strategic location
		• Difficult to find and minimal instructions to the location.
		• Image location as a place for the middle to lower
	Building	• Buildings that are not maintained
		• Unprecedented building hygiene
		• The convenience of the building is less in note
Professionality	Personnels	• Number of personnel who do not fit the needs
		• Ludruk only sideline
	Capability	• Personnel or artists who have no artistic expertise
	Schedule	• Absence of timeliness of performance
		• Cancellation of performance conducted
Supporting Equipment	Musical Instruments	• Outdated musical instruments
		• The quality is not appropriate
	Lighting	• The lighting is poor
	Stages	• Stage is not well-arranged
		• Screen background or screen is already been not good enough
	Sound system	• Sound system that does not support or does not meet the needs
Creativity Paradox	Script	• The performances that should favor creativity in improvisation become monotonous, especially from the senior group

3.2 Perspectives of consumers

The consumer's perspective on the perceived problems of players, directors and managers is more or less mutually confirmed. Location problems, professionalism, lack of creativity, equipment and equipment out of date are also felt by informants who have seen Ludruk performances. But there are additional elements from the consumer point of view that is the lack of promotional activities.

"*I do not ever see ads like banners, ban-ners and others (about ludruk) I never see it.*" (Farah, student).

Young consumers are also expecting the latest technology that can give the feel of contemporary that are mixed with traditional nuances and moral messages.

"*Ludruk as a cultural heritage must be considered as well, now we see abroad like America. everybody knows for its Broadway. Ludruk it should began it innovation, more modern background story but still giving the elements of local culture*" (Rahman, students).

Ludruk is a performance art that basically can be an icon in Surabaya City East Java, Indonesia. Ludruk abandoned his audience considering the inability ludruk to balance the progress of the times. The world of entertainment should follow the development of information technology. The world of art shows has entertainment for tourists even challenged to be able to communicate with universal language. With the ability to continue to adapt and innovate hence, the world of traditional performances will not get caught up in creativity paradox, the creativity that is stuck in monotonous behavior. Ludruk as a traditional art show experienced paradox of creativity that is a condition that demands creativity but management is run by using routine (Kacerauskas 2016). Shown every week with something new is not an easy thing, especially for a senior player. So it takes the younger generation to be able to accommodate the process of creativity. But now ludruk is not a promising thing to become a profession. It takes a bold activity cutting out the vicious circle that exists.

Ludruk will have a high potential as a show business if able to combine traditional and modern elements, and mixed with a universal language. The creativity industry requires innovations that must be tailored to the latest generation, particularly in terms of promotion. Butler (2010) "By Popular Demand: Marketing Art" explaining that: "consumers of art products are so diverse (diversity) it is necessary to have special treatment in promotion". O'Sullivan (2010) Dagling Conversations: advises the use of the internet in the marketing of the arts, and the performing arts managers are expected to be active in managing forums and web communities on the internet in different ways according to the consumer faced.

4 CONCLUSION

Based on the results of this study found some things that should be practitioners in order to better manage the art show business, among others:

1. The limitations of culturists to develop themselves require strong support from the government and corporations, in order to improve the facility of adequate show for sale to tourists.
2. As for the Ludruk cultures are expected to also be enhanced encourage Innovativeness.
3. Practitioners are also expected to be able to hold the government and private parties in terms of promotions made, for example by holding travel and travel entrepreneurs to make art performances Ludruk as one of the cultural tourism destinations when visiting Surabaya East Java Indonesia. Practitioners and related parties that the government is expected to begin to launch Ludruk art performance as one of the agendas appointed in every big event in Surabaya, this is to introduce and remind people of Ludruk art performance that has been forgotten.

ACKNOWLEDGEMENT

Thank you to the Faculty of Economics and Business Universitas Airlangga for being supportive of this research.

REFERENCES

Botti, S. 2000. What role for marketing in the arts? An anal-ysis of arts consumption and artistics value. *International Journal of Arts Management* 2(3): 14–27.

Butler, P. 2000. By popular demand. *Journal of Marketing Management* 16: 343–364.

Colbert, F. 2003. Entrepreneurship and leadership in marketing the arts. *International Journal of Arts Management* 6(1): 30–39.

Dholakia, R.R., Duan, J. & Dholakia, N. 2015. Production and marketing of art in China: Traveling the long, hard road from industrial art to high art. *Arts and the Market* 5(1): 25–44.

Evrard, Y. & Colbert, F. 2000. Arts management: A new discipline entering the millenium? *International Journal of Arts Management* 2(2): 4–13.

Hurek, L. 2015. Ludruk makin ditinggalkan penonton. http://hurek.blogspot.co.id/2015/07ludruk-makin-ditinggalkan-penonton.html. Accsessed January 2018.

Kacerauskas, T. 2016. The paradoxes of creativity management. *Business Administration and Management* 4(XIX).

Kemenperin.go.id. 2016. Era industri kreatif. http://www.kemenperin.go.id/2016/Era-Industri-Kreatif. Accsessed January 2018.

Kerrigan, F., O'Reilly, D. & Vom Lehn, D. 2009. Producing and consuming arts: A marketing perspective. *Consumption, Market & Culture* 12(3): 156–164.

Kotler, P. & Scheff, J. 1997. *Standing room only: Strategies for marketing the performing arts*. Harvard Business Review Press.

Kurnia, F. 2015. Industri kreatif Indonesia masih kalah dibandingkan negeri Jiran. *Retrieved from http://m.wartaekonomi.co.id/berita66455/industri-krestf-indonesis-masih-kalah-di-banding-negeri-jiran.html*. Accsessed January 2018.

Lange, C. 2009. Visibilty and involvement in effective art marketing. *Marketing Intelligence and Planning* 28(5): 650–668.

O'Sullivan, T. 2010. Dagling conversations: Web-forum use by a symphony orchestra's audience members. *Journal of Marketing Management* 26(7–8): 656–670.

Rejeki, S. 2012. Peluang angkat seni tradisi dalam era ekonomi kreatif. https://ekonomi.kompas.com/read/2012/12/10/07241188/Peluang.Angkat.Seni.Tradisi.dalam.Era.Ekonomi.Kreatif. Accsessed February 2018.

Schroeder, J. 2006. Aesthetic awry: The painter of light and commodification of artistic values. *Consumption Market and Culture* 9(2): 87–99.

Advances in Business, Management and Entrepreneurship – Hurriyati et al (eds)
© *2020 Taylor & Francis Group, London, ISBN 978-0-367-27176-3*

The quality of halal tourism destinations: An empirical study of Muslim foreign tourists

S. Sumaryadi, R. Hurriyati, V. Gaffar & L.A. Wibowo
Universitas Pendidikan Indonesia, Bandung, Indonesia

ABSTRACT: The purpose of this research is to measure the quality of halal tourism destinations in relation with the tourist's satisfaction. This research used quantitative methods with Muslim foreign tourists as the main respondents who travel in Bandung. The respondents involved were 123 people (n = 123). The data were collected through questionnaires. An exogenous variable in this research is the tourism destination quality, an endogenous variable is the tourist's satisfaction, and a moderating variable is Islamic attributes. Interviews were also conducted with the accelerated halal tourism development team of the tourism ministry to strengthen the understanding of the research findings. The data were analyzed by using Moderation Regression Analysis (MRA). The results provide strong support to the notion that the quality of halal tourism destination products significantly influences the Muslim foreign tourist's satisfaction and the Islamic attributes variable contributes to moderate the quality of the halal tourism destination for the Muslim foreign tourist's satisfaction. Descriptively, tourism destinations have fulfilled the expected quality of Muslim foreign tourists, although it's not optimal yet (68%). Muslim foreign tourists feel satisfied during the visit to Bandung (80%). The value of this research is that it identifies the quality of the tourism destination in a Muslim foreign tourists' perspective. It will contribute to a new model of thematic tourist destinations development, which is halal tourism destination. Other than that, it will also increase the number and variety of research on halal tourism destination and Muslim tourist behavior which is still a handful.

1 INTRODUCTION

Halal tourism is a form of tourism activity that is in line with the Islamic law (shari'ah) to protect or meet the needs of Muslim tourists, but can also provide benefits for non-Muslim tourists (Battour 2017). Islamic values can be applied to developing destinations, resorts, hotels, travel organizers, attractions, restaurants, public facilities, transportation, and others. Developing halal tourism are not restricted to the Muslim countries domain, but can also be developed by non-Muslim countries, as long as they follow the Islamic values in the development. (Battour & Ismail 2015).

The word 'Halal' is defined as "something which is permissible, with respect to which there is no restriction, and performs according to the law of God" (Battour & Ismail 2015). The term 'halal' means 'allowed' according to the teachings of Islam. Thus, halal tourism as the definition of 'halal' above is an activity in the field of tourism that is 'allowed' according to Islamic teachings (Battour & Ismail 2015). Halal tourism is not merely about religious tourism (Shakiry 2006) and the motivation to enjoy 'Islamic culture' (Henderson 2009), but also has a broader meaning as a kind of tourism activities in accordance with Islamic teachings, which include the behavior, dress code, ethics, products and services, food, and other things that are not contrary to Islamic values.

There are four basic things that Muslim tourists need in their daily life when participating in tourism activities, which are:

1. The need of water for cleanup, much better with segregation services
2. The needs of facilities for worship, much better with segregation services
3. The needs of halal food
4. The needs for tourism activities that fit with Islamic values, such as no element of pornography, immorality, and shirk. (Bahardeen 2013)

In the development of halal tourism, Islamic factors influence destination choices and tourism product preferences (Weidenfeld & Ron 2008), (RWahman 2014). Nevertheless, Battour et al. (2010) state that research on the relationship between religion, behavior, and choice of tourist destinations is still limited, especially considering the Islamic factors. The available research is mostly about religious tourism needs rather than the needs of Muslim tourists.

Morrison (2013) states that destination management is the coordination and integration of all elements of the destination mix in a region based on the destination strategy. In the context of halal tourism, the halal tourism destination is an environment which contributes to tourism destination quality in the perspective of Muslim tourists. Thus, a halal tourism destination can be interpreted as a destination that serves/meets the needs of

Muslim tourists. All components of product destinations must be aligned and able to provide quality in services for Muslim tourists. The conditions of destination products, including the presence of physical products (tourist attraction, facilities, accessibility, facilities and infrastructure), programs (events, festivals, activities), tour packages, and people (Morrison, 2013), must fit with the criteria of Muslim-friendly tourists.

In managing a tourism destination, it is required to develop products that fit the market needs. Developed themes can be cultural, natural, manmade, or some specific theme base on trends or destination strengths and characteristics, such as halal tourism. The ability to provide destination products for the Muslim market, will be a key success factor for developing halal tourism destinations.

Requirements for achieving the target is the quality of tourism destination products in applying islamic values/attributes in order to create a positive experience (Battour & Ismail 2014, Rahman et al. 2017), the linkage between the attributes of halal tourism destinations (Battour & Ismail: 2014, Battour et al. 2017), satisfaction (Battour & Ismail, 2014), tourists loyalty and effective strategies. (Namin 2013).

The ability to create compatibility between destination products and Muslim market needs (Battour et al. 2010, 2011), will be a key factor for halal tourism destinations quality. Shakona et al. (2015), Eid & El-Gohary (2014) and Sriprasert et al. (2014) emphasize the importance of fulfilment of the Muslim tourists' needs for destination products to create tourist satisfaction and tourist loyalty.

In line with the fast-growing number of Muslim travelers, halal tourism is a solution to cater their core needs. In halal tourism, destinations and facilities must be able to fulfill the various faith-based needs, such as prayer facilities, shari'ah-compatible toilets, and halal food (Battour et al. 2011, Weidenfeld & Ron 2008, Shakona et al. 2015). Fulfillment of the needs of Muslim tourists are a form of extended services in order to improve the quality of the travel experience.

As a global trend, halal tourism today is increasing rapidly, with a growth of 6% per year projected to 2020 (the number of world Muslim tourists is projected to reach 156 million in 2020, with a total of spendings up to US$220 billion; GMTI 2017 Mastercard Crescentrating 2017). This means that many countries, both Islamic and non-Islamic countries, develop halal tourism progressively to attract the Muslim travel market. Indonesia has also established halal tourism as one of the leading themes of tourism destination development. In 2017, Indonesia was in the third rank as a halal-friendly holiday destination in the world after Malaysia and the United Arab Emirates (WEF

index year 2017). As one of Indonesia's main halal tourism destination, West Java must contribute in achieving the target of 5 million Muslim foreign tourists and rank 1st in the GMTI (Global Muslim Travel Index) in 2019.

As a new theme for developing tourism destinations in Indonesia, Indonesia, with the largest Muslim population in the world, has a potential to become a world halal tourism destination (Tourism Minister of Indonesia, 2016). However, at this moment, the halal tourism development in Indonesia is still facing major challenges, namely the readiness of halal tourism destination products. The condition of tourism destination products has not been able to function optimally to meet the specific needs of Muslim tourists yet. (Sofyan, Chair of the Accelerated Halal Tourism Development Team of Tourism Ministry, 2016).

Based on the background of the research above, the problems are formulated as follows:

1. How is the tourism destination quality/TDQ?
2. How is the tourist's satisfaction/TS?
3. How is the influence of TDQ and Islamic Attributes/IA to TS?
4. How is the influence of TDQ to TS with IA as a moderating variable?

The purpose of this research is to measure the quality of halal tourism destinations in relation with the tourist's satisfaction with Islamic attributes as moderating variable.

2 METHOD

This research used quantitative methods with Muslim foreign tourists as the main respondents who travel in Bandung. The respondents involved were 123 people (n = 123). The data were collected through question-naires. An exogenous variable in this research is the tourism destination quality, an endogenous variable is the tourist's satisfaction, and a moderating variable is Islamic attributes. Interviews were also conducted with the accelerated halal tourism development team of the tourism ministry to strengthen the understanding of the research findings. The analysis used is Moderation Regression Analysis (MRA).

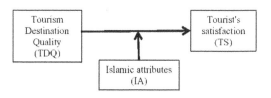

Figure 1. Research framework.

3 RESULTS AND DISCUSSION

The profile of respondents in this research is dominated by tourists from Malaysia (77.2%) and followed by Saudi Arabia (12.2%); age ranges from 25–44 years (70.7%); traveling in group with family (48.8%) and friends (39%); main purpose of the visit being recreation (42.3%) and leisure (39.8%) which visited shopping, natural, and culinary attractions (26.9%; 24,2%; 21,1%). Most of them chose Bandung Airport (74.7%) as airport entry, and the rest (25.3%) chose Jakarta because of "cheaper flights."

Table 1. Respondents' profile (n = 123).

	Profile	%
Gender	Female	57.7
	Male	42.3
Age	35–44	37.4
	25–34	33.3
	18–24	19.5
	45–54	6.5
	55–64	2.4
	Younger than 18	0.8
Marital Status	Married	60.2
	Single	39.8
Country Residence	Malaysia	77.2
	Saudi Arabia	12.2
	UAE	3.3
	Singapore	2.4
	Tiongkok	0.8
	India	0.8
	Morroco	0.8
	Pakistan	0.8
	Germany	0.8
	Russia	0.8
	Australia	0.8
Degree	Under Graduate	44.7
	Master Degree	41.5
	Bachelor	8.9
	Diploma	4.1
	Doctoral Degree	0.8
	Student	23.6
Profession	Business Owner	22
	Job Holder	19.5
	Housewife	13.8
	Academician	10.6
	Retiree	4.1
	Government	2.4
	Private Sector	2.4
	Executive	1.6

(Continued)

Table 1. (Cont.)

	Profile	%
Travel Frequency to Bandung (in the last 2 years)	First Time	54.5
	Second Time	27.6
	More than 3 Times	13
	Third Time	4.9
Whom do you go with	Family	48.8
	Friends	39
	Group Tour	10.6
	Alone	1.6
Airport Entry in In-donesia and Reason for chosing "Jakarta" as Airport Entry	Bandung	74.7
	Jakarta	25.3
	Cheaper flight	58.3
	Comfortable	8.3
	Explore Jakarta	8.3
	For job	8.3
	Simple	8.3
	Visit another destination	8.3
Main Purpose of Visit	Recreation	42.3
	Leisure	39.8
	Family Visit	7.3
	Business	7.3
	Education	1.6
	Shopping	0.8
	Volunteer	0.8
Visited Attractions in Bandung	Shopping	26.9
	Natural Attractions	24.2
	Culinary	21.1
	Cultural Attractions	17.2
	Heritage Site	7.6
	Floating Market	1.8
	Farmhouse	0.9
	Lembang	0.3

Based on the perception of Muslim foreign tourists to TDQ, the respondents score of TDQ is equal to 3.5 (in scale of 5) (= Median 45/number of questions 13), with "mode > median > mean." It is negative when the level of TDQ tends to be in the positive scale and the distribution of TDQ score tends to be on the right side of the curve. Descriptively, tourism destinations have fulfilled the expected quality of Muslim foreign tourists, although it's not optimal yet (68%). High scores on TDQ are given by respondents to the physical product (quality of tourist attraction, quality of hotel, quality of restaurant), the packages (halal packages availability), and the community (local hospitality people, local culture and lifestyle that support halal tourism); while the average score is for the program (halal events and festivals, programs of activities with specific interest) and price for halal tourism packages.

The respondents' score of IA is equal to 3.7 (in scale of 5) (median 30/number of questions 8), with 'mode > median > Mean'. This means negative skewness, that the level of IA tends to be in the positive scale and the distribution of IA score tends to be on the right side of the curve. Islamic attributes have fulfilled the expectation of Muslim foreign tourists, although it is not optimal yet (75%). High scores on IA are given by respondents to the quality of prayer facilities, availability of halal food, and availability of the qibla direction in the hotel room; while an average score is for the availability of a copy of the holy Qur'an in the hotel room, the availability of shari'ah-compatible toilets, the availability of segregated services, the availability of shari'ah-compatible television channels, and the availability of art that does not depict human forms.

Based on the perception of Muslim foreign tourists to TS, respondent score of TS is equal to 4.0 (in scale of 5) (Median 40/number of questions 10), with "mode > median > mean." It means negative skewness, that the level of TS tends to be in the positive scale and the distribution of TS score tends to be on the right side of the curve. Descriptively, Muslim foreign tourists feel satisfied during the visit in Bandung (80%).

Analysis of the influence of TDQ and IA on TS can be seen in Table 3.

From the coefficients table, the regression equation is obtained as follows:

$$Y = 12.603 + 0.346Xi + 0.4X2 \quad (1)$$

Table 2. The tourism destination quality/TDQ, Islamic attributes/IA, and tourist's satisfaction/TS.

		TDQ	IA	TS
N	Valid	123	123	123
	Missing	0	0	0
Mean		44.6098	29.7236	39.9512
Median		45.0000	30.0000	40.0000
Mode		48.00	32.00	40.00
Skewness		-.987	-.854	-1.051
Std. Error of Skewness		.218	.218	.218
Kurtosis		3.088	4.565	4.566
Std. Error of Kurtosis		.433	.433	.433

Table 3. Coefficients.

Model	B	Unstandardized Coefficients Std. Error	Standardized Coefficients Beta	t	Sig.
1 (Constant)	12.603	2.774		4.543	.000
TDQ	.346	.071	.432	4.904	.000
IA	.400	.113	.312	3.539	.001

a. Dependent Variable: TS

Table 4. Model summary.

Model	R	R Square	Adjusted R Square	Std. Error of the Estimate
1	,676a	.457	.448	4.08786

a. Predictors: (Constant), IA, TDQ

From the equation, it is concluded that the variable IA shows a greater positive coefficient to TS than TDQ.

The model summary table (Table 4) shows the correlation coefficient (R) and the coefficient of determination (R Square) variable TDQ and IA to TS.

Overall, TDQ and IA variables have an influence of 45.7% to TS.

The influence of TDQ to TS with IA as a moderating variable is shown in Table 5.

From the coefficients table (Table 5), the regression equation is obtained as follows:

$$Y = -20.37 + 0.701Xi + 0.952X2 - 0.013XL2 \quad (2)$$

From the equation, it is concluded that variable IA shows a greater positive coefficient to TS than TDQ. Each coefficient has a significance level of $P < 0.05$; which is TDQ is 0.000, IA is 0.001, and TDQ*IA is 0.28. It means that the existence of IA strengthens the influence of TDQ to TS.

Table 5. Coefficients.

Model	B	Unstandardized Coefficients Std.E	Standardized Coefficients Beta	t	Sig.
1 (Constant)	-2.037	7.113		-.286	.775
TDQ	.701	.174	.874	4.037	.000
IA	.952	.271	.741	3.509	.001
TDQ*IA	-.013	.006	-.804	-2.229	.028

a. Dependent Variable: TS.

Table 6. Coefficient of determination TDQ*IA, IA, to TS.

No.	Predictor Variables	R	R Square
1	TDQ, IA	0,675	0,457
2	TDQ*IA, TDQ, IA	0,692	0,479

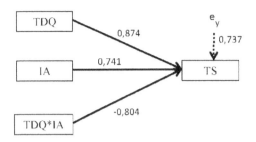

Figure 2. Result model.

The influence of TDQ to TS with IA as a moderating variable is able to increase the score of TDQ (2.2%)

The findings are formulated in the result model in Figure 2.

From Figure 2, we obtained information on the direct influence of predictor variables (TDQ, IA, TDQ* IA) on TS, namely TDQ 76.39%; IA 54.91%; and TDQ*IA 64.64%.

4 CONCLUSION

Based on the results of the discussion on the analysis of TDQ, TS, and the influence of TDQ & IA to TS, as well as the influence of TDQ to TS with IA as a moderating variable, it can be concluded that the tourism destination quality significantly influences the Muslim foreign tourist's satisfaction, and the Islamic attributes variable contributes to moderate the quality of a halal tourism destination to the Muslim foreign tourist's satisfaction.

Descriptively, tourism destinations have fulfilled the expected quality of Muslim foreign tourists, although it's not optimal yet (68%). Muslim foreign tourists feel satisfied during their visit to Bandung (80%).

The implication of this research is the need for special attention from all stakeholders. The government, through the tourism ministry, must continue to accelerate in improving the quality of halal tourism destinations. The socialization and certification of the tourism industry is also very important to ensure the standardization of products and services that are in accordance with Islamic values.

Allowing for the limitations of the research, this is expected to encourage further research in halal tourism, especially in the field of halal destination products, Muslim tourist preference and the satisfaction, which is relatively limited.

REFERENCES

Bahardeen, F. 2013. *Developing the concept of Halal Travel* Singapore: Crescentrating.

Battour, M., & Ismail, M.N. 2014. The Role of Destination Attributes in Islamic Tourism. *SHS Web of Conferences* 12: 556–564.

Battour, M., & Ismail, M.N. 2015. Halal tourism: Concepts, practises, challenges and future. *Tourism Management Per-spectives* 19: 150–154.

Battour, M. M., Ismail, M.N. & Battor, M. 2010. Toward a Halal Tourism Market. *Tourism Analysis* 15(4): 461–470.

Battour, M., Ismail, M.N. & Battor, M. 2011. The impact of destination attributes on Muslim tourist's choice. *International Journal of Tourism Research* 13(6): 527–540.

Battour, M., Ismail, M.N., Battor, M. & Awais, M. 2017. Islamic tourism: an empirical examination of travel motivation and satisfaction in Malaysia. *Current Issues in Tourism* 20(1): 50–67.

Eid, R. & El-Gohary, H. 2015. Muslim Tourist Perceived Value in the Hospitality and Tourism Industry. *Journal of Travel Research* 54(6): 774–787.

Henderson, J.C. 2009. Islamic Tourism Reviewed. *Tourism Recreation Research* 34(2): 207–211.

Mastercard & Crescentrating. 2017. Global Muslim Travel In-dex 2017 (GMTI 2017). Dubai: Mastercard & Crescentrat-ing.

Morrison, A.M. 2013. *Marketing and Managing Tourism Destinations*. New York: Routledge.

Rahman, M. K., Zailani, S. & Musa, G. 2017. What travel motivational factors influence Muslim tourists towards MMITD?. *Journal of Islamic Marketing* 8 (1): 48–73.

Namin, T. 2013. Value Creation in Tourism: An Islamic Ap-proach. *International Research Journal of Applied and Basic Sciences* 4(0): 1252–1264.

Rahman, M.K. 2014. Motivating factors of Islamic tourist's destination loyalty: An empirical investigation in Malaysia. *Journal of Tourism and Hospitality Management* 2(1): 63–77.

Shakiry, A. S. 2006. *The academy of Islamic tourism project*. *Islamic Tourism*. www.islamictourism.com. Downloaded February 25, 2018.

Shakona, M., Backman, K., Backman, S., Norman, W., Luo, Y. & Duffy, L. 2015. Understanding the traveling behavior of Muslims in the United States. *International Journal of Culture, Tourism and Hospitality Research* 9(1): 22–35.

Sriprasert, P., Chainin, O. & Rahman, H.A. 2014. Understanding Behavior and Needs of Halal Tourism in Andaman Gulf of Thailand: A Case of Asian Muslim. *Journal of Advanced Management Science* 2(3): 216–219.

Weidenfeld, A.D.I & Amos S.R. 2008. Religious Needs in the Tourism Industry. *International Journal of Tourism and Hospitality Research* 19(2): 357–361.

Advances in Business, Management and Entrepreneurship – Hurriyati et al (eds)
© 2020 Taylor & Francis Group, London, ISBN 978-0-367-27176-3

Internal marketing effects on the Islamic work commitment and nurses' performance at Islamic hospitals

R.T. Ratnasari & A.P. Pamungkas
Airlangga University, Surabaya, Indonesia

ABSTRACT: This research aims to determine the effect of internal marketing on Islam work commitment and performance of Islamic hospital nurses in East Java. This research uses quantitative approach and path analysis technique. The sampling method is probability sampling by using simple random sampling. This research uses questionnaire data from 100 employees at the Islamic hospital who work as nurses for at least three years. The result of this study showed that internal marketing has a significant effect on Islam work commitment, meanwhile internal marketing has no significant effect on the nurses' performance, and Islam work commitment has significant effect on the nurses' performance. Therefore, the Islamic hospital is expected to provide more provisions about the Islam work ethic to the employees.

1 INTRODUCTION

Human resources is seen as an asset to achieve company goals. The human resources department has an important role to play in every company activity. George said companies realize that employees are among a company's best assets and serve as effective links with external audiences (Bailey et al. 2016). Employees have an important role to the success of the company, especially in companies that offer services, because they deal directly with customers, it causes to the result that employees are very influential in improving the quality of products and services provided, and this should be monitored as well as possible. Though employees often need to be conscious of the needs of customers, they should also effectively cooperate and coordinate within their organization, as their experience of satisfaction, motivation and commitment to the organization may be greatly influenced by external service quality (Yu et al., 2017). Kotler and Keller (2016) also disclosed that highly managed service companies are believed to be the result of relationships with employees and will affect their relationship with customers. Thus, there is a high interaction between employees and customers in companies engaged in services.

Internal marketing was first mentioned in the service literature in the late 1970s as a solution to providing high-quality service (Varey & Lewis 1999). Since the mid-1980s, marketing has proved very important to be studied (Caruana & Calleya 1998). Furthermore, Caruana and Calleya (1998) have argued that marketing in service is very important to learn because it concerns the interests of internal customers, as the case where employees are made to feel that their organizations care about them and meet their needs as employees. The most fundamental thing about internal marketing is the fact that effective service to customers requires employees who are well-informed of the customer's wishes (Ghorbani & Sedeh 2014). Ernest and Young found that 68% damage to corporate relationships with customers resulting from employee attitudes and negligence (Hamidah 2012). The internal marketing of an Islamic perspective assumes that the relationship between the management of a company and its employees is like a neighborly relationship, that is, one another must respect each other (Ratnasari 2012). If it is conducted on a company, it will cause a positive attitude that employees will have high firmness in the work, and will have a high sense of responsibility to the company where they work (work commitment) (Shahab 2010).

Therefore, this study aimed to determine the effect of internal marketing on Islamic job commitment performance of Islamic hospital. The performance of employees in this study is the performance of nurses because nurses is one of the employees in an Islamic hospital. This research used questionaire data and sample criteria determined are employees of the Islamic hospital who work as nurses with a minimum of three years of service, because the respondents are considered to have had sufficient work experience and have the perception to be able to assess the policy of the Islamic hospital in East Java. The data obtained then went through various tests and then was subjected to a path analysis. After the process and the results of the analysis, it can be stated that internal marketing is significantly affect the work commitment of Islam, meanwhile internal marketing does not significantly affect the performance of the nurses, and Islam work commitment significantly influence the performance of the nurses. Based on the results of this study, the existing leaders of

Table 1. Data normality test on univariate and multivariate.

Variable	Min	Max	Skew	C.R.	Kurtosis	C.R.
Internal Marketing	2.817	3.861	0.488	1.723	−0.043	−0.054
Islam Work Commitment	2.866	4.023	−0.501	−1.589	−0.987	−1.705
Nurses' Performance	2.912	4.018	−0.276	−0.962	−1.445	−2.545
Multivariate:					−1.734	−1.409

Islamic hospital are expected to provide more education and training to the nurses so they can improve the knowledge and skills in working, and also provide the best services.

2 METHOD

The approach in this research was quantitative, through hypothesis testing, data measurement, and making conclusions. The three dimensions of internal marketing were divided into 12 indicators used as a question item. The Islamic work commitment variable consists of 7 indicators used as the question items, and the nurses' performance variable consists of 11 indicators which are also used as question item. Scale of measurement in this research is by using the modified Likert scale model and employs four answers: (SA) strongly agree = 4, (A) agree = 3, (D) disagree = 2, and (SD) strongly disagree = 1.

The data collection for this research was done through preliminary survey, literature study, and field research which then produce primary and secondary data. The number of samples used in this study was 100 respondents. Sample criteria determined are employees of the Islamic Hospital who work as nurses with a minimum of three years of service, because the respondents are considered to have had sufficient work experience and have the perception to be able to assess the policy of the Islamic Hospital in East Java.

The data obtained then going through a testing using validity test, reliability test, and then analyzed by using Path Analysis technique. Path Analysis is a statistical analysis tool to test the existence of intervention variables on the relationship between variables X to variable Y.

3 RESULT AND DISCUSSION

3.1 Normality test

The result of univariate normality test showed that all variables have CR skewness and kurtosis value, which is < 2.58 or below standard. With this result, it can be stated that the data meet the standard of univariate normality.

In addition to univariate testing of path analysis by Hair and Anderson (2006), the data must meet multivariate normality requirements. It is said to meet the normal multivariate assumption if the value of CR kurtosis < 2.58. Multivariate normality test results obtained a value of −1.409. This value is already below 2.58, so it can be said that the data has met the assumption of normal distribution multivariate.

3.2 Path analysis testing result

3.2.1 Path coefficient

The figure of path analysis test result on the next page will show the coefficient value of the path or standardize on internal marketing variable to work commitment of Islam, internal marketing to nurse performance, and work commitment of Islam to nurse performance.

Table analysis of path analysis results based on path coefficient.

Based on the value of standardized path coefficient in Table 2, it can be seen that the relationship between variables influence dominantly. Relationship between the largest variables is linking the work of Islamic commitment to the performance of nurses with a coefficient value of 0.843.

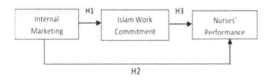

Figure 1. Analysis model.

Table 2. Effect between variables path coefficient value.

Variable		Direct Path	Indirect Path
Internal Marketing	→ Islam Work Commitment	0.476	
Internal Marketing	→ Nurses' Performance	−0.011	
Islam Work Commitment	→ Nurses' Performance	0.843	
Internal Marketing towards Nurses' Performance through Islam Work Commitment			0.387

Table 3. Determination coefficient value.

Variable	R^2
Internal Marketing→Islam Work Commitment	0.342
Internal Marketing and Islam Work Commitment→Nurses' Performance	0.605

3.2.2 Determination coefficient

The result of determination coefficient is the amount of change of Islam work commitment caused by the existence of internal marketing is 0.342 or 34.2% and the amount of change of nurse performance caused by the existence of internal marketing and Islam work commitment is 0.605 or 60.5%.

3.2.3 Coefficient of direct value path

To determine the presence or absence of exogenous variables influence on the endogen intervening, the following provision is used: comparing probability value with level of significant $\alpha = 0.05$. If the significance value ≤ 0.05, then there is a significant exogenous variables influence on the endogen intervening. Conversely, if the significance value > 0.05, then there is no significant effect of exogenous variables on the endogen intervenin. The complete result of hypothesis test can be seen in Table 4.

According to Table 4:

1. CR value magnitude of internal marketing variables to Islam work commitment is 4.652. Level of significance obtained value is 0.000, this value is smaller than 0.05. Because this level of significance is less than 0.05, internal marketing affects the Islam work commitment. So the research hypothesis is accepted.
2. CR value of the internal marketing variable on nurses' performance is –0.107. Level of significance obtained value is 0.916, this value is greater than 0.05. Therefore the significance level is greater than 0.05, then the internal marketing does not affect the performance of nurses. So, the research hypothesis is rejected.
3. CR value of the Islam work commitment variable to nurses' performance is 9.111. Level of significance obtained value is 0.000, this value is smaller than 0.05. Because of this significance level is less than 0.05, the Islam work commitment has an effect on the performance of the nurse. So the

Table 4. Direct effect hypothesis test result.

Variable		CR value	Sig.	Desc.
Internal Marketing	→ Islam Work Commitment	4.652	0.000	S
Internal Marketing	→ Nurses' Performance	–0.107	0.916	NS
Islam Work Commitment	→ Nurses' Performance	9.111	0.000	S

research hypothesis is accepted. The direct influence of internal marketing on Islam work commitment and Islam work commitment to the performance of nurses showed that internal marketing indirectly influences nurses' performance through Islam work commitment.

3.2.4 Indirect coefficient of effect path

Path analysis is tested in addition to examine the direct influence of exogenous variables on endogen, also indirect influence of exogenous variable on endogen through endogen intervention variable. There is an indirect influence of internal marketing on nurses' performance through Islamic work commitment, with indirect path coefficient value is 0.384.

3.3 Discussion

3.3.1 Internal marketing effect on islam work commitment

The results of the first hypothesis testing is in accordance with Tasmara (2006), that the vision of the company (an indicator of internal marketing variables) can generate enthusiasm, pride, and value that is considered very important for employees. This supports the seventh indicator of Islam work commitment variable, which is that nurses strive to achieve the company's objectives. Constructively, this study supports Putri (2013) study, which stated that internal marketing in the company is found to have a significant effect on employee work commitments. Specifically, Putri (2013) stated that when company leaders run policies against employees by treating employees as well as their internal customers pleasantly, such as by strengthening the Islamic sharia practice in every activity, employees' commitment to the company will increases.

This study shows that by running the Islamic Hospital's policy on nurses, such as treating employees like their internal customers well, it will affect the Islam work commitment of the nurses. The existence of internal marketing influence is reflected from the respondent's answer that has a high score on the second and fifth indicator which can be a grip for the nurses to realize the company's goals, and then it will increase the Islam work commitment of the nurses. In addition, spiritual spirits, development, knowledge and skills of nurses are conducted continuously in Islamic hospitals can also grow the Islam work commitment in Islamic Hospital, this is reflected in the nurse awareness in working with full responsibility while carrying out their duties.

On the other hand, the low indicator of the eighth and the twelfth in the internal marketing variables that is the Islamic hospital always pay attention to the welfare of the nurses by giving incentives or bonuses in return for the nurses who produce the best service. The low indicator caused by the incentive allocation for each nurse by Islamic hospital, whether for nurses who provide the best service or not. This can be an input for Islamic hospital, which considered that the Islamic hospital needs to provide more incentives or bonuses for nurses who provide the best service, with the leadership of Islamic hospital provide a secret questionnaire to the patient to choose the nurse with the best service.

3.3.2 *Effect of internal marketing on nurses' performance*

The result of this second hypothesis test are not in accordance with the result of research conducted by Magatef and Momani (2016) about the concept of internal marketing and its impact on business performance in the hospital sector stated that there is a significant relationship between internal marketing and business performance. However, the result of this study supported the result of research conducted by Putri (2013), which stated that internal marketing has no significant effect on employees' performance, because after all when the company implement its internal policy, the target given the company remains unchanged.

This research indicates that the respondent's answer to the nurses' performance variable with the second and fourth indicator shows that the nurses work in accordance with Standard Operating Procedures (SOP), in addition, the nurses are intelligent and reliable in the work, which is in accordance with the opinion of Rivai (2005), that performance is an achievement of a person in performing his duties or work in accordance with the standards and criteria established for the job.

Both of these indicators, if implemented, the nurses will be able to achieve the target above 50% in treating the patient well in accordance with the provisions of the Islamic hospital. However, these targets make the nurses' performance based on the target, not because of the internal policy of the Islamic hospital. So it can be considered that internal marketing has no significant effect on the performance of nurses. However, the internal policy of the Islamic hospital will not affect the performance of the nurses, as the respondents' answers (in the description) that internal marketing does not affect the performance, but because of the responsibilities as a nurse.

In addition, the first low indicator in the nurses' performance variable is the respondent as the nurse able to overcome the problems encountered while working. The low indicator is because nurses have not mastered deep nursing techniques and lack of knowledge about nursing. This can be an input to the Islamic hospital, which shows that the Islamic hospital needs to improve training on nursing science to every nurse, and handling techniques in patients properly. The leader of the Islamic Hospital needs to ensure that nurses really master and apply the knowledge and techniques of handling patients correctly,

by directly assess the performance of nurses once in a month.

3.3.3 *Work commitment effect on nurses' per-formance*

The result of this third hypothesis test in accordance with the opinion of Tasmara (2006) who stated that a very high commitment allow the employees to fight against challenges so that the performance of the employees will be also high. In addition, research results from Memari et al. (2013) on organizational and performance commitments in the oil and gas sector of Pakistan showed that performance arises because of organizational commitment. Respondents' answers said that employees who are comfortable with the work environment, will increase organizational commitment that will ultimately also improve their performance.

This study shows that Islam work commitment arises as a response from nurses on the policy of Islamic hospital. The influence of Islam work commitment is reflected in the respondent's answer on the fifth indicator, which shows that the nurse works with full responsibility in performing their duties. Nurses who have a high commitment to their work in an Islamic hospital will work with full responsibility in carrying out their duties. In addition, nurses who work sincerely as a form of worship to Allah SWT (on the third indicator), will also complete the task well, and not only as a form of responsibility to the Islamic hospital but also to their responsibility as a servant of Allah SWT.

On the other hand, the fourth low indicator in work commitment variable is that the nurse works independently by not relying on the leader or coworker. The low indicator is caused of the nurse still lacks of knowledge about nursing. This can be an input for Islamic hospital, which that the Islamic hospital needs to improve the provision of training on nursing science and provide provisions about the Islamic work ethic continuously, and straighten the intention of nurses in the work to the purpose to worship Allah SWT, so they will be able to have a high work commitment.

4 CONCLUSIONS

After the process and the results of the analysis, it can be obtained that internal marketing affecting Islamic work commitment, however not with nurse's performance. Meanwhile Islam work commitment significantly influence the performance of the nurses. And based on the results, the existing leaders of Islamic hospital are expected to provide more incentives or bonuses to the nurses who provide the best services, and to increase the amount of training to the nurses so they can improve the knowledge and skills in working, and finally the Islamic hospital might as well providing more provisions about the Islam work ethic to the employees, because internal marketing is needed in order to realize the company's expectations with the aim of implementing the provisions in Islamic law and achieving the blessings of Allah SWT (Ratnasari 2014).

REFERENCES

Bailey, A.A., Albassami, F. & Al-Meshal, S. 2016. The roles of employee job satisfaction and organizational commitment in the internal marketing-employee bank identification relationship. *International Journal of Bank Marketing* 34(6): 821–840.

Caruana, A. & Calleya, P. 1998. The effect of internal marketing on organizational commitment among retail bank managers. *International Journal of Bank Marketing* 16(3).

Ghorbani, H. & Sedeh, H.A. 2014. An empirical investigation on the impact of internal marketing on organizational effectiveness within human resource capabilities perspective (Case study: Islamic Azad University Branches in Region 4 of Isfahan Municipality). *International Journal of Academic Research in Business and Social Sciences* 4(1).

Hair, J.F. & Anderson, R.E. 2006. *Multivariate data analysis*. Upper Saddle River, NJ: Pearson Education Inc.

Hamidah, R.A. 2012. Pengaruh pengamalan karakter fathanah dalam pemasaran internal terhadap kesejahteraan karyawan perspektif maqashid syariah pada bank syariah mandiri di Surabaya. *Skripsi Tidak Diterbitkan*. Fakultas ekonomi dan Bisnis Universitas Airlangga Surabaya.

Kotler, P. & Keller, K.L. 2016. *Manajemen pemasaran*. Terjemahan oleh Bob Sabran. Jakarta: Erlangga.

Magatef, S.G. & Momani, R.A. 2016. The impact of internal marketing on employees' performance in private Jordanian hospital sector. *International Journal of Business and Management* 11(3).

Memari, N., Mahdieh, O. & Marnani, A.B. 2013. The impact of Organizational Commitment on Employees Job Performance: A study of Meli bank. *Interdisciplinary Journal of Contemporary Research in Business* 5 (5): 164–171.

Putri, A.F. 2013. Pengaruh istiqamah dalam pemasaran internal terhadap komitmen kerja dan kinerja karyawan bank BNI syariah Surabaya. *Skripsi Tidak Diterbitkan*. Fakultas Ekonomi dan Bisnis Universitas Airlangga Surabaya.

Ratnasari, R.T. 2012. Manajemen pemasaran Islam (jilid satu). *Modul Tidak Diterbitkan*. Departemen Ekonomi Syariah Fakultas Ekonomi dan Bisnis Universitas Airlangga Surabaya.

Ratnasari, R.T. 2014. Pengaruh kepemimpinan islam dan pemasaran internal terhadap kinerja serta kesejahteraan karyawan Bank Islam di Jawa Timur. *Disertasi Tidak Diterbitkan*. Program Pasca Sarjana Universitas Airlangga Surabaya.

Rivai, V. 2005. *Performance appraisal*. Jakarta: RajaGrafindo Persada.

Shahab, A. 2010. Implementasi kepemimpinan dan komitmen serta pengaruhnya terhadap kepuasan kerja dan kinerja karyawan pada BMT di Provinsi Jawa Tengah dalam perspektif Islam. Disertasi Tidak Diterbitkan, Universitas Airlangga Surabaya.

Tasmara, T. 2006. *Spiritual centered leadership (Kepemimpinan berbasis spiritual)*. Jakarta: Gema Insani.

Varey, R.J. & Lewis, B.R. 1998. A broadened conception of internal marketing. *European Journal of Marketing* 33(9/10).

Yu, Q., Yen, D.A., Barnes, B.R. & Huang, Y.A. 2017. Enhancing firm performance through internal market orientation and employee organizational commitment. *The International Journal of Human Resource Management*: 1–24.

Advances in Business, Management and Entrepreneurship – Hurriyati et al (eds)
© 2020 Taylor & Francis Group, London, ISBN 978-0-367-27176-3

Analysis of green marketing on purchasing decisions

D. Silvia & H. Hendrayati
Universitas Pendidikan Indonesia, Bandung, Indonesia

ABSTRACT: Green marketing is a marketing strategy that puts forward the concept of green marketing and prioritizes the sustainability of the environment as a social responsibility. In beating the competition, a company must have the right strategy in marketing its products. Innisfree is a beauty product that promotes natural conservation in every business process. This research used a quantitative approach, which aims to understand and analyze whether the green marketing mix simultaneously and partially influences the consumer decision to purchase Innisfree products at VPJ Bandung. The strategy of the green marketing mix in this research consists of the following variables: Price, Product, Place, and Promotion. Data in the study were collected through questionnaires distributed to consumers visiting the Innisfree VPJ outlet. Sampling using nonprobability technique and accidental sampling with a sample size of 100 respondents was carried out. The method of data analysis technique used in this research is multiple regression analysis. The results showed that the green marketing mix simultaneously and partially influences the purchase decision of Innisfree products in theInnisfree PVJ Bandung outlet based on the F test value of 5.592 > f table of 2.47.

1 INTRODUCTION

Human awareness of environmental damage resulting in global warming has brought a change in human lifestyle and consciousness. The change of lifestyle is based on their concerns about environmental disasters that threaten them. The aims are to reduce the effects of global warming, preserve the environment, and be more concerned towards health care. The increase of the awareness towards the environment has led people to look for a variety of eco-friendly products for their consumption. Currently many businesspeople use this issue to run their marketing strategy with the strategy being called green marketing. Many business people develop products that are environmentally friendly and provide satisfaction for the customers. The products are made with natural materials and can be renewed. Green marketing has a good chance in the future to be used in the marketing strategy in today's modern era where the community will be guided to be more conscious in taking part in preserving the environment by consuming products created with environmentally friendly technology and natural materials that can be renewed.

The American Marketing Association (Rahayu et al. 2017) explains that there are three definitions of green marketing: (1) retailing definition: the marketing of products that are presumed to be environmentally save; (2) social marketing definition: the development and marketing of products designed to minimize the negative effects on the physical environments or to improve its quality; (3) environments definition: the efforts by organizations to produce, promote, package, and reclaim products in a manner that is sensitive or responsive to logical concerns. So

green marketing becomes an activity to create products that are processed environmentally-friendly starting from product creation, packaging, and promotion activity through to advertisement. The marketing of products through the concept of the green marketing mix can be done on various elements starting from product, price, and promotion to distribution. Governments and communities in Indonesia are starting to show up with various environmental conservation activities such as government regulations on the use of plastic shopping bags, the use of recyclable packets of goods, up to the forest saving campaign.

The beauty industry is an interesting industry to discuss, because many users of beauty products want to reduce the use of chemicals in the beauty products that they use. The users of beauty products today begin to switch to products that contain natural ingredients that are environmentally friendly. One of the global beauty companies that uses the concept of green marketing is Innisfree. Innisfree Company has a brand concept that introduces the rich, natural beauty-producing island of Jeju Island in South Korea. This company is engaged in beauty and cosmetic treatments that use natural ingredients and are environmentally friendly. Innisfree also uses life movement campaigns for each of its products and the launch of outlets, such as tree planting, campaigns for handkerchiefs, empty bottle recycle campaigns, and so on.

Most of the consumers of skincare and cosmetics products, before deciding to buy a product, will look at some specific aspects, because not everyone has the same skin condition and type. So they try to find products that have good reviews from its users. Consumers of today's beauty products are also beginning to avoid the use of various chemicals in their beauty treatments to avoid the chemical dependence effects and avoid the risk of organ damage from chemical cosmetics. Innisfree is known as a brand that offers beauty and cosmetic products where they claim that in making their products, they use natural ingredients that are environmentally friendly.

Based on this background, the researcher is interested to conduct a deeper study on how big the influence of the green marketing mix strategy is, seen from product factor, price, distribution place, and promotion, to the purchase decision of Innisfree products.

1.1 *Green marketing*

The American Marketing Association has explained that green marketing is a safe and environmentally friendly process of marketing. Every activity of green marketing, starting from the production process to creating the products, can reduce the problem of global warming with a quality product. So green product offering through green marketing strategy will increase the company's ability to sell and introduce its products better. According to Kotler and Armstrong (2008), the marketing mix is a collection of tactical marketing tools, controlled byproducts, prices, places, and promotions which are combined by the company to produce the desired response in the target market. According to Singh (Paysal 2016), the 4P part marketing tools are required in an innovative way, that is:

1. Green Product: green marketing concepts emerge in several ways: first, identifying customers' environmental requirements and then develop environmental-friendly products to meet their needs.
2. Green Price: an important element in green marketing is price. Many consumers are willing to spend more on quality products. These values will improve performance, function, design, visual appeal, and taste. The importance of environmental existence will make companies that use the green marketing strategy get a determinant factor compared to the company's competitors. In the creation of products that follow the green marketing strategy, the company requires additional costs both in the form of research and creation of raw materials, so the cost incurred by consumers to obtain quality products and environmentally friendly products will be greater. In the end, an environmental-friendly product will provide a better positive value than other products in achieving satisfaction.
3. Green Place: ease in terms of time and mileage to be spent by consumers to get green products will be very influential in attracting customers. So, the location is a very significant influence. The aura of the store should also display a visualization that reflects the concept of green marketing, using materials that can be recycled.

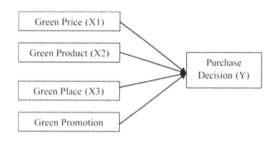

Figure 1. Hypothesis model.

4. Green Promotion: in promoting green products to the target consumers, the company can do advertising, public relations, sales promotion in the form of discounts and shopping prizes. Companies that use green marketing concepts will reinforce sustainable credibility through the environment as a means of communication and practice.

1.2 *Purchase decision*

Consumers must have done some encouragement by finding information about a product in making purchasing decisions. Kotler and Armstrong (2008) explain that there are seven components in purchasing decisions: decisions on product types, about product form, regarding sales, on product quantities, about the time of purchase, and on payment. Peter and Olson (Romadon et al. 2014) explain that purchasing decisions are a process done to combine the consumer's acquired knowledge and consideration, in order to select two or more alternatives to decide on one product.

1.3 *Research result*

The hypothesis can be explained using the image in Figure 1.

2 METHOD

The type of research used in this study is a quantitative approach, which, according to Sugiyono (2010), is used to examine the population or specific samples, data collection and using research instruments, quantitative/statistical analysis, with the aim to test the hypothesis that has been established. The sample used in this research consists of 100 respondents using the non-probability technique with accidental sampling which is s sampling technique that does not give equal opportunity or opportunity in every selected population in the sample, with the characteristic of respondents who have used Innisfree products. Because of the unknown population number, the sample is taken using the incidental sampling method, which is a sample determination technique based on anyone who happens to meet the researcher who can be used as a sample, if the person met is suitable as a data source.

This research was conducted by Store Innisfree Paris Van Java Bandung. Data collection techniques used in this study were to spread the questionnaire to Innisfree consumers, where each questionnaire answer was measured by the ordinal scale with a score of 1 to 5; the analysis techniques used in this study were descriptive analysis techniques with multiple linear regression analysis.

3 RESULTS AND DISCUSSION

In this research, there were 25 male respondents and 75 female respondents. The background of as much as 27 people was student, 2 people are civil servants, 48 people private employees, 5 entrepreneurs, 9 people IRT, and other work 23 people. Analytical techniques used are multiple linear regression analysis to determine the effect of green marketing on purchasing decisions, with the results of the analysis as seen in Table 1.

3.1 Validity test results

The validity test results can be found in Table 1.

3.2 Reliability test results

The reliability test results can be found in Table 2.

Table 1. Validity test results.

Variable	Items	R	Sig	Information
Green Price	X11	0.553	0.000	Valid
	X12	0.532	0.000	Valid
	X13	0.611	0.000	Valid
	X14	0.589	0.000	Valid
	X15	0.732	0.000	Valid
Green Product	X21	0.661	0.000	Valid
	X22	0.721	0.000	Valid
	X23	0.623	0.000	Valid
	X24	0.759	0.000	Valid
Green Place	X31	0.885	0.000	Valid
	X32	0.812	0.000	Valid
Green Promotion	X41	0.827	0.000	Valid
	X42	0.824	0.000	Valid
	X43	0.76	0.000	Valid
	X44	0.869	0.000	Valid
Purchase Decision	Y1	0.813	0.000	Valid
	Y2	0.742	0.000	Valid
	Y3	0.763	0.000	Valid
	Y4	0.463	0.000	Valid

Table 2. Reliability test results.

Variabel	Koefisien Alpha	Ket
Green Price (X1)	0.601	Reliable
Green Product (X2)	0.644	Reliable
Green Place (X3)	0.618	Reliable
Green Promotion (X4)	0.839	Reliable
Purchase Decision (Y)	0.649	Reliable

3.3 Results of multiple linear regression analysis

In measuring and testing the effect of green marketing on the Innisfree product purchase decision, data processing was done through multiple linear regression equation using SPSS data processing program 22. The results of data processing can be seen in Table 3.

A t test was conducted to find out how far the partial influence of each independent variable reached. From the result of data processing we got that the Green Price variable is t arithmetic (3.657) > t table (1.66), hence the green price can be said to significantly influence the variable purchase decision. Green Product resulted in t arithmetic (0.690) <t table (1.66), hence green product can be said not to influence the variable purchase decision much. Green Place resulted in t value (2.051)> t table (1.66), hence green place can be said to significantly affect the purchase decision variable. Green Promotion had t count (0.288) <t table (1.66), so it can be said that green promotion affects the variable purchase decision not so much.

Based on the results of the anova test, we obtained F indigo 5.592, while residual f table (df) is 95 as denominator and df Regression is 2.47. Because F calculates to 5.592> F table 2.47, H1 is accepted. Based on the results of data processing, it can be seen that there is a column significance of 0.000, which means the probability is 0.000 <0.005, and then H0 is rejected. Thus, with H1 being accepted, it can be concluded there is a significant relationship between green marketing and the purchase decision of the Innisfree product.

Table 3. T Test.

Model	Unstandardized Coefficients		Standardized Coefficients	t	Sig
	B	Std. Error	Beta		
1 (Constant)	5.386	2.440		2.208	.030
SUMX1	.343	.094	.350	3.657	.000
SUMX2	.094	.136	.076	.690	.492
SUMX3	.333	.162	.198	2.051	.043
SUMX4	−.028	.098	−.032	−.288	.774

Table 4. F test.

Model	R	R Square	Adjusted R Square	Std. Error of the Estimate	Durbin Watson
1	.437a	.191	.156	1.998	1.951

a. Predictors: (Constant), SUMX4, SUMX3, SUMX1, SUMX2
b. Dependent Variable: SUMY

Table 5. R Test.

Model		Sum of Squares	Df	Mean Square	F	Sig.
1	Regression	89.334	4	22.333	5.592	0.000b
	Residual	379.426	95	3.994		
	Total	468.760	99			

a. Dependent Variable: SUMY
b. Predictors: (Constant), SUMX4, SUMX3, SUMX1, SUMX2

In Table 5, the magnitude of R square shows the number 0.191. Determination coefficient is used to determine the magnitude of the influence of independent variables on the dependent variable, where the smaller the value of R square is, the weaker the relationship of each variable and vice versa: the greater the value of R square, the stronger the relationship of each variable. From the result of processing of the green linear regression of green marketing and the purchasing decision, it is found that the purchasing decision as influenced by green marketing is equal to 19.1% while 80.9% of the purchasing decision of Innisfree product is influenced by other factors not examined by the researcher.

4 CONCLUSION

The results of this research lead to the following conclusions:

1. Green marketing on purchasing decisions that have a real effect seen from the value of t table is Green Price with t count (3.657)> t table (1.66) and Green Place known t value (2.051)> t table (1.66).

2. While for the green product variable t arithmetic (0.690) <t table (1.66) then the green product can be said not so significantly affect the purchase decision variable. And Green Promotion with t count (0.288) <t table (1.66) so it can be said that Green promotion not so significantly influence to variable purchase decision.

REFERENCES

Kotler & Amstrong. 2008. Prinsip-Prinsip Pemasaran. Edisi 12 Jilid 1. Jakarta: Erlangga.
Paysal. 2016. Pengaruh Green Marketing Terhadap Keputusan Pembelian Pada Produk Nike di Bandung. e-Proceeding of Applied Science 2(3).
Rahayu, dkk. 2017. Pengaruh Green Marketing Terhadap Keputusan Pembelian Konsumen. Jurnal Administrasi Bisnis 43(1).
Romadon, dkk. 2014. Pengaruh Green Marketing Terhadap Brand Image Dan Struktur Keputusan Pembelian. Jurnal Administrasi Bisnis (JAB) 15(1).
Sugiyono. 2010. Metode Penelitian Kuantitatif Kualitatif dan R & D. Bandung: Alfabeta.

Advances in Business, Management and Entrepreneurship – Hurriyati et al (eds)
© 2020 Taylor & Francis Group, London, ISBN 978-0-367-27176-3

The effect of Customer Relationship Management (CRM) on customer loyalty

A. Mulyana
Universitas Pendidikan Indonesia, Bandung, Indonesia

ABSTRACT: The purpose of this study is to examine the effect of Customer Relationship Management (CRM) on customer loyalty. The unit of analysis in this study is Bank Jabar Banten (BJB), and the observation unit is the management of the bank. The population is customers of BJB Precious, which is a service from BJB Bank intended for high-net-worth individuals. Fifty respondents of BJB Precious customers in Bandung city were taken as sample. The data were analyzed using Partial Least Square (PLS). The results showed that CRM proved to have a significant effect on customer loyalty of BJB Precious in Bandung. The result of this study is expected to have an implication for the management of BJB bank, namely that the efforts to increase the implementation of CRM of BJB Precious will improve the customer loyalty. The development of CRM would increase repeat purchases, purchase across product lines, referrals of others, and priority customer immunity that would have an impact on achieving company targets.

1 INTRODUCTION

Bank Jabar Banten (BJB) is a business entity bank owned by the Government of West Java and Banten Province, and which is headquartered in Bandung. This bank was established May 20, 1961, in the form of a limited liability company (PT), and then changed the status to a Regional Owned Enterprise. Currently, BJB Bank has 63 branch offices, 311 assistant branch offices, 330 cash offices, 1202 ATMs, 103 payment points, 4 regional offices, and 473 Waroeng BJBs.

Along with the development of premium services for customers, BJB has issued a priority product called "BJB Prioritas" or BJB Priority, which later changed its name to BJB Precious in 2012. BJB Precious is the best service given to privileged persons who need maximum service in terms of hospitality, convenience, security and reliability for High Net Worth Individuals with a minimum amount of IDR 500,000,000 (five hundred million rupiah). The facilities provided for BJB Precious customers are financial statement, travel arrangement, prime service room (precious lounge) at the branch office, precious outlet, special gifts, special events, customer gatherings, free airport lounge for two persons, the BJB precious magazine, a welcoming gift, and merchandise.

The name renewal of BJB Prioritas to BJB Precious aims to strengthen the exclusive branding and services for every customer's financial transaction needs. Re-branding is an effort to increase the loyalty of premium customers whose market potential is greater, so that banking services are increasingly segmented in accordance with the needs of its customers. This service is constantly evolving. BJB Precious outlets have spread in several cities in Indonesia, such as Bandung, Serang, Jakarta, Bekasi, Tasikmalaya, Bogor, Cirebon, Kelapa Gading, Tanggerang, Surabaya, and Semarang.

However, the achievement of BJB Precious service is faced with a number of threats, such as the products of a competitor bank. A number of national banks also have priority services, so that retaining the loyalty of priority customers is the hope of every bank management to be able to realize and to achieve a superior position to its competitors. Meanwhile, according to Kuusik (2007), loyalty is a function to create repeat purchases (Cunningham 1956 and Farley 1964), and, similarly, loyalty is a function of repurchases or a pattern of repeat purchasing (Tucker 1964).

The results of the observations found that problems arise related to the implementation of Customer Relationship Management (CRM) to the priority customers. The membership card program has not been effective yet in increasing customer loyalty. In addition, there is no uniqueness of the personal program to fulfil customers' needs. On the other hand, Coulter and Coulter (2002) found that the impact of service characteristic services on trust is greater if there are longer relations. Thus, this study investigates the effect of CRM on customer loyalty BJB Precious in Bandung City.

2 METHOD

This research used a quantitative approach in which the data were collected using cross-section/one shot, meaning that the data were taken from the results of research conducted at one particular time, namely during the year 2017. The unit of analysis in this research was BJB Bank and the observation unit was the management of the bank. The research population was BJB Precious customers intended for high-net-worth individuals, and the samples were 50 respondents of BJB Precious customers in Bandung city. The data were collected through questionnaires and then processed through causality analysis which was used to obtain evidence of a causal relationship between variables using Partial Least Square (PLS).

3 RESULTS AND DISCUSSION

3.1 Measurement model analysis (inner model)

The two constructs in the structural model are exogenous and endogenous. The CRM and customer loyalty are formative constructs, so PLS is used to analyze data and assess the construct convergent and discriminant validity.

3.1.1 Convergent validity

The measurement used to assess convergent validity from reflective construct are average variance extracted, construct composite reliability, and factor loadings. Table 1 shows the average variance extracted for every reflective construct over 0.50 and construct composite reliability over 0.70 (Fornell & Larcker 1981 and Hair et al. 2006). Factor loadings and cross loadings for the reflective indicator are displayed in Table 2 and Table 3, which show that factor loadings for all of the cross loadings is over 0.70. This shows that all of the variables have a good reliability.

Table 1. Test of the outer model.

	AVE	Composite Reliability	Cronbach's Alpha
Customer loyalty	0.558	0.834	0.735
CRM	0.546	0.782	0.587

Table 2. Loading factor of latent variable indicator.

Indicator-Variable	λ	SE	t-value	Conclusion
X1 <- CRM	0.687	0.082	8.360	valid
X2 <- CRM	0.734	0.072	10.190	valid
X3 <- CRM	0.792	0.038	20.898	valid
Y1 <- Customer loyalty	0.744	0.061	12.132	valid
Y2 <- Customer loyalty	0.737	0.055	13.456	valid
Y3 <- Customer loyalty	0.787	0.051	15.530	valid
Y4 <- Customer loyalty	0.718	0.062	11.657	valid

Table 3. Cross loading of latent variable—indicator.

	Customer loyalty	CRM
X1	0.470	0.687
X2	0.514	0.734
X3	0.632	0.792
Y1	0.744	0.560
Y2	0.737	0.534
Y3	0.787	0.551
Y4	0.718	0.555

Factor loadings and cross loadings for reflective indicator are displayed in Table 2. Factor loadings of all cross loadings were over 0.70.

The results of the measurement model of dimensions by its indicators showed that the indicators were valid with a value of t value < t table (= 2.01).

3.2 Structural model analysis (inner model)

The analysis of the inner model showed the relationship between latent variables. The inner model is evaluated by the value of R Square on endogenous constructs and prediction relevance (Q square), known as Stone-Geisser's, shows the difference between the values of the observations results with the values predicted by the model. This test is indicated by the value of R Square and prediction relevance (Q square), known as Stone-Geisser's, and was used to find out the capability of prediction with the blind-folding procedure. The value obtained was 0.02 (minor), 0.15 (medium) and 0.35 (large), and only used for the endogenous construct with reflective indicator. Referring to Chin (1998), the value of R square was 0.67 (strong), 0.33 (medium) and 0.19 (weak).

The data shown in Table 4 show that the value of R2 on competitive strategy as endogenous variable is in the medium criteria range (> 0.33 = medium), and the value of Q Square is in the large criteria range (> 0.15 = medium), so that it can be concluded that the research model is fit. Figure 1 shows the complete path diagram.

Based on the research framework, a structural model is obtained, as follows:

$$Y = 0.737X1 + \zeta 1 \qquad (1)$$

where Y = customer loyalty; X1 = CR;
$\zeta 1$ = residual.

Table 4. Test of the outer and inner models.

	R Square	Q Square
Customer Loyalty	0.543	0.274
CRM		0.201

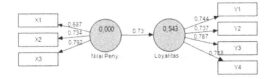

Figure 1. Complete path diagram of the research model.

3.3 Hypothesis testing

Table 5 shows the result of partial testing of the hypothesis.

Table 5 shows that CRM influenced the customer loyalty significantly with 54.3%.

3.4 Results

Based on the results of hypothesis testing, the model findings are illustrated in Figure 2:

The results of this study illustrated that CRM proved to affect the customer loyalty, so this result supports the research hypothesis. CRM is measured by continuity marketing, One-to-One Marketing, and Partnering/Co-Marketing dimensions. The result of hypothesis testing shows that the CRM aspect, which gives the highest effect in improving customer loyalty, is Partnering/Co-Marketing, followed by One-to-One Marketing, and Continuity marketing. The results of this study illustrate that efforts to increase customer loyalty should rely on increasing efforts to create a close partnership relationship between customers of BJB Precious with the bank. Customer loyalty is characterized by high repeat purchases, purchase across product lines, referrals, and immunity.

Table 5. Partial testing of the hypothesis.

Hypothesis	γ	SE (γ)	t-value	R^2	Conclusion
CRM -> Customer loyalty	0.737	0.046	15.995*	0.543	Hypothesis accepted

The results of this study support the findings of Al Dalayeens' study (2017) measuring the effect of variations in CRM on reliability, responsiveness, assurance, empathy and customer relations. Coulter and Coulter (2002) found that the impact of the service characteristic on trust was greater if there were longer existing relations.

4 CONCLUSION

The results of this study support the hypothesis that CRM has a significant effect on customer loyalty of BJB Precious in Bandung city. It is expected that the study provides benefits for BJB bank management in an effort to improve the customer loyalty of BJB Precious through the efforts of increasing the implementation of CRM, prioritized in the aspect of creating a closer partnership between BJB Precious customers and the bank. The findings in this paper can also be an input in building a framework for further research that is interested in the study of priority customer loyalty of BJB banks and in other banks that have priority customer services.

REFERENCES

Al Dalayeen, B. 2017. Impact of customer relationship management practices on customer's satisfaction in Jordan Ahli Bank and Bank Al-Etihad. *Journal of Service Science and Management* 10(01): 87.

Chin, W.W. 1998. The partial least squares approach to structural equation modeling. *Modern Methods for Business Research* 295(2): 295–336.

Coulter, K.S. & Coulter, R.A. 2002. Determinants of trust in a service provider: The moderating role of length of relationship. *Journal of Services Marketing* 16(1): 35–50.

Cunningham, R.M. 1956. Brand loyalty what, where, how much? Harvard. *Business Review* 34: 116–128.

Farley, J.U. 1964. Why does "brand loyalty" vary over products? *Journal of Marketing Research* 1(4): 9–14.

Fornell, C. & Larcker, D.F. 1981. Evaluating structural equation models with unobservable variables and measurement error. *Journal of Marketing Research* 18(1): 39–50.

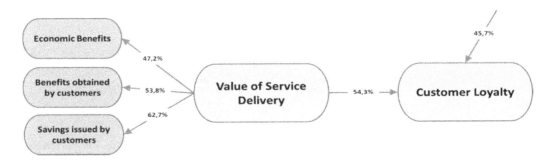

Figure 2. Research findings.

Hair, J.F., Anderson, R., Tatham, R. & Black, W. 2006. *Multi-variate data analysis 6th edition.* New Jersey: Prentice Hall.

Kuusik, A. 2007. Affecting customer loyalty: Do different factors have various influences in different loyalty levels? *The University of Tartu Faculty of Economics and Business Administration Working Paper* 58.

Tucker, W.T. 1964. The development of brand loyalty. *Journal of Marketing Research* 1(3): 32–35.

Advances in Business, Management and Entrepreneurship – Hurriyati et al (eds)
© *2020 Taylor & Francis Group, London, ISBN 978-0-367-27176-3*

Customer relationship marketing performance in Sharia banking

L. Lisnawati, D. Qibtiyah & R. Hurriyati
Universitas Pendidikan Indonesia, Bandung, Indonesia

ABSTRACT: The purpose of this study is to investigate the description of customer relationship management performance on customer loyalty to Sharia banking customers in Indonesia. The design of this study is a cross sectional method. This research uses a descriptive approach with a confirmatory survey method. A total of 110 respondents were selected using probability sampling. A questionnaire was used as a research instrument to collect data from respondents. The analysis technique was descriptive technique by using frequency distribution. The result showed that customer relationship management is in the 'good'-category, while customer loyalty is in the 'good enough'-category. The differences in this study were based on object research, time research, a measuring instrument, the literature that was used, the theory that was used and the results of the study.

1 INTRODUCTION

Maintaining customers is very important for companies in this competitive business world. One way of doing this is by maintaining customer loyalty (Ehigie 2006; McMullan & Gilmore 2008). Customer loyalty is very important for the continuity of the business world (Kassim & Abdullah 2010). Customer loyalty has a significant long-term positive impact on business profitability (Lacey & Sneath 2006, Ramaseshan 2013, Ribbink et al. 2004).

Customer loyalty has developed over time. Before the mid-1980s, customer loyalty only referred to product quality, in other words, giving promises and meeting minimum standards. In the early 1990s, the company began focusing on what customers wanted and listened to customer complaints. In the late 1990s, the market's focus shifted to competitors. The company began creating new and unique products to defeat competitors (Gonring 2008). The process of managing customer relationships is growing and customer loyalty is at the center of business success. The company's focus is on how to attract and maintain existing targets by creating good relationships through trust and commitment (McMullan & Gilmore 2008, Gonring 2008).

The development of customer loyalty has been an important focus in the field of marketing in recent years because of the benefits associated with retaining existing customers (McMullan 2005, Söderlund 2006). The four basic relationships of marketing are trust, commitment, communication, and conflict handling (Ndubisi 2007). The company continues to maintain relationships with customers to obtain valuable information about how best to serve customers so that they do not switch to other brands (Ndubisi 2007). Although service quality is important in developing customer loyalty, there must be some customers who are not served the same, for example

due to limited organizational resources. When they feel their services are differentiated, customers have perceptions that will affect their judgment about the service quality (Prentice 2013).

The company should make customers repeat and develop loyalty to the company. Customer loyalty can decrease due to many competitors. In the banking industry there are many problems with customer loyalty. Many researchers discuss customer loyalty in the banking industry; (Ndubisi et al. 2007), in their research entitled *"Supplier-customer relationship management and customer loyalty, the banking industry perspective"*, explained the problem with customer loyalty in the banking industry. Amin et al. (2013), in *"Islamic banks contrasting customers of image satisfaction, trust, and loyalty of Muslims and non-Muslim customers in Malaysia,"* explained the existence of problems regarding customer loyalty in the banking industry. The banking industry in Indonesia also continues to improve the quality of services to maintain its customers, just like conventional banks and Islamic banks continue to compete to obtain and retain customers.

Islamic banking in Indonesia is claimed to be the largest retail Islamic banking in the world. With more than 18 million customers and more than 4500 branches in 2015, Islamic banking in Indonesia has a great potential for development.

It is endeavored that Islamic banking not only attracts Muslims, because this will naturally limit the growth of the Islamic banking industry. Non-Muslims also have the potential to increase growth to a greater extent and may lead to the internationalization of the success of Islamic banking. The situation in Malaysia is quite interesting because many non-Muslims use Islamic banking products. Non-Muslim customers usually have an ongoing relationship with a conventional bank that sells cross-Sharia banking products from wholly-owned Islamic banking subsidiaries. The fact is, Malaysian Islamic banks seem to attract an increase in the number of non-Muslim customers, perhaps indicating that Malaysia is very open to diversity. The ethnic composition of the country is as follows: non-Muslims account for around 40 percent of the population and is the main segment of banking customers. However, little is known about how non-Muslims consider Islamic banks. Customer loyalty in Islamic banks deal with contrasting perceptions of Muslim and non-Muslim customers of Islamic banks (Amin et al. 2013).

Sharia banking experienced a significant decrease in loyalty compared to other Islamic banks. The decline in loyalty in some banking industries has resulted in an increasingly fierce competition. Every bank continues to improve the quality of its services in all aspects. There are many things that really influence the decline in customer loyalty, such as customer satisfaction, customer engagement, customer relationship management, customer trust, and so on. One of them is by improving the customer relationship management performance, which is important for a company as the company's survival is based on the existence of customers. Every company must be able to plan and identify what customers need; customer relationship management can increase mutual trust and commitment, which can affect the relationship between the two parties. The importance of good customer relationship management, combined with good service quality will provide the experience to the customer and lead to a feeling of trust and satisfaction (Barati et al. 2016).

Customer relationship management is not only about sellers and buyers but has a closer relationship than that. Customer relationship management focuses on maximizing customer value and certain customer segments. Customer relationship management also focuses on improving and developing strong relationships with clients and suppliers, markets created for customer needs, markets created by the influence of influential people, the recruitment market, and the domestic market (Fayazi & Gaskari 2016).

Good and proper implementation of customer relationship management can have a positive impact on the company. Customers will feel cared for and given benefits which makes customers feel satisfied and which ultimately creates loyalty to the company and provides benefits for the company. Sharia banking continues to improve customer relationship management, for instance through the presence of corporate banking programs (export transactions, collections, foreign money transfer transactions, domestic documented credit letters, import transactions, cash management, corporate internet banking and bank guarantees), small business (bank guarantee, bank statements and remittances) and consumer banking (bank notes).

The purpose of this research is to: (1) gain an overview of customer relationship management;

(2) obtain an overview of customer loyalty; and (3) get an overview of customer relationship management performance and customer loyalty.

2 METHOD

This research was conducted to analyze customer relationship management in shaping customer loyalty to Sharia banking customers in Indonesia. Independent variables or exogenous variables contained in this study are customer relationship management with bonding, trust, communication, commitment and satisfaction dimensions. The dependent variable or endogenous variable in this study is customer loyalty with the dimensions of repurchase intention, word-of-mouth, customer retention and customer value.

The object/unit of analysis in this study is Nasa-bah Sharia banking in Indonesia. This research was conducted in less than one year, so the data collection technique used in this study was a cross-sectional method. The technique used in this study is a non-probability sampling technique,accidental sampling with a total sample of 110 respondents. Data collection techniques used were library studies, field studies with questionnaires, and literature studies. Data analysis was performed with descriptive analysis using frequency distribution.

3 RESULT AND DISCUSSION

3.1 Overview of customer relationship management

Based on the results of data processing which has been done through distributing questionnaires to Sharia banking customers, it can be measured through the calculation of dimension scores from customer relationship management. The total customer relationship management score is 11,174, while the ideal score is 3850.

This comparison shows that the implementation of customer relationship management performance at Sharia banking customers in Indonesia has been

Table 1. Customer response recapitulation regarding the performance of customer relationship management.

Dimension	Score	Average Score	Ideal Score	%
Bonding	2810	562.00	770	72.99
Trust	1693	564.33	770	73.29
Communication	2220	555.00	770	72.08
Commitment	1690	563.33	770	73.16
Satisfaction	2761	552.20	770	71.71
Total	11,174	2796.87	3850	72.65

going well. Based on the results of the research from the questionnaire distributed to 110 respondents, it can be seen that customer relationship management of Sharia banking customers in Indonesia receives a score of 11,174, or 72.65% from the ideal score of 15,400. The continuum score is in the 'good'-category with an interval between 11,628.57 and 13,514.29. When viewed by dimensions, the trust dimension gets the highest achievement percentage with a score of 1693, or 73.29% from the ideal score of 2310 The satisfaction dimension gets the lowest percentage of achievement with a score level of 2761, or 71.71% of the ideal score of 3850.

The highest score is the trust dimension of 73.29%, because trust is an important component in the creation of customer relationship management, and because trust is the main thing that ensures a positive relationship arises between consumers and the company. These findings are in accordance with research (Ndubisi 2007) that trust is an important element in the company's relationship with customers and will ultimately shape customer loyalty. Therefore, companies must strive to win customer trust. This can be achieved by keeping promises to customers, showing concern for transaction security, providing quality services, showing respect for customers through frontline staff, fulfilling obligations, and acting to build customer trust in the company and its services. While the lowest score was obtained by the dimension of satisfaction of 71.71%, due to the creation of customer relationship management starting from a very high level of trust that would lead to satisfaction for consumers. Though customer satisfaction is an important antecedent to consumer retention (Wibowo & Yuniawati 2007).

3.2 Overview of customer loyalty

Based on the results of the research from the questionnaire distributed to 110 respondents it can be seen that customer loyalty in Sharia banking in Indonesia received a score of 24,064 or 72.64%, in comparison to the ideal score of 33,110. The continuum score was in the 'fairly good'-category, between 20,947.14 and 25,001.43. If viewed based on dimensions, the customer value dimension gets the highest achievement percentage with a score of 2813 or 73.06% from the ideal score of 3850 and the word-of-mouth dimension gets the lowest percentage of achievement with a score of 5005 or 72.22% of the ideal score 6930. When viewed based on the indicator of overall customer loyalty, the indicator that gets the highest score is the amount of knowledge and competence given by Sharia banking with a score of 572 or equal to 74.29%. The indicator that gets the lowest score is the desire to reference Sharia banking, Sharia products/services to the closest friends in the future with a score of 539 or 70%.

The results showed that customer value is an important dimension and the first step in achieving customer loyalty. This was stated in the study

Table 2. Recapitulation of customer responses regarding customer loyalty.

Dimension	Score	Average Score	Ideal Score	%
Repurchase Intention	7236	557	770	72.29
Word of Mouth	5005	556.11	770	72.22
Customer Retention	8992	562.00	770	72.99
Customer Value	2813	562.60	770	73.06
Total	24,046	2237	3080	72.64

(Atalık 2009) in a marketing strategy that aims to create value for customers: companies must determine services and criteria that are considered important by customers accurately and in accordance with customer expectations. Thus, the level of customer satisfaction about services and the criteria offered must continue to be measured. As well as creating benefits for customers, services and criteria that satisfy customers and are considered important by them also create value and loyalty for customers.

4 CONCLUSION

Customer relationship management study at Sharia banking in Indonesia is in the 'good'-category. This means that Sharia banking customers have good overall service/product usage experience. The dimension that gets the highest response is the trust dimension and the lowest response is satisfaction. This is because trust is an important component in the creation of customer relationship management, because trust is the main thing that ensures a positive relationship arises between consumers and companies.

The description of customer loyalty in Sharia banking customers is in a 'fairly good'-category. This means that the condition of customer loyalty in the use of Sharia banking services/products has shown good performance. The dimension that gets the highest response is the customer value dimension and the dimension with the lowest response rate is word-of-mouth. The results of the study show that customer value is a very important dimension and the first step in achieving customer loyalty. In a marketing strategy that aims to create value for customers, companies must determine the services and criteria that are considered important by customers accurately and in accordance with customer expectations. Thus, the level of customer satisfaction about services and the criteria offered must continue to be measured.

The existence of this research is expected to be able to help the next researcher in conducting research on customer relationship management and customer loyalty by using different indicators from more diverse sources of theory, and on different objects, because there are still many limitations in this study, especially those relating to research methods and data collection techniques. This research is expected to help companies pay attention to the steps of analyzing customer relationship management performance and customer loyalty.

The results show that customer relationship management performance and customer loyalty are good. But the company needs to analyze where the advantages and disadvantages of each of these variables lie, in order to improve the things that need to be improved and maintain the things that are already good.

In this study the dimensions of customer relationship management variables that must be considered by Sharia banking companies in Indonesia in particular, must pay more attention to aspects of satisfaction because they get the lowest rating from all dimensions. The principle of usefulness of services and products has not been conveyed perfectly to consumers so that the company needs to make a better effort to improve it.

Even though customer loyalty variables performance has shown good results, the level of customer loyalty from consumers is still at the initial level, which can be seen from the acquisition of research results which show that consumers mostly have high ratings on the customer value dimension. This means that companies must strive harder to increase customer loyalty levels to the highest level so they want to recommend the brand to others.

REFERENCES

Amin, M., Isa, Z. & Fontaine, R. 2013. Islamic banks. *International Journal of Bank Marketing* 31(2): 79–97.

Atalık, Ö. 2009. A Study to Determine the Effects of Customer Value on Customer Loyalty in Airline Companies Operating: Case of Turkish Air Travellers 4(6): 154–162.

Barati, M., Jafari, D. & Moghaddam, S. S. 2016. Investigating the Effect of Types of Relationship Marketing in Customer Loyalty by using Structural Equation Modeling (SEM)(Case Study Mellat Bank Branches of Tehran. *International Journal of Humanities and Cultural Studies (IJHCS)* 2(2): 632–650.

Ehigie, B. O. (2006). Correlates of customer loyalty to their bank: A case study in Nigeria. *International Journal of Bank Marketing*, 24(7),494–508. https://doi.org/10.1108/02652320610712102.

Fayazi, A. R., & Gaskari, R. (2016). The relationship between relationship marketing and customer loyalty in Mahshahr Petrochemical Special Economic Zone Companies, 2208–2215.

Gonring, M. P. 2008. Customer loyalty and employee engagement: An alignment for value. *Journal of Business Strategy* 29(4): 29–40.

Kassim, N. & Abdullah, N. A. 2010. The effect of perceived service quality dimensions on customer satisfaction, trust, and loyalty in e-commerce settings: A cross cultural analysis. *Asia Pacific Journal of Marketing and Logistics* 22(3): 351–371.

Lacey, R. & Sneath, J. Z. 2006. Customer loyalty programs: are they fair to consumers? *Journal of Consumer Marketing* 23(7): 458–464.

McMullan, R. 2005. A multiple-item scale for measuring customer loyalty development. *Journal of Services Marketing* 19(7): 470–481.

McMullan, R. & Gilmore, A. 2008. Customer loyalty: an empirical study. *European Journal of Marketing* 42(9/10): 1084–1094.

Ndubisi, N. O. 2007. Relationship marketing and customer loyalty. *Marketing Intelligence & Planning* 25(1): 98–106.

Ndubisi, N. O., Wah, C. K. & Ndubisi, G. C. 2007. Supplier-customer relationship management and customer loyalty: The banking industry perspective. *Journal of Enterprise Information Management* 20(2): 222–236.

Prentice, C. 2013. Service quality perceptions and customer loyalty in casinos. *International Journal of Contemporary Hospitality Management* 25(1): 49–64.

Ramaseshan, B. 2013. Effects of customer equity drivers on customer loyalty in B2B context. *Journal of Business & Industrial Marketing* 28(4): 335–346.

Ribbink, D., Riel, A. C. R. Van, Liljander, V. & Streukens, S. 2004. Comfort your online customer: quality, trust and loyalty on the internet. *Managing Service Quality* 14(6): 446–456.

Söderlund, M. 2006. Measuring customer loyalty with multi-item scales: A case for caution. *International Journal of Service Industry Management* 17(1): 76–98.

Wibowo, L. A. & Yuniawati, Y. 2007. The Influence of Tourist Product Attribute and Trust to Tourist Satisfaction and Loyalty A Study of Mini Vacation in Bandung. *Upi* 53(9): 1689–1699.

Advances in Business, Management and Entrepreneurship – Hurriyati et al (eds)
© 2020 Taylor & Francis Group, London, ISBN 978-0-367-27176-3

Potential development strategy of marathon sports tourism in improving the visits of tourists to Bandung

I. Yusup, S. Sulastri, A. Fauziyah & T. Koeswandi
Universitas Pendidikan Indonesia, Bandung, Indonesia

ABSTRACT: Tourism development requires creative and innovative management based on careful planning, consistent implementation, and measurable and constructive evaluation. Tourism development in Bandung is an integrated and holistic development that will bring satisfaction to all parties. In addition, the improvement of the image of Bandung can be accompanied by community-based tourism activities, such as marathon running tours. Marathons are one type of sporting event that is popular today. The purpose of this research is to conduct an in-depth study examining the potential of running sports in Bandung in order to contribute maximally, and know the success factors in organizing a marathon running event that has an impact on the people of Bandung and its impact on tourism in Indonesia. This research uses descriptive quantitative method. The data was collected through a closed questionnaire to runners or community runners and participants in the marathon race held Bandung.

1 INTRODUCTION

Building tourism requires the involvement of the community in order to spur an increase in global competitiveness and increase foreign exchange income. Increasing the image of Indonesian tourism is accompanied by community-based tourism development. Based on Law No. 10 (2009), tourism is defined as a variety of tourism activities and supported by various facilities and services provided by the community, businessmen, government and regional government. Tourism is the activity of traveling with the aim of gaining pleasure, seeking satisfaction, knowing something, improving health, enjoying sports or resting, performing tasks, making pilgrimages and others. Meanwhile, exercise is a series of organized and planned physical movements to maintain motion and improve movement ability.

In reality, sports tourism is also known as sporting events. The events engage in sports activities, so sports tourism is a tourist activity while exercising. Sports in tourism is not only for professionals, but for certain segments, which include professional and amateur communities. For example, a marathon race involves many communities of professional and amateur runners, who seem to have become the main sport tourism.

Indonesia is rich in natural and cultural resources that have the potential to sporting events as well as a factor of attraction in tourism products that are worth selling. Tourism sports events have begun to be held such as the Tour de Singkarak, Musi Triatlon, and Sail Indonesia, which highlight the vastness and beauty of marine tourism in Indonesia. Bandung also has great potential in organizing sporting events, such as rock climbing, mountain biking and marathon running. For the past few years Bandung has routinely held various marathon races, which is one of the efforts to make Bandung the new running sports destination as well as a culinary and shopping destination.

In an effort to develop a marathon running event in a big city such as Bandung, building an image as a tourist destination for participants and the wider audience is needed. Therefore, if you are successful in organizing a running event, the chance will be an on-going event every year. The success of a marathon race can be seen not only from the number of runners participating in the competition, but also from the number of visitors who watched the event as spectators.

The economic activities of the community can be increased through the visits of tourists who watch the races. The marathon running event can reach the business stage that stimulates tourism and is expected to give benefit to the community. The purpose of this research is to examine the potential of running sports in the city of Bandung in order to contribute maximally, and know the success factors in organizing a marathon running event that has an impact on the people in Bandung and its impact on tourism in Indonesia.

2 METHOD

This study uses a quantitative descriptive approach. This study uses a survey through questionnaires and the involvement of industry players in a discussion group (FGD). The purpose of this study is to describe the tourism potential of marathon running sports in Bandung and also get an overview of tourists' perceptions of running sports tourism in Bandung related to the development strategy of tourist destinations to increase tourist visits in Bandung as an international marathon running destination.

The main data collection instrument in this study was a questionnaire consisting of indicators to measure the internal and external environment, and the image of the city as a tourist destination. Interviews using closed questionnaires, observation, and documentation were also involved.

3 RESULT AND DISCUSSION

In developing the tourism potential of marathon running sports in Bandung, various efforts and strategies can also be carried out so that Bandung can be a new running sports tourism destination consistently. It must be consistence because today Bandung is better known as a shopping tourist destination, which sometimes experiences a stagnant situation. This situation can occur because the need for an item for each person depends on certain conditions, such as economic conditions, trends and others. Therefore, Bandung, known as a creative city, must be able to find another attraction as a tourist destination that will have a positive impact on the people's economy, one of which is the tourism potential of marathon running. In order for the city of Bandung to become a world-class marathon-run tourism destination, strategies and appropriate efforts are required so that the city of Bandung can develop into a new running sports tourism destination and increase tourist visits.

3.1 The role of social media

Based on the latest data marathon running events in the city of Bandung have increased every year. One of them was when Bandung had a marathon running event which was quite large, the Pocari Bandung Marathon 2017. The event was able to attract the attention of lovers of marathon running with the ability to collect 6500 participants. Seventy-five percent of participants came from outside Bandung. This reflects the sport proved to be able to have a significant impact to attract tourists to come to the city of Bandung. This marathon event contained several categories, including the Kid Dash, 5 kilometers, 10 kilometers, Half Marathon or 21 kilometers and Full Marathon which is 42 kilometers away (Subhan 2017).

Then, in the following year, the Pocari Sweat Bandung Marathon 2018 was held and was able to gather more participants, namely 8000 participants from home and abroad with the following categories.

From 8000 participants, 10% of them were foreign participants, most of them from Japan and Malaysia. From Indonesia were around 90% consisting of residents from West Java 40%, from outside West Java 50% (Wijanarko 2018).

One of the factors that made the high number of participants of the Pocari Bandung Marathon event was the massive promotional role in various media. One of them is the use of social media that can be accessed at home and abroad. Thus, choosing the right social media as a marketing communication tool is needed.

Marketing communication is a tool used by companies in an effort to inform, persuade, and remind consumers, both directly and indirectly about the products and brands (Kotler & Keller 2007).

Facebook is in the first position with the number of users about 2.1 billion. The second place is occupied by YouTube with the number of users around 1.5 billion, and the third place is occupied by WhatsApp and Facebook Messenger with the number of users about 1.3 billion. Furthermore, table 1 and 2 are the data from the top 10 countries and the top 10 cities with the highest number of Facebook users in the world (We Are Social and Hootsuite, 2018).

In addition, social media also plays an important role in the development of various running communities in the city of Bandung. Many of these communities use social media as a medium to communicate and capture the interests of their members. Instagram social media and Facebook have been chosen as priorities because of the large number of users in Indonesia. This social media also seems to be a place to inspire a wide audience with their various posts while held the running activities.

3.2 The role of government

Bandung has a potential value to be a tourist destination for marathon running. The cool air and the many city forests are supporting factor that will help to build a very comfortable ambience to run. In maintaining and developing the existing potential, the government has a very important role.

The central government through the Ministry of Tourism actively supports each region in order to create a destination or activity as a new attraction for tourism. This year's Ministry of Tourism targets to bring in 17 million foreign tourists. A total of 1.5 trillion is allocated this year to be able to promote Indonesian tourism abroad and 1 trillion in the country (Ministry of Tourism, 2017). A large promotion budget is also prepared for the European, Australian, Singapore and Malaysian markets (Pitana 2017). Because the potential of foreign tourists has increased every year.

The Ministry of Tourism will carry out promotions through eight schemes, namely online media, electronic media, print media, space media, travel fair, festival administration, sales mission, and family trip. Of all these promotion schemes, the Ministry of Tourism will be more dominant in conducting promotional activities through digital media (Pitana 2017).

Based on research by UNWTO in 2017 regarding the activities of tourists using digital platforms, it is known that around 82% of tourists prefer to search for information directly about a tourist destination using a digital platform. The remaining 53% of tourists use digital platforms to look for accommodation, 47% are used to find out transportation in tourist destinations, 36% of tourists use digital platforms to find recommendations on where to eat, and 40% use it to look for tourist activities done at destinations. This platform directly connects service providers with consumers, especially those who provide accommodation. With all the efforts of the central government through the Ministry of Tourism this year, the Bandung City Government must be proactive and synergize in order to develop new tourism, especially for marathon running sports that are in great demand by residents, both domestic and foreign. In addition, the local government also actively supports various sports activities with regular car free day events every week. The program also fosters residents' interest in running. Every week the Car Free Day program is crowded with residents, both those who are just looking for breakfast and those who run or run a bicycle.

The government actively participates in revamping sports infrastructure in order to grow the interest of the citizens such as making a running track in the Gasibu field, renovating the Saparua sports complex and also starting to renovate a number of sidewalks to make it more spacious and friendly for runners who want to enjoy the running atmosphere in urban areas.

3.3 The role of society

In building the image of being a city that deserves to be a tourist destination for marathon running, the role of the community is also very important. The warm hospitality of Bandung residents can make the participants run the marathon more comfortably when the marathon runs. Bandung residents' high interest in sports is also an important factor so that a marathon run can take place successfully, because of the high level of citizen participation both as participants and organizers.

Today, the government through the event organizer collaborates with various running communities in holding various marathon running events. The running community can play an active role both as a participant and as a committee. These communities will be happy to participate because they feel their existence is recognized, facilitated and supported by the government.

The more marathon running events held in the city of Bandung, the more natural it will build the image that Bandung is a tourist destination that is worthy of international standard marathon running. If a good image has been formed, the participants of the marathon run will make Bandung a destination for compulsory marathon running every year.

4 CONCLUSION

Sport tourism activities in the city of Bandung are part of an effort to create a new tourist attraction.

Sport tourism is considered very effective because it has a high value of media value or media branding.

The obtained media value can at least give twice the direct impact of tourists who come, because it is promoted by national and international media before, at the moment, and after the event.

The city of Bandung has the potential to become a world-class marathon-run tourism destination. The cool air with many parks and urban forests is very convenient for running.

Social media plays a huge role in promoting various marathon running events in Bandung. Besides that, social media also builds the perception of Bandung as a decent city in holding various sporting events, especially marathon running events.

The participation of the central government through the Ministry of Tourism through massive promotions at home and abroad can attract tourists to take part in marathon running activities in Indonesia.

The participation of the Bandung's government can build the image of Bandung as a new marathon run tourism destination after Bali and Jakarta.

Collaboration between running communities, the government and event organizers in organizing routine marathon events can increase people's interest in running and also maintain the existence of existing running communities.

REFERENCES

Fitriani, E. 2017. 2018 *Tourism promotion, ministry of tourism prepares IDR 2.5 trillion*. [Online]. https://www.beritasatu.com/bisnis/469772. Accessed: April 20, 2018.

Hootsuite. & We Are Social. 2018. *Digital in 2018 in Southeast Asia, essential insight into internet, social media, mobile, and ecommerce use across the region*. [Online]. https://hootsuite.com/uploads/images/stock/Digital-in-2018–001-Global-Overview-Report-v1.02-L.pdf. Accesed: June 19, 2018.

Kotler, P. & Keller, K.L. 2007. *Manajemen Pemasaran*. Jakarta: Indeks.

Ministry of Tourism. 2017. *Strategi Baru Pemerintah Datangkan 17 Juta Wisman dan 275 Juta Wisnus di Tahun 2018*. [Online]. https://kominfo.go.id/content/detail/12849/strategi-baru-pemerintah-datangkan-17-juta-wisman-dan-275-juta-wisnus-di-tahun-2018/0/artikel_gpr. Accessed: March 17, 2018.

Pitana, I.G. 2017. *Promotion schemes*. Jakarta: The Ministry of Indonesia Tourism.

Subhan, I. 2017. *Yuk ikut bandung west java marathon, total hadiah Rp 500 juta*. [Online]. https://www.pikiran-rakyat.com/olah-raga/2017/04/28/yuk-ikut-bandung-west-java-marathon-total-hadiah-rp-500-juta-400060. Accessed: April 14, 2018.

Wijanarko, Y. 2018. *July, Bandung West Java marathon 2018 will be held in Bandung City*. [Online]. https://www.pikiran-rakyat.com/bandung-raya/2018/06/01/juli-ini-bandung-west-java-marathon-2018-digelar-425286. Accessed: Juni 15, 2018.

Advances in Business, Management and Entrepreneurship – Hurriyati et al (eds)
© 2020 Taylor & Francis Group, London, ISBN 978-0-367-27176-3

Effect of online servicescape on behavioural intention online reservation hotel services

G. Razati, A. Irawati & P.D. Dirgantari
Universitas Pendidikan Indonesia, Bandung, Indonesia

ABSTRACT: This study aims to examine the influence of online servicescape against the behavioural intention of Agoda products in Indonesia. The research was verification study, and the method used was explanatory survey. Using a simple random sampling technique, 165 respondents were involved and path analysis was used. The results showed that there was a significant effect of online servicescape on behavioural intention. In an effort to improve the behavioural intention of consumers, it is recommended that companies better understand the factors online servicescape as one of the basis in providing services for users of Agoda websites in Indonesia.

1 INTRODUCTIONS

Behavioural intent or behavioural intention is a key driver for consumers to behave, and has been identified as an important construct in measuring the success of an organization (Lin 2016, Goode & Harris 2007, Liao et al. 2014) where one formulates a conscious plan to do or not to do some behaviour in the future (Foroughi et al. 2016). The concept of behavioural intention refers to the possibility of consumers returning to the company they have used or can be seen as an indicator that marks a consumer wanting to remain a consumer or leave the company that has been serving (Amoah et al. 2016; Zeithaml et al. 1996).

Research on behavioural intention has been carried out on several sectors, including the retail industry (Abubakar 2015), the hospitality industry (Palanivelrajan & Kannan 2015) and (Bhakar & Bhakar 2015), the fast food restaurant industry (Rana 2017), medical industry (Wang 2017), banking industry (Narteh 2016), manufacturing industry (Alok & Kabra 2018), hospital industry (Kim 2010, Kuruzuum & Koksal 2010), tourism industry (Karami et al. 2018), and airline industry (Lerrthaitrakul & Panjakajornsak 2014). The behavioural intention concept experiences changes along with the development of the business world, one of which is caused by the internet (Hakim & Deswindi 2015). The internet plays an important role in improving the quality of services provided by business units (Gangeshwer 2013).

The high internet users resulted in the growth of e-commerce business. The procurement of products and services has changed significantly in recent years, consumers are now using online tools and increasingly searching for online channels to research and buy products or services they need in SWA 2014. This condition resulted in one major impact: the increasing number of companies using a website to interact with potential customers (Hakim & Deswindi 2015).

The role of website websites is very important be-cause it can present the intangible elements or the physical environment of the service itself. Reputa-tion of profitable sites can reduce the uncertainty of new users and help build trust (Pan et al 2013). Through the website, the seller realizes the potential of the internet to support online ordering of the offered products (Doherty & Ellis-Chadwick 2010).

The use of the website can have an enormous impact and greatly affect the course of a business industry, one of which is the hotel services industry. Online Hotel Reservation is popping up in Indonesia, one of them is Agoda. The rapid use of the Agoda website indicates that corporate behavioural intentions are declining.

According to Harris and Goode (2010) most e-commerce fails to generate profits because online consumers are easy to change behaviour, many do not believe in companies especially on the available online payment systems. There can be some impact to the company if they ignore the hate of intention, low loyal customers, leave the company, spend less money with the company, and take legal action for the company (Saha & Theingi 2009).

Another study mentions that if companies ignore behavioural intention, consumers will have negative word of mouth (WOM). WOM is an important tool in business, used as informal communication to direct other consumers about the ownership, use or characteristics of their goods and services (Durna et al. 2015). Consumers tend to spread negative experiences to other customers and indicate intention to switch to competitors (Jandavath & Byram 2016). Consumer switching to competitors can lead to lower levels of product or service purchases and reduced corporate earnings (Kotler & Armstrong 2013).

Factors affecting behavioural intention include service interaction orientation (Liang & Zhang 2011), image perceptions (Durna et al. 2015), product quality (Foroughi et al. 2016), anticipatory emotions (Koenig-lewis & Palmer 2014), service quality (Ha & Jang 2012; Hooper et al. 2013), perceived quality (Jalil et al. 2016), customer satisfaction (Jalil et al. 2016), servicescape (Hooper et al. 2013, Jalil et al. 2016), and online servicescape (Gunawan, Wicaksono 2013, Hakim & Deswindi 2015, Lee & Jeong, 2012; Hunter & Mukerji 2011, Jeon & Jeong, 2009, Kaikkonen, 2012).

Measurement of physical services or servicescape is one of the variables that can increase or affect the behavioural intention, the concept was first proposed by Bitner (1992). In the online context, servicescape is designed indirectly or called online servicescape to create a good impression of the environment when a customer experiences website services (Kaikkonen 2012). Online servicescape has an important role in service organizations because customers can observe the physical arrangements of the parties before they consume direct services (Hakim & Deswindi 2015), as well as because the website provides customers with diverse opportunities to view products or services before they actually visited the location (Lee & Jeong 2012).

2 METHOD

This study aims to analyze the effect of online servicescape on the behavioural intention of Agoda website users in Indonesia. The independent variables in this study are online servicescape (X) as measured by aesthetic appeal, online layout & functionality, financial security and social factors. De-pendent variable is behavioural intention (Y) with dimension consisting of loyalty to company, propensity to switch, willingness to pay more, external response to problem and internal response to problem.

This study was conducted over a period of less than one year, from April to July 2017. Information gathering was conducted only once in a period of time, so this study was one-shot or cross-sectional (Maholtra 2010). Based on the type of verification research conducted through data collection in the field, the method used in this study is an explanatory survey. Data collection techniques used were literature studies, field studies with questionnaires, and literature studies. While the data analysis technique used is path analysis.

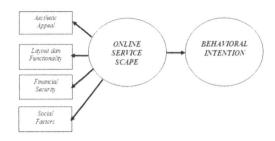

Figure 1. Research paradigm.

3 RESULT AND DISCUSSION

Based on the test of coefficient of path and correlation obtained by the correlation result in the dimension (X_1) to Y equal to 0,986, (X_2) to Y equal to 0,976, (X_3) to Y equal to 0,890 and (X_4) to Y equal to 0,987. Declares that X (online servicescape) completely affects Y (behavioural intention).

The direct and indirect influence between online servicescape's dimensions to behavioural intention on the Agoda website in Indonesia is the most influential partially that is the influence of social factors (X4) on the behavioural intention (Y) of 45.9% (45.45: 98, 98 x 100%) and the least influence is the influence of financial security (X3) to the behavioural intention (Y) of 5.3%. From these calculations we can see the dimensions of online servicescape that reinforce other dimensions of exogenous variables are social factors (X4), while the dimensions that undercut other dimensions of exogenous variables are financial security (X3). In general, the total impact of online servicescape variables on behavioural intention is 98.98%.

This means that online servicescape with behavioural intention has a significant/positive relationship. Thus, the better the servicescape owned online the higher the behavioural intention. This is in accordance with research conducted by Jeon & Jeong (2009), Lee & Li 2014; Hooper et al. 2013) that there is an influence of online servicescape on behavioural intention. Servicescape describes the perceived quality of service as an important factor for encouraging more positive

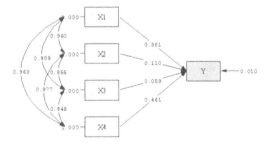

Figure 2. Chart of variable line X on variable Y.

behavioural intentions (Ha & Jang 2012), the difference with the Ha & Jang's research found in the industry taken in the food industry. But not only in the form of food products, a significant topic in the service marketing literature, including in the online context Research conducted by Williams and Dargel (2004) in the online industry shows that designing and planning virtual environments is important in cyberspace, to create an optimal browser experience as in real space.

An online consumer in using a particular product or service must have a reason why he or she continues to reuse or spread WoM positive. Research (Jeon & Jeong 2009, Harris & Goode 2010) in some industries shows that consumer intention is significantly influenced by the evaluation and interpretation of the website. The quality of the website can be represented by e-atmospherics or online servicescape, likewise according to (Kaikkonen 2012, Hunter & Mukerji 2011, Lee & Jeong, 2012) states that online servicescape has a significant effect on behaviour consumers in an online setting. Online servicescape can influence consumers in choosing products to buy or use, they can affect store image through different atmospheric attributes.

4 CONCLUSIONS

The results stated that there is a positive influence between Online Servicescape to Behavioural intention on Online Hotel Reservation Service of Agoda website in Indonesia. Some dimensions to be maintained and improved are aesthetic appeal, online layout and functionality, and social factors. Through a good combination of these dimensions will encourage consumer behaviour to use Agoda products such as by enhancing features on social media used by Agoda, enhancing branding, generating new icons or new programs so Agoda is more easily remembered by consumers and increasing the likelihood of consumers interacting with the service environment provided by Agoda.

The improvement is needed in the financial security dimension that has the lowest value of influence on behavioural intention. Agoda receives payments through PayPal or via a credit card. The payment system can be accessed by internet users from any country, as long as the consumer has a PayPal account or a credit card. However, credit card users in Indonesia are still few enough that companies need to allow other payment methods, including payment by ATM transfer, or via services such as e-pay BRI and e-pay Mandiri. This will expand the reach of consumers and facilitate the payment process which will have an impact on the increase of consumer behavioural intention.

Behavioural intention of consumers on the Agoda website as a whole is still low, so there are several things that must be improved, including willingness to pay more and propensity to switch. This condition shows that consumers are easy to switch from one company to another and are reluctant to pay more

for the products used. Consumers easily switch between them because many new companies present offer a variety of attractive programs with relatively lower prices. Agoda should view this as a profitable challenge if it can implement a marketing strategy that suits market demand over time, enabling the company to compete with other service providers.

Thus, the company especially online hotel reservation Agoda can pay attention to the needs and desires of consumers in the physical environment of the company online or online servicescape, in the form of factors of beauty, layout and function, payment system and social factors website that can be perceived by consumers, because basically the consumer's mood to use is influenced by the above situational factors.

REFERENCES

Abubakar, F.M. 2015. *Behavioural intention to adopt point of sales technology in Nigerian retail industry.* (Doctoral dissertation, Universiti Utara Malaysia).

Alok, S., Kabra, A. & Mudam, A. 2018. Predicting the behavioural intention to adopt lean practices: an empirical study in the manufacturing industry. *International Journal of Services and Operations Management* 29(4): 557–578.

Amoah, F., Radder, L. & van Eyk, M. 2016. Perceived experience value, satisfaction and behavioural intentions: A guesthouse experience. *African Journal of Economic and Management Studies* 7(3): 419–433.

Bhakar, S. S., Bhakar, S., & Bhakar, M.S. 2015. Customer Satisfaction or Service Quality–Identifying Mediating Variable and Evaluating Behavioral Intention Model in Hotel Industry: An SEM Approach. *Research Journal of Social Science and Management* 5(2): 111–124.

Bitner, M.J. 1992. Servicescapes: the impact of physical surroundings on customers and employees. *Journal of Services Marketing* 56(2): 57–71.

Doherty, N.F. & Ellis-Chadwick, F. 2010. Internet retailing: the past, the present and the future. *International Journal of Retail & Distribution Management*, 38(11/12): 943–965.

Durna, U., Dedeoglu, B.B. & Balikçioglu, S. 2015. The Role of Servicescape and Image Perceptions of Customers on Behavioural Intentions in the Hotel Industry. *International Journal of Contemporary Hospitality Management* 27(7): 1728–1748.

Foroughi, B., Nikbin, D., Hyun, S.S. & Iranmanesh, M. 2016. Impact of core product quality on sport fans' emotions and behavioral intentions. *International Journal of Sports Marketing and Sponsorship* 17(2): 110–129.

Gangeshwer, D.K. 2013. E-commerce or Internet Marketing: A business Review from Indian context. *International Journal of u-and e-Service, Science and Technology* 6(6): 187–194.

Goode, M. M. H., & Harris, L. C. 2007. Online behavioural intentions: an empirical investigation of antecedents and moderators. *European Journal of Marketing* 41(5/6): 512–536.

Gunawan, C.K. & Wicaksono, A. 2013. Analisa Persepsi Konsumen Terhadap Online Servicescape Website Bakery. *Jurnal Hospitality dan Manajemen Jasa* 1(2): 215–229.

Ha, J. & Jang, S. 2012. The Effects of Dining Atmospherics on Behavioural Intentions Through Quality Perception. *Journal of Services Marketing* 26(3): 204–215.

Hakim, L. & Deswindi, L. 2015. Assessing the Effects of e-servicescape on Customer Intention: A Study on the Hospital Websites in South Jakarta. *Procedia - Social and Behavioural Sciences* 169: 227–239.

Harris, L.C. & Goode, M.M.H. 2010. Online servicescapes, trust, and purchase intentions. *Journal of Services Marketing* 24(3): 230–243.

Hooper, D., Coughlan, J. & Mullen, M.R. 2013. The servicescape as an antecedent to service quality and behavioral intentions. *Journal of Services Marketing* 27(4): 271–280.

Hunter, Rory & Mukerji, B. 2011. The Role of Atmospherics in Influencing Consumer Behaviour in the Online Environment. *International Journal of Business and Social Science* 2(9): 118–125.

International CHRIE Conference-Refereed Track, Event 14.

Jalil, N.A.A., Fikry, A. & Zainuddin, A. 2016. The Impact of Store Atmospherics, Perceived Value, and Customer Satisfaction on Behavioural Intention. *Procedia Economics and Finance*, 37(16): 538–544.

Jandavath, R.K.N. & Byram, A. 2016. Healthcare service quality effect on patient satisfaction and behavioural intentions in corporate hospitals in India. *International Journal of Pharmaceutical and Healthcare Marketing* 10(1): 48–74.

Jeon, M. & Jeong, M. 2009. A Conceptual Framework to Measure E-Servicescape on a B&B Website. *International CHRIE Conference-Refereed Track*, Event 14.

Kaikkonen, T. 2012. The role of online store atmospherics on consumer behaviour: Literature review. *Unpublished research paper from Aalto School of Economics*: 1–27.

Karami, A., Bozbay, Z. & Arghashi, V. 2018. The influence of social media trust on consumer behavioural intention in tourism industry. 1–18.

Kim, M.G. 2010. Do interesting things increase behavioural intentions? A test of the appraisal structure of interest and relationship between interest and behavioural intention : applications in the hospitality industry. *Post presentation at The 15th Annual Graduate Education & Graduate Student Research Conference*.

Koenig-Lewis, N. & Palmer, A. 2014. The effects of anticipatory emotions on service satisfaction and behavioral intention. *Journal of Services Marketing* 28(6): 437–451.

Kotler, P. & Armstrong, G. 2013. *Principles of Marketing (15th Edition)*. New York: Pearson Education.

Kuruuzum, A. & Koksal, C.D. 2010. The impact of service quality on behavioral intention in hospitality industry. *International journal of business and management studies* 2(1): 9–15.

Lee, L.Y. & Li, L.Y. 2014. Effects of servicescape, waiting motivation and conformity on time perception and behavioral intentions. *International Journal of Marketing Studies* 6(4): 83–91.

Lee, S. & Jeong, M. 2012. Effects of e-servicescape on consumers' flow experiences. *Journal of Hospitality and Tourism Technology* 3(1): 47–59.

Lerrthaitrakul, W. & Panjakajornsak, V. 2014. The airline service quality affecting post purchase behavioural intention : empirical evidence from the low cost airline industry. *International Journal of Trade, Economics and Finance* 5(2): 155.

Liang, R.D. & Zhang, J.S. 2011. The effect of service interaction orientation on customer satisfaction and behavioural intention: The moderating effect of dining frequency. *Procedia-Social and Behavioural Sciences* 24: 10261035.

Liao, Y.W., Wang, Y.S. & Yeh, C.H. 2014. Exploring the relationship between intentional and behavioral loyalty in the context of e-tailing. *Internet Research* 24(5): 668–686.

Lin, Y. 2016. An examination of determinants of trade show exhibitors' behavioural intention. *International Journal of Contemporary Hospitality Management* 28 (12): 2630–2653.

Maholtra, K. N. 2010. *Marketing Research: An Applied Orientation Sixth Ed Pearson Education (sixth edition)*. New York: Pearson Education.

Narteh, B. 2016. Service fairness and customer behavioural intention: Evidence from the Ghanaian banking industry. *African Journal of Economic and Management Studies* 7(1): 90–108.

Palanivelrajan, B. & Kannan, A. C. 2015. Service Quality and Behavioural Intention in Hotel Industry: A Path Model Analysis. *International Conference on Inter Disciplinary Research in Engineering and Technology*: 150–157.

Pan, M.C., Kuo, C.Y., Pan, C.T. & Tu, W. 2013. Antecedent of purchase intention: Online seller reputation, product category and surcharge. *Internet Research* 23 (4): 507–522.

Rana, M.W., Lodhi, R.N., Butt, G.R., & Dar, W.U. 2017. How determinants of customer satisfaction are affecting the brand image and behavioral intention in fast food industry of Pakistan? *Journal of Tourism and Hospitality* 6(6).

Saha, G. C., & Theingi. 2009. Service quality, satisfaction, and behavioural intentions. *Managing Service Quality: An International Journal* 19(3): 350–372.

Wang, Y.H. 2017. Expectation, service quality, satisfaction, and behavioral intention-evidence from Taiwan's medical tourism industry. *Advances in Management and Applied Economics* 7(1): 1–16.

Williams, R. & Dargel, M. 2004. From servicescape to "cyberscape." *Marketing Intelligence & Planning* 22(3): 310–320.

Zeithaml, V. A., Berry, L. L., & Parasuraman, A. 1996. The Behavioural Consequence of Service Quality. *Journal of Marketing* 60: 31–46.

Advances in Business, Management and Entrepreneurship – Hurriyati et al (eds)
© *2020 Taylor & Francis Group, London, ISBN 978-0-367-27176-3*

The shape of member loyalty on cooperative enterprise and the factors to impact it

R.R. Padmakusumah
Universitas Widyatama, Bandung, Indonesia

ABSTRACT: This paper aims to determine the form and the factors that determine the loyalty of customer (member) in the cooperative enterprise (cooperative). The approach used in this paper is descriptive analysis. Descriptive analysis was conducted in the form of studies on various literature, namely international journals, national journals, international seminar results, the results of national seminars, unpublished research results, textbooks, professional experience, government regulations and legislation. It can be seen that member loyalty can be seen through two things: contributive participation and incentive participation. Factors affecting cooperative member loyalty include member value, service quality, member satisfaction, relational marketing, and loyalty program. The results of this study can used by other researchers as a basis for them to make another research in relevant field and cooperative managers can obtain references to perfect their understanding and efforts related to member loyalty.

1 INTRODUCTION

Customers are the lifeblood for every business, not least the cooperative business (cooperative). At the cooperative, customer referred to as member. Members who promote the cooperative are members who earn a lot of income for cooperative or member who transact intensively with the cooperative. Aside from being a member, customer on a cooperative also has a function as owner. These two functions are called dual identity.

One important indicator of business or marketing success of a company is seen from the level of customer loyalty. Loyal customers will generate a lot of revenue as well as profits for the company. Customer loyalty does not appear suddenly but is formed or influenced by various factors. International research related to customer loyalty has been done by various parties on various forms of companies but rarely do on the form of cooperative enterprises (cooperatives).

The customer (member) of the cooperative has a dual identity which is in addition to being the customer as well as the owner of the cooperative. This dual identity should certainly have the potential to produce a uniqueness among the concept of customer loyalty to cooperative companies compared to the concept of customer loyalty in non-cooperative companies. This study aims to identify how the uniqueness or form of customer loyalty (member loyalty) on the cooperative and what factors will affect it.

2 METHOD

The approach used in this paper is descriptive analysis, conducted in the form of studies of various literature, including international journals, national journals, international seminar results, the results of national seminars, unpublished research results, textbooks, professional experience, government regulations and legislation.

3 RESULT AND DISCUSSION

3.1 Member loyalty definition

Customer loyalty is defined as a strong commitment of a customer to subscribe to a product or service of a particular company even though it has been influenced by marketing efforts from a competitor company that has the potential to cause a change of behavior to switch to a competitor's product or service (Kotler & Keller 2016). Customer loyalty as a commitment from customers to create bonds or relationships with companies is also expressed by Szczepanska and Gawron (2011) and Mouhammadipour (2014).

Customer loyalty is simply defined as a repeat purchase behavior performed by a customer for a product or service in a company (Tsai et al. 2010, Roostika 2011, and Jain et al. 2014). The customer of the cooperative is called a member and the member of the cooperative has a double identity (National Tim LAPENKOP Nasional 2006), referring to this and referring to the above literatures can be summed up simply by the definition of member loyalty as "a strong desire or commitment of members to build long-term relationships with their cooperatives and make transactions (buying product also investing capital) and manage their the cooperative."

3.2 Elements (dimensions) of member loyalty

Repeat purchase is one of the characteristics or forms of customer loyalty. Another form of customer loyalty is buying more than one types of products that the company offers (Alma 2007), positive recommendation (Tsai et al. 2010, Hur & Kang 2012, Grigouroudis et al. 2013, Jain et al. 2014), willingness to pay (Tsai et al. 2010, Hur & Kang 2012), complain (Tsai et al. 2010), and resistance to better alternatives (Roostika 2011).

One more loyal behavior mentioned above will basically be shown by every loyal customer in a company, not to mention the cooperative company. However, the various forms of loyal behavior above only show the loyal behavior of members as customers and have not shown the loyal behavior of members as owners. In a cooperative concept, some of the earlier-mentioned loyal behaviors (repeated transactions on a cooperative product, purchasing various types of cooperative products and recommending cooperative products to others) are referred to as incentive contributions (Hendar & Kusnadi 1999).

As an owner, member loyalty is demonstrated by the strong behavior or commitment of members to capitalize cooperatives, participate in planning cooperative policies, participate in strategic decision making, and participate in controlling cooperatives (Hendar & Kusnadi 1999). The next section will explain loyal behaviors shown by members of a cooperative more clearly, including:

a. Repeat purchases: members who are loyal to the cooperative are members who make transactions on an ongoing basis with their cooperatives. This recurring transaction is either a transaction to buy a service product provided by a cooperative or a transaction to finance the cooperative (pay principal savings, mandatory savings, voluntary savings, and fund business units owned by the cooperative).

b. Purchases of more than one type of cooperative product: loyal members not only buy one type of product provided by the cooperative but also buy other products provided by the cooperative.

c. Positive word of mouth (recommendation): members who are satisfied with the products or services provided by the cooperative then will invite others to buy products or services and invite others to become members of the cooperative.

d. Willingness to pay: even if the price of service products provided by the cooperative is sometimes expensive, if members feel confident with the prospect of progress and the benefits of cooperative feeding, then they will accept the prices. Members will believe that the increase is temporary.

e. Resistance to better alternatives: competition will certainly make the product cooperative services will always be opposed by competitors in various ways and programs. Loyal members tend to be less susceptible, because they believe that their cooperative service products are better than those offered by other companies. The price, quality, and service of other companies may be better today, but members believe that this is only temporary; once the cooperatives are developed or developed they are confident that the price, quality, and service of the cooperative will be better than those offered by another company.

f. Investing capital: investing capital has become the obligation of members who have status as the owner of the cooperative. Capital that they must plant in the cooperative is the principal savings and mandatory savings. Other capital that they can also cultivate in the cooperative is voluntary savings and business funding to one more business unit owned by the cooperative. To assess the loyalty of members in cooperative capital, it can be measured by the cooperative through their database related to "the level of order of members in paying principal savings as well as mandatory savings." In addition, it can be measured through the perception of related members as "level of interest members to investing capital cooperatives outside the principal savings and mandatory savings (examples: voluntary savings, business financing on cooperative business units, and others)."

g. Participate on managing cooperative: as an owner, members must show at least three roles, namely (1) as planners, (2) as decision makers, and (3) as controllers of their cooperatives. The roles are mainly done by members through member meetings or rapat anggota (RA). The RA can be either an annual member meeting (RAT) or an extraordinary member meeting (RALB).

Planning is done by members in the RA in the formulation, discussion, and ratification of various strategic plans related to business and cooperative organizations. Decision making is done by members in RA in the form of ratification of cooperative work program plan, approval of cooperative budget plan (RAPBK), election of management, election of supervisor, and selection of manager. Control is done by members through RA in the form of approval (acceptance or rejection) on accountability report of cooperative management made by the board and on account-ability report of supervisory cooperative program made by cooperative commissioner.

This form of loyalty as a member of this owner can be measured directly by the cooperative through the "attendance level of members in various member meetings held by the cooperative" and can be measured by asking related members "how important are they present and actively participate in the RA?"

If it is classified into the concept of cooperative (Hendar & Kusnadi 1999) then the behavior of a-b-c-d-e is a category of loyal behavior "incentive participation" and f-g behavior is a category of loyal behavior "contributive participation." If it is classified into the concept of loyalty from Buttle (2007) then the behavior of a-b-c-f is "behavioral loyalty" and the

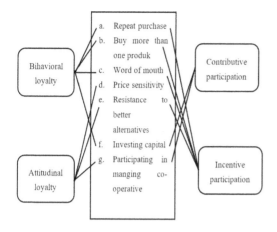

Figure 1. The linkage of loyal behavior, member participation, and loyalty calcification.

behavior of d-e-g is "attitudinal loyalty." "Behavioral loyalty" is a form of loyalty that is seen or can be observed directly by the cooperative of its members, while "attitudinal loyalty" is a form of loyalty that cannot be observed directly by the cooperative but is in the hearts and minds of its members. Figure 1 illustrates more simple and clear interrelationships between various loyal behaviors with the concepts of member participation (Hendar & Kusnadi 1999) and the concept of loyalty classification (Buttle 2007), namely:

3.3 Factors affecting member loyalty

Member loyalty does not appear suddenly, many factors influence it namely: member value (Mutasowifin 2002, Tim LAPENKOP Nasional 2006, Kotler & Keller 2016), service quality (Kasmiri 2010, Guntara 2011, Wulandari, 2016), member satisfaction (Kotler

& Keller 2016), relationship marketing (Chou 2009, Wulandari 2016, Kotler & Keller 2016), and loyalty program (Kotler & Keller 2016). Figure 2. Below shown the relationship between determine factors that affect member loyalty, namely:

4 CONCLUSION

Member loyalty is defined as "a strong desire or commitment of members to build long-term relationships with their cooperatives and make transactions (buying product also investing capital) and manage their cooperative. Forms of loyalty on cooperative are repeat purchase repurchase, buy more than one product offered by cooperative, word of mouth, willingness to pay, resistance to better alternatives, investing capital, participate in managing cooperative. That loyal behavior is classified into two categories, namely "contributive participation" and "incentive participation." Factors that impact member loyalty are determines by member value, service quality, customer satisfaction, relationship marketing, and loyalty program. Member satisfaction not appearance by itself but is determined by member value and service quality.

REFERENCES

Alma, B. 2011. *Manajemen pemasaran dan pemasaran jasa. cetakan ke-9*. Bandung: Alfabeta.
Buttle, F. 2007. *Customer relationship management. Edisi pertama*. Yogyarkata: Bayumedia Publishing.
Chou, H.J. 2009. The effect of experimental and relationship marketing on customer value: A case study of international American casual dining chains in Taiwan. *Social Behavior and Personality, ProQuest Sociology* 37(7):9 33.
Grigoroudis, E., Tsitsiridi, E. & Zopounidis, C. 2013. Linking customer satisfaction, employee appraisal, and business performance: An evaluation methodology in the banking sector. *Annals Operation Research* 205(1): 5–27.
Guntara, A. 2011. *Penguruh keemimpinan manjer dan kualitas peayanan terhadap partisipasi anggota sebagai pelanggan: studi kasus pada unit waserda koperasi karyawan PT. Pindad Bandung*. Tesis. Bandung: Institut Koperasi Indonesia.
Hendar & Kusnadi. 1999. *Ekonomi koperasi: untuk perguruan tinggi*. Jakarta: Lembaga Penerbit FE-UI.
Hur, W.M. & Kang, S. 2012. Interaction effects of the three commitment component on customer loyalty behaviors. *Social Behavior and Personality* 40(9): 1537–1542.
Jain, K., Bhakar, S. & Bhakar, S. 2014. The effect of communication and personalization on loyalty with trust as mediating variabel. *Prestige International Journal of Management & IT- Shancayan* 3(1): 1–20.
Kasmiri, A. 2010. *The service quality of waserda as a unit cooperatives in shaping the member's commitmen and loyalty as customer*. Disertasi. Bandung: Universitas Padjadjaran.
Kotler, P. & Keller, K.L. 2016. Marketing Management. 15th Edition. New Jersey: Pearson Education, Inc.
Mouhammadipour, M. 2014. Comprehensive Model of Customer Loyalty. *Advances in Natural and Applied Sciences* 8(6): 783–791.

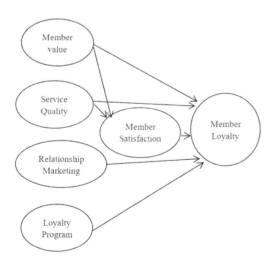

Figure 2. Factors affecting member loyalty.

Mutasowifin, A. 2002. Penerapan balanced score-card sebagai tolak ukur penilaian badan usaha berbentuk koperasi. *Jurnal Universitas Paramadina* 1(3): 245–264..

Roostika, R. 2011. The effect of perceived service quality and trust on loyalty: customer's perspectives on mobile internet adoption. *International Journal of Innovation, Management and Technology* 2(4): 286–291.

Szczpanska, K. & Gawron, P. 2011. Changes in approach to customer loyalty. *Contemporary Economics* 5(1): 60–69.

Tim LAPENKOP Nasional. 2006. *Lebih Mengenal Koperasi*. Sumedang: LAPENKOP Nasional.

Tsai, M.T., Tsai, C.L. & Chan, H.C. 2010. The effect of customer value, satisfaction, and switching costs on customer loyalty: an empirical study of hypermarkets in Taiwan. *Social Behavior and Personality* 38(6): 729–740.

Wulandari, S. 2016. The impacts of service quality, marketing communication, and relationship marketing to customer loyalty and business performance. Thesis. Jakarta: STIMA IMMI.

Advances in Business, Management and Entrepreneurship – Hurriyati et al (eds)
© 2020 Taylor & Francis Group, London, ISBN 978-0-367-27176-3

Analysis of the model of memorable tourist experience, destination image, and tourist value on the revisit intention

V. Verinita & F. Yola

Universitas Andalas, Padang, Indonesia

ABSTRACT: This study aims to analyze the linkage model of memorable tourism experience, destination image and in tention to revisit value on cultural tourism destinations, in particular Istano Basa Pagaruyung in West Sumatera. The study used an explanatory approach. The samples in this study were 160 domestic tourists visiting Istano Basa Pagaruyung at least one time. The sampling technique used a purposive sampling method with cross sectional time scope. Data were analyzed using Structural Equation Modeling (SEM) using SmartPLS version 2.0. The results showed that the variable of memorable tourism experience significantly influences the destination image. Memorable tourism experience affects the tourist value. Destination image significantly influences the tourist value. The variable of the tourist value has a significant effect on the revisit intention, and the destination image has no effect on the revisit intention. These findings imply that the destination image should be enhanced by building the reputation of Istano Basa Pagaruyung as a unique cultural destination and creating comfort and security.

1 INTRODUCTION

Indonesia is a very rich country, both in natural resources, rich in traditional culture of the region, to the diversity of tourism. Tourism is one of the fastest growing sectors in the global economic era. Based on data from the Ministry of Tourism and Creative Economy from November 2017, foreign tourist arrivals reached 1,061,055 visits, a 5.86% increase from the same month in 2016. In addition to providing a foreign exchange contribution for the country, because it is not vulnerable to the tourism sector crisis, it has the potential to become the national economy's safety belt at a time of world economic slowdown due to the global economic crisis. This is because Indonesia has natural resources as well as human resources; culture is very rich and abundant, so that Indonesia has a unique comparative advantage as a tourist destination.

The number of foreign tourist arrivals to Indonesia in November 2016 reached 1 million visits, up 19.98% compared to November 2015. Meanwhile, when compared to October 2016, it decreased by 3.68%. The number of foreign tourists visiting in November 2016 consists of foreign tourists who visit through the 19 main gates of 878,839 thousand visits, and foreign tourists outside the 19 main gates as much as 123,494 thousand visits.

West Sumatera is one of the tourist destinations for domestic tourists and foreign tourists. West Sumatera has many interesting tourist attractions ranging from cultural tourism, nature tourism, sports tours, as well as food tours. The number of foreign tourists visiting West Sumatera through Minangkabau International Airport (BIM) and Teluk Bayur Bay in July 2017 reached 4,080 people, an increase of 38.78 percent compared to June 2017, which recorded 2,940 people. When compared to July 2016, foreign tourists in July 2017 decreased by 0.34 percent. Meanwhile, the number of foreign tourists in January-July 2017 increased by 7.19 percent over the same period the previous year. Foreign tourists in July 2017 contributed to 0.35 percent of the total foreign tourists who visited Indonesia (national foreign tourists 1,180,399 people) (BPS West Sumatera 49/09/13/Th.XX, September 4, 2017).

Istano Basa Pagaruyung is located in Nagari Pagaruyung, Tanjung Emas Subdistrict Tanah Datar Regency which is 5 kilometers from Batusangkar City. Istano Basa Pagaruyung is one of the nine destinations of the world-leading halal tourism of West Sumatera (http://www.sumbarprov.go.id).

This building serves not only as a tourist destination but also as a center of custom development and culture of Minangkabau, as well as an open museum. It can be said to be a storefront or a representation of the well-known Minangkabau culture for the world. Istano Basa Pagaruyung has 11 gonjong, 72 milestones and 3 floors equipped with various carvings that respresent Minangkabau history and cultural philosophy. Istano Basa Pagaruyung is a relic of history, which used to function as the residence of the king of Minangkabau and as the center of the Minangkabau Kingdom government, led by a king known as "Rajo Alam" or king in the Minangkabau kingdom. Istano Basa Pagaruyung was built in 1976 and is a duplicate of the Istano Basa Pagaruyung which was burned by the Dutch in 1804. Istano Basa Pagaruyung also burned due to lightning strikes in 2007 and was then rebuilt. As a tourist place, Istano Pagaruyung offers a collection of relics of the Minangkabau kingdom which are currently duplicates of the original object that had been burned in 2007 (http://harianhaluan.com).

From the previous descriptions, serving as the background for the research, there are some problems in tourism in Tanah Datar regency. Visitors felt a low impressive tourist experience during the trip, which affected to create an unmemorable tourism experience. The experience of negative destinations will have an impact on the planning process and decision of tourists to visit a tourist destination.

Based on this, researchers decided to analyze the model of memorable tourist experiences, destination images, and the value of tourists to revisit intention.

The problem formulations of this research are as follows:

1. How do memorable tourist experiences influence the destination image of Istano Basa Pagaruyung tourist attraction.
2. How does a memorable tourist experience influence the tourist value of Istano Basa Pagaruyung tourist attraction,
3. How does the destination image to tourist value on Istano Basa Pagaruyung influence the tourist attraction.
4. How does tourist value influence the revisit intention to Istano Basa Pagaruyung tourist attraction, and
5. How does the destination image influence the revisit intention to Istano Basa Pagaruyung tourist attraction.

1.1 Concept of tourism

In general, Zalukhu & Meyers (2009) said that tourism is a temporary travel activity undertaken by a person from their original place to a destination for the reason not to settle or earn a living but only to fulfill their curiosity, spend their leisure time or holiday and other purposes.

Tourism is a service industry and travel industry. The successful marketing of this service is highly dependent on the entire travel industry (Kotler & James 2010). William (2006) says that the products of tourism services are very diverse and technically the tourism service comprises all the services, facilities, packages and programs offered by the service providers.

1.2 The concept of Memorable Tourism Experience (MTE)

The importance of a memorable tourist experience theory comes from the influence of past memory powers on consumer decisions. Kim et al. (2012) define memorable tourist experience (MTE) as an easy-to-remember travel experience after the event.

To deliver a unique, exceptional and unforgettable MTE, targeting tourists remains a sustainable competitive advantage (Kim et al. 2010). According to Kim et al. (2012) the components that affect memorable tourism experts are hedonism, novelty, local culture, refreshment, meaningfulness, involvement, and knowledge.

1.3 Destination image

The study of the image of the destination is expressed in the analysis of Luo and Hsieh (2013) which states that the image of the destination refers to the overall impression of a tourist towards the attraction of the destination (or host country).

Suwarduki et al. (2016), adapting the Schiffman and Kanuk concept (2007), stated that there are seven factors that can shape the brand image, including quality or reliability, reliability or reliability, usefulness or usefulness, service, risk, price, and image owned by the brand itself.

Some indicators of the destination image by Setiawan et al. (2014) are entertainment and events, natural tourism, tourist environment, historical attractions, infrastructure, accessibility, relaxation, outdoor activities, price and value.

1.4 Tourist value

The tourist value in a multidimensional perspective is an appropriate study to be used in the service context because it includes the emotional value, social and epistemic value dimensions (William & Soutar 2009) in addition to functional value.

Values can be described as overall customer ratings of the net worth of a service or product based on their assessment of what is received (benefits provided by the service) against what is offered (cost or sacrifice of time, effort, and opportunity cost of obtaining and utilizing the service) (Lai 2015).

Murphy et al. (2000) found that travelers who developed their trip quality had higher perceptions of value, which also increased the intention to return. Cronin et al. (2000) found that perceived value had the most significant impact on the consumers' buying intentions, while quality and satisfaction showed a less significant effect on repurchase intentions. Perceived value refers to the overall evaluation of the customer on the effectiveness of the service product based on the assessment of what is received and what is provided (Zeithaml 1988).

1.5 Revisit intention

Lourerio's research (2014) states that there is a rural experience economy influence on place attachment and behavioral intention with pleasant arousal and memory as intervening variable. Mehmetoglu and Engen (2011), who researched the concept of Pine and Gilmore (1999), examine the concept of the dimensions of the experience economy (education, esthetics, escapism, and entertainment) on the satisfaction of tourists at Ice Music Festival and The Maihaugen Museum.

Meanwhile, Wang and Wu (2011) measured alternative attractiveness, self-image congruity and disconfirmation of the revisit intention of museum visitors with involvement as a moderator variable. Claudia (2009) tested four indicators of tourism experience theory as a structure for the study of tourism experiences.

According to Luo and Hsieh (2013), personal experiences that are felt will affect tourists to visit these tourist destinations again. Some of the indicators that influence them to revisit, among others, are: (1) wanting to gain an impressive personal experience, (2) the destination makes the traveler relaxed, both physically and mentally, (3) the destination presents an interesting attraction, (4) revisit the destination for information obtained from social media. It was revealed that the new rating does not intend a re-visit, but they tend to recommend the MTEs to others. Having a personal experience of cultural activities at a tourist destination also encourages tourists to visit again.

2 METHOD

This research is a quantitative research, using hypothesis testing (Explanatory Research). The type of investigation used is a causal study. The time horizon used is a short one (cross-sectional) in the period 2017. The unit of analysis of this research is the visitor attraction Istano Basa Pagaruyuang.

Data source in this research is primary data. According to Sekaran (2011), primary data is data obtained from first hand for analysis and to find a solution or problem. The data collection techniques used are questionnaires.

The population in this study are all visitors of the attraction Istano Basa Pagaruyung. The sample consists of local tourists who have been to Istano Basa Pagaruyuang at least once and originated from the Province of West Sumatera. According to Hair et al. (2010) the number of the research indicator can calculate the determination of the number of samples multiplied by 5. That means the number of minimum samples in this study is 32 x 5 = 160 people. The sampling technique used is non-probability sampling with a sampling method of purposive sampling method.

The data analysis technique used is multiple regression analysis by using data processing software (SmartPLS Version 2.0 for Windows). Test the instrument that is done the test validity, reliability and hypothesis testing.

3 RESULT AND DISCUSSION

In this study, the respondents are tourists from the Province of West Sumatera, who had visited Istano Basa Pagaruyung. The questionnaires were distributed during November 2017. The samples in this study were local tourists who visited Istano Basa

Pagaruyung at least once and who were at least 17 years old. Respondents in this study amounted to 160 respondents who were grouped according to several characteristics based on demographics as sex, age, last education, occupation, monthly average income, marital status, where they got information about Istano Basa Pagaruyung, travel group, transportation mode used, and area of origin. This aspect has an important role in knowing how the effect of memorable tourist experience, destination image, and tourist value is on the revisit intention of visitors of Istano Basa Pagaruyung.

3.1 Hypothesis testing

3.1.1 The influence of memorable tourist experience on the destination image

The hypothesis 1 data analysis, Memorable Tourist Experience toward Destination Image, can be accepted. The hypothesis has been proven from the output obtained that T arithmetic (7.807113) > T table (1.65455). This research indicates overall that memorable tourist experience has a positive and significant effect on the destination image. It can be interpreted that the memory of an impressive experience perceived by tourists , will have an impact on the image of Istano Basa Pagaruyung attractions. This is in accordance with the conditions in the field, that traveling to Istano Basa Pagaruyung can provide an impressive experience for visitors because of the uniqueness-uniqueness in Istano Basa Pagaruyung, which is not held by other attractions.

Prayag and Ryan (2012) stated that the affective image (a pleasurable experience) for tourists will affect their destinations.

3.1.2 The influence of memorable tourist experience on tourist value

The result of the hypothesis 2 data analysis, Memorable Tourist Experience to the Tourist Value, is acceptable. The hypothesis has been proven from the output obtained that T arithmetic (2.463902) > T table (1.65455). This study indicates that overall memorable tourist experience has a positive and significant impact on the value of tourists. It can be interpreted that the memory of an impressive experience perceived by tourists, will have an impact on the values felt by visitors on the object of Istano Basa Pagaruyung. Zatori and Beardsley (2017), suggesting that an impressive tourist experience gives more value than a tourist experience in a tourist spot.

3.1.3 The influence of the destination image on tourist value

The result of the hypothesis 3 data analysis, Destination Image to Tourist Value, is acceptable. The hypothesis has been proven from the output obtained that T arithmetic (20.28037) > T table (1.65455). This study indicates that the overall destination image has a positive and significant impact on the tourist value. It can be interpreted that the image of a good destination felt by tourists during the visit, will have an impact on the perceived value of visitors of the object Istano Basa Pagaruyung.

This is in accordance with Allameh et al. (2015), who state that the destination image affects the value felt by tourists. Aliman et al (2014) suggest that the image of purpose affects the perceived value.

3.1.4 *The influence of tourist value on the revisit intention*

The result of the hypothesis data analysis, that is, Value of Tourists to Revisit Intention, can be accepted. The hypothesis has been proven from the output obtained that T arithmetic (2.309011) > T table (1.65455). This study indicates that overall tourist value has a positive and significant impact on the revisit intention.

This result is in line with Quintal and Polczynski's (2010) study, indicating that satisfaction with attractiveness and the value given positively influences the intention of returning to visit. It can be interpreted that the presence of positive values felt by tourists during the visit will have an impact on the intention to revisit (return visit) the tourist attraction Istano Basa Pagaruyung. It is also seen in the field with the tourists who have visited more than once.

3.1.5 *The influence of the destination image on the revisit intention*

The results of hypothesis 5, the Destination Image to Revisit Intention, was denied. The hypothesis has been proven from the output obtained that T arithmetic (0.483511) <T table (1.65455). This study indicates that the overall destination image has no effect on the revisit intention. It can be interpreted that the image of the destination does not affect the revisiting of tourists.

The research conducted by Mahdzar et al. (2015) states that the destination image positively affects the intention to revisit. This differs from research where it was found that the destination image has no effect on the revisit intention. This is due to the indicator of the destination image variable that has the average answer of the respondents are low answer from the indicators to enjoy the natural attractions behind Istano Basa Pagaruyung. This explains that the existing nature tourism in Istano Basa Pagaruyung cannot give a good impression of Istano Basa Pagaruyung. The Istano Basa Pagaruyung manager should optimize the existing tours in Istano Basa Pagaruyung, like riding, which can be done in Istano Basa Pagaruyung area.

4 CONCLUSION

Based on the results of the research and discussion, it can be concluded that:

1. Variable memorable tourist experience has a significant effect on the variable destination image of the attraction Istano Basa Pagaruyung.

2. Variable memorable tourist experience has a significant effect on the value of variables travelers.
3. Destination image variable (the image of the destination) has a significant effect on the variable value of tourists.
4. The variable of the tourist value has a significant effect on the revisit intention variable.
5. Destination image variable does not have a significant effect on the revisit intention of Istano Basa Pagaruyung

REFERENCES

Aliman, N.K., Hashim, S.M., Wahid, S.D.M. & Harudin, S. 2014. The effects of destination image on trip behavior: Evidences from Langkawi Island, Malaysia. *European Journal of Business and Social Sciences* 3(3): 279–291.

Allameh, S.M., Khazaei Pool, J., Jaberi, A., Salehzadeh, R. & Asadi, H. 2015. Factors influencing sport tourists' revisit intentions: The role and effect of destination image, perceived quality, perceived value and satisfaction. *Asia Pacific Journal of Marketing and Logistics* 27 (2): 191–207.

Claudia, Jurowski. 2009. An examination of the four realms of tourism experience theory. *International CHRIE Con-ference-Refereed Track*: 3.

Cronin, J.J., Brady, M.K., Hult, G.T.M., 2000. Assessing the effects of quality, value and customer satisfaction on consumer behavioral intentions in service environments. Journal of Retailing 76 (2), 193–218.

Hair, J.F., Black, W.C., Babin, B.J., Anderson, R.E. & Tatham, R.L. 2010. *Multivariate data analysis, 7th edition*. Englewood Cliffs: Prentice Hall.

Kim, J.H., Ritchie J.R.B & McCormick. B. 2012. Development of a scale to measure memorable tourism experiences. *Journal of Travel research* 51(1): 12–25.

Kim, J.H., Ritchie, J.R.B. & Tung, V.W.S. 2010. The effect of memorable experience on behavioral intentions in tourism: A structural equation modeling approach. *Tourism Analysis* 15(6): 637–648.

Kotler, J.B. and James Makens. 2010. *Marketing for Hospitali-ty and Tourism. Fifth Edition. International Edition*. New Jersey: Pearson Education Inc.

Lai, I.K. 2015. The roles of value, satisfaction, and commitment in the effect of service quality on customer loyalty in Hong Kong–style tea restaurants. *Cornell Hospitality Quarterly* 56(1): 118–138.

Loureiro, S.M.C. 2014. The role of the rural tourism experience economy in place attachment and behavioral intentions. International *Journal of Hospitality Management* 40:1–9.

Luo, S.J. & Hsieh, L.Y. 2013. Reconstructing revisit intention scale in tourism. *Journal of Applied Sciences* 13 (18):3638–3648.

Mahdzar, M., Shuib, A., Ramachandran, S., & Afandi, S. H. M. (2015). The role of destination attributes and memorable tourism experience in understanding tourist revisit intentions. *American-Eurasian Journal of Agricultural & Environmental Sciences (Tourism & Environment, Social and Management Sciences)* 15: 32–39.

Mehmetoglu, M. & Engen, M. 2011. Pine and Gilmore's concept of experience economy and its dimensions: An empirical examination in tourism. *Journal of Quality Assurance in Hospitality & Tourism* 12(4): 237–255.

Murphy, P., Pritchard, M. P., & Smith, B. (2000). The destination product and its impact on traveller perceptions. Tourism management, 21(1),43–52.

Pine, B. J., Pine, J. & Gilmore, J. H. 1999. *The experience economy: work is theatre & every business a stage.* Boston: Harvard Business Press.

Prayag, G. & Ryan, C. 2012. Antecedents of tourists' loyalty to Mauritius: The role and influence of destination image, place attachment, personal involvement, and satisfaction. *Journal of travel research* 51(3): 342–356.

Quintal, V.A. & Polczynski, A. 2010. Factors influencing tourists' revisit intentions. *Asia Pacific Journal of Marketing and Logistics* 22(4): 554–578.

Schiffman, L. & Kanuk, L.L. 2007. *Perilaku konsumen, edisi ke 7.* Jakarta: PT. Indeks.

Sekaran, U. 2011. *Metodologi penelitian untuk bisnis. Edisi 4 Jilid 1 & 2.* Jakarta: Salemba Empat.

Setiawan, P.Y., Troena, E.A. & Armanu, N. 2014. The effect of e-WOM on destination image, satisfaction and loyalty. *International Journal of Business and Management Invention* 3(1): 22–29.

Suwarduki, P.R., Yulianto, E. & Mawardi, M.K. 2016. Pengaruh electronic word of mouth terhadap citra destinasi serta dampaknya pada minat dan keputusan berkunjung (survei pada followers aktif akun instagram indtravel yang telah mengunjungi destinasi wisata di Indonesia). *Jurnal Administrasi Bisnis* 37(2): 1–10.

Wang, C.Y. & Wu, L.W. 2011. Reference effects on revisit intention: Involvement as a moderator. *Journal of Travel & Tourism Marketing* 28(8): 817–827.

William P. & Soutar G.N. 2009. Value, satisfaction and behavioral intention in an adventure tourism context. *Annals of Tourism Research* 36(3):413–438.

William, A. 2006. Tourism and hospitality marketing: fantasy, feeling and fun. International Journal of Contemporary Hospitality Management 18(6): 482–495.

Zalukhu, Sukawati & Meyers, Koen. (2009). Panduan Dasar Pelaksanaan Ekowisata. Jakarta: UNESCO Office.

Zatori, A & Beardsley, M. 2017. On-site and memorable tourist experiences: trending toward value and quality-of-life outcomes, in Joseph S. Chen (ed.) *Advances in Hospitality and Leisure (Advances in Hospitality and Leisure* 13: 17–45.

Zeithaml, V. A. (1988). Consumer perceptions of price, quality, and value: a means-end model and synthesis of evidence. Journal of marketing, 52(3),2–22.

Advances in Business, Management and Entrepreneurship – Hurriyati et al (eds)
© 2020 Taylor & Francis Group, London, ISBN 978-0-367-27176-3

Measuring regional working units development performance using the MDGS scorecard

R. Hurriyati, M. Mayasari, L. Lisnawati & S. Sulastri
Universitas Pendidikan Indonesia, Bandung, Indonesia

ABSTRACT: In the current era of decentralization, the regional government has a role in resolving various building problems. Although development planning that targets the priority agenda has been included as an agenda in the vision and mission of regional leaders, in its development there is often a gap between the vision and mission with the effectiveness of implementing various programs and activities in planning documents. The purpose of this activity is to prepare planning staff at each Regional Working Unit (SKPD) to develop plans that are included in budgeting in the regional. The plans are to ensure the quality and availability of data for the preparation of important regional development reports and documents, such as the mid-term regional development plan (RPJMD), MDG Regional Action Plans reports on achievement of the MDGs, human development re-ports and others. The workshop was held in Bandung City and followed by participants representing agencies, including Development Planning Board (BAPPEDA), Central Statistics Agency (BPS), Health Office, Education Agency, Agriculture/Livestock Services, Forestry/Forestry Service, Fisheries, Cooperatives, BPM, Public Works, Universities, and NGOs and other related institutions. As a follow-up program, the training pro-grams for SKPD need to be widely implemented not only in the city of Bandung but also in other cities in West Java even in Indonesia.

1 INTRODUCTION

In the current era of decentralization, the Regional Government has a role in resolving various development problems, which are the main keys to the success of regional implementation in Indonesia. One important aspect of the development efforts that prosper the community is how to create planning that focuses on priorities that occur in the community. Through PNPM Mandiri, one of the community's participatory development programs, the government, which covers all sub-districts in Indonesia, has sought to strengthen participatory planning based on community needs. This is in line with Law No. 25/2004 concerning the National Development Planning System that emphasizes that strengthening community participation in the development process has been identified as a key approach in planning achievement and focusing on the priorities needed.

Although development planning that targets the priority agenda has been included as an agenda in the vision and mission of regional leaders, in its development there is often a gap between the vision and mission with the effectiveness of implementing various programs and activities in planning documents. This gap is caused partly by: 1) lack of understanding and ability to ana-lyze regional development concepts and prob-lems; 2) weak quality of policy formulation and design of development program implementation;

and 3) the high repositions of trained officials and staff. To that end, it is necessary to strive further to improve public services and strengthen the capacity of Regional Work Units (SKPD) in planning and budgeting activities that are more in favor of priority programs and siding with the interests of society at large.

Community-based Planning and Budgeting is actually not new. Related agencies such as the Nation-al Development Planning Agency (Bappe-nas), with the support of Technical Assistance from ADB and UNDP, have developed approaches and tried various practical instruments used to improve the planning and budgeting process in 29 districts/cities, especially to strengthen pro-poor planning and budgeting.

Various references and lessons learned from the implementation of the Technical Assistance have been published in several training modules on Planning and Budgeting that are Pro-poor. However, the small number of locations that were used as piloting led to the use of various tools and analysis that could be used in the regions that had not yet been conveyed and even evenly distributed to all regions in Indonesia.

In relation to the aforementioned plan, we intend to conduct training aimed at development conceptors in the area with the approach used in the workshop approach and FGD to focus on the types of activities that can be planned in regional development.

The target of key SKPD planning staff to be trained is Bappeda as the regional development lead-ing sector and several other key SKPDs such as:

Education, Health, Agriculture, Plantation/Forestry, Fisheries, Cooperatives, BPM, Public Works, BPS, Higher Education and selected NGOs. The training participants are expected to become agents to socialize the results of analysis that can be utilized in a broader planning forum both to their institutions so that they are expected to be utilized in regional development planning forums (Musrenbang) activities.

2 METHOD

This study uses an experimental method using the Scorecard Millennium Development Goals (MDGs) Technique. The MDGs scorecard (MDGs Scorecard) is a card that can provide an assessment of the achievement of the MDGs target with a global standard of reference that has been agreed upon by UN member states. The MDG assessment cards have a focus on seeing achievements in standard targets that have been set in the MDG declaration. For quantitative building indicators Determination of color words uses indicator index value (NII) indicators with the following formula:

a) For Development Target with Mini-value
$$NII = (x-T)/T$$
Where:
NII = Indicator Index Value
x = indicator value
T = Target Value
b) For development targets with Maximum value
$$NII = (T-x)/T$$
Color determination:

NII $\leq$ 0	NII = Green
0 < NII $\leq$ 0,25	NII = Yellow
NII > 0,25	NII = Red

Figure 1. Color determination.

Determination of NII intervals for indicators that have quantitative targets using the concept of normal curve data distribution. By adopting a normal curve, it can be simply analogous to an indicator with an achievement below 75% of the target rated red which is identical to NII above 0.25, more than 75% to less than 100% rated yellow which is identical to NII between 0 to 0.25, while achieving 100% or more rated green which is identical to NII is less than 0.

The next step, after accumulating the overall achievements of the indicators, and analyzing using a scoring card, the calculation of the overall performance analysis is carried out by summing and calculating the percentage of each measurement result (red, yellow, green).

3 PROGRAM IMPLEMENTATION

3.1 Location and Target Audience

The workshop was held in Bandung City, in the office of research and development planning agency (Bapelitbang). The workshop was participated by representatives from agencies, such as Bappeda, BPS, Health Office, Education Agency, Agriculture/Animal Husbandry Service, Plantation/Forestry Service, Fisheries, Cooperatives, BPM, Public Works, Universities, and NGOs and other related institutions. Participants must meet requirements such as:

a. Having basic knowledge of computers, especially Windows and MS Office Excel, governance of MS Access is a plus.
b. Each participant must bring one laptop or computer.
c. Participants must bring sector data from their respective agencies, especially geo-graphic oriented data (data in the form of soft files is better), such as:
1) Health Office: (latest) health profile,
2) Education Agency: Education Profile (most recent),
3) Agriculture Service: Agricultural Data (most recent),
4) Bapelitbang: Data on the poor, PPLS 08.
5) Other services: Data on relevant services.

The direct influence of business management training on performance is (0,517) 2 × 100 = 26.7%. It showed that business management training gave 26.7% influence on MSMEs performance. The influence of business management training indirectly on MSMES performance in this research is not found.

The hypothesis above used t-statistic value. The t-statistic for variables X1 and X2 to Y was obtained by 2349 + 6003 = 8352. The value was greater than 1.96, so it can be concluded that Ha > 0, which means that microcredit and business management training was proven to simultaneously affect the performance of SMEs.

3.2 Steps of Activities

3.2.1 Community service
The workshop on measuring the SKPD development performance uses the MDGs scorecard.

a. Pre-training
1) Preparing and coordinating with SKPD in the city of Bandung.
2) Compilation of criteria for SKPD development performance SKPD development performance uses the MDGs scorecard by Tim.
3) Designing the Measurement of SKPD Performance by the Team. Designing the Key Performance Index (KPI) data structure of SKPD

a) Making measurable KPIs.

b) Making KPIs based on the MDGs.

c) Making KPI on the performance of SKPD development with MDGs.

4) Trial Performance Measurement System using the MDGs Scorecard.

5) Making Performance Measurement Models using the MDGs scorecard.

b. Training

1) Theory

a) Introduction

b) Making KPI

c) Model for measuring SKPD performance with MDGs Scorecard

d) Measurement results

2) Practice

After getting the theory about KPI, the performance of SKPD based on the MDGs scorecard will immediately practice. As a whole the stages of implementation can be seen in Figure 1.

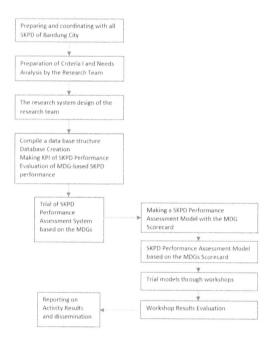

Figure 2. Workshop Implementation Stages measuring the SKPD development performance using the MDGs scorecard.

4 PROGRAM FOLLOW UP

In the digitalization era like now, the use of website technology is very important, especially as a means of measurement. Therefore, as a follow-up program, the training and training program for SKPD needs to

be carried out widely not only in Bandung, but also in other cities in West Java, even in Indonesia.

REFERENCES

Consistency of style is very important. Note the spacing A Sidik Prawiranegara. 2003. Strategi Pengembangan dan Operasionalisasi Pola Pengembangan Pengusaha Kecil yang terintegrasi dengan Pola Pembinaan Koperasi. Hasil Seminar IKOPIN. PT. PUSRI dan PT. Jasa Marga.

Agus Suroso. 2004. Peranan Lembaga Keuangan Formal dan Informal dalam Pengembangan Industri Kecil: Suatu Sur-vey di Propinsi Jawa Tengah. Desertasi UNPAD. Bandung.

Badan Pusat Statistik. 2004. Buletin Statistik Bulanan-Indikator Ekonomi. Badan Pusat Statistik. Jakarta.

Basri Faisal. 2007. Perekonomian Indonesia menjelang abad XXI: Distorsi, peluang dan kendala. Fakultas Ekonomi Universitas Indonesia. Jakarta.

Best, Roger J. 2000. Market-Based Management: Strategies for Growing Customer Value and Profitabiliy. Prentice-Hall. New Jersey.

Bina Industri. 2005. Laporan Tahunan Kantor Wilayah Departemen Perindustrian Propinsi Jawa Barat tahun 2005. Bandung.

Chang, Tung-Zong., Mehta, Rajiv., Chen, Su-Jane., Polsa, Pia., Mazur, Jolanta. 1999. The Effects of Market Orientation on Effectiveness And Efficiency: The Case of Automotive Distribution Channels in Finland and Poland. The Journal of Services Marketing; Santa Barbara.

Cravens, David W & Nigel F. Piercy. 2007. Stategic Marketing. McGraw-Hill. Boston.

Czinkota, Michael R & Masaaki Kotabe. 2001. Marketing Management. Second Editon. South-Westen College Pub. USA.

Day, George S. 1999. Market Driven Strategy: Processes for Creating Value. The Free Press. New York.

Day, G. S. 1994. The capabilities of market driven organiza-tions. Journal of Marketing. 58(October): 37–52.

Day, George. 1999b. Creating a Market-Driven Organization. Sloan Management Review. 41 (Fall): 11-22.

Doyle, P. 1994. Marketing Management and Strategy. Prentice Hall Publishing London.

Gale & Bradley T, 1994. Marketing Management. second Edi-tion. International Thomson Business Press.

Hamel, Gary & C. K. Prahalad. 1994. Competing for the Fu-ture. Harvard Business Review. 73 (July-August): 122-128.

Kotler. 2006. Marketing Management. Prentice Hall Inc. New Jersey

Kotler, Philip & Garry Amstrong. 2006. Principles of Marketing. Millenium edition. A Simon & Schucer Company, Ebngle-wood Cliff, Pretice Hall International, Inc, New Jer-sey,.

Porter, M. E. 1986. Changing patterns of international compe-tition. California Management Review. 28 (Winter), 9–40.

Porter, Michael E. 1996. The Competitive Strategy. the Free Press.

Porter, Michael E. 1990. The Competitive Advantage of Na-tions. the Free Press.

Sucherly. 2003,.Peranan Manajemen Pemasaran Strategik Da-lam Menciptakan Keunggulan Posisional Serta

Im-plikasinya terhadap Kinerja Organsasi Bisnis dan Non Bisnis (Pendekatan 5A). Pidato pengukuran Jabatan Guru Bsar dalam Ilmu Ekonomi pada Fakultas Universitas Pad-jadjaran. 22 Maret 2003.

Thomson, Athur A, & A.J Strickland. 2003. Strategic Management. Irwin- McGraw Hill.

Wheelen, Thomas L. dj Hungger. 2002. Strategic Management: Business Policy. Prentice Hall Int. New Jersey.

Yuyun Wirasasmita. 2004. Pemecahan masalah Bisnis Me-lalui Logical Framework Approach (LFA); Praktek-Praktek Bisnis Antara Teori dan realita, Program DMB UNPAD.

Canvas business model 4.0 and evaluation of the effectiveness of using the lecture system for students

R. Hurriyati, L. Lisnawati & Y. Rochmansyah
Universitas Pendidikan Indonesia, Bandung, Indonesia

ABSTRACT: In the face of the industrial revolution 4.0 era, competition will be increasingly fierce, so that policies and activities which can increase the competitiveness of UKMT in the future are needed instantly. Difficulties and obstacles in UKMT in Indonesia in developing their business are weak marketing channels, technology support and limited capital. Moreover, for beginner entrepreneurs, this problem will look bigger and become a considerable obstacle in developing their business. The purpose of this study is to look at the potential of the Business Canvas 4.0 Model. In particular, this study focuses on its potential in accordance with the learning process in the Business Study Program, at the Faculty of Economics and Business UPI. The method used in this study was the experimental method. Data collection was done by means of documentation study of the object of study both in the learning model and in the classroom. Interviews were conducted on lecturers to learn their responses to the model trials conducted. Thus, the results of this study seem to be quite effective in increasing the motivation and student learning culture of entrepreneurship as one of the subjects that have been considered less able to create educated.

1 INTRODUCTION

The current globalization has been unstoppable in Indonesia. Accompanied by increasingly sophisticated technological developments, the world is now entering the era of industrial revolution 4.0, which emphasizes a pattern of digital economy, artificial intelligence, big data, robotics, and so on, known as phenomena disruptive innovation. Facing these challenges, teaching in higher education is also required to change.

The Minister of Research, Technology and Higher Education (Menristekdikti), Mohamad Nasir, explained that based on an initial evaluation of the country's readiness to face the industrial revolution 4.0, Indonesia is estimated to be a country with high potential. Even though it is still below Singapore, on a Southeast Asian level, Indonesia's position is quite calculated. Whereas related to the global competitiveness index of the 2017–2018 World Economic Forum, Indonesia ranked 36, five ranks higher than the previous 41 position from 137 countries. Nevertheless, when compared to Malaysia, Singapore and Thailand, we are still behind. This year, Thailand's global competitiveness index ranks 32, Malaysia 23, and Singapore 3.

In the case of Indonesia, there are some causes including the weakness of higher education and training, science and technology readiness, and innovation and business sophistication. This needs to be improved to increase our competitiveness. Currently, the Ministry's strategic objectives are considered relevant so that changes are only made to service programs and models that provide more or use digital technology.

Higher education policies must also be adapted to the conditions of the industrial revolution 4.0. Strategic policies need to be formulated in various aspects ranging from institutions, fields of study, curriculum, resources, and cyber university development, risk to innovation. Five important elements must be considered and will be carried out by the Ministry to encourage economic growth and national competitiveness in the era of Industrial Revolution 4.0. The elements include:

a. Preparing more innovative learning systems in higher education, such as adjusting the learning curriculum, improving the students' abilities in data Information Technology (IT); Operational Technology (OT); Internet of Things (IoT); and Big Data Analysis, integrating physical objects, digital, and humans to produce competitive and skilled higher education graduates, especially in the aspects of data literacy, technological literacy and human literacy.

b. Reconstructing higher education institutional policies that are adaptive and responsive to the industrial revolution 4.0 in developing the required transdisciplinary study programs. In addition, Cyber University programs are being pursued, such as distance learning lecture systems, thereby reducing the intensity of lecturer and student meetings. Cyber University is expected to be a solution for the nation's children in remote areas to reach quality higher education.

c. Preparing human resources, especially lecturers and researchers, to be responsive, adaptive and reliable engineers in dealing with the industrial revolution 4.0. In addition, the rejuvenation of infrastructure and the construction of education, research and innovation infrastructure also needs to be carried out to support the quality of education, research and innovation.

d. Creating breakthroughs in research and development that supports the industrial revolution 4.0 and the ecosystem of research and development to improve the quality and quantity of research and development in Universities, R&D Institutions, LPNK, Industry, and Society.

e. Creating breakthrough innovations and strengthening innovation systems to improve industrial productivity and improve technology-based startup.

At present, most of the college graduates in Indonesia are still weak in their entrepreneurial spirit. The small proportion who has an entrepreneurial spirit generally comes from a family of entrepreneurs or trade. In fact, it shows that entrepreneurship is a state-of-mind that can be learned and taught. Someone who has an entrepreneurial spirit generally has the potential to become an entrepreneur although this is not a guarantee of being an entrepreneur, and entrepreneurs generally have an entrepreneurial spirit. Practicum Learning Model as a form of learning process through the creation of a canvas business model was designed as an effort to synergize theory (20%) and practice (80%) of the various competencies in business management sciences obtained with technology and industry.

This canvas 4.0 business model can be used as a center for learning activities with a conducive business atmosphere and supported by adequate 4.0 laboratory facilities. The purpose of implementing innovation in designing the canvas 4.0 business model is to develop an entrepreneurial spirit for students as students. The benefits obtained for the institution are the achievement of the institution's mission in building a generation of IT-based entrepreneurship and increasing the relevance between the world of education and the industrial world. The benefits for partners is the establishment of business and educational cooperation. This collaboration was developed in the form of a real business of similar products which has a high potential market economy.

1.1 Research context

A business model is a rational model that explains how an organization creates, gives, and captures a value. The Business Model is not only for a commercial enterprise but also for organizations in general. The organization can be a commercial business, a school, a government agency, a foundation, and so on. The emphasis on value added aspects is not only the goods produced or services that will be given, but what the organization contributes to its customers that are mediated by the goods or services it offers.

This canvas business model is a tool for describing, analyzing and designing business models, where the pattern of business models is based on the concept of leading businesspeople through technical techniques. The techniques help students in designing business models that translate back through the lens of business models and generic processes to help students design innovative business models and bring together all engineering concepts and tools in the business model generation.

1.2 Potential canvas 4.0 business model as a vehicle for improving learning processes

The Canvas 4.0 Business Model developed in the FPEB-UPI Business Management Education study program is an effort to facilitate the process of creating new businesses, especially those that are closely related to the competence of the study program. The canvas business model 4.0 FPEB-UPI Business Management Education study program is expected to be one of the institutions that will foster UPI's independence. Internally, the creation of innovations from inventions at UPI will continue to grow due to commercialization activities. The canvas business model 4.0 FPEB-UPI Business Management Education study program also remains one of the parts in the education process in the MU, especially learning in real terms about how to create value added creation, increase professionalism, be responsible (to be committed), create reliable entrepreneurs, and how to form a business society at UPI. In addition to the Business Model of Canvas 4.0, the FPEB-UPI Business Management Education Study Program is expected to be one of the spearheads of UPI in the efforts of the UPI to maintain its independence. Furthermore, the Canvas Business Model 4.0 FPEB-UPI Business Management Study Program must be able to become a mediator to encourage the creation and growth of social welfare around the UPI campus in particular and in Indonesia in general.

Based on this, the Canvas 4.0 business model is a vehicle for increasing the learning process for FPEB-UPI's Business Management Education Study Program. So that UPI's vision to become a pioneer and superior university can be achieved.

2 METHOD

This study used an experimental method, in which the initial stages were designed for the learning model criteria through the integration of pedagogical aspects and technological aspects that were used as experiments through need-assessment. It was then followed by the process of designing a design-based model, the teaching model that will be applied. In the next stage, an implementation test was carried out on the model using the teaching model and set subjects, then the system was tested with the same teaching model and the same subjects and students. In the process of testing the implementation of the model of all learning activities was recorded with camera video. Interviews were conducted with students and lecturers after the implementation was carried out, to learn their opinions from the trial model.

3 RESULT AND DISCUSSION

The procedure of the trial model in this study was aimed at producing a Business Canvas 4.0 Model, hereinafter abbreviated as MBC 4.0, which is a learning process model that aims to enhance the entrepreneurial spirit. Figure 1 shows the flow diagram of the learning mechanism of the MBC 4.0 model.

Based on the above trial, it appears that at this 1–2 meeting, the ability of students to express their ideas and opinions has shown improvement and reflects their ability to form hypotheses on problems that have been discussed in a classical manner. This increase seems to have been contributed by learning services provided by lecturers, both individual and classical, accompanied by the provision of several examples of the procedures for formulating a business idea. In addition, learning activities in teams or groups of students also indicate an increase, which can be seen from how they were actively and openly involved in the discussion activities within their respective work teams during the group discussions that were developed by lecturers. The comparison of student scores in the first and second trials can be seen in Table 1.

Based on Table 1, we can see that the value of t is calculated for the four times the test has been carried out, namely: 36.462; 33.548; 35.585; 34.638, and is significant with α = 0.0001. Thus it can be concluded that there is a significant difference between the pre-score and the post-test score (α = 0.0001). This means that the acquisition of student post-test scores is higher than the acquisition of pre-test scores, so

Table 1. Comparison of obtained student scores in the first and second trials.

Variable	Mean	Std.	T value	Sign.
pre-test 01	4.137	1.274	36,462	0.0001
post-test 01	7.301	1.167		
pre-test 02	3.744	1.163	33.548	0.0001
post-test 02	7,593	1.081		
pre-test 03	4.072	1.224	35.585	0.0001
post-test 03	7.906	0.818		
pre-test 04	4.224	1.105	34.638	0.0001
post-test 04	7.824	0.938		

Table 2. Mean differences in post-test scores between control group (KK) and the experimental group (KE).

Variable		Mean	Std.	T value	Sign.
post-test 01	KE	6.401	1.067	23.513	0.0001
	KK	7.707	1.316		
post-test 02	KE	6.593	1.081	51.294	0.0001
	KK	7.718	1.245		
post-test 03	KE	6.906	0.818	70.667	0.0001
	KK	7.244	0.844		
post-test 04	KE	6.824	0.938	45.381	0.0001
	KK	7.409	1.311		

that the difference is an implication of the learning activities that have been developed by the lecturers. To find out more about the difference in the mean post-test score between the control group and the experimental group, a statistical test was done, namely the t test, the results of which can be seen in Table 2.

Based on Table 2, we can see that the value of t counts from the comparison between the post-test of the experimental group and the control group, namely: 23.513; 51.294; 70.667; 45.381, is significant with α = 0.0001. Thus, the results show the significance of the difference between the experimental group post-test results (KE) and the control group (KK). Based on the results of this analysis, it can be concluded that the acquisition of post-test scores for the experimental group is better than the control group.

The comparison of the average score of the pre-test and post-test virtual market (web 2.0 application) can be described as seen in Table 3.

Based on Table 3, we can see that the value of t is calculated for the four times the tests have been carried out, namely: 39.462; 32.548; 33.585; 38.638, and is significant with α = 0.0001. These results show that, first, the creativity stage of the program implemented today enables the development of the Business Canvas 4.0 Model in entrepreneurship learning. Second, viewed from the perspective of

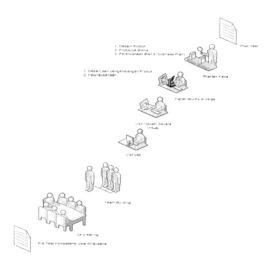

Figure 1. Flow chart of the mechanism.

Table 3. Comparison of the average score of the pre- and post-market market space.

Variable	Mean	Std.	Nilai t	Sign.
pre-test 01	5.037	1.374	39.462	0.0001
post-test 01	7.401	1.067		
pre-test 02	5.744	1.163	32.548	0.0001
post-test 02	7.593	1.081		
pre-test 03	5.072	1.224	33.585	0.0001
post-test 03	7.906	0.818		
pre-test 04	5.224	1.105	38.638	0.0001
post-test 04	7.824	0.938		

Table 4. Mean differences in post-test scores between ex experiments with control levels.

Variable		Mean	Std.	T value	Sign.
post-test 01	KE	7.401	1.067	24.513	0.0001
	KK	6.707	1.316		
post-test 02	KE	7.593	1.081	52.294	0.0001
	KK	6.718	1.245		
post-test 03	KE	7.906	0.818	71.667	0.0001
	KK	6.244	0.844		
post-test 04	KE	7.824	0.938	47.381	0.0001
	KK	6.409	1.311		

learning models with the Business Canvas 4.0 Model, existing entrepreneurship education has several formulations general learning objectives (TPU) that lead to the achievement of an educated entrepreneurial spirit. Third, it can be said that entrepreneurship education is currently in force, enabling the development of learning models through the Business Canvas 4.0 Model. To find out more about the mean differences in post-test scores between the experiment group with the control group, the statistical test was done, namely the t test, the results of which can be seen in Table 4.

Based on Table 4, the value of t is calculated from the comparison between the post-tests of the KE experimental group and the KK control group, namely: 24.513; 52.294; 71.667; 47.381, and is significant with $\alpha = 0.0001$. Thus, these results indicate the significance of the difference between the post-test results of the KE group and the KK group. The difference in the scores of the experimental group and the control group may still be influenced by the pretest.

4 CONCLUSION

Based on the testing of the results of the research that has been carried out in the activities of this research project, the following conclusions can be drawn. The design of the web 2.0 technology-based business incubator system in the Professional Skills course, namely Small and Medium Scale Business and Entrepreneurship courses has been carried out with a model approach from Fayolle Allain & Gaily (2006) Learning and Learning Processes in implementing entrepreneurship Education. This model enables the occurrence of individuation, acceleration, enrichment, expansion, effectiveness and productivity of learning, which in turn will improve the quality of education as an infrastructure of overall human resource development. Through this learning model, every student will be stimulated to attain advanced progress in accordance with the potential and skills they have. In addition, this model demands creativity and self-reliance to enable it to develop all its potential. It is expected that through this learning model students will obtain various information related to the courses developed, as well as in a wider and deeper scope to increase their perspectives. This is a conducive stimulus for the development of student independence, especially in terms of developing competence, creativity, self-control, consistency, and commitment to both oneself and others. The model makes it easy for instructors/lecturers of educational study programs to manage business in providing lecture material and assignments. Conducting trial activities of web 2.0 technology-based business incubator models through professional expertise courses and evaluating the effectiveness of the use of the lecture system for business management education study students in learning small and medium-sized business and entrepreneurship.

REFERENCES

Gibb A. 1999. "Can we Build Effective Entreupreunership Through management Development." Journal of general manager. 24 (4): 1–21.

Heinonen, J. 2007. "An Entreupreunerial Directed approach to Entreupreunership Education Mission Immposible." The Journal of Management Development. 25 (1): 80–94.

Lynch, AD. 2013. "Adolescent academic achievement and school engagement: an examination of the role of school-wide peer culture." @@@. 42 (1): 6–19.

Osterwalder, Alexander., & Pigneur Yves. 2010. "Business Model Generation: A Handbook for Visionaries, Game Changers, and Challengers."

Efendy, Onong Uchana. 2005. "Ilmu Komunikasi Teori dan Praktek," Bandung: Remaja Rosda Karya.

Lewis, David. 2002. "Tharp's third theorem" analysis, 62 (2): 95–97.

Sari Dowling, C., Godfrey, J.M., & Gyles, N. 2003. "Do hybrid flexible delivery teaching methods improve accounting students' learning outcomes?," Accounting Education, 12(4):373–391.

Cheung, L. L. W. and Kan, A. C. N. 2002. "Evaluation of Factors Related to Student Performance in a Distance Learning Business Communication Course." Journal of Education for Business, 77 (5): 257–263.

Fayolle, A., Gailly, B. t., & Lassas-Clerc, N. 2006. Assessing the impact of entrepreneurship education programmes: a new methodology. Journal of European Industrial Training.

Advances in Business, Management and Entrepreneurship – Hurriyati et al (eds)
© 2020 Taylor & Francis Group, London, ISBN 978-0-367-27176-3

The influence of reputation and Customer Relationship Management (CRM) towards the competitive advantage of airlines in Indonesia

Y. Mardani, R. Hurriyati, D. Disman & V. Gaffar
Universitas Pendidikan Indonesia, Bandung, Indonesia

ABSTRACT: One of the fastest growing transportation modes in Indonesia is air transportation. The air transportation market in Indonesia is experiencing a rapid growth. This leads to the increase in the number of airline passengers in almost all airports. However, the facts show that the competitive advantage of the industry has not been superior yet. These conditions are related to the issues in the development of reputation and Customer Relationship Management (CRM). Hence, this study was done to examine the influence of reputation and CRM towards the competitive advantage of the airline industry in Indonesia. This study employed a quantitative approach by involving 100 passengers member from five different airlines namely Garuda, Lion air, Citilink, Batik Air and Air Asia as the samples. The data was then analyzed using the PLS statistical test tool. The results showed that reputation and CRM affected the competitive advantage both simultaneously and partially, in which reputation gave greater impact than CRM. To conclude, competitive advantage in the airline industry in Indonesia was dominantly more influenced by reputation than by CRM.

1 INTRODUCTION

Since 2013, there has been a growth and shift in market share. Garuda Indonesia experienced a 23% increase in market share into 20.49% in 2008. The airline that has the largest market share increase was Lion Air (24.63% of market share in 2008) and increased more than 50% in 2013. Therefore, Lion Air controlled 43% of market share and became the leader of market share in that year. Lion Air controlled the low-cost carriers airline market share (LCC) in Indonesia. It shows a trend shift in using airline services chosen by customers. These conditions reflect that passengers tend to be disloyal to an airline. The low loyalty of customers in the aviation industry sector shows that the airlines have not developed the customers' value yet.

The phenomenon is allegedly related to the problems in maintaining customer relationship management. This phenomenon indicates that the company has not taken the incentives given to customers seriously. On the other hand, there are some issues related to the company's reputation. The current phenomenon indicates that the airline company has not been able to create a service with high credibility. In addition, the level of customer confidence in the service also tends to be low.

Based on the aforementioned issues, this study aims at examining the influence of reputation and customer relationship management (CRM) in increasing the competitive advantage of airlines in Indonesia.

2 METHOD

This study employed a quantitative approach. The unit analysis is the passenger with membership of air-lines in Indonesia. The observation was conducted in a cross section/one shot, in which the information or data obtained was the results of research conducted in 2017. The study used the primary data obtained from the questionnaire taken from 100 passengers with membership of five airlines, namely Garuda, Lion air, Citilink, Batik air and Air Asia.

The validity test used is the construct validity, while the reliability test was done by using the technique of split half. A measure of consistency where a test is split in two and the scores for each half of the test is compared with one another (Davidshofer et al. 2005). To test the hypothesis, partial least squares path modeling (PLS-PM) was used. The method to structural equation modeling allows estimating complex cause-effect relationship models with latent variables.

3 RESULT AND DISCUSSION

Based on hypothesis testing result, the Research Model Finding can be described as follow:

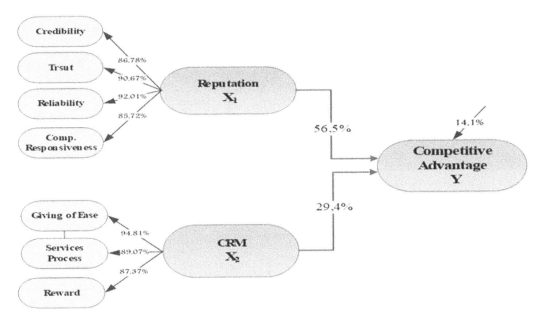

Figure 1. Research finding.

The results showed that competitive advantage in the airline industry in Indonesia was more dominantly influenced by reputation than by CRM. The test results accepted the hypothesis testing that reputation and CRM influenced the competitive advantage simultaneously or partially, where reputation gave greater influence.

Reputation is more dominant in improving competitive advantage, in which its contribution was mainly derived from reliability, trust, credibility, and responsibility. Reliability showed that the company was able to develop a work culture that led to superior performance and able to perfect their internal business process on an on-going basis. In addition, the company was also able to provide professional sales coaching to its employees, and perform a good selling process to its customers, so they were satisfied with the service of company. Meanwhile, the trust illustrated that the company had a strong company image, and passengers were confident about the airline and its products. Trust is manifested from the expertise or experience of the company in conducting its business, so it was able to have its own attraction that was liked by passenger. In addition, trust also manifested how far the employees believe in the company, so that they were able to provide the best performance.

The credibility was formed by how far the companies built a growth in their business and how far they were able to build responsibility towards passenger value. Such company will be able to build prospects in the future. Meanwhile, responsibility showed how far the company's responsibility towards their passenger, investors, social, and employees.

The hypothesis testing showed that the involved aspects (reliability, trust, credibility, and responsibility) gave dominant role compared to CRM in improving competitive advantage of airlines industry in Indonesia.

The results of this hypothesis test are in line with the previous research. Iwu-Egwuonwu (2011) found that the management of a strong reputation was the necessary basis for today's companies that intend to beat the competition, improve market prospects and financial performance and sustainable enterprise. Besides that, Lee & Roh (2012) discovered that the company's reputation was significantly and positively related to most of the company's performance measurement index. Lastly, Hasanudin & Budianto (2013) pointed out that the company's reputation had a positive effect on business performance.

Meanwhile, CRM also had an influence on competitive advantage. CRM was formed primarily by giving of ease, business development assistance, and rewards. These aspects supported the reputation in improving the competitive advantage of airlines industry in Indonesia. The expected ease by the customers were ease in ordering products, payments, along with interesting special facilities and can provide special benefits for passenger. Meanwhile, the service process was related to the handling of complaints from passengers, the provision of services during check in, and in the cabin aircraft. The rewards were in the form of giving special greeting from the company to passengers in a certain moment, conducting of an interesting customer gathering, and giving interesting prizes to passengers.

Analysis of measurement model (outer model) was used as validity and reliability test to measure latent variable and indicator in measuring dimension that is a construct. It is explained by the value of Cronbach's Alpha, in order to see the reliability of dimension in measuring variables. If the value of Cronbach's Alpha is bigger that 0.70 (Nunnaly & Berstein 1994), then the dimensions and indicators are reliable in measuring variables. Table 2showsthat Composite reliability and Cronbach's Alpha of variables > 0.70 indicated that all of variables in the model estimated fulfilled the criteria of discriminant validity. Then, it can be concluded that all of variables had a good reliability

4 CONCLUSION

Reputation and CRM affected the competitive advantage either simultaneously and partially, in which reputation gave greater influence. The results showed that competitive advantage of airlines industry in Indonesia was still more dominantly influenced by reputation than by CRM. Hence, in the effort to develop the competitive advantage, the company must rely on the development of reputation supported by the development of CRM.

Based on the results of this study, it is suggested that airlines industry in Indonesia develop reputation supported by CRM development. The reputation should primarily be developed by prioritizing the reliability aspect, supported by the development of trust, credibility, and responsibility. Meanwhile, in developing CRM, companies are advised to prioritize ease for passenger.

The results of this study is expected to be a reference for further researchers who are interested in competitive advantage of airlines industry in Indonesia, by making the results of this study as a premise in preparing the framework. This study took the population in Indonesia; therefore, it is suggested for future researchers to take a wider population such as in big cities in Indonesia.

REFERENCES

Davidshofer, K.R. & Murphy, C.O. 2005. *Psychological testing: principles and applications (6th edition)*. New Jersey: Pearson/Prentice Hall.
Hasanudin, A.I. & Budianto, R. 2017. The Implications of corporate social responsibility and firm performance with reputation as intervening variable empirical study in the manufacturing company in Indonesia. *GSTF Journal on Business Review (GBR)* 2(4): 106-109.
Iwu-Egwuonwu, D. & Chibuike, R. 2011. Corporate reputation & firm performance: empirical literature evidence. *International Journal of Business and Management* 6 (4): 197–206.
Lee, J. & Roh, J.J. 2012. Revisiting corporate reputation and firm performance link. *Benchmarking: An International Journal* 19(4/5): 649-664.
Nunnally, J.C. Berstein, I.H. 1994. *Psychometric theory (3rd edition)*. New York: McGraw-Hill.

Analysis of tourist satisfaction against tourism product at Tanjung Kelayang beach, Belitung regency

I. Khairi, F. Rahmafitria & S. Suwatno
Universitas Pendidikan Indonesia, Bandung, Indonesia

ABSTRACT: The purpose of this research is to analyze the tourists' satisfaction in Tanjung Kelayang Beach through the tourists' assessment on importance level and tourism product performance in Tanjung Kelayang Beach. There are three tourism product indicators including attractions, facilities, and accessibilities. This research used a descriptive method with quantitative approach using a simple random sampling technique. This research involved 100 respondents using questionnaire. To measure the tourists' satisfaction level, Important Performance Analysis (IPA) method was used and resulted four quadrants with priority scale. The results of this study found dissatisfaction from the tourists who came. Even though the performance level from tourism product in Tanjung Kelayang Beach was already good, in the reality, the satisfaction point cannot be reached because the expectation value from the tourist to Tanjung Kelayang Beach's tourism product was bigger than the performance value. There were three quadrants that resulted from IPA method that became the manager's reference to pay attention to tourism product's condition in Tanjung Kelayang Beach, and to maintain the tourism product performance that is already rated well by the tourists.

1 INTRODUCTION

Bangka Belitung Province is one of the provinces located in the eastern part of Sumatra Island, close to South Sumatra Province. Bangka Belitung Province is also known as the largest tin producer in Indonesia and has an amazing natural beach charm. Besides, Bangka Belitung also has a variety of tourist attractions such as water tourism, nature tourism, cultural tourism and special interest tours. This diversity can provide alternative travel options that are more varied for tourists.

At this time, tourism is the key to development, prosperity and happiness. Increasing tourism destinations and tourism investment makes tourism a major factor in export revenues, the creation of broad employment opportunities, business development and infrastructure. In this case, the government is trying to accelerate the development of 10 (ten) priority destinations to form 10 "New Bali" in Indonesia.

The Indonesian Ministry of Tourism has released 10 lists of tourist destinations in Indonesia based on the official website (http://www.kemenpar.go.id). Referring to 10 tourist destinations from the Ministry of Tourism, Tanjung Kelayang Beach becomes one of famous destinations. The beach has been made a National Tourism Strategic Area (KSPN) and Special Economic Zone (KEK) which was specifically inaugurated by the President of the Republic of Indonesia according to his decision through a Cabinet Secretariat letter Number: B 652/Seskab/Maritim/2015 dated 6 November 2015 Regarding Directives The President of the Republic of Indonesia concerning

Tourism and Presidential Direction at the cabinet meeting at the beginning of the year on 4 January 2015. The Tanjung Kelayang region has geo-economic and geostrategic potential and excellence. The geo-economic advantage of the Tanjung Kelayang region is that it has a maritime tourism object which is a white sandy beach with exotic panoramas and has the proximity of small islands which become tourism destinations. The geostrategic advantage of the Tanjung Kelayang area is to have an environmentally friendly tourism development concept with the development of sustainable tourism areas. (Indonesian Government Regulation No. 6 of 2016 concerning Tanjung Kelayang Special Economic Zone).

Based on data on tourist visits from year to year the flow of tourist visits tends to be unstable, especially for foreign tourists, this must of course be increased. Based on the results of the pre-research and interview with Mr. Alex Suryadi as the head of the Tanjung Kelayang Beach UPTD, the level of tourist visits was not stable, especially foreign tourists and there were still many complaints from tourists visiting. Middleton (2001) gives a deeper understanding of tourism products, namely: "The tourist products to be considered as an amalgam of three main components of attraction, facilities at the destination and accessibility of the destination." From this understanding we can see that tourism products in general are formed due to three main components, namely tourist attractions, facilities in tourist destinations and accessibility. Some research on tourism products that use various models shows the importance of tourism products in a tourist destination (Xu 2010).

Tourist satisfaction is a key factor for managers to be able to increase visits and attract tourists to come back (Kotler et al. 2009) define consumer satisfaction as the level of one's feelings as a result of the comparison of the reality and expectations received from a product or service. If the service is perceived to be lower than expected, the consumer will feel disappointed. If the perception meets or exceeds the expectations of consumers, the consumers will feel satisfied and there is a tendency consumers will use these service providers. To find satisfaction of tourists researchers, the IPA (Importance-Performance Analysis) method was used to facilitate explanation of data and practical proposals (Izadi et al. 2016). Some research on customer satisfaction on a product shows a direct impact on customers (Nilsson et al. 2001).

2 METHOD

Data collected in this study were taken from primary data (interviews and questionnaires) and secondary data (library studies, documentation and data search on the internet). The population in this study were tourists who visited Tanjung Kelayang Beach and samples taken with the Slovin Formula resulting in a sample of 100 respondents.

This study used a descriptive research method with quantitative approaches and the instruments used in this study consisted of questionnaires using a Likert scale approach. For data analysis, this study used data tabulation using Microsoft Excel. The data analysis technique used in this study is the Importance Performance Analysis (IPA) method with the formula below:

$$CS = \sum (Ii - Ppi) \qquad (1)$$

Information:
CS: Customer satisfaction
I: Importance (Importance)
Pp: Performance Level (Perceived Performance)
Where:
CS < 0 visitors feel very satisfied
CS = 0 visitors are satisfied
CS > 0 visitors feel dissatisfied

The Cartesian diagram is an Importance-Performance Matrix used in which a construct is divided into four quadrants which are bounded by two lines intersecting perpendicular to the point (X, Y), each calculated by the formula:

$$= \frac{\sum_i^n = x^i}{k} \qquad (2)$$

$$Y = \frac{\sum_i^n = y^i}{k} \qquad (3)$$

Where:
X = the value of the performance average of all statements.

3 RESULT AND DISCUSSION

The results of rating of tourists on the level of importance of tourism products at Tanjung Kelayang Beach.

Based on Table 1, it can be seen that the score of the tourist response recapitulation regarding the importance of tourism products in Tanjung Kelayang Beach in Belitung is 11,482. If it is changed to the continuum line, it can be seen using the following formula:

Maximum Index Value = highest scale X number of questions X respondent
= 5 x 27 x 100 = 13.500
Minimum Index Value = lowest scale X number of questions X respondent
= 1 x 27 x 100 = 2.700
Interval Distance = (maximum value - minimum value): 5
= (13.500 –2.700): 5
= 2,160

Recapitulation of tourists' responses to the level of importance of tourism products in Tanjung Kelayang Belitung Beach, the highest score was a total score of 11,482, which in the continuum line shows in very important categories and is in the range of values between 11,340-13,500. Thus it can be said that the overall response of tourists regarding the level of importance of tourism products in Tanjung Kelayang Beach in Belitung as a whole is in the

Table 1. Tourist Rating of Tourism Product Importance Levels at Tanjung Kelayang Beach in Belitung.

| No | Sub Variable | Amount | | Ideal Score |
		Total Score	%	
1	Attractions	3564	89.1	4000
2	Facilities	5209	80.2	6500
3	Accessbility	2709	90.3	3000
Total Score		11,482	85.1	13,500

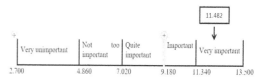

Figure 1. Continuous line of tourism product importance level.

Table 2. Tourist rating of tourism product performance levels at Tanjung Kelayang Beach in Belitung.

| No | Sub Variable | Amount | | Ideal Score |
		Total Score	%	
1	Attractions	3515	87.8	4000
2	Facilities	5181	79.8	6500
3	Accessbility	2596	86.6	3000
	Total Score	11,292	91.1	13,500

Source: Processed by researchers, 2017

very important category. This shows the level of importance or very high expectations of tourists on tourism products at Tanjung Kelayang Beach in Belitung.

Based on Table 2, it can be seen that the score of tourist response recapitulation regarding the level of performance of tourism products in Tanjung Kelayang Beach is 11,292. If it is converted to a continuum line, it can be seen using the following formula:

Maximum Index Value = highest scale X number of questions X respondent
= 5 x 27 x 100 = 13.500

Minimum Index Value = lowest scale X number of questions X respondent
= 1 x 27 x 100 = 2.700

Interval Distance = (maximum value - minimum value): 5
= (13.500 – 2.700): 5
= 2.160

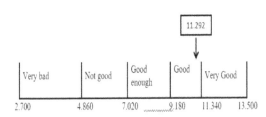

Figure 2. Continuous line of tourism product performance level.

Recapitulation of tourists' responses to the level of tourism product performance at Tanjung Kelayang Beach in Belitung Regency, the highest score was a total score of 11,292, which in the continuum line showed in the good category and was in the range between 9,180 - 11,340. Where the attraction variable sub gets the highest percentage of 87.8%. Thus it can be said that tourists' responses regarding the level of tourism product performance on Tanjung Kelayang Beach as a whole are in the good category. This shows the level of performance or condition of tourist products in Tanjung Kelayang Belitung Beach related to attractions, amenities, and accessibility is considered good according to tourists.

To measure the level of tourist satisfaction with tourism products at Tanjung Kelayang Beach can be calculated using the following formula:

$$CS = \sum (Ii - Ppi)$$

Information:
CS : Customer satisfaction
I : Importance (Importance)
Pp : Performance Level (Perceived Performance)
Where:
CS < 0 visitors feel very satisfied
CS = 0 visitors are satisfied
CS > 0 visitors feel dissatisfied

The level of tourist satisfaction is measured using the three aspects asked, namely attractions, facilities, and accessibility. Then it will be compared using the formula. If the Importance score is greater than performance (Performance), it can be said that tourism products have not reached tourist satisfaction.

The aspects that exist in the assessment of tourism products have not reached tourist satisfaction. Although in the previous discussion performance (performance) shows on the good line, but when processed using the level of satisfaction can be said the user feels dissatisfied because the performance is at a value below importance.

Then after knowing the average of each indicator, to make a diagram of Importance-Performance Analysis researchers use the help of SPSS 20 program and the following are the results.

In quadrant A there are two indicators, namely the availability of information boards at Tanjung Kelayang Beach and the ease of transportation to get to Tanjung

Table 3. Tourist satisfaction level on tourism products at Tanjung Kelayang Beach in Belitung.

Sub Variable	Total Interest Score	Average Score of Interest (%)	Performance Scale Total Score	Average Performance Score (%)	Tourist Satisfaction $\sum (Ii - Ppi)$	Information
Attractions	3564	89.1	3515	87.8	49	Not satisfied
Facilities	5209	80.2	5181	79.8	28	Not satisfied
Accessibility	2709	90.3	2596	86.6	113	Not satisfied

Source: Processed by researchers, 2017

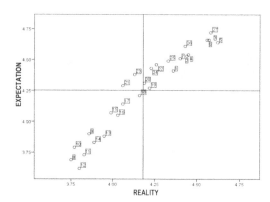

Figure 3. Efforts to increase tourist satisfaction at Tanjung Kelayang Beach in Belitung.

Kelayang Beach. Programs that can be done are related to the availability of information boards, namely:

a. Creating information boards with creative and innovative information in them to increase tourists' interest to visit Tanjung Kelayang Beach.
b. Adding information boards in several places on Tanjung Kelayang Beach to make it easier for tourists to get information about the objects and attractions in Tanjung Kelayang Beach.
c. Create information boards with unique and artistic forms so that tourists can not only add information and insight but can also be used for photo taking.

For quadrant B, it is related to items that are considered good and can meet the expectations or expectations of tourists. Programs that can be done include:

a. Making rules and conservation areas of Tanjung Kelayang Beach.
b. Regular maintenance and maintenance.

In the C quadrant there are eleven items, where the items are items that receive less priority or are considered less important by tourists, in the conditions or performance of these items are considered normal or sufficient.

4 CONCLUSION

The importance of tourism products at Tanjung Kelayang Beach in Belitung is in a very important category which shows the expectations or high expectations of tourists on existing tourism products related to attractions, facilities and accessibility to support tourism activities at Tanjung Kelayang Beach Belitung.

The performance level (Performance) of tourism products at Tanjung Kelayang Beach in Belitung is in a good category that shows the condition or performance of existing tourism products related to attractions, facilities, and accessibility of existing facilities is good to support tourism activities at Tanjung Kelayang Beach in Belitung.

The tourist satisfaction with tourism products at Tanjung Kelayang Beach in Belitung shows tourists are still dissatisfied with existing tourism products because based on the results of the importance value recapitulation is still higher than the value of the performance level in Tanjung Kelayang Belitung Beach, it shows that the level of performance or condition of existing tourism products does not meet the expectations of tourists who come.

Based on the Importance-Performance Analysis (IPA) in an effort to improve tourist satisfaction there are groupings of indicators examined in 4 quadrants, each of which explains different aspects. Where the quadrant can see what aspects should be prioritized for attention and which aspects must be maintained or developed by the manager.

REFERENCES

Indonesian Government Regulation No. 6 of 2016 concerning Tanjung Kelayang Special Economic Zone.
Izadi, A., Jahani, Y., Rafiei, S., Masoud, A. & Vali, L. 2016. Evaluating health service quality: using importance performance analysis. *International Journal of Health Care Quality Assurance* 30(7): 656–663.
Kotler P.T., Bowen, J.T., Makens, J. & Baloglu, S. 2009. *Marketing for hospitality and tourism*. London: Pearson.
Middleton, V.T.C. & Jackie Clarke. 2001. *Marketing in travel and tourism 3rd edition*. Oxford: Butterworth-Heinemann.
Nilsson, L., Johnson, M. D., & Gustafsson, A. (2001). The impact of quality practices on customer satisfaction and business results: product versus service organizations. *Journal of Quality Management* 6(1): 5–27.
Xu, J. B. (2010). Perceptions of tourism products. *Tourism Management* 31(5): 607–610.

Advances in Business, Management and Entrepreneurship – Hurriyati et al (eds)
© 2020 Taylor & Francis Group, London, ISBN 978-0-367-27176-3

Factors influencing halal cosmetic brand advocacy: The role of self-brand connection as a mediator

N. Rubiyanti
Telkom University, Bandung, Indonesia

E.S. Hariandja
Pelita Harapan University, Tangerang, Indonesia

ABSTRACT: This paper aims to determine the role of self-brand connection in mediating factors that affect brand advocacy. A conceptual model is developed from the branding literature to explain the relationship of the antecedents of self-brand connection and brand advocacy in the context of the halal cosmetic product. The model is tested using structural equation analysis on a sample of 150 women from Bandung city that have purchased the halal cosmetic product in question. The research model is empirically supported. Findings indicate that consumer who has self-brand connection may become advocates of the brand. Managers can consider using celebrity to develop self-brand connection within consumer and the brand. Additionally, other factors related to the quality and uniqueness of the brand can be considered to enhance consumer self-brand connection in shaping advocacy behavior. Hope this research provides future insight into the application of branding theory for halal cosmetic products.

1 INTRODUCTION

Branding is an important tool in marketing strategy. Building a strong brand is paramount for companies for being successful because when a consumer buys a product, they buy not only the functional benefits but also the reputation, prestige, symbols, and social meanings related with the image of the brand (Chun 2014). Today, more and more cosmetics brands try to engage connection to the consumers with the favorable, ethical, and distinctive image of their brand promoting similarity between their consumer and the brand. For that reason, scholars have devoted their efforts in studying the relationship between the brand and the consumer (Dwivedi et al. 2015, Kemp et al. 2012, Tho et al. 2016), and marketers are increasingly trying to create good relationships between their brands and consumers to reflect consumer self-brand connection. Cosmetics brands strive to create a brand image similar to the self-image of their target, thus formed self-brand connection (So et al. 2013).

Today, increasing numbers of cosmetics brands try to attract consumers by holding a celebrity to the endorser. The primary task of the endorser is to create a functional association between the endorser to a product being advertised until the resulting positive attitude within its customers (Madiawati & Pradana 2015). Other antecedents that would affect self-brand connection such as perceived quality and brand uniqueness (Kemp et al. 2012) are also discussed in this study. Therefore, this study purposed a conceptual model with a theoretical basis in the empirical branding literature to illuminate the antecedents of self-brand

connection and brand advocacy within the context of halal cosmetic branding. The construct is tested on a sample of the consumer from a local Halal cosmetic brand in Indonesia that has taken a branding effort.

1.1 *Celebrity endorser and self-brand connection*

Consumers often use brands as tools for creation and maintenance of self-identity, the self-brand connection potentially signifies a deeper level of consumer identification with a brand (Dwivedi et al. 2015). Because celebrities comprise symbolic aspirational reference group associations, this study expects celebrity endorsements to impact self-brand connections. It is important to matches between the celebrity's image and the endorsed product-brand. The congruency between the celebrity and the brand depends on the degree of match between them (Magnoni 2016). Celebrities who are well-liked and attractive tend to be active spokespersons, which helps to enhance the effectiveness of advertising. Celebrities make advertisements more believable, improve a message recall, create a positive attitude towards the brand, build a distinct personality for the endorsed brand, create an immediate identity or person for the endorsed brand (Suki 2014) Thus, this study suggests that the more likeable and attractive the celebrity, the higher connected the consumer and the brand.

1.2 *Perceived quality and self-brand connection*

Perceived quality is defined as the consumer's evaluation of a product's overall excellence or superiority

(Tsiotsou 2006). Perceived quality also refers to customer's assessment of a product or a brand that meet an individual's expectation (Vo & Nguyen 2015). In the context of the Halal cosmetic brand, the perceived quality of a brand can encompass consumers' beliefs that branding efforts imbue what is genuinely a signature and positive attribute of a cosmetic.

1.3 Brand uniqueness and self-brand connection

The perceived uniqueness of a brand is determined by consumers' assessment of features that distinguish brands from each other (Kemp et al. 2012). In the whole market there are thousands of goods and their producers, and especially many products have the similar purposes as the competitors (Vazifehdoost 2016) . Brand as a symbol identifying the source of a producer of a product that allows consumers to establish responsibility to a particular manufacturer or distributor (Keller 2003). If the consumer feels that brand is unique and get benefits from buying this brand in the long term, they will develop confidence in this brand and continue to but it. The self-brand connection will be formed (Vazifehdoost 2016).

1.4 Brand advocacy

Favorable communication about a brand from consumers can accelerate new product acceptance and adoption (Keller 2003). It is critical construct in the marketing literature as it describes the strength of the bond consumers has with the brand. This bond subsequently affects their behavior, and in turns fosters firm profitability and customer lifetime value (So et al. 2013). A brand that has established a self-brand connection with the consumer will benefit from favorable brand advocacy.

The self-brand connection is formed when the consumer has extensive knowledge and experiences about the brand and the consumer. The consumer will become an active partner in the relationship, which is likely to produce consumer advocacy behavior (Tho et al. 2016). In the context of Halal cosmetic, consumers that have developed a connection to the brand can become an "ambassador" for the brand and promote the brand to others.

Self-brand connection as the formation of stable and meaningful ties between a particular brand and a consumer's self-identity (Dwivedi et al. 2015). Con-sumers become committed to brands that help them to create or represent their desired self-concepts (Kemp et al. 2012). When a customer becomes more connected to a brand, they are likely to maintain the bond with the brand because the presence of the attachment object offers feelings of comfort, happiness, and security (So et al. 2013).

As mentioned previously, celebrity endorser, perceived quality, brand uniqueness can play a significant role in helping some consumer construct, cultivate, and express their identities. Once self-brand connection of consumer is formed, and consumer feel that

a brand personalizes who they are, they are likely to encourage positive evaluations and assessments about the brand to others. Afterward, it is proposed that in the context of Halal cosmetic branding, a self-brand connection will mediate the relationships among celebrity endorser, perceived quality, brand uniqueness and brand advocacy.

2 METHOD

This research focused on the relationship among celebrity endorsement, perceived quality, brand uniqueness and self-brand connection and brand advocacy of Wardah. Respondent were those consumers who use Wardah cosmetics. These people were chosen based on the belief that they possess a high degree of willingness to make brand advocacy. Data collection were using online survey of Google docs to 150 women.

The constructs in this study were measured using adapted scales from the previous research. Measurement scales for celebrity endorsement using eight items were developed adapted from (Dwivedi et al. 2015). Measurement scales for perceived quality using four items, brand uniqueness using four items, self-brand connection using six items, and brand advocacy using three items were adapted from (Kemp et al. 2012). This research used validated scales to operationalize the construct, measured by multi-item five-point Likert scale type format ranging from (1) strongly disagree to (5) strongly agree (Ringle et al. 2010). This research using Smart PLS 3.0 to conduct structural equation model analysis as recommended by (Hair et al. 2014), a two-step procedure was used to assess the model for construct and discriminant validity first and hereafter to test the hypotheses in the structural model.

3 RESULTS

As many as 150 questionnaires were obtained. All the respondent were women who had purchased and used Wardah thus indicated they had more willing to make brand advocacy. The findings also showed that most of the respondent (65.4%) had been used Wardah for two years and the rest were using Wardah less than two years.

3.1 Construct validity

The validity of the scales was accomplished in a confirmatory factor analysis of the measurement model. In the confirmatory factor analysis, the measurement model was assessed for construct and discriminant validity as well as reliability (Hair et al. 2010). All items of the variables must have loading s and cross-loadings significant at 0.708 or higher (Hair et al. 2010). However, several items had been deleted due to load weakly. Therefore 1 item from

Table 1. Loading and cross loading.

	Brand Advocacy	Brand Uniqueness	Celebrity Endorser	Perceived Quality	Self-Brand Connection
BA1	0.919	0.676	0.666	0.138	0.756
BA2	0.848	0.581	0.538	0.129	0.561
BUI	0.498	0.748	0.534	0.095	0.596
BU2	0.655	0.843	0.561	0.088	0.616
BU3	0.646	0.868	0.659	0.133	0.660
BU4	0.565	0.841	0.738	0.183	0.794
CE2	0.646	0.588	0.808	0.179	0.666
CE3	0.547	0.658	0.853	0.239	0.663
CE4	0.614	0.686	0.884	0.300	0.744
CE5	0.531	0.664	0.833	0.198	0.699
PQ1	0.064	0.085	0.169	0.760	0.219
PQ4	0.168	0.156	0.261	0.804	0.239
SBC1	0.514	0.721	0.656	0.185	0.783
SBC2	0.470	0.699	0.621	0.188	0.760
SBC3	0.683	0.623	0.728	0.208	0.824
SBC4	0.683	0.722	0.758	0.329	0.866
SBC5	0.708	0.723	0.706	0.262	0.902
SBC6	0.689	0.604	0.670	0.277	0.852
CE1	0.575	0.635	0.852	0.250	0.736

Note: bold values are loadings for items which are above the recommended value of 0.708.

brand advocacy, three items from celebrity endorser, and two items from perceived quality were eliminated for further analysis. Table 1 shows all the items have significant construct validity.

3.2 Convergent validity

Convergent validity measures whether variables positively correlated with other measures of the same construct (Hair et al. 2014). This study measured the composite reliability and average variance extracted method (Hair et al. 2010), which require significant values higher than 0.7 and 0.5 respectively.

Table 2 shows the composite reliability values for each construct, ranging from 0.759 to 0.931 which exceed suggested minimum value (Hair et al. 2010). The average variance extracted values are 0.612 to 0.782 which also exceed the recommended value of 0.5 (Hair et al. 2010).

3.3 Discriminant validity

It is essential to measure and test whether the correlations between the measures possibly overlap by discriminant validity, it is the opposite of convergent validity (Neuman 2014). Table 3 shows that the average variance extracted values shared between each construct were higher than the difference divided between other constructs, means people answers all the question in deferent ways (Neuman 2014). In this study has been

confirmed that all the construct have adequate discriminant validity using Fornell Larcker criterion (Fornell & Larcker 1981).

Table 2. Results of the measurement model.

Construct	Measurement Items	Loading	Composite Reliability	AVE
Brand Advocacy	BA1	0.919	0.877	0.782
	BA2	0.848		
Brand Uniqueness	BUI	0.748	0.896	0.683
	BU2	0.843		
Brand Uniqueness	BU3	0.868		
	BU4	0.841		
Celebrity Endorser	CE1	0.852	0.927	0.717
	CE2	0.808		
	CE3	0.853		
	CE4	0.884		
	CE5	0.833		
Perceived Quality	PQ1	0.760	0.759	0.612
	PQ4	0.804		
Self-Brand Connection	SBC1	0.783	0.931	0.693
	SBC2	0.760		
	SBC3	0.824		
	SBC4	0.866		
	SBC5	0.902		
	SBC6	0.852		

Table 3. Discriminant validity.

	Brand Advocacy	Brand Uniqueness	Celebrity Endorser	Perceived Quality	Self-Brand Connection
Brand Advocacy	0.884				
Brand Uniqueness	0.715	0.826			
Celebrity Endorser	0.688	0.764	0.847		
Perceived Quality	0.151	0.156	0.277	0.782	
Self-Brand Connection	0.757	0.817	0.830	0.293	0.833

Note: bold values represent the average variance extracted (AVE) while the other entries represent the squared correlations.

3.4 Hypothesis testing using bootstrapping

The structural model and hypotheses were evaluated after attaining a validated measurement model. In this study, 150 cases were used for each of the 5000 bootstrap samples. Table 4 and Figure 1 present the path coefficients resulting from bootstrapping process.

The findings show that all constructs supported in the relationship with brand advocacy.

H1-H3 predicted direct effects among celebrity endorser, perceived quality, brand uniqueness and self-brand connection. H1 proposed that celebrity en-dorser would be positively related to the self-brand connection. This hypothesis is supported. Next, H2 predicted that perceived quality would be positively associated with a self-brand connection, and the data support this hypothesis. Finally, H3 proposed that brand uniqueness would be positively related to the self-brand connection. This hypothesis is also con-firmed. H4 predicted that a self-brand connection would mediate brand advocacy. Results showed that self-brand connection is positively related to brand advocacy. The final hypothesis that are H5a, H5b, and H5c suggested that self-brand connection would be a mediator among celebrity endorser, perceived quality, and brand uniqueness to brand advocacy. Following (Barron & Kenny 1986) consideration for a mediator, the self-brand connec-tion was confirmed rolled as the partial mediator for those variables.

4 DISCUSSION

This study concluded that celebrity endorser, perceived quality, and brand uniqueness could build consumer self-brand connection. Celebrity endorser was expected to positively and significantly influence self-brand connection toward Wardah. Consistent with the previous study (Dwivedi et al. 2016, Dwivedi et al. 2015), the findings indicate a healthy relationship between the variables. Thus, Wardah marketing communications should design marketing messages that have attractive emotional appeals and use credible celebrities who can successfully express their suits efficiently in the product advertisements to encourage connection between the brand and the consumers. It is essential for not only consider the product attributes but also the broader meaning related with the endorser when selecting the celebrity.

The findings of this study are in line with the results of the survey conducted by Kemp et al. (2012) where perceived quality is found to have a positive and significant relationship with the self-brand connection. For example, the consumers who had consumed tend to evaluate the quality of the brand highly when they have the proper feeling about the product performance. If consumer satisfies with the brand, it will impact on consumer self-brand connection (Dwivedi et al. 2016). Hence, perceived quality becomes an important thing to build customer self-brand connection.

Table 4. Path coefficients and hypothesis testing of bootstrapping results.

Hypotheses	Relationships	Coefficient	T Values	P Values
HI	Celebrity Endorser -> Self-Brand Connection	0.066	6.932"*	0.000
H2	Perceived Quality -> Self-Brand Connection	0.037	2.607"*	0.009
H3	Brand Uniqueness-> Self-Brand Connection	0.066	6.S49***	0.000
H4	Self-Brand Connection -> Brand Advocacy	0.032	23.346***	0.000
HSa	Celebrity Endorser -> Self-Brand Connection -> Brand Advocacy	Partial mediation		
HSb	Perceived Quality -> Self-Brand Connection -> Brand Advocacy	Partial mediation		
HSc	Brand Uniqueness -> Self-Brand Connection -> Brand Advocacy	Partial mediation		

Note: **p < 0.05, *** < 0.01.

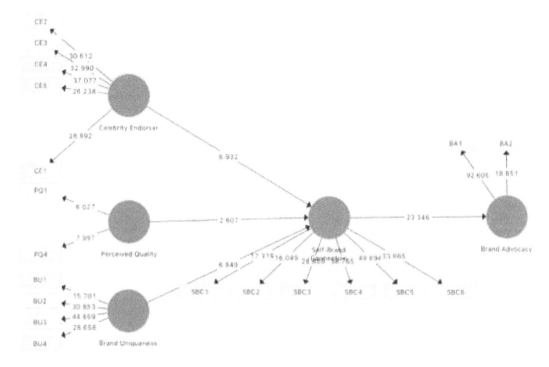

Figure 1. Results of path coefficient from bootstrapping.

The variable brand uniqueness has its role in a positive and significant relationship with the self-brand connection. These findings are consistent with the previous empirical research by (Kemp et al. 2012). By realizing the importance of brand unique-ness, the company should improve its brand attributes that associate with the consumers. Consumers tend to prefer a particular brand because they feel that the brand is consistent with their self-images and person-alities (Banerjee 2016). As long as the consumer feels a similarity between their self and the brand, the self-brand connection would be formed.

The development of self-brand connections through celebrity endorser, perceived quality, and brand uniqueness has significant implications for the company. When consumers' self-concepts are corre-lated to a brand, the company or entity behind the brand may be able to achieve a sustainable competi-tive advantage. Moreover, when individuals develop a self-brand connection, the brand becomes less of switching behavior (Kemp et al. 2012).

Establishing a self-brand connection by creating a favorable brand association with consumers can play a vital role in enhancing the branded offerings of Wardah. The consumer who has a strong brand con-nection are more likely to search more infor-mation to purchasing and consuming the brand. Additionally, if consumers develop a self-brand connection to Wardah, they can have willing to share the good things about the brand to other people. They can become an ambassador for the brand. If the brand advocacy willing has been per-formed on the consumer, the company can involve consumers in blogging and talk about the brands.

5 CONCLUSION

The findings in this study provide significant implica-tions. In spite of that, this study has some limitations and offers several directions for future research. First, this study only examines one brand (i.e. Wardah). We assumed that consumers would have direct or indirect experience with other brands to test the hypothesis. In spite of that, because of this study used the real brand (i.e. Wardah), there could be a possible influence of the brand and the company used in this study. It would be interesting for future research to examine the other brands and categories. Second, the sampling coverage is limited, which means that the study find-ings cannot be generalized. Thus, it was unable to compare different brands from the same industry and brands from various sectors. Finally, this study con-siders results from only one national context; cross-national research could be conducted to extend the validity of the findings.

REFERENCES

Banerjee, S. 2016. Influence of consumer personality, brand personality, and corporate personality on brand preference. *Asia Pacific Journal of Marketing and Logistics* 28(2): 198–216.

Barron, R.M. & Kenny, D.A. 1986. The moderator-mediator variable distinction in social psychological research: Conceptual, strategic, and statistical considerations. *Journal of Personality and Social Psychology* 51(6): 1173–1182.

Chun, R. 2014. What holds ethical consumers to a cosmetics brand: The body shop case. *Business & Society* 55(4): 528–549.

Dwivedi, A., Johnson, L.W. & McDonald, R. 2016. Celebrity endorsements, self-brand connection and relationship quality. *International Journal of Advertising* 35(3): 486–503.

Dwivedi, A., Johnson, L.W. & McDonald, R.E. 2015. Celebrity endorsement, self-brand connection and consumer-based brand equity. *Journal of Product & Brand Management* 24(5): 449–461.

Fornell, C. & Larcker, D.F. 1981. Evaluating structural equation models with unobservable variables and measurement error. *Journal of Marketing Research* 18(1): 39.

Hair, J.F., Black, W.C., Babin, B.J., & Anderson, R.E. 2010. multivariate data analysis (7th ed.). New York: Pearson.

Hair, J.F., Hult, G.T. M., Ringle, C.M., & Sarstedt, M. 2014. A primer on partial least squares structural equation modelling (PLS-SEM). Sage Publications, Inc.

Keller, K.L. 2003. Brand synthesis: The multidimensionality of brand knowledge. *Journal of Consumer Research* 29(4): 595–600.

Kemp, E., Childers, C.Y., & Williams, K.H. 2012. Place branding: Creating self-brand connections and brand advocacy. *Journal of Product & Brand Management* 21 (7): 508–515.

Madiawati, P.N. & Pradana, M. 2016. The appeal of celebrity endorsers and halal certificates on customers'

buying interest. *Journal of Administrative & Business Studies* 1(1): 28–34.

Magnoni, F. 2016. The effects of downward line extension on brand trust and brand attachment. *Recherche et Applications En Marketing (English Edition)* 31 (1): 2–25.

Neuman, W.L. 2014. Social research methods: qualitative ad quantitative approaches (seventh). Pearson Education Limited.

Ringle, C.M., Wende, S. & Will, A., 2010. Finite mixture partial least squares analysis: Methodology and numerical examples. *In Handbook of partial least squares*: 195–218.

So, J.T., Parsons, A.G. & Yap, S.F. 2013. Corporate branding, emotional attachment and brand loyalty: The case of luxury fashion branding. *Journal of Fashion Marketing and Management* 17(4): 403–423.

Suki, N.M. 2014. Does celebrity credibility influence Muslim and non-Muslim consumers' attitudes toward brands and purchase intention?. *Journal of Islamic Marketing* 5(2): 227–240.

Tho, N.D., Trang, N.T. M., & Olsen, S.O. 2016. Brand personality appeal, brand relationship quality and WOM transmission: A study of consumer markets in Vietnam. *Asia Pacific Business Review* 22(2): 307–324.

Tsiotsou, R. 2006. The role of perceived product quality and overall satisfaction on purchase intentions. *International Journal of Consumer Studies* 30(2): 207–217.

Vazifehdoost, H. 2016. The role of brand personality in consumer's decision making: A review of the literature 6(04): 15–29.

Vo, T.T.N. & Nguyen, C.T.K. 2015. Factors influencing customer perceived quality and purchase intention toward private labels in the Vietnam market: The moderating effects of store image. *International Journal of Marketing Studies* 7(4): 51–63.

The implementation of mobile commerce applications to order systems for bike-sharing programs in Bandung

R.M.A Rifki & M.A Sultan
Universitas Pendidikan Indonesia, Bandung, Indonesia

ABSTRACT: The purpose of this research was to explore the level of technology acceptance of a model mobile commerce application called Boseh bike sharing by its consumers in Bandung. This research used a verification approach with survey method. A total of 103 respondents were chosen using systematic random sampling. A survey questionnaire was used to collect the data from respondents. The result showed that perceived ease of use, perceived usefulness and attitude towards using it has an influence on information technology acceptance, both simultaneously and partially. The research provides a new concept basis to understand the information technology acceptance for the bike sharing mobile commerce application named Boseh, which is the first bike sharing in Indonesia based on information technology. The difference of this research compared to earlier research on objects that are the variable used support the theory as well as the different references used by previous researchers.

1 INTRODUCTION

In July 2017, the government launched a bike-sharing program named Boseh, which stands for Bike on Street Everybody Happy. This concept is the first bike-sharing program in Indonesia. It is not only an anticipatory program but also an innovative program because it integrates the bike-sharing process with mobile commerce application technology. Its presence is expected to decrease traffic jams in Bandung, especially during weekends and public holidays.

The innovation of the technology mobile commerce application Boseh is used by local government as a marketing strategy to attract people, especially tourists who come to Bandung, in order to switch to using the bicycle as an alternative transportation, especially for traveling on a short distance. The bike-sharing marketing strategy using mobile commerce applications targeting tourists as a potential niche in the market is supported by Tuckwell's research in Aquarita et. al. (2016), which states that an effective promotion strategy is through tourism activities. The utilization of mobile commerce applications to rent bicycles in Boseh is also used to simplify the ordering process where Boseh customers can use applications for rental registration and get information about outlets locations and bicycles availability at each outlet. In addition, mobile applications could be an effective promotion, because based on data cited from Hootsuite until January 2018, from a total population of 265 million in Indonesia, as many as 177 million Indonesians are users of smart devices such as mobile smartphone, laptop computer and tablet computer. Other data from Hootsuite discloses that throughout January 2018, as many as 40% of the user population of smart devices in Indonesia use their gadget for online transactions.

The method to learn the acceptance level of the integration of the order system by Boseh's consumers who use the mobile commerce application technology could be measured by theoretical approachment that could describe the acceptance level of technology, the method is Technology Acceptance Model (TAM). According to Destiana (2012) the concept of TAM is purposed to understand the reaction and perception of information technology and affect the user attitude in the acceptance of technology. In a case study of Boseh bike-sharing, the user is a Boseh consumer and information technology is a mobile commerce application used for bicycle order registration.

According to Sensuse and Widiatmika (2012) who reviewed TAM to understand the user's reaction and perception of technology, concluded that the technology acceptance model on using the internet technology by students could be proposed as a model of internet technology acceptance by students, especially high school students, with proper facilities and conditions. Another research that examines the technology acceptance model of BPJS online application has led to the conclusion that BPJS online application users assess the application technology fulfills the usability criteria and ease-for-use and it becomes the preference for members of BPJS to use the online application for transactions (Marini 2016). The concept of TAM is specifically about the integration of mobile commerce applications with the bike-sharing order system and has been discussed to learn the acceptance from the Chinese society of information technology for online registration to order the bike-sharing program. The results of Hazen et al. (2015) revealed there is a positive effect that significantly affects the acceptance of mobile commerce applications to order and registrate on the bike-sharing program, and consumers greatly benefited from the presence of information technology to facilitate online registration and improve safety standards when using the bicycles. The research from Yu et al. (2018) yielded two conclusions, first, that the concept of TAM as one of the methods to measure consumer acceptance in Beijing showed a positive effect from the level of public acceptance of mobile commerce application technology, and, two, that it has an impact on the public interest to re-hire bicycles as an environmentally friendly alternative to transportation in China. TAM could be used to find out the user satisfaction. Based on the conclusions, Chabibie (2012) is known that the usefulness and ease of use on information technology is closely related to user satisfaction on information technology.

In the digital era, the technology acceptance model is also used to analyze the technology acceptance model in the entertainment industry. Bakar and Bidin (2014) revealed that the use of mobile advertising technology contributes significantly to the interest of watching movies at consumers aged 15 to 29 in Malaysia. The research aims to find out whether baby boomers and millennials in Malaysia are generations that take advantage of technological roles positively especially for entertainment. TAM could be used to analyze consumer behavior in social media and its impact on online business strategy on Facebook (Rauniar et al. 2014).

In the case of Boseh bike-sharing, which integrates the registration process with mobile commerce application technology, TAM could be used to obtain the level of Boseh consumer acceptance to the successful implementation of mobile commerce application in registration process. TAM is considered important because until now there has been no research on the model of mobile technology application acceptance for bike-sharing program in Indonesia, because Boseh bike-sharing program is the first integrated bicycle rental technology concept and is still the only one in Indonesia. Therefore, based on the phenomena that occur compiled the questions to answer the encountered problems in this research is the acceptance of mobile commerce application technology by Boseh consumer with technology acceptance model method and the formulation of the problems are:

1. Does perceived ease of use affect perceived usefulness?
2. Does perceived ease of use affect the attitude toward using?
3. Does perceived usefulness have an effect on attitude toward using?
4. Does perceived usefulness affect the acceptance of mobile commerce ?

5. Does attitude toward using affect the acceptance of mobile commerce?

2 METHOD

In this research, construct behavioral intention of use and actual system usage will be changed by modified construct acceptance of mobile commerce apps (IT-ACC). This construct is adopted by Al-Gahtani's research in 2001, which examine the technology acceptance outside North America. According to Al-Gahtani (2001), the idea of construct behavioral intention of use and actual system usage is indicator to measure information technology acceptance. In this research, IT acceptance is user acceptance of mobile commerce application technology for bike-sharing registration process. The IT acceptance (IT-ACC) construct indicator is the motivation to keep using technology, influencing other user to use technology, high usage frequency and high technology satisfaction.

This research was conducted to obtain the information technology acceptance by Boseh consumer that adapted for registration using mobile commerce application. Independent variable in this research is perceived ease of use (PEOU), while dependent variable is perceived usefulness (PU), attitute toward using (ATU) and acceptance of mobile commerce (IT-ACC).

The research begins by compiling a questionnaire consisting of 21 questions divided into six questions on the perceived ease of use (PEOU), five questions on the perceived usefulness (PU), four questions on attitute toward using (ATU) and five questions on acceptance of mobile commerce (IT-ACC). The beginning of this research is examine validity and reliability of 21 questions distributed to 30 respondents generated conclusion by software AMOS 20.0 that there are three invalid questions. There are two questions on PEOU and one question on ATU.

The questionnaire was then re-distributed by using Google Form application for 10 days and found 73 respondents who are willing to fill up to a predetermined time limit. Method to get a whole sample research is using systematic random sampling technique. All respondents' answers will be combined with the initial research respondents, so that the total respondents in this research were 103 respondents.

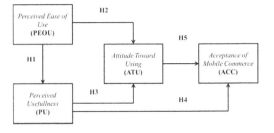

Figure 1. Research methodology flowchart.

Table 1. Goodness of fit model.

Goodness of Fit Index	Goodness of Fit	Value	Result
Chi-Square	Smaller is Better	1.447	Good
Probability	≥ 0.05	0.229	Good
CMIN (DF)	≤ 2	1	Good
AGFI	≥ 0.90	0.930	Good
GFI	≥ 0.90	0.993	Good
TLI	≥ 0.95	0.977	Good
CFI	≥ 0.95	0.996	Good

3 RESULT AND DISCUSSION

Testing the research hypothesis is one of the steps in this research to determine the level of Boseh consumer acceptance on the mobile commerce applications to order registration. Hypothesis testing is calculated simultaneously and partially using path analysis technique using software AMOS 20.0. Path analysis technique in this research will do two tests, a goodness of fit model and a hypothesis test. The result from the goodness of fit model is showed in Table 1, which presents the calculation data.

The results of data testing assisted by software Amos 20.0 showed that all parameters of the goodness of fit model fulfill the acceptance limit indicator, so that the whole equation model of path analysis used is acceptable and the hypothesis testing can proceed. To analyze the results of data processing in hypothesis testing, the variable has a significant influence if $C.R \geq 1.96$ and $P < 0.05$ and then the research hypothesis result was built is fulfill criteria. The results and calculation from software Amos 20.0 are presented in Table 2.

In part, the hypothesis of the research has a positive and significant influence. The value of direct influence from each variables could be seen in the standardized estimate column in Table 2. Simultaneously, the research hypothesis results can be seen based on variables factors PEOU which has a directly positive and significant influence in the amount of 0.9868. The simultaneous value effect of Boseh mobile commerce applications that could be said to be very high means that applications designed to provide ease for the consumer to register and information about outlets location and bicycles availability at each outlet has been well received by consumers.

As per the partial calculations and results contained in Table 2, the highest value is 0.5375 which explains that the strongest direct influence of the correlation occurs between PEOU and PU. The similar results also occurred in the research results on health-supporting mobile applications in South Korea, examined by Jeon and Park (2015) which stated that PEOU and PU have an equally positive effect on the intensity of mobile application usage, which causes the application to be well received by the user. According to Kang (2014), PEOU is a trigger for users to choose and be loyal for using a technology. The technology is present to replace a previous system that works conventionally into working programmed by computerized systems with the aim of helping humans as a users. The positive feedback of technology utilization is that human activityhas become easier and faster to complete. Therefore, the design of technology must fulfill a main functionin making it easier for humans as a user in order for the technology to be accepted by the community.

The second most important factor of a technology is that the functions of each feature work properly. Boseh mobile commerce application has some features for online registration, store information, map and timer. Based on the results of the research, consumers give a positive appreciation for this innovative technology. This could be seen from the value of the influence of PU to the IT-ACC which has the second strongest influence. The results of the research on ease of use, completeness of features, and usefulness to support the success of a mobile application technology could be accepted by a community. Roy (2017) concluded that the usefulness of applications participating in technology acceptance is supported also by Hamid et al. (2016), who stated that the strong influence of the presence of technology is accepted when the ease of use, the technology and feature completeness, as well as the features are perfectly functioning, which becomes the determinant factor of users using the application periodically. But in order to determine whether the technology received is favored by users or not, the variable ease of use could not be used as a reference. According to Olumide (2016), to determine that the users choose technology not because of compulsion, could be seen directly from the correlation between variables PU and ATU, which according to PEOU general construction theory and PU becomes the determinant variable a consumer feels comfortable

Table 2. Research hypothesis result.

ft	To	Estimate			Ft	To	Estimate
EU	PU	0.5634	0.5375	0.0875	6.4372	0.0000	Significant
EU	ATU	0.1341	0.1924	0.0657	2.0425	0.0411	Significant
HI	ATU	0.3133	0.4712	0.0626	5.0021	0.0000	Significant
ART	IT	03341	0.2141	0.1532	2.1803	0.0292	Significant
PU	IT	0.4507	0.4344	0.1019	4.4241	0.0000	Significant

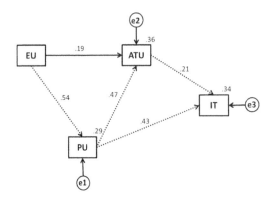

Figure 2. Path analysis diagram.

with and finds patterns of using technology. In this research, the correlation between PU and ATU found in Table 2 could be said to be relatively high and strong. A summary of the direct correlation between variables is shown in Figure 2, made with the Amos 20.0 software, and it could be said that each variable has a strong correlation and is significant both partially and simultaneously. Therefore, Boseh mobile commerce applications could be said to be accepted by consumers as a technology-based innovation for bike-sharing order registration.

In the implementation, the research hypothesis results in Table 2 explain that consumers will receive a mobile application as an information technology, if an application is easy to use and has features in the application that facilitate the consumer in the transaction; in the case of mobile applications commerce Boseh, perceived usefulness could be accepted by consumers, if the application helps simplify the order registration and provide useful information for consumers. In addition, the level of acceptance is influenced by user attitudes that assess whether or not a mobile application technology has benefits for consumers, in the implementation of mobile commerce applications Boseh, consumers will accept Boseh mobile commerce application technology, if consumers feel the application is user-friendly and helpful.

4 CONCLUSION

The empirical research purpose to obtain the acceptance level of Boseh consumers, consisting of the society of Bandung and the tourists, resulted that all hypotheses in the formulation of the problem are acceptable and the independent variables either partially or simultaneously have influence on the dependent variable. In addition, it could also be concluded that the direct influence was simultaneously present; the acceptance level of Boseh mobile commerce applications reached 0.9868. The value is considered relatively high for the acceptance level for a new technology that was launched mid-2017 and

became the only bike-sharing program that uses mobile commerce applications in Indonesia.

In the future, this research is expected to be periodically re-examined to determine the acceptance level of Boseh mobile commerce applications by consumers for other variables that have not been researched in this case. Research as a continuous improvement on Boseh mobile commerce applications is needed, especially since this application is periodically updated by the provider to improve the mobile commerce application performance and increase the number of order bike-sharing programs, which has an impact on the increasing people who are interested in using a bicycle as an alternative to other transportation.

REFERENCES

Al-Gahtani, S. 2001. The applicability of TAM outside North America: An empirical test in the United Kingdom. *Information Resources Management Journal (IRMJ)* 14(3): 37–46.

Aquarita, D., Rosyidie, A. & Pratiwi, W.D. 2016. Potensi pengembangan wisata sepeda di Kota Bandung berdasarkan persepsi dan preferensi wisatawan [The potency of bicycle tour development in Bandung based on tourist perception and preference]. *Jurnal Pengembangan Kota* 4(1): 14–20.

Bakar, M.S.A. & Bidin, R. 2014. Technology acceptance and purchase intention towards movie mobile advertising among youth in Malaysia. *Procedia-Social and Behavioral Sciences* 130(1): 558–567.

Chabibie, H. 2012. Pengaruh penerimaan teknologi dengan kegunaan web (Studi kasus portal rumah belajar http://belajar.kemdikbud.go.id). *Tesis*. Depok: Universitas Indonesia.

Destiana, B. 2012. analisis penerimaan pengguna akhir terhadap penerapan sistem e-learning dengan menggunakan pendekatan technology acceptance model (TAM) di SMAN 1 Wonosari. *Skripsi*. Yogyakarta: Universitas Negeri Yogyakarta.

Hamid, A.A., Razak, F.Z.A., Bakar, A.A. & Abdullah, W. S.W. 2016. The effects of perceived usefulness and perceived ease of use on continuance intention to use e-government. *Procedia Economics and Finance* 35: 644–649.

Hazen, B.T., Overstreet, R.E. & Wang, Y. 2015. Predicting public bicycle adoption using the technology acceptance model. *Sustainability* 7(11): 14558–14573.

Jeon, E. & Park, H.A. 2015. Factors affecting acceptance of smartphone application for management of obesity. *Healthcare Informatics Research* 21(2): 74–82.

Kang, S. 2014. Factors influencing intention of mobile application use. *International Journal of Mobile Communications* 12(4): 360.

Marini, S. 2016. Analisis model penerimaan teknologi (Technology acceptance model) aplikasi BPJS online. *Jurnal Wawasan Manajemen* 4(1): 233.

Olumide, D. 2016. Technology acceptance model as a predictor of using information system ' to acquire information literacy skills. *Library Philosophy and Practice* 1450.

Rauniar, R., Rawski, G., Yang, J. & Johnson, B. 2014. Technology acceptance model (TAM) and social media usage: an empirical study on facebook. *Journal of Enterprise Information Management* 27(1): 6–30.

Roy, S. 2017. App adoption and switching behavior: applying the extended tam in smartphone app usage. *JISTEM-Journal of Information Systems and Technology Management* 14(2): 239–261.

Sensuse, D. & Widiatmika, I.M.A.A. 2012. Pengembangan model penerimaan teknologi internet oleh pelajar dengan menggunakan konsep technology acceptance model (TAM). *Jurnal Sistem Informasi* 4.

Yu, Y., Yi, W., Feng, Y. & Liu, J. 2018. Understanding the intention to use commercial bike-sharing systems: An integration of TAM and TPB. *In Proceedings of the 51st Hawaii International Conference on System Sciences*.

Advances in Business, Management and Entrepreneurship – Hurriyati et al (eds)
© *2020 Taylor & Francis Group, London, ISBN 978-0-367-27176-3*

Influence of trust on online shopping in predicting purchase intention

P. Permatasari & D. Mardhiyah
Airlangga University, Surabaya, Indonesia

ABSTRACT: The amount of internet users in Indonesia grows quickly nowadays. The growth of internet users leads to an increase in online shopping. Today's popularity of social media becomes a chance for firms to do a business all over the world, including in Indonesia. One of the most popular online commerce media to do business is Instagram. Risks doing online shopping could decrease the consumer's trust. Therefore, trust is the most important aspect in online shopping. This research aims to analyze the effect of trust on familiarity, information seeking and purchase intention on Instagram. Data was collected through questionnaires that were given to respondents who own an Instagram account and have never purchased any fashion product from an online shop on Instagram. The amount of respondents used to analyze the hypotheses were 201 people. The results of this research show that trust has an effect on purchase intention and information seeking, information seeking has an effect on purchase intention and familiarity, and familiarity has an effect on purchase intention.

1 INTRODUCTION

Currently the growth of internet users in Indonesia is considerable. This is evident from the survey data from the Association of Internet Service Providers Indonesia (APJII) in 2016, which shows that internet users in Indonesia amount to 132.7 million, which in 2014was only 88 million people (APJII 2016). The growth of internet users is driving the creation of online shopping. In 2016, as many as 84.2 million Internet users in Indonesia have been shopping online.

The increase of internet users in Indonesia occurs because of easy access which allows the usage of the internet for various purposes, one of which is to buy products through social media. The popularity of social media is now an opportunity that is used to conduct business activities around the world, including in Indonesia. The strategy that utilizes social media sites to facilitate the trading of various products and services is called social commerce (Kim & Park 2013). Social commerce is a web 2.0 based online application, as a means of content providers, for enhanced user interaction in e- commerce (Liang et al. 2011).

Many multinational companies use social commerce for their businesses, for example; Starbucks uses Facebook and Adidas uses Instagram. Some retailers (such as Armani Exchange, Toms, and Samsung) and service providers (such as banks, insurers, and airlines) are experiencing the success of using social commerce to improve their business. There are also companies that fail to use social commerce for their business (such as Walmart) and there are many complaints about trust, security, and privacy in the exchange of information (Liang & Turban 2011, Kim & Park 2013).

One of the popular social commerce media used to do business online is Instagram. Instagram is a mobile-based app that allows users to take photos or images, apply digital filters, and share photos or pictures directly with friends on different social networking sites (Hochman & Schwartz 2012). Instagram has now provided a means to do business for its users, called Instagram business profile.

Indonesia became the country with the most Instagram users in Asia Pacific by 2017 (Bohang 2017). Of the 700 million active Instagram users worldwide, 45 million users are from Indonesia. It showed a significant increase when compared with January 2016. At that time, active users on Instagram in Indonesia were only 22 million. According to Country Director of Indonesia Facebook, Sri Widowati quoted from wartakota.tribunnews.com, currently as many as 80 percent of Instagram users follow at least one business account (Setyaningsih 2017). This proves the number of Instagram users who are interested in making a purchase through Instagram. And it also proves the role of Instagram as a container that supports the growth of various local businesses, especially SMEs in Indonesia.

Based on the Creative Economy Agency (BEKraf), Indonesia's fashion growth throughout 2015 is the second largest among creative businesses in Indonesia. The existence of opportunities for the fashion business raises many fashion brands in Indonesia, who market their products through the social commerce platform. Some fashion brands in Indonesia using the Instagram platform include Neon Lights Clothing (@neonlightsclothing), Giyomi (@giyomi.id), and Erigo (@erigostore).

Many consumers are afraid of online transactions for reasons such as fraud, product conformity, and security. Unable to meet potential customers and sellers face-to-face can directly make consumers think the condition is prone to fraud. Products that cannot be physically examined can also make consumers feel uncertain about the quality of the product, including whether the product can be delivered safely to consumers (Doolin et al. 2008, Mathur 2015). The existence of these risks can reduce consumer trust in shopping online. It is different from offline buying, where sellers and buyers meet face-to-face in transactions. Therefore, trust is an important aspect in buying and selling online.

The existence of fraud cases on the Instagram platform makes Instagram users hesitant to buy and transact via e-vendor on Instagram. Moreover, the payment system on e-vendors in Instagram is still manually checked, which is different from e-commerce systems that have been using the automatic check system, which further adds to the doubt of consumers to buy and transact via e-vendor on Instagram. In this research e-vendor is defined as an electronic seller or online shop. Therefore, consumers who believe in e-commerce necessarily believe in social commerce (Kim & Park 2013).

For a company that is building a sales chain through a social commerce platform, trust is a key concept of interaction quality and consumer purchase intention (Gefen et al. 2003). Trust is a belief in the reliability, truth, and ability of the parties concerned. The absence of trust is one of the main reasons customers refrain from making electronic purchases (Gefen 2000).

In this study, trust is studied in the context of social commerce, where users will be informed about products on social commerce and can be involved in the purchase. Thus, the trust in social commerce and embedded content provided by fellow social commerce users (such as other followers' comments on a product, the number of likes, mentioning of particular followers and the number of shares) may increase the purchase intention of users from e-vendor. Kim and Park (2013) suggest that trusts over social commerce companies (e.g. Amazon.com) instantly increase purchases and word-of-mouth intentions.

Trust can increase the customer dependence on content generated by fellow users of social commerce, reduce the uncertainty, and extend the duration of a relationship (Suh & Han 2003). Consumers decide whether they will transact with e-vendors on the platform by evaluating every aspect of their trust.

The low trust also makes consumers perform social commerce information seeking. Information seeking is a process whereby a person forms a personal point of view (Kuhlthau 2004). Social commerce information seeking is the effort of consumers to obtain information about products/e-vendors from sources available in social commercemedia such as reviews, ratings, and recommendations in the online community. In the context of social commerce, users can search information about products

through various channels, including recommendations from fellow users of social commerce, ratings and reviews, and forums and communities (Hajli & Sims 2015). Hertzum et al. (2002) on trust and information seeking, indicates that trust affects the user's information seeking behavior in relation to virtual persons, documents, and agents. McKnight et al. (2002) argue that if customers have a high confidence in the platform, they will be more willing to rely on that platform.

Social commerce information seeking enhances the individual's knowledge of a product through access to various information on the product. Increased customer knowledge of a product can facilitate decision-making processes and increase the purchase intention (Chiou et al. 2002). Products reviewed through social commerce help consumers to be able to evaluate the product thoroughly, so that consumers can make the right purchasing decisions (Heinonen 2011). Information seeking can enhance an individual's knowledge of products and e-vendors as well as their skills in using the features of the platform, so that customers will be more likely to purchase a product (Choo et al. 2000). Previous research has shown that product reviews and multimedia texts that provide the ability to interact with products before buying, have a positive effect on customer purchasing behavior and increase the purchase intention (Maria & Finotto 2008).

Social commerce information seeking can affect the familiarity with the online platform, that is, the extent to which consumers understand the procedures on the website (Gefen et al. 2003), such as familiarity with search engines of a website. Searching for information on the website deepens the user's understanding of the content and knowledge of the platform (Choo et al. 2000). For example, active users looking for information about a product on different channels, such as reviews and forums, become familiar with search tools, ratings policies, recommendations, and purchasing processes. Users who often seek information about a product will be more familiar with and understand the features used on social media.

Complexity in the online environment refrains customers from making purchases, while platform familiarity increases the customer's understanding of the shopping process and makes it easier to make decisions (Gefen et al. 2003). In Van Der Heide & Lim (2015), research shows that users who are familiar with a platform are more likely to rely on content generated by fellow social commerce users for their online purchases.

In research conducted by Kim and Park (2013) and Kaiser and Muller-Seitz (2008), it is explained how trust can affect the purchase intention, but if the mechanism of trust itself affects the purchase intention needs to be investigated more deeply (Hajli et al. 2015). Therefore this study is intended to examine the mechanism more deeply.

1.1 Trust and purchase intention

Trust is a key concept of the quality of interaction and the emergence of a purchase intention in consumers (Gefen et al. 2003). Ba and Pavlou (2002) mentioned that trust is the individual belief that an exchange will occur in a way consistent with the expectation or beliefs of a person. In an online context, trust is based on trust in exchange groups in terms of competence, integrity and goodwill (McKnight et al. 2002). Trust on the platform can be interpreted as an individual belief that a platform has competence, integrity and goodwill consistent with the expectations or beliefs of the individual. Purchase intention in the context of social commerce refers to the customer's intention to engage in online purchasing from e-vendors via social commerce. Intention itself has long been studied as a predictor of behavior and is defined as the magnitude of a person's possibility to perform certain behavior (Fishbein & Ajzen 1975).

The literature suggests that purchases from e-vendors depend on customer trust in e-vendors (Gefen et al. 2003). Trust in social commerce can increase the customer's purchase intention because it plays to reduce the risk perception in social commerce activities. The social commerce platform makes it easy for customers to contact e-vendors and provide facilities for value exchange between the parties. In social commerce, customers find ads, images/videos/news, recommendations, and likes related to e-vendors. However, trust on the social commerce platform can determine the customers' dependence on the credibility of content and e-vendor activities, thereby reinforcing the consumers' buying intentions.

Consumers decide whether they will transact with e-vendors on the platform by evaluating trusts on e-vendors and platforms used. High confidence in the platform means the user has a perception that the platform has the integrity and ability to deliver expected results, which increases the purchase intention on the platform (Kaiser & Muller-Seitz 2008). Kim and Park (2013) show that users who believe in social commerce sites are more likely to make purchases on the platform. Based on the above explanation, the following hypothesis is proposed:

H1: Trust in a social commerce platform has a positive effect on the purchase intention.

1.2 Trust, social commerce information seeking, and purchase intention

Social commerce information seeking is defined as the effort of the customer to obtain information about the product/e-vendor from the resources available in social commerce media such as reviews, ratings, and recommendations in the online community. Information gathering improves the knowledge of customers from various aspects of the product and helps in decision-making and purchasing (Turcotte et al. 2015).

The research done by Hertzum et al. (2002) on trust and information seeking indicates that trust affects the user's information seeking behavior in relation to virtual people, documents, and agents. The willingness of users to engage with a website depends on the extent to which the website succeeds in conveying accountability and trust (Hertzum et al. 2002). Thus, trust in social commerce motivates customers to engage in searching for a particular product or service (Turcotte et al. 2015) and vice versa; lack of trust prevents consumers from engaging in the information exchange (Wang et al. 2016). Based on the above explanation, the following hypothesis is formed:

H2: Trust in the social commerce platform positively affects social commerce information seeking.

Social commerce information seeking enhances the individual's knowledge of a product through access to various information on the product. Thus, increased customer knowledge of a product can facilitate decision-making processes and increase the purchase intention (Chiou et al. 2002). Products reviewed through social commerce help consumers to be able to evaluate the product thoroughly, among others related to price, product specifications, advantages over competing brands and others, so that consumers can make the right purchasing decisions (Heinonen 2011). Information seeking can enhance the individual's knowledge of products and e-vendors, as well as their skills in using features of the platform, so that customers will be more likely to purchase a product (Choo et al. 2000). With the increasing customer knowledge about a product, the customer will become more interested in the product and, ultimately, this can increase the purchase intention. Product reviews and multimedia text that enable potential customers to interact with products before buying have a positive effect on the customer buying behavior and increase the purchase intention (Maria & Finotto 2008). Murray (1991) also pointed out that information retrieval, a risk management strategy, increases the purchase intention. Based on the above explanation, the following hypothesis is proposed:

H3: Social commerce information seeking positively affects the purchase intention.

1.3 Social commerce information seeking, familiarity, and purchase intention

Familiarity with online platforms is the extent to which consumers understand the procedures on the website (Gefen et al. 2003), such as familiarity with the search engines of a website and the channel of interaction with fellow social commerce users. Understanding the procedures on the website can occur when users recognize the features listed on the website, know the function of each feature, and run each feature by following the steps on the website.

Searching for information on websites increases the understanding of customers in understanding the content and knowledge of a platform (Choo et al. 2000). This understanding improves the skills and expertise of users about the features available from an online platform. Active users seek information about a product on different channels, such as reviews and forums, and become familiar with features on online platforms such as search tools, rating policies, recommendations, and purchasing processes. Based on the above explanation, the following hypothesis is proposed:

H4: Social commerce information seeking has a positive effect on familiarity.

In the context of social commerce, familiarization with social commerce can increase the purchase intention. Online purchase intention requires certain steps such as finding the appropriate product, finding reviews/comments on products and e-vendors from other customers, selecting products and e-vendors, providing information, and placing orders. However, depending on the platform, these activities can be implemented differently and can become more complicated. The complexity of online purchases causes customers to refrain from making purchases, so in that condition, platform familiarity increases the customer's understanding of the shopping process and makes it easier to make decisions (Gefen et al. 2003). Van Der Heide and Lim's (2015) study shows that users who are familiar with a platform are more likely to rely on content generated by fellow social commerce users for their online purchases than those who are unfamiliar with it. Martinez-Lopez et al. (2015) also shows that familiarity with the platform improves the perceived ease of use, the intention to use the platform, thereby increasing the desire to purchase a product. Based on the above explanation, the following hypothesis is proposed:

H5: Familiarity positively affects the purchase intention.

2 METHOD

2.1 *Participants, procedures, and measurements*

This research uses a quantitative research approach with survey method. Data collection is done through the distribution of questionnaires to Instagram users who know of the existence of fashion products sold online through Instagram, but have never made a purchase of fashion products through Instagram. The amount of respondents used are 201 people, with a high response rate of 91%. Most of the unused questionnaires are either due to the respondents not having an Instagram account or having an Instagram account, but having shopped fashion products through Instagram already.

The measurement indicators of the variables used in this study are adapted from several previous studies.

The trust variable has 4 indicators adapted from Gefen et al. (Hajli et al. 2015). Social commerce information seeking variables have 3 indicators adapted from Hajli and Sims research (Hajli et al. 2015). Familiarity variables use 4 indicators adapted from Gefen (2000). The purchase intention variable uses 4 indicators adapted from Gefen et al. (Hajli et al., 2015). Measurement scale for all variables used 5 items of the Likert scale. Hypothesis testing in this study uses Structural Equation Modeling (SEM).

3 RESULT AND DISCUSSION

Respondents in this study amounted to 201 people, with the majority of the respondents aged 20–23 years (72.1%). More than half of the respondents have a high school education background (57.7%) and work as students (69.7%). The majority of the respondents of this study earned between IDR 1,000,000 to IDR 2,500,000 (46.3%).

According to Solimun in 2002, if there is at least one criterion that meets the criteria of goodness of fit criteria, then the model can be said to be good. The goodness of fit test indicates that the model fulfills its requirements (probability level = 0, CFI = 0.926, TLI = 0.908).

Ferdinand (2002) states that the value of standardized regression weight is acceptable when it is above 0.4 and the limit value used to assess the reliability level is 0.7. From this research, the standardized regression weight of all indicators of each research variable is valid (0.593–0.934). In addition, the reliability constructs all the research variables are declared reliable (0.853–0.943).

Based on hypothesis testing using AMOS it was shown that trust has a significant effect on the purchase intention (H1 supported). Increased trust in the fashion product online store on Instagram will have an impact on the increase of the purchase intention of fashion products. The assessment of the respondents is that the promise of reliability by the fashion product online shop on Instagram is able to increase the user's intention to buy fashion products from an online shop on Instagram. Concerns shown by online shops of fashion products on Instagram are able to make users of Instagram plan to spend some money, if there is a desired fashion product. The honesty afforded by the online fashion product shop on Instagram encourages users to make purchases from the fashion product shop online at Instagram. The more users believe in e-vendors in social commerce, the more it will make customers make purchases through social commerce. This is in line with several studies showing that trust has an influence on the purchase intention (Hajli et al. 2015, Doolin et al. 2008, Mathur 2015).

This study found that trust has a significant effect on social commerce information seeking (H2 supported). Increased trust in a fashion product online store on Instagram will have an impact on the increased information seeking about fashion products in online shops on Instagram. The absence of any doubt in the respondents regarding the honesty of an online fashion product shop on Instagram, prompted users to want to use reviews and ratings about the desired fashion product as a form of online information search. If the user has a high trust in a platform, then he/she is willing to find out information about using that platform. Several studies that also show similar results are Turcotte et al. (2015) and Wang et al. (2016).

In this study it was found that social commerce information seeking has no effect on the purchase intention (H3 is not supported). However much effort a person goes through to get information about a product, especially a fashion product, whether in the form of reviews or rating of a product, it does not make a person have the intention to make purchases of fashion products from online shops on Instagram. Because the majority of respondents in this study are aged 20–23 years, thus including the millennial generation who has the characteristic of being "literate" in technology, this makes it easy for them to collect information online. With the increasing amount of information they get, they did not rule out a lot of the information that is the opposite or not supportive of each other. It can even confuse the majority of consumers who are still students (69.7%) and who have a fairly low income level, which is between IDR 1,000,000-IDR 2,500,000 (46.3%). With limited financial resources, it increasingly makes the person have no intention of making a purchase.

In this study it was found that social commerce information seeking affects familiarity (H4 supported). Increased information seeking of fashion products in the online shop on Instagram has an impact on the increasing familiarity with the Instagram platform. The process of information retrieval of users by looking at recommendations, reviews, and ratings of fashion products that users will buy in an online shop at Instagram, can make customers more familiar with the process of searching for fashion products in online shops on Instagram. As more users actively seek information on a product from a platform, customers become more familiar with features on the platforms such as search tools, rating policies, and the buying process. This is in line with several studies showing that social commerce information seeking has an influence on familiarity (Choo et al. 2000, Hajli et al. 2015).

The test results show that familiarity affects the purchase intention (H5 supported). Increased familiarity with the Instagram platform has an impact on the increased purchase intention of fashion products on Instagram. Users who are familiar with Instagram for buying fashion products online will have a greater chance of having the intention to make a purchase of fashion products at an online shop on Instagram. The process of searching for fashion products that are familiar to the online shop of fashion products on Instagram can encourage users to plan to buy from an online shop on Instagram, if there is a desired fashion product. The more familiar users are with a certain online shop on Instagram, the more the respondent

deems the purchase as being low risk and gets familiar with the reputation of the online shop, which results in consumers having no trouble purchasing through Instagram. Users who are familiar with the platform they use will find it easy to use the platform, thereby increasing the desire to buy a product using the platform they use. Several studies that also showed similar results were Van Der Heide and Lim (2015) and Lopez et al. (2015).

Table 1. Final hypothesis results.

	Hypothesis	Conclusion
H1	Trust to Purchase Intention	Accepted
H2	Trust to Information seeking	Accepted
H3	Information seeking to Purchase Intention	Rejected
H4	Information seeking to familiarity	Accepted
H5	Familiarity to Purchase Intention	Accepted

4 CONCLUSIONS

As in some previous studies, trust and familiarization of Instagram accounts can form a person's purchase intention. In the context of online shopping, trust plays an important role because potential customers will make purchases that are quite risky because the product to be purchased is not seen directly and is only based upon the displayed image. Marketers must be able to build trust to (prospective) consumers. Trust makes people move to seek information online (social commerce information seeking), related to price, product specifications, and advantages compared to competitors' brands amongst others. The more they seek information via online shopping accounts, the more familiar they will be with the account, which will affect the familiarity of the person with online shopping as a whole. The more familiar a person is with online shopping, the more he feels at ease in making transactions and the more secure the person is in the transaction, the greater the purchase intention.

4.1 Theoretical implications

The previous study (Hajli et al. 2015) and several studies of marketing literature suggest that information seeking triggers the purchase intention, but the results of this study show an interesting fact, namely that increasing information seeking will decrease the likelihood or opportunity for purchasing products. But increasing the purchase intention, can be done by increasing trust and familiarity. According to Heinonen (2011), products reviewed through social commerce help consumers to be able to evaluate the product as a whole, among others related to price, product specifications, excellence compared to competitor brands etc., so consumers can make the right purchasing decisions. From this research it can be seen that the more users find out about the fashion products in the online shop on Instagram, desired through reviews, ratings, and recommendations provided by other users,the further it decreased the possibility or opportunity to purchase the product.

4.2 *Practical implication*

Trust and familiarity can affect the purchase intention. The author suggests that online fashion shop enthusiasts using the Instagram platform must show that they are honest with their customers, so customers will not doubt their honesty. This can be demonstrated by showing the purchase testimony, the delivery receipt, and the original product photo without being edited. Then, the online shop of fashion products on Instagram should make it easier for its customers to find information about the products they sell. This can be demonstrated by providing informative content on each uploaded product, including links related to product information that cannot be explained only by an Instagram caption, such as the online shop's website link, so that it will be easier for customers to get information in more detail. According to Heinonen (2011), products reviewed through social commerce help consumers to be able to evaluate the product as a whole, among others related to price, product specifications, excellence compared to competitor brands and others. So consumers can make the right purchasing decisions. From this research it can be seen that the more users find out about the fashion products in the online shop on Instagram, desired through reviews, ratings, and recommendations provided by other users, the further it decreased the possibility or opportunity to purchase the product.

4.3 *Limitations and future researches*

This study also has limitations, namely that this study only investigated the Instagram platform. Each platform has different features, allowing e-vendors to display their products visually, as well as through audio, in different durations and interaction capacities. So it is worth investigating whether different platforms (Twitter, Path, etc.) can affect aspects of trust, information seeking, familiarity, purchase intention or other aspects. This study also only examines only fashion products, so it is expected that subsequent research will also examine other products. The other products studied can be determined by first conducting preliminary studies in the next study population. This study found different results than others, in that social commerce information seeking has a negative effect on the purchase intention. Further research is expected to re-examine the relationship between the two variables through literature and field observation to obtain a more comprehensive explanation. Furthermore, this study did not distinguish between what online shop was being studied. So it is expected that the next research will focus more on researching an online shop.

REFERENCES

APJII. 2016. Penetrasi dan perilaku pengguna internet di Indonesia. *Retrieved from* www.apjii.or.id. Accsessed 17 January 2017.

Ba, S. & Pavlou, P.A. 2002. Evidence of the effect of trust building technology in electronic markets: Price premiums and buyer behavior. *MIS Quarterly* 26(3): 243–268.

Bohang, F.K. 2017. Indonesia pengguna Instagram terbesar se-Asia Pasifik. *Retrieved from* www.tekno.kompas. com. Accsessed 24 October 2017.

Chiou, J.S., Droge, C. & Hanvanich, S. 2002. Does customer knowledge affect how loyalty is formed?. *Journal of Service Research* 5(2): 113–124.

Choo, C.W., Detlor, B. & Turnbull, D. 2000. Information seeking on the web: An integrated model of browsing and searching. *First Monday* 5(2).

Doolin, B., Dillon, S., Thompson, F. & Corner J.L. 2008. Perceived risk, the internet shopping experience, and online purchasing behavior: A New Zealand perspective. *IGI Global* 2(13): 324–345.

Ferdinand, A. 2002. *Structural equation model dalam penelitian manajemen: Dasar-dasar permodelan (Edisi Indonesia)*. Jakarta: Elex Media Komputindo Kelompok Gramedia.

Fishbein, M. & Ajzen, I. 1975. *Intention and behavior: An introduction to theory and research*.

Gefen, D. 2000. E-commerce: The role of familiarity and trust. *Omega* 28(6): 725–737.

Gefen, D., Karahanna, E. & Straub, D.W. 2003. Inexperience and experience with online stores: The importance of TAM and trust. *IEEE Transactions on Engineering Management* 50(3): 307–321.

Hajli, N. & Sims, J. 2015. Social commerce: The transfer of power from sellers to buyers. *Technological Forecasting and Social Change* 94: 350–358.

Hajli, N., Sims, J., Zadeh, Arash H. & Richard M. 2015. A social commerce investigation of the role of trust in a social networking site on purchase intention. *Journal of Business Research* 71: 133–141.

Heinonen, K. 2011. Consumer activity in social media: Managerial approaches to consumers' social media behavior. *Journal of Consumer Behaviour* 10(6): 356–364.

Hertzum, M., Andersen, H.H., Andersen, V. & Hansen, C. B. 2002. Trust in information sources: Seeking information from people, documents, and virtual agents. *Interacting with Computers* 14(5): 575–599.

Hochman, N. & Schwartz, R. 2012. Visualizing Instagram: Tracing cultural visual rhythms. *In the Proceedings of the Workshop on Social Media Visualization in Conjunction with the 6th International AAAI Conference on Weblogs and Social Media* 12(3): 7–9.

Kaiser, S. & Muller-Seitz, G. 2008. Leveraging lead user knowledge in software development: The case of weblog technology. *Industry and Innovation* 15(2): 199–221.

Kim, S. & Park, H. 2013. Effects of various characteristics of social commerce (S-Commerce) on consumers' trust and trust performance.' *International Journal of Information Management* 33: 318–332.

Kuhlthau, C.C. 2004. *Seeking meaning: A process approach to library and information services*. Westport, CT: Libraries Unlimited Incorporated.

Liang, T.P. & Turban, E. 2011. Introduction to the special issue, social commerce: A research framework for social commerce. *International Journal of Electronic Commerce* 16(2): 5–14.

Liang, T.P., Ho, Y.T., Li, Y.W. & Turban, E. 2011. What drives social commerce: The role of social support and relationship quality. *International Journal of Electronic Commerce* 16(2): 69–90.

Maria, E. D., and Finotto, V. (2008). Communities of Consumption and Made in Italy. Industry and Innovation, 15 (2), 179–197.

Martinez-Lopez, F.J., Esteban-Millat, I., Cabal, C.C. & Gengler, C. 2015. Psychological factors explaining consumer adoption of an e-vendor's recommender. *Industrial Management and Data Systems* 115(2): 284–310.

Mathur, N. 2015. Perceived risks towards online shopping: An empirical study of Indian customers. *International Journal of Engineering Development and Research* 3(2): 296–300.

McKnight, D.H., Choudhury, V. & Kacmar, C. 2002. Developing and validating trust measures for e- commerce: An Integrative typology. *Information Systems Research* 13(3): 334–359.

Murray, K.B. 1991. A test of services marketing theory: Consumer information acquisition activities. *Journal of Marketing* 55(1): 10–25.

Setyaningsih, L. 2017. Warga pengguna aktif Instagram di Indonesia capai 45 juta terbesar di Asia Pasifik. *Retrieved from* www.wartakota.tribunnews.com. Accsessed 24 October 2017.

Suh, B. & Han, I. 2003. The impact of customer trust and perception of security control on the acceptance of electronic commerce. *International Journal of Electronic Commerce* 7(3): 135–161.

Turcotte, J., York, C., Irving, J., Scholl, R.M. & Pingree, R. J. 2015. News recommendations from social media opinion leaders: Effects on media trust and information seeking. *Journal of Computer-Mediated Communications* 20(5): 520–535.

Van Der Heide, B. & Lim, Y. 2015. On the conditional cueing of credibility heuristics: The case of online influence. *Communication Research* 3(4): 1–22.

Wang, Y., Min, Q. & Han, S. 2016. Understanding the effects of trust and risk on individual behavior toward social media platforms: A meta-analysis of the empirical evidence. *Computers in Human Behavior* 56(7): 34–44.

The influence of endorser credibility, brand credibility, and brand equity on the purchase intention in online shopping: The Instagram phenomenon

Z. Agustiansyah & D. Mardhiyah
Universitas Airlangga, Surabaya, Indonesia

ABSTRACT: The number of social media users in Indonesia is increasing every year. According to e-Marketer market research institutes, in 2017 there were approximately 82 million people in Indonesia who used social media at least once a month. This number increased from the number of 72.3 million users in 2015. That means more than one third of the Indonesian people are people who are fond of using social media. One of the popular online social media applications today is Instagram. Instagram is the most used social networking app (82%), followed by Facebook (66%) and Path (49%). The increasing users of Instagram applications create new trends of credible endorsers triggering a brand purchase intention. The purpose of this study was to analyze the influence of endorser credibility and other factors on the purchase intention on Instagram. Other factors were brand credibility and brand equity. This study used a quantitative online survey approach. There were 211 respondents, who were active in using social media, particularly, Instagram, and who knew the Indonesian artist named Raffi Ahmad, who is used as a case study to analyze the hypotheses. The data testing technique used is SEM AMOS. The results showed that brand credibility and brand equity positively affected the purchase intention; brand credibility positively affected brand equity; and endorser credibility positively affected brand credibility, while the endorser credibility didn't affect brand equity and the purchase intention.

1 INTRODUCTION

Technological developments have an impact on the development of communication which is marked by the presence of smartphones, which can make it easy for everyone to access the internet anywhere and anytime so that everyone can quickly find out the latest news. The development of internet and smartphone usage in Indonesia has become a marketer opportunity to utilize internet, social media, and the smartphone to implement a marketing strategy.

The number of social media users in Indonesia is increasing every year. According to market research institute e-Marketer, in 2017 there were about 82 million people in Indonesia who used social media at least once a month. This number increased from 72.3 million users in 2015. It meant that more than one third of the Indonesian people are people who really like using social media. According to research conducted by We Are Social, Indonesia was recorded to have 130 million active social media users in 2018. At least 49% of the total population has access to and actively uses social media, with 120 million users using mobile phones to connect with their social media accounts (Yusuf 2014).

Instagram is one of the online social media applications in demand today. According to Mastel and APJII 2016, Instagram became the most widely used social networking application (82%), leaving Facebook (66%) and Path (49%) behind. Instagram apps prioritize their service in sharing photos and videos rather than text, so it is appropriate for users who like to show off photos of their daily activities or while being on vacation. This application can also be used to buy & sell online, by using a feature called Instagram business profile. This makes it easier for users of Instagram to market their business.

The increase in Instagram app users has led to a new trend of credible endorsers in adding to the brand purchase intention. This can be proved by the increasing number of Indonesian artists who advertise certain brands on their Instagram account. Examples are Ayu Ting Ting: on her Instagram account she advertises products from Gluta; Nagita Slavina advertises products from RA Jeans; and Raffi Ahmad, the most sosmed celebrity in 2016 who has 14 million followers already verified by Instagram, also advertises products on his Instagram account (Anggraeni 2016).

The problem for most new brands is the issue of credibility that can affect the buying intentions of those brands. Companies should be able to communicate their products to potential customers or target markets, because even if the product is good, if consumers do not know of its presence in the market then consumers will not be interested in the product.

Another problem of the new brands is the lack of brand ability in establishing brand equity (Aaker 1991) so that consumers do not have enough confidence to buy the product. Endorser credibility makes it possible to introduce a product well in offline advertising (Spry et al. 2011). This has a positive effect on brand credibility, in which trust in a product (brand) increases due to the impact of brand credibility. This study aims to determine the effect of endorser credibility in introducing a product through social media. This is based on research results by Djafarova and Rushworth (2016), who stated that the purchase intention rate was higher in a product that used well-known figures in social media than those that did not. A study found that social media has a 100% higher likelihood for a close rate than non social media marketing. More followers of a brand on social media tend to increase the trust and credibility of that brand.

1.1 Endorser credibility and brand credibility

Endorser credibility is how much the endorser has the knowledge, skills, or experience of the brand they advertise, and how much the recipient of the message trusts the information that is conveyed by the endorser (Belch & Belch 2009). The greater the ability and knowledge of an endorser about the product being advertised, the greater the endorser credibility is in the eyes of consumers. It can be said that endorser credibility is the ability and characteristics of the endorser in influencing and convincing the recipient of the message through their attractiveness and expertise (Rao 2012).

The concept of brand credibility arises from the brand signaling theory proposed by Erdem and Swait (1998). Brand signals are based on the clarity and credibility of a brand (Erdem & Swait 1998). Brand clarity refers to the absence of information ambiguity conveyed by the brand to the recipient of the information so that the information provided is not exaggerated and easily understood. While brand credibility is the brand's ability to convey true and reliable information to the customer about the brand's quality and the ability of the brand to deliver what they have promised to customers. Based on this, it can be concluded that brand credibility is a belief in product information within a brand, and consumers then need to understand that the brand has the ability and willingness to deliver what has been promised (Erdem & Swait 1998).

Schiffman (2010) says that endorser credibility can affect brand credibility by communicating the image of the endorser to a brand, such as an endorser who has a good career suggesting or promoting a particular brand, then the brand is also indirectly lifted or affected by the positive impact of the success of the endorser of the brand. High endorser credibility will lead to high brand credibility as well. Based on the description, the following hypothesis can be proposed:

H1: Endorser credibility has a positive effect on brand credibility.

1.2 Brand endorser and brand equity

Brand equity occurs when consumers recognize and remember the brand as an attractive, powerful, and unique brand in their memories and when it creates added value for consumers (Kamakura & Russell 1991). According to Kotler and Keller (2012), brand equity is an added value given to products and services, such as brand awareness, brand associations, perceived quality and brand loyalty. Brand equity variables are operated as a positive perception and judgment about quality by the consumers. It is more than just a brand. It is perception and positive judgment that add value to the brand (Lassar et al. 1995).

Spry et al. (2011) argue that when consumers think about celebrities who endorse a brand, they automatically think of the brand that is endorsed by the celebrity. Characteristics of a celebrity can be an early picture of a product. This is an added value not found in other brands. If the celebrity has a good and positive credibility, then the consumer believes that that product is good and positive. Endorser credibility can increase brand equity (Keller, 1993) because well-known people with high credibility can attract an audience's attention and direct them to brands and be able to shape their own brand perceptions based on the knowledge they have about the famous person. In his research, Sivesan (2013) explains that endorser credibility can affect brand equity. The more attractive and trustworthy an endorser, the more the brand will also be seen as attractive or trustworthy. Based on the description, the following hypothesis can be proposed:

H2: Endorser credibility has a positive influence on brand equity.

1.3 Brand credibility and brand equity

Erdem and Swait (1998) show the position of brand credibility as well as a differentiating element for other brands and can add added value in the minds of consumers. Products that have a good brand credibility will be perceived positively by consumers. Conversely, on products that have a low brand credibility it will be perceived poorly by consumers.

A credible brand can reduce the fear and risk that a consumer receives when making a purchase or reduce the distrust of the brand (Erdem & Swait 1998). Furthermore, a higher brand credibility can improve consumers' perceptions of product quality (Erdem et al. 2002). Based on the description, the following hypothesis can be proposed:

H3: Brand credibility has a positive influence on brand equity.

1.4 Endorser credibility and purchase intention

The use of celebrities as endorsers representing brands in an advertisement is a very good way because celebrities are considered to have a large fan base and especially if the celebrity has a good image in the eyes of consumers (Sertoglu et al. 2014). Celebrities who have a good image and have a large fan base are credible celebrities. If the endorser is considered credible then the consumer will receive and assume the information from the endorser is accurate and will use the recommended brand (Lafferty & Goldsmith 1999).

Intention is a subjective assessment of what the future consumer will do. One form of intention is the intention to buy (Martin & Hererro 2011). The definition of purchase intention according to Aaker (2009) is the desire of consumers to own or buy a product or service resulting from some encouragement. Thus purchase intention can be interpreted as the intention or desire of individuals to make purchases of products or services. Given the impact of celebrity credibility that has a positive image, it can cause consumers to buy products that use these celebrities in their ads. Based on the description, the following hypothesis can be proposed:

H4: Endorser credibility has a positive effect on purchase intention.

1.5 *Brand credibility and purchase intention*

A credible brand can reduce the fear and risk that consumers will have when they make a purchase. With brand credibility, distrust of the brand can be reduced (Erdem & Swait 1998). Furthermore, a higher brand credibility can increase consumers' perceptions of product quality (Erdem et al. 2002). With the brand credibility, the brand distrust will decline and can increase the consumer's purchase intention of a brand.

In the Baek et al. (2010) study, a good brand credibility in the eyes of consumers can create a consumer purchase intention. Brands that can be trusted and are perceived reliable by consumers will make the consumers' perceptions positive, so they will make a purchase. Based on the description, the following hypothesis can be proposed:

H5: Brand credibility has a positive effect on the purchase intention.

1.6 *Brand equity and purchase intention*

Currently consumers are faced with a wide selection of brands so that sometimes consumers have difficulty making a brand purchase decision. In this condition, brand equity becomes an important thing because brand equity helps consumers recognize the positive brand characteristics that distinguish it from other brands. These distinctive features will create a distinct impression and enable consumers to remember a brand name as well as create additional value that will encourage the purchase intention (Tharmi & Senthilnathan 2011).

Shah and Adeel (2016) explain that brand equity has been regarded as a requirementthat affects the purchase intention. Some research indicates that once a consumer considers the differentiation of a brand and experience that other brands do not have, it increases the likelihood of buying that brand in the future. In addition, research in consumer behavior illustrates that brand equity is a key element that directly affects brand purchase intention. Based on these descriptions, the following hypothesis can be proposed:

H6: Brand equity has a positive influence on the purchase intention.

2 METHOD

This research uses a quantitative research approach with survey method. Online questionnaires were distributed via e-form links to netizens who have Instagram accounts, are active in using Instagram, and know the Indonesian celebrity named Raffi Ahmad. This study used 210 respondents (response rate 70.33%). The questionnaires were adapted from several previous studies (Dwivedi & Johnson 2013, Ohanian 1990, Erdem & Swait 1998, Ahmad & Sherwani 2015, Abzari et al. 2014, Khan et al. 2015). Researchers conducted back-translation so that the questionnaire was understood by the majority of the respondents who use bahasa. Scale measurement of 29 question items used 5 Likert scale (1 = Strongly Disagree; 2 = Disagree; 3 = Neutral; 4 = Agree; 5 = Strongly Agree). The analysis technique used in this research is Structural Model Equation by using Analysis of Moment Structures (AMOS).

3 RESULTS AND DISCUSSION

The respondents in this study amounted to 211 people, of which the majority of respondents were women (60.7%) aged 20–25 years (70.6%). More than half of the total respondents work as students (68.2%). Most respondents of this study earned between IDR 3,000,000-IDR 3,500,000 (46%).

Goodness of fit testing showed that the model meets the fit model requirements (RMR = 0.04, CFI = 0.92, TLI = 0.914). In addition, the standardized regression weight value of all indicators of each variable in this study was valid (0.767–1.090) and reliable, which was seen from the reliability construct of all variables having a Cronbach alpha value of 0.964–0.979.

Hypothesis testing using AMOS showed that endorser credibility has a significant positive effect on brand credibility (H1 is supported). This suggests that increasing endorser credibility will have a significant impact on increasing brand credibility, and vice versa. Respondents believe that the product under study is a credible brand resulting from the credibility of the endorser.

As opposed to hypothesis 1, endorser credibility had no significant effect on brand equity (H2 is unsupported). This shows that credible endorsers do not impact the increase in brand equity, and vice versa. This can happen because the endorser credibility does not quite represent the endorsed product. So, if the consumer considers the product as having a high equity, it is not because of the credibility of the endorser.

In this study it was found that brand credibility has a significant positive effect on brand equity (H3 is supported). This shows that the increasing brand credibility will have a significant impact on the increase of brand equity, and vice versa. Respondents believe that RA Jeans has a strong brand equity as a result of the brand credibility of RA Jeans.

In this study it was found that the endorser credibility had no significant effect on purchase intention (H4 is not supported). This shows that high endorser credibility does not make consumers intend to buy products that are endorsed by the celebrity, and vice versa. This can happen because the endorser credibility is not enough to represent an endorsed product. So, if a consumer buys a product it is because of a good quality product (it has a high brand equity) and not because of an endorsing celebrity.

Test results indicated that brand credibility had a significant positive effect on the purchase intention (H5 supported). This shows that increasing the brand credibility will have a significant impact on the increase of purchase intentions, and vice versa. The emergence of the desire to buy products is due to the confidence of respondents of the credibility of product brands.

In this study it was found that the brand credibility had a significant positive effect on the purchase intention (H6 is supported). This shows that increased brand equity will have a significant impact on the increase of the purchase intention, and vice versa. The emergence of the desire to buy products is due to the priviledge of the brand in forming brand equity so as to provide added value that is not owned by other brands.

Table 1. Results of the hypotheses test.

	Hypotheses	Conclusion
H1	Endorser credibility to Brand credibility	Supported
H2	Endorser credibility to Brand equity	Not Supported
H3	Brand credibility to Brand equity	Supported
H4	Endorser credibility to Purchase intention	Not Supported
H5	Brand credibility y to Purchase intention	Supported
H6	Brand equity to Purchase intention	Supported

4 CONCLUSIONS

4.1 Conclusions

Through social media, especially Instagram, endorser credibility affects brand credibility. The choice of an endorser must be in accordance with the endorsed product, otherwise, even if the endorser has a high credibility it cannot affect the brand equity of the endorsed brand. It can also affect the unaffected purchase intention by endorser credibility. However, the results showed that brand credibility affects brand equity. Consumers continue to recognize the quality of products that are endorsed by celebrities who have a good credibility, because celebrities will not sacrifice their reputation to represent a product that is not qualified. The reason why celebrities are so in demand for producers to advertise their products is that messages conveyed by interesting sources (popular celebrities) get more attention and will be very easy to remember (Royan 2004). As in some previous studies (Huang et al. 2013, Shah & Adeel 2016), this study also found that brand equity has a positive effect on the purchase intention.

4.2 Theoretical implications

This study found that endorser credibility has no effect on brand equity and the purchase intention. For building brand equity, brand credibility can be used. Meanwhile, to generate a purchase intention, one can take advantage of brand credibility. The endorser credibility of Raffi Ahmad has no effect on RA Jeans' brand equity, which is likely due to the fact that respondents saw the advertisement on Instagram only once, making the influence of Raffi Ahmad in building RA Jeans' brand equity less strong. This is supported by a concept called multiple celebrity-brand pairings that states that the effects of a celebrity on brand equity will be strengthened over time and brand placement will have a stronger impact on consumers' minds if consumers see ads twice or more.

4.3 Practical implications

Endorser credibility has no effect on brand equity. This is because the credibility of the endorser does not fit with the brand. It is important for companies to use more appropriate and representative endorsers with product characteristics. In this case the endorsed class does not fit the product class.

Since brand equity has an effect on the purchase intention, it is important for companies to make efforts to establish brand equity, which can be done through forming brand credibility. Through social media, brand credibility can be formed by utilizing testimonials on the use of the products by consumers. In addition, the company may add features or other advantages that are not owned by a similar brand, either from design, quality of materials or they can also provide free shipping when making a purchase through Instagram. Participating as a sponsor of an event attended by young people, such as music events, can also be done to further strengthen the credibility of the product.

4.4 Limitations and future research

For improving the generalization of these results, in the context of fashion products, further research can be considered to use other endorsers, i.e. competent expert endorsers in fashion, such as the young designer Barly Asmara. In addition, the level of live-liness of respondents using social media may be considered as a factor that may influence the respondent's assessment of the credibility of the endorser, such as the frequency of viewing ads. It is based on the concept of multiple celebrity-brand pairings described earlier.

REFERENCES

Aaker, D.A. 1991. *Managing brand equity: Capitalizing on the value of a brand name.* New York: The Free Press.
Aaker, D.A. 2009. *Strategic market management* (4th ed).
Abzari, M., Ghassemi, R.A. & Vosta, L.N. 2014. Analysing the effect of social media on brand attitude and purchase intention: The case of Iran Khodro Company. *Procedia-Social and Behavioral Sciences* 143: 822–826.
Ahmad, F. & Sherwani, N.U. 2015. An empirical study on the effect of brand equity of mobile phones on customer satisfaction. *International Journal of Marketing Studies* 7(2): 59.
Anggraeni, R. 2016. Daftar lengkap pemenang seleb on news awards 2016. *Retrieved from* https://lifestyle.sindonews.com. Accsessed 13 June 2017.
Baek, T.H., Kim, J. & Yu, J.H. 2010. The differential roles of brand credibility and brand prestige in consumer brand choice. *Psychology & Marketing* 27(7): 662–678.
Belch, G.E. & Belch, M.A. 2009. *Advertising and promotion: An integrated marketing communication perspective.* Boston: McGraw Hill-Irwin.
Djafarova E. & Rushworth C. 2016. Exploring the credibility of online celebrities' Instagram profiles in Influencing the purchase decisions of young female users. *Computer in Human Behaviour.*
Dwivedi, A. & Johnson, L.W. 2013. Trust-commitment as a mediator of the celebrity endorser brand equity relationship in a service context. *Australasian Marketing Journal (AMJ)* 21(1): 36–42.
Erdem, T. & Swait, J. 1998. Brand equity as a signalling phenomenon. *Journal of Consumer Psychology* 07(02): 131–157.
Erdem, T., Swait, J. & Louviere, J. 2002. The impact of brand credibility on consumer price sentivity. *International Journal of Research in Marketing* 19: 1–19.

Huang P., Wang C., Tseng Y. & Wang R. 2013. The impact of brand equity on customer's purchase intention: Taking perceived value as a moderating variable. *Journal of Information and Optimization Sciences* 32(3).
Kamakura, W. & Russel, G. 1993. Measuring brand value with scanner data. *International Journal Research ing Marketing* 10(03): 09–22.
Keller, L.L. 1993. Conceptualising, measuring and managing customer based brand equity. *Journal of Marketing* 57(01): 1–22.
Khan, N., Rahmani, S.H.R., Hoe, H.Y. & Chen, T.B. 2015. Causal relationships among dimensions of consumer-based brand equity and purchase intention: Fashion industry. *International Journal of Business and Management* 10(1): 172.
Kotler, P. & Keller, K.L. 2012. *Marketing management (14th Ed).* New Jersey: Pearson Prentice Hall.
Lafferty, B. & Goldsmith, R.E. 1999. Corporate credibility's role in consumers attitudes and purchase intention when a high versus a low credibility endorser is used in the Ad. *Journal of Business Research* 44: 109–116.
Lassar, W., Mittal, B. & Sharma, S. 1995. measuring customer-based brand equity. *Journal of Consumer Marketing* 12(04): 11–19.
Martin, H.S. & Hererro, A. 2011. Influence of the user's psychological factors on the online purchase intention in rural tourism: Integrating innovativeness to the UTAUT framework. *Tourism Management* 33(12): 341–350.
Ohanian, R. 1990. Construction and validation of a scale to measure celebrity endorsers perceived expertise, trustworthiness, and attractiveness. *Journal of Advertising* 19(3): 39–52.
Rao, R. 2012. Brand credibility and brand involvement as an antecedent of brand equity: An empirical study Kangwon National University. *Asia-Pacific Journal of Business* 27(01): 67–82.
Royan, F.M. 2004. *Marketing selebriti, selebrit dalam iklan, dan strategi memasarkan diri sendiri.* Jakarta: Alex Media Komputindo.
Sertoglu, A.E., Catli, O. & Korkmaz, S. 2014. Examining the effect of endorser credibility on the consumers' buying intentions: An empirical study in Turkey. *International Review of Management and Marketing* 04(01): 66–77.
Shah, M. & Adeel, M. 2016. the impact of brand equity on purchase intentions with moderating role of subjective norms. *Journal of Industrial and Business Management* 04(01): 18–24.
Sivesan, S. 2013. Impact of celebrity endorsement on brand equity in cosmetic product. *International Journal of Advanced Research in Management and Social Sciences* 04(12): 176–186.
Spry A., Pappu R. & Cornwell T.B. 2011. Celebrity endorsement, brand credibility and brand equity. *European Journal of Marketing* 45(06): 882–909.
Tharmi, U. & Senthilnathan, S. 2011. The relationship of brand equity to purchase intention. *The Journal of Marketing Management* 11(02): 07–26.
Yusuf, O. 2014. Pengguna internet Indonesia nomor enam. *Retrieved from* https://tekno.kompas.com. Accsessed 13 June 2017.

Advances in Business, Management and Entrepreneurship – Hurriyati et al (eds)
© 2020 Taylor & Francis Group, London, ISBN 978-0-367-27176-3

Online reviews by beauty vloggers and its impact on buying interest

H. Hendrayati, N.C. Noorfadila & M. Achyarsyah
Universitas Pendidikan Indonesia, Bandung, Indonesia

M.I. Atrisia
Chang'An University, Xi'an, China

R.K. Syahidah
Northwestern Polytechical University, Xi'an, China

ABSTRACT: Social media enables people to get information about a product more easily. One of them are online reviews by beauty vloggers on the social media channel YouTube. Purbasari is one of the local products that quite often became the object of material reviews by Indonesian beauty vloggers on YouTube. This study aims to explore online reviews by beauty vloggers, to get to know the buying interest of purple lipstick, and to learn the effect of online reviews by beauty vloggers on buying interest. Therefore, the method used is descriptive and verification. The respondents are hundred viewers of the Purbasari lipstick review video. The sampling technique of this research is purposive sampling where the criteria consist of women who know the Purbasari products and have watched Purbasari lipstick review video reviewed by an Indonesian beauty vlogger. The questionnaires are distributed through social media by using Google Form. The result of the research shows positive and significant influences of online reviews on the buying interest.

1 INTRODUCTION

The large number of cosmetic products in circulation, especially lipsticks, can indirectly trigger consumer behavior. The more products in circulation, the more selective consumers will become and the more information of the products they will use. Online review is basically used for consumer-to-consumer (C2C) communications. Therefore, they can also be referred to as "Electronic Word-of-Mouth" communication (e-WOM) (Cheung et al. 2008, Lee et al. 2013). Online reviews are considered to be one of the most powerful electronic word-of-mouth (e-WOM) types in shaping consumer behavior (Plummer 2007). In addition, online reviews become an important information channel for consumers and can have a greater impact than advertising (Jalilvand & Samiei 2012).

Nowadays, social media has revolutionized consumers' buying decisions which allow consumers to review products before making a purchase by consulting sites that offer user-generated reviews and information. This ultimately helps consumers to make the best decisions with the least risk (Riegner 2007, MacKinnon 2012). A social media site that became a favorite platform for consumers in searching or making a review of a product is the YouTube site. A total of ±369,000 reviews on cosmetic products are available on the YouTube website (YouTube.com, 2018). The number of videos on product reviews circulating on YouTube encourages interaction with viewers through comments. Ratings can

be seen from the number of thumbs up and down. This practice is called vloggin (Burgess & Green 2009). The vlog or vlogging culture is a new culture where Indonesian people use the YouTube platform. Someone involved in vlog activities is called a vlogger.

All industries seem to start glancing towards famous YouTube stars, or vloggers. Many beautician brands increasingly believe in the power a vlogger has in influencing their audience. This practice where the beauty vlogger starts to create a place for the cosmetics brand is quite prevalent. Because consumers in the digital age are saturated and tend not to believe the brand message delivered through television advertising. Women are now looking for someone who can speak honestly and really understands the world of beauty. All characters can be found on the figure of a beauty vlogger. For cosmetics companies, a beauty vlogger is considered capable to educate people about a products' usability through video tutorials and product reviews to the target markets clearly (Bachdar 2017).

The purpose of this study is to reveal the online review by a beauty vlogger and its impact on the buying interest of Purbasari lipstick.

2 METHOD

This research employed descriptive and verification methods. Descriptive research in this study aims to reveal the description of online reviews by beauty vloggers, and the description of the buying interest of Purbasari lipstick. The verification method in this research tested the hypothesis of the influence of online reviews by beauty vloggers on the buying interest of Purbasari lipstick. The independent variable (X) in this study is the online review. In the online review variable, the measured indicators are usefulness of online reviews, expertise of the reviewer, timeliness of the online reviews, valence of the online reviews, and comprehensiveness of the online reviews. The buying interest variable is measured by transactional interest, referential interest, preferential interest, and explorative interest. The population of the study are viewers of the Purbasari lipstick review video made by beauty vloggers such as Abel Cantika, Linda Kayhz, Suhay Salim, Rachel Goddard, and Tasya Farasya, totaling to $\pm 3,723,000$ viewers (YouTube.com 2018) and sampled by using the Slovin formula on 100 respondents. The purposive sampling method was used. The sample was taken by considering certain characteristics. They are, among others, women who know Purbasari products and have watched the Purbasari lipstick review video by Indonesian beauty vloggers. In this study, the questionnaire was in the form of Google forms and distributed through social media.

3 RESULT AND DISCUSSION

Online reviews can be a powerful promotional tool for marketing communications because they are considered new products of information channels that generate agrowing interest and popularity. According to Gretzel and Yoo (2008), online reviews are important sources of information for customers in generating a buying interest. In this study, there are five dimensions used to measure online review variables, as proposed by Zhao et al. (2015): usefulness of the online reviews, expertise of the reviewer, timeliness of the online review, valence of the online reviews, and comprehensiveness of the online reviews.

Based on the results of data processing and analysis on the online review variable, the dimension 'comprehensiveness of the online review' has the highest performance, which means respondents love the beauty vlogger who reviewed the Purbasari lipstick in detail, ranging from lipstick texture, color selection, and price. The indicator with the lowest performance is the dimension 'valence of the online review'. The respondents feel that the beauty vloggers convey information excess of the Purbasari lipstick product compared with the shortage of the Purbasari lipstick product itself.

Buying interest according to Mowen and Minor (2002) can be explained as the interest to search for information, to tell someone else about an experience with a product or service, and to dispose product. In this research, there are three dimensions that can be measured according to Ferdinand (2006),

namely transactional interest, referential interest, preferential interest, and explorative interest.

Based on the results of data processing and analysis on the buying interest variable, the indicator of transactional interest has the highest performance, which means that respondents have an interest to buy the Purbasari lipstick after watching the Purbasari lipstick review video reviewed by the beauty vloggers. The indicator with the lowest performance is the indicator of preferential interest. It shows that although the interest in buying the Purbasari lipstick is high, it is not necessarily an interest to buy the purple lipstick. This is because the number of lipstick products circulating in the community is large, so the desire to use the Purbasari lipstick as the main choice is not necessarily high.

The study of the influence of online reviews on the buying interest used a normality test, correlation test, simple regression test, and hypothesis test (t-test). Based on the results of the research conducted with a simple regression analysis and hypothesis test, it is obtained that online reviews have a positive and significant effect on the buying interest, with a coefficient of determination (KD) 41.7%. This shows that online reviews give the buying interest a positive effect of 41.7%. Thus, it can be identified that H_0 is rejected and H_1 is accepted.

Based on the results of this study, MacKinnon (2012), in "*Too Popular to Ignore: The Influence of Online Reviews on Purchase Intentions of Search and Experience Products*" expressed the opinion that online reviews affect the buying interest. The results of this study indicate that a credible review that contains detailed information can positively affect consumers' buying interests.

A beauty vlogger is a YouTube user who creates and publishes videos related to beauty or cosmetics on their YouTube channel, and who is not affiliated with a brand (Pixability 2014). Makeup tutorials and beauty product reviews are just a few examples of the content on the YouTube channel of a beauty vlogger. In other words, a beauty vlogger is a regular consumer who creates YouTube videos about beauty products. Consumer-generated beauty vlogs on YouTube can generally attract more viewers and subscribers than videos uploaded by authorized YouTube channels of a brand (Pixability 2014). Beauty vlogging becomes a communication medium in conveying information of a product to a consumer through the review of a product. Among a variety of online review sources that do cosmetic product reviews such as beauty bloggers, beauty vloggers, and celebrities, the beauty vloggers are considered more reliable and knowledgeable in conveying information than the celebrities (Djafarova & Rushworth 2017). Therefore, many beauty vloggers successfully affect consumers' buying interests of beauty products through the reviews made by them. That is because information submitted by a beauty vlogger can affect the buying interest, since information desired or sought by consumers is a consideration of the selection of products or services. In addition, the information obtained through the beauty vlogger can be more trusted by consumers, because they are considered to be experts in the field of beauty.

4 CONCLUSION

Based on the results of this study, it can be concluded that there is an influence of online reviews by beauty vloggers on the buying interest of purple lipstick. Therefore, the more interesting review videos are made by beauty vloggers, the higher the interest to buy Purbasari lipstick products will be.

ACKNOWLEDGEMENT

The authors would like to thank the Ministry of Research, Technology, and Higher Education for funding this research, and Universitas Pendidikan for facilitating this research.

REFERENCES

Bachdar, S. 2017. Beauty vlogger dan pengaruhnya bagi brand kosmetik. *Retrieved from http://marketeers.com/beauty-vlogger-dan-pengaruhnya-bagi-brand-kosmetik/*. Accsessed 1 January 2018.

Burgess, J. & Green, J.,2009. YouTube's popular culture. *YouTube. Online Video and Participatory Culture.*

Cheung, C., Lee, M. & Rabjohn, N. 2008. The impact of electronic word-of-mouth: the adoption of online opinions in online customer communities. *Internet Research* 18: 229–247.

Djafarova, E. & Rushworth, C. 2017. Exploring the redibility of online celebrities instagram profiles in influencing the purchase decision of young female users. *Computers in Human Behavior* 68: 1–7.

Ferdinand, A. 2006. *Metode penelitian manajemen (Edisi kedua)*. Semarang: Badan Penerbit UNDIP.

Gretzel, U. & Yoo, K. 2008. Use and impact of online travel reviews, information and communcation technologies in tourism. *Proceedings of the International Conference in Innsbruck*: 35–46.

Jalilvand, M. R. & Samiei, N. 2012. The effect of electronic word of mouth on brand image and purchase intention: An empirical study in the automobile industry in Iran. *Journal of Marketing Intelligence and Planing* 30(4): 460–476.

Jiménez, F.R. & Mendoza, N.A. 2013. Too popular to ignore: The influence of online reviews on purchase intentions of search and experience products. *Journal of Interactive Marketing* 27(3): 226–235.

Lee, M., Kim, M. & Peng, W. 2013. Consumer reviews: Reviewer avatar facial expression and review valence. *Internet Research* 23: 116–132.

MacKinnon, K.A. 2012. User generated content vs. advertising: Do consumers trust the word of others over advertisers. *The Elon Journal of Undergraduate Research in Communications* 3(1): 14–22.

Mowen, J. & Minor, M. 2002. *Perilaku konsumen (Edisi kelima, terjemahan oleh Lina Salim)*. Jakarta: Erlangga.

Pixability. 2014. "How youtube is radically transforming the beauty industry and what that means for brands. *Retrieved from* https://www.pixability.com/youtubes-radical-transformation-beauty-study/. Accsessed 5 Jnaury 2018.

Plummer, J. 2007. *Online advertising playbook*. New Jersey: John Wiley and Sons Inc.

Riegner, C. 2007. Word of mouth on the web: The impat of web 2.0 on consumer purchase. *Journal of Advertising Research* 47: 436–437.

YouTube. 2018. Review produk kosmetik. *Retrieved from https://www.youtube.com/results?search_query=review+produk+kosmetik*. Accsessed 5 January 2018.

Zhao, X., Wang, L., Guo, X. & Law, R. 2015. The influence of online reviews to online hotel booking intentions. *International Journal of Contemporary Hospitality Management* 27(6): 1343–1364.

Advances in Business, Management and Entrepreneurship – Hurriyati et al (eds)
© 2020 Taylor & Francis Group, London, ISBN 978-0-367-27176-3

The influence of product knowledge on attitude and purchase intention of *mudharabah* funding products in sharia banks in Mataram

B.V. Khairunnisa & A. Hendratmi
Airlangga University, Surabaya, Indonesia

ABSTRACT: This study aims to determine the effect of product knowledge on attitude and purchase intention for *mudharaba* financing in Islamic bank in Mataram. This study uses a quantitative approach using the test path analysis. Sources of data in this study using the primary data that is questionnaire. The characteristics of the study population are customer of Islamic bank in Mataram who have business and not use *mudharaba* financing. The number of samples in this study were 250 respondents. The sampling technique that can be used is probability sampling with random sampling. Based on the results of the study, it is indicated that product knowledge significantly influences the attitude, product knowledge significantly influences the purchase intention, and attitudes significantly influence the purchase intention. The suggestions for Islamic Bank in Mataram is to use personal promotion about *mudharaba* financing and increase the service and facilities to improve the costumer intention. The suggestions for further researcher, can be done by adding the level of purchase knowledge and usage knowledge as eksogen- variable and research can be done on other objects in the field of services such as BMT.

1 INTRODUCTION

1.1 Background

Marketing is defined as a social and managerial process, in which individuals and groups obtain what they need and want through the creation and exchange of products and values (Kotler & Armstrong 2004).

Marketing becomes one of the main activities that need to be done by a company, be it a goods or service-based company, in an effort to maintain its business survival. To win the competition, a company must be able to implement the right marketing strategy for the product or service they produce. Appropriate marketing strategies can be made by understanding consumer behavior (Assael 1992).

Engel et al. (1994) argues that consumer behavior can be defined as the actions of individuals directly involved in the business in obtaining and using economic services, including the decision-making processes that precedes and determines those actions.

In consumer behavior, there is a set of decision-making processes to determine consumer choice alternatives, in which consumers assess the various alternative options and choose one or more of the necessary alternatives based on particular considerations. Consumer knowledge is an important component affecting the decision-making process (Engel et al. 1994). When consumers have knowledge, they will act in a more efficient and more appropriate manner in processing information and be able to recall information about the product better. Engel et al. (1994) define product knowledge as a collection of various information about the product. Product knowledge includes product category, brand, product terminology or product features, product pricing, and product trust.

According to Brucks quoted from Engel (1994), there are two ways to measure product know-ledge based on previous studies, namely Subjective Know-ledge and Objective Knowledge. After con-sumers search for the product information, they begin to assess and evaluate, resulting in a positive attitude towards the product. Attitudes toward certain behaviors cannot be observed or measured directly, but can be inferred from one's evaluative responses to the attitudes of certain objects according to the consumer's memory.

After attitude of the evaluation of the product knowledge has emerged, then intention in the product will manifest. With intention, consumers can make purchase decisions after comparison and assessment of the product (Lin & Lin 2007).

The development of Islamic banks in Indonesia began in 1990, after the Workshop on Bank Usury held by the Indonesian Ulema Council (MUI) in West Java, resulting in the birth of the first Islamic bank in Indonesia named Bank Muamalat Indonesia (Antonio 2001). In the following years, Islamic banks are growing rapidly with the emergence of Sharia Commercial Bank (BUS), Sharia Business Unit (UUS), and BPR Syariah with assets that continue to increase from year to year.

The process of consumer decision-making by analyzing relationship between product knowledge and the intention of purchasing product is put into consideration in determining the right marketing strategy for companies, including Sharia Banks. Sharia banks as financial institutions have financing products with various contracts (*akad*), one of which *mudharabah* contract. Karim (2010) cites the opinion of M. Anwar Ibrahim that *mudharabah* is a joint partnership between the property of one party and the work of another, in which one party acts as the owner of the capital and entrusts the amount of capital to be managed by a second party, the business owner, for the purpose of creating profit.

Mataram as the capital of West Nusa Tenggara Province has the potential to become the target of sharia banks in developing sharia banking products. The Islamic finance industry in West Nusa Tenggara, particularly in Mataram, is experiencing rapid growth. Over the past five years, their combined assets reached IDR 144 trillion, while their total financing assets amounted to 1.955 trillion (bi.go.id). Therefore, this research designated the Bank Syariah in Mataram as the object of research to determine the influence of customer product knowledge on the attitude and intention of using *mudharabah* financing product.

1.2 Problem statement

1. Does product knowledge affect the attitude of using financing *mudharabah* in sharia banks in Mataram City?
2. Does attitude affect intention in using *mudharabah* financing in sharia bank in Mataram City?
3. Does product knowledge affect the intention in using *mudharabah* financing in sharia banks in Mataram City?

2 THEORETICAL FRAMEWORK AND HYPOTHESIS DEVELOPMENT

2.1 Consumer behavior

Engel (1994) argues that consumer behavior can be defined as the actions of individuals directly involved in the business in obtaining and using economic services, including the decision-making processes that precede and determine those actions. Loudon & Della (1993) define consumer behavior as a process of decision making and the physical activity of individuals involved in evaluating, obtaining, and using goods and services.

2.2 Consumer behavior in an Islamic perspective

Consumer behavior in Islam should reflect adherents' relationships with God. Every movement in the form of daily shopping is nothing but the manifestation of remembering (*zikir*) the name of God by not choosing forbidden (*haram*) goods, not being parsimonious, and not being greedy to survive in both the present world and the hereafter (*akhirat*) (Muflih 2006).

Islam sees human need as more than a matter of clothing, food, and housing, because they are only related to worldly businesses. As-Syatibi in Muflih (2006) said that the formulation of human needs in Islam consists of three levels, namely:

1. *Dharuriyat,* which comprises of religion (*din*), life (*nafs*), education (*'aql*), offspring (*nasl*), and property (*mal*).
2. *Hajiyat* is any complement that supports, strengthens, and protects the *dharuriyat* needs.
3. Tahsiniyat is any addition to the forms of pleasure and aesthetics other than *dharuriyat* and *hajiyat*.

2.3 Decision-making process

Kotler & Keller (2009) mentioned that the purchase decision process plays an important role in understanding how consumers actually make purchasing decisions. Marketing scholars have compiled a "phase order model" in the consumer purchase decision process.

2.4 Product knowledge

Engel et al. (1994) stated that product knowledge is a collection of various in-formation about the product. This knowledge includes product categories, brands, product terminology, product attributes or features, product prices, and product beliefs. Product knowledge also includes various information that is processed by consumers to obtain a product. Product knowledge also consists of knowledge of where to buy products and when to buy them. When a consumer decides to buy a product, he will decide where he will buy the product and when to buy it. The consumer's decision regarding where to buy the product will be largely determined by his knowledge. Brucks (1985) stated, "*Product knowledge is based on memories or known knowledge from consumers.*"

2.5 Measurement of product knowledge

Brucks in Engel et al. (1994) says that product knowledge can be assessed by two indicators, namely Objective knowledge and Subjective knowledge.

Objective knowledge are the correct information about the classes of products stored in the consumers' long-term memory.

a. Product Knowledge. The product knowledge itself is a collection of many different types of information. Product knowledge includes: Awareness of product categories and brands within product categories, product terminology, product attributes or attributes, and beliefs about product categories in general and for specific brands. Belief in this case where there is a belief that a product will provide utilities or benefits. Consumers will pay more attention to products that can generate utility, putting it into consideration in a purchase decision. Consumer trust can be formed from the conformity of a product with the principles of sharia. Product Criteria in the Islamic perspective are:

1. Products in Islam according to Abuznaid (2012) are prohibited to contain the practice of cheating, gambling (*maysir*), and usury. This is in accor¬dance with the word of God in Al-Maidah (5):4: "*lawful unto you are (all) things good and pure: and what ye have taught your trained hunting animals (to catch) in the manner directed to you by Allah: eat what they catch for you, but pronounce the name of Allah over it: andfear Allah; for Allah is swift in taking account*" (DEPAG RI 2012).
2. Prophet Muhammad, when trading, always provided information as clear as possible regarding the advantages and disadvantages of the products he sold. Islamic business ethics also requires that accurate information be provided not only to the goods sold but also in terms of their advertising (Abuznaid 2012).
3. Interest in Islam is expressed as *riba*. The usage of *riba* is strictly prohibited in the Qur'an and hadith. The word *riba* in English is often represented as usury, which in modern use means more than ordinary interest rates or stifling interest rates. Nafik (2009) argues that because Islam forbids usury, then banks that operate in accordance with sharia should implement a system of profit-sharing.

b. Purchase Knowledge. Purchase knowledge includes various pieces of information owned by consumers that are closely re¬lated to product acquisition. The basic dimension of purchasing knowledge involves information regard¬ing decisions about where the product should be purchased, when purchase should occur, and usage knowledge.

c. Subjective Knowledge is the level of understanding of consumers on a product, which is called self-assessed knowledge. Indicators of subjective knowledge according to Engel et al. (1994) are:

1. The amount of consumer knowledge towards a product.
2. The value of consumer knowledge of a product when compared to other consumers.
3. How familiar consumers are to a product.

2.6 Consumer attitude

Consumer attitude studies the tendency to respond consistently to something liked or disliked to the attention given to an object (Schiffman & Kanuk 2007). Consumer attitudes are a reflection of the overall evaluation of the association of a brand, product, object, or activity.

2.7 Attitude measurement

Schiffman & Kanuk (2007) suggested that Attitude toward object model is very suitable for expressing attitudes toward a specific product (or service) category or brand. Based on this model, consumer attitudes toward a specific product or brand of a product is a convincing presence and evaluation of specific functions-beliefs and/or attributes. According to Schiffman & Kanuk (2007), there are several indicators that can be used to measure consumer attitudes.

1. Consumers like the benefits provided by the product or brand.
2. Consumers like the ease of use of the product or brand used.
3. Consumers like the reliability of the product or brand.
4. Consumers like the service of the brand.
5. Consumers like products or brands that portray the consumer's personality.
6. Consumers like the added value of a given product or brand.

2.8 Purchase intention

Consumer intention is a form of desire (behavior intention). The desire to behave can be defined as the desire of the consumer to behave in a certain way in order to own, use, and dispose of products or services (Mowen & Minor 2002).

According to Ferdinand (2002), intention in using services or products can be identified through the following indicators:

1. Transactional intention, namely the tendency of someone to buy a product.
2. Referential intention, namely the tendency of someone to reference a product to others.
3. Preferential intention, which is an intention that describes the behavior of someone who has a primary preference for a product. This preference can only be changed if something happens with the preferred product.
4. Explorative intention, which portrays the behavior of a person who is always looking for information about the product he is intentioned in and seeking in-formation to support the positive characteristics of the product.

2.9 Sharia bank

Sharia bank is a bank that in its activities, either in collecting funds or in the framework of the distribution of funds, provides and imposes rewards on the results on the basis of sharia principles, namely trade and profit-sharing. The main principle of bank operations based on sharia principles is the Islamic law that comes from the Qur'an and Hadith. Bank operations must pay attention to orders and restrictions in the Qur'an and the Sunnah (ways) of Prophet Muhammad. Prohibition mainly relates to bank activities which can be classified as *riba* (usury) (Susilo 2000).

2.10 *Mudharabah*

Antonio (2001) quoted al-Syarbasyi's opinion as follows: "Mudharabah *is a contract of business between two parties where one* (shahibul al-mal) *provides all (100%) of the capital, and the other party becomes the manager. The profit of the business is divided according to the agreement set forth in the contract, while the loss is borne by the owner of the capital as long as the loss is not the result of the manager's negligence.*" *Mudharabah* is described also in QS.Al-Maidah verse one.

2.11 *Determination of* mudharabah *in sharia banking*

The concept of *mudharabah* in sharia banking is usually applied to funding and financing products. On the fund-raising side, *mudharabah* applies to the following products: (Antonio 2001)

a. Savings futures, namely savings intended for special purposes, such as savings for pilgrimage, and savings for cattle slaughter (*kurban*), and so forth.
b. Ordinary and special deposits, special deposits (special investment), where the funds deposited by customers, specifically for certain businesses, such as in *mudharabah* or ijarah.

Meanwhile, on the financing side, *mudharabah* applies to:

a. Work capital financing, namely financing intended to meet the needs of business funds for the purchase/procurement/provision of elements of goods in the framework of capital turnover. Examples are for the purchase of raw materials or goods to be traded.
b. Special investment, namely the financing provided to meet the needs of business facilities/infra-structure (fixed assets). For example, the purchase of production equipment.

2.12 *Relationship between the product knowledge and attitude*

According to Engel (1994), consumer knowledge consists of information stored in memory. Marketers in particular are interested in knowing consumer knowledge. Information held by consumers about the product will greatly affect the consumer pattern.

An awareness and image analysis is useful for exploring the nature of product knowledge.

Consumer knowledge is a key determinant of consumer behavior. What consumers buy, where they buy it, and when they buy it will rely on knowledge relevant to the decision. The underlying influences of consumer behavior are environmental influences, individual influences, and psychological influences. In this case, the element of knowledge includes the influence of the individual. Engel et al. (1995) divide consumer knowledge into three knowledge categories, namely product knowledge, purchasing knowledge, and knowledge of usage.

2.13 *Relationship between attitude and intention*

Attitude studies the tendency to respond consistently to something liked or disliked to the attention given to an object (Schiffman & Kanuk 2007). Thus, it can be concluded that attitude is the result of the overall evaluation of an association of a product or brand in which the feelings and emotions that form can affect one's lifestyle. A positive attitude towards a brand or product can increase consumer purchase intention in a product.

2.14 *Relationship between product knowledge and intention*

Lin & Lin (2007) says that the level of product knowledge affects consumer purchase intention. In general, consumers with higher product knowledge have better memory, recognition, analysis, and logical skills than those with lower product knowledge. Consequently, those who think that they have higher product knowledge tend to rely on intrinsic cues instead of stereotypes to make an assessment of product quality as they realize the importance of product information. A similar study was conducted by Ghalandari & Norouzi (2012), who overall stated that product knowledge affects purchase intention in a product.

Research framework and hypotheses:

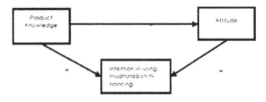

Figure 1. Analysis model.

The research used an analytical model developed from a research by Lin & Lin (2007). *The Effect of Brand Image and Product Knowledge on Purchase Intention Moderated by Price Discount.*

This research used structural equation for endogenous intervening variable (Z) as follows:

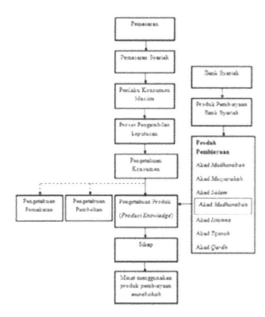

Figure 2. Framework of thinking.

$$Z = yX + e \qquad (1)$$

This study also used structural equations for endogenous variable (Y) as follows:

$$Y = PZ + yX + e \qquad (2)$$

where Y = intention to use *mudharabah* financing product in sharia banks; B = path coefficient from endogenous variables to exogenous variables; Z = attitude towards *mudharabah* financing products in sharia banks in Mataram; r = path coefficient from exogenous variable to endogenous variable; X = product knowledge E = standard error.

The framework of thinking in this study is outlined Figure 2.

3 RESEARCH METHOD

This study used some research variables, including: product knowledge as exogenous variable, attitudes as endogenous intervention variable, and pur- chase intention as an endogenous dependent variable. Product knowledge has two indicators, including: subjective knowledge and objective knowledge, consisting of 20 items of questions. Attitude variable consists of five indicators, with five question items. The purchasing intention variable has four indicators, namely transactional, referential, preferential, and explorative intentions. The total number of question items for knowledge, attitude, and intention is 29 question items.

The reliability test used Cronbach Alpha. In this study, the Cronbach Alpha value was above 0.6, so this study can be considered reliable. The validity test yields 3 items.

The scale of measurement in this research is ordinal scale with likert scale model which used four answer, namely: (SS) Strongly agree with a score of 4, (S) Agree with a score of 3, (TS) Disagree with a score of 2, (STS) agree with a score of 1. The type of data used was primary data obtained through questionnaires that had been filled by respondents and secondary data as a support in the form of corporate documents, some of which are supporting data such as journals, documentation photos, internet sources, and literature related to the problems studied.

The subjects or population in this study are the customers of sharia banks in Mataram. The characteristics of the sample used were customers of sharia banks in Mataram who own or plan to have business. The population in this study was 250 people.

Sampling technique in this research used the Probability sampling, namely sampling which gives equal opportunity for each element (member) of the population to be elected as a sample member, so the number of samples in this study is 250 respondents of sharia bank customers in Mataram. The number of questionnaires distributed was 250 questionnaires. The analysis model used to determine the effect was-path analysis using SPSS and Amos programs.

4 RESEARCH RESULT AND DISCUSSION

Data processing in the testing of path analysis assumptions used Amos and interpreted through the following Table 1.

The table above can be interpreted as follows:

1. If the product knowledge variable changes, then it will cause a change in consumer attitudes. The positive sign indicates a unidirectional change, that is, if the product knowledge variable increases then the consumer's attitude will increase, and vice versa, if the variable of product

Table 1. Coefficient values of influence paths between variables.

Variable		Direct Path Coefficient	Indirect Path Coefficient
Product Knowledge	Consumer Attitude	0.625	
Consumer Attitude	Purchase Intention	0.380	
Product Knowledge	Purchase Intention	0.394	
Product Knowledge to Purchase intention through Attitude			0.237

knowledge decreases the consumer attitude will also decrease with a pah coefficient of 0.625.

2. If the product knowledge variable changes, the purchasing intention will change. The negative sign indicates a change in the opposite direction, which is if the product knowledge variable increases then the purchase intention will decrease, and vice versa, if the variable of knowledge of the product decreases then the purchase intention will increase. The path coefficient value was -0.009. If the consumer's attitude towards variable changes, then the purchase intention will change. The positive sign indicates a unidirectional change, that is if the consumer attitudinal variable increases then the purchase intention will increase, and vice versa, if the consumer attitude variable decreases the purchase intention will also decrease. The path coefficient value was 0.380.

The coefficient of determination is a coefficient that shows the magnitude of influence or contribution of an exogenous variable to and endogenous intervening variable and exogenous and endogenous variables toward endogenous variables. The following are the test results that show the coefficient of determination:

The following is an explanation of Table 2.

1. The amount of attitude change caused by the knowledge of the product was 0.359 or 35.9%. In other words, the effect of product knowledge on attitude was 35.9%.
2. The amount of intention change caused by product knowledge and attitude was 0.360 or 36%. In other words, the effect of product knowledge and attitudes toward intention was 36%.

The next step is proving the hypotheses. The complete results of the hypothesis test can be seen in Table 3.

Table 2. Coefficient of determination value.

Variable	R2
Product knowledge to Attitude	0.359
Product knowledge to attitude towards intention	0.360

Table 3. Results of hypothesis test on direct effect.

Variable		CR count	Sig-	Notes
Product Knowledge	Consumer Attitude	11.601	0.000	Significant
Consumer Attitude	Purchase Intention	5.211	0.000	Significant
Product Knowlege	Purchase Intention	5.183	0.000	Significant

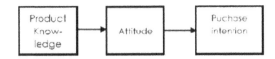

Figure 3. Indirect path coefficient.

Table 3 can be explained as follows:

1. The value of CR count of product knowledge variable to attitude was 11.605. The level of significance was 0.000. This value is less than 0.05. Since the significance level is less than 0.05, the product knowledge affects attitudes. Thus, the research hypothesis is accepted.
2. The value of CR count of consumer attitude variable to purchase intention was 5.211. The level of significance was 0.000. This value is less than 0.05. Because of this significance level is less than 0.05, then consumer attitude affects the purchase intention. Thus, the research hypothesis is accepted.
3. The value of CR count of product knowledge variable to purchase intention was 5.183. The level of significance was 0.000. This value is less than 0.05. Since the significance level is less than 0.05, the product knowledge affects attitudes. Thus, the research hypothesis is accepted. The direct influence of product knowledge on attitude and attitude towards purchase intention indicates that product knowledge indirectly influences purchase intention through attitude.

Figure 3 shows the indirect effect of product knowledge on purchase intention through attitude. The value of the indirect path coefficient was 0.237. Value is derived from the multiplication of the direct influence of product knowledge on attitude and attitude towards intention. Thus, the indirect coefficient value of product knowledge on intention through attitude was $0.625 \times 0.380 = 0.237$. This value is greater than the direct path coefficient (-0.009). With this result, product knowledge indirectly influences intention through attitude. Thus, the research hypothesis is accepted.

5 CONCLUSION

Based on the results of data processing using path analysis, the following conclusions may be drawn:

1. Product knowledge significantly affects consumer attitude. With this result, the research hypothesis is accepted.
2. Attitude significantly influences purchase intention. With this result, the research hypothesis is accepted.
3. Product knowledge significantly affects purchase in-tention. With this result, the research hypothesis is accepted.

REFERENCES

Abuznaid, S. 2012. Islamic marketing: Addressing the Muslim market. An-*Najah University Journal of Research (Humanities)* 26(6): 1473–1503.

Al Qur'an dan Terjemahnya. 2012. Departemen Agama Republik Indonesia.

Antonio, M.S. 2001. *Bank syariah, dari teori ke praktek*. Jakarta: Gema Insani.

Assael, H. 1992. *Consumer behavior and marketing action*. New York: PWS-KENT Publishing Company.

Engel, F.J., Blackwell, R.D. & Miniard, P.W. 1994. *Perilaku konsumen*. Jakarta: Binarupa Aksara.

Ferdinand, A. 2002. *Metode penelitian manajemen*. Semarang: Badan Penerbit Universitas Diponogoro.

Karim, A.A. 2010. *Bank Islam analisis fiqh dan keuangan (Edisi ke empat)*. Jakarta: PT. Raja Grafindo Persada.

Kotler, P. & Amstrong, G. 2004. *Principles of marketing (Tenth edition)*. New Jersey. Prentice Hall Inc.

Kotler, P. & Keller, K.L. 2009. *Manajemen pemasaran (Edisi 12 Jilid 1)*. Jakarta: Airlangga.

Lin. N.H. & Lin. B.S. 2007. The effect of brand image and product knowledge on purchase intention moderated by price discount. *Journal of International Management Studies*.

Loudon, D.L. & Della, A.J. 1993. *Costumer behaviour: Concept and Aapplications*. Singapore: McGraw Hill Book Company.

Mowen, J.C. & Minor. 2002. *Perilaku konsumen (Edisi ke 5. Jilid 2)*. Jakarta: Airlangga.

Muflih, M. 2006. *Perilaku konsumen muslim dalam perspektif ekonomi Islam (Edisi pertama)*. Jakarta: Raja Grafindo Persada.

Schiffman, L. & Kanuk, L.L. 2007. *Costumer behavior (Eight edition)*. New Jersey: Pearson Prentice Hall Inc.

www.bi.go.id. Accsessed 3 March 2014.

Advances in Business, Management and Entrepreneurship – Hurriyati et al (eds)
© 2020 Taylor & Francis Group, London, ISBN 978-0-367-27176-3

The influence of economy experience with Pine and Gilmore concept on customer satisfaction: Survey of Transmart consumers in Padang City

Y.P. Gubta & V. Verinita
Andalas University, Padang, West Sumatera, Indonesia

ABSTRACT: This study aims to analyze how education experience, entertainment experience, esthetic experience, and experience escapism affect customer satisfaction on Transmart consumers in Padang City. The sample of this research is people who visit Transmart Padang City at least once and are at least 17 years old. There were 170 people who participated in this study but respondents who met the criteria of this study were only 120 people. Sampling technique used is non-probability sampling with purposive sampling method. Management of research data is done by using Statistical Package for Social Science (SPSS) version 22 software. The result of the research shows that the education experience does not significantly affects the customer satisfaction, whereas the greater or enhanced education experience will not affect the customer satisfaction level of Transmart Padang City. Entertainment experience, esthetic experience, and escapism experience have a significant effect on customer satisfaction, whereas the greater or more enhanced entertainment experience, esthetic experience, and escapism experience, the higher the customer satisfaction in Transmart Padang City.

Keywords: experience economy, education experience, entertainment experience, esthetic experience, escapism experience, customer satisfaction

1 INDRODUCTION

Based on data from the Ministry of Home Affairs, the total population of Indonesia in 2017 shows more than 261 million people (www.kemendagri.go.id). From year to year per capita income Indonesia also continues to increase, in 2016 per capita income Indonesia recorded Rp 47.96 million per capita per year. This figure is higher than in 2015 (Rp 45.14 million) and in 2014 (Rp 41.92 million) (finance.detik.com).

The National Defense Agency also stated that during the second quarter of 2017, the realization of domestic investment amounted to Rp 61.0 trillion, up 16.9% from Rp 52.2 trillion in the same period in 2016 and Foreign Investment of Rp 109.9 trillion, up 10.6% from Rp 99.4 trillion in the same period in 2016 (www.bpn.go.id). One of the industries that attract investors is the retail industry. The retail industry is also increasingly contributing greatly to the national economy. According to the Central Bureau of Statistics (BPS) in 2016, the retail or trade industry contributes 15.24% to the total Gross Domestic Product (GDP) and absorbs 22.4 million workers or 31.81% of non-agricultural labor (ekbis.sindonews.com).

The retail industry makes use of the changing lifestyle of a society from the traditional pattern to the modern pattern and one part of this change is trend of shopping in traditional retail to modern retailers (Soliha, 2008). Changes in the pattern of human life trigger changes in the economic pattern of society, people go shopping place not only to buy certain products or services, but more than that unique and unforgettable experience by every consumer (Gilmore and Pine, 2002).

Kartajaya (2006) states that "today many customers are becoming more sophisticated, customers become not only need a service or a high-quality product, but also a positive experience, which is emotionally touching and." This is in order to realize consumer satisfaction that is the feeling of pleasure or disappointment of someone who derives from the comparison between his impression of the performance (or outcome) of a product and its expectations (Kotler and Keller, 2009: 138–139).

One of the fastest growing retail companies today is Transmart. Apparently Transmart provides a different and new concept for its customers, not just shopping but Transmart also provide an interesting experience for its customers, namely: education experiences, knowledge, beauty, fun, and entertainment. Therefore, Transmart is a new attraction for consumers to shop or just visit to release fatigue, calm down, looking for entertainment, or looking for new inspiration.

Based on the background described here, it is necessary to research the Effect of Economy Experience with Pine and Gilmore Concept on Customer Satisfaction (Survey of Transmart Consumers in Padang City).

The problem formulation of this research is as follows:

1. How is the influence of education experience on customer satisfaction on Transmart consumers in Padang City?
2. How does the influence of entertainment experience on customer satisfaction on Transmart consumers in Padang City?
3. How does the influence of esthetics experience on customer satisfaction on Transmart consumers in Padang City?
4. How does the influence of escapism experience on customer satisfaction on Transmart consumers in Padang City?

The purpose of this research is to analyze the influence of education experience, entertainment experience, esthetics experience, and escapism experience to customer satisfaction at Transmart consumer of Padang City.

Benefits to be achieved from this research are:

1. *Academic Benefits*

Can provide study direction about science concept in strategic management and marketing management, especially about experience economy (education experience, entertainment experience, esthetic experience, and escapism experience) and customer satisfaction which can be used as comparative material in literature for those who want to do research in strategic management or marketing management.

2. *Practical Benefits*

Practical usefulness in this research is in the form of suggestions or inputs that resulted as a research output that can be used as a guide for entrepreneurs who will open or run the retail business.

2 LITERATURE REVIEW

2.1 *Concept of experience*

The Oxford English Dictionary in Hosany and Mark, 2010 defines experience as "an event or event that leaves an impression." Experience comes from a complex set of actions between a customer and a company or an enterprise product offering. Customers can also make their own unique experiences (Prahalad and Ramaswamy 2004).

Ideas about experience included in marketing and consumption were first revealed by Holbrook and Hirshman (1982). Holbrook and Hirshman (1982) say that elements of fun, beauty, symbolic meaning, creativity, and emotion can enrich and broaden the understanding of consumer behavior.

Schmitt (1999) also introduced the concept of experiential marketing. Where experiential marketing is defined as a way to make consumers create experiences through the stimulation of the five senses: stimuli to feel (affective); stimuli for creative

thinking; stimulation to perform physical activity; behave and interact with others; and stimuli to socialize that reflect his lifestyle and culture.

Pine and Gilmore (1999) presents experience as an offer in the new economy, which emerges after commodities, goods, and services. Experience is regarded as the development of economic value. The company does not sell experience but provides tangible facilities and profitable experience contexts so that consumers can do well in order to realize a unique experience. Where there are four dimensions of experience introduced by Pine and Gilmore are education, entertainment, esthetic, and escapism.

Based on some concept of experience put forward by experts, the authors judge that the concept of experience that fits with research conducted in Transmart Padang City is a concept that was conveyed by Pine and Gilmore (1999). Where there are four dimensions of experience namely education, entertainment, esthetic, and escapism. This is because Transmart Padang City has all four dimensions are offered to consumers as one of the competitive advantage.

2.2 *Experience economy concept*

Experience economy is a concept whereby a company can start a new beginning by creating a memorable experience for each customer. Experience can be interpreted as the subjective level of mental perceived visitors during a service (Otto and Ritchie, 1996 in Amsal dan Mahardika, 2017). Customer experience helps business organizations maintain their long-term customer satisfaction and how the company gains an extra competitive edge (Andajani, 2015).

Pine and Gilmore (1998) state that economic value shifts from commodities, goods, services and the last is experience. There are four dimensions that can evoke memories or consumer experience: education, entertainment, escapism, and esthetics (Pine and Gilmore, 1998).

2.2.1 *Education experience dimension*

The dimension of education experience with the focus of creating an educational experience enables consumers to absorb an event that takes place in it, by means of intellectual interactive engagement/ physical training (Pine and Gilmore, 1999). According to Pine and Gilmore (1998) educational experiences actively engage the minds of consumers, attract them and propose their desire to learn something new. According to Mykletun and Meira (2014), education is the experience seen as active participation, for example when a person takes a parachuting course he is actively involved, but they also absorb information and not fully immersed in the activity. Other examples such as: snorkelling, ski school, seminars/talk shows, theater performances, and others.

2.2.2 *Entertainment experience dimension*

According to Mykletun and Meira (2014), entertainment is a passive participation and relationships with the event come from an absorption perspective, such as watching a performance. Entertainment developed when passive absorption is observed by the customer (Ali et al, 2014). Examples: reading books, listening to music, watching movies, and more.

Pine and Gilmore (1998) noted that the company is now a "stage" used to please and entertain customers. Some common examples of entertainment include variety shows and live concerts.

2.2.3 *Esthetics experience dimension*

The dimensions of esthetics see that consumers can drift in a particular event, event or environment, but the role of the consumer in experiencing an esthetics experience will become passive. Consumers will not be able to exert any influence on such events or events. This will lead to an experience that involves feelings. The key concepts in this aesthetic dimension are: style, taste, beauty, design, and art. (Pine and Gilmore, 1999). Examples for aesthetic dimensions include a beautiful historic setting of museums, an attractive service or atmosphere at a resort hotel or enjoying the sights of Niagara Falls (Mehmetoglu and Marit, 2011).

2.2.4 *Escapism experience dimension*

This dimension will be more or less marked with action, thrill, and adrenaline. Experience out of this reality requires greater engagement and participation. Consumers who participate in other experiences of reality not only start but also share the overall activity (Pine and Gilmore, 1999). Example: Play roll coster, paitnball, and others. According to Mykletun and Meira (2014) when one participates actively and is immersed in their activity, this experience is called escapism. Escapism that occurs when a participant affects an actual performance in a real or virtual environment. minsalnya with white water rafting or playing in the casino (Hosany and Mark, 2010).

Breakout experience can be defined as the extent to which a person is really fun and engrossed in his activities (Hosany and Mark, 2010). The escape experience is very high immersive and requires active participation. Examples of amusement parks, adventure land, simulation destinations, and themed attractions among others.

Shopping while playing provides many opportunities for breakout experience. Vacation is one means for escape, solving problems, energy sources, strengths and expectations. Escape also offers a psychological holiday from the routine of everyday life. (Hosany and Mark, 2010)

2.2.5 *Customer satisfaction*

Satisfaction is an excellent predictor of repurchase intentions (Choi and Chu, 2001), consequently understanding what causes a person to be satisfied to be one of the greatest challenges facing the business

(Mykletun and Meira, 2014). Satisfaction is usually associated with the consumer's response to supply and reseller differences between actual performance and expected product (Yoo and Park, 2016). According to Parasuraman, et al. (1988), service quality is "a global judgment or attitude related to the superiority of a service."

Oliver (1999) suggests that satisfaction may be conceptualized at two levels: micro (in response to transaction-specific experience) and global (as a result of the customer's cumulative experience with specific services). In fact, satisfaction tends to result from a global evaluation of all aspects that make up customer relationships, not just transaction-specific results with providers (Thakur, 2016).

Customer satisfaction is one of the most common surveys and marketing measures of how well a company's products or services meet or exceed customer expectations (Mykletun and Meira, 2014). Expectations often reflect many aspects of an organization's business activities including products, services, companies and how companies operate globally (Petrick, 2004).

Oliver (1999) defines satisfaction as a deep commitment by customers to consistently repurchase the same product or service, and has no direct effect on competitors' marketing strategies, which have the potential for a change in buying behavior.

2.3 *Conceptual framework*

The conceptual framework is a tentative conclusion from a theoretical review reflecting the relationships between the variables being studied. In this research, a conceptual concept model has been used as a model of empirical research and serves as a guide in conducting research, which will then show the relationship of education experience, entertainment experience, esthetics experience, and escapism experience to customer satisfaction, as can be seen in Figure 1.

2.4 *Hypothesis*

2.4.1 *The influence of education experience on customer satisfaction*

Research conducted by Mehmetoglu and Marit (2011) at the Ice Music Festival and Maihaugen Museum with 117 questionnaires spread in 2009, stated that for

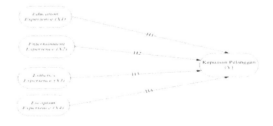

Figure 1. Research Model.

the case of Ice Music Festival: education experience has no effect on customer satisfaction, while for the case Maihaugen Museum: effect on customer satisfaction. This research also shows that the test of experience education depends on the object of research in the test.

In a study conducted by Quadri-Felitti and Ann (2014), conducted research through an online questionnaire to 9733 respondents and who sent back as many as 257 respondents. The object of this research is the Lake Erie Wine Country traveler. The results of this study indicate that the education experience affects customer satisfaction.

Also in research conducted by Ali et al (2014) on international delegations in the context of academic conferences, states that education experience affects customer satisfaction. Based on the previous research, the first hypothesis can be drawn, namely:

H1: Suspected existence of the influence of education experience on customer satisfaction on Transmart consumers Padang City.

2.4.2 The influence of entertainment experience on customer satisfaction

Research conducted by Hosany and Mark (2010) on 169 cruise ship passengers Rhapsody of the Sea, shows the result that entertainment experience affects customer satisfaction. Similarly, in the study of Ali, et al. (2014) conducted to international delegates in the context of academic conferences, where as many as 210 questionnaires distributed were returned as many as 188 questionnaires and the results show that entertainment experience positively affects customer satisfaction. This shows that the higher one's entertainment experience, the higher the customer satisfaction of a product or service.

In Mykletun and Maira's research (2014) also showed that entertainment experience has a significant effect on customer satisfaction. The study was conducted in 2011 to Extreme Sport Weeks at Voss Norway, where 292 athletes responded. Based on the previous research, the second hypothesis can be drawn, namely:

H2: Suspected of the influence of entertainment experience on customer satisfaction on Transmart consumers Padang City.

2.4.3 The influence of esthetics experience on customer satisfaction

Research conducted by Mehmetoglu and Marit (2011) states that for the case of Ice Music Festival: esthetics experience affects customer satisfaction, while for the case of Maihaugen Museum: esthetics experience also affects customer satisfaction. Quadri-Felitti and Ann (2014) conducted research on Lake Erie Wine Country travelers, where the results show that esthetics experience affects customer satisfaction.

In a Hosany and Mark study (2014), conducted on cruise ship passengers on Rhapsody of the Seas also showed similar results that esthetics experience had an effect on customer satisfaction. Mykletun and

Maira (2014); Ali, et al (2014) Also stated that esthetics experience has a positive influence on customer satisfaction. Based on the previous research, the third hypothesis can be drawn, namely:

H3: Suspected the influence of esthetics experience on customer satisfaction in Transmart consumers Padang City.

2.4.4 The influence of escapism experience on customer satisfaction

Research conducted by Ali, et al (2014) to international delegates in the context of academic conferences, from 188 respondents who returned the questionnaire showed that the escapism experience had a positive effect on customer satisfaction. In the research of Mehmetoglu and Marit (2011) at the Ice Music Festival and Maihaugen Museum in 2009 with 117 questionnaires spread, stated that for the case of Ice Music Festival: escapism experience affects customer satisfaction, while for the case of Maihaugen Museum: escapism experience has no effect on customer satisfaction. Based on the previous research, the fourth hypothesis can be drawn, namely:

H4: Suspected of the influence of escapism experience on customer satisfaction on Transmart consumers Padang City.

3 RESEARCH METHODS

This research is a quantitative research by using hypothesis testing (Explanatory Research). The type of investigation that is used is a causal study is a study where researchers want to find the cause of one or more problems. The time period (time horizon) used is one short (cross sectional) within the period of 2018. The unit of analysis of this study is consumers who visited or have been to Transmart Padang City.

Sources of data in this study are primary data and secondary data. Primary data from this research is obtained by spreading questionnaires to respondents as the object of research with questionnaire form used is closed questionnaire. Secondary data is primary data that has been processed by other parties, such as books, scientific journals, literature articles, and internet sites related to research conducted. Data collection techniques in this study are documentation and questionnaires.

The population in this study is all consumers who visited or have been to Transmart Padang City.

While the sample is a person who had visited Transmart Padang at least once. According to Hair, Black, Babin and Anderson (2010), if the population is unknown then the determination of the minimum sample quantity can be calculated by the number of indicator multiplied by 5. The minimum sample size in this study is 24 x 5 = 120 people. The sampling technique used is non probability sampling with the method of sampling purposive sampling method, that is the researcher determine the sample taken because there is a certain consideration, so the sample is not taken randomly, but determined by the researcher according to the requirement (the characteristics, characteristics, criteria). Samples are taken based on criteria that have been formulated in advance by the researcher. The criteria formulated in this study are respondents who are over 17 years old and have been to Transmart Padang city at least one visit.

Data analysis technique used is multiple regression analysis by using software SPSS version 22. Test instrument that is done is test of validity and reliability. Then also tested the classical assumption and hypothesis test.

4 RESULTS AND DISCUSSIONS

Respondents who participated in this study amounted to 170 people, while respondents who meet the criteria of this study only 120 people. This respondent can be specified by gender, marital status, age, last education, current profession, monthly income, number of visits, domicile and distance from Transmart Padang City.

4.1 Multiple linear analysis

Based on Table 1, it can be seen that the regression equation is formed as follows:

$Y = a + p\ 1X1 + p\ 2X2 + P3X3 + P4X4 + e$
$Y = 4{,}123 + 0{,}084\ X1 + 0{,}295\ X2 + 0{,}192\ X3 + 0{,}208\ X4$

1. From the multiple regression equation above shows that the constant value of 4.123 indicates that without the independent variable to customer satisfaction is fixed.
2. The value of coefficient of education experience (X1) that is 0,084. This shows that if the education experience increases by one unit then customer satisfaction will increase by 0.084 or 8.4% with the assumption that the variable of entertainment experience (X2), esthetics experience (X3), and escapism experience (X4) is fixed.
3. The value of entertainment experience coefficient (X2) is 0.295. This shows that if entertainment experience (X2) increases by one unit then

Table 1. Multiple Linear Analysis.

Model	Unstandardized Coefficients		Standardized Coefficients		
	B	Std. Error	Beta	t	Sig-
1 (Constant)	4,123	1,384		2,980	,004
XI	,084	,076	,084	1,096	,275
X2	,295	,092	,298	3,218	,002
X3	,192	,077	.215	2,504	,014
X4	,208	,059	,281	3,539	,001

Source: Primary data processed by SPSS (2018)

customer satisfaction will increase by 0. 295 or 29.5% with assumption that the variable of education experience (X1), esthetics experience (X3), and escapism experience (X4) is fixed.
4. The coefficient of esthetics experience (X3) is 192. This shows that if esthetics experience (X3) increases by one unit then customer satisfaction will increase by 0.192 or 19.2% with assumption that the variable of education experience (X1), entertainment experience (X2), and escapism experience (X4) is fixed.
5. The value of coefficient escapism experience (X4) that is 0.208. This shows that if escapism experience (X4) increases by one unit then customer satisfaction will increase by 0208 or 20.8% with assumption that the variable of education experience (X1), entertainment experience (X2), and esthetics experience (X3) is fixed.

4.2 Individual parameter significance test (Test Statistic t)

Based on Table 1 shows that the results of partial hypothesis test are as follows:

1. Based on Table 1 for education experience variable (X1), the value of t arithmetic (1,096) < t table (1,981) with significance value (Sig.) 0,275 > 0,05. Then it can be concluded that H0 is accepted and Ha is rejected, which means education experience (X1) has no significant effect on customer satisfaction (Y).
2. Variable entertainment experience (X2) shows the value of t arithmetic (3.218) > t table (1,981) with significance value (Sig.) 0,002 < 0,05. So it can be concluded H0 rejected and Ha accepted, which means entertainment experience (X2) significant effect on customer satisfaction (Y).
3. Esthetics experience variable (X3) shows the value of t arithmetic (2.504) > t table (1,981) with significance value (Sig.) 0,014 < 0,05. Then it can be concluded H0 rejected and Ha accepted, which means esthetics experience (X3) have a significant effect on customer satisfaction (Y).
4. Variable escapism experience (X4) shows the value of t arithmetic (3.539) > t table (1,981) with significance value (Sig.) 0.001 < 0.05. Then it can be concluded H0 rejected and Ha accepted, which means escapism experience (X4) have a significant effect on customer satisfaction (Y).

Based on Table 2 it can be seen that the value of linear correlation coefficient is 0.697, it means showing strong relationship between independent variable, that is Education Experience, Entertainment Experience, Esthetics Experience and Escapism Experience to Customer Satisfaction. The value of multiple linear determination coefficient (Adjusted R Square) shows the number 0.467 which means only 46.7% variable X affects variable Y and 53.3% of the rest is influenced by other variables that are not examined.

4.3 Coefficient of determination (R2)

Table 2. Determination Coefficient Analysis (R2).

Model	R	R Square	Adjusted R Square	Std. Error of the Estimate	Durbin-Watson
1	,697	,485	,467	1,950	1,956

a. *Predictors: (Constant), Education Experience, Entertainment Experience, Esthetics Experience* and *Escapism Experience*
b. *Dependent Variable:* customer satisfaction Source: Primary data processed by SPSS (2018)

4.4 Hypothesis testing

4.4.1 The influence of education experience on customer satisfaction

Hypothesis 1, the influence of education experience on customer satisfaction is rejected. Hypothesis rejected because t value (1,096) < t table (1,981) with significance value (Sig.) 0,275 > 0,05. Research indicates the overall education experience does not affect customer satisfaction. These results provide the understanding that any improvement in the education experience will not impact on customer satisfaction on Transmart consumers Padang City, meaning that the better education experience given then no effect on customer satisfaction.

This study is in line with previous researches researched by Mehmetoglu and Marit (2011) on the visitors of the Ice Music Festival and Hosany and Mark (2010) on cruise ship passengers Rhapsody of the Seas. Where in the previous research also found that the education experience does not affect customer satisfaction.

From the respondent's characteristic it can be seen that the respondents aged 17–25 years and 2635 years are the most dominant respondents are 84.1% with the dominant level of senior high school and undergraduate (83.3%), the level of work which is dominated as private employees, students, civil servants/civil servants and entrepreneurs 76.7% with average income below Rp5.000.000, - per month about 80%. Consumers have argued that education experience is not one of the factors that make them happy to visit Transmart Padang City. Because the average respondent has a young age that is under 35 years and work as a student and work so most consumers have been saturated with lectures and work, so they visit the Transmart to shop and remove the saturation of the daily busy.

4.4.2 The influence of entertainment experience on customer satisfaction

Hypothesis 2, the influence of entertainment experience on customer satisfaction is accepted. Hypothesis accepted because t value count (3,218) > t table (1,981) with significance value (Sig.) 0,002 < 0,05.

Research indicates that overall entertainment experience has an effect on customer satisfaction. These results provide the understanding that any increase in entertainment experience will impact on customer satisfaction on Transmart consumers Pa- dang City, meaning that the better entertainment experience given Transmart Padang city will be the higher the satisfaction felt by the customer.

This study is in line with previous research investigated by Hosany and Mark (2010) on cruise ship passengers Rhapsody of the Seas, Ali, et al. (2014) conducted to international delegates in the context of academic conferences, and Mykletun and Maira (2014) to Extreme Sport Weeks athletes in Voss Norway. Where in the study also found that entertainment experience have a positive effect on customer satisfaction.

From the characteristics of respondents can be seen that the average respondent has a relatively young age range of 17–35 years and comes from among students and who have worked, it shows that indeed the tendency of people in those days prefer to visit places which is fun to let go of boredom, hang out, shopping, find new inspiration, or just fill their days off. Based on marital status also shows that most of the visiting consumer is single status (53,3%) and married consumer but not yet have children (21,7%) because those who are single and not have children, have more time to be together with friends or their spouses. So the place that provides entertainment becomes their main choice and Transmart Padang City is one of their main goals in addition to entertainment is also available hangout, shopping, and so forth.

4.4.3 The influence of esthetic experience on customer satisfaction

Hypothesis 3, the effect of esthetic experience on customer satisfaction is accepted. The hypothesis is accepted because the value of t arithmetic (2.504) > t table (1,981) with significance value (Sig.) 0.014 < 0.05. Research indicates that overall esthetic experience affects customer satisfaction. This result gives an understanding that every increase of esthetic experience will have an impact on customer satisfaction on Transmart consumer of Pa-dang City, meaning that the better esthetic experience will give higher customer satisfaction.

This study is in line with previous research investigated by Mehmetoglu and Marit (2011) for the case of the Ice Music Festival, Hosany and Mark (2010) on cruise ship passengers Rhapsody of the Seas, Ali, et al. (2014) conducted to international delegates in the context of academic conferences, and Mykletun and Maira (2014) to Extreme Sport Weeks athletes in Voss Norway. Where in the study also found that esthetic experience has a positive effect on customer satisfaction.

Characteristics of respondents who are the average generation is still young and live in a massive social media era then things that have a beautiful and modern aesthetics become one of the attractions to visit and perpetuate the moment that they do the place.

Nowadays, the things that are full of aesthetics are the places that are the target of the generation of socialites and this generation is mostly the generation that is in the age range 35 years and under, where they live in the era of information technology and social media very rapidly. This is what triggers them to always be active and come to places that have high aesthetic value. Transmart Padang City is one of the excellent shopping centers in presenting the aesthetic values proficiency in order to give satisfaction to its customers.

4.4.4 *The influence of escapism experience on customer satisfaction*

Hypothesis 4, the influence of escapism experience on customer satisfaction is accepted. Hypothesis accepted because t value count (3,539)> t table (1,981) with significance value (Sig.) 0,001 <0,05. Research indicates that overall escapism experience has an effect on customer satisfaction. This result gives an understanding that every escapism experience improvement will have an impact on customer satisfaction at Transmart consumer of Padang City, meaning that the better escapism experience will give higher customer satisfaction.

This study is in line with previous studies studied by Mehmetoglu and Marit (2011) at the Ice Music Festival and the Maihaugen Museum, and Ali, et al. (2014) conducted to international delegates in the context of an academic conference. Where in the study also found that escapism experience have a positive effect on customer satisfaction.

The facts of this study indicate that the escapism experience significant influence on customer satisfaction on Transmart consumers Padang City. When visiting Transmart consumers perceive a change in their character from negative to positive as previously passive to active, quiet to be excited, and so on. Being in the Transmart makes consumers let go of all the routine activities they do to be relaxed and excited again. Transmart Kota Padang has an interesting design and concept so that consumers feel they are in a different place than their places of residence. In addition, many activities can also be done by the visitors who come to Transmart, such as games, watching the cinema, shopping, culinary tours, hanging out, and so forth, so that this can make consumers dissolve with various activities. Of the several things considered in determining the experience escapism, of the five items of the proposed statement then only one statement has an average of more than 3.50 is 3.53 while others are below that number. The statement that got the highest response was "When I was in Transmart I felt I was in a different place compared to my everyday place." Although most respondents apply neutral to the revelation, but overall escapism experience affects customer satisfaction.

5 CONCLUSION

Based on the results of research and discussion can be concluded that:

1. Education experience does not affect customer satisfaction significantly, where the greater or more improved education experience Transmart Padang then it does not affect the level of customer satisfaction Transmart Padang.
2. Entertainment experience affects customer satisfaction significantly, whereby the greater or more enhanced entertainment experience Transmart Padang city then the higher customer satisfaction Transmart Padang City.
3. Esthetic experience affects customer satisfaction significantly, where the greater or more enhanced esthetic experience Transmart Padang City then the higher customer satisfaction Transmart Padang City.
4. Escapism experience affects customer satisfaction significantly, where the greater or more enhanced escapism experience Transmart Padang city then the higher customer satisfaction Transmart Padang City.

Based on the result of the research, entertainment experience, esthetic experience, and escapism experience must be improved to get high customer satisfaction. While the education experience needs to be a concern Transmart Padang because no significant effect on customer satisfaction.

5.1 *Implication*

Based on the result of the research and the conclusion, there are some useful implications for Transmart management of Padang City as manager in optimizing education experience, entertainment experience, esthetic experience, and escapism experience in order to create customer satisfaction Transmart Padang City.

5.1.1 *Education experience*

Education experience has no significant effect on customer satisfaction. This research indicates that for every improvement of education experience variable, it does not have an impact on increasing customer satisfaction. It means that if education experience is improved by management then it will not improve customer satisfaction Transmart Padang City. This means that the increase in education experience is not so they ignore in deciding to visit Transmart Padang City. Nevertheless the education experience needs to also be managed by Transmart management of Padang City as one of the competitive advantage in today's business development. Researchers suggest that the management redesign the room and fill the empty space or provide a special place that makes consumers can enjoy the atmosphere of experience in order for customer satisfaction to education experience can be improved by increasing items of education experience such as virtual library (virtual) is full of ebooks, educational games, and more.

5.1.2 *Entertainment experience*
Based on the results of data processing, entertainment experience variable is a variable that has a significant effect on customer satisfaction. For that entertainment experience is not to be lost in Transmart Padang city for contributing in improving customer satisfaction.

Some examples of entertainment experience is watching movies, shopping, culinary tours, playing in Mini Trans Studio, hanging out, and others. Transmart Management Padang City must maintain and improve the entertainment experience given to its customers. Because it is one of the attractions and competitive advantages that must be maintained. Moreover, the tight competition in the retail sector today shopping center should provide something different and unique to attract consumers to come and shop to the shopping center. The thing that should be a concern for management is that activities in Transmart Padang City must be more interesting so that lure consumers to follow these activities, such as live shows, competitions, learning about the use of products and how to use them and other interesting activities.

5.1.3 *Esthetic experience*
Based on the results of data processing, esthetic experience variable is a variable that has a significant effect on customer satisfaction. Therefore, any increase of esthetic experience given will also increase the customer satisfaction of Transmart Pa- dang City. For that esthetic experience must be maintained, maintained, and enhanced in order to provide complete customer satisfaction to Transmart consumers Padang City.

Transmart management should make this esthetic experience one of the focus that should be the focus of attention and improvement. Improvement indicators of this esthetic experience can be done through the improvement of the design of the room, the preparation of goods, the alignment of the product sold, the arrangement of the lay out of the room, the unique items and so forth.

5.1.4 *Escapism experience*
Based on the results of data processing, the variable of experience escapism is a variable that has a significant effect on customer satisfaction. Therefore, any increase of experience escapism given will be directly proportional to the customer satisfaction of Transmart Kota Padang. Escapism experience must always be maintained and enhanced by Transmart management Padang City, because escapism experience is gained through something that adrenaline spur consumers to make them feel something

new and challenging in their life. So this makes them dissolve and forget the daily busyness.

In Padang City Transmart we can see this as shopping for unique items that make consumers busy and forget the time to shop. Other examples are challenging adrenaline games such as rool coster, adventure venues, virtual games, and so on. Things that need to be improved in creating an experience escapism is a vehicle that can make parents and children can play together so they can enjoy togetherness. This could be one of the focus in an effort to increase the experience escapism so that consumers are able to get a higher level of satisfaction and better still.

5.1.5 *Customer satisfaction*
Based on the results of data management, seen in the frequency distribution almost all respondents agreed that they are satisfied to visit Transmart Pa- dang City. However, based on four independent variables studied only three variables that have a significant effect on customer satisfaction and only 46.7% customer satisfaction influenced by the four variables. This means there are still 53.3% more other variables that affect customer satisfaction. Therefore, this should be a concern for management Transmart Padang that to get maximum satisfaction to the consumer then the management must examine other variables that affect customer satisfaction.

5.2 *Limitations of research*

This research is inseparable from the limitations and weaknesses. On the other hand, the limitations and weaknesses found in this study may be an improvement for future research. The limitations found in this study are:

1. Factors that affect customer satisfaction in this study only consist of education experience, entertainment experience, esthetic experience, and escapism experience while there are many other factors that affect customer satisfaction, such as: service quality, price, product quality, emotion, cost, and other factors not investigated in this study.
2. Research is only focused on Transmart consumers Padang City, so the results and implications only for Transmart Padang only.
3. Limitations of this study are respondents who are taken only Transmart consumers Padang city who is domiciled in the city of Padang and had visited the Transmart Padang at least once.

5.3 *Suggestion*

The results of this study and the limitations found to be a source of ideas and inputs for future research development, the researchers suggest to further researchers to be able to examine the variables outside the variable that has been researched or use the concept of experience other than that proposed by Pine and Gilmore. This is to obtain more varied results that can affect customer satisfaction, such as experiential marketing concept introduced by Schmitt consisting of the dimensions sense, feel, think, act, and related.

REFERENCES

Ali, Faizan, Kashif Hussain, dan Neethiahnanthan Ari Ragavan. (2014). Romance of Experience, Satisfaction and Behavioral Intentions: An Empirical Examination of International Delegates in Academic Conferences. *SHS Web of Conferences*, 12, 01009 (2014).

Amsal, Ares Albirru dan Harryadin Mahardika. (2017). Pendekatan experience economy pada pemasaran festival pariwisata: Pengaruh terhadap kepribadian festival yang dirasakan pengunjung dan reputasi festival. *Jurnal Ilmiah Manajemen*, Vol. VII, No. 2, Juni 2017.

Anjani, Erna. (2015). Understanding Customer Experience Management in Retailing. *Jurnal Social and Behavioral Sciences* 211 (2015) 629–633.

Choi, T.Y. dan Chu, R. (2001). Determinants of hotel guest's'satisfaction and repeat patronage in the Hong Kong hotel industry. *Hospitality Management*, Vol. 20, pp. 277297.

Gilmore, J.H. dan B.J. Pine. (2002). Differentiating Hospitality Operations via Experiences: Why Selling Services Is Not Enough. *Cornell Hotel and Restaurant Administration Quarterly.*

Hair, Jr, J. F, Black, W. C, Babin, B. J, & Anderson, R. E. (2010). *Multivariate Data Analysis: A global Perspective* (7th ed.) Upper Saddle River, NJ: Pearson Prentice Hall.

Holbrook, M.B. dan Hirschman, E.C. (1982). The experiential aspects of consumption: Consumer fantasies, feelings and fun. *Journal of Consumer Research*, Vol. 9 (2), (p.132140).

Hosany, Sameer, dan Mark Witham. (2010). Dimensions of Cruisers' Experiences, Satisfaction, and Intention to Recommend. *Jurnal of Travel Research*, 49(3), 351–364.

Kartajaya, Hermawan. (2006). *Hermawan Kartajaya on Service.* Bandung: Penerbit Mizan.

Kotler, Philip dan Kevin L. Keller. (2009). Manajemen pemasaran jilid 1, edisi Ketigabelas. Jakarta: Erlangga.

Mehmetoglu, Mehmet dan Marit Engen. (2011). Pine and Gilmore's concept of experience economy and its dimensions: An Empirical Examination in Tourism. *Journal of Quality Assurance in Hospitality & Tourism*, 12:4, 237–255.

Mykletun, Reidar J. dan Maira Rumba. (2014). Athletes' experiences, enjoyment, satisfaction, and memories from the Extreme Sport Week in Voss, Norway." *Sport,*

Business,and Management: An International Journal, *Vol. 4 Iss 4 pp.* 317–335.

Oliver, R. L. (1999). Whence consumer loyalty? *Journal of Marketing*, Vol. 63 No. 4, pp. 33–44..

Parasuraman, A., Zeithaml, A. V. dan Berry, L. L. (1988). Servqual: A multiple item scale for measuring consumer perceptions of service quality. *Journal of Retailing*, Vol. 64 No.1, pp. 12–40.

Petrick, J.F. (2004), The roles of quality, value, and satisfaction in predicting cruise passengers behavioral intentions. *Journal of Travel Research*, Vol. 42 No. 4, pp. 397–407.

Pine, B.J. II, dan Gilmore, J.H. (1998). Welcome to the Experience Economy. *Harvard Business Review*, 76 (04): 97–105.

Pine, B.J. dan Gilmore, J.H. (1999). The experience economy: Work is theatre and every business a stage. *Harvard Business School Press*, Boston.

Prahalad, C. K., dan V. Ramaswamy (2004). Co-Creation Experiences: The Next Practice in Value Creation. *Journal of Interactive Marketing*, 18 (3): 5–14.

Quadri-Felitti, Donna L. dan Ann Marie Fiore. (2013). Destination loyalty: Effects of wine tourists' experiences, memories, and satisfaction on intentions. *Jurnal and Hospitality Research*, 0(0) 1–16.

Schmitt, B.H. (1999). Experiential Marketing. *Jurnal of Marketing Management*, 15:1–3, 53–67.

Soliha, Euis. (2008). Analisis Industri Ritel di Indonesia. *Jurnal Bisnis dan Ekonomi (JBE)*,Hal., 128–142, ISSN: 1412–3126.

Takur, R. (2016). Understanding customer engagement and loyalty: A case of mobile devices for shopping. *Journal of Retailing and Consumer Services*, Vol. 32, pp. 151–163.

Yoo, J. dan Park, M. (2016). The effects of e-mass customization on consumer perceived value, satisfaction and loyalty toward luxury brands. *Journal of Business Research*. DOI: 10.1016/j.jbusres.2016.04.174.

http://www.kemendagri.go.id/news/2017/09/10/mendagri-lebih-175-juta-wni-telah-merekam-data-kependudukan. Diakses 25 Oktober 2017.

https://finance.detik.com/berita-ekonomi-bisnis/3414745/pendapatan-per-kapita-ri-naik-j adi-rp-4796- jutatahun. Diakses 25 Oktober 2017.

http://www.bpn.go.id/BERITA/Siaran-Pers/realisasi-investasi- triwulan-ii-tahun-2017-mencapai-rp-1709-triliun-68746. Diakses 25 Oktober 2017.

https://ekbis.sindonews.com/read/1202726/34/mendag-kontribusi-ritel-pada-perekonomian-nasional-makin-besar-1493963949. Diakses 25 Oktober 2017 https://finance. detik.com/berita-ekonomi-bisnis/3706734/amburadul-bisnis-ritel-pasar-tradisional-ikutan-sepi. Diakses 9 Januari 2018.

Advances in Business, Management and Entrepreneurship – Hurriyati et al (eds)
© 2020 Taylor & Francis Group, London, ISBN 978-0-367-27176-3

Analysis of brand relationship in Ganesha public speaking

P. Priambudi, F. Wijaya & S. Fakhrudin
Universitas Pendidikan, Bandung Indonesia

ABSTRACT: Bandung, the capital of West Java Province, is one of the big cities in Indonesia. Training needs for the community in this city, especially in the field of public speaking, is increasing. This can be seen from the increasing number of training institutions that open the public speaking class and managed to attract participants to attend the training. The tight competition among training institutions and the emergence of similar products at competitive prices makes every training institution work harder to be more creative and innovative in creating products that are superior to other competing products. Ganesha Public Speaking (GPS) is one of the training institutions that was established since 2009. This institution continues to improve quality and continue to innovate for getting new customers and retain old customers. Consumers will be more willing to loyal to a brand or company if they have a close relationship with the brand or company. When it comes to loyalty, it's important to win the competition and keep customers. This study aims to obtain a description of the Brand Relationship between GPS and its alumni. This study uses a marketing management science approach that focuses on Brand Relationship as the main variable. In this study, the samples used are 33 people who are all alumni of Ganesha Public Speaking training and sampling is done with nonprobability sampling technique. Based on the results of statistical tests, this study shows that the brand relationship in Ganesha Public Speaking is at a very high level.

1 INTRODUCTION

Public speaking training in Bandung is emerging more now. This can be seen from the increasing number of public speaking training classes opened and attracted participants to attend the training. Public speaking training places in Bandung include Ganesha Public Speaking, Tantowi Yahya Public Speaking School, Public Speaking Academy, Trainer Academy and Rudy Sugiono. There are other less-well-known public speaking training places in Bandung.

Ganesha Public Speaking (GPS) is one of the most popular public speaking training courses in Bandung. Starting in 2009, Ganesha Public Speaking has more than 140 batch and many universities, government agencies and companies that invite them to fill in seminars or in-house training around public speaking.

Ganesha Public Speaking has a distinctive character in every training held. GPS stands out as a training institution that has an 80%-time allocation for participant practice, limiting the number of participants with up to 10 people in each class and some experimental learning simulations and games. Strict competition among training institutions and emerging similar products at competitive prices makes every training institution available to work harder to be more creative and innovative in creating products that are superior to other competing products. GPS is trying to offer something different in terms of price, service and facilities obtained by the participants who attended the training. One way to attract the existing market, especially in the field of public speaking, is to provide services that will make the participants satisfied and then follow the existing classes and recommend it to others.

In today's world, it is not just a matter of purchase that is important. More and more marketers are focusing on an increasingly important form of brand equity: the strength of the brand's relationship with their customers. A recent study conducted by Earned Brand (2016) clearly shows that the stronger the relationship between consumers and brands, the better brands will be in the face of Disruption/disruption of existing markets. This is important in markets where there is a lot of pressure (competition between products, start-ups, peer influences, etc.) Constantly threatening to undermine relationships between consumers and brands that have formed, while at the same time, expectations consumers to the brand continues to increase.

Consumers are always influenced by sellers in various ways such as advertisements, brochures, articles or videos. This insistent information will make consumers overload with information and become saturated. Morgan and Hunt (2009) argue that consumers will be more willing to loyal to a company or they if they have a relationship with the brand or company. When it comes to loyalty, it's important to win the competition and keep customers. According to research conducted by some of the following authors turned out brand relationship plays a role in shaping loyalty.

2 RESEARCH METHOD

The research method used in this research is quantitative descriptive method. In this study, the samples used are 33 people, all alumni of GPS training. In this study used primary data. Primary data used in this study is quantitative data collected through survey method with questionnaires that are filled independently by the respondents. Quantitative data was obtained through spreading the questionnaire online through google forms in WhatsApp group of GPS alumni. While sampling was done with non-probability sampling

2.1 *Brand concept*

Kotler and Gary Armstrong (2007: 70) define brands as the names, terms, signs, symbols, designs, or combinations of these, all of which are intended to recognize the product or service of a person or a seller and to distinguish them from competing products.

2.2 *Brand relationship*

Deborah (2009) said that building brand relationships are quite complex. There are so many types of brand relationships that can be identified and each related to the difference of emotions and norms

Glossary (2011) says that brand relationship is a recurring interaction between brands and customers that begin to reflect similar characteristics of relationships among people, such as love, connection, interdependence, intimacy and commitment. Fournier (1998) suggests that brand relationship is an emotional bond resulting from the interaction between the consumer and the brand. According Syafrizal Helmi (2009), brand relationship is how companies can provide stimulus to customers to more actively interact with a brand.

The study of brand relationship stating the existence of a quality relationship between consumers with brands (Breivik and Thorbjornsen, 2008).

Furthermore, Brand Relationship's perspective can add to the understanding of the role of brands in the lives of consumers. According to Fournier's Brand Relationship has six major aspects in Brand Relationship quality: Interdependence, Self-Concept connection, Commitments, Love/Passions, Intimacy, Partner quality.

3 DATA ANALYSIS AND RESULTS

Statistical analysis was conducted based on usable responses from selected data.

3.1 *Descriptive analysis*

Of 33 questionnaires, 36,40% were from male and 63,60% were from female respondents. Most respondents are aged between 20 and 30 years (48.50%). The work of most respondents is Private (45.50%) and income above 2 million Rupiah (63.60%).

From the results of statistical tests on the attributes of Ganesha Public Speaking. The highest value is item number 1 (Whenever I need a Public Speaking, I remember the brand Ganesha Public Speaking) and number 5 (I give positive information about Ganesha Public Speaking to other people). This shows when the respondent has a top of mind of a brand then it is also possible that they will refer to others about it. In addition, from the table above can be concluded that the GPS brand relationship is high Because there is no answer that shows disagreement (Strongly disagree/Disagree) on the statement in the questionnaire.

The recapitulation of the brand relationship variable image can be seen in the overall continuum review through the calculation process. Based on the results of data processing has been presented, Ideally the expected score for the assessment of respondents to the statements 1—10 is 1650. From the calculation obtained total score of 1427 or equal to 86.48% of the ideal score of 1650. The results indicate that the assessment variable Brand Relationship can be said to be very high which means through the measurement results known that the variable brand relationship has been running well in this study.

4 CONCLUSIONS

4.1 *Conclusions*

This study shows that the brand relationship in Ganesha Public Speaking is at a very high level. However, there are still some items that still have enough responses (neutral) so that required follow-up for each attribute statement in the questionnaire so that the level of brand relationship Ganesha Public Speaking can be better again

5 LIMITATION AND FUTURE RESEARCH DIRECTION

This paper has limitations that only examine one variable only (brand relationship). Therefore, in the future in this research needs to include other variables such as Brand Loyalty, Consumer Satisfaction or Brand Image. The objective is that the variables can be tested empirically

REFERENCES

MacInnis, Deborah J., C. Whan Park, and Joseph R. Priester. *Handbook of Brand Relationships*. Armonk, NY: M.E. Sharpe, 2009.

Fournier. 1998. *Consumers and Their Brands: Developing Relationship Theory in Consumer Research*. John Wiley and Sons.

Glosarry. 2011. Literary Terms. New York: McGraw Hill.

Sugiyono. 2010. *Metode Penelitian Kuantitatif Kualitatif dan R&D*. Alfabeta, Bandung.

Trisno Mushanto, 2004. Faktor-Faktor Kepuasan Pelanggan dan Loyalitas.

Pelanggan: Studi Kasus Pada CV. Sarana Media Advertising Surabaya.

Jurusan Ekonomi Manajemen, Fakultas Ekonomi-Universitas Kristen Petra.

Jurnal Manajemen & Kewirausahaan. Vol. 6, No. 2, September 2004: 123 – 136.Nazir, M. 2003.

Wittome, James New Zealand. 2000. Understanding the Customer-brand-relationship within Positive & Negative Service Environments. Ministry of Tourism Industry Report.

Borneo, K. 2011. Pengaruh Implementasi Relationship Marketing Terhadap Loyalitas Konsumen.

Breivik, Einar, dan Helge Thorbjornsen. (2008). Consumer brand relationship: An investigation of two alternatives models. Journal of the Academic Merketing Science, 36, 443–472.

Malhotra. (2010). *Marketing Research: An Applied Orientation* 6th edition. Pearson: New Jersey.

Kotler dan Gary Armstrong. 2007. Dasar-Dasar Pemasaran. Jakarta: Erlangga.

Jung, H., & Soo, M. (2012). The effect of brand experience on brand relationship quality. *Academy of Marketing Studies Journal*, 16(1),87–99.

Papista, E., & Dimitriadis, S. (2012). Exploring consumer-brand relationship quality and identification. *Qualitative Market Research: An International Journal*, 15(1),33–56. https://doi.org/10.1108/13522751211191982

Smit, E., Bronner, F., & Tolboom, M. (2007). Brand relationship quality and its value for personal contact. *Journal of Business Research*, 60 (6), 627–633. https://doi.org/10.1016/j.jbusres.2006.06.012

Chang and Chieng. (2013). Building consumer - brand relationship. *Psychology & Marketing*, 30(6),461–469. https://doi.org/10.1002/mar

Lee, H. J., & Kang, M. S. (2012). The effect of brand experience on brand relationship quality. *Academy of Marketing Studies Journal*, 16(1),87–98.

Lee, H. J., & Kang, M. S. (2013). The effect of brand personality on brand relationship, attitude and purchase intention with a focus on brand community. *Academy of Marketing Studies Journal*, 17(2),85–98. https://doi.org/ Sirdeshmukh, Singh and Sabol (2002).

Section 2: Financial management and accounting

Advances in Business, Management and Entrepreneurship – Hurriyati et al (eds)
© *2020 Taylor & Francis Group, London, ISBN 978-0-367-27176-3*

The effect of life cycle stages on leverage in Indonesian listed companies

S.R. Nidar & I. Sugianti
Universitas Padjadjaran, Bandung, Indonesia

ABSTRACT: Capital structure policy is one of the important decisions for Indonesian listed companies. Two main approaches used in the capital structure policy are the pecking order theory and the trade-off theory. The life cycle stage experienced by a company requires it to adjust its capital structure policy. The purpose of this study is to analyze the effect of the life cycle of the company and other factors, such as profitability, growth opportunities, liquidity, non-debt tax shield, as well as the tangibility on the company's capital structure level. This study uses secondary data and panel data regression analysis as well as a fixed effect estimation model. The results of this study indicated that the life cycle stage has a negative effect on leverage. Profitability has a positive effect on leverage, growth opportunities have a positive effect on leverage, liquidity has a negative effect on leverage, non-debt tax shield has a negative effect on leverage and tangibility has a positive effect on leverage.

1 INTRODUCTION

The main objective to be achieved by a company is welfare maximization for its shareholders, which is reflected in the value or price of a company's stock. All activity within a company is meant to maximize the value of a company so that it will be able to improve the welfare of shareholders. Companies manages all of the company's operational activities well. The managerial side will be faced with decisions about how the company can increase its funding needed to fund the company's operational activities.

The increasing operational activities of acompany raises an important policy that will be taken by the managerial. One of the policies that should be considered carefully is financing. This financing policy will be recorded on the right side on the balance sheet of the company. The source of funds will be the company's investment to run the company's operational activities. This is called the investing activity and will be recorded on the left side of the balance sheet.

Capital structure policy is one of the most important decisions for the managerial of a company. The capital structure policy is concerned with the sustainability of a company's operational activities. Choosing the right composition of capital structure will maximize the value of a company. The main approaches used in capital structure policies are pecking order theory and trade-off theory; in other words, the basis for determining the capital structure for a company.

1.1 Literature review

The arising problems are many factors that can influence the policy in the capital structure decision, both external environment, and corporate characteristics.

Frank and Goyal (2009) argued that the theory of capital structure is adjusted to the circumstances of the firm. A factor that significantly affects the capital structure of a company may not be significant for other companies in different circumstances.

A company experiencing growth indicates that there is an increase in the company's operational activities. The company needs funds to finance the increasing operational activities. There is a relationship between growth and funding: funding will be used to support the company's growth.

This research was conducted on consumer goods industry, property sector and real estate, and trade, services and investment sector listed in Indonesia Stock Exchange (IDX). The selection of the sector is based on industry consideration. The selected industries had relatively increasing development from year to year for 5 years. Other sectors which were not selected fluctuated.

The problem formulation of this research is as follows: Do the life cycle stage, profitability, growth opportunity, liquidity, non-debt tax shield, and tangible assets affect company leverage?

Regarding the influence of the life cycle of a firm and leverage, Castro (2015) argued that the factors affecting capital structure such as profitability, liquidity, age, size, non-debt tax shield, asset tangible, intangible assets, and growth are affected by the life cycle of the company, leading to the establishment of the company's debt capacity.

(Adizes 1998, Mueller 1972, Myers 1977 in Omrani et al. 2011) stated that in the early stages of the company lifecycle, the company's assets are still low, operating cash flow and profitability are low, the company needs high liquidity to finance the company in achieving growth opportunity. Dickinson (2011) stated that cash flow financing at the time of introduction and growth is positive which indicates the flow of funds from debt. At the introduction and growth stage, the company usually needs more resources to buy assets used as an investment. Thus, it can increase the company's operations and increase sales, so that corporate liquidity increases, too. It can be concluded that at the early stage of the life cycle, that is, at the introduction and growth stage, there is a positive influence on leverage.

Dickinson (2011) stated that. at the decline stage, the company's cash flow operating is negative. This is due to the declining operational performance of the company. Cash flow investing is positive because there is cash receipt from the sale of assets for investment which improves operational performance. While the company's cash flow financing can be negative. It can be concluded that in the decline stage, a company will prefer to use equity because the ability of corporate liquidity decreases, evidenced by the company's operating activities which also decrease. It is not possible for the company to owe, so there is a negative influence of the life cycle at decline stage on leverage.

It can be concluded that there is an influence of the company life cycle on leverage. Corporate funding is largely determined by the development of the lifecycle stages passed by the company so that there are different capital structure patterns in the different stages of life cycle.

Regarding the effect of profitability on leverage, Brigham and Houston (2006) stated that firms with very high investment returns will use a relatively low debt. Frydenberg (2004) stated that high-profit companies use more internal funding sources. High profit generates high retained earnings that can be used as a source of funds for the company in the next year so as to reduce the composition of debt on planned investments. It can be concluded that there is a negative effect of profitability on leverage level. When profitability is high, leverage is low. Conversely, if profitability is low, leverage is high.

Regarding the influence of growth opportunity on leverage, Brigham and Houston (2006) stated that the fast-growing companies rely more on external capital, other things are considered ceteris paribus. However, the cost of emission associated with the sale of shares usually exceeds the cost when the company decides to owe. This encourages the company experiencing rapid growth to rely more on debt. Myers (1977) stated that company experiencing growth can be seen as a company with a high capital asset that will produce value. Growth opportunity is a key factor affecting corporate finance (Fama & French 2005). The company experiencing growth will tend to be aggressive in making a loan to fund the investments. It can be concluded that there is a positive influence of growth opportunity on the level of leverage.

There is an influence of liquidity on leverage. The company with high liquidity will not tend to use financing from debt. This is due to the availability of large internal funds of the company. If the company has a high liquidity, it will optimize the internal

funds to finance its investment. The statement is supported by Akdal (2011). These researchers stated that when liquidity is high, the leverage will decrease. It can be concluded that there is a negative influence of liquidity on leverage level. If the liquidity is high, the leverage level is low. Conversely, if the liquidity is low, the leverage is high.

Regarding the influence of non-debt tax shield on leverage, DeAngelo and Maulis (1980) showed an optimal model of the capital structure, which is the effect of corporate taxes, personal taxes, and non-debt related on the firm's tax shield. The study states that tax deductions on depreciation and investment tax credits generate tax benefits in debt financing. This indicates that for the companies with high non-debt tax shields, the use of the source of funds derived from debt will be lower because non-debt tax shield will eliminate the profit from the use of debt or interest tax shield. The non-debt tax shield becomes a proxy of the profit earned apart from debt. It can be concluded that non-debt tax shield negatively affects leverage.

Regarding the influence of tangible assets on leverage, Rajan and Zingales (1995) showed that there is a positive influence of collateral on debt. Tangible assets can be used as collateral to obtain loans and are generally more valuable when compared to intangible assets at the time of liquidation. This gives a positive signal to the creditor because it gives a second way out for debt payments to the creditor. Assets are part of a company's investment useful to support the company's operational activities. Companies with large assets are generally large in scale. It can be concluded that tangible assets have a positive effect on leverage.

Based on the description of the influence of the factors explained previously, the hypotheses can be formulated as follows:

H1: Life cycle stage affects leverage.

H1a: introduction life cycle has a positive effect on leverage.

H1b: growth life cycle has a positive effect on leverage.

H1c: maturity life cycle has a negative effect on leverage.

H1d: shake-out life cycle has a negative effect on leverage.

H1e: decline life cycle has a negative effect on leverage.

H2: Profitability has a negative effect on leverage based on introduction, growth, mature, shake-out and decline cycles.

H3: Growth has a positive effect on leverage based on introduction, growth, mature, shake-out, and decline cycles.

H4: Liquidity has a negative effect on leverage based on introduction, growth, mature, shake-out, and decline cycles.

H5: Non-debt tax shield has a negative effect on leverage based on introduction, growth, mature, shake-out, and decline cycles.

H6: Tangible assets have a positive effect on leverage based on introduction, growth, mature, shake-out and decline cycles.

2 METHOD

The method used in this research is a verification quantitative approach, which tests the influence of capital structure variables in the life cycles based on debt value, Ordinary least square is used for verify the regression and correlation. In addition, the test in this study also uses dummy variables to determine whether the grouping of companies based on their life cycle of introduction, growth, mature, shake-out, and decline will affect the decision of capital structure policy within the company. It is necessary to determine the limits and indicators for each variable used. This research has three variables, as follows:

1. Independent variable (X) is the variable that will affect the dependent variable or variable that explains the effect on the dependent variable. The independent variables in this research are life cycle stage (LCS), profitability (PRO), growth opportunities (GRO), liquidity (LIQ), non-debt tax shield (NDT), and tangible assets (TAS).
2. The dependent variable (Y) is the variable influenced by the independent variables. The dependent variable in this research is the capital structure of the company represented by the leverage, projected by debt to equity ratio (DER).
3. A dummy variable is a qualitative variable that indicates the presence or absence of an attribute. Ghozali and Ratmono (2013) stated that dummy or qualitative variable indicates the presence or absence of an attribute by forming an artificial variable with a value of 1 or 0.

Number 1 shows a company that is at a stage of the life cycle, while the number 0 indicates the company that is not in a stage of life cycle. There is also ordinal scale for classifications based on the life cycle. Number 1 is for companies classified at the introduction stage, number 2 is for companies classified at the growth stage, number 3 is for companies classified at maturity stage, number 4 is for companies classified at the shake-out stage, and number 5 is for companies classified at decline stage.

The statistical analysis used to test the hypothesis in this study is multiple regression analysis, both without the classification of the life cycle and with the classification of the life cycle.

1. Multiple Regression Model without Lifecycle Classification.

$$Yi, t = \alpha + \beta 1 \ i1i, t + \beta 2 \ i2i, t + \beta 3 \ i3i, t$$
$$+ \beta 4 \ i4i, t + \beta 5 \ i5i, t + \beta 6 \ i6i, t + \varepsilon i.$$

2. Multiple Regression Model with Lifecycle Classification. The model is tested five times into different life cycle stages so that there will be five models representing each company's life cycle. Thus, the regression model is as follows:

- At the time of Introduction D1.

$$Yi, t = \alpha + \beta 1 \ D1i, t + \beta 2 \ X2i, t + \beta 3 \ X3i, t$$
$$+ \beta 4 \ X4i, t + \beta 5 \ X5i, t + \beta 6 \ X6i, t + \varepsilon i, t.$$

- At the time of Growth D2.

$$Yi, t = \alpha + \beta 1 \ D2i, t + \beta 2 \ X2i, t + \beta 3 \ X3i, t$$
$$+ \beta 4 \ X4i, t + \beta 5 \ X5i, t + \beta 6 \ X6i, t + \varepsilon i, t.$$

- At the time of Mature D3.

$$Yi, t = \alpha + \beta 1 \ D3i, t + \beta 2 \ X2i, t + \beta 3 \ X3i, t$$
$$+ \beta 4 \ X4i, t + \beta 5 \ X5i, t + \beta 6 \ X6i, t + \varepsilon i, t.$$

- At the time of Shake-Out D4.

$$Yi, t = \alpha + \beta 1 \ D4i, t + \beta 2 \ X2i, t + \beta 3 \ X3i, t$$
$$+ \beta 4 \ X4i, t + \beta 5 \ X5i, t + \beta 6 \ X6i, t + \varepsilon i, t.$$

- At the time of Decline D5.

$$Yi, t = \alpha + \beta 1 \ D5i, t + \beta 2 \ X2i, t + \beta 3 \ X3i, t$$
$$+ \beta 4 \ X4i, t + \beta 5 \ X5i, t + \beta 6 \ X6i, t + \varepsilon i, t.$$

There are 96 companies as unit of analysis for this research that are listed in 2009 – 2014.

3 DISCUSSION

The profitability variable has a positive influence on leverage level and is different from the hypothesis. Variable growth opportunities have a positive effect on leverage and are consistent with the hypothesis. The liquidity variable has a negative influence and consistent with the hypothesis. The non-debt tax shield variable has a negative effect on leverage and is consistent with the hypothesis. The tangibility variable has a positive influence on leverage and is consistent with the hypothesis. Only non-debt tax shield variable has a significant effect on leverage. Based on regression result without dummy variables, the following equations are obtained:

A. Regression model without Life Cycle Stage.
   DER = 1.672720 + 0.008055PRO + 0.009522GRO - 0.002297LIQ - 1.715046NDT + 0.459522TAS.
B. Regression Model with Life Cycle Stage (Ordinal Scale).
   DER = 2.001687 - 1.21872LCS + 0.012681PRO + 0.009862GRO - 0.002102LIQ - 1.641666NDT + 0.411794TAS.

The life cycle variable at the shake-out stage is inconsistent with the hypothesis, while the hypothesis test results at introduction, growth, maturity, and decline cycle stages are consistent with the hypothesis. Only variable non-debt tax shield which has a significant effect on leverage. The coefficient result of profitability variable is not consistent with the hypothesis, while the coefficient result of growth opportunity, liquidity, non-debt tax shield, and tangibility are consistent with the hypothesis. The effect of independent variables on leverage, namely:

Regarding the effect of life cycle stage on leverage, the statistic test result without classifier of dummy variable on life cycle stage shows that life cycle stage has a negative effect, but no significant effect on leverage. This result is shown by p-value is insignificant, thus H0 cannot be rejected. That is, the life cycle has no effect on leverage.

The classification using dummy variable in the life cycle stage shows that at introduction and growth stage, the life cycle has a positive effect on leverage, but not significant. That is, at the time of introduction and growth, the leverage tends to increase.

Meanwhile, during maturity and decline, the life cycle has a negative effect on leverage but is not significant. The results showed that companies at the growth stage have a higher leverage than companies at the decline stage. In addition, the results of this study indicate that in the cycle of maturity and decline, the company's funding from debt decreases. The results of this study are in accordance with Teixeira and Santos (2014) who found that the debt ratio continues to decline during the lifetime of the life cycle. These results are also in line with Bulan and Yan (2010) who found that pecking order theory can explain the funding pattern in the company at the mature stage more than at growth stage. Meanwhile, at the shake-out stage, the results are not consistent with the hypothesis, stating that there is a positive influence but not significant.

Regarding the effect of profitability on leverage, the statistical test results showed that profitability has a positive influence on leverage, but not significant. The results of this study contrasted with Castro

(2015), who found that profitability has a negative effect on leverage.

The results of this study do not support pecking order theory but are in line with trade-off theory. Profitability has a positive effect on leverage, meaning that if profitability is high, the leverage will also increase. This shows that when the profitability of the company is high, the company will be more indebted to get the tax benefits. At the time of indebted, the trade-off theory can better explain the effect of profitability on leverage.

Regarding the influence of growth opportunities on leverage, the statistical test results showed that growth opportunities positively affect leverage, but not significant. This explains that the higher the growth rate of the company is, the higher the company's funding from debt becomes. The result of this research is in line with Castro (2015), who proved that growth opportunities positively affect leverage. Companies with high growth opportunities are high-prospecting companies that are improving their operational performance and thus require more funds. The results of this study support the pecking order theory, which states that growth opportunities have a positive effect on leverage.

Regarding the effect of liquidity on leverage, the statistical test results showed that liquidity has a negative effect on leverage, but not significant. The results of this study are in line with several previous studies, such as Castro (2015), stating that firms with high liquidity will tend to have low leverage.

The results of this study are in accordance with the pecking order theory which states that any increase in liquidity will lead to a decrease in leverage. This is because the high liquidity companies have a larger internal funding so that the funding is prioritized from the internal funds. Regarding the influence of non-debt tax shield on leverage, the statistical tests showed that non-debt tax shield has a negative and significant effect on leverage.

Non-debt tax shield has a negative effect on leverage, meaning that the bigger the non-debt tax shield is, the smaller the leverage becomes. This result is in line with the theory presented by DeAngelo and Maulis (1986) stating that for companies with high non-debt tax shields, the use funds from debt will be even smaller because non-debt tax shield will eliminate the benefits of using debt or interest tax shield. The results of this study are also in line with Castro (2015), who found that non-debt tax shield negatively affects leverage.

Non-debt tax shield is a proxy of the profit earned apart from debt. The results of this study are in accordance with the pecking-order theory which states that the higher non-debt tax shield is, the higher the free cash flow of the company becomes. Thus, the funding of the company from the free cash flow is higher. The results of this study are in line with pecking order theory which states that the company optimizes more funding from the internal capital.

Regarding the influence of tangibility on leverage, the statistical test results showed that liquidity has a positive effect on leverage, but not significant. The results of this study are in line with Rajan and Zingales (1995), who showed that there is a positive influence of tangible assets on debt. The results of this study are also in line with Castro (2015), who found that tangibility positively affects leverage.

Tangible assets can be used as collateral to obtain loans and are generally more valuable when compared to intangible assets at the time of liquidation. This gives a positive signal to the creditor because it provides a second way out for debt payment to the creditor. Assets are part of a company's investment useful for supporting the company's operational activities. Companies with high tangibility will increase funding from debt. These results further support the trade-off theory. It can be concluded that tangibility has a positive effect on leverage.

The average value of corporate leverage in the introduction cycle of 2.370622, which means that the company at this stage, the average use of corporate debt as much as 237.0622% of the overall equity it has. The average Debt to Equity Ratio (DER) at the time of introduction is greater than the average DER of the company in the growth, maturity, shake-out and decline cycles.

The highest average profitability value occurs when the shake-out cycle is 0.387700 which means that in this cycle the company is able to generate profits of 38.77% compared with the total value of assets. The average value of the highest Growth Opportunities of 5.252977 that occurs in the maturity cycle means that companies with high average prospects occur during the maturity cycle.

4 CONCLUSION

It can be concluded that the life cycle stage has a negative effect on leverage. The division of the company's life cycle has not significant effect on the company's leverage decisions. Profitability has a positive effect on leverage but not significant effect on the decision of leverage company. Growth opportunities have a positive effect on leverage, but not significant effect on the decision of leverage company. Liquidity has a negative effect on leverage, but not significant effect on the decision of leverage company. Non-debt tax shield has

a negative effect on leverage and significant effect on corporate leverage decisions. This states that the higher the value of non-debt tax shield, the lower the level of corporate leverage. Tangibility has a positive effect on leverage, but does not have a significant effect on the leverage decision of the company.

REFERENCES

Adizes, I. 1998. *Managing Corporate Life Cycle*. Prentice-Hall.

Akdal, S. 2011. How do firm characteristics affect capital structure? Some UK evidence. *Some UK Evidence*.

Brigham, E.F. & Houston, J.F. 2006. *Fundamental of Financial Management: Twelfth Edition*. Southwestern USA: Thomson.

Bulan, L.T. & Yan, Z. 2010. Firm maturity and the pecking order theory. *International Journal of Business and Economics* 9(3): 179–200.

Castro, C. 2015. Dynamic analysis of capital structure in technological firms based on their life cycle stages. *Forthcoming Spanish Journal of Finance and Accounting*.

DeAngelo, H. & Masulis, R.W. 1980. Optimal capital structure under corporate and personal taxation. *Journal of Financial Economics* 8(1): 3–27.

Dickinson, V. 2011. Cash flow patterns as a proxy for firm life cycle. *The Accounting Review* 86(6): 1969–1994.

Fama, E.F. & French, K.R. 2005. Financing decisions: Who issues stock. *Journal of Finance Economics* 76(3): 549–582.

Frank, M.Z. & Goyal, V.K. 2009. Capital structure decisions: Which factors are reliably important. *Financial Management* 38(1): 1–37.

Frydenberg, S. 2004. Theory of capital structure a review. *Journal of Economic Literature*.

Ghozali, I. & Ratmono, D. 2013. *Analisis multivariat dan ekonometrika: Teori, konsep, dan aplikasi dengan EVIEWS 8*. Semarang: Badan Penerbit Universitas Diponegoro.

Mueller, D.C. 1972. A life cycle theory of the firm. *Journal of Industrial Economics* 20(3): 199–219.

Myers, S.C. 1977. Determinants of corporate borrowing. *Journal of Financial Economics* 5(2): 147–175.

Omrani, O. & Hamed, H. 2011. Corporate life cycle and the explanatory power of risk measures versus performance measures. *Journal of Education and Vocational Research* 2: 199–206.

Rajan, R.G. & Zingales, L. 1995. What do We Know about Capital Structure? Some Evidence from International Data. *The Journal of Finance* 50(5): 1421–1460.

Teixeira, G. & Coutinho, D.S.M.J. 2014. Do firms have financing preferences along their life cycle? Evidence from Iberia. *Working Paper*.

Spillover of panel causality Granger for monetary variables in some countries' emerging markets

D. Supriyadi & E. Mahpudin
Universitas Pendidikan Indonesia, Bandung, Indonesia

ABSTRACT: This paper aims to propose the modeling of the panel data analytic framework, in which the analytical framework can be used as a simple analysis tool in panel data modeling. With the advent of modeling proposals using this panel data, it will be able to contribute to the debate issues, especially those related to monetary economic analysis in developing countries. This paper uses data analysis over the period 2012 to 2017, consisting of monthly data from six developing countries. The integrated proposal model contained in this paper shows that inter-variables together are significantly related., Only Inflation variables do not show significant numbers; the rest of the significant variables do. With the advanced test of causality, it became visible that the variables in the emerging market country have no causality relationship. This shows that the monetary economic situation in developing countries is not influenced by the monetary variables of developing countries.

1 INTRODUCTION

Investors are concerned about the existence of information related to the decline in the value of financial assets originating from developing countries over the last few years. These conditions are considered to have occurred during the monetary crises of 1997 and 2008, making economic conditions in some developing countries in Asia and America slump. The occurrence of the crisis caused trauma for almost several decades, and many encourage owners of capital to always be vigilant and want to feel the security of assets invested in the State logged. A new crisis is not upcoming and the fear of market participants is still far from reality. The year 2018 is not the year 2008 or 1997; during which many developing countries did not have a strategy and analysis model in the field to face the crisis. Some developing countries now have schemes and analyses to deal with a more complex financial crisis, and for now developing countries also already have foreign exchange reserves to protect the country's currency exchange rates (Agénor et al. 1999, Angelini et al. 2011, Aguiar et al. 2007).

Exchanges, the weakening prices in emerging markets are heavily influenced by the selling of investors, who are worried about the prospects for investments. The emerging markets index, MSCI EM, has been corrected by as much as 7.1% so far this year. Market participants assess that the conditions that are happening in developing countries today will not get better but gradually worsen. The main consideration is none other than the discourse withdrawal of monetary stimulus by the United States Bank within two years. Since the financial crisis hit the United States approximately 6 years

ago, developing country assets have been favored by investors. The Federal Reserve's monetary easing, through stimulus and cuts in interest rates to extreme levels, means that financial products of the superpower do not give a good return. The buying action of developing country assets continued at least until one year ago so that the exchange rate of some developing country currencies consistently strengthened (Aoki et al. 2016, Bank Indonesia 2004).

But now in the middle of the initial phase of the US stimulus withdrawal, the outlook for emerging markets is no longer promising. At least this assumption is used by the majority of investors. Some countries are beginning to promote ways to protect their domestic economy and exchange rates. Bank Indonesia raised interest rates to a level of 7.50% in just a few meetings. The Reserve Bank of India yesterday also raised interest rates from 7.75% to 8.00%, as its currency rate continued to decline. The RBI said in a statement that an interest rate hike would help achieve consumer inflation targets although nobody could argue that, later, the decline in capital market performance and expectations of tightening in credit markets would instead be vulnerable to undermining the economic growth while weakening the Rupee exchange rate (Pohan 2008).

In Turkey, the central bank even almost doubled its benchmark interest rate from 7.75% to 12%. Central banks are not shy anymore in taking a policy to strengthen their monetary fort. Every developing country has the same problems in terms of global economic distortions, namely the tapering of the Federal Reserve's stimulus and an indication of a medium-term economic slowdown. But on the other side, each government also has its own problems, which makes investors begin to fear: Indonesia with its current account deficit, India with a decline in its growth rate and Turkey with its corruption issue. But behind all the facts, it is almost certain that the 1997 monetary crisis will not be repeated. Shocks that can be present at any time are only sectoral rather than fundamental to the economy (Bekaerr 2003, Borensztein 2001, Broda 2004)

1.1 Literature review

In the following, the authors present some of the theories contained in monetary policy in general that are necessary and widely used in developing countries. First, it is important in a monetary policy to understand the basic concepts and understanding of Monetary Policy, then to understand what is the concept of lag effect of the monetary policy on economic development, next, to be able to explain the strategic framework of monetary policy, and to be able to explain the monetary policy transmission mechanism, to describe the operational framework of monetary policy and finally to be able to understand the Inflation Targeting Framework concept.

Monetary Policy describes the basic concepts and notions of monetary policy, the lag effect of monetary policy on economic development, the strategic framework of monetary policy, some monetary policy transmission mechanisms, among others, through the interest rate path, asset price path, credit line, and expectation path; monetary policy work is outlined through monetary policy framework, and inflation targeting concept or Inflation Targeting Framework as a new concept in monetary policy (Dooley 2000, Frankel 2003).

1 The purpose of the concept and understanding of monetary policy.
   Monetary Policy is the policy of the monetary authority or central bank in the form of control of the monetary and/or interest rate to achieve the desired economic activity development. Basically, the purpose of monetary policy is the achievement of the balance of internal (internal balance) and external balance (external balance). Internal balance is usually manifested by the creation of high employment opportunities and low inflation rate. On the other hand, the external balance is intended to balance the international payment balance.
2 What is meant by the grace period Effect of Monetary Policy.
   The monetary policy for the purpose of economic stability depends on the strength of the relationship between changes in monetary policy and economic activity and the time period between the changes in monetary policy and its effect on economic activity. This period is often called the lag.
3 The purpose of the Monetary Policy Strategic Framework.
   The strategic framework of monetary policy is basically related to the determination of the

ultimate goal of monetary policy and strategy to achieve it. The problem is that the final goal to be achieved from a monetary policy is very high and not necessarily all can be achieved simultaneously and it can even contradict each other. For example, efforts to boost economic growth and expand employment opportunities can generally lead to an increase in prices so that the achievement of macroeconomic stability is not optimal. Recognizing this, some countries have gradually shifted to applying monetary policies that focus more on single goals (Krugman & Obstfeld 1997, Mihaljek 2011).
4 In the meantime, the mechanism of Monetary Policy Transmission.
   The strategic framework of monetary policy adopted by the central bank is heavily influenced by the central bank's confidence in a particular process of how monetary policy affects the economy. This process is known as the monetary policy transmission mechanism.
5 The purpose of the Monetary Policy Operational FrameworkTo find out how a monetary policy is implemented, it is necessary to understand the operational framework of monetary policy which generally includes the instruments, operational targets, and intermediate targets used to achieve the prescribed final objectives.
6 In the intention of inflation Targeting Framework
   The Inflation Targeting Framework is a monetary policy framework that has key features, namely an official statement from the central bank and a reinforcement by law that the ultimate goal of monetary policy is to achieve and maintain a low inflation rate, and the announcement of inflation targets to the public (Mimir & Sunel 2018).

2 METHOD

In this study we use the model of panel data testing with random effect model and Panel Causality like Dumitrescu and Hurlin (2012) and Pedroni (1999, 2004). In general, bivariate regression in the context of panel data takes the following form:

$$Y_{i,t} = \alpha_{0,i} + \alpha_{1,i} y_{i,t-1} + ... + \alpha_{k,i} y_{i,-k} \\ + \beta_{1i} X_{i,} t - 1 + \beta_{ki} X_{i,-k} + \varepsilon_{i,t.} \quad (1)$$

$$X_{i,t} = \alpha_{0,i} + \alpha_{1,i} X_{i,t-1} + ... + \alpha_{k,i} X_{i,t-k} \\ + \beta_{1i} X_{i,} t - 1 + ... + \beta_{ki} X_{i,-k} + \varepsilon_{i,t.} \quad (2)$$

The t value is the time period dimension of the panel, and it shows the cross-sectional dimension. The difference in form of causality tests lies in the assumptions made about the homogeneity of cross-sectional coefficients. By treating panel data as a large set of stacked data and then performing a Granger causality test in a standard way, with the exception of not allowing data from one cross-section to include data left behind from the next cross. -part. This method assumes that all coefficients are equal across all cross sections. The second approach adopted by Dumitrescu and Hurlin (2012) makes extreme opposite assumptions, allowing all coefficients to be different cross-sectional (Phillips & Hyungsik 1999, Wooldridge 2002).

This test is calculated by simply running a standard Granger causality regression for each cross-section individually.

3 RESULTS AND DISCUSSION

Monetary policy is an effort in controlling the macroeconomic conditions in order to run as desired by regulating the amount of money circulating in the economy. The effort is done to prevent price stability and inflation, as well as increase the output of the equilibrium. The regulation of the amount of money circulating to the community is regulated by increasing or decreasing the amount of money in circulation. The results of this study for data related to monetary analysis in developing countries, using random effects and panel causality is presented in Tables 1 and 2.

In Table 1, we presented the results of panel data analysis with random effect model. What makes the difference of fixed panel effect panel data analysis model and random effect is one of them, display random cross-section and idiosyncratic random value. Probabilistic value of panel data generated by random effect test method has a significant impact using F test, but by using the t test there are only three significant monetary economic variables, while one variable is not significant, whicht is the variable of inflation. The value of R squared produced by using the regression random effect panel data test is quite high at 79%. This shows the proposed regression model with panel data modeling can be received. After the panel data test was done, the authors continued this analysis test by using granger causality panel method with stacked test model and Granger causality panel with Dumitresco- Hurlin test model. The results are presented in Tables 2 and 3.

In this study, the results are shown for paired Stacked test and Dumitresco-Hurlin test using data from "monetary variable." In this study it rejects that monetary variables do not homogenously cause monetary variables, but do not lead in the opposite direction.

This research hoped to show a causality relationship between the variables, but after doing the test by using the causality test Stacked paired test and Dumitresco-Hurlin test, only 1 variable showed a two-way relationship or mutual reciprocity with

Table 1. Result for panel data random effect.

Dependent Variable: ER

Method: Panel EGLS (Cross-section random effects)

Variable	Coefficient	Std. Error	t-Statistic	Prob.
C	-799.8659	1457.696	-0.548719	0.5835
IHSG	0.017208	0.005113	3.365758	0.0008
INF	12.28492	12.94111	0.949294	0.3430
IR	107.6687	14.90123	7.225494	0.0000
JUB	1.910006	4.78E-08	39.96103	0.0000
Effects Specification				
Rho				
Cross-section random			3551.302	0.9925
Idiosyncratic random			308.1372	0.0075
Weighted Statistics				
R-squared	0.791214	Mean dependent var		22.53587
Adjusted R-squared	0.789259	S.D. dependent var		668.8947
S.E. of regression	307.0667	Sum squared resid		40261812
F-statistic	404.5401	Durbin-Watson stat		0.173296
Prob(F-statistic)				
0.000000				
Unweighted Statistics				
R-squared	0.779463	Mean dependent var		2203.975
Sum squared resid	1.890009	Durbin-Watson stat		0.021980

Table 2. Result for panel causality Granger test type stacked test.

Panel Granger Causality Tests

Lags: 1

Null Hypothesis:	Obs	F-Statistic	Prob.
IHSG does not Granger Cause ER	426	0.00152	0.9689
ER does not Granger Cause IHSG		0.04085	0.8399
INF does not Granger Cause ER	426	1.67485	0.1963
ER does not Granger Cause INF		0.18517	0.6672
IR does not Granger Cause ER	426	0.16971	0.6806
ER does not Granger Cause IR		0.00133	0.9710
JUB does not Granger Cause ER	426	0.16574	0.6841
ER does not Granger Cause JUB		7.66272	0.0059
INF does not Granger Cause IHSG	426	0.09054	0.7636
IHSG does not Granger Cause INF		0.30285	0.5824
IR does not Granger Cause IHSG	426	1.84784	0.1748
IHSG does not Granger Cause IR		4.55324	0.0334
JUB does not Granger Cause IHSG	426	0.10536	0.7456
IHSG does not Granger Cause JUB		0.19190	0.6616
IR does not Granger Cause INF	426	2.20511	0.1383
INF does not Granger Cause IR		27.6591	2.E-07
JUB does not Granger Cause INF	426	0.03090	0.8605
INF does not Granger Cause JUB		1.24180	0.2658
JUB does not Granger Cause IR	426	0.06116	0.8048
IR does not Granger Cause JUB		1.41914	0.2342

Table 3. Result for panel causality Granger test type Dumitresco- Hurlin test.

Dumitrescu Hurlin Panel Causality Tests

Lags: 1

Null Hypothesis:	W-Stat.	Zbar-Stat.	Prob.
IHSG does not homogeneously cause ER	1.39832	0.60466	0.5454
ER does not homogeneously cause IHSG	0.96724	-0.10361	0.9175
INF does not homogeneously cause ER	1.66749	1.04692	0.2951
ER does not homogeneously cause INF	3.90414	4.72183	2.0006
IR does not homogeneously cause ER	2.84424	2.98037	0.0029
ER does not homogeneously cause IR	5.20925	6.86618	7.0012
JUB does not homogeneously cause ER	0.92499	-0.17303	0.8626
ER does not homogeneously cause JUB	0.64575	-0.63184	0.5275
INF does not homogeneously cause IHSG	1.81596	1.29086	0.1968
IHSG does not homogeneously cause INF	4.11311	5.06517	4.0007
IR does not homogeneously cause IHSG	1.34129	0.51097	0.6094
IHSG does not homogeneously cause IR	10.6463	15.7994	0.0000
JUB does not homogeneously cause IHSG	3.06892	3.34953	0.0008
IHSG does not homogeneously cause JUB	1.11777	0.14371	0.8857
IR does not homogeneously cause INF	2.50749	2.42707	0.0152
INF does not homogeneously cause IR	11.3930	17.0263	0.0000
JUB does not homogeneously cause INF	5.20421	6.85789	7.0012
INF does not homogeneously cause JUB	0.49472	-0.87998	0.3789
JUB does not homogeneously cause IR	5.39579	7.17267	7.0013
IR does not homogeneously cause JUB	1.05828	0.04597	0.9633

Dumitresco-Hurlin test, which is inflation and the interest rate in the country in which the research is conducted.

4 CONCLUSION

From the research that has been done, it can be concluded that it was done with the purpose of wanting an analysis of the influence of entity differences (individual) and/or the influence of different periods of observation, so in this study we used panel data regression analysis. But in addition there were other goals, such as the influence of different entities in this study using panel data regression and panel data causality. Thus, this research can add to the literature in determining the direction of causality between the monetary variable in developing countries, so that a complex analysis model can be obtained from the result of this study. In addition, the authors used Granger Causality Panel test to identify the direction of the relationship between the variables in the perusal. However, the results of the data show no one-way or two-way causality relationship between the monetary variables in developing countries.

REFERENCES

Agénor, A., Richard, P. & Montiel, P. 1999. *Development Macroeconomics, 2ndedition*. Princeton, NJ: Princeton University Press.

Angelini, P., Neri, S. & Panetta, F. 2011. Monetary and macroprudential policies. *Bank of Italy Economic Working Paper*: 801.

Aguiar, A., Mark, M. & Gopinath, G. 2007. Emerging market business cycles: The cycle is the trend. *Journal of Political Economy* 115(1).

Aoki, K.G., Benigno, B. & Kiyotaki, N. 2016. Monetary and Financial Policies in Emerging Markets. *Mimeo*.

Bank Indonesia. 2004. *Bank Indonesia Bank Sentral Republik Indonesia: Sebuah Pengantar, Pusat Pendidikan dan Studi Kebanksentralan*. Jakarta: BI.

Bekaert, B. 2003. Emerging markets finance. *Journal of Empirical Finance* 10(1–2): 3–56.

Borensztein, B. 2001. Monetary independence in emerging markets: does the exchange rate regime make a difference. *IMF Working Paper* 1(1).

Broda, C. 2004. Terms of trade and exchange rate regimes in developing countries. *Journal of International Economics* 63(1): 31–58.

Dumitrescu, E.I. & Hurlin, C. 2012. Testing for granger non-causality in heterogeneous panels. *Economic Modeling* 29: 1450–1460.

Dooley, M. 2000. A model of crises in emerging markets. *Economic Journal* 110: 256–272.

Frankel, J. 2003. Coping with Crises in Emerging Markets: Adjustment versus financing, in Dilip Das, ed *Perspectives in Global Finance*. London: Routledge.

Krugman, P.R. & Obstfeld, M. 1997. *International Economics Theory and Policy*. Reading, MA: Addison-Wesley.

Mihaljek, D. 2011. Domestic bank intermediation in emerging market economies during the crisis: locally-owned versus foreign-owned banks. *BIS Papers* 54: 31–48.

Mimir, Y. & Sunel, E. 2018. External shocks, banks and optimal monetary policy: A recipe for emerging market central banks. *International Journal of Central Banking, conditionally accepted for publication*.

Pedroni, P. 1999. Critical values for cointegration tests in heterogeneous panels with multiple regresion. *Oxford Bulletin of Economics and Statistics* 61: 653–670.

Pedroni, P. 2004. Panel cointegration: Asymptotic and finite sample properties of pooled time series tests with an application to the PPP hypothesis. *Econometric Theory* 20: 597–625.

Phillips, P.C.B. & Hyungsik R.M. 1999. Linear regression limit theory for nonstationary panel data. *Econometrica* 67: 1057–1111.

Pohan, A. 2008. *Kerangka Kebijakan Moneter*. Jakarta: Rajawali Press.

Wooldridge, J.M. 2002. *Econometric Analysis of Cross Section and Panel Data*. Cambridge, MA: MIT Press.

Earnings management and value relevance before and after IFRS convergency

H. Hamidah & R.D.A. Albertha
Universitas Airlangga, Surabaya, Indonesia

ABSTRACT: The IASB (International Accounting Standards Board) issued International Financial Reporting Standards (IFRS) issued to bridge the differences in the existing standards in various countries. The application of IFRS is expected to impact the decreased earnings management and increased relevance value of information in the financial statements to improve the quality of financial statements. This research is aimed to analyze the differences between earnings management and value relevance before and after IFRS convergence in Indonesia. This research used data from 77 manufacturing companies listed in the Indonesia Stock Exchange (IDX) during the period of observation before IFRS (2008–2010) and after IFRS (2012–2014). Tests of this research used Wilcoxon signed rank test and multiple linear regression. The result of this research shows that the application of IFRS in Indonesia can reduce earnings management but cannot increase the value relevance of information.

1 INTRODUCTION

The need for financial reporting and globally accessible information in international business encourages the International Accounting Standards Board (IASB) to develop an internationally understandable, accepted, and high-quality financial reporting standard. To achieve these objectives, the IASB released a standard called International Financial Reporting Standard (IFRS), which is a standard designed to bridge the differences of standards existing in different countries (Hamidah 2017).

The adoption of IFRS in Indonesia began in 2008, whereby all IFRS adopted by PSAK required infrastructure preparation; the impact of adoption by PSAK was evaluated and managed, which was valid until 2010 (Zamzami 2011). In 2011, the preparation of the supporting infrastructure for the implementation of PSAK, which had adopted IFRS, began and in 2012 the full adoption of IFRS for companies that have public accountability was a fact (Purba 2010).

IFRS is a market-based, market-oriented standard and requires greater disclosure than previous standards (Dimitropoulos et al. 2013). A level of disclosure that is closer to full disclosure will reduce the level of information asymmetry between managers and users of financial statements. Information asymmetry is a condition in which managers have superior information compared to others (Apriliani 2012). This is what causes managers to be inclined to show dysfunctional behavior by doing earnings management especially if the information is related to the performance measurement manager.

IFRS limits the existence of managerial discretion related to accounting alternatives so that it is expected to suppress opportunistic actions (Dimitropoulos et al.

2013). In addition, the use of fair value is expected to reflect the company's economic condition better. Based on this, IFRS adoption should be able to suppress the practice of earnings management. This is in line with Barth (2008), who stated that the limitation of managerial discretion associated with accounting alternatives can better reflect the performance and real company conditions. However, Barth (2008) also mentioned that flexibility in the principle-based standards can also be a greater opportunity for firms to make earnings management.

In addition to earnings management, IFRS may also affect the relevance of the value of accounting information in financial statements. According to Lev (1989), the relevance of the value of accounting information is important because it includes one of the qualities of accounting information. Francis and Schipper (1999) define the relevance of accounting information as the ability of accounting figures to summarize the information underlying stock prices so that the relevance of values is indicated by a statistical relationship between financial information and price or stock returns. IFRS puts participants in capital markets or investors as the party most in need of accounting information in the financial statements so that the implementation of IFRS is expected to provide high-quality information and is beneficial to investors in decision making. The high quality of accounting information is indicated by the strong correlation between stock price, accounting profit, and equity book value, because accounting information reflects the company's economic condition (Barth, 2008).

There are inconsistencies in the research findings on the impact of IFRS implementation on accounting quality information, both in Indonesia and abroad.

Ashbaugh and Pincus (2001), Barth (2008), Chua et al. (2012), Zeghal (2011), and Dimitropoulos et al. (2013) suggest that IFRS implementation can improve accounting quality because IFRS can limit the opportunistic behavior of management and increase the relevance of values, whereas Ball et al. (2000), Breeden (1994), Burgstahler (2006), and Lin et al. (2012) stated that the application of IFRS actually increases management aggressiveness in managing profit so that earnings management is increasing and the relevance of value decreases.

According to Cahyonowati and Ratmono (2012), the application of IFRS-based standards in Indonesia has not been able to improve the quality of accounting information proxied by value relevance. The findings of this study support the argument of Karampinis and Hevas (2011) that in code law countries (including Indonesia) with constitutional environments such as weak investor protection, lack of law enforcement, concentrated ownership and funding oriented to banking, IFRS adoption may not necessarily improve the value-relation of accounting information. The cause of the unavailability for improving the quality of information in developing countries, particularly Indonesia, is the problem of infrastructures, such as legislative conditions that are not necessarily synchronous with IFRS and a lack of human resources in education in Indonesia (Sianipar & Marsono, 2013). Edvandini (2014) states that there has been an increase in the quality of financial statement information since the adoption of IFRS than before the adoption of IFRS. This is due to IFRS standards based on fair value principles that are considered more concise and effective globally to improve transparency and information power financial statements of companies in Indonesia.

Based on the description, this study is aimed to examine the effect of the IFRS implementation on the quality of accounting information proxied by earnings management and value relevance on manufacturing companies listed in the Indonesia Stock Exchange. Consistent with previous studies, this study analyzes the overall influence of IFRS convergence and not the influence of any adopted standards. The impact of IFRS implementation in this study can be seen by comparing earnings management and value relevance between the period before- and after IFRS convergence. Thus, this study is expected to contribute conceptually to the effect of IFRS on the quality of accounting information.

1.1 Positive accounting theory

Positive accounting theory can be used to explain the relationship between accounting standards and earnings management. Scott (2012) states that positive accounting theory is a theory related to predicting actions as a choice of accounting policies by corporate managers and how managers respond to proposed new accounting standards. This theory assumes that managers are rational and will choose accounting policies that can produce the best benefits for themselves, which are not necessarily the best for the company's interests.

According to the positive accounting theory, accounting procedures used by firms do not necessarily coincide with other companies, but companies are given the freedom to choose one of the available alternative procedures to minimize contract costs and maximize corporate value. With this freedom, the manager has a tendency to perform an action which, according to positive accounting theory, is called opportunistic behavior. The opportunistic action is an action undertaken by the company in choosing a profitable accounting policy and maximizing the satisfaction of the company (Scott 2012).

1.2 Signaling theory

Theory of the signal is an explanation of the asymmetry of information. The occurrence of information asymmetry is caused by the management having more information about the prospects of the company. To avoid information asymmetry, the company must provide information as a signal to the investor where the information contained in the form of financial statements. Information asymmetry needs to be minimized so that the company goes public to inform the company's circumstances transparently to investors (Subalno 2009).

The theory of signals discusses how successors' signals or failures should be delivered to owners or investors. Management, in this case the person who runs the company, has more accurate information than the owner or investor, therefore management conveys the information to the market which will be responded to as a signal that an event has occurred and affects the value of the company.

Quality information in the financial report as well as no asymmetry is needed by the investor as the investment decision-making material. Information from each account in the financial report becomes very important to be submitted by management because it is a signal for investors and potential investors (Subalno 2009).

Information submitted by companies in the form of financial statements will show the difference between companies that have good news and companies that have not. The information will be responded to by the market through changes in the price and volume of stock trading transactions as a signal whether the company experienced a certain incident that affects the value of the company (Subalno 2009).

1.3 Hypothesis development

Positive accounting theory mentions that management tends to take opportunistic actions in determining the accounting policies applied in the company. This action arises as a result of information asymmetry, where management has superior information.

Management can take advantage of the asymmetry of this information for the sake of their personal interests.

Information asymmetry needs to be minimized so that the company goes public to inform the company in a transparent manner. Quality information in financial statements and no asymmetry is needed by stakeholders as a reference for the decision-making process. If information asymmetry can be suppressed, it is hoped that the opportunistic actions by management can also be reduced so that a more qualified financial report can be produced that reflects the real condition of the company. To achieve these objectives, the IASB has published a principle-based standard and eliminated alternative accounting estimates by requiring better accounting measures to better reflect the position and performance of a better corporate economy, IFRS (Dimitropoulos et al. 2013).

The elimination of alternative accounting and full disclosure estimates may limit the management's movement in managing reported earnings so that the implementation of IFRS is expected to suppress earnings management practices. According to Barth (2008), Zeghal (2011), Dimitropoulos et al. (2013), and Nuraini and Rahmanti (2014), the adoption of IFRS has an impact on declining earnings management practices. Based on this, the first hypothesis in this research is:

H1: Earnings management after IFRS convergence is lower than before IFRS convergence.

The relevance of values can be illustrated by the extent to which the information in the financial statements can inform the company's performance or the value of the firm through price or stock return (Beisland 2008). IFRS puts the capital market participants or investors as the ones most in need of accounting information in the financial statements, so that the implementation of IFRS is expected to provide high-quality information and will be useful for investors in decision making. This is in line with the research of Barth (2008), Dimitropoulos et al. (2013), Nuraini and Rahmanti (2014), and Edvandini (2014), that the adoption of IFRS has an impact on increasing the relevance of the value of information in the financial statements. Based on this explanation, the second hypothesis in this study is:

H2: Relevance of value after IFRS convergence is higher than before IFRS convergence.

2 METHOD

2.1 Population, samples and sampling techniques

The population used in this study are 77 manufacturing companies listed in the Indonesia Stock Ex-change in the period 2008–2010 and 2012–2014 with the division of 2008–2010 being the period before the convergence of IFRS and

2012–2014 the period after IFRS convergence. The year 2008 was chosen because at that time the IFRS adoption process in Indonesia began. The year 2012 is chosen because all companies going public listed on the BEI in that year were required to use IFRS convergence PSAK. The sample was taken by purposive sampling.

2.2 Research variables

Earnings management is a choice of accounting policies selected by managers or actions-managers to achieve specific goals in reporting earnings (Scott, 2012). The measurement of earnings management in this study refers to Dimitropoulos et al. (2013) using the Kothari (2005) model with discretionary accrual proxy. The research model is an expansion of the Jones modified model (1991), adding a firm performance of Return On Assets (ROA), as a control variable in total accrual regression. The calculation steps are as follows:

1) Calculate total accrual by using cash flow approach, that is:

$$TACC = NI - CFO \qquad (1)$$

Where:
TACC = total company accruals
NI = net income from CFO's operating activities = cash flow from the company's operating activities

2) Determine the coefficient of the accrual total regression.

Discretionary accruals are the difference between total accrual (TACC) and nondiscretionary accrual (NDACC). The first step to determine nondiscretionary accrual is by doing regression as follows:

$$TACC/TAit - 1 = \beta1 + \beta2\,(\Delta REVit/TAit - 1) + \\ \beta3\,(PPE/TAit - 1) + \beta4ROA + e \qquad (2)$$

Where:
TACC = total company accruals
TAit-1 = total assets of the company at the end of year t-1
ΔREV = change in corporate profits
ΔREC = change of net receivable of company PPE = property, plant, and equipment company
ROA = return on assets of the company at the end of the year obtained from the calculation of net income divided by total assets.
e = error

3) Determine the nondiscretionary accrual.
Regression performed at number 2 will produce the coefficients $\beta1$, $\beta2$, $\beta3$, and $\beta4$ which are used to

predict nondiscretionary accruals through the following equation:

$$NDACC = \beta 1 + \beta 2\,(\Delta REV - \Delta REC)/TAit - 1) +$$
$$\beta 3\,(PPE/TAit - 1) + \beta 4 ROA + e$$
$$(3)$$

Where:
NDACC = nondiscretionary accrual company

4) Determine the discretionary accrual.
Once the nondiscretionary accrual is obtained then the discretionary accrual can be calculated by subtracting the total accrual (calculated at number 1 with nondiscretionary accrual at number 3).

$$DACC = (TACC/TAit - 1) - NDACC \qquad (4)$$

Where:
DACC = discretionary accrual company

5) Value relevance.
According to Lev (1989), value relevance is one of the hallmarks of quality accounting information. Measurement of value relevance in this study refers to Barth's (2008) re-search by regressing stock price (P) and net income per share (NIPS) as well as book value equity per share (BVPS) as the following:

$$Pit = \alpha 0 + \alpha 1\,NIPSit + \alpha 2\,BVPSit + eit \qquad (5)$$

Where:
Pit: The stock price used in this study uses the aver-age closing price of common stock between January and March after the fiscal year ends (t + 1) for 2008, 2009, 2010, and 2012. Furthermore, for 2013 and 2014 flat common stock closing price between January and April after the book year ends (t + 1) was used. Thus, the stock price is expected to reflect the stock market price after the audited financial statements are published and reflects the market reaction to the accounting information contained in the financial statements.
NIPS: Net income per share (NIPS) or earnings per share (EPS) is derived from net income divided by the number of shares outstanding.
BVPS: Book value equity per share (BVPS) is derived from total equity divided by the number of shares outstanding.
e: error

Table 1. Descriptive statistics.

	Min	Max
DA before	.00001	.59107
DA after	.00029	.93687
P before	5.00	267950.00
P after	5.10	1096250.00
NIPS before	-2234.00	21021.00
NIPS after	-17350.00	55576.00
BVPS before	-2493.00	38032.00
BVPS after	-12182.44	47740.25

	Mean	Standard deviation
DA before	.0817745	.09151290
DA after	.0687658	.08307329
P before	6953.3489	24493.61579
P after	20476.9257	101906.19413
NIPS before	649.8928	2179.12477
NIPS after	798.8935	4529.72375
BVPS before	2636.7741	5673.35512
BVPS after	3266.9111	7014.08713

Source: Data processed 2016.

3 RESULT AND DISCUSSION

3.1 Descriptive statistics

Based on Table 1, for the period before IFRS convergence, the discretionary accrual (DA) has alowest value of 0.0001 and a highest value of 0.59107. DA average is 0.08 and the standard deviation is 0.0915. For the second model, the value relevance needed data on stock price (P), earnings per share (NIPS), and book value of equity per share (BVPS) with the following description: first, stock price (P) has a lowest value of 5.00 and a highest value of 267,950. The average stock price is 6,935.35 and the standard deviation is 24,493.6. Secondly, NIPS has a lowest value of –2.234 and a highest score of 21,021. The average NIPS is 649.89 and the standard deviation is 2,179. Thirdly, BVPS has a lowest value of -2.493 and its rated value is 38,032. The average BVPS is 2636.77 and the standard deviation of 5673.36.

Data on the period after convergence, ie: DA after IFRS convergence has a lowest value of 0.00029 and a highest value of 0.93687. DA average is 0.069 and standard deviation is 0.083. The stock price (P) has a lowest value of 5.10 and a highest value of 1,096,250. The average stock price is 20,476.93 and the standard deviation is 101,906.19. The lowest NIPS is 17,350 and the highest is 55. The NIPS average is 798.89 and the standard deviation is 4,529,72. The lowest BVPS value is –12.182.44 and the highest is

Table 2. Wilcoxon test for model I.

		N	Mean Rank	Sum of Ranks
After - before	Negative Ranks	129[a]	119.42	15405.00
	Positive Ranks	102[b]	111.68	11391.00
	Ties	0[c]		
	Total	231		

Source: Data processed 2016.
Information:
a. after < before
b. after > before
c. after = before

47,740.25. The average BVPS is 3,266.91 and the standard deviation is 7,014.09.

3.2   Wilcoxon test

Model 1: Earnings Management
The Wilcoxon test is used to analyze the results of paired observations of two data whether they are different or not. This Wilcoxon test is used only for interval or ratio type data, but the data is not normally distributed. Referring to the normality test previously discussed, it is known that the data is not normally distributed so that the Wilcoxon test is used. The results of the Wilcoxon test are shown in Table 2.

Based on the method used for the data using the Wilcoxon test, we obtained the values: negative ranks, positive ranks, and ties. Negative ranks mean the sample with the second group value (after) is lower than the first group value (before), which applied to 129 samples. Positive ranks mean the sample with the second group value (after) is lower than the first group value (before), which was equal to 102 samples. Ties mean the value of the second group (after) is equal to the value of the first group (before), which was equal to 0. The symbol N indicates the number; the mean rank is the average rating; and the sum of ranks is the sum of its rank. Proof of the hypothesis can be seen in Table 3.

Based on Table 3, it can be seen that the value of Z is -1.974 and Asymp. Sig (2-tailed) is 0.048. The value of Z equal to -1.974 indicates that the average DA value before is greater than DA after. This means that earnings management in the period after convergence is smaller than before convergence. In addition,

Table 3. Hypothesis testing results 1.

	After-before
Z	-1.974b
Asymp. Sig. (2-tailed)	.048

Source: Data processed 2016.

Asymp. Sig (2 tailed) of 0.048, where 0.048 <0.05, can be interpreted as there being a significant difference between the earnings before and after the IFRS convergence. This indicates that the hypothesis 1 of this study is evident, i.e.: earnings management after IFRS convergence is lower than before IFRS convergence.

3.3   Results of multiple linear regression

Model 2: Value Relevance
In this second model, multiple linear regression test is only done to get the adjusted R2 which will be used as comparison material for the period before the convergence of IFRS and after the IFRS convergence, which was done separately. The calculation is shown in Table 4.

Table 4 presents the results of the hypothesis testing of the relevance of the value of accounting information before and after the IFRS convergence. Test results in the table show a feasible research model with significant F value for both periods of 818,683 and 536,180 and thus statistically significant.

The relevance test focuses on adjusting the adjusted value of R2 after IFRS convergence. If the adjusted value of R2 increases significantly it can be concluded that accounting information increases the relevance of its value due to IFRS convergence. Results in Table 4 showed the value of adjusted R2 actually decreased by 0.054 from 0.877 to 0.823. These results indicate that IFRS convergence has no influence or negative effect on the combined value relevance of net income and equity book value (Cahyonowati & Ratmono 2012). This indicates that the second hypothesis in this study, that the relevance of value after IFRS convergence is higher than before IFRS convergence, is unaccepted.

Further analysis of Table 4 shows that the NIPS coefficients increase from 10,587 to 19,144. The coefficient of BVPS also increased from -0.036 to 1.036. This result indicates that there is an increase in value relevance for each dimension of accounting information.

3.3.1   The effect of IFRS convergence on earnings management
Hypothesis 1 stated that earnings management after IFRS convergence is lower than before IFRS convergence and this hypothesis was proven. The results

Table 4. Hypothesis testing results 2.

	Before IFRS (n=231)	After IFRS (n=231)
Variable	Coefficient	Coefficient
Constant	167,326	1798,027
NIPS	10,587	19,144
BVPS	-0,036	1,036
F Value	818,683	536,180
Adjusted R²	0.877	0.823

Source: Data processed 2016.

showed that there are significant differences in earnings management before and after the IFRS convergence, marked with a significance value of 0.048, smaller significance level of 0.05. In addition, the Z value of -1.974 indicates that the average DA value before is greater than the DA after.

The results of this study prove that IFRS can suppress earnings management practices. This is in line with research conducted by Barth (2008), Zeghal (2011), Dimitropoulos et al. (2013), and Nuraini and Rahmanti (2014), that the adoption of IFRS impacts the decline in earnings management practices.

The use of fair value in IFRS makes the assets and liabilities items more reflective of the true value of the financial statement date. This is in line with Barth (2008) who stated that fair value use can better reflect the company's economic condition. Thus, it has been proven that the transition from historical cost to fair value can reduce the practice of earnings management.

The results of this study are also in line with Nuraini and Rahmanti (2014), that IFRS is a standard that can reduce earnings management because accounting standards put more emphasis on the principle (principle-based standards). Arrangements at the principle level will include everything below. Such standards are consistent with the objectives of financial reporting that can reflect the actual condition of the company so that it can be used as decision-making material (Cahyati 2011).

IFRS also requires a full disclosure level so as to suppress information asymmetry between managers and stakeholders (Apriliani 2012). Information asymmetry is a condition in which managers have superior information to others. This is what drives management to show dysfunctional behavior by doing earnings management, especially if the information is related to the performance measurement manager. Condition asymmetry information is the cause of earnings management practices. Thus, with the existence of more and more detailed disclosure, information asymmetry can be suppressed and management earnings can be minimized (Cahyati 2011).

In addition to information asymmetry, the leniency in the selection of accounting methods used and accounting procedures is often utilized by management to manage earnings, therefore, Dimitropoulos et al. (2013) mentions that by applying IFRS, opportunistic management actions can be suppressed through the limitation of managerial discretion related to accounting alternatives.

3.3.2 The effect of IFRS convergence on value relevance

Hypothesis 2 stated that the relevance of values after IFRS convergence is higher than before IFRS convergence, which was not proven. The results of this study indicated that there is a decrease in the relevance of values represented by the adjusted R2 value of 0.054 from 0.877 to 0.823. These results indicated that IFRS convergence has no influence or negative

effect on the combined value relevance of net income and book value of equity.

The adjusted R2 decline in the period after IFRS convergence shows that fewer investors use the information available in the financial statements for investment decision making than in the period prior to the IFRS convention.

The results of this study are also in line with Lin (2012), where the relevance of information decreased more after IFRS convergence than when using GAAP. The results of this study also match the results of research done by Cahyonowati and Ratmono (2012), that there is no increase in the relevance of accounting information after the IFRS convergence. H1 can be caused by an institutional environment that is still not supported so that the application of IFRS cannot increase the relevance of value.

These findings support the arguments of Karampinis and Hevas (2011) that in code law countries (including Indonesia) with constitutional environment characteristics, such as weak investor protection, lack of law enforcement and funding-oriented banking, IFRS convergence may not necessarily improve relevance value of accounting information.

Karampinis and Hevas (2011) also hypothesize and provide evidence that only accounting standard factors (including IFRS) are not sufficient to improve the quality of accounting information, and the relevance of values. Karampinis and Hevas (2011) argue that it is the institutional environment of report compilers financial, non-standard, that determines the quality of accounting information. The IRRS is structured on a conceptual framework similar to the conceptual framework of common law accounting standards (Barth 2008). Therefore, IFRS benefits for countries with code law traditions are still important questions.

3.3.3 Earnings management and value relevance

This section will explain the implications of the results of research model 1 (earnings management) and model 2 (value relevance). Earnings management after IFRS convergence did decrease but the same thing also happened in value relation. This is in contrast with the existing theory, therefore regression testing was done of earnings per share and book value per share against share price separately to find out what value relevance is causing the decline after

Table 5. Test result relevance value of values and relevance of book value separately.

	Adjusted R2		
	Before IFRS	After IFRS	Changes
NIPS	0,877	0,822	-0,055
BVPS	0,315	0,546	0,231
NIPS and BVPS	0,877	0,823	-0,054

Source: Data processed 2016.

IFRS convergence. The result can be seen in Table 5.

The results of model 1 suggest that profitability after IFRS convergence has decreased. This certainly makes the reported returns more qualified, but it has not been able to increase the relevance of values. This is because the quality of profit is not a major factor affecting the relevance of value, especially the relevance of the value of accounting earnings. According to Naimah (2006), as for the factors that affect the relevance of accounting profit value, in addition to the quality of profit, there are: capital structure, auditor quality, corporate risk, company size, growth opportunities, and profit persistence.

Furthermore, referring to the information presented in Table 5 it is visible that only adjusted R2 BVPS is increasing. Adjusted R2 is a reflection of the relevance of equity book value. It can be concluded that there is a shift in the focus of investor valuation from profit to book value of equity. This is caused by the practice of earnings management which often happens so that the profit information is less relevant and then the investor also switches to the value books in making investment decisions (Kusuma 2007, Kusumo & Subekti 2014).

4 CONCLUSIONS

Based on the results of data analysis and data testing in this paper, the research can be summarized as follows: The average discretionary accrual (DA) of earnings management after IFRS convergence is lower than the average discretionary accrual (DA) after IFRS and the Wilcoxon test results show it to be significant, it can thus be concluded that there is a difference in earnings management between before and after IFRS convergence. Earnings management after IFRS convergence has decreased; Relevance value after IFRS convergence is lower than before IFRS convergence. The test was done by comparing the adjusted R2 values before and after the IFRS convergence and the result of adjusted R2 after IFRS convergence has decreased. Overall, it can be concluded that the adoption of international financial accounting standards (IFRS) in Indonesia can reduce the profit management practices but has not been able to improve the value of accounting information.

REFERENCES

Apriliani. 2012. Kajian Kualitas Pelaporan Keuangan Second Order Terhadap Asimetri Informasi. *Accounting Analysis Journal*, 1(1).

Ashbaugh, H. & Pincus, M. 2001. Domestic accounting standards, international accounting standards, and the predictability of earnings. *Journal of Accounting Research*, 39(3): 417–434.

Ball, R., Kothari, S.P. & Robin, A. 2000. The effect of international institutional factors on properties of accounting earnings. *Journal of Accounting and Economics*, 29(1): 1–51.

Barth, M.E., Landsman, W.R. & Lang, M.H. 2008. International accounting standards and accounting quality. *Journal of Accounting Research*, 46(3): 467–498.

Breeden, R.C. 1993. Foreign companies and US securities markets in a time of economic transformation. *Fordham Int'l LJ* S77, 17.

Burgstahler, D.C. Hail, L. & Leuz, C. (2006). The importance of reporting incentives: Earnings management in European private and public firms. *The Accounting Review*, 81(5): 983–1016.

Cahyati, A.D. 2011. Peluang Manajemen Laba Pasca Konvergensi IFRS: Sebuah Tinjauan Teoritis dan Empiris. *JRAK (Jurnal Riset Akuntansi dan Komputerisasi Akuntansi)*, 2(1): 1–7.

Cahyonowati, N. & Ratmono, D. 2012. Adopsi IFRS dan Relevansi Nilai Informasi Akuntansi. *Jurnal Akuntansi dan Keuangan*, 14(2): 105–115.

Chua, Y.L. Cheong, C.S. & Gould, G. 2012. The impact of mandatory IFRS adoption on accounting quality: Evidence from Australia. *Journal of International Accounting Research*, 11(1): 119–146.

Dimitropoulos, P.E., Asteriou, D., Kousenidis, D. & Leventis, S. 2013. The impact of IFRS on accounting quality: Evidence from Greece. *Advances in Accounting*, 29(1): 108–123.

Edvandini, L., Subroto, B. & Saraswati, E. 2014. Telaah Kualitas Informasi Laporan Keuangan dan Asimetri Informasi Sebelum dan Setelah Adopsi IFRS. *Jurnal Akuntansi Multiparadigma*, 5(1), 88–95.

Francis, J. & Schipper, K. 1999. Have financial statements lost their relevance? *Journal of Accounting Research*, 37(2), 319–352.

Hamidah. 2017. IFRS Adoption in Indonesia: Accounting Ecology Perspective. *International Journal of Economics and Management*, 11(1): 121–132.

Karampinis, N.I. & Hevas, D.L. 2011. Mandating IFRS in an unfavorable environment: The Greek experience. *The International Journal of Accounting*, 46(3), 304–332.

Kusuma, H. 2007. Dampak Manajemen Laba terhadap Relevansi Informasi Akuntansi: Bukti Empiris dari Indonesia. *Jurnal Akuntansi dan Keuangan*, 8(1): 1–12.

Kusumo, Y.B. & Subekti, I. 2013. Relevansi Nilai Informasi Akuntansi, Sebelum dan Setelah Adopsi IFRS Pada Perusahaan yang Tercatat dalam Bursa Efek Indonesia. Jurnal Ilmiah Mahasiswa FEB, 2(1).

Lev, B. (1989). On the usefulness of earnings and earnings research: Lessons and directions from two decades of empirical research. *Journal of Accounting Research*, 153–192.

Lin, S., Riccardi, W. & Wang, C. 2012. Does accounting quality change following a switch from US GAAP to IFRS? Evidence from Germany. *Journal of Accounting and Public Policy*, 31(6): 641–657.

Naimah, Z. & Utama, S. 2006. Pengaruh ukuran perusahaan, pertumbuhan, dan profitabilitas perusahaan terhadap koefisien respon laba dan koefisien respon nilai buku ekuitas: Studi pada perusahaan manufaktur di Bursa Efek Jakarta. *Simposium Nasional Akuntansi IX*, 1: 26.

Nuraini, H.I. & Rahmanti, W. 2014. Kualitas informasi akuntansi sebelum dan sesudah konvergensi IFRS di Indonesia studi empiris pada perusahaan non keuangan di BEI Periode 2005–2012. *Akuntansi dan Bisnis Sistem Informasi (ABSI)*. 7(13).

Purba, M.P. 2010. *IFRS: Konvergensi dan Kendala Aplikasinya di Indonesia*. Yogyakarta: Graha Ilmu.

Scott, W.R. 2012. *Financial Accounting Theory*. Canada: Pearson Prentice Hall.

Sianipar, G.A.E. & Marsono, M. 2013. Analisis Komparasi Kualitas Informasi Akuntansi Sebelum dan Sesudah Pengadopsian Penuh IFRS di Indonesia. *Diponegoro Journal of Accounting*: 350–360.

Subalno, S. 2009. Analisis pengaruh faktor fundamental dan kondisi ekonomi terhadap return saham (study kasus pada perusahaan otomotif dan komponen yang listed di Bursa Efek Indonesia periode 2003–2007) *Doctoral dissertation, Program Pasca Sarjana Universitas Diponegoro*.

Zamzami, F. 2011. Perkembangan Konvergensi International Financial Reporting Standards (IFRS) di Indonesia. *Paper presented at the Seminar dan Pelatihan IFRS Serta Penyusunan Kamus Akuntansi Indonesia P2EB UGM*.

Zéghal, D., Chtourou, S. & Sellami, Y.M. 2011. An analysis of the effect of mandatory adoption of IAS/IFRS on earnings management. *Journal of International Accounting, Auditing and Taxation*, 20(2): 61–72.

Advances in Business, Management and Entrepreneurship – Hurriyati et al (eds)
© 2020 Taylor & Francis Group, London, ISBN 978-0-367-27176-3

Performance of the government's subsidized mortgage during the period 2015–2017

L. Purnamasari & N. Nugraha

Universitas Pendidikan Indonesia, Bandung, Indonesia

ABSTRACT: The current major housing problems in Indonesia are a trend that continues significantly with a 11,8 million home backlog; 5.9 million existing houses which are uninhabitable; 4,8 million families who share roofs with other families; and 50.8% of Indonesian citizens being informal workers who have no access to loans from banks. Although Malaysia's GDP per Capita is quite similar to Indonesia, its mortgage rate is far higher (near 40%). This indicates that mortgage financing in Malaysia and Singapore is a good banking business. To reduce the backlog comprehensively, the government provides financial assistance through a mortgage subsidy scheme of Housing Financing Liquidity Facility (FLPP). Provision of subsidies on the housing sector is one of the government policies on the provision of housing, especially for low-income people. In 2015, the government allocated IDR 8.1 trillion to build 98,300 housing units and IDR 5.1 trillion in FLPP to provide MBR financing assistance and financing, which means financing with 5% mortgage rate, tenor up to 20 years, 1% down-payment and VAT exemption. OJK survey in 2016 revealed that the factors hindering the growth of housing business were mortgage interest (20.36%), down-payment needs (16.57%), taxes (16.13%), permits (14.45%), and the price of building materials (11.68%). More than 76% of the consumers still rely on bank credit (KPR/KPA) to buy a house. According to data analysis, from about 30 banks as FLPP mortgage providers, there are only 8 banks listed in IDX, and from those banks only BTN dominated the credit disbursement significantly (FLPP, SSB, and subsidized construction). Therefore, the evaluation of subsidized credit performance should focus on BTN. Compared to total mortgage amount, the subsidized scheme only takes 10%. Therefore, even if the Non-Performing Loan (NPL) from that 10% subsidized scheme is high, it will not significantly affect the total mortgage performance.

1 INTRODUCTION

The need for home provision for the people in Indonesia ranges from 820,000 to 1 million homes per year. This requirement is only fulfilled for about 40 percent of the private sector, and only about 20 percent comes from government intervention. While the remaining 40 percent cannot be fulfilled, hence the backlog. There are 40 percent of the people who can afford to buy a house with government assistance in the form of a subsidy, and another 20 percent who cannot afford to own a house.

The current major housing problems in Indonesia are a trend that continues significantly:

* 11.8 million home backlog;
* 5.9 million existing houses which are uninhabitable;
* 4.8 million families who share a roof with other families; and
* 50.8% of Indonesian citizens are informal workers who do not have access to loans from banks.

These figures will continue to grow if it is not resolved and facilitated with financing aspects.

Compared to other Asian countries, Indonesia's Mortgage Credit % of GDP is among the lowest

(2%), far below Malaysia and Singapore. Although Malaysia's GDP per capita is quite similar to Indonesia, its mortgage rate is far higher (near 40%). This indicates that mortgage financing in Malaysia and Singapore is a good banking business.

The bank's largest income is from credit interest. One of these is the existence of consumer loans, in particular mortgage loans (Home Ownership Credit), so the bank can help fulfill the people's right to provide decent housing. The existence of Home Ownership Loan (KPR) and Apartment Ownership Credit (KPA) can have a positive impact in helping the community obtain a home. Commercial banks can also run services in payment traffic in the form of general term payments of increased housing construction projects (guarantee), etc.

Since 2010, the government has developed a program to cooperate with banks in providing subsidized mortgage to reduce the large number of backlogs. Sixty percent of middle-class people need assistance and can get this through a mortgage subsidy scheme of the Housing Financing Liquidity Facility (FLPP). Housing finance liquidity facilities (FLPP) are mostly intended for workers with an income of less than IDR 5.7 million per month. Meanwhile, workers who receive 5.7 million per month income or more, should utilize BPJS as an additional housing service.

In 2018, the PUPR Ministry has partnered with 40 banks, formerly 11 banks, to disburse funds from the Housing Financing Liquidity Facility (FLPP). During the period 2010–2017, the government has disbursed FLPP funds as much as IDR 30.93 trillion for a total of 519,828 subsidized housing units. Meanwhile, FLPP funds that will be distributed in 2018 will reach 42,326 units, with a value of IDR 4.5 trillion, consisting of IDR 2.2 trillion of PUPR Ministry budget and IDR 2.3 trillion from optimization of the return of principal. Those 40 banks consist of 6 national banks and 34 regional development banks (Pertiwi et al. 2014).

This study tries to evaluate the performance of the government's subsidized mortgage during the period 2015–2017. The study was carried out in order to find significant problems and potentials to develop the program further with regard to management academic approaches.

1.1 The legal foundation of the 1 million houses program

1.2 Backlog of houses

The backlog of houses is one of the indicators used by the government as outlined in the Strategic Plan and Medium Term Development Plan (RPJMN) related to the housing sector to measure the number of homes needed in Indonesia. The backlog of homes can be measured from two perspectives, namely from the occupancy side or from the ownership side.

1.2.1 Backlog of residential homes
The home backlog from a residency perspective is calculated with reference to the concept of this ideal calculation: 1 family inhabits 1 house. The formulas used to calculate the home backlog from a residency perspective are:

$$\text{Backlog} = \sum \text{Family} - \sum \text{Home} \qquad (1)$$

In the Book 1 attachment of the Presidential Regulation of the Republic of Indonesia No. 2 of 2015 on the National Medium-Term Development Plan (RPJMN) for 2015–2019, the baseline backlog of Indonesia in 2014 was 7.6 million.

The concept of inhabitants in the calculation of the backlog means that not every family is required to own a house, but the government encourages that every family, especially those belonging to the Low-Income Community (MBR) can occupy a decent house, either in the form of lease/contract, buy/occupy own house, or live in a house belonging to relatives/family for guaranteed secure settlement (secure tenure) (Rakhmadhani 2013).

1.2.2 Backlog of home ownership
Backlog ownership is calculated based on the number of home ownership rate/percentage of a household occupying their own property. The basic data source used in this calculation is sourced from BPS data. Table 1 shows how to calculate the backlog of home ownership.

In order to utilize the data and information to support the implementation of the task of disbursement and fund management of Housing Financing Liquidity Facility (FLPP), the Center for Housing Financing Management (PPDPP) continues to utilize various strategic data of housing, one of which are data of homeownership rate/occupancy of a self-owned house in 2015 issued by the Central Bureau of Statistics (BPS) in 2016.

1.3 Concepts related to housing mortgage

1.3.1 Government subsidies
According to the public interest theory approach, regulation is presented in response to public demand for correction of inefficiencies or inadequate market prices. The main purpose of this theory is to protect the people and to realize a prosperous country (Carrigan et al. 2011).

The subsidy is a payment given by the government to business entities and households in the hope of achieving better conditions; subsidy can be direct or indirect. Direct subsidies can be in the form of cash, interest-free loans and so on, whereas indirect subsidies are in the form of exemption from depreciation, rent deductions and the like. Provision of subsidies in the housing sector is one of the government's policies on the provision of housing, especially for low-income people (Redmond 2012).

Table 1. Calculation of house ownership.

Year	Population	Number of household	House Owned by the household (%)	Total House Owned by the household	Total Owned household/Backlog house ownership
(1)	(2) = BPS	(3)=BPS	(4)=BPS	(5)=(3)X(4)	(6)=(3)-(5)
2010	237,641,326	61,390,300	78.00	47,884,434	13,505,866
2015	255,461,700	65,503,000	82.63	54,125,129	11,377,871

The adjustment of the market price of housing, which is very high, does not provide opportunities for low-income people to own a house. Provision of subsidies to low-income communities is expected to realize for the community, especially low-income people, to own a house. The government has allocated housing subsidies through the Ministry of Public Works and Public Housing (PUPR). Between 2015 and 2019, housing and housing subsidies of the Ministry of PUPR are estimated to reach Rp 74 trillion.

1.4 Housing loans (KPR and KPA)

A Home Ownership Loan (KPR) is a credit facility provided by banks to individual customers who will buy or repair houses. In Indonesia, there are currently two types of mortgages.

1.4.1 Subsidized KPR
Subsidized KPR is a credit that is intended for lower-middle-income communities in order to meet the needs of housing or home improvements for houses that are owned. A form of subsidy given to alleviate credit and subsidies increase the fund for the construction or improvement of the house. This subsidized credit is regulated separately by the government, so not every community applying for credit can be granted this facility. In general, the limit set by the government in subsidizing is the applicant's income and the maximum credit granted.

1.4.2 Non-subsidized KPR (e.g., KPA)
Non-subsidized KPR is a mortgage that is intended for the whole community. The mortgage provision is determined by the bank, so the determination of the amount of credit and interest rate is done according to the policy of the bank concerned.

The government support-program in housing financing for low-income communities is also continued as part of the People's Million Development Program launched by the government in 2015. In 2015, the government allocated IDR 8.1 trillion to build 98,300 housing units and IDR 5.1 trillion in FLPP to provide MBR financing assistance with 5% mortgage rate, tenor up to 20 years, 1% down payment and VAT exemption.

Until now, the distribution of KPR and KPA is still concentrated in Java Island because of the dense population. Based on data from the Financial Services Authority (OJK) as per February 2017, 70% of KPR and KPA values are located in Java followed by Sumatra (13%), Kalimantan (6%), Sulawesi (6%), Bali-Nusa Tenggara (4%), and Maluku-Papua (1%). The five provinces with the largest KPR and KPA distribution are DKI Jakarta, West Java, East Java, Banten and Central Java. On the other hand, the five provinces with the lowest KPR and KPA distribution are West Papua, Gorontalo, Maluku, North Maluku and West Sulawesi.

Figure 1. Proposed methodology of the study.

2 METHODS

Proposed methodology for this study to evaluate the performance of the government's subsidized mortgage during the period 2015–2017 include (see also Figure 1):

- Reviews on theories and concepts related to housing and subsidy policy, banks and finance institutions, mortgage, financial management, Credit and Installment scheme, Risk and NPL;
- Reviews on previous research on subsidized housing policy, mortgage performance, and NPL of subsidized mortgage. Possible benchmark of neighboring countries (Singapore, Malaysia, etc.) on their subsidized mortgage policy and practice;
- Analysis of hindrances, difficulties and risks on mortgage practices (subsidized and non-subsidized). Evaluation of the performance of the mortgage-provider banks listed in IDX during 2015–2017; and
- Explanation of the findings referring to financial management concepts and standards to obtain alternative practice solutions, conceptual novelties, or policy recommendations to improve the housing program.

3 RESULTS AND DISCUSSION

Economic growth is predicted to continue to show an upward trend. In the fourth quarter of 2017 it reached 5.4% year on year (yoy), in line with economic improvement, the supply and demand side growth of the housing sector, which are also expected to increase.

Bank Indonesia (BI) recorded property loans grew 15% to IDR 731.4 trillion at the end of December 2016 compared to IDR 620.4 trillion at the end of December 2015. The highest growth of property loan segment was construction with the growth of 24.2%: IDR 214.3 trillion per December 2016 compared to IDR 172.5 trillion in December 2015. Real estate credit grew 22.2 percent: IDR 130.8 trillion per December 2016 compared to IDR 107.1 trillion in December 2015.

Credit for mortgage and apartment ownership loans only grew 8.1 percent: IDR 368.3 trillion per December 2016 compared to IDR 340.8 trillion per December 2015. Bank Indonesia (BI) recorded a month-long growth in property loans accelerating from 13% yoy growth in October 2017 to 13.6% yoy in November last year. Property credit value, until November, reached IDR 791.8 trillion. The property credit acceleration is mainly driven by loans disbursed to the construction sector, as well as housing and mortgage loans (KPR).

BI data showed that mortgage and mortgages in November reached IDR 402.9 trillion or grew 11% yoy, which was higher compared to the previous month, IDR 397.4 trillion, which grew 10.8% yoy. The residential segment, consisting of apartment and housing, became the biggest contributor to the national property capitalization value, reaching 55,8%. The growth rate of the capitalization value of the residential sector reached 16.5%, from Rp 152.7 trillion to IDR 177.9 trillion in 2017. There are a number of factors that trigger the increase in property market capitalization value in 2017. These factors include the increasing interest rates of banking, which is low and are expected to reach single digits, and the loan to value (LTV) easing.

On the other hand, the policy of loan to value from BI causes the demand for mortgage loans to decline, because the debtor feels that the downpayment must be fulfilled in order to do the second, third and subsequent home loans, which are still too high. However, the supply of mortgage loans will continue to increase because of the large number of homes needed in Indonesia. Therefore, the high supply is only absorbed by the demand for credit in a certain amount, which further makes the condition of credit channeling slow down.

3.1 Performance of subsidized mortgage

In an effort to boost economic growth from the monetary side, from the beginning of 2016 to the first quarter of 2017, Bank Indonesia embarked on an accommodative monetary policy in the form of:

1. Interest rate policy, by lowering the reference rate (policy rate) of 150bps;
2. Liquidity policy, through a decrease in Statutory Reserves (GWM) of 150 bps; and
3. Macro-prudential policy, in the form of relaxation of Loan to Value (LTV) for the housing sector.

During 2015 and 2016, subsidized fund disbursed by the government was 6 and 5,5 trillion rupiah for 80.000 and 60.000 houses, while the credit for mortgage and apartment ownership loans in total were IDR 368.3 trillion per December 2016 compared to IDR 340.8 trillion per December 2015. This means that only about 10% of the Bank's mortgage fund was allocated for subsidized housing.

1. OJK survey in 2016 revealed that the factors hindering the growth of the housing business were: mortgage interest (20.36%), down payment needs (16.57%), taxes (16.13%), permits (14.45%), and the price of building materials (11.68%). More than 76% of the consumers still rely on bank credit (KPR/KPA) to buy a house; and
2. The LVT policy issued by BI in 2015 and 2016 did not improve the mortgage credit significantly, while for non-subsidizing the LVT has increased the credit.

The rapid development of KPR and KPA is also supported by a stable credit quality, with the non-performing loan ratio (NPL) below 3%. As of February 2017, the ratio of NPL, KPR and KPA was 2.85%, lower than the NPL of total credit which reached 3.16%. In more detail, the NPL ratio for a mortgage is 2.86% and KPA is 2.61%. In general, the distribution of mortgage and KPA credit has a relatively low risk, because it has collaterals in the form of houses, whose value continues to increase.

1. In fact, about 80% of the total disbursed mortgage with FLPP scheme are provided by BTN Bank, and also for the Interest Difference Subsidy (SSB) scheme mortgage is 95% provided by BTN;
2. The absence of BTN during 2017 and early months of 2018 has reduced the number of FLPP credit realization significantly, namely more than

Table 2. Realization mortgage agreement by banks in the period 2010–2017.

No	Bank	2010	2011	2012	2013	2014	2015	2016	2017	2018	Total
1.	BTN	7,775	104,646	59,833	87,079	65,397	66,563	43,821	-	-	435,114
2.	BTN Syariah	184	4,699	3,255	7,656	5,328	6,220	4,112	-	-	31,454
3.	BRI Syariah	-	-	201	1,588	3,445	1,449	2,917	5,703	3,784	19,087
4.	ASBANDA	-	22	159	670	587	782	3,950	10,238	5,648	22,056
5.	BNI	-	86	584	1,956	319	1,098	1,255	1,456	1,307	8,061
6.	Artha Graha	-	-	-	-	-	191	2,349	5,579	1,394	9,523
7.	Mandiri	-	-	376	1,693	562	26	16	166	247	3,086
8.	Bukopin	-	139	271	1,496	-	-	-	-	-	1,906

60%. This was caused by the change in government policy to reduce FLPP subsidy fund, and to reduce the burden on the government finance capacity;

3. According to data analysis, from the roughly 30 banks as FLPP mortgage providers, there are only 8 banks listed in IDX, and from those banks, only BTN dominated the credit disbursement (FLPP, SSB, and subsidized construction) significantly. Therefore, the evaluation of subsidized credit performance should focus on BTN; and

4. Compared to the total mortgage amount, the subsidized scheme only takes 10%, therefore, even if the Non-Performing Loan (NPL) from that 10% subsidized scheme is high, it will not affect the total mortgage significantly.

4 CONCLUSIONS

- The government's 1 million house program is an important program to improve the quality of life of low-income people (MBR) and government officials such as police, army, and civil government employees;
- The government provided a significant amount of subsidy funds for the program and also fiscal facilities for financial institutions (banks) that were willing to join as provider;
- The program has been running since 2010 up until now; it has developed 532.283 new houses with the FLPP scheme, 239.000 houses with the SSB scheme, and 561.046 units for the construction credit scheme in which BTN is the majority of credit providers, taking up80% of FLPP, 95% of SSB, and 100% of construction credit;
- Compared to total mortgage amount of IDR 368.3 trillion per December 2016, the subsidized scheme only takes 10%, therefore even if the

Non-Performing Loan (NPL) from that 10% subsidized scheme is high, it will not significantly affect the total mortgage performance.

- The goal of Bank Indonesia's Macro-prudential policy rally (LTV) is to encourage the channeling of Credit to Real Estate (Non-Subsidies) and mortgages for the lower-middle segment on target. When it has been implemented for 3 years and the results are not as expected, BI may issue a regulation on additional relaxation to be opened, so the mortgage providers should be able to anticipate it (Kusumastuti, 2015).

REFERENCES

Andaru, N., Nurdin, A. R., & Maulisa, N. 2015. The Implementation of risk management in house ownership credit (KPR) related to the policy of loan to value ratio (study on X Bank). *Jurnal Departemen Ilmu Hukum FHUI*, (15).

Carrigan, Christopher & Coglianese, C. 2011. *The theory of economic regulation*.

Direktorat Jenderal Anggaran, K. K. 2015. Peranan APBN dalam mengatasi Backlog perumahan bagi masyarakat berpenghasilan rendah (MBR).

Kusumastuti, D. 2015. Kajian terhadap kebijakan pemerintah dalam pemberian subsidi di sektor perumahan, *4*(3), 541–557.

Lisnawati. 2015. Pembiayaan dalam mengatasi. *Info Singkat Ekonomi Dan Kebijakan Publik*, *VII*(14), 13–16.

Pertiwi, A.O.D. & Arifianto, E.D. 2014. Loss given default (LGD) kredit pemilikan rumah (KPR) di Indonesia : analisis model industri perbankan dan bank BTN Cabang Purwokerto Tahun 2002–2013. *Journal of Management University of Diponegoro, 3*.

Rakhmadhani, I. D. 2013. Determinasi tingkat penyaluran kredit pemilikan rumah (periode setelah kebijakan loan to value). *Universitas Airlangga, Surabaya*, (September), 1–21.

Redmond, D. (2012). Housing in Indonesia: expanding access, improving efficiency, 1–88.

Advances in Business, Management and Entrepreneurship – Hurriyati et al (eds)
© *2020 Taylor & Francis Group, London, ISBN 978-0-367-27176-3*

An analysis of influence of the government health expenditure on the performance of pharmacy sector issuers in Indonesia stock exchange during 2015-2017

I. Sugianto & M. Kustiawan
Universitas Pendidikan Indonesia, Bandung, Indonesia

ABSTRACT: The big echoes of the Health Social Security Organization program initially gave hope to the pharmaceutical industry to increase sales. The program is believed to increase demands on pharmaceutical products. The drug's demand is actually increasing and types of drugs needed are also more varied. However during 2015-2017, pharmaceutical issuers grew slower and its stock prices declined. This research includes library research; the research method is mixed method research category with the procedure of data collecting activities in a span period of time (2015-2017) and the final presentation technique descriptively. Based on the financial statements of pharmaceutical companies listed on the Indonesia Stock Exchange, during the period 2015-2017, the increase in drug sales actually eroded its profit margin. For example, as the largest player in the pharmaceutical sector, Kimia Farma's profit only grew 2.2 percent in 2016. In fact, in 2015 it grew 13.15 percent and in 2014 grew 9.36 percent. So, while it is still profitable, the growth of the profits seems to be slowing down. This phenomenon is interesting to be studied further, by analyzing the influence of government health expenditure and Health Social Security Organization on pharmaceutical sector performance in Indonesia Stock Exchange period 2015-2017. The national health insurance program will still be a magnet for pharmaceutical issuers. Population growth, increasing government budgets to pay the contributions of the poor and improving public incomes can be a supplement to corporations engaged in the sector. Different characteristic of government procurement conditioned pharmaceutical companies must have sufficient liquidity and working capital because they have to import raw materials, production line investment (increase production capacity), distribution costs; that is still quite expensive in Indonesia, and the cost of money that must be considered as it can undermine the company's net profit.

1 INTRODUCTION

In order to embody people's right in receiving social security health, Law No. 40 of 2004 on National Social Security System was enacted. Through the Law, the Government established the Health Insurance Program as part of a program to build the National Health Insurance System. Referring to Law 36/2009 and Law 40/2004 in regards to providing health insurance to the public, the government should provide health facilities, health workers, medical devices and medicines that the community needs (Kadarisman 2015).

As an implementation of National Social Security System per January 1, 2014, a National Health Insurance Program instituted. The program joints all health insurance programs that have been implemented by the government are integrated into a Social Security Administering Body. Similar to the Community Health insurance program, the government is responsible for paying National health insurance fees for the poor and incapable that registered as beneficiaries of the payment support. The National health insurance organized by the Social Security Administrative Agency has become a revolution in health services in Indonesia. Previously, Indonesia's health services faced a very crucial problem, especially in terms of financing. There are still layers of society that are not covered by health

services. Through the National health insurance program, the government also regulates that drugs can be reached by people's purchasing power. Therefore, generic drugs with cheap prices to be excellent in health services (Susanti 2016).

In 2017 Ministry of Health allocated with a reasonable budget, which is IDR 58.3 trillion. The details for the fund are IDR 25.5 trillion (43.8 percent) for the allocation of the National health insurance program, through the National health insurance program, the government provides financial protection, especially for the poor in access to health services. In 2017 the number of beneficiaries of the payment support of 92.3 million people with a total contribution of IDR 25.4 trillion. As a comparison of achievements in 2016 was 91.1 million PBIs with a total contribution of IDR 24.8 trillion (Kemenkes 2013).

Indonesia's public health spending is still low, only about 3.8% of total gross domestic product (GDP). But with the National Health Insurance program, the average spending estimated will increase to 5% of total GDP. Meanwhile, with National Health Insurance Program drug procurement using e-catalogs, local pharmaceutical wholesalers lose their captive market. Meanwhile for the producers, the fixed-price of government contract has given great pressure to its profitability.

1.1 Previous researches

As starting point, previous researches related with the impact of National Health Insurance Program and the increase of government spending to support the programs are described below.

Research by Gloria is found that procurement of drugs that focus on generic drugs in large quantities brings major changes to the Indonesian pharmaceutical market. Consequently, pharmacies are losing customers, pharmaceutical wholesalers are losing the hospital market, while pharmaceutical industries are experiencing minus growth because they have to operate at a low price and low margin. The prescriptive pattern that turns into generic drugs makes the pharmacy margin only around IDR 5,000 per prescription, so the pots experience a decrease in turnover ranging from 20 to 60 percent (Gloria 2016).

Though Health Social Security Organization's generic drug policy has shaken retail drug market, according the Ilvia et al, apart from the impact of Health Social Security Organization implementation on the pharmaceutical industry in Indonesia which is temporarily perceived to be burdensome, the business prospects for health services are more attractive in the future. Shares of the health sector companies and its supporters began ogled many parties into an alternative investment (Ilvia et al. 2014).

According to research by Suhaji in 2105, pharmaceutical products sale at least are affected by two factors. The first is a public awareness of the importance of health. Second, the level of people's spending on healthcare tends to increase (Sahaji & Widiastuti 2016)

Worldwide perspective on universal health coverage practices (Health Social Security Organization like program) relationship or effects on pharmaceutical industries according to Tannoury, in developing countries the universal health coverage become the driving force for the growth of health and wellness sectors and pharmaceutical industry are forecasted to account for 49% of the market share by 2019. The market prognosis of Intercontinental Market Services estimated pharms market development to be worth $1,190 billion in 2016, with emerging markets accounting for 30% of that market (Tannoury & Attieh 2017).

Dieleman projected that the largest growth in total health spending would be in government spending from 2015 to 2040. The allocation of the national health budgets towards universal health coverage (UHC) also needs to be balanced with spending on other crucial health areas, including health emergency preparedness, health promotion, and capital investments such as hospitals (Dieleman et al. 2018).

2 METHOD

Studies on the impact of government spending on health and Health Social Security Organization, on pharmaceutical companies listed on the Indonesia Stock Exchange, need to first describe the related policies and related concepts, as a basis for conducting the in-depth and comprehensive analysis.

2.1 Policies related to national health insurance and health social security organization

2.1.1 Law and regulations

Health is the basic right of everyone, and all citizens are entitled to health services. The 1945 Constitution mandates that health insurance for the people, especially the poor and incapable, is the responsibility of the central and regional governments. In the change of 1945 Constitution, Article 34 paragraph 2 states that the government develops Social Security System for all Indonesian people. The Government enacted the 1945 Constitution by issuing Law No. 40 of 2004 on the National Social Security System to provide comprehensive social security for every person in order to fulfill the basic needs of decent living towards the realization of a just and prosperous Indonesia. In Law No. 36/2009 on Health, it is also affirmed that everyone has equal rights in obtaining access to health resources and obtaining safe, quality and affordable health services (Ministry of Health Republic of Indonesia 2013).

Prior to National Health Insurance, the government has attempted to pioneer some forms of social security in the health sector, such as Social Insurance for civil servants, pension and veteran's recipients, Health-Preserving Insurance for state and private employees, and Health Insurance for the Army and Police. For the poor and underprivileged, since 2005 the Ministry of Health has implemented a social health insurance program, originally known as the Health Insurance Program for the Poor, or more popularly known as the Health Insurance for the Poor program. Then from 2008 until 2013, the program changed its name to the Public Health Insurance program (Hamdani 2013).

2.1.2 Concepts of universal health coverage

Universal health coverage is a system that ensures everyone in a society obtains the health-care services they need without incurring financial hardship. An universal health implementation in a country depends on factors: who is covered, for which services are they covered, and with what level of financial contribution (Savedoff et al. 2012).

2.2 Theories concerning financial analysis and risk management

2.2.1 Understanding financial management

Financial management, or often called corporate finance or business finance, is the application of economic principles that are primarily concerned with financial decision making within business entities. Financial management decisions include maintaining the stability of cash balances, extending credits, acquiring other companies, borrowing from banks, and issuing shares and bonds (Fabozzi 2003).

2.2.2 Understanding financial analysis

Financial analysis is a financial management tool. It consists of evaluating the financial condition and operating performance of a business, industrial, or even economic enterprise, and forecasts of its future conditions and performance. In other words, the means to check for expected risks and returns (Fabozzi 2003).

2.2.3 The influence of government projects on industry

According to Jones in his book Financial Economics, states that projects that produce public goods, which are considered to provide direct benefits to society, as well as production costs derived from the government budget, can directly affect the demand for public goods and services. This means that through spending for the needs of public goods, the government indirectly makes investments through public goods and services providers so that the required goods can be available (through contracts/procurement tenders) and immediately utilized for the community needs (without going through various distribution channels). Conversely, for the industry, government procurement will be an attractive captive market, because it is massive, repetitive and sustainable (Alghifari 2013).

2.2.4 Understanding project risk management

Project risk management is about doing the right thing in everything, and doing it right - effectiveness and efficiency in general. The efficiency risks in the sense of some attributes embrace all this holistically and integrated. Project risk management is typically associated with the development and evaluation of contingency plans that support an activity-based plan, but effective project risk management will play a role in the development of preliminary plans and contingency plans for all six Ws, Who, What, When, Why, and Where, which should be a fundamental question in analyzing project risk before it is decided to be taken or implemented (Chapman et al. 2003)

2.3 Problem formulation, research purposes, and research methodology

2.3.1 Formulation of the problem

This research is built on previous academic researches and studies on listed companies (issuers) reports. Main variables being analyzed are trend on government spending for health sector and Health Social Security Organization, respond of the pharmaceutical industries to the trend of government health spending, factors influenced pharmaceutical industries, and performance of pharmaceutical issuers in IDX due to the spending trend.

The problem formulation in this research is whether or not trend on public health spending significant increase becomes a good and sustainable opportunities for pharmaceutical industries in Indonesia.

2.3.2 Research purposes

This research aims to identify supporting factors, immediate problems, and possible solutions to improve the performance of pharmaceutical industries in maximizing the increase on public spending for benefit both of the industry and Indonesian people.

2.3.3 Research method

This research includes library research, the main object of which is books or other literature sources, namely journals, articles, and research reports from previous researches. The data are searched and found through literature reviews that are relevant to the discussion. Research method is the mixed method research category with the procedure of data collecting activities in a span period of time (2015-2017) and the final presentation technique descriptively. According to Denzin and Lincoln, the information may be quantitative or qualitative, as long as the research uses a scientific setting, with the intention of interpreting phenomena that occur and are carried out by involving various existing methods (Johnson et.al. 2007).

The study carried out is to solve the problem, which is basically based on a critical and in-depth study of the relevant library material. Because this research is a library research, the data collection methods used in the form of library materials that are continuous (coherent) with the object of the study are pharmaceutical issuers in IDX. Sources of data obtained through library research of valid printed and online journals, academic studies, and IDX issuer reports that support the deepening and sharpness of analysis.

3 RESULT AND DISCUSSION

In January 2014, there was a major change in the healthcare sector in Indonesia. The mandate of the Law on the National Social Security System and the Law on the Social Security Administering Body are effective. The law mandates the holding of a national health insurance program for all Indonesian citizens. Health is not as much as private health insurance premiums. Guarantees are given no limit. The pre-existing condition also does not apply. So, those who are sick can still apply and enjoy health insurance.

3.1 Trend of government spending on health sector and health social security organization

In 2019, the government targets all Indonesians already registered in Health Social Security Organization. Those who are poor and incapable pay the premium, are paid by the government. The big echoes of the National health insurance program initially gave hope to the pharmaceutical industry to increase sales. The program is believed to increase drug demand. Pharmaceutical companies also put great expectations on this program. Drug demand is increasing. The types of drugs needed are also more diverse. By 2015, there are 796 items of drugs listed in the e-catalog auction. Last year, the type of drug auctioned almost doubled, i.e. 1,240 items (KPK 2016).

Until January 27, 2017, participants are National Health Insurance-Indonesia Health Card (NHI-IHC) has reached 172.9 million people. That's almost 70% of the total population of Indonesia. So the presence of NHI-IHC program brings a tremendous effect. In 2016, NHI-IHC has contributed IDR 152.2 trillion to Indonesia's economic growth. In addition, the NHI-IHC program also has a role to create employment opportunities for 1.45 million people. If projected to 2021, NHI-IHC contributes 289 trillion Rupiah and generates employment for 2.26 million people (Humas 2017b).

The National Health Insurance (NHI) has provided many benefits to the community, can be seen from the total trend of utilization NHI-IHC which continues to increase from 2014 of 92.3 million people, become 146.7 million people in 2015, by 2016 of 192.9 million people, and in October 2017 of 182.7 million people. In terms of the government's role for financial protection, especially how the poor accessing to health services, the number of poor participants paid by the government as beneficiaries of the payment support (PBI) is 87.8 million people with a total contribution of IDR. 19.8 trillion (2015), 91.1 million people with a total contribution of IDR. 24.8 trillion (2016), and in 2017 92.3 million people with a total contribution of IDR. 25.4 trillion (Ministry of Finance Republic of Indonesia, 2015).

The number of health facilities receiving NHI-IHC services is increasing every year. At a Primary level facilities are 19,969 units (2015), 20,708 units (2016), and 21,763 units (2017). At advanced health facilities 1,847 units (2015), 2,068 units (2016), and 2,292 units (2017). At pharmacy and optical facilities, there are 2,813 units (2015), 2,921 units (2016), and 3,380 units (2017) (Ministry of Finance Republic of Indonesia 2017).

3.2 Pharmaceutical companies respond to the increase of government budget in health sector and health social security organization

One of the most remarkable impacts on the pharmaceutical industry in Indonesia over the past two years is the regulation of Health Social Security Organization. On one hand, Health Social Security Organization is becoming a new opportunity, but on the other hand is very challenging. The pharmaceutical company's policy in addressing the market also changed. Generic product growth is high, while branded products are trimmed. Moreover, Health Social Security Organization currently targeted 125 millions of people. It is like Obamacare in the United States that meet many challenges.

The availability and affordability of medicines are one of the important factors that should be of concern

to the government as national drug consumption accounts for 40 (forty) percent of overall health spending. The cost according to Professor of Public Health Faculty of the University of Indonesia, Prof. Dr. Hasbullah Thabrany is one of the highest in the world. Unlike other more advanced countries such as the United States, Britain, and Germany, the cost of drugs is only around 11 percent to 12 percent of health care costs (Humas 2017). Therefore, the cost-effectiveness of medicines is important to note in order to have an impact on the costs efficiency of national health care. Generic drugs are widely used as an alternative to expensive drug prices. The high prices of drugs, mainly is caused by raw materials of drugs that still depend on imports. More than 96% of raw materials are imported, so the price depends on dollar price. In fact, some of imported drug raw materials still do not meet the standards of pharms grade. To overcome this, the solution of the pharmaceutical industry is to produce its own raw materials by utilizing local materials.

Hospitals that used branded products experienced cuts in their third-grade markets. Branded products still have a place in the hospital, but are in the upper middle segment. By 2014, government hospitals are already implementing the Health Social Security Organization policy. Since 2015 until now, private hospitals have begun to adopt this BPJS policy. Many are already using generics. This is a challenge for the pharmaceutical industry, especially producers of ethical products branded. Because previously only targeting customers for ethical product segment is not consumer OTC (over the counter) and consumer of the generic drug as an end user, but ethical consumer which is decision maker like doctors and middle class to above (Bank of Indonesia 2017).

These conditions encourage all players in the pharmaceutical industry to increase capacity because the need for drugs and pharmaceuticals will be even greater. The strategy that some pharmaceutical companies use is cost leadership because the competition is in the price. Before 2015, for example, the utilization of the national drug production capacity has reached 70%. Meanwhile, pharmaceuticals still grow about 10% -15% (Marketeers Editor 2015). Without the addition of capacity of pharmaceutical companies will be difficult to meet the quality and quantity demand demanded in this segment. Because if national production is not ready, while the demand is clear and large, other countries like India and China, may pose a threat to the Indonesian pharmaceutical industry.

3.3 Drug production is a key of competition and profit of pharmaceutical companies

The price of medicine in Indonesia is still very expensive when compared with some other countries in the world. The high price of drugs in Indonesia partially is caused by drugs sold in the market are branded drugs and not those generic drugs made by the government that commonly used by the community. In addition, high prices are caused by the absence of regulation related to Highest Retail Price (HET) for a drug with the trademark. In Indonesia, there are three types of drugs in circulation, namely generic drugs, branded generic drugs, and patent medications (Rachman 2015). In addition, the price of expensive drugs is due to high promotional costs (marketing fees) and gratuity (in the form of sponsorship and/or facilities directly) to doctors of drug manufacturers due to unhealthy competition. Another factor that keeps medicine price still expensive is the trend of healthy living began to be part of the middle income lifestyle. People needs supplement products to support their modern healthy lifestyle. People are now starting to be more preventative than curative in health problems. This trend is often used by drug mafia to trick consumers (who believe more branded drugs than generic).

According to data from the Ministry of Health, the number of pharmaceutical companies in Indonesia reached 210 companies with the number of distributors or Pharmaceutical Wholesalers (PBF) as much as 2087 (Ariati 2017). This picture shows that there is a fairly tight competition among drug providers so that unhealthy competition arises, especially in the mechanism of cooperation between pharmaceutical companies and PBF with doctors. In such cooperation, doctors will receive 10-20 percent discount on drug sales from pharmaceutical companies. The discount is given in the form of money and facilities. This is one of the factors that cause drug prices to soar and patients often get unnecessary prescriptions. This condition is possible because the people of Indonesia are now more aware of health. This is also triggered by the increasing purchasing power and reflected by the increasing number of middle class in Indonesia.

Up to now, there is no dominant player in the pharmaceutical industry that dominates all market segments. The company that is considered the number one player, Kimia Farma, only has a 7% market share. Other players are fragmented in their respective specialties, such as orthopedic products, ENT, cardio neuro, and so on. Each manufacturer possesses a competitive advantage and an innovation that continues to develop. In general, however, the pharmaceutical industry, especially drug manufacturers, is divided into three business focuses, namely OTC (over the counter), ethical branded, and generic. When it comes to competition in marketing, people still rely on both OTC and ethical branded products (Mapping Facility for Pharmaceutical Services 2016).

3.4 Performance of listed pharmaceutical sector companies (issuers) on the stock exchange during 2015-2017

An issuer is an entity/company conducting a Public Offering, namely Securities offering conducted by Issuers to sell Securities to the public based on the procedures provided for in the applicable Law (OJK 2018). Issuers may be individuals, corporations, joint ventures, associations or organized groups. In this study, publicly-listed pharmaceutical companies are public companies that focusing their efforts on the pharmacy and pharmaceutical production sectors.

Pharmaceutical issuers will give benefit from the JKN drug procurement tender. Considering the state budget revenue and expenditure for the health sector also increased. From the research of the Ministry of Finance (MoF) the state budget for health this year amounted to IDR 104 trillion compared to last year amounted to IDR 67.2 trillion. Of course, the proportion of drug procurement is also getting bigger in 2018 in the future. The National health insurance program has grown rapidly to more than 172 million members. In order to seize the opportunity of the segment of generic drugs without a brand on National health insurance program, issuers began to crowd to increase the production and type of drug. One of the listed companies that benefited was Kimia Farma Company (KFEF). KAEF is a large state-owned pharmaceutical company. As well as KAEF is increased its production up to four times at the plant of Banjaran West Java and build the first salt pharmaceutical factory in Indonesia (Public Relation- Health Social Security Organization 2017a).

In 2016, Kimia Farma's net profit grew only 2.2 percent. In fact, in 2015, the growth was about 13.15 percent and in 2014 grew 9.36 percent. So, while still profitable, the growth of the profits seems to be slowing down. In terms of sales, compared to 2014 and 2015, Kimia Farma's sales growth in 2016 is the highest, at 19 percent with total sales of IDR 5.8 trillion. Meanwhile in 2015, the growth is only 6.6 percent and in 2014 only reached 4.14 percent (Public Relation- Health Social Security Organization 2017a).

According to stock market analysts, by 2019, Indonesia's per capita income is predicted to be $ 4,100 USD. By that amount, about $ 205 USD is projected as healthcare expenditure. Meanwhile KAEF can have a positive trend in the stock because it is a state-owned company and JKN/BPJS is a government program. Thus, KAEF is predicted to be flooded with demand for drugs, especially generic drugs to support the JKN program. Another big player is Kalbe Farma Inc (KLBF), still a leader in the pharmaceutical sector, KLBF preparing special facilities for generic drugs. By the end of the third quarter of 2017, KLBF recorded revenues of more than IDR 15 trillion and a net profit of IDR 1.8 trillion.

However, although it is considered as the main booster of the growth of the national pharmaceutical market, the National Health Insurance Program which includes the procurement of generic drugs without a brand makes pharmaceutical issuers to tighten the profit margin. In the annual report of Kimia Farma Inc. citing data from Intercontinental Marketing Service (IMS) Health that in 2016 the Indonesian pharmaceutical market grew by 7.5% compared to 2015 or reached IDR 67.21 trillion.

The growth rate is expected to continue until 2017. In fact, in the annual report, the KAEF-coded company estimates the market will grow in line with government policies to support the pharmaceutical industry from upstream to downstream. This is expected to encourage the growth of the pharmaceutical industry to be able to increase the demand for medicines, so the national pharmaceutical market is predicted to grow 5% -6% to reach IDR 78 trillion (Public Relation- Health Social Security Organization 2017a).

The growth of the pharmaceutical market is not proportional to the rate of return on profit of each pharmaceutical issuer. Particularly the issuers who are the main players of the National Health Insurance program such as INAF or KAEF. Meanwhile the National Health Insurance program may double the production of drugs by the company, the profit margin level must be reduced, since government projects have special provisions, especially those concerning profit margins, which usually should not exceed 10 percent of the sale price. This has an impact on each issuer competing for government projects/tenders, although it must have adequate liquidity and working capital. Plus, the demand must bring in imported raw materials and payment deadlines from the government that has not been stable or long enough.

In 2016, the decline in profits occurred in most pharmaceutical issuers. This phenomenon causes a lot of stock market analysts feel pessimistic National Health Insurance program can be a pharmaceutical manufacturer's boon to gain significant profit growth. The reason, in addition to the margin that is limited by government regulations, the winners of project tenders must also carry distribution to remote areas. The issuers must be willing to distribute to the lowest health service network and in distant locations such as hospitals throughout Indonesia. A number of academics in the pharmaceutical field also stated that the National Health Insurance program failed to hoist the profit of the pharmaceutical industry. Procurement of drugs that focus on generic drugs in large quantities is considered to bring major changes to the Indonesian pharmaceutical market. Consequently, pharmacies lost consumers, pharmaceutical wholesalers lost the hospital market. Meanwhile the pharmaceutical industry experienced minus growth due to operating low price and low margin.

In 2016, the companies that receive the most National Health Insurance projects are KAEF and INAF, as well as other state-owned companies such as Phapros Inc. KAEF, which has achieved significant sales growth of 19.6%, has only a 2.3% profit growth. Meanwhile, Phapros assisted by the growth of sales of over-the-counter drugs and medical devices still packed a profit of 38.1%, and sales growth of 18.1%.

However, for INAF, sales growth of 3.3%, actually resulted in losses of IDR17,4 billion. The production cost of raw materials, production line investment (for production capacity increase), distribution costs are still quite expensive in Indonesia, and the cost of money that must be borne due to the smoothness of payments, erode the company's net profit such as INAF. So the National Health Insurance program has the potential to erode the profit margins of the issuers. Side costs or initial investment is not worth the margin. The market share of generic drugs with a margin that is not too large, causing the pharmaceutical industry needs to implement effective strategy in order to reduce the cost of production (Danzon & Pauly 2002).

4 CONCLUSIONS

Referring to the results of the analysis of the above discussion it can be concluded that some matters related to the analysis of the influence of government health spending and Health Social Security Organization, on the performance of the pharmaceutical issuers in the Indonesia Stock Exchange period 2015-2017, among others:

1) The National Health Insurance Program (NHIP) organized by the Health Social Security Organization has become a revolution in health services in Indonesia. By means of the NHIP program, the government also regulates that drugs can be reached by people's purchasing power. Therefore, the procurement/production of generic drugs at low prices becomes the main issue of Health Social Security Organization health implementation.

2) The national health insurance program Health Social Security Organization will still be a magnet for pharmaceutical issuers. Population growth, increasing government budgets to pay the contributions of the poor (PBI), and improving public incomes can be a supplement to corporations engaged in the sector.

3) Pharmaceutical companies must have sufficient liquidity and working capital because they have to import raw materials, production line investment (increase production capacity), distribution costs; that is still quite expensive in Indonesia, and the cost of money that must be considered as it can undermine the net profit company.

REFERENCES

Alghifari, E. S. 2013. Pengaruh risiko sistematis terhadap kinerja perusahaan dan implikasinya pada nilai perusahaan (Studi pada perusahaan food and beverage di bursa efek indonesia tahun 2007-2011). *Jurnal ilmu manajemen dan bisnis*, 04(01): 1–16.

Ariati, N. 2017. Tata kelola obat di era sistem jaminan kesehatan nasional (JKN). *Integritas*, 3(2): 231–243.

Bank of Indonesia. 2017. Maret 2017 KSK. Igarss 2017, (1): 1–5.

Chapman, Chris; Ward, S. 2003. *Project risk management* (*2nd ed.*). Chichester: John Wiley & Sons Ltd.

Danzon, P. M., & Pauly, M. V. 2002. Health insurance and the growth in pharmaceutical expenditures. *The journal of law and economics*, 45(S2): 587–613.

Dieleman, J. L., Sadat, N., Chang, A. Y., Fullman, N., Abbafati, C., Acharya, P., & Murray, C. J. L. 2018. Trends in future health financing and coverage: future health spending and universal health coverage in 188 countries, 2016–40. *The Lancet*, 391(10132): 1783–1798.

Fabozzi, F. J. 2003. *Financial management* (*2nd ed.*). New Jersey: John Wiley & Sons, Inc.

Gloria, G. 2016. BPJS kesehatan pengaruhi industri farmasi secara signifikan. Universitas Gadjah Mada. Retrieved on June 8, 2018 from https://ugm.ac.id/id/berita/11045-pjs.kesehatan.pengaruhi. industri.farmasi.secara. signifikan.

Hamdani, P. (2013). Implementasi program pelayanan jaminan kesehatan masyarakat berdasrkan Peraturan Kementrian Kesehatan Republik Indonesia Nomor 903/Menkes? per/V/2011. Fakultas Hukum Universitas Brawijaya.

Humas (2017a). Emiten Farmasi : Margin laba tergerus JKN [Online]. Retrieved http://www.jamsosindonesia. com/newsgroup/selengkapnya/emiten-farmasi-margin-laba-tergerus-jkn_10345. Accessed 25 May 2018

Humas. (2017b). JKN-KIS Tingkatkan produktivitas tenaga kerja Accessed 25 May 2018.. Retrieved https://bpjs-kesehatan.go.id/bpjs/index.php/post/read/2017/555/Lawan-Kanker-Serviks-dengan-Deteksi-Dini-IVAPap-Smear. Accessed 25 May 2018.

Ilvia, R., Desiyanti, R., & Husna, N. 2014. Faktor faktor yang mempengaruhi likuiditas saham pada perusahaan farmasi Yang Go Publik. *OJS Bung Hatta University*, 5(2): 1–14.

Johnson, R. B., Onwuegbuzie, A. J., & Turner, L. A. 2007. Toward a definition of mixed methods research. *Journal of mixed methods research*, 1(2): 112–133.

Kadarisman, M. 2015. Analisis tentang pelaksanaan sistem jaminan sosial kesehatan pasca putusan Mahkamah Konstitusi No. 07/PUU-III/2005. *Jurnal Hukum IUS QUIA IUSTUM*, 22(03): 467–488.

Kemenkes. (2013). Buku pegangan sosialisasi jaminan kesehatan nasional dalam sistem jaminan sosial nasional. departemen Kesehatan RI, 1–75.

KPK. 2016. Kajian obat dalam sistem sistem nasional kesehatan: Kegiatan pencegahan korupsi pada JKN 2013-2016. Komisi Pemberantasan Korupsi. Retrieved https://acch.kpk.go.id/id/component/bdthe mes_shortcodes/?view=download&id=4849acbc ba136b3f2ef1c35ee1752c. Accessed 25 May 2018.

Marketeers Editor. 2015. Ini dia masa depan farmasi menurut phapros marketeers. Majalah bisnis & marketing online - marketeers. Retrieved http://marketeers.com/ini-dia-masa-depan-farmasi-menurut-phapros/. Accessed 25 May 2018.

Rachman, T. 2015. Harga obat di indonesia lebih mahal dari malaysia republika online. Retrieved https://www. republika.co.id/berita/nasional/umum/15/12/14/nzcof8219-harga-obat-di-indonesia-lebih-mahal-dari-malaysia. Accessed 25 May 2018.

Savedoff, W. D., De Ferranti, D., Smith, A. L., & Fan, V. 2012. Political and economic aspects of the transition to universal health coverage. *The Lancet*, 380(9845): 924–932.

Suhaji, & Widiastuti, T. 2016. Faktor-faktor yang mempengaruhi peningkatan (Studi pada tenaga penjualan perusahaan farmasi di Semarang), 13: 157–180.

Susanti. 2016. Kualitas pelayanan pasien badan penyelenggara jaminan sosial di pusat kesehatan masyarakat Biromaru Kabupaten Sigi. *E Jurnal Katalogis*, 4(3): 47–57.

Tannoury, M., & Attieh, Z. (2017). The influence of emerging markets on the pharmaceutical industry. *Current Therapeutic Research - Clinical and Experimental*, 86 (January 2015): 19–22.

Advances in Business, Management and Entrepreneurship – Hurriyati et al (eds)
© 2020 Taylor & Francis Group, London, ISBN 978-0-367-27176-3

The role of financial constraint on the relationship between working capital management and firms' performance

R.H. Setianto & R. Hayuningdyah
Universitas Airlangga, Surabaya, Indonesia

ABSTRACT: This study aimed at analyzing the impact of working capital management on firm perform-ance and how financial constraints determine this relationship. For this purposes, the annual data of manufac-turing firm listed in Indonesian stock exchange during 2011-2015 periods were employed. The results from regression analysis indicated an inverted U-shape relationship between working capital and firm performance. Fur-thermore, it was also found that financial constraints weakened the relationship between working capital and firms' value. There were robust results for the various measurement of financial constraint.

1 INTRODUCTION

The ultimate goal of financial management is to maximize the corporate value. Corporate value can be seen through the performance of the company which is reflected in the stock price. In maximizing the value of the company, the company certainly must have a good performance.

One of the things that affects the company's per-formance is working capital management. A research conducted by Schiff & Lieber (1974) suggested that there was an influence of the decision of working cap-ital on the performance of the company. Accordingly, Smith (1980) argues that working capital management has an important role in profitability, corporate risk and corporate value.

Increased investment in working capital is expected to have a positive impact on the com-pany's performance. Large inventory levels can reduce the cost of bookings to be borne by the company, provide protection from fluctuations in the price of input goods, and prevent any con-straints in the production process due to future supply shortages or future cash depletion (Fazzari & Petersen 1993). In addition, an increase in the firm's receivables allows the company to drive sales when demand decreases and allows custom-ers to assess the quality of the company's products and services before making payments that will strengthen the relationship between the company and its customers (Wilner 2000). Early debt repay-ment to earlier suppliers also allows the company to get discounts from suppliers, thereby reducing the costs incurred by the company to supply sup-plies. This is supported by research conducted by Charitou et al. (2012) which revealed that there was a positive influence between working capital management and company performance.

However, investment in working capital can also have a negative impact on the performance of the company. Large inventories will increase storage costs. Moreover, since most of the company's funds are allocated for working capital, then investing in large working capital can hinder companies to take investment opportunities that will increase the value of the company. This is supported by research con-ducted by Vahid et al. (2012) which suggest that there was a negative influence between working cap-ital management and profitability.

The positive and negative impacts of working capital indicate that there is a tradeoff in working capital policy. The performance of the company will improve with the increase of working capital up to a certain optimum point and will decrease when working capital exceeds the optimum point. The working capital management applied by the com-pany varies. One of the things that distinguishes the use of working capital is the financial condition faced by each company. Investments made by firms depend on financial factors such as the availability of internal funding sources, access to capital markets or financing costs (Fazzari et al. 1988). Since invest-ment in current assets is less profitable and positive working capital requires funding, investment in working capital is more sensitive to firms that have financial constraints than investment in fixed assets (Fazzari & Petersen 1993). The smaller the com-pany's internal funding, the more difficult it is for companies to access the capital market and the less likely it is for firms to invest in working capital. In addition, working capital management also depends on the complexity of each company.

Companies that have a high dependence on supply, as well as companies that apply the credit system to their sales system require a tighter working efficiency of working capital than companies that do not apply the credit system to their sales. The use of the credit system at the company will increase the volume of sales made by the company. On the other hand, the use of the credit system also delays the cash coming into the company so as to disrupt the company's liquidity. To anticipate this, some companies decide to add cash reserves instead of managing working capital.

One industry that requires proper management of working capital efficiency is the manufacturing industry. The industry is dependent on the availability of raw materials and supplies in carrying out its operations. When the company supplies more raw materials, the company may get a discount from the supplier and reduce the company's likelihood of losing the revenue that should be earned due to stock shortages. Often to increase the level of sales, manufacturing companies impose a credit system so that they cannot directly earn revenue on products that have been sold. Hill et al. (2010) also revealed that the working capital policy of manufacturing companies was different from the service companies because the manufacturing sector generally has high inventory levels whereas the service sector generally has no inventory at all. If it is not properly managed, the company is at risk of losses due to the high cost borne by the company. The working capital management also differs between manufacturing companies that have and do not have financial constraints. Therefore, this study focused on companies in the manufacturing sector.

In Indonesia, research conducted to examine the effect of working capital management on the performance and risk of the company has been done by Ermawati (2011) who examined 14 manufacturing companies listed in the Indonesia Stock Exchange. The result of the research showed that the working capital management had no significant effect to profitability or performance and company risk. However, although not significant, the direction of the relationship between cash conversion cycle had a negative effect on company performance.

So far, there are not many research in Indonesia considering financial constraints in the use of working capital and its relation to firm performance. This research attempted to contribute related to the study of working capital management in at least two things: (1) this research used nonlinear model to accommodate trade off in working capital management; (2) the influence of working capital on the performance of the company may be affected by the condition of the company. This study measured the influence of financial constraints on the relationship between working capital and firm performance by using several criteria: cash flow, interest coverage, dividend payout ratio, cost of external financing, and size to obtain robust result.

1.1 Working capital and performance

The net trade cycle is the period when the cash is issued by the company to finance the inventory on the supplier until the cash is obtained from the sale to the customer indicated by the length of the sales period. The longer the sales period, greater the value of net trade cycle, which also means the longer the company can collect cash. The large net trade cycle value shows large working capital. In contrast, the small net trade cycle value shows little working capital.

Companies that put a great investment in working capital will improve the company performance. High levels of working capital allow companies to store more inventory and thus have the possibility of getting discounts on purchases from suppliers, avoiding the occurrence of stock-outs and price fluctuations in the future. In addition, the company can also provide looser credit to customers thereby encouraging customers to make purchases that will impact on increased sales at a time of declining demand (Emery 1987). However, there is also a negative impact of investment on working capital on the performance of companies at a certain level of working capital. Large inventories will increase storage costs that will increase as inventories (Kim & Chung 1990). In addition, maintaining a high level of working capital will reduce the company's ability to make other investments that can maximize the value of the company.

The positive and negative impacts of working capital indicate that there is a trade-off in working capital decisions. Specifically, the increase of working capital will improve company performance until certain optimum point of performance will decrease. Companies must apply optimal working capital levels to create a balance between cost and benefits so as to maximize the value of the company.

1.2 The role of financial constraint

Financial constraints are the company's limitations in obtaining external funding. According to Hennessy & Whited (2006), financial constraints occur when a company has a lucrative investment opportunity, but the company is experiencing limitations to fund the investment opportunity by using external funding.

Fazzari et al. (1988) argue that corporate investment depends on funding factors such as the availability of internal funding, capital market access or financing costs. The existence of information asymmetry on external funding will incur external funding costs more than internal funding.

Similarly, investments in working capital also depend on external funding access. Working capital is more sensitive than investing in fixed assets in companies with financial constraints (Fazzari & Petersen 1993). Because the level of positive working capital requires funding, then the optimal working capital level will be lower in companies with financial constraints. In the case of financial constraints, funds are owned by a limited company, so that the company must make strict management of current assets in order to remain able to meet obligations and run the company's operations by applying strict credit and managing fewer inventories. Contrary to companies with large internal funding capacity and easy access to capital markets, high levels of working capital are applied by applying loose credit and managing more inventories (Hill et al. 2010).

2 METHODS

2.1 Data

This research employed annual data of manufacturing firms listed in Indonesian Stock Exchange in 2011-2015 periods. According to Hill et al. (2010), working capital policy of manufacturing companies is clearly different from service companies since the manufacturing sectors normally have a high inventory level while the service sectors generally have no inventories at all. Thus, to avoid bias due to differences in the business characteristics of the different industries, this study used manufacturing firms as samples. Data were obtained from the financial statements provided by the Indonesian Stock Exchange website. Sampling was conducted by using purposive sampling method with the following criteria: a) companies are listed on the Stock Exchange in 2005-2014 and included in the manufacturing companies; b) the company financial statements present a complete financial data necessary for calculating the variables; c) there is no negative equity; and d) the financial statements are presented in rupiahs. Table 1 provide the descriptive statistics of the variables.

2.2 Methodology

To test the first hypothesis that there was inverted u-shaped relationship between net trade cycle and company performance the following regression model was employed:

$$Q_{i,t} = \beta_0 + \beta_1 NTC_{i,t} + \beta_2 NTC^2_{i,t} + \beta_3 SIZE_{i,t} + \beta_4 LEV_{i,t} + \beta_5 GROWTH_{i,t} + \beta_6 ROA_{i,t} + \varepsilon_{i,t}$$

(1)

To test the second hypothesis that the financial constraints moderation had an effect on

Table 1. Descriptive statistic.

Variables	Mini	Max	Mean	Std. D
TOBIN'S	0.33	7.83	1.33	0.94
NTC	-14.69	493.59	109.31	62.03
SIZE	21.66	32.94	27.93	1.56
LEVERAGE	0.04	2.66	0.48	0.24
GROWTH	-0.58	1.29	0.15	0.21
ROA	-0.51	0.56	0.08	0.09
N (samples)	435			

sharpening working capital on company performance, the following model was used:

$$Q_{i,t} = \delta_0 + \delta_1 NTC + \delta_2(NTCi_t \times DFC_{i,t}) + \delta_3 NTC^2 + \delta_4(NTC^2_{i,t} \times DFC_{i,t}) + \delta_5 SIZE_{i,t} + \delta_6 LEV_{i,t} + \delta_7 GROWTH_{i,t} + \delta_8 ROA_{i,t} + \varepsilon_{i,t}$$

(2)

Where $Q_{i,t}$ is Tobin Q as measurement of company's performance measured by ratio of market value of asset (market value of equity plus book value of debt) divided by book value of asset. It was calculated based on the following formula:

$$Q_{i,t} = \frac{MV \ of \ equity_{i,t} + BV \ of \ liability_{i,t}}{BV \ of \ assets}$$

$NTC_{i,t}$ is net trade cycle as measurement of working capital management, NTC was calculated based on the following formula:

$$NTC_{i,t} = \frac{acc.receivable_{i,t}}{sales_{i,t}} \times 365 + \frac{inventories_{i,t}}{sales_{i,t}} \times 365 - \frac{acc.payable_{i,t}}{sales_{i,t}} \times 365$$

DFC is dummy variable for financial constraint, which takes value of 1 if firm is classified as financial constraint and 0 otherwise. Various measurements of financial constraint were employed to gain robustness of the results namely; cash flow, interest coverage, dividend paying, dividend payout, cost of external financing, and size.

Control variables consisting of firm size is natural logarithm of the total sales, leverage is total debt to total asset, asset growth, and return on asset.

3 RESULTS AND ANALYSIS

Table 2 provides us with regression results based on equation 1. Based on regression result, it can be seen that working capital has inverted U-shaped relation to company performance. The inverted U-shaped relationship on the effect of working capital on firm performance can be described as follows.

The higher working capital will further improve the performance of the company to a certain opti-mum point of increase in working capital which will result in decreased company performance. The in-crease in working capital before reaching its opti-mum point can improve performance due to the in-crease of working capital related to the increase of sales which then impact on the increase of company's revenue. The company will order large inventories to suppliers so as

Table 2. The effect of working capital management on performance.

	Coefficient	P-value
Constanta	-3.842	0.000*
NTC	0.005	0.010*
NTC^2	-0.012	0.042*
SIZE	0.165	0.000*
LEVERAGE	-0.035	0.869
GROWTH	-0.107	0.619
ROA	3.036	0.000*
R^2	0.194	
F	0.000	
Durbin-Watson	2.002	

Notes: **,* indicate significance levels of 1% and 5%, respectively

to get a discount on the purchase of the inventory and can reduce the risk of future price fluctuations and provide more credit to customers to drive increased sales. This will make the company get more benefits than the cost incurred so that it will increase the company's

earnings and attract investors to buy shares that will then impact on increasing the value of the company.

On the other hand, rising working capital beyond the optimal limits of the company will cause the per-formance to decline because the use of excessive working capital will make the company bear the op-portunity cost to obtain higher profits by investing in long-term investment. In addition, increased work-ing capital beyond the optimal limits will increase storage costs as well as the risk of bad debts that could harm the company. This will make the benefits gained less than the cost incurred by the company so that it will hamper the company to gain greater profits and will impact on the declining performance of the company. The results of this study are in accordance with research conducted by Banos-Caballero et al. (2014) and Mun et al. (2015) which proved an inverted U-shaped relation on working capital and firm performance.

Table 3 provides regression results based on equation 2, examines the moderation effect of finan-cial constraint to the relationship between working capital and performance.

Table 3. The moderating role of financial constraint on the relationship between working capital and performance.

	Financial constraints criteria					
	Cash flow	Interest coverage	Dividend paying	Dividend payout ratio	Cost of External Financing	Size
NTC	0.012	0.010	0.010	0.015	0.005	0.012
	(0.003)*	(0.006)*	(0.013)*	(0.001)*	(0.001)*	(0.010)*
NTC x DFC	-0.009	-0.007	-0.007	-0.014	-0.003	-0.007
	(0.008)*	(0.016)*	(0.047)*	(0.000)*	(0.000)*	(0.042)*
NTC^2	-0.068	-0.054	-0.052	-0.076	-0.010	-0.061
	(0.001)*	(0.002)*	(0.018)*	(0.003)*	(0.041)*	(0.013)*
NTC^2 x DFC	0.069	0.055	0.044	0.078	2182.000	0.053
	(0.000)*	(0.001)*	(0.030)*	(0.002)*	(0.025)*	(0.023)*
SIZE	0.153	0.130	0.169	0.126	0.102	0.159
	(0.000)*	(0.000)*	(0.000)*	(0.001)*	(0.000)*	(0.002)*
LEVERAGE	0.429	0.501	-0.022	0.549	0.067	-0.034
	(0.067)	(0.032)*	(0.933)	(0.020)*	(0.584)	(0.898)
GROWTH	-0.501	-0.605	-0.226	-0.345	-0.050	-0.173
	(0.083)	(0.030)*	(0.412)	(0.235)	(0.678)	(0.530)
ROA	8.282	8.051	5.084	7.745	1.170	5.243
	(0.000)*	(0.000)*	(0.000)*	(0.000)*	(0.000)*	(0.000)*
R^2	0.433	0.426	0.289	0.430	0.201	0.289
F	0.000	0.000	0.000	0.000	0.000	0.000
Number of observation	435	434	372	435	335	376

Notes: **,* indicate significance levels of 1% and 5%, respectively

Regression results on all financial constraints measurements indicate that financial constraints weaken the relationship between working capital management and firm performance. Companies experiencing financial constraints tend to have lower optimal working capital levels than those without fi-nancial constraints. This is because companies that face financial constraints have limited cash so that the company must allocate the cash held to meet the obligations in advance, such as debt and interest costs. In order for the company to continue fulfilling its obligations with limited cash, the company tries not to divide the dividend or divide in small amounts and manage the current assets effectively. The company manages its current assets by managing fewer inventory for quick inventory turnover and applying strict credit policies to minimize the risk of unsophisticated receivables and extend the period of trade payables to suppliers so that cash from working capital can be obtained quickly and does not hamper operational activities, some other cash held by the company can be allocated to meet corporate liabilities.

The results of this study are in accordance with research conducted by Caballero et al. (2014) proving that companies facing financial constraints had lower optimum working capital values than firms without financial constraints.

4 CONCLUSION

This study aimed at analyzing the impact of working capital management on firm performance and how financial constraints determine this relationship. For this purposes, the annual data of manufacturing firm listed in Indonesian stock exchange for the period of 2011-2015 was employed. The results from regression analysis indicated an inverted U-shape relationship between working capital and firm performance. Furthermore, it was found that financial constraints weaken the relationship between working capital and firms' value. The results were robust for the various measurement of financial constraint.

REFERENCES

Caballero, S.B., P.J. Teruel & P.M. Solano. 2013. Working capital management, corporate performance, and financial constraints. *Journal of Business Research*, 67, 332-338.

Charitou, M., Lois, P., Santoso H.B. 2012. The relationship between working capital management and firm's profitability: an empirical investigation for an emerging Asian country. *International Business & Economic Reaserch Journal*, Vol.11 No.8.

Emery, G. W. 1984. A pure financial explanation for trade credit. *Journal of Financial and Quantitive Analysis*, 19, 271-285.

Emery, G. W. 1987. An optimal financial response to variable demand. *Journal of Financial and Quantitative Analysis*, 22, 209-225.

Ermawati, Wita Juwita. 2011. Pengaruh woking capital management terhadap kinerja dan risiko perusahaan. *Jurnal Management dan Organisasi*, Vol.2, No.1.

Fazzari, S.M., Hubbard, R. G., & Petersen, B. C. 1988. Financial constraints and corporate investment. *Brookings Paper on Economic Activity*, 1, 141-195.

Fazzari, S.M & Bruce C. Petersen. 1993. Working capital and fixed investment: new evidence on financing constraints. *The Rand Journal of Economics*, 24, 328-342.

Hennessy, Christophe A & Whited, Tony M. 2006. How costly is external financing? Evidence from structural estimation. *Journal of finance*, Vol. LXII No.4.

Hill. M. D., Kelly, G., & Hoghfield, M. J. 2010. Net operating working capital behavior: A first look. *Financial Management*, 39, 783-805.

Kim, Y. H., & Chung, K. H. 1990. An integrated evaluation investment in inventory and credit: A cash flow approach. *Journal of Business Finance & Accounting*, 17, 381-390.

Schiff, M., & Lieber, Z. 1974. A model for the integration of credit and inventory management. *Journal of Finance*, 29, 133-140.

Smith, J. K. 1980. Profitability versus liquidity tradeoffs in working capital management. *Readings on the management of working capital*, 549-562.

Vahid, T. K., Elham, G. Mohsen, A. K., Mohammadreza, E. 2012. Working capital management and corporate performance: Evidence from Iranian companies. *Procedia – Social and Behavioral Sciences*, 62, 1313-1318.

Whited, T. M., & Wu, G. 2006. Financial constraints risk. *Review of Financial Studies*, 19, 531-559.

Wilner, B. S. 2000. The exploitation of relationship in financial distress: The case of trade credit. *Journal of Finance*, 55, 153-178.

Advances in Business, Management and Entrepreneurship – Hurriyati et al (eds)
© *2020 Taylor & Francis Group, London, ISBN 978-0-367-27176-3*

Risk management in *zakat* institutions

A. Shofawati
Universitas Airlangga, Surabaya, Indonesia

ABSTRACT: Non-profit institutions, like *zakat* institutions, are one of the instruments of Islamic economy instruments. *Zakat* is one of the pillars of Islam and also one of public finance source. Recently *zakat* was categorized as a principle of Islamic social finance which is developed by the Islamic Research and Training Institute-Islamic Development Bank (IRTI-IDB). The role of *zakat* is very important to increase the welfare of society; therefore, the management of *zakat* institutions is very important to increase its accountability. One of *zakat* management policies is risk management. This research investigated and described the process of risk management in *zakat* institutions using a qualitative method based on literature review. This research described the implementation of risk management in *zakat* institutions in Indonesia. The result of this research described risk management of *zakat* institutions to increase effectiveness.

1 INTRODUCTION

Zakat is an obligation in Islam which has two dimensions, a vertical dimension and a horizontal dimension. Therefore, the role of *zakat* is very important as an instrument of the Islamic Economic system to achieve welfare of the state.

There is no doubt that the *zakat* on money is obligatory for all Muslim, as cited from the al Quran, *sunnah* and *ijma*. Allah says, "*And there are those who hoard gold and silver and spend it not in the way of Allah. Announce unto them a most grievous penalty on the day when heat will be produced out of that wealth in the fire of Hell and with it will be branded their foreheads, their flanks, and their backs. This is the treasure which ye hoarded for yourselves. Taste ye then the treasure ye buried*" (Al Tawbah, verses 34–35) (Senawi et al. 2018).

There has been much research on *zakat*. Senawi et al. (2018) examined views from classical jurists and contemporary jurists with regards to the *nisab* of *zakat*, particularly on the issue of the fluctuation of the benchmark prices of the *nisab*, which are gold and silver. The study used document analysis to discern the characteristics of the views from both eras. Basically, the findings indicated that the classical jurists made some justification on the issue, though the system of the market price was not similar to today, however, the concentration on the issue was limited. On the other hand, the contemporary showed their interest in resolving this issue. Jaafar (2017) conducted a study to examine the current challenges and criticisms of *zakat* distribution and proposed an integration between *zakat* and development impact bond as an approach to maximizing the impact from *zakat* investment. Ghani et al. (2018) conducted a study to examine whether perceived board management and governmental model may influence *zakat* payers' trust on *zakat* institutions. Using a questionnaire survey on

184 *zakat* payers, this study showed that perceived board management influenced *zakat* payers' trust on *zakat* institutions. By contrast, perceived governmental model did not influence *zakat* payers' trust on *zakat* institutions. These findings implied that *zakat* institutions should focus on strengthening the efficiency of their institutions to increase the trust level among *zakat* payers. In doing so, the *zakat* payers were driven to pay *zakat* diligently, thereby ensuring the sustainability of *zakat* institutions, particularly in developing countries such as Malaysia.

This paper emphasized risk management in *zakat* institutions. Although research on *zakat* had been undertaken by many researchers, there are fewer studies on risk management in *zakat*; therefore, this study is aimed at contributing to understanding risk management in *zakat* institutions. According to the Nation *Zakat* Institution, there are some *zakat* institutions in Indonesia which have the competency to manage *zakat*, among which is Rumah *Zakat*. Rumah *Zakat* is one of the *zakat* institutions with many achievements. According to Republika (2017), Rumah *Zakat* was audited by the British Standard Institution (BSI). BSI certified Rumah *Zakat* as an institution that deserved the ISO certificate. Rumah *Zakat* is also rated as an excellent institution in risk management. Rumah *Zakat* has obtained its ISO certificate in 2015. At that time Rumah *Zakat* was awarded as a worthy institution in the distribution of *zakat*. After two years of running, Rumah *Zakat* was audited again by the British institution, BSI. Rumah *Zakat* holds a title as an institution that has passed the risk management.

2 METHOD

This research used a qualitative method based on literature review to describe the implementation of risk management in *zakat* institutions in Indonesia. Risk management in *zakat* institutions in this research was analyzed according to Six and Kowalski (2005), who describe Five Steps to Risk Management in Nonprofit and Charitable Organizations. This paper described risk management in Rumah *Zakat*. Rumah *Zakat* is rated as an excellent institution in risk management and obtained the ISO certificate in 2015, therefore this paper described risk management according to the Annual Report of Rumah *Zakat* in 2015.

3 RESULT AND DISCUSSION

3.1 *Zakat*

Zakat is one of the five fundamental pillars of Islam, categorized under obligatory charity (Rahman et al. 2012) in (Ghani et al. 2018). The *Shariah* refers to *zakat* as a redistribution of wealth to the deserving category of people (Rahim & Kaswadi 2014) in (Ghani et al. 2018). The main purpose of *zakat* is to improve the socio-economic status of individual *asnafs* (*zakat* recipients) and the economy of the nation as a whole (Othman et al. 2015) in (Ghani et al. 2018). *Zakat* is perceived as an alternative means to reduce poverty and to improve the quality of life among *zakat* recipients (Abdul Rahman et al. 2015, Nadzri et al. 2012) in (Ghani et al. 2018). If *zakat* collection is managed efficiently, it can positively affect the socio-economic status of the . However, studies showed that *zakat* payers tend not to pay *zakat* through formal institutions (e.g. *zakat* institutions) because they believe that these institutions are untrustworthy (Ghani et al. 2018). Given that, the effectiveness of *zakat* institutions is highly dependent on *zakat* payers' support, therefore, understanding the factors influencing trust among *zakat* payers is imperative (Ghani et al. 2018).

One of the possible factors of trust is board management. *Zakat* payers expect the board management to demonstrate competence, effectiveness and fairness in their relationship with all stakeholders, including *zakat* beneficiaries (Mustafa et al. 2013) in (Ghani et al. 2018). Therefore, the *zakat* management should clearly understand the role of board management as a vital component in organizational resource management. The board management plays four critical roles, namely: monitoring, service, strategy and resource provision (Daily et al. 2003, Ong & Wan 2008, Zahra & Pearce 1989) in Ghani et al. (2018).

Zakat is one of the five pillars of Islam, the elements of the faith that are obligatory for all believers. *Zakat* is the only pillar that deals with finances and is also the only one that relates directly to Islamic governance (Bremer 2013).

3.2 *Risk management*

Risk management is the term for the procedures that an organization follows to protect itself, its staff, its clients and its volunteers. Practicing sound risk management is more than just looking out for potential problems, buying insurance and avoiding lawsuits. It is an ongoing process (Six & Kowalski 2005). Large organizations may have formalized risk management policies and managers whose job it is to oversee risk management. However, for small and medium-sized organizations without large staffing resources, risk management may simply mean ensuring that a systematic, well-planned series of steps is readily available for program managers and volunteers to follow in order to minimize risks (Six & Kowalski 2005). In the eyes of the law, it makes no difference whether staff or volunteers provide a service. If something goes wrong, the organization could be exposed to liability (Six & Kowalski 2005).

Therefore, nonprofit and charitable organizations should take as much care in screening, training and supervising volunteers as they do with staff, in order to ensure the safety of staff, volunteers and clients (Six & Kowalski 2005). Organizations must also ensure that they provide a safe workplace and take swift action to stop anything that could potentially harm or damage staff, volunteers, or clients. Failure to do so could result in liability to the organization (Six & Kowalski 2005). Five steps for dealing with risk consist of:

Step 1: Identify, assess and document program's risks. This is the crucial first step in risk management. It will take some time and effort, but all other efforts will be based upon this step. Gather a group of board members, experienced volunteers, paid staff and perhaps a few clients. Brainstorm about the risks inherent in the activities your organization undertakes (Six & Kowalski 2005).

Step 2: Establish and implement procedures for screening, supervising and evaluating. Volunteers risk can be minimized by the proper selection, screening, training and evaluation of volunteers. This may seem like a daunting task, but it becomes manageable when you break it down into smaller parts (Six & Kowalski 2005).

Step 3: Make sure to have appropriate insurance coverage Insurance cannot protect the clients or the organization from harm or loss, nor can it safeguard a board from allegations of wrongdoing. What insurance can do is help pay the cost of investigating or refuting allegations of wrongdoing and, of course, help pay for insured losses. Many organizations do not realize that there are no insurance policies or guidelines that pertain exclusively to volunteer activities. Volunteers, including board members, can be held legally liable for the same types of claims that are made against businesses and other organizations (Six & Kowalski 2005).

Step 4: Develop and use a code of ethics for volunteers. A code of ethics tells volunteers what is expected of them and communicates some of the core values of the organization. It can also help to motivate volunteers to put these values into practice as they participate in the activities of your organization (Six & Kowalski 2005).

Step 5: Develop, monitor and communicate written policies and procedures; they must be well understood and known about by everyone in the organization. This can be accomplished through education, communication and follow-up (Six & Kowalski 2005).

Zakat institutions are one of the service providers which have an intermediary function from *muzakki* and *mustahiq*. Based on this condition, in conducting a *zakat* institution there are many risks which can happen, so the role of risk management is very important. *Zakat* management also depends on the competency of *Amil*. Some of the *zakat* institutions conduct permanent employees and also volunteers.

Risk management in *zakat* institutions in this research will be analyzed according to Six & Kowalski (2005), who described five steps to risk management in nonprofit and charitable organizations. According to Herman et al. (2003), the general risks nonprofit organizations face are: property risks, income risks, liability risks, people risks, reputation and mission risks, managing volunteer risks, governance and fiduciary risks, managing risks related to serving vulnerable populations, managing the risks of transporting clients, and managing collaboration risk. According to Rumah *Zakat*, one of the best *zakat* institutions in Indonesia, there are two main risks that became the focus of the audit in 2015, namely fraud and reputation.

Risk management is very important because the health of *zakat* institutions depend on the competency of the *zakat* institution to minimize and mitigate the risk. One of the mechanisms to minimize risk is through the internal control mechanism. According to Ghani et al. (2018), internal control is considered as one of the crucial components for an organization to gain public trust. Internal control is a process affected by the entity's board of directors, management and other personnel and designed to provide reasonable assurance regarding the effectiveness of operations, reliability of financial reporting and compliance with applicable laws and regulations (COSO 1994). An effective *zakat* institution needs an effective internal control to ensure that the firm's operational and financial objectives are met. Internal control must be comprehensive and involve people at all levels in a firm (Deloitte & Touche LLP 2004 in Ghani et al. 2018).

Rumah *Zakat* also conducts internal control through the Internal Audit Department (IAD). According to the Annual Report of Rumah *Zakat* in 2015, the Internal Audit Department (IAD) is part of the SPI element (Internal Control System), which is a third line of defense that is responsible for the entire business process of internal examination of the institution through a systematic and orderly approach to evaluate and improve the effectiveness of internal control. And to maintain independence and objectivity so as to express the views and thoughts without distortion. The structure and position of the IAD is under the board of directors and it is directly responsible to the CEO. RZ has the Internal Audit SOP (standard operational procedure), which is the foundation and guides the work in conducting internal audit activities that include the purpose, function, position, scope, authority and responsibility, codes of ethics and internal audit activities for realizing an effective internal control system. In exercising its functions, it is supported by two IAD employees, consisting of one auditor leader and one auditor. Auditor competence development will be carried out through international certification programs and internal and external training with the aim to improve the efficiency, effectiveness and quality of the audit results. The development of IAD personnel committed in 2015, namely, risk management, anti-fraud seminar, and MHMMD:

a. Authorities of the IAD implement re-examining of the internal control system in all activities that exist within the institution.
b. Have access to all documents, records, personal and physical, as well as information on where objects examination is conducted to obtain data and information related to the performance of its duties.
c. Determine the scope, method, manner, technique and frequency of the audit.
d. Submit the report directly to the top management.

In response to the strategies and expectations of management, the IAD had set a priority risk that became the primary focus of the audit in 2015. This is in line with the risk-based audit approach applied, which is to ensure the effectiveness of controls, risk management, and risk assessed in the high category.

According to the Annual Report of Rumah *Zakat* in 2015, two main risks were the focus of the audit in 2015, which are fraud and reputation. Based on these risks the audit plan 2015 was prepared to cover the area exposed to such risks. In addition, the IAD also conducts audits to monitor the potential operational risk compliance and the implementation of policies that have been set. There are some risk indicators which became the main focus and will be monitored through continuous auditing mechanisms, so that the detection of indication digression can be found early. The detection of such a digression can be determined by using risk indicators that exceed specified tolerances. Results of continuous auditing will be submitted to the relevant unit for immediate correction, meanwhile, if it is required, further evaluation can be performed on site and surprised audit by internal audit. In addition to the internal audit, the 2015 IAD also carried out a coordination with the external auditors on audit activities performed by the external auditor as a form of business process monitoring of institution suitability. In line with the principle of transparency, in particular the value of integrity, related to offenses committed by including RZ management, it is necessary to build a complaint system of alleged violations, which is usually called the Whistleblowing System (WBS) that allows for all parties, both internal and external to submit the report. WBS is a management tool used to assist enforcement of business ethics and work ethic in RZ. RZ has built the web-based Whistleblowing System (WBS), allowing anyone to create and submit a report of violations of fraud and other institutions form of ethical violations that occurred.

4 CONCLUSION

Zakat institutions have intermediary functions from *Muzakki* to *Mustahiq*, so the role of Amil is very important for *zakat* institution operational to minimize the risk and to manage *zakat* properly. According to the Herman et al. (2003), the general risks faced by non-profit organizations are property risks, income risks, liability risks, people risks, reputation and mission risks, managing volunteer risks, governance and fiduciary risks, managing risks related to serving vulnerable populations, managing the risks of transporting clients, managing collaboration risk. According to Rumah *Zakat*, one of the best *zakat* institutions in Indonesia, there are two main risks that were the focus of the audit in 2015, namely fraud and reputation. The risk management of *zakat* institution as a non-profit organization, according to Six and Kowalski (2005) are (1) identify, assess and document program's risks, (2) establish and implement procedures for screening, supervising and evaluating volunteers, (3) make sure to have appropriate insurance coverage, (4) develop and use a code of ethics for volunteers, (5) develop, monitor and communicate written policies and procedures.

REFERENCES

Bremer, J. 2013. Zakat and economic justice: Emerging international models and their relevance for Egypt. *Takaful 2013 Third Annual Conference on Arab Philanthropy and Civic Engagement*. June 4–6,2013, Tunis, Tunisia.

COSO. 1994. Internal control–integrated framework. *Retrieved www.coso.org*.

Ghani, E.K., Aziz, A.A., Tajularifin, S.M. & Samargandi, N. 2018. Effect of board management and governmental model on zakat payers' trust on zakat institutions. *Global Journal Al-Thaqafah*: 73–86.

Herman, M.L., Head, G.L., Jackson, P.M. & Fogarty, T.E. 2003. *Managing risk in nonprofit organizations: A comprehensive guide*. October 2003.

Jaafar, A.Z. 2017. Integrating zakat with Development Impact Bond (DIB): Assessing possible approaches to maximizing impact from zakat investment. *Social Impact Research Experience (SIRE)* 55.

Rahim, S. & Kaswadi, H. 2014. An economic research on zakat compliance among muslim's staff in UNIMAS. *Proceeding of the International Conference on Masjid, Zakat and Waqf (IMAF 2014)*.

Rumah Zakat. *Laporan tahunan (annual report)*. 2015.

Republika.co.id. 2017. Rumah Zakat Gets ISO Certificate from British Standard Institution. *Retrived https://www.rumahzakat.org/en/rumah-zakat-dapat-sertifikat-iso-dari-lembaga-audit-inggris/*. Accessed 7 September 2017.

Senawi, A.R., Isa, M.P.M., Kamarul-zaman, M.A. & Husain, H. 2018. Assessing the classical and contemporary jurists' view on the nisab of zakat issue. *International Journal of Academic Research in Business and Social Sciences* 8(1): 794–806.

Six, K. & Kowalski, E. 2005. Developing a risk management strategy: Five steps to risk management in non-profit and charitable organizations. *Social Planning Council for the North Okanagan*. Imagine Canada.

Advances in Business, Management and Entrepreneurship – Hurriyati et al (eds)
© *2020 Taylor & Francis Group, London, ISBN 978-0-367-27176-3*

The effect of risk, growth, firm size, capital structure, and earnings persistence on earnings response coefficient

Z. Naimah & A.T. Rahma
Faculty of Economic and Business, Universitas Airlangga, Surabaya, Indonesia

ABSTRACT: This study aims to obtain empirical evidence of the association of risk, growth, firm size, capital structure, and earnings persistence to earnings response coefficient. This research used data of all manufacturing companies listed on the Indonesia Stock Exchange (IDX) in 2014 and 2015. Hypothesis testing in this research used multiple linear regression model. The results of this study found that risk has a negative significant effect on earnings response coefficient, growth has a negative but not significant effect on earnings response coefficient, firm size has a negative significant effect on earnings response coefficient, capital structure has positively but insignificantly effects on earnings response coefficient, and earnings persistence positively and significantly effects on earnings response coefficient.

1 INTRODUCTION

One of the parts in the financial statements that gets the most attention from investors is earnings. Earnings information of an enterprise may affect an investor's decision to buy, sell or hold securities issued by the company. The earnings response coefficient shows the market response to earnings information published by the observable company from stock price movements around the date of issuance of financial statements (Diantimala 2008). The earnings response coefficient (ERC) is the effect of earnings surprise on the accumulation of abnormal returns (CAR), which is shown through the slope coeficient of the regression of earnings surprise to stock abnormal return (Cho & Jung 1991). This shows that the earnings response coefficient (CAR) is the reaction to earnings announced by the company. The reactions given depend on the quality of earnings generated by the company.

The level of earnings response coefficient is an investor response to information and records in the company's financial statements. There are several things that concern the investors in dealing with information and records of financial statements that are risk, growth, company size, capitals' structure and earnings persistence. The study of earnings information was conducted by Ball & Brown (1968) that predicted that increase in unexpected earnings was followed by positive abnormal returns and decrease in unexpected earnings followed by negative abnormal returns. The results of this study indicated that accounting earnings reflects one of the factors affecting stock prices and is useful information. Beaver et al. (1987) also examined the information content of annual earnings announcement indicated that earnings announcement is an event that the investor perceives

to influence stock prices, so investors use the information to change their earnings predictions and adjust the right price.

The first factor that can affect the ERC (earnings response coefficient) is risk (beta). The riskier the future yields the lower the value of the firm in the views of investors. Collins & Kothari's (1989) and Easton & Zmijewski (1989) studies showed that risk negatively affects earnings response coefficients.

The second factor that can affect ERC is growth. One way to measure the growth of a company is to sales growth. Good News and Bad News in financial statement information indicate the prospect of future earnings growth and will change investor response, so, earnings response coefficient also increases (Scott 2015).

The third factor that can affect ERC is firm size. According to Arfan & Antasari (2008), companies with larger size have larger assets, larger sales, larger capital, and more employees. A study conducted by Chaney & Jeter (1992) found that firm size positively affects earnings response coefficients.

The fourth factor that can affect the ERC is the capital structure. The capital structure is related to leverage, which is a measure that indicates to what extent debt securities and preferred stock are used in the capital structure of the firm so that if the firm uses debt and preferred stock, the firm imposes all business risks to the common stockholders. The research done by Dhaliwal et al. (1991) showed that leverage affects earnings response coefficient.

The fifth factor that can affect the ERC is the earnings persistence. According to Scott (2015) earnings persistence is an expected revision of future earnings that can be viewed from the innovation of current earnings associated with changes in stock prices. The value of ERC is predicted to be higher if the company's earnings are more presistent in the future and higher earnings quality. Earnings persistence is an earning that has the ability as an indicator of future earnings generated by the company repetitive in the long term (sustainable) (Imroatussolihah 2013). The more persistent earnings show more informative returns, and more investors respond to companies that have predictable earnings in the future period so that ERCs will increase.

1.1 Signaling theory

There is asymmetry information between two parties, such as investor and management. Signaling theory concerned with how this information asymmetry is reducing (Spence 2002). Signaling theory explained how a company gives signals to users of financial statements stating. Managers are encouraged to signal to external parties due to asymmety information.

In market analysis with asymmetry information by Akerlof (1995), Spence (2002), and Stiglitz & Weiss (1992) it is argued that many markets indicated the existence of asymmetry information in which players on one side of the market have better information than players on the other. Borrowers knew better about their repayment prospects than lenders and managers knew more about the profitability of the company than shareholders.

In Akerlof's study (1995) it is said that an unbalance information between the seller and the buyer (asymmetry information) will lead to adverse selection due to hidden information where the buyer has risks to buy a lower quality product (lemons) but thinks the goods are high quality, which will cause the displacement of high quality goods market. Then the high quality seller needs a way to signal the quality of his goods.

A research conducted by Spence (2002) revealed that signaling is a strategy to avoid problems associated with adverse selection. Persons who have information will provide information to outside parties, the outsider will interpret the signal and adjust its buying behavior. High quality companies will deliberately signal to the market, and it is expected that the market can differentiate high and low quality companies.

Signaling theory shows that earnings response coefficient is a signal for investors that can be seen from the movement of stock prices at the time of announcement by the company. One type of information from a company that can be a signal for investors is the annual report. This information may be accounting information such as financial statements as well as non-accounting information. If the announcement is positive, it is expected that there will be a market reaction that can be seen from the stock price movement when the announcement has been received by the market.

1.2 Hypothesis development

1.2.1 The effect of risk on earning response coefficient

Risk is predicted to affect earnings response coefficient, because the firm risk level affects investors' decision making, firms with high risk level tend to be avoided by investors, especially investors who have limited funds and investors who do not want to bear risks.

Based on signaling theory, risk becomes a signal for investors that shows how much risk is beared by the company in the existing business competition. It is put into consideration for investors who tend to avoid risk. As a result, they will avoid high-risk companies. This is in accordance with Collins & Kothari (1989) and Easton & Zmijewski (1989) studies which suggested that systematic risk (beta) negatively affects earnings response coefficients. Based on the above explanation, the hypothesis in this study are:

H_1: Risk affects earnings response coefficient.

1.2.2 The effect of growth on earnings response coefficient

Growth is predicted to affect earnings response coefficient. This is because companies that have high sales growth are considered by investors as companies that have the possibility of increasing future earnings, so it will benefit investors. Continuous growth indicates that the company is able to well-manage the company therefore the earnings information announced by the growth company will be responded positively by investors (Naimah & Utama 2006).

Based on the concept of signaling theory, the sales level in the financial statements, is a signal for investors because the company's sales that grow each period shows that the company has a chance to earn earnings in the future, so this will benefit investors.

Collins & Kothari (1989) stated that growth positively affects earnings response coefficient, this study uses market value to book ratio (M/B) in calculating growth variable. The higher the M/B the higher the company's future earnings growth expectation. Naimah & Utama's (2006) research stated that the effect of accounting earnings on stock prices will be greater in firms with high growth than low-growth firms. Based on the above explanation, the hypothesis in this study is:

H_2: Sales growth affects earnings response coefficient.

1.2.3 The effect of firm size on earnings response coefficient

Firm size is predicted to affect the earnings response coefficient, this is because large firms have a lot of information available that makes it easier for investors to make decisions regarding their investment, and large companies are perceived to have high growth opportunities and high profitability than small firms so investors tend to choose large companies.

According to the signaling theory, firm size becomes a signal for investors that large companies can provide more information than small companies. Because big companies will try to give a signal to the market that this company has better capability than other companies, so in big companies, information will be widely available throughout the year in order to facilitate investors to differentiate their quality from other companies.

A study conducted by Chaney & Jeter (1992) found that firm size positively affects earnings response coefficients because large firms have many alternative sources of information, so markets can interpret information in the full financial statements and accurately predict future cash flows to reduce the level of uncertainty. Mulyani et al. (2007) shows that firm size has a positive effect on earnings response coefficient because with the large amount of information available from big companies, investors use firm size as one of the factors that can be used in investment decision making. Based on the above explanation, the hypothesis built in this study is:

H_3: Firm size affects earnings response coefficient.

1.2.4 *The effect of capital structure on earnings response coefficient*

Capital structure is predicted to affect the earnings response coefficient, because investors assume that companies with more leverage is more profitable for lenders than shareholders so the capital structure effect on decisions made by investors. Because a company that has leverage means the company has a larger debt amount than capital. Because of that, if there is an increase in corporate profits, investors feel that the profit is more profitable for debtholders rather than shareholders. Therefore, a little reaction of investors to the company's earnings information with high leverage means its earnings response coefficient is low (Scott 2015).

According to the signaling theory, the company's capital structure can be a signal for investors, because of this signal investors know which companies with high and low leverage levels, so this signal helps investors in making decisions. In this hypothesis the firm with its debt in its capital structure will be avoided by investors because the profit generated by the company is more favorable to debtholders than shareholders.

The research done by Dhaliwal et al. (1991) shows that leverage affects earnings response coefficients, where firms that have debts in their capital structure have low earnings response coefficients than firms with overall capital and firms with high leverage also have lower earnings response coefficients than firms with small or no leverage.

Imroatussolihah (2013) who measured leverage using Debt Equity Ratio, founds that DER negatively affects earnings response coefficient because high debt level of a company causes most of the profit obtained by company to be distributed to creditors, so shareholders obtain less. Based on the above explanation, the hypothesis build is as follows:

H_4: Capital structure affects the earnings response coefficient.

1.2.5 *The effect of earning persistence on earnings response coefficient*

Earnings are predicted to affect earnings response coefficients, because firms that can generate persistent earnings are considered capable of generating a stable profit in the future. Earnings persistence is an indicator of future earnings, where current earnings has the ability to act as an indicator of future earnings generated by the company repeatedly in the long term.

According to Scott (2015) earnings response coefficient will be high if good news that appears on the financial statements is due to the sale of new product companies or large cost-cutting by management rather than if good news arises because of asset sales profit. Expectations for future earnings increase are high if earnings are more persistent where earnings are generated by firms on a long-term basis (Imroatussolihah 2013).

According to signaling theory, persistent corporate earnings signaled investors the ability of firm to generate stable earnings in the future. This signal becomes the investors' consideration in determining where they will invest their capital, this earnings persistence will be a signal that there is a high chance the company will generate earnings and will be profitable to investors.

Collins & Kothari (1989) found that earnings persistence positively affect the earnings response coefficient because the greater the persistence of a company's earnings the greater the profit that can be expected by investors in the future. Meanwhile the results of research Imroatussolihah (2013) has a different result where the persistence of earnings does not affect the earnings response coefficient due to the number of transitory components in earnings that causes earnings can not be used as a measure of earnings response coefficient. Based on the above explanation, the hypothesis built is as follows:

H5: Earnings persistence has an effect on earnings response coefficient.

2 METHODOLOGY

2.1 *Population, sample, and sampling technique*

The population of this research are all manufacturing companies listed on the Indonesia Stock Exchange from 2014 until 2015. Sample selection in this study uses purposive sampling method, meaning the samples are deliberately selected based on certain criteria in order to represent the population.

2.2 Research variables

2.2.1 Earnings response coefficient

Earnings response coefficient is the coefficient obtained from the regression of accounting earnings to stock price proxy representing the market reaction. The stock price proxy used is Cummulative Abnormal Return (CAR), while the proxy for accounting earnings is Unexpected Earnings (EU). The model regression results the earnings response coefficient of each target population is used for subsequent analysis. Calculating the earning response coefficient (ERC) uses the regression between the price proxy validation CAR and accounting earnings that is EU. The amount of earnings response coefficient can be calculated by the regression equation of each company's data, the formula as follows:

$$CAR_{it} = \alpha + \beta_1 UE_{it} + \varepsilon \qquad (1)$$

Where: CAR_{it} = cumulative abnormal return of company i in t period; UE_{it} = unexpected earning of company i in t period; A = constant; β_1= Earning response coefficient (ERC)

2.3 Risk

Risk is one of the factors that influence investment decision making. The riskier a company the lower the value of the company in the views of investors. Risk can be measured using systematic risk (beta) using market model with Capital Asset Pricing Model (CAPM) formula (Scott 2015).

$$R_{it} = {}_0 + \beta_1 R_{mt} + e_{it} \qquad (2)$$

Where: R_{it} = daily returns of company i in year t; R_{mt} = daily market returns in year t; β_1 = risk

2.4 Growth

Growth is in terms of 2-year sales growth rate because sales growth is more appropriate for manufacturing companies (Charitou et al. 2001).

$$S2 = \frac{SALESit - SALESit - 1}{SALESit - 1} \qquad (3)$$

Where: SALESit = sales of company i during year t.

2.5 Firm size

Firm size is a proxy of price informativeness (Scott, 2015). According to Chaney & Jeter (1992) large companies have more information than smaller companies. The available information makes it easier for

market to interpret the information contained in the financial statements. Firm size is measured by the book value of total assets of the company at the beginning of the year (Barth et al. 1998).

2.6 Capital structure

Capital structure describes the permanent financing of a company consisting of debt and equity that is used to finance the firm's activities. Capital structure in this research is measured using financial leverage. This variable corresponds to Scott (2015) which stated the level of leverage in the capital structure of a firm will affect the earnings response coefficient.

$$Levit = \frac{Total\ Debt\ it}{Total\ Asset\ it} \qquad (4)$$

2.7 Earnings persistence

Earnings persistence is earnings that has the ability to perform as an indicator of future earnings generated by the company repeatedly in the long term (Imroatussolihah 2013). Persistence will be measured from the regression slope of the current profit gain in earnings (Scott 2015).

$$X_{it} = {}_0 + {}_1 X_{it-1} + e_{it} \qquad (5)$$

Where: X_{it} = annual profit of company i in year t; X_{it-1} = annual profit of company i in year t-1; a_1 = earnings persistence

3 RESULTS AND DISCUSSION

3.1 Descriptive statistics

Based on Table 1 the minimum value of ERC is -0.6072. For all sample companies, the average ERC is 0.033172 with standard deviation of 0.3487028. The level of distribution of ERC for all firms has a value of 1051%.

RISK has a minimum value of -1.5823 and a maximum value of 1.3626. For all sampled companies,

Table 1. Descriptive statistics.

	Minimum	Maximum	Mean	Std. Deviation
ERC	-.6072	2.8229	.033172	.3487028
RISK	-1.5823	1.3626	-.156897	.4655594
GROWTH	-.9888	1.9253	.045146	.2571759
SIZE	25.2954	33.1341	28.327477	1.6105751
LEV	.0413	.9150	.437287	.2109929
PERS	-1.4068	1.7490	.390515	.4906052

the average risk is -0.156897 with a standard deviation of 0.4655594. The level of distribution of risk data for the entire company has a value of -297%.

Sales growth (GROWTH) has a minimum value of -0.9888 and maximum 1.9253. For all the sample companies, the average sales growth is 0.045146 with the standard deviation of 0.2571759.

Firm size (SIZE) has a minimum value of 25.2954 and maximum 33.1341. For all sampled companies, the average size of the company is 28.327477 with the standard deviation of 1.6105751. The SIZE data distribution rate for all companies has a value of 6%.

The capital structure (LEV) has a minimum value of 0.0413 and a maximum value of 0.9150. For all sampled companies, the average leverage is 0.437287 with the standard deviation of 0.2109929. The level of spread of leverage data for all companies has a value of 48%.

Earnings Persistence (PERS) has a minimum value of -1.4068 and a maximum of 1.7490. For all of the sample companies, the average earnings persistence is 0.390515 with the standard deviation of 0.4906052. The level of distribution of earnings persistence data for all companies has a value of 126%.

3.2 Results

Based on the results contained in table 2, it can be seen how the influence of each independent variable to earnings response coefficient.

The results of data analysis indicate that risk and firm size negatively and significantly affect earnings response coefficient, earnings persistence positively and significantly affect earnings response coefficient. Otherwise, growth and leverage do not significantly affect earnings response coefficient.

3.3 Discussion

3.3.1 The effect of risk on earnings response coefficient

The results showed that the risk had a significant negative effect on earnings response coefficient so that hypothesis 1 is accepted. This result is in accordance with Collins & Kothari's (1989) and

Table 2. Results.

Variable	Coefficient	t-stat	Prob-value
Constant	0.445	3.157	0.002
RISK	-0.030	-1.778	0.078
GROWTH	-0.014	-0.500	0.618
SIZE	-0.017	-3.266	0.001
LEV	0.023	0.626	0.532
PERS	0.054	-2.808	0.006
Adjusted R ^{2}			0.079

* Dependent variable ; Cumulative abnormal return.

Easton & Zmijewski (1989), and Naimah & Utama (2006) studies that showed that risk negatively affects earnings response coefficients.

The negative regression coefficient indicates that the higher the risk of a company the lower the earnings response coefficient. This happens because investors view that earnings is an indicator of future returns so the riskier the company, the lower the investor's reaction to earnings shock.

Investors are trying to avoid companies that have high risk. Companies with a high risk level will be avoided by investors, especially investors who have limited funds and risk avoider. This results in decreasing investor reaction level to companies that have a high risk. The negative impact of corporate risk with earnings response coefficient indicates that investors in Indonesia are dominated by investors who do not like and tend to avoid risk.

3.3.2 The effect of growth on earnings response coefficient

Hypothesis 2 is rejected because the results of this study which shows that sales growth has a negative but not significant effect on earnings response coefficient.

Investors consider that firms with high growth rates can give confidence that the company will continue to grow in the future, as growing sales indicate the possibility of companies earning profits that will benefit investors, but the results show growth affects investor decisions negatively because the company with high sales growth rate are also accompanied by high operational costs. Usually investors view the side of net income and sales of a business, when the company's sales are high but they are followed by a high expense, so the net income generated by the company has not changed much. Another condition is investors who make short-term investments, which investors expect capital gains from the sale of shares and not long-term investors who expect a dividend yield. Short-term investors usually only see the movement of stock prices, so this condition makes this sales growth factor has no significant effect on earnings response coefficient.

The results of research are in accordance with Imroatussolihah (2013), and Sandi (2013) studies which suggested that growth is negatively significant but this may not be due to investors' motivation not to gain long-term benefits but to obtain capital gain. The growth factor of the company is usually observed by investors who have long-term perspective to obtain the yield of their investment.

3.3.3 The effect of firm size to earnings response coefficient

Hypothesis 3 is accepted as of the results of this study indicates that firm size has a significant negative effect on earnings response coefficient.

The larger the company the more public information is available about the company and the more informative the stock price, hence current earnings information content is smaller (Scott 2015). Hence, at the time of the earnings announcement, the market is less reacting because big companies have a lot of information available throughout the year. The results of this study are in line with Collins & Kothari (1989), and Murwaningsari (2008) indicating that firm size has a significant negative effect on earnings response coefficient due to high informativeness of stock price causing the information content of accounting earnings decreases.

3.3.4 *The effect of capital structure on earnings response coefficient*

The results of this study indicate that the capital structure has positive but not significant effect on earnings response coefficient so that hypothesis 4 is rejected. The results of this study are not in accordance with research Imroatussolihah (2013), Dhaliwal et al. (1991), Naimah & Utama (2006), and Murwaningsari (2008) found that leverage had a significant negative effect on earnings response coefficient.

The results of this study are consistent with the results of the research done by Sandi (2013), and Delvira & Nelvirita (2013) which states that the level of corporate leverage does not significantly influence the earnings response coefficient. The positive regression coefficient indicates that the use of debt in the source of corporate funds will not necessarily lead to bankruptcy. The use of debt in the company can provide tax protection benefits. This is because the interest payments are tax deductions, so the earnings received by the investor increases. Companies that use debt to gain more earnings than the cost of assets and sources of funds will increase shareholder earnings. Investors assess that the increase in earnings derived from the company derived from the contribution of capital structure because companies with optimal capital structure can potentially maximize the performance of the company.

3.3.5 *The effect of earnings persistence on earnings response coefficient*

The results of this study indicate that earnings persistence significantly positive effect on earnings response coefficient. Hence, hypothesis 5 is accepted and supports the research of Naimah & Utama (2006), Mulyani et al. (2007), Collins & Kothari (1989), Delvira & Nelvirita (2013), and Easton & Zmijewski (1989).

These results show that the firm's ability to generate persistent or stable earnings, provides good news to investors when the company announces earnings information. So that investors respond more by buying stock that have persistent earnings.

This result is consistent with Scott's (2015) theory that stated that the more permanent the change in earnings over time, the higher the earnings response coefficient. The earnings announcement will be reacted by investors because of the introduction of new products or the company has managed to find

a method to increase efficiency rather than earnings announcement due to the sale of fixed assets because investors assume that the increase in earnings due to the sale of fixed assets will not necessarily happen again in the next year. This suggests that market reactions are higher for information that is expected to be consistent in the long run than temporary earnings information.

4 CONCLUSION

This study aims to examine whether risk, growth, firm size, capital structure, and earnings persistence affect in manufacturing companies listed on the Indonesia Stock Exchange in 2014-2015 can affect the earnings response coefficient. Based on the results of data analysis t, it can be concluded that:

1. Risk has a significant negative effect on earnings response coefficient. The negative regression coefficient indicates that risk is an indicator for investor in choosing company to invest capital because investor has risk averse so investors avoid firms with high risk. The results of this study are in accordance with research conducted by Chaney & Jeter (1992), Delvira & Nelvirita (2013), Imroatussolihah (2013), and Mulyani et al. (2007).

2. Growth has negative effect but not significant to earnings response coefficient. The negative regression coefficient shows that firms with high sales growth are followed by high operational costs as well, and the number of investors who make short-term investments causes sales growth factors are not used by short-term investors that cause sales growth factors have no effect on earnings response coefficient. The results of this study in accordance with research conducted by Imroatussolihah (2013) and Sandi (2013)

3. Firm size has a significant negative effect on earnings response coefficient. The negative regression coefficient indicates that large companies have a lot of information available throughout the year so that when an earnings information announcement is made by the company, the market does not react much, so the earnings response coefficient becomes low. The results of this study are in accordance with research conducted by Collins & Kothari (1989), Murwaningsari (2008), and Novianti (2015).

4. Capital structure has positive but not significant effect on earnings response coefficient. It does not always have a negative effect because the use of debt can reduce the tax burden and also increase the value of the company. The results of this study in accordance with research conducted by Delvira & Nelvirita (2013) and Sandi (2013).

5. Earnings Persistence have a significant positive effect on earnings response coefficient. The market will respond to consistent long-term earnings information, so that firms with persistent

earnings each year have a higher earning response coefficient than firms with non-persistent earnings. The results of this study are consistent with research conducted by Collins & Kothari (1989), Delvira & Nelvirita (2013), Easton & Zmijewski (1989), and Mulyani et al. (2007).

REFERENCES

Arfan, M., & Antasari, I. 2008. Pengaruh ukuran, pertumbuhan, dan profitabilitas perusahaan terhadap koefisien respon laba pada emiten manufaktur di bursa efek Jakarta. *Jurnal Telaah dan Riset Akuntansi* 1(1): 50-64.

Akerlof, G. 1995. *The market for "lemons": Quality uncertainty and the market mechanism*. Springer.

Ball, R., & Brown, P. 1968. An empirical evaluation of accounting income numbers. *Journal of accounting research* 159-178.

Barth, M.E., Beaver, W.H. & Landsman, W.R. 1998. Relative valuation roles of equity book value and net income as a function of financial health. *Journal of Accounting and Economics* 25(1):1-34.

Beaver, W. H., Lambert, R. A., & Ryan, S. G. 1987. The information content of security prices: A second look. *Journal of accounting and economics* 9(2): 139-157.

Cho, J. Y., & Jung, K. 1991. Earnings response coefficients: A synthesis of theory and empirical evidence. *Journal of Accounting literature* 10(1): 85-116.

Chaney, P. K., & Jeter, D. C. 1992. The effect of size on the magnitude of long window earnings response coefficients. *Contemporary Accounting Research* 8(2): 540-560.

Charitou, A., Clubb, C., & Andreou, A. 2001. The effect of earnings permanence, growth, and firm size on the usefulness of cash flows and earnings in explaining security returns: empirical evidence for the UK. *Journal of Business Finance & Accounting* 28(5-6): 563-594.

Collins, D. W., & Kothari, S. 1989. An analysis of intertemporal and cross- sectional determinants of earnings response coefficients. *Journal of accounting and economics* 11(2): 143-181.

Delvira, M. & Nelvirita, N., 2013. Pengaruh Risiko Sistematik, Leverage Dan Persistensi Laba Terhadap Earnings Response Coefficient (ERC). *Wahana Riset Akuntansi* 1(1).

Dhaliwal, D. S., Lee, K. J., & Fargher, N. L. 1991. The association between unexpected earnings and abnormal security returns in the presence of financial leverage. *Contemporary Accounting Research* 8(1): 20-41.

Diantimala, Y. 2008. pengaruh akuntansi konservatif, ukuran perusahaan, dan default risk terhadap koefisien respon laba (ERC). *Jurnal Telaah dan Riset Akuntansi* 1(1): 102-122.

Easton, P. D., & Zmijewski, M. E. 1989. Cross-sectional variation in the stock market response to accounting earnings announcements. *Journal of accounting and economics* 11(2-3): 117-141.

Imroatussolihah, E. 2013. Pengaruh resiko, leverage, peluang pertumbuhan persistensi laba dan kualitas tanggungjawab sosial perusahaan terhadap earning response coefficient pada perusahaan high profile. *Jurnal Ilmiah Manajemen* 1(1): 75-87.

Mulyani, S., Asyik, N. F., & Andayani, A. 2007. Faktor-faktor yang mempengaruhi earnings response coeficient pada perusahaan yang terdaftar di Bursa Efek Jakarta. *Jurnal Akuntansi & Auditing Indonesia* 11(1).

Murwaningsari, E. 2008. Pengujian simultan: beberapa faktor yang mempengaruhi earning response coefficient (ERC). *Simposium Nasional Akuntansi* 11.

Naimah, Z., & Utama, S. (2006). Pengaruh ukuran perusahaan, pertumbuhan, dan profitabilitas perusahaan terhadap koefisien respon laba dan koefisien respon nilai buku ekuitas: Studi pada perusahaan manufaktur di Bursa Efek Jakarta. *Simposium Nasional Akuntansi IX* 1-26.

Nuswandari, C. (2009). Pengungkapan pelaporan keuangan dalam perspektif signalling theory. *Jurnal Ilmiah Kajian Akuntansi* 1(1).

Sandi. 2013. Faktor-faktor yang mempengaruhi earnings response coefficient. *Jurnal Analisis Akuntansi*. UNNES.

Scott, R. W. 2003. Financial accounting theory, 2003:, Toronto Ontario: Pearson Education Canada Inc.

Spence, M. 2002. Signaling in retrospect and the informational structure of markets. *The American Economic Review* 92(3): 434-459.

Stiglitz, J.E. & Weiss, A. 1992. Asymmetric information in credit markets and its implications for macro-economics. *Oxford Economic Papers* 44(4): 694-724.

The impact of tax amnesty policy in 2016 on the abnormal return and trading volume activity in banking companies

Y. Permatasari & N. Ardiyanti
Universitas Airlangga, Surabaya, Indonesia

ABSTRACT: This research aimed to demonstrate the empirical evidence on the effect of tax amnesty policy in Indonesia to abnormal return and trading volume activity in banking companies listed in Indonesia Stock Exchange during 2016-2017. Event Study of Tax Amnesty Policy method was used in this study from July 1st 2016 until December 31st 2017. The sample of this research was 35 banking companies listed in Indonesia Stock Exchange. The first hypothesis was test using Wilcoxon signed rank test. The dependent variables were abnormal return and trading volume activity of the stock, while the independent variable was the Tax Amnesty Policy. This study showed that tax amnesty policy gave trading volume activity significant effect in the first period rather than in the latter period. Meanwhile, tax amnesty policy had no significant effect to the abnormal return.

1 INTRODUCTION

In order to encourage the acceleration of development in Indonesia, the government increased the state revenue from the tax sector. In addition, tax is also able to determine the economic climate of a country. Economical tax is an important fiscal policy instrument in controlling a country's economic growth.

However, in reality, there are still many taxpayers who are not obedient in terms of payment to the state. Although tax system in Indonesia employs the self-assessment system, there are still many taxpayers reporting and paying taxes accordingly. This cer-tainly has an impact on state revenue. Therefore, in 2016, the government restated the Tax Amnesty Pol-icy.

On the 1st of July 2016, President Joko Widodo passed the Law No. 11 of 2016 on tax amnesty sup-ported by the issuance of PMK-118/PMK.03/2016, valid since July 15, 2016 on the implementa-tion of the Law No. 11 of 2016 on tax amnesty. Since then, the policy of Tax Amnesty 2016 entered into force in Indonesia, which began on the 1st of July, 2016 to March 31, 2017.

According to the article 1 (Law No. 11 of 2016), Tax Amnesty or better known in Indonesia as tax pardon or the reduction of taxes that should be paid. The implementation of Tax Amnesty Policy 2016 is considered quite success-ful in some circles. One proof of the success was the closing of Statement of Property (SPH) reached the amount of 4.855, mean-while the repatriation reached 147 T (SWA Maga-zine September 6 2016).

The success of Tax Amnesty Policy 2016 affected many sectors, including banks listed in Indonesia Stock Exchange. Some of the banks were appointed by the finance minister as the payment place for the Tax Amnesty 2016 ransom, as well as the transfer of property in the form of funds from outside of the country by the taxpayer. Given the increasing liquid-ity due to the large amount of funds coming into government appointed banks, it was assumed pos-sible to change abnormal return and trading volume activity in banking companies listed in Indonesia Stock Exchange because of the perception.

Tax Amnesty Policy 2016 is divided into 3 periods. The first period was unknown to the public, it caused a lack of effort for the community to join the program. While in the second period, community ef-fort increased until its peak at the end of period three.

Tax Amnesty Policy 2016 imposed by the gov-ernment was not an easy factor to consider for investors in making the decision to increase their investment in capital markets, which could affect the abnormal return and trading volume of existing activity. They certainly did a lot of analysis on the financial information available in the capital market along with qualitative factors from outside the capital market.

According Hartono (2013), return and risk are two things that cannot be separated because the con-sideration of an investment made by investors is a trade-off of the two existing factors.

This research was conducted with the aim to find out how the impact of Tax Amnesty Policy 2016 affected abnormal return and trading volume activity, especially in the banking sector companies listed on the Indonesia Stock Exchange. The time span used for this research is during the Tax Amnesty period I (July 1-September 30, 2016), period II (October 1-December 31, 2016), and period III (January 1-March 31, 2017).

2 METHOD

The data used in this study were taken from www. idx.com and www.finance.yahoo.com. Using 35 banking companies listed on the Indonesia Stock Exchange during the period of Tax Amnesty Policy starting July 1, 2016-March 31, 2017, this study also employed event-study method to study the reaction of capital market towards the Tax Amnesty Policy 2016.

The normality test in this research was based on Kolmogorov Smirnov test with 5% significance level. Test of difference was used in this research to know whether there was a difference of abnormal return and difference in trading volume activity between Tax Amnesty 2016 period I, period II and period III by using Paired Sample T-Test for normally distributed data, as for abnormal distributed data Wilcoxon Signed Rank Test was used.

3 RESULT AND DISCUSSION

3.1 Result

3.1.1 The first hypothesis testing

Test results from the first hypothesis using Wil-coxon Signed Rank Test method in this study was obtained from comparing between AAR in period II and period I. There were 16 companies in the se-cond period that had higher AAR than in period I and there were 19 sample companies in period II that had lower AAR than in period I. It means that there were more firms with lower AAR in period II than in period I, but this

Table 2. T-test result of average abnormal return.

	AAR Period II– AAR Period I	AAR Period III–AAR Period II	AAR Period III–AAR Period I
Z	-0.246	-0.344	-0.655
Asymp.Sig. (2-tailed)	0.806	0.731	0.512

was not statistically significant as evidenced by two-tailed significant results of 0.806, higher from α = 5% or 0.05.

Furthermore, when comparing between AAR in period III and period II, there were 18 sample firms with higher AAR in period III than in period II and there were 17 sample firms with lower AAR in period III than in period II. It meant that there were more firms with higher AAR in period III than in period II, but this was not statistically significant as evidenced by two-tailed significant results of 0.731.

In the last AAR analysis, comparison between AAR in period III and period I showed there were 20 sample firms with higher AAR in period I than in period III and there were 15 sample with lower AAR in period I than in period III. It means that there were more firms with higher AAR in period I than in period III, but this was not statistically significant as evidenced by two-tailed significant results of 0.512, higher than α = 5% or 0.05. The three analyzes can be seen from the level of significant level of comparison average abnormal return per period of Tax Amnesty 2016.

Table 1. Wilcoxon signed rank test result of average abnormal return.

		N	Mean Rank	Sum of Ranks
AAR Period II– Period I	Negative Ranks	19[a]	17.37	330.00
	Positive Ranks	16[b]	18.75	300.00
	Ties	0[c]		
	Total	35		
AAR Period III– Period II	Negative Ranks	17[d]	17.29	294.00
	Positive Ranks	18[e]	18.67	336.00
	Ties	0[f]		
	Total	35		
AAR Period III– Period I	Negative Ranks	15[g]	18.33	275.00
	Positive Ranks	20[h]	17.75	355.00
	Ties	0[i]		
	Total	35		

Table 3. Wilcoxon signed rank test result of average trading volume activity.

		N	Mean Rank	Sum of Ranks
ATVA Period II– Period I	Negative Ranks	26[a]	19.54	508.00
	Positive Ranks	9[b]	13.56	122.00
	Ties	0[c]		
	Total	35		
ATVA Period III– Period II	Negative Ranks	16[d]	15.00	240.00
	Positive Ranks	19[e]	20.53	390.00
	Ties	0[f]		
	Total	35		
ATVA Period III– Period I	Negative Ranks	21[g]	17.29	363.00
	Positive Ranks	14[h]	19.07	267.00
	Ties	0[i]		
	Total	35		

So it can be concluded that the first hypothesis that stated there was abnormal difference caused by the policy of Tax Amnesty 2016 in period I, period II, and period III in banking companies listed on the Indonesia Stock Exchange was unacceptable because it was not statistically significant. This can be caused by investors assuming that the issuance of the policy of Tax Amnesty 2016 in Indonesia does not give much profit for them therefore, abnormal return was not affected by existence of Tax Amnesty Policy 2016.

3.2 The second hypothesis testing

When comparing between ATVA in period II and period I, there were 26 companies that had higher ATVA in period I than in period II and there were 9 companies that had lower ATVA in period I than in period II. This means that there were more firms with higher ATVA in period I than in period II, and this was statistically significant as evidenced by two-tailed significant results of 0.002, lower than $\alpha = 5\%$ or 0.05.

Furthermore, when comparing ATVA in period III and period II, there were 19 companies that had higher ATVA in period II than in period III and there were 16 sample companies that had lower ATVA in period II than in period III. This means that there were more firms with higher ATVA in period II than in period III, but this was not statisti-cally significant as evidenced by two-tailed signifi-cant results of 0.219, higher than $\alpha = 5\%$ or 0.05.

In the last ATVA analysis, when comparing be-tween ATVA in period III and period I, there were 14 companies with higher ATVA in period I than in period III and there were 21 companies that had lower ATVA in period I than in period III. It means that there were more firms with lower ATVA in pe-riod I than in period III, but this was not statistically significant as evidenced by two-tailed significant re-sults of 0.432, higher than $\alpha = 5\%$ or 0.05. The three analysis can be seen from the level of significant comparison of aver-age abnormal return per period of Tax Amnesty 2016 including the significance of ATVA Period II-ATVA Period I of 0.002, ATVA Period III-ATVA Period II 0.219, ATVA Period III-ATVA Period 0.432 and Z-Score i.e. ATVA Period II-ATVA Period I -3.161, ATVA Period III-ATVA Period II -1.222, ATVA Period III-ATVA Period -0.786.

Table 4. T-test result of average trading volume activity.

	ATVA Period II–Period I	ATVA Period III –Period II	ATVA Period III–Period I
Z	-3,161	-1,228	-0,786
Asymp. Sig. (2-tailed)	0,002	0,219	0,432

Then the second hypothesis which stated that there was a difference on trading activity volume caused by Tax Amnesty 2016 in period I, II and III at the banking companies listed in Bursa Efek Indo-nesia could be accepted because it was statistically significant and also the ATVA between period I and period II were statistically significant while others were not.

The conclusion of this study was that since one of the three tests had significant results, then it can be concluded that there was a difference in trading volume activity caused by policy of Tax Amnesty 2016. This may be because some investors assumed that at the time of Tax Amnesty period I, the information in the public was sufficient and they thought following the Tax Amnesty program would benefit them. Therefore, investors flocked to sell their shares, especially shares of banking companies in the stock exchange in order to make a profit, whether from taking advantage of the sale saving it for long term investment. The offered rate of 2% was also considered more attractive than the 3% rate in period of II and 5% in the third period. This was what caused the discrepancy between trading volume activity during the Tax Amnesty period I compared with Tax Amnesty in period II. Investors assumed that in period I the banking companies achieve the highest liquidity than in period II and period III.

3.3 Discussion

3.3.1 The difference on abnormal return during tax amnesty policy period I, II and III

The calculation of abnormal return in this study was aimed to see whether the Tax Amnesty 2016 con-tained information that could affect the decision of the investor or not. It was also used as a testing tool in market efficiency.

It was known that the average level of abnormal return during the second amnesty tax period com-pared to the first period was 0.806, the average of the third period compared to the second period was 0.731, and the average of the third period compared to the first period was 0.512. Therefore, this research concluded that during period I, II and III there was no difference on abnormal return.

This was in line with Tiswiyanti & Asrini (2015) research. They argued, in their study about fuel prices announcement event, that there was no differ-ence in abnormal returns before and after the announcement of fuel prices. This means that the announcement of the event was deemed not to have sufficient information to describe the conditions and influence the investors' decisions in the capital market.

In Hartono (2010), the market was said to be inef-ficient if one or several market participants could enjoy abnormal returns in a long period of time. This means that the Tax Amnesty 2016 policy did not cause a difference in abnormal returns because at that time the market could be said to be efficient.

3.3.2 The difference on trading volume activity during the tax amnesty policy period I, II, and III

Trading volume activity reflected the behavior of investors in the banking stock market. It was discovered that the average level of average trading volume activity during the Tax Amnesty period II compared to the first period was 0.002, the third period compared to the second period was 0.219, and the third period compared to the period I was 0.432. From the data it could be seen that only when comparing between Tax Amnesty period II with period I α exceeded 5% or 0.05. So this study concluded that there were differences in trading volume activity in period I, II and III.

This may be because by some investors assumed that during Tax Amnesty period I, the information available in the public was sufficient and they thought following the Tax Amnesty program would benefit them. Therefore, investors flocked to sell their shares, especially shares of banking companies in the stock exchange in order to make a profit, whether from taking advantage of the sale saving it for long term investment.

This caused an imbalance condition between trading volume activity in period I compared with period II. Some investors assumed that in period I, the banking company reached its highest liquidity than in period II and period III.

The difference in trading volume activity in banking companies indicated that there was a change in investor expectations individually in banking stocks. As Hartono (2013) argued that testing of activity volume volumes reflected changes in investor expectations.

4 CONCLUSION

It can be concluded that there was no difference on abnormal return on banking companies listed in Indonesia Stock Exchange. This can be attributed to investors assuming that in relation with Tax Amnesty 2016 in Indonesia, there was no sufficient information to describe the condition that would influence investors' decision in capital market. However, there was a difference in trading volume activity in banking companies listed on the Indonesia Stock Exchange, but this only proved significantly in trading volume activity test in period I compared with period II. This can be attributed to investors assuming that at the time of the Tax Amnesty period I, the information available in the public was sufficient and they thought following the Tax Amnesty program would benefit them. Therefore, investors flocked to sell their shares, especially shares of banking companies in the stock exchange in order to make a profit, whether from taking advantage of the sale saving it for long term investment. This causes the discrepancy between trading volume activity during the Tax Amnesty in period I compared to Tax Amnesty in period II. Investors assumed that in period I the banking companies reached its highest liquidity than in period II and period III.

REFERENCES

Hartono, J. 2010. *Teori portofolio dan analisis investasi.* Yogyakarta: BPFE.

Hartono, J. 2013. *Teori portofolio dan analisis investasi.* Yogyakarta: BPFE.

Republik Indonesia PMK.118/PMK.03/2016. *Tata Cara Pelaksanaan Pengampunan Pajak.* 2016. Jakarta: Diperbanyak oleh Ikatan Akuntan Indonesia.

Republik Indonesia.Undang – Undang No. 11 Tahun 2016. *Pengampunan Pajak.* 2016. Jakarta: Diperbanyak oleh Ikatan Akuntan Indonesia.

SWA Magazine. 2016.

Tiswiyanti, W. & Asrini. 2015. reaksi investor atas pengumuman kenaikan harga bbm terhadap abnormal return, security return variability dan trading volume activity saham perusahaan transportasi di bursa efek indonesia tahun 2014. *Jurnal Studi Manajemen dan Bisnis* 2.

Advances in Business, Management and Entrepreneurship – Hurriyati et al (eds)
© 2020 Taylor & Francis Group, London, ISBN 978-0-367-27176-3

Comparison of model, stabilization, and finance performance of sharia commercial banks and conventional commercial banks in Indonesia in 2012–2016

Y.M. Dewi & D.F. Septiarini
Universitas Airlangga, Surabaya, Indonesia

ABSTRACT: The aim of study was to determine the difference in business model, banking stability and financial performance of Islamic banks and conventional banks around the period of 2012–2016. This study used quantitative research method. The populations of this research were Islamic banks and conventional banks in Indonesia and the sample was three Islamic banks and six conventional banks that had commercial banking business model. The data was analyzed with comparative analysis test using independent sample t-test and Mann Whitney test. The comparative analysis result in business model showed that there were differences in secondary banking in terms of fee-based income, on the other hand, there was no differences in primary banking which consisted of third-party fund ratio and financing to deposit ratio. There was no difference in banking stability measured using z-score and neither was on financial performance measured using net profit margin.

1 INTRODUCTION

The issuance of Act No. 21 year 2008 regarding Islamic banking gave Islamic banking in Indonesia a legal basis and became a competitor to conventional banks which existed beforehand. The banking industry was forced to innovate a business model in order to win competition and to attract customers.

Therefore, this study aimed to compare business model, stability and finance performance on sharia commercial banks and conventional commercial banks, to understand the differences in business models and to investigate their stability level during the period of 2012–2016.

2 METHOD

This study used a quantitative approach. Based on problem formulation, the variables in this research were (1) Fee Based Income Ratio (FBI), (2) Financing to Deposit Ratio (FDR), (3) Third Party Fund (DPK), (4) Z-score, (4) Net Profit Margin (NPM). The type of data used in this research was secondary data such as annual reports from December 31, 2012, to December 31, 2016, that were published in each bank's website. Other data resources used in this research included books, other literature, previous research, and articles from internet. This research used panel (pooled) data to collect data,

because it involved many samples on each period (cross section) and time (time series).

This research used purposive sampling to determine the samples. The selection criteria were (1) sharia commercial bank and conventional commercial bank in Indonesia those had commercial banking business model type, (2) excluding BPD and foreign bank having branch offices in Indonesia, (3) sharia commercial bank and conventional commercial bank with comparable assets around > 26 trillion and ≤ 79 trillion, (4) sharia commercial bank and conventional commercial that those published their annual report from 2012 to 2016.

Data collected was then analyzed using statistical tools. This research used two stages in analyzing data; descriptive analysis and inferential statistics.

3 RESULT AND DISCUSSION

The result of the normality test shown in Table 1 showed that DPK variable met the requirement for t-test; however, FBI, FDR, NPM and z-score variables could only be tested with a Mann-Whitney test because the data were not normally distributed.

3.1 *Fee based income ratio (FBI)*

Table 2 shows that the feebased income of a conventional commercial bank was higher than that of a sharia commercial bank in the period of

Table 1. Normality test.

Variables	Categories	Df	Statistic	Sig	Explanation
FBI	Conventional	30	0.889	0.004	Not normally distributed
	Sharia	15	0.850	0.017	Not normally distributed
FDR	Conventional	30	0.895	0.006	Not normally distributed
	Sharia	15	0.933	0.304	Normally distributed
NPM	Conventional	30	0.920	0.027	Not normally distributed
	Sharia	15	0.630	0.000	Not normally distributed
DPK	Conventional	30	0.932	0.057	Normally distributed
	Sharia	15	0.963	0.747	Normally distributed
Z-score	Conventional	30	0.924	0.034	Not normally distributed
	Sharia	15	0.689	0.000	Not normally distributed

Table 2. FBI difference test using Mann-Whitney test.

Variables	Rank	Sig	Explanation
Conventional	27.03	0000	H0 accepted
Sharia	13.13		

2012 to 2016. This was supported by research conducted by Beck et al. (2012) which showed that fee based income in Islamic bank was lower than conventional bank.

The fee-based income of Islamic bank was lower than that of a conventional bank, indicating that the conventional commercial bank had more developed secondary banking business model. This could be seen from services and products offered for example collection, credit card, ATM, garage bank, letter of credit, spot and derivative transactions. Spot and derivative transaction made bank receive increasingly high income, on the contrary, services in sharia commercial bank relatively could not be maximized due restriction in Islamic law.

This finding also showed that secondary banking business model in sharia commercial bank was less developed due its focus on primary banking; as a result, the secondary banking business was not maximized.

Table 3. FDR difference test using Mann Whitney test.

Variables	Rank	Sig	Explanation
Conventional	25.7	0.075	H0 rejected
Sharia	18.07		

3.2 Financing to deposit ratio (FDR)

Table 3 shows that there was no difference in primary banking business model between the two types of banks in the period 2012 to 2016. Beck et al. (2012) found that business models among Islamic banks and conventional banks were not too different, but Islamic banks had higher intermediation ratio compared to conventional banks.

The result meant that both banks have operated their primary banking business model as commercial banking well with the provisions set by the Bank Indonesia. Circular Letter of Bank Indonesia no. 26/5/BPPP dated May 29, 1993 stated that FDR and/or LDR should be at the level of 85%–110%. However, the LDR ratio in conventional commercial banks were over the safe limits set, indicating that conventional commercial banks are too aggressive in providing credits due to the exceeding of safe limits set by Bank Indonesia, while sharia commercial banks are more selective in providing financing so that its financing functions fall into the safe category.

3.3 Third party fund

Based on independent t-test conducted before, the result revealed that there were no differences in primary banking business model as measured by the growth of third-party funds. It meant that both types of

Table 4. Third party fund growth difference test using independent sample test.

Variables		Levene's test		Sig	Explanation
		F	Sig		
DPK	Equal variance assumed			0.756	
		9.001	0.004		H0 rejected
	Equal variance not assumed			0.689	

Table 5. Z-score difference test using Mann Whitney test.

Variable	Rank	Sig	Explanation
Conventional	25.7	0.075	H0 rejected (there is no difference)
Sharia	18.07		

Table 6. Net profit margin (NPM) difference test.

Variable	Rank	Sig	Explanation
Conventional	28.40	0.000	H0 accepted (there is difference)
Sharia	12.20		

Source: SPSS test result.

banks had maximized their primary banking model business in the financial function, this was proved by the growth of third-party funds in sharia commercial banks that were equal with the growth of conventional commercial banks. Although Islamic banks were relatively new compared with conventional banks, the growth of third-party funds in sharia commercial banks indicates a positive growth of third-party funds compared to those of conventional commercial banks and some other banks that suffer negative growth. This result also implies that the public has entrusted their funding to sharia commercial banks, which exclude systems of interest or usury.

3.4 *Z-score*

Based on the z-score test shown in Table 5, it can be stated that there was no difference in stability between the two types of banks. The result from descriptive analysis showed that sharia commercial bank had higher z-score value indicated sharia commercial banks were more stable. This meant that both types of bank have managed risk well.

Banks as intermediary institution will always factor in many risks and return in their business activity, which are related to bank stability. Bank stability is measured by using z-score; the higher z-score value indicated the more stable banks and the smaller value of z-score the closer the banks to bankruptcy. The average value of z-score in sharia commercial banks were higher than in conventional commercial banks, this condition illustrated that sharia commercial banks were more stable. Sharia commercial banks are more stable due to prohibition of haram transactions, such as spot trade and derivative transactions which contain gharar and gambling as they are risky transactions.

3.5 *Net profit margin (NPM)*

The result of Mann Whitney shown in Table 6 shows differences on finance performance. This was supported by Winar (2016), who found that there were differences on finance perfor-mance in terms of efficiency among sharia commer-cial banks and conventional commercial banks as measured by NPM ratio.

The disparity in NPM ratio was due to smaller net profit owned by sharia commercial banks thus performance efficiency is affected by amount of operational income and how far banks can use it maximally followed by emphasis on unforeseen expenses. The higher of NPM ratio proved the higher bank efficiency to gain profit from its operational activities. Therefore, sharia commercial banks should get maximum operational income, and always be based on guidelines set according to sharia.

4 CONCLUSION

It can be concluded that there was a difference between model business in secondary banking function among sharia commercial banks and conventional commercial banks in Indonesia as measured by fee based income ratio, financing to deposit ratio (FDR) and/or loan to deposit ratio (LDR) ratio, growth of third party fund. There was no difference in bank stability among sharia commercial banks among conventional commercial banks, but the value of z-score in convention commercial banks were higher. Thus, sharia commercial banks are more stable.

REFERENCES

Beck, T., Demirgüç-Kunt, A. & Merrouche, O. 2012. Islamic vs conventional banking: Business model, effeciency and stability. *Journal of Banking and Finance* 37.

Republic of Indonesia. 1998. *Act Number 10 Year 1998 about Banking*. Jakarta: Republic Indonesia.

Winar, N.I. 2016. Analisis perbandingan kinerja keuangan bank umum syariah dan bank umum konvensional periode 2013–2015. Doctoral Dissertation. Universitas Airlangga.

Advances in Business, Management and Entrepreneurship – Hurriyati et al (eds)
© 2020 Taylor & Francis Group, London, ISBN 978-0-367-27176-3

Industry growth, ownership structure, and capital structure in Indonesia

I. Harymawan, A. Arianto & Y.I. Paramitasari

Universitas Airlangga, Surabaya, Indonesia

ABSTRACT: The purpose of this study is to analyze whether industry classification has an effect on capital structure and whether foreign ownership structure has more influence on capital structure than domestic ownership structure. This study used 524 companies listed in the Indonesia Stock Exchange for three periods, from 2014 to 2016. To test the hypothesis, the Ordinary Least Square (OLS) regression method was used. This study found that industry classification had an effect on capital structure. Companies that were classified as high growth industry had a significant positive effect on capital structure, while low growth industry had a significant negative effect. Furthermore, firms with a foreign ownership structure had a significantly higher capital structure than domestic ownership. The findings from this study indicate that firms with foreign ownership and firms in high growth industries are more attractive to debt holders in Indonesia.

1 INTRODUCTION

Since Modigliani and Miller (1958) first proposed the classical MM-Irrelevant theory, firm value does not depend on its structure in a perfect financial market, capital structure has been an important research subject. For half a century, different theories have been developed to explain company's financing decisions, including the trade-off theory (Miller 1977), the hypothesis pecking order (Myers & Majluf 1984), the agency cost theory (Jensen & Meckling 1976), and equity market timing theory (Baker & Wurgler 2002). Meanwhile, research shows that capital structure is influenced by several characteristics within the company, such as profitability, firm size, value of collateral, non-debt tax shields, growth opportunities, uniqueness, industry classification, volatility (Titman & Wessels 1988), macroeconomic environment (Korajczyk & Levy 2003), and ownership structures (Bajaj et al. 1998).

Since the opening up of Indonesia's capital market in 1997 by the Indonesian government, both foreign and domestic investors can actively participate to invest in the market. This causes foreign investors to buy all of the companies in Indonesia, which results in almost no restriction of ownership by foreign investors in Indonesia. Therefore, both domestic and foreign investors can conduct free trading transactions with 99% ownership.

Various studies on the effect of foreign ownership and domestic ownership of the company's capital structure show contradictory results. Research conducted by Phung and Le (2013) showed that the foreign ownership structure had an influence on the level of corporate debt. When this perspective is viewed from the perspective of asymmetric information, according to Zou and Xiao (2006) and (Phung & Le 2013), foreign investors have no discretion in overseeing foreign ownership of listed companies in

the stock market, requiring companies with foreign ownership to use more debt as a monitoring channel to protect their investments. This is in accordance with the research of Li et al. (2009), which revealed that foreign ownership had a positive influence on all measurements of the leverage company. Another study, by Gurunlu and Gursoy (2010) put forward the initial assumption that foreign investors generally faced greater risks, such as country risk, currency risk, and business risk, compared to domestic investors. This study showed that firms with high levels of foreign ownership used less debt in the components of the company's capital structure because foreign investors faced a greater range of risks than domestic investors.

In addition to foreign ownership and domestic ownership, industry classification becomes a factor that can affect the company's capital structure. Companies, however, have different growth rates and funding needs (Hoitash et al. 2016). Research conducted by Scott et al. (1975) demonstrated that the decision of financing could be decided by a company by paying attention to the company structure of similar companies. This comparison is made by comparing firms operating in the same industry and finding that leverage variance emerges as an indication that industrial classification is a determinant of the company's financial structure. From the research, it can be said that the classification of the industry influences the decision that will be taken by a company manager, because of intra-industry information transfers. Based on this theory, the researcher suspects that the behavior of managers in deciding the capital structure of the company is influenced by the incidents that occur in similar companies. In addition, because the regulations that apply in an industry tend to be the same, it can be said that similar industries have financial ratios, for example the same leverage. Research conducted by Bauer (2004) examined the factors affecting capital structure and found that firm size, tangibility, and tax positively and significantly influenced the capital structure, profitability, and growth opportunities. They had a significant and negative effect on capital structure, non-debt tax shields. Moreover, they also had a negative and insignificant effect on capital structure, while vitality and industry classification had no effect on capital structure. The research conducted by Bauer (2004) is in contrast to that of Scott et al. (1975). Scott et al. (1975) found that industry classification had no effect on capital structure.

This study attempts to examine industry classification and ownership structure on capital structure. Specifically, it investigates whether industry classification has an effect on capital structure, and whether foreign ownership structure has more influence on capital structure rather than domestic ownership.

The funding decision (capital structure) involves the shareholder of the company or the owner of the company (principal), which is part of the ownership structure. Much of the empirical research on the relationship between ownership and capital structure has been done using data from the industrial economy (Kester 1986). Theory of ownership and capital structure emphasizes the role of debt in reducing agency problems between managers and shareholders. Hussain and Nivorozhkin (1997) say that foreign ownership of shares may have a positive effect on debt lending because firms with foreign ownership show better performance than firms with domestic ownership. In this context, Hussain and Nivorozhkin (1997) assert that average corporate foreigners show a significantly higher debt ratio than their domestic counterparts in their own country. On the other hand, a study by Sivathaasan (2013) examining manufacturing firms listed in the Colombo Stock Exchange found that correlation analysis showed a strong positive relationship between foreign ownership and leverage. In addition, there was a negative relationship between domestic ownership and leverage ratio. Using the Ordinary Least Squares (OLS) equation, it was found that the leverage ratio increased when foreign ownership was added to the equation, assuming domestic ownership was zero. In contrast, the leverage ratio decreased because domestic ownership increased, with the assumption of foreign ownership being zero. Overall, the Sivathaasan (2013) study showed that foreign ownership had more influence on leverage decisions than on domestic ownership. Based on the findings from these studies, the first hypothesis of this study is:

H1: Foreign ownership structure has more positive effect on capital structure than domestic ownership structure.

Research on the relationship of industrial classification to capital structure is done by Scott et al. (1975) and found that there was no single assessment that precisely determined the right capital structure for an enterprise. Decision of financing can be decided for a company by paying attention to the financing structure of similar companies. This comparison can be done by comparing companies in the same industry. By using parametric and nonparametric testing, Scott et al. (1975) found that leverage variance emerged as an indication of industry classification as this was the determinant of the company's capital structure. From the research, it can be said that the classification of the industry influences the decision that will be taken by a company manager, because of intra-industry information transfers. The theory of intra-industry information transfer holds the argument that information revealed by one industry member may affect the stock price of its rival (Besley & Cohers 2000). On the basis of this theory, the investigators suspect that the behavior of managers in deciding the corporate financing structure is influenced by events occurring in the same company or in the same group in which the company exists. In addition, because the regulations that apply in an industry tend to be the same, it can be said that different industries will have different financial ratios as well. Usually in an industry, there is one company that makes references by other companies in the industry because it is considered to have a good performance. Thus, companies in different industries will have different capital structures. Both theorists and empirical literature suggest that the ratio of leverage differs significantly for different industries (Schwartz & Aronson 1967, Hamada 1972, Harris & Raviv 1991, Mackay & Phillips 2005, Chen 2004, Jensen 1986).

Industry classification has an influence on corporate financing decisions. By grouping companies, the risk of the company through competition can be seen. The more members an industry group has, the greater the risk of business, and the company will be more careful in making decisions of financing. If the company is in the classification of the industry where the product or service produced is much needed by the consumer and does not have many competitors, it tends to get external sources of funds. Conversely, for companies in many competing groups it tends to be more difficult to obtain funds. Manufacturing companies tend to have higher leverage than trading companies, services and investments, because they have higher fixed assets that can be used as collateral for credit. In addition, the risk of operating the manufacturing company is higher, because of the longer cycle. As a result, the velocity of funds is longer and requires greater operational funds to run the company's operations. Based on this explanation, the second hypothesis of this study is:

H2: Industry classification has a significant effect on capital structure.

2 METHOD

2.1 Sample and data sources

The sample of this study consisted of all companies listed in the Indonesia Stock Exchange (BEI) for the period 2014–2016. Sources of data in this study were obtained through the company's financial statements,

ORBIS database, and ICMD database. There were 1665 data. Furthermore, the researchers applied sample selection criteria as follows: (1) exclude companies that belong to the finance, insurance, and real estate industries (SIC 6) because of having 426 different financial reporting data; and (2) exclude 715 incomplete data for all variables in this study. After applying the sample selection criteria, there were 524 observations obtained as the main samples in this study.

2.2 Definition of variables

The dependent variable in this study was the structure measured by the ratio of debt to total assets. The independent variables in this study were industry classification (high growth industry and low growth industry) and ownership structure (foreign ownership structure and domestic ownership structure). Industry classification and ownership structure were both measured by dummy variables. For industry classification, value of 1 was given if the company was a high growth industry. For ownership structure, a value of 1 was provided if it was a foreign ownership structure. The control variables used in this research were profitability (PROFIT), firm's size (FSIZE), growth opportunity (GROWTH), intangibility (INTANGIBLE), and business risk (RISK) (see Appendix Table A.1 for a summary of variable definitions).

3 RESULTS AND DISCUSSION

3.1 Statistics descriptive and univariate analysis

The definitions and measurements of the variables used in this study are presented in Appendix Table A.1. Table I contains the sample distributions by industry group. There was a total of 524 companies, divided into 8 sectors. The industry with the highest frequency was the manufacturing industry (36.26%) with a frequency of 190 companies. In Table 2, the 8 industry sectors were subdivided into 2 classifications based on industry growth rate. The high growth

Table 2.   Industry distribution based on industry growth.

CLASSIFICATION	NAME OF INDUSTRY	FREQ	%
HIGH GROWTH INDUSTRY	I4: Manufacturing I5: Transportation and Public Utilities I6: Wholesale Trade I7: Retail Trade	347	66.22
LOW GROWTH INDUSTRY	I1: Agriculture, Forestry, Fishery I2: Mining I3: Construction I8: Services	177	33.78

industry consisted of 347 companies, while the low growth industry consisted of 177 companies.

Table 3 shows the statistics descriptive table. The independent variable of capital structure (CAPSTRUC) had an average value of 0.514. Control variables such as profitability (PROFIT) had an average of 0.069, while business risk (RISK) and intangibility (INTANGIBLE) had an average value of 0.059 and 0.041, respectively.

Table 3.   Statistics descriptive.

	Average	Median	Minimum	Maximum
CAPSTRUC	0.514	0.506	0.064	1.898
FOREIGN-OS	0.553	1.000	0.000	1.000
DOMESTIC-OS	0.689	1.000	0.000	1.000
GROWTH	0.079	0.047	-0.887	2.661
PROFIT	0.069	0.064	-0.170	0.454
TASSETS	1.261e +10	4.126e +09	7648194.000	2.619e +11
RISK	0.059	0.071	-0.986	0.899
INTANGIBLE	0.041	0.009	0.000	0.500

Table 1.   Sample distribution based on industrial classification: Industry distribution based on SIC code.

INDUSTRY CLASSIFICATION	SIC	NAME OF INDUSTRY	FREQ	%
I1	0100–0999	Agriculture, Forestry, Fishery	34	6.49
I2	1000–1499	Mining	54	10.31
I3	1500–1799	Construction	28	5.34
I4	2000–3999	Manufacturing	190	36.26
I5	4000–4999	Transportation and Public Utilities	93	17.75
I6	5000–5199	Wholesale Trade	36	6.87
I7	5200–5999	Retail Trade	28	5.34
I8	7000–8999	Services	61	11.64
TOTAL			524	100

Table 4. Pearson correlation test.

		CAPITAL STRUCTURE							
		1	2	3	4	5	6	7	8
1	*CAPSTRUC*	1.000							
2	*FOREIGN-OS*	0.107**	1.000						
		(0.014)							
3	*DOMESTIC-OS*	-0.009	-0.230***	1.000					
		(0.834)	(0.000)						
4	*GROWTH*	-0.140***	-0.064	-0.084*	1.000				
		(0.001)	(0.143)	(0.056)					
5	*PROFIT*	-0.300***	-0.143***	-0.003	0.060	1.000			
		(0.000)	(0.001)	(0.951)	(0.173)				
6	*FSIZE*	0.059	-0.095**	-0.165***	-0.011	0.131***	1.000		
		(0.180)	(0.030)	(0.000)	(0.799)	(0.003)			
7	*RISK*	-0.010	-0.093**	-0.001	0.003	0.510***	0.155***	1.000	
		(0.824)	(0.033)	(0.975)	(0.938)	(0.000)	(0.000)		
8	*INTANGIBLE*	0.129***	0.087**	-0.114***	0.012	-0.102**	0.086**	-0.014	1.000
		(0.003)	(0.047)	(0.009)	(0.775)	(0.020)	(0.049)	(0.743)	

Table 4 shows the results of the Pearson correlation test. The relationship between capital structure variables (CAPSTRUC) and foreign ownership structure (FOREIGN-OS) was positive, meaning that firms with foreign ownership structures had higher debt levels. Meanwhile, the relationship between capital structure variable (CAPSTRUC) and domestic ownership structure (DOMESTIC-OS) was negative which means the structure of domestic ownership had a lower debt.

3.2 *Main analysis*

This research was conducted by using regression analysis with Ordinary Least Square (OLS), which controlled the year fixed effect and industry fixed effect and Ordinary Least Square (OLS) with robust. In robust, the OLS model used a cluster model approach from Petersen (2009) by grouping companies based on ticker and year to estimate the regression model.

In testing the relationship between capital structure and ownership structure, a regression model was used, as follows:

CAPSTRUC*t* = α + β1FOREIGN-OS*t* + β2DOMESTIC-OS*t* + β3GROWTH*t* + β4PROFIT*t* + β5FSIZE*t* + β6RISK*t* + β7INTANGIBLE*t* + *fixed effect year* + *fix effect industry* + ε

This study also examined the relationship between capital structure and industry classification. The regression model used is presented here:

CAPSTRUC*t* = α + β1HIGH-INDUSTRY*t* + β2LOW-INDUSTRY*t* + β3GROWTH*t* + β4PROFIT*t* + β5FSIZE*t* + β6RISK*t* + β7INTANGIBLE*t* + *fixed effect year* + ε

3.3 *First regression: Test the relationship between foreign ownership structure and domestic ownership structure on capital structure*

Table 5 shows the results of multiple linear regression analysis. In the specification (1), the FOREIGN-OS coefficient was 0.039 (t = 1.88), showing a positive and significant relation to capital structure. This means that every 1-point increases FOREIGN-OS and CAPITAL STRUCTURE. Specification (2), the coefficient of DOMESTIC-OS was -0.006 (t = -0.24) showing a negative but not significant result on capital structure. The result shows that the foreign ownership structure has high debt because it has a significant positive effect on capital structure. Meanwhile, the domestic ownership structure cannot be seen because its effect is not significant. Nevertheless, it shows a negative relation to capital structure.

Control variables such as RISK and INTANGIBLE on specifications (1) and (2) showed a positive and significant relationship to capital structure. While GROWTH and PROFIT had a negative and significant relationship to capital structure. The value of r2 indicated that the regression model was able to explain the relationship between the independent and dependent variables of 19.8% and 19.4%.

3.4 *Second regression: Examine the relationship between high growth industry and low growth industry on capital structure*

Researchers conducted a regression test by dividing the classification of industries into two groups based on the growth of industrial companies. The two groups were a group of companies with high growth industry and a group of companies with low growth industry.

Table 5. Multiple linear regression analysis.

	Prediction of Direction	CAPITAL STRUCTURE	
		(1)	(2)
FOREIGN-OS	+	0.039*	
		(1.88)	
DOMESTIC-OS	-		-0.006
			(-0.24)
GROWTH	-	-0.068**	-0.072**
		(-2.15)	(-2.22)
PROFIT	-	-1.145***	-1.166***
		(-4.57)	(-4.56)
FSIZE	+	0.009	0.008
		(1.33)	(1.16)
RISK	+	0.230**	0.228**
		(2.07)	(2.02)
INTANGIBLE	+	0.274**	0.291**
		(2.24)	(2.43)
Constant	?	0.390**	0.438***
		(2.45)	(2.61)
Year Dummies		Included	Included
Industry Dummies		Included	Included
r2		0.198	0.194
N		524	524

This table shows the result of regression using Ordinary Least Square (OLS) with robust on companies with a structure of foreign ownership and domestic ownership, and control variables. The dependent variable is the capital structure (CAPITAL STRUCTURE). The total sample of the study consists of 524 companies listed in the Indonesia Stock Exchange (BEI) in the period 2014–2016. Levels of significance are * 10%, ** 5%, and *** 1%.

Table 6 shows the regression test results from the high growth industry (HIGH-INDUSTRY). Regression results in the low growth industries were not shown because the result was the reverse of the regression results of the high growth industry. The specification (1) shows that the high growth industry (HIGH-INDUSTRY) variable had a significant positive effect on capital structure (CAPITAL STRUCTURE), with a coefficient value of 0.069 (t = 2.86) and a significance level of 1%. These results indicated that the high growth industry had a high debt level. The specification (2) shows the robust results of the specification (1).

Control variables such as FSIZE, RISK and INTANGIBLE on specifications (1) and (2) showed a positive and significant relationship to capital structure. At the same time, GROWTH and PROFIT had a negative and significant relationship to capital structure. The r2 values in the specifications (1) and

Table 6. Results of multiple linear regression: Industry classification and capital structure.

	Prediction of Direction	CAPITAL STRUCTURE	
		1	2
HIGH-INDUSTRY	+	0.069***	0.069***
		(2.86)	(2.80)
GROWTH	-	-0.082***	-0.082**
		(-2.90)	(-2.42)
PROFIT	-	-1.139***	-1.139***
		(-8.36)	(-4.68)
FSIZE	+	0.016**	0.016**
		(2.15)	(2.45)
RISK	+	0.207***	0.207*
		(3.52)	(1.94)
INTANGIBLE	+	0.351**	0.351***
		(2.51)	(2.87)
Constant	?	0.176	0.176
		(1.04)	(1.23)
Year Dummies		Included	Included
r2		0.159	0.159
N		524	524

This table shows the regression result using Ordinary Least Square (OLS) not with robust on specification (1) and with robust on specification (2) for high growth industry and control variables. The dependent variable is the capital structure (CAPITAL STRUCTURE). The total sample of the study consists of 524 companies listed on the Indonesia Stock Exchange (BEI) in the period 2014–2016. Level of significance * 10%, ** 5%, and *** 1%.

(2) show that the regression result was capable of representing 15.9% of the total of 524 samples.

4 CONCLUSION

The results of this study have two implications. First, companies with foreign ownership structures have high debts, so this can be a consideration for companies that want high external capital to recruit more foreign investors than domestic investors. Second, the capital structure of the high growth industry has high debts, so it can also be a consideration for companies that want high external capital to run a business in manufacturing, transportation and public utilities, wholesale trade, and retail trade (Mckinsey Global Growth Model 2011).

REFERENCES

Bajaj, M., Chan, Y. and Dasgupta, S. 1998. The relationship between ownership, financing decisions and firm performance: a signaling model. *International Economic Review*, 39(3): 723–744.

Baker, M. and Wurgler, J. 2002. Market timing and capital structure. *Journal of Finance*, 57(1): 1–32.

Bauer, Patrik. 2004. Determinants of Capital Structure: Empirical Evidence from the Czech Republic, *Czech Journal of Economics and Finance (Finance a uver)*, 54 (1–2): 2–21.

Besley, Scott dan Cohers, Ninon. 2000. Reactions of issuers and rivals to private placement of common equities, 1–30.

Chen, J.J. 2004. Determinants of capital structure of Chinese-listed companies. *Journal of Business Research*, 57(12): 1341–1351.

Gurunlu, Meltem and Gursoy, Güner. 2010. The influence of foreign ownership on capital structure of non-financial firms: Evidence from Istanbul Stock Exchange. *The IUP Journal of Corporate Governance*, 9(4): 21–29.

Hamada, R. 1972. The effects of the firm's capital structure on the systematic risk of common stocks. *Journal of Finance*, 27(2): 435–452.

Harris, M. and Raviv, A. 1991. The theory of capital structure. *Journal of Finance*, 46(1): 297–355.

Hoitash, R., Hoitash, U., & Kurt, A. C. (2016). Do accountants make better chief financial officers? *Journal of Accounting and Economics*, 61(2–3): 414–432.

Hussain, Qaizar and Nivorozhkin, Eugeniy. 1997. The capital structure of listed companies in Polanda, *IMF Working Paper* no. 97/175.

Jensen, M. 1986. Agency costs of free cash flow, corporate finance, and takeovers. *American Economic Review*, 76 (2): 323–339.

Jensen, M. and Meckling, M. 1976. Theory of the firm: managerial behavior, agency costs and ownership structure. *Journal of Financial Economics*, 3(4): 305–360.

Kester, W.C. 1986. Capital and ownership structure: a comparison of United States and Japanese manufacturing corporation. *Financial Management in Japan*, 15(1): 5–17.

Korajczyk, R.A. and Levy, A. 2003. Capital structure choice: macroeconomic conditions and financial constraints. *Journal of Financial Economics*, 68: 78–109.

Li, K., Yue, H., & Zhao, L. 2009. Ownership, institutions, and capital structure: evidence from China. *Journal of Comparative Economics*, 37 (3), 471–490.

Mackay, P. and Phillips, G.M. 2005. How does industry affect firm financial structure? *Review of Financial Studies*, 18(4): 1433–1466.

Mckinsey Global Institute. 2011. Urban World: Mapping the economic power of cities. Retrieved from https://www.mckinsey.com/~/media/McKinsey/Featured%20Insights/Urbanization/Urban%20world/MGI_urban_world_mapping_economic_power_of_cities_full_report.ashx

Miller, M. 1977. Debt and taxes. *Journal of Finance*, 32 (2): 261–275.

Modigliani, F. and Miller, M.H. 1958. The cost of capital, corporation finance and the theory of investment. *American Economic Review*, 48(3): 261–297.

Myers, S.C. and Majluf, N.S. 1984. Corporate financing and investment decisions when firms have information that investors do not have. *Journal of Finance*, 39(2): 575–59.

Phung, Nam Duc dan Thi Phuong Vy Le. 2013. Foreign Ownership, Capital Structure and Firm Performance: Empirical Evidence from Vietnamese Listed Firms. *The IUP Journal of Corporate Governance*, 12(2).

Schwartz, E. and Aronson, R. 1967. Some surrogate evidence in support of the concept of optimal financial structure. *Journal of Finance*, 22(1): 10–19.

Scott, Jr, David dan John D. Martin. 1975. Industry Influence on Financial Structure, *Financial Management*, spring, 67–73.

Sivathaasan, N. 2013. Foreign ownership, domestic ownership and capital structure: Special reference to manufacturing companies quoted on Colombo Stock Exchange in Sri Lanka. *European Journal of Business and Management*, 5(20).

Titman, S. and Wessels, R. 1988. The determinants of capital structure choice. *Journal of Finance*, 43(1): 1–19.

Zou, H., and Xiao Jason, Z. 2006. The financing behavior of listed Chinese firm. *The British Accounting Review*, 38: 239–258.

Appendix Table A.1 Variable Definition

Variables	Operational Definition	Data Sources
CAPSTRUC	The ratio of total debt to total assets.	ORBIS
FOREIGN-OS and DOMESTIC-OS	Dummy variable, given value 1 if it is foreign ownership, and value 0 if it is domestic ownership.	ICMD 2014–2016
HIGH-INDUSTRY and LOW-INDUSTRY	Dummy variable, given value 1 if it is high growth industry, and value 0 if it is low growth industry.	ORBIS
PROFIT	EBIT (earnings before interest expense and tax expense) divided by total assets.	ORBIS
FSIZE	*Log of total assets.*	ORBIS
GROWTH	Net sales of the current year minus sales of previous year divided by net sales of previous year.	ORBIS
RISK	Net profit to equity ratio.	ORBIS
INTANGIBLE	Intangible asset ratio to total assets.	ORBIS

Advances in Business, Management and Entrepreneurship – Hurriyati et al (eds)
© *2020 Taylor & Francis Group, London, ISBN 978-0-367-27176-3*

Investment experience to expected return: Consequences of risk behavior

F. Ismiyanti
Universitas Airlangga, Surabaya, Indonesia

P.A. Mahadwartha
Universitas Surabaya, Surabaya, Indonesia

ABSTRACT: Decision making for financial products requires scientific justification because it involves uncertainty in returns and financial risks. This study aims to examine the effect of investment experience on risk propensity, risk propensity to risk perception, risk perception to return expectation and investment experience to return expectation. Past investment experience acts as an anchor of benchmarks and the basis of individual decision making in investment. Risk propensity and risk perception are risk behaviors that show a person's attitude and behavior based on previous investment experience, while the return expectation of the financial products reflects the decision of individual investors on the chosen investment opportunities. This study expects to discover how the influence of past investment experience on variable risk propensity, risk propensity to risk perception, and the effect of investment experience on return expectation. Participants are investors and traders of financial instruments in the capital market. This study uses Structural Equation Modelling to test the hypothesis. The result showed that investment experience has positive effect on risk propensity of investors. Higher investment experience will increase the awareness of investors toward risk. Risk propensity has negative effect on risk perception, therefore will support the investors perception toward reducing risk mechanism. Risk perception has positive effect on return expectation, while investment perception has insignificant effect on return expectation. The result showed that investment perception has indirect effect to return expectation through risk perception.

1 INTRODUCTION

The subprime mortgage crisis in United States middle of 2008 was followed by the bankruptcies suffered by several major US financial firms such as Lehman Brothers, Merrill Lynch, Goldman Sachs and other financial firms which have had a global financial impact and caused a domino effect. As a result of the financial crisis, the financial products suffered a fantastic loss of value so as to give a hard blow to investors. Today's financial institutions need to pay attention to risk and understand the investor's assessment of the risks and how investor's attitude toward risk matches their risk needs and preferences. Thus, financial institutions can assist the decision-making process of investors and will affect the return expectation. Several issues arise during investors' investment process, such as experience of losses, and the benefits they gain will have a meaningful personal experience for them, affecting their behavior and attitudes toward the risks of any financial products (Chou et al. 2010). It will affect attitudes and behavior, as well as investment decision making. Investors who typically invest in high risk assets will reconsider investing in these assets even if their future returns are high enough to influence investors' investment decisions in the future. Attitudes gained through experience will have a direct effect on subsequent behavior. In this study, the variable that shows the attitude and behavior is

the risk propensity and risk perception. Risk propensity is the tendency of a person to a certain risk, while risk perception is the judgment of decision makers on the risks that exist in certain situation (Sitkin & Pablo 1992).

Chou et al. (2010) argued that past investment experience acts as anchor, the benchmark as well as the basis of individual decision making in investment. Risk propensity and risk perception are risk behaviors that show how people's attitudes and behaviors are based on previous investment experiences. The return expectation of the financial products reflects the decision of individual investors on their chosen investment. This research examines how the influence of past investment experience on risk propensity, risk propensity to risk perception and the influence of investment experience on return expectation which is the basic decision-making process of investors. The gap between studies of investment experience, risk propensity, and risk perception showed that each factor has its specific effect to financial decision of investor. However, this research tries to combine the issues of investment experience, risk propensity, and risk perception with return expectation from investor's point of view. The research problem formulated is there is a direct influence of investment experience on the return expectation; as well as the indirect effect of investment experience on the return expectation through risk propensity and risk perception.

1.1 Risk propensity

Wong (2005) defines risk propensity as a decision-makers tendency to avoid or take risks. Each individual has a different understanding of risk-taking behavior. Maturity of decision makers in terms of seniority and age will decrease risk propensity compared to young decision-makers. Thus, individuals who have high risk propensity are more risk-seeking, while individuals with lower risk propensity will be risk-averse.

1.2 Risk perception

Risk perception is defined by Sitkin and Pablo (1992) as the assessment of decision makers of the risks that exist in a situation. The definition of risk perception is consistent with the definitions of several previous studies (Douglas 1985, Dutton & Jackson 1987, Vlek & Stalen 1980). According to Sitkin and Weingart (1995), the assessment of those risks reflects the degree to which a person perceives a situation as negative, threatening, and out of control. Weber and Milliman (1997) stated that risk perception is the main factor causing behavior change in betting, because basically risk preference or risk attitude of a person is stable.

1.3 Investment experience

Osborn and Jackson (1988) and Thaler and Johnson (1990) argued that the success or failure of an event influences how decision makers assess risky situations and their actions. Thus, the events that occur beforehand will influence the tendency of decision makers to take risks. Chou et al. (2010) stated that previous investment experience and investor expertise will provide awareness of risk and be an important factor for future risk assessment. Investment experience is an important factor and has an anchor effect toward behavior.

1.4 Return expectation

According to Chou et al. (2010), in traditional financial concepts, investors assume do not like risks, but investments with high returns will have a high level of risk. Muradoglu (2005) describes professional traders supporting a positive relationship between risk and return, but novice traders and low-ability investors feel that expected returns are negatively related to risk. In Chou et al. (2010), the behavior of decision making is reflected in the expected return and how the portfolio configures by investors. Based on these considerations, in this study decision making in the investment will be represented by the return expectation that the investor expect on the amount of funds invested.

1.5 Hypothesis

Based on the theory described earlier, the hypotheses that can be drawn are as follows:

H_1: Investment experience positively influence the risk propensity.

H_2: Risk propensity negatively affects risk perception.

H_3: Risk perception has a positive effect on return expectation.

H_4: Investment experience has a positive effect on return expectation.

1.6 Research model

The structural model for the path diagram in this study is shown in Figure 1.

2 METHOD

This study obtains respondents directly through the responses of questionnaires, for example investors who invest in several financial instruments. There are 200 questionnaires distributed to potential respondents, and 197 were returned and only 153 were valid and fully completed questionnaires.

2.1 Operational definition of variables

The indicators used to measure investment experience are adapted from the research of Chou et al. (2010) and Byrne (2005). Investment experience can be known from the investment knowledge as well as the successful experience or failure of respondents in investing in several types of financial products. Indicators for investment experience include: adequacy of knowledge about financial management, ability or experience in investing, success in investing in financial products previously invested. All investment experience indicators (X1) will be measured using a Likert scale (1–5).

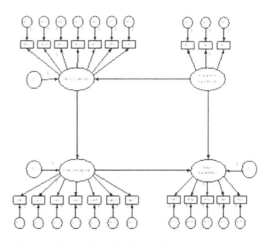

Figure 1. Structural model of path diagram.

The indicators used to measure risk propensity are adapted from the research of Byrne (2005) and Chou et al. (2010). Risk propensity can be drawn from how the investor's subjective opinion on the risk level contained in risk propensity indicators, which include: use daily income to buy lottery tickets or place them on sports betting (football, boxing, horse racing, etc.), lend money to friends whose amount is as big as income in 1 month, invest 10% of your annual income on highly speculative stocks, invest 10% of your annual income on government bonds, invest in a business that has a high risk, but prospects good and profitable, gambling using weekly earnings at the casino, take a job where you earn an exclusively commission-based income. All indicators of risk propensity will be measured using a Likert scale (1–5).

While risk perception is defined as the decision maker's judgment of the risks contained in a situation (Sitkin & Pablo 1992). According to Sitkin and Weingart (1995), the assessment reflects the degree to which individuals perceive a particular situation as negative, threatening, and out of control. People who have high risk perception will be risk averse or prefer to avoid risk. For risk perception assessments will also be measured using the Likert scale (1–5).

The return expectation of investors can be seen from the amount of expected returns from a number of investment funds. The indicators used to assess the return from investors are based on Byrne (2005) and Chou et al. (2010), where investors will be given the option of the return value they expect from their investment. The questions asked are of each type of investment below, estimate the amount you expect back within the next 3 years assuming you have $300 million to invest. Choice of answers include: (a) less than 2 hundred million, (b) 2–3 hundred million, (c) 3–4 hundred million, (d) more than 4 hundred million. The answer option (a) has a score of 1, the answer option (b) has a score of 2, the answer choice (c) has a score of 3, and the answer option (d) has the highest score of 4.

3 RESULT AND DISCUSSION

3.1 Variable description

Description begins with identifying the characteristics of respondents based on demographic factors. Respondents characteristic were gender, marital status, age, final education, occupation, and monthly income of the respondents with 153 respondents (111 men and 42 women). Goodness of fit for Goodness of index (GFI) with maximum likelihood method.

Table 1. Direct hypothesis.

Variable			CR	Significance
Investment Experience	→	Risk Propensity	2.669	0.008
Risk Propensity	→	Risk Perception	-2.692	0.007
Risk Perception	→	Return Expectation	2.445	0.014
Investment Experience	→	Return Expectation	-0.278	0.781

3.2 Direct hypothesis

CR value on investment experience to risk propensity is 2.669, which has a value greater than 1.96. Level of significance is 0.008 which has a value smaller than 0.05. Because the level of significance is smaller than 0.05 and the value of CR calculated greater than 1.96 it concludes that there is a direct signify–can't affect between investment experiences on risk propensity. Thus, the hypothesis that investment experience has a positive effect on risk propensity is hold.

CR value of investment experience against the risk propensity is –2.692, which has a value smaller than –1.96. The significance level is 0.007 which has a value smaller than 0.05. Therefore, there is a direct significant between risk propensity to risk perception. Thus, the hypothesis that risk propensity negative effect on risk perception is hold.

The CR value of risk perception to the return expectation is 2.445, which has a value greater than 1.96. The significance level is 0.014, which has a value less than 0.05. There is a direct significant influence between risk perceptions to return expectation. Thus, the hypothesis is risk perceptions positively affect the expected return expectation.

The value of CR in the investment income variable to risk propensity is –0.227, which has a value smaller than 1.96. The significance level is 0.781 where the value has value greater than 0.05. There is no significant effect between investment experiences to return expectation. Thus, the hypothesis that investment experience has a positive effect on the return expectation is rejected. In addition, because there is no direct influence, the investment experience has an indirect influence on the return expectation through risk propensity and risk perception.

4 CONCLUSION

Chou et al. (2010) suggests that investment experience is an important factor affecting behavior, in addition to the failure or success of previous investment experiences can affect investors' tendency towards risk.

An experience can certainly affect how a person behaves in his or her life, especially in investing. Experience on investment and success in investing will increase investor confidence (self-confidence). Such confidence gives optimism to investors, so optimistic investors will prefer to take any risk and bolder to take more risk.

These results support the argument of Sitkin and Weingart (1995), suggesting that someone who is risk-averse is someone who likes challenges and wants to take high risks and expects high returns on investment, therefore their perception of risk is low. The results of this study are also in line with previous research by Byrne (2005) and Chou et al. (2010) that risk propensity has a negative effect on risk perception.

The practical implication is every investor with post success investment experience will prefer riskier investment rather than post unsuccess investment experience investors or even inexperience investor. The tendency of experienced investors to have lower risk perception will increase their risk making behavior. Regulators should have more agile regulation to accommodate the needs of higher risk investment opportunity for experienced investors while protect inexperienced investors from making unfavorable investment decisions.

REFERENCES

Byrne, K. 2005. How do consumers evaluate risk in financial products?. *Journal of Financial Services Marketing* 10: 21–38.

Chou, S.R., Hsu H.L. & Huang G.L. 2010. Investor attitudes and behavior towards inherent risk and potential returns in financial products. *International Research Journal of Finance and Economics* 44: 16–30.

Douglas, M. 1985. Risk *acceptability according the social sciences*. New York: Russell Sage Foundation.

Dutton, J.E & Jackson, S.E. 1987. Categorizing strategic issues: Links to organizational actions. *Academy of Management Review* 11: 76–90.

Muradoglu, Y.G., Aslihan A.S. & Mercan M. 2005. A behavioral approach to efficient portfolio formation. Journal of Behavioral Finance. 6(4): 202–212.

Osborn, R.N. & Jackson, D.H. 1988. Leaders, riverboat gamblers, or purposeful unintended consequences in the management of complex dangerous technologies. *Academy of Management Journal* 31: 924–944.

Sitkin S.B. & Weingart, L.R. 1995. Determinants of risky decision making behavior. A test of the mediating role risk perception and propensity. *Academy of Management Journal* 38(6): 1573–1582.

Sitkin, S.B & Pablo, A.L. 1992. Reconceptualizing the determinants of risk behavior. *Academy of Management Review* 17: 9–39.

Thaler, R.H. & Johnson, E.J. 1990. Gambling with the house money and trying to break even: The effects of prior outcomes on risky choices. *Management Science* 36: 643–660.

Vlek, C. & Stallen, P.J. 1980. Rational and person aspects of risk. *Acta Psychological* 45: 273–300.

Weber, E.U. & Milliman, R.A. 1997. Perceived risk attitudes: Relating risk perception to risky choice. *Management Science* 43: 123–144.

Wong, F.E. 2005. The role of risk in making decisions under escalation situations. *Applied Psychology: An International Review* 54(4): 584–607.

Analysis of return on equity, current ratio and debt to equity ratio on Islamic stock price Jakarta Islamic Index

M. Masrizal, T. Widiastuti, I. Mawardi & W. Wisudanto
Universitas Airlangga, Surabaya, Indonesia

ABSTRACT: Before deciding to invest in certain stocks, investors are faced with a desire to get maximum return. One of the basic indicators for the assessment is financial performance. This study aimed to analyze of return on equity, current ratio and debt to equity ratio on Islamic stock prices listed in Jakarta Islamic Index (JII). The research used quantitative approach with regression analysis using panel data in EViews. As for sampling, purposive sampling method produced sample of 8 companies listed as consistent issuers recorded in JII from 2009 to 2017. The empirical results showed that independent variables of return on equity, current ratio and debt equity ratio simultaneously had a significant influence on stock prices. Partially, ROE and DER were the only variables that had a significant effect on stock prices, while current ratio had no significant influence on stock prices. The results of this study served as a basis for decision making for investors.

1 INTRODUCTION

The stock market is an important channel for the corporate sector to raise capital needed for invest-ment and business activities (Shin & Thaker 2017). Stock markets perform a key function in providing the necessary critical links between companies that need funds to start new businesses or to expand their current operations and investors who have surplus funds to invest in such companies (Avdalovic & Milenkovic 2017).

The stock price index is an indicator showing the movement of stock prices. Stock index serves as an indicator of market trends. In other word, the movement of the index describes market conditions at a time, whether the market is active on trading or vice versa. One of the most attractive sectors to invest in Indonesia's capital market is the shares that belong to the Jakarta Islamic index (JII) group. This can be seen from the capitalization of Jakarta Islamic Index, which has an increasing trend for the last 6 years.

The companies listed in the JII conduct their business activities in compliance with Islamic principles. The Jakarta Islamic Index is an index of 30 stocks of companies that put in their investments in Islamic way. JII is a sharia-based index (Hadi 2015).

A study conducted by Wardhana et al. (2011) discovered that although the sharia capital market is affected by changes in market prices during financial crises in the United States, impacts are more severe in conventional capital markets. The sharia capital market has proved to have a better ability to adapt to external changes and disruptions. Thus, it can be concluded that the sharia capital market has more resistance to crises than conventional capital markets.

Before deciding to invest in certain stocks, investors have a desire to get maximum returns. In addition, investors are exposed to the risks associated with the investments. The most common way to assess stock prices is through a fundamental analysis of the company.

According to Darmadji and Fakhruddin (2012), fundamental analysis is a way to conduct stock valuation by studying or observing various indicators related to macroeconomic and industry conditions of a company as well as various financial indicators and understanding the company management.

To assess company performance and evaluate its value, some commonly used indicators, including profit margins and other financial data, are used. A prospective investor assesses company performance gradually. One of the basic indicators for assessment is net income/profit. The higher the net profit, the higher the stock price in the market, which leads to increased demand (Utari et al. 2014). Profitability is the ability of management to earn a profit. One of the profitability analyses is return on equity ratio (ROE).

Liquidity is the ability of the company to meet all of its obligations according to their maturity. Liquidity is realized if the level of current assets is greater than the current debt or liability. One of the liquidity ratios is the current ratio (CR), defined as the ratio that compares the current assets of the company with short-term debt (Sutrisno 2017).

Leverage is the company's ability to use debt to finance investment. The ideal ratio of total debt to asset is 40%. However, in good economic conditions, leverage can be high, as the business is expected to generate higher operating profit (Utari et al. 2014). The level of corporate risk is reflected in the ratio of debt to equity (DER). This shows the ability owned by the company to pay off its obligations.

Based on previous research and the phenomenon described earlier, the authors conducted research to analyze the influence of Return on Equity (ROE), Current Ratio (CR) and Debt to Equity Ratio (DER) on share prices in the Jakarta Islamic Index (JII) between 2009 and 2016.

1.1 Return on equity

Return on Equity (ROE) is the indicator used to assess the effectiveness of investments by capital owners and to consider whether to maintain that investment in a competitive system. A high rate of return on capital incites for investment and facilitates reinvestment of profits as a source of future growth of the company, to the detriment of its immediate distribution as dividends. Return on equity expresses equity ability to create added value, after remuneration of borrowed capital, which will allow shareholders equity compensation and self-financing of the company (Vasiu 2016).

ROE is the ratio between profit after tax and equity. This is a measurement of the income available to the owner of the company for the capital invested in the company (Widayanti et al. 2009). The ratio gives a clue of company's ability to generate profits with its owned capital. This ratio shows the efficiency of capital utilization. Higher ROE indicates better company's performance. The shareholders received higher return and vice versa (Kasmir 2015). In other words, higher ROE indicates higher profits gained by the shareholders. The investor will favor this performance so that the stock price will rise (Takarini & Hendrarini 2011). Studies conducted by Shin and Thaker (2017) and Al-Omoush and Al-Shubiri (2013) stated that return on equity (ROE) provides a significant positive relationship to stock prices, supporting this statement.

1.2 Current ratio

The current ratio shows company liquidity as measured by comparing current assets to current debt (Keown et al. 2011). A high current ratio (CR) increase investors' confidences to invest their capital to the company. According to Sutrisno (2017), current ratio is a financial ratio that compares current assets owned by a company with short-term debt. Current assets include cash, trade receivables, securities and inventories. While short-term debt includes trade payables, notes payable, bank loans, payables and other debts that must be paid immediately. Thus, it can be said that a high current ratio indicates that a company is able to fulfill its obligations or short-term debt by using current assets. In addition, high current ratio increases the confidence of shareholders to the company for paying its obligations.

This statement is supported by research conducted by Kohansal et al. (2013) and Banchuenvijit (2016), which states that current ratio has a significant influence on stock prices.

1.3 Debt to equity ratio

Debt to Equity Ratio (DER) is one leverage ratio or solvency. Solvency ratio is the ratio to determine the company's ability to pay obligations if the company is liquidated.

According to Kasmir (2015), debt to equity ratio is a ratio that is used to assess debt with equity. Investors want to invest in companies with a debt equity ratio that does not exceed the debt limit and is no more than the company's capital. Debt ratio reflects the level of a company's risk, investors are not only in favor of profit, but also concerned with the level of risk exposed to the company. Investors should avoid investing in companies with high debt-to-equity ratios as they reflect a higher level of risk. This statement is supported by research conducted by Suparningsih (2017), Nordiana and Budiyanto (2017), Rahmawati and Suryono (2017), and Vedd and Yassinski (2015), which stated that the debt equity ratio (DER) has a significant effect on stock prices.

2 METHOD

This study is a quantitative research and conducted by collecting data in the form of price data obtained from Bursa Efek Indonesia (Indonesia Stock Exchange) and financial reports obtained from the official company website. The sampling technique used for the study was purposive sampling technique, which could be understood as sample determination technique with certain consideration (Sugiyono 2015). The result of sampling produced eight companies that were consistently recorded in the Jakarta Islamic index: Astra Agro Lestari Tbk, Astra Internasional Tbk, Indocement Tunggal Perkasa Tbk, Kalbe Farma Tbk, London Sumatra Indonesia Tbk, Semen Indonesia Tbk, Telekomunikasi Indonesia Tbk and Unilever Indonesia Tbk. Regression analysis of panel data was used for the data analyses. Such data could be obtained by observing a series of cross section observations over a given period (Ariefianto 2012). According to Manurung et al. (2010), a regression model of panel data can be done using three approaches: Common Effect or Pooled Least Square (PLS) method, Fixed Effect Model (FEM) method and Random Effect Model (REM) Method. The analysis was conducted using EViews 9 computer software.

3 RESULT AND DISCUSSION

Table 1 shows the multiple regression results using panel data.

Table 1. Panel data regression results.

Variable	Coefficient	Std. Error	t-Statistic	Prob.
C	-2917.137	2680.010	-1.088480	0.2807
ROE	188.4983	69.75906	2.702134	0.0089
CR	2.462611	7.578880	0.324931	0.7463
DER	117.6446	31.77425	3.702514	0.0005

Cross-section fixed (dummy variables)

R-squared	0.878607	Mean dependent var		11,568.54
Adjusted R-squared	0.858706	S.D. dependent var		10,993.32
S.E. of regression	4132.290	Akaike info criterion		19.63081
Sum squared resid	1.04E+09	Schwarz criterion		19.97864
Log likelihood	-695.7093	Hannan-Quinn criter.		19.76928
F-statistic	44.14987	Durbin-Watson stat		0.946759
Prob (F-statistic)	0.000000			

Based on the data in Table 2, the regression equation can be described as follows:

$$Y = 2917.137 + 188.4983X1 + 2.462611X2 + 117.6446X3 \quad (1)$$

where Y = share price; C = constanta; X1 = return on equity; X2 = current ratio; and X3 = debt to equity ratio.

The result of this equation gives the value of -2917.137 as a constanta value (a) indicating the absence of return on equity, the current ratio and debt to equity ratio, offer the value of stock price is -2917.137. Moreover, the value of regression coefficient of return on equity equals to 188.4983 explains, every one-unit increase for return on equity will rise the stock price to 188.4983. Furthermore, the value of 2.46261 is a regression coefficient that explains every one-unit increase for current ratio will be a rise in stock price to 2.46261. Additionally, 117.6446 is a regression coefficient that explains every one-unit increase for debt to equity will result the increase of value of stock price to 117.6446.

3.1 Discussion

Based on data processing that has been done showed, return on equity have significantly affected stock price. Profitability ratio measures the ability of a company to earn profits for its shareholders. Higher ROE indicates higher profits gained by the share-holders. The investor will favor this performance so that the stock price will rise (Takarini & Hendrarini 2011). The results are in line with studies conducted by Shin and Thaker (2017) and Al-

Omoush & Al-Shubiri (2013) which stated that ROE provides a significant positive relationship to stock prices. However, the findings also show that current ratio has no significant effect on stock price. This contradicts the research findings from Kohansal et al. (2013) and Banchuenvijit (2016), which state that current ratio has a significant effect on stock prices.

Furthermore, the results showed that debt to equity ratio has significant effect on stock price. This can be interpreted in this way: the use of company's debt will increase the company's revenue. The company uses its leverage to finance its operating costs, with the aim that the profits earned by the company will be greater than the cost of assets and sources of funds, thereby increasing the profitability of share-holders. These findings are in line with results of studies conducted by Suparningsih (2017), Vedd and Yassinski (2015), and Avdalovic and Milenkovic (2017), which state the debt equity ratio (DER) has a significant effect on stock prices.

The capital structure determines the earnings per share and the results of company owned capital. In good business conditions, using high leverage will accelerate the development of the company, for the reason that in general the operating profit is greater than the interest expense. Higher leverage contributes higher return on equity in good economic conditions (Utari et al. 2014). According to an OECD survey (OECD 2015), Indonesia's economic conditions show remarkable performance for a decade after the Asian crisis. Such performance is impressively generated from policy reforms implemented especially in terms of building robust macroeconomics frameworks.

This is also supported by the credit rating agency Standard & Poor's (S&P) that raised Indonesia's debt rating to a viable investment. The improved rating

316

indicates Indonesia has the ability to be more resilient to external shocks as geopolitics issues. A number of macroeconomic policies issued are considered capable of maintaining economic stability. The inflation rate is under control, the rupiah exchange rate is relatively stable and the government is considered successful enough to maintain the supply of food.

4 CONCLUSION

The findings show that independent variables of return on equity, current ratio and debt equity ratio simultaneously have a significant influence on stock prices. Partially, ROE and DER are the only variables that have a significant effect on stock prices, while current ratio has no significant influence on stock prices. Our recommendation is that investors should invest on shares that have high ROE and DER due to their significant effects on the return of index. Further research should look for other fundamental factors that play important roles in determining the stock price.

ACKNOWLEDGEMENT

We would like to extend a apreciation to the Faculty of Economic and Business Universitas Airlangga for giving research funding for this topic.

REFERENCES

Al-Omoush, B.H. & AL-Shubiri, F.A. 2013. The impact of multiple approaches financial performance indicators on stocks prices: An empirical study in Jordan. *Journal of Global Business and Economics* 6(1): 1–11.

Ariefianto, M.D. 2012. *Ekonometrika esensi dan aplikasi dengan menggunakan eviews*. Jakarta: Erlangga.

Avdalovic, S.M. & Milenkovic, I. 2017. Impact of company performances on the stock price: An empirical analysis on select companies in Serbia. *Economics of Agriculture* 2: 561–570.

Banchuenvijit, W. 2016. Financial ratios and stock prices: Evidence from the agriculture firms listed on the stock exchange of Thailand. *UTCC International Journal of Business and Economics*.

Darmadji, T. & Fakhruddin, H.M. 2012. *Pasar modal di Indonesia: Pendekatan tanya jawab*. Jakarta: Salemba Empat.

Hadi, N. 2015. *Pasar modal*. Yogyakarta: Graha Ilmu.

Kasmir. 2015. *Analisis laporan keuangan*. Jakarta: Rajawali.

Keown, A.J., Martin, J.D., William, P.J. & Scott, D.F. 2011. *Manajemen keuangan* (10 ed). Jakarta: PT Indeks.

Kohansal, M.R., Dadrasmoghadam, A., Mahjori, K.K. & Mohseni, A. 2013. Relationship between financial ratios and stock price for the food industry firms in stock exchange of Iran. *World Applied Programming* 3(10): 512–521.

Manurung, J.J., Manurung, A.H. & Saragih, F.D. 2010. *Ekonometrika teori dan aplikasi*. Jakarta: PT Elex Media Komputindo.

Nordiana, A. & Budiyanto, B. 2017. Pengaruh DER, ROA dan ROE terhadap harga saham pada perusahaan food and beverage. *Jurnal Ilmu & Riset Manajemen* 6(2).

OECD. 2015. *Economic surveys Indonesia*.

Rahmawati, D. & Suryono, B. 2017. Pengaruh DPR, EPS dan DER terhadap harga saham. *Jurnal Ilmu dan Riset Akuntansi* 6(6).

Shin, W.K & Thaker, H.M.T. 2017. Macroeconomic varia-bles, financial ratios and property stock prices in Malaysia. *International Journal of Business and Innovation* 3 (1).

Sugiyono. 2015. *Statistik untuk penelitian*. Bandung: Alfabeta.

Suparningsih, B. 2017. Effect of debt to equity ratio (DER), price earnings ratio (PER), net profit margin (NPM), return on investment (ROI), earning per share (EPS) in influence exchange rates and Indonesian interest rates (SBI) share price in textile and garment industry Indonesia stock exchange. *International Journal of Multiciplinary Research and Devlopment* 4(11): 58–62.

Sutrisno. 2017. *Manajemen keuangan teori: Konsep dan ap-likasi*. Yogyakarta: Ekonisi.

Takarini. N. & Hendrarini, H. 2011. Rasio keuangan dan pengaruhnya terhadap harga saham perusahaan yang terdaftar di Jakarta Islamic Index. *Journal of Business and Banking* 1(2): 93–104.

Utari, D., Purwanti, A. & Prawironegoro, D. 2014. *Manajemen keuangan*. Jakarta: Mitra Wicana Media.

Vasiu, D.E. 2016. Case study regarding the correlation between the main indicators of financial performance and the bet index, for the companies participating in the bet index. *Management and Economics*: 357–363.

Vedd, R. & Yassinski, N. 2015. The effect of financial ratios, firm size & operating cash flows on stock price: Evidence from the Latin America industrial sector. *Journal of Business and Accounting* 8(1): 15–26.

Wardhana, W., Beik, I.S. & Setianto, R.H. 2011. Pasar modal syariah dan krisis keuangan global. *Iqtishodia Jurnal Ekonomi Islam Republika*.

Widayanti, R., Ekawati, H., Dorkas, A., Rita, M.R. & Sucahyo, U.S. 2009. *Manajemen Keuangan*. Salatiga: Fakultas Ekonomi UKSW.

Advances in Business, Management and Entrepreneurship – Hurriyati et al (eds)
© 2020 Taylor & Francis Group, London, ISBN 978-0-367-27176-3

The relationship between financial accountability in a good corporate governance concept with organizational performance

D.N. Fakhriani
Universitas Pendidikan Indonesia, Bandung, Indonesia

ABSTRACT: The purpose of this research was to find the correlation between financial accountability in a good corporate governance concept with organizational performance. The method used in the research was quantitative and data analysis using nonparametric statistics. The study found that there was a positive correlation between financial accountability in the good corporate governance concept with organizational performance; hence, if it was implemented correctly, it would increase organizational performance.

1 INTRODUCTION

Private companies, public companies and government institutions are required to implement good corporate governance. The implementation of good corporate governance (GCG) concept in Indonesia universities is still a new thing and requires a process of learning that is not simple. This concept is a derivation from business world, so it could not be implemented without any academic adaptation (Wahab & Rahayu 2013).

The governance principles of transparency, accountability, responsibility, efficiency, fair and reward-punishment already is a must in higher institutions (Astina 2011). To achieve the purpose of auditing corporate governance effectively, developed conceptual models can be seen in the diagram showing the concept and scope of public sector organizations and provide a framework for thinking and structure for operating the theory into phases of the audit (Fleming & McNamee 2005). Empirical research regarding governance in the organization continues to develop, both regarding the understanding of the structure of corporate governance (Zhang 2015), and the relationship of good corporate governance with financial performance (Jinarat & Quang 2003).

In university the implementation of good corporate governance is not only an obligation but also a necessity. Along with more stringent competition situation, university must continue to strive to realize good governance a system within the dynamics of higher education (Muhi 2011).

It is important to create commitment to encourage organizational effectiveness (Pratolo 2007). Organizational commitment is the indicated with involvement in the organization.

This research refers to studies by Praset-yono and Kompyurini (2007), Suyono and Hariyanto (2012), Amelia et al. (2013) and Nurhayati (2013). This study tested the relationship of financial accountability in

good corporate governance concept with organization performance. Research on this topic is dominated with manufacturing companies and service companies that aim on profit. However, this research focused on non-profit organization (public sector), i.e., universities.

GCG is the system of regulating and controlling the companies that create added value for all stakeholders (Tjager 2003). Definition of Corporate Governance from IICG (Brazilian Institute of Corporate Governance) is a process and structure applied in managing the company, with the main goal of increasing shareholder value in the long term and remain attentive to the interests of other stakeholders (Sarafina & Saifi 2017). The application of corporate governance in University should consider the following: (1) transparency, which can be implemented through the development of information infrastructure in the form of intranets and knowledge management; (2) accountability, one example implementing this principle is by delivering clear and periodical report regarding with budget plan and its realization to the stakeholders; 3() corporate responsibility, which means that university must always prioritize on the compliance to management in accordance with the legislation and the principles of a healthy and qualified institution. Each section/unit has their own task and function and responsibilities are clearly stated in university regulations (Regulation Rector); (4) independency, which means in university, the rector and senate have independent opinions in any decision taken. In addition, it is pos-sible to obtain the advice of the independent con-sultant and legal consultants to support the smooth running of the university regulations; and (5) fairness, which can be implemented by giving equal treatment to the whole civitas academica (Daniri 2005).

Relationships with employees was also maintained, i.e., by avoiding the practice of discrimination, respecting the rights of employees, giving equal opportunities without discriminating age, race, religion and gender, treating employees as a valuable resource through the means of a system of knowledge-based management. In ensuring fairness in the execution of the remuneration system, and the principles in procurement is an efficient, effective, open and competitive, transparent, fair and accountable. Through auction of creating transparency, accountability and efficiency of the implementation of the auction.

2 METHOD

There are two variables in this study:

1) Financial accountability in good corporate governance concept as a free or independent variables (X), that is, variables whose existence is not affected by other variables.
2) Organizational performance as the dependent (Y), i.e., a variable whose existence is affected by the independent variable.

Primary data was collected through interviews with leaders and staff and via distribution of a questionnaire. In addition to primary data, secondary data sources were also obtained through related literature to obtain a theoretical foundation related to issues that were examined.

This study was aimed at determining relationships between variables. The hypothesis was that there was a positive relationship between financial accountability in good corporate governance concepts (X) and (Y) organizational performance. The hypothesis was tested using a t test with significance level (α) at 5%, which means that the likelihood of correctly (confidence level) of the withdrawal probability 95% have a conclusion or tolerance error of 5%.

3 RESULTS AND DISCUSSION

In this study, financial accountability was measured by: (1) financial integrity; (2) disclosure; (3) obedience and compliance, while organizational performance was measured by: (1) effectiveness; (2) efficiency; (3) employee growth and (4) customer satisfaction. Data was obtained from the respondent's answers in the form of a Likert scale ranging from 1 to 5.

Respondents answers regarding financial accountability had the highest score of 5, with an average of 49% of the total answers and the second highest (score 4) with an average of 24.33% stated frequently. These data indicated that that most of the respondents said always and often for the implementation of financial accountability in the organization and that the application of financial accountability complied with the expected maximum.

The answers regarding an organization's performance had the highest score of 5 with an average of 40.83% of the total and the second highest proportion with a score of 4 had an average of 33.33%.

Table 1. Responses on financial accountability in good corporate governance concept.

Variable	Indicator	No	Score Achieved	Score Max	%
Financial account-ability in	Financial Integrity	1	139	150	92.67
		2	132	150	88.00
		3	129	150	86.00
		4	130	150	86.67
		5	128	150	85.33
		6	134	150	89.33
		7	140	150	93.33
		8	109	150	72.67
		Σ	**1041**	**1200**	**86.75**
Financial account-ability	Disclosure	9	122	150	81.33
		10	98	150	65.33
		11	104	150	69.33
		12	117	150	78.00
		13	109	150	72.67
		14	116	150	77.33
		15	107	150	71.33
		16	106	150	79.67
		17	111	150	74.00
		Σ	**990**	**1350**	**73.33**
	Obedience and Compliance	18	144	150	96.00
		19	134	150	89.33
		20	116	150	77.33
		Σ	**394**	**450**	**87.56**
	TOTAL		2425	3000	80.83

This data indicated that most of the respondents said always and often in the implementation of activities that could improve the performance of the organization as expected.

Based on Table 1, the total score obtained by financial accountability variable was 2425 of the maximum score 3000 or 80.83%. It could be inferred that financial accountability has worked well.

From Table 2, the total score obtained from organizational performance variable was 2467 of the maximum score 3000 or 82.23%; it could be inferred that organizational performance was good.

Through hypothesis testing, it can be concluded that there was a positive relationship between financial accountability in good corporate governance concept with organizational performance. The relationship was discovered using correlation analysis, Spearman correlation coefficient showed a value of 0.732 which meant that financial accountability in good corporate governance concept had significant and positive relationships positive financial accountability implemented.

At a confidence level of 95%, t value was 8.34, or greater than the value of t table = 1.701, so it could be inferred that there was a significant relationship between financial accountability and organizational performance.

Table 2. Responses on organizational performance.

Variable	Indicator	No	Score Achieved	Score Max	%
Organizational Performance	Effectiveness	1	142	150	94.67
		2	131	150	87.33
		3	136	150	90.67
		4	138	150	92.00
		5	135	150	90.00
		6	135	150	90.00
		7	119	150	79.33
		Σ	**936**	**1050**	**89.14**
	Efficiency	8	117	150	78.00
		9	131	150	87.33
		10	129	150	86.00
		11	126	150	84.00
		12	126	150	84.00
		Σ	**629**	**750**	**83.87**
	Employee Growth	13	121	150	80.67
		14	102	150	68.00
		15	103	150	68.67
		16	120	150	80.00
		17	117	150	78.00
		Σ	**563**	**750**	**75.07**
Organizational Performance	Customer Satisfaction	18	128	150	85.33
		19	112	150	74.67
		20	99	150	66.00
		Σ	**339**	**450**	**75.33**
	TOTAL		**2467**	**3000**	**82.23**

4 CONCLUSION

Based on the hypothesis test result, it was concluded that the better the financial accountability, the more improved the performance of the organization would be. This meant that financial accountability had a positive relationship with organizational performance. This is one of the hallmarks of a performance accountability review in which the existence of financial liability is in accordance with the legislation.

REFERENCES

Amelia, I., Desmiyawati, D. & Azlina, N. 2013. Pengaruh good governance, pengendalian intern, dan budaya organisasi terhadap kinerja pemerintah daerah (Studi pada satuan kerja pemerintah Kabupaten Pelalawan). Universitas Riau.

Astina, I.M. 2011. Tata kelola perguruan tinggi berbasiskan dosen dan mahasiswa. Optimalisasi Peran Teknik Mesin dalam Meningkatkan Ketahanan Energi. Proseding. *Seminar Nasional Teknik Mesin X*, 2–3 November 2011 Jurusan Mesin Fakultas Teknik UB. ISBN 978–602–19028–0–6.

Daniri, M.A. 2005. *Good corporate governance: Concept and its application in the context of Indonesia, prints 1.* Jakarta: PT Ray.

Fleming, S. & McNamee, M. 2005. The ethics of corporate governance in public sector organizations. *Public Management Review* 7(1): 135–144.

Jinarat, V. & Quang, T. 2003. The impact of good governance on organization performance after the Asian crisis in Thailand. *Asia Pacific Business Review* 10(1): 21–42.

Muhi, A.H. 2011. Build good governance at universities in Indonesia.

Nurhayati, I. 2013. Pengaruh good university governance, efektivitas audit internal, komitmen organisasional terhadap kinerja manajerial dengan partisipasi penganggaran sebagai variabel intervening (Survey pada universitas negeri dan swasta di Jawa Barat). *Disertasi. Program Pascasarjana Universitas Padjadjaran*. Bandung.

Prasetyono, P. & Kompyurini, N. 2007. Analisis kinerja rumah sakit daerah dengan pendekatan balanced scorecard berdasarkan komitmen organisasi, pengendalian intern dan penerapan prinsip-prinsip good corporate governance (GCG) (Survei pada rumah sakit daerah di Jawa Timur). *Simposium Nasional Akuntansi X. Universitas Hasanuddin, Makasar.* 26–28 Juli 2007.

Pratolo, S. 2007. Good corporate governance dan kinerja BUMN di Indonesia: Aspek audit manajemen dan pengendalian intern sebagai variabel eksogen serta tinjauannya pada jenis perusahaan. *Simposium Nasional Akuntansi X. Universitas Hasanudin, Makasar.*

Sarafina, S. & Saifi, M. 2017. The influence of the good corporate governance against the financial performance and the value of the company, the University of Brawijaya, Malang. *Journal of Business Administration* 50 (3), September 2017.

Suyono, E. & Hariyanto, E. 2012. Relationship between internal control, internal audit,and organization commitment with good governance: Indonesian case. *China-USA Business Review* 11(9): 1237–1245.

Tjager, T. 2003. *Corporate governance: Challenges and opportunities for the business community of Indonesia.* Jakarta: PT Prenhallindo.

Wahab, A.A. & Rahayu, S. 2013. Pengaruh penerapan prinsip-prinsip good university governance terhadap citra serta implikasinya pada keunggulan bersaing perguruan tinggi negeri pasca perubahan status menjadi BHMN (Survei pada tiga perguruan tinggi negeri berstatus BHMN di Jawa Barat). *Jurnal Administisistrasi Pendidikan* XVII(1): 154–173.

Zhang, W. 2015. Six understandings of corporate governance structure in the context of China. *European Journal of Finance* 4364(April): 1–16.

The relationship of the income tax system to self-assessment in the Badung Bali district

N.L.P. Suastini, G. Wirata & I.W. Astawa
Universitas Ngurah Rai, Denpasar, Indonesia

ABSTRACT: This article aims to analyze and describe the practice of income tax collection system through self-assessment in Badung Regency, Bali. The method used was the qualitative method. Badung Regency is a tourism area that has the highest earning. A self-assessment system is a taxation system that gives trust to taxpayers to fulfill and carry out their own obligations and taxation rights. Power is not a matter of ownership. It is available everywhere in every social relationship. States often become autonomous and ignores society's needs. The relationship between state and society is not always harmonious. In other words, the created relationship is based on the desire to subdue the other.

1   INTRODUCTION

As a source of national income, tax has an important role in national development. Budgetary function is a function in the public sector. It functions to collect tax money in accordance with the applicable law. The money then is used to finance the state's expenditure. The aim is to realize the prosperity of the people.

Every individual or business has to pay a tax to a state. In fact, most of them do not do it. This is because of several factors, such as: the ignorance of tax rules, the lack of supervision form tax officer, and the weak law enforcement, the complex procedure in tax office, unfriendly tax officer, and so on. In addition, taxpayers' understanding of tax provision is still low. In the case of Annual Notification Letter (Surat Pemberitahuan Tahunan/SPT), for example, most of them do not know how to fill it. They also do not know the information of new tax regulation.

Taxpayers' disobedience is inseparable from imbalance power relation between society and government. Power relation is a form of social relation that shows an unequal relationship (the asymmetric relationship). Asymmetric relationship tends to create power distance. Power distance focuses on the level of equality (or inequality) between people in a country. Within a high power distance, there is inequalities of power and wealth, where the holder of power has privileges (Hofstede 1980).

Basically, there are so many powers in human relationships. This force is found in various aspects of human relationships, such as human-to-human relationships and human with their environment and situations (Deacon 2002). Power is not a possession, but a strategy. It also does not always work through oppression and repression, but mainly through normalization and regulation. In addition, power is not destructive but productive; the power to construct discourse, knowledge, objects and subjectivity (Jorgensen & Phillips 2007).

Power works at the smallest institutional level: in families, husband and wife relationships, interfriends, and so on. It can also be in dialogues (discursive practices) between two friends, children and parents, even lovers.

However, power cannot be seen simply from one event. It has to be seen from interrelated events. It also cannot be allocated in one place. It is available everywhere. It gets its form from inter-subject relationship. The idea that power must be in the form of a state or organization is outdated.

2   METHOD

This study applies descriptive method with a qualitative approach.

2.1   Tax

According to the Law No. 28 (2007) concerning General Provisions and Taxation Procedures (Article 1, paragraph 1), tax is a compulsory contribution to a country owed by an individual or a coercive body under the law, by not obtaining a direct repayment. It is used for the purposes of the state, especially for people prosperity.

2.2   Income tax

Self-assessment is a system that gives a big responsibility to the taxpayer. Taxpayers here carry out all the processes in the fulfillment of tax obligations. Waluyo and Wirawan (2013) state that self-assessment is a tax collection that gives the authority, trust, and responsibility to the taxpayers to calculate, pay, and report the amount of tax to be paid.

2.3 Power relations

A power relation is individual interest in a group or in single entity. In a power relation, groups with large capital tend to have a control over the other groups. However, this is not absolute. A weak group, for example, can apply a bargaining position if it has the ability to do so. The problem occurs when a group does not have the ability to negotiate. If people recognize this, their bargaining power in power relationships can be stronger.

The key word in this concept is "power," an important term in various discipline. Power is ubiquitous. It includes the mediation, the creation of meaning, and the control exercise (Lewis 2010). McQuail Burton (1999) mentions that the location of power is everywhere, and it varies.

The term power itself was introduced by Foucault. He adopted Nietzsche's thinking (Best 2003) about the relationship between power and knowledge. In genealogical, Foucault explains that knowledge can control an individual and other. He thinks that knowledge can produce a power, by which it can govern the people. Foucault also gave considerable attention to technology. He noticed the way technology used by various agencies to impose power on people. Although there is a link between knowledge and power, he believes there is a resistance between them (Ritzer 2005).

Foucault's powers should be viewed as diverse and scattered relationships such as networks that have strategic scope. Foucault's power is not a domination. It must be understood as a form of power relation in a place in which it operates. Power must be understood as something that maintain power relation (Mudhoffir 2013). Therefore, it is a strategy, and power relation is its impact. The issue of power is not a matter of ownership, in the context of who controls and who is powerful. It is scattered, omnipresent, and immanent in every social relation. It is always produced in every moment and every relation. The power is everywhere not because it covers everything, but because it comes from anywhere (Mudhoffir 2013).

Public awareness of the power is determined by the socio-political situation. In an authoritarian government, people are in a weak position. They do not have the courage to bargain in power relations as they will get the pressure. Conversely, in a democratic society, where people are free to express their aspirations, the bargaining power of society will be stronger. They dare to confront or negotiate with the government and/or investor to get their rights or to fulfill their interests.

3 RESULTS AND DISCUSSION

Income tax collection in Badung regency, Bali, is done by using a self-assessment system. As a tourism area, Badung is the region with the highest regional income in Bali. It has various venues, including Kuta, Nusa Dua and Ngurah Rai International Airport. In the national level, it becomes one of the taxpayers.

Tax income in Badung increased during the period studied. In 2013, it increased 17.25%. In 2014, it raised 7.08%. In 2015, it enhanced 10.00%, while in 2016, it grew 12.00%. The table also reveals that the increase in realization was above 10% each year. In addition, the number of taxpayers was in line with the increase of target and realization. It also can be seen that the increase of tax income in Badung is influenced by the Decree of Bandung Regent No No. 2366 sq/02/HK/2014 regarding the setting and awarding of the best taxpayers area in Badung Regency. Moreover, the government also apply the sanction for those who do not pay a tax. The taxpayers who do not have a tax identification number have to pay income tax that is higher 20% than those who have it.

4 CONCLUSION

Based on the findings, it can be concluded that self-assessment of the new system will work properly if there is government role. In other words, intervention from the authority is needed to make the taxpayers obedient. The intervention can be in the form of guidance, coaching, and supervision (Soemitro 1988).

REFERENCES

Best, Steven, Kellner Douglas. 2003. Postmodern Theory: Critical Interrogations. London: Macmillan Education Ltd.
Burton, S. 1999. "Evaluation of Healthy City Projects: Stakeholders Analysis of Two Project in Bangladesh." Environment and Urbanization, Vol. 11, No. 1, pp. 41–53.
Coser, Lewis A. 2010. The Function of Social Conflict. New York: The Free Press.
Deacon, Roger. 2002. "An Analytics of Power Relations: Foucault on The History of Discipline." History of The Human Science, Vol. 15, No. 1, pp. 89–117.
Hofstede, G. H. 1980. Cultures Consequences: International Differences in workrelated values. In Beverly Hills. CA: Sage Publications.
Jorgensen, Marianne W. dan Phillips, Louise J. 2007. AnalisisWacana. Terj. Suyitno. Yogyakarta: PustakaPelajar.
Mudhoffir, Abdil Mughis. 2013. "Teori Kekuasaan Michel Foucault: Tantangan bagi Sosiologi Politik," Jurnal Sosiologi Masyarakat, 3, Vol. 8, pp. 22–31.
Ritzer, George. 2005. Teori Sosial PostModern. Yogyakarta: Kreasi Wacana.
Soemitro, R. 1988. Azas dan Dasar Perpajakan 2. Bandung: PT Eresco.
Waluyo, dan Wirawan B Ilyas, 2013, Perpajakan Indonesia, Jakarta: Salemba. Empat.

Advances in Business, Management and Entrepreneurship – Hurriyati et al (eds)
© 2020 Taylor & Francis Group, London, ISBN 978-0-367-27176-3

Greed, parental influence and teenagers' financial behavior

L. Wenatri, S. Surya & M. Maruf
Universitas Andalas, Padang, Indonesia

ABSTRACT: This study aimed to investigate the relationship between personal attitude reflected by level of greed, and the individual background based on parental influence on teenagers' financial behavior with financial knowledge and financial attitude as mediating variables. This study used quantitative approached and data was obtained by distributing questionnaires to 240 high school students in Padang City. The results showed that financial behavior was mainly influenced by parents. In addition, it can be concluded that parents had weak role in the development of financial knowledge and attitude. Teenagers in Padang were strongly depended on parents in financial decision and greed was not reflected in their behavior. Implication of this research should be the concerned of stakeholders such as financial institutions, because financial personality traits and behavior during adolescence will last until they were adults and will also affect a person's financial decisions in the rest of his life.

1 INTRODUCTION

Adolescence is a period of transition of children into adulthood. Adolescence begins when a person reaches 12 years old and ends at the age of 21. Although adolescence is a period of growth and has tremendous potential, adolescence is also a time that has a high risk where the social environment provides a strong influence on behavior (WHO 2017).

Financial management by teenagers in high school is not easy, especially when they have not been able to earn their own income. The students still depended on allowance given by their parents. Moreover, adolescence is a time of self-discovery. The period in which they always want something that is up to date, both in fashion and gadgets. They always want to follow the trend in order not to be considered old-fashioned. They will always want something more and will never be satisfied with what they have.

Research shows that financial personality traits and behavior during adolescence will last until they become adults and will also affect a person's financial decisions for the rest of their life (Eccles et al. 2013). Ashby et al. (2011) also found that a person's saving habit at the age of 16 could predict his saving habit at the age of 34. So, if teenagers have good financial behavior, it will have a positive impact on their future.

Theory of Planned Behavior (TPB) helps to predict individual intention to engage in behavior at a certain time and place. The theory argues that individual behavior is driven by behavioral intention, where the intention of the behavior is a function of the three determinants, namely: individual's attitude toward behavior, Subjective norms, and perceived behavioral control (Ajzen 1991).

According to Ajzen (1991) planned behavior is determined by three factors: (a) attitude toward behavior, (b) subjective norm, (c) perceived behavioral control. This theory provides a useful conceptual framework to deal with the complexity of human social behavior. This theory incorporates several central concepts in the social sciences and behavior, and it also theory defines these concepts in a way that allows the prediction and understanding of certain behaviors in certain contexts. Attitudes toward the behavior, subjective norms on behavior, and perceived control over the behavior are usually used to predict behavioral intention with a high degree of accuracy. In turn, the intention, in combination with the perceived behavioral control, can explain a significant proportion of variance in behavior (Ajzen 1991).

Ricciardi and Simon (2002) explained financial behavior, trying to explain what, why, and how to finance and investments, from a human perspective. They also revealed three financial foundations of behavior that is psychology, sociology, and finance. One of the psychological factors that affect a person's financial behavior is greed. Greed is the desire to always get (property, etc.) as much as possible (Indonesian Dictionary Online 2017). The Merriam-Webster Online Dictionary (2017) defines greed as selfish desire and the desire to get something more (such as money) than is needed. Seuntjens et al. (2015) found that the desire to acquire more and never have enough and settle for what they have owned are two important components of greed. Greed is associated with materialism and envy and can cause behavior and disguised self-interest. Greed can affect a person's behavior, especially financial behavior. Seuntjens et al. (2016) in their research found that greed could be positive (work harder) or negative. Seuntjens et al. (2015) also mentioned that teens tend to be more greedy than adults.

In addition to psychology, the foundation of financial behavior is sociology. The first social environment of the child is the family. Family environment has an important role in the formation of a child's behavior and personality. Parental influence is one of the determining factors in financial behavior of teenagers. Shim et al. (2010) tested a variety of financial socialization process and found that parent direct teaching about personal finances had more influence on financial literacy among high school students rather than through financial education in schools.

Shim et al. (2010) used financial knowledge and financial attitude as a mediating variable. The results of research showed that financial knowledge played an important role in predicting financial attitude and financial attitude predict financial behavior. The results supported the hierarchical relationship of knowledge-attitude-behavior.

Jorgensen and Savla (2010) found that parents deemed to affect financial attitude and financial behavior of children. Parental influence was composed by two main ideas: (a) the amount of financial learning that occurs and (b) the frequency of financial learning. In their research, Jorgensen and Savla (2010) used financial knowledge and financial attitude as a mediating variable between parental influences and financial behavior. Overall, parents were considered to have a direct and significant effect on financial attitude and had an indirect influence and significant to financial behavior through financial attitude as mediating variables but did not affect financial knowledge.

These two foundations of financial behavior were the background for the author to research the influence of greed and parental influences on high school student's financial behavior in Padang with financial knowledge and financial attitude as a mediating variable. The diversity of high school in Padang City which consists of public and private high schools, public and private Vocational High School, and public and private *Madrasah Aliyah* (Islamic schools) was the reason to take this city as location of the research. With this diversity, the sample in this study was expected to have more varied characteristics. Based on this background, the author aimed to investigate the influence of greed and pa-rental influence on teenagers' financial behavior in Padang with financial knowledge and financial attitude as a mediating variable.

2 METHOD

This research used quantitative approach to test the hypothesis. The samples of this study were 240 high school students in Padang. To determine validity and reliability, the classic assumption test using SPSS 20.0 (Statistical Program for Social Science) was conducted and to test the hypotheses Structural Equation Modeling using AMOS 22.0 (Analysis of Moment Structure) was used.

This study used definition of financial behavior from Sewell (2010), that is "financial behavior is the study of the influence of psychology on the behavior of financial practitioners and the subsequent effects on the market." Dew and Xiao (2011) measured financial behavior using 14 indicators, namely: (1) compare shops; (2) pay bills on time; (3) keep financial record; (4) strict to the budget; (5) pay off credit card; (6) max out credit card; (7) make minimum payment on loans; (8) maintain or create an emergency fund; (9) make saving from every paycheck; (10) make saving from a long-term goal other than retirement; (11) save for retirement; (12) invest money; (13) obtain or maintain adequate health insurance; and (14) obtain or maintain adequate life insurance.

According to Krekels and Pandelaere (2014) *greed* was an insatiable desire for more resources, money or other. In this research author used the initial model of Dispositional Greed Scale (Seuntjens et al. 2015) which used seven indicators of measurement and added indicator "materialistic" were also measured in the study Seuntjens et al. (2015) which included: "When I eat a packet of chips, I will not stop until the chips it up," and "When I use social media (such as facebook, twitter, instagram), I would like to have a friend/follower as much as possible." The scale used is ordinal scale where 1 = strongly disagree and 5 = strongly agree. The Online Oxford Advanced Learner's Dictionary (2017) defines *parental influence* as the ability of parents to influence the development of a person's behavior. Norvilitis and MacLean (2010) measured parental influence by using four dimensions: (1) parents' instruction, the dimension that assessed parents' instructions on financial matters; (2) parents' facilitation, the dimension that assessed the involvement of parents in assisting their children in managing finances; (3) parent worries measured how often children watched their parents struggle to get money and worried about money; and (4) parents reticent when parents avoided discussing financial issues with their children.

Financial knowledge according to Parrotta (1992) was a proprietary knowledge and understanding of the information related to financial issues, particularly financial management practices recommended by experts in the field. Jorgensen et al. (2010) measured financial knowledge with 27 indicators and covered four areas of content: general financial literacy, savings and loans, insurance, and investments. Responses were given in the form of a value of 1 (true) or 0 (false) and summed. The author chose to use Jorgensen and Savla (2010) method in measuring the financial knowledge of teenagers in Padang because the questions and statements were tailored to the Indonesian condition.

Parrotta (1992) stated that financial attitude was a psychological tendency expressed by evaluating the financial management with some level of agreement or disagreement. Lim and Teo (1997) measured financial attitude with several dimensions. In this study, the authors used the indicators of ob-*session* in measuring financial attitude because the authors wanted to see exactly how parental influence affected on obsession. Indicators of obsession were: (1) I firmly believe that money can solve all of my problems; (2) I feel that money is the only thing that I can really count on; (3) Money is the most important goal in my life; (4) Money can buy every-thing; (5) I believe that time not spent on making money is time wasted; (6) I would do practically anything legal for money if it were enough; (7) I often fantasies about money and what I could do with it.

Seuntjens et al. (2016) in his research found that *greed* could be positive or negative in financial behavior. The positive impact was greed could make someone work harder to get a higher income to make ends meet. However, greed could also cause a person to have greater spending but lesser money saved and be indebted with the people around them such as family, friends, as well as financial institutions. Based on this study, the following hypothesis was proposed: H1. *Greed had an effect on adolescent financial behavior.*

Shim et al. (2010) found that direct teaching from parents about personal finance had a stronger effect on financial literacy among high school students than the acceptance level finance. Jorgensen and Savla (2010) also found that parents deemed to affect financial attitude and financial behavior of children. Overall, parents were considered to have a direct and significant effect on financial attitude and had an indirect influence and significance to financial behavior through financial attitude as mediating variables, but did not affect the financial knowledge. Therefore, the second hypothesis was: H2. *Parental Influence allegedly affect the Financial Attitude.*

A study by Shim et al. (2010) showed that parents who deliberately taught their children about financial management could provide a greater influence on children's financial knowledge rather than a combination of lessons in high school and in the workplace. Instead, Jorgensen and Savla (2010) found that young children did not think parents influence their financial knowledge. No significant relationship between perceived parental influence and financial knowledge supported the idea that many parents did not teach their children financial literacy (Lyons & Hunt 2003). Thus, it was hypothesized that: H3. *Parental influence had an effect on financial knowledge.*

Jorgensen and Savla (2010) also found that financial knowledge had a significant influence on financial attitude, which had a significant influence on financial behavior. With the increasing financial knowledge, financial attitude of young people would also increase. These results suggested that with increasing knowledge and attitude, the ability of young people to make the right financial decision could be increased. Thus: H5. *Financial Knowledge has effect on financial attitude* and H6. *Financial attitude has an effect on financial behavior.*

3 RESULTS AND DISCUSSION

Data of the respondents in this study was obtained from questionnaires (19 November to 15 December 2017), either directly or online from students of high schools, vocational schools, and Islamic schools in Padang. Respondents were grouped based on gender, age, school of origin, place of residence, parents' occupation, parent education, parents' income, allowances, expenses, saving and debt per month.

Based on the descriptive statistics on the respondents profile, it could be concluded that 63.8 percent of the respondents were women (153 respondents), while remaining 36.3 percent were men (87 respondents). From grade level, 41.3 percent of respondents were in grade 3 (99 respondents), while respondents in grades 1 was 89 respondents or 37.1 percent and from grade 2 was 52 respondents, or 21.7 percent.

86.7 percent of the respondents still lived with their parents (208 respondents), 11.3 percent (27 respondents) lived alone in renting a room and 5 other respondents, or 2.1 percent lived with their other relatives such as the grandfather/grandmother or uncle/aunt. Majority of respondent's parents work as self-employed (33.3 percent or 80 respondents). Most of the respondents' parents were high school graduates, the father by 129 respondents or 53.8 percent and the mother as 122 respondents or 50.8 percent. In terms of income, 35 percent or 84 of respondents' parents had an income between Rp 2,000,000 and 4,000,000 per month. To examine the relationship between variables, this study used structural equation modeling using AMOS application 22. Based on the result of the validity test in SPSS, the research model can be described as follows:

Figure 1 shows a conformance test model that fits properly. Although the chi-square value and the probability were not good, for the sample > 20 the RMSEA must be < 0.08.

Test of normality showed that the value of the critical ratio of the model was fit before, equal to 0.630. This value was 2.58 < 0.630 < 2.58, thus it could be said that the data were normally distributed.

Table 1 shows the result of hypothesis testing based on regression weights, which shows the relationship between variables:

Relationships between variables was significant when the value of CR was greater than 1.96 and the p-value was less than 0.05. Based on Table 1 there were only two relationships that had a significant influence, namely greed on financial attitude and parental influences on financial behavior.

Based on these results, direct effect value of greed on financial behavior amounted to –0.106. This showed that greed had negative effect on financial behavior and was not significant because its CR value was only –1.241 (less than 1.96) and p 0.106 (greater than 0.05), thefore it could be concluded H1 rejected. In a previous study, Seuntjens et al. (2016) found that greed had positive and negative impact on financial behavior. The positive impact was that greed could

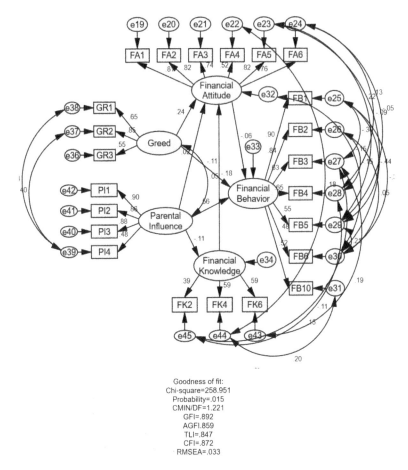

Goodness of fit:
Chi-square=258.951
Probability=.015
CMIN/DF=1.221
GFI=.892
AGFI.859
TLI=.847
CFI=.872
RMSEA=.033

Figure 1. Structural model.

Table 1. Regression weights.

	estimate	SE	CR	P	Label
FK <− PI	-0.062	0.062	-0.998	0.318	
FA <− G	0.177	0.065	2733	0.006	Sig.
FA <− PI	0.025	0.138	0.182	0.856	
FA <− FK	-0.493	0.305	-1615	0.106	
FB <− G	-0.089	0.072	-1241	0.215	
FB <− PI	1,037	0.147	7043	***	Sig.
FB <− FA	-0.068	0.091	-0.75	0.453	

Note: FA = Financial Attitude, PI = Parental Influence, FK: Financial Knowledge, G = Greed, FB = Financial Behavior

make someone work harder to get a higher income to make ends meet. However, greed could also cause a person to have greater spending but less saving, and be indebted with people around them such as family, friends, as well as financial institutions; however, in their study Seuntjens et al. (2016) did not use ordinal scale to calculate the financial behavior.

Greed did not affect financial behavior of teen-agers in Padang supposedly because of demographic factors. Almost 60% of the respondents' parents had income between IDR 2,000,000 to IDR 4,000,000 which was relatively small, and the average allow-ance of high school teenagers was IDR 547,791 per month, which restricted the teenager to spend money even though they had a great desire to buy some-thing. Moreover, the use of credit cards in Padang by teenagers was still relatively small compared to the use of credit cards by teenagers in foreign country.

Although in this study there was no hypothesis between greed and financial attitude, but results revealed that greed had significant effect on financial attitude. The direct effect value of greed on financial at-titude was 0.241. When the teenager had more greed so their obsession with money were higher. The significant effect of greed on financial attitude was supported by the theory of planned behavior from Ajzen (1991). Before a variable influenced behavior, these variables would affect the attitude of the person first. Behavior was not only determined by the attitude but also by subjective norm that society forced to perform or not perform.

The value of parental influence effect on financial attitude was 0.016. This suggested that parental influence had no significant effect on financial attitude. It could be concluded that H2 was rejected. These results contradicted the research by Jorgensen and Savla (2010) who found that parents deemed to affect financial attitude and financial behavior of children. Overall, parents were considered to have a direct and significant effect on financial attitude and had an indirect influence and significant to the financial behavior through financial attitude as a mediating variable. This was because in this study the authors wanted to see the attitude towards money obsession, so that other dimensions of teenager financial attitude was not been illustrated in this study. It was expected for the next research to add another dimension to the financial attitude so that all financial dimensions could be illustrated.

Parental influence direct effect on financial knowledge was to –0.109. This suggested that parental influence negatively affected financial knowledge. However, the influence of parental in-fluence on financial knowledge was not significant because it only had –0.998 cr value (less than 1.96) and p 0.318 (greater than 0.05), therefore, H3 was rejected. Although this result was contrary to Shim et al. (2010), who found that the direct teaching from parents about personal finance had a stronger effect on financial literacy than the acceptance of financial education, it corresponded with the results of the study by Jorgensen and Savla (2010) who found that young children did not think parents influenced their financial knowledge. Lyons and Hunt (2003) also found that many parents did not teach financial literacy to their children and more children obtained financial information from financial officer although they wished to receive financial information from their parents.

There was no significant effect of parental influence on financial knowledge, supposedly because the educational background of the parents was still relatively low. In this study, 53.8 percent of respondents' fathers had high school education, 7.9 percent of junior high school and 3.3 per cent of primary school background. Respondents' mothers also had low educational background, where 50.8 percent were senior high school graduated, 8.8 per-cent from junior high school, and 1.7 percent had a primary school educational background. It was expected for further research to choose respondents whose parents had more evenly educational back-ground so that the influence of parents on financial knowledge can be measured better.

Although parental influence did not have a significant effect on the financial attitude and financial knowledge, this study indicated that parental influence had significant effect on teenagers financial behavior. The direct effect value was 0.563 and cr value was 7.043 (greater than 1.96) and p ***, which meant the figures showed very small number

(< 0.05). It could be concluded that parental influence had significant effect on financial behavior, so H4 was accepted.

These results concurred with Norvilitis and Mac-Lean (2010), who found that parents involved in hands-on approach to teach children to handle money through actions such as teaching them how to manage allowance and how to manage bank accounts had children with lower debt in college.

Research by Shim et al. (2010) supported the proposition that the behavior of parents, work experience in middle school, and high school financial education during adolescence helped shape learning, attitudes, and behavior of students while growing up. The study also found that older people who had demonstrated a positive financial behavior and engaged in a more direct teaching during adolescence was considered to be the main model for their children for financial management during the first year of college.

Shim et al. (2010) found that seeing parents as a financial role model led students towards positive and healthy financial behavior. What's more, when parents were involved and acted as a role model, the child would grow up feeling that they can control their financial behavior better.

The direct effect of financial knowledge on financial attitude amounted to –0.175. This showed that financial knowledge negatively affected financial attitude. However, the influence of parental influence and financial behavior were not significant because its cr value was –1.615 (less than 1.96) and p value was 0.106 (greater than 0.05); therefore H5 was rejected. The results of this study were not consistent with the research by Jorgensen and Savla (2010) who found that financial knowledge had significant influence on financial attitude, and financial attitude influenced significantly on financial behavior. With increasing financial knowledge, financial attitude of young people would also increase. These results suggested that with increasing knowledge and attitude, the ability of young people to make the right financial decision for behavior can be improved.

The direct effect of financial variables attitude to-wards financial behavior amounted to –0.053. This showed that financial knowledge had no significant effect on the financial attitude, therefore H6 was re-jected. Herdijono and Damanik (2016) research showed that individuals with high financial knowledge did not necessarily have a good financial behavior and also someone with a low financial behavior did not necessarily have a bad financial behavior. They also found that the financial attitude affected the financial behavior, in other words, a person with good financial attitude would show a good mindset about money which related to (1) perception of the future (obsession); (2) wise use of money; (3) ability to control financial situation; (4) adjustment towards the use of money to make ends meet; (5) retention towards excessive spending; (6) ever-evolving view about money; (7) ability to exercise control over its consumption; (8) ability to balance expenditures and revenues held (cash flow); (9) ability to allocate money for savings and investment; and (10) ability to manage debts.

No significant effect was found between financial knowledge and financial attitude; and nor was be-tween financial attitude and financial behavior. It was presumably because the authors only wanted to measure the attitude of financial obsession with money alone. Another dimension of financial attitude was not represented in this study. It was suggested for subsequent research to measure the financial attitude by adding another dimension such as power, budget, achievement, evaluation, anxiety, and retention so that research results can be better.

4 CONCLUSION

The results showed that greed has no significant effect on financial behavior, but it had significant effect on financial attitude. Parental influence had a significant effect on financial behavior, but no significant effect was found on financial knowledge and financial attitude. Financial knowledge had no significant effect on financial attitude and financial attitude had no significant effect on financial behavior of teenagers in Padang City. The future study is expected to be able to add another psychological variable such as self-control, impulsiveness, envy, social desirability bias and other variables that affect a person's financial behavior. Future studies are also expected to add another dimension of financial attitude such power, budget, achievement, evaluation, anxiety, retention and selfishness, and choose indicators that more accurately measure variables. In addition, researchers can also add sample size of the study and try to do research on other types of respondents such as adults to find out about their financial behavior as a teenager and connect with their financial behavior today.

ACKNOWLEDGEMENTS

This research was supported by author's family and friends. We thank many high schools and students in Padang City that participated in this study.

REFERENCES

Ajzen, I. 1991. The theory of planned behavior. *Organizational Behavior and Human Decision Processes* 50: 179–211.

Ashby, J., Schoon, I. & Webley, P. 2011. Save now, save later? Linkages between saving behavior in adolescence and adulthood. *European Psychologist* 16: 227–237.

Dew, J. & Xiao, J.J. 2011. The financial management behav-ior scale: Development and validation. *Journal of Financial Counseling and Planning* 22(1): 59–53.

Eccles, D.W., Ward, P., Goldsmith, E. & Arsal, G. 2013. The relationship between retirement w–lth and householders' lifetime personal financial and investing behaviors. *Journal of Consumer Affairs* 47: 432–464.

Herdijono, I. & Damanik, L.A. 2016. Financial influence attitude, financial knowledge, parental income on financial management behavior. *Journal of Theory and Applied Management* 3: 226–241.

Indonesian Dictionary Online. 2017. Greed. *http://kbbi.web.id/tamak*. February 2017.

Jorgensen, B.L. & Savla, J. 2010. Financial literacy of young adults: The importance of parental socialization. *Family Relations* 59: 465–478.

Krekels, G. & Pandelaere, M. 2014. Dispositional greed. *Journal of Personality and Individual Differences* 74: 225–230.

Lim, V.K.G. & Teo, T.S.H. 1997. Sex, money and financial hardship: An empirical study of attitudes toward money among undergraduates in the Singapore. *Journal of Economic Psychology* 18: 369–386.

Lyons, A.C. & Hunt, J.L. 2003. The credit practices and finan-cial education needs of community college students. *Association for Financial Counseling and Planning Education* 14(1): 63–74.

Merriam-Webster Online Dictionary. 2017. Greed. *https://www.merriam-webster.com/dictionary/greed*. February 2017.

Norvilitis, J.M. & MacLean, M.G. 2010. The role of parents in college students' financial behaviors and attitudes. *Journal of Economic Psychology* 31(1): 55–63.

Online Oxford Advanced Learner's Dictionary. 2017. Greed. *https://en.oxforddictionaries.com/definition/greed*. February 2017.

Online Oxford Advanced Learner's Dictionary. 2017. Influence. *https://en.oxforddictionaries.com/definition/influence*. February 2017.

Online Oxford Advanced Learner's Dictionary. 2017. Parent. *https://en.oxforddictionaries.com/definition/parent*. February 2017.

Parrotta, J.L.M. 1992. *The impact of financial attitudes and knowledge on financial management and satisfaction*. Thesis, University of British Columbia.

Ricciardi, V. & Simon, H.K. 2000. What is behavioral finance? *Business, Education and Technology Journal Fall*.

Seuntjens, T.G., Van de Ven, N., Zeelenberg, M., & Van der Schors, A. 2016. Greed and financial adolescent behavior. *Journal of Economic Psychology* 57: 1–12.

Seuntjens, T.G., Zeelenberg, M., Breugelmans, B.C. & Van de Ven, N. 2015. Defining greed. *British Journal of Psychology* 106: 505–525.

Seuntjens, T.G., Zeelenberg, M., Van de Ven, N., & Breugelmans, S.M. 2015. Dispositional greed. *Journal of Personality and Social Psychology* 108: 917–933.

Sewell, M. 2010. *Behavioural finance*. University of Cambridge.

Shim, S., Barber, B.L., Card, N.A., Xiao, J.J. & Serido, J. 2010. Financial socialization of first-year college students: The roles of parents, work, and education. *Journal of Youth and Adolescence* 39: 1457–1470.

WHO. Adolescent development. March 3, 2017. *www.who.int/maternal_child_adolescent/topics/adolescence/dev/en/*.

Executive compensation and risk: An empirical study in Indonesia

A.R. Setiawan & B. Zunairoh
Universitas Airlangga, Surabaya, Indonesia

ABSTRACT: This study investigates the relationship between executive compensation and risk for banks in Indonesia over the period 2010–2013, with 374 observations. The risk of banks was measured by non-performing loans that implicated loan risk. Non-performing loans were the dependent variable and executive compensation was the independent variable. Firm size, firm age, and capital asset ratio were the control variables. The result of the study indicated that executive compensation, firm age, and capital asset ratio had a negative effect on risk and firm size had a positive effect on risk.

1 INTRODUCTION

The banking institution is a financial institution that acts as an intermediary between excess funds and those who need funds. Banks collect funds from excess funds in the form of savings in current accounts, savings deposits, and deposits and rechanneling funds obtained from the community in the form of credit. However, one of the major problems in banking is customers who fail to return loans and thus those become Non-Performing. The reasons may include the capacity to pay and intention to pay. Banks have to verify this prior to approval of the loans through their procedures of risk management, whether a customer may have the capacity to repay the loan in the future or if it goes default (Shahbaz 2012).

In a banking institution, the achievement of management can be shown through the way management anticipates business risks. According to (Muwardi 2004), bank risk is the potential event of an event that can cause bank losses. The rapidly expanding internal and internal environment of banks will be followed by increasing complexity of risks for banking activities. One of the risks involved in banking is credit risk which occurs when a debtor does not fulfill his obligations to the bank, such as principal repayment, interest payment, etc. that will cause loss to the bank.

Bank owners reward executives in the form of a compensation which has in recent years shown an increase. The compensation is given with the aim that executives can run the company well so as to minimize the risks. Increasing the salaries of national bankers is considered a natural thing. In the three years since 2007, remuneration received by directors and bank commissioners has increased with 65%, averaging over 21% per annum. In 2010 the increase in remuneration of directors and bank commissioners was 19%, which was much higher compared to the previous year when it was only 5% (Infobank 2011). Bankers in Indonesia receive the highest compensation compared

to bankers in other countries in the ASEAN region. The compensation of bankers in Indonesia, consisting of salaries, bonuses, regular allowances and other facilities in the form of natura and non-natura reaches Rp 12 billion per year, while in Malaysia this is only Rp 5.6 billion per year, and in Filipina it is only Rp 1.1 billion per year, or one twelfth of Indonesia (Infobank 2012). The remuneration of banks in Indonesia has been regulated by Indonesian banks with remuneration and nomination committees. However, remoderation given to banking executives is based on the Limited Liability Company Act (PT) in 2007. Salaries are set at the shareholders' meeting which becomes the highest forum of the shareholders.

Research on the effects of compensation on performance has been made in developed countries, especially the United States (US) and the United Kingdom (UK) (Murphy 1999). This is due to the growing popularity of executive compensation issues and the ease of obtaining data through the stock exchange of large private companies in the country (Ramaswamy et al. 2000). According to (Kato et al. 2006), in developing countries, especially in Asia, research on executive compensation is rarely done due to data constraints. The phenomenon also occurs in Indonesia because no regulations exist requiring companies to disclose the amount of compensation, therefore not all companies disclose the amount of compensation their executives receive. In addition, the company also greatly maintains the confidentiality of employee salaries because it expresses the salary scale, which is still regarded as something sensitive. However, organizations that follow a set of key principles can design effective incentive plans that align with the organizational strategy, motivates individuals and teams to achieve incremental performance, and incorporates appropriate risk-adjusted design safeguards. If organizations follow a set of principles, they found a positive effect of bank risk (reflected in return variability) on each component of compensation (salary, bonus, and equity-based payment) (Brewer et al. 2003).

Different from the previous research, (Chen et al. 2006) treated bank risk as the dependent variable while compensation was one of independent variables in the research model. The results are inconsistent with the conclusion of (Houston & James 1995). They showed that the equity-based payment induces risk-taking and that the executive compensation structure promotes risk in the banking industry.

Based on this background, the problem formulation in this study is whether executive compensation with capital adequacy control variables, firm size, and age of the company affects the credit risk of Banks in Indonesia.

1.1 Understanding and measurement of credit risk

Credit risk is one of the risks faced by a bank. It is the non-payment of loans that have been given to debtors. According to (Siamat 2004), credit risk is a risk due to failure or inability of customers to return the loan amount received from the bank and its interest within the specified time period. The ratio used to gauge credit risk is a non-performing loan (NPL) with the formula:

$$NPL = (Total\ problem\ loans)/(Total\ credits) \times 100\%$$

$$(1)$$

This ratio shows the total amount of nonperforming loans to the total credit provided by the bank. Based on Bank Indonesia regulations, good NPLs are below 5%. The higher the NPL ratio, the worse the performance of banks in managing credit, causing more and more problematic loans to be borne by banks. Provisions are regarded as a controlling mechanism over expected loan losses. Previous practices have shown that provisions are triggered by default incidents on loans, higher levels of non-performing loans are associated with high rates of provisioning (Hasan & Wall 2004).

1.2 Understanding and form of executive compensation

Compensation, according to (Rivai 2010), is the reward received by employees as a substitute for the contribution of services to the company. In this study, executive compensation is defined as a special compensation package designed for executive-level employees. According to (Sutrisno 2010), the compensation based on the form of payment is divided into two, including financial compensation (the compensation paid to employees in the form of money, in the form of basic salaries, bonuses, allowances, and incentive payments) and non-financial compensation (a compensation provided in kind or non-money, in the form of facilities or facilities to be owned or which cannot be owned).

In this study, executive compensation of banks can be paid in both financial and non-financial forms. The amount of compensation paid to bank executives is derived from the good corporate governance report in the annual report of each bank.

1.3 Measurement of executive compensation

In this study, the compensation in question is the compensation given to the board of directors. The total compensation of directors is obtained from remuneration in the form of salaries, bonuses, routine allowances, and other facilities in the form of housing, transportation, health, and may others which either can be or cannot be owned (Infobank 2015). The royalty compensation is a description of the size of the executive compensation that shows the amount of compensation received by executives each year. The magnitude of the prediction component is measured by the formula:

$$Compensation = Ln((Total\ Directors'\ Compensation)/\ (Number\ of\ Directors))$$

$$(2)$$

1.4 The effect of executive compensation on credit risk

In the banking sector with its high and competitive compensation to employees (especially executives), productivity, profit, and business are highly expected to continue to grow. The banking sector, in achieving good performance, cannot be separated from various risks such as credit risk.

Credit risk occurs if the debtor fails to fulfill his obligations to the bank, such as principal payments, interest payments, and so on, which lead to a loss to the bank. Credit risk must be well managed to maximize the rate of return to the bank. One way to minimize credit risk is by giving a bonus or compensation. Given the high compensation, the management owner hopes that management can run the credit system well so that it can minimize credit risk. Based on this description, this study is intended to examine the effect of compensation on credit risk. The agency theory suggests that incentive compensation can reduce the difference in risk preference of principal (shareholders) and agent (executives) by increasing the executives' risk-taking. While the influences of CEO incentives on bank risks are robust, results of research examining the ability of bonuses to affect risk have produced equivocal evidence. For example, bonuses are illustrated to induce an executive to take greater risks, according toresearch of (Salami 2009, Kaplanski & Levy 2012, & Fortin et al. 2010). However (Ayadi 2011), in studying a sample of 53 European banks from 19992009, showed evidence that annual bonuses are negatively related to bank risk.

H1: Executive compensation negatively affects credit risk.

1.5 Company size

The size of the company is the amount of assets owned by the company. Large companies are better able to achieve economies and large-size banks are more competitive to obtain low-cost funds than small-size banks as large banks are more trusted by the public. According to (Sastradiputra 2004), the assets of the bank show management strategies and activities related to fund-raising areas including cash, accounts with central banks, short-term and long-term loans, and fixed assets. The greater the assets owned by a bank, the greater the volume of credit that can be channeled by the bank. (Dendawijaya 2009) argues that a greater volume of credit provides an opportunity for the bank to reduce the spread rate, which in turn decreases the lending rate (loan interest rate) so that the bank will be more competitive in providing services to customers to accelerate credit payments and reduce the number of problem loans. The size of the company size is measured by the formula:

$$SIZE = Ln(\text{total assets}) \tag{3}$$

H2: Size negatively affects credit risk.

1.6 The age of the company

The company's age reflects the length of the company's standing in running its operations. The age of the company provides an opportunity to take advantage of the benefits associated with the company's reputation and experience. Banks with good reputations are likely to have small non-performing loan risk in the future, so the bank can run its operations smoothly. Banks that survive long periods of time with considerable experience can boost public confidence as banks are perceived as to be able to deal with the risk of losses so that the public will not be worried about saving funds in the bank. The size of the company's age is measured by the formula:

$$AGE = \text{year of research} - \text{year of establishment of company} \tag{4}$$

H3: Age has a negative effect on credit risk.

1.7 Capital adequacy

Capital adequacy is the availability of capital owned by banks to support assets that contain or generate risks. In accordance with Bank Indonesia Regulation Number 10/15/PBI/2008, banks are required to provide a minimum capital of 8% of risk-weighted assets. The bank's capital adequacy ratio is measured by the Capital Adequacy Ratio (CAR) showing how many of the bank's assets contain risks (credit, investment, securities, and claims to other banks) which can be anticipated by using its own capital. The size of a bank's CAR can be calculated by the formula:

$$CAR = \text{Capital}/(\text{Risk Weighted Assets(ATMR)}) \times 100\% \tag{5}$$

A higher CAR value of a bank shows a better capital position of the bank, because it has sufficient capital to anticipate the potential losses caused by operational activities. High CAR values can lead to increased public confidence because they will not be worried when saving funds in the bank. It means that the bank's own capital is used to finance risk-bearing assets. The higher the capital the bank has, the easier it will be for banks to finance risk-bearing assets. Conversely, if high credit is not accompanied by sufficient capital, it will potentially lead to problematic loans. Thus, it can be concluded that the higher the CAR, the lower the credit risk faced by the bank. If the credit is disbursed, the credit risk will increase. According to Bank Indonesia (Diyanti & Widyarti 2012), the capital has a negative effect on the problem condition. This gives a negative indication of the effect of CAR on NPL, which is in accordance with the results of research carried out by (Diyanti & Widyarti 2012), who, in the study, suggest CAR negatively against NPL.

H4: CAR negatively affects credit risk.

2 METHOD

2.1 Model analysis

The analysis in this study used a multiple linear analysis by an equation of a regression model as follows:

$$NPLit = \beta0 + \beta1Compit - 1 + \beta2SIZEit + \beta3AGEit + \beta4CARit + \varepsilon it \tag{6}$$

Where:
Compit-1 : Compensation of company executives i
in year t-1
CARIT : Capital Adequacy Ratio of company i in year t

SIZEit : Company size i in year t
NPLit : Credit risk of company i in year t
AGEit : Age of company i in year t
α : Constants
β (1,2,3,4) : Regression coefficient
εit : The fault factor or error of firm i in year t
i : Company sample
t : Period of time (year)

2.2 Types and data sources

The data used in this study were secondary data from banking companies for the period of 2010–2013. The data were taken from the banks' websites. For executive compensation, the data were extracted from Infobank, a research magazine, the October editions of 2011, 2012, 2013, and 2014.

2.3 Operational definition

To give a clear picture of the variables used in this study, they are defined as follows:

a Credit risk is the risk of the non-payment of loans that have been given to debtors, as measured by the ratio of non-performing loans (NPLs) using formula (1).
b Executive compensation is the average of the compensation of the debtor consisting of salaries, bonuses, and incentives paid in cash in one year. Executive compensation is obtained from Infobank from October 2012 to 2015 and is measured using formula (2).
c The size of the company is the size of the company measured by the total natural logarithm of assets using formula (3).
d The age of the company is the length of time the company operates since the company was established until the period 2010–2014, as measured using formula (4).
e Capital adequacy is the availability of capital owned by banks to support risk-bearing assets, measured using the capital adequacy ratio (CAR) by formula (5).

2.4 Sampling technique

In this study, the sample was selected using a purposive sampling method based on certain criteria, name-ly:

a A banking company that presents its annual financial statements ending on 31 December.
b Conventional commercial banks, not sharia or rural banks in 2010–2013.
c Banking companies that publish the amount of compensation received by the director in total during 2009–2012.

3 RESULTS AND DISCUSSION

3.1 Text and indenting

In Table 1, the description of each variable analyzed in this research is presented.

The average NPL of the sample banks in 2010–2013 was 1.74%, with a lowest value of 0.03% and a highest score of 8.89%. This indicates that there are still commercial banks in Indonesia that exceed the provisions of Bank Indonesia. Based on Bank Indonesia regulations, good NPLs are below 5%.

The compensation of bank executives who were sampled in 2009–2012, as measured by natural logarithm of compensation, reached an average value of 23.00, a lowest value of 20.27 and a highest value of 25.99. This indicates that the compensation given to bank directors in Indonesia is varied.

The average CAR of banks in 2010–2013 was 18.46%, with a highest value of 8.07% and a highest value of 75.04%. This indicates that commercial banks in Indonesia have complied with Bank Indonesia regulations to provide a minimum capital of 8% of risk-weighted assets.

3.2 Hypothesis analysis and testing

The results of the analysis of executive compensation effects, CAR, AGE, SIZE on NPLs, and the results are summarized in Table 2.

Table 1. Descriptive statistic.

	N	Minimum	Maximum	Mean	Std. Deviation
NPL (%)	372	0.03	8.89	1.74	1.48
COM	372	20.27	25.99	23.00	1.26
CAR (%)	372	8.07	75.04	18.46	8.61
AGE	372	20.00	87.00	46.38	13.89
SIZE	372	10.91	20.41	16.01	1.77
Valid N (listwise)	372				

Table 2. Coefficients.

	Unstandardized coefficients		Standardized coefficients		
Model	B	Std. Error	Beta	t	Sig.
(constant)	3.968	1.624		2.443	0.015
CAR	-.026	0.009	-0.152	-2.943	0.003
COM	-.212	0.104	-0.181	-2.032	0.043
AGE	0.000	0.006	-0.001	-0.024	0.981
SIZE	0.196	0.075	0.235	2.625	0.009

First, we analyze the effects of control variables on bank risks. Table 2 shows that executive compensation variables, CAR, and AGE have a negative influence on NPLs, whereas SIZE has a positive effect on NPLs. The granting of credit aims to obtain profits derived from the difference between interest income on loans with interest expense deposits. Problem loans will cause a decrease in loan interest income due to the absence of principal and interest loan principal payments from the debtor. The results of the t test show that executive compensation variable, CAR, and AGE have no significant effect on NPL at α of 0.05, while SIZE has significant effect on NPL at α equal to 0.05.

Our empirical results illustrated that the capital adequacy ratio (CAR) only has a significant negative effect on NPL%. We found no evidence of the relationship between CAR and other type of bank risks.

Turning to compensation variables, annual bonus appears to have significant negative impact on NPL%. In the relation with the other types of risk, we could not find any effect of an annual bonus on them. This result could explain that the annual bonus may encourage executives to take a higher risky portfolio correlated to the market as well as to accept riskier lending projects. Therefore, we support the point of view that the annual bonus induces myopic behavior in executives (Phu-ong 2012).

Related to the age of firms regarding bank experience, banks that were established earlier are expected to be more efficient. However, after the bank reaches a certain age, the incremental effect of experience will be negligible. It is believed that the relationship between non-performing loans and efficiency is bidirectional instead of unidirectional. Low efficiency reflects poor daily operations and loan portfolio management practices. Poor management skills in terms of credit scoring will lead to negative returns on a high proportion of loans and thus higher non-performing loans. Hence, efficiency also affects non-performing loans (Berger & De Young 1997).

4 CONCLUSIONS

Based on the results of data analysis that refers to the purpose of this research, the hypotheses, and model analysis, the following conclusions can be drawn:

a executive compensation variables, CAR, and AGE have a negative influence on NPL, whereas SIZE has a positive influence on NPL;
b the executive compensation variable, CAR, and AGE have no significant effect on NPL at α of 0.05, while SIZE has significant effect on NPL at α of 0.05.

REFERENCES

Ayadi, R., Arbak, E. & Pieter D.G.W. 2011. *Business models in European banking: a pre-and post-crisis screening.*

Berger, A.N. & DeYoung, R. 1997. Problem loans and cost efficiency in commercial banks. *Journal of Banking & Finance* 21(6): 849–870.

Biro Riset Infobank. 2011. *Remunerasi Bankir 2011.* XXXIII (39). Majalah Infobank: Jakarta.

Chen, Z. & Zhang, M. 2006. A research on the top executive's pay distributions China Account. *Scopus* 1(2006): 15–28.

Dendawijaya, L. 2009. *Banking management.* Bogor: Ghalia Indonesia.

Diyanti, A. & Widyarti, E.T. 2012. Analisis pengaruh faktor internal dan eksternal terhadap terjadinya non-performing loan (studi kasus pada bank umum konvensional yang menyediakan layanan kredit pemilikan rumah periode 2008–2011).

Fortin, R., Goldberg, G.M. & Roth, G. 2010. Bank risk taking at the onset of the current banking crisis. Financial Review 45(4): 891–913.

Hasan, I. & Wall, L.D. 2004. Determinants of the loan loss allowance: Some cross-country comparisons. *Financial Review* 39(1): 129–152.

Houston, J.F. & James, C. 1995. CEO compensation and bank risk Is compensation in banking structured to promote risk taking?. *Journal of Monetary Economics* 36 (2): 405–431.

Kaplanski, G. & Levy, H. 2012. Real estate prices: An international study of seasonality's sentiment effect. *Journal of Empirical Finance* 19(1): 123–146.

Kato, T., Kim, W. & Lee, H.J. 2006. Executive compensation, firm performance, and chaebol in Korea: Evidence from new panel data. *Pacific-Basin Finance Journal* 15: 36–55.

Murphy, K.J. 1999. Executive compensation. *Handbook of labor economics* 3: 2485–2563.

Muwardi, W. 2004. *Analisis faktor-faktor yang mempengaruhi kinerja bank umum di Indonesia.* Program Studi Magister Manajemen. Universitas Diponegoro Semarang.

Petit, R.J., Aguinagalde, I., de Beaulieu, J.L., Bittkau, C., Brewer, S., Cheddadi, R., Ennos, R., Fineschi, S., Grivet, D., Lascoux, M. & Mohanty, A., 2003. Glacial refugia: Hotspots but not melting pots of genetic diversity. *Science* 300(5625):1563–1565.

Ramaswamy, R. 2000. A study of determinants of CEO compensation in India. *Management International Review* 40: 167–191.

Rivai, R. & Veithzal, V. 2010. *Manajemen sumber daya manusia untuk perusahaan: Dari teori ke praktik.* Jakarta: Rajawali Pers Siamat, Dahlan. 2004.

Salami, H., Shahnooshi, N. & Thomson, K.J. 2009. The economic impacts of drought on the economy of Iran: An integration of linear programming and macro econometric modelling approaches. *Ecological Economics* 68 (4): 1032–1039.

Sastradiputra, S. & Komarrudin, K. 2004. *Strategi management bisnis perbankan.* Bandung: Kappa Sigma.

Shahbaz, M., Zeshan, M. & Afza, T. 2012. Is energy consumption effective to spur economic growth in Pakistan? New evidence from bounds test to level relationships and Granger causality tests. *Economic Modelling* 29(6): 2310–2319.

Sutrisno, Edi. 2010. *Manajemen sumber daya manusia. edisi pertama.* Jakarta: Prenada Media Group.

Thi Phuong, M.L. & Mireille, J. 2011. CEOinfobank.

Advances in Business, Management and Entrepreneurship – Hurriyati et al (eds)
© 2020 Taylor & Francis Group, London, ISBN 978-0-367-27176-3

Influence of microcredit and business management training to micro and small business performance in West Java, Indonesia

F. D. Trisnasih, L. Layyinaturrobaniyah & A. M. Siregar
Universitas Padjadjaran, Bandung, Indonesia

ABSTRACT: Micro, Small, and Medium Enterprises (MSMEs) are the backbone of national economy. However, there are still many problems faced by SMEs among other are the availability of capital, the difficulties to access financial institution as a lender, low understanding in business management, limited market access, and lack of science and technology mastery. This study aims to analyze the effectiveness of microcredit financing provided by banks and business management training on MSMEs performance. The method used in this research is mixed method using Sequential Explanatory Analysis. Quantitative data was analyzed with Partial Least Square (PLS) and to convert ordinal data into interval data this study used Rasch Model. The subject of this study is the customers who are getting microcredit loan from Jabar-Banten Bank (BJB) located in Regional 1. The results show that microcredit financing and business management training partially have a positive and significant impact on the performance of MSMEs with path coefficients 8.352, which is bigger than 1.96 (t-stat). It means that good quality of microcredit and the good quality of training on business man-agement would result in a better business performance for MSMEs.

1 INTRODUCTION

The society have basic rights such as the fulfillment of food, health, education, employment, housing, clean water, land, natural resources and environment, protection from threat or violence and the right to participate in social life politics, both for women and men (Supriyanto 2014). If the basic rights of society are not met, there will be poverty issues. The total poverty rate in West Java (BPS 2018) has increased slightly from 4168.11 in 2016 to 4168.44 in 2017.

To reduce the poverty rate, the West Java government launched a program that aimed to create 100 thousand new entrepreneurs (MSMEs) in 2011. This is done because MSMEs is considered a way that can be used to strengthen the economy in the community (Tambunan 2012).

The growth of MSMEs as seen from its performance makes it a source of employment opportunities by absorbing many workers, thus reducing poverty. The contribution of MSMEs to the Indonesian economy is no doubt. When the economic crisis hit the world and automatically worsened the economic conditions in Indonesia from 1997 to 1998, only MSMEs sector was able to remain relatively stable.

Data from Statistics Central Bureau showed the economic crisis did not lessened the number of MSMEs, on the contrary it increased its growth and was even able to absorb 85 million to 107 million workers until 2012. The remaining 0.01% or 4968 units were large-scale business (Suci 2017). This indicated that MSMEs in Indonesia were able to fulfill the basic needs of people for food, health, education, work, housing, clean water. These six needs were in line with the strategy of IMM (Integrated Microfinance Management) in which IMM = f (2 + 3 + 5), where 2 is the main goal of IMM that is reducing poverty and empowerment; 3 is the principles of quality, output and outcomes; and 5 is the basic service to the society which includes inclusive financial, health, education, communication and social culture (Slikkerveer 2011).

MSMEs are also referred to as the backbone of national economy because the business contribute about 59.08% of GDP or about IDR 4869.57 trillion, with a growth rate of 6.4% per year. In addition, MSMES contributed 14.06% to export volume with transaction value IDR 166.63 Trillion of total national exports, and support the establishment of Gross Fixed Capital (PMTB) nationwide of 52.33% or equivalent to IDR 830.90 trillion with the level of energy employing about 97% of the total national workforce and contributing 57% to the gross domestic product (GDP) (Bank Indonesia & LPPI 2015). Financial institutions including banks are the institutions that are able to provide access to the public financial inclusion. This is evident from the support given by Bank Indonesia. In its publication it stated that the publication of statistics on SME loans were based on the definition and criteria of business under Law No. 20 of 2008 on SMEs and was also based on the limitation of ceiling, namely: (1) Microcredit with loan ceiling up to IDR 50 million, (2) Small credit with a loan ceiling above IDR 50 million to IDR 500 million, and (3) Medium credit with loan ceiling above IDR 500 million to IDR 5 billion.

As a regional bank, BJB has a vision to become one of the ten largest banks with high performance in Indonesia and its mission is to become an eco-nomic driver West Java and as a source of the prov-ince revenue.

MSME credit types offered by BJB are Kredit Cinta Rakyat (KCR) with ceiling starting from IDR 5 million up to a maximum of IDR 50 million with an effective interest rate of 8.3% per annum, Kredit Usaha Rakyat (KUR) with ceiling ranging from IDR 5 million up to a maximum of IDR 500 million with an effective interest rate of 7% per annum and Kredit Mikro Usaha (KMU) with ceiling ranging from IDR 5 million up to a maximum of IDR 500 million and interest rate at 1% per month.

In distributing MSMEs credit, banks also have classic problem; problem loans by debtors resulting in Non-Performing Loans (NPLs). According to Mantayborbir et al. (2002) a problem credit occurs when debtors refuse to settle their obligations in accordance with the agreement, whether in terms of time limit, interest or the loan itself. The grouping of collectability consists of current credit (collectabil-ity 1), special attention (collectibility 2), substandard (collectibility 3), doubtful (collectibility 4) and loss (collectibility 5).

Other problems faced lies in the disbursement of micro and small enterprise credit, which are basically related to the profiles of MSMEs debtors that are mostly bankable (not meeting the technical banking requirements). The non-bankable debtors of MSME make the feasibility aspects of micro and small enter-prise lender neglected. Unable to meet the technical requirements of banks, prospective borrowers of micro and small businesses lose the opportunity to obtain credit facilities from banks (Mulyati 2016).

The purpose of this research is to know the effect of microcredit financing on the performance of SMEs, the influence of business management train-ing on the performance of SMEs and whether both have simultaneous effect on the performance of SMEs (research on BJB).

2 METHOD

In this study, the authors used mix method model, which combined quantitative and qualitative meth-ods, so that the data obtained were more comprehen-sive, valid, reliable and objective. The study used Sequential Explanatory Analysis.

The population of this research was 4691. They were debtors in Bank BJB Regional 1 who used mi-croloan product with amount of credit ≤ IDR 50 mil-lion. Based on Slovin formula, the sample was 98 and was rounded up to 100 people.

Data was analyzed using quantitative method in the first stage continued with qualitative analysis in the second stage. Likert measurements are presented as follows.

- 5 = Strongly agree.
- 4 = Agree.

- 3 = Doubtful.
- 2 = Disagree.
- 1 = Strongly disagree.

After the data was collected, measurements using Partial Least Square (PLS) was conducted to test the hypothesis using *p values* with alpha 5%.

The hypothesis to be tested are:

1) Microcredit has a positive effect on the perfor-mance of small micro-enterprises.
2) Provision of business management training has a positive effect on the performance of small micro-enterprises.
3) Microcredit and training on business management simultaneously have a positive effect on the per-formance of small micro-enterprises.

3 RESULT AND DISCUSSION

3.1 *Result*

The direct effect of microcredit on performance is $(0.228) \, 2 \times 100 = 5.2\%$. This number showed that without regard to other variables, microcredit gave 5.2% influence on MSMEs performance. The indi-rect effect of microcredit on performance in this study was not found.

The direct influence of business management training on performance is $(0.517) \, 2 \times 100 = 26.7\%$. It showed that business management training gave 26.7% influence on MSMEs performance. The influ-ence of business management training indirectly on MSMES performance in this research is not found.

The hypothesis above used t-statistic value pre-sented in the picture below, t-statistic for variables X1 and X2 to Y was obtained by $2349 + 6003 = 8352$. The value was greater than 1.96, so it can be concluded that $Ha > 0$, which means that microcredit and business management training was proven to simultaneously affect the performance of SMEs. The PLS calculations by performing a bootstrapping resulted in the following path analysis.

3.2 *Discussion*

From the result of statistical test it could be seen that the microcredit offered by Bank BJB and used by the debtor was included in either category. This showed that credit products up to IDR 50 million remained in demand by the MSMEs. However, the debtor must maintain the quality of the credit taken to retain trust from the banks.

The trust given by the bank is based on character, capacity, capital, guarantee and economic condition.

This study also revealed that microcredit had a di-rect influence on the performance of MSMES as much as 5.2%. This is in accordance with a research by Gustika (2016) which found that the provision of microcredit affects the income of SMEs.

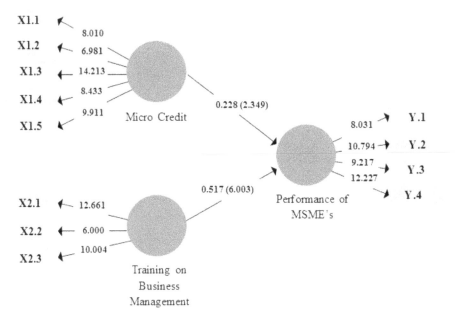

Figure 1. The result of PLS model.

Every business person has a financial record even though the recording is not done every day in the book. Financial records will facilitate them in sorting the business finances according to their needs.

As for the second variable, business management training, it was discovered that most respondents felt that training given by Bank BJB was a unique ser-vice, which was not available in other bank facilities. In fact, the training was meant to build the relationship among business people in order to expand the marketing net-work. Based on the statistics meas-urement, the effect of this second variable on the performance of MSMEs was as much as 26.7%. This is also in line with previous research conducted by Laodengkowe (2016) which stated that there is a positive and significant correlation between education and training variables with the improvement of MSMEs business.

The variable of MSMEs performance is comprised of 4 sub-variables: financial perspective, customer perspective, internal business process perspective and growth and development perspective. The statistical calculation showed that the two dependent variables, microcredit and business management training, effected the performance of MSMEs as much as 8.352 or 31.9%.

Based on the results of field observations and in-terviews, respondents had the ability to pay their ob-ligations including local contributions, arisan (a unique form of rotating savings in Indonesia), and other obligations including vehicle installments and loan installments to Bank BJB. However, some-times they forgot to pay their short-term liabilities. Some of them found it helpful to get a reminder of the credit due date through on call communication

received two days before the due date. In the finan-cial perspective, the respondents agreed that the interest rate of the bank was not a problem. The most important for the business people was a fast process and easy requirements in getting loan and the presence of more services than they expected, such as being invited to the training and involved in the exhibition or bazaar. Some respondents had been invited to attend an exhibition organized by Bank BJB. Direct sales activities during the exhib-ition was really useful to increase the respondents' income, moreover, they were also not required to pay for booth rent and lodging was provided for participants from out of town. For some products that are in demand by the market such as coffee drinks and batik, the exhibition gave them oppor-tunity to expand their marketing network.

Based on observations in the field, it was found that respondents' business place were visited by customers from time-to-time. According to the respondents who sold food, their places were usually crowded before Lebaran (the popular name for Eid al-Fitr in Indonesia) or during public holidays. Respondents also had under-standing towards the business process, since most of them had more than 2 year's experiences in their field. They were also trying to expand the marketing network and make sales online through some existing market places.

4 CONCLUSION

From this research we can conclude that micro-credit had an enormous effect on the performance of MSMEs. If the quality of credit is good, then the performance of MSMEs will also be good. This research also proved that business management training had a significant influence on MSMEs performance, which meant that business management training was required by Bank BJB's debtors in order to improve the performance of its business. Overall, the simultaneous effect of microcredit and business management training on the performance of SMEs was 31.9%, while the remaining 68.1% came from other variables that were not measured in this study. The results of this study can be used to determine the strategy of giving microcredit to the SMEs, especially in determining the products. Business management training remains as a unique package offered for micro products. Training pattern is suggested to be grouped based on debtor business scale or per product accessed, for example, debtors who apply for credit below IDR 50 million cannot be put together with debtor who apply for credit above IDR 50 million. Training providers and product marketers can collaborate in every activity so that the credit disbursement process accelerates.

The result of this study can also be taken into consideration by the government when making policy on microcredit financing especially in determining credit interest and the amount of ceiling, so that it can be easily accessed by business actors. It is also expected to create more synergy between government and non-government parties in order to help the growth of MSMEs so as to sustain the regional economy.

ACKNOWLEDGEMENTS

The authors would like to thank various parties who have helped in completing this research, especially the management of BJB who have provided valuable information on micro-credit data, micro-credit recipient respondents from BJB who were willing to respond to the questionnaire and be interviewed, as well as the integrated microfinance management master program at the Faculty of Economics and Business, Universitas Padjadjaran.

REFERENCES

BPS. 2018. *Kemiskinan kabupaten/kota di Jawa Barat 2012–2017*: Badan Pusat Statistik Provinsi Jawa Barat.

Bank Indonesia & LPPI. 2015. *Profil bisnis usaha mikro, kecil dan menengah (UMKM)*.

Gustika, Rosa. 2016. Pengaruh pemberian kredit usaha rakyat terhadap pendapatan masyarakat Ladang Panjang Kecamatan Tigo Nagari Kabupaten Pasaman. *e-Jurnal Apresiasi Ekonomi* 4(2): 107–115.

Laodengkowe, A. 2016. Analisis hubungan program pendidikan pelatihan dan bantuan permodalan usaha mikro kecil dan menengah dengan peningkatan usaha UMKM di kota Semarang. *Skripsi*. FEB Universitas Diponegoro.

Mantayborbir, S., Jauhari, I. & Widodo, A.H. 2002. *Hukum piutang dan lelang negara di Indonesia*. Medan: Pustaka Bangsa.

Mulyati, E. 2016. Kredit perbankan aspek hukum dan pengembangan usaha mikro kecil dalam pembangunan perekonomian Indonesia.

Slikkerveer, L.J. 2011. *Introduction to the first International Conference Integrated Microfinance Management for Sustainable Community Development*.

Suci, Y.R. 2017. Perkembangan UMKM (Usaha Mikro, Kecil dan Menengah) di Indonesia. *Jurnal Ilmiah Cano Ekonomos* 6(1).

Supriyanto, R.W., Ramdhani, E.R. & Rahmadan, E. 2014. *Perlindungan sosial di Indonesia: Tantangan dan arah ke depan*. Jakarta: Direktorat Perlindungan dan Kesejahteraan Masyarakat Kementerian PPN/Bappenas.

Tambunan, T. 2012. *UMKM di Indonesia, isu-isu penting*. Jakarta: LP3ES.

Profitability, dividend policy and stock price volatility: Indonesia stock exchange

I. Setyawati
Universitas Bhayangkara, Jakarta, Indonesia

D.P. Alamsyah
Universitas Bina Sarana Informatika, Bandung, Indonesia

K. Khotimah
Institut Bisnis Dan Informatika Kosgoro, Jakarta, Indonesia

ABSTRACT: The purpose of this study is to analyze profitability and dividend policy in predicting stock price volatility in the sector of property manufacturing companies with the sub-sector of real estate and property and building construction listed on the Indonesia Stock Exchange in the period 2009–2015. The data was taken in the form of corporate financial statements, while the company used as research samples taken using a purposive random sampling technique. This research found that return on asset and dividend payout ratio have a positive effect on the company stock price. Thus, as both management and investors are very concerned about changes in stock prices, research can provide a way to find the dominant factor in volatility of stock prices and should be considered by investors before making investment decisions, and by management in formulating the dividend policy for the company.

1 INTRODUCTION

Why the stock market always is moving at all times? Economists argue that almost every decade there are factors causing the volatility of stock prices. The macro-economic factor is one of the causes of stock price volatility (Schwert 1990), the business cycle also influences the volatility of stock prices (Schwert 1989). Some empirical studies have found that the volatility of stocks price is much higher when the economy is worse than when the economy is good (Brandt & Kang 2004, Schwert 1989). A company's management announcements are also an important source of short-term stock price movements (Bomfim 2003).

Financial theory and empirical evidence indicate that there are two things that affect the volatility of the stock market in the short term. First, the effect of volatility stock price is higher on the pre-announcement than on the day of the announcement itself. Second, decisions about company policy have the potential to affect stock price volatility in relation to the nature of the decision itself, meaning that policy decisions can reveal new information that was not previously included in asset prices and stock prices may move as long as market participants process newly received information (Agrawal et al. 2010, Engle et al. 2009, Hamao et al. 1990, Kim 2001, Vlastakis & Markellos 2012).

Stocks become one of the investment alternatives in capital markets that are mostly used by investors because it provides a premium return through stock price volatility (Baker & Wurgler 2007, Setyawati 2014). Investors need to choose stocks that are efficient so as to provide maximum returns with a minimum level of risk. But investors may be faced with the risk of a decline in stock prices and the possibility of issuing the company by the delist for some reason even further investors are also faced with the possibility of companies being bankrupted. Investment risks in stocks and high bonds can be minimized by conducting an in-depth analysis of the factors that are expected to affect stock prices, the risks borne by investors, which may also affect the development of capital markets (Kim 2002, Mele 2007).

Investors need adequate information about the current state of the company, which is often used to project future profits with varying levels of profitability. One of the most important types of information is the financial information that is briefly presented in the company's financial statements. A company's ability to generate profits from its operations is the main focus of a company's achievement. Investors will see the return on equity of the company because the return on equity indicating how the effectiveness of the company's management in the use of shareholder funds (Setyawati et al. 2017, Sufian & Habibullah 2010). The higher return on equity gives an indication to shareholders that the rate of return on investment is higher (Mitchell & Stannford 2000).

In addition to investment and funding decisions, dividend payout is a common problem faced by companies. Dividends are a reason for investors to invest in, where dividends represent returns to be received for investment in the company (Denis & Osobov 2008). Investors have the ultimate goal of improving prosperity by expecting dividend returns, while companies expect continued growth to maintain their survival while providing welfare to shareholders, so dividend policy is important to meet shareholders' expectations of dividends (West 1988).

Dividend payout ratio, which is a parameter to measure the amount of dividend to be distributed to shareholders as a "cash dividend." Companies with high growth rates will have a low dividend payout ratio (Amidu & Abor 2006). Conversely, companies with low growth rates will have high ratios. The greater this ratio than the slower or smaller the growth of the company (Gill et al. 2010). The large dividend payout is not undesirable by investors, but if dividend payout ratio is greater than 25% it is feared that there will be financial liquidity difficulties of the company in the future (Amidu & Abor 2006, Gill et al. 2010, Gugler 2003).

The purpose of this study is to find out how the volatility of stocks caused by the acquisition of prof-its and dividend policies made by the sector of property manufacturing companies listed on the Indonesia Stock Exchange.

This study was conducted on the sector of property manufacturing companies, as the construction market and building materials sector in Indonesia has grown significantly, driven by the rapid growth of the property/real estate market in the country, increased private investment and government spending, the price of land tends to rise, while demand will always increase along with the increase in population and the increasing of human need.

The contribution of the construction sector to the gross domestic product (GDP) of the country has grown from about 7.07% in 2009 to 13% in 2014 and has encouraged the growth of Indonesia's building and construction industry. The construction market is projected to grow by 14.26% to reach Rp. 446 trillion by 2015 and will be one of the most promising sectors due to the accelerated government infrastructure development plan, released by OJK (Financial Services Authority in Indonesia) 2017. The construction sector plays an important role in the economy of the country as it affects most sectors of the country's economy and is an important contributor to the infrastructure development process that provides the physical foundation on which development efforts and the upgrading of living standards can be realized (Widayanti & Haryanto 2013).

2 METHOD

The study used a causality design that aims to find explanations in the form of cause-effect between several variables. The data used in the form of quantitative data in the form of the data panel. The data source is secondary data derived from the financial statements of a sector of property manufacturing companies listed on the Indonesia Stock Exchange in the period 2009–2015.

The population in this study amounted to 47 companies. Selection of sample in this research using purposive sampling technique, that is a sample selection method based on certain consideration or with a certain criterion to the sample to be studied. Based on the criteria, 17 research samples were obtained.

In making an estimation, this research uses a panel data model (Wooldridge 2009). To assess the relationship between stock price movements with profitability and dividend policy, multiple regression equations are used, as it allows to explicitly control many other factors simultaneously affecting the dependent variable (Gujarati & Porter 2010), such as the following models:

$$LnSP_{it} = \beta_0 + \beta_1 ROE_{it} + \beta_2 DPR_{it} + \epsilon_{it} \quad (1)$$

Table 1 shows the variables affecting stock price volatility in the sector of property manufacturing companies in Indonesia.

Multiple regression model using panel data, as shown in equation (1), using fixed effect least square approach. Tests using ordinary least squares (OLS) are needed to determine whether there is multicollinearity, heteroscedasticity, and autocorrelation. After being tested with the Hausman test, fixed effect selected rather than random effects. In addition, the amount the research time (T) is greater than the number of individuals (N), so the user is fixed effect panel models are more appropriate (Greene 2002). Using a fixed effect panel model, individual effects of each sector of property manufacturing companies can be shown.

Table 1. Variables used in the regression model.

Variable	Variable Definitions	Hypothesis
Dependent Variable		
Logarithm natural of stock price (LnSP)	Stock price is the embodiment of the stock market value which is the ongoing stock market price in the stock exchange.	NA
Independent Variable		
Dividend Payout Ratio (DPR)	The ratio between dividends distributed to shareholders in cash dividends with earnings per share	-
Return on equity (ROE)	The ratio between profit after tax with own capital	+

339

3 RESULT AND DISCUSSION

The heteroscedasticity test was performed using Bruesch-Pagan Lagrange Multiplier (BP-LM test) and Likelihood Ratio (LR test) test (Gujarati & Porter 2010). Table 2 shows that the p-value is less than 0.05, meaning the model structure of the variance is not heteroscedastic.

As for the autocorrelation test used The Wooldridge test, in Table 3 shows that the p-value is less than 0.05, and shows no autocorrelation.

The goodness of fit test stated that the model is significant because p-value < 0.05; thus the model can be accepted in describing the dependent variable. With R2 88.27%, it means that variation stock price can be explained by ROE and DPR, while 11.73% is explained by variations of other variables, which are not included in the model. Table 5 shows the summary of the dependent and independent variable.

The estimation result of the research model is presented in Table 6.

Table 2.   Summary of dependent variables and independent variables.

Variable	Mean	Deviation Standard	Min	Max
ROE	7.621008	6.156175	.03	27.21
DPR	5.081849	3.601952	.005	18.25
LnSP	5.857149	2.357337	-3.91	9.34

Table 3.   Multicollinearity test.

	LnPS	ROE	DPR
LnPS	1.0000		
ROE	0.3863	1.0000	
	0.0000		
DPR	-0.1010	-0.0345	1.0000
	0.0566	0.5161	

Table 4.   Heteroscedasticity test.

Breusch-Pagan Lagrange Multiplier Panel Heteroscedasticity Test

Ho: Panel Homoscedasticity - Ha: Panel Heteroscedasticity	
Lagrange Multiplier LM Test	= 1141.11622
Degrees of Freedom	= 16.0
P-Value > Chi2(16)	= 0.00000

Greene Likelihood Ratio Panel Heteroscedasticity Test

Ho: Panel Homoscedasticity - Ha: Panel Heteroscedasticity	
Likelihood Ratio LR Test	= 345.49009
Degrees of Freedom	= 16.0
P-Value > Chi2(16)	= 0.00000

Table 5.   Autocorrelation test.

Wooldridge test for autocorrelation in panel data

H0: no first-order autocorrelation	
F(1, 16)	= 15.144
Prob > F	= 0.0013

ROE has a significant positive effect on stock price volatility. This shows that any increase in ROE on the sector of property manufacturing companies will significantly affect the volatility of stock prices in Indonesia Stock Exchange.

Return on equity is the sum of returns from net income to the firm's equity. It's used to measure the ability of an issuer in generating profits with capitalizing equity that has been invested by shareholders. Basically, the higher the return on equity, the better, it means that company management is able to make the company as efficient as possible, with the same capital. Thus, investors will be interested in investing funds in companies with high return on equity and ultimately will raise the stock price of the company. Research on the influence of return on equity on the stock prices volatility is still very little, some researchers found that profitability has a positive and significant effect on stock price volatility (Deitiana 2013, Purnamawati 2016). Thus required further research that discusses it.

DPR has a significant positive effect on stock price volatility. This shows that any increase in LDR on the sector of property manufacturing companies will significantly affect the stock prices volatility in Indonesia Stock Exchange.

Research on the effect of dividend payout ratio on stock price volatility has been done and most found that dividend payout ratio has the positive and significant effect to stock price volatility (Amidu & Abor 2006, Gill et al. 2010, Gugler 2003, Rehman 2012, West 1988).

The dividend policy is an integral part of the company's funding decision, so it is part of the

Table 6.   Estimation research.

Variable	Dependent variable – Stock price volatility – N = 357
INTERCEPT	+5.296***
	.201
ROE	+.023***
	.009
LDR	+.0235***
	.009
R^2	0.883
F (prob)	0.0000

*, **, *** *indicates significant at the 1 percent, 5 percent and 10 percent levels respectively*

agency theory (Denis & Osobov 2008). The dividend payout ratio determines the amount of profit that can be retained as a source of funding. The greater the retained earnings the less the amount of profit allocated for dividend payments (Amidu & Abor 2006, Rehman 2012). Information on earnings per share may be used by the management of the company to determine the dividend to be distributed. This information is also useful for investors to understand the development of the company, but it can also be used to measure the profit level of a company. Dividend payment ratio is very important for investors who are interested in short-term earnings will prefer to invest in companies whose dividend payout ratio is high while for investors who choose to have capital growth will be more interested to invest in companies that dividend payout ratio is low (Gugler 2003).

4 CONCLUSION

We found that return on equity and dividend payout ratio has the positive and significant effect on stock price volatility. This shows that the stock prices volatility is influenced by how much return received by investors from the investments made and how much acceptance of cash dividends received by shareholders. Therefore, companies need to pay attention to two things that want the share price to rise.

This study has limitations because it only examines one sector, a sector of property manufacturing companies, therefore it is advisable to examine all sectors on the Indonesia Stock Exchange. In addition, the use of return on equity and dividend payout ratio is not enough to describe the stock prices volatility. Other variables such as growth opportunities, equity mixtures obtained/contributed, capital structure, macroeconomic and political factors to enrich research on stock price volatility.

ACKNOWLEDGEMENTS

I thank the leaders of the University of Bhayangkara Jakarta Raya for the opportunity given to attend the Third Global Conference on Business, Management, and Entrepreneurship organized by Universitas Pendidikan Indonesia.

REFERENCES

Agrawal, G., Srivastav, A. K. & Srivastava, A. 2010. A Study of Exchange Rates Movement and Stock Market Volatility. *International Journal of Business and Management*, 5(12): 62–73.

Amidu, M. & Abor, J. 2006. Determinants of dividend payout ratios in Ghana. *The Journal of Risk Finance*, 7 (2): 136–145.

Baker, M. & Wurgler, J. 2007. Investor Sentiment in the stock market. *Working Paper*, 53(2): 160.

Bomfim, A.N. 2003. Pre-announcement effects, news effects, and volatility: Monetary policy and the stock market. *Journal of Banking & Finance* 27(1): 133–151.

Brandt, M.W. & Kang, Q. (2004). On the relationship between the conditional mean and volatility of stock returns: A latent VAR approach. *Journal of Financial Economics*, 72(2): 217–257.

Deitiana, T. 2013. Pengaruh current ratio, return on equity dan total asset turn over terhadap dividend payout ratio dan implikasi pada harga. *Jurnal Bisnis dan Akuntansi* 15(1): 82–88.

Denis, D.J. & Osobov, I. 2008. Why do firms pay dividends? International evidence on the determinants of dividend policy. *Journal of Financial Economics*, 89(1): 62–82.

Engle, R.F. Ghysels, E. & Sohn, B. 2013. Stock market volatility and macroeconomic fundamentals. *Review of Economics and Statistics*, 95(3): 776–797.

Gill, A., Biger, N. & Tibrewala, R. 2010. Determinants of Dividend Payout Ratios: Evidence from the United States. *The Open Business Journal*, 3: 8–14.

Greene, W.H. 2002. Econometric Analysis. *Journal of the American Statistical Association*, 97.

Gugler, K.2003. Corporate governance, dividend payout policy, and the interrelation between dividends, R&D, and capital investment. *Journal of Banking and Finance*, 27(7): 1297–1321.

Gujarati, D.N. & Porter, D.C. 2010. *Essentials of Econometrics*. New York: McGraw-Hill Education.

Hamao, Y., Masulis, R.W. & Ng, V. 1990. Correlation in Price Changes and Volatility across International Stock Market. *The Review of Financial Studies*, 3(2): 281–301.

Kim, C. 2002. Is There a Positive Relationship between Stock Market Volatility and the Equity Premium?. *Journal of Money, Credit and banking*, 339–360.

Kim, K.A. 2001. Price limits and stock market volatility. *Economics Letters*, 71(1): 131–136.

Mele, A. 2007. Asymmetric stock market volatility and the cyclical behavior of expected returns. *Journal of Financial Economics*, 86(2): 446–478.

Mitchell, M.L. & Stannford, E. 2000. Managerial Decisions and Long-Term Stock Price Performance. *The Journal of Business*, 73(3): 287–329.

Purnamawati, I.G.A. 2016. The Effect of Capital Structure and Profitability on Stock Price (Study of the Manufacturing Sector in Indonesia Stock Exchange). *International Journal of Business, Economics, and Law*, 9 (1): 10–16.

Rehman, A. 2012. Determinants of Dividend Payout Ratio: Evidence from Karachi Stock Exchange (KSE). *Journal of Contemporary Issues in Business Research*, 1(1): 21–27.

Schwert G.W. 1989. Business Cycles, Financial Crisis and Stock Volatility. *Carnegie-Rochester Conference Series on Public Policy*, 31: 83–126.

Schwert G.W. 1990. Stock market volatility. *Financial Analysts Journal*, 23–34.

Setyawati, I. 2014. Analisis Portfolio Optimal dari 10 Saham Unggulan di Bursa Efek Indonesia dengan Menggunakan Single Index Model. *Mediastima*, 20(1): 49–65.

Setyawati, I., Suroso, S., Suryanto, T. & Siti, D. 2017. Does Financial Performance of Islamic Banking Is Better? Panel Data Estimation. *European Studies Research Journal*, 20(2): 592–606.

Sufian, F. & Habibullah, M.S. 2010. Assessing the Impact of Financial Crisis on Bank Performance Empirical Evidence from Indonesia. *ASEAN Economic Bulletin*, 27 (3): 245–62.

Vlastakis, N. & Markellos, R.N. 2012. Information demand and stock market volatility. *Journal of Banking & Finance*, 36: 1808–1810.

West, K.D. 1988. Dividend Innovations and Stock Price Volatility. *Econometrica: Journal of the Econometric Society*, 37–61.

Widayanti, P. & Haryanto, A.M. 2013. Analisis pengaruh faktor fundamental dan volume perdagangan terhadap return saham (studi kasus pada perusahaan real estate and property yang terdaftar di BEI periode 2007–2010). *Diponegoro Journal of Management*: 238–248.

Wooldridge, J.M. 2009. *Introductory Econometrics*. Canada: Nelson Education, Ltd.

Advances in Business, Management and Entrepreneurship – Hurriyati et al (eds)
© 2020 Taylor & Francis Group, London, ISBN 978-0-367-27176-3

Why did Baitul Maal wa Tamwil (BMT) discontinue the linkage program with Islamic banks? A case study in Indonesia

M.N.H. Ryandono & D.A. Mi'raj
Universitas Airlangga, Surabaya, Indonesia

ABSTRACT: This research aims to determine the reason why BMT decided to discontinue the linkage program with Islamic Bank. This study used a qualitative approach and single-case study strategy. Primary data was collected by in-depth interview, while secondary data was obtained from the annual financial statements of BMT. Explanation building was used as the analytical technique: explaining the results of in-depth interviews, in order to know the reasons of BMT to stop the linkage program. The results showed that the BMT option to stop the linkage program is not just because of the expensive rate of return and other operational aspects only. The biggest point of decision is because there is a difference of *aqad* applied to the linkage program, which was not in accordance with the *mudharaba* system contract. The factors that caused BMT to discontinue the linkage program will be explained qualitatively in this research.

1 INTRODUCTION

Community access to financial institutions is one of the keys to optimize the function of the financial system. If the community can utilize financial services easily, it will encourage an increase in capital turnover (Oxtora 2015). That way, financial institutions can implement capital equity in a society, which will then encourage economic growth. However, in reality, 54today, Indonesia has still not been established in a stage of financial inclusion. World Bank's survey in the Booklet on Inclusive Finance (Bank Indonesia 2014) shows that about 52% of the Indonesians live in rural areas and 60% of them do not have access to formal financial services.

For this reason, the government, in this case OJK, who has authority of regulation and supervision in the financial services sector, began to concern itself with the strengthening of microfinance institutions, starting with Baitul Maal wa Tamwil, better known as BMT. In accordance with efforts to improve the inclusive financial system, BMTs are Islamic microfinance institutions that have the opportunity to provide the needs of the poor because the process for BMT is relatively easier than it is for banks (Sanrego & Yulizar 2008).

However, the obstacle faced by BMT is unfulfillment of funds needed by the community, because the requested amount is greater than the collected funds (Sudarsono 2007), so therefore the economic growth of the lower economic classes progresses slowly. The issuance of regulations from the Minister of Cooperatives and Small and Medium Enterprises of the Republic of Indonesia Number: 03/Per/M. KUKM/III/2009 about general guidelines of the linkage program between commercial banks and cooperatives, becomes the hope for BMT to develop and overcome obstacles faced during this time, which is to enlarge the capital capacity through partnership with Islamic Banks. This linkage program is one of the strategies to achieve a more inclusive financial system. The linkage program is divided into three types, namely executing, channeling, and joint financing. The total number of linkage funds of Islamic banking programs as of October 2012 which are channeled to BMT for one year reached IDR 439.2 billion (https://www.syariahmandiri.co.id/).

However, in reality, not all BMTs do these linkage program with Islamic Bank. There are even BMTs who previously did a linkage program with an Islamic bank, but then no longer continued their linkage program. Evidently, until September 2012, there were still BMTs who do not do a linkage with any institution. Data from the Directorate of Indonesian Islamic Banking, Bank Indonesia 2012, shows that out of 128 BMTs, only 75 BMTs (58.5%) have a linkage with Islamic Bank (BUS/UUS). The reasons why BMTs do not use linkage programs are, first, because of the convoluted linkage procedure of 29%; second, because the requirements cannot be met in 18%; third, because no benefit was felt in 12%; fourth, the association policy 4%; and , fifth, other reasons, being 37%. Another reason, which reached the highest percentage of 37%, indicates that there are some unknown linkage program weaknesses, which then cause BMTs to choose to not do the linkage with the Islamic Bank.

When viewed from the side of BMT and inclusive financial implementation, indeed linkage is a relevant strategy World Bank Document 2011. However, in terms of SME customers, it makes BMT products become more expensive. This is because the linkage loan capital also has a rate of return, which must be paid by BMT to the Islamic

Commercial Bank or Sharia Business Unit,/the link-age partner of BMT itself. So the cost of profit sharing to the Islamic Commercial Bank is charged to users of BMT financial products. The high cost of capital to be paid by SMEs to BMT can cause the growth of SMEs itself to become hampered. Whereas SME is an important sector that contributes to the national economic growth in a country. Based on the above problem, the main focus in this research is the reason why Baitul Maal wat Tamwil did not continue its linkage with Islamic Bank.

2 METHOD

The approach used in this study is a qualitative approach because the problem formulation with the "why" question requires an answer that cannot be obtained using statistical methods. According to Herdiansyah (2010), a qualitative approach is a study that intends to understand the phenomenon of what is experienced by the subject of research, such as behavior, perception, motivation, action, and so forth. The strategy used in this qualitative research is an explorative case study. Specifically, it uses an explorative single case study method, where there are only one case and one object of research, intended to find the factors causing BMT to choose to stop the linkage program with Islamic banking.

The scope of this study is limited to why BMT chose to stop the linkage program with an Islamic bank. The case study is BMT Mandiri Sejahtera Gresik, who has been running a linkage program with an Islamic bank for more than 2 years .

The type of data used in this study are primary and secondary data. Primary data is the main data derived from the results of in-depth interviews and direct observation in the field, whereas the secondary data are taken from literature such as research, financial reports, and other documents.

The author uses data triangulation techniques. Data triangulation according to Moleong (2005), is a technique of checking the validity of data by utilizing other data. According to Sugiyono (2014), triangulation of source is to test the credibility of data by checking the data to several sources.

Data Analysis Technique, the analytical technique of this research is an explanation approach that aims to analyze case study data by making an explanation of the case concerned for the sustainability of the study.

3 RESULTS AND DISCUSSION

The Indonesian government's efforts to create a more inclusive financial system has become a support of enthusiasm for the Islamic microfinance organizers. The government's partisans develop microfinance in order to improve the economic performance at SMEs level and also to provide new

hope for the middle-low-income people. Therefore, the implementation of the linkage program as a manifestation of a more inclusive financial system should also be in favor of microfinance institutions and the middle low incomes. Especially for the inclusiveness of Islamic finance, it is fit for Muslims to fight for it. Allah has described struggles based on faith and heart conviction with a business that can save mankind in his word:

أَمْ لَهُمْ مُلْكُ السَّمَٰوَٰتِ وَالْأَرْضِ وَمَا بَيْنَهُمَا فَلْيَرْتَقُوا فِى الْأَسْبَابِ
جُنْدٌ مَّا هُنَالِكَ مَهْزُومٌ مِّنَ الْأَحْزَابِ

"[They are but] soldiers [who will be] defeated there among the companies [of disbelievers]"

BMT Mandiri Sejahtera is one example of how a small Islamic microfinance institution is able to demonstrate the potential of a more inclusive financial system that enables middle-low-income people around the region get access to formal finance. The linkage program used by BMT Mandiri Sejahtera has since 2009 been one of the government's efforts to realize a more inclusive financial system. In its development, BMT Mandiri Sejahtera always made the amount of outstanding financing from the linkage program annually, especially with Bank Syariah Mandiri. However, the linkage program with Bank Syariah Mandiri was discontinued in 2013. The researcher asked the question whether there was any significant progress towards BMT Mandiri Sejahtera's performance after linkage with Bank Syariah. One informant, Mr. Ayubi said that there is still a linkage effect on performance development after the linkage with Islamic Bank, but it would be more convenient if BMT used the member funds because it is cheaper.

In previous research, the Islamic bank linkage program has a positive impact on the performance of BMT Mandiri Sejahtera in the period 2009 until 2012. However, in 2013 BMT Mandiri Sejahtera decided to discontinue the linkage program. Rationally, if the linkage program has a positive impact, then BMT Mandiri Sejahtera should continue the linkage program with Islamic Bank. Based on information from Mr. Ayubi, as the Manager of BMT Mandiri Sejahtera, there are three reasons why BMT Mandiri Sejahtera discontinued the linkage program with Islamic Syariah in 2013. Therefore, based on a board meeting's results it was decided not to continue the linkage program. First, Mr. Ayubi considered the sharia pattern of the linkage program with Bank Mandiri Syariah to be less appropriate. The executing model used by BMT Mandiri Sejahtera is by the pattern of *mudharabah* wall *murabaha* contract. The purpose of this contract pattern is the financing of Bank Syariah Mandiri to BMT Mandiri Sejahtera using a contract, while the channeling of linkage funds to the community/members of BMT Mandiri Sejahtera uses the *murabahah* scheme. This pattern was explained by Mr. Ayubi, that the return from the contract with Bank Syariah Mandiri is still flat. *Mudharabahwal murabaha aqad* that occurred between Islamic banks and BMTs can be said to be a form of cooperation that is less appropriate when viewed in terms of Islamic economic goals, even if the sharia law is legitimate. The pattern of executing that should be better suited to use the *mudharabah* contract and added with *murabaha* with the reason that the sharia bank knows the direction of channeling of the linkage fund which is sure to mitigate the financial risk.

وَاللّٰهُ فَضَّلَ بَعْضَكُمْ عَلٰى بَعْضٍ فِى الرِّزْقِ فَمَا الَّذِيْنَ فُضِّلُوْا بِرَاۤدِّيْ رِزْقِهِمْ عَلٰى مَا مَلَكَتْ اَيْمَانُهُمْ فَهُمْ فِيْهِ سَوَاۤءٌ اَفَبِنِعْمَةِ اللّٰهِ يَجْحَدُوْنَ

"And Allah has favored some of you over others in provision. But those who were favored would not hand over their provision to those whom their right hands possess so they would be equal to them therein. Then is it the favor of Allah they reject?" (Q. S.An-Nahl: 71)

In business, if any profit generated from the cooperation should occur, it should be equally divided among partners prospectively. So it was reasonable for BMT Mandiri Sejahtera to discontinue being a partner of Islamic banks due to injustice in the partnership run between the BMT and Islamic bank. It was because there is a leveling between middle-upper incomes and middle-lower income classes. This verse is also one of the foundations of and equality in Islam and which corresponds to the rights that every Muslim has.

Second, the profit and loss sharing of the linkage program with Bank Syariah Mandiri is still quite high. The effective margin that Bank Syariah Mandiri wants to get from a linkage program with BMT Mandiri Sejahtera is 14% (about 13.75%). Bank Syariah Mandiri linkage program is then thrown to members/communities with a financing effective margin of 20% -24%. Another informant, Mrs. Khusnul Khotimah, as head of the administration department of BMT Mandiri Sejahtera, also assumed that the profit sharing of Bank Syariah Mandiri was too high, therefore in 2013, BMT Mandiri Sejahtera decided to discontinue its linkage program with Bank Syariah Mandiri.

The third is the process of liquefaction, which is still too long with approximately a month. BMT Mandiri Sejahtera, which has been working together for five years with Bank Syariah Mandiri should be able to get more quick liquefaction processes because it has mutual trust. However, still Mr. Ayubi, as the manager of BMT Mandiri Sejahtera, was less satisfied with the process of liquefaction of funds linkage the fastest one month.

According to the calculation of total flat linkage program refunds for the 2015 financing year with the *murabahah* wall contract, BMT Mandiri Sejahtera must return the basic financing fund plus a margin equivalent to almost 21% of the financing fund. This caused that BMT Mandiri Sejahtera must do channel linkage funds to their members with a margin of more than 21%. Based on the calculation it can be seen that the linkage program with high-profit sharing makes the financing margins distributed to the community also higher. It is because BMT Mandiri Sejahtera must give a high enough return to the Islamic Bank. This makes the role of the linkage program limited to only create an inclusive finance but not to take the side of the middle-lower income classes who are in fact the majority of users of financial products in cooperatives and BMT. Yet it is clear what Allah says:

وَهُوَ الَّذِيْ جَعَلَكُمْ خَلٰۤىِٕفَ الْاَرْضِ وَرَفَعَ بَعْضَكُمْ فَوْقَ بَعْضٍ دَرَجٰتٍ لِّيَبْلُوَكُمْ فِيْ مَاۤ اٰتٰىكُمْ اِنَّ رَبَّكَ سَرِيْعُ الْعِقَابِ وَاِنَّهٗ لَغَفُوْرٌ رَّحِيْمٌ

"It is He who made you successors on the earth and raised some of you in ranks over others, in order to test you through what He has given you. Your Lord is Quick in retribution, and He is Forgiving and Merciful." (Q.S. Al-An'am: 165)

The Islamic banking industry can be said to be on track in its effort to achieve a more inclusive financial system. Inclusive here means that the lower income people, who previously had difficult access to formal finance, can easier access it, through the linkage program. However, the Islamic banking industry is still not altogether fair in the treatment of financing objects. Mr. Ayubi said, *"If we want to work on the lower income people, he wishes it will be great if the margin is cheaper. In our financial system, the big industries even got a cheap financing, while the middle low-income classes such as SMEs, it's actually got a higher margin rate. Well, this is the opposite."* So what Allah says, in the Qur'an Surah Al-Maidah verse 2 and Surat an-Nahl verse 71, it has still not implemented . Based on the explanation of Mr. Ayubi, the society in the middle-low classes should be helped in order to live more prosperous. He thought that the handling of banks' financing was not the same, especially for large industries it to be easier, unlike the SMEs which is a bit hard, even when SMEs failed and they no longer got financing.

The different characteristics of BMT and Islamic banks can surely cause different system and profit earnings. It also happened to Islamic banks, for which the linkage program with a BMT doesn't guarantee a better income for them. This condition is in line with research conducted by Kumara (2010), indicating that BPRs that follow the linkage program are no better than BPRs who do not follow the linkage program. Similarly, research conducted by Bimaprawira (2009) shows that, in general, linkage programs are still not successful enough in their efforts to increase the competitiveness of rural banks. It is proven that there is no better level of profitability and the higher level of risk that must be borne by linkage program BPR participants. So it can be said that the success of the linkage program depends on the system and the method used in the application.

BMT Mandiri Sejahtera wishes that the profit/ margin of linkage cooperation can be cheaper. However, sharia banks also cannot be prosecuted, because they basically depend on the revenue-based income: when the income is low, the profit margin is also low, and vice versa when the income is high, the profit margin is also high. Thus, if the profit-sharing equivalent is higher than the *murabaha* margin, then it is natural. It cannot be used as an excuse by the BMT Mandiri Sejahtera to blame Islamic banks. Especially if the reason is that the BMT wants to get more results by giving the smaller part of profit sharing to the Islamic bank, but takes a large margin of funds linkage to the SMEs. So Allah Almighty has warned in His following words: *"Woe to those who give less [than due]; who, when they take a measure from people, take in full; But if they give by measure or by weight to them, they cause loss"* (Q.S. Al-Muthaffifin: 1–3).

Islamic banks have not yet run a system for sharia-compliant results, where the profit sharing should be adjusted to the income produced, but in reality, it is not. In this case, both BMT and Sharia Banks must be careful in channeling funds to the productive sector. Because if financing for the productive sector uses a fixed margin as in *murabaha aqad*, it can lead to the *ribawi* trap, which then impacts the economy with interest. The choice of contract for financing, in this case, must be in accordance with the income to be earned. If the income is fixed, then the contract can use a fixed margin. However, if the income is uncertain, then the contracts of margin should be varied. That is why *riba* is forbidden because it seems to help but essentially it does not. As mentioned in the first stage forbidding usury in the following Word of Allah:

وَمَا آتَيْتُمْ مِّن رِّبًا لِّيَرْبُوَا۟ فِىٓ أَمْوَٰلِ ٱلنَّاسِ فَلَا يَرْبُوا۟ عِندَ ٱللَّهِ ۖ وَمَآ آتَيْتُم مِّن زَكَوٰةٍ تُرِيدُونَ وَجْهَ ٱللَّهِ فَأُو۟لَٰٓئِكَ هُمُ ٱلْمُضْعِفُونَ

"And whatever you give for interest to increase within the wealth of people will not increase with Allah. But what you give in zakat, desiring the countenance of Allah - – those are the multipliers." (Q.S. Ar-Ruum: 39)

Moreover, BMT Mandiri Sejahtera uses a lot of *murabaha* products. Even if its *mudharabah* product does not exist, just because the calculation of it is too complicated for the members, so that their SME members are not interested. Yet BMT is not only business-oriented but also the of Islamic economics. Then the task of making customers undertand Islamic financial products such as *mudaraba*, should be an absolute responsibility. And so with Islamic banks. So the substance is how they both agree in running the Islamic financial system in *kaffah* accordance with the context of welfare.

4 CONCLUSION

BMT's decision not to continue the linkage was not due to the linkage program being incriminating for BMT. The *mudharabah* system that does not comply with the contract caused the BMT to discontinue it. The BMT party turns out in terms of high-profit sharing, but still has not run the substance of sharia in *mudaraba* cooperation. At the moment they want to pay a low amount, but want a high result. Furthermore, the application of the Islamic banks system is also different when the margin or profit sharing was not variable to the income created. Thus, it turns out the understanding and practice of a profit-sharing system in accordance with the principles of *ta'awun* is still not implemented properly. So, the substance is how they both agree in implementing the sharia system in *kaffah*.

REFERENCES

Al-Qur'an. 2009. *Departemen Agama Republik Indonesia, Syamil Al-Qur'an The Miracle 15 in 1*. Bandung: PT Sygma Exemedia Arkanleema.

Oxtora, R. 2015. Komisi XI DPR Dukung Program MPS (mobile payment system) BI. [Online]. Retrieved: http://www.antaranews.com/berita/484838/komisi-xi-dpr-dukung-program-mps-bi. Acessed 30 Maret 2015.

Ascarya. 2008. *Redefinisi Klasifikasi Kredit UMKM*. Republika newspaper, 12 Mei 2008.

Bank Indonesia. 2011. *Indonesia: Credit ceiling for linkage program reaches Rp 6.4 trillion*. Bangkok: Thai News Service Group.

Bank Indonesia. 2012. *Outlook Perbankan Syariah Tahun 2013*. Jakarta: Direktorat Perbankan Syariah Indonesia —Bank Indonesia.

Bank Indonesia. 2014. *Booklet Keuangan Inklusif*. Jakarta: Departemen Pengembangan Akses Keuangan dan UMKM.

Berg, Bruce L. 2007. *Qualitative Research Methods for the Social Sciences*. Boston: Pearson Inc.

Berger, A.N. Hasan, I. & Klapper, L.F. (2004). Further evidence on the link between finance and growth: An international analysis of community banking and economic performance. *Journal of Financial Services Research*, 25(2–3): 169–202.

Bimaprawira, C. (2009). Linkage program dan kinerja bank perkreditan rakyat. *Doctoral dissertation, Universitas Airlangga*.

Challen, R. 2000. *Institutions, transaction costs, and environmental policy: institutional reform for water resources*. UK: Edward Elgar Publishing.

Herdiansyah, H. 2010. *Metodologi Penelitian Kualitatif Untuk Ilmu-Ilmu Sosial*. Jakarta: Salemba Humanika.

Kumara, R. 2010. Analisis Uji Beda Kinerja BPR Yangg Mengikuti Linkage Program Dengan BPR Yang Tidak Mengikuti Linkage Program Pada Wilayah DPC Depok.

Sanrego, Y. D. & Yulizar, D. 2008. Redefine Micro, Small and Medium Enterprises Classification and Potency of Baitul Maal wa Tamwiel as Intermediary Institutions in Indonesia. *Enhancing Islamic Financial Service for Micro and Medium Sized Enterprises (MMES)*, 2.

Sugiyono. 2014. *Metode Penelitian Kuantitatif, Kualitatif, dan R&D*. Bandung: Penerbit Alfabeta.

Advances in Business, Management and Entrepreneurship – Hurriyati et al (eds)
© 2020 Taylor & Francis Group, London, ISBN 978-0-367-27176-3

Information asymmetry in capital market: What, why and how

P.T. Komalasari & M. Nasih
Universitas Airlangga, Surabaya, Indonesia

ABSTRACT: Information is a valuable commodity and resource to maximize the utility of economic agents. Information asymmetry occurs when there is an information gap among economic actors. It has received considerable attention in both accounting and finance literature. Basically, information asymmetry is not directly observable and therefore researchers use proxy variables. However, so far there are no studies classifying the proxies of information asymmetry based on its characteristics. Furthermore, market characteristics are often overlooked in choosing a proxy of information asymmetry. This paper attempts to partially address this gap in literature by classifying proxies of information asymmetry and reviewing proxies that are more appropriate to be used in emerging market. This paper also discusses how information asymmetry takes place in capital market and why it is important in emerging capital market research. We pay more attention to emerging markets because information asymmetry is presumed resulting capital market collapse.

1 INTRODUCTION

Information can be viewed as a commodity which is sought to maximize the utility of economic agents (Allen 1990), and becomes a valuable resource (Stigler 1961). Implicitly, Grossman & Stiglitz (1976) say that individuals do not need to waste their effort to seek information because it is freely available by observing market prices. However, the incentive for seeking information is still high when no one else can access it, so that traders get benefit from expensive information acquisitions (Allen 1990). This condition implies that even though information might be public goods, but the information asymmetry amongst market participants may remain high.

The concept of information asymmetry continues to grow and becomes one of factors to be considered in decision analysis. Researchers have also used information asymmetry as a vocal construct in their research, both as antecedent and a consequence. However, basically information asymmetry is very difficult to measure, so that the majority of empirical research uses proxies to measure it. Moreover, prior empirical studies largely did not take into account market characteristics in the selection of information asymmetry proxies. Developed markets and emerging markets have different characteristics influencing the trading mechanisms in capital market. This is certainly has an impact on the selection of asymmetric information proxies, especially when using market data or market microstructure. This paper attempts to partially address this gap in literature by classifying proxies of information asymmetry and reviewing proxies that are more appropriate to be used in emerging market. It also discusses how information asymmetry takes place in the capital market and why it is important in emerging capital market research.

2 DEFINING INFORMATION ASYMMETRY

The theory concerning information asymmetry is originally developed by George Akerlof, Michael Spence and Joseph Stiglitz. Initially, Akerlof (1970) develops the concept of information asymmetry by describing that car buyers have different information concerning used cars from sellers, and therefore they encourage sellers to sell it below the market prices.

While Akerlof (1970) describes information asymmetry in the context of market for used cars, Michael Spence in 1973 identified information asymmetry between employers and employees. Employers were not convinced of productive capabilities of their employees. It triggers lowpaying job scenarios that ultimately lead to inhibition of wage bargaining mechanisms in the labor market.

The results of Stiglitz's thought have brought the theory of information asymmetry increasingly recognized. By using the theory of market screening, Stiglitz develops theory of information asymmetry in insurance market. Both Spence (1973) and Stiglitz describe that information asymmetry occurs when one party has more information than others. This definition is consistent with Akerloff's (1970).

Information asymmetry has been discussed extensively in the context of agency theory. There is no problem arises in agency relationship when information between principal and agent was symmetrical, that is principal knows all that is known by agent. However, when information is asymmetricthat is principal (agent) controls the private information which is unknown by agent (principal) agency conflict will rise. In other words, information asymmetry occurs when there is an imbalance of in-formation among participants.

3 THE IMPORTANCE OF INFORMATION ASYMMETRY IN CAPITAL MARKET

The importance of information asymmetry in the capital market is supported by some analytical and empirical research results. Some research suggests that information-based transactions in the financial market impacts on firm's cost of capital (e.g. Diamond & Verrechia 1991, Easley et al. 2004, Komalasari & Baridwan 2001, Hughes et al. 2007, Armstrong et al. 2011), underpricing IPO (Hoque 2014), lockup periods (Yung & Zender 2010) and asset prices (e.g. Kelly & Ljungqvist 2012).

Information asymmetry is not only associated with company's market performance, but it also influences market stability. Market collapse occurs because outsiders refuse to trade with insiders due to of high asymmetric information (Bhattacharya & Spiegel 1991). Insiders have more opportunities for hedging than outsiders, and this causes losses for outsiders. Fernando et al. (2008) also state that market collapse may occur because there is substantial information asymmetry concerning market fundamentals.

Allen et al. (1993) suggest that bubbles in financial markets happen when information in the market is heterogeneous. It means that every trader has private information. Bursting bubbles that occur due to information asymmetry can cause market collapse.

4 MEASUREMENT ISSUES IN INFORMATION ASYMMETRY

The attempts to construct taxonomy of information asymmetry measurements are based on the type of data sources. Based on the results of the empirical research review, proxies of information asymmetry can be classified into 4 classifications as follows.

- Measurement based on accounting.
- Measurements based on market microstructure.
- Measurements based on analyst forecast.
- Market-based measurements.

Measurements of information asymmetry based on market microstructure use trading mechanisms in securities market as a basis for estimating the level of information asymmetry. This trading mechanism encompasses the role of each market participant in order to buy or sell assets under a set of rules. In other words, market microstructure analyzes market structures and individual behavior in the process of price formation.

In contrast to market microstructures that emphasize on the process of price formation, measurements based on market data use the outcome of the trading mechanism process. The data used as the basis for the estimation of information asymmetry comprises of closing price of securities and trading volume.

Measurements based on analysts' financial use the data of financial analysts' forecasts to estimate the level of information asymmetry. Finally, measurements based accounting use the figures of financial statements to calculate information asymmetry.

4.1 *Measurement based on market microstructure*

Model of information asymmetry (e.g. Copeland & Galai 1983) assumes that there are three types of agents in the market, that are traders with excess information (informed traders), traders with little information (uninformed traders) and risk neutral specialist. Traders with excessive information will make transaction based on their private information that is not reflected in the stock price, and they are speculative. Traders with excess information come to the market because they have information regarding the value of assets in the future that have not been publicized, while traders with little information, or better known as liquidity traders, trade to adjust their portfolio in order to optimize their cash flow. Specialists are market participants who can act as a broker or dealer. Brokerage transaction is aimed to meet investor orders from its client, while a dealer has an authority to make transactions for himself. Specialist is assumed to have identical information with liquidity traders. In this condition, dealers face the problem of adverse selection and facing a potential loss when trading with informed traders. In order to cover losses from more informed traders, dealer should increase spread from liquid traders.

Based on this concept, asymmetric information can be measured by using market liquidity. Proxy variable often used as a basis for estimating information asymmetry is bid-ask spread.

Liquidity in the market has a variety of definitions and interpretations. The simplest definition of liquidity is the ability to conduct transactions without significant costs. The faster a transaction can be executed without a significant price reduction, we said that the asset has a high of liquidity.

Kyle (1985) brakes down liquidity into 3 components, namely tightness, depth and resilience. Tightness refers to deviation of stock prices from their efficient price, i.e. the price that should occur in equilibrium. Dealers often set bid and ask prices slightly above and below equilibrium value. Market is said to be liquid perfectly if the spread between bids and ask price is set to be zero, so the broker could buy and sell on the same price. Tightness component is often called as bid-ask spread.

The second component of liquidity is depth that represents trading volume at quoted prices. Technically, the bid-depth is a number of shares to be purchased by a specialist or a dealer at the current bidprice, while the ask-depth is a number of shares to be sold by a specialist or a dealer at the current ask-price. Based on the perspective of market liquidity, depth indicates the number of shares that can be traded with no effect on the market price.

The third component is the resilience, i.e. the speed of a price to return to efficient (equilibrium) price after any deviation or price jump. In a highly liquid market, stock price will return to the efficient level quickly after a temporary price jump that does not affect the value of stock. However, this component is very difficult to be measured because of continuous flow of information into the market, so that it is difficult to determine the speed of price jump to return to efficient price for specific information. In other words, it is very difficult to control other factors that go into the market.

Lee et al. (1993) also emphasize the importance of depth in the component of spread. Briefly, there are four dimensions of liquidity, namely trading time, tightness, depth and resilience. Trading time is the speed to execute transactions without any significant price changes. Trading time is measured by using the waiting time between one transaction and the next transaction, or the number of transactions per unit of time.

The literature of market microstructure states that there is a component of the spread that contributes to the losses borned by dealers when conduct transaction with an informed trader. These components are:

1) Order processing cost, consisting of fees that is charged by dealer for the readiness to fulfill buy and sales orders, and compensation for the time that the dealer spends to complete the transaction.
2) Inventory holding cost, i.e. costs associated with storing a number of shares to meet the demand from investors.
3) Adverse selection component, representing a reward for dealers to take a risk when dealing with informed trader. This component is closely related to the flow of information in the capital market.

Several measurements of information asymmetry by using the market microstructure theory are presented in Table 1.

Some investors have more information about the value of a security than dealer, and dealer know that these informed traders will only trade when they are in favorable position. On the other hand, dealer is also knows that they will benefit from trading with uninformed traders. Based on this argument, adverse selection is thought more appropriate to measure information asymmetry.

The market microstructure model states that dealer sets a bid-ask spread in such a way that benefit from trading with uninformed investors can offset the losses from informed investors. Therefore, the adverse selection component of spread will be greater when a dealer dealing with more informed traders, or when a dealer believes that informed trader has more accurate information. In this condition, the adverse selection component of bid ask spread reflects the risk of information asymmetry suffered by dealer. Thus, when dealers conduct transaction with informed traders, the transaction costs will increase, and this information asymmetry condition will lead to larger bid-ask spread.

Based on these arguments, Komalasari & Baridwan (2001) use absolute residual error of the expected spread function as a proxy for adverse selection. The higher the absolute residual error indicates the higher of information asymmetry between informed trader and dealer.

4.2 Market-based measurements

Another proxy used to measure information asymmetry is measurements based on market data. These market-based proxies rely on returns and trading volumes as a basis for estimating the level of information asymmetry. Table 2 presents proxies of asymmetric information based on market data.

4.3 Measurement based on analyst forecast

The third proxy that is also often used to measure information asymmetry is measurements-based analyst

Table 1. Proxies of information asymmetry based on market microstructure.

Proxy Variables	Initiator	Theory
Relative Spread	Bagehot (1971)	Dealers will widen spreads considerably when dealing with more informed traders that have superior information.
Adverse Selection Component of Spread	Glosten & Harris (1988), Komalasari & Baridwan (2001), Komalasari (2016)	Information asymmetry in the capital market is shown by the adverse selection problem faced by market makers (dealers) when dealing with informed investors. Adverse selection problem is refracted in adverse selection component of bid-ask spread. The high of adverse selection indicates increasing in information asymmetry.
Probability of Informed Trading (PIN)	Easley et al. (1996)	The PIN measures the availability of private information in the market by observing the amount of securities that be sold, purchased, and quoted. Abnormal buys or sells indicate the existence of information asymmetry.

Table 2. Proxies of information asymmetry based on market data.

Proxy Variables	Initiator	Theory
Illiquidity	Amihud (2002)	Illiquidity ratio increases when there is a huge price change but small trading volume changes. The higher ILLIQ indicates that stock liquidity is low and the information asymmetry is higher.
Volatility in abnormal return	Dierkens (1991)	The high positive or negative market reaction around the date of an announcement indicates that there is high information asymmetry for the company.
Residual volatility in daily stock return	Bhagat et al. (1985), Blackwell et al. (1990)	Investors and managers have the same information about the economy-wide factors that affect company's value. Investors face uncertainty about firm-specific information until such information is announced publicly. Information asymmetry of a company is high when managers have value-relevant, firm-specific information that is not distributed to the market.
Trading volume	Glosten & Milgrom (1985), Karpoff (1986)	Increasing in information asymmetry causes trading volume to be lower as uninformed investors reduce their trading in the securities.

Table 3. Proxies of information asymmetry based on analyst forecast.

Proxy Variables	Initiator	Theory
Analyst Forecast Error	Elton et al. (1984)	The higher the forecast error, higher the information asymmetry that occurs between manager and outside investor.
Standard Deviation of Forecast	Krishnaswami & Subramaniam (1999)	The dispersion among analysts is an indication of lack of available information about a firm. The higher the standard deviation of this forecast the higher the information asymmetry between managers and outside investors.
Normalized Forecast Error	Krishnaswami & Subramaniam (1999)	

forecast. Assumption underlying the use of these measurements is that security analysts always try to extract information about securities that have a good prospect. The following is presents proxy of asimmetry information using analyst forecast:

4.4 Measurement based on accounting

The fourth proxy for measuring information asymmetry is using financial statement data. The variables that are often used as a proxy for information asymmetry is presented in Table 4.

5 PROBLEMS WITH INFORMATION ASYMMETRY'S MEASURES

The researchers mostly used proxy variables based on market data and market microstructure to estimate the level of information asymmetry. The main criticism

Table 4. Proxies of information asymmetry based on accounting.

Proxy Variables	Initiator	Theory
Firm Size	Atiase (1985)	Incentives for obtaining private information are positively related to firm size. The larger size of the company, more public information is available because majority of large companies are followed by more financial analysts. Thus, the larger the size of the firm indicates a low level of information asymmetry.
Accruals quality	Lee & Masulis (2009)	The financial statement is a major source of information for investors to analyze firm's performance. The quality of financial statement relates directly to investors' uncertainty regarding corporate health and past performance. Conversely, the quality of financial statement will not affect the uncertainty that faced by managers because they have better internal information. Therefore, increasing uncertainty that is reflected in quality of information indicates high information asymmetry between managers and investor

of both proxies is that it does not reflect the magnitude of information gap between insider and outsider companies. Market microstructure emphasizes the role of dealers-broker and traders by assuming there are several traders who have private information about a firm. Traders may consist of an institutional owner, a corporate manager, or an institutional investor who has access to more accurate data and have sophisticated techniques to analyze firm performance. Variable PIN assumes that the market structure has a risk-neutral market maker. This condition is not the case in emerging market.

Likewise for market-based measurements, trading volume variable has a weakness. It suffers from an errors in variables problems such as the impact of earnings surprises (see Beaver 1968, Bamber 1987). Bartov & Bodnar (1996) tried to control it by taking residual values from regression function of trading volume variable based upon annual share turnover on earnings changes.

The use of analyst forecast to estimate information asymmetry also doesn't mean free from measurement error. One disadvantage of this proxy is the presence of optimism bias (O'Brien 1988), and it may correlate with the riskiness of firm (Krishnaswami & Subramaniam 1999) because of higher volatility of earnings.

6 MEASUREMENT OF INFORMATION ASYMMETRY IN EMERGING MARKET

Emerging markets are characterized by non-synchronous trading, i.e. securities traded passively, so there are trading days that contains zero trading volume. Market where shares are rarely traded is also referred to as thin market. The lower number of transactions over a given period of time leads to insufficient to ensure efficient price discovery (Anderson et al. 2007).

In addition, the difference between emerging markets and developed markets is that most emerging markets do not have a complete market structure such as developed markets. Most emerging markets do not have specialists in their capital market structure. In addition, there are no dealers but brokers, so the use of bid-ask offers can be misleading because there is no data on the dealer's bid-ask spread but the market spread. Spread markets are generally lower than dealer bid-ask spreads. Lesmond (2005) provides an alternative measurement of information asymmetry based on market microstructure by using firmlevel of bid-ask spread as basis for measuring liquidity.

Under these conditions, the use of information asymmetry proxies based on market microstructure becomes less accurate. The remaining alternative is a proxy based on market data, forecast analysts, and accounting. Researcher can obtain market data easily because it is available for public. However, the existence of zero transactions causes the market price to not reflect the available information. Kang &

Zhang (2014) provide alternative solutions for measuring liquidity in emerging markets by combining Amihud ratios with size of non-trading-frequency.

Analyst forecast is also an alternative, which can be used to measure information asymmetry in emerging market by controlling for optimism bias. Unfortunately, the number of companies followed by analyst securities is more limited in emerging markets compared to developed markets. Finally, accounting-based measures are a proxy that can be applied in emerging markets and more closer to the concept of information asymmetry.

7 CONCLUSION

Information asymmetry is a concept that is widely used in various fields, including finance. Empirical studies have documented the importance of information asymmetry in investment decision making. Unfortunately, it is not directly observable and measurable, so empirical research uses variety of proxies to estimate it. Based on the sources of data, proxy of information asymmetry is divided into four, i.e. based on market microstructure, market data, analyst forecast, and accounting-based proxies. Selection of proxy variables is also influenced by market characteristics. Market-based measurement is recommended for measure information asymmetry in emerging market after making adjustment to zero trading volume. Measurement based on accounting is another proxy variable that can be used in emerging market because financial statements is available for public and conceptually more relevant to measure information asymmetry.

The reliability of information asymmetry's proxy should be examined through variety of empirical researches in order to find the most stable and consistent proxies used in developed markets and emerging markets. Further researches should develop new proxy variables given the results of empirical research have not shown consistent measure yet in various contexts.

REFERENCES

Allen, B. 1990. Information as an economic commodity. *American Economic Review* 80 (2): 268-273.

Allen, F., S. Morris, M. & A, Postlewaite. 1993. Finite bubbles with short sale constraints and asymmetric information. *Journal of Economic Theory* 61: 206-229.

Akerlof, G.A. 1970. The market for 'Lemons': Quality uncertainty and the market mechanism. *Quarterly Journal of Economics* 84(3): 488–500.

Amihud, Y. 2002. Illiquidity and stock returns: Cross-section and time series effects. *Journal of Financial Markets* 5: 31-56.

Anderson, J.D., Darre, H., Ardia, H. & Stev, T. 2007. A new taxonomy of thin markets. *Working Paper. Selected Paper present-ed at the Southern Agricultural Economics Asso-ciation*, Annual Meeting.

Armstrong, C.S., John E.C., Daniel J.T. & Robert E.V. 2011. When does information asymmetry affect the cost of capital. *Journal of Accounting Research* 49(1): 1-40.

Atiase, R.K. 1985. Predisclosure information, firm capitalization, and security price behavior around earnings announcements. *Journal of Accounting Research* 23(1): 21-36.

Bagehot, W. 1971. The only game in town. *Financial Analysts Journal* 22: 12-14.

Bamber, L.S. 1987. Unexpected earnings, firm size, and trading volume around quarterly earnings announcements. *Accounting Review* 62: 510-532.

Bartov, E. & Bodnar, G.M. 1996. alternative accounting methods, information asymmetry and liquidity: Theory and evidence. *Accounting Review* 71 (3): 397-418.

Beaver, W.H. 1968. The information content of annual earnings announcements. *Journal of Accounting Research* 6 (Supplement): 67-92.

Bhagat, S.W., Marr, M. & Thompson, R. 1985. The rule 415 experiment: Equity markets. *Journal of Finance* 40: 1385-1401.

Bhattacharya, U. & Spiegel, M. 1991. Insiders, outsiders and market breakdowns. *Review of Financial Studies* 4: 255-282.

Blackwell, D.W., Marr, M. & Spivey, M. 1990. Shelf registration and the reduced due diligence argument: Implications of the underwriter certification and the implicit insurance hypothesis. *Journal of Financial and Quantitative Analysis* 25: 245-259.

Copeland, T.E. & Galai, D. 1983. Information effects on the bid-ask spread. *Journal of Finance* 38 (5): 1457-1469.

Diamond, D. & R Verrecchia. 1991. Disclosure, liquidity and the cost of capital. *Journal of Finance* 46 (4): 1325245-1325359.

Dierkens, N. 1991. Information asymmetry and equity issues. *Journal of Financial and Quantitative Analysis* 27: 181-199.

Easley, S., Hvidkjaer, H. & O'Hara, M. 2004. Information and the cost of capital. *Journal of Finance* 59 (4): 1553-1583.

Easley, D.N., Kiefer, K., M, O´Hara & Paperman, J.B. 1996. Liquidity, information, and infrequently traded stocks. *Journal of Finance* 51: 1405-1436.

Elton, E., Martin, G. & Mustafa, N.G. 1984. Professional expectations: Accuracy and diagnosis of errors. *Journal of Financial and Quantitative Analysis* 19(4): 351-363.

Engle, R. & Lange, J. 1997. Measuring, forecasting and explaining time varying liquidity in the stock market. *Working Paper.* University of California, San Diego.

Fernando, C.S. Herring, R.J. & Subrahmanyam, A. 2008. Common liquidity shocks and market collapse: Lessons from the market for perps. *Journal of Banking and Finance* 32: 1625-1635.

Glosten L.R. & Harris, L.E. 1988. Estimation the components of the bid/ask spread. *Journal of Financial Economics* 21: 123-142.

Glosten, L.R. & Milgrom, P.R. 1985. Bid, ask and transaction prices in a specialist market with heterogeneously informed traders. *Journal of Financial Economics* 14: 71-100.

Grossman, S.J. & Joseph, E.S. 1976. Information and competitive price systems. *American Economic Review Proceedings* 66: 246-253.

Hoque, H. 2014. Role of asymmetric information and moral hazard on IPO underpricing and lockup. *Journal of International Financial Markets, Institutions & Money* 30: 81-105.

Hughes, J.S., Liu, J. & Liu, J. 2007. Information asymmetry, diversification, and cost of capital. *Accounting Review* 82 (3): 705-729.

Kang, W. & Zhang, H. 2014. Measuring liquidity in emerging markets. *Pacific-Basin Finance Journal* 27: 49-71.

Karpoff, J. 1986. A theory of trading volume. *Journal of Finance* 41: 1069-1087.

Kelly, B. & Ljungqvist, A. 2012. Testing asymmetric-information asset pricing models. *Review of Financial Studies* 25 (5): 1366-1413.

Komalasari, P.T. & Baridwan, Z. 2001. Asimetri informasi dan cost of equity capital. *Jurnal Riset Akuntansi Indonesia* 4(1): 64-81.

Komalasari, P.T. 2016. Information asymmetry and herding behavior. *Jurnal Akuntansi dan Keuangan Indonesia* 13(1): 70-85.

Krishnaswami, S. & Subramaniam, V. 1999. Information Asymmetry, valuation, and the corporate spin-off decision. *Journal of Financial Economics* 53: 73-112.

Kyle, A.S. 1985. Continuous auction and insider trading. *Econometrica* 53(6): 1315-1336.

Lee, C.M., Mucklow, C.B. & Ready, M.J. 1993. Spreads, depths, and the impact of earnings information: An intraday analysis. *Review of Financial Studies* 6: 345-374.

Lee, G. & Masulis, R.W. 2009. Seasoned equity offerings: Quality of accounting information and expected floation costs. *Journal of Financial Economics* 92: 443-469.

Lesmond, D.A. 2005. Liquidity of emerging markets. *Journal of Financial Economics* 77: 411-452.

O'Brien, P. 1988. Analysts' forecasts and earnings expectations. *Journal of Accounting and Economics* 10: 53-83.

Spence, M. 1973. Job market signaling. *Quarterly Journal of Economics* 8 (3): 355–374.

Stigler, G.J. 1961. The economics of information. *Journal of Political Economy* 69(3):. 213-225.

Stoll, H. 1989. Inferring the components of the bid-ask spread: Theory and empirical tests. *Journal of Finance* 44: 115-134.

Yung, C. & Zender, J.F. 2010. Moral hazard, asymmetric information and IPO lockups. *Journal of Corporate Finance* 16: 320-332.

Advances in Business, Management and Entrepreneurship – Hurriyati et al (eds)
© *2020 Taylor & Francis Group, London, ISBN 978-0-367-27176-3*

Effect of the return on investment towards fixed assets investment at Rapih Metalindo Corporation

N.H. Yuris, F.M. Kurnia & T. Yuniarsih
Universitas Pendidikan Indonesia, Bandung, Indonesia

ABSTRACT: The purpose of this research was to understand the effect of the return on investment (ROI) towards fixed assets investment at the Rapih Metalindo Corporation. This research used a descriptive method with a quantitative approach through observation and literature study as the data collection technique. The sample taken was derived from the balance sheet and report income statements from the 2008–2012 period. The technique used to test the normality data analysis, correlation coefficient analysis of Pearson product moment, a coefficient of determination, regression, and hypotheses was through SPSS. Based on the data analysis, the results indicated that there was a close correlation between the return on investment and the fixed assets investment—as much as 0.915 in the prologue. In addition, the return on investment funds positively and significantly influenced the fixed assets investment by 83.7%. However, the remaining 16.3% was influenced by other factors that were not examined in this research, such as cash ratio, return on equity, total assets turnover, and debt to assets ratio. Companies should increase the volume of sales in order to boost profits and as decision making tool in investing, especially in fixed assets investment. This is because the total assets which continue to increase in the current assets, are not comparable to the fixed assets and net income earned each year as inventory is accumulated by innovating products such as making safes, scissors, stairs, fences, and so on, to increase profits.

1 INTRODUCTION

In conducting its operations, every company must be able to compete in order to increase profits and avoid the risk of bankruptcy. Companies are required to improve their activities and increase resources so that the company's performance can be improved that can be seen from the profit increase (Phillips 2003).

Every year, the company's financial position will continue to change according to the company's operations as well as the used assets, especially the investment in fixed assets, in which its amount and value are basically increasing from year to year (Russ-Eft & Preskill 2005).

Fixed assets are the tangible assets obtained in the form of ready-made or to be built, which are firstly used for the operation of the company and not intended for sale in the framework of the normal activities of the company that have a useful life for more than one year (Ikatan Akuntansi Indonesia 2002). It is intended to enhance the company's overall performance. Romero (2011), however, does not rule out the number and value decreased due to unfavorable corporate activity or other unfavorable conditions, such as the less conducive condition of the country's economy.

Kasmir (2008) states that the result of the return on investment (ROI) is one indicator of the profitability ratios that seeks to measure a company's ability to use all of its assets to the company's operations in generating profits. In addition, Priatinah and Kusuma (2012) state that the analysis of return on investment in financial statement analysis is important as one of the analytical techniques to measure the performance of companies.

In line with the development of a nation, manufacturing companies—both private and public—are expected to strive to improve operations in order to be able to increase the growth of the manufacturing industry (The Central Bureau of Statistics 2012). Table 1 shows the growth of manufacturing production from 2010 to 2012.

As is shown in Table 1, the growth of large and medium manufacturing industry in the first quarter of 2010 decreased by 1.59% from the fourth quarter in 2009, and the second quarter in 2010 increased by 4.00% from the first quarter in 2010. The growth of the large and medium-sized manufacturing industry in the third quarter fell by 4.13% and in the fourth quarter rose by 3.04%. The growth of large and medium-sized manufacturing industries in the first, second, and third quarters experienced an increase by 0.75%, 3.09%, and 0.52%, respectively, from the previous quarter. In contrast, the growth in the fourth quarter went down by 1.53%. In 2012, growth of large and medium manufacturing industry in the first quarter fell by 0.31 from the fourth quarter in 2011 and the growth in the second quarter of 2012 rose by 3.94% from the first quarter in 2012.

The data shows that the manufacturing industry in Indonesia has gone down from the previous period.

Increasing competition in the business is one of the factors affecting the growth of large and medium manufacturing industry.

In addition to some of the things that have been described previously, another concern regarding fixed assets is an investment that absorbs the largest portion of the capital invested in the company. This is a necessity for the production process in the enterprise, because without such assets it will be impossible to process.

Rapih Metalindo Corporation is a privately owned enterprise whose activities include the manufacturing industry, which is engaged in the manufacture of products made of metal located at Jl. Ciroyom 302/77 Bandung. The following are the acquisition of fixed assets at Rapih Metalindo Corporation during the period 2008 – 2012.

The acquisition of fixed assets in 2008 decreased by IDR (Indonesian Rupiah) 42,118,000, in 2009 decreased by IDR 28,851,000, in 2010 decreased by IDR 14,412,000, in 2011 decreased by IDR 12,951,500, in 2012 decreased by IDR 11,725,000. The problems that occurred at Rapih Metalindo Corporation is a fixed asset that always decreases every year.

The ratio indicators used in the analysis of Rapih Metalindo Corporation during the period 2008–2012 was the ratio of return on investment. Every year, the financial position of Rapih Metalindo Corporation changed according to the company's operations seen in profit as well as the book value of its fixed assets which decreased from the previous year. Apart from these problems, this study was conducted based on the previous research regarding financial performance, return on investment, and fixed asset investment (Karnawiredja et al. 2013). Research conducted by Simanjuntak, Akuntansi, and Lampung, (2011) indicated that "the Return on Investment (ROI) and Total Assets Turnover (TATO) variables simultaneously showed a significant effect on the investment of fixed assets." In addition, Darminto (2007) revealed, "The result of F-test or ANOVA showed that current ratio, long-term debt to equity, return on assets, fixed asset turnover and inventory turnover had a significant influence on investment in fixed assets." Lily (2005) also stated that "variable rate assets investment, the composition of the funding, and management of assets simultaneously showed significant effect on the level of 1% on the financial performance of the path coefficients (path) 0.327 to variable ratio of the company's financial performance." Results of research conducted by Manullang (2005) showed that there was a significant difference between the financial performance based on ROI and TATO on investment in fixed assets amounted to 79.3%. The decisions regarding investment in fixed assets had either a good or bad influence on the company profitability (ROI). Javitz et al. (2004) suggested that "employers can receive competitive returns on investment from sponsoring smoking cessation programs that 150 mg of bupropion doses yield better returns than 300-mg

doses, and that PTC treatments should be preferred to TM if smoking cessation rates in the targeted employee population are lower than those in the study population." Another study conducted by Goetzel et al. (2005) showed that "[t]he results of 44 studies investigating financial impact and return on investment (ROI) from disease management (DM) programs for asthma, congestive heart failure (CHF), diabetes, depression, and multiple illnesses were examined. A positive ROI was found for programs directed at CHF and multiple disease conditions." Jones et al. (2017) found a positive ROI after the large storm events when fire mitigation treatments were placed in priority areas with diminishing marginal returns after treating > 50–80% of the forested area.

Based on the description of the problem, then it is necessary research the "Effect of The Return on Investment towards Fixed Assets Investment at Rapih Metalindo Corporation."

Based on these problems, the purposes of this study are to find out:

1. The condition of return on investment Rapih Metalindo Corporation in the period 2008–2012.
2. Fixed asset investment in Rapih Metalindo Corporation in the period 2008–2012 during the running operation of the company.
3. The magnitude of the effect of the return on investment on fixed asset investment in the Rapih Metalindo Corporation in the period 2008–2012.

2 METHODS

Sugiyono (2012) states that quantitative research is based on the philosophy of positivism. Quantitative research consists of descriptive and associative problem formulations. Descriptive problem formulation is a formulation of the problem with regard to the question of the existence of independent variables, both just at one or more variables.

2.1 Research instruments

The independent variable is the ROI, while the dependent variable is the fixed asset investment. The figure of the framework can be displayed as follows:

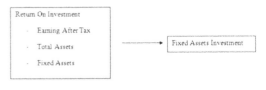

Figure 1. Conceptual model and research hypotheses.

2.2 Operational variable

Table 1. Operational variable.

No	Variable	Definition Variable	Indicator	Scale
1	Return on Investment (X1)	Return on Investment is the ratio that shows the results of the amount of assets used in the company	EAT, Total Asset, Fixed Assets	Ratio
2	Fixed Asset Investment (Y)	Investment in fixed asset is a form of investment with the hope that the company can generate profits through its operations	Book Value of fixed assets	Ratio

2.3 Population and sample

The population in this study was the financial statements of Rapih Metalindo Corporation consisting of balance sheet and profit/loss during the period 2008–2012. The samples were Rapih Metalindo Corporation financial statements, consisting of the balance sheet and profit/loss during the 2008–2012 periods.

3 RESEARCH AND DISCUSSION

As is shown in Table 2, the return on investment experienced by Rapih Metalindo Corporation in 2008 amounted to 6.47%. In the year 2009, it decreased by 0.67% to 5.80%. In 2010, it decreased by 0.54% to 5.26%. In 2011, it decreased by 0.50% to 4.76%. In 2012, there was an increase by 0.31% to 5.07%. During the period 2008–2012, the average ROI was equal to 5.47% with the highest value which in 2008 amounted to 6.47% and its lowest value which in 2011 amounted to 4.76%. During the period 2008–2012, it stayed below the industry standard; therefore, it is in the poor ratings category.

Ratings were given by comparing the median value of the percentage of the book value of fixed assets by 17%. In 2008, 2009 and 2012, it received a good rating. In 2010–2011, it received unfavorable

rating because it was below the median percentage value. On average, in the years 2008–2012, it received unfavorable ratings because it was below the median percentage value. The percentage of the book value of the largest occurred in 2008 by 25% and the lowest is in 2010 by 15%, as is shown in Table 3.

The results from a Pearson Product Moment Correlation Analysis showed significant value < 0.05 or $0.029 < 0.05$, meaning that there is a very strong relationship, significant and direction between ROI and IAT at 0.915, as is shown in Table 4.

Based on Table 5, the coefficient of determination (R^2 or R Square) is 0.837 or 83.7% which means that ROI affect the IAT. This suggests that the effect of ROI on is by 83.7%, and the remaining 16.3% is influenced by other factors not examined, such as the ratio of cash, TATO, and the accounts receivable turnover ratio.

A simple linear regression equation is:

$$Y = a + b. x$$
$$IAT = -0.142 + 6.102. ROI$$

With the regression equation above means constant IAT value. If there is no increase for the ROI, the IAT value will experience shrinkage by -0.142. The value of 6.102 is the coefficient of ROI. ROI will get lower once the IAT decreases by 6.102 as show in Table 6.

With the t value of 3.930, while t table with $\alpha = 0.05$ and n = 5, the two-party test, df = 5–2 = 3, the t table obtained is 3.182. It turned out that it is greater than t-table or 3,930> 3,182. Hence, Ho was refused and Ha was accepted which means that there was an influence of ROI on the IAT.

The results of the analysis showed that the average percentage value of return on investment was 5.47%, in which the highest value was in 2008 amounted to 6.47% and the lowest value was 4.76%. Only in 2012, it experienced a significant increase by 0.31%. In 2008–2012, it was far below the BI (Central Bank of Indonesia) rate industry standard, so it received a poor rating category. This showed that the average percentage value of investment in the fixed assets sector was 14%, including in the unfavorable category because it was below the median percentage value, with the highest value was in 2008 by 25% and the lowest value in 2010 was 15%. In 2008, 2009, and 2012, it received a good rating category. In contrast, during the

Table 2. Return on investment, Rapih Metalindo Corporation.

No	Year	Total Asset (IDR)	EAIT (IDR)	Score (%)	Up/Down (%)	BI Rate (%)	Assesment
1	2008	319,966,383	20,702,803	6.47	-	9.25	Bad
2	2009	361,792,263	20,981,275	5.80	-0.67	6.50	Bad
3	2010	402,013,310	21,133,027	5.26	-0.54	6.50	Bad
4	2011	447,560,744	21,289,568	4.76	-0,50	6.00	Bad
5	2012	458,600,310	23,230,967	5,07	0,31	5.75	Bad
Average		458,600,310	21,467,528	5.47		6.80	Bad

Table 3. Return on investment Rapih Metalindo Corporation.

No	Year	Book Value of Fixed Asset (IDR)	Up/down	%	Assesment
1	2007	165,470,000			
2	2008	123,352,000	42,118,000	25%	Good
3	2009	94,501,000	28,851,000	23%	Good
4	2010	80,089,000	14,412,000	15%	Good
5	2011	67,137,500	12,951,500	16%	Bad
6	2012	55,412,500	11,725,000	17%	Bad
Average		97,660,333	22,011,500	14%	Bad
Median		87,295,000	14,412,000	17%	
Max		165,470,000	42,118,000	25%	
Min		55,412,333	11,725,000	15%	

Table 4. Pearson product moment correlation analysis.

		ROI	IAT
ROI	Pearson Correlation	1	.915
	Sig. (2-tailed)		.029
	N	5	5
IAT	Pearson Correlation	.915	
	Sig. (2-tailed)	.029	
	N	5	5

Table 5. Coefficient of determination.

Model	R	R Square	Adjusted R Square	Std. Error of the estimate
1	.915*	.837	.783	.02093

* Predictors: (Constant). ROI
Dependent Variable: IAT

Table 6. Coefficients.

	Unstandardized Coefficient		Standardized Coefficient		
Model	B	Std. Error	Beta	t	Sig.
(Constant)	−.142	.085		−1.660	.195
ROI	6.102	1.553	.915	3.930	.029

* Dependent variable: IAT

2008–2012 period, there was no increase in fixed asset investment. The results of hypothesis testing showed that return on investment had a strong, significant and unidirectional effect on fixed asset investment by 83.7 %, with the regression equation IAT = -0.142 + 6.102 ROI.

The problems occurred in return on investment at Rapih Metalindo Corporation was the total assets that continue to increase in current assets were not comparable to fixed assets and net income earned each year. Whereas the problems occurred in the investment of fixed assets at the Rapih Metalindo Corporation was the book value of fixed assets, which always decreases every year. The declining value of this book shows that the Rapih Metalindo Corporation did not take action related to fixed assets that had a long economic life during the period 2008–2012 (Ichsani & Suhardi, 2015). Efforts must be made by the Rapih Metalindo Corporation to overcome the problem of return on investment, namely, increasing sales volume and utilizing the amount of current assets such as inventory that accumulates by innovating products such as making safes, scissors, stairs, fences, and so on, so as to increase profits. The Rapih Metalindo Corporation also needs to put more effort into overcoming problems in fixed asset investments, which always decreases every year.

4 CONCLUSION

Based on the data analysis result using SPSS and the discussion in previous chapter, it can be concluded that the average percentage of the return on investment during the period 2008–2012 were far below the industry standard of BI rate, so it got unfavorable rating category. During the 2008–2012 period, it had never experienced an increase in fixed asset investment. The hypothesis testing results showed that there was a strong and significant influence on the return on investment towards fixed asset investment amounted to 83.7%, with the regression equation IAT = −0.142 + 6.102 ROI.

REFERENCES

Badan Pusat Statistik. 2012. *Perkembangan indeks produksi industri manufaktur 2010–2012.*

Darminto. 2007. Pengaruh investasi aktiva pendanaan dan pengelolahan terhadap kinerja keuangan. *Jurnal Ilmu-Ilmu Sosial, 19.*

Goetzel, R. Z., Ozminkowski, R. J., Villagra, V. G., & Duffy, J. 2005. Return on investment in disease management: A review. *Health Care Financing Review, 26*(4), 1–19.

Ichsani, S., & Suhardi, A. R. (2015). The effect of return on equity (ROE) and return on investment (ROI) on trading volume. *Procedia – Social and Behavioral Sciences, 211*, 896–902.

Ikatan Akuntansi Indonesia. 2002. *Standar akuntansi keuangan.* Jakarta: Salemba Empat.

Jack J. Phillips. 2003. *Return on investment in training and performance improvement programs* (Second). New York: Routledge.

Javitz, H. S., Swan, G. E., Zbikowski, S. M., Curry, S. J., Mcafee, T. A., Decker, D., Jack, L. M. 2004. Return on investment of different combinations of bupropion sr dose and behavioral treatment for smoking cessation in a health care setting : an employer ' s perspective. *Value in Health*, 7(5).

Jones, K. W., Cannon, J. B., Saavedra, F. A., Kampf, S. K., Addington, R. N., Cheng, A. S., Wolk, B. 2017. Return on investment from fuel treatments to reduce severe wildfire and erosion in a watershed investment program in Colorado. *Journal of Environmental Management*, 198, 66–77.

Karnawiredja, R. A. M., Hidayat, L., & Effendy, M. 2013. Pengaruh kinerja keuangan terhadap investasi aktiva tetap, 1(3), 263–272.

Kasmir. 2008. *Analisis laporan keuangan*. Jakarta: Rajawali Pers.

Lily. 2005. *Pengaruh kinerja keuangan berdasarkan return on investment dan total asset turnover terhadap investasi aktiva tetap*. Universitas Widiyatama.

M.Manullang 2005. *Dasar dasar manajemen*. Yogyakarta: Gadjah Manada University Press Bulaksumur.

Priatinah, D., & Kusuma, P. A. 2012. Pengaruh return on investment (ROI), earning per share (EPS), dan dividen per share (DPS) terhadap harga saham perusahaan pertambangan yang terdaftar di bursa efek indonesia (BEI) periode 2008–2010. *Jurnal Nominal*, 1(1), 15.

Romero, N. L. (2011). ROI. Measuring the social media return on investment in a library. *Bottom Line*, 24(2), 145–151.

Russ-Eft, D., & Preskill, H. (2005). In search of the holy grail: return on investment evaluation in human resource development. *Advances in Developing Human Resources*, 7(1), 71–85.

Simanjuntak, M., Akuntansi, J., & Lampung, U. (2011). *Aktiva tetap pada perusahaan food and beverages yang go publik di bursa efek*.

Sugiyono. (2012). *Metode penelitian kuantitatif kualitatif dan R&D*. Bandung: Alfabeta.

Advances in Business, Management and Entrepreneurship – Hurriyati et al (eds)
© 2020 Taylor & Francis Group, London, ISBN 978-0-367-27176-3

Understanding taxpayers' attitudes towards tax amnesty policy

I.F.A. Prawira
Universitas Pendidikan Indonesia, Bandung, Indonesia

ABSTRACT: This paper reported on a project which was designed to increase the participation of individual taxpayers on tax amnesty policy. The focus of the paper was on an Indonesian-based project which overcomes the identified barriers to offer tax amnesty policy. This research project was carried out due to the lack of taxpayers' response to take advantage of tax amnesty policy in Indonesia. This paper drew on the social psychology, taxpayers' experience, and tax accounting education literature to design a connective model of taxpayers' attitude towards tax amnesty policy for individual taxpayer. The project adopted an action research methodology which engaged professional tax consultant, tax officers, and the university sector to explain the benefits of tax amnesty policy for taxpayers. The results showed that 83% of respondents decided to join the tax amnesty program after their participation. The taxpayers felt that they had received information clearly and completely, and felt well-served, so that they practically participated in this program.

1 INTRODUCTION

Tax evasion is a global problem that occurs in all societies and economic systems in both developed and developing countries. It is more prevalent in developing countries given the rapid growth of investment in their economies and the lack of adequate experience in dealing with tax evasion (Junpath et al. 2016). Tax evasion reduces the collection of taxes, thereby affecting taxes that compliant taxpayers pay and the public services that citizens receive. Furthermore, tax evasion creates misallocations in resource use when individuals cheat on their taxes, thereby altering the distribution of income in unpredictable ways (Alm & Gomes 2008).

To address the problem of tax evasion, some countries have reformed their taxation policies and implemented tax amnesty programs over the years (Uchitelle 1989). Similarly, after the Sunset Policy in 2007 which did not give substantial revenues (Prawira 2015), Indonesia re-issued a tax amnesty policy in July 2016. This policy aims to 1) repatriate or withdraw the funds from Indonesian citizens who are outside the country; 2) boost national growth; 3) improve the national tax base, i.e the assets submitted in the application of tax amnesty can be utilized for future taxation; and 4) increase tax revenues.

Tax amnesty is a government policy to eliminate taxes payable as well as the administrative burden in the form of tax penalties and criminal sanctions for taxpayers that do not currently report their property and assets held by paying ransom in accordance with the tax laws and regulations. Taxpayers can take advantage of Tax Amnesty in the form of Enterprises Taxpayers, Personal, and Entrepreneurs who wish to report a majority of the property that have been reported through the Annual Tax Return. This program can be reserved even for those who do not

have Taxpayer Identification Number (TIN) (Yanirahmawati 2016). This policy is motivated by; 1) Moderate Global Economic Growth; 2) The US economy which has not been stable; 3) Slow Chinese Growth; 4) Uncertain Monetary Policy; 5) Declining Commodity Prices; and 6) Geopolitical risk: Middle East Country & Brexit.

A lot of studies have been done to see the taxpayer's response to the tax amnesty policy measured by the amount of tax revenue from tax amnesty. The study conducted by James et al. (2009) showed that tax amnesty had a short- or long-term impact on revenues. They concluded that the Russian amnesties, like most other amnesties, seemed unlikely to have significant and positive - or negative - impacts on the revenues of the Russian Federation, a conclusion that questioned their usefulness as a policy instrument. Junpath et al. (2016) presented the results of a survey on the attitudes of taxpayers towards tax compliance and tax amnesties in South Africa. The findings from their study indicated that taxpayers saw that the offering of multiple tax amnesties might not generate additional revenue, as non-compliant taxpayers would continue to evade taxation in anticipation of additional future amnesties. Torgler et al. (2003) analyzed the effect of multiple tax amnesties on tax compliance in Switzerland and Costa Rica. Their experimental design included a second amnesty contrary to the participants informed that no further amnesties would be offered. They found that the first amnesty had a significant positive effect on tax compliance. However, the second amnesty did not significantly increase compliance in the post-amnesty period. These findings supported the view that multiple tax amnesties should not be conducted at short intervals, since individuals anticipate future tax amnesties, thus causing non-compliance. Torgler et al. (2003) noted that the offering of a second amnesty gave the impression that the government could not be trusted, which increased the expectations of additional tax amnesties. A study regarding the Turkish tax amnesty was conducted by Saraçoğlu & Çaşkurlu (2011) who revealed that honest taxpayers perceived tax amnesty as a reward for tax delinquents. Furthermore, frequent tax amnesties complicated tax compliance, as routine tax amnesty executions affected honest taxpayers in a negative manner in which the tax amnesty was considered as a privilege for tax offenders. The frequency of the tax amnesties made it impossible to convince taxpayers that a tax amnesty is a one-time opportunity to become tax compliant, as it has become part of the legislation. Thus, the repetition of tax amnesties in Turkey made compliant taxpayers believed that they were disadvantaged because their honesty caused these groups of taxpayers not to fulfil their tax obligations (Saraçoğlu & Çaşkurlu 2011). Alabede et al. (2011) found that the perceptions of taxpayers in Nigeria on the fairness of the tax system were seen as an important factor that could have a significant influence on individuals' tax compliance behavior. The above studies have yielded mixed results, confirming the comments of Christian, Gupta, and Young (2002) that it is unclear from the existing empirical research whether amnesties have achieved the goal of encouraging those who are not currently compliant with the law to become compliant in the future (Alm 1998).

The objective of this paper was to present the taxpayers' attitude of implementation of tax amnesty in West Java, as currently there is no information available on taxpayers' perceptions towards tax amnesty in Indonesia. The remainder of the paper is organized as follows. The following section provides the background of the enactment of tax amnesty policy and includes an overview of the tax system in Indonesia and a discussion on tax avoidance and tax evasion. Next, the research methodology and a discussion of the results are then presented. Finally, the conclusions, limitations and recommendations for further research are presented.

1.1 Tax avoidance and tax evasion

The most interesting feature of the modern system of income tax is that it is fundamentally voluntary in nature. Individual taxpayers always face some chances of being audited by the revenue authorities. However, most taxpayers' actual assessment depends on the income chosen to report (Junpath et al. 2016). This may not necessarily be the same as their actual income (Reinganum & Wilde, 1985). Feld & Frey (2002) found that tax compliance is determined by a psychological tax contact between citizens and tax authorities. Rechberger et al. (2010) add that in order to keep the contract in balance, taxpayers must pay their taxes honestly and tax authorities must recognize and support

the contract with taxpayers by treating them in a respectful manner, but also preventing "honest taxpayers from being exploited in the process" (Feld & Frey 2002).

According to Brown & Mazur (2003), tax compliance is a multi-faceted measure consisting of payment, filing and reporting compliance which requires the taxpayer to meet all his or her tax obligations in full. Feld & Frey (2007) suggest citizens to honestly declare their income even if they do not receive a full public assets equivalent to tax payments as long as the political process is perceived to be fair and legitimate. Alm & Gomez (2008) found that an individual's tax morale is significantly and positively associated with a taxpayer's perception of the benefits perceived by the society from the public delivery of goods and services.

The level of taxation, however, is also related to tax morale and income inequality, which enhances the decision to participate in the informal sector. The lower the degree of tax morale, the more unequal the distribution of income and wealth is. In addition, the larger the informal sector, the lower the tax effort is (Steenekamp 2007). Therefore, tax compliance has caused international concerns for tax authorities and public policymakers, as tax evasion seriously threatens the capacity of government to raise public revenue (Chau & Leung 2009).

In an ideal world, all taxpayers would voluntarily pay their taxes and comply with all their tax obligations (Junpath et al. 2016). At the end of each financial year, some individuals plan on how they can arrange their tax affairs strategically in order to pay as little tax as possible (Prawira 2015). This may cause some taxpayers make use of strategies which are legal in order to minimize their taxes (Murphy 2004) referred as tax avoidance.

Tax evasion and fiscal corruption have been a major problem throughout the world, with significant economic consequences (Fjeldstad 2006). Bailey (2006) states that tax planning and tax avoidance are the methods used entirely legally to reduce one's tax. However, tax evasion is an offence which always involves dishonesty in some form or another. Tax avoidance can be described as the legal utilization of the Income Tax Act and other related acts to one's own advantage to reduce or defer one's liability for tax by means that are legal and within the provisions of the law (Killian & Kolitz 2004). Tax evasion, on the other hand, can be described as an illegal, dishonest activity that usually involves taxpayers deliberately misrepresenting or hiding the true status of their tax affairs from the tax authorities in order to reduce their tax liability. This may either occur by the taxpayer not declaring income accrued or by claiming disallowed deductions against income. Tax evasion is simply a fraud against the revenue officer, for which appropriate penalties are usually provided in the tax legislation (Killian & Klitz 2004).

1.2 Indonesia tax amnesty

Tax amnesty in Indonesia has been implemented since 1964 through Presidential Decree No. 5 of 1964. It contends that for the sake of the Indonesian National Revolution and National Development Planning in general and to facilitate the implementation of the Economic Declaration of March 28, 1963 and the mobilization of all funds, power and strength in particular, tax amnesty is necessary for the capital in a society that has never been taxed. The capital in a society that has never been subject to corporate taxes, income taxes, and property taxes, registered in the Directorate of Taxes before August 17, 1965, are not used as the grounds for fiscal or criminal government agencies to conduct any inquiry, investigation, and examination of its origins. The capital at registration is subject to a one-time levy of 10 (ten) percent as a ransom of the amount of taxes, which is based on the actual fiscal policy, is owed to the State.

Furthermore, through Presidential Decree No. 26 dated April 18, 1984, the tax amnesty was granted to an Individual Taxpayer or an institution under any name and form which has or has not been registered as a Taxpayer. The tax amnesty is given to the taxes that have not been or have not been fully imposed or levied in accordance with applicable laws and regulations. The tax amnesty is subject to ransom by:

1. 1% (one percent) of the amount of wealth used as the basis for calculating the amount of taxes requested for amnesty, for the Taxpayer in which on the date of stipulation of this Presidential Decree has submitted the Notice of Income Tax/Corporate Tax in 1983 and Wealth Tax in 1984.
2. 10% (ten percent) of the amount of wealth used as the basis for calculating the amount of taxes requested for amnesty, for the Taxpayer in which on the date of stipulation of this Presidential Decree has not submitted the Notice of Income Tax/Corporate Tax in 1983 and the Wealth Tax in 1984.

However, the implementation is not effective because Taxpayers are less responsive and not followed by the reform of the tax administration system as a whole.

In 2008, Directorate General of Taxes (DGT) issued a tax amnesty policy namely Sunset Policy. This policy provides an opportunity for taxpayers to make corrections of their tax return and be exempted from fines. However, this policy is less successful in increasing tax revenues (Prawira 2012). One of the reasons is because this policy is considered to be less attractive and partial because the tax calculation payable in sunset policy does not get special rate. In addition, the one-year time frame given by the government in sunset policy is not effective enough to make its objectives achieved.

In 2016, the government re-rolled tax amnesty policy with the aim to increase tax revenue, especially in the case of repatriation of taxpayers' assets. This policy is motivated by the wealth of Indonesian citizens who are kept abroad to avoid taxes in the country so as to harm the country in a number that is not small and low public awareness in paying taxes. This policy is expected to return funds deposited abroad in the state authority and attract new taxpayers and greater tax revenues in the future by taking the risk of not receiving the tax payment as it is supposed to. On the contrary, the state only receives a ransom based on the rate set in accordance with the period of its implementation.

However, there is one risk that cannot be underestimated; i.e the unfairness for the obedient taxpayers who report taxes every year. Hence, it is expected that the tax amnesty is not offered in the future because it might cause tax avoidance as it is expected that there will be a tax amnesty again, which cause greater loss for the State.

2 METHODS

This research project adopted an action research methodology which engages professional tax consultant, tax officers, and the university sector to explain how tax amnesty policy gives many benefits for taxpayers. The action research process undertaken was iterative, involving several discrete stages, followed by a re-evaluation of the project and the re-commencement of the process. The action research plan for the project was adapted from the model used by Paisey and Paisey (2005). The action research process may be diagrammatically formulated differently (Doran et al. 2011, Kaplan 1998, Lewin 1946, Paisey & Paisey 2005). It is the systematic and iterative process of collaborative reflective evaluation which is at the heart of the action research process. This paper reported on each of the action research stage.

2.1 Defining the problem

The first action research cycle involved identifying the research problem, conducting the literature review, undertaking an initial study, and framing the research questions inherent in the falling numbers of tertiary tax amnesty involvements. Tax compliance has been defined in various ways. For example, Andreoni et al. (1998) claim that tax compliance should be defined as taxpayers' willingness to obey tax laws in order to obtain the economy equilibrium of a country. Kirchler (2007) perceives a simpler definition in which tax compliance is defined as the most neutral term to describe taxpayers' willingness to pay their taxes. A wider definition of tax compliance is described by Song and Yarbrough (1978) who suggest that due to the remarkable aspect of the operation of the tax system in the United States and that is largely based on self-assessment and voluntary compliance, tax compliance should be defined as taxpayers' ability and willingness to comply with tax laws which are determined by ethics, legal environment and other situational factors at a particular time and place. Alm (1991) and Jackson and Milliron (1986) define tax compliance as the report of all incomes and payment of all taxes by fulfilling the provisions of laws, regulations and court judgments. Furthermore, tax compliance has also been categorized into two perspectives, namely compliance in terms of administration and compliance in terms of completing (accuracy) the tax returns (Chow 2004, Harris 1989).

Attitudes represent the positive and negative evaluations that an individual holds of objects (Jayawardane & Low 2016). It is assumed that attitudes encourage individuals to act accordingly to them. Thus, a taxpayer with positive attitudes toward tax evasion is expected to be less compliant than a taxpayer with negative attitudes. Attitudes towards tax evasion are often found to be quite positive (Kierchler et al. 2008). Many studies on tax evasion found significant, but weak relationships between attitudes and self-reported tax evasion (Trivedi et al. 2004). The attitude is important for both the power and the trust dimension. On the other hand, favorable attitudes will contribute to the authorities trust and consequently will enhance voluntary tax compliance. In contrast, attitudes towards the authorities will be relevant with the interpretation of the use of power as generous or malevolent (Jayawardane & Low 2016). According to Ajezen's (1991), attitudes relates to one's own personal views about a behavior. Attitudes are generally assumed to influence compliance behavior because they represent taxpayer's propensity to respond positively or negatively to a particular situation (Eagly & Chaiken 1993, Ajzen 1993). There is a numerous ways to operationalize and measure attitudes towards taxation starting from general judgments of the government and state (Schmolders 1960), subjective assessments of tax evasion (Porcano 1988), ended with moral attitudes towards tax evasion (Orviska & Hudson 2002). Several empirical studies revealed a statistically significant link between attitudes and self-reported behavior (Chan et al. 2000, Trivedi et al. 2004). Nevertheless, because the link was weak in most of the studies, i.e., attitudes could not be fully perceived as a convincing proxy for behavior, the results have to be cautiously interpreted (Kirchler 2007).

Due to the low role of taxpayers in the tax amnesty, the purpose of issuing this policy is difficult to achieve. The taxpayers assume that the benefits of participating in the tax amnesty program are small, so they become reluctant to participate. Besides, they also assume that there will be the same program next time (Prawira 2015). To resolve this problem, an action research project involving professional tax consultant, tax officers, and the university sector to explain how tax amnesty policy gives many benefits for taxpayers.

2.2 Framing the research questions

The very nature of action research makes the research process both iterative and chronological. In such an iterative research process, findings from earlier stages were used to answer the research questions asked in subsequent stages.

The initial question that needed to be resolved was the taxpayer's perception of the benefits of this policy and how easy it is to meet the requirements.

Research question:

What do you know about this tax amnesty program?

This question was addressed to know the taxpayer knowledge about the tax amnesty program, given that the population in this study was an individual taxpayer domiciled in Bandung regency which is located far from the tax office service. Thus, from the respondents' answers, it can be identified whether the information conveyed through electronic media is effective or not, and what constrains occured if it is not effective.

2.3 Initial findings

This initial study was aimed at measuring the level of taxpayer's personal knowledge about the tax amnesty program. The study also included meetings with tax officials and tax consultants who handled individual taxpayers in the area.

The results showed that the socialization made by the tax office to taxpayers located in the area was still lacking, even some taxpayers never received a notification letter from the tax office regarding the tax amnesty program. These facts revealed a few positive responses made by taxpayers that most of the taxpayers ignored the program.

However, there were also taxpayers who tried to find the information about this tax amnesty program. Their curiosity about the program was triggered by their desire to become a responsible taxpayers and some were triggered because they were worried about the sanctions in this policy. Both triggers were caused by the lack of taxpayers' knowledge about this tax amnesty program. They just wanted their job or business run well that was not disturbed by tax affairs. These preliminary findings were conveyed and discussed to the tax office and tax consultants to identify the barriers that make it happen.

2.4 Barriers identified by tax offices

One of the barriers faced by the government was the inadequate number of tax employees. In fact, the ratio between tax officials and the number of people in Indonesia was much smaller than in other countries such as Australia and Japan. One tax officer must handle 800 taxpayers. Compare to Australia, with 18 million taxpayers and 22 thousand tax employees, in Indonesia there were only 37 thousand with a total population of 250 million people.

Besides the lack of employee taxes, the application period of the tax amnesty program was also narrow. This can be seen at the time of the issuance of the tax amnesty law and when it was enacted. Hence, it becomes a problem for the tax office to promote to the areas far from the tax office.

2.5 Barriers identified by tax consultants

In order to add information and understand the implementation and participation in a tax amnesty program, seeking assistance to a tax consultant often done by taxpayers who are financially stable and indeed look for an assistance in terms of Tax Return reporting or tax amnesty policy applied in Indonesia. Hence, tax consultants will surely know the prospects of the tax amnesty whether it will be successful or not in the future in order increase tax revenue seen from the enthusiasm of the taxpayers in the tax amnesty program.

The role of tax consultants in delivering tax amnesty programs to taxpayers seemed very low, even taxpayers felt that the tax consultants did not play their role well in informing this tax amnesty program. No one from a tax consulting firm came to their area to provide information about the program, let alone provide consultation on how to use the program. There were some taxpayers who used the services of tax consultants, and even it was because the taxpayers are asking, not because the tax consultant informed about this tax amnesty program voluntarily. It was allegedly due to a lack of coordination between the tax's office and the tax consultant's office.

2.6 An exposure program development and data collection

The program was designed by adopting the conceptual research by Sari et al. (2016), where the program can be seen in Figure 1.

This program was designed to inform the taxpayers as respondents, where previously the taxpayer has provided an answer to the question of initial study. The purpose of this program was to encourage taxpayers to be able to take advantage of the tax amnesty program issued by the tax office.

2.7 Implementation of the program

The tax officer together with tax consultants and academics provided taxpayers with knowledge on tax amnesty programs, in which short-term and long-term benefits are presented to taxpayers. In addition, technical implementation of taxpayer participation in tax amnesty also was guided so as to facilitate taxpayers in the implementation.

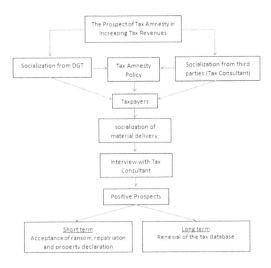

Figure 1. Conceptual frame of tax amnesty prospect.

2.8 Evaluation of the program

Following the various implementations of the program, the data were collected and analyzed to gauge the effectiveness of the intervention. The analysis focused on the capacity of the program to increase the awareness level of the taxpayers in fulfilling their tax obligations and to encourage taxpayers' interest in this tax amnesty program. The data were collected and analyzed from interviews with taxpayers, tax officers, and tax consultant involved in this program. The findings from these sources were discussed in relation to the identified research questions.

3 RESULTS AND DISCUSSION

3.1 Background information

Table 1 below shows the distribution of respondents based on their income, gender, and age. Table 1 shows that the majority of the respondents in this study were male (25 out of a total of 30 or 83.3%) and 5 respondents were female (16.7%). As a result, the database used in this study involved more males who were qualified and met the criteria to apply for amnesty. Although the majority of the respondents was male, this did not influence any parameters examined in the study.

The age group analysis of the respondents (male and female) showed that all economically active age groups were represented by the respondents. Other analyses of the respondents (not shown) revealed that most of the respondents were self-employed (25 or 86.3%). Of the remainder, 3 respondents (10%) were employees, and 2 respondents (6.7%) were retired.

The next part of the program probed the respondents' attitudes towards tax compliance by focusing on tax amnesty policy.

Table 1. Distribution of income, gender and age of respondents.

Income bracket (Rp'000.000)	Age group						Total	%
	<20	20–29	30–39	40–49	50–59	60+		
Male res- pondents								
<R60	-	1	3	9	2	-	15	60
60–120	-	-	5	2	-	-	7	28
121–180	-	-	1	1	-	-	2	8
181 – 240	-	-	-	-	1	-	1	4
241 - 280	-	-	-	-	-	-	0	0
Total male	0	1	9	12	3	0	25	
% of total	0	4	36	48	12	0		
Female respondents								
<R60	-	-	-	2	-	-	2	40
60–120	-	-	1	1	-	-	2	40
121–180	-	-	-	-	-	-	0	0
181 – 240	-	-	-	-	1	-	1	20
241 - 280	-	-	-	-	-	-	0	0
Total male	0	0	1	3	1	0	5	
% of total	0	0	20	60	20	0		
Total respondents								
<R60	-	1	3	11	2	-	17	56
60–120	-	-	6	3	-	-	9	30
121–180	-	-	1	1	-	-	2	7
181 – 240	-	-	-	-	2	-	2	7
241 - 280	-	-	-	-	-	-	0	0
Total	0	1	10	15	4	0	30	
% of total	0	3	34	50	13	0		

3.2 Research findings

Research question 2:

What is the taxpayer's perception of the benefits of this policy?

After going through the stages in this program, participants were interviewed on how they perceived the tax amnesty program. The result showed that 100% of participants had enough information about the tax amnesty policy, benefits and how to get involved in it. All of them agreed that tax amnesty policy had many benefits, such as they could report their assets list that had not been submitted in Tax Return. However, they were also constrained because the details of the assets must be submitted while the proof of ownership of the assests has not been completed.

It showed that this program had successfully overcome the lack of information received by taxpayers regarding the tax amnesty policy. Direct and directed information delivery to taxpayers will be more effective than submitting it in the form of a notification letter or invitation letter for the socialization because there will be only a few taxpayers who will respond to it.

Research question 3:

How easy is it to meet the requirements?

Several different answers revealed and among the answers are:

"The tax amnesty program is good, but we are constrained because our domicile is far from the tax office, so we have to leave our business activities for a few days. We recommend that the tax office open special services in remote areas like ours, a kind of cash office or mobile tax office to service taxpayers who want to join this tax amnesty program."

"We do not have other assets that have not been reported to Tax Return, so we have nothing to report"

"It is very easy to follow this tax amnesty policy, we are only asked to submit a list of assets that have not been reported in tax returns with fair value or market value. We do not need to ask the appraiser to conduct an asset assessment."

"It looks like I will take part in this tax amnesty program, because if not then I will be examined. It will be very troublesome for me because I have to spend a lot of time and energy on it"

These answers show a variety of taxpayer responses on how easy for them in participating in the tax amnesty program. Participants involved in this action research tended to see that it was easy and beneficial to participate in the tax amnesty program. They also saw more benefits, even though it was not easy for them to participate in this program for several different reasons they stated above. This is in accordance with the results of research conducted by Junpath et al. (2016) which showed that taxpayers who saw many benefits of this policy will use it more than taxpayers who saw fewer benefits.

Research question 4:

Table 2. Research question 4.

	Yes %	No %	Unsure %
After completing this program, will you take advantage of the tax amnesty policy?	83	10	7

Participants who decided to take advantage of the tax amnesty policy reasoned that they got more benefits, especially because there was a possibility of being examined when not using this policy. However, participants who decided not to use the tax amnesty policy reasoned that they thought that the government would issue a tax amnesty policy again later on, so that they would not utilize tax amnesty policy this year. Participants who answered that they were not sure that they would use this policy were those who felt they did not have other assets that had to be reported in tax amnesty.

4 CONCLUSION

Overall, taxpayers who participated in this program agreed that the program fulfilled their expectations and managed to overcome the barriers in utilizing all tax policies issued by the government. The cooperation between tax employees, tax consultants, and academics in educating taxpayers was very effective in increasing the understanding of taxpayers and then could increase the voluntary compliance from taxpayers.

The study was geographically limited to one region; however, there is no reason to believe that the positive results achieved in this case will not be applicable in other regions. Similarly, a pre-existing close relationship between the researchers, tax employees, and tax consultants, might have contributed to the success of the implementation. The evidence of the potential success of the program at promoting tax amnesty policy for personal taxpayers on a larger scale was demonstrated by its successful delivery, which resulted in the high interest of taxpayers to utilize the tax amnesty policy. This research action has limitations including not involving corporate taxpayers. Therefore, future research is expected to involve corporate taxpayers in a wider range, not only in one region.

REFERENCES

Alabede, J. O., Ariffin, Z. Z., & Idris, K. M. 2011. Individual taxpayers' attitude and compliance behaviour in Nigeria: The moderating role of financial condition and risk preference. Journal of Accounting and Taxation 3(5): 91–104.

Alm, J. & W. Beck 1991. Wiping the Slate Clean: Individual Response to State Tax Amnesties, Southern Economic Journal 57: 1043–1053.

Alm, J., & Gomez, J. L. 2008. Social capital and tax morale in Spain. Journal of the Economic Society of Australia 38: 73–87.

Alm, J., Martinez-Vazquez, J., & Wallace, S. 2009. Do tax amnesties work? The revenue effects of tax amnesties during the transition in the Russian Federation. Economic Analysis & Policy 39: 235–253.

Bailey, J. 2006. The difference between tax planning, tax avoidance, and tax evasion. Tax Insider.

Brown, R. E., & Mazur, M. J. 2003. IRS's comprehensive approach to compliance measurement.

Chau, G., & Leung, P. 2009. A critical review of Fischer tax compliance. Journal of Accounting and Taxation 1(2): 34–40.

Chow, C.Y. 2004. Gearing up for the self-assessment tax regime for individuals. Tax National (2): 20–23.

Christian, C., Gupta, S., & Young, J. 2002. Evidence on subsequent filing from the state of Michigan's income tax amnesty. National Tax Journal 55: 703–721.

Dickens, L., & Watkins, K. 1999. Action research: Rethinking Lewin. Management Learning, 30 (2): 127–140.

Doran, J., Healy, M., McCutcheon, M., & O'Callaghan, S. 2011. Adapting case-based teaching to large class settings: An action research approach. Accounting Education 20(3): 245–263.

Feld, L. P., & Frey, B. S. 2002. Trust breeds trust: How taxpayers are treated. Economics of Governance, 3, 87–99.

————. (2007). Tax compliance as the result of a psychological tax contract: The role of incentives and responsive regulation. Law & Policy, 29, 102–120.

Fjeldstad, O. H. 2006. Tax evasion and fiscal corruption: Essays on compliance and tax administrative practices in East and South Africa (Unpublished doctoral dissertation). Norwegian School of Economics and Business Administration, Bergen, Norway.

Greenwood, D. J., & Levin, M. 2007. Introduction to action research. Thousand Oaks, CA: Sage.

Hand, L. 1998. Tackling an accounting coursework assignment – Action research on the student perspective. Accounting Education, 7(4), 305–323. doi:10.1080/096392898331090.

Hazelton, J., & Haigh, M. 2010. Incorporating sustainability into accounting curricula: Lessons learnt from an action research study. *Accounting Education*, 19(1–2): 159–178.

Kaplan, R. S. 1998. Innovation action research: creating new management theory and practice. *Journal of Management Accounting Research* 10: 89–118.

Killian, S., & Kolitz, M. 2004. Revenue approaches to income tax evasion: A comparative study of Ireland and South Africa. *Journal of Accounting, Ethics and Public Policy* 4: 235–255.

Junpath, S.V., Kharwa, M.S.E. & Stainbank, L.J. 2016. Taxpayers' attitudes towards tax amnesties and compliance in South Africa: An exploratory study. *South African Journal of Accounting Research* 30 (2), 97–119.

John Hutagaol, 2007, *Perpajakan: isu-isu kontemporer.* Yogyakarta: Graha Ilmu.

Lewin, K. 1946. Action research and minority problems. *Journal of Social Issues* 2(4): 34–46.

McSweeney, B. 2000. Comment: 'Action research': Mission impossible? Commentary on 'towards the increased use of action research in accounting information systems' by C. Richard Baker. *Accounting Forum*, 24(4), 379–390. doi:10.1111/1467-6303.00047

Murphy, K. 2004. *An examination of taxpayer's attitudes towards the Australian tax system: Findings from a survey of tax scheme investors (Working Paper 46)*. Canberra: The Australian National University.

Paisey, C., & Paisey, N. J. 2003. Developing research awareness in students: An action research project explored. *Accounting Education*, 12(3): 283–302.

Paisey, C., & Paisey, N. J. 2004. An analysis of accounting education research in accounting education: an international journal – 1992–2001. *Accounting Education* 13(1): 69–99.

Paisey, C., & Paisey, N. J. 2005. Improving accounting education through the use of action research. *Journal of Accounting Education*, 23(1): 1–19.

Prawira, I.F.A. 2012. Upaya Pemerintah dalam Meningkatkan Rasio Kepatuhan Pajak, Sudah Efektifkah?, *Jurnal KIAT AL Khairaat*, 6(1).

―――――――. 2015. The Effect of Granting Tax Amnesty to Tax Revenues, *Research Journal of Finance and Accounting (RJFA)* 6(4): 2222–2847.

Rechberger, S., Hartner, M., Kirchler, E., & Hämmerle, F. (2010). Tax amnesties, justice perceptions, and filing behavior: A simulation study. *Law & Policy* 32: 214–225.

Reinganum, J. F., & Wilde, L. L. 1985. Income tax compliance in a principal-agent framework. *Journal of Public Economics*, 26, 1–18..

Saraçoğlu, O. F., & Çaşkurlu, E. 2011. Tax amnesty with effects and effecting aspects: Tax compliance, tax audits and enforcements around; the Turkish case. *International Journal of Business and Social Science* 2(7): 95–103..

Sari, Novita & Khairani, Siti. (2016). *Prospek tax amnesty dalam meningkatkan penerimaan pajak dari sudut pandang konsultan pajak (studi kasus pada konsultan pajak di Palembang).*

Steenekamp, T. J. 2007. Tax performance in South Africa: A comparative study. *South African Business Review* 11(3): 1–16..

Torgler, B. 2003. *Tax morale: Theory and empirical analysis of tax compliance* (Unpublished doctoral dissertation). University of Basel: Switzerland.

Uchitelle, E. 1989. The effectiveness of tax amnesty programs in selected countries. *FRBNY Quarterly Review*, 48–53.

Yanirahmawati, N. 2016. Sudah tau cara ikut program tax amnesty? *Easy Accounting System* 3.

Advances in Business, Management and Entrepreneurship – Hurriyati et al (eds)
© 2020 Taylor & Francis Group, London, ISBN 978-0-367-27176-3

Family ownership and control in dividend and leverage decision making

M.G.A. Aryani, S.M. Soeharto & I. Ariyani
Universitas Airlangga, Surabaya, Indonesia

ABSTRACT: All of the businesses were started as a family business. Data has shown that 35% of the Fortune top 500 companies around the world in 2016 were family firms. This research aims to find a linkage between family ownership and control to dividend and leverage decision making. The use of both family ownership and control was to separate the different nature of family firms. This research was a quantitative one and performed on non-financial companies listed on the Indonesia Stock Exchange during 2011–2016, analyzed with the multiple regression method. The results of this study showed that family ownership and family control had a significant and positive effect on dividend policies and leverage had a significant positive effect on dividend policies. While family ownership did not affect leverage significantly, family control, dividend and firm size had a significant positive effect on leverage. This research has confirmed that Indonesian family firms chose to retain control over ownership, because of its direct impact to protect the family interest by making strategic decisions.

1 INTRODUCTION

There are various ownership structures in companies and one of them is family ownership. Family ownership refers to a company that is founded, owned, controlled and managed by a member of the founder's family. Family ownership tends to have a constant managerial role to lessen the potential conflict between shareholder and manager.

A survey conducted in 2014 by Price Waterhouse Cooper revealed that more than 95% of companies in Indonesia are in the form of family business. Family business is defined by the majority of the voting right being in the hand of the founder or its descendants, at least one family member is involved whether in management or administration (Supriadi 2014). This data confirms the importance of family business in the economy of Indonesia.

The distinguishing character of a family-controlled firm is that the majority stock is owned by the founder or his family. The owner takes an active part in running the company as a part of the management team or as block holder. The common agency problem in family-controlled firms is that conflict arises between majority and minority stockholders. Majority stockholders will have utmost control over the minority, hence the minority stockholders face potential expropriation.

Agency problems lead to agency costs for the the family-controlled firms; there are a few alternatives to minimize the aforementioned costs such as increasing the dividend payment and the proportion of debt in the capital structure. These two options can be used as bonding mechanism by the

management to synchronize the interest of management and shareholders (Megginson 1997).

A higher dividend payment to the stockholder will lessen the company's internal fund for working capital or expansion. Hence, the company will have to look for new sources of funding in the form of debt or new equity. The dividend decision as bonding mechanism will minimize agency costs of capital because it will lessen the manager's opportunity to use the company cash flow for his own interest. (Megginson 1997).

External funding by debt will increase supervision and control from the stockholder and creditors (whether banks or bondholder). The leverage decision acts asa bonding mechanism so that the manager can convey positive intentions to the outside stockholder. Megginson (1997) has stated that the leverage decision as a bonding mechanism will decrease agency costs of capital. The company is burdened to repay the debt and its interest costs periodically, hence the manager must perform well to ensure the continuity and punctuality of the payment.

Another solution is by increasing insider ownership to solve the agency problem with higher supervision to managers. It is caused by insider ownership and can align the interest of the principal and agent, hence a lower agency problem. (Jensen and Meckling, 1976).

This research also used three control variables which are company size, profitability and cash ratio. Previous research by Harjito (2015) found that insider ownership and dividend payment affected leverage negatively. In contrast, Arifin (2003) deduced that family ownership affected leverage positively.

Indonesian firms chose to retain control rather than ownership during the research period. The founder and his descendants tend to place a member of their family in the board of directors to uphold the family control of the firm. The family member as board of director will also choose external funding of debt rather than issuing new equity to avoid dilution of family control.

Family firms have been the subject of various researches before, but it was mostly proxied with family ownership. Family control has been overlooked or used separately. This research uses both family control and ownership to separate the different nature of the family business. With the background elaborated above, this research will analyze the effect of family ownership and family control in dividend and leverage decision making in non-financial firms listed in the Indonesia Stock Exchange during 2011 – 2016.

Family ownership, according to Chen et al. (2010), is defined as a company with its founding family members keeping the position of upper level management, board of directors and/or block holder. Vilalonga and Amit (2006) explained that a family company should have family ownership, whether its founder or family members are connected through blood or marriage and have strategic positions or stock ownership by at least 5%, individually or in a group.

Based on these definitions, a family member will be the key to control the company in running its business. Family participation will strengthen the company because of the loyalty and high dedication of the family members. However, family ownership likely faces agency conflicts between the family and other non-family shareholders. Family ownership as block holder may pressure the management to provide higher profit, and an even bigger pressure may arise if the family runs the company from within the management. This pressure will hurt minority shareholders because their interest is neglected.

The presence of family members in strategic positions will add positive value. Based on previous research by Anderson and Reeb (2003), a company that is dominantly managed by family members will have a higher value of the firm. More specifically, a family company will increase the profit by 15.9%, compared to a non-family company.

Even though the family firm has a higher profit, it also tends to minimize the dividend payment. Family ownership will prefer the dividend to be allocated to projects with positive NPV so that the company will benefit in the long run. The family ownership structure will tend to choose external financing with debts instead of issuing new stocks to avoid the dilution of ownership, voting rights and profit within the family.

Family control reflects some or most of the family members who have direct control in the company and influences the company's management policy. According to Peraturan Bursa Efek Jakarta (Indonesia Stock Exchange Rulebook number 305/BEJ/07–2004), family control can also be said to be like a majority shareholder because it has the ability to affect the management and or its policy in any way, even though the number of shares owned is less than 25%.

Previous research by Tabalujan (2002) stated that a family firm can control the firm through it shared ownership. The founder will retain its control by placing its family members in strategic positions within the company and tend to govern the board of directors. Atmaja (2010) specified that family-controlled firms should own at least 20% of their ownership from outstanding shares. According to Villalonga and Amit (2007), the family-controlled firm often chooses the indirect ownership method to retain secrecy of business ownership, especially in Asia in order to get around the foreboding rules.

Siregar (2007) has stated that less dividend in family-controlled firms shows a sign of expropriation because the firm cash flow is used to satisfy the interest of the majority stockholders. Family firms also tend to avoid issuing new equity and choose debt instead to lessen the higher cost of capital associated with a secondary offering and evade dilution of family control in strategic decision making.

The proportion of debt in capital structure is defined as leverage. The research conducted by Rozeff (1982) concluded that companies with high financial leverage tend to lower their dividend payout to lessen transaction costs related to external funding. This finding was supported by Desmukh (2005), who found that managers choose to lower the dividend payment when facing financial distress to avoid being unable to pay the dividend in the upcoming years.

Hence, based on the previous literature, the hypotheses for this research are as follows:

H1: Family ownership affects dividend negatively.
H2: Family control affects dividend negatively.
H3: Leverage affects dividend negatively.
H4: Family ownership affects leverage positively.
H5: Family control affects leverage positively.
H6: Dividend affects leverage positively.

2 METHODS

2.1 Research population and sample

This research was conducted to all non-financial family firms listed in the Indonesia Stock Exchange during 2011–2016. A sample is determined as family firm if it has at least a member of the board of directors (BOD) in a family relationship with the founder of the company, which is characterized by the same surname. By employing a purposive sampling method, 201 companies were chosen as samples.

2.2 Research model

1) $DPR_{it} = \alpha_i + \beta_1 FO_{it} + \beta_2 FC_{it} + \beta_3 Lev_{it}$
$+ \beta_3 Size_{it} + \beta_4 Prof_{it} + \beta_5 Cash_{it} + \varepsilon_{it}$

2) $Lev_{it} = \alpha_i + \beta_1 FO_{it} + \beta_2 FC_{it} + \beta_3 DPR_{it}$
$+ \beta_3 Size_{it} + \beta_4 Prof_{it} + \beta_5 Cash_{it} + \varepsilon_{it}$

2.3 Operational definition of variables

3 RESULTS AND DISCUSSION

3.1 Statistic descriptive

All of the data were obtained from published financial reports of each company and from the Indonesia Stock Exchange during 2011–2016.

Table 2 confirms that the majority of family owned firms in Indonesia chose to retain control rather than ownership, as seen by the difference in the average number in family ownership, which is 5.25% while the family control's mean is 34.56%. By focusing on control, the family will be able to

Table 1. Operational definition of variables.

Variables	Operational Definition
Dividend Pay-out (Dpr)	Dividend per share / earnings per share
Leverage (Lev)	long term debt / total asset
Family Ownership (FO)	Family owned stocks / Outstanding stocks
Family Control (FC)	$\sum$family members in BOD / $\sum$members in BOD
Size	Ln Total Assets
Profitability (Prof)	Earning after taxes / Total Assets
Cash	Cash+Marketable securities / Total Assets

Table 2. Statistic descriptive results.

	N	MIN	MAX	MEAN	STDEV
FO	344	0,00	0,83	0,0525	0,1223
FC	344	0,07	0,71	0,3456	0,1458
DPR	344	-0,07	0,77	0,1709	0,1890
LEV	344	0,002	0,61	0,1429	0,1061
SIZE	344	113,46	183,35	14,530,9	16,041,5
ROA	344	-0,13	0,30	0,0566	0,0602
CR	344	0,001	0,48	0,1203	0,1137
Valid N	344				

Table 3. Regression results using Model 1 and 2.

Variables	Model 1 Dependent: DPR	Model 2 Dependent: LEV
FO	-0.186* (0.002)	0.000 (0.789)
FC	-0.003* (0.000)	0.211* (0.000)
DPR	-	0.135* (0.011)
LEV	0.328* (0.011)	-
SIZE	0.246** (0.000)	0.024* (0.000)
ROA	0.032* (0.048)	-0.281* (0.011)
CR	0.231* (0.037)	-0.006 (0.088)
R Square (R^2)	0.434	0.247

(*: significant at 5%).

steer the company towards family interest especially in leverage and dividend decision making.

3.2 Regression result and discussion

3.2.1 Model 1

Based on Table 3 from the first model, family ownership affected dividend negatively, hence the first hypothesis was accepted. This finding shows that family members find it easier to build close relationships with corporate managers to improve the company's performance. This easiness provides an opportunity for family-related shareholders to influence the manager's decision to allocate dividend payments. They choose to increase retained earnings to fund future expansion, hence decreasing the dividend payout. This finding is similar to Arifin (2003), who stated that family ownership affected dividend negatively.

Based on the results of this study, it can be concluded that family control had a significant negative effect on dividends, hence the second hypothesis was accepted. This result indicated that the existence of family control within management ranks will put pressure on other managers to diminish dividend payout to minority shareholders. Family-related managers will encourage other managers to allocate profits generated by the company to pay interest expenses from debt, boost earnings as a source of internal funding, and increase working capital and expansion. Family members in the company's management structure will seek to increase retained earnings, reduce debt obligations, long-term debt, and expand by minimizing dividend payout to shareholders. This finding agrees with Arifin

(2003) that family ownership affected dividend negatively.

The third finding is that leverage had a significant positive effect on dividend; hence the third hypothesis was rejected. Shareholders hold the residual claim, which will decrease with a higher leverage, and it harmed their rights for profit. Thus, with higher risk coming from higher leverage, shareholders will ask for higher returns in the form of higher dividend, thus confirming the positive effect.

The three control variables used in this research are size, profitability and cash ratio, and they all have a positive effect towards dividend payout policy. Bigger firms will be able to pay more dividends as they can easily access external funding (either banks or the capital market) due to their size, so they will retain less profit. This is supported by previous research from Jensen et al. (1992), Fama and French (2001) and Mulyani et al. (2016). When companies earned a higher profit, they will convey a positive signal to investors by also increasing the dividend payout. This is also an attempt to reduce agency conflicts between management and shareholders. This is supported by previous research from DeAngelo et al. (1992), Fama and French (2001), and Mulyani et al. (2016). Higher cash should not be left idle when no prospective projects available to the company will result in a higher dividend payout. This is supported by previous research from Mulyani *et al.* (2016), Gill et al. (2010), and John and Muthusamy (2010).

3.2.2 *Model 2*

The next model with leverage as the dependent variable shows that it was not affected by family ownership, hence the fourth hypothesis is declined. It showed that one form of policy that cannot be intervened by shareholders is related to debt policy. Shareholders have no power to influence management to increase or decrease the amount of debt held by the company. Family ownership expects the management of the company to increase its debt in the event of additional capital for operational activities run by the company, rather than issuing new shares. Moreover, leverage decision making is based on various factors such as internal investment plans, availability of collateral, existing debt capacity, and macroeconomic factors like interest rate, inflation and central bank policy so this one cannot be intervened easily by the stockholders. This result is similar to Peilow (2017) and Abor (2008), who stated that family ownership does not affect leverage.

Family control proved to have a significant and positive effect on leverage, hence the fifth hypothesis was accepted. This suggests that family members within management ranks will use their position to protect the interests of the owner. Family related managers seek to maintain control of the company, hence resulting in an effort to restrict new ownership in the company. The addition of non-family ownership in the company deters the influence of family members in making important decisions for the sustainability of the company. Family control in the management ranks will strive to safeguard the interests of the family as one of the owners of the company by not increasing the amount of equity owned by the company. Managers of family members prefer debt as an external source of funding rather than issuing new shares. This finding is supported with previous research by Mulyani et al. (2016), who found that family control affected leverage positively.

The last hypothesis was declined as the result showed that dividend had a significant positive effect on leverage. It shows that family ownership attempts to set a certain proportion of dividends payout and retain it for a few years. It is important that dividend payout should not decrease for this will send a negative signal towards investors. When the company's free cash flow is lower than the amount of dividend payout, the manager will decide to increase the amount of debt owned by the company. High dividends are followed by high investment so that the company must add debt to fund the company's investment activities. This result is supported by Mahhadwartha and Hartono (2002), that dividend affects leverage positively.

The three control variables employed in this research had different effects towards leverage. Size affected leverage positively while profitability affected it negatively and cash did not affect leverage significantly. The bigger the company, the more resources are needed to run it; hence big companies are supported with a higher leverage. Bigger companies also have more collaterals available to apply for loans and are more trusted by banks or other creditors. This result is similar to Mulyani et al (2016). A higher profitability enables the company to retain internal funds needed for future expansion; thus, lower leverage. Higher profit also enables a company to pay back the loan faster and pay it off sooner. Lastly, a high cash ratio indicates higher liquidity within the company. Cash is used mostly for operational purposes and not in the long term. Meanwhile leverage is one of the components of capital structure to fund projects and investments, hence cash availability did not affect the leverage decision since it has different purposes. This result is parallel with Piaw and Jais (2013), that cash ratio does not affect leverage.

4 CONCLUSION

This research examined the relationship between family ownership and -control for different natures of the family business with leverage and dividend payout decision making. There was a higher degree for family control than ownership in Indonesia, because in typical Asian countries the pyramid structure is used to veil the family intrusion within the company. Hence the founder family chose to retain the power by placing family members within the management to make strategic decisions and protect family interests rather becoming a majority shareholder.

This research result showed that family ownership and control indeed affected dividend payout negatively to ensure that the company profit will suffice for future expansion rather than issuing new stocks. Different result for leverage decision was where family ownership increased the leverage to avoid the dilution of voting rights, control and profit from family members. Meanwhile leverage decision making seems immune to family control since this is a complex one with various factors to be considered, from internal firm condition to external macroeconomic factors.

For future research, it is advised to add more observations than the limitations of the sampling procedure in this study. A higher observation number will increase the model's reliability to explain family related influences inthe company.

REFERENCES

Abor, J. 2008. Agency theoretic determinants of debt levels: evidence from Ghana. *Review of Accounting aAnd Finance* 7(2): 183–192.

Arifin, Z. 2003. Efektifitas Mekanisme Bonding Dividen dan Hutang untuk Mengurangi Masalah Agensi pada Perusahaan di Bursa Efek Jakarta (The Effectivity Of Bonding Mechanism Of Dividend And Leverage To Lessen Agency Problem In Public Companies). *Jurnal Siasat Bisnis* 1(8).

DeAngelo, H., DeAngelo, L., and Skinner, D. J. 1992. Dividends and losses. *The Journal of Finance* 47(5): 1837–1863.

Gill, A., Biger, N. and Tibrewala, R. 2010. Determinants of Dividend Payout Ratios: Evidence from United State. *The Open Business Journal* 3: 8–14.

Jensen, M. C. and Meckling, W. H. 1976. Theory of the firm: Managerial behavior, agency costs, and capital structure. *Journal of Financial Economics* 3(4): 305–360.

Jensen, G. R., Solberg, D. P., and Zorn, T. S. 1992. Simultaneous determination of insider ownership, debt, and dividend policies. *Journal of Financial and Quantitative Analysis* 27(2): 247–263.

John, Franklin S and Muthusamy, K. 2010. Leverage, growth and profitability as determinants of dividend payout ratio – evidence from Indian paper industry. *Asian Journal of Business Management Studies* 1(1): 26–30.

Megginson, W. L. 1997. *Corporate finance theory.* Reading, MA: Addison-Wesley Educational Publishers.

Mulyani, E., Singh, H., and Mishra, S. 2016. Dividends, leverage, and family ownership in the emerging Indonesian market. *Journal of International Financial Markets, Institutions and Money* 43:16–29.

Mahadwartha, P.A. and Hartono, J. 2002. Uji teori keagenan dalam hubungan interdependensi antara kebijakan utang dengan kebijakan dividen (test of agency theory in interdependence relations of leverage and dividend decision making). *Proc. in Simposium Nasional Akuntansi 2002*, Universitas Diponegoro, Semarang.

Piaw, LLT and Mohamad J. 2013. The Capital Structure of Malaysian Firms in the Aftermath of Asian Financial Crisis 1997; *Proc. 2nd International Conference on Management, Economics, and Finance* 455–474.

Siregar, B. 2008. Pengaruh Pemisahan Hak Aliran Kas dan Hak Kontrol Terhadap Dividen (the effect of separation of cash flow and control rights to dividend). *Jurnal Riset Akuntansi Indonesia*, 11(2): 158–185.

Sudana, I Made. 2011. *Manajemen keuangan perusahaan: teori dan praktik (financial management, theory and practical) 2nd ed.* Jakarta: Erlangga.

Effect of profitability, investment opportunity set, free cash flow and collateralizable assets to dividend

A. Juliarti & S. Sumani
Universitas Katolik Indonesia Atma Jaya, Jakarta, Indonesia

ABSTRACT: In the development and progress of the business world, competition encourages companies to have a competitive advantage. One of these is reflected in the company's dividend policy. This study aims to examine the effect of profitability, investment opportunity set (IOS), free cash flows (FCF), and collateralizable assets (CA) on dividend policy. The population were consumer goods companies listed on the Indonesia Stock Exchange from 2013–2015. The number of samples used in the study were 19 companies. The results showed that profitability and IOS variables had a significant negative effect on corporate dividend policy, while FCF and CA had no significant effect on company dividend policy. This implies that a residual theory of dividend has to be applied in managing the dividend decision.

1   INTRODUCTION

In the context of financial management, firms are seen as a set of contracts between corporate managers and shareholders. They are working together to achieve the company's goal, which is maximizing corporate value for the owners. To achieve this goal, the financial manager must be able to perform its function well. The function of financial managers is to make decisions related to financial decisions, investment decisions, and dividend decisions (Karnadi 1993)

Appropriate management considerations are needed in dividend decisions as it involves two different stakeholders and company management. The company's dividend payout policy is reflected in the dividend payout ratio. The amount of corporate dividend policy can be influenced by several factors: profitability, investment opportunity set (IOS), free cash flows (FCF), and collateralizable assets (CA).

Several studies reveal the effect of profitability which is seen through ROA on the company's dividend policy. Research conducted by Suharli (2007), Dewi (2008), and Pradana and Sanjaya (2013) demonstrated that profitability had a positive effect on dividend policy. In contrast, Nuringsih (2005) found that the profitability of the company had no positive effect on the dividend policy.

Regarding IOS, Gaver and Gaver (1993) mention that companies with a high IOS will reduce the share of dividends. This is in line with research conducted by Pujiati (2015) and Giriati (2016), discovering that IOS had no positive effect on dividend policy. On the other hand, the study conducted by Handriani and Irianti (2015) and Prasetio and Suryono (2016) found that IOS had a positive effect on corporate dividend policy.

Some studies have been done to investigate the effect of free cash flows on dividend policy. Arfan

and Maywindlan (2013), Suci (2016), Giriati (2016), and Prasetio and Suryono (2016) revealed that free cash flows had a positive effect on dividend policy. Meanwhile, Pradana and Sanjaya (2013) argued that free cash flows had no significant effect on dividend policy.

Related to collateralizable assets, the study conducted by Arfan and Maywindlan (2013) as well as Lie and Astuti (2015) revealed that collateralizable assets had a positive influence on dividend policy. While different things are stated by (Suci 2016) who found that collateralizable assets had no positive effect on the company's dividend policy.

2   METHOD

The above studies show different results. This prompted the researcher to further investigate the effect of profitability, IOS, free cash flows, and collateralizable assets on dividend policies. The conceptual hypotheses of this study are:

H1: Profitability has a significant negative effect on dividend policy.
H2: Investment Opportunity Set (IOS) has a significant negative effect on dividend policy.
H3: Free cash flows have a significant positive effect on dividend policy.
H4: Collateralizable assets has a significant positive effect on dividend policy.

The variables used in this research are dependent and independent variables. The dependent variable was dividend policy, while the independent variables were profitability, IOS, free cash flow, and collateralizable assets. The description of the variables and their measurement techniques is given in Table 1.

Table 1. Definition of variable operations and measurement techniques.

Variable		Formula
Dependent	Dividend policy (DPR)	$DPR = \dfrac{Dividend\ per\ Share\ (DPS)_{t+1}}{Earning\ per\ Share\ (EPS)_t}$
Independent	Profitability (ROA)	$ROA = \dfrac{Earning\ after\ tax}{Total\ assets}$
	IOS (MTBVER)	$MTBVER = \dfrac{Closing\ price\ of\ stock\ \times\ Outstanding\ C/S}{Total\ Equity}$
	FCF (FCF)	$FCF = \dfrac{\begin{array}{c}Cash\ flow\ from\ operations\ - \\ Net\ capital\ expenditure\ +\ Change \\ in\ working\ capital\end{array}}{Total\ assets}$
	CA (CA)	$CA = \dfrac{Net\ Fixed\ Assets}{Total\ Assets}$

Source: Processed Data, 2018.

The data used in this study was secondary data. Data related to the closing stock price and the number of cash dividends distributed were obtained through the Indonesia Capital Market Directory (ICMD) and www.ksei.co.id. Meanwhile, other data were obtained through the company's financial reports published by the Indonesia Stock Exchange, which were accessed through www.idx.co.id and each of the company's website pages.

The population in this study were consumer goods companies listed on the Indonesia Stock Exchange in the years 2013–2015. The total number of listed companies was 37 companies. The author then set the following population limits:

a) The companies listed on the Indonesia Stock Exchange in the years 2013–2015 had to run consumer goods industry sector.
b) The companies did consistent payments of dividends during the study period (2013–2015).
c) The companies fully published their audited financial statements on December 31.
d) The companies had a net profit and positive free cash flow.

From the total population of 37 companies in the consumer goods industry during 2013–2015, there were only 19 companies that met the criteria.

The method of data analysis used was panel data analysis. It was the combination of time series and cross-section data. The data processing tool used was E-Views program version 9.0 with specified significance value (α) of 5%. Some approaches were used as the estimation method of the variables, namely:

1. Descriptive statistics analysis
2. Selection of panel data estimation model
   • Pooled Model (Pooled Least Square)
   • Fixed Effect Model (FEM)
   • Random Effect Model (REM)
3. Inference analysis
4. The coefficient of determination test (R2)

3 RESULTS AND DISCUSSION

Table 2 shows the descriptive statistics of the variables.

The results of the Chow Test and Hausman Test conducted using E-Views 9.0 with a significance level of 5% is described in Table 3.

The probability of the Chi-square cross-section listed in Table 3 was 0.0000. Since the significance level (α) was greater than the probability of the cross-section Chi-square (0.0500> 0.0000), H1 failed to be rejected. This means that there are the differences in parameters in each individual studied. This confirms that the FEM method is better than the PLS in this study. Because the FEM method is better, the Hausman Test has to be conducted.

Table 2. Descriptive statistics.

	DPR	ROA	MTBVER	FCF	CA
Mean	0.567630	0.179278	8.507316	0.524488	0.315821
Median	0.497000	0.109245	3.999305	0.492568	0.258553
Maximum	1.459200	0.715090	58.48124	1.082175	0.918363
Minimum	0.099900	0.037882	0.263810	0.082365	0.088424
Std. Dev.	0.328665	0.144292	13.06281	0.246750	0.175908

Source: Processed Data, 2018.

Table 3. Results of the Chow Test and Hausman Test.

Results	Prob.
Cross-section Chi-square	0.0000
Cross-section random	0.0000

Source: Processed Data, 2018.

The probability of the random cross-section listed in Table 3 was 0.0000. Since the level of significance (α) was greater than the probability of cross-section random (0.0500> 0.0000), H1 failed to be rejected. Thus, it can be concluded that the FEM method is better than REM in this study.

Inference analysis was conducted to determine whether the profitability, investment opportunity set (IOS), free cash flow (FCF), and collateralizable assets (CA) variables simultaneously or partially affected the company dividend policy. A hypothesis test was done by using F statistic test (F-test) and t-test (t-test) with a significance level (α) equal to 5%.

The probability value of t-statistic after dividing by two (p-value/2) of profitability variable was 0.02975 with coefficient –0.949637. As the p-value was less than the level of significance (0.02975 < 0.0500), H0 was rejected and H1 was accepted. This means that profitability variables have a significant influence on the company dividend policy. A coefficient value which is less than zero indicates that there is a negative influence of ROA on the company dividend policy. The profitability variable shows a significant negative effect on the dividend policy. Hence, the first hypothesis (H1) which states that profitability has a significant negative effect on dividend policy is accepted.

This result is in accordance with the research conducted by Nuringsih (2005), in which the profitability of the company has no positive effect on dividend policy.

The dividend policy is concerned with the use of profits that shareholders are entitled to, so this is closely related to the agency theory. For management, cash dividends represent cash outflows which reduce the company's cash. Profitability owned by the company can be divided as a dividend or retained for reinvestment. Research data taken from the period 2013–2015 also showed the trend of an increase in investment opportunities that could be done by the company. This is an indication that the company chooses to use the profitability that it receives for more retained earnings (RE), which reduces the amount of dividend payout to shareholders.

In the long-term expenditure policy approach, all post-tax profits earned by the firm are long-term sources of funds. The announcement of the dividend distribution as a dividend means a reduction to the long-term funding sources that can be used to finance the business development needs. Therefore, the dividend payout will result in an emphasis on business development or coercion on the disbursement of external funds. If the company has a business development plan which is rather encouraging in the future, it needs sources of funds from within the company. In addition, if based on the consideration that low capital costs are required to form the desired capital structure, the source of capital funds is from themselves. The higher level of expansion planned by the company results in reducing the dividend payout ratio because the profit earned is prioritized for additional activities.

In addition, the state of the company's shareholders also affects dividend. If the company's shareholders are in the high tax category and prefer to obtain capital gain, the company will tend to maintain a low dividend payout ratio. If a low dividend payout ratio will increase the amount of profit, the company can hold it for the internal company. This is relevant to the theory of Tax Difference Theory.

Regarding Investment Opportunity Set (IOS), Table 4 shows that the obtained t-statistic probability after being divided by two (p-value/2) was equal to 0,0263. The p-value was less than the significance level (0.0263 < 0.0500), so H0 was rejected and H2 was accepted. Thus, the IOS variable has a significant influence on the company dividend policy. The IOS variable coefficient value was –0.019821. A coefficient value which was less than zero showed that there was a negative influence of IOS on dividend policy. This implies that the second hypothesis (H2), which states that IOS has a significant negative effect on the policy dividends, is accepted.

This result is in line with Gaver and Gaver (1993), who state that companies with high IOS will reduce the share of dividends. This is also in line with studies conducted by Pujiati (2015) and Giriati (2016), that IOS had a significant negative effect on dividend policy. However, this result is not in

Table 4. Regression analysis results.

Variable	Coefficient	Prob.
C	0.536090	0.0878
ROA	-0.949637	*0.0595
MTBVER	-0.019821	*0.0526
FCF	0.316448	0.3842
CA	0.647326	0.2449
R-squared		0.829715
Adjusted R-squared		0.719531

Source: Processed Data, 2018*= research variables are partially significant to the dividend policy at the level of significance 5%

accordance to those disclosed by Handriani and Irianti (2015) as well as Prasetio and Suryono (2016), that IOS has a positive effect on company dividend policy.

The residual theory of dividend states that dividend payments are seen as residual after all useful investment opportunities have been made. Therefore, cash owned by the company is preferred for investment activities, and then the remaining cash used for new investments is distributed to shareholders. If the company has a high IOS, then the company will be more inclined to invest, by which the investment made by the company will reduce the availability of cash that can be used to pay dividends. The funds that should be paid to the shareholders as dividends will be used to make a profitable investment so that the dividend payout ratio (DPR) of the company will also decrease.

Conversely, if the company is experiencing a slow growth with low IOS, it tends to share higher dividends to overcome the issue of overinvestment occurring in the company. In addition, by dividing higher dividends, it will signal to investors that the company has better prospects for the future. Even for a growing company, the increase in dividend payments that occur will be assumed to be bad news because it implies that the company will reduce its future investment plan. This supports the signaling hypothesis theory, which implies the current state of the company as a signal to describe the state of the company in the future.

Table 4 shows that the probability value of t-statistic after being subdivided (p-value/2) of the free cash flows variable was 0.1921, with the coefficient 0.316448. In this study, the p-value was higher when compared to the level of significance (0.1921 > 0.0500), so H0 failed to be rejected. This indicates that the variables of free cash flows do not have a significant effect on the company dividend policy. The free cash flows variables show an insignificant positive effect on dividend policy. Therefore, the third hypothesis (H3), which states that free cash flows had a significant positive effect on dividend policy, is rejected.

This result supports previous research conducted by Endang and Minaya (2005) and Pradana and Sanjaya (2013), which also found that free cash flows had no significant effect on dividend payout ratio (DPR). The results of this study, however, contradicts the research conducted by Arfan and Maywindlan (2013), Suci (2016), Giriati (2016), and Prasetio and Suryono (2016), which all found that free cash flows had a positive effect on dividend policy.

This study also found that the size of free cash flows did not affect the high dividend distribution. Companies that have free cash flows may not pay dividends to shareholders, but it is used to pay the corporate debt (Ross et al. 2010). Lenders often set conditions to ensure a smooth payment of receivables. One of them is the limitation of dividend payments. Free cash flows of excess companies can also be used by companies to make investments rather than having to pay dividends. By investing, the company can increase the company's growth and the company gets a return on the investment.

In addition, macroeconomic conditions are considered capable of affecting the discount rate, the company's ability to generate cash flows, and future dividend payout. This is because the market risk is closely related to changes in stock prices of certain types or certain groups caused by investors' anticipation of expected changes in expected returns. Macroeconomic factors such as inflation rate, interest rate, and exchange rate, including the value of JCI in 2013 until 2015 are also considered to be a consideration of the company's decision in the use of free cash flows owned by the company. Fluctuations in the JCI value are likely to increase in 2013 (4.274), 2014 (5.226), and 2015 (4.593). This illustrates that market conditions improve in the period 2013–2015. Thus, the company chose not to allocate its free cash flows to be distributed as dividends to shareholders.

Another finding that also supports that free cash flows are not a dividend policy consideration is that firms apply fixed dividend payouts. Therefore, even if free cash flows are owned by large companies, the amount of dividends the company shares is worth. Conversely, despite the free cash flows of small companies, dividends to be distributed will remain. This can be seen in Indofood CBP Sukses Makmur Tbk and Indofood Sukses Makmur Tbk.

Related to collateralizable assets (CA), Table 4 shows that the value of the t-statistic probability after being divided by two (p-value/2) of the collateralizable assets variable was 0.12245, with the coefficient 0.647326. In this case, the p-value was higher than the level of significance (0.12245 > 0.0500), so H0 failed to be rejected. This indicates that the collateralizable assets variable does not have a significant effect on the company's dividend policy. The fourth hypothesis (H4) which states that collateralizable assets (CA) have a significant positive effect on dividend policy was rejected.

This result is in accordance with the research conducted by Suci (2016), who found that collateralizable assets had no significant effect on the company dividend policy.

Collateralizable assets (CA) have a close relationship with the company's debt policy. When the company decides to make a loan, there are a number of additional conditions that have been negotiated and approved by both parties, company and creditor. One is a negative covenant in which the creditor forbids the debtor to perform an action including dividends.

Debt policy is a policy taken by the management in order to obtain the source of financing for the company so that it can be used to finance the company's operational activities. Debt policies are also seen as internal control mechanisms that can reduce agency conflicts between management and shareholders. Through loans with collateralizable assets, the company has an obligation to repay loans and pay interest expenses on a periodic basis.

The absence of any significant effect of the collateralizable assets variable on dividend policy can be due to a large number of fixed assets owned by a company. The assets here do not reflect the availability of cash to be considered in determining the dividend policy of a company, while the ability of a company to obtain a number of loans by pledging its fixed assets to its creditors does not increase the availability of cash to be distributed to investors as dividends. However, most companies tend to use the cash for internal funding which is more important than distributing cash dividends. This causes collateralizable assets to be insignificant in the company's dividend policy.

After analyzing the data by using the t-test, the next step conducted was the F-test. The F-test was performed to find out whether the dependent variable affected the independent variables. The simultaneous significance test in this research was done by comparing the probability statistic value of F (p-value) to a significance level (α) used, which was 5%. The data in the table shows that the probability value of F statistic (p-value) was less than the level of significance (0.0000 < 0.0500), so H0 was rejected. This can be interpreted that, simultaneously, the dependent variables in the study significantly influence the dependent variable.

The magnitude of the determination coefficient can be seen through the magnitude of Adjusted R-Square. The data in the table shows that the magnitude of Adjusted R-Square was 0.719531. This showed that 71.9531% of the variations that occurred in corporate dividend policies were explained by profitability, IOS, free cash flow, and collateralizable assets. Meanwhile, 28.0469% was explained by other variables that were not discussed in this study.

4 CONCLUSION

Based on the results of this study, it can be concluded that profitability of the company and investment opportunity set (IOS) have a significant negative effect on the company's dividend policy, whereas free cash flow (FCF) and collateralizable assets (CA) have no significant effect on dividend policy. Thus, high FCF and CA levels do not affect the amount of dividend payout. This implies that the residual theory of dividend has to be applied in managing the dividend decision.

ACKNOWLEDGMENT

We would like to thank the Atma Jaya Catholic University of Indonesia for giving us the opportunity to present this paper.

REFERENCES

Arfan, M and Maywindlan, T. 2013. Pengaruh arus kas bebas, *collateralizable assets*, dan kebijakan utang terhadap kebijakan dividen pada perusahaan yang terdaftar di *jakarta islamic index. Jurnal Telaah and Riset Akuntansi, 6*, (2), 194–208.

Dewi, S. C., 2008. Pengaruh kepemilikan managerial, kepemilikan institusional, kebijakan hutang, profitabilitas dan ukuran perusahaan terhadap kebijakan dividen, *Jurnal Bisnis dan Akuntansi, 10*, (1), 48–57.

Endang, and Minaya. 2003. Pengaruh insider ownership, dispersion of ownership, free cash flow, collaterizable assets dan tingkat pertumbuhan terhadap kebijakan dividen. *Jurnal Ekonomi dan Bisnis, 14*, (21), 281–301.

Gaver, Jeniffer J., and Gaver, K.M. 1993. Additional evidence on the association between the investment opportunity set and corporate financing, dividen, and compensation policies. *Journal of Accounting and Economics, (16)*, 125–160.

Giriati. 2016. Free cash flow, dividend policy, invesment opurtunity set, opportunistic behavior and firm's value. *Procedia-Social and Behavioral Sciences, 219*, 248–254.

Handriani, E., and Irianti, T.E. 2015. Investment opportunity set (ios) berbasis pertumbuhan perusahaan dan kaitannya dengan upaya peningkatan nilai perusahaan. *Jurnal Ekonomi Dan Bisnis, XVIII*, (1), 83–99.

Karnadi, S.H. 1993. *Manajemen Pembelanjaan* (Jilid 1). Jakarta: Yayasan Promotio Humana.

Lie, Mei Fong and Astuti, D. 2015. Pengaruh agency cost terhadap kebijakan dividen perusahaan keuangan dan non-keuangan. *FINESTA, 3*, (2), 18–22.

Nuringsih, K. 2005. Analisis pengaruh kepemilikan manajerial, kebijakan utang, roa dan ukuran perusahaan terhadap kebijkan dividen: studi 1995–1996. *Jurnal Akuntansi dan Keuangan Indonesia, 2*, (2), 103–123.

Pradana, S.W.L., dan Sanjaya I.P.S., 2013, Pengaruh Profitabilitas, Free Cash Flow, dan Investment Opportunity Set Terhadap Dividend Payout Ratio (Studi Empiris Pada Perusahaan Perbankan Yang Terdaftar Di BEI), Simposium Nasional Akuntansi (XVII), Lombok.

Prasetio, D. A., and Suryono, B. 2016. Pengaruh profitabilitas, free cash flow, investment opportunity set terhadap dividen payout ratio. *Jurnal Ilmu dan Riset Akuntansi, 5*, (1), 1–19.

Pujiastuti, T. 2008. Agency cost terhadap kebijakan dividen pada perusahaan manufaktur dan jasa yang go public di indonesia. *Jurnal Keuangan dan Perbankan, 12*, (2), 183–197.

Pujiati. (2015). Faktor-Faktor Yang Mempengaruhi Kebijakan Dividen Pada Sektor Industri Barang Konsumsi. *Jurnal Nominal, IV.Universitas Negeri Yogyakarta.*

Suci, R. I. W. 2016. Pengaruh arus kas bebas, kebijakan pendanaan, profitabilitas, collateralizable assets terhadap kebijakan dividen. *Jurnal Ilmu dan Riset Akun-tansi, 5*, (2), 1–17.

Suharli, M. 2007. Pengaruh profitability dan investment opportunity set terhadap kebijakan dividen tunai dengan likuiditas sebagai variabel penguat (studi pada perusahaan yang terdaftar di bursa efek jakarta periode 2002–2003). *Jurnal Akuntansi dan Keuangan, 9*, (1), 9–17.

Tax expense and bonus mechanism on transfer pricing

G.S. Manda, H.M. Zakaria & A. Rakhman
Universitas Singaperbangsa Karawang, West Java, Indonesia

ABSTRACT: The aim of this research is to examine the influence of tax expense and bonus mechanism on transfer pricing by using a sample manufacturing company at LQ 45 in the period 2012–2016. The variables of this research consist of two independent variables and one dependent variable. The first independent variable is tax expense; the second independent variable is bonus plan, and transfer pricing. The research consists of quantitative methods and analysis techniques, using logistic regression analysis. This is because the dependent variable in this research is a dummy variable: the companies which are using transfer pricing is coded by 1 and the companies which are not using transfer pricing is coded by 0. This research uses financial statements as data sources. The results of the research showed that tax expense and bonus plan do not have an influence on transfer pricing of the manufacturing company at LQ 45 in the period 2012–2016.

1 INTRODUCTION

Multinational companies are companies whose business activities are not only conducted in their own country, but also extend to foreign countries through subsidiaries and branches. Multinational companies have many transactions between their branch companies, two of which are sale and purchase transactions. The price of transactions occurring in a multinational company is known as transfer pricing, which is the amount sold and bought by a subsidiary or branch (Utari et al. 2016).

Transfer pricing is often used by multinational companies to avoid tax costs. Tax is a contribution for citizens, both private and business entities to the state, reciprocity of such contributions in the form of public facilities for the welfare of citizens. The types of taxes consist of income tax; value added tax; sales tax, etc. The type of tax that affects transfer pricing practice is income tax. Earnings on business entities represent the company's gross profit multiplied by the tax rate (Anthony & Govindarajan 2007).

Each country has different tax rates, multinational companies tend to earn profit management and tax planning by opening branches in countries with low tax rates. A branch company in a country with high tax rates will pass transfer pricing to a branch company in a country with low tax rates and low transfer rates, so that the company's profit in the country with high tax rates is decreasing because of its low transfer price. Branch companies with low tax rates can sell products at high prices without having to worry about the tax cost (Mardiasmo 2011).

The income tax rate in Indonesia is included in the top 10 countries with the highest tax rates in Asia Trading Economics 2017. Therefore, it can be potentially profitable for companies in

Indonesia to practice transfer pricing in order to avoid taxes.

A company that allegedly does transfer pricing to avoid taxes is PT Toyota Motor Manufacturing Indonesia (PT TMMI), as reported in an article which said that there are a number of findings indicating that Toyota Indonesia sells their production cars to Singapore at an unnatural price. An example is found in Toyota's tax report document from 2007. It is noted that Toyota Motor Manufacturing in Indonesia exported 17,181 units of car type Fortuner to Singapore. An employee tax investigation in Toyota's financial statements discovered that the Fortuner's cost of goods sold (COGS) is Rp 161 million per unit, while Toyota's internal documents show that all Fortuner's are sold 3.49 percent cheaper than that. Therefore, Toyota Indonesia bears the loss of selling the cars to Singapore (Tempo 2014).

Based on the article, it can be said that if PT TMMI sells its production car to Toyota Motor Asia Pacific Pte. Ltd. in Singapore before it is sold to other branches, the transfer pricing practice is performed because the corporate income tax rate in Singapore is lower than the corporate income tax rate in Indonesia. The unfair practice of transfer pricing makes the profit of PT TMMI very low and the tax cost should be low. Toyota Motor Asia Pacific Pte. Ltd. can resell those cars with a high sales volume and fair market prices. Sales at such fair prices make Toyota Motor Asia Pacific Pte., Ltd earn a high profit, and the tax cost to be borne by the company remains low because the company stands in a country with low corporate tax rates. Based on the example case, it can be concluded that tax can be one factor that influences the practice of transfer pricing (Tempo 2014).

Another factor that may affect transfer pricing is the bonus mechanism. Management tends to utilize transfer pricing practices to maximize the bonuses that they will receive if the bonuses are based on earnings of the company (Agnes et al. 2010).

A bonus is a reward for the performance of the board of directors on the basis of the calculation of corporate profits. The higher the profit, the more likely the bonus will be received, so the directors in the branch company tend to make earnings management to maximize net profit so that the bonus will be maximized (Nurjanah et al. 2016).

Previous research relating to transfer pricing includes research conducted by Nurjanah et al. (2016) under the title "*Determinant factors of corporate decision in transfer pricing*" which shows the result that taxes, bonus mechanisms and firm size influence transfer pricing. Another study was conducted by Kusuma (2016) under the title "The influence of tax expense, tunneling incentive, and bonus mechanism to transfer pricing of multinational enterprises listed on Indonesian stock exchange (IDX)". The results show that taxes influence transfer pricing, incentive tunneling, and that a bonus mechanism has no effect on transfer pricing.

Based on the phenomenon and the differences of previous research results, the authors are interested in re-examining the effect of taxes and bonus mechanisms on transfer pricing. This study uses different data from previous studies. The case study in this research is a manufacturing company which is included into LQ 45 in IDX that is 10 companies. The data of this research is sourced from the financial statements of the years 2012 until 2016.

1.1 Literature review

1.1.1 Theoretical basis
This study will discuss the influence of tax and bonus mechanism on transfer pricing. The basis of this research theory is.

1.1.2 Tax
According to Waluyo (2013), tax is contribution to the state (which may be enforced) owed by obligatory payers according to the rules, with no immediate return, which may be appropriately appointed, and are useful for financing general expenses related to the duty of the State which administers them.

According to Mardiasmo (2011), tax is a contribution of the people to the state treasury under the law (which can be enforced) with no received lead services (counter-achievement) that can be directly shown and used to pay public expenditures.

The definition of tax is also contained in the Law of Article 1 No. 28 of 2007 on General Provisions and Procedures of Taxation, which states that: Tax is a compulsory contribution to a country that is indebted by an individual or an institution which is coercive under the act, by not being directly reciprocated and used for state purposes for the greatest possible prosperity of the people.

Based on these definitions, it can be concluded that tax is a contribution of citizens, both of private and business entities, to the state which is mandatory; reciprocity of the fee is in the form of public facilities for the welfare of citizens.

1.1.3 Bonus mechanism
A bonus can be a motivation for employees to work better for the progress of the company. The bonus mechanism according to Nurjanah et al. (2016) is the greater profit generated by the company, the better image of the directors in the eyes of the company owner. The board of directors is considered to have a good performance for the company, therefore the owner of the company will give awards to the directors who have managed the company well. The award is called a bonus.

Another understanding of the bonus mechanism according to Kusuma (2016) is one of the strategies or calculation motives in accounting whose purpose is to give a reward to the board of directors or management by looking at the company's overall profit.

1.1.4 Transfer pricing
According to Utari et al. (2016), transfer pricing is the value unit of a product or service charged by one division to another within an organization.

According to Anthony and Govindarajan (2007), transfer pricing is the value awarded to a transfer of goods or services in a transaction where at least one of the two parties involved is the profit center.

The purpose of transfer pricing, according to Anthony and Govindarajan (2007), is to:

a. Provide relevant information to each business unit to determine the optimum feedback between cost and revenue.
b. Generate decisions that are in harmony with ideals, which means that the system must be designed in such a way so that decisions that increase the profit of the business unit will also increase the company's profit.
c. Help to measure the economic performance of individual business units.
d. Make sure that the system is easy to be understood and managed.

The current business development ensures transfer pricing is not only done between divisions in one domestic company. Transfer pricing can also occur in multinational companies. The price of transactions occurring within a multinational company is better known as transfer pricing. So, transfer pricing can be interpreted as a number of prices sold and purchased by fellow subsidiaries or branches (Utari et al. 2016).

Therefore, it can be said that transfer pricing is the price of the sale and purchase transactions between the parent company and the subsidiary company or its branch company with the method determined by the company.

1.1.5 Results of previous research
This research uses textbook as material of theory study. However, the authors also use previous

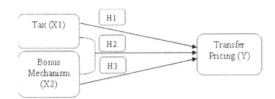

Figure 1. Research framework.

researchers' reference material in order to get a better understanding of the phenomena associated with transfer pricing.

1.1.6 Research framework

Based on the framework that has been described previously, the hypotheses proposed in this study are:

a. Tax expenses have an influence on transfer pricing.Multinational companies avoid the tax burden by making earnings management between branch companies. Therefore, the tax burden can be an influence factor for companies in conducting transfer pricing practices.
H1: tax expenses have an influence on transfer pricing.
b. Bonus mechanism has an influence on transfer pricing.Management trends utilize transfer pricing transactions by making earnings management to maximize the bonuses they will receive. Therefore, the bonus mechanism can be an influence factor for the company in transfer pricing practices.
H2: bonus mechanism has an influence on transfer pricing.
c. Tax expenses and bonus mechanism simultaneously have an influence on transfer pricing. Earning management which is done by multinational companies to avoid taxes can also be done to maximize the bonus. Therefore, taxes and bonus mechanisms can simultaneously influence the practice of transfer pricing.
H3: Tax expenses and bonus mechanism simultaneously have an influence on transfer pricing.

2 METHOD

2.1 Research method

According to Sugiyono (2015), the method of quantitative research is a method based on the philosophy of positivism, used to examine a population or a particular sample; data collection uses research instruments, quantitative/statistical data analysis, with the aim to test the hypothesis that has been established.

This research uses the quantitative method. Research data is in the form of numbers and analyzed by using SPSS (Statistical Package for the Social Sciences).

2.2 Research variable

2.2.1 Conceptual definitions

Based on the theoretical basis that has been described previously, the conceptual definitions in this study include:

a. Tax is a contribution of citizens, both of private and business entities, to the state, which is mandatory, reciprocity of the fee is in the form of public facilities for the welfare of citizens.
b. Bonus mechanism is the company's strategy in rewarding the performance of the board of directors on the basis of the company's profit calculation.
c. Transfer pricing is the price of the sale and purchase transactions between the main company and the subsidiary company or its branch company with the method determined by the company.

2.2.2 Operational definitions

The operational definition in this research is to calculate the influence of tax and bonus mechanism on transfer pricing at a manufacturing company in IDX in the period 2012–2016 by using the specified sample. Therefore, it can be seen that taxes and bonus mechanisms are factors that influence companies in transfer pricing practices in business activities.

According to Yuniasih et al. (2012), tax can be measured with the effective tax rate which is the ratio of tax expense minus differed tax expense divided by taxable income. So, in the calculation later, the value of the tax expenses is converted into natural logs.

The bonus mechanism can be measured based on the net profit trend index, the percentage of net income achievement in year t against the net income of year t-1 (Yuniasih et al. 2012).

Transfer pricing is calculated by a dichotomous approach, that is, by looking at the existence of the sale to a related party. For companies that make a sale to a related party a value of 1 is assigned and for a not-related party a value of 0 is given (Yuniasih et al. 2012).

2.2.3 Research instrument

According to Sugiyono (2015), a research instrument is a tool used to measure the natural and social phenomena which are observed. Specifically, all of these phenomena are called research variables.

Based on this understanding, the instrument of research in each of these research variables are as follows.

2.3 Method of collecting data

2.3.1 Population, sample, and sampling technique

Population in this research is a manufacturing company which is included into company LQ 45, in the period 2012 until 2016. There are 10 companies in this study and we used a saturated sample in which all members of the population were regarded as research samples.

2.3.2 *Data resources*

The data used in this study is secondary data, data obtained through intermediary media, such as archives, both published and unpublished. The secondary data in this study referred to the form of company's financial statements of 2012 to 2016 that can be accessed through the official website of the Indonesia Stock Exchange.

2.3.3 *Data collection technique*

Data collection techniques used in this study are documentation study. Documentation study is a type of data collection that examines various documents useful for research material.

This study uses documents in the form of previous research journals, relevant books, and financial statements of sample research available on the official website of the Indonesia Stock Exchange.

2.3.4 *Method of data analyses*

2.3.4.1 Inferential statistics analyses

Inferential statistical analysis is used for hypothesis testing. Hypothesis testing in this research is using the logistic regression model. Logistic regression is a regression used to test whether the probability of occurrence of dependent variables can be predicted with independent variables. This analysis technique no longer requires the test of normality and heteroscedasity, it means the dependent variable does not require homoscedacity for each independent variable. The purpose of the normality and heteroscedasity test is that the regression analysis model used in the study yields a sophisticated parametric value. Logistic regression is used to test the analysis of the influence of tax expenses and the bonus mechanism on transfer pricing. This test is performed at the level of significance (α) 5%.

The use of logistic regression analysis is because the dependent variable is dichotomous. Analytical techniques in processing this data no longer require the test of normality and test the classical assumption on the independent variables (Ghozali 2011).

2.3.5 *Hypothesis testing research*

2.3.5.1 Assessing the overall model

This test is used to assess the model that has been hypothesized to be fit or not with the data. The hypothesis for assessing the fit model is:

Ha: The hypothesized model fits the data.

Ho: The hypothesized model does not fit the data.

A model can be said fitting with the data if Ha is accepted. Statistics are used by Likelihood. Likelihood L of the model is the probability that the model is hypothesized describing the input data. The existence of a value reduction between -2LogL, initial (initial - 2LogL, function) with the -2LogL value in the next step indicates that the model that is hypothesized fits the data (Ghozali 2011). Log likelihood on logistic regression is similar to the definition "Sum of Square Error" in the regression model, so the decrease in Log Likelihood shows a better regression model.

2.3.5.2 Assessing the eligibility of the regression model

The feasibility of the regression model is using the Hosmer and Lemeshow Goodness of Fit Test. The hypothesis for assessing the feasibility of the regression model is:

Ha: There is no difference between the model and the data.

Ho: There is a difference between the model and the data.

If the statistical value of Hosmer and Lemeshow goodness of fit is less than 0.05, then the null hypothesis is rejected, which means there is a significance difference between the model and the observed value so that the goodness of fit model is not good because the model cannot predict the observation value. If the value of Hosmer and Lemeshow goodness of fit is greater than 0.05 then the null hypothesis cannot be rejected and means the model is able to predict the observed value or it can be said that the model is acceptable because it is in accordance with the observational data (Ghozali 2011).

2.3.5.3 Coefficient of determination

Nagelkerke R Square is a test conducted to find out to what extent independent variables are able to explain and influence the dependent variable. The value of Nagelkerke R Square varies between 1 (one) and 0 (zero). The closer the value is to 1, then the more the model is considered goodness of fit, while closer to 0 means the model is not considered goodness of fit (Ghozali 2011).

2.3.5.4 Multicollinearity test

According to Ghozali (2011), the multicollinearity test aims to test whether the regression model found a correlation between independent variables.

A good regression model is a regression that does not experience strong correlation symptoms among its independent variables. To facilitate statistical calculations, this study is assisted by SPSS software in the data processing.

2.3.5.5 Classification matrix

The classification matrix shows the predictive power of the regression model to predict the possibility of transfer pricing by the firm.

2.3.5.6 Simultaneous testing

Simultaneous testing in logistic regression uses the Chi-Square value of the difference between -2 Log likelihoods before the independent variables enter the model and -2 Log likelihoods after the independent variables enter the model. This test is also called Maximum likelihood testing.

2.3.5.7 Parameter estimation and interpretation

The analysis used in this research is logistic regression analysis, by analyzing of the effect of tax burden and bonus mechanism to transfer pricing at a manufacturing company. Hypothesis testing is done by comparing the probability value (sig) to the significance level (α). If the significant number is greater than α (0.05) then Ha is accepted and Ho is rejected, which means the independent variable has no significant effect on the occurrence of the dependent variable. The model used in this research is as follows:

$$Ln \frac{p(transferpricing)}{1 - p(transferpricing)} = \alpha + \beta 1 + \beta 2 + e$$

Information:
α = Constants
β1 = Tax Expenses
β2 = Bonus Mechanism
e = Error

3   RESULT AND DISCUSSION

3.1   Company profile

The research samples are companies which appear on the LQ 45 list in the period 2012–2016. LQ 45 is an index in the Indonesia Stock Exchange consisting of 45 emitters with different sectors of business and have a high value selected through several criteria of voters. The Indonesia Stock Exchange is routinely supervised with issuers included in the calculation of the LQ 45 index and evaluated on those shares every 3 (three) months.

The list of shares that becomes the reference for the index LQ 45 in the Indonesia Stock Exchange has been evaluated every 6 (six) months, that is, the period of February-July, and August until January. This period occurred since the first stock divestment LQ 45 index was launched, that is, in February 1997. Various issuers to be included in the LQ 45 calculation index are the following factors:

a. Listed on the BEI at least 3 months
b. Transaction activities in the regular market are value, volume and frequency of transactions.
c. The number of trading days in the regular market.
d. Market capitalization over a period of time.
e. In addition, to considering the above-mentioned liquidity criteria and market capitalization, the financial condition and growth prospects of the company will also be seen.

This study uses the data of the company's financial statements sample period 2012–2016. The data can be obtained from the official website of the Indonesia Stock Exchange. The Indonesia Stock Exchange is often shortened to BEI, or Indonesia Stock Exchange (IDX) and which is the result of the merger of the Jakarta Stock Exchange (JSE) with the Surabaya Stock Exchange (BES). For the sake of operational effectiveness and transactions, the government decided to merge the Jakarta Stock Exchange as a stock market with the Surabaya Stock Exchange as a bond and derivative market. This merged stock started operations on December 1, 2007.

BEI uses a trading system called Jakarta Automated Trading System (JATS) since May 22, 1995, replacing the manual system used previously. Since March 2, 2009, JATS system itself has been replaced with a new system called JATS-NextG provided OMX. The Indonesia Stock Exchange is centered on Indonesia Stock Exchange Building, Niaga Sudirman Area, Jalan Sudirman 52–53, Senayan, Kebayoran Baru, South Jakarta.

IDX provides more complete information about the development of the stock to the public by disseminating stock price movement data through print and electronic media. One indicator of such stock price movements is the stock price index. IDX has several index types, plus ten sectoral index types.

3.2   Results of research and discussion

According to Sugiyono (2015), descriptive statistics are used to analyze data by way of describing or describing the data collected as it is without intending to make conclusions that apply to the public. Descriptive statistics can also be done to look for strong relationships between variables through correlation analysis.

Based on this understanding, the descriptive statistical analysis in this study is intended to describe the data of each variable, namely variable Tax Expense (X1), Bonus Mechanism (X2), and Transfer Pricing (Y). Descriptive statistical analysis is also intended for the writer to know in advance about the description of the three variables in the sample company to be examined before testing the effect of taxes and bonus mechanism on transfer pricing.

3.2.1   Tax expense

The first independent variable in this study is the tax expense symbolized by X1. Taxes are proxied with the effective tax rate which is the ratio of tax expense minus the differed tax expense divided by taxable income. Table 1 is the result of the descriptive analysis of tax variables.

Based on Table 1, it can be seen that the sample company has the lowest tax burden of 275 billion rupiah, while the company has the highest tax burden of 5,226 billion rupiah. The average value derived from the tax burden is 1,882.58 greater than the standard deviation of 1,305,770.

Table 1. Descriptive analysis of tax expense.

	N	Min.	Max.	Sum	Mean	Std. Dev.	Var.
BM	50	275	5226	94129	1882.58	1305.77	1705036.289
Valid N (listwise)	50						

Table 2. Descriptive analysis of the bonus mechanism.

	N	Min.	Max.	Sum	Mean	Std. Dev.	Var.
BM	50	-0.306	0.322	2.343	0.047	0.136	0.019
Valid N (listwise)	50						

3.2.2 Bonus mechanism

The second independent variable in this research is the mechanism symbolized by X2. The bonus mechanism can be measured based on the net profit trend index, the percentage of net income achievement in year t of the t-1 net income. Table 2 shows the descriptive variable analysis of the bonus mechanism.

Based on Table 2, it can be seen that the sample company has the lowest bonus mechanism of − 0.306%. For companies with negative bonus calculations it can be assumed that the company does not provide bonuses, because it is not possible if the bonus is given in a negative amount. The company with the highest bonus mechanism is 0.32%. The average value obtained from the bonus mechanism is 0.047%. The standard deviation and variance values of the mechanisms are 0.136% and 0.019%, respectively.

3.2.3 Transfer pricing

The dependent variable in this research is transfer pricing symbolized by Y. Transfer pricing is the price of sale and purchase transaction between the main company and subsidiary company or its branch company. Transfer pricing is calculated by dichotomous approach, that is, by looking at the existence of the sale to a related party. Companies that make a sale to a related party are rated 1 and not 0.

Information about companies that make a sale to a privileged party can be seen in the Notes to

Financial Statements; how much sales are for external parties and how much of a sale for a related party. Table 3 shows the results of the descriptive analysis of transfer pricing variables.

Based on Table 3, it can be seen that the sample companies that do transfer pricing are as many as 9 companies or, in other words, 90% of the sample companies do transfer pricing and as much as 1 company, or 10% of the sample companies, do not do transfer pricing.

3.3 Inferential statistics analysis

Sugiyono (2015) said that inferential statistics (often called inductive statistics or probability statistics) is an analytical technique used to analyze sample data and the results are applied to the population.

Inferential statistical analysis is used for hypothesis testing. Hypothesis testing in this research was done by using logistic regression model, then by analyzing the influence of tax burden and bonus mechanism to transfer pricing at manufacturing company.

Hypothesis testing in this research consisted of: Assessing Overall Model, Assessing Feasibility of Regression Model, Coefficient of Determination, Multicollinearity Test, Classification Matrix, Simultaneous Test and Parameter Estimation and Interpretation.

3.3.1 Assessing the entire model

This test is used to assess the model that has been hypothesized to be fit or not with the data. The hypothesis for assessing the fit model is:

Ha: The hypothesized model fits the data.

Ho: The hypothesized model does not fit the data.

The model can be considered fit if there is a subtraction value between -2LogL, initial (initial-2LogL, function) with the -2LogL value in the next step indicating that the model is hypothesized fit with the data (Ghozali 2011).

Table 3. Descriptive analysis of transfer pricing.

		Freq.	Percent	Valid Percent	Cumulative Percent
Valid	NTP	5	10.0	10.0	10.0
	TP	45	90.0	90.0	100.0
	Total	50	100.0	100.0	

381

Table 4. Comparison analysis of value -2LogL.

No	Keterangan	Nilai -2LogL
1	Value of -2LogL at beginning (*Block number* 0)	32.508
2	value -2LogL last (*Block number* 1)	32.356
	Nilai Penurunan -2LogL	152

Table 4 shows the comparison between indigo -2LogL initial and -2LogL end.

Based on Table 4, it can be seen that the decrease of -2LogL value of 152 indicates that the hypothesized model has a fit with the data, so Ha is accepted.

3.3.2 *Assessing the eligibility of the regression model*

The feasibility of the regression model was assessed using the Hosmer and Lemeshow Goodness of Fit Test. The hypothesis for assessing the feasibility of the regression model is:

Ha: There is no difference between the model and the data.

Ho: There is a difference between the model and the data.

The model can be considered feasible if the sig value in Hosmer and Lemeshow Goodness of Fit is greater than 0.05, so that Ha is accepted showing that the model is able to predict the observed value or it can be said that the model is acceptable according to the observation data. Table 5 shows the result of the feasibility test of the regression model in SPSS output.

Based on Table 5, it can be seen that the sig value obtained is 0.325. The value is greater than 0.05, which means Ha is accepted and the regression model is feasible for use in further analysis because the model is in accordance with the observational data.

3.3.3 *Coefficient of determination*

Nagelkerke R Square is a test conducted to find out how much independent variables are able to explain and influence the dependent variable. The value of Nagelkerke R Square varies between 1 (one) and 0 (zero). The closer it is to the value of 1 the more the model is considered a goodness of fit while closer to 0 means the model is not considered goodness of fit (Ghozali 2011).

Table 6 shows the results of the coefficient of determination analysis.

Table 5. Feasibility analysis of the regression model Hosmer and Lemeshow test.

Step	Chi square	Df	Sig.
1	9.205	8	0.325

Table 6. Coefficient determination analysis.

Step	-2 Log likelihood	Cox & Snell R Square	Nagelkerke R Square
1	32.356[a]	.003	.006

Based on Table 6, it can be seen that the value of Nagelkerke R Square is equal to 0.006 which indicates that the independent variable is able to explain and influence the dependent variable of 0.6% while the other 0.3% is explained and influenced by other variables outside this study.

3.3.4 *Multicollinearity test*

According to Imam Ghozali (2011), the m multicollinearity test aims to test whether the regression model found a correlation between the independent variables. A good regression model is a regression that does not experience strong correlation symptoms among its independent variables. Table 7 shows the result of the multicollinearity test.

The correlation values between the variables is greater than 0.8. This shows that there are no multicollinearity symptoms that occur between the variables.

3.3.5 *Classification matrix*

The classification matrix shows the predictive power of the regression model to predict the possibility of transfer pricing by the firm. Table 8 shows the classification matrix test results in SPSS.

Based on Table 8, it can be seen that the percentage rate of companies that are in the practice of

Table 7. Multicollinearity test.

		Constant	X1	X2
Step 1	Constant	1.000	-.791	-.222
	X1	-.791	1.000	-.068
	X2	-.222	-.068	1.000

Table 8. Classification matrix analysis.

	Observed		Predicted Transfer Pricing		Percentage Correct
			Non TP	TP	
Step 0	Transfer	Non TP	0	5	.0
	Pricing	TP	0	45	100.0
	Overall Percentage				90.0

transfer pricing is 100% and companies that are not is 0%. Overall models, tax variables and bonus mechanisms can be statistically estimated at 90%.

3.3.6 *Simultaneous test*
Simultaneous testing in logistic regression uses the Chi-Square value of the difference between -2 Log likelihood before the independent variables enter the model and -2 Log likelihood after the independent variables enter the model. This test is also called maximum likelihood testing. Table 9 shows the results of omnibus test analysis to determine the value of Chi-square count.

Based on Table 9, it can be seen that the value of the calculated chi square is smaller than the value of the chi square table (value df 2 on chi square table) being equal to 0.152 and smaller than 5.99. Therefore, it can be concluded that Ho is rejected, and Ha accepted, which means there is no significant simultaneous influence between tax and bonus mechanism on transfer pricing.

3.3.7 *Parameter estimation and interpretation*
Hypothesis testing is done by comparing the probability value (sig) with the significance level (α). If the significant number is smaller than α (0.05) then Ha is accepted and Ho is rejected, which means the independent variable has a significant effect on the occurrence of the dependent variable. Table 10 shows the results of logistic regression analysis in SPPS output v.23.

Based on Table 10, it can be seen that the logistic regression equation in this study are

This study shows that the tax symbolized by X1 is the ratio of the tax burden minus the deferred tax burden with taxable profit; this variable is significant in probability 0.939. The bonus mechanism is

Table 9. Omnibus test analysis.

		Chi-square	Df	Sig.
Step 1	Step	.152	2	.927
	Block	.152	2	.927
	Model	.152	2	.927

Table 10. Logistic regression analysis.

	B	S.E.	Wald	Df	Sig.	Exp (B)
Step 1[a] X1	0.00	0.00	0.01	1	0.94	1.00
X2	-1.38	3.63	0.15	1	0.70	0.25
Constant	2.22	0.87	6.61	1	0.01	9.22

symbolized by X2 and measured by the percentage of net income achievement in year t of the t-1 net income; this variable is significant at prob 0.703.

3.4 *Discussion*

3.4.1 *The effect of tax expense*
The result of the logistic regression analysis in Table 10 shows that the load has a significant value of 0.939 which is greater than the value of α that is equal to 0.05. The results of this study are not consistent with the proposed hypothesis where Ha is rejected because the value of significance at the tax burden is greater than the value of α.

This study also does not support previous research from Nurjanah et al. (2016) and Kusuma (2016). Differences in research results may be due to differences in object research. Previous research focused on manufacturing companies in BEI, while this research focused on manufacturing companies from LQ 45. LQ 45 company itself is a company that has high liquidity value, where the current debt can be repaid with current assets. From the low tax burden in the LQ 45 corporation it cannot be assumed that the company does earnings management with transfer pricing. Transfer pricing that exists within the company also shows that the characteristics of LQ 45 manufacturing companies is different from the characteristics of multinational companies that become the phenomenon of this research.

3.4.2 *The influence of the bonus mechanism*
The result of the logistic regression analysis in Table 10 shows that the bonus mechanism does not significantly influence the transfer pricing. The bonus mechanism has a significant value of 0.703 which is greater than the value of α which is 0.05, so Ha is rejected and Ho accepted. The results of this study are not consistent with the proposed hypothesis. This study does not support previous research from Nurjanah et al. (2016) but supports research from Kusuma (2016), which states that the bonus mechanism has no effect on transfer pricing. Differences in research results may be due to the owners of the companies in giving bonuses to the directors and not only see the earnings of one company that performs transfer pricing, but also see the overall profit of all companies that conduct transfer pricing within a group of companies.

3.4.3 *Simultaneous effect of tax expense and bonus mechanism*
Based on Table 9, it can be seen that the value of calculated chi square is smaller than the value of the chi square table (value df 2 on the Chi-square table), beingequal to 0.152, which is smaller than 5.99. Therefore, it can be concluded that Ha is rejected and Ho accepted, which means there is no significant simultaneous influence between tax and the bonus mechanism on transfer pricing. The results of this

study do not support the transfer pricing theory of Utari et al. (2016), which states that foreign companies generally increase profits by avoiding taxes in countries where they invest through the transfer pricing system. Differences in research results may be due to the characteristics of the manufacturing companies LQ 45 which are different from the characteristics of foreign companies in their intent according to Utari et al. (2016). The theory will be the same if the characteristics of the company are the same, as described previously. LQ 45 Company is a company that has a high liquidity value, which means that current liabilities can be immediately settled using current assets. From the low tax burden and corporate earnings in LQ 45 it cannot be assumed directly that the company is transferring pricing.

4 CONCLUSION

Based on the results of the data analysis and the discussion that has been described previously, it can be concluded that the tax burden has no significant effect on the transfer pricing. From the low tax burden in a manufacturing company LQ 45 it cannot be assumed directly that the company is transferring pricing, since the characteristics of a manufacturing company LQ 45 can be different from the characteristics of multinational corporations which are the phenomenon of this research. The bonus mechanism has no significant effect on transfer pricing. The result may be that the owner of the company, in giving bonuses to the directors, not only saw the earnings of one company transferring pricing but saw the overall earnings of all companies that transfer pricing within a group of companies. Tax expense and bonus mechanism have no significant effect simultaneously on transfer pricing. The low tax burden and profit of manufacturing companies LQ 45 cannot be assumed directly to state that the company is

transferring pricing, because the characteristics of manufacturing companies LQ 45 can be different from the characteristics of multinational companies which became the phenomenon of this research.

REFERENCES

Agnes, W.Y.L., Raymond M.K., Wong,W. & Michael F. 2010. Tax, financial reporting, and tunneling incentives for income shifting: An empirical analysis of the transfer pricing behavior of Chinese-listed companies. *JATA American Accounting Association* 32(2): 1–26.

Anthony, N., Robert, R. & Govidarajan, V. 2007. *Management control systems, 12th edition.* New York: McGraw-Hill Higher Education.

Ghozali, I. 2011. *Application of multivariate analysis with SPSS program.* Semarang: Diponegoro University.

Kusuma, N.A. 2016. Analysis of the influence of tax expenses, tunneling incentives, and bonus mechanisms on multinational corporate transfer pricing listing on the Indonesia stock exchange. *Thesis Faculty of Economics.* Semarang: State University.

Mardiasmo, M. 2011. *Taxations, revised editions.* Yogyakarta: Andi Publishing.

Nurjanah, N., Ika, I., et al. 2016. Determinants factors of the company's decision to transfer pricing. Full paper of lambung mangkurat university.

Sugiyono, S. 2015. *Qualitative and quantitative research methods and R & D.* Bandung: Alfabeta.

Tempo Inti Media. 2014. *Prahara pajak raja otomotif lika-liku transfer pricing.*

Trading Economics. 2017. *Corporate Tax Rate.*

Utari, U., Dewi, D., Ari, P. & Prawironegoro D. 2016. *Management accounting, fourth editions.* Jakarta: Mitra Wacana Media.

Waluyo, W. 2013. *Indonesian taxation.* Jakarta: Salemba Empat.

Yuniasih, Y., Wayan, W., Rasmini N.N.K. & Wirakusuma M.G. 2012. The effect of tax and tunneling incentive on the transfer pricing decision of manufacturing companies which listing on the Indonesia stock exchange. *Journal of Udayana University.*

Advances in Business, Management and Entrepreneurship – Hurriyati et al (eds)
© 2020 Taylor & Francis Group, London, ISBN 978-0-367-27176-3

Forecasting volatility stock price using the ARCH/GARCH method: Evidence from the Indonesia stock exchanges

L.P. Anggita, N. Nugraha & I. Waspada
Universitas Pendidikan Indonesia, Bandung, Indonesia

ABSTRACT: This study aims to predict stock price volatility using the ARCH/GARCH model in Indonesia. The research method used is a quantitative method on the Indonesia stock price index for the period 2011–2017. The analysis technique used the ARCH/GARCH model with data processing using Eviews 9 software. The results show the best volatility model in predicting stock price is the EGARCH model and the level of accuracy of forecasting shows an error rate of less than 5 percent, so the model is quite accurate in predicting stock prices.

1 INTRODUCTION

The volatility of stock price movement is an increase or decrease in the stock price on the stock exchange and can be used as a measuring tool to measure the risk of a stock. Investments in stocks are known as being on high-risk high-return terms, meaning that if an investor wants a higher rate of return, then he should dare to take higher risks (Edelen et al. 2016). Investing in stocks can make shareholders gain huge amounts of profit in a short period of time but it can also bring them big losses in a short time. This is meant with return and risk in investing shares, providing a unidirectional relationship: when high returns are to be obtained by shareholders, the higher the risk will have to be for shareholders. Stock returns are an important indicator of investing to provide satisfaction for investors (Mclean & Pontiff 2016). The movement of returns or stock returns on the composite stock price in Indonesia on average shows considerable movement, released by OJK (Financial Services Authority in Indonesia) 2017. The occurrence of large volatility in stock prices at the return of the composite stock price index in Indonesia indicates a lack of stability in rising and falling stock prices and may impac the level of returns or the returns that will be received by investors and other stakeholders. In order to get profit in shareholder investment one needs to predict the stock price that will occur and when the stock should be sold and purchased by the company, using a particular model (Padmakumari & Maheswaran 2017). Information is needed for stakeholders in predicting firm pricing, thus some pioneering research on price volatility has been done since the 1980s in the United States Composite Stock Price Index (Shiller 1980). In addition, in the period 1926–1987 research on stock price volatility was performed on the S & P500 Composite Index (French et al., 1987). Furthermore, the development of stock price

volatility patterns in developed countries is done by analyzing the comparison of root test results on previous researchers (West 1988). The volatility of the stock market varies greatly on important issues for the international investment process (Trivedi & Birău 2017). The response of variables in calculating volatility is not only influenced by exogenous but also influenced by its own variables from behavior of the past. On the basis of the foundation of the theory, the autoregressive model was developed. Box Jenkins time series modeling is essential in analyzing stochastic processes (Box et al. 2016). Forecasting stock price volatility has evolved the usual volatility models used from simple to complicated calculations. Volatility models include Standard Deviation (SD), Simple Moving Average (SMA), Exponential Weighted Moving Average (EWMA), Autoregressive Conditional Heteroscedasticity (ARCH) or Generalized Autoregressive Conditional Heteroscedasticity (GARCH) (Brailsford & Faff 1996).

Engle (1982) introduced the model of Autoregressive Conditional Heteroscedasticity (ARCH), which serves to model the residual volatility that often occurs in the financial data. When testing the ARCH model there is often variance which depends on the volatility of several periods in the past. This raises the number of parameters in conditional variance which must be estimated. Estimating these parameters is difficult to do with precision. In 1986 the ARCH model developed into the more general form of Generalized Autoregressive Conditional Heteroscedasticity (GARCH); this model is considered to have symmetrical volatility response characteristics to shocks (Bollerslev 1986). As long as the intensity is the same, the volatility response to a shock as well as positive and negative shocks is aligned with a more specific theory of a market equilibrium model, which is a model that shows the nature of the equilibrium market when the price "fully reflects" the available information (Fama 1969). The GARCH model of conditional variance is not only influenced by the past residual but by the conditional variant's lag itself. The development of the GARCH model further accommodates the possibility of an asymmetric volatility response when negative information has a greater impact than positive information. The techniques in modeling GARCH responses are the following two modeling techniques: Threshold GARCH (TGARCH) and Exponential GARCH (EGARCH) model. Previous studies have suggested that the GARCH (1.1) model is used to forecast monthly US stock market data (Akgiray 1989). The American stock market in the period 1834–1925 shows that conventional ARCH and GARCH cannot capture stock return behavior because there is asymmetric data, so the best model is EGARCH (Pagan & Schwert 1990). EWMA and ARCH/GARCH models can predict Japanese and Singapore stocks (Tse 1991), (Kuen & Hoong 1992). Japan's capital markets show that EGARCH is a model that most accurately predicted the stock price (Engle & Ng 1993). The results show the GARCH-M model (1.1) is the best model in explaining stock price volatility in China (Song et al. 1998). Aggarwal et al. (1999) observed the GARCH, EGARCH, GJR, GQARCH models on the Hang Seng index. Research on Hongkong, Indonesia, Japan, Korea, Malaysia, Philippines, Singapore, Taiwan, Thailand, and the US with samples from 1990–1999 shows that the GARCH (1.1) model can predict stock return (Chiang et al. 2000). The EGARCH asymmetric model performance is better compared to other GARCH models (Giot & Laurent 2004), (Karanasos et al. 2014). Research on stock prices of Singapore, Indonesia, Philippines and Malaysia shows the GARCH (1.1) model can predict (Wong & Kok 2005). The results show that the EGARCH model in KOSPI Korea Exchange and TASE Index is the best model in forecasting the volatility of stock price index (Roh 2007) (Alberg et al. 2008). The stock exchange in India in 2003–2008 showed that the GARCH model provides the most accurate estimates and is the best model compared to other models (Srinivasan 2017). Other studies on the S&P 500 stock index in the US from 31 March 1964 to 31 December 2014 show the estimated parameters of the GARCH model can determine the volatility of stock prices resulting in better forecasts (Byun 2016).

The daily closing index of stocks in the United States, Britain, Japan, France, Hungary, Poland, Romania and Slovakia using the GARCH model shows that developed countries are more stable in stock price movements than developing countries (Trivedi & Birău 2017). The Nordic Stock Exchanges of the Norwegian index, the Danish index, the Swedish index and the Finnish index of 1983–2016 show the result that each market is well described in the ARCH/GARCH model except the Swedish index (Dritsakis & Savvas 2017). The research conducted on the Stock Index in Malaysia,

Indonesia, Japan and Hongkong in 1998–2015 shows the EGARCH model can predict stock prices in Malaysia, Indonesia, and Japan, while the EWMA model can predict stock prices in Hongkong (Lee et al. 2017). The result of the research on the volatility of the Indonesian stock price from July 2007 to September 2015 shows that the EGARCH model can be used to predict stock prices (Awalludin et al. 2018). Based on research studies that have been conducted by previous researchers, Indonesia is a country that includes emerging markets and similarly sharp price volatility also occurred in the price of joint stocks in Indonesia. This study will be conducted to assess the forecasting of stock prices by observing the stock price volatility at stock prices. The combination in Indonesia of using the ARCH/GARCH model is in line with previous research (Roh 2007, Alberg et al. 2008, Ou & Wang 2010, Srinivasan 2017, Byun 2016, Trivedi & Birău 2017, Dritsakis & Savvas 2017).

2 METHOD

This study uses a quantitative method and descriptive analytic method that aims to obtain an overview of the level of stock price volatility and forecasting in the stock prices in Indonesia by using a model of ARCH/GARCH. The variable studied in this research is about the volatility of the stock price of IHSG in Indonesia in 2011–2017. Types of data used are secondary data with data collection techniques, documentation and literature study. Figure 1 shows the steps of the data analysis techniques using the ARCH/GARCH model and using software Eviews 9.

3 RESULT AND DISCUSSION

The result of the descriptive statistics for composite stock price indices in Indonesia can be seen in Table 1. The result of the data stationarity test, using the ADF test, shows that the composite stock price index in

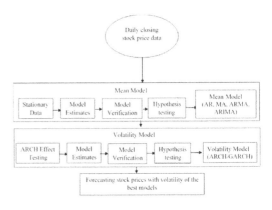

Figure 1. ARCH/GARCH model workflow

Indonesia obtained a probability value of 0.00; because the probability value < 0.05, it can be said that the data are stationery and can be used to make a mean model. The mean model of the composite share price index in Indonesia is MA (3). The selection of this model means is done by using the smallest AIC and SC values. After the selection of the model the hypothesis was tested through the partial test and simultaneous test, with the simultaneous test results being 17.25144 and the probability 0.00, while results for the partial test are -10.31503 and a probability of 0.00. The partial and simultaneous test results show that the resulting model mean is the best model that can be used for modeling the volatility.

Table 2 shows the verification of the best model carved out of the smallest AIC and SC values so that the volatility model for the composite stock price index in Indonesia is obtained by the EGARCH model with the following equation:

Volatility Model EGARCH:

$$\sigma_t^2 + 0.000152 \ e_{t-1}^2 - 0.090943 \ \sigma_{t-1}^2 \qquad (1)$$

This research shows the EGARCH model to be the best model to predict stock prices at the composite stock price index in Indonesia. The EGARCH model is able to capture the phenomenon asymmetric effect which occurs when negative information has a greater impact than positive information. In addition, the existence of asymmetric information in the form of differences in information obtained by investors causes the difference in return.

The data is said to be bad news when the volatility decreases while the condition is said to be good news when the volatility increases periodically. Asymmetrical effects often occur in financial data, especially in stock trading because it is usually a response in the sense that market turmoil is more violent when news is negative rather than positive.

The financial return has three characteristics: big changes tend to be followed by big changes and

Table 1. Testing of data stationarity and mean model of stock price index in Indonesia.

	IDX Composite in Indonesia
Mean	0.000353
ADF	−16.90221
Probability	0.00
Decision	Stationer
Mean Model	MA 3
T-test	−10.31503
Probability	0.00
F-test	17.25144
Probability	0.00

Source: Data processing results 2018.

Table 2. Verification of the volatility model of the stock price index in Indonesia.

Stock	Mean Model	Volatility Model	AIC	SC
IHSG	ARIMA (3,1,3)	ARCH (1)	−6.384838	−6.372451
		GARCH (1,1)	−6.55804	−6.542556
		ARCH-M	−6.383721	−6.368238
		TGARCH	−6.563278	−6.541601
		EGARCH	−6.567389	−6.545711

Source: Data processing results 2018.

minor changes; fat tailedness in financial return often displays a fat tail greater than the normal distribution of stands and the leverage effect is the result of negative return resulted from positive return on the same size (Mehmet 2008).

The results show that the EGARCH model becomes the best model in predicting stock prices not only in Indonesia but in developing countries as well, and even in developed countries. In line with previous research conducted on the American, the strongest data attributes so that the best model is EGARCH (Pagan & Schwert 1990).

The Japanese capital market shows that EGARCH is the most accurate model for forecasting stock prices (Engle & Ng 1993). The performance of the EGARCH asymmetric models is better compared to other GARCH models (Giot & Laurent 2004, Karanasos et al. 2014). The results show that the EGARCH model in KOSPI Korea Exchange and TASE Index is the best model in forecasting the volatility of the stock price index (Roh 2007, Alberg et al. 2008).

4 CONCLUSION

The EGARCH model shows to be a more accurate model in making model volatility; consistency of the research result can be seen from the result of previous research which has been done from 1998 until now. Knowing stock price movements will give an idea to investors. High volatility causes relatively high risks and uncertainties and vice versa. The best volatility model obtained using EGARCH will help the investor in establishing the policy of investing.

REFERENCES

Aggarwal, R., Inclan, C. & Leal, R. 1999. Volatility in emerging stock markets. *Journal of Financial and Quantitative Analysis*, 34(1): 33–55.

Akgiray, V. 1989. Conditional heteroscedasticity in time series of stock returns: Evidence and forecasts. *Journal of Business*: 55–80.

Alberg, D., Shalit, H. & Yosef, R. 2008. Estimating stock market volatility using asymmetric GARCH models. *Applied Financial Economics*, 18(15): 1201–1208.

Awalludin, S.A., Ulfah, S. & Soro, S. 2018, January. Modeling the stock price returns volatility using GARCH (1.1) in some Indonesia stock prices. *Journal of Physics: Conference Series*, 948(1): 012068.

Bollerslev, T. 1986. Generalized autoregressive conditional heteroskedasticity. *Journal of Econometrics*, 31(3), 307–327.

Box, G.E., Jenkins, G.M., Reinsel, G.C. & Ljung, G.M. 2015. *Time series analysis: Forecasting and control* (fifth ed.). New Jersey: John Wiley & Sons.

Brailsford, T.J. & Faff, R.W. 1996. An evaluation of volatility forecasting techniques. *Journal of Banking & Finance*, 20(3): 419–438.

Byun, S.J. 2016. The usefulness of cross-sectional dispersion for forecasting aggregate stock price volatility. *Journal of Empirical Finance*, 36: 162–180.

Chiang, T.C., Yang, S.Y. & Wang, T.S. 2000. Stock return and exchange rate risk: evidence from Asian stock markets based on a bivariate GARCH model. *International Journal of Business*, 5(2): 97–117.

Dritsakis, N. & Savvas, G. 2017. Forecasting Volatility Stock Return: Evidence from the Nordic Stock Exchanges. *International Journal of Economics and Finance*, 9(2): 15–31.

Edelen, R.M., Ince, O.S. & Kadlec, G.B. 2016. Institutional investors and stock return anomalies. *Journal of Financial Economics*, 119(3): 472–488.

Engle, R. F. 1982. Autoregressive conditional heteroscedasticity with estimates of the variance of United Kingdom inflation. *Econometrica: Journal of the Econometric Society*, 50(4): 987–1007.

Engle, R.F. & Ng, V.K. 1993. Measuring and testing the impact of news on volatility. *The Journal of Finance*, 48(5): 1749–1778.

Fama, E.F. 1969. Efficient Capital Markets: A Review of Theory and Empirical Work. *The Journal of Finance*, 25(2): 28–30.

French, K.R., Schwert, G.W. & Stambaugh, R.F. 1987. Expected Stock Returns and Volatility. *Journal of Financial Economics*, 19: 3–29.

Giot, P. & Laurent, S. 2004. Modelling daily value-at-risk using realized volatility and ARCH type models. *Journal of Empirical Finance*, 11(3): 379–398.

Karanasos, M., Paraskevopoulos, A.G., Ali, F.M., Karoglou, M. & Yfanti, S. 2014. Modelling stock volatilities during financial crises: A time varying coefficient approach. *Journal of Empirical Finance*, 29: 113–128.

Kuen, T.Y. & Hoong, T.S. 1992. Forecasting volatility in the Singapore stock market. Asia Pacific *Journal of Management*, 9(1): 1–13.

Lee, S.K., Nguyen, L.T. & Sy, M.O. 2017. Comparative study of volatility forecasting models: the case of Malaysia, Indonesia, Hong Kong and Japan stock markets. *Economics*, 5(4): 299–310.

McLean, R.D. & Pontiff, J. 2016. Does academic research destroy stock return predictability? *The Journal of Finance*, 71(1): 5–32.

Mehmet, A. 2008. Analysis of the Turkish Financial Market with Markov Regime Switching Volatility Models. *The Middle East Technical University*.

Ou, P. & Wang, H. 2010. Financial volatility forecasting by least square support vector machine based on GARCH, EGARCH and GJR models: evidence from ASEAN stock markets. *International Journal of Economics and Finance*, 2(1): 51–64.

Padmakumari, L., & Maheswaran, S. 2017. A new statistic to capture the level dependence in stock price volatility. *The Quarterly Review of Economics and Finance*, 65: 355–362.

Pagan, A.R. & Schwert, G.W. 1990. Alternative models for conditional stock volatility. *Journal of econometrics*, 45(1–2), 267–290.

Roh, T.H. 2007. Forecasting the volatility of stock price index. *Expert Systems with Applications*, 33(4), 916–922.

Shiller, R.J. 1980. Do Stock Prices Move Too Much To Be Justified By Subsequent Changes in Dividends? *NBER Working Paper* (456).

Song, H., Liu, X. & Romilly, P. 1998. Stock returns and volatility: an empirical study of Chinese stock markets. *International Review of Applied Economics*, 12(1), 129–139.

Srinivasan, K. 2017. Modeling the Symmetric and Asymmetric Volatili ty for Select Stock Futures in India: Evidence from GARCH Family Models. *Ushus-Journal of Business Management*, 12(1), 61–82.

Trivedi, J. & Birău, R. 2017. Investigating international transmission patterns of stock price volatility 2 Literature review. *Journal of Financial Economics*, 126–131.

Tse, Y.K. 1991. Stock returns volatility in the Tokyo Stock Exchange. *Japan and the World Economy*, 3(3), 285–298.

West, K.D. 1988. Bubbles, fads and stock price volatility tests: a partial evaluation. *The Journal of Finance*, 43(3), 639–656.

Wong, Y.C. & Kok, K.L. 2005. A comparison forecasting models for ASEAN equity markets. *Sunway Academic Journal*, 2, 1–12.

Advances in Business, Management and Entrepreneurship – Hurriyati et al (eds)
© 2020 Taylor & Francis Group, London, ISBN 978-0-367-27176-3

Peer group model as a reference for bank performance assessment

S. Sugiarto
Universitas Pelita Harapan, Jakarta, Indonesia

S. Karnadi
Sekolah Tinggi Ilmu Ekonomi Wiyatamandala, Jakarta, Indonesia

ABSTRACT: Banks peer group is a group of banks that have common conditions. The plotting of a bank peer group is based on the homogeneity of the banks condition relative to the other banks. The purpose of this study was to test the bank peer group model formulated by Sugiarto using cross section data per December 2016. To simulate the performance of the bank peer group model, Bank Jasa Jakarta will be used as an anchor bank. The test was carried out on several variables; Non-Performing Loans, Return on Assets and Return on Equity, all of which have an important role in bank sustainability. Data was taken from the website of the Indonesia Financial Services Authority. The test results indicated that the bank peer group model was a reliable model since the results obtained were able to represent the actual conditions. The test results showed that it was possible to establish an objective bank peer group determination model using a statistical point of view and estimation in the form of data distribution was also possible to do. The decision making regarding the conclusions made are based on a calculated probability.

1 INTRODUCTION

A Bank's rating condition is the result of the Bank's assessment according to their risk and performance (SE BI. No. 13/24/DPNP-Bank Indonesia Regulation number 13/1/PBI/2011). The assessment of a bank operating in Indonesia is measured by considering the peer group, which is carried out by considering various aspects such as similarities of level, trend, structure, and stability, taking into account the performance of peer groups. Bank's peer group is a group of banks that have similarities, although there are also differences among them. In determining the peer group, banks need to pay attention to the business scale, characteristics, and/or complexity of the bank's business (Sugiarto 2018).

In order to carry out the responsibility of the bank's business continuity, the board of directors and commissioners of the banks are responsible for maintaining and monitoring the bank's condition as well as taking steps to maintain and improve the bank's condition. The bank's peer group can be used by banks to evaluate the progress of their performance over time, compared to the performance of banks in their peer group. According to De Aghion and Gollier (2000), peer group systems can be viewed as an effective risk pooling mechanism, which is able to enhance efficiency, not just in the full information set up. As stated by Glewwe (1997), if the peer group effects exist, governments may be able to manipulate them to better achieve policy objectives.

According to the important role of a bank's peer group, the determination of the bank's peer group should be done objectively. Empirical data showed that many banks operating in Indonesia establish their peer groups subjectively, which is not scientifically accountable (Sugiarto 2018). By viewing the needs of banks operating in Indonesia, several important issued needed to be addressed; first, the possibility of establishing an objective bank peer group determination model using a statistical point of view; second, an analysis of a bank's relative position compared to its partners in a peer group based on the bank peer group model; and third, the variables that a bank needs to emphasize to improve its performance.

Sugiarto (2017a) created a model of a bank peer group using statistic probabilistic approach which can be used to analyze the relative position of a bank compared to other banks in the peer group. This research was conducted to test the reliability of the bank peer group model by using cross section data. To simulate the performance of the bank peer group model, Bank Jasa Jakarta was used as an "anchor" bank on the basis of consideration that throughout its history, Bank Jasa Jakarta has proved itself a bank with consistent performance and achievement.

2 METHOD

The data was taken from a bank publication report published by Otoritas Jasa Keuangan (OJK, Financial Services Authority).

Data derived from 71 banks (Commercial Banks, Foreign Exchange Banks) was analyzed using cross section data per December 2016. For the purpose of testing the performance of the peer group model, the performance analysis of Bank Jasa Jakarta was compared to the peer group, especially the variables that significantly affected the performance of Bank Jasa Jakarta such as NPL (Non-Performing Loan), ROA (Return on Assets) and ROE (Return on Equity).

3 RESULT AND DISCUSSION

The 17 selected banks were banks belonging to BUKU 2 and 3. Banks classified as BUKU 1 when its core capital was close to IDR 1 Trillion. After considering the similarity of core capital, the filtering continued by looking at the similarity of the bank's business focus and the anomalies of the banks. In order to obtain a more affiliated peer group member, it was necessary to eliminate Stanchart and BRI Agro considering the business suitability of those banks.

After the advanced screening was carried out, the banks were included in the peer group. Based on the results of the processing, we obtained 15 banks as Peer Group of Bank Jasa Jakarta and measured the ranking on the acquisition of bank performance results based on the data of financial statements of bank publications. Bank Jasa Jakarta's core capital is ranked seventh.

After the peer group model was formed, their performances were analyzed based on NPL (Non-Performing Loan), ROA (Return on Assets) and ROE (Return on Equity).

A major challenge faced by banks is to ensure that a debtor uses the funds wisely so that the likelihood of repayment is enhanced (Stiglitz 1990). In connection with this, the performance of the institution is reflected by the management of the Non-Performing Loan (NPL) variable (Sugiarto 2018). Regarding to NPL, the ranking is done by an ascending model from small to large. A rating of 1 is given to a bank with the best NPL performance, which after the capital equity equalization, generates the smallest NPL, thereby declaring a low credit default opportunity. In this case, although the core capital of Bank Jasa Jakarta (BJJ) is not the largest among the existing banks in the peer group, Bank Jasa Jakarta was able to be in the second-best position, according to the management of NPL. The NPL of Bank Jasa Jakarta of 0.51% was far below the NPL limit set by the Indonesian banking authorities. In accordance with the provisions stipulated by Bank Indonesia, a good bank should keep its NPL below 5%. Thus, Bank Jasa Jakarta is a bank that has an excellent ability in managing credit risk.

Regarding ROA, the ranking was done by an ascending model from small to large. Rank 1 was given to a bank that had the best ROA performance, which after the capital equity equalization, generated the largest ROA, thereby declaring the bank's ability to generate profits. The greater the ROA generated by a bank, the better the bank's position. In this case, although Bank Jasa Jakarta (BJJ)'s core capital was not the largest among the existing banks in the peer group, Bank Jasa Jakarta was able to show that they were the best in ROA management. After capital equity equalization, ROA of BJJ of 2.74% which was the highest compared to other banks in the peer group. The value of ROA of Bank Jasa Jakarta exceeded 1.5% which was the best standard for ROA. Thus, Bank Jasa Jakarta is a bank that has an excellent ability in generating profit with regard to ROA.

Regarding the ROE variable, the ranking was done by an ascending model from small to large. Rank 1 was given to a bank with the best ROE performance, i.e. generated the largest ROE, thereby declaring the bank's ability to generate profits. The greater the ROE generated by a bank, the better the bank's position. In this case, although Bank Jasa Jakarta (BJJ)'s core capital was not the largest among them, Bank Jasa Jakarta was able to show that they were in the first place in ROE management. After equalization of capital equity, the ROE of BJJ (10.43%) was highest compared to the other banks in the peer group. The value of ROE from Bank Jasa Jakarta exceeded 7%, which was the best standard of ROE. Thus, Bank Jasa Jakarta is a bank that has an excellent ability in generating profits regarding the ROE.

ROA and ROE are variables that represent the bank's profitability. By looking at both variables showing a consistent performance, the results obtained prove that Bank Jasa Jakarta is a bank that performs well and deserves an award. The result of this analysis based on the peer group model was one of the proponents of BJJ's award for various awards such as the award of Indonesian Banking Awards 2016 for Best Bank In Financial Aspects award category: Non Public Bank with The Best Rank, Bisnis Indonesia Banking Award 2016 for the category of the Most Efficient Bank awards Category Bank BUKU 2, Indonesia Banking Award 2016 for the category of The Most Reliable Bank, The Most Efficient Bank, The Best Bank In Productivity, Titanium Trophy Infobank Awards 2016 Award for "Excellent" Performance Bank Award for 19 Years successively. In addition, the peer group bank model formulated by Sugiarto (2017a) can be expressed as a reliable peer group bank model which is capable in representing actual conditions.

4 CONCLUSION

The test results, using cross section data, indicate that the model of the bank peer group formed is a reliable model because the results obtained from the model are able to represent the actual conditions. The test results also show that it is possible to establish an objective bank peer group determination model using a statistical point of view. The decision making regarding the conclusions made are based on a calculated probability.

REFERENCES

De Aghion, B.A. & Gollier, C. 2000. Peer group formation in an adverse selection model. *The Economic Journal* 110(465): 632–643.

Glewwe, P. 1997. Estimating the impact of peer group effects on socioeconomic outcomes: Does the distribution of peer group characteristics matter? *Economic of Education Review* 16(1): 39–43.

Peraturan Bank Indonesia. 2011. *Nomor 13/1/PBI/2011 Penilaian tingkat kesehatan bank umum.*

SE BI. 2011. No.13/24/DPNP. Bank Indonesia perihal: Penilaian tingkat kesehatan bank umum.

Stiglitz, J.E. 1990. Peer monitoring and credit markets. *The World Bank Economic Review* 4(3): 351–366.

Sugiarto, S. 2016. Performance evaluation of Indonesian banks and foreign banks operating in Indonesia related to classification of capital. *Advances in Economics, Business and Management Research* 15. 1st Global Conference on Business, Management and Entrepreneurship (GCBME–16).

Sugiarto, S. 2017. Core capital performance evaluation of banks operating in Indonesia. *International Journal of Applied Business and Economic Research* 15(6).

Sugiarto, S. 2017a. Formulation of bank peer group model. *Workshop of bank peer group model, Bank Jasa Jakarta.* Jakarta, June 2017.

Sugiarto, S. 2018. Time series analysis of performance consistency of bank peer group model. *2nd International Conference on E-Business and Internet* (ICEBI 2018). 16–18 May 2018, National Taipei University of Business, Taiwan.

Sugiarto, S. & Nursiana, A. 2016. Determinants identification of public banks stock prices in Indonesia based on fundamental analysis. *International Journal of Applied Business and Economics Research* 14(6): 4705–4712.

Advances in Business, Management and Entrepreneurship – Hurriyati et al (eds)
© *2020 Taylor & Francis Group, London, ISBN 978-0-367-27176-3*

Mandatory financial accounting standard based international financial reporting standard adoption and audit delay

R. Mawardi & H. Hamidah
Universitas Airlangga, Surabaya, Indonesia

ABSTRACT: The institute of Indonesia Chartered Accountants (IAI) in 2012 issued a policy decision for mandatory International Financial Reporting Standard (IFRS) adoption applied by companies on the Indonesia Stock Exchange (IDX). This paper aims to examine whether mandatory IFRS adoption has impacted increasing or decreasing audit delay caused by increasing complexity of accounting practices. This research was designed by multiple regression model. The result of descriptive analysis shows that the average of audit delay between the period of voluntary and mandatory IFRS adoption are 77 days. The result of coefficient of determination shows $R^2 = 0.194$ in voluntary IFRS adoption and $R^2 = 0.264$ IFRS adoption mandatory. This result gives empirical evidence that audit delay increases in mandatory IFRS adoption but does not make the company late in submitting financial reports to the capital market.

1 INTRODUCTION

Mandatory IFRS adoption has impacted the increasing complexity of accounting practices, especially for countries whose accounting systems are influenced by capital market and government as the standard setter (Ball et al. 2000). It also can reduce the value of the relevance of incremental information for investors in the capital market for investment decisions (Houqe et al. 2012, Kousenidis et al. 2010, Ball et al. 2000). However, the positive impact of mandatory IFRS adoption is reducing the practice of earnings management activities by corporate executives by increasing the effectiveness and efficiency of corporate governance (Zéghal et al. 2011, Marra et al. 2011).

Habib (2015) empirically examined the effects of audit report lag and the new accounting standards in China that were implemented in 2007 based on a fair value accounting system. He documented significant empirical evidence in audit report lag in China after the adoption of new accounting standards. This leads to delays in issuing audit opinions over financial statements while companies are required to present earnings information in timeliness (Givoly & Palon 1982).

Recent research on determination factors of audit delay are corporate governance, corporate executive characteristics (CEO and directors), gender and earnings management (Alfraih 2016, Harjoto et al. 2015, Suryanto 2016). This present research continues (Alfraih's 2016) which discusses the influence of corporate governance on audit delay by adding duality indicators of CEO independent and directors affecting the auditing process. Gender differences of CEO and directors have an effect on audit delay (Harjoto et al. 2015. Cotter et al. 2001) interpreted the theory of ceiling glass that female

executives as minorities tend to maintain their reputation and have a strong impetus to keep their careers in the company. The role of the audit committee assists the effectiveness and accuracy of the company's internal controls. (Ettredge et al. 2006) examine the quality of internal controls qualitatively in which good internal controls are able to mitigate risks so that public accountants no longer require thorough audits and accelerate the audit process. An analysis of external factors were initiated from the side of audit opinions and corporate relations relationships with public accounting partners (Ashton et al. 1987, Ashton et al. 1989, Carslaw & Kaplan 1991).

This paper is the first study to provide an overview of the problem of audit delay in financial reporting related to changes in accounting standards and ex-amine corporate governance and gender executives in voluntary and mandatory IFRS adoption. Our re-search has been motivated by the impact of problems arising from changes in accounting standards on accounting practices leads to increased complexity to be audited during the auditing process may cause the value of financial information to decrease as audit lag reports increase. Audit delay is interesting to be studied because previous research also revealed the similarity of determination factors with the quality of financial statement information in the research based on of IFRS adoption from voluntary to mandatory.

Finally, hypotheses to be tested empirically in this study are based on the above exposure and previous research by (Habib 2015) regarding IFRS adoption and audit delay, including:

H: There is an increase of audit delay in the period of mandatory IFRS adoption.

2 METHOD

This study used a sample from the Indonesian Capital Market Directory (ICMD) and Data Center FEB Airlangga University, with a total of 159 manufacturing companies listed on the Indonesia Stock Ex-change (IDX) in 2008–2016. Company year end must be December 31, and this has been shown to have an influence on audit reporting (Leventis et al. 2005). A further seventy companies were excluded due to lack of financial reporting and 44 were excluded due to lack of annual report. The final sample was 405 observation data.

This research is a quantitative research that uses a descriptive approach and conducts hypothesis testing. This research also describes and compares the occurrence of a social phenomenon related to the research hypothesis (Sekaran 2003).

The dependent variable is audit delay, measured as the period of completion of the audit of the annual financial statements from December 31 to date on the independent auditor's report issued by the public accounting firm.

In this study, the independent variables are internal corporate attributes are role duality, CEO financial expertise, board size, independent director, women CEO, women director, women audit committee, audit committee size, audit committee meet and external factor attributes are audit firm size and audit opinion. Control variables are profitability, leverage and assets. The strength of the association between audit delay and corporate governance and executive gender is measured using a linear regression model (Al-Ghanem & Hegazy 2011, Cohen & Leventis 2013, Pourali et al. 2013, Wan-Hussin & Bamahros 2013, Khlif & Samaha 2014, Pizzini et al. 2014, Harjoto et al. 2016, Alfraih 2016). Here, the following multiple regression model is used:

$$AUD = \beta 0 + \beta 1(CEODUAL) + \beta 2(CEOFINEXP) +$$
$$\beta 3(BODSIZE) + \beta 4(DIRIND) + \beta 5(WMNCEO) +$$
$$\beta 6(RWMNBOD) + \beta 7(RWMNCA) + \beta 8(CASIZE) +$$
$$\beta 9(CAMEET) + \beta 10(KAP) + \beta 11(OPINION) +$$
$$\beta 12(ROA) + \beta 13(LEV) + \beta 14(ASSETS)$$

3 RESULT AND DISCUSSION

3.1 Descriptive statistics

Table 1 shows descriptive statistics for independent, dependent, and control variables in this study. The minimum of audit delay in the period of IFRS adoption voluntary are 39 days while mandatory IFRS adoption manufacturing company longer audit delay process has increased to 44 days. The average of audit delay in the period of voluntary IFRS adoption (2008–2011) are 77 days, while in the period of

Table 1. Descriptive statistics.

Variables	Voluntary (2008–2012)		Mandatory (2012–2016)	
	Min.	Average	Min.	Average
AUD	39.00	77	44.00	77
CEODUAL	0.00	0.73	0.00	.81
CEOFINEXP	0.00	0.87	0.00	.95
BODSIZE	2.00	5.38	2.00	5.51
DIRIND	0.00	0.02	0.00	.11
WMNCEO	0.00	0.29	0.00	.37
RWMNBOD	0.00	0.10	0.00	.11
RWMNCA	0.00	0.10	0.00	.13
CASIZE	0.00	2.94	2.00	3.14
CAMEET	0.00	7.35	0.00	6.19
KAP	0.00	0.56	0.00	.61
OPINION	0.00	0.54	0.00	.63
ROA	0.00	0.13	0.00	.09
LEV	0.00	0.61	0.00	.47
ASSETS	13174	8028503.3	135849	146434988
Valid N	180		225	

mandatory IFRS adoption (2012–2016) has the same average audit delay 77 days.

These results indicate that the entire manufacturing company delivers timely financial report information to the capital market based on Bapepam and Financial Institution no: KEP-431/BL/2012 regulation regarding the submission of the company's annual report to the maximum capital market 120 days.

These results also provide inconsistent with Habib (2015), who found that there was an increase in the audit report lag when the company experienced accounting standards. Overall, mandatory IFRS adoption in Indonesia does not make the company late in submitting financial reports to the capital market so that shareholders continue to receive financial information in timeliness.

3.2 Multiple regression analyses

Using multiple regression analyses model to investigate the effect of the internal and external company factors on audit report lag and control variables was presented in Table 2.

Table 2 shows the results of estimating the audit delay model. It is evident that variables representing corporate governance and executive gender in combination are highly significant in explaining audit delay in voluntary IFRS adoption (F 2.823, p 0.001) and mandatory IFRS adoption (F 5.360, p 0.000), as the adjusted R2 indicates that these corporate governance mechanisms explain about 19 percent of the variation in voluntary IFRS adoption and 26 percent of the variation in mandatory IFRS adoption. It can be concluded that Hypothesis is not rejected because audit delay has increased in the period of mandatory

Table 2. Multiple regression result.

Variables	Voluntary (2008–2012)		Mandatory (2012–2016)	
	t	Sig.	t	Sig.
(Constant)	7.847	0.000	11.006	0.000
CEODUAL	1.049	0.296	−0.144	0.885
CEOFINEXP	−2.328	0.021**	−3.180	0.002***
BODSIZE	−1.246	0.214	1.205	0.230
DIRIND	−.937	0.350	2.551	0.011**
WMNCEO	2.366	0.019**	0.146	0.884
RWMNBOD	−0.329	0.743	0.911	0.363
RWMNCA	−1.393	0.166	−0.759	0.449
CASIZE	2.246	0.026**	−1.893	0.060*
CAMEET	−0.490	0.625	−2.292	0.023**
KAP	1.880	0.062*	−1.945	0.053*
OPINION	−2.109	0.036**	−1.345	0.180
ROA	−1.007	0.315	−2.174	0.031**
LEV	3.118	0.002***	3.758	0.000***
ASSETS	−2.066	0.040**	−2.758	0.006***
F Statistik	2.823	0.001***	5.360	0.000***
R	0.441		0.514	
R^2	0.194		0.264	

Notes: *, **, *** is significant at 0.10, 0.05 and 0.01 levels, respectively (two-tailed).

IFRS adoption caused by internal and external factor of company.

These results are supported by several studies (Yaacob & Che-Ahmad 2012, Hitz et al. 2013, Habib 2015), all of which found significant increases in audit report lag following the adoption of new accounting standards. Company characteristics and market forces become more influential in the calculation of the time span for auditors to perform their professional obligations in the era of potentially increasing regulation of the audit market. The convergence of IFRS impacts on audit work becomes complicated for auditors in performing audit procedures due to a number of changes in standards. The level of complexity of IFRS adoption is mainly due to certain standards that have received much criticism from compilers and auditors due to the measurement and recognition of ambiguous new accounting standards.

This study aims to prove whether there is an increase or decrease audit delay caused by internal and external factors in the period of mandatory IFRS adoption. This has a direct impact on additional audit procedures that result in companies requiring additional time to submit financial reports to capital markets and investors.

The result of descriptive analysis shows the first finding is the comparison of average and maximum audit delay between the period of voluntary IFRS adoption (2008–2011) and mandatory IFRS adoption (2012–2016) proves that manufacturing companies in Indonesia as a whole provide the newest financial information faster and timely to capital market. This finding is rejected the argument that mandatory IFRS adoption leads to increased complexity in accounting practices for firms in developing countries. However, the minimum value of audit delay during the period of mandatory IFRS adoption audit delay manufacturing firms has increased. This result also illustrates that there is an increase in the lag of the audit report while corporates have changed the accounting standard. This result has implication for support to the Indonesian Accounting Standards Association, the Indonesian Institute of Accountants (IAI), on the policy of implementing the IFRS adoption mandatory for all listed companies listed on the Indonesia Stock Exchange. This finding leads directly to the impact of implementing IFRS based financial standards proven to increase the relevance of information values based on audit timeliness.

Second, finding based on the result of determination coefficient (R^2) shows that audit delay has increased in the period of IFRS adoption mandatory caused by the influence of internal and external factors of the company. However, based on the analysis of the coefficient of determination IAI should still pay attention to the increase lag audit report that occurs when the company undergoes changes in accounting standards.

4 CONCLUSION

This study discusses corporate governance and gender executive on audit delay at manufacturing companies in Indonesia Stock Exchange related to changes in voluntary and mandatory IFRS adoption. This study aims to prove whether there is an increase or decrease audit delay caused by internal and external factors in the period of mandatory IFRS adoption. This has a direct impact on additional audit procedures that result in companies requiring additional time to submit financial reports to capital markets and investors.

The result of descriptive analysis shows the first finding is the comparison of average and maximum audit delay between the period of voluntary IFRS adoption 2008–2011 and mandatory IFRS adoption 2012–2016 proves that manufacturing companies in Indonesia as a whole provide the newest financial information faster and timely to capital market. This finding is rejected the argument that mandatory IFRS adoption leads to increased complexity in accounting practices for firms in developing countries. However, the minimum value of audit delay during the period of mandatory IFRS adoption audit delay manufacturing firms has increased. This result also illustrates that there is an increase in the lag of the audit report while corporates have changed the accounting standard. This result has implication for support to the Indonesian Accounting Standards

Association, the Indonesian Institute of Accountants (IAI), on the policy of implementing the IFRS adoption mandatory for all listed companies listed on the Indonesia Stock Exchange. This finding leads directly to the impact of implementing IFRS based financial standards proven to increase the relevance of information values based on audit timeliness.

Second, finding based on the result of determination coefficient (R2) shows that audit delay has increased in the period of IFRS adoption mandatory caused by the influence of internal and external factors of the company. However, based on the analysis of the coefficient of determination IAI should still pay attention to the increase lag audit report that occurs when the company undergoes changes in accounting standards.

ACKNOWLEDGEMENT

This research was supported/partially supported by Perbanas Institute. We thank our colleagues from Airlangga University who provided insight and expertise that greatly assisted the research, although they may not agree with all of the interpretations/conclusions of this paper.

We thank Editorial GCBME-UPI 2018 for assistance with (particular technique, methodology), and Reviewe GCBME-UPI 2018 for comments that greatly improved the manuscript.

REFERENCES

Al-Ghanem, W. & Hegazy, M. 2011. An empirical analysis of audit delays and timeliness of corporate financial reporting in Kuwait. *Eur-Asian Business Review* 1(1): 73–90.

Alfraih, M.M. 2016. Corporate governance mechanisms and audit delay in a joint audit regulation. *Journal of Financial Regulation and Compliance* 24 (3): 292–316.

Ashton, R.H., Graul, P.R. & Newton, J.D. 1989. Audit delay and the timeliness of corporate reporting. *Contemporary accounting research* 5(2): 657–673.

Ashton, R.H., Willingham, J.J. & Elliott, R.K. 1987. An empirical analysis of audit delay. *Journal of Accounting Research*: 275–292.

Ball, R., Kothari, S. & Robin, A. 2000. The effect of international institutional factors on properties of accounting earnings. *Journal of Accounting and Economics* 29(1): 1–51.

Carslaw, C.A. & Kaplan, S.E. 1991. An examination of audit delay: further evidence from New Zealand. *Accounting and Business Research* 22(85): 21–32.

Cohen, S. & Leventis, S. 2013. Effects of municipal, auditing and political factors on audit delay. *Paper Presented at the Accounting Forum*.

Cotter, D.A., Hermsen, J.M., Ovadia, S. & Vanneman, R. 2001. The glass ceiling effect. *Social Forces* 80(2): 655–681.

Ettredge, M.L., Li, C. & Sun, L. 2006. The impact of sox section 404 internal control quality assessment on audit delay in the SOX era. *Auditing: A Journal of Practice & Theory* 25(2): 1–23.

Fee, C.E. & Hadlock, C.J. 2004. Management turnover across the corporate hierarchy. *Journal of Accounting and Economics* 37(1): 3–38.

Givoly, D. & Palmon, D. 1982. Timeliness of annual earnings announcements: Some empirical evidence. *Accounting Review* 486–508.

Habib, A. 2015. The new Chinese accounting standards and audit report lag. *International Journal of Auditing* 19(1): 1–14.

Harjoto, M.A., Laksmana, I., & Lee, R. 2015. The impact of demographic characteristics of CEOs and directors on audit fees and audit delay. *Managerial Auditing Journal* 30(8/9): 963–997.

Hitz, J.M., Löw, P. & Solka, M. 2013. Determinants of audit delay in a mandatory ifrs setting/bestimmungsfaktoren der prüfungsdauer von verpflichtend aufzustellenden ifrs-abschlüssen. *Die Betriebswirtschaft* 73(4): 293.

Houqe, M.N., van Zijl, T., Dunstan, K. & Karim, A.W. 2012. The effect of IFRS adoption and investor protection on earnings quality around the world. *The International Journal of Accounting* 47(3): 333–355.

Jensen, M.C. 1993. The modern industrial revolution, exit, and the failure of internal control systems. *The Journal of Finance* 48(3): 831–880.

Khlif, H. & Samaha, K. 2014. Internal control quality, Egyptian standards on auditing and external audit delays: Evidence from the egyptian stock exchange. *International Journal of Auditing* 18(2): 139–154.

Kousenidis, D.V., Ladas, A.C. & Negakis, C.I. 2010. Value relevance of accounting information in the pre-and post-IFRS accounting periods. *European Research Studies* 13(1): 143.

Leventis, S., Weetman, P. & Caramanis, C. 2005. Determinants of audit report lag: Some evidence from the athens stock exchange. *International Journal of Auditing* 9(1): 45–58.

Marra, A., Mazzola, P. & Prencipe, A. 2011. Board monitoring and earnings management pre-and post-IFRS. *The International Journal of Accounting* 46(2): 205–230.

Pizzini, M., Lin, S. & Ziegenfuss, D.E. 2014. The impact of internal audit function quality and contribution on audit delay. *Auditing: A Journal of Practice & Theory* 34(1): 25–58.

Pourali, M.R., Jozi, M., Rostami, K.H., Taherpour, G.R. & Niazi, F. 2013. Investigation of effective factors in audit delay: Evidence from Tehran stock exchange (TSE). *Research Journal of Applied Sciences, Engineering and Technology* 5(2): 405–410.

Sekaran, U. 2003. *Research methods for business a skilling-building approach, fourth edition.* New York: John Wiley and Sons.

Suryanto, T. 2016. Audit delay and its implication for fraudulent financial reporting: a study of companies listed in the Indonesian stock exchange. *European Research Studies* 19(1): 18.

Wan-Hussin, W.N. & Bamahros, H.M. 2013. Do investment in and the sourcing arrangement of the internal audit function affect audit deblay. *Journal of Contemporary Accounting & Economics* 9(1): 19–32.

Yaacob, N.M. & Che-Ahmad, A. 2012. Adoption of FRS 138 and audit delay in Malaysia. *International Journal of Economics and Finance* 4(1): 167.

Zéghal, D., Chtourou, S. & Sellami, Y.M. 2011. An analysis of the effect of mandatory adoption of IAS/IFRS on earnings management. *Journal of International Accounting, Auditing and Taxation* 20(2): 61–72.

Advances in Business, Management and Entrepreneurship – Hurriyati et al (eds)
© 2020 Taylor & Francis Group, London, ISBN 978-0-367-27176-3

Analysis of constant correlation optimal portfolio model

I. Yunita
Telkom University, Bandung, Indonesia

ABSTRACT: The purpose of this research was to perform and analyze the optimal portfolio by using a Constant Correlation Model. The sample of this research was all securities in Jakarta Islamic Index Period 2018. This research used time series data from 2013 to 2018. The results showed that the optimal portfolio selection consisted of two securities that were TPIA and BPRT. The monthly return of portfolio was above the individual return and the risk of portfolio was below the individual risk. The portfolio performance index was positive and above the market.

1 INTRODUCTION

Investment represents a delay in current consumption which is incorporated into productive assets or productive processes which result for future consumption (Jogiyanto 2016). The investment process includes understanding the basics of investment decisions and how to organize activities in the investment decision process (Tandelilin, 2017). The fundamental thing in the investment process is understanding the relationship between the expected return and the risk of an investment. The expected risk and return relationship of an investment is linear, the greater the expected return, the greater risk to be considered (Tandelilin 2017). The steps of the investment process (Husnan, 2005) include: (1) determination of investment policy, (2) securities analysis, (3) portfolio formation, (5) portfolio revision, and (5) portfolio performance evaluation.

In the formation of investment portfolios in capital markets such as stocks, investors often face problems in choosing from many stocks in the capital market. As it is known that there are approximately 500 shares in the capital market. Investors can choose stocks that are on the index in the capital market. Shares incorporated in the index selected based on certain criteria, such as having a high liquidity, that meet certain criteria, such as sharia, can be used as a reference in stocks investment.

Jakarta Islamic Index (JII) is one of capital Market Index in Indonesia Stock Exchange that can be used as a preference for investor to invest their money. JII is an index of 30 stock companies that meet the cri-teria of investment based on Islamic sharia. During the period from 2013 to 2018, the development of JII index has increased.

The trend lines of Jakarta Islamic Index increased from 2013 to 2018. This condition can be a basis for investor to select securities in order to form the optimal portfolio. Portfolio is a combination of various investment instruments (Zubir 2011). Each investment

has different return and risk characteristics. However, in general the concept of rate of return and risk is "High Risk, High Return" means that the higher the return, the greater the risk is.

A stock portfolio is the combination of more than one stock of a company with the expectation that if one share price decreases, while the other increases, the investment does not incur losses (Zubir 2011). There are two ways in stock investment process (Bodie et al. 1993): (1) analyze the return and risk of individual stock which will form the portfolio and (2) form the optimal portfolio including fund allocation, calculating the return and risk of portfolio and selecting the best portfolio.

The optimal portfolio is a combination of several assets that provide the rate of return and risk adjusted to the characteristics of the investor. The optimal portfolio will be different for each investor. Investors who prefer risk will choose high returns by paying a higher risk compared to less risk-seekers investors (Jogiyanto 2016). According to Markowitz (1952), rational investors will seek efficient portfolios because these portfolios are optimal based on expected return and risk.

There are various techniques that can help investors in forming an optimal portfolio such as the constant correlation model. The constant correlation model assumes that the correlation between all pairs of securities is the same (Elton et al. 2007). The selection of optimal portfolio using Constant Correlation Model is based on excess return to standard deviation (ERSD).

In evaluating the performance of the portfolio that has been established, there are several methods that can be used: (1) Reward to Variability (Sharpe Measure) in which portfolio performance is measured by dividing the excess return with the variability of portfolio return, (2) Reward to Volatility (Treynor Measure) by dividing the excess return with volatility portfolio and (3) Jensen's Alpha, that measures the angle or slop of the portfolio (Jogiyanto 2016).

2   METHOD

This research employed descriptive and quantitative research which used time series data of stock including in Jakarta Islamic Index (JII Index) period 2018. This research used secondary data, such as monthly closing price of securities, closing market price index (IHSG), and Bank Indonesia (BI) rate as risk free risk. Data were obtained from www.yahoofi nance.com, www.idx.co.id and www.BI.go.id. The population was all securities in Jakarta Islamic Index 2018 by using non purposive sampling. The criterion was securities having a complete data of closing price in 2013–2018.

To form the optimal portfolio, constant correlation model was used based on excess return to standard deviation ratio. This ratio is (Elton et al. 2007):

$$ERSD_i = \frac{E(R_i) - Rf}{\sigma_i}$$

ERBi = Excess return to standard Deviation
    securities-i
E(Ri) = Expected return
Rf = Risk Free Return
σ_i = Standard deviation of securities i
The steps to find securities on optimal portfolio using Constant Correlation Model are (Jogiyanto 2016):

1. Calculate Individual Return
    $Rit = \frac{P_t - P_{t-1}}{P_{t-1}} + \frac{D_t}{P_{t-1}}$
    R_{it} = Return in current period
    P_{t-1} = Share price previous period
    P_t = Share price current period
    Rate of return is a measure of the outcome of an investment (Zubir 2011). In making an investment, a person will choose an investment that provides a high rate of return.
2. Calculate Individual Expected Return
    $E(Ri) = \frac{\sum_{i=1}^{n} Rit}{n}$
    Expected return is the average value of stock returns over a given period (Zubir 2011).
3. Calculate Individual Variance
    $\sigma i^2 = 1/n\text{-}1 \sum(R_{it} - E(Ri))^2$
    Risk (varians) is the diference between expected return and realized return and the differences can be positive, negative or zero (Zubir 2011).
4. Calculate Individual Standard Deviation
    $\sigma_i = \sqrt{\sigma_i^2}$
5. Calculate Covarians of 2 Stock
    $$\sigma_{12} = \sum_{i=1}^{n} \frac{[(R_{1i} - E(R_1)) \cdot (R_{2i} - E(R_2)]}{n}$$
    R_{1i} = Stock return 1 period-i
    R_{2i} = Stock return 2 period-i
    $E(R_1)$ = Expected return stock 1

$E(R_2)$ = Expected return stock 2
n = Historical Observation Number

Covariance is a linear independence between two random variables. In stock, covariance measures the magnitude of changes in returns of one share and other shares together, in which the greater the covariance, the stronger the relationship and the interplay of the two stock returns is (Zubir 2011). According to Zubir (2011), if covariance is positive, the rate of return of a stock and other shares changes in the same direction. If covariance is negative, shares will move in opposite directions with others.

6. Calculate Coefficient Correlation of 2 stock
    $$\rho_{12} = \frac{\sigma_{12}}{\sigma_1 \cdot \sigma_2}$$
    σ_{12} = Covarians of stock 1 and 2
    σ_1 = Standard Deviation stock 1
    σ_2 = Standard Deviation stock 2

    The correlation coefficient moves between +1 and –1. The point –1 indicates perfect negative correlation, which means that the return varians of 2 stocks will move perfectly in the opposite ways. The point +1 indicates perfect positive correlation which means that the return varians of 2 stocks perfectly will move in the same direction (Zubir 2011).
7. Rank the securities based on the basis of excess return to standard deviation from the largest to the smallest. Securities with the largest ERSD are candidates to be included in optimal portfolio.
8. Set the Cut off Rate. First, calculate the values Ci, where I represents the fact that the first I securities are included in the computation of Ci. Ci can be found from:
    $$C = \frac{\rho}{1 - \rho + i\rho} \sum_{j=1}^{i} \frac{R_j - R_f}{\sigma_j}$$
    where ρ is the correlation coefficient - assumed constant for all securities. The subscript I indicates that Ci is calculated using data on the first i securities.
9. Determine the appropriate level of the cut off rate C* when the ci has been found such that: (1) All stock ranked 1 to i have a value of ERSD lower than Ci. (2) All stock ranked i+1 through N have a value of ERSD lower than Ci.
10. The amount of Cutoff Point (C *) is the Ci value where the last ERSD value is still greater than Ci value.
11. Securities that form the optimal portfolio are securities that have ERSD values greater than or equal to the value of ERSD at point C *. Securities that have smaller ERSD with ERSD point

C * are not included in optimal portfolio formation.

12. Calculate the optimum amount to invest in any security:

$$X_i = \frac{Z}{\sum\limits_{j=1}^{N} Zj}$$

$$Z_i = \frac{1}{(1-\rho)\sigma i}\left[\frac{E(Ri) - Rf}{\sigma i} - C^*\right]$$

The formula to calculate the return and risk of portfolio is:

Return of portfolio:

$$E(Rp) = Rf + \sum\limits_{i=1}^{s} wi.[E(Ri) - Rf]$$

Risk of portfolio (Jogiyanto 2016):

$$\alpha_p^2 = \beta_p^2 . \sigma_M^2 + \left(\sum_{i=1}^{n} wi . \sigma ei^2\right)^2$$

The portfolio performance analysis used Sharpe index, Treynor Index and Jensen Index. The higher the index, the better the portfolio performance (Jogiyanto 2016).

a. Sharpe Index; $Spi = (R_p - R_f)/SD_{pi}$

R_p = Return of Portofolio
R_f = Risk Free Rate
SD_{pi} = Standard Deviation of Portfolio

The Sharpe index measures the performance by dividing excess return with portfolio return variability (Jogiyanto 2016). The Sharpe index measures the slop or angle of the portfolio drawn from the risk-free return point. The best portfolio is a portfolio that has the largest angle. This method was introduced by William F. Sharpe in 1966.

b. Treynor Index; $S_{ti} = (R_p - R_f)/\beta_p$

The Treynor index measures the performance by dividing excess return with portfolio return volatility (beta of portfolio) (Jogiyanto 2016). According to Treynor, the portfolio that is formed should be the optimal portfolio, the unique risk can be ignored and that is left

Table 1. The Result of ERSD.

Code	Expected Return E(Ri)	Excess Return E(Ri) - Rf	βi	Varians, σ_i^2	σi	ERSD
AKRA	0.02060	0.0154	0.7478	0.0129	0.11367	0.1354
ADRO	0.02071	0.0155	0.9421	0.0134	0.11598	0.1337
ASII	0.00280	-0.0024	1.4131	0.0042	0.06552	-0.0366
BPRT	0.06042	0.0552	1.8335	0.0419	0.20477	0.2697
BSDE	-0.00012	-0.0053	1.8602	0.0068	0.08250	-0.0644
CTRA	0.00382	-0.0014	2.3874	0.0162	0.12745	-0.0109
EXCL	-0.00674	-0.0119	0.7148	0.0098	0.09912	-0.1204
ICBP	0.01532	0.0101	0.6231	0.0276	0.16618	0.0609
INCO	0.01365	0.0084	0.7554	0.0253	0.15929	0.0531
INDF	0.00189	-0.0033	0.8769	0.0045	0.06729	-0.0492
KLBF	0.00134	-0.0038	0.9696	0.0035	0.05921	-0.0652
LPKR	-0.01505	-0.0202	0.0672	0.0091	0.09555	-0.2119
LPPF	0.00152	-0.0036	0.8752	0.0090	0.09517	-0.0387
LSIP	0.00184	-0.0033	0.3148	0.0173	0.13153	-0.0255
MYRX	0.00689	0.0016	0.3982	0.0091	0.09546	0.0177
PGAS	-0.00810	-0.0133	1.4403	0.0146	0.12088	-0.1100
PTBA	0.01224	0.0070	1.3017	0.0184	0.13577	0.0519
PTPP	0.01411	0.0089	1.6393	0.0127	0.11295	0.0788
PWON	0.01546	0.0102	0.0735	0.0099	0.09982	0.1027
SMRA	-0.00004	-0.0052	2.3582	0.0144	0.12009	-0.0436
SCMA	0.00188	-0.0033	0.7874	0.0072	0.08534	-0.0389
SMGR	-0.00688	-0.0120	1.3720	0.0051	0.07139	-0.1693
TPIA	0.05366	0.04846	0.6337	0.0235	0.15352	0.3156
UNTR	0.01320	0.0080	0.7487	0.0048	0.06992	0.1145
UNVR	0.00944	0.0042	0.1037	0.0027	0.05265	0.0806
WIKA	-0.00131	-0.0065	1.6529	0.0128	0.11341	-0.0574
WSBP	-0.00689	-0.0120	0.1195	0.0092	0.09595	-0.1260
WSKT	0.02364	0.0184	2.0249	0.0170	0.13042	0.1414
TLKM	-0.00131	-0.0065	0.7739	0.0035	0.05947	-0.1095
IHSG	0.00407			0.0012	0.03490	

behind is the systematic risk as measured by beta (Jogiyanto 2016). This method was introduced by Jack L. Treynor in 1966.

c. Jensen Index; $J_{pi} = (R_{pi} - R_f) - (R_m - R_f) \beta_{pi}$

R_p = Return of Portfolio
R_f = Risk Free Rate
β_{pi} = Beta Portfolio

$$= \left(\sum_{i=1}^{n} Xi \cdot \beta i \right)$$

Xi = weight security i;
βi = Beta Security i.
$\beta i = \sigma im/(\sigma^2 m)$

The Jensen Index measures the angle of the portfolio in which the greater the angle or slop of the portfolio, the better the portfolio performance is. This method was introduced by Michael C. Jensen in 1968.

3 RESULTS AND DISCUSSION

Excess return to standard deviation measures the excess return premium to a single unit of diversifiable risk as measured by standard deviation. The value of ERSD is used as the basis for determining the stock that includes the optimal portfolio. A high value ERSD is a candidate for optimal portfolio. The Rf is the average monthly BI Rate from 2013–2018, that is 0.52%.

Table 1 shows that there are securities that have positive and negative expected return. Securities that have negative value will not be included in optimal portfolio calculations because investors will prefer more to less (Markowitz, 1952). The security that has the highest return is BPRT and the lowest return (loss) is LPKR. In Table 1, it can be seen that the beta of

individual securities is from 0 to 2.4. The beta of a stock is the slope of the characteristic line between returns for the stock and those for the market Horne (2002). The higher the beta, the more volatile the price of securities is than the market. Conversely, the lower the beta, the less volatile the price of securities is than the market. According to Titman et al. (2011), βs for common stocks are typically between 0.6 and 1.6. The

Table 3. The result of Ci and C*.

No	Code	ERSD	$\rho/(1-\rho + i\rho)$	$\sum_{j=1}^{i} \frac{R_j - R_f}{\sigma_j}$	C
1	TPIA	0.3156	0.3827	0.3156	0.1208
2	BPRT	0.2697	0.2768	0.5853	0.1620, C*
3	WSKT	0.1414	0.2168	0.7267	0.15753
4	AKRA	0.1354	0.1782	0.8621	0.15360
5	ADRO	0.1337	0.1512	0.9959	0.15060
6	UNTR	0.1145	0.1314	1.1104	0.14585
7	PWON	0.1027	0.1161	1.2131	0.14085
8	UNVR	0.0806	0.1040	1.2937	0.13458
9	PTPP	0.0788	0.0942	1.3725	0.12933
10	ICBP	0.0609	0.0861	1.4334	0.12343
11	INCO	0.0531	0.0793	1.4865	0.11786
12	PTBA	0.0519	0.0735	1.5384	0.11301
13	MYRX	0.0177	0.0684	1.5561	0.10649

Table 4. The result of Xi and Wi.

No	Code	C	Zi	Xi
1	TPIA	0.1208	0.3156	0.5393
2	BPRT	0.1620	0.2697	0.4607
		Σ	0.5853	1

Table 5. The result of return portfolio.

No	Code	Rf	Xi	E(Ri)	Xi. (E(Ri) - Rf)
1	TPIA	0.0052	0.5393	0.05366	0.02613
2	BPRT	0.0052	0.4607	0.06042	0.02544
					0.05158
	E(Rp)				0.05678

Table 2. The result of ERSD.

No	Code	Expected Return E(Ri)	Excess Return E(Ri) - Rf	Standard Deviation, σi	ERSD
1	TPIA	0.0537	0.0485	0.1535	0.3156
2	BPRT	0.0604	0.0552	0.2048	0.2697
3	WSKT	0.0236	0.0184	0.1304	0.1414
4	AKRA	0.0206	0.0154	0.1137	0.1354
5	ADRO	0.0207	0.0155	0.1160	0.1337
6	UNTR	0.0132	0.0080	0.0699	0.1145
7	PWON	0.0155	0.0103	0.0998	0.1027
8	UNVR	0.0094	0.0042	0.0527	0.0806
9	PTPP	0.0141	0.0089	0.1129	0.0788
10	ICBP	0.0153	0.0101	0.1662	0.0609
11	INCO	0.0137	0.0085	0.1593	0.0531
12	PTBA	0.0122	0.0070	0.1358	0.0519
13	MYRX	0.0069	0.0017	0.0955	0.0177

Table 6. The result of risk portfolio.

No	Code	X_i	σ_{ei}^2	$(\sum X \cdot \sigma_{ei}^2)$	β_p^2	σ_M^2
1	TPIA	0.5393	0.0240	0.0129		
2	BPRT	0.4607	0.0427	0.0197	0.001	0.0012
$(\sum Wi \cdot \sigma ei2)^2$				0.0010		
α_p^2					0.00106	

399

Table 7. Portfolio performance.

Rp	σp	βp	Performance	Sharpe index	Treynor Index	Jensen Index
0.0568	0.0326	1.186	Optimal Portfolio	0.0435	1.5819	0.0529
σm	Rm	Rf	Market	-0.0011	-0.0324	0.0000
0.0349	0.00407	0.0052				

highest beta is CTRA and the lowest beta is LPKR. The next step to form the portfolio selection by using constant correlation model is selecting securities that has positive excess return to standard deviation (ERSD).

Table 2 shows that there are 13 securities that have positive excess return to standard Deviation. The securities are: TPIA, BPRT, WSKT, AKRA, ADRO, UNTR, PWON, UNVR, PTPP, ICBP, INCO, PTBA and MYRX. The rank of the securities is based on the basis of excess return to standard deviation from the largest to the smallest. Securities with the largest ERSD are candidates to be included in optimal portfolio.

Table 3 shows that the Cut of Rate (C*) is 0.1620, that is, BPRT with ERSD 0.2768. The stocks on optimal portfolio that have ERSD above and same with C*is TPIA and BPRT. Thus, those two stocks will be in-cluded in the optimal portfolio. The correlation is 0.3827.

Table 4 shows that the proportion of each stock is TPIA (53.93%) and BPRT (46.07%).

Tables 5 and 6 show that the monthly return of portfolio is 5.7 % and the risk of portfolio is 0.106%, that is, below the risk of all individual stock in portfolio. The risk of portfolio is below the individual risk (variance, α_i^2) of TPIA and BPRT. Thus, it can be said that individual risk can be minimized by forming a portfolio due to diversification into several types of securities.

Table 7 shows that the portfolio performance is positive and better than the market. Thus, the portfolio's performance has been well diversified. This result supports Pratiwi et al. (2015), which proved that the optimal portfolio had better performance when using the Sharpe, Treynor and Jensen indexes. This result is also in line with the modern portfolio theory by Markowitz, that portfolio diversification will reduce the overall risk.

4 CONCLUSIONS

The optimal portfolio selection using Constant Correlation involved two companies in Jakarta Islamic Index such as TPIA with the proportion for each security was 52.93% for TPIA and 46.07% for BPRT. The monthly return of portfolio was 5.7 % and the risk of portfolio was 0.106% that was below the risk of all individual stock in portfolio. Based on the Sharpe, Treynor and Jensen indexes, the performance portfolio was positive and better than the market. It can be said that the performance of portfolio was well diversified because the value of all performance index was positive and above the market index.

ACKNOWLEDGEMENTS

We would like to thank PPM Telkom University for providing the grant by the scheme of Research Publication (Publikasi Penelitian) 2018.

REFERENCES

Bodie, Zvi, Kane, Alex & Marcus A. J. 1993. *Investments*. Illinois: Richard Irwin, Inc.

Elton, J.Edwin, Gruber, J.Martin, Brown, J.Stephen, & Goetzmann N.William. 2007. *Modern portfolio theory and investment analysis*. New York: John Wiley & Sons, Inc.

Horne, Van C James 2002. Financial management and policy. 12th edition. Prentice Hall.

Husnan, Suad. 2005. *Dasar-dasar teori portofolio dan analisis sekuritas*. Edisi Kelima. Yogyakarta: UPP STIM YKPN.

Jogiyanto. Hartono. 2016. *Teori portofolio dan analisis investasi*. Edisi kesepuluh. Yogyakarta: BPFE-Yogyakarta.

Markowitz, Harry. 1952. Portfolio selection. *The Journal of Finance*. Vol. 7, No.1, pp. 77–91.

Pratiwi, Dhea Ayu and Yunita, Irni. 2015. Optimal portfolio constraction (a case study of LQ45 index in Indonesia stock exchange). *International Journal of Science and Research (IJSR)*. Vol.4 Issue 6.

Tandelilin, Eduardus. 2017. *Pasar Modal: Manajemen portofolio & investasi*. Yogyakarta: PT. Kanisius.

Titman, S., Keown, A. J., & Martin, J. D. 2011. Financial management: Principles and applications (11th ed.). Upper Saddle River, NJ: Pearson/Prentice Hall.

Zubir, Zalmi. 2011. Manajemen portofolio:penerapannya dalam investasi saham. Jakarta: Salemba Empat.

Advances in Business, Management and Entrepreneurship – Hurriyati et al (eds)
© *2020 Taylor & Francis Group, London, ISBN 978-0-367-27176-3*

Gender responsive planning and budgeting implementation in Indonesia: Historical review and lessons learned

F. Fithriyah
Universitas Airlangga, Surabaya, Indonesia

ABSTRACT: This paper is a historical review and lessons learned from the Indonesia experience in implementing Gender Responsive Planning and Budgeting (GRPB). It is very important for Indonesia since it addresses the strategy towards reducing inequality between men and women in development, at national and district levels. This paper aims to review and share the lessons learned, towards better implementation of GRPB in Indonesia in the future. The methodology used in this paper is a qualitative approach, based on literature review. The results showed that GRPB should be mainstreamed into the existing planning and budgeting process. Sex-disaggregated data and gender statistics are very imminent for conducting gender analysis and understanding gender perspective is the first and the most important aspect in conducting GRPB. Therefore, gender awareness should be conducted regularly. GRPB have successfully narrowed the gender gap in Indonesia's development, nationally and regionally.

1 INTRODUCTION

1.1 *Gender* Mainstreaming *(GM)*

The Gender Mainstreaming (GM) strategy in Indonesia's national development planning and policies means the integration of gender perspective into every stage of the national development process from planning and budgeting, coordinating and monitoring the implementation, to the evaluation. Therefore, the gender perspective is integrated into the existing national development cycle (from planning until the evaluation) with small adjustments, not creating the new/special cycle. In order to do the gender analysis, the Ministry of National Development Planning (MoNDP/Bappenas)—assisted by the Canadian International Development Agency (CIDA) and supported by the Ministry of Women Empowerment (MoWE) have developed the Gender Analysis Pathway (GAP) in 1998. This GAP is designed as the analysis tool for engendering the development programs/activities/projects. Therefore, the GAP expected output is gender responsive programs/activities/projects. This tool is revised many times due to the changing of the governance system.

The last version was distributed in 2007. GAP in terms of planning consists of two parts: (1) Gender Responsive Policy Analysis and (2) Policies, Action Plan. The result of this planning process will be implemented, monitored and evaluated, as the input for the next planning process.

1.2 *Gender Responsive Planning and Budgeting (GRPB)*

The GRPB initiative is formulated as the Ministry of National Development Planning's Decree on GRPB Steering Committee and Technical Teams, which was officially signed and implemented in 2009. This decree has included the Ministry of Finance (MoF) on the board of GRPB's drivers, together with the MoNDP/Bappenas and Ministry of Women Empowerment (MoWE). These teams also included the Ministry of Home Affairs (MoHA), Ministry of National Education (MoNE), and Ministry of Health (MoH).

This decree was followed by the issuance of the Ministry of Finance's Regulation Number 105/PMK.02/2008, on Guidelines for the Preparation and Review of Ministry/Agency Work Plan and Budget and the Preparation, which mentioned GRPB for the first time as one of the new budgeting mechanisms. This was also in line with the restructuring and reorganization of all LMs in Indonesia at the end of 2009, as a result of implementing new approaches of planning and budgeting: Unified Budget, Medium Term Expenditure Framework (MTEF) and Performance Based Budgeting (PBB). The detailed guidelines on preparation and reviewing the LMs work plan and budget according to GRPB was explained in the following year's circular letter, the Ministry of Finance's Regulation Number 119/PMK.02/2009. However, the restructuring of all LMs organizations, which took place at the end of 2009, has resulted in the postponing of this GRPB implementation to the following year. In 2010, GRPB has been implemented in seven pilot LMs: MoNDP/Bappenas, MoF, MoWE, MoNE, MoH, Ministry of Agriculture (MoA) and Ministry of Public Works (MoPW); assisted by three drivers: MoNDP/Bappenas, MoF and MoWE.

The PBB required each LMs to create its unique performance indicators, which could not be replicated in other directorate/bureau within and/or inter-LMs.

This brought a significant change in the planning and budgeting structure.

For example, after PBB, these multi-executing-agencies programs are not allowed. Therefore, gender mainstreaming was classified as one of the mainstreaming strategies in NMTDP 2010–2014, together with good governance and sustainable development. Besides that, the NMTDP also stated that GM is one of the three other national priorities, under the national priorities of people welfare. Moreover, NMTDP 2010–2014 consists of three books: (1) National Priorities; (2) Field Priorities; and (3) Regional Priorities (Bappenas 2010). In order to support this, all of the drivers and implementing LMs should conduct GRPB in line with their existing duties and functions.

Furthermore, to support the gender analysis, we should have the gender analysis tool first, which is supported by the sex-disaggregated data, especially for development indicators and statistics, which is addressing the gender issues on access, participation, control (on resources) and benefit. According to NMTDP 2010–2014, GM here covers three priorities: (1) increasing the quality of life and the role of women; (2) increasing the protection from discrimination and violence against women and (3) increasing the capacity of GM institutions (including the systemized and updated sex-disaggregated data).

GRPB does not mean: (1) special budget allocation for women/men; (2) budget allocation of 50% for men and 50% for women; or (3) extra budget for integrating GM (for example: dissemination of GRPB should be integrated in the planning and budgeting dissemination, which is allocated annually for all line ministries and local government). Based on PBB, the national government should set the 'shopping list' for GM, in alignment with the NMTDP (RPJMN) and Government Annual Work Plan (RKP), which is supported by the local government and related stakeholders, in order to ensure the sustainability of related GM policies/programs/activities. The GM efforts should be conducted by all of the stakeholders, not only the government. However, the government should set firm policies and activities related to GM, which will be supported by other stakeholders. This attempt was then formalized by the National Strategy to Accelerate Gender Mainstreaming through Gender Responsive Planning and Budgeting, a circular letter signed by four GRPB drivers: Bappenas, Ministry of Finance, Ministry of Home Affairs, Ministry of Women Empowerment and Child Protection (Sardjunani 2012).

In 2013, the implementation of GRPB ex-panded to 37 LMs and 33 provinces. In addition to that, the preparation of the third phase of RPJMN 2015–2019 has also brought in-depth gender stocktaking of GRPB in 36 LMs. This was expected to support the preparation of the Background Study of the RPJMN 2015–2019 related to Gender Equality.

Besides the policies, GM also needs a change of perspective/mindset from the grass root level, starting from the households. Therefore, the Ministry of National Education (MoNE) has conducted the revision of some school textbooks, which were gender biased. In terms of the household level, this change of perspective will be more effective for the husbands and wives, if addressed by community figures, such as the community leaders, the religious organizations' figures, and other community-based organizations.

The other critical points in implementing GM and its acceleration through GRPB is the common understanding of GM itself. When GM and GRPB are becoming national and local government efforts simultaneously, understanding gender perspective and translating it into GRPB in their national and local government development planning will be very important. Therefore, integrating the GM and GRPB aspects into the curricula of government planners' education and training is also crucial. Last but not least, this effort needs a consistent, regular and never-ending coordination within all related stakeholders; not only in terms of substance but in funding as well.

The acceleration of GM in Indonesia through GRPB from 2010 to 2013 has shown positive results. This can be seen from the improvement of the Gender Related Development Index/GDI (Indeks Pembangunan Gender/IPG) and Gender Empowerment Measurement/GEM (Indeks Pemberdayaan Gender/IDG), which is calculated annually by the Central Board of Statistics (CBS/Statistics Indonesia), assisted by MoWE. The calculation by CBS-MoWE is slightly different from the calculations by UNDP in Human Development Reports (UNDP 2014).

GDI calculated the basic human capabilities on education, health and economy, focusing on the inequality between men and women. GDI is a compo-site index which is calculated based on several variables: Life Expectancy at Birth, Literacy Rate, Mean Years of Schooling and Income in Non-Agriculture Sectors (which used the GDP/capita calculated based on Purchasing Power Parity/PPP).

Meanwhile, GEM is focusing on women's roles in economy, politics and decision making. GEM is also a composite index, which covers the following variables: Women's Representative in Parliament, Female Labor Force, Professional Female Workers, High Level Female Officials and Managers and the wage of female workers in non-agricultural sectors. The increase of GDI and GEM in Indonesia during the period 2010–2013.

This paper aims to review and share the lessons learned of GRPB implementation in Indonesia, towards its improvement in the future. The review method referred to the previous GRPB evaluation report (Sanjoyo et al. 2011). The implication of improved GRPB is expected to reduce the gender gaps and as a result, reduce the inequality between men and women.

2 METHOD

The method used in this paper is a qualitative approach, based on literature reviews (such as government reports and evaluation results). This paper focuses on the historical review of the GRPB implementation in Indonesia during the period of 2010 to 2013, which is the groundbreaking period of GRPB in Indonesia.

3 RESULTS AND DISCUSSIONS

3.1 *Results*

Based on the analysis, there are several lessons to be learned from the implementation of GRPB in Indonesia. First, the progress of GRPB's implementation up to now was supported by the integration of GM and GRPB into the existing planning and budgeting system, instead of developing the exclusive new system in the planning and budgeting cycle separately. This has effectively proven to minimizing the resistance of the drivers and line ministries/regional institutions in implementing GRPB, which is what usually happens when facing new strategies/policies. Besides, the well preparedness of its tools/documents and the indicators for the whole cycle at the early stage of implementing GRPB have helped them to conduct it much easier.

Second, the implementation of GRPB by the drivers (Ministry of National Development Plan-ning/Bappenas, Ministry of Finance, Ministry of Home Affairs and Ministry of Women Empower-ment and Child Protection) on the national level, as well as its transla-tion by the regional drivers, has been assigned according to their existing tasks and scope of works. This is conducted by integrating the gender perspective into the existing process, which resulted in no "new tasks" for the related institutions.

Third, the importance of awareness and education to the entire staff regarding this matter, especially in LMs and regional government offices, is to obtain a mutual level of understanding at the implementation level, as well in the way of institutionalizing and systemizing it. This is due to the rapid promotion, rotation and mutation of government officers, especially in the regional levels. If the new staff appointed for GRPB does not have any understanding regarding that matter, it will result in the setback of GRPB implementation in that institution. Therefore, integrating the related materials of GM and GRPB into the curricula of the government officers' education and training becomes very imminent. It should be integrated for all levels, from the highest to the lowest levels of structural and functional educations and trainings. Its expected result is that whoever assigned for GRPB in the future will have equal understanding. This is the importance of integrating GM-GRPB related materials into the curricula for government officers' education and training which is developed by the State Administration Institution (LAN)/LMs/Regional Government Offices, and is included in the National Strategy for Accelerating GM through GRPB.

Fifth, the institutionalization of sex-disaggregated data and gender statistics needs to be encouraged and supported in each LMs/regional office. Sex-disaggregated data and gender statistics will support the more accurate and well targeted development, which will result in reducing the gender gap in Indonesia.

Sixth, all of the related stakeholders must be involved. The gender gap is one of the reasons of development inequality in general. Therefore, reducing the gender gap could not be conducted only by a LM/regional office. It should be supported and executed together with the religious, social and community leaders, NGOs, CSOs, university, as well as the community itself.

Seventh, the most important thing in having a similar understanding in gender perspective is the standardization of gender responsive guidelines/modules/teaching materials by the LMs who are authorized in providing and distributing them. Therefore, whoever teaches it, the concept will be the same, and no confusion and resistances will occur due to the inconsistency of delivering these concepts.

Eighth, the mechanism of delivering these materials should be tailor-made/customized to the audience. This is taking the related culture, location, LMs/regional offices duties/functions and relevant issues into account in delivering the gender perspective concepts and practices in development. For example, gender issues in the Ministry of Social Welfare will be different to the Ministry of Public Works. Also, the Province of West Sumatra has different gender issues than West Papua Province.

Ninth, considering that the understanding of gender perspective is the first and the most important aspect in conducting GRPB, the gender awareness and technical assistance/consultation in internal LMs/regional offices must be conducted regularly. There-fore, Gender Working Groups (GWGs) which have been founded based on the President's Instruction No. 9/2000 should be revitalized and supported by involving the planners in that institution. It is also ensuring that the acceleration of GM through GRPB is relevant for the planning and budgeting process and regulations, and not going the other way, outside the system.

Tenth, the coordination within the drivers must always be maintained. When there are any changes in regulations/laws/policies on planning and budgeting, they should communicate and discuss it altogether. This is to ensure that GRPB will always be in the corridor of the existing planning and budgeting system, both nationally and regionally.

3.2 *Discussions*

Despite the positive progress, there are some existing obstacles in implementing GRPB, which need further discussion, for example the different levels of understanding among the planners on national and regional levels, on GM and its acceleration through GRPB. Staff rotation or promotion which hap-pens very often followed by a low understanding of GM by the replacement staff has resulted in GM progressing very slowly.

Besides, the different levels of understanding among the parliament members on national and regional levels on this is also a significant obstacle. When they assume that GM is the same as Women in Development (WID) or subject to women empowerment, they will not prioritize this GM related programs, activities, or policies, and switch it into other programs, such as infrastructure development (Sanjoyo et al. 2011).

The last is the limited availability of updated and systemized sex disaggregated data, especially related to development achievements or gender statistics. Despite the recording process this data has in the grass root level, somehow it is lost during the reporting process to the upper level (Sanjoyo et al. 2012).

4 CONCLUSION

It can be concluded that the acceleration of GM through GRPB has been developed much faster than GM itself since it has been directly banded to development budget documents. In the early period of GM implementation, gender responsive programs/ activities were integrated in the development planning documents only, then "vanished" in budgeting documents, which resulted in non-executed related programs and activities. Therefore, gender responsive programs and activities stated in the budgeting documents should be followed up by allocating its budget. In addition, as a planning cycle, the related programs and activities should always be monitored and evaluated, as inputs for a better development planning in the future.

Therefore, the emergent recommendations for improving GM and its acceleration through GRPB are: First, the dissemination of GM and its acceleration through GRPB should be conducted systematically and regularly, for all government officials, starting from the freshmen. Therefore, it should be integrated into the curricula of the government officials' systematic and regular trainings and educations, both for national and local government officials.

In order to conduct these trainings and educations, we should have standardized trainers or instructors who share the same gender perspective with the government in terms of national and local development, to avoid confusion with the trainees, which could result in the ineffective and inefficient gender responsive development planning and policies. It is also important to increase the awareness of parliament members through the advocacy in the importance of GM in the national and local development process. In order to support the advocacy to the parliament members and other high-level officials, the awareness materials should include the updated and systemized sex-disaggregated data and gender statistics.

Second, evaluating the programs and activities that have been stated in the Gender Budget Statement so far. Have those programs and activities supported pursuing the national priority goals and SDGs significantly, have they been significantly proven to reduce the gender gaps, have they been gender responsively conducted, have they been sex-disaggregated data and/or gender statistics institutionalized as the basis for implementing the updated GRPB accurately? These are to ensure that the chosen programs and activities are the highest leveraged ones in reducing the gender gap. As stated in the National Strategy of GRPB, it is better to focus on one highly-leveraged/significant program in reducing the gender gap, instead of conducting many programs and activities which are not significant or low-leverage. Therefore, the next important step is developing the updated and systemized sex-disaggregated data, in order to measure and analyze the indicators or statistics more accurately, for formulating effective, efficient and gender-responsive development planning and policies.

ACKNOWLEDGMENT

The author thanks Universitas Airlangga for funding the participation in GCBME 2018 and Bappenas for the references.

REFERENCES

Bappenas. 2010. *Peraturan Presiden RI No. 5/2010 tentang Rencana Pembangunan Jangka Menengah (RPJMN) 2010–2014 Buku II: Memperkuat Sinergi Antarbidang Pembangunan.* Jakarta: Kementerian PPN/Bappenas.

Bappenas. 2010. PPRG. Accessed in 2015.

Bappenas-Direktorat keluarga, anak, perempuan, pemuda, dan olahraga, 2015. *Kajian pendalaman penyusunan indeks keadilan dan kesetaraan gender (IKKG) dan indikator kelembagaan pengarusutamaan gender (IKPUG).* Jakarta.

PPN/Bappenas, Kementerian, dan KNPP. 2007. Gender analysis pathway (GAP): Alat analisis gender untuk perencanaan pembangunan. Jakarta.

RI, Presiden. 2007. "Undang-undang Nomor 17 Tahun 2007 tentang rencana pembangunan jangka panjang nasional (RPJPN) 2005–2025."

Sanjoyo, Fithriyah, Aini Harisani, & Puspasari S, 2011. Evaluasi terpadu pelaksanaan uji coba perencanaan dan penganggaran yang responsif gender (PPRG) tahun anggaran 2009–2010. directorate of population, women empowerment, and child protection-Bappenas.

Sanjoyo., Fithriyah., Aini, H., Yohanna M.L., Gultom, P.S. & Lilis H.M. 2012. *Indeks kesetaraan dan keadilan gender (IKKG) dan indikator kelembagaan pengarusutamaan gender (IKPUG): Kajian awal.* Jakarta: Direktorat Kependudukan, Pemberdayaan Perempuan, dan Perlindungan Anak-Kementerian PPN/BAPPENAS.

Sardjunani, N. 2012. *National strategy to accelerate gender mainstreaming through gender responsive planning and budgeting. bappenas, ministry of finance, ministry of home affairs, ministry of women empowerment and child protection, assisted by un women* Jakarta: MoNDP/Bappenas.

UN. 2001. *Gender mainstreaming: strategy for promoting gender equality (rev.).* Office of the special advisor on gender issues and advancement of women.

UNDP. 2014. *Human development report 2014*: *Sustaining human progress*: *Reducing vulnerabilities and building resilience*. UNDP.

Advances in Business, Management and Entrepreneurship – Hurriyati et al (eds)
© 2020 Taylor & Francis Group, London, ISBN 978-0-367-27176-3

Corporate governance mechanisms and their performance

W.P. Setiyono
Universitas Muhammadiyah Sidoarjo, Sidoarjo, Indonesia

ABSTRACT: The purpose of this study is to investigate the impact of internal governance structures in relation to managerial performance in Indonesian companies. In this research, we employed three proxy variables to measure internal governance structures such as: board, ownership, and compensation structure as independent variables, and Tobins Q as a proxy of managerial performance. The sample of the study were the companies whose stocks are actively traded on the Indonesia Stock Exchange. The data used were panel data, namely, the data of cross section and time series from the period of 2006 to 2011. The sampling was simple random sampling, and the analytical techniques were logistics regression analysis. The findings generally suggest a strong effect of internal governance structure measures in this data set. First, it is found that independent commissioners are effective in monitoring managerial performance, but the impact is negative. Secondly, the small proportion of managerial ownership is also found to be important result in this analysis. A higher proportion of insider ownership leads to decrease in managerial performance. Finally, there is a strong support for the view that the provision of executive bonuses has a positive impaction of managerial performance. The results suggest that granting incentive compensation to managers is an appropriate way to increase their performance.

1 INTRODUCTION

The emergence of the global financial crisis from 2007 to 2008 has changed the way stakeholders of a company should be managed (Setia 2009). Furthermore, the quality of corporate governance has now become a major consideration in managing a company's survival in times of crisis. Therefore, the definition and model of corporate governance have evolved, as many scandals arise (Kasey et al. 2005). Indonesia, although one of the countries not directly affected by the crisis is facing similar problems. As research on corporate governance in the Indonesian context is still rare (Patrick 2001), this study investigates the ways in which corporate governance is conducted in the Indonesian context. This paper is expected to expand the knowledge of corporate governance mechanisms issues in emerging countries, including Indonesia.

Theoretically, corporate governance consists of two mechanisms: internal and external (Fiorucci's 2008). In addition, the external mechanisms are associated with conditions outside the company, which are particularly controlled by the capital market. The internal mechanisms, on the other hand, are not under the control of the capital market. These mechanisms are not only of concern to the effectiveness of management in order to achieve the corporate goals, but also play an important role in the monitoring of investment decisions (Fiorucci's 2008). In addition, the internal mechanisms that exist in companies typically consist of the governance structure which includes boards, ownership and compensation structure (Dong & Gou 2010). Some

scholars claimed that the internal governance structure has an important role in ensuring the ability of the internal corporate governance mechanisms to achieve the objectives of a company (Singh 2003, Fiorucci's 2008).

Poor internal governance structures have led companies to face many difficulties, including agency problems. In a crisis, these agency problems often occur more frequently because of weak corporate governance (Purmerend 2012). For example, the US government has rescued some failed companies by bailing them out; however, these companies still gave huge bonuses to management (as agents) (Fiorucci's 2008). Provision of incentives and bonuses is actually reasonable because it falls within the corporate governance structure. But, considering that companies were hit by the crisis, incentives and bonuses have given rise to agency problems and the US government as the owner of funds has declined to continue with this policy. Such situations might also occur in many other countries, including Indonesia (Purmerend 2012).

The discussion regarding the corporate governance in Indonesia is relatively new. According to Purmerend (2012), intense discussion about the corporate governance in Indonesia has taken place since the financial crisis began in late 1997. Poor corporate governance of Indonesian companies at that time has led Indonesia to face difficulties in recovering from the crisis. Many cases of bank collapses were caused by the intervention of the owners who forced managers to give excessive credits to their affiliated groups, thus reflecting poor control. This was reinforced by the findings of a study by the Asian

Development Bank in 2001, which revealed that poor corporate governance was one of the major factors contributing to Asian countries being vulnerable to the crisis (Siregar & Utama 2008). Indonesia, as one of the most affected countries, has been forced to consider issues of corporate governance at the fore front of its national agenda. Compared to other countries that were also affected by the financial crisis such as Thailand, South Korea and Malaysia, the situation in Indonesia was worse. As an illustration, Indonesia had only recovered by the end of 2002, four years later than the other countries mentioned.

Therefore, the Indonesian government took an initiative to improve regulations of the corporate governance. The initiative was presented in the form of "Codes for Good Corporate Governance," which was established by the National Committee on Corporate Governance. This initiative was also followed by recommendations for law reform and legislation to support the implementation of this code (Indonesia 2006). The Committee believed that the importance of an institutional framework and further development of policies for the code at the institutional level should be applied in the context of Indonesia. The corporate governance reforms in Indonesia are also aimed at strengthening the current institutional structure (Siregar & Utama 2008).

The agency theory was developed from the original work of Jensen and Mackling in 1976 (Fama et al. 1983). This theory explains various issues that arise with regard to the separation of corporate ownership, control and management. The effect of the separation of ownership and management has been the subject of debate since the study was presented by Berle and Means in 1932. Jensen and Meckling define an agency relationship as a contract made by one or more persons as owners (the principals) who invite the participation of another person (the agent) to perform some services on the former's behalf which would include some delegations of decision-making authority. This contract is made as a reflection of the owners' efforts to improve corporate value by delegating the authority to managers.

For instance, the high level of managerial ownership as an internal mechanism of corporate governance is important in aligning managerial ownership (Fama et al. 1983). By holding a certain level of ownership, managers are motivated to generate profits and improve firm value for their own interests. Hence it is argued that a high level of managerial ownership would increase their performance.

Furthermore, a unique phenomenom is found in the Indonesian setting, especially when Indonesia adopted a two-tier board system with two separate functions, namely, the Supervisory Board and Executive Board. A Board of Commissioners acts as a Supervisory Board while a Board of Directors acts as an Executive Board. The existence of a clear separation of functions between the Board of Commissioners and Board of Executives has become an important issue in research on managerial performance (Patrick, 2001). Therefore, the research in internal governance structures has inconclusive results in relationship with the managerial firm performance.

Some scholars found that there are significant results in the relationship between board structure and managerial performance (Dong & Gou 2010, Victoria 2006). However, (Fiorucci's 2008) does not find any significant effect of commissioner board size on managerial performance. A study by Florakis (2008) found that the board independence significantly enhances the board's effectiveness and increase managerial performance. However, Fama and Jensen 1983 acknowledged that the effectiveness of the board independence depends on the capacity to control and make decisions.

Other variables in this research, including ownership concentration, have a positive impact on the managerial performance. This relationship is in accordance with the logic of the agency theory. Furthermore, the managerial ownership has a negative influence but is not significant on managerial performance (Fama et al. 1983). Finally, some scholars state that the amount of compensation given to the executive directors has a significant positive influence on the managerial performance (Dong & Gou 2010).

2 METHOD

The research was carried out through the construction of a positive empirical model (Damodaran Gujarati, 2014). Data were collected from the Indonesian Capital Market Directory (ICMD) and annual financial reports of firms listed on the Indonesian Stock exchange (IDX) from 2006 to 2011. Thirty-four companies listed on the IDX were selected based on random sampling.

The relationship between internal governance structures and managerial performance was tested by correlation analysis (Hair et al. 2010). Several statistical and econometric tests were used to test the effect of internal governance structures on managerial performance. The data for the tests included a combination of cross-sectional and time series observations and are termed "panel data." Furthermore, this study employs the managerial performance by using a binary variable as a proxy. Because of the binary variable, this study employed the logit regression models for the hypothesis analysis (Gravette & Wallan, 2007).

3 RESULTS AND DISCUSSION

The regression results are reported in Table 2. This study employs the managerial performance by using a binary variable as a proxy. Because of the binary variable, this study employs the logit regression models for the analysis. The logit regression model

Table 1. Variables measurements.

Variable	Definition	Source
The Dependent Variables		
Tobinsq	A dummy variable, which take 1 if {The ratio of (market capitalisation of equity + Book value of preference shares + Book value of long-term debt)/ Book value of total assets} ≥ 1 and 0 otherwise	Author's calculations based on annual report
The Internal Governance Structures Variables		
Comsize	The total number of commissioners on the board	Author's calculations based on annual report
Comind	The total number of independent board members of commissioners on the board	Author's calculations based on annual report
Concert	The percentages block holders' shares.	Author's calculations based on annual report
Execown	The percentage of equity ownership held by executive directors.	Author's calculations based on annual report

Table 2. The effect of IGS on managerial firm performance (logit regression).

VARIABLES	Predict Sign	Internal Governance Structures Effect on tobinsq	
		Coef. Regression	z-statistics
Comsize	+	–0.2922	(–1.55)
Comind	+	–5.7851*	(–1.87)
Constr	+	0.4151	(0.39)
Execown	+	–24.4321**	(–2.17)
exbonus	+	106.3162***	(2.66)
firmsize	+	0.4860***	(3.68)
Constant	–	0.1066	(0.91)
Observations		204	
Pseudo R2		0.1202	
LR chi2(6)		33.46***	

equations in the empirical model are presented in Table 1.

Tobinsq = 0.1066 – 0.2922 (comsize) – 5.7851 (comind) + 0.4151 (constr) – 24.4321 (execown) + 106.3162 (exbonus) + 0.4860 (firmsize) $= \pi r^2$

The first step is to perform a logit regression model between the internal governance structures and the managerial performance (Tobinsq). The result indicated that the explanatory power of the regression (pseudo (R^2)) is 12.02 per cent. Meanwhile, the LR

chi square value is 36.46 and is significant at a level of 0.01. The board of commissioner size (Comsize) has a negative, but not significant effect on the managerial firm performance. The variable of the commissioner independence (Comind) and the executive ownership (Execown) have negative and statistically significant effects at a level of 10% and 5% respectively. Furthermore, block holder ownership (Constr) has a positive influence, but no significant impact on the managerial firm performance, whereas the executive bonuses (Exbonus) have a positive influence and significant effect on the managerial firm.

4 CONCLUSION

Firstly, this study provides an empirical analysis of the model, comprising of the internal governance structures and managerial performance which have different measures. The results from this analysis provide the model relationship between the internal governance structure and the managerial performance, and the model is fit. Secondly, this study also extends the investigation of internal governance structure on managerial performance. The result from this analysis shows that the independent commissioners are effective in monitoring management but they lack the power to impactthe managerial performance. The block holders are also important in analyzing this relationship. A higher proportion of block holders leads to an increase in managerial performance. Finally, this study shows a strong support for the view that the executive bonuses have a positive impact on the managerial performance. The result suggests that granting compensation to managers is an appropriate way to higher their performance.

REFERENCES

Damodaran, G. 2014. Econometrics by example (first). Palgrave Macmillan.
Dong, J. & Gou, Y.N. 2010. Corporate governance structure, managerial discretion, and the R&D investment in China. International Review of Economics and Finance 19(2): 180–188.
Fama, E.F., Jensen, M.C., Law, J. & Conference, P.P.A. 1983. Separation of ownership and control. Control 26 (2): 301–325.
Fiorucci's, C. 2008. Agency costs and corporate governance mechanisms: Evidence for UK firms. International Journal of Managerial Finance 4(1): 37–59.
Gravette, F. & Wallan, L. 2007. Statistics for the behavioral sciences (First). Belmont, CA: Thompson Learning.
Hair, J.F., Anderson, R.E., Tatham, R.L. & Black, W. 2010. Multivariate data analysis (7th ed.). New Jersey: Prentice-Hall International.
Indonesia, F.C.G. 2006. Corporate governance (II). Jakarta.
Kasey, K., Short, H. & Wright, M. 2005. The development of corporate governance codes in the UK. Corporate Governance: Accountant Ability, Enterprise, and International Comparisons, John Wiley.

Patrick, H. 2001. Corporate governance and the Indonesian financial system: A comparative perspective, 1 paper for CSIS-Columbia University Joint Research Program on Indonesian Economic Institution Building in a Global Economy.

Purmerend, A. 2012. *Good corporate governance pays off for Indonesia*. The Jakarta Post.

Setia, A.L. 2009. Governance mechanism and firm value: the impact of ownership concentration and dividends. *Corporate Governance: An International Review* 17(6): 694–709.

Singh, A. 2003. Competition, corporate governance and selection in emerging markets. *The Economic Journal* 113(491): 443–464.

Siregar, S.V & Utama, S. 2008. Type of earnings management and the effect of ownership structure, firm size, and corporate-governance practices: Evidence from Indonesia. *The International Journal of Accounting* 43 (1): 1–27.

Victoria, K. 2006. Ownership, board structure, and performance in continental Europe. *The International Journal of Accounting* 41(2): 176–197.

Advances in Business, Management and Entrepreneurship – Hurriyati et al (eds)
© *2020 Taylor & Francis Group, London, ISBN 978-0-367-27176-3*

Impact of political risk, financial risk and economic risk on trading volume of Islamic stocks exchange in Indonesia

M.U.Al Mustofa, I. Mawardi & T. Widiastuti
Universitas Airlangga, Surabaya, Indonesia

ABSTRACT: Risk is an important factor that should be considered in every investment decision, especially for international business. Higher risk tends to decrease the motivation of investors to invest their wealth in a stock market and vice versa. The purpose of this study is to examine the impact of political risk, financial risk and economic risk on investment decision to trade in Islamic stocks in Indonesia. Quantitative research method was used in this study, and data were analyzed using regression analysis using views. Islamic Sharia Stock Index, ISSI, listed in Indonesia Stock Exchange would be used to best represent the transaction of Islamic Stocks in Indonesia's capital market. Empirical results showed economic risk negatively and significantly affect the trade volume of ISSI. While political risk insignificantly influenced the trading volume with a negative direction. On the other hand, the financial risk positively and significantly influenced the trade volume. Simultaneously, all risks show significant influence on the dependent variable. Few papers discuss how risks explicitly influenced the decision to invest in Islamic Stocks, especially in Indonesia. Our suggestion for investors is to assess the risks and put under control to minimize the potential loss.

1 INTRODUCTION

There is no universally agreed-upon definition of risk. It is widely known as the possibility that an actual return will differ from the expected return. Risk for a single asset represents the spread of the distribution of historical return. It is the dispersion of how much a particular return deviates from the mean return. The most common measure of variability or dispersion is the variance of returns and its square roots, the standard deviation. The higher the standard deviation, the riskier a particular asset would be.

Every manager, especially for a multinational company, conducts country risk analysis before investing for a business in specific country. The assessment of county risk is one of important factors to consider in the decision-making process, such as whether a company invest new projects in certain foreign countries. Rationally, the managers start divesting company's businesses in a country with increasing risk and evade investing in a country with extreme risk. The managers must understand the assessment of country risk in order to maximize potential cash flows and minimize the potential loss of the company. Most multinational companies will not be affected by every event that occur, but they will pay attention to any events that may affect the industry or the country. Madura (2010) defines country risk as the potential adverse impact of a country's environment on companies' cash flows and divided it into political and financial risks.

Madura (2010) further discussed common forms of political risk. Firstly, attitude of consumers in a particular country is a tendency for residents to buy and acquire specific products. Secondly, government action, law enforcement and regulations may affect company cash flows. Next, war and terrorism attacks make business cycle more volatile, exposing threats to the safety of the employees and the asset of the corporations. Moreover, inefficient government bureaucracy and corruption can complicate the business process for expanding the business and investing in new projects. Some irresponsible government employees expect gifts before giving approval for application submitted by the companies. They make the business competition unhealthy by giving contract or project to companies that bribed the government officials. The extreme form for a political risk is that the government would take over a business without any compensation.

Along with political risk, financial risk represents the current and potential state of the country's economy. The demand for product, service and commodity strongly depend on the economy of the country. Economic growth is one of the variables used for the assessment of financial factor affecting the country's businesses. In some case, the use of forecasting future economic growth is necessary for the evaluation. Madura (2010) believes that three factors influence the growth of the economy: inflation, currency exchange rate and interest rates. Inflation affects the purchasing power as higher inflation declines the consumption for goods and services, and vice versa. Further, currency exchange rate influences the demand for a product in the country. Strong currency reduces the demand for country's export while the opposite is true. Interest rate affects the growth of the country as higher rate tends to slowdown the economic growth while the low rate stimulates faster growth.

Hoti and McAleer (2003) define country risk as the capability of a country to pay off such international obligations. Country risk includes the credit obligations in a certain country or all the risks that depend upon economic, financial and social conditions that are likely to have an effect over the investments made in that particular country. Further studies on risk discovered that the systematic risk of stock is represented in the context of country risk that come from economic, political, financial and other environmental conditions. These macroeconomic variables affect the economy simultaneously but to different degrees. They also occur beyond the control of the investors (Kara & Karabiyik 2015, Trabelsi 2017) argues that political uncertainty seems to generate unstable financial markets and more pronounced stock market cycles. The shock caused by Tunisian Revolution increased the volatility of stock markets, leading to a deviation of the trend from its original path. Amihud and Wohl (2004) revealed a rise in the probability to start and end the war affected the movement of stock price. In addition, Vortelinos and Saha (2016) discovered that political risks explain the high volatility and discontinuity in international stock markets.

The relationship of risks with the stock prices have always become a popular topic of discussion of the financial literatures and discourses. Few papers discussed how country risks explicitly influenced the decision to invest in Islamic stocks and sharia indices. The purpose of this study is to examine the impact of country risk on investment decision to trade in Islamic stocks using a representative of an Islamic index.

Several country risk assessments are provided by international rating agencies such as the Euromoney and Economic Intelligence Unit. Further, a study by (Asiri & Hubail 2014) attempted to replicate risk valuation by using few numbers of political and macroeconomic variables affecting the GDP per Capita. The object of risk in this study is derived from the risk assessment of International Country Risk Guide (ICRG). It is a model used to forecast and analyze the risk exposed in a certain country. It gives risk indicators for investment evaluation, especially in the background of international business. The editors of International Reports, a widely respected weekly newsletter on international finance and economics, created this database in 1980, which comprises of financial, economic, and political risk. It differs the division of country risk by Madura, which focus only on political and financial risks. A separate index is created for each of the subcategories. According to the ICRG Methodology 2016, the Political Risk index is based on 100 points, Financial Risk on 50 points, and Economic Risk on 50 points. The total points from the three indices are divided by two to produce the weights for inclusion

in the composite country risk score. The composite scores, ranging from zero to 100, are then broken into categories with lower value indicating the high level of risk in a particular country and vice versa. In 1992, the International Reports' editors and analysts moved from International Reports to the PRS Group, becoming an integral part of the company's services to the international business community. below is explained the variables used for the assessment of ICRG risk components. Each variable owns different scores and weight of calculation, depending on its importance and influence on the country.

- Political risk: Government Stability, Socioeconomic Conditions, Investment Profile, Internal Conflict, External Conflict, Corruption, Military in Politics, Religious Tensions, Law and Order, Ethnic Tensions, Democratic Accountability, and Quality of Government Bureaucracy.
- Economic Risk: GDP per Head for a given year, Real GDP Growth, Annual Inflation Rate, Government Budget Balance as a Percentage of GDP, and the balance of Current Account as a Percentage of GDP.
- Financial Risk: Foreign Debt as a Percentage of GDP, Foreign Debt Service as a Percentage of Exports of Goods and Services, Current Account as a Percentage of Exports of Goods and Services, Net International Liquidity as Months of Import Cover, Exchange Rate Stability.

Sharia index is an index based on Islamic sharia. Stocks listed in the sharia index are listed companies whose business activities are not contradictory to sharia laws, such as first, gambling business or prohibited trade. In addition, the business of conventional financial institutions, business that runs with the use of interest rate, includes conventional banking and insurance. Further, businesses that produce, distribute and provide food and beverages are categorized as haram goods or services that are morally destructive and harmful (Manan 2012). To assist achieving the goal of the study and represent trading of Islamic Stocks in Indonesian stock market, an index of Islamic Stocks was used, in this case, Indonesia Sharia Shares Index (ISSI). ISSI is a composite stock index that reflects the total sharia stocks listed in the Indonesia Stock Exchange (IDX) and included in List Islamic Stock (DES) which issued by Financial Service Authority (OJK). It is an indicator of trading performance for Indonesia Islamic stock market. ISSI constituents are reselected twice in every year, May and November, following the DES review schedule and published at the beginning of the following month. The purposes of reselection are to evaluate and adjust if there are newly listed sharia stocks and remove stocks that no longer qualified as Islamic stock according to the assessment method conducted by OJK. Weighted average of market capitalization is used for the calculation of ISSI. The base year used for ISSI calculations is the start of the issuance of the List of Sharia Securities (DES) that is December 2007. ISSI was launched on May 12, 2011. According to the January 2018 update, there are 366 Islamic stocks in ISSI.

2 METHOD

The research method used for the study is quantitative research method. The application of this method is to look for the influence and of one variable to another. The study would apply classical tests of data sampling and multiple regression analysis using EViews software. To best represent the exchange of Islamic Stocks in Indonesia's stock market, the study used the data of trading volume of Islamic Sharia Stock Index, ISSI. While the assessment of risk is derived from International Country Risk Guide, ICRG, established by the PRS Group. The data used for the analysis is a time horizon with monthly frequency of trading volume in Indonesian Rupiah, starting from May 2011 to March 2016.

3 RESULT AND DISCUSSION

Several macroeconomic and monetary factors affected the performance of ISSI. A study conducted by Suciningtias and Khoiroh (2015) showed that macroeconomic variables such as inflation and currency exchange rate have negative and significant impact on Indonesia Sharia Stock Index (ISSI). However, the Bank Indonesia Sharia Certificates (SBIS) and world oil prices do not have a significant influence on the dependent variable. The independent variables influenced significantly and simultaneously the performance of ISSI. Furthermore, the model built in their study is considered weak for the reason of low R2. Soon after, Saputra et al. (2017) studied some macroeconomics and monetary variables that affect ISSI and conclude that, simultaneously, BI Rate, Inflation, Rupiah Exchange Rate and SBIS affect the performance of ISSI significantly and positively. While partially, only the exchange rate significantly affects the price movement of ISSI. Earlier studies conducted with the focus of considering macroeconomic variables; however, few papers discussed how country risks explicitly influenced the decision to invest in Islamic Stocks, especially in Indonesia. This may due for the availability of the data.

The analyses started with the classical test of data sampling with the following results. The data is normally distributed since the probability value of Jarque-Bera is 0.115 and is higher than the significance rate. The model is free from the issue of Heteroscedasticity as the probability of Glejser test equals to 0.7341 and is higher than significance rate. The model suffers from no multicollinearity issue. This can be seen at correlation matrix in Table 1, as the correlation value of one variable to other shows no more than 0.90.

Table 1. Correlation matrix.

Matrix	Y	Polit_Risk	Eco_Risk	Finc_Risk
Y	1.0000			
Polit_Risk	−0.1024	1.0000		
Eco_Risk	−0.0130	−0.3516	1.0000	
Finc_Risk	0.2282	−0.0175	0.1785	1.0000

Based on the multiple regression result in Table 2, discussion over some analysis and its finding is:

Table 2. Independent variable: Monthly average trading volume of.

Variable	Coefficient	t-Statistic	Prob.
ECO_RISK	−286174554918.8522	−3.0912	0.0031
FINC_RISK	188434007849.9115	2.8066	0.0069
POLIT_RISK	−86212304455.18683	−1.6851	0.0976
C	11108502886246.64	2.7618	0.0078

3.1 Economic risk

The regression analysis shows a negative relationship between economic risk and monthly trading volume of ISSI. The negative relationship come for the reason of negative value of coefficient regression. The value of coefficient gives the clue if the as-assessment of economic risk increases by one unit, the trading volume will decrease by 286 billion Rupiah. This correlation is statistically significant at all significance levels since the probability (p-value) is 0.003 and it is less than 0.1, 0.05, and 0.01. Statistically, economic risk has a significant effect on trading volume in ISSI with a negative direction.

3.2 Financial risk

In opposition to the negative relationship of economic risk, the regression analysis shows a positive relationship between financial risk and monthly trading volume of ISSI for the reason of positive value of coefficient regression. The value of coefficient gives the clue if the assessment of financial risk increases by one unit, the trading volume will increase by 188 billion Rupiah. This correlation is statistically significant at all significance levels since the probability (p-value) is 0.007 and it is less than 0.1, 0.05, and 0.01. Statistical finding conclude that financial risk has a significant effect on trading volume in IS-SI with a positive direction.

3.3 Political risk

An interesting finding come from the regression analysis that shows a negative relationship between

412

political risk and monthly trading volume of ISSI for the reason of negative value of coefficient regression. The value of coefficient gives the clue if the assessment of political risk increases by one unit, the trading volume will decrease by 862 billion Rupiah. On the other hand, this correlation is statistically insignificant at 5% significance rate since the Probability (p-value) is 0.098 and it is higher than 0.05. This means that political risk has an insignificant effect on trading volume of ISSI with a negative direction. Nonetheless, statistically at 10% significance rate, political risk would affect the dependent variable significantly at a negative direction.

The three variables of risk consisting of economic, financial and political risk can jointly influence the trading volume of ISSI. This come from the result of F-test. It represents joint hypothesis of independent variables and test whether the independent variables influence simultaneously the dependent variable. The finding shows that the probability of F-Test is 0.0000, which is less than statistical value of significance levels. Therefore, the finding concluded that the independent variables could conjointly influence the dependent variable. The degree of how far these independent variables could affect the dependent variable could be identified by assessing the coefficient of determination. It is defined as the proportion of the total variation of the dependent variable in a multiple regression model that is explained by its relationship to the independent variables (Groebner et al. 2014). It is widely called as R-squared and is denoted as R2. The multiple coefficient of determination is 0.3314, which means that more than 33.14% of the variation in Monthly Average Trade Volume of ISSI can be explained by the variation in the independent variables of risks consisting of political, financial and economic risks. The model built in this study considered weak to explain the variation of the dependent variable due to low coefficient of determination. Early study conducted by Suciningtias and Khoiroh (2015) resulted with the weak model, too. The reason comes from the selection of macroeconomic variables, which failed to explain the main variability of ISSI performance. The only factor used in this model is country risk, excluding other macroeconomic variables that affect the movement of stock index. The addition of other macroeconomic variables might give better R-square.

Considering country risk as one of macroeconomic variables, country risk affects the performance of trading in stocks listed in ISSI. The finding results supported the studies conducted by Suciningtias and Khoiroh (2015) and Saputra et al. (2017), which stated that macroeconomic variables affect the movement of stock price and return volatility.

Further analysis showed that negative relationships for economic and political risks are in line with the research conducted by Kara and Karabiyik

(2015) that discussed how country risks affect negatively to the movement of a BIST100 index. Moreover, Political risk has an insignificant influence on trading volume of ISSI. The finding opposed the research conducted by several scholars (Amihud & Wohl 2004, Trabelsi 2017, Vortelinos & Saha 2016), which concluded that political uncertainty affect the volatility of stock at significant level. It also competes against the argument of Madura, which stated that political risk contributes to high consideration of investment decisions. However, at 10% significance rate, the political risk will influence negatively and significantly to the dependent variable.

The positive impact of financial risk on trading volume of ISSI opposes the finding result of study conducted by Kara and Karabiyik (2015), which stated that financial risk affect the stocks movement negatively. This positive influence comes from the positive correlation of 0.2282 between financial risk and the monthly average trading volume of ISSI. In addition, the correlation between the financial risk and the return of ISSI is showed to be a positive value of 0.003. Positive correlation revealed to be in line with the general perspective of high risks is compensated with greater return and opposed the paradigm of higher risk tend to lower the motivation of investors as stated by Madura (2010). These results give evidences of existing risk-taker investors in Indonesian Islamic stock exchange, granting the consideration of risk factor, especially financial risk. During the period of the study, the total average of monthly return for the ISSI index is a positive value of 0.03%.

4 CONCLUSION

ISSI acts as an index for the performance of Indonesia sharia stock market. This study aimed to analyze the impact of political risk, financial risk, and economic risk on trading volume of ISSI. To conclude, there is a negative and significant relationship between economic risk and the monthly average trading volume. While the political risk resulted the same direction but insignificantly. Positive and significant correlation of financial risk and the dependent variable give evidence of the presence of risk takers investors in Indonesian Islamic stocks market. All risks influence cooperatively the monthly average trading volume. To recommend, every manager shall have taken care the risk wisely. Periodic assessment of country risk is sought after to determine value at risk and predict what kind of risk determinants would give the most impact. Managers can formulate several hedging strategies after conducting risks valuation and assessment before plotting for investment decision. Further research is needed to discover risk determinants that have major influences on the trading of ISSI.

ACKNOWLEDGEMENT

We would like to show our great appreciation to the Faculty of Economic and Business Universitas Airlangga for providing research funding for this topic.

REFERENCES

Amihud, Y. & Wohl, A. 2004. Political news and stock prices: the case of Saddam Hussein contracts. *Journal of Banking and Finance* 28(5): 1185–1200.

Asiri, B.K. & Hubail, R.A. 2014. An empirical analysis of country risk ratings. *Journal of Business Studies Quarterly* 5(4): 52–67.

Groebner, D.F., Shannon, P.W., Fry, P.C. & Smith, K.D. 2014. *Business statistics a decision-making approach.* Edinburgh: Pearson Education Limited.

Hoti, S. & McAleer, M. 2003. An empirical assessment of country risk ratings and association models. *Journal of Economic Surveys* 18(4): 539–550.

Kara, E. & Karabiyik, L. 2015. The effect of country risk on stock prices: An application in Borsa Istanbul. *The Journal of Faculty of Economics and Administrative Sciences* 20(1): 225–239.

Madura, J. 2010. International *corporate finance.* joe sabatino.

Manan, A. 2012. *Hukum ekonomi syariah: dalam perspektif kewenangan peradilan agama.* Jakarta: Kencana Prenada Media Group.

Saputra, R., Litriani, E. & Akbar, D.A. 2017. Pengaruh BI rate, inflasi, nilai tukar rupiah, dan sertifikat bank indonesia syariah (SBIS) terhadap Indeks saham syariah Indonesia (IS-SI). *I-Economic* 3(1): 51–72.

Suciningtias, S.A. & Khoiroh, R. 2015. Analisis dampak variabel makro ekonomi terhadap indeks saham syariah Indonesia (ISSI). *Conference in Business, Accounting, and Management (CBAM)* 2(1): 398–412.

Trabelsi, M.A. 2017. Political uncertainty and behavior of tunisian stock market cycles: Structural unobserved components time series models. *Research in International Business and Finance* 39: 206–214.

Vortelinos, D.I. & Saha, S. 2016. The impact of political risk on return, volatility and discontinuity: Evidence from the international stock and foreign exchange markets. *Finance Research Letters* 17: 222–226.

Demographic analysis for the selection of an investment type for amateur golfers

H. Sulistiyo & E. Mahpudin
Universitas Pendidikan Indonesia, Bandung, Indonesia

ABSTRACT: This research examined the demographic factor affecting the choice of investment for amateur golfers in Karawang City. This research divides investment into two groups: real asset and financial asset investment. Data in this research was obtained by distributing a questionnaire to employees working in governmental and private offices. The sample was collected using a purposive sampling method. Then the data were analyzed using cross-tabulation and chi-square analysis. The research found that demographic factors influenced the choice of investment among young professionals in Karawang City. Out of the six demographic factors observed, there were five factors that significantly influenced the investment selection. These factors were gender, profession, education, the number of family members and income. However, the one factor that did not influence the investment selection was age.

1 INTRODUCTION

The background of golfers, among others, came from businessmen, bureaucrats, political elites, expatriates, or others who can play in the area Karawang. They use golf as an important tool for hobby sports as well as expanding their business network. Aside from being a useful sport for body health, another important benefit of golf is that it provides a unique window into the personality and behavioral values of others.

According to Nurmayanti (2017), the industrial area is currently one of the development priorities in West Java. Based on the data of the Ministry of Industry, there are about 2,381.97 hectares of land being developed into ten new industrial areas of national and international standards in West Java. Of these, 851.97 hectares or about 35% are in the Karawang area.

Sutedja (2017) stated that fforeign companies, of course, employ many foreign employees. The golf course is an ideal venue for the meeting of two different cultures namely, foreign culture and local culture. The number of golf courses in Indonesia which is under the member of APLGI (Association of Entrepreneurs of Golf Course of Indonesia), from the total range of golf courses in Indonesia about 140 courses.

It is interesting to note how the demographic conditions of different groups of people influence the preferences of choosing an investment. That's why the author tries to bring this theme to be the theme of this research.

In some studies, it shows that demographic factors are one of the factors that influence the decision-making to invest.

Age, education and income factors as demographic factors, in Kusumawati's research (2013), have an influence on decisions in investing. Subsequent research was conducted by Saputra and Anastasia (2013) with the title "Type of Investment-Based on Risk Profile". This research is carried out on heads of households who live in Surabaya, are married and have children who are in school. The research variables used were gender, age, level of education, number of children, work and income. Conclusions from the results of this study are that four of the six demographic variables tested have a relationship with the grouping of the respondents' risk profiles, namely gender, number of children, work and income. Meanwhile, Rudyanto (2014) in his research entitled "Preference for the Selection of Types of Young Professional Investments in Surabaya" got the following results: there is a relationship between demographic factors including gender, marital status, education, employment and income with the preference of choosing the type of investment in Surabaya, except for the number of family members.

Based on the description of the background, the formulation of the problem in this study is as follows: Are there any influences of demographic factors (sex, age, and education, number of family members, occupation and income)on the preferences of investment type selection among amateur golfers?

2 METHOD

2.1 *Population and sample*

The population used are amateur golfers who play in Karawang. The total number of amateur golfers enrolled in several golf clubs in Karawang is 100. Therefore, this study used the number of samples taken using saturated samples, where the whole population is taken as a sample.

2.2 Research variable

In this study variables that are considered closely related to the research objectives will be observed and analyzed. The research variables are the demography factors as the independent variables and the investment type as the dependent variables.

Demographic factors as independent variables in this study include gender, age, education, number of family members, employment and income. The dependent variables in this study are: Investment type, which is divided into Real Asset and Financial Asset investments.

2.3 Data analysis technique

To answer the hypothesis this study used Chi-Square Test analysis techniques (X2). This study uses the nominal measurement scale. The nominal measurement scale only makes it possible to compare the observed results into specific categories. To analyze the relationship of these categories the freedom of the categorical variables will be seen. Chi-square based on the cross table is also relatively popular in measuring the correlation between variables (Junaidi 2010)

Procedure X2 Test (Chi-Square test) based on this cross-tabulation is tabulating (arranging in a table form) a variable in the category and testing the hypothesis, whether the observed frequency (observed data) is not different from the expected frequency (theoretical frequency). The goodness-of-fit test of chi-square compares the observed frequencies and expected frequencies in each category to test that all categories contain the same proportion of values or test that each category contains a certain proportion of values.

3 RESULT AND DISCUSSION

In this study, a cross tabulation test was conducted to see the relationship between demographic variables and investment types. Table 1 presents the results of the chi-square cross tabulation test in this study.

And Table 2 describes the results characteristics of respondents in relation to the investment decision they choose.

Table 1. The chi-square cross tabulation test.

	Result
Gender with Investment	0.840
Age with Investment	0.002
Education with Investment	0.004
Number of Family members with Investment	0.049
Work with Investment	0.007
Revenue with Investment	0.045

Table 2. Characteristics of respondents.

Variables	Type of investment		Total
	Real Asset	Financial Asset	
Gender			
Male	83	5	88
Female	8	4	12
Total	91	9	100
Age			
< 21 Years Old			
21–30 Years Old	15	2	17
31–40 Years Old	45	5	50
41–50 Years Old	22	1	23
51–60 Years Old	9	1	10
Total	91	9	100
Education			
Senior High School	4	3	7
Bachelor	77	6	83
Post Graduated	10	0	10
Total	91	9	100
Numbers of Family Members			
1	32	2	34
2	52	4	56
≥ 3	7	3	10
Total	91	9	100
Work			
Bureaucrats	6	3	9
Private	85	6	91
Total	91	9	100
Income			
≤ Rp 25.000.000	1	1	2
Rp 25.000.000–Rp 50.000.000	67	4	71
≥Rp 50.000.000	23	4	27
TOTAL	91	9	100

In this article, we have presented data about the relationship between demography and decision making in investing. The results show that the demographic factors are gender, occupation, education, number of family members and income, and which all have an influence on investment decisions, except the age factor, which is the only factor in this study that has no effect. This finding contributes to the decision to invest for amateur golfers, based on demographic factors. There are many other factors that can be used as a discussion about amateur golfers, for example in terms of the products used, as well as other things, so as to provide a comprehensive perspective on amateur golfers.

4 CONCLUSION

We have presented the influence of demography factors on the selection of investment type among amateur golfers in Kabupaten Karawang. Of the six demographic factors studied, there are five factors that have a significant effect on the type of investment selection. These factors are gender, occupation, education, number of family members and income. But there is one factor that is not remotely relevant to the choice of investment type, i.e. age.

Of the 100 respondents studied, 91 respondents chose real estate investment, which is, gold, houses, and land, while 9 respondents chose deposit/saving, bond, and stock.

REFERENCES

Junaidi. 2010. Statistika Non-Parametrik. Fakultas Ekonomi Universitas Jambi. Jambi.

Kusumawati, Melisa. 2013. Faktor Demografi, Economic Factors dan Behavioral Motivation Dalam Pertimbangan Keputusan Investasi Di Surabaya. *Finestas*, 1(2): 30–35.

Nurmayanti. 2017. Kota Industri Karawang Bisa Jadi Kantong Ekonomi Baru. [Online]. Retrieved: http://www.liputan6.com/bisnis/read/3187844/kota-industri-karawang-bisa-jadi-kantong-ekonomi-baru. Accessed 8 Maret 2018.

Rudyanto, D. 2014. Preferensi Pemilihan Jenis Investasi Profesional Muda di Surabaya. *Finestas*, 2(1): 103–108.

Saputra, H.I & Anastasia, N. 2013. Jenis Investasi Berdasarkan Profil Risiko. *Finesta*, 4(2): 47–52.

Sutedja. 2017. Nuansa Konsep Golf di Hunian Vertikal. [Online]. Retrieved: http://www.koran-jakarta.com/berita-detail.php?id=7973. Accessed 4 Maret 2018.

Advances in Business, Management and Entrepreneurship – Hurriyati et al (eds)
© *2020 Taylor & Francis Group, London, ISBN 978-0-367-27176-3*

Intended use of initial public offerings proceeds, underpricing, and long-term market performance of stocks in Indonesia

H. Meidiaswati & D. Novita
Universitas Kartini, Surabaya, Indonesia

N. Sasikirono
Universitas Airlangga, Surabaya, Indonesia

ABSTRACT: The aim of this study is to examine the effect of the intended use of Initial Public Offerings (IPO) proceeds on the market performance of stocks in Indonesia Stock Exchange (IDX). The test was conducted by multiple linear regression. The purpose was to find out whether the information in the prospectus of stock offerings related to IPO proceeds utilization for acquisition, group financing, long-term investment, debt repayment, and working capital, affect both the initial and the long-term stock returns. The sample consisted of 115 companies that made initial public offerings in Indonesia Stock Exchange during the period of 2006-2013. The results showed that the disclosure of specific information about the use of funds for acquisitions negatively affected the initial returns and long-term market performance. This result is expected to provide a foundation for capital market authorities to further encourage the quality of information disclosure of issuers in order to minimize information asymmetry in the primary market.

1 INTRODUCTION

Information asymmetry influences transaction incentives in the capital market. Healey & Palepu (2001) argue that the lack of information can encourage undervalued capital markets against good investment opportunities and overvalued on poor investment opportunities. The greatest information asymmetry occurs in the initial public offerings (IPO) event. In the initial public offering of stocks, information about the company is relatively limited in terms of publication. It also tends to focus on sophisticated investors. The lack of public information in IPO causes common investors facing uncertainty in high investment.

One solution to reduce information asymmetry is through the regulation of the capital market authority on the necessity of a company to disclose all private information. Disclosure increases not only in efficiency and incentives, but also the endogeneity of the market process. It however involves multiple market participants (Verrechia 2001). Based on the disclosure of information, investors, especially unsophisticated investors, can moderate the risks through investment in information and utilize the services of capital market supporting institutions, both information intermediaries and financial intermediaries institutions (Healey & Palepu 2001).

Rock (1986) states that, at the time of the IPO, the information gap between investors precludes is-suers from selling stocks at their fair value. The imbalance of information, irrespective of good or bad quality of issuers, causes sophisticated investors to always outperform unsophisticated investors. This process will ultimately eliminate unsophisticated investors and become a disincentive for investment in the capital market. Issuers and underwriters overcome this problem by underpriced their stock. The impresario hypothesis (Ritter 1998) states that the issuers and underwriters systematically fix the IPO price below the intrinsic value in order to leave the good taste for investors in the form of a high initial returns. Based on the impresario hypothesis, in the long run, as more information about the company reveals, the market will understand the true value of the company. This understanding encourages correction of stock prices and results in lower long-term market performance of IPO stocks; known as the phenomenon of long-term IPO underperformance.

One form of disclosure in the IPO process is about the issuer's plan for the use of proceeds in the stock offering prospectus. Regulation of Indonesian Capital Market Supervisory Agency (Bapepam) Number IX. C. 2 of 1996, stipulates the obligation of issuers to disclose material information in the initial public offering (IPO) prospectus; one of them is the plan of the use of proceeds. Through the disclosure of the plan to use the IPO proceeds, investors are not only informed about the proportion of the use of funds for various common purposes, but they can obtain information about the use of proceeds for specific purposes such as acquisitions and financing of a subsidiary (group financing). In other words, the disclosure of the proceeds plan can reduce the information asymmetry in the primary market, and on both the underpricing and long-term market per-formance of IPO stocks.

Leone et al. (2007) mention that disclosure of plans to use IPO proceeds can reduce ex-ante uncertainty and assist investors in estimating values in the secondary market. Due to the disclosure nature that minimizes information asymmetry, its impact on underpricing as well as long-term market perfor-mance is that IPO stocks can be pre-dicted. Specifi-cally, if the intended use of IPO proceeds provides positive information, the issuer is motivated to set the IPO stock price not much different from the in-trinsic value, and vice versa. According to (Jeanneret 2005), the company that plans to utilize IPO proceeds for debt repayment is basically mak-ing improvements to its financial flexibility and cap-ital structure. Therefore, such companies will exhibit low underpricing level followed by the improved long-term market perform-ance. Companies planning to use IPO proceeds for long-term investments basi-cally also convey infor-mation of the increase of asymmetry information and agency problems in the future. Therefore, they should offer a high level of underpricing and be addressed with poor long-term market performance. The relationship between dis-closure of the use of proceeds for investment to un-derpricing and long-term market performance can be moderated if the long-term investment objective is the acquisition and the financing of companies in the group. This moderation is the impact of potential synergies and value enhancement post-acquisition activities and group financing. In the case of intend-ed use of IPO proceeds for working capital, Gumanti et al. (2015) note that the utilization of funds for work-ing capital is perceived as risky due to low investor control over operational activities. Therefore such intention will increase the level of underpricing. Ben Amor & Kohli (2016) in their study found no relationship between disclosure of the main use of IPO funds for investment and long-term market performance of IPO stocks. On the con-trary, there is a strong negative relationship between the dis-closure of the main uses for debt repayment and long-term market performance. Autore et al. (2009) found that disclosure of purpose-specific use of proceeds in SEO demonstrated higher long-term market performance than non-specific ones.

This study aims to examine the effect of the dis-closure of intended use of IPO proceeds on under-pricing and long-term market performance of IPO in the Indonesian Stock Exchange (IDX).

2 METHOD

2.1 Sample and data

The population was the entire IPO in Indonesia Stock Exchange (formerly the Jakarta Stock Ex-change) during the period 2006-2013. The sample was 115 IPO companies obtained by purposive sam-pling technique based on research objectives.

The number of samples obtained covering 70% of IPOs during the period 2006-2013. Secondary data were obtained by documentation techniques, from Indo-nesia Stock Exchange (IDX), Indones-ian Capital Market Directory (ICMD), Indonesian Capital Mar-ket Library (Icamel), Yahoo Finance and The Wallstreet Journal. The data being used was the companies conducting IPOs (i.e. the plan of use of proceeds, age, asset value, sales, total proceeds ob-tained through IPO) described in the initial offering prospectus, stock price, as well as the Composite Stock Price Index 2006-2016 (adjusted to the IPO year).

2.2 Measure

Market Adjusted Initial Returns (MAIR) was used as an underpricing proxy and was calculated by adjusting the initial returns of a stock with the returns of the market on the same day. Initial returns was the first day returns of stocks traded in the primary market. It was calculated as the differ-ence in closing stock price on the first trading day on the secondary market with IPO stock price div-ided by IPO stock price. Abnormal returns esti-mated using market adjusted abnormal returns; that was, daily stock returns minus market returns. CAR36 was the accumulation of abnormal returns of 1st to 36th month, after the first trading day of IPO stocks.

Other variables used in this study included: DACQUI= dummy variable for acquisition, which was 1 if part or all of the IPO funds were used for the acquisition of both assets and stocks, and 0 otherwise; DINFIN= dummy variable for group financing which was 1 if part or all of IPO funds used to finance subsidiaries or other company in the group either in the form of capital investment, debt repayment, or working capital investment; INVEST= proportion of IPO proceeds was used for long-term investments in the form of capital goods, investments in subsidiaries, or acquisitions; DEBREP= proportion of IPO proceeds was used for short-term and long-term repayment of both parent and subsidiary debts; WORCAP= propor-tion of IPO proceeds was used for investments in working capital and financing operating costs for both parent and subsidiary companies; LNAGE= natural logarithm of company age at IPO; LNTAS= natural logarithm of total company assets at IPO; LNSE= natural logarithm of com-pany sales at IPO; and LNPROC= natural loga-rithm of IPO proceeds.

2.3 Model of analysis

This study was an explanatory study that aimed to analyze the relationship between variables and to explain the influence between variables through hy-pothesis testing. The tests were performed by linear regression, with the following models:

$$MAIR = \beta_0 + \beta_1 DACQUI + \beta_2 DINFIN + \beta_3 INVEST$$
$$+ \beta_4 DEBREP + \beta_5 WORCAP + \beta_6 LnTAS$$
$$+ \beta_7 LnSLE + \beta_8 LnPROC + \beta_9 LnAGE + e$$

$$(1)$$

$$CAR36 = \beta_0 + \beta_1 MAIR + \beta_2 DACQUI + \beta_3 DINFIN$$
$$+ \beta_4 INVEST + \beta_5 DEBREP + \beta_6 WORCAP$$
$$+ \beta_7 LnTAS + \beta_8 LnSLE + \beta_9 LnPROC$$
$$+ \beta_{10} LnAGE + e$$

$$(2)$$

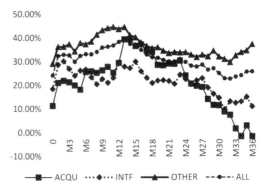

Figure 1. CAAR36 of acquisition, inter-group financing, other using IPOs.
Source: Processed Data, 2018.

3 RESULTS AND DISCUSSION

3.1 Descriptive

Table 1 shows that there were 15 companies (13%) that disclosed the acquisition plan using IPO pro-ceeds. This included the companies that acquired as-sets through the purchase of stocks for the purpose of acquiring external business entities. Thirty-six compan-ies (31%) disclosed the intended use of IPO proceeds for funding companies in the same busi-ness group, either subsidiaries or other companies in the same parent company. The average underpricing, measured using MAIR, was 23.7%. IPOs in Indone-sia were gen-erally underpriced, with 92 issuers (80%) producing MAIR positive. Forty-two compa-nies (37%) produced CAR36 that was higher than MAIR. The CAR36 aver-age value was 1.59%; lower long-term market perform-ance of stocks could not be said to be common in Indonesia. This condition resulted in an average cumu-lative average abnormal returns over 36 months or CAAR36 (i.e. 25.8%) was higher than MAIR.

Table 1. Descriptive.

	N	Min	Max	Mean	Std. Deviation
DACQUI	115	.00	1.00	.11	.32
DINFIN	115	.00	1.00	.30	.46
INVEST (%)	115	.00	100.00	58.11	32.76
DEBREP (%)	115	.00	100.00	17.04	25.10
WORCAP (%)	115	.00	100.00	24.15	29.10
MAIR (%)	115	-63.81	122.85	24.21	31.11
CAR36 (%)	115	-203.60	333.57	1.59	94.31
T. ASSETS (million IDR)	115	9.580	44.990.000	3.241.000	6.386.000
SALES (mil-lion IDR)	115	2.942	17.860.000	1.643.000	3.118.000
PROCEEDS (million IDR)	115	22.500	6.292.000	700.700	1.014.000
AGE (years)	115	1.10	90.42	19.96	14.77
Valid N (listwise)	115				

Source: Processed Data, 2018.

Compared to other specific use plans of the pro-ceeds, the acquisition yields the lowest initial returns (Figure 1). The low underpricing of the acquisition IPO is due to perceptions about the relatively low level of investment risk through the purchase of as-sets and other companies' stocks, caused by post-acquisition synergies. IPOs with plans for the use of proceeds for internal financing of the group are also perceived to be less risky than other fund usage plans. However, there are the differences in long-term per-formance characteristics of stocks for both plans of the use of the proceeds. Meanwhile, long-term stock performance for the purpose of using proceeds for group financing shows a downward trend beginning in the second post-IPO month; this is not the case of acquisition IPOs. The increase in abnormal returns in the period to one year after the IPO occurs. This is related to the synergy of acquisition results that can be formed in a relatively short time. However in the long run, the impact of synergy begins to fade and issues arising from the acquisition shows an impact on stock performance (Davidson 1988, Datta 1991).

Figure 2 shows that the characteristics of the use of IPO proceeds, which were dominant in debt repayment purposes (DEB) were similar to those used for acquisitions. This can be seen in the low underpicing and market performance that tends to increase in the period of one year after the IPO. This can be attributed to an increase in reserve borrowing capacity as well as a decrease in corporate financial risk that drives stock-market performance. However, unlike the acquisition, in the long-term the accumu-lation of returns is not much different from the initial returns. This condition is different from the IPO with the main goal of long-term investment (INV). Char-acteristics of long-term investment and high risk of Indonesian business are interpreted by investors as the high risk of investing in issuers with a primary motive for long-term investment. Interests in buying stock is encouraged by companies doing high under-pricing (26.46%). However, as more information of the company's intrinsic value is revealed, long-term

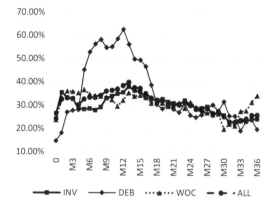

Figure 2. CAAR36 by dominant use of proceeds.
Source: Processed Data, 2018.

IPO market performance with the primary goal of long-term investment is declining (24.09%).

The grouping of issuers into high and low underpricing, based on the average initial returns (as in Figure 3), shows the difference in cumulative aver-age abnormal returns (CAAR) in both. Long-run stock market performance of IPO stocks in the high underpricing group shows an average increase in av-erage abnormal returns (AAR) of 0.46% within 3 years after IPO. The phenomenon of IPO groups with high underpricing is in contrast to the low un-derpricing group which actually shows the decrease of 0.38% in AAR. CARs from the high underpricing group have an increasing tendency, but not for low underpricing groups. Consequently, there are signifi-cant differences in long-term market performance of high underpricing IPO groups (with CAAR 68.01%) and low underpricing groups (with CAAR -6.74%) at α 1%.

3.2 Intended use of proceeds and underpricing

The results show that the disclosure of information on the use of IPO proceeds in the prospectus, is sufficiently considered by the investor (Table 2). Alt-hough other plans for the use of proceeds have no

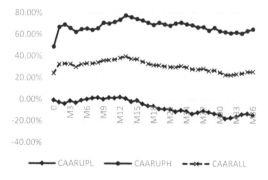

Figure 3. CAAR 36 of high and low underpricing groups.
Source: Processed Data, 2018.

Table 2. Regression result on MAIR.

	FULL SAMPLE		SAMPLE EXCLUDING ECONOMIC CRISIS	
	Model 1	Model 1A	Model 2	Model 2A
(Constant)	21.726***	137.425**	23.157***	133.585**
	(2.805)	(2.157)	(2.883)	(2.049)
DACQUI	-21.628**	-20.874**	-21.455**	-20.768**
	(-2.379)	(-2.245)	(-2.353)	(-2.223)
DINFIN	-9.801	-7.183	-9.839	-7.320
	(-1.561)	(-1.118)	(-1.563)	(-1.132)
INVEST	.170	.129	.153	.123
	(1.598)	(1.225)	(1.401)	(1.142)
DEBREP	-.117	-.215	-.133	-.222
	(-.886)	(-1.538)	(-.989)	(-1.561)
LNAGE		-9.243		-9.152
		(-2.436)		(-2.394)
LNTAS		2.402		2.602
		(.744)		(.787)
LNSLE		-1.566		-1.608
		(-.792)		(-.808)
LNPRO		-4.233		-4.241
		(-1.332)		(-1.329)
R^2	0.101	.186	.101	.183
F	3.093***	3.036***	3.059***	2.948***

Table 2 reports the results from OLS regressions of MAIR for 2006 - 2013 of full sample and sample excluding data on the year of economic crisis. In each group, the analysis was performed for data with and without control variables. T-statistics are in parentheses. *, **, *** Indicate significance at the 10, 5, and 1% levels, respectively.
Source: Processed Data, 2018

effect, information on the acquisition plan is negatively correlated with MAIR. These results are consistent both in tests with control variables and without control variables. The results also show consistency in the overall sample and samples that do not consider the data during an economic crisis. Investors consider the company's acquisition plan from IPO proceeds as a positive signal related to risk. This make investors interested in investing and encourage issuers to sell stocks at prices that are not too much different from its intrinsic value.

Initial returns, in addition to being influenced by information disclosure related acquisition plans, are also influenced by the age of the company. This means that the older the company's age when doing an IPO, the smaller the initial stock returns. The old-er the IPO company's age, the more information the market had, the less likely it is to get a high price difference during the first day of stock trading. Age is also related to quality. The longer the life of the company, the longer the company survives in its business environment, the higher the ability of management in managing their business. It defines issu-er's quality. Signaling theory of Welch (1996) ex-plained that a good quality IPO issuer can bear the greater loss of money on the table and will therefore establish a low underpricing. Meanwhile, information on the utilization of IPO funds for group financing and general long-term investment has no effect on MAIR on testing with or without control variables. These results indicate that the proportion of IPO fund utilization for general investment and group financing activities, in general, is not a major consideration of IPO stock pricing.

3.3 Intended use of proceeds and long-term market performance

Testing of long-term stock returns, measured by CAR36, indicates that the proportion of IPO funds to be utilized for: long-term investments, debt repayment, and investment in working capital consistently does not affect long-term returns either with or without control variables (Table 4). The results of the analysis are also consistent in the full sample and sample group that do not take into account the economic crisis.

The effect of disclosure intended use of IPO proceeds for acquisition purposes arises because of the positive consequences of acquisitions, which are instant. The synergies that result from the acquisition can reduce the uncertainty of investment and encourage issuers to sell stocks at prices close to intrinsic value. Even so, issuers do not seem to consider the negative consequences of long-term acquisitions. Investors initially respond to positive signals of acquisitions in a positive way but over the course of time, investors' understanding of the true value of the acquisition project pushed the CAR down. It appears that it takes at least one year for investors to realize real project value.

The relationship between the proportion of the use of proceeds with underpricing and long-term market performance indicates that underpricing is a common mechanism used by issuers and underwrit-ers in Indonesia to increase the interest in investing in the primary market. This is done without consid-ering the impact of disclosure of the issuer's use plan and underwriter's underpricing in most of the IPO. This action can be understood as the level of inves-tor confidence in the Indonesian capital market is still low. This condition is then exacerbated by the low interest of Indonesians to invest. Different things happen in the IPO with the aim of acquisition that is considered to provide a signal of potential growth that is relatively instant. In the long run, along with a better understanding of the real value of the acquisition project, there is also a decline in long-term market performance.

Similar to underpricing, the only control variable that shows the effect on long-term market performance is the age of the firm. The negative relation-ship between age and long-term market performance suggests that companies that are already at maturity tend to have difficulty in finding new investment opportun-ities and focus more on maintaining opera-tional

Table 3. Regression Result on CAR36.

	FULL SAMPLE		SAMPLE EXCLUDING ECONOMIC CRISIS	
	Model 3	Model 3A	Model 4	Model 4A
(Constant)	-2.916	-289.899	-5.363	-286.715
	(.899)	(-1.477)	(-.225)	(-1.430)
DACQUI	-31.508	-47.349*	-31.803	-47.424*
	(.245)	(-1.651)	(-1.174)	(-1.645)
DINFIN	-11.267	-18.984	-11.202	-18.860
	(.547)	(-.974)	(-.599)	(-.961)
INVEST	.045	.092	.074	.097
	(.886)	(.289)	(.228)	(.298)
DEBREP	.229	-.083	.256	-.077
	(.560)	(-.194)	(.643)	(-.177)
MAIR		-.155		-.155
		(-.531)		(-.525)
LNAGE		-23.546**		-23.617**
		(-2.003)		(-1.995)
LNTAS		12.767		12.594
		(1.309)		(1.260)
LNSLE		6.461		6.499
		(1.080)		(1.079)
LNPRO		-6.323		-6.312
		(-.655)		(-.650)
R^2	0.022	.095	.022	.094
F	0.629	1.228	0.625	1.199

Table 3 reports the results from OLS regressions of CAR36 for 2006 - 2013 of full sample and sample excluding data on the year of economic crisis. In each group, the analysis was performed for data with and without control variables. T-statistics are in parentheses. *, **, *** Indicate significance at the 10, 5, and 1% levels, respectively.
Source: Processed Data, 2018

capabilities. This will lead to the growth of companies that tend to weaken long-term market performance.

4 CONCLUSION

This study aims to see the relationship between the intended use of IPO proceeds and the market performance of IPO stocks, both initial and long-term returns. The results show that there is a relationship between intended uses of proceeds for the purpose of acquisition and underpricing. The same relationship does not occur in the other plan of use of proceeds. Even though disclosure of intended use of proceeds is capable of minimizing information asymmetry between issuers and investors, it is not considered as important in IPO stock pricing. Although IPOs in Indonesia generally experience underpricing, disclosure of the use of proceeds for acquisitions is capable of reducing post-IPO risks and thereby reducing underpricing levels. However, the realization of the real value of the acquisition project and the delayed adverse impact of acquisitions, degrades the long-term market performance of such IPO stocks.

ACKNOWLEDGEMENT

Authors would like to thank Universitas Pendidikan Indonesia for providing the opportunity to present research results in GCBME 2018.

REFERENCES

Autore, D. M., Bray, D.E., & Peterson, D.R. 2009. Intended use of proceeds and the long-run performance of seasoned equity issuers. *Journal of Corporate Finance*, 15, 358–367.

Ben Amor, S., & Kooli, M. 2016. Intended use of proceeds and post-IPO performance. *The Quarterly Review of Economics and Finance*, 65, 168–181.

Datta,D.K. 1991. Organizational Fit and Acquisition Performance, Effects of Post-Acquisition Integration. *Strategic Management Journal*, 12(4), 281–297.

Davidson, K.M. 1988. The Acquisition Risk. *Journal of Business Strategy*, 9(3), 56–58.

Gumanti, T.A., Nurhayati, and Maulidia, Y. 2015. Determinants of Underpricing in Indonesia Stock Market. *Journal of Economics, Business, and Management*, 3(8), 802–806.

Healey, P., & Palepu, K. 2001. Information asymmetry, corporate disclosure, and the capital markets, A review of the empirical disclosure literature. *Journal of Accounting and Economics*, 31, 405–440.

Jeanneret, P. 2005. Use of proceeds and long-term performance of French SEO firms. *European Financial Management*, 11, 99–122.

Leone, A. J., S. Rock, &M. Willenborg. (2007). Disclosure of intended use of proceeds and underpricing in initial public offerings. *Journal of Accounting Research* 45, 111–153.

Ritter, J.R.1998. *Initial Public Offerings*. Contemporary Finance Digest, 2(1), 5–30.

Rock, K., 1986. Why new issues are underpriced. *Journal of Financial Economics*, 15, 187–212.

Verrechia, R. 2001. Essays on disclosure. *Journal of Accounting and Economics*, 32, 97–180.

Welch, I. 1996. Equity Offering Following the IPO Theory and Evidence. *Journal of Corporate Finance*, 2, 227–259.

Advances in Business, Management and Entrepreneurship – Hurriyati et al (eds)
© *2020 Taylor & Francis Group, London, ISBN 978-0-367-27176-3*

Influence of financial literacy and financial attitude on individual investment decisions

N. Hasanuh & R.A.K. Putra
Universitas Pendidikan Indonesia, Bandung, Indonesia

ABSTRACT: The purpose of this research was to examine the influences of financial literacy and financial attitudes on individual investment decisions. The samples in this study were street vendors in Rengasdengklok Traditional Market. This study used questionnaires administered to 96 people as respondents. This study used multiple regression analysis, using SPSS IBM 22 to analyse the data. The analysis results showed that the financial literacy and financial attitude influenced individual investment decisions. Simultaneously financial literacy and financial attitude influence individual investment decisions. This research has two implications. Financial literacy is the key factor on individual investment decisions, with the indicators are knowledge and financial concepts, ability to communicate about financial concepts, talent in managing personal finances, skills in making the right financial decisions; and confidence in effective financial planning for future needs. In addition, financial attitude influenced on individual investment decisions, with indicators are confidence, self-development, security of condition finance and attitude itself.

1 INTRODUCTION

There is a tendency that the street vendors are not well ordered in the financial management, Kompas 2011. Every day they receive and spend money with no apparent arrangement. In managing finances, the attitude of street vendors depend on their behaviour Kompas 2011. Some street vendors spend their income in a day, while some even feel the lack of income after selling. There is a group of street vendors who partly set aside their income for savings and investment.

In (Aminatuzzahra 2014), there are a number of problems in financial management in society, caused by personal household debt, a growing credit consulting business, and dependence on credit card use. Everyone having debt problems has four options to distribute their current income, such as: paying debts, adjusting costs, satisfying desires and saving for the future. This practice is called financial behaviour.

According to (Nofsinger 2005) behavioral differences will make the determination of finance different from each other. Choosing a bad financial determination will have a negative impact and will continue in the long term (Suryanto 2017).

A lack of understanding of finance will make financial decisions worse (Aminatuzzahra 2014). Public must understand the financial system appropriately because every person needs basic financial knowledge and skills to effectively manage financial resources (Shaari et al. 2013). Personal financial management is one of the most basic competencies needed by modern society as consumer choice from day to day affects one's financial security and standard of living (Howell 1993). Financial literacy can be interpreted as financial knowledge, with the aim of achieving prosperity (Lusardi & Mitchell 2009).

Welfare will be obtained in a good investment decision. Lintner in (Aminatuzzahra 2014) says that financial behaviour is the study of how humans interpret and react to information to make decisions in investment. Investment is a sacrifice, which is done today, with the aim of getting greater benefits in the future (Haming & Basalamah 2010).

Mugo (2016) found that financial literacy and financial behaviour influence the decision to save funds. Strengthened by (Al-Tamimi & Kalli 2009) financial literacy has a significant relationship with personal investment decisions. (Singh et al. 2016) concluded that the level of financial literacy plays an important role in investment decisions as evidenced by the level of correlation between financial knowledge and personal investment decisions of 68.1%. (Rasuma & Rahyuda 2017) argue that financial literacy has a significant effect on individual investment decision and understanding of personal financial management is a key factor in determining investment decisions. There was an influence between financial attitudes and financial management behaviour, with the chi square test yielding significant values of 0.001 <0.05 (Herdjiono et al. 2016). Susdiani (2017) study shows that financial knowledge does not affect a person's behaviour to invest.

Based on the description above, there are several factors that can influence financial behaviour in personal investment decisions. A number of studies investigated financial behaviour to some respondents, such as students, investors in the stock exchange or the owners of funds in banks. This study is a replicated study involving street vendors as the respondents. Thus, this study fills the gap on financial literacy, financial attitude and individual investment decisions in street vendors.

1.1 Behavioural finance concept

Behaviour is the way in which a person acts or self-behaviour (Zahroh 2014). Personal finance is an individual's financial management that is undertaken to obtain budgets and savings by taking into account the various financial risks and future life events. When planning personal finances, a person will consider the suitability of his needs from various banking products (demand deposits, savings, credit cards and consumer credit) or investments (stock market, bonds, mutual funds), insurance products (life insurance, health insurance) and pension plans.

Financial behaviour is the involvement of existing behaviors in a person that includes emotions, attributes, preferences and various things inherent in human beings as intellectual and social creatures that interact and underlie the emergence of the decision to perform an action (Ricciardi et al. 2000). It is how humans actually behave in a financial determination particularly how psychology influences financial decisions (Nofsinger 2005).

Olson (2001) provides a financial behaviour perspective of the decision making process. Financial decision-making preferences tend to be multifaceted, open to change and often formed during the decision-making process itself. Financial decision makers are satisfiers and not optimizers. Adaptive financial decision makers mean that the nature of the decision and the environment in which it makes the influence of the type of process used. Neurological financial decision makers tend to combine influences (emotions) into the decision-making process.

According to Suryanto (2017), financial behaviour related to how people treat, manage, and use the available financial resources. Individuals who have responsible financial behaviour tends to be effective in using money, such as making a budget, saving money and controlling spending, investing, and paying their obligations on time. It can be concluded that financial behaviour is a study of how people determine their finances and make financial investments that are influenced by psychological factors.

1.2 Personal investment decisions

Pandey (2004) defines an investment as an activity that is engaged by people who have savings by committing their funds in capital assets/goods and services, with an expectation of some positive rate of return.

The personal investment decision concept is that the person allocates limited resources between competing opportunities (investments products) in a process known as capital budgeting. Making this investment, or capital allocation, decision requires estimating the value of each opportunity or project, which is the function of the size, timing and predictability of future cash flows (Hodge 2000).

1.3 Financial literacy

Theoretically, financial literacy can be interpreted as financial knowledge, which has a goal to achieve prosperity (Lusardi & Mitchell 2007). Knowledge is power and when the power is in the financial sphere it will be a tool that can generate returns and minimize financial risk (Mugo 2016). Megawangi (1989) proposed to have financial knowledge then needs to develop financial skills and learn to use financial tools. Financial tools are a form of financial behaviour in decision making.

Remund (2010) explains that the concept of financial literacy has focused on five domains, including 1) knowledge and financial concepts; 2) ability to communicate about financial concepts; 3) talent in managing personal finances; 4) skills in making the right financial decisions; and 5) confidence in effective financial planning for future needs.

1.4 Financial attitude

Zahroh (2014) argues that attitude is the mental state and level of readiness which are governed by experiences that give dynamic influence or directed towards the individual response to all objects and situations associated with it. Financial attitude refers to that state of mind or opinion and judgment about one's finances reflecting a position one has taken (Pankow 2012). First aspect related to personality is self-confidence investors in financial behaviour as it is detached from his approach to his career, his health and the money (Pompian 2006). It is an emotional situation and investor confidence about some things or how many investors tend to feel worried. Both approaches are related to whether investors think methodically, carefully, and analytically in their financial behaviour or investors are emotional, intuitive, and patient. The two hypotheses in this study include:

a. Financial literacy affects personal investment decisions.
b. Financial attitude affects personal investment decisions.

2 METHOD

The population in this study are the street vendors registered as member of the Association of street vendors in the traditional market Rengasdengklok, Karawang. The sampling technique used incidental sampling, and the selection of sample members based on the population who were willing to fill out the questionnaires. Data were collected using

Table 1. Operational definition.

No	Variables	Indicator	Source
1	Financial Literacy (X1)	Basic Personal Finance Money Management Credit and Debt Management Saving and Investment Management Risk Financial Management	Ni Made Dwiyana Rasuma Putri and Henny Rahyuda (2017)
2	Financial Attitude (X2)	1.Confidence 2.Self-Development 3.Security of Conditions Finance 4.Financial Attitude	Aminatuzzahra (2014) Michael M Pompian (2006)
3	Individual Investment Decision (Y)	1.Calculating Security in an investment 2.Risk Factors and Investment Value Changing 3.Revenue in cash and defined 4.Increase in value	Ni Made Dwiyana Rasuma Putri and Henny Rahyuda (2017)

questionnaires distributed to 96 street vendors. Some definitions in the study are presented in Table 1.

The collected data were analysed to examine the effect of financial literacy and financial attitude on individual investment decisions with multiple regression model.

$$Y = a + b1X1 + b2X2 + e$$

Information:
Y = Individual Investment Decision
a = Constants
b1 = Regression coefficient of Financial Literacy
b2 = Regression coefficient of Financial Attitude
X1 = Financial Literacy
X2 = Financial Attitude
e = Error

3 RESULTS AND DISCUSSION

The results of analysis show the influence of financial literacy and financial attitude on individual investment decisions for the street vendors. Results of data processing as follows:

Based on the descriptive statistics shown in Table 2, the financial literacy variables have an average value of 3.5990 and are categorized quite well. Meanwhile,

Table 2. Descriptive statistics.

	N	Minimum	Maximum	Mean	Std. Deviation
FinLit	96	2.50	4.50	3.5990	0.42891
FinAtt	96	2.25	4.75	3.6302	0.49335
BeFi	96	2.25	4.50	3.5547	0.49231
Valid N (listwise)	96				

financial attitude average value was 3.6302 also categorized quite well. The variable of individual investment decision has an average value of 3.5547 which is good category.

3.1 Partial test results (t test)

Multiple regression was used to test the influence analysis of financial literacy and financial attitude on personal investment decisions. This test is used to determine whether there is influence of attachment between financial literacy with individual investment decisions and financial attitude with individual investment decisions, which can be seen from t count to t table with 1 side test. In this test it is known that n = 96 at the 5% significance level. At the error rate ($\alpha = 0.05$) using the 1 side test obtained t table value (95,00) for 1.661.

From the calculation of multiple linear regression using SPSS for windows program then obtained the regression equation as follows:

$$Y = 1.909 + 0.254X1 + 0.202X2 + e$$

From the Equation above, it can be explained that Financial Literacy (X1) has an effect on Personal Investment Decisions. The result of hypothesis testing is obtained by probability value of 0,028 < 0.05. The results of calculations on multiple regression obtained value of t arithmetic is 2.233. Whereas t table is (df = n-1, two sides/0.025) or (95,025) is 1.985. Thus t arithmetic reside in area H0 rejected

Table 3. T-test results.

Model	Unstandardized Coefficients		Standardized Coefficients		
	B	Std. Error	Beta	t	Sig.
1 (Constant)	1.909	.546		3.497	.001
FinLit	0.254	0.114	0.221	2.233	.028
FinAtt	0.202	0.101	0.197	1.993	049

Dependent Variable: PersInvestDec.

and Ha accepted, hence the number indicates significant value. This means that there is an influence between the Financial Literacy with Individual Investment Decision. Positive influence of 0.254, meaning that the higher the Financial Literacy is, the higher Individual Investment Decision is. The influence of financial literacy variables is directly proportional to the behaviour of individual investment decisions, meaning that the higher the financial literacy of a person, the better the behaviour of individual investment decisions. This supports the results of research conducted by (Al-Tamimi & Kalli 2009, Herdjiono et al. 2016, Mugo 2016). who develop a good level of financial knowledge (financial literacy). The financial behaviour tends to be better than someone with a lower level of financial knowledge. The results of research conducted by (Lusardi & Mitchell 2007, Zahro 2014, Putri et al. 2017, Shaari et al. 2013, Singh, Ajay & Rahul 2016) said that an adequate level of financial literacy would make a person do planning including planning anticipation with investment since productive age.

In addition, Financial Attitude (X2) has an effect on Individual Investment decision. Hypothesis Test results obtained error probability value of 0.049 < 0.05. The results of calculations on multiple regression obtained t arithmetic is 1.993. Whereas t table is (df = n-1, two sides/0.025) or (95,025) is 1.985. Thus t arithmetic reside in area H0 rejected and Ha accepted meaning there is an influence between Financial Attitude to Individual Investment Decisions. Positive influence of 0.202 means if the Financial Attitude is higher than Individual Investment Decision is also the same. Street vendors with better financial attitudes tend to be wiser in their financial behaviour, compared to Street vendors with poor financial attitudes. Financial attitudes influence to determine one's financial behaviour. Financial attitudes direct people to manage various financial behaviours. With a good financial attitude, people will be better at making various decisions related to their financial management. A person with a good level of financial attitude will show a good mindset about money, namely his perception of the future (obsession), not using money for the purpose of controlling other people or as a problem solver (power), able to control the financial situation (effort), adjusting the use of money so that it can meet its living needs (inadequacy). Besides, this person do not want to spend money (retention) and have a view that is always developing about money not with an old-fashioned view (securities) so that it will be able to control its consumption, be able to balance expenses and income owned (cash flow), set aside money for savings and investment, and manage the debt held for its welfare. Hypothesis 2 statement is still accepted. The results of this study are consistent with the research conducted by (Amminatuzahra 2014,

Table 4. F Test result.

ANOVA[a]

Model	Sum of Squares	df	Mean Square	F	Sig.
1 Regression	2.066	2	1.033	4.583	.013 [b]
Residual	20.960	93	.225		
Total	23.025	95			

a. Dependent Variable: PersInvestDec.
b. Predictors: (Constant), FinAtt, FinLit.

Table 5. Coefficient of determination (R2).

Model Summary

Model	R	R Square	Adjusted R Square	Std. Error of the Estimate
1	.300[a]	.090	.070	.47473

a. Predictors: (Constant), FinAtt, FinLit

Herdjiono at el. 2016, Zahroh 2014) suggesting that there is a significant relationship between one's financial attitudes will tend to be wiser investment decisions.

Based on Table 4, with a significant level of 5% and degrees of freedom df1 = 2 and df2 = 93, then the table obtained F (3; 93) is 2.703. In the calculation obtained the F arithmetic > F table, or 4.583> 2.703, it means H0 rejected. Meanwhile, if seen from the probability is 0.13 < 0.05, then the decision also rejects H0 showing that simultaneously there is the influence of Financial Literacy and Financial Attitude on Personal Investment Decisions.

The result of multiple linear regression analysis can be seen from Adjusted R Square 0,070 (see Table 5). It shows that Financial Behavior in Individual Investment Decisions is influenced by Financial Literacy and Financial Attitude variable 7%. The other 93% is influenced by other variables that have not been studied in this research.

4 CONCLUSION

It can be concluded that there is an influence between Financial Literacy on Personal Investment Decisions. There is influence between Financial Attitude on Personal Investment Decisions. Simultaneously Financial Literacy and Financial Attitude influence on Individual Investment Decisions. This research has limitations especially in the case of respondents which are limited to Street vendors who joined in the Association of Street Vendors in the traditional market Rengasdengklok. For further research, more street vendors could be involved.

REFERENCES

Al-Tamimi, H.A.H., Kalli, A.A. 2009. Financial literacy and investment decisions of UAE investors. *The Journal of Risk Finance* 10(5): 500–516.

Aminatuzzahra, A. 2014. Perception effect of financial knowledge, financial attitudes, social demography against financial behavior in individual investment decision making. *Journal of Business Strategy* 23(2).

Courchane, M. 2005. Consumer literacy and creditworthiness. *Proceedings, Federal Reserve Bank of Chicago*.

Haming, M. & Basalamah, S. 2010. *Feasibility study on investment project and business*. Jakarta: Earth Script.

Herdjiono, H., Irine, I. & Lady A.D. 2016. Pengaruh financial attitude, financial knowledge, parental income terhadap financial management behavior. *Jurnal Manajemen Teori dan Terapan* 9(3).

Howell, J.M. & Avolio, B.J. 1993. Transformational leadership, transactional leadership, locus of control, and support for innovation: Key predictors of consolidated-business unit performance. *Journal of Applied Psychology*.

Lusardi, A., Mitchell, O.S. & Curto, V. 2009. Financial literacy among the young: evidence and implication for consumer policy. *National Bureau of Economic Research*.

Megawangi, R., Zetlin, M.F. & Garman, D. 1989. Structural models of family social health theory. strengtening the family.imp.

Michael M.P. 2006. *Behavioral finance and wealth management*. Canada: John Wiley & Sons, Inc., Hoboken.

Mugo, E. 2016. Effects of financial literacy on investment decisions among savings and credit co-operative societies in nairobi membership, KCA University: School of Business and Public Management.

Nofsinger, J.R. 2005. *Psychology of investing second edition new jersey*. Precentice-Hall Inc.

Olson, J.M. & Hafer, C.L 2001. *Tolerance of personal deprivation*. In *JT Jost & B. Major (Eds .), The psychology of legitimacy: Emerging perspectives on ideology, justice, and intergroup relations*. Cambridge, UK: Cambridge University Press.

Pandey, I.M. 2004. *Financial management*. Vikas Publishing House.

Pankow, D. 2012. *Financial values, attitudes and goals*. North Dakota State University, Fargo, North Dakota.

Putri, P., Rasuma, N.M.D. & Rahyuda,H. 2017. Influence of financial literacy level and sociodemographic factors against behavior of individual investment decision. *E-Journal of Economics and Business of Udayana University* 6(9): 3407–3434.

Remund, D.L. 2010. Financial literacy explicated: The case for a clear definition in an increasingly complex economy. *The Journal of Consumer Affairs* 44(2): 276–295.

Ricciardi, V. & Simon, H.K 2000. What is behavioral finance. *Business Education and Technology Journal* 2(2): 1–9.

Robb, C.A. & Woodyard, USA. 2011. Financial knowledge and best practice behavior. *Journal of financial counseling and planning* 22(1).

Shaari, et al. 2013. Financial literacy: A study among the university students. *Interdisciplinary Journal of Cotemporary Research in Business* 5(2).

Siahaan, M.D.R. 2013. The influence of financial literacy on financial management behavior in university students In Surabaya. *Sekolah Tinggi Ilmu Ekonomi Economics Surabaya*.

Singh, S., Ajay, A. & Rahul S. 2016. Financial literacy & its impact on investment behavior for effective financial planning. *International Journal of Research in Finance and Marketing (IJRFM)* 6(8).

Suryanto, S. 2017. Student financial behavior patterns in higher education. *Journal of Political Science and Communication* 7(1).

Susdiani, L. 2017. Influence of financial literacy and financial experience to behavior of investment planning of pns in Padang city. *Journal of Nagari Development* 2(1).

Zahroh, F. 2014. Testing the financial knowledge level, personal finance attitudes, and personal financial behavior students management department faculty of economics and business semester 3 and semester 7. Semarang. faculty of economics and business Diponegoro university.

Advances in Business, Management and Entrepreneurship – Hurriyati et al (eds)
© 2020 Taylor & Francis Group, London, ISBN 978-0-367-27176-3

Capital structure and investment opportunity set on the value of the company

A. Suwandhayani & N. Fitdiarini
Universitas Airlangga, Surabaya, Indonesia

ABSTRACT: This research aims to examine the impact of capital structure and investment opportunity set on firm's value of 25 consumer goods companies listed in Indonesian Stock Exchange during 2009-2012. Capital structure proxies in this research were debt to equity ratio, debt to assets ratio and long term debt to assets ratio. Investment opportunity set proxies in this research were market to book value of assets, market to book value of equity, capital expenditure to book value of assets and firm's value proxies in this research were price to book value and price earnings ratio. The empirical result of Partial Least Square showed that both capital structure and investment opportunity set had positive significant impact on firm's value. All indicators on capital structure variables proved to be valid for measuring capital structure while CAPBVA and PER were not valid used in measuring each variables. This findings showed that funding policies and investment opportunity set of a firm profoundly influenced the way a firm is viewed by investors and creditors.

1 INTRODUCTION

In this era of globalization, many companies are investing to support economic growth. The investment financing required a large funding source in the capital structure of the company, either by issuing new shares or by lending.

The long-term goal of the company is to optimize the value of the firm (Fama & French 1978), by planning to acquire and use the funds (Weston & Copeland 1997). Corporate value does not only reflects the company's current performance but also describes the company's future prospects. The value of the firm can be measured by the stock market price reflecting the investor's assessment of the overall equity held.

Determining the proportion of debt and capital in using as a source of corporate funds is closely related to the term capital structure. The company can create a beneficial combination of the use of the source of capital funds derived from external funds. An increase in the usage of debt can help the company in carrying out activities, but on the other hand, the excessive use of debt can increase the financial distress ratios experienced by firms as increasing the cost of bankruptcy.

The focus of the current performance appraisal is not only limited access to the financial statements, but many people think that the value of the company is also reflected in the investment and the value of the investment to be issued by the company in the future. Investment is an activity to get some funds during a certain period in the hope of earning and increasing the value of investment in the future. The additional funds are used to increase the company's growth. The prospect of growing companies for investors is a lucrative project because investments are expected to provide high returns (Nugroho & Hartono 2002). If the investor expects to obtain a high-profit rate, then it must be willing to bear a high risk as well.

Investment Opportunity Set (IOS) is a choice of future investment opportunities that can affect the growth of company assets or projects that have a positive net present value (NPV). IOS provides a broader clue where the value of the company as the primary goal depends on the company's spending in the future (Myers 1977).

In the previous research, Carpentier (2006) has examined the effect of changing in capital structure to firm value. The result indicated that changes in the capital structure did not affect the firm value. While research conducted by Chowdhury & Chowdury (2010) showed that the use of debt in the company's capital structure could increase the value of the company. The research conducted by Yuliani et al (2012) showed that the IOS is peroxided by Market to Book of Asset (MKTBA) and Market to Book of Equity (MKTBE) affect the firm value. These findings supported the findings of McConnell & Muscarella (1985), Fama & Frech (1998), Chung et al (1998) and Hasnawati (2005) suggesting that investment opportunities have an effect on corporate value. This research try to find more deeply about (1) the effect of the capital structure on the value of the company (2) the effect of the IOS on firm value.

1.1 Capital structure

The capital structure is related to long-term expenditures of a company as measured by the ratio of its long-term debt and its own capital. The purpose of capital structure management is to integrate the permanent source of funds used by the firm by maximizing the firm's stock price and minimizing the firm's capital cost as well as the balance between risk and return (Keown et al. 2005)

There are several theories about capital structure. First is the theory of Modigliani and Miller (MM). According to MM, the total value of the firm is not influenced by the company's capital structure, but it is influenced by the investment made by the company and its ability to generate profit, assuming: perfect capital market, expected value from the probability distribution for all investors are the same, companies can be grouped in the same class of risk, and there is no corporate income tax. Second, Pecking Order Theory states that companies with high profitability level are low debt level, because the company has abundant internal fund resources (Myers 2001). In Pecking Order Theory, there is no optimal capital structure, but there is a sequence of preferences (hierarchy) in the use of funds and preferring internal financing (retained earnings) rather than external funding (issuing new shares). Third, Trade-Off Theory explains the relationship between taxes, the risk of bankruptcy, and the use of debt due to capital structure decisions are taken by the company. According to Myers (2001), the company will owe up to certain debt levels, where the tax shield of additional debt equals the cost of financial distress. Finally, Signaling Theory where capital structure is a signal delivered by managers to the market. If the manager has confidence that the prospect of the company is good, then the manager uses more debt as a sign that the company is confident with the prospect of where to come.

1.2   *Investment opportunity set*

Investment opportunity set (IOS) was first introduced by Myers (1977), where the value of the company is influenced by owned assets and options for future investment. IOS is more emphasis on future investment. Corporate investment options can be obtained if the company has a project with a positive Net present value (Kallapur & Trombley 1999). According to Gaver & Gaver (1993), future investment options are not solely indicated by projects directed by research and development activities alone, but also with the ability of firms to more exploit the opportunity to take advantage than other equivalent companies an industry group. This ability is unobservable.

1.3   *The value of the company*

The main goal of the company is to maximize the value of the company. For the company go public this is marked by the market price of their shares because the stock market price reflects the overall valuation of the investor on the equity of the company (Sartono 2001). The value of the company becomes very important because the high corporate value will be followed by high shareholder wealth. The value of the firm is the value that the stock market gives to the management of the company.

1.4   *Development of hypotheses*

There are two views on funding decisions. The first view holds that capital structure does not affect the value of firms proposed by Modigliani and Miller. This theory is evidenced by research conducted by Carpentier (2006), which states that the capital structure does not affect the value of the company.

While the second view, based on trade-off theory and pecking order theory, states that maximizing shareholder wealth requires the right combination of debt and equity. This is evidenced by research conducted by Chowdury & Chowdury (2010), thereby changing the composition of the company's capital structure can increase the value of the company in the market.

According to Brigham & Houston (2009), the use of debt can also affect the company's price range. The greater the debt, it will increase the value of the company. Firms with high levels of debt will be able to increase their earnings per share which will eventually increase the company's stock price which means it will increase the value of the company.

H1: Capital structure has a positive effect on firm value.

According to Myers (1977), investment decisions represent the value of a company whose size depends on future management expenditures, in which case investment options are expected to generate greater profits. According to signaling theory, investment spending gives a positive signal about future company growth, so as to increase the stock price used as an indicator of corporate value (Wahyudi & Pawestri 2006).

IOS is the relationship between current and future expenditures with value or return as a result of investment decision to generate company value (Hasnawati 2005). Yuliani et al (2012) also said that the IOS variable affects the firm's value. This indicates the existence of investment opportunities owned by the company will bring influence to the value of company because investment opportunities will be selected will bring a change in the company and will affect the investor's view of the company.

H2: Investment opportunity set positively affects the value of the firm.

2   METHOD

The approach used in this research is the quantitative approach. The data used in this study is data from the company's financial statements in the consumer goods industry sector listed on the Indonesia Stock Exchange period 2009-2012,

Table 1. Definition of variable.

Variables	Definition
PBV	Market price per share/book value per share
PER	Market price per share/earnings per share
DER	Total debt/total equity x 100%
DAR	Total debt/total assets x 100%
LTDA	Total long-term debt/total assets x 100%
MKTBA	(Asset - equity) + (shares outstanding x stock closing price))/total asset x 100%
MKTBE	(Stocks outstanding x stock closing price)/total equity) x 100%
CAPBVA	(Δ Book value of fixed assets/total assets) x 100%

Source: Output SPSS 18

with criteria companies do not have negative earnings. There are 25 companies that meet the defined criteria with details as follows:

This research has one dependent variable that is company value. The value of the firm is charged with the price to book value (PBV) and price earnings ratio (PER) or overpriced (Sawir 2001).

The independent variables in this research are capital structure and IOS. The capital structure in this study is measured using the debt to equity ratio (DER), debt to asset ratio (DAR), and long-term debt to asset ratio (LTDA). Based on research conducted by Smith & Watts (1992), Gaver & Gaver (1992), Kallapur & Trombley (1999), and Adam & Goyal (2007), IOS is measured using MKTBA, MKTBE, and capital CAPBVA.

This research uses Partial Least Square (PLS) method. The research model to test the hypothesis as follows:

$$\text{Corporate Value}_{it} = \alpha + \beta 1 \text{ Capital Structure}_{it} \\ + \beta 2 \text{ IOS}_{(it-1)} + e_{it} \quad (1)$$

Table 2. Descriptive Statistics.

	N	Min	Max	Mean	Std. Dev.
DER	75	0.03	2.28	0.730	0.528
DAR	75	0.02	0.70	0.372	0.166
LTDA	75	0.02	0.79	0.114	0.130
MKTBA	75	0.40	23.65	2.935	4.371
MKTBE	75	0.17	38.97	4.398	7.558
CAPBVA	75	−0.11	0.36	0.028	0.065
PBV	75	0.36	38.97	4.372	7.168
PER	75	2.93	109.87	21.321	17.758

Source: Output SPSS 18

3 RESULT AND DISCUSSION

3.1 Descriptive statistics

On Table 2, a number of 75 research data declared valid for research.

3.2 Testing model

The testing model uses two stages, namely testing outer model and inner model. Outer model is used to validate the research model built. Convergent validity of measurement model with reflective indicators can be seen from the correlation between score item or indicator with construct score. Individual indicators are considered reliable if they have a correlation value above 0.5 (Ghozali 2011)

Based on the Table 3, the indicators of the capital structure have the overall loading factor value above 0.5, so the three indicators are considered valid. Indicators of IOS that have a loading factor score above 0.5 i.e. MKTBA and MKTBE are considered valid, while the CAPBVA indicator has a loading factor value below 0.5 so it is considered invalid and is eliminated from the IOS measurement indicator. The indicator of firm value that has a loading factor score above 0.5 is PBV so it is considered valid, whereas PER which has scored below 0.5 is eliminated from company value measurement indicator.

AVE measurements in the discriminant validity test are used to ensure that the indicators used in the measurement of each variable are consistent and valid. The AVE value is valid if it is greater than 0.5.

Table 3. Results of loading factor indicator variable.

Variable	Indicator	Loading Factor	Notes
CAPITAL STRUCTURE	DER	0.984	Valid
	DAR	0.989	Valid
	LTDA	0.557	Valid
IOS	CAPBVA	0.204	Not Valid
	MKTBA	0.974	Valid
	MKTBE	0.986	Valid
CORPORATE VALUE	PBV	0.982	Valid
	PER	0.355	Not Valid

Source: Output SPSS 18

Table 4. AVE Test results.

Variable	AVE
IOS	0.968
Corporate Value	1.000
Capital Structure	0.750

Source: Output SPSS 18

Table 5. Bootstrapping test results.

	Ori. Sample	Mean	Std. Dev	Std. Error	T Stat
Capital Structure → Company Value	0.269	0.262	0.078	0.078	3.455
IOS → Company Value	0.733	0.733	0.071	0.071	10.288

Source: Output SPSS 18

Reliability test on Table 5 shows the accuracy, consistency, and accuracy of a measuring instrument in measuring with values must be above 0.6. Based on the above table, it can be seen that the three constructs have composite reliability above 0.6.

Table 5 shows that, based on the parameter coefficients, capital and IOS structures both have a positive effect. And by looking at t-statistics, it can be seen that the two constructs, both capital, and IOS structures, have a t-statistic value of more than 1.96, which in turn both have a significant positive effect on firm value.

4 CONCLUSION

The existence of a funding policy may affect the value of the company. This is consistent with the theory of signals that an increase in debt can be interpreted by outsiders as the company's ability to pay future liabilities or low corporate business risks, so the addition of debt can provide a positive signal. More debt usage is a signal that the company is more credible and seen more confident in its prospects in the future. But the higher the debt risks that the company, the greater the cost of bankruptcy that will be borne by the company. Therefore, the company is required to use the funding policy optimally.

In addition, the existence of various investment opportunities owned by the company will affect the value of the company. Based on signal theory, firms with high investment opportunities will have bright future prospects and will have an effect on increasing stock prices, so that the value of the company increases. In addition, if the company can utilize its capital well in running the business, the greater the opportunity for the company to continue to grow, so that the value of the company also increases and can attract many investors.

REFERENCES

Adam, T. & Goyal, V.K. 2007. The investment opportunity set and its proxy variables: theory and evidence. Working paper. The Journal of Financial Research 31(1): 41–63.

Carpentier, C. 2006. The valuation effects of long-term changes in capital structure. International Journal of Managerial Finance 26(2): 129–142.

Chowdhury, A. & Chowdhury, S.P. 2010. The impact of capital structure on firm's value: evidence from Bangladesh. Business and Economic Horizons 3(3): 111–122.

Chung, K.H., Wright, P. & Charoenwong, C. 1998. Investment opportunities and market reaction to capital expenditure decisions. Journal of Banking & Finance 22(1): 41–60.

Copeland, E.T. & Fred W. 1997. Manajemen Keuangan, Jilid 2. Jakarta: Binarupa Aksara.

Fama, E.F. & French, K.R. 1998. Taxes, financing decision, and firm value. The Journal of Finance 53(3): 819–843.

Fama, E.F. 1978. The effect of a firm's investment and financing decision on the welfare of its security holders. American Economic Review 68(3).

Gaver, J.J. & Gaver, K.M. 1993. Additional evidence on the association between the investment opportunity set and corporate financing, dividend, and compensation policies. Journal of Accounting and Economics 16: 125–160.

Ghozali, I. 2011. Analisis multivariate dengan program IBM SPSS 19. Semarang: Badan Penerbit Fakultas Ekonomi Universitas Diponegoro.

Hasnawati, S. 2005. Dampak set peluang investasi terhadap nilai perusahaan publik di Bursa Efek Jakarta. Jurnal Akuntansi dan Auditing Indonesia 9(2): 117–126.

Houston, J.F. & Brigham, E F. 2009. Fundamentals of Financial Management. Nashville: South-Western College Pub.

Kallapur, S. & Trombley, M. 2001. The investment opportunity set: determinants, consequence and measurement. Manajerial Finance 27(3).

Keown, A.J., Martin, J.D., Petty, J.W. & Scott Jr, D.F. 2005. Financial Management: Principle and Applications. Tenth Edition. New Jersey: Prentice Hall.

McConnell, J.J., & Muscarella, C.J. 1985. Corporate capital expenditure decisions and the market value of the firm. Journal of financial economics 14(3): 399–422.

Myers, S.C. 1977. Determinant of Corporate Borrowing. Journal of Financial Economics 5(2): 147–175.

Myers, S.C. 2001. Capital Structure. Journal of Economic Perspective 15(2): 81–102.

Nugroho, J.A. & Hartono, J. 2002. Analisis hubungan antara gabungan proksi investment opportunity set dan real growth dengan pendekatan confirmacy factor analysis. Jurnal Riset Akuntansi Indonesia 6(1): 69–92.

Sartono, A. (2001). Manajemen keuangan teori dan aplikasi. Yogyakarta: BPFE.

Sawir, A. 2001. Analysis of Financial Performance and Corporate Financial Planning. Jakarta: Gramedia Pustaka Utama.

Wahyudi, U., & Pawestri, H.P. 2006. Implikasi struktur kepemilikan terhadap nilai perusahaan: dengan keputusan keuangan sebagai variabel intervening. Simposium Nasional Akuntansi 9: 1–25.

Yuliani, Y.L., Djumilah Zain, D.Z., Sudarma, M. & Solimun, S.M. 2012. Diversification, investment opportunity set, environmental dynamics and firm value (empirical study of manufacturing sectors in Indonesia Stock Exchange). Journal of Business and Management (IOSR-JBM) 6(4): 01–15.

Advances in Business, Management and Entrepreneurship – Hurriyati et al (eds)
© *2020 Taylor & Francis Group, London, ISBN 978-0-367-27176-3*

Effect of market, profitability and solvability ratio to Market Value Added (MVA) companies listed in the Jakarta Islamic Index (JII)

P.S. Sukmaningrum & H.K. Prawira
Universitas Airlangga, Surabaya, Indonesia

ABSTRACT: This study aims to investigate the effect of Market Value Ratio, Profitability Ratio, and Solvability Ratio to Market Value Added Companies Listed in the Jakarta Islamic Index partially and simultaneously. The sample used in this study are companies registered in the Jakarta Islamic Index (JII) from 2012 to 2016. The results of the best estimation model based on the Random Effect Model shows that the variables of Earning Per Share, Return on Equity and Debt to Equity Ratio affect the Market Value Added simultaneously and significantly. Variable of Return on Equity has a positive and significant influence and Debt to Equity has a significant negative effect to market value added. Companies should strengthen financial ratios, in particular the profitability of the company by keeping the level of profit, in order to raise high market value added. The use of debt must be minimized, because it will impact low market value added.

1 INTRODUCTION

In today's competitive world, the main business goal is to create value and wealth for shareholders. Businesses that focus on shareholder value are much healthier and improve the overall economy. As a result, companies need reliable and accurate activity measures. Market Value Added (MVA) can predict the value created by a company by deducting from the current market value of its capital (equity and debentures) (Zafiris & Bayldon 1999). MVA can be measured with the difference between the firm's market value (including equity and debt) and the overall capital invested in the firm. Market Value Added is technically obtained by multiplying the difference between the stock price per share and the book value per share. Market Value Added increases only if the invested capital gets a payback rate larger than the cost of capital (Young & O'Byrne 2000; Banerjee 2000).

MVA is the external indicator that is able to measure how much wealth the company has which had been created for investors or, in other words, the MVA states how great the prosperity is that has been achieved or is eliminated by a single company. To see how big the wealth and prosperity of the company is can be seen from the performance of a company. Assessing the performance of companies can be done using profitability ratio, market value ratio, and solvency ratio (Van Horne & Wachowicz 2008).

Return on Equity Ratio (ROE) as an indicator of profitability compares net income after tax with equity invested by the company shareholders (Brigham & Houston 2015). This ratio shows the power to generate return on investment based on shareholder book value and is often used for comparing two or more companies for good investment opportunities and cost-effective management.

The first essential component which must be noticed in the analysis of companies' performance is the Earning Per Share ratio (EPS) as an indicator of market value (Tandelilin 2010). EPS becomes the indicator of market value. Investors often focus on the amount of EPS in conducting stock analysis. In general, the EPS is the level of net profit for each of the shares of a company at the time it is able to run its operation.

Debt to Equity Ratio (DER) as an indicator of solvability is a ratio used to assess the debt with equity. This ratio is sought to compare the entire total debt with the equity. This ratio is used to understand the amount of funding provided by the borrower (creditors) and owner of the company. In other words, this ratio serves to know every capital that was made to guarantee the debt. It can be concluded that the debt to equity ratio is a ratio that measures how far a company is financed by debt and the company's ability to meet its liabilities with equity owned (Brigham & Houston 2015).

The research question in this research is whether there is a partial and simultaneous influence of the market value ratio (EPS), the ratio of Profitability (ROE), and Solvency Ratio (DER) against Market Value Added (MVA) in companies listed in the Jakarta Islamic Index (JII) in the period 2012–2016.

2 METHOD

The approach used in this research is a quantitative approach, which analyzes the measurement of an economic phenomenon. It is a combination of economic theory (financial report information) and a statistical model, which is classified in a certain category by using tables to facilitate in analyzing and by using panel data regression (Gujarati & Porter 2009).

The type of data used in this study is secondary data, namely in the form of figures from annual financial statements of companies listed in the Jakarta Islamic Index (JII) in 2012–2016.

The population in this study is companies registered in the Jakarta Islamic Index (JII) in 2012–2016. Criteria for the samples in this study include companies that are listed continuously, publish annual financial statements, and do not experience loss in JII during the period 2012–2016. There is a total of 70 samples from 14 companies.

3 RESULT AND DISCUSSION

3.1 Result

Based on the results of testing on the panel data regression analysis, the equation regression model can be arranged as follows:

$$\text{LogMVAit} = 30.9822 - 3.15E - 05\text{EPSit} \\ + 3.981805\text{ROEit} - 0.843708\text{Derit} \quad (1)$$

The constant value indicates if the earning per share (EPS), return on equity (ROE) and debt to equity ratio (DER) is zero or constant; in that case, the market value added (MVA) is 30.9822. The variable coefficient of Earning per Share is equal to –3.15E-05 which shows that with every increase of one unit of earnings per share, the market value added shares also decrease by –3.15E-05, assuming that the other variable is constant. The variable coefficient of Return on Equity is equal to +3.981805 which shows that with every increase of one unit of ROE, the market value added shares also increase by +3.981805, assuming that the other variable is constant. The value coefficient variable of Debt to Equity Ratio is equal to –0.843708. This shows that with every increase of one unit of debt to equity ratio, the market value added shares also decrease by 0.843708, with the assumption that other variables are constant.

By using the analysis of Panel Data Regression Test conducted with the result of the Hausman Test, the best model for this research is the Random Effect Model (REM); the probability value is presented in Table 1.

The Earnings Per Share (EPS) variable has a probability value of 0.9493, greater than the 5% significance level (α = 0.05), so H0 is accepted and H1 is rejected. This shows there is no significant influence between EPS and MVA. From the value of the coefficient it can be concluded that EPS has a negative insignificant effect to market value added.

The Debt to Equity Ratio (DER) variable has a probability value equal to 0.0243, smaller than the

Table 1. Random effect model (REM) result.

Variable	Coefficient	Std. Error	t-Statistic	Prob.
C	30.9822	0.498722	61.97485	0.0000
EPS?	-3.15E-05	0.000493	-0.063874	0.9493
ROE?	3.981805	0.920520	4.325604	0.0001
DER?	-0.843708	0.365250	-2.309950	0.0243
Weighted Statistics				
R-squared	0.286225	Mean dependent var		8.114251
Adjusted R-squared	0.250536	S.D. dependent var		0.908284
S.E. of regression	0.594646	Sum squared resid		21.21622
F-statistic	8.020033	Durbin-Watson stat		1.636006
Prob (F-statistic)	0.000141			
Unweighted Statistics				
R-squared	0.407449	Mean dependent var		31.20394
Sum squared resid	76.23844	Durbin-Watson stat		0.455280

Dependent Variable: LOGMVA?
Method: Pooled EGLS (Cross-section random effects)
Date: 01/01/18 Time: 00:47
Sample: 1 5
Included observations: 5
Cross-sections included: 14
Total pool (unbalanced) observations: 64
Swamy and Arora estimator of component variances

5% significance level (α = 0.05), so H0 is rejected and H3 is accepted. This shows that there is significant influence between DER and MVA company registered in the Jakarta Islamic Index in the period 2012–2016. From the value of the coefficient it can be concluded that the Debt to Equity Ratio has a negative significant effect to the market value added.

The Return on Equity Ratio (ROE) variable has a probability value equal to 0.0001, smaller than the 5% significance level (α = 0.05), so H0 is rejected and H3 is accepted. This shows that there is a significant influence between ROE and MVA company registered in the Jakarta Islamic Index in the period 2012–2016. From the value of the coefficient it can be concluded that the return on equity ratio has a negative significant effect to the market value added.

The F-statistics probability value is 0.000141; this value is smaller than 0.05 and significant at a 95% confidence level (α = 0.05), so H0 is rejected and H1 is accepted. Thus, it can be concluded that the independent variables, namely earnings per share (EPS), return on equity (ROE), and debt to equity ratio (DER) simultaneously have a significant effect on the dependent variable market value added in the listed companies from the Jakarta Islamic Index in the period 2012–2016.

The value of Adjusted R-Squared is 0.2862. It can be concluded that the independent variables are Earning Per Share, Return On Equity, and Debt to Equity Ratio and they can explain 28.62% dependent variable which is the market value added companies listed in the Jakarta Islamic Index in the period 2012–2016. The rest, 71.38%, was influenced by other factors not used in the regression model in this study.

3.2 Discussion

The Earnings Per Share (EPS) variable has no significant effect on the market value added to companies listed in the Jakarta Islamic Index in 2012–2016. This result contradicts the research by Alipour and Pejman (2015), which stated that EPS has a significant influence to the MVA. A higher value of EPS will attract investors, because a higher EPS means high profit provided to the shareholders.

Figure 1 shows the development of EPS against MVA companies registered in the JII for 2012–2016. It was seen in every period when the EPS in-creased, the company's MVA not always followed the increase, which showed that there was no influence between EPS and the company's MVA registered in the Jakarta Islamic Index.

The variable Return on Equity (ROE) has a significant effect on the Market Value Added (MVA) in companies listed in the Jakarta Islamic Index in 2012–2016. This is in line with research conducted by Alipour and Pejman (2015), which states that return on equity has a significant positive effect on firm value. However, it contradicts the research by Akgun et al. (2018), which states ROE has no significant impact on Return on Equity (ROE).

ROE is a profitability ratio that measures a company's ability to generate net income based on certain capital. This ratio is a measure of profitability seen from the perspective of shareholders. A high figure for ROE shows a high level of profitability (Eugene F. Brigham, 2015). The high level of corporate profitability is the success of the company's performance and is a positive signal for investors to invest in the company.

Companies that have good quality will get a good level of corporate profitability, because only companies that have high profitability are able to pay dividends. A company with poor performance and low profitability will not be able to imitate by paying a large amount

of dividends because it does not have enough cash or if it still pays dividends, the funds for investment will not exist and this will worsen the company's performance. Thus, increasing ROE means increasing the company's market value as reflected in the MVA.

Figure 2 shows the development of ROE in MVA companies registered in the JII in 2012–2016. It was seen in 2016 when the ROE increased, the company's MVA increased as well, which showed that there was a positive influence of ROE against the company's MVA registered in the Jakarta Islamic Index.

The variable debt to equity ratio has a negative and significant effect on the market value added in companies registered in the Jakarta Islamic Index in 2012–2016. This is in line with the research of Alipour and Pejman (2015) and Olivia (2017). According to Van Horne and Wachowicz (2008), debt to equity ratio is the ratio used to assess debt with equity. This ratio is sought by comparing the entire debt, including current debt with all equity. This ratio is used to determine the amount of funds provided by the borrower (creditors) with the company owner. In other words, this ratio serves to find out every amount of its own capital that is used as collateral for debt.

If the Debt to Equity Ratio (DER) is high, there is a possibility that the MVA of the company will be low because if the company has a high DER level, the company will be more at risk in default and bankruptcy. Investor confidence also declined, which resulted in a decline in stock prices.

Figure 3 shows the development of DER against MVA companies registered in the JII for 2012–2016.

Figure 2. Return on equity to market value added.

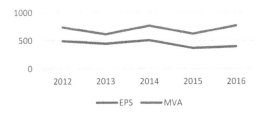

Figure 1. Earnings per share against market value added.

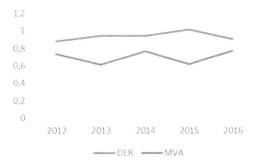

Figure 3. Debt to equity ratio on market value added.

It was seen in 2013 when DER increased, the influence of MVA declined, which showed that there was a negative influence between DER on the company MVA registered in the Jakarta Islamic Index.

4 CONCLUSION

Based on the results of the research and discussion, we can conclude that Earning Per Share (EPS) has no significant influence on market value added. The Return on Equity (ROE) has a significant positive effect on the market value added, thus companies with a high return on equity have a high market value added as well. The Debt to Equity Ratio (DER) has a significant negative effect on the market value added, thus the low level of debt to equity ratio can increase the market value added of the company. Furthermore, Earning Per Share (EPS), Return on Equity (ROE) and Debt to Equity Ratio (DER) simultaneously have a significant effect on the market value added companies listed in the Jakarta Islamic Index in 2012–2015.

REFERENCES

Akgun, A. I., Samiloglu, F. & Oztop, A.O. 2018. The impact of profitability on market value added: Evidence from Turkish Informatics and Technology Firms. *International Journal of Economics and Financial Issues* 8(4): 105–112.

Alipour, M. & Pejman, M.E. 2015. The impact of performance measures, leverage and efficiency on market value added: Evidence from Iran. *Global Economics and Management Review* 20(1): 6–14.

Banerjee, A. 2000. Linkage between economic value added and market value: An analysis. *Vikalpa* 25(3): 23–36.

Brigham, E.F. & Houston, J.F. 2015. *Fundamentals of Financial Management (14th ed.).* Boston: Cengage Learning.

Gujarati, D.N. & Porter, D.C. 2009. *Basic Econometrics (5th ed.).* New York: McGraw-Hill/Irwin.

Olivia, S.G. 2017. Pengaruh economic value added dan leverage terhadap market value added pada perusahaan properti dan realestate yang terdaftar di bursa efek Indonesia. Unpublished, Universitas Sumatera Utara.

Tandelilin, E. 2010. *Portofolio dan Investasi: Teori dan Aplikasi.* Yogyakarta: Kanisius.

Van Horne, J.C. & Wachowicz, J.M. 2008. *Fundamentals of financial management (13th ed.).* Harlow: Prentice Hall.

Young, S. D., & O'Byrne, S. F. 2000. *EVA and Value-Based Management: A Practical Guide to Implementation (1st ed .).* New York: McGraw-Hill Education.

Zafiris, N. & Bayldon, R. 1999. Economic value added and market value added: A simple version and application. *Journal of Applied Accounting Research* 5(2): 84–105.

Financial literacy and retirement savings ownership

A.M. Adiandari, N.D.A. Amrita & H. Winata
Universitas Ngurah Rai, Denpasar, Indonesia

ABSTRACT: This study aims to determine the relationship of financial literacy and retirement savings owner-ship. This study also tests whether socio-economic characteristics, such as marital status, number of children, tenure of employment, and amount of monthly income can affect a person's decision to own retirement savings for future prosperity. The research method is quantitative, using Spearman correlation test with the sample number of 170 employees from a state-owned bank in Payakumbuh, West Sumatra, Indonesia. Results showed that only number of children and tenure of employment of the socio-economic characteristics have relationship with retirement savings ownership, while marital status and amount of monthly income have no relationship with retirement savings ownership. Moreover, results show that financial literacy has a relationship with retire-ment savings ownership. Hence, improving financial knowledge, especially directly related to retirement savings is necessary, to increase penetration and density of retirement savings in Indonesia.

1 INTRODUCTION

Retirement savings is one of the products of non-bank financial institutions whose penetration in Indonesian society is still very low. Data from the results of the national survey of financial literacy and inclusion in 2016 in Indonesia indicate that the inclusion rate of retirement savings or the level of access of Indone-sians to retirement savings products is only 1.53% in 2013 and 4.66% in 2016. The lowest figure when compared to other non-bank financial products, namely insurance 12.08%, financing institutions 11.85% and pawnshops of 10.49%. Meanwhile, when compared to the inclusion rate of banking products in Indonesia, the figure is still far behind where the inclusion of banking products reached 57.28% in 2013 and 63.63% in 2016 (Financial Services Author-ity/OJK 2016a).

OJK (2016b) survey results related to the under-standing and participation of the Indonesian society to the retirement fund above is still very low. This condition indicates at least two depictions of the con-dition of Indonesian society. The first indicates that many Indonesians do not understand that life after retirement is long, even if they still have a dependent cost for children's education or dependents of a sick family. Based on data in the book of the People's Welfare Indicator, BPS catalog (Bureau of Statistics 2015), women life expectancy in Indonesia is 72.8 years and men life expectancy is 68.9 years. Based on the number of 72.8 years, we could get people longest period of productive working time for about 33 years obtained from the difference of retirement age at 55 years and the age of 22 years for the starter, it is necessary to make saving for deposit for 17.8 years after not being productive. In the productive period, the average Indonesian worker is able to meet the needs of his life and his family, but in the retirement period he becomes dependent on others, which is caused by the lack of understanding of the length of life that will be lived after retirement.

The second indicates that the number of retired Indonesians who are secured from pension funds is so small that many of the retired communities are dependent on their children or others. This is what makes it difficult for a family to break the chain of dependence or poverty, let alone to achieve a prosperous life and to achieve financial freedom, this condition will be very difficult to achieve.

Understanding retirement will be different for people of different backgrounds. For certain profes-sions such as lawyers, doctors, musicians, retirement is the time when they intentionally stop doing their profession to enjoy their life, which means they can stop working whenever they want. As for employ-ees, retirement is defined as stopping work due to an age-related rule to work in their respective compan-ies. For employees, there are retirement age rules in 55 years, 58 years or even up to 65 years. From this overview of the understanding of pensions, it gener-ally indicates that employees have tighter time limits which cause these employees to be disciplined in preparing their early retirement funds.

In terms of corporate retirement savings, retire-ment savings is one of the key points offered by companies in the work environment to attract, retain, motivate, and ultimately retire their employees (Clark et al. 2017). In addition, the management of pension funds is a key to success for a family who wants financial freedom. This financial freedom can be obtained with the discipline of expenditure and discipline of financial income so that the target of retirement income of 80% of current income can be achieved. From a case study conducted in the United States against Federal Reserve workers, information can be obtained that the workers were asked to watch a video entitled "Your Retirement Plan". From a video of no more than 30 minutes, one of its explanations is related to suggest that retirees should save enough to replace about 80% of their final earnings. This figure will certainly be an independent financial source for one's pension age in particular and as a pool of funds that can be used for development within a country in general.

Furthermore, a research conducted in some countries around the world shows that not only do individuals indicate low levels of financial literacy, but also that low financial knowledge can be attributed to lack of financial planning and lack of resources in retirement period (Lusardi & Mitchell 2011). This is in line with the conditions in Indonesia, where the level of understanding and membership of retirement funds is still very low. The author conducted a study in one of state-owned enterprises in Payakumbuh West Sumatra aimed to see the relationship between financial literacy with retirement savings ownership. This is due to the phenomenon of the very low penetration and density of pension products in Indonesia, especially in West Sumatra province, while on the other hand, the membership of the retirement fund is actually obliged to all employees in Indonesia by the Government. Meanwhile, unpreparedness in facing retirement will be a fatal mistake that results in a very long term, causing dependence and welfare for the individual and leading to an increase in the poverty rate and the level of crime of poverty in a country. Therefore, this study will also look at the relationship between socio-economic factors such as marital status, number of children, tenure of person's employment and the amount of monthly income with the ownership of retirement funds. The objective is to strengthen the analysis of the findings by determining whether there are differences in the decision of ownership of retirement savings based on marital status, number of children, tenure of person's employment and the amount of individual monthly income.

1.1 The life-cycle hypothesis theory

The Life-cycle Hypothesis theory of Modigliani (1963) states that the individual or household in the economy will delay consumption by making saving. The savings will be collected until the individual/ household reaches retirement age. After retirement, new individuals use their retirement savings to meet retirement needs. The author uses the theory of Modigliani as a basis in doing this research where, whether it is realized or not, the individual will do saving activities to deal with his retirement. Furthermore, whether the level of individual financial literacy has any relationship with the ownership of retirement savings will be researched in this study.

1.2 Socio-economic characteristics

Based on a research conducted by several experts in several countries in the world, most found researches are related to the relationship between economic demographic characteristics with pension preparation, where empirical differences are found among some countries. Based on age demographic factors, Clark et al. (2017) found that, in the United States, each additional 10 years of age was associated with 1.2 percentage points more pay being contributed to the DC plan, the findings in the Netherlands from the research conducted by Alessie et al. (2011), show that respondents do not tend to think much about retirement when they are young and retirement is a distant concept. Similar findings from a study conducted in Canada by Boisclair et al. (2015), that age also has a significant effect, with older individuals being more likely to hold retirement savings. Different findings are found in Germany by Koenen & Lusardi (2015) which reveal that there is no significant difference in retirement planning between age groups, and there are no large differences in planning across education. Nevertheless, the author still has not found a study that specifically examines the relationship between socio-economic characteristics with the ownership of retirement savings.

Therefore, this study will investigate the relationship of socio-economic characteristics including marital status, number of children, tenure of employment and the amount of monthly income of individuals with retirement savings ownership. The hypothesis shall be as follows:

H1. There is a significant relationship between socioeconomic characteristics and the ownership of retirement savings.
H1a. There is a significant difference in Retirement Planning ownership based on marital status
H1.b. There is a significant difference in ownership of Retirement Planning based on the number of children
H1.c. There is a significant difference in ownership of Retirement Planning based on the tenure of employment.
H1.d. There is a significant difference in Retirement Planning ownership based on the amount of monthly income.

1.3 Financial literacy

Several literacy-related studies in relation to retirement planning have been conducted in several countries such as America, Canada and the Netherlands. The results of studies in these countries show some issues related to the relationship of financial literacy with the level of management of personal retirement funds. A case study conducted on Federal Reserve employees in America found that the most financially knowledgeable employees are more likely to participate in their pension plans, contribute a higher percent of their pay, and hold more equity in their account requirements. (Clark et al. 2017). Findings from a Canadian study found that retirement planning is strongly associated with financial literacy. Typically, in Netherlands, most of the retirement saving decision, are beyond employees' control and the pension system is perceived as quite generous, and this might contribute to the low number of households (one in eight) claiming to have thought a lot about retirement. (Alessie et al. 2011).

Considering the low penetration and density of pension products in Indonesia, researcher is interested in investigating the relationship between a person's financial literacy and the ownership of the person's retirement savings. Whether someone who has higher financial literacy will better show a higher level of retirement savings ownership will be known from the results of this study. The hypothesis tested shall be follows:

H2. There is a significant relationship between financial literacy and retirement savings ownership.

2 METHOD

2.1 Research approach

In this study, the approach used is a quantitative approach through the use of hypotheses to find the relationship between variables.

2.2 Sample and data collection

Methods of data collection is using questionnaires that are directly distributed to all employees of one of the state-owned commercial banks (BUMN) located in Payakumbuh, West Sumatra Province, Indonesia. The total population was 274 people and the number of questionnaire returns was from 170 people who subsequently author set as the number of samples in this study. For information, as requested by the BUMN leadership board where the research is conducted, the name of SOE is not mentioned in this research. If you need for more information, you may contact the author directly.

2.3 Questionnaire design

Questionnaire design in this research consists of 2 parts: (1) Section of information related to the background and socio-economic characteristics of the respondents, (2) The question section consists of 5 questions to determine the level of basic financial literacy of the respondents where the question standard refers to Babiarz & Robb (2014). The author adjusts some questions solely for ease of understanding of respondents in Indonesia and adjustment of the economic condition of the respondent without removing the original meaning of the question.

2.4 Measurement of variables

Measurements of variables in this study are described in Table 1 as follows:

Table 1. Measurement of variables.

Variable	Phenomenon
Dependent Variable	
Retirement savings ownership	Retirement savings ownership
Independent Variable	
Financial Literacy	
	Interest
	Inflation
	Bond Price
	Mortgage
	Portfolio
	Score of Financial Literacy
Social Ec. Characteristic	
Marital Status	Marital Status Category
Number of Children	Number of Children Category
Tenure of Employment	Tenure of Employment Category
Monthly Income	Monthly Income Category

2.5 Data analysis technique

The analysis of the relationship between socioeconomic characteristics with the ownership of retirement savings and the relationship between financial literacy and retirement savings ownership using Spearman correlation test is using SPSS 20 statistical tool.

3 RESULT AND DISCUSSION

Based on cross tabulation between marital status and retirement savings ownership, it can be seen that in unmarried respondents and did not have retirement savings of 26 people or 15.3%, while the respondents were unmarried and had retirement savings of 23 people or 13.5%. For married respondents who do not have retirement savings are 46 people or 27.1%, and married respondents who have retirement savings of 74 people or 43.5%. While from a divorced respondent, she did not have retirement savings. From the data above it can be noted that married respondents and has a retirement savings have the largest percentage, which is 43.5%. To find out whether there is a relationship between marital status with retirement savings ownership, it can be seen on spearman correlation test result. Based on Spearman correlation test between marital status and retirement saving ownership, it can be seen that the significance value is 0.126 > 0.05, so it can be concluded that there is no relationship between marital status and retirement savings ownership.

Furthermore, based on cross-tabulation between the number of children and retirement savings ownership, it can be seen that the respondents who do not have children actually have the highest retirement savings ownership rate, which is 35 people or 20.6% of the total respondents. This figure is almost comparable with respondents who have no children and no retirement savings, i.e. as many as 33 people or 19.4% of the total respondents.

Table 2. Descriptive statistic.

Marital Status		
Single	49	28.8%
Married	120	70.6%
Divorce	1	0.6%
Number of Children		
0	68	40.0%
1	49	28.8%
2	37	21.8%
3	16	9.4%
Tenure of Employment < 5 years		
< 5 years	98	57.6%
5 - 10 years	49	28.8%
> 10 years	23	13.5%
Monthly Income		
< 5 million	145	85.3%
5 - 10 million	23	13.5%
> 10 million	2	1.2%
Financial Literacy		
Very Low	21	12.4%
Low	24	14.1%
Medium	56	32.9%
High Enough	55	32.4%
High	13	7.6%
Very High	1	0.6%
Retirement Saving Ownership		
Do not have	73	42.9%
Have	97	57.1%

To determine whether there is a relationship be-tween the number of children and retirement savings ownership, it can be seen on spearman correlation test where the significance value is 0.020 < 0.05 so it can be concluded that there is relationship be-tween number of children and retirement savings ownership.

For the factors of tenure of employment and retirement savings ownership can be seen in cross tabulation where the tenure of employment is div-ided into 3 classes and most respondents are in the first class with the duration of work less than 5 years with a total of 98 respondents. For the largest number of retirement savings ownership in respond-ents with tenure of employment is less or equal to 5 years, that is 49 people or 28.8% from respondent. This amount is exactly equal to the tenure of employment less than or equal to 5 years and that has no retirement savings.

Based on Spearman correlation test between tenure of employment and retirement savings owner-ship, it can be seen that the significance value is 0.014 < 0.05 so it can be concluded that there is a relationship between tenure of employment and retirement savings ownership.

The next factor is monthly income which can be seen in cross tabulation result where monthly income is divided into 3 classes and most

respondents are in first class with income below IDR 5 million with the number of retirement savings ownership is 79 people or 46.5% of the total respondents and who have no retirement savings ownership is 66 people or 38.8% of the total respondents. Then when further examined, respond-ents with income greater than or equal to 10 million who do not have retirement savings are 2 or 1.2%.

To find out whether there is a relationship between income with retirement savings ownership, it can be seen on spearman correlation test where the significance value is 0.125 > 0.05 meaning no rela-tionship between monthly income with retirement savings ownership.

The last factor observed is the financial literacy and retirement savings ownership that can be seen in cross tabulation. The division of the financial literacy level is divided into 5 classes, namely very low, low, medium, high enough and high. For the largest number of respondents are in the medium class of financial literacy with a total of 56 respondents. For the largest number of retirement savings ownership in respondents with financial literacy is high enough, as many as 40 people or 23.5% of the total respondents.

Furthermore, based on Spearman correlation test results between financial literacy and retirement sav-ings ownership, it can be seen that the significance value of 0.002 < 0.01 so it can be concluded that there is a relationship between financial literacy and retirement savings ownership.

From the results of the number of children with retirement savings ownership there is a significant relationship. Cross tabulation results can be seen that those with children less than or equal to 1, the ratio between those who have retirement saving with those who otherwise is almost the same. But once the indi-vidual has more than or equal to 2 children, the own-ership of the retirement savings is two to four times higher than those who have no retirement savings. It is possible that the need for the cost of Child Educa-tion still exists when the individual has retired.

From the results, there is significant relationship between the tenure of employment with retirement savings ownership there is a significant relationship. Cross tabulation results show that when an individ-ual enters an employment period of less than or equal to 5 years, the amount of retirement savings equals to those with no retirement savings. However, when entering an employment period above 5 years, the percentage of retirement savings ownership is higher than the percentage with no retirement sav-ings of 1.5 to 3 times. It is possible that individuals are increasingly aware that retirement is getting closer so they prepare their retirement savings.

From the results, there is a significant relationship as well between the financial literacy with retirement savings ownership. Cross tabulation results show that individuals with very low financial literacy have retirement savings rate of 4.7%. It is lower than those without retirement savings with the rate of 7.6% from the total of 12.4% of respondents that have very low rate of literacy. Furthermore, the percentage comparison of the rate of ownership of retirement savings is higher than the percentage that has no retirement savings when the level of individual literacy is higher. The most distinguishable percentage difference is in the high level of financial literacy where individuals with retirement savings reach almost 3 times that of no retirement savings. This suggests that financial literacy is a driver of retirement savings. This is also in line with the results of previous studies which found that financial literacy has a relationship or effect on retirement planning or retirement savings.

In relation to the results of this study with the low penetration rate and the density of retirement fund products in Indonesia, it can be said that there are still incomplete programs to improve the financial literacy of Indonesian society especially for retirement products, either by the government or other stakeholders. In addition, it should also be noted that there is cultural community who make pension fund preparation by using traditional instruments such as investing in land, building land, gold or pledge of gold. Traditional investment is not recorded in the system but prevalently done by the community, especially people in West Sumatra who became the object of research. This is expected to cause the penetration rate and the density of retirement fund products to be low.

4 CONCLUSION

From the statistic test result in regards to socioeconomic characteristic relationship to retirement savings ownership, it can be seen that there is a significant correlation between factor of number of children, tenure of employment and financial literacy with retirement savings ownership. Meanwhile, there is no significant relationship to marital status and the amount of monthly income. For future research, it is very interesting to do the development of a vibrant society culture perform traditional ways of preparing for retirement. In addition, the development of research can also be done by using qualitative research methods with the intent to find more in depth how the true meaning of the relationship or the influence of financial literacy on ownership of retirement savings in Indonesia.

ACKNOWLEDGEMENT

We acknowledged to the state owned bank in Payakumbuh, West Sumatera Indonesia for supporting this research.

REFERENCES

Alessie, R., Van Rooij, M. & Lusardi, A. 2011. Financial literacy and retirement preparation in the Netherlands. *Journal of Pension Economics & Finance* 10(4): 527–545.

Babiarz, P., & Robb, C.A. (2014). Financial literacy and emergency saving. *Journal of Family and Economic Issues* 35(1): 40–50.

Boisclair, D., Lusardi, A. & Michaud, P.C. 2017. Financial literacy and retirement planning in Canada. *Journal of Pension Economics & Finance* 16(3): 277–296.

Bureau of Statistics. 2015. Indikator Kesejahteraan Rakyat 2015 katalog BPS.

Clark, R., Lusardi, A. & Mitchell, O.S. 2017. Employee financial literacy and retirement plan behavior: a case study. *Economic Inquiry* 55(1): 248–259.

Koenen, T.B. & Lusardi, A. 2011. Financial literacy and retirement planning in Germany. *Journal of Pension Economics & Finance* 10(4): 565–584.

Lusardi, A., & Mitchell, O. S. (2011). Financial literacy around the world: an overview. *Journal of pension economics & finance* 10(4): 497–508.

Otorisasi Jasa Keuangan (OJK). 2016a. *Dana Pensiun. Seri Literasi Perguruan Tinggi*. Jakarta: OJK.

Otorisasi Jasa Keuangan (OJK). 2016b. *Survey Literasi dan Inklusi Keuangan 2016*. Jakarta: OJK.

Advances in Business, Management and Entrepreneurship – Hurriyati et al (eds)
© 2020 Taylor & Francis Group, London, ISBN 978-0-367-27176-3

Dynamics relationship between the composite and Islamic index in the capital market of Indonesia

S.A. Rusmita, I.N. Muharam, L.N. Rani & E.F. Cahyono
Universitas Airlangga, Surabaya, Indonesia

ABSTRACT: This paper tries to analyze the relationship between the Islamic and composite index by using the VECM method. The result shows that there is a co-integration among the variables of the Jakarta Composite Index (JCI) and Jakarta Islamic Index (JII). Furthermore, the results show JII and JCI indicate having a negative correlation. JII significantly affects JCI, while JII was significantly influenced by JCI in lag 2 and JII in lag 1. The results of this study indicate a negative correlation between two variables. In the context of diversification, this measure explains the extent to which the return index of a security is related to others. If the correlation was negative, investors could make portfolio between the Islamic and conventional but they could not completely eliminate this portfolio risk. For investors, the results will give them choices for their investment portfolio.

1 INTRODUCTION

Capital markets have an important role in the economy of a country. A capital market can be a medium to transfer funds from excess funds to the party who lacks funds in the hope of a profit (return). The funded party can utilize the funds to develop the business. According to Tavinayati and Yulia (2009), the stock market, also called stock exchange, is a place to trade securities of public companies that in the process involves institutions and professions associated with securities.

The capital market as the market in general is the meeting place between the seller and buyer. In the capital market the object of sale is capital or funds in the form of securities, in order to obtain sources of financing for owners of the business and as a container of investment for the investor community (Hartono 2009).

Tandelilin (2010) states that the capital market is a place to buy securities that are generally more than one year old, such as stocks and bonds. The physical form of the capital market is the stock exchange, which is where the sale and purchase of securities take place. The capital market is where various parties meet, especially companies that sell stocks and bonds, with the purpose of the sale to be used as additional funds or to strengthen the company's capital.

Based on the Capital Market Law Number 8 of 1995, the capital market has a strategic role in the national development as a source of financing for business and investment vehicles for investors. UUPM also provides capital market restrictions, namely activities related to public offerings and securities trading, public companies related to

securities issued, as well as institutions and professions related to securities (Rokhmatussa & Suratman, 2010).

The Indonesian Stock Market (IDX/JCI) is one of the best-performing stock markets lately. At the end of 2017, the Indonesia Stock Exchange had 566 listed companies with a combined market capitalization of IDR 7,052.39 trillion. In December 2017, based on the Single Identification Number, there were 628,346 domestic investors; 51.33% were foreign investors and 48.67% were domestic investors. JCI is one of the guidelines for an investor in investing in the Capital Market because JCI indicates the stock price movement in the Indonesia Stock Exchange.

As the world's most populous Muslim country, Islamic-based investments are growing rapidly in Indonesia. For investors who believe Islamic principles it is more suitable to invest in Islamic investment instruments, including stocks.

Within 5 years, the value of the Islamic capital market capitalization in Indonesia increased 42%. In 2012, Islamic capital market capitalization was recorded at Rp2.451 trillion. In September 2017 it increased to Rp3.473 trillion. The development of Islamic market capitalization is caused by Islamic Sharia Stock Indonesia (ISSI), the new index that is issued by the Indonesian Stock Exchanges (IDX).

In the first period of 2010, the total of Islamic shares amounted to 210 shares and the number increased until 2014. At the end of the two-year period until 2014, the total of Islamic shares amounted to 336 shares, so it can be said that the number of Islamic shares in Indonesia increased 37% over 4 years.

The presence of this rapidly growing portfolio of Islamic stocks is interesting to choose and scrutinize, especially if the stock portfolio provides a promising return. An investor in investment is confronted by the return on investment and the risk of investment (Tandelilin 2010); therefore, portfolio selection is necessary.

The existence of Islamic-compliant stocks has been recognized since the emergence of the Jakarta Islamic Index (JII). JII is an index of shares traded on the Indonesia Stock Exchange. The selection process of JII is based on the performance of Islamic stock trading conducted by BEI, among others: Shares selected are sharia stocks included in the DES issued by Bapepam and LK; Islamic shares are then selected, 60 shares based on the largest capitalization sequence during the last year and 60 shares that have the largest capitalization. Then 30 shares are selected based on liquidity in the order of the largest transaction value in the regular market during the last year.

The difference between the Islamic capital market and conventional capital market lies fundamentally in the fact that the Islamic capital market does not recognize any kind of short selling trading activity. It is a sell or buy activity in a very short time. Islamic shareholders will pay high attention and ownership, so it will be an effective control for the management of the company.

In Indonesia the capital market system still uses a dual system, where the Islamic and conventional are still in one market. Therefore, there are still many investors who doubt or assume that investment in sharia instruments or conventional instruments is the same thing.

The cointegrated test between variables is interesting since on the long run perhaps every variable has a correlation. However, to test co-integration is not simple; it needs a large and multiplier test, like Dickey Fuller and Johansen Tests (Engle & Granger 2001).

Previous research done by Majdoub et al. (2016), assesses the market integration between conventional and Islamic stock prices from the long- and short-run perspectives for France, Indonesia, the United Kingdom and the United States from September 8, 2008, to September 6, 2013, using various econometric approaches. The results show long-run relationships for all countries, except for the United Kingdom, where there is no co-integration between conventional and Islamic stock prices. There is evidence of weak linkages between the Indonesian market and the developed markets for both conventional and Islamic stock prices, thus suggesting that investors can diversify their portfolios at the international level to minimize risk (Majdoub et al. 2016). Majdoub and Mansour (2014) also wrote about the correlation between the US market and a sample of five Islamic emerging markets, namely Turkey, Indonesia, Pakistan, Qatar, and Malaysia. The estimation results of

the three models show that the US and Islamic emerging equity markets are weakly correlated over time (Majdoub & Mansour 2014).

Beik and Wadana (2011) also analyzed Indonesia's Islamic Stock Market, namely the Jakarta Islamic Index (JII), in relation to other Islamic as well as conventional stock markets in Malaysia and the United States, especially during the subprime crisis that started in early 2006.

The results show that there is no long run relationship between Indonesia's capital market and both Malaysia and the US markets. For investors, the results will give them choices for their investment portfolio. Meanwhile, in Indonesia's perspective, this should be an opportunity for promoting its capital market as the potential destination for profitable investment. In the short run, the JII is significantly affected by the shock or disturbance taking place in the other markets. However, the results also indicate that the JII is the least volatile and a more stable market.

This research will use cointegration analysis, just as the previous study. Co-integration analysis was used by Beik and Wardana (2011) on the Jakarta Islamic Index and another Islamic index in the world.

Guyot (2011) analyzed the market quality and price dynamics of a sample group of Islamic indexes. The results highlight that efficient investment allocation is not compromised by the application of Shariah criteria. However, although few indexes impose an additional liquidity cost on investors, a vast majority of indexes present degrees of liquidity that are similar to conventional indexes.

El Khamlichi et al. (2014), used cointegration analysis for exploring the efficiency and diversification potential in comparison with the conventional benchmarks. The sample includes Islamic and mainstream indices. The Islamic indices of Dow Jones and S&P do not have co-integrating relations with their respective benchmarks, which suggests the existence of long-run diversification opportunities.

Alexakis et al. (2017) explored long-run relationships between Islamic and conventional equity indices for the period 2000–2014. They adopted a hidden co-integration technique to decompose the series into positive and negative components. The result is that relationships for the negative components retain their significance, indicating that the Islamic index is the least responsive during bad times. Also the robust nature of Islamic investments and a possible differentiated investor reaction to financial information during market downtrends.

Al-Khazali et al. (2014) used stochastic dominance to analyze the performance of Islamic and conventional indexes. The result is that all conventional indexes stochastically dominate the Islamic indexes at second and third orders in all markets except the European market. However, the European, US, and global Islamic stock indexes dominated the conventional ones during 2007–2012. The results indicate that Islamic indexes outperformed their conventional peers during the recent global financial crisis. Thus, Islamic investing performs better than conventional investing during a meltdown economy.

Miniaoui et al. (2015) wrote about the performance of Islamic and conventional indices of the Gulf Cooperation Council (GCC) countries in the wake of the financial crisis of 2008 and tested whether Islamic indices were less risky than conventional indices by using GARCH models.

According to Lean and Teng (2013), other countries also have integration to the local index. They have analyzed two emerging powers (China and India) into the Malaysian stock market. The result is that a strong financial integration between the stock markets in India and Malaysia was observed. In contrast, the volatility spillover effect from the United States to Malaysia disappeared, especially on the short term. Nevertheless, the study suggests that in the long run, investors in Malaysia could gain by diversifying their portfolios in China and Japan, relative to India and the United States.

Recently, previous studies focused on the cointegration between Islamic and conventional; this research focused more on whether the Islamic market is influenced by the conventional or vice versa.

Looking at the facts, the research this time will dig deeper on the existence of a linkage between the sharia index with conventional stock markets in Indonesia only, with data over a longer time, to prove the previous study about a relation between the conventional and Islamic index.

The need for this linkage research in the Islamic financial market is to convince investors that if there is a linkage, for the investment in the Islamic or conventional, there is no difference. The hypothesis in this study is that the Islamic Issues (Jakarta Islamic Index) are influenced by the Jakarta Composite Indonesia (JCI) index or joint stock price index (IHSG) and vice versa.

This paper is structured as follows. After the introduction, theb literature review related to the long-run relationship and short-run dynamics between the Islamic and composite index will be presented. Then, the next part is an explanation about the method used in this research. In the second to last part, the finding and analysis of the test results will be described. The final part provides a conclusion and suggestion for future research.

2 METHOD

2.1 Data

The data used in this research is -first- the Jakarta Composite Index (JCI). This index comprises all equities listed in the Indonesia Stock Exchange Market; therefore, its movements show an upward and downward trend of the overall Indonesian Market.

$$IHSG = sum \frac{(p \times q)}{d} \times 100 \qquad (1)$$

p = Closing Price at Regular Market
q = Number of Shares
d = Share Value

100 = the total market value of total shares recorded on August 10, 1982, making the base day of the computation of IHSG (the base value of 100 out of a total of 13 shares listed at that time)

The second data used in this research is the Jakarta Islamic Index (JII). It represents liquid selected equities that comply with the Islamic principles. Thirty Islamic-compliant companies were selected based on the assessment that was conducted by the regulator every 6 months.

The sample consists of 1347 daily observations from the period March 2012 to December 2017. All data were collected from the historical prices of the Indonesia Stock Exchange Market. Daily data were used to get robust results, while the time period selection was intended to provide the most recent illustration of the Indonesian market condition.

2.2 Empirical framework

To assess the long-run relationship and short-run dynamics among variables, this research conducted Cointegration and Vector Error Correction Model (VECM) tests. The former was used to observe the existence of long-run relationships, while the latter was used to estimate the short-run dynamics among variables. The software used to conduct all tests was Eviews 10.

2.3 Dickey-Fuller (DF-test)

The Augmented Dickey Fuller Test (ADF) is a unit root test for stationary. Unit roots can cause unpredictable results in your time series analysis. Unit root test must be performed because the requirement to perform a Cointegration test is that all variables must be stationary at first difference. Furthermore, the presence of Cointegration is a requirement for performing the VECM test. If the value of t-ADF is less than the critical value of MacKinnon, then it can be concluded that the data used is stationary (does not contain the unit root). The level of significance used in this research is 5%.

2.4 Cointegration test

In order to find out the relationship between variables in the long run, it is necessary to do a cointegration test. If there is co-integration, then it indicates a long-term relationship between the variables. In this research, the method used in testing the existence of co-integration is the Johansen Cointegration method (Engle & Granger 2001).

Table 1. Dickey-Fuller (DF-test).

	Test Statistic	1% Critical Value	5% Critical Value	10% Critical Value	p-value for Z(t)
dfuller dJCI	-61.174	-3.430	-2.860	-2.570	0.0000
dfuller djii	-63.277	-3.430	-2.860	-2.570	0.0000

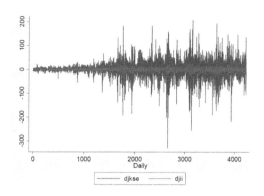

Figure 1. DF-test statistics.

2.5 Granger causality test

The Granger Causality Test needs to be done because the influence between variables is unknown. Such tests are useful when we know that two variables are related, but we do not know which variable causes other variables to move. By doing this test, it will be known whether a variable influences another variable, or otherwise, or whether there is no relationship between the variables

2.6 Empirical model of VECM

The final step is to test the VECM model. VECM is a terrestrial VAR model due to the existence of non-stationary but cointegrated data. VECM is often referred to as a VAR design for non-stationary series, which has cointegration relationships. The VECM specification restricts the long-term relationships of the endogenous variables to converge into their cointegration relationships, while still allowing the existence of short-run dynamics.

2.7 Impulse response function analysis

This analysis will explain the impact of shocks which occur on one variable against another variable. With this method, the dynamics that occur in each variable when there is a certain shock of 1 standard error in each equation can be seen. Analysis of Impulse Response Function also serves to see how long the influence occurs. The horizontal axis represents the period (in this case how many days, since the data is on a daily basis), while the vertical axis shows the response value in percentage.

3 RESULT AND DISCUSSION

3.1 Dickey-Fuller (DF-test)

Based on the Augmented Dickey Fuller test, the results show that all variables are stationary at first-difference.

At the first different Test Statistical level shows greater than the Critical Value of 5% then the tested variables are probability (Prob *) or p-value is below 0.05. This means that the data is valid to be used in further tests in this research.

3.2 Optimum lag

This optimal lag length test is very useful to eliminate autocorrelation problems. The results are based on the optimum lag test with a selection of order criteria: Likehood Ratio (LR), Final Prediction Error (FPE), Akaike Information Crition (AIC), Schwarz Information Crition (SC), dan Hannan-Quin Crition (HQ). Determination of the optimal lag in this study is based on the criterium of sequential modified LR test statistics (LR).

The result is that the optimum lag is the fourth lag, which contains the most star sign. After the optimum lag is known, the next step is cointegration testing using Johansen Test.

Table 2. Optimum lag.

lag	LL	LR	df	P	FPE	AIC	HQIC	SBIC
0	- 33435.8				25471.5	15.82	15.82	15.82
1	-32807	1257.7	4	0.0	18952	15.525	15.528	15.5344
2	-32741	131.4	4	0.0	18404.3	15.496	15.501	15.5111
3	-32680.2	121.58	4	0.0	17916.4	15.469	15.477	15.4903
4	-32627.1	106.28*	4	0.0	17504.7*	15.446*	15.455*	15.473*

3.3 Cointegration test

This test is to find out whether there is influence between the variables we will examine. If it is proven that there is co-integration, then the VECM stage can be continued, but if it is not proven, then VECM cannot be continued.

In Table 3, the co-integration test using Johansen Test showed the results of cointegrated variables in the long term, which is indicated by trace statistic > critical value 5%. It means that in the long run, JCI and JII have a relationship and similarity of movements. In other words, all variables tend to adjust to each other to achieve their long-term equilibrium. The existence of co-integration means that the VECM test is possible.

3.4 Granger causality test results

The Granger Causality Test is performed to see if two variables have a reciprocal relationship. From the results obtained in Table 4, it can be seen that those who have a causality relationship are those that have a probability value smaller than alpha 0.05 so that H0 will be rejected, which means that a variable will affect other variables.

The hypotheses used:

H0: The dependent variable is not significantly affected by the independent variable.
H1: The dependent variable is significantly affected by independent variables.

Table 3 demonstrates that the causality relationship that occurs between variables is one way. JII significantly affects JCI with a probability less than 0.05. JCI also significantly affects JII with a probability of 0.001. Thus, it is concluded that there is unidirectional causality between the JCI and JII variables.

3.5 VECM results

VECM estimation results will be obtained between short and long-term relationships between JCI and JII. VECM estimation results used to analyze the effects of short-term and long-term effects of the dependent variable on the independent variables can be seen in Table 5.

Table 4. Granger causality tests.

Equation	Excluded	Obs	Chi2	Probability
JII	JCI	4227	839.78	0.000
JCI	JII	4227	15.021	0.001

Table 5. VECM short run.

Variable	coefficient	z-Statistic partial	Probability
D(JII(-1),2)	-.1461522	-14.78	0.0000
D(JII(-2),2)	-.0785808	-10.59	0.0000
D(JII(-3),2)	-.0351748	-8.21	0.0000
D(JCI (-1),2)	-.2.132608	-5.90	0.0000
D(JCI (-2),2)	-1.88272	-7.09	0.0000
D(JCI (-3),2)	-1.246825	-8.63	0.0000

From Table 5 it can be seen that two variables have a reciprocal relationship; JII is significant to JCI on the short run and long run, also JCI is significant to JII. For JII, the significant variables at a 1% significance level are JCI in lag 1, 2, 3 and JII in lag 1, 2, 3. So if there is an increase in index in JII, it will cause JCI to decrease by −0.1461522%.

According to a theory by Markowitz about diversification in portfolio, portfolio formation could be formatted by considering covariance and negative correlation coefficients between assets in order to reduce portfolio risk. The results of this study indicate a negative correlation between two variables. In the context of diversification, this measure explains the extent to which the return index of a security is related to others. The result shows that the correlation is negative but is not more than −1 so that the correlation is not perfect. So, investors could make portfolio between Islamic and conventional but they cannot completely eliminate this portfolio risk. Sharia and conventional index have a negative correlation so that if the sharia index rises, the conventional index falls. This is a good diversification for investors to form a portfolio.

Table 3. Co-integration test.

maxium rank	parms	LL	eigenvalue	trace statistic	5% critical value
0	14	-34077.358	.	2900.5723	15.41
1	17	-33148.741	0.35556	1043.3383	3.76
2	18	-32627.072	0.21872		

Table 6. VECM long run.

Variable	Chi2	P > Chi2
D_JIII	4211.731	0.0000
D_JCI	2091.365	0.0000

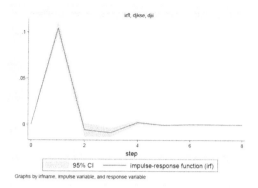

Figure 3. IRF Jakarta Islamic index.

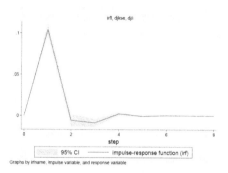

Figure 2. IRF Jakarta composite index.

The correlation in the long term, as seen in Table 6, exists between sharia and conventional, because it is in the same market in Indonesia by using the dual system.

3.6 Impulse response function analysis

Impulse Response Function (IRF) is used to describe the shock rate level of the variables used in the study. The dynamic behavior of the VECM model can be seen through the response of each variable to the shock of the variable and other endogenous variables. The response is in the short term usually quite significant and tends to change. In the long run, responses tend to be consistent and keep getting smaller. Impulse Response Function provides an overview of how the response of a variable in the future acts if there is interference with another variable.

Figure 2: The horizontal axis is the time in the next day after the shock occurs, while the vertical axis is the response value.

In this model the response of changes in each variable in the presence of new information is measured by 1-standard deviation. In this analysis, a positive or negative response from a variable against other variables will be known. Figure 2 shows that JCI starts responding to the shock with a negative trend (−) after the first period until entering the third period.

Figure 3 shows that if there is any shock in the market, JII starts responding to the shock with a positive trend (+) in the first period and then declines after the first period until entering the second period, but it is still positive. The second period curve is declined to negative, but it is not going so far, after that it increases and is stable. So, this means that sharia could be a buffer if there are any shocks in the economy.

4 CONCLUSION

From this analysis, between JII and JCI is a relation and co-integration. The relation concludes on the short and long run and has a reciprocal relationship for both JCI and JII. Sharia and conventional index have a negative correlation so that if the sharia index rises, the conventional index falls. This is a good diversification for the investors' base of theory portfolio.

For investors, the results will give them choices for their investment portfolio, because both Islamic and composite index have a negative relation. Also, Sharia could be a buffer if there are any shocks in the economy, because when the market shock happens, the response of the Islamic market is still positive, and more stable.

REFERENCES

Alexakis, C., Pappas, V. & Tsikouras, A. 2017. Hidden cointegration reveals hidden values in Islamic investments. *Journal of International Financial Markets, Institutions and Money* 46, 70–83.

Al-Khazali, O., Lean, H. H. & Samet, A. 2014. Do Islamic stock indexes outperform conventional stock indexes? A stochastic dominance approach. *Pacific-Basin Finance Journal* 28: 29–46.

Beik, I. S. & Wardhana, W. 2011. The relationship between Jakarta Islamic Index and other selected markets: evidence from impulse response function. *Journal of Economics and Business Airlangga* 21(2).

El Khamlichi, A., Sarkar, K., Arouri, M. & Teulon, F. 2014. Are Islamic equity indices more efficient than their conventional counterparts? Evidence from major global index families. *Journal of Applied Business Research* 30(4): 1137.

Engle, R. & Granger, C. 2001. Co-integration and error-correction: Representation, estimation, and testing. *Econometric Society Monographs* 33: 145–172.

Guyot, A. 2011. Efficiency and dynamics of Islamic investment: evidence of geopolitical effects on Dow Jones Islamic market indexes. *Emerging Markets Finance and Trade* 47(6): 24–45.

Hartono, J. 2009. *Teori Protofolio dan Analisis Investasi. (6*ed.*).* Yogyakarta: BPFE.

Lean, H.H. & Teng, K.T. 2013. Integration of world leaders and emerging powers into the Malaysian stock market: A DCC-MGARCH approach. *Economic Modelling* 32: 333–342.

Majdoub, J. & Mansour, W. 2014. Islamic equity market integration and volatility spillover between emerging and US stock markets. *The North American Journal of Economics and Finance* 29: 452–470.

Majdoub, J., Mansour, W. & Jouini, J. 2016. Market integration between conventional and Islamic stock prices. *The North American Journal of Economics and Finance* 37: 436–457.

Miniaoui, H., Syani, H., & Chaibi, A. (2015). The impact of financial crisis on Islamic and conventional indices of the GCC countries. *The Journal of Applied Business Research* 31(2): 357–370.

Rokhmatussa, & Suratman. 2010. *Hukum Investasi dan Pasar Modal.* Jakarta: Sinar Grafika.

Tandelilin, E. (2010). *Portofolio dan Investasi teori dan aplikasi.* Yogyakarta: Kanisius.

Tavinayati, & Yulia. 2009. *Hukum Pasar Modal di Indonesia.* Jakarta: Sinar Grafika.

Advances in Business, Management and Entrepreneurship – Hurriyati et al (eds)
© *2020 Taylor & Francis Group, London, ISBN 978-0-367-27176-3*

Empirical analysis of non-performing financing in the case of Indonesian Islamic banks

L.N. Rani, E.F. Cahyono & S.A. Rusmita
Universitas Airlangga, Surabaya, Indonesia

ABSTRACT: Non-Performing Financing is the main problem faced by Syariah banks in running their operations. This study examines the specific factors of Islamic banks that affect the Non-Performing Financing of Islamic banks in Indonesia in the period 2008–2016. The analysis technique used in this study was multiple regression (OLS) to measure the effect of the variables Efficiency, Financing, Capital Adequacy Ratio and Size of Bank against Non-Performing Financing (NPF). The results showed that the Financing (FIN) and Capital Adequacy Ratio (CAR) variables significantly influence the Non-Performing Financing of Islamic Banks, while the Efficiency variable obtained from the calculation of DEA (Data Envelopment Analysis) and SIZE Islamic banks has no significant effect on Non-Performing Financing (NPF). Financing that positively affects NPF of Islamic banks shows that the increase in financing variables leads to an increase in Islamic bank NPF, but, on the other hand, the CAR variable, which has a negative influence on NPF, shows that the financing risk can still be managed well by the Islamic bank and Islamic bank management can still sufficiently capital minimum banking.

1 INTRODUCTION

Islamic banks have developed in Indonesia since 199, beginning with the establishment of Bank Muamalat Indonesia (BMI). BMI is the first Islamic bank in Indonesia. Islamic banks are banks operating on principles based on the Qur'an and Hadith.

Financing is one of the bank's intermediation functions that determines the income of Islamic banks. The bank's financing risk is Non-Performing Financing (NPF). An NPF indicator is when the failure of payment from customers to the bank can cause the bank to lose.

Research on non-performing financing in Islamic banks is less abundant than non-performing loans of conventional banks. Thus, this research can provide an input for policies related to risk management of financing at Islamic banks. This study aims to examine how much influence the internal factors of Islamic banks have on the NPF. This study refers to Bahrini's (2011) research on the problem of financing at banks in Tunisia; this is because financing is a bank's performance and the main factor affecting the occurrence of problematic financing is an internal bank factor, which is bank efficiency and capitalization. Bahrini (2011) showed that there is a significant causality relationship between Non-Performing Loan and Cost Efficiency.

Karim et al. (2010) examined the NPL (Non-Performing Loan) relationship with bank efficiency in Malaysia and Singapore. The results showed that there is no significant difference in efficiency costs in Malaysia and Singapore, even though efficiency costs in Singapore are higher than Malaysia.

The onset of the global financial crisis that began in 2007 in the United States also affected the economy in Indonesia, as well as banks including the Non-Performing Financing of Islamic banks. Thus, this study aims to analyze the Non-Performing Financing of Islamic banks in Indonesia after the occurrence of the global financial crisis in the period 2008-2016. Non-Performing Financing for Islamic banks is a disadvantage because banks lose the opportunity to get refunds from financing customers and if this happens continuously, it can cause loss and operational failure for Islamic banks.

This is similar to Bahrini (2011), and shows that it is evident that a substantial amount on Non-Performing Loans can lead banks to bank failures.

Zaib et al. (2014) explained that NPLs cause serious problems for the banking sector itself and the economy as well. NPLs affect the banks' performance and finances. This has led to bank management always focusing on planning for NPLs. NPLs reduce the profit margins of banks and can even cause banks to lose money and thus reduce the size of bank assets and their lending capacity can be limited. In the worst case, banks may face problems due to failure of financing. Failure of large amounts of financing can lead to a financial crisis.

NPF can lead to disruption of liquidity of Islamic banks. Liquidity has an important role for bank operations, especially when facing a financial crisis. According to Wagner (2007), when in a crisis condition, if a bank more liquid bank asset then this can give the bank profit in the form of stability because the crisis costs incurred by the bank are fewer.

Troubled financing influenced by bank efficiency is found in Karim et al.'s research (2010) showing that based on the regression result, the low level of efficiency increases the occurrence of problem loans. In the opinion of Berger and De Young (1997), the inefficiency of banks can cause problems in the performance of bank loans, examples of banks that have management deficiencies in monitoring costs and total borrowing from borrowers so that when a loss occurs the loan has the potential for failure.

Keeton (1999) explains that loan growth can increase problem loans because banks lower the minimum standards for lending instances; banks lower the number of collateral borrowers that make borrowers repay loans; they accept borrowers with bad credit history, and overly believe that borrowers will be able to repay loans. A decrease in lending standards gives the borrower the opportunity to make his loan problematic.

The difference of this study with Bahrini (2011) is the presence of the SIZE variable. The determination of the SIZE variable is based on Rajan and Dhal's (2003) research, which shows that there is a positive relationship between bank size and NPL; the result shows that larger banks have larger NPLs than smaller banks. It is based generally that bank managers provide loans assessed less competent in assessing the ability to pay customers so that it can lead to NPLs.

2 METHOD

This study used multiple regression Ordinary Least Square (OLS) to test the hypothesis to identify the relationship between Non-Performing Financing of Islamic Banks with bank specific variables, namely: Efficiency (EFF), Total Financing (FIN), Capital Adequacy Ratio (CAR) and SIZE (Bank Size). The equation model of the research is as follows:

$$NPF_{it} = \beta_1 + \beta_2 EFF_{it} + \beta_3 FIN_{it} + \beta_4 CAR_{it} + \beta_5 SIZE_{it} \quad (1)$$

This research data is sourced from Financial Services Authority: Islamic Bank Financial Report in Indonesian for Five Bank Islam period 2008–2015. The data type used in this research is panel data. The variables used in this research are:

- NPF (Non-Performing Financing): measured by comparison ratio of non-performing financing with total financing.
- EFF (Efficiency): Measured the level of efficiency of Islamic banks using the DEA method.
- FIN (Financing): total Islamic bank financing
- CAR (Capital Adequacy Ratio): Measured from the ratio of bank capital to the earning assets of a bank that is weighted according to the risks. CAR demonstrates the capability of bank capital to cover productive risks.

- SIZE: measured from natural logic data of total Bank Assets.

DEA (Data Envelopment Analysis) was first introduced by Banker et al. (1984). Researchers in several areas recognize that DEA is an excellent and relatively easy-to-use methodology in the operational modeling process for performance evaluation. In this study, DEA is used as a tool to measure the efficiency of Islamic banks in the period 2008-2016.

Banker et al. (1984) explain that DEA is used to evaluate the efficiency of planned management, using mathematical programs. DEA measures techniques and scale of inefficiency comprehensively, through input and output.

The measurement of bank efficiency refers to Bahrini's research (2011). For output variables use data from the bank's financial statements of total financing and investment assets. As for the input variables, we used data from the bank financial statements of labor (total labor costs), fixed assets (Book Value of Property, Plant and Equipment), Bank Management Quality (measured from non-operational costs divided by total bank assets). The measurement of efficiency in this study used MaxDEA.

3 RESULT AND DISCUSSION

3.1 Haussman test

Panel data testing must start with the Haussman test to determine whether panel data is tested with FEM (Fixed Effect Model) or REM (Random Effect Model), using the eviews tool 6. The Haussman test results are listed in Table 1.

Table 1 is the result of the Haussman test which shows a probability of 0.0619. This suggests to researchers to do the hypothesis testing with Ordinary Least Square (OLS) using REM (Random Effect Model).

3.2 Result

Based on the test results, Haussman hypothesis testing was done using REM (Random Effect Model). Table 2 shows the results of hypothesis testing of Islamic bank data in Indonesia from 2008 to 2016 using REM.

The results of the multiple regression of OLS in Table 2 can be formulated in a model equation as follows:

$$NPF_{it} = 3.374928 - 0.655909 EFF_{it} + 1.04E$$
$$-07FIN_{it} - 0.045860 CAR_{it} - 8.40E - 08SIZE_{it} \quad (2)$$

Table 1. Haussman test.

Test Summary	Chi-Sq. Statistic	Chi-Sq. d.f.	Prob.
Cross-section random	8.96981	4	0.0619

Table 2. Multiple regression results OLS REM Islamic bank.

Variable	Coefficient	Std. Error	t-Statistic	Prob.
C	3.374928	0.6504	5.189007	0
EFF?	−0.655909	0.66167-5	−0,991287	0.3275
FIN?	1.04E-07	5.98E-08	1.739955	0.0896
CAR?	−0.04586	0.02614-5	−1.754075	0.0871
SIZE?	−8.40E-08	5.30E-08	−1.583533	0.1212
Random Effects (Cross)				
_BMI–C	0			
_BSM–C	0			
_BMGS–C	0			
_BKPNS–C	0			
_BRIS–C	0			

Effects Specification		
	S.D.	Rho
Cross-section random	0.000000	0.0000
Idiosyncratic random	1.049015	1.0000

Weighted Statistics			
R-squared	0.150541	Mean dependent var	2.438889
Adjusted R-squared	0.065595	S.D. dependent var	1.150654
S.E. of regression	1.112276	Sum squared resid	49.48628
F-statistic	1.772201	Durbin-Watson stat	0.962428
Prob (F-statistic)	0.15349		

Unweighted Statistics			
R-squared	0.150541	Mean dependent var	2.438889
Sum squared resid	49.48628	Durbin-Watson stat	0.962428

Based on Table 2, the OLS results in Islamic banks show that the variables FIN (Financing) and CAR with a significance level of 10% significantly affect the NPF (Non-Performing Financing). There is a difference: where the FIN variable positively affects the NPF, it indicates that any increase in FIN will cause an increase of the Islamic bank NPF, but when the CAR variable has a negative effect on NPF, it shows that any increase in CAR will cause the decrease of Islamic bank NPF.

Based on Table 2, the results of the multiple regression (OLS) indicate that the variable EFF (Efficiency) with a probability of 0.3725 does not significantly affect NPF of Islamic banks. In the opinion of Berger

& De Young (1997) the occurrence of problem loans can also be caused by events outside the bank as the economic conditions are declining. The results of this study are similar to Bahrini's (2011) research results.

Financing (FIN) variables significantly and positively affect the NPF. This suggests that an increase in the number of financing in Islamic banks led to an increase in NPF. The results of this study differ from Bahrini (2011), which does not use Financing as a factor affecting Non-Performing Financing. The use of financing variables refers to Keeton's (1999) research that the reasons for financing can also lead to problem financing when lowering the financing standards.

The CAR (Capital Adequacy Ratio) variable significantly and negatively affect NPF. This suggests that an increase in the number of CARs in Islamic banks led to a decrease in NPF. The results of this study are similar to Bahrini's (2011) research, namely that capitalization has a significant and negative influence on the NPL.

The effect of the SIZE variable on NPF is not significant. This is different from the results of Rajan and Dhal's research (2003) which showed there is a positive relationship between the SIZE variable and NPL. That SIZE variable does not significantly affect the NPL was also experienced by Zaib et al. (2014). The not-significant influence of SIZE on the NPL shows that both large banks and small banks have no difference in determining the level of bank efficiency in handling NPLs.

3.3 Discussion

Non-Performing Financing (NPF) represents the risk of Islamic bank financing and illustrates the risk of non-acceptance of borrowing costs from borrowers in order to reduce profits and disrupt the liquidity of Islamic banks.

The results of the test show that the significant variables affecting the Non-Performing Financing are Financing and Capital Adequacy Ratio. The effect of Financing on NPF is positive, indicating that the increase in financing variables leads to an increase in Islamic bank NPF. This result is in accordance with Keeton's (1999) research which shows that the increase in financing the bank conducted indicates a decline in the standard of credit, which can lead to an increased risk of credit so that the bank can suffer losses. In addition, research conducted by Rajan and Dhal (2003) also showed a significant effect of credit with NPF. Hence the recommended policy for Islamic banks should further improve the standard in providing financing to reduce the risk of problem financing.

CAR variables that negatively affect NPFs indicate that an increase in CAR can decrease the NPF. This shows that risk management of Islamic bank financing is good, so that the increase in CAR can decrease the NPF level of Islamic banks. The results of this study are in line with research by Bahrini (2011), indicating that a connection between the reverse capitalization banks with financing failed to pay. The results showed that the lower the capital owned bank opportunities in the higher loan loss that the bank managers prefer utilizing the financing risk to earn income. Suggestion for banking management is to keep improving CAR because it can help lower NPF.

The EFF (Efficiency) variable has no effect on NPF of Islamic banks, indicating that efficiency does not have an impact on the increase and/or decrease of Islamic bank NPF. The results of this study are similar to Bahrini's (2011) research results which became the reference in this study.

The SIZE variable, which is an additional variable based on Zaib et al. (2014), indicated that the test results show no significant SIZE variables (Islamic bank size) to an increase or decrease in Islamic bank NPF. The results of this study are similar to those of Zaib and Kamran (2014).

4 CONCLUSION

The conclusion of this study is that FIN (Financing) and CAR are variables that influence NPF in Islamic Bank partially. The results showed that FIN had a significant positive effect on NPF, while CAR had a significant negative effect on NPF., while EFF and SIZE are both variables that do not influence NPF.

The results of the OLS multiple regression testing showed that the variables Efficiency, Financing, CAR and Size simultaneously and together only give an effect of 15% ($R2 = 0.15$). This suggests that 85% of NPF levels are influenced by other variables which may be the development of this study. This is similar to Bahrini's (2011) research result which also has R (Adj) less than 0.20 or 20 % which is 12.45 %.

The recommendation of this study is that Islamic banking management should focus on improving the standard of financing so as not to run the risk of financing and Islamic banks should keep increasing the CAR (Capital Adequacy Ratio) so that the NPF value can be lowered.

REFERENCES

Bahrini, R. 2011. Empirical analysis of non-performing loans in the case of Tunisian banks. *Journal of Business Studies Quarterly* 3(1): 230.

Banker, R.D., Charnes, A. & Cooper, W.W. 1984. Some models for estimating technical and scale inefficiencies in data envelopment analysis. *Management science* 30(9): 1078–1092.

Berger, A.N. & DeYoung, R. 1997. Problem loans and cost efficiency in commercial banks. *Journal of Banking & Finance* 21(6): 849–870.

Karim, M.Z.A., Chan, S.G. & Hassan, S. 2010. Bank efficiency and non-performing loans: Evidence from Malaysia and Singapore. *Prague Economic Papers* 2(2010): 118–132.

Keeton, W.R. 1999. Does faster loan growth lead to higher loan losses? *Economic review Federal Reserve Bank of Kansas City* 84: 57–76.

Rajan, R. & Dhal, S.C. 2003. Non-performing loans and terms of credit of public sector banks in India: An empirical assessment. *Reserve Bank of India Occasional Papers* 24(3): 81–121.

Wagner, W. 2007. The liquidity of bank assets and banking stability. *Journal of Banking & Finance* 31(1): 121–139.

Zaib, A., Farid, F. & Khan, M.K. 2014. Macroeconomic and Bank-Specific Determinants of Non-Performing Loans in the Banking Sector in Pakistan. *International Journal of Information, Business and Management* 6(2): 53.

Advances in Business, Management and Entrepreneurship – Hurriyati et al (eds)
© 2020 Taylor & Francis Group, London, ISBN 978-0-367-27176-3

Is overconfidence and herding in Ponzi scheme investors influenced by demographic factors?

M. Sari & N. Nugraha
Universitas Pendidikan Indonesia, Bandung, Indonesia

ABSTRACT: This study aims to analyze the influence of demographic factors on investment bias in Ponzi scheme investors. The study analyzes demographic factors such as gender, age, marital status, level of education, occupation, working experience and income level, and investment biases such as herding and overconfidence. The respondents in this study were 187 Ponzi scheme investors in Bandung. They were selected by using purposive random sampling. The data were collected by using a questionnaire and they were analyzed by using multiple regression analysis with dummy variables. The results of the study showed that the demographic factors had a significant effect on investment bias. Gender, age, marital status and occupation had a negative effect on both the herding and overconfidence bias. The working experience and income level had a positive effect on herding and overconfidence. The level of education had a negative effect on the herding and overconfidence bias. The findings of this study contribute to developing financial education models in the community.

1 INTRODUCTION

Investors often make irrational decisions (Agarwal et al. 2016). An investment decision on Ponzi schemes is one example of irrationality in making decisions. Several findings in various countries, including Indonesia, show that high-risk Ponzi schemes are still interesting for many investors, while clearly not all investors can make a profit from it. During the period of 1975–2015, the practice of Ponzi schemes in Indonesia caused 1.3 million people to lose their invested funds up, making up IDR 126 Billion. (https://www.hukumonline.com/berita/baca/lt5925388dbec70)

Irrational financial decisions are caused by the psychological factors of investors (Tverskey & Kahneman 1973). The impact of these factors leads to an investment bias such as overconfidence and herding. Overconfidence is a belief that someone has above average abilities and knowledge (Nofsinger 2002). Investors who have high overconfidence tend to overestimate their knowledge and underestimate the risks. Elan and Goodrich (2010) found that overconfidence investors can be found among the victims of fraudulent investments, such as Ponzi schemes.

Several studies show that there is an effect of demographic factors on overconfidence. Schaefer et al. (2004) found that female investors were more overconfident compared to male investors and there was positive relationship between age and the level of overconfidence. Jamshidinavid et al. (2012) also found a positive relationship between the level of education and the level of overconfidence. Sun et al. (2011) showed that the married investors group had higher levels of overconfidence. The opposite result about the

relationship of demographic factors with overconfidence was found in other studies. Bengtson et al. (2005) found male investors were more overconfident and Lin (2011) found there was a negative relationship between age and the level of overconfidence.

Herding behavior in investment is described as a behavioral tendency for an investor to mimic others, thereby ignoring important substantive information (Scharfstein & Stein 1990). When the individual investors deal with uncertainty, they will make decisions not only by looking at the fundamentals of the risk asset, but also following the actions of other investors on the same circumstances (Waweru et al. 2008).

The effect of demographic factors on herding biases were found in several studies. Baddeley (2010) found differences in herding biases among the group of investors with different gender and age. Jamshidinavid et al. (2012) showed women investors had higher herd biases than male investors and there was a negative relationship between age and the herding bias level. Menkhoff et al. (2006) showed that there was no difference in the herding bias in the investor group with different gender and a negative relationship between education and herding biases.

This study aims to analyze the influence of demographic factors, including gender, age, marital status, education, occupation, working experience and income level, towards overconfidence and herding bias in Ponzi scheme investors. A demographic factor analysis research on Ponzi scheme investors is relatively rare, so the findings of this study are expected to provide information to develop a financial education model in accordance with the characteristics of investors.

2 METHOD

2.1 Research variable

The research variables consisted of demographic factors as independent variables which include gender, age, marital status, occupation, years of work and income rate, and investment bias as dependent variable consisting of overconfidence and herding. The overconfidence was measured by (1) the level of confidence in the skill and investment knowledge, (2) the level of financial success expectancy, (3) the level of confidence to offer investment advice and (4) the level of confidence to make a right decision. The herding variables were measured using the indicators (1) the investment encouragement due to the success of others investing in the Ponzi scheme, (2) the investment encouragement because friends/relations make the same decisions, and (3) the investment encouragement due to the invitation of friends/relations.

2.2 Participants

Participants in the study were 187 Ponzi scheme investors in Bandung, selected through purposive random sampling. They were women (51.9%), aged 41–50 years (61%), their marital status being married (77.5%) and with undergraduate education (65.8%). Based on occupation, the majority of the participants were private employees (61%) with a working experience of over 10 years (56.1%). The total income of the respondents, on average, is around IDR 3 million & 6 million.

The instrument in collecting the data were questionnaires that were distributed by sending an email and responding directly to participants.

2.3 Analysis ttechnique

The relation analysis of demographic factors and investment bias used multiple regression analysis with a dummy variable, with the equation model as follows:

$$Y1 = B_0 + b_1 \text{GEN} + b_2(\text{AGE1}) + b_3(\text{AGE2})$$
$$+ b_4(\text{STATUS}) + e_1 B_5(\text{EDU1}) + b_2(\text{EDU2})$$
$$+ b_3(\text{EDU3}) + b_4(\text{OCCUP1}) B_5(\text{OCCUP2})$$
$$+ b_6(\text{OCCUP3}) + b_3(\text{WRKEXP1})$$

$$Y2 = B_0 + b_1 \text{GEN} + b_2(\text{AGE1}) + b_3(\text{AGE2})$$
$$+ b_4(\text{STATUS}) + e_1 B_5(\text{EDU1}) + b_2(\text{EDU2})$$
$$+ b_3(\text{EDU3}) + b_4(\text{OCCUP1}) B_5(\text{OCCUP2})$$
$$+ b_6(\text{OCCUP3}) + b_3(\text{WRKEXP1})$$
$$+ b_4(\text{WRKEXP2}) + \text{INCOME} + e_1$$

Y1 = Herding
Y2 = Overconfidence
GEN = 1 if participants are man, 0 if woman

AGE 1 = 1 if participants are 31 – 40 years old, 0 if not
AGE 2 = 1 if participants 41 – 50 years old, 0 if not
STATUS = 1 if participants are married, 0 if not
EDU1 = 1 if participants have a Diploma, 0 if not
EDU2 = 1 if participants are an Undergraduate, 0 if not
EDU3 = 1 if participants are a Magister, 0 if not
OCCUP1 = 1 if participants are a Government employee, 0 if not
OCCUP2 = 1 if participants are an entrepreneur, 0 if not
OCCUP3 = 1 if participants are a private employee, 0 if not
WKEXP1 = 1 if years of work of the participants is < 5 years, 0 if not
WKEXP2 = 1 if years of work of the participants is 5-10 years, 0 if not
INCOME = Income per year

3 RESULT AND DISCUSSION

Table 1 shows the results of the cognitive bias measurement which consists of overconfidence and herding.

The result of the study shows that Ponzi scheme investors have a high overconfidence, mainly derived from the perception that investors have a high level of knowledge about the investment they choose and they have the confidence that the investment will provide a great advantage. The results of this study are in line with research of Elan and Goodrich (2010) who found there was an investment overconfidence among victims of investment crimes such as Ponzi schemes.

Table 1 also explains that the cognitive biases of herding among Ponzi scheme investors is high. The results indicate the decision to invest in Ponzi

Table 1. Description of variables.

Variabel/sub variable	Average Score	Ideal Score	Conclusion
Overconfidence	**3.88**	5	high
a. Skill and knowledge	4.00	5	high
b. Hope for Success	4.04	5	very high
c. Ability to offer investment advice	3.58	5	high
Herding	**3.9**	5	high
Investment drive from the success of other investors	3.94	5	high
Investment drive because they follow friends/relations that make the same decisions	4.05	5	very high
Investment drive due to other people	3.71	5	high

schemes are made of the encouragement of friends/ relations who make the same decision. These findings suggest that Ponzi's investment decisions are not based on economic considerations, but rather "mimic" behavior in the hope of gaining high profits in a short time. This finding is relevant to studies suggesting that the investment decision in Ponzi schemes is due to social relations (Shafi 2014).

Table 2 shows the relationship of demographic factors with herding and overconfidence of the Ponzi scheme investor.

The female investors group has higher levels of overconfidence compared to male investors. Overconfidence has a negative significant relationship with age, marital status, and education but has a positive significant relationship with occupation, working experience and level of income.

Young investors will consider various factors in making investment decisions. Conversely, the increasing age of an investor will do the opposite, because of the experiences that they have (Cronqvist et al. 2015). This result is supported by the research of Lin (2011), Jamshidinavid et al. (2012), and Bhandari and Deaves (2006) saying that educated investors make investments based on their own knowledge, abilities and confidence. Mishra and Metilda (2015) stated that the higher the level of education, the higher the level of self-confidence, because a higher education makes an investor feel more competent in almost all fields, including finance. This means that the higher the education of investors, the more susceptible they are to overconfidence.

The relationship between demographic factors and herding bias is also described in Table 2. Female investors have higher levels of herding compared to male investors. Herding has a negative significant relationship with age and marital status, but has a positive significant relationship with education, occupation, working experience and level of income.

The results of this study showed female investors have a higher level of overconfidence and herding biases compared to male investors. The results of this study are relevant to findings from Jamshidinavid et al. (2012). The gender's effect related to optimism has usually been viewed as the result of Type I errors and being uninterpretable (Felton et al. 2003).

The more sociable individuals will be more responsive to social influence. They will be more likely to herd, thus the personality traits such as socialization and extraversion will correlate positively with the propensity to herd (Baddeley et al. 2012). This finding is relevant for the personality traits of women because they are more extravert than man, so they are more prone to herding behavior than men.

The negative relationship between age and overconfidence and herding indicates that with the age increasing, the level of herding and overconfidence will decrease. These results are in line with studies by Baddeley (2010), Jamshidinavid et al. (2012) and Lin (2011).

Education is one factor that influences investment biases. This study shows that education has a positive relationship with herding bias but has a negative relationship with overconfidence. This result is supported by the research of Pompian (2012) and Lai et al. (2013).

Table 2. The relation of demographic factors with herding and overconfidence.

Variable Independent	Herding (Y1)		Overconfidence (Y2)	
	Unstandardized Coefficients	Sig	Unstandardized Coefficients	Sig
(Constant)	2.255	.000	4.428	.000
GEN	-.080	.012	-.122	.000
AGE1	-.453	.000	-.687	.000
AGE2	-.134	.028	-.312	.000
STATUS	-.110	.059	-.269	.000
EDU1	1.524	.000	-.568	.000
EDU2	1.646	.000	-.350	.000
EDU3	1.360	-000	-1.011	.000
WORK1	.445	.000	.865	.000
WORK2	.120	.015	.256	.000
WORK3	-.170	.004	-.273	.000
LW1	.101	.216	-.087	.320
LW2	.152	.001	.265	.000
INCOME	.078	.007	.131	.000
Adjusted R Square	.868		.653	
F Statistic	3.625 (sig = 0.000)		27.437 sig = (0.000)	

The marital status negatively affects the level of herding and overconfidence. In the group of the unmarried, investors herding have higher overconfidence. This finding is relevant with the research of Shie and Chang (2017).

The groups of investors who have a job as entrepreneur have a higher herding and overconfidence bias than others. Working experience positively affects the attitude of herding, where the longer they are working, the herding attitude increases and they are being associated with an increasing overconfidence working bias. The results also show that income has a positive effect on herding and overconfidence. It means the higher the level of the investor's income, the more herding and overconfidence they show. This result is supported by the research of Forbes (2005) and Bhandari and Deaves (2006).

4 CONCLUSION

The purpose of this study was to analyze the relevance of demographic factors against investment bi-as. The high level of overconfidence and herding found in the Ponzi scheme investor group showed that rational economic factors were not the main factors considered in making investment decisions. In the Ponzi scheme investor group, the belief in the knowledge and the investment skills and the follow-up behavior were the dominant factors affecting the investment decision. The research result showed that the demographic factor of the investor Ponzi scheme had a significant effect on the investment bias. Gender, Age, Marital Status and Employment had a negative effect on both herding and overconfidence biases. Work's duration and income level have a positive effect on herding and overconfidence bias. Educational level has a positive effect on herding bias but negatively affects overconfidence. The findings of this study are expected to give a contribution to policy makers in developing financial education models in the community, so that investors can make investment decisions wisely.

REFERENCES

Agarwal, A., Verma, A. & Agarwal, R.K. 2016. Factors influencing the individual investor decision making behavior in India. *Journal of Applied Management and Investments* 5(4): 211–222.

Baddeley, M. 2010. Herding, social influence and economic decision-making: socio-psychological and neuroscientific analyses. *Philosophical Transactions of the Royal Society B: Biological Sciences* 365(1538): 281–290.

Baddeley, M., Burke, C., Schultz, W., & Tobler, P. 2012. Herding in financial behaviour: A behavioural and neuroeconomic analysis of individual differences. *Journal of Economic Behavior and Organisation.*

Bengtsson, C., Persson, M. & Willenhag, P. 2005. Gender and overconfidence. *Economics Letters* 86(2): 199–203.

Bhandari, G.R. & Deaves. 2006. The Demographics of Overconfidence. The Journal of Behavioral Finance 7: 5–11.

Cronqvist, H., Siegel, S. & Yu, F. 2015. Value versus growth investing: Why do different investors have different styles? *Journal of Financial Economics* 117(2): 333–349.

Elan, S.L., & Goodrich, M.K. 2010. Behavioral patterns and pitfalls of US investors. In Federal Research Division, Library of Congress.

Felton, J., Gibson, B. & Sanbonmatsu, D.M. 2003. Preference for risk in investing as a function of trait optimism and gender. *The journal of behavioral finance* 4(1): 33–40.

Forbes, D.P. 2005. Are some entrepreneurs more overconfident than others? *Journal of business venturing* 20(5): 623–640.

Jamshidinavid, B., Chavoshani, C. & Amiri, S. 2012. The Impact of Demographic and Psychological Characteristics on the Investment Prejudices in Tehran Stock. *European Journal of Business and Social Sciences* 1(5): 41–53.

Lai, M.M., Tan, S.H. & Chong, L.L. 2013. The behavior of institutional and retail investors in Bursa Malaysia during the bulls and bears. *Journal of Behavioral Finance* 14(2): 104–115.

Lin, H.W. 2011. Elucidating the influence of demographics and psychological traits on investment biases. *World Academy of Science, Engineering and Technology* 77: 145–150.

Menkhoff, L., Schmidt, U. & Brozynski, T. 2006. The impact of experience on risk taking, overconfidence, and herding of fund managers: Complementary survey evidence. *European Economic Review* 50(7) 1753–1766.

Mishra, K.C. & Metilda, M.J. 2015. A study on the impact of investment experience, gender, and level of education on overconfidence and self-attribution bias. *IIMB Management Review* 27(4): 228–239.

Nofsinger. 2002. Overconfidence. [Online} Retrieved: http://www.overconfidence.behavioralfinance.net/. Accesed December 1, 2011.

Pompian, M. M. 2012. *Behavioral finance and investor types: managing behavior to make better investment decisions.* New Jersey: John Wiley & Sons.

Schaefer, P.S., Williams, C.C., Goodie, A.S. & Campbell, W.K. 2004. Overconfidence and the big five. *Journal of research in Personality* 38(5): 473–480.

Scharfstein, D.S. & Stein, J.C. 1990. Herd behavior and investment. *American Economic Review* 80(3): 465–479.

Shafi, M. 2014. Determinants influencing individual investor behavior in stock market: a cross country research survey. *Arabian Journal of Business and Management Review* 2(1): 60–71.

Shie, F.S. & Chang, C.Y. 2017. The herding behaviour and announcement of insider transfer trading: *A study in Taiwan. Investment Analysts Journal* 46(4): 249–262.

Sun, C.F., Ching, C.P, Huey-Cher, J.Q. & Hwa, M.C. 2011. The existence of overconfidence among individual and institutional investors in Malaysia (Doctoral dissertation, Universiti Tunku Abdul Rahman).

Tversky, A. & Kahneman, D. 1973. Availability: A heuristic for judging frequency and probability. *Cognitive psychology* 5(2): 207–232.

Waweru, N.M., Munyoki, E. & Uliana, E. 2008. The effects of behavioural factors in investment decision-making: a survey of institutional investors operating at the Nairobi Stock Ex-change. *International Journal of Business and Emerging Markets* 1(1): 24–41.

Advances in Business, Management and Entrepreneurship – Hurriyati et al (eds)
© 2020 Taylor & Francis Group, London, ISBN 978-0-367-27176-3

Effects of corporate governance and barriers to entry on financial performance with intellectual capital as a mediating variable

N. Soewarno, B. Tjahjadi & R.D Istiqomah

Universitas Airlangga, Surabaya, Indonesia

ABSTRACT: The objective of this research is to test empirically the mediating effect of intellectual capital on the relationship of corporate governance and barriers to entry to financial performance. This study was conducted with secondary data, which were the financial statements of mining sector companies in the period 2013–2015, and 93 samples were obtained. The hypotheses of this study were analyzed by means of partial least squares using WarpPLS 5.0 software. The results showed that corporate governance has a direct significantly negative influence on financial performance. Barriers to entry also showed a direct significantly negative effect on financial performance. This study proves the role of intellectual capital as a partial mediator in the relationship of corporate governance and barriers to entry to financial performance.

1 INTRODUCTION

Intellectual capital affects a company's financial performance because it focuses not only on monetary but also on knowledge assets, so that intellectual capital can be an important component in supporting corporate wealth for both profit and non-profit-oriented firms (Roodposhti et al. 2011; Suhendah 2012).

A significant negative relationship exists between barriers to entry and intellectual capital because having fewer competitors makes employees unmotivated to innovate (El-Bannany 2008; Sefidgar et al. 2015). A significant negative relationship between the two variables can also be created because a larger amount of assets owned and used in the production process will increase production costs, reducing the profitability of the company. In addition, the authors wish to prove the relationship of corporate governance to financial performance by using intellectual capital as a mediator because of differences from the results of previous research (research gap) (Nkundabanyanga et al. 2014).

Mining companies were selected as the subject of this research because they have high barriers to entry, requiring large funds and resources of intellectual capital with specific capabilities as disclosed by Delta Mining Corporation CEO Bob Kamandanu in the final article at www.okezone.com.

Donaldson and Preston (1995) explain that corporate stakeholders can consist of many parties, such as suppliers, customers, employees, political groups, communities, trade associations, governments, and even investors or shareholders. Stakeholder theory can be a tool for linking ethics and strategy because it focuses on ways to treat stakeholders well and manage their interests, which can help companies create value and improve company performance. This theory is appropriate to explain the importance of corporate governance in improving financial performance. The agency relationship arises between two (or more) parties when either party is designated as an agent (management), acts, becomes part of, or as representative for another party, principal (shareholder), within a particular area for decision issues (Ross 1973).

Barriers to entry can be defined as socially undesirable restrictions, which are caused by resource protection by existing owners in the industry (Weizsacker 1980). A potential competitor seeks a market where barriers to entry are relatively insignificant. The absence of entry barriers increases the likelihood that a newcomer can operate profitably.

In the current knowledge-based economy, the importance of intellectual capital as a competitive advantage factor is undeniable. Some intellectual capital components are developed using various definitions of intellectual capital as a base. Bontis et al. (2000) state that there are three elements that build intellectual capital: human capital, structural capital, and customer capital. Performance management is defined as a systematic process to improve organizational performance by developing individual and team performance (Armstrong 2006).

Intellectual capital in the financial statements also has a relationship with corporate governance. This view is also expressed by Makki and Lodhi (2014), who state that corporate governance does not improve financial performance directly but through intellectual capital. Intellectual capital significantly acts as a mediator in corporate governance relationships with financial performance (Nkundabanyanga et al. 2014).

Hypothesis 1: Corporate governance significantly influences financial performance with intellectual capital as a mediating variable.

Several previous studies have shown that there is a relationship between barriers to entry and intellectual capital (El-Bannany 2008; Sefidgar et al. 2015), intellectual capital and financial performance (Roodposhti et al. 2011), and barriers to entry and financial performance (Listyawati & Sirine 2007). It can be assumed that intellectual capital can act as mediator in the relationship between barriers to entry and financial performance.

Hypothesis 2: Barriers to entry significantly influence financial performance with intellectual capital as a mediation variable.

The following is a conceptual framework that will explain the influence of corporate governance (CG) and barriers to entry (BE) on financial performance (FP) with intellectual capital (IC) as a mediating variable.

2 METHOD

2.1 Sampling procedure

The population of this study is 43 mining companies listed on the Stock Exchange in 2013 to 2015. Sample selection criteria in this study include

(1) Mining companies listed on the Indonesia Stock Exchange in the period 2013–2015 and up to 30 April 2016.
(2) Mining sector companies that publish complete annual financial statements relating to variables in the study. Based on the criteria, the number of samples that meet the criteria of 31 samples each year of mining companies in 2013–2015.

2.2 Operational definition of variables

2.2.1 Corporate governance
Proxies used to measure corporate governance are board size (BS), female directors (FD), and managerial ownership (MO). BS is the number of seats on the board of directors (Wang 2015). FD is the number of women on the board of directors (Rambo 2013).

2.2.2 Barriers to entry
The variable barriers to entry is measured using the Capital Intensity Ratio (CIR) (Kim & Lyn 1987; Listyawati & Sirine 2007).

2.2.3 Financial performance
The financial performance indicators used are Return on Assets (ROA), Return on Equity (ROE), and Tobin's Q Ratio (Amba 2014; Maditinos et al. 2011).

2.2.4 Intellectual capital
Intellectual capital is proxied with the Value Added Intellectual Coefficient (VAIC) and Modified-Value Added Intellectual Coefficient (M-VAIC). M-VAIC is a model developed by Ulum et al. (2014) based on the VAIC model by Pulic (1999).

3 RESULT AND DISCUSSION

3.1 Analysis of the outer model

3.1.1 Convergent validity
Based on the results of analysis and data processing, two indicators do not meet the criteria because they have loading < 0.4, that is, BS and Tobin's Q. Both indicators are then eliminated. In the second iteration, all indicators have loads above 0.7.

3.1.2 Discriminant validity
Discriminant validity is used to measure the quality of measurement instruments. Comparative analysis of AVE square root values and quadratic correlation (R^2) is used to check the value of AVE. Table 1 shows the results of the AVE calculation.

Based on Table 1, it can be concluded that all latent variables have met the discriminant validity requirements.

3.1.3 Comparison analysis of loading and cross-loading
The results of the analysis show that each indicator has a loading factor value that exceeds 0.7 and is larger than the loading indicator for other existing constructs, so it can be concluded that all indicators of latent variables meet the criteria of discriminant validity.

3.1.4 Composite reliability
A construct is said to be reliable if it has a composite reliability value >0.70. Results of construct research data processing in this research show a composite reliability value above 0.7.

3.2 Measurement analysis of the inner model and discussion

3.2.1 Direct relation between corporate governance and financial performance
The results of analysis show that corporate governance has an effect on financial performance, with a β value of −0.38 and $p < 0.01$. Laible (2013) states that the presence of women in management can negatively affect the performance of companies in certain contexts. Ruan et al. (2011) adds that managerial ownership, which is one of the proxies of corporate governance, can improve the company's performance to a certain point; when the point is

Table 1. Correlations among latent variables with square roots of AVEs.

	CG	BE	IC	FP
CG	(0.758)	−0.056	−0.217	-0.240
BE	−0.056	(0.921)	-0.287	-0.611
IC	−0.217	−0.287	(1.000)	0.539
FP	−0.240	−0.611	0.539	(0.992)

exceeded, managerial controls to align the interests of managers and other stakeholders with the addition of ownership management structure will no longer improve the company's performance. R^2 analysis indicates that corporate governance is able to explain a financial performance of 14.8%, while the other 85.2% is explained by other variables. The value of Q2 of 0.131 indicates that the above model has qualified predictive relevance because it has a value above zero.

3.2.2 *Direct effect of barriers to entry on financial performance*

The result of the analysis shows that barriers to entry have an effect on financial performance, with a β value of −0.64 and $p < 0.01$. This indicates that barriers to entry have a significant negative direct impact on financial performance. When the capital intensity ratio, which is the proxy of barriers to entry, is high, then firms tend to be inefficient because firms are issuing larger assets for production (high production costs). The high cost of production will result in price increases per unit of product and decrease of product sales so that the profitability of the company will decline (Listyawati & Sirine 2007). The value of R^2 based on Table 1 shows that financial performance can be explained by barriers to entry of 40.7% while 59.3% is explained by other variables. The above model has qualified predictive relevance because it has a positive Q2 value.

3.3 *Hypothesis testing and discussion*

The structural model analysis is shown in in Figure 1.

The direct relationship of corporate governance to financial performance before and after the intervening variable has a negative β value, down from −0.38 to −0.21. Figure 1 shows a significant p value (0.02 < 0.05). In addition, the relationship of corporate governance and intellectual capital also has a significant p value at < 0.01 with β −0.027, while the relationship of intellectual capital to financial performance has a β value of 0.36 with a significant p value < 0.01. This proves that intellectual capital acts as a partial mediator in the relationship of

corporate governance relation to financial performance. Thus, hypothesis 1 states that corporate governance significantly influences financial performance with intellectual capital as an acceptable mediation variable.

Negative correlations between corporate governance and intellectual capital are possible because the industry used in this research is mining, which prioritizes value in operational activities. The expertise of female directors according to Swartz and Firer (2005) as role models, mentors, and supporters in recruitment, training, and promotion of other female employees becomes less meaningful. Managerial ownership can negatively affect intellectual capital as described by Mohd-Saleh et al. (2009), who state that when managerial ownership exceeds a certain level, it may have adverse effects on minority shareholders. Managers who have control over corporate financial decision making may make adverse decisions such as raising the company's debt for the purpose of getting more cash to make investment policies that are not optimal or build "empire management" (Ruan et al. 2011).

The relationship of intellectual capital and financial performance is significantly positive, in agreement with Nazari (2010), who states that recognizing the most influential intellectual capital elements of corporate performance will help the company to understand more clearly and develop better ways of managing the capabilities they have. The study of intellectual capital as a mediating variable on corporate governance and financial performance relationship is in line with Nkundabanyanga et al. (2014) stating that by improving the quality of board governance and intellectual capital, directors will show relevance to corporate objectives and improvement of company performance for the benefit of all stakeholders.

The relationship of barriers to entry to the financial performance before and after the mediation variable has a significant fixed value at $p < 0.01$ and the β value falls from −0.64 to −0.46. The relationship of barriers to entry to intellectual capital was significant, with $p < 0.01$ and β of −0.34. The relationship of intellectual capital to financial performance has a significant value of $p < 0.01$ and β of 0.36. The result of the analysis shows that intellectual capital partially mediates the relationship of barriers to entry to financial performance. Thus, hypothesis 2 states that barriers to entry significantly influence financial performance, with intellectual capital as an acceptable mediation variable.

Research on the relationship between barriers to entry and financial performance with intellectual capital as a mediation variable has never been conducted before. However, several researchers have conducted research on the relationshipbetween barriers to entry, intellectual capital, and financial performance. Sefidgar et al. (2015) states that companies with high barriers to entry tend not to take action to encourage and motivate employees to innovate and such circumstances can

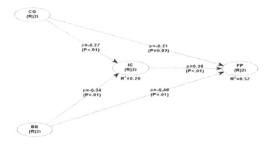

Figure 1. Analysis results.

negatively affect human capital that is part of intellectual capital. Conversely, intellectual capital proved to have a positive significant relationship to financial performance, in line with Chen et al. (2005) stating that investors will place a higher value on firms with better intellectual capital efficiency because the company will generate greater profitability as well as revenue growth both in the current year and the following years.

The results of R^2 analysis show that for corporate governance and barriers to entry, 19.5% is explained by intellectual capital while the remaining 80.5% is explained by other variables. The value of R^2 for financial performance illustrates that the three variables that influence it are corporate governance, barriers to entry, and intellectual capital, and financial performance explains 57.2% while 42.8% is explained by other variables. Intellectual capital and financial performance both have positive Q2 values of 0.2 and 0.574. This means that the model used has qualified predictive relevance.

The results of partial least squares analysis show that the influence of corporate governance on financial performance indirectly through intellectual capital as a mediation variable has $p = 0.093$, or significant at 10%, while the influence of barriers to entry on financial performance indirectly through intellectual capital as a mediation variable has $p = 0.043$, or significant at the level of 5%. Intellectual capital is therefore proved as a mediating variable.

4 CONCLUSION

Based on the results of the analysis and discussion that has been described previously, it can be concluded that significant negative impacts are evident on corporate governance and financial performance relationships, barriers to entry and financial performance, corporate governance and intellectual capital, as well as barriers to entry and intellectual capital. Conversely, a significantly positive relationship was found between intellectual capital and financial performance. In addition, intellectual capital is proven to be a mediator of corporate governance and financial performance relationships and barriers to entry and financial performance relationships. The form of mediation that occurs in the model is partial mediation.

ACKNOWLEDGMENTS

We would like to thank to the Dean of the Faculty of Economics and Business for sponsoring us to the Conference.

REFERENCES

Amba, S.M. 2014. Corporate governance and firms' financial performance. *Journal of Academic and Business Ethics* 8(1): 1.

Armstrong, M. 2006. *Performance management: Key strategies and practical guidelines*, 3rd edition. London and Philadelphia: Kogan Page.

Bontis, N., Keow, W.C.C. & Richardson, S. 2000. Intellectual capital and business performance in Malaysian industries. *Journal of Intellectual Capital* 1(1): 85–100.

Chen, M., Cheng, S. & Hwang, Y. 2005. An empirical investigation of the relationship between intellectual capital and firms' market value and financial performance. *Journal of Intellectual Capital* 6(2): 159–176.

Donaldson, T & Lee, E. P. 1995. The stakeholder theory of the corporation: Concepts, evidence, and implications. *Academy of Management Review* 20(1): 65–91.

El-Bannany, M. 2008. A Study of determinants of intellectual capital performance in banks: The UK case. *Journal of Intellectual Capital* 9(3): 487–498.

Kim, W.S. & Lyn, E.O. 1987. Foreign direct investment theories, entry barriers, and reverse investments in US manufacturing industries. *Journal of International Business Studies* 18(2): 53–66.

Laible, M. 2013. Gender diversity in top management and firm performance: an analysis with the IAB-establishment panel. Paper presented at the CAED Conference, Atlanta.

Listyawati, T.M. & Sirine, H. 2007. Analisis pengaruh profitabilitas industri, rasio leverage keuangan tertimbang, rasio intensitas modal tertimbang dan pangsa pasar terhadap roa dan roe perusahaan manufaktur yang go-public di Indonesia. *Jurnal Ekonomi dan Bisnis* 2(3).

Maditinos, D., Chatzoudes, D., Tsairidis, C. & Theriou, G. 2011. The impact of intellectual capital on firms' market value and financial performance. *Journal of Intellectual Capital* 12(1), 132–151.

Makki, M.A.M. & Lodhi, S. A. 2014. Impact of corporate governance on intellectual capital efficiency and financial performance. *Pakistan Journal of Commerce and Social Sciences* 8(2): 305–330.

Mohd-Saleh, Norman, Rahman, C.A., & Ridhuan, M. 2009. Ownership structure and intellectual capital performance in Malaysia. *Asian Academy of Management Journal of Accounting and Finance* 5(1): 1–29.

Nazari, J.A. 2010. An investigation of the relationship between the intellectual capital components and firm's financial performance. *Thesis. Haskayne School of Business.*

Nkundabanyanga, S.K., Ntayi, J.M., Ahiauzu, A. & Sejjaaka, S. K. 2014. Intellectual capital in Ugandan service firms as mediator of board governance and firm performance. *African Journal of Economic and Management Studies* 5(3): 300–340.

Pulic, A. 1999. *Basic information on VAIC™*. [Online]. Retrieved: www.vaic-on.net. Accessed: August 23, 2016.

Rambo, C.M. 2013. Influence of the capital markets authority's corporate governance guidelines on financial performance of commercial banks in Kenya. *The International Journal of Business and Finance Research* 7(3): 77–92.

Roodposhti, F. R. 2011. The effect of any relation between intellectual capital based on financial patterns and economic value added for measuring business of accepted companies in Iranian stock exchange organization. *African Journal of Business Management* 5(27): 11022–11033.

Ross, S.A. 1973. The economic theory of agency: the principal's problem. *The American Economic Review* 63(2): 134–139.

Ruan, W., Tian, G. & Ma, S. 2011. Managerial ownership, capital structure and firm value: evidence from china's civilian-run firms. *Australasian Accounting Business & Finance Journal* 5(3): 73–92.

Sefidgar, M., Minouei, M. & Maleki, S. 2015. Studying factors that affect intellectual capital performance in listed banks in Tehran stock exchange. *Indian Journal of Fundamental and Applied Life Sciences* 5(S1): 769–776.

Suhendah, R. 2012. Pengaruh intellectual capital terhadap profitabilitas, produktivitas, dan penilaian pasar pada perusahaan yang go public di Indonesia pada tahun 2005–2007. *Jurnal dan Prosiding Simposium Nasional Akuntansi* 15.

Swartz, N. & Firer, S. 2005. Board structure and intellectual capital performance in South Africa. *Meditari Accountancy Research* 13(2): 145–166.

Ulum, I,. Ghozali, I., Chariri, A. 2014. Intellectual capital performance of Indonesian banking sector: A Modified VAIC (M-VAIC) Perspective. *Asian Journal of Finance & Accounting* 6(2): 103–123.

Wang, M. 2015. Value Relevance of Tobin's q and corporate governance for the Taiwanese tourism industry. *Journal of Business Ethics* 130(1): 223–230.

Weizsacker, C.C.V. 1980. A welfare analysis of barriers to entry. *The Bell Journal of Economics* 11(2): 399–420.

Advances in Business, Management and Entrepreneurship – Hurriyati et al (eds)
© *2020 Taylor & Francis Group, London, ISBN 978-0-367-27176-3*

The effect of adoption of the International Financial Reporting Standard on earning management

H. Hamidah & A. Rahmah
Universitas Airlangga, Surabaya, Indonesia

ABSTRACT: Corporate governance is a system that directs and controls the company (FCGI, 2001). By implementing corporate governance, it is expected that the manipulation of financial statements by managers will be reduced. The purpose of this study is to provide empirical evidence of the influence of the adoption of the International Financial Reporting Standard (IFRS) on earnings management and the Good Corporate Governance (GCG) mechanism in moderating the relationship between IFRS adoption and earnings management. This research used data of a total of 120 banking companies listed on the Indonesia Stock Exchange (IDX) during 2007 to 2014. Hypothesis testing in this research was conducted using a moderated regression analysis test. The results show a decrease in the level of earnings management at the time of the adoption of IFRS in banking companies in Indonesia. However, GCG mechanisms cannot moderate the influence of IFRS adoption on earnings management.

1 INTRODUCTION

In 2012 Indonesia began to use the globally accepted financial reporting standard, the International Financial Reporting Standard (IFRS; Hamidah 2017). According to Barth et al. (2008) the convergence of Statements of Financial Accounting Standards with IFRS is expected to enhance the functioning of the global capital market by providing more comparable and high-quality information to investors. Adoption of international accounting standards into domestic accounting standards aims to produce financial statements that have a high degree of credibility, and the requirements of disclosure items will be higher so that the value of the company will be higher also. As management has a high degree of accountability in running the company, the company's financial statements will produce more relevant and accurate information and the financial statements will be more comparable and generate valid information for assets, debt, equity, income, and expenses (Petreski 2005).

The adoption of IFRS also has an impact on the reclassification of financial instruments. In the early period of IFRS adoption, it is still permissible to reclassify previously reclassified financial instruments, recognizing profits or losses. It is this alternative that enables management to carry out the profit-making (Pratama & Ratnaningsih 2013). The practice of earnings management in the financial statements occurs because there is a difference of interest between the management (agent) and the principal in managing the company, causing a conflict between agencies. The role of managers in administration of the company can increase the productivity of the company (Bertners 2000). However, the role of the manager cannot remove the aspect of the principal– agent relationship in which there is a difference of interests between the entrepreneur and the manager of the company. This is a consequence of the delegation of work from principal to agent (Jensen & Meckling 1976).

The action of earnings management can be suppressed by the implementation of a good corporate governance mechanism (Nasution & Setiawan 2007).The goal of applying corporate governance, which is a concept based on agency theory, is to reduce the impetus for manipulation of financial reports by managers. In relation to Good Corporate Governance (GCG), Nasution and Setiawan (2007) show that the corporate governance mechanism has been effective in reducing earnings management practices in banking companies in Indonesia. Similarly, the results of the study by Ajina et al. (2013) of 145 companies in France show that the GCG mechanism has really narrowed the level of earnings management during the IFRS adoption period. Based on the background description, this study aims to examine the influence of IFRS adoption on earnings management with a good corporate governance mechanism as a moderating variable in banking in Indonesia.

1.1 Agency theory

According to agency theory, agency relationships are a contract between one or more persons who are regarded as principals and others who are perceived as agents to act on behalf of the principal involving delegation of authority in the decision-making of the principal to the agent. When both parties work to maximize their interests, the agent will not always act in accordance with the interests of the principal. This is because there is a difference of interest between the principal and the agent in maximizing their respective interests (Jensen & Meckling 1976). According to Scott (2009), however, agency theory is the appropriate contract design to align the interests of principal and agent in the event of a conflict of interests.

According to Messier (2006), this agency relation-ship results in two problems: (1) information asymmetry, in which management in general has more information about the actual financial position and position of the entity's operations; and (2) pursuit of the same goal, where management does not always act in accordance with the interests of the owner. The agency problem can lead to the practice of earnings management in the company as a result of information asymmetry and differences of interest between the principal and the agent. Asymmetry in information can occur because the management (agent) has more knowledge of the condition of the company than the principal.

According to Healy and Wahlen (1999), earnings management occurs when managers use judgment in financial reporting such as estimated economic life and residual value of fixed assets, liabilities for pensions, deferred taxes, loss of accounts receivable, and impairment assets and managers have a choice of accounting methods, such as depreciation and costing methods. The purpose is to manipulate the amount of earnings in information to stakeholders about the economic performance of the company or to influence the outcome of the contract, which depends on the accounting figures reported. The manager has the authority to manage the company, with a strong interest in selecting accounting policies for the company. This is the result of the delegation of work from principal to the agent.

According to Watts (2003) one of the ways to monitor contract issues and limit opportunistic management behavior is corporate governance. Shleifer and Vishny (1997) argue that corporate governance deals with how investors believe that managers will benefit them and will not steal or invest in unprofitable projects with funds from investors, and how investors control the manager.

1.2 Development of hypotheses

The adoption of IFRSs in PSAK 50 and 55 revised 2006 derived from IAS 32 and 39 is related to changes in the measurement, recognition, disclosure, and presentation of financial instruments, one of which is the classification of loans and receivables. The latter contains a change in the method of calculating the Allowance for Impairment of Earning of Credit Reduction in Bank Indonesia Regulation no.11/2/PBI/2009 using the determination of credit quality (current, substandard, doubtful, and loss) and the required reserve percentage. After the implementation of IFRS, the calculation of the allowance for impairment losses is no longer based on expectations but on objective evidence of impairment of credit values, making it difficult for managers to practice earnings management through the CKPN account of such credits. Further, the strict provisions on the reclassification of financial instruments mean

that managers cannot exploit any gain or loss due to reclassification of financial instruments.

According to Anggraita (2012) and Nurazmi (2015), IFRS adoption negatively affects earnings management, proving that implementation of IFRS can lower the level of earnings management. This may occur because prior to the implementation of IFRS in the banking industry, especially in the PSAK 50 and 55 2006 revision, the banking company calculated the provision of credit loss known as the Allowance for Earning Assets Losses (PPAP) based on the rules issued by Bank Indonesia, making it easier for managers to state reserves are higher or lower because they are still expectations. Based on the aforementioned exposure, the research hypothesis can be formulated as follows:

Hypothesis 1: IFRS adoption has a negative effect on earnings management in banks in Indonesia.

Implementation of GCG is expected to reduce the level of earnings management after the adoption of IFRS at a company because with good corporate governance the company will have a high level of control in lowering the tendency of managers to practice earnings management. It can also it can limit opportunist actions of managers in judging or assessing the treatment of accounting policies in corporate financial reporting. Nasution and Setiawan (2007) prove that GCG has effectively reduced earnings management practices in banking companies in Indonesia. Based on this exposure, the research hypothesis can be formulated as follows:

Hypothesis 2: GCG can moderate the influence of IFRS adoption on earnings management in banking in Indonesia.

2 METHOD

2.1 Population, samples, and sampling techniques

All of the constituents of the population in this study are commercial banks listed on the Indonesia Stock Exchange (BEI). The total population during the study period was 26 commercial banks. The research period was 2007–2014. The year 2007 was mandatory year for GCG self-assessment by banks, 2007–2009 is the period in which IFRS had not been adopted, and 2010–2014 is the period in which adoption of IFRS began. Purposive sampling representative of the criteria specified was conducted.

2.2 Research variables

This research used dependent, independent, moderating, and control variables. The dependent variable used is earnings management, which is proxied by discretionary accruals. The independent variable used is IFRS. The moderating variable in this research is GCG, which is proxied with a composite value obtained from the results of self-assessment by banking. The control variables in this research are firm size and growth.

2.3 Profit management

The earnings management in this study was proxied by discretionary accruals using the models of Beaver and Engel (1996). Profit management includes both accounting policy choices and concrete actions performed by management. The accounting options are discretionary accruals such as allowance for failed credits, warranty costs, and options discretionary accruals whose amount depends on the management policy (Scott 2009). The steps in measuring earnings management with the Beaver and Engel (1996) models are as follows:

1. Establish coefficients α, $\alpha 1$, $\alpha 2$, $\alpha 3$, and $\alpha 4$ by performing regression,

$$TA_{it} = \alpha + \alpha 1 CO_{it} + \alpha 2 LOAN_{it} + \alpha 3 NPL_{it} \\ + \alpha 4 \Delta NPL_{it} + 1 + e \quad (1)$$

Before the regression is performed, first all the variables are deflated by total equity and reserves of credit losses.

2. After the coefficients of α, $\alpha 1$, $\alpha 2$, $\alpha 3$, and $\alpha 4$ are obtained, nondiscretionary accrual (NDAit) is calculated:

$$NDA_{it} = \alpha + \alpha 1 CO_{it} + \alpha 2 LOAN_{it} + \alpha 3 NPL_{it} \\ + \alpha 4 \Delta NPL_{it} + 1 + e \quad (2)$$

3. Once the value of NDA_{it} is known, the value of the discretionary accrual (DA_{it}) can be calculated by subtracting the nondiscretionary accruals from total accruals and can be formulated as follows:

$$DA_{it} = TA_{it} - NDA_{it} \quad (3)$$

where

TA_{it}	= total accrual, calculated based on CKPN of credit company i in year t
COIT	= loan charge off (company write-off loan) for i in year t
$LOAN_{it}$	= outstanding loan (outstanding loan) of company i in year t
NPL_{it}	= nonperforming loan of company i in year t consists of (1) special attention, (2) substandard, (3) doubtful, and (4) freeze
$\Delta NPL_{it} + 1$	= difference between non performing loan year $t+1$ and year t
NDA_{it}	= nondiscretionary accrual of company i in year t
DA_{it}	= discretionary accrual of company i in year t
$\alpha \ldots \alpha 4$	= coefficients obtained from the regression equation
e	= error term

2.4 IFRS adoption

The adoption of IFRS in the banking sector was one of the revised SFAS 50 and 55 revisions of 2006, which was effectively adopted in 2010. IFRS adoption was measured using a dummy variable rated 0 for banks that had not yet adopted IFRS and rated 1 for those who had adopted IFRS.

2.5 Good corporate governance

GCG implementation is regulated by Bank Indonesia Regulation No. 8/14/PBI/2006 on the implementation of GCG in commercial banks. Therefore, in this study GCG is proxied by a composite index resulting from self-assessment by each bank. The GCG composite index categories are shown in Table 1.

Following the research of Anggraita (2012), the bank's GCG mechanism is measured using a dummy variable. It is given a value of 1 if the composite index of GCG is predicated very well and 0 if other.

2.6 Company size (firm size)

Company size can be measured using natural logs of total assets. Total assets are more stable and representative in showing the size of the firm compared to market capitalization and sales, which are strongly influenced by demand and supply (Sudarmandji & Sularto 2007).

2.7 Company growth

According to Kallapur and Trombley (1999), company growth is the company's ability to increase size. Kallapur and Trombley (1999) measure growth of the company by using asset growth according to the formula:

$$Growth = \frac{Total\ asset - Total\ asset_{t-1}}{Total\ asset} \quad (4)$$

Table 1. Category of GCG composite index.

Composite value	Predicate composite
<1.5	Very good
1.5–2.5	Good
2.5–3.5	Pretty good
3.5–4.5	Poor
4.5–5	Not good

3 RESULTS AND DISCUSSION

3.1 Descriptive statistics

Based on Table 2, the maximum value of DA is 0.7969, with the mean of 0.0991 indicating that the bank practices earnings management by increasing profit, which can be seen from the mean of the positive value of discretionary accrual. A mean value of 0.0991 and a standard deviation of 0.1699 indicates that the level of DA data distribution has a variation rate of 171.44%. This shows that the earnings management that occurs in the sample company is relatively fluctuating because the amount of DA owned by the interbank in the sample has various values.

The average of dummy variables, which is the measurement of IFRS adoption in the period be-fore and after adoption, is equal to 0.62 or 62%. This indicates that 38% of the sample companies had not adopted IFRS in the period 2007–2009 and the rest of the sample companies had already adopted IFRS with the enactment of PSAK 50/55 revised 2006 in the period 2010–2014. The standard deviation for this IFRS variable is 0.486, which means it has a data distribution rate of 78.3%. This indicates that the sample company has applied IFRS evenly in accordance with a circular issued by the Financial Accounting Standards Board (DSAK) No. 1705/DSAK/IAI/XII/2008 regarding the effective date of IFRS adoption for banking.

Furthermore, 45% of sample companies have conducted GCG with a composite index category of very good and the rest of the sample companies conducted GCG with composite index categories of good, good enough, and less good. 111%. This shows that the assessment of GCG by self-assessment of the sample banks in this study has a relatively diverse value with a very good, good, good enough, and less good predicates.

Firm Size, which is a control variable calculated using the natural logarithm of total assets. The minimum value of the total natural asset logarithm is 27.7661, while the maximum value of the total natural logarithm of assets amounts to 34.3822. The average size is 31.9276 and the standard deviation of 1.5582 indicates that the firm size data distribution level has a 4.8% variation rate. This shows that the firm size that occurs in the sample company does not fluctuate because of the large natural algorithm of the total assets held between the banks that are sampled relatively the same.

Growth has a minimum value of −0.0808 and its maximum value is 1.3786. The average for the growth variable is 0.1776 and the standard deviation is 0.1649, indicating that the data distribution rate is 92.8%. This shows that the growth of companies that occur in the sample is relatively diverse.

3.2 Results of regression analysis

In this research there are three models of regression analysis to determine the influence of IFRS adoption on earnings management, with GCG as the moderating variable. Table 3 shows the results of a multiple linear regression analysis test.

Model I:

$$DA = -0.364 - 0.045 \ IFRS + 0.022 \ Size + 0.018 \ Growth + e \quad (5)$$

Based on Table 3, the results of multiple regression tests can be interpreted as indicating that the value of the constant (α) in the above equation is −0.364, which means that if the IFRS, Size, and Growth variables are unchanged or constant then the value of earnings management will decrease by −0.364 times.

The value of the regression coefficient (β1) of the IFRS variable is −0.045, which means that the negative sign indicates uneven or reversed change in the IFRS variable relationship with the variable DA, so if the company has implemented IFRS then earnings management will decrease by 0.045 times, with the assumption that other variables remain constant.

The value of the regression coefficient (β2) of variable size is 0.022, which means that a positive value indicates a unidirectional change or a change in the relation between variable size and the variable of earnings management, so if the size of the firm increases by one unit then profit management will increase by 0.022 times, with the assumption that other variables remain constant.

Table 2. Descriptive statistics.

Variable	Min	Max	Mean	Standard dev.
DA	0.0009	0.7969	0.0991	0.1699
IFRS	0	1	0.62	0.486
GCG	0	1	0.45	0.500
Size	27.7861	34.3822	31.9276	1.5582
Growth	−0.0808	1.3786	0.1776	0.1649

Source: Data processed, 2016.

Table 3. Results of multiple linear regression analysis model I and II.

Model	Variable	Coefficient	T	Sign.
I	Constant	−0.364	−22.989	0.000
	IFRS	−0.045	−2.685	0.009
	Size	0.022	1.403	0.164
	Growth	0.018	0.802	0.425
II	Constant	−0.365	−22.857	0.000
	IFRS	−0.044	−2.600	0.011
	GCG	0.009	0.494	0.622
	Size	0.018	0.971	0.334
	Growth	0.014	0.626	0.533

Source: Data processed, 2016.

The value of the regression coefficient (β3) of the growth variable is 0.018, which means that if the growth of the firm increases by one unit then the profit segment will increase by 0.018 times, with the assumption that other variables remain constant.

Model II:

$$DA = -0.365 - 0.044\ IFRS + 0.009\ GCG + \\ 0.018\ Size + 0.014\ Growth + e \quad (6)$$

This regression model includes the independent variable IFRS adoption as measured by the dummy variable and the moderation variable, which is considered a second independent variable that is GCG. The result of this multiple regression test can be interpreted as indicating that the value of the constant (α) in the equation above is equal to -0.365, which means that if IFRS is the independent variable, and the GCG moderation variable and control variable size and growth are constant or not changed, the value of earnings management is equal to -0.365.

The value of the regression coefficient (β1) of the IFRS variable is -0.044, which means that the negative sign indicates unaligned or reversed change in the relationship between the IFRS variable and the variable DA, so if the company has implemented IFRS then earnings management will decrease by 0.044 times, with the assumption that other variables remain constant.

The value of the regression coefficient (β2) of the GCG variable is 0.009, which means that the positive value indicates a direct change or a change in the correlation between the GCG variable and the variable of earnings management, so that if GCG increases by one unit then profit management will increase by 0.009 times, assuming other variable remain constant.

The value of the regression coefficient (β3) of variable size is 0.018 which means that a positive value indicates a unidirectional change or a change in the correlation between the size variable and the variable of earnings management, so that if the firm size increases by one unit then profit management will increase by 0.018 times, with the assumption that other variables remain constant.

The value of the regression coefficient (β4) of the growth variable is 0.014, which means that if the growth of the firm increases by one unit then the earnings management will increase by 0.014 times, with the assumption that other variables remain constant.

Furthermore, the regression model used in model III is Moderated Regression Analysis (MRA). Its purpose is to determine whether the moderating variables used in this study (GCG) will strengthen or weaken the relationship between the independent variable (IFRS) and the dependent variable (DA).

This moderation regression test was used to examine the effect of GCG variables in moderating

Table 4. Moderated regression analysis model III results.

Model	Variable	Coef.	T	Sign.
III	Constant	-0.364	-22.989	0.000
	IFRS	-0.045	-2.685	0.009
	GCG	0.009	0.485	0.629
	IFRS*GCG	-0.002	-0.101	0.920
	Size	0.022	1.403	0.164
	Growth	0.018	0.802	0.425

Source: Data processed, 2016.

the relationship between IFRS adoption as measured by dummy variables and profit management proxies with discretionary accruals. The results of the MRA test in Table 4 can explain the influence of the relationship between independent variables and moderation variables on dependent variables, and the influence of moderation variables on the relationship between independent variables and the dependent variable. The mod-regression equation is as follows:

Model III:

$$DA = -0.365 - 0.044\ IFRS + 0.009\ GCG - \\ 0.002\ IFRS * GCG + 0.018\ Size + \quad (7) \\ 0.015\ Growth + e$$

The results of this moderation regression test can be interpreted as indicating that the constant value (α) of the regression model is -0.365, which means that if all the variables are unchanged or constant, the value of earnings management is -0.365.

The value of the regression coefficient (β1) of the IFRS variable is -0.044, which means that the negative sign indicates an unrelated or a reversed change between the IFRS variable and DA variable, so if IFRS has been implemented the profit management will decrease by 0.044 times, with the assumption of other variables remaining constant.

The value of the regression coefficient (β2) of the GCG variable is 0.009, which means that the positive value indicates a direct change in the correlation between the GCG variable and the variable of earnings management, so that if GCG increases by one unit then profit management will increase by 0.009 times assuming other variables remain constant.

The value of the regression coefficient (β3) of the IFRS * GCG variable is -0.002. The negative sign indicates a nondirectional or inverse change in the relationship between the IFRS * GCG variable with the DA variable, so if IFRS * GCG increases by one unit then earnings management will decrease by -0.002, assuming other variables remain fixed.

The value of regression coefficient (β4) of variable size is 0.018, which means that a positive value indicates a unidirectional change or a change in the relationships between the variable of size and the

Table 5. Determination coefficient test.

Model	R	R^2	Adjusted R^2
I	0.274	0.075	0.046
II	0.278	0.077	0.039
III	0.278	0.077	0.029

Source: Data processed, 2016.

variable of earnings management, so if the size of the company increases by one unit then the profit management will increase by 0.018 times assuming other variables are constant.

The value of the regression coefficient ($\beta5$) of the growth variable is 0.015, which means if the growth of the firm increases by one unit then the earnings management will increase by 0.015 times with the assumption that other variables remain constant.

Based on the results shown in Table 5, the value of the adjusted R^2 from regression model I was 0.573 indicates that the influence of the independent variable of IFRS, size control variable, and growth on the dependent variable of earnings management was 4.6% while the remaining 95.4% influence of other variables was not examined.

In the regression models II and III to test the moderation variables, the adjusted R^2 value for model II is 3.9% and adjusted R^2 for model III is 2.9%, which is a regression model to test hypothesis 2 in this research. These results indicate that the magnitude of the influence of the IFRS independent variable, size control variable, and growth on the dependent variable of earnings management with GCG as moderating variable is 96.1% while the remaining 97.1% is explained by other variables not examined.

3.3 Individual parameter significance test r-test)

The statistical *t*-test basically indicates the effect of one independent variable on the dependent variable by assuming that the other independent variable is constant. If the significance value is < 0.05 then the relationship between the dependent variable and the independent is significant.

Based on the result of regression testing using MRA, the independent, moderating, and control variables are included in the regression model. The independent variable of IFRS adoption is significantly below the 0.05 mark. The other variables of variation-bel GCG, size control variable, and growth in this study are not significant because they are above the level of significance of 0.05.

The result of the first hypothesis testing using regression I in this research is that IFRS adoption has a negative effect on earnings management (DA) as measured by the discretionary accruals proxies, with a *t* value equal to −2.685 and significance equal to 0.009. Because the significance value is below 0.05 and the β value has a negative (−) direction, hypothesis 1 is accepted.

The result of the second hypothesis test using regression models II and III indicating that in this study the variable of GCG cannot be considered a moderating variable in the relationship between IFRS adoption and earnings management because the GCG variable has a *t* value equal to 0.494 with a significance of 0.622 and the IFRS * GCG interaction variable has a *t* value of −0.101 with significance of 0.920. As the value of significance is far above 0.05, hypothesis 2 is rejected.

3.4 Discussion

The results of the first hypothesis testing in this study indicate that the adoption of IFRS in banking companies in Indonesia marked by the implementation of PSAK 50 and 55 revised 2006 on the effective date of January 1, 2010 in accordance with a circular issued by the Financial Accounting Standards Board (DSAK) No. 1705/DSAK/IAI/XII/2008 tends to decrease the level of earnings management. This is evident from the results of hypothesis testing that shows the significance value of the IFRS variable is 0.009 and *t* arithmetic shows the value of −2.685, which means earnings management decreased with the application of IFRS. The results of this study are in line with previous research conducted by Oosterbosch (2009), Anggraita (2012), and Nurazmi (2015).

The adoption of IFRS in PSAK 50 and 55 revised 2006 caused some accounting policies to be made more stringent. The increased stringency in the policy of calculating the value of the reserves of credit decline can not only narrow the movement of managers in the business manipulation of data financial statements, but can also have an impact in terms of reclassification of financial instruments. Before the adoption of IFRS companies were still allowed to reclassify previously reclassified financial instruments, recognizing profit or loss so that bank managers could use the accounting policy to refine their financial statements through earnings management.

Adoption of a fair value IFRS that reflects the actual value at the date of financial reporting and disclosure of information in the financial statements in more detail will reduce the level of information asymmetry between managers and users of financial statements. This will make it difficult for managers to practice earnings management and make the quality of information in the financial statements reflect high earnings quality with reduced profit management actions undertaken by bank managers. Financial reporting is therefore presented in accordance with the truth (representation faithfulness).

The result of the second hypothesis testing in this research, on the moderation variable of the GCG mechanism, indicates GCG is not a moderating variable in the relationship between IFRS adoption and earnings management. GCG cannot moderate the influence of IFRS adoption on earnings management.

The results of this research are in line with those of Siregar and Utama (2006), Anggraita (2012), and Nurazmi (2015). The assessment of GCG implementation through self-assessment conducted by each bank has not dampened the decline in earnings management practices after the adoption of IFRS in Indonesian banks. This is because there are still many banks that have composite values above 1.5, indicating that the implementation of GCG banks needs further improvement. Descriptive statistical results also indicate there are only 45% of banks that have a GCG composite index below 1.5, which is with very good predicate. This means there are still as many as 55% of banks with predicates that are only good, good enough, and less good.

Implementation of the GCG mechanism in accordance with Indonesian banking regulation No. 8/4/PBI/ 2006 also exists in the regulation assessing the soundness of banks, namely Bank Indonesia regulation No. 6/10/PBI/2005 and the new bank regulation No. 13/1/ PBI/2011. If banks do self-assessment of GCG in accordance with the regulations issued by Indonesian banks then they can avoid the sanction of decreasing bank soundness. This means the implementation of GCG in banking is limited to the fulfillment of regulation alone to make it possible to avoid sanctions. Hence the implementation of GCG has not been done effectively and in this study GCG cannot moderate the decline in profit management practices after the adoption of IFRS in banking in Indonesia. Implementation of GCG mechanisms used to monitor contract issues and limit opportunistic management behavior cannot monitor banks in suppressing earnings management actions. Indonesian banking implements corporate governance only to comply with law or regulation (Anggraita 2012). Hence banks cannot realize their responsibility to holders (stakeholders) through the implementation of GCG mechanisms based on Bank Indonesia regulations.

4 CONCLUSION

Based on the results of the research discussed in this article, the following conclusions can be drawn:

1. The results of testing and analysis show that IFRS adoption is empirically proven to reduce earnings management practices in the banking sector. The adoption of IFRS in banking, which is marked by the enactment of PSAK 50 and 55 revised 2006 on January 1, 2010, can decrease earnings management practice because of a new policy of calculation of CKPN credit based on objective evidence derived from historic data on the existence of credit default from the debtor in the preceding 3 years and the existence of new rules on the strict reclassification of financial instruments. With the adoption of IFRS, companies are also required to use fair value and disclosure of more detailed information in financial statements, thus narrowing the actions of bank managers in manipulating financial statements.

2. Test results and analysis show that the corporate governance mechanism using the GCG composite self-assessment based on Bank Indonesia regulation has not been able to moderate the decline of earnings management. This is because the implementation of GCG in banking is limited to regulatory fulfillment alone to make it possible to avoid sanctions. GCG implementation therefore has not been done effectively and yet can monitor banking in suppressing earnings management action

REFERENCES

Ajina, A., Bouchareb, M. & Souid, S. 2013. Corporate governance mechanisms and earnings management after and before the adoption of IFRS. *The Business & Management Review* 3(4): 147.

Anggraita, V. 2012. Dampak penerapan PSAK 50/55 (revisi 2006) terhadap manajemen laba diperbankan: peranan mekanisme corporate governance, struktur kepemilikan, dan kualitas audit. *Jurnal Simposium Nasional Akuntansi* XV.

Barth, M.E., Landsman, W.R. & Lang, M.H. 2008. International accounting standards and accounting quality. *Journal of accounting research* 46(3): 467–498.

Beaver, W.H. & Engel, E.E. 1996. Discretionary behavior with respect to allowances for loan losses and the behavior of security prices. *Journal of Accounting and Economics* 22(1–3): 177–206.

Bertens, K. 2000. Pengantar etika bisnis. Yogyakarta: Kanisius.

Hamidah. 2017. IFRS Adoption in Indonesia: accounting ecology perspective. *International Journal of Economics and Management* 11 (1): 121–132.

Healy, P.M. & Wahlen, J.M. 1999. A review of the earnings management literature and its implications for standard setting. *Accounting horizons* 13(4): 365–383.

Jensen, M.C. & Meckling, W.H. 1976. Theory of the firm: Managerial behavior, agency costs and ownership structure. *Journal of financial economics* 3(4): 305–360.

Kallapur, S. & Trombley, M.A. (1999). The Association between Investment Opportunity Set Proxies and Realized Growth. *Journal of Business & Accounting* 26(3-4): 505–519.

Messier, W.F., Glover, S.M. & Prawitt, D.F. 2008. *Auditing & assurance services a systematic approach (6th edition)*. New York: McGraw-Hill Irwin.

Nasution, M. & Setiawan, D. 2007. Pengaruh corporate governance terhadap manajemen laba di industri perbankan Indonesia. *Simposium Nasional Akuntansi X* 1(1): 1–26.

Nurazmi, Handajani, L. & Effendy, L. 2015. Dampak adopsi ifrs terhadap manajemen laba serta peran mekanisme corporate governance pada perbankan Indonesia. *Jurnal Simposium Nasional Akuntansi* 18.

Oosterbosch, R. 2009. Earnings management in the banking industry. *Economics*.

Petreski, M. 2005. The Impact of international accounting standards on firms. *Financial Accounting and Reporting Section (FARS) Meeting Paper.*.

Pratama & Ratnaningsih. (2013). Perbedaan kualitas laba sebelum dan sesudah adopsi IAS 39 pada perusahaan perbankan yang terdaftar di Bursa Efek Indonesia. *Jurnal Universitas Atma Jaya Yogyakarta*.

Scott, R.W. 2009. *Financial accounting theory*, ;5th ed. Upper Saddle River, NJ: Prentice Hall.

Shleifer, A. & Vishny. R.W. 1997. A survey of corporate governance. *Journal of Finance* 52(2): 737–783.

Siregar, S.V.N. & Utama, S. 2006. Pengaruh struktur kepemilikan, ukuran perusahaan, dan praktik corporate gorvernance terhadap pengelolaan laba (earnings management). *The Indonesian Journal of Accounting Research* 9(3).

Sudarmadji, S. & Sularto, L. 2007. Pengaruh ukuran perusahaan, profitabilitas, dan tipe kepemilikan perusahaan terhadap luas voluntary disclousure laporan keuangan tahunan. *Jurnal Akuntansi dan Bisnis* 2(2): 149–155.

Watts, R.L. 2003. Conservatism in Accounting part I: Explanations and Implications. *Accounting Horizon* 17 (3): 207–221.

The Sharia microfinancial institution as an option to social investment decisions in a disruptive era

A.P.B. Eka & N. Nugraha
Universitas Pendidikan Indonesia, Bandung, Indonesia

ABSTRACT: Islamic financial practices have become the main motor of Islamic economics, which is believed to be a potential alternative for the development of a more applicable and sustainable economic system in this disruptive era. One of financial practices based on Sharia or Islamic Compliances was Sharia Microfinance Institutions (MFIs), which had been appearing throughout Indonesia. Indonesia's Sharia financial industry style is driven by community preferences or market forces from the demand side, making the industry grow bottom-up, which provides a strong foundation for further development. Since its financial activities are registered and overseen at the Financial Services Authority (OJK), the community feels secure and comfortable to conduct financial transactions, especially for Muslim communities, in this disruptive era, owing to the impact of globalization in response to a rapidly changing business and economic environment (dynamic). The purpose of this study was to measure the correlations among factors that influenced the investment decisions of the community in Islamic MFIs in Indonesia, with funding placement activities, financing, and receipt of funds determined by financial performances of Sharia MFIs as the intervening variables. This study used quantitative methods of research. By collecting secondary data on a group of Islamic MFIs in Indonesia listed in the OJK, this research analyzed the effects of funding placements, financing, and receiving funds in Sharia MFI financial statements on the financial performance of Islamic MFIs that affected investment decisions over three years. This study used path analysis to measure the effect of the financial performances of Islamic MFIs on investment decisions so as to provide an overview to investors who will make financial investments in Sharia MFIs.

1 INTRODUCTION

The current period in which uncertainties and dynamic changes in circumstances have affected the business world is considered a disruptive era and has also influenced the development of investment. Islam is a religion that continues to grow, not only in the country of origin but also in Indonesia, as the country with the largest Muslim population in the world. The development of investment in Islam, or known as Sharia investment, had been growing very quickly in this disruptive era.

Investment in Islam was, as highly recommended, a muamallah activity Because of the investment, the property owned became productive and also brought benefits to others. In the surah At-Taubah (9:35) of the Qur'an, that stockpiling of property and assets was forbidden in Islam. The basic principle of Islamic investment referred to the Qur'an and hadith as the legal sources for Muslims. Islamic (Sharia) principled investment placed more emphasis on profit sharing and prohibited usury (Tatiana et al. 2015; Zainal et al. 2016).

Islamic (Sharia) investment is a privilege. In making investment decisions, Islam practices its individual and social visions, which are about keeping property, distributing property, developing the economy, developing society, and justice. Although

the investment decision-making activities are aimed toward obtaining a certain rate of return and risk, in Islam Sharia compliance with Islamic rules must also be considered. Investment activities must be based on individual responsibility and social effects.

The Islamic belief in responsibility transcended the goal of profitability alone and coincided with a new perception of the business that had recently been at stake in the advanced sectors of Western business and civil society. Far from the limits set by neoclassical thinking, this new wave implied a new kind of responsibility on behalf of the company falling under the rubric of corporate social responsibility (Segrado 2005).

Because the ultimate goal was to maximize social benefits compared to maximizing profits, through the creation of healthier financial institutions that could provide effective financial services as well as at the grassroots level, some authors (Al-Harran 1996) argued that Islamic finance, if inserted in a new paradigm, could have been a viable alternative to a socioeconomic crisis associated with a Western paradigm (Segrado 2005).

As stated earlier, relatively few studies and experiences in the field concentrated on Sharia microfinance or Islamic microfinance institutions (IMFIs). Among the most comprehensive studies on this topic, Dhumale and Sapcanin (1998) developed technical notes in which they attempted to analyze how to combine Islamic banking with microfinance. They considered the three main Islamic financial instruments (mudharabah, musharakah, and murabaha) to try to use them as tools for designing successful microfinance programs (Segrado 2005).

The Islamic approach placed great emphasis on micro-development efforts through financial and nonfinancial assistance and adherence to the principles of transparency, empathy, and cooperation. At the same time, diverse microfinance products and services should not violate Sharia norms and restrictions. The Islamic approach to poverty reduction was more inclusive than the conventional one. It provided basic and ongoing conditions for successful microfinance, integrating wealth creation with empathy for the poorest. It also followed that the Islamic approach was merged (Obaidullah 2008).

In this study, although Islamic MFIs were inseparable from all the rules that are in accordance with principles of Islamic law, and in accordance with the guidance of the Qur'an and hadith, the general picture did not view Islam as a religion in ritual form. Rather it held a broader perspective of the financial and investment actors in carrying out an understanding of perspectives that brought more goodness and higher value than simply carrying out financial activities and conventional investment. These includedIslamic finance principles such as empowering the people to use the economy not only for worldly purposes (Abdul Rahman 2008; Obaidullah 2008) and viewing Islamic LKM as one of the ways to alleviate poverty and provide investment to benefit the afterlife (Zainal et al. 2016).

Globally, the relevance of MFIs had been accepted in practice for Islamic banks. The Islamic banking system had an internal dimension that promoted financing activities for the poor, because it was in a financial trajectory that was supported by the power of Sharia's command. These Sharia orders established Islamic financial transactions with genuine concern for eradicating poverty, and at the same time promoting social justice and wealth distribution by prohibiting involvement in illegal acts or activities that were detrimental to social and environmental welfare (Rahim & Rahman 2007; Wajdi Dusuki 2008).

The financial performance shown in the financial statements of Islamic financial practices was thought to be important in making investment decisions. Various Islamic finance practices in Indonesia had grown rapidly. One that attracted attention was the Islamic MFIs, the transactions of which were mostly in the microcommunity environment, for both microfinance and investment in micro-businesses in the form of fund-raising products, namely deposits and investments, as well as product distribution of funds in the form of profit sharing, buying and selling, rents, and loans (Darsono & Sakti 2017).

Several approaches were used to fund Islamic MFIs that were carried out by collecting and using funds sourced from the MFIs' own institutions, the community, and from financial or other institutions (Ahmed 2007; Sharia Professional Education 2007), such as those originating from waqf, zakat, shadaqah, and others that would have been used and allocated to Funding Placement, funding, and Fund Receipts (IKIT 2018).

As one of the financial institutions, Islamic MFIs cannot be separated from financial performance assessment, which showed efficiency and effectiveness as a basis for assessing financial successes and failures in accordance with institutional objectives. Financial Performance was described in the Financial Report. In Indonesia, Financial Performance Reports of Sharia MFIs presented a Statement of Financial Positionthroughout Indonesia that is combined into one report that presents the amount of Fund Placement, Financing, and Fund Receipts. In some countries, financial performance of Islamic MFIs was still better than that of conventional banks and Islamic banks, because the benefits and values offered were more than just profits, namely benefits for the common good (Ahmed 2002, 2007), and institutional finance was managed with low costs compared to non-Sharia financial institutions (Wajdi Dusuki 2008). Developments in the State of Pakistan show that the growth rate of the number of borrowers in Islamic MFIs is approximately 1,200,000 active borrowers and continues to increase (Siddiq 2008).

From the results of empirical studies, the decision to invest in Islamic finance was influenced by investor behavior in enforcing compliance with Sharia principles (Tahir & Brimble 2011; Adam & Shauki 2012; Azmat et al. 2016).

From the scientifically and empirically supported facts about the conditions of Islamic MFIs in Indonesia, the authors were interested in conducting research on investment decisions that are influenced by the financial performance of Islamic MFIs by considering the use of financial resources, so that in the future people will make an alternative Sharia investment, in accordance with Islamic norms and rules.

We expected that this study would appear as novel not only in terms of factors that influence investment decisions in terms of product type and contract in financial transactions in investment, as well as investor behavior, but also in that investment could be done by all people in the community, especially small communities, through information obtained from the financial performance of Islamic MFIs through transactions in fund placements, financing, and receipt of funding so that investment decisions based on Sharia principles could be taken.

1.1 Sharia finance

The concept of Sharia financial institutions, explicitly not mentioned in the Qur'an but if financial institutions had elements, such as structure, management, functions, rights and obligations, they were mentioned in words such as peoples, ummah, balad, suq, and so on that indicated the existence of certain functions and roles in society. They were also present in the economic concept that referred to the words zakat, shadaqah, fa'i, ghanimah, mal, dain, and so on (Zainal et al. 2016). The following were types of contracts in Islamic financial institutions contra:

1. Mudharabah, was a microfinance program and micro-business partner, with a program of investing money and micro-entrepreneurs early in its establishment. Micro-entrepreneurs were rewarded for their work and share profits while the program only shares profits. Of course this model presented a series of difficulties, largely caused by the fact that micro-entrepreneurs usually did not maintain accurate accountability. It was more difficult to determine the correct share of profits. As stated earlier, these models were complex to understand, manage, and handle, which implied that those involved need special training on these issues. For this reason, and for the easier management of revenue-sharing schemes, the mudharabah model might be more direct for businesses with longer profit cycles.
2. The murabahah model operated under a contract in which the microfinance program bought goods and resold them to a micro-enterprise for the cost of goods plus markup for administrative costs.
3. Borrowers often paid for items in equal amounts of installments, and microfinance programs kept the goods until the last installment was paid.
4. Musharakah contracts could be used to provide additional working capital for companies or for joint investment, for example, in real estate or agriculture. This type of contract was often used to finance long-term investment projects (Tatiana et al. 2015).

1.2 Sharia investment

Investments were made using funds that were not consumed at the present time. Islamic investment was a part of investment activities that took into account Islamic ethical values. Islamic Ethical Investment was often paralleled by Social Responsible Investment because it combined the consideration of profit and morality (Zainal et al. 2016).

According to Chapra, there are several ways suggested in Islam to carry out capital development:

1. Sole proprietorship
2. A combination of personal ownership and cooperation
3. Joint ventures
4. Syirkah (Zainal et al. 2016)

Principles of Sharia investment were halal (lawful) and maslahah (Sharia Professional Education 2007; Muhammad 2016).

1.3 Financial performance

Fund Placement consisted of placement of funds in the Sharia MFIs in the form of savings, and the funding for this deposit was sourced from the community (IKIT 2018). Financing was the provision of funds or bills equivalent to the financing derived from mudharabah and musyarakah contracts. The source was the community.

Receipt of funds or Fund Receiving, wasrecognized as debt or liability. The usual source of funds in the form of those collected from the community could also be obtained from other financial institutions (IKIT 2018).

Financial performance was the result achieved by financial institutions in managing their resourcesto achieve their goals. The ratio used Return on Assets (ROA). ROA was the ratio between earning before tax (EBT) to average total assets (Mersland & Strom 2007).

1.4 Investment decisions

Investment decisions are the decisions to invest funds made by a company into an asset in the hope of obtaining income in the future in accordance with Islamic legal norms. The income is measured by ROE, achieved by a useful financial institution for shareholders by determining profit generated from financial institutions using their average capital (Prastowo 2014; Zainal et al. 2016).

Among theories about investment decisions, one was based on signaling theory. According to Brigham and Houston the signalis an action taken by the company to provide clues to investors about how management views the prospects of the company. This signal provided information about what had been done by the management to realize the objectives of the owner. Information issued by the company is important because the effect on investment decisions comes from outside the company. Such information is important to investors and business people because it essentially presents data, notes or figures, whether for the past, present, or future circumstances regarding the sustainability of the company and its effect on the company. Signaling theory explained why companies had an incentive to provide financial statement information to external parties.

The company is encouraged to provide information because there is asymmetry of information between the company and external actors. The company knows more about the company itself (internal) and prospects than external actors would (Connelly 2011).

In relation to the theory of Islamic investment, the fact that the investment expenditure was made, signals, especially to the investor or borrowers, that the institution will grow in the future, by considering the return to be received and certainly the most profitable and appropriate compliance with Sharia. In this case the ROA was a good signal to determine the company's financial performance and Return on Equity (ROE) was one of the good signals on investment decisions. Likewise with proxy theory to measure investment decisions, using a proxy indicates the existence of options or investment opportunities, because the decision could not be observed. The proxy that was commonly used was the Investment Opportunity Set, and referred to a rate of return that was assumed by taking into account the assets owned or the capital (Zainal et al. 2016).

1.5 Sharia microfinance institutions

Microfinance was constituted by a range of financial services for people who are traditionally considered non-bankable (unfit bank), mainly because they lack the guarantees that could protect a financial institution against a loss risk.

The real revolution of microfinance was that this tool provided an opportunity for people who has been deniedaccess to the financial market, opened new perspectives, and empowered people who could finally carry out their own projects and ideas with their own resources, and outside of assistance, subsidies, and dependence. Microfinance experiences all around the world had definitely proved that the poor demanded a wide range of financial services, were willing to bear the expenses related to them, and were absolutely bankable.

The target group of microfinance was not constituted by the poorest of the poor, who need other interventions such as food and health security, but by those poor who lived at the border of the so-called poverty line. These individuals could more easily attain a decent quality of life and had entrepreneurial ideas but lacked access to formal finance (Segrado 2005).

The difference between a conventional MFI (non-Sharia with a Sharia LKM is shown below.

1. Conventional MFI
   – Liabilities (sources of funds): external funds, and savings of clients
   – Assets (mode of financing): interest-based
   – Financing the poorest: poorest are left out
   – Funds transfer: cash given
   – Deductions at inception of contract: part of the funds deducted at inception
   – Target group: women
   – Objective of targeting women: empowerment of women
   – Liability of the loan (when given to women): recipien
   – Work incentive of employees: monetary
   – Dealing with default: group/center pressure and threats
   – Social development program: secular (or un-lslamic) behavioral, ethical, and social development

2. Islamic MFI
   – Liabilities (sources of funds): external funds, savings of clients, and Islamic charitable sources
   – Assets (mode of financing): Islamic financial instruments
   – Financing the poorest: poorest can be included by integrating -akaft with microfinancing
   – Funds transfer: good transferred
   – Deductions at inception of contract: jSlo deductions at inception
   – Target group: family
   – Objective of targeting women: ease of availability
   – Liability of the loan (when given to women): recipient and spouse
   – Work incentive of employees: monetary and religious
   – Dealing with default: group/center/spouse guarantee, and Islamic ethics.
   – Social development program: religious (includes behavior, ethics, and social)

2 METHOD

2.1 Location and time of research

The study was conducted on the Financial Services Authority's website (OJK) from the Sharia Non-Bank Financial Institutions (IKNB) data over40 working days from April to May 2018.

2.2 Population and sample

The population and sample for this research comprised all financial statements of Sharia Micro Finance Institution registered in OJK for which permission had been obtained. This study used purposive sampling of data from 167 Sharia IMFIs over the period October 2015 to March 2018.

2.3 Data collection and data types

The data in this study were collected
using nonparticipant observation, and the type of data was the type of secondary data, namely data in the form of ratios and numerical values. Data were obtained from financial reports that have been published on the official website of OJK. Fund Placement Transactions, Financing, and Fund Receipts in 167 Sharia LKMs are registered in OJK.

Data regarding Fund Placement, Fund Receipt, and Financing transactions were obtained during the period of October 2015 to March 2018 in billions of rupiah and processed in graphical form to see how growth and numbers are increasing for fund placements and financing. As for the receipt of funds obtained from funds raised despite having experienced a decline, the trend will increase at the end of the period.

Financial performance data were calculated by the ROA while the investment decision was calculated using the proxyROE.

In financial performance, there were significant increases over time, while investment decisions were more stable at the level of 8%, above the average rate of Indonesian bank deposits for the 2015–2018 period of 4–6%.

2.4 Research design and data analysis

The data analysis used quantitative descriptive analysis, by using path analysis to create a multiple regression model. There is no classical assumption test because data acquisition is secondary data (given), so that the validity, reliability, and normality of test data are not provided (Pardede & Manurung 2014; Sugiyono 2016).

Multiple regression analysis was used to determine the effect of partial and simultaneous. Related to regression analysis, by paying attention to the regression equation, prediction value, coefficient of determination, standard error of estimation, standard error of regression coefficient, value of F arithmetic and t-count value, conclusions could be drawn on whether an influence existed between the independent variables and the dependent variable.

Correlation analysis was used to learn the degree of linear relationship between one variable with other variables. If the change was unidirectional then the variables had a positive correlation; if the change was in the opposite direction the correlation was negative; and if the change of one variable was not followed by a change in the other variable then the variables were not correlated (Pardede & Manurung 2014). This research used product moment or Pearson correlation to understand the relation between variables when data had a scale interval or ratio, as shown with the formula

$$r_{xy} = \frac{n\Sigma X_i Y_i - (\Sigma X_i)(\Sigma Y_i)}{\left\{n\Sigma X_i^2 - (\Sigma X_i)^2\right\}\left\{\Sigma Y_i^2 - (\Sigma Y_i)^2\right\}} \quad (1)$$

where r_{xy} = correlation coefficient; n = number of observations; ΣX = number of observations from X value; ΣY = number of observations from the value of Y.

The correlation category was used to determine the category of correlation coefficients, but could not be used to determine the correlation's significance, and it also depended on the sample size and tolerance level used.

Correlation analysis was seen in terms of the level of significance to test whether there was a significant relationship between independent variables with each other.

The criteria for determining the weakness of a correlation are as follows:

1. 0.00–0.29 = Very weak
2. 0.30–0.49 = Weak
3. 0.50–0.69 = Enough
4. 0.70–0.79 = Strong
5. 0.80–1.00 = Very strong

The coefficient analysis of determination was used to test the goodness of fit, denoted by R^2. Hypothesis testing was carried out by using an F test and t test. An F test was used to test the coefficient (slope) regressions together, and a t test was used to test the regression coefficients including individual intercepts.

2.5 Operationalization of variables

The variables used in this research were $X1$= Pd = funding placement, consisting of fund placement in the Sharia MFIs in the form of savings and time deposits; $X2$ = Pb = financing was financing derived from mudharabah and musharakah contracts; $X3$ = Pn = receipt of funds was received funds recognized as debt/liability; Y = Kinerja = financial performance, measured by ROA; Z = KepInvest = investment decisions, which were the decisions by a company to invest funds into an asset asset in the hope of obtaining future revenue in accordance with Islamic legal norms, measured by ROE (Zainal et al. 2016).

2.6 Hypothesis

Based on the formulation of the research hypothesis, problems were:

1. Ho1 = There was no simultaneous influence of Fund Placement, Financing, and Revenue on Financial Performance.
2. Ha1 = There was a simultaneous influence of Fund Placement, Financing, and Acceptance on Financial Performance.
3. Ho2 = There was no simultaneous influence of fund placement, financing, receipt of funds, and financial performance on investmentdecision.
4. Ha2 = There was a simultaneous influence of fund placement, funding, receipts of fund, and financial performance on investment decision.

3 RESULTS AND DISCUSSION

Based on the results of data processing and statistical data analysis, the results of the findings are shown in Table 1.

Based on the formulation of the problem designed into the research objectives and framework of thought, the model of this study is illustrated in Figure 1.

Table 1. Statistical data.

Variable	N	Min	Max	Mean	Std. deviation
Pd	30	3.000	40.120	22.73267	9.680571
Pb	30	1.000	12.460	9.00733	2.738780
Pn	30	1.330	10.000	4.33967	2.725287
Performance	30	0.000	0.013	0.01007	0.003073
Investment decision	30	0.000	0.081	0.06293	0.018682
Valid N (listwise)	30				

From the structural equation model:

$$Y = pYx1 + pYx2 + pYx3 + elZ = PxZlx1 + pZx1 + pZx1 + e2 \qquad (2)$$

The results of analysis of each structure of equation using regression analysis and ANOVA test are shown in Table 2.

This showed that a simultaneous influence of Fund Placement, Financing, and Receipts of Fund on Financial Performance had an effect of of 0.774 (R^2) with determination coefficient of 77.4%, while the remaining 22.6% was influenced by other factors. The simultaneous influence of Fund Placement, Financing, Revenue, and Financial Performance had an effect of 0.702 (R^2) on investment decisions with a determination coefficient of 70.2%, while the remaining 29,8% was influenced by other factors. A hypothesis test was done by testing the following criteria:

1. Test F: if F arithmetic > F table, then Ho was rejected, Ha accepted; if F arithmetic < F table, then Ho accepted, Ha rejected.
2. Test T: if t arithmetic > t table, then Ho rejected, Ha accepted; if t arithmetic < t table, then Ho accepted, Ha rejected.

The hypothesis test results are shown in Table 3.

This showed the feasibility of the regression model of the hypothesis in the structural model of

Table 2. Regression analysis ANOVA.

Model	R	R^2	Determination	F	Sig
Coefficient					
Model 1	0.880	0.774	77.4%	29.641	0.000
Model 2	0.838	0.702	70.2%	14.719	0.000

Table 3. Hypothesis testing analysis.

Model	Unstandardized coefficients B	Std. error	Standardized β coefficients	t	Sig.
1					
(Constant)	6.41E-17	0.092		0	1
(pd)	0.642	0.205	0.642	3.125	0.004
(pn)	0.132	0.096	0.132	1.381	0.179

a. Dependent variable: financial performance

Model	B	Std. error	β coefficients	t	Sig.
2					
(Constant)	−1.42E-15	0.107			
(pd)	−0.945	0.282	−0.945	0	1
(pn)	−0.236	0.116	−0.236	−3.353	0.003
(pb)	0.871	0.248	0.871	−2.037	0.052
Kin	0.776	0.23	0.776	3.516	0.002

b. Dependent variable: investment decision

equation (1), which was obtained from F arithmetic = 29.641. In an F table for 3 degrees of freedom 3 (df = 3) and significance level 0.05, F table = 2.98, and F arithmetic > F table, which meant that the feasibility of the regression model answered the hypothesis, there was an influence, and the Ho hypothesis was rejected. (Ho1 was rejected and Ha1 accepted) and in the second equation model F arithmetic = 14.719. In an F table for 4 degrees of freedom 4 (df = 4) at the significance level of 0.005 was F table = 2.74, F arithmetic > F table, meaning that the model 2 regression equitably answered the

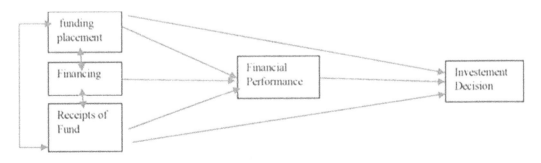

Figure 1. Research model.

hypothesis, there was influence, and rejected the Ho hypothesis (Ho2 rejected and Ha2 accepted).

To test partially influence from t test analysis from SPSS 24 data processing result that, Placement of fund to performance obtained t count equal to 3.125 meant of t table at level of significance 0.05 obtained t table = 2.048. This meant t arithmetic > t table, indicating that there was influence Placement of funds with financial performance, which amounted to 0.642, meaning if, the Placement Funding rose or fell 1 unit then the Financial Performance will rise or fall by 64.2%. While for Financing and Revenue from t count equal to 1.381 and 1.361, where t table equal to 2.048, meaning t arithmetic < t table, so accept hypothesis Ho, which means there is no influence Financing to Financial Performance and no Influence of Revenue (Receipt of Fund) to Financial Performance.

To test the partial influence between Fund Placement with Investment Decision obtained t count equal to -3.353, meant t arithmetic < t table, so there was no influence between placement of fund to investment decision, as well as receipt of fund, obtained t count equal to -2.037, t arithmetic < t table of 2.048. Fund receipts have no effect on investment decisions.

The test of partial influence between financing to investment decisions and financial performance on investment decisions obtained t calculations totaling 3.516 and 3.381, indicating that t calculated financing > t table, and t arithmetic Financial performance > t table, proving there was a partial influence between financing with Investment Decision, and Financial Performance with Investment Decision. This meant that every increase or decrease 1 unit of financing, investment decisions would rise or fell by

87.1% and each increase or decrease 1 unit of financial performance, investment decisions would rise or fell by 77.6%.

The criteria of correlation analysis to determine the level of significance of influence between independent variables with the others was defined as follows: if Sig. 2-tailed > 0.05 then there was no significant effect; if Sig. 2-tailed < 0.05 then the level of influence was significant. Results of correlation analysis indicated that:

1. The correlation between Fund Placement and Financing, 0.890, was very strong and unidirectional, meaning that if the Fund Placement increased then the financing would also increase. The relationship between Fund Placement and Financing was significant, at 0.00 < 0.05.
2. The correlation between Fund Placement and Receipt of Funds, wasof -0.214, was very weak and not directional, and the level of significance was also not significant, 0.256 > 0.05.
3. The correlation of Financing and Clerical Admission of Funds of -1.96 was very weak and not directional, nor did it have a significant relationship (Sig. 2-tailed 0.3 > 0.05).
4. The correlation of the placement of funds with financial performance, 0.86, was very strongly positive and unidirectional. Sig. 2-tailed 0.00 < 0.05 means the relationship of influence was significant.
5. The correlation of Financing with Financial Performance, of 0.823, was very strongly positive and unidirectional. Sig, research 0.00 < 0.05 indicates a significant influence.
6. Correlation of Fund Receipts with Financial Performance, amounting to -0.06, showed the relationship was not unidirectional and very weak. Sig. study 0.7 > 0.05 meant the relationship was not significant. From the analysis results the following equation was obtained:

$$\text{Final Performance} = 0.642\text{Pd} + 0.132\text{Pn} \\ + 0.278\text{Pb} + 0.226 \tag{3}$$

$$\text{Investment Decision} = 0.871\text{Pb} - 0.945\text{Pd} \\ - 0.236\text{Pn} + 0.298 \tag{4}$$

Calculation of the influence in total showed the following correlations:

1. Receipt of Funds to Financial Performance to Investment Decision, of 1.418
2. Receipts of Funds to Financial Performance to Decision Investment amounted to 0.908.
3. Financing to Financial Performance to Investment Decision of 1.054.
4. Funding Placement to Investment Decision of -0.945
5. Financing to Investment decision of 0.871
6. Financial Performance to Investment Decision of 0.776

Table 4. Correlation analysis.

		Pd	Pb	Pn	Financial performance
Pd	Pearson correlation	1	0.890**	-0.214	0.861**
	Sig. (2-tailed)		0	0.256	0
	N	30	30	30	30
Pb	Pearson correlation	0.890**	1	-0.196	0.823**
	Sig. (2-tailed)	0		0.3	0
	N	30	30	30	30
Pn	Pearson correlation	-0.214	-0.196	1	-0.06
	Sig. (2-tailed)	0.256	0.3		0.753
	N	30	30	30	30
Financial performance	Pearson correlation	0.861**	0.823**	-0.06	1
	Sig. (2-tailed)	0	0	0.753	
	N	30	30	30	30

** Correlation is significant at the 0.01 level (2-tailed).

The results of the analysis and the findings that the correlation is very strong between the factors of Fund Placement, Financing, and Fund receipts and performance and investment decisions also respond to the discussion and development of research ideas based on empirical studies conducted by previous researchers demonstrating that Islamic MFIs effectively support the economic activities of small communities (Alhifni & Huda 2015). As studies show that Islamic MFIs have the potential to raise public funds to be invested in Islamic MFIs (Siddiq 2008; Wajdi Dusuki 2008; Murwanti & Sholahuddin 2013) and have become topics forfor future research it would be beneficial for the development of Sharia MFIs to receive special attention from the government because of their potential to improve financial performance, which has also been proven by Mersland and Strom (2007).

4 CONCLUSIONS

From the results of analysis and discussion that answered the formulation of the problem, the study concluded that the effects of Fund Placement, Financing, and Fund Receipts on the financial performance of Islamic MFIs partially was that placement of funds in Islamic MFIs affects the financial performance in a unidirectional and positive direction, while for the Receipt of Funds and Financing, there was no influence on the financial performance.

However, when Fund Placement, Financing, and Fund Admission were combined simultaneously, the effect on financial performance becomes strong and significant.

Partially, Fund Placement and Fund Admission had no effect on investment decisions, and the amount of influence was not significant, whereas Financing had a very significant positive and unidirectional influence. Fund Placement, Financing, and Admission of funds simultaneously had a strong influence on Sharia financial performance and investment decisions.

The effect of financial performance on investment decisions was strong and significant, with a correlation of 77.6%.

This strong level of influence indicated that if the Sharia MFI performed well then the investment decision would be on target, and this could be one of the alternative investment options, because with a good level of performance, the ability of Islamic MFIs to produce expected returns or profits would increase.

Limitations of the study were that it did not measure the common factors in investment decision making as in conventional financial theory, but rather the elements contained in the Sharia MFIs Financial Statements that were subject to Sharia compliance, as well as investor behavior factors that were not investigated in this study.

Islamic MFIs should develop products and contracts in accordance with the market and the vision of developing the Indonesian Islamic financial industry, which focuses on achieving optimal benefits for greater economic growth to the public through Islamic financial products and the benefits that Islamic financial institutions can provide in microfinance institutions

The choice to invest through Sharia MFIs, which are directly related to the community and investors in the company or the micro-enterprise and connected to the economic joints of the lower layers, is a good one in times of an uncertain and highly dynamic economy, and can not only create benefits locally but can also contribute to the social welfare of other societies and communities.

REFERENCES

Abdul Rahman, A.R. 2008. Islamic microfinance: an ethical alternative to poverty alleviation. *ECER Regional Conference.*

Adam, A.A. & Shauki, E. 2012. Socially responsible investment in Malaysia: Behavioural framework to evaluate investors' decision-making process. *International Graduate School of Business Journal* 80(2): 224–240.

Ahmed, H. 2002. Financing microenterprises: An analytical study of islamic microfinance institutions. *Islamic Economic Studies* 9(2): 27–64.

Ahmed, H. 2007. Waqf-based microfinance: Realizing the social role of islamic finance "Integrating awqaf in the islamic financial sector." *Integrating Awqaf in the Islamic Financial Sector*: 1–22.

Al-Harran, S. 1996. Islamic finance needs a new paradigm. *New Horizon* 48: 7–9.

Alhifni, A. & Huda, N. 2015. Kinerja LKMS dalam mendukung kegiatan ekonomi rakyat berbasis pesantren (Studi pondok pesantren darut tauhid dan BMT Darut Tauhid). *Jurnal Aplikasi Manajemen* 13(4): 597–609.

Azmat, S., Jalil, M.N., Skully, M. & Brown, K. 2016. Investor's choice of Shariah compliant replicas and original Islamic instruments. *Journal of Economic Behavior and Organization* 132: 4–22.

Connelly, B. 2011. Signaling theory: A review and assessment.

Darsono, D. & Sakti, A. 2017. *Dinamika produk dan akad keuangan syariah di Indonesia (1st. ed)* . Depok: Raja-grafindo Persada.

Dhumale, R. & Sapcanin, A. 1998. An Application of Islamic Banking Principles to microfinance. *Study by the Regional Bureau for Arab States, United Nations Development Programme, in Cooperation with the Middle East and North Africa Region*: 1–14.

IKIT. 2018. *Manajemen dana bank syariah (1st ed.).* Yogyakarta: Gava Media.

Mersland, R. & Strom, R.O. 2007. Performance and corporate governance in microfinance institutions. *MPRA Paper* (3887): 37.

Muhammad, M. 2016. *Manajemen keuangan syariah (2nd ed).* Yogyakarta: UPP STIM YKPN.

Murwanti, S. & Sholahuddin, M. 2013. Peran keuangan lembaga mikro syariah untuk usaha mikro di Wonogiri. *Call Paper Sancall*: 300–309.

Obaidullah, M. 2008. *Introduction to Islamic microfinance. New Delhi.* India: IBF Net.

Pardede, R. & Manurung, R. 2014. *Analisis jalur teori dan aplikasi dalam riset bisnis (1st ed.).* Jakarta: PT Rinea Capta.

Prastowo, D. 2014. *Analisis laporan keuangan konsep dan aplikasi (3rd ed.).* Yogyakarta: Unit Penerbit danPercetakan STIM YKPN.

Rahim, A. & Rahman, A. 2007. Islamic economics: Theoretical and practical perspectives in a global context. Islamic Microfinance: A Missing Component in Islamic Banking. *Kurenai* 1(2): 38–53.

Segrado, C. 2005. Case study "Islamic microfinance and socially responsible investments." *MEDA Project*: 1–20.

Sharia Professional Education. 2007. *Keuangan dan investasi syariah (2nd ed.).* Jakarta: Renaisan.

Siddiq, K. 2008. Potential of islamic microfinance in Pakistan. *Dissertation.* Management.

Sugiyono, S. 2016. *Metode penelitian manajemen (5th ed.).* Bandung: Alfabeta.

Tahir, I. & Brimble, M. 2011. Islamic investment behaviour. *International Journal of Islamic and Middle Eastern Finance and Management* 4(2): 116–130.

Tatiana, N., Igor, K. & Liliya, S. 2015. Principles and Instruments of Islamic Financial Institutions. *Procedia Economics and Finance* 24: 479–484.

Wajdi Dusuki, A. 2008. Banking for the poor: The role of Islamic banking in microfinance initiatives. *Humanomics* 24(1): 49–66.

Zainal, V.R., Waluyanto, R., Veithzal, A. P. & Handoko, D. 2016. *Manajemen investasi Islami* (1st ed.). Yogyakarta: BPFE-Yogyakarta.

Advances in Business, Management and Entrepreneurship – Hurriyati et al (eds)
© 2020 Taylor & Francis Group, London, ISBN 978-0-367-27176-3

Corporate governance and finance pattern

W. Windijarto & G. Gestanti
Airlangga University, Surabaya, Indonesia

ABSTRACT: This research aimed to analyze the effect of corporate governance on the finance pattern for nonfinancial going public firms that were listed on the Indonesia Stock Exchange for 2010–2014. Linear multiple regression was used in this study. The dependent variables are the finance pattern (FP) Retained Earnings (RE) and FP Debt. The independent variables are board of director (BOD) share ownership, independent commissioner, board size, and block holder. BOD ownership and profitability had a significant positive effect on FP RE, whereas block holder and the size of the company had a significant negative effect. Independent commissioner, board size, and tangibility had no significant effect on FP RE; firm size and tangibility had a significant positive effect on FP Debt; board size and profitability in the previous period had a significant negative effect on FP Debt; BOD ownership, as well as independent commissioner and block holder, had no significant effect on FP Debt.

1 INTRODUCTION

According to Corbett and Jenkinson (1996), a company has a finance pattern within a certain time. According to Turkalj and Srzentic (2011), the finance pattern shows the use of one source of funds (retained earnings, debt, or stock) as the primary source or that has the highest composition for several consecutive periods. According to Moeinaddin and Karimianrad (2012), there are three types of finance pattern (FP): FP Retained Earnings (RE), Debt, and Equity.

Based on the results of data processing of financial statements of going public firms on the Indonesia Stock Exchange during 2009–2014, there are 21 firms with FP RE, 32 firms with FP Debt, and 4 firms with FP Equity.

This study is interesting because the literature on finance patterns is limited. The finance pattern is explained using research related only to financial structure and capital structure because thethe formation of the finance pattern begins with the option of financial structure combination. However, because of limited sample size, the finance pattern of equity is not analyzed.

According to Myers (1984), the finance pattern demonstrates a tendency of managers to feel comfortable choosing a particular funding source. This convenience could lead to the occurrence of agency problems because it does not maximize the shareholder value and prioritizes the utility of a comfortable position for the manager. According to Raza et al. (2013), the use of a false finance pattern can cause the firm to be worse off and cause financial distress and even bankruptcy.

According to Berger et al. (1997), Abor (2007), and Moeinaddin and Karimianrad (2012), corporate governance can influence the option of funding sources, thus forming a corporate finance pattern. The application of good corporate governance to the FP Debt will be able to restore debt to the optimum level (Berger et al. 1997; Abor 2007). This mechanism will align the interests of shareholders and managers (through managerial ownership) and oversee managerial decision-making (through the number of independent commissioners and block holder ownership), which will affect the proportion of retained earnings or debt. In this study the variables board of director share ownership, independent commissioner, board size, and block holder ownership are used as a proxy of corporate governance.

Furthermore, the firm has flexibility in selecting a different finance pattern according to the company's condition (Ross et al. 2008). The characteristics of the company will be able to explain the emergence of the finance pattern. According to Berger et al. (1997), Wen et al. (2002), and Abor (2007), firm characteristics will affect the company's capital structure. In this study, firm characteristics are used as control variables such as firm size, tangibility, and profitability.

Based on this explanation and phenomenon, the researcher is interested in knowing the influence of corporate governance on finance pattern in the going public firms of the nonfinancial sector in Indonesia during 2010–2014.

1.1 Finance pattern retained earnings

FP RE shows retained earnings have been used as the main funding source; it is also called pecking order pattern (Singh & Hamid 1992). According to Myers (1984) in Singh (1996) the rational manager will adopt this pattern. Such actions will make managers more manageable in obtaining financial sources. According to Myers (1984), when managers adopt pecking order theory, they will adapt the amount of dividend payments, so that profits can be reinvested. According to

Ross et al. (2008), this pattern will lead to financial slack, or the tendency of not optimizing the option of funds.

1.2 Finance pattern debt

Moeinaddin and Karimianrad (2012) show that debt represents the largest proportion used as a source of funds compared with other funding sources This condition raises a trade off, because finance pattern financing firms bear a higher risk of bankruptcy, as well as pay higher interest, but on the other hand the company can also save taxes and obtain a bonding mechanism.

$$\text{FP Debt}_{it} = \textbf{Total Debt}_{it} \textbf{Total Asset}_{it} \quad (1)$$

1.3 Board of director share ownership

A board of directors (BOD) can be a risk taker or risk averse. When risk averse the BOD will cause a problem called managerial entrenchment (Berger et al. 1997). When the BOD is a risk taker it wants to continue to expand the company or engage in empire building, so that a BOD that owns shares is expected to position itself with and have the same goals as shareholders, especially minorities. The number of BOD shareholdings in Indonesia tends to be small because large holdings of BOD shares no longer represent minorities (Shleifer & Vishny 1997). BOD shareholding aims to make managers also position themselves as shareholders.

Total Share Ownership of Board of Directors

it

Number of Outstanding Shares $_{it}$

$$(2)$$

According to Shleifer and Vishny (1997), a BOD that own shares will tend to maximize capital structure. Initially, the BOD as a manager tends to take the decision of low debt level, under the optimal point (Berger et al. 1997). After the BOD owns the stock, positioning itself like a shareholder, it wants to achieve maximum profit by increasing the debt to the optimum point. In this research, if the debt level of FP Debt is high, then the BOD that owns the shares will try to reduce the debt to the optimum level.

Hypothesis 1: BOD share ownership negatively affects FP Debt.

According to Jensen et al. (1992), insider (BOD) ownership has a negative influence on dividends. According to Florackis et al. (2015), when BOD ownership is <10%, it will negatively affect dividends and increase retained earnings. The small number of directors' shares causes the BOD to

position itself as a manager rather than a shareholder. As managers, the BOD tend to use profits for reinvestment rather than as dividends. Thus, the increase of share ownership of directors will increase retained earnings.

Hypothesis 2: BOD shareholding positively affects FP RE.

1.4 Independent commissioners

Independent commissioners are commissioners from outside the firm. Independent commissioners become part of the board of commissioners (Wen et al. 2002). Independent commissioners are expected to be mechanisms that protect minority shareholders. The minimum number of independent commissioners is 30% of the total number of board of commissioner members.

According to Berger et al. (1997), the managerial entrenchment of independent commissioners will improve by increasing the debt. This happens because the independent commissioner provides close supervision to the directors. Different research results are provided by Jensen (1986), Wen et al. (2002), and Abor (2007). Independent commissioners have a negative effect on debt. FP debt is already high, so the increase in the supervision by independent commissioners encourages directors to reduce the use of debt to the optimum point. Independent commissioners will negatively affect FP debt.

Hypothesis 3: Independent commissioners negatively affect FP Debt.

Independent commissioners will protect shareholders, especially minorities, by encouraging directors to use profits as dividends (Atmaja 2010) rather than reinvesting them in the form of retained earnings. This reduces the amount of funds managers can manage, especially when in the firm there are majority shareholders who will tend to lead the BOD to make riskier decisions through the RUPS.

$$\text{INDP}_{it} = \textbf{Total Member of Independent}$$
$$\textbf{Commissioner it}$$
$$\textbf{Total Member of Board} \quad (3)$$
$$\textbf{Commissioner it}$$

Hypothesis 4: Independent commissioners negatively affect to FP RE.

1.5 The size of the board of commissioners

Boards of commissioners cause free riders, conflicts, and agreement difficulties, so supervision of the BOD declines (Pelt 2013). The size of the board of commissioners is measured by

$$UDK_{it} = \text{£ Total Member Board of Commissioner}_{it}$$
$$(4)$$

where UDK = board of commissioners.

Jensen (1986) and Abor (2007)argue that board size (UDK) has a positive effect on the firm's debt when the amount of debt is high. An increase of UDK will have a negative effect on the performance of the board of commissioners because it causes a decrease in supervision of the BOD. In firms with a very high FP Debt ratio, through the supervision of the commissioner, the use of debt will fall. On the contrary, when the number of members of the board of commissioners is too high to be any longer effective and causes oversight of the BOD to be low, the board of commissioners will allow the debt level to increase.

Hypothesis 5: The size of the board of commissioners positively affects FP Debt.

According to Rozeff (1982), dividends are a mechanism to overcome agency problems. According to Pelt (2013), dividend positively affects board size. Dividend payments reduce the amount of money managed; if the block holder takes steps that aretoo risky, the dividend secures minority shareholders. The existence of the board of commissioners optimally encourages directors to use profit as a dividend payment rather than a reinvestment through retained earnings, especially in FP RE which already has enough internal funding. Conversely, UDK that is too high will cause suboptimal supervision of directors, thus allowing a decrease in the return of dividends that can increase the amount of retained earnings.

Hypothesis 6: The size of the board of commissioners has a positive effect on FP RE.

1.6 The concentration of ownership (block holder)

According to Kep-305/BEJ/07/2004, a block holder (controlling shareholder) is a shareholder with the number of shares ≥ 25% or the largest number of shares.

$$\text{Block holder}_{it} = \text{Y} \frac{\textbf{Controlling Shares}_{it}}{\textbf{Total of Outstanding Shares}_{it}}$$
$$(5)$$

The results of Berger et al. (1997) show that block holders have a significant positive effect on debt. When there is an agency problem between shareholders and managers, block holders can be used to address them. Block holders will guide the BOD to increase debt through voting rights in the RUPS, while La Porta et al. (1999) found a negative correlation between the concentration of ownership and the mechanism of investor protection. A block

holder who has a high risk profile will cause harm to minority shareholders (Alba & Djankov 1998), so that the increasing ownership of block holder shares will still increase the yet high debt level.

Hypothesis 7: Concentration of ownership positively affects FP Debt.

According to Harada and Nguyen (2011), concentration of ownership positively affects dividend payments and decreases retained earnings. This result shows that the controlling shareholder wants to enjoy the investment result because it is perceived that the firm has internal capital. This is good for minority shareholders, as it also achieves a refund. By contrast, according to La Porta et al. (1999), concentrations of ownership decrease dividends. When there is poor protection for a minority, the majority shareholder is in control and wants a return profit. In FP RE the controlling shareholder will want dividends and reduce the allowance for retained earnings, as it feels that internal funding is sufficient.

Hypothesis 8: Concentration of ownership negatively affects FP RE.

The larger the size of the firm the greater its wealth and the greater its ability to pay its obligations. According to Rajan and Zingales (1995), large firms tend to be more diversified and have a lower probability of bankruptcy because they are "too big to fail."

$$\text{Firm size}_{it} = \text{lnTotal Assets}_{it} \qquad (6)$$

According to Rajan and Zingales (1995) there is a negative influence in that the greater the firms the more transparent they are than the small firms and tend to like internal funding because other funding costs are more expensive. Meanwhile, according to Berger et al. (1997), Wen et al. (2002), and Abor (2007), size has a significant positive effect on debt. The size shows the wealth of the company; as seen by the creditor, the richer the firm, the greater the company's ability to pay off debt, so the greater the chance of the company getting a loan.

Based on the results of research by Megginson (1997) and Flo0rackis et al. (2015), firm size has a positive effect on dividends. The bigger the company, the richer it is and the greater is its ability to pay bigger dividends compared to small firms. However, based on the results of Harada and Nguyen (2006), large firms tend to pay lower dividends because large firms tend to fit the pecking order prediction.

Hypothesis 10: Firm size negatively affects FP RE.

Tangibility shows the amount of fixed assets the company can use for production. According to Jensen and Meckling (1976), when there is no guarantee there is a possibility of moral hazard, or in other words the BOD replaces a safe project with a risky project.

$$Tangibility_{it} = \frac{Fixed\ Asset\ or\ (\textbf{property, plant, and equipment})\ it}{\textbf{Total Assets}\ it}$$

$$(7)$$

Titman and Wessel (1988) state that tangibility has a positive effect on debt. Tangibility reflects the value of the firm's fixed assets. Assets are still used as collateral when the firm has debt. Increased assets as collateral will increase the amount of funds the firm can lend.

Hypothesis 11: Tangibility positively affects FP Debt.

According to Vo and Nguyen (2014), tangibility positively affects retained earnings. When tangible assets are part of high assets, the remaining short-term assets are low, so to secure the firm's condition, the company needs funds and retained earnings will be increased. In addition, an increase in tangibility or fixed assets will increase production and impact the increase in the amount of profits generated. When the firm embraces pecking order theory, there is an increase in profit used to pay off debt, and so it is recorded as a rise in retained earnings.

Hypothesis 12: Tangibility positively affects FP RE.

Profitability shows the firm's ability to generate profits using the source of funds owned (Sudana 2011). According to Wen et al. (2002) and Abor (2007), the greater the profitability, the greater the firm's capacity to finance itself and fulfill its obligations. In this study two kinds of Return on Assets (ROA) were used: ROA_{it} and ROA_{it-1}. ROA_{it} describes the use of retained earnings in FP RE firms, as retained earnings and dividends are residual decisions. Debt decisions are set at the beginning of management, so to get the debt, the firm sees the profitability of the previous period or ROA_{it-1} (Sartono 2001).

According to Titman and Wessels (1988) and Abor (2007), profit in the previous period had a negative effect on debt. A profitable firm may have easy access to credit to increase the amount of debt, but the BOD does not like the bonding mechanism and the risks it incurs when the company has debt, and the profitable company will tend to replace its debt with retained earnings. This management behavior is in accordance with pecking order theory; that is, the BOD as manager prefers internal rather than external funding sources. When the firm's ability to generate profits at this time increases, the the BOD will tend to use a profit increase that is retained earnings. Meanwhile, the greater the profitability in the previous period, the more the firm will reduce or pay off the firm's debt. Such actions alter the composition of liabilities, resulting in a reduced proportion of debt usage and increased retained earnings.

Hypothesis 13: The profitability of the previous period negatively affects FP Debt.

In addition to the preceding explanation, the increase in profitability will also increase retained earnings, because although according to Harada and Nguyen (2006) and Vo and Nguyen (2014), high profitability enables the firm to pay high dividends, when the payment is not up to 100% earnings, there is still a profit listed as an increase in retained earnings.

Hypothesis 14: Profitability positively affects FP RE.

2 METHOD AND ANALYSIS

2.1 Framework

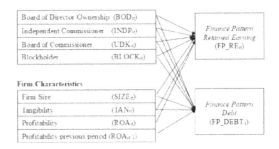

Figure 1. Research framework.

2.2 Analysis model

The analysis model in this research is as follows:

$$FP\_RE_{it} = a + b_1BOD_{it} + b_2INDP_{it} + b_3UDK_{it} \\ + b_4Block_{it} + b_5Size_{it} + b_6TAN_{it} \\ + b_7ROA_{it} + S_{it}$$

$$(8)$$

$$FP\_Debt_{it} = a + brBOD_{it} + b_2INDP_{it} + b_3UDK_{it} \\ + b_4BLOCK_{it} + b_5SIZE_{it} + b_6TAN_{it} \\ + b_7ROA_{it} - 1 + S_{it}$$

$$(9)$$

where FP RE = finance pattern retained earnings; FP DEBT = finance pattern debt; BOD = board of directors share ownership; INDP = independent commisioner; UDK = board of commissioner; BLOCK = block holder; SIZE = size; TAN = tangibility; ROA_{it} = profitabilty of firm i in year t; ROA_{it-1} = profitability of firm i in year $t − 1$; £ = error term; it = firm i in year t.

2.3 Research approach

This study uses a quantitative approach. It uses historical data to determine the effect of corporate governance on the finance pattern of non-financial going public firms during 2010–2014 in Indonesia.

Sampling using purposive sampling method with criteria:

1. Companies including the nonfinancial industry on the Indonesia Stock Exchange during 2009–2014.
2. The Firm has financial statements for each period of 31 December.
3. No loss on net income, or retained earnings.
4. In FP Debt firms: a long-term bank debt composition.

3 RESULTS AND DISCUSSION

3.1 Description of research results

Based on Table 1, retained earnings on corporate FP RE are 62.6%, and debt in an FP Debt company is 58.7%. The minimum values are 38.1% and 37%, which indicates a firm with a specific finance pattern but one that is still low, allowing for different directions to source optimal funds.

The total share ownership of BOD is at least zero and averages 0.001 for the FP RE, and 0.07 for the FP Debt. The number of BOD shares is small because if it is too large then this mechanism cannot be used to protect minority shareholders. INDP shows the number of independent commissioners from outside and is expected to be independent in the overall board of commissioners. In addition, the value shows that it has met the 30% requirement, but there are still firms that have not complied with the regulation.

3.2 Model analysis and hypothesis testing

Based on Table 2, the constant value of 1.173 means that if all independent variables and control value are zero, the value of FP RE is predicted to be 1.173. An Adj Value R^2 of 0.401 shows all independent variable and control can explain FP RE equal to 40.1% and the remaining 59.9% influenced by other variables not observed.

BOD share ownership has a significant positive effect on FP RE. Increased ownership of BOD shares will increase the use of retained earnings in FP RE firms. A positive influence shows the BOD prefers the use of profits to be reinvested instead of getting a dividend return. These results indicate that BOD tends to use its voting rights as a manager rather than as an owner. The results of this study are consistent with the research of Madaschi (2010) and Florackis et al. (2015). The number of small shareholdings (under 10%) will keep the directors behaving as managers rather than as shareholders. In addition, BOD also wants to realize empire building; that is, the temptation to grow the company becomes larger because it is more prestigious and generates higher payouts related to its size(Madaschi 2010). Thus, the presence of BOD shareholders will be an inappropriate corporate governance mechanism for FP RE firms.

Table 2. Results of multiple linear regression test.

	FP RE		FP Debt	
Variable	B	Sig.	B	Sig.*
Constant	1.173	0.000	0.305	0.053
BOD	9.333	0.012*	−0.014	0.966
UNDP	0.146	0.059	−0.022	0.781
UDK	0.023	0.335	−0.090	0.000*
BLOCK	−0.192	0.000*	0.054	0.12S
SIZE	−0.020	0.003*	0.013	0.027*
IAN	−0.004	0.921	0.113	0.004*
ROA	0.409	0.000*	—	–
ROA(.j	—	—	−0.352	0.006*
R^2	0.441		0.192	

Table 1. Statistical description of FP RE and Debt.

Variable	FP RE				FP Debt			
	Min	Max	Mean	SD	MId	Mai	Mean	SD
FP	0.5S1	0.004	0.020	0.07ft	0.570	0.S18	0.587	0.089
BOD	0.000	0.010	0/10]	0.002	0.000	0.032	0.007	0.021
INDP	0.112	0.750	0.507	0.077	0.2SQ	0.667	0.190	0.036
UDK	2	1!	5.51	2.14	2	12	4.8!	2.01
BLOCK	0.102	0.057	0.657	0.107	0.167	0.972	0.521	0.195
SIZE	26.440	41.695	2S.S94	1.412	25.659	55.005	2S.2S1	].551
TAN	0.05!	0.346	P.JJ0	0.175	0/J0J	0.700	0.277	0.172
KOA	0.o n	0J9G	0.155	0.070	—	*	—	—
ROA	—	—	0.155	*	0.001	0.508	0.059	0.054

BOD share ownership has no significant negative effects on the FP Debt. The results of this study are in line with those of Moeinaddin and Karimianrad (2012). The results of research are not significant, which means not all directors who have shares in the FP Debt firm will take the same decision to reduce debt, as a BOD manager does not like a high level of debt because it can be risky and tie up capital. According to Berger et al. (1997), the intensive form of shares given to BOD should give BOD a purpose and encourage it to behave like a shareholder by directing debt to the optimum level. Although the firm has an FP Debt, there are firms with a low level of debt, thus optimizing the use of different debt levels (Berger et al. 1997), such as MPPA, INDF, and INDR companies, so the company's BOD will increase the level of corporate debt. In addition, this is also because BOD is likely to use its rights as a shareholder to influence the use of other funding sources, such as retained earnings. BOD as a manager tends to favor internal or external funding sources as expressed by pecking order theory (Florackis et al. 2015).

Independent commissioners have no significant effects on FP REand FP Debt. The results of this study indicate characteristics of independent commissioners who do not have a particular interest in the company, so that some Commissioners cannot affect the source of corporate funds.

The size of the board of commissioners (UDK) has no significant effect on FP RE. This shows that not all members of the board of commissioners will increase the amount of high corporate retained earnings. These results are inconsistent with those of Pelt (2013) and Florackcis et al. (2015). In this study, the number of UDK averages five people in accordance with Pelt (2013); the number is optimal for supervision so there is no conflict of opinion when giving advice to the directors. Differences in giving advice and approval for the decision of the BOD due to the level of use of corporate retained earnings vary.

The size of the board of commissioners has a significant negative effect on FP Debt. The increase in the number of boards succeeded in reducing the high level of debt. It is inconsistent with Wen et al. (2002), Abor (2007), and Vakilifard et al. (2011). Based on research datain Table 2, the number of board of commissioners in a finance pattern firm, in accordance with Pelt (2013), is optimal, so it can provide supervision, suggestions, and approvals to reduce debt levels in an FP Debt firm. Although previously described there are companies that have an optimal level of debt usage, but when firms are making decisions about high debt then the board of commissioners will tend to reduce the risk of companies to protect shareholders whose positions are lower than those of creditors who can take guarantees and liquidate the company. Thus, this mechanism works well in finance pattern companies in Indonesia.

Block holders have a significant negative effect on FP RE. This result shows that the block holder as a professional decision maker realizes that the company's internal fund usage rate is yet very high, requiring a decrease in the use of retained earnings through dividend payments.

Based on data compiled by DLTA in 2010 and 2011, INCO 2012 and 2013, TINS and BRANDS in 2012 have paid dividends higher than earnings generated during that period to reduce the amount of retained earnings that are too high. In addition, the proportion of retained earnings may also be seen to decrease due to an increase in other external funding sources. Figure 2 shows FP RE companies having a trendline of retained earnings decline, as investors want dividends.

According to Harada and Nguyen (2006) it is done to reduce monitoring costs, because there is already a return on investment. This reduces the funds owned by shareholders who are still managed managers (Jensen 1986) and maximize the financial structure. Thus, the block holder may serve as a protection against minority shareholders, when there is an agency problem between the owner and the manager. These results are consistent with the research of Shleifer and Vishny (1986), but differ from those of Harada and Nguyen (2006) and Moeinaddin and Karimianrad (2012).

Block holders have no significant effects on FP debt. These results are consistent with the study of La Porta et al. (1998) showing that block holder in FP Debt firms have different views in using their voting rights. The use of the voting rights of the block holder is quite important, because in this position, the block holder is the shareholder with concentrated ownership characteristics.

Block holders can have other views, such as the current debt level is too high and needs to be lowered. Basically, shareholders do not like the existence of debt, because the creditor has a greater capacity to get a refund before the shareholders. According to Ross et al. (2008), an increase in debt raises debt costs and decreases the ability of firms to provide returns to shareholders, increasing bankruptcy claims.

The effects of firm characteristics on finance pattern
Size has a significant negative effect on FP RE. Size shows the firm's wealth. The results indicate an

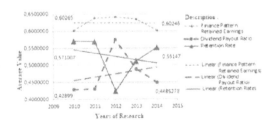

Figure 2. Dividend payout ratio and retention rate of FP RE.

484

increase in corporate wealth will encourage the firm's FP RE to reduce the proportion of the firm's use of retained earnings. In addition, it can also be interpreted as indicating that firms that have a more dominant retained earnings pattern are smaller. These results are consistent with the findings of Beck et al. (2007) and Moeinaddin and Karimianrad (2012), as well as inconsistent with research by Harada and Nguyen (2006).

According to Beck et al. (2007), large firms tend to have wider access to financial resources, and will tend to be less dependent on internal funding sources or retained earnings. In addition, the advantages of other sources of funding such as debt can be a tool to overcome agency problems and increase profits when in use.

Size has a significant positive effect on FP Debt. Size shows the company's wealth. These results indicate an increase in wealth owned by finance pattern companies will encourage companies to increase the proportion of debt usage. In other words, these results also show companies that become FP Debt firms tend to be large. The results of this study are consistent with those of Berger et al. (1997), Abor (2007), Beck et al. (2007), and Moeinaddin and Karimianrad (2012).

According to Beck et al. (2007), large firmsdiffer from small firms when in need of additional funds. Large firms tend to prefer the use of external funding sources. In the views of creditors, large firms have a greater potential to pay off their liabilities. According to Abor (2007) and Vo and Nguyen (2014), creditors consider large firms to be diversified, thus having less risk, and possibly lower probability of bankruptcy, to be more trustworthy, and to have cheaper debt fees. Then, from a firm perspective, the larger the firm, the greater the number of employees and the capital to be managed, so that the use of debt will be a corporate governance mechanism in a firm that can oversee its performance, while also generating tax savings (Beck et al. 2007).

Tangibility has no significant negative effects on FP RE. This result shows that tangibility is not absolute in affecting FP RE. This result is not in accordance with previous research such as that by Moeinaddin and Karimianrad (2012) and Vo and Nguyen (2014). This insignificant result indicates that tangibility is less precise in explaining retained earnings, and tangibility is more accurate in explaining the use of other sources of funds, such as debt in a firm. This is because obtaining sources of funds in the form of retained earnings does not require any tangibility as collateral, such as when trying to get a loan (Sudana 2011). Tangibility results have a significant positive effect on FP Debt. This indicates that increased tangibility will increase the proportion of debt usage in finance pattern companies. These results are consistent with the findings of Moeinaddin and Karimianrad (2012) and are inconsistent with those of Berger et al. (1997) and Wen et al. (2002) but the difference is not significant. Tangibility can be used as

a debt guarantee, so increasing debt guarantees will cause creditors to feel secure and lend more. The creditor's attitude is in accordance with the 5C's characteristics that need to be seen from the debtor (Sudana 2011), one of which is fixed asset as collateral. The higher the value of the collateral provided, the higher the amount that the lender can lend.

Profitability has a significant positive effect on FP RE. These results indicate that an increase in corporate profitability will lead to an increase in the proportion of retained earnings. Thus, the firm's ability to generate profits is one factor that causes the company to have an FP RE. The results of this study are in accordance with Mulama (2014). These results indicate that the firm embraces the pecking order theory; as the firm's ability to generate profits increases, the firm will try to keep the funds set aside to fund the company. In addition, even an increase in profitability will increase the amount of dividend payments, but when the payments are not more than 100% of the profits generated during that period, there will still be retained earnings.

The profitability of the previous period had a significant negative effect on FP Debt. These results show that the greater the ability of firms to generate profits, the smaller the proportion of debt used for FP Debt. These results are consistent with those of Titman and Wessel (1988), Berger et al. (1997), and Abor (2007) and different from those of Wen et al. (2002), which showed negative results are not significant. The results of this study showed that finance pattern firms tend to embrace pecking order theory. Under pecking order theory, firms will tend to prefer internal rather than external funding. In corporate FP Debt, the debt usage rate is too high and risky, so when the company has the ability to fund itself such as in pecking order theory, it will do so to reduce the already high debt level. This is especially true when the ability of the firm to generate profits increases, so it can at the same time use the profit to pay off debt and reduce the already high level of debt.

4 CONCLUSION

1. BOD ownership and profitability (ROA) have a significant positive effect on FP RE.
2. Block holder and firm size have significant negative effects on FP RE.
3. Institutional shareholding, independent commissioner, board size, and tangibility have no significant effect on FP RE.
4. Block holder and tangibility have a significant positive affect on FP Debt.
5. Institutional shareholding, board size, and profitability in the previous period (ROA$_{t-1}$) have a significant negative effect on FP Debt.
6. BOD ownership, independent commissioner, and firm size have no significant effect on FP Debt.

REFRENCES

Abor, J. 2007. Corporate governance and financing decisions of Ghanaian listed firms. *Corporate Governance: The International Journal Of Business In Society* 7(1): 83–92.

Alba, P.C. & Djankov, S. 1998. Thailand's corporate financoing and governance structures: Impact on firms' competitives. Conference on Thailand's Dynamic Economic Recovery and Competitive*ness*. UNCC, Bangkok.

Atmaja, L.S. 2010. Dividend and debt policies of family controlled firms: The impact of board independence. *International Journal of Managerial Finance* 6: 128–142.

Beck, T., Demirguc-Kunt, A. & Maksimovic, V. 2008. Financing pattern around the world: Are small firms different. *Journal of Financial Economics* 89: 467–487.

Berger, P., Ofek, E. & Yermack, D.L. 1997. Managerial entrenchment and capital structure decisions. *The Journal of Finance* 52: 1411–1438.

Corbett, J. & Jenkinson, T. 1996. The financing of industry, 1970–1989: An international comparison. *Journal of The Japanese and International Economies* 10:71–96.

Florackis, C., Kostakis A. & Ozkan, A. 2015. Managerial ownership and performance. *Journal of Business Research* 62: 1350–1357.

Harada, K. & Nguyen, P. 2006. Ownership concentration, agency conflicts, and dividend policy in Japan. *Working Paper*:1–25.

Harada, K. & Nguyen, P. 2011. Ownership concentration and dividend policy in Japan. *Managerial Finance* 37 (4): 362–379.

Jensen, G.R., Solberg, D.P. & Zorn, T.S. 1992. Simultaneous determination of inside ownership, debt and dividend policies. *Journal of Financial and Quantitative Analysis* 27: 247–263.

Jensen, M.C. 1986. Agency cost of free cash flow, corporate finance, and takeovers. *American Economic Review* 76(2): 323–329.

Jensen, M.C. & Meckling, W.H. 1976. Theory of the firm: managerial behavior, agency costs, and ownership structure. *Journal of Financial Economics* 3(4): 305–360.

La Porta, R., L'opez-de-Silanes, F. & Shleifer, A. 1999. Corporate ownership around the world. *Journal of Finance* 54(2): 471–517.

Madaschi, A. 2010. *On corporate governance.*

Megginson, W.L. 1997. *Corporate finance theory.* Addison-Wesley.

Moeinaddin, M. & Karimianrad, M. 2012. The relationship between corporate governance and finance patterns of the listed companies. *Interdisciplinary Journal of Contemporary Research in Business* 4(7): 489–500.

Mulama, L.W. 2014. The determinants of retained earnings in companies listed at Nairobi Securities exchange. *A Research Project.* University of Nairobi.

Myers, S.C. 1984. The capital structure puzzle. *Journal of Finance* 34: 575–592.

Pelt, T.V. 2013. The effects of boar characteristic on dividend policy. Tilburg University.

Rajan, R. & Zingales, L. 1995. What do we know about capital structure? Some evidence from international data. *Journal of Finance* 50: 1421–1460.

Raza, H., Aslam, S. & Farooq, U. 2013. Financing pattern in developing nations empirical evidence from Pakistan. *World Applied Sciences Journal* 22(9): 1279–1285.

Ross, S.A., Westerfield, R. & Jordan, B.D. 2008. *Fundamentals of corporate finance.* Tata McGraw-Hill Education.

Rozeff, M.S. 1982. Growth, beta and agency costs as determinants of dividend payout ratios. *Journal of Financial Research* 5(3): 249–259.

Sartono, A. 2001. *Manajemen keuangan teori dan aplikasi.* Yogyakarta: BPFE.

Shleifer, A. & Vishny, R.W. 1986. Large shareholders and corporate control. *Journal of Political Economy* 94(3, Part 1): 461–488.

Shleifer, A. & Vishny, R.W. 1997. A survey of corporate governance. *The Journal of Finance* 52(2): 737–783.

Singh, A. & Hamid, J. 1992. *Corporate financial structures in developing countries (Vol. 22).* World Bank Publications.

Singh, A. 1996. Pension reform, the stock market, capital formation and economic growth: A critical commentary on the World Bank's proposals. *International Social Security Review* 49(3): 21–43.

Sudana, I.M. 2011. *Manajemen keuangan perusahaan teori dan praktik.* Jakarta: Erlangga.

Titman, S. & Wessel, R. 1988. The determinants of capital structure choice. *The Journal of Finance* XLIII (1).

Turkalj, K.G. & Srzentić, N. 2011. Financing patterns of firms in transition countries and its implications: Evidence from Croatia.

Vakilifard, H.R., Gerayli, M.S., Yanesari, A.M. & Ma'atoofi, A.R. 2011. Effect of corporate governance on capital structure: Case of the Iranian listed firms. *European Journal of Economics, Finance and Administrative Sciences* 35: 165–172.

Vo, D.H. & Nguyen, V.T.Y. 2014. Managerial ownership, leverage and dividend policies: Empirical evidence from Vietnam's listed firms. *International Journal of Economics and Finance* 6(5): 274–284.

Wen, Y., Rwegasira, K. & Bilderbeek, J. 2002. Corporate governance and capital structure decisions of the Chinese listed firms. *Corporate Governance: An International Review* 10(2): 75–83.

Advances in Business, Management and Entrepreneurship – Hurriyati et al (eds)
© 2020 Taylor & Francis Group, London, ISBN 978-0-367-27176-3

Tick size change and market liquidation in the Indonesia stock exchange

W.M. Soeroto, T. Widiastuti & L. Cania
Airlangga University, Surabaya, Indonesia

ABSTRACT: One action taken by the Indonesia Stock Exchange (IDX) in order to compete with members of the World Federation of Exchange (WFE) is changing the tick size. It is hoped that an increase to a five tick size in the price group will increase liquidity. This research used the bid–ask spread and depth was estimated using stock volume in closing price before and after a new tick size policy was applied in each price group. We used the nonparametric test to examine the mean difference in two related samples. An increase in the tick size leads to increases in the spread. Bid depth and ask depth also increased; however, ask depth did not show any difference. Therefore, to eliminate the ambiguity this study used the depth to relative spread ratio, which resulted in a broader spread. The IDX needs to consider a tick size that can increase liquidity in each stock price group, which therefore becomes more attractive for investors.

Keywords: bid–ask spread, market depth, tick size, market liquidity

1 INTRODUCTION

Stock markets around the world are striving to provide a liquid market. Toward increasing market liquidity, Indonesian stock exchanges are improving capital market regulation. One of the regulations used to increase liquidity is a change in price fraction. Many of the world's stock exchanges have changed the price fraction. In the United States, both the American Stock Exchange (AMEX) and The New York Stock Exchange (NYSE) lowered the price fraction on September 3, 1992 and June 24, 1997. In Asia, the Stock Exchange of Singapore (SES) lowered the price fraction on July 18, 1994 (Ahn, Cao, and Choe, 1996). In Australia, the Australian Stock Exchange (ASX) lowered the price fraction on December 4, 1995. In New Zealand, the New Zealand Exchange (NZX) lowered the price fraction in 2011.

The numerous stock exchanges making price-fraction changes pushed the Indonesia Stock Exchange (BEI) to follow the changes to compete with members of the World Federation of Exchange (WFE). The BEI price fraction has changed several times. Before July 3, 2000, BEI implemented a Single Fraction system of Rp25. As of July 3, 2000, the price fraction decreased to Rp5. Furthermore, since October 20, 2000, BEI has implemented the multifraction system up to now. BEI imposed a new price fraction based on SK Direksi PT BEI No. Kep-00023/BEI/04–2016 effective on May 2, 2016. The new price fraction applies five price fraction groups from the previous three groups of price fractions. The application of this new price fraction is expected to be better suited to the needs of both retail and institutional investors to increase the value and volume of transactions.

Based on Harris (1994), it remains unclear whether a smaller price fraction will improve market liquidity. Harris (1994)) finds a trade-off in the fractional decline in prices, in that a smaller fraction of the price could lead to a decrease in bid–ask spreads, thereby lowering trading costs. However, the decline in price fraction also has the potential to decrease liquidity because it causes a decrease in depth. Although in general the lower price fraction led to an increase in liquidity Aitken and Comerton-Forde (2005) found that stocks with a high fraction of relative prices experienced the highest increase in liquidity, while stocks with relatively low price fractions and trading volumes experience a low decreased liquidity.

Several previous studies have found different results regarding the effect of price fraction changes on market liquidity. This research therefore aimed to determine whether the change of price fraction in the Indonesian capital market increases market liquidity. The study examined the change of price fraction dated May 2, 2016 by analyzing the difference in liquidity in the period two months before and after the change of price fraction as measured by bid–ask spread and depth. This study used a control group consisting of stock price groups that had not changed the price fraction to ensure the difference in liquidity is due to a change in price fraction. The change in the price fraction should not have any effect on the control group, so there will be no difference of liquidity in the period before and after the change.

2 THEORY AND HYPOTHESIS DEVELOPMENT

2.1 Influence of price fraction on bid–ask spread

In a competitive market, a decrease in price fraction increases the bid–ask spread (Aitken & Comerton-Forde 2005). Aitken and Comerton-Forde (2005) suggested that the decrease in price fraction is crucial in stocks where the previous spread was limited by the minimum fraction and where the relative tick size is high. However, even though the stock is not limited, the price may also decrease the spreads as investors can place orders at prices previously unavailable. As the price fraction is the lowest price increase at which the investor can place a limit order, the minimum price fraction causes the stock to be traded on a narrower spread (Anderson & Peng 2014).

On May 2, 2016, BEI increased the price fraction in the price group of Rp200 to <500 and at the price group Rp2000 to <5000. In the price group Rp200 to <500, the fraction increased from Rp1 to Rp2, while in the price group Rp2000 to < Rp5000, the fraction increased from Rp 5 to Rp10. An increase in price fraction causes investors to place orders on wider price increases.

Similar to Aitken and Comerton-Forde (2005) this study used a relative spread, as it allows liquidity to be compared between stocks with different prices (Aitken & Comerton-Forde 2003). The relative bid–ask spread can be calculated by the formula

$$Relative\ Spread\ \% = \frac{\left(Ask_{j,t} - Bid_{j,t}\right)}{\left(Ask_{j,t} + Bid_{j,t}\right)/2}$$

Ask Depth, t is the best ask volume of stock j on day t. Bid Depth, t is the best bid volume of shares j on day t. Relative spread, t is the relative bid–ask share spread j on day t (Ekaputra & Ahmad 2006).

Hypothesis 1: An increase in the price fraction increases the bid-ask spread in the Indonesian capital market.

2.2 Influence of an increase in price fraction on depth

A decrease in the price fraction will reduce the premium paid to the limit order to provide market liquidity (Aitken & Comerton-Forde 2005). As a result, investors and traders who previously placed a limit order on the best bid–ask price chose to move some or all of their orders away from the best bid and ask price in order to earn a higher premium (Aitken & Comerton-Forde 2005). Another possibility is that the

impatient investor chooses to use the market order rather than the limit order as the cost of demanding liquidity decreases. As a result, the depth offered at the bid and ask best price will be reduced.

On May 2, 2016, BEI increased the price fraction in the price group of Rp200 to <500 and at the price group Rp2000 to <5000. In the price group Rp200 to <500, the fraction increased from Rp1 to Rp2, while in the price group Rp2000 to < Rp5000, the fraction increased from Rp 5 to Rp10. An increase in the price fraction increases the order of the best bid and ask prices and increases the use of limit orders.

Similar to Ekaputra and Ahmad (2006), this study uses bid closure and ask depth (volume) data. In fact, both the price and bid–ask volume change continuously during trading hours.

Hypothesis 2: An increase in price fraction increases the bid–ask depth in the Indonesian capital market.

2.3 Influence of an increase in price fraction on liquidity

Harris (1994) tested the trade-offs associated with a minimum fraction reduction. Harris (1994) states that a smaller price fraction will lead to a decrease in bid–ask spreads, thereby lowering transaction costs and increasing trading volume. However, the decline in the faction also has the potential to reduce liquidity if investors use the opportunity to free ride on other investors. If the price fraction is too small, the time priority rule becomes negligible because of quote-matcher or front-runner problems (Ekaputra & Ahmad 2006). The quote-matcher will try to put the order slightly better than the order queue, which will be more advantageous if the fraction of the price is small. Thus, although a small fraction of the price causes lower bid–ask spreads, it may also make investors less willing to expose orders, thereby reducing depth. The influence of price fraction on liquidity in terms of spread and depth still cannot be determined.

In order-driven markets, the benefits of decreasing price fractions also vary. Aitken and Comerton-Forde (2005) found that stocks with higher relative price fractions actually experienced the highest increase in liquidity, while stocks with relatively small price fractions and low trading volume decreased liquidity. Aitken and Comerton-Forde (2005) argue that price fractions are more important in order-driven markets, as are their own limit orders that provide the only source of market liquidity. Therefore, it is important to set the price fraction at the level that will encourage the placement of limit orders and provide protection from free-riders (Aitken & Comerton-Forde 2005).

The Indonesia Stock Exchange is an order-driven market, facilitating trade through a system called JATS (Jakarta Automated Trading System). The order is automatically executed based on price priority rather than time priority. The effect of the decline in price fractions on spread and depth explains the adverse effect on liquidity. For example, regarding spreads, it shows a decrease in liquidity when the price fraction is increasing. In contrast, increasing depth indicates that liquidity increases following an increase in price fraction. For that reason, we suspect that the price increase makes a difference in liquidity, but we do not predict its direction.

Ekaputra and Ahmad (2006) measured the trade-off between relative spread and depth by calculating Depth to Relative Spread Ratio (DRS), defined as:

$$DRS = \frac{\left(Ask\ Depth_{j,t} - Bid\ Depth_{j,t}\right)}{\left(Relative\ Spread_{j,t}\right)}$$

Ask Depth$_{j,t}$ adalah volume *ask* terbaik saham j pada hari t. *Bid Depth*$_{j,t}$ adalah volume *bid*terbaik saham j pada hari t. *Relative spread*$_{j,t}$adalah *relative bid ask spread* saham j pada hari t (Ekaputra & Ahmad 2006).

Hypothesis 3: An increase in price fraction causes changes in liquidity in the Indonesia capital market,

3 METHOD

3.1 *Population and sample*

The population of this study consists of all stocks traded on the BEI period two months before and after the price fraction changes. The sample selection was made using purposive sampling, with the following criteria:

1. Shares with a closing price of Rp200 to <Rp500 and Rp2000 to <Rp5000 will be the tested group and the stock with the closing price <Rp200, Rp500 to <2000, and Rp ≥ 5000 will be the control group.
2. Shares that move from one fraction to another during the study period as a result of changes in stock prices will be excluded from the sample.
3. Issuers do not conduct corporate actions, such as the announcement of dividend, rights issue, stock split, merger or acquisition, during the observation period.
4. Shares trading at less than one transaction per day will be excluded from the sample.

3.2 *Research data*

The data used in this research are secondary data sourced from the Indonesia Stock Exchange. Secondary data needed in this research are daily stock data based on closing data in the form of stock price data, ask price, bid price, ask volume, and bid volume.

3.3 *Variables*

Dependent variable:

1. The bid–ask spread is the difference between the lowest ask price and the highest bid price of the relevant quotes. Bid–ask spreads reflect trading costs and market liquidity.
2. Depth is the ability of securities to absorb buy orders and sell orders without dramatic stock price movements. Bid–ask depth is estimated using stock volume at the bid–ask closing price.

Dependent variable: The price fraction is the minimum allowable price variation in a sequence, usually determined by the exchange in which securities are traded.

3.4 *Analysis model*

To test the above hypothesis, we used these models:

1. After *Relative Spread* $_{(Rp\ 200\ to\ <500)}$ > Before *Relative Spread*$_{(Rp\ 200\ to\ <500)}$
2. After *Relative Spread* $_{(Rp\ 2000\ to\ <5000)}$ > Before *Relative Spread* $_{(Rp\ 2000\ to\ <5000)}$
3. After *Bid–Ask Depth* $_{(Rp\ 200\ to\ <500)}$ > Before *Bid–Ask Depth* $_{(Rp\ 200\ to\ <500)}$
4. After *Bid–Ask Depth* $_{(Rp\ 2000\ to\ <5000)}$ > Before *Bid–Ask Depth* $_{(Rp\ 2000-\ to\ <5000)}$
5. After Liquidity $_{(Rp\ 200-<500)}$ ≠ Before Liquidity $_{(Rp\ 200\ to\ <500)}$
6. After Liquidity $_{(Rp\ 2000\ to\ <5000)}$ ≠ Before Liquidity $_{(Rp\ 2000\ to\ <5000)}$

4 RESULTS AND DISCUSSION

4.1 *Description of research results*

The results show research data before and after the fractional changes of each stock price group on each variable studied, which include the relative spread and depth. Table 1 shows the relative spread which includes the mean and standard deviation values in the period before and after.

Based on Table 1 it can be seen that there is a difference in the highest (maximum) and the lowest (minimum) relative spread value in the period before and after. The lowest (minimum) difference between the pre- and post- period is greater in the price group of Rp200 to <500 and the price group Rp2000 to <5000, while in the price group <Rp200, Rp500 to <2000, and ≥ Rp5000, the lowest value (minimum) did not experience a big difference. This can be due to the ranking of the price fraction in the price group of Rp200 to <500 and the price group of Rp2000 to <5000 resulting in the investor placing the order on a wider price increase in the period after. Conversely, in the price group <Rp200, Rp500 to <2000, and

Table 1. Descriptive statistics of *Relative Bid–Ask Spread*, before and after May 2, 2016.

Group of price	N	Mean	Std. deviation	Min	Max
Sebelum					
<Rp200	1.428	2.151	1.177	0.63	5.5
Rp200 to < 500	924	1.44	1.923	0.21	7.52
Rp500 to < 2000	2.436	1.272	1.405	0.33	8.98
Rp2000 to < 5000	840	1.169	1.452	0.18	5.08
≥Rp5000	1.092	0.983	1.173	0.23	4.59
Sesudah					
<Rp200	1.428	2.364	1.423	0.62	5.45
Rp200 to < 500	924	1.647	1.744	0.42	7.17
Rp500 to < 2000	2..436	1.368	1.491	0.3	8.55
Rp2000 to < 5000	840	1.468	1.709	0.34	6.14
≥Rp5000	1.092	1.119	1.458	0.2	6.46

≥Rp5000, there is no change in price fraction resulting in the lowest value (minimum) and the group did not experience a big difference between the period before and after.

Based on Table 2 it can be seen that in the groups of Rp200 to <500 and Rp2000 to <5000 there is an increase in the mean (average) on bid and ask depth (volume) which indicates an increase in selling orders and purchase orders in the period after the change in price fraction. An increase in the price fraction of May 2, 2016 in the price group of Rp200 to <500 and Rp2000 to <5000 leads to an increase in premiums paid to the limit order to provide market liquidity and

increased use of limit orders rather than market orders as the cost of demanding liquidity increases, thus increasing the depth offered in the group.

In the price group <Rp200 there is also an increase in the mean (average) on bid and ask depth (volume). Based on the analysis, BKSL shares experienced a large increase in bid and ask depth in the period after. In the group of Rp500 to <2000 there was an average decrease (bid) on bid and ask depth (volume). Based on the analysis, CTRA stocks experienced a large decline in the ask depth and ADRO shares decreased greatly in bid depth in the period after. The decrease could have been caused by the investors in the group of Rp500 to <2000 choosing to invest in the price group of Rp200 to <500 and Rp2000 to <5000 which has increased the price fraction, so that in the group of Rp200 to <500 and Rp2000 to <5000 there is an increase of bid and ask depth. The increase in price fraction causes an increase in premiums paid to investors, making the stock more profitable, while in the group ≥Rp5000 there was a decrease in average (mean) on ask depth but increase in bid depth. Based on the analysis, ASII shares experienced the highest decrease in ask depth, but also experienced the highest increase in bid depth over the period after. The drop in ask depth could also have been caused by investors choosing to invest in price groups that have increased the price fraction.

4.2 Analysis of research results

The analysis tool used in this study was the nonparametric sign test to test the average difference in two related samples. The dependent variable will be compared to know the difference between the

Table 2. Descriptive statistics of *Bid–Ask Depth*, before and after May 2, 2016.

Group of price	N	Before (%) Mean	Before (%) Std. deviation	After (%) Mean	After (%) Std. deviation
Bid Depth					
< Rp200	1.428	773748.79	2349506.213	798636	2699882.975
Rp200 to < 500	924	396561	772020.855	673161.36	1182412.668
Rp500 to < 2000	2.436	333715.31	583931.692	270706.28	397178.823
Rp2000 to< 5000	840	244132.7	387196.496	360047.1	621017.859
≥Rp5000	1.092	111492.65	158169.22	154247.19	233478.285
Ask Depth					
<Rp200	1.428	555362.85	1637039.819	598952.68	2225824.575
Rp200 to < 500	924	306279.86	414700.532	548870.09	842372.576
Rp500 to < 2000	2.436	319491.6	486410.051	235268.28	359491.17
Rp2000 to < 5000	840	266522.35	486792.49	289663.55	453155.236
≥Rp5000	1.092	264968	434842.715	189290.31	302805.13

periods before and after. The dependent variable in this research is bid–ask spread and bid depth and ask depth. The independent variable in this research is price fraction.

4.3 Model analysis and hypothesis test

The following are the results of the nonparametric sign test to test the average difference in two related samples.

4.3.1 Hypothesis 1

The following is the result of the nonparametric sign test to test the average difference in two samples Table 3 shows the difference of the average of the bid–ask spread for each stock price group before and after the price fraction change. The results show that the stocks in the price group Rp200 to <500 (group 2) and the stocks in the price group Rp2000 to <5000 (group 4) experienced a significant increase in spreads. The average spread in group 2 increased from 1.44% to 1.647% and in group 4 increased from 1.169% to 1.468%. This result supports Hypothesis 1 that an increase in the price fraction causes an increase in the related spread.

There was no significant change in bid–ask spread in the control group, which indicated that changes in the test group were due to changes in the price fraction rather than caused by other unrelated factors.

4.3.2 Hypothesis 2

Table 4 shows the difference in bid–ask depth for each stock price group before and after the price fraction change. The results show that stocks in the price group of Rp200 to <500 (group 2) experience a significant increase in bid depth. The average bid depth in group 2 increased from 396,561 to 673,161.36. The average ask depth in group 2 also increased from 306,279.86 to 548,870.09. Using a nonparametric sign test, the increase in ask depth is not significant at the 5% level. In the price group Rp2000 to <5000 (group 4), there is an increase in bid and ask depth. The average bid depth in group 4 increased from 244,132.7 to 360,047.1, while the ask depth increased from 266,522.35 to 289,663.55. Using a nonparametric sign test, the bid increase and ask depth are not significant at the 5% level.

Table 3. Average of *Bid–Ask Spread*, before and after May 2, 2016.

Group	N	Before	After	Difference	Sig.
Bid Depth					
1(Control)	1.428	2.15	2.364	0.213	0.417
2	924	1.44	1.647	0.207	0.046
3(Control)	2.436	1.272	1.368	0.095	0.138
4	840	1.169	1.468	0.299	0.002
5(Control)	1.092	0.983	1.119	0.135	0.382

Table 4. Average of *Bid–Ask Depth*, before and after May 2, 2016.

Group	N	Before	After	Difference	Sig.
Bid Depth					
1(Control)	1.428	77,3748.79	798636	24,887.21	0.417
2	924	39,6561	673161.36	276,600.36	0.046
3(Control)	2.436	33,3715.31	270706.28	−63,009.03	0.138
4	840	24,4132.7	360047.1	11,5914.4	0.002
5(Control)	1.092	11,1492.65	154247.19	42,754.54	0.382
Ask Depth					
1(Control)	1.428	55,5362.85	598952.68	43,589.83	0.417
2	924	30,6279.86	548870.09	242,590.23	0.046
3(Control)	2.436	31,9491.6	235268.28	−84,223.32	0.138
4	840	26,652 2.35	289663.55	23,141.2	0.002
5(Control)	1.092	26,4968	189290.31	−75,677.69	0.382

There were significant differences in ask depth in the control group, i.e., the price group of Rp500 to <2000 (group 3) and the price group ≥ Rp5000 (group 5). Companies with large market capitalization in Indonesia typically have high family ownership rates and often perform stock transactions that can affect liquidity. Based on the decision of the directors of PT BEJ No. Kep-305/BEJ/07–2004, in Indonesia the controlling shareholder owns 25% or more of the company's shares. The family companies include BCA, Gudang Garam, Unilever, Astra, etc.

4.3.3 Hypothesis 3

The result of relative bid–ask spread in Table 3 and bid–ask depth in Table 4 shows that in the test group Rp200 to <500 (group 2) there was a significant increase in spreads. Bid depth in group 2 also increased significantly, while ask depth increased insignificantly at level 5%. In the test group Rp2000 to <5000 (group 4) there was a significant increase in spreads. Bid and ask depth in group 4 increased insignificantly at the 5% level.

Whether the increase in spread and bid depth in group 2 significantly led to the effect of increasing the price fraction on liquidity was inconclusive. So to calculate the trade-off between relative spread and depth, this study calculates depth to relative spread ratio (DRS). The DRS ratio measures whether the increase in depth is greater or less than the relative spread increase.

Table 5 shows the average difference of DRS for the price group Rp200 to <500 before and after the change of price fraction. The results show that after the new price fraction, the average DRS decreased by 16,260,723.27. The decrease in DRS means that the spread increase is greater than the increase in depth. Although the DRS average decreased, the decrease was not significant at the 5% level, which means that the increase in

Table 5. DRS average, before and after May 2, 2016.

Group	N	Before	After	Difference	Sig.
Rp200 to < 500 (Kelompok 2)	924	107,101,196.09	90,840,472.82	−16,260,723.27	0.101

the price fraction did not decrease the overall stock liquidity in the Rp200 to <500 price group.

4.4 Discussion

Based on the hypothesis tests, we could explain the following.

4.4.1 Influence of price fraction on bid–ask spread

The test results for Hypothesis 1 show that there is a significant increase in the relative spread, which indicates a greater transaction cost after the increase of the price fraction. As Anderson and Peng (2014) argue, the price fraction is the lowest stock price increase at which investors can place a limit order, so a decrease in the price fraction will cause the stock to be traded on a narrower spread. On the contrary, the increase of price fraction in the Indonesia Stock Exchange on May 2, 2016 will increase the spread width. Results of this test indicate lower liquidity because it will be more expensive for investors to make transactions immediately.

4.4.2 Influence of price fraction on bid–ask depth

The test for Hypothesis 2 shows different results for each group of tested prices. In the price group Rp200 to <500 (group 2) there was a significant increase in bid depth, but an insignificant increase in ask depth. The price group Rp2000 to <5000 (group 4) shows an increase in bid and ask depth, but the bid and ask depth increase is not significant. The increase in bid and ask depth indicates that after the change of price faction of May 2, 2016, the greater the depth offered in the price group Rp200 to <500 and Rp2000 to <5000. Aitken dan Forde (2005) suggests that a decrease in the price fraction will reduce the premiums paid to the limit order to provide market liquidity. Another possibility is that impatient investors choose to use market order rather than limit order as the cost of demanding liquidity decreases. On the contrary, an increase in the price fraction of the Indonesia Stock Exchange on May 2, 2016 will increase the premiums paid to the limit order to provide market liquidity and increase the use of limit orders rather than market orders as the cost of demanding liquidity increases, thus increasing the depth offered in the price group Rp200 to <500 and Rp2000 to <5000.The rising depth indicatesincreasing liquidity, as there is more supply and demand in the price group.

4.4.3 Influence of price fraction on liquidity

Harris (1994) states that a smaller fraction of the price will lead to a decrease in bid–ask spreads, thereby lowering transaction costs and increasing trade volume. However, the decline in the faction also has the potential to reduce liquidity if investors use the opportunity to free ride on other investors. Thus, a small fraction of the price causes lower bid–ask spreads, and it may also make investors less willing to expose orders, thereby reducing depth. On the contrary, the increase of price fraction on the Indonesia Stock Exchange on May 2, 2016, viewed from the spread shows a decrease of liquidity when the price fraction is increased, while viewed from depth it shows increased liquidity following the increase in price fraction.

The test results for Hypothesis 3 show that in the price group of Rp200 to <500 there is an increase in spread higher than the increase of depth. Although the spread is higher, the overall increase in the price fraction does not decrease the stock liquidity in the Rp200 to <500 price group. In the price group Rp2000 to <5000, however, there is an increase in spread without being followed by a significant increase of depth. Thus, overall liquidity becomes lower in the price group of Rp2000 to <5000.

5 CONCLUSION

Based on the results of data analysis related to the purpose of the research, hypotheses and model analysis, the following conclusions can be drawn:

1. The increase in price fraction on May 2, 2016 at the Indonesia Stock Exchange in the price group Rp200 to <500 and Rp2000 to <5000 causing a bid–ask spread in the group was significant. The higher spread indicates lower liquidity, as it will be more expensive for investors to make transactions immediately.

2. The increase in price fraction on May 2, 2016 at the Indonesia Stock Exchange in the price group Rp200 to <500 and Rp2000 to <5000 caused an increase in the bid and ask depth. The increase in bid and ask depth indicates that after the change of price fraction of May 2, 2016, a larger depth is seen in the price group of Rp200 to <500 and Rp2000 to <5000. The increasing depth indicates increased liquidity, as there is more supply and demand in the price group.

3. The price group Rp200 to <500 has a massive spread increase compared with the depth increase. Although the spread is the highest, the overall increase in the price fraction does not decrease stock liquidity in the Rp200 to <500 price group. Conversely, in the price group Rp2000 to <5000 there is an increase of spread without being followed by a significant increase in depth. Thus, overall liquidity decreases in the price group of Rp2000 to <5000.

REFERENCES

Ahn, H.J., Cao, C.Q., & Choe, H. 1996. Tick size, spread, and volume. *Journal of Financial Intermediation* 5(1): 2–22.

Aitken, M. & Comerton-Forde, C. 2003. How should liquidity be measured? *Pacific Basin Finance Journal.*

Aitken, M. & Comerton-Forde, C. 2005. Do reductions in tick sizes influence liquidity? *Accounting and Finance.*

Anderson, H. & Peng, A. Y. 2014. From cents to half-cents and its liquidity impact. *Pacific Accounting Review.*

Chung, K. H. & Chuwonganant, C. 2002. 'Tick size and quote revisions on the NYSE' *Journal of Financial Markets* 5(4): 391–410.

Ekaputra, I.A. & Ahmad, B. 2006. 'The Impact of Tick Size Reduction on liquidity and order strategy: Evidence from the Jakarta Stock Exchange (JSX).

Harris, L.E. 1994. Minimum price variations, discrete bid–ask spreads, and quotation sizes. *The Review of Financial Studies.*

Lau, S.T. & McInish, T.H. 1995. Reducing tick size on the stock exchange of Singapore. *Pacific-Basin Finance Journal*, 3(4): 485–496.

Risk profile, good corporate governance, profitability, capital, and third-party funds interest rate of Indonesian banking

I.M. Sudana & K. Yuvita
Airlangga University, Surabaya, Indonesia

ABSTRACT: This research aims to determine the effect of risk profile, good corporate governance, profitability, and capital to third-party fund interest rate of Indonesian banking. This research using multiple linear regression analysis to determine the effect of independent variables consisting of risk profile, Good Corporate Governance (GCG), profitability (ROA), and capital (CAR) towards dependent variable which is third-party fund interest rate. Based on the results of analysis, it could conclude that risk profile and GCG significantly applies a positive influence to third-party fund interest rate, ROA significantly applies a negative influence to third-party fund interest rate, while CAR applies a negative effect but not significant to third-party fund interest rate.

1 INTRODUCTION

In January 2012, the bank assessed its health level with the latest criteria through Bank Indonesia regulation no. 13/1/PBI/2011 known as RGEC (Risk Profile, Good Corporate Governance, Earnings, Capital). The rating of the health of a commercial bank replaces the previous Bank Indonesia regulation, CAMEL. This is due to the increasingly complex and diverse banking industry. In addition, the international assessment approach also leads to a risk-based supervision approach.

Healthy banks are assumed to have low risk so that the bank does not need to increase interest rates of third party funds to attract customers, and vice versa. Most researchers in Indonesia (Tasyia 2014, Larasaty 2014) use fundamental factors (CAMEL) to determine the effect on third party fund interest rates. However, not all fundamental factors of the bank have a significant effect on the interest rate of third party funds in both studies. According to the results of research related to market discipline (Peria & Schmukler 2001), depositors in Argentina, Chile and Mexico, demanded high-risk banks by requiring higher interest rates, because high-risk banks can lower depositor's confidence and make them becomes unsecure to invest in a bank. Based on the background described, the problems studied in this research are: are the risk profile, good corporate governance, profitability, and capital affecting the interest rate of third party funds in Indonesian banking?

1.1 Definition and measurement interest rates of third party funds (IR)

The deposit interest rate is the interest rate obliged to be paid by the bank to its customers for deposits made. High risk makes the customers afraid to put their funds in the bank, so as compensation to customers who are willing to keep funds in banks with high risk, the bank will try to raise interest rates of savings so that customers become interested. Therefore, the focus of this research is the deposit interest rate, i.e. the interest rate of third party funds. The formula for calculating the interest rate of third party funds is:

$$IR = \frac{\text{Interest Expense}}{\text{Total Third Party Funds}} \times 100\% \quad (1)$$

1.2 Risk profile (RP)

A risk profile is a risk assessment of a bank at a certain time in which it contains a set of interconnected risks using a consistent method (Polk 2014). Assessment of risk profile factors is an assessment of the inherent risk and quality of risk management implementation in bank operations performed on 8 (eight) risk components, including credit risk, market risk, liquidity risk, operational risk, legal risk, strategic risk, compliance risk, and reputation risk. The composite rating of bank risk profile is determined based on a comprehensive and structured analysis of each factor's ranking, rating 1 (very low risk), rank 2 (low risk), rank 3 (risk is high enough), rank 4 (high risk), ranked 5 (risk is very high).

If the bank is at high risk and the public savings are not guaranteed, the depositor feels insecure and loses confidence in placing the funds in the bank, thus reacting to the risky banks by requiring higher interest rates or withdrawing deposits. In order to avoid the possibility of withdrawal of funds by the customers, the bank seeks to reclaim its customers by increasing the interest rate of third party funds. It can be concluded that the risk profile positively affects the interest rate of third party funds, in other words the greater the risk profile rating (bad risk management), the greater the interest rate of third party funds, and viceversa.

H1: The risk profile positively affects the interest rate of third party funds.

1.3 Good corporate governance (GCG)

Corporate governance is a series of mechanisms that direct and control a company to run the company's operations in accordance with the expectations of the stakeholders. Good corporate governance is a concept that emphasizes the importance of shareholder and other stakeholder rights to obtain correct, accurate, timely, and corporate liability to disclose a company's disclosure. Banks must conduct periodic self-assessment which includes at least 11 (eleven) integrated and integrated GCG implementation assessment factors. The GCG composite rank matrix comprises, ranked 1 (excellent GCG implementation), ranks 2 (good GCG implementation), ranks 3 (GCG implementation is good enough), ranks 4 (poor GCG implementation), ranks 5 (very poor GCG implementation). According to Greuning & Bratanovic (2011) bad corporate governance can increase the possibility of a bank failure. Therefore, bad corporate governance causes customers to be afraid to put their funds in the bank, because they have a high risk of failure, so the bank will attract customers by increasing the interest rate of third party funds to return to believe. It can be concluded that good corporate governance positively affects third party fund interest rate, in other words the greater the composite of GCG (the worse the application of GCG) the greater the interest rate of third party funds, and vice versa.

H2: Good corporate governance positively affects the interest rate of third party funds.

1.4 Profitability

Earning is one of the bank's health assessment in terms of profitability. Profitability factor used as an assessment of a bank's performance associated with the ability of banks in generating profits. The ratio often used to measure the profitability of a bank is the return on assets (ROA). The higher the ROA of a bank means the more efficient the bank is in using its assets. The formula for calculating ROA:

$$ROA = \frac{Earning\ Before\ Taxes}{Total\ Assets} \times 100\% \quad (2)$$

The ROA ratio measures the company's ability to generate profit before taxes by using the total

assets owned by the bank. The higher the percentage ROA means the better and efficient ability of the bank in managing all its assets to generate profit. On the other hand, the credibility of the bank also increases as customers feel secure when saving their funds in banks that have high profitability, so customers still entrust their funds to the bank even though third party fund interest rate is low. Thus, ROA negatively affects the interest rate of third party funds.

H3: Return on asset negatively affects third party fund interest rate.

1.5 Capital (CAR)

Capital, which is proxied with capital adequacy ratio (CAR) is also one of the approach to assess the soundness of a bank. Capital is to ensure the adequacy of capital and reserves to bear the risks that may arise.

$$CAR = \frac{Bank\ Capital}{Risk\ Weighted\ Assets} \times 100\% \quad (3)$$

The higher the CAR ratio means the better the bank's capital position because it indicates the bank has sufficient capital to anticipate the potential losses caused by its operational activities, so that this can increase public confidence in placing funds in the bank. With the increase of public confidence in the bank, the bank tends to lower the interest rate of third party fund. Accordingly, the CAR negatively affects the interest rate of third party funds.

H4: Capital adequacy ratio negatively affects the interest rate of third party funds.

1.6 Model analysis

The model of analysis used is multiple regression model.

$$IR_{it} = \beta_0 + \beta_1 PR_{i,t-1} + \beta_2 GCG_{i,t-1} \\ + \beta_3 ROA_{i,t-1} + \beta_4 CAR_{i,t-1} + \varepsilon_{i,t} \quad (4)$$

where IR_{it} = interest rate of third party fund of bank i in year t; $PR_{i,t-1}$ = bank risk profile i in year t-1; $GCG_{i,t-1}$ = good corporate governance bank i in year t-1; $ROA_{i,t-1}$ = return on asset of bank i in year t-1; $CAR_{i,t-1}$ = capital adequacy ratio of bank i in year t-1; β_0 = constants; β_1, β_2, β_3, β_4 = regression coefficient; $\varepsilon_{i,t}$ = error.

2 METHOD

The sample in this research is determined by using purposive sampling method.

2.1 Operational definition of variables

1. Interest rate is the interest rate of third party funds which is the ratio between interest expense and total third party funds, equation (1).
2. The risk profile is an assessment of the inherent risk and quality of risk management implementation in bank operations measured by the composite risk profile rating calculated by each bank concerned based on PBI no. 13/1/PBI/2011 on the annual report of each bank.
3. Good corporate governance is an assessment of bank management on the implementation of GCG principles measured based on the composite GCG ratings calculated by each bank concerned by PBI no. 13/1/PBI/2011 on the annual report of each bank.
4. Profitability is the bank's ability to generate profit before tax with all assets owned by banks or return on assets (ROA), equation (2).
5. Capital or capital proxy with capital adequacy ratio (CAR) is the ratio between total bank capital with risk-weighted assets, equation (3).

3 RESULT AND DISCUSSION

3.1 Description of research results

In the description of research results, will be presented the data of banking research variables in Indonesia. The minimal number of observations due to the official BI regulations relating to the research variables was only published in 2011. Not all banks active in Indonesia in the period 2012-2013 include risk and GCG profile reports in their annual reports. On the other hand, not all banks publish such data, especially foreign ownership banks. Table 1 shows descriptive variables that include minimum values, maximum values, average of each variable of observation and standard deviation. Dependent variable in this research is the interest rate of third party fund (IR), while independent variable consist of risk profile (PR), good corporate governance (GCG), ROA, and CAR.

3.2 Model analysis and hypothesis testing

The results of analysis and hypothesis testing are summarized in Table 2 below.

Table 1.   Statistical description of research variables.

Varme	N	Min	Max	Mean	SD
PR*	91	1.00	3.00	2.0659	0.48995
GCG*	91	1.00	3.00	1.9780	0.57693
ROA	91	-0.18	5.46	2.2610	1.30873
CAR	91	10.09	48.75	19.6513	7.33717
IR	91	1.81	10.39	5.5790	1.72405

PR*, GCG* ordinal data.

Table 2.   Results of risk profile regression analysis, GCG, ROA, and CAR on interest rate of third party funds.

	Unst Coef	Stand Coef		
Var	β	β	t	Sig
Const	3.866		4.939	0.000
PR	0.682	0.194	2.008	0.048
GCG	1.001	0.335	3.561	0.001*
ROA	-0.659	-0.501	-5.719	0.000*
CAR	-0.009	-0.040	-0.471	0.639
R^2		0.402		
DW		1.655		
N		91		

* significant $\alpha = 5\%$.

3.3 Discussion

3.3.1 The effect of risk profile on third party fund interest rate

Risk profile variables (PR) have a significant positive effect on third party fund interest rate indicates that the higher the risk of a bank causing the customer's trust to decrease to place the funds in the bank, so that the customer tend to withdraw their funds. Therefore, to withdraw the interest of customers, the bank raised the interest rate of third party funds in order to replace their funds in the bank that has a high risk, and vice versa. The greater the risk of a bank is marked by the greater the composite rating of the risk profile, which means that the interest rate of third party funds offered is also high, and vice versa. The results of research are in line with the results of research conducted abroad, such as Peria & Schmukler (2001), and Demirguc-Kunt & Huizinga (2004) stating that the depositor demanded high interest rates as long as the bank's risk is also high.

3.3.2 The influence of GCG on third party fund interest rates

The GCG variable has a significant positive effect on the interest rate of third party funds indicating good corporate governance will make customers believe to place their funds in the bank, because the bank tends to have a low risk of failure, so the bank does not need to attract customers by increasing the interest rate of third party funds. The better the GCG implementation is marked by the smaller GCG composite ratings, which means the interest rate offered is also low, and vice versa.

3.3.3 The effect of ROA on third party fund interest rate

Return on assets has a significant negative effect indicates that the higher the value of ROA, the more efficient the bank in managing its assets to generate profit. On the other hand, the credibility of banks also increases as customers feel secure if they save their funds in banks that have high profitability. Increased public confidence in banks, banks tend to lower interest rates third party funds.

3.3.4 The effect of CAR on third party fund interest rates

The CAR variable has a negative effect on the interest rate of third party funds, but not significant. The effect of CAR is not significant because the majority of banks have CAR value that has exceeded the provisions of Bank Indonesia, which is above 8%, so the bank is considered to have sufficient capital to anticipate the potential losses caused by operational activities, so this can increase public confidence in placing and does not need to pay attention to the CAR variable as a determinant of bank soundness. In addition, the guarantee of IDR 2 billion makes customers feel safe. The insignificant effect of CAR on interest rates on third party funds is in line with Tasyia's (2014) study in the period under which the guarantee and Larasaty (2014) are in full guarantee.

4 CONCLUSIONS AND SUGESSTIONS

4.1 Conclusions

1. Risk profile variables, GCG have a significant positive effect, and ROA has a significant negative effect on third party fund interest rate, because risk profile, GCG, and ROA are important factors considered by customers in placing their funds in banks.
2. CAR variable has no significant negative effect on third party fund interest rate, because the average CAR ratio of banks in Indonesia has exceeded the provisions of Bank Indonesia.
3. The value of R^2 in this study is 40.2 percent, which means that most of the variations in third

party fund interest rates (IR) are explained by other variables outside this study.

4.2 Sugesstion

1. Bank management should pay attention to risk profile, GCG, and ROA variables in the determination of third party fund interest policy, because these three variables have a significant effect on the interest rate of third party funds.
2. Customers should pay attention to variable variables of risk profile, GCG, and ROA as consideration in the determination of the bank when it will place its funds in the bank.
3. Further research is expected to add another variable outside this study, because the value of R square in this study is low.

REFERENCES

Bank Indonesia. 2011. *Peraturan Bank Indonesia Nomor 13/1/PBI/2011 tentang penilaian tingkat kesehatan bank umum*. Jakarta.

Demirguc-Kunt, A. & Huizinga, H. 2004. Market discipline and deposit insurance. *Journal of Monetary Economics*.

Greuning, H.V. & Bratanovic, S.B. 2011. *Analyzing banking risk: Analisis risiko perbankan*. Jakarta: Salemba Empat.

Larasaty, A. 2014. fundamental bank tingkat bunga deposito, dan perubahan penjaminan simpanan bank di Indonesia. *Doctoral Dissertation*. Universitas Airlangga.

Peria, M.M.S. & Schmukler, S.L. 2001. Do depositors punish bank for bad behaviour? Market discipline, deposit insurance, and banking crisis. *Journal of Finance* 56(3): 1029–1051..

Polk, D. 2014. *Risk governance visual memorandum on guidelines proposed by the OCC*. New York.

Tasyia, F. 2014. Pengaruh fundamental bank terhadap suku bunga deposito pada periode sebelum dan saat penjaminan. Universitas Airlangga.

Advances in Business, Management and Entrepreneurship – Hurriyati et al (eds)
© 2020 Taylor & Francis Group, London, ISBN 978-0-367-27176-3

The influence of current ratio, debt-to-equity ratio, inventory turnover, and return on investment on price-earnings ratio of cement industry companies listed at Indonesia stock exchange

N.A. Hamdani, A. Solihat & G.A.F. Maulani
Universitas Garut, Garut, Indonesia

ABSTRACT: A funding gap may be reduced with debt instruments or by issuing shares to the capital market. When the latter option is preferred, the company ownership is no longer private but dispersed among the general public. A publicly listed company's business performance can be measured through its financial statement. Before taking any decision to buy a company's shares in the capital market, investors will first analyze them to see the potential profit. This research is aimed at analyzing if the Current Ratio (CR), Debt-to-Equity Ratio (DER), INventory Turnover (INTO), and Return On Investment (ROI) have influence on Price-Earnings Ratio (PER) of cement industry companies listed at Indonesia Stock exchange. To measure the relationships between variables, this study employed descriptive and verification methods. The results of an F-test show that CR, DER, INTO, ROI has significant influence on PER simultaneously as the Sig. F value is lower than 5%. It is also revealed that partially CR and DER do not have significant influence on PER, but INTO and ROI do.

1 INTRODUCTION

As a developing country, Indonesia is incessantly building infrastructure. This development requires a lot of raw materials, one of which is cement. In this case, cement is indispensable. Therefore, to meet domestic needs of cement in the next few years, many cement companies are expanding their businesses. But to this end, relatively big funds are required.

A funding gap may be reduced with debt instruments or by issuing shares to the capital market. When the latter option is preferred, the company ownership is no longer private but dispersed among the public. According to Djuhana (2009), "Capital market is a means to obtain capital. Through the capital market, companies can strive to improve their professionalism and performance. Obtaining capital can be done by issuing shares (going public)."

A publicly listed company's business performance can be measured through its financial statement. As for the investors, before taking any decision to buy a company's shares in the capital market, they will first analyze them to see the potential profit. Generally, there are two approaches to analyzing shares, technical analysis and fundamental analysis. Technical analysis is carried out by analyzing statistics gathered from external factors such as economic, social, political factors. Fundamental analysis is carried out by evaluating internal state of a company, such as the company's financial statements. Lusiana (2010) points out, "Fundamental analysis is always used as a reference for investors in making investment decisions in the capital market." Fundamental analysis looks at the financial ratios taken from the company's financial statement. One of the most important ratios is price-earnings ratio

(PER) because it is easily understood by investors and prospective investors. According to Jones, in Siagian (2004), "Price-earnings ratio is used to estimate the value of shares by dividing the current share price by earnings per share (EPS)."

There are three cement companies listed at Indonesia Stock exchange: PT Holcim Indonesia Tbk, PT Indocement Tunggal Prakarsa Tbk, and PT Semen Gresik Tbk (www.idx.co.id). Table 1 shows data on prices, earnings, and price-earnings ratios over the past nine years.

Previous studies suggest that PER is influenced by some variables other than price per share and earnings per share. Kholid (2006) states that PER is influenced by return on equity (ROE), dividend payout ratio (DPR), return on investment (ROI), debt-to-equity ratio (DER), and earnings per share (EPS). While according to Kurniawan (2013), PER is influenced by loan-to-deposit ratio (LDR), loan-to-assets ratio (LAR), DER, operating expense ratio (OER), return on assets (ROA), ROE, DPR and price-to-book value (PBV). Setyorini (2005) states that PER is influenced by DPR, earnings growth (EG), variance of earnings growth (VEG), ROE, and financial leverage (FL).

This study seeks to answer if the current ratio (CR), debt-to-equity ratio (DER), inventory turnover (INTO), and return on investment (ROI) have influence on price-earnings ratio (PER) of cement industry companies listed at Indonesia Stock exchange.

2 LITERATURE REVIEW

Munawir (2007) states the financial statements are the result of an accounting process that can be used as a communication tool between financial data or the activities of a company with those who have an interest in the data or activities of the company in question. Financial statement is an important means for investors to monitor the company's development periodically (Samsul, 2006). The faster issuers issue the periodic financial statement, either audited (by Certified Public Accountant) or unaudited, the more useful is to investors. Harahap (2007) points out financial ratios are numerical values resulted from the comparison of one financial statement post with another that has a relevant and significant relationship. While according to James C Van Horne, cited in Kasmir (2012), financial ratios are indices that connect two accounting numbers and are obtained by dividing one number by another. Financial ratios are used to evaluate a company's financial conditions and performance. Current ratio is the most common and frequently used ratio to measure a company's liquidity (Munawir, 2007). According to Djuhana (2009) current ratio shows the current assets of the issuer in paying short-term liabilities; the higher the ratio, the better. According to Harahap (2007:304), "Solvency ratios describe a company's capabilities in paying long-term liabilities and obligations in case the company is liquidated." While accroding to Djuhana (2009), solvency ratios refer to a company's capabilities in paying long-term obligations including debt ratio, debt-to-equity ratio, long-term debt-to-equity ratio, long-term debt-to-capitalization ratio, times interest earned, cash flow to interest coverage, cash flow net income, and cash return on sales ratio. Kasmir (2012:158) states, "Debt-to-equity ratio provides a general description of a company's feasibility and financial risks." According to Djuhana (2009:105), "Debt-to-equity ratio demonstrates the issuer's the capital structure compared to liabilities. While Harahap (2007:303) says that DER describes the extent to which the owner's capital can cover debts to outsiders. The lower the ratio is, the better. This ratio is also called a leverage ratio. For the security of external parties, the best ratio is if the amount of capital is more than the amount of debt or at least the same. However, for the shareholders or advantage ratio management, the capital should be more than the debt.

According to Sawir (2005), inventory turnover is a popular indicator for assessing operational activities that show how well management controls the capital in the inventory. Djuhana (2009) states inventory turnover shows how quickly supplies turn into cash. The fastest the company sell inventory, the fastest the company cashes in their investment. While according to Harahap (2007), inventory turnover shows how fast a company has sold and replaced inventory during a normal production cycle. The higher the ratio, the better. It means that sales activities run fast.

Tandelilin (2010:2) states, investment is the allocation of some amount of funds or other resources in the expectation of some benefit in the future.

According to Financial Accountancy Standards Notice (*PSAK, Indonesian: Pernyataan Standar Akuntansi Keuangan*) no 13 since October 1, 2004, investment is an asset used by a company for the accretion of wealth through the distribution of investment returns (such as interest, royalties, dividends, and rents), for investment value appreciation, or for other benefits. Inventories and fixed assets are not part of investment.

When an investor buys shares, he becomes the owner and be called company shareholder (Anoraga, 2006). According to Djuhana (2009), shares show proof of a company ownership and shareholders have claim rights to company income and assets. Likewise Samsul (2006) states, shares are company ownership proof, and the holders are called shareholders. One or a party can be considered as a shareholder if he is registered in the shareholder list. For some, buying shares is a way to get capital gains in a relatively quick manner. For some others, shares bring about current income in the form of dividends (Kertonegoro, 1995). The basic nature of share investment is that it allows investors' participation in the company profit. This is the value of shares.

PER approach is often used as a reference by investors for decision-making. According to Harmono (2009), price-earnings ratio is the price per share; this indicator has practically been applied in the final profit and loss financial statements and is a form of financial reporting standards for public companies in Indonesia. Therefore, understanding PER is important an can be an indicator of company value in a research model. Samsul (2006) states price-earnings ratio approach is very popular and commonly used in many countries to estimate price per share. Likewise, Sunariyah points out price-earnings ratio approach is the most frequently used approach by investors and security analysts. This approach is based on the expectation of estimated earnings per share in the future so that how long investment will return can be projected. According to Sawir (2005), investors in the well-established capital markets use price-earnings ratios to measure whether a share is underpriced or overpriced. Shares with low price-earnings ratio are underpriced or priced lower than their intrinsic values, and those with high price-earnings ratio are overpriced or priced higher than their intrinsic values. According to Suariyah (2000), the intrinsic value is the true value of a share, determined by some company's fundamental factors. Suad Husnan, cited in Lusiana (2010), suggest that overvalued shares should be avoided or be sold because there may be a market correction in the near future. On the contrary, undervalued shares should be bought or hold because chances are that there will be a surge in market prices in the future." According to PER approach, shares are influenced by two factors: earnings per share (E) and multiplier (R).

Table 1. State of the Art Studies.

Researcher, Title, and Research Period	Studied Variables (X)	Analysis Methods	Result
Abdul Kholid (2006), Analisis Faktor-Faktor yang Mempengaruhi *Price Earning Ratio* Saham-Saham Perusahaan yang Terdaftar di Bursa Efek Jakarta, Research period 2000-2003	Sales growth, ROE growth, DPR, Bank Indondesia Certificates (SBI) rate, DER growth, and ROI growth.	Multiple linear regression, F-Test, and t-Test.	Simultaneous analysis shows that all studied variables (X) have influence on PER. Partial analysis shows that Sales growth, ROE growth, DPR, Bank Indondesia Certificates (SBI) rate, and ROI growth have significant influence on PER. But DER growth has no significant influence on PER.
Farida Wahyu L (2010), Analisis Pengaruh Rasio Likuiditas, Rasio Solvabilitas, Rasio Aktivitas dan Rasio Profitabilitas terhadap *Price Earning Ratio,* Research Period 2006-2009	CR, DER, INTO, and ROE.	Multiple linear regression, F-Test, t-Test, and coefficient of determination.	Simultaneous analysis shows that all studied variables (X) have influence on PER. Partial analysis shows that CR, INTO, and ROE have significant influence on PER. But DER growth has no significant influence on PER.
Parwati Setyorini (2005), Faktor-Faktor yang Mempengaruhi *Price Earning Ratio* Pada Saham LQ 45 di Bursa Efek Jakarta Tahun 2000-2002	DPR, EG, VEG, ROE, and FL.	Multiple linear regression, F-Test, t-Test, and coefficient of determination.	Simultaneous analysis shows that DPR, EG, VEG, ROE, and FL have influence on PER. Partial analysis shows that DPR and VEG have significant influence on PER. And EG, ROE, and FL have no significant influence on PER.
Radot Ruben S (2004), Analisis Faktor-Faktor yang Mempengaruhi *Price Earning Ratio* pada Perusahaan Manufaktur di Bursa Efek Jakarta, Research Period 2001-2002	ROE, PBV, CR, DER, and INTO.	Multiple linear regression, F-Test, and t-Test.	Simultaneous analysis shows that ROE, PBV, CR, DER, and INTO have significant influence on PER. Partial analysis shows that ROE and PBV have significant influence on PER. And CR, DER and INTO have no significant influence on PER.
Rofi Kurniawan (2013), Analisis Faktor-Faktor yang Mempengaruhi *Price Earning Ratio* pada Perusahaan Perbankan yang Terdaftar di Bursa Efek Indonesia, Research Period 2009-2011	LDR, LAR, DER, BOPO, ROA, ROE, DPR, and PBV.	Factor analysis, one-way equation analysis, F-Test, and t-Test.	Of the eight variables (X) studied, only five variables passed the factor analysis test that was compiled into the profitability factors; LDR, BOPO, ROA, ROE, and PBV. Simultaneous analysis shows that all of these variables have significant influence on PER. Partial analysis shows that ROE negatively influences PER and that PBV positively influences PER.
Nizar Hamdani and Asri Solihat (2018), Pengaruh Rasio Likuiditas, Rasio Solvabilitas, Rasio Aktivitas dan Rasio Profitabilitas terhadap *Price Earning Ratio* pada Perusahaan Industri Semen yang Terdaftar di Bursa Efek Indonesia, Research Period 2009-2011	CR, DER, INTO, and ROI.	Multiple linear regression, F-Test, t-Test, and coefficient of determination.	Simultaneous analysis shows that CR, DER, INTO, and ROI have significant influence on PER. Partial analysis shows that INTO and ROI have significant influence on PER. And CR and DER have no significant influence on PER.

3 RESEARCH METHOD

This study employed descriptive and verification methods. The independent variables were current ratio (X_1), debt to equity ratio (X_2), inventory turnover (X_3) and return on investment (X_4). In addition, the dependent variable was price-earnings ratio. The data used in this study is secondary data obtained by downloading the financial statements of cement industry companies listed at Indonesia Stock Exchange from the official website of the Indonesia Stock Exchange (BEI). Multiple linear regression method was used to analyze the influence of independent variables on the dependent variable, after of course normality, multicollinearity, heteroscedasticity, and autocorrelation testing. To obtain more accurate results and facilitate the data processing, data were analyzed by means of SPSS 20.

4 RESULT AND DISCUSSION

Based on Table 2, the average current ratio (CR) of cement industry companies in 2009-2017 is 277.91. There was an increase in average CR in 2009-2012 but a decline in 2013-2017.

Based on Table 3, the average debt-to-equity ratio (DER) of cement industry companies is 56.60% in 2009, 32.75% in 2010, 31.79% in 2011, and 83.70% in 2017. It can be concluded that the DER experienced a decline with an average of 50.58%.

Table 4 shows that the average inventory turnover (INTO) of cement industry companies is 6.21X in 2009, 5.91X in 2010, 6.44X in 2011, and 7,34X in 2017. It can be concluded that the INTO in 2009-2017 experienced fluctuations with the average of 7.15X.

Table 5 shows that the average return on investment (ROI) of cement industry companies is 19.57% in 2009, 17.49% in 2010, 16.56% in 2011, and 2.23%

Table 2. Calculation of Current Ratio in 2009-2017.

| No | Kode | Current Ratio (%) | | | | | | | | |
		2009	2010	2011	2012	2013	2014	2015	2016	2017
1	INTP	300,19	555,37	698,54	602,76	614,81	493,37	488,66	452,50	370,31
2	SMCB	126,99	166,19	146,58	140,46	63,92	60,17	65,67	45,94	54,36
3	SMGR	358	291,79	264,65	170,59	188,24	220,95	159,70	127,25	156,78
Jumlah		785,60	1.013,35	1.109,77	913,81	866,96	774,50	714,02	625,69	581,44
Rata-rata		261,87	337,78	369,92	304,60	288,99	258,17	238,01	208,56	193,81

Table 3. Calculation of Debt-to-Equity Ratio.

| No | Kode | Debt to Equity Ratio (%) | | | | | | | | |
		2009	2010	2011	2012	2013	2014	2015	2016	2017
1	INTP	24,03	17,14	15,36	17,18	15,80	16,54	15,81	15,35	17,54
2	SMCB	120,34	52,9	45,48	44,55	69,78	96,33	105,58	145,18	172,70
3	SMGR	25,43	28,2	34,53	46,32	41,75	37,30	39,04	44,65	60,86
Jumlah		169,80	98,24	95,37	108,06	127,33	150,17	160,42	205,18	251,10
Rata-rata		56,60	32,75	31,79	36,02	42,44	50,06	53,47	68,39	83,70

Table 4. Calculation of Inventory Turnover.

| No | Kode | Inventory Turn Over (X) | | | | | | | | |
		2009	2010	2011	2012	2013	2014	2015	2016	2017
1	INTP	3,93	4,36	5,69	6,45	6,82	6,95	6,21	5,47	5,31
2	SMCB	9,61	8,41	8,73	9,01	9,91	11,30	11,00	13,57	10,45
3	SMGR	5,10	4,97	4,90	4,80	5,50	5,65	6,25	6,41	6,25
Jumlah		18,64	17,74	19,32	20,26	22,22	23,90	23,46	25,45	22,01
Rata-rata		6,21	5,91	6,44	6,75	7,41	7,97	7,82	8,48	7,34

Table 5. Calculation of ROI.

No	Kode	Return On Investment (%)								
		2009	2010	2011	2012	2013	2014	2015	2016	2017
1	INTP	20,69	21,01	19,84	20,93	19,61	17,84	15,41	12,60	6,37
2	SMCB	12,33	7,96	9,71	11,10	6,39	3,89	1,01	-1,44	-3,86
3	SMGR	25,68	23,51	20,12	18,54	17,37	16,22	11,86	10,25	4,17
Jumlah		58,70	52,48	49,67	50,57	43,37	37,95	28,28	21,42	6,68
Rata-rata		19,57	17,49	16,56	16,86	14,46	12,65	9,43	7,14	2,23

Table 6. Result of Multiple Linear Regression.

coefficients[a]

Model	Unstandardized coeficients		Standardized coeficients		
	B	Std. Error	Beta	t	Sig.
1					
(Constant)	40.988	5.725	.242	7.160	.002
CR	.003	.003	.003	.975	.385
DER	.000	.027	-1.682	.008	.994
INTO	-1.958	.621	-1.880	-3.154	.034
ROI	-.741	.160		-4.635	.010

a. Dependent Variable: PER

Table 7. Result of Simultaneous F-test.

ANOVA[b]

Model		Sum of Squares	Df	Mean Square	F	Sig.
1	Regression	42.782	4	10.696	6.969	.043[a]
	Residual	6.139	4	1.535		
	Total	48.921	8			

a. Predictors: (Constant), ROI, CR, DER, INTO
b. Dependent Variable: PER

in 2017. It can be concluded that the ROI experienced a decline in 2009 – 2017 with the average of 12.93%.

What follows is the result of data analysis using SPSS 20.

The above multiple linear regression results in the following regression equation:

PER = 40.988 + 0.003 CR + 0.000 DER – 1.958 INTO – 0.741 ROI   (1)

Based on the above equation, it can be seen that the constant value of the equation is 40 40.988. It means that if CR, DER, INTO, and ROI = 0, the PER is 40.988.

The regression coefficient of Variable X_1 (CR) is 0.003. It means that if the values of other independent variables remain the same and the CR experiences an increase of 1%, the PER will experience an increase of 0.003%. The positive coefficient indicates positive correlation between the CR and the PER. The higher the CR, the higher the PER.

The regression coefficient of Variable X_2 (DER) is 0.000. It means that if the values of other independent variables remain the same and the DER experiences an increase of 1%, the PER will experience a change. The percentage of the change cannot

be made certain because the SPSS analysis shows a value with three decimal digits. Thus, the authors conclude that the change in the PER is very small (near 0).

The regression coefficient of Variable X_3 (INTO) is -1.958. It means that if the values of other independent variables remain the same and the INTO experiences an increase of 1%, the PER will experience a decline of 1.958%. The negative coefficient indicates negative correlation between the INTO and the PER. The higher the INTO, the lower the PER.

The regression coefficient of Variable X_4 (ROI) is -0.741. It means that if the values of other independent variables remain the same and the ROI experiences an increase of 1%, the PER will experience a decline of 0.741%. The negative coefficient indicates negative correlation between the ROI and the PER. The higher the ROI, the lower the PER.

Table 7 presents the result F-Test calculation using SPSS 20.

The SPPS calculation results in the observed F of 6.969 with the significance level of 5% with df1=number of variables -1 or 5-1 = 4 and df2=n-k-1 or 9-4-1=4 where n is the number of samples and k is the number of independent variables. Based on Ms. Excel computation equation =finv(0.05,4,4) the critical F value is 6.388. Since the observed F> critical F (6.969 > 6.388), H_o is rejected and H_a is

Table 8. T-test reults.

		Coefficients			
	Unstandardized Coefficients		Standard-ized Coef-ficients	T	Sig.
Model	B	Std. Error	Beta		
1 (Constant)	40.98-8	5.725		7.160	.002
CR	.003	.003	.242	.975	.385
DER	.000	.027	.003	.008	.994
INTO	-1.958	.621	-1.682	-3.154	.034
ROI	-.741	.160	-1.880	-4.635	.010

a. Dependent Variable: PER

accepted. It means that current ratio, debt-to-equity ratio, inventory turnover, and return on investment together influence price-earnings ratio (PER) significantly. The statistical calculation results in the probability (significance) value of 0.043. Since this significance value of 0.043 is lower than 0.05, H_o is rejected and H_a is accepted. It means that current ratio, debt-to-equity ratio, inventory turnover, and return on investment together influence price-earnings ratio (PER). Table 8 presents the result of t-Test using SPSS 20.

With the significance level of 0.05 and df = n-k-1 or 9-4-1, it was obtained the critical t of 2.776 using Ms. Excel computation equation = tinv(0.05,4).

Table 9 presents the calculation result of determination coefficient (R^2) using SPSS 20.

The SPSS output shows that the determination coefficient (R^2) is 0.749. It means that the influence of all independent variables; i.e., current ratio, debt-to-equity ratio, inventory turnover, and return on investment, simultaneously on the dependent variable, price-earnings ratio (PER), that can be explained by this equation model is 74.9%, and the rest (100% - 74.9% = 25.1%) is influence by other factors.

Table 9. Calculation Result of Determination Coefficient (R^2).

Model Summary[b]

Model	R	R Square	Adjusted R Square	Std. Error of the Estimate	Durbin-Watson
1	.935[a]	.875	.749	1.23886	1.847

a. Predictors: (Constant), ROI, CR, DER, INTO
b. Dependent Variable: PER

5 CONCLUSION

Based on the result of analysis and discussion, the authors come to the following conclusions:

1. The SPSS processing results in the observed F of 6.969. With significance level of 5% the critical F is 6.388. Since the observed F > critical F (6.969 > 6.388), H_o is rejected and H_a is accepted. It means that current ratio, debt-to-equity ratio, inventory turnover, and return on investment simultaneously have significant influence on the price-earnings ratio (PER).
2. The statistical calculation results in the probability (significance) value of 0.043. Since this significance value of 0.043 is lower than 0.05, Ho is rejected and Ha is accepted. It means that current ratio, debt-to-equity ratio, inventory turnover, and return on investment simultaneously have significant influence on the price-earnings ratio (PER).
3. The partial influence of current ratio, debt to equity ratio, inventory turnover, return on investment on the price-earnings ratio (PER) is as follows:
   a. The result of the test shows that current ration (CR) has no significant influence on the price-earnings ratio (PER). The influence of CR on PER is merely 0.242 or 24.2%.
   b. The result of the test shows that debt-to-equity ratio (DER) has no significant influence on the price-earnings ratio (PER). The influence of DER on PER is merely 0.003 or 0.3%.
   c. The result of the test shows that inventory turnover (INTO) has significant influence on the price-earnings ratio (PER). The influence of INTO on PER is 1.682 or 168.2%.
   d. The result of the test shows that return on investment (ROI) has significant influence on the price-earnings ratio (PER). The influence of ROI on PER is 1.880 or 188%.

It is recommended that further studies take into account of other ratios that are not discussed in this study such as quick ratio, total assets turnover, net profit margin, dividend payout ratio, and other ratios that are assumed to have influence on the price-earnings ratio.

REFERENCES

Anoraga, Pandji dan Piji Pakarti. (2006). Pengantar Pasar Modal. Edisi Revisi. Jakarta: Penerbit Rineka Cipta.
Djuhana, Djudju. (2009). Pasar Uang dan Pasar Modal. Garut: Penerbit Universitas Garut.
Ghozali, Imam. (2011). Aplikasi Analisis *Multivariate* dengan Program IBM SPSS 19. Cetakan ke Lima. Semarang: Badan Penerbit Universitas Diponegoro.
Hamdani, N. A., Abdul, G. and Maulani, F. (2018) 'The influence of E-WOM on purchase intentions in local

culinary business sector', International Journal of Engineering & Technology, 7, pp. 246–250.

Harahap, Sofyan Syafri. (2007). Analisis Kritis Atas Laporan Keuangan. Edisi ke Lima. Jakarta: Penerbit Raja Grafindo Persada.

Harmono. (2009). Manajemen Keuangan. Jakarta: Penerbit Bumi Aksara.

Hasan, Iqbal. (2006). Analisis Data Penelitian Dengan Statistik. Jakarta: Penerbit Bumi Aksara.

Ikatan Akuntan Indonesia (2004), Standar Akuntansi Keuangan, Penerbit Salemba Empat, Jakarta.

Indonesian Stock Exchange (2008). Investasi di Pasar Modal. Jakarta: Penerbit: Bursa Efek Indonesia.

Indonesian Stock Exchange (2018). Laporan Keuangan (online). Tersedia:http://www.idx.co.id/id-anda/perusahaantercatat/laporankeuangandantahunan.asp xunduh tanggal 15 Juli 2018.

Indonesian Stock Exchange (2017). Sejarah, visi, misi dan struktur pasar modal (online). Tersdia:http://www.idx.co.id/idanda/perusahaantercatat/sejarah,visi,misidanstrukturmodal.aspxunduh tanggal 15 Jul 2018.

Kasmir. (2012). Analisis Laporan Keuangan. Edisi Revisi. Jakarta: Penerbit Raja Grafindo Persada.

Kertonegoro, Sentanoe. (1995). Analisa dan Manajemen Investasi. Jakarta: Penerbit Widya Press.

Advances in Business, Management and Entrepreneurship – Hurriyati et al (eds)
© *2020 Taylor & Francis Group, London, ISBN 978-0-367-27176-3*

Effects of accounting information and environmental information on investor's decisions: An experimental study

A. Ardianto & F. Farhanah
Faculty of Economic and Business, Airlangga University, Surabaya, Indonesia

ABSTRACT: This study aimed to examine the effects of accounting information, environmental information, and the interaction between accounting information and environmental information on investors' decisions. An experimental research method using 2 × 3 factorial designs (between-subject) was adopted. T covariance (ANCOVA) as an additional analysis to estimate the value of confounding variables that could potentially affect investors' decisions. The study subjects were 86 undergraduate students in the Accounting Faculty of Economics and Business, Airlangga University. The results of this research showed that accounting information and environmental information have an effect on investors' decisions, whereas the interaction between accounting information and environmental information does not have an effect on investors' decisions.

1 INTRODUCTION

Interest in learning more about the relationship between a company's business activities and environmental conditions has been increasing . In Indonesia, environmental pollution is a serious problem that has received special attention and should be addressed immediately. Forest fires, a water crisis, and an increase in the volume of wastes are among the many factors causing environmental problems in Indonesia. The importance of managing and preserving the environment for companies in Indonesia is evident in the government's stipulation of Law No. 32 of 2009 on the Protection and Management of the Environment.

Nugroho (2013) states that it is not uncommon for companies to ignore the social and environmental impacts of the company's economic activities in sole pursuit of material-oriented goals. Companies that exploit natural resources and society uncontrollably potentially cause damage to the natural environment such as deforestation; soil, air, and water contamination; and climate change that will ultimately interfere with human life.

Previous studies that tested the use of environmental information on corporate values in the business world included Al-Tuwaijri, Christensen, and Hughes (2004), Anderson and Frankle (1980), Belkaoui (1976), Chan and Milne (1999), Cormier et al. (2011),Cormier and Magnan (2014), Guidry and Patten (2010), Hassel and Nilsson (2005), Hendricks (1976), Ingram (1978), Madein and Sholihin (2015), Madein and Sholihin (2015), Moneva and Ortas (2010), Murray et al. (2006), Rikhardsson and Holm (2006), and Teoh and Shiu (1990). Most of these studies show that disclosure of environmental issues is important because it contains information relevant to decision-making. These conditions indicate that stakeholders consider environmental information in decision-making. Previous studies have explained much about the effects of accounting and environmental information on investor decisions, but no specific research has explained the influence of each of the main effects, and the interaction effect between accounting information and environmental information investor decisions using experimental methods where the internal validity obtained is relatively high compared with the methods that have been used in previous research.

This research tested whether there are effects of the accounting information variable (main effect) on investor decisions, of the environmental effect variable (main effect) on investment decisions, and of the interaction between accounting information and environmental information on investor decisions of whether to invest or do not invest using experimental methods. The researchers examined all of the effects to confirm any significant effect of in decision- making related to the independent variables of accounting information and environmental information. The research subjects were undergraduate accounting students in the Faculty of Economics and Business at Airlangga University who had taken a management accounting course, financial management 1, and financial management 2. Students were chosen as research subject in an effort to eliminate the experience factor, which can interfere with this experimental test.

2 THORETICAL REVIEW

2.1 *Stakeholder theory*

Robert Edward Freeman is a pioneer of the corporate stakeholders approach with a book entitled *Strategic*

Management: A Stakeholder Approach, published in 1984 (Donaldson & Preston, 1995). The main idea of this theory is the need for organizations to manage relationships with stakeholder groups or individuals who can affect or be influenced by the achievement of corporate goals (Freeman 1984). The success of an organization depends on how well organizations manage relationships with key groups such as customers, employees, suppliers, societies or communities, financiers, and others that can affect the achievement of their goals (Freeman & Phillips 2002), including the environment (Starik 1995; Gibson 2012). Freeman (1984) noted that this systematic stakeholder theoretical approach a rational one for organizational managers if they want to manage the organization effectively. This theory predicts that managers will consider stakeholders, which in this case are the environment.

2.2 *Theory of reasoned action*

The Theory of Reasoned Action was developed by Ajzen and Fishbein (1975) and explains that behavior is performed because the individual has the intention to do so and is related to volitional activities. Volitional behavior is based on the assumption, first, that humans do things in a reasonable way, and second, that humans consider all information. Third, either explicitly or implicitly human beings take into account the implications of their actions. The intent to act is a function of two basic determinants, one related to personal factors and the other to social influences.

The theory of intent to behave (Ajzen & Fishbein 1975), just basing and declaring one's intention to behave, is influenced only by two factors: attitude and subjective norms. It is therefore still widely open for constructive developments in a specific behavior. Individual behavior is also indirectly influenced by external variables that then also interact with environmental factors in determining behavior. These external variables are demographics, personality characteristics, beliefs about objects, attitudes toward objects, task characteristics, and situational. Attitudes interact with each other.

3 HYPOTHESIS DEVELOPMENT

Stakeholder theory states that managers as administrators of the company should pay attention to all stakeholders including investors, creditors, governments, communities, employees, and the environment. Profit is no longer the main focus of the company, but rather the sustainability of the company (the going concern assumption) in the future, which in this case depends on environmental factors. Companies that do not pay attention to environmental stakeholders will tend to make decisions that are not on the side of the environment, resulting in pol-

lution and environmental damage. In the event of pollution and environmental damage, the potential for the company to not be trusted by stakeholders or even suffer loss and bankruptcy is higher, because facing the demands of the people affected by pollution and addressing the damage require relatively large funds.

The Theory of Reasoned Action states that a behavior is performed because the individual has the intention to engage in that behavior based on self-belief and will. In addition, if the individual will make a decision, then all information will be considered. This study discusses the decisions of investors when given accounting and environmental accounting information on whether to invest or not to invest. The decision to make an investment based on the Theory of Reasoned Action is influenced by investor confidence and willingness, and consideration of all available information. Where accounting information and environmental information are provided, according to this theory, both types of information will be considered by the decision maker, in this case the investor.

Hypothesis 1: Accounting information affects investor decisions.
Hypothesis 2: Environmental information affects investor decisions.
Hypothesis 3: The interaction between accounting information and environmental information affects investor decisions.

4 RESEARCH METHODOLOGY

This experimental study used a 2 × 3 factorial design between subjects (between-subject). The subjects of this study were undergraduate students of accounting in Faculty of Economics and Business of Airlangga University, who are or have taken the course in management accounting and financial management 1 and financial management 2.

The experiment consisted of a control group and an experimental group. Each group received different treatment. There are two kinds of manipulation: accounting information and environmental information. Manipulation of information is done by providing information in good and bad forms. The control group is given only accounting information, while the experimental group is given both accounting information and environmental information.

The research subjects were asked to behave like actual investors who will make a decision to invest or not. The decision is represented by a scale of 1 to 10. The scale 1–5 (the left) shows the decision not to invest, while the 6–10 scale (right) shows the decision to invest.

Before choosing to invest or not, the subject will be given a material case containing (1) current condition of the research subject, (2) company condition (company profile), (3) company financial analysis (financial report attached) accounting, and (4) environmental information described in the form of a summary of sustainability reports (for experimental groups). For manipulation of no environmental information, the researcher directly gives a sheet to decide investment decisions based on accounting information only.

5 RESULTS

5.1 Influence of accounting information on investors' decisions

The first hypothesis in this research is that there is a difference in decision-making between investors

Table 1. Descriptive statistics.

Accounting Information	Environmental Information	Main	Std. deviation	N
Good	Good	6.8235	1.46779	14
	Not Good	4.7273	2.28433	14
	None	6.9333	1.90738	15
	Total	6.3256	2.04382	43
Not Good	Good	4.9286	1.73046	14
	Not Good	2.8667	1.64172	15
	None	3.4286	1.86936	14
	Total	3.7209	1.91890	43
Total	Good	5.9677	1.83455	28
	Not Good	3.6538	2.11551	29
	None	5.2414	2.57259	29
	Total	5.0233	2.36632	86

Table 2. Test of between-subjects effects (ANOVA).

Source	Type III sum of squares	df	Mean square	F	Sig.
Corrected Model	216.277[a]	5	43.255	13.326	0.000
Intercept	2072.116	1	2072.116	638.369	0.000
Accounting Information	123.760	1	123.760	38.128	0.000
Environmental Information	61.225	2	30.613	9.431	0.000
Accounting Information * Environmental Information	12.620	2	6.310	1.944	0.150
Error	259.676	80	3.246		
Total	2646.000	86			
Corrected total	475.953	85			

$R^2 = 0.454$ (Adjusted $R^2 = 0.420$)

Table 3. Test of between-subjects effects (ANCOVA).

Source	Type III sum of squares	df	Mean square	F	Sig.
Corrected Model	232.023[a]	10	23.202	7.134	.000
Intercept	14.926	1	14.926	4.589	.035
Sex	2.211	1	2.211	.680	.412
Age	3.787	1	3.787	1.164	.284
GPA	.037	1	.037	.011	.916
Semester	.508	1	.508	.156	.694
Experience	7.799	1	7.799	2.398	.126
Accounting Information	106.276	1	106.276	32.676	.000
Environmental Information	50.835	2	25.418	7.815	.001
Information Accounting * Environmental Information	9.540	2	4.770	1.467	.237
Error	243.931	75	3.252		
Total	2646.000	86			
Corrected total	475.953	85			

$R^2 = 0.487$ (Adjusted $R^2 = 0.419$)

who are given accounting information (good and not good) and those not given the information. The result of data analysis shows that the significance value of the main effect of accounting information is 0.000 (<5%). It shows that there is a significant effect of accounting information on investors' decisions.

The implication of this effect is that investors will consider accounting information to determine whether or not to invest in a company. Accounting information is of great concern to all stakeholders including investors, As accounting information is an overall portrait of company performance over a certain period. Therefore, good accounting information and in accordance with the interests of stakeholders will make stakeholders choose to invest, but if the opposite occurs, that is, accounting information that is not good, then investors will choose not to invest. This is evidenced by researchers by manipulating accounting information variables with good and not good levels. The results of this study also support the previous research of of Farj et al. (2016), Hai et al. (2015), and Nagy and Obenberger (1994). These studies explain the influence of accounting information on decision-making by stakeholders, especially investors.

5.2 The influence of environmental information on investor decisions

The second hypothesis in this study is that there is a difference in decision-making between investors who are given environmental information (none, good, and not good) and those who are not. The result of data analysis shows that the significance value of the main effect of environmental information is 0.000 (<5%). This indicates that there is a difference in investors' decisions between those who are given environmental information (none, good, or not good) and those who are not, or there is significant effect of environmental information on investors' decisions.

The influence of environmental information on decisionmaking by stakeholders, especially investors, is also explained in previous studies by Cormier et al. (2011), Cormier and Magnan (2014), Hussainey and Salama (2010), Milne and Patten (2002), Murray et al. (2006), Said et al. (2014), Rikhardsson and Holm (2006), and Teoh and Shiu (1990). These studies explain that environmental information in the form of Corporate Social Responsibility (CSR) contained in the Sustainability Report is one of the important sources of information that investors consider in making decisions to invest.

5.3 The influence of the interaction between accounting information and environmental information on investors' decisions

The third hypothesis in this research is that there is an effect of the interaction between accounting

information and environmental information on investors' decisions. This hypothesis examines the effect of interaction between accounting information variables and environmental information on investor decisions. The result of data analysis shows that the significance value of the interaction effect is 0.150 (>5%). It shows that the the interaction of accounting information and environmental information has no significant effect on investors' decisions. That the influence of the interaction shows an insignificant value can also be interpreted as indicating that the environmental information variable is the moderation variable of the accounting information variable, which means that the environmental information variable only strengthens or weakens the variable of accounting information. If environmental information is good, it will strengthen the relationship between accounting information and investor decisions, but if the environmental information is bad, it will weaken the relationship between accounting information and investor decisions.

6 CONCLUSION

Based on experimental research on the effect of accounting information and environmental information on investors' decisions, it can be concluded that accounting information (main effect) and environmental information (main effect) have significant effects on investor decision. Thus, the use of accounting information and environmental information at both good and not good levels has an effect on investors' considerations in making investment decisions. Meanwhile, an interaction effect between accounting information with environmental information does not significantly influence investors' decisions. The use of accounting information and environmental information in combination (interaction) has no effect on investor decision making.

REFERENCES

Ajzen, I. & Fishbein, M. 1975. Understanding Attitudes and Predicting Social Behavior. Englewood Cliffs, NJ: Prentice-Hall.

Al-Tuwaijri, S.A., Christensen, T.E. & Hughes, K.E. 2004. The relations among environmental disclosure, environmental performance, and economic performance: a simultaneous equations approach. Accounting, Organizations and Society 29(5): 447–471. doi:https://doi.org/10.1016/S0361–3682(03)00032–1

Anderson, J.C. & Frankle, A.W. 1980. Voluntary social reporting: an iso-beta portfolio analysis. The Accounting Review 55(3): 467–479.

Belkaoui, A. 1976. The impact of the disclosure of the environmental effects of organizational behavior on the market. Financial Management 5(4): 26–31.

Chan, C.C.C. & Milne, M.J. 1999. Investor reactions to corporate environmental saints and sinners: an experimental analysis. Accounting and Business Research 29(4): 265–279.

Cormier, D., Ledoux, M.J. & Magnan, M. 2011. The information contributionn of social and environmental disclosure for investors. *Management Decision* 49(8): 1276–1304.

Cormier, D. & Magnan, M. 2014. The impact of social responsibility disclosure and governance on financial analysts. *Corporate Governance* 14(4): 467–484.

Donaldson, T. & Preston, L.E. 1995. The stakeholder theory of corporation: concepts: evidence and implications. *The Academy of Management Review* 20(1): 65–91.

Farj, R.M.H., Jais, M.B. & Isa, A.H.B.M. 2016. Importance of accounting information to investors in the stock market: a case study of Libya. *IOSR Journal of Economics and Finance* 7(1): 70–79.

Freeman, R.E. 1984. *Strategic Management: A Stakeholder Approach*. Boston: Pitman.

Freeman, R.E. & Phillips, R.A. 2002. Stakeholder theory: a libertarian defense. *Business Ethics Quarterly* 12(3).

Gibson, K. 2012. Stakeholders and sustainability: an evolving theory. *Journal of Business Ethics* 109(1): 15–25.

Guidry, R.P. & Patten, D.M. 2010. Market reactions to the first-time issuance of corporate sustainability reports: evidence that quality matters. *Sustainability Accounting, Management and Policy Journal* 1(1): 33–50.

Hai, T.T.T., Diem, N.N. & Binh, H.Q. 2015. The relationship between accounting information reported in financial statements and stock returns: empirical evidence from Vietnam. *International Journal of Accounting and Financial Reporting* 5(1): 229–238.

Hassel, L. & Nilsson, H. 2005. The value relevance of environmental performance. *European Accounting Review* 14(1): 41–61.

Hendricks, J.A. 1976. The impact of human resource accounting information on stock investment decisions: an empirical study. *The Accounting Review* 51(2): 292–305.

Hussainey, K. & Salama, A. 2010. The importance of corporate environmental reputation to investors. *Journal of Applied Accounting Research* 11(3): 229–241.

Ingram, R.W. 1978. An investigation of the information content of (certain) social responsibility disclosures. *Journal of Accounting Research* 16(2): 270–285.

Madein, A. & Sholihin, M. 2015. The impact of social and environmental information on managers' decisions. *Asian Review of Accounting* 23(1): 156–169.

Milne, M.J. & Patten, D M. 2002. Securing organizational legitimacy: an experimental decision case examining the impact of environmental disclosures. *Accounting, Auditing & Accountability Journal* 15(3): 372–405.

Moneva, J.M. & Ortas, E. 2010. Corporate environmental and financial performance: a multivariate approach. *Industrial Management & Data Systems* 110(2): 193–210.

Murray, A., Sinclair, D., Power, D. & Gray, R. 2006. Do financial markets care about social and environmental disclosure? Further evidence and exploration from the UK. *Accounting, Auditing & Accountability Journal* 19(2): 228–255.

Nagy, R.A. & Obenberger, R.W. 1994. Factors influencing individual investor behavior. *Financial Analysts Journal* 50(4): 63–68.

Rikhardsson, P. & Holm, C. 2006. The effect of environmental information on investment allocation decisions: an experimental study. *Business Strategy and the Environment* 17(6).

Said, R.M., Sulaiman, M. & Ahmad, N.N.N. 2014. Environmental information usefulness to stakeholders: empirical evidence from Malaysia. *Social Responsibility Journal* 10(2): 348–363.

Starik, M. 1995. Should trees have managerial standing? Toward stakeholder status for non-human nature. *Journal of Business Ethics* 14: 207–217.

Teoh, H.Y. & Shiu, G.Y. 1990. Attitudes towards corporate social responsibility and perceived importance of social responsibility information characteristics in a decision context. *Journal of Business Ethics* 9(1): 71–77.

Comparison of the quality between net income and total comprehensive income in an IFRS implementation context in Indonesia: Empirical study on companies going public that are listed on the Indonesia stock exchange in the period 2011–2014

A. Rizki & O.D. Megayanti
Universitas Airlangga, Surabaya, Indonesia

ABSTRACT: This research aimed to investigate the quality of total relative comprehensive income compared with the quality of the net income arranged according to the International Financial Report Standard (IFRS) for companies listed on the Indonesia Stock Exchange in the period 2011–2014. To compare the quality between net income and total comprehensive income, this study focused on the two attributes of profit, value relevance and predictive value. This research used a combination of time series and cross-sectional data. The tests used in this study were Ordinary Least Squares (OLS) and multiple linear regression. The results show that net income has a more relevant value than the total comprehensive income in explaining the market value and return on organization shares return. The net income also can better predict the future operating cash flow and profit than the total comprehensive income can.

1 INTRODUCTION

The global economic environment, where a company conducts business activities not only in a regional territory, requires that the company's financial statements are presented in compliance with internationally accepted accounting standards. The International Financial Reporting Standards (IFRS) have been adopted by Indonesia Financial Accounting Standards (IFAS) since 2011, and the latest Financial Accounting Standard Guidelines (FASG) revision has been in effect since January 1, 2012. GAAP-based FASG convergence with FASG-based IFRS has an impact on the application of the principle-based concepts of fair value, professional judgment, and full disclosure. One form of IFRS adoption by FASG, IAS 1, has been adopted in FASG 1. It concerns the presentation of elements of financial statement is the shift of income statements to total comprehensive income with profit classification change in net income and total comprehensive profit, which is the sum of other comprehensive income and net income. Owing to the use of total comprehensive income in the IFRS-based financial statements, the controversial issue of total comprehensive income is debated among academics and stakeholders for the best performance measurement of a firm in total comprehensive income, net income, or other accounting elements (Devalle 2012).

According to Goncharov and Hodgson (2011), the disclosure of total comprehensive income provides insight to investors about the future prospects of the company, as well as the future profitability predictions and cash flow advances. Tsujiyama (2007) stated that total comprehensive income can provide objective and more useful information related to changes in net assets, given that assets and liabilities possessed by the entity objectively observed. Previous research conducted by Jaweher and Mounira (2014) stated that net income has more relevant value than total comprehensive income in predicting future operating cash flow and revenue, is more consistent, more timely, and has better accrual quality. Devalle (2012) used the variable value relevance in evaluating whether total comprehensive income is more relevant than net income and showed that total comprehensive income is better in terms of value relevance compared to net income.

Pronobis and Zülch (2011) stated in their research that there is no evidence to suggest that total comprehensive income has an advantage in predicting the company's future operating performance compared to net income. Research conducted by Krismiaji et al. (2013) showed that IFRS adoption positively affects the relevance of the value of accounting information and faithful representation. Kusumo and Subekti (2013) analyzed the relevance of the value of accounting information prior toand after the adoption of IFRS to companies listed in the Indonesia Stock Exchange. The study showed that standardized IFRS implementation in Indonesia has not been able to improve the quality of accounting information. Research on the comparison of the ability between net income and total comprehensive income in presenting the attribute of quality of profit named value relevance and predictive value in Indonesia have never been done before. The authors therefore used some of the aforementioned profit quality factors to determine the quality of net income and total comprehensive income after the enactment of IFRS in Indonesia.

1.1 Agency theory

Jensen and Meckling (1976) argued that agency relationships are a contract between the management (agent) and the investor (principal). Separation between management and owners can lead to potential agency conflicts that may affect the quality of financial statement information. It may be assumed that management and owners have different interests and that it can lead to a conflict of interest, whereby management seeks to maximize profits for the purpose of obtaining bonuses and other personal benefits while the owner wants high profits. Management actions that seek to maximize profits by not reporting the actual information to the owner result in the emergence of information asymmetry. Information asymmetry is a condition in which there is an imbalance of information acquisition between management as a provider of information (preparer) with shareholders and stakeholders in general as the users of information (user) (Ujiyantho & Pramuka 2007).

2 HYPOTHESIS DEVELOPMENT

2.1 The ability of net income vs total comprehensive income in presenting the profit quality attribute named value relevance in the context of IFRS implementation in Indonesia

Separation between management and owners can lead to potential agency conflicts that may lead to management reporting earnings opportunistically in order to maximize its personal interests (profit management), and if that happens it can lead to poor earnings quality (Rachmawati & Triatmoko 2007). Gu, in Pratiwi and Eddy (2012), stated the relevance of the value of profit is the ability to explain (explanatory power) the effect of the accounting information on the price or return of stock. Jaweher and Mounira (2014) stated that they evaluate the ability of net income and total comprehensive income to summarize the company's performance reflected in stock prices and stock returns.

Several previous studies on the value relevance of net income and comprehensive profits showed different results, such as research conducted by Jaweher and Mounira (2014) and Devalle (2012) showing that total comprehensive income is not superior in value relevance compared to net income. Conversely, research conducted by Groen (2010) showed that total comprehensive income is better in value relevance compared with net income. In Indonesia previous similar research conducted by the Krismiaji et al. (2013) showed that IFRS positively influences the value relevance of accounting information as measured by the ability to predict and to provide faithful

representation. Based on the theoretical basis and previous research, the following hypothesis is proposed:

Hypothesis 1: Net income has the ability to present the quality of profit that is the value relevance better than the total comprehensive income in the context of IFRS implementation in Indonesia.

2.2 The ability of net income vs. total comprehensive income in presenting the quality attribute of profit that is predictive value in the context of IFRS implementation in Indonesia

Pronobis and Zülch (2011) suggest that the predictive power of income or profit figures is a high value attribute of relevance to the analysis because it reduces their forecasting risk.

Predictability implies that the profit presented should provide information that can be used as a predictor in the performance appraisal process for both cash and revenue (Jaweher & Mounira 2014). The main purpose of profit reporting is to assist investors and creditors in predicting future cash flows. If the information is relevant to investors, this information will be used to analyze and interpret the value of the shares of the company concerned (Vestari 2012).

Several previous studies related to predicted values quality between net income and total comprehensive income. Pronobis and Zülch (2011) and Jaweher and Mounira (2014) showed that total comprehensive income is no better than net income in predictive power quality. Different results are shown by Krismiaji et al. (2013), that is, the application of IFRS in Indonesia has been shown to increase the relevance of values as measured by predicted values. Based on the theoretical basis and previous research, the following hypothesis is proposed:

Hypothesis 2: Net income has the ability to present a better quality of profit predictive value compared to total comprehensive income in the context of IFRS implementation in Indonesia.

3 METHOD

3.1 Population and sample

The population used in this study was a nonfinancial company listed on the Indonesia Stock Exchange in 2011–2014. The sampling method used was purposive sampling with the aim to obtain representative samples in accordance with the criteria specified. There are 125 of companies that meet the specified criteria.

3.2 Research variables

3.2.1 Value relevance
Jaweher and Mounira (2014) stated that there are two choices in measuring the relevance of earnings

value: measurement perspective and signaling perspective. In measuring the value relevance of profit, this study used the measurement perspective, namely the ability of profit in explaining market data with the following formula:

$$MVEit/MVEi, (t-1) = a0 + a1NIit/ \\ MVEi, (t-1)+ \\ a2BVit/MVEi, (t-1)+ \\ roit \quad (1)$$

$$MVEit/MVEi, (t-1) = P0 + P1TCIit/ \\ MVEi, (t-1)+ \\ P2BVit/MVEi, (t-1)+ \\ kit \quad (2)$$

$$RETURNit = a0 + a1NIit/MVEi, (t-1)+ \\ a2VARNIit/MVEi, (t-1) + roit \quad (3)$$

$$RETURNit = P0 + p1TCIit/MVEi, (t-1)+ \\ P2VARTCIit/MVEi, (t-1) + kit \quad (4)$$

where MVEit/MVEi, (t – 1) = market value equity of company i 3 months after the fiscal (t) year ended divided with company i's MVE in year t – 1; BVit/MVEi, (t – 1) = book value equity company i in period (t) divided with company i's MVE in year (t – 1); Nit/MVEi, (t – 1) = net income company i in period (t) divided with company i's MVE in year t – 1; TCIit/MVEi, (t – 1) = total comprehensive income company i in period (t) divided with company i's MVE in year t – 1; RETURNit = the return rate of company i's stock in period (t); VARNit/MVEi, (t – 1) = annual change in company i's net income in period t divided with company i's MVE in year t – 1; VARTCIit/MVEi, (t – 1) = Annual change in the total comprehensive income of company i in period (t) divided with company i's MVE in year t – 1.

3.2.2 Predictive value
Based on previous research conducted by Brochet et al. (2008), Gordon et al. (2010), and Jemaa et al. (2015), where predictive value variables were assessed by assessing the strength of predictions and the relationship of earnings, both net income and total comprehensive income in the previous year, to the current operating cash flow and corporate earnings. The research model for comparison of quality predictive value between net income and total comprehensive income is as follows:

$$OCFi, t/TAi, (t-1) = a0 + a1NIi, (t-1)/ \\ TAi, (t-1) + si, t \quad (5)$$

$$OCFi, t/TAi, (t-1) = P0 + p1TCIi, (t-1)/ \\ TAi, (t-1) + si, t \quad (6)$$

$$NIi, t/TAi(t-1) = a0 + a1NIi(t-1)/TAi (t-1) \\ + si, t \quad (7)$$

$$TCIi, t/TAi(t-1) = P0 + p1TCIi, (t-1)/ \\ TAi(t-1) + si, t \quad (8)$$

where OCFi, t/TAi, (t – 1) = operating cash flows of company i in year t divided with the total asset of company i in year t – 1; NIi, (t – 1)/TAi, (t – 1) = net income of company i in year t divided with the total asset of company i in year t – 1; TCIi, (t – 1)/TAi, (t – 1) = total comprehensive income of company i in year t divided with the total asset of company i in year t – 1; NIi, t/TAi, (t – 1) = net income of company i in year t – 1 divided with the total asset of company i in year t – 1; TCIi, t/TAi, (t – 1) = total comprehensive income of company i in year t – 1 divided with the total asset of company i in year t – 1.

4 RESULTS AND DISCUSSION

Table 1. Descriptive statistic result of predictive value variable in period 2011–2014.

	N	Min	Max	Mean	Std. deviation
OCF (t)/ TA (t – 1)	500	–0.4908	3.0401	0.126892	0.2072042
NI (t – 1)/ TA (t – 1)	500	0.0015	1.0837	0.097956	0.0889154
TCI (t – 1)/TA (t – 1)	500	0.0005	1.0837	0.101239	0.0896321
NI(t)/TA (t – 1)	500	0.0001	0.7031	0.109930	0.0948397
TCI (t)/TA (t – 1)	500	0.000043	0.7031	0.113988	0.0970411
MVE (MAR CH)/MVE (t – 1)	500	0.2036	6.5068	1.482689	8034906
NI/MVE (t – 1)	500	0.000040	0.8601	103161	0.838882
BV/MVE (t – 1)	500	0.0004	4.8030	853154	6878713
TCI/MVE (t – 1)	500	0.000038	0.8601	110518	0.987172
Return	500	–0.96	3.49	0.2314	0.53179
VARNI/ MVE (t – 1)	500	–0.6859	0.6330	013392	0.716856
VARTCI/ MVE (t – 1)	500	–0.6963	0.8401	029750	1038275

512

Table 2. Multiple regression test result (adj R^2) value.

Model								M 1.1						
Price model	a1	t	Sig.	a2	t	Sig.	F	Sig.	Adj.. R^2
	446	9.290	000	(.137)	2.865	004	45. 565	000	152
Nbr.					500				

Model								M 1.2						
Price model	B1	t	Sig.	B2	t	Sig.	F	Sig.	Adj.. R^2
	392	7.983	000	(.111)	-2.270	024	34. 185	000	117
Nbr.					500				

Model								M 2.1						
Return model	a1	t	Sig.	a2	t	Sig.	F	Sig.	Adj. R^2
	244	4.506	000	0.135	2.498	013	32.551	000	115
Nbr.					486				

Model								M 2.1						
Return model	a1	t	Sig.	a2	t	Sig.	F	Sig.	Adj. R^2
	244	4.506	000	0.135	2.498	013	32.551	000	115
Nbr.					486				
111 Model	313	5.435	000	(0.024)	(0.425)	671	23.431	000	085
Nbr.					486				

The results of the hypothesis test were evaluated according to the previously described model where the decision is based on the regression result (Adj. R^2) being larger and more significant among the research models. Research models (M 1.1) and (M 2.1) representing net income were compared with research models (M 1.2) and (M 2.2) representing total comprehensive income for their ability to explain the market price of equity and stock return of the firm. The comparisons show the regression value of Adj. R^2 (M 1.1) has a significantly greater value than obtained with (M 1.2), thus indicating a better ability of net income in presenting the attribute of earning value quality compared to the total comprehensive income. Therefore hypothesis 1 is accepted.

Decision comparison of earnings ability in presenting attribute of earnings quality, that is value relevance between net income and total comprehensive income, is based on the comparison result of regression between price model (M 1.1) and (M 1.2) and return model (M 2.1) and (M 2.2). The decision that can be taken is that it is not proven that total comprehensive income has a better quality of value relevance compared with net income. The results obtained in this test are in accordance with previous research conducted by Devalle (2012), Kusumo and Subekti (2013), and Jaweher and Mounira (2014) stating that the relevance of total comprehensive income value is no better than the net value of net income and there was no increase in the relevance of earnings value after IFRS was adopted. This study does not state the same results as those of Groen

(2010) and Krismiaji et al. (2013)that stated otherwise. The difference of the result is caused by the difference of sample and the research done by the author compared with the research of Groen (2010) and Krismiaji et al. (2013) and there is a difference of measuring value relevance value while the research conducted by Krismiaji et al. (2013) used the proxy of predictive and faithful representation in measuring the quality of value relevance.

Thus, it can be stated that the quality of value relevance of net income is better than total comprehensive income. Net income is a profit that is still pure and free from the revenue and expenses that are not the main operational results of the company or based on managing performance, so the quality of the resulting profit is better. On the other hand, total comprehensive income is a profit that has been added with revenue and expenses other than the performance of management or an income and expenses beyond the main operational control of the company. Jaweher and Mounira (2014)stated that the results of research that has been done do not support the statement that total comprehensive income gives a better attribute quality than net income. Cahyonowati and Ratmono (2012) stated that the findings of their study support the hypothesis that an institutional environment that is still not supportive can lead to the adoption of IFRS has no effect on the quality of accounting information. The research findings support the Karampinis and Hevas (2011) argument that in code law countries (including Indonesia), with the characteristics of institutional environments such as weak investor protection, lack of law enforcement, concentrated ownership, and banking-oriented funding, IFRS adoption may not necessarily increase the relevance of the value of accounting information.

Table 3. Multiple regression test result (Adj. R^2) predicitive value.

Model	M 3.1					
Price	a1	t	Sig.	F	Sig.	Adj. R^2
model	0.566	15.192	0.000	230.812	0.000	0.319
Nbr.				492		

Model	M 3.2					
Price	B1	t	Sig.	F	Sig.	Adj. R^2
model	0.556	14.789	0.000	218.711	0.000	0.307
Nbr.				492		

Model	M 4.1					
Price	a1	T	Sig.	F	Sig.	Adj. R^2
model	0.804	29.151	0.000	849.809	0.000	0.646
Nbr.				466		

Model	M 4.2					
Price		t	Sig.	F	Sig.	Adj. R^2
model	0.775	26.395	0.000	696.674	0.000	0.599
Nbr.				466		

Decision-making hypothesis testing in this study was based on test results that have been done on two proxy measures of predictive value variable, the ratio of the ability between net income and total comprehensive income of the previous year in predicting operational cash flow and company earnings for the current year based on the value of larger and more significant Adj. R^2 of the models of (M 3.1) and (M 4.1) compared with models of (M 3.2) and (M 4.2). The test results show that Adj. R^2 (M 3.1) is significantly larger than Adj. R^2 (M 3.2), where Adj. R^2 (M 3.1) represents net income prediction ability in predicting a company's operational cash flow, while Adj. R^2 (M 3.2) represents the total predictive ability of comprehensive income in predicting the company's operating cash flow. In other words, it can be stated that net income is significantly better in predicting the company's operating cash flow compared to the total comprehensive income, so that hypothesis 2 is accepted.

The test results of the models (M 4.1) and (M 4.2) show that the value of Adj. R^2 (M 4.1) is significantly greater than the value of Adj. R^2 (M 4.2). Based on these results, it can be stated that the model (M 4.1) is better than the model (M 4.2), where the model (M 4.1) represents the net income ability of the previous year against net income of the current year. The model (M 4.2) represents the ability of the total comprehensive income forecast for the previous year to total comprehensive income of the current year. In other words, the ability of net income is better in predicting earnings compared to total comprehensive income.

Based on the model testing results of (M 3.1) and (M 4.1) representing net income capability in predicting operating cash flows and corporate earnings compared to models of (M 3.2) and (M 4.2) representing total comprehensive income capability in predicting operational cash flows and profit company, it can be stated that net income has a better prediction capability compared with total comprehensive income in predicting operating cash flow and corporate earnings.

The results of this study are in conformity with the results of previous research conducted by Jaweher and Mounira (2014), and Pronobis and Zülch (2011), stating that total comprehensive income is not proven compared to net income. However, this study has different results from the research conducted by Krismiaji et al. (2013), which stated that IFRS affects the improvement of the quality of predictive ability. Thus, it can be stated that the predictive value of net income is better than that of the total comprehensive income. The results are supported by statements in several previous studies, including Pronobis and Zülch (2011), which stated that the predicted strength of net income and total comprehensive income for future operating performance of the company appears to have deteriorated as a consequence of IASB's new initiatives and actions. Thus, the deterioration in predictive power appears to be driven by a standard change. Jaweher and Mounira (2014) stated that the results of research that has been done do not support the statement that total comprehensive income gives a better attribute quality than net income. This finding appeals to standard-setters because the results indicate that the IASB should focus on the items included in other comprehensive income, in order to improve the total comprehensive comprehension quality.

5 CONCLUSIONS

The results of the value relevance test between net income and total comprehensive income of hypothesis 1 show that the ability of net income to present the attribute of earnings quality of value of relevance is statistically better and significant compared to total comprehensive income, either by measuring price regression models or return regression models. The reason is that net income is a pure profit that comes from the performance of management and is free of revenue and expenses that are not derived from the performance of management. The findings are in accordance with previous research by Devalle (2012), Kusumo and Subekti (2013), and Jaweher and Mounira (2014). This results also implies that total comprehensive income has not been proven to have a better quality of value relevance compared to net income, and standards-based IFRS implementation in Indonesia has not been able to improve the quality of accounting information.

The predictive value test result between net income and total comprehensive income inHypothesis 2 shows that it is statistically proven that net income has a significantly better ability in presenting the attribute of profit quality of predictive value compared with total comprehensive income, by measurement ability in predicting the company's operating cash flow in the future as well as the ability to predict future corporate earnings. This is because net income is a pure profit that comes from the performance of management and is free of revenue and expenses that are not derived from the performance of management. The findings are in accordance with previous research by Pronobis and Zülch (2011) and Jaweher and Mounira (2014), which stated that total comprehensive income is not proven to have a better prediction ability and significance compared to net income.

REFERENCES

Brochet, F., Nam, S. & Ronen, J. 2008. The role of accruals in predicting future cash flows and stock returns. *Social Science Research Network*.

Cahyonowati, N. & Ratmono, D. 2012. Adopsi IFRS dan relevansi nilai informasi akuntansi. *Jurnal Akuntansi dan Keuangan* 14(2): 105–115.

Devalle, A. 2012. Assessing the value relevance of total comprehensive income under IFRS: An empirical evidence from European stock exchanges. *International Journal of Accounting, Auditing and Performance Evaluation* 8: 43–68.

Goncharov, I. & Hodgson, A.C. 2011. Measuring and reporting income in Europe. *Journal of International Accounting Research* 10(1): 27–59.

Gordon, E., Jorgensen, B. & Linthicum, C. 2008. Could IFRS replace US GAAP? A comparison of earnings attributes and informativeness in the US market. Manuscript, Temple University.

Groen, A. 2010. The value relevance of other comprehensive income under IAS 1: An empirical investigation. Working Paper, Amsterdam Business School.

Jaweher, B. & Mounira, B.A. 2014. Quality of net income vs total comprehensive income in the context of IAS/IFRS regulation. *International Journal of Finance and Accounting Studies* 1(2): 17–14.

Jemaa, O.B., Toukabri, M. & Jilani, F. 2015. Accruals and the prediction of future operating cash flows: Evidence from Tunisian companies. *International Journal of Accounting and Economics Studies* 3(1): 1–6.

Jensen, M. & Meckling, W. 1976. Theory of the firm: Managerial behavior, agency, and ownership structure. *Journal of Financial Economics* 3: 305360.

Karampinis, N.I. & Hevas, D.L. 2011. Mandating IFRS in an unfavorable environment: The Greek experience. *The International Journal of Accounting* 46(3): 304–332.

Krismiaji, A.Y., Aryani, A. & Suhardjanto, D. 2013. Pengaruh adopsi International Financial Reporting Standard terhadap kualitas informasi akuntansi. *Jurnal Akuntansi dan Manajemen (JAM)* 24(2): 63–71.

Kusumo, Y.B. & Subekti, I. 2013. relevansi nilai informasi akuntansi, sebelum adopsi IFRS dan setelah adopsi IFRS pada perusahaan yang tercatat dalam Bursa Efek Indonesia. *Jurnal Ilmiah Mahasiswa FEB* 2(1).

Pratiwi, E. & Eddy, S. 2012. Relevansi nilai informasi laporan keuangan dan komponen laba rugi komprehensif dalam menjelaskan harga dan return saham. *Forum Bisnis & Keuangan* I: 289–307.

Pronobis, P. and Zülch, H. 2011. The predictive power of comprehensive income and its individual components under IFRS. *Problems and Perspectives in Management* 9(4).

Rachmawati, A. & Triatmoko, H. 2007. Analisis faktor-faktor yang mempengaruhi kualitas laba dan nilai perusahaan. *Simposium Nasional Akuntansi X Makassar* 26–28 Juli.

Tsujiyama, E. 2007. Two concepts of comprehensive income. *Accounting and Audit Journal (JICPA)* 19(11): 30–39.

Ujiyantho, M.A. & Pramuka, B.A. 2007. Mekanisme corporate governance, manajemen laba dan kinerja keuangan. *Simposium Nasional Akuntansi X* 10(6).

Vestari, M. 2012. pengaruh earnings surprise benchmark terhadap prediktabilitas laba dan return saham. *Prestasi* 9(1): 62–84.

Advances in Business, Management and Entrepreneurship – Hurriyati et al (eds)
© 2020 Taylor & Francis Group, London, ISBN 978-0-367-27176-3

Evaluation of forensic auditor role as corruption eradicator

A.W. Mardijuwono & F. Daniyah
Airlangga University, Surabaya, Indonesia

ABSTRACT: This study contains the evaluation of the role of forensic auditors in implementing their roles, namely to eradicate corruption by using forensic audit procedures. Seeing the growing corruption crime rate, this research examines the effectiveness of the role of forensic auditors, especially at BPK & BPKP Representative of East Java region. This research uses a qualitative research method of case study by using primary data. The results show that the role of forensic auditors is to find evidence for cases in court, as well as to prevent corruption, but prevention is limited to internal agencies only. Forensic auditors still cannot cause deterrent effects for corruptors because they are beyond the control of the forensic auditor. Forensic auditors cannot impose criminal penalties on corruptors. The effectiveness indicator of the presence of forensic auditors in combating fraud is the realization that exceeds the planned target.

1 INTRODUCTION

The last three decades of the accounting world have been thrown into a fraudulent financial reporting scandal by big companies like Enron, WorldCom, Parmalat, Satyam, and sub-prime mortgage company, Olympus. According to the Association of Certified Fraud Examinations (ACFE) fraud can be classified into three categories, namely: (1) Fraud Financial Statements, (2) Misuse of Assets, (3) Corruption. These forms of fraud can have an adverse effect on an entity such as, change of management personnel, falling stock prices, delisting, bankruptcy, decreased productivity, and punishment (Debnath 2017).

In Indonesia, fraud cases also occur frequently. One of the most commonly occurring fraud cases are corruption cases such as the case of Bank Bali, Hambalang Project, Century Bank, and others. Corruption cases reduce public confidence in accounting, resulting in increased forensic techniques, one of which is forensic auditing (Astuti 2013). During the year of 2017, East Java has resolved at least 112 cases of corruption. Although the eradication of corruption has been carried out as it should, a deterrent effect against corruptors is still lacking. One reason is because law enforcers in Indonesia are bribed and there exists the influence of power or position. Strategies have come from government agencies such as BPK, Inspektorat, and KPK, to NGOs such as MTI and ICW which have been formulated to combat corruption; however, these are still not enough to eradicate corruption cases (Wiratmaja 2010).

Astuti (2013) states the success of forensic auditing in Indonesia is seen where Price Waterhouse and Cooper managed to show a number of fund flows from people who were involved in the case of Bank Bali. According to Matson (2016), a forensic audit is a discipline that combines knowledge and expertise in accounting, audit, loss prevention, and law enforcement. The Government of Indonesia supports forensic auditors, as evidenced by BPKP conducting competence training for forensic auditors in preparation to become forensic auditors following certification of competence of forensic auditors held by Certification of Professional Auditor (LSPAF) (Astuti 2013). Government support for forensic accountants is still not enough to combat corruption in Indonesia. According to Wiratmaja (2010), there needs to be a strategy to eradicate corrupted acts that are socialized to the public and to strengthen institutions that have the authority to combat corruption.

1.1 Strain theory

The theory of strains is the theory of sociology and criminology developed in 1938 by Robert K. Merton. The theory states that societies put pressure on individuals to achieve socially acceptable goals even though they lack the means, causing pressure that can cause individuals to commit crimes. An example would be selling drugs or committing criminal acts of corruption (Merton 1957). Merton (1957) recalled his theory that corruption is a violation of norms, from a human behavior caused by social pressure. The Strain Theory of Merton shows the answer to how a culture that emphasizes economic success but limits the opportunities to achieve it makes the level of corruption even higher.

1.2 Fraud triangle theory

The elements of a fraudulent triangle developed by Donald Cressey in his book titled Other People's Money argue that a person cheats because of (Cressey 1953):

1. Pressure: the act of cheating behavior depends on the individual's conditions, such as experiencing financial problems, bad habits such as gambling and drinking, greed, or high expectations. These conditions can make the individual commit a fraud.
2. Opportunities: Weaknesses of an authorization and approval procedure from management; low information in personal finance information (fraud in banking); there is no separation between the assignor and the custody of assets. These things create the opportunity for individuals to commit fraud.
3. Rationalization: Justification occurs when a person or group of people builds the truth of the fraud committed. The perpetrators of fraud will look for justification that they are not thieving or cheating.

1.3 Gone theory

Gone Theory proposed by Jack Bologne in 1993 discusses fraudulent behavior. Bologne explains the factors that encourage individuals to engage in fraudulent deviant behavior as follows:

1. Greed: an attitude that potentially exists within each individual. The presence of disgruntled attitudes that make individuals commit fraud
2. Opportunity: is a situation where an organization or public agency opens an opportunity for individuals to commit fraud.
3. Needs: factors needed by the individual to fulfill the necessity of his/her life, which according to the individual are reasonable.
4. Disclosure: disclosure or a consequence obtained by the perpetrator of fraud if the offender is involved in fraudulent behavior.

2 METHOD

The approach used in this study is a descriptive qualitative approach. The data used in this study is the primary data. This approach will explain the role of forensic auditors as eradicators of corruption.

This research uses the Interview as the data collection technique. This study uses an in-depth interview type in which researchers and informants are face-to-face and directly involved in the interview conducted. Informants came from BPK and BPKP Representative of East Java region and only one informant from each agency was used. By fulfilling the criteria that have been determined to be an informant which are: (1) forensic auditor, (2) has CFE or CFrA certificate, (3) has ever been an expert informant in a corruption criminal proceeding. The next data collection type is Documentation. The documents referred to in this study are documents such as standards or procedures for conducting forensic audit techniques.

In testing the validity of data, this study uses triangulation of sources. An informant from BPK Representative of East Java region was compared with an informant from BPKP Representative of East Java region. Literature studies that have been described in the previous chapter are also considered.

3 RESULT AND DISCUSSION

3.1 The role of forensic auditors in eradicating corruption

The BPK Representative of East Java region has a role as an auditor of the region's financial statements. The role of CPC is crucial to find fraud those results in state losses. As explained by Mr. Arief in the interview:

"Generally we conduct an audit of Financial Statements. There are two segments of the audit, financial statement of central government and financial statement of regional government. We have three types of audit reports: Local Government Financial Statements (LKPD) semester one, performance audits, and Checks with Specific Purposes (PDTT). There are two types of audits that can result in losses, namely statement on compliance, PDTT on spending generated by BPK itself. So the role of BPK is to see the performance and whether financial statement is in accordance with the actual situation in general over the assets or inventory. PDDT is specifically consisted of two also: PDTT revenue and PDTT regional spending. Whether there is fraud in shopping PDTT, because shopping area is very vulnerable of fraud".

The role of the BPKP representative of East Java region has been regulated in the regulation of the head of BPKP No. 1 year 2016 on the code of conduct of BPKP mentioned in Article 3 and Article 4. Mr. Roeddy stated that:

"The role of BPKP can be seen in the regulation of the head of BPKP No. 1 year 2016, one of its roles is to supervise internal financial accountability in which activities are cross-sectoral, then we usually also carry out coaching on the implementation of internal control system of government in its working area. Sometimes we will also provide assistance to the management of local finance in State-Owned or Regional Government-Owned Enterprises".

Auditors have a very important role, especially in eradicating corruption. Forensic auditors should be able to disclose state losses caused by corruptors. Forensic auditor's duties also include preventing fraud. The forms of fraud include corruption, misuse of assets, and fraud to financial statements. Basically, both BPK and BPKP have conducted prevention both internally and externally on the parties concerned. This was expressed by Mr. Arief as follows:

"For BPK, there are two ways to prevent fraud in our internal and external operation. Internal prevention leads to the prevention from the perspective of auditor for example there is a project worth 1 M, can it be reduced? If so, the auditor will be rewarded, so in our internal is more on preventing the auditor to do fraud because one form of corruption is bribery to our auditors. For external prevention, we conduct checks in accordance with existing procedures".

If BPK is trying to prevent fraud, BPKP has its own method of detecting fraud. BPKP representatives perform first detection of fraud by doing Fraud Risk Assessment. The methods were described by Mr. Roeddy very clearly as follows:

"We have two types of fraud detection methods: (1) the development of findings from routine audit results such as (Operational audits or Performance audits), (2) conducting the Fraud Risk Assessment in East Java Provincial Government on the capital expenditure planning process in 2017. In FRA, the method used is Control Self-Assessment in which BPKP only acts as facilitator. We also cooperate with KPK in conducting Fraud prevention program called Coordination program, Supervision and Prevention of Corruption with object of Regency/City Government, the result of this program is recommendation which lead to coaching and improvement of existing system and procedure".

The prevention and detection of fraud is now done maximally; the role of corruption eradication institutions has been maximised as much as possible in carrying out their role. However, corruption continues to exist and corruption eradication agencies seek to prevent corruption. Basically, the efforts of BPK and BPKP still do not cause a deterrent effect for the perpetrators of corruption, because after submitting the report of examination, the task of a forensic auditor is completed. This is unlike Law Enforcement Officials that provide administrative penalties and imprisonment punishment. This was conveyed by Mr. Arief:

"In fact, corruption still happens a lot, specifically among law enforcement officials. BPK or BPKP only make the report without any severe sanctions, administrative sanctions only. For example if you work in the financial management agency, then there is the budget to buy office supplies, when we check it there is a fictitious note. We just report the result of the inspection and finish the job. Unlike law enforcement officials when they find fictitious notes, they will process it to court, giving the sentence of confinement and return the loss of state. You may have heard of

KPK, KPK has caused a deterrent effect in local government because executives and legislatives are targeted and also processed to trial and imprisonment".

It is almost the same as Mr. Roeddy's opinion on whether the role of corruption eradication institution has caused a deterrent effect:

"This question is difficult to answer clearly (relative) because the cause of Fraud is Fraud Triangle, therefore eradication of corruption program is done in several approaches that is Preventive, Detective and Repressive. These three approaches have been done by KPK, BPKP and BPK".

The statement from Mr. Roeddy is in accordance with the Fraud Triangle theory which explains the causes of fraud which are pressure, opportunity, and rationalization (Cressey 1953). From the above statement it can be concluded that the role of BPK and BPKP has not caused a deterrent effect for corruptors because their limits mean they cannot give penalties or sanctions which would be able to cause a deterrent effect. If assessed whether results of corruption eradication agencies have been effective or not, basically, the nature of corruption is very close to greed because these institutions have conducted anti-corruption prevention activities. The theory of Bologne in 1993 suggests that factors causing deviant behavior are factors of greed, opportunity, need, and disclosure. But the biggest factor in deviant behavior is due to the motive of need and greed. This was stated by Mr. Roeddy:

"Based on the motive, one of the causes of corruption is because of the need and greed. In accordance with this motive actually each agency often do spiritual guidance to improve morals, in addition the government has made income improvements such as providing Performance Allowance to civil servants. To narrow the opportunity of corruption by BPKP, we have conducted Fraud Risk Assessment Evaluation, Evaluation SPIP (Government Internal Control System), and conduct socialization, Diagnostic and Implementation of Fraud Control Plan (Internal control system with Fraud prevention base) in several agencies and also routine audits under the Annual Supervisory Work Program. BPKP also conducts Audit of Investigation and Audit of State Financial Losses in order to assist investigators to solve Corruption cases (Repressive action)".

In carrying out his/her assignment, the forensic auditor is guided by existing standards and codes of ethics. BPK has the same code of ethics as the central

Table 1. Preventive approach conducted by BPKP representative of East Java Region.

| No | Description | 2016 (Per 31-12-2016) | | 2017 (Per 30-11-2017) | |
		Target	Realization	Target	Realization
1	Fraud control plan (FCP) program	1	2	5	5
2	Socialization of anti-corruption program	1	7	1	8

BPK to standardize. The code of ethics is contained in Article 29 Paragraph (1) of Law no. 15 of 2006 on the execution of functions, assignments, and authorities shall refer to the Regulations of the Supreme Audit Board No. 4 of 2016 concerning the Honorary Council of the Code of Ethics of the Supreme Audit Board. The regulation has a more detailed and technical basis; it is intended that there is a unity of view and action in carrying out its duties. The standard used by BPK is the State Financial Review Standard (SPKN). This was stated by Mr. Arief:

"For the code of ethics, we follow the internal center, while the standard we have SPKN (State Auditing Standards) has the technical guidance and is for internal only. Then the derivative of the technical guidance resulted in the Inspection Procedure (P2). Currently we are developing this technical guidance since it is still draft".

Also, BPKP has a government internal audit standard issued by AAIPI and the regulation of the Head of Finance and Development Supervisory Agency Number: PER-1314/K/D6/2012 About Guidelines for Assignment of Investigation Sector, as well as a Pocket book containing Code of Conduct of BPKP.

Standards are clearly defined for carrying out assignments and existing codes of ethics. Cases of corruption still cannot be reduced, although the case has been disclosed in mass media with the existence of corruption eradication agencies such as KPK, BPK and BPKP. This is because basically corruption is only seen once it has occurred, and the role of these institutions is as a deterrent to corruption. The statement from Mr. Arief is in harmony with Gone Theory, a theory put forward by Jack Bologne in 1993 stating that fraud behavior is derived from Greed, Opportunity, Needs, and Exposure. This was conveyed by Mr. Arief in his interview:

"Corruption occurs because it is more likely in each individual or group. In principle humans entering professional work life will never be satisfied (greedy). Despite everything that has been received, individual will want more things. Despite anything that has been received, humans will continue to feel less. Once when I attended my instructor certification I worked for a foreign company in a developed country, then he discovered that fraud has occurred and the suspect was a German citizen which is a developed country. So in any country corruption will continue to exist, in the most advanced country, corruption still exists. It may even now have a larger and modern corruption crime due to the development. We have prevented or mitigated the corruption, but in reality corruption is discovered because it happens".

Mr. Arief's statement is similar to the description given by Mr. Roeddy as follows:

"The case of Corruption is a complex problem mainly related to the income of civil servants in large cities not in harmony with their daily needs that encourage unscrupulous civil servants to corrupt. Besides that there are also civil servants who do corruption because of the greed factor, usually associated with civil servants with strategic positions. It is also influenced by the direct election system of the Head of Region which requires enormous campaign costs so as to encourage the regional head to commit corruption in lieu of the campaign costs he has spent".

Forensic auditors act as a preventive, detective, and repressive act, meaning that forensic auditors prevent fraud, and as detective or investigator in solving the corruption crime problem, as well as performing repressive actions to handle or process the act of corruption in accordance with applicable legislation. The role of forensic auditors in Indonesia is that they are expected to be able to prevent, disclose, and solve corruption cases in a preventive, detective, and repressive manner effectively, efficiently and economically (Wiratmaja 2010). Suppressive prevention is used and implemented to prevent the causes of corruption, because the criminal act of corruption cannot be eliminated, only minimized by preventive means. Detectives are involved in solving criminal acts of corruption that have occurred or corruption indication cases; the case can be disclosed in a short period of time and more accurately in order to prevent any greater loss. Repressive acts are investigations which reveal the criminal act of corruption.

3.2 Indicators of successful forensic auditors in handling corruption crime

In Indonesia, the suitability of forensic auditors is seen in the emergence of a fraud case at Bank Bali, where PwC as a forensic auditor is able to show a certain amount of funding flow from certain individuals through a series of forensic audits that have been conducted. The subsequent successes at discovering corruption were achieved by the BPK, namely the corruption case in the General Election Commission, the BPK as a forensic auditor was able to complete up to the KPK court stage. (Jumansyah et al. 2011). The success of the forensic auditor is marked by the appearance of the Report on the Result of Examination, and the appearance of the element of state loss. This is expressed by Mr. Arief in his interview:

"LHP (Report of the Audit Result), the existence of elements of the problem, and the element of state loss".

This is similar to the statement from Mr. Roeddy:

"The indicator is the Report of the Investigative Audit Result/PKKN which has been published through a tiered review so that it has been in accordance with the applicable Audit Standards and in the hearing of Corruption usually always defendant guilty of corruption. Moreover, it can be seen from the success of the target that has exceeded the realization. Then we find the state losses".

Table 2 explains the realization that occurred has exceeded the target planned in the previous year. This indicates that there is success in eradicating

Table 2. Implementation of audit of investigation and audit calculation of financial losses as an audit of assistance to police investigator and attorney office, with realization.

Description	2016 (Per 31-12-2016)			2017 (Per 30-11-2017)		
	Target	Realization	Loss Amount	Target	Realization	Loss Amount
Investigation Audit	1	1		6	5	
State Budget Loss Audit (PKKN)	31	65	IDR 45,534,627,687	13	29	IDR 36,189,733,884
Provision of Expert Statement at The Corruption Crime Trial	38	192		26	153	

corruption by the BPKP Representative of East Java region. It can be concluded that the role of forensic auditor is quite effective in eradicating corruption; it is indicated by the element of loss found each year and the realization that exceeds the planned target.

Barriers when conducting forensic audits sometimes cause auditors to experience audit failures, then from prior failures, a forensic auditor will take action mitigation before doing his/her assignment. At least forensic auditors have minimized the occurrence of such failures by way of mitigation. Basically audit failure is inherent when conducting the auditing process, because forensic audits are very risky and related to state losses. It is very necessary to do mitigation which is useful to minimize audit failure. As expressed by Mr. Arief:

"Failure occurs if there is no preventive mitigation process in the absence of failure, such as lack of documents. Maybe it's called a failure. Like the case of Pertamina tankers, Pertamina sells tankers to export oil at too low a price. One way to determine the price is expensive or cheap by having to find price comparison to a several third parties in the area. In this case there is no comparison because the ship is special. If they mention the ship is overpriced then how much is fair market at that time. In fair market calculation using depreciation method from age of useful life was not possible because the price is still in normal normative so that BPK stated that examination cannot be conducted, because state losses cannot be measured. Maybe that's why audit fails".

In regards to the BPKP Representative, they have never experienced an audit failure, because it is rare that the results of an investigative audit are not proven. This was described by Mr. Roeddy:

"BPKP Representative of East Java region has never experienced an audit failure, but it is very rare that the results of the Investigative Audit that the conclusion is not proven and for the Audit calculation of the State Financial Losses the result was there is no loss of state finances or the loss cannot be calculated".

Mitigation is a process to reduce the risk that occurs during the auditing process. With mitigation,

it is likely that audit failure will be avoided. If the forensic audit process has occurred and in the process it cannot continue, then it is a waste of time, energy, and cost because the examination process requires considerable cost. One effective method before doing a forensic audit is mitigation. This is the same as what was stated by Mr. Arief:

"Currently doing mitigation, my own example is in PKN section, we will request exposing the evidence completely or partially, which is important that state losses can be captured and calculated. Why completely or partially? We can calculate in advance and this can cost efficiency eventually because the cost of inspection is very high. So before doing the process, further examination is done by calculating the evidence that existed first. If the result is not there, normatively the process has been done, the result is not there, and the output is not there. The run will be a waste. Only the real happening and there must be a loss although not yet can be calculated but already have the picture then assessment will be continued".

In overcoming the failure of the audit, BPKP chose its own way in managing the failure, by conducting a minimum standardized review of 3W + 2H and then an investigative assignment will be done. The criterion for assessing whether or not a reason for forensic audit is sufficient is to answer questions 3W + 2H (Kayo 2013). This was expressed by Mr. Roeddy:

"The process to minimize the occurrence of the problem is done after the Public Complaint Letter that has been formerly reviewed with a minimum standard to meet 3W + 2H, then the investigative audit assignment is conducted. For Investigative audit the request of the investigator should be done then reviewed firstly with minimum standard to meet 3W + 2H then investigative audit assignment is conducted. For the PKKN Audit, exposure and assessment is conducted whether the case has been investigation stage, it is clear that the action is against the law and has been supported with sufficient evidence".

Audits fail because there has been no mitigation process previously to reduce it. In order to realize the success of the forensic auditor's role in conducting forensic audits, it is necessary to mitigate to reduce the risk of the impact of audit failure. With the mitigation, forensic auditors can run efficiently, effectively, and, economically in performing their role. Thus, the role of the forensic auditor can be successful if the forensic auditor performs a mitigation action.

4 CONCLUSIONS

The role of forensic auditors in preventing and detecting is quite effective in finding corruptive acts, through accounting powers with auditing procedures, but the role of forensic auditors still has not resulted in a deterrent effect for corruptors, since forensic auditors do not impose criminal penalties. The success indicators of the forensic auditor are indicated by the Report of Examination Results and the discovery of the element of loss. This is further marked by the realization that exceeds the target that should be achieved.

REFERENCES

Astuti, N.P.S. 2013. Peran audit forensik dalam upaya pemberantasan korupsi di Indonesia. *Jurnal Akuntansi Unesa* 2(1).

Cressey, D.R. 1953. *Other people money, study in the social psychology of embezzlement*. Montclain: Patterson Smith.

Debnath, R.K.T.J. 2017. Forensic accounting: A blend of knowledge. *Journal of Financial Regulation and Compliance* 25(1).

Jumansyah, J., Dewi, N.L. & En, T.K. 2011. *Akuntansi forensik dan prospeknya terhadap penyelesaian masalah-masalah hukum di Indonesia*. Maksi.

Kayo, A.S. 2013. *Audit forensik: Penggunaan dan kompetensi auditor dalam pemberntasan tindak pidana korupsi*. Yogyakarta: Graha Ilmu.

Matson, D.M. 2016. independent studies in forensic accounting: Some practical ideas. *Journal of Forensic & Investigative Accounting* 8(2).

Merton, R.K. 1957. *Social theory and social structure (Rev. and enlarged ed)*. New York: The Free Press.

Wiratmaja, I.D.N. 2010. Akuntansi forensik dalam upaya pemberantasan tindak pidana korupsi. *Artikel*. Universitas Udayana.

Advances in Business, Management and Entrepreneurship – Hurriyati et al (eds)
© 2020 Taylor & Francis Group, London, ISBN 978-0-367-27176-3

The impact of asset, mudharabah time deposit and Non Performing Financing (NPF) to profitability Islamic Banking in Indonesia

R. Sukmana & N. Junun
Airlangga University, Surabaya, Indonesia

ABSTRACT: The purpose of this research is to analyze the impact of Islamic bank's internal factors, such as size, Mudharabah time deposit and non performing financing (NPF) on return on asset (ROA) and return on equity (ROE) in Indonesia. This research uses quantitative approach such as data stationarity test and cointegration test. The data used are obtained from the Bank of Indonesia, which cover the period during September 2003 until September 2014. The results of this study show that the Mudharabah time deposit and the non performing financing (NPF) affect ROA negatively. These results suggest that the sample of Islamic banks have been successfully maintained its credit risks and have been able to get the optimum profitability.

Keywords: asset, Mudharabah time deposit, npf, profitability, roa and roe.

1 INTRODUCTION

The rapid development in the financial industry, including the emerging of sharia banks in Indonesia, is the result of the fact that the majority of Indonesia citizens are Muslim and its government also supports the dual banking system by making the regulation that allows every conventional bank to open a branch using Sharia service system (dual banking system).

Sharia Banking as an intermediary institution which collects and distributes funds requires the role of an accounting system to manage its finance, and the one thing that must be carefully managed is bank's assets. The asset is the most important indicators used in the sharia banking industry because asset is the source of banking funds that can be used in order to carry out the intermediary function of the mentioned banks.

Currently, the number of assets owned by Sharia Commercial Bank (SCB) and Sharia Business Unit (SBU) have kept increasing. This means that the resources owned by the Sharia Commercial Bank (SCB) and the Sharia (Islamic) Business Unit (SBU) are increasing every year.

Sharia banking resources that keep increasing every year is related with the collection of funds entrusted to the Sharia Bank to be managed by its managers in order to not make it idle and to produce benefits for people. The collection of funds by the Sharia Bank is known as the Third Party Fund, which consists of 3 items, which are current accounts, savings accounts and deposit accounts.

There are several causes that can decrease the amount of net income of the Sharia Commercial Bank (SCB) and Sharia (Islamic) Business Unit (SBU) in a certain period. If associated with the assets owned by Sharia banks, some Sharia banks are indeed confronted with a non-performing financing condition. This condition is experienced by the Sya- ria Bank when some of their customers have problems to perform its obligation to the bank, either in the form of having difficulty to return the principal or profit sharing, margin, fee or wage. This will result in a decrease in the earnings of the Sharia bank. The study of Arninda's (2014) also states that assets, Mudharabah time deposits and NPFs have significant negative relationships with net income.

Based on the previous explanation and research, this research will discuss about "The Impact of Assets, Mudharabah Time Deposit and Non Performing Financing (NPF) on Profitability on Islamic Bank Industry in Indonesia".

2 THEORETICAL BASIS AND HYPOTHESES DEVELOPMENT

2.1 *Function of Sharia Bank*

Huda (2010: 38) stated in her research that the functions of the establishment of Islamic banking industry are:

1. In order to direct the Muslims to conduct their muamalah activities based on the Islamic principles and to avoid the practice of usury.
2. In order to create justice in the economic framework by distributing the income equally
3. In order to improve the quality of life of people by opening up bigger business opportunities.
4. In order to help overcome poverty problems.

5. In order to maintain the level of economic and monetary stability and to avoid the unhealthy competitions that may occur between financial institutions.

2.2 Asset

Reeve (2013: 57) stated that assets and properties are the resources owned by a business entity.

Antonio (2001: 203) also stated that assets are something that have the capability of generating positive cash flow or other economic benefits, either by itself or with other assets, whose rights are acquired by the Islamic banks as a result of past transactions or events.

2.3 Mudharabah time deposit

Al Khofdho (2009: 50) stated that the Mudharabah time deposit is the deposit by third parties to the bank, where its withdrawal can only be done after a certain time period based on the agreement.

2.4 Non performing financing (NPF)

According to the Bank of Indonesia Form Letter No. 3/30/DPNP dated December 14, 2001, it is stated that the Non Performing Financing (NPF) is a financial ratio used in the banking industry that calculates the percentage of the troubled financing to the total of financing.

2.5 Profitability

Arifin (2005: 71) stated that there are two ratios that are usually used to measure the performance of banks, namely return on assets (ROA) and return on equity (ROE). ROA is the ratio between net income to average assets. ROE is defined as the ratio of net income to average equity or bank owners' investments.

2.6 Relationships between variables

According to Al-Qudah (2013), it stated that assets can influence the size of the bank (Bank Size), because large banks have a greater advantage to provide their financial services to certain customers with large values. So, the growth in total assets will signify the profitability of the banks, and by the nature of the relationship between assets and return on equity (ROE), banks with high capital will achieve higher profits because they can use cheaper sources of financing with lower costs and lower risk financing.

According to Arninda (2014), she stated that the amount of deposits or raised fund has a significant negative impact on profitability. As a result, the number of fund depositors also has a significant negative relationship with profitability. This shows that the intense competitions between banks have made it difficult for banks to lower their rate of return that must be paid to the depositors.

According to Arninda (2014), Al-Qudah (2013), Curak (2012), Akhtar (2011), and Alper and Adem Anbar (2011), they stated that Non Performing Financing (NPF) is significantly and negatively influencing bank's profitability. The higher is the percentage of bank's NPF, the bank's health level will be lower.

Hypothesis 1: Asset has a significant impact on return on assets (ROA) of Sharia banks in Indonesia.
Hypothesis 2: Asset has a significant impact on return on equity (ROE) of Sharia banks in Indonesia.
Hypothesis 3: Mudharabah time deposit has a significant impact on return on assets (ROA) of Sharia banks in Indonesia.
Hypothesis 4: Mudharabah time deposit has a significant impact on return on equity (ROE) of Sharia banks in Indonesia.
Hypothesis 5: Non Performing Financing (NPF) has a significant impact on return on assets (ROA) of Sharia banks in Indonesia.
Hypothesis 6: Non Performing Financing (NPF) has a significant impact on return on equity (ROE) of Sharia banks in Indonesia.

3 METHODOLOGY

3.1 Research approach

The approach used in this research is the quantitative approach and the statistical tests used in the analysis include the stationary test, unit root test, and cointegration test. The data used in this study are gathered from the Sharia Banking Statistics (SBS) in a monthly time period starting from September 2003 to September 2014, thus that the total observation time is 133 (N = 133).

3.2 Variables

The independent variables of this study consist of:

1. Assets
2. Mudharabah Time Deposit
3. Non Performing Financing (NPF)

The dependent variable of this study is the profitability of Sharia Banks in Indonesia, which are depicted by return on asset (ROA) and return on equity (ROE).

3.3 Analysis model

Based on several previous theoretical studies and hypotheses that are supported by theories, this research analysis will explain about the specific relationship between the exogenous variable and the endogenous variable as its research framework.

1. ROA = (Net Income/Total Assets) x 100%. The data are collected from the Sharia Banking Statistics (SBS) in a monthly type of data with the measurement scale in percentages.
2. ROE = (Net Income/Common Equity) x 100%. . The data are collected from the Sharia Banking Statistics (SBS) in a monthly type of data with the measurement scale in percentages.
3. TA = Total Assets. The data are collected from the Sharia Banking Statistics (SBS) in a monthly type of data with the measurement scale in nominal Rupiah.
4. MTD = (Mudharabah Time Deposit/Total Third Party Fund) x 100%. The data are collected from the Sharia Banking Statistics (SBS) in a monthly type of data with the measurement scale in percentages.
5. NPF = (Total Troubled Financing/Total Revenue) x 100%. The data are collected from the Sharia Banking Statistics (SBS) in a monthly type of data with the measurement scale in percentages.

4 RESULTS AND DISCUSSIONS

4.1 *Description on research results*

4.1.1 *Correlation tests*
Based on the correlation test results in Table 1, it can be seen that the there is no multicollinerity problem on the bank samples. We get this conclusion because the total value of the correlation matrix of all variables is less than 0.8.

4.1.2 *Tests on data stationarity*
The number of data used in this study amounts to 131 observations. As the data cover the period between September 2003 until September 2014, this data is called "time series" data. The underlying of the time series data is stationary, because if the time series data is non stationary, then the results of the study analysis will be spurious (deceptive or unreasonable) and can not be used as a reference for behavioral forecasting.

The results show that some variables tested in the unit root test (sta- sionerity test) using Augmented

Dickey Fuller (ADF) method at the intercept level, such as: NPF and ROE have insignificant results, ADF t-statistics are smaller than Mackinnon Critical Value. Due to the significant asset is at the level of a = 5%, and mudharabah time deposit is significant at the level of a = 1%, and in this study is using the level of a = 5%, then it is concluded that variable of asset and mudharabah time deposits have significant results.

In the unit root test (stationary test) using the Augmented Dickey Fuller (ADF) method at the trend level and intercept level, some variables such as asset, NPF, ROA and ROE have insignificant. These results indicate that the ADF t-statistic value is less than the Mackinnon Critical Value. Due to mudharabah time deposit is significant at the level of a = 1% level and in this study is using the level of a = 5% level, then it is concluded that variable of mudharabah time deposit have significant result.

In the unit root test (stationary test) using the Augmented Dickey Fuller ADF method at first (1st) difference level both at the intercept level and the trend and intercept level, all significant variables are showed at a = 1%, a = 5% and a = 10 %, meaning the value of ADF t-statistic is higher than the Mackinnon Critical Value.

The second unit root test (stationary test) in this study is using the Philips Perron (PP) method. At the intercept level, the results show that ROE is not significant, however, at the trend & intercept level, the variable of asset seems to be significant. Meanwhile, the results of mudharabah time deposit, NPF, ROA and ROE variables at the intercept level and the trend and intercept level seem to be all significant at the level of a = 1%, a = 5% and a = 10%. This results show that the value of ERS t- statistics are higher than the Mackinnon Critical Value.

The third unit root test (stationary test) in this study is using the Kwiatkowski Philips Schmidt Shin (KPSS) method. At the intercept level, the variable of asset, mudharabah time deposits, NPF and ROA seem to be insignificant, indicating that the KPSS t-statistic is higher the Kwiatkowski Philips Schmidt Shin (KPSS) Critical Value. Meanwhile, other variable such as ROE shows a significant result, which is shown by the value of KPSS t-statistic, which is lower than the Kwiatkowski Philips Schmidt Shin (KPSS) Critical Value.

Table 1. Results of the correlation test.

	ROE	ROA	DM	LNASET	NPF
ROE	1	0.4845	-0.5279	-0.0476	0.5898
ROA	0.4845	1	-0.1633	0.5076	0.2396
DM	-0.5279	-.1633	1	0.2878	-0.5515
LNA-SET	-0.0476	0.5076	0.2848	1	-0.2261
NPF	0.5898	0.2396	-0.5515	-0.2261	1

	ROE	NPF	LNASET
		-25.01081	-32.14937
t-std.Error		(6.50384)	(66.0213)
t-statistics		3.8455	0.4869

The fourth unit root test (stationary test) in this study is using the Kwiatkowski Philips Schmidt Shin (KPSS) method. At the trend and intercept level, the variable of asset shows a significant result, indicating that the value of KPSS t-statistics is less than the Kwiatkowski Philips Schmidt Shin

(KPSS) Critical Value. Meanwhile, the results of other variables, such as mudharabah time deposits, NPF, ROA and ROE seem to be not significant, indicating that the value of KPSS t-statistics is higher than the Kwiatkowski Phillips Schmidt Shin (KPSS) Critical Value.

The fourth unit root test (stationary test) in this study is using the Kwiatkowski Philips Schmidt Shin (KPSS) method. At the first (1st) difference at the intercept level, all variables except of the asset are significant at the level of a = *1%, a = 5%* and a = 10%. It can be concluded that the value of KPSS t-statistics is lower than the Kwiatkowski Phillips Schmidt Shin (KPSS) Critical Value. Meanwhile, at the first (1st) difference in the trend and intercept level, the results show that only asset and mudharabah time deposits seem to be insignificant, indicating that the value of KPSS t-statistic is greater than the Kwiatkowski Philips Schmidt Shin (KPSS) Critical Value.

4.1.3 Optimal lag determination

The optimal lag testing in this study uses the Final Prediction Error (FPE), the Akaike Information Criterion (AIC), the Schwawrz Information Criterion (SIC) and the Hannan-Quinn Information Criterion (HQ) criteria. Based on the optimal lag testing using EViews 8, it is known that ROA, based on the Final Prediction Error (FPE), the Akaike Information Criterion (AIC), the Schwawrz Information Criterion (SIC) and the Hannan-Quinn Information Criterion (HQ) criteria, is lag 1. Meanwhile, ROE, based on the Final Prediction Error (FPE), the Akaike Information Criterion (AIC) and the Hannan-Quinn Information Criterion (HQ) criteria is lag 2.

Based on optimal log test result, the lag used for ROA is lag 1 and the lag used for ROE is lag 2.

4.1.4 Cointegration tests

The purpose of conducting thecointegration tests is to see whether long-term relationships exist between the dependent and independent variables. Cointegration can easily occur in the time series data in a long period of time. The approach used in testing the cointegration between variables is the Johansen method, as this method can be used to determine the credibility of a number of variables.

The long term equations are known from the results of the cointegration testing,: -0.283020 NPF(0.14046)2.0149

	-0.272301	1.527315
ROA	DM	LNASET
t-std.Error	(0.03907)	(1.49687)
t-statistics	6.9695	1.0270

Based from the results of the cointegration tests, the long term equation in which the non performing financing (NPF) has an impact of 0.28% against ROA and its t-statistics value is 2.0149, which is greater than the t-table significance level of 0.05 (1.98) are known. These results mean that the NPF variable has a significant relationship with ROA. In addition, the NPF variable's coefficient mark shows a negative value, meaning that the NPF has a negative impact in the long term equation. Any increase in the amount of NPF by 1% will decrease the ROA by 0.28%.

Based from the results of the cointegration tests, Mudharabah time deposit (MD) has an impact of

1. 27% against ROA and its t-statistics value is 6.9695, which is greater than the t-table significance level of 0.05 (1.98). These results mean that the DM variable has a significant relationship with ROA. In addition, the Mudharabah time deposit variable's coefficient mark shows a negative value, meaning that the Mudharabah time deposit variable has a negative impact in the long term equation. Any increase in the amount of Mudharabah time deposit by 1% will decrease the ROA by 0.27%.

Based from the results of the cointegration tests, the variable of asset (LNASET) has an impact of 1.53% against ROA and its t-statistics value is 1.0270, which is lower than the t-table significance level of 0.05 (1.98). These results mean that the asset variable has an insignificant relationship with ROA. In addition, the asset variable's coefficient mark shows a positive value, meaning that the asset variable has a positive impact in the long term equation. Any increase in the amount of assets by 1% will decrease the ROA by 1.53%.

The long term equations are known from the results of the cointegration testing:

-14.10772
DM

(1.89905)
7.4288

Based from the results of the cointegration tests, the long term equation in which the non performing financing (NPF) has an impact of 25% against ROE and its t-statistics value is 3.8455, which is greater than the t-table significance level of 0.05 (1.98) are known. These results mean that the NPF variable has a significant relationship with ROE. In addition, the NPF variable's coefficient mark shows a negative value, meaning that the NPF has a negative impact in the long term equation. Any increase in the amount of NPF by 1% will decrease the ROE by 25%.

Based from the results of the cointegration tests, Mudharabah time deposit (MD) has an impact of 14.1% against ROE and its t-statistics value is 7.4288, which is greater than the t-table significance level of 0.05 (1.98). These results mean that the DM variable has a significant relationship with ROE. In addition, the Mudharabah time deposit variable's coefficient mark shows a negative value, meaning that the Mudharabah time deposit variable has a negative impact in the long term equation. Any increase in the amount of Mudharabah time deposit by 1% will decrease the ROE by 14.1%.

From the results of the cointegration tests, the variable of asset (LNASET) has an impact of 32.1% against ROE and its t-statistics value is 0.4869, which is lower than the t-table significance level of 0.05 (1.98). These results mean that the asset variable has an insignificant relationship with ROE. In addition, the asset variable's coefficient mark shows a negative value, meaning that the asset variable has a negative impact in the long term equation. Any increase in the amount of assets by 1% will decrease the ROE by 32.1%.

4.2 Analysis on research results

The hypothesis in the previous chapter by obtained by analyzing the relationship of independent variables that consist of asset, Mudharabah time deposit, and non performing financing (NPF) that have been estimated to impact the dependent variables, consisting of Return On Assets (ROA) and Return On Equity (ROE).

The first hypothesis, where it is assumed that the asset variable has a significant positive impact on the return on asset (ROA), is rejected because the result of its t-statistic value is 1.02, which is lower than the t-table significance level of 5% (1,98). Thus, it is concluded that the variable of asset is not significantly influencing the return on asset (ROA).

The second hypothesis, where it is assumed that the asset variable has a significant positive impact on the return on equity (ROE), is rejected because the result of its t-statistic value is 0.48, which is lower than the t-table significance level of 5% (1.98). Thus, it is concluded that the variable of asset is not significantly influencing the return on equity (ROE).

The third hypothesis, where it is assumed that the Mudharabah time deposit variable has a significant negative impact on the return on asset (ROA), is accepted because the result of its t-statistic value is 6,96, which is higher than the t-table significance level of 5% (1.98). Thus, it is concluded that the variable of Mudharabah time deposit is significantly influencing the return on asset (ROA).

The fourth hypothesis, where it is assumed that the Mudharabah time deposit variable has a significant negative impact on the return on equity (ROE), is accepted because the result of its t-statistic value is 7.42 which is higher than the t-table significance level of 5% (1.98). Thus, it is concluded that the variable of Mudharabah time deposit is significantly influencing the return on equity (ROE).

The fifth hypothesis, where it is assumed that the Non Performing Financing (NPF) variable has a significant negative impact on the return on asset (ROA) is accepted, because the result of its t-statistic value is 2.01, which is higher than the t-table significance level of 5% (1.98). Thus, it is concluded that

the variable of Non Performing Financing (NPF) is significantly influencing the return on asset (ROA).

The sixth hypothesis, where it is assumed that the Non Performing Financing (NPF) variable has a significant negative impact on the return on equity (ROE) is accepted, because the result of its t-statistic value is 3.84, which is higher than the t-table significance level of 5% (1.98). Thus, it is concluded that the variable of Non Performing Financing (NPF) is significantly influencing the return on equity (ROE).

4.3 Discussions

4.3.1 The impact of non performing financing on ROA

Based on the statistical data analysis in this study, Non Performing Financing (NPF) variable has a negative significant impact on return on asset (ROA).

Slamat (2001: 174) said that the tendency for losses arising from bank's credit disbursements are caused by the lack of bank interest after the credit runs and the lack of further analysis conducted by the bank in the event of changes in business cycles. Therefore, the real issue here is the problem's early detection, and it must be done properly.

This result is in line with some previous statistical analysis studies done by Arminda (2014), Al-Qudah (2013), Curak (2012), Akhtar (2011), and Alper (2011).

Thus, it can be concluded that Non Performing Financing (NPF) represent the level of problematic financing, where higher Non Performing Financing (NPF) will eventually decrease the value of Profitability, which is proxies by banks' return on asset (ROA).

4.3.2 The impact of Mudharabah time deposit on ROA

Based on the statistical data analysis in this research, Mudharabah Time Deposit variable has a significant negative impact on return on asset (ROA), or in another term, Mudharabah time deposit gives a negative influence on Return On Assets (ROA)

The Bank of Indonesia (2005: 188) stated that in the event where Third Party Fund is increasing, it appears that the risk of displacement (the risk of transfer of funds from Sharia bank to conventional banks) is also increasing, leading to the increase in bank's interest rates. This condition is expected to affect the performance of Islamic banking fund accumulation activities.

The Bank of Indonesia (2013: 151) also stated that the unbalanced financing structure between credit and Third Party Fund can lead to the increase of competitions in the banking industry to obtain third party funds. Such cases will encourage some banks to raise their interest rates and to provide a special rate to large depositors, which can lead to the shrinking of interest rate spreads. This condition is also evident from the impact of the event when the rate of the Bank of Indonesia is increasing, it will be directly responded by the raising of interest rate of

Third Party Fund in the banking sector, even though this event is not entirely and directly affecting the loan interest rate. This is in line with the behavior of banks that generally adjust theier loan interest rate between 5-6 months after the rate of the Bank of Indonesia is increasing.

Based on the results of statistical data analysis, it can be concluded that the relationship between the Mudharabah time deposit and bank's return on asset (ROA) is in accordance with the proposed hypothesis, which states that Mudharabah time deposit has a significant impact on return on assets (ROA) of Sharia banks in Indonesia.

The results of this study's statistical analysis is also in accordance with previous finding by Arninda (2014) in a journal entitled "Islamic Rural Bank Profitability: Evidence from Indonesia, Journal of Islamic Economics, Banking and Finance, Vol. 3, July-Sep 2014 ".

In conclusion, higher given rate from the Bank of Indonesia will result in a high risk of displacement, and the Sharia banking industry will also participate by raising their equivalent rates, resulting in an increase in bank costs of funding, thus reducing the profitability of the banks.

4.3.3 *The impact of asset on ROA*
Based on the statistical data analysis in this research, asset variable has an insignificant positive impact on the return on asset (ROA).

The asset of the bank illustrates the size of the bank, so as the asset of the bank is increasing, the size of the bank is also increasing because there are highly productive assets that can be used by the bank to do their operational activities and to function as an intermediary institution. However, management's ability to perform its functions as the professional investment managers is also highly influencing the quality of the assets they manage.

Based on the results of the statistical data analysis in this research, it can be concluded that the relationship between asset with return on asset (ROA) is not in accordance with the proposed hypothesis, which stated that asset has a significant impact on return on assets (ROA) of Sharia banks in Indonesia.

The results of statistical analysis is in line with the opinion of Alper (2011) in the journal of "Bank Specific and Macroeconomic Determinant of Comercial Bank Profitability: Empirical Evidance from Turkey, Business and Economics Research Journal, Vol 2 No. 2 Year 2011 PP. 139-152. ISSN: 1309-2448 "

In conclusion, having more assets may not necessarily provide more benefits for the bank because there may be diseconomies of scale, such as longer bureaucracy processes or managerial inefficiencies that can deter the asset's positve contribution to

bank's profitability, and as a result, it is insignificantly impacting the bank's profitability.

4.3.4 *The impact of non performing financing on ROE*
Based on the results of statistical data analysis in this research, Non Performing Financing (NPF) variable has a significant negative impact on the return on equity (ROE).

Slamat (2001: 103) has already stated that if the bank is confident that there will be no losses due to the existing non performing loans, they likely will reduce the amount of its capital to increase their equity multiplier, which in turn will increase the return on equity (ROE).

Still based on the results of statistical data analysis, it can be concluded that the relationship between Non Performing Financing (NPF) and return on equity (ROE) is in accordance with the proposed hypothesis, which state that Non Performing Financing (NPF) has a significant impact on return on equity (ROE) of Sharia banks in Indonesia.

The results of statistical analysis is in accordance with the opinion expressed by Al-Qudah (2013), Akhtar et al (2011) and Alper (2011), stating that Non Performing Financing (NPF) is significantly and negatively influencing bank's profitability.

4.3.5 *The impact of mudharabah time deposit on ROE*
Based on the statistical data analysis in this research, Mudharabah Time Deposit variable has a significant negative impact on return on equity (ROE).

Similar with the return on assets (ROA), the reasons Mudharabah time deposit has a negative impact to bank's return on equity (ROE) are because higher Mudharabah time deposit will increase the risk of displacement the risk of transfer of funds from Sharia bank to conventional banks), leading to the increase in bank's interest rates, and the impact of the increase in the Bank of Indonesia's rate will be directly responded by the banks by raising their interest rates of Third Party Fund, even though it is not affecting directly into their loan interest rates. This phenomena can lead to an unbalanced credit and deposits financing structure of the banks, which will impact to the increasing cost of funds incurred by the banks.

Based on the previous study by Slamat (2001: 150), the ability of banks to lower their deposit rates depend on the reaction of certain groups of customers towards the changes in interest rate. The more sensitive is the group of clients to the deposit rate changes, the more difficult it will be for the banks to to lower their deposit rates . The success of the bank to minimize its deposit rates will affect its cost of finance.

Based on the results of statistical data analysis, it can be concluded that the relationship between Mudharabah time deposit and return on equity (ROE) is in accordance with the proposed hypothesis that states Mudharabah time deposit has a significant impact on return on equity (ROE) of Sharia banks in Indonesia. This results of statistical analysis is also in accordance with the opinions expressed byninda Arninda (2014).

In conclusion, higher rate given by the Bank of Indonesia will result in higher risk of displacement in the banking industry including in the Sharia banking industry, which will lead these banks to raise their equivalent rate, resulting in an increase in the costs of banks' finances, thus reducing the profitability of the banks.

4.3.6 *the impact of asset on ROE*

Based on the statistical data analysis in this research, the asset variable has an insignificant negative impact on bank's return on equity (ROE).

Slamat (2001: 102) has already stated that bank owners tend to be interested in the ability of the bank to gain profits from capital they have invested. These owners often use return on equity (ROE) to measure the ability of banks to obtain profits, because it calculates the ratio of bank's net income to their total equity (capital) in percentage.

Slamat (2001: 103) said that the benefits of utilizing bank capital include the security on investment by minimizing the possibility of insolvency or bankruptcy. But capital itself is actually expensive because the higher amount of bank capital will lead to lower ROE for a certain amount of ROA in a certain period. Therefore, in determining the amount of capital, bank management must decide how much profit that can be gained by the bank when the bank increase its capital amount, because increasing the capital will also decrease the bank's ROE at the same time.

This correlation is a trade off between security and profit for the bank shareholders. In some circumstances where the possibility of the banks to experience difficult conditions due to the presence of non performing assets is high, then the management should increase their bank's capital. However, if the management believes that no losses will arise due to the non performing loans, then the management reduce the amount of bank's capital to increase its equity multiplier, which in turn will increase the ROE. Based on the results of statistical data analysis in this study, it can be concluded that the relationship between the assets and the return on equity (ROE) is not in line with the proposed hypothesis that state the asset has a significant impact on return on equity (ROE) of Sharia banks in Indonesia.

The results of the statistical analysis in this study is in accordance with the opinion of Akhtar (2011) published in the journal of "Factors Influencing the Profitability of Islamic Banks of Pakistan International Research Journal of Finance and Economics-Issue 66 (2011)"

Thus, it can be concluded that higher non performing assets onwed by banks causes the bank to increase its capital in order to decrease the profitability ob return on equity (ROE).

5 CONCLUSSIONS

Based on the analysis of data and results that have been done, it can be concluded as follows:

1. The variable of asset has no significant impact on return on asset (ROA).
2. The variable of asset has no significant impact on return on equity (ROE).
3. The variable of Mudharabah Time Deposit Variable has a significant impact on return on asset (ROA).
4. The variable of Mudharabah Time Deposit Variable has a significant impact on return on equity (ROE).
5. The variable of Non Performing Financing (NPF) has a significant negative impact on return on asset (ROA).
6. The variable of Non Performing Financing (NPF) has a significant negative impact on return on quity (ROE).

REFERENCES

Ajija, Sochrul Romatul, et al. 2011. Cara Cerdas Menguasai Eviews. Jakarta:Penerbit Sa- lemba Empat.

Akhtar, Muhammad Farhan. 2011. Factors Influencing the Profitability of Islamic Banks of Pakistan International Research Journal of Finance and Economics-Issue 66 (2011).

Alper, Deger & Adem Anbar.2011. Bank Specific and Macroeconomic Determinant of Commercial Bank Profitability: Empirical Evidence from Turkey Business and Economics Research Journal, Vol 2 No 2 2011 PP.139-152.ISSN:1309-2448.

Al-Qudah, Ali Mustafa & Mahmoud Ali Jaradat. 2013. The Impact of Macroeconomic Va- riabels and Banks Characteristics on Jordanian Islamic Banks Profitability: Empirical Evidence, International Business Research, Vol. 6, No 10; 2013.

Antonio, Muhammad Syafi'i. 2001. Bank Syariah Dari Teori ke Praktik. Jakarta: Gema Insani.

Arifin, Zainul. 2005. Dasar-Dasar Manajemen Bank Syariah. Jakarta: Pustaka Alvabet Anggota IKAPI.

Arifin, Zainul.. 2006. Dasar-Dasar Manajemen Bank Syariah. Jakarta: Pustaka Alvabet Anggota IKAPI.

Arninda, Titi Dewi. 2014. Islamic Rural Bank Profitability: Evidence from Indonesia, Journal of Islamic Economics, Banking and Finance, Vol. 10 No. 3, July - Sep 2014.

Bank Indonesia. 2005. Laporan Perekonomian Indonesia Tahun 2005. Jakarta: Bank Indonesia.

Bank Indonesia. 2013. Laporan Perekonomian Indonesia Tahun 2013. Jakarta: Bank Indonesia.

Bank Indonesia. 2014. Statistik Perbankan Syariah Edisi September 2003 sampai den- gan September 2014. www.bi.go.id.

Curak, Marijana, et al. 2012. Profitability Determinants of the Macedonian Banking Sector in Changing Environment, Procedia- Social and Behavioral Sciences, 44 (2012) 406-416.

Huda, Nurul & Mohammad Heykal. 2010. Lembaga Keuangan Islam, Tinjauan Teori- tis dan Praktis. Jakarta: Kencana.

Javaid, Saira, et al. 2011. Determinants of Bank Profitability in Pakistan: Internal Factor Analysis MEDITERRANEAN JOURNAL OF SOCIAL SCIENCES, Vol. 2, No. 1, January 2011, ISSN 2039-2117.

Muhammad. 2004. Manajemen Dana Bank Syariah. Yogyakarta:EKONISIA.

Reeve, James M. et al. 2013. Pengantar Akuntansi Adaptasi Indonesia Buku I. Jakarta: Salemba Empat.

Slamat, Dahlan. 2001. Manajemen Lembaga Keuangan (Edisi Ketiga). Jakarta: Lembaga Penerbit Fakultas Ekonomi Universitas Indonesia.

Winarno, Wing Wahyu. 2011. Analisis Ekonome- trika dan Statistika dengan Eviews 3. Yo-gyakarta: UPP STIM YKPN.

Section 3: Green business

Advances in Business, Management and Entrepreneurship – Hurriyati et al (eds)
© *2020 Taylor & Francis Group, London, ISBN 978-0-367-27176-3*

The impact of good corporate governance on firm value with corporate social responsibility as a mediating variable: Empirical study of publicly listed mining companies on the Indonesia stock exchange

B. Tjahjadi, N. Soewarno & H. Vitus
Universitas Airlangga, Surabaya, Indonesia

ABSTRACT: This study examined the association among good corporate governance, corporate social responsibility, and firm value. Using purposive sampling, 35 mining companies were selected from the Indonesia Stock Exchange (IDX) for the period 2013–2015 and resulted in 105 observed data. The results indicate that good corporate governance affects firm value and corporate social responsibility, corporate social responsibility does not affect firm value, and corporate social responsibility does not mediate the relationship between corporate governance and firm value. This study contributes to agency theory and the signaling theory in terms of providing empirical evidence on the mining industry in Indonesia.

1 INTRODUCTION

Good corporate governance is a critical aspect in improving firm value. The stronger good corporate governance is, the higher the firm value. A study by Guest et al. (2009) revealed that good corporate governance affects firm value. According to stakeholder theory, a firm is responsible for a triple bottom line: economy, society, and environment (Elkington & Rowlands 1999). A firm with good governance tends to have a greater corporate social responsibility. A study by Khan et al. (2013) in Bangladesh reported that good governance affects corporate social responsibility disclosure, and furthermore is related to financial performance or firm value.

Studies on the impact of good corporate governance on firm value show inconclusive results. Swastika (2013) reports that corporate governance is represented by a board of directors (BOD), audit quality, and independent board of commissioners. The results of this study do not show that corporate governance has a significant influence on firm value, and it cannot ascertain that corporate governance practices have an impact on performance. Firms that implement good corporate governance do not necessarily perform better than others, because of other factors, such as length of corporate governance implementation in a firm. Guest et al. (2009) state that corporate governance affects firm value through board size. Some arguments state that a good board size consists of five or six people. If the board is too large, it will have less effective communication and lead to poor decision making. When a firm grows larger, the board size tends to be larger and it will have a negative impact. Ararat et al. (2016) in their research in Turkey show that good corporate governance can increase firm value. Black et al. (2006)

in Korea, Leal and Silva (2005) in Brazil, and Black et al. (2006) in Russia also show the same effect even with different figures because different types of firms in each country led to different practices.

Disclosure is one of the most dominant factors influencing firm value. Chan et al. (2014) state that corporate governance improves a firm's value. This is because the BOD can play a significant role in controlling agency issues (García-Meca 2009). Wright (1996) finds that the composition of the audit committee is closely related to good financial reporting. McMullen and Raghunandan (1996) provide support for a more reliable relationship between audit presence and financial reporting. Based on the discussion, the first hypothesis is formulated as follows:

Hypothesis 1: Good corporate governance affects firm value.

Corporate responsibility is based on the triple bottom line, which is financial, social, and environmental (Elkington & Rowlands 1999). Khan et al. (2013) report that through board independence and audit committees corporate governance has a positive influence on corporate social responsibility disclosure. In family firms, independent commissioners and other independent counsels play an important role in increasing the desire of companies to disclose corporate social responsibility information and assume more responsible for the social environment. Chan et al. (2014) in their research stated the key to disclosure of corporate social responsibility information is contained in the annual report.

Jizi et al. (2014) state that an effectively designed board structure can contribute to safeguarding all stakeholder interests, and not just contributing to shareholders appointing them. Corporate governance mechanisms can affect the effectiveness of the BOD and do not limit their attention to corporate governance related to corporate social responsibility. Independent commissioners have an attitude that encourages companies to engage in corporate social responsibility and voluntary disclosure.

The corporate social responsibility report can be considered in establishing mutually beneficial long-term relationships with stakeholders. Esa and Ghazali (2012) in their research in Malaysia provide evidence that the size of the BOD has a positive influence on corporate social responsibility disclosure. The larger the size of the BOD the wider and more beneficial is the contribution to corporate social responsibility. Based on the previous discussion, the second hypothesis is formulated as follows.

Hypothesis 2: Good corporate governance affects corporate social responsibility disclosure.

Each firm carries out corporate social responsibility in different ways depending on the size of the business, overall environment and culture in which it is located, and owners' willingness to assume corporate social responsibility. In this intensely competitive era, it is strategic for firms to increase their value and image among their stakeholders through social activities.

Scholars reveal a positive relationship between corporate social responsibility and corporate value. Firms with higher income quality tend to engage in more corporate social responsibility activities. Supriyono et al. (2015) also stated a positive influence of corporate social responsibility disclosure on corporate value. Disclosure of corporate social responsibility information can be used by investors to invest and to provide information related to the sustainability of investment security.

Corporate social responsibility is part of the business strategy in forming a good image for a firm. If corporate social responsibility activities are viewed as part of a strategic management plan to meet the stakeholders' demands, it is reasonable to expect a positive relationship between stakeholders and corporate social responsibility (Roberts 1992). Based on the discussion, the third hypothesis is formulated as follows.

Hypothesis 3: Corporate social responsibility disclosure affects a firm's value.

The shareholders as the principal want the greatest return on their investments, while the manager as the agent who is authorized by the principal to manage the company expects a large incentive for performance. One way is via a system of oversight through good corporate governance.

Disclosure of information becomes an important factor in the influence of corporate governance on corporate value. One form of disclosure is the disclosure of corporate social responsibility information (Ararat et al. 2006). The information disclosed is a form of communication between the company and its stakeholders.

Based on stakeholder theory, corporate social responsibility can be used as a strategy to improve

the company's reputation. Improved good corporate reputation achieves a good response from stakeholders so it can be a stimulus in increasing company value. The activity and disclosure of corporate social responsibility depend on the attitude of management toward stakeholders. When management is more in favor of shareholder interests to maximize profit the company will not sacrifice its profit to conduct corporate social responsibility activities. However, when management uses the concept that shareholders are one of the other stakeholder groups, it is of interest not only to shareholders but also to customers, suppliers, and communities and the environment (Heath & Norman 2004). In addition, the information disclosed can also have a negative effect because it will be used by stakeholders in making decisions. When the information disclosed does not match what the stakeholder expects, it will not have a positive impact on firm value. The imposition of social activities will negatively affect firm value because it is considered unprofitable to stakeholders. Therefore, the fulfillment of the wishes or needs of all stakeholders through corporate social responsibility activities will be borne by shareholder profits (Moser & Martin 2012). Based on previous discussions, the fourth hypothesis is formulated as follows.

Hypothesis 4: Corporate social responsibility mediates the relationship between good corporate governance and firm value.

2 METHOD

2.1 *Research model*

Based on previous discussions, a research model is presented in Figure 1.

2.2 *Design*

This study was designed to use a quantitative approach at an explanatory level of. It explains the relationships among the variables, namely good corporate governance, corporate social responsibility disclosure, and firm value.

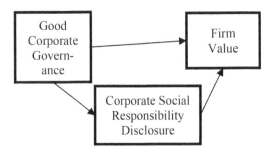

Figure 1. Research model.

2.3 Measurement

Firm value is measured using Tobin's Q with the following formula (Herawaty 2009):

$$FV = \frac{(CP \times \text{Number of Stocks} + TL + I) - CA}{TA} \quad (1)$$

where FV = firm value; CP = closing price; TL = total liabilities; I = inventory; CA = current assets; TA = total assets.

Based on Swastika (2013), good corporate governance is measured using the proxies of BOD, independent board of commissioners, and audit quality. Size of the BOD is determined according to Guest et al. (2009) as follows: $< 3 = 0; 3 - 5 = 1; 6 - 8 = 2; > 8 = 3$. Size of independent commissioners is determined using the BAPEPAM regulation stating that at least 30% of the board must be independent. Therefore, the size of independent commissioners is determined as follow: $< 30\% = 0; > 30\% = 1$. Audit quality is the dummy variable: 1 if audited by one of the Big Four, and 0 if audited by a non-Big Four accounting firm. Corporate social responsibility is based on GRI 4 Corporate Social Responsibilities Disclosure Index (CSRDI): 1 for disclosed items and 0 for undisclosed items. Haniffa and Cooke (2005) state that CSRDI is calculated using the following formula:

$$CSRDIj = \frac{\sum_{t=1}^{nj} Xi.j}{nj} \quad (2)$$

where CSRDIj is the corporate social responsibility disclosure index for company j; n_j is total items disclosed by firm j, $nj \leq 90$; Xij is the dummy variable, and is 1 if item i is disclosed and 0 if item i is not disclosed.

2.4 Sample

Using purposive sampling on publicly traded mining companies listed on the IDX, we get 35 companies that have financial statements for the period 2013–2015, and all data needed for this study are available on their financial statements.

2.5 Analytical technique

This study employed path analysis to test the hypotheses. The equations used for this study is the following:

$$FV = f(GCE)$$
$$FV = a + b(GCG) + c \quad (3)$$

$$CSRD = f(CG)$$
$$CSRD = a + b(CG) + c \quad (4)$$

$$FV = f(CSRD)$$
$$FV = a + b(CSRD) + c \quad (5)$$

$$FV = f(CG, \; CSRD)$$
$$FV = a + b(CG) + \; CSRD + c \quad (6)$$

3 RESULTS AND DISCUSSION

3.1 Results

3.1.1 Descriptive statistics

Table 1 shows the firm value has an average of 1.0064 and standard deviation of 0.97531. This reveals that firm value data varies at 96.91%. Because the level of variation is <100%, it can be concluded that the data are quite homogeneous, which also means that firm value is relatively the same between companies.

Table 1. Tobin's Q.

Test	N	Min	Max	Mean	Std. deviation
Tobin's Q	105	0.0300	8.4600	1.006350	0.9753039
Valid N (listwise)	105				

Table 2 shows that as many as 11 companies have a BOD of fewer than 3 people, 66 companies have BOD between 3 and 5 people, and as many as 25 companies have 6–8 directors. Guest et al. (2009) argue that a good composition of the BOD is between five and six people or at least three people. Therefore, it can be concluded that 91 companies already have BODs of the appropriate size.

Table 2. BOD size.

Valid	Frequency	Percent	Valid percent	Cumulative percent
0	11	10.5	10.5	105
1	66	62.9	62.9	73.3
2	25	23.8	23.8	97.1
3	3	2.9	2.9	100.0
Total	105	100.0	100.0	

Table 3 shows that 97 companies already have a board of commissioners of more than 30% independent members, while the other 8 companies still do not meet the BAPEPAM regulation stating that at least 30% of members of the board of commissioners must be independent.

Table 3. Size of independent BOC.

Valid	Frequency	Percent	Valid percent	Cumulative percent
0	8	7.6	7.6	7.6
1	97	92.4	92.4	100.0
Total	105	100.0	100.0	

Table 4 shows that as many as 52 companies have been audited by one of the Big Four, and 53 companies were audited by non-Big Four accounting firms. Although this measurement is debatable, at least those 52 companies already have a good audit quality.

Table 4. Audit quality.

Valid	Frequency	Percent	Valid percent	Cumulative percent
0	53	50.5	50.5	50.0
1	52	49.5	49.5	100.0
Total	105	100.0	100.0	

3.1.2 Hypotheses testing

Hypothesis 1: Table 5 shows that the value of beta (β) is 0.196 with $p = 0.045$ ($p < 0.05$). It can be concluded that hypothesis 1 stating that good corporate governance affects firm value is supported. The value of coefficient of determination (R^2) is 0.038, meaning that only 3.8% variation in firm value data can be explained by variation in good corporate governance data, and 96.2% is explained by other variables.

Hypothesis 2: Table 5 also shows that the value of beta (β) is 0.585 with $p = 0.000$ ($p < 0.05$). It can be concluded that hypothesis 2 stating that good corporate governance affects corporate social responsibility disclosure is also supported. The value of the coefficient of determination (R^2) is 0.342, meaning that 34.2% variation in corporate social responsibility

Table 5. Results of path analysis.

Path	R^2	Beta (β)	p-value
Good Corporate Governance > Firm Value	0.038	0.196	0.045
Good Corporate Governance > Corporate Social Responsibility Disclosure	0.342	0.585	0.000
Corporate Social Responsibility Disclosure > Firm Value	0.004	−0.064	0.518
Good Corporate Governance > Corporate Social Responsibility Disclosure > Firm Value	0.087	GCG: 0.355 CSRD: −0.271	GCG: 0.003 CSRD: 0.022

data can be explained by good corporate governance data, and 65.8% is explained by other variables.

Hypothesis 3: Table 5 also shows the value of beta (β) is −0.064 with $p = 0.518$ ($p > 0.05$). Since $p > 0.05$, it can be concluded that hypothesis 3 stating that corporate social responsibility disclosure affects firm value is not supported.

Hypothesis 4: Because one of the paths is not significant, hypothesis 4, stating that corporate social responsibility mediates the relationship between good corporate governance and firm value, is not supported.

3.2 Discussion

From the previous results, it can be concluded that good corporate governance affects firm value. This is consistent with studies conducted by Black et al. (2006) in Korea, Ararat (2016) and Leal and Silva (2005) in Brazil, and Black et al. (2006) in Russia. Disclosure is one of the most dominant factors in influencing firm value. A BOD can play a significant role in controlling agency issues as stated by García-Meca (2009). The practice of corporate governance does not appear to be dominant in this study although it has a significant positive effect. This variable cannot represent the practice of corporate governance because there are many other considerations, such as foreign ownership, government or family, size, and type of firms.

This study proves that good corporate governance affects corporate social responsibility disclosure. This result is consistent with studies conducted by Esa and Ghazali (2012), Khan et al. (2013), Chan et al. (2014), and Jizi et al. (2014). . Independent commissioners have an attitude to encourage companies to engage in corporate social responsibility and voluntarily disclose and advise stakeholders to pay more attention to corporate social responsibility disclosure. With a larger board size, corporate social responsibility information disclosure is more significant than others are. A BOD with more members and diverse experiences and backgrounds will be more conducive to having more lively discussions about social activities and more investments in those activities. As stated by Khan et al. (2013), independent commissioners and other independent boards play an important role in increasing the desire to disclose corporate social responsibility information conducted by a company and assume more responsibility for the social environment.

This study fails to support that corporate social responsibility disclosure affects firm value. This result is in line with previous research by Freire et al. (2011) in Brazil. The relationship with employees and the environment creates a social burden on companies. Alotaibi and Hussainey (2016) share the same opinion that corporate social responsibility disclosure has no effect on firm value. This is because every company carries out corporate social responsibility in different ways because of different sizes of the business, the environment and culture, and owners' willingness. The impact of disclosure also requires a long period of time and is sometimes less important for investors. In their study on Saudi Ariabia, Alotaibi and Hussainey (2016) also prove that corporate social responsibility disclosure has no effect on firm value.

4 CONCLUSION

This study has proved or disproved the following points. First, good corporate governance affects firm value. Second, good corporate governance affects corporate social responsibility disclosure. Third, corporate social responsibility disclosure does not affect firm value. Lastly, corporate social responsibility disclosure does not mediate the relationship between good corporate governance and firm value. This study has the following limitations. It focused on publicly listed mining companies in the IDX, which therefore limits the generalization. Future studies need to include other sectors such as manufacturing and trading companies for better generalization. This study employed regression analysis to test the hypotheses. Future studies need to use other methods, such as experiments to test causal relationships. This study is limited by only three variables. Future studies can explore other variables to build a more comprehensive model.

REFERENCES

Alotaibi, K. & Hussainey, K. 2016. Quantity versus quality: The value relevance of CSR disclosure of Saudi companies. *Corporate Ownership and Control* 13(2).

Ararat, M., Black, B.S. & Yurtoglu, B.B. 2017. The effect of corporate governance on firm value and profitability: Time-series evidence from Turkey. *Emerging Markets Review* 30: 113–132.

Black, B.S., Jang, H. & Kim, W. 2006. Does corporate governance predict firms' market values? Evidence from Korea. *The Journal of Law, Economics, and Organization* 22(2): 366–413.

Black, B.S., Love, I. & Rachinsky, A. 2006. Corporate governance indices and firms' market values: Time series evidence from Russia. *Emerging Markets Review* 7(4): 361–379.

Chan, M.C., Watson, J. & Woodliff, D. 2014. Corporate governance quality and CSR disclosures. *Journal of Business Ethics* 125(1): 59–73.

Elkington, J. & Rowlands, I. H. 1999. Cannibals with forks: The triple bottom line of 21st century business. *Alternatives Journal* 25(4): 42.

Esa, E. & Ghazali, N.A.M. 2012. Corporate social responsibility and corporate governance in Malaysian government-linked companies. *Corporate Governance: The International Journal of Business in Society* 12(3): 292–305.

García-Meca, E. & Sánchez-Ballesta, J. P. 2009. Corporate governance and earnings management: A meta-analysis. *Corporate Governance: An International Review* 17(5): 594–610.

Guest, J.S., Skerlos, S.J., Barnard, J.L., et al. 2009. A new planning and design paradigm to achieve sustainable resource recovery from wastewater. *Environmental. Science & Technology* 43(16): 6126–6130.

Haniffa, R.M. & Cooke, T.E. 2005. The impact of culture and governance on corporate social reporting. *Journal of Accounting and Public Policy* 24(5): 391–430.

Heath, J. & Norman, W. 2004. Stakeholder theory, corporate governance and public management: What can the history of state-run enterprises teach us in the post-Enron era? *Journal of Business Ethics* 53(3): 247–265.

Jizi, M.I., Salama, A., Dixon, R. & Stratling, R. 2014. Corporate governance and corporate social responsibility disclosure: Evidence from the US banking sector. *Journal of Business Ethics* 125(4): 601–615.

Khan, A., Muttakin, M.B. & Siddiqui, J. 2013. Corporate governance and corporate social responsibility disclosures: Evidence from an emerging economy. *Journal of Business Ethics* 114(2): 207–223.

Leal, R.P. & Silva, A.L. 2007. Corporate governance and value in Brazil (and in Chile). *Investor protection and corporate governance: Firm-level evidence across Latin America.* 213–287.

McMullen, D.A. & Raghunandan, K. 1996. Enhancing audit committee effectiveness. *Journal of Accountancy* 182(2): 79.

Moser, D.V. & Martin, P.R. 2012. A broader perspective on corporate social responsibility research in accounting. *The Accounting Review* 87(3): 797–806.

Roberts, R.W. 1992. Determinants of corporate social responsibility disclosure: An application of stakeholder theory. *Accounting, Organizations and Society* 17(6): 595–612.

Supriyono, E., Almasyhari, A.K., Suhardjanto, D. & Rahmawati, S. 2015. The impact of corporate governance on corporate social disclosure: Comparative study in South East Asia. *International Journal of Monetary Economics and Finance* 8(2): 143–161.

Swastika, D.L.T. 2013. Corporate governance, firm size, and earning management: Evidence in Indonesia stock exchange. *IOSR Journal of Business and Management* 10(4): 77–82.

Wright, D.W. 1996. Evidence on the relation between corporate governance characteristics and the quality of financial reporting. University of Michigan.

Advances in Business, Management and Entrepreneurship – Hurriyati et al (eds)
© 2020 Taylor & Francis Group, London, ISBN 978-0-367-27176-3

Greed, parental influence, and adolescent financial behavior

L. Wenatri, S. Surya & Maruf[1]
Jurusan Manajemen, Universitas Andalas, Padang, Sumatera Barat

ABSTRACT: This study aimed to investigate the relationship between personal attitude as reflected by level of greed and individual background based on parental influence on adolescent financial behavior, with financial knowledge and financial attitude as mediating variables. This study used a quantitative approache and data were obtained by distributing questionnaires to 240 high school students in Padang City. The results show that financial behavior of adolescents is mainly influenced by parents. In addition, it can be concluded that parents have weak role in the development of financial knowledge and attitude. Adolescents in Padang City are strongly dependent on parents for financial decisions and greed is not reflected in their behavior. Implication of this research should be the concern of stakeholders such as financial institutions, because financial personality traits and behavior during adolescence will last until adulthood and will also affect a person's financial decisions over the rest of his or her life.

Keywords: Adolescent, Financial Behavior, Greed, Parental Influence

1 INTRODUCTION

Adolescence is a period of transition of children into adulthood, beginning at age 12 years and ending at age 21. Although adolescence is a period of growth and self-discovery and has tremendous potential, it is also a time of high risk where the social environment exerts a strong influence on behavior (WHO 2017).

Financial management by teenagers in high school is not easy, especially when they have not been able to earn their own income. The students still depend on pocket money given by their parents. They will always want something that is up to date, both in fashion and gadgets, following trends in order not to be considered old-fashioned. They will always want something more and will never be satisfied with what they have.

Research shows that financial personality traits and behavior during adolescence will last until adulthood and will also affect a person's financial decisions over the rest of his or her life (Eccles et al. 2013). Ashby et al. (2011) also found that a person's saving habits at the age of 16 can predict his or her saving habits at the age of 34. So, if someonehas good financial behavior as a teenager, then it will have a positive impact on the person's financial behavior as he or she grows up.

The Theory of Planned Behavior (TPB) helps to predict an individual intention to engage in behavior

at a certain time and place. The theory argues that individual behavior is driven by behavioral intention, where the intention of the behavior is a function of three determinants: an individual's attitude toward behavior, subjective norms, and perceived behavioral control (Ajzen 1991).

This theory provides a useful conceptual framework to deal with the complexity of human social behavior. It incorporates several central concepts in the social sciences and behavioral psychology and defines them in a way that allows the prediction and understanding of certain behaviors in certain contexts. Attitudes toward the behavior, subjective norms on behavior, and perceived control over the behavior are usually found to predict behavioral intention with a high degree of accuracy. In turn, the intention, in combination with the perceived behavioral control, can explain a significant proportion of variance in behavior (Ajzen 1991).

Ricciardi and Simon (2002) analyze financial behavior, trying to explain what, why, and how to finance and investments, from a human perspective. They also revealed three financial foundations of behavior: psychology, sociology, and finance. One of the psychological factors that affect a person's financial behavior is greed: the desire to always get (property, etc.) as much as possible (Indonesian Dictionary Online 2017). Merriam-Webster Online Dictionary (2017) defines greed as a selfish desire and the desire to get something more (such as

---
[1] Corresponding author: maruf@eb.unand.ac.id

money) than is needed. Seuntjens et al. (2015) found that the desire to acquire more and feeling one never has enough and cannot settle for what one owns are two important components of greed. Greed is associated with materialism and envy and can cause self-interested behavior in disguise. It can affect a person's behavior, especially financial behavior. Seuntjens et al. (2016) in their research found that greed could be positive (work harder) or negative (greed). Zeelenberg et al. (2015) also mentioned that teens tend to be greedier than adults.

In addition to psychology, the next foundation of financial behavior is sociology. The first social environment of the child is the family. Family environment has an important role in the formation of a child's behavior and personality. Parental influence is one of the determining factors in the financial behavior of adolescents. Shim et al. (2010) tested a variety of financial socialization processes and found that direct teaching by parents about personal finances has a greater influence on financial literacy among high school students than financial education in schools.

Shim et al. (2010) used financial knowledge and financial attitude as a mediating variable. Theirresearch results show that financial knowledge plays an important role in predicting financial attitude and financial attitude predicts financial behavior. The results support the hierarchical relationship of knowledge–attitude–behavior.

Jorgensen and Savla (2010) found that parents were deemed to affect financial attitude and financial behavior of children. Parental influence had two main components: (1) the amount of financial learning that occurs and (2) the frequency of financial learning. In their research, Jorgensen and Savla (2010) used financial knowledge and financial attitude as mediating variables between parental influences and financial behavior. Overall, parents are considered to have a direct and significant effect on financial attitude and an indirect and significant influence on financial behavior through financial attitude as a mediating variables but do not have an effect on financial knowledge.

These two foundations of financial behavior were the background for the authors researchgreed and parental influences on high school students' financial behavior in Padang with financial knowledge and financial attitude as mediating variables. Padang City was chosen as the location for this research because of the diversity of high schools, which consists of public and private, public and private vocational high schools, and public and private Madrasah Aliyah (Islamic schools). The sample in this study was therefore expected to have more diverse characteristics. Based on this background, the authors aimed to investigate the influence of greed and parental influence on adolescent financial behavior in Padang with financial knowledge and financial attitude as mediating variables.

2 METHODS

This research used a quantitative approach to test the hypothesis. The samples of this study were 240 high school students in the city of Padang. In this study, the test data were analyzed using descriptive statistics, validity and reliability, the classic assumption test using SPSS 20.0 (Statistical Program for Social Science), as well as structural equation modeling using AMOS 22.0 (analysis of moment structure) to test the hypothesis.

This study uses the definition of *financial behavior* of Sewell (2010), which is the study of the influence of psychology on the behavior of financial practitioners and the subsequent effects on the market.

Dew and Xiao (2011) measured financial behavior using 14 indicators: (1) comparison shop, (2) pay bills on time, (3) keep a record of financial transactions, (4) stay within budget, (5) pay off credit card, (6) max out your credit card, (7) make the minimum payment on loans, (8) maintain or create an emergency fund, (9) save from every paycheck, (10) save for a long-term goal other than retirement, (11) save for retirement, (12) invest money, (13) obtain or maintain adequate health insurance, and (14) obtain or maintain adequate life insurance.

According to Krekels and Pandelaere (2014), *greed* is an insatiable desire for more resources, money, or other goods. In this research the authors used the initial model of the Dispositional Greed Scale (Seuntjens et al., 2015), which uses seven indicators of measurement and added the materialistic indicators which were also measured in the study of Seuntjens, Zeelenberg, et al. (2015) and included statements such as: "When I eat a packet of chips, I will not stop until the chips it up," and "When I use social media (such as Facebook, twitter, Instagram), I would like to have a friend/follower as much as possible." The scale used is an ordinal scale where 1 = strongly disagree and 5 = strongly agree. The Online Oxford Advanced Learner's Dictionary (2017) defined *parental influence* as the ability of parents to affect the development of a person's behavior. Norvilitis and MacLean (2010) measure parental influence by using four dimensions: (1) parental instructionassesses the instructions parents give on financial matters; (2) parental facilitation assesses the involvement of parents in assisting their children in managing finances; (3) parental worries measure how often children watched their parents struggle to get money and worried about money; and (4) parental reticence measured situations when parents avoid discussing financial issues with their children.

According to Parrotta (1992), *financial knowledge* is a proprietary knowledge and understanding of the information related to financial issues, particularly financial management practices recommended by experts in the field. Jorgensen and Savla (2010) measure financial knowledge with 27 indicators and cover four areas of content: general financial literacy, savings and loans, insurance, and investments. Responses were given a value of 1 (true) or 0 (false) and summed. The authors chose to use Jorgensen and Savla (2010) method in measuring the financial knowledge of teenagers in Padang because the questions and statements are tailored to the Indonesian state.

Parrotta (1992) mentioned that *financial attitude* is a psychological tendency expressed by evaluating financial management with some level of agreement or disagreement. Lim and Teo (1997) measured financial attitude with several dimensions. In this study, the authors will only use the indicators of *obsession* in measuring financial attitude because the authors wanted to see exactly the parental influence on the dimensions of obsession. Indicators of obsession were: (1) I firmly believe that money can solve all of my problems; (2) I feel that money is the only thing that I can really count on; (3) Money is the most important goal in my life; (4) Money can buy everything; (5) I believe that time not spent on making money is time wasted; (6) I would do practically anything legal for money if it were enough; and (7) I often fantasies about money and what I could do with it.

Seuntjens et al. (2016) in their research found that *greed* could have a positive or negative effect on financial behavior. The positive impact is that greed can make someone work harder so as to earn a higher income to make ends meet. However, greed can also cause a person to spend more, save less, and cause indebtedness to the people around them such as family, friends, as well as financial institutions. Based on this study, the following hypothesis is proposed:

Hypothesis 1: Greed has an effect on adolescent financial behavior.

Shim et al. (2010) found that direct teaching from parents about personal finance has a stronger effect on financial literacy among high school students than financial education in schools. Jorgensen and Savla (2010) also found that parents were deemed to affect financial attitude and financial behavior of children. Overall, parents are considered to have a direct and significant effect on financial attitude, and have an indirect and significant influence on the financial behavior through financial attitude as a mediating variable, but do not affect financial knowledge. Therefore, the second hypothesis is:

Hypothesis 2: Parental influence allegedly affects financial attitude.

A study by Shim et al. (2010) showed that parents who deliberately teach their children about financial management can provide a greater influence on children's financial knowledge rather than a combination of lessons in high school and learning in the workplace. Conversely, Jorgensen and Savla (2010) found that young children do not think parents influence their financial knowledge. A nonsignificant relationship between perceived parental influence and financial knowledge supports the idea that many parents do not teach their children financial literacy (Lyons & Hunt 2003). Thus, it is hypothesized that:

Hypothesis 3: Parental influence has an effect on financial knowledge.
Hypothesis 4: Parental influence has an effect on financial behavior.

Jorgensen and Savla (2010) also found that financial knowledge has a significant influence on financial attitude, which has a significant influence on financial behavior. With increasing financial knowledge, the financial attitude of young people will also increase. These results are encouraging and suggest that with increasing knowledge and attitude, the ability of young people to make the right financial decisions can be increased. Thus:

Hypothesis 5: Financial knowledge has an effect on financial attitude.
Hypothesis 6: Financial attitude has an effect on financial behavior.

3 RESULTS AND DISCUSSION

Data were obtained from questionnaires (November 19 to December 15, 2017), either directly or online from students of SMA, SMK, MA in Padang. Characteristics of respondents in this study were divided by gender, age, school of origin, place of residence, work of parents, parent education, parental income, allowances, expenses, saving, and debt per month.

Based on the descriptive statistics of the identity of respondents it can be concluded that 63.8% of the respondents were women (153 respondents), while the remaining 36.3% were men (87 respondents). With regard to grade level, 41.3% of respondents (99) were from grade 3, 37.1% (89) were from grade 1, and 21.7% (52) were from grade 2.

Most of the respondents (208, or 86.7%) still lived with their parents, 11.3% (27) lived alone in a rented room, and 5 other respondents, or 2.1%, lived with other relatives such as the grandfather/grandmother or uncle/aunt. The majority (33.3%, or 80) of respondents' parents are self-employed. Most of the respondents' parents are high school graduates: 53.8% (129) of fathers and 50.8% (122) of mothers. In terms of income, 35% (84)of respondents' parents have an income of Rp 2,000,000–4,000,000 per month. To examine the relationship between variables, this study used structural equation modeling using AMOS application 22.The results of the validity test in SPSS research models can be described as follows.

Figure 1 shows a conformance test model that fits properly. Although the chi-square value and the

probability were not good, for the sample > 20 the RMSEA must be < 0.08.

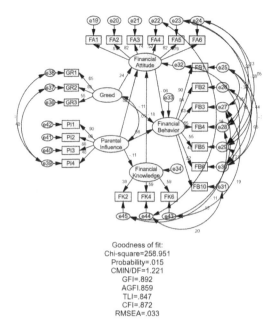

Goodness of fit:
Chi-square=258.951
Probability=.015
CMIN/DF=1.221
GFI=.892
AGFI.859
TLI=.847
CFI=.872
RMSEA=.033

Figure 1. Structural model.

Test of normality shows that the value of the critical ratio of the model was fit before that is equal to 0.630. This value is 2.58 < 0.630 < 2.58; thus it can be concluded that the data are normally distributed.

Table 1 shows the result of hypothesis testing based on regression weights which shows the relationship between variables.

Table 1 shows the CR and p values for the relationship between variables. Relationships between variables are significant whenCR > 1.96 and a p <0.05. Based on Table 1, only two relationships had a significant influence: greed on financial attitude and parental influences on financial behavior.

Based on these results, the value of the direct effect of greed on financial behavior amounted to −0.106. This shows that greed has a negative effect on financial behavior. But the influence of greed on financial behavior was not significant because its CR value was only −1.241 (<1.96) and p = 0.106

Table 1. Regression weights.

	Estimate	SE	CR	p	Label
FK <− PI	−0.062	0.062	−0.998	0.318	
FA <− G	0.177	0.065	2733	0.006	Sig.
FA <− PI	0.025	0.138	0.182	0.856	
FA <− FK	−0.493	0.305	−1615	0.106	
FB <− G	-0.089	0.072	−1241	0.215	
FB <− PI	1,037	0.147	7043	***	Sig.
FB <− FA	−0.068	0.091	−0.75	0.453	

FA, financial attitude; FB, financial behavior; FK, financial knowledge; G, greed; PI, pa-rental influence.

(>0.05). Hypothesis 1 can therefore be rejected. In a previous study Seuntjens et al. (2016) found that greed has both a positive and a negative impact on financial behavior. The positive impact was that greed can make someone work harder to earn a higher income to make ends meet. However, greed can also cause a person to spendmore, rarely save money, and can can also cause indebtedness to the people around them such as family, friends, as well as financial institutions. But in this study Seuntjens et al. did not use an ordinal scale to calculate the financial behavior.

Greed did not affect financial behavior of teenagers in Padang allegedly because of demographic factors. In terms of respondents' incomes, 24.6% of parents had a low income of IDR 2,000,000 by 24.6 percent and 35% had an income ofIDR 2,000,-000–4,000,000 income. The average allowance of high school teenagers is IDR 547,791 per month so as to restrict teenagers from spending money even though they have a great desire to buy something. Moreover, the use of credit cards in Padang by adolescents was still relatively low compared to the use of credit cards by adolescents in other countries.

Although in this study there was no hypothesis relating greed and financial attitude, our results indicate that greed has a significant effect on financial attitude. The value of the direct effect of greed on financial attitude was 0.241. When the teenager had more greed his or her obsession with money was higher. The significant effect of greed on financial attitude was supported by the theory of planned behavior (Ajzen 1991). These variables will affect the attitude of the person before the other variables will. Behavior is determined not only by attitude but also by subjective norms of social pressure to perform or not perform the behavior and perceived behavioral control refers to the perceived ease or difficulty of performing the behavior.

The value of the direct effect of parental influence on financial attitude was 0.016. This suggests that parental influence has no significant effect on financial attitude. Hypothesis 2 can therefore be rejected. These results contradict the research by Jorgensen and Savla (2010), which found that parents were deemed to affect financial attitude and financial behavior of children. Overall, parents are considered to have a direct and significant effect on financial attitude and have an indirect and significant influence on financial behavior through financial attitude as a mediating variable. This is because in this study the authors wanted to see the financial dimensions of adolescent attitude regardingobsession with money, and therefore other dimensions of financial adolescent attitude have not been illustrated in this study. It is expected that in the next research another dimension will be added to the financial attitude so that all financial dimensions can be illustrated with a good attitude.

The value of the direct effect of parental influence on financial knowledge was −0.109. This suggests that parental influence negatively affects financial knowledge. However, the influence of parental influence on financial knowledge is not significant because it is only has a −0.998 CR value (<1.96) and $p = 0.318$ (>0.05). Hypothesis 3 can therefore be rejected. However, this result is contrary to research by Shim et al. (2010), who found that the direct teaching by parents about personal finance has a stronger effect on financial literacy among secondary school students than financial education in schools. These results corresponded with the results of the study by Jorgensen and Savla (2010), which found that young children do not think parents influence their financial knowledge. Lyons and Hunt (2003) also found that many parents do not teach financial literacy to their children and more children obtain financial information from financial officers, despite wishing to receive financial information from their parents.

The effect of parental influence on financial knowledge allegedly caused by the educational background of the parents of respondents wasstill relatively low and not significant. In this study, 53.8% of respondents' fathers had a high school education, 7.9% of junior high school, and 3.3% of primary school background. Respondents' mothers also had low educational background, with 50.8% senior high school graduates, 8.8% junior high school graduates, and 1.7% with a primary school educational background. It is expected that in future researchrespondents will be chosen more evenly with regard to educational background of parents, so that the influence of parents on financial knowledge of the child can be better measured.

Although parental influence does not have a significant effect on financial attitude and financial knowledge, this study indicated that parental influence had a significant effect on adolescent financial behavior. The value of the direct effect was 0.563 and CR was 7.043 (>1.96) and p ***, which means the figures show very small number (< 0.05). It can be concluded that parental influence has a significant effect on financial behavior, so hypothesis 4 is accepted.

These results concur with those of Norvilitis and MacLean (2010), who found that parents who have a hands-on approach to teaching children to handle money such as teaching them how to manage pocket money andbank accounts have children with lower debt levels and credit card usage in college.

Research by Shim et al. (2010) supports the proposition that the behavior of parents, work experience in middle school, and high school financial education during adolescence helped shape learning, attitudes, and behavior of students while growing up. The study also found that older people who have demonstrated a positive financial behavior and engage in a more direct teaching during adolescence are considered to be the main model for their children on financial management during the first year of college.

Shim et al. (2010) found that seeing parents as financial role models had a positive association with the expectations of parents and showed a good attitude toward healthy financial behavior on the part of adolescents. What's more, when parents are involved and have acted as role models, children grow up feeling that they can have better control of their finances.

The value of the direct effect of financial variables on financial knowledge attitude amounted to −0.175. This shows that financial knowledge negatively effect on financial attitude. However, the influence of parental influence and financial behavior is not significant because its CR value was −1.615 (<1.96) and $p = 0.106$ (>0.05). Hypothesis 5 can therefore be rejected. The results of this study are not consistent with the research of Jorgensen and Savla (2010), which found that financial knowledge has a significant influence on financial attitude, which has a significant influence on financial behavior. With increasing financial knowledge, the financial attitude of young people will also increase. These results are encouraging and suggest that with increasing knowledge and attitude, the ability of young people to make the right financial decisionscan be improved.

The value of the direct effect of financial variables attitude toward financial behavior amounted to −0.053. This shows that financial knowledge had no significant effect on financial attitude. Hypothesis can therefore be rejected. The research of Herdijono and Damanik (2016)showed that individuals with high financial knowledge do not necessarily have good financial behavior and also someone with low financial knowledge does not necessarily have a poor financial behavior. They also studied the effect of financial attitude. A person with a good financial attitude will show a good mindset about the money that he or she perceives will be present in the future (obsession), does not use the money for the purpose of controlling others or as a solution to a problem (power), is able to control his or her personal finances(effort), can adjust the use of money to be able to make ends meet (inadequacy), does not want to spend money (retention), and has an ever-evolving view about money or does not rely on the past (securities) so as to be able to exercise control over its consumption, balance expenditures and revenues held (cash flow), set aside money for savings and investment, as well as manage the debts owed to welfare.

There was no significant effect of financial knowledge and financial behavior on financial attitude as the authors chose to measure the attitude of financial obsession with money alone.Other dimensions of financial attitude are thus not represented in this study. To improve subsequent research, it is suggested that financial attitude be measured by adding another dimension such as power, budget, achievement, evaluation, anxiety, retention, and lack of generosity.

4 CONCLUSION

The results showed that greed has no significant effect on financial behavior but a significant effect on the financial attitude of adolescents. Parental influence has a significant effect on the financial behavior of teenagers, but no significant effect on financial knowledge and financial attitude. Financial knowledge had no significant effect on financial attitude and financial attitude has no significant effect on adolescent financial behavior in Padang City. A future study is expected to be able to add other psychological variables such as self-control, impulsiveness, envy, social desirability bias, and others that affect a person's financial behavior. Future studies are also expected to add another dimension of financial attitude variable such as power, budget, achievement, evaluation, anxiety, retention, and lack ofgenerosity and to choose indicators that more accurately measure variables. Researchers can also add to the sample size and try to conduct research on other respondents such as students and adults to ask about their financial behavior as teenagers and relate it to their present financial behavior.

ACKNOWLEDGMENTS

This research was supported by the authors' families and friends. We thank the many high schools and students in Padang City who participated in this study.

REFERENCES

Ajzen, I. 1991. The theory of planned behavior. *Organizational Behavior and Human Decision Processes* 50: 179–211.

Ashby, J., Schoon, I. & Webley, P. 2011. Save now, save later? Linkages between saving behavior in adolescence and adulthood. *European Psychologist* 16: 227–237.

Dew, J. & Xiao, J.J. 2011. The financial management behavior scale: Development and validation. *Journal of Financial Counseling and Planning* 22 (1): 59–53.

Eccles, D.W., Ward, P., Goldsmith, E. & Arsal, G. 2013. The relationship between retirement wealth and householders' lifetime personal financial and investing behaviors. *Journal of Consumer Affairs* 47: 432–464.

"Greed." Indonesian Dictionary Online. February. 2017. http://kbbi.web.id/tamak

"Greed." Merriam-Webster Online Dictionary. February. 2017. https://www.merriam-webster.com/dictionary/greed

"Greed." Online Oxford Advanced Learner's Dictionary. February. 2017. https://en.oxforddictionaries.com/definition/greed

Herdijono, I. & Damanik, L.A. 2016. Financial influence attitude, financial knowledge, parental income on financial management behavior. *Journal of Theory and Applied Management* 3: 226–241.

"Influence." Online Oxford Advanced Learner's Dictionary. August. 2017. https://en.oxforddictionaries.com/definition/influence

Jorgensen, B.L. & Savla, J. 2010. Financial literacy of young adults: The importance of parental socialization. *Family Relations* 59: 465–478.

Krekels, G. & Pandelaere, M. 2014. Dispositional greed. *Journal of Personality and Individual Differences* 74: 225–230.

Lim, V.K.G. & Thompson, S.H.T. 1997. Sex, money and financial hardship: An empirical study of attitudes toward money among undergraduates in the Singapore. *Journal of Economic Psychology* 18: 369–386.

Lyons, A.C. & Hunt, J.L. 2003. The credit practices and financial education needs of community college students. *Association for Financial Counseling and Planning Education* 14(1): 63–74.

Norvilitis, J.M. & MacLean, M.G. 2010. The role of parents in college students' financial behaviors and attitudes. *Journal of Economic Psychology* 31(1): 55–63.

"Parent." Online Oxford Advanced Learner's Dictionary. August. 2017. https://en.oxforddictionaries.com/definition/parent

Parrotta, J. & McFarlane, L. 1992. The impact of financial attitudes and knowledge on financial management and satisfaction. Thesis, University of British Columbia.

Ricciardi, V. & Simon, H.K. 2000. What is behavioral finance? *Business, Education and Technology Journal* Fall.

Sekaran, U. 2006. *Research methods for business*. Book 1. Jakarta: Salemba Empat.

Seuntjens, T.G., Van de Ven, N., Zeelenberg, M. & Van der Schors, A. 2016. Greed and financial adolescent behavior. *Journal of Economic Psychology* 57: 1–12.

Seuntjens, T.G., Zeelenberg, M., Breugelmans, B.C. &s Van de Ven, N. 2015. Defining greed. *British Journal of Psychology* 106: 505–525.

Seuntjens, T.G., Zeelenberg, M., Van de Ven, N. & Breugelmans, S.M. 2015. Dispositional greed. *Journal of Personality and Social Psychology* 108: 917–933.

Sewell, M. 2010. Behavioural finance. University of Cambridge.

Shim, S., Barber, B.L., Card, N.A., Xiao, J.J. & Serido, J. (2010). Financial socialization of first-year college students: The roles of parents, work, and education. *Journal of Youth and Adolescence* 39: 1457–1470.

WHO. Adolescent Development. March 3, 2017. www.who.int/maternal_child_adolescent/topics/adolescence/dev/en/,

Advances in Business, Management and Entrepreneurship – Hurriyati et al (eds)
© *2020 Taylor & Francis Group, London, ISBN 978-0-367-27176-3*

Green skills for green industries: Meeting the needs of the green economy

L.C. Sern
Universiti Tun Hussein Onn, Johor, Malaysia

ABSTRACT: Climate change and environmental pollution have negatively affected the sustainable development of the society, economy, and environment in every country around the world, especially the developing and underdeveloped ones. In order to diminish the impacts caused by environmental problems, many countries have started to shift the existing economy to a green economy model which requires green industries, creates green jobs, and demands green skills. It is expected that in the near future green skills will be imperatively needed by employers along with the conventional hard and soft skills. Therefore, this article focuses on the green skills needed by the various green industrial sectors. Specifically, there are several common green skills that are critical for the various green industries: design, communication, waste management, energy, city planning, management, leadership, management, financial, and procurement. Therefore, institutions of higher learning should play a more significant role in producing graduates who are equipped with green skills in order to cater for the needs of the green industry. As a suggestion, the existing curricula have been revised and the green skills elements should be embedded in the curricula.

1   INTRODUCTION

There is a chain reaction between environment, economy, and social development. Economic activities such as manufacturing of products, deforestation for farming, and transportation, will bring about devastating impacts on the environment, as they produce air pollution and release more greenhouse gases, which induce environmental pollution and climate change. Environmental pollution causes changing weather patterns, more extreme droughts and monsoons, more frequent tropical cyclones, and rising sea level. These disasters will, in turn, affect social as well as economic development (Chinowsky et al. 2011). Owing to low technical and technological levels, many developing and undeveloped countries have paid a very high cost of environmental pollution and climate change. If this problem is not resolved, many countries such as Maldives, Tuvalu, Micronesia, and so on could sink and disappear in the future (Pariona 2017).

Some countries, such as Germany and Japan, have taken action to mitigate the environmental problems by changing the existing economic model to a more environmentally friendly or a greener one (CEDEFOP 2010). A green economy model means the economic activities related to manufacturing, commercial, service, and other industries produce lower carbon emissions, consume lesser nonrenewable energy, and are friendlier to the environment, which leads to improved human well-being (UNEP 2011). The industries that are less energy efficient and less environmentally friendly are transforming into ones that are less polluting and more efficient. This transformation creates more green occupations and greening of existing jobs in industry. For

instance, within the construction industry, technical experts are needed to invent environmentally friendly building materials, and architects are required to integrate natural elements into the building design in which natural light and ventilation are optimally utilized. In the automobile industry, designers have to produce engines with less fuel consumption and that use alternative sources of energy, and engineers have to work with new fuel-efficient technologies.

Beginning a few decades ago, people started paying more attention to the issues related to the environment because everyone looks for a better living quality, safer living environment, and sustainable living condition. There is therefore a growing political and public awareness of the need to reduce the negative impacts on the environment due to economical and industrial activities. In order to counteract the issues pertaining to environmental pollution and climate change, the European Union has developed the "EU 2020 Strategies" with a target of increasing renewable energy consumption by 20% and at the same time reducing greenhouse gas emissions by 20% in the EU member countries (CEDEFOP 2010). Likewise, China has also taken drastic action to cut down greenhouse gas emissions by launching the emission trading system, which is the biggest such mechanism in the world (Harvey 2017).

Although the contents might be different, similar policies have been formulated and endorsed in other countries. For example, the NPE (National Policy on the Environment) has been enforced in Malaysia since 2002, which aims to reduce environmental pollution and conserve nature (Sharom 2002). Additionally, in the 11th Malaysia Plan (11MP) that focuses on sustainable development, there are six thrusts to ensure the vision of achieving a sustainable economy and becoming a developed nation by 2020 (Asohan 2015). One of the them stresses green growth (4th thrust: pursuing green growth for sustainability and resilience) that specifically aims at achieving a resilient, low-carbon, and resource-efficient economy model that leads to higher quality of living and promotes well-being of the population. To achieve this noble goal, both civil and private sectors have their role to play. Government agencies should institute more efficient policies and the enforcement of rules and regulations related to ecological environments should be strengthened. From the private sector side, economic and industrial activities should go green by creating more environmentally friendly products and generating less pollution that causes climate change.

From an economical perspective, those environmental policies have a great impact on the industrial sector, where a large pool of employees with green skills are needed for developing a greener economy model and building up an environmentally friendly society. However, how the industries respond to government policies and the impact of environmental policies on the job market and skill requirements are largely unknown.

1.1 The need for green skills

Green industries create green jobs. This change has a great impact on the labour market and skills demanded by the green industry. The green industry that underpins a green economy requires workers with green skills to perform the tasks. This change of skill requirements is pervasive and calls for a major effort to revise the existing training frameworks because the element of green skills is almost invisible in the existing training curricula and framework.

From the perspective of employment, two types of skills are always mentioned: hard skills and soft skills. A hard skill is generally regarded as a technical skill needed to perform a task related to the occupation (Laker & Powell 2011). Examples of hard skills are programming, welding, and bookkeeping. Soft skills, or, as some researchers call them, generic skills or employability skills, typically are associated with personal ability in dealing with other people or jobs, for instance, teamwork, communication, and problem solving (Schulz 2008; Zhang 2012). Previously, an individual who had both hard and soft skills could secure a job without much difficulty. However, the scenario has been changing gradually in which hard and soft skills are no longer sufficient for an individual to compete in the job market. Apart from hard and soft skills, employers have started looking for manpower with green skills, which are much needed for promoting sustainable development in social, economic, as well as environmental areas.

The green industry has caused a change in employment demand and worker skill requirements.

In other words, significant skill gaps will be created in the labor market due because of the needs of the green industry. Neglecting to fill these skills gaps could impede economic and employment growth and would also represent an obstacle to broader efforts to fight climate change. As mentioned by the UNEP (2008), "Shortages of skilled labour could put the brakes on green expansion ... it is important both to prepare the workforce at large for the skills requirements inherent in green jobs and to ensure that green industries and workplaces do not face a shortage of adequately skilled workers."

This statement appears to reflect the fact that a large number of workers with green skills are needed in order to succeed in the transition to a green economy. In order to avoid a shortage of workers with green skills, immediate action needs to be taken by training institutions to provide the adequate workforce training. Several surveys concerning this issue have been conducted in different countries. The outcomes have revealed difficulties in finding well-trained employees for green industries, and the workforce is still lacking green skills.

Training institutions must go beyond equipping graduates with hard and soft skills because green skills are equally important and they must be embedded in all levels of education. In this aspect, TVET institutions have an essential role to play in producing workers with green skills since TVET is closely and directly connected with the development of the economy and society. It is meaningless and useless to talk about quality if the skills mastered by graduates do not meet the needs of the green industries.

1.2 Definition of green skills

A green skill is referred to as a skill for sustainable development. According to McDonald et al. (2012), green skills are regarded as skills for sustainability that are related to the technical skills, knowledge, values, and attitudes needed in the workforce to develop and support sustainable social, economic, and environmental outcomes in business, industry, and the community. This definition is partially in line with the ideas of Vona et al. (2015), who stated that green skills are technical know-how associated with management, production, and control over technology that ultimately leads to technological advancement and progress, and improvement of the well-being of people.

Likewise, CEDEFOP (2012) defines green skills as the knowledge, abilities, values, and attitudes needed to live in, develop, and support a society that reduces the impact of human activity on the environment. Taken together, green skills generally are composed of three dimensions: knowledge (cognitive dimension), skills/abilities (psychomotor dimension), and attitudes/values (affective dimension) needed by workers to promote sustainable development in the society, economy, and environment. From the cognitive dimension, knowledge concerning environmental protection can be regarded as an element of green skills. From the psychomotor perspective, green skills refer to the ability to, for instance, minimize energy consumption or reduce greenhouse gases. Green skills also refer to the affective aspect, for example, the motivation of an individual to conserve natural resources.

Some people argue that green skills are such a new set of skills that many of them are still absent within the existing labor market. According to a report published by the European Commission (ECORYS, 2008), apart from the traditional skills (e.g., construction, programming), some green skills will be needed for the growth of the green economy such as knowledge of sustainable materials and waste management and landscaping skills. In this case, the challenge of filling the skills gap created by the green economy should be resolved by ensuring the existence of a qualified workforce well equipped with traditional skills as well as green skills. These new skills will have to be incorporated into training and education programs.

1.3 Green skills for the green industry

Several green skills have been identified from the existing literature, such as design, leadership, management, city planning, landscaping, energy, financial, waste management, and communication skills. These skills are commonly required in accomplishing the duty and tasks depending on the sectors.

Many sectors (e.g., building, machine, and circuit design) need people with good design skill. Nowadays, design skill is considered as one of the green skills because designers should be able to embed green elements into a design that is friendlier to the environment (Ragheb et al. 2015). Likewise, many small cities in the world are gradually evolving to be metropolitans, and the existing metropolitans are transforming into smart cities. This transformation of cities needs proper planning in order to make the cities conducive to live in and sustainable (Adhya et al. 2010). Therefore, city planning and landscaping skills are critical green skills in this modern society.

In green industries, leadership and management skills are important to support green activities, such as lean production or life-cycle management (UNEP 2012). Leadership and management skills are needed to bring changes to an organization in terms of its structure, function, and operation in relation to green activities. In addition leadership and management skills, financial skill and procurement skill are also needed by green industry. These skills play an important role in ensuring the implementation of economic responsibility that leads to sustainable social and environmental development. Specifically, financial skill is very much needed to monitor and control the organization's expenditure in order to balance the revenue and responsibility for environmental conservation (Krechovská 2015). Similarly, procurement personnel have to deal with many internal departments of an organization and external agencies to manage, coordinate, and purchase materials. Within the green industrial context, procurement skill is imperative to ensure the materials purchased are harmless to the environment, thereby reducing the environment impact during their life cycle (Bohari & Xia 2015).

Energy skill is seen as another critical green skill. To date, energy production in the world derives mostly from fossil fuel, such as coal, natural gas, and petroleum. The by-products of fossil fuel consumption during energy generation processes are very detrimental to the environment. Therefore, it is essential to equip workers with energy skills that help minimize the use of nonrenewable energy, and at the same time replace those nonrenewable resources with the ones that are safer to use and more environmentally friendly. Apart from energy skill, waste management skill is also becoming increasingly significant for the green industry. Waste management skill means the ability to reduce, reuse, and recycle waste through proper planning, implementation, and coordination of the waste management system (Bozkurt & Stowell 2016). Waste management skill is in high demand today, especially in the waste management sector, which contributes enormously to the sustainability of the environment and prevention of pollution.

Within the context of green skills, communication skill does not refer to verbal and nonverbal communication between people. Green communication skill means communication between people using technology that consumes lesser energy and is a more environmentally friendly type of communication (Buntat et al. 2012).

2 METHOD

Skills are the abilities and capacities people have to perform tasks that are in demand in the workforce. These skills can be generic or specific regarding functions at work such as managing people, computing, collaborating or dealing with risk and uncertainty, or developing a new product or service (Green 2011).

The training framework is composed of several skill components that include basic skills, hard/core skills (Nikitina & Huruoka 2012), and soft skills (Ngang et al. 2015). These skills are usually acquired through education and training and comprise the following:.

1. Basic skill, generic
- Linguistic skill
- Numeracy and literacy
- General IT user skills

2. Hard core skill, technical (dependent on the field of training)
- Knowledge specific (e.g., welding, computer programming, technical drafting skill)

3. Soft skills that are associated with personal ability in dealing with other people or jobs
- Communication skills
- Critical thinking and problem-solving skills
- Team-working skills
- Lifelong learning and information management skills
- Entrepreneurial skills
- Moral and professional ethics
- Leadership skills

4. Green skills, skills for sustainability that are related to technical skills, knowledge, values, and attitudes
- Design, leadership, and management skills
- City planning, landscaping, energy, financial, procurement, waste management, communication skills, and so on

The basic skills are the general skills that must be mastered by students across the world. They include effective use of linguistic, arithmetical, and basic IT skills. Apart from the basic skills, students are trained in hard skills. Hard skills are regarded as technical skills that are domain specific and require consistent training and practice in order to be acquired. Hard skills are various and depend on the field of training. For example, in the field of welding, the hard skills are related to welding techniques, welding inspection, and skill in handling welding machines. In addition to the basic and hard skills, soft skills are also a part of the training components. In brief, soft skills are nonacademic skills such as leadership, teamwork, communication, and lifelong learning (Selamat et al. 2013). In line with the advent of the green economy, it is advisable to include green skills in the training framework and curricula. There are a number of green skills related to the green economy, and therefore the relevancy of green skills is dependent on the industrial sector. For instance, in the waste management sector, waste management skill is very relevant, whereas in the construction sector, the design, landscaping, city planning, and procurement skills are more relevant to the sector.

3 CONCLUSION

In a nutshell, green skills are very much needed by the green industries. However, the existing skill training framework for TVET does not include green skill elements. Therefore, the existing framework must be revised and restructured in order to cater for the needs of green industries. In addition, the policies related to environmental protection have been generated and enforced by many countries. There are

several impacts of those environmental policies on industrial sectors and skill requirements. Employers not only look for employees with conventional skills (hard and soft skills), but they also need someone with green skills in order to perform the tasks efficiently. This information is important for redesigning the training curricula in TVET. Since green industries are growing more and more dominant within the economic sphere, green skills that are needed by the green industries should be given emphasis by training institutions in order to produce high-quality graduates to cater for the needs of the green economy.

ACKNOWLEDGMENTS

This research project (vot 1554) was financially supported by the Ministry of Education Malaysia under the Fundamental Research Grant Scheme (FRGS).

REFERENCES

Adhya, A., Plowright, P. & Stevens, J. 2010. Defining sustainable urbanism: Towards a responsive urban design. In *Proceedings of the Conference on Sustainability and the Built Environment*, 2010, pp. 17–38.
Asohan, A. 2015. 11th Malaysia Plan: Broadband gets some love. *Digital News Asia*. Retrieved from https://www.digitalnewsasia.com/digital-economy/11th-malaysia-plan-broadband-gets-some-love.
Bohari, A.A.M. & Xia B. 2015. Developing green procurement framework for construction projects in Malaysia. In *The Proceedings of the 6th International Conference on Engineering, Project, and Production Management* (EPPM2015), pp. 282–290. Association of Engineering, Project, and Production Management (EPPM).
Bozkurt, Ö. & Stowell, A. 2016. Skill in the green economy: Recycling promises in the UK e-waste management sector, *New Technology, Work, and Employment* 31(2): 146–160.
Buntat, Y. & Othman, M. 2012. *Penerapan kemahiran insaniah 'hijau' (green soft skills) dalam pendidikan teknik dan vokasional di Sekolah Menengah Teknik, Malaysia. Journal of Social Science* 5: 32–41.
CEDEFOP. 2010. *Skills for green jobs: European synthesis report.* Luxembourg: Publications Office of the European Union.
CEDEFOP. 2012. *Green skills and environmental awareness in vocational education and training: Synthesis report.* Luxembourg: Publications Office of the European Union.
Chinowsky, P., Hayles, C., Schweikert, A., Strzepek, N., Strazepek, K. & Schlosser, C.A. 2011. Climate change: Comparative impact on developing and developed countries. *The Engineering Project Organization Journal* 1: 67–80.
ECORYS 2008. *Environment and labour force skills: Overview of the links between the skills profile of the labour force and environmental factors.* Retrieved from: http://ec.europa.eu/environment/enveco/industry_employment/pdf/labor_force.pdf.
Green, F. 2011. *What is skill? An inter-disciplinary synthesis.* Centre for Learning and Life Chances in

Knowledge Economies and Societies. Retrieved from www.llakes.org.

Harvey, F. 2017. China aims to drastically cut greenhouse gas emissions through trading scheme. *The Guardian*. Retrieved from.

Krechovská, M. 2015. Financial Literacy as a Path to Sustainability. *Business Trend*, 2, 3–12.

Laker, D.R. & Powell, J.L. 2011. The differences between hard and soft skills and their relative impact on training transfer. *Human Resource Development Quarterly* 22(1): 111–122.

McDonald, G., Condon, L. & Riordan, M. 2012. *The Australian Green Skills Agreement: Policy and industry context, institutional response and green skills delivery.* NSW: TAFE.

Ngang, T.K., Hashim, N.H. & Mohd Yunus, H. 2015. Novice Teacher perceptions of the soft skills needed in today's workplace. *Procedia: Social and Behavioral Sciences* 177, 284–288.

Nikitina, L. & Huruoka, F. 2012. Sharp focus on soft skills: A case study of Malaysian university students' educational expectations. *Education Research Policy and Practice* 11: 207–224.

Pariona, A. 2017. 10 countries that could disappear with global warming. Retrieved from https://www.worldatlas.com/articles/10-countries-that-could-disappear-with-global-warming.html.

Ragheb, G., El-Shimy,H., Ragheb, A. 2015. Green architecture: A concept of sustainability. *Procedia: Social and Behavioral Science*. 2: 324–333.

Schulz, B, 2008. The importance of soft skills: Education beyond academic knowledge. *Journal of Language and Communication* 2(1): 146–154.

Selamat, J., Ismail, K. Ahmad,A., Hussin, M.H. & Seliman, S. 2013. Framework of soft skills infusion based on learning contract concept in Malaysia higher education. *Asian Social Sciences* 9(7): 22–28.

Sharom, A. 2002. Malaysian environmental law: Ten years after Rio. *Singapore Journal of International & Comparative Law* 6: 855–890.

UNEP. 2008. *Green jobs: Towards decent work in a sustainable, low-carbon world.* Retrieved from: http://www.unep.org/annualreport/2011/docs/UNEP_ANNUAL_REPORT_2011.pdf.

UNEP 2011. *UNEP year book: Emerging issues in our global environment.* Retrieved from www.unep.org/yearbook/2011.

Vona, F., Marin, G., Consoli, D., Popp, D. (2015). Green skills. National Bureau of Economic Research. Working Paper No. 21116. JEL No. J24,Q52. April 2015.

Zhang, A. 2012. Peer assessment of soft skills and hard skills. *Journal of Educational Technology Education* 11: 155–168.

Standard energy management system PDCA cycle of ISO 50001 to minimize energy consumption in service operation

I. Usman & E. Sopacua

Universitas Airlangga, Surabaya, Indonesia

ABSTRACT: The purpose of this research is to explore the green operations activities, especially in energy consumption, and to develop a PDCA cycle framework that can be used as a guide to minimize energy consumption, using Delphi Method. The PDCA cycle ISO 50001 framework is developed through (1) Identifying risks using Delphi method and mean score ranking from the activities found in existing empirical case study in beauty clinic. (2) Analyzing the most crucial risks found in the field. (3) Developing the PDCA cycle as a guide to minimize the energy consumption. The research found twelve risks, and find the crucial risk. PDCA cycle ISO 50001 developed, and found that the most crucial risk is in the act of PDCA cycle, which indicate that there was no continuance in order to solve the problem of energy usage in the business process. This research provides a simple practical guideline of PDCA cycle ISO 50001 template as a critical thinking to make a continuous improvement process.

1 INTRODUCTION

Environmental management is a topic of mutual concern for the business community, government and consumers, in line with the increasing industrialization (New, Green & Morton 2002, Azzone & Manzini 1994, Azzone & Bertele 1994, Azzone & Noci 1996). Growing concerns in global markets over "green" issues and the scarcity of natural resources are forcing executives to look at business strategies from an environmental perspective. The increasing energy used in general has a broad impact on all sectors of life, therefore it needs to build awareness of the importance of an energy management system. The International Organization for Standardization (ISO) cares about the importance of energy for life, thus issuing standards for energy management systems contained in the ISO 50001: 2011 standard -Energy Management System (ENMs). This study aims to build a PDCA template based on ISO 50001 to solve the energy management problems of a company. The research object is a clinic of health and beauty, which is facing the problem of waste in the use of electric energy. Electricity consumption is increasing from time to time that is not in line with business growth.

The existence of ISO 50001 energy management system is an opportunity for companies in creating energy efficiency to increase their income. Based on the reasons above it is necessary to conduct research on improvement of energy management through the PDCA cycle in accordance with ISO 50001 which aims to improve the operation process so as to improve clinical performance to increase efficiency in the use of electrical energy.

1.1 Conceptual background

The energy management system (ENMs) ISO 50001 is a tool that can help to create energy efficiency improvements in the organization, and in the operations process of the business. ISO 50001 is intended to avoid wasteful use of energy and provide precise instructions for building more efficient and economical operations and organizational processes. Energy Management System ISO 50001: 2001 is an international standard energy management system issued by the International Standardization Agency. In Indonesia the standard is adopted to SNI ISO 50001: 2011. The management model of EnMS ISO 50001 is P-D-C-A-Continual Improvement emphasizing on significant Energy Uses and how its management to keep energy performance always increase continuously.

The PDCA ISO 50001 process:

1. Plan

- Responsibility of top management.
- Energy policy.
- Management representative.
- Energy review.
- Objectives and action plan.

2. Do

- Implementation and realization.
- Communication.
- Training.
- Awareness.
- Operational control.

3. Check

- Monitoring.

- Analysis.
- Corrective action.
- Preventive action.
- Internal audit.

4. Act

- Management review.
- New strategic goals.
- Optimizations.

Given the issue of energy costs, it is important for companies to implement energy management systems. By implementing of such systems then energy efficiency efforts can produce optimal results. Implementing this system means that all elements of the organization must be concerned and play a role in energy management.

1.2 Delphi method and mean score ranking

The Delphi method is a "systematic, intuitive" forecasting procedure (Rayens & Hahn 2000), based on expert judgment, derived from a series of questionnaires, and interspersed with feedback (Skulmoski et al. 2007). This method requires the knowledge and contributions of individual experts to respond the questions and send the results to the coordinating center. Delphi method also has the characteristics of anonymity and iteration. Experts are given freedom of association and freedom of participation (Kher et al. 2010). To achieve this goal requires experienced practitioners and experts who are experienced in energy management and have a good understanding of the company's operational managerial concepts. In addition, to ensure reliable quality and information, the tools used to collect data openly to obtain diverse opinions from experts and to seek the most appropriate topic-based feedback (Schmidt et al. 2001).

The Delphi technique, designed in such a way that unbiased results from experts through statistics are repeated and controlled feedback (Kher et al. 2010), adopted as a primary data collection tool. The feedback gathered from the delphi survey was analyzed using the mean score ranking methods. Average score is used because it is a measure of central tendency and widely used in construction management studies (Ke et al. 2010, Chan et al. 2001) to determine the significance of the variable list. Using such an analysis allows to determine the relative significance of each risk factor.

2 METHOD

This research uses case study approach, conducted in a health care and beauty clinic which have an energy usage problem. A case study is a research strategy, which is done through empirical review by investigating the symptoms of problems that occur in the object of research. This strategy can find qualitative evidence obtained from various sources and previous developments of theoretical propositions. (Yin 2003). The purpose of this case study is to understand the object under investigation. Nevertheless, case study research has a specific purpose for explaining and understanding the object it examines specifically as a 'case'. Yin 2003, Yin et al. 2009 states that the purpose of using case study research is not just to explain what the object is being studied, but to explain how it exists and why it can occur. This research uses two tools in identifying risk and designing operational process (PDCA) for research object by Delphi method and mean score ranking.

2.1 Delphi

The identification of risk and design of the PDCA is based on the opinions of experts and referred to ISO 50001, conducted by interviewing and distributing questionnaires to 3 delphi from internal clinics as well as external clinics (operations management experts and business practitioners). The identification of risk and design of the PDCA is based on the opinions of experts and referred to ISO 50001, conducted by interviewing and distributing questionnaires to 3 delphi from internal clinics as well as external clinics (operations management experts and business practitioners).

2.2 Mean score ranking

After obtaining the risk list we have weighted by using a scale of 1-4 to calculate the mean score ranking in order to find the crucial risk correlated to the electrical energy cost issues. Furthermore, should be sought the solution to the problem. The solution will be given is the PDCA design that is contained in the form of PDCA templates.

3 RESULT AND DISCUSSION

The research process conducted interviews for risk identification through the Delphi method, in some beauty clinic experts and operations management experts. Implementation of identification is done through two stages. The first stage is by conducting interviews with experts and clinic practitioners to classify activities into the PDCA process. The second phase identifies the risk and determines the mean score for each identified risk. Risk identification is done by involving beauty clinic experts and also operations management experts. The data were collected by semi structured interview and determination of mean score was done by distributing questionnaire to the experts in order to determine the score for each identified risk. The results of risk identification and analysis in PDCA ISO 50001 process indicate 12 risks spread over PDCA ISO 50001 process. There are four risks in the PLAN phase: Risk number-1.

Table 1. Mean score ranking.

Risk.no	1	2	3	4	5	6
M.Score	2.67	3.33	2.67	3.33	2.67	3

Risk.no	7	8	9	10	11	12
M.Score	2.33	2.67	3	2.67	2.67	4

The company's goal has not been fixed. Risk number-2, no energy policy standard is used. Risk number-3, error indicator in the energy assessment. Risk number-4, the level of awareness to green operation has not been high. There are three risks in the DO stage: Risk number-5, standard Operating procedure is not yet complete, especially to handle energy problem. Risk number-6, has no expertise. Risk number-7, the implementation of staff training is minimal. There are three risks in the CHECK stage: Risk number-8, the reference standard for monitoring is incomplete and less valid. Risk number-9, monitoring to employees is less stringent. Risk number-10, there is an error causing the bias. There are two risks in the ACT phase, namely: Risk number-11, evaluation is too general. Risk number-12, the feedback is not immediately followed up by the top leader.

The next step is weighted risk and mean score ranking to determine the most crucial risks that are the main cause of waste in the use of electrical energy. Each expert participant in delphi is asked to provide a score to determine the weight of each potential risk. Then calculate the mean score for each risk, to determine the highest weight, which is the most crucial risk. Table 1 showing the result of mean score ranking for 12 identified risks.

Based on the result of weight calculation and the mean score ranking, indicate that the most crucial risk with the highest mean score (4) is risk number 12, there is no immediate follow-up by top management of feedback. Two other risks considered crucial are risk number-2 and risk number-4. Further steps are taken to mitigate risks and their solutions in the form of PDCA templates. The process of preparing PDCA templates is done by involving four expert delphi participants. Delphi-1 focuses on the analysis of strategies on energy management. Delpi-2 focuses on increasing revenue with consideration of electricity utilization cost fix. Delphi-3 focuses on communication with top management. The design of operational processes presented in the form of PDCA templates is used to formulate the critical thinking process for the related problems encountered by answering questions that are intended to facilitate the clinic in doing continuous improvement. Based on the successful set of PDCA templates, it is known that the clinic should start

doing new things, such as starting a strategy analysis related to energy management to tackle the existing electrical energy cost issues. The analysis of the strategy includes new policies that must be made and taken either by the head of operations or by top management in terms of human resources, controlling, and clinical tactical strategies to increase revenue.

4 CONCLUSION

Based on the result of risk identification and weighting of mean score ranking, it is found risk in act stage, that is, there is no follow up of top management for the given feedback. This risk is considered very important by the delphi, because in fact, top management has a great influence on the success or failure of the clinic. In the future, clinics need to communicate and change in managerial and clinical operations by applying critical thinking process through the application and use of PDCA templates. The first step to do is to immediately draft a strategy of energy management and controlling more tightly to the entire range of clinic employees either by giving rewards or controlling on the existing operational process.

ACKNOWLEDGEMETS

I would like to thanks to Dean of faculty of economic and business of Airlangga Unversity, especially management department facilitating us in research process and opportunity to present in GCBME International conference as well.

REFERENCES

Azzone, G. & Bertelè, U. 1994. Exploiting green strategies for competitive advantage. *Long Range Planning*, 27(6): 69-81.

Azzone, G. & Manzini, R. 1994. Measuring strategic environmental performance. *Business Strategy and the Environment*, 3(1): 1-14.

Azzone, G., Noci, G., Manzini, R., Welford, R. & Young, C. W. 1996. Defining environmental performance indicators: an integrated framework. Business Strategy and the Environment, 5(2): 69-80.

Chan, A.P.C., Yung, E.H.K., Lam, P.T.I., Tam, C.M. & Cheung, S.O. 2001. Application of Delphi method in selection of procurement systems for construction projects. *Construction Management and Economics*, 19(7): 699-718.

Ke, Y.J., Wang, S.Q., Chan, A.P.C. & Lam, P.T.I. 2010. Preferred risk allocation in China's public–private partnership (PPP) projects. *International Journal of Project Management*, 28(5): 482–492.

Kher, S.V., Frewer, L.J., De Jonge, J., Wentholt, M., Howell, D.O., Luijckx, N.L. & Cnossen, H.J. 2010. Experts' perspectives on the implementation of traceability in Europe. *British Food Journal*, 112(3): 261-274.

New, S., Green, K. & Morton, B. 2002. An analysis of private versus public sector responses to the environmental challenges of the supply chain. *Journal of Public Procurement*, 2(1): 93-105.

Rayens, M.K. & Hahn, E.J. 2000. Building consensus using the policy Delphi method. *Policy, politics, & nursing practice*, 1(4): 308-315.

Schmidt, R., Lyytinen, K., Keil, M. & Cule, P. 2001. Identifying software project risks: An international Delphi study. *Journal of management information systems*, 17(4): 5-36.

Skulmoski, G.J., Hartman, F.T. & Krahn, J. 2007. The Delphi method for graduate research. *Journal of Information Technology Education*, 6: 1-21.

Yin, R.K. 2003. Design and methods. *Case study research, 3*.

Yin, Y. Zhang, X. Peng, D. & Li, X. 2009. Model validation and case study on internally cooled/heated dehumidifier/regenerator of liquid desiccant systems. *International journal of thermal sciences*, 48(8): 1664-1671.

Advances in Business, Management and Entrepreneurship – Hurriyati et al (eds)
© 2020 Taylor & Francis Group, London, ISBN 978-0-367-27176-3

Value creation for competitive advantages of vegetable and dairy farmers through an integrated farming system in rural Bandung of West Java, Indonesia

K. Saefullah, R. Sudarsono, Y. Yunizar, L. Layyinaturrabbaniyah & A. Widyastuti
Universitas Padjadjaran, Bandung, Indonesia

ABSTRACT: This study attempts to investigate whether a competitive or cooperative strategy is more preferable in the creating value in agriculture sector, by which the integrated farming systems is applied. This study used a non-participative observation. Interview with some key informants in the Rural Bandung of West Java, Indonesia was conducted to collect the data needed for this study. The findings revealed that the ability of the farmers to reach competitive advantages in farming depended on their ability to cooperate among existing agents and local institutions at the community level.

1 INTRODUCTION

Agriculture sectors have been struggling in maintaining their sustainability in Indonesia. The statistical data on the Indonesian Economy in 2017 shows that agriculture sectors contributed only 14% to the national income, while services and industry contributed by 46% and 40 % consecutively. The percentage shown a decrease proportion in comparison to the earlier period of Indonesian development, in which the agricultural sector was the major contributed sector to the economy in 1965 (Indonesia-Investments 2018). At a regional level, the competitiveness of Indonesian agriculture sector is also considered less than the other ASEAN-5 countries. Although each of the member countries is complementarily supporting each other, the competitiveness of Indonesia's agriculture exports in 2001 to 2014 is under Malaysia and Thailand, and is only slightly above Vietnam and Philippines (Hamid & Aslam 2017).

Agriculture sector, farm and livestock breeds relate to a wide range of other economic agents in the economy. It also altered by the various policy settings that is relevant to national economy. While there are different agriculture policies, it will affect to how agricultural firms operate. The relevant factors that determine the competitiveness in agriculture within a sector of a national economy are very complex. This complexity leads to the difficulties for policy makers to understand the impact of different policy measures on agriculture sectors (Massini et al. 2013)

Competitive Advantage Theory (CAT) and Integrated Farming System (IFS) are approaches which have been acknowledged to support sustainability. Sustainability in this context refers to the ability of a business entity or development sector to survive in

the long run. While competitive advantage supports business to sustain in any changing environment and outperform among its industries (Porter 1985a, 1990b, 2001c, 2011d, Boehlje et al. 2004). Integrated farming supports the sustainability of agriculture sector through innovative systems. It sustains the system operations in a complex environment which requires any farmers to create value-added opportunities, despite of the fact that there are plenty of constraints. It enhances the productivity and competitiveness of agriculture sectors, *i.e.* by increasing the efficiency in agriculture production (Rodriguez et al. 1998, Dekker 2003, Devendra 2011).

Agriculture sector mainly deals with the communities. Unfortunately, the study at a community level is often being neglected, despite of the fact that various economic agents play their important roles in the local economy at the community. Many economists mostly emphasize on the macro and micro-level of the economic performance (Shaffer 2006).

Porter's framework for a competitive strategy has been used numerously in various sectors of the economy. Porter's model of five-forces model and the diamond framework have been used to identify the most relevant input resources by considering certain circumstances among the business and economic environment. The model also suggests the most appropriate strategies to enhance the competitive advantages (Porter 2001, Maudi & Kusnadi 2011).

The objective of this study is to investigate the practice of an integrated farming system (IFS) in the rural area of Bandung and how Porter's framework of competitive strategy can be used to create value-added in the agriculture sector between vegetable farming and livestock breeds.

Furthermore, a community-based research is urgently needed to understand the competitiveness of a nation, by looking at the competitiveness and sustainability of a community. By focusing the study in a community of rural area in Bandung, West Java, this paper attempts to answer the following questions: 1) How can integrated farming systems (IFS) be applied to increase the competitiveness in agriculture sectors, particularly between economic agents of cattle breeding and vegetable farmers?; 2) Should the economic agents in the integrated farming systems do cooperative or competitive actions among themselves to increase the competitiveness of the agriculture sectors?

2 METHODS

2.1 Research methods

This study was a part of a fundamental research on integrated farming systems in rural area of Bandung, West Java, which was supported by the research grant from Universitas Padjadjaran (HIU). This paper used a qualitative approach through a non-participative observation, in-depth interview, and focus group discussion (FGD). For the theoretical approach, this paper used Porter's framework to reach competitive advantage and Porter's general strategy. The strategy offered two approaches to optimize the integrated farming system through value chain analysis: by reducing the cost of the whole business process or by providing product differentiation. To implement these approaches, main and supporting activities of the business were distinguished. Main activities focused on optimizing the use of resources to have efficient business processes, while supporting activities incorporated any possible innovation to create value-added in the whole processes, including a better way of doing the activities (Maudi & Kusnadi 2011).

In using Porter's Value Chain Analysis, integrated farming systems was suggested to combine between main activities and supporting activities in agriculture, by integrating each of identified sub-systems in the agriculture business activities. There were two challenges to ensure that the approach supported the value-added achievement: competition and cooperation among existing local institutions, which played the role of each sub-systems (Ostadi et al. 2013). Both competition and cooperation supported the achievement of competitive advantage through efficiency and innovation.

2.2 Research location

The study was conducted in Suntenjaya village, the Northern part of Bandung, particulary in *Kampung* Pasirangling (Pasirangling sub-village).

The sample is located in the northern area of West Bandung, which can be reached for about 30-60 minutes from the center of Bandung. The village consists of 150 households and 600 inhabitants. Farming and livestock breeding (particularly dairy farmers) are the main occupation of the people in Pasirangling. Since 2015, *Yayasan Walungan* or Walungan Foundation conducted a community based program, which focused on the improvement of farming and breeding systems through an integrated farming systems.

The area has been struggling with farming and breeding inefficiency, where the input resources in the farming and breeding activities spend most of the cost. The farmer, for instance, should spend 100 thousand rupiah per day and walk for more than 10 km to collect grasses (animal feeds). In addition, the dairy farm produces a residual output of 20 kg of feces per cow per day or 3-4 ton in total per day. This is for the whole sub-village that have 248 cows in the dairy farms. The feces are thrown directly to the Cikapundung River. As a result, it creates environmental problems as the river supplies the water for the people in the lower area of Bandung.

2.3 Data collection and research analysis

The data was collected from May to June 2018. It was the combination of non-participative observation of the research area, a depth interview with the local farmers, and focus group discussion with the foundation and the representative of the people in the village. Descriptive analysis was used to discuss the findings from the data collection, combined with the elaboration from the related literatures.

3 RESULTS

3.1 General activities of dairy and vegetable farmers

Based on the non-participative observation and focus group discussion (FGD), it shows the chain activities of the vegetable farmers and dairy farmers.

According to the interview with the dairy farmers, the breeder had to collect the animal feeds every day. It took 4-5 hours to search, cut and collect the grasses for the animal feeds. The breeder could walk more than 2-3 km to search the source for the feeds. Alternatively, the farmer could buy the animal feeds from others. However, they had to spend IDR 500 to 800/kg of grass. Each cow consumed around 20 kg of grasses every day. It meant that the cost of animal feeds were around IDR 10,000 to 16,000/day or for about IDR 3000 to 48,000/dairy farmer, assuming every dairy farmer had 3 cows. Every day the dairy farm produced 15-20 liter of fresh milk for one cow, apart from the feces of the cows which was also for about 20 kg/cow or 60 kg/dairy farmer. The dairy farmer also had to spend IDR 50,000 for other expense (meal for the breeder, etc). Hence, the total cost of dairy farm was around IDR 100,000/day.

In terms of gross income, each farmer, who had two or three cows, could get IDR 211,500–282,000/day (1 liter fresh milk was sold at 4700 rupiah). By using a simple cost-benefit analysis, the farmer could generate income around 111.500 to 182,000/day or 2,8 to 4,5 million/month. However, considering the farmer usually borrowed capital from a local cooperative institution (for food consumption, for buying the feeds and equipment, or for paying the breeder or animal food gatherer, etc) generally a farmer could only get IDR 23.000/day or IDR 600,000/month (*pers.comm* with the dairy farmer, 2018). This level of income is quite low comparing to the minimum wage of labor in Bandung Barat district, which is about 2,7 million/per month (West Java Governor decree No. 561/2017). To get additional income, the dairy farmer often combined their dairy farming with other farming, *i.e.* vegetable farming.

The researchers interviewed the farmers during the observation. It shows the daily activity of vegetable farming in the research area. Most of the inhabitants had no land for farming. They rented some private or government lands and operated the farming based on profit-loss sharing, known as *maro* in local term (the farmer pays rent to the landlord and share the revenue with landlord based on specific agreement, usually 50-50 to 40-60). The cost of farming input, such as seeds, fertilizer, equipment, and the production processes from planting to harvesting, were under the responsibility of the farmers. The peasant were paid on a daily basis with a wage IDR 40,000 to IDR 50,000/day for male workers and IDR 30,000 to Rp.35,000/day for female workers. In a monthly basis, the peasant received IDR 900,000 IDR 1,5 million at maximum. Similar to dairy farmer, the wage of the peasant is under the standard of minimum wages.

There were various vegetables which the farmer planted, *i.e.* tomatoes, cauliflower, cabbage, chilies, eggplant and spring onions. However, broccoli and cabbage were the most planted ones as it could be harvest more often in a year than other vegetables. The revenue for the farming were varied. It depended on the season. The tomatoes could be sold at IDR 2500/kg. However it could be decreased into IDR. 1000 when result of plantation resulted was oversupplied. In terms of the market, the farmer had certain market. The vegetable were sold to some particular buyers who already made an agreement to buy in every harvest season.

have been proposed to the villagers: 1) applying an integrated farming system, combining between dairy farming and vegetable farming; and 2) applying Porter's generic strategy of cost leadership by minimizing the cost of farming inputs. Although the objective of this study is to increase the competitive advantage of the farming, however, both proposals prompted the farming system to accommodate 'cooperation' approach than 'competition' approach among the existing farming agents in the village.

Figure 1 shows the integrated farming system model. The integrated farming system was proposed to accommodate the activities of the villagers which were categorized into two: a) dairy farming and b) vegetable farming. As the villagers were doing both farming to have a better income, it would be more efficient if they combined the two main activities into one chain of farming.

The Figure 2 shows the proposed integrated farming system, which combined between dairy farming and vegetable farming. By using Porter's framework to minimize the cost of production (farming's input), two innovative proposals were added in the value of chain model: self-grown grass to supply the animal feeds through grass planting and feces-based fertilizer to transform the 'farming's waste' from the dairy farming into an organic fertilizer using the *kascing* method (with the help of worms).

According to the interviews, the self-grown grass plantation could reduce the cost of gathering animal feeds almost one-third of the cost. Meanwhile, the production of self-organic fertilizer from the cow's feces through a *kascing* method could not only reduce the cost of fertilizer for vegetable farming by half, but also could generate additional income. The organic fertilizer could be produced about 1500 kg/month and sold IDR 3000/kg. This generated additional gross income for the farmer with Rp 4,5 million/month, excluding the labor cost that spent for about half of the gross income (*pers.comm* with the farmers 2018).

The integrated farming systems basically had been proposed by the *Walungan* Foundation to the villagers since 2015. Unfortunately, less villagers were interested to do it. There were only two out of 200 farmers that had implemented the system. Most of the farmers still collected the grasses for the animal feeds by cutting it from the ground. They

3.2 Integrated farming system and porter's framework to increase the competitive advantage of farming

Based on the data collection and analysis towards the daily and vegetable farming activities, an integrated farming system had been proposed by a local institution. It was established by a *Walungan* Foundation, a community-based development institution. The foundation had identified the problems in the agriculture process, which were done by the farmers in the village. There were two suggestions which

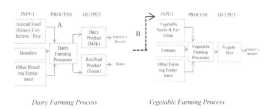

Dairy Farming Process   *Vegetable Farming Process*

Figure 1. Integrating Dairy & Vegetable Farmings.
Source: Authors' Logbook (2018)

A. Self-Grown grass for Animal Feeds

B. Organic-Fertilizer from Cow's Feces

Figure 2. Two-Innovative programmes in integrated farming system to reduce inefficiency.

Source: Authors' Logbook (2018)

also preferred to buy a fertilizer than made organic fertilizer from the cow's feces. They said that it just gave them additional time to work.

*"kumaha bade midamel nu kitu, kangge mayunan anu parantos aya wae tos kacida cape-na "(*transl: *how can I do that approach? I have already tired with the usual one).*

(*pers.comm* with the farmer, 2018)

In conclusion, regular habits become an obstacle to implement new approach. It becomes the most difficult factor in the implementation of the integrated farming system in the research area.

The study shows that an integrated farming system introduces a new approach in farming activities. By using Porter's framework of general strategy and value added analysis, the integrated farming system can reduce the inefficiency in the farming's

input. It also can generate additional income on the dairy farming by reducing the cost of fertilizer for the vegetables farming. Unfortunately, less farmers are willing to shift their way of farming to the integrated system. This result is similar to the study of Baide (2005) in Georgia. The study found that despite having a support from agents to change, farmers were rarely adopting an integrated farming system, which were more supporting the sustainable development and increasing farming's competitive advantages. In this case, farmers are still unsure that the new approach will produce a better result. They also experience various challenges to obtain accurate information regarding the new system.

To solve this problem, the farmers need to do cooperation among the farming agents. The cooperation includes creating a good relationship with the government, as the farmer will need a land to rent for the self-grown grass as well as the network for selling the organic fertilizer. In addition, the farmers should also make good connection with the peasant, the dairy milk buyers, as well as the local people who wish to implement the integrated system approach. In this context, a cooperation to increase the competitive advantages is preferable than a competition among farming agents. The farmers should cooperate among themselves to get better welfare without neglecting sustainable agriculture. In this context the existence of local cooperative can provide any assistance to support the behavioral changes of the farmers in implementing the integrated farming systems.

4 CONCLUSION

This study shows that an integrated farming system can support the sustainable agriculture and increase the competitive advantages of the dairy and vegetable farming. By using Porter's framework of a cost reduction strategy and value-added analysis, the integrated farming system can reduce inefficiency in the dairy farming and vegetable farming. Moreover, it can also generates additional income for the farmers. The implementation of the integrated system can be achieved if the farmers do cooperative actions with the existing economic and farming agents in the community.

REFERENCES

Baide, J.M.R. 2005. Barriers to Adoption of Sustainable Agriculture Practices in the South: Change Agent Perspectives. *Thesis.* Auburn University. 2005.

Bohlie, M., Gray, A., Dobbins, C. 2004. Strategi Development For the Farm Business: Options and Analysis Tools. Staff Paper No 4:12. *Department of Agricultural Economics*, September2004.

Dekker, H.C. 2003. Value Chain Analysis in Interfirm Relationship: A Field Study. *Management Accounting Research* 14 (1): 1–23.

Devendra, C. 2011. Integrated tree-crops ruminants systems in Southeast Asia: Advances in Productivity Enhancement and Environmental Sustainability. *Asian-Australasian Journal of Animal Sciences*. 24 (5) 587–602.

Hamid, M.F.S., Aslam, M. 2017. The Competitiveness and Complementarities of Agriculture Trade among ASEAN-5 Countries: An Empirical Analysis. *International Journal of Economics and Finance*; *Vol. 9. No. 8*. Canadian Center of Science and Education.

Indonesia-Investments. 2018. Gross Domestic Product of Indonesia. https://www.indonesia-investments.com/finance/macroeconomic- [online] Retrieved from: indicators/gross-domestic-product-of-indonesia/item 253.

Massini, R., Sudira, P., Mawardi, M., Darwanto, D.H. 2013. Pengembangan Konsep Agroindustri Berbasis Sistem Usahatani Terpadu di Wilayah Pasang Surut Bagian I: (Konsep Pemikiran). *Agritech*, Vol. 33 No. 1 February 2013.

Maudi, F., Kusnadi, N. 2011. Model Usaha Tani Terpadu Sayran Organik-Hewan Ternak: Studi Kasus Gapoktan Pandan Wangi, Desa Karehkel, Kecamatan Leuwiliang Kabupaten Bogor, Propinsi Jawa Barat. *Forum Agribisnis*.

Ostadi, H., Hortamani, A., Mojoudi, S. 2013. The Determination of Competitive Advantage in the Agricultural Sector of Iran Based on TM Index. *International Journal of Academic Research in Progressive Education and Development*. October 2013. Vol 2. No. 4.

Porter, M.E. 1985. Competitive Advantage. Free Press. ISBN 0-684-84146–84140.

Porter, M.E. 1990. The Competitive Advantage of Nations. In *Harvard Business Review*, March-April 1990. https://hbr.org/1990/03/the-competitive-advantage-of-nations.

Porter, M.E. 2001. The Value Chain and Competitive Advantages. In Barnes, D. 2001. *Understanding Business: Processes: 50–66*.

Porter, M.E. 2011. The Competitive Advantage of Nations, States and Regions. *National Council of Professor.* Kuala Lumpur Malaysia. 7 July 2011.

Rodríguez J. L., Preston, T.R., Lai, N.V. 1998. Integrated farming systems for efficient use of local resources. http://www.ias.unu.edu/proceedings/icibs/rodriguez/In: Editors: Eng-Leong Foo & Tarcisio Della Senta. Integrated Bio-Systems in Zero Emissions Applications. Proceedings of the Internet Conference on Integrated Bio-Systems. http://www.ias.unu.edu/proceedings/icibs/wang

Shaffer, R., Deller, S., Marcouiller, D. 2006. Rethinking Community Economic Development. *Economic Development Quarterly*, 20 (1): 59–74. *SAGE Publications*.

Sudarsono, R., Saefullah, K., Yunizar. Widyastuty, A., Layyinaturrabbaniyah. 2018. Field Research on "Penerapan Value Chain pada Pengembangan Integrated Farming System untuk memberikan Nilai Tambah Produk Pertanian dan Peternakan ". *Log Book. Fundamental Research*. Padjadjaran University. 2018.

Sunterjaya Village. 2013. Profile Book 2013. Kantor Desa Sunterjaya. Kabupaten Bandung Barat.

The influence of research and development intensity, firm size, and family ownership on green product innovation

D. Meicistaria & I. Isnalita

Airlangga University, Surabaya, Indonesia

ABSTRACT: The purpose of this research is to examine the influence of Research and Development (R&D) intensity, firm size, and family ownership on Green Product Innovation (GPI). Former research has shown mixed results. The data of this research was 145 manufacturing companies listed on the Indonesia Stock Exchange in 2012-2016. The R&D intensity was measured by dividing R&D expense into sales, firm size, family ownership, and GPI. Sales and firm size was measured by the natural logarithm of total assets. Family ownership was measured by percentage of share proportion, and GPI was measured by the GPI disclosure index. The results of this study showed that R&D intensity and firm size positively and significantly influenced GPI, while family ownership negatively influenced GPI. The research model in this study has been adjusted according to the conditions of Indonesian companies in order to obtain more relevant research results.

1 INTRODUCTION

Green Product Innovation (GPI) is the creation of new products or the improvement of products that can reduce the impact on the environment throughout the product life cycle (Dangelico & Pujari 2010, Kammerer 2009, Lin et al. 2013). GPI also means the process in which three key focus types (matter, energy, and pollution) are highlighted based on their major environmental impacts at different stages of the product's physical life cycle (manufacturing process, product use, and disposal). It is important to note that not all products have significant environmental impacts at each stage of the life cycle of physical products. The impact of the product also does not come from all aspects (matter, energy, and pollution), but almost all products have significant environmental impacts in at least one of its stages.

GPI encourages companies to improve product development, as well as existing processes, while maintaining environmental sustainability. According to Chen et al. (2012), GPI can be an advantage for companies in implementing differentiation strategies to improve green image, as well as a competitive advantage for the company. This innovation can encourage the companies to understand the consumer choice in order to improve the product quality, and to create and produce new products by using more efficient resources and energy, thereby reducing environmental impact.

Some studies related to GPI provide inconsistent results. The study conducted by Huang et al. (2016) showed that the company's internal factors had a positive effect on GPI and family ownership had a negative effect on GPI. Horbach (2008) found that the improvement of R&D technology

capabilities triggers environmental innovation. In addition, the study also revealed that in general, environmental stewardship, environmental management tools, and organizational change also encouraged environmental innovation. Dangelico & Pujari (2010) revealed that strong motivations as well as environmental policies and targets affected the GPI. Meanwhile Lin et al. (2013) found that market demand was positively associated with GPI and firm performance.

The studies done by Cleff & Rennings (1999) and Rehfeld et al. (2007) indicated that firm size had a positive effect on GPI. In contrast, another studies also found that firm size had no significant positive effect on GPI (Engels, 2008); (Nogareda 2007) and (Craig & Dibrell's 2006) study showed that family influences led to more flexible processes and decision-making structures. However, Konig et al. (2013) argued that family influences reduced aggressiveness and flexibility.

The results of previous studies show inconsistency. This finally spur the researcher to explore and test empirically the influence of Research and Development (R&D) intensity, company size, and family ownership on the GPI in Indonesia's manufacturing companies. This article consists of several sections, namely method, results and discussion, and conclusion. Method presents an explanation of the research approach used in this study, the operational definition, and how to measure the variables of R&D intensity, firm size, family ownership, and GPI. Furthermore, results and discussion describes statistical results of hypothesis testing and its interpretation. The final section of this article discusses the test results, implications, and recommendations for further research.

2 METHOD

Data was obtained by collecting information from annual reports published by the Indonesia Stock Exchange. Sampling of this research was done by using a purposive sampling method. From the 723 manufacturing companies that published their financial statements in IDX from 2012 to 2016, there were only 145 companies that met the criteria.

The independent variables of this research were the intensity of R&D, company size, and family ownership. Meanwhile, dependent variable was GPI. GPI was measured by looking at the disclosure of ten items related to GPI of manufacturing companies in Indonesia. The ten items of this indicator were: a) substitution of materials that caused contamination; b) substitution of hazardous materials; c) product designs that focused on reducing the resource constraints in the production stage; d) the product design that focused on the reduction of waste generation in the production stage; e) the design that focused on the improvement and decomposition products; f) the designs that focused on improving product usage; g) the designs that focused on product recycling; h) product life-cycle analysis that aimed to improve product design; i) expansion of market coverage of green products; and j) improving manufacturing technology for green products. GPI was calculated by using the formula below:

$$GPI = \frac{Total\ Company\ Disclosure}{10} \quad (1)$$

The intensity of research and development (R & D) was measured by the following ratio;

$$IRND = \frac{RND\ Charge}{Sales} \quad (2)$$

The determination of firm size was based on the total assets of the company. It was measured by the total natural logarithm of the asset.

$$SIZE = Ln(total\ asset) \quad (3)$$

Family ownership in this study was measured using dummy variables. Companies belonging to family firms were given number 1, others were given num-ber 0. Next, for an existing company, the family ownership was calculated by the following formula:

$$fown = \frac{shares\ amount\ owned\ by\ the\ family}{number\ of\ shares\ outstanding} \quad (4)$$

This research used multiple linear regression. The regression model used to test the hypothesis was formulated as follows:

$$GPI = \alpha + \beta 1 RND + \beta 2 SIZE + \beta 3 FOWN + e \quad (5)$$

Descriptions:
α : *intercept*
$\beta 1$, $\beta 2$, $\beta 3$: Coefficient
GPI: GPI
IRND: R&D Intensity
SIZE: Firm Size
FOWN: Family Ownership
e: Residual Error

3 RESULTS AND DISCUSSION

3.1 *Resuls*

Table 1 presents the descriptive statistics of the variables used in the research. The table shows that the lowest value of GPI was 0.1 and the highest was 1.00. The average GPI of all sample companies was 0.528966 with a standard deviation of 0.2521896 and a variation of 0.4767. This indicates that GPI that occurs in the sample company is relatively even. The company has the same relative capability to create new products based on the environment.

The lowest value of R&D intensity (IRND) was 0.0004, while the highest was 24,4174. The average of R&D intensity of all sample companies was 1.179089 with a standard deviation of 3.6852088 and the level of variation was 3.1254. This implies that the intensity of R&D that occurs in the sample company is relatively uneven. The ability of companies to conduct R&D is different.

The lowest values of firm size (SIZE) was 25,6348 and the highest was 33,1988. The average size of the company owned by the entire sample company was 28.527158 with the standard deviation of 1.816646 and the variation level of 0.06368. The assets held between the companies sampled were relatively the same.

The lowest values of family ownership (FOWN) was 0.00 and the highest was 0.4752. The average family ownership of all sample companies was 0.029534 with a standard deviation of 0.0910746 and the rate of variation was 3.0837. The family ownership that occurs in the sample company is relatively even.

Table 1. Descriptive statistic.

Variable	Mean	Standard deviation	Min	Max
GPI	.528966	.2521896	.1000	.2521896
IRND	1.179089	3.6852088	.0004	3.6852088
SIZE	28.527158	1.8166464	25.6348	1.8166464
FOWN	.029534	.0910746	.0000	.0910746

Source: Processed Data, 2018.

559

Table 2. Determination coefficient test.

Model	R	R Square	Adjusted R Square
1	.530[a]	.281	.266

Source: Processed Data, 2018.

Table 3. Multiple Linear Regression Analysis Result.

Model	Unstandardized Coefficients		T	Sig.
	B	Std. Error		
(Constant)	-1,089	,297	-3,665	.000
IRND	,014	,005	2,835	.005
SIZE	,057	,010	5,470	.000
FOWN	-,426	,206	-2,065	.041

Source: Processed Data, 2018.

Table 2 presents the results of determination coef-ficient test. The table shows that the deter-mination coefficient (adjusted R2) for the built model was 0.266. This implies that 26.6% of GPI can be ex-plained by the intensity of R&D, firm size, and family ownership. Meanwhile, 73.4% can be explained by other variables which are not used in this research.

The result of multiple linear regression testing of the research model is presented in Table 3.

Hypothesis 1, which states that the intensity of R&D has a positive effect on GPI, is accepted. Stat-istical test shows that t value was 2.835 and signifi-cant value was 0.005 with significant level 0.05. This indicates that the higher the intensity of R&D, the higher the company performed GPI.

Hypothesis 2, which states that firm size has a positive effect on GPI, is accepted. From Table 3, it can be seen that t value was 5.470 and significance value was 0.000 with significant level 0.05. This means that the larger the size of the company, the more likely it was to perform GPI.

Hypothesis 3, which states that family ownership negatively affects GPI, is also accepted. The t value was -2.065 and significance value was 0.041 with significant level 0.05. Thus, the higher the share of family ownership in the company, the lower the company's concern for GPI would be.

3.2 Discussion

The results show that companies that actively conduct R&D will find it easier to make product development as part of innovation. Companies that are active in conducting R&D will also pay attention to the problems arising in the commu-nity. One of the problems in today's society is related to the companies' ability in preserving the environment. Regarding this, companies should grow their efforts to maintain environmental sus-tainability. The intensity of R&D conducted by the companies will lead them to create new prod-ucts that are more environmentally friendly (GPI). As a consequence, public trust to the com-panies will be maintained. Based on the theory of legitimacy, the companies attempt to absorb the values and norms in society. One of the ways is through GPI. GPI is the companies' effort to create more environmentally friendly products that can of-fer the benefits for environment sus-tainability. To have this effort, companies should intensively con-duct R&D. Huang et al. (2016) found that the inten-sity of R&D had a significant effect on GPI. This is supported by Hegarty & Hoffman (1990), and Rehfeld et al. (2007) who say that R&D activities tends to have a positive effect on environmental product innovation.

The results also show that the size of the com-pany, which is measured by the number of assets, will have an impact on GPI. Companies that have a lot of assets have the ability to create new products. Companies with large resources also will produce large operational activities that enable them to contribute more to the environ-ment. Therefore, these companies will strive to create new innovations on products that are environmental friendly. Based on the theory of legitimacy, the total assets owned by the compan-ies is impacted to the increase of the companies' operations. As a result, environmental impact will be greater. Damanpour (1992) found a positive influence between firm size and innovation. Child & Tsai (2005) argue that large companies can more easily deal with environmental stress. They are more likely to be environmentally respon-sible. Thus, large companies is expected to improve GPI and reduce its unfavorable impact.

The results show that the greater the composition of shares owned by the family in the company, the less they want to create the products that are more environ-mental friendly. Companies with family shared-ownership tend to be apathetic to the environmental problems that are caused by their opera-tions. These companies will focus more on creating attractive prod-uct for the consumers, than creating environmental friendly products. Thus, they do not implement GPI as their corporate strategy. Based on stakeholder theory, the company should pay close attention to stakeholder interests, so that GPI activities are supported by stake-holders. This research shows that the owner becomes the focus of the companies. The companies will try to increase the welfare of the family as the owner. Thus, they tend to not pay attention to the environmental aspect as a corporate strategy. This is in line with DeMassis et al. (2015) who revealed that family firms tended not to engage in innovation activities, especially collaborative innovation projects, because innovation required professional expertise, which the company may not have. In other words, family members risk resisting and being reluctant to invest in radical in-novation.

4 CONCLUSION

The results show that GPI is influenced by the intensity of R&D, firm size, and family ownership. This can be used as a consideration for management in determining the strategy of creating innovative products that are environmental friendly. Stakehold-ers also should give more attention to environmental friendly products. Thus, environmental sustainability can be created. Further researches are suggested to concern on green process innovation that also con-tributes to environmental sustainability.

REFERENCES

Chen, C. C., Shih, H. S., Shyur, H. J. and Wu, K. S. (2012). A business strategy selection of green supply chain management via an analytic network process. *Computers and Mathematics with Applications*, 64 (8), 2544–2557.

Child, J. and Tsai, T. 2005. The dynamic between firms' environmental strategies and institutional constraints in emerging economies: Evidence from China and Taiwan. *Journal of Management Studies*, 42 (1), 95–125.

Cleff, T., and K. Rennings. 1999. Determinants of environmental product and process innovation. *European environment*, 9(5), 191–201.

Craig, J. and Dibrell, C. 2006. The Natural environment, innovation, and firm performance: A comparative study. *Family Business Review*, 19 (4), 275–288.

Damanpour, F. (1992). Organizational size and innovation. *Organization Studies*, 13(3), 375–402.

Dangelico, R. M. and Pujari, D. 2010. Mainstreaming GPI: Why and how companies integrate environmental sustainability. *Journal of Business Ethics*, 95 (3), 471–486.

De Massis, A., Frattini, F., Pizzurno, E. and Cassia, L. 2015. Product innovation in family versus nonfamily firms: An exploratory analysis. *Journal of Small Business Management*, 53 (1), 1–36.

Engels, S. V. 2008. *Determinants of environmental innovation in the Swiss and German food and beverages industry: What role does environmental regulation play?*. ETH Zurich, Zurich, Switzerland.

Hegarty, W. H., and R. C. Hoffman. 1990. Product/market innovations: a study of top management involvement among four cultures. *Journal of Product Innovation Management*, 7(3), 186–199.

Horbach, J. 2008. Determinants of environmental innovation: New evidence from German panel data sources. *Research Policy*, 37 (1), 163–173.

Huang, Y. C., Wong, Y. J. and Yang, M. L. 2016, The effect of internal factors and family influence on firms'adoption of GPI. *Management Research Review*, 39 (10), 1167–1198.

Kammerer, D. 2009. The effects of customer benefit and regulation on environmental product innovation: Empirical evidence from appliance manufacturers in Germany. *Ecological Economics*, 68 (8-9), 2285–2295.

Konig, A., Kammerlander, N. and Enders, A. 2013. The family innovator's dilemma: How family influence affects the adoption of discontinuous technologies by incumbent firms. *Academy of Management Review*, 38 (3), 418–441.

Lin, R. J., Tan, K. H., and Geng, Y. (2013). Market demand, green innovation, and firm performance: Evidence from Vietnam motorcycle industry. *Journal of Production*, 40, 101–107.

Nogareda, J. S. 2007. *Determinants of Environmental Innovation in the German and Swiss Chemical Industry*. University of Zurich.

Rehfeld, K. M., Rennings, K. and Ziegler, A. 2007. *Integrated product policy and environmental product innovations: An empirical analysis*. Ecological Economics, 61 (1), 91–100.

Section 4: Innovation, information and technology, operations

and supply chain

Advances in Business, Management and Entrepreneurship – Hurriyati et al (eds)
© 2020 Taylor & Francis Group, London, ISBN 978-0-367-27176-3

The influence of information technology and entrepreneurial orientation on competitiveness and business performance

N.A. Hamdani & S. Nugraha
Universitas Garut, Garut, Indonesia

ABSTRACT: SMEs are demanded to be sensitive to the rapid changes in technology. These changes can be anticipated with an understanding of the concept of market-based theory and resource-based theory. This study is aimed at analyzing the influence of information technology, knowledge sharing, and entrepreneurial orientation on competitiveness and SME business performance. This study used an explanatory research design. The samples were 115 SMEs selected randomly. Data were analyzed using SEM-PLS. Based on the data analysis result, it can be concluded that information technology, knowledge sharing, and entrepreneurial orientation have influence on competitiveness and business performance. The result also reveals that information technology and competitiveness will not be optimal without an entrepreneurial orientation of SMEs and that competitiveness can boost SME business performance.

1 INTRODUCTION

In Indonesia, small and medium-sized enterprises (SMEs) are the largest business group in number. This group is proven not to be affected by economic crisis. The performance of SMEs in Indonesia has received special attention because SMEs are economic supporting sectors and job creators. Some studies, for example Hamdani, Abdul and Maulani (2018) and Hamdani (2018) emphasize the importance of competitiveness to improve business performance. Furthermore, Wu (2015) explains empirically how SMEs in Taiwan can survive by relying on competitiveness. While according to Mohebi and Farzollahzade (2014), competitiveness and business performance is established upon entrepreneurial competence and orientation.

It is necessary for SMEs to improve their quality and competitiveness considering the tight business competition in today's free market era. It is not easy to win both local and global competition. SMEs need to maintain their business stability and anticipate the arrival of new products. They need to adapt to the technological changes to have global competitiveness (Carr, 2005). Technological adoption is required to address the impact of globalization (Ongori and Migiro, 2010).

In a developing country like Indonesia, SMEs are one of the backbones of the country's economy. Their growth has reached 6%. Government and private sectors continue to provide them with business supports.

The problem frequently associated with SMEs is that they are reluctant to learn Internet technology to develop their businesses despite the fact that Internet is getting easier to access and has significant impact on the SME development (Kula and Tatoglu, 2003).

Internet is very useful as a marketing medium (Andriyanto, 2010).

Indonesian people are very creative, and creativity is all that is needed to develop highly competitive SMEs. The problem is that they are yet to learn to use Internet to penetrate the market. This condition is an opportunity for technology activists to help SMEs to adapt technology-based services to run online marketing (Windrum and De Berranger, 2010).

Garut's dodol (sweet jelly-like snack) is widely known for its distinctive taste and chewiness. Dodol industry in Garut was pioneered by a businessperson named Mrs. Karsinah in 1926. Garut's dodol can survive due to its (1) distinctive taste, (2) affordable price, (3) simple manufacturing process and raw material, (4) being free from preservatives and synthetic food materials, and (5) relatively long expiration period (up to 3 months) (Pemda, 2015).

The purpose of this study is to analyze the influence of information technology, knowledge sharing, and entrepreneurial orientation on competitiveness and SME business performance.

2 LITERATURE REVIEW

SMEs may not think that investing in information technology (IT) is not urgent, especially when their businesses can run well without it. However, very few small businesses are able to operate effectively. The use of IT can significantly boost the success of business (Wahid *et al.*, 2015), (Ashrafi and Murtaza, 2008). The question is how to implement IT effectively and efficiently in SMEs (Albesher, 2012). IT is a means to facilitate, create, change, store, and share information. IT can combine high-speed computing and communication and plays an important role in facilitating businesses (Apulu and Latham, 2010).

Entrepreneurship is a dynamic process in which people gain additional wealthy gradually by taking risks, such as capital risk, time risk, and career risk, in providing values for particular products or services (Hisrich D.Robert, Peters P. Michael, 2008).

Entrepreneurial orientation is characterized by three aspects: innovativeness, risk-taking and pro-activeness (Covin and Slevin, 1989). This is what has been discussed in some studies; for example, Effendi and Hadiwidjojo, 2013; Awang, 2010). Entrepreneurial orientation is associated with the use of technology to generate new products or services. Business people with entrepreneurial orientation are those with dynamic behaviors, bravery to take risks, creative ideas, visions (Nursalida, 2013; Duru, Ehi-diamhen and Chijioke, 2018).

Most SMEs in Indonesia are yet to improve their competitiveness. Their low competitiveness leads to the low export intensity. In the domestic market, domestic SME products are still inferior to imported SME products. There are several contributing factors to the low competitiveness of domestic products such as: low quality due to minimum use of technology, non-optimal production efficiency, and Indonesia's macroeconomic sector policies that discourage SMEs (Beselly, 2017). In relation to business performance, SME competitiveness includes several dimensions such as educational level, innovation level, capital level, and business strategy (Anton, Muzakan and Muhammad, 2015).

Performance is level of success of a person during a certain period in carrying out a task in terms of work standards and achievement of predetermined and mutually agreed upon targets (Duru, Ehidiamhen and Chijioke, 2018). Performance can been from financial and non-financial perspective (Lumpkin and Dess, 2001).

3 RESEARCH METHOD

This study used an exploratory research method. Data were collected through questionnaires addressed to 115 dodol SMEs in Garut selected using simple random sampling technique. The gathered data were analyzed using SmartPLS software. The research variables included information technology with three indicators: cultural aspect (X11), organizational aspect (X12) and technical aspect (X13); entrepreneurial orientation with four indicators aggressiveness (X21), pro-activeness (X22), innovativeness (X23) and risk-taking (X24); competitiveness with four indicators: product (Y1), service (Y2), image (Y3), and cost (Y4); and performance with the indicators: product (Z1), financial and non-financial performance (Z2).

4 RESULT AND DISCUSSION

Data analysis using SmartPLS software consists of several stages. The first stage in PLS analysis is outer model analysis. This analysis is to ensure that all indicators for every variable are valid and reliable. Outer model analysis also can show the relationship between every indicator and its variable.

What follows is the resulting modeling of SmartPLS:

The model that represents all data, indicators, and variables is then analyzed using PLS Algorithm.

Figure 2 shows values that represents the relationship between indicators and their variables. These values can be interpreted as follows:

1. The path coefficient between information technology (X1) and competitiveness (Y) is 0.484. It means that the latent variable X1 have influence on Y as much as 0.484.
2. Loading of cultural aspect (X11) is 0.955. It means that the latent variable information technology (X1) contribute to cultural aspect (X11) as much as 0.955.
3. Loading of organizational aspect (X12) is 0.962. It means that the latent variable information technology (X1) contribute to organizational aspect (X12) as much as 0.962, and so on.
4. The path coefficient between entrepreneurial orientation (X2) and competitiveness (Y) is 0.379. It means that entrepreneurial orientation have influence on competitiveness as much as 0.379.
5. Loading of aggressiveness (X21) is 0.653. It means that entrepreneurial orientation (X2) contribute to aggressiveness (X21) as much as 0.653, and so on.

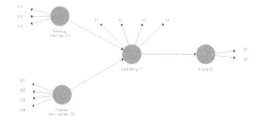

Figure 1. PLS Research Modeling.

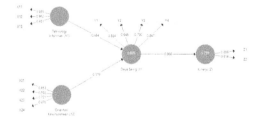

Figure 2. Result of PLS Algorithm Analysis.

6. The path coefficient between competitiveness (Y) and the latent variable performance (Z) is as much as 0.860. It means that competitiveness have influence on performance as much as 0.860.
7. Loading of product (Y1) is 0.824. It means that competitiveness (Y) contribute to product (Y1) as much as 0.824, and so on.

The result of analysis reveals that technical aspect (X13) has the loading factor value of 0.451 (lower than 0.5) so that it should be excluded from the model. What follows is the resulting model after excluding X13 (Figure 3).

PLS also shows parameters to analyze data. These parameters include average variance extracted (AVE), composite reliability, Cronbach's alpha, and cross loading.

The result of data analysis shows that the variance value of each latent variable has met the requirement that says the AVE value should be higher than 0.50. The higher the AVE value is, the better and the more information could be obtained of each latent variable. Since the values of information technology (X1), entrepreneurial orientation (X2), competitiveness (Y) and performance (Z) are higher than 0.50, all variables can be said to have met the requirement.

Composite reliability (CR) and Cronbach's alpha (CA) show the reliability level of every latent variable. The recommended value is 0.700. The higher, the better. The result of analysis reveals the following CR values: 0.979 for X1, 0.805 for X2, 0.895 for Y, and 0.860 for Z. The result also reveals the following CA values CA: 0.956 for X1, 0.708 for X2, 0.767 for Y, and 0.804 for Z. It could then be concluded information technology (X1), entrepreneurial orientation (X2), competitiveness (Y) and performance (Z) are valid and reliable.

In this study, it is also necessary that the correlation values between indicators are higher than other constructs as seen in the Figure 5.

Figure 5 shows that the correlation between cultural aspect (X11) and the latent variable information technology (X1) is 0.978 and that between cultural aspect (X11) and the latent variable in its block (X1) is higher than the correlation between X11 and latent variables in other blocks (X2, Y, and Z). It could then be concluded that X11 is valid. Using the same logics, all indicators could be said to be valid.

The outer model analysis is followed by inner model analysis. Parameters used in this analysis include t-statistic, R-squared, effect size, and path coefficient. To perform inner model testing and obtain these parameters, PLS should perform bootstrapping. What follows (Figure 6) is the result of PLS modeling using bootstrapping method:

To test the significance of hypothesized path, t-statistic is used. Using alpha level of 5%, the critical t-statistic value is 1.96 and stated as valid. The correlation can be said to be significant since P-Value < 0.05.

Figure 7 shows that the path correlation between information technology (X1) and competitiveness (Y) is significant because its t-statistic value is 3.088 (>1.96) and its P-Value is 0.002 (<0.05). It could then be concluded that information technology has

Discriminant Validity

	Fornell-Larcker Criteri...	Cross Loadings	Heterotrait-Monotrait R...	Heterotr
	Daya Saing (Y)	Kinerja (Z)	Orientasi Kewir...	Teknologi Info...
X11	0.684	0.836	0.587	0.978
X12	0.703	0.798	0.533	0.979
X21	0.387	0.439	0.654	0.314
X22	0.239	0.370	0.594	0.224
X23	0.415	0.544	0.720	0.404
X24	0.718	0.886	0.869	0.572
Y1	0.825	0.915	0.676	0.908
Y2	0.643	0.407	0.275	0.136
Y3	0.790	0.520	0.444	0.408
Y4	0.847	0.628	0.563	0.402
Z1	0.718	0.886	0.869	0.572

Figure 5. Cross Loading Criteria.

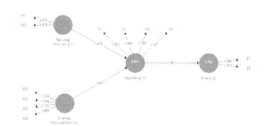

Figure 3. Result of PLS Algorithm Analysis without X13.

Construct Reliability and Validity

	Matrix	Cronbach's Alpha	rho_A	Composite Reliability	Average Variance Extracted (...
		Cronbach's Al	rho_A	Composite Rel...	Average Varian...
Daya Saing (Y)		0.404	0.409	0.360	0.473
Kinerja (Z)		0.767	0.727	0.895	0.411
Orientasi Kewirusahaan (X2)		0.708	0.810	0.305	0.534
Teknologi Informasi (X1)		0.956	0.957	0.979	0.958

Figure 4. Construct Reliability and Validity.

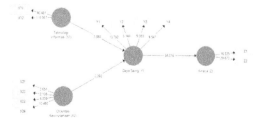

Figure 6. Output of Path Testing

Path Coefficients

	T Statistics (|O/STDEV|)	P Values
Daya Saing (Y) -> Kinerja (Z)	34.016	0.000
Orientasi Kewirusahaan (X2) -> Daya Saing (Y)	2.392	0.017
Teknologi Informasi (X1) -> Daya Saing (Y)	3.088	0.002

Figure 7. Path Coefficients.

significant influence on the competitiveness of dodol SMEs in Garut. Figure 7 also shows that the path correlation between entrepreneurial orientation (X2) and competitiveness (Y) is significant because its t-statistic value is 2.392 (>1.96) and its P-Value is 0,017 (<0.05). It could then be concluded that entrepreneurial orientation has significant influence on the competitiveness of dodol SMEs in Garut. The path correlation between competitiveness (Y) and performance (Z) of dodol SMEs in Garut is also significant because significant because its t-statistic value is 34.016 (>1.96) and its P-Value is 0.000 (<0.05). It could then be concluded that competitiveness has significant influence on the performance of dodol SMEs in Garut.

The R-squared value represents the influence of every latent variable on its indicators. The R-squared value of variable competitiveness (Y) is 0.614. It means that the latent variables information technology (X1) and entrepreneurial orientation (X2) with each one of their indicators have significant influence on competitiveness (Y). The R-squared value of variable performance (Z) is 0.740. It means that the latent variable competitiveness (Y) with its four indicators have significant influence on performance (Z).

5 CONCLUSION

The use of information technology and entrepreneurial orientation become influential factors in improving the competitiveness of Garut's dodol and competitiveness improves business performance. Adapting to new information technology developments and taking risks are the key to the development of dodol SMEs in Garut. The fact that Garut is dubbed as the City of Dodol also improves the competitiveness of Garut's dodol besides its affordable price. To most Dodol SMEs in Garut, financial performance is more tangible than non-financial performance. Non-financial performance is associated with their sustainability.

REFERENCES

Albesher, A. A. (2012) 'The impact of IT resources on SMEs innovation performance', in *DRUID Academy*. Londo: Druid Society, pp. 1–35.

Andriyanto, R. D. (2010) 'Analisis Pengaruh Internet Marketing terhadap Pembentukan Word of Mouth dan Brand Awareness untuk Memunculkan Intention to Buy', @@@@ 9(1).

Anton, S. A., Muzakan, I. and Muhammad, W. F. (2015) 'An Assessment of SME Competitiveness in Indonesia', *Journal of Competitiveness*, 7(2), pp. 60–74. doi: 10.7441/joc.2015.02.04.

Apulu, I. and Latham, A. (2010) 'Benefit of Information and Communication Technology in Small and Medium Enterprises', in *UK Academy for Information Systems*. AISel, pp. 1–19.

Ashrafi, R. and Murtaza, M. (2008) 'Use and Impact of ICT on SMEs in Oman.', *Electronic Journal of Information Systems Evaluation*, 11(3), pp. 125–138. doi: ISSN 1566–6379.

Awang, A. (2010) 'Study of Distinctive Capabilities and Entrepreneurial Orientation on Return on Sales among Small and Medium Agro-Based Enterprises (SMAEs) in Malaysia', *International Business Research*, 3(2), pp. 34–48.

Beselly, D. and M. K. (2017) 'KEBIJAKAN PENGEMBANGAN DAYA SAING GLOBAL USAHA KECIL MENENGAH (UKM) DI KOTA BATU MENGGUNAKAN SME DEVELOPMENT INDEX', *Jurnal Administrasi Bisnis (JAB)*, 47(1), pp. 25–31.

Carr, J. (2005) 'The Implementation of Technology-Based SME Management Development Programmes Diffusion of management learning technology from HE to SMEs', *Educational Technology & Society*, 8(3), pp. 206–215.

Covin, J. G., & Slevin, D. P. (1989) 'Strategic management of small firms in hostile and benign environments', *Strategic Management Journal*, 10(2), pp. 75–87.

Duru, I. U., Ehidiamhen, P. O. and Chijioke, A. N. J. (2018) 'Role of Entrepreneurial Orientation in the Performance of Small and Medium Enterprises: Evidence from Federal Capital Territory, Abuja, Nigeria', *Asian Journal of Economics, Business and Accounting*, 6(1), pp. 1–21. doi: 10.9734/AJEBA/2018/39748.

Effendi, S. and Hadiwidjojo, D. (2013) 'The Effect Of Entrepreneurship Orientation On The Small Business Performance With Government Role As The Moderator Variable And Managerial Competence As The Mediating Variable On The Small Business of Apparel Industry In Cipulir Market, South Jakarta', *IOSR Journal of Business and Management*, 8(1), pp. 49–55.

Hamdani, N. A. (2018) 'Building Knowledge Creation For Making Business Competition Atmosphere in SME of Batik', *Management Science Letters*, 8, pp. 667–676. doi: 10.5267/j.msl.2018.4.024.

Hamdani, N. A., Abdul, G. and Maulani, F. (2018) 'The influence of E-WOM on purchase intentions in local culinary business sector', *International Journal of Engineering & Technology*, 7, pp. 246–250.

Hisrich D.Robert, Peters P.Michael, S. A. D. (2008) *Entrepreneurship*. Internatio. Mc Graw Hill.

Kula, V. and Tatoglu, E. (2003) 'An exploratory study of Internet adoption by SMEs in an emerging market economy', *European Business Review*, 15(5), pp. 324–333. doi: 10.1108/09555340310493045.

Lumpkin, G. T. and Dess, G. G. (2001) 'Linking two dimensions of entrepreneurial orientation to firm performance: The moderating role of environment and industry life cycle', *Journal of Business Venturing*, 16 (5), pp. 429–451. doi: 10.1016/S0883-9026(00)00048-3.

Mohebi, M. M. and Farzollahzade, S. (2014) 'Improving Competitive Advantage and Business Performance of SMEs by Creating Entrepreneurial Social Competence', *management Reserach*, 2(spesial Issue), pp. 20–26.

Nursalida (2013) *THE INFLUENCE OF ENTREPRE-NEURIAL ORIENTATION AND RESOURCES ON FIRM PERFORMANCE THROUGH DISTINCTIVE CAPABILITIES AND COMPETITIVE ADVANTAGE (Study of micro firm of Aceh specific food industry in Aceh Province)*. Universitas Padjajaran.

Ongori, H. and Migiro, S. O. (2010) 'Information and communication technologies adoption in SMEs: literature review', *Journal of Chinese Entrepreneurship*, 2(1), pp. 93–104. doi: 10.1108/17561391011019041.

Pemda (2015) *Profil Ekonomi Kabupaten Garut, Ekonomi*. Available at: http://www.garutkab.go.id/ (Accessed: 16 March 2017).

Wahid, K. A., Numprasertchai, H., Sudharatna, Y., Laohavichien, T. and Numprasertchai, S. (2015) 'the Impact of Knowledge Sources on Knowledge Creation : a Study in Thai Innovative Companies', in *Joint International Conference*. Bari Italy: TIIM, pp. 237–250. doi: http://dx.doi.org/10.1590/1982-451320150305.

Windrum, P. and De Berranger, P. (2010) 'The adoption of e-business technology by SMEs', *University Business*, pp. 177–201. Available at: http://hdl.handle.net/2173/93160.

Wu, K. (2015) 'An Empirical Study of Small- and Medium-Sized Firms in Taiwan : Entrepreneurship, Core Competency and Market Performance An Empirical Study of Small- and Medium-Sized Firms in Taiwan : Entrepreneurship, Core Competency and Market Performance', 18(3).

Advances in Business, Management and Entrepreneurship – Hurriyati et al (eds)
© *2020 Taylor & Francis Group, London, ISBN 978-0-367-27176-3*

QFD as a tool for improvement of transportation services in Bandung City

M.A. Sultan, R.R. Ahmad & A. Ciptagustia
Universitas Pendidikan Indonesia, Bandung, Indonesia

ABSTRACT: This study aims to measure the voice of customers in the mode of transportation in the city of Bandung by using the method of Quality Function Deployment (QFD) to improve the performance of service quality in conventional transportation modes in the city of Bandung. The research method used is survey method, using questionnaire as data collection tool. The subject of research is a conventional taxi company in Bandung. Observations use "one shoot"/cross sectional time coverage. Data analysis was done by resource-based view, SWOT analysis, and visual mapping then analyzed descriptively. The research method used is survey method. Observations use "one shoot"/cross sectional time coverage. Triangulation is used as a data collection technique. Data were analyzed by the method of quality function deployment, Pareto diagram, fishbone analysis, and resource-based view, with a reference to the data analysis model from Miles and Huberman.

1 INTRODUCTION

Convenient public transportation mode is now an absolute thing, based on the number of consumer complaints of the existing transportation. Discomfort in the current mode of public transportation has an impact on the increasing number of people considering private vehicles as compared to public transport modes.

The decline in the number of users of the public transportation mode would have an impact on the imbalance between the number of existing modes of transportation and the number of passengers. So that some public transportation modes become more stop and impact the traffic jam.

Calculated in the period of 2009 to 2014 the types of passenger vehicles experienced a growth of 59.5%, which certainly affected the reduced capacity of the road. The impact of the reduced capacity of the road made the public feel inconvenient to drive privately, so that some users started thinking to choose public transportation.

People's desire for a comfortable public transportation would be an absolute thing. Therefore, this study tried to measure the convenience value felt by the public transport users. This research focuses on taxi mode users as research subjects.

There is a diversity of methods that can be used to measure comfort values. Previous research revealed the quality of public transport services using the SERVQUAL method to measure consumer expectation and perception based on service attributes of 5 dimensions namely reliability, responsiveness, assurance, empathy, tangibles (Pramesti, 2013). Another study revealed the use of the Quality Function Deployment (QFD) method with House of Quality (HOQ). Angelina (2012) uses the Importance Performance Analysis (IPA) method to determine the priority of various attributes to measure the quality of Citi Trans travel services. While Haryanto (2012) uses factor analysis and Importance Performance Analysis (IPA) to provide information about important service factors according to the consumer and still needs improvement to improve service quality.

This research uses the method of Quality Function Deployment (QFD) with the consideration that this method is able to explore the consumer's expectation and requirement and compare it with the ability of the service company to fulfill the expectation (Francheschini, 2002). QFD is a method capable of providing continuous improvement which involves consumers from the beginning of the product development process (Goetsch and Davis, 2010). QFD is also able to determine customer needs and translate these into the attributes of each functional area (Heizer and Render, 2013). Another expression states that QFD is a structured methodology used in product design and development processes to define specifications according to customer needs (Cohen, 1995).

2 RESEARCH METHODOLOGY

This research uses qualitative and quantitative research approach, by understanding the "customer voice" using QFD method. The research population focuses on the taxi users in Bandung. Here are the steps performed in the customer research of taxi users. The steps of data processing method using QFD method include (Couhen Lou, 1995):

1. House Of Quality (HOQ)
   The application of QFD methodology in product design process begins with the formation of a product planning matrix.

2. Part of Deployment Matrix

   In this second house the technical requirements chosen to be developed are transformed into a more technical concept design called the critical part. In determining the critical section, it is necessary to draw up a conceptual analysis first.
3. Process Planning Matrix

   Before determining the process matrix, the stages of the process must be considered through which the raw materials are to be finished products and ready for market. At this stage the analysis begins with making a product development process map. The map is then connected to the resulting critical part and the previous matrix.
4. Product Planning Matrix

   After going through the planning and process stage, for the last stage the actions can be seen that need to be taken for quality improvement.

3 RESULTS AND DISCUSSION

Consumer desires list based on priority. Quality service aspects are shown below.

1. Taxi fleets make you comfortable.
2. The AC in the car works.
3. The driver initiates to bring the goods into the trunk.
4. The entertainment facility works. (tape, radio)
5. Taxi fleet is easily available.
6. The driver asks you to wear a seat belt before leaving. (especially if you sit in the front)
7. The driver keeps your luggage left behind and immediately informs him.
8. The driver greets you before and after the trip.
9. The driver knows an alternative way if needed. (when stuck)
10. The driver has an insight into the sights.
11. The driver mastered the technical stuff. (understanding of car engines)
12. The taxi fleet comes according to the time of the order.
13. The argo count corresponds to the stated rate.
14. Taxi fleet is an output car (minimum) 2014.
15. The company provides consumer services. (to provide complaints and suggestions)
16. The driver speaks politely and is polite to you.
17. The company provides the latest info about the latest service in social media or attaches it to the taxi fleet.
18. There is a driver's identification in every taxi fleet. (for your safety)
19. The driver prioritizes traffic safety. (driving according to traffic rules)
20. There is an identity (taxi fleet number) on every car, outside or inside.
21. The driver checks the luggage cabin if all the passenger goods have been unloaded.
22. The driver is neat and clean

As can be seen in Table 1, almost every aspect of service quality that has been given by the taxi company X is as good as the quality of service provided by the online taxi provider, but the taxi company X outperformed the online taxi provider in the presence of the identity of the Fleet and the driver's identification.

Taxi X company strategy to satisfy the wants and needs of consumers include:

1. Outlets are spread out in various strategic areas in Indonesia.
2. Taxi looking for passengers also operate on the highway. (350 fleets in Bandung Raya area)
3. Taxi X company provides e-payment for consumers who do not carry cash.
4. There is customer care service.
5. There are driver identities and fleets on the taxi fleet.
6. Did lots of promotion
7. The price set fits the quality that the customer receives.
8. Conducted training in accordance with customer service SOP on driver and all staff.
9. The taxi fleet is used for an average of at least 2015.
10. Provides fleet booking service over the phone and through applications
11. Collaborates with other online taxi companies.

From the analysis of the interrelationship between corporate strategy the conclusion can be drawn that: The strategy of the X taxi company outlets scattered in various strategic areas is closely related to the 40,000 operating fleet strategy. Pricing strategy tailored to the quality of service is related to payment service strategy using e-payment. The operating fleet of 40,000 is closely related to the operated fleet as of 2015. Strategies that require drivers to wear their identities and each fleet have an identity number is closely related to the SOP strategy that X taxi company runs. The strategy of having a booking service using phones and apps is closely related to X taxi company collaborating with other online taxi providers.

Based on the analysis of the relationship between the needs and desires of consumers with corporate strategy the conclusion can be drawn:

The taxi fleet makes the consumer comfortable with regard to the fleet used in 2015, the taxi company provides e-payment service, the X taxi company provides customer care services and the latter in relation to the price set according to the service.

The functioning of air conditioning and other facilities such as tape, radio and others in the car is closely related to the fleet used being a new fleet from 2015.

The taxi fleet is easy to get as it relates to the X taxi company providing outlets in various strategic

Table 1. Analysis of service quality with competitors.

Aspects of Service Quality Company Taxi X	Answer		
	Better	Equaly good	Less
Taxi fleets make you comfortable.	25%	65%	10%
The AC in the car works.	21%	74%	5%
The driverinitiates to bring the goods into the trunk.	8%	64%	28%
The entertainment facility works. (tape, radio)	28%	46%	26%
Taxi fleet is easily available.	41%	50%	9%
The driver asks you to wear a seat belt before leaving. (especially if you sit in the front)	23%	57%	20%
The driver keeps your luggage left behind and immediately informs him.	24%	62%	14%
The driver greets you before and after the trip.	30%	44%	26%
The driver knows an alternative way if needed. (when stuck)	32%	59%	9%
The driver has an insight into the sights.	26%	54%	20%
The driver mastered the technical stuff. (understanding of car engines)	25%	68%	7%
The taxi fleet comes according to the time of the order.	16%	61%	23%
The argo count corresponds to the stated rate.	24%	56%	20%
Taxi fleet is an output car (minimum) 2014.	14%	73%	13%
The company provides consumer services. (to provide complaints and suggestions)	33%	61%	6%
The driver speaks politely and is polite to you.	21%	54%	25%
The company provides the latest info about the latest service in social media or attach it to the taxi fleet.	36%	56%	8%
There is a driver's identification in every taxi fleet. (for your safety)	27%	71%	2%
The driver prioritizes traffic safety. (driving according to traffic rules)	31%	66%	3%
There is an identity (taxi fleet number) on every car, outside or inside.	58%	41%	1%
The driver checks the luggage cabin if all the passenger goods have been unloaded.	56%	42%	2%
The driver is neat and clean	26%	63%	11%

areas, taxi company X provides 40,000 fleets and special areas of Bandung Raya and provides 350 fleets, phone and app booking services, and is closely related to taxi company X collaborating with online taxi providers.

The driver's initiative to bring the goods into the trunk, the driver asks the consumer to use the seat belt before leaving (especially when sitting in the front), the driver greets the customer before and after the trip, the driver has an insight into the place of the tour, the driver knows an alternative way if needed when the traffic jams) The driver mastered technical matters (understood the car engine), the driver was neat and clean, the driver spoke polite and was polite to the customer, the driver put traffic safety first (driving in accordance with the traffic rules) and checked the luggage cabin when all passenger goods had been lowered. All of the needs and desires of consumers above are closely related to the taxi company X which conducts customer service SOP

training especially to the driver and to all staff and employees in general.

The driver keeps the consumer items left behind and immediately informs it is closely related to customer care service and customer service SOP training held by the taxi company X for the driver.

Consumers who want a taxi fleet from the year 2014 is closely related to the strategy of taxi company X which has megopersikan fleet which averages from the year 2015.

Taxi fleet arrives according to the time of the order is closely related to the booking service using the phone and the application as well as being closely related to taxi company X which collaborates with the online taxi provider.

Consumers want taxi companies to provide consumer service, and it is very much related to customer care services that have been provided by taxi companies X.

The company provides info on the latest service info and it is very related to the taxi company X has held many promotions and also related to customer care services.

There is an identity of drivers and fleets inside and outside, which is very much related to the taxi company X which requires the driver to carry identity and the fleet being plastered with identity numbers, and is closely related to customer service SOP that taxi company X had run so far.

4 CONCLUSION

The quality of service provided by the taxi company X to the service users, showed that the performance of service in both categories based on the voice of the customer indicates that the taxi company x has tried to facilitate the service user's desire, the dominant value that needs to be improved is the improvement activity of operational implementation according to the operational standard procedure.

Recommendations for further research include it to be applied to mass transportation modes, so as to solve the problems of urban transportation.

REFERENCES

Angelina, Arresty Theresia. 2012. *Service Quality Improvement at CV. Citra Tiara (citi trans) Bandung*, Thesis, Bandung: Program Business Administration-SBM Institut Teknologi Bandung.

Cohen, Lou. (1995). *Quality Function Deployment: How to make QFD Work for You*. Boston: Addison-Wesley Publishing Company.

Franceschini F. (2002), *Advanced Quality Function Deployment*. Boca Raton, FL: CRC Press.

Goetsch, D. L., & Davis, S. (2010). *Quality Management for Organizational Excellence*. Pearson.

Pramesti, Mia Indria. 2013. *Quality of Commuter Line Train Service Jabodetabek*. Thesis. Jakarta: Program Magister Perencanaan dan Kebijakan Publik Fakultas Ekonomi Universitas Indonesia.

Heizer, J. and Render, Barry. (2011). 10th Edition. *Operations Management*. New jersey: Pearson Education, Inc.

Advances in Business, Management and Entrepreneurship – Hurriyati et al (eds)
© *2020 Taylor & Francis Group, London, ISBN 978-0-367-27176-3*

Agility logistics service providers performance model

R. Nurjaman & L.A. Wibowo
Universitas Pendidikan Indonesia, Bandung, Indonesia

ABSTRACT: Environmental monitoring is assessed on the importance of success on the six forces of threats of newcomers; competition between existing companies, product threats or replacement services, buyer bargaining power, supplier bargaining power, and relative strengths of other stakeholders. One of the activities that can affect the performance of competing companies in meeting the customers' demands is the supply chain system (supply chain), in which part of this system is operational logistics. Logistics management has become part of its own business. This article discusses a model of agility logistics system based on the transformation of agility that offers chain performance improvements to logistics service providers. Model design use Structural Equation Models (SEM). Logistics performance can affect a company's performance. In fact, each indicator can define a strong or low impact against other indicators. Agility logistic service variable attributes are environment (geographic, industrial structure, economic condition, and political features), strategic response, business practice (order cycles, inventory levels, profitability, volume of sales, and share price), innovation, competition (alliance, restructuring legalistically arrangement, and activity base costing), and technology (computer integrated, leanness, and flexibility).

1 INTRODUCTION

In an environment characterized by strong and sophisticated competitors, each organization tries to develop sustainable competitive advantage, many organizations have recognized that logistics competency holds the key to developing or maintaining continued business success. As international competitive pressures continued, and with advances in the field of strategic management, attention shifted to logistics management as a competitive weapon and as an important dimension of competitive strategy. In the last few years, an additional shift in orientation has taken place (Newton 2007). Agility in a supply chain is the ability of the supply chain and its members to rapidly align the network and its operations to dynamic and turbulent requirements of the customers. The main focus is on running businesses in network structures with an adequate level of agility to respond to changes as well as proactively anticipate changes and seek new emerging opportunities (AMBE 2010).

The scope of the supply chain begins with the source of supply and ends at the point of consumption. It extends much further than simply a concern with the physical movement of material and is just as much concerned with supplier management, purchasing, materials management, manufacturing management, facilities planning, customer service and information flow as with transport and physical distribution (Stevens 1989).

Operating system in companies grows rapidly. Supply chain will follow the model of the operating system. Most logistic activities are handed over to the third parties. The company that takes care of its

logistics, also known as logistics provider. The competition to get the market in the sector of logistic service providers is very tight. Logistics service providers do not only operate on the area of the management of goods, but also in its development, a company has to meet expectations of consumers in providing and delivering the goods. With more competitors and the company wooing customers, agile logistic system has emerged as a new mantra. Those who can meet the demands of customers are considered more successful. With consumers' preferences which always change, it is even more important to bring the new products at a good pace. This condition indicates the company has to move dynamically to follow the changes in the environment. The ability of the internal dynamic resources can make the company become flexible and agile in fulfilling the wishes of consumers and environmental change. Skills are required in order for the company to get the expected dynamic performance. Agility is all about customer responsiveness, people and information, cooperation within and between firms and fitting a company for change. To be truly agile, a supply chain must possess several distinguishing characteristics which include: market sensitivity, virtuality, process integration, and networking (Khan K, et al., 2009). That is, supply chains need to have an agile approach to deal with all these changes. In addition to changes in customer preferences, supply chains are vulnerable to disruptions, and consequently, the risk to business continuity has increased (Carvalho et al. 2012). In the design of a supply chain, the customer order decoupling point is of importance as it separates the part of the supply chain where planning is based on forecast (leanness) to where it is based on actual customer orders (agility). As a result, agile supply chain design has been very intriguing towards achieving logistics performance (Muh, 2008).

Agility models are discussed more in the supply chain system in public companies, in which part of the supply chain system is operational logistics. Logistics management has become part of its own business. This article discusses a model of agility logistics system based on the transformation of agility that offers chain performance improvements to logistics service providers.

2 METHOD

Structural Equation Models (SEM) is a methodology which uses empirical evidence to confirm a set of hypotheses representing a widely accepted theory. The structural model proposed for this research can be found within the framework of strategic management theory described in almost any modern-day textbook in strategic management such as Wheelen & Hunger (2010). The basis for this theoretical structure lies within the framework of strategic management theory. The process of classical strategic management begins with environmental scanning (identifying strategic factors), followed by strategy formulation (creating mission statement, objectives, strategies, and policies). The next stage is strategy implementation (developing programs, budgets, procedures) and finally evaluation and control (monitoring objectives). These activities proceed in a sequential, yet interactive, progression where previous steps may be modified due to feedback from subsequent steps (Kohn et al. 2011).

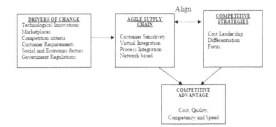

Structural model design is required to review previous models. The model that will be formed in this article is the system logistic agility model based on the adaptation of the agility supply chain to improve the performance of logistics service providers. Several studies have been conducted with Structural Equation Models (SEM). Research with the SEM model includes attributes as a tool for measuring the relationship between variables.

3 RESULT AND DISCUSSION

Based on the results of the previous research review, the agility logistic attribute model is obtained as shown in Figure 1. The quantitative model was used as a tool for measuring the performance of logistics. Logistics performance can affect a company's performance. In fact, each indicator can define a strong or low impact against other indicators. Therefore, in the process of increased performance can be used as a scale of priorities.

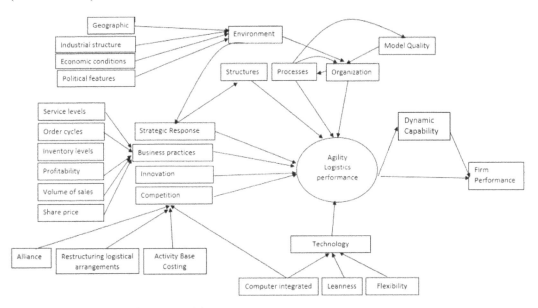

Figure 1. Agility logistics performance model.

4 CONCLUSION

In an environment characterized by strong and sophisticated competitors, each organization tries to develop sustainable competitive advantage, many organizations have recognized that logistics competency holds the key to developing or maintaining continued business success. Supply chains consist of different structures: business processes and technological, organizational, technical, topological, informational, and financial structures. Agility in a supply chain is the ability of the supply chain and its members to rapidly align the network and its operations to dynamic and turbulent requirements of the customers. Likewise, for logistic service provider company, changes in the environment and the consumers 'desires happen rapidly. Thus, operation of the logistic service provider can adapt to the supply chain agility in the company's innovation performance.

Agility logistic service variable attributes are environment (Geographic, industrial structure, economic condition, and political features), Strategic response, Business practice (order cycles, inventory levels, profitability, volume of sales, and share price), innovation, Competition (alliance, restructuring legalistically arrangement, and activity base costing), and Technology (computer integrated, leanness, and flexibility).

REFERENCES

AMBE, I. M. 2010. Agile supply chain: strategy for competitive advantage. *Journal of Global Strategic Management* 1(4): 5–5.

Carvalho, H., Azevedo, S. G., & Cruz-Machado, V. 2012. Agile and resilient approaches to supply chain management: Influence on performance and competitiveness. *Logistics Research* 4(1–2): 49–62.

Ivanov, D., Sokolov, B., & Kaeschel, J. 2010. A multi-structural framework for adaptive supply chain planning and operations control with structure dynamics considerations. *European Journal of Operational Research, 2002*, 409–420.

Khan K, A., Bakkappa, B., Metri, B. A., & Sahay, B. S. 2009. Impact of agile supply chains' delivery practices on firms' performance: cluster analysis and validation. *Supply Chain Management: An International Journal* 14(1): 41–48.

Kohn, J. W., McGinnis, M. A., & Kara, A. 2011. A structural equation model assessment of logistics strategy. *The International Journal of Logistics Management* 22(3): 284–305.

Muh, F. N. 2008. *A Framework supporting the design of a lean-agile supply chain towards improving logistics performance.*

Newton, I. 2007. *Conceptual Frameworks for Supply Chain.*

Qrunfleh, S., & Tarafdar, M. 2013. Lean and agile supply chain strategies and supply chain responsiveness: the role of strategic supplier partnership and postponement. *Supply Chain Management: An International Journal, 186*, 571–582.

Soltan, H., & Mostafa, S. 2015. Lean and agile performance framework for manufacturing enterprises. *Procedia Manufacturing, 2*: 476–484.

Stadtler, H., & Kilger, C. 2008. *Supply Chain Management and Advanced Planning.*

Stevens, G. C. 1989. Integrating the supply chain. *International Journal of Physical Distribution & Materials Management* 19(8), 3–8.

Tarafdar, M., & Qrunfleh, S. 2017. Agile supply chain strategy and supply chain performance: complementary roles of supply chain practices and information systems capability for agility. *International Journal of Production Research* 55(4): 925–938.

Teece, D. J., Pisano, G., & Shuen, A. 1997. Dynamic capabilities and strategic management. *Strategic Management Journal* 18(7), 509–533.

Yusuf, Y. Y., Gunasekaran, A., Adeleye, E. O., & Sivayoganathan, K. 2004. Agile supply chain capabilities: Determinants of competitive objectives. *European Journal of Operational Research* 159(2): 379–392.

The design of integrated information system bakery resource planning using an enterprise resource planning system approach

A.A.G.S. Utama & D.W. Putra
Universitas Airlangga, Surabaya, Indonesia

ABSTRACT: This research is related to the design of the integrated information system Bakery Resource Planning. The design adopts the Enterprise Resource Planning System (ERP), which is a combination of the Manufacturing Resource Planning (MRP) system and accounting, finance, and resource systems. Bakery Resource Planning integrates divisions within Dea Cake and Bakery. The aim is to improve the company's effectiveness and profitability. The approach used was qualitative with a case study research method. The data were obtained through observation, documentation, and interviews of several parties related to the object of this study. The input subsystems in this research were Operational Module, Financial Accounting Module and Human Resources Module. The subsystem was processed using a Database Management System (DBMS). It generated reports that provided information for Bakery Resource Planning users. The result of this research showed that the operational procedure at Dea Cake and Bakery used a manual system. Costs related to production also could be misstated by management. In addition, employee's absence used check sign system on paper. Based on these results, Bakery Resource Planning is expected to be a solution in conducting business activities at Dea Cake and Bakery.

1 INTRODUCTION

Small and medium enterprises (SMEs) are one of the most important parts of a country's economy, including in Indonesia. Their development is becoming more extensive and complex. Data from the Central Bureau of Statistics showed that the total value of Indonesia's Gross Domestic Product (GDP) in 2014 reached IDR 3957.4 trillion. At this point, SMEs contributed 2212.3 trillion, or 53.6% of Indonesia's total GDP.

The food industry, especially the bread industry, is one of the SMEs that is growing rapidly at present. Dea Cake and Bakery is an example. From 2009 to 2014, it had 22 outlets consisting of 21 outlets and 1 bakery outlet. In running its business, Dea Cake and Bakery still uses a manual system. As a result, surveillance is difficult, as well as very time consuming and costly.

In response to this problem, Dea Cake and Bakery needs to upgrade the system from a manual to a computerized system. The system required by Dea and Cake Bakery is Bakery Resource Planning, which adopts an integrated information system that is widely used by the manufacturing company Enterprise Resource Planning (ERP). ERP integrates the Manufacturing Resource Planning (MRP) system with accounting, finance, and resources systems.

Kremzar and Wallace (2001) define ERP as a management tool that employs proven business processes for decision making and provides integration among high-level functions of sales, marketing, manufacturing, operations, logistics, purchasing, finance, product development, and human resources. Broadly speaking, ERP can be described as a management tool that balances the company's inventory and demand to connect customers and suppliers within a single supply chain, to adopt proven business processes in decision making, and to integrate the entire functional parts of the company including sales, marketing, production processes, operations, delivery, purchasing, finance, and resources. Thus, businesses can run with a high level of customer service and productivity, lower costs and inventory, and effective e-commerce.

As integrated information system, Bakery Resource Planning, is expected to provide benefits and convenience in the process of bread production. Supported by an effective and efficient manufacturing management process, it can provide a competitive advantage for the company and solutions for management in improving production effectiveness and management performance with systems that integrate data and business processes of all departments and functions into a single computer system.

2 METHOD

The methodology used in this research was an empirical case study that investigated phenomena in a real-life context when the boundaries of the phenomena and contexts were not clearly visible and where multiple sources of evi-dence were used (Yin 2011).

A case study was used because it was in accordance with the purpose of the research to find new knowledge, explained in detail the object of study, studied various management and accounting processes, and analyzed the object of study. Hence, solutions and views for the management of information systems were provided. Data analysis was basically done before, during, and after the process in the field. However, it focused more on the time during the process in the field while the data were collected. The study needed to be limited to make the research more focused and directed. This research was limited to Bakery Resource Planning as an integrated information system that could help Dea Cake and Bakery achieve a high level of effectiveness and profitability.

3 RESULTS AND DISCUSSION

3.1 General description

Dea Cake and Bakery is an up-and-coming company in the food industry that produces bread and cake. It is located in Malang, East Java.

The financial performance of Dea Cake and Bakery is good, as seen from its income of Dea Cake and Bakery in 2009–2013. In 2009, its income was 3 million Rp per month, which increased to 156 million Rp per month in 2013.

In producing bread and cake, Dea Cake and Bakery uses new equipment and machines. It has technicians who operate those tools. The operational hours are divided into two shifts. The first shift is from 06.30 to 16.00, and the second one is from 10.00 to 20.00.

3.1.1 Analysis of running systems

The operational procedures at Dea Cake and Bakery are described in the subsections that follow.

3.1.2 Raw material order and receipt procedures

The warehouse receives a request for a raw material from the production department. It then reports the stock of the available material through the stock card to the purchasing department. The purchasing department authorizes the raw material order form that is provided by the warehouse to be submitted to the accounting department. The accounting department processes the order of raw materials, whether by cash or credit, to the supplier through the purchase division. The supplier sends raw materials. It is received by the warehouse. The warehouse then checks the suitability of the order with the raw materials received and provides a purchase invoice for the accounting department. The aim is for a standard acceptance report.

3.1.3 The procedure of using and producing raw materials

The production department (cake and bakery trainer) makes a list of the requested raw material. This list is then handed over to the warehouse. If the requested raw material is available, the warehouse will report it

as production demand. On the contrary, if the requested material is not available, the warehouse will ask the purchasing department to process the order and provide a recap of the order to the finance department.

The available raw material is then processed in accordance with production needs. After this is completed, the production department will make authorized production list. This list is then archived.

To check the quality of the finished product, the production department delivers the product to the warehouse. When the product does not meet the quality standards, it is separated from the products that meet the quality standards. After the sorting process is finished, the production cost and selling price of the good quality product are calculated.

3.1.4 Sales procedure of production result

Products that meet quality standards are recorded and authorized by the warehouse to be sold or shipped to the consumer. The products sold are guaranteed to be quality products fresh from the oven. There are two kinds of sales at Dea Cake and Bakery, direct sales and order sales. For direct sales in each outlet, a consumer can select the products on display and pay by cash, debit, or credit card. After making a payment, the consumer will get a payment slip. For order sales, a consumer may order the products several days in advance and pay on the day when the product is delivered. The products are delivered to a consumer according to the agreed date. At the end of the operating hour, the cashier receives cash receipts based on sales of the day. Sales and cash receipts are submitted to the finance department for a daily cash record.

3.1.5 Employee payment procedure

Every month, the human resources department makes a recapitulation of employees' absences and working hours. The recapitulation is given to the general manager to be authorized. It is then handed over to the finance department. The finance department then makes the salary recapitulation and salary slip for each employee. The department distributes salaries to employees via bank transfer. Pay slips are received by each employee via email on the day after the payroll process.

3.2 Analysis of system needs

Bakery Resource Planning basically adopts the concept of the Enterprise Resource Planning System that integrates and automates business processes for the operational and production aspects of a company. The ERP system is a development of Manufacturing Resource Planning (MRP), which is integrated with the financial system and staffing. The module developed in Bakery Resource Planning adapts to the needs of Dea Cake and Bakery and consists of

1. User module: To manage password changes, logout, login, and administrator menu in managing user permissions

2. Production module: To record product data (bread), production management, production demand, and reports related to production
3. Warehouse module: To manage the supplier data, raw material data, raw material data on the division, raw material data entry and exit in the warehouse including information related to the ordering of suppliers, as well as reports relating to the warehouses and purchases
4. Customer module: To record customer data and information
5. Supplier module: To record supplier data and information
6. Financial module: To record account data used, existing financial transactions, and financial reports
7. Sales module: To record orders for bread, cash sales, and sales reports
8. Employment module: To record employees' data, employees' abilities, and employees' absence reports

Bakery Resource Planning will provide benefits for Dea Cake and Bakery such as

1. Integration of financial data. With the existence of Bakery Resource Planning, all processes related to finance can be well integrated. Thus,

management can control the company's finances easily.
2. Integration and standardization of operational processes. With the use of Bakery Resource Planning, the company will be able to minimize mistakes in operational processes. As a consequence, the company's efficiency, productivity, profitability, and quality will improve.
3. Integration or standardization of data and information. With Bakery Resource Planning, misunderstandings in terms of information exchanged between employees and the company will be reduced.

3.3 Bakery resource planning as an integrated information system design

Based on the results of system analysis, it can be concluded that Dea Cake and Bakery needs to upgrade its existing systems into the new one. To explain how the new system is applied, the Data Flow Diagram (DFD) is designed to facilitate the understanding. DFD is a graphical representation that describes the data flow in a system, starting from sources, processes, objectives, to data storage.

Figure 1 shows seven entities that are associated with the integrated information system, Bakery

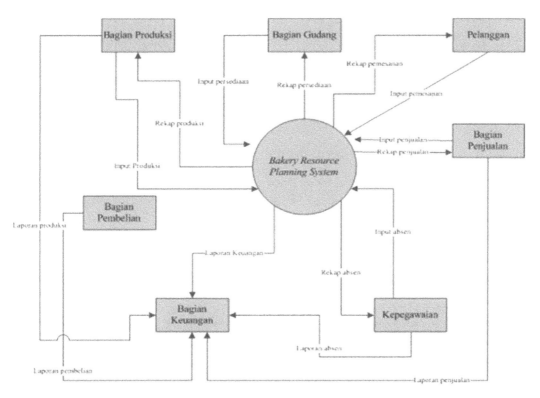

Figure 1. Context diagram of integrated information system, bakery resource planning.
Source: Data Processed, 2014

Resource Planning: production, warehouse, purchasing, sales, customer, staffing, and accountimg. The seven entities are run by the corresponding sections and each section holds the module associated with each of its functions.

3.4 Database management system

The Database Management System (DBMS) functions to efficiently store, centralize, and process data into outputs to meet the needs of information users. The databases collected are Operational Module, Financial Accounting Module, and Human Resources Module that are integrated and stored into a centralized database. They are then tested in a web-based computerized system.

3.5 Output subsystem of integrated information system, bakery resource planning

The output subsystem serves to transmit the processed information (input subsystem) to people or activities that will use the information. Information generated through Bakery Resource Planning is in the form of reports that are processed from the data in the database that has been integrated, namely Operational Module, Financial Accounting Module, and Human Resources Module. The results of the reports are as follows:

1. Request Order Report of Raw Materials
2. Purchasing Report
3. Receipt Report of Raw Materials
4. Request Report of Raw Materials
5. Usage Report of Raw Materials
6. Production Report
7. Product Stock Report
8. Sales Report
9. Absence Recording Report
10. Payroll Report
11. Cash Expenditure Report
12. Cash Inflow Report
13. Financial Position Report (Balance Sheet)
14. Income Statement Report

4 CONCLUSION

Based on the results of analysis and discussion, Dea Cake and Bakery is still using a manual operational procedure system. The calculation of production cost is inaccurate, as it is done manually. Production processes cannot be well maintained because duplicate inventory data, production procedures, and financial data are unconnected. The control of employees' absence is also less effective as it still uses a paper signature system. As a result, these cause difficulties in making a report and decision. To overcome all of these problems, an integrated information system such Bakery Resource Planning can be used in running business activities at Dea Cake and Bakery. This system provides complete input subsystems consisting of operational modules, financial accounting modules, and human resources modules that are processed using a database and in accordance with the needs of the company. In addition, subsystem output produced in the form of reports is in real time, which is certainly useful in making management decisions.

REFERENCES

Wallace, T.H. & Kremzar, M.H.. 2001. *ERP: Making IT happen: The implementer's guide to success with enterprise resource planning*. Hoboken, NJ: John Wiley & Sons.
Yin, R.K. 2011. *Case study: Design and methods*. Jakarta: Raja Grafindo Persada.

Advances in Business, Management and Entrepreneurship – Hurriyati et al (eds)
© *2020 Taylor & Francis Group, London, ISBN 978-0-367-27176-3*

Determinants of innovation among manufacturing firms in a developing country: Insights from Indonesia

Y. Isnasari & P. Prasetyoputra
Indonesian Institute of Sciences, Jakarta, Indonesia

ABSTRACT: Many studies have examined the drivers of innovation among firms in developed settings. However, such studies in the Indonesia context is limited. This study aims to investigate the determinants of innovation among firms in Indonesia. The data was taken from World Bank Enterprise Survey (WBES), particularly the Indonesia dataset. The focus was on 996 manufacturing firms. Set product and process innovation became the dependent variables. Multivariable probit regression models was constructed. The results showed that the firms which had higher product innovation were the following firms: part of a larger group; had patent; man-aged by female; located in a large city; faced informal competitors; and conducted export activities. Mean-while, firms which were more likely to introduce the process innovation were the following firms: part of a larger group; had patent; conducted training, research, and development; had the access to the internet; and faced competition from informal firms. This information enriches the knowledge of innovation among firms. It is also useful for the Indonesia government to make policy.

1 INTRODUCTION

Innovation is one of the well-recognized drivers of economic growth (Solow 1956, Schumpeter 1983, Aghion & Howitt 1997). One of its role is in the expansion of the manufacturing sector and the general economy of a country. Hence, it is necessary to comprehend the determinants of innovation at the country level.

The endogenous growth model proposed by Romer (1986, 1990) notes that technological innovation is generated by the research and development (R&D) in the country that uses the existing stock of knowledge and human capital in the country to produce finished goods and services which then drives the economic growth. R&D is one of the forms of innovation. National innovation system (NIS) has been shown to positively affect the economic growth of a country (Sesayet al. 2018).

Despite the pertinent role of innovation in a country's economic growth, little attention has been given to studies that explore the drivers of innovation and Indonesia. Mahendra et al. (2015) cleverly combined data from the 2010 Indonesian World Bank Enterprise Survey (WBES) and the 2007 Local Economic Governance Surveys to investigate the determinants of firm innovation in Indonesia. However, they mainly focused on access to finance and institutional quality. Although they consider the age of firm, the size of firm, human capital, and foreign ownership, there are still many factors that may determine innovation at the firm level.

Therefore, this study aims to investigate the determinants of process and product innovation among manufacturing firms in Indonesia. Compared to the study by Mahendra et al. (2015), the most recent WBES data for Indonesia was used. Regarding organization of the paper, it is structured as follow: Section 2 discusses the theoretical basis of this study and previous empirical studies; section 3 describes briefly the data source, variables, and statistical techniques used in this study; section 4 presents main empirical findings; and section 5 concludes the findings.

2 METHODS

2.1 Data source

The data from the World Bank Enterprise Survey (WBES), a firm-level survey of a representative sample of private non-agricultural sector, was used. The Indonesian WBES collected the data from both manufacturing and service sector firms. It collected information on various topics such as innovation, technology, human resource, and performance measures of firms. More detailed information can be read elsewhere (World Bank 2011, 2014).

In analyzing the data, this study used cross-section of manufacturing firms in Indonesia, which covered 1069 firms of a total of 1320 surveyed firms. However, the final analytic sample for the multivariable regression analysis was further reduced to 996 due to missing observations (listwise deletion process).

2.2 Dependent variables

There were two dependent variables, namely product innovation and process innovation. The respondents were given two questions: 1) during the last three years, has this establishment introduced any new or significantly improved products or services? and 2) during the last three years, has this establishment introduced any new or significantly improved methods of manufacturing products or offering services? These variables were then coded into binary variables (1 = 'Yes', 0 = 'No'). The time span of three years was used to prevent bias from accidental or one-off innovation effort (Van et al. 2017). Perception-based innovation measures have been shown to be correlated with the objective of innovation measures (Hagedoorn & Cloodt, 2003).

2.3 Explanatory variables

The potential determinants were chosen based on extant literature (Mahendra et al. 2015, Adeyeye et al, 2016, Crespi et al. 2016, Mohan et al. 2016, Van et al. 2017, Abdu & Jibir 2017).

In general, there were common determinants of innovation among firms. These determinants covered firm's age, size, and strategic characteristics, such as the export activity, the obstacles in access to finance, the level of competition in the market, the economic and political situation of the country, and the government's incentive in research and development (Abdu & Jibir 2017).

A recent study by Adeyeye et al. (2016) using data from the 2008 Nigerian Innovation Survey reported that intramural R&D and investment in machinery and equipment positively affected the propensity to conduct product innovation, process innovation, organizational innovation, and marketing innovation. This finding is in line with other previous studies conducted by (Mairesse & Mohnen 2004, Carvalho et al. 2013).

The size of the firm, one of the characteristics of a firm, has been shown to significantly affect innovation behavior of the firm. For instance, using the Community Innovation Survey (CIS) collected in the Zemplinerová et al. (2012) found that the bigger the firm was, the higher the probability of investing in innovation activities, although with diminishing returns.

Competition in the market has also been shown to influence a firm's propensity to innovate. Carlin, et al. (2004) found evidence that firms with no competition at all (monopolies) did fewer innovation activities and had less productivity compared to the firms with higher degree of competition. Moreover, Alder (2010) suggests that the increase of competition results in higher innovation if the competition is low at the start. In the longer run, competition has also been found to be positive for innovation (Artés, 2009).

Furthermore, the financial aspect of a firm was also important in determining innovation. For example, Lonrenz (2014) analyzed data from nine African countries and found negative effect of financial constraint on innovation activities. Better access to financial resources also has been shown to enable firm conduct more innovation activities (Mahendra et al. 2015, Fombang & Adjasi 2018).

Lastly, the human capital factor of a firm also played a pivotal role in determining a firm's propensity to innovate. Van et al. (2017) analyzed data from 13 sub-Saharan African countries and demonstrated a significant relationship between human capital endowment and innovation activities.

Description of variables:

1. Dependent
   - Product innovation: (0/1) if the firm introduced a new or significant product in the past three years.
   - Process innovation: (0/1) if the firm introduced new or significant process in the past three years.

2. Explanatory
   - Age: The number of years a firm has been in operation (natural logarithm).
   - Size: The natural logarithm of total number of firm's full-time employees.
   - Group: (0/1) the firm is part of an economic group or subsidiary of a multinational corporation.
   - Knowledge stock: (0/1) if the firm has any patents abroad.
   - R&D: (0/1) if the firm conducts any formal research and development activity.
   - Female owner: (0/1) if there are any females amongst the firm owners.
   - Female top manager: (0/1) if the top manager of the firm is a female.
   - Formal training: (0/1) if during the last three years the firm provide formal training specifically for the development and/or introduction of new or significantly improved products.
   - Large city: (0/1) if the firm is located in a city with more than 1 million of population.
   - License: (0/1) if the firm uses licensed technology from a foreign-owned company.
   - Internet: (0/1) if the firm has broadband access.
   - Foreign owner: (0/1) if at least 10% of the firm is owned by foreign individual or company.
   - Informal competitor: (0/1) if the firm faces competition from informal establishments.
   - International market: (0/1) if the firm's main market is the international market.
   - Export status: (0/1) if the firm conducts exporting activities.
   - Access to finance is an obstacle: (0/1) if the access to finance is perceived as a major obstacle for the firm.

2.4 Estimation strategy

Quantitative method was used to answer the objective of this study. Considering dependent variables were binary dummies; probit regression models (Long & Freese 2014) were applied. Average marginal effect (AME) was used as the measure of association between the explanatory variables and the dependent variables. All the analyses were performed by using Stata version 13.1 (StataCorp 2013).

3 RESULTS AND DISCUSSION

3.1 Descriptive analysis

Table 1 presents the distribution of Indonesian firms. It can be seen that the majority of firms (36.67 firms) operated a medium scale. Those operated large scale was 33.12% (354 firms) and small was 30.22 % (323 firms).

Table 2 reveals the incidence of innovation by firm type. It is observed that innovation incidence was the largest among firms that operated at the large scale. Incidence of product innovation was higher than incidence of process innovation in small and medium firms. Meanwhile, incidence of process innovation was higher in large firms.

3.2 Drivers of innovation

The final multivariable probit models, containing sixteen explanatory variables, were statistically significant ($p < 0.001$). Table 3 presents the marginal effects of probit models on determinants of innovation. Age of firm corresponded to higher probability of product innovation. This finding is different from the study conducted by Abdu & Jibir (2017) where they found a negative relationship. Moreover, firms that were part of a larger group were more likely to introduce product and process innovation. This is in line with Crespi et al. (2016) who demonstrated such positive relationship between group and product innovation. They did not, however, find a significant association between group and the introduction of process innovation. Furthermore, with regard to location, firms that were located in a large city were more likely to introduce product innovation.

As with knowledge stock, firms that had registered a patent abroad were more likely to introduce product and process innovation. The research by Crespi et al. (2016) also showed the positive effect

Table 3. Marginal effects of probit models on determinants of innovation (N = 996).

Variable	(1) Product		(2) Process	
Ln(age)	0.0381	**	-0.0033	
Ln(size)	0.0029		0.0053	
Group	0.1292	***	0.0735	*
Knowledge stock	0.1960	***	0.1635	**
Research & development	0.0902		0.1067	*
Female owner	-0.0462	*	0.0094	
Female top manager	0.0875	***	0.0303	
Formal training	0.0962	**	0.1171	**
Large city	0.0515	**	0.0253	
License	-0.0419	*	-0.0044	
Internet	-0.0148		0.0482	*
Foreign owner	-0.0686	**	-0.0283	
Informal competitor	0.0533	***	0.0392	**
International market	0.0219		0.0062	
Export	0.0863	**	0.0473	
Access to finance an obstacle	-0.0379	*	-0.0198	
Pseudo R2	16.58		21.28	
Prob > chi2	p<0.001		p<0.001	
Observation	996		996	

Note: ***p<0.01, **p<0.05, *p<0.1
Source: Authors' calculation of WBES data.

of knowledge stock on process innovation, but not on product innovation. Moreover, firms that conducted training were more likely to introduce product and process innovation. Furthermore, it was observed that firms that conducted formal R&D activities had higher probability of introducing process innovation. This confirms the previous research done by Van et al. (2017) in the context of sub-Saharan countries.

Gender also was found to be associated with the firms' propensity to innovate. Firms that were owned by females were observed to perform less product innovation. However, those which were led by females are more likely to introduce product innovation, but not with process innovation.

Firms that had licensed foreign technology and foreign-owner were less likely to introduce product innovation. This is similar to a previous study by Van et al. (2017) where they showed a negative relationship between foreign ownership and innovation efforts in firms located in sub-Saharan countries. Moreover, firms that had access to internet had higher propensity to introduce process innovation.

Alder (2010) demonstrated the positive effect of competition on the propensity of a firm to innovate. We found a similar relationship where firms that faced competition with informal or unregistered firms were more likely to introduce product and process innovation. Moreover, firms that exported their products were also observed

Table 1. Distribution of Indonesian firms sampled by their size.

Firm-type	Frequency	Percentage
Small	323	30.22
Medium	392	36.67
Large	354	33.12

Source: Authors' calculation of WBES data.

Table 2. Incidence of innovation by firm type.

Type of innovation	Small	Medium	Large	Full sample
Product	8.05	12.24	16.95	12.54
Process	6.19	11.22	19.77	12.54

Source: Authors' calculation of WBES data.

to have higher probability of introducing product innovation. Lastly, firms that considered the access to finance as a major obstacle were less likely to introduce product innovation

3.3 Study limitations

This study has two limitations. First, the sample does not include micro enterprises. Second, other types of innovation have not been considered in the analysis.

4 CONCLUSION

The main objective of this study is to examine the major determinants of a firm's innovation in Indonesia using the WBES dataset. Probit regression models was used to answer this objective. The study produced the determinants of process and product innovation. This information will enrich the knowledge on innovation among firms. This is also useful for the Indonesia government to make the policy.

ACKNOWLEDGEMENTS

We would like to thank the World Bank for providing the Indonesia WBES dataset. We also thank the Center for Innovation, Indonesian Institute of Sciences, for providing funding for us to present this paper at the 3rd Global Conference on Business, Management and Entrepreneurship. The funder, however, did not have a part in the design and analysis in this study. Hence, the views expressed in this paper are solely of the authors and do not represent that of the Indonesian Institute of Sciences.

REFERENCES

Abdu, M., & Jibir, A. 2017. Determinants of firms innovation in Nigeria. Kasetsart *Journal of Social Sciences*.

Adeyeye, A. D., Jegede, O. O., Oluwadare, A. J., & Aremu, F. S. 2016. Micro-level determinants of innovation: analysis of the Nigerian manufacturing sector. *Innovation and Development*, 6(1),1-14.

Aghion, P., & Howitt, P. 1997. *Endogenous Growth Theory*. Cambridge, Massachusetts: The MIT Press.

Artés, J. 2009. Long-run versus short-run decisions: R&D and market structure in Spanish firms. *Research Policy* 38(1),120-132.

Carlin, W., Schaffer, M., & Seabright, P. 2004. A minimum of rivalry: Evidence from transition economies on the importance of competition for innovation and growth. The B.E. *Journal of Economic Analysis & Policy*, 3(1).

Carvalho, L., Costa, T. & Caiado, J. 2013. Determinants of innovation in a small open economy: a multidimensional perspective. *Journal of Business Economics and Management* 14(3), 583-600.

Crespi, G., Tacsir, E., & Vargas, F. 2016. *Innovation Dynamics and Productivity: Evidence for Latin America. In B. Inter-American Development*, M. Grazzi & C. Pietrobelli *(Eds.), Firm Innovation and Productivity in Latin America and the Caribbean: The Engine of Economic Development* (pp. 37-71). New York: Palgrave Macmillan US.

Fombang, M. S., & Adjasi, C. K. 2018. Access to finance and firm innovation. *Journal of Financial Economic Policy*, 10(1), 73-94.

Hagedoorn, J., & Cloodt, M. 2003. Measuring innovative performance: is there an advantage in using multiple indicators? *Research Policy*, 32(8), 1365-1379.

Long, J. S., & Freese, J. 2014. *Regression Models for Categorical Dependent Variables using Stata (3rd ed.)*. College Station, Texas: Stata Press.

Lorenz, E. 2014. Do Credit Constrained Firms in Africa Innovate Less? A Study Based on Nine African Nations.

Mahendra, E., Zuhdi, U., & Muyanto, R. 2015. Determinants of Firm Innovation in Indonesia: The Role of Institutions and Access to Finance. *Economics and Finance in Indonesia*, 61(3), 149-179.

Mairesse, J., & Mohnen, P. 2004. The Importance of R&D for Innovation: A Reassessment Using French Survey Data. *The Journal of Technology Transfer*, 30(1), 183-197.

Mohan, P., Watson, P., & Strobl, E. 2016. *Innovative Activity in the Caribbean: Drivers, Benefits, and Obstacles. In B. Inter-American Development*, M. Grazzi & C. Pietrobelli *(Eds.), Firm Innovation and Productivity in Latin America and the Caribbean*: The Engine of Economic Development (pp. 73-101). New York: Palgrave Macmillan US.

Romer, P. M. 1986. Increasing Returns and Long-Run Growth. *Journal of Political Economy*, 94(5),1002-1037. Retrieved from http://www.jstor.org/stable/1833190

Romer, P. M. 1990. Endogenous technological change. *Journal of Political Economy*, 98(5),S71–S102. Larch, A.A. 1996a. Development.

Schumpeter, J. A. 1983. The Theory of Economic Development: An Inquiry into Profits, Capital, Credit, Interest, and the Business Cycle: Transaction Publishers.

Şeker, M. 2012. Importing, Exporting, and Innovation in Developing Countries. *Review of International Economics*, 20(2), 299-314.

Sesay, B., Yulin, Z., & Wang, F. 2018. Does the national innovation system spur economic growth in Brazil, Russia, India, China and South Africa economies? Evidence from panel data. *South African Journal of Economic and Management Sciences*, 21(1).

Solow, R. M. 1956. A Contribution to the Theory of Economic Growth. *The Quarterly Journal of Economics*, 70(1), 65-94.

StataCorp 2013. *Stata Statistical Software: Release 13*, College Station, TX, StataCorp LP.

World Bank. 2011. *World Bank*'s Enterprise Survey: Understanding the Questionnaire. Washing-ton, D.C.: World Bank. Retrieved from.

World Bank. 2014. *World Bank Enterprise Surveys: Indonesia 2015 Country Profile*. Washington, DC: The World Bank. Retrieved from.

Van Uden, A., Knoben, J., & Vermeulen, P. 2017. *Human capital and innovation in Sub-Saharan countries: a firm-level study*. Innovation, 19(2), 103-124.

Zemplinerová, A. & Hromádková, E. 2012. *Determinants of firm's innovation*. Prague Economic Papers 4(487-503).

Advances in Business, Management and Entrepreneurship – Hurriyati et al (eds)
© *2020 Taylor & Francis Group, London, ISBN 978-0-367-27176-3*

An integrated framework to support the process supply chain in the tourism sector

L.C. Nawangsari & A.H. Sutawijaya
Universitas Mercu Buana, Jakarta, Indonesia

ABSTRACT: Tourism today has developed into a sector that can potentially increase the economy of a country. One of the strategies of the tourism industry that can improve competitiveness is tourism supply chain management. The purpose of this research was to analyze the supply chain model of the tourism sector. The research methodology used in this research was a qualitative method with case study design. The data were gathered through deep interviews and detailed observations on key information sources. The study focused mainly on the supply chain framework model in the tourism sector. The framework has three levels: strategic, tactical, and operational. Some other factors that were affecting the effectiveness of the supply chain were shareholders, tourism destinations, agents, tour operators, accessibility, attractiveness, and facilities. These components were suggested to be well managed by the stakeholders. Thus, they can increase the strategic values of business sustainability.

1 INTRODUCTION

The tourism sector is one of the main sectors that can promote economic activity. The shifting of community needs from primary needs to secondary and tertiary needs leads to an increase in consumer demand for tourism services. Consequently, the government's steps to boost the development of Indonesia's tourism industry is considered the right strategy. The World Tourism Organization (WTO) predicts that the tourism industry will continually develop, with an average growth of the number of international tourists of about 4% per year (Pitana & Gayatri 2005). Meanwhile, domestic tourists are estimated to reach ten times the number of international tourists, which also plays a major role in the economic development of tourist destinations.

The number of foreign tourist arrivals in Indonesia has grown steadily from 2013 to 2017. The Travel & Tourism Competitiveness Report of the World Economic Forum statess that the growth of the tourism sector in Indonesia is due to the arrival of foreign tourists, as well as the role of the local government in the development of tourism industry and improved investment in infrastructure.

Tourism is one of the fastest growing sectors of the economy in countries throughout the world. This sector can contribute significantly to social and economic development and to reduction of poverty. Despite the occasional decline, international tourist arrivals increased every year by an average of 4.3% between 2013 and 2017.

The tourism sector needs a large amount of manpower. It is a significant source of development and employment, especially for those with limited access to employment such as women, youth, migrant workers, and rural communities. The relationship between tourism and supply chains is inseparable within this framework. The relationship between the tourism sector and the other related sectors is a precondition for improving good service for tourists. If there is no improvement in supply chain management (SCM), it will affect the business of tourism.

Based on this description, an integrated supply chain is needed to increase the advantages of the product. The problem raised then is how the concept of tourism Supply Chain Management can be integrated in West Java so it can increase tourist visits.

SCM is a set of approaches used to efficiently integrate suppliers, manufacturers, warehouses, and sellers, so that products (merchandise) are produced and distributed in exact quantities to the right locations at the right time. SCM is also used to minimize system-wide costs, to maximize the total supply chain surplus, and to satisfy the service levels.

According to Heizer and Render (2014), SCM describes the coordination of all supply chain activities, starting with raw materials and ending with a satisfied customer. Thus, a supply chain includes suppliers, manufacturers, and/or service providers, distributors, wholesalers, and/or retailers who deliver the product and/or service to the final customer.

Furthermore, Chopra and Meindl (2013) define supply chain as all parties involved, directly or indirectly, in fulfilling a customer request. The supply chain includes not only the manufacturer and suppliers, but also transporters, warehouses, retailers, and even customers themselves.

1.1 Tourism

Tourism is a very important aspect of economic activity. However, the exact measurement of travel and tourism is not always easy. Tourism is an activity that takes place when, in international terms, people cross borders for leisure or business and stay at least 24 hours but less than one year (Mill & Morrison 1998). Clearly, tourism activities cover a variety of sectors, including accommodation, attractions, the travel trade, and transport.

1.2 *Tourism Supply Chain*

The Tourism Supply Chain (TSC) can be defined as a network of tourism organizations that supplies various components of tourism products/services, such as transportation and accommodation, for the distribution and marketing of final tourism products at a particular tourism destination. The distribution and marketing involve various parties, both private and public.

2 METHOD

The method describes the type or design of the research, variables and measurements, population and sampling technique, data type and collection technique, and data analysis.

3 RESULTS AND DISCUSSION

3.1 *Supply chain evaluation*

There were three important aspects of the supply chain that needed to be evaluated, which were strategies relating to the individual performance of each supply element, a supply chain design, and supply chain model simulation. The first and second aspects were related to the performance of communal similar supply elements (intragroup performance of supply elements) or forecasting and capacity planning. The third operational element related to communal performance of various elements of supply (intergroup performance of supply elements), such as transportation, inventory management, and production scheduling. Regarding this, the authors thought that the performance of the tourism supply element in West Java was not optimal. In the context of supply chain elements, the authors offered the framework of the supply chain process in the tourism sector. The framework dealt with the government as a regulator, service provider, or the owner of the facility and BUMN as a service provider, large private business, medium private business, and small private enterprise. Government, however, played a major role in defining development strategies, and programs, policies and legal requirements related to security, safety, sanitation, working conditions, infrastructure, education, and training. As a regulator, the government had to make clear policies, rules, and strategies. These became the basis for sustainable tourism development, large-scale poverty reduction,

Figure 1. Framework for a continuous process of partition through supply chain implementation.

natural resource protection and lifestyle, and the promotion of economic development. The policy framework shown in Figure 1 was the basis for a continuous process of partition through supply chain implementation.

3.2 *Tourist destination*

The visiting decision in this research was determined from the purchase decision in the stages tourists passed in determining the choice of destinations. This was equal to the stage of purchase decision in general. The visiting decision was a process by which the tourists decide to choose the products or services through various alternative options that were provided by the company supported by agents, tour operators, tourism and accessibility, attraction, amenities, and ancillary factors.

3.3 *Government (culture and tourism)*

Local government through the tourism department was expected to support the tourism supply chain in West Java. The government had to give more attention and guidance to the suppliers of tourism.

3.4 *Sustainable tourism*

The supply chain was committed to improve local prosperity by maximizing the contribution of the tourism industry to economic prosperity. The tourism industry had to generate income and suitable work without affecting the environment and culture of tourist destinations. The tourism industry also had to provide long-term benefits. In short, the tourism industry can simultaneously impact all aspects of human life, such as economic, social, cultural, and environmental.

4 CONCLUSION

The tourism industry has many positive effects, including A very significant impact on the economy. It provides job opportunity and increases regional income. Some tourist destinations even depend on the tourism industry for their earnings.

The main objectives of the tourism supply chain are to improve tourists' satisfaction and to create sustainable tourism. These objectives can be achieved by using specific tactics as well as analyzing the market and its components as a whole. This research showed that the tourism supply chain in West Java was run well. It could be managed by three levels of supply chain management: strategic, tactical, and operational.

REFERENCES

Chopra, S. & Meindl, P. 2013. *Supply chain management: Strategy, planning, and operation*, 5th ed. Harlow: Pearson Education.

Heizer, J. & Render, B. 2015. *Operations management, continuous sustainability.* Jakarta: Edisi sebelas, Salemba Empat.

Mills, J.E. & Morrison, A.M. 2003, *Expanding and re-testing E-SAT: An instrument and structural model for measuring customer satisfaction with travel websites.* Travel and Tourism Research Association Conference, St. Louis.

Pitana, I.G. & Gayatri, P.G. 2005. *Sosiologi Pariwisata.* Yogyakarta: Penerbit ANDI.

Advances in Business, Management and Entrepreneurship – Hurriyati et al (eds)
© 2020 Taylor & Francis Group, London, ISBN 978-0-367-27176-3

Service quality analysis of the outpatient section at public hospitals

T.A. Auliandri & R.M. Wardani
Universitas Airlangga, Surabaya, Indonesia

ABSTRACT: The objective of this research was to observe the gap in the level of service quality in the outpatient section (Instalasi Rawat Jalan) at Dr. Mohammad Soewandhi Public Hospital in Surabaya, Indonesia, according to patients' expectations and perceptions. The data was analyzed by the Servqual method, furthermore, evaluation and recommendation to improve the quality of the service was formulated by using a Fishbone Diagram. Respondents in this research were patients in the outpatient section. Variables which have been observed were tangible, reliability, responsiveness, assurance and empathy. The research resulted in gap values between expectation and reality of variable tangible -0,35, reliability -0,35, responsiveness -0.29, assurance -0.28 and empathy -0.23, which indicated that the services have not met the customers' expectation. Therefore, some improvements need to be implemented in priority variables such as tangible and reliability. Academic and managerial implications will be explained further in the final part of this paper.

1 INTRODUCTION

The outpatient section is the most frequently visited in every hospital, where most of its services and performances are given directly to the patients from their arrival to their leaving. Generally, according to the Decree number 36 by the Ministry of Health of the Republic Indonesia (2009), the number of patients is used as the hospital's performance indicator in determining its service quality. However, it is no longer considered representative enough and needs to be assessed further.

In Surabaya, Indonesia, one of the most recommended hospitals is Dr. Mohammad Soewandhie Regional Public Hospital (Rumah Sakit Umum Daerah) located in Tambakrejo, Tambaksari District. It provides a number of health services such as outpatient service, inpatient service, an emergency unit and other supporting services. It also engages services for patients holding JAMKESMAS (Indonesia's government-funded health insurance program for the poor), BPJS Kesehatan (Health Public Insurance), and other public health insurance such as Asabri, ASKES, Jamsostek, and so on (Ministry of Health 1995).

The number of patients at RSUD Dr. Soewandhie fluctuated during 2010–2012. There is an increasing trend of 15,460 to 17,380 new patients in 2011–2012. Meanwhile, a decreased number of patients was seen previously in 2010, as many as 16,719 new patients. It is shown that during 2010–2011 there was a decreasing number of patients in the hospital (Mayor of Surabaya 2013).

Service quality is defined as the quality perceived by the customer (Gronroos 1990). According to Parasuraman et al. (1985) service quality is a measurement of how properly the services are delivered according to the customer's expectation. Service quality confirms the customer's expectation

of being consistent in basic services. Additionally, service quality is also defined as an overhaul evaluation of a corporate by comparing the company's performance with the customer's general expectation of the service industry. Hoffman & Bateson (2001) explained that service quality is the customer's appraisal of the service providing process.

Servqual is an instrument of service quality measurement developed by Parasuraman, Zeithaml and Berry. Servqual measures expectation and perception of the customers and occurred gaps in the service quality model. In its development, Parasuraman et al. (1985:47) divide Servqual components into five aspects: (1) tangibles, (2) reliability, (3) responsiveness, (4) assurance and (5) empathy.

Related studies conducted previously include:

a. Sari and Harmawan (2012), titled "Service Quality Improvement Suggestions for Outpatient Facility using Servqual and TRIZZ methods (A Case Study at Muhammadiyah Roemani Hospital)," ("Usulan Perbaikan Kualitas Pelayanan pada Instalasi Rawat Jalan dengan Metode Servqual dan TRIZZ (Studi Kasus di RS Muhammadiyah Roemani)."

b. Asyati (2007), titled "The quality analysis of services in the Public Hospital at Wonogiri in Central Java using Fuzzy-Servqual Method."

2 METHOD

This study used a quantitative approach in achieving the objectives of the study by using the Servqual analysis method. Variables in this research are defined as follows:

a. Tangibles, which includes physical facilities, equipment and performance of staff

b. Reliability, which describes an ability to deliver reliable and promised services since the first visit of the patient without making any mistakes and in a timely manner as mutually agreed upon.

c. Responsiveness, which shows a willingness to assist participants and perform quick, responsive and accurate services for patients as well as to clearly inform the time of service if necessary.

d. Assurance, which is also commonly known as certainty, referring to the staff's knowledge, attitude and ability to grow the patients' sense of safety and confidence in the company.

e. Empathy, which means the company's genuine concern for each individual customer, shown by understanding the patients' needs and acting out for their best interests.

This research used a questionnaire to respondents using a five-level rating scale (Likert scale).

2.1 Population, sample and sampling technique

The population included in this study were the patients in the Outpatient Section of RSUD Dr. Mohammad Soewandhie Surabaya. The samples were patients who had been undergoing medical treatments or had visited OPD of RSUD Dr. Mohammad Soewandhie four times or more. The purposive sampling method was applied to take samples, by using a certain sampling technique based on certain criteria as the study required. The number of samples to use was determined based on the Bernoulli model, with a confidence interval of 95% and a 5% margin of error. The samples that have been obtained in this study are seen in Equation 1.

$$N \geq \frac{(1.96)^2 0.95 * 0.05}{0.05^2}$$

$$N \geq 72.99 \approx 73 \text{ respondents} \qquad (1)$$

3 RESULT AND DISCUSSION

3.1 Questionnaire data

The distributed questionnaire and the survey analysis resulted in 119 respondents to take part in this study.

The distribution of respondents is explained in Table 1.

3.2 Calculation of service quality value

The next process was to find the gap score between the expected value and reality value. The gap score was calculated based on the following formula:

$$\text{Gap Score} = X \text{ reality} - X \text{ expectation} \qquad (2)$$

Table 2 shows the gap score between the expectation value and the reality value of services in OPD of RSUD Dr. Mohammad Soewandhie Surabaya.

3.3 Service gap score calculation analysis

The gap of the service value shows the patients' expectation of service and evaluation of the real services received during medication. The previous calculation of the gap score shows a negative result, which means that the services in the hospital have not satisfied the patients' expectation. Nonetheless, the score given by patients was on a scale of 4, which indicates the patients' high level of satisfaction of the hospital's services. The result is described in the Table 3, showing the average gap score for all dimensions.

As can be seen in Table 3, all dimensions have a negative gap score. It shows that patients still feel that the service quality in OPD of RSUD Dr. Soewandhie has not exceeded their expectations, even though the average score shows 4 points, which indicates their satisfaction.

The table also shows that the average gap score of all dimensions is -0.30, which will be used in determining improvement suggestions by using a fishbone diagram for determining the priorities of the dimensions that will be improved.

From the Servqual analysis conducted previously and based on the average gap score, in each dimension there are two dimensions with an under average score: tangible dimension and reliability dimension, both of which have the same gap score of -0.35. Although the other three dimensions also have a negative score, they are acceptably above average. Therefore, priority for improvement should be implemented in the tangible dimension and reliability dimension.

Table 1. Distribution of respondents.

Gender		Age				Frequency of medication		
F	M	15–24	25–40	41–55	>55	≤ 2x	3x- 4x	≥ 4x
62	57	5	24	35	55	0	0	119
52%	48%	4%	20%	29%	46%	0%	0%	0%
119		119				119		

589

Table 2. Expectation–reality gap score.

Dimension	Attribute	X Expectation	X Reality	Gap Score
TG1	Having a well-organized and clean building	4.49	4.09	-0.40
TG2	Having modern medical supplies and equipment	4.44	4.12	-0.32
TG3	Having a neat and clean waiting room	4.38	4.01	-0.37
TG4	Having a clear and detailed information board and directional signs	4.35	4.04	-0.31
TG5	Having a clean and comfortable consulting room	4.39	4.05	-0.34
RB1	Quick and on time service	4.36	3.86	-0.50
RB2	Convenient and less confusing service procedures for patients	4.34	4.05	-0.29
RB3	Quick and great services from friendly staff	4.33	4.02	-0.31
RB4	Reliable doctors and medical personnel to give accurate treatment	4.38	4.08	-0.30
RP1	Timely new patients' reception based on the right procedures	4.35	4.03	-0.32
RP2	Quick payment service	4.45	4.19	-0.26
AS1	Employees' knowledge and qualification in serving patients	4.28	4.01	-0.27
AS2	Doctors and medical personnel's commitment to keep the patients' information confidential	4.35	4.08	-0.27

(Continued)

Table 2. (Cont.)

Dimension	Attribute	X Expectation	X Reality	Gap Score
AS3	Availability of medical personnel ready to serve patients on every nurse station	4.34	4.07	-0.27
AS4	Clear information of every medical treatment to be applied (e.g. medicine prescription)	4.40	4.08	-0.32
EP1	Personal attention to each patient	4.31	4.08	-0.23
EP2	Responsive feedback toward every patient's complaint	4.34	4.08	-0.26
EP3	Equal treatment for all patients without discriminating on social status	4.45	4.26	-0.19

Table 3. Average gap score of all dimensions.

Dimension	Average Gap Score of Each Dimension	Average Gap Score of all Dimensions
Tangibles	-0.35	
Reliability	-0.35	
Responsiveness	-0.29	-0.30
Assurance	-0.28	
Empathy	-0.23	

4 CONCLUSION

In the result of the Servqual equation it can be observed that the gaps of all reality and expectation values of the examined variables have a negative value. It means that the service quality of the Outpatient Section of RSUD Dr. Mohammad Soewandhie has not satisfied the patient's expectation. Suggested improvements are increasing the number of medical personnel to balance the ratio of medical personnel and patients; Expanding the hospital public area, thus allowing the hospital to add more reception counters in the Outpatient Section;

Encouraging Hospital management to intensively socialize service procedures to all staff.

ACKNOWLEDGMENT

The research is funded by Section of Management, Universitas Airlangga, Indonesia.

REFERENCES

Asyati, A. 2007. *The quality analysis of services in the Public Hospital at Wonogiri in Central Java using Fuzzy-Servqual method.* Jakarta: Gema Tekhnik Publisher.

Gronroos, C. 1990. Service management: A management focus for service competition. *International Journal of Service Industry Management* 1(1): 6–14.

Hoffman, K.D. & Bateson, E.G. 2001. *Essential of service marketing.* 2nd ed. South-Western.

Mayor of Surabaya. 2013. *The 2014 planning of regional planning and development.* Surabaya: City of Surabaya Regulation Book.

Ministry of Health of Republic Indonesia. 2009. *Rules number 36 about Health Regulation.* Jakarta: Ministry of Health Indonesia Regulation Book.

Ministry of Health of Republic Indonesia. 1995. *The regulation of Sanitation for Hospital in Indonesia.* Jakarta: Ministry of Health Indonesia Regulation Book.

Parasuraman, A., Zeithaml, V.A. & Berry L.L. 1985. A conceptual model of service quality and its implications for future research. *Journal of Marketing*: 41–50.

Sari, D.P. & Harmawan, A. 2012. The quality improvement at the outgoing patient section using servqual method and TRIZ. *Journal of IT Universitas Diponegoro* VII(2).

Evaluation of hospital management information systems: A model success through quality, user satisfaction, and benefit factors

V. Pujani, R.F. Handika, H. Hardisman, R. Semiarty & R. Nazir
Universitas Andalas, Padang, Indonesia

ABSTRACT: The study investigated the influencing factors of using Hospital Management Information Systems including quality of system, information, ser satisfaction, and net benefit from the perspective of the medical staff in a hospital. Medical staffs that have operated the system to support hospital activities were selected as respondents using accidental sampling and were asked to participate through drop-and-collect surveys. Four hypotheses were tested and analyzed using structural equation modeling and the PLS program. The results of three of the four hypotheses showed significant influences of information and system quality on user satisfaction in using HMIS and user satisfaction on net benefit. The rest of the hypotheses showed no significant influence of human quality on user satisfaction. Finally, the discussion and conclusion describe contributions, limitations, and ideas for further research.

1   INTRODUCTION

The utilization of information technology in health sectors has been able to deliver both in-dividual and organizational benefits (Lawson-Body et al. 2017l Noh & Park 2017; Rosdini & Ritchi 2017). A Health Management Information System (HMIS) is used not only in developed countries but also in developing countries including Indonesia. The implementation of an HMIS in Indonesia is facilitated by the penetration of Indonesian internet users, who reach 53.7% of the country's population or 143,260,000 people, which has placed Indonesia among the top five countries with the highest number of internet users (Internet Worldstat 2018). However, despite its potential, the use of an information system in hospital activities remains low and not optimal.

The Indonesian Health Ministry has issued a regulation on the implementation of a e-health system in PERMEN No. 36 of 2009 and No. 192/MENKES/SK/VI/2012 through integrated collaboration between hospitals, governments, universities, private companies, and telecom providers. In prior studies in Indonesia Khoja et al. (2007) believed that the idea of e-healthcare is relatively new and undeveloped although numerous hospitals, clinics, and primary health care facilities (Puskesmas) have been using health systems in their routine activities. In addition, changes of the medical system paradigm and health industry including future social developments in the fourth industrial revolution are imminent and should be addressed with high urgency (Noh & Park 2017).

Healthcare organizations across the world believe in the necessity of evaluation of information systems to ensure delivery of high-quality service to patients, rapid information retrieval, and efficient data management (Ojo & Popoola 2015) as a tool for lower cost care (Farzandipour et al. 2017). Therefore, from the management perspective, evaluating the implementation of information systems is identified as an important agenda item to provide suitability and sustainability of the system operations in contributing to various users and organizations (Noh & Park 2017). The contributions to healthcare programs of the information system in relation to are measured through user satisfaction and benefit provided through quality factors (Hossain 2016; Noh & Park 2017). However, studies related to the evaluation of the information systems in healthcare in Indonesian hospitals as a developing country are currently underdeveloped.

Accordingly, the current study aimed to investigate the impact of HMIS in one province of Indonesia as an initial study. The objective of this study is to evaluate the impact of HMIS through the factors quality, technology infrastructure, user satisfaction, and net benefits.

This article is organized as follows. In the first part some concepts available in the literature related to the implementation of an information system in a hospital's healthcare program are reviewed and used to develop variables and a research model. The research methodology, data analysis, and results are discussed and conclusions drawn in the last part of this article.

1.1   *Hospital management information systems*

The utilization of information systems (IS) and information technology (IT) is evident in many hospitals to support some parts or the entire activity in delivering good public services and improving internal performance (Subiyakto et al. 2015). The HMIS can use both integrated and separate software applications depending on certain units and activities, for instance electronic health records (EHR) and health information exchange (HIE) (Strudwick & McGillis Hall 2015; Pietro & Francetic 2017) and, hospital information systems (HIS) (Noh & Park 2017). This study tends to utilize the DeLone and McLean (1992) IS success model as the basic measurement and was revised in hospital contexts as the HMIS success model.

1.2 Quality

The success factor for HMIS is how qualified the system is in information and human aspects. System quality is believed to be a success factor in some IS contexts First, system quality, such as technical usability and processing including ease of use, accessibility, speedy, flexibility, usefulness, and integration, is identified as an important factor during operation of the system (Jamal et al. 2009; Ojo & Popoola 2015; Mou & Cohen 2017; Noh & Park 2017). Second is the quality of information related to measurement of IS output in the form of reports, including reliability, timeliness, precision, meaningfulness, and relevance (DeLone & McLean 1992; Halawi et al. 2008; Hossain 2016). Third, the human quality is related to capabilities, knowledge, skills, behavior, attitude, and talents (Swanson & Holton 2001; Nair 2005). In this research, the human quality is measured by top management supports, managerial IT knowledge, management style, and organization culture (Chaveesuk & Hongsuwan 2017).

1.3 User satisfaction

User satisfaction is identified as users' perspectives and responses toward the effectiveness of the systems (Rosdini & Rictchi 2017) and the output system had been tested. The prior study of an IS success model by DeLone and McLain (1992) used a generic model in evaluating IS contexts, while user satisfaction was used as a significant factor to measure the IS success model besides quality, use, individual, and organizational impact.

In terms of information systems in health contexts there are numerous empirical studies that use user satisfaction as a success factor that has a high impact on both personal and hospital performance (Danielsen et al. 2010; Strudwick & McGillis Hall 2015; Farzandipour et al. 2017; Lawson-Body et al. 2017Noh & Park 2017,). User satisfaction is believed to be an influencing factor on performance individually and net benefits of using HMIS in different hospitals and medical services (Lankton & Wilson 2007; Mou & Cohen 2017).

Some studies attempt to test the quality factor in using the IS success model as an independent variable. The prior studies proved that system, information, and human quality influenced user satisfaction in using HMIS (Saputro et al. 2015; Strudwick & McGillis Hall 2015; Chaveesuk & Hongsuwan 2017; Noh & Park 2017; Rosdini & Ritchi 2017).

Hypothesis 1: System quality has an effect on user satisfaction in using HMIS.

Hypothesis 2: Information quality has an effect on user satisfaction in using HMIS.

Hypothesis 3: Human quality has an effect on user satisfaction in using HMIS.

1.4 Net benefits

Net benefits are identified as performances with consequences of working with an information system according to the DeLone & McLean model (1992). In this study, net benefits presented as individual impacts on IS uses that were influenced by user satisfaction. The current research employed net benefits as perceived usefulness, task accomplishment, job performance, effectiveness, ease of job, and usefulness in work (Ojo & Popoola 2015; Saputro et al. 2015; Ojo 2017).

Accordingly, the examination in this research aims to describe the relationships between user satisfaction and net benefits in the next hypothesis.

Hypothesis 4: User satisfaction has an effect on net benefits in using HMIS.

1.5 Research model

Evaluation of the HMIS success model employing the factors of system quality, information quality, human quality, user satisfaction, and net benefits are presented in this research. The framework of the current research is based on the DeLone and McLean model (1992) and was modified using human quality as an additional factor in this research model of four hypotheses in Figure 1.

2 METHOD

2.1 Research design

This study examined the HMIS success model in quantitative and explanatory research through a drop-and-collect survey conducted on 104 nurses in Padang, Indonesia. The research respondents were selected using accidental sampling in three hospitals to find out their perceptions in relation to the operation of information systems. The data obtained were then analyzed

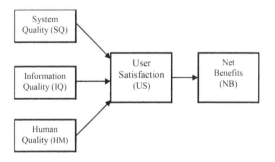

Figure 1. The research model.

using structural equation modeling (SEM), especially the SmartPLS program. Partial least squares analysis using SEM/PLS was conducted to test two different aspects of the research model. In the first stage the model was measured through validity and reliability tests. In the second stage the structural model was measured through hypothesis testing as empirical research on the HMIS success model.

The operational variables measured are net benefit (6 items), user satisfaction (3 items), human quality (4 items), information quality (4 items), and system quality (6 items) using a 5-Likert scale.

3 RESULTS

The research population consisted of the end users who operated the information system/application in doing their jobs in hospitals. The data analysis was conducted using SEM/PLS software; as a result, the measurement model and the structural model were tested.

3.1 Measurement model

The data obtained were tested using the measurement of the research model in terms of validity and reliability testing. Accordingly, validity was measured by convergent and discriminant validity. The validity test was identified based on the value of outer loading through the algorithmic process. The indicator is considered valid if it has a value of outer loading up to 0.70. However, for loading 0.50–0.70 is still accepted (Cooper & Schinder 2013).

The validity test used construct and discriminant validity. The construct validity is measured by identifying the value of cross-loading, by comparing the indicator correlation of that construct with the other constructs. Table 1 shows all loading scores are adequate in testing the construct. Again, the cross-loading score shows all of the indicators already have a higher score correlated to the own variable tested rather than other variables, and consequently the validity model was presented.

The discriminant validity is considered the value of cross-loading and also can be seen by comparing the square root of AVE (Average Variant Extracted) in terms of correlation between latent variables. To examine the AVE value of every construct, it should be >0.5 (Table 2) as the convergent validity. In the meantime, discriminant validity presented by the value of one construct must be higher loading to its construct rather than correlated with other constructs in latent variables correlation (Table 3).

The comparison of the square roots of AVE of every variable in the diagonal element to correlate with other variables. Table 3 shows that each square root of AVE for the variable is larger than the correlation among other variables. Accordingly, it can be concluded that the latent variable has a good discriminant validity.

Table 1. Cross-loading of HMIS model.

	HQ	IQ	NB	SQ	US
HQ1	**0.846221**	0.631813	0.701947	0.742165	0.581556
HQ2	**0.921651**	0.61459	0.687044	0.667167	0.574935
HQ3	**0.897572**	0.645856	0.603565	0.61163	0.497449
HQ4	**0.871691**	0.682402	0.563692	0.566256	0.543083
IQ1	0.574146	**0.82168**	0.560342	0.578115	0.433849
IQ2	0.650944	**0.881134**	0.612129	0.642619	0.585439
IQ3	0.668577	**0.853168**	0.622888	0.568551	0.579562
IQ4	0.611734	**0.904671**	0.518186	0.506905	0.50387
NB1	0.672976	0.653578	**0.900557**	0.753494	0.687286
NB2	0.683894	0.584457	**0.889974**	0.717034	0.609493
NB3	0.689072	0.613352	**0.940933**	0.698849	0.698234
NB4	0.659525	0.633366	**0.935972**	0.725214	0.709626
NB5	0.639635	0.597824	**0.910036**	0.67866	0.631617
NB6	0.639635	0.597824	**0.910036**	0.67866	0.631617
SQ1	0.49144	0.334182	0.5016	**0.716382**	0.491751
SQ2	0.532057	0.460188	0.592809	**0.689038**	0.386588
SQ3	0.664535	0.626377	0.742297	**0.880393**	0.618081
SQ4	0.438495	0.292426	0.435972	**0.626989**	0.417247
SQ5	0.64897	0.680002	0.704046	**0.890032**	0.644632
SQ6	0.620628	0.623773	0.591525	**0.82018**	0.496373
US1	0.55555	0.586604	0.641858	0.62564	**0.916982**
US2	0.620175	0.622166	0.715773	0.641346	**0.949002**
US3	0.572126	0.513711	0.674151	0.610087	**0.940176**

Table 2. Overview of HMIS measurement.

	AVE	Composite reliability	R^2	Cronbach's alpha
HQ	0.782753	0.935055	—	0.907232
IQ	0.74947	0.922785	—	0.888951
NB	0.836798	0.968506	0.525433	0.960935
SQ	0.603506	0.899838	—	0.864488
US	0.87513	0.954588	0.509286	0.928587

Table 3. Latent variable correlation.

	HQ	IQ	NB	SQ	US
HQ	**0.884733**				
IQ	0.727461	**0.865719**			
NB	0.725884	0.671318	**0.914767**		
SQ	0.734826	0.664945	0.774967	**0.776856**	
US	0.623691	0.614901	0.724868	0.669081	**0.935483**

Diagonal elements (bold) are the square roots of AVE.

A reliability test was conducted on the measurement tools to assess the consistent accuracy and precision of measurement. The instrument reliability is determined by the value of Cronbach's alpha and composite reliability for each block constructs. As a rule of thumb, value of Cronbach's alpha and composite reliability must be >0.7 though 0.6 is still accepted (Table 2).

Cronbach's alpha and composite reliability of each variable were over the rule of thumb since the value was >0.7.

4 DISCUSSION

The results of the research are shown in Table 4. In operating HMIS, system and information quality significantly influence the satisfaction of users (nurses) (Noh & Park 2017). This result concerns about HMIS indicates usability (system) quality provided for instance flexibility, integration, and responsiveness of system (Chaveesuk Hongsuwan 2017; Noh & Park 2017). The item of system quality was believed to be user satisfaction with HMIS. In line with system quality, the item of information quality was being able to be user satisfied during operating HMIS, for instance; accuracy, meaningful information, and reliability of system contents. Unfortunately, the item of human quality was insignificant empirically to be user satisfied using HMIS, which was in contrast with system and information quality.

The benefit of HMIS was influenced by user satisfaction. This study predicted working with HMIS has benefits including reduced time needed for people to do their jobs, job performance, effectiveness, ease of doing their jobs, and helpfulness, in line with the studies of Saputro et al. (2015) and Ojo (2017). These benefits would be achieved when HMIS users are quite satisfied with the systems.

5 CONCLUSION

In conclusion, the study tested four hypotheses of HMIS model success factors. Three of the hypotheses have a significant influence: a system quality and information quality which can predict user satisfaction in using HMIS and have an impact on net benefits of working with HMIS. On the contrary, human quality has an insignificant influence in predicting user satisfaction using HMIS.

ACKNOWLEDGMENTS

The authors are grateful to The Indonesian Ministry of Higher Education (MenristekDikti-DP2M) for its financial support of this research through International Research Collaboration grant and Scientific Publication 2018 scheme.

REFERENCES

Chaveesuk, S. & Hongsuwan, S. 2017. A structural equation model of ERP implementation success in Thailand. *Review of Integrative Business & Economics* 6 (3): 194–204.

Cooper, D.R. & Schindler, P.S. 2013. *Business research methods*, 12th ed. New York: McGraw-Hill.

Danielsen, K., Bjertnaes, O.A., Garratt, A., Forland, O., Iversen, H.H. & Hunskaar, S. 2010. The association between demographic factors, user reported experiences and user satisfaction: Results from three casualty clinics in Norway. *Biomed Central* 11(73): 1–8.

DeLone, W.H. & McLean, E.R. 1992. Information systems success: The quest for the dependent variable. *Information Systems Research* 3(1): 60–95.

Farzandipour, M., Meidani, Z., Gilasi, H. & Dehghan, R. 2017. Evaluation of key capabilities for hospital information system: A milestone for meaningful use of information technology. *Annals of Tropical Medicine and Public Health* 10(6): 1579–1586.

Halawi, L.A., McCarthy, R.V. & Aronson, J.E. 2008. An empirical investigation of knowledge management systems success. *Journal of Computer Information Systems* 121–135.

Hossain, M.A. 2016. Assessing m-Health success in Bangladesh: An empirical investigation using IS success models. *Journal of Enterprise Information Management* 29(5): 774–796.

Internet World Stat. 2018. Asia internet use, population data and Facebook statistics-June 30, 2018. Retrieved from www.internetworldstats.com/stats3.htm.

Jamal, A., McKenzie, K. & Clark, M. 2009. The impact of health information technology on the quality of medical and health care: A systematic review. *Health Information Management Journal* 38(3): 26–37.

Khoja, S., Scott, R.E., Casebeer, A.L., Mohsin, M., Ishaq, A.F.M. & Gilani, S. 2007. e-Health readiness assessment tools for healthcare institutions in developing countries. *Telemedicine and e-Health* 13(4): 425–432.

Lankton, N.K. & Wilson, E.V. 2007. Factors influencing expectations of e-health services within a direct-effects model of user satisfaction. *E-Service Journal* 5(2): 85–112.

Lawson-Body, A., Willoughby, L., Lawson-Body, L. & Logossah, K. 2017. Developing and validating a cultural user satisfaction instrument in developing countries. *Journal of Computer Information Systems* 57(4): 319–329.

Mou, J. & Cohen, J.F. 2017. Trust and online consumer health service success: A longitudinal study. *Information Development* 33(2): 169–189.

Nair, K.D.D. 2005. Development of an instrument to asses' human resource quality (HRQ) and measuring the impact of TQM efforts on HRQ using the instrument. Thesis, Doctor of Philosophy in Management. Faculty of Social Sciences. Chocin University of Science and Technology.

Noh, M.J. & Park, H.H. 2017. The quality analysis on the satisfaction and performance of HIS. *International Information Institution* 20(12): 8365–8372.

Ojo, A.I. 2017. Validation of the DeLone and McLean information system success model. *Health Information Research* 23(1): 60–66.

Ojo, A.I. & Popoola, S. 2015. Some correlates of electronic health information management system success in Nigerian teaching hospitals. *Biomedical Informatics Insights* 23(1): 60–66.

Pietro, C. & Francetic, I. 2017. E-health in Switzerland: The laborious adoption of the federal law on electronic health records (EHR) and health information exchange (HIE) networks. *Article in press*: 1–6.

Rosdini, D & Ritchi, H. 2017. Examining accounting information system in a shared environment: The measure of system adoption efficacy- Indonesia case. *Review of Integrative Business and Economics Research* 6(2): 326–342.

Saputro, P. H., Budiyanto, D. & Santoso, J. 2015. Model Delone and Mclean success measurement *E-Government* in Pekalongan. *Scientific Journal of Informatics* 2(1): 1–8.

Strudwick, G. & McGillis Hall, L. 2015 Nurse acceptance of electronic health record technology: A literature review. *Journal of Research in Nursing* 20(7): 596 607.

Subiyakto, A.A., Ahlan, A.R., Putra, S.J. & Kartiwi, M. 2015. Validation of information system project success model: A focus group study. *Creative Commons* 1–14.

Swanson, R.S. & Holton, E.F. 2001. *Foundations of human resource development*. San Francisco: Berrett-Koehler.

Advances in Business, Management and Entrepreneurship – Hurriyati et al (eds)
© 2020 Taylor & Francis Group, London, ISBN 978-0-367-27176-3

The linkage between co-creation and soft innovation in firm performance: A survey of Indonesian hijab fashion creative industries

E. Astuty
Universitas Widyatama, Bandung, Indonesia

A. Rahayu, D. Disman & L.A. Wibowo
Universitas Pendidikan Indonesia, Bandung, Indonesia

ABSTRACT: The performance of creative industries can be triggered by soft innovation putting forward an aesthetic that appeals both to the senses and the intellect. Co-creation as a new paradigm of product development has improved firm performance for at least two decades. This research aimed to study more deeply the interrelationship between them in the creative industries in Indonesia. The data in this research were collected by using a survey method. A questionnaire was distributed to 30 creative businesses of hijab fashion in Bandung. Research confirmed that co-creation did not show a significant influence on creative industry performance, but it could directly improve soft innovation. Therefore, soft innovation had a positive effect on performance, and it was also confirmed that soft innovation could be a good mediator for co-creation to improve firm performance. Furthermore, it was found that partnerships between private sectors were related to successful new products as one of the measurements of the performance of creative industries.

1 INTRODUCTION

Firm performance is conceptualized as the result of the business activity in a company (Wheelen & Hunger 2015). The measurement tool for the performance is based on the elements to be assessed and the objectives to be achieved. The predetermined objectives in strategy formulation as part of the strategic management process have to be used in measuring firm performance once the strategy is implemented (Wheelen & Hunger 2015). The performance of creative industries in Indonesia, which comes from creativity and innovation (Kemenparekraf 2014), will be greatly influenced by several distinguished concepts that encompass creativity as a result of the utilization of creative ideas, science, cultural heritage, and technology (Kemenparekraf 2014). Several studies related to the improvement of firm performance have been carried out previously, some of which confirmed that innovation positively improved firm performance (Shan et al. 2014; Zhang et al. 2016; Jogaratnam 2017).

Currently, the creative industry market is so competitive that more and more companies use innovation as a tool to achieve growth and performance improvement. However, a number of new products fail to achieve market share because they do not meet the needs or desires of customers. This has led many companies to empower customers to participate actively in the process of creating new products. Recent research on innovation, competitive advantage, and company performance found that customer-driven innovation has now become a core business practice for some companies. A great deal of analysis and research

began to explore the concept of customer-driven innovation, better known as co-creation, which is related to the improvement of firm performance (Prahalad & Ramaswamy 2000; Sharma et al. 2002; Von Hippel & Katz 2002; Vargo & Lusch 2004; Etgar 2008).

Some of these studies have raised theoretical gaps associated with improving firm performance, which suggest that the performance can be triggered by innovation, while other studies suggest that performance can be triggered by co-creation activities between a company and a customer or a partner. Therefore, this theoretical gap is the underlying factor that has encouraged the authors to learn more about the relevance and influence of co-creation, innovation, and firm performance on the creative industries in Indonesia, especially hijab as a subsector of fashion that is currently growing rapidly in Indonesia. The purpose of this research is to gain deeper insights into the interrelationship between them in the creative industries in Indonesia.

2 METHOD

The data in this research were collected through surveys examining

three objects of the research: co-creation, soft innovation, and firm performance on the subject of the hijab creative industry as a subsector of fashion in Bandung. The authors used random sampling of 30 hijab creative industries in Bandung by distributing questionnaires. The questionnaire was divided into two: those that explored the demographic data of the unit of analysis and questionnaires related to the data of the research object using a numerical scale of seven points by using the bipolar adjective on both ends (Sekaran & Bougie 2009). In processing and analyzing the data, the authors used descriptive statistics methods to identify some specific characteristics of observed units of analysis, while methods of inferencing statistics used path analysis in order to answer some of the hypotheses proposed in this study. The statistical tool used was SPSS 22.

3 RESULTS AND DISCUSSION

The data collected from 30 hijab fashion creative industries in Bandung resulted from 13% medium-scale hijab fashion units, 47% small scale, and 40% microscale. The respondent position representing each business unit comprised 87% as the owner, 10% as manager, and 3% as the management team in the business unit surveyed. Of the 30 units of business, 80% conducted online marketing, while 20% did not. Distribution of marketing reaching 30 business units consists of 17% on a local scale (Bandung and surrounding

area), 77% on a national scale, and 7% on an international scale. Demographic data are presented in Figure 1.

Based on the results of validity and reliability testing, it was found that there were 8 indicators of co-creation, 12 indicators of soft innovation, and 2 indicators of firm performance that were valid and reliable. The variables, dimensions, and indicators that were successfully identified as valid and reliable, and then further processed in this study in order to obtain research objectives, are presented in Table 1.

Furthermore, the classical assumption test was carried out in order to find out the data characteristics of on each variable. It was found that the value of asymp. Sig. from one sample Kolmogorov Smirnov test on co-creation and soft innovation variables was >0.05, so it was concluded that the data in the two variables were normally distributed, ex-cept for the data on the firm performance variable due to the asymp value. The Sig. one sample Kolmogorov Smirnov test was 0.010 (Asymp. Sig < 0.05). The sig. of the chi-square test in Table 2 is 1.000 and 0.857 for each co-creation and soft innovation variable. It means that all of the data collected are homogeneous, except for the data on heterogeneous firm

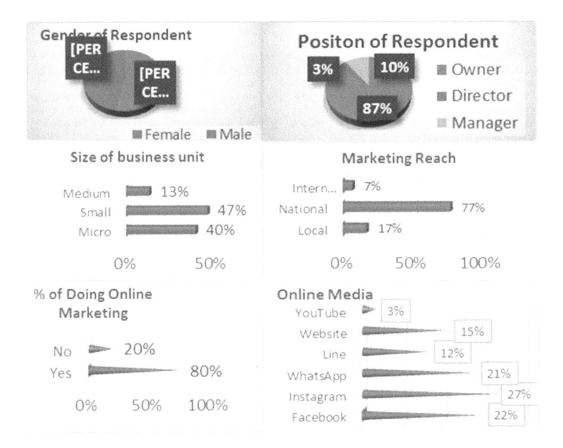

Figure 1 Demographic data

Table 1. Validity and reliability results.

	Validity test results	
CO-CREATION	SUB	Product development ideas from customers
		Product development ideas from partners
	CD	Product innovation designed with customers
		Product innovation designed with partners
	TI	Product innovation modified with customers
		Product innovation modified with partners
		Organizational development enhances employee innovativeness
	NI	Companies are active in community with fashion entrepreneurs
		Companies actively interact with the customer community
	LE	High value products come from local resources
	CO	Product innovation as a result of collaboration with customers
		Product innovation as a result of collaboration with partners
SOFT INNOVATION	TPP	Company's product innovation > competitor's
		Company's process innovation > competitor's
	AES	The company's ability in offering aesthetics of products
	IP	Ability to create intellectual property products
	ORG	Organizational development enhances employee creativity
		The unique product/hard to imitate comes from local resources
		High-value products come from local cultural
		The unique product/hard to imitate comes from local culture
	NFP	Percentage of successful new product
	FP	**Sales growth**
	Reliability test result	
	CC	Cronbach's alpha = 0.941
	SI	Cronbach's alpha = 0.877
	FP	Cronbach's alpha = 0.790

Table 2. One-sample Kolmogorov–SmirnovS and chi-square.

	cc avg	si avg	fb avg
Kolmogorov-Smirnov Z	0.502	0.769	1.632
Asymp. Sig. (2-tailed)	0.963	0.595	0.010
Chi-Square	4.200	11.800	11.067
Df	18	18	3
Asymp. Sig.	1.000	0.857	0.011

Source: The results of processed research, 2017.

performance variables due to asymp values. Sig. chi-square test value is 0.011 (Asymp Sig < 0.05).

The Durbin–Watson value on path analysis test is 2.000 and VIF value is 1.153, which proved that the data on co-creation and soft innovation variables obtained from 30 units of Islamic fashion creative business were free from multicollinearity and autocorrelation. Judging from the preceding description, it can be concluded that the quality of the data is qualified to be used as the research data and can be further processed in the effort to obtain research objectives. The results of path analysis are presented in Table 3.

Table 3 shows that co-creation did not have a significant impact on the performance of the creative

Table 3. Path analysis results.

	Path coef.	t_{count}	p-value	
Direct effect				
Cc to SI	0.364	2.066	0.048	Sig.
CC to FP	0.249	1.494	0.147	Not sig.
SI to FP	0.454	2.728	0.011	Sig.
Indirect effect				
CC to FP with SI as mediator	0.165			
Total effect				
CC to FP	0.414			
SI to FP	0.454			

Source: Data processing using SPSS 22.

business in Bandung. This was contrary to the results of previous research stating that co-creation could play an important role in improving company performance (Prahalad & Ramaswamy 2000; Von Hippel & Katz 2002; Lakhani & Wolf 2005; Von Hippel 2005; Grewal et al. 2006; Shah 2006). This was likely because Islamic fashion business players had not fully comprehended the meaning, the process (including submitting, tinkering, co-designing, and collaboration), and the output that can be generated from the practice of co-creation between the industry players and customers or industry partners, as this theory is still relatively new to them.

Another result showed that co-creation had a significant effect on the emergence of soft innovation in the business by as much as 36.4%. This was in line with previous research confirming that intensive company co-creation was expected to improve the company's capability to innovate(Sharma et al. 2002; Von Hippel & Katz 2002; Von Krogh et al. 2003; Leadbeater & Miller 2004; Prahalad & Ramaswamy 2004; Vargo & Lusch 2004; Evans & Wolf 2005; Grewal et al. 2006; Huston & Sakkab 2006; Pitt 2006; Shar 2006; Etgar 2008;;, Malhotra 2010; Parmentier & Mangematin 2014; Anningdorson 2017).

The soft innovation ability in influencing the performance of Islamic fashion creative business units in Bandung is directly, positively, and significantly able to contribute 45.4% in strengthening the results of previous research. It confirmed that continuous soft innovation could improve firm performance (Marques & Ferreira 2009; Stoneman 2010; Parkman et al. 2012; Kemenparekraf 2014; Shan et al. 2014; Utami & Lantu 2014; Batsakis & Mohr 2016; Zhang et al. 2016; Anningdorson 2017; Jogaratnam 2017). In addition, soft innovation also makes it possible for co-creation to influence the performance of creative business units of hijab fashion, enabling it to increase by 41.4% (from 24.9%). This clearly proves that soft innovation is a good mediator for co-creation in improving the performance of creative businesses in Bandung.

Furthermore, the authors also explored the factors that were considered able to improve the performance of the creative business units in the fashion sector in Bandung. Based on the chi-square test on the crosstab menu in SPSS, it can be said that the performance of hijab creative business units in creating a successful new product in the market was influenced by partnership with private companies or private sectors that accommodated it. In contrast, the performance was not the same when the units formed a partnership with the government. The results of chi-square tests are shown in Table 4.

From Table 4, it can be seen that the p-value of the creative business unit partnership with the private sector was 0.049 ($p < 0.05$). Thus, it indicated that partnerships with private parties had a positive relationship with creation of new products that were successful in the market. Some forms of partnership that have been done by several units of the creative business with private parties are in the form of business training, product marketing assistance, expansion of business network, capital/loan assistance, raw material procurement, and machinery/equipment procurement. Conversely, partnerships with the government had a p-value of 0.083 ($p- > 0.05$), which meant that they had no positive relationship with the creation of a successful new product in the market.

A further result obtained was the success of new product creation in the market supported (having high relation with) the increase of sales growth from each creative business unit of hijab that was analyzed in this research. This is the evident from the results of crosstab testing, as outlined in Tables 5 and 6.

4 CONCLUSION

The results confirmed that the improvement of creative business performance in Indonesia, especially in the hijab creative industries as a subsector of fashion in Bandung, was strongly supported by innovation in development that was devoted to soft aspects (soft innovation), such as the aesthetics of a product that appeals to customers because it comes from high

Table 4. Partnership with the private sector and government: Successful new product cross tabulation.

		Succesful new product				Chi-square tests	
		21%-40%	41%-60%	61%-80%	Total	Value	Sig.
Partnership with the private sector	Y	2	8	5	15		
	N	6	2	7	15		
Total		8	10	12	30		
Pearson chi-square						5.933[a]	0.049
Partnership with government	Y	3	5	6	14		
	N	5	5	6	16		
Total		8	10	12	30		
Pearson chi-square						0.368[b]	0.832

Source: Data processing using SPSS 22.

Table 5. Successful new product × sales growth cross tabulation.

		Sales Growth				
		21%-40%	41%-60%	61%-80%	81%-100%	Total
Successful	21%-40%	5	3	0	0	8
new product	41%-60%	1	3	4	2	10
	61%-80%	0	2	7	3	12
Total		6	8	11	5	30

Source: Data processed with SPSS 22.

Table 6. Chi-square tests.

	Value	df	Asymp. sig.
Pearson chi-square	17.369[a]	6	0.008
Likelihood ratio	21.239	6	0.002
Linear-by-linear association	12.928	1	0.000
N of valid cases	30		

Source: Data processed with SPSS 22.

local values derived from Indonesia's natural wealth and cultural richness, as well as the distinctiveness of Indonesian heritage that makes it difficult to replicate. Soft innovation was able to directly and significantly (45.4%) improve the performance of creative efforts in Indonesia. This is in line with the results of previous studies that confirmed that soft innovation was continuously improving firm performance (Marques & Ferreira 2009; Stoneman 2010; Parkman et al. 2012;Kemenparekraf 2014; Shan et al. 2014; Utami & Lantu 2014; Batsakis & Mohr 2016; Zhang et al. 2016; Anningdorson 2017; Jogaratnam 2017).

Another result was that co-creation as a new paradigm of product development was not well adopted by Indonesian creative business actors. The practice of submitting, co-designing, tinkering, and collaborating, which in fact is the dimension of co-creation, had not been fully understood by the hijab fashion creative industry players in Indonesia. This resulted in the implementation of co-creation that has not been able to directly improve the performance of creative business in Indonesia.

However, co-creation was able to influence soft innovation (36.4%) positively and significantly, which in turn had a positive implication in improving the performance of the creative industry of hijab fashion in Indonesia. In addition, the research revealed another factor that was considered to contribute positively to the improvement of performance such as partnership between creative business actors and private sectors. On the other hand, partnerships between business actors and the government did not show an actual contribution in improving the performance of hijab fashion as a creative industry in Indonesia.

REFERENCES

Anning-dorson, T. 2017. Customer involvement capability and service firm performance: The mediating role of innovation. *Journal of Business Research* 86: 269–280.

Batsakis, G. & Mohr, A.T. 2016. Revisiting the relationship between product diversification and internationalization process in the context of emerging market MNEs. *Journal of World Business* 52(4): 564–577.

Etgar, M. 2008. A descriptive model of the consumer co-production process. *Journal of the Academy of Marketing Science* 36(1): 97–108.

Evans, P. & Wolf, B. 2005. Collaboration rules. *IEEE Engineering Management Review* 33(4): 50.

Grewal, R., Lilien, G.L. & Mallapragada, G. 2006. Location, location, location: How network embeddedness affects project success in open source systems. *Management Science* 52(7): 1043–1056.

Huston, L. & Sakkab, N. 2006. Connect and develop inside Procter & Gamble's new model for innovation. *Harvard Business Review* 84(3): 58–67.

Jogaratnam, G. 2017. How organizational culture influences market orientation and business performance in the restaurant industry. *Journal of Hospitality and Tourism Management* 31: 211–219.

Kaplan, R.S. & Norton, D.P. 1991. The balanced scorecard–measures that drive performance the balanced scorecard-measures. *Harvard Business Review* 1: 1–11.

Kemenparekraf R.I. 2014. *Ekonomi Kreatif: Kekuatan Baru Indonesia menuju 2025*. Jakarta: Kementerian Pariwisata dan Ekonomi Kreatif.

Lakhani, K.R. & Wolf, R.G. 2005. Why hackers do what they do: Understanding motivation and effort in free/open source software projects. *Perspectives on Free and Open Source Software* 1–27.

Leadbeater, C. & Miller, P. 2004. The proam revolution: How enthusiasts are changing our economy and society. *Demos* 4(1): 5–74.

Malhotra, N.K. 2010. *Review of marketing research*, 6th ed. New York: M.E. Sharpe.

Marques, C.S. & Ferreira, J. 2009. SME innovative capacity, competitive advantage and performance in a "traditional" industrial region of Portugal. *Journal of Technology Management and Innovation* 4(4): 53–68.

Parkman, I.D., Holloway, S.S. & Sebastiao, H. 2012. Creative industries: Aligning entrepreneurial orientation and innovation capacity. *Journal of Research in Marketing and Entrepreneurship* 14(1): 95–114.

Parmentier, G. & Mangematin, V. 2014. Orchestrating innovation with user communities in the creative industries. *Technological Forecasting and Social Change* 83(1): 40–53.

Pitt, L.F. 2006. The penguin's window: Corporate brands from an open-source perspective. *Journal of the Academy of Marketing Science* 34(2): 115–127.

Prahalad, C.K. & Ramaswamy, V. 2000. Co-opting customer competence. *Harvard Business Review* 78(1): 79–90.

Prahalad, C.K. & Ramaswamy, V. 2004. Co-creation experiences: The next practice in value creation. *Journal of Interactive Marketing* 18(3): 5–14.

Sekaran, U. & Bougie, R. 2009. *Research methods for business: A skill building approach*, 5th ed. Hoboken, NJ: John Wiley & Sons.

Shah, S.K. 2006. Motivation, governance, and the viability of hybrid forms in open source software development. *Management Science* 52(7): 1000–1014.

Shan, P., Song, M. & Ju, X. 2014. Entrepreneurial orientation and performance: Is innovation speed a missing link? *Journal of Business Research* 69(2): 683–690.

Sharma, S., Sugumaran, V. & Rajagopalan, B. 2002. A framework for creating hybrid–open source software communities. *Information Systems Journal* 12: 7–25.

Stoneman, P. 2010. *Soft innovation: Economics, product aesthetics and the creative industries*. Oxford: Oxford University Press.

Utami, R.M. & Lantu, D.C. 2014. Development competitiveness model for small-medium enterprises among the creative industry in Bandung. *Procedia: Social and Behavioral Sciences* 115: 305–323.

Vargo, S.L. & Lusch, R.F. 2004. Evolving to a new dominant logic for marketing. *Journal of Marketing* 68(1): 1–17.

Von Hippel, E. 2005. *Democratizing innovation*. Cambridge, MA: MIT Press.

Von Hippel, E. & Katz, R. 2002. Shifting innovation to users via toolkits. *Management Science* 48(7): 821–833.

Von Krogh, G., Spaeth, S. & Lakhani, K.R. 2003. Community, joining, and specialization in open source software innovation: A case study. *Research Policy* 32(7): 1217–1241.

Wheelen, T.L. & Hunger, J.D. 2015. *Strategic management and business policy globalization, innovation and sustainability*. Upper Saddle River, NJ: Pearson.

Zhang, J.A., Edgar, F., Geare, A. & O'Kane, C. 2016. The interactive effects of entrepreneurial orientation and capability-based HRM on firm performance: The mediating role of innovation ambidexterity. *Industrial Marketing Management* 59: 131–143.

Advances in Business, Management and Entrepreneurship – Hurriyati et al (eds)
© *2020 Taylor & Francis Group, London, ISBN 978-0-367-27176-3*

Identifying a defect's cause using the Six Sigma method and designing an improvement: A case study

D.I. Nisa & F. Wurjaningrum
Universitas Airlangga, Surabaya, Indonesia

ABSTRACT: The purpose of this study was to examine the process of producing bottled mineral water, to detect the cause of a production defect, and to propose an improved production design using the Six Sigma method. This study used a descriptive qualitative approach with several steps in collecting both primary and secondary data. The result of the research was that in the production process of bottled mineral water, three types of defect emerged that became critical to quality. Based on a Pareto diagram, the most common cause of defects was established, and a cause–effect diagram was employed to determine the cause's factors. Furthermore, an FMEA analysis was used to look for the source of the defect and to determine which improvement should be prioritized in order to reduce the failure. The company management is expected to implement the improvement design in order to achieve zero defects in the production process.

1 INTRODUCTION

Good quality results from a good production process in accordance with quality standards that have been determined by the company based on the wants and needs of the market (Crosby 1979). Schultz et al. (1998) defined quality as fitness for use, or suitability across functions and needs. In quality, two important things must be considered: First, does the product suit consumers' needs and give them satisfaction? Second, is the product free from errors or defects? Every company needs to apply quality control that reduces any defects that have occurred in its products. A defective product is a failure in the production process, and it will affect the company both in rework and in financial costs. Therefore, the company must be able to produce its product with the smallest possible defect (Gasperz & Vincent 2007).

The purpose of this research was to determine the cause of defects in bottled water production and packaging at PT. Surya Prima Cipta Mandiri, Madiun, East Java, Indonesia, using the DMAIC (Define, Measure, Analyze, Improve, Control) Six Sigma method, and to find the best solution.

2 METHOD

The research method used was a qualitative approach. Researchers triangulated sources to compare the data obtained between one source and another and to match them so that the data were qualified and reliable. Data analysis, according to Moelong and Lexi (2007), is a process of sequencing data and organizing those data into a pattern, category, or basic description unit. The unit or category gives significant meaning to

the analysis, explains the description pattern, and establishes the relationship between the description dimensions. Data analysis is used to organize data, which consist of documents, field records, photos, reports, and other information sources.

The Six Sigma method can be utilized to identify the cause of defects in production processes. Six Sigma is a tool used to perform quality control by determining the level of disability so that steps to improvement can be formulated. Six Sigma employs data analysis to achieve a defect-free production process and to reduce variation. Six Sigma corrects the problem by focusing on its cause. Since its introduction by Motorola in the 1980s, Six Sigma has been widely adopted by various companies to improve product quality (Gasperz & Vincent 2007). Six Sigma improves product quality by reducing a product's defects through five stages: define, measure, analyze, improve, and control (Prashar 2014). This research was conducted from December 2016 to June 2017 at PT. Surya Prima Cipta Mandiri.

3 RESULTS AND DISCUSSION

In order to identify the occurrence of defects in its product and to glean suggestions for improvement on its bottled drinking water production process, PT. Surya Prima Cipta Mandiri used the Six Sigma concept, which applies the DMAIC method. The steps taken to get the results of a Six Sigma analysis are as follows.

3.1 *Define*

The initial stage is to clearly identify the problem by creating a project statement, identifying the process to be improved via a SIPOC (Supplier, Input, Process, Output, Customer) diagram, and knowing the CTQ (Critical to Quality) of the product or process that needs improvement. The CTQ needs to be defined based on the input of the customer concerning the desired quality of the product. The focus of this research was to reduce defects in the packaging of bottled drinking water products. This is because the quality of the water produced by this company has reached the standard specified by the health body on Drinking Water in Packaging. To ensure that the water produced meets the feasibility standard as bottled drinking water, the company tests in pH, TDS (Total Dissolved Solid), or so-called water-soluble substances. The standard for pH of the finished product is 7.2 and the standard for TDS is 0.6–0.20. This test is performed in the laboratory to make absolutely sure that the water produced is hygienic, healthy, and safe for human consumption. The next step after the project statement is to identify the key processes of the company. This is usually illustrated using SIPOC diagrams. Based on the SIPOC diagram in the previous stage, there are three potential CTQs on bottled drinking water production: (a) The cup lid is perfectly closed (either no leakage is visible or the leak is very small). (b) The product has a standard shape. (c) The product maintains a clean image (e.g., the product is not exposed to ink so the cup lid is clean and the image is clearly visible). The production data and defect data discussed in this analysis are the production and defect data on good glass products of 220 ml and 120 ml. For example, in January 2016, the number of bottled water products produced was 6,498 pieces. From the total number of 4,998 glasses identified among those products, 309 pieces were defective.

3.2 Measure

The second stage is to measure the level of Defect per Million Opportunities (DPMO) and sigma capabilities and to collect some baseline information from the product or process. The performance of bottled drinking water processes in December 2016 was selected as a baseline performance with a sigma capability of 3.65 and a DPMO of 15,851. In January, there was a decrease in sigma capability (3.65 sigma to 2.98 sigma), and in February, there was an incremental increase from the baseline performance (3.65 sigma to 3.79 sigma). March saw a decline in returns (3.65 sigma to 3.16 sigma), and April saw a decrease from the baseline performance (3.65 sigma to 3.08 sigma). In May, again there was a decline (3.65 sigma to 3.17 sigma). Then, in June, there was an increase (3.65 sigma to 3.73 sigma). July brought a decrease from 3.65 sigma to 3.57 sigma. In August, there was an increase (3.65 sigma to 3.92 sigma). In September, again there was a decrease (3.65 sigma to 3.56 sigma). Similarly, October saw a decrease (3.65 sigma to 2.91 sigma), and in November, there was also a decline (3.65

sigma to 2.93 sigma). Performance in the production process always increases and decreases for each month. This affects the capability of Six Sigma, which means that the performance of the drinking water production process in the packaging has not been maximal, as seen from the number of bottled drinking water products that have defects. Therefore, it is necessary to detect the factors causing the failure of drinking water production process in the packaging so that the company can improve its performance and value.

A Pareto diagram is a data analysis that aims to prioritize problems based on the number of events, and it is embodied through a bar graph. This diagram is used to rank the causes of product defects from the smallest to the largest. The results of this analysis showed that the biggest cause of disability (58%) comes from a cup lid that is not closed perfectly, while the second largest (24%) is imperfect products or glasses. The last (18%) is the gross cup lid product. From these disability-level data, the disability to be solved first is the cup lid that is not sealed perfectly.

3.3 Analyze

The third stage is to analyze and look for the root cause of the problem by identifying the dominant defect with the help of the Pareto diagram, identifying the root cause of the problem with cause–effect diagrams, and using the FMEA (Failure Mode Effect Analysis) method to prioritize improvements. In this case, the tool used to find the root cause of the problem was a cause–effect diagram. This diagram identifies the overall cause of failure in a process. In this study, the largest number of defects in products produced during the production process of bottled drinking water was classified into three defect criteria. The identification stage only emphasized the problems that occur in the packaging of cup and cup lid products because many disabilities occur in the raw materials packaging, not in the water produced. The water produced met the specified standard for Drinking Water in Packaging products. A cause–effect diagram facilitated finding the cause of these defects. Some of the defects were caused by material (poor-quality raw materials), by human error (less-conscientious workers, lack of work experience, alternating workers, lower work discipline), by method (no clear standard operating procedure [SOP]), and by machine (less engine maintenance, fewer engine checks, less-stable engines). The FMEA was based on interviews with company production managers and employees. The results of the FMEA revealed that the failure of this imperfect product form has a severity value of six, which means the severity level is moderate. Production can still be maintained at a later stage, but may cause unsatisfied customers. Categories that cause imperfect product forms include materials, human error, method, and machine with an occurrence value of between three and six, which means the frequency of occurrence of these factors is low to moderate. The company's current efforts to detect disability were high, as evidenced by the detection value of three to four. The highest RPN value was a machine factor of 144, and that can be interpreted as meaning that the machine factor is the first priority for improvement. The cause of disability with the highest RPN value requires immediate treatment and affects the subsequent production process. Therefore, it should be prioritized for improvement so as not to cause more defects in the future. The highest RPN value of the two factors causing the defect was caused by the human factor with an RPN value of 522. This shows that the human factor needs to be handled first. The second highest value was caused by the machine factor with an RPN value of 216. The third highest value came from the method factor with an RPN value of 192, and the last RPN value considered here derived from the material factor with an RPN value of 108.

3.4 Improve

The fourth stage is to present proposed improvements obtained from the interpretation of the results of the cause–effect diagram and the FMEA. At this stage, the researcher proposes improvements that the company may apply as a form of quality control. Based on the results of the analysis in the previous stage, the proposed improvements are sorted based on the RPN value as follows.

3.4.1 Man
The solutions provided are: (a) award rewards and bonuses to workers; (b) provide employees opportunities to raise problems that occur during the production process; (c) give instructions before production begins and reviewing the work after production is completed, with the aim that the production process can continue to be reevaluatedslaughtered continuously so that if there is an abnormality, it can be found quickly; (d) require supervisors to conduct strict and continuous monitoring and inspection of the work stations under their responsibility; (e) ensure mechanics sufficiently check any machinery that is easily damaged or needs special care; (f) create SOPs on issues, prevention, and improvement; (g) conduct training to improve the skills of workers; and (h) build good coordination between all parts of the process and the quality control section.

3.4.2 Machine
Treatment includes preventive maintenance and corrective maintenance. Repair solutions that can be implemented include: (a) preventive maintenance – that is, inspection, planning and scheduling, recording, and analysis, as well as training for care personnel; and (b) corrective treatment once damage has occurred.

3.4.3 Method
Currently, some companies operate without being supported with a standard system. Such companies' operations are based largely on habits that have been practiced for years. Employees work by following the habits of previous workers. Companies need to implement SOPs for daily employee performance. The SOPs are a set of standard operating procedures that are used as guidance in the company to ensure that the employees are performing effectively and consistently, and that they are meeting the company's standards and systematics (Tambunan 2013).

The departments within the company use SOPs as guidelines for job implementers, i.e., employees from the lowest level to managers. Managers use SOPs to ensure that everyone follows the same steps each time they perform procedures and that employees understand their tasks and responsibilities. This SOP covers both the work process and the operating procedures of the production machinery.

3.4.4 Material
The solutions provided were (a) selecting suppliers that have good cup or cup lid quality to minimize defects; (b) choosing and buying cups with raw materials above the standard specification (< 135 ml); and (c) establishing the standard criteria of raw materials that the company wants. Standard criteria of raw materials can be established by making a database of suppliers that have a good track record of supplying raw materials to the company, and then evaluating the selected supplier periodically during the year of the contract.

3.5 Control

The fifth stage is to continue to evaluate the results of improvements over time, which can be done by standardizing the process for the company. In the previous stage, an improvement plan was prepared, so at this stage, the focus is to maintain the improvement. Expected countermeasures prevent the same problem from recurring. The steps that must be done at the control stage include: (a) Perform machine checks and maintenance on a regular basis, considering that the machine is old. (b) Monitor employee performance, provide good direction, and conduct regular training so that employees can become skilled in using the machine. (c) Conduct production processes in accordance with existing procedures and evaluate them for efficiency and effectiveness. (d) Ensure Eensure that the quality of raw materials such as the cup and the cup lid is in accordance with the company standard.

4 CONCLUSION

This research examined the disabilities that occur in the process of producing packaged drinking water.

They are: (1) The cup lid is not sealed perfectly (perforated and leaking). (2) The product shape is incomplete (tilted tumbler glass and cup lid). (3) The cup lid appears as a dirty product (the cup lid shows exposed ink and the picture is not clear).

Several methods of analysis revealed that the cup lid does not close perfectly (it is perforated and leaking) and that the product form is not ideal. The disability is due to human factors, machinery, methods, and materials. The managerial implications from this research are: (1) The company needs to do more intensive maintenance of the machines it owns, considering they are quite old. In addition, long-term proposals can be prepared to make new machinery procurement a priority so that in the future, when the demand for products has increased, the company can meet all of its targets. (2) The company may consider the recommendations for improvement proposed by the writer.

ACKNOWLEDGMENTS

This research was supported fully by the Department of Management, Faculty of Economics and Business, Universitas Airlangga Surabaya. We would also like to show our gratitude to the GCBME 2018 committee for comments that greatly improved the manuscript.

REFERENCES

Crosby, F. 1979. Relative deprivation revisited: A response to Miller, Bolce, and Halligan. *American Political Science Review* 73(1): 103–112.

Gasperz, G. & Vincent, V. 2007. *Lean Six Sigma for manufacturing and service industries*. Jakarta: Gramedia Pustaka Utama.

Moelong, M. & Lexy, J. 2007. *Metodologi Penelitian Kualitatif Edisi Revisi. Cetakan Kedua puluh empat*. Bandung: PT Remaja Rosdakarya.

Prashar, A. 2014. Adoption of Six Sigma DMAIC to reduce cost of poor quality. *International Journal of Productivity and Performance Management* 63(1): 103–126.

Schultz, K. L., Juran, D. C., Boudreau, J. W., McClain, J. O., & Thomas, L. J. 1998. Modeling and worker motivation in JIT production systems. *Management Science* 44(12-part-1): 1595–1607.

Tambunan, R. M. 2013. *Standard operating procedures (SOP) edisi 2*. Jakarta: Maeistas Publishing.

Section 5: Organizational behavior, leadership and human

resources management

The impact of the implementation of a quality management system on teacher productivity

T.S. Rahayu & R. Rasto
Universitas Pendidikan Indonesia, Bandung, Indonesia

ABSTRACT: This study aimed to determine the influence of the ISO 9001:2015 quality management system (QMS) on teacher productivity. The research method used was an explanatory survey. Data collection techniques included questionnaire rating scale models. Respondents were 54 teachers at a vocational high school in Bandung. Data were analyzed using regression. The results of the study revealed that the ISO 9001:2015 QMS has had a positive and significant influence on teacher productivity. Therefore, the productivity of teachers can be improved through improving the effectiveness of the ISO 9001:2015 QMS implemented in schools.

1 INTRODUCTION

Teacher productivity is an interesting thing to study. This is inseparable from the position of teachers as education personnel (Suhardan 2006). Teachers are the determinants of the quality of education (Broder & Dorfman 1994; Kárpáti 2009; Akareem & Hossain 2016) and play a role in ensuring the development and viability of a nation (Juangsih 2014). Education has the main task of preparing human resources as a foundation for the advancement of a nation, which certainly requires a variety of mutually supportive and sustainable resources, especially teachers (Sugiarti & Rasto 2017).

Teachers are at the forefront of education (Nakpodia 2011; Okorie 2016), which plays an important role in producing qualified graduates, and in turn contributes significantly to the achievement of organizational goals (Siburian 2013), and to the ability of national education to carry out quality teaching (Ndugu 2014). Therefore, educational institutions should make quality plans based on national and international standards (Anwar & Yusuf 2008). This is in line with the government's policy that every educational unit across formal and informal channels is obliged to ensure education quality (Anwar & Yusuf 2008).

Teachers have several duties, such as to educate, teach, guide, direct, train, assess, and evaluate students (Shabir 2009). As professional educators, teachers must always increase their productivity (Sutikno 2009) in order to build quality educational institutions (Agustina 2012).

The results of preliminary studies have indicated that teacher productivity is not yet optimal. The immediate question is, "Why is teachers' productivity not yet optimal?" Referring to the perspective of behavior theory, many factors can affect teacher productivity. The management system applied in school is one factor that allegedly influences teacher productivity.

Based on this, the problem statement of this research is, "Does the ISO 9001:2015 quality management system influence teacher productivity?"

1.1 Quality management system

The QMS is one of the constructs that affect teacher productivity. The ISO 9001:2015 QMS is described as a basic international standard (Kaziliūnas 2010) management system (Paunescu 2002) used to ensure conformity and quality control (El-Morsy 2014) of processes and products (Hatane & Zulkarnain 2016).

The implementation of the QMS provides efficiency and effectiveness within an organization (Donald et al. 2015), thus making a difference in quality (Heras, Casadesús, & Dick 2002; Donald et al. 2015). Quality improvement is needed to continuously enrich all activities in the education process (Dahil & Karabulut 2013) because quality improvement requires a process (Tigani 2011).

The ISO 9001:2015 QMS is appropriate for schools because it motivates teachers to increase their productivity (Sedarmayanti 2001), due to a clear explanation of the tasks of each element in the school (Supriyadi 2012).

1.2 Teacher productivity

Productivity is a construct that shows a commitment to achieve professional goals (Nakpodia 2011). The concept of productivity evolves from a technical sense to behavior (Tjutju & Suwatno 2013). Productivity in the technical sense refers to the degree of effectiveness and efficiency in the use of various resources, whereas in the sense of behavior, productivity is a mental attitude that constantly strives to grow.

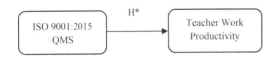

Figure 1. Theoretical framework.

Note: *H = influence of the ISO 9001:2015 QMS on teacher work productivity.

In simple terms, productivity is the ratio of output to input (Daft 2003; Darra 2006; McLaughlin & Coffey 2006; Hanushek & Ettema 2017) – how efficiently a set of resources is used to achieve a particular set of goals (Okorie 2016). Teacher productivity is the result of work on what the teacher does (Wenglinsky 2001), reflected in planning, implementing, and assessing the learning process, which is based on work ethic and on professional discipline in the learning process (Sutikno 2009).

Teacher productivity can be seen from the performance of teachers (Ndugu 2014), which is a manifestation of the understanding and application of teacher competency (Sutikno 2011) and the output of teacher tasks that are contained in the main tasks and functions of teachers (Sutikno 2009, 2013). Teachers with high productivity can better perform their various duties and functions (Peretomode & Chukwuma 2012; Omebe 2001) so as to provide quality education services (Ahmad, Rasi, & Zakuan 2017) and advance the school (Shamaki 2015) in producing qualified graduates.

Factors affecting productivity include wages, education (Ndugu 2014), management (Sedarmayanti 2001; Sinungan 2008; Ndugu 2014), work ability, work discipline (Tanto et al. 2012), mental attitude, nutrition and health, social security, environment and work climate, production facilities, technology, achievement opportunities, professional training, and educational infrastructure (Ndugu 2014).

The theoretical framework is as shown in Figure 1.

2 METHOD

This research used an explanatory survey method. This method is considered appropriate for this study to gather factual information through questionnaires. Respondents were 54 teachers from a vocational high school in Bandung, West Java, Indonesia.

Descriptive statistics used the average score, which was used to obtain a description of the level of respondents' perception of the implementation of the quality management system (QMS) and teacher productivity. Inferential statistics used regression analysis to test the hypothesis.

3 RESULTS AND DISCUSSION

3.1 Implementation of ISO 9001:2015 QMS

The ISO 9001:2015 QMS can be measured from five indicators (ISO, 2007). First, the development of systems related to education design in the planning and implementation of the QMS, teachers' understanding of and involvement in planning the QMS, and document control by the school personnel are important to run the system and procedures well. Second is the responsibilities related to the commitment of top school management to make continual improvement in order to provide quality education services. Third is the management of resources related to the provision of educational services in order to ensure the availability of resources that the QMS might require. Fourth is the realization of services that educational organizations should plan, including design and development of teaching methods, assessment of learning and follow-up, support of service activities, resource allocation, and evaluation criteria. Fifth is the measurement, analysis, and improvement related to the school's ability to take corrective and preventive action for continuous improvement.

Implementation of the ISO 9001:2015 QMS, according to the perception of respondents, is in the effective category, as indicated by the average score of the respondents' answers of 3.58. Table 1 presents the average scores of each indicator as the measure of the ISO 9001:2015 QMS.

The highest score is on the indicator of educational service realization. This indicates that the teacher constantly strives to undertake education services in accordance with the system and available provisions. The lowest indicator is on the development of systems with categories of understanding and the involvement of teachers in system development and planning

3.2 Teacher productivity

Teachers who have high productivity have six indicators (Mulyasa 2013). First, a productive teacher must plan and implement the learning process, which means that the teacher must arrange something that will be implemented in order to achieve predetermined goals,

Table 1. Implementation of the ISO 9001:2015 QMS.

Indicators	Average	Category
System development	3.25	Almost effective
Management responsibilities	3.65	Effective
Resource management	3.63	Effective
Education service realization	3.73	Effective
Measurement, analysis, and improvement	3.66	Effective
Average	3.58	Effective

in accordance with educational targets. Teachers making learning plans should be able to arrange various teaching programs according to the approach and method that will be used. Carrying out the process of learning means that teachers interact with teaching and learning through the application of various strategies and learning techniques, as well as the use of a set of media. Second, a productive teacher must assess the learning outcomes and perform measurements as well as assessment, processing, and interpretation in order to make decisions about the learning outcomes achieved by students and to analyze students' weaknesses and strengths. Third, a productive teacher provides mentoring and training, because the teacher has the duty, the responsibility, and the authority to give guidance and counseling services to students who have problems. Coaching can also be given to colleagues, especially beginner teachers. In addition, teachers should participate in activities or training to develop their skills and knowledge. Fourth, a productive teacher does research in order to advance the profession. Fifth, a productive teacher assists in the development of school programs and should be actively involved in helping schools implement such programs. Sixth, productive teachers focus on professional development.

Teacher productivity, according to the perception of respondents, is in the high category, as indicated by the average score of 3.74. Table 2 presents the average scores of each indicator that measures the productivity of teachers.

The highest score is in the indicator of planning and implementing the learning process. This indicates that teachers can work well in planning and implementing the learning process in schools. Assessing learning outcomes, coaching and training, helping to develop and manage school programs, and developing professionalism are in the high category with lower scores. The lowest score is the indicator to conduct research.

3.3 Influence of ISO 9001:2015 QMS on teacher productivity

The linear regression equation shows that the effect of the ISO 9001: 2015 on teacher productivity is

Table 2. Teacher work productivity.

Indicators	Average	Category
Planning and implementing learning process	3.92	High
Assessing learning outcomes	3.87	High
Conducting mentoring and training	3.66	High
Conducting research	3.42	High
Helping with the development and management of school programs	3.69	High
Developing professionalism	3.86	High
Average	3.74	High

$\hat{Y} = 83,071 + 0,272X$. The positive sign (+) indicates the relationship between variables goes one way, meaning the more effective the ISO 9001: 2015 QMS is, the higher the productivity of teachers. A hypothesis test shows that the t-stat is bigger than the t-table value (4,134 > 4,027), with df1 = 1, df2 = 2 = n−2, and α = 0.05. Thus, the ISO 9001: 2015 QMS has a significant effect on teacher productivity. Based on the coefficient of determination, the effect of the ISO 9001: 2015 QMS on teacher productivity is 7%.

In line with this research, several previous theories and research have stated that the ISO 9001: 2015 QMS has many advantages such as competitiveness, consistency in providing quality service, productivity improvement, work spirit, and job satisfaction (Kumar & Balakrishnan 2011). The QMS is one of the factors affecting productivity (Donald et al. 2015); because the QMS provides a clear explanation of work procedures and responsibilities (Donald et al. 2015), it can ensure consistency in the quality of service (Rahman, Mohd Rahim, & Mahyuddin 2014).

A quality-oriented work system (Zahari & Zakuan 2016) and consistent systematic management (Khan 2015) indirectly encourage school personnel to make changes to improve performance (Kanji & Yui 1997; Albulescu, Drăghici, Fistiş, & Truşculescu 2016). More optimal teacher performance can lead to continual improvement (Sucipto 2017) in the teaching and learning process that produces qualified graduates. Thus, the implementation of the ISO 9001: 2015 QMS empirically can improve organizational performance (Hendricks & Singhal 1996) and have a significant positive effect on productivity (Zulfah, Suwandono, & Luthfianto 2004; Indraswari, 2007).

4 CONCLUSION

The ISO 9001:2015 QMS that includes system development, management responsibilities, resource management, the realization of educational services and measurement, analysis, and improvement is in the effective category. Teacher productivity that includes planning and implementing learning, assessing learning outcomes, conducting mentoring and training, conducting research, helping in the development and management of school programs, and developing professionalism is in the high category. The ISO 9001:2015 QMS has a positive and significant effect on teacher productivity. This shows that the ISO 9001:2015 QMS is a predictor of teacher productivity improvement. Thus, the ISO 9001:2015 QMS should be improved in order to increase teacher productivity.

REFERENCES

Agustina, B. D. & D. A. 2012. Guru Profesional Sebagai Faktor Penentu Pendidikan Bermutu.

Ahmad, M., Rasi, R., & Zakuan, N. 2017. Effect of total quality management on the quality and productivity of human resources.

Akareem, H. S. & Hossain, S. S. 2016. Determinants of education quality: What makes students' perception different? *Open Review of Educational Research* 3(1): 52–67.

Albulescu, C. T., Drăghici, A., Fistiş, G. M., & Truşculescu, A. 2016. Does ISO 9001 quality certification influence labor productivity in EU-27? *Procedia: Social and Behavioral Sciences* 221: 278–286.

Anwar, M. A. & Yusuf, M. 2008. Pengaruh Sistem Manajemen Mutu ISO 9001:2008 dan Kompensasi Terhadap Kinerja Guru di SMA Darul Ulum 2 Unggulan BPPT Jombang. *Jurnal Manajemen dan Pendidikan Islam* 3(1): 17–38.

Broder, J. M. & Dorfman, J. H. 1994. Determinants of teaching quality: What's important to students? *Research in Higher Education* 35(2): 235–249.

Daft, R. L. 2003. *Manajemen Sumber Daya Manusia*. Jakarta: Erlangga.

Dahil, L. & Karabulut, A. 2013. Effects of total quality management on teachers and students. *Procedia: Social and Behavioral Sciences* 106: 1021–1030.

Darra, M. 2006. Productivity improvements in education: A replay. *European Research Studies* IX.

Donald, G. I., Otieno, O. C., Obura, J. M., Abong, B., & Ondoro, C. 2015. Effect of implementing quality management system on the performance of public universities in Kenya: A case of Maseno University, Kenya. *American Journal of Business, Economics and Management* 3(3): 145–151.

El-Morsy, G. S. 2014. Implementation of quality management system by utilizing ISO 9001:2008 model in the emerging faculties. *Life Science Journal* 11(8): 81–84.

Hanushek, E. A. & Ettema, E. 2017. Defining productivity in education: Issues and illustrations. *American Economist* 62(2): 165–183.

Hatane, S. & Zulkarnain, J. 2016. Pengaruh Sistem Manajemen Mutu ISO Terhadap Kinerja Karyawan Melalui Budaya Kualitas Perusahaan (Studi Kasus PT Otsuka Indonesia Malang), 3.

Hendricks, K. B. & Singhal, V. R. 1996. Quality awards and the market value of the firm: An empirical investigation. *Management Science* 42: 415–436.

Heras, I., Casadesús, M., & Dick, G. P. M. 2002. ISO 9000 certification and the bottom line: A comparative study of the profitability of Basque region companies. *Managerial Auditing Journal* 17(1/2): 72–78.

Indraswari. 2007. Pengaruh Penerapan ISO 9001: 2000 Terhadap Produktivitas Kerja Karyawan PTPN VIII Gunung Mas Bogor.

ISO. 2007. International Workshop Agreement (IWA) 2: Guidelines for the application of quality management systems.

Juangsih, J. 2014. Peran lptk dalam menghasilkan guru yang profesional. *Wahana Didaktika* 12(2): 72–83.

Kanji, G. K. & Yui, H. 1997. Total quality culture. *Total Quality Management* 8(6): 417–428.

Kárpáti, A. 2009. Teacher training and professional development. In *Green book: For the renewal of Public education in Hungary*, 203–226.

Kaziliūnas, A. 2010. The implementation of quality management systems in service organizations. *Public Policy and Administration* 2603(34): 71–82.

Khan, J. H. 2015. Impact of total quality management on productivity (December 2003).

Kumar, D. A. & Balakrishnan, V. 2011. A study on ISO 9001 quality management system certifications: Reasons behind the failure of ISO certified organizations. 11(9).

McLaughlin, C. P. & Coffey, S. 2006. Measuring productivity in services. *International Journal of Service Industry Management* 1(1): 46–64.

Mulyasa, E. 2013. *Standar Kompetensi dan Sertifikasi Guru*. Bandung: Remaja Rosdakarya.

Nakpodia, E. D. 2011. Work environment and productivity among primary school teachers in Nigeria. *International Multidisciplinary Journal, Ethiopia* 5(22): 367–381.

Ndugu, M. M. 2014. Quality and productivity of teachers in selected public secondary schools in Kenya. *Mediterranean Journal of Social Sciences* 5(5): 103–116.

Okorie, A. 2016. Teachers' personnel management as determinant of teacher productivity in secondary schools in Delta State, Nigeria. *British Journal of Education* 4 (8): 13–23.

Omebe, B. 2001. Management strategies and secondary school teacher job performance in Akwa Ibo South Senatoria District Asia. *Journal of Management Science and Education* 4(2): 13–20.

Paunescu, C. 2002. *Commitment to quality education services through ISO 9000: A case study of Romania*, 1–15.

Peretomode, V. F. & Chukwuma, R. A. 2012. Manpower development and lecturers' productivity in tertiary institutions in Nigeria. *European Scientific Journal* 8(13): 16–28.

Rahman, H. A., Mohd Rahim, F. A., & Mahyuddin, N. 2014. Implementing a quality management system for built environment programs: University of Malaya's experience. Centre for Project & Facilities Management.

Sugiarti, R. & Rasto, R. 2017. *Keterlibatan Kerja dan Komitmen Organisasi Sebagai Determinan Kepuasan Kerja Guru*.

Sedarmayanti. 2001. *Sumber Daya Manusia dan Produktivitas Kerja*. Bandung: Mandar Maju.

Shamaki, E. B. 2015. Influence of leadership style on teachers' job productivity in public secondary schools in Taraba State, Nigeria. *Journal of Education and Practice* 6(10): 200–204.

Siburian, T. A. 2013. The effect of interpersonal communication, organizational culture, job satisfaction, and achievement motivation to organizational commitment of state high school teacher in the District Humbang Hasundutan, North Sumatera, Indonesia Tiur Asi Siburian. *International Journal of Social Science* 3(12): 247–264.

Sinungan, M. 2008. *Produktivitas Apa dan Bagaimana*. Jakarta: Bumi Aksara.

Sucipto, E. H. 2017. Analisis Implementasi ISO 9001:2008 Terhadap Produktivitas Dan Kepuasan Pelanggan Studi Kasus Manufaktur Kompor Gas, 1.

Suhardan, D. 2006. Pengaruh Kemepimpinan Kepala Sekolah Terhadap Produktivitas Kerja Guru SMP Se-Gugus 08 di Kabupaten Bandung (1).

Supriyadi, E. 2012. *Pengaruh Penerapan Sistem Manajemen Mutu ISO 9001:2008 Terhadap Kinerja Guru di SMK Negeri 1 Sedayu Bantul*. Yogyakarta.

Sutikno, T. A. 2009. Indikator produktivitas kerja guru sekolah menengah kejuruan. 32(1): 107–119.

Sutikno, T. A. 2011. Studi produktivitas kerja guru pada sekolah menengah kejuruan negeri di malang raya. 34(1): 1–12.

Sutikno, T. A. 2013. Pengaruh Persepsi Tentang Sertifikasi Guru, Strategi Penyelesaian Konflik, dan Motivasi Kerja

Terhadap Produktivitas Kerja Guru SMKN. *Cakrawala Pendidikan* (1): 150–160.

Tanto, D., Dewi, S. M., Budio, S. P., Teknik, J., Fakultas, S., & Brawijaya, U. 2012. Faktor-faktor Yang Mempengaruhi Produktivitas Pekerja Pada Pengerjaan Atap Baja Ringan di Perumahan Green Hills Malang. 6 (1): 69–82.

Tigani, O. 2011. The impact of the implementation of the ISO 9000 Quality Management System upon the perception of the performance of the organization's worker. *Global Journal of Management and Business Research* 11(8): 11–22.

Tjutju, Y. & Suwatno. 2013. *Manajemen Sumber Daya Manusia*. Bandung: Alfabeta.

Wenglinsky, H. 2001. Teacher classroom practices and student performance: How schools can make a difference. *Educational Testing Service* (September): 1–17.

Zahari, M. K. & Zakuan, N. 2016. The effect of total quality management on the employee performance in Malaysian manufacturing (October): 1–6.

Zulfah, Suwandono, & Luthfianto, S. 2004. Analisa Penerapan Sistem Manajemen Mutu ISO 9001:2000 Terhadap Kualitas Produk dan Produktivitas Kerja Karyawan di PT Unilon Bandung. *Teknik Industri*.

Advances in Business, Management and Entrepreneurship – Hurriyati et al (eds)
© 2020 Taylor & Francis Group, London, ISBN 978-0-367-27176-3

Development of a human resources management capacity-strengthening model for village government

M.O. Fauzan & D. Disman
Universitas Pendidikan Indonesia, Bandung, Indonesia

ABSTRACT: Every year, the government has budgeted a fund for villages. This fund is used to improve the quality of villagers' life. The data show that this village fund decreased rural equality from 0.34 in 2014 to 0.32 in 2017. With a good fund management system, this achievement is predicted to continually increase in the forthcoming years. The key to success in developing the village is a good relationship between the community in a village and its local government apparatus. The apparatus, here, should have the ability to manage the affairs of the village, including the fund given by the government. The fact shows that most such apparatuses do not have this ability. Therefore, it is necessary to make a capacity-building model for these apparatuses in managing sustainable, transparent, and accountable village funds, so that villagers' welfare can be protected.

1 INTRODUCTION

Law No. 6/2014 has placed the village as the spearhead of the development and the improvement of people's welfare. The village is given authority and resources in order to improve the economy and community welfare. To assist the village, the government has budgeted an annual fund. In 2017, the budget was Rp 60 trillion with an average of Rp 800 million per village (Ministry of Finance 2017).

Based on the evaluation of three years' implementation, the fund is truly used for the development of villages, including building facilities or infrastructure (roads, bridges, drainage and irrigation units, early childhood development programs, markets, etc.). The fund is also used for economic development, including developing tourist areas, fisheries, and livestock, or holding training seminars. As a result, the quality of villagers' life has been improved. The data show that the ratio of rural inequality decreased from 0.34 in 2014 to 0.32 in 2017. The number of poor areas also declined from 17.7 million in 2014 to 17.1 million in 2017. Moreover, poverty dropped from 14.09% in 2015 to 13.93% in 2017. This success is inseparable from the strong initiation, innovation, creation, and cooperation between the apparatus and the community in achieving common goals (Ministry of Finance 2017).

This success spurs on the government to continue the program. The government has even added up to Rp 120 trillion to the fund, twice the previous fund. The focus of this program is creating job opportunities and providing the ability for the village to manage itself (Ministry of Finance 2017). To realize this program, it is necessary to strengthen the institutions and the human resources capacity of a village. This includes the government apparatus and community and village advisors. Furthermore, the transparency, accountability, and supervision of the fund are also needed (Regional 2015).

Human resources are people who work in an organization. They are called employees, personnel, leaders/managers, workers, laborers, employers, and other titles. In educational organizations, human resources are all administrative personnel, educators/teachers, lecturers, and other education staff members (Hasbi 2016). In human resources development, there is also monitoring. Employees' performance will be observed to see the improvements. If there is no improvement in the employees' performance after the introduction of a new skill, guidance or discipline will be conducted (Armstrong 2006).

Human resources development differs from training in terms of objective and scope. It deals not only with strategic focus but also with initiatives (Boxall, Purcell, & Wright 2007). It helps employees to do their jobs more easily. This will definitely contribute to an organization's performance (Mayadi 2016). This is supported by Suryani and Agusdin (2017), who say that organizations need to prepare human resources that can compete in the future through behavior and competence.

2 METHOD

According to the Ministry of Village Development of Disadvantaged Regions and Transmigration, the lack of quality of human resources in rural areas is due to the lack of an educational background. Most laborers are primary school graduates (57.79%). Only 18.87% are junior high school graduates and 13.07% are senior high school graduates. To handle this problem, vocational education can be an option to increase the quality of the labor force (Directorate General of Village Development of Disadvantage Regions).

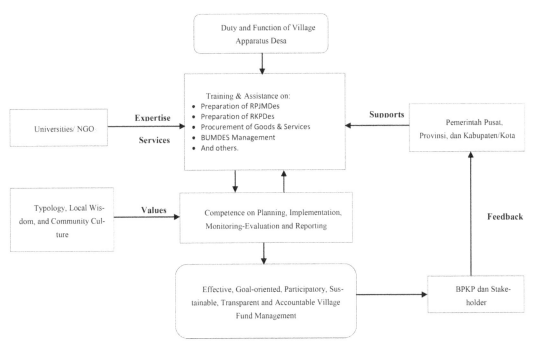

Figure 1. Model of the empowerment of human resources as accomplished through the village apparatus.
(Source: Analysis result)

Regarding this, it is necessary to change the current paradigm that equates a village counselor with a village officer. Thus, village counselors can effectively assist the village apparatus, including helping with financial reports as requested by the mayor or regent (Kemendesa 2015; Zakia & Maksum 2017).

3 RESULTS AND DISCUSSION

The administrative problem can be solved by using information technology as a means of managing village functions (planning, administration, and financial management).

The Finance and Development Supervisory Agency along with MoHA has made an application to solve the village funding issue – namely, the Village Information and Management System (SIMDA). With this application, it is expected that the village government can easily organize village financial management, from the preparation of the income budgeting expenditure of the village, its administration, its implementation, and its accountability. The use of this system, however, will be directly supervised by a finance and development supervisory agency with the role of an internal auditor. Thus, it is recommended that each head office encourage its officers to learn to use this system.

The need to strengthen village human resources is a matter that cannot be negotiated. Strengthening activities should refer to the main tasks and functions set by

regulations. The formed competencies can be translated into a model that can bolster the apparatus' capacity. The model is developed based on the following:

1. Regulation of the minister of the village on the prioritization of village funds usage;
2. Regulation of the Minister of Home Affairs Law No. 84 of 2015 on Organizational Structure and Work Procedure (SOTK) of Village Government'
3. Preparation for the operation of SIMDA;
4. Procedures to confirm the content of village regulations and the mechanisms for enacting them;
5. Procedures for the preparation of accountability reports on the realization of the income budgeting expenditure of the village, village property wealth reports, and monitoring and evaluation to prevent misuse of village funds; and
6. Regulation of the Minister of Home Affairs Law No. 1/2017 on Village Arrangement.

Based on this model, village-level human resources empowerment can focus on coordinating the main tasks and on improving the function of the village apparatus by taking into account the typology, the local wisdom, and the community culture. The model can be applied in the form of leadership training programs, management, and finance. It can cover the following areas:

a. Apply discretion in the management of policies and programs. Enhance the staff's ability to translate, adapt, and implement generic government policies and programs according to local circumstances.

b. Focus on problem-solving and decision-making. Improve the staff's capability to manage day-to-day services and responsibilities; to identify problems, constraints, and challenges; and to respond effectively within a reasonable time.

c. Provide effective coordination. Improve the staff's capabilities in vertical and horizontal coordination with internal and external stakeholders in order to ensure better local service.

d. Provide effective facilitation. Increase facilitation of local stakeholders and their needs, including community empowerment.

e. Provide accountability. Build the staff's capacity in providing accountable financial administration and management.

f. Identify and apply cutting-edge innovations. Improve the staff's ability to identify and create creative solutions through service innovation.

g. Strengthen village planning and budgeting. Improve the staff's capacity to facilitate participatory planning and budgeting processes in coordination with village facilitators based on village mid-term planning (RPJMDes) and district regulations/policies on village laws.

h. Mainstream gender and social inclusion. Increase the capacity of village staff and assistants on gender and inclusion in RPJMDes and village work plans (RKP).

i. Supervise the use of the village fund. Improve the staff's capacity to review the implementation, administration, and financial report on the uses of the village fund, particularly in response to the needs of the poor community.

4 CONCLUSION

Village apparatuses, such as village consultative bodies and village heads, are required to be transparent in the use of funds. The weakness of the village apparatus concerns a lack of ability in planning and implementing. Accountability in finance is low. A competent apparatus that can identify the condition of the village and understand the needs of the community can run the program independently. The training for the village apparatus should be prioritized in line with the increasing village funds. Training activities should focus on improving the capacity to manage the funds. The funds have to be used for the development of village. Village heads and apparatuses are the key to achieving this objective. The training is not the socialization of the government program. It should directly teach how to manage the fund comprehensively.

REFERENCES

Armstrong, M. 2006. *Strategic human resources management. A guide to action.*

Boxall, P. F., Purcell, J., & Wright, P. M. 2007. *The Oxford handbook of human resource management.* Oxford: Oxford University Press, xv.

Daerah, P. K. D. dan O. 2015. Implementasi uu desa dan tantangan pengembangan kapasitas pemerintah desa, 1–4.

DIRJEN-PDT. 2018. Percepatan pembangunan desa.

Hasbi, M. and Y. 2016. Kinerja Guru Aqidah Akhlak, Ski, Al-Qur'an Hadits, Fiqih di Madrasah Tsanawiyah (Mts) Al-Ikhlas Keban Ii Kec. Sanga Desa Kab. Muba. *Journal of Islamic Education Management* (1): 1–7.

Kemendesa. 2014. Sistem keuangan desa (Siskeudes), 85910031(6),2014–2015.

Kemendesa. 2015. SIMDA Desa (Vol. 85910031).

Kemendesa. 2016. *Forum Pertides.* 1st edn. Jakarta: Kementerian Desa, Pembangunan Daerah Tertinggal dan Transmigrasi.

Kementerian Keuangan Republik Indonesia. 2017. Buku Pintar Dana Desa, 125. Retrieved from www.kemenkeu.go.id/media/6749/buku-pintar-dana-desa.pdf.

Kementerian Keuangan RI. 2017. Buku Saku Dana Desa, 1–103.

Mayadi, N. L. 2016. Model Pengembangan Sdm Berbasis Kearifal Lokal Dan Regional (Analisis Kritis Dari Kesiapan Sdm Dalam Menghadapi Masyarakat Ekonomi Asean) Mea, 13(April): 96–117.

Peraturan Menteri Dalam Negeri Nomor 82Tahun 2015 Tentang Pengangkatan dan Pemberhentian Kepala Desa.

Peraturan Menteri Dalam Negeri Nomor 83Tahun 2015 Tentang Pengangkatan dan Pemberhentian Perangkat Desa.

Peraturan Pemerintah Nomor 43 tahun 2014 tentang Peraturan Pelaksanaan Undang-Undang Nomor 6 Tahun 2014 tentang Desa.

Rachmawati, I. K. 2010. *Manajemen Source Daya Manusia. Cetakan Pertama.* Yogyakarta: Penerbit CV. Andi Offset.

Suryani, E. & Agusdin, A. (2017). Analisis Strategi Pengembangan Sumber Daya Manusia Aparatur Pemerintah Kota Mataram dalam Mewujudkan Kota Mataram Sebagai Kota Layak Anak. *Jurnal Magister Manajemen Universitas Mataram.*

Undang-Undang Republik Indonesia Nomor 6 tahun 2014 tentang Desa.

Zakia, Maksum, I. R. 2017. Kapasitas Pemerintahan Desa dalam Menghadapi Implementasi Undang-Undang No. 6 Tahun 2014 Tentang Desa. *Reformasi Administrasi* 4(2).

Advances in Business, Management and Entrepreneurship – Hurriyati et al (eds)
© 2020 Taylor & Francis Group, London, ISBN 978-0-367-27176-3

Millennial generation employment: The impact of millennial characteristics on performance achievement

E. Siahaan
Universitas Sumatera Utara, Medan, Indonesia

ABSTRACT: Employment is undergoing enormous changes across the globe. The presence of the Millennial Generation (Millennials) has greatly altered the working environment. The characteristics of Millennial workers affect the behavior, mind-set, and performance of the workforce. The research examined here aimed at analyzing the impact of Millennial characteristics on performance achievement. The research population comprised Millennial employees from PT. Telkom Regional I and included a total of 306 employees born during the 1980–2000 period. The total sample in this research consisted of 173 employees. The data were analyzed using logistic regression, which can predict the level of performance accurately up to 82.1% based on the baseline performance of the average employee. The results indicated that various Millennial characteristics encouraged improved performance. Millennials' fondness for technology, independence, team-oriented attitude, and desires can lead to higher performance.

1 INTRODUCTION

Performance is a key issue in addressing organizational achievements. An organization's performance is accumulated from the performance of its individual members. This performance is achieved by optimizing individual capabilities. Basically, a company always expects its employees to provide their best performance in order to enhance organizational performance. For this reason, it is important to understand a worker both as an individual and as a part of the organization.

The current workforce is grouped into three main generations: the Baby Boomers, born between 1946 and 1964; Generation X, born between 1965 and 1981; and Millennials – also called Generation Y – born after 1982. Each generation has certain characteristics, traits, and values. The difference is due to evolving circumstances and changing times (Shragay & Tziner 2011). By 2014, Generation Y comprised 36% of the global workforce, and it is estimated that that proportion will grow to 46% by 2020 (Oktariani, Hubeis, & Sukandar 2017). Therefore, it is interesting to pay attention to Generation Y in the world of work. PT Telkom Regional Access I North Sumatra, a government company engaged in telecommunications, until now has hired a workforce that is 97% Millennials.

Currently, the world is facing the Industrial Revolution 4.0 era characterized by the development of technological sophistication and by the Internet of/for things followed by new technology in data design, artificial intelligence, robotics, cloud computing, three-dimensional printing, and nanotechnology that have distorted previous innovations. All companies, especially telecommunications companies such as PT Telkom, must adapt to the changing times in order to survive. Changes start from the work plan, the work design, the equipment used, the funds budgeted, and the human resources trained to master the technology.

The key to managing Generation Y is to understand its characteristics (Luscombe, Lewis, & Biggs 2013). The previous study did not conclude that the characteristics can be used to evaluate Generation Y's performance level. This study aimed at examining the characteristics of Generation Y in the telecommunications industry. This study also analyzed the relation between Millennials' characteristcis and their performance level. After understanding Millennials' characteristics, management can create a working system to improve the performance of individual members of the Y Generation. This research also discussed the relationship between gender and Millennial characteristics in the level of achievement of employee performance. This study sought to address whether the emphasis on understanding Millennial characteristics can help a company to drive employee performance in general.

1.1 Conceptual framework

Members of Generation Y have grown up with the Internet and mobile communications. They are the most connected and technology-friendly generation in the workforce (Sharma 2012). They are always connected digitally and globally via improved access to e-mail and mobile phones. Generation Y (or the Millennials) was born during an era of global development of information technology and education, so it has different characteristics than the previous generation. For example, Generation Y is more concerned about aspects of work–life balance than Generation X (Meier, Austin, & Crocker 2010). The current phenomenon is that many of the workers belonging to Generation Y want a flexible, demanding work schedule, and an opportunity to have an impact on the company itself. Generation Y is also known as the "Next Generation" or the "'Net Generation" (Luscombe et al. 2013) and the "Millennium Generation" (Kultalahti & Viitala 2015). In other words, this generation was born during the rapid advancement of information and the era of networking that has changed their lifestyle. This generation displays unique positivism as its members love open communication, work efficiently in teams, look for challenging work, value autonomy and work flexibility, are highly independent, and possess strong knowledge and skills in information technology (Kultalahti & Viitala 2015).

As a younger generation in the workforce, employees in the Millennium Generation are often regarded as more demanding, less experienced in practice, overconfident, and expecting unrealistic salaries (Chang & Lee 2011). Generation Y members are technology enthusiasts, self-centered, and ambitious, and they want meaningful work (Kilber, Barclay, & Ohmer 2015). The most prominent features of Generation Y are high self-esteem, a sense of entitlement, and self-centeredness (Laird, Harvey, & Lancaster 2014). *Entitlement* is a person's tendency to judge and feel that he or she deserves praise or appreciation, regardless of his or her actual performance (Laird et al. 2014).

The key in managing Generation Y is to understand its characteristics (Luscombe et al. 2013). Millennials appreciate a pleasant work environment that provides opportunities for creative communication, as well as recognition. They seek access to up-to-date technology, career development opportunities, and workplace flexibility (Cates, Cojanu, & Pettine 2013). Generation Y members often expect reciprocity and attention so that they can feel the satisfaction they expect from their workplace. This does not mean that they do not like challenges; according to Cates et al. (2013), providing a significant challenge for Generation Y is important.

Members of Generation Y usually have a more participative style of work. This generation is quite expert in computer usage and has grown along with technology. Millennials thrive in virtual connections with others through social media. They are very comfortable using technology in everyday life and view technology as a way to facilitate life and to connect with friends and family (Taylor & Keeter 2010).

Self-esteem is when someone desires to be appreciated more and more. Generation Y employees want their contributions and efforts to be respected and recognized so that they can accept challenging work and develop their knowledge, skills, and abilities (Puteh, Kaliannan, & Alam 2015). Generation Y believes that it does not have to sacrifice or work as hard as the generation of its predecessors to get the promotion or flexibility its members think they need (Kelly & McGowen 2011). Generation Y sees itself as a necessary commodity and desires special treatment when it enters the workforce (Lancaster & Stillman 2010).

Independent can refer to freedom or can mean the ability to stand alone. This generation is an independent generation and applies a "work my way" mind-set. Millennials are loyal to their work, not to a company (Chang & Lee 2011). They want to do things their own way and want meaningful work (Boston College 2011). Generation Y prefers working in teams to complete independent tasks while using the skills, knowledge, and resources of team members to satisfy individual needs. However, when interacting with managers, they are more appreciative on an individual level (Sharma 2012).

This generation loves collaboration and working with new people. Generation Y appreciates organizations that uphold justice, transparency, and decent work practices (Puteh et al. 2015). Organizations are encouraged to adopt a collaborative rather than a totalitarian approach to Generation Y (Luscombe et al. 2013).

Brown, Griffith, and Johnson (2012) state that Generation Y wants career and job flexibility. With the introduction of technology, this generation believes that it can work more efficiently. More specifically, it can reduce the activities it thinks are a waste of time, such as face-to-face interactions that occur in the typical office setting.

2 METHOD

2.1 *Participants*

At the time of this study, PT Telkom Regional Access I North Sumatra had 306 Generation Y employees, 13 Generation X employees, and 3 Baby Boomer employees. This study included as many as 173 of the Generation Y employees. The selected sample was contacted directly and asked to participate in the research. The sample participation rated from selected samples reached 100%, or all selected samples participated in this study.

2.2 *Research procedure*

The employee data of PT Telkom Regional Access I North Sumatra were used as the basis of sampling. Samples were randomly selected using the *Microsoft Excel 2016* randomization system. The researcher directly contacted the selected samples and asked them to participate in the study. The samples who agreed to participate in this research were given a questionnaire sheet that measured the research variables based on the employee's point of view. Meanwhile, when samples refused to participate, additional samples were taken at random until the minimum sample requirement in this study was met.

2.3 Instrument validity and reliability

To test the validity and reliability of the questionnaire, 30 employees of PT Telkom Regional Access I North Sumatra outside the research sample were involved. The test was conducted to evaluate whether the statement items submitted on the questionnaire were correct and consistent in measuring the research variables. The results of the validity and reliability test of the research questionnaire are summarized in Table 1.

Table 1 indicates that all of the statements submitted in the research questionnaire had accurately measured the research variables with a correlation value toward the total score on each variable > r-table (N = 30; 0.361). Cronbach's alpha on each research variable also indicated a reliable value with a value > 0.7. This indicated that the research questionnaire was consistent in measuring perceptions of respondents and deserved to be used as a research instrument.

2.4 Data-processing technique

Data processing was done using the logistic regression method. Logistic regression analysis was chosen by looking at employee performance achievement based on Millennial characteristics in PT Telkom Regional Access I North Sumatra. By using this method, it was expected that the steps in an effort to improve employee performance could be portrayed. Employee performance was converted to a binary based on an average score of employee performance. Employees who achieved above-average scores were converted to a value of 1 and those with the same or below-average score were converted to a value of 0. The predictor variables in this study were set using the interval scale based on the average score of the participants' assessment of each variable research.

2.5 Participant

The sample was taken as proportion of the total number of employees of the company based on the proportion of the number of employees in each working position. The composition of employees is summarized in Table 2.

Table 2 indicates that the composition of employees at PT Telkom Regional Access I North Sumatra was dominated by the Customer Service and Helpdesk sections. PT Telkom Access is a telecommunications company that provides communication services to its customers. In the service context, customers often reported complaints or inquiries to service providers. Naturally, the company multiplied employees on

Table 1. Validity test results and instrument reliability.

Variables	Number of Items	Index	Pearson Correlation	Cronbach's Alpha
Gender (X6)	1	c1	Single Construct	
Self-Appreciation (X1)	7	q1	0.937	0.928
		q2	0.936	
		q3	0.545	
		q4	0.904	
		q5	0.960	
		q6	0.527	
		q7	0.958	
Tech Enthusiast (X2)	9	q8	0.913	0.922
		q9	0.657	
		q10	0.913	
		q11	0.914	
		q12	0.604	
		q13	0.530	
		q14	0.657	
		q15	0.914	
		q16	0.856	
Self-Independence (X3)	6	q17	0.991	0.943
		q18	0.991	
		q19	0.962	
		q20	0.449	
		q21	0.958	
		q22	0.963	
Team Oriented (X4)	7	q23	0.677	0.900
		q24	0.902	
		q25	0.882	
		q26	0.660	
		q27	0.882	
		q28	0.902	
		q29	0.565	
Flexible (X5)	4	q30	0.695	0.760
		q31	0.633	
		q32	0.888	
		q33	0.812	
Employee Performance (Y)	11	q34	0.967	0.947
		q35	0.967	
		q36	0.561	
		q37	0.960	
		q38	0.967	
		q39	0.531	
		q40	0.960	
		q41	0.960	
		q42	0.561	
		q43	0.960	
		q44	0.967	

Table 2. Participant by division.

Position	Frequency	Percent	Cumulative Percent
Commerce & Billing Collection	2	01.02	01.02
Consumer Service	100	57.08.00	59.00.00
Finance Service	4	02.03	61.03.00
Helpdesk	27	15.06	76.09.00
HR Services	5	02.09	79.08.00
Inventory	7	04.00	83.08.00
Inventory & Asset Management	2	01.02	85.00.00
Operation & Mainten-ance Performance	2	01.02	86.01.00
Operation Perform-ance Support	3	01.07	87.09.00
Procurement & Commerce	2	01.02	89.00.00
Project Admin.	4	02.03	91.03.00
Warehouse SO	11	06.04	97.07.00
Wi-Fi	4	02.03	100.00.00
Total	173	100.00.00	

customer handling so that all customers can be served well.

3 RESULTS AND DISCUSSION

3.1 Current situation

The analysis of the current situation was generally taken from the responses of the participants. The maximum and minimum values provided an overview of the range of variable values. The situation analysis is summarized in Table 3.

Table 3 indicates that the average score for each variable was excellent with an average score > 3.40. On average, the employee performance highlighted in this research was 4.192. In the classification, most of the employees were categorized below the average performance with a proportion of 58.96%, or equivalent to 102 out of 173 participants. This simultaneously

Table 3. Current situation analysis.

Variables	Minimum	Maximum	Average
Self-Appreciation	3.571	5.000	4.267
Tech Enthusiast	3.556	5.000	4.270
Self-Independence	3.167	5.000	4.130
Team-Oriented	3.571	5.000	4.300
Flexible	3.000	5.000	4.221
Performance	3.273	5.000	4.192

indicated a leaning to the left distribution of the employee performance. Employee performance could be optimized more. The minimum score on employee performance was 3.273 in the "enough" category. The performance appraisal conducted through the questionnaire indicated that self-evaluation of employees still existed, which only ranked in the "enough" category.

3.2 Logistic regression results

The data obtained in this study were evaluated using logistic regression methods to see the overall picture in predicting and encouraging employee performance based on the characteristics of the Millennial Generation. Employee performance was classified into above-average and equal-to or less-than-average categories. Predictors in this study were gender and Millennial Generation characteristics. Gender was codified as a dummy variable with a value of 0 for women and 1 for men. Other predictor variables used scale data based on measurement results through a research questionnaire.

Table 4 indicates that models in logical regression could accurately predict the employee performance classifications up to 82.1%. More specifically, the model could accurately evaluate employees with equal-to or less-than-average performance levels up to 85.3%. The model could accurately predict employees with above-average performance by 77.5%. The false-positive error rate in the proposed logistic regression model was 14.7% while the false-negative was 22.5%. The overall model could account for 58.6% variance in the data. The result of

Table 4. Classification table.

Classification Table*

	Observed		Predicted Performance		
			Less Than Average	Above Average	Percentage Correct
Step 1	Performance	Less Than Average	87.0	15.0	85.3
		Above Average	16.0	55.0	77.5
	Overall Percentage				82.1

* The cut value is 0.500.

Table 5. Logistic regression predicting employees' performance.

		B	S.E.	Wald	Sig.	Exp(B)
	X1	1.016	0.508	3.996	0.046	2.763
	X2	1.660	0.605	7.532	0.006	5.259
	X3	1.083	0.417	6.742	0.009	2.955
Step 1[a]	X4	1.570	0.477	10.846	0.001	4.809
	X5	1.043	0.420	6.160	0.013	2.837
	X6	0.439	0.435	1.019	0.313	1.552
	Constant	−29.948	4.478	44.722	0.000	0.000

logistic regression of predictor variables in this study is summarized in Table 5.

Table 5 shows the logistic regression results, the Wald tests, and the odd-ratio ratios of each predictor variable in predicting employees' performance. The significance level used in this study was 5%. At that level, self-esteem predictors, enthusiasm for technology, self-independence, team orientation, and self-flexibility positively and significantly affected employee performance. While gender in the logistic regression model did not have a significant effect on employee performance, in general, female employees had the potential to achieve a higher performance up to 1.5 times that of male employees.

The odd-ratio level on self-esteem characteristics gave a value of 2.763, indicating that in each increase of one unit, self-esteem characteristics will increase 2.763 if the employee's probability resulted in above-average performance. Thus, employees who are increasingly able to appreciate themselves and their potential will increasingly be able to achieve better job performance. Cowen and Glazer (2007) perceive self-esteem as quest for status. This condition results in an increased motivation to achieve more and bring forth better performance (Kuhnen & Tymula 2012). Sajuyigbe, Bosede, and Adeyemi (2013) report similar findings.

The odd-ratio rate on the characteristics of technological savvy gave a value of 5.259. This value was the strongest predictor in evaluating potential employee performance. The more employees like the latest technology, the greater their potential to achieve better performance. It was suggested that people who love technologies will find it easier to adopt given technologies (Simuforosa 2013). As in Industry 4.0, many more techologies will be adopted, and those who possess the ability to understand technologies will perform better than those who do not. Binuyo and Aregbesola (2014) give an example of how technologies support overall performance.

In self-independence predictors, the odd-ratio value of employee performance achievement was 2.955. This indicated that the more employees were able to freely express themselves, the more likely they would be to provide higher performance. Self-independence related to employees' flexibility and self-management. As Kalkaja (2015) states, the flexibility given to

employees encourages their performance. Yet flexibility should be under control as it risks giving too much freedom to employees.

The odd-ratio value in the team orientation predictor was 4,809, which indicated that this predictor played an important role in evaluating employee performance. The more employees were able to do work with a team orientation, the greater the chances that those employees would give above-average performance. Team orientation helped employees to finish their job easier. Fapohunda (2013) contends that a team can reduce workload and finish the given job faster. A similar result was also revealed by Boayke (2015), who concludes that team orientation could help employees to improve their performance.

The odd-ratio rate on the flexibility predictor gave a value of 2.387, which indicated that the more an employee was granted self-flexibility on the job, the greater his or her potential for performance. Thus, the more the employees could be flexible in carrying out their duties and responsibilities, the greater their potential to achieve higher performance. A similar result emerges in Fantazy, Kumar, and Kumar (2009), which suggests that employees who hold flexibility will be tougher and better when adapting to problems. A sword should be not only tough but also flexible; otherwise, it will break when faced with a strong stimulus.

Gender-based predictors used dummy variables in binary methods. Although it was not statistically significant, female employees were more likely to perform above average than male employees. Pfair and Warner (2014) argue that it remains unclear whether there was a significant difference in performance based on gender.

3.3 Discussion

The fondness for technology is essentially a favorable characteristic for a company because the development of technology is a certain thing, and a company must be able to adapt quickly to existing changes. The Milienial Generation could cultivate the technologies to support their work, thus improving their performance. Apulu and Latham (2011) demonstrated how important the role of fondness for technology is in technology adaptation, which had a positive impact on employee performance. The fondness for technology encourages one to adapt more easily to technological changes that aid in improving employee performance.

Kalkaja (2015) states how individual self-management encouraged the improvement of his performance. The Millenial Generation feels unease with too much instruction; they want to be more trusted at work. Employees who are given freedom in self-management have more flexibility in managing their time and completing their tasks. This will help employees achieve higher performance. However, note that self-management independently deals with risk and requires individual capability.

Teamwork is one of the important components in improving employee performance both individually and in groups (Fapohunda 2013). Boayke (2015) reveals how team orientation could aid performance. Millennials can position themselves in group work, enabling them to learn and improve in achieving the required performance and competencies. Team-oriented work also helps them grow by learning from the team's capabilities.

Self-flexibility promotes the efficiency of job execution. Work done efficiently will drive employee performance. Fantazy et al. (2009) shows how flexibility could drive organizational performance. The key to raising the Millenial Generation is not to be bound by current activities.

There is no significant effect of gender on employee performance. Both male and female employees are given jobs in accordance with their capabilities.

4 CONCLUSION

The characteristics of the Millennial Generation in the work environment can encourage employee performance. Behind the negative Millennial characteristics, the generation can optimize individual characteristics. Those who are confident encourage themselves to achieve higher performance. Those who are fond of technology easily adapt. Those who are independent can be assisted in self-management. Those who are team-oriented are helped in work optimization and group learning. Those who are flexible can be assisted in prioritizing work. All characteristics of the Millennial Generation can drive better performance.

This research is contributing to knowledge development, especially in the science of organizational behavior and in human resources. This study demonstrated that Millennials are different, so they need to be treated differently from previous generations in management and policy in order to produce superior performance.

Based on the research results, it is suggested that management should support, understand, and guide the Generation Y characteristics in order to stimulate and optimize Millennials' performance. The better that management understands Generation Y, the higher potential that Millennials will perform better in the workplace.

REFERENCES

Apulu, I. & Latham, A. 2011. An evaluation of the impact of information and communication technologies: Two case study examples. *International Business Research* 4(3): 3–9.

Binuyo, A. O. & Aregbesola, R. A. 2014. The impact of information and communication technology (ICT) on commercial bank performance: Evidence from South Africa. *Problems and Perspectives in Management* 12(3): 59–68.

Boayke, E. O. 2015. The Impact of Teamwork on Employee Performance. doi: 10.13140/RG.2.1.4959.8804.

Boston College. 2011. *The multi-generational workforce: Management implications and strategies for collaboration*. Boston, MA: Boston College Press.

Brown, S., Griffith, R., & Johnson, E. 2012. *Generation Y in the workplace*. College Station: Texas A&M University Press.

Cates, S., Cojanu, K., & Pettine, S. 2013. Can you lead effectively? An analysis of the leadership styles of four generations of American employees. *International Review of Management and Business Research* 1025–1041.

Chang, Y., & Lee, M. 2011. *Factors that influence job performance in Generation Y*.

Cowen, T., & Glazer, A. 2007. Esteem and ignorance. *Journal of Economic Behavior & Organization* 63: 373–383.

Fantazy, K. A., Kumar, V., & Kumar, U. 2009. An empirical study of the relationships among strategy, flexibility, and performance in the supply chain context. *Supply Chain Management: An International Journal* 14(3): 177–188.

Fapohunda, T. M. 2013. Towards effective team building in the workplace. *International Journal of Education and Research* 1–12.

Kelly, M. & McGowen, J. 2011. *BUSN: Student edition*. Mason, OH: Nelson Education Ltd.

Kilber, J., Barclay, A., & Ohmer, D. 2015. Seven tips for managing Generation Y. *Journal of Management Policy and Practice* 80–91.

Kuhnen, C. & Tymula, A. 2012. Feedback, self-esteem and performance in organizations. *Management Science*.

Kultalahti, S. & Viitala, R. 2015. Generation Y: Challenging clients for HRM? *Journal of Managerial Psychology* 1–27.

Laird, M., Harvey, P., & Lancaster, J. 2014. Accountability, entitlement, tenure, and satisfaction in Generation Y. *Journal of Managerial Psychology* 87–100.

Lancaster, L. C. & Stillman, D. 2010. *The M-factor: How the Millennial Generation is rocking the workplace*. New York: Harper Collins.

Luscombe, J., Lewis, I., & Biggs, H. C. 2013. Essential elements for recruitment and retention: Generation Y. *Education and Training* 272–290.

Kalkaja, M. 2015. Self-management and its part in knowledge workers' experiences of high performance. Master's thesis. Oulu Business School.

Meier, J., Austin, S., & Crocker, M. 2010. Generation Y in the workforce: Managerial challenges. *Journal of Human Resource and Adult Learning* 68–78.

Oktariani, D., Hubeis, A. V., & Sukandar, D. 2017. Kepuasan kerja generasi X dan Y terhadap komitmen kerja di Bank Mandiri Palembang. *Jurnal Aplikasi Bisnis dan Manajemen* 3(1): 12–22.

Puteh, F., Kaliannan, M., & Alam, N. 2015. Assessing Gen Y's impact on organizational performance: An analysis from top management perspective. *Journal of Administrative Science* 47–59.

Sajuyigbe, A. S., Bosede, O. O., & Adeyemi, M. A. 2013. Impact of reward on employees' performance in a company in Ibadan, Oyo-state, Nigeria. *International Journal of Arts and Commerce* 2(2): 27–32.

Sharma, L. 2012. Generation Y at the workplace. *SIBM Journal* 74–78.

Shragay, D. & Tziner, A. 2011. The generational effect on the relationship between job involvement, work satisfaction, and organizational citizenship behavior. *Journal of Work and Organizational Psychology* 143–157.

Simuforosa, M. 2013. The impact of modern technology on the educational attainment of adolescents. *International Journal of Education and Research* 1(9): 1–8.

Taylor, P. & Keeter, S. 2010. *Millennials: A portrayal of Generation Next*. Washington, DC: Pew Research Center.

Advances in Business, Management and Entrepreneurship – Hurriyati et al (eds)
© 2020 Taylor & Francis Group, London, ISBN 978-0-367-27176-3

Developing strategies to stimulate employees' performance: The case of an Indonesian banker

E. Siahaan
Faculty of Economic and Business, Universitas Sumatera Utara, Medan, Indonesia

ABSTRACT: Rapid changes in technology, information, and deregulation create tight competitions in all sectors. Today, national banks compete not only with local banks but also with foreign banks. To be competitive, banks need to maximize their employees' performance. This study aimed to evaluate two kinds of strategies: (1) strategies based on self-quality – that is, employees' competencies and emotional intelligence – and (2) strategies based on motivation – that is, performance appraisal and opportunity for promotion. Each strategy was evaluated using regression analysis. Data were collected through structured questionnaires. The sample was taken from employees at BNI USU Branch. This sample was chosen using saturation sampling. Fifty-six employees participated in this study. The result indicated that both strategies significantly influenced employees' performance. However, building competencies was more effective to stimulate their performance. Thus, banks need to bolster employees' emotional intelligence so they can better deal with their jobs.

1 INTRODUCTION

1.1 *Building strategies*

In order to survive and have the competitive advantage, the ability to create and implement a strategy is highly necessary for a company. Almost all company resources can be easily duplicated by competitors. Nevertheless, human resources are something that competitors cannot imitate. Having tough-minded, wise, intelligent, and honest employees is difficult for rivals to copy. Therefore, it is very important to strategize in human resource planning, managing, fostering, and development so that the performance of bankers increases significantly.

1.2 *Importance of building strategies in improving banking employees' performance*

Strategy is the steps that must be taken in achieving a goal. In improving a company's performance, strategy is very necessary in managing the owned resources, primarily human resources. Human resources play an important, major, and strategic role in the success of banking, starting from planning, organizing, implementing, and supervising. Due to the important role of human resources (HR), the Indonesian government emphasizes national development policy to improve the quality of human resources.

The role of banks as financial institutions is very important for Indonesia's economic growth. Banks act as mediators. Banks disburse credits in the form of loans to people who are deficit. Banks keep the cash for customers who are surplus. They stabilize finance as well as control inflation, payment systems, monetary authorities, and so on. Because of the very important and strategic role of the banking system

for the overall condition of Indonesia, it is necessary to build a strategy for improving the performance of bankers in order to improve banking performance.

Even though strategy is important to improve the performance, a previous study (Koeseoglu, Barca, & Karayormuk 2009) has shown that implementing strategies did not necessarily guarantee success. Higgins (2005) states that the key success factor for implementing strategy is building the strategy itself. Studies have yet to explore how to build strategy, especially among banking employees.

Currently, banks compete to create effective and efficient payment transactions for customers. Indonesia is heading to the Industrial Revolution 4.0 and a more advanced economy. Banks need to support this. Banks, however, have a comprehensive, cheap, efficient, and secure payment system. Consequently, banks need employees who have the ability to run the system. Training and fair reward are indispensable to improve employees' spirit and performance.

Based on the foregoing discussion, it is important to formulate a strategy to improve bankers' performance. Regarding this, Patricia and Emilia (2013) state that there is no general strategy to improve all employees' performance. However, the strategies can be generalized, while the implementation can be personalized for each employee. This study covers the general strategies.

This study aims to improve bankers' performance by formulating a priority strategies. There are two main proposed strategies: (1) a strategy based on self-quality, which deals with competencies and emotional intelligence, and (2) a strategy based on employees' motivation, which deals with regulation for evaluation and promotion. The focus of these two strategies is to build productive organizational behavior and excellent human resources. Thus, qualified and competitive bankers are created.

1.3 Employees' performance

Employees' performance is the result of work achieved by the employees in performing their duties in accordance with the responsibility given to them (Mangkunegara, 2013).

Employees' performance will impact business profits, company success, and business sustainability. Therefore, management needs to know the factors that affect employees' performance.

1.4 Competence

Competence is the ability to carry out work that is driven by skills and knowledge and that is supported by a positive attitude toward the work (Wibowo 2017). Wu (2008) says that employees' competence has to be maintained and improved in order to encourage professionalism among workers.

Much research shows the significant effect of competence on performance. This means that the more superior the competence of the employees, the greater their performance (Kumar & Sharma 2001; Hsu 2008; Rahmadhani 2009; Qamariah & Fadli 2011; Lotunani et al. 2014; Siahaan 2016).

1.5 Emotional intelligence

Emotional intelligence is the ability to control oneself; to appropriately deal with problems, impulses, motivations and moods; and to display empathy in relations with others (Goleman, 2009). Emotional intelligence is needed for all business activities, mainly those related to services and direct relations with clients and customers. Employees who have high emotional intelligence will behave wisely (Carmelia & Josman 2006; Hopkins & Bilimorui 2008; Mohamad & Jais 2016; Siahaan 2017).

1.6 Performance appraisal

Decisions made by the leader will be appropriate if they come from the right information. One way to find out employees' performance is to carry out a performance appraisal. A true performance appraisal process will result in the right performance appraisal information and appropriate HR decisions. *Performance appraisal* is a process undertaken by the leader to determine whether an employee performs his work in accordance with his duties and responsibilities (Mangkunegara, 2013).

1.7 Job promotion

Promotion is a move that brings the authority and responsibility of employees to a higher position within an organization, so that the obligations, rights, status, and income of that employee are greater (Hasibuan 2014). The promoted employee will have displayed a high performance (Lumbantoruan 2008; Qamariah & Fadli 2011; Siahaan 2016).

2 METHOD

2.1 Type of research

This study aimed to investigate the potential application of banking employees' performance improvement strategies based on employees' development and motivation. To get the answer, a quantitative method with a descriptive design was applied.

2.2 Place and time of research

The research was carried out at Bank Negara Indonesia KCU USU, which is located at Jl. Universitas 50 Padang Bulan. The study was conducted from August 2017 to February 2018.

2.3 Variables and model of research

This study evaluated the factors that affected the improvement of employees' performance. There were two predictor variables. First was competence and emotional intelligence as employees' self-improvement factor. Second was performance appraisal and promotion of position as the motivator of employees' work. The models in this study were evaluated based on both factors. Model I dealt with the effect of employees' self-development, while Model II was concerned with giving boosts to employees to work sincerely so they produced a qualified and effective performance.

2.4 Population and Sample

The population who participated in this research consisted of 56 employees at Bank BNI KC USU. The entire population was used as a sample.

2.5 Data analysis technique

Secondary and primary data were used. Data analysis techniques were descriptive and inferential. Descriptive analysis was used to analyze the characteristics and the respondents. Inferential statistical analysis was used to test the hypothesis with multiple regression analysis.

3 RESULTS AND DISCUSSION

3.1 Analysis of respondents' characteristics

The majority of employees at Bank BNI KC USU were male. They were 26–36 years old. Regarding their education, most of them held bachelor's degrees. They had worked for more than four years.

3.2 *Analysis of respondents' answers*

The employees' competence was very good. They had superb skills, knowledge, and attitudes. They could communicate well with colleagues, operate working equipment, and solve problems quickly.

The employees were satisfied with the performance appraisal system conducted by the leader because it was implemented fairly, transparently, and objectively.

Job promotion ran well because the criteria considered in promoting an employee were experience, personality, skills, and education.

The emotional intelligence of BNI employees was good. For example, they could speak politely to customers and colleagues despite being angry. Knowing the factors causing anger could strengthen the bond between employees.

The employees' performance was very good. For example, employees could utilize existing resources in the office for things that were useful for their job. They could also complete the job properly in accordance with standard procedures. Furthermore, they were punctual in the completion of work and loyal to the orders given by superiors.

3.3 *Discussion*

The first model shows that by improving employees' emotional intelligence and competence, employees' performance could be improved up to 48.2%.

From Table 1, it can be seen that the bankers' competence and emotional intelligence positively and significantly affected their performance. The higher the competence of the bankers, the greater their performance. The higher the emotional intelligence of the bankers, the greater their performance. This result corroborates previous studies conducted by Siahaan (2017), Lotunani et al. (2014), Ainon (2003), Qamariah and Fadli (2011), Rahmadhani (2009), Hsu (2008), Mohamad and Jais (2016), Carmelia and Josman (2006), and Hopkins and Bilimorui (2008).

However, bankers who have working competence, such as knowing the technicalities of and the procedures for completing their work, will produce a good performance. Employees who are skilled in completing their work, in communication, and in using working equipment will quickly resolve problems and produce high-quality and effective work. Bankers who have a good working attitude, such as being creative in completing work and solving problems, and having a positive working spirit and strong motivation, will produce a superior performance. Besides, employees who have emotional intelligence will be wiser in thought and behavior. They can finish the job well. They will also have a qualified and effective working performance. Employees who have emotional intelligence are confident. They are able to control their emotion, able to motivate themselves to achieve goals, able to feel what is felt by others (empathy), and able to establish harmonious relationships with others. Thus, if a company continues to build employees' emotional intelligence, employee performance will increase significantly. This will have positive impacts on banking performance and reputation.

In the second model, it can be seen that employees' performance increased 42.7%. This was due to the two following factors: (1) promotions based on experience, personality, skill, and educational background; and (2) fair, objective, and transparent performance appraisal.

Fair, objective, and transparent performance appraisal makes employees feel appreciated. As a result, high performance is created. This is in line with the studies conducted by Khan, Khan, and Khan (2017), Wanjala and Kimutai (2015), Mangkunegara (2013), and Iqbal et al. (2013).

Besides, promotions based on experience, personality, skill, and educational background make the qualified employees have greater responsibility. This builds high motivation and good performance. This statement is supported by Siahaan (2016), Wayan (2016), Qamariah and Fadli (2011), Lumbantoruan (2008), Peter (2014) and Hasibuan (2014).

Table 1. Regression result of Model I.

Model I	Unstandardized Coefficients		Standardized Coefficients		
	B	Std. Error	Beta	t	Sig.
(Constant)	0.930	0.462		2.01	0.049
Competence	0.343	0.106	0.347	3.23	0.002
Emotional Intelligence	0.437	0.096	0.487	4.54	0.000

a. Dependent variable: Performance
b. Adjusted r-Squared = 0.482

Table 2. Regression result of Model II.

Model II	Unstandardized Coefficients		Standardized Coefficients		
	B	Std. Error	Beta	t	Sig.
(Constant)	0.771	0.533		1.448	0.154
Job Evaluation	0.353	0.096	0.397	3.687	0.001
Job Promotion	0.471	0.119	0.428	3.976	0.000

a. Dependent variable: Performance
b. Adjusted r-Squared = 0.427

4 CONCLUSION

Based on these findings, it can be concluded that both self-equity (employees' competency and emotional intelligence) and motivation (performance appraisal and promotion) strategies can be used to improve employees' performance. Both strategies have a significant effect on bankers' performance. However, the most significant contribution comes from self-equity, especially emotional intelligence. These findings support the previous literature that argues that performance can be improved through better competencies, emotional intelligence, effective performance appraisal, and promotion.

To have the competitive advantage, banks should focus on developing the quality of their employees. The employees here should have emotional intelligence. This can be built by providing a workshop that invites experts and experienced speakers. Thus, optimal service for customers can be created. Furthermore, to improve employees' competence, banks can offer scholarships. They can also allow the employees to join both national and international seminars that can improve their knowledge, skill, and attitude.

REFERENCES

Ainon, M. 2003. *Psikologi Kejayaan*. Pahang: PTS Publications.

Carmeli, A. & Josman, Z. E. 2006. The relationship among emotional intelligence, task performance, and organizational citizenship behaviors. *Human Performance* 19: 403–419.

Dessler, G. 2013. *Human resource management*. 13th edn. Upper Saddle River, NJ: Pearson Education.

Goleman, D. 2009. *Kecerdasan Emosional*. Terjemahan oleh T Hermaya. Jakarta: Gramedia Pustaka Utama.

Hasibuan, M. 2014. *Manajemen Sumber Daya Manusia, Edisi Revisi*. Jakarta: Bumi Aksara.

Higgins, J. M. 2005. The eight "S's" of successful strategy execution. *Journal of Change Management* 5(1): 3–13.

Hopkins, M. M. & Bilimoria, D. 2008. Social and emotional competencies predicting success for male and female executives. *Journal of Management Development* 27(1): 13–15.

Hsu, I.-C. 2008. Knowledge sharing practices as a facilitating factor for improving organizational performance through human capital: A preliminary test. *Expert System with Applications* 35(1): 316–326.

Iqbal, N., Ahmad, N., Haider, Z., Batol, Y., & Ain, Q. 2013. Impact of performance appraisal on employees' performance involving the moderating role of motivation. *Arabian Journal of Business and Management Review* 3(1).

Khan, Z., Khan, A. S., & Khan, I. 2017. Impact of performance appraisal on employees' performance including the moderating role of motivation: A survey of commercial banks in Dera Ismail Khan, Khyber Pakhtunkhwa, Pakistan. *Universal Journal of Industrial and Business Management* 5(1): 1–9.

Koeseoglu, M. A., Barca, M., & Karayormuk, K. 2009. A study on the causes of strategies failing to succeed. *Journal of Global Strategic Management* 3(2): 77–91.

Kumar, A. & Sharma, R. 2001. *Personnel management theory and practice*. Washington, DC: Atlantic Publishers.

Lotunani, A., Idrus, S. M., Afnan, E., & Setiawan, M. 2014. The effect of competence on commitment, performance and satisfaction with reward as a moderating variable (a study on designing work plants in Kendari City Government, Southeast Sulawesi). *International Journal of Business and Management Invention* 3(2): 18–25.

Lumbantoruan, R. R. 2008. *Analisis Pengembangan Karir Karyawan dalam Meningkatkan Kinerja Karyawan Pada PT (Persero) Pelabuhan Indonesia I*. Medan: Skripsi, Fakultas Ekonomi Universitas Negeri Medan.

Mangkunegara, A. P. 2013. *Manajemen Sumber Daya Manusia Perusahaan*. Bandung: Remaja Rosdakarya.

Mohamad, M. & Jais, J. 2016. Emotional intelligence and job performance: A study among Malaysian teachers. *Procedia Economics and Finance* 35: 674–682.

Patricia, R. & Emilia, S. C. 2013. Performance improvement strategies used by managers in the private sector. *Economic Science* 1: 1613–1624.

Peter, C. G. 2014. *Impact of promotion on employees' performance at Dar Es Salaam City Council*. Dar Es Salaam: MZUMBE University Digital Research Repository.

Qamariah, I. & Fadli. 2011. Pengaruh Perencanaan SDM dan Kompetensi Karyawan terhadap Kinerja Karyawan pada PT. Indonesia Asahan Aluminium Kuala Tanjung, *Jurnal Ekonomi* 14(2).

Rahmadhani, S. 2009. Analisis Penempatan Terhadap Kinerja Karyawan Pada PT Perkebunan Nusantara III Medan. Medan, *Skripsi*, Fakultas Ekonomi, Universitas Sumatera Utara.

Siahaan, E. 2017. Evaluating the effect of work–family conflict and emotional intelligence in the workplace: Review to increase employees' performance. *IOP Conferences Series: Earth and Environmental Science* (126): 2018.

Siahaan, E. 2016. Improvement of employee banking performance based on competency improvement and placement working through career development (case study in Indonesia). *International Business Management* 10 (3): 255–261.

Wanjala M. W. & Kimutai, G. 2015. Influence of performance appraisal on employee performance in commercial banks in Trans Nzoia County, Kenya. *International Journal of Academic Research in Business and Social Sciences* 5(8): 332–343.

Wibowo. 2017. *Manajemen Kinerja (Edisi Ke 5)*. Jakarta: Rajawali Pers.

Wu, W. 2008. Exploring core competencies for R&D technical professionals. *Expert System with Applications* 36(5): 9574–9579.

Advances in Business, Management and Entrepreneurship – Hurriyati et al (eds)
© *2020 Taylor & Francis Group, London, ISBN 978-0-367-27176-3*

The mediating role of psychological empowerment on the effect of person-organization fit on innovative work behavior

M.S. Melina & C.W. Sandroto
Universitas Katolik Indonesia Atma Jaya, Jakarta, Indonesia

ABSTRACT: Innovative work behavior is needed to remain sustainability in an organization. Previous studies have suggested that person-organization fit affects the innovative work behavior and psychological empowerment mediates the influence between them. This study aims to prove further whether the influence of person-organization fit on innovative work behavior is mediated by the psychological empowerment of the employees. Population in this study is all employee CHC, Co and sampling technique used is convenience sampling. Innovation work behavior is assessed through supervisor-rated and self-rated. Results of the study indicate that person-organization fit has a significant effect on innovative work behavior (supervisor-rated), person-organization fit has a significant effect on psychological empowerment, psychological empowerment has a significant effect on innovative work behavior (supervisor-rated), and person-organization fit has a significant effect on employee's innovative work behavior (supervisor-rated) with psychological empowerment as mediator. This study showed that the direct effect of person-organization fit on innovative work behavior (supervisor-rated) is greater than its indirect effect. While assessed through self-rated, person-organization fit has no significant effect on the innovative work behavior. Assessment by supervisor-rated is considered more objective than self-rated.

1 INTRODUCTION

Organizations need innovative employees. There are several factors to cultivate innovative employees, such as person-organization fit, which is defined as the degree to which individuals are matched to the culture and values of the organization (Snell & Bohlander 2013) as well as psychological empowerment, that is employee's confidence to the degree where they affect work environment, competence, diversity of work, and work autonomy.

Employees with high psychological empowerment are perceived to have more confidence and motivation so that they can affect the work environment. According to Bruce (2003), an employee can influence the motivation of other employees and improve productivity by trying the new and innovative techniques. According to Gruman and Saks (2011), one of the most important factors to demonstrate innovative work behavior (IWB) is a work involvement.

Cigading Habeam Center, Co (CHC) is welded H-beam factory. CHC operates in three divisions: steelwork construction, manufacturing, and service. Driven by a strong vision to become a leading steel structure manufacturer in Indonesia, CHC requires employees who have strong capabilities. One of CHC's strategy to improve the quality of the workforce is through conducting training programs that refer to company values, such as innovative.

1.1 *The relationship between person-organization fit and innovative work behavior*

Person-organization fit is a theory that people are attracted to and selected by organizations that match their values and leave when there is no compatibility (Robbins & Judge 2017). According to (Sekiguchi 2004), person-organization fit can be grouped into four dimensions: value congruence, goal congruence, the employee need fulfilment, and culture personality congruence. Person-organization fit brings motivation to support the overall success of an organization (Amy & Jon 2013). Verquer et al. (2003) state that person-organization fit affects positive work and promotes innovative work behavior, where innovative work behavior is voluntary and is not a formal part of job description. Correspondingly, Vilela et al. (2008) revealed that person-organization fit positively influences innovative work behavior. Innovative work behaviors could be assessed as self-rated, but also supervisor-rated, to see the differences in outcomes, because in general self-rated experience basic problem where employees tend to assess themselves higher than the assessment of supervisors or peers (Dessler 2017). Therefore, on the basis of these arguments, it can be hypothesized:

Hypothesis 1a: Person-organization fit has an effect on innovative work behavior (supervisor-rated).

Hypothesis 1b: Person-organization fit has an effect on innovative work behavior (self-rated).

1.2 The relationship between person-organization fit and psychological empowerment

High person organization fit gives a better understanding of expectations organization realization, that behaviors and attitudes can be tailored to meet expectations, fostering a unique feeling of psycho-logical empowerment (Gregory et al. 2010). An individual who shares the organization's values and fits better processes the behavioral expectations more cognitively and proactively than an individual with the low person organization fit (Schein 1985). Spreitzer (2016) defines construct psychological empowerment with four cognitions: meaning, competence, self-determination, and impact. Employees who have better person-organization fits are expected to identify their organization and, therefore, are expected to have greater feelings that they can have an impact on their organizations. This impact influences important behaviors and attitudes towards work roles such as intrinsic motivation, creativity, and innovation, through enhancement to their psychological empowerment (Pieterse et al. 2010, Seibert et al. 2004). Based on this description, the following hypothesis is proposed:

Hypothesis 2: Person-organization fit has an effect on psychological empowerment.

1.3 The relationship between psychological empowerment and innovative work behavior

Generally, an individual with psychological empowerment feels autonomy and freedom to engage in "trial and error" and develop new ideas of carrying out organizational processes efficiently and effectively (Ramamoorthy et al. 2005). According to Spreitzer et al. (1999), the perceptions of higher psychological empowerment lead to increased inspiration, innovation and upward influenced. Laschinger et al. (2004) stated that employees with high psychological empowerment are committed to their work and generate innovative work behaviors. Based on that argument, it can be hypothesized:

Hypothesis 3: Psychological empowerment has an effect on innovative work behavior.

1.4 The relationship between person-organization fit and innovative work behavior with psychological empowerment as a mediator

Afsar and Badir (2016) stated the relationship of person organization fit on innovative work behavior with psychological empowerment as a mediator. They argue that congruency perception with an organization has an impact on the employee's journey about the working environment and the organization's scheme and then that feeling affects employee engagement in innovative work behavior. Based on this argument, the following hypothesis is proposed:

Hypothesis 4: Person-organization fit has an effect on innovative work behavior with psychological empowerment as a mediator.

2 METHOD

2.1 Population and sampling

The population in this research is 72 employees of CHC. The researchers used the Slovin formula and obtained a minimum number of samples of 61 respondents, and the researchers distributed 68 questionnaires, and all returned and filled completely. Sampling technique with non-probability sampling method, using convenience sampling, which is sampling based on conveniently available (Sekaran 2016).

2.2 Data and measurement

There are three variables in this study: person-organization fit (independent variable), psychological empowerment (intervening variable), and innovative work behavior (dependent variable). Respondents were asked to indicate their level of approval of the item questions on a 5-point Likert scale, ranging from strong disagreement (1) to strong agreement (5).

The innovative work behavior (IWB) variable is the employee's work behavior that can generate, introduce, and apply new things from within which can be useful to his work within the company. Measurements of innovative work behavior variable was developed by De Jong & Den Hartog (2010) as referenced in Afsar & Badir (2016). The innovative work behavior questionnaire is supervisor rated and self-rated (we used two assessment to obtain more objective results).

Person-organization fit (P-O Fit) is the appropriateness of value between the individual with the company and vice versa. Person-organization fit was measured using Cable & Judge's (1996) scale (Afsar & Badir 2016).

Psychological empowerment (PE) is an employee's belief in the ability to perform work activities related to their skills and competencies. Psychological empowerment was measured through scale developed by Afsar & Badir (2016).

3 RESULT AND DISCUSSION

Data analysis used reliability and validity test, normality test, descriptive statistic, mean score, and Preacher Hayes test analysis the simple mediation model. Validity and reliability test results show the scale is valid variable ($p < 0.01$) and reliable

(Cronbach's Alpha > 0.7) (Correlation coefficient P-O Fit ranged from 0.889 to 0.915 and Cronbach's Alpha 0.878, correlation coefficient IWB supervisor-rated ranged from 0.453 to 0.895 and Cronbach's Alpha 0.925, correlation coefficient IWB self-rated ranged from 0.643 to 0.941 and Cronbach's Alpha 0.969, correlation coefficient PE ranged from 0.641 to 0.828 and Cronbach's Alpha 0.931). Based on the normality test it is found that the data comes from a normally distributed population where the residual has significance > 0.05 (Ghozali 2011).

3.1 Descriptive statistics

The characteristics of the respondents are presented in the following table:

The majority of respondents are male, in the 20–29 year old age range, level of education is an undergraduate degree, and they had to work for more than 5 years.

3.2 Mean score

An overall mean score of person-organization fit is in the high category (mean 4.02), innovative work behavior (self-rated) score higher than (supervisor-rated) (mean 3.99 > 3.96), both are in the high category, and psychological empowerment is in the high category (mean 4.18).

3.3 Preacher-hayes test analysis the simple mediation model

To determine whether there is an effect between independent variable on intervening variable and intervening variable on the dependent variable or indirect effect independent variable on dependent variable through intervening variable path analysis test was conducted using SPSS 23.00 by Preacher - Hayes The Simple Mediation Model (Preacher & Hayes 2008).

Table 2 shows that direct effect of person-organization fit on innovative work behavior (self-rated) is insignificant because it has Sig. (0.3102) > 0.05. Hypothesis 1b is rejected.

Table 3 shows that person-organization fit significantly affects the innovative work behavior assessed by the supervisor (R square = 0.8788, Sig value. (0.0000) < 0.05). Hypothesis 1a is accepted.

Table 4 shows that person-organization fit significantly affects the psychological empowerment (R square = 0.6145, Sig. (0.0000) < 0.05). Hypothesis 2 is accepted.

Table 5 shows that psychological empowerment significantly affects the innovative work behavior (supervisor-rated). (R square = 0.9020, Sig. (0.0002) < 0.05). Hypothesis 3 is accepted.

Table 1. Characteristics of respondents.

	Frequency	Percentage
Gender		
Male	46	67.65
Female	22	32.35
Total	68	100
Age		
< 20	3	4.41
20 – 29	22	32.35
30 – 39	21	30.88
40 – 49	12	17.65
50	10	14.71
Total	68	100
Education		
Senior high school	17	25
Diploma	12	17.65
Undergraduate degree	37	54.41
Postgraduate degree	2	2.94
Total	68	100
Number of years working		
<1	6	8.82
1 – 2	4	5.88
2 – 3	3	4.41
3 – 4	11	16.18
5	44	64.7
Total	68	100

Table 2. Effect of person-organization fit on innovation work behavior (self-rated) (direct effect).

	Coeff	Se	t	P
P-O Fit	0.5493	0.5371	1.0228	0.3102

Table 3. Total effects of person-organization fit on innovation work behavior (supervisor-rated) (c path).

	Coeff	Se	t	R-sq	P
P-O Fit	2.6978	0.1233	21.8771	0.8788	0.0000

Table 4. Effect of person-organization fit (independent variable) on psychological empowerment (a path).

	Coeff	Se	T	R-sq	P
P-O Fit	2.5549	0.2491	10.2563	0.6145	0.0000

Table 5. Direct effect of psychological empowerment on innovative work behavior (supervisor-rated) - b path.

	Coeff	Se	T	R-sq	P
PE	0.2166	0.0552	3.9227	0.9020	0.0002

Table 6. Direct effect of person - organization fit on innovative work behavior (supervisor-rated) (c 'path).

	Effect	Se	T	P
P-O Fit	2.1444	0.1800	11.9165	0.0000

Table 7. Indirect effect of person-organization fit on innovative work behavior (supervisor-rated) with mediator psychological empowerment (ab).

	Effect	se	Z	P
PE	0.5533	0.1517	3.6487	0.0003

Table 6 shows that person-organization fit significantly effects directly on innovative work behavior (Sig. (0.0000) < 0.05).

Table 7 shows that the person-organization fit significantly affects directly on innovative work behavior (supervisor-rated) with psychological empowerment as a mediator. (Sig. (0.0003) < 0.05). In the Preacher Hayes test, there is a product of coefficient strategy used to determine indirect effects. The indirect effect is proved if z > 1.96. Table 7 shows that z is (3.6487) > 1.96. In another word, this result indicates that person-organization fit has an indirect effect on the innovative work behavior (supervisor-rated) with psychological empowerment as a mediator. Hypothesis 4 is accepted.

The results of simple mediation model Preacher-Hayes shows that person-organization fit can affect directly on innovative work behavior and also affect indirectly from person-organization fit on psychological empowerment (as an intervening variable) and on innovative work behavior. The coefficient of direct effect is 2.1444 while indirect effect is 0.5533. Because the coefficient of direct effect is greater than the indirect effect, it can be concluded that the actual effect is direct.

Person-organization fit and psychological empowerment are effect positively on innovative

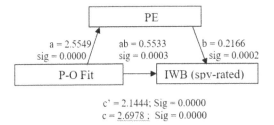

c' = 2.1444; Sig = 0.0000
c = 2.6978 ; Sig = 0.0000

Figure 1. Simple mediation test: effect model per-son-organization fit on innovative work behavior (supervisor-rated) with psychological empowerment as a mediator.

work behavior (supervisor-rated). This supports the theory that innovative work behavior requires a high level of engagement, long-term commitment, courage, peer support, which arises if it has conformity between the value of the employee and the organization (Afsar & Badir 2016). The direct effect of person-organization fit and psychological empowerment on innovative work behavior (supervisor-rated) shows that the improvement of innovative work behavior can be achieved by improving person-organization fit or psychological empowerment. These findings correspond with studies that employees with high-fit person-organizations feel empowered psychologically because their needs, skills, and abilities are match with organizational demands and resources (Gregory et al. 2010). Psychological empowerment significantly affects the innovative work behavior (supervisor-rated). These correspond with studies by Spreizer et al. (1999) which shows that psychological empowerment positively affects the creative process, when employees have strong psychological empowerment, then employees are creatively engaged in work, managerial effectiveness and intrinsic motivation. Person-organization fit affected by innovative work behavior (supervisor-rated) with psychological empowerment as a mediator. These findings correspond with studies by Afsar and Badir (2016). In this research, the coefficient value of direct effect is greater than indirect effect. Possibly this can be caused by the role of psychological empowerment as suppressor variable. The suppressor variable will cause psychological empowerment increase the effect of person-organization fit on innovative work behavior.

Psychological empowerment has an R-sq of 90.2% on innovative work behavior (supervisor-rated), while person-organization fit has R-sq 87.88% on innovative work behavior. These results show that the degree variability of the innovative work behavior (supervisor-rated) can be explained more by the psychological empowerment than the person-organization fit variable.

Person-organization fit can be improved by redefining the vision, mission and corporate goals clearly, and then socialized them to all CHC's employee. The values of the organization must be imparted, trusted, and followed by all employees. In the case of recruitment and selection, to obtain new employees, it is necessary to ensure that the new hires are highly competent, disciplined, and goal-oriented and adopt the same values as the company. Training programs were undertaken to equalize and recall corporate values. Mean score employees' person-organization fit is high, therefore should be maintained and improved for maximum result. Training can be held to fulfilled job demand that more challenging in technology and science field. Employees must continue to improve knowledge and adjust with the latest innovation that is needed by the company. Psychological empowerment can be improved by giving autonomy and assigning responsibilities to employees. Mean score psychological empowerment is high, therefore must be maintained.

4 CONCLUSION

This study finds that in CHC, person-organization fit has an effect on innovative work behavior (supervisor-rated). Person-organization fit has no effect on innovative work behavior (self-rated). Person-organization fit has an effect on psychological empowerment. Psychological empowerment has an effect on innovative work behavior. Person-organization fit has an effect on innovative work behavior with psychological empowerment as a mediator. This study shows that person-organization fit can affect directly on innovative work behavior and also affect indirectly from person-organization fit on psychological empowerment (as an intervening variable) and on innovative work behavior, but the actual effect is direct.

REFERENCES

Afsar, B. & Badir, Y. 2016. The mediating role of psychological empowerment on the relationship between person-organization fit and innovative work behavior. *Journal of Chinese Human Resource Management* 7(1): 5–27.

Amy, K. & Jon, B. 2013. *Organizational fit: Key issues and new directions*. New York: John Wiley and Sons.

Bruce, A. 2003. *How to motivate every employee*. Jakarta: PT Bhuana Ilmu Populer.

Dessler, D. & Gary, G. 2017. *Human resource management (15th ed global edition)*. Boston: Pearson Education.

Ghozali, G. & Imam, I. 2011. *Aplikasi analisis multivariate dengan program SPSS*. Semarang: BP Universitas Diponegoro.

Gregory, B.T., Albritton, M.D. & Osmonbekov, T. 2010. The mediating role of psychological empowerment on the relationships between p-o fit, job satisfaction, and in-role performance. *Journal of Business and Psychology* 25(4): 639–647.

Gruman, J.A. & Saks, A.M. 2011. Manage employee engagement to manage performance. *Journal Industrial and Organizational Psychology* 4(2): 204–207.

Laschinger, H.K.S., Finegan, J.E., Shamian, J. & Wilk, P. 2004. A longitudinal analysis of the impact of workplace empowerment on work satisfaction. *Journal of Organizational Behavior* 25: 527–545.

Pieterse, A.N., Van Knippenberg, D., Schippers, M. & Stam, D. 2010. Transformational and transactional leadership and innovative behavior: the moderating role of psychological empowerment. *Journal of Organizational Behavior* 3(4): 609–623.

Preacher, K.J. & Hayes, A.F. 2008. Asymptotic and resampling strategies for assessing and comparing indirect effects in multiple mediator models. *Behavior Research Methods* 40:879–891.

Ramamoorthy, N., Flood, P.C., Slattery, T. & Sardessai, R. 2005. Determinants of innovative work behavior: Development and test of integrated model. *Creativity and Innovative Management* 14(2): 142–150.

Robbins, S.P. & Judge, T. 2017. *Organizational behavior (17th ed.)*. Upper Saddle River, NJ: Pearson Prentice Hall.

Schein, E.H. 1985. *Organizational culture and leadership, Josey-Bass*. San Francisco.

Seibert, S.E., Silver, S.R., & Randolph, W.A. 2004. Taking empowerment to the next level: A multiple-level model of empowerment, performance, and satisfaction. *Academy of Management Journal* 47(3): 332–349.

Sekaran, U. 2006. *Research methods for business a skill building approach*. New York: John Wiley and Sons.

Sekiguchi, T. 2004. Person-organization fit and person-job fit in employee selection: A review of the literature. *Osaka Keidai Ronshu* 54(6): 179–196.

Snell, B. 2013. *Principles of human resource management (16th ed.)*. South Western: Cengage Learning.

Spreitzer, G.M., De Janasz., S.C. & Quinn, R.E. 1999. Empowered to lead: the role of psychological empowerment in leadership. *Journal of Organizational Behavior* 20(4): 511–526.

Sukrajap, M.A. 2016. Pengaruh kepemimpinan transformasional terhadap kepuasan kerja dan komitmen organisasinal dengan dimediasi oleh pemberdayaan psikologis pada rumah sakit PKU Muhammadiyah Yogyakarta.

Verquer, M.L., Beehr, T.A. & Wagner, S.H. 2003, A Meta-Analysis of the relations between person-organization fit and work attitudes. *Journal of Vocational Behavior* 63(3): 473–489.

Vilela, B.B., Gonzalez, J.V. & Ferrin, P.F. 2008. Person–organization fit, OCB and performance appraisal: Evidence from matched supervisor-sales person data set in a Spanish context. *Industrial Marketing Management* 37(8): 1005–1019.

Advances in Business, Management and Entrepreneurship – Hurriyati et al (eds)
© 2020 Taylor & Francis Group, London, ISBN 978-0-367-27176-3

Decision support system in determining an outstanding employee based on employee performance assessment with analytical hierarchy process method

U. Mulyana, A.M. Siddiq & K. Kusnendi
Universitas Pendidikan Indonesia, Bandung, Indonesia

ABSTRACT: The problem that occurred in the process of selecting outstanding employees in Faculty of Sports and Health Education (Fakultas Pendidikan Olahraga dan Kesehatan) in Universitas Pendidikan Indonesia (UPI) was in the decision-making that had not fully used objective criteria or assessment instruments and was based only on subjective judgments from the highest leadership, in this case, the dean. The purpose of this study was to create a support system to provide objective consideration for the dean in the selection of outstanding employees. The decision support system determined the employee's performance-based achievement using the Analytical Hierarchy Process (AHP) method based on the criteria/assessment instruments applied to a civil servant (Pegawai Negeri Sipil), namely the List of Work Implementation Assessments (DP3) consisting of five assessment criteria. The AHP process was carried out to determine the score of employees who excelled (a good performance), which was the basis of recommendations for decision makers to choose outstanding employees at the faculty level. This application was made by using Xampp 2.5 and Adobe CS 5 as its tools. The results of the implementation of this system showed that the faculty had clear criteria in selecting employees to perform further and the system objectively facilitated and assisted the decision-making process.

1 INTRODUCTION

In an educational institution or agency, both government and private, employees are important to the sustainability of the organization. High quality employees will facilitate the organization in achieving its goals.

The performance of good employees can improve the image of institutions/agencies. To encourage employees to work and achieve better, the organization can reward employees who are considered outstanding. Some research has shown that rewards can have an impact on employee performance (Ibrar & Khan 2015, Murphy 2015, Salah 2016).

The award for employee performance can be in the form of promotion, class change, or other things that can encourage employees to work better, and is given to employees who excel. The Faculty of Sports and Health Education (FPOK), Universitas Pendidikan Indonesia (UPI) annually decides the employees, lecturers, and outstanding students to be sent to the selection of university-level employees, lecturers, and students. In determining the outstanding employees in FPOK, objective instruments or criteria have not been fully used, so the subjective assessment of decision-makers often greatly influences the decision to determine outstanding employees.

This problem needed further research on systems in human resources affairs. Some research showed that the existence of technology could facilitate the work of personnel or Human Resources Development (HRD) (Faliagka et al. 2014, G & O 2016, Abdullah et al. 2017, Boşcai 2017). From that result,

it was shown that the technology could make the HRD employee's job easier.

The process of selecting outstanding employees is a major problem in this study. Outstanding employees are those who can provide excellent performance. There has been a lot of research on employee performance (Azka et al. 2011, Bedarkar & Pandita 2014, J. 2014, Menguc et al. 2017). Ishizaka & Pereira 2016, for example, explained the ease of use of criteria included in the system to help assess employee performance.

Registered employees are civil servants who work in the Faculty of Sports Education and Health, UPI. The criteria used were those set by the staffing service and divided into five hierarchical sections, namely loyalty, performance, responsibility, obedience, and honesty, and entered into the Analytical Hierarchy Process (AHP) to determine the ranking. One study showed that AHP could be used in evaluating employee performance (Rafikul & Shuib 2005). The purpose in this study was to make a decision support system that could help faculty leaders in determining outstanding employees who delivered excellent performance at the faculty level.

2 METHODOLOGY

There were five methods used in this research:

1. Literature review
2. Observation
3. Analysis and design

4. Implementation
5. Testing

Literature review was used to get the criteria needed because objectivity is a decision that does not come from a personal view that has been mixed with emotions. Observation was used to determine platforms and interfaces that were easy to use by faculty leaders, namely the dean. The results were analyzed. The next step was the implementation of the system and system testing.

Progress hierarchy analysis is the analytical technique used to create the hierarchy of criteria used. The analysis processes were:

1. Hierarchy development
2. Determining comparative value
3. Determining priority value
4. Measuring consistency

3 RESULTS AND DISCUSSION

The display used in building this system is like on web pages in general, namely hypertext. In the construction of this application, there are two main pages that can be accessed directly by the user. One is the page of the process displayed after the user has logged in, and another page contains the process of selecting outstanding employees.

The data used/processed in the decision support system for outstanding employee selection based on performance with the AHP method were as shown in Table 1.

The work procedures described here are used to support the selection of an outstanding employee based on performance with the AHP method as seen in Table 2.

For the process, a document is needed to assess employee performance. Documents involved in

Table 1. Data analysis.

No	Data Name	Detail
1	Criteria	Id criteria
		Name
		Description
		Score priority
2	Sub-criteria	Id criteria
		Id Sub criteria
		Name
		Score priority sub criteria
3	Weight sub criteria	Id importance sub criteria
		Id criteria
		Id sub criteria 1
		Id sub criteria 2
		Weight
4	Weight criteria	Id importance criteria
		Id criteria 1
		Id criteria 2
		Weight
5	Employee	Id employee
		ID number
		Name
		Unit
		Title group
		Gender
		Education
		Work length
		Training
6	Employee score	Id employee
		Id criteria
		Id sub criteria
		Score
7	Admin	Id admin
		Name
		Username
		Password

Table 2. Procedure and process.

No	Procedure Name	Process Name	Function
1.	Criteria management	Changes in criteria data	To add, enter and delete criteria data
		Changes in criteria values	Enter and change criteria values
2.	Sub-criteria management	Sub data changes	To add, enter and delete sub-criteria data
		Sub-criteria value changes	Enter and change sub-criteria values
3.	Employee data management	Adding employee data	To add and enter employee data
		Changes in employee data	To change employee data if there is a change in employee data.
		Changes in employee value	To change employee value in the event of changes in employee data.

Table 3. Document analysis.

No	Information Name	Function
1	Each part of an out-standing employee document	An employee score document based on criteria
2	AHP results report of an outstanding employees	To know the scoring data for an outstanding employee based on AHP method

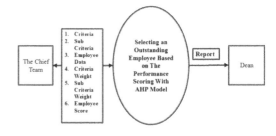

Figure 3. Context diagram.

decision support system for selecting outstanding employee selection based on performance with the AHP method included those shown in Table 3.

In this system, the first step is to create a hierarchical structure consisting of main objectives, criteria and sub-criteria. The following is a picture for the hierarchy contained in this system:

In Figure 1, there are five criteria used: loyalty, performance, responsibility, obedience, and honesty. From these five criteria, an assessment will be carried out with five rating scales: very good, good, enough, medium, and bad.

In Figure 2, the first process in this system is employee data input. After this, the next thing to do is to enter the weight criteria into the system.

Figure 1. Hierarchy.

Figure 2. System flow.

If the weight value is equal to or below 0.01 then the system will proceed to the next stage, but if it is above 0.01, the system will reject and will return to the data input. The next step is to enter the weight sub-criteria. The weight sub-criteria has the same conditions, namely that it must be equal to or below 0.01. Next, the system will begin to enter the employee's value and begin the process to generate a report that will be given to the leader, the dean of the Faculty of Sports and Health Education in UPI.

In Figure 3, the use of this system is given to a team formed by the dean to assess outstanding employees. The team leader will determine the outstanding employees using this system. After the system issues a report, the team leader will provide the report to the dean to be taken into consideration in making a decision to choose outstanding employees.

4 CONCLUSION

The system provides an alternative consideration for the dean in determining an outstanding employee in Faculty of Sports and Health Education, UPI. The system can give another perspective, or an objective perspective, for the dean before taking any decision to annouce who the outstanding employees are.

The weakness of this system is that the system was made based on statistical data and in the case of two or more employees who excel with the same final score, the system cannot determine which one is better.

The system has been built in the form of a web-based application so that it can be further developed for the internet network use. Looking at the development of today's technology, an android-based application could be built, and it will make the input process faster and easier.

For more objective results of the decision support system determination of outstanding employees based on employee performance assessment, objective assessment data is needed, and the recording of the value for each criterion from each employee by each section head within the assessment period is also necessary.

REFERENCES

Abdullah, M.S., Hadi, A., Manaf, A., Asyraf, M., Kassim, M., & Salahuddin, S.N. 2017. Graduate students' perception towards e-recruitment, 6(August), 51–59.

Azka, G., Tahir, M.Q., M, A.K., & Syed, T.H. 2011. Transformational leadership, employee engagement and performance: Mediating effect of psychological ownership. *African Journal of Business Management* 5(17), 7391–7403.

Bedarkar, M., & Pandita, D. 2014. A study on the drivers of employee engagement impacting employee performance. *Procedia - Social and Behavioral Sciences* 133, 106–115.

Boşcai, B.G. 2017. The evolution of e-recruitment: the introduction of online recruiter. *Management and Organization: Concepts, Tools and Applications* 161–170.

Faliagka, E., Iliadis, L., Karydis, I., Rigou, M., Sioutas, S., Tsakalidis, A., & Tzimas, G. 2014. On-line consistent ranking on e-recruitment: Seeking the truth behind a well-formed CV. *Artificial Intelligence Review 42*(3), 515–528.

G, A.D., & O, O.M. 2016. Development of efficient e-recruitment system for university staff in Nigeria. *Circulation in Computer Science 1*(1), 10–14.

Ibrar, M., & Khan, O. 2015. The impact of reward on employee performance (a case study of Malakand private school). *International Letters of Social and Humanistic Sciences 52*(2005), 95–103.

Ishizaka, A., & Pereira, V.E. 2016. Portraying an employee performance management system based on multi-criteria decision analysis and visual techniques. *International Journal of Manpower 37*(4), 628–659.

J., A. 2014. Determinants of employee engagement and their impact on employee performance. *International Journal of Productivity and Performance Management 63*(3), 308–323.

Menguc, B., Auh, S., Yeniaras, V., & Katsikeas, C.S. 2017. The role of climate: Implications for service employee engagement and customer service performance. *Journal of the Academy of Marketing Science 45*(3), 428–451.

Murphy, B. 2015. *The impact of reward systems on employee performance*. Dublin Business School.

Rafikul, Islam; Shuib, M.R. 2005. Employee performance evaluation by AHP: A case study. *Proceedings of the 8th International Symposium on the Analytic Hierarchy Process Multi-Criteria Decision Making*.

Salah, M. 2016. The Influence of rewards on employees performance. *British Journal of Economics, Management & Trade 13*(4), 1–25.

Impact of leadership style and organizational culture on employee performance

Y.R. Widjaja, D. Disman & S.H. Senen
Universitas Pendidikan Indonesia, Bandung, Indonesia

ABSTRACT: Leadership and organizational culture have a very close relationship that is highly interdependent because every aspect of leadership will ultimately shape an organizational culture. The purpose of this research is to know the influence of leadership style and organizational culture on employee performance at Duta Rama Company, Surabaya. In this study, the population consists of 92 people selected using saturated samples. The results indicate that leadership style has an effect, but not a significant effect, on employee performance. Meanwhile, organizational culture has a positive and significant influence on employee performance. Together, leadership style and organizational culture have a positive and significant effect on employee performance.

1 INTRODUCTION

In this era of disruptive innovation, various industries are growing very rapidly. The idea of disruptive innovation in the business world that was initiated by Christensen (1997), the innovator's dilemma, reveals that thinking of successful companies not only meets the needs of today's customers but also anticipates their future needs. The theory explains how small companies with fewer resources are able to enter the market and replace the established systems. Certainly in this disruptive era every organization is required to make changes in order to develop and defend itself amid the competition. A fundamental change must of course start with a leader within the organization. According to Burns (1978), leadership is a phenomenon that receives the most attention. The role of basic leaders in the organization is as a regulator in every activity that motivates the group, and to set the task to achieve targets. The survival of an organization depends largely on its leader. The leader is seen as the one who is able to create orderly conditions out of chaos, and direct the organization out of the turmoil in its environment. As described by Yukl (1989), leaders are capable of influencing followers in many ways, including coordination, communication, training, motivation, and benefit. Leadership and organizational culture have a very close relationship, and we have seen that every leader has different leadership styles that will shape the organizational culture. Thus, in an innovative organizational culture, authentic leaders foster innovative behavior in followers. Therefore, organizational culture is focused on innovation as well as authentic leadership relationships, giving positive effects to employees as key to human resource management. In other words, if the style of leadership is applied well and can give

good direction to subordinates, then it creates confidence and motivates employees to work, thus increasing employee morale and having a positive impact on employee performance.

The measure of human performance can only be determined by the performance of leaders. In human affairs, the distance between the leader and the average is constant. If leadership performance is high, group performance will go up (Drucker 1996). In other words, organizational culture can reflect leadership within the organization as two sides that are interconnected with each other. This study aims to determine the influence of leadership style and organizational culture on employee performance at Duta Rama Company, Surabaya.

1.1 *Leadership*

According to Robin et al. (2003), leadership is the ability to influence a group to achieve a goal. Leadership differs from management because formal managerial rank in an organization does not always allow a manager to effectively lead and the ability to influence can emerge outside the formal structure of the organization. In general, there are three categories of leadership theories, namely trait theory, behavioral theory, and contingency theory (Robbins & Coulter 2007). Meanwhile, according to DuBrin (2004), leadership is the achievement of goals through communication between one party and another party. Leadership is a dynamic condition, which is the ability to influence groups to achieve goals (Koontz & Weihrich 1990), which is in harmony with other opinions that leadership is a process of interaction between leaders and followers in which leaders seek to influence followers to achieve common goals (Northouse 2010, Yukl 2005). According to Chen and Chen (2008), previous studies on leadership have identified the different leadership styles that leaders adopt in managing organizations (Hirtz et al. 2007, House et al. 2004). Based on these definitions, leadership is a person's ability to influence individuals or other groups in order to work together to achieve goals that have been established together.

1.2 *Organizational culture*

Organizational culture, according to Trefry (2006), is divided into two levels:

1. Behavior, attitude.
2. Underlying belief in values, customs, customs.

Deshpande and Webster (1989) define organizational culture as a set of values, customs, norms, and beliefs generally accepted by individuals in an organization. On the other hand, Chatman and Jehn (1994) believe that organizational culture will be different in each organization, depending on the values, habits, and also environmental factors of the organization. McDonald (1992) suggests four types of cultural categories: (1) culture of adaptation, (2) achievement culture, (3) clan culture, and (4) bureaucratic culture. Meanwhile, according to Daft (2005), the organizational culture is a collection of a values, assumptions, understanding, and norms believed by individuals within the organization and spread to other individuals as something that is considered true. Furthermore, according to Cole (1997), organizational culture is a set of shared values, norms, and beliefs within an organization, and she distinguishes cultures in two categories, namely explicit culture and implicit culture. Based on these opinions, organizational culture is a collection of values, customs, and norms, believed and considered correct and thus implemented in an organization.

1.3 *Employee performance*

Employee performance shows in both financial and nonfinancial results that relate to its success in relation to organizational performance. Another opinion suggests that performance is a group of behaviors generated based on a person's technical knowledge in the form of skills, adaptability, and interpersonal relationships, resulting in increased productivity, customer satisfaction, organizational growth and development, and so on. Employee performance is also one of the elements that can be assessed through level of productivity, i.e., through quality, quantity, knowledge, and creativity over a predetermined period based on an assessment system with reliable standard parameters. Employee performance is a group of behaviors or attitudes related to the success of the organization's performance and related to levels of productivity based on a standards assessment system set in the organization. According to Armstrong (2009), the factors that affect the level of

individual performance are the motivation, ability, and opportunity to participate.

1.4 *Hypothesis*

- Leadership style and organizational culture are suspected to together affect employee performance.
- Leadership style affects employee performance.
- Organizational culture affects employee performance.

2 METHOD

This study uses an approach with quantitative methods by looking at phenomena to measure the influence of leadership style and organizational culture on employee performance. The study used explanatory research that explains the causal relationship of variables through hypothesis testing. In this study, the authors use multiple regression analysis using SPSS program version 17. The number of employees in Duta Rama Company was 102 people, consisting of 10 people at the leadership level and 92 people at the employee level. The population is 92 employees, and the whole population is sampled, so it can be said that this research uses a saturated sample. Operational variables consist of:

1. Leadership style ($X1$).
2. Organizational culture ($X2$).
3. Employee performance (Y).

3 RESULTS AND DISCUSSION

3.1 *Characteristic of respondents research*

Table 1. Characteristic of respondents research.

No	Characteristic of Respondents	Circumstances	
		Amount	Percentage (%)
1.	Gender		
	a. Male	51	55.4
	b. Female	41	44.6
2.	Age of Respondent		
	a. 20 – 30 years	25	27.1
	b. 31 – 35 years	8	8.6
	c. 36 – 40 years	13	14.1
	d. 41 – 45 years	23	25
	e. 46 – 50 years	20	21.7
	f. $\geq$ 50 years	3	3.6

(Continued)

Table 1. (Cont.)

No	Characteristic of Respondents	Amount	Percentage (%)
		Circumstances	
3.	Level of Education		
	a. High school	21	22.8
	b. Diploma	44	47.8
	c. Bachelor	20	21.7
	d. Postgraduate	7	7.60
4.	Group/Level		
	a. Leader (top&middle)	25	27.1
	b. Supervisor	57	61.9
	c. Staff	10	10.8
5.	Marriage Status		
	a. Married	69	75
	b. Not Married	20	21.7
	c. Widow/widower	3	3.2

Source: Processed Results of Respondent Data.

3.2 Analysis of multiple correlation coefficient (R) and coefficient of determination (R2)

Multiple correlation coefficient analysis and determination coefficient analysis were done to determine the strength of the relationship and the amount of contribution influence between variables. The model summary is presented below.

3.3 Results of data analysis research

Table 2. Model summary.

Model	R	R Square	Adjusted R Square	Std. Error of the Estimate
1	0.631[a]	0.398	0.385	7.68716

Source: Output SPSS 17, 2018.

Data analysis research shows that the correlation coefficient of 0.631 (63.1%) means that leadership style (X1) and organizational culture (X2) have a strong relationship (over 50%) with employee performance variable (Y). The determination coefficient analysis result of 0.398 (39.8%) shows that leadership style (X1) and organizational culture (X2) influence (39.8%) employee performance variable (Y): the remainder is accounted for by other variables not included in equations of this model.

1. Leadership style and organizational culture are expected to together affect employee performance.

First Hypothesis Test Results.
From the regression analysis results of testing the first hypothesis as contained in Table 3, it can be interpreted that leadership style and organizational culture together have a significant and positive effect

Table 3. Anova[b].

Model		Sum of Squares	Df	Mean Square	F	Sig.
1	Regression	3480.724	2	1740.362	29.451	0.000[a]
	Residual	5259.232	90	59.092		
	Total	8739.957	92			

on employee performance, as evidenced by the f-test with a result of 29.451 with a significance level of 0.000 or less than 0.05. Thus the first hypothesis in this study is accepted.

2. The leadership style affects employee performance.

Second Hypothesis Test Results.

Table 4. Coefficients[a].

Model	Unstandardized Coefficients B	Std. Error	Standardized Coefficients Beta	T	Sig.
1 (Constant)	12.079	30.307		0.399	0.691
Leadership Style	0.670	0.878	0.063	0.764	0.447

Looking at the results of regression analysis testing the second hypothesis as shown in the table above, it can be interpreted that leadership style. This is supported by the standardized beta coefficient value obtained of 0.063, the value of regression coefficient (a) leadership style variable of 0.670 and the value of t-test 0.764 with a significance value of 0.447. For the value of regression coefficient (b) and t-test using a significant level of 0.05, it can be concluded that this result shows that the leadership style affects the employee performance but not significantly because the significance value of 0.447 is greater than the significance level of 0.05. Thus, the second hypothesis in this study was rejected.

3. Organizational culture affects employee performance.

Third Hypothesis Test Result.

Table 5. Coefficients[a].

Model	Unstandardized Coefficients B	Std. Error	Standardized Coefficients Beta	T	Sig.
1 (Constant)	12.079	30.307		0.399	0.691
Organizational Culture	1.020	.136	.621	7.500	.000

It can be interpreted that organizational culture is significant to employee performance. This is supported by the value of the standardized coefficient beta obtained, 0.621. The value of the regression coefficient (a) the organizational culture variablen is 1.020 and the t-test value is 7.50 with a significance value of 0.000. For the value of regression coefficient (b) and t-test using a significance level of 0.05, it can be concluded that this result shows that organizational culture has a positive and significant effect on employee performance because the significance value of 0.000 is smaller than the significant level of 0.05. Thus, the third hypothesis in this study is accepted.

4 CONCLUSION

Based on the results of the study, the authors draw the following conclusions:

Leadership style might partially influence employee performance but not significantly. Organizational culture has a positive and significant effect on employee performance.

Based on the results of the discussion, the following suggestions can be put forward for use by researchers, academics, and Duta Rama Company:

1. Duta Rama Company's leadership style should be adjusted to the situation and conditions that exist at this time.
2. Other research can be developed in the future by developing other variables such as work motivation, learning organization and organizational commitment as intervening variable.

REFERENCES

Armstrong, M. 2009. *Armstrong's handbook of management and leadership: A guide to managing for results.* UK: Kogan Pade Limited.

Ball, R., Robin, A. & Wu, J. S. 2003. Incentives versus standards: Properties of accounting income in four East Asian countries. *Journal of Accounting and Economics* 36(1–3): 235–270.

Burn, J. M. 1978. *Leadership.* New York: Harper & Row.

Chatman, J.A. & Jehn, K.A. 1994. Assessing the relationship between industry characteristics and organizational culture: How different can you be. *Academy of Management Journal* 3(3): 522–553.

Christensen, C. 1997. Patterns in the evolution of product competition. *European Management Journal* 15(2): 117–127.

Chen, X., Ba, Y., Ma, L., Cai, X., Yin, Y., Wang, K., Guo, J., Zhang, Y., Chen, J., Guo, X. & Li, Q. 2008. Characterization of microRNAs in serum: A novel class of biomarkers for diagnosis of cancer and other diseases. *Cell Research* 18(10): 997.

Chen, X., Lenhert, S., Hirtz, M., Lu, N., Fuchs, H. & Chi, L. 2007. Langmuir-Blodgett patterning: A bottom-up way to build mesostructures over large areas. *Accounts of Chemical Research* 40(6): 393–401.

Cole, S. 1997. Cultural heritage tourism: The villagers' perspective. A case study from Ngada, Flores. *Tourism and Heritage Management.*

Daft, R. L. 2005. *The leadership experiences.* Mason, OH: Thomson South-Western.

Deshpande, R. & Webster, J.F.E. 1989. Organizational culture and marketing: Defining the research agenda. *The Journal of Marketing*: 3–15.

Drucker, H. & Cortes, C. 1996. Boosting decision trees. *Advances in Neural Information Processing Systems* 8: 479–485.

House, R.J., Hanges, P.J., Javidan, M., Dorfman, P.W. & Gupta, V. 2004. *Culture, leadership, and organizations: The globe study of 62 societies.* Sage Publications.

Koontz, H. & Weihrich, H. 1990. *Essential of management,* international edition.

McDonald, P. & Gandz, J. 1992. Getting value from shared values. *Organizational Dynamics* 20(3): 64–77.

Northouse, L.L., Katapodi, M.C., Song, L., Zhang, L. & Mood, D.W. 2010. Interventions with family caregivers of cancer patients: Meta-analysis of randomized trials. *CA: A Cancer Journal for Clinicians* 60(5): 317–339.

Robbins, S.P. & Coulter, M. 2007. Foundation of planning. *Management.*

Trefry, M.G., 2006. A double-edged sword: Organizational culture in multicultural organizations.

Yukl, G. 1989. Managerial leadership: A review of theory and research. *Journal of Management* 15(2): 251–289.

Yukl, G. & Lepsinger, R. 2005. Issues & observations: Improving performance through flexible leadership. *Leadership in Action: A Publication of the Center for Creative Leadership and Jossey-Bass* 25(4): 23–24.

Intellectual capital and knowledge sharing linkages for enhancing institutional performance: Indonesia colleges case

M. Maryani
Universitas Winayamukti, Bandung, Indonesia

H. Djulius
Universitas Pasundan, Bandung, Indonesia

ABSTRACT: In recent years, the Indonesian government has established various policies to improve the performance and scientific publications of lecturers in Indonesia, in an effort to improve the performance of higher education institutions. Based on this phenomenon, this research examined the relationship between intellectual capital and knowledge sharing and its impact on the knowledge management performance of the higher education institutions. This research used the quantitative method by conducting a survey using questionnaires given to the decision makers in college institutions involving 156 respondents. Its analysis unit was the management department of various colleges in West Java and Banten, which is one of the largest clusters of college management in Indonesia. The survey results data were processed using path analysis to elaborate the research hypothesis. The result of the study indicated that there was partially significant influence either directly or indirectly from intellectual capital and knowledge sharing on commitment to the study program.

1 INTRODUCTION

The policy to equalize the performance of the lecturers in Indonesia to be equivalent to the performance of international lecturers has had implications on the importance of higher education institution performance even to the smallest entity, namely, the department. The main issue of institutional performance in this research is knowledge management, considering that the universities possess a high role in forming human capital through the transfer of technology and science to the graduates (Djulius 2017a). This research focuses on the management departments at private universities under Kopertis Region IV covering West Java and Banten provinces. The National Accreditation Board of Higher Education released data indicating only six of the 100 management departments of the universities in West Java and Banten hold A-level accreditation.

To address this problem, a strong employee commitment is required to support the performance of the departments within the university environment (Haftkhavani et al. 2012, Jafri & Lhamo 2013, Zafeiti & Noor 2017), as well as other factors such as intellectual capital and knowledge sharing activities. Organizations with higher levels of intellectual capital (i.e., elements of human capital, structural capital, and relationship capital) will result in higher performance than the competitors that have lower levels of intellectual capital (Kamal et al. 2012, Yi et al. 2014, Cloud 2015, Todericiu & Şerban 2015).

Considering that part of the core business of the university is knowledge management, then one of its elements is knowledge sharing. Without knowledge sharing, the learning process, which is the process of improving knowledge, will be hampered. Individuals who share knowledge with others will not lose their knowledge but will multiply the value of that knowledge or science. Through knowledge sharing there will be an increase in the value of knowledge possessed by the organization. The addition of this value will affect employee performance (Zack et al. 2009, Ha et al. 2016).

In addition to causality between intellectual capital, knowledge-sharing activities, and commitment to institutional performance, this research also analyzes the relationship among the three variables. Knowledge management is the basis of intellectual asset management, which contributes to the improvement of the organization. One important application of knowledge management is knowledge sharing. Hsu (2008) has also concluded that knowledge sharing plays an important role in improving performance. The role will be more important through human capital with a better quality. Intellectual capital consists of human capital, relationship capital, and structural capital, which exists in the organization and will grow and expand if it is shared or distributed to employees (Radaelli et al. 2011, Reza & Mehrabian 2015, Djulius 2017b). Moreover, the application of knowledge management must be supported by the commitment of employees to build a prime company that is able to compete with other

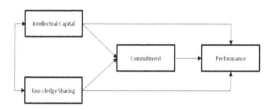

Figure 1. Logical framework.

companies. If the employees are not committed to the company then the implementation of knowledge management will not lead to significant progress in terms of knowledge sharing and will also affect the knowledge of the organization (Li-Ying 2015, Djulius et al. 2018). Finally, this research also hypothesizes the relationship between intellectual capital and commitment. Through high commitment, the knowledge gained from the institution can be transformed into new knowledge. The main issue of knowledge management is not just information technology but also the behavior that enables members of the organization to increase their knowledge, and that of their work unit as well, by combining their existing knowledge with new knowledge (Yi et al. 2012, Zeinoddiniet al. 2015, Wilkins et al. 2017).

The logical framework formed from the in-depth elaboration of the theoretical phenomena and the above empirical phenomenon is shown in Figure 1.

2 METHOD

This research used descriptive and verification methods. The population in this study was the overall S1 management department of universities in West Java and Banten. The sampling technique was done using stratified random sampling with a population of 273 and a sample number of 156. The populations and samples consisted of heads of the study program, secretary, and staff of the S1 management study programs.

Samples were selected through the proportional stratified random method as shown in Table 1.

The respondents of the research were the heads, secretaries and staff of the departments, for a total of 156 persons.

Table 1. Sampling frame.

No	Accreditation Status	Population	Sample
1	A	6	4
2	B	30	22
3	C	64	47
Total		100	73

Table 2. Concepts and variables.

Variable	Concept Variable	Sub Variables
Intellectual Capital	Intangible assets comprising of human capital, structural capital, and capital relations	Human capital / Structural capital / Relationship capital
Knowledge Sharing	The process by which the individuals exchange their knowledge, both tacit knowledge and explicit knowledge, and create new knowledge in an integrated manner	Ability to acquire knowledge / Ability to assimilate knowledge / Ability to transform knowledge / Ability to exploit knowledge
Organizational Commitment	A sense of attachment such that an employee or member of an organization will stay within the organization to achieve the vision, mission, and goals of the organization	Affective commitment / Sustainable commitment / Normative commitment
Performance	The work achieved by a person or a group in the organization that is influenced by various factors to achieve organizational goals within a certain period of time	Work knowledge / Work quality / Innovation / Personal ability / Efficiency / Effectivity / Cooperation

The variables and concepts used in this research are shown in Table 2.

3 RESULTS AND DISCUSSION

The result of the path analysis estimation of the research model depicting relationships among the research variables is shown in Table 3.

The direct influence of the intellectual capital variable on departmental commitment is 17.31%, while its indirect influence, that is, through knowledge sharing variables, is 13.74%. The direct influence of intellectual capital variables on departmental commitment is greater than its indirect influence. The direct influence of knowledge sharing variables to departmental commitment is 18.75%, while the indirect influence, which is through the intellectual capital variables, is 13.74%. The indirect influence of knowledge sharing on departmental commitment is greater than its direct influence. Based on these results, it is obvious that the relationship between intellectual capital and knowledge sharing has a smaller impact than each variable directly.

641

Table 3. Path analysis estimation.

| Variable | Direct Influence (1) | Indirect Influence through | | | Total Influence |
		Intellectual Capital	Indirect Influence (2)	Knowledge Sharing	
Intellectual Capital	17.31%		13.74%	13.74%	31.05%
Knowledge Sharing	18.75%	13.74%		13.74%	32.49%
Sub Structure 1	36.05%	13.74%	13.74%	27.49%	63.54%
Commitment					4.76
Sub Structure 2					68.3%

Knowledge sharing had a greater impact on departmental commitment than did intellectual capital. This indicates that the sharing of knowledge produces increased commitment more than does intellectual capital, because sharing knowledge from different parties can make the chairman and secretary of the department increase their commitment in order for the institution to gain public trust. Knowledge sharing facilitates learning techniques and solving problems; knowledge sharing can also encourage exchange and creation within the organization to improve competitive advantages, as when people share news, experiences, knowledge, and information with other employees. Hsu (2008) has also concluded that knowledge sharing plays an important role in improving performance and that this role will be more important through human capital with a better quality. In fact, people with skills, abilities, knowledge, and an improved attitude, will have the ability to share implicit and explicit knowledge. Nevertheless, to have a high commitment still also requires intellectual capital of the chairman and secretary of the department for achieving the aspired-to goals.

4 CONCLUSION

There was a partially significant influence, either directly or indirectly, from intellectual capital and knowledge sharing on the commitment of the study program. In addition, there was a significant influence on the study program's commitment to the performance of the study program. The better commitment of the study program will improve the performance of the study program.

REFERENCES

Djulius, H. 2017a. Foreign direct investment and technology transfer: knowledge spillover in the manufacturing sector in Indonesia, *Global Business Review* 181: 57–70.

Djulius, H. 2017b. How to transform creative ideas into creative products: learning from the success of batik fractal. *International Journal of Business and Globalisation* 192.

Djulius, H., Juanim, J. & Ratnamiasih, I. 2018. Knowledge spillover through foreign direct investment in textile industry. *International Journal of Economic Policy in Emerging Economies* 111(2): 12–25.

Ha, S.-T., Lo, M.-C. & Wang, Y.-C. 2016. Relationship between knowledge management and organizational performance: a test on SMEs in Malaysia. *Procedia - Social and Behavioral Sciences. The Authors* 2: 184–189.

Haftkhavani, Z.G., Faghiharam, B. & Araghieh, A. 2012. Organizational commitment and academic performance case study: students at secondary schools for girls. *Procedia - Social and Behavioral Science* 69: 1529–1538.

Jafri, H. and Lhamo, T. 2013. Organizational commitment and work performance in regular and contract faculties of Royal University of Bhutan, *Journal of Contemporary Research in Management* 82: 47–58.

Kamal, M.H.M. et al. 2012. Intellectual capital and firm performance of commercial banks in Malaysia. *Asian Economic and Financial Review* 24: 577–590.

Li-Ying, J., Li, J., Yuan, L. and Ning, L. 2015. Knowledge sharing and affective commitment: the mediating role of psychological ownership. *Journal of Knowledge Management* 196.

Radaelli, G. et al. 2011. Intellectual capital and knowledge sharing: The mediating role of organisational knowledge-sharing climate. *Knowledge Management Research and Practice. Nature Publishing Group* 94: 342–352.

Reza, M. & Mehrabian, M. 2015. Study the effect of intellectual capital and knowledge sharing on organizational innovation of software companies. *Introduction* 310: 45–59.

Shehzad, U. et al. 2014. The impact of intellectual capital on the performance of universities. *European Journal of Contemporary Education* 104: 273–280.

Todericiu, R. & Șerban, A. 2015. Intellectual capital and its relationship with universities. *Procedia Economics and Finance* 2715: 713–717.

Wilkins, S., Butt, M.M. & Annabi, C.A. 2017. The effects of employee commitment in transnational higher education: the case of international branch campus. *Journal of Studies in International Education* 214: 295–314.

Yi-Ching Chen, M., Shui Wang, Y. & Sun, V. 2012. Intellectual capital and organizational commitment. *Personnel Review* 413: 321–339.

Zack, M., McKeen, J. & Singh, S. 2009. Knowledge management and organizational performance: an exploratory analysis, *Journal of Knowledge Management* 136: 392–409.

Zafeiti, S.M.B. & Noor, A.M. 2017. The influence of organizational commitment on Omani public employees' work performance. *International Review of Management and Marketing* 72: 151–160.

Zeinoddini, S., Esfahani, S.A. & Soleimani, H. 2015. The effect of intellectual capital on organizational commitment: a case study of the ministry of economic affairs and finance of Kermanshah province. *International Journal of Organizational Leadership* 4: 324–341.

Advances in Business, Management and Entrepreneurship – Hurriyati et al (eds)
© *2020 Taylor & Francis Group, London, ISBN 978-0-367-27176-3*

The influence of leadership style on employee performance in construction company

P.D.H. Ardana, N.K. Astariani & I.G.M. Sudika
Universitas Ngurah Rai, Denpasar, Indonesia

ABSTRACT: To make an enhancement of employee performance in Construction Company, the human resources play an important role where the leadership factor is inside in. In this research, the role of leadership in Construction Company becomes the main topic of discussion. The study used quantitative approaches and a questionnaire was designed. A five-point Likert scale questionnaire was used to determine the impact of leadership style on employee performance. SPSS software was used in analyzing the questionnaires. Validity test, reliability test, descriptive analysis and multiple regression analysis were presented. Descriptive statistics show that the most significant value associated with employee performance is democratic leadership style. The value of democratic leadership style can be shown from multiple regression testing, t-test, F-test, and coefficient determination which produce the best result. The subsequent findings of this study show the leadership behaviors are influencing amount 32.9% for the employee performance in X Construction Company.

1 INTRODUCTION

Construction is a widely diverse industry brimming with innovation and change. In general, the construction industry plays a key role for governments in both developed and developing economies. The sector creates a new job, drives economic growth, and provides solutions to address social, climate and energy challenges. The construction industry has important linkages with other sectors so that its impact on GDP and economic development goes well beyond the direct contribution of construction activities.

In Indonesia, infrastructure construction becomes a main priority by the government because can make an equity throughout the region of Indonesia. The purpose of equity is to achieve economic independence by moving the strategic sectors of the domestic economy. So the province of Bali, infrastructure development has developed very rapidly. Therefore, the large or small contractor as a construction company has an important role in infrastructure development.

Many factors that influence the performance of construction companies (contractor). These factors are work experience, salary, loyalty, and equally important is the leadership style applied to a construction company which all connecting with human resources. According to (Maddepunggeng et al. 2016), (Putri 2014), & (Nurdin 2014) was shown that work experience, amount of salary, a loyalty which combined with leadership style provided significant results in improving the performance of the human resources in companies. Human resource is a key element in construction companies. Failure in managing human resources can be a disruption in achieving objectives, whether in the performance of human resources or the performance of the company. Organizational or company success in achieving its goals and objectives depends on the leaders of the organization and their leadership styles (Voon et al. 2011). Huemann et al. (2007) in (Nauman & Khan 2011) suggests that the project is a social system, and includes several areas focused on organizational behavior, leadership, communication, team building, and human resource management.

According to (Buba & Tanko 2017), leadership has proven to be an important factor in creating a successful project which comes from different leadership theories. Neither from contingency or universal theory dealing with the fact that leaders influence their team members in a way to fulfill their desired goals and the result show the directing leadership styles better than another leadership style in improving construction performance.

Leadership styles have significant effects not only on small construction companies but also in the world's largest construction companies or other organizations. Different leadership styles may affect organizational or human resources effectiveness or performance. Leadership style is the most prevalent factors that influence employee's attitudes and behaviors including organizational commitment (Veliu et.al 2017). Leaders have adopted various styles when they lead others in the organization as according to (Kaihatu & Rini 2007) which raises the transformational leadership what was applied by the principal gave satisfaction to the high school teachers in Surabaya; (Khan et al. 2014) shows the combination both the features of transformational and transactional leadership was influentially for project success; according to (Jiang et al. 2017) was reveal that employee sustainable performance is positively influenced by transformational leadership; (Kurzydlowska 2016) showed the transformational leadership styles better than transactional leadership in project management. In this case, there has not been seen the use of an autocratic, democratic and delegative or laissez-faire leadership styles that are analyzed to see its influence on employees in a company especially construction company. The autocratic, democratic and laissez-faire leadership styles are three major leadership, particularly in decision making.

So for this paper focuses on three leadership styles that are autocratic, democratic and laissez-faire (delegative) styles and how styles of leadership can influence on the employee performance of company X in Bali which engaged in construction.

2 METHOD

2.1 *Leadership*

There is no single definition of leadership. It is a complex, emergent process that can be described in many different ways, using different components, styles, and traits. In general terms, it is a process by which a person influences others to accomplish an objective. According to (Naoum 2001), leadership is aligning people towards common goals and empowering them to take the actions needed to reach them. While (Blanchard & Hersey 1993) said leadership as the process of influencing the activities of an individual or group in efforts toward goal achievement in a given situation.

Leadership in project construction is very important in the implementation of the construction work especially to lead the human resources. There is a possibilities cause of bad leadership style is a decrease in the performance of human resources will impact the performance of the project construction.

2.2 *Leadership styles*

Theories about leadership styles seek to examine the behavior or actions of leaders in influencing and/or moving their followers to achieve a goal. These behaviors and actions can basically be understood as two distinct but interlocked things, there is focus on task completion (work) or task/production-centered; focus on coaching efforts on personnel performing the task/job (job/employee centered). According to Lippit and White (Clark 2015), leadership style is divided into three: Authoritarian or autocratic: the leader tells his or her employees what to do and how to do it, without getting their advice; Participative or democratic: the leader includes one or more employees in the decision-making process, but the leader normally maintains the final decision-making

authority; Delegative or laissez-faire (free-rein): the leader allows the employees to make the decisions, however, the leader is still responsible for the decisions that are made.

2.3 *Leadership styles and employee performance*

Performance is understood as the achievement of the organization in relation with its set goals. It includes outcomes achieved or accomplished through the contribution of individuals or teams to the organization's strategic goals. Performance is a multi-component concept and on the fundamental level one can distinguish the process aspect of performance, that is, behavioral engagements from an expected outcome (Pradhan & Jena 2017). The performance-driven objective is expected to be aligned with the organizational policies so that the entire process moves away from being event-driven to become more strategic and a people-centric perspective

In the organizational context, performance is usually defined as the extent to which an organizational member contributes to achieving the goals of the organization. Employees are a primary source of competitive advantage in service-oriented organizations. In addition, a commitment performance approach views employees as resources or assets and values their voice. Employee performance plays an important role in organizational performance. Employee performance is originally what an employee does or does not do.

A good leader understands the importance of employees in achieving the goals of the organization, and that motivating the employees is of paramount importance in achieving these goals. Different leadership styles bring about different consequences, which have a direct or indirect impact on the attitude and behaviors of the employees (Veliu et al. 2017). The employee performance is positively influenced by leadership styles (Jiang et al. 2017).

2.4 *Conceptual framework*

The following conceptual framework was developed after a review of the existing literature to investigate the research questions. The framework shows leadership styles (autocratic, democratic, and laissez-faire) as the independent variables used to explain

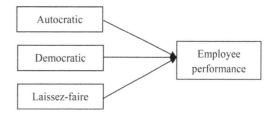

Figure 1. Conceptual framework.

employee performance as the dependent variable. The research model is illustrated in Figure 1.

3 RESULT AND DISCUSSION

The main purpose of this study was to identify the influence between leadership style and employee's performance. To conduct the study it is essential to plan and formulate appropriate study area and period, research design, research methodology includes sampling design, target population, source of data, data collection instrument, data analysis, reliability, and validity test were incorporated. The research strategy is based on quantitative research. A five-point Likert scale questionnaire was used to determine the impact of leadership style on employee performance. SPSS software was used in analyzing the questionnaires. Validity test, reliability test (Cronbach's Alpha), descriptive analysis and multiple regression analysis were presented.

3.1 Sample

The sample comprised of 47 employees who work for company X. Based on the age of employees, as much 55.32% of respondents are 26-35 years; the respondents which ageless from 26 years are 14.89%, and other (> 35 years) are 29.79%. From gender, 85.11% respondents were males and 14.8% were females. From an educational level of at least show 53.19% a senior high school degree; 23.4% a graduate degree, and other educational is 23.4%. Most respondents had been with the company between one and five years (85.11%) and six to ten years (6.38%) and others more than ten years 8.51%.

3.2 Determination of range

In this research, the instrument measured by a Likert scale with the highest point in each question is 5 and the lowest point is 1. The response from respondents show for the autocratic style the average value is 101.4, democratic style is 198.8, negative or Laissez-faire is 108.8, and the average value for the performance of employees is 201.

3.3 Validity test

Validity indicates the extent an accuracy of a measuring instrument in performing its measuring function. This validity test can be done by looking at the value of significance between the score of each item in the questionnaire with the total score to be measured. If the significance value < 0.05 then valid and if the value significance > 0.05 then is not valid. In this research, the analysis used SPSS 21 (Statistical Package for Social Science 21) program.

From the analysis obtained that all items of the question have a significance because of the value below 0.05. So it can be said that all items are a valid instrument and all included in the measurement.

3.4 Reliability test

Reliability test is used to know the consistency of the measuring instrument, whether the measuring instrument can be relied upon for further use. Reliability test results in this study using coefficient Cronbach's alpha, which according to (Ghozali 2011) that the instrument is said reliable if it has coefficient Cronbach's alpha equal to 0.60 or more. Results of data reliability test through SPSS 21 can be seen in the following below.

Based on the reliability test results, the dependent and independent variables produce the value which greater than 0.6. So it is mean all of the indicators in this research were said to be reliable.

3.5 Hypothesis test

Hypothesis testing is an analysis to test the influence of leadership style (autocratic, democratic, and delegative) by using multiple regression analysis and coefficient of determination (R^2).

3.6 Multiple regression analysis

Multiple regression equation can be used to predict how high the value of dependent variable when the value of independent variable is manipulated or modified (Sugiyono, 2016). The creation of multiple regression equations can be done by implanting the numbers in the unstandardized beta coefficient using the SPSS 21 program, the equations are generated:

$$Y = 14.432 - 0.006X1 + 0.430X2 - 0.179X3 \quad (1)$$

Where:
Y = employee performance;
b_0 = constants;
X1 = autocratic style;
X2 = democratic style;
X3 = delegative style; and
b_1, b_2, b_3 = regression coefisient.

From the regression equation it can be interpreted several things, among others: The number 14.432 is a constant value, indicating the level of employee performance gained obtained when the leadership style variable is ignored; The number 0.0061 indicate that the autocratic leadership style variables have a negative effect or not have a significant effect on employee performance; The number 0.430 indicates that the variable of democratic leadership style has a positive effect or significant on employee performance; The number 0.179 indicates that the *delegative* leadership style variables do not have a significant effect on employee performance.

Based on the standardized coefficients beta (scb) test result of the regression equation above it can be seen that the most influential independent variable on employee performance is a democratic leadership style. A second hypothesis states that the factor of a democratic leadership style has the most dominant influence on employee performance is acceptable.

3.7 T-test

The t-test is performed to determine the influence of each or partially independent variable (autocratic, democratic, and delegative leadership style) on the dependent variable (employee performance). Meanwhile, the partial effect of these three independent variables on employee performance is shown in the following table.

The influence of each leadership style variables autocratic. democratic and *laissez-faire* or *delegative* on the employee's performance can be seen from the direction of sign and level of significance (probability). If the level of significance < 0.05 and t-count is greater than t-table then it can be said that the variable is influential and significant.

Effect of autocratic and *laissez-faire* leadership style on employee performance can show from the t-test. The result of the partial test (t-test) between autocratic and *laissez-faire* leadership style variable to employee's performance variable shows t-count $<$ t-table. this means that autocratic and laissez-faire leadership style has negative and no significant effect to employee performance.

The result of the partial test (t-test) between the variables of democratic leadership style to employee performance variable shows t-count equal to 3.668 while t-table equal to 1.68023 or t-count $>$ t-table, this indicates that democratic leadership style has a positive and significant influence to employee performance.

3.8 F-test

The F-test basically shows whether all the independent variables included in the model have a mutual influence on the dependent variable. The result of F-test can be seen in Table 2.

F-test (simultaneous test) used to determine the effect of probability of whether or not it can be used to forecast the value of the independent variable to the dependent variable. it must look at F-count and F-table. With F-count equal to 8.535 and F-table equal

Table 2. F-test results.

Model	Sum of Squares	df	Mean Square	F	Sig.
Regression	64.605	3	21.535	8.535	.000b
Residual	108.501	43	2.523		
Total	173.106	46			

Table 3. The coefficient of the determination result.

Model	R	R Square	Adjusted R Square	Std. The error of the Estimate
1	0.611a	0.373	0.329	1.588

to 2.82 then it can be said that together free variable can explain a dependent variable simultaneously because of F-count $>$ F-table. Hence the hypothesis that leadership styles have a positive and significant effect on employee performance is acceptable.

3.9 The coefficient of determination test

The coefficient of determination test is to know how closely the influence of leadership style on employee performance. It is shown in Table 3.

Based on the results of the above data using SPSS got the coefficient of determination (R^2) is 0.329 this shows that as much as 32.9% employee performance on the X construction company influenced by independent variable that is leadership styles while the rest that is equal to 67.1% influenced by other factors which not included in this model.

4 CONCLUSION

The relationship between the leadership styles with the employee performance in the construction company was discussed. In this study found. the democratic leadership style has positive and significant relationships with employee performance. The value of democratic leadership style can be shown by multiple linear testing. T-test. F-test. and coefficient determination which produce the best result. While autocratic and laissez-faire leadership style showed a negative and no significant relationship for employee performance. And in this study. the leadership was given influence on employee performance.

ACKNOWLEDGMENT

We would like to thank the members of the Scientific Committee of the 3rd Global Conference on Bussiness. Management and Entrepreneurship for their

Table 1. T-test results.

Model	t_{count}	t_{table}	Sig
Autocratic (X1)	-0.067	1.68023	0.947
Democratic (X2)	3.668	1.68023	0.001
Laissez-faire (X3)	-1.994	1.68023	0.053

contributions as reviewers of abstracts and full papers. We would like to thank too for the Ngurah Rai University especially Department of Civil Engineering Faculty of Engineering for support in this research.

REFERENCES

Blanchard, P. & Hersey, K. 1993. *Management of Organizational Behavior: Utilizing Human Resources.* Englewood Cliffs: Prentice Hall.

Buba, S. & Tanko, B. 2017. Project Leadership and Quality Performance of Construction Projects. *International Journal of Built Environment and* Sustainability: 63-70.

Clark, D. 2015. *From The Performance Juxtaposition.* [Online]. Retrieved: http://www.nwlink.com/~donclark/leader/leadstl.html. Accesed May 8. 2018.

Ghozali, I. 2011. *Aplikasi Analisis Multivariate dengan Program IBM SPSS 19.* Semarang: Badan Penerbit Universitas Diponegoro.

Jiang, W., Zhao, X. & Ni, J. 2017. The Impact of Transformational Leadership on Employee Sustainable Performance: The Mediating Role of Organizational Citizenship Behavior. *Sustainability*: 1-17.

Kaihatu, T. & Rini, W. 2007. Kepemimpinan Transformasional dan Pengaruhnya Terhadap Kepuasan atas Kualitas Kehidupan Kerja. Komitmen Organisasi. dan Perilaku Ekstra Peran: Studi pada Guru-Guru SMU di Kota Surabaya. *Jurnal Manajemen dan Kewirausahaan*: 49-61.

Khan, M.S., Khan, I., Akhtar, B.Y., Abbasi, Z., Khan, F., Jan, F. & Ahmad, R. 2014. Styles of Leadership and Its Impact upon the Projects Success. *Public Policy and Administration Research*: 48-52.

Kurzydlowska, A. 2016. Leadership in Project Management. *Humanities and Social Sciences*: 103-112.

Maddepunggeng, A., Abdullah, R. & Mustika, T.F. 2016. Pengaruh Pengalaman Kerja Dan Gaya Kepemimpinan Terhadap Kinerja Sumber Daya Manusia (SDM) Konstruksi. *Konstruksia*: 99-108.

Naoum, S. 2001. *People and Organizational Management in Construction.* London: Thomas Telford Publishing.

Nauman, S. & Khan, A. 2011. Patterns of Leadership for Effective Project Management. *Journal of Quality and Technology Management*, (20): 1-14.

Nurdin. 2014. Pengaruh Gaya Kepemimpinan Dan Besarnya Gaji Terhadap Efektivitas Kerja Karyawan PT. Telkom Bekasi. *Transparansi*, (6)1.

Pradhan. R. & Jena. L. 2017. Employee Performance at Workplace: Conceptual Model and Empirical Validation. *Business Perspectives and Research*: 69-85.

Putri, S. 2014. Pengaruh Gaya Kepemimpinan dan Loyalitas Karyawan Terhadap Kinerja Karyawan Pada PT. Kurnia Alam Perista Kudus. *Doctoral dissertation, Fakultas Ekonomika dan Bisnis.* Semarang.

Sugiyono. 2016. *Metode Penelitian Kuantitatif Kualitatif dan R & D.* Yogyakarta: Alphabeta.

Veliu, L., Manxhari, M., Demiri, V. & Jahaj, L. 2017. The Influence Of Leadership Styles On Employee's Performance. *Vadyba Journal of Management*: 59-69.

Voon, M., Lo, M., Ngui, K. & Ayob, N. 2011. The influence of leadership styles on employees' job satisfaction in public sector organizations in Malaysia. *International Journal of Business. Management and Social Sciences*: 24-32.

Advances in Business, Management and Entrepreneurship – Hurriyati et al (eds)
© *2020 Taylor & Francis Group, London, ISBN 978-0-367-27176-3*

Impact of knowledge management on career development: A study among lecturers in private college

R. Widyanti & B. Basuki

Universitas Islam Kalimantan MAB, Banjarmasin, Indonesia

ABSTRACT: This study aims to test the relationship between knowledge management and career development of lecturers in private colleges in Kopertis Region XI Kalimantan Indonesia. Data were collected from lecturer civil servants; specifically, 103 questionnaires were collected and analyzed using the statistical technique of partial least squares (PLS). Some the open questions are presented to explain career development programs that a lecturer needed. This study concludes by measuring the impact of knowledge management as a very important tool for lecturers to develop a career in an effective way.

1 INTRODUCTION

Human resources have an important role in the progress of a nation. Moreover, fourth-generation Indonesia must be ready to face the era of industry. In this era of science and technology, changeis occurring rapidly so it must be anticipated in producing human resources.

In addition, global economic and business changes are fast, unpredictable, surprising, and complex, leading to conflicts within an organization (Bahaudin 2001). In such an environment of change, innovative networks and global economic knowledge movements add momentum, enabling companies and organizations to realize that competition and methods of information-based enterprise management from the previous century have fundamentally changed toward knowledge-based, where strategic collaboration is important mindset (Leibold et al. 2005). Organizations or companies are required to have the ability to detect the tendency of environmental change and able to take decisions quickly.

Private colleges are one sector of organizations producing quality human resources. Qualifications for private colleges or universities in South Kalimantan (based on university accreditation rankings) indicate mostly qualified B accreditation; that is, 11 colleges/universities, and another two qualified colleges C accreditation (web: kopertis11.net).

This is closely related to the qualification of education level and lecturer functional level. According to Miarso (2006), the development of lecturers, which includes personal development, professional development, organizational development, facility development, and career development, is a core part of the development of the institute. Evident from career development data of lecturers at private higher education in South Kalimantan, no one has become a professor yet. Some challenges have emerged in universities in this era of disruptive innovation, among them the extinction of science due to the emergence of various artificial intelligence, the entry of the era of abundance of science, and the application of nanotechnologies into another threat to science and humanity. Developments in digital technology of artificial intelligence (AI) that converts data into information has made if possible for people to easily and inexpensively obtain it.

This change affects the workings of the college as one of the sources of these conveniences, including changes in learning and teaching procedures. Likewise, the career development of lecturers has been done with the online system through PAK online higher education. Knowledge management resources are closely related to human resources (HR).

Differences or gaps in the results of the study are possible because the study uses different measurements and different test equipment as well. In this research, we will fill the gap by using career development variables by adding the indicators organizational needs and qualifications (Hedge et al. 2006, Law 2005).

Based on theoretical phenomenon, empirical phenomenon, and the difference of findings from previous research, the research question posed is: Does knowledge management affect the career development of lecturers?

The aim of this research is to test and empirically examine the influence of knowledge management on the career development of private college lecturers in South Kalimantan.

1.1 *Knowledge management*

According to Alavi and Leidner (2001), data is a representation of raw figures and facts. Once the data is processed systematically and structures are organized or given, data turns into information. When individuals have such information and apply to take action or make decisions, then it will be considered as knowledge (Rao 2003). Knowledge management is the creation, extraction, transformation, and storage of the correct knowledge and information to design better policies, modify actions, and deliver results (Horwitch & Armacost 2002).

Knowledge management (KM) was initially defined as a process of applying a systematic approach to capture, structure, manage, and disseminate knowledge across organizations in order to work faster, reproduce best practices, and reduce costly reworking of projects (Nonaka & Takeuchi 1995, Pasternack & Viscio 1998, Pfeiffer & Sutton 1999, Ruggle & Holtshouse 1999). Knowledge management solutions have proven to be most successful in the capture, storage, and dissemination of explicitly provided knowledge – especially lessons learned and best practices. Effective KM needs a continuous knowledge conversion process.

1.2 Career development

Career development during the post-graduate period is one of the prerequisites of producing graduate engineers (Hall 2005).

Career development is defined as a lifelong process, whereby an individual defines and refines a job role. These include individual awareness, skills, attitudes, talents, and abilities, especially as they change and develop during experiential education. This process helps students to explore diverse employment opportunities, learning the reality in the workplace, and identifying both the technical skills and the individual qualities they must possess to succeed (Hall 2005). Individuals typically experience four career stages: exploration, establishment, maintenance, and disengagement. In his studies, Noe (2006) defines management careers as "career exploration, career development goals, and implementation of career strategy" (Gary 2010).

The career development program is basically an integration of individual career planning and organizational career management activities comprised of individuals, elements of leadership, and the organization (Igrabia et al. 1991, Hall 1995, Hedge et al. 2006, Iskandar 2007).

1.3 Operational variable definition

The knowledge management (X) variables referred to in this research are the capability/ability of lecturers to acquire, store/document, disseminate/share, and use/apply knowledge related to opportunities and the external environmental challenges facing the lecturer and the knowledge of the internal resources (strength or weakness) of the lecturer. This variable is measured from four indicators: knowledge acquisition (X1.1), knowledge sharing (X1.2), knowledge storing (X1.3), and knowledge application (X1.4).

Career development is an important organizational effort in design and implementation, to achieve a balance between individual career needs and job skills required by the organization. The factors that make up an organizational career development program according to this study are (1) organizational needs, (2) the role of the superior and/or leader in developing career of lecturer, (3) qualification, and (4) organizational reward system.

2 METHOD

This research is explanatory research with the causality approach, seeking to explain, in the form of cause and impact (cause-effect), the relationship between some concepts, variables, or strategies developed in the management. Totally population is 162 lecturer civil servants, but the sample of respondents in this research is 103 people be used as criteria as become a lecturers at more five year.

3 RESULT AND DISCUSSION

3.1 Knowledge management (X)

Knowledge management illustrates the capability/ability of lecturers to acquire, share, document (storing), and utilize (application), knowledge related to opportunities and external environmental challenges faced by lecturers and knowledge of internal resources (strength or weakness). This variable is measured from four indicators: knowledge

Table 1. Knowledge management description variable (X).

Indicator	Perception Proportion Respondents (%)					
	Very bad	Bad	Fair	Good	Better	Average
Knowledge acquisition (X1)	0	10.53	55.26	34.21	0	3.75
Knowledge sharing (X2)	0	7.89	73.68	18.42	0	3.60
Knowledge storing (X3)	0	11.53	55.26	34.21	0	3.70
Knowledge application (X4)	0	5.26	47.37	47.37	0	3.91
Knowledge Management (X)						3.74

Source: Primary data, process, 2018.

acquisition (X1), knowledge sharing (X2), knowledge storing (X3) and knowledge application (X4). In general, the overview of the results of these research variables can be seen in Table 1, below.

The information presented in Table 1 indicates that knowledge management is working well enough for private college lecturers, as the overall perception of respondents is concentrated on good and fair value with an overall average score of 3.74. It describes organizational knowledge about the opportunities and challenges of the external environment, and knowledge of internal resources (strengths or weaknesses) are adequately managed.

3.2 Career development

The information presented in Table 2 indicates that career development is working well enough for private college lecturers, as the overall perception of respondents is concentrated on good and fair value with an overall average score of 4.43. It describes career development as about qualification and ability of the person.

3.3 Impact of knowledge management on career development

Knowledge management positively affects career development programs, with a path coefficient of 0.279. This means that the better the knowledge management, the better the career development of the lecturer civil servant at private colleges/universities. Conversely, less good knowledge management will have a negative impact on career development lecturers (Figure 1).

The findings of this study reinforce the theory of Nonaka and Takeuchi (1995) which states that knowledge management (KM) is a process of applying a systematic approach to capture, structure, manage, and disseminate knowledge throughout the organization in order to work faster, use best practices, and avoid expensive reworking for the project (Nonaka & Takeuchi 1995, Pasternack & Viscio 1998, Pfeiffer & Sutton 1999, Ruggle & Holtshouse 1999). Knowledge management solutions have

Figure 1. Impact of knowledge management on career development.

proven most successful in further capture, storage, and dissemination of explicitly given knowledge, particularly the more effective and efficient learning process. Thus, lecturers can work on other obligations related to the Tridharma of higher education. Effective knowledge management is required in the process of continuous knowledge conversion.

3.4 Discussion

Knowledge management is measured from four indicators: knowledge acquisition, knowledge sharing, knowledge storing, and knowledge application.

Based on the value of outer loading, it appears that knowledge acquisition is the most important indicator of knowledge management. This shows that, conceptually, the ability of lecturers to gain knowledge is the most important aspect for the application of good knowledge management. Knowledge gained from various sources, both internal and external, includes tacit (unspoken) and explicit knowledge. Knowledge is derived from information on the development of socio-economic conditions of society and population development, publication of various print and electronic media, and through careful observation of the development of science. Knowledge from external sources such as knowledge of reference sources, college students, knowledge of educational experts/consultants, and knowledge gained through cooperation/partnership with relevant external institutions, are also of great importance. Meanwhile, knowledge from internal sources, can be obtained from the knowledge of college leaders and employees, and by carefully identifying internal resources and college routines.

Table 2. Career development description variable (Y).

| Indicator | Perception Proportion Respondents (%) | | | | | |
	Very bad	Bad	Fair	Good	Better	Average
Organizational need (Y1)	0	.0	65.0	34.21	30.0	4.25
The role of superior (Y2)	0	1.7	51.7	18.42	30.0	4.10
Qualification (Y3)	0	0	55.26	34.21	48.3	4.43
System reward (Y4)	0	0	47.37	47.37	36.7	4.33
Career Development (Y)						4.28

Source: Primary data, process (2018).

In practice, some colleges (10.53%) are still poor in knowledge acquisition. These results illustrate that knowledge of external and internal college conditions, whether tacit or explicit, has not been explored. Based on the results of this study, it appears that the lecturers are more perceptive regarding the use of knowledge (knowledge application) as the most important aspect reflecting knowledge management. This indicates that the college leadership places more emphasis on the ease, accuracy, and effectiveness of the use of knowledge about opportunities, challenges, strengths and weaknesses in implementing Tridharma, as a reflection of the application of knowledge management.

These results prove that, conceptually, knowledge acquisition is the most important aspect. Therefore, in order for lecturers in private colleges to facilitate career development, they should start with the effort to get the correct and appropriate knowledge as necessary. The quality of the knowledge that will be the basis of decision-making or other uses depends heavily on the accuracy of the acquired knowledge, be it explicit or tacit knowledge, which can be obtained from internal or external sources.

Sharing knowledge between managers, between leaders and employees, or between units or sections in private colleges, either through formal or informal meetings, can be done effectively. This includes sharing knowledge with management consultants or online using an intranet that may already be available in all areas of the private college. Furthermore, knowledge will be accessible again easily and quickly by storing it neatly and securely in the storage media. By doing so, knowledge can be used easily, opportunely, and effectively for lecturer career development.

Organizations should pursue individual career planning and development tailored to their career needs (Igrabia, Greenhaus & Parasuraman 1991, Chen et al. 2004). As the gap between career development programs designed by organizations widens, individual career needs will be of lower priority to the organization (Nachbaggeur & Ridle 2002, Hedge et al. 2006).

The qualification indicator was adopted from the findings of (Hedge et al. 2006) in the form of propositions and did not perform empirical testing found the result of positive and significant influence in career development. This is in accordance with PPRI No. 37/2009 on Lecturers that lecturers are required to have academic qualifications, competence, Certificate of Educator, be physically and mentally healthy, and meet other required qualifications (Chapter II article 2).

Associated with the respondents' answer, *qualification* is the strongest indicator in forming a career development variable. This shows that respondents consider that lecturers should have high-level education, try to improve their learning system by evaluating themselves and their students at the end of each semester, and need to create professionalism in the learning process for lecturers, and to improve research quality by cooperating with other parties. This is shown from the fact that the result of the average score of items indicator question is 4.08. This indicates the tendency of respondents to agree that lecturers are professional educators and scientists who are considered superior than students. A professional in the field of lecturers' knowledge will be tied to professional ethics and academic ethics, so that lecturer qualification can be seen from the practice of teaching ethics. In accordance with the opinion of Sigit (2003) the higher the education level of a person, the better their morals as well.

This explanation provides an understanding that the lecturer always committed positively to the work and institution, then the institution, in this case one of higher education, should always improve and likewise improve the career development program for the lecturer. The career development of lecturers known as the profession development of lecturers (Miarso 2006) points to a broad effort in improving learning and performance in universities. Citing the opinion of Bergquist and Philips (Miarso 2006), the development of lecturers includes the development of facilities, career development as part of organizational development, and welfare development as an important part of personal development. Improving the career development of lecturers can be done through improving placement systems, improving human resources skills, and providing facilities and infrastructure in implementing Tridharma college activities. In addition, based on respondents' answers to open questions, lecturer civil servant wanted the opportunity to obtain training such as research methodology, data analysis, and the opportunity to obtain higher education.

4 CONCLUSION

Based on the discussion of research results, it can be concluded as follows: Implementation of good knowledge management can improve the accuracy of strategic planning. These results indicate that the implementation of knowledge acquisition, knowledge sharing, knowledge storing, and knowledge application is proven to improve the organization's ability to identify opportunities, challenges, strengths, and weaknesses, so that the formulation, implementation, and evaluation of strategies can be carried out appropriately.

Knowledge management can also improve the ability of lecturers in planning their career development. This indicates that the private college personnel who are actively involved as lecturer partners can improve the accuracy of online loan formulation, implementation, and evaluation of credit proposals (PAK).

REFERENCES

Alavi, M. & Leidner, D.E., 2001. Knowledge management and knowledge management systems: Conceptual foundations and research issues. *MIS Quarterly*: 107–136.

Beer, M., Voelpel, S.C., Leibold, M. & Tekie, E.B. 2005. Strategic management as organizational learning: Developing fit and alignment through a disciplined process. *Long Range Planning* 38(5): 445–465.

Cairns, G. 2010. Concept of prioritization and capture of tacit knowledge.

Hall, D.T. & Chandler, D. 2005. Psychological success: When the career is a calling. *Journal of Organizational Behavior* 26: 155–176.

Hedge, J.W., Borman. W.C. & Bourne. M.J. 2006. Designing a system for career development and advancement in the US Navy. *Human Resource Management Review* 16: 340–355.

Horwitch, M. & Armacost, R. 2002. Helping knowledge management be all it can be. *Journal of Business Strategy* 23(3): 26–31.

Igrabia, M., & Greenhaus, J. 1991. Career orientations of MIS employees: An empirical analysis. *Journal of MIS Quarterly* 15(2): 151–169.

Iskandar. 2007. Pengaruh jangkar karir dan sistem pengembangan karir terhadap kepuasan kerja dan kinerja pns pada pemkab. Kartanegara.

Miarso, Y. 2006. Action research di perguruan tinggi. *makalah disampaikan pada seminar penelitian ibii*. Jakarta.

Mujtaba, B.G. & Cavico, F.J. 2013. Corporate social responsibility and sustainability model for global firms. *Journal of Leadership, Accountability and Ethics* 10(1): 58–75.

Noe, R.A., Hollenbeck, J.R., Gerhart, B. & Wright, P.M. 2006. *Human resource management: Gaining a competitive advantage*, third edition. USA.

Nonaka, I. 2006. Creating sustainable competitive advantage through knowledge-based management.

Rao, R.N. 2003. The role of computer technology in knowledge acquisition. *Journal of Knowledge Management Practice* 11(3).

Sigit & Soehardi. 2002. *Esensi perilaku organisasional, BPFE – UST.* Yogyakarta.

Wang, H., Law, K.S., Hackett, R.D., Wang, D. & Chen, Z.X. 2005. Leader-member exchange as a mediator of the relationship between transformational leadership and followers' performance and organizational citizenship behavior. *Academy of Management Journal* 48(3): 420–432.

Yusufhadi. M. 2006. Menyemai benih teknologi pendidikan, pemutakhiran.

Advances in Business, Management and Entrepreneurship – Hurriyati et al (eds)
© 2020 Taylor & Francis Group, London, ISBN 978-0-367-27176-3

The role of job demands and teamwork effectiveness in a harmonizing relationship within police organizations

P. Yulianti, I.M. Rohmawati & N.A. Arina
Universitas Airlangga, Surabaya, Indonesia

ABSTRACT: Burnout is a condition which should be prevented and minimized in an organization, whereas engagement is one of the essential issues for an organization. The police, as the forefront agent of the state, hold an important role in maintaining public order and controlling crime. Therefore, the understanding of burnout and engagement is considered crucial for all police members. This study aimed at identifying the effects of job demands on burnout and engagement of police officers, and also to recognize the effects of job demands on burnout moderated with teamwork effectiveness. The data were collected through a survey with questionnaire involving 131 police officers at the Police Resort of Tanjung Perak Port Surabaya, Indonesia. The findings showed a significant effect on three hypotheses: the effect of job demand on burnout, job demands on engagement, and teamwork effectiveness on engagement. However, there was no significant effect of teamwork effectiveness on burnout, and job demands on burnout moderated with teamwork effectiveness.

1 INTRODUCTION

In carrying out their duties, the police are faced with a great responsibility. People still expect an increase in the role and duties of police as protectors, guards, and public servants, as well as the worthy law enforcers. The existence of a substantial responsibility to the community can sometimes cause the police officers to experience work stress and it can lead to physical and mental problems. Stress with the job as a police officer not only affects individuals but also influences the relationships between team members in carrying out their duties. Studies show that work stress is associated with mental health, tension, job satisfaction, and burnout (Maslach et al. 2012).

Burnout is the psychological response to prolonged work stress (Maslach et al. 2012). Burnout is the physical and/or psychological exhaustion of an employee or professional due to a work situation that does not support their expectations and needs. Under these conditions, employees or professionals may experience health problems psychologically or physically, and this can affect task performance. The incidence of burnout conditions can be affected by several factors, one of which is job demands.

Job demand is defined as an aspect of work that requires long-term physical, emotional, or cognitive effort because job demand is related to physical and psychological conditions (Demerouti & Bakker 2011). In this case, the police have enormous job demands, as frequent conflicts can come up when completing their duties, for example, when the police are managing traffic or criminal-related problems. Therefore, teamwork is expected to overcome

the problems faced by the police, because teamwork will reduce the burden of each individual police officer. The team also allows employees to collaborate, enhance individual skill, and provide constructive feedback by escaping the conflict between individuals (Jones et al. 2007, Manzoor et al. 2011)

Effective teamwork is expected to moderate job demands and burnout (Bakker & Demerouti 2008, Montgomery et al. 2015). Effective teamwork is also used to strengthen or weaken the influence of job demands on burnout. Thus, effective teams have a major role in reducing high job demands. The police will feel that the influence of effective teamwork within the organization or the police resort will affect the work demands so that it can reduce burnout.

An effective team also encourages the police officer to build engagement with his work. Work engagement is defined by Kahn (1990) as the organization members' effort to bind themselves to job roles. Under these conditions, members will actively engage and express themselves physically, cognitively, and emotionally as long as they are playing their role. The cognitive aspects of work engagement involve employees' trust in the organization, their leaders, and their working conditions. The emotional aspect involves the employees' feelings toward these three points, and whether the employees behave positively or negatively toward the organization and their leaders.

The physical aspect involves how much energy the employees use to accomplish their task within the organization. Employees with engagement values are workers with full involvement and enthusiasm to do their job (Harter et al. 2002, Tritch 2003).

1.1 The effects of job demands on burnout

According to a research conducted by Montgomery et al. (2015), there is a strong correlation between job demands and burnout. In nursing staff, job demands such as time pressures or required emotional interactions is predicted to cause a burnout. In addition, research conducted by Rothmann et al. (2007) explains that the dimensions of burnout (emotional exhaustion) are positively associated with job demands due to workload (which includes physical, emotional, and cognitive workload), and is negatively related to job resources due to inadequate organizational support (focused mainly on management support, communications, performance feedback, participation in decision making, work autonomy, and clarity of roles). When employees feel their job burden is very large and the resources given are very narrow or limited, then employees will experience physical and psychological fatigue. When employees are feeling tired or stressed, they will no longer feel a sense of commitment to the organization. Meanwhile, the higher the job demand perceived by the employees, the higher their burnout rate. In other words, job demands play a role in increasing burnout rate for employees.

Hypothesis 1: Job demands have a significant effect on burnout of police members.

1.2 The effects of job demands on engagement

In a study of professional nurses conducted by Montgomery et al. (2015), it was found that job demand negatively affects engagement: in addition to the role played in job development, job demands also have a negative relationship with the motivation of individuals that have been engaged with their job. This is also supported by other researchers (Bakker 2006, Mauno 2007). Job resources (e.g., autonomy, support from peers and supervisors, feedback, training and development opportunities) play a role to achieve the goals of work, to facilitate learning and growth, and to reduce the negative effects of job demands (Schaufeli & Bakker 2004). Job resources also increase employees' motivation to be engaged in work, and reduce the burnout rates caused by job demands (Schaufeli & Bakker 2004, Bakker et al. 2006). From the above explanation, can be concluded that when job demand is high, it will result in low engagement.

Hypothesis 2: Job demand has a significant effect on engagement of police members.

1.3 The effects of teamwork effectiveness on burnout

Montgomery et al. (2015) argue that a person will improve the quality of their work if they get support from the surrounding environment. The feeling of mutual respect and cooperation with each other can slightly relieve the feelings of fatigue and the feelings of presence of energy

between employees. So, it can be concluded that employees who can cooperate with each other well in an effective team can reduce the occurrence of burnout.

Hypothesis 3: Teamwork effectiveness has a significant effect on the burnout of police members.

1.4 The effects of teamwork effectiveness on engagement

Employees will feel involved and motivated by the job as they enjoy their work in the presence of support among team members and mutual contribution to carry out the responsibilities (Montgomery et al. 2015). Montgomery et al.'s (2015) study also presented a model that proposes teamwork effectiveness positively affecting employee engagement. This means that teamwork effectiveness has the ability to influence employee work motivation. Job resource is a potential motivation when employees are faced with high job demands that can cause burnout. According to the JD-R model, when employees are faced with high emotional demands, social support from peers may become more visible and more instrumental. Thus, the effectiveness of teamwork can affect employee motivation between the tension in work environment, which contributes to evidence that burnout and engagement are not just opposites. These results are consistent with the studies of burnout and engagement that indicate that fatigue and involvement have different patterns that are likely to cause consequences, suggesting that team involvement strategies should be used when high burnout occurs (Schaufeli & Bakker 2004). Thus, it can be concluded that high teamwork effectiveness among employees will also raise employee engagement. Therefore, in this research, teamwork effectiveness will affect the engagement on the police officers of Tanjung Perak Port Police Surabaya.

Hypothesis 4: Teamwork effectiveness has a significant effect on the engagement of police members.

1.5 The effects of job demands against burnout moderated by teamwork effectiveness

Teamwork is a work process within the scope of a group with participative leadership, shared responsibility, goal equality, intensive communication, focusing on the future, focusing on tasks, talents and creativity, and rapid responses to organizational goals (Buchholz 2000). When employees can adapt to their job and the environment of the team in overcoming difficulties and challenges, it is expected that they also can overcome the problems of job demands that trigger burnout. Montgomery et al.'s (2015) study describes effective teamwork as a very important predictor, so that as long as the person works under pressure they can cope proactively and resolve their responsibility well without emotional negativity arising. Many studies have suggested that job demands have a significant effect on burnout, but with an effective team within the work environment, it is expected that employees will be more motivated to engage themselves in their work and reduce the impact of burnout. So, it can be concluded that teamwork effectiveness can reduce the occurrence of burnout caused by high job demands.

H5: Job demand has a significant effect on burnout moderated by teamwork effectiveness of police members.

2 METHOD

The method of data collection is a survey method, by distributing a questionnaire. This research uses PLS (Partial Least Square) analysis technique. The population in this research is members of the Police Resort of Tanjung Perak Port Surabaya. The corresponding population number was 161 respondents, consisting of 100 respondents from "Sabhara and 61 respondents from traffic units. The sampling technique used is the census method, where each of the returned and qualified data will be used as a sample. From 161 questionnaires that had been dispersed, only 132 (response rate 81.99%) questionnaires were returned, with 1 unqualified questionnaire, thus making the total number of questionnaires examined 131 questionnaires. The measurement of job demands, teamwork effectiveness, burnout, and engagement variables is based on the respondents' answers or ratings on the statements in the questionnaire whose value is determined on a Likert scale, with the choices of "Strongly Disagree," "Disagree," "Neutral," "Agree" and "Strongly Agree."

3 RESULTS

The result of research hypothesis testing in PLS analysis using the result of inner weight estimation is described in Table 1.

3.1 The effect of job demand on burnout

The result of PLS value representing the effect of job demands on burnout is equal to 0.5563, and the t-statistics value is equal to 2.7318. This shows that there is significant influence on burnout at the Police Resort of Tanjung Perak Port Surabaya. Any increased job demands will result in instances of burnout, because the influence of job demands against burnout is proven true. Based on the results of the highest indicators, it can be seen that the members of the police force feel that they have high job demands that make them feel that they have many unfinished tasks. Because of this pressure of many unfinished tasks, many members don't feel unenthusiastic with their work anymore. On the other hand, burnout can occur because the work is always the same and increases over time. Job demands is perceived by the members of the Police Resort of Tanjung Perak Port Surabaya as at a moderate level, and burnout is also at a moderate level. Job demands can act as opportunities that offer a potential advantage, such as the opportunity for police officers to learn and adapt to the environment, thus the problems that come from burnout can be resolved.

The results of this study are in line with Montgomery et al. (2015), who proposed that job demands (workload and emotional demand) positively affect burnout (emotional exhaustion and depersonalization). A job demand is an aspect of work that requires long-term physical or emotional effort and is therefore associated with physiological and psychological costs (Demerouti & Bakker 2011). Job demands (irregular hours, time pressures,

Table 1. Result.

	Original Sample (O)	Standard Error (STERR)	T Statistics (|O/STERR|)	Result
Job demand →Burnout	0.5563	0.2036	2.7318	S (+)*
Job demand →Engagement	-0.5321	0.0773	6.8820	S (-)
Teamwork Effectiveness → Burnout	-0.2107	0.1474	1.4300	NS (-)
Teamwork Effectiveness → Engagement	0.2787	0.0898	3.1021	S (+)
Job demand → Teamwork Effectiveness → Burnout	0.0482	0.1712	0.2818	NS (+)
Variable			R Square	
Engagement			0.4754	
Burnout			0.4807	

* Note: S = Significant, NS = Not Significant

demanding interactions with the community) will turn into job stress if they require excessive effort from which one fails to get better. When this is done on a long-term basis, it can cause some physical and psychological disorders such as burnout (Demerouti et al. 2011) or depression (Hall et al. 2011). Evidence of a positive relationship between job demands and burnout was found in some organizational settings (Schaufeli & Bakker 2004, Bakker et al. 2006).

3.2 The effect of job demand on engagement

The results of PLS analysis show that job demands have a significant negative effect on the engagement of police officers of the Police Resort of Tanjung Perak Port Surabaya, as proven by the value of t-statistics of 6.8820 > 1.96. It means that when job demands increase, engagement will decrease. Based on results of the highest job demands and engagement indicator, it can be seen that members of the Police Resort of Tanjung Perak Port Surabaya actually have high engagement with their job if they do not feel high job demand. This can be an opportunity for members of the police force to be motivated by their work, in order to make job demands a motivation to work. High job demands significantly and negatively affect engagement, but job resources have a significant positive effect on engagement. An excess job demand is one of the best predictors of engagement. High job demands as perceived by the employees will result in low engagement. However, with strong motivation, the job demands can be resolved.

3.3 The effect of teamwork effectiveness on burnout

The result of PLS analysis representing the influence of teamwork effectiveness on burnout is −0.2107, with t-statistics value of 1.4300. This indicates that there is no significant influence of teamwork effectiveness on burnout of members of Police Resort of Tanjung Perak Port Surabaya. In other words, improving teamwork effectiveness will not result in a decreased burnout, because the influence of teamwork effectiveness on burnout is not real. Based on these results, the third hypothesis in this study that suspects a significant effect of teamwork effectiveness on the members of the Police Resort of Tanjung Perak Port Surabaya is unacceptable. The description of our respondents' answers shows that the members of the Police Resort of Tanjung Perak Port Surabaya have good or high effectiveness teamwork, and this high effectiveness teamwork has no significant or insignificant effect on moderate or neutral burnout of the members. This can be caused by other factors, such as conflict with the community, feeling depressed, emotional, and others. This is in line with the theory described by Maslach and Jackson (1981), which explains that the beginning of emotional

occurrence is when police personnel deal with individuals with some severe problems, and long-term emotional interactions with the community are the main source of burnout for the police officers. The members of the police force are expected to manage their emotions to present a neutral, solid, and controlled physical expression. The work of members of the police expects an emotional emphasis when handling an event like conflict, manipulation, and aggression. On the other hand, the members of the police are also asked to show affection and understanding, for example, for the victims of a crime. Thus, members of the police need to master the art of changing attitudes according to circumstances and controlling their emotional expression. This insignificant negative result is not surprising, as Montgomery et al.'s study of professional nurses suggests that researchers have found no evidence that teamwork effectiveness in the medical department affects the individual nurses' perception of job demands and burnouts (Montgomery et al. 2015).

3.4 The effect of teamwork effectiveness on engagement

The result of a statistical test with the PLS indicates the influence of teamwork effectiveness on the engagement of the police officers from Police Resort of Tanjung Perak Port Surabaya. The result shows that teamwork effectiveness has a significant positive effect on engagement, proven by the value of the PLS test results representing the effect of teamwork effectiveness to engagement of 0.2787, and the value of t-statistics of 3.1021. This means that the existence of good and effective teamwork will result in increased engagement. Based on this result, the fourth hypothesis of the study, namely that the influence of teamwork effectiveness is significant to the engagement on the police officers of Police Resort of Tanjung Perak Port Surabaya, can be accepted.

3.5 The effects of job demands against burnout moderated by teamwork effectiveness

The result of PLS regarding the effects of job demands against burnout moderated by teamwork effectiveness is 0.0482, and with t-statistics value of 0.2818. This shows that there is no significant influence from the moderation of teamwork effectiveness to job demands and burnout to members of Tanjung Perak Port Surabaya. This also means that the moderation of teamwork effectiveness on job demands and burnout will not result in a reduction of burnout. So, in this study, the hypothesis of moderation of teamwork effectiveness on job demands and burnout on members of the Police Resort at Tanjung Perak Port Surabaya has to be rejected. This study is in line with Montgomery's study that indicated no significant moderation of teamwork effectiveness on job demands and burnout (Montgomery et al. 2015). Due to the strong influence of job demands on burnout, an effective team is unable to overcome this problem. There are other factors that trigger burnout prevention. This proves that high teamwork effectiveness cannot overcome burnout caused by job demands on police officers. The burnout that occurs in police members has a different context than teamwork effectiveness. The burnout experienced by members of the police force occurs because of problems with the community and members of the police feel the work done together and grow from time to time.

4 CONCLUSION

Referring to results related to the burnout of police officers, tension and feeling uninspired at work does exist. This indicates a negative reaction that impacts work stress. The Police Resort of Tanjung Perak Port Surabaya should provide some training or counseling related to stress management that aims to reduce work stress and to improve the work motivation of police personnel. In addition, it is necessary to thoroughly evaluate the job demands owned by police officers at the Police Resort of Tanjung Perak Port Surabaya, because the value of job demands perceived by the members of the police force appears to be low in some aspects and medium in another aspect. Task rotation should be considered in order to provide police members a fresh perspective on day-to-day operational work situations and also in order to achieve future workload alignment for police officers at the Police Resort of Tanjung Perak Port Surabaya.

REFERENCES

Bakker, B.A. & Ellen, H. 2006. Emotional dissonance, burnout, and in-role performance among nurses and police officers. *International Journal of Stress Management* 13(4): 423–440.

Bakker, B.A. & Demerouti, E. 2008. Towards a model of work engagement. *Career Development International* 13(3): 209–223.

Buchholz, S. 2000. *Creating the High Performance Team*. Canada: John Wiley and Sons.

Demerouti, E. 2001. The job demands-resources model of burnout. *Journal of Applied Psychology* 86(3): 499–512.

Demerouti, E. & Pascale, L.B.M. 2009. Present but sick: A three- wave study on job demands, presenteeism and burnout. *Career Development International* 14 (1).

Demerouti, E. & Bakker, B.A. 2011. The job demands-resources model: Challenges for future research. *SA Journal of Industrial Psychology* 37(2): 1–9.

Griffin, P.S. & Thomas, J.B. 2003. Angry aggression among police officers. *Police Quarterly* 6 (1): 3–21.

Harter, J.K. Schmidt, F.L., & Hayes, T.L. 2002. Business-unit-level relationship between employee satisfaction, employee engagement, and business outcomes: A meta-analysis. *Journal of Applied Psychology* 87: 268–279.

Hall, G.B. et al. 2010. Job demands, work-family conflict, and emotional exhaustion in police officers: a longitudinal test of competing theories. *Journal of Occupational and Organizational Psychology* 83: 237–250.

Jones, A., Richard, B., Paul, D., Sloane K. & Peter, F. 2007. Effectiveness of teambuilding in organization. *Journal of Management* 5(3): 35–37.

Kahn, W.A. 1990. Psychological conditions of personal engagement and disengagement at work. *Academy of Management Journal* 33(4): 692–724.

Manzoor, S.R, Hussain, H.U.M. & Zulqarnain, M.A. 2011. Effect of teamwork on employee performance. *International Journal of Learning and Developmen*. 1(1): 110–126.

Maslach, C. & Susan, E.J. 1981. The measurement of experienced burnout. *Journal of Occupational Behaviour*. 2: 99–113.

Maslach, C., Leiter, M.P. & Jackson, S.E. 2012. Making a significant difference with burnout interventions: researcher and practitioner collaboration. *Journal of Organizational Behavior* 33(2): 296–300.

Mauno, S. 2007. Job demands and resources as antecedent of work engagement: A longitudinal study. *Journal of Vacation Behavior* 70 (1): 149–171.

Montgomery, A et al. 2015. Job demands, burnout, and engagement among nurses: A multi-level analysis of ORCAB data investigating the moderating effect of teamwork. *Burnout Research* 2: 71–79.

Naude, J.L.P. & Rohman, S. 2004. The validation of the maslach burnout inventory-human services survey for emergency edical technicians in Gauteng. *SA Journal of Industrial Psychology* 30(3): 21–28.

Rothmann, S. & Joubert, J.H.M. 2007. Job demands, job resources, burnout and work engagement of managers at a platinum mine in the north west province. *South African. Journal of Business Management* 38(3).

Schaufeli, W.B. & Bakker A.B. 2003. The measurement of work engagement with a short questionnaire: A cross-national study. *Education and Psychological Measurement* 66 (4): 701–716.

Schaufeli, W.B & Bakker A.B. 2004. Job demands, job resources, and their relationship with burnout and engagement: a multi-sample study. *Journal of Organizational Behavior* 25: 293–315.

Schaefuli, W.B & Bakker A.B. 2010. Defining and measuring work engagement: bringing clarity to the concept. *Journal of Organizational Behavior* 10–25.

Schaufeli, W.B & Taris T.W. 2014. *A critical review of the job demands-resources model: Implications for improving work and health*. Springer Science and Business Media, Dordrecht.

Tritch, T. 2003. Engagement drives results at new century. *Gallup Management Journal*.

The influence of clan culture on Organizational Citizenship Behavior (OCB) with affective organizational commitment as intervening variable of employees

M.D.T. Pamungkas & P. Yulianti
Universitas Airlangga, Surabaya, Indonesia

ABSTRACT: Organizational Citizenship Behavior (OCB) is an individual's free behavior that can improve effective functioning and organizational efficiency. This study aims to examine the effect of clan culture on OCB with affective organizational commitment as an intervening variable. This study involved 54 people as respondents. Data were obtained through observation and questionnaire distribution, which were then analyzed using Partial Least Square (PLS) and Sobel tests to test the mediating effect. The results show that the clan culture has a significant positive effect on affective commitment, and affective commitment has a significant positive effect on OCB both OCB-I and OCB-O. Affective commitment proved to have a full mediating effect on the influence of clan culture on organizational citizenship behavior. This implication shows that affective commitment is one of the important elements to be considered in an effort to improve the OCB of employees in an organization.

1 INTRODUCTION

The success of a company to achieve organizational goals is determined by the success of the organization's human resources in carrying out their duties (Mathis & Jackson 2001). The involvement of not only teachers as educators, but also of other parties such as administrative staff, library staff, accounting staff, human resources staff, multimedia staff, facilities maintenance and infrastructure staff, information technology staff, and CHB-Centre staff is also very important in supporting operation activities of a school. Therefore, there is a need for encouragement that motivates each individual to try to show their best and go beyond what is their duty and responsibility, often called the extra-role behavior or organizational citizenship behavior (OCB).

OCB is influenced by several interrelated and complex factors. Culture, as one of the external factors, can trigger OCB because it is always related to the life of an organization. The existence of an organizational culture, such as clan culture that prioritizes teamwork and the participation of members of the organization, will be able to cause an affective commitment that makes employees have the desire to remain in an organization. Titisari (2014) states that affective commitment is a predictor of OCB-I and OCB-O.

Previous research conducted by Organ et al. (2006) also states that clan culture cannot be directly related to OCB, but must be related through affective commitment. This is because affective commitment can arise if it starts from a relatively high family atmosphere and strong teamwork. With the formation of affective commitment, employees will feel emotionally attached, try

to always be involved, and provide the best for the organization. Employees will tend to do positive things for voluntary organizations without a formal reward system. Therefore, this research is needed to prove the problem.

School X Surabaya is a private school which has been established since 1994 and has levels from PG kindergarten to elementary, junior, and senior high school. The results of interviews, observation, and the distribution of questionnaires – Organizational Culture Assessment Instrument (OCAI) – conducted in School X Surabaya, obtained information that:

a Culture in School X Surabaya is clan culture.
b Reasons employees keep working is because employees want to work in School X Surabaya.
c Employees of School X Surabaya tend to not be willing to work beyond working hours.
d Employees tend not to replace the work of colleagues who are not entering. Employees choose to help by simply recording and delivering the necessary work when a coworker is signed in.

Based on these issues and the importance of the OCB role, this research aims to find out how far clan culture and affective commitment of employees of School X Surabaya can make a positive contribution to OCB.

1.1 *Organizational culture*

Organizational culture is a system of shared meanings adopted by members that distinguishes one organization from anther (Robbins 2001). According to Cameron and Quinn (2006), there are four organizational cultures:

1. Clan culture is similar to a family outside the home characterized by a leader as a facilitator or mentor, teamwork, involvement of members of the organization, and consensus.
2. Advocacy culture has a dynamic and creative environment, innovator leaders, entrepreneurs, and visionaries.
3. Market culture focuses on the completion of work. The leader is a rigorous director, producer, and rival all at the same time.
4. Hierarchy culture has a formal and structured work environment with a leader as a coordinator, monitor, and organizer.

Cameron and Quinn's (2006) typology of organizational culture is considered most suitable for this study when compared with other organizational culture typologies. The typology from Cameron and Quinn (2006) is not only able to diagnose a number of elements related to the internal conditions of an organization, such as the relationship of socialization between organizational members and the limitations of organizational management functions, but also able to diagnose external organizational conditions that can affect the sustainability of the organization.

1.2 *Organizational commitment*

Organizational commitment is seen as a strong attitude, belief, trust, and willingness of employees to remain employed with and become members of the organization. Organizational commitment has three dimensions according to Allen and Meyer (1996):

1. Affective commitment relates to members' emotional connection to their organization to continue working in the organization (want to).
2. Continuance commitment relates to the awareness of members of the organization so that they will lose if leaving the organization.
3. Normative commitment relates to feelings of attachment to stay in the organization because they feel they must be in the organization.

1.2.1 *Affective commitment*
Newstrom (2015) states that affective commitment is a positive emotional state where employees want to exert effort and choose to stay with the organization. Another opinion was expressed by Hartmann and Bambacas (2000), that affective commitment is a feeling of belonging and a bond to the organization, and has a relationship with personal characteristics, organizational structure, and work experience, such as salary, supervision, role clarity, and various skills. Based on Gautam, Dick, and Wagner (2004), affective commitment consists of three aspects: emotional attachment, identification, and involvement.

1.3 *Organizational citizenship behavior*

Pierce (2002) states that OCB is a behavior shown by employees and not part of their task, which is a form of sacrifice and commitment to organizational prosperity. Based on the conceptual work of Organ et al. (2006), OCB is identified as five dimensions, namely, altruism, civic virtue, conscientiousness, courtesy, and sportsmanship, whereas according to William and Anderson (1991), OCB is divided into two dimensions, namely, the OCB-O and OCB-I dimensions, where OCB-O is a behavior that benefits the organization in general, and OCB-I is a behavior that directly benefits employees (individuals) and indirectly benefits the organization. The results of previous studies conducted by Organ and Konovsky (1989) and Smith et al. (1983) concluded that OCB-I was altruism and OCB-O was general compliance.

1.4 *Hypothesis and analysis model*

Hypotheses in this research can be formulated as follows:
   Hypothesis 1: Clan culture has a significant effect on OCB.
   Hypothesis 2: Clan culture has a significant effect on OCB through affective organizational commitment.

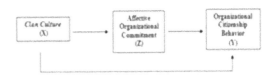

Figure 1. Analysis model.

2 METHOD

This research used a quantitative approach by formulating and testing hypotheses to produce conclusions that can be generalized. This research is included in the category of applied research because it is not necessarily a new invention but is looking for solutions to the problem (Nazir 2005).
   The sample determination technique in this study used a saturated sampling or census sampling technique. The sample of this study was 59 employees of School X Surabaya (the School), at PG-kindergarten, elementary, middle, and high school institutions. This research data was obtained from preliminary surveys, literature studies, and questionnaires. The data analysis technique used in this study is the test of partial least square (PLS) and a Sobel test of the mediation effect.

3 RESULT AND DISCUSSION

3.1 Hypothesis testing

The results of the analysis are shown in Table 1. The coefficient of the direct influence between clan culture and OCB is 0.175 with the value of t-statistics of $1.128 < 1.96$. So, there is no significant and positive influence directly between clan cultures and OCB employees of the School. Based on this result then Hypothesis 1 research is rejected.

In addition, the coefficient of the direct influence between clan culture to OCB is 0.175 with the value of t-statistics of $1.128 < 1.96$. So, there is no significant and positive influence directly between clan cultures against OCB employees of the School. Based on this result then Hypothesis 1 research is rejected.

The coefficient of the direct influence between clan cultures and affective commitment is 0.516 with t-statistics value is $5.640 > 1.96$. So, there is a positive influence directly between clan cultures on affective commitment of employees of the school.

The coefficient of the direct influence between affective commitments to OCB is 0486 with t-statistics value of $2.617 > 1.96$. So, there is a significant direct positive influence between affective commitments and OCB employees of the school.

Table 1 shows that the coefficient of influence between clan culture and OCB indirectly through affective organizational commitment is 0.251 with t-statistics value of $2.374 > 1.96$. These results conclude that there is an indirectly significant positive influence between clan cultures and organizational citizenship behavior in the school

Table 1. Hypothesis testing.

Hypothesis	Coefficient of direct influence	T Statistics
Clan Culture → Organizational Citizenship Behavior	0.175	1.128
Clan Culture → Affective Commitment	0.516	5.640
Affective Commitment → Organizational Citizenship Behavior	0.486	2.617

through affective organizational commitment. Based on these results Hypothesis 2 research is accepted.

3.2 The influence of clan culture (X) on organizational citizenship behavior (Y)

The results showed that clan culture had no significant effect on organizational citizenship behavior with a coefficient of influence of 0.175 with a t-statistics value of $1.128 < 1.96$ at a 5% error level. It means that individual initiative and free choice is not related to the formal system of an organization. The results of this study support previous research conducted by (Kim 2014) in the public sector in South Korea which states that clan culture has no significant direct relationship with organizational citizenship behavior because the clan culture characteristics created by local governments in Korea still reflect the highly structured element of bureaucracy that shows high power spacing and cultural collectivism. In terms of the school, although the existing culture is a clan culture it is still not able to influence the existence of OCB due to the low level of understanding of employees about clan culture in the school, especially for employees who have been working there less than two years.

3.3 The influence of clan culture (X) on organizational citizenship behavior (Y) through affective organizational commitment (Z)

From the results of the hypothesis testing, it is known that clan culture has a significant positive effect on organizational citizenship behavior through affective organizational. Clan culture has a significant positive effect on affective commitment on the school employees with a regression coefficient of 0.51 and a t-statistic value of 5.64, which is greater than the critical limit of 1.96 at a 5% error level. The stronger the clan culture at the school, the stronger the employee's desire to work. This study was also able to prove that affective commitment also has a significant positive effect on OCB with a regression coefficient of 0.48 and at t-statistic value of 2.61, which is greater than the critical limit of 1.96 at the 5% error level. That is, affective commitment and OCB have unidirectional changes. If the affective commitment of the school

Table 2. Sobel test.

| Hypothesis | Sobel Test | | | | | |
	A	B	Sa	Sb	t-statistics	p value
H2 Clan culture → Affective commmitment → organizational citizenship behavior	0.516 A x b = 0.251	0.486	0.091	0.186	2.374	0.017

employees becomes stronger, the desire of employees to provide their best work will also increase.

The stronger the clan culture in the school, the stronger the desire of employees to keep working. The results of this study are in accordance with research conducted by several scholars (Zain et al. 2009, Carter 2013, Momeni et al. 2012, Nongo & Ikyanyon 2012), that some components of clan culture such as teamwork, involvement, and adaptability have significant relationships with organizational commitment. Affective commitment and OCB have a similar exchange. If affective commitment of the employees of the school is getting stronger, the desire of the employees to provide their best will also increase. This study supports previous research conducted by Salehi and Gholtash (2011) and Lavelle et al. (2007) that affective commitment is positively related to OCB because affective commitment is capable of generating a sense of desire to remain in the organization, thereby encouraging employees to be involved in providing everything that is best for their organization. It can be concluded that organizational commitment acts as a good mediation variable to understanding the relationship between organizational culture and OCB.

4 CONCLUSION

Based on the results and discussion of the study it can be concluded that clan culture does not have a significant influence on OCB for the school employees. However, clan culture has a significant influence on OCB through affective. Clan culture has a significant positive effect on affective commitment and affective commitment has a significant positive effect on OCB. That is, affective commitment is proven to have the full mediating effect of clan culture on OCB.

This study has implications that affective organizational commitment is one of the important elements that must be considered in improving the organizational citizenship behavior (OCB) of employees in an organization.

REFERENCES

Allen, N.J. & Meyer, J.P. 1996. Affective, continuance, and normative commitment to the organization: An examination of construct validity. *Journal of Vocational Behaviour* 49(3): 252–276.

Cameron, K. & Quinn, R. 2006. *Diagnosing and changing organizational culture: Based on a competing values framework*. San Francisco: Josey-Bass.

Gautam, G., Thaneswor, T., Van Dick, R. & Wagner, U. 2004. Organizational identification and organizational commitment: Distinct aspects of two related concepts. *Asian Journal of Social Psychology* 7(3): 301–315.

Hartmann, L.C. & Bambacas, M. 2000. Organizational commitment: A multi method scale analysis and test of effects. *International Journal of Organizational Analysis* 8(1): 89–108.

Kim, H. 2014. Transformational leadership, organizational clan culture, organizational affective commitment, and organizational citizenship behavior: A case of South Korea's public sector. *Journal of Public Organize Review* 14(3): 397–417.

Lavelle, J.J., Rupp, D.E. & Brockner, J. 2007. Taking a multifoci approach to the study of justice, social exchange, and citizenship behavior: The target similarity model. *Journal of Management* 33(6): 841–866.

Mathis, M., Robert, L., John, H. & Jackson, J. 2001. *Manajemen sumber daya manusia*. Jakarta: Salemba Empat.

Momeni, M., Babak. M. A. & Saadat. V. 2012. The relationship between organizational culture and organizational commitment in staff department of general prosecutors of Tehran. *International Journal of Business and Social Science* 3(13): 217–221.

Nazir, M. 2005. *Metode penelitian*. Bogor: Ghalia Indonesia.

Newstrom, J.W. 2015. *Organizational behavior: Human behavior at work, (4th* edition). New York: McGraw-Hill Education.

Nongo, E.S. & Ikyanyon D.N. 2012. The influence of corporate culture on employee commitment to the organization. *International Journal of Business and Management* 7(22): 21–28.

Organ, D.W., Podsakoff, P.M. & MacKenzie, S.B. 2006. Organizational citizenship behavior: Its nature, antecedents, and consequences. Thousand Oaks, CA: Sage Publications.

Pierce, J.L. & Garden, D.G. 2002. *Management and organizational behavior: An integrated perspective*. Mason, OH: South-Western.

Robbins, S.P. 2001. *Organizational behavior*, (9th ed). New Jersey: Prentice-Hall.

Salehi, M. & Gholtash. A. 2011. The relationship between job satisfaction, job burnout and organizational commitment with the organizational citizenship behavior among members of faculty in the Islamic azad university first district branches, in order to provide the appropriate model. *Procedia, Social and Behavioral Science* 15: 306–310.

Short, E.C. 2013. The relationship between clan culture, leardership member exchange, and affective organizational commitment. Dissertation, Trevecca Nazarene University.

Tititsari, P. 2014. *Peranan organizarional citizenship behavior (OCB) dalam meningkatkan kinerja karyawan*. Jakarta: Mitra Wacana Media.

Williams, L.J. & Anderson, S.E. 1991. Job satisfaction and organizational commitment as predictors of organizational citizenship and in-role behaviors. *Journal of Management* 17: 601–617.

Zain, Z.M., Ishak, R. & Ghani, E.K. 2009. The influence of corporate culture on organizational commitment: a study on a Malaysian listed company. *European Journal of Economics, Finance and Administrative Sciences* 1: 1450–2275.

Advances in Business, Management and Entrepreneurship – Hurriyati et al (eds)
© *2020 Taylor & Francis Group, London, ISBN 978-0-367-27176-3*

Transactional leadership in public sector

I.A.P.S. Widnyani & G. Wirata
Universitas Ngurah Rai, Denpasar, Indonesia

ABSTRACT: This study determined the value and meaning of transactional leadership in the public sector. This study used a limited qualitative descriptive approach. Results suggested that transactional leadership styles in the public sector are not always negative. Leaders give priority to the people who support it, compared to people who do not support it. This is a form of reward and punishment by elected leaders in the public sector. Reward and punishment as a result of the transactional leadership model in order to always get support from the community. Transactional models as one of the leadership styles are still needed in the public sector, because there must be a reward between what you have received, so you need to support to reach the goal optimally.

1 INTRODUCTION

The ideal type of a leader depends heavily on his personality and commitment. The leader moves ahead, walks ahead, takes the first step, does the very first, pioneers, exerts thoughts and opinions, actions of others, guides, demands, moves others through his influence. In this case a leader is supported by the personality of self, position and role, ability and skill, whereas followers or subordinates need to have trust, obedience, critical thinking, individual needs which are different in their way of fulfillment, pressure or stressing, the environment, work culture, and also involving other interest groups served by the organization in which a leader is in power. According to (Kartakusumah 2006) the formation of leadership has two main factors namely internal factors concerning personality, vision, capabilities and achievements, while individual external factors include learning processes learned in all learning environments in everyday life.

However, a leader is a person whose personal qualities, with or without an official appointment, may influence the group he leads to move the joint effort toward achieving certain targets which have been previously set. According to Ackerman's lively leadership means finding yourself into yourself and showing yourself (find yourself, be yourself and show yourself). Leadership as a person's identity when entering society and organization, not only in his ability to occupy the top position, but has become a color in everyday life "life-centered leadership itself" or buffer the responsibility to lead yourself, transformationed himself in the form of leadership, being able to do this will resulted in leadership towards others (Ackerman 2000). This is required in the realization of public in public organizations, whose service interests are based on genuine devotion to the organization and the interests of the community to be the main thing, rather than the

effort to gain advantage or interest to gain power and respect from the leadership that is above it.

According to (Senn & Children 1999 in Hartman et al. 2005) a leader must have nine factors needed in the organization in creating and sustaining cultural sustainability, including: 1) creating a shared vision, 2) ensuring that senior leaders are committed to the process and an agreed-upon model of behavior, 3) define sample behaviors that support organizational values, 4) performing gap analysis through cultural audit, 5) connecting the need for changes to business cases and outcomes, 6) remember that the correct behavior change takes place on the emotional level, not the intellectual level, 7) create a culture of coaching and feedback, 8) Remember that change requires critical mass to eliminate old behaviors, 9) align the support system to strengthen the desired culture (e.g. performance management, hiring/firing, training, recognition, etc.).

Furthermore, Jason Miletsky mentioned some examples of creating a positive culture in the organization: strengthening the culture of positive praise and never injure, freedom to do one's job with a little direction and a lot of trust, flexible work schedule but with creative process schedule which ensure high quality result, the absence of dominance of a particular person, allowing employee to work according to their contracts and obligations, let one's passion bring creativity in their work.

In public sector, public organization can run well and resulted in good service quality if it can fulfill 6 dimension including 1) voice and accountability: freedom of expression, free association and dissemination of performance through media, 2) political stability and absence of pressure from anyone, comfort in performing service duties, consistent rules and free from the influence of political change, 3) government effectiveness, is concerned with the achievement of the quality of public services and civil service, the quality of decision-making and implementation of decisions, the credibility of the executors and freedom from political pressures, 4) regulatory quality, the ability of governments to prepare policies and legal products to improve and ensure high quality public sector services, 5) rule of law, supported by legal and regulatory frameworks for guaranteeing the public and the public sector from the threat of crime and irregularities, 6) control of corruption, developing the power to deal with corruption and the interests of certain groups that allow deviations. These six dimensions are presented by Pippa Norris in his writing "measuring governance" (Bevir 2011).

However, political issues, freedom of expression, the quality of rules and legislation, law enforcement and government oversight in the public sector are still of high concern within the public sector, in which this has an effect on the level of leadership in the region, which will also affect performance of public organizations in the region.

2 METHOD

This paper is written by descriptive qualitative limited methods, data collection through interview and documentation to obtain primary data and secondary data. Informants were determined purposively. Methods of data analysis by reducing data, presentation based on describing and drawing conclusions.

3 RESULTS AND DISCUSSION

The concept of transactional leadership was first formulated by (Burns 1978 in Yukl 1994), based on his descriptive research on political leaders and subsequently refined and introduced into the organizational context by (Bass 1985). Transactional leadership according to (Burns in Yukl 1998) motivates followers by showing self-interest. Political leaders exchange jobs, subsidies, and profitable government contracts for votes and contributions to the campaign. Corporate leaders often exchange wages and status for employment. Transactional leadership involves values, but in the form of values relevant to the exchange process, such as honesty, fairness, responsibility and exchange. The term transactional comes from how this type of leader motivates followers to do what they want to do. The transactional leader determines the followers' desires and gives something in exchange because the followers perform certain tasks or find specific goals. Thus, a transaction or exchange process between a leader and a follower occurs when a follower receives a reward from a job performance and the leader benefits from the completion of the tasks. In transactional leadership, the leader-follower relationship is based on a series of exchanges or agreements between leaders and followers (Howell & Avolio 1993). Transactional

leadership is the leader who guides or motivates their followers toward defined goals by clarifying the terms of roles and tasks (Robbins 2008). According to (Gibson et al. 1997) Transactional leaders identify the desires or choices of subordinates and help them achieve reward-generating performance that can satisfy subordinates. (Bass 1990) defines transactional leadership as a leadership model that involves an exchange process in which followers get immediate and real rewards after executing leaders' orders. Furthermore, (Mc Shane & Von G 2003) define transactional leadership as leadership that helps people achieve their present goals more efficiently such as linking job performance with assessed rewards and ensuring that employees have the resources needed to complete work. Burns in (Usman 2009) defines transformational leadership as "a process in which leaders and followers raise to higher levels of morality and motivation". This style of leadership will be able to bring followers consciousness by generating productive ideas, synergistic relationships, responsibility, educational awareness, and shared ideals.

As expressed by experts on transactional leadership models, empirically it does so in the public sector. When political positions such as elected regional heads by elections by popular vote, the so-called rewards and punishments occur. Some informants stated about transactional leadership in the public sector. Wirawan stated that: "the definition of punishment is not a punishment, but as a coaching to a society that has not been supportive, so that they can change the mind from not supportive to being supportive". This is also reinforced by Darma that it is important to reward supporters for maintaining the trust given by the community through the choice of the leader, and thus, supporters need to be prioritized. Transactional leadership style as a motivation in improving the performance of subordinates, maintaining public trust to leaders.

4 CONCLUSION

Transactional models as one of leadership styles are still required in the public sector, because there must be a reward between what has been received, so that take-and-give remains to sustain the support to achieve the goal optimally. Suggestion: although transactional in the style of leadership is still required to pay attention to the needs of the community, as well as to be fair without discrimination. Because fair treatment as a social investment to the community to attract sympathy support in achieving the goal optimally.

ACKNOWLEDGEMENT

We would like to thank the informants for their willingness to provide information for the completion of this paper.

REFERENCES

Ackerman, A. & Laurence, L. 2000. *Identity is destiny leadership and the roots value creation. diterjemahkan oleh probo suasanto*. Jakarta: PT Gamedia Pustaka Utama.

Bass, B.M. 1985. *Leadership and performance beyond expectations*. New York: Free Press.

Bass, B.M. 1990. *Bass and stogdill's handbook of leadership: theory, research, and managerial application. Third Edition*. New York: Free Press.

Bevir, B. & Mark, M. 2011. Governance. *Sage Publication*.

Bond, L.A., Holmes, T.R., Byrne, C., Babchuck, L. & Kirton-Robbins, S. 2008. Movers and shakers: How and why women become and remain engaged in community leadership. *Psychology of Women Quarterly* 32(1): 48–64.

Cokroaminoto, C. 2007. *Membangun kinerja (memaknai kinerja karyawan)*.

Hartman, H., Phyllis G., SPHR & John T.H 2005. Creating a positive culture.

Howell, J.M. & Avolio, B.J. 1993. Transformational leadership, transactional leadership, locus of control, and support for innovation: Key predictors of consolidated-business-unit performance. Journal of applied psychology, 78(6), p.891.

Kartakusumah, B. 2006. *Pemimpin adiluhung geneologi kepemimpinan kontenporer*. Jakarta: Teraju PT Mizan Publika.

Gibson, J.L., Ivancevich, J.M. & Donnelly, J.H. 1997. *Organisasi: perilaku–struktur–proses*. Penerbit: Binarupa Aksara.

Yukl, G. 1994. *Leadership in organizations. Third Edition*. New Jersey. Prentice-Hall, Inc. Englewood Cliffs.

Yukl, G. 1998. *Kepemimpinan dalam organisasi. Edisi Ketiga. Edisi Bahasa Indonesia*. Jakarta: Prenhallindo.

Usman, H. 2009. *Manajemen: teori, praktik, dan riset pendidikan. edisi ketiga*. Jakarta: Penerbit PT. Bumi Aksara.

Advances in Business, Management and Entrepreneurship – Hurriyati et al (eds)
© 2020 Taylor & Francis Group, London, ISBN 978-0-367-27176-3

The influence of perceived organizational support and job characteristics on organizational citizenship behavior with employee engagement as intervening variable on pharmacists

D.M. Machfud & P. Yulianti
Universitas Airlangga, Surabaya, Indonesia

ABSTRACT: This study was aimed at examining the influences of perceived organizational support, job characteristics, employee engagement, and organizational citizenship behavior. This research was quantitative research using questionnaires. The data were collected through surveys and filled by 52 pharmacists from 25 community franchise-pharmacies in Surabaya, East Java, Indonesia. Employee engagement was influenced significantly by perceived organizational support and job characteristics. Organizational citizenship behavior was influenced significantly by perceived organizational support, job characteristics, and employee engagement. There was a significant positive influence indirectly between perceived organizational support and organizational citizenship behavior through employee engagement, and an indirectly significant positive influence between job characteristics and organizational citizenship behavior through employee engagement. The implication of this study is that perceived organizational support of pharmacists is lacking but pharmacists have engagement to their profession, so OCB-I is greater than OCB-O in pharmacists.

1 INTRODUCTION

The pharmaceutical industry merits a much higher level of importance because its often-life-saving products are an essential part of health care around the world (Taher et al. 2012). The number of community pharmacies continues to grow from year to year resulting in hyper-competition (Rabbanee et al. 2015), so community pharmacies must be able to provide good service, including the management of human resources (Na-Nan et al. 2016). A community franchise pharmacy is the original Indonesian pharmacy which achieved the title TOP Brand 2016 in pharmacy version. Community franchise pharmacies in Surabaya amounted to 25 outlets spread across Surabaya, with an average of two pharmacists in each outlet. In developing countries, the pharmacist plays a key role in the health care system and the medicines patients ultimately purchase, so marketing activities directed toward the pharmacist may decreace cost and maximize the pharmaceutical firms' profit (Taher et al. 2012). The success of an organization in an increasingly competitive environment depends on the quality of service that can be further enhanced by extra-role behavior, better known as organizational citizenship behavior (OCB) (Hadjali & Salimi 2012).

OCB is a positive behavior that affects all parties and it is necessary for the organization to encourage the emergence of OCB in every employee (Podsakoff et al. 2009). Recent research shows that one of the factors influencing the quality of service of a company that creates a competitive advantage is

the OCB of employees, especially employees who interact directly with customers (Hadjali & Salimi 2012). The community franchise pharmacies treat patients with different cases 24 hours a day, so OCB is needed in pharmaceutical services because patients come to pharmacies requiring special attention and positive behavior from the pharmacists handling their cases.

As members of a pharmacy organization, pharmacists are not only bound to fulfill their professional obligations but are also affected by different personal and organizational factors, such as support received from the organization (perceived organizational support or POS), that may influence their behavior and, consequently, the quality of the services provided to patients (Urbonas et al. 2015). POS generates greater commitment to the organization, an increased desire to help the organization succeed, and greater psychological well-being (Kurtessis et al. 2015). POS may affect OCB, as shown in Peele's (2007) research that POS is found as an OCB antecedent and POS impact and also involves some employee engagement (Sulea et al. 2012).

Employee engagement is an important aspect in organizations that deal with organizational effectiveness and the achievement of competitive advantage (Kataria et al. 2013). Employee engagement has a positive and significant relationship with OCB-I, but the relationship between employee engagement and OCB-O is positive and non-significant (Wickramasinghe & Perera 2014, Abd-Allah 2016). One common antecedent of employee engagement that has been found is job characteristics (Sulea et al. 2012).

Pharmacists working in community pharmacies are not only dispensing medicine but also frequently find themselves taking on more responsibility for running the business and may be involved in marketing, financing, logistics, human resource management, and information management (Lin et al. 2007). A job characteristics theory (Hackman & Oldham 1980) describes the relationship between job characteristics and individual responses to work (Mukul et al. 2013). The relationship between job characteristics with OCB when engagement works as a mediator shows that the highest and most significant relation was between employee engagement and OCB (Shantzet et al. 2013, Abd-Allah 2016).

This study was aimed at examining the influences of perceived organizational support, job characteristics, employee engagement, and organizational citizenship behavior.

2 METHOD

2.1 Sampling and data collection

This research was a quantitative research using questionnaires. The data were collected through surveys and filled by 52 pharmacists from 25 community franchise pharmacies in Surabaya, East Java, Indonesia.

2.2 Measures

2.2.1 Perceived organizational support (POS)

The construct POS was measured using the Survey of Perceived Organizational Support (SPOS) instrument developed by Urbonas et al. (2015). Items of the construct were: (1) the organization cares about employee welfare benefits; (2) the organization cares about the development of employee knowledge; (3) the organization considers the values and purpose of the employee's existence; and (4) the organization considers the opinions of employees. Participants indicated their response on a 5-point Likert-type scale with anchors (1) strongly disagree to (5) strongly agree.

2.2.2 Job characteristics

Job characteristics were measured using indicators developed by Hackman and Oldham (1980) in Rai et al. (2017), so that the measurement indicators of job characteristics in this study were: (1) the work requires a variety of skills; (2) the work has a clear and comprehensive definition from beginning to end; (3) the work is related to the work of others; (4) the work provides freedom to use skills and define work procedures; and (5) the job provides information about the resulting performance. The participants

responded by using a 5-point Likert-type scale with anchors (1) strongly disagree to (5) strongly agree.

2.2.3 Employee engagement

One of the best models that measures employee engagement through vigor, dedication, and absorption is the UWES (Utrecht Work Engagement Scale) instrument developed by Schaufeli & Bakker (2004). The measurement indicators of employee engagement used in this study were vigor, dedication, and absorbtion. Vigor can be determined by three items: (1) At my work, I feel bursting with energy; (2) At my work, I am very resilient, mentally; and (3) At my work, I always persevere, even when things do not go well. Dedication can be determined by four items: (1) My work that I do is full of meaning and purpose; (2) My job is inspiring other people; (3) I am proud on the work that I do; and (4) My job is challenging to me. Absorption can be determined by three items: (1) Time flies when I'm working; (2) I am immersed in my work; and (3) It's difficult to detach myself from my job. The participants responded by using a 5-point Likert-type scale with anchors (1) strongly disagree to (5) strongly agree.

2.2.4 Organizational citizenship behavior (OCB)

The construct OCB was directed to the individual (OCB-I) and organization (OCB-O), and was each measured by The Organizational Citizenship Behavior Checklist (OCB-C) developed by Fox and Spector (2011). Participants responded using a 5-point Likert-type scale with anchors (1) strongly disagree to (5) strongly agree. A sample item from the OCB-I scale is, "Helping co-workers who are less skilled in solving work problems" and a sample item from the OCB-O scale is, "Assisting new employee orientation on the job."

2.3 Hypotheses

The hypotheses proposed in this study are:
Hypothesis 1: POS will significantly influence employee engagement.
Hypothesis 2: Job characteristics will significantly influence employee engagement.
Hypothesis 3: POS will significantly influence OCB.
Hypothesis 4: Job characteristics will significantly influence OCB.
Hypothesis 5: Employee engagement will significantly influence OCB.
Hypothesis 6: POS will significantly influence OCB through employee engagement.
Hypothesis 7: Job characteristics will significantly influence OCB through employee engagement.

Our proposed model is shown in Figure 1.

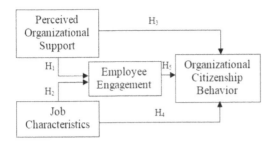

Figure 1. Research model.

3 RESULTS

Demographic information in Table 1 showed that among the sample of 52 respondents, 80.8% of them fell in age category 21–30 years, 17.3% fell in the category of 31–40 years. Meanwhile 1.9% respondents were in the age category of more than 40 years. The gender division of the respondents. 84.6% of the respondents were female, while 15.4% were male. In terms of respondents job tenure, out of 52 respondents, the shortest and longest tenure were 36.5% and 7.7%.

The direct impact hypothesis testing results indicate that *Hypothesis 1* through *Hypothesis 5* in the study are acceptable.

In Table 3, there is an indirectly significant positive influence between POS and OCB through

Table 1. Demographic profile of respondents.

Demographic information	Frequency	Percent
AGE		
21 to 30 years	42	80.8
31 to 40 years	9	17.3
More than 40 years	1	1.9
GENDER		
Male	8	15.4
Female	44	84.6
JOB TENURE		
1 to 12 months	19	36.5
13 to 60 months	29	55.8
More than 60 months	4	7.7

Table 2. Hypotheses testing.

Hypotheses	Coefficient of direct influence	T-Statistics	Significant
H_1 POS → EE	0.3521	3.9408	Significant
H_2 JC → EE	0.5176	5.9893	Significant
H_3 POS → OCB	0.2797	2.7130	Significant
H_4 JC → OCB	0.3294	3.4166	Significant
H_5 EE → OCB	0.4048	3.9881	Significant

Table 3. Indirect hypotheses testing.

Hypotheses	Coefficient of indirect influence	T-Statistics	P Value
H_6 POS → EE → OCB	0.143	2.816	0.005
H_7 JC → EE → OCB	0.210	3c.338	0.001

employee engagement. There is an indirectly significant positive influence between job characteristics and OCB through employee engagement. Based on those results, *Hypothesis 6* and *Hypothesis 7* are acceptable.

4 DISCUSSION

4.1 The influence of perceived organizational support on employee engagement

From the results of the research, it is found that the POS variable shows a sufficient average value, so the perception that emerged from the pharmacists as regards the concern of the organization is still sufficient. According to research conducted by Rigoni and Belson (2016), only 29% of millennials are engaged at work, 6 in 10 millennials say they're open to different job opportunities, and only 50% plan to be with their company one year from now.

In this study, respondents who are millennial generation pharmacists have high engagement, even though the POS value is sufficient. Respondents had an average age of 23–30 years, 80.8% were the millennial generation and 82.7% had a working period of one to three years. According to Urbonas et al. (2015), pharmacists have responsibility as professionals in health services so each pharmacy can create its own culture and organizational climate.

4.2 The influence of job characteristics on employee engagement

Characteristics of the millennial generation, such as knowledge, skills, and abilities, are in accordance with job characteristics of the community franchise pharmacy in Surabaya. According to Gilbert (2011), the millennial generation has a low level of engagement, but this generation is very confident, has a high lovel of social life, likes to challenge, is creative in completing work, and is interested in feedback on its performance. In addition, the fit between individual characteristics and job characteristics refers to person-job fit that poses a positive work attitude and high employee engagement (Doyle 2014). Therefore, job characteristics that are more enriched and adapted to the pharmacist's profession are essential to building a positive environment and creating higher employee engagement.

4.3 The influence of perceived organizational support on organizational citizenship behavior

Organizational awareness of the welfare and opinions of pharmacists is lacking. Pharmacist participation in decision making and peer support has an impact on motivation to engage in OCB activities. Organizational support for pharmacists from the is still lacking, but the level of OCB is high among pharmacist because, according to Urbonas et al. (2015), pharmacists have responsibility as professionals in health services so each pharmacy can create its own culture and organizational climate.

Pharmacists not only bound to fulfill their professional obligations but are also affected by different personal and organizational factors, such as support received from the organization, that may influence behavior and, consequently, the quality of the services provided to patients (Urbonas 2015). Therefore, the support provided by the organization is essential for pharmacists so that increasing POS will have an impact on their improved OCB.

4.4 The influence of job characteristics on organizational citizenship behavior

The results of this study are in accordance with the results of research by Pohl et al. (2013) that there is a significant direct relationship between job characteristics and OCB; this study is also compatible with Podsakoff et al.'s (2000) findings that job characteristics have a consistent relationship with OCB.

4.5 The influence of employee engagement on organizational citizenship behavior

Engaged employees are involved in OCB as they efficiently achieve their professional goals and feel able to do OCB (Christian et al. 2011). Therefore, when the pharmacist experiences a positive affective motivational state in the workplace (i.e., employee engagement), then the pharmacist is willing to engage in voluntary activities at his workplace.

4.6 The influence of perceived organizational support on organizational citizenship behavior through employee engagement

A significant positive POS effect on OCB through employee engagement showed that employee engagement is a good intervening variable, as it can be relied upon to support the positive POS effect on OCB. According to Sulea et al. (2012), employees with POS will demonstrate engagement to the organization and conduct a profitable OCB at work. Meanwhile, in the

Wickramasinghe and Perera study (2014), employee engagement is positively and significantly correlated with OCB-I, but it is positively and insignificantly associated with OCB-O (Abd-Allah 2016).

The POS in this study is sufficient, so the pharmacists' OCB is more inclined to OCB-I than OCB-O. The results show that agreement with Wickramasinghe and Perera (2014) conducted in Abd-Allah (2016), so it can be concluded that POS proved to have a significant positive effect, either directly or indirectly, on OCB, and it can be successfully used for community franchise pharmacy management in Surabaya to improve OCB and pharmacists' engagement. In addition, employee engagement is also an important variable that is urgently needed to improve the OCB of pharmacists working at pharmacies.

4.7 The influence of job characteristics on organizational citizenship behavior through employee engagement

A significant positive effect of job characteristics on OCB through employee engagement showed that the employee engagement is a good and reliable intervening variable to support the positive effect of job characteristics on OCB. According to study of Shantzet et al. (2013), the relationship between job characteristics and OCB when employee engagement as a mediator showed the highest and most significant relationship (Abd-Allah 2016). Thus, job characteristics are proven to have a significant positive effect, either directly or indirectly, on OCB and can be used by community franchise pharmacy management in Surabaya to improve OCB and pharmacist engagement. In addition, employee engagement is also an important variable that is urgently needed to improve pharmacists' OCB.

5 CONCLUSION

Based on the analysis and discussion of the results obtained, employee engagement was influenced significantly by perceived organizational support and job characteristics. Organizational citizenship behavior was influenced significantly by perceived organizational support, job characteristics, and employee engagement. A significant positive influence was found indirectly between perceived organizational support and organizational citizenship behavior through employee engagement, and an indirectly significant positive influence was found between job characteristics and organizational citizenship behavior through employee engagement. The implication of this study is that although perceived organizational support of pharmacists is lacking, pharmacists have engagement to their profession, so OCB-I is greater than OCB-O in pharmacists.

ACKNOWLEDGEMENTS

This research was supported by Airlangga University, Indonesia, and community franchise pharmacies in Surabaya. Big thanks are given to all pharmacists for supporting this research.

REFERENCES

Abd-Allah, O.Z. 2016. The relationship between organizational citizenship behavior and employee engagement in cement industry in Egypt. *International Journal of Management and Commerce Innovations* 4(1): 362–376.

Christian, M.S., Garza, A.S. & Slaughter, J.E. 2011. Work engagement: A qualitative review and test of its relations with task and contextual performance. *Personnel Psychology* 64: 89–136.

Doyle, A.L. 2014. Person-job fit: Do job characteristics moderate the relationship of personality with burnout, job satisfaction, and organizational commitment? [Online]. Retrieved https://etd.auburn.edu/bitstream/handle/10415/4281/AndreaDoyleDissertation_Final.pdf.txt?sequence=3&isAllowed.=y. Accessed 2 May 2018.

Fox, S. & Spector, P.E. 2011. Organizational citizenship behavior checklist (OCB-C) [Online]. Retrieved http://shell.cas.usf.edu/~pspector/scales/ocbcpage.html. Accessed 2 May 2018.

Gilbert, J. 2011. The millennials: A new generation of employees, a new set of engagement policies [Online]. Retrieved https://iveybusinessjournal.com/publication/the-millennials-a-new-generation-of-employees-a-new-set-of-engagement-policies/. Accessed 2 May 2018.

Hadjali, H.R. & Salimi, M. 2012. An investigation on the effect of organizational citizenship behaviors (OCB) toward customer-orientation: A case of nursing home. *Procedia - Social and Behavioral Sciences* 57: 524–532.

Kataria, A., Garg, P. & Rastogi, R. 2013. Employee engagement and organizational effectiveness: The role of organizational citizenship behavior. *International Journal of Indian Culture and Business Management* 6(1): 102–113.

Kurtessis, J.N., Eisenberger, R., Ford, M.T., Buffardi, L.C., Stewart, K.A. & Adis, C.S. 2015. Perceived organizational support: A meta-analytic evaluation of organizational support theory. *Journal of Management* 20(10): 1–31.

Lin, B.Y., Yeh, Y.C. & Lin, W.H. 2007. The influence of job characteristics on job outcomes of pharmacists in hospital, clinic, and community pharmacies. *Journal of Medical Systems* 31 (3): 224–229.

Mukul, A.Z.A., Rayhan, S.J., Hoque, F. & Islam, F. 2013. Job characteristics model of Hackman and Oldham in garment sector in Bangladesh: A case study at Savar area in Dhaka district. *International Journal of Economics, Finance and Management Sciences* 1(4): 188–195.

Na-Nan, K., Panich, T., Thipnete, A. & Kulsingh, R. 2016. Influence of job characteristics, organizational climate, job satisfaction and employee engagement that affect the organizational citizenship behavior of teachers in Thailand. *Medwell Journals, The Social Sciences* 11(18): 4523–4533.

Podsakoff, N.P., Whiting, S.W., Podsakoff, P.M. & Blume, B.D. 2009. Individual and organizational level consequences of organizational citizenship behaviors: A meta-analysis. *Journal of Applied Psychology* 94(1): 122–141.

Podsakoff, P.M., MacKenzie, S.B., Paine, J.B. & Bachrach, D. G. 2000. Organizational citizenship behavior: A critical review of the theoretical and empirical literature and suggestions for future research. *Journal of Management* 26(3): 361–278.

Pohl, S., Battistelli, A. & Librecht, J. 2013. The impact of perceived organizational support and job characteristics on nurse's organizational citizenship behaviours. *International Journal of Organization Theory and Behavior* 16(2): 193–207.

Rabbanee, F.K., Burford, O. & Ramaseshan, B. 2015. Does employee performance affect customer loyalty in pharmacy services? *Journal of Service Theory and Practice* 25(6): 725–743.

Rai, A., Ghosh, P., Chauhan, R. & Mehta, N.K. 2017. Influence of job characteristics on engagement: does support at work act as moderator? *International Journal of Sociology and Social Policy* 37(1/2): 86–105.

Rigoni, R. & Nelson, N. 2016. Few millennials are engaged at work. Business journal. Retrieved http://news.gallup.com/businessjournal/195209/few-millennials-engaged-work.aspx. Accessed 10 May 2018.

Schaufeli, W.B. & Bakker, A.B. 2004. *Test manual for the Utrecht Work Engagement Scale*. Utrecht University, The Netherlands.

Shantzet A., Alfesb K., Trussc C. & Soaned, E. 2013.The role of employee engagement in the relationship between job design and task performance, citizenship and deviant behaviours, *The International Journal of Human Resource Management* 24(13): 2608–2627.

Sulea, C., Virga, D., Maricutoiu, L P., Schaufeli, W., Dumitru, C.Z. & Sava, F.A. 2012. Work engagement as mediator between job characteristics and positive and negative extrarole behaviors. *Career Development International* 17(3): 188–207.

Taher, A., Stuart, E.W. & Hegazy, H. 2012. The pharmacist's role in the Egyptian pharmaceutical market. *International Journal of Pharmaceutical and Healthcare Marketing* 6(2): 140–155.

Urbonas, G., Kubilienė, L., Kubilius, R. & Urbonienė, A. 2015. Assessing the effects of pharmacists' perceived organizational support, organizational commitment and turnover intention on provision of medication information at community pharmacies in Lithuania: A structural equation modeling approach. *BioMed Central Health Services Research* 15(82): 1–10.

Wickramasinghe, V. & Perera, S. 2014. Effects of perceived organisation support, employee engagement and organisation citizenship behaviour on quality performance. *Total Quality Management & Business Excellence* 25(11–12): 1280–1294.

Advances in Business, Management and Entrepreneurship – Hurriyati et al (eds)
© *2020 Taylor & Francis Group, London, ISBN 978-0-367-27176-3*

The dynamics of community political participation in direct elections of Gianyar Regency in the reformation era

I.M. Artayasa, G. Wirata & I.W. Astawa
University of Ngurah Rai, Denpasar, Indonesia

ABSTRACT: The democratic dynamics of the representative system in the election of regional heads in Gianyar Regency is the implementation of policies issued in the reformation era. Direct elections in Gianyar Regency were carried out in 2008 and 2012. The problem raised in this study is the political participation of people of the Gianyar district in the direct election in the reformation era. The objective of the study is to identify and describe the political participation of the community and the influential power relations in the direct elections of the reformation era. This study used qualitative research methods. The data collected through observation, interview, document study and literature study, are presented through the stages of reduction, presentation and conclusions drawing. The results of the analysis showed that he political participation is still high because of the public's knowledge and local figure, especially the rural community.

1 INTRODUCTION

General elections of the regional head and deputy head of the region started after the fall of the new order regime called the reformation era, followed by the revision of the 1945 Constitution. It led to many changes in the state order, including in the election of regional heads both at a provincial and region/city level. The elections in the reformation era began with changes in election regulations for the presidential election, the elections of the Regional Representative Council, the Parliament, the Provincial People's Representative Assembly, the Regency/City People's Representative Assembly. The regulations have been directly implemented in accordance with the mandate of the law as well as the demands of the society in the reformation era.

Gianyar Regency is one of the nine regencies/cities in Bali that held elections twice from 2004 to 2014. The political participation is, on average, above 80%, although in the 2012 presidential election reached only 79%. The high political participation of the society was especially visible in the elections in 2008 with a percentage of 86% and decreased to 82% for the elections of 2012. Based on the data, it is important to find out about the dynamics of community political participation in direct elections in Gianyar Regency during the reformation era.

2 METHOD

2.1 Political participation

The second direct regional election resulted in a decrease in political participation of the society to 81%, from 86% in the first direct election. A decrease in participation occurred in all districts.

Implementation of direct local elections as a local political arena was strongly influenced by society participation and whether the society knows how to use their political rights by electing their leaders directly in accordance with the mandate of the law. There are some powers who can mobilize people's choice in direct elections, community figures such as Kelihan Dinas and Adat, Bendesa Adat, Village chief/Perbekel. These figures play an important role and can mobilize and direct political choices of people, especially in rural areas, because they are respected in the society. In addition to the mentioned figures, there are also many other figures such as Dadia, Kelihan subak, and Kelihan sekehe gong in the structure of Banjar and village organizations.

Another influence comes from government programs that can stimulate young people to make choices. With the development of communication technology, it is very easy to influence other young friends. This is in line with what Foucault said about power that tends to be hidden.

In summary, political participation of the Gianyar society is influenced by their local elite and peer people will feel ashamed of not voting because it has often been agreed in their Banjar/hamlet.

2.2 Discourse of power and knowledge

Foucault in his theory of discourse of power and knowledge (Barker 2005) emphasized the mutual reciprocal relationship between power and knowledge so that knowledge cannot be separated from power. Knowledge is formed within the context of relations and power practices, and further contributes to the development, improvement and maintenance of new power techniques. Foucault (Lubis 2014) saw that in the history of culture, the physical and mental (intellectual) forces always have strong minorities to impose their ideas onof true or good for the majority. The power or will of power is universal in politics and is exercised by all living beings. For Foucault there is no single source of power that becomes the origin of power. Power is understood as a dispersed one, and conflict is always specific, "distinctive" according to certain cultures' territories and cultural contexts. Power according to him is not something that has been there just like that. Power is relations that work in a certain space and time. Power is no longer in the hands of one person or institution but is widespread in society and tends to be hidden. Foucault's theory of power discourse is used to analyze the power relation of forces that influence the various forces in direct regional elections involving political society, economic society and civil society.

3 RESULT AND DISCUSSION

3.1 The political participation of the society in the direct regional election of the reformation era in Gianyar regency

Gianyar is one of the nine regencies/cities in Bali province and has an area of 36,800 hectares. Gianyar has held elections with candidates dominated by Puri areas; Puri Ubud, Puri Gianyar and also from Puri Peliatan.

In the first direct election in 2008 there was a battle between the candidate pairs for regional head and the deputy head of the region position, A.A. Gde Beratha from Puri Gianyar paired with Putu Yudany Tema against the couple Tjokorda Oka Artha Ardana Sukawati from Puri Saren Ubud in pair with Dewa Sutanaya. In the first direct election the battle was won by Tjokorda Artha Ardana Sukawati with Dewa Sutanaya as a newcomer with 138,182 votes (50.67%) while the incumbent A.A. Gde Beratha with Putu Yudany Tema earned 134,527 votes (49.33%).

Society's participation reached 86% and was the accumulation of seven districts with a fairly varied participation. The highest participation was in Ubud District reaching 91% during the election of Tjok Oka Artha Ardhana Sukawati.

The second direct election was held by the Election Committee of Gianyar Regency on 4 November 2012 in a fight between the candidate pair Cokorda Gde Putra Nindia from Puri Peliatan Ubud with A.A. Ngurah Agung from Puri Gianyar who went against the candidate pair A.A. Gde Beratha from Puri Gianyar paired with I Made Mahayastra from Payangan District. In this regional election A. A. Gde Beratha and I Made Mahayastra won with a vote of 193,643 (69.79%) while the pair Cokorda Gde Putra Nindia and A.A. Ngurah Agung obtained 83,838 votes (30.21%).

The second direct regional election resulted in a decrease in political participation of the society to 81%, from 86% in the first direct election. A decrease in participation occurred in all districts.

3.2 The power relations that affected the political participation of the people in the direct election of the reformation era in Gianyar regency

Implementation of direct local elections as a local political arena was strongly influenced by society participation and whether the society knows how to use their political rights by electing their leaders directly in accordance with the mandate of the law. There are some powers that can mobilize the people's choice in direct elections, community figures such as Kelihan Dinas and Adat, Bendesa Adat, Village chief/Perbekel. These figures have an important role in mobilizing and directing political choices of people, especially in rural areas, because they are respected in the society. In addition to the mentioned figures there are also many other figures, such as Dadia, Kelihan subak, and Kelihan sekehe gong in the structure of Banjar and village organizations.

As Foucault said, power is widespread in society and tends to be hidden, and candidates or their winning team/volunteers must approach local leaders to be able to mesimakrama to their territories to introduce their programs.

Another influence comes from government programs that can stimulate young people to make choices. With the development of communication technology, it is very easy to influence other young friends. This is in line with what Foucault said, that power tends to be hidden.

In summary, political participation of the Gianyar society is influenced by their local elite, and peer people will feel ashamed of not voting because it has often been agreed in their Banjar/hamlet.

4 CONCLUSION

The political participation of Gianyar society in the reformation era is still high because of the public's knowledge about the importance of the elections, although there are still influential local figures, especially in the rural community, who mobilize and direct the political choice of the community, since the rural community still respects them. In order to win the battle in regional elections in Gianyar regency, the candidate pair and the winning team must be able to embrace community leaders besides creating good programs so that it can attract the attention of rational voters and young voters.

REFERENCES

Barker, C. 2005. Cultural studies teori dan praktik. Yogyakarta: PT. Bentang Pustaka.
Lubis, A.Y. 2014. Teori dan metodologi ilmu pengetahuan sosial budaya kontemporer. Jakarta: PT. Rajagrafindo Persada.

Advances in Business, Management and Entrepreneurship – Hurriyati et al (eds)
© *2020 Taylor & Francis Group, London, ISBN 978-0-367-27176-3*

The influence of discipline coaching and physical working environment on employee motivation in Kertas Padalarang company

S. Sedarmayanti
Universitas Dr. Soetomo, Surabaya, Indonesia

S. Gunawan
STIA LAN, Bandung, Indonesia

B.W. Wibawa
Universitas Kristen Maranatha, Bandung, Indonesia

ABSTRACT: Discipline coaching and provision of a safe, comfortable working environment generate employee motivation to work optimally and to achieve the goals of the company. This research aims to show the effect of discipline coaching and working environment on employee motivation at Kertas Padalarang Company. This research is taking 121 respondents and using saturated sampling. Data for this research are collected through a questionnaire, using semantic differential scale. The research methods applied are multiple linear regression analysis, and simultaneous and partial determination of coefficient analysis. Meanwhile, the hypotheses testing is applying the *f*-test and *t*-test. The count result and the descriptive analysis show that the implementation of discipline coaching of the production section of Kertas Padalarang Company and conditions of the physical working environment of the production section of Kertas Padalarang Company are categorized as excellent, while employee working motivation is categorized as high. Positive effect of discipline coaching on employee motivation is shown by coefficient value 0.259 and 0.039 significance and *t* count < *t* table (2.084 > 1.981) and positive effect of physical environment on employee motivation is shown by coefficient value 0.360 and 0.000 significance and *t* count < *t* table (7.360 > 1.981). Discipline coaching and physical work environment effect employee motivation simultaneously with 0.000 significance and *F* count > *F* table (31.943 > 3.07). The effect of discipline coaching and physical work environment on employee motivation is shown simultaneously by 35.1%. The partial effect of discipline coaching is shown by 3.6% and the partial effect of physical work environment is shown by 31.5%. The next research will be the effect of organizational culture, compensation, or leadership on employee motivation in another entity such as a clinic, school, or hospital.

1 INTRODUCTION

One aspect of the power of human resources can be reflected in disciplinary attitudes and behaviors, because discipline has a strong impact on an organization's ability to achieve success in pursuing its planned goals (Thaief et al. 2015). According to Rivai (2004), work discipline is a tool used by managers to communicate with employees so they are willing to change their behaviors and make efforts to raise their awareness and willingness to comply with all corporate rules and prevailing social norms.

Based on interviews conducted at Kertas Padalarang Company, some information was obtained to identify employees' work discipline as based on their level of attendance and details of their delayed start times during 2016.

During 2016, employees were absent from work with permission 67 times. It is necessary to discern whether these absences were due to something that could not be postponed so that employees legitimately required permission or whether the employees only used an excuse to not come to work.

There are 11 out of 121 employees from the Production Bureau who came late to work 24 times during 2016. Some employees reasoned they were late to work due to oversleeping, taking children to school, traffic jam, and bad weather. With the existence of the absenteeism phenomenon and production bureau employees coming to work late in 2016, the researchers are interested in conducting research about employee work motivation at Production Bureau, Kertas Padalarang Company (Persero).

In addition to disciplinary coaching factors, another aspect that is considered as very important and needing attention in an effort to improve the employees' spirit and passion for their work is the physical work environment within a company. According to Simaremare and Isyandi (2015), a good physical work environment is expected to encourage employees to work as well as possible. According to Sedarmayanti (2011), there are several factors that can affect the formation of a physical work environment, including conditions such as the light in the workplace, temperature, air circulation, noise, mechanical vibration, odor, color, decoration and music, and security.

By conducting disciplinary coaching and facilitating the availability of a safe and comfortable physical work environment, the company expects to enhance its employees' motivation to work optimally so as to achieve the company's goals. Work motivation is a condition that is able to move employees to achieve goals owned by the company, because work motivation is closely related to the success of employees and the company itself.

2 METHOD

In order to clarify the influence of disciplinary coaching and the provision of work facilities toward work motivation, we can take a look at the framework described below.

This research uses a qualitative and quantitative approach and the population used is 121 employees of Production Bureau in Kertas Padalarang Company, five of whom are the supervisors of production bureau and the other 116 people are the operational staff members. Techniques used in data collection are library research and field research in the form of interview, questionnaire, observation, and documentation.

The quantitative descriptive method is used to answer the identification of problems 1, 2, and 3,

Table 1. Criteria of respondents' attitude on each variable.

| Interval | Variable Criteria | | |
	Disciplinary coaching	Physical work environment	Work Motivation
1,00 – 1,79	Very Bad	Very bad	Very low
1,80 – 2,59	Bad	Bad	Low
2,60 – 3,39	Fair	Fair	Fair
3,40 – 4,19	Good	Good	High
4,20 – 5,00	Very good	Very good	Very high

and the analysis is conducted based on Table 1 as follows:

Meanwhile, double regression is used to answer problem identification number 4, which aims to know how big the influence is from independent variable to variable dependent.

Several tests were performed on the data normality test using a one-sample Kolmogorov-Smirnov (KS) test, a multicollinearity test to test whether the regression model found a correlation with the independent variable, and a Glejser test which proposed to regress the residual absolute value against the independent variable.

Hypotheses testing was conducted to know the significance of the influence between the independent variable (X) and the dependent variable (Y). Hypothesis testing can be done by applying a simultaneous test and a partial test. Meanwhile, the coefficient of determination is used to measure how far the ability of the model goes in explaining the variation of the dependent variable

3 RESULTS AND DISCUSSION

Based on the validity test results, all items for all statement variables are valid. While the reliability test carried out to the valid questions to determine whether the result of measurement remain consistent done again with the measurement of symptoms are still the same. The variables of discipline coaching, physical work environment, and work motivation have Cronbach's Alpha value of > 0.70 or greater than the criterion value. Then it can be concluded that the instrument for the variables discipline, physical work environment, and work motivation is said to be reliable.

3.1 Descriptive analysis of discipline coaching

To understand the implementation of discipline coaching, questionnaires were distributed to 121 employees by using a saturated sampling method: five people are the supervisors of the production

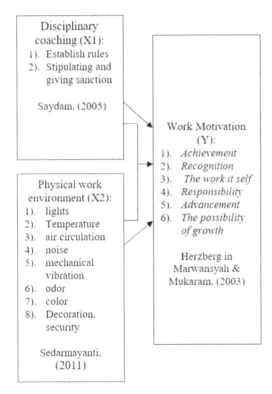

Figure 1. Framework scheme.

bureau and 116 are the operational staff members. The following is the result of the questionnaire distribution of discipline coaching variable.

Based on Table 2, after the calculation of the average score on the implementation of discipline coaching was conducted, it is known that the sub variable that has the smallest average score is the 'et and sanction violators' (3.30) and that it is included in the 'fair' category. Meanwhile, the 'set the rules and discipline that must be implemented by employees' sub variable has average score of 4.26 and falls into the 'good' category. It can be seen that the average score for discipline coaching is 3.78 and it falls into 'good' category. This indicates that the respondents assume that the implementation of discipline coaching to employees of Production Bureau, Kertas Padalarang Company is good.

3.2 Descriptive analysis of physical work environment

To determine the state of the physical work environment, questionnaires were distributed to 121 employees of the Production Bureau (five supervisors and 116 operational staff members) by using a saturated sampling method. The following is the result of the questionnaire distribution of the physical work environment variable.

Based on Table 3 above, after the calculation of the average score on the implementation of physical work

Table 2. Average total of discipline coaching variable.

No	Sub Variable	Average Score	Category
1.	Establish rules and regulations to be implemented by employees	4.26	Very good
2.	Set and sanction violators	3.30	Fair
	Discipline Coaching	3.78	Good

Sources: Questionnaires, tabulated data, 2017.

Table 3. Average total of physical work environment.

No	Sub Variable	Average Score	Category
1	Lights	2.32	Bad
2	Temperature	4.09	Good
3	Air Circulation	4.19	Good
4	Noise	3.99	Good
5	Mechanical Vibration	4.26	Very Good
6	Odor	4.26	Very Good
7	Color	3.96	Good
8	Decoration	4.22	Very Good
9	Security	4.50	Very Good
	Physical Work Environment	3.98	Good

Source: Questionnaires, tabulated data 2017.

environment was conducted, it is known that the smallest average score is the 'light' sub variable (2.32) and it falls into the 'bad' category. Meanwhile, the largest average score is 'security' with an average score of 4.50 and it is included in the 'very good' category. For an average score of 3.98, physical work environment is included in either category. This indicates that the state of physical work environment of employees of Production Bureau, Kertas Padalarang Company is good.

3.3 Descriptive analysis of work motivation variable

To find out the employees' work motivation, questionnaires were distributed to 121 employees of the production bureau (five supervisors and 116 operational staff members) using a saturated sampling method. The following is the result of the questionnaire distribution of work motivation variable.

Based on Table 4, after calculation of the average score on the implementation of work motivation to the production bureau employees of Kertas Padalarang Company, it is known that the smallest average score is in the sub variable of advancement of 2.32 and falls into to the 'high' category. The largest average is the possibility of growth sub variable with an average of 4.50 and is included in the 'high' category. For an average score of 3.98, physical work environment is included in either category. This indicates that the state of work motivation of production bureau employees, Kertas Padalarang Company is high.

3.4 Normality test results

The results of the data calculation for normality test were calculated by using the SPSS V.24 program, and showed that the value of asymp.sig (2-tailed) for variables of discipline, physical work environment, and motivation at 0.200. It means that the asymp.sig (2-tailed) value is above the significant level of 5% (0.05) or 0.200> 0.05. So, it can be concluded that the variables of discipline coaching, physical work environment and motivation are normally distributed.

Table 4. Average total of work motivation variable.

No	Sub Variable	Average score	Category
1	Achievement	4.18	High
2	Recognition	4.13	High
3	The work itself	3.97	High
4	Responsibility	4.33	Very High
5	Advancement	3.88	High
6	The possibility of growth	4.35	Very High
	Work Motivation	4.14	High

Sources: Questionnaires, tabulated data, 2017.

3.5 Multicollinearity test results

The results of data calculations for the multicollinearity test calculated by using the SPSS V.24 program shows a tolerance value for the discipline and physical work environment variables of 0.981 greater than 0.10 or 0.981 <0.10. As for the VIF value for the discipline coaching variable and the physical work environment variable is 1.019 (condition <10). So it can be concluded that there is no multicollinearity in the independent variable in the regression model.

3.6 Heteroscedasticity test results

The results of data calculations for heteroscedasticity test calculated by using SPSS V.24 program shows that the value of significance for the discipline coaching variable is 0.211 and the physical work environment variable is 0.663. Both values of significance of the variable are above 5% or 0.05. So it can be concluded that the regression model does not contain heteroscedasticity.

3.7 Results of multiple linear regression analysis

Data processing using SPSS V.24 obtained the results in Table 5.

Based on Table 5, the prediction of model equation of discipline coaching (X1) and physical work environment (X2) to work motivation (Y) can be drawn as follows:

$$\tilde{Y} = 9.999 + 0.259X_1 + 0.360X_2 + e \quad (1)$$

Based on the equation of the prediction model, regression coefficients and variables can be interpreted as follows:

a = 9.999   This means that if the variable of discipline coaching (X1) and Physical work environment variable (X2) is zero (0), then the work motivation (Y) variable will be worth 9999. In other words, the regression lines will intersect the Y axis at point 9.999

b_1= 0.259   This means that if the discipline coaching variable (X1) increases by 1 and the other variable remains, then the work motivation (Y) variable will increase by 0.259

b_2= 0.360   This means that if the physical work environment variable (X2) increases by 1 and the other variable remains, then the work motivation (Y) variable will increase by 0.360.

Based on the data, it can be seen that the physical work environment variable has greater influence when compared with the variable of work discipline.

3.8 Simultaneously test results (F)

Testing the hypothesis using the f-test statistic with the calculation results as follows:

$$\text{Value of Table } F = F(K : N - K)$$
$$= F(2 : 119) = 3.07 \quad (2)$$

Based on Table 6, it shows that the value f count > f table or 31.943 > 3.07, where f table obtained from the f test table with the above calculation and f significance 0.000 < 0.05. Then H0 is rejected and HA is accepted, which means discipline coaching and

Table 6. Simultaneous test results (F).

ANOVA[a]

Model		Sum of Sq	df	MS	F	Sig.
1	Reg	55.961	2	27.980	31.943	0.000[b]
	Res	103.362	118	0.876		
	Total	159.323	120			

a. Dependent variable: work motivation
b. Predictors: (constant), LKF, PD
Source: SPSS V24 result, processed, 2017.

Table 5. Multiple linear regression test results.

Coefficients

Model		Unstandardized Coefficients		Standardized Coefficients		
		B	Std. Error	Beta	t	Sig.
1	(Constant)	9.999	1.874		5.336	.000
	Discipline coaching	.259	.124	.156	2.084	.039
	Work environment	.360	.049	.551	7.360	.000

a. Dependent variable: work motivation
Source: Result of SPSS V24, processed, 2017.

physical work environment simultaneously have a positive or unidirectional effect on employee work motivation at Production Bureau, Kertas Padalarang Company.

3.9 *Partial test results (t)*

Testing the hypothesis using the t-test statistic with calculation results as follows:

$$t_{table} = t\left(\frac{a}{2} : n - k - 1\right)$$
$$t_{table} = t(0.025 : 118) = 1.981 \qquad (3)$$

Based on Table 7, for the discipline coaching variable, the value of t count $> t$ table or 2.084> 1.981, where t table is derived from the above calculation and t significance is 0.039 < 0.05. Then H0 is rejected and HA is accepted which means discipline coaching partially has a positive or unidirectional influence to employee work motivation at Production Bureau, Kertas Padalarang Company.

Table 1.9 shows that the physical work environment variable is t count $> t$ table or 7.360 > 1.981, where t table is obtained from the above calculation and t significance is 0.000 < 0.05. Then H0 is rejected and HA is accepted which means the physical work environment partially has a positive or unidirectional influence to employee work motivation at Production Bureau, Kertas Padalarang Company.

3.10 *Determination coefficient*

$$Determination\ Coefficient = R\ square$$
$$= (0.351 \times 100\%) \qquad (4)$$
$$= 35.1\%$$

Based on these calculations, and using the determination coefficient of 35.1%, the influence of discipline coaching and physical work environment to

Table 8. Determination coefficient.

Model Summary

Model	R	R Square	Adjusted R Square	Std. Error of the Estimate
1	0.593[a]	0.351	0.340	0.93592

a. Predictors: (constant), work environment, discipline coaching
Source: SPSS V24 result, processed, 2017.

employee work motivation is 35.1%, while the rest of 64.9% influenced by other factors which are not discussed in this research.

The result of beta and zero-order multiplication is an important component in determining the determination coefficient of discipline and physical work environment. From the influence of 35.1% of variables simultaneously, it can be known partially that discipline coaching has an influence of 3.6% (0.156 × 0.231 × 100) and the physical work environment has an influence of 31.5% (0.551 × 0.572 × 100). Thus, it can be seen that the physical work environment variable has greater influence than discipline coaching on employee work motivation at Production Bureau, Kertas Padalarang Company.

4 CONCLUSIONS

Based on the results of research and discussion conducted to determine the influence of discipline coaching and physical work environment on employee work motivation, both simultaneously and partially, it can be concluded as follows:

1) The discipline coaching applied to the employees of Production Bureau at Kertas Padalarang Company is in the 'good' category.
2) Physical work environment at Production Bureau, Kertas Padalarang Company, is included in the 'good' category.

Table 7. Partial test tesults (t-test).

Coefficients[a]

Model		Unstandardized Coefficients		Standardized Coefficients		
		B	Std. Error	Beta	t	Sig.
1	(Cons)	9.999	1.874		5.336	0.000
	PD	0.259	0.124	0.156	2.084	0.039
	LKF	0.360	0.049	0.551	7.360	0.000

a. Dependent variable: work motivation
Source: SPSS V24 result, process, 2017.

3) Employee work motivation at Production Bureau, Kertas Padalarang Company is included in the 'high' category.
4) Discipline coaching and physical work environment simultaneously has a positive or unidirectional influence on employee work motivation at Production Bureau, Kertas Padalarang Company.
5) Discipline coaching has a partial positive or unidirectional influence on employee work motivation at Production Bureau, Kertas Padalarang Company.
6) Physical work environment partially has a positive or unidirectional influence on employee work motivation at Production Bureau, Kertas Padalarang Company.
7) Based on the result of the determination test, discipline and the physical work environment have a positive and unidirectional influence to work motivation simultaneously by 35.1%, while the other 64.9% is influenced by other factors not discussed in this research. From the percentage of the influence (35.1%), it can be known partially that discipline coaching has an effect of 3.6% and the physical work environment has an influence of 31.5%.

REFERENCES

Marwansyah & Mukaram. 2003. *Manajemen Sumberdaya Manusia. Bandung.* Pusat Penerbit Administrasi Niaga Politeknik Negeri Bandung.

Rivai, V. 2004. *MSDM untuk perusahaan: Dari Teori ke Praktek.* Jakarta: PT. Raja Grafindo Persada.

Saydam, G. 2005. *Manajemen Sumberdaya Manusia, Suatu Pendekatan Mikro.* Jakarta: Djambatan.

Sedarmayanti. 2011. *Tata Kerja dan Produktivitas Kerja. Bandung.* Mandar Maju.

Simaremare, C. & Isyandi, H. 2015. Pengaruh Pelatihan, Lingkungan Kerja Fisik Dan Kepemimpinan Terhadap Kepuasan Kerja Dalam Meningkatkan Kinerja Karyawan Pada PT. Federal International Finance Wilayah Riau. *Jurnal Tepak Manajemen Bisnis* 7(3).

Thaief, I., Baharuddin, A. & Idrus, M.S.I. 2015. Effect of training, compensation and work discipline against employee job performance (Studies in the Office of PT. PLN (Persero) Service Area and Network Malang). *Review of European studies* 7(11): 23.

Advances in Business, Management and Entrepreneurship – Hurriyati et al (eds)
© 2020 Taylor & Francis Group, London, ISBN 978-0-367-27176-3

The influence of ability on employee performance

M. Masharyono, S.H. Senen & D.A. Dewi
Universitas Pendidikan Indonesia, Bandung, Indonesia

ABSTRACT: Companies that engage in the manufacturing sector rely on high quality and high employee performance. Employee performance is still an important problem in manufacturing companies, especially at PT. Indorama Synthetics Tbk. Polyester Division in Purwakarta. One of the efforts made by the company to improve employee performance is by improving employee skill. The present study uses descriptive analysis technique and verification. The method of the study is an explanatory survey of 103 respondents, while the data are obtained through the questionnaires. Then the analysis of the study is by using a simple linear regression analysis. The finding of the study indicates that the employees' work ability is very good, but there are weaknesses in the knowledge dimension. Other findings state that the employees' performance is very high, especially on the dimensions of ethics and communications. The working ability is influence 70, 3% to employee performance. If the company does not pay attention to the performance, the company will have difficulty to achieve any goals and the productivity of work.

1 INTRODUCTION

Human resources are essential to organizational excellence and act as a key factor in achieving organizational goals (Muda et al. 2014). Organizations need human resources that work efficiently and effectively in order to take over the management of employees well (Sumiyati et al. 2016). The main objective of human resource management is to improve employee performance. In other words, human resource management is expected to produce high-performing employees (Wayan & Surachman 2016).

Employee performance is a concern; therefore, it is at the core of human resource management issues (Schaefer et al. 2015). Low employee performance in this competitive era is still a major problem in human resource management (Senen et al. 2016). Employee performance issues are encountered by companies in various sectors in many countries, both public and private, in health, education, banking, state-owned enterprises, and small-scale enterprises (Senen et al. 2016). Employee performance is very important to a company's efforts to achieve its goals and can become one of the main factors of success in the company (Nuryanti & Rahmawati 2016). High employee performance is a priority, so that the company can maximize the performance of its employees that become the main challenge for the organization (Osa 2014).

Organizations require their employees to demonstrate good performance and dedicate themselves to the organization in their work (Senen et al. 2017). Performance can affect the activities of an organization or company; better performance shown by the employees will be very helpful for any development. (Wardhana et al. 2016). In order to survive in the competitive international market, the company needs human resources that have high performance, because, it is one of the success factors within an organization or company (Senen & Wahyuni 2016).

The quality of the individual's performance depends on understanding and ability to reach the target (Khan et al. 2016). Employee performance can be viewed in terms of productivity and output that affect and help the organization to work effectively and efficiently in achieving its goals (Emelia & Darko 2017).

A study conducted by Odunlami and Matthew (2014) on employee performance in the manufacturing industry, stated that employees are an inseparable part of human resource management. This is reinforced by a study by Emeka et al. (2015) that in Nigerian companies the manufacturing sector is listed as one of the growth engines in jobs that can create wealth for development, but the sector is unable to overcome the challenges reflected in its poor performance over the years.

A company in the manufacturing sector is a company engaged in the production sector based on the efficiency and effectiveness of employees. Therefore, companies that engage in the manufacturing sector rely on the quality and performance of high-functioning employees (Senen & Wahyuni 2016). Based on a study by Supriyani and Mahmud (2013) on employee performance in the manufacturing industry, it is concluded that the decline of employee performance is indicated by the number of employees who are absent, the target of production, and the employees' punctuality.

There are several factors that affect employee performance such as work quality, skills, responsiveness, speed, initiative, ability, and communication (Torang 2012). However, the key determinants of poor employee performance and their lack of involvement in the workplace are the salary structure, lack of opportunities, weak incentive schemes, seniority-promotional codes, delays in promotion, and the lack of a motivational strategy (Zheng & Lamond 2010, Assan 2015).

One thing that can play an important role for a company to improve the performance of its employees is ability (Burns 1993). Performance can be improved by using the ability to generate new ideas for building relationships and work processes (Sajid 2017). Employee performance issues occur because knowledge and skills between employees have a gap so that the work cannot be completed on time (Chen et al. 2014).

Based on these problems, the purpose of this study is to obtain findings on (1) description of work ability, (2) the employee performance picture, and (3) the influence of work ability on employee performance.

2 LITERATURE REVIEW

Human resource management can be interpreted as a science and an art that regulate relationships and the role of labor to be effective and efficient in the use of human capabilities in order to achieve goals in each company (Mathis & Jackson 2011). Organizations need human resources that can work efficiently and effectively in order to be able to take over the management of employees well (Sumiyati et al. 2016).

One of the resources that is the most important asset of the organization is the human resources because human resources play a role in achieving the goals of the organization (Masharyono & Sumiyati 2016).

Mathis and Jackson (2011) suggests seven functions of human resource management: (1) strategic HR management, (2) equal employment opportunity, (3) staffing, (4) talent management and development, (5) total rewards, (6) risk management and worker protection, and (7) employee and labor rRelations.

Based on the human resource function, there is a talent management function focusing on individuals who are ready to work when needed and talented people being developed for future organizational needs (Mathis & Jackson 2011). One way to improve talent management is by recruiting and selecting employees through training, succession planning, career planning, development, and performance management (Mathis & Jackson 2011).

Training seeks business success and enables an organization to make periodic improvements to produce planned output. With regular training, the organization will produce employees with the knowledge, skills, and abilities needed to compete effectively (Mathis & Jackson 2011). In the function of human resource management, work ability has a role to be one of the factors that support the achievement

of the vision and mission of the organization in order to facilitate successful development (Kumar 2016). Whether a company will succeed or even fail is largely determined by its ability (Thoha 2010). Ability can generate employees who can grow and achieve performance levels beyond what is expected by the company; in addition, ability is considered effective in any situation and culture (Yukl 2010).

Ability is the personal potential that allows a person to be able to do the job or not. One of the most important and influential factors to employees' success in performing a job is the ability to work (Bolli & Farsi 2015). Each type of work demands certain knowledge and skills and attitudes in order that the task be performed well. Therefore, the job is done effectively or the employee performs well in the workplace in certain situations (Moeheriono 2010).

Sutermeister's (1976) theory states that ability shows employees' potential to perform tasks or jobs in which each type of work requires certain knowledge, skills, and attitudes in order to carry out the task well and to emphasize understanding as outcomes of a job and their contribution to the organization.

Working ability reflects a balance between individual capacity, and mental and job demands (Calatayud et al. 2015). In line with Robbins and Judge (2013), ability is an individual capacity to perform various tasks in a job. According to Robbins and Judge (2013), characteristic abilities are similar to intellectual abilities and physical abilities (intellectual physical). According to Sutermeister (1976), dimensions of ability consist of (1) knowledge based on education, experience, practice, and interests, and (2) skills consisting of attitude and personality.

Working ability is influenced by several factors. Brasileira et al. (2014) mention the following factors that influence abilities: (1) dysfunction, which limits workers' ability to work; (2) elderly and female workers show an inverse relationship with loss of productivity in the workplace; (3) education level, participation in activities; and (4) the mental and physical demands of doing the work.

Human resources are essential to organizational excellence and act as a key factor in achieving organizational goals (Muda et al. 2014). The main objective of human resource management is to improve employee performance. In other words, human resource management is expected to produce employees with high performance (Wayan & Surachman 2016). Employee performance can be viewed in terms of productivity and output, and employees that affect and help the organization to work effectively and efficiently in achieving its goals (Emelia & Darko 2017).

Performance is a real behavior that is displayed by each individual as work performance generated by employees in accordance with their role in the company (Rivai & Sagala 2014). Wibowo (2014: 62) views performance as a way of ensuring that

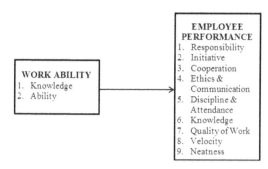

EMPLOYEE PERFORMANCE
1. Responsibility
2. Initiative
3. Cooperation
4. Ethics & Communication
5. Discipline & Attendance
6. Knowledge
7. Quality of Work
8. Velocity
9. Neatness

WORK ABILITY
1. Knowledge
2. Ability

Figure 1. Paradigm of the study.

individual workers or teams know what to expect and that they stay focused on effective performance by paying attention to objectives, measures, and assessors. According to Albrecht et al. (2012) performance is work achieved by individuals in performing tasks assigned to them on the basis of experience, sincerity, and measurement.

Performance indicators are used for behavioral activities that can be observed by prioritizing future perspectives rather than looking backwards. The company has nine assessment standards to measure employee performance: (1) responsibility, (2) initiative, (3) cooperation, (4) ethics and communication, (5) discipline and attendance, (6) knowledge, (7) quality, (8) speed and (9) neatness.

The five performance indicators of employee performance indicators according to Gomez-Mejia (2012) are (1) quality of work, (2) quantity of work performance, (3) interpersonal effectiveness, (4) competencies, and (5) job knowledge. Employee performance can be measured through (1) timeliness, (2) achievement of targets, (3) understanding of work, (4) understanding of operational standards, (5) submission of, (6) cooperation, (7) responsibility, and (8) leadership (Zachary 2017).

Based on the explanation of the effect of work ability on employee performance, we then compiled a paradigm of the study of work ability to employee performance as clearly described in Figure 1.

3 METHOD

The unit analysis of the study is employees of PT. Indorama Synthetics Tbk. Polyester Division in Purwakarta in less than one year, using a cross-sectional data collection technique. A sampling technique in the study is a saturated sample of 103 employees of PT. Indorama Synthetics Tbk. Polyester Division in Purwakarta.

The data collection techniques used are literature study and field study with direct questionnaire distribution. Meanwhile, the technique of data analysis used descriptive and verificative analysis. The analysis of verificative data used simple linear regression analysis by software SPSS 24.0 for Windows.

4 RESULT AND DISCUSSION

4.1 Descriptive analysis

4.1.1 Work ability
Based on the results of 103 questionnaires, a description of the work ability of PT. Indorama Synthetics Tbk. Polyester Division in Purwakarta is presented in Table 1 below.

Based on Table 1, the highest dimension is ability at about 3556 or 82%, while the lowest is knowledge at about 5181 or 80%. Overall, skill variables obtained 8737 scores or 81%. Therefore, it can be said that almost all respondents expressed their skill or ability at PT. Indorama Synthetics Tbk. Polyester Division as in the "very good" category. A good work ability will improve prosperity and support in the work field. Whether a company will succeed or even fail largely determined by its ability (Thoha 2010). Ability can generate employees who can grow and achieve performance levels beyond what is expected by the company. Moreover, ability is considered effective in any situation and culture (Yukl 2011).

4.1.2 Employee performance
Based on the results of 103 questionnaires, there is a picture of employee performance at PT. Indorama Synthetics Tbk. Polyester Division in Purwakarta in Table 2.

Based on Table 2, the highest performance aspect is found in the ethics and communication dimension which scored 1853 or 86%, while the lowest dimension is initiative which scored 2264 or 79%. Overall, the score of performance variables is about 17,176 or 82%, it can be said that almost all respondents stated that their performance at PT. Indorama Synthetics Tbk. was in the very high category. Ideal performance score is 20,909 for 29 statement items. Based on the score, the result of data processing on the performance variable is about 17,176. Performance can affect the activities of a company or organization; better performance by employees will be very helpful in the development of the organization or company (Wardhana et al. 2016). In the competi-

Table 1. Recapitulation of work capability response.

No	Dimension	Total Score	Ideal Score	%
1	Knowledge	5181	6489	80%
2	Ability/Capability	3556	4326	82%
Total		8737	10,815	81%

Table 2. Recapitulation of employee performance feedback.

No	Dimension(s)	Total Score	Ideal Score	%
1	Responsibility	2345	2884	81%
2	Initiative	2264	2884	79%
3	Cooperation	2380	2884	83%
4	Ethics & Communication	1853	2163	86%
5	Discipline & Attendance	1792	2163	83%
6	Knowledge	1735	2163	80%
7	Quality of work	1808	2163	84%
8	Speed	1737	2163	80%
9	Neatness	1226	1442	85%
	Total	17,176	20,909	82%

tive international market, companies need high-performance human resources, because this is one of the success factors in an organization or company (Senen & Wahyuni 2016).

4.2 Verificative analysis

In the present study, there is a free variable, the ability (X), while the dependent variable is the employee's performance (Y). To test whether there is any independent influence on the dependent variable, we used a simple linear regression test.

The model of multiple regression equation that will be formed in this study follows:

$$Y = a + bX \qquad (1)$$

Source: Sugiyono (2010).

where Y = employee performance; a = constants; b = regression coefficient; and X = ability.

Based on the results of data processing by SPSS 24.0 for Windows, we obtained the regression coefficient as follows:

Based on Table 3, column B, the constants and the values of simple linear regression coefficients for independent variables are listed. Based on the values, it can be determined that a simple linear regression model can be expressed in terms of equations as follows:

$$Y = a + bX$$

$$Y = 38.372 + 1.510X \qquad (2)$$

The above equation can be interpreted as follows:
a = 38.372 it means that if the variable X, is zero (0), then the variable Y will be worth 38.372.
b_1 = 1.510 it means that if the working ability (X) increases by one unit and the other variable is constant, then the variable Y will increase by 1.510.

To find out the percentage of influence X to Y, the coefficient of determination can be known by Sugiyono (2013). The formula as follows:

$$KD = r2 \times 100\% \qquad (3)$$

where KD = coefficient of determination; r = correlation coefficient; and 100% = constants.

The effect of ability on employee performance can be seen from the result on Table 4 below:

$$KD = R^2 \times 100\%$$
$$= (0.821)^2 \times 100\% \qquad (4)$$
$$= 67.5\%$$

The obtained value of KD was 67.5%, which indicates that the ability to work has a partial influence of 67.4% on employee performance. The remaining 32.6% is influenced by factors that are not studied in the present study, such as work quality, skills, responsiveness, speed, initiative, ability, and communication (Torang 2012).

Table 3. Regression coefficient.

Coefficients[a]

Model	Unstandardized Coefficients		Standardized Coefficients	T	Sig.
	B	Std. Error	Beta		
(Constant)	38.372	8.907		4.308	0.000
Work ability	1.510	0.104	0.821	14.477	0.000

a. Dependent variable: employee performance.

Table 4.   Partial determination coefficient.

Model Summary[b]

Model	R Square	Adjusted R Square	Std. Error of the Estimate
1	0.821[a]	0.675	0.672	10.238

a. Predictors: (constant), ability.
b. Dependent variable: employee performance.

4.3   Partial hypothesis testing (t-test)

The statistical *t*-test basically indicates how far the influence of an individual explanatory/independent variable can explain the dependent variable. To find out the extent of the influence of ability on employee performance, the present study is using the program SPSS 24.0 for Windows. The resulting output is as follows:

Based on Table 5, *t*-test results of the influence of ability to performance with SPSS obtained t count = 14.477 with a significance level of 0.000. By using the 0.05 significance limit, the significance value is below 5% and the t count is 14.477 > t table of 1660 (df = 103 − 3 = 100, α = 5%). Thus, ability can affect performance. Thus Ho1 is rejected and Ha1 is accepted, it means that if ability is good, it will improve the performance of employees at PT. Indorama Synthetics Tbk. Polyester Division in Purwakarta. The magnitude of influence of ability to performance is to 0.821 or 82.1%.

4.4   Simultaneous hypothesis testing (f-test)

The statistical *f*-test basically shows whether all independent variables included in the model have a mutual influence on the dependent/bound variable. The *f*-test is used to test the significance of the effect of ability and the social work environment on employee performance of *f*-test formula or ANOVA test described in Table 6.

Based on Table 6, the influence of independent variables together on the dependent variable is done by using the *f*-test. The result of statistical calculation determines the value of F count = 209.584 > F table = 3.09 (df1 = 3–1 = 2, df2 = 103–2 = 101, α = 5%) and significance of 0.000 < 0.05. It can be concluded that the hypothesis in the present study shows that H0 is rejected and HA is accepted. Therefore, at the same time there is influence between work ability to employee performance in PT. Indorama Synthetics Tbk. Polyester Division in Purwakarta.

Table 5.   Significant value *t*-test.

Coefficients[a]

Model	Unstandardized Coefficients		Standardized Coefficients	T	Sig.
	B	Std. Error	Beta		
(Constant)	38.372	8.907		4.308	0.000
Work ability	1.510	0.104	0.821	14.477	0.000

Dependent variable: employee performance.

Table 6.   Significant value *f*-test.

ANOVA[a]

Model		Sum of Squares	df	Mean Square		Sig.
1	Regression	21968.818	1	21968.818	209.584	0.000[b]
	Residual	10586.910	101	104.821		
	Total	32555.728	102			

a. Dependent variable: employee performance.
b. Predictors: (constant), ability work.

5 CONCLUSION AND RECOMMENDATION

The result of the study stated that (1) the description of employee work ability in PT. Indorama Synthetics Tbk. Polyester Division is in the "very good" category; (2) an overview of employee performance at PT. Indorama Synthetics Tbk. Polyester Division is in a very high category; (3) ability has an effect on employee performance. This shows that the better the ability of employees in the company, the higher the performance of employees of PT. Indorama Synthetics Tbk. Polyester Division in Purwakarta.

Based on the results of the study, the authors suggest improving employee performance by improving the knowledge aspects of the company, by accepting employees to work in accordance with their field of education, and that the company considers an employees' year of service, and conducts training.

REFERENCES

Albrecht, W.S., Albrecht, C.O. & Zimbelman, C.C. 2012. *Fraud examination* (Fourth) Edi. USA: South-Western.

Bolli, T. & Farsi, M. 2015. The dynamics of productivity in Swiss Universities. *Journal of Productivity Analysis* 44(1): 21–38.

Brasileira, R., Coelho, J.S., Salmaso, C., Universidade, T. & Coelho, J.S. 2014. Factors associated with work ability in the elderly. *Revista Brasileira de Epidemiologia* 17: 830–841.

Burns R. B. 1993. *Konsep Diri: Teori, pengukuran, perkembangan dan perilaku.* Jakarta: Arcan.

Calatayud, J., Jakobsen, M.D., Sundstrup, E., Casaña, J. & Andersen, L.L. 2015. Dose-response association between leisure time physical activity and work ability: Cross-sectional study among 3000 workers. *Scandinavian Journal of Public health* 43(8): 819–824.

Chen, S., Zhu, X., Welk, G.J., Kim, Y., Lee, J. & Meier, N. F. 2014. Science direct using sensewear armband and diet journal to promote adolescents' energy balance knowledge and motivation. *Journal of Sport and Health Science* 3(4): 326–332.

Emeka, H., Ifeoma, J. & Emmanuel, I. 2015. The international journal of business & management an evaluation of the effect of technological innovations on corporate performance: A study of selected manufacturing firms in Nigeria Abstract 3 (1): 248–262.

Emelia, A.A. & Darko, T.O. 2017. Leadership, employee engagement and employee performance in the public sector of Ghana. *Journal of Business and Management Sciences* 5(2).

Gomez-Mejia, L.R. 2012. *Managing human resources.* New York: Prentice Hall.

Khan, A.A., Abbasi, S.O.B.H., Waseem, R.M., Ayaz, M. & Ijaz, M. 2016. Impact of training and development of employees on employee performance through job satisfaction: A study of telecom sector of Pakistan. *Business Management and Strategy* 7(1): 29–46.

Kumar, S. 2016. Innovative motivational techniques and their impact on performance and job satisfaction: 393–397.

Masharyono, M. & Sumiyati, S. 2016. Physical work environment effect on employee productivity of textile industry. *Advances in Economics, Business and Management Research* 15: 630–632.

Mathis, R.L. & Jackson, J.H. 2011. *Human resource management (13th ed.).* United States of America: Cengage Learning.

Moeheriono, M. 2010. *Pengukuran kinerja berbasis kompetensi.* Surabaya: Ghalia Indonesia.

Muda, I., Rafiki, A. & Harahap, M.R. 2014. Factors influencing employees' performance: A study on the Islamic banks in Islamic Science University of Malaysia University of North Sumatera. *International Journal of Business and Social Science* 5(2): 73–81.

Nuryanti, B.L. & Rahmawati, R. 2016. The influence of situational leadership and work environment towards employees' performance 15: 540–543.

Odunlami, I.B. & Matthew, A.O. 2014. Compensation management and employees performance in the manufacturing sector: A case study of a reputable organization in the food and beverage industry. *International Journal of Managerial Studies and Research* 2(9): 108–117.

Osa, I.G. 2014. Monetary incentives motivates employee's on organizational performance. *Global Journal of Arts Humanities and Social Sciences* 2(7).

Rivai, V. & Sagala, E.J. 2014. *Manajemen sumber daya manusia untuk perusahaan dari teori ke praktik.* Jakarta: Rajawali Pers.

Robbins, S.P. & Judge, T.A. 2013. *Organizational behavior.* (15th ed.). New Jersey.

Sajid, H. 2017. Impact of organizational politics on employee performance in public sector organizations. *Pakistan Administrative Review* 1(1): 19–31.

Sattar, T., Ahmad, K. & Hassan, S.M. 2015. Role of human resource practices in employee performance and job satisfaction with mediating effect of employee engagement. *Pakistan Economic and Social Review* 53(1): 81–96.

Schaefer, C.P., Cappelleri, J.C., Cheng, R., Cole, J.C., Guenthner, S., Fowler, J. & Mamolo, C. 2015. Health care resource use, productivity, and costs among patients with moderate to severe plaque psoriasis in the United States. *Journal of the American Academy of Dermatology* 74(4): 585–593.

Senen, S.H., & Wahyuni, Y. 2016. Pengaruh gaya kepemimpinan dan budaya organisasi terhadap kinerja karyawan PT Sugih Instrumendo Abadi di Padalarang. *Journal of Business Management Education* 1(2): 59–69.

Senen, S.H., Sumiyati, S. & Masharyono, M. 2017. Employee performance assessment system design based on competence. *Innovation of Vocational Technology Education* 13(2): 68–70.

Senen, S.H., Sumiyati, S., Masharyono, M. & Triananda, N. 2016. The employee performance influenced by communication: A study of BUMD in Indonesia 15: 596–598.

Sugiyono, S. 2010. *Metode penelitian kuantitatif kualitatif & R&D.* Jakarta: Alfabeta.

Sugiyono, S. 2013. *Metode penelitian pendidikan (Pendekatan kuantitatif, kualitatif, dan R&D).* Bandung: Alfabeta.

Sumiyati, S., Masharyono, M., Purnama, R. & Pratama, K. F. 2016. The influence of social work environment on employee productivity in manufacturing in Indonesia 15: 649–652.

Supriyani, S. & Mahmud, M. 2013. Membangun kinerja karyawan melalui motivasi kerja, kepuasan kerja dan komitmen organisasi pada PT. Astra Internasional di Semarang. *Skripsi*. Fakultas Ekonomi Bisnis.

Sutermeister, R.A. 1976. *People and productivity*. New York: McGraw-Hill.

Thoha, M. 2010. *Kepemimpinan dan manajemen*. Jakarta: PT. Raja Grafindo Persada.

Torang, S. 2012. *Perilaku organisasi*. Bandung: Alfabeta.

Wardhana, R.M.D.H., Tarmedi, E. & Sumiyati, S. 2016. Upaya meningkatkan kinerja dengan cara memberikan motivasi kerja dan menumbuhkan komitmen organisasional pegawai dinas perhubungan provinsi jawa barat. *Journal of Business and Management Education* 1(2): 91–96.

Wayan, S.N. & Surachman, H.D.R.F. 2016. Performance based compensation effect on employee motivation satisfaction of employees and performance of employees (Study on private universities in the Province of Bali conceptual). *International Journal of Business, Economics, and Law* 11(2): 62–70.

Wibowo, W. 2014. *Manajemen kinerja*. Jakarta: Rajawali.

Yukl, G. 2010. *Kepemimpinan dalam Organisasi (kelima)*. Jakarta: PT. Indeks.

Zachary, L. 2017. *Creating a mentoring culture: The organization guide (John Wille)*. San Francisco: John Wiley & Sons.

Zheng, C. & Lamond, D. 2010. Organisational determinants of employee turnover for multinational companies in Asia. *Asia Pacific Journal of Management* 27(3): 423–443.

Advances in Business, Management and Entrepreneurship – Hurriyati et al (eds)
© 2020 Taylor & Francis Group, London, ISBN 978-0-367-27176-3

The relationship of the factors that motivate nurses to provide complete nursing care documentation

A. Jaelani, A.M. Siddiq & K. Kusnendi
Universitas Pendidikan Indonesia, Bandung, Indonesia

W. William
Immanuel College of Health Sciences, Bandung, Indonesia

ABSTRACT: Research related to nursing care documentation is important to know the extent of documentation completeness in the Inpatient Prima I room of Immanuel Hospital and find out what factors influence the completeness of nursing care documentation. The purpose of this study is to identify the factors that influence nurses in achieving the completeness of nursing documentation in the primary inpatient room of the Immanuel Bandung Hospital. The population in this study was 44 nurses, consisting of both permanent nurses and contract nurses in the Inpatient Room I Prima Immanuel Hospital Bandung. The tools used are questionnaires and observation sheets. The implementation of nursing care documentation in the Prima Inpatient Room I Immanuel Hospital in Bandung has not been implemented well. The workload of nurses at Immanuel Hospital is in the light category. The nursing care documentation format at Immanuel Hospital has support categories. The leadership policy on achieving completeness of nursing care documentation is in the low category. The results of chi-square analysis showed a significant relationship between workload factors, documentation format factors, and leadership policy factors for the implementation of complete nursing care documentation in the Inpatient Prima I Room of Immanuel Hospital in Bandung with values of 0.041, 0.004 and 0.007.

1 INTRODUCTION

Documentation as discussed here is a medical record that describes the comprehensive status and needs of a client that will be given, is being given, or has been given care by the nurse The documentation is written or printed on the patient's status record and can be relied upon as evidence of court records if there is a patient or family demand. Good documentation reflects the quality of good care and proves the responsibility of each member providing care (Potter & Perry 2005). Nursing care documentation is a way of communicating nursing care among other health teams and is a patent document in providing nursing care that includes assessment, identifying problems, interventions, implementation, and nurses' observations and evaluations of client responses to interventions that have been given by nurses (Potter & Perry 2005).

Many studies have explained the importance of nurse motivation as one of the problems to be studied (Jooste & Hamani et al. 2017, Bernardino 2018). In this research, the influence of the work is the workload, documentation format, and leader's policy. These three factors are the problems identified after an interview and observations.

Documentation of important nursing care is carried out as a means of communication between nurses and other health professionals. In addition to being a communication tool, nursing care documentation is also proof of actions taken in regard to patients and will be evidence of accountability for nurses. Therefore, research related to nursing care documentation is important. The purpose of this study is to identify the factors that influence nurses in achieving completeness of nursing documentation in the prime inpatient room of the Immanuel Bandung Hospital. The results of this study can be input for the administrators of Immanuel Hospital to improve the quality of service of Immanuel Hospital in Bandung.

2 METHOD

This research is analytic quantitative research with a cross-sectional approach, to determine the prevalence or effect of a phenomenon by connecting independent variables and dependent variables. Data is collected once and scored at the same time (point time approach). That is, each research subject is only observed once and measurements are taken on the character status or subject variables at the time of the examination. The population in this study was 44 nurses (Philip 14 nurses, Gideon 15 nurses, and Beria 15 nurses), which consisted of permanent nurses and contract nurses in the Inpatient Room I Prima

Immanuel Hospital Bandung. The tools used are questionnaires and observation sheets.

3 RESEARCH AND DISCUSSION

3.1 Univariate analysis

The results of the univariate analysis present four tables that will be analyzed for their effect on achievement of completeness of nursing care documentation conducted by nurses. These variables are variable nurse workload, documentation format, leader's policy, and completeness of nursing care documentation. Univariate analysis results are presented in the following table form:

a. Nurse workload factors in the implementation of nursing care documentation in the Prima Inpatient Room I Immanuel Hospital Bandung.

Table 1. Workload of nurses.

Workload	Frequency	Percentage
Light	15	41.67
Medium	10	27.78
Heavy	11	30.56
Total	36	100

Analysis of Table 1 shows that most respondents, 15 people or 41.67%, have a light workload.

b. Documentation format factor in the implementation of nursing care documentation in the Prima Inpatient Room I Immanuel Hospital Bandung.

3.2 Bivariate analysis

Bivariate analysis was performed to obtain the variables that affect nurses in achieving completeness of documentation of nursing care using a chi square, with the help of a computer program.

Table 2. Documentation.

Documentation Format	Frequency	Percentage
Supporting	19	52.78
Not Supporting	17	47.22
Total	36	100

Analysis of Table 2 shows that most respondents, 19 people or 52.78%, stated that the documentation format was supporting.

a. Leadership policy factor in the implementation of nursing care documentation in the Prima Inpatient Room I Immanuel Hospital Bandung.

Table 3. Leadership policy leadership.

Policy	Frequency	Percentage
High	17	47.22
Low	19	52.78
Total	36	100

Analysis of Table 3 shows that the majority of respondents, 19 people or 52.78%, rated the leadership policy as low.

b. Implementation of nursing care documentation in the Inpatient Prima I Nursing Room Bandung Immanuel Hospital.

Table 4. Implementation of nursing care documentation.

Implementation	Frequency	Percentage
Complete	13	36.11
Not Complete	23	63.89
Total	36	100

Analysis of Table 4 shows that most (63.89%) of the nursing care documentation is incomplete.

Analysis of Table 5 shows that out of 11 respondents who have a heavy workload most (9 or 81.8%) have incomplete nursing care documentation. Of the 15 respondents who have a light workload most have complete implementation of complete nursing care documentation as many as 9 documentation (60%). Based on the chi square test, obtained p value was 0.041. Due to the value of $p < 0.05$, then H0 is rejected, which means there is a significant relationship between workload factors and the completeness of nursing care documentation, and confirms results from the above that the heavier workload will lead to incomplete nursing care documentation, but the lighter the workload, the more complete the implementation of nursing care documentation. The results of this study are supported by Martini (2007) which indicates a relationship between the workload of respondents and the practice of documenting nursing care.

Analysis of Table 6 shows that of the 17 respondents who stated the documentation format that did not support most (15 or 88.2%) had incomplete nursing care documentation. And of the 19 people who stated that the documentation format supported most (11 or 57.89%) had complete implementation of nursing care documentation

Table 5. Relationship between workload factors and nursing documentation implementation.

| Workload | Completeness | | | | | | P value |
| | Complete | | Not Complete | | Total | | |
	Freq	%	Freq	%	Freq	%	
Heavy	9	81.8	2	18.2	11	100	
Medium	8	80	2	20	10	100	0.041
Light	6	40	9	60	15	100	

Table 6. Relationship between documentation format factors and nursing care documentation implementation.

| Documentation Format | Completeness | | | | | | P value |
| | Complete | | Not Complete | | Total | | |
	Freq	%	Freq	%	Freq	%	
Not Support	15	88.2	2	11.8	17	100	0.004
Support	8	42.1	11	57.89	19	100	

Table 7. Relationship between leadership policy factors and implementation of nursing care documentation.

| Leader Policy | Completeness | | | | | | P value |
| | Complete | | Not Complete | | Total | | |
	Freq	%	Freq	%	Freq	%	
Low	15	84.2	3	15.8	19	100	
High	7	41.2	10	58.82	17	100	0.007

The chi square test obtained a p value of 0.004. Due to the value of $p < 0.05$, then H0 is rejected, which means there is a significant relationship between documentation format and the completeness of nursing care documentation, and results from the above that if the documentation does not support the format it lead to incomplete implementation of nursing care documentation, but if the documentation format is more supportive, there will be more complete nursing care documentation.

The results of this study are supported by one of the results of the Martini (2007) study indicates a relationship between the availability of format facilities and facilities for nursing care standards with documentation of nursing care.

Based on the chi square test, the obtained p value value was 0.007. Due to the value of $p < 0.05$ then H0 is rejected, which means there is a significant relationship between policy factors and the completeness of nursing care documentation, and results

from the above that the lower the level of leadership policy, the more likely there will be incomplete nursing care documentation, but the higher the level of leadership policy, the more complete the implementation of documentary nursing care.

The results of this study are supported by one of the results of Yanti and Warsito's (2013) study which indicates a relationship between the supervision of the head of the room with the quality of documentation.

REFERENCES

Bernardino, A.O., Coriolano, M., Santos, A.H., Linhares, F. M.P., Cavalcanti, A.M.T. & Lima, L.S. 2018. Motivation of nursing students and their influence in the teaching-learning process. *Texto & Contexto—Enfermagem*.

Jooste, K, & Hamani, M. 2017. The motivational needs of primary health care nurses to acquire power as leaders in a mine clinic setting. *Health SA Gesondheid*.

Martini. 2007. Relationship between nurse characteristics, attitudes, workload and availability of facilities, supervision with the practice of documenting nursing care in Salatiga City. *BPRSUD Inpatients*. Indonesia: Salatiga.

Pandey, R., Goel, S. & Koushal, V. 2018. Assessment of motivation levels and nursing factors among nursing staff of tertiary-level government hospitals. *International Journal of Health Planning and Management*.

Potter, A.P. & Perry, G.A. 2005. *Fundamental nursing*. St. Louis, Missouri: Mosby Company.

Rebecca, M., Margaret, N. & Scott, M. 2005. Nursing careers: What motivated nurses to choose their profession.?. *Australian Bulletin of Labor*.

Sajjadnia, Z., Sadeghi, A., Kavosi, Z., Zamaru, M. & Ravangard, R. 2015. Factors affecting the nurses' in the training motivation for participating courses: Study. *Health Man & Info*.

Sheikhbardsiri, H., Khademipour, G. & Aminizadeh, M. 2018. Motivation of nurses in pre-emergency and educational hospitals, hospitals, emergency in the Southeast of Iran. *International Journal of Health Planning and Management*.

Toode, K., Routasalo, P. & Suominen, T. 2011. Work motivation of nurses: A literature review. *International Journal of Nursing Studies*.

Usher, K., West, C., Macmanus, M., Waqa, S., Stewart, L. & Henry, R. 2013. Motivations to nurse: An exploration of what motivates students in Pacific Island countries to enter nursing. *International Journal of Nursing Practice*.

Yanti, R.I. & Warsito, B.E. 2013. Relationship between nurse characteristics, motivation, and supervision with quality documentation of nursing care processes. *Journal of Nursing Management*.

Advances in Business, Management and Entrepreneurship – Hurriyati et al (eds)
© 2020 Taylor & Francis Group, London, ISBN 978-0-367-27176-3

Work-life balance and work stress as antecedents of employee turnover intention in private food processing organizations

R. Saragih, A.P. Prasetio & I.Z. Naufal
Telkom University, Bandung, Indonesia

ABSTRACT: Voluntary turnover can harm organizations in achieving their objectives. Therefore, organizations need to keep the turnover as low as possible. In order to tackle the problems, they need to identify the current turnover intention level. The aim of this study is to find out the direct influence of work-life balance on employee turnover intention and the mediation of work stress in the relationship. A questionnaire was used to gather data from seven private food processing companies in Tasikmalaya, west Java. The sample for this study are 120 employees. A quantitative method was applied by using correlation and process macro to analyze the results. Work-life balance shows negative correlation with stress and turnover intention, while positive relation is found in the relation of stress and turnover. The study also finds that work stress mediates the influence of work-life balance on turnover intention. This means that employees who have balance in their work-life will less expose to stress which in the end do not have the intention to leave the organization. Management should focus on their human resources policy in order to maintain such balance and the low stress level.

1 INTRODUCTION

The performance of organizations is affected by the ability to manage human resources (Bartel 2004). As we all understand, management needa to adapt the human ability with the business that will become key factor that differentiate organization from its competitors (Sun 2017). In fact, according to (Thomas 2014), employees play an important role as a value proposition. In competitive strategy, human resources have the advantages compare to other resources. People can be developed according to the new challenges and changes in business environment (Rahardjo 2016, Saragih et al. 2017). When the organizations take care of its people, they create employees with the physical energy, mental focus, and emotional drive necessary to power their businesses (Boyce 2015). Indeed, the link between human resources practices and performance is acknowledged because great HR practices will impact on employee attitude and behavior toward works (Torrington et al. 2017). Implementing human resources practices require the ability of organizing human resources department to fulfill their function properly. Oliveira et al. (2017) argued for the importance of human resources function for the survival, growth, and competitiveness of the organization. Organizations are facing new challenges to develop program that can attract and retain valuable employees (Burke, 2006). The low employee turnover is one of the success measurement of the human resources program. Turnover intention is serious problem in managing human resources (Fah et al. 2010). (Grissom et al. 2012) explain the impact of high levels of employee turnover for organization.

Significant costs related to turnover can range from spending for recruitment and filling vacant positions, resources devoted to training, and, loss of knowledge and talented human capital. According MENA Workplace poll employees retention level for recent generation is lower (Al-Masri 2015). More recent finding (Mercer 2016) showed average global voluntary turnover was 7.2%. Turnover has become a hot topic in today's HR and this lead for recognition by the organizations regarding the importance of measuring, monitoring, and maximizing the level of loyalty and engagement amongst their employees. This study examines the work-life balance (WLB) and work stress as antecedents of employee turnover intention (TI) in the private-owned food processing organizations in West Java. Food processing companies need experience employees to support their business. The strict method, procedures, safety and health regulation can only be maintained with experienced employees. It is important for the organizations to retain their best employees to ensure the smoothness of their daily operation. Since the competitors also need experienced and knowledge employees, each organization should develop program which could prevent the employees to move out or resign. Excellent and experience employees will keep more money inside organization and gives fewer problems. We maintain human resources practices which could drive the WLB and lower stress among the employee. When there is balance in work and private life, they tend to produce less stress. This would decrease the thought of leaving the organization. The purpose of this study is to identify the direct effect of WLB on TI and also to identify the mediation of work stress in that relation. Such model could help organization to design their human resource program in order to manage work stress level and reducing turnover intention.

1.1 Work-life balance

Clark (2000) defined WLB as satisfaction and good functioning at work and at home, with a minimum of role conflict. Greenblatt (2002) viewed WLB from the absent of conflict between work and family demand. Meanwhile, Greenhaus and Beutell (1985) explained WLB as a form of inter-role conflict in which the role pressures from the work and family domains are mutually incompatible in some respect. The participation in the work (family) role is made more difficult by virtue of participation in the family (work) role. WLB is a state experienced by individual regarding his or her perception towards the work and private life where there is a little interference between both of them. Or, the condition which an individual can minimizes the level of conflict between work and non-work demands. WLB is affected by various factors such as individual, organizational, social and others (Moorehead & Griffin 2013). Factors such as role overload, dependent care issues, quality of health, problems in time management and lack of support are consider major influences for WLB (Mathew & Panchantham 2011). Meanwhile, Haar et al. (2018) found that job autonomy, family demands, work demands, work hour, overtime, and supervisor support have correlation with WLB. A study in New Zealand by Deery and Jago (2009) found that flexible work strategy including flexible schedule, work from home, easy access to pay and unpaid leave and sharing job between coworkers will help employees to achieve WLB. Organization should care for employees WLB since its affect the individual and organization performance. AlHazemi and Ali (2016) concluded that WLB can create positive consequences. Aryee et al. (2005) related WLB with job satisfaction and organizational commitment. Das and Khusawa (2013) argued that WLB practices could minimize the work family conflict. Other consequences revealed by various study were low productivity (Kamran et al. 2014, Bloom & Van Reenen 2006, Fapohunda 2014), employee turnover (Fapohunda 2014, Cao et al. 2013, Giau-que et al. 2016), employee performance (Anwar & Shahzad 2011), and stress (Schiemann et al. 2003, Sheraz et al. 2014). It is clear why organization should manage their employee work-life balance. They need to do that in order to excel their human resources to perform and achieve organization's objectives.

1.2 Work stress

Schermerhorn et al. (2012) defined stress as the tension from extraordinary demands, constraints, or opportunities. They differentiated stress in two parts, eustress which is a stress that has a positive impact on both attitudes and performance, and distress which is a negative impact on both attitudes and performance. Robbins and Judge (2017) explained that stress is a dynamic condition where individual faces with opportunities, demands, or resources which relate with their interest, while the outcomes are still uncertain. Such conditions can cause psychological inconvenient because of pressure from environment. Of course, stress will influence employee behavior at work. Determinants of stress can be identified from previous literatures. Several scholars (Al-Hosam et al. 2016 in Yemen & Gill et al. 2010) in India argued that leadership style can affect employee stress, while Wani (2013) in India found that work motivation can become source of employee stress. Other factors affecting stress was explained by Schermerhorn et al. (2012) including task or job demands, role ambiguities, role conflicts, ethical dilemmas, interpersonal problems, career developments, environmental setting (noise, lack of privacy, pollution), personal lives (new child, economic difficulties, separation or divorce). Furthermore, Schermerhorn et al. (2012) also described the consequences of stress in the workplace. Employees who suffer too much stress and work overload could experience a break down in physical and mental systems. Stress can result in higher absenteeism, turnover, mistakes and errors, work accidents, dissatisfaction, reduced performance, unethical behavior, and even illness. Similar finding from Robbins and Judge (2017) showed several consequences of stress such as illness, chronic health conditions, anxiety, low emotional well-being, low job satisfaction, low job performance, high absenteeism and turnover. In the end these outcomes will significantly affect profitability, productivity, effectiveness, and efficiency of organizations.

1.3 Turnover intention

Torrington et al. (2017) regarded employee turnover rates as the measurement of the employee departures from an organization. Employees may resign, retire, or be dismissed. Turnover is costly and organization try hard to minimize the turnover levels. Dessler (2013) described turnover as the rate of employee resignation from the organization. Snell and Bohlander (2013) defined turnover as the movement of employees out of the organization. From these definitions, it comes out the term of Turnover Intention (TI). TI is the tendency from employees to seek out new job opportunities outside the organization and to resign (Branham 2012). The difference of TI from turnover is that TI is still limited on the tendency or intention. By identifying TI management could predict how far the seriousness of employee to leave the organization in certain time frames. The term turnover intention discussed here is the voluntary turnover which means that employee resign because of their own desire. Various factors can affect TI. Organizational commitment, job satisfaction, work environment, work stress, workload, motivation, organizational support are several of them (Anton 2009, Dawley et al. 2010, Valentine et al. 2011; Qureshi et al. 2013, Sajjad et al. 2013, Mxenge et al. 2014). Employees who feel uncomfortable while working for the organization will have stronger tendency to exit. Turnover can harm organization. Study from Spain (Anton 2009) found the negative correlation between TI and employee performance. (O'brien-Pallas et al. 2006) found that minimum level of nurse's turnover will influence patient satisfaction and safety, and also impact on the overall turnover costs experienced by institutions. (Brandt et al. 2016) lower staff consistency (high turnover) which is defined as the consecutive time without staff turnover will result in lower patient treatment quality. A study in South Korea (Kim & Han 2018) also found the relation between nurse's TI with the patient outcome (patient health condition after having treatment). These find-ings were supported by (Nuhn et al. 2017) who ex-plained that turnover intentions in both organiza-tional forms negatively affect performance on the individual, temporary organizational and the perma-nent organizational. They also suggested that man-agement should prevent the development of turnover intentions as early as possible by applying various policies which attract employee to stay longer. It is important for organizations to identify the determi-nants of turn-over so they can predict future turnover more accur-ately and manage their human resources policy to prevent the voluntary resignation. Man-agement should extend their effort to achieve opti-mal staff retention.

1.4 Work-life balance and work stress

Employees need to maintain a healthy balance between work and their private lives with less stress. Report from Lowe (2006) indicated that there is inverse relation between WLB and stress. Yadav & Yadav (2014) in India found the negative relation between WLB and stress. Both findings indicated that WLB and stress showed opposite direction. Giauque et al. (2016) also found negative relation between WLB and stress when they conducted study using United Nations employees. When WLB is high, work stress tends to become lower. Meanwhile, Nart and Batur (2014) and Karabay et al. (2016) conducted study in Turkey using the opposite of WLB, which is work-life conflict (WLC) and showed the positive relation between WLC and stress. Those studies were supported by Yun et al. (2012) in Korea which also found when employees experience higher conflict between work and private domain they developed higher stress level. The same direction also found by several scholars (Jamadin et al. 2015, Malaysia & Karkoulian et al. 2016) in Lebanon. (Chiang et al. 2010) found the positive relation between WLB and stress, but since it was quite small, there was no effect of WLB on stress. Based on these findings, since we use WLB, we propose the first hypothesis as below;

H1: work-life balance will have a significant negative effect on work stress.

1.5 Work-life balance and turnover intention

Oosthuizen et al. (2016) conducted a study in South Africa and found that WLB did not have significant effect on turnover intention. Javed et al. (2014) in Pakistan also found no significant effect of WLB on turnover intention. Meanwhile, study in Malaysia from (Noor 2011) indicated that perceived work-life balance was correlated negatively with intention to leave the organisation among academics. When employees perceive balance, their intention to quit will reduced. Lee et al. (2013) revealed in his study in Taiwan hospital that seven dimensions of quality of work-life were significantly negatively correlated with intention to leave the organization. Finding from study in Iran also supported that WLB had a significant negative relationship with turnover inten-tion (Fayyaz & Aslani 2018). Other studies discussing WLB and turnover intention and found the negative direction came from several scholars (Cao et al. 2013, Malik et al. 2010, Giauque et al. 2016). Contrary to WLB, work-life conflict was known to have positive relationship with turnover intention. Employees who experienced greater conflict in their work and private life would have higher turnover to leave the organiza-tion. Suifan et al. (2016) in Jordan proved and sup-ported the notion when they found significant and positive relation between work-life conflict and turn-over intention. Sang et al. (2009) also revealed that there was a very high correlation (0.973) between overall work-life conflict and turnover intentions. This means that increases in work-life conflict was associated with an increase in turnover intention. Fapohunda (2014) and Noor and Maad (2008) con-ducted studies regarding work-life conflict and turn-over and found positive relationship. Since this study using WLB, then our second hypothesis is;

H2: work-life balance will have significant nega-tive effect on turnover intention.

1.6 Work stress and turnover intention

Previous literatures discussing the relation of work stress and turnover intention mostly reveal the positive direction. Higher level of stress tends to increase the intention to leave the organization. Research studying the influence of stress on TI has often been done. Most of the result suggest that stress positively affect TI. (Naufal & Prasetio 2017) in Indonesia, (Mxenge et al. 2014) in South Africa, (Lin et al. 2013, Liu & Onwuegbuzie 2012) in China, study in western culture from (Jaramillo et al. 2006) in Korea, (Tongchaiprasit & Ariyabuddhiphongs 2016) in Thailand, (Sewwandi & Perere 2016) in Srilanka, and three studies from

Iran (Mosadeghrad 2013, Ahanian et al. 2016, Arshadi & Damiri 2013) supported findings that work-related stress can forecast employee intentions to quit.

H3: work stress will have significant positive effect on turnover intention.

One of the objectives of this study is to measure the mediation effect from work stress in the relation of WLB and TI. Therefore, our is;

H4: work stress will mediate the relationship of WLB and turnover intention.

2 METHOD

2.1 Participants and research instrument

The location of this study is in Tasikmalaya, Indone-sia with seven private companies who run food pro-cessing especially using chicken meat. Total em-ployees in this organization are 168. The authors dis-tribute the questionnaires to 120 employees ranging from officers to managers as participants.

The research intrument is questionnaire that consists of 40 items represent WLB, work stress, and turnover intention. Each items has 5 answers based on Likert's scale option from 1 - strongly disagree to 5 - strongly agree.

WLB is measured with 9 item adapted from (Hayman 2005). The internal consistency reliability was α= .748. Work stress is assessed with 19 items develop based on the dimension from (Leung et al. 2007) including personal, interpersonal, task, and physical stressors. The Cronbach's Alpha for this varible is .887. Finally, turnover intention is measured with 12 items developed from (Abelson 1987) and the Cronbach's Alpha is .918.

2.2 Data analysis

SPSS is used to analyze the correlation between variables. Then, we use regression analysis to identify the significant of the effect between variables. And to determine the mediation of work stress, boot-strapping approach using the PROCESS Macro for SPSS is used (Hayes, 2013). This approach has been believed as a good tool since it avoids normality as-sumptions of the sampling distribution (Preacher et al., 2007). Using PROCESS Macro in SPSS makes the imple-mentation of the bootstrapping easy. The result from this tool directly shows the significance of mediation effects. Work stress will have media-tion role if the value of Upperlevel and Lowerlevel Confidence Interval contain no zero value (either positive or negative).

3 RESULT AND DISCUSSION

Table 1 displays means, standard deviation, and correl-ation between variables. It shows that WLB has nega-tive correlation with stress and TI, while stress

Table 1. Mean, standard deviation, & correlation.

	Mean	Std. Deviation	WLB	Stress	Turnover
WLB	3.5862	.55946	1		
Stress	2.9343	.62077	-.490**	1	
Turnover	2.2675	.65960	-.476**	.762**	1

**Correlation is significant at the 0.01 level (2-tailed).
*Correlation is significant at the 0.05 level (2-tailed).

positively correlates with TI. WLB Correlation with work stress is -.490 and with TI is -.476 classified as moderate (Rumsey 2011). Meanwhile, the correlation between work stress and TI is consider strong (.762). Since the variables used in this study are correlated each other, we may continue our investigation to find out the effect of WLB on TI with the mediation of work stress.

Tabel 2 displays the regression result other statistics information resulted from the calculation.

WLB has significant negative effect on work stress and TI (p-value .000 and .048). on the other hand, work stress has significant positive effect on TI (p-value .000). thus, hypothesis H1, H2 and H3 are supported. The balance in work and life will minimize the stress level and inten-tion to leave. The low of work stress will reduce the intention to leave. Whereas the effect is small which mean there are other factors which might also affect the turnover intention. Figure 1 exhibits the effect model.

Table 3 shows the mediation analysis result and indicate that work stress mediates the relationship between WLB and turnover intention. The value of Lower Level Confidence Interval (LLCI) and Upper Level Confidence Interval (ULCI) are negative. It means that if employees perceive higher balance in work and life they will experience less work stress and in the end they do not develop the intention to leave the organizations. Thus, H4 are supported.

The result reflects the condition in human resources management in the organization. Nor-mally, employees inside organization with better work-life balance policy will experience less

Figure 1. The mediation model.

Table 2. Regression coefficient, standard error, & model summary.

	Work Stress			Turnover		
	Coeff	SE	p-value	Coeff	SE	p-value
Work-life Balance	-0.544	0.089	0.000	-0.160	0.079	0.048
Work Stress	-	-	-	0.739	0.072	0.000
Contant	4.885	0.321	0.000	0.669	0.432	0.124
	R Square		0.490	R Square		0.771
	F=		37.315	F=		85.672
	P=		0.000	P=		0.000

Table 3. Indirrect effect.

Indirect Effect of Work-life Balance on Turnover Intention

	Effect	Boot SE	Boot LLCI	Boot ULCI
Through Work Stress	-0.402	0.088	-0.609	-0.254

stress from work. Flexibility, organizational support, remote work, reasonable sick leave, and educational leave could help employees manage their work and life. They do not have to be affraid of losing their job or face heavy consequences while taking care of personal/family problems because organization give flexibility which focus on the final result. Compare with traditional organization which implement strict work hours (from 9 to 5 for example), employees with flexible work schedule should develop less stress. With low job-related stress, employees feel comfortable in the organization. They will do great and perform better. In fact, they usually do more than expected because they want to keep the positive condition which allow them to manage their life and work at the same time without sacrifacing eachothers. With this in mind, do they plan to quit? No wonder the level of TI will decrease because employee already fit with the organization and their jobs. Referring to previous studies, the result of this study promotes that WLB has negative relation with work stress and TI. Thus, supports the studies from (Yadav & Yadav 2014, Giauque et al. 2016, Nart & Batur, 2014, Karabay et al. 2016, Yun et al. 2012, Jamadin et al. 2015) It also corroborates with (Noor, 2011, Lee et al. 2013, Cao et al. 2013, Malik et al. 2010, & Giauque et al. 2016). Finally, regarding the correlation between work stress and turnover intention, this study is in line with several studies (Mxenge et al. 2014, Lin et al. 2013, Lu et al. 2017, Li & Onwuegbuzie

2012, Mosadeghrad 2013, Ahanian et al. 2016; Jaramillo et al. 2006, Tongchaiprasit & Ariyabuddhiphongs 2016, Arshadi & Damiri 2013), all of whom argued that work stress had significant positive effect on turnover intention.

Turnover intention in these particular organizations are low. We find that employees perceive high WLB and moderate level of work stress. This is a perfect combination for organizations. At least they do not have problems to retain their employees. But, low turnover also could become a problem if the cause of low turnover is only come from the comfort to stay in the organizations. Considering this reason, organizations can suffer from unproductive employees who only stay because they feel relax. Or, they just wait for the right time to move out when there is opportunity. Organization could take first step by identifying the reason of the low TI. The result will guide management in developing program to retain the best perform employee. In the meantime, they can implement the various action which support WLB such as providing flexible hour, special work leave permission for taking care of the elders or children (sickness, educational), give training on how to manage time for work and family/personal, provide better tools which help them to do the job faster and better, and provide medical and education allowance.

4 CONCLUSIONS

All four hypotheses are supported. WLB significantly negative affects work stress and turnover intention and work stress significantly positive affect turnover intention. Related to mediation analysis, the relation between WLB and turnover intention is mediated by work stress. Both variables can directly predict the turnover. The limitations of this study by using of cross-sectional data, self-reported surveys, and sample limitation that could be improved for next research. Future study can use longitudinal data, using various

694

different varibales, increase the participant diversity, and greater samples.

REFERENCES

Abelson, M.A. 1987. Examination of avoidable and unavoidable turnover. *Journal of Anplied Psychology* 72 (3): 382–386.

Ahanian, E., Mirzae, A. & Fardi, A.S. 2016. The study of correlation between job sress and turnover intentions among the operating room nurses in selected hospitals of tehren university medical science. *Acta Medica Mediterranea* 32: 1045.

AlHazemi, A.A. & Ali, W. 2016. The notion of work-life balance, determining factors, antecedents and consequences: a comprehensive literature survey. *International Journal of Academic Research and Reflection* 4 (8): 74–85.

Al-Hosam, M.A.A., Ahmed, S., Ahmda, B.F. & Joarder, R. H.M. 2016. Impact of transformational leadership on psycological empowerment and job satisfication relationship: a case of yemeni banking. *Binus Business Review* 7(2): 109–116.

Arshadi, N. & Damiri, H. 2013. The relationship of job stress with turnover intention and job performance: moderating role of OBSE. *Procedia-Social and Behavioral Sciences* 84: 706–710.

Aryee, S., Srinivas, E.S. & Tan, H.H. 2005. Rhythms of life: antecedents and outcomes of work–family balance in employed parents. *Journal of Applied Psychology* 90 (1): 132–146.

Bartel, A.P. 2004. Human resource management and organizational performance: Evidence from retail banking. *Industrial and Labor Relations Review* 57(2): 181–203.

Bloom, N. & Van Reenen, J. 2006. Management practices, work-life balance, and productivity: a review of some recent evidence. *Oxford Review of Economic Policy* 22 (4): 457–482.

Brandt, W.A., Bielitz, C.J. & Georgi, A. 2016. The impact of staff turnover and staff density on treatment quality in a psychiatric clinic. *Frontiers in Psychology* 7(457): 1–7.

Burke, R.J. 2006. The human resources revolution.

R.J., Burke & C.L., Cooper (Eds.), The human resources revolution: Why putting people first matters: 3–11.

Cao, Z.T., Chen, J.X., & Song, Y.X. 2013. Does Total Rewards Reduce the Core Employees' Turnover Intention? *International Journal of Business and Management* 8 (20): 62–75.

Chiang, F.F.T., Birtch, T.A., & Kwan, H.K. 2010. The moderating roles of job control and work-life balance practices on employee stress in the hotel and catering industry. *Inter-national Journal of Hospitality Management* (29): 25–32.

Clark, S.C. 2000. Work-family border theory a new theory of work-family balance. *Human Relations* 53(6): 747–770.

Das, S.C. & Kushwaha, S. 2013. Identifying critical factors of work-life balance and its impact on insurance employees in India–an exploratory factor analysis. *Time's Journey* 2(1): 1–13.

Dawley, D., Houghton, J. D. & Bucklew, N.S. 2010. Per-ceived organizational support and turnover intention: the mediating effects of personal sacrifice and job fit. *The Journal of Social Psychology* 150(3): 238–257.

Deery, M. & Jago, L. 2009. A framework for work-life balance practices: Addressing the of needs tourism industry. *Tourism and Hospitality Research* 92(2): 97–108.

Dessler, G. 2013. *Human resource management, 13th ed.* New Jersey: Pearson.

Fah, C.B., Foon, S.Y., Leong, C. & Osman, S. 2010. An exploratory study on turnover intention among private sector employee. *International Journal of Busniness and Management* 5(8): 55–64.

Fapohunda, T.M. 2014. An exploration of the effects of work-life balance on productivity. *Journal of human resources management and labor studies* 2(2): 71–89.

Giauque, D., Anderfuhren-Biget, S. & Varone, F. 2016. Stress and turnover intents in international organizations: social support and work–life balance as resources. *The In-ternational Journal of Human Resource Management.*

Greenblatt, E. 2002. Work/life balance: wisdom or whining. *Organizational Dynamics* 31(2): 177–193.

Greenhaus, J.H. & Beutell, N.J. 1985. Sources of conflict between work and family roles. *The Academy of Management Review*: 76–88.

Grissom, J.A., Nicholson-Crotty, J. & Keiser, L. 2012. Does my boss's gender matter? Explaining job satisfaction and employee turnover in the public sector. *Journal of Public Administration Research and Theory* (22): 649–673.

Haar, J.M., Sune, A., Russo, M., & Ariane, O.M. 2018. A Cross-national study on the antecedents of work–life balance from the fit and balance perspective. *Social Indicators Research, Springer Netherlands.*

Hayes, A.F. 2013. PROCESS SPSS Macro (Computer software and manual).

Hayman, J. 2005. Psychometric assessment of an instrument designed to measure work-life balance. *Research and Practice in Human Resource Management* 13(1): 85–91.

Huang, W.L., Tung, C.W., Huang, H.L., Hwang, S.F. & Ho, S.Y. 2007. ProLoc: prediction of protein subnuclear localization using SVM with automatic selection from physicochemical composition features. *BioSystems* 90 (2): 573–581.

Indebetouw, R., Mathis, J.S., Babler, B.L., Meade, M.R., Watson, C., Whitney, B.A., Wolff, M.J., Wolfire, M.G., Cohen, M., Bania, T.M. & Benjamin, R.A. 2005. The wavelength dependence of interstellar extinction from 1.25 to 8.0 μm using GLIMPSE data. The Astrophysical Journal 619(2): 931.

Jamadin, N., Mohamad, S., Syarkawi, Z. & Noordin, F. 2015. Work-Family Conflict and Stress: Evidence from Malaysia. *Journal of Economics, Business and Management* 3(2): 309–312.

Jaramillo, F., Mulki, J. & Solomon, P. 2006. The role of ethical climate on salesperson's role stress, job attitudes, turnover intention, and job performance. *Journal of Personal Selling and Sales Management* 26(3): 271–282.

Javed, M., Khan, M. A., Yasir, M., Aamir, S. & Ahmed, K. 2014. Effect of role conflict, work-life balance and job stress on turnover intention: evidence from Pakistan. *Journal of Basic and Applied Scientific Research* 4(3): 125–133.

Kamran, A., Zafar, S. & Ali, S.N. 2014. in J. Xu et al. (eds.). Impact of work-life balance on employees productivity and job satisfaction in private sector universities of pakistan. *Proceedings of the Seventh International*

Conference on Management Science and Engineering Management (2): 242.

Karabay, M.E., Akyüz, B. & Elçi, M. 2016. Effects of family-work conflict, locus of control, self confidence and extraversion personality on employee work stress. *12th International Strategic Management Conference ISMC, Antalya, Turkey, Procedia-Social and Behavioral Sciences* (235): 269–280.

Kim, Y.S. & Han, K.Y. 2018. Longitudinal associations of nursing staff turnover with patient outcomes in long-term care hospitals in Korea. *Journal of Nursing Management*: 1–7.

Lee, Y.W., Dai, Y.T., Park, C.G. & McCreary, L.L. 2013. Predicting quality of work-life on nurses' intention to leave. *Journal of Nursing Scholarship* 45(2): 160–168.

Leung, M., Sham, J. & Chan, Y. 2007. Adjusting stressors - job demand stress in preventing rustout/burnout in estimators. *Surveying and Built Environment* 18(1): 17–26.

Lin, Q.H., Jiang, C.Q., & Lam, T.H. 2013. The relationship between occupational stress, burnout, and turnover intention among managerial staff from a Sino-Japanese joint venture in Guangzhou China. *Journal of Occupational Health* 55: 458–467.

Liu, S. & Onwuegbuzie, A.J. 2012. Chinese teachers' work stress and their turnover intention. *International Journal of Educational Research* 53: 160–170.

Lowe, G. 2006. Implications of Work-Life Balance and Job Stress. *Human Solutions Report*.

Lu, Y., Hu, X.M, Huang, X.L., Zhuang, X.D., Guo, P., Feng, L.F., Hu, W., Chen, L., Zou, H. & Hao, Y.T. 2017. The relationship between job satisfaction, work stress, work–family conflict, and turnover intention among physicians in Guangdong, China: A cross sectional study.

Malik, M.I., Gomez, S.F., Ahmad, M. & Saif, M.I. 2010. Examining the relationship of work-life balance, job satisfaction, & turnover In Pakistan. *International Journal of Sustainable Development* 02(01): 27–33.

Mathew, R.V. & Panchanatham, N. 2011. An exploratory study on the work-life balance of women entrepreneurs in South India. *Asia Academy of Management Journal* 16(2): 77–105.

Memish, Z.A., Assiri, A., Turkestani, A., Yezli, S., al Masri, M., Charrel, R., Drali, T., Gaudart, J., Edouard, S., Parola, P. & Gautret, P. 2015. Mass gathering and globalization of respiratory pathogens during the 2013 Hajj. *Clinical Microbiology and Infection* 21(6): 571–5e1.

Mirzaee, F., Aslani, H., Nourbakhsh, S.T., Fayyaz, M.R., Zafarani, Z. & Sazegari, M.A., 2018. Platelet-rich plasma for frozen shoulder. *The Archives of Bone and Joint Surgery* 6: 15–15.

Moorhead, G. & Griffin, R.W. 2013. *Perilaku organisasi: Manajemen sumber daya manusia dan organisasi.* Jakarta: Salemba Empat.

Mosadeghrad, A.M. 2013. Occupational stress and turnover intention: implications for nursing management. *International Journal of Health Policy and Management* 1(2): 179–186.

Mxenge, S., M.Dywili. & Bazana, S. 2014. Organisational stress and employees' intention to quit amongst administrative personnel at the university of fort hare, eastern cape, South Africa. *International Journal of Research in Social Sciences* 4(5): 13–29.

Nart, S. & Batur, B. 2014. The relation between work-family conflict, job stress, organizational commitment and job performance: A study on turkish primary teachers. *European Journal of Research on Education* 2(2): 72–81.

Naufal, I.Z. & Prasetio, A.P. 2017. Pengaruh stres kerja terhadap turnover intention pada karyawan CV. sukahati pratama. *Journal SMART STIE STEMBI* 14(3): 57–64.

Neufer, P.D., Bamman, M.M., Muoio, D.M., Bouchard, C., Cooper, D.M., Goodpaster, B.H., Booth, F.W., Kohrt, W.M., Gerszten, R.E., Mattson, M.P. & Hepple, R.T. 2015. Understanding the cellular and molecular mechanisms of physical activity-induced health benefits. *Cell metabolism* 22(1): 4–11.

Nishimura, R.A., Otto, C.M., Bonow, R.O., Carabello, B.A., Erwin, J.P., Guyton, R.A., O'gara, P.T., Ruiz, C.E., Skubas, N.J., Sorajja, P. & Sundt, T.M. 2014. 2014 AHA/ACC guideline for the management of patients with valvular heart disease: a report of the American college of cardiology/ American heart association task force on practice guidelines. *Journal of the American College of Cardiology* 63(22): e57–e185.

Noor, K.M. 2011. Work-life balance and intention to leave among academics in Malaysian public higher education institutions. *International Journal of Business and Social Science* 2(11): 240–248.

Noor, S. & Maad, N. 2008. Examining the relationship be-tween work-life conflict, stress and turnover intentions among marketing executives in Pakistan. *International Journal of Business and Management* 3(11): 93–102.

Nuhn, H.F.R., Heidenreich, S. & Wald, A. 2017. Performance outcomes of turnover intentions in temporary organi-zations: a dyadic study on the effects at the individual, team, and organizational level. *European Management Review*.

O'Brien-Pallas, L., Griffin, P., Shamian, J., Buchan, J., Duffield, C., Hughes, F., Laschinger, H.K.S., North, N. & Stone, P.W. 2006. Nurse, and system outcomes: a pilot study and focus for a multicenter international study. *Policy, Politics, & Nursing Practice* 7(3) 169–179.

Oliveira, L.B., Cavazotte, F. & Dunzer, R.A. 2017. The interactive effects of organizational and leadership career management support on job satisfaction and turnover intention. *The International Journal of Human Resource Management*: 1–21.

Oosthuizen, R.M., Coetzee, M. & Munro, Z. 2016. Work-life balance, job satisfaction and turnover intention amongst information technology employees. *Southern African Busi-ness Review* (20): 446–476.

Preacher, K.J., Rucker, D.D. & Hayes, A.F. 2007. Addressing moderated mediation hypotheses: Theory, methods, and prescriptions. *Multivariate Behavioral Research*, 42(1): 185–227.

Qureshi, M.I., Iftikhar, M., Abbas S.G., Hassan, U., Khan, K., & Zaman, K. 2013. Relationship between job stress, workload, environment and employees turnover intentions: what we know, what should we know. *World Applied Sciences Journal* 23(6): 764–770.

Rumsey, D.J. 2011. How to interpret a correlation coefficient. Statistics for dummies.

Sajjad, A., Ghazanfar, H. & Ramzan, M. 2013. Impact of motivation on employee turnover in telecom sector of pakistan. *Journal of Business Studies Quarterly* 5(1).

Sang, K.J.C., Ison, S.G. & Dainty, A.R.J. 2009. The job sat-isfaction of UK architects and relationships with work-life balance and turnover intentions. *Engineering, Construction and Architectural Management* 16(3): 288–300.

Saragih, R., Rahayu, A. & Wibowo, L.A. 2017. External environment impact on business performance in digital creative industry: Dynamic capability as mediating variabel. *International Journal of Advanced and Applied Sciences* 4(9): 61–69.

Schermerhorn, Jr. J.R., Osborn, R.N., Uhl-Bien, M., Hunt, J.G. 2012. *Organizational behavior, 12th edition*. New Jersey: John Wiley & Sons.

Schiemann, S., McBrier, D.B. & Gundy, K.V. 2003. Home-to work conflict, work qualities, and emotional distress. *So-ciological Forum* 18(1): 137–164.

Sewwandi, D.V.S. & Perere, G.D.N. 2016. The impact of job stress on turnover intention: a study of reputed apparel firm in Sri Lanka 3(1): 223–229.

Sheraz, A., Wajid, M., Sajid, M., Qureshi, W.H. & Rizwan, M. 2014. Antecedents of job stress and its impact on employee's job satisfaction and turnover intentions. *International Journal of Learning & Development* 4(2): 204–226.

Snell, S.A. & Bohlander, G. 2013. *Managing human resources, sixteenth edition*. USA: South-Western Cengage Learning.

Suifan, T.S., Abdallah, A.B. & Diab, H. 2016. The influence of work-life balance on turnover intention in private hospitals: the mediating role of work-life conflict. *European Journal of Business and Management* 8(20): 126–139.

Sun, R.A.W. 2017. Human ability to adapt is key to success.

Tongchaiprasit, P. & Ariyabuddhiphongs, V. 2016. Creativity and turnover intention among hotel chefs: The mediating effects of job satisfaction and job stress. *International Journal of Hospitality Management* (55): 33–40.

Torrington, D., Hall, L., Taylor, S. & Atkinson, C. 2017. *Human resources management, 10th edition*. United Kingdom: Pearson Education Limited.

Wani, K,S. 2013. Job stress and its impact on employee motivation: a study of a select commercial bank. International *Journal of Business and Management Invention* 2(3): 13–18.

Yadav, R.K. & Yadav, S.S. 2014. Impact of work-life balance and stress management on job satisfaction among the working women in public sector banks. *International Letters of Social and Humanistic Sciences* 26: 63–70.

Yun, H., Kettinger, W.J. & Lee, C.C. 2012. A New open door: the smartphone's impact on work-to-life conflict, stress, and resistance. *International Journal of Electronic Commerce* 16(4): 121–152.

Advances in Business, Management and Entrepreneurship – Hurriyati et al (eds)
© *2020 Taylor & Francis Group, London, ISBN 978-0-367-27176-3*

The effects of environmental work and individual characteristics on job stress among hospital nurses

S. Sumiyati, M. Masharyono, H. Yuliadi & R. Purnama
Universitas Pendidikan Indonesia, Bandung, Indonesia

ABSTRACT: Job stress is a significant problem in manufacturing and service companies, especially hospitals. One effort made by companies to reduce the level of job stress among employees is to improve the work environment and individual characteristics. The present study used a descriptive analysis technique. The explanatory survey consisted of 73 respondents and data were collected through the questionnaires. Data analysis in the study used multiple linear regression analyses. The findings show how important the work environment of a company is. However, weaknesses in the physical work environment, such as an inadequate working room temperature, and weaknesses such as in the nurses' knowledge contribute to job stress. Other findings show that the job stress experienced by nurses is quite high, particularly with regard to workload. Influence of the work environment on job stress amounts to 40.20%, while individual characteristics amount to 32.72%. If work environment, and if individual characteristics among nurses lead to an increase in the absenteeism level, this will result in a decrease in the productivity level, tolerance, and performance.

1 INTRODUCTION

Employees are one of the most important parts of a company, helping to maintain sustainability, development, competitiveness, and profitability (Senen et al. 2017). Employee performance is very important in the company's efforts to achieve business goals (Senen & Triananda 2016). Increasing employee performance is a priority for companies nowadays, and maximizing performance is a key challenge for any organization (Osa 2014, Senen & Triananda 2016).

Job stress is one of the main causes of declining employee performance in an organization (Shivendra & Kumar 2016). Approximately 30% of the employees in developed countries suffer from job stress; it is likely that developing countries have an even higher job stress level (Hosseini et al. 2016). As reported in *Fortune* magazine, job stress among workers in America is increasing, and job stress is considered to be one of the main problems facing labor in this century (Sager 1994).

Studies of job stress has been done in various types of companies, including the service industry and manufacturing. Studies in India have been conducted on various groups, including the education sector, the banking sector, and information technology sector (Shukla et al. 2016), as well as in the healthcare sector, particularly nursing (Dagget et al. 2016).

Job stress among nurses is one of the problems faced by human resources management in hospitals worldwide. The results of a national survey in France found that approximately 74% of nurses experienced work stress (NIOSH 2008). A study in

the United Arab Emirates consisting of at 216 nurses at Al-Zahra Hospital found a link between nurse's work stress and occupational environmental factors; 44.4% had a low stress level, 55.1% had an intermediate stress level, and 0.5% had a high stress levels (Mozhdeh et al. 2008).

As of February 2012, Indonesia had 120.4 million people in work, an increase of 1.0 million people from February 2011. This shows the potential for huge losses as a result of job stress (Fitri 2013). A study by Martina stated that nurses working in the RSPG Cisarua Bogor Hospital experienced moderate stress (86%). Nurses at the Yogyakarta Islamic Hospital PDHI are known to experience moderate work stress (82.70%). Nurses with average stress levels are still able to do their job properly (Hariyono et al. 1978).

The high level of job stress among nurses will cause problems in the hospital. Consequences such as increased absenteeism, decreased productivity, and tolerance lead to a high rate of turnover (Zainal et al. 2014). Individuals tend to experience job-related stress when there is a perceived threat and an inability to solve a problem (Chetty et al. 2016).

The work environment is identified as the most important factor affecting job stress among nurses (Loo-See Beh 2016). Suwatno and Priansa (2011) suggest that both the physical and psychological aspects of the work environment influence job stress among employees. Social work affects job satisfaction and productivity (Purnama & Pratama 2016). A work environment located around the workplace may affect employees either directly or indirectly (Masharyono et al. 2016). A good social or psychological work environment can encourage productivity and can increase benefits for the organization (Pratama & Purnama 2016).

Another factor that affects job stress is individual characteristics: these include psychological, physiological, and individual reactions and specific personality traits and behavioral patterns based on attitudes, needs, values, past experiences, living conditions, and abilities (Zainal et al. 2014). Individual characteristics also consist of individual skills, experiences, knowledge, and demographics (Gibson et al. 2012). These individual characteristics are important elements. The existence of good relationships in a company help employees to perform their duties properly (Juliandiny et al. 2016) because the progress of a company is determined by the quality of its human resources (Senen 2008).

Based this, the purpose of the present study is to determine (1) the influence of work environment on job stress, and (2) the influence of individual characteristics on job stress.

2 METHOD

The subjects of the study were nurses at Avisena General Hospital in Cimahi. The period under study was less than one year, and the data collection technique used was a cross-sectional method. The sampling technique in the study is a saturated sample of 73 nurses.

Data collection techniques used included literature study, field study, and questionnaires. Data analysis was done through descriptive and verification analysis. Analysis was done using SPSS 24.0 for Windows.

3 RESULTS AND DISCUSSION

3.1 Descriptive analysis

3.1.1 Work environment
The description of the work environment at Avisena Public Hospital in Cimahi can be seen in Table 1.

According to Mangkunegara (2014), the social work environment covers all aspects of the physical and psychological environment as well as any work regulations that affect job satisfaction and productivity achievement. The scores shown in Table 1 reflect the responses regarding opinions on both the physical and psychological work environment for nurses at Avisena

General Hospital. Based on the item statement (indicator), the highest value is in the psychological work environment, 3.851 or 75.36%. The lowest value is for the physical work environment, 2.253 or 73.48%. Empowerment in the work environment is a concept that can be used to generate and sustain workers' job satisfaction prosperity (Dirik & Intepeler 2017).

3.1.2 Individual characteristics
The scores shown in Table 2 reflect the responses regarding the description of individual characteristics of nurses working at the Avisena General Hospital.

According to PPNI (2005), nurses must exhibit knowledge, skills, and good attitudes, and there are standards to be met. Skills means the ability to handle a job in an easy and careful manner using basic abilities (Robbins & Judge 2015). Based on the item statement (indicator), the highest score is in skills, 3.111 or 76,10%. The lowest category is knowledge, 1.553 or 75.98%.

3.1.3 Job stress
is the scores shown in Table 3 reflect the responses regarding the description of the work environment for nurses at Avisena General Hospital.

If demands on employees gets greater and it becomes difficult to complete a job, this will lead to stress and poor employee performance (Stres et al., 2018). Based on the item statement (indicator), the highest score is for workload, 1.550 or 75.83%. The lowest score is for role ambiguity, 2.010 or 65.56%. According to Robbins and Judge (2015), role ambiguity arises when expectation for a worker are not clearly understood, so that employee is unsure of what has been, is being, or should be done.

3.2 Verification analysis

This study consists of two independent variables, namely work environment (X1) and individual

Table 2. Description of individual characteristics.

No	Dimension	Ideal Score	Total Score	%
1	Skills	4.088	3.111	76.10
2	Experience	3.066	2.333	76.09
3	Knowledge	2.044	1.553	75.98
TOTAL		9.198	6.997	76.07

Table 1. Description of the work environment.

No	Dimension	Ideal Score	Total Score	%
1	Physical work environment	3.066	2.253	73,48
2	Psychological work environment	5.110	3.851	75,36
TOTAL		8.176	6.104	74,66

Table 3. Response to questions regarding job stress.

No	Dimension	Ideal Score	Total Score	%
1	Workload	2.044	1.550	75.83
2	Working pressure	3.066	2.138	69.73
3	Work conflicts	2.555	1.884	73.74
4	Role ambiguity	3.066	2.010	65.56
TOTAL		10.731	7.582	70.66

characteristics (X2), while the dependent variable is job stress (Y). To find out the influence of work environment and work characteristic on job stress, a double linear regression test was needed.

The model of multiple regression equation that was performed in this study was as follows:

$$Y = a + b_1 X_1 + b_2 X_2 + e$$

Source: (Sugiyono 2017)
where:
Y = Job Stress
X1 = Work Environment
X2 = Individual Characteristics
a = Constant Numbers
b1,2 = Regression Coefficient
e = Standard Error

The results based on the data processing by SPSS 24.0 for Windows are shown in Table 4.

As can be seen in Table 4, the correlation coefficient (R) of the working environment and individual characteristics on job stress are about 0.673 and are included in the strong category (Sugiyono 2017).

To find out the contribution of work environment and individual characteristics on job stress, the following formula is used:
KD = r^2 x 100% (Riduwan 2013)
where:
KD : Determination Coefficient
r^2 : Correlation Coefficient
100% : Constants

Table 5 shows the result of the coefficient determination calculation from X1 to Y using SPSS 24.0 for Windows.

$$
\begin{aligned}
KD &= r^2 \times 100\% \\
&= (0,634)^2 \times 100\% \\
&= 40,2\%
\end{aligned}
$$

The result of determination correlation is about 40.2%, which means that the work environment gives a partial influence of 40.2% on job stress of the

Table 4. Coefficient of correlation model summary.

Model Summary[b]

Model	R	R Square	Adjusted R Square	Std. Error of the Estimate
1	.673[a]	.452	.437	10.366

* Predictors: (Constant), individual characteristics, working environment.
** Dependent variable: job stress.

Table 5. Coefficient of determination of work environment to job stress.

Model Summary[b]

Model	R	R Square	Adjusted R Square	Std. Error of the Estimate
1	.634[a]	.402	.394	10.752

* Predictors: (Constant), work environment.
* Dependent Variable: job stress.

Table 6. Determination of coefficient of individual characteristics on job stress.

Model Summary[b]

Model	R	R Square	Adjusted R Square	Std. Error of the Estimate
1	.572[a]	.327	.317	11.410

* Predictors: (Constant), individual characteristics.
** Dependent Variable: job stress.

remaining 59.8% was influenced by factors not examined in this study, including extra-organizational stressors, organizational stressor, stressor groups, and individual stressors (Luthans 2011). The result shown in Table 6 is the result of calculation determination coefficient from X2 to Y.

$$
\begin{aligned}
KD &= r^2 \times 100\% \\
&= (0,572)^2 \times 100\% \\
&= 32,7\%
\end{aligned}
$$

Thus, it can be said that the coefficient determination value of 32.7% means that individual characteristics have a partial effect of 32.7% on job stress. of the remaining 67.3% was influenced by factors not examined in this study, including extra-organizational stressors, organizational stressors, stressor groups, and individual stressors (Luthans 2011).

3.3 Simultaneous hypothesis testing (test F)

Test F in this study used an Anova test, which was used to determine the effect of simultaneously independent variables (work environment and individual characteristics) on the dependent variable (job stress). Table 7 shows the results of hypothesis testing simultaneously.

Based on the output shown in Table 7, it can be seen that the value of Fcount is 28.901 with p-value (sig) 0.000. With $\alpha = 0, 05$ and degrees of freedom df2 = n - k = 73 - 3 = 70 and df1 = k - 1 = 3 − 1 = 2,

Table 7. Test F.

ANOVA^a

ANOVA[a]

Model	Sum of Squares	df	Mean Square	F	Sig.
1 Regression	6210.809	2	3105.405	28.901	.000[b]
Residual	7521.520	70	107.450		
Total	13732.329	72			

* Dependent Variable: job stress.
** Predictors: (Constant), individual characteristics, work environment.

then Ftable 3.13 is obtained. Due to the value Fcount > Ftable (28,901 > 3, 13), H0 is rejected and Ha accepted. This means that the work environment and individual characteristics have a significant effect on job stress.

Thus, it can be concluded that the present study shows that H0 is rejected and Ha is accepted. In other words, the work environment and individual characteristics have an influence on job stress among nurses at Avisena General Hospital in Cimahi.

3.4 Partial hypothesis testing (T test)

The T test can be used to determine the effect of independent variables (work environment and individual characteristics) on job stress. Table 8 shows a partial result of hypothesis testing.

As can be seen in Table 8, the result of a T test regarding the influence of work environment on job stress shows that t count is about 6,913 for the work environment and 5,872 for individual characteristics. Significance level (α) is about 5%, and degrees of freedom df = n – k = 73 – 3 = 70 obtained tcable value 1.667. Due to thitung > ttabel or 6,913 > 1,667, Ha is accepted. This means that work environment has a significant effect on job stress. The thitung value of individual characteristics is about

5,872. Due to thitung > ttabel or 5,872 > 1,667, Ha is accepted. This means that individual characteristics have a significant effect on job stress.

Based on these test results, it can be concluded that there is a significant effect of work environment and a significant influence of individual characteristics on job stress among nurses at Avisena General Hospital in Cimahi.

4 CONCLUSION

4.1 Conclusion

The result of the present study indicates that (1) the work environment has a significant effect on job stress; the more conducive the work environment, the lower the job stress of the nurses at Avisena General Hospital in Cimahi; and (2) individual characteristics have a significant effect on job stress; the better the individual characteristics, the lower the job stress of nurses at Avisena General Hospital in Cimahi.

4.2 Recommendation

Based on the results of this study, the authors have several suggestions to help to reduce the level of job stressincluding:

a. A good work environment is one way to reduce stress and create a comfortable workplace for nurses. The work environment is identified as the most important factor affecting job stress in nurses (Loo-See Beh 2016). Excessive work demands can lead to "poor health" between individual ability and work environment (Zhao et al. 2016). Thus, the authors recommend that companies improve the physical and psychological work environments in order to create a conducive working environment, work comfort, and prosperity.

b. Educated and knowledgeable nurses will help to create better and more skilled performance in the

Table 8. The significance value of a T test.

Coefficients[a]

Model	Unstandardized Coefficients		Standardized Coefficients		
	B	Std. Error	Beta	t	Sig.
(Constant)	34.329	10.143		3.385	.001
Work environment	.832	.120	.634	6.913	.000
(Constant)	39.887	10.983		3.632	.001
Individual characteristic	.661	.113	.572	5.872	.000

* Dependent Variable: job stress.

workplace. For instance, individual characteristics such as Type A personality patterns, personal control, learned helplessness, and psychological resilience can affect job stress. In addition, there is the level of the intrinsicdividual conflict stems from frustration, goals, and roles (Luthans 2011). Thus, the authors recommend that companies may continue to improve and take into account individual characteristics of each nurse, leading to an improvement in their knowledge and understanding of individual nurses in the work environment.

REFERENCES

Chetty, P.J., Coetzee, M. & Ferreira, N. 2016. Sources of job stress and cognitive receptivity to change: the moderating role of job embeddedness. *South African Journal of Psychology* 46(1): 101–113.

Dagget, T., Molla, A. & Belachew, T. 2016. Job related stress among nurses working in Jimma Zone public hospitals, South West Ethiopia: A cross sectional study. *BMC Nursing* 15(1): 39.

Dirik, H.F. & Intepeler, S.S. 2017. The work environment and empowerment as predictors of patient safety culture in Turkey. *Journal of Nursing Management* 25(4): 256–265.

Fitri, A.M. 2013. Analisis faktor-faktor yang berhubungan dengan kejadian stres kerja pada karyawan bank (Studi pada karyawan Bank BMT). *Jurnal Kesehatan Masyarakat* 2013 2(1).

Gibson, J.L., Ivancevich, J.M., Donnelly, J.H. & Konopaske, R. 2012. *Organizations: Behavior, structure, processes (14th ed.).* New York: McGraw-Hill.

Hariyono, W., Suryani, D. & Wulandari, Y. 1978. Hubungan antara beban kerja, stres kerja dan tingkat konflik dengan kelelahan kerja perawat di rumah sakit Islam Yogyakarta Pdhi Kota Yogyakarta. *Kesmas* 3(3): 186–197.

Hosseini, S., Habibi, E., Barakat, S., Ahanchi, N., Fooladvand, M. & Khorasani, E. 2016. Investigating the relationship of mental health with job stress and burnout in workers of metal industries. *International Journal of Educational and Psychological Researches* 2(2): 111.

Juliandiny, T., Senen, S. H. & Sumiyati. 2016. Kompensasi serta motivasi kerja pada kinerja keperawatan kontrak. *Journal of Business Management Education* 1(2): 81–90.

Loo-See Beh. 2016. Job stress and coping mechanisms among nursing staff in public health services leap-han loo. *International Journal of Academic Research in Business and Social Sciences* 6(5): 131–176.

Luthans, F. 2011. *Organizational behavior: An evidence-based approach (12th Ed.).* New York: McGraw-Hill /Irwin.

Masharyono, M., Sumiyati, S. & Toyib. 2016. Physical work environment effect on employee productivity of textile industry. *Atlantis Press* 15: 630–632.

Mozhdeh, S., Sabet, B., Irani, M., Hajian, E. & Malbousizade, M. 2008. Relationship between nurse's stress and environmental-occupational factors. *Iran J Nurs Midwifery* 13(1): 1–5.

NIOSH. 2008. *Exposure to stress occupational hazards in hospital.* NIOSH.

Osa, I.G.M. 2014. Incentives motivates employees on organizational performance. *Global Journal of Arts Humanities and Social Sciences* 2(7).

Pratama, K.F. & Purnama, R. 2016. The effect of social work environment on employee productivity in manufacturing company in Indonesia. *Atlantis Press* 15: 574–575.

Purnama, R. & Pratama, K.F. 2016. The influence of social work environment on employee productivity in manufacturing in Indonesia. *Atlantis Press* 15: 649–652.

Riduwan, R. 2013. *Cara menggunakan dan memakai analisis jalur (path analysis).* Bandung: Alfabeta.

Robbins, S.P. & Judge, T.A. 2015. *Perilaku organisasi (edisi enam).* Jakarta: Salemba Empat.

Sager, J.K. 1994. A structural model depicting salespeople's job stress. *Journal of the Academy of Marketing Science* 22(1): 74–84.

Senen, S.H. 2008. Pengaruh motivasi kerja dan kemampuan kerja karyawan pada PT. Safilindo Permata. *Jurnal Pendidika Manajemen Bisnis* 7(14): 1–15.

Senen, S.H., Sumiyati, S. & Masharyono, M. 2017. Employee performance assessment system design based on competence. *Innovation of Vacational Technology Education* 2(8): 68–70.

Senen, S.H. & Triananda, N. 2016. The employee performance influenced by communication: A study of BUMD in Indonesia 15: 596–598.

Shivendra, D., & Kumar, M.M. 2016. A study of job satisfaction and job stress among physical education teachers working in government, semi-government and private schools. *International Journal of Sports Sciences & Fitness* 6(1): 89–99.

Shukla, A., Srivastava, R. & Eldridge, D. 2016. Development of short questionnaire to measure an extended set of role expectation conflict, coworker support and work life balance: The new job stress scale. *Cogent Business & Management* 3(1): 1.

Stres, P., Dan, K., Kerja, K., Kinerja, T., Studi, K., Karyawan, P. & Internusa, C. 2018. Pengaruh stres kerja dan konflik kerja terhadap kinerja karyawan (studi pada karyawan pt. surya cipta internusa gresik). 6.

Sugiyono. 2017. *Metode penelitian kuantitatif, kualitatif dan r & d.* Bandung: Alfabeta.

Zainal, V.R., Hadad, M.D. & Ramly, M. 2014. *Kepemimpinan dan perilaku organisasi.* Depok: PT. RajaGrafindo Persada.

The linking of knowledge management enabler and employee's performance

A. Silvianita, O.P. Pramesti & M. Fakhri
Telkom University, Bandung, Indonesia

ABSTRACT: In this era, to face competition in business, organizations need to improve strategy and innovation. In addition to human resources and capital, knowledge is an important resource. Through knowledge, people share their experiences and apply that knowledge in the organization. Hence, the company is a concern about the importance of knowledge management as an essential asset. In order to manage it, the organization has to concern with enabler's factors. The aim of this research is to identify the influence of organizational structure and information technology (IT) on employee performance in the Processing Post Unit, Pos Indonesia Corporation, Bandung. Based on the results, both organizational structure and IT affect employee performance. It is also known that still many other factors can accelerate knowledge management enabler.

1   INTRODUCTION

Human factors can determine the organizational performance. According to Pawirosumarto et al. (2017), performance is an achievement in one period compared to a predetermined working target. Basically, employee performance is the result of the achievement of the work undertaken (Anitha 2014). In other words, employee performance is closely related to an organization's overall performance (Muda et al. 2014).

Pos Indonesia Corporation is required to have a higher organizational performance compared to competitors. In terms of quantity from 2015 and 2016, in Pos Processing Unit, there is a decrease in a number of production of mail and package items received and sent by SPPos. This is caused by manual processing process. During 2015, the number of complaints increased. Most complaints are either regarding damage or loss of packages. In addition, there were problems involving quality and timeliness. In August 2016, the SPPos unit introduced cost-effective tools in order to speed up processing packages and letters. The new machine has simplified processes, but employees have to be trained to be able to use it. Before using the new machine, five to six people were needed to do the job that now only requires two to three people.

With regard to knowledge as the major resource, there are several efforts made by Pos Indonesia Corporation related to knowledge management. To create knowledge becomes a resource in an organization, it needs a mechanism to build a sharing habit and protect knowledge in organization as a process stimulus in creating knowledge, or Knowledge Management (KM) Enabler (Silvianita & CL Tan 2015). KM Enabler is needed in an organization to actualized knowledge management into organizational aspects.

This research will focus on organizational structure and information technology (IT). According to Fu-Ho et al. (2014), there are two types of organizational structure: centralization and decentralization. Centralized organizational structure is when employees must obey all of the rules issued by the top management, whereas decentralized organizational structure refers to the type of decision making that can be made through policy.

In general, IT is a tool that creates to helping people in create, change, store, communicate or share information (Mulyadi 2015). In terms of organizational structure, Pos Indonesia Corporation is a pioneer in the field of mailing delivery service in Indonesia. Along with the growth in mailing and package service, organizational structure in Pos Indonesia Corporation has to change and innovate. Related to the IT, Pos Indonesia Corporation already has a site as a supported tool for knowledge management implementation. Through that site, is expected that all the employees will share knowledge.

2   METHOD

According to this explanation, the framework of this research can be seen in Figure 1.

According to the research framework, as the following hypotheses can be formulated:

a. H1: Organizational structure will be positively related to employees' performance.
b. H2: IT will be positively related to employees' performance.
c. H3: Organizational structure and IT will be related to employees' performance.

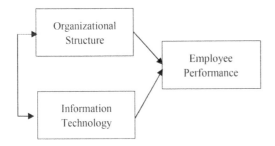

Figure 1. Research framework.

Based on 95 respondents who work in SPPos unit obtained some results. To answer the hypothesis, this research using path analysis as a statistical tool. Based on Table 1, a relationship between variable Xi (Organizational structure), X2 (IT) and Y (Employees) can be seen.

As can be seen in Table 1, we found that the path coefficient from each X variable to Y variable is 0.287 for Xi to Y. Variable X2 to Y variable is 0.377. The significance value of Xi is $0.003 < 0.005$, while the significance value for X2 is $0.000 < 0.005$.

Table 2 shows the path coefficient of the outside model. The path coefficient of the outside model is 0.840, where it can calculate by: $Vl - 0.294 = 0.840$.

Means that there are still many factors that can be measured to determine the KM Enabler to employees' performance. In path analysis it also calculates the indirect relationship between variable X and Y. Because it is known that both organizational structure and IT can work independently to form KM Enabler. Here is the result of the indirect relationship between Xi, X2 and Y.

Table 3 shows that the total influence from variable Xi to Y, either direct or through variable X2, is 0.116. Table 4 shows that the influence from variable X2 to Y, either direct or through variable Xi, is 0.176.

3 RESULTS AND DISCUSSION

In organizational,KM can be defined as various effort that can be performed by an organization in a systematic process to get, capture, share and use knowledge productively for the purpose enriched knowledge and organizational performance (Karim et al. 2012). In general, organizational structure was formed based on

Table 2. Model summary.

Model	R	R Square	Adj.R Square	Std. Error of the Estimate
1	0.542	0.294	0.278	0.21784

Table 3. Indirect influence among Xi, X_2 and Y.

Direct influence X!	0.287 x 0.287	0,082
Influence through X?	0.287x0.319x0.377	0,034
Total Influence X! to Y		0.116

Table 4. Indirect influence among X_2, Xi and Y.

Direct influence X_2	0.377 x 0.377	0.142
Influence through Xi	0.377x0.319x0.287	0.034
Total Influence X_2 to Y		0.176

operational or organizational needs; if these are not properly formulated, they can create a barrier for the organization (Fu-Ho et al. 2014). Moreover, organizational structure becomes a supporting factor in knowledge creation (Jeng & Dunk 2013).

Based on result, it found that variable Xi (Organizational structure) directly influence variable Y (Em-ployees performance) for 0.082 (8.2%) and through variable X2 (IT) for 0.034 (3.4%). Therefore, variable Xi totally determines the change of variable Y for 11.6%. It appropriates with previous research that mentioned that each factor is together created KM enabler in an organization. In SPPos unit, found that organizational structure itself will influence the employees' performance.

There are other factors that can be used to measure KM's supported factors, however, basically KM Enabler can encourage employees to share knowledge and experience with others in order knowledge in organization can be evolved systematically (Allameh et al. 2011). KM is a complex process supported by factors such as strategy and leadership, culture, appropriate measurement, and technology. From many factors, Information and Communication

Table 1. Coefficients.

Model		Unstand. Coef.		Std. Coefficient		
		B	Std. Error	Beta	t	Sig.
1	Constant	2.866	0.185		15.476	0.000
	Xi	0.155	0.050	0.287	3.108	0.003
	X,	0.184	0.045	0.377	4.075	0.000

Technologies (ICTs) is a main factor in KM Enabler (Okunoye & Karsten 2002). Consistent with previous research, variable X2 (IT) directly influence variable Y (em-ployee performance) for 0,142 (14.2%) and through variable X1 (organizational structure) for 0.034 (3.4%). Therefore, variable X2 totally determine the change of variable Y for 17.6%. In SPPos unit found that IT itself has a bigger influence to employees' performance compared to organizational structure.

KM Enabler constitutes a mechanism that used by organization in build a knowledge, supporting knowledge creation within organization, to then share to all employees and protected as an organizational knowledge (Allameh et al. 2011). Variable X1 and X2 together influence variable Y for 11.6% + 17.6% = 29.2%. While, the amount proportional influence caused by another variable beside X1 and X2 claimed by p2 ye = $(0.840)2$ = 0.705 or 70.5%. From this cal-culation found that there are many factors that have to elaborate in SPPos Unit to improve the employees' performance through KM Enabler.

The influence accepted by variable Y from both X1 and X2 is 29.2% + 70.5% = 100%. From this cal-culation, it can be seen that organizational structure and IT in the SPPos unit have an influence on employees performance. Therefore, in order for an organization to be able to achieve a high level of performance, it has to see from many things, such as employees' performance. Employee performance is influenced by many factors, including appropriate organizational structure and IT readiness.

4 CONCLUSION

Based on the literature review and statistical ana-lysis, this research has some of conclusions is organizational structure and Information Technol-ogy as factors of KM Enabler have an important role for improving the employees' performance in SPPos unit Pos Indonesia Corporation Bandung. According to the statistical analysis, there is still research to be done.

REFERENCES

Allameh, S. M. & Zare, S. M. & Davoodi, S.M.R. 2011. Examining the Impact of KM Enablers on KM Process. *Procedia Computer Science*, Vol. 3, pp. 1211–1223.

Anitha, J. 2014. Determinant of Employee Engagement and Their Impact on Employee Performance. *International Journal of Productivity and Performance Management*, Vol. 63, Issue 3, pp. 308–323.

Fu Ho, C. & Hsuan Hsieh, P. & His Hung, W. 2014. Enablers and Processes for Effective Knowledge Management. *Industrial Management and Data Systems*, 114 (5), pp. 734–754.

Jeng, D. & Dunk, N. 2013. Knowledge Management Enablers and Knowledge Creation in ERP System Success. *International Journal of Electronic Business Management*, Vol. 11, Issue 1, pp. 49–59.

Karim, N. S. A. & Razi, M. J. M. & Mohamed, N. 2012. Measuring Employee Readiness for Knowledge Management using Intention to be Involve with KM SECI Processes. *Business Process Management Journal*, Vol. 18, Issue 5, pp. 777–791.

Kaswan. 2016. *Peak Performance*. Bandung: Indonesia.

Lee, H. & Choi, B. 2003. Journal of Management Enablers, Processes, and Organizational Performance: An Integrative View and Empirical Examination, *Journal of Management Information Systems*, Vol. 20, Issue 1, pp. 179–228.

Muda, I. & Rafiki, A. & Harahap, M.R. 2014. *Factor's Influencing Employees' Performance: A Study of the Islamic Banks in Indonesia*. International Journal of Business and Social Science, Vol. 5 No 2; February 2014.

Mulyadi, D. 2015. *People in Organization*. Bandung: Indonesia.

Okunoye, A. & Karsten, H. 2002. Where the Global Needs the Local: Variation in Enablers in the Knowledge Management Process. *Journal of Global Information Technology Management*, pp. 12–33.

Pawirosumarto, S. & Sarjana, P.K. & Muchtar, M. 2017. Factors Affecting Employee Performance of PT. Kiyokuni Indonesia. *International Journal of Law and Management*, Vol. 5 No.4, p. 602–617.

Pradhan, R.K. & Jena, L. K. 2017. Employee Performance at Workplace: Conceptual Model and Empirical Validation. *Business Perspective and Research*, 5(1), pp. 1–17.

Silvianita, A. & Tan, C.L. 2017. A Model Linking the Knowledge Management (KM) Enabler, KM Capability and Operational Performance in Indonesian Automobile Industry. *Advance Science Letters*, Vol. 23, Issue 1, pp. 640–642.

The impact of coaching and person-job fit on self-efficacy and performance

A.S. Hidayat
Universitas Swadaya Gunung Jati, Cirebon, Indonesia

ABSTRACT: The present study aims to determine the impact of coaching and person-job fit on self-efficacy and performance. It uses a quantitative research. The population involves 45 employees of marketing division at Lambang Putra Perkasa Motor Company. This study uses saturation sampling, the population is also considered as research sample. This study then uses questionnaire by using likert scale as instrument measurement scale. In analysing the data, the path analysis is supported by SMART PLS application. The results show that (1) coaching doesn't have any significant effect on self-efficacy; (2) coaching doesn't have any significant effect on performance; (3) person-job fit has significant effect on self-efficacy; (4) person-job fit has significant effect on performance; and (5) self-efficacy has significant effect on performance.

1 INTRODUCTION

In Indonesia, the competition in the automotive industry is getting tighter as Toyota, Honda, Daihatsu, and Mitsubishi are fighting over market share. The increasing number of companies requires each company to have great organizational performance. It is necessary to maintain company's competitiveness. The main indicator of organizational performace, in the industry, is the number of selling, therefore high selling performance represents high quality of company's performance.

The discussion on automotive product selling, specifically cars, cannot be separated from employees of marketing departement. The employees of marketing department are the spearhead of realizing the target. Therefore, the development of their performance will have influence on product selling.

It is a fact that there are some companies with low-level product selling. It is resulted from the low performance of marketing division. Lambang Putra Perkasa Motor Company is recently dealing with the respective problem. As an automotive company of Honda, based on selling report from 2016–2017, there has been a steady decrease in the number of selling.

In 2016, there were 1277 sold products, then, in 2017, there were 1227 sold products. It means that there was a decrease; 50 units. This decrease reflects the low performance of employees of marketing departement as they cannot meet the target.

The decline of employees' performance is a crucial problem. Therefore, the solution should be soon discovered to improve their performance. By leaning on theories and studies, performance is underpinned by some factors.

Regarding marketing division performance, self-efficacy is an interesting issue to discuss. Some scholars believe that self-efficacy is important for enhancing performance. Self-efficacy is personal belief in his ability to accomplish a job (Bandura 1982).

Being employees in the marketing department is challenging. The reason is that a member will be forced to achieve the target. Personal belief to meet the target is an important factor to boost the achievement. Personal belief is the foundation of making the target happen. Ayundasari et al. (2017) stated that employees, with high self-efficacy, will have better performance.

The fact directs the importance of discovering solutions to improve self-efficacy. A factor of self-efficacy is coaching. Coaching is a process where individuals obtain skills, abilities, and knowledge which they need to develop themselves professionally and become more effective in their work (Stone, 1999). Employees who are given coaching by superiors have higher self-efficacy (Baron & Morin 2010).

In addition to the coaching factors, person-job fit is also a factor that is considered to improve self-efficacy.

Kridtof-Brown, Zimmerman and Johnson (2005) define person-job fit as suitability of ability-demands, where employees' knowledge, skills, and abilities are commensurate with what is demanded by work and the suitability between needs, desires, or employee preferences are met by the work they do. Employees who have the harmony between the abilities they have and the demands of the work have high self-efficacy (Peng & Mao 2015).

2 METHOD

The research method, used in this research, is quantitative method. The method is used to discover the influence of coaching and person-job fit variables on self-efficacy, as well as its impact on perfromance.

The population was 45 employees of the marketing division at P Lambang Putra Perkasa Motor Company, Cirebon. This study used saturation sampling and the population was also considered to be the research sample.

This study used questionnaire by using a Likert scale as an instrument measurement scale. In analysing the data, the path analysis was supported by SMART PLS application. The data analysis is performed by using validity and reliability tests, path analysis, and hypothesis test.

3 RESULT AND DISCUSSION

The result of path analysis shows the influence of coaching and person-job fit on self-efficacy. The result is as follows.

The figure shows that (1) coaching has influnce on self-efficacy, it is 9.5%, (2) person-job fit has influence on self-efficacy, it is 62.1%, (3) coaching and person-job fit has simultaneous influence on self-efficacy of 45.4%, (4) coaching has influence on performance, it is 21.1%, (5) person-job fit has influence on performance, it is 42.6%, (6) self-efficacy has influence on performance, it is 28, 3%, and (7) coaching, person-job fit, and self-efficacy have simultaneous influence on performance, it is 64,1%.

To discover the result of hypothesis test, the analysis is shown by path coeficient analysis. The results are seen in Table 1.

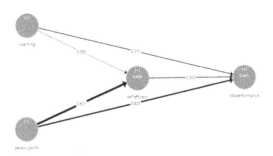

Figure 1. The result of path analysis.
Source: data analysis, 2018.

Table 1. Path coeficient.

	T Statistic (IO/STDEV)	P Values
Coaching - > self-efficacy	0.546	0.58
Coaching - > performance	1.670	0.09
Self-efficacy - > performance	2.427	0.01
Person-job fit - > self-efficacy	3.400	0.00
Person-job fit - > performance	3.053	0.00

Source: data analysis, 2018.

Table 1 shows that (1) there is no significant influence of coaching on self-efficacy, as the P value is $0.58 > 0.05$, (2) there is no significant influence of coaching on performance, as the P value is $0.09 > 0.05$, (3) there is significant influence of person-job fit on self-efficacy, as the P value is $0.01 < 0.05$, (4) there is significant influence of person-job fit on performance, as the P value is $0.00 < 0.05$, and (5) there is significant influence of self-efficacy on performance, as the P value is $0.01 < 0.05$.

Based on the analysis, coaching has no significant influence on self-efficacy. It means that, although, coaching has positive influence on self-efficacy, the influence is not sufficient to improve self-efficacy. The finding is different from Ammentorp and Kofoed (2009), Baron and Morin (2010), and Moen and Allgood (2009), all of whom state that coaching is significant to self-efficacy.

The analysis shows that coaching has no significant influence on performance. It means that coaching has positive influence on performance, but the influence is not sufficient to improve employee's performance.

The finding is different from Kalkavan and Katrinli (2014), Latham et al. (2012), and Utrilla et al. (2015), all of whom state that coaching is significant to performance. The result is interesting to be discussed by the future researchers.

The analysis also shows that person-job fit has positive (and significant) influence on self-efficacy. It means that higher person-job fit will increase self-efficacy. The findings corroborate the results of DeRue and Morgeson (2007) and Ru and Hsu (2012), who state that person-job fit has positive (and significant) influence on self-efficacy.

Furthermore, the analysis shows that person-job fit has positive (and significant) influence on performance. It means that higher person-job fit will improve employee's performance. The findings corroborate the results of previous studies proposed by Caldwell and O'Reilly (1990), Farooqui and Nagendra (2014), Iqbal et al. (2012), and June and Mahmood (2015), all of whom state that person-job fit is a determinant of performance. Employees who have suitability between personal characteristics and characteristics of their work have higher performance, compared to employees who do not have suitability between personal characteristics with job characteristics.

Farooqui and Nagendra (2014) and Iqbal et al. (2012 used job satisfaction variables as intervening variables that correlate person-job fit to performance. In addition, Caldwell and O'Reilly (1990) and June and Mahmood (2015) examined only the direct influence of person-job fit on performance. In the studies, the researchers examined the influence of person-job fit on performance through self-efficacy variable as intervening variable.

The result then shows self-efficacy has positive (and significant) influence on performance. It means that employees' self-efficacy will improve their performance. The findings corroborate the results of Ayundasari et al. (2017), Carter et al. (2016), and Cherian and Jacob (2013), all of whom state that self-efficacy has positive (and significant) influence on performance

4 CONCLUSION

The present study concludes that (1) coaching has no significant influence on self-efficacy, (2) coaching has no significant influence on performance, (3) person-job fit has significant influence on self-efficacy, (4) person-job fit has significant influence on performance, and (5) self-efficacy has significant influence on performance.

The managerial implication of this study is, for the company, to improve employee's person-job fit, especially salary alignment with employee's expectations, as well as the suitability of the training provided with the employee's needs.

The limitation of this study is the object of research that is only carried out one company. Therefore, the sample size is small, so that generalization of the problems in the industry cannot be made. It can only be applied to the respective company.

There are some suggestions for further research: (1) expanding the object of research, (2) re-examining the influence of coaching on self-efficacy and performance, and (3) conducting research on the influence of self-efficacy on performance and its impact on competitive advantage.

REFERENCES

Ammentorp, J., & Kofoed, P. E. (2009). Coach training can improve the self-efficacy of neonatal nurses: A pilot study. *Patient Education and Counseling* 79(2): 258–261.

Ayundasari, D. Y., Sudiro, A., & Irawanto, D. W. (2017). Improving employee performance through work motivation and self-efficacy mediated by job satisfaction. *Jurnal Aplikasi Manajemen* 15(4): 587–599.

Bandura, A. 1982. Self-efficacy mechanism in human agency. *American Psychologist* 37(2): 122–147.

Baron, L., & Morin, L. (2010). The impact of executive coaching on self-efficacy related to management soft-skills. *Leadership & Organization Development Journal* 31(1): 18–38.

Caldwell, D. F., & O'Reilly III, C. A. 1990. Measuring Person – Job Fit Using a Profile Comparison Process Measuring Person-Job Fit with a Profile-Comparison Process. *Journal of Applied Psychology* 75(6): 648–657.

Carter, W. R., Nesbit, P. L., Badham, R. J., Parker, S. K., & Sung, L. K. 2016. The effects of employee engagement and self-efficacy on job performance: A longitudinal field study. *The International Journal of Human Resource Management*: 1–20.

Cherian, J., & Jacob, J. 2013. Impact of Self Efficacy on Motivation and Performance of Employees. *International Journal of Business and Management* 8(14): 80–88.

DeRue, D. S., & Morgeson, F. P. 2007. Stability and change in person-team and person-role fit over time: the effects of growth satisfaction, performance, and general self-efficacy. *Journal of Applied Psychology* 92(5): 1242–1253.

Farooqui, M. S., & Nagendra, A. 2014. The Impact of Person organization Fit on Job Satisfaction and Performance of the Employees. In *Procedia Economics and Finance* 11: 122–129.

Iqbal, M. T., Latif, W., & Naseer, W. 2012. The Impact of Person Job Fit on Job Satisfaction and its Subsequent Impact on Employees Performance. *Mediterranean Journal of Social Sciences* 3(2): 523–530.

June, S., & Mahmood, R. 2015. The Relationship between Person-job Fit and Job Performance: A Study among the Employees of the Service Sector SMEs in Malaysia. *International Journal of Business, Humanities and Technology* 1(2): 95–105.

Kalkavan, S., & Katrinli, A. 2014. The effects of managerial coaching behaviors on the employees' perception of job satisfaction, organisational commitment, and job performance: case study on insurance industry in Turkey. In *Procedia-Social and Behavioral Sciences* 150: 1137–1147.

Kristof-Brown, A. L., Zimmerman, R. D., & Johnson, E. C. 2005. Consequences of Individuals'fit at Work: A Meta-Analysis of Person–Job, Person–Organization, Person–Group, and Person–Supervisor Fit. *Personnel Psychology* 58(2): 281–342.

Latham, G. P., Ford, R. C., & Tzabbar, D. 2012. Enhancing employee and organizational performance through coaching based on mystery shopper feedback: A quasi-experimental study. *Human Resource Management* 51(2): 213–229.

Moen, F., & Allgood, E. 2009. Coaching and the effect on self-efficacy. *Organization Development Journal* 27(4): 69.

Peng, Y., & Mao, C. 2015. The impact of person–job fit on job satisfaction: the mediator role of Self efficacy. *Social Indicators Research* 121(3): 805–813.

Ru, Y., & Hsu. 2012. Mediating roles of intrinsic motivation and self-efficacy in the relationships between perceived person-job fit and work outcomes. *African Journal of Business Management* 6(7): 2616–2625.

Stone, F. 1999. *Coaching, Counseling and Mentoring*. New York: AMA Publication.

Utrilla, P. N., Grande, F. A., & Lorenzo, D. 2015. The effects of the coaching in employees and organizational performance: The Spanish case. *Intangible Capital* 11(2): 166–189.

Advances in Business, Management and Entrepreneurship – Hurriyati et al (eds)
© *2020 Taylor & Francis Group, London, ISBN 978-0-367-27176-3*

Polychronicity in the hotel industry in Bandung city

A.R. Andriani & D. Disman
Universitas Pendidikan Indonesia, Bandung, Indonesia

ABSTRACT: Polychronicity is important to service companies, as its implementation has a great impact on improving efficiency, quality of work and employee satisfaction. Employee satisfaction in multitasking ultimately contributes to both themselves and the company. The research was conducted to analyze the factors of polychronicity and the impact of polychronicity on the level of job satisfaction of hotel employees. The research was con-ducted through a survey to 64 employees of star hotels in Bandung Indonesia as the respondents. The study found that polychronicity can improve employee job satisfaction. The dominant factors formed in polychronicity are related to the ability of employees to do many tasks, multitask, and complete a task before starting another task. Another finding is that the level of employees' position affects polychronic behavior, so does their gender. This research is beneficial for the service industry identical with multitasking but still prioritizing excellent services.

1 INTRODUCTION

Hospitality industry is one service company that faces human resource challenges, for example in recruiting qualified employees (Arasl et al. 2014). Hotel employees should be able to respond to hotel guests' requests on time, provide high-quality services, and manage to overcome complaints of hotel guests (Karatepe & Kilic 2007). The success of hotel employees in serving hotel guests can be caused by the hotel employees' job satisfaction (Arasl et al. 2014). Satisfaction judgments capture the extent to which an employee feels pleased, happy, and rewarded, or displeased, unhappy, and exploited. Further, satisfied employees deliver the company's promise by creating a favorable image and striving to provide better services than their competitors (Malhotra & Mukherjee 2004).

Hotel is a work environment where employees' activities are increasingly complex, so that the individual gets involved in many jobs at the same time (Kayaalp 2014). Every industry will certainly appreciate employees who can handle many tasks at the same time (Kantrowitz et al. 2012, Kayaalp 2014). Employees must be able to set the time to achieve their work goals. While some employees may prefer to deal with multiple jobs simultaneously, others choose to focus on one task before engaging with another (Kayaalp 2014). With increasing competition in the hospitality industry, employees are expected to be involved in the various tasks, activities, and additional roles they must handle simultaneously (Persing et al. 1998).

The hospitality industry is a workplace where time is of paramount importance and employee behavior to switch to multiple tasks is often necessary within a certain timeframe (Jang & George 2012). This work environment introduces polychronic behavior, which requires employees to work on two or more activities over the same period (Bluedorn et al. 1999). Several studies have shown that poly-chronicity has a significant relationship with job satisfaction (Arasl et al. 2014, Arndt et al. 2006, Jang & George 2012).The main purpose of this research is to analyze the polychronicity factor and its implementation on the level of job satisfaction of hotel employees. This study then proposes a hypothesized relationship between variables. The research findings and implications for the practice are then suggested. The article ends with limitations and direction for future research.

1.1 Polychronicity

Polychronic can be understood as behavior to per-form multiple jobs simultaneously, whereas mono-chronicity is a behavior for doing work sequentially (Adams & Eerde 2012). Polychronicity refers to the extent to which people prefer to engage in two or more jobs simultaneously; and believe that their behavior is the best way to do something (Bluedorn et al. 1999). A person who chooses to complete a task, activity, or project before engaging with another is called a monochron, whereas people who prefer to engage with multiple tasks, activities or projects are called polychrons (Bluedorn et al. 1999). Polycronic is con-sidered a relatively stable nature or culture over time (Slocombe 1999).

A polychronic culture is a culture in which a person engages in multiple activities and events over the same period (Bluedorn 1998). This definition in-cludes behavioral elements (i.e., multitasking behavior) and evaluation elements as well as the relation-ship between the two. Polycronic people do some activities at the same time, and the reason for doing so is because of judgment. Hall incorporated several other phenomena under the polychronic concept (König & Waller 2010). For example, polychronic people are oriented, have complex information net-works, pay less attention to formal time limits, and can be more easily distracted than monochronic people (König & Waller 2010).

Polycronic refers to the extent to which a person prefers to switch between multiple tasks over the same period (Bluedorn et al.1999). Polycronic individuals feel comfortable engaging in some activities. People with polychronic orientation will anticipate engagement with some activities over a period of time, intending to move quickly between multiple tasks and projects. Unplanned events will be interpreted as part of normal activities rather than as a disturbance or deviation from a plan or work schedule (Arndt et al. 2006). For example, some individuals may be able to divert their attention from one task to another within a period of time. Therefore, the concept of multi-tasking is conceptually different from the concept of task shifting. In the context of hospitality, the task-shifting behavior is appropriate because it makes sense to assume that employees can engage in various activities by moving back and forth between multiple jobs within the same time frame. In this study, we use task shifting as a polychronic concept (Jang & George 2012).

Employees with higher levels of polychronicity may find it more comfortable to work in the hospitality industry where employees are expected to engage with multiple tasks within a short span of time and simultaneity to move back and forth in order to meet customer demands that are consistent between the individual personality and the surrounding environment (Jang & George 2012). Polycronic employees have a high tendency to display extra role behaviors and experience greater job satisfaction (Daskin 2015). Because of their natural habit to focus on many jobs, polychronic employees are involved in their work (Karatepe et al. 2013). Job involvement refers to a positive, satisfying, and work-related state of mind characterized by strength, dedication, and absorption (Schaufeli et al. 2002).

Polychronic as a fundamental strategy, involves consistent choice-whether consciously or unconsciously about how to involve tasks and events (Bluedorn 1999). Each job will involve many activities with qualitatively different skills, a pattern of behavior characterized by a relatively high degree of polychronicity would be more suitable for this type of task. A high level of polychronicity would be more appropriate because some skills will be used as well as many activities performed while doing one job (Bluedorn & Time 2015). People who are highly polychronic may find it difficult to focus on work, which can hamper their ability to complete. Conversely, people with low polychronicity (monokronic) may hold on to one job so they fail to show the flexibility needed to take advantage of opportunities (Bluedorn & Time 2015).

(Cotte & Ratneshwar 1999) found that some people seem to have a relatively natural, stress-free

ability to move back and forth over large distances along the polychronic sequence in context. Positive impact of polychronic orientation is related to job performance and negatively impacts role ambiguity (Fournier et al. 2013). However, this is different from multi-tasking that refers to "situations in which individuals are asked to divert their attention between multiple independent but simultaneous tasks (Mattarelli et al. 2015).

In short, polychron is the quality that results from a person's character but multi-tasking is to obey orders in order to fulfill the different tasks required by the supervisor or working conditions (Arasli et al. 2018). In particular, in the hospitality industry, significant polychronicity for frontline employees as they can shape the form of service experience (Paek et al. 2015) and act as the face of the company. Frontline employees, who lack these skills, have a tendency to provide poor service, hinder effective service processes and service recovery efforts.

1.2 Job satisfaction

Job satisfaction is defined as "a pleasant emotional state resulting from the assessment of one's work as achieving or facilitating the achievement of one's work values" (Locke 1969) Job satisfaction is the center of employee work life and for effective use of personnel within the organization (Koeske et al. 1994) Employee job satisfaction can be predicted by employees' evaluation of work climate, level of organizational support and employment situation (Patah et al. 2009) Employee job satisfaction can be judged by how far they feel happy, happy, and appreciated, or unhappy, unhappy, and exploited, satisfied employees in their work will create a good image and strive to provide better service (Malhotra & Mukherjee 2004).

When an employee is satisfied in the workplace, he tends to be more stable, productive and achievable toward organizational goals (Jessen 2010). Job satisfaction is defined as the overall affective orientation of the individual part of the occupational role they currently occupy (Kalleberg 1977). Job satisfaction as a function of a certain range of satisfaction and dissatisfaction resulting from the assessment of the various dimensions of the work he or she is experiencing. Assessment of the various dimensions including the work itself, supervision, payments, promotional policies and co-workers (Efraty & Sirgy 1990). When employees express their feelings about their work, whether positive or negative, they tend to refer to their job satisfaction (Pool 1997). Thus, job satisfaction implies subjective and emotional re-actions to various aspects of work, which are perceived as emotional states resulting from the assessment of one's situation, related to the characteristics and demands of one's job (Jessen 2010).

In the literature of organizational behavior, job satisfaction has been shown to have a positive relation-ship with many variables. Hospitality studies on job satisfaction can be shared with the types of respondents and measurements used in the study, whether general or specific job satisfaction (Chuang et al. 2009). For example, (Hancer & George 2003) explored job satisfaction rates with a sample of 798 restaurant employees, using the short version of Minnesota Satisfaction Questionnaire (MSQ) and they found that payday satisfaction showed the lowest average job satisfaction score, while job security was rated highest job satisfaction

Polychronicity refers to the extent to which people prefer to engage in two or more occupations or events simultaneously and believe that their behavior is the best way to do something (Bluedorn et al. 1999). Polycronic individuals prefer to be involved in some tasks over a period of time, while mono-chronic individuals prefer to complete one task at a time before taking on another task (Conte & Gintoft 2005). Only a few studies have attempted to investigate the relationship between polychronicity and job satisfaction. (Arndt et al. 2006) conducted an empirical study using 313 employees to examine the relationship between polychronicity and job satisfaction. They found that front-line employees with polychronic orientation had a direct and positive effect in predicting job satisfaction. In particular, employees in service organizations such as hotels not only constantly face uncertain and unexpected situations such as angry customers and special requests (Bitner et al. 1990), but are also expected to work faster in a limited time. Similarly, (Jang & George 2012) reported a positive relationship between polychronic and job satisfaction for a sample of hotel employees. Here is a conceptual model of research (Figure 1)

2 METHOD

Research respondents consist of 5-star hotel employees in Bandung, Indonesia. Hotel employees are required to be able to handle the various needs and complaints of hotel guests (Karatepe et al. 2006). Based on information from the Tourism Department there are as many as nine five star hotels in Bandung, Indonesia. Of the 100 questionnaires distributed, only 64 were returned. Questionnaire is divided into 3 parts, namely the demographic part as the control

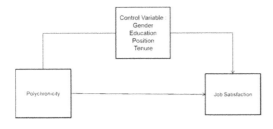

Figure 1. Conceptual model of job satisfaction.

variable, polychronicity and employee job satisfaction. Part A consists of control variables such as job positions, gender, employment, and recent employee education. Part B consists of 8 items used to measure polychronicity and part C consists of 10 items used to measure employee job satisfaction. This study was measured using SPSS to analyze data from questionnaires.

3 RESULTS AND DISCUSSION

Respondents involved in this study were 64 people. Based on the results of the data, there are 58 respondents (90.6%) work as hotel staff, while six other respondents (9.4%) served as hotel managers. Based on sex, the majority of the questionnaires were women as many as 36 respondents (56.3%) and the rest were men as many as 28 respondents (43.8%). Based on the period of work, there are 53 respondents (82.8%) whose working period is below five years and 11 respondents (17.2%) whose working period is between 5-10 years. Based on education level, 37 respondents (57.8%) are graduates of SMA/SMK, 16 respondents (25%) diploma graduates, 9 respondents (14.1%) are graduates of bachelor degree and the rest are 2 respondents (3.1%) is a graduate of Graduate.

Based on Table 1 the greatest influence on Polychronicity variables is item number six, that they believe and are able to do a good job even though they are facing some work. While for the lowest effect is item statement number five, that they prefer to do one job at a time.

The implication of this experimental study resulted in the finding that polychronicity was able to improve employee job satisfaction. Factors in the dominant polychronicity are formed regarding the ability of employees to do a lot of work, work at

Tabel 1. Polychronicity.

	Extraction
1. We like to juggle several activities at the same time	0.414
2. We would rather complete an entire project every day than complete parts of several projects	0.581
3. We believe people should try to do many things at once	0.616
4. When we work by ourselves, we usually work on one project at a time	0.417
5. We prefer to do one thing at a time	0.142
6. We believe people do their best work when they have many tasks to complete	0.723
7. We believe it is best to complete one task before beginning another	0.608
8. We believe it is best for people to be given several tasks and assignments to perform	0.565

once and able to complete the task before starting another task. Another finding is that the level of office affects polychronicity behavior, as well as the gender of the employee. This research is useful for the service industry that spans the multi-tasking job, but still prioritizes excellent service.

4 CONCLUSION

The study found that polychronicity can improve employee job satisfaction. The dominant factors formed in polychronicity are related to the ability of employees to do many tasks, multitask, and complete a task before starting another task. Another finding is that the level of employees' position affects polychronic behavior, so does their gender. This research is beneficial for the service industry identical with multitasking but still priori-tizing excellent services.

REFERENCES

Adams, S.J. & Eerde, W.V. 2012. Polychronicity in modern Madrid: An interview study. *Time & Society*, 21(2): 175–202.

Araslı, H., Daşkın, M. & Saydam, S. 2014. Polychronicity and intrinsic motivation as dispositional determinants on hotel frontline employees' job satisfaction: do control variables make a difference?. *Procedia-Social and Behavioral Sciences*, 109: 1395-1405.

Arasli, H., Hejraty, B. & Abubakar, A.M. 2018. Workplace incivility as a moderator of the relationships between polychronicity and job outcomes. *International Journal of Contemporary Hospitality Management*, 30(3): 1245-1272.

Arndt, A., Arnold, T.J. & Landry, T.D. 2006. The effects of polychronic-orientation upon retail employee satisfaction and turnover. *Journal of Retailing*, 82(4): 319–330.

Bitner, M.J., Booms, B. H. & Tetreault, M.S. (1990). The Service Encounter: Diagnosing Favorable and Unfavorable Incidents. *Journal of Marketing*, 54(1): 71.

Bluedorn, A.C. 1998. An interview with anthropologist Edward T. Hall. *Journal of Management Inquiry*, 7(2): 109-115.

Bluedorn, A.C., Kalliath, T.J., Strube, M.J. & Martin, G. D. (1999). Polychronicity and the Inventory of Polychronic Values (IPV): The development of an instrument to measure a fundamental dimension of organizational culture. *Journal of Managerial Psychology*, 14(3/4): 205–231.

Chuang, N., Yin, D. & Dellmann-Jenkins, M. 2009. Intrinsic and extrinsic factors impacting casino hotel chefs' job satisfaction. *International Journal of Contemporary Hospitality Management*, 21(3): 323–340.

Conte, J.M. & Gintoft, J.N. 2005. Polychronicity, Big Five Personality Dimensions, and Sales Performance. *Human Performance*, 18(4): 405–407.

Cotte, J. & Ratneshwar, S. 1999. Juggling and hopping: what does it mean to work polychronically?. *Journal of Managerial Psychology*, 14(3/4): 184–205.

Daskin, M. 2015. Antecedents of extra-role customer service behavior: polychronicity as a moderator. *Anatolia*, 26(4): 521–534.

Efraty, D. & Sirgy, M.J. 1990. The effects of quality of working life (QWL) on employee behavioral responses. *Social Indicators Research*, 22(1): 31–47.

Fournier, C., Weeks, W.A., Blocker, C.P. & Chonko, L.B. 2013. Polychronicity and Scheduling's Role in Reducing Role Stress and Enhancing Sales Performance. *Journal of Personal Selling & Sales Management*, 33(2): 197–209.

Hancer, M. & George, R.T. 2003. Job Satisfaction of Restaurant Employees: An Empirical Investigation Using the Minnesota Satisfaction Questionnaire. *Journal of Hospitality and Tourism Research*, 27(1): 85–100.

Jang, J. & George, R.T. 2012. Understanding the influence of polychronicity on job satisfaction and turnover intention : A study of non-supervisory hotel employees. *International Journal of Hospitality Management*, 31(2): 588–595.

Jessen, T.J. 2010. Job satisfaction and social rewards in the social services. *Journal of Comparative Social Work*, 1: 1-18.

Kalleberg, A.L. 1977. Work values and job rewards: A theory of job satisfaction. *American sociological review*, 42(1): 124-143.

Kantrowitz, T.M., Grelle, D.M., Beaty, J.C. & Wolf, M.B. 2012. Time Is Money: Polychronicity as a Predictor of Performance Across Job Levels. *Human Performance*, 25(2): 114–137.

Karatepe, O.M. & Kilic, H. 2007. Relationships of supervisor support and conflicts in the work-family interface with the selected job outcomes of frontline employees. *Tourism Management*, 28(1): 238–252.

Karatepe, O.M., Karadas, G., Azar, A.K. & Naderiadib, N. 2013. Does Work Engagement Mediate the Effect of Polychronicity on Performance Outcomes? A Study in the Hospitality Industry in Northern Cyprus. *Journal of Human Resources in Hospitality and Tourism*, 12(1): 52–70.

Karatepe, O.M., Uludag, O., Menevis, I., Hadzimehmedagic, L. & Baddar, L. 2006. The effects of selected individual characteristics on frontline employee performance and job satisfaction. *Tourism Management*, 27(4): 547-560.

Kayaalp, A. 2014. The octopus approach in time management: Polychronicity and creativity. *Military Psychology*, 26(2): 67–76.

Koeske, G.F., Kirk, S.A., Koeske, R.D. & Rauktis, M.B. 1994. Measuring the Monday blues: Validation of a job satisfaction scale for the human services. *Social Work Research*, 18(1): 27–35.

König, C.J. & Waller, M.J. 2010. Time for reflection: A critical examination of polychronicity. *Human Performance*, 23(2): 173–190.

Locke, E.A. 1969. What is job satisfaction?. *Organizational Behavior and Human Performance*, 4(4): 309–336.

Malhotra, N. & Mukherjee, A. 2004. The relative influence of organisational commitment and job satisfaction on service quality of customer-contact employees in banking call centres. *Journal of Services Marketing*, 18(3): 162–174.

Mattarelli, E., Bertolotti, F. & Incerti, V. 2015. The interplay between organizational polychronicity, multitasking behaviors and organizational identification: A mixed-methods study in knowledge intensive organizations. *International Journal of Human Computer Studies*, 79: 6–19.

Paek, S., Schuckert, M., Kim, T.T. & Lee, G. 2015. Why is hospitality employees' psychological capital important? The effects of psychological capital on work engagement and employee morale. *International Journal of Hospitality Management*, 50: 9–26.

Patah, M., Zain, R., Abdullah, D. & Radzi, S. 2009. An Empirical Investigation into the Influences of Psychological Empowerment and Overall Job Satisfaction on Employee. *Journal of Tourism, Hospitality & Culinary Arts*, 1(3): 43–62.

Persing, D.L., Supe, Â. & Recherche, E.S.C. 1998. Managing in polychronic times venues. *Journal of Managerial Psychology*, 14(5): 358–373.

Pool, S.W. 1997. The Relationship of Job Satisfaction with Substitutes of Leadership, Leadership Behavior, and Work Motivation. *The Journal of Psychology*, 131(3): 271–283.

Schaufeli, W.B., Salanova, M., González-Romá, V.A. & Bakker, A.B. 2002. The Measurement of Engagement and Burnout: a Two Sample Confirmatory Factor Analytic Approach. *Journal of Happiness Studies*, 3(1): 71–92.

Slocombe, T.E. 1999. Applying the theory of reasoned action to the analysis of an individual's polychronicity. *Journal of Managerial Psychology*, 14(3/4): 313–324.

Advances in Business, Management and Entrepreneurship – Hurriyati et al (eds)
© 2020 Taylor & Francis Group, London, ISBN 978-0-367-27176-3

Learning culture and technology acceptance as predictors of employee performance in public electrical companies

A.P. Prasetio, B.S. Luturlean & G. Riyadhi
Telkom University, Bandung, Indonesia

ABSTRACT: The aim of this study is to examine the effects of learning culture and technology acceptance on employee performance. Technology is believed to have a fundamental importance in every business. Ever-changing technology, especially in electrical companies, demands constant improvement from employees. Employees need to enhance their knowledge and skills constantly. Organizations must have the ability to become a learning organization. A learning culture needs to be welcomed among employees. Previous studies suggest that a strong learning culture can bring about positive output from individuals. The data in this study was gathered via questionnaire sent to 67 employees from one branch of a public electrical company in Bandung. Data were analysed using multiple regressions and bootstrap confidence interval. The results indicate that a learning culture has a significant positive effect on employee performance and the technology acceptance mediating this relations.

1 INTRODUCTION

The study analysed the impact of the learning culture and technology acceptance on the employee' performance in PLN Majalaya Distribution Office. The benefit of the study is to provide empirical discussion for the management to promote the employee performances by identifying important factors which can be developed to boost the performance. Employee performance an important factors in organizational success. Ferres (2015) explained that better understanding of the human element will help the profitability of the organization. That is why managing employee performance becomes an important goal for all human resources departments. Organizations should evaluate employee achievement and apply proper care when identifying decreased performance. Organization learning culture is one of factors which had effect on performance (Rahman et al. 2015). The role of the learning process in an organization is also believed to have an impact on work performance (Rose et al. 2009). Meanwhile, Imran et al. (2014) found that technology and employee performance had a significant relationship. As the sole provider of electricity for the community, PLN bears a great responsibility. PLN relies on their human resources to provide the service. Therefore, they need to maintain their employee performance. The PLN Majalaya Distribution Office also needs to improve performance. Recent evaluation indicates that they achieve standard performance which, however, is still lower than expectations. The organization needs to know what factors affecting the performance. Implementing new technology in the workplace sometimes has its own challenges. One of them is the possibility of rejection from the employees. One of the reasons to resist is because the future is uncertain and may adversely affect people's competencies, worth and abilities (Agboola & Salawu 2010). If they cannot adjust, they feel that the organization will force them to quit or at least they could be stuck in their career. Whatever reason and no matter how much resistance, organizations need to adapt to new technology if they want to survive.

Previous research from Mutuku and Nyaribo (2015) found the significant positive influence of implementation of technology on employee productivity of the banks. Al-Nashmi and Amer (2014) argued that that the low level of information technology adoption in non-government entities, organizations, and societies are strongly related to the low level of employee productivity in these government entities. The effort to encourage employees to adopt and use technology is aimed to enhance the effectiveness of human resources and reinforce corporate performance in general. Fujino and Kawamoto (2013) revealed that employee performance improved after using computers to communicate and interact. This result is supported by Dauda and Akingbade (2011). Other findings regarding the relationship between technology acceptance and employee performance came from Pakistan and Malaysia (Imran et al. 2014, Abbas et al. 2014, Hasan & Nadzar 2014). They reckoned that there was a positive link between technological advancement and changes on employee performance. Employees experienced an increase in their efficiency and productivity which in the end affect their overall work performance and the competitiveness of the organization. Therefore, our first hypothesis, H1, is that technology acceptance will have a significant positive effect on employee performance.

The importance of OLC for a company's survival and effective performance has been emphasized in the literature (Zahay & Handfield 2004). Many studies focus on the effect of learning culture (learning organization) on organizational performance (Binuyo & Adewale 2014, Kontoghiorghes et al. 2005, Yang et al. 2004, Yu & Chen 2015). In the following discussion, we present previous studies which analysed the relation between learning culture and individual performance. According to Joo Song et al. (2009), organizational learning culture plays an important role as an antecedent for many dependent variables of HRD: learning, performance, satisfaction, change, creativity, productivity, and effectiveness. Most studies regarding learning culture and job performance resulted support the positive relation. Cerne et al. (2012) found a very strong positive relationship between organizational learning culture and innovative culture, while study from St (Maria & Watkins 2005) reported that the dimensions of the learning organization explained 31.5 percent of the variance in use of the innovation. Innovativeness is one of the ingredients for individuals to perform better in their jobs. Rahman et al. (2015) revealed that organizational learning culture helps in promoting the career planning and career management skills of the employee which further promotes their career development. These findings are supported by several scholars (Dekoulou & Trivellas 2014, Gorelick & Tantawi-Monsou 2005, Ali 2012, Abassi & Zamani-Miandashti 2013) which found that learning organizational culture improves performance. Therefore, our second hypothesis, H2, is that learning culture will have a significant positive effect on employee performance.

Next, we discuss the relationship of two variables which are relatively rare. This notion came follows Tuan (2010), who proposed the model of relation between organizational culture and technological innovation. Tuan used continuous learning culture which part of the innovation stimulator. Shameem (2016) proposed a model to explain the connection between micro culture (school culture) and teachers' attitudes toward technology. He examined the role of culture in shaping the mindset of individuals toward something. We explore various sources in order to find articles which discuss and measure the relation of learning and technology acceptance but not much to be had. It seems such study is still rare and difficult to obtain. However, we found several studies which connect those variables. In the first and most direct one, Reardon (2010) explicitly expresses that learning culture has a strong positive relationship on the incorporation of technological innovations into day-to-day work activities. The second study, by Palis (2006), reported that the success of the Farmer Field School associated with cultural norms that en-courage experiential and collective learning which in the end lead to the adoption of integrated pest management

(agricultural technology used to reduce pesticide use) methods among the farmers. Both studies propose a positive relationship between culture (of learning) and technology acceptance. Dasgupta and Gupta (2017) also discovered that organizational culture has an impact on individual acceptance and use of technology. Based on limited literature in the study of learning culture and technology acceptance, our third hypothesis, H3, is that learning culture will have a significant positive effect on technology acceptance.

In examining the mediation role of technology acceptance in the relation of learning culture and employee performance, we have our fourth hypothesis, H4, that technology acceptance will have a significant positive mediation in the relation of learning culture and employee performance.

The contribution of this study is the complete analysis of the relation among learning culture, technology acceptance, and performance. Currently, study regarding those variables even though important, was not widely conducted.

2 METHOD

2.1 Participants

For the purposes of data collection, we used non-probability sampling consist of all employees from the PLN Majalaya Distribution Office (67 employees). The authors distributed the questionnaire to all employees. Detailed information regarding the demographic aspects of this sample is presented in Table 1.

2.2 Data analysis

Bootstrapping confidence interval using the PROCESS Macro for SPSS 23 was used to test the significance of mediation (Hayes 2013). This approach has been advocated as appropriate in such cases because it avoids normality assumptions of the sampling distribution through the application of bootstrapping confidence intervals (Preacher et al. 2007). Using macros in SPSS facilitates the easy implementation of bootstrapping. The important thing is that directly show the significance of mediation effects. For the purpose of measurement, technology acceptance is said to have a mediation role if the value in Upper Level and Lower Level Confidence Interval contain no zero value.

2.3 Measurement

Technology acceptance, 17 for learning culture, and 13 for employee performance. All items met the validity requirement as well as reliability. Participants were asked to give their response using 5-point scale ranging from 1 (Strongly Disagree) to 5 (Strongly Agree). Technology acceptance with 22 items was adapted from Davis (1989), learning organization measurement was developed based on Dimension of Learning Organization Questionnaire from Watkins,

Table 1. Demographic aspects.

Gender	Male	47	70%
	Female	20	30%
Age	< 25	5	7%
	21–25	10	15%
	> 25–30	8	12%
	> 30–35	5	7%
	> 35–40	4	6%
	> 40–45	0	0%
Gender	Male	47	70%
	Female	20	30%
	> 45	35	52%
Education Background	High School	29	43%
	Diploma	19	28%
	Bachelor	18	27%
	Post Graduate	1	1%
Marital Status	Married	49	73%
	Single	18	27%
Length of service	< 1 year	2	3%
	1–3 years	10	15%
	> 3–5 years	4	6%
	> 5–10 years	12	18%
	> 10 years	39	58%
Position	Staff	54	81%
	Supervisor	11	16%
	Manager	2	3%

Larasati and Anggoro (2016), and employee performance was assessed with 13 items developed based on the four dimensions of performance (Edison et al. 2016). The Cronbach's Alpha for the each of the variables is .965 for technology acceptance; .945 for job performance; and .882 for learning culture.

3 RESULT AND DISCUSSION

Organizations filled with relatively old employees (over 45 years old) usually will experience obstacles regarding learning and accepting new things. Senior employees often had their own way and quite diffi-cult to change the habit. The data gather from the self-assessment showed that employees regard their organization to have learning culture. Another surprising finding is that employees consider them-selves open to new technology. They also perceived that their performance is quite satisfying. But management was not happy about the way that employees reacted to new technology implementation. They felt that employees have no motivation to learn which in the end affect their willingness to accept new technology. These differences may be caused by the different points of view between employees and management. To overcome these problems, both parties should conduct routine discussions regarding the process of learning and implementing technology.

Process macro for SPSS was used to test the significance of conditional indirect effects (Hayes 2013). This method is been considered to be an appropriate approach because it avoids the normality assumptions (Preacher et al. 2007) and has become popular as it is widely used in other articles (Prasetio et al. 2017, Gu et al. 2016). Table 2 will help us to answer the hypotheses. The relationship between technology acceptance and employee performance reveals the significant and positive relationship, just as predicted in H1 (0.534, p < 0.000). Thus, the result confirm and support previous literature (Mankins 2016, Mutuku & Nyaribo 2015, Al-Nashmi & Amer 2014, Vogiatzi 2015, Fujino & Kawamoto 2013, Dauda & Akingbade 2011, Imran et al. 2014, Abbas et al. 2014, Hasan & Nadzar 2014). Those studies showed significant and positive relationship between information technology acceptance with employee performance. Technology can do a lot to help organizations to achieve their goals. Thus, organizations need to think about the right technology for their business and how to make their employees willing and ready to use this.

Next, we present the result for the relation of learning culture and employee performance that a significant and positive relationship (0.483, p < 0.000). This result supported H2. In order to face new challenges, organizations need to adapt and learn, organizations need to develop their employees' capabilities. Continuous learning will enhance knowledge and ability. With new knowledge, employees become smarter and more efficient. Previous literature had the same out-come (e.g., Cerne et al. 2012, Sta Maria & Watkins 2003, Rahman et al. 2015, Dekouloua & Trivellas 2014, Ali 2012).

Table 2. Regression coefficient, standard error, and model summary.

	Technology Acceptance			Employee Performance		
	coeff	SE	p-value	coeff	SE	p-value
Learning Culture	0.823	0.123	0.000	0.483	0.105	0.000
Technology Acceptance	-	-	-	0.534	0.082	0.000
Constant	0.877	0.382	0.025	0.143	0.261	0.587
	R Square = 0.408			R Square = 0.730		
	F = 44.780			F = 86.640		
	p =0.000			p =0.000		

Table 2 also presents significant and positive relations between learning culture and technology acceptance (0.823, p < 0.000), which supported H3. Studies regarding the relationship between both variables are still rare. In fact, the previous study did not explicitly use learning culture as a variable. For example, a study from Tuan (2010) explained the relation between organizational culture and technological innovation. Meanwhile, Shameem (2016) discussed the connection between micro culture (school culture) and teachers' attitudes toward technology. Dasgupta and Gupta (2017) used organizational culture as preditors for the technology acceptance. Studies from Reardon (2010) and Palis (2006) explicitly expressed the term of learning culture which had a strong positive relationship with the implementation of technology at work. The results of this study support the significant and positive relation between learning culture and technology acceptance, aligned with previous researches. The finding becomes one of the important contributions from the study because it provides a clear explanation of the relationship between the dynamics of learning culture and technology acceptance. The correlation of 0.823 considers quite high or strong.

Table 3 showed that Lower Level and Upper Level of confidence interval did not contain zero. The result confirms that technology acceptance acts as a mediator between learning culture and employee performance.

The finding will benefit future research because it provides a solid ground for scholars to study learning culture and technology acceptance as per-formance predictors. We have not yet found other studies which performed similar analysis regarding these three variables. Therefore, we cannot compare this result with previous literature.

Figure 1. Displayed the indirect effect model of learning culture on employee relation through technology acceptance.

Table 3. Indirect effect.

	Effect	Boot SE	Boot LLCI	Boot ULCI
Through Technology Acceptance	0.44	0.117	0.230	0.693

Management should take immediate action to improve or develop such culture which accelerate the acceptance of new technology. They must encourage their employees to permeate the learning culture to in-crease their knowledge. Continuous learning should become an important part of an organization's culture. The organization should hire employees who are eager to learn or have certain knowledge and the ability to use technology. The organization also needs to look after current employees. Older people usually takes more time to learn and to adapt. Fortunately, in this organization this is not the case. According to the survey, employees perceived that they did not have difficulties in adopting new technology. Management still needs to conduct assessments regarding this behavior. Once they identify that employees are eager to adopt new things, they can proceed with next strategy on how to put the classroom lessons into practices. A series of training programs can help an organization to ensure that the learning culture will run smoothly and will provide a positive impact on employee and organizational performance.

4 CONCLUSION

We found evidence that if employees consider that they have learning culture, they will become faster in accepting new technology which eventually increase their performance. Furthermore, the study regarding the relationship of learning culture, technology acceptance, and employee performance is still relatively new, and therefore further research is needed to confirm the interconnection between these variables.

ACKNOWLEDGEMENT

The writer would like to express appreciation to LPPM Telkom University, who provided part of the funds to complete this report.

REFERENCES

Abbas, J., Muzaffar, A., Mahmood, H.K., Ramzan, M.A. & Rivzi, S.S.U.H. 2014. Impact of technology on performance of employees (A case study on Allied Bank Ltd, Pakistan). *Sciences Journal* 29(2): 271–276.

Abbasi, E. & Zamani, M.N. 2013. The role of transformational leadership, organizational culture and organizational learning in improving the performance of Iranian agricultural faculties. *High Educ* 66:505–519.

Agboola, A.A. & Salawu, R.O. 2010. Managing deviant behavior and resistance to change. *International Journal of Business and Management* 6(1).

Ali, A.K. 2012. Academic staff's perceptions of characteristics of learning organization in a higher learning institution. *International Journal of Educational Management* 26(1): 55–82.

Al-Nashmi, M.M. & Amer, R.A. 2014. The impact of information technology adoption on employee productivity in nongovernmental organizations in Yemen. *International Journal of Social Sciences and Humanities Research* 3(2): 32–50.

Binuyo, B., Adekunle, A., Adewale, A. & Aregbeshola R. 2014. The impact of information and communication technology (ICT) on commercial bank performance: Evidence from South Africa. *Research Gate*: 1–25.

Cerne, M., Jaklic, M., Skerlavaj, M., Aydinlik, A.U. & Polat, D.D. 2012. Organizational learning culture and innovativeness in Turkish firms. *Journal of Management and Organization* 18(2): 193–219.

Dasgupta, S. & Gupta, B. 2017. Organization culture dimensions as antecedents of internet technology adoption.

Yogesh, K., Dwivedi, D., Helle, Z.H., David, W. & Rahul D. 2013. International working conference on transfer and difusion of IT (TDIT). *IFIP Advances in Information and Communication Technology, AICT-402*: 658–662.

Dauda, Y.A. & Akingbade, W.A. 2011. Technological change and employee performance in selected manufacturing industry in Lagos state of Nigeria. *Australian Journal of Business and Management Research* 1(5): 32–43.

Davis, F.D., Bagozzi, R.P. & Warshaw, P.R. 1989. User acceptance of computer technology: A comparison of two theoretical models. *Management Science* 35(8): 982–1003.

Dekoulou, P. & Trivellas, P. 2014. Measuring the impact of learning organization on job satisfaction and individual performance in greek advertising sector. International conference on strategic innovative marketing, IC-SIM 2014. *Procedia-Social and Behavioral Sciences* 175(2015): 367–375.

Edison, E., Anwar, Y. & Komariyah, I. 2016. *Manajemen sumber daya manusia: Strategi dan perubahan dalam rangka meningkatkan kinerja pegawai dan organisasi (1st* ed). Bandung: CV Alfabeta Bandung.

Ferres, Z. 2015. *The human element: Your most important business resource.*

Fujino, F., Yuriko, Y., Kawamoto, K. & Rieko, R. 2013. Effect of information and communication technology on nursing performance. *Feature Article* 31(5): 244–250.

Gorelick, C. & Tantawy, M.B. 2005. For performance through learning, knowledge management is the critical practice. *Learning Organization* 12(2): 125–39.

Gu, J., Song, J. & Wu, J. 2016. Abusive supervision and employee creativity in China: departmental identification as mediator and face as moderator. *Leadership & Organization Development Journal* 37(8): 1187–1204.

Hasan, H. & Nadzar, F.H.M. 2010. Acceptance of techno-logical changes and job performance among administrative support personnel in the government offices in Maran, pahang darul makmur. *Gading Business and Management Journal* 14.

Hayes, A.F. 2013. *Introduction to mediation, moderation, and conditional process analysis: A regression-based approach*. New York: The Guilford Press.

Imran, M., Maqbool, N. & Shafique, H. 2014. Impact of technological advancement on employee performance in banking sector. *International Journal of Human Resource Studies* 4(1): 57–70.

Kontoghiorghes, C., Awbrey, S.M. & Feurig, P.L. 2005. Ex-amining the relationship between learning organization characteristics and change adaptation, innovation and organizational performance. *Human Resource Development Quarterly* 16: 185–211.

Larasati, K. & Anggoro, Y. 2016. Learning culture assessment and its influence on knowledge creation process (Case Study: PT Pindad (Persero)). *Proceedings on International Seminar and Conference on Learning Organization*. Bandung, Indonesia.

Mankins, M. 2016. Is technology really helping us get more done?.

Mutuku, M.N. & Nyaribo, W.M. 2015. Effect of information technology on employee productivity in selected banks in Kenya. *Review of Contemporary Business Research* 4(1): 49–57.

Palis, F.G. 2006. The role of culture in farmer learning and technology adoption: A case study of farmer field schools among rice farmers in central Luzon, Philippines. *Agriculture and Human Values* 23:491–500.

Prasetio, A.P., Yuniarsih, T. & Ahman, E. 2017. The direct and indirect effect of three dimension of work life inter-face towards organizational citizenship behavior. *Polish Journal of Management Studies* 15(1): 174–184.

Preacher, K.J., Rucker, D.D. & Hayes, A.F. 2007. Addressing moderated mediation hypotheses: theory, methods, and prescriptions. *Multivariate Behavioral Research* 42(1): 185–227.

Rahman, H., Saeed, T., Rahman, W. & Ali, N. 2015. Organi-zational learning culture as the antecedent of employee career and job performance as its outcome. *International Journal of Management Sciences* 5(12): 771–784.

Reardon, R.F. 2010. The impact of learning culture on worker response to new technology. *Journal of Workplace Learn-ing* 22(4): 201–211.

Rose, R.C., Kumar, N. & Pak, O.G. 2009. The effect of organizational learning on organizational commitment, job satisfaction and work performance. *The Journal of Applied Business Research* 25: 55–66.

Shameem, A. 2016. Influence of culture on teachers' attitudes towards technology.

Song, J.H., Joo, B.K. & Chermack, T.J. 2009. The dimensions of learning organization questionnaire (DLOQ): A valida-tion study in a Korean context. *Human Resources Devel-opment Quarterly* 20(1): 43–64.

Maria, R.F. & Watkins, K.E. 2003. Perception of learning culture and concerns about the innovation on its use: a question of level of analysis. *Human Resource Development International* 6(4): 491–508.

Tuan, L.T. 2010. Organizational culture and technological innovation adoption in private hospitals. *International Business Research* 3(3): 144–153.

Yang, B., Watkins, K. & Marsick, V. 2004. The construct of the learning organization: Dimensions, measurement, and validation. *Human Resource Development Quarterly* 15: 31–55.

Yu, T. & Chen, C.C. 2015. The relationship of learning culture, learning method, and organizational performance in the university and college libraries in Taiwan. *LIBRI* 65(1): 1–14.

Zahay, D.L. & Handfield, R.B. 2004. The role of learning and technical capabilities in predicting adoption of B2B technologies. *Industrial Marketing Management* 33: 627–641.

The role of job satisfaction and organizational citizenship behavior in developing knowledge sharing behavior in private universities

A.M. Nurdiaman & G.G. Akbar
Garut University, Garut, Indonesia

A.P. Prasetio, A. Rahmawati & R.P. Yasmin
Telkom University Bandung, Indonesia

ABSTRACT: The current study discusses the effect of job satisfaction on knowledge sharing intention and behavior. It also examines the role of Organizational Citizenship Behavior (OCB) as mediation in the relationship. This study was conducted in private universities in Bandung and Garut by choosing 133 lecturers as participants. The results showed that job satisfaction had a significant positive influence on knowledge sharing. It was also found that OCB had a mediating role in the relationship. It was identified from the Lower Level Confidence Interval (LLCI) and Upper Level Confidence Interval (ULCI) which had positive value and contained no zero. Research that examined the relationship of job satisfaction, OCB, and knowledge sharing is still limited so that this work can contribute in expanding such study in the future. Organizations need employees who have broader insight and knowledge to face future challenges. This can be achieved by implementing policies that encourage employees to share knowledge. In order for employees to share with colleagues, organizations need to improve their level of job satisfaction. Satisfied employees tend to have an intention to produce better results for the company. One of the contributions of satisfied employees is the increase of OCB spirit. This study confirms that the management of a business organization needs to build both elements so that the application of knowledge sharing can be successful.

1 INTRODUCTION

One of the key factors in organization success is the effective use of knowledge resources. In the era of knowledge economy, where everything based on intangible assets, organizations and their managers face enormous challenges to manage these important resources. Knowledge as an intellectual capital can become a competitive advantages to face future competition. Organization needs to create, nurture, enhance, modify, and transfer the knowledge. Knowledge should be made available for all employees. To distribute knowledge, organization need to develop the process of knowledge sharing. People in organization should share their expertise to one another. In doing so, they spread the best practices in the business.

Saeed (2016) found the significant and positive relation of sharing with employee performance. Muqadas et al. (2016) strengthen the finding in their study which found the positive relation of sharing behavior with work performance and employees' creativity.

A study using university senior students in Taiwan revealed that knowledge sharing gave meaning in life for those seniors (Chang et al. 2018). Further study from China reinforced the imprtance of knowledge sharing in supporting firms performance (Law & Ngai 2008). The

result from Allameh et al. (2014) using sport managers in Iran also supported the premise that knowledge sharing positively affected organizational performance. Though it is crucial, still most individual do not willing to share their knowledge. One of the reasons is that they feel such knowledge will help them to remain valuable in an organization. So, instead of enforcing employees to share their knowledge, an organization should emphasize how to motivate them to share. According to Rehman et al. (2011), the change of behavior becomes great challenges for the manager to drive sharing behavior. It becomes essential for managers or leaders to understand what factors influencing employees' willingness to share knowledge.

Previous studies provide several causes of employees' unwillingness to share their knowledge. Lin (2008) found that the characteristics of the organizational structure, organizational culture, and the level of trust and commitment affected knowledge sharing. Meanwhile study in Dubai Police Force (Seba et al. 2012) found organizational structure, leadership, time allocation, and trust could become barriers for knowledge sharing. This mean that if the organizations fail to develop strong culture and good leadership as well as to increase the level of trust, then they cannot implement knowledge sharing practices effectively.

Another finding from Palo and Charles (2016) indicated that individual norms and their attitude toward knowledge sharing influence the sharing behavior. They also found the tendency from employees to weight the benefit and cost of the sharing. Finally, the study from Whiterspoon et al. (2013) found that organization culture and reward system had positive effect on knowledge sharing intention and behavior. They argued that knowledge sharing practices would be easier to be implemented in the collectivism culture than in individual culture.

Some studies find positive relation between job satisfaction and knowledge sharing (Almahamid et al. 2010, Rehman et al. 2011, Suliman & Al-Hoshani 2014, Pruzinsky & Mihalcova 2017, Tarigh & Nezhad 2016, Batenburg 2017, Cheema & Javed 2017). Organizational Citizenship Behavior (OCB) can become an important link in the job satisfaction and knowledge sharing relation. Regarding OCB, previous studies (Foote & Tang 2008, Salehi & Gholtash 2011, Mohammad et al. 2011, Sesen & Basim 2012, Swaminathan & Jawahar 2013, Talachi et al. 2014, Pavalache-Ilie 2014, Qamar 2012 and Prasetio et al. 2017) revealed the positive relation between OCB and job satisfaction.

This study aim to analyze the relation of job satisfaction with knowledge sharing and the mediation of OCB in that relation. The study was conducted in two private universities in Bandung and Garut. Educational institution faced continous challenges to provide better education. Providing education means they have to invest in people as key resources to deliver the content.

2 METHOD

2.1 Participants

We had 133 participants from two private universities in Garut and Bandung. The comparison between male and female employees were quite balance (49% male and 52% female). All participants were lecturers; most of them were held a Master's degree (74%) and 90% already served more than three years in the instituion. This fact shows that both universities met the regulation requirements for lecturer's basic education which was master. Regarding the age distribution, most of participants were 40 years old and over, but young lecturers were strategically scattered. Both institutions had a well plan. They had to prepare young lecturers to learn from the seniors and replaced them when the time came.

2.2 Measurement

Items used in this study developed from Teh and Sun (2012). The measurement of job satisfaction used eight items, OCB four items, and knowledge sharing five items. The survey items were scored on a 6-point Likert scale, which ranged from 1 (strongly disagreed) to 6 (strongly agreed).

2.3 Data analysis

To identify whether there was a mediation effect from OCB in job satisfaction relation with knowledge sharing, bootstrap approach using the PROCESS Macro for SPSS was used. The result directly shows the significance of mediation effects if Upperlevel and Lowerlevel Confidence Interval contained no zero value. The macro facilitated the implementation of the recommended bootstrapping approach and readily assessed the significance of conditional effects at different values of a mediation.

Based on the previous studies, the hypotheses of this study are:

H_1: *Job satisfaction will have significant positive effect on OCB*

H_2: *Job satisfaction will have significant positive effect on knowledge sharing*

H_3: *OCB will have significant positive effect on knowledge sharing*

H_4: *OCB will mediate the relation between job satisfaction and knowledge sharing*

3 RESULTS AND DISCUSSION

The researchers first conducted a correlation in SPSS, then performed multiple regression analyses to test the expected relationships as presented in hypotheses 1, 2 and 3. For hypothesis 4, a mediation analysis based on Hayes (2013) was performed. Table 1 shows the result.

Correlation value between knowledge sharing and job satisfaction was 0.681 and considered as moderate positive. Meanwhile, the correlation between knowledge sharing and OCB (0.494) and job satisfaction and OCB (0.424) were considered weak positive.

Table 2 revealed that job satisfaction had significant positive effect on both OCB and knowledge sharing (p-value 0.000). Meanwhile, OCB also had significant positive effect on knowledge sharing (p-value 0.000). Thus, these results supported the hypothesis H_1, H_2, and H_3. Simultaneously and partially, job satisfaction and OCB can predict the level of employees to share knowledge in both

Table 1. Mean, standard deviation, and correlation.

	Mean	Std. Deviation	sharing	Statis faction	OCB
Sharing	4.5850	0.60132			
Statisfaction	4.5893	0.5433	0.681		
OCB	4.3158	0.64762	0.494	0.424	

Source: procced data 2018.

Table 2. Regression coefficient, standard error, and model summary.

Coeff	OCB SE		p-value	Knowledge Sharing		
				Coeff	SE	p-value
Job statisfaction	0.503	0.094	0.000		0.074	0.000
OCB	-	-	-		0.063	0.000
constant	2.006	0,433	0.000		0,335	0.000
	0.18				0.515	
	F=28.779				F=68.97	
	P=0.000				P=0.000	

Source: procced data 2018.

organizations. Lecturer is considered as a person who always have to enhance and improve their knowledge. In order to do so, they can do it individually or in group. Sharing session in educational institution is a routine activities. In fact, one of the performance measurement for lecturer is to participate in a conference and write a scientific papers. Both is routine activities since at least once a year they had to do it. The level of job satisfaction and OCB will have impact on the willingness to share. If they perceived organization treat them well (providing funds for research, send them to participate in learning, build positive work environment, and provide adequate financial rewards) then lecturers felt satisfied. Satisfy people according to the theory of social exchange will act positive to their organization. Social exchange theory argues that if an individual treated well, they will reciprocate to the other party. Satisfy lecturer will develop higher OCB which in the end increase their intention and behavior to share knowledge. Table 3 shows the mediation analysis result. OCB mediated the relation between job satisfaction and knowledge sharing. The Lower Level Confidence Interval (LLCI) and Upper Level Confidence Interval (ULCI) were both positive, which can be explained as the increase in satisfaction will affect OCB and then gave impact on lecturer's intention and behavior to share knowledge. Thus, hypothesis H_4 was also accepted.

After identifying the mediation of OCB in the relation of job satisfaction and knowledge sharing, the model of mediation is presented in Figure 1. It shows that job satisfaction had direct positive effect on OCB, OCB also had direct positive effect on knowledge sharing, and job satisfaction had direct positive effect on knowledge sharing.

The present study examined the effect of job satisfaction on knowledge sharing intention and behavior. The focus was on exploring the mediating role of OCB commitment in connecting job satisfaction and knowledge sharing. It was found that if employees felt satisfy with their job, they were more willing to demonstrate OCB toward their organizations, which eventually increasing their intentions and behavior to share knowledge to others.

This study corroborates previous literature that discussed about the relation of job satisfaction and OCB on knowledge sharing and also the relation of job satisfaction and OCB. The positive relation between job satisfaction and OCB support the previous literatures from Malaysia (Mohammad et al. 2011), India (Swaminathan & Jawahar 2013), Iran (Zeinabadi 2010); (Salehi & Gholtash 2011), and Turkey (Sesen & Basim 2012). They also used participants from educational intitution. Study from other industries such as banking, mining, manufacturing, hotel, and public sector also found the same positive relation (Foote & Tang 2008, Qamar 2012, Intraprasong et al. 2012, Talachi et al. 2014, Pavalache-Ilie 2014, Prasetio et al. 2017). The result is also in line with previous studies regarding the relation of job satisfaction and knowledge sharing. Cheema & Javed (2017) in Pakistan, Batenburg (2017) in Holland, Rehman et al. (2011), and Almahamid et al. (2010) in Jordan Suliman & Al-Hoshani (2014) in UAE, Husain & Husain (2016) in Indonesia, Pruzinsky & Mihalcova (2017) in Slovenia, Tarigh & Nezhad (2016) in Iran found positive relation between both variables. They were using participants

Table 3. Indirect effect.

	Effect	Boot SE	BootLLCI	BootULCI
Through OCB	0.117	0.042	0.049	0.214

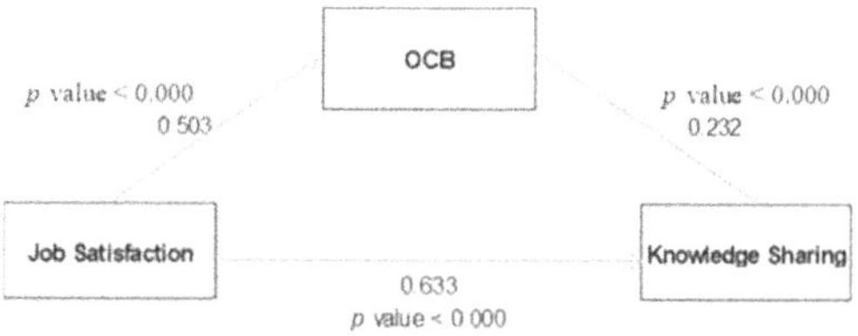

Figure 1. The mediation model.

from various industries, education, health care, oil, public sector, and manufacturing. The positive relation between OCB and knowledge sharing obtained from this study also supports previous literatures from Murtaza et al. (2014) in Pakistan, Ramasamy & Thamaraiselvan (2011) in India, Dehghani et al. (2015) in Iran, Lin (2008) in Taiwan. They conducted studies using participants from educational institutions. Other studies from other industries (hotel, IT, pharmacy, public sector, food service) also showed similar results (Al-Zu'bi2011, Jo & Joo 2011, Hsien et al. 2014, Shadeg, 2015, Teh & Sun 2012, Sivasakthi & Selvarani, 2016, Tuan, 2016, Husain & Husain, 2016). The current study addressing the gap in the organizational research by identifying that OCB can be more effective in enhancing the knowledge sharing intention and behavior when it is combined with the job satisfaction.

4 CONCLUSION

The current study aims to analyze the effect of job satisfaction on knowledge sharing and the mediation of OCB in the relation. There are four hypotheses and findings show that all are accepted. Hypotheses H1, job satisfaction have significant positive effect on OCB; Hypotheses H2, job satisfaction have significant positive effect on knowledge sharing; Hypotheses H3, OCB have significant positive effect on knowledge sharing; and Hypotheses H4, OCB mediate the relation between job satisfaction and knowledge sharing. Facing growing competition, organizations should develop competitive advantages that are hard to beat.

REFERENCES

Allameh, S.M., Pool, J.K., Jaberi, A., & Soveini, F.M. 2014. Developing a Model for Examining the Effect of Tacit and Explicit Knowledge Sharing on Organizational Performance Based on EFQM Approach. *Journal of Science & Technology Policy Management*, 5 (3), 265–280.

Almahamid, S., McAdams, A. & Kalaldeh, T. 2010. The Relationships among Organizational Knowledge Sharing Practices, Employees' Learning Commitments, Employees' Adaptability and Employees' Job Satisfaction: An Empirical Investigation of the Listed Manufacturing Companies in Jordan. *Interdisciplinary Journal of Information, Knowledge, and Management*, Vol. 5, pp. 327–356.

Al-Zu'bi, H.A. 2011. Organizational Citizenship Behavior and Impacts on Knowledge Sharing: An Empirical Study. *International Business Research*, Vol. 4, No. 3, 221–227.

Batenburg, A. 2017. Healthcare Workers Sharing Knowledge Online: Intrinsic Motivations and Well-Being Consequences of Participating in Social Technologies at Work. *Communication Management Review*, 2, 104–124.

Chang, I.C., Chang, C.H., Lian, J.W., & Wang, M.W. 2018. Antecedents and Consequences of Social Networking Site Knowledge Sharing by Seniors: *A social capital perspective*. *Library Hi Tech*.

Cheema, S. & Javed, F. 2017. Predictors of Knowledge Sharing in the Pakistani Educational Sector: A Moderated Mediation Study. *Cogent Business & Management*, 4, 1–16.

Dehghani, M.R., Hayat, A.A., Kojuri, J., & Esmi, K. 2015. Role of Organizational Citizenship Behavior in Promoting Knowledge Sharing. *Journal of Health Management & Informatics*, Vol. 2, Issue 4, 126–131.

Foote, D.A. dan Tang, T.L. 2008. Job Satisfaction and Organizational Citizenship Behavior (OCB) Does Team Commitment Make a Difference in Self-Directed Teams? *Management Decision*, Vol. 46 No. 6, 933–947.

Hayes, A.F. 2013. *Introduction to Mediation, Moderation, and Conditional Process Analysis: A Regression-Based Approach*. New York: The Gilford Press.

Hsien, L.C., Pei, N.F., Yung, P.C., & Sheng, C.T. 2014. A Study on the Correlations between Knowledge Sharing Behavior and Organizational Citizenship Behavior in Catering Industry: The Viewpoint of Theory of Planned Behavior. *The Anthropologist*, 17:3, 873–881.

Husain, S.N. & Husain, Y.S. 2016. Mediating Effect of OCB on Relationship between Job Attitudes and Knowledge Sharing Behavior. *International Journal of Science and Research*, Volume 5 Issue 1, 1008–1015.

Intaraprasong, B., Dityen, W., & Krugkrunjit, P.T. 2012. Job Satisfaction and Organizational Citizenship Behavior of Personnel at One University Hospital in Thailand. *Journal Med Assoc Thai*, 95 (6), 102–108.

Jo, S.J. & Joo, B.K. 2011. Knowledge Sharing: The Influences of Learning Organization Culture, Organizational Commitment, and Organizational Citizenship Behaviors. *Journal of Leadership & Organizational Studies*, 18(3) 353–364.

Law, C.C.H. & Ngai, E.W.T. 2008. An Empirical Study of the Effects of Knowledge Sharing and Learning Behaviors on Firm Performance. *Expert Systems with Applications*, 34, 2342–2349.

Lin, C.P. 2008. Clarifying the Relationship Between Organizational Citizenship Behaviors, Gender, and Knowledge Sharing in Workplace Organizations in Taiwan. *J Bus Psychol*, 22:241–250.

Lin, W.B. 2008. The effect of knowledge sharing model. *Expert Systems with Applications*, 34, 1508–1521.

Mohammad, J., Habib, Farzana Q. dan Alias, M.A. 2011. Job Satisfaction and Organization Citizenship Behavior: An Empirical Study at Higher Learning Institutions. *Asian Academy of Management Journal*, Vol. 16, No. 2, 149–165.

Muqadas, F., Ilyas, M., Aslam, U., & Rehman, U.U. 2016. Antecedents of Knowledge Sharing and Its Impact on Employee's Creativity and Work Performance. *Pakistan Business Review*, October 2016, 655–674.

Murtaza, G., Abbas, M., Raja, Roques, O., Khalid, A., & Mushtaq, R. 2014. Impact of Islamic Work Ethics on Organizational Citizenship Behaviors and Knowledge-Sharing Behaviors. *Journal of Business Ethic*, DOI 10.1007/s10551–014–2396–0.

Palo, S. & Charles, L. 2016. Investigating Factors Affecting Knowledge Sharing Intention of Salespeople. *Management and Labour Studies*, 40 (3&4), 1–23.

Pavalache-Ilie, M. 2014. Organizational Citizenship Behaviour, Work Satisfaction and Employees' Personality. *Procedia—Social and Behavioral Sciences* 127, 489–493.

Prasetio, A.P., Yuniarsih, T., Ahman, E. 2017. Perceived Work-Life Interface and Organizational Citizenship Behavior: Are The Job Satisfaction & Organizational Commitment Mediates The Relations? (Study on Stars Hotel Employees in Indonesia). *International Journal of Human Resource Studies*, Vol. 7, No. 2, 122–135.

Pruzinsky, M. & Mihalcova, B. 2017. Employee Satisfaction and Knowledge Management. *12th International Workshop on Knowledge Management*, 12–13 October 2017, Trenčín, Slovakia.

Qamar, N. 2012. Job Satisfaction and Organizational Commitment as Antecedents of Organizational Citizenship Behavior (OCB). *Interdisciplinary Journal of Contemporary Research in Business*. Vol. 4 No. 7, 103–122.

Ramasamy, M. & Thamaraiselvan, N. 2011. Knowledge Sharing and Organizational Citizenship Behavior. *Knowledge and Process Management*, Vol. 18 No. 4, 278–284.

Rehman, M., Mahmood, A.K.B., Salleh, R., & Amin, A. 2011. Review of Factors Affecting Knowledge Sharing Behavior. *2010 International Conference on E-business, Management and Economics*, Hongkong.

Sadegh, T. 2015. Introducing a Model of Relationship between Knowledge Sharing Behavior, OCB, Psychological Empowerment and Psychological Capital: A Two-Wave Study. *American Journal of Applied Psychology*, 4(4): 95–104.

Saeed, M.S. (2016). The Impact of Job Satisfaction and Knowledge Sharing on Employee Performance. *Journal of Resources Development and Management*, Vol. 21, 16–23.

Seba, I., Rowley, J. & Delbridge, R. 2012. Knowledge sharing in the Dubai police force. *Journal of Knowledge Management*, Vol. 16, No. 1, pp. 114–128.

Sesen, H., & Basim, N.H. 2012. Impact of Satisfaction and Commitment on Teachers' Organizational Citizenship. *Educational Psychology: An International Journal of Experimental Educational Psychology*. Vol. 32, No. 4, 475–491.

Sivasakthi, K. & Selvarani, A. 2016. Mediating Effect of OCB on Relationship Between Job Attitude and Knowledge Sharing Behavior in Add Soft Technologies, Bangalore. *International Journal of Management*, Vol. 7, Issue 2, 437–442.

Suliman, A. & Al-Hosani, A.A. 2014. Job Satisfaction and Knowledge Sharing: The Case of the UAE. *Issues in Business Management and Economics*, Vol. 2 (2), 024–033.

Swaminathan, S. & Jawahar, P. D. (2013). Job Satisfaction as a Predictor of Organizational Citizenship Behavior: An Empirical Study. *Global Journal of Business Research*, Vol. 7, No. 1, 71–80.

Talachi, R.K., et al. 2014. The Role of Job Satisfaction in Employees' OCB. *Coll. Antropol*, Vol. 38 No. 2, 429–436.

Tarigh, M.P. & Nezhad, F.R. 2016. The Impact of Job Satisfaction and Work Environment Friendly on Services Innovation with Mediator Role of Knowledge Sharing in Companies under Supervision of Parstousheh in Gilan Province. *Bulletin de la Société Royale des Sciences de Liège*, Vol. 85, 2016, pp. 1017–1025.

Teh, P.L. & Sun, H.Y. 2012. Knowledge Sharing, Job Attitudes and Organisational Citizenship Behaviour. *Industrial Management & Data Systems*. Vol. 112, No. 1, pp. 64–82.

Tuan, L.T. 2016. Knowledge Sharing in Public Organizations: The Roles of Servant Leadership and Organizational Citizenship Behavior. *International Journal of Public Administration*.

Witherspoon, C.L., Bergner, J., Cockrell, C., & Stone, D.N. 2013. Antecedents of Organizational Knowledge Sharing: A Meta-Analysis and Critique. *Journal of Knowledge Management*, Vol. 17 Issue 2, pp. 250–277.

Zeinabadi, H. 2010. Job Satisfaction and Organizational Commitment as Antecedents of Organizational Citizenship Behavior (OCB) of Teachers. *Procedia Social and Behavioral Sciences*. Vol. 5, 998–1003.

Advances in Business, Management and Entrepreneurship – Hurriyati et al (eds)
© 2020 Taylor & Francis Group, London, ISBN 978-0-367-27176-3

The driving factors of *wakif*'s intention to pay *waqh* with cash at Bantul Regency Yogyakarta

K.C. Kirana, S. Hermuningsih & R. Widiastuti
Universitas Sarjanawiyata Tamansiswa, Yogyakarta, Indonesia

ABSTRACT: The purpose of this study is to investigate the driving factors of Wakif's intention to pay *waqh* with cash at Bantul Regency Yogyakarta. Theory of planned behavior was used with quantitative method. Three variables were tested as driving factors of Wakif's intention to their attitude and intention to pay *waqh* with cash. The variables were subjective norms, government support, and self-efficacy. The object of research are people who is going to pay *waqh*. People that are members of Bazwa with 98 respondents, divided into four village are 35 from Banguntapan, 23 from Bantul City, 11 from Dlingo, and 29 from Kasihan. The type of data that have been collected is primary and secondary data. The methods of data analysis that were used in this research included validity test, reliability test and path analysis. The results of this study indicate that driving factors of intention to *Wakif* are subjective norms, government support and self-efficacy. The result showed that the effect of subjective norms toward intention is attitude. Subjective norms positive significant to attitude. subjective not significant intention, Self-Efficacy positive significant to Attitude, Self-Efficacy not significant to Intention, Government Support not significant to Attitude and Intention, and Attitude significant to Intention.

1 INTRODUCTION

Bantul Regency is part of the province of Special Region of Yogyakarta, with a Muslim majority population. Unstable and barren land conditions make Bantul an area with a high rate of poverty. Poverty coupled with lack of education has led the Bantul people to work as in low-skill jobs in other cities. Poverty and unemployment have made the working-age population wander in big cities, and not infrequently there are desperate to be beggars and homeless. Some Bantul people have been economically successful, but many have failed. This situation must of course be a serious concern not only for the government, but also for intellectuals and public figures. More efforts are needed to overcome poverty in various regions of Indonesia, and in particular in Bantul Regency. In connection with efforts to alleviate poverty in Bantul, the government has actually made many programs. Whether set in the policy of the Central Government, through the Regional Government, or in cooperation with the private sector. However, the problem of poverty still cannot be solved completely. This is indicated by DIY statistics showing that Bantul ranks second in DIY as the district with the poorest population (BPS 2016).

So far, Indonesians know *waqf* in the form of land or other immovable objects that the utilization is limited only for the construction of houses of worship or orphanages and boarding schools. Along with the many studies on Islamic economics in the world, began to recognize the existence of cash

waqf, the meaning of cash *waqf* is the kind of *waqf* that is fulfilled both before being given and after it with the consent of the endowment. Through *waqf*, is expected to occur the process of distributing benefits for the community more broadly, from private benefits to private benefits (Kusrini et al. 2018).

In relation to poverty eradication efforts, the government has made many programs. Whether set in the policy of the Central Government, through the Regional Government, or in cooperation with the private sector. However, the problem of poverty in Indonesia still cannot be solved completely. This is shown by the number of regions that are poor and lag behind other regions. The main problem in the effort to utilize cash *waqf* funds in Bantul is the lack of understanding of the Muslim community regarding cash *waqf*. Therefore, the researchers feel the need to conduct research with a primary focus on knowing factors wakif (people who are going to pay *waqf*), to choose a cash *waqf* as a vehicle of charity. The factors in question will be analyzed through Planet Behavior Theory (TPB) and Planet Action Theory (TPA) following Azjen (2005). In addition, this study also uses other theories about poverty and economic independence following Swasono (2003), so the definition of cash *waqf* based on UU No. 41 Tahun 2004 as for the previous research that refined in this study are, *Waqf* as a Source of Productive Economic Empowerment (Perspective on *Waqf* Benefits Based on Islamic Financial System), GIS-Based Decision Support System for Cash *Waqf* Distribution (Kusrini et al. 2018).

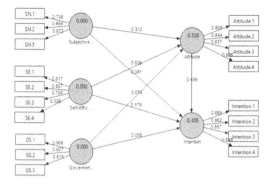

Figure 1. Research method.

2 METHOD

The model that is proposed in this research can be seen in Figure 1.

The methods in this study used surveys and in-depth interviews. This research is quantitative; the data was collected using sampling methods and the samples were taken from the population of existing research subjects (Arikunto 2010). The location of the research wass Bantul Regency of Special Region of Yogyakarta, taken from three subdistricts: Dlingo, Bantul and Banguntapan. The research time frame was six months. The population in this study is the wakif or the family who is an adults and able to fulfill wakaf in cash, which is domiciled in Bantul district of Yogyakarta. The sample used to test the theory in this study was taken from a selected population of 98 people.

3 RESULT AND DISCUSSION

Discriminant validity test results are obtained that, the loading factor value for each indicator of each latent variable has a value of loading factor between 0.660235 and 0.95823. This means that each latent variable has good discriminant validity. The criteria of validity and reliability can also be seen from Composite Reliability and Average Variance Extracted (AVE) values of each construct. The construct is said to have a high reliability if the value of composite reliability is above 0.70 and AVE is above 0.50. Based on the data analysis, it can be seen that the entire construct meets the composite reliability because the value is in the range of 0.863–0.930. Therefore, the value of AVE generated is in accordance with the recommended value, which is more than 0.5. Another way to test discriminant validity is by comparing the AVE root values of each construct with the correlation between constructs.

Based on the results of data analysis can be seen that hypothesis testing using a significance level of 5% with t table of 1.985 (N = 96). The results of

hypothesis testing for each variable are: At the level of significance (α) = 0.05, the value of t arithmetic (4.102) is greater than the value of t table of 1.985. This shows that *Subjective Norms* significant posi-tive to Attitude, at the level of significance (α) = 0.05, the value of t arithmetic (0.637) is smaller than the value of t table of 1.985. This shows that *Subjective Norms* not significant to Intention, at the level of significance (α) = 0.05, the value of t arithmetic (95.255) is greater than the value of t table of 1.985. This shows that *Self-Efficacy* significant positive to *Attitude*, at the level of significance (α) = 0.05, the value of t arithmetic (1.799) is smaller than the value of t table of 1.985. This shows that *Self-Efficacy* not significant to Intention, at the level of significance (α) = 0.05, the value of t arithmetic (0.962) is smaller than the value of t table of 1.985. This shows that *Government Support not significant to attitude,* at the level of significance (α) = 0.05, the value of t arithmetic (0.751) is smaller than the value of t table of 1.985. This shows that Government Sup-port not significant to Intention, and at the level of significance (α) = 0.05, the value of t arithmetic is (4.979) greater than the value of t table of 1.985. This shows that Attitude significant positive to Intention.

Based on the results of data analysis of respond-ents can be described from gender, a men's *wakif* who become respondent equal to 71% and women is 29%. Wakif ages between 20 and over 68 years, with a level of education equivalent to Senior High School (SMA) 34%, Equivalent Universities 59% and the rest of 7% educated under high school. Based on the location of their territory located, spread in 3 District of Bantul namely; Dlingo 26%, Bantul 43%, and Banguntapan 27%.

Based on the results of data analysis and discus-sion it can be concluded the implications of this research that are; the potential of cash *waqf* in Bantul is quite high, just need intensive socialization from related parties, Muslim attitudes in Bantul over the cash *waqf* are influenced by subjective norms and self-efficacy, the intention of cash *waqf* in Muslim society in Bantul is influenced by self-efficacy and attitude, while government support does not affect the attitude and intention of the wakif to perform cash *waqf*. This is because during this time the role of related institution less to do socialization about cash *waqf* in Bantul.

ACKNOWLEDGE

Thank you to the KEMENRISTIKDIKTI (Kemen-terian Riset, Teknologi dan Pendidikan Tinggi) which sponsored and funded this research; thank you also to the leadership of UST Yogyakarta who have provided support facilities; lastly, we thank BAZNAS (Badan Amil Zakat) Bantul Regency as a partner in this research.

REFERENCES

Arikunto, S. 2010. *Prosedur penelitian suatu pendekatan praktek*. Jakarta: PT Renika Cipta.

Azjen, I., & Fishbein, M. 2005. The influence of attitudes on behavior. In D. Albarrcin, B.T.

BPS. 2016. Badan Pusat Statistik D.I. Yogyakarta. Retrieved from https://yogyakarta.bps.go.id/. Accessed 17 August 2018.

Kusrini, K. C. K., Purwanto, I & Laksito, A. 2018. GIS-based decision support system for cash *waqh* distribution. *Journal of Theoretical and Applied Information Technology*. Vol. 96, no. 3, issue 15 (February) pp. 71–77.

Swasono, S. E. (2003). Kemandirian Ekonomi: Menghapus Sistem Ekonomi Subordinasi Membangun Ekonomi Rakyat. *Jurnal Ekonomi Rakyat (in Indonesian)*.

Undang-Undang No. 41 Tahun 2004 Tentang Wakaf.

Advances in Business, Management and Entrepreneurship – Hurriyati et al (eds)
© 2020 Taylor & Francis Group, London, ISBN 978-0-367-27176-3

Work-family conflict: With antecedents of job involvement, role ambiguity and job demand with social support moderation

P. Yulianti & A.P. Sari
Universitas Airlangga, Surabaya, Indonesia

ABSTRACT: This study examined work–family conflict with antecedent of job involvement, role ambiguity and job demand in moderation by social support. The sample of this study was 134 nurses in private hospitals. Data analysis was done with Smart PLS. The result of this study is job involvement has negative and significant effect to work–family conflict reinforced by social support as moderator. Job demand positively affects work–family conflict but is weakened by social support, while role ambiguity has no significant effect on work–family conflict. The conclusions of this study are that social support from supervisors, colleagues and families can reduce work–family conflict.

1 INTRODUCTION

Nurses have an important role to play in hospitals. In the performance of their duties, nurses deal not only with patients, patient's families, and co-workers but also with physicians as well as adhering to workplace regulations. Nurses' workloads are often judged to be incompatible with their physical, psychological and emotional situations. Work–family conflict is defined as a person's inability to divide their time and commitment to job and family roles equally (Greenhaus & Powell 1985, in Tharmalingam & Bhatti 2014).

Nurses in Indonesia are predominately women. This is a major factor in work–family conflict. This conflict will hamper job performance. Female nurses tend to complain about the difficulty of fulfilling their multiple roles of motherhood at home and their role as nurses working in hospitals; they feel pressure to be able to balance their time, energy and mind between their family and their job.

Some studies explain the antecedents of work-family conflict that are job involvement, work overload, and role ambiguity (Razak et al. 2011). The Tharmalingam and Bhatti study (2014) also discovered that job involvement, role ambiguity, and job demands influence work–family conflict with social support as moderator.

Lodahl and Kejner (1965) in Seo (2013) described job involvement as for a major factor in one's self-image. Individuals who have high job involvement consider work to be very important in their life. Job involvement has an influence on work–family conflict. As explained by Greenhaus and Beutell (1985) in Tharmalingam and Bhatti (2014), if one is heavily involved in work, this will lead to work–family conflicts.

Role ambiguity can occur when the individual experiences an uncertainty about some matters relating to his work such as the scope of his responsibilities, expectations on him, and how to do various jobs. Job demand includes time demands, emotional demands and demands physically related to workloads, where job demand causes negative influences such as tension or fatigue while working (Cabrera 2013). Supported by Bakker et al. (2003) states job demand refers to high work pressures, role overload, and poor working environment conditions.

Conflict aroused can be suppressed by support from the sorroundings. Almasitoh states that the support and assistance provided by the husband and other family members will provide an opportunity for the wife to develop his career. Social support provided by family members will provide a sense of security for women. The results of a study by Newsom and Schultz (1996) in Almasitoh found that a person's physical weakness was associated with minimal family support, peer support and reduced feelings of belonging. Social support from the workplace can especially contribute to productivity and employee benefits. The purpose of this study is to explore the role of job involvement, role ambiguity, job demand as antecedent work–family conflict with moderator social support.

2 METHOD

Sampling using probability sampling with proportionate stratified random sampling technique. The sample was 134 private hospital nurses in Surabaya. Analysis technique using SEM with partial least square with SmartPLS program. The job involvement measuring instrument uses four indicators by Yoshimura (1997), role ambiguity using four indicators according to Rizzo, House and Lirtzman (1970) (Harris & Bladen 1994), job demand using instruments developed by Bakker et al. (2004) and

Xanthopoulou et al. (2013) with five indicators. Work-Family Conflict uses four indicators developed by Netemeyer et al. (1996) and social support using three indicators developed by Caplan et al. (1975) modified by Lee et al. (2004)

3 RESULT AND DISCUSSION

Outer loading values of each indicator on job involvement variables, role ambiguity, job demand, social support and work–family conflict are P value > 0.05. Value of composite reliability of job involvement variables, role ambiguity, job demand, social support and work family conflict are all above 0.70. The results of this study hypotheses 2 and 5 are not proven. Job involvement has a negative and significant influence on work–family conflict and job demand has a positive and significant influence on work–family conflict (T-value > 1.96). Social support strengthens the negative relationship of job involvement with work–family conflict and, by contrast, social support weakens the positive relationship of job demand with work–family conflict.

3.1 The influence of job involvement on work family conflict

The results of the hypothesis show that there is a significant negatively relationship between job involvement and work–family conflict. Individuals who have low job involvement are individuals who view work as a less important part of their lives, and therefore they will have less pride in the company, participate less, and be less satisfied overall with their work. In contrast, individuals who have a high level of involvement in implementing their role or occupation will increase the conflict between roles of work and the role of their families (Tharmalingam & Bhatti

Table 1. Path coefficient.

Hypothesis	T-Statistic	Original sample	Sig.
Job Involvement → Work-Family Conflict	-.225	2.322	Sig.
Role Ambiguity → Work-Family Conflict	1.012	.115	Not Sig.
Job Demand → Work-Family Conflict	2.330	2.233	Sig.
Job Involvement * Social Support → Work-Family Conflict	2.898	.305	Sig
Role Ambiguity * Social Support → Work-Family Conflict	.730	.088	Not Sig.
Job Demand * Social Support → Work-Family Conflict	3.213	-.428	Sig.

2014). The results of a study by Greenhaus and Beutell (1985) suggests that high job involvement can improve work–family conflicts. This study is consistent with those of Tharmalingam and Bhatti (2014) and Razak et al. (2011) that show a negative relationship between job involvement and work–family conflict.

3.2 The influence of role ambiguity on work–family conflict

The results of the hypothesis does not show a significant positive relationship between role ambiguity and work–family conflict. The degree of individual discomfort in anticipating their work is called role ambiguity (Tharmalingam & Bhatti 2014). Lack of clear information about jobs and tasks for employees in job roles can lead to conflict and stress. Several studies have explained the relationship between role ambiguity and work–family conflict. Ryan et al. (2009) explained that there is a significant and positive relationship between role ambiguity and work–family conflict. This study is consistent with Tharmalingam and Bhatti (2014), which stated that the relationship between role ambiguity and work–family conflict a positive and insignificant.

3.3 The influence of job demand on work–family conflict

The results of the hypothesis show that there is a significant positive relationship between job demand and work–family conflict. Tharmalingam and Bhatti (2014) concluded that there is a strong relationship between job demand and work stress in work–family conflict. Working for a long period of time causes employees to reduce time with their families, which they should have a lot to spend time with. With this, they can easily engage in work–family conflict (Yildirim & Aycan 2008).

There are several previous studies that shows a positive relationship between job demand and work–family conflict. Tharmalingam and Bhatti (2014) explain that job demand is positively and significantly correlated with work–family conflict. Supported by research by Cabrera (2013) and Shimozu et al. (2011) who conducted the study and found that there is a significant and positive relationship between job demand and work-family conflict.

3.4 Relationship of job involvement, role ambiguity, and job demand with work–family conflict with social support as moderating variables

The results of the hypothesis show that social support moderated the relationship between job involvement and work–family conflict, also social support moderated the relationship between job demand and work–family conflict. Social support is not moderated the relationship between role ambiguity and work-family conflict. The theoretical point of view based on social support research shows that the existence of social support will provide comfort (Tharmalingam & Bhatti 2014). Social support is divided into supervisor support, peer support, organizational support and family support. Social support used for social sources where there is an influence on work–family conflict (Yildirim 2008, Selvarajan et al. 2013).

Social support from the workplace and family can help a person to reduce work–family conflict, and peer support can improve psychological well-being and reduce role conflict. Support from leaders can reduce stress in the workplace, thus contributing to a successful relationship between work and family (Hamid & Amin 2014).

Tharmalingam and Bhatti (2014) found that social support can be used as a moderator and the results are significant in job involvement, role ambiguity, and job demand for work–family conflict.

4 CONCLUSION

The results of this study found that nurse jobs have high job involvement in their profession and assume that work has an important role in life and as an actualization of themselves. Nursing is a high-demand job, both due to the amount time required to treat patients as well as due to demands of the patient's family. Social support from leaders, colleagues and families plays a very important role to reduce work–family conflict.

ACKNOWLEDGEMENT

This work was supported by Faculty of Economics and Business, Universities Airlangga research grants.

REFERENCES

Bakker, A. B et al. 2005. Job resources buffer the impact of job demands on burnout. *Journal of Occupational Health Psychology* 10(2): 170–180.

Cabrera, E. F. 2013. Using the Job Demands-Resources Model to Study Work-Family Conflict in Women. *The International Journal of Management and Business* 4(1): 112–130.

Hamid, R. A., & Amin, S. Mohd. 2014. Social support as a moderator to work-family conflict and work-family enrichment: a review. *Journal of Advanced Review on Scientific Research* 2(1): 1–18.

Harris, M.M. and Bladen, A., 1994. Wording effects in the measurement of role conflict and role ambiguity: A multitrait-multimethod analysis. *Journal of Management* 20(4): 887–901.

Lockley, S.W., Cronin, J.W., Evans, E.E., Cade, B.E., Lee, C.J., Landrigan, C.P., Rothschild, J.M., Katz, J.T., Lilly, C.M., Stone, P.H. and Aeschbach, D., 2004. Effect of reducing interns' weekly work hours on sleep and attentional failures. *New England Journal of Medicine* 351(18): 1829–1837.

Netemeyer, R. G., Boles, J. S., & McMurrian, R. 1996. Development and validation of work–family conflict and family–work conflict scales. *Journal of Applied Psychology* 81: 400–410.

Razak A.Z.A, Yunus N.K.Y, & Nasurdin M. 2011. Impact of work overload and job involvement on work-family conflict among Malaysian doctors. *Labuan e-Journal Maumalat Soc.* 5:1–10.

Ryan, B., Ma, J.E. and Ku, M.C., 2009. Role conflict, role ambiguity and work-family conflict among university foodservice managers.

Selvarajan, T. T., Cloninger, P. A., & Singh, B. 2013. Social support and work-family conflict: A test of an indirect effects model. *Journal of Vocational Behavior* 83: 480–499.

Seo, J.Y. 2013. Job involvement of part-time faculty: exploring associations with distributive justice, underemployment, work status congruence, and empowerment. *Iowa Research Online*.

Shimozu, A., Bakker, A.,B., Demerouti., E., & Peeters., M.,C.,W 2011. Work-family conflict in Japan: How job and home demands affect psychological distress. *Industrial Health* 48: 766–774.

Tharmalingam, S. D & Bhatti, M. A. 2014. Work-family conflict: An investigation on job involvement, role ambiguity and job demand: Moderated by social support. *Journal of Human Resource Management* 2(2): 52–62.

Xanthopoulou, D., et al. 2007. The role of personal resources in the job demands-resources model. *International Journal of Stress* 14(2): 121–141.

Yildirim, D & Aycan, Z. 2008. Nurses work demands and work-family conflict: A questionnaire survey. *International Journal of Nursing Studies*: 1–12.

Yoshimura, A. 1997. A review and proposal of job involvement. *Keio Business Review* (33):175–184.

Advances in Business, Management and Entrepreneurship – Hurriyati et al (eds)
© 2020 Taylor & Francis Group, London, ISBN 978-0-367-27176-3

The role of mediation: Creative self-efficacy on the relationship between creative role identity, job creativity requirements and supervisor creative expectation to creative performance

P. Yulianti & M. Mutiara
Universitas Airlangga, Surabaya, Indonesia

ABSTRACT: This study examines the influence of creative role identity, job creativity requirements, and creative expectation supervisors on creative performance. The study also examined the role of creative self-efficacy mediation on the relationship between creative role identity, job creativity requirements, and supervisor creative expectation to creative performance. The samples in this study are employees who work in marketing, engineering, program, production & news division in television media. The number of employees is 157 employees. Analysis technique using SEM partial least square with Smart PLS program. The result of this study is creative self-efficacy fully mediated on the relationship between creative role identity with creative performance and creative self-efficacy partially mediated on the relationship between job creativity requirements, creative support supervisor and creative performance. The implication of this study is that creative self-efficacy is an important factor in building performance and creativity.

1 INTRODUCTION

The development of technology makes the business environment becomes more dynamic, more complex and full of uncertainty so that every company is required to be able to change and adapt to face the condition (Anantan 2010). Companies are required for creativity as a source of innovation in an organization (Amabile et al. 1996) and as a key organizational performance in a competitive environment (Choi 2004).

Creative performance is a manifestation of the behavior of a person's creative ability, as an individual's ability to provide new ideas for solving problems encountered (Oldham & Cummings 1996). Creative performance can be built from the existence of creative role identity, job creativity requirements, and creative expectation supervisor with creative self-efficacy mediation.

Creative performance can be influenced by the creative role identity, because the stronger the role of identity possessed by a person, the more likely that the individual will display behavior consistent with his identity (Stryker 1980). Role identity theory asserts that one way that influences when one sees their own ability is through role identity (Stryker 1980). Employees will determine their own abilities (Gist & Mitchell 1992) whether they can do creativity in the workplace or not. Creative identity provides motivation for individuals to engage in creative behavior (Petkus 1996). Research on creative identity shows that individuals with creative identity have individual involvement in creative behavior (Jaussi et al. 2007).

Creative performance can be influenced by job creativity requirements which are the creativity

needs of a job that is the main requirement in carrying out the work specified (Tierney & Farmer 2011). Job creativity requirements can affect a person's creativity through individual characteristics, but creativity also involves taking risks, so it is not easy for individuals to meet high job creativity requirements (Zhou & George 2001). Job creativity requirements are an aspect of job design that encourages complexity, and the task of creativity that is specifically defined as the part of the job description (Shalley & Gilson 2004). Job creativity requirements are positively linked with intrinsic motivation and creative performance.

Supervisor's support for employee creativity refers to the extent to which employees perceive the support of superiors about creativity to their juniors, such as providing feedback and information related to creativity (Madjar et al. 2002). Supervisor creative expectation serves to provide support and inculcate attitudes to employees in order to be able to finish their work in a creative way, thus establishing employee confidence (Liao et al. 2010). In addition, employees will gradually meet the expectations of their superiors (Liao et al. 2010) so that creative expectation supervisors can be used as strong emotional social support, encouragement of the creative ability of employees and assist employees in overcoming unpleasant situations while doing complex tasks and more challenging, so as to enhance their creative abilities (Liao et al. 2010).

Creative self-efficacy is an important pioneer of creative performance by (Carmeli & Schaubroeck 2007). Overall, leaders who convey the creativity expectations to employees can enhance their creative self-efficacy, and can improve the creativity performance of the employee, enables creative self-efficacy to mediate the effect of creative self-efficacy on employee creative performance (Liao et al. 2010)

Creative self-efficacy has a role as a mediator between individual and contextual factors as well as the creative performance of employees (Gong et al. 2009). Tierney & Farmer (2002) explain that one promising mediator is creative self-efficacy that reflects the beliefs that are formed of one's creativity knowledge and skills and reflects the intrinsic motivation to engage in a creative activity so that creative self-efficacy becomes a strong foundation for employee creativity (Tierney & Farmer 2004). This study explores the role of creative self-efficacy in mediating the influence of creative role identity, job creativity requirements, and creative expectation supervisors on the creative performance of employees.

2 METHOD

The samples in this study are employees working in the marketing, engineering, program, production & news division of the television media. The number of employees is 157 employees. Analysis technique using SEM with partial least square with SmartPLS program. The measuring instrument Creative role identity was measured using 3 modified indicators from Farmer et al. (2003). Job creativity requirements used four modified indicators from Axtell et al. (2000), creative expectation supervisors using 6 modified indicators from Farmer et al. (2003) Creative self-efficacy using 3 indicators from Tierney & Farmer (2002) and Employee creative performance (Y) were measured using 5 modified indicators from Tierney et al (1999).

3 RESULT AND DISCUSSION

3.1 Result

Outer loading values of each indicator on the creative role identity variable, job creativity requirement, supervisor creativity expectation, creative self-efficacy and employee creative performance are P value > 0,05. Composite reliability value of creative role identity, job creativity requirement, supervisor creativity expectation, creative self-efficacy and employee creative performance are all above 0.70. The results of this study hypothesis 2 are not supported. Creative role identity, job creativity requirement, supervisor creativity expectation have a positive and significant effect on creative self-efficacy. Creative self-efficacy has a significant effect on employee creative performance. Job creativity requirement and creative expectation supervisor have a significant effect on employee creative performance. (T-value> 1.96). Tabel 1 describes the path coefficient the relationship among variables.

Table 1. Path coefficient.

Hypothesis	T-statistic	Original sample	Sign
Creative Role Identity -> Creative Self-Efficacy	2.637	0.275	Sign
Creative Role Identity ->Employee Creative Performance	0.580	-0.044	Not sign
Creative Self-Efficacy -> Employee Creative Performance	3.190	0.310	Sign
Job Creativity Requirements -> Creative Self-Efficacy	3.024	0.317	Sign
Job Creativity Requirements ->Employee Creative Performance	2.961	0.263	Sign
Supervisor Creative Expectation ->Creative Self-Efficacy	3.145	0.344	Sign
Supervisor Creative Expectation -> Employee Creative Performance	2.957	0.319	Sign

3.2 Discussion

3.2.1 The influence of creative role identity on the creative self-efficacy

The results of the hypothesis show that there is a significant positive relationship between creative role identity and creative self-efficacy. Creative role identity is an identity of the individual role as a creative person and makes the identity of a creative role as the main thing in carrying out its work (Tierney & Farmer 2011). Role identity motivates the emergence of role performance in an individual (Markus & Wurf 1987). Employees with high creative role identity will display creative behaviors in the workplace to gain recognition from their peers (Farmer et al. 2003). Tierney & Farmer (2011) explain that creative role identity and creative self-efficacy have a positive relationship.

3.2.2 The influence of creative role identity on employee creative performance

The results of the hypothesis show that there is not a significant negative relationship between creative role identity and employee creative performance. Creative role identity is an attitude and behavior of an individual who is consistent with the identity of a creative role in carrying out his work (Farmer et al. 2003). Creative role identity will motivate creative role performance in individuals (Markus & Wurf

1987). The results of the study inconsistent with a study by Grube & Piliavin (2000).

3.2.3 *The influence of creative self-efficacy on employee creative performance*

The results of the hypothesis show that there is a significant positive relationship between creative self-efficacy and employee creative performance. Creative self-efficacy is defined as a belief in someone who has the ability to generate new ideas (Tierney & Farmer 2002). Bandura (1997) argues that self-efficacy is a necessary condition for creative activities and the discovery of new knowledge or ideas. This happens because self-efficacy can affect the motivation and ability to engage in creative behavior (Bandura 1997). Creative self-efficacy is a dominant factor in building creativity in individuals (Gong et al. 2009). Having a high level of confidence required employees to be able to engage in creative behavior (Gong et al. 2009; in Tierney & Farmer 2011). Self-efficacy is one of many elements in the formation of creativity in an individual.

3.2.4 *The influence of job creativity requirements on the creative self-efficacy*

The results of the hypothesis show that there is a significant positive relationship between job creativity requirements and creative self-efficacy. Job creativity requirements are jobs that require creativity as a key requirement in carrying out the work specified (Farmer & Tierney 2011). Job creativity requirements have an efficacy-building element because by living a job that requires creativity, employees will have the confidence to be a creative employee.

3.2.5 *The influence of job creativity requirements on employee creative performance*

The results of the hypothesis show that there is a significant positive relationship between job creativity requirements and employee creative performance. Job creativity requirements affect creative self-efficacy because it serves as a trigger for creativity for employees (George 2007) and creative self-efficacy can encourage creative performance in employees (Bandura & Locke 2003). Research conducted by Gong et al. (2009) also supports the statement that creative self-efficacy has a positive influence with creative performance, thus both conceptual and empirical support shows that job creativity requirements and creative self-efficacy are succeeded, so creative performance is established (Tierney & Farmer 2011).

3.2.6 *The influence of supervisor creative expectation on the creative self-efficacy*

The results of the hypothesis show that there is a significant positive relationship between supervisor creative expectation and creative self-efficacy. Supervisor creative expectation is defined as the head's expectations of creative behaviors displayed by employees (Carmeli & Schaubroeck 2007). In particular, creative expectation supervisors are expected to enable employees to understand the expectations of employers and firms on employee creativity, thus motivating employees to build their creativity (Jiang & Gu 2015). Several studies have suggested the important role of superiors in shaping one's behavior in the workplace (Gong et al. 2009), one of which is a creative expectation supervisor who is recognized as an approach to encouraging employee creativity (Carmeli & Schaubroeck 2007).

3.2.7 *The influence of supervisor creative expectation on employee creative performance*

The results of the hypothesis show that there is a significant positive relationship between supervisor creative expectation and employee creative performance. Research conducted by Liao et al. (2010) suggests that creative expectation supervisors are an effective social persuasion in shaping their trustworthiness, as employees get the strong social-emotional support that can encourage their self-confidence to be creative when performing new tasks and challenging, thus improving their creative performance (Liao et al. 2010).

4 CONCLUSION

Employees with creative role identity will have a creative performance if they have confidence in their ability to generate creative ideas because creative self-efficacy is the dominant factor to build creativity in individuals (Gong et al. 2009). Job creativity requirements are jobs that require creativity as a key requirement in carrying out the work specified (Farmer & Tierney 2011). Job creativity requirements are also known to have an efficacy-building. Supervisor creative expectation is expected to make employees understand the expectations of leaders and companies on employee creativity, so as to motivate employees to build creativity (Jiang & Gu 2015).

ACKNOWLEDGMENT

This work was supported by the Faculty of Economics and Business, Universities Airlangga research grants.

REFERENCES

Amabile, T.M. 1988. A model of creativity and innovation in organizations. *Organizational Behavior*, 10(1): 123–167.

Anatan, L. 2010. Coorporate Social Responsibility (CSR): tinjauan teoritis dan praktik di Indonesia. *Jurnal Manajemen Maranatha*, 8(2): 66.

Axtell, C.M., Holman, D.J., Unsworth, K.L., Wall, T.D., Waterson, P.E. & Harrington, E. 2000. Shopfloor innovation: Facilitating the suggestion and implementation of ideas. *Journal of occupational and organizational psychology*, 73(3): 265–285.

Bandura, A. & E.A. Locke. 2003. Negative self-efficacy and goal effects revisited. *Journal of Applied Psychology*, 88(1): 87–99.

Bandura, A. 1997. Editorial. *American Journal of Health Promotion*, 12(1): 8–10.

Carmeli, A. & Schaubroeck, J. 2007. The influence of leaders' and other referents' normative expectations on individual involvement in creative work. *The Leadership Quarterly*, 18(1): 35–48.

Choi, J.N. 2004. Individual and contextual predictors of creative performance: The mediating role of psychological processes. *Creativity Research Journal*. 16(2-3): 187–199.

Farmer, S.M., Tierney, P. & Kung-Mcintyre, K. 2003. Employee creativity in Taiwan: An application of role identity theory. *Academy of Management Journal*, 46 (5): 618–630.

George, J.M. 2007. 9 Creativity in organizations. *The academy of management annals*, 1(1): 439–477.

Gong, Y., Huang, J.C. & Farh, J.L 2009. Employee learning orientation, transformational leadership, and employee creativity: The mediating role of employee creative self-efficacy. *Academy of Management Journal*, 52(44): 765–778.

Grube, J.A. & Piliavin, J.A. 2000. Role identity, organizational experiences, and volunteer performance. *Personality and Social Psychology Bulletin*, 26(9): 1108–1119.

Jaussi, K.S., Randel, A.E. & Dionne, S.D. 2007. I am, I think I can, and I do: The role of personal identity, self-efficacy, and cross-application of experiences in creativity at work. *Creativity Research Journal*, 19(2-3): 247–258.

Jiang, W. & Gu, Q. 2015. A moderated mediation examination of proactive personality on employee creativity: A person-environment fit perspective. *Journal of Organizational Change Management*, 28(3): 393–410.

Liao, H., Liu, D. & Loi, R. 2010. Looking at both sides of the social exchange coin: A social cognitive perspective on the joint effects of relationship quality and differentiation on creativity. *Academy of Management Journal*, 53(5): 1090–1109.

Madjar, N., Oldham, G.R. & Pratt, M.G. 2002. There's no place like home? The contributions of work and nonwwork creativity support to employees' creative performance. *Academy of management journal*, 45(4): 757–767.

Markus, H. & Wurf, E. 1987. The dynamic self-concept: A social psychological perspective. Annual review of psychology, 38(1): 299–337.

Oldham, G.R. & Cummings, A. 1996. Employee creativity: Personal and contextual factors at work. *Academy of Management Journal*, 39(3): 607–634.

Petkus Jr, E.D. 1996. The creative identity: Creative behavior from the symbolic interactionist perspective. *The Journal of Creative Behavior*, 30(3): 188–196.

Shalley, C.E. & Gilson, L.L. 2004. What leaders need to know: A review of social and contextual factors that can foster or hinder creativity. *The leadership quarterly*, 15(1): 33–53.

Stryker, S. 1980. *Symbolic interactionism: A social structural version*. Benjamin-Cummings Publishing Company.

Tierney, P. & Farmer, S.M. 2002. Creative self-efficacy: Its potential antecedents and relationship to creative performance. *Academy of Management journal*, 45(6): 1137–1148.

Tierney, P. & Farmer, S.M. 2004. The Pygmalion process and employee creativity. *Journal of Management*, 30(3): 413–432.

Tierney, P. & Farmer, S.M. 2011. Creative self-efficacy development and creative performance over time. *Journal of Applied Psychology*, 96(2): 277.

Tierney, P. Farmer, S.M. & Graen, G.B. 1999. An examination of leadership and employee creativity: The relevance of traits and relationships. *Personnel psychology*, 52(3): 591–620.

Zhou, J. & George, J.M. 2001. When job dissatisfaction leads to creativity: Encouraging the expression of voice. *Academy of Management journal*, 44(4): 682–696.

Advances in Business, Management and Entrepreneurship – Hurriyati et al (eds)
© 2020 Taylor & Francis Group, London, ISBN 978-0-367-27176-3

Study of minimum wage comparison in Indonesia and Malaysia: An Islamic economics perspective

I. Jauhari, S. Herianingrum & T. Widiastuti
Universitas Airlangga, Surabaya, Indonesia

ABSTRACT: The minimum wage policy in Indonesia has been set since 1970, undergoing several changes to the latest government rule number 78 of 2015. The minimum wage in Malaysia is set to begin in 2011, implemented in 2012, and amended by 2016. The purpose of this study is to analyze minimum wage in Indonesia and Malaysia and examine how it is viewed through the perspective of Islamic economics. This study used a qualitative approach using literature analysis techniques. The results of this study indicate that both countries in providing wages are in acaccordance with Islamic economics. First, wages for workers in both countries are fair. Second, wages in both countries are feasible because in the wage-related policy-making process the needs of the workers are always considered. Third, wages are paid with no delay.

1 INTRODUCTION

Working is an important factor in human life because it directly affects the level of prosperity. Labour is an important factor in the process of production, and labourers, or workers, are anyone who works and receives wages or other forms of reward (Hardijan 2011). But there is often a disharmony between employers and workers caused by mistakes in making policy within a company especially in terms of wage provision.

Wages are the right of workers or labourers who are received and expressed in the form of money in return for employers to workers or labourers who are established and paid under an employment agreement or law. Minimum wage is a minimum standard used by employers or industry players to provide wages to workers in a business or work environment. Implementation of minimum wage policy directs to two consequences, labour welfare and employment (Rohadi 2015).

In Indonesia, minimum wages often change in the direction of their policies starting from the minimum requirements minimum requirements (KFM) minimum wage policy, after which it is revamped with the minimum living needs (KHM), then the reference for the determination of the minimum wage is based on the living need (Hakim 2006). The minimum salary in Malaysia has been established by the State Payroll Consultative Board approved on 30 June 2011. The minimum wage in Malaysia has actually been in effect since 1949, but the level is still too low but continues to apply in today's life. And the Malaysian Trade Union Congress (MTUC) has struggled to demand a fairer private sector minimum wage policy. MTUC in its demands demanded RM1200 for the minimum salary of the private sector in Malaysia. Under this 2011 State Payroll

Negotiation Act, employers who fail to comply with the Minimum Salary Command may be subject to a maximum fine of RM10,000 per employee.

Carpio et al. (2015) investigated the employment and wage impacts of minimum wage changes in Indonesia, differentiating the effects on production workers and non-production workers, by educational category, and by gender. The results indicate that assistance to vulnerable groups is necessary when minimum wage increases are implemented. In the Indonesian context, they are low-skilled, female and non-production workers, and they tend to lose jobs in manufacturing when minimum wages are raised.

Senasi and Samihah (2015) revealed that the employers face minimal challenges in implementing the minimum wage policy in Malaysia. Most of the employers' concerns centre on adjusting the profit and cost in their respective firms. This result supports the notion that employers impose the minimum wages as top-down pressure which can lead to investment cost.

This study examined the wage system in Indonesia and Malaysia. This research compared the model of remuneration in both countries after it was analyzed using the concept of wages in Islam, whether in both countries in terms of wages are in line with the existing wage system in Islam. This research has never been done before that it only concentrates on the theme of a particular theme that is not exactly the same as this one research.

1.1 *Definition wages in Islam*

In Islam, wages are also called ujrah: resulting from Ijarah contract. According to ulama Hanafiyah, Ijarah is a transaction against a benefit with certain allowable benefits. So wages (ujrah) is a form of compensation for services that have been provided by the workforce.

Al-Qur'an Surah Al-Tawbah verse 105 explains that according to the concept of Islam, wages consist of two forms, namely, the wages of the world and the rewards of the hereafter. In other words, this verse defines wages for the rewards that one receives for his work in the form of material reward in the world and the reward of a reward in the afterlife. The material rewards that a worker receives in the world must be just and worthy, while the reward of the afterlife is a better reward received by a Muslim from his God.

1.2 Principles of wages in Islam

Islam has mentioned many basic principles of wages as workers' rights, whether mentioned in the Qur'an or hadith. From the verses of Al-Quran and hadith mentioned above, it can be concluded that there are four principles in terms of employment especially in terms of wages. The four principles are (following Salim 2013): the principle of human freedom, the principle of the glory of the human degree, the principle of justice and anti-discrimination, and the principle feasibility of wages.

1.3 Minimum wages in Indonesia

Wage-setting mechanism is regulated in Law Number 13 the Year 2003 regarding Manpower with systematic as follows:

a Establishment of the minimum wage at provincial and regency or city level (Article 88).
b Determination of wages through agreements or collective bargaining (Article 91).
c Application of structure and wage scale (article 9 2, paragraph 1).
d Periodic wage review (Article 92 paragraph 2).

In addition to macro regulatory regulations (in the form of law), the government also makes the rules of execution in the form of government regulations, ministerial decrees, and ministerial regulations.

1.4 Minimum wages determination mechanism

A mechanism in the determination of the minimum wage in Indonesia with the following process:

a The District or City Wage Council establishes a survey team whose membership consists of Wage Council members from tripartite elements, college or expert elements, and includes the local Central Bureau of Statistics.
b For the District or City that has not yet been established the Wage Council, the survey is conducted by a Survey Team formed by the Regent or Mayor. The Survey Team is membership on a tripartite basis and with the inclusion of the local Central Bureau of Statistics.

c The survey team then conducts a price survey based on the components of the living necessities of the worker or single worker as set forth in the attachment Permenakertrans (Minister of Labour and Transmigration Republic Indonesia) number 13 of 2012.
d The survey is done once a month from January to September, while for the month of October to December is to do the prediction using the least square method. The survey results each month are then taken to get the value of KHL.
e Based on the results of the price survey, the Regency or Municipal Wage Council then submits the KHL score and proposes the value of MSE to the Regent or Mayor who is then presented to the Governor. Upon hearing the advice and consideration of the Provincial Wage Council, the Governor also considers the balance of the minimum wage value among districts or cities in the province; then set the Minimum Wage Value of the Regency or city concerned.
f The determination of the Regency or City Minimum Wage is set not later than 40 days prior to 1 January.
g The regulated Minimum Wage of the city must be greater than the provincial Minimum Wage.

In addition to the minimum wage set by the government through legislation, wage fixing can also be made through agreements. Determination through this agreement is usually made for workers with a working period of more than one year as regulated in the provisions.

1.5 Minimum wages in Malaysia

In Malaysia, wages are also referred to as wages in which salaries are paid to all employees who have used their labour in doing any work to fulfil the ob-ligations and needs of a master. The national minimum wage is a salary set by the government or a tripartite party (representatives of government, employers 'and workers' representatives). It is the most widely used model around the world and is divided into two categories: one minimum wage level for all countries, or minimum wage rates by sector or region. The minimum wage is stipulated by the State Payroll Consultative Council, approved on 30 June 2011. The Council is responsible for conducting research on all matters related to minimum wage implementation, including making recommendations to the government to establish minimum wage rules by sector, occupation, and region and to impose conditions for matters relating to it.

The minimum wage rate is not static and is to be reviewed from time to time, depending on the level of state prosperity, national competitive-ness, and productivity, to ensure that the minimum wage will always be above the national poverty line. The minimum wage rate in Malaysia paid to employees governed by the 2016 minimum wage order is valid from 1 July 2016, this rule in accordance with the 1955 Working Act (Deed 265).

2 METHOD

The research used a qualitative approach using literature analysis techniques. Literature analysis technique methodologically analyses and synthesizes quality literature, provides a firm foundation to a re-search topic and to the selection of research methodology, and demonstrates that the proposed research contributes something new to the overall body of knowledge or advances the re-search field's knowledge-base (Levy 2006). The main sources in this study were obtained from academic journals relevant to wage themes in Islamic economics as research themes, then sources of written regulations such as books, scriptures, Islamic scholars, internet media (websites), and other reading material related to research. Then, in accordance with the approach and type of research used than in this study.

3 RESULT AND DISCUSSION

Indonesia and Malaysia are two countries with different legal systems. However, general principles concerning remuneration also apply to both Malaysia and Indonesia as a member of the International Labour Organization. Regulations on employment in Malaysia are the authority of the Ministry of Human Resources Management under the Prime Minister, similar to the Ministry of Manpower and Transmigration in Indonesia as an assistant to the president in carrying out government functions. As a federal state, employment is a direct federal authority, not a state authority, so in case of employment problems resolved in federal courts specifically dealing with labour or industrial disputes, in contrast to the Indonesian state, there is an industrial relations court in the provinces below the Supreme Court for which some of the disputes are final.

The explanation of this remuneration can be seen by how wage conditions exist in Indonesia and Malaysia.

First, the remuneration in Indonesia has been going on for a long time as discussed here, to see the condition if the wage model applied in Indonesia is in fact in accordance with Islamic principles, has put forward a justice, has seen the basic needs of a worker. But in reality, the existing wages in Indonesia are still not in line with the rules already in place. Many entrepreneurs are deliberately paying their employees under the minimum wage set by the government. Under such conditions should the government also give firm punishment to entrepreneurs who intentionally paid wages below the prescribed minimum wage. To date the government is very

unfair in taking action against entrepreneurs who pay wages under minimum wage rules Baydoun (Willet 2000).

Secondly, the minimum wage in Indonesia and Malaysia in practice is also influenced by the existing model of government. Indonesia uses a system of democratic government while in Malaysia using the system of the kingdom. The existing governmental model is very influential with the implementation of existing wages, for example in Indonesia. Many entrepreneurs are not obedient to the government but in Malaysia none of the entrepreneurs who dare to fight the government policy. This happens because the model of royal government like in Malaysia is authoritarian so that entrepreneurs do not dare to give wages under the rules set by the government, different from the existing democratic model in Indonesia. This makes the freedom so that the entrepreneurs as if not afraid of the rules made by the government.

Thirdly, wages in Indonesia and Malaysia are also influenced by the geographical region of the region. For example, in Malaysia, the wage is only divided with two regions, both central Malaysia and eastern Malaysia. Unlike in Malaysia, the wage in Indonesia is divided by many existing areas and has various differences so that it makes the government somewhat difficult in controlling in the implementation of existing management. Thus, fraud rate is very possible, and the government should be more careful in supervising the payment of wages of employers to employees or workers.

The rate of wages paid to labourers should be sufficient to finance the needs of proper food, clothing, and shelter (Beekun & Badawi 2005). The concept of wages in Islamic wages is as described in the following paragraph.

First, wages must be fair for both parties (entrepreneurs and workers). The process of providing wages to workers through several processes in the determination of wages, such as in Indonesia, is very fair. In Malaysia, the determination of the minimum wage is eligible related to the amount of wages received by workers. The feasibility of wages received by workers can be seen from three aspects, namely food, clothing, and boards (shelter) (Hafidhudin & Tanjung 2003). Islam also sets the concept of the highest wage in paying the workers. That is, workers should not charge their work beyond the limits of the company's ability to pay for it. In Islam, the pre-determined wages in the contract can be corrected by the management of the company, whether at the time of profit or loss. However, these improvements should first be discussed with workers. Qardhawi states, "It should not be for workers to demand wages above their rights and above the ability of their services (the company) through pressure by way of strikes, labor organization or other means" (Hafidhudin & Tanjung 2003). This concept emphasizes a very important point. Entrepreneurs are required to meet the needs of their employees. On the other hand, workers are asked not to charge high fees until the employer can't afford it. In this regard, Islam has laid the groundwork for protecting the rights of employers and workers. If entrepreneurs are fully aware of their obligations to the workers, they will most likely to pay their workers with enough wages to cover basic needs. This happens if they really believe and hope Allah SWT in his devotion to humanity.

Second, Wage Feasibility The limit on wages according to the wage council is as follows: the wage is an acceptance in return for the employer to the recipient for a work or service that has or will be per-formed, which serves as a guarantee of viable sur-vival for humanity and production, declared and valued in the form of money stipulated in accordance with an agreement of the act and regulations paid on the basis of an employment agreement be-tween the employer and the employee. The setting of wages for the workforce should reflect justice, and consider various aspects of life so that the view of Islam on labour rights in receiving wages is realized. As in the Qur'an, it is also advisable to be fair by explaining justice itself. In Surat an-Nisa' verse 135, Allah affirms: O those who believe, be you truly enforcers of justice, be witnesses For Allah even to yourself or your father and relatives (Surah an-Nisa': 135). Wages paid to a person other than should be proportional to the activities that have been issued should also be quite beneficial for the fulfilment of the needs of a reasonable life. In this case either because of differences in the level of needs and abilities of a person or due to environmental factors and so on (Kartasaputra 1994).

Third, entrepreneurs (musta'jir) pays tribute to workers who have finished his work. Whether it's daily, weekly, monthly, or other. Islam advocates to accelerate the payment of wages when the work is perfect or at the end of the work as agreed, without delay. If it is terminated without any obstacles, then including acting wrongfully. Allah SWT speaks of a child who is divorced by a divorced wife which means "Then if they nurse (your children) for you then give them their wages" (Surah At-Talaq: 6). The above verse commands to pay the wages as soon as possible after the completion of the work. Prophet Muhammad SAW said, "Give to a worker his wages before his sweat dry" (HR. Ibn Majah). The purpose of the hadith is to hasten workers' rights after completion of the work, as well as if there has been a salary agreement every month. Delays of payment of wages in Islam arbitrarily to workers is prohibited, unless the delay has been set in the contract (agreement). Likewise, the suspension of payment of wages by employers must first be arranged in the contract. If it is not regulated, then the employer is obliged to pay the worker's wage after completing his work. Delaying the salary to the employee is a crime.

In order for such a partnership to work properly and all parties involved to benefit from each other, Islam regulates it clearly and in detail with laws relating to ijaratul ajir (contract of employment) (Possumah et al. 2013). Islam asserts that ijara transactions that are still obscure points of agreement are transactions that are facade (broken). It is expected that each party can understand their rights and obligations, respectively. This will be able to prevent tyrannical employers in hiring workers outside his working hours such as cases of workers this domestic. The actual wage is the value of services (benefits) given by the workers (ajir) to employers (employers/musta'jir). Wages in the Islamic view are an agreement between ajīr (workers) and musta'jir (entrepreneurs). The standard used to establish it is the benefit of labor (manfa'at al juhd) provided by the marketer, not the lowest cost of living. Therefore, there will be no exploitation of labour by employers. Workers and civil servants are the same because workers earn their wages in accordance with the provisions of the corresponding wage prevailing in society.

Determining wages is an agreement between workers and employers by making the benefits of energy as a standard determination (Diamond 1982). The burden of living necessities, health costs and other dependents of workers is not a determinant of wages (Gindling 1991). There is no exploitation of labour because everything is already known to each other. Nor will it burden the authorities because they bear the burden of costs that do not affect the production such as health insurance, educational benefits, and pension funds. Thus, the state does not need to set a minimum wage (regional minimum wage). In fact, such a stipulation is not allowed, similar to a ban on price fixing. Both price and wage are compensations received by someone. The difference, the price is the compensation of goods, while the wage is a compensation of services.

4 CONCLUSION

Indonesia and Malaysia are two countries with different legal systems. However, general principles concerning remuneration also apply to both Malaysia and Indonesia as a member of the International Labour Organization. Regulations on employment in Malaysia are the authority of the Ministry of Human Resources Management under the Prime Minister, similar to the Ministry of Manpower and Transmigration in Indonesia as an assistant to the president in carrying out government functions.

In Islamic wages should be done as follows; first, wages must be fair for both parties (entrepreneurs and workers). Second, wage feasibility, the wage setting for the workforce should reflect justice, and consider various aspects of life, so that the view of Islam on labor rights in receiving wages is more manifest. Third is no delayed payments, and employers (musta'jir) are obliged to pay wages to workers who have completed their work.

The problems in the wages of workers will always be there, especially in the global era where business competition is so tight that entrepreneurs minimize losses by earning big profits. Some suggestions given in this research are:

a For workers, since wages are a reward provided and as a source of income for them-selves and their families, it requires seriousness and professionalism in working. This is a measure of how much it will receive.
b For the government, wages are the protection of workers and employers in sustaining national development strategies. Therefore, it is necessary to create and enact rules that are not biased but are in favour of workers and employers. The government must also take firm action against the negligent rulers in paying the wage minimum, in addition, the government must also more often oversee the implementation of wages in the field.
c For the community, wages are nothing but a means to increase the dignity of workers and their families, and are useful in alleviating poverty.

Thus, it is expected that the good cooperation between employees and the company will be established in order to realize their respective objectives. An adequate wage for workers and a satisfactory profit for the company could be achieved.

REFERENCES

Beekun, R.I. & Badawi, J.A. 2005. Balancing ethical responsibility among multiple organizational stakeholders: The Islamic perspective. *Journal of Business Ethics* 60(2):131–145.

Carpio et al. 2015. Do minimum wages affect employment? Evidence from the manufacturing sector in Indonesia. *IZA Journal of Labor & Development, Springer* 4(17): 1–35.

Didin, H. & Hendri, T. 2003. *Manajemen syariah dalam praktek*. Jakarta: Gema Insani.

Gindling, T.H. 1991. Labor market segmentation and the determination of wages in the public, private-formal, and informal sectors in San Jose, Costa Rica. *Economic Development and Cultural Change* 39(3): 585–605.

Hardijan, R. 2011. Hukum ketenagakerjaan. ghalia Indonesia. Bogor.

Orman, W.H. 2006. Giving it away for free? Motivations of open-source software developers. University of Arizona.

Levy, Y. & Ellis, T.J. 2006. A systems approach to conduct an effective literature review in support of information systems research. *Informing Science* 9.

Possumah, B.T., Ismail, A.G. & Shahimi, S. 2013. Bringing work back in Islamic ethics. *Journal of Business Ethics* 112(2): 257–270.

Rohadi, R. & Ahmad, A. 2015. Sectoral implications of Indonesia minimum wage policy: Empirical evidence using panel data. *International Institute of Social Studies*.

Senasi, S. Vally, V. & Samihah, K. 2015. Implementation of minimum wage policy in Malaysia: Manufacturing employers' perceptions of training provision and fringe benefits. *International Journal of Humanities and Social Science* 5(12).

Advances in Business, Management and Entrepreneurship – Hurriyati et al (eds)
© 2020 Taylor & Francis Group, London, ISBN 978-0-367-27176-3

Nurses service quality in community health service using Analytic Hierarchy Process (AHP)

A.Z. Abidin, S. Suwatno, Y. Yuniarsih & D. Disman
Universitas Pendidikan Indonesia, Bandung, Indonesia

ABSTRACT: The health service quality plays a strategic role in improving public health quality. Therefore, this study was conducted to determine the nursing service quality and to figure out the most important criteria in nursing service quality based on the priority scale of five dimensions of service quality. This study was conducted by involving 40 patients of Community Health Service in Tangerang Selatan City, who had used the health service for at least three years. The data obtained were analyzed using Analytic Hierarchy Process (AHP) method and calculated using Super Decision software. The results showed that accordingly the most important criteria in nursing service qualities were Assurance, Empathy, Responsiveness, Reliability, and Tangibility.

1 INTRODUCTION

Service is as any intangible act that one party can offer to another party and does not result in the ownership of anything. To obtain quality service, there are five dimensions that must be fulfilled: Tangibility, empathy, reliability, responsiveness, assurance (Youssef 1996, Kotler 2012).

According to (Andaleeb 2001), patient's perception of health care appears to have been neglected by healthcare providers in developing countries. Such perceptions, especially regarding the service quality, may establish confidence and subsequent behavior with respect to the choice and use of health care facilities. Whereas, accountable quality service is vital in health service quality. It is in accordance with (Lawrence & Olesen 1997) stating that a quality assurance system is required, partly to advance professional services and development to provide accountability. This is considered important as the consumers tend to choose better performing and responsive health plan and health-care providers that provide quality information (Kolstad & Chernew 2009, As Zastowny et al. 1995) point out that the use of patient satisfaction and personal health care experience are the measurement of health care quality.

Studies on the health care quality have been previously done. (Sofaer & Firminger 2005, Van Campen et al. 1998) measured the patient's perception of the health care quality. Meanwhile, (Anderson 1995) conducted a study to assess the services quality provided by the public health clinics of state universities. His study used SERVQUAL instruments that were administered to patients at the University of Houston Health Center, in order to evaluate customer perceptions of service quality. (Oermann & Templin's 2000) study was conducted to identify consumer interests of attributes of

quality health care and nursing care, and examine the relationship be-tween consumer perspective and health status and demographic variables selected. Besides, a study conducted by (Headley & Miller 1993) adapted SERVQUAL scales for health care services and examined them from reliability, dimensions, and validity aspects in primary care clinic settings.

Furthermore, a study done by Wisniewski (1996) analyzed the use of various measurement of service quality in the private sector as an important indicator of organizational performance and customer satisfaction. The study has resulted in considerable empirical research. Furthermore, a study of Sadiq (2003) examined and measured the services quality provided by private hospitals in Malaysia. This study employed empirical research that was used to deter-mine patients' expectations and perceptions about service quality. The comprehensive scale used in his study was adapted from SERVQUAL, which was empirically evaluated in Malaysian hospital settings. Oermann (1999) study found that consumer-oriented health report cards have emerged as a strategy to disseminate information to consumers about the health plans quality and relative costs, with the aim of enabling them to make informed choices. Another study, which was done by Carman (2000) used empirical investigation, using conjoint methodology to weight the importance of providing quality attributes for acute care services in hospitals.

Based on aforementioned findings of previous studies, this study under the theme of nursing service quality in community health service became the fo-cus of this study Analysis Hierarchy Process (AHP) was chosen to assess the indicator ranks of service quality. It is expected that the results will be beneficial to provide excellent nursing service quality.

2 METHOD

Analytic Hierarchy Process (AHP) is a mathematical theory that is able to analyze the effect of the as-Sumption approach to solve a problem. AHP was first introduced by (Saaty 1977). This method provides a measuring scale to obtain a priority value of each indicator.

AHP is a qualitative method approach in the decision-making process. It starts with a goal then criteria, and ends with an alternative. There are various forms of decision hierarchy adapted that can be resolved with AHP.

2.1 Hierarchy preparation

Preparation of hierarchy is done by identifying the knowledge or information being observed. The hierarchy consists of goals, criteria and alternatives. In this study, this stage was done by preparing list of interview questions and questionnaire before being administered to the respondents. The respondents assessed the questionnaire items using the following scale developed by (Saaty 2006).

2.2 Priority setting

For each level of hierarchy, pairwise comparison is needed to determine the priority. A pair of elements is compared based on certain criteria and weighs the intensity of preference among elements. The relationship between elements of each hierarchical level is determined by comparing the elements in pairs. In this study, the priority setting was done by weighting the indicators of quality service, in or-der to determine which indicators should be included in the interview and questionnaire.

2.3 Logical consistency

All elements are grouped logically and consistently upgraded according to a logical criterion. A high consistency assessment is necessary in decision-making issues for accurate decision-making. AHP measures the overall consistency of various considerations through a consistency ratio. The value of the consistency ratio should be 10% or less, if more than 10% then the random assessment needs to be fixed. In this study, the overall logical consistency of questionnaire items and interview questions had been assessed.

This study implemented AHP by using primary and secondary data. The primary data was obtained from in-depth interview and filled questionnaire. Moreover, the secondary data was obtained from textbooks and books issued by related institution.

In employing AHP approach, the numbers of respondents involved is not really matter as long as they are representative enough to share their experience in using certain service. According to (Kenagy et al. 1999), the principle of service quality is seen from a service quality point of view, which can be analyzed from regular meetings to health care. Therefore, this study involved purposely chosen patients who had experienced direct benefits from service quality in Community Health Service. There were 40 patients of Community Health Service in Tangerang Selatan City involved in this study, who have used the service for at least three years.

In this study, the elements of the problem were formed in the following hierarchy:

a. Target to be achieved: Nursing service quality.
b. Selection criteria: Using quality service dimensions, namely Reliability, Responsiveness, Assurance, Empathy, and Tangibility.

3 RESULTS AND DISCUSSION

The obtained data was processed using super Decision software. Following are the results obtained:

a. Reliability = 0.1365
b. Responsiveness = 0.1410
c. Assurance = 0.3700
d. Empathy = 0.2648
e. Tangibility = 0.0877

Assurance dimension amounted to 0.3700. It means that it was considered as the most important dimension in health quality service. Accordingly, Empathy was considered as the second important dimension in health quality service and was amounted to 0.2468. Then, Reliability was considered as the third important dimension in health quality service and was amounted to 0.1365. Next, Responsiveness was considered as the fourth important dimension in health quality service, which was amounted to 0.1410. Lastly, Tangibility was considered as the fifth important dimension in health quality service, which was amounted to 0.0877. Therefore, improving nursing service quality needs to consider these five dimensions.

Table 1. Definition of assessment and numerical scale.

Definition	Intensity of Importance
Equal Importance	1
Weak	2
Moderate importance	3
Moderate plus	4
Strong importance	5
Strong Plus	6
Very strong or demonstrated importance	7
Very, very strong	8
Extreme importance	9

4 CONCLUSION

The nursing service quality plays an important role in producing quality public health. Its quality has to be assessed from the five dimensions of quality service, namely: Tangibility, Empathy, Reliability, Responsiveness, Assurance. The findings showed that accordingly the most important criterion of nursing service in the community health service is Assurance, Empathy, Responsiveness, Reliability and Tangibility. Moreover, the use of Analysis Hierarchy Process method has significantly beneficial to the studies on health service quality, since previous related studies have not been found using this method.

This study suggests the nurses of public health centers, especially Community Health Service, who wish to achieve the best service quality, to not only provide the same level of service quality for all service criteria, but also they are suggested to provide quality services based on the level of preference (how important it is) based on each of the service quality criteria.

REFERENCES

Andaleeb, S.S. 2001. Service quality perceptions and patient satisfaction: A study of hospitals in a developing country. *Social Science & Medicine* 52(9): 1359-1370.

Anderson, E.A. 1995. Measuring service quality at a university health clinic. *International Journal of Health Care Quality Assurance* 8(2): 32-37.

Carman, J.M. 2000. Patient perceptions of service quality: Combining the dimensions. *Journal of Services Marketing* 14(4): 337-352.

Headley, D.E. & Miller, S.J. 1993. Measuring service quality and its relationship to future consumer behavior. *Journal of Health Care Marketing* 13(4).

Kenagy, J.W., Berwick, D.M. & Shore, M.F. 1999. Service quality in health care. *JAMA* 281(7): 661-665.

Kolstad, J.T. & Chernew, M.E. 2009. Quality and consumer decision making in the market for health insurance and health care services. *Medical Care Research and Review* 66(1): 28S-52S.

Kotler, Philip. & Keller, K.L. 2012. Marketing management, 14th edition. United State of America: Pearson.

Lawrence, M. & Olesen, F. 1997. Indicators of quality in health care. *The European Journal of General Practice* 3(3): 103-108.

Oermann, M.H. 1999. Consumers' descriptions of quality health care. *Journal of Nursing Care Quality* 14(1): 47-55.

Oermann, M.H. & Templin, T. 2000. Important attributes of quality health care: Consumer perspectives. *Journal of Nursing Scholarship* 32(2): 167-172.

Saaty, T.L. & Vargas, L.G. 2006. Decision making with the analytic network process. economic, political, social and technological applications with benefits, opportunities, costs and risks. Pittsburgh USA: Springer. Rws publication.

Saaty, T.L. 1977. A scaling method for priorities in hierarchical structures. *Journal of Mathematical Psychology* 15: 234-281.

Sadiq, S.M. 2003. Service quality in hospitals: More favorable than you might think. *Managing Service Quality: An Inter-National Journal* 13(3): 197-206.

Sofaer, S. & Firminger, K. 2005. Patient perceptions of the quality of health services. *Rev. Public Health* 26: 513-559.

Van Campen, C., Sixma, H.J., Kerssens, J.J., Peters, L. & Rasker, J.J. 1998. Assessing patients' priorities and perceptions of the quality of health care: the development of the QUOTE-Rheumatic-Patients instrument. *British Journal Of Rheumatology* 37(4): 362-368.

Wisniewski, M. 1996. Measuring service quality in the public sector: The potential for SERVQUAL. *Total Quality Management* 7(4): 357-366.

Youssef, F.N. 1996. Health care quality in NHS hospitals. *International Journal of Health Care Quality Assurance* 9(1): 15-28.

Zastowny, T.R., Stratmann, W.C., Adams, E.H. & Fox, M. L. 1995. Patient satisfaction and experience with health services and quality of care. *Quality Management in Healthcare* 3(3): 50-61.

Advances in Business, Management and Entrepreneurship – Hurriyati et al (eds)
© *2020 Taylor & Francis Group, London, ISBN 978-0-367-27176-3*

Trade liberalization and labor demand in Indonesia

R.D. Handoyo & F. Rabbanisyah
Universitas Airlangga, Surabaya, Indonesia

ABSTRACT: Trade liberalization agreement has made great opportunities to enhance economic growth for those participating countries. Trade liberalization would then inevitably affect factors of production within countries. The aim of study is to analyse the impact of trade liberalization on labour demand in manufacturing industries in Indonesia. This study used industrial manufacturing firm level data over the period 2008–2013 and estimates using panel data regression analysis following a fixed effect method. The results showed that tariff rates, value added, imports of raw materials, and export significantly affect the labour demand. Furthermore, labour wages negatively affect labour demands of the Indonesia manufacturing industry.

1 INTRODUCTION

Trade liberalization has become a blueprint for countries in the world to establish an open market with very few barriers to trade. It aims to stop or reduce state intervention in the form of entry barriers (tariffs and quotas) on trade and international economy (Mayana 2004). Theories regarding trade liberalization were first put forward by the classical economists, such as Adam Smith in 1776, who emphasized that international trade occurs when there is an absolute advantage in a country. David Ricardo (1817) stated that international trade is based on the comparative advantage of each country.

Currently, trade liberalization has spread to the whole world, including the community countries of Southeast Asia, ASEAN (Association of Southeast Asian Nations). On January 28, 1992, in Singapore, several ASEAN countries, including Indonesia, Singapore, Brunei Darussalam, Malaysia, the Philippines, and Thailand signed a trade liberalization agreement called AFTA (ASEAN Free Trade Area). The goal was to make the ASEAN region a competitive production base thereby attracting more foreign direct investment in ASEAN as well as increasing trade among members of the ASEAN Regional and Bilateral Policy Center 2012.

In Indonesia, years after the signing of the AFTA trade liberalization agreements, declines could be observed in the value of the average tariff on all trade products, from 12.51% in 1993 to 4.66% in 2013. In the same time period, there was also an increase in the total workforce in Indonesia. Since 1993, the number of workers in Indonesia has amounted to 76.72 million and continued to rise to the year 2013 with 110.8 million. But it is unclear if the effect of increase in variable labour is associated with trade liberalization.

Based on previous research conducted by Mitra and Shin (2012), Mouelhi and Ghazali (2013), and Njikam (2016), the survey data of manufacturing industry are used to see the impact of trade liberalization on labour demand. This present research uses industry-level data because generally during the liberalization of trade, the level of the tariff for the manufacturing sector is the highest compared to other sectors (Mouelhi & Ghazali 2013), indicating the manufacturing sector as the category of "government controlled" (Njikam 2016). On the other hand, the manufacturing industry in Indonesia accounts for 65% of employment (Ministry of Industry, 2016). The manufacturing industries has played a strategic role in employment in Indonesia. Thus, manufacturing industry data can show the contribution of the industry sector on employment in Indonesia during the liberalization trade period.

Based on this, the present study uses the variables that have been used by previous studies. This study seeks to examine the impact of trade liberalization through the variable tariff rate on imported raw materials, value added, imports of raw materials, labour costs and the value of exports to the labour demand using firm-level data in the Indonesian manufacturing industry. In addition to supporting, strengthening, and adding to the literature, this study can serve as a reference and create a benchmark for the extent to which these policies affect the level of market openness in Indonesia and its relation to labour demand.

1.1 *Liberalization of the labour demand*

Trade liberalization according to Rodrik (1997) is a condition in which trade was at a more open market conditions (international trade) through the reduction of related barriers that exist in the market.

Rodrik (1997) mentions that there are two paths that show the effect of trade liberalization leading to the demand for labour. The first path, trade liberalization allows import of raw materials, inputs of capital and work-in-process goods for processing in the importing country. The second path, associated with the Marshallian demand, through trade liberalization, the company will increase output efficiently by reducing the amount of labour. The linkage between trade liberalization and labour demand expressed by Rodrik (1997) is closely associated with international trade theory propounded by Hechscher-Ohlin (1991), in which international trade will be carried out by the industry that are labour-intensive and also capital-intensive (capital intensive). Trade liberalization will be a positive influence on the addition of labour for labour intensive industries. However, trade liberalization also negatively affects the demand for labour because it will be a process of substitution between the almost-finished raw material and the use of domestic labour.

1.2 Analysis model

A panel data regression model was used to examine the relationship of trade liberalization variable with labour demand. The econometric equations used can be formulated as follows:

$$logTK_{it} = \beta_1 + \beta_2 logVA_{it} + \beta_3 logIM_{it} \\ + \beta_4 persentarif_{it} + \beta_5 logw_{it} \\ + \beta_6 logEX_{it} + \varepsilon_{it}$$

Information:

TK_{it}	: Total employment in the industry i the period (years) t
VA_{it}	: The value added of industry i in period (years) t.
IM_{it}	: Imported raw material in industry i in period (years) t
$persentarif_{it}$	: The tariff rate of imported raw material in industry in period (years) t
w_{it}	: Labor wages in industry i the period (years) t
EX_{it}	: The export value of the industry i the period (years) t
ε_{it}	: Error term

2 METHOD

The dependent variable used is total labour force manufacturing industry that illustrates the demand for labour. Indicators of total employment are seen from the number of workers at each classification of the manufacturing industry in Indonesia. Independent variables used are the sum of value added, the tariff rate on imported raw materials, imported raw materials, labour wages, and the value of exports. The cross section data includes 17.677 Indonesian companies in the manufacturing industry, in nine

industry classifications: food, beverages, and tobacco; textiles and textile processed products; furniture and result of forest; footwear and leather products; industrial metals and metal products; paper and paper products; chemical and pharmaceutical industries; rubber and plastics industry; and other industries, as well as time-series data for the period 2008–2013. Data used were from Company Annual Survey of Big and Medium Manufacturing Industry sourced from BPS and WITS in World Bank.

Variable labor demand or total labor is derived from the number of production workers and other workers working in every company of the manufacturing sector which is divided into each classification in each year. Tariffs variable on imported raw materials in this study is Tariffs Most-Favored Nation (MFN) dutiable imports for each sector of the manufacturing industry. The variable of value-added production is measured by dividing the value of output per manufacturing company in a different class to the Consumer Price Index (CPI). The variable of imported raw materials obtained from dividing the total of imported raw materials with the CPI. The variable of wage is the average wage of workers obtained from the result of total labor costs divided by the total workforce in any manufacturing industries company divided by CPI. The variable of export value derived from multiplying the percentage of exports in the manufacturing sector in a different year with a total output of companies in each sector which then divided by CPI.

Table 1. Effect of variable in trade liberalization on labour demand in Indonesian manufacturing industry 2008–2013.

Variable	Coefficient
Dependent variable: logTK *(Labour Demand)*	
Constants	2.339
	(0.000)
Value Added (logVA)	0.218
	(0.000)
Imported Raw Material (logIM)	0.007
	(0.000)
Tariff of Imported Raw Material	(0.000)
Labor Wages (logw)	-0.028
	(0.000)
Exported Values (logEX)	0.003
	(0.000)
Observation	106.062
R-squared	0.6378
F-statistic	3547.66
Prob > F	0.0000

*Source: Data processed.
*Description: Figures in brackets [] stated p-value.

3 RESULTS AND DISCUSSION

This research uses panel data. In addition, regression is done by using Pooled Least Square, Random Effect Model and Fixed Effect Model. According to Hausman, obtained best model which is the Fixed Effect Model. So that the model estimates using Fixed Effect Model obtained the results shown in Table 1.

Table 1 shows Prob > F indicates the number 0.0000 less than the level of significance α by 1 percent, and the p-value for each variable which states that either simultaneously or partially, the independent variable in our model significantly affect the dependent variable.

The coefficient in Table 1 shows the different number and notations on each variable. The variable of value added (logVA), raw material imports (logIM), and exports (logEX) has significant positive effect on labor demand variables (logTK), with a coefficient value respectively 0.218, 0.007 and 0.003. Furthermore, the variable of tariffs on imported raw materials (persentarif) and labor wages (logw) has significant negative effect on the labor demand (logTK), with a coefficient value, respectively, 0.039 and 0.028.

These findings indicate that an increase of 1 percent in each of the added value, raw material import and export value, it will significantly increase the demand for labor in the manufacturing industry amounting to 0.218 percent, 0.007 percent and 0.003 percent. Conversely, a negative relationship shown by variable of tariffs on imported raw materials and labor wages to labor demand, indicate that an increase of one per cent on imported raw materials, tariffs, and labor wage, then it will significantly reduce the demand for labor, respectively, 0.039 and 0.028 percent.

Figure 1 shows the results of the estimation of each variable that describe the sectoral trade liberalization on the manufacturing industry in Indonesia during the period 2008–2013 using the Fixed Effect Method. Table 2 significance all variables. Food, beverages, and tobacco sectors significantly affect

the demand for labor in the industry. Influence shown in the sector of the food, beverages, and tobacco industry as a whole has the same effect with the variables of trade liberalization on labor demand in the Indonesian manufacturing industry.

All variables showed trade liberalization also affects the demand for labor significantly in the manufacturing of textile and apparel, furniture and processed forest products, as well as the chemical and pharmaceutical industries, although there are several variables that show different results of the analysis in the manufacturing industry as a whole. Additionally, the variable of imported raw material and export value in the industrial sector of footwear and leather products, the variable of imported raw materials and tariffs on imported raw materials in the industrial sector of metal and metal products, the variable of import raw materials, tariffs of imported raw material and exports value for the paper and paper products industry, and the variable of tariffs on imported raw materials did not significantly affect the demand for labor in the manufacturing sector of rubber and plastic in Indonesia.

4 DISCUSSION AND CONCLUSION

This study aimed to analyze the impacts of trade liberalization on labor demand in Indonesia. The value added has positive and significant impact on the demand for labor. This is consistent with the theory expressed by Cobb-Douglas, that demand for labor will arise because of the demand for the output. From these results, it can be said that the existence of trade liberalization agreements in Indonesia positively managed to trigger value-added manufacturing to be increased so that it is accompanied by an increase in the demand for labor.

Imported material raw has a positive and significant impact on the demand for labor in the manufacturing industry in Indonesia. The existence of trade liberalization agreements in Indonesia lowered both tariffs and non-tariffs for goods imported into the country.

The existence of trade liberalization agreements in Indonesia positively triggered some increases in imported raw materials which further led to an increase in the demand for labor to process the raw materials.

Tariff of imported raw material has a significant negative effect on the demand for labor in the manufacturing industry in Indonesia. This research obtains results that the decrease in tariff rates managed to increase the demand for labor, aligned with previous research.

The findings of this study stated that trade liberalization agreements have an impact on the increase in demand for imported raw materials in the manufacturing industry Indonesia during the period of implementation of the agreement, so the it encourages an increase in demand for labor. This positive

	Food, Beverage, and Tobacco	Textile and apparel	Furniture and Forest Product	Footwear and Leather Product	Metal and Metal Porduct	Paper and Paper Product	Chemistry and Pharmacy	Rubber and Plastic	Other industry
Constanta	2.359	9.733	1.025	0.058	2.621	3.432	2.951	2.392	1.824
Value Added (log VA)	0.182	0.288	0.317	0.354	0.19	0.162	0.179	0.172	0.153
Imported Raw Materials (log IM)	0.014	0.003	0.005	0.004	0.002	0.003	0.008	0.006	0.007
Imported Raw Materials Taeff (Persen Tariff)	-0.425	97.866	3.485	9.591	-1.79	-10.087	-0.411	-0.26	10.085
Wages (log W)	-0.022	-0.021	-0.031	-0.035	-0.043	-0.051	-0.042	-0.027	-0.27
Export Value (log EX)	-0.005	0.007	0.002	0.003	0.014	0.004	0.006	0.003	0.003
Observation	4735	3363	1791	524	731	410	853	1127	3841
R-squared	0.624	0.755	0.714	0.805	0.622	0.719	0.468	0.567	0.652

Figure 1. Variables influence on trade liberalization on labor demand manufacturing industry sectoral Indonesia in 2008–2013.

*Source: Data processed. *Description: The yellow color indicates significant figures.

relationship between trade liberalization with labor demand is evident in this study.

The findings of this study stated that trade liberalization agreements applied in Indonesia, on the other hand have an impact on wage increases, so it impacts on the decrease of the demand for labor. Despite the increase in wages due to trade liberalization has a negative effect on the demand for labor, it describes the level of competition in the labor market is increasing. Someone will increase the competence in working through an increase in productivity, this is indicated by the higher value-added in companies, therefore trade liberalization agreements through the decreasing of wages in other hand also have positive impact on the labor productivity.

The value of exports have a positive and significant impact on the demand for labor in the manufacturing industry in Indonesia. Liberalization of trade lead to an increase in greater output. Companies that produce labor-intensive goods will tend to increase the workforce so that it has a positive impact on the demand for labor in the manufacturing industry in Indonesia. The condition of export value in Indonesian manufacturing industry showed an upward trend during the period of implementation of trade liberalization.

In the food, beverages, and tobacco industry, the demand for labor in the food and beverage industry and tobacco is strongly influenced by the increase in value added, where the food beverage and tobacco industries have considerable opportunities to enhance its role in the condition of market openness as one of the sectors that absorb labor.

The use of imported raw materials significantly affect the demand for labor in the food, beverages, and tobacco industry, textile and apparel industry, the furniture and forest product industry, chemical and pharmaceutical industry, rubber and plastics industry, and other industries. This shows that with the implementation of trade liberalization will increase the intensity of the use of imported raw materials, which in turn will increase the demand for labor in the production process, due to the increase of raw materials, will lead to an increase in the demand for labor to produce these materials. The different conditions indicated by the processing industries of footwear and leather products, industrial metals and metal products, as well as paper and paper products industry in which the value of imports of raw materials for a third of the industrial sector did not significantly affect the demand for labor. This shows that the trend of using imported raw materials has no effect on employment, because intensity of the use of imported raw materials is relatively small for these industries or there is a probability that all three of the industry is a capital-intensive industry (capital intensive).

Tariffs on imported raw materials have a significant effect on the demand for labor in the manufacturing industries of food, beverages, and tobacco, textile and apparel industries, the furniture and the forest products industry, footwear and leather products, chemicals and pharmaceuticals, and other industrial sectors. On the other hand, tariffs on imported raw materials have no significant effect on the metal processing and metal products industries, paper and paper products industry, and rubber and plastics industries. This shows that the food, beverages, and tobacco industries as well as other industrial sectors that have significant influence through variable rates of imported raw materials, decreases in tariff of imported raw materials will decrease companies' expense to pay for imported raw material cost which result in fund allocation to improve production through increase in demand for labor.

Overall, the study found that during the period of implementation of the trade liberalization agreement in Indonesia, there was an increase in the demand for labor through the reduction of tariff rates, the increase in value added, an increase in raw material imports, and an increase in the value of exports in the aggregate in the manufacturing industry in Indonesia. On the other hand, trade liberalization agreements also have a negative impact on the demand for labor through an increase in labor costs impact on the decrease of the overall demand for labor in the manufacturing industry in Indonesia.

The conclusion is that trade liberalization significantly affected the demand for labor in the manufacturing industry in Indonesia. During the period of implementation of trade liberalization agreements, the demand for labor was increased through the variables of reduction of tariffs on imported raw materials, an increase in value added, an increase in imports of raw materials and an increase in the value of overall exports in the manufacturing industry in Indonesia. On the other hand, trade liberalization agreements also have a negative impact on labor demand through increased labor wages which then impact on the decrease of the overall demand for labor in the manufacturing industry in Indonesia.

REFERENCES

Heckscher, E.F. & Ohlin, B.G. 1991. Heckscher-Ohlin trade theory. *The MIT Press*.
Mayana. R.F. 2004. *Protection of industrial design in Indonesia*. Jakarta: Grasindo.
Mouelhi, R.B.A. & Ghazali, M. 2013. Impact of trade reforms in Tunisia on the elasticity of labour demand. *International Economics* 134: 78–96.
Njikam, O. 2016. Trade liberalization, labormarket regulations and labor demand in Cameroon. *International Review of Economics and Finance* 30: 01–17.
Ricardo, D. 1817. On the principles of political economy and taxation.
Rodrik, D. 1997. *Has Globalization Gone Too Far?* Washington, DC: Institute of International Economics.
Smith, A. 1776. *An inquiry into the nature and causes of the wealth of nations: Volume One*. London: printed for W. Strahan; and T. Cadell, 1776.

Advances in Business, Management and Entrepreneurship – Hurriyati et al (eds)
© 2020 Taylor & Francis Group, London, ISBN 978-0-367-27176-3

The effect of job satisfaction and Organizational Citizenship Behavior (OCB) on employee engagement at star-rated hotels in Indonesia

S. Sofiyah, E. Ahman & S.H. Senen
Universitas Pendidikan Indonesia, Bandung, Indonesia

ABSTRACT: This study aims to determine and analyze the effect of job satisfaction and Organizational Citizenship Behavior (OCB) on employee engagement in star-rated hotels in Indonesia. The research method used is descriptive and verification analysis. The data collection used was interviews using questionnaires accompanied by observation and literature techniques. The data analysis technique used was Path Analysis. The study population was 251 employees of 3-star hotels in Indonesia and 154 employees were taken as samples. Based on the results of the study we found that job satisfaction and Organizational Citizenship Behavior (OCB) partially had a significant influence on employee engagement.

1 INTRODUCTION

The successful organizations are founded on 3 important qualities that must be present in all employees, namely: competence, commitment or engagement, and contribution. In other words, it is important to realize that work performance and organizational success do not only depend on the competencies or cognitive skills possessed by employees, but also depend on how employees respond to the work and organization where they work. (Ulrich 2007)

According to Robinson et al. (2004), employee engagement is an individual's positive attitude towards the organization and organizational value. An employee who has a high level of attachment to the organization has an understanding and concern for the operational environment of the organization, is able to work together to improve the achievement of the work unit or organization through cooperation between individual employees and management.

According to a Gallup research, around the world, only 13% of employees work for an organization feel engaged (Gallup 2013). Research on employee engagement was also carried out by Gallup in Southeast Asian countries. The results state that the Philippines is the country with the highest employee engagement, with 29% of employees who can be said to be actively involved and engaged in the organization where they work. On the other hand Indonesia is in the 5th position where the total workers who feel engaged in the organization is only around 8%, meaning that there are still many Indonesians who do not feel engaged in their organization, which is 77%.

On several occasions, Kahn was declared as the first researcher to bring the concept of engagement to work (Avery et al. 2007). In 1990, Kahn developed a basic theory of engagement and disengagement in the workplace. Kahn used a number of data and data collection methods, especially in in-depth interviews to examine statements about engagement and disengagement. Based on a research conducted by Kahn (1990), employee engagement is defined as "*The harnessing of organization members' selves to their work roles; in engagement people employ and express themselves physically, cognitively, and emotionally during role performances*".

According to Gallup (2013), there are three levels of engagement for employees, namely:

1) Engaged. An engaged employee is a builder. They always show a high level of performance. These employees will be willing to use their talents and strengths in working every day and always work with passion and always innovate so that the company develops.

2) Not Engaged. Employees of this type tend to focus on tasks rather than achieving the goals of the job. They are always waiting for orders and tend to feel their contributions are ignored.

3) Actively Disengaged. This type of employee is a cave dweller. They consistently show resistance to all aspects. They only see the negative side on various occasions and every day, this type of being actively disengaged weakens what is done by engaged workers.

Job satisfaction is an interesting problem because it has been proven to have a great influence on employees and companies. For employees, job satisfaction will create a pleasant feeling at work. Meanwhile for companies, job satisfaction is useful in efforts to increase production, change attitudes and increase employee engagement. According to Osborn (1995), "*Job satisfaction is the degree to which an individual feels positively or negatively about the various facets of the jobs tasks, the work setting and relationship with co-worker*".

Sopiah (2008) mentioned several definitions of job satisfaction, including: Luthans (1995), who suggests that, *"Job satisfaction is a pleasurable or positive emotional state resulting from the appraisal of one's job or job experience"*. Job satisfaction is an emotional expression that is positive or pleasant as a result of an assessment of a job or work experience. Mathis & Jackson (2000) suggested, *"job satisfaction is a positive emotional state resulting from one's job experience"*. Job satisfaction is a positive emotional statement that is the result of evaluation of work experience.

Job satisfaction has many dimensions, which can represent the attitude as a whole or refers to the part of someone's work. According to Robbins (2001), in general job satisfaction consists of 5 (five) dimensions, namely:

a. The work itself

   According to Robbins, employees tend to like jobs that provide opportunities for them to prove their skills and abilities, and provide varied tasks, freedom and feedback about the work they do.

b. Wages and promotions

   According to Robbins, employees usually want a wage and promotion system that suits their expectations. Basically what employees want is not the amount of wages they receive but justice in terms of compensation. Similarly, they expect justice in the promotion system.

c. Working conditions

   Robbins & Luthans (2002) both argue that employees usually pay great attention to their workplace environment for personal comfort and to support their work.

d. Colleagues, supervisors and bosses

   Robbins stated that for most employees, the workplace is also a place for socialization so it is very important for them to have colleagues who support them and can work together well with them. They usually expect superiors who pay attention to their welfare, provide lots of guidance and assistance in work, communicate well, and want to involve themselves in work.

e. Unity between work and personality

   Robbins adds an important element in job satisfaction, namely employees tend to feel satisfied if there is a match between their personality and work.

   Organ (2006) suggest that, *"Organizational citizenship behavior (OCB) as individual behavior that is discretionary, not directly or explicitly recognized by the formal reward system, and that in the aggregate promotes the effective functioning of the organization"*.

Organ (1995) and Sloat (Zurasaka 2008), suggested several factors that influence OCB as follows: Organizational culture and climate, personality and mood, perception of organizational support,

perception of the quality of subordinate relationships/interactions, work period, and gender.

According to Organ et al (Meilina 2017) the OCB dimensions are as follows:

a. Altruism: A behavior where employees help other employees without coercion on tasks that are closely related to organizational operations.

b. Courtesy: A behavior where employees alleviate problems related to work faced by others

c. Conscientiousness: A behavior where employees deliver performance and preconditions for roles that exceed minimum standards.

d. Sportmanship: A behavior where employees abstain from making destructive issues even if they feel annoyed

e. Civic Virtue: A behavior where employees demonstrate voluntary participation and support for organizational functions both professionally and socially.

Today the hotel and resort business is growing rapidly in Indonesia, especially in the Sumedang district as a "Sundanese cultural center". Tourists come and stay based on several things including distance, costs that need to be spent, as well as services provided, where service is the most valued and highlighted thing. Member of Commission V Nurhasan Zaidi and the field of tourism marketing development said, that they hope the government will endeavor to prepare the community morally, skillfully, and entrepreneurially to be able to take advantage of the Kertajati Majalengka International Airport and the Cisundawu toll road, so Sumedang can become a magnet for good tourist attraction locally and internationally because Sumedang is preparing to become a tourism destination (www.sumedangonline.com).

Given the importance of employee engagement, the researchers will analyze how starred hotels in Sumedang district as objects of research provide support for employees in carrying out their duties and responsibilities and how this can improve employee engagement.

Based on the thinking framework outlined, our hypotheses are namely:

a. Job Satisfaction (X1) has an effect on Employee Engagement (Y).

b. Organizational Citizenship Behavior (OCB) (X2) has an effect on Employee Engagement (Y).

2 METHOD

The research method used is descriptive and verification analysis. The data collection technique used was interviews using questionnaires accompanied by observation and literature techniques. The data analysis techniques used was Path Analysis. The study population was 251 employees of 3-star hotels in Sumedang district and 154 employees were taken as samples.

3 RESULTS AND DISCUSSION

Based on the results of descriptive analysis, variable X1 gives an overall description of the answers regarding job satisfaction of 3-star Hotel employees in Sumedang Regency. Recapitulation of answers provides an average value of employee job satisfaction variables of 4.18 in the value interval of 3.41 - 4.20 with a percentage of 83.64%, which means the job satisfaction of 3-star Hotel employees in the District is in the good or satisfied criteria. When compared between dimensions, the dimension of promotion provides the best description while the dimensions of supervision provide the least description.

The result of descriptive research of variable X2 gives an overall description of the answers about organizational citizenship behavior of 3-star Hotel employees in Sumedang Regency. The answer recapitulation gives the average value of employee organizational citizenship behavior variable of 4.20 in the interval value of 3.41 - 4.20 with a percentage of 83.52%, which means that organizational citizenship behavior of 3-star Hotel employees in the District is in the criteria of good. When compared between dimensions, the dimension of conscientiousness gives the best description while the dimension of courtesy gives the least description.

The results of descriptive analysis of variable Y provide an overall description of the answers regarding employee engagement of 3-star Hotel employees in Sumedang Regency. The recapitulation of the answers gives the employee engagement variable an average value of 4.33 in the interval value of 4.21 - 5.00 with a percentage of 86.60%, which means that employee engagement of 3 Star Hotel employees in the Sumedang District is in the criteria of very good. When it is compared between the dimensions, the dimension of vigor and dedication gives the best description while the absorption dimension provides the least description.

The first beta coefficient is 0.269. the tcount is 3.916 with an α significance level of 5%. Which means the value of t table or t0.05.151 =

Table 1. Coefficients[a].

| Model | Unstandardized Coefficients | | Standardized Coefficients | | |
	B	Std. Error	Beta	t	Sig.
(Constant)	.185	2.375		-.078	.938
1 X1	.170	.043	.269	3.916	.000
X2	1.176	.124	.652	9.507	.000

* Dependent Variable: Y.
Source: Processed data. SPSS 21.0 for Windows. 2018.

1.833. So. because tcount = 3.916 is greater than t table = 1.833. then we reject H0 or in other words job satisfaction (X1) affects employee employee engagement (Y). Based on the significancelevel. if the significance is <0.05 then H0 is rejected and if the significance is> 0.05 then H0 is accepted. Because the significance of job satisfaction (X1) is 0.000 <0.05. H0 is rejected. This means that job satisfaction has an effect on employee engagement of employees of 3-star hotels in Sumedang Regency.

The second beta coefficient is 0.652. The tcount is 9.507 with an α significance level of 5%. Which means the value of t table or t0.05.151 = 1.833. So. because tcount = 9.507 is greater than t table = 1.833. We reject H0 or in other words organizational citizenship behavior (X2) affects employee engagement (Y). Based on the significance level. if the significance is < 0.05 then H0 is rejected and if the significance is > 0.05 then H0 is accepted. Because of the significance of (X2) which is 0.034 < 0.05. H0 is rejected. This means that organizational citizenship behavior has an effect on employee engagement of employees of 3-star hotels in Sumedang Regency.

Based on the table above. the regression equation model is as follows:

$$Y = \rho YX1 + \rho YX2 + e1 \qquad (1)$$

The value of R2 or R square is 0.792. This shows that the contribution of the influence of Job Satisfaction (X1) and OCB (X2) on Employee Engagement (Y) is 79.2% while the remaining 20.8% is the contribution of other variables not included in the study. Meanwhile. The value of e1 can be found with the formula e1 = $\sqrt{(1-0.792)}$ = 0.456. The path diagram obtained is as follows:

3.1 Summary of model parameter results

That job satisfaction and organizational citizenship behavior (OCB) can encourage employee engagement is reinforced by the results of research conducted by (Garg & Mishra 2018, Wahyu 2013, Tahir 2014).

Table 2. Model summary.

Model	R	R Square	Adjusted R Square	Std. Error of the Estimate
1	.890[*]	.792	.790	3.34561

* Predictors: (Constant). X2. X1.
Source: Processed data. SPSS 21.0 for Windows. 2018.

Table 3. Summary of results.

Model	Path Coefficient	t	p	R^2
Structural 1 (X1, X2 - Y)				
X1 (ρYX1)	0.269	3.916	0.000	0.792
X2 (ρYX2)	0.652	9.507	0.000	

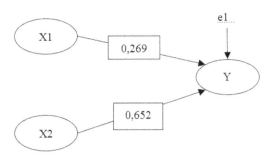

Figure 1. Path diagram.

4 CONCLUSIONS

Based on the results of the research that has been done. It can be concluded that:

a. The results showed that job satisfaction had a significant effect on employee engagement.
b. Research results show that OCB has a significant effect on employee engagement.

ACKNOWLEDGEMENTS

First of all. Thank to ALLAH S.W.T for his grace and guidance in giving us full strength to complete this paper. We would like to thank Mr. Prof. Dr. H. Eeng Ahman. M.Si for all support and guidance. Because without his guidance our paper cannot be done properly. Besides that. Acknowledgments are conveyed to all those who have assisted in the making of this paper. May all guidance. Help and support from all parties are rewarded by Allah SWT.

REFERENCES

Avery. D. R.. McKay. P. F.. & Wilson. D. C. 2007. Engaging the aging workforce: the relationship between perceived age similarity. satisfaction with coworkers. and employee engagement. *Journal of Applied Psychology* 92(6): 1542–1556.

Gallup. 2013. State of the Global Workplace. Employee Engagement Insights for Business Leaders Worldwide. [Online] Retrieved *http://www.gallup.com/file/services/ 176735/ State of the Global Workplace Report 2013.pdf \npapers2://publication/uuid/4F576D34-017E-4BC6- 8B6E-E3760C5FCD5E.* Accessed 18 June 2018.

Garg. K., Dar. I. A. & Mishra. M. 2018. Job Satisfaction and Work Engagement: A Study Using Private Sector Bank Managers. *Advances in Developing Human Resources* 20(1): 58–71.

Kahn, W. A. 1990. Psychological conditions of personal engagement and disengagement at work. *Academy of Management Journal* 33(4): 692–724.

Luthans, F. 2006. *Perilaku organisasi (10 cetakan).* Yogyakarta: Andi Offset.

Mathis, R. & Jackson, J. 2000, Job analysis and the changing nature of jobs. in human resource management (Cincinnati, OH: South-Western College Publishing), 9th Edition.

Meilina, R. 2017. Kepuasan kerja. komitmen organisasi. organizational citizenship behaviour (OCB) dan pengaruhnya terhadap kualitas layanan publik.

Organ, D. W., Podsakoff. P. M. & MacKenzie, S. B. 2006. Organizational citizenship behaviour: Its nature. antecedents. and consequences. Sage.

Ratanjee, V. & Emond, L. 2013. Why Indonesia must engage younger worker [Online]. Retrived https://news. gallup.com/businessjournal/166280/why-indonesia-engage-younger-workers.aspx,%E2%80%9D. Accessed 5 May 2018.

Robbin, S.P. 2001. *Perilaku organisasi: konsep. kontroversi. aplikasi.* Jakarta: Prenhallindo.

Robinson, D., Perryman, S. & Hayday, S. 2004. The drivers of employee engagement.

Schermerhorn, J.R., Hunt, J.G. & Osborn, R.N., 1995. *Basic organizational behavior.* J. Wiley.

Sopiah, D. 2008. Perilaku organisasional (1st ed.). Yogyakarta: CV. Ani Offset.

Tahir, R. 2014. Kualitas kehidupan kerja. perilaku kewargaan organisasional. dan keterikatan karyawan.

Ulrich, D. 2007. The talent trifecta. *Workforce Management* 86(15): 32–33.

Wahyu, A.D. 2013. The relationship between employee engagement. organizational citizenship behavior. and counterproductive work behavior. *International Journal of Business Administration* 4(2): 46–56.

Zurasaka. A. 2008. Teori Perilaku Organisasi.

Advances in Business, Management and Entrepreneurship – Hurriyati et al (eds)
© 2020 Taylor & Francis Group, London, ISBN 978-0-367-27176-3

Person-organization value fit and employee attitude: A study of bluecollar employees

R.A. Pebriani & R. Hurriyati
Universitas Pendidikan Indonesia, Bandung, Indonesia

ABSTRACT: Most employee-attitude and behavior is determined by both personal and situational characteristics. Person Organization (PO) value fit, which is defined as the congruence between individual and organizational values, supports this assumption. The higher person-organization value fit perceived, the higher the employee attitude showed to be positive. PO fit has been extensively investigated in the empirical literature. This study focused on blue collar employees to examine the relationship between the fit of person- and organizational values with the employee attitude towards the organization. The overall findings suggested that perceived person-organizational value fit is positively associated with the employee attitude towards job satisfaction and employee commitment to the organization.

1 INTRODUCTION

An organization can run effectively if it is supported by the quality of human resources. However, the main problem is how to manage human resources properly to reach the goal of the organization. In line with this challenge, person organization fit is considered to be able to answer the problem. Thus, the fit between individual values and organizational values represents one of the main factors to maintain employee commitment in competitive business environments. Due to this fact, an organization is required to create person-organization value (P-O fit) accurately through the process of hiring, communication and socialization to find a high degree of P-O fit, which in the future will have a high compatibility between employees and the organization.

Person-organization fit deals with the congruence between the employee's personal values and the values of the organization (Kristop 1996). Further, Kristop (1996) defined PO fit as the compatibility between people and the organizations in which they work. Essentially, PO fit theory posits that there are characteristics of organizations that have the potential to be congruent with characteristics of individuals, and that an individuals' attitudes and behaviors will be influenced by the degree of congruence or "fit" between individuals and organizations.

The concept of P-O value fit argued that if an employee fits well with an organization, they are likely to perform more positive work-related attitudes and behavior such as higher satisfaction with and higher commitment to organizations. This relation is supported by the literature and the studies that have found the relationship between PO fit and

attitudes and behavior (Amos 2008, Kim et al. 2010, and Park 2013).

The study of PO fit has been conducted widely and has been an area of interest among both researchers and managers in recent years, whose concern is the congruence of employee and organization related to work attitudes. Amos and Weathington (2008) examine the relationship between value congruence of an employee and organization in graduate and undergraduate students from a midsize university in the southern United States, which reveals that there is a positive relation among the values dimension in the congruence of an employee and organization. In addition, Park (2013) focusing on the Korean Public sector shows that the value of person and organization fit is positively associated with employee attitude such as job satisfaction, career satisfaction and organizational commitment. However, this study tries to modify the study of Amos and Weathington (2008) which focuses more on blue-collar employees in Islamic Foundation. Moreover, this study is conducted to explore the relationship between Person Organization fit and employee attitude of blue collars employees and the impact on their attitude. In the present study we were interested to examine the person organization value fit dimension and their relation of P-O fit to job satisfaction and organizational commitment in order to understand how this value fit dimensions affect blue collar employee attitude. The value is said to be fit when the value of an employee matches those of an organization and their coworkers are also colleagues in the organization. We suggest that value fit between employee and organizations is positively related to attitude such as job satisfaction and organizational commitment. Therefore, this study investigated this relation by using the seven values of Peters and Watermen.

1.1 Person organization fit

P-O fit refers to the match between employees' values and their organizations values (Kristop 1996) and (O'Reilly et al. 1991). It is one of the person-environment studies that received much attention in many research areas such as organizational psychology, leadership, counseling, organizational behavior and management and development of human resources (Edwards et al. 2006).

Some researchers argued that the congruence of the value in employees and organization influenced the employee selection process and employee attitude to choose the job (Chatman 1991). Peters and Waterman (1982) identified seven value congruences between an employee and organization. These values are (a) superior quality and service, (b) innovation, (c) importance of people as individuals, (d) importance of details of execution, (e) communication, (f) profit orientation, and (g) goal accomplishment. Based on those values, we propose that employees who have a high level of fit on these value dimensions will also report higher levels of employee attitude towards job satisfaction and organizational commitment.

1.2 P-O fit and job satisfaction

Job satisfaction refers to one's feelings or condition of mind according to the nature of the work. According to Gibson et al. (1985), it is a person's attitude towards their work. This attitude comes from their perception of work. Job satisfaction is influenced by various factors such as the kind of organization policies, supervision, administration, salary and quality of life.

Generally, employees showed a higher level of satisfaction when their values matched those of their coworkers (Adkins et al. 1996). In addition, employees who perceived their supervisor's value to be similar to their own values, were found to have more satisfaction in their job (Meglino et al. 1989). Based on the studies that were conducted earlier, it is demonstrated that there was a positive relation of P-O fit on job satisfaction (O'Reilly et al. 1991, Edwards et al. 2006, Boxx et al. 1991, Erdogan et al. 2004, Amos & Weathington 2008, Kim et al. 2010, and Park 2013).

1.3 Person organization fit and organizational commitment

Organizational commitment is reflected as the feeling of like or dislike towards the organization in which the employee works and refers to the employee 's loyalty towards the organization. In other words, organizational commitment is the employee's attitude to the values of the organization. It is showed by an individual acceptance of the vision, mission and value organization, without a willingness to leave the organization. This is in line with Allen and Meyer (1996), who define organizational commitment as a psychological link between the employee and his or her organization that makes it less likely that the employee will leave the organization. Alan and Meyer also suggest that there are three components of organizational commitment;

1. Affective commitment in which employees stayin their job because they want to stay.
2. Continuance commitment in which employees stay in an organization because it would cost them more to leave it.
3. Normative commitment in which employees remain a part of their organization because they feel obligated to stay.

Some literature identified several studies that examined the relation between PO value fit and organizational commitment (Adkins et al. 1996, Meglino et al. 1989, Posner 1992, Ugboro 1993, Verquer et al. 2003, Rosete 2006, and Lawrence 2009). Those studies describe the positive relationship between PO Fit and organizational commitment.

2 METHOD

This study is a descriptive method. Through this way, the study describes the relation of person organization value fit, job satisfaction and organizational commitment. The sample consisted of 30 bluecollar employees; the data was analyzed by conducting regression analysis with the help of SPSS. This study has two variables; person organization value fit and employee attitude. The PO value fit variable has seven dimensions; (a) superior quality and service, (b) innovation, (c) importance of people as individuals, (d) importance of details of execution, (e) communication, (f) profit orientation, and (g) goal accomplishment. In addition, employee attitude has two dimensions; job satisfaction and organizational commitment. The hypothesis of this research is that a positive relation exists between person organization value fit and job satisfaction and organizational commitment.

3 RESULT AND DISCUSSION

The result of the study supports the hypothesis. Total PO value fit dimension is significantly associated with job satisfaction ($r = 0.90$) and organizational commitment ($r = 0.90$). Besides, each dimension of value fit significantly correlated with job satisfaction and organizational commitment. We also calculated correlations for congruence of each value dimension.

Superior quality and service were significantly cor-related with job satisfaction (r = 0.20); organiza-tional commitment (r = 0.15); Innovation was associated significantly with job satisfaction (r = 0.19); organizational commitment (r = 0.17). The value of the importance of people as individuals sig-nificantly correlated with job satisfaction (r = 0.35); organizational commitment (r = 0.30). The import-ance of details of execution correlated significantly with job satisfaction (r = 0.19); organizational com-mitment (r = 0.14). The value of communication cor-related significantly with job satisfaction (r = 0.28); organizational commitment (r = 0.22). In addition, the value of goal accomplishment significantly cor-related with job satisfaction (r = 0.27); organiza-tional commitment (r = 0.18). The value of profit orientation significantly negatively correlated with organizational satisfaction (r = 0.16). Profit orienta-tion significantly correlated with job satisfaction (r = 0.21); organizational commitment (r = 0.15). More-over, based on the regression analysis there was a relation between person organization value fit and employee attitude. We found that the total value fit of person and organization had a significant relation (p < .001) with job satisfaction (β = .21); organiza-tional satisfaction (β = .21). Much research has shown that Person Organization value fit is signifi-cantly related to job satisfaction and organizational commitment (Farooqi & Nagendra 2014, Astakhova 2015, and Sarac et al. 2017) and the present results further support these previous findings. In this research, the dimension of person organization fit among blue collar employees in Islamic Foundation shows that the value of importance of people as indi-viduals is the highest value dimension which affects the employee attitude towards job satisfaction and organizational commitment.

4 CONCLUSION

The results of the present study support the idea that employees who fit well or perceive themselves as fit-ting well in an organization will likely be more satis-fied with their job and more committed to remaining with the organization. In addition, it is important for an organization to understand how the value of person and organization could affect employees' atti-tudes, mainly in satisfaction and commitment. Based on the explanation above we conclude that values can influence employees' attitudes. Specifically, either the fit of person and organizational values can relate to job satisfaction and organizational commit-ment. When individual employees' values match those of their organization, they are likely to report a higher level of satisfaction and commitment. Fur-ther research is needed to expand on the relation between person – organization value fit and differing employee attitudes between blue collar and white-collar employees.

REFERENCES

Adkins, C.L., Ravlin, E.C. & Meglino, B.M. 1996. Value congruence between co-workers and its relationship to work outcomes. *Group & Organization Management* 21 (4): 439–460.

Allen, N.J. & Meyer, J.P. 1996. Affective, continuance, and normative commitment to the organization: An examin-ation of construct validity. *Journal of Vocational Behav-ior* 49(3): 252–276.

Amos, E.A. & Weathington, B.L. 2008. An analysis of the relation between employee—Organization value con-gruence and employee attitudes. *The Journal of Psych-ology* 142(6): 615–632.

Astakhova, M.N. 2016. Explaining the effects of perceived person-supervisor fit and person-organization fit on organizational commitment in the US and Japan. *Jour-nal of Business Research* 69(2): 956–963.

Boxx, W.R., Odom, R.Y. & Dunn, M.G. 1991. Organiza-tional values and value congruency and their impact on satisfaction, commitment, and cohesion: An empirical examination within the public sector. *Public Personnel Management* 20(2): 195–205.

Chatman, J.A. 1991. Matching People and Organiza-tions: Selection and Socialization in Public Account-ing Firms. *Administrative Science Quarterly* 36(3): 459–484.

Edwards, J.R., Cable, D.M., Williamson, I.O., Lambert, L. S. & Shipp, A.J. 2006. The phenomenology of fit: link-ing the person and environment to the subjective experi-ence of person-environment fit. *Journal of Applied Psychology* 91(4): 802.

Erdogan, B., Kraimer, M.L. & Liden, R.C. 2004. Work value congruence and intrinsic career success: the com-pensatory roles of leader-member exchange and per-ceived organizational support. *Personnel Psychology* 57 (2): 305–332.

Farooqui, M.S. & Nagendra, A. 2014. The impact of person organization fit on job satisfaction and perform-ance of the employees. *Procedia Economics and Finance* 11: 122–129.

Gibson, J.L., Ivancevich, J.M. & Donnely, J.H. 1985. *Organizations 5th Ed/Organisasi: Perilaku, Struktur, Proses Jilid I*. Jakarta, Erlangga, 1985.

Kim, J.S., Park, Y.K. & Yim, H.C. 2010. A study on the effects of person-organization fit on the job attitudes using OCP. *Korean Journal of Industrial and Organiza-tional Psychology* 23 (4): 711–731.

Kristof, A. L. Person-organization fit: An integrative review of its conceptualizations, measurement, and implications. *Personnel Psychology*, 49, 1–49, 1996.

Lawrence, P. 2009. Values congruence and organizational commitment: P-O fit in higher education institutions. *Journal of Academic Ethics* 7(4): 297–314.

Meglino, B.M., Ravlin, E.C. & Adkins, C.L. 1989. A work values approach to corporate culture: A field test of the value congruence process and its relationship to individ-ual outcomes. *Journal of Applied Psychology* 74(3): 424–432.

O'Reilly, C. A., Chatman, J. & Caldwell, D. F. 1991. People and organizational culture: A profile comparison approach to assessing person-organization fit. *Academy of Management Journal*, 34(3): 487–516.

Park, S. 2013. The relationship between person-organization value fit and job satisfaction: Toward identifying the moderating roles of individual and organizational characteristics. *Journal of Human Resource Management Research* 20(2): 1–26.

Peters, T.J. & Waterman, R.H. 1982. *In search of excellence: Lessons from America's best-run companies.* New York: Harper and Row.

Posner, B.Z. 1992. Person-organization values congruence: No support for individual differences as a moderating influence. *Human Relations* 45(4): 351–361.

Rosete, D. 2006. The impact of organizational values and performance management congruency on satisfaction and commitment. *Asia Pacific Journal of Human Resources* 44(1): 7–24.

Saraç, M., Meydan, B. & Efil, I. 2017. Does the relationship between person–organization fit and work attitudes differ for blue-collar and white-collar employees? *Management Research Review* 40(10): 1081–1099.

Ugboro, I.O. 1993. Loyalty, value congruency, and affective organizational commitment: an empirical study. *American Journal of Business* 8(2): 29–36.

Verquer, M.L., Beehr, T.A. & Wagner, S.H. 2003. A meta-analysis of relations between person–organization fit and work attitudes. *Journal of Vocational Behavior* 63 (3): 473–489.

Advances in Business, Management and Entrepreneurship – Hurriyati et al (eds)
© *2020 Taylor & Francis Group, London, ISBN 978-0-367-27176-3*

The impact of implementation of financial rewards and occupational safety and health on job satisfaction at X Company

B. Widjajanta, S. Sumiyati, M. Masharyono, N. Fadhlillah & H. Tanuatmodjo
Universitas Pendidikan Indonesia, Bandung, Indonesia

ABSTRACT: The purpose of this study is to see the influence of financial reward and occupational health and safety on job satisfaction. The design of this study is cross sectional. This research used a verification approach with an explanatory survey method with a total of 60 respondents. A questionnaire was used as a research instrument to collect data. The analytical technique used was a multiple linear analysis. Based on the results of the research, it can be seen that the magnitude of the effect of financial reward and occupational health and safety is 53.8%, the financial reward has an effect toward job satisfaction is equal to 32.9%%, and the influence of occupational health and safety toward job satisfaction is equal to 38.1%.

1 INTRODUCTION

The success of an organization must be supported by qualified employees (Umar 2015) because there is no more important resource than human beings (DeNisi & Grifin 2008). Organizations need employees who can work effectively and efficiently, and an organization needs to be able to manage employees well (Sumiyati et al. 2016). The effectiveness and efficiency in achieving company goals will not work without the support from qualified employees (Senen 2008). Quality improvements and labor utilization are important factors in increasing the rate of economic growth of a company (Manoj & Das 2016).

Problems regarding job satisfaction are interesting for academics and practitioners to study. Job satisfaction has become one of most studied variables in organizational behavior and human resources (Wan-Yih & Htaik 2011). Job satisfaction level is an interesting issue from the point of view of employees, managers, and scientists (Bakotić & Tomislav 2013). Keeping employees satisfied with their job is a main priority for every company (Gregory 2011). Companies with relatively high employee satisfaction levels show better performance rates (Armstrong 2014). Workers with high satisfaction levels tend to be more productive and have higher job involvement as well (Qasim & Sayeed 2012).

Job satisfaction affects the desirable turnover by the organization. Job satisfaction can be an important indicator of how employees feel about their work. Managers need to be interested in discussing employee job satisfaction as employee behavior can provide warnings about potential problems (e.g., intentions to quit working, decreased productivity, absence, and organizational turnover). Managers, supervisors, human resource specialists, employees, and citizens generally should be concerned about ways to improve job satisfaction (Tessema et al.

2013). Job satisfaction can be seen as a machine for organization to bring better change. Changes in improvement of services to the employees will also improve services to customers (Ariani 2015). Satisfied employees will produce quality performance and increase company revenue (Odembo 2013).

Workers who are dissatisfied with their work will not be motivated to work better so that organization will not achieve success (Dobre 2013). Low employee job satisfaction will result in a corresponding low level of commitment and might affect performance and achievement of overall organizational goals (Rast & Tourani 2012).

Problems in employee job satisfaction levels occur in industries ranging from telecommunications (Khartik & Saratha 2012), hotels (Soni & Rawal 2014), banking (Hoshi 2014), catering (Lin 2015) manufacturing (Ravichandran & Rajan 2015), education (Saeed & Nasir 2016), hospitals (Nemmaniwar & Deshpande 2016), service industry (Zaidi & Iqbal 2012), and electricity companies (Utami et al. 2016). Important factors can help provide sustainable long-term results for organizations in the service sector if the company makes employee satisfaction in the center of the management process (Nedeljkovic 2012).

A low level of employee job satisfaction is a very important concern in companies in the service industry (Ch et al. 2015). Many studies have shown the results of low employee satisfaction in service industries occurring in several countries, namely China (Yee et al. 2008), Canada (Sledge et al. 2011), Pakistan (Daniel 2012), and Ghana (Sarwar & Abugre 2013).

Based on several studies, there are factors that can affect employee satisfaction, such as compensation (Yaseen 2013), salary (Tessema et al. 2013), quality of work life (Soni & Rawal 2014), career development and organizational commitment (Kaya & Ceylan 2014), rewards (financial or non-financial) and recognition (Imran et al. 2014), and motivation (Umar 2015). Other factors that may affect employee satisfaction are work environment (Saeed & Nasir 2016), job safety and health (Huang et al. 2016), and cynicism and organizational commitment (Khan et al. 2016).

Solutions taken to improve employee satisfaction are through financial rewards and job safety and health. The first solution, financial reward, was studied by Ali and Akram (2012), who proved that financial rewards offering will motivate and make employees feel satisfied. The second solution is job safety and health; according to a study conducted by Huang et al. (2016), there is a relationship between job safety and health and employee job satisfaction.

The purpose of this present study is to examine the impact of financial rewards and job safety and health on job satisfaction.

2 METHOD

This study was conducted to find the influence of financial rewards and job safety and health on job satisfaction. The independent variable of this research was financial rewards: pay, bonuses, and insurance; and job safety and health: safety and health inspections, accident prevention, safety and health communication, and safety and health training. The dependent variable in this research was job satisfaction: personal dispositions, task and roles, and supervisors and co-workers.

The object of analysis in this study was Company X in Bandung. This research was conducted for less than a year, and the data collection technique used in this research was cross-sectional method. The saturated sample used in this study was 66 employees. Data collection used in this research included literature review, observation, interview, and questionnaire. The verification data analysis used multiple linear regression analysis with SPSS 22.0 software for Windows.

3 RESULT AND DISCUSSION

Based on the multiple linear regression equation, the constant value of 47.299 states that if the variable of financial reward and job safety and health is zero (0), then the job satisfaction variable will be worth 47.299. The regression coefficient on the variable of financial reward is 1.051, meaning that if the financial reward increases by one unit and the other variable is constant, then the job satisfaction variable will increase by 1.051. Regression coefficient on job safety and health variable is 0.561, meaning that if safety and health work increase by one unit and other variable constant, hence the job satisfaction variable will increase equal to 0.561 unit.

To understand the influence of the variable of financial reward on the job satisfaction variable

shows the existence of R-Square value equal to 0.329. Based on the calculation, the coefficient of determination for financial reward towards job satisfaction is 32.9%. In other words, the effect of financial reward on job satisfaction is 32.9%.

To determine the influence of job safety and health variables on job satisfaction indicate a value of R-Square is 0.381. Based on the calculation result, the coefficient of determination for job safety and health on job satisfaction is 38.1%. In other words, the effect of job safety and health on job satisfaction is 38.1%.

To determine the influence of variable of financial reward and job safety and health on the job satisfaction variable, the result shows the value of R-Square equal to 0.538. Based on this calculation, the coefficient of determination for financial reward and job safety and health on job satisfaction is 53.8%. In other words, job satisfaction is influenced by 53.8% by financial reward and job safety and health while 46.2% is influenced by other factors than financial rewards and job safety and health.

Hypothesis test in this research is by using F_{count} equal to 36,662 with significance level of 5%. Compared with the F_{table} value, obtained the value of 3.14. Due to the value $F_{count} > F_{table}$ (36.662 > 3.14) then Ho is rejected and Ha accepted, meaning that financial reward and job safety and health together have effect on job satisfaction of employees of X Company in Bandung.

Hypothesis test in this research is by using t test (t-Test), we acquired the value of t_{count} equal to 5,596 for financial reward and 6,281 for job safety and health. A significant level (α) of 5%, and degrees of freedom (v) = 63 = (n - (k + 1)) obtained table value 3.452. According to Ghozali (2011), t test is done by comparing the significance of t_{count} with t_{table} with condition:

1. H0 is accepted and Ha is rejected if $t_{count} \leq t_{table}$ to α = 0.052.
2. H0 rejected and Ha accepted if $t_{count} > t_{table}$ for α = 0.05.

Since $t_{count} > t_{table}$ or 5.596 > 3.452, then Ha is accepted, meaning that financial rewards have an impact on job satisfaction. The value of t_{count} for job safety and health is 6.281. Since $t_{count} > t_{table}$ or 6.281 > 3.452, then Ha is accepted, meaning that job safety and health have an impact on job satisfaction of employees of X Company in Bandung.

This means that job satisfaction is influenced by financial rewards and job safety and health. The value of the direct influence of financial rewards and job safety and health towards job satisfaction is 53.8%. It can be concluded that the financial rewards and job safety & health affect the job satisfaction. Usman (2011) says that the factors impacting job satisfaction are (1) service rewards, (2) sense of security, (3) interpersonal influences, (4) working environment conditions, (5) opportunities for development, and (6) self-improvement.

Organizations reward employees as a form of reciprocity given to the performance provided by employees. Employees expect financial and nonfinancial rewards to improve their service and performance so they will be satisfied (Yousaf 2014). Providing financial rewards is not a major factor to satisfy employees, but if compensation is inadequate or uneven, it can lead to discontent (Armstrong 2015). The rewards needed in the employees depend on how each individual needs. Management should pay attention to financial rewards levels so as to not affect employee perceptions negatively (Nazir et al. 2015). In addition to improving the performance of the company, financial and non-financial rewards can also increase employee job satisfaction. Job satisfaction as a dependent variable can be influenced by independent variables, namely rewards consisting of financial rewards (salary and benefits) and non-financial rewards (employee recognition) (Tessema et al. 2013).

Itika (2011) argues that financial reward is a tool in the form of money that used to gain good commitment from employees as business partner, to then improve employee satisfaction in fulfilling their living needs. Salary and benefits, financial incentives, and employee recognition are all a part of financial and non-financial rewards that can affect the level of employee job satisfaction (DeCenzo & Robbins 2010). The rate of financial rewards such as salary will affect job satisfaction which should contribute to better performance (Mondy & Martocchio 2016). Financial rewards offered by organizations might affect employee attitudes toward their job (job satisfaction) as well as the organizations they work for, so then financial rewards are used as a key tool to record behaviors and activities in order to attract and retain the most competent employee and make them feel satisfied and motivated (Nazir et al. 2015).

Employee job satisfaction can be created from the safety and health of a good working environment. A healthy worker is a worker free from illness, injury, and mental and emotional problems that can interfere with normal work activities. When workers understand the importance of work safety and health regulations and procedures and tools used for work, this can result in better employee performance (Dwomoh et al. 2013). Unsafe work conditions, too heavy a work-load and other work will have an impact on employee dissatisfaction, and a sense of dissatisfaction may encourage employees to look for better employment opportunities and ineffective management and safety or health procedures will cause poor performance of the employees, therefore enjoyable work and good safety and health will create employee job satisfaction (Salman et al. 2016).

Job satisfaction is determined by how well the job or the organization can fulfill certain employee needs such as job safety and health. Employee who has positive perception towards job safety and health condition tends to support the organization, has high commitments, leads to beneficial result like increased employee job satisfaction (Huang et al. 2016). Job

safety and health policies play important roles in reducing accidents and injuries at work. Effective job safety and health system may reduce employee absence, improve commitment, satisfaction, employee physical and mental health, which will directly impact on productivity improvement that will improve organizational profitability (Dwomoh et al. 2013).

Job safety and health may affect employee retention with the role of motivational mediation and job satisfaction. Improvement in job safety and health may improve employee satisfaction and motivation which will affect employee retention level (Sal-man, et al., 2016). Employees who feel that the management cares about job safety and health system management will have a positive impact on employee work result, such as in motivation, involvement, commitment, and employee job satisfaction (U & Tom 2016).

4 CONCLUSION

Based on the result of research conducted by using a verification analysis through multiple linear regression analysis, we can conclude that financial rewards have a positive impact on job satisfaction; job safety and health have positive impacts on job satisfaction; and both altogether have positive impacts on job satisfaction.

REFERENCES

Ali, A. & Akram, N.M. 2012. Impact of Financial Rewards on Employee's Motivation and Satisfaction in Pharmaceutical Industry, *Pakistan. Global Journal of Management and Business Research* 12(17): 44–50.

Ariani, D.W. 2015. Employee Satisfaction and Service Quality: Is There Relations? *International Journal of Business Research and Management* 6(3): 33–44.

Armstrong, M. 2014. *Armstrong's Handbook of Human Resource Management Practice. Human Resource Management (13th Edition)*. London: Kogan Page.

Armstrong, M. 2015. *Armstrong's Handbook of Reward Management Practice (5th Edition)*. London: Kogan Page.

Bakotić, D. & Tomislav, B. 2013. Relationship between Working Conditions and Job Satisfaction: The Case of Croatian Shipbuilding Company. *International Journal of Business and Social Science* 4(2): 206–213.

Utami, P.S., Hubeis, M. & Affandi, M.J. 2016. The Impact of Working Climate and Motivation towards Job Satisfaction That Implies the Employee Performance in PT Indonesia Power Generation Business Unit of Suralaya Banten. *International Journal of Scientific and Research Publications* 6(7): 26–32.

Ch, J. I., Tasleem, M., & Iqbal, R. (2015, March). The impact of employee satisfaction and service quality on perceived firm's performance in high contact service industry of Pakistan. *2015 International Conference on Industrial Engineering and Operations Management (IEOM)*: 1–8.

Daniel, A. (2012). An impact of employee satisfaction on customer satisfaction in service sector of Pakistan. *Journal of Asian Scientific Research* 2(10): 548–561.

Decenzo, D.A., & Robbins, S.P. 2010. *Fundamentals of Human Resource Management (10th Ed.)*. New Jersey: John Wiley & Sons, Inc.

DeNisi A.S. & Griffin R.W. 2008. *Managing Human Resources (3rd Edition)*. Boston: Houghton-Mifflin.

Dobre, O. 2013. Employee motivation and organizational performance. *Review of Applied Socio-Economic Research*, 5(1),53–60.

Ravichandran, A., Rajan, L.S., & Kumar, G.B. 2015. A study on job satisfaction of employees of manufacturing industry in Puducherry, India. *International Journal of Innovative Research & Development (IJIRD)* 4(2): 344–349.

Dwomoh, G., Owusu, E.E. & Addo, M. 2013. Impact of Occupational Health and Safety Policies on Employees' Performance in the Ghana's Timber Industry: Evidence from Lumber and Logs Limited. *International Journal of Education and Research* 1(12): 1–14.

Gregory, K. (2011). The importance of employee satisfaction. *The Journal of the Division of Business & Information Management*: 29–37.

Hoshi, S. (2014). Employee Satisfaction of Commercial Banks : The Case of North Cyprus. Thesis, Eastern Mediterranean University.

Huang, Y.-H., Lee, J., McFadden, A.C., Murphy, L.A., Robertson, M.M., Cheung, J.H. & Zohar, D. 2016. Beyond safety outcomes: An investigation of the impact of safety climate on job satisfaction, employee engagement and turnover using social exchange theory as the theoretical framework. *Applied Ergonomics* 55: 248–257.

Imran, A., Ahmad, S., Nisar, Q. A. & Ahmad, U. 2014. Exploring Relationship among Rewards, Recognition and Employees' Job Satisfaction : A Descriptive Study on Libraries in Pakistan. M*iddle-East Journal of Scientific Research* 21(9): 1533–1540.

Itika, J. 2011. Fundamentals of human resource management: Emerging experiences from Africa. *African Public Administration and Management Series*.

Kaya, C. & Ceylan, B. 2014. An Empirical Study on the Role of Career Development Programs in Organizations and Organizational Commitment on Job Satisfaction of Employees. *American Journal of Business and Management* 3(3): 178–191.

Khan, R., Naseem, A. & Masood, S.A. 2016. Effect of Continuance Commitment and Organizational Cynicism on Employee Satisfaction in Engineering Organizations. *International journal of innovation, management and technology* 7(4),141–146.

Khartik, R.S, Saratha, M.S. 2012. A Study on Job Satisfaction in Iti Limited, Bangalore. *International Journal of Management, IT and Engineering* 2(6): 402–416.

Lin, T.H. 2015. Factors That Influence Employees' Job Satisfaction in Hotel and Catering Industry. Factors That Influence Employees' Job Satisfaction in Hotel and Catering Industry. *Unpublished, Universiti Malaysia Sarawak*.

Manoj, N. & Das, V.T. 2016. Management Human Resource Management Practices Impact on Employee Job Satisfaction in The Public Sector Undertakings—An Empirical Study Guntur: 209–212.

Mondy, R.W. & Martocchio, J.J. 2016. *Human Resource Management (Fourteenth)*. London: Pearson Education Limited.

Nedeljkovic, M. 2012. Organizational Changes and Job Satisfaction in the Hospitality Industry in Serbia. *UTMS Journal of Economics* 3(2): 105–117.

Nemmaniwar, A.G. & Deshpande, M.S. 2016. Job Satisfaction among Hospital Employees: A Review of Literature. *Journal of Business and Management (IOSR-JBM)* 18(6),27–31.

Odembo, S. A. (2013). Job satisfaction and employee performance within the telecommunication industry in Kenya : a case of Airtel Kenya Limited. *Ph. D. thesis, Kenyatta University.*

Qasim, S. & Sayeed, F.C. 2012. Exploring Factors Affecting Employees' Job Satisfaction at Work. *Journal of Management and Social Sciences* 8(1): 31–39.

Nazir, S., Qun, W., Akhtar, M. N., Shafi, A., & Nazir, N. (2015). Financial rewards climate and its impact on employee attitudes towards job satisfaction in the retail organizations. European Scientific Journal, ESJ, 11(1): 351–364.

Rast, S. & Tourani, A. 2012. Evaluation of Employees' Job Satisfaction and Role of Gender Difference: An Empirical Study at Airline Industry in Iran. *International Journal of Business and Social Science* 3(7): 91–100.

Saeed, H.M.I. & Nasir, N. 2016. Work Environment on Job Satisfaction with mediating effect of Motivation among School Teachers in Lahore, Pakistan. *Journal of Management Engineering and Information Technology* 3(6).

Sarwar, S. & Abugre, J. 2013. The Influence of Rewards and Job Satisfaction on Employees in the Service Industry and Key words. *The Business & Management Review* 3(2): 22–32.

Senen, S. H. 2008. Pengaruh Motivasi Kerja dan Kemampuan Kerja Karyawan terhadap Produktivitas Kerja Karyawan pada PT. Safilindo Permata. *Strategic* 7(14): 1–15.

Sledge, S., Miles, A.K. & Van Sambeek, M. 2011. A Comparison of Employee Job Satisfaction in the Service Industry: Do Cultural and Spirituality Influences Matter? *Journal of Management Policy and Practice* 12 (4): 126–145.

Soni, H. & Rawal, Y.S. 2014. Impact of quality of work life on employee satisfaction in hotel industry. *IOSR Journal of Business and Management* 16(3): 37–44.

Sumiyati, S., Masharyono, M., Pratama, K. F. & Purnama, R. 2016. The Effect of Social Work Environment on Employee Productivity in Manufacturing Company in Indonesia. *2016 Global Conference on Business, Management and Entrepreneurship* 15: 574–575.

Tessema, M.T., Ready, K.J. & Embaye, A.B. 2013. The Effects of Employee Recognition, Pay, and Benefits on Job Satisfaction : Cross Country Evidence. *Journal of Business and Economics* 4(1): 1–12.

Maryjoan, I.U. & Tom, E.E. 2016. Effects of industrial safety and health on employees' job performance in selected cement companies in cross river state, Nigeria. *International Journal of Business and Management Review* 4(3): 49–56.

Umar, A. 2015. The Effect of Motivation and Career Development against Employees' Performance and Job Satisfaction of the Governor Office South Sulawesi Province, Indonesia, *International Journal of Management Sciences* 5(9): 628–638.

Usman, H. 2011. *Manajemen teori praktek dan riset pendidikan (Edisi 3)*. Jakarta: Erlangga.

Wan-Yih, W., & Htaik, S. 2011. The Impacts of Perceived Organizational Support, Job Satisfaction, and

Aorganizational Commitment on Job Performance in Hotel Industry. *The 11th International DSI and the 16th APDSI Joint Meeting, Taipei, Taiwan* 12 (16).

Yaseen, A. 2013. Effect of Compensation Factors on Employee Satisfaction-A Study of Doctor's Dissatisfaction in Punjab. *International Journal of Human Resource Studies* 3(1): 142–158.

Yee, R.W.Y., Yeung, A.C.L. & Cheng, T.C.E. 2008. The impact of employee satisfaction on quality and profitability in high-contact service industries. *Journal of Operations Management* 26(5): 651–668.

Yousaf, S. 2014. Impact of Financial and Non-Financial Rewards on Employee Motivation.

Middle-East Journal of Scientific Research 21(10): 1776–1886.

Zaidi, F.B. & Iqbal, S. 2012. Impact of career selection on job satisfaction in the service industry of Pakistan. *African Journal of Business Management* 6(9): 3384–3401.

Ghozali, I. 2011. *Analisis multivariate dengan program IBM SPSS 19*. Semarang: Badan Penerbit Fakultas Ekonomi Universitas Diponegoro.

Salman, S., Mahmood, P. A., Aftab, F. & Mahmood, A.2016. Impact of Safety Health Environment on Employee Retention in Pharmaceutical Industry: Mediating Role of Job Satisfaction and Motivation. *Journal Of Business Studies* 12(1):185–197.

Effectiveness of HR departments' roles implementation in hospitality industry

R. Wahyuningtyas
Telkom University, Bandung, Indonesia

ABSTRACT: This study will provide relevant information related to HR departments' roles implementation effectiveness. The method of this research is descriptive through survey. The sample consisted of 592 employees in five-star hotels on Java Island. Variables in this study cover areas related to Employee Relations, Functional Specialist, Strategic Integrator, and Innovator. Almost of the indicators are in the "effective" category, which means that employees determined that HR departments in the companies in which they work have performed their role effectively. Employee expectation for all indicators are higher than the current condition score. The existence of a gap between employees' expectations regarding the importance of the HR department's role implementation with the current condition and also the significant differences in almost all indicators shows that companies still need to improve their HR departments' role's effectiveness through improvement of existing indicators in order to meet employee expectations.

1 INTRODUCTION

The current organization, especially in the hotel sector, is being geared towards finding better ways to acquire, train and retain employees who help to provide a competitive advantage for their organizations. Human resource (HR) management is the key for organizations to maintain continuity in today's competitive environment. Therefore, human beings should be viewed as assets that can add value to an organization. Based on preliminary survey results of 189 employees at five-star hotels on Java Island, there are some conditions in the companies that are still less appropriate related to human resource management:

1. job demands, long working hours and requests for completion of work even during periods of leave: 60.53%.
2. unclear job descriptions: 11.64%.
3. the promotion system is not yet intended for high-performing employees: 8.47%.
4. lack of response and organizational assistance in resolving employee issues outside of work: 6.88%.
5. lack of employee development and rewards: 11.64%.

Saleem and Perwez (2012) stated that especially in the hotel sector, HR departments focus primarily on administrative activities. Edgley-Pyshorn and Huisman (2011) found that HR departments have difficulty in implementing culture change because most employees are resistant to change. Cohen and Karatzimas (2011) found that HR departments need to use budgeting for motivating, communicating, evaluating and controlling employees. HR function in modern business organizations was not originally conceived as an employee-facing department (Kaufman, 2014). Based on Sheehan et al. (2016), HR departments have difficulties in balancing the roles of strategic decision making with other HR department roles. Ingham and Ulrich (2016) said that better HR departments will create better organizations. Choi and Dickson (2010) found that there is a lot of research regarding HR management practices. However, there are only a few studies that measure the effectiveness of HR management practices in the hospitality industry.

According to Robbins and Judge (2013), a role is a set of behavioral patterns expected of someone who occupies a particular position within a social unit. According to Ardana et al. (2012), an HR department is defined as a working group formed by an organization or company to help manage human resources (hence the title) to work effectively and efficiently for the good of the workforce, company and community. According to Ulrich and Brockbank (2005), the role of an HR department is to define identity seen as the fulfillment in channeling value as an HR professional. Some of the roles of an HR department are shown here (based on Amstrong [2009]):

1. John Storey
 – *Change masters*. Advisers is playing as an internal consultant, while HR practices are carried out by line managers.
 – *Regulator* is playing a role in the formulation and monitoring of rules in the work.

– *Handmaidens* is serving as a service provider to meet the demand of line managers.

2. Peter Reilly
– *Strategist/integrator* is a role in contributing to make long-term strategies.
– *Administrator/controller* is a role in making short term tactical contributions.
– *Adviser/consultant is the role between the two previous roles.*

3. Dave Ulrich and Wayne Brockbank
– *Employee advocate* is a role that focuses on the needs of today's employees through understanding, listening and empathy
– *Human capital developer* is a role in managing and developing human capital (both individuals and teams), focus on preparing employees to succeed in the future
– *Functional expert* is a role that focus on administrative efficiency practices (such as technology or process design) and others through policies and interventions
– *Strategic partner* is a role that aligns the HR system to achieve the vision and mission of the organization, helping managers to be able to walk and disseminate learning in the organization
– *Leader* is leading the HR function, collaborating with other functions, creating and improving standards in strategic thinking and ensuring organizational governance

With regard to the various roles of an HR department, some of the roles expressed by some experts have similarities to the role expressed by other experts. The company should increase the role of the HR department to manage all issue-related to employees and their functional roles. It is important to review HR department performance on a regular basis as a part of control system (Wahyuningtyas et al. 2015). In this study, the roles of HR department are measured through employee assessment toward implementation of the HR department's roles in their organization. Individual employees will assess the implementation of the HR department's roles based on their experience and thinking with regard to the roles that the HR department has been undertaking. The dimension of HR department role used in this research are:

a. Employee relations, namely the role of an HR department in motivating, resolving problems related to work and out of work and improve employee competency and employee preparation to occupy key positions in the future.
b. Functional specialists, namely the role of an HR departments in leading the implementation of functional activities, supervision of policy implementation in human resource management and efficiency in administrative activities.
c. Strategic integrators, namely the role of an HR department in integrating policies in HR management with other departmental policies including

initiating changes that align with the company's business strategy.
d. Innovator, namely the role of an HR department in creating innovative ideas and policies in enhancing effectiveness and efficiency in human resource management.

The purpose of this study is to examine the effectiveness of HR departments' roles implementation in five-star hotels on Java Island. The results of this study will provide input for HR managers in the hotel sector to maximize managing human resources and help employees provide the best possible contribution to the organization.

2 METHOD

The type of research is descriptive. The researcher obtained a picture of the effectiveness of HR department's role implementation in five-star hotels in Java Island. Variables in this research are the HR department's roles consisting of dimensions employee relations, functional specialists, strategic integrators and innovators. Indicators of employee relations consist of motivation, competence, completion and preparation. Indicators of functional specialists consist of lead, oversight and administrative. Indicator of strategic integrator consists of integrating and initiating. Indicator of innovator consists of creating and efficiency.

Population in this study is all employees who work in five-star hotels in Java. The number of samples were 592 employees. This research uses rank order mean analysis and Paired Samples T Test for different test.

Interpretation of category:

1. 20%–36% = HR Department Role implementation is very ineffective.
2. > 36%–52% = HR Department Role implementation is ineffective.
3. > 52%–68% = HR Department Role implementation is moderate.
4. > 68%–84% = HR Department Role implementation is effective.
5. > 84%–100% = HR Department Role implementation is very effective.

3 RESULT AND DISCUSSION

A validity test was done with Confirmatory Factor Analysis (CFA). Validity is done by checking the standardized check factor loadings ≥ 0.50 (Wijanto 2008). A reliability test is also needed, and this was done by using composite reliability measurement and variance extracted measurement. A construct has good reliability when the value of Construct Reliability (CR) ≥ 0.70 and the value of Variance Extracted (VE) ≥ 0.50.

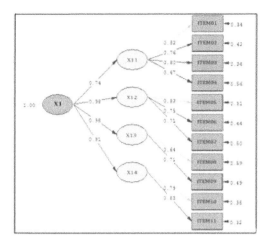

Figure 1. CFA test for variable of HR departments' roles.

The result of the Validity Test for Variable of HR departments' roles show that all indicators are valid. In Table 1, it can be seen that the variable of HR departments' roles has a Construct Reliability (CR) value ≥ 0.70 and a Variance Extracted (VE) value ≥ 0.50. This indicates that instrument is reliable.

A descriptive analysis was performed on each dimension (sub-variable) of the HR department's role. Table 2 is the result of the measurement for effectiveness in current condition and the level of importance as per employee expectations.

Most of the 11 indicators are in the "effective" category means employees assume that HR department in the company where they work has performed its role effectively. The indicator with the highest score is indicator 4, which states, "HR department is able to provide employees to fill key position in the future," with a score of 84.05%. Here the HR department coordinates with heads of other departments to jointly explore potential and improve employee competence. In addition to the implementation of training activities, there is also the provision of jobs to work with the level of difficulty and responsibility one level above its current position. This is done when the head of department sees that their subordinate has the potential to be developed. This

Table 1. Reliability test for variable of HR departments' role.

Calculation	Result	Condition
$(\sum \lambda)^2$	13.03	
$\sum \lambda^2$	3.30	
$\sum e_j$	0.70	Reliable
CR	0.95	
VE	0.82	

Table 2. Current condition and expectation score for indicator of HR department role.

		Score	
No	Indicator	Current	Expectation
1	HR department is able to motivate employees	65.61%	83.16%
2	HR department is able to solve employee's problem	63.18%	83.47%
3	HR departmen is able to increase employee's competence	81.99%	84.28%
4	HR department is able to provide employees to fill key position in the future	84.05%	84.78%
5	HR department is able to lead the implementation of HR functions	82.36%	83,20%
6	HR department is able to supervise the implementation of human resource management policies	76.05%	84.68%
7	HR department is able to carry out administrative activities efficiently	83.31%	84.11%
8	HR department is able to synergize HR management with other departments	80.98%	82.49%
9	HR department is able to initiate changes within the organization	79.93%	83.16%
10	HR department is able to create innovative human resource management policies	66.42%	83.97%
11	HR department is able to create efficient operational activities of the organization	75.44%	83.13%

condition is usually communicated to the potential employee that employer provides additional employment at a higher level. The boss also communicates the purpose of this situation for preparing employees if at any time there is a vacancy in a particular position above it.

As can be seen in Table 2, there are some indicators that are considered in medium category for its implementation effectiveness, namely "HR department is able to motivate employees" with 65.61%, indicator "HR department able to solve employee's problems" with value 63.18% and indicator "HR department is able to create innovative human resource management policies" with a value of 66.42%. In relation to the process of motivating and solving employee's problems, HR departments have prioritized the completion of their respective departments. Motivation to employees is provided by the HR department by facilitating formation activities such as discussion forums every few months that bring together superiors with subordinates to discuss organizational achievement and employee complaints submission. Increased personal motivation is more focused implemented by the head of each department unless the condition is very serious. In this case, the head of department have an important role as the HR department partners to solve employee's problems. The HR department will help solve employee's problems if the problem is already disrupting employee and environmental performance or there are any requests from the department heads to help solve the problems. Meanwhile, the indicator "The HR department is able to create an innovative human re-source management policy" is also rated moderately. Employees judge that often new policies issued by HR department are emphasized on sanctions rather than rewards for employees. Such as the implementation of new rules regarding delay sanctions through employee HKE deductions.

Based on the result of measurement for employee expectation regarding the implementation of the HR department role, shows that for all indicators, employee expectation is higher than the current condition score. The overall score for employee expectation is 83.95%, while the score for current conditions is 76.31%. Figure 2 shows a comparison and the gap between the current condition and employee expectations related to implementation of the HR departments' roles.

To find out whether the current condition score with the employee's expectation has significant differences or not, a different test using Paired Samples T Test as shown in Table 3 was conducted. If $t_{count} > 1.96$, then there is a significant difference statistically between the current conditions with employee expectations on indicators assessed. Table 3 shows the existence of gap between employee's expectations regarding the importance of the HR department's role implementation with the current condition, and also the significant differences in almost all of the indicators.

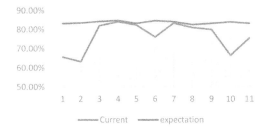

Figure 2. Comparison between current and expectation score.

Table 3. The result of paired samples t-test.

No	T_{count}	Result
1	-22.245	There is significant difference on indicator "HR department is able to motivate employees"
2	-20.828	There is significant difference on indicator "HR department is able to solve employee's problem"
3	-2.967	There is significant difference on indicator "HR department is able to increase employee's competence"
4	-1.029	There is no significant difference on indicator "HR department is able to provide employees to fill key position in the future"
5	-1.243	There is no significant difference on indicator "HR department is able to lead the implementation of HR functions"
6	-11.583	There is significant difference on indicator "HR department is able to supervise the implementation of human resource management policies"
7	-1.213	There is no significant difference on indicator "HR department is able to carry out administrative activities efficiently"
8	-2.085	There is significant difference on indicator "HR department is able to synergize HR management with other departments"
9	-4.191	There is significant difference on indicator "HR department is able to initiate changes within the organization"
10	-21.624	There is significant difference on indicator "HR department is able to create innovative human resource management policies"
11	-9.432	There is significant difference on indicator "HR department is able to create efficient operational activities of the organization"

4 CONCLUSION

There is still a gap between the effectiveness of an HR department's role in comparison to an employee's expectations. Higher employee expectations indicate that there is still opportunity for organization to make improvement in order to meet employee expectations on an HR department's role. Indicators that need to get main attention related to HR department role is the role in "employee relation" that include ability to motivate and solve the problems of employees and role of "innovator." An HR department should take an initiative to help resolve employee issues even if it has not impacted to employee's work and not entrusted entirely to the head of department. It is important to do so that employee problems can be resolved and employee motivation is not interrupted. Finally, this condition will not decrease

Table 4. Priorities and operational steps increasing the role of an HR department.

Role	Operational strategies
Employee Relation (Priority: High)	Personal counseling to find out constraints, complaints, satisfaction and employee inputs for organizational improvement
	Coordinate with employee union leaders to discuss and solve employee complaints
	Coordinate with head of department for employee development and employee career movement
Functional Specialist (Priority: Moderate)	Socialize policies and discipline rules to all employees
	Conduct monitoring process of HR management policy implementation by all employees
	Formulate the consequences of disciplinary offenses in writing by employees
Strategic Integrator (Priority: Moderate)	Coordinate with other departments related to company's HR management
Innovator (Priority: High)	Creating innovative programs in the implementation of HR management policies
	Giving rewards to employees who have contributed to give creative ideas
	Implement employee satisfaction measurement of its work regularly with an appropriate measuring instrument

employees' performance. Then with regard to the role of innovator, innovative policies should not always focus on the adoption of rules and punishment that will de-grade employee motivation in work.

Priorities and operational steps to improve the effectiveness of an HR department's role implementation is shown in Table 4.

ACKNOWLEDGEMENT

I would like to express my appreciation to Telkom University for supporting the completion of this research.

REFERENCES

Amstrong, M. 2009. *A handbook of human resources management practices. Eleventh edition.* London: Kogan Page.

Ardana, I.K., Mujiati, N.W & Utama, I.W.M. 2012. *Manajemen Sumber Daya Manusia.* Bandung: Alfabeta.

Choi, Y. & Dickson, D.R. 2009. A case study into the benefits of management training programs: Impacts on hotel employee turnover and satisfaction level. *Journal of Human Resources in Hospitality & Tourism* 9(1): 103–116.

Cohen, S., & Karatzimas, S. 2011. The role of the human resources department in budgeting: Evidence from Greece. *Journal of Human Resource Costing & Accounting* 15(2): 147–166.

Edgley-Pyshorn, C., & Huisman, J. 2011. The role of the HR department in organisational change in a British university. *Journal of Organizational Change Management* 24(5): 610–625.

Ingham, J. & Ulrich, D. 2016. *Building better HR departments. Strategic HR Review* 15(3): 129–136.

Kaufman, B.E. 2014. History of the British Industrial Relations Field Reconsidered: Getting from the Webbs to the New Employment Relations Paradigm. *British Journal of Industrial Relations* 52(1): 1–31.

Robbins, S.P. & Timothy A.J. 2013. *Organizational behavior. 15th edition.* San Diego, CA: Prentice Hall.

Saleem, S.M. & Perwez, S.K. 2012. The human resources role and challenges in the hotel sector in Kanyakumari, Tamil Nadu. *International Journal of Management Research and Reviews* 2(10): 1758–1763.

Sheehan, C., De Cieri, H., Cooper, B. & Shea, T. 2016. Strategic implications of HR role management in a dynamic environment. *Personnel Review* 45(2): 353–373.

Ulrich, D. & Brockbank, W. 2005. The HR value proposition. Cambridge, MA: Harvard Business Press.

Wahyuningtyas, R., Sule, E.T., Kusman, M. & Soemaryani, I. 2015. Employee Turnover Intentions in Hotel: How to reduce it? *Advanced Science Letters* 21(4): 719–722.

Wijanto, S. H. (2008). *Structural equation modeling dengan Lisrel 8.8.* Yogyakarta: Graha Ilmu.

Advances in Business, Management and Entrepreneurship – Hurriyati et al (eds)
© *2020 Taylor & Francis Group, London, ISBN 978-0-367-27176-3*

Collaborative-based academic supervision for principals

B. Bahrodin, J. Widodo, M. Rachman & A. Slamet
Universitas Negeri Semarang, Semarang, Indonesia

ABSTRACT: The study aims to analyze and synthesize a collaborative-based academic supervision model feasible among principals in Banyumas Regency. This research and development study employed data from the results of academic supervision, need analysis, hypothetical model testing, and model validation. Questionnaires, interviews, documentation studies, observations, and FGDs were used to collect the data. Test validity used model validation from experts and practitioners, and the qualitative descriptive analysis used data display, reduction, verification, and conclusion. The results showed that (1) the factual model of collaborative-based academic supervision model for principals was categorized into good category; (2) hypothetical model of collaborative-based academic supervision model for principals was categorized into very important; and (3) the final model of collaborative-based academic supervision model was feasible. The collaborative-based academic supervision model for principals contributed to the improvement of teachers' performance. The results concluded that the model is good, but it is not yet optimal. It is more essential, more practical, effective and efficient by emphasizing the steps namely report building, basic determination of collaboration, principles, problem sets, directing settings, participation, team discussions, cooperation, and follow-up. The academic supervision model is feasible to improve teachers' performance.

1 INTRODUCTION

Headmaster has an important and strategic role in school supervision. Supervision is a quality control to supervise the process of education in schools (Bahrodin 2017). Therefore, the principal has the duty, authority and responsibility to conduct good oversight in the form of monitoring, supervision, evaluation and reporting. The Regulation of the Minister of National Education Number 13 Year 2007, on Headmaster Standard in School/Madrasah, has stipulated that there are five dimensions of principal competence, and one of which is academic supervision.

Academic supervision is a series of activities to help teachers develop their skills in managing the teaching and learning process (Glickman 2007, Daresh 2001, and Dharma 2009). It is also used to help improve the competence, professionalism and performance of teachers in classroom learning (Rahayu 2015). Moreover, it also becomes coaching venues for quality improvement, and improvements in teacher learning performance in the classroom (Marzano et al. 2011) and (Guntoro et al. 2016)

Although academic supervision is important, the performance of principals in conducting academic supervision is still low and not optimal. Competence and skills of principals in conducting academic supervision is still low (Cahyono 2014). The principal is still formalistic which emphasizes only the instructional administration tool in carrying out academic supervision (Clark & Olumese 2013). The principals carry out academic supervision of new

class visits (77.5%), and democratic (75%) (Behlol 2011). The result of measurement of Minimum Education Service Standard indicator in Banyumas shows principals and supervisors who perform routine, scheduled and new sustainable supervision (56.25%). Performance assessment of junior high school principals was conducted in Banyumas Regency (2015/2016) and it is on process component from new supervisor aspect 63.4%. Although academic supervision is important, the performance of principals in conducting academic supervision is still low and not optimal. Competence and skills of principals in conducting academic supervision are still low (Cahyono 2014). The principal in carrying out academic supervision is still formalistic which emphasizes only the instructional administration tool (Clark & Olumese 2013). The principals carry out academic supervision of new class visits (77.5%) and democratic (75%) (Behlol 2011). The result of Minimum Education Service Standard (SPM) indicator in Banyumas shows that principals and supervisors who perform routine, scheduled and new sustainable supervision were 56.25%. Performance assessment of junior high school principals in Banyumas Regency (2015/2016) was on process component from new supervisor aspect of 63.4%.

The principal academic supervision problems in Banyumas Regency are caused by several factors, such as the culture and low performance of the principal in performing academic supervision, low competence of the principal as an academic supervisor, inadequate conceptual, interpersonal, and technical skills of the principal, and low quality of academic supervision performance in program preparation, observation instrument, analysis and evaluation of observation result, reporting and follow up.

Based on the facts of various efforts to improve the performance of principals, academic supervision has been widely implemented, including workshops, trainings, performance assessment principals (PKKS), coaching principals through self-study materials on sustainable development activities and activities of Principal Working Congress. However, these efforts have not produced results to improve the ability of the principal in academic supervision so that it needs to look for one solution by implementing collaborative-based supervision.

Collaborative supervision is a jointly planned supervision between the supervisor and the person being supervised (Bahrodin 2017). Collaborative supervision is capable of generating changes in cooperative attitudes, innovative behavior, character and emotional skills of teachers (Fakhruddin 2017) and (Taoefik 2016). Collaborative-based academic supervision can improve the effectiveness of improving the ability of teachers to implement learning in the classroom (Fakhruddin 2017).

Based on the background, it is necessary to develop collaborative principal-based supervisory school supervision model in Banyumas District. The study aims to analyze the factual model, hypothetical model design, and synthesize the final model that deserves the collaborative principal supervision of the collaborative principal from state junior high schools in Banyumas District. The results of the research are useful for Banyumas District Education Office, supervisors, principals and teachers as a reference and input in the implementation of academic supervision in schools.

2 METHOD

The location was conducted in public junior high schools in Banyumas. The research design employed Research and Development (R&D) method, and the research procedure used was adopted from Borg and Gall (2007) with three main stages: preliminary stage, development and validation. The population in this studies were 300 people, and the sample of research was 10% i.e. 30 people. The Source of data used in this research included data of principal's performance documentation, questionnaire, observation, interview, and Focus Group Discussion (FGD). Data were collected using questionnaire, interview, observation, document study and FGD. The validity of data uses the validity of the constructs and the validity of the items. Data analysis techniques used quantitative and qualitative.

3 RESULT AND DISCUSSION

The results of preliminary study of academic supervision of principal in Banyumas Regency was good with the average score of 420.07 (74.87%). The good category distributed for the system performance aspect was obtained with the average score of 192.70 (range of 165-214) with 72.64%; the average system component score of 227.37 with percentage of 75.99%.

In the aspect of the performance, the highest value of the system is the implementation (74.53%). This means that 74.53% of respondents stated that the implementation is good, but there are still 25.47% which is less good. Indicators that need to be improved were the reverse meeting stage. The reversal meeting is a vehicle of reflection on success and weakness (Ernawati 2014). The activities include analyzing observations of teaching behavior of teachers in teaching (Makawimbang 2013). The lowest value of system performance was evaluation (70.71%). This means that the evaluation needs to be improved because there are still 29.29% which states less good. Indicators that need to be improved include solutions to improve teacher performance and reporting.

The aspects of system components have been implemented in both categories average score 227.37 (75.99%). The good category was distributed to the supervisors average score of 101.17 (76.64%); the people being supervised obtained mean score 46.67 (72.92%); average material score of 18.90 (78.75%). The highest score is the material percentage of 78.75% which is in good category. This means the material is relevant and has usefulness. The lowest score is supervise acquisition value (72.92%). This means teachers' performance needs to be improved, and supervision is as part of the object and subject of target (Syukri 2015). Performances of teachers that need to be improved are ability to prepare administration of teaching and learning materials, mastery of learning component and learning steps ranging from apperception to reflection (Surachman 2014).

The findings of the factual model of academic supervision of the principals showed some aspects of system performance including planning, implementation, evaluation and follow-up. The results of the average score system performance has scored 192.7 meaning that the performance of the system has been implemented according to procedures/phases starting from planning, implementation, evaluation, reporting and follow-up (Gultom 2014).

Meanwhile, the components of the academic supervision system include aspects of supervisor, supervision, materials, methods, facilities and infrastructure and time. The result of the average system component score was 227.37 within the range of good category (201-262). This means that the system components have supported the implementation of activities according to their roles and functions. Supervision works well if supported by several components, such as supervisors, materials, methods, facilities and infrastructure (Manggar 2011).

In addition to this, the result of requirement analysis of collaborative principal is very important with mean score of 612.51 (85.82%). The category is very important, distributed for aspect analysis needs system performance average score of 227.47 (85.94); the average system component score of 260.57 (84.91%); the average collaboration strategy scores of 124.47 (86.39%).

In the aspect of system performance, the highest score is: the implementation with percentage 87.22%. It means that the implementation is very important starting from the initial meeting, core activities and discussion back. Implementation activities include: program socialization, classroom observation, evaluation and follow up (Prihatin 2017). Implementation includes: preparation, execution, method determination, implementation of collaboration strategy and brainstorming (Bahrodin 2017). The lowest score is: evaluation, the percentage of 84.78% category is very important. Evaluation serves as a feedback (Prihatin 2017). Evaluation activities include: output, feedback, reporting and follow-up (Widodo 2007) and (Himdani 2017).

In the aspect analysis requirement of component of system, the highest value is material with percentage 88.19%. It means that the matter is very important. Required materials include the creation of working program, schedule, observation instrument, evaluation, follow up; implementation (initial meeting, core and feedback), data analysis, and reporting (Bahrodin 2017). The lowest value is the method with percentage 85.29% with category very important. This means that the principal needs to pay attention to the effectiveness and usefulness of the method.

In the aspect of requirement analysis of the highest value collaboration strategy, the determination of collaborative basic values with percentage 89.30%. Collaborative values include the development of mutual respect, cooperation, commitment in completing the task (Bahrodin 2017). Collaborative values can improve principal synergy (Wasitohadi 2016). The lowest value of the collaborative needs analysis is directing setting percentage 81.25%. This means that directing settings need to be improved starting from: basic rules determination, agenda setting, organizing, group determination. This is in line with the characteristics of collaborative supervision: communicating, sharing, brainstorming, consensus, negotiation, supervisory teamwork, goal setting together, and division of tasks between (Silck 2000) and (Pidarta 2009).

Academic supervision models were validated by experts, practitioners and feasibility tests through FGD-1 through hypothetical models. Further it was tested in a limited and expanded, and evaluated test. The results are listed in Table 1.

Table 1 described that the results of limited trial test of collaborative principal-based school principal supervision obtained averages of 71.91 with good category. The good category is distributed for the average performance quality of 69.33, the average collaborative strategy is 73.33. The highest score of the trial is limited on the performance of the principal with average score 72.50. This means that the principal's performance quality is good from initial meetings, core activities, and reverse meetings. The execution of academic supervision activities is given priority to the provision of guidance in the teaching and learning process of teachers (Rahayu 2015). In order for teacher development to run effectively and efficiently, it is necessary to prepare an observation instrument. Classroom observation activities include checking learning tools, teacher observations in teaching, and feedback meetings (Prihatin 2017). The lowest value of the trial is limited, namely analyzing the average data value of 66.67. This means the principal needs to improve his ability to analyze data. Data processing of monitoring results of academic supervision can be records, recordings, and documentation of the real condition of the results of supervision, observation (Ernawati 2014) and (Prihatin 2017). The results of the analysis are then interpreted to be used to define priority objectives, targets, strategies and subsequent supervisory program measures (Rahayu 2015).

In a limited trial, the results obtained the category of good with the average value of 73.33. The highest score is report cards, cooperation between supervisors, and discussions with an average of 80. Rapport is fostering good relationships between principals as supervisors with supervisors/supervised teachers. The lowest score is the determination of baseline collaborative values and the mean follow-up of 65. This means that the determination of collaborative and follow-up basic values still needs to be improved. Collaborative base value is very important, because it is able to create a conducive, fun, mutually appreciative and open cooperation (Djumara 2008). Collaborative supervision is able to change teacher behavior, and principal synergy (Fakhruddin 2017).

The expanded trial obtained an average of 86.25 which is categorized into very good. The excellent category, distributed for the average performance quality aspect 87.66, the average collaboration strategy is 87.66 which is in very good category. The highest score for the test is expanded, namely: the average report card score of 95.00 which is very good. The lowest value is determination of the basic value of collaborative average of 80 which is in good category. The basic value of collaboration is beneficial for improving teacher character development and emotional skill (Fakhruddin 2017). Collaborative values developed are respecting for people, honor and integrity, ownership and alignment, consensus, full responsibility and accountability, trust-based relationship recognition and growth (Djumara 2008).

The results of limited and expanded trial there is an increase in the average value of 14.40 points. It

Table 1. Limited and expanded trial results in school-based academic supervision.

No	Aspect	Limited Trial Average Score	Category	Pilot testing Average score	Category	Score Improvement
A.	Principals" Performance Quality					
	Creating working program. schedule. and arranging observation instrument	70.00	Good	88.33	Very Good	18.33
	Conducting academic supervision	72.50	Good	84.00	Very Good	11.50
	Analyzing observation results data	66.67	Good	87.00	Very Good	20.33
	Writing academic supervision report	70.00	Good	89.00	Very Good	19.00
	Feedback and follow up	67.50	Good	85.00	Very Good	17.50
	Number of Indicators	346.67	Good	Good	Very Good	Very Good
	Average of Indicators	69.33	Good	Good	Very Good	Very Good
B.	Collaborative Strategy					
	6. Rapport	80.00	Good	95.00	Very Good	15.00
	7. Determination of collaboration of Basic Values	65.00	Good	80.00	Good	15.00
	8. Collaborative Principles	70.00	Good	82.50	Very Good	12.50
	9. Problem Setting	70.00	Good	87.50	Very Good	17.50
	10. Directing Setting	75.00	Good	85.00	Very Good	10.00
	11. Cooperation among Supervisor	80.00	Good	85.00	Very Good	5.00
	12. Participation	75.00	Good	85.00	Very Good	10.00
	13. Discussion	80.00	Good	87.50	Very Good	7.50
	14. Follow Up	66.00	Good	87.50	Very Good	22.50
	Number of Indicators	660	Good	775	Very Good	Very Good
	Average of Indicators	73.33	Good	86.11	Very Good	Very Good
	Number of all indicators	1006.67	Good	1200.83	Very Good	Very Good
	Average of all indicators	71.91	Good	86.25	Very Good	Very Good

means that collaborative supervision is able to contribute to improving teacher performance quality by 14.40% from other dimensions. The results of the increase are distributed for performance quality which obtained an increase of 17.34%, and collaborative strategy 12.78%. This means that collaborative strategies can contribute to the quality of school principals in building rapport, the value of collaboration, problem setting, directing settings, participation, team discussions, cooperation, and follow-up. Strategy optimization is able to give synergy of main task and function of headmaster (Djumara 2008).

Assessment of the collaborative supervision model that is tested is limited or expanded, consulted with experts and discussed through FGD-2 for model improvement, manuals and guidance; the results are presented in Table 2.

Table 2 shows the results of expert and practitioner validation on the feasibility of models, manuals and guidebooks. The results showed that they are in very reasonable categories with the percentage of 83.99%. The highest score is mean value supervision model of 85.94%. This means that the development model design is eligible for use in enhancing collaborative-

Table 2. Expert and expert validation results on appropriateness of model, guidebook and guidance on academic supervision by collaborative headmaster.

No	Aspect	Score	Average Score	Score Range	Conversion (%)	Category
1.	Model design	220	27.50	26-32	85.94	Very Feasible
2.	Reference Book	800	100	98-120	83.33	Very Feasible
3.	Guide Book	688	86	84-104	82.69	Very Feasible
Total		1708	213.5	208-256	251.96	Very Feasible
Average		213.5	71.17	70-74	83.99	Very Feasible

based principal supervision. The lowest value is the manual with a percentage of 82.69% so it needs to be fixed. Handbooks that need to be improved are: clarity of benefits, techniques, and sufficiency of the bibliography. The findings of the final model of collaborative principal-based principal supervision in Banyumas District are presented in Figure 1.

Figure 1 shows the final model of collaborative-based principal supervision as a refinement of the factual model and the hypothetical model. Factual and hypothetical models after the FGDs are expert validation and practitioners for input and refinement to become a viable final model. Expert and practitioner validation results on system performance are planning, implementation, evaluation, reporting, follow-up, and control. Aspects of planning include goals, work programs, time, steps, instruments and teams. Implementation includes: pre-observation, observation and post-observation. The evaluation includes: the principal's performance qualities in academic supervision, including making working program, schedule, preparation of observation instrument, practice of academic supervision, analysis of observation data, reporting, and feedback and follow-up. Collaborative strategies include report cards, basic determination of collaboration, principles, problem setting, directing settings, participation, team discussions, cooperation, and follow-

up. Aspects of reporting include: reporting systematics, and document completeness. Follow up principal academic supervision with workshop, training and mentoring. Controlling includes: monitoring and evaluation of principal supervision. System components include supervisors, supervisors, materials, methods, facilities and infrastructure and time.

Practical implications of the collaborative principal supervisory principal supervision model are able to improve the quality of principals' performance in implementing supervision so that it can be an alternative model to improve teacher performance in learning.

It is suggested that the principals of junior high school in Banyumas District provide guidance to the principal in academic supervision. The principal should improve the competence, qualification, and track record in conducting academic supervision. School supervisors as partners can improve the performance of academic supervision. Teachers should participate actively in academic supervision activities and to improve learning performance.

4 CONCLUSION

Based on the results of the research that has been described previously, it is suggested that the model of principal supervision that has been implemented is in good category; design of collaborative-based principal supervision model in Banyumas District is very important and necessary; the final model of collaborative-based principal supervision is well-suited for improving teacher performance in learning.

REFERENCES

Bahrodin. 2017. Academic Supervision Model for Headmasters by Applying Collaboration Strategy for Teachers in State Junior High School in Banyumas Regency. *Proceedings of National Seminar and Call for Paper. Untidar* 1(1): 35–44.

Behlol, M.G. 2011. Concept of Supervision and Supervisory Practices at Primary Level in Pakistan. *International Education Studies Journal* 4(4):28–35.

Borg.G. 2007. *Educational Research*. New York: logman.

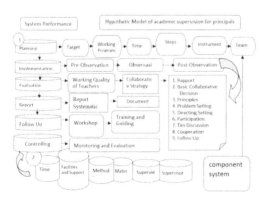

Figure 1. Final model of academic supervision for principals collaborated with teachers from state Junior high schools in Banyumas Regency

Cahyono, B. 2014. Development of Academic Supervision Model Based on Total Quality Management at State Junior High School in Tulungagung Regency East Java Province, *EM* 3(2): 114–118.

Clark, A.O. & Olumese, H.A. 2013. Effective Supervision as a Challenge in Technical and Vocational Education Delivery: Ensuring Quality Teaching/Learning Environment and Feedback Mechanism. *Basic Research Journal of Education Research and Review* 2(1): 6–15.

Daresh, J.C. 2001. *Supervision as proactive leadership. 3rd ed.* Illinois: Waveland Press.

Dharma, S. 2009. *Supervision Module for Educational Working Conference of Headmasters*. Jakarta: Director General of Quality Improvement of Education and Education Personnel of the Ministry of National Education.

Djumara, N. 2008. *Negotiation, Collaboration and Networking*. Jakarta: National Adminsitration Institute.

Ernawati. 2014. Development of Academic Supervision Model by Applying Senior-Teacher-Based Classroom Visitation for ICT Teachers in Senior High Schools in Semarang. *Educational Management* 3 (1): 41–46.

Fakhruddin, 2017. Effect of Mediation of Emotional Intelligence on the Effects of Collaborative Supervision and Leadership on Master's Innovative Behavior. *Educational Management* 6 (1):71–79.

Glickman, C.D. 2007. *Supervision of Intruction: A Developmental Approach*. Boston: Allyn and Bacon. Inc.

Gultom. S. 2014. *Teaching Material Supervising Headmaster Training*. Jakarta: Ministry of Education and Culture.

Guntoro, D., Totok, S. & Ahmad R. 2016. Development of a Web-Based Esupervision Assisted Supervision Model. *EM5* 2(2): 22–28.

Himdani. 2017. Development of Clinical Supervision Model of Group Counseling Technique at Counseling Teachers from Senior High School of East Lombok Regency. *EM* 6 (1): 1–8.

Makawimbang, J.H. 2013. *Clinical Supervision Theory and its Measurement (Analysis in the Field of Education.* Bandung: Alfabeta.

Manggar, Y, 2011. *Headmaster Academic Supervision.* Jakarta: Ministry of Education and Culture Institute for School Development and Empowerment Indonesia.

Marzano, R., Frontier, T & Livingston, D. 2011. *Effective Supervision: Supporting the Art and Science of Teaching.* Alexandria: ASCD.

Pidarta, M. 2009. *Contextual Educational Institution.* Jakarta: Rineka Cipta.

Prihatin, T. 2017. Development of Academic Supervision Model with Mentoring Method in Educational Learning at Vocational Schools in Kupang District. *EM* 6 (1): 9–19.

Rahayu, P. 2015. The Role of the Principal in Academic Supervision to Improve Teacher Professionalism. *Jurnal Unila* 3(3): 1–13.

Silck.M. 2000. *Handbook Educational Supervion. A. Guide for the Practition.* Boston: Ailyn & Bacon Inc.

Surachman, M. 2014. Development of Monitoring Model for Implementation of National Standards for Education Based on Information and Communication Technology. *EM* 4 (2): 125–133.

Syukri, 2015. Implementation of Academic Supervision by the Principal to Improve the Performance of Elementary School Teachers in Cluster I UPTD Dewantara North Aceh. *Jurnal* 3 (2): 79–80.

Taoefik, M. 2016. Collaborative Supervision and Principal Leadership on Master's Innovative Behavior. *EM* 5(3): 1–78.

Wasitohadi, 2016. Collaboration and Inter Institutional Synergy in Principal Competency Improvement. *Jurnal UKSW* 3 (2): 230–245.

Widodo, 2007. Supervision of Economics Teachers in Indonesia between Theory and Reality. *Journal of Economic Education* 2 (2): 291–313.

Advances in Business, Management and Entrepreneurship – Hurriyati et al (eds)
© *2020 Taylor & Francis Group, London, ISBN 978-0-367-27176-3*

The effect of organization culture, leadership style and personality on job satisfaction of auditors and supervisors: A case study of provincial and district/city inspectorates in West Java-Indonesia

E. Sudaryanto
Universitas Pendidikan Indonesia, Bandung, Indonesia

R.W. Kurniasari
Universitas Pakuan, Bogor, Indonesia

ABSTRACT: The Regional Inspectorate as the Internal Control Official Government (APIP) plays a role in controlling and supervising the management of regional finance. This can spur the implementation of good governance and the development of regions, which that can impact on improving the welfare of the community in the area concerned. Control and supervision by the local inspectorate (provincial/district/city) is capable of immediately implementing good governance, which is one of the indicators of free corruption, collusion, and nepotism among local government. Therefore, there should be APIP internal controls in each local government. A change of supervision system by APIP is needed to strengthen the capacity and capability of human resources, especially for the functional officials such as the auditor and supervisor (P2UPD) as the front guard in controlling and supervising the heads of regions (Governor, Regent, and Mayor) in decision-making. The role of auditors and supervisors in driving the tasks and functions carried out by the regional inspectorates surely cannot be separated from the auditor and supervisors as human figures who would expect job satisfaction; this will influence the organization culture and the personality of the auditors or supervisor. Due to these problems, this research uses a quantitative method and random sampling, focused on whether the existence of two functional positions of auditors and supervisors, who perform almost similar tasks in one institution/inspectorate, will affect the job satisfaction. This is viewed from the aspect of organization culture, leadership style and personality.

1 INTRODUCTION

The findings of the Audit Board of Indonesia (BPK-RI) on the performance of APIP in 2013 are that the appointment of the Functional Position of Supervisors for the Execution of Regional Government Affairs (Jabatan Fungsional Pengawas Penyelenggaraan Urusan Pemerintahan di Daerah/JF-P2UPD) by the Ministry of Administrative and Bureaucratic Reform was not executed properly. The causes of this problem were that the Head of BKN gave technical considerations that were just a formality. The Ministry of Administrative and Bureaucratic Reform did not give enough consideration to the appointment of JF-P2UPD, which in turn meant the appointment of JF-P2UPD did not have a legal basis; the professionalism of APIP was not realized; the effectiveness of the supervision was not realized; and the quality of existing supervision was not guaranteed (source: Report of BPK-RI Audit on the Performance of APIP in 2013).

The lack of professionalism by the supervisors in APIP, as stated in the findings of BPK-RI, is due to the existence of two functional positions doing internal auditing, namely auditors and supervisors. Employees carrying out their duties as internal auditors in APIP, especially in the regional government (Provincial and Regency/City Inspectorate), are

functional officials, namely auditors and supervisors who are regulated by the Ministerial Regulation of Administrative and Bureaucratic Reform No. PER/220/M.PAN/7/2008 regarding the Functional Position of Auditor (JFA) and Its Credit Points and the Ministerial Regulation of Administrative and Bureaucratic Reform No. 15 Year 2009 regarding JF-P2UPD and Its Credit Points.

Auditors and supervisors as internal auditors are not functioning as the quality assurance, consultants and catalyst for their agency. They are supposed to be professional in every aspect, so that the internal audit is not seen in a negative light by society and the auditees, especially regarding their moral attitude and behavior.

Auditors and supervisors have the main responsibility of determining whether policies or procedures stipulated by the top management have been followed, determining the quality of bookkeeping for the organization, the efficiency and effectivity of the organization activity procedures, and also the reliability of the information generated by various parts of the organization. In order that APIP can function as the agent of quality assurance and consulting, and for its parent organization (regional government), human resource support is needed in the form of professional auditors and supervisors (P2UPD).

Two functional positions in one inspectorate institution, with no regulations about the requirements of the functional position levels, for instance who can be the team leader, technical controller or quality controller, or about the duties/responsibilities that both functional positions would do, create a dilemma. This is because naturally the auditors and supervisors (P2UPD) would seek what is beneficial for themselves, be it ease of promotion or obtaining credit scores for monetary gains (functional allowance). It is no wonder that in several regional APIP there were mutations from one functional position to another and this is very disruptive to the performance of the APIP (Sudaryanto, 2015).

This research is focused on whether the existence of two functional positions, namely the auditor and supervisor, with similar supervisory duties in one institution/agency, would have impacted the job satisfaction of auditors/supervisors in Inspectorates of the Provincial and City/Regency in West Java.

2 METHOD

The design applied in this research is a quantitative using descriptive approach. In quantitative research, the main focus is positive principles and the research used variables and hypotheses that are tested accurately through calculations.

The qualitative method applied by the writers also uses the correlation research method, which Purwanto (2010) described as "a research which involves the relation of one or more variables to another set of one or more variables. The relation happens in one group." This correlation method is applied to discover how far the impact of organization culture, effectivity of leadership style, the strength of employee personality, and the amount of job satisfaction would influence the effectivity of employee performance as described by the model developed by Colquitt et al. (2009).

The data in this research consists of two types of data: primary and secondary data. Primary data were obtained using instruments measuring the research variables via questionnaires and by conducting interviews with several respondents. The secondary data was supporting data from the auditors and supervisors in the Inspectorates of the Provincial and City/Regency in West Java, and data obtained from the fostering agency for auditors (BPKP) or for supervisors of regional government affairs/P2UPD (training agency of the Ministry of Internal Affairs/ Badan Diklat Kementerian Dalam Negeri).

There are 832 auditors and supervisors in Inspectorates of the Provincial and City/Regency in West Java, according to the fostering agency above, consisting of 418 auditors and 414 supervisors. These auditors and supervisors are distributed in Inspectorates of the Provincial and City/Regency in West Java, which are divided into four Coordinating Agencies for Governance and Development (BKPP/Badan Koordinasi Pemerintahan dan Pembangunan). BKPP Area I covers the

City of Bogor, Regency of Bogor, City of Depok, City of Sukabumi, and the Regency of Cianjur. BKPP Area II covers the City of Bekasi, Regency of Bekasi, Regency of Purwakarta, Regency of Subang, Regency of Majalengka, Regency of Indramayu, City of Cirebon, Regency of Cirebon, and Regency of Kuningan. BKPP Area III covers the City of Banjar, Kab. Ciamis, City of Tasikmalaya, Regency of Pangandaran, Regency of Tasikmalaya, Regency of Sumedang, and Regency of Garut. BKPP Area IV covers the City of Bandung, Regency of Bandung, Kota Cimahi, and Regency of West Bandung.

The sampling technique applied in this research was random sampling, performed by taking samples proportionately using a randomizing technique. Due to the consideration of location and time, the samples were taken in the Inspectorate of West Java Province and in BKPP Area I and Area IV, with 202 respondents consisting of 120 auditors and 82 supervisors.

The research model is described in the problem constellation in Figure 1.

3 RESULT AND DISCUSSION

3.1 Result

3.1.1 Data description
The data description in the form of mean, median, modus, distribution and histogram from each variable are as follows.

3.1.1.1 Organizational culture
The data distribution of the organization culture variable can be seen in Table 1.

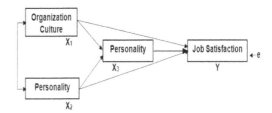

Figure 1. Model research.

Table 1. Frequency distribution of organization culture.

Organization culture	Category	Frequency	Percentage	Cum.
51–65	Very weak	12	5.94	5.94
66–80	Weak	40	19.80	25.74
81–95	Moderate	77	38.12	63.86
96–110	Strong	60	29.70	93.56
111–125	Very strong	13	6.44	100
Total		202	100	

For frequency distribution, the data findings are classified in five categories: very weak, weak, moderate, strong, and very strong. The score of organizational culture with the frequency/total of 77 respondents is around 81–95 (third interval class), which shows that the strength of organization culture is needed to improve job satisfaction of auditors and supervisors.

3.1.1.2 Leadership style

The data distribution of the leadership style variable can be seen in Table 2.

For frequency distribution, the data findings are classified in five categories: ineffective, less effective, moderately effective, effective, and very effective. The leadership style score with a frequency/total of 65 respondents is around 83–98 (third class interval). This shows that the effectiveness of leadership style is needed to improve job satisfaction of auditors and supervisors.

3.1.1.3 Personality

Data distribution of the personality variable can be seen in Table 3.

For the frequency distribution, the data findings are classified in five categories: not good, less good, moderately good, good, and very good. The personality score with a frequency/total of 69 respondents is around 114–126 (third class interval). This shows that a good personality is needed to improve job satisfaction.

3.1.1.4 Job satisfaction

The data distribution of the job satisfaction variable can be seen in Table 4.

Table 2. Frequency distribution of leadership style.

Leadership Style	Category	Frequency	Percentage	Cum.
51–66	Ineffective	15	7.43	7.43
67–82	Less effective	43	21.29	28.71
83–98	Moderately effective	65	32.18	60.89
99–114	Effective	44	21.78	82.67
115–130	Very effective	35	17.33	100
Total		202	100	

Table 3. Frequency distribution of personality.

Personality	Category	Frequency	Percent	Cum.
88–100	Not good	18	8.91	8.91
101–113	Less good	62	30.69	39.60
114–126	Moderately good	69	34.16	73.76
127–139	Good	29	14.36	88.12
140–152	Very good	24	11.88	100
Total		202	100	

Table 4. Frequency distribution of job satisfaction.

Job satisfaction	Category	Frequency	Percent	Cum.
56–66	Very low	9	4.46	4.46
67–81	Low	46	22.77	27.23
82–96	Moderate	65	32.18	59.41
97–111	High	51	25.25	84.65
112–126	Very high	31	15.35	100
Total		202	100	

For the frequency distribution, the data findings are classified in five categories: very low, low, moderate, high, and very high. The job satisfaction score with a frequency/total of 65 respondents is around 82–96 (third class interval). This shows that the value of job satisfaction for auditors and supervisors is affected by other variables.

3.1.2 Normality test

A normality test is conducted to test if the data has a normal distribution. The tool to test the data normality in this research is the Kolmogorov Smirnov test. The summary of the normality test can be seen in Table 5.

3.1.3 Path analysis

Based on the causal model created theoretically, a diagram of path analysis will be obtained and coefficient values or r for each path will be calculated. Simple correlation coefficient between the variables job satisfaction and organization culture, leadership style, and personality is 0.000 or smaller than 0.01. This means the inter-variable correlation between independent and dependent variables is very significant. The calculation of inter-variable correlation is shown in Table 6.

3.1.4 Correlation model of job satisfaction path

The calculation of coefficient values for paths pYX1, pYX2, and pYX3 can be seen in Table 7.

Table 7 shows that $\rho YX1 = 0.384$, $\rho YX2 = 0.289$, and $\rho YX3 = 0.142$, which means that the dependent variable (job satisfaction) is affected by the independent variable (organization culture) with a value of 0.384, the independent variable (leadership style) with a value of 0.289, and the independent variable (personality) with a value of 0.142.

Table 5. Summary of normality test analysis.

Variable	n	D_{max}	p-value	Legend
Organization culture	202	0.0668	0.328	Normal
Leadership style	202	0.0677	0.312	Normal
Personality	202	0.0887	0.069	Normal
Job satisfaction	202	0.0781	0.109	Normal

Table 6. Simple inter-variable correlation.

Variable	Organization culture	Leadership style	Personality	Job satisfaction
Organization culture	1.000			
Leadership style	0.462	1.000		
Sig	0.000			
Personality	2.276	0.331	1.000	
sig	0.000	0.000		
Job satisfaction	0.557	0.514	0.344	1.000
Sig	0.000	0.000	0.000	

Table 7. Calculation of path coefficient ρYX1, ρYX2, and dependent variable: job satisfaction .

Variable	Coef.	Stad. Err.	t	ρ	Beta
Organization culture	0.428	0.069	6.180	0.000	0.384
Leadership style	0.260	0.057	4.570	0.000	0.289
Personality	0.166	0.068	2.430	0.016	0.142
Constant	10.052	8.413	1.190	0.234	

Other than the variables organization culture, leadership style, and personality which affect job satisfaction (Y), there is an influence of (e)/other variables, whose calculation uses the formula $e = \sqrt{(1-R \text{ square})}$ which produced e = 0.767463. The residual path coefficient is obtained with the formula $P4e = \sqrt{(1 - R2)}$. The value $R2$ is obtained from the STATA as follows:

Therefore the value of the residual coefficient is $PY = \sqrt{(1-0.411)} = \sqrt{0.589} = 0.767$

3.1.5 Decomposition
Calculation of direct and indirect influence between independent variables and dependent variables can be seen in Table 8.

Table 8 shows that the path with the biggest value is the path X2Y (pYX2), with a total value of 0.689. This means that to increase the job satisfaction of auditors and supervisors, the fastest path is to

Table 8. Decomposition of direct and indirect between variables.

Influence	Direct	Indirect X₃	Total
X₁ toward Y	0.384	0.299	0.683
X₂ toward Y	0.258	0.400	0.689
X₃ toward Y	0.142	-	0.142

increase the effectiveness of leadership style through the strength of personality.

3.1.6 Structural model form
The structural model after the analysis is shown in Figure 2.

3.1.7 Hypothesis test
The summary of the hypothesis test can be seen in Table 9.

Table 9 shows that organization culture, leadership style, and personality have direct impact toward job satisfaction of the auditors and supervisors.

3.2 Discussion

Job satisfaction of auditors and supervisors are affected by the variables organization culture, leadership style, and personality. The impact of organization culture, leadership style toward job satisfaction are described below.

Organization culture is part of organization mechanisms which has direct and indirect impacts (through personality) toward job satisfaction, which is part of individual mechanisms. Organization culture related with values and behavioral norms of a good work place, such as cooperative working, attentive toward details, being result-oriented, and personality, which is characterized as being reliable, tenacious, and trustworthy are proven to affect the

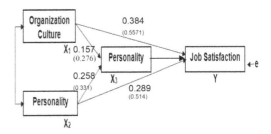

Figure 2. Structural model.

Table 9. Summary of hypothesis test results.

No	Hypothesis	Hypothesis Statistic	Statistic Test Result	Conclusion
1.	Organization culture has direct impact toward job satisfaction	H₀: ρYX₁ = 0 H₁: ρYX₁> 0	H₀ rejected	Has direct impact
2.	Leadership Style has direct impact toward job satisfaction	H₀: ρYX₂ = 0 H₁: ρYX₂ > 0	H₀ rejected	Has direct impact
3.	Personality has direct impact toward job satisfaction	H₀: ρYX₃ = 0 H₁: ρYX₃ > 0	H₀ rejected	Has direct impact

job satisfaction of auditors and supervisors in Inspectorates of the Provincial and City/Regency in West Java. This is characterized by the auditors and supervisors feeling safe in carrying out their duties, feeling that there is a career development opportunity, and feeling that there are rewards.

Leadership style is part of group mechanisms that have direct and indirect impacts (through personality) toward job satisfaction, which is a part of individual mechanisms. Leadership style is characterized by how a superior officer maintains a good relationship with their subordinates and delegates authority. Personality is characterized as being reliable, tenacious, and trustworthy. Both are proven to affect the job satisfaction of auditors and supervisors in Inspectorates of the Provincial and City/Regency in West Java. This is characterized by the auditors and supervisors feeling safe in carrying out their duties, feeling that there is a career development opportunity, and feeling that there are rewards.

Personality is part of individual characteristics that have a direct impact toward job satisfaction, which is a part of individual characteristics. Personality, which is characterized as being reliable, tenacious, and trustworthy is proven to affect the job satisfaction of auditors and supervisors in Inspectorates of the Provincial and City/Regency in West Java. This is characterized by the auditors and supervisors feeling safe in carrying out their duties, feeling that there is a career development opportunity, and feeling that there are rewards.

However, other than the fact that the job satisfaction of auditors and supervisors in Inspectorates of the Provincial and City/Regency in West Java is affected by the factors of organization culture in the related agency, leadership style of the superior officers, and a good personality, there are other factors with a value of 76.75% (e = 0.767463) that can improve the job satisfaction of auditors and supervisors in Inspectorates of the Provincial and City/Regency in West Java.

4 CONCLUSION

Organization culture has an impact toward job satisfaction. The strength of organization culture developing in the Inspectorates of the Provincial and City/Regency in West Java can influence the job satisfaction of auditors and supervisors in the agencies involved, making it advantageous or disadvantageous.

Leadership style has an impact toward job satisfaction. The effectiveness of leadership style of a superior officer in the Inspectorates of the Provincial and City/Regency in West Java can influence the job satisfaction of auditors and supervisors in the agencies involved, making it advantageous or disadvantageous.

Personality has an impact toward job satisfaction. The strength of the personality of auditors or supervisors in the Inspectorates of the Provincial and City/Regency in West Java can influence the job satisfaction of auditors and supervisors in the agencies involved, making it advantageous or disadvantageous.

REFERENCES

Colquitt, J., Lepine, J.A., Wesson, M.J. & Gellatly, I.R. 2009. *Organizational behavior: Improving performance and commitment in the workplace*. New York: McGraw-Hill Irwin.

Ministerial Regulation of Administrative and Bureaucratic Reform No. PER/220/M.PAN/7/2008 regarding the Functional Position of Auditor (JFA) and Its Credit Points.

Ministerial Regulation of Administrative and Bureaucratic Reform No. 15 Year 2009 regarding JF-P2UPD and Its Credit Points.

Purwanto, N. 2010. *Prinsip-prinsip dan teknik evaluasi pengajaran*. Bandung: PT. Remaja Rosdakarya.

Sudaryanto, E. 2014. Policy effectivity of government internal supervision body (apip) to actualize clean, and free of corruption, collusion, and nepotism government in Indonesia. *International Journal of Science and Research (IJSR)* 3(9): 306–310.

Advances in Business, Management and Entrepreneurship – Hurriyati et al (eds)
© *2020 Taylor & Francis Group, London, ISBN 978-0-367-27176-3*

The effectiveness of a mobile-based reward system on performance: A case study on Go-Jek online transportation drivers in Garut, Indonesia

R. Muttaqin & S. Suwatno
Universitas Pendidikan Indonesia, Bandung, Indonesia

ABSTRACT: The purpose of this study is to see the effectiveness of a mobile-based reward system on performance. The research method was a descriptive and verificative with quantitative approach and the method is used is an explanatory survey with a simple random sampling technique. The study population is 4238 drivers of Go-Jek in Garut, with a sample of 356 respondents. A questionnaire was used as a research instrument to collect data from respondents. Path analysis was used to analyze the data, with SPSS software as the tool. The result shows that the effectiveness of a mobile-based reward system has positive impact on performance of drivers Go-Jek in Garut.

1 INTRODUCTION

Reward systems are one of the strategies used by human resource managers for attracting and retaining suitable employees as well as facilitating them to improve their performance through motivation and to comply with employment legislation and regulations (Rugami et al. 2016). A reward system is important for employee performance, and a high-quality reward system will be more effective. That how to be high performance is base high reward vice versa. Job performance is also part of human resources management. Performance and employees' good work is important for an organization's success and achieving the goals. Employees provide good effort for achieving goals and that good effort depends on rewards they receive (Ibrar & Khan 2015).

One company that improves service quality by improving employee performance is PT Go-Jek Indonesia. Go-Jek is one of the largest online motorcycle transportation companies in Indonesia and their utilization of the internet and smartphones also helps provide employment. This company is able to absorb large numbers of workers who operate in as many as 63 cities in Indonesia.

Garut, one of the regencies in West Java Province, has a Go-Jek branch office. Job openings are one of the benefits of the presence of Go-Jek in Garut. The size of the Garut community needs online motorcycle taxi drivers to be the background of Go-Jek to penetrate the market in Garut regency.

The fundamental problem in company organization is how an institution or agency can achieve superior performance and maintain competitive advantage (Radulovich et al. 2016). Strategy is the key to success that influences organizational performance (Widodo 2011). This is because a strategy is an overall plan that explains the position of a company's competitiveness (Widodo 2009).

Factors that influence drivers' performance include the reward system, job design, training and development opportunities, leadership, personality, communication, trust, and recognition (Nzuve & Njambi 2015, Peters et al. 2017). This is consistent with Wilson's study, which shows that rewards include systems, programs and practices that can influence a person's actions and make a positive contribution to the desired performance (Rugami et al. 2016).

The coefficient of a reward system shows that it can significantly affect performance (Rugami et al. 2016). The ideal combination of reward systems helps organizations maximize employee commitment, motivation, and job satisfaction, which results in improved employee performance and organizational productivity. Applying a reward system can significantly improve employee performance (Khan & Afzal 2016).

According to Karami et al. (2013), the reward system is accompanied by several actions, both from a corporate and individual point of view. Research on human resource management shows that organizational compensation plays an important role in ensuring employees work hard and take responsibility, because rewards can positively affect employee performance. As mentioned in Khan and Afzal (2016), the reward can significantly affect employee performance.

According to Ndungu (2017), the indicator profile of the reward system is used to understand the characteristics of employees and this knowledge is used to improve employee performance through (1) extrinsic rewards, (2) intrinsic rewards, (3) recognition (which involves non-cash rewards and social benefits), (4) financial rewards (such as performance bonuses), (5) work environment, and (6) leadership style taken as an independent variable.

According to Nzuve and Njambi (2015), employee performance is success shown by certain individuals or individuals, determined by the supervisor or organization, to set standards that can be used for an environment that changes in a changing environment. Employee performance uses positive rewards from bonuses, awards, wider interpersonal awards, promotion, achievement, satisfaction, and blessings.

Research and experience on the concept of performance strength have shown that effective and timely feedback is a motivation to increase employee productivity. This feedback means the rewards given by the company to employees as an award and dedication. Generally, appreciation of rewards for all employees to improve performance. In addition, in return will improve employee performance to be better with the contributions made. Employee performance is very important in improving organizational performance and improvement (Nazir & Islam 2017).

According to Ndungu (2017), the indicators of performance used to measure this dimension, with relevant sources from which they were adopted, are (1) work quality, (2) initiative, (3) team work, (4) problem-solving, (5) response against stress and conflict, (6) productivity, and (7) development of employee performance. The recognition model is shown Figure 1.

In this way, awards and recognition will elicit good work from Go-Jek drivers. The company provides many mobile-based rewards for all drivers in Garut regency. The goal is to make it easier for companies to reward drivers directly and quickly, to reward drivers who are shown to depend on human resources, which can increase company profits in the long run. For the driver, a sense of appreciation and trust will immediately improve their work performance.

The mobile-based reward system can be used as an application tool to encourage employees to work properly in accordance with their responsibilities. This method can encourage the actions of Go-Jek drivers and improve their attitudes so they always feel happy. In addition, it aims to make employees more active at work so that they can improve and enhance presentations, which Go-Jek hopes can foster good performance by its drivers in Garut regency. Thus, it can aim for the level of performance that is the company's goal. The quality of human resources will be fulfilled if the mobile-based rewards system can be implemented properly, which supports the effectiveness of drivers performance that can be made perfect.

2 METHOD

This study is a quantitative research with a cross sectional approach. The research population was 4238 drivers in Garut, Indonesia. The formula used to measure the sample was by Al-Rasyid (1994), namely the size of the sample which is a comparison of the size of the population with the presentation of leniency inaccuracy, because the sampling can be tolerated or desirable. This sampling used an error rate of 5%. So that the sample used was representative, the study determined a sample of 356 respondents. The analytical technique used was path analysis. The data were collected through an explanatory survey with a simple random sampling technique. Collection was done at a meeting in the basecamp and on the street, from 8 a.m. until 11 a.m., not during the work day.

The descriptive verification method was used to describe the effectiveness of the mobile-based reward system (X), which has sub-variables consisting of extrinsic rewards (X1), intrinsic rewards (X2), recognition rewards (X3), financial rewards (X4), work environment (X5), and leadership styles (X6) toward performance (Y), a case study on Go-Jek online transportation drivers in Garut. The data were analyzed using path analysis supported by SPSS 25.0 software.

3 RESULTS AND DISCUSSION

The results of this study have had a significant impact. Therefore, to know which exogenous variables are affected individually by endogenous variables, this can be followed by partial testing.

Figure 2 shows the coefficients of the mobile-based reward system, with sub-variables including extrinsic rewards (X1), intrinsic rewards (X2), recognition rewards (X3), financial rewards (X4), work environment (X5), and style leadership (X6), that have a positive influence on the performance of (Y) Driver Go-Jek in Garut. The results of this hypothesis are clarified by the calculation data from partial testing, which is shown in Table 1.

From Table 1, it can be seen that X1 (2.420), X2 (2.866), X3 (2.592), X4 (3.143), X5 (2.282), and X6 (2.560) greater than 1.96 significantly influence the variable Y, where t score > t table the level of significant test (0.000) < (0.05), therefore H0 was

REWARD SYSTEM (X)	PERFORMANCE (X)
1. Extrinsic Rewards	1. Quality work
2. Intrinsic Rewards	2. Initiative
3. Recognition Rewards	3. Team work
4. Financial Rewards	4. Problem Solving
5. Work environment	5. Response to stress
6. Leadership styles	6. Conflict
	7. Productivity
	8. Employee performance Development.

Figure 1. Conceptual model and hypothesis research

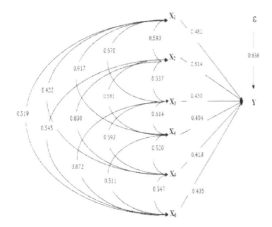

Figure 2. Hypothesis path diagram of reward system toward performance of Go-Jek drivers in Garut.

Table 1.

Alternative hypothesis	t Hitting	t Tibet	Significant	Decision
XL affects Y	2.420		0.015	H; rejected
X affects Y	2.866		0.004	H; rejected
Xj affects Y	2.592		0.007	H; rejected
X4 affects Y	3.143	1.96	0.002	H; rejected
X affects Y	2.282		0.023	H; rejected
X affects Y	2.560		0.011	H; rejected

rejected and there was significant effectiveness of the mobile-based reward system (X) on drivers' performance (Y).

This is in accordance with the opinion of Freedman, as cited in Sajuyigbe et al. (2013), that when effective rewards and recognition are implemented within an organization a favorable working environment is produced, which motivates employees to excel in their performance. Employees take recognition as their feelings of value and appreciation and, as a result, this boosts the morale of the employee, which ultimately increases the productivity of organizations. Rewards play a vital role in determining the significant performance in a job and is positively associated with the process of motivation (Sajuyigbe et al. 2013).

4 CONCLUSION

This study concludes that there is a significant influence of the mobile-based reward system on drivers' performance. The policy of providing benefits through smartphones that are implemented by the company is a strong one, so the drivers are able to improve performance and this has an impact on profits for the company. The

atmosphere in giving rewards directly through the mobile platform mediates the experience and provides emotional stimulation to the drivers. The mobile-based rewards system can motivate employees and more presentations directly in the smartphone application. A gift system has been used in this business for a long time, but every company has a different way of implementing it. In the era of disruption, a mobile-based reward system is an application that can be easily used by all companies directly, without being face-to-face with employees, aiming to increase motivation and performance by the employees to meet the goals that the company has set. It is very important for companies to maintain the relationship system reward variables and this performance in human resource practices. In addition, this study also supports previous research, which suggests that the mobile-based reward system significantly affect drivers' performance. The results of this study also imply a theory that explains from recommendation results of this study are:

1. In order for a mobile-based reward system to run well, the company must first design the gift program and driver performance carefully, so that programs and policies are not only written in the Go-Jek application, but are implemented for the driver.
2. The better the implementation of the reward-based mobile platform, the better the performance of drivers for the company. Thus, the company should focus on the application of the reward-based mobile platform to improve driver performance further.
3. Contributing to knowledge development, especially in the science of the organization behavior and management human resources.

REFERENCES

Al-Rasyid, H. 1994. *Teknik penarikan sampel dan penyusunan skala*. Bandung: Universitas Padjajaran.
Karami, A., Dolatabadi, H.R. & Rajaeepour, S. 2013. Analyzing the effectiveness of reward management system on employee performance through the mediating role of employee motivation case study: Isfahan Regional electric company. *International Journal of Academic Research in Business and Social Sciences* 3(9): 327.
Ndungu, D.N. 2017. The effects of rewards and recognition on employee performance in public educational institutions: A case of Kenyatta University, Kenya. *Global Journal of Management and Business Research*.
Khan, H.G.A. & Afzal, M. 2016. The effect of reward systems, organizational commitment and experience on job satisfaction with respect to employee's perceived performance. NUML *International Journal of Business & Management* 11(2): 35–49.
Ibrar, M. & Khan, O. 2015. The impact of reward on employee (A case study of Malakand Private School) 52: 95–103.

Nzuve, S.N.M. & Njambi, M.P. 2015. Factors perceived to influence employees' performance: A case of the independent electoral and boundaries commission. *Scienta Socialis Journals* 10(2): 88–99.

Nazir, O. & Islam, J.U. 2017. Enhancing organizational commitment and employee performance through employee engagement: An empirical check. *South Asian Journal of Business Studies* 6(1): 98–114.

Peters, J., Basit, A. & Hassan, Z. 2017. Influence of team related factors on employee performance: A case study on Malaysian context.

Radulovich, L., Javalgi, R. & Scherer, R.F. 2016. Intangible resources influencing the international performance of professional service SMEs in an emerging market: Evidence from India. *International Marketing Review* 35(1): 113–135.

Rugami, I., Wambua, P., Mwatha, S. & Rugami, I. 2016. Reward systems and employee performance in the print media sector in Kenya. *European Journal of Business and Strategic Management* 1(1): 100–116.

Sajuyigbe, A.S., Olaoye, B.O. & Adeyemi, M.A. 2013. Impact of reward on employees' performance in a selected manufacturing companies in Ibadan, Oyo state, Nigeria. *International Journal of Arts and Commerce* 2(2): 27–32.

Widodo, W. 2009. Research on Smart Work Patterns and Coordination to Improve Organizational Performance 2(1): 25–45.

Widodo, W. 2011. Building strategy quality. *International Journal of Business and Management* 6(8): 180–192.

Advances in Business, Management and Entrepreneurship – Hurriyati et al (eds)
© *2020 Taylor & Francis Group, London, ISBN 978-0-367-27176-3*

Influence of ISO 9001: 2015 Quality Management System implementation on employee performance: A case study of UPI Academic Directorate

D. Lavianti & E. Ahman
Universitas Pendidikan Indonesia, Bandung, Indonesia

ABSTRACT: The purpose of this study was to determine the effect of ISO 9001: 2015 Quality Management System (QMS) on employee performance, with the Universitas Pendidikan Indonesia (UPI) Academic Directorate research object. ISO 9001: 2015 Quality Management System is viewed from three processes: quality management plan, organizational commitment, and ISO 9001: 2015 Implementation Procedure. The analysis unit is all employees in the Academic Directorate at UPI. The research data was obtained from the survey results in the form of questionnaires with a Likert 5 scale distributed at one time. The results of the study found that when planning for ISO 9001: 2015 certification, a quality management plan, organizational commitment, and implementation of work procedures/work instructions had a significantly positive impact on employee performance.

1 INTRODUCTION

Currently the development of quality management covers various aspects, such as quality management ISO 9001 (Heras-Saizarbitoria & Boiral 2013). Since ISO was first introduced, it has become a global benchmark for quality management systems and has issued more than one million certificates in 178 countries (Ngai et al. 2013). The latest version of ISO 9001: 2015, is also predicted to improve the performance and targets of an organization/company. However, some studies have found no relationship between performance with ISO (Ramadhany 2015). ISO 9001 is one of the benchmarks for quality management systems that have been widely implemented in various companies.

In 2013, ISO certificates were issued for 184 countries in all sectors, both manufacturing and services (Allur et al. 2014). Universitas Pendidikan Indonesia (UPI) is one of the Legal Entity Higher Education Institutions (PTN-BH) given autonomous rights to manage institutions independently. The Autonomous Management of the State Entity Legal Entity (PTN-BH) is carried out in accordance with the standards set by the government and covers the academic and non-academic fields. In this case study, one of PTNBH's management autonomies in the non-academic field includes operational policies and quality management of the activities carried out. Quality improvement becomes increasingly important for institutions and is used to get better control. According to Widodo (2011), quality is a dynamic idea that is difficult to uniform.

To improve quality and competitiveness, it is necessary to do planning and set out processes that are in accordance with the established rules. A good management system is needed that can encourage the performance of employees, one of which is imple-menting the Quality Management System ISO 9001 and currently the newest version is ISO 9001: 2015 Quality Management System. Many studies show that ISO 9001 implementation improves operational performance and work culture and product/service quality (Psomas & Pantouvakis 2015).

ISO 9001: 2015 is an advanced process and audits/evaluations are carried out every year. ISO 9001: 2015 is the latest version that has been upgraded, in the implementation of ISO 9001: 2015 this requires organizational commitment. The benefits and processes of the quality management plan and implementation of ISO 9001: 2015 certification differ in each organization/company but still have the same rules and objectives.

The purpose of this study was to determine the effect of the implementation of ISO 9001: 2015 Quality Management System on employee performance, based on the results of a survey conducted with employees in the research object.

2 LITERATURE REVIEW

According to Gasperz (2002), "A Quality Management System is a set of documented procedures and standard practices for system management that aims to ensure the suitability of a process and product (goods/services) to the requirements or requirements determined or specified by customers or organizations."

Referring to the two meanings mentioned earlier, it can be said that a quality management system is a tool that is applied in an organization to provide a transparency about the activities in the organization. This activity is expected to provide satisfaction, and can meet the needs of customers and markets.

According to the International Organization for Standardization, a quality management system is the way in which a company controls activities that are interconnected (both directly and indirectly) to achieve the desired results. Hadiwiardjo and Wibisono (1996) state that companies that run a quality management system tend to show the following traits: (1) the existence of a philosophy that prevents better than detecting, correcting, and results; (2) communication that is consistent in the process and between production, suppliers and buyers; (3) careful maintenance of documents and controlling them efficiently; (4) quality awareness of all employees.

According to Gasperz (2002), the purpose of the quality management system is to ensure the suitability of a process and product to specific needs or requirements and provide satisfaction to consumers through meeting the needs and requirements of processes and products determined by customers and organizations.

Rothery (2000) states that "ISO 9000 series is a first and foremost quality management system, a global system to optimize the quality effects of an organization or company, by creating a framework for continuous improvement".

ISO 9000 series sejak diterbitkan pertama kali pada tahun 1987, standar ini sudah mengalami empat kali perubahan yang mencakup beberapa standar yaitu ISO 9001, ISO 9004, dan ISO 19011. Semua standar ISO selalu ditinjau dan direvisi secara berkala untuk memastikan persyaratan di dalamnya tetap relevan terhadap kondisi pasar. Versi terbaru ISO yang saat ini berlaku adalah ISO 9001:2015, menggantikan versi sebelumnya yaitu ISO 9001:2008. Revisi ISO 9001:208 bertujuan agar standar ISO 9001:2015 bisa diterapkan pada semua jenis perusahaan hal tersebut terlihat ISO 9001:2015 dibandingkan dimana strukturnya disesuaikan dengan struktur di dalam An-nex SL, yaitu High Level Structure (HSL) yang menjadi acuan dasar bagi semua struktur sistem manajemen mutu yang diterbitkan ISO, sehingga memudahkan perusahaan untuk menggunakan sistem manajemen yang lain.

Pasal-pasal yang terdapat di dalam ISO 9001:2015 berfokus pada berpikir berdasarkan risiko (risk based thinking), di mana perusahaan diharapkan mampu memahami cara berpikir berdasarkan risiko secara lebih rinci sehingga dapat mewujudkannya dalam pelaksanaan dan peningkatan sistem manajemen mutu serta proses bisnisnya.

The International Organization for Standardization is an international body that expresses itself in terms of standardization. ISO standards are reviewed and revised periodically to ensure the requirements therein remain relevant to market conditions. Since the first publication in 1987, ISO has undergone four changes. The latest version of ISO currently in effect and used in the UPI is ISO 9001: 2015, replacing the previous version of ISO 9001: 2008 (Ramadhany 2015).

This last revision was published and passed in September 2015 and is expected to be a stable standard, at least for the next 10 years.

ISO 9001: 2015 contains several main points, such as leadership, where the role of the company is expected to become more active in taking responsibility for implementing the management system. The application of risk management in each line of business allows the company to always calculate the risks that will be faced from each action or decision taken.

ISO states that there are seven principles that underline the ISO 9001: 2015 Quality Management System: customer focus, leadership, people involvement, process approach, motivation, evidence-based decision-making, and relationship management, (Fonseca & Domingues 2016).

Armstrong & Baron (2005) explain performance management as ensuring that managers effectively manage this, which means that they ensure the people or teams they lead (1) know and understand what is expected, (2) have the skills and abilities to meet expectations, (3) have the support of the organization to develop their capacity to meet expectations, and (4) have the opportunity to discuss and contribute to the individual and team goals and objectives (Jain 2014).

3 METHOD

This study uses descriptive quantitative methods with survey analysis techniques. Data was obtained by distributing questionnaires to all staff in the academic environment. The questionnaires took into account the indicators that exist on ISO, with a Likert 5 scale. The research data obtained 35 samples of employee respondents who were randomly selected. Data were analyzed by performing multiple linear regression analyzes with the help of SPSS. This study has an X variable that is ISO 9001: 2015 with three independent variables of quality management plan (X1), organizational commitment (X2), and implementation of work instructions (X3) yang terlihat pada gambar paradigma penelitian. See Figure 1.

H0: Variable quality management plan, organizational commitment, and implementation of work

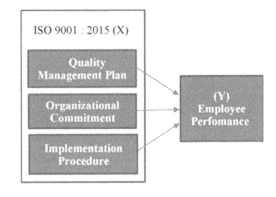

Figure 1 Reasearch paradigm

instructions has a significant effect on employee performance.

H1: Variables of the quality management plan, organizational commitment, and implementation of work instructions have no significant effect on the performance of the stewards' (X1), organizational commitment (X2) to employee performance (Y).

4 RESULT AND DISCUSSION

The Academic Directorate of UPI has been certified to ISO 9001: 2015 Quality Management System since Ok-2017, after previously being ISO 9001: 2000 certified. The implementation of the ISO 9001: 2015 Quality Management System at the Academic Directorate of UPI aims to improve the management system and documentation of unstructured activities. This is done so that institutional management is more structured, and agency costs are more easily achieved.

The research location was the Academic Directorate of Universitas Pendidikan Indonesia Bandung, West Java. This unit was chosen because the Academic Directorate is a central unit that is strategic and serves all UPI students. The Academic Directorate has the data needed to support this research, which is planning data up to the implementation of ISO 9001: 2015 Quality Management System. The data collection technique was questionnaires with a Likert 5 scale, given at one time to all employees (staff) in the academic environment. It is the responsibility of each employee to get involved and understand the importance of the quality management system, which makes clear the proced-

ures and accountability of work in every part of the organizational structure of the company unit. Thirty-five questionnaires were returned and declared valid. The questionnaire data obtained were processed with SPSS assisted data analysis techniques. The results of the processing are shown in Tables 1–3.

Table 1 shows that the R figure is 0.604 and there is a very strong relationship between work instruction (X3) and quality management plan.

Table 2 shows that R (R square) is 0.365 or (36.5%). There is a percentage contribution of 36.5% influence on the independent variable quality management plan (X1), organizational commitment (X2), and work instructions(X3) to employee performance (Y).

Standard error of the estimate is a measure of the number of errors in the regression model in predicting the value of Y. From the regression results, this is a value of 2.70992.

The steps for conducting the F test are as follows:

1. Hypothesis
   Ho: There is no significant effect between the quality management plan (X1), organizational commitment (X2), and work instructions (X3) on employee performance.
   Ha: There is a significant influence between the quality management plan (X1), organizational commitment (X2), and work instructions (X3) on employee performance.
2. The level of significance
   The level of significance uses a = 5% a = 0.05. Thus, from the results of the analysis it can be concluded that the quality management plan (X1),

Table 1. Multiple correlation analysis results.

Model summary[b]

| Model | R | R Square | Adjusted R square | Std. error of the estimate | Change Statistics | | | | | |
					R square change	F change	df1	df2	Sig. F change	Durbin Watson
1	0.604[a]	0.365	0.304	2.70992	0.365	5.942	3		0.03	1.757

Table 2. Determination analysis (R2).

Model summary[b]

| Model | R | R Square | Adjusted R Square | Std. Error of The Estimate | Change Statistics | | | | | |
					R Square Change	F Change	df1	df2	Sig. F Change	Durbin Watson
1	0.604[a]	0.365	0.304	2.70992	0.365	5.942	3		0.03	1.757

a Dependent variable: Kineria Karyawan (Y)

Table 3. Test regression coefficients together (test F).

Model 1 Regression	Sum of squares	df	Mean square	F	Sig.
	130.918	3	43.639	5.942	0.003[a]
Residual	227.653	31	7.344		
Total	358.571	34			

[a] Predictors: (Constant), Intruhsi Kerja (X3), Rencana Manajemen Mutu (XI), Komitmen Organisasi (X2)

organizational commitment (X2), and work instructions (X3) jointly affect the performance of employees.
3. F count ≥ based on the table, F count is 5.942.
4. F table ≥ menggunakan using a 95% confidence level, a = 5%, the results obtained for F table are 2.9113.
5. Test criteria: -Ho is accepted if F counts < F table or Ho is rejected if F counts > F table.

Then: F value is calculated > F table (5.942 > 2.9113), then Ho is rejected. It can be concluded that there is a significant influence of the quality management plan (X1), organizational commitment (X2), and work instructions (X3) on employee performance.

5 CONCLUSION

The results of this study support the hypothesis that the implementation of ISO 9001: 2015 has a positive influence on employee performance. Thus, it is important for institutions and employees to understand and implement ISO 9001: 2015 in the assigned work process, from the planning process to documentation. In this case the implementation of ISO 9001: 2015 has an impact on the management of work units where the quality of service documents improves.

REFERENCES

Allur, E., Heras-Saizarbitoria, I. & Casadesus, M. 2014. Internalization of ISO 9001: A longitudinal survey. *Industrial Management and Data Systems* 114(6): 872–885.
Armstrong, M. & Baron, A. 2005. *Managing performance: Performance management in action.* CIPD publishing.
Fonseca, L. & Domingues, J.P. 2016. ISO 9001:2015 edition-management, quality and value. *International Journal for Quality Research* 11(1): 149–158.
Gasperz, V. 2002. ISO 9001 and continual quality improvement.
Hadiwiardjo, B.H. & Wibisono, S. 1996. *Memasuki pasar internasional dengan ISO 9000: Sistem menajemen mutu.* Ghalia Indonesia.
Heras-Saizarbitoria, I., & Boiral, O. 2013. ISO 9001 and ISO 14001: Towards a research agenda on management system standards. *International Journal of Management Reviews* 15(1): 47–65.
Jain, S. & Gautam, A. 2014. Performance management system: A strategic tool for human resource management. *Prabandhan Guru* 5: 28–31.
Ngai, E.W., Moon, K.L.K., Lam, S.S., Chin, E.S. & Tao, S. S. 2015. Social media models, technologies, and applications: An academic review and case study. *Industrial Management & Data Systems* 115(5): 769–802.
Psomas, E. & Pantouvakis, A. 2015. ISO 9001 overall performance dimensions: An exploratory study. *TQM Journal* 27(5): 519–531.
Ramadhany, F.F. 2015. Analisis penerapan sistem manajemen mutu ISO 9001: 2015 dalam menunjang pemasaran (Studi pada PT Tritama Bina Karya Malang) 55(1): 31–38.
Rothery, B. 2000. *ISO 9000 & 14000.*
Widodo, 2011. *Manajemen mutu pendidikan.* Jakarta: PT. Ardadizya Jaya.

The mediating role of academic motivation in the influence of the Big Five personalities on academic performance

I.B.G.A. Permana
Universitas Airlangga, Surabaya, Indonesia

ABSTRACT: This study aims to seek the influence of the Big Five Personalities on academic performance, using academic motivation as the intervening variable. This research is classified as quantitative research, which is research using hypothesis testing with a statistical test tool. The population in this study are 750 students of the Faculty of Economics and Business at Airlangga University. The findings show that openness partially influences the academic performance after being mediated with intrinsic motivation; agreeableness and openness respectively influence fully and partially the academic motivation. The research findings suggest that the faculty can create a comfortable learning atmosphere both in the process of teaching and learning facilities that can be used by students

1 INTRODUCTION

Qualified Human Resources (HR) are one of the important capital in supporting a nation to be able to compete and balance the progress of both technology and the economy of other countries. The challenges of global competition that companies must face in the coming decade will increase the importance of human resource management (Noe, Hollenbeck, Gerhart, & Wright, 1994). Others argue that at present there is only one cornerstone of success for a strong competitive advantage for firms, which is how to organize and manage the human factor within the firm (Pfeffer, 1996).

An indicator in measuring the quality of human resources of a country can be seen from the Human Development Index (HDI). Based on the HDI Report from 2012, there are 4 categories: Very High Human development, High Human Development, Normal Human Development, and Low Human Development. Indonesia is in the normal category of human development, ranking 121 out of 185 countries. Compared to the previous year, Indonesia has increased its ranking by 3 levels from 124 to 121 out of 186 countries. This data shows that Indonesia should try harder in order to achieve a better category.

Both college and university are the most common formal paths of education in Indonesia. The number of universities in Indonesia reaches 100 or more. One of them is Airlangga University which is expected to develop work-ready graduates who are ready to contribute in advancing the country through companies. To be able to reach the international standard for graduates, one of the requirements is that any organization or educational institution must be certified by the International Organizational for Standardization (ISO).

Airlangga University has been certified ISO 9001: 2008 since 2009, which means Airlangga University is an educational organization that already has international quality management standards. This research focuses on the Faculty of Economics and Business (FEB) of the Airlangga University management program. In an effort to maintain the various awards that have been achieved by the University of Airlangga and especially the Faculty of Economics and Business, the Department of Economics and Business held an annual evaluation that aims to see and control how the progress has been achieved, especially the quality of the graduates.

One assessment of the ISO in the world of education is to see the academic performance of students who are participating in or who have graduated from the Department of Management FEB Airlangga University. If averaged by looking at the number of graduates based on the duration of graduation, the graduation rate generated by management courses in the academic year of 2011/2012 is 4.57. This result increased in the academic year of 2010/2011 to a score of 4.62. This illustrates that there is an improvement in the quality of graduates generated by the Airlangga University management program.

Based on the background that has been described, the formulation of the problem is as follows:

1. Does intrinsic motivation have any direct and significant effects on the academic performance of the students of the Faculty of Economics and Business Management Program of Airlangga University?

2. Does extrinsic motivation have any direct and significant effects on the academic performance of the students of the Faculty of Economics and Business Management Program of Airlangga University?

3. Does amotivation have any direct and significant effects on the academic performance of the students of the Faculty of Economics and Business Management Program of Airlangga University?

1.1 The Big Five Personality theory

The Big Five Personality theory is an approach used by psychologists in assessing or viewing the personality of a person. Initially, this theory was first introduced by Goldberg in 1981. However, in its development, the Big Five theory originated from Cartell (Srivastava & John, 1999). It uses a multidimensional model of the personality structure of Allport and Odbert. Costa and McCrae call the Big Five personality theory the Five Factor Model. This theory is researched or made based on a simple approach, so that not only researchers understand the basic parts of personality or units used, but also people in general. The Big Five theory is not derived from a theoretical perspective on personality but comes from an analysis of everyday language used by a person in describing himself and others. The Big Five theory is classified by Costa and McCrae into five dimensions: extroversion, agreeableness, conscientiousness, neuroticism, and openness to experience (Brehm , 2002).

1.2 Academic motivation

Motivation is a term that describes something that makes a person take action, keeps them going, and helps them accomplish their tasks (Pintrich, 2003).

Motivation is an important factor in many fields, especially in education. This is especially important for subjects that are considered by students difficult subjects. According to Pintrich and Zusho (2002), academic motivation is defined as an internal process that incites and maintains an activity that aims to achieve certain academic goals. The self-determination theory itself says that academic motivation is multidimensional and consists of three types of global motivation: intrinsic motivation, extrinsic motivation, and amotivation (Deci & Ryan, 2002).

According to Ryan and Deci (2000), motivation is divided into three types based on the locus of causality: Extrinsic Motivation, Intrinsic Motivation, and Amotivation.

1.2.1 Intrinsic motivation
Intrinsic motivation is the motivation that arises from within the individual and concerns the pleasure for himself and doing activities for himself. This motivation can increase if there are psychological needs of autonomy, relatedness, and competence and when supported by the surrounding social environment (Ryan & Deci, 2000).

1.2.2 Extrinsic motivation
Extrinsic motivation is an encouragement from within the individual in doing something that is not for his own sake (Deci & Ryan, 2002). Individuals

who are extrinsically motivated act by obtaining external impulses (Ryan & Deci, 2000).

1.2.3 Amotivation
Amotivation arises when a person gets unsuitable results from the effort that someone has spent on the results obtained and they are not motivated intrinsically and extrinsically (Ryan & Deci, 2000).

1.3 Academic performance

Academic performance is the result of learning or education obtained by a student. In Komarraju et al.'s study from 2009, it is concluded that academic performance was influenced by the Big Five theory. This shows that academic performance can be influenced by the Big Five theory. Students who have neuroticism, extraversion, openness to experience, agreeableness, and conscientiousness more than other students have a higher level of value.

Other research states that academic performance is influenced by the existence of academic motivation as concluded by Asghar Hazrati et al. The higher the academic motivation of the students, the value obtained will be proportional to the level of the academic motivation. With a high academic motivation in students, it indirectly causes increased curiosity and easier comprehension of a concept.

2 METHOD

This research is classified as quantitative research, which is research using hypothesis testing with a statistical test tool. The population in this study are 750 students of the Faculty of Economics and Business at Airlangga University. The respondents in this study are students who are taking their bachelor (S-1) in the management program, which is ultimately used to generalize the population. The sampling technique used in this research is by using convenience sampling with 87 respondents.

Hierarchical regression analysis is the further development of multiple regression analysis. Hierarchical regression analysis was performed to determine the effect of the Big Five Personality on academic performance with academic motivation as intervening variable (Sekaran & Roger, 2013).

3 RESULTS AND DISCUSSION

3.1 The effect of the Big Five Personality on academic performance with intrinsic motivation as the intervening variable

The result of the regression analysis shows that the Big Five Personality of openness has an effect on intrinsic motivation. This means that in this study, only the dimension of openness has a significant influence on intrinsic motivation. The value of beta shows that the higher the students' openness in teaching and learning activities, the higher the intrinsic motivation or motivation inside the students' selves. The result also shows a significant influence between the Big Five personality and the academic performance of the students. In this study, the dimensions of the Big Five Personality that have a significant influence on student academic performance are extraversion, agreeableness, neuroticism, and openness.

Therefore, it can be concluded from the above results that intrinsic motivation mediates the relationship between openness to partial mediation of student performance as described by Baron and Kenny (1986). According to Baron and Kenny (1986), variables that have a partial mediation influence mean that these variables can directly affect the dependent variable or indirectly through intervening or mediation variables. In this study, openness has a partial influence of mediation, meaning that openness can directly affect the academic performance or can affect the academic performance through intrinsic motivation.

3.2 The effect of the Big Five Personality on academic performance with extrinsic motivation as the intervening variable

Based on the results of the regression analysis that has been done, it was shown that the dimensions of the Big Five Personality that have significant effects on extrinsic motivation are extraversion and openness. In the hierarchy regression analysis, there is an influence between extrinsic motivation on students' academic performance. With a significance level below 0.05, namely 0.030, and a beta value equal to 0.117, it can be interpreted that extrinsic motivation has a positive effect on students' academic performance. The higher the level of extrinsic motivation of a student, the higher the academic performance obtained will be by the students. By having extrinsic motivation, students can get motivation from the surrounding environment such as motivating friends, teachers, and so forth. External support will certainly make students feel eager to undergo the teaching and learning process on campus.

The hierarchy regression table also shows the influence of agreeableness, neuroticism, and openness. The significance level of the three are less than 0.05, namely 0.000 in agreeableness with a beta value of 0.069, neuroticism has a significance level of 0.040 with beta of -0.077, and openness has a significance of 0.378 and beta of 0.378.

Based on these results, it can be concluded that based on the idea of Baron and Kenny (1986), in this study, extrinsic motivation mediates the full extraversion dimension (full mediation) and partial mediation to the students' academic performance. According to Baron and Kenny (1986), variables that have a full mediation effect means that these variables can affect the dependent variable through intervening or mediation only. For the variables that

have partial mediation influence, it means that these variables can directly affect the dependent variable or indirectly through intervening or mediation variables. In this study, openness which has a partial influence of mediation means that openness can directly affect the academic performance or can affect the academic performance through intrinsic motivation.

3.3 The effect of the Big Five Personality on academic performance with amotivation as the intervening variable

Based on the results of the hierarchical regression, it can be seen that none of them have a significance level less than 0.05. In addition to the regression result, none of the influences of the Big Five Personality on amotivation has a significance level less than 0.05. Thus, it can be concluded that amotivation cannot mediate the relationship between Big Five Personalities and students' academic performance. It can be interpreted that someone who is not motivated either from within himself or from outside, cannot improve his academic performance even though the student already has a good personality.

4 CONCLUSION

Based on this discussion, the following conclusions can be made. First, not all of the dimensions of the Big Five Personalities have an influence on academic performance, but overall, it can be said that there is a relationship between the Big Five Personality and the students' academic performance. These results indicate that the student's personality plays a role in a high or low academic performance of the student.

Second, this research found that academic motivation has a significant influence on academic performance. Students who have internal motivation both internally and externally have a high level of academic performance as well. Thus, it can be said that the academic motivation of students participates in the acquisition of a high or low academic performance of a student.

Third, it was found that intrinsic motivation only mediates the influence of openness on partially mediated student's performance; extrinsic motivation only mediates the influence between extraversion and openness on the academic performance of students who each have fully mediated and partially mediated; and amotivation does not mediate the relationship of the Big Five Personality toward students' academic performance. This test refers to Baron and Kenny (1986) on mediation.

Suggestions submitted based on the results of this study are: In the process of teaching and learning, lecturers are expected to provide a participative and comfortable learning process to the students in order to make the students' academic potential be honed and improved. As to which the results of this study state that the Big Five Personality has a significant influence on student academic performance. With an interesting learning process, the personality of students who are not really good will be minimized because a comfortable condition, besides increasing student motivation, it can also make students more eager to learn and discuss (Caers 2012).

We recommend that the faculty can create a comfortable learning atmosphere both in the process of teaching and learning facilities that can be used by students. As was found in this study, extrinsic motivation from students has an effect on students' academic performance (Vallerand 1992). Therefore, by creating a learning atmosphere or a comfortable learning environment, this can make the students' motivation increase to learn or gain knowledge on campus.

ACKNOWLEDGMENTS

This study was funded by Universitas Airlangga and made possible with a grant from the Faculty. The authors would like to thank Nidya, Deva, and interns at the Faculty of Economics and Business at Universitas Airlangga.

REFERENCES

Baron, R.M., & Kenny, D.A., 1986. The Moderator-mediator variable distinction in social psychological research: conceptual, strategic, and statistical consideration. *Journal of Personality and Social Psychology.*

Brehm, S., Miller, Rowland S., Perlman, D., & Campbell, Susan M. 2002. Intimate Relationship (3rd ed.). Boston: McGraw-Hill.

Noe, R.A., Hollenbeck, J.R., Gerhart, B., Wright, P.M. (1994). Human Resource Management: Gaining a Competitive Advantage. Illnois : Austen Press.

Caers, R., Cladia, V., Dries, B. 2012. Unraveling the impact of the Big Five personality traits on academic performance: The moderating and mediating effects of self-efficacy and academic motivation. Learning and Individual Differences. Vol 22 pg 439–439448.

Deci, E.L. & Ryan, R.M. 2002. Handbook of self-determination research. Rochester, UK: University of Rochester Press.

Deci, E. L. & Ryan, R. M. 2000. The "What" and "Why" of Goal Pursuits: Human Needs and the Self-Determination of Behavior. *Psychological Inquiry* 11 (4), 227-268.

Komarraju, M., Karau, S. J., & Schmeck, R. R. 2009. *Role of the Big Five personality traits in predicting college students' academic motivation and achievement. Learning and Individual Differences.* 19, 47–52.

Pfeffer, M. 1996. The Art to Maintance Human Resources. New York: McGraw-Hill.

Pintrich, P. R. 2003b. Motivation and classroom learning. In W. M. Reynolds & G. E. Miller (Eds.), *Handbook of psychology: Educational psychology* (Vol. 7, pp. 103–122). Hoboken, NJ: Wiley.

Pintrich, P. R., & Zusho, A. 2002. The development of academic self-regulation: the role of cognitive and motivational factors. In A. Wigfield & J. S. Eccles (Eds.), *Development of achievement motivation* (pp. 249–284). San Diego, CA: Academic Press.

Sekaran, U. & Roger, B. 2013. *Research Methods for Business.* Chichester: John Wiley & Sons Ltd.

Srivastava, S, & John, O. P. 1999. The Big Five Traits Taxonomy: History, Measurement, and Theoritical Perspectives. Berkeley: University of California.

Vallerand, R. J., Pelletier, L. G., Blais, M. R., Briere, N. M., Senecal, C., & Vallieres, E. F. 1992. The Academic Motivation Scale: A measure of intrinsic, extrinsic, and amotivation in education. *Educational and Psychological Measurement* 52, 1003–1017.

Advances in Business, Management and Entrepreneurship – Hurriyati et al (eds)
© 2020 Taylor & Francis Group, London, ISBN 978-0-367-27176-3

The influence of transformational leadership on the affective commitment mediated by inclusive organizational culture in an elementary school in Zainuddin, Sidoarjo

M.A. Rozzak & A. Eliyana
Airlangga University, Surabaya, Indonesia

ABSTRACT: The subject of this study is Zainuddin Elementary School in Sidoarjo. The article discusses affective commitment, which is influenced and mediated by an inclusive organizational culture. Issues about commitment in the organization can be seen from the results of an initial observation, which shows a tendency of too many job desks for teachers in the last five years. This study uses SEM-PLS as the analysis method. The number of the population and sample consists of 45 teachers and uses the census as the technique to help identify the amount of respondents. The results of this research find that there is a direct influence of transformational leadership and affective commitment, and a significant result for inclusive organizational culture as the mediation between transformational leadership and affective commitment.

1 INTRODUCTION

The success or failure of an organization or company in achieving good goals and performance is determined by the leader (Ashikali and Groeneveld, 2015). In performing its activities, a company is required to have a reliable leader who is able to anticipate the future of the organization and take opportunities for change, so as to lead the organization toward its objectives. Good leadership in an organization is supported by good organizational culture, and a strong leadership style applied and supported by good organizational culture, will improve company performance. The establishment of a good organizational culture will cause employees to be more motivated in working to achieve higher performance.

Transformational leadership is a leadership style that fosters awareness of the vision and mission of the organization, with the processing of work activities through using the skills of employees, so that every employee feels involved and responsible in completing the work (Fasola et al., 2013). To achieve the ideals of an organization, in addition to having a reliable leader they must also be supported by human resources existing in the organization. With competent human resources, the ideals and goals of the organization will be more easily achieved. In fact, it is difficult for this to happen if the people in the organization are less committed in carrying out the tasks within the organization.

The diversity of human resources in an organization can become one of the causes of the lack of commitment to each individual (Ashikali and Groeneveld, 2015). In Zainuddin Elementary School (SD) there is not a level of high diversity, but every person has different desires and different backgrounds. Thus, an organization will create sustainable behavior that eventually forms a culture that is often called as an inclusive organizational culture.

Inclusive organizational culture is the openness and appreciation of employees toward the organizational environment (Ashikali and Groeneveld, 2015). If more employees are open and appreciative, it will be easier for transformational leadership to direct and control the organization. Inclusive organizational culture is like receiving new things that feel good and in accordance with the culture of the organization.

Education becomes a portrait as well as the main highlight of the implementation of inclusive organizational culture because it is in direct contact with the community. The diverse backgrounds of employees, teachers, and students make organizational management a great concern. Cultures that lead to understanding, acceptance, and respect for others of different ethnicity, culture, values, personality, and physical or psychological functioning must be maintained. However, a positive culture without a foundation of organizational commitment can lead to the erosion of organizational commitment, due to the negative influence of various parties that come and go.

In education settings such as elementary schools, in addition to requiring a transformational leader, the affective commitment of employees is also necessary to achieve the goals of the school. In fact, this affective commitment is still hard to find in an employee. Affective commitment helps employees pay attention to the compensation of material that can be given by the organization, because employees who have affective commitment will have an emotional attachment to the organization and survive within it.

Kent and Chelladurai (2001) show that there is a significant relationship between transformational leadership and organizational commitment. Research by Fasola et al. (2013) indicates that transformational leadership has a significant positive effect on organizational commitment. Boon and Arumugami (2006) explain organizational culture as a management philosophy, how to manage the organization to improve the overall effectiveness and performance of the company, and that organizational culture has a significant effect on employees' commitment to the organization. The four dimensions of organizational culture—consisting of communication, training and development, respect and recognition, and cooperation—are positively related to employee commitment. Communication becomes a dominantly influential dimension to organizational commitment.

2 THEORETICAL BASIS

2.1 Transformational leadership

According to Wiguna (2012: 94), transformational leadership motivates followers to work toward a goal, not for short-term personal gain, and to achieve self-actualization, not for security. Transformational leadership can attract highly motivated employees, while improving the quality of life for members of the organization (Fasola et al., 2013). Transformational leadership is able to motivate followers who struggle in achieving organizational goals, without the use of power and authority but through deep passion and thought.

Gumusluoglu and Ilsev (2009) explain that transformational leadership inspires awe, respect, loyalty, and emphasizes the importance of a collective sense of corporate mission. This involves individual considerations, as leadership builds relationships with subordinates and assumes that each has different needs/skills.

Tschannen-Moran (2003) describes the following dimensions of transformational leadership:

1. Individualized influence emphasizes the type of leadership that demonstrates belief, admiration and praise of followers.
2. Inspirational motivation emphasizes how to motivate and inspire subordinates in task challenges. The influence is expected to enhance the spirit of the group.
3. Intellectual stimulation emphasizes the type of leadership that seeks to encourage subordinates to think innovatively and creatively in new ways.
4. Individualized consideration emphasizes the type of leadership that shows attention or appreciation to the achievement, development and achievement needs of subordinates.

2.2 Inclusive organizational culture

Organizational culture according to Tobari (2015: 46) is a collection of habits, values, wisdom, beliefs, attitudes that are easily absorbed for everything that is done and thought in an organization. Adewale and Anthonia (2013) describe organizational culture as a driving force that contributes to providing a thorough understanding of what and how to achieve, how goals are interrelated, and how each employee can achieve goals.

Inclusive organizational culture is a climate that values equality, and positive acknowledgment of culturally diverse and disadvantaged social and institutional responses do not create barriers to positive work experience (Shore et al. 2011). Ashikali and Groeneveld (2015) explain that inclusive organizational culture is openness and appreciation of employees to the work environment. Organizational culture can strengthen bonds among employees in an organization by behaving with mutual respect and high tolerance. Inclusive culture shows that employees will have a commitment to the organization (Shore et al., 2011).

2.3 Affective commitment

According to Allen and Meyer (1991), affective commitment is an emotional attachment, identification and involvement in an organization. In this case the individual resides within an organization because of their own desires or the existence of an emotional attachment to the organization. Affective commitment can occur in an employee when they want to be part of the company based on emotional attachment.

3 HYPOTHESES

3.1 The relationship between transformational leadership and affective commitment

Ashikali and Groeneveld (2015) argue that transformational leadership has a direct positive effect on employee affective commitment. The higher the supervisor has transformational leadership, the more employees feel and have an organization. When employees feel they have an organization then the percentage of employees that commit to the organization will be higher, which is also supported by how much employees know the organization. The quality of transformational leadership relationships with employees should be considered because this can predict affective commitment, which describes the extent that employees are committed to the organization.

H1: Transformational leadership affects the affective commitment.

3.2 The relationship between transformational leadership and inclusive organizational culture

Transformational leadership increases the employee's affective commitment in line with the increasing inclusive organizational culture. Ashikali and Groeneveld (2015) state that the success of inclusive organizational culture also influences the extent of employees' perspective on their transformational leadership. These results also showed that supervisors or leaders apply transformational leadership styles so that employees will increasingly support inclusive organizational culture.

H2: Transformational leadership affects inclusive organizational culture.

3.3 The relationship between inclusive organizational culture and affective commitment

According to Meyer and Allen (1991) affective commitment in the form of emotional attachment by employees to the organization, is expressed by the identification and involvement of activities within the organization, where openness and appreciation of employees to the work environment is a form of inclusive organizational culture (Ashikali and Groeneveld, 2015). The higher the employer lifts the inclusive organizational culture, the higher the employees will be committed to the organization. Inclusive organizational culture is also a reinforcement for employees to be more committed to the organization, and Ashikali and Groeneveld (2015) stated that inclusive organizational culture strengthens affective commitment.

H3: Inclusive organizational culture affects the affective commitment.

3.4 The relationship between transformational leadership and affective commitment mediated by inclusive organizational culture

Transformational leadership will approach and understand its employees through four dimensions. Transformational leadership is mediated by an inclusive organizational culture, which has an effect on instilling affective commitment to what has been conveyed and exemplified by transformational leadership. Ashikali and Groeneveld (2015) stated that transformational leadership increases the employee's affective commitment by increasing the inclusive organizational culture.

H4: Transformational leadership improves the affective commitment of employees by increasing inclusive organizational culture.

4 RESEARCH METHODOLOGY

This study uses a quantitative approach. The population in this research is 45 Zainuddin elementary teachers. The sampling technique conducted was probability sampling with a census method. Sampling, or sampling saturation census, is a sample determination technique where all members of the population are used as a sample (Sugiyono, 2012: 96). Determination of the sample by census method is performed on the basis of a relatively small population size in the hope of minimizing errors. In this research, the data collection method is interview and questionnaire. The analytical technique used in this research is PLS-Path analysis, because the data do not have to have

multivariate normal distribution and the sample size should not be large and uses a Sobel Test to test the mediation effect.

5 RESULTS AND DISCUSSION

5.1 Evaluation of outer model

Table 1 shows the result of the convergent validity test, that all indicators have met the loading value limit above 0.5. So, it can be concluded that the overall data collected is valid and has good convergent validity.

The result of construct validity test in Table 2 shows that the AVE value for each variable in this research analysis model is greater than 0.5. So it can be concluded that the overall data collected is valid and has good construct validity.

The result of cross loading test in shows that each indicator on this research variable has the

Table 1. Test results convergent validity (value loading factor).

Variables	Indicators	Loading factor	Remarks
Transformational leadership	TL1	0.711	Valid
	TL2	0.663	Valid
	TL3	0.733	Valid
	TL4	0.652	Valid
	TL5	0.503	Valid
	TL6	0.777	Valid
	TL7	0.811	Valid
	TL8	0.852	Valid
Inclusive organizational culture	IC1	0.808	Valid
	IC2	0.824	Valid
	IC3	0.679	Valid
	IC4	0.785	Valid
	IC5	0.814	Valid
	IC6	0.610	Valid
Affective commitment	AC1	0.778	Valid
	AC2	0.835	Valid
	AC3	0.805	Valid

Source: Data processing with PLS

Table 2. Construct validity test results (average variance extracted value).

Variables	Average variance extracted (AVE)
Affective commitment	0.651
Inclusive organizational culture	0.574
Transformational leadership	0.518

* Source: Data Processing with PLS

Table 3. Discriminant Validity Test Results (Cross Cross Rate).

Variables	Composite reliability	Remarks
Affective commitment	0.848	Reliable
Inclusive organizational culture	0.889	Reliable
Transformational leadership	0.894	Reliable

Source: Data processing with PLS

largest crossload value on the variable it formed compared to the value of cross loading on other variables. So it can be stated that the indicators used in this study have had good discriminant validity in preparing their respective variables.

The composite reliability test results in Table 3 show that the composite reliability value of all research variables is > 0.7. This indicates that each variable has met the composite reliability, so it can be concluded that the overall variable has a high level of internal consistency reliability.

5.2 Evalution of inner model

The result of the determination coefficient test in Table 4 shows that the R-square value for affective commitment is 0.572. The value acquisition shows that the percentage of affective commitment can be explained by transformational leadership by 50.2%. The value of R-square for inclusive organizational culture is 0.469. This result shows that inclusive

Table 4. Composite reliability test results.

Indicators	Variables		
	AC	IC	TL
AC1	0.778	0.510	0.495
AC2	0.835	0.665	0.612
AC3	0.805	0.539	0.495
IC1	0.665	0.808	0.554
IC2	0.499	0.824	0.564
IC3	0.329	0.679	0.394
IC4	0.663	0.785	0.573
IC5	0.607	0.814	0.585
IC6	0.368	0.610	0.382
TL1	0.499	0.329	0.711
TL2	0.505	0.428	0.663
TL3	0.483	0.596	0.733
TL4	0.526	0.492	0.652
TL5	0.385	0.231	0.503
TL6	0.349	0.506	0.777
TL7	0.572	0.613	0.811
TL8	0.494	0.610	0.852

* Source: Data Processing with PLS

organizational culture can be explained by transformational leadership by 46.9%.

Test of goodness of fit:

$$Q - Square = 1 - [(1 - 0.469) \times (1 - 0.572)]$$
$$= 1 - (0.531 \times 0.428)$$
$$= 1 - 0.227268$$
$$= 0.772732 = 0.773$$

The above calculation results obtained a Q-Square value of 0.773. This shows that the amount of diversity of research data that can be explained by the research model is 77.3%, while the remaining 22.7% is explained by other factors outside this research model. Based on these results, the model in this study can be stated to have a good result of goodness of fit.

The result of the path coefficient test in Table 5 shows that the effect of the transformational leadership variable toward affective commitment is 0.335, with a t-statistic value 2.232 > 1.96, at significance level = 0.05 (5%) and p-value of 0.026 < 0.05. These results suggest that there is a significant influence of transformational leadership on affective commitment in teachers at Zainuddin Elementary School, Waru, East Java. A positive value on the meter coefficient indicates that with an increasing influence of transformational leadership the affective commitment will also increase. Thus hypothesis 1 is proved true and accepted. That is, higher transformational leadership by the principal will result in increased affective commitment of teachers at Zainuddin Elementary School. The findings are in line with research conducted by Ashikali and Groeneveld (2015), that the more bosses or headmasters have transformational leadership, the more employees feel to own and recognize the organization.

The result of the hypothesis 1 test shows that there is a positive influence of the transformational leadership construct toward affective commitment in Zainuddin Elementary School teachers. This illustrates that the four dimensions of transformational leadership fit perfectly with an approach that involves emotions, given that an affective commitment assists survival in the organization because of the emotional ties. Master deeply feels given special attention by transformational leadership with the four dimensions of transformational leadership of the head seokalah. The level of appropriateness of individual dimensions of influence, inspirational

Table 5. Coefficient determination test result (r-square value).

Variables	R-square
Affective commitment	0.572
Inclusive organizational culture	0.469

Source: Data processing with PLS

Table 6. Hypothesis test results (path coefficient value).

	Original Sample (O)	T-Statistics	P-values	Remarks
Inclusive Organizational Culture -> Affective Commitment	0.486	3.446	0.001	Accepted
Transformational Leadership-> Affective Commitment	0.335	2.232	0.026	Accepted
Transformational Leadership-> Inclusive Organizational Culture	0.685	9.466	0.000	Accepted

Source: data Processing with PLS

motivation, intellectual stimulation, individual consideration, the employee will feel an emotional attachment to the organization that has been channeled by the principal.

The Haisl test path coefficient in Table 5 also shows the influence of transformational leadership relationship on inclusive organizational culture, obtained by the path meter coefficient 0.685, with t-count value equal to 9.466 > 1.96 at significance level = 0.05 (5%) and a p-value of 0.000 < 0.05. These results suggest that there is a significant influence of transformational leadership on inclusive organizational culture. A positive value on the meter coefficient means that increasing transformational leadership has an effect on increasing inclusive organizational culture. It can be concluded that transformational leadership has a positive influence on inclusive organizational culture in the teachers at Zainuddin Elementary School, Sidoarjo. Thus hypothesis 2 is accepted, which means that the higher level of conformity between the transformational leadership of the principal and their influence as the power owner can make it easier for the organization to create inclusive organizational culture.

There is a positive influence given by the transformational leadership constructing indicators on the inclusive organizational culture. This shows that if the teacher has a level of trust toward the principal then an inclusive organizational culture will gradually be created, which at this stage of leadership collects supporters to create a culture of ketebukaan. After receiving a good response from the teacher, the headteacher takes action by introducing a new idea or policy, for example employing female teachers as permanent staff and creating a culture of openness in the organization (inclusive organizational culture).

The Haisl path coefficient test in Table 5 also shows that the influence of inclusive organizational culture on affective commitment, obtained by the path meter coefficient 0.486, with t-count value 3.466 > 1.96, at significance level = 0.05 (5%) and a p-value of 0.001 < 0.05. These results suggest that there is a significant influence between inclusive organizational culture on affective commitment. A positive value on the

meter coefficient means that increasing inclusive organizational culture has an effect on increasing affective commitment. It can be concluded that inclusive organizational culture has a positive influence on the affective commitment of teachers in Zainuddin Elementary School. Hence hypothesis 3 is proved true and accepted.

There is a positive influence given by constructive indicators of inclusive organizational culture toward the affective commitment of teachers in Zainuddin Elementary School. It shows that as teachers experience openness within the organization their level of commitment will be higher. The sense of kinship that the teacher possesses is the greatest possibility of the creator of a feeling that can not be explained by the sentence and becomes the trigger that it is that makes the teacher to remain in the organization or often called affective commitment.

The result of the test in Figure 1 shows that the p-value is 0.00119243 smaller than the critical limit of hypothesis acceptance of 0.05 at the 5% error level. This means that inclusive organizational culture has a significant mediation effect on how transformational leadership influences affective commitment. Thus hypothesis 4 can be accepted as true.

The relationship of transformational leadership can be through inclusive organizational culture. This means that when transformational leadership is felt by high employees, then what has been exemplified and directed by transformational leadership will be followed and implemented without any objection because the teacher already believes in transformational leadership, exemplifies and directs each teacher to receive an inclusive organizational culture in the organization. Inclusive organizational culture in the organization can help teachers have affective commitment to themselves, because the openness is

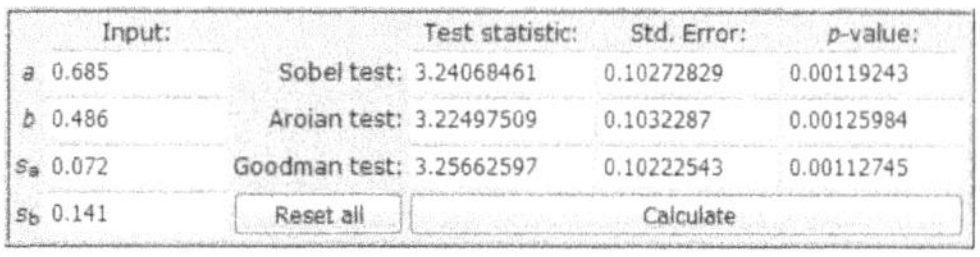

Input:	Test statistic:	Std. Error:	p-value:
a 0.685	Sobel test: 3.24068461	0.10272829	0.00119243
b 0.486	Aroian test: 3.22497509	0.1032287	0.00125984
Sₐ 0.072	Goodman test: 3.25662597	0.10222543	0.00112745
Sᵦ 0.141	Reset all	Calculate	

Figure 1. Sobel test result (mediation effect).

linked to the leadership. From this openness the teacher learns to appreciate and tolerate others and the environment is important, as has been exemplified by the leadership. Ashikali and Groeneveld (2015) point out that transformational leadership increases the employee's affective commitment by increasing the inclusive organizational culture.

6 CONCLUSIONS AND SUGGESTIONS

After the process and the results of the analysis, the conclusions from this research are as follows:

1. Transformational leadership has a significant positive effect on affective commitment of teachers at Zainuddin Elementary School.
2. Transformational leadership has a significant positive effect on inclusive organizational culture for teachers at Zainuddin Elementary School.
3. Inclusive organizational culture has a significant positive effect on affective commitment of teachers at Zainuddin Elementary School.
4. Inclusive organizational culture positively mediates transformational leadership's relationship with the affective commitment of teachers at at Zainuddin Elementary School.

Suggestions based on the results of this study are as follows:

1. Increase teachers' awareness of school-organized activities and teaching-learning activities, by increasing the intensity of communication between school leaders and teachers to provide shared perceptions and ownership of the organization.
2. The leadership of at Zainuddin Elementary School should prioritize freedom of opinion and appreciate each input given by teachers.
3. Based on the results of questionnaires, respondents provide a description of the answers for inclusive organizational culture variables higher than transformational leadership. This indicates that inclusive organizational culture has an important role in mediating transformational leadership toward affective commitment, where the culture of mutual understanding of other people's opinions really needs to be maintained. The highest value of transformational leadership that teachers brand as highly credible leadership can also directly influence affective commitment, but the survey results show that the role of mediation is greater than the direct influence of transformational leadership on affective commitment.

REFERENCES

Adewale, Osibanjo, O., Anthonia, Adeniji, A. 2013. Impact of Organizational Culture on Human Resources Practices: A Study of Selected Nigerian Private University. *Journal of Competitiveness* 5, 4, 115–133.

Aydogdu, S., Asikgil, B.. 2011. The Effect of Transformational Leadership Behavior on Organizational Culture: An Application in Pharmaceutical Industry. *International Review of Management and Marketing* 1, 4, 65–73.

Ashikali and Groeneveld. 2015. Diversity Management in Public Organizations and Its Effect on Employees' Affective Commitment: The Role of Transformational Leadership and the Inclusiveness of the Organizational Culture. *Review of Public Personnel Administration* 35, 2, 146–168.

Boon, Ooi Keng., Arumugam, Veeri. 2006. The Influence of Corporate Culture on Organizational Commitment: Case Study of Semi Conductor Organizations In Malaysia. *Sunway Academic Journal* 3, 99–115.

Chou, P. 2013. The Effect of Transformational Leadership on Follower's Affective Commitment to Change. Minghsin University of Science and Technology, Taiwan (ROC). Minghsin University of Science and Technology, Taiwan (ROC). 3, 1, 38–52.

Creswell, J.W. 2010. Research design: pendekatan kualitatif, kuantitatif, dan mixed. Yogyakarta: Pustaka pelajar.

Ershi, Q., Jiang, Shen., Runliang Dou. 2013. The 19th International Conference on Industrial Engineering and Engineering Management: Management System Innovation. Springer Science & Business Media.

Fasola, O.S., Adeyemi., M.A., Olowe, F.T. 2013. Exploring the Relationship between Transformational, Transactional Leadership Style and Organizational Commitment among Nigerian Banks Employees. *International Journal of Academic Research in Economics and Management Sciences* 2, 6.

Ghozali, I. 2002. Aplikasi Analisis Multivariate Dengan Program SPSS. Semarang: Badan Penerbit Universitas Diponegoro.

Gumusluoglu, Lale., Ilsev, Arzu. 2009. Transformational Leadership, Creativity, and Organizational Innovation. *Journal of Business Research* 62, 461–473.

Hamdi, A.S., & Bahruddin, E. 2015. Metode Penelitian Kuantitatif Aplikasi Dalam Pendidikan. Deepublish.

Ivancevich, J.M., Konopaske, R., Matteson, M.T. 2011. Perilaku dan Manajemen Organisasi. Jakarta: Penerbit Erlangga.

Jain, P, Dugga, T. 2015. The Role Of Transformational Leadership In Organizational Commitment. International *Journal Of Busines Quantitative Economic And Applied Management Research* 2, 5.

Kent, A., Chelladurai, P. 2001. Perceived Transformational Leadership Organizational Commitment, and Citizenship Behavior: A Case Study in Intercollegiate Athletics. *Journal of Sport Management* 15, 135–159.

Kusumaputri, E.S. 2015. Komitmen pada Perubahan Organisasi. Deepublish: Grup Penerbit CV Budi Utama.

Meyer, J.P, Allen, N.J. 1991. A Three-Component Conceptualization Of Organizational Commitment. *Human Resource Management Review* 1, 1, 61–89.

Moeljono, D. 2005. Budaya Organisasi dalam Tantangan. Jakarta: PT Gramedia Pelled.

Pelled, L.H., Ledford G.E., Jr., Mormhan, S.A. 1999. Demograpic Dissimilarity And Workplace Inclusion. Blackwell Publishers, Oxford.

Rhoades, L., Eisenberger, R., Armeli, S. 2001. Affective Commitment to the Organization: The Contribution of Perceived Organizational Support. *Journal of Applied Psychology* 86, 5, 825–836.

Robbins, S.P., Judge, T.A. 2008. Perilaku Organisasi. Jakarta: Salemba Empat Silalahi, U. 2012. Metode Penelitian Sosial. Bandung: Refika Aditama.

Roberson, Q.M. 2006. Disentangling the Meanings of Diversity and Inclusion in Organizations. *Group & Organization Management*, 31, 2, 212–236.

Sabri Pirzada Sami Ullah., Ilyas Muhammad., and Amjad Zahra. (2013). The Impact of Organizational Culture on Commitment of Teachers in Private Sector of Higher Education. *Pakistan Journal of Social and Clinical Psychology* 11, 69–76.

Smith, K.L., McCracken, J.D, dan Turiman Suandi, 1983, Agent's Organizational Commitment. *Journal of Extention* 21–26.

Sugiyono. 2007. Statistika untuk penelitian. Bandung: Alfabeta.

Sugiyono. 2012. Metode Penelitian Pendidikan, Pendekatan Kuantitatif, Kualitatif, dan R&D. Bandung: Alfabeta.

Sugiyono. 2013. Statistika untuk penelitian. Bandung: Alfabeta.

Tobari, H. 2015. Membangun Budaya Organisasi pada Instantsi Pemerintahan. Deepublish: Grup Penerbit CV Budi Utama.

Tsai, Ming-Ten. Chung-Lin Tsai. & Yi Choung Wang. 2011. A study on the relationship between leadership style, emotional intelligence, self-efficacy and organizational commitment: A case study of the Banking Industry in Taiwan. *African Journal of Business Management* 5, 13, 5319–5329.

Wanjiku, Njugi Anne., Agusioma, Nickson L. 2014. Effect of Organisation Culture on Employee Performance in Non Govermental Organizations. *International Journal of Scientific and Research Publications* 4, 11.

Wiguna, A. 2012. Isu-isu Komtemporer Pendidikan Islam. Deepublish: Grup Penerbit CV Budi Utama.

Yulk, Gary. 2009. Leadership dalam Organisasi. Jakarta: PT. Indeks Kelompok Gramedia.

Advances in Business, Management and Entrepreneurship – Hurriyati et al (eds)
© 2020 Taylor & Francis Group, London, ISBN 978-0-367-27176-3

Effect of work–family conflict on job performance through emotional exhaustion as a mediating variable on nurses in the Emergency Department of the Regional Public Hospital (RSUD) Dr. Iskak Tulungagung

M.B. Habibi & A. Eliyana
Airlangga University, Surabaya, Indonesia

ABSTRACT: This study aims to test whether there is influence of work–family conflict on job performance through emotional exhaustion as a mediating variable on nurses in the Emergency Department of the Regional Public Hospital (RSUD) Dr. Iskak Tulungagung. Research was conducted on nurses with a sample size of 70 people, and the analytical techniques used was path analysis using SPSS 21. The independent variable in this study is work–family conflict (X1) and family–work conflict (X2), the mediating variables are emotional exhaustion (Z), and the dependent variable is job performance (Y). The results of this study indicate that work–family conflict and family–work conflict have direct and significant impacts on job performance. The result shows that emotional exhaustion is the mediating influence between work–family conflict and family–work conflict on job performance.

1 INTRODUCTION

The performance of a company is inseparable from the performance of every individual or employee in it. An employee's performance can be influenced by conflicting factors that can come from the family or work (Karatepe, 2013). Therefore, companies need to pay attention to issues related to the conflict of family and work so that performance in the company can be maximized.

The conflict that needs attention is the work–family conflict, which has two directions: work–family conflict and family–work conflict. Greenhaus and Beutell (1985) suggest that work–family conflicts can arise from demands that should be run on different roles and an individual may experience emotional exhaustion while trying to meet the demands of responsibilities of family and work. Emotional exhaustion is the first stage of burnout (Karatepe et al., 2008). Burnout is a state of stress experienced by individuals for long periods of time at high intensities, characterized by physical fatigue, mental, emotional and low self-esteem, resulting in an individual feeling separate to their environment.

Margaret Posig and Fill Kickul (2004) in Sari (2010) suggest that work–family conflict has a direct relationship to emotional exhaustion. This suggests that a work–family conflict will have an impact on emotional exhaustion, and the worker will feel irritable about their surroundings, resulting in them being not maximal in doing the work they are responsible for.

Research conducted by Karatepe (2013) found that emotional exhaustion mediates fully between

work–family conflict and job performance. This means that when a worker has a work–family conflict they will experience emotional exhaustion, resulting in an irritability that will ultimately affect the worker's performance. A worker who has experienced emotional fatigue will not be able to work maximally, because they will have no energy for their professional life.

Emotional fatigue is manifested in self-pity, irritability, loneliness and loss of morale (Anah, 2003). A worker who cannot work optimally will have loss of performance, because emotional fatigue cannot be solved by rest alone, in contrast with physical fatigue that can. Anah (2003) stated that emotional fatigue cannot be lost by rest or sleep.

Work–family conflict is also experienced by nurses. Sari (2011) stated that nurses are required to always work professionally within the time specified by the hospital where they work. It is this demand that can affect the time for their family, so that between the demands of family and the demands of work there can be a conflict. Not only required to have professional skills in work, a nurse is also required to be able to control their emotions.

A nurse experiencing emotional exhaustion is not expected by patients, because patients expects a nurse to work with a professional manner. The nurse faces a variety of emotional burdens in her work, such as dealing with uncooperative clients, dealing with the suffering of patients, as well as other obligations outside her work involving family. Less than optimal body conditions will result in an easily irritable nurse, easily discouraged and even emotionally exhausted.

2 THEORETICAL BASIS

2.1 Work–family conflict

Work–family conflict is defined as a form of two-way conflict where the conflict arises from the domain of work and family (Netemeyer and Boles, 1996). Greenhaus and Beutell (1985) stated that work–family conflict is a form of dual-role conflict where the pressure of the role of the family domain and work are conflicting in various ways. If the role of work and family is difficult, it will lead to work–family conflicts, where family affairs will disrupt the work and/ or work affairs disrupt the family (Christine, 2010).

According to Armstrong et al. (2015), work–family conflict can be experienced in two directions: work–family conflict (WFC) and family–work conflict (FWC). Work–family conflict (WFC) occurs when the responsibility of the job affects family responsibilities (Karatepe, 2013), and family–work conflict (FWC) occurs when the responsibilities of the family affect the responsibilitis at work (Karatepe, 2013).

2.2 Job performance

Jackofsky in Christine (2010) argues that performance is the result of individual work that affects the whole organization with the various rewards received related to its performance. Christine (2010) also states that performance is the achievement of an outcome characterized by the expertise of one's task or group on the basis of a predetermined goal. Motowido and Van Scorter in Roboth (2015), state that performance (job performance) is the work that refers to the results obtained from substantive tasks that differentiate one's work from another and include more technical aspects of performance. Bernadin and Rusell in Roboth (2015) state that performance is a record of the consequences that resulted in a particular work function or activity within a given time period.

2.3 Emotional exhaustion

Emotional exhaustion (emotional exhaustion) is the first level of burnout syndrome (Netemeyer, 1996). This means that someone with burnout syndrome will experience an emotional phase of fatigue. In contrast to Netemeyer, Churiyah in Princess (2012) stated that emotional exhaustion is an emotional state where the individual feels helpless and depressed.

Halbesleben and Buckley in Halbesleben et al. (2011) suggest that emotional exhaustion is generally associated with negative employee attitudes and behaviors and ultimately affects employee performance and employee turnover. This shows that emotional fatigue is very important to diagnose because it is closely related to attitudes and behavior of employees. When an employee has a high level of emotional fatigue this may also cause a negative attitude and behavior by the employee.

According to Maslach and Jackson (1980) in Sari (2010), emotional exhaustion is one of the levels in burnout. Burnout is a psychological syndrome of emotional fatigue, depersonalization, and reduced personal attainment that can occur among individuals working with others in some capacity.

3 HYPOTHESIS

3.1 The relationship between work–family conflict job performance

The research conducted by Karatepe and Bekteshi (2007) shows that a decrease in performance is caused by work–family conflict, when it is difficult for employees to balance between job responsibilities and family responsibilities. Research conducted by Karatepe (2013) also shows that work–family conflict (WFC) and family–work conflict (FWC) have relationships with performance. Work–family conflict (WFC) can reduce the performance of an employee, as the conflict will mean their performance is not maximal.

H1: Work–family conflict (WFC) has a significant effect on job performance.

3.2 The relationship between family–work conflict job performance

Research conducted by Karatepe and Bekteshi (2007) shows that a decrease in performance is caused by family–work conflict (FWC), when employees find it difficult to balance job responsibilities and family responsibilities. Research conducted by Karatepe (2013) also shows that family–work conflict (FWC) has a relationship with performance. Work–family conflict (WFC) can reduce the performance of an employee, as the conflict will mean their performance is not maximal.

H2: Family–work conflict (FWC) has a significant effect on job performance.

3.3 The relationship between work to family conflict (WFC) and emotional exhaustion

Work–family conflict (WFC) can prevent a person from meeting the demands of their job and family role and it is easy for someone with WFC to experience an increase in emotional fatigue (Karatepe et al., 2008). This shows that when WFC increases it will increase emotional fatigue. The influence of WFC is more likely to affect emotional exhaustion in comparison with FWC, because job-related domains have more influence on emotional exhaustion (Karatepe, 2013).

3.4 The relationship between family–work conflict (FWC) and emotional exhaustion

Family–work conflict is a significant predictor of predicting emotional exhaustion (Karatepe, 2013).

Research conducted by Karatepe also found that FWC is a weaker predictor than WFC, because the domain of work is more powerful than the emotional exhaustion cause by family. Family–work conflict may prevent a person from meeting the demands of their job and family role and it is easy for someone with FWC to increase emotional fatigue (Karatepe et al., 2008). This shows that when FWC increases it may increase the emotional fatigue of an employee.

3.5 The relationship between emotional exhaustion and job performance

The relationship between emotional exhaustion and performance appears in a study conducted by Cordes and Dougherty (1993) in Karatepe (2013), stating that high levels of emotional exhaustion will an employee's performance to be low. An employee who has a high level of emotional fatigue will feel helpless and depressed, which will affect their focus in work. The employee will not be performing their responsibilities maximally and their performance will decrease.

3.6 The relationship between work–family conflict, emotional exhaustion and job performance

Work–family conflict has a relationship with job performance and emotional exhaustion, as demonstrated by Karatepe (2013) who found that emotional exhaustion fully mediates between work–family conflict and job performance. The role of emotional exhaustion has a negative effect on job performance, because individuals who no longer have the appropriate resources will face emotional exhaustion, which is indicated by performance decline. Emotional exhaustion affects employees whose work is oriented with many people (Karatepe, 2006), and those who constantly experience work–family conflict have an increased tendency to experience emotional exhaustion. Ahmad (2010) also suggested that there is a relationship between work–family conflict and emotional exhaustion. This can be shown from the results of research that work–family conflict has a significant relationship with emotional exhaustion. He also argued that work–family conflict partially mediates emotional exhaustion and role overload.

H3: Emotional exhaustion mediates the relationship between work–family conflict (WFC) and job performance.

H4: Emotional exhaustion mediates the relationship between family–work conflict (FWC) and job performance.

4 RESEARCH METHODOLOGY

This study uses a quantitative approach. The population used in this research was nurses from IGD hospital, Iskak Tulungagung regency, with sample size of 70. The sampling technique in this research used a census method. The census method collects the entire data from the population associated with the research to extract the required information (Firdaus, 2012). This was used because the population is limited and can be reached by researchers. Date was collected to provide a set of written questions or questionnaires to nurses from IGD Hospital Dr. Iskak Tulungagung, as respondents, and from research journals, as well as data from the website of BPS Kabupaten Tulungagung. The analysis technique used in this research was path analysis with SPSS 21 software, and the Sobel test to test the mediation effect.

5 RESULTS AND DISCUSSION

Table 1 shows that all statement items in the work–family conflict (WFC) variable have Pearson correlation significance value < 0.05, thus it can be concluded that all statement items measuring work–family conflict (WFC) variable are valid.

Table 2 shows that all statement items in the family–work conflict (FWC) variable have a correlation significance value of Pearson < 0.05, thus it can be concluded that all statement items that measure the family–work conflict (FWC) variable are valid.

Table 3 shows that all statement items on job performance variables have a Pearson correlation significance value < 0.05, thus it can be concluded that all statement items that measure job performance variable are declared valid.

Table 4 shows that all statement items in the emotional exhaustion variable have a Pearson correlation

Table 1. Validity test results for variable work–family conflict (WFC).

Items	Pearson correlation	Significance value	Remarks
Job demands interfere with my family life	0.799	0.000	Valid
The time of my job makes family responsibilities difficult to fulfill	0.861	0.000	Valid
Something I want to do at home cannot be done because of job demands	0.777	0.000	Valid
My work makes family tasks difficult to fulfill	0.818	0.000	Valid
Job-related tasks, made me change plans for family activities	0.754	0.000	Valid

Source: Appendix (data processed)

Table 2. Validity test results for variable family–work conflict (FWC).

Items	Pearson Correlation	Significance Value	Remarks
The demands of the family affect the activity at work	0.770	0.000	Valid
Things I want to do at work I cannot do because of family demands	0.567	0.000	Valid
Sometimes I ignore work to meet family responsibilities	0.685	0.000	Valid
The family environment affects my job responsibilities, such as completing work on time, completing daily tasks and doing overtime	0.784	0.000	Valid
Family-related tensions interfere with my ability to perform tasks in a job	0.706	0.000	Valid

Source: Appendix (data processed)

Table 3. Job performance validity test results.

Items	Pearson Correlation	Significance Value	Remarks
Able to complete the number of jobs according to the time specified	0.884	0.000	Valid
Able to complete the quality of work according to the conditions specified	0.850	0.000	Valid
Have a positive work attitude	0.871	0.000	Valid
Be able to complete the task with creative ideas	0.927	0.000	Valid
Can be trusted in completing the work according to the conditions specified	0.739	0.000	Valid

Source: Appendix (data processed)

significance value < 0.05, thus it can be concluded that all statement items that measure emotional exhaustion are declared valid.

Table 5 shows that all variables in the study have cronbach alpha value > 0.6 critical value, thus all

Table 4. Test validity of emotional exhaustion variables.

Items	Pearson Correlation	Significant Value	Remarks
I feel tired of work	0.823	0.000	Valid
I feel frustrated with the job if it does not work well	0.889	0.000	Valid
I feel the job I'm doing is too heavy	0.810	0.000	Valid
I am depressed because I work with people all day long	0.791	0.000	Valid
I feel tired when I wake up in the morning to face work again	0.783	0.000	Valid

Source: Appendix (data processed)

Table 5. Test reliability in variable research.

Variables	Cronbach Alpha	Critical Value	Remarks
Work–family conflict (WFC)	0.862	0.6	Reliable
Family–work conflict (FWC)	0.738	0.6	Reliable
Job performance	0.908	0.6	Reliable
Emotional Exhaustion	0.876	0.6	Reliable

Source: Appendix (data processed)

statement items that measure the five research variables are declared reliable.

Table 6 shows that the overall tolerance value is > 0.10 and VIF < 10, thus there is no multicollinearity, which means that all independent variables used are not correlated.

Table 7 shows that work–family conflict has a direct positive and significant effect on job performance: a coefficient value 0.845 indicates that if the work–family conflict variable increase one unit, it will cause a job performance increase equal to

Table 6. Multicollinearity test results.

	Collinearity statistics	
Model	Tolerance	VIF
Work–family conflict	1.000	1,000
Family–work conflict	1.000	1,000
Emotional exhaustion	1.000	1,000

Source: Output from SPSS

Table 7. Simple linear regression test result.

Variables		Coeffi-cient path	t-calcula-tion	Sig
Work–family con-flict (X1)	→ Job Per-formance (Y)	0.568	0.845	0.000
Family–work con-flict (X2)	→ Job Per-formance (Y)	0.488	0.943	0.000
Work–family con-flict (X1)	→ Emotional exhaustion (Z)	0.461	0.676	0.000
Family–work conflict (X2)	→ Emotional exhaustion (Z)	0.388	0.747	0.000
Emotional exhaustion (Z)	→ Job Per-formance (Y)	0.746	0.972	0.000

Source: Output from PSS

Table 8. Results of direct effect hypothesis test.

Variables	Unstand-ardized coeffi-cients	F signific-ance	T signific-ance	Remarks
Work–family conflict (WFC) → job per-formance	0.845	0.000	0.000	Accepted
Family–work conflict (FWC) → job per-formance	0.943	0.000	0.000	Accepted

Source: Output from SPSS

0.845. The value of R2 indicates that work–family conflict affects job performance by 56.8%, while the rest are variables outside the model.

Family–work conflict has a direct positive and significant effect on job performance: a coefficient value 0.943 indicates that if the family–work conflict variable increased by one unit, it will cause a job performance increase equal to 0.943. The value of R2 indicates that family–work conflict affects job performance by 48.8%, while the rest are variables outside the model.

Work–family conflict has a direct positive and significant effect on emotional exhaustion: a coefficient value of 0.676 shows that if work–family conflict variable increases by one unit, it will cause an emotional exhaustion increase equal to 0.676. The value of R2 shows that work–family conflict affects emotional exhaustion by 46.1%, while the rest are variables outside the model.

Family–work conflict has a positive and significant direct effect on emotional exhaustion: a coefficient value of 0.747 indicates that if the family–work conflict variable increases by one unit, it will cause an emotional exhaustion increase of 0.747. The value of R2 shows that family–work conflict affects emotional exhaustion by 38.8%, while the rest are variables outside the model.

Emotional exhaustion has a direct positive and significant effect on job performance: a coefficient value of 0.972 indicates that if emotional exhaustion variable increases by one unit it will cause a job performance increase equal to 0.972. The value of R2 indicates that emotional exhaustion affects job performance by 74.6%, while the rest are variables outside the model.

Table 8 shows that the coefficient of work–family conflict (WFC) → job performance is 0.845 with significance F and t < 0.05, so there

is a work–family conflict (WFC) influence on job performance equal to 84.5%. A coefficient marked positive indicates that a high work–family conflict (WFC) will increase job performance. The first hypothesis that work–family conflict (WFC) has a significant effect on job performance on emergency department nurses at RSUD Dr. Iskak Tulungagung is accepted.

Based on the questionnaire data, the average value of 3.81 for nurses' responses about work–family conflict (WFC) is included in the high category. This indicates that in general the nurses of the emergency department of RSUD Dr. Iskak Tulungagung experience high work–family conflict, while the average value of nurses' responses about job performance was 3.69, in the high category.

Nurses consider their job as a dedication to be done well and professionally in accordance with Law number 38 of 2014 on nursing, where it is mentioned that the nurse must do their work professionally. In an interview, one of the heads of red room IGD RSUD Dr. Iskak Tulungagung described that many applicants who apply for jobs at this hospital show that although nurses have a high work–family conflict they keep working professionally. The age of nurses who are majority still at the age of 25–35 years make them aware of the importance of professionalism. This is also supported by superiors who always maintain the professionalism of the nurses, so that RSUD Dr. Iskak Tulungagung is a hospital that is renowned in the surrounding area.

The coefficient of family–work conflict (FWC) → job performance is 0.943, with significance F and t < 0.05. Work–family conflict (WFC) influence on job performance is 94.3%. Coefficients with positive signs indicate that a high family–work conflict (FWC) will improve job performance. The second hypothesis, that family–work conflict (FWC) has

Table 9. Test result for the influence of work–family conflict (WFC) on job performance through emotional exhaustion.

Influence	Coeff	B	Std. error	T	Sig.	R2
X1 → Y	c	0.845	0.089	9.464	0.000	0.568
X1 → Z	a	0.676	0.089	7.628	0.000	0.461
Z → Y	b	0.972	0.069	14.145	0.000	0.746
X1, Z→Y	c'	0.348	0.084	4.154	0.000	0.798

Source: Output from SPSS

a significant effect on job performance on emergency department nurses at RSUD Dr. Iskak Tulungagung, is accepted.

Based on the questionnaire data, the average value of 3.68 for nurses' responses about family–work conflict (FWC) are included in the high category. This indicates that, in general, the nurses of the emergency department at RSUD Dr. Iskak Tulungagung have a high family–work conflict (FWC), while the average score of nurses' responses on job performance is 3.69, included in the high category. Nurses at IGD RSUD Dr. Iskak Tulungagung must be able to work professionally in accordance with Law number 38 year 2014 about nursing, which states that nurses have to do their work professionally. The results of this study indicate a positive influence between family–work conflict (FWC) and job performance, when family–work conflict (FWC) has increased then job performance has also increases. The facts on the ground show that this positive effect is caused by the demand to work professionally as a nurse in accordance with hospital regulations. Although conflicts are high, hospital nurses are able to manage conflict and improve performance. This is in accordance with the opinions expressed by Robbins and Judge in research conducted by Christine (2010), which states that the level of conflict positively and negatively affect performance.

Despite the fact that the conflict of the family and work is high, it actually makes the nurse perform maximally. The demand for high nursing responsibilities leads them to fulfill responsibilities in their work. The number of nurses who apply for work in RSUD Dr. Iskak Tulungagung causes them to perform well in order not to be periodically evaluated by hospitals. Nurses who do not comply with the provisions in the hospital hmay be dismissed from their work, so nurses will try to perform as well as possible despite the conflict with family. In an interview with the IGD head of RSUD Dr. Iskak Tulungagung, it was mentioned that because there are many applications for nurse posts at the hospital, there is a demand that must be considered by the nurse if they do not want to be dismissed. The good performance of the nurse is what contributes to the overall performance of the hospital,, so that RSUD Dr. Iskak Tulungagung is a hospital that is renowned in the surrounding area.

The Sobel Test is used to test the significance of the reduction between the c-c' regression coefficients with the following equation:

$$Z \text{value} =$$

$$\frac{0.676 \times 0.972}{\sqrt{(0.972^2 \times 0.089^2) + (0.676^2 \times 0.069^2) + (0.089^2 \times 0.069^2)}}$$

$$Z \text{ value} = \frac{0.657072}{\sqrt{0.009697}}$$

$$Z \text{ value} = 0.657072/0.098473 = 6.672$$

Based on the Sobel Test, the value of z is equal to 6.672 > 1.96 indicating that there is significant decrease between regression coefficient c compared with coefficient c'. Both of these test results show that emotional exhaustion mediates the influence of work–family conflict on job performance through emotional exhaustion. This shows that the third hypothesis, that emotional exhaustion mediates the relationship between work–family conflict (WFC) and job performance for emergency hospital nurses at Dr. Iskak Tulungagung, is accepted to mediate partially.

The results of this study are consistent with a study conducted by Karatepe in 2013, which that found that emotional exhaustion mediated the relationship between work–family conflict (WFC) on job performance at frontline hotels and managers in Romania. The equation of research on nurses at IGD RSUD Dr. Iskak Tulungagung with Karatepe's research is that emotional exhaustion variables mediate the work–family conflict (WFC) relationship with job performance.

This study agrees with Karatepe (2013) that emotional exhaustion mediates work–family conflict (WFC) variables with job performance, but there is a difference in the effect of mediation. This study found that emotional exhaustion mediates partially, in contrast to Karatepe who found that emotional exhaustion mediates fully. Evidence that emotional exhaustion mediates partially is the influence of variable X to Y through Z, which is significant; the coefficient, which decreased significantly from coefficient value 0.845 to 0.348; and the result of the Sobel Test, with z value equal to 6.672 > 1.96 with a mean significant decreasing from path c versus line c'. A nurse who has a conflict in their work environment will constantly experience emotional exhaustion that will impact on family life, This is because the responsibility and role as a nurse may create conflict in the family environment if the problems are brought home and consequently will impact on performance decline.

Table 10. Results of family–work conflict (FWC) influence tests on job performance through emotional exhaustion.

Influence	Coeff	B	Std.error	T	Sig.	R2
X2 → Y	c	0.943	0.117	8.047	0.000	0.488
X2 → Z	a	0.747	0.114	6.571	0.000	0.388
Z → Y	b	0.972	0.069	14.145	0.000	0.746
X2.Z → Y	c'	0.353	0.097	3.639	0.000	0.788

Source: Output from SPSS

The results of the tests in this study show that the direct influence of work–family conflict (WFC) on job performance through emotional exhaustion decreased, which means that when someone experiences emotional exhaustion then the direct influence of work–family conflict of emotional exhaustion variables are greater. The characteristics of respondents who are mostly women and married make the nurses feel more tired of the work performed, which is indicated by the highest emotional fatigue item chosen by the nurses. This shows that the high sense of failure can have a higher effect on performance compared to work–family conflict (WFC), because work–family conflict (WFC) is more directly influential on performance without involving emotional exhaustion.

The Sobel Test is used to test the significance of the reduction between the c-c' regression coefficients with the following equation:

$$Z\,\text{value} =$$

$$\frac{0.747 \times 0.972}{\sqrt{(0.9722 \times 0.1142) + (0.7472 \times 0.0692) + (0.1142 \times 0.0692)}}$$

$$Z\,\text{value} = \frac{0.726084}{\sqrt{0.014997}}$$

$$Z\,\text{value} = 0.726084/0.122462 = 5.929$$

Based on the Sobel Test, the value of z is equal to 5.929 > 1.96 which shows that there is a significant decrease between regression coefficient c compared with coefficient c'. Both of these test results show that emotional exhaustion mediates the influence of family–work conflict on job performance through emotional exhaustion. This proves the fourth hypothesis that emotional exhaustion mediates the relationship between family–work conflict (FWC) and job performance for emergency department nurses at RSUD Dr. Iskak Tulungagung is accepted to mediate partially.

The results of this study are in accordance with research conducted by Karatepe in 2013, who found that emotional exhaustion mediates the relationship between family–work conflict (FWC) and job performance in frontline hotels and managers in

Romania. The equation of research on nurses at IGD RSUD Dr. Iskak Tulungagung with Karatepe's research is a variable emotional exhaustion mediate the relationship of family–work conflict (FWC) to job performance.

This study agrees with Karatepe that emotional exhaustion mediates the family–work conflict (FWC) variable to job performance, but there is a difference in the influence of mediation. This study found that emotional exhaustion mediates partially, in contrast to Karatepe who found that emotional exhaustion mediates fully. Evidence that emotional exhaustion mediates partially is the influence of variable X to Y through Z, which is significant; the coefficient, which decreased significantly from coefficient value 0.943 to 0.353; and the result of the Sobel Test with z value equal to 5.929 > 1.96, which means a significant decrease from path c versus line c'. A nurse who has a role as a parent will have a burden or pressure in carrying out their duties as a nurse, and a nurse who is unable to control family–work conflict (FWC) will have an increased tendency to experience emotional exhaustion. Fulfilling duties as parents and important figures in the family and the obligations as a nurse, at the same time, will affect performance decline.

The results of the tests in this study show a direct the influence of family–work conflict (FWC) on a decrease in job performance through emotional exhaustion, which means that when a person experiences emotional exhaustion then the direct influence of family–work conflict diminishes of emotional exhaustion variables are greater. The characteristics of the respondents – mostly married and women – make the nurses feel emotional fatigue against the work they have to do. The test results show that the nurses feel tired of their work as the highest item chosen, indicating that high fatigue can affect higher performance compared to family–work conflict (FWC). This is because family–work conflict (FWC) is more influential directly on performance without involving emotional exhaustion.

6 CONCLUSIONS

Based on the results of data analysis and discussion, the conclusions in this study are as follows:

a. Work–family conflict (WFC) has a significant effect on job performance for emergency nurses of RSUD Dr. Iskak Tulungagung.
b. Family–work conflict (FWC) has a significant effect on job performance for emergency nurses of RSUD Dr. Iskak Tulungagung.
c. Emotional exhaustion mediates the relationship between work–family conflict (WFC) and job performance for emergency department nurses at RSUD Dr. Iskak Tulungagung.
d. Emotional exhaustion mediates the relationship between family–work conflict (FWC) and job

Performance for emergency department nurses at RSUD Dr. Iskak Tulungagung.

Based on the discussion and conclusions in this study, the researchers provide suggestions as follows:

a. Dr. RSUD. Iskak Tulungagung provides time programs for families, such as not providing additional tasks on holidays, as well as controlling the workload of nurses to reduce the level of work–family conflict (WFC), family–work conflict (FWC), and emotional exhaustion.

b. Dr. RSUD. Iskak Tulungagung needs to create performance-based performance management systems, as well as reward nurse achievement to maintain high job performance levels.

c. Subsequent research is expected to further improve the sample of research, not only the nurses at IGD RSUD Dr. Iskak Tulungagung, and distinguish between female and male nurses.

REFERENCES

Adam, Giri Aulia Fauziah. 2013. Pengaruh Work Family Conflict dan Work Family Facilitation Terhadap Kinerja Melalui Kepuasan Kerja Pada Karyawan JW Marriot Hotel Surabaya. Skripsi. Universitas Airlangga.

Advani, Jay Yashwant, et al. 2005. Antecedents And Concequences of "Burnout" In Services Personnel A Case of Indian Software Professionals.

Aminah Ahmad. 2010. work-family conflict among junior physicians: its mediating role in the relathionship between role overload and emotional exhaustion. *Journal of Social Science*, 6, 2.

Baron and Kenny. 1986. The Moderator-Mediator Variable Distinction in Social Psychological Research: Conceptual, Strategic, and Considerations. *Journal of Personality and Social Psychology*, 51, 6.

BPS Kabupaten Tulungagung. 2016. https://tulungagung kab.bps.go.id/linkTabelStatis/view/id/800, accessed 22 October 2016.

Christine, W.S., Oktorina, Megawati, &Mula, Indah. (2010). Pengaruh konflik pekerjaan dan konflik keluarga terhadap kinerja dengan konflik pekerjaan keluarga sebagai intervening variabel (studi pada dual career couple di Jabodetabek). *Jurnal Manajemendan Kewirausaan*, 12 (2), 121–132.

Golden Timothy D. 2011. "Altering the Effect of Work and Family Conflict on Exhaustion: Telework During Traditional and Nontraditional Work Hours." *Springer Science+ Business Media*, 27, 255–269.

Greenhaus, Jeffrey H, and Beutell, Nicholas J. 1985."Soures of Conflict between Work and Family Roles." *Journal of the Academy of Management Review*, 10.

Greenhaus, Jeffrey H. (2002). Work-family conflict. *Journal of the Academy of Management Review*, 45, 1–9.

Halbesleben et. al. (2011). The cost and benefits of working with one's spouse: A two-sample examination of spousal support, work-family conflict, and emotional exhaustion in work-linked relationship. *Journal of Organizational Behavior*, 33.

Karatepe, O.M. (2013). The effects of work overload and work-family conflict on job embeddedness and job performance: The mediation of emotional exhaustion. *International Journal of Contemporary Hospitality Management*, 25 (4), 614–634.

Netemeyer, R. G., James, S., McMurrian, R. (1996). Development and validation of work-family conflict and family-work conflict scales. *Journal of Applied Psychology*, 881 (4), 400–410.

Putri, Anggia. (2012). Pengaruh kelelahan emosional terhadap perilaku belajar pada mahasiswa yang bekerja. *Jurnal Ilmiah*.

Rahadiono, Alldino Putra. 2015. Pengaruh kepuasan kerja terhadap kinerja karyawan dengan komitmen organisasional sebagai variabel intervening pada karyawan pt. Jaya abadi energy. Skripsi. UniversitasAirlangga.

Rakhmawati, Annisa. (2015). Pengaruh work-family conflict dan lingkungan kerja terhadap kinerja perawat rumah sakit Ghrasia Yogyakarta. Skripsi.Universitas Negeri Yogyakarta.

Roboth, Jane Y. (2015). Analisis Work-family conflict, stress kerja dan kinerja wanita berperan ganda pada yayasan Copmassion East Indonesia. *Jurnal Riset Bisnis dan Manajemen*, 3 (1), 33–46.

Firdaus, M. Aziz. 2012. Metode Penelitian. Edisi pertama. Tangerang Selatan: Jelajah Nusa.

Maghfiroh, Ursula. 2013. Pengaruh Enterpreneurial Orientation Terhadap Kinerja Usaha Mikro, Kecil, Dan Menengah Di Kabupaten Kediri Korcam Ngadiluwih Melalui Innovation Capacity.Skripsi tidak diterbitkan, Surabaya Fakultas Ekonomi dan Bisnis Universitas airlangga.

Sekaran, Uma. 2006. Metodologi Penelitian Untuk Bisnis. Jakarta: Salemba Empat.

Ghozali, Imam. 2013. AplikasiAnalisis Multivariate dengan Program IBM SPSS 21. Edisi tujuh. Semarang: Badan Penerbit Universitas Diponegoro.

Sugiyono (2009). Metode Penelitian Bisnis (Pendekatan Kuantitatif, Kualitatif, dan R&D), Bandung: Alfabeta.

Undang-Undang Republik Indonesia No.13 Tahun 2003 tentang Ketenagakerjaan.

Wijayanti, Rahardian. (2012). Analisis kinerja RSUD Dr. ISKAK Tulungagung dengan metode Balanced Scorecard. Tesis. Universitas Indonesia.

Advances in Business, Management and Entrepreneurship – Hurriyati et al (eds)
© 2020 Taylor & Francis Group, London, ISBN 978-0-367-27176-3

Influence of perceived organizational support for creativity and creative self-efficiency on job satisfaction toward individual creativity in members of non-profit organizations engaged in domestic stray animal care in Surabaya and Sidoarjo

T.S. Agustina
Universitas Airlangga, Surabaya, Indonesia

ABSTRACT: Creativity is one of the main aspects that individuals need to continuously develop and solve problems. This study focused on perceived organizational support (POS) for creativity, creative self-efficacy, job satisfaction, and individual creativity, and how the variables affect each other. The subjects were 50 members from five non-profit organizations engaged in domestic stray animal care in Surabaya and Sidoarjo. Using the partial least square (PLS) analyses method showed significant influences on several hypothesis, namely the influence of POS for creativity toward job satisfaction, creative self-efficacy toward job satisfaction, POS for creativity toward individual creativity, and creative self-efficacy toward individual creativity. The variables with no significant effect were job satisfaction toward individual creativity caused by lack of intrinsic and extrinsic motivation in members, POS for creativity toward individual creativity with job satisfaction as the intervening variable, and creative self-efficacy toward individual creativity with job satisfaction as the intervening variable. The results of this study can be used as an input in decision-making related to quality and organizational development in order to increase job satisfaction and creativity of non-profit organizations engaged in domestic stray animal care in Surabaya and Sidoarjo.

1 INTRODUCTION

Employee creativity is acknowledged as essential for an organization in creating innovation (Amabile, 1998). In a rapidly changing and competitive business environment, an organization might regard their employees as a creative solution to crucial problems (Thompson and Choi 2006). Things that help creativity development are also important for the organization to know in order to help improve their employees' skills.

Perceived organizational support (POS) is the degree to which employees believe that their organization appreciates their contribution and care about their well-being (Robbins and Judge, 2008; 103). Zhou and George, and a number of other theorists, define perceived of organizational support for creativity as "the extent to which an employee perceives that the organization encourages, respects, rewards, and recognizes employees who exhibit creativity" (DiLiello et al., 2011). Another important aspect of the development of creative mindset is the development of self-efficacy (Ghafoor et al., 2011). Bandura's research (1997) shows that self-efficacy is a driver to increase work performance. According to Bandura, creative self-efficacy is a self-assessment of one's creative potential that specifically involves seeing oneself as being good at

creative problem-solving and novel idea generation (DiLiello et al., 2011). Several studies have also shown a positive relationship between creative self-efficacy and creativity (Gong et al., 2009; Ghafoor et al., 2011).

In addition to the two variables, this study also examines job satisfaction as an intervening variable in its relation to creativity. Locke (1976) describes job satisfaction as a positive emotional state as the result of a job evaluation or a person's work experience. Job satisfaction is very important because an employee's positive attitude in an organization is an important factor to determine their ability to work properly and provide optimal performance, which is the company's goal.

Related to this matter, this research aimed to determine the influence of POS for creativity and self-efficacy on job satisfaction toward individual creativity, in non-profit organizations engaged in domestic stray animal care in Surabaya and Sidoarjo, as this type of animal-care organization has grown in Indonesia over the last five years. However, non-profit organizations engaged in domestic stray animal care in Indonesia are still very poorly supported by society because the majority of people are not aware of how often animals in the streets are treated badly. Such organizations also need creative ideas to help their work program and solve problems.

2 THEORITICAL CONTEXT AND RESEARCH HYPOTHESIS

2.1 Literature study

2.1.1 Perceived organizational support for creativity
POS for creativity is defined as an employee's perception or belief about the organization's ability to support, motivate, direct, and appreciate the employee's contribution to creating and producing products, ideas, and thoughts that are different from the usual and are reflected through creative performance (Amabile, 1983). Zhou and George (2001) examined whether the support of organizations and colleagues can help dissatisfied employees to produce creative ideas. This can happen if the employee's continuance commitment is high and when a colleague can provide useful and organizational support for employee creativity (Novellea, 2014).

POS for creativity is an employees's trust, assessment, and perceptions of how the organization can appreciate and provide support to their creativity. In addition, POS for creativity is built to create and maintain an organizational environment that supports creativity and innovation.

2.1.2 Creative self-efficacy
Creative self-efficacy is a self-assessment of one's creative potential that specifically involves seeing oneself as being good at creative problem-solving and novel idea generation (Bandura in Diliello et al., 2011). Bandura also said the magnitude of self-efficacy possessed by individuals can influence their competence development and liveliness. From the above explanation, it can be concluded that when someone has a high level of self-efficacy, then they will try harder when encountering difficulties at work. This relates to creative self-efficacy because employees who can work creatively usually also love the challenges in their work.

2.1.3 Job satisfaction
Job satisfaction is an individual thing because everyone has a different level of job satisfaction. The high aspect of work that fits the expectations of the individual can increase the level of job satisfaction. Job satisfaction is a feeling of satisfaction from the outcome of a work achievement that has an important value (Noe et al., 1997: 23). The existence of such an achievement is an awareness of each employee through positive perceptions of the organization that supports their performance in utilizing their creative potential. This can also be reflected through the attitude and the feeling of the individual about their work (Armstrong, 2006). Thus, it can be said that positive behavior can be an indication of job satisfaction.

2.1.4 Individual creativity
Creativity is the ability to do a job in a new and precise way (McLean, 2005). Creativity is one of the human needs for self-actualization (Novellea, 2014) and is also regarded as a thought, idea, or a way that satisfies two conditions: the first is a thought, idea or way of doing something that is new and original, and the second has the potential to be used in organizations (Oldham and Cummings 1996). Basically, creativity is the creation of new ideas that are beneficial to all aspects of human activity and everyday life (Amabile, 1997). A useful new idea is considered one that has not been thought of before that can solve a problem in an organization. Amabile also stated that creativity is the first step of innovation, which means success in implementing new ideas, and is an important success for the company in the long run. Broadly speaking, creativity cannot be separated from novelty and usefulness. Therefore, creativity is a result or idea that is useful and precise for solving problems effectively (Novellea, 2014).

2.2 Research hypothesis

Emerson (2013) shows that POS serves as a mediator between organizational culture and both turnover intentions and job satisfaction. There are twelve hypotheses in this study, and the first hypothesis states that there is a positive relation between POS and job satisfaction, in which this hypothesis is accepted.

H1: POS for creativity has a significant influence on the work satisfaction of members of non-profit organizations engaged in domestic stray animal care in Surabaya and Sidoarjo.

Research conducted by Karatepe et al. (2006) on employees at a Turkish-owned Northern Cyprus hotel discussed competitiveness, self-efficacy, and effort as a significant predictor of frontline employee performance. The effect of competitiveness variables on performance is stronger than the effect of the business variable. It is also known that the direct influence of self-efficacy on job satisfaction is stronger than the influence of business. Zorlu (2011) examines the relationship between the role of sources of stress (role of ambiguity-role conflict) and job satisfaction, and to determine whether the role of self-esteem and self-efficacy perceptions of employees can be assumed to make this relationship positive.

H2: Creative self-efficacy has a significant influence on job satisfaction at non-profit organizations engaged in domestic stray animal care in Surabaya and Sidoarjo.

Aschenbrener et al. (2007) examined the effect of creativity and job satisfaction on second-year secondary agricultural teachers in Columbia. The relationship in this study revolves between creativity and job satisfaction, but some in-category creativity used can be a reference in determining teacher's job satisfaction, such as fluency, originality, elaboration, and flexibility. In this study, there is a positive but weak relationship between job satisfaction and creativity.

Miao (2011) examined the interrelationships between perceptions of organizational support, job performance, and job satisfaction at the Department of the Steel Corporations in Liaoning, China. The study shows a correlation between the perception of organizational support for work-worker behavior, which states that high levels of support can affect employee work behavior and will impact on the formation of effective performance and job satisfaction.

H3: Job satisfaction has a significant influence on the individual creativity of members of non-profit organizations engaged in domestic stray animal care in Surabaya and Sidoarjo

Houghton and DiLiello (2010) examine the role of leadership development as a modulator relationship between POS for creativity, creative self-efficacy, and individual creativity. The result of this study is that POS for creativity and creative self-efficacy has a positive effect on individual creativity. In his research, Tierney and Farmer (2010) also discussed factors that can influence a change of creativity in the self-efficacy of employees. There are three main objectives in this study: 1) to assess whether employees' creative self-efficacy can change and the type of factors that might account for such development; 2) to explore whether employee creative performance levels change and whether this corresponds with changes in creative self-efficacy, thereby providing greater insight into the nature of the creative self-efficacy-creativity association; 3) to test the study model in a not-for-profit organization, in which both a sense of creative efficaciousness and creative performance would be particularly salient. The study took place in a child welfare organization, characterized by a dynamic work context where employees faced the emergence of novel and unpredictable issues that needed to be creatively resolved in the face of limited resources. Finding creative ways to solve problems related to this work had been emphasized by their chief executives, supervisors and HRD members. The results show that increases in creative self-efficacy corresponded with increases in creative performance. Therefore, the researchers put forward the hypotheses as follows:

H4: POS for creativity has a significant influence on the individual creativity of members of non - profit organizations engaged in domestic stray animal care in Surabaya and Sidoarjo.
H5: Creative self-efficacy has a significant influence on the individual creativity of members of non-profit organizations engaged in domestic stray animal care in Surabaya and Sidoarjo.

Mardhatillah (2014) on the "Influence of POS for Creativity on Creativity with Job Satisfaction as intervening variable at PT. Kompas Gramedia

Surabaya Circulation" indicated that POS for creativity influences creativity, with job satisfaction as an intervening variable. Therefore, the researchers put forward the hypothesis as follows:

H6: POS for creativity has a significant influence on individual creativity, with job satisfaction as an intervening variable, in members of non-profit organizations engaged in domestic stray animal care in Surabaya and Sidoarjo.

Anggarwati (2014) found that creative self-efficacy has a significant influence on creativity through job satisfaction as an intervening variable. Therefore, the researcher put forward the hypothesis as follows:

H7: Creative self-efficacy has a significant influence on individual creativity, with job satisfaction as an intervening variable, for members of non-profit organizations engaged in domestic stray animal care in Surabaya and Sidoarjo.

There are several similarities between this study and some of the previous studies described above, namely the influence of POS for creativity with job satisfaction and credibility, and the influence of creative self-efficacy with job satisfaction and creativity, and job satisfaction as an intervening variable. In addition, there is also a positive influence between job satisfaction and creativity. The difference between previous studies and this study is that there are several variables from the other studies that are then combined in different models in this study, but mainly still based on the previous results.

3 DATA AND METHOD

3.1 Research approach

This study uses a quantitative method, designed to explain the influence of variables through hypothesis testing (explanatory research).

3.2 Variable identification

1. An exogenous variable influences an endogen variable positively or negatively (Sekaran, 2003: 89). In this study, the exogenous variable is POS for creativity (X1) and creative self-efficacy (X2).
2. An endogenous variable has changes predicted by an exogenous variable (Sekaran, 2003: 88). In this study, the endogenous variable is individual creativity (Y).
3. An intervening variable exists as an operational function to an exogenous variable in various situations and helps to conceptualize and explain the influence of exogenous variable on endogenous variable (Sekaran, 2003:94). The intervening variable in this study is job satisfaction (Z).

Organization Name	Number of members	Members > 1 year	Sample
Surabaya Sidoarjo Peduli Kucing Domestik (S2PKD)	36	20	19
Surabaya Animal Care Community	15	8	8
(SACC) Surabaya Dog Lovers (SDL)	23	6	4
Surabaya Save Paws (SSP)	13	8	8
Komunitas Pecinta Kucing Sidoarjo (KOMPAKS)	27	11	11
Total	114	53	50

Source: Data of non-profit organizations engaged in domestic stray animal care in Surabaya and Sidoarjo.

3.3 Type and data source

1. Primary data is collected from empiric observations, interviews with concerned parties, and questionnaires.
2. Secondary data is collected from books, journals, articles, and other supporting documents.

3.4 Sampling method

3.4.1 Research population
According to Sugiyono (2010:115) population is a "generalization region consisting of objects or subjects that have certain qualities and characteristics set by the researcher to be studied and then drawn conclusions of." The population is important, and should be carefully selected to be suitable for the expected characteristics. In this study, the population is non-profit organizations engaged in domestic stray animal care in Surabaya, and Sidoarjo's members who have joined the organization for at least a year, so the influence of POS for the creativity variable can be studied in members that have actually experienced support given by the organization.

3.4.2 Research sample
Since the population of the study site is 53 members only, the entire population is sampled. The sampling technique used in this study is saturated sampling. According to Sugiyono (2008: 124), a saturated sampling technique is when all members of the population are used as a sample, and it is also called saturated samples or census.

Related to POS for creativity, creative self-efficacy and individual creativity variables, the population is

determined by the basis on which creativity has a relationship in individual performance. Members of the domestic caring organization in Surabaya and Sidoarjo who have joined for a minimum of one year were selected as samples because members of the organization are required to think creatively in solving problems and running their own work program. There were 50 questionnaires collected that were usable to test the validity and reliability. Table 2.1 shows data of the members of non-profit organizations engaged in domestic stray animal care in Surabaya and Sidoarjo.

3.5 Analysis method

3.5.1 Data processing method
PLS is an analysis of variance-based structural equations that can simultaneously test the measurement model and structural model (Ghozali, 2008: 17). The measurement model is used to test the validity and reliability of the measuring tools, while the structural model is used to test causality. Ghozali (2006) explains that evaluation of the PLS model is done by evaluating the inner model and outer model:

1. The outer model evaluation is used for measuring validity and reliability model. The self-measurement model is used to test the validity of the construct and the reliability of the tools using three criteria: convergent validity, composite reliability, and discriminant validity.
2. The inner model evaluation is a structural model in evaluated PLS used to see the relationship between construct, value significance, and R-square. The structural model is evaluated by using R-square for dependent constructs and t-test as well as the significance of the structural path of parameter coefficients (Ghozali, 2006). R-square is used to assess the effect of a latent dependent variable and whether it has substantive influence. Testing of the proposed hypothesis can be seen from the magnitude of the t-statistics. The criteria for rejecting and accepting the proposed relationship can be seen from the comparison between t-arithmetic and t-table values. If the value of t-arithmetic > t-table, i.e., 1.96, then the hypothesis is accepted.

3.5.2 Validity test
In PLS, the validity test is measured through convergent validity and discriminant validity. Convergent validity is judged by the correlation between item score (component score) and construct score. According to Chin (1998) in Ghozali (2008: 24), the size of validity is considered to meet the valid criteria if the measuring scale of the loading value is at least 0.7. The discriminant validity of the indicator is judged based on cross-loading measurements with construction. If the correlation of the construct with the measurement item is greater than the size of the other construct, then this indicates that the latent

construct predicts the block size better than the other block size. The indicator is considered valid if the loading factor value is at least 0.7 and has a t-statistic above the t-table value of 1.96 (at 5% significance level).

3.5.3 *Reliability test*

Ghozali (2008: 25) states that composite reliability is an indicator block that measures a construct, which can be evaluated by an internal consistency measure developed by Werts, Linn, and Joreskog (1974). Internal consistency is a closer approximation, with the assumption that the parameter's estimation is accurate. If the value of composite reliability is above 0.70 then it can be said that the construct is reliable (Ghozali, 2008: 43). In addition, reliability can also be tested by seeing if the value of Cronbach's alpha is above 0.70. Thus, it can be said that the variable has good reliability.

4 RESULT

4.1 *Respondents profile*

The information in Table 4.1 shows that most respondents in this study are female.

The information in Table 4.2 shows that the majority of respondents in this study are members aged between 20–30 years.

The information in Table 4.3 shows that the majority of members of non-profit organizations engaged in domestic stray animal care in Surabaya and Sidoarjo have a diploma or bachelor level education.

The information in Table 4.4 shows that the majority of respondents in this study have been with the organization for 1–3 years.

Tabel 4.1 Description of respondent's characteristics by sex.

Sex	Members	Percentage
Male	8	16%
Female	42	84%
Total	50	100%

Source: Primary data, 2016

CSE -> JB	0.6086	0.615	0.064	0.0643	9.4722
CSE -> IC	0.3065	0.2995	0.1511	0.1511	2.0286
JB -> IC	0.0247	0.0209	0.1015	0.1015	0.2429
POSC -> JS	0.2432	0.2426	0.073	0.073	3.3317
POSC -> IC	0.4763	0.482	0.1193	0.1193	3.9935

Source: PLS

Table 4.2. Description of respondent's characteristics by age

Age	Members	Percentage
< 20 years	1	2%
20–30 years	33	66%
> 30 years	16	3 2 %
Total	50	100%

Source: Primary Data, 2016

Table 4.3. Description of respondent's characteristics by highest level of education

Highest level of education	Members	Percentage
High school	14	28%
Diploma/Bachelor	36	72%
Total	50	100%

Source: Primary Data, 2016

Table 4.4. Description of respondent's characteristics by participation time

Participation Time	Members	Percentage
1–3 years	38	76%
3–5 years	9	18%
> 5 years	3	6%
Total	50	100%

Source: Primary data, 2016

4.2 *Respondents' answers*

Description of the answers was performed by calculating the mean value of respondent's answer to each question and all questions as a whole. To categorize the average respondents' answers, class interval is calculated by the following formula:

$$\frac{\textit{Highest Value — Lowest Value}}{\textit{Number of Classes}}$$

(1)

With a class interval of 0.8, the average criteria of respondents' answers are presented Table 4.5.

4.2.1 *POS for creativity*

Table 4.6 shows the description of respondents' answers about POS for creativity.

Based on Table 4.6 it can be concluded that the average answers from respondents to variable POS for creativity is 4.29, with a very high category, which means that organizational support to

Table 4.5. Category of average respondents answer

Interval	Category
4,20 < a =< 5,00	Very High (VH)
3,40 < a =< 4,20	High (H)
2,60 < a =< 3,40	Intermediate (I)
1,80 < a =< 2,60	Low (L)
1,00 < a =< 1,80	Very Low (VL)

Source: Calculation

Table 4.6. Respondents' responses about POS for creativity

Indicator	Statement	Means	Category
POSC1	The organization supports me in doing creative things to solve the organization's problems.	4.28	VH
POSC2	The organization's leader appreciates my creative ideas related to the organization's program and problem-solving.	4.34	VH
POSC3	There is teamwork in the form of working groups in program and problem-solving.	4.30	VH
POSC4	The organization provides facilities to help the member work creatively.	4.24	H
POSC5	I have freedom in which program I want to do.	4.24	VH
POSC6	In doing this work I feel motivated to think creatively.	4.36	VH

Source: Primary data, 2016

creativity is very high. Respondents perceive organizational support in creativity mostly when their members feel the urge to think creatively with an average score of 4.36. However, respondents feel that the organization is least supportive in providing facilities for the respondent to perform creatively, with an average value of 4.24, which falls into the high category.

4.2.2 Creative self-efficacy
Table 4.7 shows respondents' answers about creativity self-efficacy.

Based on Table 4.7 it can be concluded that the average answers from respondents to the creative self-efficacy variable is 4.17, in the high category, which means the respondent has high creative self-efficacy. The highest level of creative self-efficacy that respondents feel is when proposing an idea yet they are not sure they can complete the job well, with an average score of 4.30. The lowest level of creative self-efficacy is when they belief that they can provide the right creative solutions in relation to problems that occur in the organization, with an average value of 3.98.

4.2.3 Job satisfaction
Table 4.8 details the respondents' answers about job satisfaction.

This information shows that the work environment has supported the development of self-creativity of the respondents, with an average value of 4.26. On the contrary, the respondents feel the lowest level of job satisfaction when they received an award from the head of the organization for their performance and efforts, with an average value of 3.68.

4.2.4 Individual creativity
Table 4.9 details respondents' answers about individual creativity.

Based on Table 4.9 it can be concluded that the average answers from respondents about the individual credibility variable is 4.02, in the high category, which means that respondents have high individual creativity. Their highest level of individual creativity, with an average score of 4.06, is when they can produce original ideas of self-thinking that have never been raised by other members. The lowest level of individual creativity, with an average value of 3.98, is felt by respondents at a level capable of generating new ideas that are unique and different.

Table 4.7. Respondents' responses about creative self-efficacy

Indicator	Statement	Means	Category
CSE1	Have many creative ideas to solve problems	4.16	H
CSE2	Thinking about multiple ideas to solve problems	4.30	VH
CSE3	Have idea that other members do not have	4.22	H
CSE4	Able to convey unconventional ideas		

Source: Primary data 2016

Table 4.8. Description of respondents' responses about job satisfaction

Indicator	Statement	Means	Category
JS1	There is a good relation between the organization's leader and fellow members.	4.16	H
JS2	Have the chance to use the organization's facilities.	3.76	H
JS3	The organization's environment encourages the development of creativity of its members.	4.26	VH
JS4	Achieve award from the organization's leader for their performance and effort in program and problem-solving.	3.68	H

Source: Primary data. 2016

Table 4.9. Description of respondents' responses about individual creativity

Indicator	Statement	Means	Category
IC1	Able to give new, different and unique ideas.	3.98	H
IC2	Able to give idea in the organization's problem-solving pro-	4.02	H
IC3	Able to develop ideas into detailed acts.	4.02	H
IC4	Able to give original idea that never been given by other members.	4.96	H

Source: Primary data. 2016

4.3 Data analysis

Data processing techniques using the structural equation modeling (SEM) method based on partial least square (PLS) requires two stages to assess the fit model of the research model. The stages are as follows:

4.3.1 Measurement model

In the data analysis method using SmartPLS there are three criteria to assess the outer model of convergent validity, discriminant validity, and composite reliability.

4.3.1.1 Convergent validity

Based on the structural model of the first validity test there are two items of questions that do not meet convergent validity because the value of loading is below 0.5, so for further analysis the items need to be reduced. Figure 4.3 shows the structural model after reducing two items that do not meet the convergent validity:

Based on Figure 4.3, each indicator on all varieties of research has a value of loading more than 0.5. This means the item questions that exist in the structural model after the reduction have met convergent validity.

4.3.1.2 Discriminant validity

Discriminant validity tests are assessed based on cross-loading measurements with their constructs. An in-catalog is said to meet the discriminant validity if the value of the cross-loading indicator of the variable is the largest compared to the other variables.

the indicators that compose each variable in this study has met the discriminant validity because it has the largest cross-loading for the variables, thus all indicators in each variable in this study have met the discriminant validity.

4.3.1.3 Composite reliability

The reliability test is used to measure the reliability of the constructs, which can be done using a Cronbach's alpha test and or by looking at composite reliability. A construct can be said to be reliable if the composite reliability value is greater than 0.7. Table 4.11 is the composite reliability of each variable:

Table 4.11 shows that the value of the composite variable and Cronbach's alpha at each variable in this study is more than 0.7. Thus, it can be concluded that each variable has met the composite reliability.

4.3.2 Inner model or structural model

The structural model in PLS was done to see the significance of influence between construct and R-square. The structural model is evaluated using a t-test to know the significance and coefficient of

Table 4.10. Cross loadings.

	Creative Self-Efficacy	Job Satis-faction	Individual Creativity	POS for Creativity
CSE1	0.688	0.5669	0.6164	0.5712
CSE2	0.8015	0.5639	0.3905	0.4606
CSE3	0.7883	0.6531	0.3815	0.6348
CSE4	0.8418	0.6419	0.5244	0.5186
KI1	0.6297	0.4905	0.7897	0.6559
KI2	0.4339	0.4254	0.8379	0.4864
KI3	0.5076	0.479	0.7731	0.5619
KI4	0.3415	0.2501	0.8228	0.3828
KK2	0.4018	0.7178	0.1259	0.2691
KK3	0.716	0.8052	0.5121	0.5625
KK4	0.6227	0.823	0.475	0.6314
POSC1	0.3766	0.2915	0.5207	0.694
POSC2	0.5289	0.5736	0.4224	0.7965
POSC3	0.6283	0.5953	0.5242	0.7477
POSC4	0.5956	0.571	0.5105	0.8618
POSC5	0.6063	0.5693	0.6705	0.8389

Source: Based

808

Source: PLS

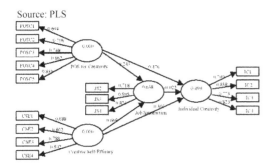

Figure 4.3. PLS structural model after reduction

Source: PLS

structural path parameter, and R-square to know the influence between variables. Table 4.12 is the result of R-square estimation using SmartPLS.

The higher the R-square the greater the independent variable can explain the dependent variable, and the better the structural equation. Job satisfaction variable has an R-square value of 0.6379, which means the percentage of the diversity of job satisfaction data that can be explained by POS for creativity and creative self-efficacy is equal to 63.79%. while the individual creativity variable has a R-square value of 0.4993, which means the percentage of the diversity of individual creativity data that can be explained by POS for creativity, creative self-efficacy, and job satisfaction is 49.93%.

After the inner model was evaluated with R-square, the the inner model was evaluated by

Table 4.11.   Composite reliability

Construct (variable)	Composite reliability value	Cronbach Alp ha value
Creative Self-Efficacy	0.8623	0.7859
Job Satisfaction	0.826	0.7038
Individual Creativity	0.8813	0.8244
POS for Creativity	0.892	0.8485

Source: PLS

Table 4.12.   R-square

Construct (variable)	R-square
Job satisfaction	0.6379
Individual creativity	0.4993

Source: PLS

looking at the significance of the influence between constructs using, the t-test of path coefficient. The influence of the path between these variables is considered significant at a significance level of a 5% that has t-statistics over 1.96.

Table 4.13 Path coefficient (Mean, STDEV, STERR, t-values)

Original Mean Standard Standard T-stats Sample Sample Deviation Error (| O (O) (M) (STDEV) (STERR) STERR|)

Based on the above path coefficient table, creative self-efficacy has a significant influence on job satisfaction at the level of a 5%, with t-statistics of 9.4722 and original sample 0.6086. Creative self-efficacy against individual creativity showed a significant influence with t-statistics of 2.0286 and original sample 0.3065. POS for creativity to job satisfaction showed a significant influence with t-statistics of 3.3317 and original sample 0.2432. In addition, POS for creativity for individual creativity also showed a significant effect with t-statistic of 3.9935 and original sample 0.4763. The variable job satisfaction had no significant effect on individual creativity because it has t-statistics below 1.96, with t-statistics of 0.2429 and original sample 0.0247.

The result of structural model analysis or inner model with PLS method can be seen more clearly in Figure 4.4.

4.4 Hypothesis test

To test the proposed hypothesis, we examined the value of t-statistics. The limit to reject and accept the proposed hypothesis is ± 1.96, in which if the value of t is in the range of values -1.96 to 1.96 then the hypothesis is rejected. The result of t-statistic estimation can be seen in the path coefficient (t-statistic) in Table 4.13.

4.4.1 Hypothesis 1 (H1)

The first hypothesis (H1) states that POS for creativity has a significant influence on job satisfaction in members of non-profit organizations

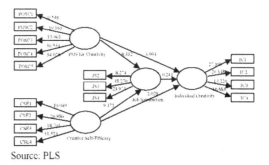

Source: PLS

Figure 4.4. Inner model analysis

Source: PLS

809

engaged in domestic stray animal care in Surabaya and Sidoarjo. The result of path coefficient between POS for creativity and job satisfaction shows that the t-statistic has positive value of 3.3317 > with a t-table value of 1.96, and indicates that POS for the creativity variable has significant effect on job satisfaction. Thus, the first hypothesis (H1) in this study is accepted.

4.4.2 *Hypothesis 2 (H2)*
The second hypothesis (H2) states that creative self-efficacy has a significant influence on job satisfaction in members of non-profit organizations engaged in domestic stray animal care in Surabaya and Sidoarjo. The result of the path coefficient between creative self-efficacy and job satisfaction shows that the t-statistic has a positive value of 9.4722 > t-table value of 1.96, which shows that the creative self-efficacy variable has significant effect on job satisfaction. Thus, the second hypothesis (H2) in this study is accepted.

4.4.3 *Hypothesis 3 (H3)*
The third hypothesis (H3) states that job satisfaction has a significant influence on the individual creativity in members of non-profit organizations engaged in domestic stray animal care in Surabaya and Sidoarjo. The result of the path coefficient between job satisfaction and individual creativity shows that the t-statistic has a positive value of 0.2429 < t-table value of 1.96, which shows that the variable of POS for creativity has no significant effect on job satisfaction. Thus, the third hypothesis (H3) in this study is rejected.

4.4.4 *Hypothesis 4 (H4)*
The fourth hypothesis (H4) states that POS for creativity has a significant influence in members of non-profit organizations engaged in domestic stray animal care in Surabaya and Sidoarjo. The result of path coefficient calculation between POS for creativity and individual creativity shows that the value of t-statistic has a positive value of 3.9935 > t-table value of 1,96, which shows that the variable of POS for creativity has a significant effect to individual creativity. Thus, the fourth hypothesis (H4) in this study is accepted.

Input:	Test statistic:	p-value:
fa 9.4722	Sobel test: 0.24282018	0.8081447
fb 0.2429	Aroian test: 0.24147909	0.80918382
	Goodman test: 0.24418386	0.80708843
	Reset all	Calculate

Source: Sobel test calculation

Figure 4.5. Sobel test result calculation of creative self-efficacy on individual creativity with job satisfaction as intervening variable

Source: Sobel test Calculation

4.4.5 *Hypothesis 5 (H5)*
The fifth hypothesis (H5) states that creative self-efficacy has a significant influence in members of non-profit organizations engaged in domestic stray animal care in Surabaya and Sidoarjo. The result of path coefficient between creative self-efficacy and individual creativity shows that the value of t-statistic has a positive value of 2.0286 > t-table value of 1.96, which shows that creative self-efficacy variable has a significant influence on creativity individual. Thus, the fifth hypothesis (H5) in this study is accepted.

4.4.6 *Hypothesis 6 (H6)*
The sixth hypothesis (H6) states that POS for creativity has a significant influence on individual creativity with job satisfaction as an intervening variable in members of non-profit organizations engaged in domestic stray animal care in Surabaya and Sidoarjo. A Sobel test can be done by looking at the t-statistic. If the t-statistic result from the Sobel test is above 1.96 then the hypothesis accepted, but if the t-statistic value is below 1.96 then the hypothesis is rejected.

The results showed that POS for creativity (X1) had significant effect on job satisfaction (Z) and individual creativity (Y), but job satisfaction (Z) had no significant effect on individual creativity (Y), so that job satisfaction (Z) as intervening variable has no significant effect on individual creativity (Y). In addition, the Sobel test calculation also showed insignificant results between POS for creativity on individual creativity with job satisfaction as intervening variable

This means that the sixth hypothesis is rejected, which states that POS for creativity has a significant effect on individual creativity with job satisfaction as the intervening variable in members of non-profit organizations engaged in domestic stray animal care in Surabaya and Sidoarjo.

4.4.7 *Hypothesis 7 (H7)*
The seventh hypothesis (H7) suggests that creative self-efficacy has a significant influence on individual creativity with job satisfaction as an intervening variable in members of non-profit organizations engaged in domestic stray animal care in Surabaya and Sidoarjo. Referring to Baron and Kenny (1986), in Frazier et al. (2004), the results show that creative self-efficacy (X2) has a significant effect on job satisfaction (Z) and individual creativity (Y), but job satisfaction (Z) has no significant effect on individual creativity (Y), so job satisfaction (Z) as the intervening variable has no significant effect on individual creativity (Y).

In addition, the Sobel test calculation also shows insignificant results between creative self-efficacy toward individual creativity and job satisfaction as intervening variable, with statistic of 0.24 < t-table 1.96. The test results are presented in Figure 4.5

This means that the seventh hypothesis is rejected, which states that creative self-efficacy does not have a significant influence on individual creativity with job satisfaction as intervening variable in in members of non-profit organizations engaged in domestic stray animal care in Surabaya and Sidoarjo.

5 CONCLUSION

The aim of this study is to see how POS for creativity, creative self-efficacy, job satisfaction, and individual creativity affect each other in 50 members from five non-profit organizations engaged in domestic stray animal care in Surabaya and Sidoarjo. There are seven hypotheses to this study: POS for creativity and creative self-efficacy has a significant influence on the work satisfaction, while job satisfaction. POS for creativity, and creative self-efficacy has significant influence on the individual creativity. This study also hypothesizes that there is a significant influence of both POS for creativity and self-efficacy on individual creativity with job satisfaction as an intervening variable. Using PLS analyses method, the result of this study proved that POS for creativity has a significant influence on job satisfaction, creative self-efficacy has significant influence on job satisfaction, POS for creativity has a significant influence on individual creativity, and creative self-efficacy has a significant influence on individual creativity. However, job satisfaction has no significant effect on individual creativity, and POS for creativity toward individual creativity with job satisfaction as intervening variable, and creative self-efficacy toward individual creativity with job satisfaction as intervening variable. Non-profit organizations engaged in domestic stray animal care in Surabaya and Sidoarjo can use this study as an input in decision-making related to quality development and organizational development to increase job satisfaction and creativity

REFERENCES

Anggarwati, Adita. 2014. *Pengaruh Creative Self-Efficacy Ter- hadap Kreativitas Dengan Kepuasan Kerja Sebagai Var- iabel Intervening Pada PT. Smile Island Surabaya.* Surabaya: Fakultas Ekonomi dan Bisms Universitas Airlangga.

Amabile, T.M. 1997. Motivating Creativity in Organization: On Doing What You Love and Loving What You Do. *California Management Review* 40, 11.

Aschenbrener, M. *et al.* 2007. Assessment of Creativity and Job Satisfaction of Second Year Agriculture Education Teachers. *Proceedings of the 2007 AAAE Research Conference*, 34.

Bandura, A. 1997. *Self Efficacy: The Exercise of Control.* New York: W.H. Freeman.

Diliello, T.C., *et al.* 2011. Narrowing the Creativity Gap: The Moderating Effects of Perceived Support for Creativity. *The Journal of Psychology* 145, 3, 151–172.

Emerson, D.J. 2013. *Organizational Culture, Job Satisfaction and Turnover Intentions: The Mediating Role of Perceived Organizational Support.* Richmond: Virginia Commonwealth University.

Ghafoor, A. *et al.* 2011. Mediating Role of Creative Self Efficacy. *African Journal of Business Management* 5, 27, 11093–11103.

Ghozali, I. 2006. *Structural Equation Modelling: Metode Alternatif dengan Partial Least Square (PLS).* Semarang: Badan Penerbit UNDIP.

Ghozali, I. 2008. *Structural Equation Modelling: Metode Alternatif dengan Partial Least Square (PLS).* Semarang: Badan Penerbit UNDIP.

Ghozali, I. 2012. *Aplikasi Analisis Multivariate Dengan Program SPSS.* Semarang: Badan Penerbit UNDIP.

Houghton, J.D & Trudy C. DiLiello. 2009. Leadership Development: The Key to Unlocking Individual Creativity in Organizations. *Leadership & Organization Development Journal* 31, 3.

Karatepe, Osman M, et al. 2006. The Effects of Selected Individual Characteristics on Frontline Employee Performance and Job Satisfaction. *Tourism Management* 27, 547–560.

Mardhatillah, Shariati. 2014. *Perngaruh Perceived for Organizational Support for Creativity Terhadap kreativitas dengan Kepuasan Kerja sebagai Variabel Intervening pada PT. Sirkulasi Kompas Gramedia (SKG) Surabaya.* Surabaya: Fakultas Ekonomi dan Bisms Universitas Air- langga.

Novellea, R. 2014. *Pengaruh Perceived of Organizational Support for Creativity dan Creative Self-Efficacy Terhadap kreativitas Individual pada Penyiar Radio di Surabaya.* Surabaya: Fakultas Ekonomi dan Bisnis Universitas Airlangga.

Oldham, G.R. & Cummings, A. 2007. Employee Creativity: Personal and Contextual Factors at Work. *The Academy of Management Journal* 39, 607–634.

Robbins, S.P & Judge, T.A. 2008. *Organization Behaviour.* (Diana Angelica, trans). 2008. Jakarta: Salemba Empat.

Sugiyono. 2006. *Metode Penelitian Kuantitatif, Kualitatif, dan R&D.* Bandung: Ganesha.

Sugiyono. 2008. *Metode Penelitian Bisnis.* Bandung: Alfabeta.

Sugiyono. 2010. *Metode Penelitian Bisnis.* Bandung: Alfabeta.

Sugiyono. 2012. *Metode Penelitian Kuantitatif, Kualitatif, dan R&D.* Bandung: Alfabeta.

Tierney, P. & Farmer, S.M. 2011. Creative Self-Efficacy Development and Creative Performance Over Time. *Journal of Applied Psychology* 96, 277–293.

Zhou, K.Z. 2006. Innovation, Imitation, and New Product Performance: the case of China. *Industrial Marketing Management* 35, 3, 394–402.

Zorlu, K. 2011. The Perception of Self-esteem and Self-efficacy as Transforming Factors in the Sources of Role Stress and Job Satisfaction Relationship of Employees: A Trial of a Staged Model Based on the Artificial Neural Network Method. *Africa Journal of Business Management* 6, 8, 3014–3025.

Advances in Business, Management and Entrepreneurship – Hurriyati et al (eds)
© *2020 Taylor & Francis Group, London, ISBN 978-0-367-27176-3*

Motivation and leadership on the performance of private higher education lecturers

N.A. Hamdani & G.A.F. Maulani
Universitas Garut, Garut, Indonesia

ABSTRACT: The lecturer has the main task to transform, develop and disseminate his knowledge through education, research and community service. At this time, lecturers only focus on educational activities, while research activities and community service have not been optimal. The purpose of this study is to examine the extent of the role of achievement motivation and leadership on the performance of lecturers. The method used in this research is quantitative, with sample of all lecturers spread in Private Higher Education of East Priangan, with proportional random sampling by using path analysis. The results of research show that achievement motivation with some indicators affect on the performance of lecturers. This shows that the role of achievement motivation, friendship and socializing is very important to improve the performance of lecturers. The study also reveals that lecturers have not been optimal in conducting research and community service among others.

1 INTRODUCTION

The improvement and empowerment of the quality of human resources is an indispensable requirement to adjust to the future changes. One of the key elements in the human resource quality improvement and empowerment is education. Through education, human resources are expected to become the development driving force (Burma, 2014). Furthermore, Abu Teir (2016) asserts the importance of human resources in higher education to bring about competitiveness.

The concept of motivation is important in individual performance studies. A highly motivated individual is one who carries out efforts for the sake of the purpose of his work unit in the organization. In contrast, a demotivated individual will not give his all when performing his duties (Mathew, Faith and Edward, 2016). Motivation deals with self-drive that cannot be observed. It drives an individual to work hard to achieve a particular objective. Some studies suggest that motivation can improve employee performance (for example State and Victor, 2014).

A leader is responsible for the improvement of his subordinates' motivation in his organization in order to improve their performance (Fisher, 2009) and (Alimo-metcalfe *et al.*, 2008). The main problem is that how the leader figures out a way to motivate his employees to work towards the goals and improve their performance (Fisher, 2009; Chipunza and Matsumunyane, 2008). He is also responsible for creating a good organizational climate so that the employees can work effectively (Zhang, 2010; Bryman, 2007). Some studies suggest that organizational climate can improve employee performance. For example, Permarupan, Mamun and Ahmad (2013) explains that the organizational climate with dimensions of flexibility,

rewards, standards, clarity, responsibility and team commitment has an impact on employee performance.

Competition in education, especially in Eastern Priangan (Garut Regency, Tasik Regency, Tasik City, Ciamis Regency, and Banjar Regency) is marked by the proliferation of private higher education institutions. There are 30 higher education institutions in Eastern Priangan, each with its own competitiveness (Ministry of Higher Education, 2018). However, they face a very similar problem when it comes to their lecturers' performance: educational, research, and community service performance. A good education program is to produce competent output that is acceptable in the labor market, and lecturers' performance is one of the indicators for the improvement of the output quality (Xiao and Wilkins, 2015; Chen, 2015).

Starting from a preliminary literature review and observation, to see a more realistic situation, the author is interested in studying the influence of motivation and organizational climate on employee performance at the Universitas Garut despite the fact that lecturers are not the only influential factor worth scrutiny. As a case in point, lecturer's performance can be influenced by the lecturer in question per se, students, and organizational climate, culture, rules, curriculum, leadership, communication, and so on. However, due to the time limitation, all of these contributing factors cannot be studied. The author only focuses on what is deemed significant.

2 LITERATURE REVIEW

Motivation is of internal and external. Motivation can trigger sincerity, encouragement, and change in an individual. All people have motivation and it is the basis for them in performing their activities.

Motivation is a sincerity to go all out to achieve organizational objectives (Robbins, Stephen P and Judge, 2013). Motivation can influence organizational performance (Joseph, 2015) and its employees (Shahzadi, 2014; Tella, 2007). Furthermore, Luthans (2011) points out that a motivated individual is one who puts a lot of effort into productive objectives of his work unit and organization where he works. Motivation is crucial for employee performance.

It can be concluded that motivation is a potential force that exists in an individual; this individual can improve his motivation by himself or with help of a number of external factors, both monetary and non-monetary rewards. The motivation can influence his performance.

According to McClelland (1987), learning from a culture of a society, there are many types of human needs such as need for achievement (n Ach), need for affiliation (n-aff) and need for power (n-pow). McClelland further suggests that the stronger the need, the higher the motivation for one to put effort into the fulfillment of that need. For example, when one has a strong need for achievement, he will be motivated to fulfill that need. These needs are not the same from one person to another. Some may have a strong need for affiliation (n-Aff) but low need for achievement (n-Ach) or low need for power (n-Pow) (Mcclelland et al., 1980).

Organizational climate refers to "an environment within an organization in which the employees work" (Davis, 1996) or a set of working environment that is deemed the main influential element. The quality of an organizational climate is on a continuum. This could be very good, good, poor, and very poor. The organizational climate, good and poor, can really affect employee motivation, performance and satisfaction. Some studies (for example Shaminah et al., 2013; Sokol et al., 2015), reveal that organizational climate in higher education can improve lecturers' creativity. According to Davis (1996), climate dimension consists of seven factors: leadership, motivation, communication, interaction, decision making, objective formulation, and control. This dimension has been discussed by several previous studies; for example, Yang (2015) who studies the organizational climate in higher education and Zhang (2010) who investigates the effect of organizational climate on the whole organization performance including its human resources.

Performance has many dimensions; each one of them comes with its own significance. One dimension is not more significant than another. Therefore, in the process of performance measurement, each one of these dimensions is measured and treated equally. Performance dimensions can be different from one to another job, depending on job description; however, generally performance, as Mitchell (1987) puts it, has five dimensions: quality of work, promptness, initiative, capability, and communication. This is in line with what has been discussed by Schwartz and Te, 1998; Allan, Clarke and Jopling, 2009).

3 RESEARCH METHOD

This study used an explanatory research design, in which all variables were explained through a hypothesis testing. The questionnaire was distributed to 95 respondents, selected using a simple random sampling technique. The questionnaire was measured using validity, reliability, and runs tests. Data were analyzed using a path analysis. This analysis technique was chosen because its characteristics are suitable for analyzing the effect of independent variables on the dependent variables. The motivation variable consists of the following driving motivators: need for achievement ($X11$), need for affiliation ($X12$), need for power ($X3$), and organizational climate variable consists of motivation ($X21$), control ($X22$), communication ($X23$), leadership ($X24$), decision making ($X25$), objective formulation ($X26$) and interaction ($X27$). The performance variable consists of quality of work, ability, initiative, communication and timeliness. These five dimensions are then formed into one performance factor variable using the Summated Rating method.

4 RESULT AND DISCUSSION

All questionnaire items of motivation variable, organizational climate variable, and lecturers' performance variable are said to be valid if their correlation coefficient equals or is higher than 0.30.

Reliability testing is done by looking at the consistency and stability of the measuring instrument. The reliability coefficient was calculated using Cronbach's alpha. The reliability coefficient of each variable is presented in Table 1.

The measuring instrument can be said reliable if all items' reliability coefficient is higher than 0.70. The answer to each item is measured using

Table 1. Reliability coefficient.

Variable	Reliability Coefficient Alpha	Note (reliable if $\alpha > 0.7$)
Need for Achievement ($X_{1.1}$)	0.7549	Reliable
Need for Affiliation ($X_{1.2}$)	0.9081	Reliable
Need for Power ($X_{1.3}$)	0.7219	Reliable
Motivation ($X_{2.1}$)	0.7629	Reliable
Control ($X_{2.2}$)	0.8844	Reliable
Communication ($X_{2.3}$)	0.7943	Reliable
Leadership ($X_{2.4}$)	0.8880	Reliable
Decision Making ($X_{2.5}$)	0.8912	Reliable
Objective Formulation ($X_{2.6}$)	0.8830	Reliable
Interaction ($X_{2.7}$)	0.7415	Reliable
Lecturers' Performance (Y)	0,8334	Reliable

a runs-test. The result is presented in the Table 1. The table shows that the scores of all answers to questionnaire items of all variables, motivation ($X1$), organizational climate (X_2), and lecturers' performance (Y) are randomly distributed as the p-value (Asymp. Sig) is higher than α 0.05.

As only need for achievement, need for affiliation, need for power, motivation, control, communication, and leadership factors showed significant influence on lecturer's performance, a trimming process was carried out. This is a recalculation of influence of factors whose hypothesis testing result was significant by excluding non-influential factors.

Based on the test results on the influence of motivation and organizational climate on the performance of the lecturers, a statistical conclusion was drawn that the factors with significant influence included need for achievement ($X_{1,1}$), need for affiliation ($X_{1,2}$), need for power ($X_{1,3}$), motivation ($X_{2,1}$), control ($X_{2,2}$), communication ($X_{2,3}$), and leadership ($X_{2,4}$).

Based on the test results on the influence of motivation and organizational climate on the performance of the lecturers, a statistical conclusion was drawn that the factors with significant influence included need for achievement (X1,1), need for affiliation (X1,2), need for power (X1,3), motivation (X2,1), control (X2,2), communication (X2,3), and leadership (X2,4).

The influence of need for achievement (X1,1), need for affiliation (X1,2), need for power (X1,3),

motivation (X2,1), control (X2,2), communication (X2,3), and leadership (X2,4) simultaneously on the performance of lecturers is presented in Table 2:

Based on Table 2, need for achievement (X1,1), need for affiliation (X1,2), need for power (X1,3), motivation (X2,1), control (X2,2), communication (X2,3), and leadership (X2,4) have simultaneous influence on the performance of lecturers as much as 80.17%, and the influence of factors other than these seven factors were 19.83%.

The influence of need for achievement (X1,1), need for affiliation (X1,2), need for power (X1,3), motivation (X2,1), control (X2,2), communication (X2,3), and leadership (X2,4) asynchronously on the performance of lecturers is presented in Table 3:

Based on the above table, the need-for-achievement factor (X1,1) positively influences the performance of lecturers as much as 8.54%. The stronger the need for achievement is, the better the performance of lecturers is. This influence shows that most lecturers like challenges and are accountable at work. It can be concluded that most of the lecturers always perform their duties very well in accordance with organizational rules.

The need-for-affiliation factor ($X_{1,2}$) positively influences the performance of lecturers as much as 9.15%. The stronger the need for affiliation is, the better the performance of lecturers is. Among McClelland's three driving motivators, this factor contributes the most to the performance of lecturers. The author deems that most of the lecturers are interactive, especially when explaining and identifying the students' needs.

The need-for-power factor ($X_{1,3}$) positively influences the performance of lecturers as much as 7.48

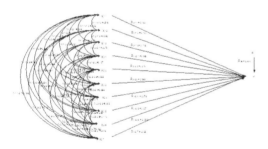

Figure 1. Result of path coefficient analysis of motivation and organizational climate.

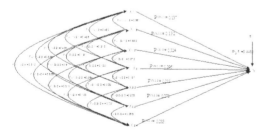

Figure 2. Result of path coefficient analysis of motivation and organizational climate (after trimming process).

Table 2. Simultaneous influence and other influence.

Types of Influence	Contribution (%)
Simultaneous Influence ($X_{1.1}$, $X_{1.2}$, $X_{1.3}$, $X_{2.1}$, $X_{2.2}$, $X_{2.3}$, $X_{2.4}$) on Y (R^2)	80.17
Other influence on Y ($P^2_{Y\epsilon}$)	19.83

Table 3. Influence of each factor.

Driving Motivators and Organizational Climate	Total Direct and Indirect Effect on Y
Need for Achievement ($X_{1,1}$)	8.54 %
Need for Affiliation ($X_{1,2}$),	9.15 %
Need for Power ($X_{1,3}$)	7.48 %
Motivation ($X_{2,1}$)	14.61 %
Control ($X_{2,2}$)	9.11 %
Communication ($X_{2,3}$)	15.09 %
Leadership ($X_{2,4}$)	16.19 %

%. The stronger the need for power is, the better the performance of lecturers is. It goes to show that lecturers in Eastern Priangan are: 1) very active in determining the purpose of organizational activities, 2) very sensitive to the structure of interpersonal influence in groups or within organization, 3) care about prestige, and 4) like to help others even if they are not asked to.

As for organizational climate factor, motivation $(X_{2,1})$ positively influences the performance of lecturers as much as 14.61%. The higher the motivation is, the better the performance of lecturers is. Motivation in this case is different from McClelland's motivation. It deals with direct motivation (supports from superiors and fellow lecturers) and indirect motivation that is associated with remuneration, job description, and infrastructure. In this case, supports from superiors and fellow lecturers are very motivating. The superiors in this case refer to Head of Department and Dean of Faculty. Infrastructure is also important for the instructional process, but remuneration is not too significant.

5 LIMITATION OF THE RESEARCH

This study has some methodological limitations. The measurement was carried out by questionnaires whose respondents also served as the measurement objects. In other words, there is very high potential for bias in self-assessment.

6 CONCLUSION

1. Need for achievement (X1,1), need for affiliation (X1,2), need for power (X1,3), motivation (X2,1), control (X2,2), communication (X2,3), and leadership (X2,4) have simultaneous influence on the performance of lecturers as much as 80.17%, and the influence of factors other than these seven factors were 19.83%.
2. Based on the data analysis result, some factors in the organizational climate variable do not have influence on performance. These factors include decision making $(X_{2,5})$, objective formulation $(X_{2,6})$, and interaction $(X_{2,7})$. Rensis Likert's theory on the influence of organizational climate is not perfect and not applicable to all types of organization.
3. From the highest to the lowest, the driving motivators that influence performance can be ordered as (1) need for affiliation, (2) need for achievement, and (3) and need for power. Using the same logic, the order of factors in the organizational climate variable is (1) leadership, (2) communication, (3) motivation, and (4) control.
4. The order in point 3 may be different in different organizations, so this conclusion cannot be generalized. Even so, the management needs to take into

account of development/strategic priorities based on the motivational and organizational climate factors in relation to their influence on performance.

REFERENCES

Abu Teir, R. and Z. Q. (2016) 'Journal of Human Resources Management and Labor Studies', *Journal of Human Resources Management and Labor Studies*, 4(1), pp. 65–83. doi: 10.15640/jhrmls.v4n1a3.

Alimo-metcalfe, B., Alban-metcalfe, J., Bradley, M., Samele, C. and Alimo-metcalfe, B. (2008) 'The impact of engaging leadership on performance, attitudes to work and wellbeing at work A longitudinal study', *Journal of Health Organization and Management*, 22(6), pp. 586–598. doi: 10.1108/14777260810916560.

Allan, J., Clarke, K. and Jopling, M. (2009) 'Effective Teaching in Higher Education : Perceptions of First Year Undergraduate Students', *International Journal of Teaching and Learning in Higher Education*, 21(3), pp. 362–372.

Bryman, A. (2007) 'Effective leadership in higher education : a literature review', *Studies in Higher Education*, 32(6), pp. 693–710. doi: 10.1080/03075070701685114.

Burma, Z. A. (2014) 'Human Resource Management and Its Importance for Today ' s Organizations', *International Journal of Education and Social Sciences*, 1(2), pp. 85–94.

Chen, C. Y. (2015) 'A Study Showing Research has been Valued over Teaching in Higher Education', *Journal of the Scholarship of Teaching and Learning*, 15(3), pp. 15–32.

Chipunza, C. and Matsumunyane, L. L. (2008) 'Motivation sources and leadership styles among middle managers at a South African university', *Administration in Social Work ISSN:2071-078X*, 3, pp. 1–13.

Davis, K. and N. J. (1996) *Organizational Behavior*. New York, USA: Mc-Graw Hill, International Book Company.

Fisher, E. A. (2009) 'Administration in Social Work Motivation and Leadership in Social Work Management : A Review of Theories and Related Studies Motivation and Leadership in Social Work Management : A Review of Theories and Related Studies', *Administration in Social Work ISSN:*, 3107(33), pp. 347–367. doi: 10.1080/03643100902769160.

Joseph, B. (2015) 'The effect of employees ' motivation on organizational performance', *Journal of Public Administration and Policy Research*, 7(May), pp. 62–75. doi: 10.5897/JPAPR2014.0300.

Luthans, F. (2011) *Organizational Behavior*. 12th edn. New York, USA: McGraw-Hill.

Mathew, S., Faith, K. and Edward, G. (2016) 'Motivational Issues for Lecturers in Tertiary Institutions : A Case of Bulawayo Polytechnic', *International Journal of Scientific and research Publication*, 6(4), pp. 167–175.

McClelland, D. C. (1987) *Human Motivation*. New York, USA: Cambridge University Press.

Mcclelland, D. C., Davidson, R., Saron, C. and Floor, E. (1980) 'THE NEED FOR POWER, BRAIN NOREPLNEPHRINE TURNOVER AND LEARNING David', *Biolpgical Psychology*, 10(2), pp. 93–102.

Ministry of Higher Education (2018) *Higher Education Data Base*. Indonesia.

Permarupan, P. Y., Mamun, A. A.- and Ahmad, R. (2013) 'Organizational Climate on Employees ' Work Passion : A Review', *Canadian Social Science*, 9(4), pp. 63–68. doi: 10.3968/j.css.1923669720130904.2612.

Robbins, Stephen P and Judge, T. (2013) *Organizational Behaviour*. 15th Editi. New Jersey, USA: Pearson.

Schwartz, D. G. and Te, D. (1998) 'Intelligent agent behavior based on organizational image theory', *Intelligent agent behavior*, 2(30), pp. 166–178.

Shahzadi, I. (2014) 'Impact of Employee Motivation on Employee Performance', *European Journal of Business and Management*, 6(23), pp. 159–167.

Shaminah, N., Kamalu, M., Education, F., Mara, U. T. and Alam, S. (2013) 'THE IMPACT OF ORGANIZATIONAL CLIMATE ON TEACHERS ' JOB PERFORMANCE Nurharani Selamat Nur Zahira Samsu', *Educational Research*, 2(1), pp. 71–82. doi: 10.5838/erej.2013.21.06.

Sokol, A., Gozdek, A., Figurska, I. and Blaskova, M. (2015) 'Organizational climate of higher education institutions and its implications for the development of creativity', *Procedia - Social and Behavioral Sciences*. Elsevier B.V., 182, pp. 279–288. doi: 10.1016/j.sbspro.2015.04.767.

State, O. and Victor, A. A. (2014) 'Motivation and Effective Performance of Academic Staff in Higher Education (Case Study of Adekunle Ajasin', *International Journal Of Innovation and resesrach in Educational Sciences*, 1 (2), pp. 157–163.

Tella, A. (2007) 'Work Motivation, Job Satisfaction, and Organisational Commitment of Library Personnel in Academic and Research Libraries in Oyo State, Nigeria Work Motivation, Job Satisfaction, and Organisational Commitment of Library Personnel in Academic and Resea', *Library Philosophy and Practice*, 1(April).

Xiao, J. and Wilkins, S. (2015) 'The effects of lecturer commitment on student perceptions of teaching quality and student satisfaction in Chinese higher education', *Journal of Higher Education Policy and Managemen*, 37(1), pp. 98–110.

Yang, C. (2015) 'The Effects of Higher Education ' s Institutional Organizational Climate on Performance Satisfaction : Perceptions of University Faculty in Taiwan', *International Business Research*, 8(8), pp. 103–117. doi: 10.5539/ibr.v8n8p103.

Zhang, J. (2010) 'Organizational Climate and its Effects on Organizational Variables : An Empirical Study', *International Journal of Psychological Studies*, 2(2), pp. 189–201.

Advances in Business, Management and Entrepreneurship – Hurriyati et al (eds)
© *2020 Taylor & Francis Group, London, ISBN 978-0-367-27176-3*

Planning for the development of village-owned enterprises (BUMDes)

N.A. Hamdani & D. Yudiardi
Universitas Garut, Garut, Indonesia

ABSTRACT: This research is based on the planning issues regarding the development of Village-Owned Enterprise (BUMDes) in Garut District, the research is aimed to find new concepts regarding the development plans for the development of BUMDes in Garut District. This research is derived from the proposition by the Department of Social and Village Empowerment which stated that the BUMDes development could increase village economics through the mobilization of village potentials and resources, as well as the external factors supporting the villages. The problem in this research is the fact that the creation of BUMDes has not been utilized as a tool for village development with the end goal of bringing prosperity to the people. Based on the characteristics of the problems analyzed, this research is done using qualitative method, data collected by observation, in-depth interview to DPMD chief, secretary, chief of society economic empowerment, village head, and BUMDes director. In testing the data validity, the writer used triangulation technique, the researcher analyzed the problem through planning approach by Stoner' theories (2006). Based on the researched, BUMDes development plans by Garut DPMD has to be able to identify supporting and inhibiting factors.

Keywords: BUMDes development planning, Dinas Pemberdayaan Masyarakat Desa (DPMD), Garut District

1 INTRODUCTION

Strategic planning is very important to be implemented to improve corporate performance (Haythem, 2015). It is also critical for the development of strategic management (Furrer and Thomas, 2008; Wolf and Floyd, 2013). The government is trying to develop the Indonesian economy by encouraging the establishment of village-owned enterprises (BUMDeses). Srirejeki (2018) points out that the existence of village-owned enterprises is very important for the economic improvement of rural communities. Likewise, according to Lin and Meulder (2012), village owned enterprise may reduce the urbanization wave.

The Regency of Garut has many comparative advantages thanks to its relatively fertile soil suitable for farming various categories of crops, vegetables and rice. However, the growth rate of the agricultural category is still not optimal yet because the linkages between agricultural subsystems and other categories are not yet fully synergized. This is reflected in the development of agro-industries that have not been optimal both in processing and marketing. In the period of 2011-2015, along with economic growth and increased purchasing power, Indonesian Bureau of Statistics (BPS) notes that the rate of macro poverty in Garut Regency has decreased significantly, especially in 2015. In 2014, Garut's population

living below the poverty line is estimated to be approximately 410,564 people, a decrease of 24,896 people from the previous year. Thus, the poverty rate in Garut in 2014 is 17.51 percent, a decrease of 1.34 percent from the previous year. The poverty number in Garut experiences another decline of 4,728 people or 0.41 percent in 2015. The total number of poor people in Garut in 2015 is 330,905 or 13.53 percent of its population. This number; however, is still deemed high taking account of the development goals of the Regency of Garut (Goverment, 2016).

Strengthening the economy through BUMDes is one strategy to improve the community welfare by exploring the rural potential. BUMDes are owned and managed by village community and government for the benefit of their own. BUMDes are established in accordance with the needs and potentials of each village. BUMDes are projected to emerge as new economic powers in rural areas. (*Act Law Number 6, years 2014 about Village*, 2014) provides a legal umbrella under which BUMDes manage village potential collectively to improve their community welfare. Substantially, Law No. 6 of 2014 encourages villages to be emancipatory development subjects to provide basic services to their communities, including managing local economic assets. BUMDes become economic centers in villages with a collective economic spirit. Schianetz, Kavanagh and Lockington (2007) suggest that BUMDes can be the driving forces of the regional tourism sector. Furthermore, according to Setyobakti (2017), BUMDes can become economic and social driving forces. Meanwhile, according to Chen (2005), privatization of village-owned enterprises can contribute to the rural economy. This study aims to design strategic planning for the development of BUMDes to improve the rural economy.

2 GETTING STARTED

One of Indonesia's main problems is the wave of urbanization triggered by several factors such as economic factors (Douglass, Tacoli and Earthscan, 2006). Rural-based economic development is very important to overcome urbanization-related problems. Some studies, for example Hamdani, Abdul and Maulani (2018) and Hamdani (2018), suggest that SMEs can improve economy. Meanwhile Karel, Adam and Radomír (2013) point out that strategic planning is necessary to improve performance. Therefore, BUMDes should have the principles of Village Business Unit, either in the form of SME or joint venture.

Strategic planning is part of strategic management (Ibraimi, 2014; Elimimian, 2013). Some studies, for example Sosiawani et al., (2015), suggest that strategic planning can improve corporate performance. In his study, Ramli describes several dimensions that can be objects of strategic planning research including (1) objective identification, (2) employer participation, (3) time span, (4) implementation), and (5) control. Meanwhile, Simcic and Br (2002) discuss strategic planning as an issue of strategic orientation.

One of important aspects in planning is decision making, which is a process of taking actions to solve particular problems. The decision should be made at the planning process (Hourani, 2017). Managers should make an economic projection and potential rivalry in the market. They should analyze organization resources and decide how best to use them in order to achieve the goals.

Furthermore, according to James F. Stoner, there are four big steps in planning that can be used in all planning activities on all levels of organization. They are: (1) defining the objectives, (2) defining the current situation, (3) identifying supporting and inhibiting factors, and (4) developing a plan or designing a series of actions. *Step 1:* Planning begins with defining what to achieve or what is needed by the organization or a subunit. Without a clear definition, the organization would waste too many resources. By setting priorities and defining objectives, an organization will put their resources into use effectively. *Step 2:* how far is the organization or its subunit from their objectives? What resources are available to achieve the objectives? Only after analyzing the current situation and problems, a plan could be made to project what to achieve. An open communication channel within the organization and

among its subunits will provide information, especially financial and statistical data needed for this second step. *Step 3:* What factors in the external and inner environments can help the organization achieve its goals? What factors may cause problems? It is easy to see what is happening now, but the future is never clear. However, anticipating situations, problems and opportunities in the future is an important part of planning. *Step 4:* The last step in the planning process deals with the designing of alternatives to achieve desired goals, evaluate these alternative, and decide which one of them is the most suitable (or at least suitable enough) alternative to achieve the goal. It is a step to make decisions about future actions and is most relevant to effective decision-making guidelines.

Planning has something to do with the process of the process of selecting strategies and policies in order to maximize the goals of the organization. Strategic management encompasses all activities that lead to the formulation of organizational goals, strategies and the development of plans, actions and policies to achieve these strategic goals for the organization (Hunger, 2017; Martínez-lópez, 2014; and Thompson and Martin, 2005).

3 RESEARCH METHOD

This study used a qualitative approach (Creswell, 2008). The qualitative approach was selected because it allows the researchers to collect detailed, comprehensive, valid, reliable, and relevant research data. This study was conducted at the Community Empowerment and Village Government Office (DPMPD) of Garut Regency. The informants in this study included (1) Head and staff of DPMPD, (2) DPMPD related officials, (3) Head of village, (4) BUMDes managing board, and (5) community prominent figures. Data were collected through observations, interviews, and documentations and analyzed through data reduction, triangulation, and conclusion drawing.

4 RESULT

BUMDes in Garut Regency have encountered so many problems. They have become the strengthening instrument of village autonomy that can encourage the village governments to develop the potential of their villages in accordance with the ability and authority of the villages. They are instrument of community welfare to involve the community participation in managing BUMDes to improve economy and reduce the unemployment rate in the village.

4.1 *Defining objective(s)*

Prior to planning the BUMDes development, it is necessary to define what to do in the form of activity planning to minimize risks. All activities, actions, and policies should be in line with the desired goals of the BUMDes. In planning, goals are the most fundamental and important things to defined in order for the planning to keep on track. What follows are the main goals of the BUMDes development planning by Garut's DPMPD:

1. To improve village own-source revenue (PADes).
2. To provide services to the community.
3. To provide entrepreneurship opportunity and reduce unemployment in the village.
4. To improve village community income.
5. To reduce poverty rate.

The goals of BUMDes have been determined in the existing legislation as explained by the informants: that BUMDes in Garut must become the backbone of the community economy that can fulfill their needs. The need fulfillment should not burden the community considering that BUMDes will become the most dominant enterprise to drive the village economy. BUMDes is also expected to provide services to the community by setting prices and services that apply market standards. This means that there must be a mutually agreed institutional/regulatory mechanism so as not to cause economic distortions in the village caused by businesses run by the BUMDes.

The initial objective of BUMDes is to encourage or facilitate all community income-generating activities, both those developed according to local customs and culture and economic activities handed over to the community through a program or project of the central and local government. As a village business unit, the establishment of BUMDes is actually to maximize the potentials of the village community whether it is of economic, natural resources, or human resource potentials. Specifically, the establishment of BUMDes is to open employment opportunities and improve low-income community creativity and productive economic enterprise.

The target of economic empowerment of rural communities through BUMDes is to serve the village community in developing their productive enterprises. Another aim is to provide a variety of business media to support the economy of rural communities in accordance with the villages' potentials and community needs.

Other points discussed include the recruitment process and payroll system and remuneration. The BUMDes managing board staffing can done through a consensus. But the selection should be based on certain criteria in order to make sure that the selected people are qualified to carry out their duties properly. Therefore, the requirements should be formulated by Board of Commissioners. The requirement formulation is then discussed in the village discussion forum to be later socialized and offered to the community. The next step is recruitment and staffing process based on the predetermined criteria. In addition, the remuneration should also consider how far the managing board achieves the target in a particular period and the profit rate. The managing board should be informed of this system before they accept the job in order for them to carry out their duties wholeheartedly.

4.2 Defining the current situation

In planning, planners must of course see how the current situation is and define it. The environment is dynamic and can change at any time in the blink of eye; planners should be able to assess the current situation. To this end, they should find relevant information in order for the plan suitable with the needs.

The information will be analyzed and used to help make decisions in the planning process. That is why the information should be accurate and relevant. Being accurate means that the information is real and not fabricated, and being relevant means that the information is in line with what is needed. The identification of current situation can also be done through the opening of communication channels within the organization and the provision of information made between sub-units (mainly financial data and statistical data). When the current situation has been analyzed and the required information has been obtained, the planning can be done to see further progress.

In the BUMDes development planning, the Community Economy Empowerment Agency of DPMPD assess the current state of the BUMDes, what information is needed for the BUMDes development planning, and open communication channels both within the BUMDes and with other related parties. Informants provide information needed by the Community Economy Empowerment Agency of DPMPD to design the road map of the BUMDes management.

BUMDes is a village business unit managed by community and village government to improve village economy and is established based on the village's needs and potentials.

As an economic institution operating in the village, BUMDes must have differences with economic institutions in general. This is to ensure that the existence and performance of BUMDes can contribute significantly to the community welfare and to prevent the development of rural capitalistic business system that can lead to disruption of community social values.

BUMDes according to Law No. 32 of 2004 on Regional Government is established, among others, to improve revenue. The development of BUMDes by Garut's DPMDP is one way to improve community economy, reduce poverty, and improve village own- source revenue as described in the BUMDes establishment and management manual.

BUMDes development and management should be done in accordance with well-planned and integrated steps in order to achieve the predetermined goals. BUMDes is regulated through a Regional Regulation (Perda) taking into account more superior regulations.

Through its self-help and member-based mechanism, BUMDes is also a manifestation of the participation of the village community as a whole, as opposed to rural capitalistic system. This means that the regulation is manifested in a solid institutional mechanism. Strengthening institutional capacity will lead to the existence of rules that bind all community members.

BUMDes should be managed using cooperative, participatory, emancipatory, transparency, accountable, and sustainable principles using a member-based and self-help mechanism that is run professionally and independently.

4.3 *Identifying supporting and inhibiting factors*

In planning, the planners should be able to project what the problems and possibilities will be; i.e., the inhibiting and supporting factors. That way, they can minimize ineffective plan.

The identification of supporting factors could be beneficial in the planning process. The inhibiting factors should also be identified and be dealt with. Therefore, these inhibiting factors will not hamper the planning process. The supporting and inhibiting factors can come from internal and external organization. Garut's DPMPD needs to identify these internal and external factors, i.e., the strengths, the weaknesses, opportunities, and threats in the BUMDes development planning in order to achieve the goals.

4.4 *Developing a plan*

The last step is developing a plan or taking actions to achieve the goals. At this last stage, the Community Empowerment and Village Government Agency (BPMDP) develop alternatives in identifying and solving problems. Along with Village Consultative Agency (BPD) and village government, they evaluate the progress in the BUMDes development planning.

What follows are several findings:

1. Defining Objective(s). Objectives are important in planning because it may serve as a guidelines and basis for planning. All actions taken and decisions made in planning are to achieve the desired objectives. This is what is said of what Garut's BPMDP do. However, this is not the case in reality. One of the objectives or BUMDes is to improve community efforts in managing the village's economic potentials, and what happens is that BUMDes's business lines do not have anything to do with culture and village's potentials so that BUMDes have not yet managed to improve community creativity. What is expected is: one village, one product.
2. Defining the Current Situation. In order for the planning to be effective and keep on the track, Garut's BPMPD needs to define the current

situation: the current state of BUMDes, information needed, and what BUMDes needs.
3. Identifying Supporting and Inhibiting Factors. Supporting and inhibiting factors should be identified in the BUMDes development planning. The identification of supporting factors could be beneficial in the planning process. The inhibiting factors should also be identified and be dealt with. Therefore, these inhibiting factors will not hamper the BUMDes development planning process.
4. Developing a Plan or a Series of Actions to Achieve Goals in the BUMDEs Development Planning. This refers to all actions to be taken and decisions to be made in order for the goals to be achieved. BPMDPD, BPD, and village government are expected to figure out what is best to achieve the desired goals. In this respect, BUMDes determines what future actions are to be taken for effective decision making. Strategies and alternatives will be needed to deal with all potential problems. Wrong decisions and actions may hamper the BUMDes development.

5 LIMITATION OF THE RESEARCH

The limitations of this study among others are that BUMDeses are heterogeneous – the strategic planning will likely to succeed only when applied to well-established BUMDeses – and that objective formulation and evaluation are not properly done due to the non-optimal involvement of regional government.

6 CONCLUSION

It can be concluded that in order for the BUMDes strategic planning to succeed, the following must be done: (1) defining objective(s), defining the current situation, (3) identifying supporting and inhibiting factors, and (4) developing a plan or designing a series of actions to achieve the objectives. Conceptually, BUMDes focuses on community intelligence, independence, and welfare and environmental sustainability. In practice, this is built around three pillars of good governance: community (KSM, self-help groups), government (of village or regency), and third parties (private sectors). This is to ensure that the objectives can be achieved through a democratic mechanism. With regards to the establishment and operationalization of BUMDes in Garut regency, the involvement of DPMPD in mentoring and strengthening community groups can optimize the role of KSM as the main actors, thereby accelerating the achievement of BUMDes's objectives in the context of social entrepreneurship.

REFERENCES

Act Law Number 6, years 2014 about Village (2014).

Chen, C. J. (2005) 'The Path of Chinese Privatisation: a case study of village enterprises in southern Jiagnsu', Corporate Governance, 13(1), pp. 72–80.

Creswell, J. W. (2008) Research Design: Qualitative, Quantitative and Mixed Method Approaches. Third Edit. Edited by V. Knight. London: SAGE Publications Inc.

Douglass, M., Tacoli, C. and Earthscan, R. L. (2006) A Regional Network Strategy for Reciprocal Rural-Urban Linkages: An Agenda for Policy Research with Reference to Indonesia, Third World Planning Review.

Elimimian, J. and E. N. (2013) 'Issues Management: Managerial Tools for Effective Strategic Planning and Implementation Jonathan U. Elimimian Albany State University Manager Operational Definitions of Terms Used in the Paper ':, International Journal of Humanities and Social Science, 3(15), pp. 34–39.

Furrer, O. and Thomas, H. (2008) 'The structure and evolution of the strategic management field: A content analysis of 26 years of strategic management research', 10(1), pp. 1–23. doi: 10.1111/j.1468-2370.2007.00217.x.

Goverment, G. D. (2016) Performarce Report on Garut. Garut. Available at: www.garut.go.id.

Hamdani, N. A. (2018) 'Building Knowledge Creation For Making Business Competition Atmosphere in SME of Batik', Management Science Letters, 8, pp. 667–676. doi: 10.5267/j.msl.2018.4.024.

Hamdani, N. A., Abdul, G. and Maulani, F. (2018) 'The influence of E-WOM on purchase intentions in local culinary business sector', International Journal of Engineering & Technology, 7, pp. 246–250.

Haythem, A. (2015) 'The Role of Strategic Planning in Performance Management', International Journal of Multi Disciplinary Research, 2(3).

Hourani, M. (2017) 'Conceptual Frameworks for Strategy Implementation: A Literature Review', 9(3), pp. 12–30. doi: 10.5296/jmr.v9i3.11222.

Hunger, D. and W. L. T. (2017) Strategic Management and Business Policy Toward Global Sustainability. 13th Editi. New Jersey: Pearson Prentice Hall.

Ibraimi, S. (2014) 'Strategic Planning and Performance Management: Theoretical Frameworks Analysis', International Academic Research Journal of Business and Technology, 4(4), pp. 124–149. doi: 10.6007/IJARBSS/v4-i4/789.

Karel, S., Adam, P. and Radomír, P. (2013) 'Strategic Planning and Business Performance of Micro, Small and Medium-Sized Enterprises', Journal of Competitiveness, 5(4), pp. 57–72. doi: 10.7441/joc.2013.04.04.

Lin, Y. and Meulder, B. De (2012) 'A conceptual framework for the strategic urban project approach for the sustainable redevelopment of "villages in the city" in Guangzhou', Habitat International. Elsevier Ltd, 36(3), pp. 380–387. doi: 10.1016/j.habitatint.2011.12.001.

Martínez-lópez, ancisco J. (2014) Handbook of Strategic. New York: Springer.

Schianetz, K., Kavanagh, L. and Lockington, D. (2007) 'The Learning Tourism Destination: The potential of a learning organisation approach for improving the sustainability of tourism destinations', Forthcoming in Tourism, 1(12). doi: 10.1016/j.tourman.2007.01.012.

Setyobakti, H. O. B. E. B. (2017) 'Identification of Business Enterprises BUMDES Bades on Social and Economic Aspect', Jurnal Ilmiah Bidang Akuntansi dan Manajemen, 14(2), pp. 101–110.

Simcic, P. and Br, C. (2002) 'Issues management as a basis for strategic orientation', Journal of Public Affairs, 2(4), pp. 247–258.

Sosiawani, I., Ramli, A. Bin, Mustafa, M. Bin, Zein, R. and Yusoff, B. (2015) 'Strategic Planning and Firm Performance: A Proposed Framework', International Academic Research Journal of Business and Technology, 1(2), pp. 201–207.

Srirejeki, K. (2018) 'Empowering the role of village owned enterprises (BUMDes) for rural development: case of Indonesia', Journal of Accounting, Management, and Economics, 20(1), pp. 5–10.

Thompson, J. and Martin, F. (2005) Strategic Management. 5Th Editio. New York: Mc Graw Hil.

Wolf, C. and Floyd, S. W. (2013) 'Strategic Planning Research: Toward a Theory- Driven Agenda', Journal of Management, 20(May 2012), pp. 1–35. doi: 10.1177/0149206313478185.

The influence of staffing on work performance: A study on employees in the production division of PT Samick Indonesia

P. Purnamasari & E. Mahpudin
Universitas Pendidikan Indonesia, Bandung, Indonesia

ABSTRACT: Employee performance in the Production Division at PT Samick Indonesia is indicated to be low, as seen from the lack of a sense of responsibility of employees, which resulted in production levels that do not meet targets. It is possible that this is because the placement of employees in the Production Division is not effective due to a mismatch between abilities, skills, and expertise with job specialization, as well as inappropriate placement of academic knowledge and skill. This research uses a quantitative method with an associative statistical approach to determine the relationship between employee placement as independent variable with work performance as dependent variable. The data collection was done by distributing questionnaires to the employees in the Production Division at PT Samick Indonesia, with as many as 60 respondents. Data analysis and statistical hypothesis testing was performed by simple linear regression analysis. The conclusion of this study was that employee placement's influence on work performance obtained a correlation coefficient value of $r = 0.49$. The results of statistical hypothesis testing was that there is a significant effect of employee placement on job performance in the Production Section at PT Samick Indonesia. The amount of influence of employee placement on work performance was based on the coefficient of determination, at 24.01%; and the remaining 75.99% was influenced by other factors not discussed in this research.

1 INTRODUCTION

Human resources is an important factor in an organization or company to achieve goals and objectives through cooperative efforts of a group of people. Thus, it can be said that the existence of human resources is as a determinant of th success or failure of a company in achieving its objectives. The human resources in question here are employees, so achieving the goals of the company depends on how employees can develop the skills, skills, and expertise. It is necessary for employees to have accuracy in performing their duties and they can work together, doing different jobs, with the knowledge, skills, and good attitudes that will work optimally, so that companies in managing human resources can be more effective, efficient, and improve employee performance

One important aspect in managing human resources iis employee placement. Employee placement is a series of action steps undertaken to decide whether or not an employee is placed in a particular position within a company by optimizing their knowledge, skills and attitudes toward job performance.

Sastrohadiwiryo quoted by Badriyah (2015) explains that employee placement is placing an employee, as the executing element of the job, in a position consistent with their abilities, skills and expertise. Placement of employees in the right position is important because it is closely related to employee performance for providing great benefits for the company.

Errors in employee placement will result in conflicts with employees associated with the work, which can lead to decreased enthusiasm and excitement of work that ultimately results in decreased employee performance. Evaluating the company's behavior and employee performance can determine whether the process of placement is implemented successfully or not.

Job performance can be interpreted as a result of the quality and quantity of work achieved by an employee during a certain period of time in carrying out their work duties in accordance with the responsibilities given to them, as stated in Mangkunegara (2007).

This is shown also from some previous research results:

– Galuh (2010) indicated that staffing at a site included both categories, with an average value of 3.97 and its employees' performance included high category with an average value of 3.84. The amount of influence of employee placement on employee performance in the research location was 52.70% while the remaining 47.30% was influenced by other variables not included in this study.
– Saleh (2014) showed that the placement of employees and job performance in the category was quite effective and high enough. The variable of employee placement included indicators with the lowest percentage of indicators of work skills and the highest indicator of working knowledge. The job performance variable included the lowest indicator of timeliness and the highest indicator of work quantity. The

results of this research also proved that employee placement has a significant effect on job performance.

– Yuliana (2014) showed that the placement of employees at the location of his research is good. P restore work shows most have performance in either category. This is seen from the answer to a questionnaire with a total percentage of 95.2%. The results of this research also proved that employee placement has a significant effect on job performance.

Based on these three research articles, it can be concluded that there is an employee placement effect on employee performance. Associated with this previous research and by taking the object of research work as performance employees at PT Samick Indonesia, there is a phenomenon that the placement of employees on the production line that is not effective. This indicates that there are still problems, such as a mismatch between abilities, skills, and specialist job skills.

Non-conformity of expertise is categorized in employees who experience similar problems of inappropriate placement of academic science knowledge and skills, resulting in low achievement levels of production employees of at PT Samick Indonesia, and a lack of a sense of responsibility for work that results in production levels that do not meet the target.

Based on the research problem, the purpose of this research is:

1. To understand the placement of employees in the production department at PT Samick Indonesia.
2. To understand the work performance of employees in the production department at PT Samick Indonesia.
3. To know the effect of employee placement on employee performance in the production department at PT Samick Indonesia.

1.1 Employee placement

Employee placements are part of the procurement concept, and the success of the entire employee procurement program lies in the accuracy of employee placement. However, it is rare for an employee to be directly placed in the right position.

Marihot TE Hariandja in Mila (Badriyah 2015), defines placement as follows: "placement is the process of assignment/assignment of office or reassignment of employees to new task/occupation or different positions." Meanwhile, according Badriyah (2015), "Placement is putting one's position in the right job position, how well an employee in doing his job will affect the quantity and quality of work."

Sastrohadiwiryo quoted by Badriyah (2015), explains job placement as follows: "Employee placement is placing an employee as the executing element of the job in a position consistent with his or her skills, abilities and skills."

According to Rivai et al. (2014), placement is allocating employees to certain working positions and this is specific to new employees. This confirms that placement of employees is not just placement but a company must match and compare the qualifications of employees with the needs and requirements of a particular position, putting the right person in the right place. More placement decisions are made by line managers and employee supervisors within a particular division

Employee placement procedures are closely related to the systems and processes used. In connection with the Sastrohadiwiryo placement system, Badriyah (2015) states, "There should be aims and objectives in plotting an employee placement system." Employee placement procedures must meet the following requirements:

1. To place personnel who come from a list of personnel developed through labor analysis.
2. The standards used to compare job candidates.
3. Job applicants who will be in the selection to be in place.

Success in the procurement of labor lies in precision in employee placement, either placement of new employees or current employees in new positions. Employee re-placement is done with a variety of reasons related to human resource planning or utilizing manpower more effectively and efficiently, which can be caused by organizational challenges, supply, and availability of employees internally and externally, career enhancement in the aspect of development of resources of human power, job satisfaction, and work motivation. The process of placement is a decisive process in getting competent employees into the company because the right placement in the right position can help companies in achieving their expected goals.

According to Sastrohadiwiryo quoted by Badriyah (2015), employee placement intention is to place the employee as the executing element of work in a position which is in accordance with the following criteria: (a) ability; (b) proficiency; (c) expertise. Ability is innate or learned and means a person can complete the work, both mentally and physically.

1.2 Work performance

Job performance has significance in the achievement of corporate or organizational goals, therefore management must try to encourage employees to always work well and achieve their best performance.

1. Mangkunegara (2007) describe this as "job performance (job performance) is the work of quality and quantity achieved by someone in doing their duties in accordance with the responsibilities given to him".
2. According to Badriyah (2015), the quality of work is the quality of an employee or employee in terms of carrying out their duties, including suitability, tidiness, and completeness.

3. Wungu & Brotoharjoso (2003) state that quantity of work is any form of unit of measure associated with the amount of work (output) that can be paired with numbers in a use time in order to achieve the targets of the company.
4. Wajdi & Abdullan (2007) explain responsibility as the ability of a person to carry out obligations because there is encouragement in him so that the creation of self-quality.
5. According to Hasibuan (2007), "work performance is a result of work accomplished by a person in carrying out the tasks assigned to him based on his skills, experience and sincerity".
6. Handoko (2007), describe that "work performance is a result of work achieved by the organization to evaluate or assess its employees".

Thus, it can be concluded that work performance is a result achieved in a job both in the quality and quantity of carrying out tasks given, based on the suitability of knowledge, skills and attitudes for the achievement of corporate objectives.

According Sunyoto (2015), work achievement by a person is carrying out and complete the work assigned to them. The working achievement factor that will be examined includes the quality of work, quantity of work, reliability and work attitude, Heidjracman quoted in Sunyoto (2015).

According to Sutrisno (2014), a definition of achievement is a record of the results obtained from certain job functions or specific activities over a period of time. Sutrisno defines achievement as the level of one's ability on tasks included on one's work. This definition indicates the weight of individual ability in fulfilling the provisions that exist in the work. The achievement of work is the result of the efforts by a person determined by their ability through personal characteristics and perceptions of their role in the job.

In general, work performance is given limitations as a person's success in performing a job, more firmly again is (Sutrisno 2014), which states that job performance is "succesful role achievement" obtained by a person doing his job assignment and this is called level of performance.

There are other terms that mean similar to or overlap with job performance, for example "proficiency" has a broader meaning because it includes both the aspects of the work effort. According to Sutrisno (2014), giving merit meaning is more a general aspect than the proficiency. Work productivity is the ratio between input and output.

It seems clear that the notion of job performance is more narrow in character, and is only with regard to what results in a person's behavior. Usually a person with a high level of performance is referred to as a productive person, and conversely a person whose level is not up to standard is said to be unproductive or have low performance.

To know the level of one's work performance, it is necessary to do a work performance appraisal. According to Handoko (2007: 135),

"performance appraisal is a process through which organizations evaluate or assess employee performance". This activity can improve personnel decisions and provide feedback to employees about their work implementation. Mangkunegara (2007) states that "the elements assessed from work performance are work quality, quantity of work, reliability and attitude. Quality of work consists of accuracy, precision, skill, cleanliness. The quantity of work consists of output and work completion with extras. Reliability consists of following instruction, initiative, caution, craft. While the attitude consists of attitudes toward the company, other employees and work and cooperation.

Thus, the overall elements/components of this performance assessment must exist in the implementation of the assessment so that the results can reflect the work performance of the employees.

According Mangkunegara (2007), there are two factors that affect the determination of work performance: ability and motivation.

– Ability factor: Psychologically, the ability of employees consists of the potential ability (IQ) and ability (knowledge + skill). An employee who has an above-average IQ (IQ 110–120) with adequate education for their position and is skilled in doing their day-to-day work will find it easier to achieve the expected performance. Employees therefore need to be placed in work that suits their expertise.
– Motivation: This grows from the attitude of an employee who faces the situation of work. Motivation is a self-directed driving condition of a targeted employee to achieve company goals.

According to Sutrisno (2014), individual work performance is a combined function of three factors:

– The ability, temperament, and interest of a worker.
– Clarity and acceptance of the explanation of a worker's role.
– Level of work motivation.

Each factor can individually have important meanings, but the combination greatly determines the level of outcomes of each worker, which in turn helps the organization's overall performance. Sutrisno (2014), reveals the existence of two factors that affect job performance: the individual and environmental factors. Individual factors are:

– Efforts that show a number of physical and mental synergies used in organizing the tasks.
– Abilities, which are the personal qualities required to perform a task.
– Role/task perception, i.e. any behavior and activities that are felt necessary by the individual to complete a job.

Sutrisno (2014) mengemukakan bahwa prestasi kerja merupakan hasil dari gabungan variabel individual dan variabel fisik, pekerjaan, variabel organisasi dan sosial.

Sutrisno (2014) mengatakan pengukuran prestasi kerja diarahkan pada enam aspek yang merupakan bidang prestasi kunci bagi organisasi, yaitu:

Sutrisno (2014) argues that work performance is the result of a combination of individual variables and physical variables, occupations, organizational and social variables. Sutrisno states that work performance measurement is directed to six key areas of achievement for the organization:

– Work result. The level of quantity and quality that has been generated and the extent to which supervision is done.
– Knowledge of work. The level of knowledge associated with job tasks that will directly affect the quantity and quality of the work.
– Initiative. The level of initiative during carrying out the work tasks, especially in terms of handling problems that arise.
– Mental attachment. Level of ability and speed in accepting work instructions, and adjusting to work situations that arise.
– Attitude. Level of morale and positive attitude in carrying out the job task.
– Discipline of time and attendance. Level of punctuality and attendance.

From the description above, it can be concluded, that the discussion of the problem of success or work performance must be viewed from two perspectives:

– Aspects that relate to the criteria of success measurement work, which is the ultimate goal of the implementation of a job.
– The behavior of the individual in their efforts to achieve success in accordance with predetermined standards. Behavior itself is influenced by two main variables: individual and situational.

Sutrisno (2010) describes the indicators of work performance as follows:

– Quality of work is the level of perfection of work processes or fulfillment of ideal and expected work activities.
– Quantity is the amount generated in the context of the value of money, the number of units, or the number of completions of an activity cycle.
– Timeliness is the level of work done or a result achieved within the shortest time expected so as to maximize the utilization of time for other activities.
– Cost effectiveness is the maximization of organizational resources to obtain the best results or reduce losses.

According to Sunyoto (2015), work performance can be measured through:

– Quality of work, related to the timeliness, skills and personality in doing the job.
– Quality of work, relating to doing additional tasks assigned by superiors to subordinates.

– Toughness, related to attendance, holiday time, and schedule delays are present in the workplace.
– Attitude that shows the level of cooperation in completing the work.

Performance appraisal is the process through which organizations evaluate or assess employee performance (Sunyoto 2015). The process of achievement assessment is intended to understand the performance of someone's work. This objective requires a process that is a series of interrelated activities. These activities consist of identification, observation, measurement, and development of the work of employees in an organization (Sutrisno 2010).

1.3 Theoretical thinking framework

One of the factors that affect employee performance is correct employee placement. Work placement is part of the following concept, according to Sastrohadiwiryo quoted in Badriyah (2015): employee placement intention is to place the employee as the executing element of work in a position that matches the following criteria: (a) ability; (b) proficiency; (c) expertise.

The ability of an employee in a particular job will mean they perform the tasks well, the skills will make the tasks easy to do, and the expertise will provide the best results in accordance with the expertise. These three criteria are important factors for employees to have good work performance.

According to Young et. al. (2010) regarding the liaison between the effect of job placement on job performance, "job placement has a significant effect on job performance so that employees will be productive".

Framework of thought refers to the object and the problem of research that describes the relationship between variables. The framework of thought will direct the research process according to the objectives to be achieved and will be the research thought flow. Based on the theoretical framework, the indicator variables x and y are as follows:

– Badriyah (2015) describes the definition of job placement as follows: "The placement of work is to place the employee as an element of the job pelakasana in a position in accordance with ability, competence and keahliaanya".
– Mangkunegara (2007) states that work performance is the work of quality and quantity achieved by an employee in performing their duties in accordance with the responsibilities given to them.

In this research, there are two variables: job placement as X variable and employee performance as Y variable. Figure 1 shows more detail of the the variables and indicators.

Based on Figure 1 above, the authors will conduct research on the sub variable influence of employee placement on work performance.

Figure 1. Framework.

1.4 Development of research hypothesis

Development of the research hypotheses conducted in this study refers to previous studies related to the effect of job placement on employee performance:

– Based on the results of hypothesis testing in Rahyang Galuh's thesis at Widyatama university, in 2010 with the title of the influence of employee placement on the work performance of employees in Sahabat sejati based on the calculation of correlation obtained results rs of 0.726 indicating a strong relationship between employee placement with work performance. This belongs to the category 0.60–0.799. The amount of influence of employee placement on employee work performance is 52.70%, and the remaining 47.30% is influenced by other factors not examined. From the calculation of the t-test statistic, the value of t_ (count)> t_ (table) or 5.59 > 1.701, means Ho is rejected and Ha is accepted. Thus, there is an influence of employee placement on job performance.

– Based on the result of hypothesis testing of the students in the Faculty of Administration of Universitas Brawijaya-Malang, it can be known that the variable of employee placement (X) simultaneously has a significant effect on work performance variable (Y). This is based on the calculation of multiple regression analysis F_hitung > F_table (53.946 > 2.92). Thus, the decision is Ho rejected and Ha accepted. The results of this hypothesis testing shows that there is an influence of the placement of employees on job performance at PT Telkom Kendatel Malang.

– According to the results of the hypothesis of university students at pamulang-tanggerang South, the variable employee placement (X) simultaneously has a significant effect on job performance variable (Y). This is based on the calculation of simple regression analysis F_hitung > F_table (54.82 > 2.92). Thus, Ho is rejected and Ha is accepted. The results of this hypothesis testing show that there is an influence of the placement of employees on job performance at hotel Acacia Jakarta.

Based on the previous framework and research, the hypothesis of research as a temporary estimate in this study is as follows:

II: Placement of employees affects the performance of the Production Division at PT Samick Indonesia.

2 METHOD

2.1 Research design

The research design is quantitative research, using statistical procedures based on the results of quantification (measurement). The quantitative approach used in this study to analyze the interrelationship between variables refers to an objective theory that will be tested based on empirical data.

Quantitative research according to Sujarweni (2015) is a process of finding knowledge using numerical data as a tool to analyze information about what we want to know. Quantitative research is a research process that starts from theory, hypothesis, and research design, then chooses subject, collects data, processes data, analyzes data, and writes conclusions.

According to Sunyoto (2015), research design describes the planning that will be done in research and refers to the problems that have been set previously. In addition to quantitative research, the research design used in this study is a descriptive method. According to Sugiyono (2014), a descriptive method aims to describe the nature of something that is going on at the time of research and examines the causes of a particular symptom. This descriptive method can be used with more facets and to a further extent than other methods. According to Consuelo in Sunyoto (2015), there are several types of descriptive method research, including case study, survey, development research, advanced research, document research, research tendency, and correlation research.

Instruments used in survey research are in the form of interviews, observation, literature study, and questionnaires. In compiling this questionnaire, the researcher uses a closed-ended questionnaire system so respondents could answer more easily and quickly, and different answers from respondents would be easier to compare and statistically analyze. The resulting data that will be processed and analyzed and finally a conclusion will be drawn. The conclusions made will apply to the entire population who become the object of research.

2.2 Population, sampling and sampling techniques

The population in this research is the employees of assembling upright pianos at PT Samick Indonesia, amounting to 70 people, with a minimum period of 2 years' employment. Determining the sample according to the hypothesis refers to the Slovin formula quoted in Sunyoto (2015), and the sample can be determined based on the total population of 70 employees assembling upright pianos at PT Samick Indonesia. This will tolerate the error of 5% and then many samples will be taken is 60 respondents.

In the research, sampling is done in such a way that the samples taken can represent the overall characteristics of the existing population. This is done in order to minimize errors that may occur during the questionnaire.

According to Sunyoto (2015), the sample is part of the number and characteristics of the population. Samples were taken to represent the population. Samples will be generalized (applied in general) to the population. The larger the number of samples, the smaller the probability of generalization error to the population.

2.3 Operational definition

Operational definition is the elements of research that measure a variable. Thus, these measurements determine the indicators that support the variables to be analyzed. The variables that will be measured in this study are placement employees (X) and job performance (Y). The operational definition used here is presented in the form of tables, as follows.

The date was collected through three stages of research: 1) library studies, 2) field studies through interviews and observations, and 3) questionnaires. At the stage of data collection, the instrument to be used as a data collection tool was first tested via 1) test validity and 2) reliability test.

2.4 Data analysis method

The data already obtained was then collected for later analysis or to test the accuracy of the data. Various techniques can be used to test the accuracy of data, and in this case the author used frequency distribution, simple linear regression analysis, Pearson correlation coefficient analysis, coefficient of determination (kd), and a t-test. To facilitate this and to achieve a high level of accuracy of calculation, this study used IBM SPSS 24.

3 RESULTS AND DISCUSSION

The results of questionnaire data collection were a picture of the characteristics of respondent employees of PT Samick Indonesia, totaling 60 people. The results are as follows.

– Sex: 46 men (77%), and 14 women (23%).
– Age: 25 people age 18–29 years (42%), 17 people age 30–39 years (28%), 15 people age 40–49 years (25%), 3 people age > 50 years (5%). The largest number of respondents are aged between 18–29 years while the fewest respondents are aged > 50 years.
– Education: elementary school 0%, junior high 20 people (33%), high school 38 people (63%), Diploma 1 person (2%), and strata one 1 person (2%). The fewest respondents are educated to Diploma and stratum one level, while the largest number of respondents have high school education.
– Years of employment: 5 years (19%), 6–10 years 26 people (43%), 11–15 years 10 people (17%), and > 15 years 5 people (8%). The fewest respondents have a working life of > 15 years, while the largest number of respondents have working life of 6–10 years.

3.1 Assessment of employee placement and job performance

The data used for the analysis was obtained from a questionnaire disseminated to 60 employees at PT. Samick Indonesia. The terms of scoring each respondent's answer are as follows:

1. Answer "strongly agree" (SS) is worth 5 points.
2. The answer "agree" (S) is worth 4 points.
3. Answer "doubtful" (RG) is worth 3 points.
4. Answer "disagree" (TS) is worth 2 points.
5. Answer "strongly disagree" (STS) is worth 1 point.

For free variables (employee placement) there were 27 statements, five of which are not valid numbers (5, 6, 8, 18, 21) so that only 22 statements were used. For the dependent variable (work achievement) there were 27 statements, one of which is not valid (45) so that only 26 statements are used.

To know the value of placement of employees as independent variables, the authors analyzed the primary data from the results of questionnaires. The analysis began by counting the number of classes and class intervals based on the Sturges formula proposed by Sugiyono (2014).

The relationship of employee response to employee placement amounted to 4.17, which can be

Table 1. Operasionalisasi varabel penelitian.

Variable	Sub Variable	Indiator	No instrument
Job placement (X)	Ability	Work accuracy	1, 2, 3
		Motivation	4, 5, 6
		Work experience	7, 8, 9
	Skills	Initiative	10, 11, 12
		Responsiblity	13, 14, 15
		Creativity	16, 17, 18
	Expertise	Understanding the work	19, 20, 21
		Skillls	22, 23, 24
		Community	25, 26, 27
Work achievement (Y)	Quality	Work accuracy	28, 29, 30
		Thoroughness	31, 32, 33
		Standard of work	34, 35, 36
	Qantity	Speed of work	37, 38, 39
		Production target	40, 41, 42
		Inreased out put	43, 44, 45
	Responsible	Time efficiency	46, 47, 48
		Production process	49, 50, 51
		Maintenance of production facilities	52, 53, 54

interpreted as the level of employee placement at PT Samick Indonesia being very good but needs improvement in various aspects to be better. The level of relationship of employee perceptions about job performance is 4.08, which can be interpreted as employee work performance is very good, but there needs to be some improvements from various aspects, in order to be better, and there is no difference of perception between employees and management.

3.2 Regression analysis

Simple linear regression analysis was used to analyze the effect of employee placement on job performance in the Production Section at PT Samick Indonesia. The following shows the results of calculations by simple linear regression analysis using IBM SPSS 24

The calculation result obtained by the model of simple linear regression equation as follows.

$$Y = 53.74 + 0.58.X$$

This regression equation model shows that if there is no change in employee placement (X = 0), then the work performance will be worth an average of 53.740 units. However, if there is an increase of 1 unit in variable employee placement (X = 1), it will affect the increase in work performance (Y) of 0.579 units.

3.3 Correlation coefficient analysis

Correlation coefficient analysis is used to determine the relationship of employee placement relationship to job performance in the Production Division at PT Samick Indonesia. Results of calculating the correlation coefficient using IBM SPSS 24 are as follows.

The correlation coefficient value between employee placement (X) and work achievement (Y) is e r = 0.490, and hence it can be stated that employee placement has a positive influence and level of relation to work performance.

3.4 Determination of coefficient analysis

Analysis of the coefficient of determination is a statistical method used to determine the influence of employee placement on work performance in the Production Division at PT Samick Indonesia. The formula used is as follows:

$$Kd = r^2 \times 100\%$$
$$Kd = (0.49)^2 \times 100\%$$
$$Kd = 0.2401 \times 100\%$$
$$Kd = 24.01\%$$

Based on the coefficient of determination above, the influence of employee placement on job performance is 24.01% and the remaining 75.99% is the influenced of other factors not discussed in this study.

3.5 Hypothesis test/Test (t)

The research hypothesis to be tested by using statistical hypothesis is as follows.

Ho: β = 0 (no significant effect of employee placement on work performance)

Ha: β ≠ 0 (there is significant influence of employee placement on job performance)

Based on the calculation of t count, the value obtained was 4.287. Using the error rate α = 5% (0.05) and the two-sided test (α/2 = 0.05/2 = 0.025) with df = n-2 = 60–2 = 58, then the t_table is.,002. From the calculation results obtained for t_hitung of 4.29 this shows that the calculation results are outside the acceptance area, which means that Ho is rejected and Ha is accepted. The value t_hitung (4.29) > value t_table (2.002) so Ho is rejected and Ha is accepted. This means there is a significant effect of employee placement on work performance in the Production Section at PT Samick Indonesia.

4 CONCLUSION

The authors have deduced the following regarding the influence of employee placement on work performance:

– Employee placement value applied at PT Samick Indonesia is very good, where the average score of employee perceptions of this variable is 4.17 (after consulting the calculations in the table interval 3.92 – 4.22).
–The value of work performance in PT Samick Indonesia is very good, where the average score of employee perceptions of this variable is 4.08 (after consulting the calculations in the table interval 4.08 – 4.31).
– The effect of employee placement (X) on employee performance (Y) can be denoted by simple linear regression equation Y = 53.74 + 0.58X.
– The relationship between employee placement (X) on work performance (Y) based on correlation coefficient value (r) = 0.49. This shows the relationship between employee placement and work performance in the production department at PT Samick Indonesia is positive and moderate. Positive means there is a direct relationship between the two variables, so if there is a good employee placement influence then the work performance will increase, and vice versa. While the relationship is because the value of 0.49 is on the rank 0.40 - 0.599 (Sugiyono 2014).
– The contribution of variable (X) influence of employee placement to variable (Y) employee performance can be seen from the determination coefficient value of 24.01%, and the remaining 75.99% is influenced by other factors not examined here, such as communication and socialization between superiors and subordinates, leadership styles, conflicts,

organizational culture, etc. Employee placement is influenced by work performance and this can be seen from result of hypothesis testing obtained for t arithmetic > t table, which is 4.29> 2.002. Thus, Ho is rejected and Ha is accepted, meaning there is significant influence between employee placement effect (X) (Y) on the Production Division at PT Samick Indonesia.

Thus, the conclusion of this research is that variable placement of employees significantly contributes to the employee performance variable and the hypothesis of this research is proven.

5 RECOMMENDATION

Based on the conclusion of the research result, the recommendation that can be the reference in decision-making in the effort to improve the work performance can be seen from the lowest response to the variable of employee placement, that is the ability indicator with the statement able to give creative idea for the company progress. It is recommended that the company increases employee motivation in contributing creative ideas by opening a suggestion box or inviting ideas for the company's progress, and if the creative idea is taken forward then the company will give a bonus or incentive.

The lowest response for job performance variables on the quantity indicator, for the statement "achieving individual production targets every day", means there are still many employees who have not been able to reach individual targets. Thus there are still some areas that have not met production targets because of many obstacles, both from material mismatch or from employees who are less eager to work. Recommendations for the management are to always be ready to provide the appropriate materials to make the production process run smoothly and thus employees can be more eager to carry out the work that is their responsibility.

REFERENCES

Badriyah, M. 2015. *Manajemen Sumber Daya Manusia.* Bandung: Pustaka Setia.

Galuh, R. 2010. Pengaruh penempatan karyawan terhadap prestasi kerjapada sahabat sejati. *Skripsi Universitas Widyatama.*

Handoko, H.T. 2007. *Mengukur kepuasan kerja.* Jakarta: Erlangga.

Hasibuan, M. 2007. *Manajemen Sumber Daya Manusia.* Jakarta: Bumi Aksara.

Mangkunegara, A.P. 2007. *Manajemen Sumber Daya Manusia.* Bandung: Remaja Rosdakarya.

Rivai, R. 2014. *Manajemen Sumber Daya Manusia untuk Perusahaan.* Jakarta: PT Rajagrafindo Persada.

Saleh, N.F. 2014. Pengaruh penempatan karyawan terhadap prestasi kerja di PT. Dhanar Mas Concern. *Skripsi Universitas Pendidikan Indonesia.*

Sugiyono, S. 2014. *Metode Penelitian Kuantitatif Kualitataif dan R&D, Cetakan ke 20.* Bandung: Alfabeta.

Sujarweni, W. 2015. *Metodologi Penelitian Bisnis dan Ekonomi.*

Sunyoto, D. 2015. *Penelitian Sumber Daya Manusia: Teori, kuesioner, Alat Statistik, dan Contoh Riset.* Yogyakarta: CAPS (Center of Academic Publishing Service).

Sutrisno, E. 2014. *Manajemen Sumber Daya Manusia.* Jakarta: Kencana Prenada Media Grup.

Suwatno, D.J.P. 2016. *Manajemen SDM Dalam Organisasi Publikdan Bisnis.* Bandung: Alfabeta.

Wajdi, D.A. & Irwani, A.N. 2007. Why do Malaysian customers patronise Islamic banks. *International Journal of Bank Marketing* 25(3): 142–160.

Wungu, J. & Brotoharsojo, H. 2003. *Tingkatan Kinerja Perusahaan.* Jakarta: PT. Raja Grafindo.

Yuliana, I. 2015. Pengaruh penempatan karyawan terhadap prestasi kerjapada hotel acacia jakarta. *Skripsi Universitas Pamulang.*

Young, W., Hwang, K., McDonald, S. & Oates, C.J. 2010. Sustainable consumption: Green consumer behaviour when purchasing products. *Sustainable development* 18(1): 20–31.

Section 6: Strategic management, entrepreneurship

and contemporary issues

Advances in Business, Management and Entrepreneurship – Hurriyati et al (eds)
© 2020 Taylor & Francis Group, London, ISBN 978-0-367-27176-3

Barriers and sukuk solutions in Indonesia

M. Kurniawati, N. Laila, F.F. Hasib & S.N.
Mahmudah
Universitas Airlangga, Surabaya, Indonesia

ABSTRACT: The world of Islamic finance has shown rapid growth in recent decades. Indonesia as a country with the largest Muslim population in the world must be a pioneer in the development of Islamic finance. However, that potential is not reflected in reality. In the last two decades, since the first appearance, Islamic finance in Indonesia has shown very slow progress, especially for sukuk or sharia bonds. Since its first appearance in 2002, sukuk market share has not been able to penetrate 5% of total sukuk and bond market. Basically, there are some obstacles that hinder the development of sukuk in Indonesia. Therefore, to cultivate public interest in Indonesia to invest in sukuk, proper and large strategy is absolutely necessary. This study aims to determine the obstacles and solutions in the development of sukuk in Indonesia. This study tries to use a strategic management approach to formulate the most appropriate strategy in developing sukuk in Indonesia. The method used in this research is qualitative exploratory. Technical analysis used is Analytical Hierarchy Process (AHP) by Miles and Hubemain. Obstacles in the development of sukuk in Indonesia from the aspect of issuers are pricing refers to labor, cost efficiency so less competitive disadvantages, lack of sukuk variant, and low regulation while for sukuk development solution in Indonesia from issuer aspect ie reduced cost, asset identification, sukuk contract variation, perspective on sukuk.

1 INTRODUCTION

Sukuk is used as an alternative to investing, many countries that issue sukuk such as Western and Asian countries with minority Muslims also choose this investment. There are several countries that issue sukuk namely Canada, UK, Germany, United Arab Emirates, Malaysia, Singapore, Qatar, Dubai, Kuwait, Pakistan. The following countries are China, India, Japan, Korea, and Indonesia.

According to (Lotunani 2014) since the economic crisis in 2008 showed the recovery of materials that are still relatively low and fragile. In the Americas, Europe, China, and Japan are still showing a slow economy. The slow growth has made it a challenge and a concern for the Islamic finance industry to continue to grow. In 1997 and 2008 in the global financial crisis it was shown that the financial system of Islamic finance is considered to be relatively stable and that the real sector based can provide a sense of security because it is believed to be higher endurance in the economic crisis.

Unlike the State Sukuk, although corporate Sukuk shows an increasing growth, but the growth of corporate sukuk is not as big as the development of the State Sukuk. Corporate Sukuk is a sukuk issued by a private company. This sukuk reaches wider investors than bonds, because these sukuk can be bought by all circles both Islamic and conventional financial institutions, while bonds can only be purchased by conventional financial institutions. In Indonesia, many private companies are issuing corporate sukuk with various business fields, namely banking, multifinance, telecommunications, construction to consumer goods.

In 2013, corporate sukuk circulating in Indonesia amounted to 9.7% and reached Rp. 9.5 trillion. In early 2014, the number of corporate sukuk decreased to Rp. 7 trillion, which initially amounted to 9.7%, due to several sukuk maturing and no new corporate sukuk issuance. From 2001 to July 2014, globally, 28 countries have published sukuk with a Muslim majority population. The dominance of the sukuk issuer country is Malasyia by 67%, then Saudi Arabia and United Arab Emirates (UAE) which owns 8% issuance of sukuk from the existing global sukuk. Then the 4th order of Indonesia which reached 4% followed by Sudan, Turkey, Qatar. The development of sukuk globally shows that more and more countries are issuing sukuk.

From the above explanation illustrated that Indonesia with the largest Muslim population in the world only ranks fourth with global percentage (4%) issuance of global sukuk very far behind from Malasyia (67%). It becomes a question, why the development of sukuk in Indonesia cannot progress or develop as in Malasyia? There are several studies on sukuk, including (Ansari 2004) who examines "Managing Financial Risks of Sukuk Structures" which addresses the challenges in the development of sukuk instruments. Ascarya & Yumanita (2007) conducted a research entitled "Comparing the Development of Islamic Financial/Bond Market in Malaysia and Indonesia" which discusses the issue of sukuk development in Indonesia.

Although there have been a lot of research on sukuk, but there is no deep and comprehensive research to answer why sukuk in Indonesia, especially Sukuk Negara, has a relatively small growth compared to sukuk in other Muslim majority countries. Given the magnitude of the role of sukuk in a country's economy, it is very important to do the research in question.

1.1 Formulation of the problem

Based on the above background, then the formulation of the problem in the following research is what obstacles and solutions of sukuk development in Indonesia.

1.2 Theoretical basis

1.2.1 Sukuk

Sukuk in its practical understanding means proof of ownership (Iggi 2003). A sukuk represents an interest, whether full or proportional in an asset or a set of assets (Elenkov et al. 2005). Based on Rule Number IX.A.13 of 2009 concerning issuance of Sharia Securities, sukuk is a Sharia effect in the form of certificates or proof of ownership of equal value and represents an unspecified part (undivided or undivided share) of Tangible Assets certain value of certain existing or existing tangible assets, existing or

existing services, certain project assets, predetermined investment activity (Kuriniawati et al. 2013).

1.2.2 Types of sukuk

Sukuk has various types, it is intended to meet the various needs of investors. According to AAOIFI standards in (Salim 2011) there are 14 types of contracts in sukuk, but there are some of the most widely used in the market, namely:

1. Sukuk ijara
   Sukuk issued under an ijara agreement or agreement whereby one party acts alone or through its representative sells or leases the beneficial interest of an asset to another party on the basis of the agreed rental price and lease period, without being followed by the transfer of ownership of the asset itself. (Ibrahim 2009).
2. Sukuk Mudharabah
   Sukuk issued under a mudaraba agreement or agreement whereby one party provides capital and the other provides labor and expertise, the benefits of such cooperation shall be divided on the basis of a pre-agreed comparison. losses incurred will be borne entirely by the party who became the provider of capital (Rusydiana 2012).
3. Musharaka sukuk
   Sukuk issued under a musyarakah agreement or contract whereby two or more parties work together to incorporate the capital used to build a new project, develop an existing project, or

Table 1. Harmonic average score criteria for developing sukuk in Indonesia.

	The most important obstacles				
Criteria	Informant 1	Informant 2	Informant 3	Informant 4	Average Harmonics
Aspects Of Investors	0.240	0.272	0.235	0.326	0.264
Aspects of Issuers	0.496	0.362	0.182	0.356	0.306
Market Aspects	0.186	0.111	0.105	0.124	0.125
Legal Aspects	0.078	0.255	0.478	0.194	0.167

Table 2. Harmonic average value of investor aspect barriers criteria.

	The most important obstacles				
Sub Criteria	Informant 1	Informant 2	Informant 3	Informant 4	Average Harmonics
Minim Investors	0.169	0.320	0.163	0.222	0.203
Pricing Referring to Libor	0.443	0.558	0.540	0.667	0.541
Buy and Hold Strategy	0.387	0.122	0.297	0.111	0.173

Table 3. Harmonic average value of aspect issuer obstacle criteria.

Sub Criteria	The most important obstacles				Average Harmonics
	Informant 1	Informant 2	Informant 3	Informant 4	
Adverse selection	0.540	0.105	0.625	0.084	0.161
Limited asset securitization	0.163	0.637	0.136	0.705	0.243
Cost efficiency is less competitive disadvantages	0.297	0.258	0.238	0.211	0.247

Table 4. Harmonic average value of market barriers criteria.

Sub Criteria	The most important obstacles				Average Harmonics
	Informant 1	Informant 2	Informant 3	Informant 4	
Minim SDM	0.163	0.079	0.097	0.143	0.111
Lack of education and socialization	0.540	0.263	0.570	0.286	0.367
Lack of sukuk variants	0.297	0.659	0.333	0.571	0.415

Table 5. Harmonic average value of legal aspect barriers criteria.

Sub Criteria	The most important obstacles				Average Harmonics
	Informant 1	Informant 2	Informant 3	Informant 4	
Low regulation	0.276	0.250	0.297	0.185	0.244
Lack of harmony opinion scholars on sukuk	0.595	0.095	0.540	0.156	0.195
Lack of academic	0.128	0.655	0.163	0.659	

Table 6. Harmonic average score criteria solution in sukuk development in Indonesia.

Sub Criteria	The most important solution				Average Harmonics
	Informant 1	Informant 2	Informant 3	Informant 4	
Aspects of Investors	0.138	0.087	0.104	0.132	0.111
Aspects of Issuers	0.363	0.615	0.510	0.526	0.485
Market Aspects	0.320	0.085	0.226	0.085	0.129
Legal Aspects	0.179	0.212	0.159	0.257	0.195

Table 7. Harmonic average value criteria aspect solution investor.

| Sub Criteria | The most important solution | | | | |
	Informant 1	Informant 2	Informant 3	Informant 4	Average Harmonics
Promote investors	0.136	0.300	0.163	0.238	0.190
Reduce costs if entering Islamic financial markets	0.625	0.600	0.540	0.625	0.595
Change the investment mindset	0.238	0.100	0.297	0.136	0.161

Table 8. Harmonic average value criteria aspect solutions issuer.

| Sub Criteria | The most important obstacles | | | | |
	Informant 1	Informant 2	Informant 3	Informant 4	Average Harmonics
Education	0.625	0.136	0.614	0.097	0.191
Identification assets	0.238	0.625	0.268	0.570	0.354
Evaluate the contracts used	0.136	0.238	0.117	0.333	0.173

Table 9. Harmonic Average Value Criteria for Market Aspect Solutions.

| Sub Criteria | The most important obstacles | | | | |
	Informant 1	Informant 2	Informant 3	Informant 4	Average Harmonics
Provision of human resources who understand sukuk	0.140	0.079	0.163	0.122	0.117
Education	0.333	0.263	0.297	0.320	0.301
Variation of sukuk contract	0.528	0.659	0.540	0.558	0.567

Table 10. Harmonic average value criteria legal aspect solution.

| Sub Criteria | The most important obstacles | | | | |
	Informant 1	Informant 2	Informant 3	Informant 4	Average Harmonics
Revised regulation on sukuk	0.249	0.243	0.286	0.185	0.235
Uniform perspective on sukuk	0.594	0.669	0.571	0.659	0.620
Cooperate with academics	0.157	0.088	0.143	0.156	0.128

finance business activities. Gains or losses incurred will be borne together in accordance with the amount of participation of each party's capital (Umer 2000).

4. Sukuk Murabahah
   Murabahah is the sale and purchase of goods at the original price with an additional profit agreed. (Huda 2008).
5. Istisna
   Istisna is an end buyer's sales contract with the supplier, in this contract the supplier receives an order from the buyer that the price or specification of goods ordered in accordance with the wishes of both parties (Sofiniyah 2005).

1.2.3 *Sukuk risk on state sukuk*

The term High Risk High Return is a term that is very often heard in the general public including in the management of sukuk. Based on contracts and relationships between parties contracting (Wahid 2010) classify sukuk risk to several risks, namely:

1. Market risk
   Market risks are two forms: systematic market risks caused by overall market price movements and less systematic market risks occur when the price of an asset changes.
2. Liquidity risk
   Risks arising from the ability to get cash on debt are prohibited in Islam
3. Credit risk
   Risks associated with the quality of assets or loans that may not be available again in the event of negligence of the parties in the settlement.
4. Asset Risk
   Risk of assets may occur due to the process of sanctuary and redemption of sukuk, where assets consisting of land, buildings, and other forms of tangible assets have been designated as contractual objects and guarantees for the expenditure of sukuk.
5. Country Risk
   The risk in this form is also called the risk of legal provisions and the problems of legislation of a country, where the certificate of sukuk is traded in a market involving various countries.
6. Counterparty Risk
   Contracts involving counterparty will typically face a moral hazard risk, in which the contracting party is required to carry out the responsibilities correctly and honestly, as it is a trust contract such as murabahah, salam, istishna ', musyarakah and mudaraba contracts. Counterparty risk may arise due to negligence of partnership on the management of sukuk management.
7. Risk of Sharia Appropriateness
   The risks in this form arise due to the theoretical understanding of the fiqih results of mujtahid

formulations understood and practiced differently by each contracting party, consequently will give effect to the form of sukuk contracts that are practiced.

1.3 *Previous research*

Aam & Jakarsih (2010) examines the corporate sukuk (corporate sukuk) in Indonesia. The purpose of this research is to know what are the development problems faced and the fiber development solution how the strategy implication. The approach used is Analytic Network Process (ANP). The result of his research is the challenge of corporate sukuk development in Indonesia dominated by market participants and regulation. Among the minimum understanding of investors about sukuk, as well as tax uncertainty factors on income from sukuk that makes the development of sukuk korporsai become less optimal.

Khairunnisa (2009) examines the use of sukuk as a source of state funding and its effect on the level of welfare State. The purpose of this research is to know how big influence of sukuk to state prosperity level. The results of his research indicate that there is a significant difference between the use of sukuk and the welfare of a State.

Karimzadeh et al. (2013) examines the role of Sukuk in the Iranian Capital Market in the Arabian Journal of Business and Management Review (Oman Chapter). Which has the purpose of research to know the role of sukuk in the Capital Market in Iran. This study resulted in the conclusion that the development of both Islamic and corporate sukuk sukuk very rapidly. Karimzadeh recommends that Iran adopt some policies to encourage sukuk development in the Iranian capital market such as promoting sukuk promotion policy, sukuk taxation policy and environmental provision policy conducive to the development of sukuk associated with monetary and fiscal markets in Iran.

Ben (2014) examines the relationship between Islamic banks and sukuk markets and their effects on economic development, a case study on Tunisia's post-revolutionary economy published in the journal of Islamic Accounting and Business Research. This study analyzes the dominance of investment in the form of savings in Islamic banks and compared with the existence of sukuk market in Tunisia. An important finding of this research is the ease provided by sharia banks to finance economic development and solving the problems of poverty and unemployment. Another finding is the increasing role of Sharia intermediary institutions with the sukuk market, as this can overcome the problems related to the mobilization of Islamic bank financing, liquidity management and long-term investment.

Nazar (2011) examines the sukuk and its implications for financial regulation to promote the development of sukuk in syariah capital markets in a country. An important finding of Nazar research is the discrepancy between the western legal system if it is used to protect the rights of sukuk holders in the event of a default sukuk. The results of the Nazar research were presented in the 8th International Conference on Islamic Economics and Finance in Qatar.

2 METHOD

This research uses the approach of Qualitative Eksploratoris Malhotra (2010) said that explorative research aims to get a picture and understanding of the problems faced by researchers.

Determination of Informant in this research using purposive technique. Information obtained by doing depth interviews to find a Sukuk process to be published soon, and cause public enthusiasm. In this research the informants are BI, OJK, Fiqh Expert, Ministry of Finance, Practitioners, and Academics.

Analytical techniques used are AHP and belong to (Miles and Hubermain 1992). A qualitative approach is used with the intention of observing sukuk which is an alternative for the source of APBN funds. Analytical Hierarchi Process (AHP) technique was used in this study to weight criteria and sub criteria to find out the main criteria most considered in decision making in terms of sukuk and how to make sukuk better known to the public (Fahrurrozi 2008). AHP techniques in this research are:

a. AHP to choose about the important things that can be used to develop sukuk. In this research the informants are BI, OJK, Fiqh Expert, Ministry of Finance, Practitioners, and Academics.
b. Develop strategies to increase public interest in sukuk. In this research, the informants are Islamic Marketing experts both practitioners and academics.

Based on the results of the two AHPs, depth interviews were conducted to Islamic marketing experts, Islamic finance experts, expert of Islamic finance institutions to be analyzed to obtain a strategy to generate community to assist the sukuk development process. This result is obtained from descriptive analysis. Descriptive analysis is used to look at the importance of sukuk in the expenditure budget in a country, therefore this information can be obtained by conducting interviews.

In this study the analytical techniques also use the theory of (Miles & Huberman 1992) in the book Yin entitled Design Case Studies and Methods consisting of three processes:

a. Data Reduction
   Data reduction is the process of selection, focusing on simplification, abstraction and rough data transfers arising from written records in the field (Miles & Huberman 1992).
b. Display Data
   The presentation of data is a set of arranged

information and provides the possibility of conclusion and subsequent action taking.
c. Conclusion Drawing/Verification
   At this stage, researchers look for patterns, themes, relationships, equations, things that often arise, hypotheses and so forth to obtain a conclusion (Nasution 1998). The conclusion must be verified during the course of the research in various ways so that its credibility can be accounted.

In this research, triangulation used to test the validity of triangulation data used in this research is triangulation of source and triangulation technique. (Sugiyono 2007):

a. Triangulation of Resources
   Triangulation of sources to test the credibility of data is done by checking the data that has been obtained through several sources.
b. Triangulation Technique
   Triangulation techniques to test the credibility of data is done by checking the data to the same source with different techniques

3 RESULTS AND DISCUSSION

This research data is taken by interview to the speakers about the obstacles and solutions of sukuk development in Indonesia. Based on the results of interviews with some informants from the side of the barriers consist of 4 criteria ie investors, issuers, markets, and legality.

Based on Table 1, it explained that of the four existing barrier criteria in the development of sukuk in Indonesia, the issuer aspect barrier is the most important barrier criterion with an average harmonic of 0.306.

Based on Table 2, it explained that the sub criteria included in the criteria of constraints aspect of investors who have the most influence and impact if it happens is sub criteria pricing refers libor with harmonic average of 0.541.

Based on Table 3, it explained that the sub criteria included in the criteria of constraints aspects of issuers that have the greatest impact and impact if it occurs is a sub criteria of cost efficiency so leess competitive disadvantages with average harmonic 0.247.

Based on Table 4, it explained that the sub criteria included in the criteria of barriers of market aspects that have the greatest impact and impact if it occurs is a sub criteria of the lack of sukuk variants with an average harmonic of 0.415.

Based on Table 5, it explained that the sub criteria included in the criteria of legal barrier that has the greatest impact and impact if it happens is sub criteria of low regulation with average harmonic sebeesar 0,244.

Based on Table 6, it explained that the criteria included in the criteria of the solution that has the greatest impact and impact if it happens is the criteria aspects of the issuer with an average harmonic of 0.485.

Based on Table 7, it explained that the sub criteria included in the criterion solution aspect of investors who have the greatest impact and impact if it happens is sub criteria to reduce costs with the average harmonics of 0.595.

Based on Table 8, it explained that the sub criteria included in the criteria solution issuer aspects that have the greatest impact and impact if it happens is the sub criteria of asset identification with harmonic average of 0.354.

Based on Table 9, it explained that the sub criteria included in the criteria of market aspect solutions that have the greatest impact and impact if it happens is a sub criteria of variation of sukuk contract with an average harmonic of 0.567.

Based on Table 10, it explained that the sub criteria included in the criteria of the solution of the legal aspects that have the greatest impact and impact when it occurs is a sub-criteria for uniform perspective on the sukuk with an average harmonic of 0.620.

4 CONCLUSION

From the above explanation can be concluded that the most influential obstacles to the development of sukuk in Indonesia is the criteria aspects of issuers and sub criteria that can inhibit the development of sukuk in Indonesia is pricing refers labor, cost efficiency so less competitive disadvantages, lack of sukuk variants, and low regulation. While the solution to develop sukuk in Indonesia should be done is from the criteria aspects of issuers and sub criteria to reduce costs when entering Islamic financial markets, asset identification, variations of ethnic contracts, and uniform perspective on sukuk.

REFERENCES

Aam, S.R. & Jakarsih, M. 2010. Menguarai masalah pengembangan sukuk korporasi di Indonesia: Pendekatan metode analitic network process (ANP).

Ansari, T.M., Marr, I.L. & Tariq, N. 2004. Heavy metals in marine pollution perspective a mini review. *Journal of Applied Sciences* 4(1): 1–20.

Ascarya, A. & Yumanita, D. 2007. The profile of micro, small and medium enterprises in Indonesia and the strategy to enhance Islamic financial services through baitul maal wa tamwiel. *In The 2nd Islamic Conference (iECONS2007) Kuala Lumpur.*

Ben, J.K.D. 2014. Islamic banks-Sukuk markets relationships and economic development: The case of the Tunisian post-revolution economy. *Journal of Islamic Accounting and Business Research* 5(1): 47–60.

Fahrurrozi. 2008. Penerapan Analytical Hierarchy Process Dalam Sistem Penunjang Keputusan Pemilihan Obat (SIPEBAT). Skripsi. Depok: Fakultas Ilmu Komputer Universitas Indonesia.

Huda, H., Nurul, N. & Mustafa, E. 2008. *Investasi pada pasar modal syariah, cet.II.* Jakarta: Perdana Media Group.

Ibrahim, W. 2009. *Islamic Finance, Keuangan Islam dalamperekonomian Cet I.* Yogyakarta: Pustaka Pelajar.

Iggi H. 2003 Investasi Syariah di Pasar Modal: Menggagas Konsep dan Praktek.

Khairunnisa, M. 2009. Rentannya Implementasi Sukuk di Indonesia.

Kuriniawati, K., Devi, D. & Dwi, D. 2013. Analis perkembangan sukuk (obligasi syariah) dan dampaknya bagi pasar modal syariah. Surabaya: Universitas Negeri Surabaya.

Karimzadeh, S., Cakir, Z., Osmanoğlu, B., Schmalzle, G., Miyajima, M., Amiraslanzadeh, R. & Djamour, Y. 2013. Interseismic strain accumulation across the North Tabriz Fault (NW Iran) deduced from InSAR time series. *Journal of Geodynamics* 66: 53–58.

Lotunani, A., Idrus, M.S., Afnan, E. & Setiawan, M. 2014. The effect of competence on commitment, performance and satisfaction with reward as a moderating variable (a study on designing work plans in Kendari City government, Southeast Sulawesi). *International Journal of Business and Management Invention* 3(2): 18–25.

Malhotra, M. & Naresh, K. 2010. *Risetpemasaran: Pendekatan Terapan. Jilid 1.* Jakarta: PT. Indeks.

Miles, B.B. & Huberman, A.M. 1992. *Analisa Data Kualitatif.* Jakarta: UI Press.

Nasution, A. 1998. The meltdown of the Indonesian economy in 1997–1998: Causes and responses.

Rusydiana, A. 2012. Analisis penguraian masalah pengembangan sukuk korporasi di Indonesia pendekatan metode ANP (Analytic Network Process).

Salim, F. 2011. *Konsep dan Aplikasi Sukuk Negara Dalam Kebijakan Fiskal Di Indonesia.* Jakarta: UIN.

Sofiniyah, G. 2005. *Briefcase Book Edukasi Profesional Syariah Cara Mudah Memahami Akad- Akad Syariah.* Jakarta: Renainsan.

Sugiyono, S. 2007. *Metode Penelitian Kuantitatif Kualitatif dan R&D.* Bandung: Alfabeta.

Umer, C. 2000. *Sistem moneter Islam.* Jakarta: Gema Insani Press.

Wahid, N.A. 2010. *Memahami & Membedah Obligasi Pada Perbankan Syariah.* Yogyakarta: Ar-Ruzz Media.

Advances in Business, Management and Entrepreneurship – Hurriyati et al (eds)
© *2020 Taylor & Francis Group, London, ISBN 978-0-367-27176-3*

Realized strategies and financial performance in the Indonesian banking industry

A. Kunaifi, M.S. Hakim & B.M. Wibawa
Institut Teknologi Sepuluh Nopember, Surabaya, Indonesia

ABSTRACT: This study aimed at explaining and describing the typology of realized strategies and financial performance in the Indonesian banking industry. The sample was 30 banks characterized as Strategy Business Units (SBU). Although descriptive analysis showed that banking with a pure strategic typology (pure defender, pure analyzer, and pure prospector) performed better than banking with a hybrid strategy (DA-like and PA-like), the one-way ANOVA analysis result showed that a realized strategy performed equally well in an oligopolistic competition market.

1 INTRODUCTION

The banking industry has an important role in the financial system in Indonesia. The intermediating function of banking in the financial system supports the country's development program. To harmonize the banking industry and the Indonesian vision development, the Financial Services Authority (*Otoritas Jasa Keuangan*) released a roadmap policy for 2015–2019, with a policy is focusing on antici-pating banking competitiveness in ASEAN countries.

Competitiveness in the banking industry must be controlled and evaluated to measure the impact of banking competition on national banking stability. The loss of market share and margin will be a threat in the banking industry, and therefore banks must have good strategies to compete.

The impact of the realized strategy and the performance of a firm must be evaluated to show the effectiveness of the strategy. Some researchers have studied the impact of a realized strategy on performance.

Kotha & Nair (1995) explained that a firm's strategy plays a significant role in influencing profitability and growth. Berman et al. (1999) argued that the relationship between strategy (efficiency and capital intensity) and financial performance was negative. Spanos et al. (2004) found that a mixed strategy was more profitable than a pure strategy. Claver-Cortés et al. (2012) revealed that a hybrid competitive strategy positively influenced a firm's performance. Anwar and Hasnu (2016) explained that the performance of a firm was significantly differentiated under different strategy typology. This research aimed at explaining the effect of strategy typology on a firm's financial performance under an oligopoly market like the banking industry.

Realized strategy is a part of a strategic management process (Mišanková & Kočišová 2014).

A strategic management process includes strategic planning, strategic implementation, and strategic control. Successful strategic management brings all processes in strategic management together in the everyday decision-management process. Realized strategy needs tools to be implemented, such as company culture, organizational structure and control, and rewards and support from administration functions (Mišanková & Kočišová, 2014). Before a realized strategy, a company must arrange strategic planning by considering and evaluating its past strategy. Strategy typology has been used to simplify the strategy classified in realized strategy.

Different strategic typologies have been explained by researchers. Miles et al. (1978) analyzed four strategy types: prospector, analyzer, defender, and reactor. A defender strategy is persistent reliance on the continued viability. A prospector strategy is opposite to a defender strategy, and responds to the chosen environment. An analyzer strategy is a combination of prospector and defender strategies: an analyzer strategy is conservative with risk and a reactor strategy always adjusts to the business environment, which makes the business inconsistent and unstable.

Porter (1980) classifies generic strategy types into cost leadership, differentiation, and focus. A cost leadership strategy promotes lower costs than competitors. A differentiation strategy requires unique products or services to have a higher price than competitors. A focus strategy concentrates on the segment of the market (Dess & Davis, 1984).

Treacy and Wiersema (1993) divide strategy into three categories: operational excellence, product leadership, and costumer intimacy. Operational excellence focuses on delivering products and services that excel in price and convenience. Product leadership produces a continuous product and service within their competency. Costumer intimacy develops customer loyalty for the long term.

Strategy and performance as a form of market competition, especially in oligopoly, have been explained by researchers (Bulow et al. 1985, Levin et al. 2009). In Indonesia, the competition market form of the banking industry tends to be oligopoly (Kunaifi et al. 2016). This research tries to explain the effect of strategy typology on financial performance in an oligopoly market like the banking industry.

2 METHOD

The sample for this research was 30 banks in the Indonesia banking industry in 2015. The criteria for selection was that the bank should only have one operating system, either sharia banking or conventional banking. The mechanism to measure the strategy was using proxy and scoring mechanisms to assess Miles and Snow's strategic types using archived financial data. The measurements are as follows. (a) COGTA: cost of goods sold to total assets (Kotha & Nair 1995). This measurement was used to assess cost-efficiency. The relationship between the measurement and dimensions of the Miles and Snow typology was high for prospector and low for defender banks. (b) CI, capital intensity ratio: ratio of the total assets to total employed each year to measure the firm's technological focus (Kotha & Nair 1995). The relationship between the measurement and dimension of the Miles and Snow typology is high for defender and low for prospector banks.

The strategy measurement is constructed by combining the score of the two proxies (COGTA and CI). The composite score is calculated by giving the highest percentiles a score of 4 and the data in the lowest percentile i a score of 0. The reverse score is calculated for capital intensity, where defender has the highest score and prospector has the lowest score. The total score is 8 and the lowest score is 0. For the proxy of the strategy, the typology is categorized as pure defender (0–1.6), DA-like (1.7–3.2), pure analyzer (3.3–4.8), PA-like (4.9–6.4), and pure prospector (6.5–8). Financial performance measured by ROA was taken as the dependent variables in this analysis. Cross-tab analysis was used to explain the banking performance in the context of the realized strategy.

3 RESULTS AND DISCUSSION

The descriptive analysis shown in Table 1 explains that the mean of the ROA is -0.1237, showing that financial performance of banks in Indonesia is relatively low and range from the maximum 5.24 to the minimum -20.13. It shows that a diverse variability in the financial performance of the Indonesian banking industry. The variable cost-efficiency indicates that, generally, the banks' performance has been efficient, and they have controlled their direct cost of sale. The capital intensity proxied by total asset to the number employed shows that banking in Indonesia has grown.

The result in Table 2 shows that there are very few pure prospectors and pure defenders in Indonesian banking. There is 43% of the realized strategy in Indonesia banking is DA-like with negative performance. However, pure analyzer have been realized by 20% bank in Indonesia with positive performance. Pure defender showed highest performance, and prospector showed positive performance.

Defender-like (DA-like) refers to hybrid strategy in the banking industry and shows negative ROA. This finding is in contrast to the result from previous research (Claver-Cortés et al. 2012, Spanos et al. 2004). Table 2 shows that pure defender has the highest performance of ROA. This finding explains that a pure strategy performs better than a hybrid strategy in an oligopoly market, because a firm action in one market can change a competitor's strategy (Bulow et al. 1985). Strategic behavior by consumers can have a serious impact on revenue in oligopolistic competition (Levin et al. 2009).

The one-way ANOVA in Table 3 indicates that there were no statistically significant differences in financial performance between the means of strategy typology. This finding explains that strategic types

Table 1. Descriptive statistics.

	Minimum	Maximum	Mean	Std. Dev
ROA	-20.13	5.24	-.1237	4.28852
COGTA	0.00	2.60	0.6911	0.54060
CI	408	27285	10352	6520
N	30			

Table 2. Strategy typology in Indonesia's banking industry.

	Strategy typology					
	Pure defender	DA-like	Pure analyzer	PA-like	Pure prospectors	Total
Percent	17%	43%	20%	17%	3%	100%
ROA	0.164286	-0.35929	0.120357	-0.08607	0.028214	-0.1237

Table 3. One-way ANOVA test on realized strategy and financial performance.

Sources	Sum of squares	df	Mean square	F	Sig.
Between groups	15.237	4	3.809	.184	.945
Within groups	518.113	25	20.725		
Total	533.350	29			

will perform equally well for financial performance. This finding is consistent with previous research conducted by Anwar and Hasnu (2016) and Saraç et al. (2014). It is important for management to not only choose an appropriate strategy type but also be consistent and focus on customer satisfaction. The successful application of realized strategies is not only in choosing the strategy type but also in the strategic management process, including strategic planning, strategic implementation, and strategic control. Every process in strategic management is interconnected and must be evaluated to assess the effectiveness. When the realized strategy was performed equally, it was well implemented in the banking industry. This implies that banks in Indonesian have been consistent in implementing their strategies. Indonesian banking within an oligopolistic market can focus on strategic management and strategic control. The success of strategic management can be realized by managers, employees, their organization and by transforming culture in the organization (Mišanková & Kočišová 2014).

4 CONCLUSION

The purpose of the research was to investigate strategy typology and performance in the Indonesian banking industry. This research applied Miles and Snow's strategic typology using archived financial data. The research in this paper described the realized strategies in the banking industry. Most of the banks in Indonesia have realized a DA-like strategy. However, the performance of ROA in DA-like banking is relatively poor. The highest performance was acquired by a pure defender strategy. This result explained that realized strategic planning was effective in an oligopolistic competition like the banking industry.

Further study could develop performance in non-financial firms to evaluate the realized strategies. Performance as the outcome from a strategic management process can be defined in a comprehensive view, not only in financial but also in non-financial performance, such as customer satisfaction.

REFERENCES

Anwar, J., & Hasnu, S. 2016. Business strategy and firm performance: a multi-industry analysis. *Journal of Strategy and Management*, 93, 361–382.

Berman, S. L., Wicks, A. C., Kotha, S., & Jones, T. M. 1999. Does stakeholder orientation matter? The relationshp between stakeholder management models and firm financial performance. *Academy of Management Journal.*, 425, 488–506.

Bulow, J. I., Geanakoplos, J. D., & Klemperer, P. D. 1985. Multimarket oligopoly: Strategic substitutes and complements. *Journal of Political Economy*, 933, 488–511.

Claver-Cortés, E., Pertusa-Ortega, E. M., & Molina-Azorín, J. F. 2012. Characteristics of organizational structure relating to hybrid competitive strategy: Implications for performance. *Journal of Business Research*, 65, 993–1002.

Dess, G. G., & Davis, P. S. 1984. Porter's 1980 Generic strategies as determinants of strategic group membership and organizational performance. *The Academy of Management Journal*, 273, 467–488.

Kotha, S., & Nair, A. 1995. Strategy and environment as determinants of performance: evidence from the Japanese machine tool industry. *Strategic Management Journal*, 16.7, 497–518.

Kunaifi, A., Hakim, M. S., & Wibawa, B. M. 2016. Financial competitiveness analysis in the Indonesian islamic banking. *Proceedings Seminar Nasional Ilmu Manajemen 2016*.

Levin, Y., Mcgill, J., & Nediak, M. 2009. Dynamic pricing in the presence of strategic consumers and oligopolistic competition. *Management Science*, 551, 32–46.

Miles, R. E., Snow, C. C., Meyer, A. D., & Coleman, H. D. 1978. Organizational strategy, structure, and process. *The Academy of Management Review*, 33, 546–562.

Mišanková, M., & Kočišová, K. 2014. ScienceDirect Strategic implementation as a part of strategic management. *Procedia -Social and Behavioral Sciences*, 110, 861–870.

Saraç, M., Ertan, Y., & Yücel, E. 2014. How do business strategies predict firm performance? An investigation on borsa Istanbul 100 index. *Journal of Accounting & Finance*, 121–135.

Spanos, Y. E., Zaralis, G., & Lioukas, S. 2004. Strategy and industry effects on profitability: evidence from greece. *Strategic Management Journal*, 25, 139–165.

Treacy, M., & Wiersema, F. 1993. Customer intimacy and other value disciplines. *Harvard Business Review*, 1, 84–93.

The effect of customer demand and supplier performance on competitive strategy

R.D. Pasaribu, A. Prasetio & O.O. Sharif
Universitas Telkom, Bandung, Indonesia

ABSTRACT: The performance of fixed broadband access penetration in Indonesia is far below the average. This poor performance is an indication that operators have not implemented competitive strategies optimally. This study aims to investigate the effect of customer demand and supplier performance on competitive strategy. A survey technique was used to collect data. The analysis units were 38 business units of the operators and the observation units were the senior management and customers of these business units. The data was then analyzed using partially least square SEM. The result showed that customer demand had a negative significant effect on competitive strategy, while supplier performance had a positive significant effect on competitive strategy.

1   INTRODUCTION

In theory, the quality and quantity of fixed broadband is better than wireless broadband, because a radio frequency is limited while an optical fiber has a very wide frequency. However, when this is viewed by performance, the fixed broadband telecommunication industry in Indonesia is still poor.

In the early 2000s, Indonesia, Malaysia, Vietnam, Philippines, and Thailand were at a relatively similar level of broadband coverage, with penetration rates below 1%. However, in later years, while the other countries grew rapidly (in 2012 Malaysia had reached about 55% and Vietnam had reached about 40% coverage), Indonesia was still at single-digit coverage of 8%. The latest ITU issue of 2015 still showed similar symptoms of lag In 2014, fixed broadband national penetration was at 10.14% in Malaysia and 6.48% in Vietnam has reached, while Indonesia was only at 1.19%.

Another phenomenon can be seen from mobile and fixed broadband comparison data. Mobile broadband meets the largest portion of the broadband market share, with a total subscriber base of 3.4 million, while fixed broadband only has 1.5 million subscribers. From this data, it is clear that fixed broadband has not been able to attract subscribers. In other words, fixed broadband services have not been able to meet customer needs. This indicates a gap in which the terms 'requirement' of the customer on the value has not been matched with the companies capabilities (Cravens & Piercy 2013).

On the other hand, the role of the broadband access network is very strategic for improving the nation's competitiveness. Therefore, the role of telecommunication operators must be better, and growth is required in all areas, including in fixed broadband, to maintain the continuity of State existence.

The preliminary study of secondary data and an interview with the officers of the operators, clearly identified that the performance of the fixed broadband industry has not grown optimally (Pasaribu et al. 2016). The alleged reason for this is that the operator still has weaknesses in their competitive strategy.

Competitiveness strategy can be improved via several inputs, such as customer demand and supplier performance, and also other factors like the collaboration of the actors (Anggadwita et al. 2016). Unfortunately, it is clear that fixed broadband operators have not been able to fulfill customer demand. Of the many characteristics of the fixed broadband service, the first priority of the customer is the quality of the network. The quality of network is the responsibility of various parties. One of the main processes is developing suppliers and access network maintenance suppliers. A preliminary study of dominant operator management also identified that other weaknesses were on the side of an access network supplier, where partners used cascading patterns without creating value added in their work chains. As a result, there was a shortage of skilled employees/technicians for building and maintaining the fixed networks broadband. Based on Griffin (2013), every company/organization can only operate properly if it has complete resources, namely human capital, financial resources, and physical resources. As suppliers are partners who fulfill the customer's request, the relationship dimensions between suppliers and companies (supplier relationship) needs to be examined.

Based on this description and the fact that there is no research yet on this subject, it is considered important to investigate the influence of customer demand and supplier performance on competitive strategy and business performance of the fixed broadband industry in Indonesia. Therefore, this study aims to investigate the effect of customer demand and supplier performance on competitive strategy.

Chen and Yang (2002) showed that the parameters used in planning and building a supplier performance rating system focus on satisfying customers, both internal and external, leading to the best value for money when acquiring goods and services. Boon-itt and Chee (2011) found that supplier integration and internal integration positively is associated with customer delivery performance. Terpend and Krause (2015) found that competitive incentives could be an effective approach to improve delivery, quality, innovation and flexibility, for purchases where buyer–supplier relations were characterized by a balanced and moderate number of interdependencies.

According to the concept of three-value discipline, the company should choose one focus, based on their strengths, which suits the needs of the customer (Treacy & Wieserma 1999). It is clear that customer requests should be one of the cornerstones of importance, and inevitably have an impact and influence on the enterprise to formulate its competitive strategy. Mao (2009) suggests that the survival and development of a company not only depends on the resources owned but also rely on consumer value. Therefore, customer demand strategy has become an important part of strategic development strategies. Bernardo (2001) considered that the role of customer demand creation in the bank's strategy is a response to the ecstatic changes in their growth opportunities. Veliyath and Fitzgerald (2000) conclude that sustainability can be enhanced by combining customer preferences (as well as the type and value provided) into strategy and by taking into account the availability of isolated mechanisms within the market environment or local business.

Al-Abdallah et al. (2014) indicated that two practical matters of SRM, the development of supplier partnership and lead time improvement suppliers, have a significant positive impact on the company's business performance. Gummesson et al. (2013) revealed that in the current conditions where consumers were becoming more active, the role of suppliers that traditionally controlled consumers was no longer appropriate. Kanagal (2009) revealed that a competitive competition strategy needed to have relationship marketing (RM) as one of the key functions in improving business performance. RM's role in the competitive marketing strategy involves the moment of truth guidance, improving profitability, building partnerships, demonstrating 'better customers', buying in customer attention, protecting emotions, understanding consumers' lives, and building trust with customers. Koufteros et al. (2012) identify strategic suppliers as a source of promising competitive advantage in resource-based views.

2 METHOD

The research method used was a quantitative method. The data analysis was used partially least square structural equation modeling (PLS-SEM). The operationalization of variables in this study was constructed on the conceptual model discussed in the previous chapters. The variable operational table describes the variable configuration, dimensions, indicators and measures used for the research variables. These variables consisted of independent variables (customer demand and supplier performance), and intervening variables (competitive strategy).

The research analysis unit was the business unit of the fixed broadband operator with the management and customers of each business unit as the observation unit. Management became the respondent for the performance variable.

3 RESULTS AND DISCUSSION

The test result shows that F arithmetic was higher than 3.267 (F table at $\alpha = 0.05$), which meant that MP and PP had simultaneously significant effects toward SB. The coefficient of determination R^2 shows that MP and PP had an influence of 28% and the remaining 72% was influenced by other factors. This finding indicates that H0 is rejected, while H2 is accepted. Thus, customer demand and supplier performance simultaneously affect the competitive strategy.

Based on this test, it can be seen that MP had a negative and significant influence on SB. This condition represents a factual condition in which the industry has not responded to MP preferences with the formulation and implementation of an appropriate SB focus. MP prioritizes product quality, while SB focuses on cost leadership, which should be more likely to lead to product differentiation. This is in accordance with the descriptive test of H1, where MP has a preference on product quality and SB tends to accentuate cost leadership.

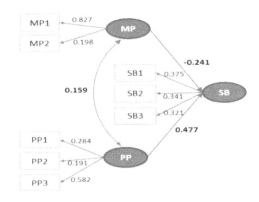

Figure 1. PLS-SEM result.

844

Conditions in the telecommunications industry as a whole encourage this to happen. Beside the pressure from competitors, a much stronger pressure comes from complementary mobile broadband. This industry is already at the stage of high price, where price competition reduces the entire telecommunication market, which has also affected the fixed broadband market. The significant negative influence of MP on SB indicates that the competing fixed broadband SB industry has not been formulated and implemented because it has not responded to MP. This needs solutions, which will be one of the programs for problem solving.

The test result shows that F arithmetic was higher than 3.267 (F table at α = 0.05), which meant that MP and PP had simultaneously significant effects toward SB. The coefficient of determination R2 shows that MP and PP had an influence of 28% and the remaining 72% was influenced by other factors. This finding indicates that H0 is rejected, while H2 is accepted. Thus, customer demand and supplier performance simultaneously affect the competitive strategy.

Based on this test, it can be seen that MP had a negative and significant influence on SB. This condition represents a factual condition in which the industry has not responded to MP preferences with the formulation and implementation of appropriate SB focus. MP prioritizes product quality, while SB focuses on cost leadership, which should be more likely to lead to product differentiation. This is in accordance with the descriptive test of H1, where MP has a preference on product quality and SB tends to accentuate cost leadership.

Conditions in the telecommunications industry as a whole encourage this to happen. Beside the pressure from competitors, a much stronger pressure comes from complementary mobile broadband. This industry already exists at the stage of high price, where price competition reduces the entire telecommunication market which is also affected on the fixed broadband market. The significant negative influence of MP on SB indicates that the competing fixed broadband SB industry has not been formulated and implemented because it has not responded to MP. This condition needs solutions. This will be one of the programs for problem solving.

On the other hand, PP also had a positive influence on SB. This was more significant than MP. This implies that the formulation of the SB fixed broadband industry is more influenced by PP. In the telecommunications industry, there are three important aspects, device/device, network, and application with network, which is heavy on process development and maintenance. Until now the technology that is believed to meet the needs of the fixed broadband industry is a fiber optic network. This technology is relatively capital-intensive and labor-intensive as network development is still physical and operators are heavily dependent on the development and maintenance of networks by their partners, not to mention the

application partners/content and device partners either on the operator side or customer premises equipment (CPE).

In this regard, network suppliers need to get serious attention from operators to be able to support proper SB formulation. Therefore, PP must be supported by its dimension level. A supplier must have better resources, work performance, and quality relationships. Thus, it can be concluded that the business unit does not respond to customers (and competitors) but it is more affected by the PP.

A partial test of MP to SB shows that the effect was significant because the t-count was higher than 1.688 (t table at α = 0,05), but with negative sign. This finding shows that H0 is rejected and H2a proves that customer requests are partially influential on the competing strategy with a negative sign.

This finding is important as it relates to the customer satisfaction, by which the competitive strategy applied by the business unit does not accommodate the customer demand preferences. A business unit is basically aware of customer preferences, but it is contested by competitors. Thus, it is driven to perform an effective and efficient process in achieving more work targets, which focuses on sales with a narrow target time. In conclusion, business unit in the application of competitive strategy is not market-oriented (customer and competitor) or market-based oriented.

The effect of PP on SB was positive and more significant. T-arithmetic was higher than 1.688 (t table at α = 0.05) and the quantity of t calculate to PP-SB was 7.813, which was greater than MP◊SB (2680). This finding implies that H0 is rejected and H2b proves that supplier performance partially influences competitive strategy positively. In general, it can be said that customer requests and supplier permanence both significantly affect the competition strategy in fixed broadband industry in Indonesia.

4 CONCLUSION

Customer demand and supplier performance are both significantly influencing the competition strategy in the fixed broadband industry in Indonesia. However, supplier performance has a more significant impact compared to customer demand. Conversely, customer demand negatively affects the competitive strategy. This is because the competitive strategy does not accommodate customer requests or is not customer/market-oriented (non-market-based oriented) and more likely to be resource-based oriented.

REFERENCES

Al-Abdallah, G.M. 2014. The Impact of Supplier Relationship Management on Competitive Performance of Manufacturing Firms. *International Journal of Business and Management*, 9, 2: 192–202.

Anggadwita, G., Amani H., Saragih R., Alamanda D.T. 2016. Competitive Strategy of Creative Application Content in the ASEAN Economic Community: Software Development using SWOT Analysis in Indonesia. *International Journal of Economics and Management*, 10 (S1): 95–107.

Bernardo Bátiz-Lazo. 2001. Customer Value Creation and Corporate Strategy In Financial Services. In: 17th Annual IMP Conference, 9–11 Sep 2001, Oslo, Norway.

Boon-itt, S., Chee, Y. W. 2011. The moderating effects of technological and demand uncertainties on the relationship between supply chain integration and customer delivery performance. *International Journal of Physical Distribution & Logistics Management*, 41, 3: 253–276.

Chen, C., Yang, C. 2002. Customer-oriented evaluation system of supplier quality performance. IIE Annual Conference Proceedings: 1–7.

Cravens, D.W and Piercy, N.F. 2013. *Strategic Marketing*. 10th ed. New York: McGraw-Hill.

Griffin, R.W. 2013. *Management* 13th ed. South-Western Cengage Learning, Australia.

Gummesson, E. Kuusela H., Narvanen, E. 2014. Reinventing Marketing Strategy by Recasting Supplier/Customer Roles. *Journal of Service Management*, 2, 2: 228–24.

Kanagal, N. 2009. Role of Relationship Marketing in Competitive Marketing Strategy *Journal of Management and Marketing Research* 2, 1–17.

Koufteros, X., Vickery, S.K., Dröge, C. 2012. The Effects of Strategic Supplier Selection On Buyer Competitive Performance In Matched Domains: Does Supplier Integration Mediate The Relationships? *Journal of Supply Chain Management*, 48, 2: 93–115.

Maohua Li. 2009. The Customer Value Strategy in the Competitiveness of Companies. *International Journal of Business and Management* 4, 2: 136–141.

Pasaribu, R.D., Dwi Kartini, O., Yevis M., Padmadisastra, S. 2016. The Effect of Customer Demand and Supplier Performance in Competitive Strategy and Business Performance (Case of Fixed Broadband Operator in Indonesia). *International Journal of Scientific & Technology Research*, epublication, 5, 2: 123–127.

Terpend, R., Krause, D.R. 2015. Competition or Cooperation? Promoting Supplier Performance with Incentives under Varying Conditions of Dependence. *Journal of Supply Chain Management*, 51, 4: 29–53.

Treacy, M. & Wieserma, F. 1995. *The Discipline of Market Leaders*. London: HarperCollins.

Veliyath, R. & Fitzgerald, E. 2000. Firm capabilities, business strategies, customer preferences, and hypercompetitive arenas: The sustainability of competitive advantages with implications for firm. *Competitiveness Review*, 10, 1: 56–82.

Advances in Business, Management and Entrepreneurship – Hurriyati et al (eds)
© *2020 Taylor & Francis Group, London, ISBN 978-0-367-27176-3*

Developing human capital industry through public private partnership: Is it profitable?

I. Helvetikasari, I.D.A. Nurhaeni & D.G. Suharto
Universitas Sebelas Maret, Surakarta, Indonesia

ABSTRACT: The industrial sector plays a strategic role in enhancing the development of the national econ-omy and community welfare. The textile and garment industry sector are a non-oil and gas sector that contrib-utes greatly to Indonesia's GDP (Ministry of Industry, 2017). Partnership between the government and the private sector in creating skilled and competent workforce is expected to help the industry to meet the needs of the workforce in accordance with the needs of the industrial world. This article discusses the results of partnerships between the government and the private sector in human capital investment in industry partners AK Tekstil Solo. This research uses descriptive qualitative method. The primary data were obtained through focus group discussion and observation. The research was conducted on the AK Tekstil Solo with the textile and garment industry partners. The result of this research is the ongoing partnership between the government and the private sector has not been able to meet the human capital needs of the textile and garment industry. The workforce that is an investment in a company does not have the expected soft skills even though the hard skills are adequate. To further the improvement of the workforce productivity, it is expected to do soft skill coaching simultaneously with hard skill quality improvement.

1 INTRODUCTION

In the era of industrialization, where there has been an increasingly high level of competition in the pro-duction of goods and services, the industrial sector plays a strategic role in the improvement of national economic development and community welfare. The industrial sector is the largest contributor to the GDP of Indonesia, which is 21.35% (Rencana Kerja Kementerian Perindustrian, 2017). The industrial sector plays an important role as a leading sector for economic growth, sources and stimulus of value-added economic creation, which has a wide and high level of relevance to almost all economic activities.

Textile products industry is one of the mainstay and priority sectors, as well as a motor of national economic development. The value of textile and tex-tile sector exports in 2013 reached US $ 12.68 bil-lion, equivalent to 8.5% of Indonesia's non-oil and gas exports (Cas, 2018). With the value of exports, Indonesia is able to meet about 1.8% of world de-mand for textile products. The Textile and textile industry sector is expected to continue to strengthen as it is a workforce-intensive industry that absorbs a lot of manpower. The textile industry absorbs approximately 2.79 million workers and is able to meet 70% of the domestic clothing needs (Studi Kelayakan Akademi Komunitas Industri Tekstil dan Produk Tekstil Surakarta 2015). Based on these con-ditions, the improvement of competitiveness is a key factor to be considered in order that the national tex-tile industry can increase its presence in both domes-tic and international markets.

One form of implementation of the 'revolusi mental', the Ministry of Industry encourages the enhancement of human resource capability through vocational higher education organized by the (AK Tekstil Solo). AK Tekstil Solo produces and pro-vides industrial workforce that has been equipped with foreman competence level. AK Tekstil Solo partnered with the textile and garment industry in learning activities both theory and practice by utiliz-ing manpower from an industry that has been already qualified, experienced and has a special skill to be able to give the lecture material. Other activities that require a partnership with the textile and garment industry is the implementation of practical lectures conducted at the factory and directly in contact with the production process, in addition to the absorption of workforce also requires a partner-ship with the textile and garment industries. These activities are carried out to create qualified human resources so that the textile and garment industry can increase its productivity if supported by a work-force that has competencies that suit the needs or demand driven.

Human Resources is one of the important and valuable factors in supporting the performance of an organization (Pella & Inayati 2011), especially in the industry sector. Because if you have good human resources, then it will bring success in the future. Not only that, the presence of good human resources will greatly affect the process of achieving organiza-tional goals. HR that has quality and added value or better is known as human capital. Human capital refers to the knowledge, attitudes, and skills needed to be developed and is an asset that determines the company's success (Ongkorahardjo 2008, Burr & Girardi 2002). On the other hand, human capital focusses on the level of workforce education as a source of productivity (Carmeli 2004) which means that education is inherent in the most important workforce is not a knowledge, attitude, and skill develop within the company. Human capital development is part of efforts to achieve more effective performance (Marimuthu et al, 2009). Therefore, companies need to understand more about the development of human capital. Human capital development can be done through education and training in terms of knowledge and skills (Marimuthu et al. 2009). According to (Schuller 2000), the core of human capital is the skills, knowledge, and competencies that are key factors that determine the welfare of the organization. In contrast, (Davenport 1999) has a different view about the core of human capital that employees should not be treated as a passive asset that can be bought, sold and replaced by the owner of the organization, but it should be noted that employees also actively have control over their working life. Employees, especially educated employees, can decide on their own. In human capital as a strategic asset, there are four dimensions (Barney 1991) that are usefulness, behavioral uncertainty, firm-specificity, and spread of strategic human capital.

This article discusses the partnership between government and private sector to the development of human capital in textile and garment industry. The development of human capital of the industry is done through a partnership between AK Tekstil Solo and its textile and garment industry partners located in Central Java. Central Java Province is a province whose economic activity is driven by one of them by textile and garment industry. Currently, Central Java Province has a textile and garment industry of 516 industries with a workforce of 223,583 people (Studi Kelayakan Akademi Komunitas Industri Tekstil dan Produk Tekstil Surakarta 2015).

2 METHOD

This study uses a qualitative method. Primary da-ta were obtained through a focus group discussion organized by AK Tekstil Solo and followed by representatives of HRD/HRM companies that became partners of AK Tekstil Solo. The sampling technique uses purposive sampling technique. Observations were made at PT Sri Rejeki Isman, Tbk., PT Dan Liris, PT Kusumahadi Santosa, PT Pan Brothers, Tbk., PT Sinar Surya Indah Lestari, PT Bintang Asahi Textile Industries, and PT Kusoema Nanda Putra. While for data analysis techniques used interactive analysis of Miles and Huberman models.

The variables that used are human capital dimensions: usefulness, behavioral uncertainty, firm-specificity, and spread of strategic human capital (Barney 1991) to see the public private partnership

between the ministry of industry through AK Tekstil Solo with industrial partners in creating human capital on AK Tekstil's industry partner.

According to (Barney 1991), there are four dimensions of human capital that affect the success of the company that is the dimension of usefulness is when the workforce has the potential to contribute to innovation and renewal strategy so it can help companies to take advantage and opportunities. The uncertainty's behavioral dimension relates to the actions of the employee and how the actions affect the results achieved by the firm. The firms specify dimension shows the specific knowledge and skills possessed by the workforce that the company can use. The dimension of the spread of strategic human capital according to (Barney & Wright 1998) is a collection of competencies owned by the workforce synergized by the company so as to form a good teamwork so that the mission of the company's missions can be achieved.

3 RESULT AND DISCUSSION

3.1 Profile of AK tekstil Solo graduates

The Ministry of Industry established a special skills-based higher education for the development of human resources in Central Java. AK Tekstil Solo is a textile and garment vocational college that focuses on developing human resources in accordance with industry needs. AK Tekstil Solo organizes a diploma (D2) which equivalent to level 4 on the Indonesian National Qualification Framework. Referring to the fourth level of the Indonesian National Qualification Framework, graduates of AK Tekstil Solo must be able to carry out specific work that is routine and in accordance with the requirement of work and quality standards, able to solve work problems, cooperate, communicate and have initiative and be able to be given responsibility for quality and the work of others. Graduates of AK Tekstil Solo is not only get a certificate as a sign of graduation but also given a certificate of competence issued by the Badan Nasional Standarisasi Profesi (BNSP).

AK Tekstil Solo since its presence in 2015 has graduated 102 people and has been absorbed entirely into the textile and industries, namely PT Sri Rejeki Isman, Tbk., PT Dan Liris, PT Kusumahadi Santosa, PT Pan Brothers, Tbk., PT Sinar Surya Indah Lestari, PT Bintang Asahi Textile Industries and PT Kusoema Nanda Putra. Based on the data that obtained from AK Tekstil Solo partner, currently graduates occupy the position of Middle-Down, which is the level of foreman, head shift and also operator.

3.2 Usefulness

Usefulness is the belief of managers regarding the use of human capital (Barncy 1991). In human capital investment conducted by AK Tekstil industry partners is in the form of absorption of AK Tekstil

Solo graduates. Graduates can be a very potential resource and contribute innovative ideas for the progress of the company. In the research found that some graduates have been able to provide ideas, theoretical and technical knowledge, and organizational experience that is very useful to provide new colors of the quality of human resources provided for the progress of the company. The human resources that come from AK Tekstil Solo contribute because they give inputs in the production process and create innovation. There is also a work-force from the graduates of AK Tekstil Solo is still limited to do field work in accordance with and has not been able to contribute to the provision of innovative ideas and innovations. The research also found that AK Tekstil graduates do not have good mental endurance and are less prepared with textile and garment production activities that are full of pressure and target, marked by several graduates applying for moving work site and resignation, whereas they have been carrying out industry practice for one month in each semester. One of the obstacles is the schedule of practice or information that is not delivered on time by both parties, so the communication and information are slightly inhibited.

3.3 Behavioral uncertainty

The behavioral uncertainty dimension in human capital is the actions undertaken by employees and how such actions affect the outcomes that the company will achieve. The results of the research show that human investments by AK Tekstil industry partners by absorbing human resources from graduates have not given the idea of reducing pro-duction and service costs, they only play a role in providing input on efficiency in running the production. Graduates of AK Tekstil Solo still occupy the position of Middle-Down in the organizational structure, so they only have limited authority, and they have no authority to decide a step of efficiency or special measures that play a ma-jor role for the progress of the company.

3.4 Firm-specificity

Firm-specificity shows specific knowledge and skills that can be used for the company's progress. Not all graduates of AK Tekstil Solo have special skills in the production department, but still have to undergo the training process in advance so hopefully there will be improvisation especially for certain parts that require competencies and special skills that become core competencies compared with expertise owned by other companies and can add the selling value of the company's products.

3.5 Spread of strategic human capital

Competence that owned by each individual and synergized to produce a new thing, in this research the

spread of strategic human capital is defined as teamwork. The result of the research on the spread of strategic human capital is the alumni of AK Tekstil Solo in building teamwork always put togetherness in work and as a partner to go towards the goal in production and can quickly build team work with other employees well. However, it was also found that some individuals behaved that they were "special cases". So, they feel the need to get a certain position and preferential treatment. This attitude can form an unfavorable stereotyping for the graduates of the next Solo Textile AK. The attitude also indicates that the soft skills possessed by the graduates are not good, the attitude that privileges the group indicates that the interpersonal soft skill needs to be improved.

3.6 Discussion

We live in complex societies where policies often seem to fail to provide satisfactory solutions to problems, so partnerships with the private sector and the community are needed to solve problems. The business world and the industrial world also need a partnership with the government and society to further productivity improvement.

Partnerships are very important because companies and governments must be concerned with the development, dissemination, and management of a more sophisticated way to solve social, economic and political problems (Clement et al. 2004, Onojaefe & Marcus 2007). Partnership is an important concept in the paradigm of state administration, namely the governance paradigm. According to UNDP in 1997 the governance paradigm can be interpreted as governance organized by multistakeholders. Partnerships with private institutions and civil society are one of the characteristics of governance. In governance, the state is no longer the only dominant actor in the administration of public affairs and public services. The state requires other parties, in this case, the private sector to carry out public services.

In implementing partnerships, uniting all actors with the same vision and mission is not easy, partners must have the same rights in a partnership (Grundey & Daugėlaitė 2009). Managing a partnership is not an easy thing, moreover, the government and the private sector have different visions and missions. The government carries out partnerships to improve community welfare in advance, while the private sector partnerships is to increase income. So that a bridge is needed to accommodate the interests and objectives of each party. As is the case with the Ministry of Industry through AK Tekstil Solo which partnerships with the textile and garment industry to create a competent workforce. The partnership that is established is expected to contribute in the form of the birth of a competent workforce and can become an investment asset of human capital in the textile and garment industry. While the partnership is established, the government can also reduce the unemployment rates.

Karim & Shabbir (2012) explained that in developing countries, human capital can improve industrial progress for sustainable development, explained that the higher the level of training and education, the higher the skills that will be owned by workers from any country, which in turn will have implications for overall economic development. Kathuria et al. (2010) in his study found that the level of education has a significant positive effect on the growth of total factor productivity (TFP) in several processing industries in India.

Based on empirical research findings, when viewed from the four dimensions of human capital namely usefulness, behavioral uncertainty, firm-specifity, spread of strategic human capital there are still some shortcomings of graduates of AK Tekstil Solo such as graduates who have not fully provided the idea of innovation that is useful for the progress of the company, expertise Specifics that are expected to be the key to get success are also still in the training and coaching stage. Soft interpersonal skills possessed by some graduates are good, but some are still privileging themselves. The three main components of human capital are initial capability qualifications; knowledge is obtained through formal education and skills; competence and expertise are obtained through workplace training. Three components contained in four dimensions of human capital can be used as human capital investments in the future. The implementation of the partnership currently carried out by AK Tekstil Solo with the textile and garment industry partners is still not optimal, because some aspects such as usefulness, behavioral uncertainty, firm-specificity cannot be fulfilled by graduates of AK Tekstil Solo. If the dimensions of human capital can be met and owned by graduates of AK Tekstil Solo, the textile and garment industry partners can use the graduates of AK Tekstil Solo as an alternative investment in human capital. Textile and garment industry partners who have partnered with AK Tekstil Solo have high hopes for graduates of AK Tekstil Solo.

4 CONCLUSION

Based on the theoretical foundation and empirical findings derived from the research, revealing that the partnership between the government and the private sector in this article is AK Tekstil Solo with the textile and garment industry partners perceived to be profitable in terms of increasing the number of workers, but for human capital investment it is still necessary to strengthen the partnership, equality of perception and communication that is more effective in the delivery of information and material given to prospective graduates. AK Tekstil Solo as a government agency tasked with educating and building competent industrial workers is expected to provide more educational material that stimulates special and unique abilities possessed by AK Tekstil Solo students, who after graduating will become industrial workers in the textile and garment industry partners.

Textile and garment industry partners as partners of AK Tekstil Solo and as a place of employment for prospective workers from AK Tekstil Solo graduates are expected to assist in the implementation of education and the development of the ability of prospective workers by providing concessions to prospective workers. Textile and garment industry partners are expected to provide access to prospective workers so that they can assist in solving problems that arise during the production process by increasing the knowledge and abilities of prospective workers who are conducting industrial practices.

In order that the impact of the partnership on industrial human capital development can gain benefit both parties who are partnering, it can be carried out the sharing of risks, responsibilities, resources and expertise of partners managed properly, and the participation of partners in making decisions that have a broad impact.

REFERENCES

Barney, J. 1991. Firm resources and sustained competitive advantage. *Journal of Management* (17): 99-120.

Barney, J. & Wright. P. 1998. On becoming a strategic partner: the role of human resources in gaining competitive advantage. *Human Resources Management* (37): 31-46.

Burr, R. & Girardi, A. 2002. Intellectual capital: More than the interaction of competence x commitment. *Australian Journal of Management* 27(1): 77-87.

Carmeli, A. 2004. Strategic human capital and performance of public sector organizations. *Scandinavia Journal Management* (20): 372-395.

Cas, C. 2018. Peluang tekstil terbuka. *Kompas*: 7.

Clement, A., Gurstein, M., Longford, G., Luke, R., Moll, M., Shade, L.R. & DeChief, D. 2004. The canadian research alliance for community innovation and networking (CRACIN): A research partnership and agenda for community networking in Canada. *The Journal of Community Informatics* 1(1): 7-20.

Darlington O. & Marcus L. 2007. The importance of partnerships: the relationship between small bussiness, ict and local communities issues in informing science and information technology 4.

Davenport, T & De Long, D. 1999. Successful knowledge management projects. *The Knowledge Management Yearbook 1999-2000.*

Grundey, G., Dainora, D. & Daugėlaitė I. 2009. Developing business partnership on the basis of internal marketing. *Economics & Sociology* 2(1):118-130.

Karim, K. & Shabbir, S. 2012. Human capital and the development of manufacturing sector in Malaysia. *International Journal of Sustainable Development* (04):04.

Kathuria, et al. 2010. Human capital and manufacturing productivity in India. *International Conference on: Human Capital and Development.*

Kementerian Perindustrian Republik Indonesia. 2015. *Studi kelayakan akademi komunitas industri tekstil dan produk tekstil Surakarta*. Jakarta: Kementerian Perindustrian Republik Indonesia.

Kementerian Perindustrian Republik Indonesia. 2017. *Rencana kerja kementerian perindustrian*. Jakarta: Kementerian Perindustrian Republik Indonesia.

Marimuthu, M. et al. 2009. Human capital development and its impact on firm performance: evidence from development economics. *Journal of International Social Research* 2(8): 256-272.

Pella, D.A. & Inayati, A. 2011. *Talent management mengembangkan sdm untuk mencapai pertumbuhan dan kinerja prima*. Jakarta: PT. Gramedia Pustaka Utama.

Rachmawati, D., Susanto, A. & Ongkorahardjo, M. 2008. Analisis pengaruh human capital terhadap kinerja perusahaan. *Jurnal Akuntansi dan Keuangan* 10(1): 11–21.

Advances in Business, Management and Entrepreneurship – Hurriyati et al (eds)
© 2020 Taylor & Francis Group, London, ISBN 978-0-367-27176-3

Strategic management and entrepreneurship in the disruption era

A. Chauhan
President University, Bekasi, Indonesia

ABSTRACT: Human history has gone through various era of evolutions. The first era was social and political evolution, the second era was industrial, business and management evolution. The third was the era of information and communication technology revolution. What has started, post 2010, is the era of disruption where the future can no longer be based on the conventional theories and historical precedence. The countries as well as business organizations are in the middle of a paradigm shift where the best prognosis failed. The best political analyzers could not predict Brexit or Trump's victory. The best economists could not predict the economic crises of 2008-9 or recent crises in Haiti, Zimbabwe and Venezuela which were on verge of bankruptcy. The best management experts, even in Harvard, could not predict the downfall of Kodak or Nokia. The conventional theories of strategy and entrepreneurship fail. It is worth analyzing how the business would shape up – who will succeed and who will fail. It is worth contemplating if the conventional theories like Porter's generic business strategies need to be relooked and modified. Of main interest of analysis is what the small and medium entrepreneurs (SME) should do because big organization can withstand the failure but SME do not have financial muscle. This paper presents an analysis of strategies and skills required to survive in the disruption era.

1 INTRODUCTION

Is disruption new to the human society? In every era, humans have faced challenges which were overcome by the mankind by finding new solutions, sometimes disruptively innovative. The result was revolution in all context i.e. social, economic, technological and political. Over time, the speed of occurrence of disruption is increasing. A dig into history could explain that disruption was always part of human evolution and will be, in future.

1.1 *First three disruptions*

Initially, there were various species of humans like Homo Erectus (East Asia), Homo Sapiens (East Africa), Homo Neanderthalensis (Europe and western Asia) and Homo Floresiensis (Indonesia) with ability to use tools and fire. By 10,000 BC, most of these vanished because of their inability to migrate and change. The Homo Sapiens were the only survivors because they migrated all over the world (Harari 2014). Great civilizations like Egyptians vanished. Even though they built great pyramids around 2500 BC, they were easily conquered by Romans, because their movement was slow and laborious in absence of wheel. Wheel could have assisted fast movement and war capabilities. There is a mention of agriculture by Egyptians but no mention of wheels. Then how were the pyramids built? The stones were tied onto sledges and pulled up on wooden tracks which were anchored on the pyramid flanks (Löhner & Zuberbühler 2006).

Western history says that archaeologists spotted potter's wheel in Mesopotamia (now Syria) which belongs to 3500 BC (Gambino 2009). But the disruption started when axle was added to wheel around 2000 BC. It was used in many civilizations in Asia and Europe. Buddhist religion mentions about the Wheel of Life. All ancient Indian religions preceding Buddha dating back to about 2,000 BC talk about Chakra, meaning wheel. The wheel brought a revolution because humans could move around fast, carry and lift heavy loads, hence opening the possibilities of socio-economic exchange, trading, business and, ultimately, war capabilities. It led mankind on a path of progress. The result was need for currency, trades, industry and banking and the creation of empires and religions. In short, invention of wheel initiated the socio-politico-economic revolution, which was the first disruption.

The second disruption, made possible by Renaissance, started in Britain during 1760-1780. It lasted till the middle half of the last century. The invention of spinning jenny by James Hargreaves in 1764 and steam engine by James Watt in 1770 lead to invention of power looms by Edmund Cartwright in 1780. The steam engine went on to power machinery, locomotives and ships initiating the Industrial Revolution (history.com, n.d.). In short, the first disruption created socio-economic revolution that resulted in what is now called mass. The second disruption started by invention of jenny and steam engine created an industrial revolution that resulted in mass production and mass transportation. The textile was the first mass manufacturing industry. This was followed by a series of inventions and innovations. The power shifted to the capitalists who created large private organizations. The theories of Marxism, capitalism and bureaucracy originated in this era. The second disruption shifted the focus from kings, religions and miracles to organization, entrepreneurship and management. Harvard Business School was established in 1906. Just after 5 years, theory of scientific management was published by Frederick Taylor in1911 followed by principles of organization and management by Henry Fayol in 1916. The stock exchanges were established. The economic thrust of capitalism was borne by the nations. The wars were fought and colonization was done to get the raw materials. It all ended with the World War 2 (WW2). The second disruption led mankind on path of progress and technology but also to wars and business competition, that needed strategic management.

A need for peace and collaboration was felt after the devastating impact of WW1, Great Depression and WW2. Most of the colonies were freed between 1945 and 1970. Global organizations like United Nations (UN), World Bank, International Monetary Fund (IMF), World Trade Organization (WTO) and International Labor Office (ILO) were established. This international peace and collaboration resulted in the third disruption. The third disruption is sometimes misunderstood as information revolution but it was information and communication technology (ICT) revolution. After the WW2 was over, scaling up of organizations became possible and their operations became multinational and complex. Finally, the technologies created for WW2 like airplanes, radio, television, nuclear power and rockets were all put to use for business and peaceful purposes. All these developments led to the need of a high computing power, information and communication management. Even though the computer was invented long back in 1833 by Charles Babbage and programmed by Ada, also known as Lady Lovelace, the first commercial computer was deployed in 1951 in University of Pennsylvania (Zimmermann 2017). The rockets became more powerful and powerful nations started eyeing sources of raw material outside the earth using rockets, after free sourcing from colonies disappeared.

Finally, man landed on moon in 1969. Using the same space technology, the geostationary satellites could be placed in earth orbit starting with the year 1963. In 1945, Arthur Clarke postulated that three satellites, placed some 35,789 kilometers (Km) above earth equidistant from each other and traveling at a speed of 11,069 Km per hour, would appear stationary from the ground and would cover the earth with continuous radio and television coverage (History Channel n.d.). As on today, there are about 402 such satellites providing geo positioning system (GPS) that we all use, weather monitoring, remote sensing earth resources, monitoring ship movements, internet data, telephony, radio and TV channels (Howell 2015). Today, it is estimated that an average American spends 70% of the waking time in front of electronic media (Richter, 2015). The whole world

became a global village, where it was possible to communicate, connect and do business instantly with any part of the world. The present day disruption could not have been possible without the connectivity, business, knowledge, technological possibilities and collaborations as a result of ICT revolution

1.2 Rise of strategic management

In WW2, the importance of strategy was realized and the ex-servicemen not only reached various industrial sectors but also reached the citadels of education popularizing theories of leadership and strategic management. The parallels were drawn between war and competition. The tactical and technological advantage over enemy in war became the competitive advantage in business. Harvard business school, established in 1906, had a capstone subject called Business Policy. After WW2, the concept of *business policy* got replaced by *strategy* and Strategic Management became the prestigious capstone subject in place of Business Policy. The strategic management was also popularized by Michael Porter, who is the most cited author in management research. His main contributions are strategic analysis (such as 5 force analysis, value chain analysis, and Porter's diamond) and the generic business strategy planning (Wheelan & Hunger 2014). The most popular was generic business strategies, which was a two-way analysis of market scope and market strategy – cost based or differentiation based.

Albert Humphrey, led a research project at Stanford University in the 1960s and 1970s using data from Fortune 500 companies. The goal was to identify why corporate planning failed. He used strength weakness opportunity and threat (SWOT) analysis, though he originally used word fault instead of weakness and called it SOFT analysis. SWOT analysis replaced 5 force analysis, in importance, and is routinely used till now in management teaching in some form (Thompson & Strickland, 2015). The experts of strategic management advocated numerical IFAS (Internal Factor Analysis Summary), EFAS (external factor analysis summary) or TOWS - a two way SWOT table, which are just different versions of SWOT analysis. Even till now Strategic Management is a capstone subject in almost all business schools. The strategic plan based on strategic analysis is part and parcel of all corporate plans in almost all organizations in business as well as government. One of the offshoots of strategic management was Blue Ocean strategy in 2005 (Kim & Mouborgnc, 2005), which became a buzz word but is, actually, about entrepreneurship and innovation rather than strategic management.

1.3 Rise of entrepreneurship

From the historical perspective, Schumpeter opined that the French economist Richard Cantillon, was the first to introduce the concept "entrepreneur" in his work in 1755. He viewed the entrepreneur as a risk taker. Many textbooks and authors have written extensively on the origin of entrepreneurship. What is interesting is that most of the authors who wrote about the origin of entrepreneurship were either economists or historians (Drucker 1985). Another renowned economist, Alfred Marshall formally recognized entrepreneurship as an important factor of production in 1890 and believed that entrepreneurship is the driving force behind organizations (Schumpeter 1951). Now, entrepreneurship is synonymous with small business and startups.

The post WW2 era provided the perfect settings for entrepreneurship because of the existence of collaborations, connectivity, institutional and economic development. Entrepreneurship identified with small and medium enterprises (SME), got regarded as the tool for a higher growth rate of nations. The economic growth, employment and inflation have been at the core of national strategies across nations and the entrepreneurship and SME have been the best means to achieve national goals. The most valuable companies today are all borne of entrepreneurship like Apple, Microsoft, Facebook, and Amazon driven by entrepreneurs with passion, guts and risk taking abilities.

2 METHOD

The start of 21st century is witnessing so many disruptions that it is easier to label it as Disruption 4.0. Disruptions have started to become a way of life. So much so that websites have sprung up to advise and warn about disruptions. It is worth exploring what is disruption and how does it affect economies, businesses and individuals. A website called *disruption-hub* conducted a survey among 1063 CEOs of various companies about what are the future disruptive trends (Prevett 2018). The technology trend that were voted are

- Artificial intelligence (AI) 42%
- Block chain 25%
- Internet of things (IOT) 16%
- 5G 10%
- Augmented reality 5%
- CRISPER, etc. 2%

Companies are demonstrating a willingness to use AI and related tools like machine learning to automate processes, reduce administrative tasks, and collect and organize data. Understanding vast amounts of information is vital in the age of mass data, and AI is proving to be a highly effective solution. The wider use of automated processes powered by AI, advanced robotics, and IOT connectivity are steps towards Industry 4.0 and the factory of the future. Industry 4.0 promises a more connected world in which machines carry out mundane tasks. Many companies, such as Amazon, have taken tangible steps to implement this in automated warehouses.

The Intelligent Electronic Agents using natural language voice commands will become common e.g. Apple's Siri. Even mundane tasks at home will be performed by AI equipped robots.

Organizations across a wide range of sectors are already experimenting with Block Chain technology to establish trusted networks, to improve transparency, and to reduce friction and costs. Despite fierce debate, interest in crypto-currencies powered by block chain remains strong. More commercial businesses are accepting crypto-currency payments, starting of course with Bitcoin and Ethereum. IOT along with AI and big data is allowing new possibilities such as smart machines, homes, cities, and cars thanks to embedded and networked sensors combined with other technologies such as GPS. Machine-to-Machine Communications using chips and micro sensors, combined with networked sensors using 5G, will create a rapidly growing IOT sharing real-time data, performing diagnostics, and making virtual repairs all without human intervention. By 2020, there will be well over a billion machines talking to each other, performing tasks, and making decisions based on predefined guidelines using artificial intelligence (Prevett, 2018).

Customer demand for resources has triggered innovation in the reuse and remanufacture of goods, reflecting the trend of environmental awareness and a preference for access over ownership. For businesses, as-a-service solutions cut costs by simplifying IT infrastructure. Between 2016 and 2020, the global XaaS (anything as-a-service) market is forecasted to grow by 40% each year (Prevett, 2018). Big Data as-a-Service (BDaaS) has emerged, as cloud providers offer midsize and smaller organizations access to much larger streams of relevant data they could not tap into otherwise. On Demand Services will increasingly be offered to companies needing to rapidly deploy new services. Advanced Cloud Services will be increasingly adopted by business of all sizes, as this represents a major shift in how organizations obtain and maintain software, hardware, and computing capacity. As consumers, we have already experienced public clouds e.g. Google Docs or Apple's iCloud.

The augmented reality is the result of cross industry disruption. Augmented Reality (AR) Apps will become more common, adding just-in-time information to our physical world. Simply aim your smart phone camera at a crowded street to find the stores who have the exact products you are looking for. Or, when you are in a store, use your phone's camera and AR app to quickly locate the products you need. Put on a pair of Google Glasses and see the information you need about how to service your lawn mower or install a water filter. Business and education could find a great use for this powerful tool. CRISPR is an acronym for Clustered Regularly Interspaced Short Palindromic Repeat, which is an editable DNA sequence. Growing interest and investment into gene editing technologies based on CRISPR will allow research teams to precisely alter, delete, and rearrange the DNA of nearly any living organism. The impact will be in battling disease and world hunger. All the disruptions discussed so far can be called digital revolution. It will have impact on societies, industries, nations and ultimately all the consumers, living beings, animals and this planet.

3 RESULTS AND DISCUSSION

3.1 *Shape of industry 4.0*

Over the next five short years game-changing technologies will *transform* how we sell, market, communicate, collaborate, educate, train, innovate, and much more. The futurists agree that industries that will be impacted most, in the same order, are as follows:

a) Technology product companies
b) Media and Entertainment
c) Retailing
d) Financial Services
e) Telecommunication
f) Education
g) Hospitality and tourism
h) Automotive and manufacturing
i) Healthcare & medicine
j) Utilities

Discussing industry by industry could be long. Hence just education industry is discussed here. The education and training industry could turn in to gaming applications. This will accelerate a fast-moving hard trend of using advanced simulations and skill-based learning systems that are self-diagnostic, interactive, game-like, and competitive, all focused on giving the student an immersive experience thanks to a augmented reality or 3D interface. Already, Online Learning and Massive Open Online Courses (MOOC) have been launched by highly recognized and traditional educational institutions, putting them in a position to challenge all educational systems by making *Location* and *Tuition* far less of a barrier to receiving the information, training, and knowledge people need to know in order to succeed in a rapidly changing world. Further, interactive eBooks may soon be crossing the tipping point due to the abundance of smart phones and tablets. The publishers would be providing apps that give a better-than-paper experience by including cut, copy, paste, print, and multimedia capabilities. Probably, the universities in the present forms may become outdated. Anyone can put on a pair of Google Glasses and see the information needed, such as how to service lawn mower or install a water filter. Every business and school could find a great use for such powerful tool.

3.2 *Impact on strategic management*

The traditional mega corporations like General Motors were conceived by Alfred Sloan as hierarchical. The strategy followed from the top and the middle and lower management just had execution responsibility. Compensation was based on performance against the predetermined key performance indicators (KPI) or objectives. No wonder strategic plan was more like a plan to win a war with thick books filled with scores of market intelligence, strategic analysis, description of tactics and procedures. Jack Welch was the first to rebel saying the books are thicker, sophisticated, with thick hard cover and better drawings but could not help the performance (McGrath 2013). I have been a top level positions in many companies. The vast majority of strategic plans that I have seen over 35 years in both government, corporate and even education sectors are simply complicated budgets with pages of justification. The real plans are more complicated compared to the simple cost leadership or differentiation strategies suggested by Porter (1980). This may be because the finance function is deeply involved in the strategy process in most organizations. But it is also the cause of the deep disconnect and almost all hate the planning cycle (Martin 2013). Same disconnect is found in education. The students fill up their assignments with meaningless IFAS, EFAS, SWOT or TOWS. They are evaluated on how they memorized each element of business and corporate strategies. None of the assignments are able to connect the strategies with the analysis. It should be realized that the strategic analysis has become irrelevant in the disruption era. The business cycles are becoming shorter than the corporate planning cycles. Clearly, syllabus of strategic management should be relooked at.

The circumstances do not go as planned e.g. the first helicopter of commandos, who arrived at Osama bin Laden in Abbottabad, crash landed destroying their plan of entry. So the strategy had to be changed. This is what Henry Mintzberg, called an emergent strategy, a strategy which optimizes on the changing circumstances. The world as well as organizations are getting flat. There is no need to be lonely at the top in the disruption era (McGrath, 2013). The digital disruption makes possible connect with the ground reality and the staff at ground level. Narendra Modi prime minister made an elaborate use of social media and set up direct channel with 1.3 billion Indians.

According to the textbooks the strategic management starts with the vision and mission statements. The vision and mission statements can be found on the doors of not only all corporations but the government offices and academic institutions also, across countries. The vision has a shelf life. For example, vision of GE in the Jack Welch days was that every business should be number one or two in its category. But the vision could not sustain after some time. Sometimes, it blinds and becomes an illusion. Jeffrey Skilling, CEO of notorious Enron believed that securitization and a quantitative approach could make the company unstoppable. Unfortunately, that same vision obscured serious problems that led to one of the great financial meltdowns in history. Therefore, vision and strategy are for long term and in the disruption era there is no long term.

3.3 *Impact on entrepreneurship*

Disruption era has changed the entrepreneurship. Earlier, when entrepreneur had a new product or service idea, it was customary to make a business plan, get some funding, get some government incentives and then launch the company. In disruption era, the business plan is no more relevant. The failure rate of new ideas, especially the product ideas, is high. The traditional funding agencies like banks have no idea about how to judge. Therefore, should entrepreneurs bother about the loans, funds or business plan? Is this the way, Go-jek became successful? Not through the business plan. The venture capital is the new nature of funding and it is not one time, but goes on in several rounds. The venture capitalists have criteria other than business plan. At the Bloomberg Global Business Forum in 2017, Masayoshi Son, the Chairman and CEO of one of the largest venture capitalists, said that he was drawn to Alibaba not because of the business plan, but because of the passion and traits of founder Jack Ma (Balakrishnan 2017). Alibaba was formed in 1999 and Softbank invested $20 billion in 2000. There are so many such stories.

Another impact is that digital disruption is that the most successful entrepreneurs are in the technology or IT based companies, especially service oriented companies. The five most valuable companies in 2001 were General Electric, Microsoft, Exxon Mobil, Citibank and Walmart in the same order but the five most valuable companies in 2018 were Apple, Amazon, Google, Microsoft and Facebook (Statista, n.d.). Except Microsoft all were brick and mortar in 2001 but click and mortar in 2018.

The third impact is that focus has shifted from product to services, user experiences and engagement. New entrepreneurship business model based on just services have sprung up. Spotify found success with their streaming music subscription services, which provide listeners with access to millions of songs on demand. AirBnB became a disruptor in the industry by improving that experience, offering local rooms in homes or apartments at significant savings for the traveler and profit for the host. Gamification involves integrating elements of gaming into an application or interface, such as points, rewards, competition or an overall engaging experience. Established companies, like Starbucks, have introduced rewards systems that encourage users to visit the store and buy certain items to earn points for future purchases.

The last impact on entrepreneurship is that exit plan has become more important than business plan. The existing order of things will find a way to stifle the entrepreneur if poses an immediate threat because disrupting force attracts attention and attracts resistance from incumbents (Thiel & Masters 2014). It takes time to become a disruptive force like Tesla. In any case, it has been noticed that 99% innovative ideas fail and only 25% entrepreneurs survive beyond 10 years. Therefore, the best idea could be to find an angel and sell off in time before the size becomes unwieldy.

4 CONLUSION

The conclusion that emerges from the earlier discussions is that the disruption era has changed the focus from planning to action. The facts speak better. The difference between strategic management and the entrepreneurship is that the first one belongs to large companies whereas the second one to small and medium enterprises (SME). Large companies have strategic plan whereas the SMEs have business plan but both are rendered useless by the disruption era. Then, what is the winning solution? First of all, it should be noticed that the objective of strategic plan is to get competitive advantage and the objective of business plan is profits. Both objectives are unviable in the disruption era.

Therefore, the first conclusion is that large companies as well as SMEs should focus on execution capabilities rather than planning capabilities. The organizations can hire consultants for a superior plan but can't hire the experience in execution. Plans are always fool proof but it is the execution which ensures success.

The essential elements of execution capability are strong leadership, cohesive team, high performance culture, best practices approach, digitization, use of disruption tools, and customer centric approach.

The second conclusion is focus on customer value. Actually, Porter had suggested this but not fully and he talked about the Best Cost strategy (Porter, 1980) which is nothing but the value for money strategy. Even talking in the Porter's framework, the generic strategies are not four but only two strategies applied to two types of market. He was actually suggesting three strategies, which is cost leadership, differentiation and value strategy, but could not make value strategy explicit. However, ultimate winner is customer value strategy whether for SME or large enterprise. IMD proposes that there are three value drivers with their own business models. The strongest disruptors like Amazon, Google, and others employ value combination strategies because the values are mutually reinforcing. First value is cost value, which tries to minimize the cost impact on the customers. The second is experience value by offering a superior experience. The third one is platform value by creating forums, associations and networks. For example, Starbucks not only sells better coffee with a superior experience at affordable prices but also provides platform value with a pre-pay mobile app, which has $1.4 billion of coffee drinkers' cash earning interest. (Wade et al. 2016).

Last conclusion is that because there is no long run in disruption era, it is better to plan for the short run. In a typical product life cycle is market development, growth, maturity and decline. The entrepreneurs and intrepreneurs could follow the same steps:

1. Start: If the value is the driver in the disruption era, the first step should be to find the value gap, which is essentially a market opportunity based on product or service enhancements, process enhancement, nearby markets, new segments, new applications, new engagement style, and new services e.g. Go-jek started with transport but filled the value gap by Go-pay, or Go-food. After finding the value gap, a concrete start is required by finding a solution and a business model.

2. Scale: Think big but start small. Then, scale as fast as possible. The success breeds success. Entrepreneurs should focus on small successes rather than business plan. If the proof of concept is good, the funding will follow. With falling interest rates everyone wants to ensure a good return on investment. Angel investors are always looking for good projects. They will make the business plan. First round of funding normally comes from internal resources of the company or in case of entrepreneurs from close friends and relatives who do not need proof of concept. The second round onwards comes from venture capitalists. The large companies have various avenues.

3. Optimize: Having grown, there should be a constant evaluation and optimizing of value chain. Because, this is the time when competition will catch up e.g. iTunes value supremacy got challenged by Spotify and couple of other similar ventures.

4. Exit: The best time to exit is when the venture is at its best because at that time the business can command the maximum value. This is the time for finding another value gap. This is the inflexion point as described by Andy grove. There are many exist strategies like public offering or sale to a large competitor. The success depends on the type of business and industry. Whatsapp was purchased at whopping USD 19 billion in Feb 2014, when its revenue for 2013 were just USD 10 million and losses USD138 million (Wagner 2014).

Hence, the best strategy in this disruption era is SSOE (start, scale, optimize and exit) strategy, which involves thinking big, finding value gap, starting small, scaling fast, and exiting when at peak of success and, then, starting the cycle all over again. The strengths and skills required for this are leadership, reliable teams, high performance culture, ability to change and IT/technology skills. Future will be technology driven even in culture, arts, media and performances.

REFERENCES

Balakrishnan, A. 2017. *Masayoshi Son is shaking up Silicon Valley $1 billion at a time — here's how he became one of tech's most powerful people*. Retrieved from cnbc.com: https://www.cnbc.com/2017/12/31/who-is-masayoshi-son.html

David, F. R. 2014. *Strategic Management: Concept and Cases (15th Ed)*. New York: Pearson.

Drucker, P. F. 1985. *Innovation and Entrepreneurship: Practice and Principles*. New York: Harper & Row.

Gambino, M. 2009. *A Salute to the Wheel*. Retrieved from Smithsonian: https://www.smithsonianmag.com/science-nature/a-salute-to-the-wheel–31805121/

Harari, Y. N. 2014. *Sapiens: a brief history of humankind*. Toronto: Signal Books.

History Channel. (n.d.). *World's first geosynchronous satellite launched 23 July 1963*. Retrieved from historychannel.com.au: https://www.historychannel.com.au/this-day-in-history/worlds-first-geosynchronous-satellite-launched/

history.com. (n.d.) *Industrial Revolution*. Retrieved from history.com: https://www.history.com/topics/industrial-revolution

Howell, E. 2015. *What Is a Geosynchronous Orbit?* Retrieved from www.space.com: https://www.space.com/29222-geosynchronous-orbit.html

Kim, W. C., & Mouborgne, R. 2005. *Blue Ocean Strategy: How to Create Uncontested Market Space and Make the Competition Irrelevant*. Boston: HBR Press.

Löhner, F., & Zuberbühler, T. 2006. *Building the Great Pyramid*. Retrieved from Cheops-pyramid: https://www.cheops-pyramide.ch/pyramid-building.html

Martin, R. L. 2013. *Don't let strategy become planning*. Retrieved from hbr.org: http://blogs.hbr.org/2013/02/dont-letstrategy-become-plann/

McGrath, R. G. 2013. *The End of Competitive Advantage: How to Keep Your Strategy Moving as Fast as Your Business*. Boston: HBR Press.

Porter, M. E. 1980. *Competitive Strategy*. New York: Free Press.

Prevett, R. 2018. *18 Disruptive Technology Trends For 2018*. Retrieved from disruptionhub.com: https://disruptionhub.com/2018-disruptive-trends/

Richter, F. 2015. *Americans Use Electronic Media 11+ Hours A Day*. Retrieved from www.statista.com: https://www.statista.com/chart/1971/electronic-media-use/

Schumpeter, J. A. 1951. *Essays of J. A. Schumpeter*. Cambridge, MA: Addison Wesley.

Statista. (n.d.). *The 100 largest companies in the world by market value in 2018 (in billion U.S. dollars)*. Retrieved from www.statista.com: https://www.statista.com/statistics/263264/top-companies-in-the-world-by-market-value/

Thiel, P. A., & Masters, B. 2014. *Zero to one: Notes on startups, or how to build the future*. New York, NY: Crown.

Thompson, J. A., & Strickland, A. J. 2015. *Crafting executive strategy: Text and cases (19th ed)*. New York, NY: McGraw-Hill.

Wade, M. R., Shan, J., & McTeague, L. 2016. *Strategies for responding to digital disruption*. Retrieved from imd.org: https://www.imd.org/research/insightsimd/strategies-for-responding-to-digital-disruption2

Wagner, K. 2014. *Facebook Paid $19 Billion for WhatsApp, Which Lost $138 Million Last Year*. Retrieved from recode.net: https://www.recode.net/2014/10/28/11632404/facebook-paid-19-billion-for-whatsapp-which-lost-138-million-last-year

Wheelan, T., & Hunger, J. 2014. *Strategic Management and Business Policy (14th ed)*. New York: Prentice Hall.

Zimmermann, K. A. 2017. *History of Computers: A Brief Timeline*. Retrieved from livescience.com: https://www.livescience.com/20718-computer-history.html

Advances in Business, Management and Entrepreneurship – Hurriyati et al (eds)
© *2020 Taylor & Francis Group, London, ISBN 978-0-367-27176-3*

How Information Technology (IT) firms survive in the disruptive technology era

J. Achmadi
Telkom Sigma, Tangerang, Indonesia

A. Rahayu, E. Ahmad & L.A. Wibowo
Universitas Pendidikan Indonesia, Bandung, Indonesia

ABSTRACT: Companies must implement disruptive concepts in order to continue to lead. The key IT company can improve its performance in this disruptive era through the uniqueness of human resources and management of innovation. HR is part of the strategic planning process and becomes part of the development of organizational policy, organizational expansion planning, organizational mergers and acquisitions processes. The ability of innovation is the company's ability to introduce new products to the market or open up new markets through a combination process and strategic orientation with innovative behavior and processes. Innovation ability becomes an important competitive weapon for small companies operating in dynamic environments. The uniqueness of resources provides a competitive advantage. All business owners and managers understand that a company needs a unique selling proposition. This strategy has been successfully applied to one of the IT companies in Indonesia. IT companies have been able to implement the concept of disruptiveness to print IT expertise in the field of IT and has become a leader in several large companies in Indonesia.

1 INTRODUCTION

Understanding the sources of sustainable competitive advantage has become a major area of research in strategic management (Barney 1991). Strategic planning can help companies achieve goals, progress and develop, and increase their market share in the midst of an increasingly sharp business competition (Allison & Kaye 2013). To win the business competition today, the company must have a good strategy. Companies that implement strategies quickly and accurately are companies that have competitive advantages to win the competition (Thompson et al. 2001). Strategic planning is complex and consists of several aspects (Wheelan & Hunger 2012) which have an influence on company goals, learning, innovative management, competitive positioning and sustainable competitive advantage (Lee et al. 2015).

In the early 1980s, Porter (1998) gave the hypothesis of sustainable competencies that were very well established as the main values of sustained superior performance (Boulianne 2007). Sustainability is the focus of the company's agenda and researchers emphasize the company's need to develop a sustainability strategy. Porter's five-forces model (1998) has been widely criticized, including mentioning that the model does not help the company in identifying and maintaining unique and sustainable advantages (Barney 1991). Companies tend to shift in their sustainability strategies in innovative ways that allow them to do more by increasing the company value without making significant cuts (Barnett et al. 2015).

Previously, the company only focused on linking cost advantages to organizational performance. Companies that are established with the Resource-Based-View (RBV) of the company's competitive advantage are one of the keys to strategic management theory. This is also part of a larger management theory that has been developed to meet the managerial needs of the organization. Competitive advantage is a related concept (Majeed 2011).

Most previous studies have investigated factors that influence the company's sustainable competitive advantage (SCA), such as intellectual capital (Hsu & Wang 2012), innovation (Barrett & Sexton, 2006) or dynamic abilities (Bowman & Ambrosini 2003, Easterby et al. 2008, Macher & Mowery 2009, Pandza & Thorpe 2009. While, on the other hand, other studies look into different traits, such as temporary competitive advantage (TCA) and SCA, with a few exceptions (Huang et al. 2015). RBV researchers show that sustained competitive advantage arises from the ownership of company resources and abilities with certain characteristics (Barney 1991). The RBV emphasizes that resource-based competitive advantage and capability are more sustainable than those based on product/market positioning (Hitt et al. 2007).

Cross-sector competition in the information and communication technology sector is a strategic challenge for telecommunications companies. Because of increasing convergence, value creation results in a greater level of interaction. The potential for diversification of the telecommunications business is very rapidly changing the related ICT sector, such as hardware, software and media (Wulf & Zarnekow 2011). Global competition and technology continue to improve with a dynamic business environment (Dwivedi et al. 2015).

There is no denying that one of the main causes of the globalization era which came faster than expected by all parties was due to the rapid development of information technology. The implementation of the internet, electronic commerce, electronic data interchange, virtual office, telemedicine, intranet, and so on has broken through the physical boundaries between countries. The combination of computer technology and telecommunications has resulted in a revolution in the field of information systems. The IT industry in Indonesia is a very competitive industry. Many new companies emerged through Start Up IT companies, but many also closed because they were unable to compete. The role of information systems in an organization is needed to support business competition strategies so that profits can be achieved. The use of information systems in an organization can be optimal, if planned properly in a strategic plan. One strategy of improving competitiveness is through the development of the role of information systems in companies. If earlier the role of information systems was only as a supporting process in obtaining data with the emphasis on operational cost efficiency and operational risk minimization of various company functions, nowadays its role has changed into a strategic tool in the company to increase its competitiveness.

2 METHOD

The development of IT companies in the world about the achievement of the value of the software industry related to the use of information technology. The highest software revenue share was achieved by Activision Blizard with a percentage of 100%. There are several business models in the IT software industry, including selling one-time licenses, monthly maintenance with distribution and one-to-one support, selling professional services in the form of application development, and software as a services (SaaS) systems through cloud computing. The software system as a services allows software users to rent systems rather than buy expensive one-software licenses, so this system is predicted to continue to experience an increase in income.

The largest revenue position of the IT sorghware industry was achieved by North America, while the Asia/Pacific region held the third lowest position. The need for information and communication technology in Indonesia is very large; where 80% is hardware requirements, 8% software, and the remaining 12% is for services. The use of hardware

is information and communication technology that is more widely used in Indonesia, indicates that the use of software in Indonesia is quite low. Top 5 UTIs Kearney's Global Service Location Index for 2016 shows that Indonesia is ranked 5th in terms of business performance potential and generally shows that in Indonesia there is a high potential in terms of high financial attractiveness, good people skills and availability and a supportive business environment.

IT startup companies in Indonesia in 2016 were recorded to be the highest number in Southeast Asia, with as many as 2,000 companies. Startups in Indonesia are projected to continue to grow to 6.5 times or as much as 13,000 in 2020 (Bisnis.tempo.co 2017). The challenge of digital transformation requires companies to create new innovations and creativity in a number of sectors to support traditional methods and overcome the various problems they face, such as human resource requirements, data management, technology infrastructure, connectivity, and improving customer experience (Arenalte.com 2017).

Competing in the IT Industry means product diversification by integrating IT with other fields (Wulf & Zarnekow 2011). There are five important aspects of the IT industry that are very dynamic both from service products and interrelated networks (Glass & Zheng 2012). Travis (1991) highlighted policy problems in the telecommunications industry and competition between countries.

All of these studies support the relationship between competitive advantage and company performance in a positive way. The nature and impact of organizational competencies empirically recommend productive opportunities to take advantage of other organizations and also help in further research to find methods to improve company performance. Most researchers on competitive advantage against business performance recommend conducting research with different sample sizes and different pribumi sources (Morioka & Carvalho, 2016).

3 RESULTS AND DISCUSSION

The results show that innovation has a positive effect on the company's business performance (Bayraktar et al. 2016). Innovation can be used as one of the factors that must be considered by companies to improve their business performance. Companies that have an innovation strategy are able to survive and increase efficiency, and long-term opportunities can provide better business performance (Al-Ansari et al. 2013).

Improving business performance can also be done by using the unique resources owned by the company. Having valuable and unique resources and abilities will be difficult to imitate by others (Al-Ansari et al. 2013). Based on resource analysis, John Kay (Lynch, 2015) suggests that the unique capabilities of organizational resources are very important in providing competitive advantage. Resources that are difficult to replicate and cannot be substituted can improve performance through more sustainable management and improve technological aspects (Brasil et al. 2016). Technology is the catalyst of strategic planning that is combined with market insights to generate innovative ideas (Berman & Hagan 2006).

4 CONCLUSION

Technology is one part of the global trend in the digital era today; through digital technology, industry players can easily build relationships with various stakeholders, including fans, customers or consumers. The use of technology allows companies to develop systems which are used to share information with their business partners around the world. The uniqueness of resources and innovation management affects business performance in the digital IT industry in Indonesia. The uniqueness of resources is a corporate strategy so that it is not easily imitated by competitors.

REFERENCES

Al-ansari, Y., Pervan, S., & Xu, J. 2013. Innovation and Business Performance of SMEs : The Case of Dubai. *Education, Business and Society: Contemporary Middle Eastern Issues*, 6(3/4), 162–180.

Allison, M., and J. Kaye. 2013. *Strategic planning for non-profit organizations: A practical guide and workbook*. John Wiley & Sons.

Arthur A. Thompson, Jr & A.J.Strickland III, "Crafting and Executing Strategy"-Text and Reading, McGraw-Hill Irwin, 2001.

Arenalte.com. 2017. Prediksi Tren Industri Teknologi Informasi dan Manajemen Data Tahun 2017. *Arenalte. com*. Retrieved from https://arenalte.com/life/hope/teknologi-informasi-dan-manajemen-data/

ATKearney.com. 2015. *The Rising Stars of IT Outsourcing*. Retrieved from https://www.atkearney.com/documents/10192/5185127/The+Rising+Stars+in+IT+Outsourcing.pdf/cd3ba9d1–3d84–4a12-a5be-e0162b3fa73c

Barnett, M. L., Darnall, N., & Husted, B. W. (2015). Sustainability strategy in constrained economic times. *Long Range Planning*, 48(2), 63–68.

Barney, J. (1991). Firm Resources ans Sustained Competitive Advantage. *Journal of Management*, 17(1), 1991.

Bayraktar Bayraktar, C. A., Hancerliogullari, G., Cetinguc, B., & Calisir, F. (2016). Competitive Strategies, Innovation, and Firm Performance: An Empirical Study in a Developing Economy Environment. *Technology Analysis & Strategic Management*.

Berman, S. J., & Hagan, J. 2006. How Technology-Driven Business Strategy Can Spur Innovation and Growth. *Strategy & Leadership*, 34(2), 28–34.

Bisnis.tempo.co. 2017. Kinerja 2017, Pelaku E-Commerce Bidik Target Fantastis. *Bisnis.tempo.co*. Retrieved from https://bisnis.tempo.co/read/news/2017/02/08/090844274/kinerja-2017-pelaku-e-commerce-bidik-target-fantastis

Boulianne, E. 2007. Revisiting fit between AIS design and performance with the analyzer strategic-type. *International Journal of Accounting Information Systems*, 8 (1), 1–16.

Brasil, M. V. de O., Abreu, M. C. S. de, Filho, J. C. L. da S., & Leocadio, A. L. (2016). Relationship Between Eco-Innovations and The Impact on Business Performance: An Empirical Survey Research on The Brazilian Textile Industry. *Revista de Administração*, 51(3), 276–287.

Dwivedi, R., Agarwal, A., & Chakraborty, S. (2015. Application of Activity based Costing and Balanced Scorecard Models in a Biscuit Industry for Sustainable Competitive Advantage. *Journal of Advanced Research in Management*, VI(1), 5–14.

Gartner 2011 Gartner Press Release: Gartner Says Android to Command Nearly Half of Worldwide Smartphone Operating System Market by Year-End 2012. http://www.gartner.com/it/page.jsp?id=1622614. Accessed 2012–05–01.

Glass, B. V., & Zheng, Y. 2012. Challenges in Predicting a Declining Product : Case Study of a Telecommunications Product. *Journal of Business Forecasting*, 31(1), 26–31.

Hitt, M. A., Ireland, R. D., & Hoskisson, R. E. (2007). *Strategic Management: Competitiveness and Globalization (Concepts and Cases)*. Thomson Learning, Inc.

Huang, K.-F., Dyerson, R., Wu, L.-Y., & Harindranath, G. 2015. From Temporary Competitive Advantage to Sustainable Competitive Advantage. *British Journal of Management*, 26(4), 617–636.

Kay, J. 1993. The Structure of Strategy. *Business Strategy Review*, 4(2), 17–37.

Lee, A. H. I., Chen, H. H., & Chen, S. 2015. Suitable organization forms for knowlede management to attain sustainable competitive advantage in the renewable energy industry. *Energy*, 89, 1057–1064.

Lynch, R. 2015. *Strategic Management* (7th ed.). Pearson Education Inc.

Majeed, S. (2011). The Impact of Competitive Advantage on Organizational Performance. *European Journal of Business and Management*, 3(4), 48–62.

Mehra, S., & Coleman, J. T. 2016. Implementing Capabilities-Based Quality Management and Marketing Strategies to Improve Business Performance. *International Journal of Quality & Reliability Management*, 33(8).

Morioka, S. N., & Carvalho, M. M. de. 2016. A Systematic Literature Review Towards a Conceptual Framework for Integrating Sustainability Performance into Business. *Journal of Cleaner Production*, 1–28. http://doi.org/10.1016/j.jclepro.2016.01.104

Porter, M. E. 1998. Clusters and the New Economics of Competition. *HArvard Business Review*, (November-December).

Thompson, A. A., Jr., & Strickland, A. J. 2001. Strategic management : concepts and cases.

Wheelen, T. L., & Hunger, J. D. 2012. *Strategic Management and Business Policy - Toward Global Sustainability. Pearson* (Vol. Thirteenth). New Jersey: Pearson Education, Inc.

Wulf, J., & Zarnekow, R. 2011. Cross-sector competition in telecommunications: An empirical analysis of diversification activities. *Business and Information Systems Engineering*, 3(5), 289–298.

Advances in Business, Management and Entrepreneurship – Hurriyati et al (eds)
© 2020 Taylor & Francis Group, London, ISBN 978-0-367-27176-3

Entrepreneurship promotion in the context of sustainable development goals achievement

M. Ali

Universitas Pendidikan Indonesia, Bandung, Indonesia

ABSTRACT: Sustainable Development Goals (SDGs) achievement is a UNESCO Global Action Program that every member country should implement. There are 17 goals every member country should achieve, and poverty eradication is goal number one. The achievement of this particular goal is expected to have multiplier effects on the achievement of other goals. For this regard, entrepreneurship extensification is considered a reasonable effort in achieving this particular SDG and entrepreneurial firms are the essential mechanism by which millions of people enter the economic and social mainstream, which may stimulate economic growth. Therefore, it is necessary to strengthen the condition for cultivating the quality of the entrepreneurial environment in order to promote a large number of people to become entrepreneurs.

1 INTRODUCTION

National development does not only have positive impacts, such as economic growth and the improvement of people's welfare, but if it is done unwisely it also has negative ones, such as the deterioration and destruction of the environment, and social as well as cultural degradation. Environmental destruction is a concern of the United Nations (UN), which led to the establishment of the United Nations Conference on Human Environment (UNCHE), in Stockholm, Sweden, in June 1972. This was the first UN-initiated conference discussing the environment, and 114 countries participated (UNESCO 2005). It was also the initial step in environment conservation. The conference identified the connection between development, poverty, and lower education level. Poverty and lower education levels are the key factors in environmental damage, so the forum agreed to connect national development policy in every member country to the policy of environment conservation. With respect to fulfilling the needs of the population and improving their welfare at the present time, while taking into account the needs of the future generations, national development should apply this concept. In these attempts, it is not unusual for the government to explore the use of existing natural resources, but if this is done in a reckless manner, the natural resources will be depleted sooner (Ali 2015). This, in turn, will destroy the environment if it is done without renewable resources for sustainability. Thus, the outcome of development will only be enjoyed by the current generation who exploit and consume the natural resources in the short-term. Meanwhile, the future generations, who are in reality also entitled to all existing natural resources, will not be able to reap the benefit of them. Development is usually related

to efforts toward improving the population's welfare. From the economic perspective, at the national level, it is designed for seeking economic growth as measured by certain indicators, such as the improvement of gross domestic product (GDP), gross national product (GNP), and income per capita, which often include output from exploiting the natural resources at the cost of environment. A proper concept of national development needs to be applied in regard to implementing national development that guarantees the equilibrium between improving the present generation's welfare and the right of the future generation to reap the benefit of the country's resources. Involving a large number of people in national development is among efforts for increasing economic growth. This is done by attempting to extend entrepreneurship in the country, which will not only strengthen the country's economic foundation, but will also significantly decrease poverty, which is in line with UNESCO's program of SDGs achievement that every member country should implement. This paper addresses the issues concerning what particular concept must apply to the national development and the role of entrepreneurship in the context of achieving SDGs.

1.1 Sustainable development

The concept of sustainable development (SD) was initially formulated for forestry, especially in the efforts of conserving the forests without hindering their usage but sustaining them through replantation. According to this concept, logging must be carried out wisely by selecting trees that fulfill the criteria for logging and replanting the logged trees to prevent environmental damage to the forest and avoid future catastrophes such as flood and erosion on critical land. In the 1980s, the World Nature

Protection for Conservation of Nature and the World Wide Fund for Nature used this to denote the protection of the environment and biosphere, with the objective of protecting the biological system from the change in its essential characteristics so the future generation can see a benefit.

A related issue is the contamination of nature by waste products. Many irresponsible businesses dispose of their toxic industrial waste in unsuitable places for private cost-saving reasons. In addition, there is thick smoke from factory chimneys or turbid solutions disposed of in rivers. These are examples of environmental damage caused by irresponsible behavior in the development of the economy with no regard for the sustainability of development or the environment. Even the existence of some industries is a potential threat to the safety and life of all living creatures on Earth. The Bhopal tragedy in India in 1984 was an example of how an industry can jeopardize human safety. The tragedy happened when a chemical substance container leaked and killed hundreds of people in the area (Ali, 2015).

In the context of national development, the types of industries prone to causing damage to the aquatic environment include vegetable oil, chemicals, textiles, bottled drinks, canned meat, pulp and rayon, soybean sauce, canned fruits, and the wood industry. Industries that often cause noise pollution include metallurgy, zinc, and iron, etc. Dust and ashes circulated around factory areas, like dust from the cement and chalk industry and toxic gas from aluminum processing factories can cause air pollution. Toxic gas waste is absorbed by local plants that are consumed by humans.

The conventional development model, which focuses on economic growth with no regard for the preservation of the environment, must be discarded. This is due to the real impact it has on the environment and for the future generation. Awareness of the negative consequences of the conventional development model has existed since the birth of the SD concept. The world realized that the management of natural resources and the preservation of the environment must be integrated with social issues such as poverty. It is agreed that if a country's economy is weak and many of its people are poor, the environment will also deteriorate. If the environment is damaged and resources are exploited and depleted in an excessive manner, the people will suffer and the economy will worsen.

These issues challenged the global community to find an appropriate solution, which led to formulating the Concept of Sustainability, from Theory to Application, which emphasized the conservation and improvement of social, ecological and economic needs, as the main aims of SD. This refers to three sustainability pillars: ecology, economy and society (Ali 2015). The concept was made in a reference to the development program of the World Commission on Environment and Development (1987), which defines sustainable development as "a type of development which combines the fulfillment of present needs without risking the future generations' ability to cater their own needs."

In July 2001, European Union member countries held a conference in Gothenburg, Sweden, to discuss the future of Europe and the general guidelines of the related policy. The conference recommended that development should take into account the principles of sustainability, to fulfill the present generation's needs without risking the fulfillment of the future generations. Therefore, all forms of the economic, ecological, and social policies must be made synergically so they can reinforce each other. If the development is unable to stop the tendencies that risk the future quality of life, society's costs will increase dramatically and the negative tendencies will reach a point of no return.

The current threats to the environment are related to uncontrolled population growth, urbanization, and development of infrastructures triggered by the expansion of development in urban and industrial areas. In fact, these have a significant effect on a country's macroeconomy, particularly with respect to the economic growth.

2 RESULTS AND DISCUSSION

This has resulted in the increase of land and air pollution and problems related to volumes of refuse or industrial and household waste, which can be hazardous to health. It has also increased society's consumption at a global scale, causing a rise in demand for raw materials, energy, and water. This has implications for the degradation of humanity, cultural and spiritual values.

The world change has certainly had an impact on the environment, which must be anticipated well. The rise of the world's population has triggered an escalating increase in the need for food, clothes, housing, and energy. Countries are competing with each other to carry out development for fulfilling these increasing needs. However, many still base their standards on the conventional development model, focusing only on economic growth. In the conventional development model, achievements in the form of increased production of goods and services, which are the elements of GDP and gross regional domestic product (GRDP), do not accommodate the environmental aspect. Social development is also overlooked so that in many aspects poor people are marginalized during development.

Overcoming the negative effects of the conventional development model requires implementing the SD concept. This includes a vast scope, while its parameters and indicators continue to grow according to the understanding of each organization involved. However, in view of the above background and concepts developed by the UN, we have to be aware of its importance because the damage inflicted on the environment as a result of economic development is already at an alarming level. This can be observed by the phenomena of climate change, global warming, drought, and natural disasters, etc. This paradigm is intended to be the collective awareness of all nations of the world when implementing their development programs.

In order to intensify awareness, the UN mandated UNESCO to include the concept in the education program. This led to a proclamation concerning an international implementation strategy for the Decade of Education for Sustainable Development (DESD), which focused on achieving SD goals (UNESCO 2005). It also can be viewed from socio-cultural, environmental, and economic perspectives (Summers 2013, Pradham & Mariam 2014):

a) From a socio-cultural perspective, it is seen as an attempt to fulfill human rights and achieve national defense, world peace, national life survival, gender equality, cultural diversity, intercultural understanding, healthcare, and prevention and management of harmful diseases such as HIV/AIDS.
b) From an environmental perspective, it becomes an attempt to utilize the natural resources in a balanced manner. This is in regard to the future generations' needs, anticipated climate change, and changes in the natural environment in both urban and rural areas due to urbanization, and prevention of the disasters, which is triggered by human activities such as forest exploitation causing floods and droughts.
c) From the economic perspective, it is seen as an effort to reduce poverty, improve welfare, achieve sustained economic growth, and establish economic independence and national competitiveness.

Education for sustainable development (ESD) means taking into consideration the SD pillars, as the independent and interrelated dimensions, that is, society, culture, economy and environment, in an attempt to improve the quality of life. Jenkins and Jenkins (2015) explained about the position of the three pillars in the implementation of sustainable development: "ESD, which includes economics and social dimensions as well as environmental ones, is viewed as being less constrained and more holistic than traditional forms of education 'about' the environment." This is a dynamic concept as well as a collective attempt to look into the future, when everyone will reap benefits from the opportunity to obtain an education, learn about important lifestyles, behavior, and values for the creation of a sustainable future.

In regard to implementing the concept, the 58th UN General Assembly in December 2005 agreed on the implementation of the DESD starting in the year 2005 through to 2014. As development itself, by its nature, is the process of learning, every education institution both formal (from preschool through college and university) and informal (various training

and adult education institutions) plays an important role in helping learners in this process (Lenglet, 2014). The basic vision of DESD is a world where everybody has equal opportunities to obtain benefits of education for the sake of social transformation. Kahriman et al. (2012) explained this as follows: Goals of the UN Decade of Education for Sustainable Development (2005–2014 DESD) (2005) recommended that principles, values, and practices of sustainable development should be integrated into all aspects and levels of education and learning within the three pillars. In summary, ESD aims to work with all levels of formal education on local and global issues and to develop appropriate ideas, attitudes, values and behaviors with respect to sustainability in all levels of the formal curriculum, starting from the early childhood. The objectives of DESD are:

a) To promote education as the basis of sustainable social life and to strengthen international cooperation for the development of policy innovations, programs and the implementation of ESD.
b) To integrate SD into the education system at all levels of education.
c) To provide funds and support for the education, research and public awareness programs as well as development institutions in developing countries and countries undergoing economic transition.

The application of this concept involves socio-cultural and socio-political issues, including equality of rights, poverty, democracy, and quality of life. Theoretically, ESD can be integrated into all school subjects (Hofman 2015). The integration is done in the form themes, and each includes various subjects such as education for the eradication of poverty, human rights, gender equality, democracy and good governance.

From 2015 through 2030 every UNESCO member country will participate in the decade of the Global Action Program (GAP) for achieving SD goals (SDGs). According to Taylor (2014), the GAP focuses on five priority action areas:

1. Advancing policy.
2. Integrating sustainability practices into education and training environments (the whole institution approaches).
3. Increasing the capacity of educators and trainer.
4. Empowering and mobilizing youth.
5. Encouraging local communities and municipal authorities to develop community-based ESD programs (p.135).

There are 17 goals for the SD during the Global Action Program, 2015–2024: 1) no poverty, 2) zero hunger, 3) good health and well-being, 4) quality education, 5) gender inequality, 6) clean water and sanitation, 7) affordable and clean energy, 8) decent work and economic growth, 9) industry innovation and infrastructure, 10) reduced inequality, 11) sustainabilities and communities, 12) responsible consumption and production, 13) climate action, 14) life below water, 15) life on land, 16) peace, justice, and a strong institution, 17) partnership for the goals (UNESCO, 2015).

In fact, SD by itself is unsustainable unless all stakeholders are committed to its implementation. Sterling (2014) explained that "Sustainable development is not itself sustainable (i.e., lasting and secured) unless relevant learning among all stakeholders is central to the process" (p. 93). This is due to the fact that education improves competencies. Therefore, in its implementation it requires: "long-term and systems thinking, dealing with complexities and working in partnerships. It also entails specific knowledge related to areas of ones' personal and professional life that impact local and global communities and ecosystems" (Fadeeva et al. 2012).

As stated above, among the 17 goals of SD is the elimination of poverty (no poverty) as goal number 1. In the broader sense, this particular goal achievement will have multiplier effects with respect to other SD goals: goal number 2, zero hunger; goal number 3, good health and well-being; goal number 4, quality education; and goal number 8, decent work and economic growth. As the country with the world's fourth-highest population, and predicted to be third-highest in 2025, it is not easy for Indonesia to eliminate the poverty issue. However, an attempt to extensify entrepreneurship among the population is considered a most reasonable attempt in achieving this particular SD goal.

2.1 Entrepreneurship promotion

The concept of entrepreneurship has various definitions. Kuratko (2004), defines entrepreneurship as an integrated concept that permeates an individual business in an innovative manner. Beckman and Cherwitz (2009), define entrepreneurship as an intrinsic human right to change the status quo. This means the concept "entrepreneurship" is concerned with business and innovation with regard to changing the status quo. Acs (2006), defines entrepreneurship as a process through which an individual or group of individuals identify opportunities, allocate resources, and create value. This creation of value is often through the identification of unmet needs or through the identification of opportunities for change. Entrepreneurs see "problems" as "opportunities," then take action to identify the solutions to those problems and the customers who will pay to have those problems solved.

In an entrepreneurial business, an entrepreneur is someone who specializes in making decisions about the coordination of scare sources. Entrepreneurial activity involves the discovery, evaluation and exploitation of opportunity within the framework of an individual-opportunity nexus. The UN Conference on Trade and Development (2015) stated:

There are three of the most frequently mentioned functional roles of entrepreneurs associated with major schools of thought on a entrepreneurship, namely:

1. Risk seeking: the Cantillon or Knightian entrepreneur willing to take the risk associated with uncertainty.
2. Innovativeness: the Schumpeterian entrepreneur accelerating the generation dissemination and application of innovative ideas.
3. Opportunity seeking: the Kiznerian entrepreneur perceiving and seizing new profit opportunities.

With respect to SDG number 1 goal achievement, it is considered important to promote entrepreneurship. In this regard, education plays an important role in entrepreneurship promotion. Of course, that should be concurrently accompanied by the government support. Through the education process we can cultivate entrepreneurial motivation and the related competencies in the country's young generation. In this pursuit, we can learn from the US and Japanese cases, for instance. In these countries, within 10 years the number of well-educated youngsters, particularly university graduates in the United States, could contribute around 42% to the country's GDP per capita%, whereas in the Japanese case this could increase 35% of the country's GDP per capita (Ali 2015).

The USA is one of the risk-taker nations and the dominant force in entrepreneurial activities. Entrepreneurial firms are the essential mechanism by which millions enter the economic and social mainstream of American society. In the US, entrepreneurship enables millions of people, including women, minorities, and immigrants, to access the American Dream. In the evolutionary process, entrepreneurship plays a crucial and indispensable role of providing the "social glue" that binds together both high-tech and "main-street" activities (Drucker 2005). Kuratko (2009) noted that there are as many as 5.6 million Americans below the age of 34 actively trying to start their own businesses. One-third of new entrepreneurs are 30, more than 60% of those aged 18–29 say they want to own their own business, and nearly 80% of those who would become entrepreneurs in the United States are between the ages of 18 and 34.

At the microeconomic level, only certain activities and functions of entrepreneurs may stimulate economic growth. Acs (2006) identified that development economists distinguish three major stages of development. In the first stage, the economy specializes in the production of agricultural products and small-scale manufacturing. In the second stage, the economy shifts from small-scale production toward manufacturing. In the third stage, with increasing wealth the economy shifts away from manufacturing toward services. Entrepreneurship has a large impact on economic growth (Aidis 2003).

An increasing number of entrepreneurs in a country is important for its economic growth and stability. According to Ali (2015), the ideal ratio between the number of entrepreneurs and the population should be ≥ 2%. This means that if the number of entrepreneurs in a country reaches 2% or more, the economic growth should be high and the unemployment rate should stay low. Extensifying entrepreneurship can have the impact of not only increasing the country's economic growth and strengthening its economic foundation but also reducing or even eliminating poverty. However, according to Kerr & Nanda (2009), financial constraints are among the reasons for a lack of entrepreneurship. In an attempt to extensify entrepreneurship a policy is required related to accessibility of financial support to new and existing entrepreneurs.

Hence, developing countries like Indonesia might need to strengthen the conditions for and improve the quality of the entrepreneurial environment for major established firms, including the rule of law, labor market flexibility, infrastructure, financial market efficiency and management skills. These are needed to attract investment that in return will provide employment, export and tax revenues. This implies the need for strong commitment to education at both the secondary and tertiary levels. According to Acs (2006), the nation's economic development depends on successful entrepreneurship combined with the force of established corporations.

3 CONCLUSION

National development has positive impacts, as indicated by improvements in the population's welfare, but economic growth also has negative impacts indicated by environmental destruction and social/cultural degradation.

In order to minimize the negative impacts, the concept of sustainable development should be applied, which is formulated by UNESCO and mandated to every member country to implement in its national development.

2015 through 2030 is the period for implementing the Global Action Program (GAP) of achieving Sustainable Development Goals (SDGs). Of these, the eradication of poverty, or no poverty, is goal number one, and it is assumed this will have a multiplier effects on the other SD goals. Entrepreneurial firms are the essential mechanism by which millions of people enter the economic and social mainstream. At the microeconomic level, entrepreneurship activities may stimulate economic growth. For this regard, extensifying entrepreneurship is considered a reasonable attempt at achieving this goal. For entrepreneurship extensification, it is necessary to strengthen the condition for and to improve the quality of the entrepreneurial environment in order to encourage a larger number of people to become new entrepreneurs.

REFERENCES

Acs, Z., 2006. How is entrepreneurship good for economic growth. *Innovation/winter 2006*, 97–107.

Aidis, R., 2003. Entrepreneurship and economic transition. *Tinbergen Institute Discussion Paper*. Available online at http://www.timbergen.nl

Ali, M., 2015. *Education for national development: A case study of Indonesia*. Bandung: UPI Press.

Beckman, G.D., and Cherwitz, R.A., 2009. *Intellectual entrepreneurship: An authentic foundation for higher education reform*. Society for College and University Planning (SCUP). Available online at www.scup.org/phe.html.

Drucker, P.F., 2005. *Management challenges for the 21st century*. Oxford: Butterworth-Heinemann.

Fadeeva, Z., Petry, R., and Payyappallimana, U., 2012. Learning and innovation for greener and socially just societies. In Fadeeva, Z., Payyappallimana, U., and Petry, R., (Editors). *Towards More Sustainable Consumption and Production Systems and Sustainable Livelihood*. Yokohama: United Nations University, Institute of Advanced Studies, pp. 8–27.

Hofman, M., 2015. What is an education for sustainable development suppose to achieve—a question of what, how and why. *Journal of Education for Sustainable Development*, 9, 2, 213–2228.

Jenkins, K.A., and Jenkins, B.A., 2005. Education for sustainable development and the question of balance: Lesson from the Pacific. *Current Issues in Comparative Education*, 7, 2, 114–129.

Kerr, W., and Nanda, R., 2009. Financing constraints and entrepreneurship. *National Bureau of Economic Research (NBER) Working Paper Series*. Available online at http://www.nber.org/papers/w15498

Kuratko, D.F., 2004. Entrepreneurship education in the 21st century: From legitimization to leadership. *A Coleman foundation white paper USASBE national conference*. January 16, 2004.

Lenglet, F., 2014. Can ESD reach the year 2020? *Journal of Education for Sustainable Development*, 8, 2, 121–125.

Pradhan, M., and Mariam, A., 2014. The global universities partnership on environment and sustainability (GUPES): Networking of higher educational institutions in facilitating implementation of the UN decade of education for sustainable development 2005–2014. *Journal of Education for Sustainable Development*, 8, 2, 171–175.

Sterling, S., 2014. Separate tracks or real synergy? Achieving a closer relationship between education and SD, post-2015. *Journal of Education for Sustainable Development*, 8, 2, 89–112.

Summers, D., 2013. Education for sustainable development in initial teacher education: From compliance to commitment—Sowing the seeds change. *Journal of Education for Sustainable Development*, 7, 2, 205–222.

Taylor, J., 2014. Shaping the GAP: Ideas for the UNpost—2014 ESD agenda. *Journal of Education for Sustainable Development*, 8, 2, 133–141.

UNESCO, 2005. *UN decade of education for sustainable development*. Paris: UN Headquarters.

UNESCO, 2015. *UN roadmap for implementing the global action program on education for sustainable development*. Paris: UN Headquarters.

United Nations Conference on Trade and Development, 2005. *Entrepreneurship and economic development: The empress showcase*. Geneva.

Advances in Business, Management and Entrepreneurship – Hurriyati et al (eds)
© 2020 Taylor & Francis Group, London, ISBN 978-0-367-27176-3

Management model of research clinic-based classroom action research assistance according to the need of social science teachers in junior high school

S. Swidarto, J. Widodo, F. Fahrudin & T. Sumaryanto
Universitas Negeri Semarang, Semarang, Indonesia

ABSTRACT: This study aims to analyze the management model of research clinic-based action research assistance based on the need for social science teachers in junior high school. The method used in this paper was Research and Development design. Data were collected from the action research, need analysis, model test, validation. Data collection techniques used questionnaires, interviews, documentation, observations, FGDs. This paper used model from experts and practitioners to test the validity. The qualitative descriptive analysis in this paper used data display, reduction, verification, conclusion. The result of research showed the factual model is "less good"; hypothetical model was essential; the model clinical research approach was very feasible. The conclusion is that assistance management model was not good enough. It was still theoretical and it had not been based on the need. The research clinic-based assistance model was essential, as it was supported by systemic performance, system components, clinical ways of identifying problems, diagnosis, prognosis, synthesis, treatment, and reflection.

1 INTRODUCTION

Classroom Action Research plays an important role for social science teachers because it becomes a miniature and prototype of learning in the classroom. Moreover, it is useful for empowerment, Teacher Performance Assessment (TPA), and Continuous Profession Development (CPD) (Swidarto 2017). A classroom action research is a process of studying a planned learning problem through a continuous cycle action starting with planning, action, observation, evaluation, and reflection (Sanjaya 2013, Hine 2013). The ongoing cycle is used for improving student's learning outcomes and teacher's performance in meaningful reflective learning practices (Morales 2016).

Although classroom action research is very important, the culture of researching and writing among social science teachers from Junior High School in Pati is low. It is proven by the lack of classroom action research they produce. Social Science Teachers ability in conducting classroom action research is in low average (65.22%). It means that 34.78% of teachers still have difficulties in conducting action research (Lisnawati 2016). Preliminary research results show that system performance management and classroom action research assistance system components so far are still not good. Effectiveness and efficiency of system performance management activities ranging from planning, organizing, coordinating, cooperation between parties, evaluation, and follow-up have not been optimal. The component of the system as a supporter of the implementation of Classroom Action Research

(CAR) assistance is still weak, ranging from competence, qualification, track record, companion and participants; facilities and infrastructure, methods, material specifications of the social science, and the sufficiency of time (Swidarto 2017).

Based on field observations, various efforts to improve teachers' skills in conducting action research have been done either through workshop, Technical Guidance, training and IHT (In-House Training) as well as assistance, but the result has not enabled teachers in arranging CAR with "ONSC" (Original, Necessary, Scientific, Consistent), and "SMART" (Specific, Manageable, Acceptable, Realistic, Time-bound).

The difficulties faced by social science teachers from Junior High School in Pati District in conducting classroom action research were caused by factors such as the competence and qualifications of teachers which are still low. Subject Teacher Forum (STF) of Social Science is not yet optimal in facilitating teachers in conducting classroom action and the teacher's track record in compiling the conducting classroom action is still inadequate.

These conditions required solution through classroom action research assistance with Research Clinic (RC) approach. RC is a model of providing research assistance in clinical ways starting with problem identification, diagnosis, prognosis, synthesis, therapy, evaluation, and follow-up (Surya 2003). RC education is adopted and adapted from health sciences. The results of these adoptions led to clinical supervision, clinical psychology, clinical sociology, and teaching clinics (Schein 2005). RC education is the provision of technical and non-technical assistance to educational personnel who have difficulties in conducting research through clinical means, starting problem identification, diagnosis, prognosis, synthesis, and intervention (Samad 2012).

Assistance with the RC approach has theoretical and practical advantages. Its theoretical advantages are directing toward experience, knowledge, concepts, and practice, while the practical advantages are teacher empowerment, diagnosis of teacher difficulties, directing during the research process, and bridging the practice gap and credibility of educational research (Bos 2008, Muniati 2013).

Based on the background, it is necessary to develop an accompaniment management model with Research Clinic approach. The objective of this research is to analyze and synthesize the classroom action research mentoring management model with RC approach based on decent needs for social science teachers of state junior high school in Pati Regency.

This research is expected to be useful both theoretically and practically for the Education Office, State Junior High School in Pati District, STF and social science teachers as input for the determination of policy in TPA, and CPD.

2 METHODS

The location of this research was in State Junior High Schools in Pati Regency. The research design used Research and Development (R & D) method. The research procedure was adopted from Borg & Gall (Borg 2007), with three stages, namely introduction, development, and validation. The population in this study were 200 people. The samples were 20 people (10%). Data sources included teacher performance documentation data, factual questionnaire results, and needs analysis, observation, interviews, Focus Group Discussion (FGD), and CAR mentoring model results. Data collection techniques used in this research were questionnaires, interviews, observations, Focus Group Discussion (FGD). The validity of data used were agined throuh the validity of the constructs and the validity of the items. Data were presented quantitatively and qualitatively.

3 RESULTS AND DISCUSSION

The result of preliminary research of classroom action research teacher training for social science teachers in junior high school generally was "not good". The average score is 206.40 in the range of 167-240 or 52.32%. The poor category was distributed for the performance of the average score system is 82.85 in the range of 66-94 or 54.01%, the system component average score is 123.55 in the range of 102-146 or 51.30%.

In the aspect of the performance in the management system, the highest score is the implementation which is 56.64%. It means that 56.64% of the respondents said the implementation was "good", but 43.36% of them said it was "less good". Implementation of assistance is effective, supported by coordination and cooperation between parties (Masrukhi 2015). Cooperation between the parties fosters is useful collegial interactions to help teachers construct knowledge, and improve their teaching practices (Zulfiani 2016, Aldridge et al. 2012). The lowest value of system performance was evaluation which is 48.92%. It means that 48.92% of the respondents stated that the evaluation was "good", but 51.08% of them stated that it was "less good". An indicator which achieved "good" category is the reflection, while the accuracy of the evaluation method achieved "less good". Reflection is the assessment of success/deficiency expressed through impressions, messages, hopes and useful criticisms which are useful for improving the quality of teachers in understanding the material and practice (Sukanti 2008, Rosiani et al. 2014). The evaluation method was used to determine effectiveness and efficiency, to measure the level of achievement of objectives, materials, stakeholder involvement, and task execution, satisfaction, material understanding, and attitudes (Masrukhi 2015, Suparlan 2013, Zarkasy 2005). Methods are in (theory), on (practice), and in, on (theory and practice) (Supriano 2016).

On the aspect of system components, the highest score is companion which was 69.17%. It means that 69.17% of the respondents stated that the companion was "good", but there is still 30.83% of them which was "poor". The inadequate indicators that need to be improved are competence, qualification, strategy, and giving feedback. Companions are required to have competence, qualifications, and track record, and are able to guide from beginning to end (Masrukhi 2015, Surapranata 2015). The lowest value of the system components was participants which was 44.70%. It means that 47.70% of the respondents stated that the participants were "good", and 55.30% of them stated that they were "poor". Inadequate indicators that need to be improved are background, literature review, methodology, scientific writing, and data analysis. The mentoring participants are required to have competence, qualification, track record, and commitment to the task (Widodo 2017). The factual model of mentoring classroom action research for social science teachers in Junior High Schools in Pati Regency has been implemented as well as has embraced the concept of system performance management and system components. The performance of the management system is a managerial process that reflects the performance of organizational devices (Widodo 2017). System performance includes aspects of planning, organizing, implementation, and supervision (Setiadi 2018). The system component is organizing and supporting element in the implementation of assistance which includes elements: companion, participants, facilities and infrastructure, material and time (Mckimm 2007),

The hypothetical model design was obtained from the results of analysis of system performance requirements, system components, and RC. The results of need analysis of classroom action research assistance, in general, showed that the assistance was "very important" with average score 474.75 in the range of 432-532 or 88.05%. The category was distributed for the analysis of the performance of needs of the system with average score 135.25 in the range 124-152 or 88.88%; average system component score is 209.80 in range of 192-236 or 88.22%; RC obtained the average score of 126.70 in the range of 117-144 or 85.59%.

Value of analysis results of the highest system performance requirements was planning with 89.46%. This means that 89.46% of the respondents stated that planning was "very important", only 10.54% who stated that it was "less important". Planning is the initial design that becomes the reference and benchmark of the success of classroom action research assistance that includes the elements of goals, objectives, programs, and the involvement of the parties (Masrukhi 2015, Suparlan 2013, Widodo 2017). The result of the analysis of the need for the lowest system performance was evaluation which was 88.44%. Evaluation is an assessment of the success or ineffectiveness of mentoring (Suparlan, 2013). Aspects of evaluation include the accuracy and ability of evaluation methods in providing a reaction, learning behavior, and skills (Masrukhi 2015, Suparlan 2013).

The result value of the highest system component requirement analysis was companion which is 90.57%. It means that 90.57% of the respondents considered the companion as "very important", and only 9.43% of them stated that it was less important. Based on the result, indicators which are considered very important include competence, qualifications, commitment in carrying out the task, while the indicator which was less important was academic qualifications. Assistance activities require qualified, professional, and competent resources (Sumaryanto 2017). A companion is someone who is in charge of helping, directing, solving work problems, and the career of the accompanist (Kaswan 2012). The role of counselors are guides, counselors, instructors, sharers, and encouragers (Ambrosetti & Dekkers 2010). Companion criteria are professional in their field, interpersonal skills (Waring 2013). The lowest value of system components was participants with 84.88%. It means that 84.88% of the respondents considered the participants as "very important", only 15.12% of them considered that participants were "less important". Indicators which were considered important are competence, qualification, and the ability of participants in completing the tasks on time.

Value of requirement analysis of highest RC approach was treatment with 86.88%. It means that the treatment was "very important" in the assistance of CAR. Treatment is an alternative form of assistance based on the background of the cause (Wiramihardja 2012). The treatment is in the form of technical and nontechnical. Treatment which is done through clinical ways means therapy, intervention, and consultation (Muniati 2013, Andaryani 2013). The lowest result value of RC requirement analysis was identification problem with 82.08%. Problem identification is a necessary action to know the root cause of the problem, as well as to find the solution (Schein 2005).

Based on the design of a hypothetical model of CAR assistance management by referring to RC needs, its implementation pattern used system performance flow, system components, and RC approach. System performance includes planning, execution, reporting, evaluation, follow-up, and monitoring. System components include anassistant, participants, facilities and infrastructure, teaching materials and time. The RC approach includes identification, diagnosis, prognosis, synthesis, treatment, and evaluation/reflection.

A validated CAR assisted expert, practitioner, and feasibility model through FGD-1 to hypothetical model is trialed in a limited way and then is expanded and evaluated, the results are presented in Table 1.

Table 1 describes the value of the results of the limited trial categorized as unfavorable with average score 6.58 or 54.82%. The highest score of the limited trial is the quality of classroom action research proposal submission average with score 7.58 or 63.13% which is in "good" category. It means that social science teachers has no difficulty in preparing a proposal of classroom action research because they have obtained workshop, In-House Training (IHT), and Teacher Professional Education and Training (PLPG) (Masrukhi 2015). The lowest score is scientific article writing with average score 6.00 or 50.00% which falls into "less good" catagory. It means that social science teachers are still experiencing difficulties, so they need to be given assistance in preparing scientific articles. The scientific articles of classroom action research result are very important because it is one of the requirements of CLA (Widodo, 2017). Good scientific articles include title, abstract, introduction, method, results and discussion, conclusions and suggestions, as well as bibliography (Supardi & Suhardjono 2013).

Evaluation of the practice of the test results expanded to a very good category with average score 27.85 in the range (26-32) or 87.02%. The highest score is CAR reporting results with average score 29.08 or 90.87% which falls into "very good" category. It means that social science teachers has improved their ability to compile reports from the beginning; Chapter I-V, list of centers, and attachments. The lowest score is making the book results of classroom action research with average score 25.12 or 78.50% which is in "good" category. It means that social science teachers need to be given assistance in arranging books of classroom action research results.

Table 1. Assessment practical result of social science teachers of state junior high schools in Pati regency at limited and expanded trial.

No.	Aspect	Limited Trial					Expanded Trial					Percentage of Improvement
		Score	Range	Average	Conversion %	Category	Score	Range	Average	Conversion %	Category	
1.	Proposal of Classroom action research	182	179-233	7.58	63.19	Good	688	624-768	28.67	89.58	Very Good	26.39
2.	Instrument	57	40-58	7.12	59.38	Poor	228	208-256	28.50	89.06	Very Good	29.68
3.	Reports of Action Research	234	202-290	6.00	50.00	Poor	1134	1014-1248	29.08	90.87	Very Good	40.87
4.	Scientific Paper	42	34-50	6.00	50.00	Poor	196	182-224	28.00	87.50	Very Good	37.5
5.	Scientific Facilities	45	34-50	6.42	53.57	Poor	194	182-224	27.71	86.61	Very Good	33.04
6.	The book of the result of classroom action research	209	172-246	6.33	52.,78	Poor	829	659-857	25.12	78.50	Good	25.72
	Total	769	616-882	39.45	328.92	Poor	3269	3068-3776	167.08	522.12	Very Good	193.2
	Average	256.33	203-292	6,58	54.82	Poor	408,63	383-472	27.85	87.02	Very Good	32.20

Based on the assessment result of preparing classroom action research in the limited and expanded trial, there is an average increase of 32.20%. It means that RC-assisted assistance can contribute to improvement the quality of IPS teachers in conducting classroom action research started from proposals, instruments, reporting, articles, scientific, and books. The improvement is 32.20% of other dimensions. The enhancement of the capacity of the certified teacher is trained through the professional development clinic at STF, by clinical means problem identification, diagnosis, prognosis, synthesis, treatment, and reflection (Muniati 2013).

Assessment of a classroom action research assistance with RC approach which has been trialed in limited and expanded way is consulted with experts and discussed through FGD-2 for model improvement, guidebooks, guides, classroom action research teaching materials. The results are presented in Table 2.

Table 2 describes the results of the expert validation of the feasibility of the model book, guidance and teaching materials of classroom action research with RC approach which is very feasible. The average score is 20.36 or 84.58%. The highest score is companion manual and retention participants with score 20.80 or 86.67%. It means that the guidebook is very worthy of use for mentoring. The guidebook is very important because it contains instructions for operational techniques for the team regarding tasks, functions, roles, and order (Surapranata 2015). The lowest score is the average model scorebook which is 19.70 or 82.08%. This means that the model book needs to be refined. The model book is very important, as it becomes the PTK's mentoring workflow. The flow of RC operations is based on system performance and system components (Swidarto 2017).

The findings of the final model of CAR assistance with RC teachers approach for social science teachers from Junior High Schools in Pati Regency is a refinement of the factual model and hypothetical model. Factual and hypothetical models after FGDs with experts and practitioners are validated for input

and refinement to become a viable final model. Expert validation and final model validation results include aspects of system performance and system components. System performance aspects that have been validated and enhanced include planning, implementation, evaluation, reporting, follow-up, monitoring and monitoring (monitoring and evaluation). Aspects of components include an assistant, participants, facilities and infrastructure, teaching materials, and time.

The findings of the research results of CAR's RC assistance approach indicated an improvement in the results ranging from factual conditions, hypothetical model design development, and final model. The factual condition of CAR assistance that has been implemented was categorized as a "less good" with percentage of 52,32%. Development of hypothetical model obtained the result in the category of "very important" with percentage of 88.05%. The final model obtained a result in the category of "very decent" with percentage of 84.58%. Based on these findings, mentoring of RC approach can improve the quality of social science teacher performance in conducting action research is 32.20% from other dimensions.

Based on these findings, the RC-assisted mentoring model for classroom action research can be implicated in the development and professional development of social science teachers. Professional development related to improving the quality of teachers' performance in conducting classroom action research is useful to overcome the problems and improvements in learning in the classroom. Professional development for career enhancement, and scientific publications as CPD requirements.

It is suggested that (1) social science teachers improve their competence and qualifications in conducting classroomaction research for CPD; (2) STF of social science subject is a forum for organizations to facilitate CPD activities. (3) The Core Teacher as a companion carries out the mentoring thoroughly; (4) Principal grant ease of permit, accommodation to teachers following CPD; (5) supervisors of the clump improve the guidance, monitoring, and evaluation of teacher activities, STF of social science in CPD.

Table 2. Recapitulation of expert validation results on appropriateness of model. Handbook, guide and teaching material for classroom action research assistance with RC approach for social science teacher in junior high school.

| No. | Evaluated Aspect | Results of Experts' Assessment/Expert | | | | |
		Score	Range	Average	Conversion (%)	Category
1.	Model Book	394	390-480	19.70	82.08	SL
2.	Reference Book	405	390-480	20.25	84.38	SL
3.	Guide Book	416	390-480	20.80	86.67	SL
4.	Teaching Materials	3434	3237-3984	20.69	85.19	SL
Number of all Indicators		4649	4317-5424	81.44	338.32	SL
Average of All Indicators		774.83	734-904	20.36	84.58	SL

4 CONCLUSION

Based on the results of the research that has been described previously, it can be stated that: (1) the factual model of management of classroom action research assistance for social science teachers so far is categorized as less good; (2) the design of the development of a classroom action research mentoring management model with a need-based RC approach is very important and needed by social science teachers in junior high school; (3) mentoring model with RC approach is suitable as a vehicle for improving the quality of social science teachers' performance in Pati District in conducting classroom action research, TPA, and CPD.

REFERENCES

Aldridge, J., Barry J. F., Lisa, B., 2012. Using a new learning environment questionnaire for reflection in teacher action research. *International Journal Science Teacher Education* 3(23): 259–290.

Ambrosetti, A. & Dekkers, J. 2010. The interconnectedness of the roles of mentors and mentees in pre-service teacher education mentoring relationships. *Australian Journal of Teacher Education* 35(6): 42–55.

Andaryani, G. 2013. Learning clinic. UNS. *IJERN* 2(3): 13–22.

Borg.G. 2007. *Educational research*. New York: logman.

Bos, J. A. B. 2008. Relevance in education research, will a clinical approach make education research more relevant for practice? *Educational Researcher* 37 (7): 412–420.

Hine, G.S 2013. The importance of action research in teacher education programs. *University of Notre Dame Australia Journal* 23(2): 151–163.

Kaswan, 2012. *Coaching dan mentoring untuk pengembangan SDM dan peningkatan kinerja organisasi.* Bandung: Alfabeta.

Lisnawati, T. 2013. The roles of subject teacher forum of social science from KOMDA PATI in improving teachers' professionalism of social science education. *Journal of Educational Social Studies* 2(1): 16–21.

Masrukhi, 2015. Developing supervision-based classroom action research to enhance mathematics teachers' professionalism in Brebes regency. *Journal Cakrawala Pendidikan* 9 (1): 59–65.

Mckimm, J. 2007. Mentoring: theory and practice. Developed from preparedness to practice. *Mentoring Sch. NHSE/Imperial College School of Medicine.* 3(4): 1–24.

Morales, M.P.E. (2016). Participatory action research (PAR) cum action research (AR) in teacher professional development: A literature review. *International Journal of Research in Education and Science (IJRES)* 2(1): 156–165.

Muniati, N. A. 2013. The effectiveness of supervising science teachers' professionalism through subject teacher forum with clinic teaching. *IKIP PGRI Semarang* 3 (2): 20–31.

Rosiani, D. Martono., Kardoyo. 2014. Developing training model to improve teachers' competence in developing classroom action research-based learning. *Journal of Educational Social Studies*: 1(2): 69–79.

Samad, B.S. 2012. *Types of research based on function.* http://educationesia. blogspot.com

Sanjaya, H.W. 2013. *Classroom action research.* Jakarta: Kencana.

Schein, S. 2005. Connecting clinical teaching practice with instructional leadership. *Journal of Education University of Melbourne Australia* 57(3): 225–236.

Setiadi, G. 2018. The development of blended learning-based self-learning on classroom action research training material to improve teachers professionalism. *IJERN* 4(9): 213–224.

Sukanti, 2008. Improving teacher's competence through classroom action research implementation. *Jurnal Akuntansi Indonesia* 6(1): 1–11.

Sumaryanto. T. 2017. The implementation of among-asuh method in guidance and mentoring management to shape the cadets' noble character (a case study in the military academy). *JED* 5 (3): 445–459.

Supardi & Suhardjono. 2013. *Strategies in planning classroom action research based on the rules of ministry of education and bureucrat reformation number 16 year 2009.* Yogyakarta: Andi Offset.

Suparlan. 2013. *School-Based Management Theory to Practice.* Jakarta: Bumi Aksara.

Supriano. 2016. *The guidance of conducting training and supervision of the 2013 curriculum.* Jakarta: Kementerian Pendidikan dan Kebudayaan.

Surapranata, S. 2015. The *guidance of career supervision in writing paper for educators and educational staffs.* Jakarta: Kementerian Pendidikan.

Surya, D. 2003. *Supervising teachers in classroom action research.* Jakarta: Kementerian Pendidikan Nasional.

Swidarto, 2017. Development of action research supervison model with clinical research-based approach for social science teachers in junior high school in Pati regency. *Proc. in Untidar* 1(1): 21–26.

Waring, H.Z. 2013. Two mentor practices that generate teacher reflection without explicit solicitations: some preliminary considerations. *RELC Journal* 44 (1): 203–119.

Widodo, J. 2017. A model for developing soft skill training management oriented toward service quality for shs counselors. *JED* 5(1): 39–49.

Wiramihardja, S. 2012. *Introduction to clinical psychology.* Bandung: Refika Aditama.

Zarkasy, A. S. (2005). *The guidance of BMPM monitoring and evaluation.* Jakarta: Departemen Agama RI.

Zulfiani, 2016. The Study of Supervision Implementation in Collaborative Classroom Action Research between University and Schools. 35 (2): 273–283.

Does central policy influence performance of small and medium-sized industry? A case study of Footwear Central Industry in West Java, Indonesia

R. Wahyuniardi & H. Djulius
Universitas Pasundan Bandung, Bandung, Indonesia

I. Sudirman
Institut Teknologi Bandung, Bandung, Indonesia

ABSTRACT: This study aims to show the influence of central industrial policy on the performance of Small and Medium Industry (SMI). Central policy by the Government of Indonesia since 2007 aims to strengthen the performance of SMI. This research discusses the influence of external collaboration, technology, and research and development on innovation, as supported by the central policy on the performance of SMI. The research analyses two things: (1) literature review of SMI performance research in various countries; and (2) empirical phenomena about SMI policies and Central Industry's performance in Indonesia. A qualitative model and research hypotheses are also used. The research is conducted at Footwear Central Industry in West Java including Cibaduyut-Bandung, Ciomas-Bogor, and Sukaregang-Garut. The results of this research show the influence of the factors mentioned, recommending the development of SMEs in the Central Industry, especially the Footwear Central Industry in West Java, Indonesia.

1 INTRODUCTION

The policy of industrial centers in Indonesia started in 2007 through the regulation of Ministry of Industry number 78/M-IND/9/2007 (Kemenperin 2007). The aim of this policy is to promote creative, innovative, unique, and local products and enhance their competitiveness. This policy is similar to the concept of the governor of Oita Province in Japan, called One Village One Product (OVOP) (Triharini et al. 2014). It is in line with the concept of clusters, which Mazur (2006) states are an innovative-active concentration in a country.

Larsen and Lewis (2007) state that a company will not last long without innovation. This is due to the needs, desires and requests of customers for change. Therefore, innovation must continue in order to ensure the company's performance (Hadiyati 2011).

In Indonesia, the footwear industry's contribution to Gross Domestic Product (GDP) rose from Rp. 31.44 trillion in 2015 to IDR 35.14 trillion in 2016, or 0.28% of state revenues. This is in line with the positive export growth of USD 4.85 billion in 2015, which rose 3.3% to USD 5.01 billion in 2016. In the footwear industry sector, Indonesia was the fifth-largest exporter worldwide, after China, India, Vietnam, and Brazil, with international market share reaching 4.4% (Ministry of Industry 2017).

Nevertheless, conditions in Indonesia, especially in the development of the footwear sector in West Java, are not very encouraging. The number of SMEs in Sentra Footwear Cibaduyut, Bandung,

"only" amounted to 177 units. This number dropped dramatically from the initial data of 400–450 units. Similar conditions occurred in the sentra of footwear in Bogor and Garut (Wahyuniardi et al. 2014).

For these reasons, the study aims to explain the importance of research on the existence of sentra in Indonesia. This study also involves literature that supports the research.

1.1 *External collaboration*

In their research, Powell et al. (1990) state that external collaboration is a key instrument in completing activities that lead to the creation of value inside organizations and enhance their competitive advantage, because, in knowledge and networking economies, innovation often comes not from within the firm but from an external partner. Collaboration can be characterized by interactivity, open and direct relationships, support for experience and innovation, and common interests and purposes for all partners (Rahbari & Jalali 2016). Research shows that foreign investment and transfer of technology affect the transfer of knowledge in Indonesia's manufacturing sector.

1.2 *Technology*

In Capability Approach (CA), technology is an input that enables capability, shared with other resources. Technology is like other inputs, but it can drive the relationship between technical objects and other inputs as well as conversion factors for more complex ideas (Haenssgen & Ariana 2018). Here technology is defined as "something created through an organizational process that shows its aspects functioning with a purpose that can provide some benefit" or "something organized and implies the process of creation" whose aspects work with goals that can provide multiple benefits" (Carroll 2017, Djulius 2017a).

1.3 Research and development

R&D is a complex concept. As defined by the OECD's Frascati Manual (OECD 2002), it is "creative work undertaken on a systematic basis in order to increase the stock of knowledge, including knowledge of man, culture and society, and the use of this stock of knowledge to devise new applications" (Prihadyanti & Laksani 2015). R&D is categorized into basic research, applied research, and experimental development.

Kim in his research on South Korea's service industry divides a company's research and development activities into three activities: internal, joint, and external (Kim et al. 2016, Djulius 2017b).

1.4 Innovation

The term "innovation" was defined by Schumpeter in the early 20th century as a product, process, and organizational change that does not always come from new scientific discoveries but may arise from a combination of existing technologies and their applications in a new context (Urbancova 2013).

Lala, Preda, and Boldea (2010) state that innovation is the ability to develop ideas. Innovation also comes from public research (Lala et al. 2010). Therefore, it is possible to conclude that, according to this definition, innovation includes not only technical and technological changes and improvements but also practical applications, especially those derived from research. A critical view of the definitions provided in the existing literature should involve three important considerations. The first consideration shows that innovation is not something that should be defined singly and in an integrated way. Innovation can come in the form of a new product, a new service, a new technology, or a new administrative practice (Razavi & Attarnezhad 2013). Areas of innovation can take different general forms, including: (1) diversification of an existing collection of products and services; (2) new additions and versions of existing types; (3) a completely new introduction of goods; (4) improved techniques and presentation styles; and (5) development of participation models.

Innovation is a costly and risky activity (Ortega-Argilés et al. 2009, Rehman 2006). From a company perspective, innovation can be defined as a complex process involving new ideas—development, transformation, and application—using technology,

capabilities, and knowledge resources (Rehman 2016, Djulius et al. 2018).

1.5 Center for Small and Medium Industries (SMI)

The Center for Small and Medium Industries is a central location for small and medium industries producing similar products, using similar materials and/or working on the same production process; it is equipped with supporting facilities and infrastructure designed based on the development of local resource potential, and it is managed by a professional board.

The development of the Center of Small and Medium Industry is a mandate in law no. 3 of 2014 on Industry, where article 14 mentions the role of the Central Government and/or Regional Government to accelerate the spread and equity of industrial development throughout the territory of the Unitary Republic of Indonesia through industrial zoning (Undang-Undang 2014). The industrial territory is implemented via the development of the Industrial Growth Center Area, the Industrial Allotment Zone, the Industrial Zone, and the Sentra of Small and Medium Industries.

In addition, Article 74 mandates the enhancement of central capability in order to strengthen the institutional capacity of small and medium industries.

At this time, the SMI Central generally grows informally within its limitations, without direct contact or intervention from government, so development is difficult. The central and/or local governments are expected to undertake the development and empowerment of SMEs in order to boost competitiveness. They are also expected to play a significant role in strengthening the national industrial structure, alleviating poverty, expanding employment opportunities, and producing industrial goods and/or services for export (Kemenperin 2016).

1.6 Performance

Performance can be measured through financial indicators (profit, market share) or non-financial indicators (customer satisfaction, production, cost improvement, sales improvement) (Demirbag et al. 2006, Sondakh & Ellitan 2017). In addition, Ainin et al. (2015) state that the use of Facebook in SMI has a strong influence on financial and non-financial indicators.

2 HYPOTHESIS

2.1 External collaboration and innovation

Rahbari and Jalali (2016) state that the ability to implement effective external collaboration will have an impact on the capabilities of innovation. Johnson et al. (2009) also show that collaboration, both internal and external, is very influential on innovation. The advantages of external collaboration come in speeding up the time necessary to develop a product. Brettel and Cleven (2011) and Findik and Beyhan (2015) state that a company's ability to use external partnerships can strengthen its innovative performance.

Hypothesis 1: External collaboration has an influence on innovation.

2.2 Technology and innovation

Conte et al. (2013) state that the acquisition of technology has a significant influence on innovation, whether innovation to process or product innovation. Meanwhile, Hao and Yu (2011) state that the selection of technology has no effect on innovation but that capabilities, and managerial technology capabilities, are what affect innovation. This is in accordance with Ndesaulwa and Kikula (2016), who state that China is more innovative than other countries because of its acquisition of technology as well as research and development. China is expected to make more than 9 billion varieties of production—more than the world's human population.

Hypothesis 2: Technology has an influence on innovation.

2.3 Research and development and innovation

Akinci and Utlu (2015), in a study conducted at an industrial center in Turkey, show that 53% of SMI have an R&D unit and conduct R&D activities. Meanwhile, Martin (2015) states that R&D activities affect innovation. In addition, Baumann and Kritikos (2016) state unequivocally that R&D activity has a considerable influence on a company's ability to introduce innovation. Rehman (2016) states that research and development is a significant intangible asset related to corporate innovation.

Hypothesis 3: Research and development influence research on innovation.

2.4 External collaboration with technology

External collaboration provides many technological advantages for SMEs. With external collaboration, SMI can ignore its internal resources in favour of improving company efficiency by collaborating with IT support companies (Noh & Lee 2015). In that same year, an online survey conducted in three countries, with a total of 753 workers in sixteen business sectors, yielded data that 83% of workers rely heavily on technology in their collaboration.

2.5 Technology and research and development

Non-tech firms are being gradually replaced by high-tech firms that emphasize added value with high investment, skilled labor, advanced machinery, and R&D (Hansen & Winther 2014, in Booltink & Saka-

Helmhout 2018). It is clear that technology plays the same role as R&D itself.

2.6 External collaboration with research and development

Olmos-Peñuela et al. (2017), in their research at 610 SMI Spain, conclude that companies with previous cooperative experience gained more benefits in terms of strengthening their innovation culture; repeated collaboration over time produces a higher ability to absorb knowledge and skills provided by fellow teachers.

2.7 Innovation and performance

According to Bocquet et al. (2017), innovation is the main factor in a company's performance. In line with that, Atalay, Anafarta, and Sarvan (2013) explain that the type of innovation affects the company's performance. To add, Rajapathirana and Hui (2017) conclude in their research on the insurance industry of Sri Lanka that innovations made by a company have an effect on that company's performance.

Hypothesis 4: Innovation has an influence on performance.

Hypothesis 5: Innovation with Sentra's support influences SMI performance.

3 METHODOLOGY

This research will be conducted in three footwear Sentra in West Java: (1) Sentra Cibaduyut Bandung; (2) Sentra Ciomas Bogor; and (3) Sentra Sukaregang Garut.

The paradigm of this research can be shown in Figure 1.

The data collected is be processed by Structural Equation Modeling (SEM) approach. The representative sample size using SEM analysis is in the range of 100 to 200 (Hair, 1998). Each variable and its indicator is to be tested for validity and reliability.

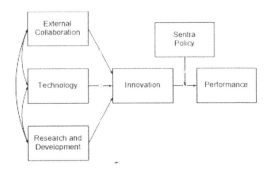

Figure 1. Paradigm of Research.

Source: Researcher, 2018.

The measurement scale to be used is a score of 1 to 5 (1 = Strongly Disagree; 5 = Strongly Agree). Data obtained will be processed using the Statistical Package for the Social Sciences (SPSS) version 22.

4 RESULTS

SMI is a major contributor in Indonesia's economy because of its flexibility. It can survive in turbulent environments better than large companies: SMI companies are more flexible and adaptable to change, and, being labour intensive, they also support job creation (Oliveira & Guarulhos 2015).

This research contributes information on how SMI innovates in the central industry and what should be done by the government in encouraging SME development, especially in three Footwear Centers in West Java.

REFERENCES

Ainin, S. et al. 2015. Factors influencing the use of social media by SMEs and its performance outcomes. *Industrial Management & Data System* 115(3): 570–588.

Akinci, G. & Utlu, Z. 2015. The research & development and innovation capacity of small and medium sized enterprises in IMES. *Procedia—Social and Behavioral Sciences* 195: 790–798.

Atalay, M., Anafarta, N. & Sarvan, F. 2013. The relationship between innovation and firm performance: an empirical evidence from Turkish automotive supplier industry. *Procedia—Social and Behavioral Sciences* 75: 226–235.

Baumann, J. & Kritikos, A. S. 2016. The link between R&D, innovation and productivity: Are micro firms different? *Research Policy* 45(6): 1263–1274.

Bocquet, R. et al. 2017. CSR, innovation, and firm performance in sluggish growth contexts: A firm-level empirical analysis. *Journal of Business Ethics* 146(1): 241–254.

Booltink, L. W. A. & Saka-Helmhout, A. 2018. The effects of R&D intensity and internationalization on the performance of non-high-tech SMEs. *International Small Business Journal: Researching Entrepreneurship* 36(1): ?.

Brettel, M. & Cleven, N. J. 2011. Innovation culture, collaboration with external partners and NPD performance. *Creativity and Innovation Management* 20(4): 253–272.

Carroll, L. 2017. A comprehensive definition of technology from an ethological perspective. *Social Sciences* 6(4): 126.

Conte, A. et al. 2013. Succeeding in innovation: key insights on the role of R&D and technological acquisition drawn from company data. Discussion paper 7671.

Demirbag, M. et al. 2006. An analysis of the relationship between TQM implementation and organizational performance. *Journal of Manufacturing Technology Management* 17(6): 829–847.

Djuliul, H. 2017a. Foreign direct investment and technology transfer: Knowledge spillover in the manufacturing sector in Indonesia. *Global Business Review* 18(1): 57–70.

Djulius, H. 2017b. How to transform creative ideas into creative products: Learning from the success of batik fractal. *International Journal of Business and Globalisation* 19(2): 183–190.

Djulius, H., Juanim, J. & Ratnamiasih, I. 2018. Knowledge spillover through foreign direct investment in textile industry. *International Journal of Economic Policy in Emerging Economies* 11(1–2): 12–25.

Hadiyati, E. 2011. Kreativitas dan inovasi berpengaruh terhadap kewirausahaan usaha kecil. *Jurnal Manajemen dan Kewirausahaan* 13: 8–16.

Fındık, D. & Beyhan, B. 2015. The impact of external collaborations on firm innovation performance: Evidence from Turkey. *Procedia—Social and Behavioral Sciences* 195: 1425–1434.

Haenssgen, M. J. & Ariana, P. 2018. The place of technology in the Capability Approach. *Oxford Development Studies*. Routledge 46(1): 98–112.

Hair, J., Anderson, R., Tatham, R. & Black, W. 1998. *Multivariate data analysis (5th ed.)*. Upper Saddle River, NJ: Prentice Hall.

Hao, S. & Yu, B. 2011. The impact of technology selection on innovation success and organizational performance. *iBusiness* 3(4): 366–371.

Johnson, W. H., Piccolotto, Z. & Filippini, R. 2009. The impacts of time performance and market knowledge competence on new product success: An international study. *IEEE Transactions on Engineering Management* 56(2): 219–228.

Kemenperin. 2007. Peraturan Menteri Perindustrian nomor 78/M-Ind/Per/9/2007. Jakarta: Kementerian Perindustrian.

Kemenperin. 2016. Kinerja industri [Online]. Retrieved from http://www.kemenperin.go.id/kinerja-industri. Accessed 12 July 2018.

Kim, S. et al. 2016. The effect of service innovation on R&D activities and government support systems: The moderating role of government support systems in Korea, *Journal of Open Innovation: Technology, Market, and Complexity*. Journal of Open Innovation: Technology, Market, and Complexity 2(1): 5.

Lala, P. I., Preda, G. & Boldea, M. 2010. A theoretical approach of the concept of innovation. *Managerial Challenges of the Contemporary Society* (1):151–156.

Larsen, P. & Lewis, A. 2007. How award-winning SMEs manage the barriers to innovation. *Creativity and innovation management* 16(2): 142–151.

Martin, M. 2015. Effectiveness of business innovation and R&D in emerging economies: The evidence from panel data analysis. *Journal of Economics, Business and Management* 3(4): 440–446.

Mazur, V. V. et al. 2016. Innovation clusters: Advantages and disadvantages. *International Journal of Economics and Financial Issues* 6(S1): 270–274.

Ndesaulwa, A. P. & Kikula, J. 2016. The impact of innovation on performance of small and medium enterprises (SMEs) in Tanzania: A review of empirical evidence. *Journal of Business and Management Sciences* 4(1): 1–6.

Noh, H. & Lee, S. 2015. Perceptual factors affecting the tendency to collaboration in SMEs: Perceived importance of collaboration modes and partners. *Journal of Technology Management and Innovation* 10(3): 18–31.

OECD. 2002. Guidelines for collecting and reporting data on research and experimental development [Online]. Frascati Manual OECD. Retrieved from http://www.oecd.org/sti/inno/frascati-manual.htm#data. Accessed 12 July 2018.

Oliveira, D. & Guarulhos, F. 2015. South American Development Society Journal. *South American Development Society Journal* 1(2): 134–148.

Olmos-Peñuela, J. et al. 2017. Strengthening SMEs' innovation culture through collaborations with public research organizations. Do all firms benefit equally? *European Planning Studies* 25(11): 2001–2020.

Ortega-Argilés, R., Vivarelli, M. & Voigt, P. 2009. R&D in SMEs: A paradox? *Small Business Economics* 33 (1): 3–11.

Prihadyanti, D. & Laksani, C. S. 2015. R&D dan inovasi di perusahaan sektor manufaktur Indonesia. *Jurnal Manajemen Teknologi* 14(2): 187–198.

Powell, W. W., Koput, K. W., Smith-Doerr, L. & Owen-Smith, J., 1999. Network position and firm performance: Organizational returns to collaboration in the biotechnology industry. *Research in the Sociology of Organizations* 16(1): 129–159.

Rahbari, D. & Jalali, S. M. 2016. Investigation of the effects of the company's capabilities on attracting external collaboration and company's performance. 8(4): 45–55.

Rajapathirana, R. P. J. & Hui, Y. 2017. Relationship between innovation capability, innovation type, and firm performance. *Journal of Innovation & Knowledge* 3(1): 44–55.

Rehman, N. U. 2016. Does internal and external research and development affect innovation of small and medium-sized enterprises? Evidence from India and Pakistan. Asian Development Bank Institute: Working Paper 577.

Sondakh, O., Christiananta, B. & Ellitan, L. 2017. Measuring organizational performance: A case study of food industry SMEs in Surabaya-Indonesia. *International Journal of Scientific Research and Management (IJSRM)* 5(12): 7681–7689.

Triharini, M., Larasati, D. & Susanto, R. 2014. Pendekatan One Village One Product (OVOP) untuk mengembangkan potensi kerajinan daerah studi kasus: kerajinan gerabah di Kecamatan Plered, Kabupaten Purwakarta,' *ITB Journal of Visual Art and Design* 6(1): 29–42.

Undang-Undang 2014. Undang-Undang No. 3 Tahin 2014 Tentang Perindustrian. Republik Indonesia.

Urbancova, H. 2013. Competitive advantage achievement through innovation and knowledge, *Journal of Competitiveness* 5(1): 82–96.

Wahyuniardi, R. et al. 2014. Penyusunan sistem informasi berbasis web untuk monitoring dan evaluasi sentra IKM Alas Kaki di Cibaduyut – Jawa Barat. *Seminar Nasional Teknik Industri BKSTI 2014* (February): 25–30.

Advances in Business, Management and Entrepreneurship – Hurriyati et al (eds)
© *2020 Taylor & Francis Group, London, ISBN 978-0-367-27176-3*

Queue teller performance analysis: Case study in a government bank

Y.D. Lestari & L. Jamilah
Universitas Airlangga, Surabaya, Indonesia

ABSTRACT: Banking services need to provide excellent services to their customers. The important service actors are tellers who are often rated as professional standards by customers. The purpose of this research is to analyze the queue teller performance and identify the queue characteristic. These problems are important to investigate as it can help banks conduct the right policy regarding the decrease of queue time. This policy is definitely beneficial as it can increase customers' satisfaction. This study used a descriptive qualitative approach. The primary and secondary data were taken from a government bank. The results showed that the addition of a new teller officer made the teller service performance more optimal, even though it led to higher service costs. The most efficient costs can be achieved when there were 6 tellers.

1 INTRODUCTION

Improving performance in the service industry, where services are performed by humans, is a complex decision (Ullah et al. 2014). One factor that decreases customer satisfaction in this industry is the queue. Queues occur due to poor scheduling, poor system design, irregular customer arrivals, and unknown service time (Hollins et al. 2006). A queuing system consists of three main components: population and the pattern of the consumers' arrival, service systems, and consumer conditions in the system (Heizer & Render 2008).

All operating activities related to cash, such as opening accounts, deposits and withdrawing savings, checking transactions, withdrawing checks, money orders, funds transfers, and so forth, and will be served by a teller (Romadhona 2008). According to Hasibuan (2009), a teller is an officer who can work fast, accurate, honest, and friendly. They are also able to work even under heavy pressure (Hasibuan 2009). They serve as the bank's front line which is often rated as a professional standard, and their attitude reflects the bank.

A common phenomenon that often happens today is that customers have to queue and wait before receiving service. According to Ariani (2009), a waiting line (queue) is one or more customers who are waiting to be served. The intended customer, in this context, may be a person or object, such as a machine requiring maintenance, an order waiting to be sent, or a stock of materials to be used. The waiting line occurs because of a temporary imbalance between the service request and the capacity of the system that provides the service. According to Hasan (2011), a customer bank often assesses the quality of the bank's operating system based on the length of waiting time or the teller speed in providing services.

Previous research on queuing at banks only calculated queue system performance (Jumaily & Jobori 2011, Berhan 2015). Agyei et al. (2015) used simulation to compare the performance of the queue by adding teller numbers, but there was no daily queue condition analysis and comparison between performance and cost efficiency. According to Hollins et al. (2006), queue management can be done by increasing the number of services provided or making the waiting time of the queue to be interesting. Therefore, this study conducts an analysis of queues performance. It also attempts to find efficiency through increasing the number of tellers in order to reduce queues.

Queuing is the science of forming queues of people or goods waiting in line to be served, including how the company can determine the best time and facilities in order to serve customers efficiently (Heizer and Render 2006).

The system is a set of interrelated and interdependent elements in performing joint activities to achieve a goal, while the queue is the waiting line of people or goods waiting to be served (Heizer & Render 2008). Russell & Taylor (2005) explained that there are three elements that exist in the queuing system, which are:

a) Customer
   The customer is the person or item waiting to be served. The meaning of the customer is not always a person. It can be the queue at the counter payment at the supermarket, people who wait their turn to pay including customers, or the items waiting to be counted by the cashier.
b) Queue
   Queue is a collection of people waiting to be served. A queue does not have to be a long waiting line. For example, a queue on a phone call.
c) The provider
   The provider is the person or something that provides service to the customer. Just like the customer, the provider does not have to be a person.

According to Aquilano et al. (2014) queue structure is divided into four models;

1. Single Channel - Single Phase
   Single channel means there is only one path to enter the service system or there is one service facility. Single phase means there is only one stage to complete the service. Single channel - Single phase is a queuing system that has only one service facility with one stage to complete the service. An example is a post office that has only one counter with one queue line, a supermarket that has only one cashier for payment, and so on.
2. Single Channel – Multiphase
   Single channel means there is only one path to enter the service system or there is one service facility. Multiphase means that there are two or more service facilities that must be implemented in sequence. Single channel - Multiphase indicates that there is one path to enter the service system but there are two or more services executed in sequence. For example, car wash, car paintman, and so on.
3. Multichannel – Single phase
   Multichannel means there are two or more paths to enter the service system. Single Phase means there is only one stage in the service facility. Multichannel - Single Phase system takes place where there are two or more paths to enter the service facility with one stage to complete the service. An example is a queue at a bank with several tellers, ticket purchases or ticket served by several counters, payment by some cashiers, and others.
4. Multichannel – Multiphase
   Multichannel means there are two or more paths to enter the service system. Multiphase means that there are two or more service facilities that must be implemented in sequence. Multi-channel system - multiphase shows that each system has several service facilities with several stages done in sequence so that there is more than one customer that can be served at the same time. Examples of this model are the services provided to patients in hospitals ranging from enrollment, diagnosis, medical treatment, to the stage of payment.

The formula for calculating the Double Line Queuing Model

1. Probability no costumer in system = Po
2. The average number of customers who are in the system = Ls
3. The average amount of time a customer spends in the system (waiting time plus service time). Ws
4. The average number of customers waiting in the queue = Lq
5. Average time spent by a customer to wait in queue. Wq

According to Siswanto (2007), service cost arises because the organization must make an additional investment in order to increase the service facility for increasing the service level (μ). This cost consists of fixed costs for additional facilities and operational costs. (Mulyono 2004). According to Mulyono (2004), the waiting cost can be assumed simply as the cost of loss of profit for the entrepreneur or the cost of productivity decrease for the worker. Meanwhile, Siswanto (2007) says that the waiting cost is the cost that appears on a customer side because he has to waste time queuing. This cost is measured through the opportunity cost of a customer. Opportunity costs for a manager in the waiting line is different than for a housewife or student; for example, the cost of a loss because the customer must be out of line.

2 METHOD

This study used a descriptive qualitative approach with a case study in Bank X. This approach was chosen to understand the phenomenon experienced by the research subject holistically, by a description in words and language, in a special context that was natural and by utilizing various scientific methods (Moleong 2007). The definition of descriptive research, according to Sukmadinata (2011), is a research method that describes the existing phenomena, which take place at this time or the past. Denscombe (2007) states that case studies focus on one or more examples of specific phenomena in order to conduct an in-depth review of an event, relationship, experience or process that occurs in the case. The respondents were bank X customers.

3 RESULTS AND DISCUSSION

Bank X service used the single phase-multichannel queue structure. The teller number was 4. They served the customers with one stage service passed. However, the teller officer performance in bank X was not good because the service time required to serve the customer was slower than the standard time (3 minutes). The number of customers who came at certain hours made the queue longer and it caused complaints. To solve this problem, a queue theory was used to overcome queuing problems, so the customers got better service and the bank could determine the optimal number of tellers.

Based on an observation over 20 days, customers' data of arrival (Table 1) was very busy. The highest total customer arrivals occurred on Monday, commonly at 10:00 to 10:59 am, being about 40 people. Whereas the average capacity of four teller officers to serve customers was 37 customers/hour (221 arrivals/6 hours), or on average 9 customers/teller. This made the queue longer because the arrival rate of customers was larger than the teller service level.

From the calculation of the Bank X queue system performance, the busiest day was Monday. Customer arrival per hour and the serving capacity per teller was 15 customers/hours. The average possibility that there were no customers in the queue system on

Table 1. Queuing system performance at Bank X.

Day	Performance calculation				
	P0	Ls	Lq	Ws (hour)	Wq (hour)
Monday	49.65%	12	10	0.302	0.235
Tuesday	52.19%	8	5	0.202	0.135
Wednesday	56.74%	4	2	0.122	0.055
Thursday	52.19%	8	5	0.202	0.135
Friday	53.97%	6	4	0.161	0.094

Monday was the highest. Every hour there were 13 customers in the queuing system and 10 customers were waiting in line. Every customer spent 18,13 minutes to finish the transaction and 14,1 minutes waiting in the queue. To reduce customer queues, the number of tellers should be expanded (Agyei et al. 2015), in order to increase the performance. In other words, an additional teller was provided to get optimal and efficient service. It was done by adding a cost comparison for every scenario.

Table 2. Cost comparison for scenario at Bank X.

Teller number	Ls	cost		total
		Service fee Rp	Waiting Rp.	
4	13	91.172	169.062	260.241
5	5	113.974	65.024	178.998
6	3	136.768	39.014	175.783

The scenario for adding a teller at bank X allowed 6 providers. The queuing system performance calculation results at Bank X shows that the addition of a new teller officer made the teller service performance become more optimal. Even though it led to higher service costs, the result in waiting costs was lower. Therefore, it is necessary to calculate service costs and waiting costs that occur in the queuing system, where the lowest total cost is the optimal cost that will provide benefits for the company.

The lowest value in waiting costs was compared to the increase of service costs. The result shows that the total queuing costs became lower. Calculation above shows the optimal teller in Bank X, by adding two teller officers.

4 CONCLUSIONS

The calculation of the queue system performance at Bank X shows that the busiest day occurred on Monday. The average rate of customer arrival per hour and each teller capacity for serving was 15 customers/hour. From the calculation, it can be seen that the teller officers' performance is not good, because the service time needed to serve the customers is slower than the standard time set by Bank X, which is 3 minutes. The cause of the queue problems in Bank X is high levels of customer arrivals. Long queues, however, can make customers feel disappointed. The total costs calculation after the addition of tellers shows that the most efficient cost can be achieved when there are 6 tellers.

ACKNOWLEDGMENT

Thanks to the dean of the faculty of economics and business and the chair of management department for the support and assistance provided.

REFERENCES

Agyei, Wallace. C, A Darko. F Odilon 2015. Modeling and Analysis of Queuing Systems in Banks: (A case study of Ghana Commercial Bank Ltd. Kumasi Main Branch). International Journal of Scientific & Technology Research. Vol 4, Issue 07, July.

Ariani, D. Wahyu. 2009. Service Operation Management. Yogyakarta: Graha Ilmu.

Aquilano, Nicholas J., Chase, Richard B dan Jacobs, F Robert 2014, Operations and Supply Chain Management. 14th edition. Singapore. McGraw-Hill Education.

Berhan, Eshetie. 2015. Bank Service Performance improvements using Multi-Server Queue system. IOSR Journal of Business and Management, Vol 17, pp 65–69.

Denscombe, Martyn. 2007. The good research guide: for small-scale social research projects. Maidenhead: Open University Press.

Hasan, Irmayanti. 2011. Model Optimasi Pelayanan Nasabah Berdasarkan Metode Antrian (Queuing System). Jurnal Keuangan dan Perbankan, Vol 15 No. 1.

Hasibuan, Malayu S.P 2007. Dasar-Dasar Perbankan Cetakan keenam. PT Bumi Aksara, Jakarta.

Heizer, Jay dan Render, Bary. 2008. Operation Management. Edisi ketujuh. Diterjemahkan oleh Dwianoegrahwati Styoningsih dan Indra Almahdy Jakarta: Salemba Empat.

Hollins, Bill, and Sadie Shinkins, 2006. Managing Service Operations, design and operations. Sage Publications. London.

Jumaily, Ahmed & H.K.T Jobori. 2011. Automatic queuing model for banking applications. International Journal of Advanced Computer science and applications, Vol 2, No 7.

Moleong, Lexy J. 2007 Metodologi Penelitian Kualitatif, Penerbit PT Remaja Rosdakarya Offset, Bandung.

Mulyono, Sri 2004. Riset Operasi. Jakarta FE-UI.

Romadhona, R. 2008. Kalkulasi dan Analisa Model Antrian M/M/1/I/I pada Bagian Customer Teller Service Bank Syariah Mandiri Bogor. Skripsi. Universitas Gunadarma. Jakarta.

Russel, Roberta S & Taylor III, Bernard W 2003. Operations Management. Fourt Edition. New Jersey: Pearson Education International.

Siswanto. 2007. Operations Research. Jilid 2. Jakarta: Erlangga.

Sukmadinata, N.S. 2011. Metode Penelitian Pendidikan. Bandung: Remaja Rosadakarya.

Advances in Business, Management and Entrepreneurship – Hurriyati et al (eds)
© 2020 Taylor & Francis Group, London, ISBN 978-0-367-27176-3

Entrepreneurial characteristics and business performance: A study of the "Suci" (holy) T-shirt production region in Bandung

C.I. Setiawati & N.A.M. Sihombing
Telkom University, Bandung, Indonesia

ABSTRACT: With the development of SME business in Bandung, the T-shirt producing or "Suci" region is no longer customers the main choice due to the strong bargaining power of buyers. The author has conducted interviews to determine the condition of that region. Some entrepreneurs said that a lack of motivation and undisciplined business can lead to unsuccessful business attempts. To face the intense competition, proper entrepreneurial characteristics are required. This study sets out to determine the relationship between entrepreneurial characteristics and business performance in the "Suci" region. This research uses a quantitative method with 80 respondents and applies the probability sampling technique. Based on the results of this study, there is a very strong relationship between self-confidence, results-oriented, risk-taking, leadership, originality, and future-oriented outlook and business success (49.7%), while the remaining 50.3% of successful cases are explained by other variables which are not examined in this research, such entrepreneurial behavior, entrepreneurial interests and innovation.

1 INTRODUCTION

The growth rate of micro, small, and medium-sized enterprises (MSMEs) in Indonesia is quite promising. As of August 2016, there are about 57.8 million MSMEs, enough to employ 97.22% of the workforce and also contribute 60.3% of the nation's Gross Domestic Product (GDP). GDP is a measure of the amount of goods and services produced in a certain area at a certain time. It is estimated that MSMEs (in Indonesian, Usaha Mikro Kecil Menengah, or UMKM) will increase in 2017 as the population of Indonesia grows (Herman 2016). In 2013 the number of UMKM units increased by 2.41% compared to 2012, while the Beat Bish (UB) in 2013 also experienced an increase of 1.97% from 2012. The workforce employed by UMKM increased by 6.20% in 2013 as well. This is a rather sharp difference compared to the number of workers employed by Large Enterprise in 2013, which increased by just 2.84%, as UMKM has more business units than large-scale businesses do.

The industrial center of UMKM shirt-making is located in the sacred area of Bandung. T-shirt centers in the sacred area began started in 1978 and began to flourish around 1982. Industrial units in the T-shirt area are supported by various types of secondary industries, such as design services, screen printing services, embroidery services, and tailor-made sewing services. The sacred area itself is located along Jl. PHH Mustafa up to Jl. Surapati, about 3 km in length, with about 400 kiosks or outlets and 500 T-shirt entrepreneurs located there. The industry was unaffected by the economic crisis of 1998. Instead, more and more new kiosks have continued to

emerge since then. Consumers come not only from the region but also from outside West Java, outside Java Island, and also foreign countries such as Malaysia (Munamah 2017). The location of the T-shirt center is strategic in terms of both location and accessibility. It is located in Surapati Street, Bandung, which is connected directly with Gasibu or Pasupati Layang Bridge to the west.

In the last three years, the number of shirt-making business units has increased, with 365 in 2013 and 415 in 2015. Workers employed in 2013 numbered 2103, while in 2015 it was 2455. While production generated in 2015 decreased by 1,725,000 pieces compared to 2014, production was recorded at 1,771,428 pieces, generating sales turnover of 69.02 billion rupiah per year.

High production means buyers have strong bargaining power because there are many choices for consumers. To face the intense competition, businesses require good entrepreneurial characteristics. The author conducted interviews on February 13, 2017, to find out more about the current conditions at the Bandung T-shirt center. One interview was with Mrs. Sundari from Pratama Production. The T-shirt business owned by Mrs. Sundari is quite successful; her entrepreneurial spirit has kept the business surviving for nine years. Among surviving and successful T-shirt companies, however, there are also some businesses that are less developed. Based on the preliminary survey, 22 respondents, or 73.3%, said that they have confidence. In addition, 66.7% of entrepreneurs said they are motivated. Of respondents, 60% claimed that their businesses were successful, and 40% said that their businesses were less successful. Some respondents also said that lack of motivation and undisciplined behavior in business can cause businesses not to run successfully.

According to the head of the T-shirt cooperative, Marnawie Munamah, turnover in the T-shirt industry is decreasing because it cannot compete with imported products. Marnawie said in 2014 the turnover of production reached Rp 50 billion, a drop compared to 2013's IDR 60 billion. In addition, the majority of artisans in sacred shirts are having difficulty innovating with current technology. The interview also highlighted the lack of promotion support from the Bandung authorities. Promotion is very important in maintaining the competitiveness of holy shirt industries (Septarini 2017).

Entrepreneurs are those who make creative and innovative efforts by developing ideas and gathering resources, leading to opportunities and improvements in life; this tendency encourages someone to take advantage of opportunities to make something profitable. Putri et al. (2013) call this the entrepreneurial characteristic.

According to Barringer and Ireland (2016), "An inventor creates something new. An entrepreneur assembles and then integrates all the resources needed—the money, the people, the business model, the strategy, the risk-bearing ability—to transfer the idea into a viable business."

The process of developing a new business is in the entrepreneurial process, which involves more than just problem-solving in a management position. An entrepreneur must find, evaluate, and develop an opportunity by overcoming the forces that hinder the creation of something new (Hisrich & Shepherd 2013). This process has four distinct phases: the identification and evaluation of opportunities; the development of business plans; the determination of required resources; and the management of the resulting enterprise. According to Suparyanto (2012), the characteristics of entrepreneurship can be possessed as a person's innate nature since birth. They can also be established through the process of education and experience. The characteristic of entrepreneurship can evolve when one wishes to earn income from unsupportive economic factors. Entrepreneurs must have some of the following characteristics to succeed: (1) the confidence to work independently, work hard, and understand the risks as part of the effort to succeed; (2) the ability to organize, determine purpose, be results-oriented, and have responsibility for the results, no matter if good or bad; (3) ability to be creative and always look for loopholes for creation; and (4) to love challenges and gain personal satisfaction when it comes to achieving ideas (Wijayanto 2011).

Zimmerer and Scarborough (2014) suggest eight entrepreneurial characteristics: (1) sense of responsibility, which means being committed and be self-aware; (2) preference for moderate risk, which means avoiding risks that are either too low or too high; (3) confidence in ability to succeed; (4) requires immediate feedback, meaning someone

wants to quickly succeed; (5) high levels of energy, meaning who has spirit and desires to work hard is more likely to realize his or her desire for a better future; (6) future-oriented, which means having perspective and insight looking far ahead; (7) skills to organize resources to create added value; (8) respecting the value of achievement more than the value of money.

Wijayanto (2011) suggests that there are three factors that affect the performance of small businesses, especially for new businesses. Sorted by increasing level of influence, these factors are industrial structure, business strategy, and entrepreneurial spirit. According to Suryana (2013), a business's success or failure is influenced by the entrepreneur's nature and personality. Wijayanto (2011) suggests that business success is influenced by several factors, including having vision and business objectives; taking risks; being able to organize business planning, power, and implementation; being able to work hard; being able to build relationships with customers, laborers, suppliers, etc.; and taking responsibility for both success and failure.

From the interviews conducted, researchers found that the characteristics of entrepreneurship affect the success of a business. Basically, every business does not always result in what the entrepreneurs expected. Many of the entrepreneurs suffer from loss, even until the company's bankruptcy. Wisdom in management can be measured by certain benchmarks and parameters. Therefore, researchers looked to conduct in-depth research on the characteristics of entrepreneurship which affect the success of a business.

2 METHOD

Based on the research objectives, this research uses quantitative methods. This study integrates descriptive and causal research methods. The data obtained from the interview is in the form of statistics, which will then be studied so as to explain the pattern of relationship between independent variables (affecting variables) and the dependent variable (that which is influenced). The population being studied is the businesspeople of the Suci (holy) shirt industry in Bandung City. There are 415 businesspeople in the Suci T-shirt industry residing in the holy area in Bandung, of whom 80 served as a sample for this study. The technique used in this research is probability sampling technique by applying Simple Random Sampling.

Primary data used in this research includes: (1) direct interviews with businesspeople in the Bandung T-shirt area to get an idea of their entrepreneurial spirit and to explore the problems they encounter; and (2) a questionnaire: a list of written questions previously formulated that respondents need to answer, usually with clearly defined alternatives. Secondary data refers to information gathered by other people and not the recent studies conducted by

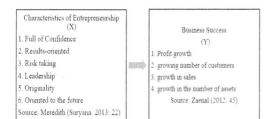

Figure 1. Framework of thinking applied this research.

the researcher. The data is accessed via the internet, tracking documents, or publication. The technique analysis used in this study is simple linear regression analysis to predict how far the change in the value of the dependent variable is, whether the value of the independent variable change is up or down. The thinking framework of this study is shown in Figure 1.

3 RESULTS

Entrepreneurship characteristics have several sub variables: being full of confidence; results-oriented; risk-taking; leadership; originality; and future-oriented. The responses of the respondents on sub-variables can be seen from the data collection through questionnaires given to the 80 T-shirt entrepreneurs in Bandung. The statements related to the variables of entrepreneurial characteristics consisted of 18 items and 6 sections, namely: 5 items of statements on confidence sub-variables, 5 items of statements about the results-oriented sub-variables, 1 item of a statement on the sub-variable of retrieval risks, 3 items of statements on leadership sub-variables, 3 items of statements on sub-originality variables, and 1 item of a statement on future-oriented sub-variables.

Table 1. Variable of X.

Sub-Variable	Average
Full of confidence	76.34%
Oriented on results	73.68%
Risk taking	71%
Leadership	72.9%
Originality	72.9%
Oriented to the future	76%
Average total value	73.80%

The average percentage for the sub-variable "full of confidence" is 76.34%, categorized in the high category. Meanwhile, item 1 (full of confidence) becomes the statement with the highest percentage with a total value of 78%.

The average percentage for the "results-oriented" variable is 73.68%, falling into the high category.

Meanwhile, the statement with the highest percentage is item 2 (results-oriented) with a total value of 80.7%.

The average percentage for the sub-variable of "risk-taking" is 71%, or high.

The percentage of average for the sub-variable of "leadership" is 72.9%, or high.

Meanwhile, the statement with the highest percentage is item 3 (open to suggestions and criticism) with a total value of 74.2%. Based on the calculation, the average percentage for the sub-variable sub is 72.9%, or high.

Meanwhile, the statement with the highest percentage is item 3 that is flexibility with a total value of 75%.

The average percentage for the "future-oriented" sub-variable is 76%, or high.

The average percentage for the "business success" variable is 73.15%, or high. Meanwhile, the statement with the highest percentage is item 3 ("growth in sales") with the total value equal to 74.2%.

To work out the influence of entrepreneurship characteristic variable (X) on the success of business (Y) for Bandung T-shirt businesses, a simple linear regression model is used.

Table 2. Table coefficient of determination test.

Model Summary[b]				
Model	R	R Square	Adjusted R Square	Std. Error of the Estimate
1	.705[a]	.497	.488	.560400

* Predictors: (Constant), X
** Dependent Variable: Y

Table 4:12 shows that the value of R = 0.705 and R Square = 0.497. The influence magnitude of entrepreneurial characteristic on business success is shown by the determination coefficient (DC) using the following formula:

$$KD = 0.497 \times 100\% = 49.7\% \qquad (1)$$

The table shows that R Square = 0.497, which means 49.7% of business success can be explained by independent variables (entrepreneurial characteristics), while the remaining 50.3% is explained by other variables which are not examined in this study.

In this study, "has a strong impetus" becomes the indicator of the independent variable that has the highest value. Statement item 7 ("motivated in running a business") has a percentage of 80.75%. The indicator of independent variable with the lowest value is profit-oriented indicators. In item 6, the statement of "owned profit continues to increase" is at 64%. For the dependent variable, the indicator with the highest value is the "increase of sales numbers," at 74.25%. The indicator that has the lowest value is "increase in business profits," at 71.75%.

4 DISCUSSIONS

Respondents divided by sex results in women at 34% and men at 66%, illustrating that Bandung has more male T-shirt entrepreneurs than female. Respondents sorted by largest to smallest age group break down as follows: 31–40 years old (57%); 21–30 years old (22%); 41–50 years old (19%); and over 50 years old (2%). The diverse ages reported indicate the family businesses being passed on through generations. The entrepreneurs' education break down as follows: SMA (64%), Diploma (34%), and S1 (2%).

Based on the validity test conducted, 22 statements that represent the variable are declared as valid. This is proofed by value of Corrected Item-Total Correlation counted. A reliability test obtained Cronbach's alpha value of 0.866. According to Suhartanto (2014), if the value of Cronbach's alpha is greater than 0.70, the 22 statements are reliable and can be used for research.

Based on the results of descriptive analysis conducted on the entrepreneurial characteristic, the entrepreneurs at the T-shirt industry have good entrepreneurship characteristics, falling into the high category with an average score of 73.80%. But there are still some issues that need to be fixed so that entrepreneurs can develop and make their businesses run better. The majority of entrepreneurs do not give priority to the profits gained; they report only running the business because it has been inherited from their family. The entrepreneurs seem to run their businesses without considering the benefits to be gained in the long term. Some entrepreneurs said that the profits obtained from their business are uncertain due to the lack of financial management. Nonetheless, with a business success variable of 73.15%, categorized as high, entrepreneurs still afloat in the Bandung T-shirt business are continuing to increase their sales and customers.

From the analysis conducted on the variable of entrepreneurial characteristic, the result from the highest statement shows that the indicator has a strong impetus of 80.75%. Based on this indicator, those entrepreneurs are motivated in running their business. Entrepreneurs said that families are one of their motivations to continue running their business. The lowest analytical result is the "profit-oriented" indicator (64%); entrepreneurs perceive that the profit held does not increase continuously but also does not decrease significantly. Businesses sometimes obtain profit, but advantages are unpredictable so profits cannot always be improved. The existence of elections, holidays, and other special occasions, however, benefit entrepreneurs greatly, with an increase in the number of the business sales (74.25%). Sales in T-shirts rise ahead of holidays to meets consumer needs for parties or other activities. The result of the lowest analysis shows that the increase in business profits is 71.75%, which means that the entrepreneurs are less able to increase profits in the shirt business.

These results show entrepreneurial characteristic variables to have a significant effect on the success of business entrepreneurs in the Bandung T-shirt center. The value of R Square = 0.497; this means that 49.7% of business success can be explained by independent variable (entrepreneurial characteristics), while the rest, 50.3%, is explained by other variables not examined in this research, such as entrepreneurship behavior, entrepreneurship interest, and innovation.

5 CONCLUSIONS

This research concluded that: (1) based on the total score of each dimension for the variable "entrepreneurial characteristics," out of an ideal score an average score of 73.80% is obtained; (2) the variable "success of the business" gives an average score of 73.15% of the ideal score; and (3) entrepreneurial characteristics influence 49.7% of business success among T-shirt entrepreneurs in Bandung, while the rest (50.3%) of success is explained by other variable not examined in this research, such as entrepreneurship behavior, entrepreneurship interest and innovation.

ACKNOWLEDGEMENT

We would like to thank Telkom University as well as the respondents who filled out the questionnaire and the companies the owners who were very patient explaining their answers to us. In addition, thanks also to the Research Center for the opportunity to publish this research.

REFERENCES

Barringer & Ireland. 2016. *Entrepreneurship: Successful launching new ventures*. 5th edition. India: Pearson Education, Inc.

Bygrave & Zacharakis. 2014. *Entrepreneurship*, 3rd ed. Wiley.

Herman, Y. U. D. 2016. Kontribusi UKM Indonesia dalam GDP sanggup kalahkan AS. Retrieved from: https://www.beritasatu.com/iptek/352112-kontribusi-ukm-indonesia-dalam-gdp-sanggup-kalahkan-as.html. Accessed January 28, 2018.

Hisrich, P. & Shepherd. 2013. *Entrepreneurship*, 9th ed. McGraw-Hill.

Munamah, Marnawie. 2017. Kampung UKM digital kaos Suci. Retrieved from https://www.kampungukmdigital.com/directory/profil/kampung-ukm-digital-kaos-suci. Accessed January 28, 2018.

Putri, K., Pradhanawarti, A. & Prabawani, B. 2013. Pengaruh karakteristik kewirausahaan, modal usaha dan peran business development service terhadap pengembangan usaha (Studi pada sentra industri kerupuk desa Kedungrejo Sidoarjo Jawa Timur). *Jurnal Ilmu Administrasi Bisnis*. Universitas Diponegoro.

Septarini, E. C. 2017. Kemampuan ini diperlukan oleh industri kreatif. Retrieved from https://kabar24.bisnis.com/read/20170915/78/690357/kemampuan-ini-diperlukan-oleh-pengusaha-industri-kreatif. Accessed 28 January 2018.

Suhartanto. 2014. *Metode riset pemasaran*. Bandung: Alfabeta.

Suparyanto. 2012. *Kewirausahaan konsep dan realita pada usaha kecil*. Bandung: Alfabeta.

Suryana. 2013. *Kewirausahaan kiat dan proses menuju sukses*. Jakarta: Salemba Empat.

Wijayanto, A. 2011. Pengaruh karakteristik wirausahawan terhadap tingkat keberhasilan usaha. *Jurnal Ilmu Sosial* 12(1): 16–28.

Zimmerer, T. W. & Scarborough, M. 2014. Essentials of entrepreneurship and small business management, 7th ed. *Instructor*.

Advances in Business, Management and Entrepreneurship – Hurriyati et al (eds)
© *2020 Taylor & Francis Group, London, ISBN 978-0-367-27176-3*

Personal attitudes, family backgrounds, and contextual elements as antecedents of students' entrepreneurial intentions: The case of Indonesian higher education

H.A. Rivai, H. Lukito & A. Morhan
Universitas Andalas, Padang, Indonesia

ABSTRACT: This study is aimed to analyzing the antecedents of entrepreneurial intentions of students in a higher education context. Those antecedents can be viewed from personal attitudes, family background, and contextual elements (i.e., capital access, availability of information, social network). This study was conducted by using quantitative approach to answer proposed hypotheses. Four universities from West Sumatra Province, Indonesia, participated as research objects. The total number of samples in the study is 240 respondents, all registered university students. Data analyses was performed using multiple linear regression. The study found that personal attitudes significantly affect entrepreneurial intentions of university students, with family background a significant factor. The study supported that contextual elements as significant determinant of entrepreneurial intentions in the universities. It can be concluded that three variables—personal attitudes, family background, and contextual elements—have a significant impact on the entrepreneurial intentions of the students in a higher education context. Implications of the study are also discussed in the paper.

1 INTRODUCTION

Entrepreneurship plays an important role in promoting national economic growth. It has become a priority for several countries, mainly developing countries, to accelerate economic development by generating new ideas and utilizing them in business activities (Rivai 2012). It can be said that entrepreneurship generates competitiveness in national and international markets by encouraging entrepreneurs to emerge with innovations. Entrepreneurs are also considered to be drivers of business development, providing jobs and generating social advancement within their environment (Liñán et al. 2005). According to the data released by the Bank of Indonesia (2015), micro, small and medium enterprises (MSMEs) contribute 60% of the total Growth Domestic Product (GDP) of Indonesia.

Entrepreneurial activity is something that is intentionally planned (Krueger et al. 2000). In general, the intention is a cognitive state, existing prior to executing the behavior (Krueger 2005). The intentions are a determining key, moderated by external variables such as family background, position in the family, parents' occupation, education, and training to create new ventures (Bird & Jalinek 1988). Intentions in an entrepreneurial context are concerned with the inclination of a person to start an entrepreneurial activity in the future. Fitzsimmons and Douglas (2011) state that understanding entrepreneurial intentions helps to predict the behavior of entrepreneurs. Entrepreneurial intentions have been described as a conscious state of mind that directs attention toward a specific object or the pathway to achieving it (Bird 1988). Therefore, it could be argued that someone might not have thought to engage in entrepreneurial activities if there were not any entrepreneurial intentions. In the context of higher education, with an understanding of higher education students' entrepreneurial intentions, we can better predict whether they will take real action to start a new business. Promoting entrepreneurial intentions of university students can effectively increase the possibility that the students will engage in entrepreneurship after graduating from university. Entrepreneurial intentions are the crucial factors in predicting entrepreneurial behaviors. Nevertheless, there is no agreed theory to explain the intentions of people to become entrepreneurs (Shook et al. 2003). In particular, the influence of entrepreneurial motivation on goal-specific intentions needs to be explored (Solesvik 2011). The current research attempts to explain the impact of personal attitudes, family background, and contextual elements on intentions to be an entrepreneur. The contextual elements can be viewed as access to capital, availability of business information, and social networks.

Thus, in this research, the researchers propose the following hypotheses:

Hypothesis 1: Personal attitudes significantly affect entrepreneurial intentions.

Hypothesis 2: Family background significantly affects entrepreneurial intentions.

Hypothesis 3: Contextual elements significantly affect entrepreneurial intentions.

Table 1. Mean, standard deviation, and Cronbach's alpha (N=240).

Variables	Item	Cronbach's Alpha	Mean	SD
Entrepreneurial intentions	4	0.815	3.82	.46
Personal attitudes	5	0.776	3.96	.55
Family background	4	0.741	4.03	.61
Contextual elements	6	0.841	4.13	.51

2 METHOD

A total of 300 questionnaires were distributed to the universities. Participants were involved on a voluntary basis, and responses were treated with confidentiality. In total, 240 were returned, comprising a response rate of 80%. All variables were measured using items which have been utilized in previous studies with the 5-point Likert scale. The entrepreneurial intentions questionnaire was adopted from Liñán and Chen (2009). Personal attitudes were measured by using questionnaires taken from Yurtkoru et al. (2014). Family background was measured by using questionnaires adopted from Ohlsson et al. (2016) and Arrighetti et al. (2016). Contextual elements were measured by using questionnaires taken from Kristiansen and Indarti (2004). Data analysis was conducted using multiple regression after checking for data entry. The reliability of each construct was assessed using Cronbach's alpha. Hair et al. (1998) suggest that the usual lower limit for Cronbach's alpha is 70, but, in exploratory research, this limit may decrease to 60. All constructs demonstrated good reliability. The psychometric properties of scales are reported in Table 1.

3 RESULT AND DISCUSSION

3.1 Result

Female respondents comprised 47.1%and males 52.9%. The survey was conducted during the odd

Table 2. Regression analysis.

Model	R	R square	Adjusted R square	S.E of the Estimate
1	0.610 (a)	0.361	0.331	0.369

a *Predictors: (constant), Personal attitude, Family background, Contextual elements*

semester. Majority of respondents registered in seventh semester (43.8%) and the rest registered in the ninth, fifth, and third semesters (18.3%, 14.8%, and 8.3%, respectively). All had taken courses in entrepreneurship. Most respondents were aged between 20 to 25 years (55.6%). A large number of respondents spent their time as full-time students without working or running a business (73.1%), while 9.9% worked a part-time job and 20.0% ran a business (20.0%). The survey found that approximately 37.3% of the students intended to be an entrepreneur once they graduated, and the rest stated they would continue to study for a master's degree or would look for a job.

3.2 Discussion

3.2.1 The influence of personal attitudes on entrepreneurial intentions

The result of the study found that personal attitudes have a positive and significant influence on entrepreneurial intentions. The hypothesis testing result demonstrates that the *p*-value of personal attitudes is .000 (less than .05). This indicates that personal attitudes impact greatly on students' willingness and propensity to be entrepreneurs. It can be argued that one's personal attitudes are really important, supportive of the statements of a study conducted by Kolvereid (1996). Krueger et al. (2000) state that attitudes are a good predictor of intentions to be an entrepreneur. The more that students have a positive view about the outcome of getting a business, the more favorable their attitude will be toward the behavior, which in turn strengthens their intentions to take the initiative to start up the business. Otherwise, the students that perceived entrepreneurship will not bring any advantages for them will have

Table 3. Coefficients correlation.

Model	Unstandardized Coefficient		Standardized Coefficient		sig
	B	Std error	B	t	Std error
1.constant	1.231	0.162		7.588	0.000
Personal attitude	0.180	0.038	0.308	4.767	0.000
Family background	0.226	0.068	0.259	3.316	0.001
Contextual elements	0.153	0.046	0.198	3.304	0.001

a negative attitude toward the behavior and will not have any intentions to pursue it. Personal attitude (e.g., self-confidence, risk-taking, innovativeness) can affect the students' future career choice. Those attitudes reflect entrepreneurial characteristics that might differentiate the students between those with low and high intentions to be an entrepreneur (Rivai 2012). Furthermore, in a review of the previous study, Yurtkoru et al. (2014) demonstrated that personal attitude and perceived behavior control have a significant influence on someone's intentions to be an entrepreneur.

3.2.2 *The influence of family background on entrepreneurial intentions*

The result of hypotheses testing supported that the students who have intentions to start up businesses because of their family background or family environment feel psychological support to be an entrepreneur. In line with several previous studies conducted by Akanbi (2013), Carr and Sequira (2007), McElwee and Al Riyami (2003), and Mueller (2006), another recent study noted family background as a determinant of someone's entrepreneurial intentions, stating that, when people are unsure in the midst of ambient normative influences and attitudes, their entrepreneurial intentions will be affected by their prior knowledge and experiences (Ajzen 2002). In this case, family background can influence someone's entrepreneurial intentions because they have grown up with such knowledge and experiences. The influence of family on an individual's propensity to be self-employed was shown to be significant, which in turn affects a student's decision once graduated (Ohlsson et al. 2016). It can be argued that self-employed individual will pursue their entrepreneurial intentions based on their family background.

3.2.3 *The influence of contextual elements on entrepreneurial intentions*

The determinant of entrepreneurial intentions can be viewed from contextual elements, including access to capital, availability of business information, and social networks (Kristiansen & Indarti 2004). The current findings note that entrepreneurial intentions are significantly influenced by availability of access to capital, availability of business information, and social networks; these facilitate running a business for entrepreneurs, mainly by allowing them access to sources of materials and market opportunities. The social network has an impact on desired career paths and the likelihood of successful entrepreneurial endeavors (Kristiansen & Indarti 2004). The studies conducted by Steel (1994) and Meier and Pilgrim (1994) demonstrate that difficulties of capital access and credit schemes can be a big obstacle for the potential entrepreneur in starting up their own business. Furthermore, access to a social network and access to supporting information cannot be regarded as unimportant. Eagerness to seek information is one of the major entrepreneurial characteristics (Singh & Krishna 1994).

4 CONCLUSION

These research findings show all three variables (i.e., personal attitudes, family background, and contextual elements) to have a significant effect on students' entrepreneurial intentions. Personal attitudes have positive and significant influences on students' entrepreneurial intentions. Family background likewise has a positive and significant influence on students' entrepreneurial intentions, as potential entrepreneurs see family as role models when making career-path decisions. Finally, contextual elements have positive and significant influence on student's entrepreneurial intentions. The findings show the influence of three situations (i.e., access to capital, availability of business information, and social networks) result in increased entrepreneurial intentions among students.

The study noted limitations discovered during the process of conducting the research: these limitations might provide platforms for future studies. This study was limited to a single industry (i.e., the educational industry), with a non-probability sampling method used to obtain the data. The current study was also limited to three determinants of entrepreneurial intentions; meanwhile, there are many factors that contribute to entrepreneurial intentions but that were not investigated in the current study. The study noted implications for developing entrepreneurship in higher education. To help establish those young people who are ready to begin entrepreneurial activities, the universities should focus on developing personal attitudes toward entrepreneurial intentions and provide a medium of business information which can provide access to sources of capital. Nurturing entrepreneurship in higher education, universities might consider accepting more students with entrepreneurial family backgrounds. They could also influence other students to choose successful family businesses as a role model to motivate them to become entrepreneurs.

REFERENCES

Ajzen, I. 2002. Perceived behavioral control, self-efficacy, locus of control, and the theory of planned behavior. *Journal of Applied Social Psychology*, 32(4): 665–683.

Akanbi, S. 2013. Familian factors, personality traits, and self-efficacy as determinants of entrepreneurial intentions among vocational based college of education students in Oyo State, Nigeria. *The African Symposium*, 13 (2), 3–13.

Arrighetti, A., Caricati, L., Landini, F. & Monacelli, N. 2016. Entrepreneurial intention in the time of crisis: a field study. *International Journal of Entrepreneurial Behavior & Research*, 22(6): 835–859.

Bird, B. & Jalinek, M. 1988. The operation of entrepreneurial intentions. *Entrepreneurship Theory and Practice*, 13 (2): 21–29.

Bird, B. 1988. Implementing entrepreneurial ideas: The case of intentions. *Academy of Management Review*, 13 (3): 442–454.

Carr, J. C. & Sequeira, J. M. 2007. Prior family business exposure as an intergenerational influence and entrepreneurial intent: A theory of planned behavior approach. *Journal of Business Research*, 60(10): 1090–1098.

Fitzsimmons, J. R. & Douglas, E. 2011. Interaction between feasibility and desirability in the formation of entrepreneurial intentions. *Journal of Business Venturing*, 26(4): 431–440.

Hair, J. F. Tatham, R. L., Anderson, R.E., & Black, W. 1998. *Multivariate Data Analysis, Fifth Edition*. Upper Saddle River, NJ: Prentice Hall.

Kolvereid, L. 1996. Organizational employment versus self- employment: Reasons for career choice intentions. *Entrepreneurship: Theory and Practice*, 20(3): 47–57.

Kristiansen, S. & Indarti, N. 2004. Entrepreneurial intentions among Indonesian and Norwegian students. *Journal of Enterprising Culture*, 12(1): 55–78.

Krueger, N. F. 2005. The cognitive psychology of entrepreneurship. *Handbook of Entrepreneurship Research*, 105–140.

Krueger, N.F. Jr., Reilly, M. D. & Carsrud, A. L. 2000. Competing model of entrepreneurial intentions. *Journal of Business Venturing*, 15(5–6): 411–432.

Liñán, F., & Chen, Y. W. 2009. Development and cross-cultural application of a specific instrument to measure entrepreneurial intentions. *Entrepreneurship Theory and Practice*, 33(3): 593–617.

Liñán, F., Rodriguez-Cohard, J.C. & Rueda-Cantuche, J. M. 2005. Factors affecting entrepreneurial intentions levels. *45th Congress of the European Regional Science Association, Amsterdam, 23–27 August 2005*.

McElwee, G. & Al-Riyami, R. 2003. Women entrepreneurs in Oman: some barriers to success. *Career Development International*, 8(7): 339–346.

Meier, R. & Pilgrim, M. 1994. Policy-induced constraints on small enterprise development in Asian developing countries. *Small Enterprise Development*, 5(2): 66–78.

Mueller, P. 2006. Entrepreneurship in the region: Breeding ground for nascent entrepreneurs? *Small Business Economics*, 27(1): 41–58.

Ohlsson, H., Broomé, P. & Schölin, T. 2016. Self-employment: the significance of families for professional intentions and choice of company type. *International Journal of Entrepreneurial Behavior & Research*, 22(3): 329–345.

Rivai, H. A. 2012. Factors influencing students' intentions to be entrepreneurs: Evidence from Indonesian Higher Education. *ASEAN Entrepreneurship Journal*, 1(1): 67–80.

Shook, C. L., Mcgee, J. E. & Priem, R. L. 2003. Venture creation and the enterprising individual: A review and synthesis. *Journal of Management*, 29 (3): 379–399.

Singh, K. A. & Krishna, K. V. S. M. 1994. Agricultural entrepreneurship: The concept and evidence. *The Journal of Entrepreneurship* 3(1): 97–111.

Solesvik, M. 2011. Attitudes towards future career choice. *ICSB World Conference Proceedings*, 1–18.

Steel, W.F. 1994. Changing the institutional and policy environment for small enterprise development in Africa. *Small Enterprise Development*, 5(2): 4–9.

Yurtkoru, E. S., Kuşcu, Z. K. & Doğanay, A. 2014. Exploring the antecedents of entrepreneurial intentions on Turkish university students. *Procedia—Social and Behavioral Sciences*, 150: 841–850.

Highest and Best Use (HBU) analysis as an alternative strategy of assessing asset utilization

M. Listyohadi, S. Sinulingga & S. Sugiharto
Universitas Sumatera Utara, Medan, Indonesia

ABSTRACT: PLN (Perusahaan Listrik Negara) is a government-owned corporation that provides electricity for the industrial and community sectors. In accordance with Article 1 of the Regulation of the Minister Number: PER-13/MBU/09/2014 concerning the utilization guideline of State-owned fixed assets—stating that that the Board of Directors must prepare a list of fixed assets that are lacking and/ or not optimally utilized—PLN reported an asset in the form of vacant land that has not been optimally utilized. In order that the vacant land can be used optimally, it is necessary to carry out Highest and Best Use (HBU) analysis on the land. HBU analysis aims to identify the most profitable and competitive use for the land using four criteria: physically possible; legally permitted; financially feasible; and maximally productive. The result shows that the most feasible investment would be the construction of gas-fired power plants, with a payback period of 2 years and 7 months, indicating a positive value for NPV with an IRR value of 28.69% and PI of 1.91.

1 INTRODUCTION

Currently PLN of North Sumatera Region has assets in the form of vacant land that has not been optimally utilized and requires operational costs to perform maintenance on it. Amin (2015) explains that Highest and Best Use (HBU) study activities ideally are carried out as part of a work package in order to optimize assets. This is to determine which assets have been optimized and which have not, which are the most optimum utilization options, and which are the most beneficial cooperation patterns for the asset. PLN has several alternative uses for vacant land, categorized as follows: PLTD (diesel-powered power plant) development; PLTG/MG (gas-fired power plant) development; other power plant construction; office building; warehouse development; flats; and substation.

HBU analysis is a well-known concept in property asset management, in terms of both asset optimization and asset valuation. With several alternatives for land use, it is necessary to carry out best-use analysis to determine which type of use is the most suitable, feasible, and profitable.

1.1 *Understanding highest and best use concepts*

HBU analysis of land vacant or considered vacant includes four main parts: physical feasibility analysis; legal permissibility analysis; financial feasibility analysis; and maximal productivity analysis. A property's use is said to have met the criteria of HBU where it is physically possible, permitted by regulations, financially feasible, and can provide maximum results. Dachyar (2012) conducted a feasibility analysis of investment and risk of the Indramayu power plant development project by considering the interest rate of loans in Japanese currency. The calculation of capital budgeting gave an NPV value of IDR 36 trillion and IRR of 9.03%. In addition, based on the Monte Carlo Simulation, it produced an average NPV of IDR 29 trillion, with a probability of producing NPV with a minus value of 26.62%.

1.2 *Appraisal approach*

According to Fanning (2005), analytical tools to be used in HBU analysis include: (1) market data approach; (2) cost approach; and (3) income approach.

Pomykacz et al. (2014) conducted an assessment of power generation because power plants can be worth billions of dollars and play an important role in infrastructure. Different technologies are applied at the plant. Uses of plant valuation are also diverse, from acquisitions, financing purposes, regulation, litigation, or property taxes to Internal Revenue Service (IRS) or Securities and Exchange Commission (SEC) reporting.

1.3 *Standard value in asset rating*

Market value is defined as the estimated amount of money on the date of valuation, which can be obtained from a sale and purchase transaction or the exchange of a property (Fanning 2005).

The cost approach is based on the amount of costs incurred to create or build a new version of each major component, including building materials and facilities. The value of the property is obtained from the multiplication of the building area by the construction cost per square meter. The property value is the sum of the value of the land, and the value of the building obtained from the new construction is reduced by depreciation.

The revenue approach, also called the investment approach, is another approach that can be used in the valuation of a property.

1.4 Indicators for financial incentive variables

Indicators for financial incentive variables are electricity cost savings, water cost savings, home value, rental value, and the willingness to pursue green home investment for long-term benefits. For the green behavior variable, indicators are the use of eco-friendly materials, energy-efficient transportation, preference for open green spaces, energy savings, efficient use of goods, preference for use of electronic documents rather than paper, electricity savings, protection of the environment, purchase of goods with less packaging, and participation in protecting the environment (Fachrudin 2017).

2 METHOD

In practice, the value of vacant land is determined using the market approach method. This comparison approach considers the sale of similar or substitute property and related market data and generates the estimated value through a comparison process. In general, the assessed property (the valuation object) is compared to a comparable property transaction, either existing or a property still in the offer stage of a buying and selling process (MAPPI 2015).

The framework can be divided into several parts, as seen in Figure 1.

Problem identification is the first step in starting the study, which can be done in many ways, such as reviewing literature, acquiring information online, observing real-life conditions, and holding discussions with a project officer, lecturer, or tutor. After that, the researcher can identify the problem, which would then be formulated into the research question. Formulation of the research question is highly dependent on the background research.

2.1 Location and time of study

The location of this research was land owned by PLN in Paya Pasir, Rengas Island, Medan Marelan, Medan, North Sumatera. The land is currently not used maximally, so research needs to be done to optimize use of the asset.

2.2 Data sources

The data used was in the form of primary and secondary data. Primary data was obtained from the survey results, observation, and direct interviews with the PLN management as well as stakeholders concerning PLN's core business.

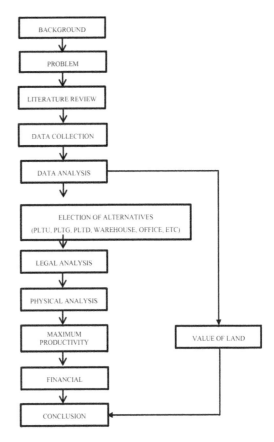

Figure 1. Research flow chart.

2.3 Data analysis

Data analysis of the alternatives chosen in this study were:

a. Analysis of alternatives in terms of legal aspects. Things to review are zoning, building codes, and environmental regulations, such as those concerning clean air, clean environment, clean water, and security.
b. Analysis of alternatives in terms of physical aspects. This analysis looked at the size of the area, its shape, its public utility, and accessibility of the site location.
c. Analysis of alternatives in terms of maximum productivity. Of the financially feasible uses, the highest-yielding residual usage consistent with the market-guaranteed rate of return for such use is the highest and best use. An alternative is said to have maximum productivity if it has the highest land value.
d. Financial analysis (capital budgeting). This analysis will determine whether the alternatives can be implemented or not and whether their profit or rate of return will be sufficient so as not to cause a loss if the project is carried out. This investment

feasibility assessment is also called the capital budgeting technique. Capital budgeting is the process of planning and decision-making regarding the expenditure of funds where the return period of funds exceeds one year (capital expenditure). These expenses include the purchase of plant investments, i.e., buildings, machinery, and expenditures for long-term advocacy projects, and research and development.

3 RESEARCH AND DISCUSSION

3.1 Analysis of market price forecast for the undeveloped land

The value of the vacant land in this study was determined based on a market price survey: IDR 1,535,592 per m².

3.2 Analysis of alternative uses based on the perception of potential users

Based on a survey of stakeholders related to PLN's core business, the two most popular possible uses were: (1) for the land to be used as a diesel-fired power plant (in Indonesian, Pembangkit Listrik Tenaga Diesel, or PLTD); (2) for the land to be used as a gas-fired power plant with gas engine (in Indonesian, Pembangkit Listrik Tenaga Gas den Mesin Gas, or PLTG/MG). These two options were then reviewed using HBU analysis, to compare the two uses in terms of maximizing legal, physical, financial and productivity factors.

3.3 HBU analysis of legal aspects

Local Regulation of Medan City No. 13 of 2011 on Spatial Planning of Medan City Year 2011–2031 allows the Medan City area to contribute to the Energy System Plan. This is in accordance with the contents of Chapter III, First Section of Article 13 Paragraph (1), part c of the rule. In Part Four of Chapter III (System of Energy Networks), Article 25 Paragraph (2) states that the energy network system consists of electric power and oil and gas pipelines. In Paragraph (3) of Article 25, the electric power network consists of electricity generation and transmission lines, whereas in Paragraph (4) it is mentioned that power plants established can be gas/gas engine power plants (MG) as well as steam- or diesel-driven power plants. Paragraph (4) on Transmission Network outlines the form of substation that can be built on the area.

The conformity of the allotment described is also mentioned in the Letter of the Head of the Office of Spatial Planning of Medan City to the General Manager of PLN Development Master Units I, numbered 640/4423 and dated 28–05–2015, regarding Explanation on Spatial Information (attached). Point 2 mentions the appropriate land allocation according to RSSW. Because the construction site is located on land owned by PLN itself, the construction of a power plant is still possible at that location, while point 1 in the letter states the land can also be used as housing.

3.4 HBU analysis of physical aspects

From field observation, the land is square, and the contour of the land is even and slightly wavy. The land is empty, with shrubs or weeds, and there are puddles when it rains. Land maturation is required for the site. About 50m from the entrance gate of Medan Power Sector (Paya Pasir), there is a road, Titi Pahlawan Street, which is rather crowded with traffic.

Regarding a gas-fired power plant (PLTG/MG), in accordance with above layout conditions, a PLTG can be built with maximum capacity of 241 MW. This land layout is obtained from an existing power plant in Arun area, with a total area of 4.8 Ha, including a reforested area. Natural gas from PT Perta Gas can be carried from the Belawan Installation to PLTG/MG MPP Paya Pasir via a gas pipeline installation.

Based on the reference of PLTU Tanjung Balai Karimun Kepulauan Riau, other power plant alternatives cannot be realized because the area of land owned is only 5 Ha (50,000m²), whereas the development of a steam power plant (2x7) MW requires at least 12 Ha of land, due to the allocation of coal storage and coal dust. As 40% of the land will be used for mechanical equipment, 60% is available for warehouse storage use, equal to 27,000m². As the area of land available for the substation alternative is ± 10,000m², with an observed area of 50,000m², a substation does not allow optimal utilization of space.

3.5 HBU analysis of maximum productivity

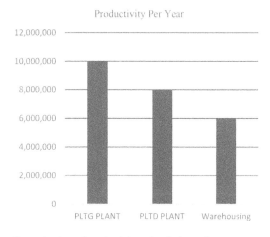

Figure 2. Annual productivity of each alternative.

3.6 HBU analysis of financial aspects

a. Investment costs

The land can be built into a PLTG with capacity of 240 MW with the following cost details:

b. Income

The revenue for a PLTG/MG development comes from the Electricity Sales Tariff applied to customers at a rate of 1467.28 IDR/kWh (January 2017) with a periodic increase of 10% annually for a period of 20 years.

c. Expense spending

With an 3% increase due to operations and maintenance costs per year over a 20-year lifespan, the plan for expenditure is illustrated in Table 2.

3.7 Financial analysis

The cost of investment, income, and business expenditure for each utilization option was calculated by capital budgeting and the LEGC (Levelized Electricity Generating Cost) calculation scheme. Taking into account the discount rate of 10.44% produced the following results.

a. Payback period

The period of return on investment capital for use as a gas-fired engine power plant (PLTG) is shown in Table 3.

As the payback period for use of the land for a gas-fired power plant is calculated as 2 years 7 months, compared to the duration of use for gas power plants (20 years), the payback period appears comparatively short.

Table 1. Investment cost for gas power plant.

Description of Work	Cost EPC (Rupiah)
Civil works	293,342,533,386
Mechanical works	1,210,523,716,019
Electrical works and control instruments	689,909,663,705
Procurement of operations and maintenance	547,809,886
Miscellaneous expenses	8,829,035,026
Total investment cost (not including tax)	2,203,152,758,023

Table 2. Operational cost of PLTG/MG per KwH.

Description of cost	Rp/kWh
Fuel costs	790.12
Operation and maintenace cost (fixed)	80.50
Operation and maintenace cost (variable)	34.50
Total	905.12

Source: Results of data processing.

Table 3. Data payback period.

Description	PLTG Plant
Payback Period	2 years 7 months

Source: Results of data processing.

b. Net Present Value (NPV)

NPV calculation results show that a gas-fired power plant has a positive value of NPV IDR 2,818,837,070,855, which means that the development of a gas-fired engine power station is feasible.An NPV greater than zero means benefits for the company and that such an undertaking can be carried out feasibly. The advantages of this NPV method include taking into account the time value of money, taking into account cash flow during the project's economic life, and taking into account the value of the remaining project.

c. Internal Rate of Return (IRR)

IRR is an indicator of the level of efficiency of an investment. A project or investment can be carried out if the rate of return is greater than the rate of return when investing elsewhere (interest on bank deposits, mutual funds, etc.). The IRR for a PLTG/MG is shown in Table 5.

The Profitability Index score for a gas-fired power plant can be seen in Table 6.

Based on these calculations, a Profitabilty Index (PI) greater than 1 indicates that gas-fired power plant alternative could feasibly be implemented.

Table 4. NPV data for 20 years.

Description	PLTG Plant (Rp)
NPV for 20 years of operation	2,818,837,070,855

Source: Results of data processing.

Table 5. Internal Rate of Return (IRR).

Description	PLTG Plant
IRR Value	28.69%

Source: Results of data processing.

Table 6. Value of Profitability Index (PI).

Description	PLTG Plant
PI value	1.91

Source: Results of data processing.

4 CONCLUSION

Based on financial analysis, alternative utilization of the vacant land as a site for a gas-fired power plant with an investment value of IDR 2,203,152,758,023 gives an NPV result of IDR 2,818,837,070,855 for a period of 20 years, with a payback period value of 2 years and 7 months, indicating a positive value for NPV, an IRR value of 28.69%, and PI of 1.91.

Local Regulation of Medan City states that the area is meant for industrial uses, so power-plant development is possible. The location has good potential and is easy to reach, has public utilities, and allows development as a commercial property.

From research conducted by the author on vacant land owned by PLN located in Paya Pasir, Rengas Village, Marelan District, Medan City, North Sumatra Province, it can be concluded via HBU analysis that development of PLTG/MG can be carried out.

REFERENCES

Adji, A.R. 2015. *Use of Highest and Best Use Analysis and Capital Budgeting as an analysis tool for investment in the property sector*. Jakarta: Indonesia.

Amin. 2015. *Revenue Approach on Property Invesatsi*. Jakarta: Indonesia.

Bolstad, W. C. 2006. *The MIT Endicott House: A Study in Highest and Best Use*. San Diego: Massachusetts Institute of Technology: San Diego, USA.

Dachyar. M. 2012. Analysis of investment and risk feasibility of the Indramayu PLTU PLN construction project. *Faculty of Economics, Master of Management*, University of Indonesia. Depok: Indonesia.

Damodaran, A. 2002. *Investment Valuation Tools: and Techniques for Determining the Value of Any Asset*. New York: Wiley.

Fachrudin, K. & Fachrudin, H.T. 2017. *The Effect of Green Home, Green Behavior, and Livability on the Financial Incentive in Medan City*. Medan: Indonesia.

Gittinger. J.P. 1986. Economic analysis of agricultural projects. Second edition. *University of Indonesia*. Jakarta: Indonesia.

Luce. Anthony J. 2012. Highest and Best Use Analysis for a site in Arlington VA. Baltimore, MD.

MAPPI. 2015. Indonesian Appraisal Ethics Code (KEPI) and Indonesian Standard Assessment (SPI). 6th edition. Jakarta: Indonesia.

Muhammad, A. 2015. Financial Feasibility Analysis of Gumantimicro Hydro Power Plant Project. *Journal of Business and Management*. Bandung: Indonesia.

Parwoto, Agus. 2014. *Theory and Practice of Property Valuation. 2nd edition*. Yogyakarta: Indonesia.

Pomykacz, M. & Olmstead, C. 2014. The appraisal of power plants. *The Valuation Journal* 9(2): 90–121.

The Appraisal Institute. 1993. *The Appraisal of Real Estate*. Chicago, IL.

Identification of e-government indicators for measuring smart governance in Bandung City, Indonesia

I. Indrawati, M.Y. Febrianta & H. Amani
Telkom University, Bandung, Indonesia

ABSTRACT: As Bandung has become the center of urbanization in Indonesia, questions have arisen as to how to manage and solve problems that result. A so-called "Smart City" is one solution for enhancing the functions of Bandung City government in favour of good city governance. To measure whether Bandung has implemented smart governance, variables and indicators must be identified. The steps in obtaining variables and indicators are, first, searching the literature and, second, conducting interviews and focus-group discussions among respondents who have expertise and experience in the field of Smart Cities. Through those steps, this study seeks to determine whether Smart Governance has been implemented in Bandung. The study reveals a new proposed model and statements to measure readiness of Smart Governance implementation in Bandung and other cities in Indonesia.

1 INTRODUCTION

In 2013 the Global Standards Initiative on the Internet of Things (IoT-GSI) defined the Internet of Things (IoT) as the infrastructure of the information society. IoT allows an object to be detected and con-trolled remotely through an existing network infra-structure to be integrated directly from the real world with a computer-based system that will result in increased efficiency, accuracy, and economic benefits. Gartner claims that by 2020 there will be more than 26 billion devices connected (Termanini 2016). Research from Ericsson (2016) stated that 93% of telecom operators mentioned that IoT would play a significant role in monetizing 5G technology, which will be developed after the implementation of current 4G LTE technology. This means IoT technology will become a trend and is set to grow in the future.

IoT is expected to offer further connectivity between devices, systems, and services to allow machine-to-machine (M2M) communication and linking of multiple protocols, domains, and applications. The interconnection of these connected devices (including smart objects) is expected to be a mediator that can automatically control all areas of the urban community and can be further developed for applications such as smart grids and covering a Smart City area. IoT is one of the platforms of today's growing Smart City.

According to the United Nations, the number of people moving to urban areas in Indonesia is increasing and will continue to do so in the coming years (Utoyo 2015). The urbanization rate in Indonesia is high compared with other countries in Asia and the Southeast Asia region, and this is set to cause problems in Indonesia's major cities.

According to data from Internet World Statistics in 2015, Indonesia ranks eighth among countries with the most Internet users. According to a survey conducted by APJII (an Indonesian Internet service provider), as many as 132.7 million Indonesians are internet users (APJII 2016). West Java, and especially the capital city of Bandung, is becoming an area with a high potential number of Internet users.

A survey of Internet user behavior by APJII Indonesia states that 25.3% of Indonesia's population or about 31.3 million Indonesians use the Internet to find information. Further, 27.6 million Indonesians have Internet-related jobs; the rest use the Internet to fill leisure time, for socialization, for education, for entertainment, and to do business or look for goods. Based on the survey, approximately 90.4% of Internet users or about 119.90 million Indonesians who use the Internet agree that social media can be used to disseminate government policies to the public (APJII 2016).

As mentioned in Pikiran Rakyat newspaper (Amaliya 2016), especially in the city of Bandung, the Internet can be used as way to improve service to the community. This matches Bandung's vision of being a comfortable and prosperous city; also, it matches the Bandung City mission, which is presenting accountable and clean government services. Bandung's mayor, Ridwan Kamil, said that he would like to show off the progress made in areas where Smart City initiatives are doing a lot to improve services to society, bureaucratic reformation, and budgetary savings.

Bureaucratic reforms conducted by Kamil (Muhammad 2015) include: an open-office bid for two heads of service departments; launching report cards for heads of villages; the launch of head of the village performance reports; launching a Government Information System; launching an online system for distributing social grants; removing tax collection team in the field; launching New Student Acceptance for elementary, juniors and high school; establish building expert teams; launching anti-corruption programs; repairing performance reports on public services; launching e-Musrenbang; launching a citizen report system called LAPOR; becoming the first city to use Twitter in all departments; launch of a quick road-patching reaction unit; launching a 24-hour community health center for the poor; launching clinics for chronic illness; increasing the goods/services procurement system via e-catalog; improving the doctor's queue mechanism; and, lastly, launching Smart City initiatives.

One of the bureaucratic reform initiatives of Kamil (2016) is to pursue five Smart City milestones: infrastructure, capacity building, open and smart government, citizen engagement, and Bandung Technopolis.

The Government of Bandung initiated the establishment of Indonesia Smart City Forum, a place to collaborate and share inter-district/municipal software for Indonesia to accelerate information- and technology-based development. "Almost 70% of the problems could not being reached before, but now this can be solved," said Kamil, meeting with reporters after the opening of the forum (Amaliya 2016). According to Kamil, with technology-based public services it is very easy to determine and resolve complaints from the public. Hence, every division should use social media to handle community reports immediately. He said infrastructure projects can be monitored directly through the implementation of the Smart City system. In addition, the performance of local government can be evaluated to optimize their performance. The implementation of the existing Smart City system can monitor the performance of the division, head district, and head villages all over Bandung City; as a result, two sub-districts and four heads of villages have been dismissed for poor performance (birokrasinews.com).

Software sharing was confirmed in a memorandum of understanding witnessed by Minister of Administrative Reform and Bureaucratic Reform Asman Abnur. "The technology can make Indonesia into a great country," said Kamil.

He states as an example that Bandung City has 320 software programs, adapted to the problems of the city. Cooperation with other parties to produce the software cost around IDR 40 billion. "But through e-Budgeting Software, we've been able to detect wasteful budgets, hidden behind the names of long, dense nomenclature, with 1,200 of our activities cut with a value of IDR 1 trillion."

Abnur said that he would issue a decree that the Smart City forum is mandatory for all districts and will be held annually. "It costs a lot to come to Japan, to Singapore etc. This movement cuts half the length of the journey." The Smart City is in line with two presidential instructions focusing on improving public services and implementing e-Government and e-Budgeting Software.

According to Ardisasmita (2015), the approach taken by Ridwan Kamil as mayor was community-based and involved mutual cooperation. Many parties are collaborating to create the Bandung Smart City, from the community, university, and private sectors to foreign cities becoming sister cities—cities invited to cooperate intensively in various sectors.

A study by Frost and Sullivan (Singh 2014) on many Smart City projects and initiatives globally identified eight aspects of a Smart City: smart governance, smart energy, smart building, smart mobility, smart infrastructure, smart technology, smart healthcare, and smart citizens.

Kardos states that government has traditionally connoted "the act or process of governing," but modern theories have expanded the connotation, focusing on a large variety of instruments "designed to alter and channel the behavior of individual and collective actors." Kardos also cites Lafferty (2004), who states that "governance has been available for influential social influences in preordained directions" (Kardos 2012). Smart governance is the most important part of the Smart City because the city government is a major stakeholder with many instruments designed to direct the behavioral habits of individuals or groups of actors who play a role in urban society.

A city is known as a Smart City if it develops open and user-based innovation ecosystems to improve innovation and people's lives (Komninos et al. 2013). A Smart City focuses on efforts to transform both rural and urban areas (Scuotto et al. 2016).

Implementation of a Smart City framework in Indonesia is needed because this platform is not only for urban services but also for business, administration, and management in the field of security, so that with this platform all activities can be improved effectively and efficiently, as mentioned by the Smart City Indonesia Community (smartcity.id).

According to the Smart City ID Community, a Smart City impacts the quality of citizens' lives by making the city more efficient, safe, and comfortable. Communities need to participate in the management and administration of the city and become active city users (smartcity.id).

Based on the formulation of these problems, the purpose of this study can be determined as follows: exploring and identifying the most appropriate variables and indicators of smart governance measurement in Bandung City; obtaining interview results and focus group discussions about the variables and indicators that can be used to measure smart governance in Bandung; listing items which can be used to measure smart governance in Bandung city.

Issues to be raised as research questions for this study are as follows. Based on the literature review, what variables and indicators are appropriate to measure smart governance in Bandung? Based on interviews or focus-group discussions, what variables and indicators can be used to measure smart governance? What can be used to measure implementation?

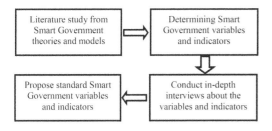

Figure 1. Smart government research stages.

2 METHOD

The method used to determine variables and indicators of smart governance was qualitative exploratory research. Explorative research results rely on secondary data such as reviewing available literature and/or qualitative data and primary data that can be collected through discussions with consumers, employees, and competitor companies and through in-depth interviews through case studies or focus group discussions (FGD) with business players and management experts.

In this study, the authors explore the variables and indicators formulated by various parties. These variables and indicators become the protocol for the authors to conduct FGD or interviews with respondents from the government, academics, business figures, and members of society at large. This research is conducted in accordance with the research stages shown in Figure 1.

Literature studies can be obtained from various sources such as journal proceedings, expert opinions (e.g. consultants), opinions from stakeholders, textbook, and so on (Indrawati 2015). From previous research and after the grouping and selection of Smart Governance variables, the authors decided to choose 15 variables for this study.

3 RESULT AND DISCUSSION

At this stage, which involves screening and grouping of definitions derived from the literature search results, the determination of variables and indicators should be conducted carefully so as not to cause a bias that can confuse the authors themselves. The variables and indicators from the literature review are collected and selected based on those most connected to Smart Governance; see Table 1.

In-depth interviewing involves obtaining information for the purpose of research by means of questions and answers between the interviewer and the respondent or the person being interviewed. In qualitative research, the in-depth interview technique is often used to confirm data already obtained from previous literature studies. The interview is also a tool to check or verify information obtained previously and also a direct communication technique between researchers

Table 1. Variables and indicators of smart governance, primarily related to e-governance.

	Variable and indicators of Smart Governance primarily related to e-Governance
1	Openness and public participation variables
1.1	Availability of online city information and feedback mechanisms
1.2	Online civic engagement
1.3	Online support for new city inhabitants
1.4	The existence of strategies, rules, and regulations to enable ICT literacy among inhabitants
2	e-Governance variables
2.1	Provision of online systems for administering public services and facilities
2.2	Application of services to support persons with specific needs
3	New public-private collaboration framework variables
4	Smart applications variables
4.1	Support information resources on decision-making
5	Support system variables
5.1	The complete rate of information standards
6	Guarantee system variables
6.1	Smart City plan and implementation scheme
6.2	Organization guarantee
7	Participatory public management variable
7.1	Public reporting sessions per year
8	Modern processes of public management of the municipal budget variables
8.1	The existence of a multi-annual budget
8.2	Remuneration of personnel based on a system of performance indicators
9	Modern systems of public management of the municipal government variables
9.1	The existence of electronic systems for tracking the municipality's management
9.2	Existence of electronic procurement system
10	Transparency and auditing of the Government's public management variables
10.1	Transparency index
10.2	Municipal government accounts audited
10.3	Municipal companies' accounts audited by a third party
11	Efficiency variables
12	Coordination and integration variables
12.1	Policy implementation in multiple dimensions variables
13	Participation and co-production variables
13.1	Information exchange
14	Local government staff variables
15	Bureaucracy variables
15.1	Accountability
15.2	Fairness

and respondents. In this case, the in-depth interview method is carried out with a list of pre-prepared questions.

The population surveyed should be determined by the researcher based on the criteria and in accordance with the topic of discussion in the study. In qualitative research, sample determination is done with the aim of selecting respondents who know the topic and problems in depth. This study uses a purposive sampling method in which the sample is selected from government, business figures, users/customers, and researchers/observers/experts. Table 2 shows the breakdown of the respondents.

Table 2. Respondent criteria.

Respondent Criteria	Number of Respondents
1 Business figure/user	7
2 Government	8
3 Researchers/observers/experts	9

3.1 Smart governance (e-governance) criteria: Literature review result

After conducting interviews and focus group discussions, the results are coded to the list of the variables and indicators, as shown in Table 3.

Variables and indicators mentioned by fewer than 50% of respondents are then eliminated (for instance, item 1.3, online support for new city inhabitants, was supported by only 4% of respondents).

3.2 Smart governance (e-governance) criteria: Based on interviews and FGD results

The results after elimination of unconfirmed variables and indicators are shown in Table 4.

Table 3. Variable based on the literature coded.

	Variable/Indicator	Coded	% confirmation
1	Openness and public participation variables		
1.1	Availability of online city information and feedback mechanisms	21	80%
1.2	Online civic engagement	24	96%
1.3	Online support for new city inhabitants	1	4%
1.4	The existence of strategies, rules, and regulations to enable ICT literacy among inhabitants	27	80%
2	e-Governance variables		
2.1	Provision of online systems for administering public services and facilities	40	100%
2.2	Application of services to support persons with specific needs	13	40%
3	New public-private collaboration framework variables	28	92%
4	Smart applications variables		
4.1	Support information resources on decision-making	11	64%
5	Support system variables		
5.1	The complete rate of information standards	6	48%
6	Guarantee system variables		
6.1	Smart City plan and implementation scheme	46	100%
6.2	Organization guarantee	19	96%
7	Participatory public management variable		
7.1	Public reporting sessions per year	2	36%
8	Modern processes of public management of the municipal budget variables		
8.1	The existence of a multi-annual budget	10	84%
8.2	Remuneration of personnel based on a system of performance indicators	5	16%
9	Modern systems of public management of the municipal government variables		
9.1	The existence of electronic systems for tracking the municipality's management	33	100%
9.2	Existence of electronic procurement system	9	60%
10	Transparency and auditing of the Government's public management variables		
10.1	Transparency index	12	52%
10.2	Municipal government accounts audited	4	40%
10.3	Municipal companies' accounts audited by a third party	1	8%
11	Efficiency variables	16	60%
12	Coordination and integration variables		
12.1	Policy implementation in multiple dimensions variables	40	100%
13	Participation and co-production variables		
13.1	Information exchange	24	80%
14	Local government staff variables	36	100%
15	Bureaucracy variables		
15.1	Accountability	4	28%
15.2	Fairness	2	12%

Table 4. Variable and indicators based on interview & FGD results.

	Variable/indicator	Coded	% confirmation
1	Openness and public participation variables		
1.1	Availability of online city information and feedback mechanisms	21	80%
1.2	Online civic engagement	24	96%
1.4	The existence of strategies, rules, and regulations to enable ICT literacy among inhabitants	27	80%
2	e-Governance variables		
2.1	Provision of online systems for administering public services and facilities	40	100%
3	New public-private collaboration framework variables	28	92%
4	Smart applications variables		
4.1	Support information resources on decision-making	11	64%
6	Guarantee system variables		
6.1	Smart City plan and implementation scheme	46	100%
6.2	Organization guarantee	19	96%
8	Modern processes of public management of the municipal budget variables		
8.1	The existence of a multi-annual budget	10	84%
9	Modern systems of public management of the municipal government variables		
9.1	The existence of electronic systems for tracking the municipality's management	33	100%
9.2	Existence of electronic procurement system	9	60%
10	Transparency and auditing of the Government's public management variables		
10.1	Transparency index	12	52%
11	Efficiency variables	16	60%
12	Coordination and integration variables		
12.1	Policy implementation in multiple dimensions variables	40	100%
13	Participation and co-production variables		
13.1	Information exchange	24	80%
14	Local government staff variables	36	100%

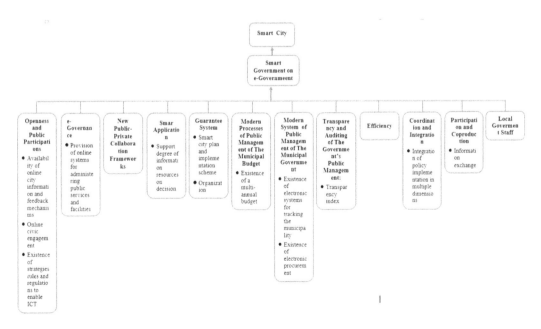

Figure 2. Schematic model for measuring smart governance.

4 CONCLUSION

This study conclude that, in the implementation of Smart Governance (specifically for e-Governance), there are 12 variables and 13 indicators. Figure 2 presents a schematic model for measuring smart governance.

This research has determined variables and indicators of Smart Governance as well as items to measure Smart Government implementation. These results can be followed by carrying out a pilot test of the items produced by this research.

The idea of Smart Cities generally and Smart Governance in particular is still very dynamic. Further research is needed to continue to accommodate parameters in the ecosystem, including parameters of people, processes, and technologies that will continue to evolve according to ever-changing demands.

ACKNOWLEDGMENT

The authors of this research would like to express their gratitude to the Ministry of Research, Technology and Higher Education of Indonesia, for financial support in doing this research.

REFERENCES

Amaliya. 2016. Forum Smart City, Ajang Berbagi Teknologi. [Online]. Retrieved: https://www.pikiran-rakyat.com/. Accessed November 20, 2017.

APJII. 2016. *Survey Penetrasi dan Perilaku Pengguna Internet Indonesia*: Jakarta.

Ardisasmita, A. 2015. Langkah Bandung Dalam Mengimplementasikan Kota Cerdas (Smart City). [Online]. Retrieved: https://id.techinasia.com/. Accessed November 20, 2017.

Ericsson. 2016. 5G Readiness Survey 1–8. [Online]. Retrieved: https://www.ericsson.com/en/press-releases/2016/6/internet-of-things-to-overtake-mobile-phones-by-2018-ericsson-mobility-report. Accessed September 17, 2017.

Indrawati. 2015. *Metode Penelitian Manajemen dan Bisnis*. Bandung: Refika Aditama.

Muhammad, D. 2015. 19 Inovasi Ridwan Kamil Kala Mereformasi Birokrasi Bandung. [Online]. Retrieved: https://www.republika.co.id/berita/nasional/daerah/15/05/29/np419h-19-inovasi-ridwan-kamil-kala-mereformasi-birokrasi-bandung. Accesed February 20, 2018.

Kamil, R. 2016. *Bandung Smart City*. Bandung: Global Center of Competence Cities.

Kardos, M. 2012. The Reflection of Good Governance in Sustainable Development Strategies. *Procedia-Social and Behavioral Sciences*, 58: 1166–1173.

Komninos N., Pallot, M. & Schaffers, H. 2013. Special Issue on Smart Cities and the Future Internet in Europe. *Journal of the Knowledge Economy*, 4(2): 119–134.

Scuotto, V., Ferraris, A. & Bresciani, S. 2016. Internet of Things: Applications and challenges in smart cities: a case study of IBM smart city projects. *Business Process Management Journal*, 22(2): 357–367.

Singh, S. 2014. Smart cities—a $1.5 trillion market opportunity. Forbes. Retrieved, 4.

Termanini, R. 2016. *The Cognitive Early Warning Predictive System Using the Smart Vaccine: The New Digital Immunity Paradigm for Smart Cities and Critical Infrastructure*. Florida: CRC Press.

Utoyo, S. 2015. Statistik Perusahaan Informasi dan Komunikasi 2015. Badan Pusat Statistik.

Reformulation in the basic of legal considerations about the basic in filing for bankruptcy in Indonesian commercial court (The legal comparison with Malaysia and Singapore)

P.E.T. Dewi & I.W.P.S. Aryana
Universitas Ngurah Rai, Bali, Indonesia

ABSTRACT: In the middle of 1997, monetary crisis hit Indonesia. The business sector was the one that got the effects of the crisis the most. Many of the businesses went bankrupt. As the result, there were many obligations of debts and accounts receivable due that were not fulfilled. Considering those conditions, the quick, open, and effective regulations are needed in order to give the chance for both creditors and debtors to attempt the fair settlement. The attempts can be done through the bankruptcy law. However, in the formulation of Article 2 paragraph 1 Law number 37 of 2004 about the Bankruptcy and Suspension of Obligation for Payment of Debts (UUK-PKPU), it can be identified that how easy a debtor, that is an individual or a company, is declared bankrupt. This study was a normative legal research in which, legal materials were collected through library research. The analysis was conducted qualitatively. The neglect of their rights in the regulation of UKK-PKPU is the important thing that needs to be considered. The debtors' rights, although they do default of risk, must be protected because their rights are the constitutional rights of every citizen.

1 INTRODUCTION

Basically, a business activity is always based on the aspects of commercial law in which those business activities need to consider the contract law underlying the occurrence of all business activities. The legal awareness of the businessmen usually appears when there are problems in the contract. These problems are usually due to the bad faith of one party. Therefore, there will be default of risk or broken promises of one party causing losses on agreements that have been made and agreed.

The concept of default of risk is a domain in civil law (private law). Article 1234 of the Civil Code (hereinafter referred to as the Civil Code) states that the purposes of the contract is to give something, do something or do nothing. Default of risk will be a more complicated problem if the party performing default of risk is the debtor whose business is failing and or already in a state of unable to pay its debts. This condition makes the creditors as the owner of the capital worried. Moreover, if the debtor has more than one creditor and his personal or company assets are estimated not enough to pay the debts to the creditors.

This situation was mostly found in mid-1997 in Indonesia when the monetary crisis hit Indonesia. The business sector was the one that got the effects of the crisis the most. Many businesses went bankrupt resulting in non-fulfillment of obligations of debts and accounts receivable due. Regarding this matter, it is necessary to arrange the quick, open, and effective regulations in order to give opportunity to the parties both creditor and debtor to attempt the fair settlement. The attempts can be done through the bankruptcy law.

Bankruptcy is defined as anything related to bankrupt (Badudu 2001). In Article 1 Sub-Article 1 of Law Number 37 of 2004 about the Bankruptcy and Suspension of Obligation for Payment of Debts (UUK-PKPU), bankruptcy shall mean general confiscation of all assets of a bankrupt debtor that will be managed and liquidated by a curator under the supervision of supervisory judge as provided for herein. The bankruptcy which is as a fair settlement for the parties is when the bankruptcy request can be filed by the debtor and the creditor. The debtor performing default of risk is the party requesting or having filed for bankruptcy to the Commercial Court. Based on Article 2 paragraph 1 UUK-PKPU, the debtor that can be filed for bankruptcy is a debtor that has two or more creditors and fails to pay at least one debt which has matured and become payable.

Formulation of this Article shows how easy a debtor who is an individual or a company shall be declared bankrupt. It will be unfair when the default risk debtor still has a good business prospect and the ability to pay its debts but then file for bankruptcy in the Commercial Court. There is a legal vacuum regarding the amount limit of the debt and ignorance of the criteria of the debtor's financial ability that can be filed for bankruptcy in the Commercial Court. UUK-PKPU does not adhere to the insolvent principle. It is a problem because debtors who can still pay a half or even all the debts can be declared bankrupt.

This empty of norms has become an obstacle to bankruptcy law enforcement in Indonesia. In fact, one of the important stages in the bankruptcy process is the insolvency stage. For this reason, it is necessary to amend the UUK-PKPU. Reformulating the basic in Filing for Bankruptcy in Indonesian Commercial Court is very necessary to realize the legal purpose itself (legal certainty, utility and justice) for all parties.

2 METHODS

This study is normative legal research, this research examined library materials or secondary data that include primary legal materials and secondary legal materials. Analysis of legal materials was made qualitatively. The discussion was presented in the form of descriptive analysis.

This study also used three approaches:

1. Statute Approach
   In the method of legislation approach that needs to be understood is the hierarchy, and the principles of the legislation (Nasution 2008). The legislative approach is an approach using legislation and regulation (Marzuki 2016).
2. Conceptual Approach.
   Conceptual approach can be used to describe and analyze research problems that move from the existence of an empty norm. This means that in the current legal system, there is no norm from a law that can be applied to legal events or concrete legal events (Diantha 2016).
3. Comparative Law Approach
   Comparative law is the study of the principles of legal science by the comparison of various systems of law. Comparative law is done by comparing the law of one country with the law of another country or the decision of a judge of one country with another country concerning the same case. Comparative law can also compare laws that applies now with laws that had been applied in the past (Susanti & Efendi 2013).

3 RESULTS AND DISCUSSION

Bankruptcy should be the last resort in the debt settlement of the debtors. The role of bankruptcy institution is basically very important to ensure the parties - the debtor and the creditor – get justice from the bankruptcy process. According to Haman (2005): "Bankruptcy is a legal procedure that allows you to get out of oppressive debt and get a fresh start financially". Other opinions according to Ventura (2004): "Bankruptcy is a constitutional right of protection against creditors". In addition to argue that bankruptcy is the constitutional right of creditor, John Ventura also explains the "fresh start financial" in bankruptcy proceedings as proposed by Edward A. Haman. According to Ventura (2004):

Bankruptcy allows consumers and businesses to make a fresh start when they owe so much money relative to their income that they cannot afford to pay their debts. Depending on the specific type of bankruptcy that is filed, most of the debts owed by a consumer or a business will be wiped out or the debts will be reorganized so that the consumer or business can afford to pay them. In exchange for the financial fresh start that bankruptcy provides, however, the consumer or business may have to give some of their assets back to their creditor. Furthermore, the bankruptcy will remain in the consumer's credit history for up to ten years and during that time, if the consumer is approved for new credit, the cost of that credit will be high.

In Malaysia, the bankruptcy gets serious attention from the government. This condition is proven by several amendments to bankruptcy law aiming to improve the insolvency system based on changes in the level of community life as well as the challenges of globalization. In the system prevailing in this neighboring country, there is a distinction and separation of understanding and regulation between the individual and company bankruptcy. Individual bankruptcy is governed by a bankruptcy law often referred to the 1967 Malaysian Bankruptcy Act or the "Akta Kebankrapan 1967", while corporate insolvency is governed by the Malaysian Limited Liability Company Act 1965 or "Akta Syarikat 1965" (Wijayanta 2016).

Until 2004, the "Akta Kebankrapan 1967" has been amended five times by the Malaysian Parliament. The last amendment was in 2003 as set in the "*Akta Kebangkrapan (Pindaan) 2003*". The changes contained in the "*Akta Kebangkrapan (Pindaan) 2003*" are: 1) the change of the name of the Institution that manages the bankruptcy property of the Treasury Board into Jabatan Insolvency Malaysia (JIM), 2) the stipulations on the distinction between social and ordinary guarantor, 3) the minimum amount of the debt is amended from RM 10,000 to RM 30,000 and the maximum amount that may be borrowed by the bankrupt debtor is amended from RM 10,000 to RM 100,000, 4) and several other amendments.

The Malaysia Companies Act 1965 regulates the winding up of insolvent companies. The statute is based on the Australian Uniform Companies Act of 1961 and the British Companies Act of 1948. In terms of concept and statutory language, the insolvency law of both these countries have influenced the Malaysian legislation. If a company were hopelessly insolvent it could be wound up by court on the ground that it is unable to pay its debts. If, however, a company is in financial difficulty but has the potential to be solvent in the long term, the law provides for it to be reconstructed under a compromise or scheme of arrangement (Abeyratne 2000).

Bankruptcy arrangements in Malaysia can be said providing legal protection not only for the creditors but also for the debtor which is embodied in the clear and firm arrangements in the Bankruptcy Act and Company Law. In Malaysia, there is a clear separation between individual and company bankruptcy and the standard of how a person or company can be declared bankrupt. The minimum amount of debt as a requirement for bankruptcy as regulated in the section 5 (1) letter a of *Akta Kebankrapan* (*Pindaan*) 2003 is RM 30,000 as well as the maximum amount to be borrowed by the insolvent is RM 100,000. In addition, Malaysia also adheres to the principles of insolvency which is reflected both in the Bankruptcy Act and Company Law where the solvent debtor should not be filed bankrupt and if the debtor is in a state of financial difficulties but still has the potential to pay, then the debt should be restructured first. That rule reflects the existence of the protection of the debtor's rights although he has done default of risk.

Similar to Malaysia, in Singapore the individual and corporate bankruptcy is set differently. As quoted from the web of Singapore Government (2018):

> Bankruptcy is a legal status of an individual who cannot repay debts of greater than $15,000 and is declared a bankrupt by the High Court. The High Court usually appoints the Official Assignee to administer the bankrupt's affairs in bankruptcy. These include the selling off of the bankrupt's assets to repay his creditors, the registration of the creditors' claims and the distribution of dividends to the bankrupt's creditors.

The Singapore bankruptcy rules are governed by two separate rules, namely the Corporations Act 2017 and the Bankruptcy Act 1995 which were amended twice in 2009 and 2015 (Bankruptcy (Amendment) Act 2015). The amendment by the government of Singapore is also a governmental concern related to the ability of people's purchasing power and also the condition of global economy especially to keep the investment climate in Singapore.

If an individual is unable to pay their debts as and when they fall due, the debtor or their creditors may apply for a bankruptcy order. It will be presumed that a person is unable to pay their debts if they do not settle or set aside a statutory demand within 21 days of it being served. The debts, however, have to be reach the statutory threshold, which is $10,000 since the Bankruptcy Act 1995 was passed. An individual with debts less than this amount will not qualify for bankruptcy. After the Bankruptcy (Amendment) Bill 2015 was passed by Parliament. The threshold for bankruptcy was increased from $10,000 to $15,000. This means that a debtor must have at least $15,000 in debts to become bankrupt. As outlined by the Ministry of Law, this figure is in line with the income-related benchmarks from the Bankruptcy Act 1995. The purpose of the threshold is to encourage both debtors and creditors of smaller loans to sort out payment without needing to resort to the bankruptcy regime. It does however exclude a larger number of people –mostly lower-income individuals – from obtaining the relief associated with bankruptcy (Gardner 2016).

In the beginning of their independence, bankruptcy in Indonesia was regulated in Faillissements-verordening (S. 1905-217) which was a bankruptcy law inherited from the Dutch colonizers. In the middle of 1997 a monetary crisis hit Indonesia. To overcome monetary turmoil and its severe consequences on the economy, one of the most pressing and urgent issues that needed to be resolved was the settlement of company debts, and thus bankruptcy regulations and delays in payment obligations that could be used by Debtors and Creditors fairly, quickly, openly and effectively became very necessary to be realized immediately. In connection with the above issues, Perpu Number 1 of 1998 was established which was then passed into Law Number 4 of 1998 about Bankruptcy (UUK). It turned out that UUK also had weaknesses, so on October 18, 2004 the UUK was changed to Law Number 37 of 2004 about Bankruptcy and Suspension of Obligations for Payment of Debts (UUK-PKPU).

Since the enactment of the UUK-PKPU, bankruptcy issues have not been resolved completely. One of the fundamental problems is that individual or corporate debtors are easily filed and bankrupted with very little debt compared to the amount of assets and may even be declared bankrupt simply because they do not pay a debt to the creditor although the debts from other creditors are not due and run well. Therefore, the bad intentions in this case are not only from the debtor who performs default of risk but also from the creditor in terms of utilizing bankruptcy institutions to drop the reputation of someone or business by filing for bankruptcy.

According to Rudhi Prasetya, the existence of bankruptcy institutions serves to prevent the arbitrariness of the creditors who force in various ways to make the debtors repay their debts (Prasetya 1996). The regulations on bankruptcy is expected to overcome the problems in the national economy and give a sense of justice to both creditors and debtors. Finally, there are some important things that must be considered in developing reformulation the basic in filing for bankruptcy applications in the future (ius constituendum):

1. The importance of limiting the amount of debt as a condition to be petitioned for bankruptcy becomes important to be regulated in Indonesian bankruptcy law.
2. Separation between individual and corporate bankruptcy due to the allocation of debt, the ability to pay the debts, and the amount of debt between individuals and a company must be different.
3. In bankruptcy, the principle of insolvency is certainly very important in determining whether someone can be said bankrupt or not.

According to The Law Offices of Tyler, Bartl, Ramsdell & Counts, P.L.C, insolvency is what many people think bankruptcy is that is to declare a state of complete inability to pay debt. Further it is explained "Now, insolvency is the "state of being" that would cause a person to file for bankruptcy. If a person cannot pay back his/her lenders on time, or if the cash flow falls, or if income is simply too low to cover expenses, then insolvency is the proper term. Of course, the first step of an insolvent person or company would be to take steps towards resolution." Therefore, in fact, insolvent state here is important as one measure in determining whether or not the debtor performing default of risk is petitioned for bankruptcy in the Commercial Court.

4 CONCLUSION

The neglect of default debtors' rights in the arrangement in UUK-PKPU is important to be noted. The rights of the debtor, even though he performs default of risk, must be protected since these rights are the constitutional rights of every citizen. Human rights arrangements in Indonesia have a place in the Constitution, namely Chapter X.A, the 1945 Constitution, from Articles 28A to 28J. Insolvency stage is important because in this stage, the fate of debtors performing default of risk is determined. Whether the debtor will be declared doing debt restructure, peace, or bankruptcy by sharing his property with the creditors equally. However, the absence of the amount limit of the debt, discrepancy between individual and corporate insolvency, and the insolvency principle in bankruptcy regulations in Indonesia will open loopholes for naughty creditors. These conditions must be the government's concern to create a bankruptcy regulation that reflects justice, ensures legal certainty, legal protection and gives benefit to the parties involved. Bankruptcy rules should be adjusted to the development of globalization. This concern is expected to create a good investment climate and more advanced economy in Indonesia.

REFERENCES

Abeyratne, Sonali. 2000. Corporate insolvency in Malaysia, *International Insolvency Review* 9(3).

Badudu, Y. 2001. *Kamus besar bahasa Indonesia*. Jakarta: Balai Pustaka.

Diantha, I.M.P. 2016. *Metodologi penelitian hukum normatif dalam justifikasi teori hukum*. Jakarta: Prenada Media Group.

Gardner, J. 2016. *Bankruptcy reforms in Singapore: what can we learn?* Singapore: Centre for Banking & Finance Law, Faculty of Law National University of Singapore.

Haman, E.A. 2005. *how to file your own bankruptcy (or how to avoid it) (6ᵗʰ ed.)*. United States of America: Sphinx Publishing.

Marzuki, P.M. 2016. *Penelitian hukum edisi revisi*. Jakarta: Prenada Media Group.

Nasution, B.J. 2008. *Metode penelitian ilmu hukum*. Bandung: Mandar Maju.

Prasetya, R. 1996. Likuidasi sukarela dalam hukum kepailitan, Jakarta: *Seminar Hukum Kebangkrutan, Badan Pembinaan Hukum Nasional Departemen Kehakiman.*

Singapore Government "Ministry of Law Insolvency Office Singapore" 2018. Bankruptcy & Debt Repayment Scheme, https://www.mlaw.gov.sg/content/io/en/bankruptcy-and-debt-repayment-scheme/bankruptcy.html, Accessed on 23th May 2016.

Susanti, D.O. & Efendi, A. 2013. *Penelitian hukum (legal research)*. Jakarta: Sinar Grafika.

Ventura, J. 2004. *The bankruptcy kit, (3ʳᵈ ed.)*. USA: Dearborn Trade Publishing A Kaplan Professional Company.

Wijayanta, T. 2016. *Undang-undang dan praktik kepailitan perbandingan Indonesia dan Malaysia*, Yogyakarta: Gadjah Mada University Press.

www.TBRCLaw.com, Accessed on 23th May 2016.

Advances in Business, Management and Entrepreneurship – Hurriyati et al (eds)
© 2020 Taylor & Francis Group, London, ISBN 978-0-367-27176-3

Analysis of small medium enterprise business performance in Indonesia

A.E. Herlinawati, A. Machmud & S. Suryana
Universitas Pendidikan Indonesia, Bandung, Indonesia

ABSTRACT: As the largest group of economic actors in Indonesia, Small and Medium Enterprises (SMEs) face various problems. It, for example, has limited access to get the capital. It also has difficulties in marketing products and having low management capabilities. This research aims to analyze the business performance of SMEs in Indonesia in terms of sales growth, number of workers, market share, profitability and additional assets, by exploring financial assistance provided by the government. This research used a quantitative method. The population were SMEs in Bandung, West Java, Indonesia, which got financial assistance from microfinance institutions. The sample were 150 the owners of SMEs. The sample was selected by using simple random sampling technique. The data was collected through a questionnaire. It was then analysed by using a Structural Equation Model (SEM). The results showed that sales growth had the strongest contribution to SMEs business performance. By contrast, expansion of market share had a smaller contribution to the SMEs business performance. This finding implies that SMEs have to apply marketing strategies; strengthen the capital, financial management, and management capabilities; and expand the relationship network.

1 INTRODUCTION

Small and Medium Enterprises (SMEs) are interested to be investigated. It, however, plays a significant role in economic development and job creation (Richardson et al. 2004). This is supported by some experts (Storey 1994); (Wardi et al. 2017); (Machmud 2017) who say that SMEs are very important for the economic growth of a country. In Indonesia, SMEs dominate the business unit. It can also absorb 97% of total national workforce. SMEs also contribute 61.41% to Indonesia's Gross Domestic Product. The growth of SMEs in Indonesia is predicted to continually increase in the years 2017–2020 (Central Bureau of Statistic 2017).

Even though SMEs make a great contribution to the country's economic development, the fact is that they still face many problems. The classic problem is lack of access to formal financial institutions, such as banking. Thus, it has limited financial assets and tends to rely much on informal sources (Niode 2009, Sudaryanto & Hanim 2012, Ramdhansyah & Sondang 2013, Rifa, 2013). Another problem is that SMEs in Indonesia have low managerial ability, marketing, production and financial management (Herlinawati & Riyandi 2017). Furthermore, most SMEs do not have a clear vision and mission. It also less competitive. The quality, design, and price of SMEs product does not match the consumers' needs. As a result, it is difficult to sell their products in domestic or international markets (Mahmud 2009, Mahmud 2017). The other problem is that SMEs still run traditionally and hereditary (Astrachan 2008, Tambunan 2014), so it is lack of innovation (Stewart & Roth 2001), lack of initiative, as well as competence and capability

(Hisrich & Peter 2002, Garengo & Bernadi 2007, Astrachan & Zellweger 2008).

The performance of small companies can be seen from various aspects. Neely et al. (2004) state that the performance of small companies generally can be seen from the profit and the growth of sales level (Neely et al. 2004). In addition, it can also be seen from three perspectives: financial perspective with profit and asset as the indicator; operation perspective with productivity of employees as the indicator; and marketing perspective with sales growth and frequency of product change as the indicator (Neely et al. 2004).

The aim of this research is to analyze the business performance of SMEs in Indonesia with a focus on sales growth, number of workers, market share, profit and total assets. The finding is expected to be helpful to policy makers when developing SMEs.

2 METHOD

This research used a quantitative method. The purpose was to investigate the contribution of each variable on SMEs' business performance. The population was SMEs that got financial assistance from microfinance institutions in Bandung and Sumedang. The sample was 150 owners/managers of SMEs. This sample was selected by using simple random sampling.

The variables in this research were measured by using a 5-point Likert Scale. The primary data was collected through questionnaire. The questionnaire was tested priory to 30 owners of SMEs in order to check its validity and reliability. The validity test was done by using Pearson's Product Moment

Correlation. The result showed that all questions were valid. The score was less than the significance level of 0.05. Meanwhile, the reliability test was done by using Cronbach's alpha. The result showed that all questions were reliable. The coefficient value, in this case, was greater than 0.7.

In analyzing the data, this research used Structural Equation Modeling (SEM). SEM was used to test mode 1 in the causal form to the performance variable of SMEs' business. In addition, normality test, outliers test and multicollinearity were also conducted.

The result of multivariate test shows that the CR value for all indicators (−1.999) was less than 2.58. This indicated that all indicators were normally distributed. The test result also shows that the value of d^2 (13,791) was less than the value of X^2 (167,6102). This meant that there were no cases of outliers. Meanwhile, the value of condition number was $7,388 > 0$. In addition, the output of AMOS SEM indicated no multicollinearity problem.

Thus, it can be concluded the sample met the main statistical assumption and could be used in further analysis. Fit test was then conducted to check the level of compatibility between data and model, validity and reliability, the measurement model and the coefficient significance of the structural model. The suitability of the measurement model was carried out to each construct by looking at the relationship between latent variables and some indicators through validity and reliability test of the measurement model.

The result of validity and reliability test indicated that the indicators used in measurement model had adequate internal consistency in measuring the constructs studied. The overall model of fit test was performed to evaluate the goodness of fit between the data and the model.

The sample was characterized based on the business type, educational level, length of business, and financial assistance agency. Regarding the business types, the result shows that most of the sample (27.33%) run service and trading business; 24.67% of sample operated craft and processing business; 14,67% had convection business; 6.77% operated livestock business; and 2% had agribusiness.

The education level of most of the sample (60%) was senior high school graduate; 16.67% was junior high school graduate; 14.66% was undergraduate; and 8.67% was diploma graduate.

Time in business for most of the sample (58.67%) was almost 5 years; 25.33% more than 10 years; and 16% was 6–10 years. Most of the sample received the financial assistance from a microfinance institution.

3 RESULTS AND DISCUSSION

The result of the analysis is shown in Table 1.

According to Maholtra (2010), there are three requirements for the goodness of fit test: (1) use at least one measure, one of which is absolute (for example GFI, AGFI); (2) use at least one very bad

Table 1. Summary SEM Model.

Test Statistic	t-test	Critical Point	Conclusion
Cmin/DF	1.1920	≤ 2.00	Fit
P-value	0.3101	≥ 0.05	Fit
AGFI	0.9525	≥ 0.90	Fit
GFI	0.9842	≥ 0.90	Fit
CFI	0.9939	≥ 0.90	Fit
TLI	0.987	≥ 0.90	Fit
RMSEA	0.0359	≤ 0.08	Fit

Source: SEM AMOS Processed Data, 2018.

measure (e.g., Chi-square, RMSR, SRMR, RMSEA); and (3) use at least one comparison measure (e.g., NFI, NNFI, CFI, TLI, RNI). The result in Table 1 shows that GFI was higher than 0.90. RMSEA is less than 0.08. CLI, TLI and GFI is higher than 0.90. This indicates that the model is in a fit condition.

Regarding the validity and reliability test, the result shows that CR value for business performance was 0.08906. This value is higher than 0.79. In addition, AVE values were 0.6204, which was higher than 0.59. This result indicates that the measurement model is valid and reliable to measure business performance.

There are three indicators that shows SMEs performance after getting financial assistance from microfinance institution. The first is related to sales growth and profit. The result reveals that SMEs sales growth and profit increased 2%–15%. The second indicator is capital or company asset. Similar to the first one, this also increased 2%–10%. Next was number of workers. Related to this, the result shows that there were 22 SMEs that hired new labor. The percentage was 20%. Meanwhile, 128 SMEs did not hire new labor after getting financial assistance from a microfinance institution. The last indicator is market share. The result indicates that of 150 SMEs, there were only 2 that were able to increase market share from local to regional convection business.

Table 2 shows that the value of sales growth (X1), number of workers (X2), profit (X4) and total assets

Table 2. Regression Weights.

			Estimate	S.E.	C.R.	P	Consl
X2	←—	Business Perf	1.482	0.210	7.049	***	par_1
X1	←—	Business Perf	1.442	0.201	6.895	***	par_1
X3	←—	Business Perf	.143	0.125	1.13	0.255	par_2
X4	←—	Business Perf	.378	0.137	2.74	0.006	par_3
X5	←—	Business Perf	1.242	0.174	7.14	***	par_4

Source: SEM AMOS Processed Data, 2018.

Table 3. Standardized Regression Weights.

			Estimate
X2	<—	Business Perf	0.907
X1	<—	Business Perf	0.619
X3	<—	Business Perf	0.101
X4	<—	Business Perf	0.247
X5	<—	Business Perf	0.751

Source: SEM AMOS Processed Data, 2018.

(X5) was less than 0,05. In contrast, the value of market share (X3) was 0.255, which was higher than 0.05. This means that all of the indicators except market share have a significant effect on SMEs business performance.

From Table 3, it can be seen that all indicators had an influence on SMEs business performance. The significant contribution was given by the number of worker (90.7%). This was then followed by asset (75.1%), sales growth (61.90%); profits (24.7%), and market share (10.1%). This result corroborates previous studies Suci (2009), Abaho (2014), Aisyah (2017), Shehu and Mahmood (2014) and Machmud (2017) which found the number of workers, sales growth, assets, profit, and market share to be indicators that improved SMEs business performance.

4 CONCLUSION

Based on the result, it can be concluded that sales growth makes a strong contribution to SMEs' business performance. In contrast, market share is an indicator with less of a contribution. This strengthens the previous research, that SMEs have difficulty in expanding their market share.

This result indicates that SMEs should give more attention to their marketing strategies, capital, financial management, management capabilities, and network. To strength all those aspects, the synergy among government, academician, and SMEs is needed.

REFERENCES

Abaho, et.al. 2014. Firm Capabilities, Entrepreneurial Competency and Performance of Uganda SMEs. *Business Management Review*. 105–125 ISSN 0856–2253.

Aisyah, St., et. al. 2017. Effect of Characteristics and Entrepreneurial Orientation towards Entrepreneurship Competence and Crafts and Arts Small and Medium Enterprises Business Performance in Makassar. *International Review of Management and Marketing* ISSN: 2146–4405.

Astrachan, J. and Zellweger T. 2008. Performance of family firms: a literature and guidance for future research. *ZfKEZeitschrift für KMU und Entrepreneurship*. 56(1–2):1–22.

Garengo, P & Bernardi, G. 2007. Organisational Capability in SMEs, Performance Measurement as a Key System in Supporting Company Development. *International Journal of Productivity*. 56(5/6): 518–532.

Herlinawati, Erna and Riyandi. 2017. The Role of Sharia Microfinance in SMEs Business Development. International Conference on Economic Education and Entrepreneurship. *Proceeding Agustus 3rd*, 2017.

Hisrich, R.D. & Peters, M.P. 2002. *Entrepreneurship. Fifth Edition*. New York: McGraw-Hill Irwin.

Machmud, Amir. (2017). Strategi Kemiskinan Berbasis Ekonomi Islam. *Penelitian yang tidak dipublikasikan*.

Machmud, Amir. 2009. Model Kemitraan Pelaku Usaha Mikro Kecil Menengah Kota Bandung. *Buletin Ekuitas*. 2 (2). ISSN 1778–1466.

Maholtra, N.K. 2010. *Marketing Research and Applied Orientation. 6th edition*. New Jersey: Pearson Prentice Hall.

Neely, Andi. 2004. *Business Performance Measurement: Theory & Practice*. Cambridge: Cambridge University Press.

Niode, Idris Yanto. 2009. Sektor UMKM Di Indonesia: Profil, Masalah, Dan Strategi Pemberdayaan. *Jurnal Kajian Ekonomi dan Bisnis Oikos-Nomos*. 2 (1). January 2009. ISSN 1979–1607.

Ramdhansyah, dan Sondang Aida Silalahi. 2013. Pengembangan Model Pendanaan UMKM Berdasarkan Persepsi UMKM. *Jurnal Keuangan dan Bisnis*. 5(1), Maret, 2013.

Idealisa Masyrafina, Republika.Co.Id. [online] 18 Agustus 2017. UMKM Di Indonesia. Retrieved from: https://www.republika.co.id/berita/ekonomi/makro/17/08/18/ouvlqv382-kontribusi-umkm-untuk-pertumbuhan-ekonomi-diprediksi-turun Accessed 25/08/2017.

Reynolds, P., Storey, D. J., & Westhead, P. (1994). Cross-national comparisons of the variation in new firm formation rates. *Regional Studies*, 28(4), 443–456.

Richardson, P., Howarth, R., & Finnegan, G. (2004). The challenges of growing small businesses: Insights from women entrepreneurs in Africa. *Geneva: International Labour Office*.

Rifa, Bachtiar. 2013. The Importance of Microfinance for Development of MSMEs in ASEAN: Evidence from Indonesia. *Jurnal Kebijakan dan Manajemen Publik*. ISSN 2303. 1 (1). January 2013.

Shehu and Mahmood. 2014. The Relationship between Market Orientation and Business Performance of Nigerian SMEs: The Role of Organizational Culture. *International Journal of Business and Social Science*. 5(1):9.

Stewart, WH & PL. Roth. 2001. Risk Propensity Differences between Entrepreneurs and Managers: A Meta-Analytic Review. *Journal of Applied Psychology*. 86(1):145–153.

Suci, Rahayu Puji. 2009. Peningkatan Kinerja melalui Orientasi Kewirausahaan. Kemampuan Manajemen dan Strategi Bisnis. (Studi pada Industri Kecil Menengah Bordir di Jawa Timur. *Jurnal Manajemen dan Kewirausahaan*. 11 (1): 46–58.

Sudaryanto dan Hanim, Anifatul. 2012. Evaluasi kesiapan UKM Menyongsong Pasar Bebas Asean (AFTA): Analisis Perspektif dan Tinjauan Teoritis. *Jurnal Ekonomi Akuntansi dan Manajemen*, 1(2).

Tambunan, Tulus TH. 2014 The Importance of Microfinance for Development of MSMEs in ASEAN: Evidence from Indonesia. *Journal of ASEAN Studies*. 2 (2):80–102.

Wardi, Yunia dkk. 2017. Orientasi Kewirausahaan pada Kinerja Usaha Kecil dan Menengah (UKM) Sumatera Barat: Analisis Peran Moderasi dari Intensitas Persaingan, Turbulensi *Pasar dan Teknologi*. *Jurnal Manajemen Teknologi*, 16 (1): 46–61.

Advances in Business, Management and Entrepreneurship – Hurriyati et al (eds)
© *2020 Taylor & Francis Group, London, ISBN 978-0-367-27176-3*

The relationship between innovation capacity and company performance of creative industries in Indonesia

R.N. Sumawidjadja
STIE Indonesia Membangun, Bandung, Indonesia

A. Machmud, S. Suryana & E. Ahman
Universitas Pendidikan Indonesia, Bandung, Indonesia

ABSTRACT: The aims of this research were to find out and analyze the relationship between innovation capacity and creative industry performance in Indonesia. This study is based on the contribution of the creative industry in the national economy. The method used in this study was quantitative method through causal explanatory survey research. The innovative capacity measurements included product innovation and process innovation, while company performance was measured through financial perspective, customer perspective, internal business process perspective, as well as learning and growth perspective. The population of this research was 368 businesspeople in the leather industry, leather goods and footwear in Bandung City, West Java, Indonesia, with a sample size referring to Slovin's Formula: as many as 192 respondents. All research variables were measured using Likert scale, and the data were collected through questionnaires on the SMEs (Small-to-Medium Enterprises). This research used Structural Equation Modeling (SEM) for the data analysis. The results showed that innovation capacity had a significant influence on the company performance. This study implies that to improve the performance of creative industry, businesspeople need to optimize the innovation capacity through product innovation and process innovation by producing superior and innovative products as well as becoming leaders of process innovation using the latest technology.

1   INTRODUCTION

Performance is the work achieved by a person or group of people within an enterprise in accordance with their respective powers and responsibilities in achieving the objectives of the enterprise legally, not violating the law and not contradictive to morals or ethics (Veithzal & Ahmad 2005). It is also described as the achievement level of task implementation in an organization, in an effort to realize the goals, mission and vision of the organization (Bastian 2001). Performance is also defined as a record of the results obtained from certain job functions or activities over a certain time (Bernardin & Russell 2002).

Company performance can be measured through sales growth and market share (Pelham & Wilson 1996). Company performance is a result made by the management on a continuous basis (Helfert 1996). One of the ways to measure company performance is by using the concept of a balance scorecard, which considers the balance between financial performance and non-financial performance with four perspectives: financial perspective, customer perspective, internal business process perspective, and learning and growth perspective (Kaplan & Norton 2004).

Innovation capacity is the ability to constantly transform knowledge and ideas into new products, processes and systems for the benefit of the company

and its stakeholders (Szeto 2000, Chen & Xu 2009). The ability to innovate is not only a way to succeed in running a new business flow, or to manage its main stream, but also to synthesize these two operating paradigms (Lawson & Samson 2001). Innovation capacity involves the interaction between key factors of a company's operations, such as different types of resources such as knowledge, processes and products/services, external relationships with the community, market changes and individual creative inputs in the enterprise (Balan & Lindsay 2010).

Studies concerning the relationship between innovation capacity and company performance have been conducted by Jayani and Hui (2017), Fatemeh and Naser (2017), Tugba and Safak (2016), Marques and Fereira (2009), and Zahra et al. (1988). The results of these studies showed that the innovation capacity with input product innovation and process innovation had an effect on company performance. However, a study conducted by Ardyan et al. (2015) showed that product innovation had no effect on business/company performance.

The aims of this study were to find out and analyze the relationship between innovation capacity and company performance of the creative industry in Indonesia. The study focused on the contribution of the creative industry to the national economy of Indonesia. This study showed a significant growth, but it was not followed by the contribution. The contribution of the creative industry for the national economy was only 7.38 percent. Nevertheless, the creative economy is able to employ a workforce of 15.9 million people (Bekraf 2016). The development of the creative industry in Indonesia is still dominated by small- and medium-scale companies. During 2011–2015, the concern was the leather, footwear and leather goods sector with the average growth was by 0.27%. This condition shows that the business performance of this sector is stagnant. This should be a concern of the stakeholders because the growth and sustainability of the industry will involve many entrepreneurs and engage a lot of labor.

2 METHOD

This research used a quantitative method through causal explanatory survey research to test the correlation betweem innovation capacity and the company performance of creative industry in Indonesia. The measurement of innovation capacity refers to opinions (Silva 2003, Roberts & Amit 2003), namely product innovation, process innovation, investment in research and development, and new distribution channels. However, in this study, there were only two dimensions using comprising product innovation dimensions and process innovation. To measure the company performance, it involved four perspectives: financial perspective, customer perspective, internal business process perspective, and learning and growth perspective (Kaplan & Norton 2004).

The research was conducted by involving businesspeople in leather industry, leather goods and footwear in Bandung City with a population size of 368 units and a sample size referring to Slovin's Formula of 192 respondents. The respondents were 93% male and 7% female. The majority of respondent ages ranged from 41–50 years old (47%), 20–30 years (21%), 31–40 years (22%) and 51 years (12%). The highest level of education was amounted to 55%, Basic by 11%, High School by 22%, Diploma by 3% and Bachelor Degree by 9%. Based on the longest establishment, most business which has run between 11–20 years was 41%, 10 years was 30%, and 21 years was 29%. Based on the number of workers, the highest number was 10 people by 85%, 11–20 people by 11% and 21 people by 4%.

All research variables were measured using a 5-point Likert scale where 5 is Strongly Agree and 1 is Strongly Disagree. Data collection was conducted through a questionnaire, and the data analysis model used Structural Equation Modeling (SEM). SEM analysis was used to test the model in causal form between Innovation Capacity and Company Performance of creative industry in Indonesia. To answer that problem, the first stage performed was normality test, test outliers, and multicollinearity, and then fit tests were carried out to check the level of compatibility between data and model, validity and reliability, the measurement model and the coefficient significance of the structural model. The suitability of the measurement model was carried out against each construct by looking at the relationship between latent variables and some indicators through validity and reliability test of the measurement model.

3 DISCUSSION AND CONCLUSION

Normality tests were performed using critical ratios skewness value and kurtosis at a 0.01 level of esignificance. The data used were said to be normally distributed when CR skewness and kurtosis all manifest variables ≤ 2.58 and Multivariate test results are smaller than 2.58. This indicates that all indicators were normally distributed. Furthermore, a Mahalanobis distance (d2) test was used to test the possibility of multivariate outliers at 0,001 and df = number of observed variables. It is said that there is no case of outliers if the value of d2 < X2. The test results showed that d2 (16,919) < X2 (20,515), which means that there was no case of outliers. The last was Multicollinearity that can be seen through the determinant of covariance matrix. The determinant value must be greater than 0. A very small determinant value indicates that there is a multicollinearity problem, so the data cannot be used for research. The AMOS SEM output results in a value condition number = 7,342 >0, so it can be concluded that there was no multicollinearity problem. Based on the test results, it can be seen that the data was normally distributed, there were no cases of outliers, and the data set of the sample empirically met the main statistical assumption that there was no multicollinearity problem. Thus, it can be concluded that the sample data sets could be used in further analysis.

The result from the validity test was significant and reliable, indicating that CR value for product innovation was 0.82 > 0.70 and AVE value was 0.57 > 0,50. CR for process innovation was 0.89 > 0.70 and AVE value was 0.62 > 0.50. CR for SMEs performance was 0.92 > 0.70 and AVE value was 0.69 > 0.50. It can be concluded that the measurement model had adequate validity and reliability to measure innovating capacity and company performance.

The overall model fit test was performed to generally evaluate the goodness of fit between the data and the model. The examination result of Model Fit: RMSEA (0,04) < 0.08; CFI (0.995) > 0.90; TLI (0.991) > 0.90 and GFI (0.984) > 0.90, overall the model is fit as stated by Maholtra (2010: 733), that: (1) Use at least one size that is absolute (e.g GFI, AGFI). In this model, GFI was above 0.90, thus it can be interpreted that the model was on Fit condition. (2) Use at least one size that is absolutely bad (e.g Chi-square, RMSR, SRMR, RMSEA). In this model, RMSEA & lt was 0.08 which means that the model under Fit condition. (3) Use at least one comparative size (e.g NFI, NNFI, CFI, TLI, RNI). In this model CFI and TLI & gt; 0.90, so it can be interpreted that the model was on Fit condition.

Table 1. Estimation and parameter test of structural model.

Hypothesis		Estimate Standard	S.E.	C.R.	P	Concl.
C.Performance	Innovation	0.851	0.165	5.150	***	Sign.
C.Performance	Innovation	0.690	0.102	6.752	***	Sign.

Table 2. Estimation and equation of structural model.

Model	Structural Equation	R2
Performance	= 0.851 X1 + 0.690 X2 +z	0.724

Table 3. The impact of innovation capacity on performance.

Construct	Product Innovation	Process Innovation
Performance	0.851	0.690

As it can be seen in Table 2, both hypotheses are significantly (p-value < 0.05) acceptable, and the test results show: (1) 72.4% of variations that occur in company performance can be explained jointly by product innovation and process innovation. The remaining 27.6% is the influence of other variables not described in the model; (2) Judging from the estimation of the coefficient of R^2, it was indicated that the proposed performance model was effective in explaining the phenomenon under study ($R^2 > 50\%$). However, there are other variables that need to be investigated to explain the phenomenon of company performance variation. The test results show: (1) The amount of influence of product innovation on the company's performance is 85.1%, while the process innovation was affect by 69%; (2) The rest is determined by other variables not described in this study; (3) Judging from the influence of each variable, product innovation has the strongest contribution to company performance.

These findings reinforce the results of previous studies conducted by Tugba and Safak (2016) and Marques and Fereira (2009), which proved that both product innovation and process innovation in entrepreneurship had a significant effect on company performance.

The results showed that innovation capacity had a significant influence on company performance. This study implies that to improve the performance of creative industry, businesspeople need to optimize innovation capacity through product innovation and process innovation by delivering superior and innovative products and leading innovation processes using the latest technology.

REFERENCES

Ardyan, Elia and Olivia T. Putri. 2015. Dampak positif seorang wirausaha yang memiliki kompetensi kewirausahaan pada kesuksesan inovasi produk dan kinerja bisnis. *Jurnal Kewirausahaan dan Usaha Kecil Menengah*, 1 (1), 11–19.

Badan Pusat Statistik Kota Bandung. 2017. *Kota Bandung dalam angka 2017* Katalog BPS: 1102001.3273.

Badan Pusat Statistik Provinsi Jawa Barat. 2017. *Hasil pendaftaran (listing) usaha/perusahaan sensus ekonomi 2016.*

Balan, P. and Lindsay, N. 2010. Innovation capabiliy, entrepreneurial orientation and performance in Australian hotels: An empirical study: Sustainable tourism. *Cooperative Research Centre.*

Bekraf. 2016. *Database peta kegiatan dalam negeri.* Jakarta.

Bastian, Indra. 2001. *Akuntansi sektor publik di indonesia.* yogyakarta: BPFE.

Chen, L. and Xu, Q. 2009. The mechanism of innovation capability leveraging via strategy in SMEs. *Paper presented at the Industrial Engineering and Engineering Management (IEEM), International Conference on,* Hong Kong.

Fatemeh, H. and Naser G. 2017. Impact of co-creation on innovation capability and firm performance: A structural equation modeling. *AD-minister no. 30 Januari-Juni p. 73–90.*

Helfert, Erich A,1996. *Teknik analisis keuangan: petunjuk praktis untuk mengelola dan mengukur kinerja perusahaan, edisi kedelapan,* Erlangga, Jakarta.

Jayani R.P. and Yan Hui. 2017. Relationship between innovative capability, innovation type, and firm performance. *Journal of Innovation & Knowledge 3: 44–55.*

Kaplan, Robert S. & David P. Norton. 1996. *The balanced scorecard: translating Strategy into Action.* Harvard Business Press.

Lawson, B., Samson, D. 2001. Developing innovation capability in organizations: a dynamic capabilities approach. *International Journal of Innovation Management*, 2(4), p. 154–178.

Marques, C and Ferreira J.2009. SME innovative capacity, competitive advantage and performance in a traditional industrial region of Portugal. *Journal of Technology Management & Innovation* 4(4).

Pelham, A.M. and Wilson, D.T.1996. A longitudinal study of the impact of market structure, firm structure, strategy and market orientation culture on dimensions of small-firm performance. *Journal of the Academy of Marketing Science* 24(1): 27–43.

Robert, P. and Amit, R. (2003). The dynamics of innovative activuty and competitive advantage: The Australian retail Banking, 1981 to 1995, *Organizatioanl Science,* 14(2): 107–122.

Ruki. 2002. *Penggunaan sistem manajemen kinerja.* Jakarta: Gramedia Pustaka Utama.

Silva, M. 2003. Capacidade inovadora empresarial—Estudo dos factores impulsionadores e limitadores nas empresas industriais portuguesas. *Unpublished doctoral thesis, Universidade da Beira Interior, Covilhã, Portugal.*

Szeto, E. 2000. Innovation Capacity: Working towards mechanism for improving innovation within an interorganizational network. *The TQM Magazine* 12(2): 149–158.

Tugba G.Y. & Safak Akssoy. 2016 The impact of innovation capacity on firm performance: evidence from the Turkish industrial cluster. *Proceding of the Annual Academic Research Conference on Global Business, Economics, Finance & Social Science.*

Veithzal, R. & Basri, M. 2005 *Performance appraisal; sistem yang tepat untuk menilai kinerja karyawan dan meningkatkan daya saing perusahaan Edisi 1.* Jakarta: Raja Grafindo Persada.

Zahra, S., Belardino, S., Boxx, W. 1988. Organizational innovation, its correlates and its implications for financial performance. *International Journal of Management,* 5:133–142.

Advances in Business, Management and Entrepreneurship – Hurriyati et al (eds)
© 2020 Taylor & Francis Group, London, ISBN 978-0-367-27176-3

The role of formal education on entrepreneurial intention among students

O.R. Yustian & H. Mulyadi
Universitas Pendidikan Indonesia, Bandung, Indonesia

ABSTRACT: This study examines the effect of formal education in developing entrepreneurial intentions among widyatama university students. Questionnaires distributed to students who have received entrepreneurship courses. The data collected were analyzed through SPSS and Structural Equation Modelling techniques using AMOS. Confirmatory factor analysis through AMOS is used to test the goodness of fit of the model and the hypotheses developed for the study. An interesting finding from this study revealed that formal education does not affect entrepreneurial intentions, while other findings reveal that attitude has a relationship with entrepreneurial intention. Since attitude is the mediation variable in this research, this research will examine the role of attitude as a mediation between formal education and entrepreneurial intention.

1 INTRODUCTION

A country can be said to prosper if at least 2% of the total population becomes entrepreneurial, per McClelland (1961) in his book *The Achieving Society*. This theory is often used as an indicator in measuring the prosperity of a country.

Based on the Central Bureau of Statistics (BPS) in 2016, Indonesia's entrepreneurship ratio has increased, but when looking at data in 2016, this ratio is still small compared to developing countries and other developed countries, such as Malaysia (5%), Singapore (7%), China (10%), Japan (11%), and the United States (12%). In theory, it finds the truth if we see the data described earlier. It is evident that the number of entrepreneurs can affect the prosperity of the State.

Most entrepreneurs in Indonesia is a group of entre-preneur necessity. Which means, the interest of these groups to build business due to family economic fac-tors. An unstable family economic condition leads to individual efforts in building their own business and the difficulty of getting a job becomes one of the reasons why they build their own business. This group also tends to be careless in business management, when in fact many entrepreneurs who have the ability to develop a better business (Pujoal-wanto 2014).

According to the Central Bureau of Statistics (2016), in recent years the increase in unemployment was dominated by educated unemployment, although by 2017 the level open unemployment decreased by 0.11%. Still, according to Central Bureau of Statistics, as many as 856,644 people who graduated from diploma and university education are still unemployed. From the data can be seen that college graduates are still dominated by the mindset of a job seeker, not of a job creator.

Entrepreneurial intentions are not inherited but can be trained and developed through education (Azjen 2005). Entrepreneur attributes can be shaped positively by educational programs that build student awareness about entrepreneurship as a career choice (Athayde 2009). Universities have an important role in increasing the number of entrepreneurs, so college graduates can reduce the number of unemployed by opening a new field of jobs.

In the previous study, there was a discovery difference between the relationship of entrepreneurial intention and formal education. The research findings indicate that participation in formal education has a positive effect on students' entrepreneurial intentions (Rengiah & Sentosa 2016. Kuttim et al. 2014. Kim-sun et al. 2016). The research findings also show that formal education has no positive effect on students' entrepreneurial intentions (Rengiah & Sentosa 2016, Mahendra et al. 2017). Rengiah and Sentosa (2016) reveal the factors of formal education such as curriculum, teaching methods, and role of the university are the factors that determine entrepreneurial intention.

Previous study about the relationship between attitude and entrepreneurial education which in this study named formal education has been investigated by Khalili et al. (2014), Farhangmehr et al., Jakubiak and Buchta (2016), and Mahendra et al. (2017). The previous studies revealed there is a relationship between entrepreneurial education and attitude.

Previous research has discussed the relationship between entrepreneurial intention and entrepreneurial attitudes (Luthje & Franke 2003, Linan et al. 2007, Fitzsimmons & Douglas 2005, Rengiah 2016, Ma-hendra et al. 2017). Many previous studies positioned research models based on the Theory of Planned Behavior (Ajzen 1991), which suggests that individuals can assess entrepreneurial intentions (Krueger et al. 2000). Ajzen (1991) states three antecedents of intention, that is, attitudes, social norms, and self-efficacy. Empirical studies generally support the relationship between entrepreneurial intention

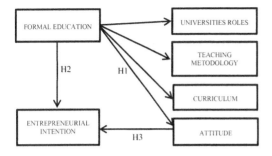

Figure 1. Conceptual framework.

and attitudes, subjective norms, and perceived behavioral control (Kolvereid 1996, Kruger et al. 2000, Douglas & Shepherd 2002, Souitaris et al. 2007). In this study, the only attitude that will be used as a mediation variable because based on previous studies by Krueger et al. (2000) and Autio et al. (2001) social norms and self-efficacy have a weak relationship to the intentions of entrepreneurship.

Based on previous research, this research hypothesis will examine the relationship between variables that have been discussed earlier. The research hypothesis will be poured in the conceptual model in Figure 1.

From the results of previous studies, researchers developed a number of hypotheses, as shown in Figure 1. There are four hypotheses developed in this study: the relationship between formal education and attitude, the relationship between formal education and entrepreneurial intention (H2), the relationship between attitude and entrepreneurial intention (H3), since attitude is mediation variable in this research, this research will examine the role of attitude as mediation between formal education and entrepreneurial intention.

2 METHOD

This research used Structural Equation Modeling (SEM) analysis. SEM is a multivariate technique of multiple regression analysis (Hair et al. 1998). The population of this research is university students by using convenience sampling as a sampling method. A total of 396 cases were analyzed for the study using structural equation modeling (SEM).

3 RESEARCH

3.1 Measurement model

The evaluation of reliability and validity of the scales is the main task in the measurement analysis. The process was performed using an iterative procedure. After data entry, the raw data were assessed for missing data, outliers, skewness and kurtosis.

Some items were deleted as have low loading coefficient and applying proposed modification indices to improve the fit of the model (Hu & Bentler 1999). The process resulted in seven reliable items for curriculum, eight reliable items for teaching methodology, seven reliable items for universities roles, six reliable items for attitude and four reliable item for entrepreneurial intention (Table 1). In the subsequent stage, the validity of constructs for the structural model was evaluated by the confirmatory factor analysis (CFA). Second-order confirmatory factor analysis (CFA) was used for measuring the model (Figure 2). The overall model fit as indicated by the j2 statistic ($x2 = 558.625$; df = 392; $p < 0.001$). However, given the x2 test's sensitivity to sample size (Hair et al., 2010) was focused on x2/df ratio and incremental fit measures other model fit indices confirmed by the adjusted goodness of fit index (AGFI), the comparative fit index (CFI), and Tucker-

Table 1. Factors loadings for curriculum, teaching methodology, universities roles, attitude and entrepreneurial intention.

	C M	TM	UR	EI	AT
CM9	0.799				
CM8	0.761				
CM7	0.602				
CM6	0.914				
CM5	0.787				
CM2	0.867				
CM1	0.623				
TM8		0.748			
TM7		0.710			
TM6		0.665			
TM5		0.824			
TM4		0.561			
TM3		0.530			
TM2		0.610			
TM1		0.700			
UR8			0.741		
UR7			0.860		
UR5			0.578		
UR4			0.782		
UR3			0.788		
UR2			0.717		
UR1			0.603		
EI2				0.873	
EI3				0.824	
EI4				0.687	
EI5				0.733	
AT2					0.845
AT3					0.758
AT5					0.599
AT6					0.649
AT8					0.699
AT9					0.783

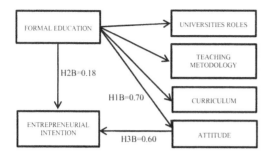

Figure 2. Model paths.

Notes: *p < 0.05; **p< 0.01; ***p < 0.001

Table 2. Composite reliabilities (CR) and average variance extracted (AVE) value CRAVE.

FE	0.683	0.511
EI	0.862	0.613
AT	0.869	0.528

Table 3. Standardized values.

$X2 = 532.432$; df = 365; $p < 0.001$ R^2 of constructs		
x2/df=	1.549	$Vi = 0.49$
CFI=	0.935	$V2 = 0.55$
TLI=	0.928	
AGFI=	0.811	
RMSEA=	0.056	

Lewis Index (TLI). On the other hand, root mean square error of approximation (RMSEA) index has been mostly preferred by researchers because it's independent from model parsimony (Simsek, 2007). The RMSEA, AGFI, CFI and TLI values were 0.054; 0.809; 0.939 and 0.932; It is approximately an acceptable level. The model's fit as indicated by these estimates although AGFI values were low but tolerable (Hu & Bentler 1999). It can be assumed that the model can be accepted due to other estimates were satisfactory.

Given that all standardized factor loadings were ranged from 0.53 to 0.91 and average variance extracted (AVE) values exceeded 0.5, which is of acceptable magnitude (Bagozzi et al. 1991, Fornell & Larcker 1981), convergent validity was supported (Table 2).

The composite reliabilities (CR) of the measures indicated acceptable internal reliability, ranging from 0.862 (Entrepreneurial Intention) to 0.869 (attitude), exceeding the threshold of 0.70 (Nunnally & Bernstein 1994) except for formal education. Results of measurement analysis show evidence of convergent validity and discriminant validity for our measurement model (Table 2).

3.2 Structural model

Structural equation modeling was performed to test the conceptual model and the hypothesized paths. Although the %2 value for the conceptual model was significant (%2 = 532.432; df = 365; p < 0.001), other model fit indices revealed that the hypothesized struc-tural relationships were explained by the data well (x2/df = 1.549; CFI = 0.935; TLI = 0.928; AGFI = 0.811; RMSEA = 0.056) (Table 3) by

satisfying ac-ceptable thresholds. Results of H indicated that formal education (P = 0.70, p < 0.05) had a positive effect on attitude, while in H2 indicated that formal education (P = 0.18, p > 0.05) did not had a positive effect on entrepreneurial intention, therefore H2 did not support. As predicted in H3, a positive influence of attitude on entrepreneurial intention (P = 0.6, p < 0.001) was found, therefore H3 was supported. Figure 2 presents the result of the structural equation model in testing hypotheses.

4 DISCUSSION

The first objective of this study was to investigate the relationship between formal education and attitude. The second aim was to investigate the relationship between formal education and entrepreneurial intention. Third, the aim is to investigate the relationship between attitude and entrepreneurial intention and the last objective of this study was to investigate the interaction of attitude, mediating the relationship between formal education and entrepreneurial intention. An interesting finding is revealed from this study that there is no direct effect between formal education and entrepreneurial intentions among students in the Department of Management Faculty of Business and Management, Widyatama University. This empirical research supports the finding revealed by Rengiah and Sentosa (2016) and Mahendra et al. (2017). Formal education such as curriculum, teaching methods, and the role of universities are the factors that determine entrepreneurial intention.

The existing entrepreneurship curriculum at universities in Indonesia should be more innovative. The curriculum focuses in the classroom and only makes the business plan without practice must be changed with features of business startup practices, visits to business areas, inviting successful entrepreneurs, working with partnership programs and including curricula that can make students more creative and innovative. Entrepreneurship teaching should also not only focus in the classroom because the focus is to become an entrepreneur, then at the beginning, it is very important to practice starting a business.

Combining theory with field practice is a way of introducing students into entrepreneurs that will encourage them to become true entrepreneurs. Resource planning, decision making, managing resources, communicating, experiencing successes and failures are ways to practice for students to explore areas they've never experienced before. Including how to create business plan competitions and experimental games that make the curriculum more interesting. The role of universities is vital in generating entrepreneurial spirit to the students, curriculum evaluation should be undertaken if it is to create an entrepreneurial environment at the university. Universities must participate actively in entrepreneurial activities by including their students. Creating business incubation within the university is one way to create an entrepreneurial environment on campus. Facilitating partnership programs can also be done to stimulate students more interested in entrepreneurship.

One of the objectives of this study was to investigate the relationship between formal education and attitude. This study finding that there is a relationship between formal education and attitude. This finding supports the study of Khalili et al. (2014), Far- hangmehr et al., Jakubiak and Buchta (2016), and Mahedra et al. (2017). Curriculum, teaching methodologies, and universities roles are part of formal education in this research. Formal education that includes an entrepreneurship curriculum with practice-based teaching methods and the role of universities in supporting the entrepreneurial environment to students will enhance students' entrepreneurship skills in attitude. In this study, the effective teaching of formal education will stimulate students' attitudes in completing their entrepreneurial study, which will be expected students become entrepreneurs in the future.

Another objective of this study was to investigate the relationship between attitude and entrepreneurial intention. This study finding that there is a relationship between attitude and entrepreneurial intention. This finding supports the study of Luthje and Franke (2003), Linan et al. (2007), Fitzsimmons and Douglas (2005), Rengiah and Sentosa (2016) and Mahendra et al. (2017). This indicates that high student entrepreneurial attitudes enable them to achieve high entrepreneurial intentions.

Since no significant influence of formal education on entrepreneurial intentions, it can be concluded that there is no mediating effect of attitude between formal education and entrepreneurial intentions.

5 CONCLUSION

In previous studies, entrepreneurial intentions were considered an important variable that contributed to the students' desire to make new ventures as career paths. This study focuses on the role of formal education on entrepreneurial intentions, with attitudes as a mediating variable. The results of this study indicate that formal education can improve student attitudes, while formal education does not affect student entrepreneurial intentions. Another result of this study indicates that the entrepreneurial attitude of students can affect student entrepreneurial intentions.

The better the entrepreneurship education will improve student entrepreneurship attitude and the higher the entrepreneurship attitude of the students, the higher the student entrepreneurial intention in making new business. Formal education can create student character. The role of the university to facilitate in developing the entrepreneurial character of the students by giving students the opportunity to acquire entrepreneurial education in theory and practice. Providing entrepreneurial experience to students is essential to anticipate and predict future business possibilities. It is suggested that entrepreneurship formal education is more innovative to stimulate student entrepreneurial intentions. For further research, other variables that may affect entrepreneurial intentions can be further investigated.

REFERENCES

Ajzen, I. 1991. The Theory of Planned Behavior. *Organizational Behavior and Human Decision Processes*, 50(2), 179–211.

Ajzen, I. 2005. Attitude, Personality, and Behavior (2nd). *Open University Press*. Poland.

Athayde, R. 2009. Measuring Entreprise Potential in Young People. *Entrepreneurship Theory & Practice*, 33(2), 481–500.

Autio, E. & Klofsten, M. & Keeley, R. 2001. Entrepreneurial Intent among Students in Scandinavia and in the USA. *Enterprise and Innovation Management Studies Vol. 2*, No. 2. DOI: 10.1080/14632440110094632.

Bagozzi, R.P, Yi, Y. & Phillips, L.W. 1991. Assessing Construct Validity in Organizational Re-search. *Administrative Science Quarterly*, 36(3), 421–458.

Douglas, E.J. & Shepherd. 2002. Self -employment as a career choice: Attitudes, entrepreneurial intentions, and utility maximization. *Entrepreneurship: Theory and Practice* 26 (3): 81–90.

Farhangmehr, M., Goncalves, P., & Sarmento, M. 2016. Predicting Entrepreneurial Motivation among University Students. 861–881.

Fitzsimmons, J. R. & Douglas, E. J. 2005. Entre-preneurial Attitudes and Entrepreneurial Intentions: A Cross-Cultural Study of Potential Entrepreneurs in India, China, Thailand and Australia. *Babson-Kauffman Entrepreneurial Research Conference, Wellesley, MA.*

Fornell, C. G. & Larcker, D. F. 1981. Evaluating structural equation models with unobservable variables and measurement error. *Journal of Marketing Research*, 18(1), 39–50.

Hair et al., 1998. Multivariate Data Analysis, Fifth Edition, Prentice Hal. *Upper Saddle River*. New Jersey: USA.

Jakubiak, M. & Buchta, K. 2016. Determinants of Entrepreneurial Attitudes in Relation to Students of Economics and Non-Economics. *Studia i Materi- aly*, 2(1), 17–30.

Khalili, B. Tojari, F. & Rezaei, M. 2014. The Impact of Entrepreneurship Training Course on the Development of Entrepreneurial Features. *European of Academic Research*, 2(9), 11942–11953.

Kim-Sun, N. & Ahmad, A. R. & Ibrahim, N. N. 2016. Theory of Planned Behavior: Undergraduates Entrepreneurial Motivation and Entrepreneurship Career Intention at a Public University. *Journal of Entrepreneurship: Research & Practice*, 1–14.

Kolvereid, L. 1996. Prediction of employment status choice intentions. *Entrepreneurship Theory and Practice*, 21, 47–57.

Krueger Jr. & N. F. & Reilly, M. D. & Carsrud, A. L. 2000. Competing Models of Entrepreneurial Intentions. *Journal of Business Venturing*, 15 (5–6), 411–432.

Kuttim, M. & Kallaste, M. & Venesaar, U. & Kiis, A. 2014. Entrepreneurship Education at University Level and Students Entrepreneurial Intentions. *Journal Procedia—Social and Behavioral Sciences*, 110, 658–668.

Linan, F. & Urbano, F. & Guerrero, M. 2007. Entrepreneurial intentions of university students in Spain: A regional comparison. *Paper presented at the XVII National Conference of ACEDE*, Seville: Spain.

Litze Hu & Peter M. Bentler. 1999. Cutoff criteria for fit indexes in covariance structure analysis: Conventional criteria versus new alternatives, *Structural Equation Modeling: A Multidisciplinary Journal*, 6:1, 1–55, DOI: 10.1080/10705519909540118.

Luthje, C. & Franke, N. 2003. The "making" of an entrepreneur: Testing a model of entrepreneurial intent among engineering students. *R&D Management*, 33(2), 135. doi:10.1111/1467–9310.00288.

Mahendra, A.M. & Djatmika, E.T. & Hermawan, A. 2017. The Effect of Entrepreneurship Education on Entrepreneurial Intention Mediated by Motivation and Attitude among Management Students, State University of Malang, Indonesia. *Canadian Center of Science and Education. International Education Studies*; Vol. 10, No. 9.

McClelland. D. C. 1961. *The achieving society.* Princeton, NJ: Van Nostrand.

Nunnally, J. C. & Bernstein, I. H. 1994. *Psychometric theory* (3rd). New York: USA.

Rengiah, P. & Sentosa, I. 2016. The Effectiveness of Entrepreneurship Education in Developing Entrepreneurial Intentions among Malaysian University Students: (a Research Findings on the Structural Equation Modeling). *European Journal of Business and Social Sciences*, 5(2), 30–43.

Simsek, O. F. 2007. Introduction to Structural Equation Modeling Basic Principles and LISREL Applications. *Ankara: Ekinox.*

Souitaris, V. & S. Zerbinati, and A. Al-Laham. 2007. Do entrepreneurship programmes raise entrepreneurial intention of science and engineering students? The effect of learning, inspiration and resources. *Journal of Business Venturing* 22 (4): 566–591.

Identification of student attitudes toward entrepreneurship

S. Doriza & E. Maulida
Universitas Negeri Jakarta, Jakarta, Indonesia

ABSTRACT: Entrepreneurship education involves educating students to think creatively and to be innovative and competitive when doing business. This research aims to identify students' attitude toward entrepreneurship, using a scale of interest measurement ranging from "very important" to "not important." Two hundred students who have already attended entrepreneurship classes were selected as subjects for the research survey. The results of this research can be used to create entrepreneurial education materials to enhance students' entrepreneurial attitudes.

1 INTRODUCTION

What kinds of attitudes toward entrepreneurship are needed for students who wish to be entrepreneurs? Positive attitudes to entrepreneurship do not merely emerge without entrepreneurship education (Oosterbeek et al. 2010). Indeed, entrepreneurship education can boost insight, creativity, confidence and work performance in order to drive someone to make more timely decisions.

Entrepreneurial attitude in students can be formed by entrepreneurship education. This is consistent with the study by Stamboulis and Barlas (2014), which shows that there are significant differences in students' attitudes before and after entrepreneurship education.

The result of a needs analysis carried out with 50 students by means of an open questionnaire and interviews shows that students need information about things related to entrepreneurship and principles of doing business, which will enable them to be on the right track to becoming successful entrepreneurs (Doriza & Maulida 2017). The results of this study can be used as in the creation of entrepreneurial education materials to enhance students' entrepreneurial attitudes.

Entrepreneurship education teaches students to face obstacles and take risks (Lyon 2008) in running a business. It can help them develop interpersonal, creative, innovative, organizational, time management, and leadership skills.

2 METHOD

This study distributed questionnaire surveys to undergraduate students during the 2017 academic year at Family Welfare Education Universitas Negeri Jakarta. A descriptive quantitative method was employed in this study. Data analysis was conducted to identify attitudes to entrepreneurship, and the scale of interest was measured using a scale of "very important (scale 4)," "important (scale 3)," "not important (scale 2)," and "very unimportant (scale 1)." Two hundred students who have already attended classes in entrepreneurship were selected as subjects for the study.

3 RESULTS AND DISCUSSION

The result of a validity test shows that only 10 out of 20 initial questions were valid to use. They asked whether students were adept at: (1) building an example of trust; (2) making a timely decision; (3) being flexible unless dealing with crucial matters; (4) controlling ego; (5) being persistent with work performance; (6) getting back on your feet immediately in response to encountering obstacles; (7) making a good plan; (8) being able to happily accept critiques; (9) being smart in utilizing available resources, and (10) embedding in yourself a fearlessness when it comes to dreaming big.

3.1 Building an example of trust

Out of 200 students, 134 students considered building an example of trust "very important," 64 considered it "important," and 2 considered it "not important" (Figure 1).

3.2 Making a timely decision

Out of 200 students, 118 considered making a timely decision "very important," 80 considered it "important," and 2 considered it "not important" (Figure 2).

3.3 Being flexible unless dealing with crucial matters

Out of 200 students, 107 considered being flexible "very important," the highest score achieved on the scale of interest for this statement (Figure 3).

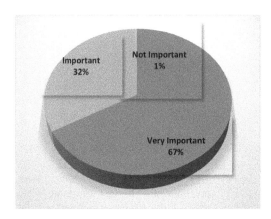

Figure 1. The scale of interest in building an example of trust..

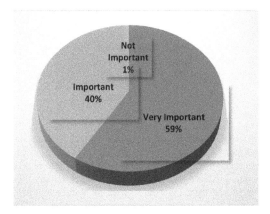

Figure 2. The scale of interest in making a timely decision.

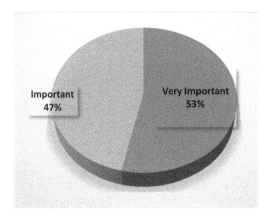

Figure 3. The scale of interest in being flexible unless dealing with crucial matters.

3.4 Controlling ego

Out of 200 students, 135 students considered controlling one's ego "very important," 62 students considered it "important," 2 students considered it "not important," and 1 student considered it "very unimportant" (Figure 4).

3.5 Being persistent with work performance

Out of 200 students, 117 students considered being persistent with work "very important" and 83 students considered it "important" (Figure 5).

3.6 Getting back on your feet immediately in response to encountering obstacles

Out of 200 students, 135 students considered getting back on one's feet immediately in response to encountering obstacles "very important" and 65 students considered it "important" (Figure 6).

3.7 Making a good plan

Out of 200 students, 134 students considered making a good plan "very important," 64 students considered it "important," and 2 students considered it "not important" (Figure 7).

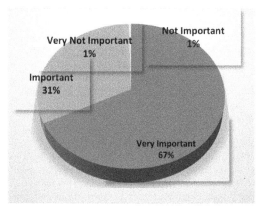

Figure 4. The scale of interest in controlling ego.

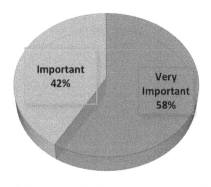

Figure 5. The scale of interest in being persistent with work performance.

918

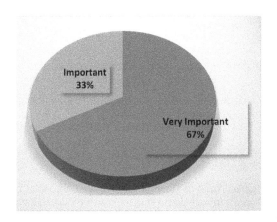

Figure 6. The scale of interest in getting back on your feet immediately in response to encountering obstacles.

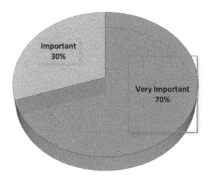

Figure 7. The scale of interest in making a good plan.

3.8 Being able to happily accept criticism

Out of 200 students, 76 students considered being able to happily accept criticism "very important," 122 students considered it "important," and 2 students considered it "not important" (Figure 8).

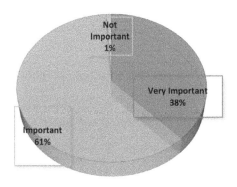

Figure 8. The scale of interest in being able to happily accept criticism.

3.9 Being smart about utilizing available resources

Out of 200 students, 107 students considered being smart about using available resources "very important," 64 students considered it "important," and 2 students considered it "not important" (Figure 9).

3.10 Embedding in yourself fearlessness toward dreaming big

Out of 200 students, 126 students considered embedding in oneself a fearlessness toward dreaming big "very important" and 74 students considered it "important" (Figure 10).

3.11 Building an example of trust

According to Kauffman (2015), "to build trust, we recommend that entrepreneurs be educated or educate themselves about the process." Building an example of trust is helps promote students' confidence. People will also have confidence in someone whom they can trust and will want to cooperate with them in a culture of integrity.

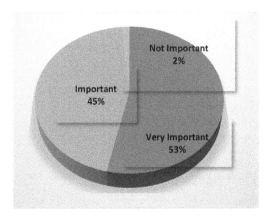

Figure 9. The scale of interest of being smart in utilizing available resources.

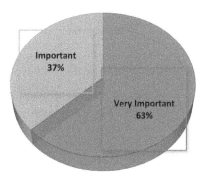

Figure 10. The scale of interest in embedding yourself fearlessness toward dreaming big.

3.12 Making a timely decision

Making a timely decision was considered "very important" by 59% of respondents, making it the most popular option on the scale of interest for this statement.

3.13 Being flexible unless dealing with crucial matters

Flexibility is one of the characteristics of a true entrepreneur (Paje 2016). In this study, the highest score achieved on the scale of interest for this statement was "very important." Flexibility enabling quick changes is one advantage a small business has over a big one.

3.14 Controlling ego

A positive attitude requires mental or emotional readiness to carry out right actions (Stamboulis & Barlas 2014). It is something we can learn from and refer to when reacting to a situation and deciding on what is important in life. A right attitude is often equated with maturity.

3.15 Being persistent with work performance

Persistence may help defeat competitors by encouraging them to work harder, especially when the product or service offered closely resembles those of competitors.

3.16 Getting back on your feet immediately in response to existing obstacles

A business may suffer from obstacles. However, one can learn from the situation and make plans to move on because the past cannot be changed.

3.17 Making a good plan

Good planning should be based on data that aims to improve business opportunities (Zimmerer et al. 2008). A good plan is required to help map a path past problems that are likely to be encountered and to set up the right steps to handle them.

3.18 Being able to happily accept criticism

An entrepreneur has to persistently convince his or her employees of the best way to run things, although there may be differences on opinion. Giving one's opinion or feedback and referring to the missions of the company should not be a matter of concern. This is supported by research by Paje (2016) that suggests changes in personal attitudes to boost respect.

3.19 Being smart about utilizing available resources

Finding an appropriate business to implement is difficult. Therefore, one should not hesitate to adapt any existing business models, as making use of appropriate business examples will be far easier and take less time than putting all systems into place from scratch.

3.20 Embedding in yourself a fearlessness when it comes to dreaming big

Over time, challenges will emerge when doing business, and having big dreams of success can help a businessperson remain strong in overcoming them.

4 CONCLUSION

Ranked from the highest to the lowest, the order of interest in attitudes to entrepreneurship is as follows: (1) making a good plan; (2) getting back on your feet immediately in response to existing obstacles; (3) controlling ego; (4) building an example of trust; (5) making a timely decision; (6) being persistent with work performance; (7) being smart about utilizing available resources; (8) embedding in yourself a fearlessness when it comes to dreaming big.; (9) being able to happily accept criticism; and (10) being flexible unless dealing with crucial matters.

ACKNOWLEDGEMENTS

The authors wish to thank: the Directorate General of Research and Development Strengthening at the Ministry of Research, Technology and Higher Education of the Republic of Indonesia, as well as theresearch Institution and community service of Universitas Negeri Jakarta, the Engineering Faculty of Universitas Negeri Jakarta, and the students of the Family Welfare Edocation Programme of Universitas Negeri Jakarta.

REFERENCES

Doriza, S. & Maulida, E. 2017. The way of entrepreneur subject. *Pandangan terhadap mata kuliah kewirausahaan.* Unpublished independent research report.

Kauffman, E. 2015. Building trust in entrepreneurial outsourced relationships. Retrieved https://www.entrepreneurship.org/articles/2005/06/building-trust-in-entrepreneurial-outsourced- relationships. Accessed May 2, 2017.

Lyon, E.M. 2008. Entrepreneurship education at a crossroads: Towards a more mature teaching field. *Journal of Enterprising Culture* 16(4): 325–337.

Oosterbeek, H., Van Praag, M. & Ijsselstein, A. 2010. The impact of entrepreneurship education on entrepreneurship skills and motivation. *European Economic Review* 54(3): 442–454.

Paje, M.J.J. 2016. Effectiveness of set in enhancing the entrepreneurial competencies of the students belonging to special population. *International Journal of Arts & Sciences* 9(3): 9–26.

Stamboulis, Y. & Barlas, A. 2014. Entrepreneurship education impact on student attitudes. *International Journal of Management Education* 12(3): 365–373.

Zimmerer, T.W., Scarborough, N.M. & Wilson D. 2008. *Essential of Entrepreneurship and Small Business Management-Kewirausahaan dan Manajemen Usaha Kecil (5th edition).* Jakarta: Penerbit Salemba Empat.

Advances in Business, Management and Entrepreneurship – Hurriyati et al (eds)
© 2020 Taylor & Francis Group, London, ISBN 978-0-367-27176-3

Impact of entrepreneurial skills and innovations on business success

R. Indriarti, H. Mulyadi & H. Hendrayati
Universitas Pendidikan Indonesia, Bandung, Indonesia

ABSTRACT: Small Medium Enterprises (SMEs) is an entrepreneurship activity that became one of Indonesia's economic development priorities because of its large contribution towards national income. The success of SMEs is expected to increase national income continuously. Therefore, identifying factors that affect business success is important. Two of these factors are entrepreneurial skills and innovations. The purpose of this study is to investigate the impact of entrepreneurial skills and innovation on business success in knitting industry. The population of this study are 225 knitting industry entrepreneurs, while the sample are 70 knitting industry entrepreneurs. The sample technique used is simple random sampling. The result shows that entrepreneurial skills and innovation have significant impact on business success.

1 INTRODUCTION

Entrepreneurship has grown as a research topic for a long time because it has positive implications for economic growth (Quadrini 2000). The city of Bandung, known as a fashion hotspot has a big enough opportunity to develop the clothing industry. Especially if the industry has a contribution towards the society and the surrounding area. There is a knitting industry that contributes to the society and the surrounding area in Bandung. The industry gave a contribution to the regional income either from the tax or sales side, and therefore the economy of the surrounding community is also lifted.

Unfortunately, since 2010 the industry has decreased productivity and profit. This is marked by a significant decrease in production and business income from year to year. Production decrease will result in a decrease in the number of sales, which also affects income decrease. The size of business income ultimately affects the size of profits earned by the business concerned. If a business succeeds in getting profit, then the business has achieved business bene-fits. Achieving business benefits suggests business success (Maxwell 2003, Lechner & Dowling 2003). A business is said to succeed when it gets profit be-cause profit is the goal of people doing business (Noor 2007). The problem cannot be allowed to drag on because it will have an impact on the business success of the knitting industry.

There are many factors that can improve and influence the business success. An analysis of the factors that influence business success found that business success is influenced by motivation, age, experience, training, education, commitment, finance, marketing, human resource management, business location, social relationships, business skills, managerial skills, family support, environment, project management, technical skill, intellectual, leadership, communication, and building business network (Uddin & Bose 2013, Gupta & Mirchandani 2018, Agbim 2013, Chatterjee & Das 2016, Pollack & Adler 2016, Alam et al. 2011). Beside these opinions and analyzes, there are other studies that tested the impact of a certain factor on business success, including entrepreneurial skills and innovations.

An entrepreneur needs to have entrepreneurial skills (Mamabolo et al. 2017, Nwarieji et al. 2017) because entrepreneurial skills are one of the significant contributors to achieving business success (Agbim 2013, Irawan & Mulyadi 2016, Unger et al. 2011, Pollack & Adler 2016, Chatterjee & Das 2016). Although some opinions assess entrepreneurial skills as ambiguous and complex (Morales & Marquina 2013, Chell 2013), research has proved that business success is the result of attitudes, traits, environments, and entrepreneurial skills (Shaw et al. 2010). Other studies have concluded that entrepreneurs with entrepreneurial skills contribute to the change and growth of their businesses (Dafna 2008).

Beside entrepreneurial skills, innovation becomes an important thing that entrepreneurs must have because entrepreneurial strength lies in innovation. Through innovation, competitive advantage and business sustainability can be guaranteed (Azar & Drogendijk 2014, Damanpour & Aravind 2012, Price et al. 2013). An enterprise's success will be aligned with the level of innovation made by the company (Leenders & Chandra 2013, Gunday et al. 2011, Duckworth 2014). In Small and Medium Enterprises (SMEs) it is more possible to adopt innovation sourced from the employees' knowledge and skills (Tan et al. 2009). A business that adopts innovation in enhancing its product, process, marketing and organizational advantages has a greater chance of overcoming uncertainty and changing competitive global business environments (Carpenter & Petersen 2002, Damanpour & Aravind 2012, Azar & Drogendijk 2014). The empirical studies proved that there is positive influence of innovation on the company success (Azar & Drogendijk 2014, Srivastava et al. 2017, Mamun et al. 2018, Ojo et al. 2017,

Gunday et al. 2011, Damanpour & Aravind 2012, Prajogo & Sohal 2006, Ko & Lu 2010).

Based on the description above, the author is interested in examining the problem with this research, titled "Impact of Entrepreneurial Skills and Innovations on Business Success (Survey on Entrepreneurs in Knitting Industry Bandung)". The research aims are: To know the description of entrepreneurial skills, innovations, and business success in the knitting industry; to find out the impact of entrepreneurial skills and innovation on business success in the knitting industry.

2 METHOD

This study used descriptive and verification methods. The descriptive method's aim in this study is to describe the entrepreneurial skills, innovations, and business success of the knitting industry. While the verification method is used to test the impact of entrepreneurial skills and innovations on business success in the knitting industry. The population of this study consist of all knitting industry entrepreneurs, 225 entrepreneurs were selected, while 70 knitting industry entrepreneurs were selected as samples in simple random sampling.

To collect data, questionnaires were used: Entrepreneurial skills questionnaire, consisting of three dimensions: technical skills, business management skills, personal entrepreneurial skills (Heflin 2011); Innovations questionnaire, consisting of three dimensions: product innovations, process innovations, marketing innovations (Manual 2005); and entrepreneurial success questionnaire, consisting of five indicators: profitability, productivity and efficiency, competitiveness, competence and business ethics, good image (Noor 2007).

3 RESULT AND DISCUSSION

3.1 Result

3.1.1 Variables description
Based on the questionnaire results for 16 items of questions about entrepreneurial skills variable, a score of 3958 or 70.68% was obtained. These results indicate that the entrepreneurial skills of entrepreneurs in the knitting industry are on the medium criteria.

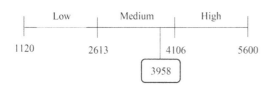

Figure 1. Entrepreneurial skills' criteria line.

Figure 2. Business success' criteria line.

Based on the questionnaire results for 12 items of questions about business success variable, a score of 2732 or 65.05% was obtained. These results indicate that business success in the knitting industry is on the medium criteria.

Overall, the score of response on entrepreneurial skills, innovations, and business success variables can be seen in Table 1.

3.1.2 Correlation analysis

(a) Product moment correlation
Based on Table 3, the value of the coefficient correlation between entrepreneurial skills with business success is 0.874 which means there is a significant relationship between entrepreneurial skills with business success. The value of 0.874 on the coefficient correlation table indicates a completely strong level of relationship.

Then, the value of the coefficient correlation between innovations with business success is 0.810 which means there is a significant relationship be-tween innovations with business success. The value of 0.810 on the coefficient correlation table indicates a completely strong level of relationship.

A positive correlation value indicates that the higher the level of entrepreneurial skills and innovations, then the higher the rate of business success will tend to be. Because the correlation value be-tween entrepreneurial skills and business success is higher than the correlation value between innovations and business success, then entrepreneurial skills variable has more influence on business success than innovations.

Table 1. Respondents response's recapitulation on entrepreneurial skills, Innovations, and business success.

Variables	Lowest Score	Highest Score	Score	Percentage
Entrepreneurial Skills	1120	5600	3958	70.68%
Innovations	700	3500	2535	72.43%
Business Success	840	4200	2732	65.05%

Source: Processed data, 2018.

922

Table 2. Correlations output.

Correlations

		Entrepreneurial Skills	Innovations	Business Success
Entrepreneurial Skills	Pearson Correlation Sig. N	1 70	.886** .000 70	.874** .000 70
Innovations	Pearson Correlation Sig. N	.886** .000 70	1 70	.810** .000 70
Business Success	Pearson Correlation Sig. N	.874** .000 70	.810** .000 70	1 70

Source: Processed data, SPSS 17.0, 2018.

Table 3. The output of impact of entrepreneurial skills and innovations on business success.

Model Summary[b]

Model	R	R Square	Adjusted R Square	Std. Error of the Estimate
1	.877[a]	.769	.762	2.137

a. Predictors: (Constant), Innovations, Entrepreneurial Skills
b. Dependent Variable: Business Success
Source: Processed data, SPSS 17.0 for Windows, 2018.

(b) Multiple correlation analysis
Based on Table 4 the value of the coefficient correlation between entrepreneurial skills and innovations in business success collectively is 0.877. This means that entrepreneurial skills and innovations have a completely strong correlation with business success. The value of R-Square 0.769 shows the business success in the knitting industry is in the high category.

Adjusted R Square or coefficient determination of 0.762 means that entrepreneurial skills in the knitting industry are influenced by entrepreneurial skills and innovations by 76.2%. Furthermore, 23.8% is impacted by factors that are not examined by the author.

(c) Multiple linear regression analysis
Based on Table 5 there are three regression coefficients: a = 5.915; b1 = 0.481; b2 = 0.164. Then multiple linear regression equations for entrepreneurial skills, innovations, and business success are as follows:

Table 4. Regression coefficients output.

Coefficients

Model		Unstandardized Coefficients		Standardized Coefficients	t	Sig.
		B	Std. Error	Beta		
1	(Constant)	5.915	2.235	.726	2.646	.000
	Entrepreneurial Skills	.481	.084	.166	5.730	.000
	Innovations	.164	.125		1.311	.194

a. Dependent Variable: Business Success
Source: Processed data, SPSS, 2018.

Table 5. Significant value of F test.

ANOVA[b]

Model	Sum of Squares	df	Mean Square	F	Sig.
1 Regression	1019.850	2	509.925	111.616	.000[a]
Residual	306.093	67	4.569		
Total	1325.943	69			

a. Predictors: (Constant), Innovations, Entrepreneurial Skills
b. Dependent Variable: Business Success
Source: Processed data, SPSS 17.0 for Windows, 2018.

$$Y = a + b_1X_1 + b_2X_2$$
$$Y = 5.915 + (0.481) X_1 + (0.164) X_2$$

Business Success = 5.915 + (0.481) Entrepreneurial Skills + (0.164) Innovations.

The regression coefficient of entrepreneurial skills variable is 0.481, it means that when the value of entrepreneurial skills is raised one, then the value of business success will increase by 0.481. While the regression coefficient of innovations variable is 0.164, which means if the value of innovations in-creased by one, then the value of business success will increase by 0.164.

(d) Hypothesis test
Hypothesis test used in this study is the F test and t-test. To test the impact of entrepreneurial skills and innovations on business success simultaneously, the F_{count} is done as follows:
Based on Table 6 the value of F_{count} is 111,616, while the F_{table} value at degrees of freedom (df) = n-2 = 68 with 70 respondents is 3.13. From the hypothesis test it is known that $F_{count} > F_{table}$ that is 111,616 > 3,13 which means H0 is rejected and H1

is accepted, it means there is a significant impact from entrepreneurial skills and innovations on business success.

Then to know the partial impact of entrepreneurial skills on business success, and innovations in business success, then the t-test is done as follows:

Based on the result value of T_{count} for entrepreneurial skills variable is 2.646 and T_{count} for innovations variable is 1,311, while T_{table} value at degrees of freedom (df) = n-2 = 68 with 70 respondents is 1.668. From hypothesis test, it is known that for entrepreneurial skills variable $T_{count} > T_{table}$ that is 2.646 > 1.668 then H_0 is rejected and H_1 is accepted, it means there is a significant impact of entrepreneurial skills on business success. Meanwhile, for the innovations variable $T_{count} < T_{table}$ that is 1.311 < 1.668 then H_0 is accepted and H1 is rejected, it means innovations has no significant impact on business success.

3.2 Discussions

The results of this study show that there is a significant impact on entrepreneurial skills and innovations on business success in the knitting industry simultaneously.

Meanwhile, partially the results showed that there is a significant impact of entrepreneurial skills on business success in the knitting industry. These results are in line with previous research that state entrepreneurial skills have a positive and significant impact on the business success (Agbim 2013, Irawan & Mulyadi 2016, Unger et al. 2011, Pollack & Adler 2016, Chatterjee & Das 2016). Therefore entrepreneurs' skill in the knitting industry is expected to develop their business in achieving business success, in accordance with previous research conducted by Dafna (2008) and Shaw et al (2010).

Table 6. Significant value of t-test.

Coefficients

Model	Unstandardized Coefficients		Standardized Coefficients		
	B	Std. Error	Beta	t	Sig.
(Constant)	5.915	2.235	.726	2.646	.000
Entrepreneurial Skills	.481	.084	.166	5.730	.000
Innovations	.164	.125		1.311	.194

a. Dependent Variable: Business Success
Source: Processed data, SPSS 17.0 for Windows, 2018.

The results also showed that there was no significant influence of innovations on business success in the knitting industry. This result contradicts previous studies that suggest that innovations have a positive and significant impact on business success (Azar & Drogendijk 2014, Srivastava et al. 2017, Mamun et al. 2018, Ojo et al. 2017, Gunday et al. 2011, Damanpour & Aravind 2012, Prajogo & Sohal 2006, Ko & Lu 2010). Yet the insignificance that occurs in the knitting industry does not mean that innovations are not important. However, the adoption of innovations is needed (Carpenter & Petersen 2002, Damanpour & Aravind 2012, Azar & Drogendijk 2014), because every change that occurs in the market must be balanced with innovation. If the knitting industry entrepreneurs do not innovate, then their market will be left behind. That is why innovation guarantees competitive advantage and business sustainability can be guaranteed (Azar & Drogendijk 2014, Damanpour & Aravind 2012, Price et al. 2013).

4 CONCLUSION

Based on the research results and discussions, it can be concluded as follow: entrepreneurial skills, innovations, and business success in the knitting industry are on medium criteria; simultaneously, entrepreneurial skills and innovations have a significant impact on business success in the knitting industry by 76.2%. Partially, entrepreneurial skills have a significant impact on business success in the knitting industry, while innovations have no significant impact on business success in the knitting industry.

REFERENCES

Agbim, K.C. 2013. The relative contribution of management skills to entrepreneurial success: A survey of small and medium enterprises (SMEs) in the trade sector. *Journal of Business and Management*, 7(1): 8–16.

Alam, S.S., Jani, M.F.M. & Omar, N.A. 2011. An Empirical Study of Success Factors of Women Entrepreneurs in Southern Region in Malaysia. *International Journal of Economics and Finance*, 3(2): 166–175.

Azar, G. & Drogendijk, R. 2014. Psychic Distance, Innovation, and Firm Performance. *Management International Review*, 54(5): 581–613.

Carpenter, R.E. & Petersen, B.C. 2002. Capital market imperfections, high-tech investment, and new equity financing. *Economic Journal*, 112(477): 54–72.

Chatterjee, N. & Das, N. 2016. A Study on the Impact of Key Entrepreneurial Skills on Business Success of Indian Micro-entrepreneurs: A Case of Jharkhand Region. *Global Business Review*, 17(1): 226–237.

Chell, E. 2013. Review of skill and the entrepreneurial process. *International Journal of Entrepreneurial Behavior & Research*, 19(1): 6–31.

Dafna, K. 2008. Managerial performance and business success. *Journal of Enterprising Communities: People and Places in the Global Economy*, 2(4): 300–331.

Damanpour, F. & Aravind, D. 2012. Managerial Innovation: Conceptions, Processes, and Antecedents. *Management and Organization Review*, 8(2): 423–454.

Duckworth, R. 2014. *Examining Relationships Between Perceived Characteristics of Innovation and Adoption Intentions of Small and Medium Enterprises*. Northcentral University.

Gunday, G., Ulusoy, G., Kilic, K. & Alpkan, L. 2011. Effects of innovation types on firm performance. *International Journal of production economics*, 133(2): 662–676.

Gupta, N. & Mirchandani, A. 2018. Investigating entrepreneurial success factors of women-owned SMEs in UAE. *Management Decision*, 56(1): 219–232.

Heflin, Z.F. 2011. *Be An Entrepreneur (Jadilah Seorang Wirausaha) Kajian Strategis Pengembangan Kewirausahaan*. Yogjakarta: Graha II.

Irawan, A. & Mulyadi, H. 2016. Pengaruh Keterampilan Wirausaha terhadap Keberhasilan Usaha (Studi Kasus pada Distro Anggota Kreatif Independent Clothing Kommunity di Kota Bandung). *Journal of Business Management and Entrepreneurship Education*, 1(1): 213–223.

Ko, H. & Lu, H. 2010. Measuring innovation competencies for integrated services in the communications industry. *Journal of Service Management*, 21(2): 162–190.

Lechner, C. & Dowling, M. 2003. Firm networks: External relationships as sources for the growth and competitiveness of entrepreneurial firms. *Entrepreneurship and Regional Development*, 15(1): 1–26.

Leenders, M.A.A.M. & Chandra, Y. 2013. Antecedents and consequences of green innovation in the wine industry: the role of the channel structure. *Technology Analysis & Strategic Management*, 25(2): 203–218.

Mamabolo, M.A., Kerrin, M. & Kele, T. 2017. Entrepreneurship management skills requirements in an emerging economy : A South African outlook. *The Southern African Journal of Entrepreneurship and Small Business Management*, 9(1): 1–10.

Mamun, A.A., Kumar, N., Ibrahim, M.D. & Yusoff, H. 2018. Establishing a valid instrument to measure entrepreneurial knowledge and skill. *Business Perspectives and Research*, 6(1): 13–26.

Manual, O. 2005. *Guidelines for Collecting and Interpreting Innovation Data* 3rd ed. Paris: OECD PUBLISHING.

Maxwell, T.P. 2003. Considering Spirituality: Integral Spirituality, *Deep Science, and Ecological Awarness*, 38 (2): 257–276.

Morales, C. & Marquina, P.S. 2013. Entrepreneurial skills, significant differences between Serbian and German entrepreneurs. *Journal of Centrum Cathedra: The Business and Economics Research Journal*, 6(1): 129–141.

Noor, H.F. 2007. *Ekonomi Manajerial*, Jakarta: Grafindo Persada.

Nwarieji, F.E., Obi, B.C. & Tochukwu, E. 2017. The study of entrepreneurial skills required by small scale poultry farmers in Orumba North and South Local Government Area of Anambra State, Nigeria. *African Journal of Agricultural Research*, 12(33): 2586–2597.

Ojo, O.D., Petrescu, M., Petrescu, A.G. & Bilcan, F.R. 2017. Impact of innovation on the entrepreneurial success: Evidence from Nigeria. *African Journal of Business Management*, 11(12): 261–265.

Pollack, J. & Adler, D. 2016. Skills that improve profitability: The relationship between project management, IT skills, and small to medium enterprise profitability. *International Journal of Project Management*, 34(5): 831–838.

Prajogo, D.I. & Sohal, A.S. 2006. The integration of TQM and technology/R&D management in determining quality and innovation performance. *Omega*, 34(3): 296–312.

Price, D.P., Stoica, M. & Boncella, R.J. 2013. The relationship between innovation, knowledge, and performance in family and non-family firms: an analysis of SMEs. *Journal of Innovation and Entrepreneurship*, 2(1): 14.

Quadrini, V. 2000. Entrepreneurship, Saving, and Social Mobility. *Review of Economic Dynamics*, 3(1): 1–40.

Shaw, E., Gordon, J., Harvey, C. & Henderson, K. 2010. Entrepreneurial Philanthropy: Theoretical Antecedents and Empirical Analysis of Economic, Social, Cultural and Symbolic Capital. *Frontiers of Entrepreneurship Research*, 30(7): 6.

Srivastava, S., Sultan, A. & Nasreen, C. 2017. Influence of innovation competence on firm level competitiveness: an exploratory study. *Asia Pacific Journal of Innovation and Entrepreneurship*, 11(1): 63–75.

Tan, K.S., Eze, U.C. & Chong, S.C. 2009. Factors influencing internet-based information and communication technologies adoption among Malaysian small and medium enterprises. International *Journal of Management and Enterprise Development*, 6(4), 397.

Uddin, R. & Bose, T.K. 2013. Motivation, Success Factors and Challenges of Entrepreneurs in Khulna City of Bangladesh. *European Journal of Business and Management*, 5(16): 148–156.

Unger, J. M., Rauch, A., Frese, M. & Rosenbusch, N. 2011. Human capital and entrepreneurial success: A meta-analytical review. *Journal of Business Venturing*, 26(3): 341–358.

Absorptive capacity as a strategy to improve business performance of automotive companies in Indonesia

S. Syahyono & L.A. Wibowo
Universitas Pendidikan Indonesia, Bandung, Indonesia

ABSTRACT: Indonesia has become an important pillar of the manufacturing sector since many of the world's famous car companies have established their factories there and increased their production capacity in Indonesia. Indonesia has also experienced an extraordinary transition because it has changed from a place for the production of imported cars in the South East Asia region to a large car sales market due to the increase in Gross Domestic Product (GDP). We use the definitions of absorption from Cohen and Levinthal and several other definitions from other researchers such as Zahra and George. This study combines quantitative and qualitative approaches, which involve complex research designs, usually with research stages that might be iterate. By combining these methods, this study aims to achieve a more comprehensive understanding of the implementation of absorption in business performance strategies, experiences and how organizations adopt, use and manage innovation to meet their strategic needs.

1 INTRODUCTION

In a more complex and competitive business environment, the company's business performance plays an important role in its survival (Indarti 2010). Absorptive capacities, which may contribute to business performance and innovation of companies, has recently gained more attention in developing competitive advantages. Over the years, researchers believed that companies with a high level of absorption capacity would be able to achieve competitive advantage and outperform their competitors. Absorption capacity of a company is closely related to business performance and subsequently affects the company's financial performance (Kostopoulos et al. 2011). In a sector where everything changes rapidly and intensively, like technology, prior knowledge is believed to be an important element which determines the level of the enterprise (absorptive capacity) according to many researchers. (Cohen & Levinthal 1990, Kim 1998).

In this research, the author examines absorptive capacity as a strategy to improve business performance in automotive companies. Indonesia has the second largest car manufacturing industry in Southeast Asia (Thailand controls about 50 percent of the car production in the ASEAN region) and due to its fast growth in recent years, Indonesia will increasingly threaten Thailand's dominant position in the next decade. However, to overtake Thailand as the largest car producer in the ASEAN region will require major efforts and breakthroughs. At present, Indonesia is highly dependent on foreign direct investment, especially from Japan, to establish car manufacturing facilities. Indonesia also needs to develop a car component industry that can support the car manufacturing industry. In the meantime, the total production capacity of cars assembled in Indonesia is approximately at two million units per year (Indonesia Investments 2018).

In 2017 the total capacity of cars installed in Indonesia was 2.2 million units. However, the capacity utilization was estimated to drop to 55 percent in 2017 because the expansion of the domestic car production capacity was not in line with the growth of domestic- and foreign demand for cars made in Indonesia. However, there are no big concerns about this situation because the domestic market demand for cars has plenty of room for growth in the next few decades, with Indonesia's car ownership per capita still at a very low level (Indonesia Investments 2018).

In terms of market size, Indonesia is the largest car market in Southeast Asia and the ASEAN region, controlling about one-third of the total annual car sales in ASEAN, followed by Thailand in second place. Not only does Indonesia have a large population (258 million people), but it is also characterized by having a rapidly growing middle class. Together, these two factors create a strong consumer power (Indonesia Investments 2018).

The Indonesian government also has high hopes for car exports in this country, since it can generate additional foreign exchange earnings, especially ahead of the implementation of the AEC that will transform the ASEAN region into a single market- and production area. MEA will open up opportunities for exporters to increase the regional trade. The cars made in Indonesia that have been exported include Toyota Avanza and Toyota Fortuner, Nissan Grand Livina, Honda Freed, Chevrolet Spin and

Suzuki APV. The most important export markets are Thailand, Saudi Arabia, the Philippines, Japan and Malaysia (Indonesia Investments 2018).

1.1 Research overview

This research deliberately draws and cuts with two prominent areas of research: Absorption Capacity and Business Performance of Automotive Companies in Indonesia. The research questions are:

a. What are the factors that determine business effectiveness to increase absorption capacity in automotive companies in Indonesia?
b. How do these determinants interact with each other in increasing the capacity of innovation and competitiveness of automotive companies in Indonesia?

1.2 Business performance strategy

According to Eckerson (2006), Business Performance Management (BPM) is a top-down approach that helps executives understand the processes needed to achieve strategic goals and then measure the effectiveness of these processes to achieve the results desired. Therefore, it can be concluded that BPM collects processes that help companies optimize their business performance to ensure the achievement of company objectives.

1.3 Absorptive capacity

Cohen and Levinthal (1990) argued that prior knowledge determines the level of absorption in a company. According to Kim (2001), previous knowledge bases refer to individual units that have knowledge available in the organization, hence, aggregation of individual knowledge. Large companies, having a larger number of employees, are therefore a more diverse and larger knowledge base. Theoretically, it is easier for large companies to obtain new knowledge from the external environment. Research conducted by O'Dwyer and O'Flynn (2005) showed that companies were willing to contribute more to the development of the absorptive capacity of the ability to acquire knowledge. Jane Zhao and Anand (2009) stated that "structural aspects and organizational culture are strong determinants of the organization's ability to adapt to other new practices, rather than the amount of prior knowledge and experience of individuals," which is contrary to the theory that large companies have better absorption because of abundant previous knowledge. In addition, large companies have a mature market position and technology, which reduces the need to obtain external knowledge (Barge-Gil 2010). Thus, there is less need to learn from the environment, in other words, absorptive capacity will be difficult for large companies to develop or maintain, since their needs decrease. The

different views above cannot conclude whether it is easy or difficult for large companies to develop their absorption capacity. However, it is certain that the absorptive capacity in large companies is different from small ones. Therefore, it is interesting to know whether the development of large companies will slow down or increase the development of absorptive capacity over a period of time. Based on the original concept of absorptive capacity (Cohen & Levinthal 1990), researchers have expanded the concept through a conceptualization process, which includes a four-dimensional model by Zahra and George (2002) and the absorptive model by Todorova and Durisin (2007). Both models argue on routine characteristics and absorptive processes, which leaves room for exploring deeper through observing antecedents and processes. At present, much of the existing literature focuses on the antecedents of absorptive capacity from both the resource-based view and the organizational perspective (Van den Bosch et al. 2003, Kamien & Zang 2000, Kinoshita 2000, Vega-Jurado et al. 2008), while lacking in studies from a process perspective in studying the development of absorptive capacity. Therefore, the uniqueness of this research lies in the lack of literature from longitudinal studies on the development of absorptive capacity from the perspective of a time-span process, let alone empirical case studies. This research will show how the absorption of a company develops over time and investigates more deeply to find the reasons that might cause differences in development. In addition, it is expected that companies can have a clear picture on the interrelation between their absorptive capacity and innovative performance over the years. This may help management teams in companies to better understand the company's capabilities and help to further develop their strategies by changing their absorption level over a certain time span to maintain their competitive advantage.

2 RESEARCH CONTRIBUTION

This research aims to contribute to the theory and practices of innovation and social networking. The contributions are:

1. For the management of automotive companies to be used as a reference and contribution of thought in enhancing the adaptation of the strengths of the external environment, corporate resources, strategic orientation, absorptive capacity, supply chain, organizational culture, leadership and improvement of current business performance.
2. For the government and regulators, to be an input for the improvement of the regulation and supervision of the automotive industry in Indonesia. For stakeholders, it is one of the sources of information that is useful for all parties in observing decisions that are in line with the dynamics of the development of the automotive industry in Indonesia.

3 APPROACH

Taking into account that the nature of this research is primarily explorative, both quantitative and qualitative methods can become inadequate on their own. Quantitative studies may be enough to map and analyze statistics, numbers and trends show the benefits, uses, problems and difficulties in implementing absorption in automotive companies. However, it cannot directly explain why the use of absorptive capacity and its implementation in certain businesses suffers failure. Qualitative research, on the other hand, can provide detailed views and perspectives on business performance and how absorbency manages innovation and social interaction, but it is very difficult to achieve generalizations and derive broader characteristics.

Therefore, this study combines quantitative and qualitative approaches (Gilbert 1992), which involve complex research designs, usually with research stages that might be iterate (Danemark et al. 2002), like in this research. By combining these methods, this study aims to achieve a more comprehensive understanding of the implementation of absorption in business performance strategies, experiences and how organizations adopt to use and manage innovation to meet their strategic needs. This research is also able to explain the process as both approaches prevent any missing picture.

REFERENCES

Barge-Gil, A. 2010. Open, semi-open and closed innovators: Towards an explanation of degree of openness. *Industry and Innovation* 17(6): 577–607.

Cohen, W.M. & Levinthal, D.A. 1990. Absorptive capacity: A new perspective on learning and innovation. *Administrative Science Quarterly*: 128–152.

Danemark, B., Ekstrom, M., Jakobsen, L. & Karlsson, J.C. 2002. Explaining society. *Critical Realism in The Social Sciences* 2.

Eckerson, W.W. 2006. Performance dashboards: Measuring. monitoring, and managing.

Gilbert, M. 1992. *On social facts*. Princeton, NJ: Princeton University Press.

Indarti, N. 2010. *The effect of knowledge stickiness and interaction on absorptive capacity*. Groningen: University of Groningen, The Netherlands.

Indonesia Investment. 2018. Automotive manufacturing industry Indonesia. *Retrieved*: https://www.indonesia-investments.com/business/industries-sectors/automotive-industry/item6047?searchstring=Automotive%20manufacturing%20industry%20Indonesia. Accsessed 5 June 2018.

Jane Zhao, Z. & Anand, J. 2009. A multilevel perspective on knowledge transfer: Evidence from the Chinese automotive industry. *Strategic Management Journal* 30(9): 959–983.

Kamien, M.I. & Zang, I. 2000. Meet me halfway: Research joint ventures and absorptive capacity. *International Journal of Industrial Organization* 18(7): 995–1012.

Kim, L. 1998. Crisis construction and organizational learning: Capability building in catching-up at Hyundai Motor. *Organization Science* 9(4): 506–521.

Kim, L. 2001. Absorptive capacity, co-opetition, and knowledge creation. *Knowledge Emergence: Social, Technical, Evolutionary Dimensions of Knowledge Creation*: 13–29.

Kinoshita, Y. 2000. R&D and technology spillovers via FDI: Innovation and absorptive capacity. *Working Paper* (349): 1–24.

Kostopoulos, K., Papalexandris, A., Papachroni, M. & Ioannou, G. 2011. Absorptive capacity, innovation, and financial performance. *Journal of Business Research* 64 (12): 1335–1343.

O'Dwyer, M. & O'Flynn, E. 2005. MNC–SME strategic alliances-A model framing knowledge value as the primary predictor of governance modal choice. *Journal of International Management* 11(3): 397–416.

Todorova, G. & Durisin, B. 2007. Absorptive capacity: Valuing a reconceptualization. *Academy of Management Review* 32(3): 774–786.

Van den Bosch, F., Van Wijk, R. & Volberda, H.W. 2003. Absorptive capacity: Antecedents, models and outcomes.

Vega-Jurado, J., Gutiérrez-Gracia, A. & Fernández-De-Lucio, I. 2008. Analyzing the determinants of firm's absorptive capacity: Beyond R&D. *R and D Management* 38(4): 392–405.

Zahra, S.A. & George, G. 2002. Absorptive capacity: A review, reconceptualization, and extension. *Academy of Management Review* 27(2): 185–203.

Advances in Business, Management and Entrepreneurship – Hurriyati et al (eds)
© *2020 Taylor & Francis Group, London, ISBN 978-0-367-27176-3*

Analysis of business model canvas to increase competitiveness

S.D. Pratiwi, H. Mulyadi & H. Hendrayati
Universitas Pendidikan Indonesia, Bandung, Indonesia

ABSTRACT: The purpose of this research is to describe the level of competitiveness of the small medium entreprise Boneka Bandung, and to describe the business model canvas used by Boneka Bandung as well as to ascertain the influence of business model canvas on competitiveness by analyzing what strategy can be used in the doll making business based on SWOT analysis. The type of research used is descriptive and verificative. Data were collected through interviews, observations and questionnaires. Data analysis is done through EFAS/IFAS matrix, SWOT matrix and strategy combination matrix and presentation technique. The population and sample in this research is the small medium entreprise Boneka Bandung and the research subject is the owner of Boneka Bandung. This study concludes that Boneka Bandung has a low level of competitiveness compared to competitors with similar products. The business model canvas illustrates that Boneka Bandung lacks physical re-sources in the form of labor and storage. Strategy analysis using business model canvas is very influential in improving the competitiveness and growth of Boneka Bandung by prioritizing the SO (strength-opportunity) strategy which has the highest score of 3.52.

1 INTRODUCTION

In the face of business competition, companies must be able to read their customers' desire because what the customers purchased are not merely products in their physical form, but the benefits offered by the products in fulfilling the customers' needs and wants (Utari & Sri 2015). Strategic decision making is one of the most important things in achieving the company's success and survival. The company's strategy establishes a comprehensive master plan that states how a corporation will achieve its missions and objectives, this maximizes competitive advantages and minimizes competitive losses (Papulova & Gazova 2016). Other than strategic decision making, companies also require the formulation of strategies, which is the process of preparing future steps intended to build the company's vision and mission, establish strategic and financial goals of the company, and design strategies to achieve these objectives in order to provide the best customer value (Amalia, Hidayat & Budiatmo 2012).

Strategy development plays an important role in im-proving the competitiveness of enterprises (Singh, Garg & Deshmukh 2009). To minimize losses or any other negative effects then the business require a strategy of competitiveness based on competitive advantages such as setting lower prices, attracting customers by offering better service than competi-tors and paying close attention to the quality of the goods (Malysheva et al. 2016). Increasing business competition, changes in consumer tastes, and economic changes bring about challenges and opportunities in business. Companies should be able to make the best choices about what customers need and how to meet those needs or demands at the lowest possible price. To overcome these problems, companies need a way to achieve competitive advantage (Ellitan 2008).

The problem faced by Boneka Bandung is decreasing growth in sales. This is due to the abundance of competitors who sell similar products of interest to customers.

The low level of competitiveness of small businesses in Indonesia is caused by many factors, including marketing, finance, management, technology, location, human resources and economic structure (Hadiati 2008).

Every company is required to have a business model to improve competitiveness. Competitiveness is a concept that refers to a company's ability to compete with other companies to create value (Nuvriasari 2015). A business model illustrates the rationale of how organizations create, deliver and capture value (Osterwalder 2010). The problem faced by small business in the doll making industry in Bandung is a declining sales growth rate. This is due to the many competitors who sell similar prod-ucts that are also in demand by consumers. The low level of competitiveness of small businesses in In-donesia is caused by many factors, including mar-keting, finance, management, technology, location, human resources and economic structure (Hadiati 2008).

1.1 Problem definition

Based on the phenomenon described above, the core problem in this research is to analyze the business model canvas to improve competitiveness. Thus the research questions for this study are as follows:

a What is the general level of competitiveness of the small medium entreprise Boneka Bandung

b What is the business model canvas Boneka Bandung.
c How can the business model canvas planning influence or improve the competitiveness Boneka Bandung.

1.2 Literatur review

1.2.1 The concept of entrepreneurship
States that entrepreneurship is the process of creating something new (new creations) and making something different from what currently exists (innovation), the goal is the achievement of individual welfare and add value for society. Simply stated that an entrepreneur is a person who has the courage to take risks to open a business using various opportunities (Kasmir 2006). According to (Simpeh 2011) there are disciplines in the study of entrepreneurship that can be classified into six categories of entrepreneurship theory: economic entrepreneurship theory, psychology entrepreneurship theory, sociological entrepreneurship theory, anthropological entrepreneurship theory, opportunity-based entrepreneurship theory and resource-based entrepreneurship theory.

1.2.2 SWOT analysis
The SWOT analysis aims to identify the strengths and weaknesses of an organization as well as potential opportunities and threats in the environment around it. After identifying the strategies used, these factors are developed that can build strengths, eliminate weaknesses, take advantage of opportunities or counter threats. Strengths and weaknesses are identified by assessing the organization's internal mechanisms while the threats and opportunities are identified by judging the organization's external surroundings. Internal assessments examine all aspects of the organization including e.g facilities, products and services, to identify organizational strengths and weaknesses. External assessment scans the political, economic, social, technological and competitive environment with a view to identify opportunities and threats (Dyson

2004). The SWOT matrix is a tool for constructing strategic factors of a company. Here is the SWOT matrix:

1.2.3 Analysis of the external enviroment
An external environment consists of opportunities and threats. Analysis of the company's external environment primarily aims to identify a number of opportunities and threats that reside in the company's. external environment. The external environment includes (Akdon 2009):

a Opportunity is the positive circumstances and external factors that help the organization achieve or surpass the achievement of its vision and mission.
b Challenges/Threats are negative external factors that can result in the organization failing to achieve its vision and mission.

To refine the strategies regarding the organization's opportunities and threats they are transcribed into the EFAS matrix in the table below. The EFAS matrix is structured to formulate external strategic factors within the enterprise's opportunities and threats framework. The results of EFAS will show the greatest opportunities and threats facing the company.

1.2.4 Analysis of the internal enviroment
The internal environment includes the strengths and weaknesses of the company. Identifying strengths and weaknesses is essential in studying the general environment of an organization. Strength is an organization's potential that can be used to achieve its objectives. The strength of the company is hidden and unrevealed. An organization's weak-ness can cause failure in maximizing opportunities and neutralizing threats (Asmarantaka 2013). The Internal Environment includes (Akdon 2009):

a Strengths are positive internal situations and abilities that allow organizations to fulfill strategic advantage in achieving their vision and mission.
b Internal Weaknesses are situations and factors outside the organization that are negative and hampers the organization in achieving or exceeding the achievement of its vision and mission.

To refine strengths and weaknesses strategy it is mapped out to the IFAS matrix in the table below. IFAS matrices are structured to formulate internal strategic factors within the framework of the company's greatest strengths and weaknesses.

1.2.5 Business model
The business model summarizes the company's logic, how it works and how it creates value for stakeholders. The business model is essentially a formula that captures important elements of the organization's chosen strategy with the goal of providing corporate profits (an added value of the organization) within a specific strategic context. Business model canvas

Table 1. The SWOT matrix.

IFAS EFAS	Strengt (S)	Weakness (W)
Opportunity (O)	Strategy SO (Create a strategy here that uses the power to take advantage of opportunities)	Strategy WO (Create a strategy here that takes advantage of opportunities by overcoming weaknesses)
Threat (T)	Strategy ST (Generate strategies that use force to avoid thearts)	Strategy WT (Generate strategies that minimize weakness and avoid thearts)

(Wheelen & Hunger 2012).

931

serves as a relationship between creating value for customers while capturing value for companies using corporate behavior (Lynch 2013). A business model illustrates the rationale of how organizations create, deliver and capture value (Osterwalder 2010).

This canvas divides the business model into nine main components, which are then separated again into nine components, ie customer segments, value propotitions, channels, customer relationships, revenue streams, key resources, key activities, key partnerships and cost structure.

1.2.6 *Competitiveness*

Competitiveness can be defined as the ability to maintain market share. This ability is largely determined by factors such as timely supply and competitive pricing. In the long run, timely supply and competitive pricing are influenced by two other important factors, namely flexibility (ability to adapt to customer desires) and product differentiation management (Rahmana 2009). Competitiveness is a measure of competitive ability that is formulated as a company's ability to generate profits in a sustainable manner. There are various indicators that can be used to measure competitiveness: productivity, output growth, market share, rapid adaptation and marketing effectiveness (Joewono & Handito 2006).

2 METHOD

2.1 *Research type*

The type of research is done in Boneka Bandung using descriptive and verificative research. The purposes are to determine the general picture and business condition of Boneka Bandung. Surely this will make it easier to find strategies that need to be executed to improve competitiveness.

2.2 *Sampling technique*

The sampling technique used in this research is the total sample technique. The population and sample of this research are the small medium entreprise Boneka Bandung.

2.3 *Types and data collection techniques*

The type of data used in this study are primary and secondary data. Data were collected through a structured interview using previously prepared questions and conducted in the production house of doll in Bandung.

2.4 *Data analysis technique*

Data analysis in this research were done through the following stages:

a Data collection phase
  At this stage data and information related to internal and external factors affecting the company were collected.
b Analysis phase
  Each internal and external factors are transcribed into EFAS and IFAS matrices by assigning weight and rating to get the subtotal of each strengths, weaknesses, opportunities and threats. Then each factor is applied into the SWOT matrix.
c The approproate strategic decision-making phase
  This stage reviews the strategies that have been formulated in the analysis phase, then the decision about strategies that felt most appropriate, effective, efficient and profitable for the company was taken. Then it is mapped out into a business model canvas to capture the current state of the company.

3 RESULTS AND DISCUSSION

3.1 *Overview of boneka Bandung*

Boneka Bandung is a company engaged in the manufacture of fabric-based dolls. It manufactures dolls of different shapes, sizes, and materials. Boneka Bandung is currently under the auspices of CV. Grow Training Organizer and ready to serve customers professionally. The dolls produced are the result of the diligence and expertise of housewives who live around the operational site of the company. The manpower involved in the production process of Boneka Bandung consists of 3 doll making tailors; 1 doll accessories maker (shirt, hat, scarf); 1 doll stuffer; personnel who stitches puppet eyes on the dolls, and 5 people working for the finishing division.

3.2 *SWOT analysis of Boneka Bandung*

3.2.1 *Analysis of internal factors*

3.2.1.1 Strengths

a Boneka Bandung uses internal human resources, ie housewives living around the company's operational site who have sewing skills to produce dolls, doll accessories, doll stuffings, doll eyes, and do the finishing work.
b Boneka Bandung offers affordable and competitive prices.
c Boneka Bandung promotes its products in several "market places" to boost sales.
d Boneka Bandung has a website that is easy to be navigated.
e The graduation ceremony dolls produced by Boneka Bandung have been sold in all universities accross Indonesia.

3.2.1.2 Weaknesses

a Boneka Bandung does not yet have a business license.
b The vision and mission of the company are not perfect.
c Does not use technology in the production of its dolls.
d The informations available on its competitors' websites are more comprehensive.
e Management is still conducted independently.

3.2.2 *Analysis of external factors*

3.2.2.1 Opportunities

a The rapid development of technology can be utilized by Boneka Bandung either in the production process or in marketing its products.
b There is high demand from customers located outside of Bandung.
c Business opportunities in Indonesia are widely available.
d Partnering with customers to increase opportunities to innovate.

3.2.2.2 Thearts

a Competitors are also offering competitive prices.
b Some competitors have superior rates.
c The emergence of new competitors in the doll industry.
d Social and cultural factors that can change people's buying interest.

3.2.3 *Matrix of internal factor analysis summary (IFAS) and external factor analysis summary (EFAS) of Boneka Bandung*

The first step in compiling the IFAS and EFAS matrices is to determine the factors that are the strengths, weaknesses, opportunities and threats for Boneka Bandung. Each factor is given a weight ranging from 0.0 (very unimportant) to 1.0 (very important), in which the total weight of each matrix is not more than 1.0. Then each factor that has been given weight is rated (in a scale of 4) from 1 (below average) to 4 (very good). The ratings of strengths and weaknesses are always contradictory, so are the opportunity ratings and threats.

Based on the IFAS matrix, the subtotal score of strengths is 1.74, while the subtotal score of weaknesses is 0.67, so the total IFAS score is 2.41.

Based on the EFAS matrix, the subtotal score of opportunities is 1.78, while the subtotal score of threats is 0.77, so the total EFAS score is 2.55.

Based on the Strategies Combination Matrix of SWOT/TOWS, there are some strategies that needed to be applied by Boneka Bandung, the strategies consist of ST, SO, WT, and WO strategies as described below:

Table 2. The IFAS matrix.

Internal Factor

Strengths	Weight	Rating	Score
Using internal human resources	0.09	3	0.27
Offers affordable prices	0.11	4	0.44
Condusting promotion in market place	0.13	4	0.52
The graduation doll is absorbed by the entire university	0.07	3	0.21
Have a website that is easy on navigation	0.10	3	0.30
Sub Total	0.50		1.74

Weaknesses	Weight	Rating	Score
Doesn't have a business license yet	0.10	1	0.10
Vision and mission is not perfect (not according to its component)	0.14	1	0.14
Not using technology in doll production process	0.08	2	0.16
Information on competitor's website is more complete	0.09	1	0.09
Management is managed independently	0.09	2	0.18
Sub Total	0.50		0.67
Total IFAS	1.00		2.41

Table 3. The EFAS matrix.

Internal Factor

Opportunities	Weight	Rating	Score
The rapid development of technology for production and marketing process	0.15	4	0.60
Request doll outside the city of Bandung	0.12	3	0.36
Business opportunities in Indonesia wide open	0.10	3	0.30

Opportunities	Weight	Rating	Score
Partner with customers to enlarge innovation opportunities	0.13	4	0.52
Sub Total	0.50		1.78

Thearts	Weight	Rating	Score
Competitors offer equally competitive prices	0.15	2	0.30
Competitors have superior rates	0.13	1	0.13
The emergence of new competitors in the doll industry	0.12	2	0.24
Social culture that can change people's buying interest	0.10	1	0.10
Sub Total	0.50		0.77
Total EFAS	1.00		2.55

Table 4. Matrix combination of SWOT strategy.

IFAS EFAS	Strength (S) Using internal resources Offers affordable prices Conducting promotion in "market place" Has a website that is easy on navigation Graduation puppets are absorbed by all universities in Indonesia	Weakness (W) Does not have a business license yet The vision and mission of the company is not perfect Not involving technology in the production process of dolls Information on competitor's website is more complete Management is still managed independently
Opportunities (O) The rapid development of technology The demand for dolls outside Bandung Business opportunities in Indonesia wide open Partner with customers to enlarge innovation opportunities	**Strategies SO:** Add promotions through market place and website Partner with customers to enlarge innovation opportunities Perform market expansion/expand distribution network	**Strategies WO:** Add promotions through market place and website Apply for a business license and create a vision of the company's mission
Threat (T) Competitors alike offer competitive prices Some competitors have a superior rate The emergence of new competitors in the doll industry Social culture that can change people's buying interest	**Strategies ST:** Make continuous improvements, maintain quality, and keep prices competitive Complete important information on the company website	**Strategies WT:** Complete important information on the company website

a. SO (Strength-Opportunity):
   (S3, S5, O1, O2, O3) The widely available business opportunities in Indonesia and the rapid development of technology can absorb the demand of dolls from outside of Bandung through market places and the company website. (S5, O4) Partnering with customers to expand innovation opportunities (e.g listening to customers as a company value or receiving criticisms and suggestions) either through direct interaction/ through the company's website.
   (S4, O2, O3) The widely available business opportunities in Indonesia, the demand for dolls from outside of Bandung and the high sales rate of well known graduation doll products can facilitate Boneka Bandung to expand its market/distribution network.

b. ST (Strength-Threat):
   (S2, T1, T3) Make continuous improvements, maintain quality, and keep prices competitive to face new competitors in the doll industry and to cope with socio-cultural changes that can change people's buying interest.(S5, T2) Important informations should be available on the company website (e.g payment information)

c. WO (Weakness-Opportunity):
   (W3, O1, O2, O3) Utilize the rapid development of technology to absorb a wider market.(W1, W2, O3) Apply for a business license and create vision and mission statements to make the achievement of company goals more focused.

d. WT (Weakness-Threat):
   (W4, T2 = S5, T2) Complete important informations on the company website (e.g payment information). This strategy is the same strategy as the ST strategy that is a combination of strategies from S5 and T2.

Table 5. Matrix planning combination of quantitative strategies.

IFAS EFAS	Strengths (S)	Weaknesses (W)
Opportunities (O)	Strategy SO: 1.74+1.78= 3.52	Strategy WO: 067+1.78= 2.45
Threat (T)	Strategy ST: 1.74+0.77=2.51	Strategy WT: 0.67+0.77=1.44

The quantitative strategies combination planning matrix shows that Boneka Bandung needs to prioritize the SO strategy that has the highest total score (3.52). Then followed by ST strategy (2.51); WO (2.45); and WT (1.44). To reiterate, these strategies are: increasing promotional activities through marketplaces and website; partnering with customers to increase innovation opportunities (e.g listening to customers as corporate value or receiving criticisms and suggestions) either through direct interactions or through social media and the corporate website; expanding market/expanding distribution network; making continuous improvements, maintaining quality, and keeping prices competitive; completing important information on the company's website; applying for a business license and creating vision and mission statements for the company.

3.2.4 *The business model canvas of Boneka Bandung*

In this article strategy canvassing is done to provide a snapshot of the position of Boneka Bandung products based on nine components, namely:

3.2.4.1 Customer segments

The Customer Segments Boneka Bandung is divided into groups that have certain behaviors criteria, namely adults aged 23 to 35 years both female and male, residing not only in the city of Bandung, but also all over Indonesia. For now, after knowing what customers need, Boneka Bandung is increasing the production of teddy bears (without graduation ceremony attributes) to meet the demand of the market.

3.2.4.2 Value propositions

The Value Propositions offered by Boneka Bandung as the right reasons for customers to choose their products are: make quality products from good raw materials, fast response, competitive prices, and refundable products.

3.2.4.3 Channel

Boneka Bandung does not have stores, distributors, and or other parties who can help communicate and convey Value Propositions to the Customers other than through several resellers and direct marketing.

3.2.4.4 Customer relationship

How doll in Boneka Bandung maintain a good relationship with its customers through interactions either on the company website or through social media.

3.2.4.5 Revenue stream

Currently doll in Boneka Bandung is beginning to increase revenue growth by selling raw materials for dolls.

3.2.4.6 Key resources

The most important resources Boneka Bandung needs currently are physical resources in the form of warehouses, human resources in the form of 11 more workers, and venture capital.

3.2.4.7 Key activties

Activities that Boneka Bandung needs to do for its business to run smoothly include making patterns, sewing dolls, finishing, and marketing their products.

3.2.4.8 Key partnership

Boneka Bandung's current business associates are resellers and suppliers of raw materials. Currently doll in Bandung the company does not cooperate with outside investors in terms of capital, 100% of its business capital comes from internal sources.

3.2.4.9 Cost structure

Costs incurred by boneka Bandung curently are the cost of raw materials, salaries for workers, and promotional fees.

3.3 *Discussions*

From the competitiveness indicator it can be seen that Boneka Bandung has a low level of productivity and does not adapt quickly to changes in the environment/market. So competitors are superior in terms of creating and selling products. Highly competitive entrepreneurs are characterized by a tendency to increase production volume, an ever increasing domestic market or export market share, a tendency to serve not only the local but also the national market and for the export market not only serving one country but multiple countries (Puryono 2017). Competitiveness can be created or enhanced by applying the right competitive strategies, one of them is by applying effective and efficient resource management. In addition, the process of determining the right strategy should be tailored to all activities of the company's function, so that the company will perform as expected or even exceeding expectations and can create value (Nuvriasari 2015).

In the business model canvas indicator (1) customer segment, the target market of doll in Bandung is everyone, especially students throughout Indonesia. The customer segment describes a group of people or

Table 6. Business model canvas of boneka Bandung.

Key Partnership	Key Activities	Value Propotition	Customer Relatinship	Customer
Reseller	Making doll pattcns	Selecting of quality raw	Website and social media	segment
Supplier	Sewing dolls	materials	Channels	Man and
	Attributing installs	Maintaining the quality of the	Reseller and direct	woman
	Finishing	doll	marketing	All the people
	Key Resources	Competitive price		Indonesia
	Increase the number of	Goods can be returned		
	workers	Quick response to customers		
	Expand the repository			
Cost structure			Revenue Streams	
Staff salaries,			Raw material dolls	
raw material				
costs and adver-				
tising costs				

organizations that the company wants to reach or serve, (2) value proposition, Boneka Bandung has provided quantitative values such as affordable prices and speed in serving/responding to customer and qualitative values such as the doll designs and customers' experience with the products so that consumers are interested in the products, (3) channels, in this indicator Boneka Bandung only have several resellers and direct marketing, this indicator describes how the company communicates with and reaches out to customers to provide value, (4) customer relationship, Boneka Bandung maintains a good relationship with customers through interactions either through the company website or social media. The customer relationship indicator when applied in the business model of a company greatly affects the overall customer experience, (5) revenue streams, Boneka Bandung is increasing income growth by selling raw materials needed to make dolls, (6) key resources, Boneka Bandung needs to increase physical resources by expanding storage warehouses and increasing the number of workers, these resources enable the company to create and offer values, reach the market, maintain relationships with customers and earn revenue, (7) key activities, activities or product manufacturing process undertaken by doll in Bandung such as choosing raw materials of high quality, making doll patterns, sewing dolls, attribute assembling and finishing. These activities are related to the design, manufacture and delivery of products in large quantities and of superior quality (8) key partnership, currently Boneka Bandung only cooperates with reseller and suppliers of raw materials using personal capital. Companies form partnerships with a variety of reasons and partnerships become the cornerstone of various business models. Companies create alliances to optimize their business model, reduce risks or gain resources, (9) Cost structure, current costs are raw material costs, employee salaries and advertising costs. By looking at these indiactors it is clear that creating and delivering value, maintaining customer relationships and generating revenue, leads to costs (Osterwalder 2010). Business model canvas serves as the link between creating value for customers and capturing value for companies using corporate actions (Lynch 2013), as well as helping to connect companies with external business environments such as customers, competitors and the general public (Ahokangas 2014).

4 CONCLUSION

a. The competitiveness of Boneka Bandung is relatively low compared to competitors with the same product.
b. The business model canvas shows that Boneka Bandung lacks physical resources in the form of labor and storage space.

c. Results obtained from the business model canvas analysis can improve competitiveness, as can be seen from the result of the SO strategy combination matrix which has highest total matrix score (3.25).

4.1 Suggestions

From the results of the analysis and our conclusions we put forward these suggestions for Boneka Bandung:

a. After analyzing the competitiveness variable it can be seen that Boneka Bandung has a lower level of competitiveness compared to competitors with the same product. Therefore it is suggested that Boneka Bandung should create more variations for its products. Some that have been planned are food characters, bouquets, etc.
b. From the business model canvas variable it can be seen that the problem is on the key resources components, namely those are the shortage of labor and storage space. It is advisable for Boneka Bandung to increase the number of manpower by recruiting skilled workers in the field of doll pattern making, sewing and doll designing and by expanding its storage warehouses by renting rooms or by building their own warehouses.
c. Analysis of business model canvas can improve the competitiveness of Boneka Bandung by prioritizing SO strategies (strenght-opportunity), by increasing promotions through marketplaces and the company website, partnering with customers to incrase innovation opportunities, doing expansions (market expansions), making continuous improvements, maintaining the quality of raw materials, and keeping prices competitive and by completing important informations on the company website.

REFERENCES

Ahokangas, P. 2014. The practice of creating and transforming a business model. *Journal of Business Models* 2 (1): 6–18.
Akdon, A. 2009. Strategi *management for education management (manajemen strategik untuk manajemen pendidikan). (Riduwan, Ed.) (Ketiga).* Bandung: Alfabeta.
Amalia, A., Hidayat, W. & Budiatmo, A. 2012. Analisis strategi pengembangan usaha pada UKM batik semarangan di Kota Semarang. *Ilmu Administrasi Bisnis* 1–12.
Asmarantaka, A. dkk. 2013. Analisis strategi pengembangan usaha bandrek Lampung pada unit usaha THP herbalist. *Jurnal Ilmu-Ilmu Agribisnis* 1(3): 210–217.
Dyson, R.G. 2004. Strategic development and SWOT analysis at the University of Warwick. *European Journal of Operational Research* 152(3): 631–640.
Ellitan, L. 2008. *Manajemen strategi operasi: Teori dan riset di Indonesia (cetakan pe).* Bandung: Alfabeta.
Hadiati, S. 2008. Perilaku wirausaha industri keramik berskala kecil untuk meningkatkan daya saing produk di Malang. *Jurnal Manajemen dan Kewirausahaan* 10: 115-123.

Joewono, J. & Handito, H. 2006. *7n1 strategy toward global competitiveness, seri manajemen, pustaka bisnis indonesia dan arrbey.*

Kasmir, K. 2006. *Kewirausahaan (Edisi Satu).*

Lynch, R. 2013. *Strategic management.* Always learning person.

Malysheva, T.V., Shinkevich, A.I., Kharisova, G.M., Nuretdinova, Y.V., Khasyanov, O.R., Nu-retdinov, I.G. & Kudryavtseva, S.S. 2016. The sustainable development of competitive enterprises through the implementation of innovative development strategy. *International Journal of Economics and Financial Issues* 6(1): 185–191.

Nuvriasari, A. 2015. Model strategi peningkatan daya saing ukm industri kreatif berbasis orientasi pasar dan orientasi kewirausahaan: 138–154.

Osterwalder, O. 2010. *Business model generation.*

Papulova, Z. & Gazova, A. 2016. Role of strategic analysis in strategic decision-making. *Procedia Economics and Finance* 39(2015): 571–579.

Puryono, D.A. 2017. Penerapan model green supply chain management untuk meningkatkan daya saing UMKM batik bakaran 9(3).

Rahmana, A. 2009. Peranan teknologi informasi dalam peningkatan daya. Seminar nasional aplikasi teknologi informasi 2009 (SMATI 2009). *2009 (Snati)*: B11–B15.

Simpeh, K.N. 2011. Entrepreneurship theories and empirical research : A summary review of the literature. *European Journal of Business Management* 3(6): 1–9.

Singh, R.K., Garg, S.K. & Deshmukh, S.G. 2009. The competitiveness of SMEs in a globalized economy: Observations from China and India. *Management Research Review* 33(1): 54–65.

Utari E. & Sri. 2015. Analisis strategi keunggulan bersaing dalam rangka memenangkan persaingan bisnis perbankan (studi pada bank nagari cabang utama kota Pekanbaru). *Jom FISIP* (2): 1–15.

Wheelen, T.L. & Hunger, J.D. 2012. *Strategic management and business policy.*

937

Advances in Business, Management and Entrepreneurship – Hurriyati et al (eds)
© 2020 Taylor & Francis Group, London, ISBN 978-0-367-27176-3

The influence of entrepreneurial attitude toward entrepreneurship intention of female students

H. Mulyadi, R.D.H. Utama, Y.M. Hidayat & W. Rahayu
Universitas Pendidikan Indonesia, Bandung, Indonesia

ABSTRACT: The purpose of this study is to describe the entrepreneurial attitude and intention and find out the influence of autonomy and authority, the economics opportunity, self-realization, perceived confidence, and entrepreneurial attitude on entrepreneurship intention. The type of research is descriptive and verification research. The method used was an explanatory survey with simple random sampling technique. The number in the sample was 145 students at Universitas Pendidikan Indonesia. The data analysis technique is a path analysis with computer software SPSS 23.0 for Windows. The outcome of this study indicated that the description of entrepreneurial attitude was in the medium category, the description of entrepreneurship intentions was in the medium category. Entrepreneurial attitude variables that had the highest influence on entrepreneurship intentions was the dimensions of autonomy and authority, while the lowest influence was the dimensions of perceived confidence, and entrepreneurship intention was affected by entrepreneurial attitude. Based on these results, it is suggested that increasing perceived confidence can increase the entrepreneurship intention. So hopefully, university can maintain the entrepreneurial attitude in order to improve students' entrepreneurship intentions.

1 INTRODUCTION

Entrepreneurship intentions are still an issue discussed in international research and entrepreneurial forums, one of the research published in December 2016 by Matthew Mayhew who states if research on entrepreneurship intentions has improved over the past two decades (www.entrepreneur.com). Introduced by Ajzen (1991) in Theory of Planned Behavior, the entrepreneurial intentions will form a productive life habits that attract other experts to develop research on the birth of new entrepreneurs (Krueger & Carsrud 1993).

Some research on entrepreneurship intentions has been largely done in Small and Medium Enterprises (Ali et al. 2011, Anggadwita & Dhewanto 2016), secondary education (Mulyadi et al. 2016a) and higher education or university (Hussain 2015, Khuong & An 2016, Farani et al. 2017). The findings of the development of entrepreneurship intentions at the university level in several countries are: (1) the personal characteristics of the individual can affect the intention to entrepreneurship and entrepreneurship programs must be included in the policy of economic reform of the country of Vietnam (Khuong & An 2016) and (2) the development of new digital-based business and an understanding of entrepreneurship related to the intention to entrepreneurship students in Iran as a career field in the future (Farani et al. 2017).

Various attempts were made to foster entrepreneurship intentions, especially changing the minds of youths who had only intended as job seekers after completing their school or college in order to become job creators.

Entrepreneurship learning is the process of shaping the mindset, attitudes, and entrepreneurial behaviors to realize a real effort as a driver of career choice to become an entrepreneur in the future (Sumanjaya et al. 2016). College teaching model is generally based on theory or research course. It makes the efforts to cultivate the intention of entrepreneurship in college runs slow (Love et al. 2013).

The results of research in Indonesia related to the intention of entrepreneurship in higher education or university found that, understanding the potential and introduction of the field of entrepreneurship as a student career can assist in the establishment of entrepreneurial intentions (Kadiyono 2014).

To understand the attitude of entrepreneurship and entrepreneurship intentions owned by UPI Bandung students as a benchmark, pre-research was conducted through the questionnaires to 30 female students from each faculty at UPI Kampus Bumi Siliwangi who have taken and graduated from entrepreneurship courses. The results of pre-research showed that entrepreneurial intentions of these students were still low.

The problem-solving approach of entrepreneurship intentions used in this study through Theory of Planned Behavior (TPB). The theory states that the intention of entrepreneurship is influenced by attitude toward behavior, subjective norm, and perceived behavioral control (Ajzen 1991). Some other factors based on the research, which may affect entrepreneurial intent include: entrepreneur education (Shah & Ali 2013, Pratiwi & Wardana 2016, Farani et al. 2017), personality traits (Cantner et al. 2016, Niranjan et al. 2016), adversity quotient (Mulyadi et al. 2016b), personal attitude and social perception (Anggadwita & Dhewanto, 2016), entrepreneurial skill and family occupation (Farooq et al. 2016), entrepreneurial attitude (Shah & Ali, 2013, Botsaris & Vamvaka 2014, Mcnally et al. 2016), demographic factors, entrepreneurial personality (Chaudhary 2017), social networking, emotional intelligence and psychosocial characteristic (Javed et al. 2016), self-efficacy (Tsai et al. 2014), and entrepreneurial knowledge and personal attitudes (Tshikovhi et al. 2015).

The principle of the intention model is determined by the attitude, the individual who views the entrepreneurial attitude profitable will increase the intention to carry out entrepreneurial activities (Botsaris & Vamvaka 2014). Research that has been done on entrepreneurial intentions among Indonesian students shows the same problem of individual characteristics, namely entrepreneurial attitude factors that are influenced by interest in a career in the field of entrepreneurship or business that still needs to be improved (Sutanto 2014, Santoso & Oetomo 2016).

The research problems that can be formulated are: (1) How should the entrepreneurship attitude and entrepreneurship intention of student of class of 2014 at UPI Bandung be described? and (2) How does the entrepreneurship attitude influence the entrepreneurship intent of the student class of 2014 at UPI Bandung?

2 METHOD

The method used in this research is explanatory survey. The type of research was descriptive and verificative. The population in this study was the totality of female students of 2014 in UPI Bandung about 2832 people from several courses or majors in 7 faculties that provide courses in entrepreneurship. The formula used to take a sample from a population was from Rasyid (1994).

$$n = \frac{n_0}{1 + \frac{n_0}{N}} \quad (1)$$

while n_0 can be searched using the following formula:

$$n_0 = \left[\frac{Z\left(1 - \frac{\alpha}{2}\right)S}{\delta} \right] \quad (2)$$

where:
N = Population;
n = Sample;
n_0 = Number of samples taken from all units;

S = Standard deviation for the variables studied in the population with using Deming's Empirical Rule:
δ = Bound of errors that can be tolerated by 5%

So the sample of this study set 145 female students. The sampling technique was probability technique with Simple Random Sampling.

3 RESULT AND DISCUSSION

The results of the variables of entrepreneurial attitude are shown in Table 1.

The entrepreneurial attitude variable can be known based on the score from the data from Table 1 where the values are compared with the standard score criteria. It is obtained through the calculation of the ideal score (citerium) and the smallest score, so that through the standard score, it can be known area continuum that shows the ideal region of entrepreneurial attitude (Sugiyono 2014).

Obtaining scores based on the results of data processing on the dimensions of entrepreneurial attitude is 8752 or 63.5%, the scores on a continuum can be described as shown in Figure 1.

Based on Figure 1, continuum value of entrepreneurship attitude variable that is located at the value of 8752 or 63.5% can be said that most respondents said the entrepreneurial attitude variable is in the medium category. However, it still needs to be improved again in the aspect of self-realization dimension, especially on the indicator of the ability to work hard to reach the job goals that still have the lowest score.

Based on Table 1, on the attitude dimension that got the highest score or percentage is on autonomy and authority dimension with score 2689 or 61.8%. While the dimensions that get the lowest score is self-realization with a score of 1918 or 66.1%. Where the overall entrepreneurial attitude variable scores 8752, if presented to the ideal score then obtained a percentage of 63.5%. It can be said that most of the respondents expressed entrepreneurship attitude student of class 2014 in UPI Bandung is in medium category (Narimawati 2008). This study supports previous research conducted by Gurbuz Aykol (2008) and Suharti (2011) on the dimensions of entrepreneurship attitude where autonomy and authority have high perception.

Variable of entrepreneurship intentions consisting of desire, planning, and action factors can be identified based on the scores obtained from the data

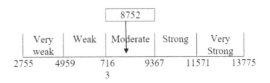

Figure 1. Continuum line of entrepreneurship attitude.

Table 1. Recapitulation of respondent response scores about attitude of entrepreneurship.

No	Statement	Scores	Ideal Scores	Average (%)
Entrepreneurial Attitude				
Autonomy & Authority				
1	The power to make work decisions	454	725	62.5
2	The ability to choose your own work	426	725	58.8
3	The ability to become a leader for themselves and others	449	725	61.9
4	The ability to choose a job that has freedom	409	725	56.4
5	Influence on others	393	725	54.2
6	Relationships with others	558	725	77
Total Scores		2689	4350	61.8
Economics Opportunity				
7	Compensation is based on achievements	472	725	65.1
8	Innovation for business success	397	725	54.8
9	Opportunities found to do business	535	725	73.8
10	Challenge of the selected job	408	725	56.3
11	The motivation of doing business	390	725	53.8
Total Scores		2202	3625	60.7
Self-Realization				
12	Creativity to meet the needs of others	474	725	65.4
13	Ability to follow the rules work task from beginning to end	551	725	76
14	Ability to work hard to achieve work goals	440	725	60.7
15	Involvement in the work process	453	725	62.4
Total Scores		1918	2900	66.1
Perceived Confidence				
16	Responsibility for the selected decision	490	725	67.6
17	Confidence to succeed by running your own business idea	480	725	66.2
18	Ability possessed for success as an entrepreneur	522	725	72
19	Ability to accept insecurity from work to be performed	451	725	62.2
Total Scores		1943	2900	67
Score Total Attitudes Of Entrepreneurship		8752	13775	63.5

recapitulation. Recapitulation of dimensions of the entrepreneurship intention variable can be seen in Table 2.

Through the standard score can be seen continuum area that shows the ideal region of entrepreneurial intentions (Sugiyono 2014)

Obtaining scores based on the results of data processing on the variable of entrepreneurship intention is 8052 or 65.3%, the scores on a continuum can be described as seen in Figure 2.

Based on Figure 2, values continuum variables intent entrepreneurship is located at the value of 8752 or 63.5% can be said that most respondents said the entrepreneurship intention variable is in the medium category. However, it still needs to be improved especially on the act dimension with the indicator on

the statement of interest in following the business community to get the lowest scored business experience.

Based on Table 2, dimensions of entrepreneurship intentions that get the highest score or percentage of desire or desire to start a business with the acquisition score of 2783 or 64% (2783/4350 x 100%), while the dimensions that get the lowest score is the act or the act of starting a business with score 2500 or 69% (2500/3625 x 100%). Where the overall entrepreneurship intention variable scores 8052, if presented to the ideal score then obtained a percentage of 65.3% (8052/12325 x 100%). It can be said that most of the respondents stated the intention of entrepreneurship student class of 2014 in UPI Ban-dung is in medium category (Narimawati 2008).

Table 2. Recapitulation of respondent response scores about intention of entrepreneurship.

No	Statement	Scores	Ideal Scores	Average (%)
Entrepreneurial Attitude				
Autonomy & Authority				
1	The power to make work decisions	454	725	62.5
2	The ability to choose your own work	426	725	58.8
3	The ability to become a leader for themselves and others	449	725	61.9
4	The ability to choose a job that has freedom	409	725	56.4
5	Influence on others	393	725	54.2
6	Relationships with others	558	725	77
Total Scores		2689	4350	61.8
Economics Opportunity				
7	Compensation is based on achievements	472	725	65.1
8	Innovation for business success	397	725	54.8
9	Opportunities found to do business	535	725	73.8
10	Challenge of the selected job	408	725	56.3
11	The motivation of doing business	390	725	53.8
Total Scores		2202	3625	60.7
Self-Realization				
12	Creativity to meet the needs of others	474	725	65.4
13	Ability to follow the rules work task from beginning to end	551	725	76
14	Ability to work hard to achieve work goals	440	725	60.7
15	Involvement in the work process	453	725	62.4
Total Scores		1918	2900	66.1
Perceived Confidence				
16	Responsibility for the selected decision	490	725	67.6
17	Confidence to succeed by running your own business idea	480	725	66.2
18	Ability possessed for success as an entrepreneur	522	725	72
19	Ability to accept insecurity from work to be performed	451	725	62.2
Total Scores		1943	2900	67
Score Total Attitudes Of Entrepreneurship		8752	13775	63.5

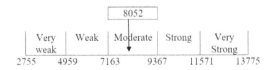

Figure 2. Continuum line of entrepreneurship intention.

4 CONCLUSION

The description of entrepreneurship attitude can be seen from the dimensions consisting of autonomy and authority, economics opportunity, self-realization, and perceived confidence are in the medium category. The perceived confidence dimension has the highest perception appraisal while the economics opportunity dimension has the lowest overall perception appraisal. The description of entrepreneurship intentions can be seen from the dimensions of the desire, plan, and act. The act dimension or the act of starting a business has the highest perceptual judgment while the plan dimension or start-up plan has the lowest overall perception appraisal. Entrepreneurship attitude positively and significantly influence entrepreneurship students of class of 2014 at UPI Bandung. This shows the positive effect that the stronger entrepreneurial attitude possessed by individuals, the higher the intention of the individual to decide the career as an entrepreneur.

REFERENCES

Ali, A., Topping, K.J. & Tariq, R.H. 2011. Entrepreneurial attitudes among potential entrepreneurs. *Pakistan Journal of Commerce and Social Sciences* (PJCSS) 5(1): 12–46.

Anggadwita, G. & Dhewanto, W. 2016. The influence of personal attitude and social perception on women entrepreneurial intentions in micro and small enterprises in Indonesia. *International Journal of Entrepreneurship and Small Business* 27(2–3): 131–148.

Ajzen, I. 1991. The theory of planned behavior. *Organizational Behavior and Human Decision Processes* 50:179–211.

Botsaris, C. & Vamvaka, V. 2014. Attitude toward entrepreneurship: Structure, prediction from behavioral beliefs, and relation to entrepreneurial intention. *Journal Knowledge Economy* 1(11): 1–28.

Cantner, U., Goethner, M. & Silbereisen, R.K. 2016. Schumpeter's entrepreneur—A rare case. *Journal of Evolutionary Economics* 1(1): 1–28.

Chaudhary, R. 2017. Demographic factors, personality and entrepreneurial inclination a study among Indian University students. *Education & Training*, 59(2), 171–187.

Farooq, M.S., Jaafar, N., Ayupp, K., Salam, M., Mughal, Y. H., Azam, F. & Sajid, A. 2016. Impact of entrepreneurial skill and family occupation on entrepreneurial intentions. *Journal of Entrepreneurship* 28(3): 3145–3148.

Gurbuz, G. & Aykol, S. 2008. Entrepreneurial intentions of young educated public in Turkey. *Journal of Global Strategic Management* 2(2): 47–56.

Hussain, A. 2015. Impact of entrepreneurial education on entrepreneurial intentions of Pakistani students. *Entrepreneurship and Business Innovation* 2(1): 43–53.

Kadiyono, A. L., Psikologi, F. & Padjadjaran, U. 2014. Efektivitas pengembangan potensi diri dan orientasi wirausaha dalam meningkatkan sikap wirausaha effectiveness of self-development and entrepreneurial orientation in improving entrepreneurial attitude. *Jurnal Intervensi Psikologi* 6(1): 25–38.

Krueger, N.F. & Carsrud, A.L. 1993. Entrepreneurial intentions: applying the theory of planned behaviour. *Entrepreneurship & Regional Development* 5(4): 315–330.

Javed, F. (2016). Role of social networks, emotional intelligence and psychosocial characteristics in developing entrepreneurial intentions of students. *Journal of Business and Management* 3(1): 54–81.

Mcnally, J.J., Martin, B.C., Honig, B. & Bergmann, H. 2016. Toward rigor and parsimony: A primary validation of Kolvereid's (1996) entrepreneurial attitudes scales. *Entrepreneurship & Regional Development* 1(4): 1–23.

Mulyadi, H., Tarmedi, E. & Buhari, R.Q. 2016a. Adversity quotient effect of achievement and its impact on student entrepreneurship intentions. *Journal Business and Management Research* 15(1): 912–914.

Mulyadi, H., Tarmedi, E. & Ruslandi, G. 2016b. Analysis of factors influencing the student's interests to participate in entrepreneurial student program. *Economics, Business, and Management Research* 15: 694–699.

Narimawati, U. 2008. *Metodologi Penelitian Kualitatif dan Kuantitatif, Teori dan Aplikasi*. Bandung: Agung Media.

Khuong, M.N. & An, N.H. 2016. The factors affecting entrepreneurial intention of the students of Vietnam national university—a mediation analysis of perception toward entrepreneurship. *Journal of Economics, Business and Management* 4(2), 104–111.

Niranjan, S. & Krishnakumare, B. 2016. Personality traits and entrepreneurial intention among management and horticultural students of a public university—comparative analysis. *Journal of Scientific Research* 5(10): 330–331.

Pratiwi, Y. & Wardana, I.M. 2016. Pengaruh Faktor Internal dan Eksternal terhadap Niat Berwirausaha Mahasiswa Fakultas Ekonomi dan Bisnis Universitas Udayana. *Jurnal Manajemen* 5(8): 5215–5242.

Rasyid, H. Al. 1994. *Teknik Penarikan Sampel dan Penyusunan Skala*. Bandung: Universitas Padjajaran.

Santoso, S. & Oetomo, B.S.D. 2016. Pengaruh karateristik psikologis, sikap berwirausaha, dan norma subyektif terhadap niat berwirausaha. *Jurnal Psikologi Kewirausahaan* 20(03): 338–352.

Shah, N. & Ali, B. 2013. Investigating attitudes and intentions among potential entrepreneurs of a developing country: A conceptual approach. *Journal of Business*: 217–220.

Sugiyono. 2014. *Metode Penelitian Manajemen*. Bandung: Alfabeta.

Suharti, L. (2011). Faktor-faktor yang berpengaruh terhadap niat kewirausahaan (entrepreneurial intention) (studi terhadap mahasiswa Universitas Kristen Satya Wacana, Salatiga). *Jurnal Manajemen Dan Kewirausahaan Dan Kewirausahaan* 13(2): 124–134.

Sumanjaya, W., Widajanti, E. & Lamidi. 2016. Pengaruh karakteristik kewirausahaan terhadap niat berwirausaha pada mahasiswa fakultas ekonomi unisri dengan motivasi berwirausaha sebagai variabel moderasi. *Jurnal Ekonomi Dan Kewirausahaan* 16(3): 433–441.

Sutanto, E. M. (2014). The study of entrepreneurial characteristics with achievement motivation and attitude as the antecedent variables. *Journal of Arts, Science & Commerce* 5(4): 125–134.

Tsai, K., Chang, H. & Peng, C. 2014. Extending the link between entrepreneurial self-efficacy and intention: A moderated mediation model. *International Entrepreneurial Management Journal* 1(1): 119.

Farani, A., Karimi, S. & Motaghed, M. 201). The role of entrepreneurial knowledge as a competence in shaping Iranian students' career intentions to start a new digital business. *European Journal of Training and Development* 41(1): 83–100.

Tshikovhi, N. & Shambare, R. 2015. Entrepreneurial knowledge, personal attitudes, and entrepreneurship intentions among South African Enactus students. *Problems and Perspectives in Management* 13(1): 152–158.

Love, B., Hodge, A., Grandgenett, N. & Swift, A.W. 2014. Student learning and perceptions in a flipped linear algebra course. *International Journal of Mathematical Education in Science and Technology* 45(3): 317–324.

Advances in Business, Management and Entrepreneurship – Hurriyati et al (eds)
© 2020 Taylor & Francis Group, London, ISBN 978-0-367-27176-3

The influence of attitude, subjective norm, self-efficacy, family environment towards entrepreneurial intention, mediated by entrepreneurship education, on students of faculty of economy of Ngurah Rai University

N.D.A. Amrita, I.M. Kartika & P.G.D. Herlambang
Universitas Ngurah Rai, Denpasar, Indonesia

ABSTRACT: According to the Indonesian Central Bureau of Statistics, the unemployment rate in 2014 was 7.24 million people and around 30 to 40% were young people, of which 10% hold a Bachelor's degree. Indonesia has a serious problem of high unemployment especially among those holding an undergraduate degree. Creating independent employment is a new breakthrough to deal with unemployment. To support the development of entrepreneurship, in 1995 the government launched the National Movement on Entrepreneurship Socialization and Development at the university level. This research was focused on observing factors that support and encourage students' interest in entrepreneurship based on their attitude, subjective norms and perceived behavioral control. This study used quantitative research methodology with multiple linear regression data analysis. Data were collected by means of observation, interview, documentary and questionnaire. The respondents of the research were the sixth-semester students of Ngurah Rai University, who had passed the class on entrepreneurship.

1 INTRODUCTION

The unemployment issue is a problem for every country. The unemployment rate has been increasing in the past decades. According to BPS (Central Bureau of Statistics) data, the unemployment rate of Indonesia in 2014 was 7.24 million, 30 to 40%of which are of young age unemployed and 10% of which are under graduated. Indonesia has been heavily dealing with this young age and under graduated unemployment issue.

No nation is prosperous and respected by other nations without economic progress. Economic progress will be achieved if the society has a strong entrepreneurial intention. According to McClelland (1987), one of the factors that support the growth of a country is when the number of entrepreneurs in the country amounts to at least 2% of the population. Indonesia urgently needs to increase the number of new entrepreneurs.

Growing entrepreneurship is a collective undertaking of the society, the government, the private sector and universities. Examining the relationship between entrepreneurial education and entrepreneurial action and compare the results of the various approaches to teaching entrepreneurship education (Heuer 2014). Creation of college graduates who become entrepreneurs is not necessarily easy to implement. Based on empirical evidence in the field, there is a tendency for college graduates prefer to seek comfort, security and reliability in a short time. This can be seen by the number of civil service (PNS) applicants. The limited absorption of university graduates in the government sector causes

attention to turn to opportunities for employment in the private sector.

Given the importance of entrepreneurship education for the community, especially for students, the Directorate General of Higher Education (DIKTI) as the institution that oversees the university level of education implements entrepreneurship courses that must be followed by students of all faculty of study. This has been in effect since 2017 (Murdjianto 2006).

Attitudes and interests toward entrepreneurship are influenced by various aspects of career choice as an entrepreneur. Entrepreneurial intention has a major role in start-up businesses (Heuer 2014).

Demographic factors such as gender, parental education background and work experience, may influence career choice to become entrepreneurs (Karimi et al. 2014). According to Sels et al. (2009), one's tendency to do or not to do something like choosing entrepreneurship as a career option, can be predicted by the Theory of Planned Behaviour (TPB). TPB uses three pillars as antecedents of the intention, that is, attitudes toward behaviour, subjective norms and self-efficacy.

Theory of Planned Behaviour (TPB) is used to explain the intention of someone who then explains the person's behaviour. This theory was originally called Theory of Reasoned Action (TRA), developed in 1967, subsequently revised and expanded by Fishbein and Ajzen (2005), who suggested that attitudes toward this behaviour are determined by beliefs about the consequences of a behaviour. According to Baron and Byrne (2003), Subjective Norms are the individual's perception of the expectations of influential people in his life regarding the conduct or omission of certain behaviours. Self-efficacy is at the core of the theory of Social Cognitive proposed by Bandura (1997), which emphasizes the role of observational learning, social experience and mutual determinism in the development of personality.

In addition to education in the school and community environments, the family, especially parents, play an important role as a direction for the future of their children. Thus, indirectly parents can also affect interest of the working area for children in the future including entrepreneurship (Soemanto 2008). Gunarsa (1991) states that the family environment is the first environment that gives the child an in-depth influence.

According to Soemanto (2006), entrepreneurship education helps Indonesian in learning process so that they have the personal power of dynamic and creative to run its business in accordance with the personality of the Indonesian nation based on Pancasila.

Kasmir (2011) mentions that there are three stages to change the mindset both mentally and motivation in entrepreneurship. The first stage is to establish a school with entrepreneurship insight in order to create an entrepreneurship mindset for students. Secondly, in entrepreneurship education it is necessary to emphasize the courage needed to start entrepreneurship. The third stage is to emphasize more that with entrepreneurship the future is definitely in our own hands, not in the hands of others, so that the motivation grows stronger.

According to Fishbein and Ajzen (2005), "intentions reflect the individual's desire to try to establish behaviour, consisting of three determinants: attitudes toward behaviour, subjective norms, and conscious behaviour control." Intention is not only controlled by the intention of someone to do or not to do a behaviour but also is influenced by everything that motivates him.

Entrepreneurship education can shape the mindset, attitudes and behaviors of students into a true entrepreneur, thus, directing them to choose entrepreneurship as a career choice. But the influence needs to be studied further whether the existence of entrepreneurship courses can encourage entrepreneurship interest for students. Therefore, research needs to be done to identify the factors that encourage entrepreneurship interest in students considering the importance of entrepreneurship for economic and social welfare. This research is to find out whether the entrepreneurship education that has been implemented in the Faculty of Economics of Ngurah Rai University can significantly generate interest in entrepreneurship of the students. Based on the background, the research problem is formulated as follows: Can entrepreneurship education affect students' entrepreneurial intention?

Hypotheses

Based on the review of the literature, we expect the following hypotheses to hold:

H1: Attitudes have a positive and significant influence on entrepreneurship education.

H2: The subjective norm has a positive and significant influence on entrepreneurship education.

H3: Self-efficacy has a positive and significant influence on entrepreneurship education.

H4: The family environment has a positive and significant impact on entrepreneurship education.

H5: Entrepreneurship education has a positive and significant influence on entrepreneurship interest.

2 METHOD

2.1 Population and sample

The population in this study were students who took Advanced Entrepreneurship courses in the even semester of academic year 2017/2018 as many as 238 students while the sample was calculated using Slovin formula. Hence, the number of sample needed was 70 students of the sixth semester.

2.2 Research sites

This research was conducted at the Faculty of Economics Ngurah Rai University of Denpasar. This location was selected due to an indication of low interest in entrepreneurship among the students.

2.3 Research instruments

The research instrument consists of four independent variables, one mediation variable and one dependent variable. Instrument Attitudes consist of 5 questions, Subjective Norm consists of 5 questions, Self-efficacy consists of 5 questions, Family Environment consists of 6 questions, Entrepreneurship Education consists of 3 questions and entrepreneurial intention consists of 5 questions. The six instruments are in the form of a checklist using a 5-point Likert scale.

2.4 Data analysis technique

The validity and reliability of the research instruments will be analyzed. The validity test is conducted using validity test product moment correlation test. The reliability test is done using Cronbach's alpha coefficient formula. The analysis technique used in this research is path analysis by using regression. Statistic test F will identify whether all independent variables have the influence of each independent variable to the dependent variable. The coefficient of determination test is used to find out how far the ability of the model in explaining the variation of the dependent variable. Path analysis is used to explain the direct and indirect effects of a set of variables as the cause of other variables that are the result variable.

3 RESULT AND DISCUSSION

3.1 Characteristics of respondents

The analysis revealed that the respondents were male (50%) the same as female respondents (50%). Most of the respondents (95.7%) were between 20 to 29 years old, age 30 to 39 years were 2.9%, and age 40 to 49 years were 1.4%.

3.2 Validity test

The result of validity test shows all components in this research variable is valid because its significance value is less than 0.05.

3.3 Reliability test

Reliability test results show that all variables in this study have Cronbach's alpha value of greater than 0.6, so it can be said that all indicators of each variable is reliable (Ghozali 2011).

3.4 Multiple regression analysis

The objective of this analysis is to figure out relationship between independent variables and dependent variable, either positive or negative and to predict the value of the dependent variable in response to the increase or decrease of the independent variable (Sugiyono 2016).

Multiple linear regression analysis model 1 is used to determine the effect of attitude variables, subjective norms, self-efficacy and family environment on entrepreneurship education. To simplify the calculation, the authors used SPSS (Statistical Program Social Science) version 21. The calculation results show the regression equation as follows:

$$Y = 5.792 + 0.242 \, X1 + 0.098 \, X2 \\ - 0.012 \, X3 + 0.032 \, X4 \tag{1}$$

where:
Y = Entrepreneurship Education,
X1 = Attitude,
X2 = Subjective Norm,
X3 = Self Efficacy,
X4 = Family Environment.

Multiple linear regression analysis model 2:

$$Y = 4.719 + 0.175 \, X1 + 0.294 \, X2 \\ + 0.007 \, X3 + 0.050 \, X4 + 0.448 \, X5 \tag{2}$$

where:
Y = Entrepreneurship Education,
X1 = Attitude,
X2 = Subjective Norm,
X3 = Self Efficacy,
X4 = Family Environment,
X5 = Entrepreneurial Intention.

From result of analysis by using SPSS above obtained that significance value F = 0,000 < 0,05 this mean H0 rejected, and Hi accepted meaning simultaneously have significant influence between Attitude, Subjective Norm, Self-efficacy, Family Environment and Entrepreneurship Education to Entrepreneurial Intention, then the hypothesis proposed that the variable Attitudes, Subjective Norms, Self-efficacy, Family Environment and Education Entrepreneurship positive and significant impact on Entrepreneurial Intention is acceptable.

3.5 T-test

The t-test is performed to test the significance of each regression coefficient, that is, attitudes, subjective norms, self-hypnosis, and family environment to entrepreneurship education.

From Table 1, it can be concluded that attitude (X1) has a positive and significant influence on entrepreneurship education variable because t count > t table, while the subjective norm variable (X2) has positive but not significant effect. Self-efficacy (X3) has negative and insignificant influence and family environment (X4) has no significant positive effect.

3.6 F-test

F-test is performed to understand the impact of the variables all together towards entrepreneurship education.

The result of this F-test using SPSS shows that significance value F = 0.001 < 0.05. This means that H0 is rejected, and H1 is accepted. It shows that simultaneously there are significant influence of attitude, subjective norm, self-efficacy and family environment to entrepreneurship education. Hence, the hypothesis proposed that attitudes, subjective norms, self-efficacy and family environment have a positive and significant influence on entrepreneurship education is accepted.

Table 1. T-test results.

Model	t_{count}	t_{table}	Sig
Attitude (X_1)	2.406	1.66691	0.019
Subjective Norm (X_2)	1.132	1.66691	0.262
Self-efficacy (X_3)	- 0.044	1.66691	0.710
Family Environment (X_4)	0.085	1.66691	0.460

Table 2. F-test results.

Model	Sum of Squares	df	Mean Square	F	Sig.
Regression	42.784	4	10.696	5.035	.001b
Residual	138.088	65	2.124		
Total	180.871	69			

Table 3. Coefficient of the determination result.

Model	R	R Square	Adjusted R Square	Std. Error of the Estimate
1.	.486[a]	0.237	0.190	1.45754
2.	.588[a]	0.345	0.294	2.40831

3.7 Coefficient of determination test

This analysis is used as a tool to understand the contribution of the independent variables towards the dependent variable, which is indicated in a percentage.

In Table 3, the determination model 1 (R2 = 0.237) shows that the attitude, subjective norm, self-efficacy and family environment have 23.7% effect towards entrepreneurship education. The rest of 76.3% is affected by other factors outside this research. The determination model 2 (R2=0,345) shows that the attitude, subjective norm, self-efficacy, family environment, entrepreneurship education all have a 34.5% effect towards entrepreneurial intention.

3.8 Path analysis

Table 4. Correlation between independent variables.

Variable	Attitude	Subjective Norm	Self Efficacy	Family Environment
Attitude	1.000	0.643	0.311	0.217
Subject Norm	0.643	1.000	0.357	0.331
Self-Efficacy	0.311	0.357	1.000	0.148
Family Environment	0.217	0.331	0.148	1.000

From the R square, we can calculate the other variable path coefficients outside the model, namely:
Entrepreneurship Education

$$py1\varepsilon = \sqrt{(1 - 0.237)} = \sqrt{0.763} = 0.8734 \quad (3)$$

Entrepreneurial Intention

$$py2\varepsilon = \sqrt{(1 - 0.345)} = \sqrt{0.655} = 0.8093 \quad (4)$$

From the graph above it can be explained that:
- The correlation coefficient of attitude toward subjective norm is 0,643.
- The coefficient of attitude correlation to self-efficacy is 0.311.
- Coefficient of attitude correlation to family environment is 0.217.
- The correlation coefficient of subjective norm in self-efficacy is 0.357.
- The correlation coefficient of subjective norm in family environment is 0.331.
- Coefficient of self-efficacy correlation to family environment that is equal to 0.148.
- The coefficient of attitude beta on entrepreneurship education is 0.343.
- The beta coefficient of subjective norm in entrepreneurship education is 0.169.
- Self efficacy beta coefficient for entrepreneurship education is equal to - 0.044.
- The coefficient of family environment beta for entrepreneurship education is 0.085.
- The beta coefficient of attitudes towards entrepreneurial intention is 0.140.
- The beta coefficient of subjective norms on entrepreneurial intention is 0.286.
- The beta self-efficacy coefficient of entrepreneurial intention is 0.015.

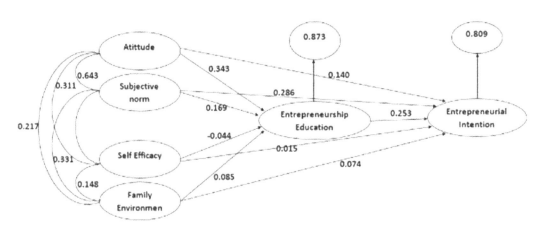

Figure 1. Path coefficient.

Table 5. Coefficient of correlation.

Variable	→ EE	EE→EI	→ EI
Attitude	0.343	0.343 + 0.253 = 0.596	0.140
Subjective Norm	0.169	0.169 + 0.253 = 0.422	0.286
Self-Efficacy	-0.044	-0.044 + 0.253 = 0.209	0.015
Family Environment	0.085	0.085 + 0.253 = 0.338	0.074

- The coefficient of beta family environment to entrepreneurship intention is 0.074.
- The beta coefficient of entrepreneurship education on entrepreneurial intention is 0.253.
- For values e1 and e2, respectively, 0.873 and 0.809.

4 CONCLUSION

From the data and hypothesis analysis using SPSS as an analytical tool to test the five research hypotheses, it can be concluded that:

1. The result of hypothesis 1 testing shows that attitude have positive and significant effect to entrepreneurship education. Hence, hypothesis 1 is accepted.
2. The result of hypothesis 2 testing shows that subjective norms have positive but not significant effect on entrepreneurship education. Hence, hypothesis 2 is rejected.
3. The result of on hypothesis 3 testing shows that self-efficacy has no significant negative effect on entrepreneurship education. Hence, hypothesis 3 is rejected.
4. The result of hypothesis 4 testing shows that family environment has a not significant positive effect on entrepreneurship education. Hence, hypothesis 4 is rejected.
5. The result of hypothesis 5 testing shows that entrepreneurship education has a positive and significant impact on entrepreneurship interest. Hence, hypothesis 5 is accepted.

ACKNOWLEDGEMENT

We offer thanks to the students of the Faculty of Economic Ngurah Rai University who have supported this research and also thanks to the friends of lecturer

REFERENCES

Bandura, A. 1997. *Self-efficacy: The exercise of control.* New York: Freeman.
Baron, R.A. & Byrne, D. 2003. *Social psychology.* Boston: Allyn & Bacon.
Fishbein, M., & Ajzen, I. 2005. Theory-based behavior change interventions: comments on Hobbis and Sutton. *Journal of Health Psychology,* 10(1), 27–31.
Ghozali, I. 2011. Aplikasi Analisis Multivariate dengan Program IBM SPSS 19. Semarang: Badan Penerbit Universitas Diponegoro.
Gunarsa, S. D. (1991). *Psikologi praktis: anak, remaja dan keluarga.* Jakarta: BPK Gunung Mulia.
Heuer, A. & Kolvereid, L. 2014. Education in entrepreneurship and the Theory of Planned Behaviour. *European Journal of Training and Development* 38 (6): 506–523.
Karimi, S., JA Biemans, H., Lans, T., Chizari, M. & Mulder, M. 2014. Effects of role models and gender on students' entrepreneurial intentions. *European Journal of Training and Development* 38(8): 694–727.
Kasmir. 2011. *Kewirausahaan.* Jakarta: PT. Rajagrafindo Persada.
McClelland, D.C. 1987. *Human motivation.* Cambridge: Cambridge University Press.
Murdjianto & Aliaras W. 2006. *Membangun Karakter Kepribadian Kewirausahaan.* Yogyakarta: Graha Ilmu.
Sels, H.L.J.M.L., Debrulle, J. & Meuleman, M. 2009. Gender effects on entrepreneurial intentions: A TPB multi-group analysis at factor and indicator level. In Paper presented at the Academy of Management Annual Meeting 7:11.
Soemanto, W. 2008. *Pendidikan Wiraswasta.* Jakarta: PT. Bumi Aksara.
Soemanto, Wasty. 2006. *Sekuncup Ide Operasional Pendidikan Kewirausahaan.* Jakarta: PT. Bumi Aksara.
Sugiyono. (2016). Metode Penelitian Kuantitatif Kualitatif dan R & D. Yogyakarta: Alphabeta.

Upgrading the business performance: The role of the community

R. Rofaida, S. Suryana & A.K. Yuliawati
Universitas Pendidikan Indonesia, Bandung, Indonesia

ABSTRACT: The digital creative industry is an industry that combines creative and digital elements in products and services. West Java is one of the provinces in Indonesia which becomes the center of the digital creative industry. The objective of this study is to explore the role of community in upgrading digital creative industry's performance. This article was based on research conducted on digital creative industries in West Java. The research was conducted using survey method and the data was analyzed by descriptive analysis. The role of the community in improving the business performance of the digital creative industry were cooperating with educational and government institutions to develop entrepreneurial capacity and provide business coaching and mentoring. Communities become a means for establishing and developing business entities and sharing knowledge amongst the digital creative industry businesses and the medium of knowledge sharing amongst business actors of the digital creative industry.

1 INTRODUCTION

In Indonesia's Creative Industry Blueprint 2015-2025, the digital creative industry becomes the focus of development for improving the competitiveness of creative industries in Indonesia. Especially this creative industry is based on technology, which creates of higher value. The digital creative industry combines creative and digital elements in their products and services. The digital creative industry is growing rapidly in Indonesia especially in West Java Province. This is evidenced by the growing digital business in West Java and the emergence of digital communities that support the development of the digital industry. West Java is also one of the provinces in Indonesia which becomes the center of digital creative industry in Indonesia.

The digital creative industry has enormous development potential supported by the development of technological infrastructure, the ease of access to information centers through the internet, social innovation processes in the community such as the rapidly expanding e-commerce phenomenon, and various applications that enable people to conduct interaction and transaction digitally. In this case, West Java Province has its own potential in supporting the development of digital creative industry. The diversity of local culture and the openness of society to modern culture can support the dynamics of creativity and innovation. West Java also has many sources of knowledge, a number of leading universities, which produce qualified human resources. The combination of these factors brings together the aesthetic and technological elements to produce a unique digital creative industry product.

However, the potential also has problems in its development. The problems are related to unsatisfactory performance in entrepreneurship aspects which

consist of the application aspect of new technology, innovation, formulation of the right business strategy, application of ICT in production process, business network, and limited talented human resource. For that reason, it is inevitable to formulate strategies to increase the entrepreneurship of corporate corporations by using pentahelix model (business, academia, government, community, and media). The focus of this article is the development of the digital creative industry by further examining the role of the digital creative industry community.

The community plays an essential role in improving the business performance of the digital creative industry, the Community being a means for the establishment and development of business entities and the means of sharing knowledge amongst the digital creative industry businesses. Sharing knowledge can be a key tool for developing quality performance among members.

1.1 Digital creative industry

The digital creative industry is a creative industry that has become a top priority. In 2015-2019, the national creative economy is in the third phase focusing on the development of creative industries that have the advantage of human resources and the utilization of science and technology. The digital creative industry combines creative and digital elements in its products and services. This industry can produce higher value because it provides technology products that are creative and unique so it can be a solution of business life or daily life. The scope of the digital creative industry consists of games, animation, applications, software, social media and digital music.

In its development, the digital creative industry has great potential for further development. These potentials are the rapid development of information technology which becomes the main factor supporting the creativity of business actors, ease of access to the information center through the internet which is also supported by the government in the form of better infrastructure improvements which improve the market potential of this industry, social innovation processes, such as e-commerce phenomenon and the number of existing applications, trigger interaction and digital transactions, the uniqueness and diversity of local cultural potentials of West Java and openness to modern culture support the dynamics of creativity and innovation, sources of knowledge, leading universities, can be a producer of qualified human resources that have high creativity and innovation and combination of creativity/innovation factors and technologies can produce aesthetic and technological elements that produce distinctive digital creative industry products.

The existence of the digital creative industry is not entirely up to expectations. There are challenges to be faced, such as bureaucratic challenges, transfer of knowledge and technology and challenges in improving human resource competence and entrepreneurship. The digital creative industry did not have appropriate competitive strategy and still showed low performance and limited human resource talent (Rofaida & Krishna 2016). Therefore, the policy of developing the digital creative industry leads to the implementation of interventions in the aspects of creative business infrastructure, human resource management, technology development and procurement, determination of the excellence and uniqueness of the product, market expansion through government support, improved technology infrastructure, and the strengthening of value chain elements, interconnected ecosystem of related digital creative industries, superior product ownership at the global market level, the compilation of policies supporting the digital creative industry, and the creation of a high-performing digital creative industry cluster. The existence of information technology, which is very essential in the digital creative industry, is closely related to the creative arts resulting from creative human resources in the industry (Liu & Xiang 2009).

The rapid development of information technology requires a digital company to be able to improve the quality of its creative resources. One process that can support the improvement of creative power and individual innovation is corporate entrepreneurship. Innovation technology are developed by copying and developing based technology through continuous improvement process (Lu & Mu 2011), (Hotho & Champion 2010). The results of research on innovations carried out in the creative industry in China show that innovation is able to provide significant influence related to indicators of business continuity indicators such as profit levels and speed of innovation (Keane & Hartley 2006), (O'Connor 2011).

1.2 Digital creative industry community

Efforts to improve business performance in the digital creative industry through corporate entrepreneurship also involves several interested parties. Among them are intellectuals/academics, business, government, community, and media. The parties are incorporated in Penta Helix. The purpose of Penta Helix is to create a circulation of science that leads to innovation, namely the ownership of economic potential, or knowledge capital. Penta Helix as the main actor must always move in circulation to form a space of knowledge that will direct these five parties to form consensus space where these five parties begin to make agreement and commitment to something that will ultimately lead to the formation of innovation space of creative products that has economic value.

Penta Helix has its own role according to its capacity. Academics play a role to produce competent human resources in accordance with industry needs. Business/industry acts as a driver of creative endeavor. The government acts as a policy maker for the growth of the creative industry. The community acts as a capacity building of creative industries through various cooperation with various parties. Media acts as a partner of creative business development through access to promotion and marketing.

The community as the closest party to the digital business becomes the medium to bring industry closer to other stakeholders. The role of the community in the development of digital business is very essential. Moreover, nowadays, the growing number of digital communities has contributed to the strengthening of the digital business. One of the largest digital communities in West Java is Bandung Digital Valley which is a means of appreciation formed by the cooperation between PT Telekomunikasi Indonesia Tbk. (TELKOM) and MIKTI (Society of Creative Industries of Technology, Information and Communication Indonesia). Bandung Digital Valley serves as a place for appreciation and development of local digital business as well as an intermediary between digital business and digital application users.

In an effort to support digital business, Bandung Digital Valley performs its role as a collector and organizer of potential developers' communities spread across Indonesia, fund managers from internal and external companies, education providers, business and commercial products advocates, and managers of product that have been produced by developers to be ready for commercialization. The result of the role of this community in connecting the industry with Penta Helix has been evidenced essential as seen from the growing number of digital businessmen, the growing scale of digital business, the wider range of marketing of digital creative products, the growing number of digital startup programs, and the increasing quality of human resources in the digital creative industry.

2 METHOD

This research is a combination of quantitative and qualitative research. Mixed method is a research approach that combines qualitative research with quantitative research. Quantitative methods with qualitative methods were employed together in a research activity in order to obtain more comprehensive, valid, reliable and objective data. The research method employed was survey method (Creswell 2014). The survey method used questionnaires as a data collection instrument. Data collection was conducted through observation, interview, in depth interview and FGD. The observations was "one shoot"/cross sectional time coverage. Data analysis was conducted descriptively.

The sample were chosen using purposive sampling method. Purposive sampling is a sampling technique used when a researcher has some considerations. It is also called a sample for a particular purpose. The selected respondents were those who had better knowledge about the variables studied. Qualitative research usually uses non probability sampling (including the purposive sampling). The sample size of qualitative research is relatively small compared to the sample size for the study using representation to increase the sample population. Sample in this research were 50 creative industries in West Java.

3 RESULT AND DISCUSSION

The development of the digital creative industry is not without problems. The results show that the existing problems in the industry are related to some challenges such as bureaucratic challenges, the challenges of knowledge transfer and technology as well as challenges in enhancing human resource competence and entrepreneurship. Based on the problems and challenges faced by the digital creative industry, the development strategy of digital creative industry is aimed at developing the aspect of creative business infrastructure, human resource management, technology development and procurement, determination of product excellence and uniqueness, market expansion through government support, as well as a complete infrastructure. Related to the strategy, the targeting of digital creative industry development is aimed at measuring and strengthening value chain elements, interconnected digital creative industry business ecosystems, superior product ownership at global market level, compilation of policies supporting the digital creative industry, and the creation of high-performing creative industry clusters. Increasing the capacity of the digital creative industry is determined by conducive ecosystems such as creative industry associations/communities, educational institutions that

prepare creative human resources, and public policies that support the growth and development of this industry (Ibrahim 2013). The form of organization, business practices carried out and the industrial environment will have a huge influence in determining the level of growth and sustainability of this industrial business (Ma et al. 2012).

In order to support the continuous improvement of the business performance of the digital creative industry, contribution from those closest to the digital business world, the community, is needed. The community can play its role in facilitating the problems faced by the digital creative industry by establishing cooperation with various stakeholders. In terms of human resources, the community plays a role in helping to increase the competence of creative industry players. In an industry context, communities can improve overall industry performance. In terms of technology, communities can use and develop technologies that support the operation of the digital creative industry. In terms of corporate resources, communities can maintain the availability of the resources needed in the digital creative industry. In the case of finance, the community can work with financial institutions as fund managers. In terms of marketing, the community can help in improving the ability in promotion and sales.

The creative community within the context of this research is a collection of individuals who share a common vision and move on their own, from creating an exchange of interacting knowledge, experience, techniques and tactics to fostering a project and eventually hatching into a robust, innovative business entity. The community can support the development policy/program of digital business capacity through various ways, such as cooperation with various parties (penta helix) for the development of business performance and entrepreneurship, and providing business coaching and mentoring. The community is also a means for the establishment and development of business entities through knowledge sharing among business actors of digital creative industry.

Creative community in West Java plays an active role in improving the performance of digital business. One of the largest communities is Bandung Digital Valley which has two strategic roles as resource pools and hubs that collect and manage potential developers' communities spread across Indonesia, managing internal and external funding, providing education, business advocacy and product management, which developers have generated for commercial use. Bandung Digital Valley has a vision to be an excellent innovation ecosystem that has a mutually beneficial partnership with several technology providers and support systems. The goal to be achieved by this community is to encourage and accelerate self-sufficiency in Information and Communication Technology (ICT), especially applications and content. Thus, in the future, it is expected that all application and content needs can be fulfilled by domestic developers.

This community plays its role by collaborating with various parties among developers from the Indonesian Creative Information Society (MIKTI), various universities, as well as some other application development community. In addition, this creative community has a strategic step to support the improvement of digital business performance through several programs, including Indigo Creative Nation (appreciation program for startup/entrepreneur who is considered successful to create ideas, products, and innovative digital business that customers want), the Daily Event commonly done is Wirabusaha (a periodic event on Wednesday that will discuss aspects of Business from startup) and technical Thursday (technical event sharing on Thursday). Other activities undertaken by Bandung Digital Valley in supporting the development of digital business is the provision of facilities for startups in the form of working space in Bandung Digital Valley in cooperation with DILO (Digital Innovation Lounge), and play an active role in organizing digital business development events in cooperation with MIKTI (Society of Creative Industries, Information and Communication Technology).

4 CONCLUSION

Community as the closest party to digital business has an essential role to improve the performance of digital business. The community has a role as a business intermediary with related stakeholders. Communities can work together to develop entrepreneurial capacity, business coaching and mentoring. Communities become a means for establishing and developing business entities and sharing knowledge amongst the digital creative industry businesses. Sharing knowledge can be a key tool for developing quality performance among members.

REFERENCES

Creswell, J.W. 2014. *Research design: Qualitative, quantitative, and mixed methods approaches*. Sage Publications, Inc.

Hotho, S., & Champion, K. 2010. We are always after that balance: Managing innovation in the new digital media industries. Journal of technology management & innovation, 5(3),36-50.

Ibrahim, N., Shariman, T.N T. & Woods, P. 2013. The concept of digital literacy from the perspective of the creative multimedia industry. *International Conference on Informatics and Creative Multimedia*: 259-264.

Jingchen, L. & Zhongping, X. 2009. The Study of Film and TV Cultural Creative Industry Based on Digital Technology. *International Conference on Environmental Science and Information Application Technology* 1: 705-707.

Keane, M.A. & Hartley, J. 2006. Creative industries and innovation in China. *International Journal of Cultural Studies* 9(3): 259-264.

Lu, F. & Mu, L. 2011. Learning by innovating: lessons from China's digital video player industry. *Journal of Science and Technology Policy in China* 2(1): 27-57.

Ma, X., Li, S. & Dai, W. 2010. Modeling and Simulating of Ecological Community in Digital Creative Industry. *Third International Symposium on Intelligent Information Technology and Security Informatics*: 708-712.

O'Connor, J. (2011). Economic development, enlightenment and creative transformation: creative industries in the New China. *Ekonomiaz: Basque Economic Review* 78(3):108-125.

Rofaida, R & Krishna, A.Y. 2016. Competitive Advantages of a Fashion Creative Industry in a Post Asean-China Free Trade Agreement (Acfta): A Case of Bandung in Indonesia. *Man in India Journal* 96 (11), 5011-5026.

Advances in Business, Management and Entrepreneurship – Hurriyati et al (eds)
© *2020 Taylor & Francis Group, London, ISBN 978-0-367-27176-3*

Development strategy of small medium enterprise in increasing the image of Tasikmalaya as a halal culinary tourist destination

S. Sulastri, A. Fauziyah, I. Yusup & E. Surachman
Universitas Pendidikan Indonesia, Bandung, Indonesia

ABSTRACT: The city of Tasikmalaya is one of the cities that is the destination of tourist attraction in the West Java region, Indonesia. Culinary is one of the industries that develops in the city of Tasikmalaya that supports the development of the city of Tasikmalaya as a tourist destination. As one of the cities known as the city of santri, the city of Tasikmalaya is currently building the image of the city of Tasikmalaya as a halal culinary destination. The purpose of this study is to explain and analyze internal and external factors that influence the image of the City of Tasikmalaya as a halal-based culinary tourism destination. This study used a descriptive verification method, and data was collected through an interview process using a closed questionnaire to culinary SMEs in the City of Tasikmalaya as many as 84 people. The results of the study indicate that the highest internal factors are marketing strategies. While the lowest is the financial aspect. The external factor of the development of MSME in Tasikmalaya City that was considered the highest was the role of the relevant institutions. While the lowest is the socio-economic aspect. The image of Kota Tasikmalaya as a halal-based culinary tourism destination that is considered the highest is the image of the city as a culinary tourism destination which is part of the cognitive image, while the lowest is the image of the city with unique food as an indicator that measures unique image. External factors affect the internal factors of Tasikmalaya City MSMEs, while internal and external factors jointly influence the image of the City of Tasikmalaya as a halal-based culinary tourism destination, where from these two factors external factors have a higher influence than internal factors.

1 INTRODUCTION

Tourism is one of the national strategic sectors as a contributor to foreign exchange, an instrument of equitable development and improvement of people's welfare for several countries in the world. In Figure 1, it can be seen the contribution of the tourism sector to the global economy from 2006 to 2016. The contribution of tourism to the global economy both directly and totally experienced a continuous increase. Spain and Singapore, which each occupy the first place in the Travel & Tourism Competitiveness Index in the World and ASEAN, contribute to tourism to GDP respectively at 5.3 percent and 4.7 percent in 2017. In Indonesia, the tourism sector has a potential to be a sector that drives future's economic growth, although this contribution is still 1.9 percent of Gross Domestic Product (GDP).

Meanwhile, in terms of direct contribution to employment opportunities, Indonesia's tourism has only reached 1.6% which is below Spain 4.8% and Singapore 4.6%. Meanwhile the contribution of tourism to do investment in Indonesia, this year reached USD 14 billion under Spain USD 18.1 billion but above Singapore USD 13.8 billion (Atmoko 2017).

In order to increase tourism contribution, one of the strategies that can be done is a tourism destination development strategy. Tourist destinations are an attraction for tourist arrivals. Tourist attraction to

visit tourism is different from each other. The choice of destination is influenced by the motivation and tastes of tourists. In Indonesia, tourist destinations that are a favorite for both domestic and foreign tourists are still owned by the island of Bali with the exotic views.

Indonesia is known for its beautiful natural scenery with a combination of mountainous terrain and stunning beaches. Priangan is an area that has many mountains located in the province of West Java which covers the areas of Bogor, Cianjur, Sukabumi, Bandung, Sumedang Garut, Tasikmalaya, and Ciamis to the border of Central Java. The priangan area is divided into four namely West Priangan, East Priangan, South Priangan, and North Priangan. The East Priangan region is more familiar to the public than the West Priangan, North Priangan and South Priangan. The cities included in East Priangan start from Sumedang, Garut, Tasikmalaya and Ciamis (including Banjar and Pangandaran).

The city in eastern Priangan region which has the greatest tourism potential is Tasikmalaya because almost 70 percent of the business centers, trade and service centers, and industrial centers in eastern Priangan are placed in this city. Tasikmalaya is located right in the middle of the heart of the East and South Priangan, flanked by Ciamis with its legendary Pangandaran tourist attraction, Sumedang with museum attractions that keep a history of the development of the earth priangan, and Garut with

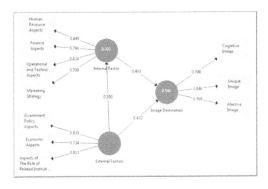

Figure 1. Model analysis of SME development factors in the city of Tasikmalaya in establishing the image of the city of Tasik as a culinary-based travel destination.

its famous Cipanas attractions. With the strategic position of Tasikmalaya, this city is the largest meeting, incentive, conference and exhibition (MICE) center in West Java after Bandung and Bogor.

Tasikmalaya has a varied tourism potential, including natural tourism, crafts, shopping, religious tourism, art, culture and others. Tasikmalaya crafts are very well known both at home and abroad. Tasikmalaya's unique crafts include Embroidery, Payung Geulis which has become an icon of West Java, Kelom Geulis, Tasikmalaya batik which is no less than other batik in Java with its distinctive features, and other crafts.

The development of the tourism industry is also closely related to the development of the creative industry. The creative industry is a strategic sector in supporting national economic growth and development and making an important contribution to the national economy. Various sectors in the creative industry have become increasingly popular tourism attractions, one of which is culinary. The city of Tasikmalaya, in addition to being famous for its craft tours, is also increasingly crowded with the presence of a number of cafes and restaurants and even karaoke venues. There are currently around a hundred cafes and restaurants in Tasikmalaya. The hotel business is also quite developed marked by the increasing number of star hotels. This shows the level of tourist visits also increases. Up to 2016 the number of foreign tourists visiting Tasikmalaya was around 501 people while domestic tourists were 207,853 people.

The growth of the cafe and restaurant business is accompanied by innovation in the culinary field. Various types of food and drinks are offered with a variety of uniqueness. There are those serving typical menus of Tasikmalaya such as Tutug Oncom Rice, Cilok Goang, and Tutut. However, there are also those who open franchise restaurants that come from within and outside the country.

One of Tasikmalaya's work programs is promoting the City as a Halal Culinary City. This is related to the nickname of the City of Tasikmalaya as a santri city. The nickname appeared around 1970 because there were around 1,200 Islamic boarding schools scattered in Tasikmalaya. Nicknamed the City of Santri, Tasikmalaya became an area known for being religious. Especially then, the local government issued Regional Regulation No. 12 of 2009 concerning Values Based on Religious Teachings, or Islamic Sharia Regional Regulations. The regulation provoked controversy so that the Tasikmalaya Government revised it, known as Regional Regulation No. 7 of 2014 concerning the Values of Life of Religious Communities in the City of Tasikmalaya.

As a tourist destination city, Tasikmalaya City also has the potential of SMEs in supporting tourism development. Tasikmalaya has the largest number of SMEs after Bandung Raya (Bandung City, Bandung Regency, West Bandung Regency) in West Java.

SMEs in the processed food industry (culinary) ranks third. The problem that arises is that from all SMEs in the culinary industry most of them have not registered their products in order to have halal certification. This is due to the lack of knowledge of SME actors regarding the procedures and benefits of halal certificates. This is important because one of the efforts that can be made to convince consumers that the food they consume is halal is by the existence of the certification. In addition, another problem is the lack of knowledge of SMEs in the culinary industry with the concept of halal culinary tourism so that in an effort to support the work program of the Tasikmalaya government, it is necessary to develop SME development strategies in improving the image of Tasikmalaya as a halal culinary tourism destination. Based on the background described, the researcher needs to conduct research on "Development strategy of small medium enterprise in increasing the image of Tasikmalaya as a halal culinary tourist destination"

1.1 Internal and external analysis

Analysis of the internal environment aims to identify a number of strengths and weaknesses contained in the internal business resources and processes of an industry. Internal business resources and processes are said to have strength when internal resources and business processes have the capability to create distinctive competencies so that the company will gain a competitive advantage. Some of the analyze used to measure the company's internal resource capabilities include: SWOT Analysis (Strength, Weakness, Opportunities, and Threat) and value chain analysis and resource-based view (RBV). Each analysis tool has advantages and disadvantages in analyzing the company's internal environment.

According to Duncan (1972), that what is meant by the company's external environment (external business environment) are various factors that are outside the organization that must be taken into account by the company organization when making decisions. The company's external environment is all events outside the company that have the potential to influence the company (Williams 2001). Pearce and Robinson (2013) define the external environment as factors out of control that affect the company's choices regarding the direction and actions, which in turn also affect the organizational structure and internal processes. Kartajaya et al. (1999) state that the current competitive conditions are in the private sector, where companies that compete in one industry and even across industries have relatively similar access to the availability of technology to produce new products. Meanwhile, Fitzroy and Hulbert (2005) describe the current corporate environment as a turbulent world, a world filled with change and difficult to pattern. In these circumstances, various changes run so fast and cannot always be predicted accurately. External environmental analysis needs to be done to identify opportunities and major threats faced by an organization against changes in the external environment of the company so that managers can formulate strategies to take advantage of these opportunities and avoid or minimize the impact of potential threats that arise. David (2010) says the main external force factors.

2 METHOD

This study uses a descriptive approach verification with survey methods through questionnaires. The population in this study was 525 culinary SMEs in the City of Tasikmalaya with sampling using the Slovin formula as many as 84 SMEs. The technique of collecting data through a questionnaire consisting of indicators to measure the internal environment, the external environment, and the image of the city as a tourist destination. The data analysis technique used is Partial Least Square.

3 RESULT AND DISCUSSION

3.1 Evaluation of measurement models (outer models)

Measurement model evaluation is done to evaluate the relationship between indicators with latent variables (constructs). The overall latent variables involved in this study both for latent variables of internal factors, external factors and business performance are measured by reflective indicators so that to evaluate the measurement model includes individual items of reliability, internal consistency, or construct reliability, average variance extracted, and discriminant validity. The first three measurements are grouped in convergent validity.

3.1.1 Convergent validity

Convergent validity consists of three tests, namely reliability items (validity of each indicator), composite reliability, and average variance extracted (AVE). Convergent validity is used to measure how much an indicator can explain its construct. This means that the greater the convergent validity, the greater the ability of the indicator in explaining the construct:

a Reliability Item: reliability items or commonly referred to as indicator validity. Testing of the item reability (indicator validity) can be seen from the value of the loading factor. The value of this loading factor is the magnitude of the correlation between each indicator and its construct. The factor loading value above 0.5 can be said to be ideal. The value of loading factor for PLS model can be seen in Table 1.

According to Ghozali (2008), an indicator is considered valid if it has a correlation value above 0.70. However, for loading values 0.50 to 0.60 it can still be accepted by looking at the output correlation between indicators and their construct. Based on Table 1, it can be seen that each indicator that forms its construct both indicators on internal aspects, external aspects and destination image mostly already have a loading value above 0.7 even though there are two indicators (Marketing Aspects and Employee Growth) below 0, 70 but still above 0.60 so that the indicator can be said to be valid.

b Composite Realiability: the statistics used in composite reliability or construct reliability are cronbachs alpha and D.G rho (PCA). The value of cronbachs alpha and D.G rho (PCA) above 0.7 indicates that the construct has a high reliability or reliability as a measuring tool. A limit value of 0.7 and above means that it can be

Table 1. Value of the initial loading factor model.

	Internal Factor	External Factor	Destination Image
Human Resource Aspects	0.851		
Finance Aspects	0.786		
Technical And Operational Aspects	0.820		
Marketing Strategy Aspects	0.701		
Government Policy Aspects		0.839	
Socio-Economic Aspects		0.734	
Aspects Of The Role Of Related Institutions		0.808	
Cognitive Image			0.798
Unique Image			0.846
Afective Image			0.769

Table 2. Composite reliability.

	Cronbach Alpha	Composite Reliability
Internal Factor	0.800	0.810
External Factor	0.708	0.714
Destination Image	0.728	0.727

Table 4. Discriminant validity.

	Internal Factor	External Factor	Destination Image
Human Resource Aspects	**0.849**	0.421	0.589
Finance Aspects	**0.784**	0.516	0.503
Technical and Operational Aspects	**0.824**	0.422	0.514
Marketing Strategy Aspects	**0.700**	0.372	0.402
Government Policy Aspects	0.462	**0.835**	0.586
Socio-Economic Aspects	0.506	**0.734**	0.435
Aspects of The Role of Related Institutions	0.334	**0.813**	0.530
Cognitive Image	0.566	0.452	**0.798**
Unique Image	0.571	0.452	**0.846**
Afective Image	0.413	0.662	**0.769**

accepted and above 0.8 and 0.9 means it is very satisfying (Nunnally and Bernstein, 1994 in Sofyan Yamin and Heri Kurniawan, 2011). The composite reliability value can be seen in Table 2.

Based on Table 2, it shows that the Cronbach alpha value and composite reliability value for constructs of internal factors, external factors and destination image are above 0.7. Then the construct has high reliability or reliability as a measuring tool.

c Convergent Validity, average variance extracted (AVE) describes the amount of the variance construct that can be explained by the items/indicators compared to the variance caused by measurement errors. The default is if the AVE value is above 0.5 then it can be said that the construct has good convergent validity. This means that the latent/construct variable can explain more than half the variance of the indicators. The results of convergent validity described through Average Variance Extracted (AVE) can be seen in Table 3.

Table 3 shows that the AVE values for each construct are internal factors, external factors and destination image above 0.5, which means that the construct is able to explain an average of more than half the variance of the indicators.

3.1.2 Discriminant validity

Checking discriminant validity from the reflective measurement model that is rated based on the cross loading value. The size of cross loading is comparing the correlation of indicators with their constructs and constructs from other blocks. Good discriminant validity will be able to explain the indicator variables higher than explaining variants of other construct indicators. Table 4 shownis the value of discriminant validity for each indicator.

Based on Table 4, it can be seen that the value of disrimanant validity or loading factor for aspects of human resources, financial aspects, technical and

Table 3. Convergent validity.

	AVE
Internal Factor	0,588
External Factor	0,648
Destination Image	0,622

operational aspects and aspects of marketing strategies are higher with internal factor variables compared to other variables. This shows that internal factor variables can explain higher variants with indicators of aspects of human resources, financial aspects, technical and operational aspects and aspects of marketing strategies.

Likewise on aspects of government policy, social and economic aspects and aspects of the role of relevant institutions where the value of the discrimant validity is higher with external factors so that the variable value of experience is able to explain the variant higher with the indicators mentioned.

In the destination image variable, the cognitive image, unique image, and affective image so that destination image variables are also able to explain a higher variance with these indicators.

3.2 Evaluation of structural models

There are several stages in evaluating structural models. The first is to see the significance of the relationship between the constructs. This can be seen from the path coefficient, which describes the strength of the relationship between the constructs.

3.2.1 The coeffecient path

Seeing the significance of the relationship between constructs can be seen from the path coefficient. Signs in path coefficient must be in accordance with the hypothesized theory, to assess the significance of path coefficient can be seen from the T-Statistic obtained from the bootstrapping (resampling method) process.

Hypothesis 1: External Factors Influence Internal Factors in the Development of SMEs in the City of Tasikmalaya

Based on Table 5, it can be seen that the t-value value for external factors on internal factors is 6.691

Table 5. Path coefficient model.

	T Statistics	P-value
Internal Factor → External Factor	6,691	0,000
Internal Factor → Destination Image	4,157	0,000
External Factor → Destination Image	4,199	0,000

with p-value 0.000 <0.05 so that it can be said that the influential external factors are influenced by internal developments in Tasikmalaya City UKM. This means that the good or not internal factors in the development of Tasikmalaya City SMEs will be influenced by whether or not the condition of external factors.

Hypothesis 2: Internal factors influence the image of the City of Tasikmalaya as a Halal Culinary-Based Travel Destination

Based on Table 5, it can be seen that the value of t-value for external factors on the image of the city of Tasikmalaya as a culinary-based tourist destination amounted to 4.157 with p-value of 0.000 < 0.05, so that it can be said that internal factors influence the image of the city as a culinary based destination. This means that the good image of the city of Tasikmalaya as a culinary-based tourist destination will be influenced by the good condition of internal factors.

Hypothesis 3: External factors influence the image of the city of Tasikmalaya as a culinary-based tourist destination

Based on Table 5, it can be seen that the value of t-value for external factors on the image of the city of Tasikmalaya as a culinary-based tourist destination is 4,199 with a value of 0,000 <0,05 so that external factors can influence the image of the city as a culinary based destination. This means that the good image of the city of Tasikmalaya as a culinary-based tourist destination will be influenced by the good condition of external factors.

The contribution of internal and external variables to the image of the city of Tasikmalaya as a culinary-based tourist destination can be seen from the path coefficient.

3.2.2 Evaluate R2

To see how well the influence of external factors on internal factors and the two factors affect business performance, we can see the R2 value in Table 6.

Based on Table 6, it can be seen that the construct of external factors is able to explain the internal factor construct of 0.302 or equal to 30.2%, while

Table 6. R2 for PLS model.

	R Square	R Square Adjusted
Internal Factor	0,302	0,294
Destination image	0,540	0,528

the remaining 69.8% is influenced by other constructs which are not included in the research model.

Constructs of internal factors and external factors simultaneously can explain the construct of destination image 0.540 or 54%, while the remaining 46% is influenced by other constructs that are not included in the research model.

4 CONCLUSION

Internal factors in the development of SMEs in the City of Tasikmalaya are measured through aspects of human resources, financial aspects, technical and operational aspects, and marketing strategies. The highest rated aspect is the marketing strategy, while the lowest is the financial aspect.

External factors in the development of SMEs in the City of Tasikmalaya are measured through aspects of government policy, socio-economic aspects, and aspects of the role of related institutions. The highest aspect is the role of related institutions, while the lowest is the socio-economic aspect.

The image of the City of Tasikmalaya as a halal-based culinary tourism destination is measured through cognitive image, unique image and affective image. The indicator that experienced the highest increase was the image of the city as a kulner tourist destination which is part of the cognitive image, while the lowest is the image of the city with unique food as an indicator that measures unique image.

External factors influence the internal factors of SMEs in the City of Tasikmalaya, while internal and external factors jointly influence the Image of the City of Tasikmalaya as a halal-based culinary tourist destination, where from both factors external factors have a higher influence than internal factors.

REFERENCES

Atmoko, C. 2017. BI: sektor pariwisata berpotensi dorong pertumbuhan ekonomi. [Online]. Retrieved: https://www.antaranews.com/berita/658816/bi-sektor-pariwisata-berpotensi-dorong-pertumbuhan-ekonomi. Accessed: April 20, 2017.

David, Fred R. (2010). Manajemen Strategis Konsep, Buku 1, Edisi 12. Jakarta.Salemba Empat.

Duncan, R.B. 1972. Characteristics of organizational environments and perceived environmental uncertainty. Administrative Science Quarterly 17: 313–327.

Fitzroy, P., dan Hulbert, J. (2005). StrategiyManagement: Creating Value in Turbulent Times. New Jersey: John Wiley.

Ghozali, I. 2008. Structural equation modeling: Metode alternatif dengan partial least square (pls). Semarang: Badan Penerbit Universitas Diponegoro.

Kartajaya, H., Yuswohady, & Taufik. (1999). Bridging to the network company. Gramedia Pustaka Utama.

Pearce, J.A. & Robinson Jr, R.B. 2013. Strategic management: Formulation, implementation, and control. New York: McGraw-Hill.

Williams, C. (2001). Manajemen. Edisi Pertama. Jakarta: Salemba Empat.

Advances in Business, Management and Entrepreneurship – Hurriyati et al (eds)
© 2020 Taylor & Francis Group, London, ISBN 978-0-367-27176-3

Influence of entrepreneurial learning on self-esteem and its impact on motivation of student entrepreneurship

E. Tarmedi, F.A. Setiadi, A. Surachim & L. Lisnawati
Universitas Pendidikan Indonesia, Bandung, Indonesia

ABSTRACT: The purpose of this study is to find out the description of confidence level and its impact on motivation of entrepreneurship that is influenced by entrepreneurship learning. The type of research is verification. The method used was explanatory survey with simple random sampling techniques and sample of 228 respondents. Data analysis technique used is path analysis with computer software SPSS 21.0. Based on the results of the research, it can be seen that simultaneously and entrepreneurship learning influence on confidence level and have an impact on motivation of entrepreneurship.

1 INTRODUCTION

Motives are one aspect that has considerable influence in attitude and can encourage someone to perform an action (Meyer & Landsberg 2015). In relation to education, motive has an important role to the young generation in entrepreneurial activity, an important motive in the life of the younger generation because it involves energy, direction, and perseverance and intention (Iqbal et al. 2012). The entrepreneurial motive is not the same as the unique personality traits of the entrepreneur, but this entrepreneurial motive is an important topic in entrepreneurship education because of the importance and significant thrust of a country's economic development (Kim-Soon et al. 2011).

Each country is increasingly recognizing the importance of entrepreneurial motives and adopting it with the aim of increasing employment and economic development (Sondari 2014). According to Hussain (2015), the entrepreneurial motives in developing countries are seen as important for increasing employment opportunities. This impact is seen from several regions in Pakistan reporting a decrease in the unemployment rate, as it has a higher level of entrepreneurial initiative. Students need to grow their entrepreneurship motive because students will be encouraged to make creative and innovative efforts by developing ideas to find business opportunities (Sodikin & Widodo 2014).

According to Ayu et al. (2016), the increasing unemployment rate in Indonesia is the number of workers in the formal sector and does not seek to create self-employment or self-employment. This is due to their preference for working in the formal sector rather than being an entrepreneur (Susetyo & Lestari 2014). The high unemployment rate helps the in-creasing poverty line in Indonesia.

The data of the Central of Statistics shows that the number of unemployed in Indonesia increased by 320,000 in August 2015. This problem is caused by the widespread termination of employment due to the economic slowdown and the limited number of job opportunities that are not able to absorb the maximum number of job seekers. The need for large employment, but available employment is limited. Indratno (2012) describes the efforts that can reduce the unemployment rate is by increasing entrepreneurship spirit as early as possible, because a nation will progress if the number of entrepreneurs amount to at least 2% of the population. Currently, Indonesia has only 1.5% of entrepreneurs of a population of approximately 252 million people.

The existence of vocational schools in preparing the current workforce is still less than optimal. This can be seen in some graduates who cannot be absorbed in the job field because the competence they have are not in accordance with the demands in the world of work. Ideally nationally vocational graduates who can directly enter the workforce of about 80–85% is being absorbed the new 61%. These results prove that there are many vocational graduates who have not worked. The reason for this is because the competence they have is not in accordance with the demands of the world of work, but they also have not been able to create their own jobs. This means that vocational graduates have not been fully acknowledged by the world of work to apply the science that has gained in school (Samsudin 2010).

Nguyen and Phan (2014) have said that the motive of entrepreneurship is the outcome directed by education so that it is considered important, by improving the entrepreneurial motive will have high responsibilities in work and life. Promoting young entrepreneurship can help reduce unemployment (Sharma & Madan 2014). Factors influencing entrepreneurial motives according to Kalyani and Kumar (2011) are ambitions to become an entrepreneur desire for self-reliance, self-confidence, prior experience, technical qualifications, good market potential, small investment, economic needs, high profitability, availability raw materials, government concessions and family businesses.

Less optimal support to the student entrepreneurial can be seen from the lack of confidence. This will have an impact on one of the students' self-encouragement to perform entrepreneurial or other actions. Confidence needs to be created from an early age. The circumstances that encourage, mobilize and direct the individual's desire to engage in entrepreneurial activities, independently, self-confident, future-oriented, risk-taking, creative and high-valued innovative desires are known as entrepreneurial motivation (Ratnawati & Kuswardani 2007).

2 METHOD

This study was conducted to determine the effect of entrepreneurial learning on self-confidence and its implications for entrepreneurial motives. Variable X is entrepreneurial learning with dimension of educational and teaching objectives, learners or students, teachers especially teachers, teaching planning, learning strategies, instructional media, and evaluation of teaching. Meanwhile, the variable Y is self-confidence that includes the belief in self-ability, optimistic, objective, responsible, and rational and realistic.

As well as variable Z is the dimension of entrepreneurship motive includes need for achievement, locus of control, vision, desire independence, egoistic passion, drive, goal setting, self-efficiency. Object/unit of analysis in this study were students of class XI Public Vocational School 1 Ciamis.

This research was conducted in less than one year, so the data collection technique used in this research is a cross-sectional method. Technique used in this research is probability technique that is simple random with sample of 228 respondents. Data collection techniques were literature studies, field studies with questionnaires, and literature studies. Meanwhile, the technique of data analysis conducted is descriptive and verificative analysis. The data were analyzed by using SPSS 21.0 for Windows.

3 RESULT AND DISCUSSION

The description of learning entrepreneurship can be seen from the dimensions consisting of educational and teaching objectives, students, teachers, teaching planning, learning strategies, teaching media, and teaching evaluation are in high enough category. This shows that entrepreneurial learning can be quite good. Based on the recapitulation of responses of respondents on the aspect of entrepreneurship learning there is the highest score that is on the dimensions of teaching evaluation by obtaining a score of 1,828 or 80.17% while the lowest dimension is the

strategy for teaching media with a score of 1.679 or 73.64%. According to Suryana (2011), entrepreneurship is not only innate talent from birth or field affairs, but can be learned and taught. A person who has an entrepreneurial talent can develop through education. Entrepreneurship education is a discipline that studies values, abilities and behaviors in addressing life's challenges.

Lestari and Wijaya (2012) states that entrepreneurship education can form the mindset, attitude, and behavior of students into a true entrepreneur thus it is possible to direct them to choose entrepreneurship as a career choice. According to Nursito and Nugroho (2013), the goals of entrepreneurship education are: to acquire knowledge that is closely related to entrepreneur-ship; acquire skills in using techniques, analysis of business situations, and developing work plans; identify motivation, potential, talent and entrepreneurial skills and developing them; eliminate the risks inherent in analytical techniques; develop empathy and support for unique aspects of entrepreneurship, change wrong attitudes and thoughts toward change; encourage the emergence of new ventures; and stimulate the element of affective socialization. Entrepreneurship learning is a process of gaining cognition and structures knowledge and gains meaning from experience (Rae 2000). Entrepreneurial learning is the result of a dynamic social process of reason-making, which is not only cognitive or behavioral but also affective and holistic (Cope 2005). Entrepreneurship learning as a dynamic process of both consciousness, reflection, association and application involves the transformation of experience and knowledge into learning outcomes (Rae 2006).

Self-confidence can be seen from the dimensions consisting of the belief in self-ability, objective, optimistic, responsible, and rational and realistic in the category of high enough. This shows that students' self-confidence is good enough. Recapitulation of respondents' responses on the self-confidence aspect is the highest score or percentage of confidence in the ability of self to get score as much as 1933 or 84.78%. According to Parsons et al. (2011), taste is a belief while confidence, according to Ghufron and Risnawati (2010), is the belief to do something on the subject as a personal characteristic in which there is self-ability, optimistic, objective, responsible, rational and realistic. Confidence is the main trait that must be owned by an entrepreneur. A person is said to have a high sense of self-confidence if he does not vacillate the opinions and suggestions of others (Alma 2013). According to Ghufron and Risnawati (2010), aspects of self-confidence consists of the belief in self-ability, optimistic, objective, responsible, and rational and realistic.

Confidence can affect the motivation of entrepreneurship. Confidence will encourage someone to be brave with self-confidence, dare to express its existence, dare to channel an unpopular and will to sacrifice for the sake of truth and firm, able to make good decisions in spite of uncertainty and pressure. Confidence will tend to encourage someone to exercise their best potential in achieving their goals; in other words, confidence will encourage one's motivation in realizing his work (Ilyas & Gumilar 2012).

Meanwhile the lowest dimension is responsible by obtaining the score 1.558 or 68.33%. Furthermore, in the recapitulation of respondents on the aspect of entrepreneurship motive there is a score or the highest percentage is on the vision dimension by obtaining a score of 1.784 or 78.24%, while the lowest dimension is self-efficiency with a score of 1.780 or 70.07%.

Based on the results of descriptive test on entrepreneurship learning variables get a score of 13,937 or 76,41%, including into high category, so it can be said that entrepreneurial learning can improve self-confidence and encourage students to do entrepreneurship but it still needs to be improved because all dimensions are below the ideal score. Obtaining scores on confidence variables is 7035 or 77.14% including into the high category, thus it can be said that self-confidence and need improvement as all dimensions are below ideal score.

The description of students' entrepreneurship motives measured through need for achievement, locus of control, vision, desire independence, egoistic passion, drive, goal setting, and self-efficacy as a whole are in fairly high categories. This implies the students' entrepreneurship motive has been running quite well. The results of the variables of entrepreneurship motive is 16.193 or 74.75%, including into the high category, so it can be said that the motive of entrepreneurship is high enough but still needs to be improved because all dimensions are below ideal.

4 CONCLUSION

The dimension of evaluation of teaching is the dimension that has the highest percentage of assessment in influencing self-confidence and student entrepreneurship motives, while the dimension that has the lowest rating in influencing the self-confidence and entrepreneurial motive is the learning media. Dimensions Belief in self-efficacy is a dimension that has the same percentage of assessment in affect affecting self-confidence and student entrepreneurial motives, while the dimensions that have the lowest judgment in affect the self-confidence and entrepreneurship motives of students is responsible. The overall vision dimension has the same high percentage value, while for the dimension that has the lowest percentage value is self-efficiency. According to Alma (2011), motivation is the ability to do some-thing, while the motive is the need, desire, encouragement or impulse. Motives are influenced by the level of physical needs, security, self-esteem and quality of self (Maslow 2003). The motive or

impetus of each individual in creating and developing his business, either from the internal influence of each individual or the external (Shane et al. 2003). Dimensions of entrepreneurship motive measurement by Shane et al. (2003) can be measured with need for achievement, locus of control, vision, desire independence, egoistic passion, drive, goal setting and self-efficacy to open self-inducement into an entrepreneur.

REFERENCES

Alma, B. 2011. *Kewirausahaan*. Bandung: Alfa Beta.
Alma, B. 2013. *Kewirausahaan*. Bandung: Alfa Beta.
Ayu, D., Anggraeni, L. & Nurcaya, I. N. 2016. Peran Efikasi Diri dalam Memediasi Pengaruh Pendidikan Kewirausahaan Terhadap Niat Berwirausaha. *E-Jurnal Manajemen Unud* 5(4): 2424–2453.
Cope, J. 2005. Toward a Dynamic Learning Perspective of Entrepreneurship. *Entrepreneurship Theory and Practice* 29(4): 373–397.
Ghufron, M.N. & Risnawati, R. 2010. *Teori-Teori Psikologi*. Yogjakarta: Ar Ruzz.
Hussain, A. 2015. Impact of Entrepreneurial Education on Entrepreneurial Intentions of Pakistani Students. *Journal of Entrepreneurship and Business Innovation* 2(1): 43–53.
Ilyas, S. & Gumilar, I. 2012. Pengaruh Persepsi Mahasiswa Tentang Pendidikan Kewirausahaan Terhadap Kepercayaan Diri Dan Motivasi Mahasiswa Program Studi Akuntansi (Studi Kasus Pada Fakultas Ekonomi Universitas Widyatama). *Universitas Widyatama*: 1199–1206.
Indratno, A.F. 2012. *Membentuk Jiwa Wirausaha*. Jakarta: Kompas.
Iqbal, A., Melhem, Y. & Kokash, H. 2012. Readiness of the university students towards entrepreneurship in Saudi Private University: An exploratory study. *European Scientific Journal* 8(15): 109–131.
Kalyani, B. & Kumar, D. 2011. Motivational factors, entrepreneurship and education: Study with reference to women in SMEs. *Journal of Psychology and Business* 3 (3): 14–35.
Kim-Soon, N., Ahmad, A.R. & Ibrahim, N.N. 2014. Entrepreneurial motivation and entrepreneurship career intention: Case at a Malaysian Public University. *Journal of Entrepreneurship: Research & Practice*: 1001–1011.
Lestari, R.B. & Wijaya, T. 2012. Pengaruh Pendidikan Kewirausahaan Terhadap Minat Berwirausaha Mahasiswa di STIE MDP, STMIK MDP, dan STIE MUSI. Forum Bisnis dan Kewirausahaan. *Jurnal Ilmiah STIE MDP* 1(2): 112–119.
Maslow, A. 2003. *Motivasi dan Kepribadian*. Jakarta: Midas Surya Grafindo.
Meyer, N. & Landsberg, J. 2015. Motivational factors influencing women's entrepreneurship: A case study of female entrepreneurship in South Africa. *International Scholarly and Scientific Research & Innovation* 9(11): 3738–3743.
Nguyen, M. & Phan, A. 2014. Entrepreneurial traits and motivations of the youth—an empirical study in Ho Chi Minh City, Vietnam. *International Journal of Business and Social Science* 5(1): 53–63.
Nursito, S. & Nugroho, A.J.S. 2013. Pengembangan Model Student Based University Reputation Terintegrasi Sebagai Strategi untuk Mencapai Keunggulan

Kompetitif Perguruan Tinggi Swasta di Jawa Tengah. *Kiat Bisnis* 5(2): 137–148.

Parsons, S., Croft, T. & Harrison, M. 2011. Engineering students' self-confidence in mathematics mapped onto Bandura's self-efficacy. *Engineering Education* 6.

Rae, D. 2000. Understanding entrepreneurial learning: A question of how ? *International Journal of Entrepreneurial Behaviour & Research* 6: 145–159.

Rae, D. 2006. Entrepreneurial learning: A conceptual frame-work for technology-based enterprise. *Technology Analysis & Strategic Management* 18(1): 39–56.

Ratnawati, D. & Kuswardani, I. 2007. Kematangan Vokasional dan Motivasi Berwirausaha Pada Siswa Sekolah Menengah Kejuruan (SMK). *Jurnal Psikohumanika*.

Samsudin, S. 2010. *Manajemen Sumber Daya Manusia*. Bandung: Pustaka Setia.

Shane, S., Locke, E.A. & Collins, C.J. 2003. Entrepreneurial motivation. *Human Resource Management Review* 13(2): 257–279.

Sharma, L. & Madan, P. 2014. Effect of individual factors on youth entrepreneurship—a study of Uttarakhand State, India. *Journal of Global Entrepreneurship Research* 2(3): 1–17.

Sodikin, S. & Widodo, J. 2014. Pengaruh Praktik Kerja Industri dan Lingkungan Keluarga Terhadap Motivasi Berwirausaha Siswa Kelas XII Pemasaran SMK Negeri 2 Semarang. *Economic Education Analysis Journal* 3(2): 391–398.

Sondari, M.C. 2014. Is entrepreneurship education really needed? Examining the antecedent of entrepreneurial career intention. *Procedia-Social and Behavioral Sciences* 115, 44–53.

Suryana, S. 2011. *Kewirausahaan : Pedoman Praktis, Kiatdan Proses Menuju Sukses*. Jakarta: Salemba Empat.

Susetyo, D. & Lestari, P. S. 2014. Developing Entrepreneurial Intention Model of University Students (an Empirical Study on University. *International Journal of Engineering and Management Sciences* 5(3): 184–196.

Advances in Business, Management and Entrepreneurship – Hurriyati et al (eds)
© *2020 Taylor & Francis Group, London, ISBN 978-0-367-27176-3*

The effect of the use of project based learning model on entrepreneurship intensity based on adversity quotient

A. Fauziyah, S. Sulastri, I. Yusup & S. Ruhayati
Universitas Pendidikan Indonesia, Bandung, Indonesia

ABSTRACT: This study aims to investigate entrepreneurial intentions of students through Project Based Learning (PBL) method in Entrepreneurship subjects. Apart from the learning method, this study also takes the internal factors of students, namely the ability to face challenges (adversity quotient). The method was a quasi-experimental study using factorial Between-Subject design using Two Ways ANOVA. The results of the study show that there are differences in entrepreneurial intentions in the class using PBL models with classes that use conventional learning models. There are also differences in the increase in entrepreneurial intentions with high, medium and low adversity quotient. An interaction between the project based learning, adversity quotient, and entrepreneurial intention models is found in the study.

1 INTRODUCTION

Entrepreneurship is an important spear in the economy of a country. Beside a long-term economic growth, entrepreneurship can also increase economic and social prosperity through increasing state revenues (Alfonso & Cuevas 2012). Because of its important role, many countries propose their citizens to have their own business (entrepreneurship) rather than working under the leadership of others (Gelderen 2010), including in Indonesia.

Today, the Indonesian people who own their business are only 1.56% of the total population of Indonesia (www.republika.co.id). Although the minimum number of entrepreneurs in a country is only 2% of the total population (GMI, 2011), this number is very small compared to some other countries. China has 12% of its population choosing to be entrepreneurs, followed by the United States (11%), Singapore (7.2%), and Malaysia (5%) (www.pikiran-rakyat.com). With the number of entrepreneurs higher than the minimum number of entrepreneurs suggested by McClelland (1987), these countries are able to increase state revenues better and able to make their communities more prosperous compared to Indonesia. This clearly indicates that the population of Indonesia is not capable of entrepreneurship or rather is said to not be able to use the resources and opportunities that exist to manage the company itself compared to working in companies owned by other people whose numbers are not in accordance with the number of existing labor force.

One of the problems is the lack of entrepreneurial intentions among people in many cities. For example, based on statistics in 2017, the City of Tasikmalaya is included in the third poorest city in West Java. Currently, the per capita income of the city of Tasikmalaya is only IDR 397,215 even though the city has a lot of business potential including craftsmen, culinary and tourism centers. If it is associated with an entrepreneurial program promoted by the government at the secondary education level with entrepreneurial potential in the city, ideally the graduates of vocational schools are encouraged to become independent young entrepreneurs.

According to the results of field observations in high school and vocational school students in the city, around 55.60% of students choose to work in the business or in the industries rather than choosing to have their own business or choose to continue their education. The choice taken by students after completing their next education is to continue their studies with 22.68% and the remaining only 21.72% of students who choose to have their own business.

Entrepreneurship education is often referred to one of the key instruments to enhance entrepreneurial behavior (Farashah 2013, Shinnar et al. 2014, and Maresch et al. 2016), but the procurement of entrepreneurship education cannot directly make someone become an entrepreneur. Entrepreneurship education is often criticized for being too rational, managerial oriented, but not directly teaching relationships between individuals and creating values in an environment (Penaluna & Penaluna 2009, Farhangmehr et al. 2016). Therefore, we need the right learning method in delivering entrepreneurship education, one of which is through a project based learning (PBL) model.

The PBL model is a learning model that uses problems as a first step in gathering and integrating new knowledge based on their experience in actual activities. The model is designed to be used on complex problems that students need to investigate and understand. It is an in-depth investigation of a real-world topic, and this is valuable for the attention and efforts of students.

In addition to entrepreneurship education, the value of personality as one of the internal factors that influences entrepreneurial tendencies is shaped by motivation and individual optimism. Motivation, optimism, and intelligence to overcome difficulties, the ability to survive, and continue to strive diligently are needed by individuals to face difficulties, where Stolzt & Stolzt (2000) mentions it as Adversity Quotient (AQ) which is a concept that can see how far a person is able to endure difficulties.

AQ in entrepreneurship is not only a person's ability to respond to obstacles or difficulties through his intelligence, but also ability to be an entrepreneur or prospective entrepreneur to take advantage of the opportunities available. An entrepreneur who has a high AQ is able to react appropriately and be able to face difficulties, whereas an entrepreneur who has a low level of AQ does not react well in overcoming difficulties (Stoltz & Stolzt 2000). Based on the background described, this study investigates the effect of PBL model implementation on entrepreneurship intensity based on AQ.

In the theory of planned behavior of Ajzen (1991), the performance of a behavior can be predicted from the plan and the intention of someone to do a behavior. The main idea of this planned theory of behavior is that one's behavior depends on the intention before the behavior is carried out. Intention is not only able to capture motivational factors that will influence behavior, but intentions are also important to understand in terms of understanding a behavior and the consequences that accompany that behavior. Bandura (1977) also emphasized that to determine a behaviour, a person's perceptions are needed regarding the activities they want to do, including entrepreneurial behavior.

Based on previous research, there are many factors that can help increase entrepreneurial success through improving entrepreneurial behavior. Several factors are classified into internal attributes, such as social-psychological state, ability, personality, intention, and attitude. Meanwhile, the external attributes consist of environmental factors, family background and situations that occur in their environment (Sasu & Sasu 2013, Solesvik et al. 2014, Kirkley 2015, and Shirokova et al. 2015).

Entrepreneurial behavior cannot be maintained if it is not supported by knowledge, skills and experience. Increased knowledge, skills and experience through the existence of entrepreneurship education are not only considered as a prerequisite for providing basic skills and abilities, but also entrepreneurship education can increase awareness and can facilitate the entrepreneurial process further.

In the process of entrepreneurship further there is a need for social-psychological state as well as the existence of entrepreneurial intentions previously cultivated through entrepreneurship education. Through social-psychological state that

refers to the theory of Entrepreneurial Event from Shapero & Sokol (1982), one's entrepreneurial behavior begins with the process of perceived desirability, perceived feasibility, and the tendency to act (propensity to act) can lead to entrepreneurial intentions that lead to entrepreneurial behavior. The stronger the individual's beliefs about personal desires, the greater the likelihood that they will behave in a certain way.

2 METHOD

The study used a quasi-experimental method to determine the effect of treatment. This study used factorial design 3 x 2. The research variables include X1 (Project Based Learning Model as an independent variable as treatment), and X2 (Adversity Quotient/High, Medium and Low as an independent variable as a factor), and Y (Entrepreneurship intention as the dependent variable). To find out in more detail how the factorial design can be seen in the following Table 1:

Table 1. Factorial experimental design.

| | | Method(A) | |
		Project Based Learning (Experiment Class) (A1)	Conventional (Control Class) (A2)
Factor (B)			
Adversity Quotient	High (B$_1$)	A1B1	A2B1
	Medium (B$_2$)	A1B2	A2B2
	Low (B$_3$)	A1B3	A2B3

3 RESULT AND DISCUSSION

The hypothesis testing used analysis of variance between subject designs. Anova in this experimental study was used to test the main and interaction effects of one or more nonmetric or categorical independent variables whose categories are more than two against one independent variable metric (interval, ratio). The independent variable is called a factor and this study involves two factors, namely Model and Adversity Quotient. Table 2 shows the results of the calculation of Anova between Subject Design.

Based on the results shown in Table 2, the significance value is < 0.05 so that the three proposed hypotheses are accepted, and the interaction between learning methods and emotional intelligence in relation to student financial literacy is clearly seen in the following profile picture plot of entrepreneurial intention (Figure 1)

Table 2. Partial results of testing anova between subjects.

Source	F	Sig.
Corrected Model	78.551	0.000
Intercept	1.082E5	0.000
Model	283.320	0.000
Adversity Quotient	31.770	0.000
Model * Adversity Quotient	3.466	0.038
Error		
Total		
Corrected Total		

a. R Squared = ,873 (Adjusted R Squared = ,862)

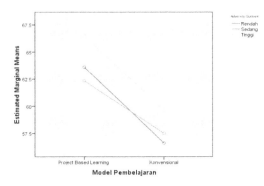

Figure 1. Profile plots interaction model and adversity quotient.

The influence of adversity quotient as one of the individual factors that influence entrepreneurial intentions aside from the learning model is a factor to optimize the increase in entrepreneurial intention.

Based on the results of hypothesis testing regarding the interaction of learning methods and adversity quotient on entrepreneurial intentions, the value of F = 3.466 and p = 0.038 <0.05 meaning that the hypothesis is accepted, and that there is interaction of project based learning models with adversity quotient on students' entrepreneurial intentions in subjects Entrepreneurship Basic Competency in Running a Small Business.

Entrepreneurial behavior cannot be maintained if it is not supported by knowledge, skills and experience. Through the use of project based learning models, the learning process not only results in increased knowledge, but is accompanied by students' entrepreneurial behavior and skills when compared to the learning process that uses conventional learning models. The existence of entrepreneurship education delivered with the right learning model can increase awareness and can facilitate the entrepreneurial process further.

In the process of entrepreneurship further there is a need for social-psychological state as well as the existence of entrepreneurial intentions previously cultivated through entrepreneurship education. Through social-psychological state that refers to the theory of Entrepreneurial Event from Shapero & Sokol (1982), one's entrepreneurial behavior begins with the process of perceived desirability, perceived feasibility, and the tendency to act (propensity to act) can lead to entrepreneurial intentions that lead to entrepreneurial behavior. The stronger the individual's beliefs about personal desires, the greater the likelihood that they will behave in a certain way

For this reason, the effect of using PBL model based on AQ on entrepreneurial intentions in Entrepreneurship subjects is one of the important factors in an effort to improve students' entrepreneurial intentions and behavior.

4 CONCLUSION

In can be concluded that there are differences in entrepreneurial intentions of students in the experimental class who use project based learning models with control classes that use conventional learning models. Entrepreneurial intention in the class that uses the project based learning model is higher than the class that uses conventional learning models.

Besides, there is a difference in increasing entrepreneurial intentions using project based learning models in students who have high, medium and low adversity quotient. Students with high AQ have a higher level of entrepreneurial intention than students with a moderate and low AQ.

In addition, there is an interaction between the PBL model, AQ, and entrepreneurial intention. Thus, in an effort to improve students' entrepreneurial intentions, in addition to the use of project based learning models, it is expected that students have adversity quotient with the aim of increasing entrepreneurial intentions and fostering entrepreneurial behavior that lasts longer.

REFERENCES

Ajzen, I. 1991. The theory of planned behavior. *Organizational behavior and human decision processes* 50(2): 179-211.

Alfonso, C.G. & Guzman, C.J. 2012. Entrepreneurial intention models as applied to Latin America. *Journal Organizational Change Management* 25(5): 721-735.

Bandura, A. 1977. Self-efficacy: toward a unifying theory of behavioural change. *Psychological Review* 84(2): 191-215.

Farashah, A.D. 2013. The process of impact of entrepreneurship education and training on entrepreneurship perception and intention: Study of educational system of Iran. *Education and Training* 55(8/9): 868-885.

Farhangmehr, M., Goncalves, P. & Sarmento, M. 2016. Predicting entrepreneurial motivation among university students. *Education and Training* 58(7/8): 861-881.

Gelderen, M.V. 2010. Autonomy as the guiding aim of entrepreneurship education. Education and Training 52(8/9): 710-721.

Kirkley, W.W. 2015. Entrepreneurial behaviour: the role of values. *International Entrepreneurial Behavior & Research* 22(3): 290-328.

Maresch, D., Harms, R., Kailer, N. & Wimmer-Wurm, B. 2016. The impact of entrepreneurship education on the entrepreneurial intention of students in science and engineering versus business studies university programs. *Technological forecasting and social change* 104: 172-179.

McClelland, D.C. 1987. *Human motivation*. CUP Archive.

Nielsen, S.L., & Stovang, P. 2015. DesUni: university entrepreneurship education through design thinking. *Education and Training* 57(8/9): 977-991.

Penaluna, A. & Penaluna, K. 2009. Creativity in business/business in creativity: ttransdisciplinary curricula as an enabling strategy in enterprise education. *Industry & Higher Education* 23(3): 209-219.

Sasu, A.L. & Sasu, B. 2013. On the asymptotic behavior of autonomous systems. *Asymptotic Analysis* 83(4): 303-329.

Shapero, A. & Sokol, L. 1982. The social dimensions of entrepreneurship. *Encyclopedia of entrepreneurship*: 72-90.

Shinnar, R.S., Hsu, D.K. & Powell, B.C. 2014. Self-efficacy, entrepreneurial intentions, and gender: Assessing the impact of entrepreneurship education longitudinally. *The International Journal of Management Education* 12(3) 561-570.

Shirokova, G., Tsukanova, T. & Bogatyreva, K. 2015. University environment and student entrepreneurship: The role of business experience and entrepreneurial self-efficacy. *Educational Studies* (3): 171-207.

Solesvik, M., Westhead, P. & Matlay, H. 2014. Cultural factors and entrepreneurial intention: The role of entrepreneurship education. *Education and Training* 56(8/9): 680-696.

Stoltz, P.G. & Stoltz, P. 2000. *Adversity Quotient@ Work: Make Everyday Challenges the Key to Your Success–Putting the Principles of AQ Into Action*. New York: William Morrow.

Advances in Business, Management and Entrepreneurship – Hurriyati et al (eds)
© *2020 Taylor & Francis Group, London, ISBN 978-0-367-27176-3*

How does financial constraint affect SOE's performance in Indonesia?

A. Ridlo & C. Sulistyowati
Universitas Airlangga, Surabaya, Indonesia

ABSTRACT: This research investigates the effect of state ownership on financial constraints, and the effect of financial constraints on firm performance. This study uses data of non-financial companies listed on the Indonesia Stock Exchange with the period 2013 - 2016, using purposive sampling method. In this research, we use two models, with two dependent variables, the first is financial constraint, and the second is firm performance. Both models are also controlled with variable age and firm size. Measurement of financial constraint in this research using KZ Index model proposed by Kaplan and Zingales (1997). Firm performance is measured using the amount of return on assets. The results show that state ownership negatively affect financial constraint and there is a negative effect between financial constraint on firm performance.

1 INTRODUCTION

State-Owned Enterprises (SOEs) was established as a form of intervention of the state in the economic and social sector. Aside from being a source of state revenue, SOEs are also supporting state programs in building social welfare. As the asset of state, the state gives special treatment to SOEs, which should make SOEs better in funding and performance. However, assurance and management as a state asset, does not guarantee SOEs will always benefit. In Indonesia, during the year 2012 to 2017, there are always state-owned losers.

In reality, there are poor performance of the company despite having political power from the state. Supported by political theory and social theory which concludes that the ownership of state shares in a company will lead to a worse performance than that of no state shareholding. In the study of Heider et al. (2017), social theory illustrates that the state set up SOEs to correct market failures in achieving social welfare. Political theory explains that the establishment of SOEs because of the objectives of state's political goals, one of which is to support programs made by the state (Shleifer and Vishny, 1998).

State ownership has an impact on the company's ability to obtain funding (Heider, 2017). Borisowa and Megginson (2011) argue that more debtors are not worried about getting their money back from SOEs, because of state guarantees .In an emergency case, the state will provide relief funds to state-owned enterprises that are losers. Therefore, SOEs grow differently from non-state enterprises in the face of financial constraints (Heider et al., 2017).

However, the external determinant of financial constraunt is underdeveloped (Heider et al., 2017). Access to external funding is essential to the company's investment and development. The limitation of companies in obtaining capital from available funding sources is a phenomenon of companies experiencing financial constraints. Kaplan and Zingales (1997) argue that financial constraints occur when firms face a difference between the cost of capital from internal funding sources and the cost of capital from external funding sources.

Research on state ownership and financial constraint is rare in Indonesia. One reason is because there is no definition and standard measurement of financial constraint. One of the most commonly used measurements, especially in developing countries is the KZ Index. KZ index is a measurement used by Kaplan and Zingales (1997) in measuring the financial constraint in his research.

Various studies have also examined how the impact of state ownership, and resulting in various outcomes, including easier access to debt (Claesens et al., 2008; Fan et al., 2008), get subsidies and tax benefits (Wu et al., 2011), has a poorer quality of financial statements than is not owned by the state (Chaney et al., 2011), is less likely to experience no financial constraint (Heider et al., 2017), and at certain financial constraint levels, firms with state share ownership, does not require assistance to reduce its financial constraint level (Lin and Bo, 2010).

Research from Haider et al. (2017) found that firms with higher state shareholdings, will tend to experience lower financial constraints, and firms with lower financial constraints will result in better performance. Lin and Bo, (2017) in his paper found that state ownership does not always reduce financial constraints. They also added that companies with state ownership did not reduce the financial constraints of a company based on the state-controlled banking sector. Nguyen et al. (2017) found empirical evidence of a negative relationship between state ownership and financial constraints in China. They also found low performance from state-controlled companies. Furthermore, they claim that different types of ownership will result in different levels of financial constraints.

Based on this background, this study intends to conduct further studies on the effect of state ownership on financial constraints and the influence of financial constraints and on the performance of non-financial listed companies on the Indonesia Stock Exchange. In addition, researchers also used the KZ Index measurement method to see how financial constraints in Indonesia are based on the sample year 2013-2016.

1.1 State ownership

In Government Regulation No. 12/1998, it states that state-owned enterprises are entities that are wholly or mostly owned by the state through direct participation of state assets. Companies with government ownership will grow differently because their attitudes toward financial constraint differ from other companies (Heider et al., 2017). With the power of government, they can negotiate with the rules of the government itself. In addition, as a government asset, companies with government ownership will have other benefits, such as subsidies, assistance with the state budget, government project allotment, even tax cuts. With these advantages, should the company with government ownership will have ease in terms of funding and also good performance. But in reality, in Indonesia many state-owned enterprises have lost.

The government invests in the company and takes over the power of a company, to engage itself in market mechanisms in improving social welfare. In Indonesia, there are various parties that try to damage the balance of the market, by hoarding and selling it when prices rise. The government has an obligation to participate in controlling the market. Social theory in believe that, the government set up State-Owned Enterprises to improve market failure in creating social welfare(Heider et al., 2017). This theory argues that SOEs are established not only to perform business functions, but also social functions. So in theory it concludes that government ownership will negatively affect firm performance.

Political theory explain that the establishment of SOEs is intended as a political tool of the government. Beck et al. (2005) says that government ownership in a company can provide benefits in the face of regulation, but also as a political tool. Government as the holder of control of a company, has many political interests. This makes the business functions of the company to be ruled out, so that will affect the company's performance. In political theory, the government is perceived to have a goal of prioritized political objectives rather than social welfare objectives. Just like social theory, political theory assumes that government ownership will have a negative impact on performance.

In this study, it uses a sample of listed companies that have government ownership in their capital structure. The measurement of the level of government stock ownership in a company is as follows:

GovOwni, t = Number of shares owned by the government$_{i,t}$/number of outstanding shares$_{i,t}$

Financial Constraint

There is no standard definition of financial constraint. According to Lamont et al. (2001), financial constraint is what prevents the company from getting financing for its investment. Kaplan and Zingales (1997) argue that financial constraints occur when firms face different capital costs from internal and external funding sources. White and Wu (2006) argue that the condition of a company experiencing financial constraint when there is a problem in obtaining external sources of funds. According to Hannessy and Whited (2006) Financial constraint is defined when a company has access to profitable investment opportunities but the company is experiencing limitations to fund such investment opportunities with external financing.

Kaplan and Zingales (1997) have expressed their criticisms of Fazzari et al. (1988) on corporate investment and cash flows to examine the existence and importance of financial constraints. Fazzari et al. (1988) found that the higher the investment cash flow sensitivity, the higher the distance between internal and external funding costs, or can be said to have a high level of financial constraint. In this study, using the KZ Index measurement proposed by Kaplan and Zingales (1997).

Companies with a high KZ value, mean that the company will have difficulty in finding external funding. Bavarsad et al. (2013) in their research to calculate the level of financial constraint using KZ index modified by Chen et al. (2012) and Lin and Bo (2017). This KZ index is calculated using the model:

$$KZ_{i,t} = -1.002(CF_{i,t}/TA_{i,t-1}) - 39.368(DIV_{i,t}/TA_{i,t-1}) - 1.315(CA_{i,t}/TA_{i,t-1}) + 3.139LEV_{i,t} + 0.283 Q_{i,t}$$

$CF_{i,t}/K_{i,t-1}$ = Cash flows divided by previous year's assets.

Dividend$_{i,t}/K_{i,t-1}$ = Dividend divided by previous year's assets.

Cash$_{i,t}/K_{i,t-1}$ = Total cash divided by previous year's assets.

Lev$_{i,t}$ = Total debt divided by total assets at the end of the year.

$Q_{i,t}$ = The ratio of market to book value divided by total assets

1.2 Firm performance

The study of performance appraisal has been widely discussed among academics and practitioners. The development of corporate performance assessment from accounting based, market performance, to synergize performance appraisal through balance scored card. However, in this study, firm performance is measured based on the company's financial performance.

In this study, the indicator used to measure performance measurement is Return on Assets (ROA). Abbasid and Malik (2015), Yammessri and Loth (2004) examine the company's financial performance with ROA as an indicator of its performance measurement. According to Sudana (2015: 25) ROA shows the company's ability to use all its assets in generating profit after tax. This ratio is important for management to evaluate the efficiency and effectiveness of the company's management in managing all of the company's assets. ROA calculations by Sudana (2015: 26), are as follows:

$$ROA_{i,t} = \text{Earning after tax}_{i,t}/\text{Total asset}_{i,t}$$

1.2.1 The effect of government ownership on financial constraints

In the study of Nguyen et al. (2017) found that there is a strong positive relationship to government ownership with financial constraints. However, some studies produce the opposite result, ie there is a negative relationship between government ownership and financial constraint (Heider et al., 2017; Cull et al., 2015)

Companies with high levels of ownership will gain political power in getting funding or income. Generally, the government will use its owned companies to work on projects from government programs, so companies will earn more revenue from other companies. In the area of funding, government-owned companies will be more reliable, because of the implicit protection of the government. In conclusion, in the area of funding companies with higher government ownership will benefit more. Thus, the first hypothesis proposed is:

H1: Government ownership negatively affects financial constraint

1.3 The effect of financial constraints on firm performance

Heider et al. (2017), found that firms with low levels of financial constraints had better performance. Muso and Schiviavo (2008) point out in his research that access to external funding increases the growth of firms using a certain period of time in measuring financial constraints. Chen et al. (2009) find that different types and levels of government ownership, will result in performance differences.

Companies with lower financial constraints will have a better chance of getting funding in running their business. Furthermore, the company is easier in investing in business projects or in its assets. Thus, with a low level of financial constraint, will have an impact on higher performance. This becomes the second hypothesis formulation, namely:

H2: financial constraint has a negative effect on firm performance

2 RESEARCH METHODOLOGY

2.1 Population and sample

The data used are secondary and quantitative data. The data is sourced from the company's financial statements of the period 2013-2016 which can be obtained from the Indonesian stock exchange website that is www.idx.co.id. The population of this study is all non-financial companies in Indonesia. The sampling technique in this research is purspo-sive sampling. The technique of purposive sampling is the technique of determining the sample with the consideration of certain criteria.

2.2 Model

To test the above hypotheses, this study uses the model as following:

Model 1:
$$FinCons_{i,t} = \beta_0 + \beta_1 GovOwners_{i,t} + \beta_2 Size_{i,t} + \beta_2 Age_{i,t} + \varepsilon_{i,t}$$

Model 2:
$$Performance_{i,t} = \beta_0 + \beta_1 FinCons_{i,t} + \beta_2 Size_{i,t} + \beta_2 Age_{i,t} + \varepsilon_{i,t}$$

Where performance is measured using Return On Assets (ROA) which is the probability ratio of the company (Abbasid an Malik, 2015). ROA can be measured by the division between income after tax with total assets (Sudana, 2015: 25) Financial constraint measurement using the merode of KZ Index (Lin and Bo, 2010). The control variables used in this study are firm size and age. Company size is measured by the log of total assets. The bigger the companies show better profitability than smaller companies. The age of the company shows how long the company was formed, operating, and surviving in business competition. The age of the company shows the company's experience in operating.

3 RESULT AND DISCUSSION

Below is the result of multiple linear regression test with two models. The first model is to examine the influence of government ownership (GovernOwn), on financial constraints, controlled by size and age of the firm. The second model is to examine the effect of government ownership, financial constraint, on firm performance (ROA) and controlled by age and firm size. Table 1 present the summary of descriptive statistics for those two models.

Based on regression test results (Table 2 and 3), the relationship between government ownership and financial constraint on the performance of significant negative value. This can be seen from the sig numbers below 0.05. This means that the results of the summary of multiple linear analyzes in the above table, all independent variables have, in model 1,

Table 1. Descriptive Statistics.

Var	N	Min	Max	Mean	Std. Dev
Model 1					
FCKZ	994	-3.50	3.46	0.159	1.254
GovOwn	994	0	0.9	0.024	0.126
Age	994	4	115	31.98	15.11
Size	994	22.97	33.2	28.28	1.655
Model 2					
ROA	969	-0.25	0.36	0.034	0.072
FCKZ	969	-12.1	8.35	0.091	1.687
Age	969	5	109	31.73	12.51
Size	969	23	33	28.28	1.67

Table 2. Regression Result Model 1.

Variable	Coef	Sig	Result
Constant	1.885	0.007	
GovOwn	-0.719	0.027	Significant*
Age	-0.010	0.000	Significant*
Size	-0.049	0.044	Significant*

Dependend Variabel: FCKZ

Table 3. Regression Result Model 2.

Var	Coef	Sig	Con*
Const	-0.215	0.000	
FCKZ	-0.024	0.000	Significant*
Age	0.0003	0.097	Significant**
Size	0.009	0.000	Significant*

Dependend Variabel: ROA

there is a significant influence of government owner-ship variable, age and firm size to financial con-straint as measured by KZ index.

Table 2 shows the influence of government own-ership on the financial constraint has a negative coef-ficient of -0.719. This negative relationship indicates that any increase in government ownership in a company, it will lower the level of financial con-straint on the company. Similarly, if there is a decrease in the level of government ownership, then there will be an increase in financial constraints on the company. This explains that in terms of exter-nal funding, firms with higher levels of government ownership will benefit more.

For variable age and firm size also have negative relation with financial constraint. The coefficient of age and company size is -0.01 and -0.049 which means, the older the company and or the larger the company, the financial constraint level will be smal-ler, and vice versa. So it can be concluded that the

experience and size of the company's assets have an influence on the company's external funding.

According to Table 3, there is influence of financial constraint variable, government ownership, controlled by age and firm size to firm performance measured by ROA. This influence is significant, can be seen from the level of significance that is above 0.05 for all variables. Financial constraints have a negative effect on government performance. Coefficient -0.215, explains that if level financial constraint up one unit, then there will be a decrease in ROA of -0.215. That is, that financial constraints faced by the company will negatively affect the company's performance.

The age and firm size control variables have a significant influence on firm performance. The older the company, the more experienced, and the better its performance. This is in accordance with the results of this study which states the relationship between firm age and performance is positive. Com-pany size also has a positive effect on performance. With a coefficient of 0.009, this positive effect means that the size of the firm will improve the per-formance of the company.

3.1 Government ownership of the financial constraint

In the results of this study, found a negative effect on the relationship of government ownership to financial constraint. This result is in accordance with the proposed hypothesis. Supported by research conducted by Heider et al. (2017), who found that companies with government-owned levels, would have a lower financial constraint. In Indonesia, companies with government ownership, especially those with BUMN status, will receive special treatment from the government. Starting from the management and supervision of the gov-ernment, to the injection of funds from the State Budget, merupakah a profit for the company. State-owned enterprises have easy access because they are state assets, and the state will not let them fall into financial difficulties. In accordance with the opinion of Borisova and Megginson (2011) that the lender would be less worried about getting his money back in a state-owned company because of implicit government guarantees.

In research of Lin and Bo (2010), found that gov-ernment ownership levels do not always reduce financial constraints. Research conducted in China produces inconsistent results, but the conclusion is that at a certain level of financial constraint, the level of government ownership can not reduce a company's financial constraints. This distinction can be generated due to different political and gov-ernance elements between companies in Indonesia and in China. Blancard and Shleifer (2000) explain that companies with government as one owner, will have political access and lobbying power, let alone the legitimacy of institutions, will have an effect on corporate finance.

The result of this empirical study, using a sample of non-financial companies with sample periods 2013-2016, indicates that one of the advantages of being a State-Owned Enterprise or at least of government ownership in the company is related to obtain external financing.

3.2 Financial constraints to firm performance

Firm performance is measured by ROA because in this study would measure how the financial performance is assessed from the company's ability to generate profits. In the second hypothesis, test whether there is influence of financial constraint on firm performance measured by ROA.The result is that there is a significant negative effect between the financial constraint on performance. Which means, the higher the financial constraint will be the lower the company's performance. This is also supported by research from Heider et al. (2017), which states that companies with low financial constraints, will perform better. They add that companies with high levels of financial constraints will have limited options in maximizing their value.

Financial constraint is an external financial constraint, where companies have difficulty in obtaining external funding, such as bank loans, issuing letters or issuing shares. The difficulty in obtaining this funding could be the high interest rate on the banking credit and or the high interest on the bonds that must be offered. If the company has difficulty in getting funding, then the company will have difficulty in making investments, so it will reduce the company's ability to generate profits.

3.3 Influence of control variable to financial constraint and firm performance

The control variables of the first and second models, using the age and size of the firm. For the first model, the influence of age and firm size on financial constraint is negatively significant. That is, the older the company and/or the larger the company, the financial constraint will be smaller, and means more and no financial constraints. The older the company, the more companies have the experience, and the published information related to the company more and more, the performance record is also more and more, so it will help in reducing the asymmetry of information from companies and outsiders. The less information asymmetry, the more experienced a company will be, the more it will help in obtaining external funding.

The size of a large company can be a guarantee if at any time the company goes bankrupt. Therefore, creditors and investors will be more confident in companies that have a large size compared with small companies. Therefore, the size of the company will negatively affect the financial constraint. These results are consistent with research by Heider et al.

(2017) and Bougheas et al. (2006) that young companies or small companies are significantly more likely to face financial constraints and older companies will have less difficulty due to low information asymmetries.

For the second model, there is a significant positive influence between the variable age and firm size on ROA. This means that the older and larger the size of the company, the ability to generate their profits the better, in other words, have a good performance.The older the company, the more experienced in dealing with business problems. Therefore, the company will be more appropriate in analyzing and making decisions, and will be better performance. Companies that are large, in general, have a large market capitalization and influence on its ability to generate profits. Large companies have large funding access and will affect performance. Therefore, it is consistent with the results of this study, that firm size will have a positive effect on performance.

4 CONCLUSIONS

Based on the results of research and analysis that have been described previously, the first conclusion is Government ownership negatively affects the financial constraint. This explains that firms with high levels of government ownership will gain political power that helps in external financing, thus having low levels of financial constraints. Second, financial constraint has a negative effect on firm performance. Company's external financial constraints significantly affect performance. The more a company is constrained in external financing, it will affect the company's performance will be lower. Third, the age and firm size control variables negatively affect the financial constraint, and have a positive effect on performance. The older and bigger the company, the lower the financial constraint level and the better performance.

REFERENCES

Abbasid, A., & Malik, Q. A. 2015. Firms' Size Moderating Financial Performance in Growing Firms: An Empirical Evidence from Pakistan. *International Journal of Economics and Financial Issues*, 5(2),334-339.

Bavarsad, B., Sinei, H., & Delavaripour, J. 2013. Study on the Relationship Between Financial Constraints and Stock Return in Tehran Stock Exchange. *International Journal of Economy, Management, and Social Sciences* 2(5),166-173.

Beck, T., Demirguc-Kunt, A., & Maksimovic, v. 2005. Financial and Legal Constraintto growth: Does firm size matter? *Journal of Finance* 60, 137-177.

Bhougeas, S., Mizen, P., & Yalcin, C. 2006. Access to external finance: Theory and evidence on the impact of monetary policy and firm-specific characteristics. *Journal of Banking and Finance*, 30, 199-227.

Blancard, Olivier, & Shleifer, A. 2000. Faderalism with and without political centralization: China vs Russia. *NEBR Working Paper 7616, National Bureau of Economic Research.*

Borisova, G., & Megginson, W. L. 2011. Does State Ownership affect the Cost of Debt? Evidence from Privatization. *Review of Financial Studies*,24, 2693-2737.

Campbell, J. L., Dhalliwal, D., & Schwartz, W. 2011. Financing constraints and the cost of capital: Evidence from the funding of corporate pension plans. *Review of Financial Studies*, 25, 868-912.

Chaney, P. K., Faccio, M., & Parsley, D. 2011. The quality of accounting information in politically connected firm. *Journal Accounting and Economics*, 58-76.

Chen, G., Firth, M., & Xu, L. 2009. Does the Type of Ownweship Control Matter? Evidence from China's Listed Companies. *Journal of Banking and Finance*, 33, 171-181.

Chen, H., & Chen, S. 2012. Investment-cash flow sensitivity cannot be a good measure of financial constraint: Evidence from the time series. *Journal of Financial Economics* 103, 393-410.

Claessens, S., Feijen, E., & Laeven, L. 2008. Political connection and preferential access to finance: The role of campaign contribution. *Journal of Finance and Economic* 88, 554-580.

Cull, R., Li, W., Sun, B., & Xu, L. C. 2015. State Connection and financial constraints: Evidence from large representative sample of Chinese firm. *Journal of Corporate Finance* 32, 271-294.

Fan, J., Wong, T. J., & Zhang, T. 2008. Politically connected CEOs, corporate governance, and Post IPO performance of China's newly partially privatized firm. *Journal of Financial and Economic* 84, 330-357.

Fazzari, S., Hubbard, R. G., & Petersen, B. 1988. Financing Constraint and Corporate Investment. *Brooking Papers on Economic Activity* 1, 41-95.

Fazzari, S., Hubbard, R. G., & Petersen, C. B. 1987. Financing Consrtraints and Corporate Investment. *NEBR Working Paper*, 2387.

Guariglia, A., Liu, X., & Song, L. 2008. Internal Finance and Growth: Microeconomics Evidenceon Chinese Firms. *IZA Discussion Paper* 3808. Institute for the Study of Labor (IZA).

Headd, B., & Kirchhoff, B. A. 2007. Small Bussiness Growth: Searching for Stylized Facts. *SSRN Electronic Journal.*

Heider, Z. A., Liu, M., & Wang, Y. 2017. State Ownership, Financial Constraint, Corruption and Corporate Performance: International Evidence. *Journal of International Financial Market, Instirution, and Money.*

Hennessy, C., & Whited, T. 2006. How Costly is External Finance? Evidence from a Structural Estimation. *Journal of Finance* 22, 1705-1745.

Indonesia Stock Exchange. (n.d.). *Retrieved from* www.idx.co.id

Jiang, B., Laurenceson, J., & Tang, K. 2008. Share Reform and The Performance of China's Listed Companies. *China Economic Review*, 489-501.

Kaplan, S. N., & Zingales, L. 1995. Do Financing Constraints Explain Why Investment is Correlate with Cash Flow? *NEBR Working Paper*, 5267.

Kaplan, S., & Zingales, L. 1997. Do Investment cash flw sensitivities profide useful measure of financial constraint?. *Quarterly Journal of Economics* 34, 169-215.

Lamont, O., Polk, C., & Saaa-Requejo, J. 2001. Financial Contraint and Stock Return. *Review Financial Study* 14, 529-554.

Lin, H.-C. M., & Bo, H. 2012. State Ownership and Financial Constraints on Investment of Cchinese Listed Firm: New Evidence. *European Journal of Finance* 18, 497-513.

Muso, P., & Sciavo, S. 2008. The impact of financial constraint on firm survival and growth. *Journal of Evolution Economics* 18, 135-149.

Nguyen, G., Nguyen, M., & Li, L. 2017. Does State Ownership Mitigate Financial Constraints? New Evidence. 30th Australasian Finance and Banking Conference 2017. *Brisbane: Financial Research Network (FIRN).*

Riyadi. 2006. *Banking Aset and Lability Management Edisi 3.* Jakarta: Penerbit Fakultas Ekonomi Universitas Indonesia.

Sapienza, P. 2004. The Effect of State Ownership on Bank Lending. *Journal of Finance and Economic* 72, 357-384.

Shleifer, A., & Vishny, R. 1998. *The Grabbing Hand: State Pathologies and Their Cures.* Cambridge, MA: Harvard University Press.

Sudana, I. M. 2015. *Manajemen Keuangan Perusahaan Teori dan Praktek*, Edisi 2. Jakarta: Penerbit Erlangga.

Wasiuzzaman, S. 2015. Working Capital and Firm Value in An Emerging Market. International *Journal of Managerial Finance*, 60-79.

Wei, G. 2007. Ownership structure, corporate governance, and company performance in China. *Asia Pacific Business Review*, Vol. 13, No. 4, 519-545.

Whited, T., & Wu, G. 2006. Financial constraint risk. *Review of Financial Studies* 19, 531-559.

Wibowo, S. S., & Lolyta, S. 2016. Cash and Financial Constraints: An Empirical Analysis of Non-Financial Firms Listed in Indonesian Stock Exchange 2005-2014. *Sebelas Maret Business Review*, Issue 1.

Wu, W., Wu, C., & Rui, O. M. 2012. Ownership and the Value of Political Connections: Evidence from China. *European Financial Management*, 695-729.

Yammesri, J., & Loth, D. C. 2004. Is Family Ownership Pain or Gain to Firm Performance. *Journal of American Academy of Business*, Cambridge(4), 263-270.

Yi, & Tzu. 2005. Relationship beween multinationally and performance of Taiwan Firms. *Journal of American Academy of Bussiness.* 130-134.

Zhao, W. 2016. Corporate Governance, Financial Constraint, and Value of Cash Holding: Research fromPErspective of Ultimate Controllers. *Modern Economy*, 1096-1119.

Advances in Business, Management and Entrepreneurship – Hurriyati et al (eds)
© 2020 Taylor & Francis Group, London, ISBN 978-0-367-27176-3

Organization culture and organization effectiveness

D. Ekowati, J. Sulistiawan & E.R. Triksina
Universitas Airlangga, Surabaya, Indonesia

ABSTRACT: This study examines the effect of organizational culture dimensions which consists of four cultural dimensions, adaptation, mission, consistency, and involvement on the company's performance effectiveness, with the use of data collected from the go-public company. Using 11 go-public companies being paired with three objective measures of organizational performance effectiveness (ROA, market to book, sales growth). Relationships among variables were analyzed using Multilevel Regression method. The results show that the combination of adaptation culture with consistency has significant effect on company's performance effectiveness (ROA). Interestingly, the consistency dimension did not affect organizations' performance due to the environmental conditions at the go-public company being so dynamic that it will ignore the flexibility and changes in the market that are important to improve profitability. Overall, the most influential on the effectiveness of the company's performance is the combination of cultural adaptation with consistency.

1 INTRODUCTION

Organizations often face a variety of issues when there is interaction with the environment, especially if the environment is dynamic. Organizations need to adjust to these changing circumstances in order to address the problems that occur and to gain competitive advantage (Schein, 1992). In addition, at the same time the organization also faces internal problems, which require the organization to cope so that there remains an alignment in the functioning of the organization. Organization attempts to overcome external and internal problems such as, organizations need to establish strong organizational culture, if they want to survive, even if they want to expand and gain competitive advantages (Schein, 1992). Basically organizational culture is a set of shared values, beliefs, and norms that affect employees' way of thinking, feeling, and behavior at work (Schein, 2011). Culture is important because culture has four functions that lead the organization to success as an organizational identity, maintaining organizational stability, tools for achieving goals and shaping commitments (Kreitner and Kinichi, 2009). In addition, organizational culture is able to control the interaction of its members with each other and also interactions with suppliers, customers and elements outside the organization (Jones, 1995). However, Fey and Dennison (2003: 198) argued that the dimensions of organizational culture (involvement, consistency, adaptability and mission) are important factors in affecting the effectiveness of an organization. Therefore, every organization needs to create shared values to build an organizational system to unite thoughts and actions and transform individual behavior into organizational behavior.

Based on the main dimensions of organizational culture has different mechanisms that may improve organizational effectiveness. For example, the dimension of involvement helps organization expand the number of alternatives to take into account when making decisions. Strong involvement may strengthen group dynamics in solving complex problems, increase employees' commitment and enthusiasm in decision execution so as to enhance internal integration, increase the flexibility of an organization and increase creativity. In other words, organizations with high involvement tend to be more effective (Denison and Mishra, 1995).

Furthermore, Denison and Mishra (1995) emphasized the importance of a combination of organizational cultural dimensions that may increase its effectiveness. Dimensions which are oriented toward adaptability and engagement will be more diversity, more input, and more solutions to a particular situation. As for the dimensions of consistency and mission may give higher emphasis on control and stability. This orientation towards control and stability is probably best in situations where the organization provides a limited set of responses that are limited but precise and adapted to a stable environment. Organization tends to apply different techniques, eliminate tension, gain compromise, balance internal and external focus, and maintain flexibility and stability control. The relationship between culture and organization effectiveness is seen as an organization's success due to a combination of values and beliefs, rules and practices, and the relationship between the two. The logical consequence of this is the need to identify organizational dimensions in supporting organizational effectiveness.

2 LITERATURE REVIEW AND HYPHOTESIS

2.1 *Organization culture*

Kreitner and Kinicki (2009) defined organizational culture as a social binder to remind members of an organization that a different personality or personality can be incorporated into an organizational strength. Organizational culture is often defined as values, symbols that are understood and obeyed together, owned by an organization so that members feel to be one family and create a condition of members of the organization feel different from other organizations (Alvesson, 2011).

Denison and Mishra (1995) classified organization culture into four dimensions

a. Mission. It reflects the extent to which an organization has a direction and clarity of purpose. Effective organizations pursue missions that provide meaning and direction for their employees (Denison and Mishra, 1995). These organizations have clear goals and direction, goals and objectives, and visions for the future (Fey and Denison, 2003).
b. Involvement. It focuses on the extent to which employees commit to their work, feel a sense of ownership, and feed into decisions that affect their work. As Fey and Denison (2003) argue, effective organizations empower their employees, use teamwork, and continue to develop the capacity of their employees (Becker, 1964, Kennedy, 1982, Lawler, 1996; Likert, 1961; Peters and Waterman, 1982).
c. Consistency. It refers to the integration regarding values and norms. Behavior roots on a set of core values and well-integrated organization activities (Kotter and Heskett, 1992; Saffold, 1988).
d. Adaptability. It reflects the employees' ability to understand and accommodate consumers' needs, change in response to new demands by learning new skills and technologies (Momot and Litvinenko, 2012).

2.2 *Organization effectiveness*

Dunn (2000: 429) defined effectiveness as whether an alternative achieves expected outcomes or results, or achieves the objectives of the action. Furthermore, Dunn (2003: 601) adds that effectiveness is an evaluation criterion that whether the desired outcome has been achieved. Robbins (2013) argued that the notion of organizational effectiveness is the extent to which the organization can achieve the various objectives (short term) and objectives (long term) that have been established, where the determination of goals and objectives that reflect the strategic constituency, subjective interests assessment and stage of organizational growth.

Organizational effectiveness is measured by suitability between organizational goals and observed results. This measurement is important in determining the degree of conformity between objectives and outcomes. Effectiveness is measured as how well it works and achieves the desired outcome. In particular, this effectiveness emphasizes the difference in subjective perceptions of sales growth, market share, profitability, quality, new product development, employee satisfaction, and overall performance (Denison et al., 2003) as well as with objective indicators of return-on-assets (ROA) (Denison and Mishra, 1995). Measuring the performance of both subjective and objective organizations is equally important, but this study focuses on the objective indicator of the return on asset ratio (ROA). As an independent parameter of effectiveness, ROA can overcome the deficiencies of common methods using the effectiveness indicators commonly used in the study. Each of these metrics provides a slightly different perspective and together, provides a comprehensive picture of effectiveness. ROA is a company's ability to use all assets owned to generate profit after tax. This ratio is important for the management to evaluate the effectiveness and efficiency of the company's management in managing all the assets of the company in achieving the effectiveness of the company.

2.3 *Hypothesis*

The results of prior studies implied that the cultural dimensions of involvement, consistency, adaptability, and mission are important variables in predicting organizational performance. The internal integration focuses on the dimensions of engagement and consistency generally as a better predictor of the measurement of operating performance such as quality and profitability, while externally focused on mission dimensions and adaptability generally as a better predictor of sales growth. Similarly, mission-focused strategic directions and intentions are the only significant predictors of market share. Another important trend is that new product development is highly correlated to consistency dimension, whilst engagement is the strongest predictor of employee satisfaction. These findings suggest that cultural aspects assessed in the DOCS (Denison Organizational Culture Survey) will contribute to overall organizational effectiveness.

Adaptation may accommodate inputs from external factors and customer expectations with internal change as it enhances the organization's ability to cope with the dynamics of growth and environmental uncertainty (Denison and Mishra, 1995). Adaptive organizations are influenced by the customer, take risks and learn from mistakes, have the capability and experience in creating change. Organizations are constantly changing in order to increase value to customers. Internal integration and internal adaptation are often becomes the obstacles for the organization. Therefore, adaptability is generally regarded as a specific indication of a organization's ability to adapt with new areas (through innovation and product development, market expansion etc.) and the ability to overcome unexpected external threats (Momot and Litvinenko, 2012). Thus,

H1: Adaption is positively related to organization effectiveness

Organization effectiveness will be achieved as long as the organization has clear direction and strategy which declare the organization's goals and vision (Fey and Denison, 2003). As the mission of the organization changes, change also occurs in other aspects of organizational culture. Based on the external focus of an organization which focused on stability that can enable organizations to maintain the value of internal integration as objectives and values that are perceived as strong factors leading to increased market share, improving financial performance and improving the overall efficiency of the organization (Denison and Mishra, 1995; Momot and Litvinenko, 2012). Thus,

H2: Mission is positively related to organization effectiveness

Organizations are effective because they have a strong, consistent, coordinated and well-integrated culture (Saffold, 1988). Behavior is perceieved as a set of core values, where leaders and followers have the skills to gain mutual agreement, even when they have different points of view. This type of consistency is a powerful source of internal stability and integration as a result of the common mindset and high degree of conformity (Kotter and Heskett, 1992). The dimension of consistency culture will be able to increase organizational efficiency and productivity significantly by improving coordination and communication. Thus, consistency can be treated as a predictor that can increase economic efficiency (Momot and Litvinenko, 2012). Thus,

H3: Consistency is positively related to organization effectiveness

Organization is effective when it involves its members to build team which will develop its members' ability. Executives, managers and staffs presume that they may influence the decision which will affect their job. High involvement implies may strengthen groups dynamic in solving complex problems, increase internal integration, flexibility and also creativity. Thus,

H4: Involvement is positively related to organization effectiveness

3 METHODS

3.1 Sample and procedures

We collected our data in two stages. In stage 1, we emailed organizations to invite them in our survey. In the end, only 11 organization responded our email. In stage 2, we use questionnaires to organization who willing to join our survey which consist of manufactures, services, distributors and mining. Total respondents in our study was 207 respondents which consists of top management (43%) and middle management (57%).

3.2 Measures

All measures used a response scale from 1 was "strongly disagree" to 7 was "strongly agree. The organization culture was measured by using Denison Organizational Culture Survey (DOCS) (Denison, 1990) which consists of five dimensions with 15 item for each dimensions, Adaptation (α=0.823), Mission (α=0.894), Consistency (α=0.819), Involvement (α=0.879). Organization performance was measured using objective organization performance which is ROA for the last 3 years.

4 RESULTS

We used Hierarchical Linear Modelling (HLM) to test our hypothesis. We regressed the dimension of organization culture on organization effectiveness (ROA) to test our hypothesis. Table 1 shows the regression between each dimension on ROA as dependent variable.

As shown in Table 1, the adaptation was statistically significant in predicting organization effectiveness ($p < 0.05$), thus supporting hypothesis 1. Mission and involvement were also statistically significant in predicting organization effectiveness ($p < 0.10$ and $p < 0.05$), thus supporting hypothesis 2 and 4. On the contrary, the result show that consistency was not statistically significant in predicting organization effectiveness, thus hypothesis 3 was not supported.

Table 1. Hierarchical linear modeling.

Dependent Variable	Parameter	Estimate
Organization Effectiveness (ROA)	Intercept	-0.03118
	Adaptation	0.042001* (0.01)
	Intercept	0.018463
	Mission	0.028365** (0.051)
	Intercept	0.043966
	Concistency	0.022375
	Intercept	0.009769
	Involvement	0.030639* (0.042)

*p < .05, **p < .10

973

5 DISCUSSION

This study examined how the dimension of organization culture which consist of four dimension (adaptation, mission, consistency and involvement) affect organization effectiveness.

Our study support prior study form Momot and Litvinenko (2012) that organization culture has positive effect on organizational performance. The adaptation may accept and accommodate inputs from external environment, such as customer expectation, which may enhance organization's ability to cope with its dynamic and uncertain environment. Therefore, adaptability plays critical role as a specific indication of company's ability to address external problems in a new region through innovation and product development, market expansion and the ability to cope with unexpected external threats.

The mission culture dimension has benefits and directions that may clarify organizational goals and objective as well as demonstrate the organization's vision of the future as goals and values, considered a strong factor leading to increased market share, improving financial performance and improving the overall efficiency of the organization Momot and Litvinenko, 2012).

Fey and Denison (2003) argued that effective organizations are empowering their employees, using teamwork, and continuing to develop their employee capacity will create increased market share, improve financial performance and improve overall organizational efficiency leading to the company's performance effectiveness (Becker, 1964; Kennedy, 1982; Lawler, 1996; Likert, 1961; Momot and Litvinenko, 2012; Peters and Waterman, 1982).

Prior study from Momot and Litvinenko (2012) also support our finding that consistency has a weak impact on organization performance. This is due to internal integration issues, which are less able to coordinate work in the field (division) and different skills within an organization. In addition, the lack of alignment in obtaining consensual agreements of complex problems by ignoring corporate values will make the condition more complicated (Momot and Litvinenko, 2012). Moreover, previous study by Ritchie, et al (2012) argued that organizational capabilities are internally focused on stability in the use of consistency value, which would ignore the flexibility and changes in the market including important things to improve profitability. In the long run, this strategy is likely to lead to a loss of market value and market share. However, this strategy will work for the short term. This is supported by Kotter and Heskett (1992) which suggest that a consistent culture is useful as long as its business environment is relatively stable. Therefore, this situation may not be effective in the long term because the business environment tends to change over time.

REFERENCES

Alvesson, M. 2011. Organizational Culture: Meaning, discourse, and identity. In: Ashkanasy N, Wilderom C and Peterson M (eds) The Handbook of Organizational Culture and Climate, 2nd edn. Thousand Oaks, CA.

Becker, G. 1964. Human Capital: A Theoretical and Empirical Analysis with Special Reference to Education. New York: Columbia University.

Denison, D.R., and Mishra A.K. 1995. Toward a Theory of Organizational Culture and Effectiveness, Organization Science. Vol. 6, pp. 204-223.

Denison, D.R, Haaland, S., and Goelzer, P. 2003. Corporate Culture and Organizational Effectiveness: Is Asia Different From The Rest of The World? Organizational Dynamics 33(1): 98–109.

Dunn, William N. 2003. Pengantar Analisis Kebijakan Publik. Yogyakarta: Gadjah Mada University Press.

Fey, C.F., and Denison, D.R. 2003. Organizational Culture and Effectiveness: Can American theory be applied in Russia? Organizational Science 14(6): 686–706.

Jones, G.R. 1995. Organizational Theory Texts and Cases, USA, Addison Wesley Publishing Company, Inc.

Keenedy, Allan A., and Deal, Terrence, E. 2003. Strong Culture: The New "Old Rule" for Bussiness Success. Wesley Publishing Company, P: 327-335.

Kotter, J., and Heskett, J. 1992. Corporate Culture and Performance. New York: Free Press.

Kreitner, Robert and Kinicki, Angelo. 2009. Organizational Behavior. McGrawHill International Edition. 8th Edition: New York USA.

Lawler, EE III. 1996. From the Ground Up: Six Principles for Building the New Logic Corporation. San Francisco, CA: Jossey-Bass.

Likert, R. 1961. New Patterns of Management. New York: McGraw-Hill.

Momot, Volodymyr E., and Litvinenko, Olena M. 2012. Relationship Between Corporate Culture and Effectiveness of an Organization. Alfred Nobel University, Dnepropetrovsk, Ukraine 1(8): 5-20.

Peters, TJ., and Waterman, RH. 1982. In Search of Excellence: Lessons from America's Best-Run, Companies. New York: Harper dan Row.

Ritchie, S. A., Kotrba, L. M., Gillespie, M. A., Schmidt, A. M., & Smerek, R. E. 2012. Do Konsistent Corporate Cultures Have Better Business Performance? Exploring The Interaction Effects. Sage Journal.

Saffold, G. 1988. Culture traits, strength, and organizational performance: Moving beyond 'strong'culture. Academy of Management Review. 13(4): 546–558.

Schein, Edgar H. 1992. Organizational Culture and Leadership, 2nd edn. San Francisco, CA: Jossey-Bass.

Schein, Edgar H. 2010. Organizational Culture and Leadership, 4th edn. San Francisco, CA: Jossey-Bass.

Schein, Edgar H. 2011. Organizational Culture and Leadership, edisi. Terjemahan, YPTK. Padang.

Business ethics among sellers of imported used clothes in Royal Plaza Surabaya

A. Armuninggar & A. Aris
Universitas Airlangga, Surabaya, Indonesia

ABSTRACT: The increasing competition in the business world often requires businesspeople to think out-side the box. However, some businesses are only concerned about making profits without regard to business ethics. After seeing financial gains, some businesspeople act contrary to the law, and this trend seems to keep increasing. Therefore, it is necessary to institute a more polite business environment that addresses business ethics in the world of commerce, including in the imported second-hand garment industry. Sales of imported second-hand clothes today are increasing and have started to break into large shopping centers. Breaches of laws and ethics has became an issue in, among other places, Royal Plaza Surabaya, where many imported second-hand clothes sellers conduct their business, leading to questions of whether young entrepre-neurs are more likely to ignore ethics when doing business.

1 INTRODUCTION

The number of entrepreneurs in Surabaya has increased by 30% in recent years, and this has had a significant effect on the economy, especially in the reduction of unemployment rate. Entrepreneurship can represent an alternative for young people looking to earn income despite the lack of jobs available. An entrepreneur is a pioneer in their business, an innovator, and a risk insurer that has a vision for the future and for achieving excellence in their field of business (Swasono). According to Suryana, an entrepreneur is someone who combines self-motivation, vision, communication, optimism, encouragement, and the ability to take advantage of business opportuneties. Entrepreneurship skills are even taught in many schools and colleges so that graduates can create new jobs and decrease the unemployment rate. Becoming an entrepreneur is con-sidered to be best started from a young age, because young people are perceived to still have the idealism, skills for thinking out of the box, and energy. Young entrepreneurs also enjoy the follow advantages:

1. Length of career
   Young entrepreneurs usually have more time to devote to planning, implementation, and evalu-ation from previous performances, which results in their businesses having more opportunities to improv and change to achieve desired results.
2. Learning to achieving goals
   Young entrepreneurs are considered to be better at absorbing knowledge, which is very important for sharpening business acumen and bringing new innovation and creativity.
3. Opportunity to take on a mentor to help you better understand your business.

Young entrepreneurs should always be under the guidance of a mentor, so they can learn what strat-egies should be taken for their business to run smoothly and to achieve their goals. Mentors can also help control the ego of the young entrepre-neurs, so they can focus more on their business.

4. Ideas for success
   Young generations are often perceived to be good at generating ideas. This is a huge advantage, to be able to find ideas outside of the box. The younger generation also has more courage in taking risks when doing business. Although they are often found to be less stable with their emo-tions, this issue can minimalized through taking on mentors as discussed previously. In addition, parents and others can offer moral support.

1.1 Innovation and creativity

Entrepreneurship requires creative power, innovation, strategies, and resources in order to pursue opportun-ities and achieve goals. Entrepreneurs tend to try to create something new and different by thinking cre-atively and acting innovatively (Wibisono 2006, Zar-kasi 2012). This could be done by doing something new or by doing something old in a new way.

That more and more young entrepreneurs are emerging is proof that being an entrepreneur allows for greater income than being an employee. But to be an entrepreneur, one must have courage, especially when it comes to taking risks. However, the risk borne by businesspeople can be predicted and calcu-lated from the beginning through a contract or agree-ment. An agreement is a form of risk management for young entrepreneurs. And risk can be managed in a contract.

In addition to having courage to take risks, entrepreneurs also need to be innovative, creative, and responsive to change. Creativity requires focus, ability, willingness, hard work, and perseverance (Shadily 1987). According to Yeni Rachmawati in 2005, creativity is the ability to create something new, in the form of either ideas or real products that are relatively different from what already exists. According to Sahlan and Masman (1988), creativity requires ideas and creative thinking skills. According to Zimmer, creativity is the ability to develop new ideas and new ways to find opportunities and solve problems (Suryana 2001).

The second factor necessary for an entrepreneur is innovation. As said by Peter Drucker, innovation is a typical tool for entrepreneurs. Innovation is the process of converting business opportunities into business ideas, but these opportunities are not always able to be accurately captured. Thus, entrepreneurs should do some research about their perceived opportunities. Through innovation, an entrepreneur can create new production resources and process existing resources to increase their business value (Suryana 2001).

There are four types of innovation:

1. Discovery. Creation of a product, service, or a process that has never been done before. Examples are the invention of airplanes by the Wright brothers, the telephone by Alexander Graham Bell, etc.
2. Development. Development of a product, service, or process that already exists through different ideas. The example is the development of McDonald's by Ray Kroc.
3. Duplication. Impersonation of a product, service, or process that already exists. To duplicate is not simply to imitate but to add some creative touches to fix the concept to make it more capable of winning against the competition. The example is Dentaland dental care.
4. Synthesis. Combination of concepts and factors that already exist into a new product with a new formulation. This process includes getting a number of ideas or products that have been discovered and shaped so that it becomes a product that can be applied in new ways. For example, the synthesis of the watch by (Petter 1985).

Innovative ideas can originate from external and internal forms of creativity. External creativity can be stimulated by being curious about ideas developing around someone. By doing this, a person can build up information about a variety of things and obtain an idea of what can be achieved and utilized. Internal creativity, however, can suddenly appear when someone is busy engaging in external creativity. One example is the effort to use personal experience as a source for knowledge, which can be obtained through learning processes.

Desire for inexpensive and quality clothing has led to a proliferation of imported second-hand clothes businesses in Indonesia. Many people, especially young entrepreneurs, see this as a business opportunity, because the need for clothes at a low price but in good quality represents a very promising opportunity for them. Indonesia has one of the largest second-hand clothes markets in Southeast Asia. Several major cities in Java, including Bandung, Jakarta, and Surabaya, have became centers of business in this industry. In Surabaya, imported second-hand clothes were once mostly sold around the area of Kingpin Market and Tugu Pahlawan. But currently the business has penetrated into the mall area, including Royal Plaza Surabaya, where various types of used clothes have been neatly arranged in windows with a variety of eye-catching concepts, with themes of casual, shabby chic, or vintage, these second-hand settings are not much different from shops selling new clothes.

1.2 Implementation of ethics in the business of imported secondhand clothes in royal plaza Surabaya

The imported second-hand clothes industry has been rapidly growing in the recent years, and the implementation of business ethics is important, both from micro and macro perspectives. The macro perspective sees that the growth of a country is highly dependent on its market systems, because a market system is more effective and efficient than a command system in allocate goods and services. Ethics viewed from a micro perspective trusts businesspeople themselves on whether they want to have good ethics.

Sonny Keraf describes five ethical business principles:

1. Autonomy, the attitude and ability to make decisions and act upon the awareness of what is considered to be a good thing to do.
2. Honesty, including fulfillment of the entire contents of a contract or agreement, offering good quality and fair prices for goods and services, and ensuring internal working relationships within a company.
3. Fairness, where everyone should be treated equally without any distinction.
4. Mutual benefit, where business needs to be mutually beneficial for all parties.
5. Moral integrity, meaning that, in running a business, all involved should uphold the good name of the company and of its owner or head.

Unfortunately, many entrepreneurs trading in imported second-hand clothing in Surabaya violate these ethical principles. According to Law No. 7 Year 2014, importing second-hand clothes to be resold is forbidden by the government. In addition, germs have been found in used clothes that can potentially harm buyers because many traders often find it unnecessary to wash the clothes before they sell them. Buyers are often not being told explicitly that the products they buy are used. Sellers even sometimes label used clothes with their shop brand and mislead the buyers into thinking that the items are new. The business of selling imported second-hand clothes, especially in the area of the mall, has also proven detrimental to tenants who sell new clothes.

2 METHOD

This paper uses qualitative method with triangulation of data. The authors interviewed imported second-hand clothes sellers, their customers, and the other tenants selling new clothes in the area of Royal Plaza Surabaya.

3 RESULT AND DISCUSSION

Based on interviews with imported second-hand clothes sellers, buyers, and other tenants who sell new clothes at the mall, the study found most sellers of imported second-hand clothes are young entrepreneurs. Many of them learned from entrepreneurship courses and already know that selling imported second-hand clothes violated regulations. They found it more difficult to get products under this law, because many imported second-hand clothes were confiscated by the state, and this led to increased prices for imported second-hand clothes. But this industry still appears profitable for sellers. For example, on average, a seller can buy a bag of second-hand clothes at a price of Rp. 5,000,000, to be sold at various price points depending on the condition and brand of the items. The fact that many buyers do not know that the articles of clothing are previously used has also helped to increase sales. Many people still assume that all goods sold in mall must be new. The fact that many imported second-hand clothes are re-labelled to match the store's brand makes the products seem new. This violates the principles of business ethics.

From the point of view of buyers, however, this business is very helpful for them. Because Royal Plaza Surabaya targets lower- and middle-income customers, second-hand clothing shops help customers find products available more cheaply than else-

where. Many buyers do not seem to care about the origin of the clothes they have purchased—even after learning that the clothes they have purchased are second-hand.

Interviews with tenants selling new clothes showed that these shops often receive complaints from prospective buyer because their price range is higher than those at stores selling second-hand clothes. Proliferation in imported second-hand clothes stores also impact sales for stores selling new clothes stores, which have seen profits drop about 20% on average since the arrival of the used-clothing vendors in the mall.

Lastly, the issue of used clothing sales can also taint the name of Royal Plaza Mall, as the mall may come to have a stigma due to the selling of second-hand clothes, which would be detrimental to all vendors.

4 CONCLUSION

The selling of used clothes at the Royal Plaza is detrimental to the other tenants in the same area. The practice turns out to be very profitable for consumers and vendors, however, as each benefit from the demand for cheap clothes. The business carried out by young entrepreneurs in the imported second-hand clothes industry at Royal Plaza Surabaya is an unethical one, however, because it violates many principles of business ethics. Doing business is not just about gaining profit and beating the competitors; in running a business, one must also consider the ethics and applicable regulations, so that a polite business environment can be achieved.

REFERENCES

Wibisono, W. *Generous. Performance Management: Concepts, Design, and Engineering Increase*
Zarkasi, Z. 2012. How Creativity In Achieving Success.
Suryana, S. 2001. *Entrepreneurship*. Jakarta: Salemba Four.
Petter, D. 1985. *Innovation and Entrepreneurship*. New York: Harper and Row.
Shadily, H. 1987. *Encyclopedia Indonesia* 4: 29.
Sahlan, S. & Misman, M. 1988. *Multi-Dimensional Human Berkratifitas*. New York: New Light.
Suryana, S. 2011. *Entrepreneurship*. Jakarta: Salemba Four.

Improving teacher capability through clinical supervision with the Peer Coaching Grow Me (PCGM) approach

M. Hanif, J. Widodo, J. Sutarto & W. Wahyono
Universitas Negeri Semarang, Semarang, Indonesia

ABSTRACT: This research analyzes the clinical supervision model of authentic assessment with the Peer Coaching Grow Me (PCGM) approach for MT (Madrasah Tsanawiyah, or Islamic school) teachers in Brebes District. The research method used was research and development. Data was obtained from the results of clinical supervision, needs analysis, model testing, and model validation. Data collection techniques used were questionnaires, interviews, documentation studies, observations, and Focus Group Discussions (FGDs). Test validity includes model validation by experts and practitioners. Qualitative descriptive analysis employed display data, reduction, verification, and conclusion. The results of the research are: (1) the latest factual model of clinical supervision scored low, with 56.83%; (2) the hypothetical model of clinical supervision scored high, with 84.45%; and (3) the final model of clinical supervision proved very feasible, with an 85.00% score of model validation results, model book, and guidance. Clinical supervision contributed to 26.67% of teachers' capability in authentic assessment. The conclusion is that clinical supervision has been categorized as mediocre, theoretical, and has not reached practical level. The model of clinical supervision, however, is very important, as it effectively and efficiently emphasizes the following steps: goals, reality, options for what's next, monitoring, and evaluation. Clinical supervision via PCGM can therefore be used by principals and supervisors to help teachers who have difficulties carrying out authentic assessment of the 2013 curriculum.

1 INTRODUCTION

The 2013 curriculum was implemented as a refinement of KTSP (Kurikulum Tingkat Satuan Pendidikan, or School-Based Curriculum, SBC), but this has implications for assessment standards. Assessment is an evaluation tool that serves to give a picture of the achievement of National Education Standards (Nizam 2015). Assessment is essential in determining the direction of learning and the quality of education (Sugiyanto 2015). One form of assessment of the 2013 curriculum that is considered appropriate for assessing students' learning outcomes is "authentic assessment." Authentic assessment refers to activities carried out to assess students according to competency requirements in the Competency Standard (CS), Core Competency (CC), and Basic Competency (BC) (Kunandar 2014).

Authentic assessment was applied to the 2013 curriculum because it is more comprehensive, covering aspects of attitude (affective), knowledge (cognitive), and skills (psychomotor) (Gulton 2014, Enggarwati 2015). Authentic assessment has the advantage of measuring students' ability, focusing on the evaluation of the process, portfolio, and overall output evaluation (Muller 2005, Mulyasa 2011).

Despite having many advantages, however, some teachers are still having difficulties implementing the 2013 curriculum assessment, as they are still used to the KTSP assessment system. Half (50%) of SMP (Sekolah Menengah Pertama, or junior high school) MT (Madrasah Tsanawiyah, or Islamic school) teachers are still having difficulties in formulating indicators, compiling items of attitude, skills, and knowledge assessment instruments, and applying authentic assessment software (Suharjono 2014). This means that teachers' ability to carry out an authentic assessment of the 2013 curriculum is still low and needs to be improved.

Efforts to improve the skill of MTs Negeri teachers in Brebes District have been carried out by the Ministry of Religious Affairs, Brebes District Education Office, MGMP (Musyawarah Guru Mata Pelajaran, or subject/teacher interviews), and schools through training, workshops, bintek, and supervision. One activity considered an effective solution for improving teachers' capability in the 2013 curriculum's authentic assessment is clinical supervision.

Clinical supervision refers to professional assistance given to teachers who have problems with the teaching-learning process (Himdani 2017). Clinical supervision emphasizes the personal relationship between supervisors and teachers with regard to solving the learning problem (Gibson & Mithell 2011). Awareness and initiative of teachers in utilizing clinical supervision to address problematic learning is still low (Purwaningsih 2016). Teachers should be aware that clinical supervision must stem from the initiative and desire of the teacher rather than of the principal (Gulton 2014). Teachers with difficulties conducting learning processes are likened to ill patients, who need to see a doctor to recover from illness (Dharma 2009). This means that teachers who have difficulties carrying out authentic assessment need to come to supervisors for help to resolve the problem.

Clinical supervision specializing in assisting teachers having difficulties with authentic assessment has not yet been implemented (Krajewski 2012). As a result, clinical supervision cannot provide problem-solving solutions according to each teacher's difficulty (Masrukan 2016). Clinical supervision is, however, routinely conducted at the initiative of supervisors and principals initiative. Therefore, it has not been able to provide solutions to teacher difficulties (Kaufman et al. 2003). A form of clinical supervision that can help teachers overcome problems in authentic assessment, however, is the Peer Coaching Grow Me (PCGM) method. PCGM is a form of professional coaching for teachers which involves reflecting on the role of supervisors and principals in promoting self-development through the steps of goals, reality, options for what's next, monitoring, and evaluation (Pak 1999).

Clinical supervision with PCGM can serve as a solution for teachers' difficulties in carrying out authentic assessment since it has both theoretical and practical advantages. Theoretical advantages include emphasizing the learning process of self-development; therefore, it has the potential to improve teacher performance in authentic assessment (Parsloe 1999). Practical advantages include prioritizing partnership activities between supervisor and supervisee, so that teachers are given the opportunity to share knowledge and professional skills (Hayes 2003, Wesly 2007).

With this in mind, it is necessary to develop a model of clinical supervision using the PCGM approach to improve the capability of MTs teachers in Brebes District when it comes to authentic assessment. The study aims to analyze and synthesize the clinical supervision model with the appropriate PCGM approach to improve the capability of MTs Negeri teachers in Brebes District in authentic assessment. The results of the study are expected to be useful for the Ministry of Religious Affairs, supervisors, heads of MTs, and teachers, serving as input into the implementation of clinical supervision in the 2013 curriculum's authentic assessment.

2 METHOD

The research was conducted in MTs Negeri, Brebes District. The research design used was the research and development (R&D) method. Research procedures were adopted from Borg (2007) with three main stages: preliminary, development, and validation. The population in this study was 200 people, with a research sample of 10%, i.e. 20 people. The data sources were: data documentation of teachers' performance quality in authentic assessment, questionnaire data, observation, interview, and focus-group discussions (FGD). Data collection techniques applied were: questionnaire, interview, observation, documentation study, and FGD. The

data validity employed constructs and item validity. The data analysis techniques used were quantitative and qualitative.

3 RESULTS AND DISCUSSION

Preliminary research on the clinical supervision model in authentic assessment at MTs in Brebes Regency showed a mediocre score of 306.15 (58.83%) in a score range of 217 to 312, seen from the aspect of performance and component systems. This score was distributed for the performance system aspect, with the average score of 136.05 (58.45%) in the score range of 97 to 140 and for the component system aspect with the average score of 170.09 (59.02%) in the score range between.

The results of clinical supervision system performance evaluation in authentic assessment had the highest score on mean planning with 68.40 (63.54%) in a score range of between 67 and 87. This means that planning was good but still needed improvement, especially on the indicator of clinical supervision stages in authentic assessment. Planning is one of the key stages in clinical supervision activities, as it aims to identify the needs and problems of teachers (Tonkin 2005). Planning can determine the standard of success in the implementation of clinical supervision from the early stages, execution, and follow-up (Himdani 2017, Fakruddin 2017). Indicators that need to be considered in the planning stage are: goals, work program, time, steps, instruments, and team planning (Hanif 2017). The lowest score was evaluation, with a mean score of 23.90 (54.29%) in the score range of 18 to 26. This means that only 54.29% of the respondents stated that the evaluation was good, but there were still 45.71% who stated that it was mediocre. Thus, this score needs to be improved. The evaluation of the initial meeting included rapport development and brainstorming (Widodo 2007). Evaluation is the ability to follow up on the supervisor's advice on the disadvantages of carrying out an authentic assessment (Nurohim et al. 2016).

Assessment of the clinical supervision system component in the authentic assessment showed a mediocre score, with the average score of 170.09 (59.02%) in a score range of between 118 and 170. The highest score of the system components was the supervisors' average score of 55.85 (66.57%) in the score range of between 51 and 67. This indicates that the supervisors did well in carrying out the clinical supervision. The lowest score of the system components was the supervisees' average score of 23.40 (53.13%) in the score range of between 18 and 26, indicating that the supervees needed to improve their ability in authentic assessment.

The factual models implemented of clinical supervision in authentic assessments include the management system, such as planning, implementation, and evaluation. Thee system components include supervisors, supervisees, materials, methods, facilities and infrastructure, and time of clinical supervision. This obtained a new value of 58.83% for the factual model of clinical supervision implementation in the authentic assessment. This showed that the implementation of clinical supervision that has been categorized as mediocre reached 58.83%, needing improvement.

Needs analysis results for the clinical supervision development model with PCGM approach was generally categorized as very important, with an average score of 506.60 (84.45%) in a score range of between 478 and 588. This category was distributed for system performance with the average score of 194.50 (85.15%) in the score range of between 185 and 228; the component system had an average score of 245.75 (85.04%) in a score range of between 224 and 276. In addition, PCGM's average score was 69.35 (83.32%) in the score range of between 68 and 84.

For the requirement analysis of system performance, the highest score was on the implementation stage, with an average score of 65.30 (86.00%) in a score range of between 62 and 76. This indicates that 86% of respondents considered the implementation of clinical supervision very important, and only 15% of them stated it was less important or not important. Implementation is the core of clinical supervision activities; thus, supervisors and supervisees need to understand the procedures and phases, including initial meeting, feedback, and follow-up (Urwaningsih 2016, Makawimbang 2013, Sugiyo 2015).

The lowest value of the system performance was evaluation, with the average score of 36.85 (83.73%) in a score range of between 26 and 44. This means that evaluation of clinical supervision—such as the accuracy and ability of the applied methods to measure competence (cognitive, affective and psychomotor), implementation, feedback, and evaluation of observational instruments—were considered very important by 83.73% of respondents, with 16.27% ranking these aspects as less important. Evaluation is a systematic and continuous process, and the key to success in the implementation of clinical supervision (Riyanto 2015).

The highest score of the system management component was for the supervisor (86.32%). This means that the competency, attitudes/principles, and duties and roles of supervisors were considered very important and needed by 86.32% of respondents; only 13.68% considered it less important. The supervisor

indicator needing improvement was interpersonal skills. Interpersonal skills here refer to a supervisor's ability to maintain a harmonious relationship with the supervisee and understand his or her desires, attitude, behavior, or feelings (Azizy 2013). Supervisors act as coordinators, consultants, group leaders, and evaluators (Sahertian 2008). The lowest value of the system components was for facilities and infrastructure, at 82.85%, meaning facilities and infrastructure (especially information and communications technology, or ICT) were still considered very important and needed by 82.85% of respondents; only 17.15% of them stated they were less important.

On the aspect of PCGM approach, this was considered important by 90.00% of the respondents, and only 10.00% considered PCGM less important. Goals that still needed to be improved were the coaches/supervisors' collaboration with the coachees/teachers in formulating the goal itself. The lowest value of the PCGM approach aspect was for reality (an assessment of the self-ability of coachees—Pak 2005), at 80.00%. This shows that reality was considered very important by 80.00% of respondents, and only 20.00% of them stated it less important.

Clinical supervision using a PCGM approach in the authentic assessment of the 2013 curriculum for MTs teachers in Brebes District was developed based on a management context consisting of two main activities: system performance and system components. The performance of the management system developed included planning, implementation, evaluation, reporting, follow-up, and monitoring. The system components included supervisors, supervisees, materials, methods, facilities and infrastructure, and time. In addition, PCGM included goals, reality, options for what's next, monitoring, and evaluation.

The clinical supervision model has validated the expert, practitioner, and feasibility test through FGD-1 to become a hypothetical model. Further tested in a limited and expanded trial and evaluated, the results are listed in Table 1.

Table 1 describes the results of a limited trial of clinical supervision with the PCGM approach in authentic assessment. It obtained an average score of 56.83 for the mediocre (M) category. The category was distributed for the quality aspect of instrument production, with the average of 63.33 for the good category; the

Table 1. The results of limited trial and expanded trial in extending clinical supervision in authentic assessment.

No	Aspect	Limited Trial		Expanded Trial		Increasing Score
		Average score	Category	Average score	Category	
1.	Instrument preparation	63.33	G	86.67	VG	23.34
2.	Determining PAP and PAN	55.00	M	85.00	VG	30.00
3.	Designing scoring system	58.33	M	83.33	VG	25.00
4.	Applying assessment software	52.50	M	80.00	G	27.50
5.	Processing rapport value	55.00	M	82.50	VG	27.50
Total		284.17	KB	M	VG	133.33
Average		56.83	KB	M	VG	26.67

determining of PAP and PAN resulted in an average score of 55.00 and was categorized as mediocre (M); the designing of scoring system resulted in an average score of 58.33, or mediocre. Moreover, the average software application received an average score of 52.50 and was categorized as mediocre.

The highest score of the limited trial went to the designing of authentic scoring instruments, with an average score of 63.33. This shows that the quality of teachers in creating assessment instruments, including attitude, knowledge, and skills, has reached 63.33% and was categorized as good, but 36.67% ranked it as mediocre and poor. Instruments are very important because they serve as a test tool; in making the instruments, teachers should pay attention to competency indicators, inputs, processes, and students' outputs (Nurohim et al. 2016, Zarkasy 2005). The lowest score of teacher performance test results in authentic assessment was the ability of teachers to apply software, with an average score of 52.50. This proves that the performance of teachers in applying software assessment was still low. Teachers who can apply software well equalled 52.50%, while the other 47.50% were mediocre. Hence, teachers need to be given clinical supervision services on how to apply software assessment.

The expanded trial obtained an average value of 83.50, categorized as very good. The category of instrument preparation had an average of 86.67; determining PAP and PAN had an average of 85.00; designing scoring system had an average of 83.33; software application had an average of 80.00; and managing rapport had an average of 82.50. The highest value of the expanded trial was for instruments with the average of 86.67, categorized as very good. The lowest value was for software application with an average of 80.00, categorized as good. The results of limited and expanded trials showed an increased average of 26.67 points. That is, the PCGM approach to clinical supervision contributed to improvement in authentic assessment of 26.67%.

Clinical supervision with a PCGM approach was then consulted on with experts and discussed through FGD-2 for the improvement of the model, manual, and guidelines (see Table 2).

Table 2 shows the results of validation done by experts and practitioners on the feasibility of models, manuals, and guidelines categorized as very feasible. The highest score (manuals) was 85.00, indicating that they are worth using to improve the ability of teachers in authentic assessment. The lowest score was 83.08 on guidelines; thus, these need improvement. Aspects needing improvement are mechanisms, evaluations, and material clarity. The findings of the final model of clinical supervision in authentic assessment with PCGM approach in State MTs of Brebes are presented in Figure 1.

In Figure 1, the final model of clinical supervision was a refinement of the factual model and the hypothetical model. Factual and hypothetical models after FGDs with experts and practitioners and after validation were refined to become worthy-use final

Table 2. Expert and practitioner validation results on model's feasibility, clinical supervision manual, and guidelines with PCGM approach.

No	Aspect	Score	Range	Conversion (%)	Category
1.	Model Design	33.80	32–40	84.38	Very feasible
2.	Manual	99.7	97–120	83.08	Very feasible
3.	Guidelines	88	84–104	85.00	Very feasible
Total		221.50	208–256	252.50	Very feasible
Average		73.83	72–96	84.15	Very feasible

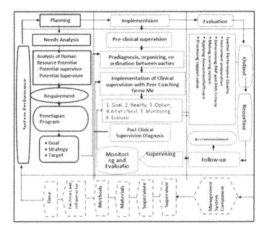

Figure 1. Final model of clinical supervision in authentic assessment with PCGM for MTs teachers in Brebes District.

models for clinical supervision in authentic assessment. Experts and practitioners provided suggestions and inputs to add a monitoring and evaluation (monev) component. Monitoring and evaluation is an activity for observing, checking, and supervising (Hanif 2017). Monitoring is carried out by a team of supervisors, principals, and senior teachers. The monitoring team is assigned to monitor and coordinate each stage of clinical supervision. The system components included supervisors, supervisees, materials, methods, facilities and infrastructure, and time.

The findings of the study showed that improvements in factual conditions, hypothetical model design development, and final model. The factual condition was categorized as mediocre, at 58.83%. The development of hypothetical models had a score of 84.45%. The final model obtained a very feasible result, with a percentage of 84.58%. Based on these findings, clinical supervision with PCGM approach contributed to improvement of teachers' ability in authentic assessment of 26.67%.

Based on these findings, clinical supervision by the PCGM approach may be implemented by school principals and supervisors to assist MTs teachers in Brebes District to overcome difficulties in authentic assessment of the 2013 curriculum, including instrument preparation, PAP and PAN criteria determination, scoring system, appraisal software applications, and processing report cards.

Suggestions given are: (1) MTs teachers can use clinical supervision as a means to overcome difficulties in authentic assessment; and (2) principals and supervisors may use PCGM clinical supervision as an alternative for assisting teachers in authentic assessment.

4 CONCLUSION

Based on the results and discussion, it can be concluded that the clinical supervision model for the authentic assessment of the 2013 curriculum is still: theoretical, unnecessarily designed, and a form of authentic assessment that emphasizes the cognitive domain and rules out the affective and psychomotor domains. The hypothetical model design of academic supervision is essential and necessary because it has a more practical, effective, and efficient advantage, emphasizing goals, reality, options for what's next, monitoring, and evaluation steps. Clinical supervision with a Peer Coaching Grow Me approach is worthy of use by school principals and supervisors to assist teachers of MTs State in Brebes District who have difficulties in authentic assessment such as instrument preparation, PAP and PAN criteria determination, scoring, applying assessment software, and processing report cards.

REFERENCES

Azizy, A. 2013. *Pedoman Pelaksanaan Supervisi Pendidikan Agama*. Jakarta: Kementerian Agama RI Direktorat Kelembagaan Agama Islam.

Borg, G. 2007. *Educational Research*. New York: Logman.

Dharma, S. 2009. *Modul Supervisi Pendidikan Musyawarah Kerja Kepala Sekolah*. Jakarta: Dirjen Peningkatan Mutu Pendidikan dan Tenaga Kependidikan Departemen Pendidikan Nasional.

Enggarwati, N.S. 2015. Kesulitan guru sd negeri glagah dalam mengimplementasikan penilaian autentik pada kurikulum 2013. *UNY* 2(3): 31–42.

Fakhruddin, F. 2017. Efek mediasi kecerdasan emosi pada pengaruh supervisi kolaboratif dan kepemimpinan terhadap perilaku inovatif guru. *EM* 6(1): 71–79.

Gibson R.L. & Mitchell, M.H. 2011. *Bimbingan dan Konseling*. Yogyakarta: Pustaka Pelajar.

Gultom, S. 2014. *Materi Pelatihan Guru Implementasi Kurikulum 2013 Tahun Ajaran 2014/2015*. Jakarta: Kementerian Pendidikan dan Kebudayaan.

Hanif, M. 2017. Model supervisi klinis dengan pendekatan peer coaching grow me dalam penilaian autentik pada guru MTs Negeri di Kabupaten Brebes. *Jurnal Untidar* 1(1): 99–110.

Hayes, H. 2003. *Leadership Coaching: A Practical Guide*. Frenchs Forest, NSW: Pearson Education.

Himdani, H. 2017. Pengembangan model supervisi klinis teknik konseling kelompok pada guru BK SMA Kabupaten Lombok Timur. *Journal of Educational Management* 6(1): 1–8.

Kaufman, K., Judith, J., & Schwartz, T. 2003. Models of supervision: Shaping professional identity. *The Clinical Supervisor* 22(1): 201–211.

Krajewski, K. 2012. Clinical supervision: A conceptual framework. *Journal of Research and Development in Education* 15(3): 94–95.

Kunandar, K. 2014. *Penilaian Autentik (Penilaian Hasil Belajar Peserta Didik Berdasarkan Kurikulum 2013) Suatu Pendekatan Praktis*. Jakarta: Raja Grafindo Persada.

Nizam, N. 2015. *Pedoman Penilaian Kelas oleh Pendidikan*. Jakarta: Pusat Penilaian Pendidikan Kementerian Pendidikan dan Kebudayaan.

Makawimbang, J.H. 2013. *Supervisi Klinis Teori dan Pengukurannya (Analisis di Bidang Pendidikan)*. Bandung: Alfabeta.

Masrukan, M. 2016. Peningkatan layanan guru smk melalui supervisi klinis. *Jurnal EM* 5(2): 1–11.

Muller, J. 2005. The authentic assessment toolbox: Enhancing student learning through online faculty development. *Journal of Online Learning and Teaching* 1(1): 1–10.

Mulyasa, M. 2011. *Kurikulum Tingkat Satuan Pendidikan Sebuah Panduan Praktis* Bandung: Remaja Rosdakarya.

Nurohim, A., Bain, B. & Andy. S. 2016. Pelaksanaan penilaian autentik dalam pembelajaran sejarah kurikulum 2013 di SMA Negeri 1 Purwareja Klampok tahun pelajaran 2015/2016. *IJHE* 4(2).

Pak, T. 2005. *Grow Me!—Coaching for Schools*. Singapore: Prentice Hall.

Parsloe, E. 1999. *The Manager as Coach and Mentor (Management Shapers)*. Oxford: Chartered Institute of Personnel & Development.

Purwaningsih, D. 2016. Supervisi klinis berbasis komunikasi efektif (SKBE) untuk meningkatkan layanan supervisi guru SMK. *Jurnal EM* 5(1): 1–11.

Riyanto, A. 2015. Model supervisi klinis berbasis "ojt" sebagai layanan peningkatan kompetensi guru dalam melaksanakan evaluasi pembelajaran praktik produktif. *Jurnal EM* 2(2): 51–66.

Sahertian, P.A. 2008. *Konsep Dasar dan Teknik Supervisi Pendidikan dalam Rangka Pengembangan Sumber Daya Manusia*. Jakarta: Rineka Cipta.

Sugiyanto, F.N. 2015. Penggunaan penilaian autentik dalam pembelajaran biologi dengan inkuiri terbimbing dan pengaruhnya terhadap hasil belajar peserta didik. *Journal of Biology Education* 4(3).

Sugiyo, S. 2015, Supervisi klinis untuk guru sosiologi SMA. *Jurnal EM* 4(2): 134–142.

Suharjono, S. 2014. Data monitoring dan evaluasi kepengawasan pendidikan, dinas pendidikan dan kebudayaan kabupaten brebes.

Tonkin, B. 2005. Clinical preparation and supervision of professional School Counselors. *Journal of School Counseling* 8(30): 30–41.

Wesly, W. 2007. Authentic assessment of social studies, Journey. *Michigan Department of Education Curriculum Development Program Unit*.

Widodo, W. 2007. Supervisi guru mata pelajaran ekonomi di indonesia antara teori dan realita. *Jurnal Pendidikan Ekonomi* 2(2): 291–313.

Zarkasy, A.S. 2005. Panduan Monitoring dan Evaluasi BMPM (Bina Mitra Pemberdayaan Madrasah). Jakarta: Departemen Agama RI.

Advances in Business, Management and Entrepreneurship – Hurriyati et al (eds)
© 2020 Taylor & Francis Group, London, ISBN 978-0-367-27176-3

Human capital to competitive advantage of micro industry coffe in Garut Regency through distinctive

N.A. Hamdani & G.A.F. Maulani
Universitas Garut, Garut, Indonesia

T. Tetep
Institut Pendidikan Indonesia, Garut, Indonesia

D. Supriyadi
Universitas Pendidikan Indonesia, Bandung, Indonesia

ABSTRACT: This study aims to reveal how the influence of human capital to competitiveness through the capability distinctive. The method used is descriptive survey method and explanatory survey method. Unit of analysis in this research is coffee industry effort in Garut Regency Samples taken in research are perpetrator micro coffee industry with a total sample size of 34, with the technique used is Partial Least Square (PLS). The results of the hypothesis testing showed that the human capital and distinctive capability is good. It is evidenced by the amount of coffee Garut for national consumption and even exports. Testing hypothesis shows that human capital has an influence on competitive advantage through the capability distinctive. Distinctive capabilities have an influence on competitiveness so it can be said that the distinctive as a partial intervening variable.

1 INTRODUCTION

The success of winning competition in the future will likely be determined by the creation of hidden differences such as in resources, organizational functions, and strategic intent. The competitive environment in this globalization era is marked by rapid and dynamic changes, a rapid and wide diffusing capacity, growth of new global competitors, and regionalization of trade (Hampshire and Wiley, 2003). In order to compete in the global market, a company must prepare itself in terms of its transformation, condition, and competition arena today and in the future.

Some studies suggest the importance of competitiveness (Mohebi and Farzollahzade, 2014) discussing the role of entrepreneurial human resources in competitive advantage and business performance. Kraja and Osmani (2013) reveal that the key to competitiveness of SMEs is tangible assets and intangible assets, and one of which is human capital. Meanwhile Mahmood (2013) discusses the importance of competitive advantage as a mediator improve competitiveness. Ruskov, Haralampiev and Georgiev (2012) assert how important it is to go online to create competitive advantage. Some studies reveal that the development of SMEs can improve performance and competitiveness (for example Hamdani, Abdul & Maulani, 2018; Hamdani, 2018).

Drinking coffee has now been very popular all over the world, including in some coffee producing countries. Sousa *et al.*, (2016) suggest that coffee has become a part of culture, and consumers have their own preferences about the coffee they like to drink. Generation Y even drink coffee to reduce tension of a very demanding workload in their workplace (Hashim, Mamat and Nasarudin, 2017). The growth of world market share of coffee is believed to always increase from year to year. The number of coffee shop businesses is increasing in every country. Consequently, the supply of coffee is also increasing. World coffee production in 2017/ 2018 reached 159.7 million bags in crop, an increase of 1.2% from the previous year (Sänger, 2018).

In terms of coffee production, despite having 1.2 million hectares of plantation land, Indonesia is still inferior to Vietnam, which only has an area of 630,000 hectares of coffee. Indonesian coffee production is 500 kilograms of coffee per hectare while Vietnamese coffee production reaches 2.7 million tons of coffee per hectare. Indonesia is the 4th largest exporter of coffee in the world (Pitoko, 2018).

Human capital is an important factor in building the competitive advantage. Some studies discuss the importance of human capital the 21st century. According to Kuehn (2014), human capital is a company asset. Human capital is also a source of knowledge to determine the competitiveness of a company (Ndinguri, Eratis, Prieto, Leon and Machtmes, 2012). Distinctive capability is a company's standout and unique competence to win the competition in terms of satisfying customers. This capability can improve company performance (Darsono, Yahya and Amalia, 2016; Rahim *et al.*, 2009). According to Awang (2010), distinctive capability is very important for SMEs.

The purpose of this study is to explore the relationship between human capital as a source of knowledge and distinctive capability to enhance competitive advantage with the locus of some Coffee SMEs in Garut.

2 LITERATURE REVIEW

Human capital reflects a company's collective capability to find the best solution with reference to the knowledge of people in the company in question. Human capital will improve provided that the company can make the best use of the knowledge of its employees (Schiemann, William and Seibert, 2013; Burma, 2014). Therefore, human capital is very important for the sustainability of the company because human capital is a combination of intangible resources of the members of the organization (Burma, 2014).

Furthermore Astuti (2005) points out that investment in the form of training to improve human capital is important. Skills, experience, knowledge, and health have economic value for the organization because they are beneficial not only for individuals but also for organization. Like other assets in general, human capital can wholly be realized only with the cooperation of each individual. It requires investment in human capital to boost productivity. Human capital is a source of competitive advantage (Memon, Muhammad Aslam, Mangi, Ahmed and Rohra, 2014).

In relation to human capital measurement, Chrysler-Fox, Pharny and Roodt (2014) mention some indicators including individual capability, individual motivation, the organizational climate, workgroup effectiveness, and leadership. Meanwhile, Campbell and Coff (2012) discuss human capital based on competitive advantage.

Capability is the capacity to best exploit the resources and potentials of individuals to perform particular activities within an organization. Darsono, Yahya and Amalia (2016) discuss the distinctive capability measurement using some indicators including development process, financial information, capital structure, general administration, entrepreneurship management, and information technology.

Competitive advantage results from the strategy employment by a company to seize the market opportunities. To win the competition, the company should choose between low-end and high-end market niche, not both (Zeebaree and Siron, 2017; Kraja and Osmani, 2013).

Coffee competitiveness is measured through comparative competitiveness, the export position of Indonesian, Brazilian, Colombian and Vietnamese coffee and the competitive advantage of the Indonesian coffee industry (Suprayogi *et al.*, 2017). Some studies (for example Sachitra, 2016) discuss competitive advantage measurement using some indicators including price, quality, delivery dependability, product innovation, and time to market.

3 RESEARCH METHOD

This study used an explanatory research method with an aim to test the theory of human capital, distinctive capability, and competitive advantage. The samples were 34 coffee industries selected using simle random sampling technique. Data were analyzed using SEM-PLS. The research variables were human capital (X1) with individual capability, individual motivation, the organizational climate, workgroup effectiveness, and leadership as the indicators, distinctive capabilities (X2) with development process, financial information, capital structure, entrepreunership management, and information technology as the indicators, and competitive advantage (Y) with price, quality, delivery dependability, product innovation, time to market as the indicators.

4 RESULT AND DISCUSSION

Data analysis was carried out by means of SmartPLS software. PLS features bootstrapping or random multiplication method so the data are assumed to be normally distributed. The samples are 34 coffee micro industries in Garut, West Java, Indonesia. The SmartPLS modeling results was shown in Figure 1:

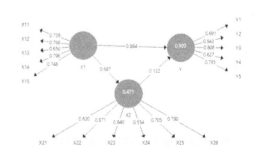

Figure 1. Result of analysis using PLS algorithm calculation.

The above PLS algorithm calculation reveals value of each indicator or variable as interpreted below:

1. The path coefficient of human capital (X1) to the latent variable competitive advantage (Y) is 0.864. It means that the X1 has an influence on Y as much as 0.864.
2. Loading for indicator X11 is 0.739. It means that the latent variable X1 contributes to X11 as much as 0.739.
3. Loading for indicator X12 is 0.768. It means that the latent variable X1 contributes to X12 as much as 0.768 and so on.
4. The path coefficient of variable X2 to the latent variable Y is 0.132. It means that the X2 has influence on Y as much as 0.132.
5. Loading for indicator X21 is 0.630. It means that the latent variable X2 contributes to X21 as much as 0.630 and so on.
6. The path coefficient of variable X1 to the latent variable X2 is 0.687. It means that the X1 has influence on X1 as much as 0.687 and so on.

The PLS Algorithm also shows that general administration (X24) has a loading factor lower than 0.6. Therefore, Item X24 is excluded. The resulting model is shown in Figure 2:

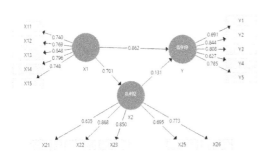

Figure 2. Result of analysis using PLS algorithm calculation without X24.

SmartPLS also results in the following matrix (Figure 3):

Construct Reliability and Validity

	Cronbach's Al	rho_A	Composite Reliability	Average Variance Extracted (AVE)
X1	0.795	0.802	0.859	0.551
X2	0.831	0.912	0.877	0.592
Y	0.958	0.824	0.868	0.579

Figure 3. Construct reliability and validity matrix.

Referring to each loading factor in Figure 4, the value of each construct is higher than 0.6. It means that all indicators used in this research are convergently valid.

The discriminant validity could be seen in the value of square root of average variance extracted (AVE). The recommended value is higher than 0.5. The AVE values in this research are 0.551 for X1; 0.592 for X2, and 0.579 for Y. It can be concluded that all of AVE values have met the requirement.

The next step is conducting the reliability test on each variable to figure out the composite reliability value of construct measuring indicator block.

The composite reliability value is satisfying if it is higher than 0.7. The output composite reliability values are as follows: 0.859 for X1, 0.877 for X2, and 0.868 for Y – all of them are higher than 0.7. The reliability test result is strengthened by the following Cronbach's alpha values: 0.795 for X1, 0.831 for X2, and 0.808 for Y. The recommended value is higher than 0.6, and Figure 5 shows that Cronbach's Alpha values of all constructs are higher than 0.6.

R Square

	R Square	R Square Adjus
X2	0.492	0.476
Y	0.919	0.914

Figure 4. R square.

In SmartPLS analysis, the first inner model testing is conducted on the model, which is R-square. The result shows that distinctive capability (X2) has the R-square of 0.492 and that competitive advantage (Y) has the R-square of 0,919. It can be concluded that distinctive capability can explain the variance of competitive advantage of coffee micro industries in Garut as much as 91%.

The hypothesis testing in SEM PLS was done by way of outer model testing using bootstrapping method. The result is as is shown in Figure 6:

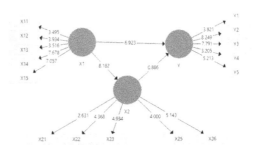

Figure 5. Bootstrapping model.

The bootstrapping also results in the following matrix (Figure 6):

Path Coefficients

	Original Sampl...	Sample Mean (	Standard Devia...	T Statistics (|O...	P Values
X1 -> X2	0.701	0.716	0.086	8.182	0.000
X1 -> Y	0.862	0.862	0.125	6.923	0.000
X2 -> Y	0.131	0.128	0.147	0.886	0.376

Figure 6. Path coefficients.

Figure 6 shows that there is a significant correlation between human capital (X1) and distinctive capability (X2) since the statistical T value is 8.182 (>1.66). In addition, the original sample estimate value shows the positive number of 0.701. This indicates that the correlation between X1 and X2 is positive. It could then be concluded that human capital has significant influence on the distinctive capability of coffee micro industries in Garut.

It can also be seen that there is a significant correlation between human capital (X1) and competitive advantage (Y) because the statistical T value is 6.923 (>1.66) and that the original sample estimate value shows the positive number of 0.862. This indicates that the correlation between X1 and X2 is positive. It could then be concluded that human capital has significant influence on the distinctive capability of coffee micro industries in Garut.

However, distinctive capability (X2) has no significant influence on the competitive advantage of coffee micro industries in Garut since the statistical T value is 0.886 (<1.66). Even so, since the original sample estimate value 0.131, the correlation between them is positive. It could then be concluded that distinctive capability as mediator variable between human capital and competitive advantage is still influential but not significantly.

5 CONCLUSION

Human capital has significant influence on the competitiveness. It goes to show that coffee business people in Garut demonstrate a good teamwork and leadership in managing their business. The fact that human capital has significant influence on the competitive advantage shows that human capital plays an important role in the distinctive capability.

The findings reveal that the quality of Indonesian coffee has yet to meet the market demand. This is because the influence of distinctive capability is not significant due to the non-optimal managerial factor of coffee SMEs in Garut. Therefore, it is necessary to improve distinctive capability through supports of government and private sectors in order to be able to compete in global market.

This study has sampling limitation. Despite the fact that the sample is sufficient for data processing, but cannot be generalized to the population in other setting. It is recommended that further studies develop innovative factor to improve competitive advantage.

REFERENCES

Astuti, P. D. (2005) 'Hubungan Intellectual Capital dan Business Performance', *Jurnal Maksi*, 5, pp. 34–37.

Awang, A. (2010) 'Study of Distinctive Capabilities and Entrepreneurial Orientation on Return on Sales among Small and Medium Agro-Based Enterprises (SMAEs) in Malaysia', *International Business Research*, 3(2), pp. 34–48.

Burma, Z. A. (2014) 'Human Resource Management and Its Importance for Today ' s Organizations', *International Journal of Education and Social Sciences*, 1 (2), pp. 85–94.

Campbell, B. and Coff, R. (2012) 'Rethinking Sustained Competitive Advantage from Human Capital', *Academy Management Review*, 37(3), pp. 376–395.

Chrysler-Fox, Pharny and Roodt, G. (2014) 'Changing Domains in Human Capital Measurement', *Journal of Human Resources Management*, 12(1), pp. 1–12.

Darsono, N., Yahya, A. and Amalia, R. (2016) 'Analysis of Distinctive Capabilities and Competitive Advantage on Business Performance of Tourism Industry in Aceh', *Journal of Economics, Business and Management*, 4(3), pp. 4–7. doi: 10.7763/JOEBM.2016.V4.395.

Hamdani, N. A. (2018) 'Building Knowledge Creation For Making Business Competition Atmosphere in SME of Batik', *Management Science Letters*, 8, pp. 667–676. doi: 10.5267/j.msl.2018.4.024.

Hamdani, N. A., Abdul, G. and Maulani, F. (2018) 'The influence of E-WOM on purchase intentions in local culinary business sector', *International Journal of Engineering & Technology*, 7, pp. 246–250.

Hampshire, N. and Wiley, J. (2003) 'Scanning Dynamic Competitive Landcapes', *Strategic Management Journal*, 24(20), pp. 1027–1041. doi: 10.1002/smj.325.

Hashim, N. H., Mamat, N. A. and Nasarudin, N. (2017) 'SOCIAL SCIENCES & HUMANITIES Coffee Culture among Generation Y', *Pertanika J. Soc. Sci. & Hum*, 25 (Spesial Issue), pp. 39–48.

Kraja, Y. and Osmani, E. (2013) 'Competitive Advantage and Its Impact In Small and Medium Enterprises : Case of Albania', *European Scientific Journal*, 9(16), pp. 76–85.

Kuehn, D. (2014) 'Human Capital in the Twenty First Century', *The European Journal of Comparative Economics*, 15(1), pp. 3–9.

Mahmood, R. (2013) 'Entrepreneurial Orientation and Business Performance of Women-Owned Small and Medium Enterprises in Malaysia : Competitive Advantage as a Mediator', *International Journal of Business and Social Science*, 4(1), pp. 82–90.

Memon, Muhammad Aslam, Mangi, Ahmed and Rohra, L. (2014) 'Human Capital a Source of Competitive Advantage " Ideas for Strategic Leadership " Human Capital a Source of Competitive Advantage " Ideas for Strategic Leadership "', *Australian Journal of Basic and Apllied Sciiences*, 4(3), pp. 4182–4189.

Mohebi, M. M. and Farzollahzade, S. (2014) 'Improving Competitive Advantage and Business Performance of

SMEs by Creating Entrepreneurial Social Competence', *management Reserach*, 2(spesial Issue), pp. 20–26.

Ndinguri, Eratis, Prieto, Leon and Machtmes, K. (2012) 'Human Capital Development Dynamics : The Knowledge Based Approach', *Academy of Strategic Management Journal*, 11(2), pp. 121–136.

Pitoko, R. A. (2018) 'Lika-Liku Produksi Kopi Indonesia dan Strategi Meningkatkan Produktivitasnya Artikel ini telah tayang di Kompas.com dengan judul "Lika-Liku Produksi Kopi Indonesia dan Strategi Meningkatkan Produktivitasnya", https://ekonomi.kompas.com/read/2018/04/27/101', *Kompas*, 27 April. Available at: https://ekonomi.kompas.com/read/2018/04/27/101500626/.

Rahim, A., Bakar, A., Hashim, F. and Ahmad, H. (2009) 'Distinctive Capabilities and Strategic Thrusts of Malaysia ' s Institutions of Higher Learning', *International Journal Marketing Studies*, 1(2), pp. 158–164.

Ruskov, P., Haralampiev, K. and Georgiev, L. (2012) 'Online Investigation of SMEs Competitive Advantage', in *MEB 2012 – 10th International Conference on Management, Enterprise and Benchmarking June*. Budapest, pp. 143–160.

Sachitra, V. (2016) 'Review of Competitive Advantage Measurements : Reference on Agribusiness Sector', *Journal of Scientific Research & Reports*, 12(6), pp. 1–11. doi: 10.9734/JSRR/2016/30850.

Sänger, C. (2018) *State of the global coffee market*. Genewa.

Schiemann, William and Seibert, J. (2013) 'Optimizing Human Capital', *Poeple and Startegy*, 36(1), pp. 33–61.

Sousa, A. G., Maria, L., Machado, M., Freitas, E. and Helena, T. (2016) 'Personal characteristics of coffee consumers and non-consumers, reasons and preferences for foods eaten with coffee among adults from the Federal District, Brazil', *Food Science and Technology*, 36(3), pp. 432–438.

Suprayogi, B. M., Arifin, Z., Mawardi, M. K., Bisnis, I. A., Administrasi, F. I. and Brawijaya, U. (2017) 'Konsumsi kopi dunia', *Jurnal Administrasi Bisnis (JAB)*, 50(2), pp. 190–194.

Zeebaree, M. R. Y. and Siron, R. (2017) 'The Impact of Entrepreneurial Orientation on Competitive Advantage Moderated by Financing Support in SMEs', *International Review of Management and Marketing*, 7(1), pp. 43–52. Available at: file:///Users/haniruzila/Desktop/3157-10307-1-PB.pdf.

Advances in Business, Management and Entrepreneurship – Hurriyati et al (eds)
© 2020 Taylor & Francis Group, London, ISBN 978-0-367-27176-3

Early-warning systems for property price bubbles in Indonesia and strategies to prevent business failures

R. Purwono, P.L. Permata & R.D.B. Herlambang
Universitas Airlangga, Surabaya, Indonesia

ABSTRACT: The Global Financial Crisis of 2008 is a prominent example of how we should be aware of potential property price bubbles. In this paper, we estimate the chances of an Indonesian housing price bubble and build a probability model to seek any possible determinant. Results from these estimations then help in developing a strategy to be used to prevent business failure. As predictors of price bubbles, we used macro-economic factors and macroprudential policy, as these are easily monitored. Results from this paper suggest that GDP growth could serve as an early warning indicator of price bubbles, implying a counter-cyclical rela-tion with the bubble. In addition, some LTV (Loan-To-Value) implementation is too strict, inducing a bubble instead of stabilizing prices. This information, along with the demonstrated procedure, could help firms plan early reaction to prevent upcoming bubbles.

1 INTRODUCTION

Over the past few decades, the global economy has faced several crises. One phenomenon these crises had in common was the presence of an asset price bubble. The recent subprime mortgage crisis also specifically pointed to the role of the housing market in creating a devastating impact on global econom-ics. As suggested by Tokic (2012), firms should stra-tegically react to this bubble, starting with acknowledging how bubbles develop.

There is some evidence that an asset price bubble could be related to a loosening of monetary policy (McDonald & Stokes 2013, Shiller 2012; Yang et al. 2010). However, during the recent Global Financial Crisis (GFC), Bernanke (2010) suggested that a good regulatory oversight was needed more than a conventional monetary move to respond to the housing bubble. To be specific, he suggested that the response should be aimed dir-ectly at the main problem, such as the underwriting process and risk management. Indonesia, as one of the world's most populous countries, adopted this policy by implementing loan-to-value (LTV) regu-lation. Implemented in 2012 Q2, LTV regulation in Indonesia has since undergone some changes, par-ticularly the adjustment by BI (Bank Indonesia, the central bank) in 2013 Q3 and 2015 Q2 and the implementation of a regressive scheme and loosen-ing of the LTV ratio.

In this paper, we estimate the presence of a housing bubble in Indonesia using PWY, the approach of Phillips et al. (2015) (PWY) approach. The estimated bubble presence i s then analyzed fur-ther with related determinants, using a dynamic probit approach as demonstrated by Huang and Shen (2017).

Results from this paper show several housing bubble episodes in Indonesia. The first was in 2009 Q2 through 2009 Q3, followed by another bubble in 2012 Q4 to 2013 Q4. Interestingly, the determinant analysis shows LTV regulation significantly encour-aged the bubble. Another interesting finding is the macroeconomic factors, which leave GDP growth as the only significant factor related to the housing bubble. The optimal approach that should be taken by firms is then discussed.

The remainder of the paper is organized as fol-lows: Section 2 contains a short review of the lit-erature related to the problem investigated; Section 3 describes the data and methodology; Section 4 presents the result of bubble estimation, estimation of bubble determinants, and discussion of business strategy; and Section 5 concludes the findings.

2 LITERATURE REVIEW

2.1 *House price bubble*

From theoretical perspective, an asset bubble heavily depends on the present value and the rational bubble assumption. Using rational expectation framework to detect a bubble, such as Lucas and Sargent (1981) and Clayton (1996), a bubble is defined to exist when the trading price of an asset (P_t) has a persistent positive misalignment trend from its expected fundamental value $(E_t P_t^*)$. The misalign-ment (ε_t) then could be calculated as:

$$\varepsilon_t = P_t - E_t P_t^* \qquad (1)$$

Assuming that ε_t is similar to the bubble component, this equation is still in line with one derived in Capozza et al. (1997) and Capozza et al. (2002).

The econometric approach to estimating this bubble, however, does not progress as smoothly as the theoretical foundation. The early approach developed for estimating the presence of a bubble is the variance bound test by Shiller (1981). This approach suggests a bubble is developed if it exceeds the bound value from the fundamental variance. Another approach that directly estimates the misalignment is the two-step test proposed by West (1987). This approach requires a detailed model to estimate the fundamental value of an asset, then used the estimated fundamental value to analyze the misalignment. However, these two approaches heavily depend on the goodness of fit in modeling the fundamental value and hence might be subject to misspecification.

Campbell and Shiller (1987) suggest estimating the bubble's progress as explosive. This approach assumes that the trading and fundamental value cannot have cointegration, since it is contrary to the explosiveness process in the gap. Diba and Grossman (1988), then, point out that the unit root and cointegration test should serve as a tool to estimate the presence of an explosive bubble. However, as demonstrated by Evans (1991), this approach could not effectively estimate the presence of a bubble if there is a periodically collapsing bubble in the sample. To this end, Phillips et al. (2011) and Phillips et al. (2015) developed a sup Dickey-Fuller (DF) test to date the presence of a bubble by using a forward recursive algorithm to handle the explosive process. This test proved to be effective in dating the presence of multiple bubbles and gained popularity in recent literature of housing bubbles (Huang and Shen 2017).

2.2 Housing price bubble determinants

Several bubbles in recent decades that prominently led to economic crises encouraged researchers to seek out bubble determinants. According to earlier housing bubble literature, the causes of housing bubbles often involve demand-side factors (Shiller 2008). This factor is highly related to the macroeconomic condition. Following Huang and Shen (2017) and Pavlidis et al. (2016), the selected macroeconomic factors include GDP growth (GR), inflation (inF), and interest rate (LR). Another determinant that could affect a housing bubble is local regulation, such as macroprudential regulation that could affect the local market (Glaeser et al. 2010).

2.3 Reacting to a housing price bubble

As suggested by Tokic (2012), firms should know how bubbles develop to help them in planning hedge programs. Acknowledging the presence of a bubble could help firms refrain from contributing to a bubble's development. In addition, Tokic points out that contributing to the bubble could hurt the firm's hedging strategy when the bubble bursts. However, the hedging strategy using future contract has also proven to be ineffective when the bubble starts to burst (Schorno et al. 2014). This leaves the safest option as minimizing the bubble's impact by having it burst earlier.

To understand how a bubble develops, one could refer to the informational friction theory that underlies bubble development. In the work of De Long et al. (1990), traders are assumed to be faced with two risks when a bubble develops. The first is the fundamental risk to the state of the economy. This risk encourages the investor to trade against the bubble, as it logically turns out as a loss. The second is the risk that arises if traders choose to ride the bubble, which turns out also to mean a loss for the trader if the bubble bursts.

These two risks cause of information friction. Even long-time risk-neutral traders therefore often choose to ride the bubble because they find it to be optimal (Abreu & Brunnermeier 2003). The information friction could be explained as the uncertainty of other rational traders that allegedly would do the same to optimize the presence of the bubble. This key element of informational friction states that only a group of traders can burst a bubble early, not an individual. It is plausible for a trader, then, to make an agreement with others to burst the bubble earlier. This turns out to be safer, as the possible outcome of a larger bubble bursting results in the worst outcome for all, as in the recent subprime mortgage crisis.

3 DATA AND METHODOLOGY

3.1 Data

For economic growth and inflation rate, we use the year-on-year growth rate of constant Gross Domestic Product (GDP) and Consumer Price Index (CPI). For interest rate, we use the lending rate for consumption as a proxy for the discount rate for housing price. All macroeconomic data was obtained from BI statistics.

To estimate the house price bubble, we used the nominal and real house-price index data from 2002 Q1 to 2016 Q4. The housing price index can be obtained from the Organization for Economic Cooperation and Development (OECD) database. However, the estimated bubble data starts from 2005 Q3 due to the rolling window approach, hence the following looks for the bubble determinant using the period using 2005 Q3 to 2016 Q2 as the sample.

3.2 Estimating house price bubble

To estimate the housing price bubble, we follow the PWY approach. The PWY approach basically uses Generalized sup Augmented Dicky-Fuller (GSADF) test to assess single or multiple episodes of a bubble. The GSADF test for an asset price (y) could be conducted by the following equation:

$$\Delta y_t = \hat{\alpha}_{r_1, r_2} + \hat{\beta}_{r_1, r_2} y_{y-1} + \sum_{i=1}^{k} \hat{\psi}^i_{r_1, r_2} \Delta y_{t-1} + \hat{\varepsilon}_t$$

$$(2)$$

where r_1 and r_2 are the first and last window for recursive estimation. The Augmented Dickey-Fuller (ADF) statistics then could be obtained from the t-statistics for $\hat{\beta}_{r_1, r_2}$. The advantage of using this test is the selection of window for estimation, which is based on recursive estimation to obtain sup ADF statistics (SADF). This approach is proven to be effective in testing the presence of a bubble even there is an explosion progress.

This approach is then converted to GSADF using the recursive rolling window approach, which is useful for obtaining best ADF statistics. To date the bubble (B_t), we then estimate the same procedure using a backward sequence to obtain backward sup ADF (BSADF) statistics. The obtained statistics are then compared to the critical value (CV), which is estimated using Monte Carlo simulation with 2000 replication.

3.3 Estimating house price bubble determinant

To estimate the housing price bubble determinant, we follow Huang and Shen (2017), who suggest using dynamic probit to estimate the impact from all determinants to the probability of housing price bubble. We define the bubble as:

$$B_t = \begin{cases} 1, & BSADF_t \geq CV_t \\ 0, & BSADF_t < CV_t \end{cases} \quad (3)$$

The probability of a bubble is then estimated using dynamic probit model, written as:

$$Prob(B_t = 1) = \Phi\left(\alpha + \sum_{i=1}^{m} \gamma_i B_{t-i} + \beta X_t\right) \quad (4)$$

where X represent the vector of independent variables, which includes macreconomics factor (GDP growth, inflation, and interest rate) and dummies for the first two LTV regulation. The notable difference in using dynamic probit is the parameter γ, which represents the autoregressive (AR) impact. Since there is no agreement in choosing the amount of m, we consider a parsimonous model of AR(1).

4 RESULTS AND DISCUSSION

4.1 House price bubble estimation and event study

The results from BSADF estimation shows that we can see a similar result in the nominal and real housing price indexes. However, if using the nominal price index as a proxy, the aftermath impact from Global Financial Crisis is not captured. Thus, using the real price index as a proxy should give more convincing results.

The observed bubble for nominal prices started in 2012 Q4 and ended in 2014 Q2. For the real price index, there are two observed bubbles. The first is a short-lived bubble in 2009 Q2 that ended in 2009 Q3, and the second started in 2012 Q4, as in the nominal price results, but ended earlier, in 2013 Q4.

One interesting point that stems from the two results is the impact of the first LTV implementation in 2012 Q2, which starts to push the BSADF value to rise in the following period. This is in line with the work of Glaeser et al. (2010), who suggest that a too-strict macroprudential policy could encourage a bubble instead. The dating of the bubble's bursting, however, show a different outcome between nominal and real housing prices. The improvement in 2013 LTV regulation shows a successful impact to reduce BSADF statistics if using nominal price index but a further rise to exceed CV after the 2015 LTV regulation change. A different result is observed if using real price index as a proxy, which shows the 2013 LTV regulation change induced an impact similar to that of the first implementation but successfully burst the bubble earlier as compared to using the nominal price as a proxy. The impact from the 2015 LTV implementation also shows counterpart results. These different results encourage the use of the real price index instead, as suggested by Pavlidis et al. (2016), for a more reliable bubble estimation.

4.2 Estimation of housing price bubble determinants

Results of the bubble estimation are further analyzed using a dynamic probit model to analyze the possible macroeconomics and regulation determinants. In Table 1, the result for the real price index is presented.

Among macroeconomics variables, the economic growth rate is the only significant determinant. In this period, the growth rate prominently reduces the probability of a bubble. This result differs from earlier studies, as in Pavlidis et al. (2016), which show that a GDP growth rate encourages a bubble instead. Still, the results from this paper are still in line with the relation of financial market and macroeconomics, suggesting the financial market price volatility to be counter-cyclical with macroeconomics conditions (Engle et al. 2013, Engle & Rangel 2008).

Table 1. Estimation results for fundamental price model.

Variable	Coefficient	Standard error
B_{t-i}	3.3754[**]	1.6447
GR_t	-2.8701[**]	1.6552
INF_t	-0.0074	0.2273
LR_t	4.9977	2.6555
$LTV2012$	0.8232[**]	1.5525
$LTV2013$	0.6403	2.1199
$Cons$	11.2367	13.9176
Period	2005 Q3–2016 Q4	
Pseudo R^2	0.6733	
Chi-squared stat	26.42[***]	

Notes: *, **, and *** indicate significance levels respectively of 10%, 5%, and 1%.

Turning to the impact of LTV regulation, the results from dynamic probit estimation show a similar conclusion to that of the previous event study from BSADF estimation. The first LTV regulation significantly encourages the bubble's development instead of stabilizing the housing price. However, the 2013 LTV regulation does not significantly reduce the probability as in the previous event study.

4.3 Proposed strategy related to Indonesian housing bubble determinants.

As explained in the previous section, the optimal reaction to a bubble is first to acknowledge its presence. The previous event study and dynamic probit estimation could help in acknowledging how the bubble developed.

The first factor that was consistently concluded in the event study and dynamic probit estimation is the role of LTV regulation. As suggested in Glaeser et al. (2010), strict asset price regulation should be made known as a way to create a bubble instead of stabilizing prices. A well-formulated policy, however, could act as a stabilizer to housing prices, as this paper's results suggest. Thus, firms should study the future macroprudential policy and monitor the market response to the policy as an early warning for a housing price bubble.

Another strategy that could be directly implemented by firms is regarding the development process of the bubble. As suggested by Tokic (2012), firms should avoid participating in positively reinforcing the bubble. This could only be achieved by creating an agreement among traders, as suggested by Abreu and Brunnermeier (2003). In addition, the previous event study and dynamic probit estimation show that an economic recession and a too-strict regulation should be an early warning to start considering early communications among traders to prevent bubbles in the future.

5 CONCLUSION

Indonesia has already undergone some improvements in macroprudential policy to prevent the housing bubble, particularly with the implementation of LTV regulations. However, results from this paper suggest that LTV regulation is too strict in the first implementation, which suggests the market ought to be made aware of the need for another macroprudential regulation in the future. In addition, results from bubble determinants in Indonesia suggest that bubbles behaves counter-cyclically to macroeconomic conditions.

The BSADF estimation and bubble determinants in this paper could help firms develop an early-warning system for housing price bubbles. The BSADF procedure could be updated with real-time data, which could help the firms to monitor any changes in housing bubble behavior. Firms could also consider monitoring the macroeconomic projectory, as it shows a counter-cyclical relation with housing bubble probability. In short, an upcoming recession should be acknowledged by firms as an early warning of an upcoming bubble.

REFERENCES

Abreu, D. & Brunnermeier, M.K., 2003. Bubbles and crashes. Econometrica 71, 173–204.

Bernanke, B.S., 2010. Monetary policy and the housing bubble: a speech at the Annual Meeting of the American Economic Association, Atlanta, Georgia, January 3, 2010.

Campbell, J.Y. & Shiller, R.J., 1987. Cointegration and tests of present value models. Journal of Political Economics 95, 1062–1088.

Capozza, D., Mack, C., & Mayer, C. 1997. The dynamic structure of housing markets. Ann Arbor 1001, 48109–48109.

Capozza, D.R., Hendershott, P.H., Mack, C., & Mayer, C.J. 2002. Determinants of real house price dynamics. National Bureau of Economic Research.

Clayton, J., 1996. Rational expectations, market fundamentals and housing price volatility. Real Estate Economics 24, 441–470.

De Long, J.B., Shleifer, A., Summers, L.H., & Waldmann, R.J. 1990. Noise trader risk in financial markets. Journal of Political Economics 98, 703–738.

Diba, B.T. & Grossman, H.I., 1988. Explosive rational bubbles in stock prices? American Economic Review 78, 520–530.

Engle, R.F., Ghysels, E., & Sohn, B. 2013. Stock market volatility and macroeconomic fundamentals. Review of Economic Statistics 95, 776–797.

Engle, R.F. & Rangel, J.G., 2008. The spline-GARCH model for low-frequency volatility and its global macroeconomic causes. Review of Financial Studies 21, 1187–1222.

Evans, G.W., 1991. Pitfalls in testing for explosive bubbles in asset prices. American Economic Review 81, 922–930.

Glaeser, E.L., Gottlieb, J.D., & Gyourko, J., 2010. Did credit market policies cause the housing bubble? Cambridge: MA: Harvard Kennedy School.

Huang, J. & Shen, G.Q., 2017. Residential housing bubbles in Hong Kong: identification and explanation based on GSADF test and dynamic probit model. Journal of Property Research 34, 108–128.

Lucas, R.E. & Sargent, T.J., 1981. *Rational expectations and econometric practice.* Minneapolis: U of Minnesota Press.

McDonald, J.F., Stokes, H.H., 2013. Monetary policy and the housing bubble. *Journal of Real Estate Finance Economics* 46, 437–451.

Pavlidis, E., Yusupova, A., Paya, I., Peel, D., Martínez-García, E., Mack, A., & Grossman, V. 2016. Episodes of exuberance in housing markets: in search of the smoking gun. Joural of Real Estate Finance Economics 53, 419–449.

Phillips, P.C., Wu, Y., & Yu, J., 2011. Explosive behavior in the 1990s Nasdaq: When did exuberance escalate asset values? *International Economic Review* 52, 201–226.

Phillips, P.C.B., Shi, S., Yu, J., 2015. Testing for multiple bubbles: Historical episodes of exuberance and collapse in the S&P 500. International Economic Review 56, 1043–1078. https://doi.org/10.1111/iere.12132

Schorno, P.J., Swidler, S.M., & Wittry, M.D. 2014. Hedging house price risk with futures contracts after the bubble burst. *Finance Research Letters* 11, 332–340.

Shiller, R.J. 2012. *The subprime solution: how today's global financial crisis happened, and what to do about it.* Princeton, NJ: Princeton University Press.

Shiller, R.J., 2008. Historic turning points in real estate. *Eastern. Economic Journal* 34, 1–13.

Shiller, R.J. 1981. Do stock prices move too much to be justified by subsequent changes in dividends? *American Economic Review* 71, 421–436.

Tokic, D. 2012. When hedging fails: what every CEO should know about speculation. Journal of Management Development 31, 801–807.

West, K.D. 1987. A specification test for speculative bubbles. *Quarterly Journal of Economics* 102, 553–580.

Yang, Z., Wang, S., & Campbell, R., 2010. Monetary policy and regional price boom in Sweden. *Journal of Policy Modeling* 32, 865–879.

Advances in Business, Management and Entrepreneurship – Hurriyati et al (eds)
© *2020 Taylor & Francis Group, London, ISBN 978-0-367-27176-3*

Impact from loan-to-value to housing price bubble in Indonesia and strategy to prevent business failure

R. Purwono, Dessy Kusumawardani & R. Dimas Bagas Herlambang
Universitas Airlangga, Surabaya, Indonesia

ABSTRACT: The recent subprime mortgage crisis showed the need for housing price bubble monitoring. The central bank of Indonesia, Bank Indonesia (BI), responded to the recent crisis by implementing Loan-to-Value (LTV) for housing credit. However, for Indonesia, the research for strategies to prevent business failure regarding the housing price bubble and its outcome is very limited. In this paper, we will estimate the housing bubble in Indonesia to give more perspective in property business strategy. Using information from the property price index for Indonesia, we also check whether the LTV policy changes the price bubble condition, hence, change the strategy for the prevention of firms' failures. Results from this paper suggest that a misconduct in LTV could lead to a boom in housing prices instead of acting as a counter-cyclical buffer. However, recent changes in LTV have been effective in reducing housing price misalignment, and therefore it should be safe for firms to respond with optimism in future housing price stability.

1 INTRODUCTION

There is some evidence that an asset price bubble could be related to a loosening monetary policy (McDonald and Stokes, 2013; Shiller, 2012; Yang et al., 2010). However, for recent Global Financial Crisis (GFC) experience, Bernanke (2010) suggested that a good regulatory is more needed than a conventional monetary move to respond to the recent housing bubble. To be specific, he suggested that the response should be aimed directly at the main problem, such as the underwriting process and risk management. This macroprudential approach, rather than a conventional monetary policy, then approved by many central banks and been adopted in many countries.

Indonesia is one of the countries that uses LTV as macroprudential regulation to prevent a bubble. The implementation of LTV in Indonesia has undergone some changes. The first LTV regulation was implemented in 2012Q2 to housing and vehicle credit with a maximum of 70 percent in LTV. This regulation is undergone further adjustment by BI in 2013Q3 and 2015Q2, particularly by implement a regressive scheme and loosening the LTV ratio.

In this paper, we estimate the bubble as the misalignment between trading price and its fundamental value, as recently demonstrated in Igan and Loungani (2012) and Shi et al. (2014). The estimated bubble then used in event study to analyze the impact of LTV implementation.

Results in this paper are in line with most views on the housing market in Indonesia, that is, that there is no hazardous bubble. A short series of bubbles after first LTV implementation, however, show us on how a counter-cyclical buffer could also bring a disruption instead of more stable market. These findings affect how the trader should react, which is discussed further in one of the following sections.

The remainder of the paper is organized as follows: Section 2 contains a short review of the literature related to the problem investigated; Section 3 describes the data and methodology; Section 4 presents the result of bubble estimation and event study for LTV implementation, and a discussion for business strategy; and Section 5 concludes the findings.

2 LITERATURE REVIEW

2.1 House price bubble

In this paper, will use the rational expectation framework to detect bubbles that developed in Lucas and Sargent (1981) and Clayton (1996). In this framework, the bubble is defined to be exist when the trading price of an asset (P_t) is having a persistent positive misalignment trend from its expected fundamental value $(E_tP_t^*)$. The misalignment (ε_t) then could be written as:

$$\varepsilon_t = P_t - E_tP_t^* \tag{1}$$

Assuming that ε_t similar to the bubble component, this equation is still in line with one derived in Capozza et al. (1997) and Capozza et al. (2002).

2.2 Housing fundamental price determinants

As explained in equation (1), we need to estimate the fundamental price to further estimate its deviation with the trading price. In the growing literature of housing price bubble, the fundamental value is determined using macroeconomic variable to represent the current economic state.

For macroeconomic factor, we consider using economic growth as the most basic proxy. This factor is widely considered as the fundamental determinant for demand factor for housing that representing the state of income level (Campbell et al., 2009; Wheaton and Nechayev, 2008). To further control the macroeconomic condition, we use inflation and lending rate to elaborate the impact from price level and discount rate, as in the work of Igan and Loungani (2012) and Shi et al. (2014).

2.3 Macroprudential policy to prevent bubble

As reported in the work of Galati and Moessner (2013), the macroprudential policy to prevent bubble could be conducted in some different approach but should be in a distinction with financial stability policy. This feature underlines the policy should be a counter-cyclical buffer and LTV served as more direct policy among other reviewed in the seminal paper of Borio et al. (2001).

The implementation of LTV is proven to be effective in cross-country (Brockmeijer et al., 2011; Crowe et al., 2013). Igan and Kang (2011) also reviewed the LTV implementation in the Republic of Korea and found that the overall transaction and appreciation in housing price declined. He also underlined the role of that decline to control expectations, which helps to prevent bubble formation.

2.4 Reacting to a housing price bubble

A specific business strategy to react in the presence of a bubble is lack of detailed research. However, one could refer to the informational friction theory to understand what to react in the presence of bubble (recently reviewed by Brunnermeier & Oehmke 2013).

First of all, as suggested in De Long et al. (1990), traders are assumed to be faced with two risks when the bubble is present. The first is the fundamental risk, suggesting the investor trade against the bubble, as it logically turns out as a loss. The second is the risk that arises if the traders choose to ride the bubble, that turns out also made a loss for the trader if the bubble burst.

The two risks mentioned here are the fundamental cause of the information friction, that as depicted in Abreu and Brunnermeier (2003) turns out made the even long-lived risk-neutral traders choose to ride the bubble because they find it to be optimal. The information friction, then further explained in their work as the uncertainty of other rational traders, that allegedly would do the same move to optimize the presence of the bubble. This key element of informational friction explained that only a group of traders could burst the bubble earlier.

This theory could help businesses work out how to react. First of all, we should accept the belief of riding a bubble could make an optimal profit, which turns out to be a more plausible scenario in the reality (see Brunnermeier & Nagel 2004 and Temin & Voth 2004, for example). But, the theory of informational friction also suggests the trader could make an agreement among other to burst the bubble earlier. This turns out to be more safely, as the possible outcome of a bigger bubble burst evidently turns out to make a worst outcome for all, as in recent subprime mortgage crisis.

3 DATA AND METHODOLOGY

3.1 Data

The data used in this paper is publicly available in the Bank for International Settlement (BIS) and BI. We use the new house price index provided by BIS as housing trading price proxy. For economic growth and inflation rate, we use year-on-year growth rate of constant Gross Domestic Product (GDP) and Consumer Price Index (CPI). For the interest rate, we use lending rate for consumption as a proxy for discount rate for housing price. All of the macroeconomic data obtained from BI statistics. Due to the availability of the data, the period used in the estimation is limited to 2004Q1–2017Q1.

3.2 Estimating house price bubble

The estimation of misalignment price in equation (1) requires the value of $E_tP_t^*$ to be estimated first. For estimation of $E_tP_t^*$, we follow the approach of Igan and Loungani (2012) and Shi et al. (2014), that is using first differenced equation with trading price as the dependent variable and considering the effect of up to 3 lags in autoregressive (AR) and independend variables. The first differenced equation could be written as:

$$\Delta P_t = \beta_0 + \sum\nolimits_{k=1}^{3} \beta_{1,k}\Delta P_{t-k} + \sum\nolimits_{j=0}^{3} \beta_{X,j}\Delta X_{t-j} + \upsilon_t$$

$$(2)$$

where X is the vector of macroeconomic variable explained before and υ is the error unexplained in the model.

Assuming that the fitted value from equation (2) is equal to the fundamental housing price, the $E_t\Delta P_t^*$ could be calculated using:

$$E_t\Delta P_t^* = \hat{\beta}_0 + \sum\nolimits_{k=1}^{3} \hat{\beta}_{1,k}\Delta P_{t-k} + \sum\nolimits_{j=0}^{3} \hat{\beta}_{X,j}\Delta X_{t-j}$$

$$(3)$$

The estimated $E_t\Delta P_t^*$ then could be used to restore $E_tP_t^*$ using equation:

$$E_t P_t^* = P_{t-1} + E_t \Delta P_t^* \qquad (4)$$

The coefficients in equation (2) are estimated using OLS. The optimal lag is determined using SIC. It is also should be noted that the fundamental housing price is not estimated as forecast value as in Shi et al. (2014) due to a limitation in data availability.

4 RESULTS AND DISCUSSION

4.1 Estimation results

Result for equation (2) estimation is presented in Table 1. From this estimation result, we can conclude that not all of the macroeconomic variables could well describe the fundamental value of housing price. The difference in the level of discount rate, growth in GDP, and inflation rate are expected to have a positive impact to the difference in housing price (Glindro et al., 2011; Igan & Loungani, 2012; Shi et al., 2014). The GDP growth and interest rate parameter are insignificant in $j = 0$; therefore, we use general-to-specific approach as in Glindro et al. (2011), that is, by removing GDP growth first from the equation as it is seems not fit in both economics and statistical terms. The

Table 1.   Estimation results for fundamental price model.

Variable	(1)	(2)
ΔP_{t-1}	0.5110***	0.3213**
	(0.1075)	(0.1329)
ΔGR_t	-0.3808	
	(0.2294)	
ΔINF_t	0.1527**	0.1514**
	(0.0671)	(0.0649)
ΔLR_t	0.2917	0.9593**
	(0.3016)	(0.3895)
ΔLR_{t-1}	0.8232**	1.5525***
	(0.3066)	(0.4258)
ΔLR_{t-2}		0.8765*
		(0.4657)
ΔLR_{t-3}		1.0902***
		(0.3918)
Cons	0.8881***	1.3748***
	(0.2316)	(0.3047)
Period	2004Q3–2017Q1	2004Q3–2017Q1
Adj. R^2	0.4191	0.4624
F -stat	8.50***	8.45***

Notes: Standard errors in parentheses. *, **, and *** indicates significance level respectively at 10, 5, and 1 percent level.

removal of GDP growth turns out to make all of the variables to be significant, so the variable removal could be stopped.

Estimation using SIC as criterion choosing an AR (1) model as an optimal impact from housing price past value. In overall, this model then could be safely used to estimate the fundamental price.

4.2 Event study for LTV implementation

From the result from equation (2) estimation, we then calculate the fundamental value using equation (4). The calculated fundamental value then was compared to the trading price to analyze the misalignment as in equation (1).

The housing price bubble is rarely visible in Indonesia. A very short series of persistence and relatively big positive misalignment could be seen in after the first LTV implementation in 2012Q2, specifically around period of 2012Q4–2013Q1. This bubble is also confirmed by BI, that later encouraged BI to respond further by the 2013 adjustment in LTV policy.[1] The bubble, however, burst before the 2013 adjustment implemented in September 2013. From this estimation we can conclude that the 2013 adjustment are not well-dated, but if we compare to the first LTV version, it is evidently more promising in maintaining the stability of housing price bubble in the following period, which also maintained after the 2015 adjustment.

4.3 Proposed strategy related to the Indonesian housing bubble condition

As we can see from the estimation results, the housing price bubble is rarely formed in Indonesia. For example, a series of bubble in 2012Q4–2013Q1 is very short to represent a real bubble but meaningful enough to be an object of study for Indonesian case.

First, we could agree that most of the misalignment price are lived shortly, suggesting that the bubbles are mostly burst not long after the trading price start to positively deviate. This is showing that the informational friction weakly occurs in Indonesian case, thus, an arrangement among trader is not urgently needed if this condition still takes place in the future.

Second, we suggest giving attention to the counter-cyclical buffer policy instead. The most visible bubble occurs in 2012Q4–2013Q1, which happens after the first LTV implementation in 2012Q2. The fist LTV implementation, as mentioned before, is followed with a riskier consumer that forcing to buy property by issuing more leverage.[2] Contrasted to the results from Igan and Kang (2011), this event shows us on how the LTV could disrupt the market and encouraging a bubble instead, as suggested by

---

[1] See press-release by BI for the 2013 adjustment in LTV: https://www.bi.go.id/en/ruang-media/siaran-pers/Pages/SP_153113_dkom.aspx.
[2] See the same press release referred to in footnote 1.

Glaeser et al. (2010). For the trader, this disruption should be anticipated in a future counter-cyclical policy by monitoring any change in consumer behavior outcome, particularly in the risk profile.

5 CONCLUSION

Recent LTV regulation in Indonesia shows a promising role as a counter-cyclical buffer to prevent a housing bubble. In addition, the estimated misalignment price shows that informational friction is very minimum in housing bubble context.

Results from this paper suggest that the trader in the housing market should maintain the *status quo* of minimum informational friction in order to keep the housing price stable. The trader, instead, should be aware of any changes in the policy environment that could disrupt the consumer behavior.

REFERENCES

Abreu, D., & Brunnermeier, M. K. 2003. Bubbles and Crashes. *Econometrica*, Vol. 71(1),173–204.

Bernanke, B. S. 2010. *Monetary Policy and the Housing Bubble: A Speech at the Annual Meeting of the American Economic Association, Atlanta, Georgia, January 3, 2010*.

Borio, C., Furfine, C., & Lowe, P. 2001. Procyclicality of the Financial System and Financial Stability: Issues and Policy Options. *BIS Papers*, Vol. 1(March), 1–57.

Brockmeijer, J., Moretti, M., Osinski, J., Blancher, N., Gobat, J., Jassaud, N., ... Nier, E. 2011. *Macro Prudential Policy: An Organizing Framework*. IMF.

Brunnermeier, M. K., & Nagel, S. 2004. Hedge Funds and the Technology Bubble. *The Journal of Finance*, Vol. 59(5),2013–2040.

Brunnermeier, M. K., & Oehmke, M. 2013. Bubbles, Financial Crises, and Systemic Risk. In *Handbook of the Economics of Finance* (Vol.2, pp. 1221–1288). https://doi.org/10.1016/B978–0–44–459406–8.00018–4

Campbell, S. D., Davis, M. A., Gallin, J., & Martin, R. F. 2009. What Moves Housing Markets: A Variance Decomposition of the Rent–price Ratio. *Journal of Urban Economics*, Vol. 66(2),90–102.

Capozza, D., Mack, C., & Mayer, C. 1997. The Dynamic Structure of Housing Markets. *Ann Arbor*, Vol. 1001, 48109.

Capozza, D. R., Hendershott, P. H., Mack, C., & Mayer, C. J. 2002. *Determinants of Real House Price Dynamics*. National Bureau of Economic Research.

Clayton, J. 1996. Rational Expectations, Market Fundamentals and Housing Price Volatility. *Real Estate Economics*, Vol. 24(4),441–470.

Crowe, C., Dell'Ariccia, G., Igan, D., & Rabanal, P. 2013. How to Deal with Real Estate Booms: Lessons from Country Experiences. *Journal of Financial Stability*, Vol. 9(3),300–319.

De Long, J. B., Shleifer, A., Summers, L. H., & Waldmann, R. J. 1990. Noise Trader Risk in Financial Markets. *Journal of Political Economy*, Vol. 98(4),703–738.

Galati, G., & Moessner, R. 2013. Macroprudential Policy–a Literature Review. *Journal of Economic Surveys*, Vol. 27(5),846–878.

Glaeser, E. L., Gottlieb, J. D., & Gyourko, J. 2010. Did Credit Market Policies Cause the Housing Bubble?.

Glindro, E. T., Subhanij, T., Szeto, J., & Zhu, H. 2011. Determinants of House Prices in Nine Asia-Pacific Economies. *International Journal of Central Banking*, Vol. 7(3),163–204.

Igan, D., & Kang, H. 2011. Effects of Loan-to-Value and Debt-to-Income Limits on Housing and Mortgage Market Activity: Evidence from Korea. *International Monetary Fund (IMF) Working Paper WP/11/297*.

Igan, D., & Loungani, P. 2012. Global Housing Cycles.

Lucas, R. E., & Sargent, T. J. 1981. *Rational Expectations and Econometric Practice* (Vol. 2). U of Minnesota Press.

McDonald, J. F., & Stokes, H. H. 2013. Monetary Policy and the Housing Bubble. *The Journal of Real Estate Finance and Economics*, Vol. 46(3),437–451.

Shi, S., Jou, J.-B., & Tripe, D. 2014. Can Interest Rates Really Control House Prices? Effectiveness and Implications for Macroprudential Policy. *Journal of Banking & Finance*, Vol. 47, 15–28.

Shiller, R. J. 2012. *The Subprime Solution: How Today's Global Financial Crisis Happened, and What to Do about It*. Princeton University Press.

Temin, P., & Voth, H.-J. 2004. Riding the South Sea Bubble. *American Economic Review*, Vol. 94(5),1654–1668.

Wheaton, W., & Nechayev, G. 2008. The 1998–2005 Housing "Bubble" and the Current "Correction": What's Different This Time? *Journal of Real Estate Research*, Vol. 30(1),1–26.

Yang, Z., Wang, S., & Campbell, R. 2010. Monetary Policy and Regional Price Boom in Sweden. *Journal of Policy Modeling*, Vol. 32(6),865–879.

Advances in Business, Management and Entrepreneurship – Hurriyati et al (eds)
© 2020 Taylor & Francis Group, London, ISBN 978-0-367-27176-3

Psycho-economic phenomena, opportunistic behavior, and impacts on entrepreneurial failure

H. Rahman & E. Besra
Universitas Andalas, Padang, Indonesia

N. Nurhayati
Padang State Polytechnic, Padang, Indonesia

ABSTRACT: This paper aims to investigate and discuss psycho-economic phenomena contributing to entrepreneurial failures. Several research studies have indicated that deterministic, emotive, and voluntaristic factors are the most prominent aspects influencing entrepreneurial failure. However, this paper proposes that causes of failure are not limited to these categories only; the opportunistic behavior of individuals can also contribute to entrepreneurial failure. The study is a quantitative study; it carries out causal analysis relating existing arguments regarding psycho-economic phenomenon to entrepreneurial failure. The study further adds to and analyzes the construct of opportunistic behavior as another possible factor in entrepreneurial failure. The sample for the study is 1541 nascent entrepreneurs in West Sumatra Province, Indonesia, who have experienced failures in their business. Analysis was undertaken by using multiple and partial regression analysis in which the statistical protocol was carried out. The study found that psycho-economic factors, together with individuals' opportunistic behavior to a greater or lesser degree, cause entrepreneurial failure. The study also argues that opportunistic behavior may not only be viewed as a source of entrepreneurial success but should also be considered as a cause of entrepreneurial failure. This finding clearly demonstrates the originality and value of this study.

1 INTRODUCTION

Some of the most important entrepreneurial lessons that can be learned by entrepreneurs come from failure. In a study of entrepreneurs, Wadhwa et al. (2009) find that successful entrepreneurs experience two to three failures on average for every new successful venture. Failure triggers entrepreneurs to become tougher, more resilient, and, most importantly, able to learn from mistakes.

The existence of failure as an event in the entrepreneurial journey has led to the identification of a particular concept in entrepreneurship: serial entrepreneurship. Lafontaine and Shaw (2014) say that serial entrepreneurship can be understood as a stage in the entrepreneurial process in which the following elements exists: (1) learning from the failure; (2) a process of change in entrepreneurial behavior after the failure; and (3) experience in managing a new business. Opinions differ as to the number of failed ventures that need to happen before an entrepreneur reaches success. The number of venture failures before achieving entrepreneurial success is situational, conditionally and contextually related to an entrepreneur as a person. This means that most entrepreneurs will experience entrepreneurial failure—and that the number of failures varies among them. However, one common argument made by scholars is that entrepreneurial success is very seldom achieved in one single venture.

Research on determinants of entrepreneurial success clearly identifies a number of influences, as formulated by Rahman and Day (2014): (1) internal and external environments of the entrepreneur; (2) psychological condition and situation of the entrepreneur; and (3) sociological situation of the entrepreneur. Each factor has its own specific determinants, with opportunism considered one psychological determinant that can lead to entrepreneurial success (see, for example, Hills & Shrader 1999, Herath 2014, Wasdani & Matthew 2014, Chang et al. 2014, Khamudeen, Keat, & Hassan 2017).

An interesting question therefore arises: what would be the other sensible psycho-economic constructs, causing entrepreneurial failure, apart from the already identified deterministic, voluntaristic, and emotive factors? Further, can any determinant of entrepreneurial success factors play intersecting roles with determinants of entrepreneurial failure? This study investigates and discusses the possibility of opportunistic behavior as a determinant not only in entrepreneurial success but also in entrepreneurial failure.

This study takes nascent entrepreneurs in West Sumatra, Indonesia, as the subject of analysis. West Sumatrans are known in Indonesia for their entrepreneurial spirit. At the same time, many West

Sumatran entrepreneurs experience failure when starting a new business. Our earlier investigation through a pilot study found that entrepreneurs in West Sumatra fail between three and four times before they successfully establish a business. This study attempts to identify reasons for failure experienced by nascent entrepreneurs and to serve as a reliable foundation to support government policies regarding nascent entrepreneurs.

2 LITERATURE REVIEW

Entreprenerial failure can be defined as the cessation of the entrepreneurial process as a result of failures that occurred during the preparation, implementation, and management of the venture. One main possible cause of failure is an inability to manage financial matters, resulting in the cessation of business operation and, worse, bankruptcy. However, even though an inability to manage financial matters is closely linked to bankruptcy, it cannot be viewed as the sole reason for entrepreneurial failure. Instead of an inability to manage financial matters, entrepreneurial failure is mainly viewed as being the result of a combination of psychological and economic factors. Psychological factors may lead individuals to make errors of judgement—which over time may result in the failure of their business.

As Smida and Khelil (2010) suggest, entrepreneurial failure is a psycho-economic phenomenon in which entrepreneurs undertake errors of action in allocating resources and can result in further consequences in the form of disappointment. The psycho-economic phenomenon may relate to:

(1) individual situations and conditions (in particular behavior and personality);
(2) the organization as a business entity, and where anentrepreneur undertakes the entrepreneurial process;
(3) the social environment that directly and/or indirectly relates to individuals;
(4) the entrepreneurial process as it happens to individuals. However, these four factors individually cannot influence entrepreneurial failure, as each does not have sufficient power alone. It is a combination of factors and elements that leads a business run by an entrepreneur to fail.

Knowledge of why businesses fail is an important way to minimize the failure rate of new businesses. Even though entrepreneurs are often described as individuals who can take risks, we argue that knowledge about business failures can also help guide nascent entrepreneurs in preparing their businesses and help reduce the rate of serial entrepreneurship. Failures can be learning events, with experience gained in *post-mortem assessment*. One form of post-mortem assessment is *cognitive structural analysis*, in which entrepreneurs analyze types of failure and re-motivate themselves to get back in business,

gaining experience from cases, a new ability to face and tackle failures, and, most important, determining how failures can transform into opportunities.

Most studies regarding failures of new ventures consider: (1) how and why a new venture fails (Artinger and Powell 2015, van Gelder et al. 2007); and (2) what consequence needs to be borne by entrepreneurs as the result of their failure (Singh et al. 2015, Yamakawa & Cardon, 2015, Jenkins et al. 2014, Ucbasaran et al. 2010). Contextually, past research has mostly analyzed the consequences of failure to the business belonging to the entrepreneur—which means the analysis has mainly looked at the business as an entity. Considering this, research contexts were mostly centered on the question of why a business fails. Wennberg et al. (2010) and Delmar et al. (2006) conclude that a business mainly fails because of the following reasons: (1) low business performance; (2) the problem of resources; and (3) unachievable positive goals and growth of the business. On a broader scale, Wennberg and DeTienne (2014) and Hammer (2014) identified three conditions leading to failure: (1) business environment; (2) the business itself—mainly inappropriate resources and incompetence; and (3) the entrepreneur as an individual/personal—mainly less commitment from entrepreneurs in managing the business.

Observed objectively, business failure, in particular new venture failure, is closely related to the individual who operates the business. The entrepreneur has a prominent role in determining whether a business succeeds or fails. Therefore, unit analysis cannot always be focused on the business as an entity. This is why we centered our analysis on the entrepreneur as an individual. According to Mellahi and Wilkinson (2004), business failure can be viewed from the angle of the entrepreneur as an individual, and it is sourced from (1) deterministic factors and (2) voluntaristic factors. A *deterministic factor* is defined as the failure of new ventures due to the entrepreneur's environment, which cannot be entirely controlled by the entrepreneur. Cardon et al. (2011) also state that the deterministic factor of business failure refers to the entrepreneur's environment, which cannot be avoided by entrepreneurs in their business operation. Mellahi and Wilkinson (2004) state that voluntaristic factors, in contrast, relate to business failure due to errors by entrepreneurs in making decisions and conducting actions. Mellahi and Wilkinson (2004) therefore show that deterministic factors are not the only one factors resulting in business failure—voluntaristic factors also play a role.

Cardon et al. (2001) further state that business failure is not only a result/consequence of uncontrollable situations and conditions such as the environmental factor. One possible source of business failure rather comes from a stigma that is embedded inside of entrepreneurs, which makes them try as hard as they can to avoid it—but unfortunately they still end up with making false decisions and wrong actions, and, as the result of these, their businesses can fail (Singh et al. 2015). Khelil (2016) further mentions that, apart from deterministic and voluntaristic factors—as in Mellahi and Wilkinson (2004) and Cardon et al. (2011)—there is also an emotive factor, which is another decisive factor in entrepreneurial failure. Emotive factors as described by Khelil (2016) can show a relationship between business performance achieved by entrepreneurs and disappointment in unachieved business performance. In relation to this, Hammer (2014) also shows that *goal setting bias* is a major cause of business failure. Goal setting bias happens because of the mismatched situation between expectations and performance. This emotive factor, as Khelil (2016) states, is a combination of deterministic and voluntaristic factors. Therefore, failure not only comes from uncontrollable or difficult-to-control environments; it also derives from errors in decision-making and wrong actions made by the entrepreneur in managing his or her business.

Khelil (2012) states that failure in new ventures mostly happens because the entrepreneur falls into one of the following patterns or types: (1) *gambler* —those who start a business without resources or clear orientation; (2) *supported at arm's length*— those with limited resources; (3) *bankrupt*—ones with limited competencies; (4) *megalomaniac*— those with too much self-confidence; and (5) *dissatisfied with lord*—ones who do not fully rely on God's will. We further can categorize those reasons into two categories: (1) individual personality as a source of failure (*gambler, megalomaniac,* and *dissatisfied with lord*); (2) environmental situation as a source of failure (*supported at arm's length* and *bankrupt*). Meanwhile, Hammer and Khelil (2014) state that, based on consideration of input, process, and output in the new venture creation process at the individual and enterprise levels, entrepreneurial failure can take the form of an individual leaving the business as a result of: (1) an *individual level* decision, which has *competency* and *security* as its dimensions; and (2) a *firm level* decision, which has *support* and *business model* as its dimensions. The existence of both factors can lead entrepreneurs to stop their business. Following this exit decision, entrepreneurs can try to find other jobs, restart a new venture, or sell their business with the hope that they will get some profits or, worse, declare bankruptcy.

Studies regarding opportunism mostly relate to the opportunistic behaviour of individuals. Opportunistic behaviour can bring a negative impact to a business (Cordes et al. 2010). Williamson (1993 and 1999) has previously revealed that opportunistic behaviour is closely related to the transaction cost concept and will create conflicts and situations in which individuals need to choose one appropriate choice from many alternative possibilities. Both concepts are clearly related to the topic of this paper and

study—they demonstrate an entrepreneur will prefer a situation which brings more benefits to him- or herself personally, whether it's a better job compared with a career in entrepreneurship or the desire to start other new businesses when the earlier business is still unstable/immature. This situation will put entrepreneurs in a conflict that requires them to make their own decision: enter a more secure job, start another new venture, or maintain the earlier venture until maturity. Sometimes the decision made is a good one, but it can also be a wrong decision, leading to failure in the earlier venture. We can see here that entrepreneurs can be trapped at an individual level (as Hammer & Khelil 2014) in a psychological situation where they must secure their future based on which career path they choose.

In a study of the impact and consequences of entrepreneurial failure, Mantere et al. (2013) describe it as a social construction in which emotional and cognitive processes exist to justify every action. Entrepreneurial failure has a psychological impact on failed entrepreneurs— as a result, they do their best to maintain their self-esteem and to avoid losing their own business. Psychological impacts can also be seen in efforts to reduce and, moreover, eliminate stigma that arises from the failure (Singh et al. 2015). An interesting phenomenon regarding the existence of stigma lies in the fact that it can also become a source of the next failure. As Singh et al. (2015) state, fear of failure and efforts to avoid the stigma of failure have contradictively resulted in negative situations for an entrepreneur —as he/she makes a wrong decision or action in the business. As the result, this makes entrepreneurs sink into failure (again). Stigma around failure is a psychological reason why an entrepreneur fails in the business.

3 DESIGN AND METHODOLOGY

The study is an explanatory study and uses quantitative methodology as its research approach. Causal analysis is used to investigate the simultaneous relationship of, and the influence of, psycho-economic factors and opportunistic behaviour as variables in entrepreneurial failure. Cross-sectional cohort data and information were collected by using questionnaires as the research instrument.

The study uses 1541 nascent entrepreneurs in West Sumatra, Indonesia, as its research sample. Samples are chosen based on certain criteria: (1) age; (2) minimum level of formal education; and (3) experience of failure in business. The study considers its samples as gender neutral; gender is not considered for data analysis. Furthermore, the study does not consider a particular business branch as background for the samples. Operation of the variables in the research uses the following guidelines (see Table 1).

Table 1. Operation of variables.

tab	Variables	Dimension	Indicators	Measurement
1	Entrepreneurial failure	a. Failure to allocate resources b. Failure in decision making c. Failure in actions	a. Ineffectiveness and inefficiency in allocating resources b. Decision making is not based on data, information and facts c. Actions do not fit with what supposed to be done	Likert scale 1–5 9 questions
2	Deterministic factors	a. The availability of supports b. Social environment of entrepreneurs c. Competition d. High operational cost of the business	a. The existence of social support from the nearest social environment b. Response of social environment to the choice of becoming an entrepreneur c. Degree of competition d. Level of operational cost	Likert scale 1–5 11 questions
3	Voluntaristic factors	a. Competencies b. Individual orientation c. Behaviour related to customers	a. Level of knowledge, skills, Motivation, and personal characteristics b. Personal orientation of the entrepreneurs c. Response to customers complaints and needs	Likert scale 1–5 14 questions
4	Emotive factors	a. Psychological pressures to get income b. Necessity motives	a. Level of psychological pressure to get income b. Necessity based motive in entrepreneurship	Likert scale 1–5 8 questions
5	Opportunistic behaviour	a. The possibility of getting a better job b. Job security in a longer period c. Desire and passion to start other businesses	a. Level of income that would be received continuously b. High level of social status c. Opportunity to get a better career d. Level of intention to start other businesses	Likert scale 1–5 8 questions

Source: Authors' own conception

Opportunistic behaviour is measured based on the conceptual foundation that transactional circumstances will appear to individuals regarding their choice of future life (Williamson 1993 and 1999). In this study, it is reflected in the possibility of a person: (1) choosing a better job rather than entrepreneurship; (2) choosing to secure a job for a longer period; and (3) following his/her desire to immediately establish other new ventures while an earlier business is still unstable or immature. These possibilities are then connected to the possibility of failure for the earlier business. For this purpose, findings and results of the study were analyzed by using multiple regression analysis and supported by the use of SPSS as the statistical tool. Design of the research framework in this study is shown in Figure 1.

Based on the research framework developed in Figure 1, the following statistical equation will be used as the multiple regression model of the study.

Yef = a + biXv + b2Xd + b3Xe + b4Xo + e

where:
Yef = entrepreneurial failure
Xv = voluntaristic factors
Xd = deterministic factors

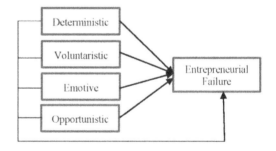

Figure 1. Research framework.
Source: Adapted from Mellahi and Wilkinson (2004) and Williamson (1993 and 1999)

Xe = emotive factors
Xo = opportunistic behaviour
e = error
a = constant

The statistical model developed for this study measures the simultaneous influence of (1) voluntaristic, (2) deterministic, (3) emotive, and (4) opportunistic behavior on entrepreneurial failure. We use the *F*-test statistic to test this model. To measure the influence of each independent variables on entrepreneurial failure as the dependent variable, we use the *t*-test statistic.

4 RESULTS AND FINDINGS

Profiles in our sample are divided into two categories: (1) personal profiles; and (2) business profiles. Our intention is to prove that, descriptively, our samples fit with the sampling criteria previously stated in the methodology part of this paper. The personal profile of our sample is shown in one integrated table (see Table 2).

Our second task in the study is to measure whether questions in our research instrument (questionnaire) are valid and reliable. For this purpose, we use *r*-table with the value of 0.1308 as the basis for the validity

Despite showing the gender profile of our sample in Table 2, as already mentioned we consider our sample gender neutral. We believe that both genders (male and female) may experience the same rates of business failure. We are more interested in discussing our sample profiles from the perspectives of current age, level of education, and age when first starting a business, because we think those factors are more related to why entrepreneurs fail in business.

Table 2. Sample profile (personal profile).

Category	Characteristic	Amount of Sample	Percentage (%)	Valid Percent
Gender	Male	840	54.50	54.51
	Female	701	45.49	100.00
Age (years)	18–20	62	4.23	4.23
	21–25	716	46.46	50.69
	26–30	763	49.51	100.00
Level of education	Elementary & Junior High School	147	9.60	9.60
	Senior High School	724	47.00	56.60
	Undergraduate	669	43.40	100.00
Age when	18–20	491	31.86	31.86
first	21–23	726	47.11	78.97
starting a busi-	24–26	252	16.35	95.32
ness (years)	27–30	72	4.68	100

Source: Survey data, processed.

The majority of our sample are male, and the sample is dominated by individuals who are 26–30 years old, attended a higher education institution, and were 21–23 years old when they first started their business. As justification for using demographic characteristics in our study, we consider Talas et al. (2013), who previously argued that gender, age, education, and type of school previously attended by individuals are the demographic factors that can influence entrepreneurship.

Meanwhile, the profile of the businesses belonging to our sample is shown in Table 3.

Table 3 suggests that around one-third of our sample ran businesses previously which failed. Most of our samples' current businesses are the second business, which means that the previous one failed: 1101 experienced business failure once before. Table 2 indicates serial entrepreneurship at work in our sample. Our samples still have the courage to start a business (either in the same line of business as the previous one or a completely new one). As Lafontaine and Shaw (2014) argue, serial entrepreneurship is a journey in which an entrepreneur experiences failures in order to achieve success, with processes of (1) learning from the failure, (2) change in entrepreneurial behavior after the failure, and (3) experience in managing the business taking place during the failure. Experiences in business operations that result in failures impact the creation of entrepreneurial resilience. Entrepreneurs can benefit from this process by improving their personal capacity to make realistic business plans, have self-confidence and a positive self-image, possess communication skills, and have the capacity to manage strong feelings and impulses (American Psychological Association [APA] 2010). As De Vries (1977) and Gratzer (2001) argue, learning from failure is an important characteristic of entrepreneurs—and once an

Table 3. Profile of business.

Category	Characteristic	Amount of Sample	Percentage (%)	Valid Percent
Current business is the …	second	990	64.24	64.24
	third	426	27.64	91.89
	fourth	84	5.45	97.34
	fifth	13	0.84	98.18
	above fifth	28	1.82	100.00
Number of failures	1	1101	71.45	71.45
	2	353	22.91	94.35
	3	57	3.70	98.05
	4	16	1.04	99.09
	5	1	0.06	99.16
	>5	13	0.84	100.00

entrepreneur learns from failures he or she has experienced, there is a greater possibility of rebuilding the business and achieving success.

analysis. Our validity measurement shows that the value of Corrected Item-Total Correlation for deterministic, voluntaristic, emotive, and opportunism factors are bigger than the value of r-table (0.1308). We then conclude that questions in our research instrument are valid. The measurement of reliability in our study uses the reliability statistics table—in which the rule says that variables are reliable if they have Cronbach's alpha value of more than 0.7. The measurement of reliability statistics from our research instrument is shown in Table 4.

In Table 4, our reliability measurement shows that the value of Cronbach's alpha for the variables of entrepreneurial failures (EF), deterministic (DEF), voluntaristic (VEF), emotive (EEF), and opportunistic behavior (OEF) factors are above 0.7—which means that all points in our questionnaire are reliable.

We then measured the simultaneous influence of the variables DEF, VEF, EEF, and OEF on entrepreneurial failure using multiple regression analysis (see Table 5).

Multiple regression analysis shows the value of F is 217.249 with the Sig. 0.000. As this value is less than $a = 5\%$, the regression model can be used to predict entrepreneurial failure. Our finding indicates that the four independent variables in this study—namely, deterministic, voluntaristic, emotive,

and opportunistic behaviors— simultaneously and significantly influence entrepreneurial failure.

The final task in our study is measuring the result of t-test for a possible relationship between each independent variable and entrepreneurial failure (see Table 6).

Partial regression analysis using t-test measures the relationship between EF and: DEF, VEF, EEF, and OEF. The t-test of each relationship shows the values 3.470 (EF and DEF), 7.396 (EF and VEF), 16.168 (EF and EEF), and 9.236 (EF and OEF). We also found that the Sig. values for all relationships are between 0.000 and 0.001. Since all the Sig. value are less than $a = 5\%$, the partial regression model from each variable (DEF, VEF, EEF, and OEF) have a significant relationship with entrepreneurial failure.

We specifically put our attention to the construct of opportunistic behavior (OEF) and its relationship to entrepreneurial failure. From the partial regression model above, it has the t-value of 9.236 with Sig. 0.000—which means that there is a significant relationship between opportunistic behaviors as a construct that can cause entrepreneurial failure. We argue that opportunistic behavior can be predicted as another factor (apart from the deterministic, voluntaristic, and emotive factors) behind the failure experienced by nascent entrepreneurs.

5 DISCUSSION

Our study proved that psycho-economic phenomena —voluntaristic, emotive, and deterministic factors— are influencing entrepreneurial failures experienced by nascent entrepreneurs in West Sumatra. If we look at the descriptive results gathered from our sample, the most deterministic factor is the inability of respondents to meet operational costs for their business. In emotive factors, the majority of our

Table 4. Reliability statistics of the variables.

Variables	Cronbach's Alpha
Entrepreneurial Failure (EF)	0.843
Deterministic Factors (DEF)	0.783
Voluntaristic Factors (VEF)	0.828
Emotive Factors (EEF)	0.823
Opportunistic Behavior (OEF)	0.766

Source: Survey data, processed.

Table 5. The result of multiple regression analysis (f-test) to measure the simultaneous influence of DEF, VEF, EEF, and OEF on entrepreneurial failure.

Model	Sum of Squares	Df	Mean Square	F	Sig.
¹ Regression	30720.010	4	7680.003	217.249	.000ª
Residual	106795.989	3021	35.351		
Total	137515.999	3025			

Source: Primary data analysis.

Table 6. The result of partial regression analysis (t-test) to measure the influence of DEF, VEF, EEF, and OEF on entrepreneurial failure.

Model	Unstandardized Coefficients		Standardized Coefficients		
	B	Std. Error	Beta	t	Sig.
(Constant)	7.393	1.049		7.045	0.000
EF DEF	0.068	0.019	0.062	3.470	0.001
EF VEF	0.126	0.017	0.126	7.396	0.000
EF EEF	0.323	0.020	0.300	16.168	0.000
EF OEF	0.195	0.021	0.167	9.236	0.000

Predictors: (Constant), DEF, VEF, EEF, OEF
Dependent Variable: EF
Source: Primary data analysis.

respondents revealed that the pressure to meet personal and family needs have left them stressed and contributed to their failure in business. In the descriptive analysis of the construct of opportunistic behavior, we found that the majority of our sample think that getting into entrepreneurship is only a way to wait for a formal job—and that, once an opportunity comes, they will simply cease and leave their business. The main findings of the study confirm the opinion of Mellahi and Wilkinson (2004), that organizational failure comes from these three factors (deterministic, voluntaristic, and emotive). Similar to Mellahi and Wilkinson (2004), and as the result of our findings, we also argue that the interaction and combination of psychological and economic circumstances of individuals in terms of deterministic, voluntaristic and emotive factors cause and contribute to business and organizational failure.

Our study also argues that proxies of (1) choosing a better job rather than entrepreneurship, (2) choosing to secure a job in a longer period, and (3) the possibility of individuals following their desire to immediately establish other new ventures while an earlier business is still unstable or immature represent the construct of opportunistic behavior. We argue that opportunistic behavior of individuals, combined with their psycho-economic circumstances as an entrepreneur, will positively contribute to entrepreneurial failure.

As in the partial regression analysis, we found that opportunistic behavior of individuals could partially cause entrepreneurial failure. In essence, the correlation between the opportunistic behavior of our sample and entrepreneurial failure is explained as follows.

First, those in our sample who already had a business but then had an opportunity to enter the formal job market tended to leave his/her own business and concentrate on working for an employer. Further investigation shows this is mostly related to securing a better future life. Considering the culture of Indonesian people in general, one particular cultural dimension is a high level of uncertainty avoidance, according to Hofstede (2017) and Mangundjaya (2010). Indonesian people prefer to choose a stable situation in their life—for them, this stability can only be achieved when having a formal job with a regular monthly salary and a pension. We can understand why members of our sample prefer to have a job rather than working to maintain themselves by staying in the business. Feelings of insecurity regarding future life due to limited opportunities and strong competition in the job market, as well as an unstable business environment—including severe business competition, inconsistent government rules, regulation, and laws about business environments (especially those related to small- and medium-scale enterprises)—and the insecurity that comes with trying to launch a sustainable business operation have psychologically influenced our

sample to choose to have a job once the opportunity appears. These psychological circumstances have made some members of our sample leave their businesses; as a consequence, the amount of attention devoted to their business diminishes and the business fails. Our finding is also consistent with the opinion of Hammer and Khelil (2014), who note that entrepreneurs can leave their businesses (causing them to fail) in favour of taking another job.

Williamson (1993 and 1999) state that during conflict individuals tend to choose the most appropriate option of many alternatives. In case of our sample, this is shown by the preference to choose a formal job rather than staying in the business that they have started. We consider this psychological circumstance as a sign of opportunistic behavior, which is one cause of entrepreneurial failure.

Second, our study found that some of our samples have an uncontrollable passion for business that led them to be aggressive in creating new ventures. They believe that business is about catching opportunities—but they forget they also need to concentrate on existing businesses started earlier. Some of our sample tend to follow their desire to establish new ventures while their previous/earlier business is still unstable or immature. We also view this as opportunistic behavior. In contrast, if an entrepreneur creates a new venture after his/her earlier business has reached the maturity stage, we view this as a consequence of a desire to develop and improve the scale of the business (whether through diversification, acquisitions, or mergers). Unlike most scholars who argue that opportunistic behavior is an entrepreneurial success factor, we believe it may be a factor in entrepreneurial success and entrepreneurial failure.

6 SUMMARY AND IMPLICATIONS

Our study investigates the psychological and economic phenomenon of individuals that may cause entrepreneurial failure. We have found that deterministic, voluntaristic, and emotive factors, together with the construct of opportunistic behavior, contribute to failure experienced by nascent entrepreneurs. We further argue that opportunism should be considered as both an entrepreneurial success factor and an entrepreneurial failure factor.

The creation of resilient entrepreneurs follows entrepreneurial processes, one of which can be the failure process, in which nascent entrepreneurs learn from mistakes they have made. Thus, government intervention to strengthen entrepreneurial personalities and focus on psychological aspects regarding nascent entrepreneurs would be a sensible and reasonable policy. Capacity-building schemes for nascent entrepreneurs would help strengthen their psychological aspects: motives, maturity, logical consideration in choosing alternatives, decision-making processes, dealing with social pressures, and so on. Other capacity-building schemes may help boost business knowledge, with nascent entrepreneurs learning to manage their businesses in more effective and efficient ways.

However, we also realize that our study has particular limitations, and the issue still needs further exploration to broader contexts. As our study only discussed one particular context (nascent entrepreneurs in one province in Indonesia), it would be worthwhile to expand the topics of this study into a comparative study regarding entrepreneurial failures between cities or regions and to consider causes of entrepreneurial failure in particular business branches. We also think that it is worthwhile to relate the content of this study in entrepreneurial failure to the study of entrepreneurial resilience and serial entrepreneurship. We believe that those three particular fields are interrelated—and this could be avenue for research related to this topic.

REFERENCES

American Psychological Association (APA). 2010. The road to resilience: Resilience factors & strategies. Retrieved 2 September 2017, from http://www.apa.org /helpcenter/road-resilience.aspx.

Artinger, S., & Powell, T., C. 2015. Entrepreneurial failure: Statistical and psychological explanations. *Strategic Management Journal* (DOI: 10.1002/smj.2378).

Cardon, M. S., Stevens, C. E., & Potter, D. R. 2011. Misfortunes or mistakes? Cultural sensemaking of entrepreneurial failure. *Journal of Business Venturing*, 26: 79–92.

Chang, W., W.G.H. Liu, & Chiang, S. 2014. A study of the relationship between entrepreneurship courses and opportunity identification: An empirical survey, *Asia Pacific Management Review*, 19(1): 1–24.

Cordes, C., Richerson, P., McElreath, R., & Strimling, P. 2010. How Does Opportunistic Behaviour Influence Firm Size? An Evolutionary Approach to Organizational Behaviour. *Journal of Institutional Economics* 7(1): 1–21.

Delmar, F. K., Hellerstedt, K., & Wennberg, K. 2006. The evolution of firms created by the science and technology labor force in Sweden 1990–2000. In *Managing Complexity and Change in SMEs: Frontiers in European Research*, 69–102.

De Vries, M. F. R. K. 1977. The entrepreneurial personality: A person at the crossroads. *Journal of Management Studies* 14(1): 34–57.

Gratzer, K. (2001). The fear of failure: Reflections on business failure and entrepreneurial activity. In M. Henrekson, M. Larsson, & H. Sjoegren (Eds.), *Entrepreneurship in business and research*. Stockholm: Institute for Research in Economic History.

Hammer, M.H.M., (2012). How Business Management benefit from Entrepreneurship. In Proceedings of *10th Conference of Management, Enterprise and Benchmarking*, Budapest, Hungary, 175–182.

Hammer, M.H.M. & Khelil, N., (2014). Exploring the Different Patterns of Entrepreneurial Exit; the Causes and Consequences. In *Proceedings of 59th ICSB World Conference on Entrepreneurship*, Dublin, June 10–15.

Herath, H. M. T. S. 2014. Conceptualizing the role of opportunity recognition in entrepreneurial career success. *International Journal of Scientific Research and Innovative Technology* 1(3): 73 82.

Hills, G.E., & Shrader, R.C. 1999. Successful entrepreneurs' insights into opportunity recognition. *Frontiers of Entrepreneurship Research*. Wellesley, MA: Babson College.

Hofstede, G. 2017. National Culture of Indonesia. Retrieved from http://geert-hofstede.com/indonesia.html.

Jenkins, A. S., Wiklund, J., & Brundin, E. 2014. Individual responses to firm failure: Appraisals, grief, and the influence of prior failure experience. *Journal of Business Venturing* 29: 17–33.

Khelil, N., (2012). What are we talking about when we talk about entrepreneurial failure? *RENTXXVI—Research in Entrepreneurship and Small Business*, Lyon.

Khelil, N. 2016. The many faces of entrepreneurial failure: Insights from an empirical taxonomy. *Journal of Business Venturing* 31: 72–94.

Lafontaine, F., & Shaw, K. 2014. Serial Entrepreneurship: Learning by Doing? *NBER Working Paper No. 20312*. Cambridge, MA.

Mangundjaya, W.L.H. (2010). Is There Cultural Change In The National Cultures Of Indonesia? Paper Presented at International Conference on Association of Cross Cultural Psychology (IACCP), Melbourne, Australia.

Mantere, S., Aula, P., Schildt, H., & Vaara, E. 2013. Narrative attributions of entrepreneurial failure. *Journal of Business Venturing* 28: 459–473.

Mellahi, K., & Wilkinson, A. 2004. Organizational failure: A critique of recent research and a proposed integrative framework. *International Journal of Management Reviews* 5: 21–41.

Rahman, H. & Day, J. 2014. Involving the Entrepreneurial Role Model: A Possible Development for Entrepreneurship Education. *Journal of Entrepreneurship Education* 17(2), 163–171.

Singh, S., Corner, P, & Pavlovich, K. 2015. Failed, not finished: A narrative approach to understanding venture failure stigmatization. *Journal of Business Venturing* 30 (1): 150–166.

Smida, A. & Khelil, N. (2010). Repenser l'echec entrepreneurial des petites entreprises emergentes. Proposition d'une typologie s'appuyant sur une approche integrative. *Revue Internationale P.M.E.: economie et gestion de la petite et moyenne entreprise*, 23(2): 65–106.

Talas, E., Celik, A.K., & Oral, I.O. 2013. The Influence of Demographic Factors on Entrepreneurial Intention among Undergraduate Students as a Career Choice: The Case of a Turkish University. *American International Journal of Contemporary Research* 3(12), pp. 22–31.

Ucbasaran, D., Westhead, P., Wright, M., & Flores, M. (2010). The nature of entrepreneurial experience, business failure and comparative optimism. *Journal of Business Venturing* 25: 541–555.

van Gelder, M., Thurik, R., & Bosma, N. 2007. Succes and Risk Factors in the Pre-Startup Phase. *Small Business Economics*: 319–335.

Wadhwa, V., Aggarwal, R., Holly. H., & Salkever, A. 2009. Anatomy of an Entrepreneur: Family Background and Motivation. The Ewing Marion Kauffman Foundation Research Report, July 2009.

Wasdani, K.P., & Mathew, M. (2014). Potential for opportunity recognition: Differentiating entrepreneurs. *International Journal of Entrepreneurship and Small Business* 23(3): 336–362.

Wennberg, K. & DeTienne, D.R. 2012. What do we really mean when we talk about 'exit'? A critical review of research on entrepreneurial exit. *International Small Business Journal* 32(1): 4–16.

Wennberg, K., Wiklund, J., DeTienne, D. R., & Cardon, M. S. 2010. Reconceptualizing entrepreneurial exit. Divergent exit routes and their drivers. *Journal of Business Venturing* 25(4): 361–375.

Williamson, O.E. 1993. Opportunism and its Critics, *Managerial and Decision Economics* 14: 97–107.

Williamson, O.E. 1999. Strategy Research; Governance and Competitive Perspectives. *Strategic Management Journal* 20: 1087–1108.

Yamakawa, Y., & Cardon, M.S. 2015. Causal ascriptions and perceived learning from entrepreneurial failure. *Small Business Economics* 44: 797–820.

Advances in Business, Management and Entrepreneurship – Hurriyati et al (eds)
© 2020 Taylor & Francis Group, London, ISBN 978-0-367-27176-3

How do we perceive failure? Introducing the integrative model of entrepreneurial failure

E. Besra & H. Rahman
Universitas Andalas, Padang, Indonesia

Nurhayati
Padang State Polytechnic, Padang, Indonesia

ABSTRACT: This paper aims to introduce and to discuss an integrative model of entrepreneurial failure experienced by nascent entrepreneurs. Previous models on entrepreneurial failure mainly described reasons and factors that caused entrepreneurial failure from individual and firm levels with less elaboration about possible follow-ups of that failure. Contrarily, this paper tries to add to our understanding regarding entrepreneurial failure by elaborating its process, the possible consequence and follow up of that failure as well as its relationship to the concept of serial entrepreneurship. The study is a qualitative study and uses exploration of a possible integrative model that can figure out the process of entrepreneurial failure experienced by nascent entrepreneurs. The model is developed based on a previous quantitative study with 1541 nascent entrepreneurs in West Sumatra, Indonesia who have experienced entrepreneurial failure. The model introduces entrepreneurial failure as a series of processes which stems from an individuals' internalities and externalities. As a process, it has also brought consequences and follow-up of that failure to nascent entrepreneurs. The introduction of this integrative model of entrepreneurial failure demonstrates the originality and value of this study. Whereas other models of entrepreneurial failure concentrate their elaboration on the sources and reasons of failure, this paper concentrates more on entrepreneurial failure as a process, in which consequences and follow-up of failures would also be elaborated as part of the process.

1 INTRODUCTION

Nascent entrepreneurs are parties who are very fragile to experience entrepreneurial failure. As they have limited experience, networks, skills and knowledge, access to finance etc. failure always haunts them. No wonder we found many facts about business failure that hit nascent entrepreneurs. Some people view failureas a negative condition and as a consequence, they have no interest in having businesses anymore. However, some people will see failure as a positive event, from which they can take benefits and advantages. For them, failure is the most important entrepreneurial lesson that can be absorbed in the business, and they can use it as part of the preparations to restart a new business and to achieve entrepreneurial success.

In a study about the anatomy of entrepreneurs, Wadhwa et al. (2009) came out with the conclusion that a successful entrepreneur has experienced -on average- two to three failures in every new venture that they previously established before he/she reached success. This situation and the condition of failure have triggered entrepreneurs to become tougher, more resilient, and the most important thing, make them able to learn from failures. The existence of failure as an event in the entrepreneurs' entrepreneurial journey—which is followed by

mental and learning processes and an experiential process, has raised a particular concept in entrepreneurship, which is introduced as the concept of serial entrepreneurship.

Lafontaine and Shaw, (2014) mentioned that serial entrepreneurship can be understood as an entrepreneurial process which happens to an entrepreneur in achieving success in new ventures after he/she experienced failures, and in which the following elements exist: [a] the learning process from the failure, [b] the process of change in entrepreneurial behavior after the failure, and [c] the experience in managing the business. There are no uniform arguments and opinions from scholars that state the number of venture failures that need to happen to entrepreneurs before he/she reaches entrepreneurial success. It is believed that the number of venture failure before achieving entrepreneurial success is situational, conditionally and contextually related to the entrepreneur as a person. This means that most entrepreneurs will experience entrepreneurial failure —and the number of failure varies among them. However, one common argument raised and approved by the scholars is that entrepreneurial success is very seldom achieved in only one single venture creation/establishment.

Studies and research revealing the determinants of entrepreneurial success factors have clearly identified that entrepreneurial success is influenced by: [a] internal and external environments of the entrepreneur, [b] psychological condition and situation of the entrepreneur, and [c] sociological situation of the entrepreneur, Rahman and Day (2014). Each factor has its own specific determinants, in which opportunism is considered a psychological determinant that can lead to entrepreneurial success (see, for example, the studies of Hills and Shrader, 1999; Herath, 2014; Wasdani and Matthew, 2014: Chang et al., 2014; Khamudeen, Keat and Hassan, 2017).

An interesting question therefore is: How can we figure the process of the entrepreneurial course that could happen to and be experienced by nascent entrepreneurs? How do we understand and perceive failure experienced by nascent entrepreneurs? Most studies and research in entrepreneurial failure only considered failure as an 'event' in the entrepreneurial process and journey of entrepreneurs. In fact, entrepreneurial failure should be considered in a broader way, because it has many events that could cause failure and can be its consequences. Analyzing entrepreneurial failure from a smaller part of this process could raise our misunderstanding about it. This situation, of course, is a challenging one——as people normally undertake necessary steps and efforts to reduce the failure rate. Therefore, our paper tries to fill the gap in the topic of entrepreneurial failure, i.e., by offering an integrative model that could figure out and explain the process of entrepreneurial failure experienced by entrepreneurs. We believe this study is interesting and valuable as it tries to reveal the process of entrepreneurial failure experienced by nascent entrepreneurs, which can be used as a reliable resource and foundation to set up government policies regarding nascent entrepreneurs.

2   LITERATURE REVIEW

In general, entrepreneurial failure can be understood as the cessation of an entrepreneurial process undertaken by entrepreneurs as a result of failures that occurred during the preparation, implementation and management of the venture. One main possible sign of this failure can be seen in the inability of the entrepreneur to manage financial matters, which has further resulted in the cessation of business operations, and in the worst case,—bankruptcy. However, even though the inability to manage financial matters is closely linked to bankruptcy, it cannot be viewed as the one and only reason for entrepreneurial failure. Instead of an inability to manage the financial matters of the venture, entrepreneurial failure is mainly viewed as a result of the combination and interaction of psychological and economic factors of an entrepreneur. It is viewed that psychological factors of individuals will lead them to decide further, and to undertake error actions——

which will further result in entrepreneurial failure. As Smida and Khelil (2010), entrepreneurial failure is a psycho- economic phenomenon which will lead entrepreneurs to undertake error actions, to allocate resources and will result in further consequences, in terms of their psychological situation in the form of disappointment.

Knowledge and observation regarding entrepreneurial failure is considered important and one of the main concerns in entrepreneurship is to minimize the failure rate of new businesses. Even though entrepreneurs are described as individuals who can tackle risks (including business risks), we view that the information and knowledge regarding failures can also be used as a guidance by nascent entrepreneurs in preparing their business. The information and knowledge regarding failures can also be used to reduce the rate of serial entrepreneurship process that happens to entrepreneurs. Apart from those, failures will also be a learning event and experience from *postmortem assessment* to analyze the reasons of entrepreneurial failure. One of this post-mortem assessment is in the form of *cognitive structural analysis* that can be used by entrepreneurs to analyze the type of failure and re-motivate themselves to be back in business, getting experiences from the cases, a new ability to face and tackle failures and most importantly, the ability to determine the transformation process of failures into opportunities.

Study and research regarding new venture failures are normally emphasized on analysis of the following topics: [a] what and why a new venture fails? (Artinger and Powell; 2015; van Gelder et al., 2007) and, [b] what is the consequence that needs to be born by entrepreneurs as the result of their failure? (Singh et al., 2015; Yamakawa and Cardon, 2015; Jenkins et al., 2014, Ucbasaran et al., 2010). Contextually, studies and research that have been completed mostly analyzed the consequence of failure to the business as belonging to the entrepreneur,— which means that the analysis has been done mainly to the business as an entity. Considering this, research contexts were mostly centered on the question 'why cana business fail?.' Wennberg et al., (2010) and Delmar et al., (2006) concluded that a business mainly fails because of the following reasons: [a] low business performance, [b] the problem of resources, and [c] unachievable positive goals and growth of the business. In a more broader scale, Wennberg and DeTienne (2014) and Hammer (2014) further identified the existence of three conditions that can lead to a failure in business, which is: [a] the business environment, [b] the business itself— mainly inappropriate resources and competence, and [c] the entrepreneur as an individual/person—— mainly meaning less commitment from entrepreneurs in managing the business.

If we objectively observe business failure, in particular new ventures, failure is closely related to the analysis on those who are operating the business. This means that the entrepreneur as a person is considered an individual who is having prominent roles to determine whether a business can be a success, or, a failure. Therefore, the unit analysis of the research cannot always be focused on the business as an entity. It is why we centred our focus and analysis merely on the entrepreneur as an individual. According to Mellahi and Wilkinson, (2004) the concept of business and organizational failure can be viewed from the entrepreneur as an individual, and it stems from: [a] deterministic factors, and [b] voluntaristic factors. *Deterministic factor* is defined as the failure of new ventures stemming from the entrepreneur's environment, which existence cannot minimally be controlled by the entrepreneur. As Cardon et al., (2011) mentioned, the deterministic factor of business failure comes from the entrepreneur's environment which cannot be avoided by entrepreneurs in their business operation. In reverse, Mellahi and Wilkinson (2004) mentioned that voluntaristic factors are related to business failure which comes from errors done by entrepreneurs in making decisions and conducting actions. It can be inferred from Mellahi and Wilkinson, (2004) that deterministic factors are not the only factors that resulted in business failure—but they rather are voluntaristic factors.

Cardon et al., (2001) further stated that business failure is not a result/consequence of an uncontrollable situation and condition for the entrepreneur, such as the environment factor. One possible source of the business failure rather comes from a stigma that is embedded inside of entrepreneurs, which makes them try as hard as they can to avoid it,—but unfortunately, they ended up with making wrong decisions and wrong actions and as a result of these, their business failed. (Singh et al., 2015). Khelil (2016) further mentioned that apart from deterministic and voluntaristic factors, as Mellahi and Wilkinson (2004) and Cardon et al. mentioned, (2011)—there is also an emotive factor inside of entrepreneurs which is a decisive factor in entrepreneurial failure. Emotive factors as mentioned by Khelil (2016), are described as a factor that can show the interaction and combination between business performance achieved by entrepreneurs (in terms of failure) and the disappointment to that unachieved business performance. In relation to this, the findings of Hammer (2014) also showed us the indication of *goal setting bias* in entrepreneurs as a major source of business failure. Goal setting bias happens because of the unmatched situation between expectation and real business performance experienced by entrepreneurs. This emotive factor as Khelil (2016) mentioned, is an interaction and a combination between deterministic and voluntaristic factors in entrepreneurial failure. Therefore, failure is not only sourced from an uncontrollable/difficult-to-control environment but it is also sourced from errors in decision-making and wrongdoing by the entrepreneur in managing his/her business.

In previous studies, Khelil (2012) mentioned that failure in a new venture mostly happened because of the following patterns and types that are embedded in entrepreneurs as an individual: [a] *gambler:*—ones who would like to start a business without resources and without any clear orientation, [b] *supported at arm's length:*—ones who have limited resources, [c] *bankrupt:*—ones who have limited competencies, [d] *megalomaniac:*—ones who have too much self-confidence, and [d] *dissatisfied with the lord:*—ones who do not fully rely on God's will. We can further categorize those reasons into three categories: [a] individual personality as a source of failure [consists of *gambler, megalomaniac, dissatisfied with the lord*], [b] environmental situation as a source of failure [consists of *supported at arm's length* and *bankrupt*]. Meanwhile, Hammer and Khelil (2014) strengthened the analysis about entrepreneurial failure by saying that based on the consideration of input, process and output in the new venture creation process on individual and enterprise levels, entrepreneurial failure can be in the form of an exit decision from the business, drawn as the result of: [a] the individual level which has competency and security as its dimensions and, [b] the firm level which has support and business model as its dimensions. The existence of both factors will lead entrepreneurs to a decision to quit their business—and therefore, the business fails. Further impact of this exit decision can lead entrepreneurs to try to find other jobs, restarting a new venture, or selling their business with the hope that they will get some profit or in the worst case, end up bankrupt.

Concept and studies regarding opportunism mostly related to the opportunistic behavior of individuals and it mentioned that this opportunistic behavior is a hidden will of an individual that can bring negative impact to the business and is shown by efforts to achieve that hidden will, (Cordes, et al., 2010). Williamson (1993 and 1999) has previously revealed that the existence of opportunistic behavior is closely related to the transaction cost concept and will create conflicts and bargain situations in which individuals need to make the most appropriate choice from many alternative decisions. Those both concepts are clearly related to the topic of this paper and study—and they demonstrate an indication that an entrepreneur will prefer a situation that brings more benefits to him/herself personally. This situation can be in terms of an alternative to secure for a better job compared to a career in entrepreneurship, or the desire to start other new businesses, but on the condition that the earlier business is still unstable/immature. This situation will put entrepreneurs in a conflict or a bargaining position which requires them to make their own decision:—securing and entering the job, or to start another new venture or maintaining the earlier venture to reach its maturity. Sometimes the decision is a good and powerful decision, but it can also be a wrong decision which will lead to failure in the earlier venture. We can see here that entrepreneurs can be trapped on an individual level (according to Hammer and Khelil, 2014), a psychological situation where they are demanded to secure their future life which means a possible future career that they need to follow.

In the study of impacts and consequences of entrepreneurial failure, Mantere et al. (2013) mentioned that entrepreneurial failure is a social construction where the process of individual psychology in terms of [a] the existence of emotional processes and [b] cognitive process to justify every action, exists. We can conclude from this opinion that entrepreneurial failure brings psychological impacts to the failed entrepreneurs—and as a result, they are trying to do their best to maintain their self-esteem and to avoid losing their own business. Psychological impacts can also be seen in efforts to reduce, and moreover, to eliminate the stigma that arises from the failure, Singh et al., (2015). An interesting phenomenon regarding the existence of stigma lies in the fact that it can also be a source for the next failure. As Singh et al. mentioned, (2015) the fear of failure and the efforts to stay away from the stigma of failure have contradictively resulted in negative situations for an entrepreneur—as he/she will probably make a wrong decision and conduct a wrong action regarding the business. As a result, this will make entrepreneurs sink into failure (again). Therefore, the stigma of failure can be seen as a psychological reason and impact entrepreneurial failure.

3 DESIGN AND METHODOLOGY

The study is an exploratory study which aims at revealing the integrative model of how the process of entrepreneurial failure together with its possible consequences and follow-up of the failure take place within nascent entrepreneurs. The study is based on the foundation of a previous quantitative study, undertaken by the authors, on 1541 nascent entrepreneurs who have experienced entrepreneurial failure in West Sumatra, Indonesia. The previous quantitative study mainly investigated the internal individual reasons of why entrepreneurs fail in their business. It came up with the result that the most plausible reasons that caused entrepreneurial failure are psycho-economic factors and opportunistic behavior of individuals.

Results of the previous quantitative study were then developed into a model of entrepreneurial failure—and it further added the possible consequences and follow-up taken by entrepreneurs after they have experienced failure.

4 THE INTEGRATIVE MODEL OF ENTREPRENEURIAL FAILURE

We began our task by drawing and introducing the figure of our model, which we call the Integrative Model of Entrepreneurial Failure. In principle, the emphasis of the model should highlight the process of entrepreneurial failure experienced by individuals, together with its consequence and follow-up of that failure and the creation of a process of serial entrepreneurship. As there is the possibility of serial entrepreneurship circumstance within this process, there could be two chances occurring for nascent entrepreneurs, which are [a] the possibility to fail again and restart the business, and [b] the possibility to achieve entrepreneurial success.

We further introduce our model as in Figure 1.

4.1 Factors identification and constructs

One major element of our model is to identify the main factors and constructs that cause entrepreneurial failure. In general, we argue that there are two major factors within individuals that can cause entrepreneurial failure, namely [a] individuals' internalities and, [b] individuals' externalities. Each of those factors is elaborated below.

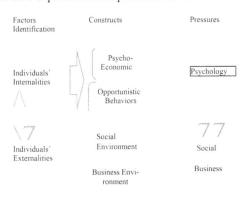

Figure 1. The Integrative Model of Entrepreneurial Failure.
Source: Authors' Own Conception.

4.1.1 *Individual internalities*

We define individuals' internalities as every construct or circumstance within individuals that raises psychological pressures on them and would push to make wrong decisions, which at the end, results in failures in business. As the basis and foundation of our understanding, we view individuals' internalities as the construction of psycho-economic circumstances and the level of opportunistic behavior belonging to individuals.

We have elaborated this in our previous empirical study with 1541 nascent entrepreneurs and found that voluntaristic, deterministic and emotive factors are the most plausible psycho-economic circumstances for causing entrepreneurial failure. If we look at the descriptive results gathered from our previous study, the most deterministic factor that causes failure is the inability of nascent entrepreneurs to meet the most efficient operational cost in their business. In emotive factors, the majority of nascent entrepreneurs revealed that the pressure to fill-up their personal and family needs, stressed them and contributed to their failure in business. Results of our empirical study supported the argument and opinion from Mellahi and Wilkinson, (2004) who said that voluntaristic, deterministic and emotive factors are the psycho-economic phenomenon of individuals that can cause entrepreneurial failures. We understand opportunistic behavior in the context of entrepreneurial failure as a psychological circumstance where individuals are led by their uncontrollable passion and desire to reach ambitions for the interest of their future life but leaves the logical consideration. Individuals tend to make a decision that will please them or their social environment and fits their ambitions. Thus, individuals will always consider alternatives which could be the answer to their passion and desire at the most—rather than thinking about logical alternatives. This psychological construction will lead them to choose the more opportunistic decision which is showed by the tendency and preference to [a] enter into a formal job (once the opportunity comes) rather than maintaining themselves to stay in the business, [b] choose to secure a job by keeping it for a longer period, and [c] follow his/her desire to immediately establish other new ventures but on the condition that the earlier business is still unstable or immature. These are confirmed by the result of descriptive analysis during our previous study which showed that the majority of nascent entrepreneurs think that getting into entrepreneurship is only a way to wait for a formal job and once the opportunity comes, they simply leave their business. Our further investigation clarifies the reasons of individuals to concentrate on the job rather than on the business. It is mostly related to the motive and interest for securing a future life. If we consider the culture of Indonesian people in general, one particular cultural dimension and value of Indonesian people is the high uncertainty avoidance, Hofstede (2017) and Mangundjaya (2010). This culture and value have made Indonesian people prefer to choose a stable situation in their life—and in

their perception, this can only be achieved if they have a formal job where they can get a regular monthly salary and pension in the future. The feeling of insecurity regarding the future life as the impact of [a] limited chances and severe competition in the job market and, [b] insecure business environment, which includes severe business competition, inconsistent government rules, regulation and laws about business environments (especially those that are related to small and medium scale enterprises) as well as insecure sustainable business operation have influenced individuals to choose to have a job once the opportunity appears (Rahman, 2016).

We found this individual phenomenon, as well as Williamson (1993 and 1999), who revealed that the existence of opportunistic behavior will create conflicts and bargain situations in which individuals need to choose the most appropriate choice from many alternative decisions. Individuals, in fact, tend and prefer to choose the most appropriate alternative in accordance with their personal situation/condition—and this becomes the basis of opportunistic behavior raised within individuals.

4.1.2 *Individual Externalities*

We understand individual externalities as every construct and/or circumstance surrounding individuals that relies from social and business environments and pressures them to make the wrong decision out of many alternatives in their daily activities. We view individual externalities as having a direct and/or reciprocal relationship with individual internalities as the foundation of our understanding regarding failure. We consider business and social environments of individuals as part of individual externalities which will be further elaborated as the cause of failure.

The business environment can be understood as a challenging business climate with various actors who will create pressures for entrepreneurs. This can be in the form of, for example, a severe business competition, unsupportive business climate and surroundings, unclear business rules and regulations, weak access to finance, weak access to the market, and so on. These have put a particular pressure on nascent entrepreneurs, since they cannot meet the most efficient cost for the operation of their business—which is one of the major complaints from nascent entrepreneurs and is directly associated with the deterministic factor in entrepreneurial failure. Unsupportive business environments have created business pressures for nascent entrepreneurs who find many barriers, starting from day one of their business and onward. Some scholars used the terminology of "barriers to entry" (either strategic or structural entry barriers) to show this situation (Martins and Orlando, 2004; Dijkstra et al., 2006; Lutz et al., 2010; Bartlett et al., 2013; and Afraz et al., 2014)

The second construct in individual externalities is social environment. We view social environment as the perception of individuals' social environment regarding entrepreneurship which affects the creation of pressure to entrepreneurs. It can be in the form of, for example, the level of entrepreneurial/business culture of a society, the social view and perception regarding entrepreneurship as a choice of career and future life, local culture and value, demand and needs from the nearest social environment, etc. The factin the previous empirical study revealed that nascent entrepreneurs who have experienced entrepreneurial failure, previously experienced social pressure when they operated their business. The majority of them revealed that filling-up personal and family needs is a major social pressure which they received from their social environment. This finding is closely associated with the fact that the emotive factor in terms of efforts fill-up the personal and family needs which have stressed nascent entrepreneurs in undertaking their business and hasfurther contributed to their failure in business. The finding also shows a reciprocal relationship between individual externalities and internalities as the cause of entrepreneurial failure. Thus, we consider individual internalities and externalities (with all their constructs) as having a direct and/or indirect association with the foundation that causes entrepreneurial failure.

4.2   Individual pressure to nascent entrepreneurs

As we argued, the mixture between individual internalities and externalities will significantly raise pressures for nascent entrepreneurs. We simply defined this pressure as a situation surrounding individuals, whether they voluntarily admit it or not, that creates a pressurized circumstance that makes them need efforts to solve that circumstance. These can be in terms of: [a] psychological pressure, [b] business pressure, and [c] social pressure.

4.2.1 Psychological pressure
We viewed psychological pressure as a pressure which comes from inside of an individual as a result of his/her psycho-economic condition and his/her uncontrolled desire and passion to get something of his/her interest. Psychological pressure will lead nascent entrepreneurs to illogical considerations when they need to make decisions and undertake actions in business. As a consequence of this illogical consideration, many decisions and actions that have been made are considerably wrong or not suitable for their business. This will be the source of failure and if it goes on for a longer term regularly, it will result in entrepreneurial failure.

4.2.2 Business pressure
Business pressure is a form of pressure that is created (by design or not) by the business environment to which may be mistakenly responded by nascent entrepreneurs. It can be for example, competition, barriers, exclusion, discrimination, fraudulence, blockage, alignments, difficulties, access to finance, distortion of information, the unbalance of power

etc. that are created and done by business partners and stakeholders of nascent entrepreneurs. Business pressure can also be unconsciously created and carried out by the government through its rules, regulations and other legal issues that can hamper existing businesses that are being run by nascent entrepreneurs. Rules and regulations regarding taxes, access to finance, certification, issues to administer the business, interest rate, foreign currency rate, competition, export-import policies etc. are just some examples of how the government may unconsciously hamper the development and empowerment of nascent entrepreneurs.

The existence of an unfriendly business environment will directly and/or indirectly put pressure on nascent entrepreneurs. Since nascent entrepreneurs are relatively new entrepreneurs in the business, and they will be more fragile to this experience. It makes them more vulnerable rather than more experienced entrepreneurs. The ability to respond to challenges and pressures will determine whether they can survive for a longer period of time or not.

4.2.3 Social pressure
Social pressure experienced by nascent entrepreneurs is mainly related to the perception of their social environment or their community regarding entrepreneurship as a choice for their future career, as a source of income, personal and social demands and needs, etc. Our empirical study has found that the demand to fill-up personal and family daily needs is the main pressure felt by nascent entrepreneurs from their nearest social environment. This creates a particular pressure for them to immediately run the business and make it profitable. On one side, this would be a good sign that entrepreneurs must work harder to make their business profitable in a short time. However, on the other side, this will also put more pressure on them when running the business with the bigger possibility to make wrong decisions and do wrong in the business.

Social pressure is also related to a regular income which would be earned by individuals. In many social environments and communities, individuals are requested to earn a regular income (mostly a monthly income) to fund their daily needs and because of that, they will get more respect from the people around them. This is a contradictory circumstance for individuals who choose to become an entrepreneur. Nascent entrepreneurs are those who are still new in the business—so there is no guarantee for them to earn a stable regular income. As a consequence, there is a cynical perception regarding the job of entrepreneur and the ability of nascent entrepreneurs to earn money. This circumstance will put nascent entrepreneurs at a particular pressure and will allow them to make wrong decisions and actions in undertaking their business.

Our understanding regarding the factors, constructs and pressure experienced by nascent entrepreneurs have led us to understand entrepreneurial failure as a result of the process of the entrepreneurial journey. The process of entrepreneurial journey is not yet a successfully guaranteed process in which nascent entrepreneurs will receive a satisfying result from their efforts in business and gain benefits from it. There is a greater possibility of failure than success during the entrepreneurial journey of an entrepreneur. As Wadhwa et al., (2009) mentioned, a successful entrepreneur has experienced, on average, two to three failures during their entrepreneurial journey before he/she reaches success. This means there is only a small guarantee that nascent entrepreneurs will succeed in a first venture establishment.

Thus, we believe and argue that attention and emphasis should actually be given to the way of how to avoid failure rather than to disseminate/spread success stories of an entrepreneur. Dissemination of the way of avoiding failure will increase awareness and alertness of individuals to choose alternatives, to make decisions and to take actions. Disseminating entrepreneurial failure will also stimulate preventive efforts and preparedness of individuals to face their psychological burdens, social- and business environments which may cause failures. Individuals can also allow a learning process that takes place inside themselves as a positive response to previous failures which will be useful in avoiding further failures. We build our argument on this overview by saying that entrepreneurial failure should not only be seen from a negative point of view,—but it should also be considered from the positive point of view which will be of benefit to nascent entrepreneurs.

4.4 *The consequence of entrepreneurial failure*

We view that every event brings its own consequence to every individual who has experienced it. This also occurs in

entrepreneurial failure where possible consequences would be felt by nascent entrepreneurs. The consequence of entrepreneurial failure could positively and/or negatively impact nascent entrepreneurs. Using previous studies and research from Singh et al., (2015); Yamakawa and Cardon, (2015); Jenkins et al., (2014); Mantere et al., (2013); Ucbasaran et al., (2010), as the foundation of our overview, we derived several possible consequences of

entrepreneurial failure of nascent entrepreneurs, namely [a] learning process, [b] stigma, [c] cognitive process, [d] emotional process, and [e] self-esteem.

Learning process is the most plausible consequence received from entrepreneurial failure. Nascent entrepreneurs who have experienced failure usually gain advantages from and learn from the failures that they have experienced. Some possible

positive outcomes from this learning process are in terms of awareness of failures, prudentiality, logical consideration and calculation, selection of partners and branch of industry, financial management and accounting, business planning etc. The process of learning will make nascent entrepreneurs tougher, more resilient, more experienced, more knowledgeable about the branch of business and their partners and able to cope with uncertain circumstances. These advantages and benefits would of course be useful if they would like to re-start their business in the future.

According to Singh et al., (2015), the positive psychological impact from entrepreneurial failure can also be seen in the efforts undertaken by nascent entrepreneurs to reduce, and moreover, to eliminate the stigma that arises from failures. Some individuals have a tendency for negative stigma that cannot be easily erased from their memories and feelings. In reverse, entrepreneurial failure would be a way to tackle, reduce and to eliminate that negative stigma. There would arise a psychological mechanism where nascent entrepreneurs are willing to prove that they are tough individuals by eliminating that negative stigma.

Efforts to eliminate negative stigma regarding entrepreneurship also relate to individuals' self-esteem. Nascent entrepreneurs will try to do the best to maintain their self-esteem by avoiding losing their own business. The feeling of self-esteem and claim as the business owner would be an extra energy for nascent entrepreneurs to avoid mistakes in the future. Thus, the outcome of failure could be in the form of improving the individuals' self-esteem as the business owners, which could improve the feeling of self-belonging to the business.

Emotional process can also be seen as a positive outcome of entrepreneurial failure. Nascent entrepreneurs will be more patient in facing and tackling bad situations that hamper their business. The event of failure is a perfect way to train and to examine the level of patience of nascent entrepreneurs. There is also a possibility of using more logical considerations rather than illogical concerns in business, such as passion, desires, hasty, recklessness and so on. Thus, in this essence, failures will give particular benefit to nascent entrepreneurs in reducing their own psychological bad side/nature and switch it to a better one.

One consequence that can be received by nascent entrepreneurs through failure is related to the cognitive process as a response to the failure. We believe that nobody can accept failure voluntarily. If individuals experienced failure as a result from their previous decisions and actions, then there is little possibility for them to repeat the same failure in the future. They will try to avoid the same failure to happen again. This situation will stimulate the consciousness and logic of nascent entrepreneurs to be more aware, preventive and anticipative to every circumstance in the future. The cognitive process is closely associated with the ability of nascent entrepreneurs to learn from failures—and this will be useful to anticipate next failures which could possibly happen in the future.

4.5 *The follow-up of entrepreneurial failure*

The consequence of the failure has led to a possible follow-up that could be taken by nascent entrepreneurs. We argue that there are twopossibilities of follow-up after the failure that can be taken by nascent entrepreneurs, namely: [a] the possibility of business cessation, and [b] the possibility of business continuation.

Business cessation is a simple decision that would be made by nascent entrepreneurs after they experienced failure in their previous business. They simply cease the opportunity and possibility of restarting the business and will find other chances for their future life, which mainly consists of finding a job. Business cessation could be a result of bankruptcy, very low and slow business progress and expansion, very small business scale, inability to maintain core customers, inability to adapt to business dynamics, inability to respond to harsh competition etc. Apart from the earlier-mentioned reasons, business cessation can also be a result of other opportunities, such as job offering.

There are also nascent entrepreneurs who are still willing to continue their business. This type of entrepreneurs normally chooses to stay in the business in two possible forms: [a] restarting a new business within the same business branch as the previous one that failed, and [b] restarting a brand-new business which is totally different compared to the previous one. The majority of nascent entrepreneurs who stay in business are those who are restarting a new business that is totally different compared to the previous one. We argue that this is a disadvantage for them, because we think that the more nascent entrepreneurs get involved in a brand-new business unrelated to the previous one, the more likely it is for them to fail again.

We argue that if nascent entrepreneurs start a brand-newbusiness unrelated to their previous business, they will always be categorized as a "newbie"—who potentially fails again. As a newbie, they need to totally learn a new thing and this has made the advantages of the learning process gained from previous failures to be minimum. In other words, maximum benefits of the learning process from failures will only be received by nascent entrepreneurs if they restart their new business in the same business branch or industry to the one which has failed. We argue that advantages taken from learning processes from entrepreneurial failure will be maximum if nascent entrepreneurs restart their new business in the same business branch/industry as the previous one.

Since there is no guarantee of success regarding the success rate of new businesses by nascent entrepreneurs after the failure, our model further figured that there will be a process of serial entrepreneurship that takes place within this situation.

This means that thereare two more possibilities that may happen to nascent entrepreneurs who choose to continue and stay in business, which are: [a] the process of failure again, and [b] entrepreneurial success.

5 SUMMARY AND IMPLICATIONS

Our paper introduces the possible integrative model to show the process of entrepreneurial failure experienced by nascent entrepreneurs. The model suggests that entrepreneurial failure needs to be viewed integratively, starting from factors that cause failure, constructs that create it, pressures to individuals, consequences and possible follow-up taken by individuals after the failure. The model also relates the process of entrepreneurial failure with the possibility of entrepreneurs to undertake the serial entrepreneurship journey, where there will be the possibilities of success and failure again.

Our model implies that studying and researching entrepreneurial failure should be undertaken integratively. Thus, we cannot completely elaborate on entrepreneurial failure as an event in the entrepreneurial journey,—but it should rather be considered as a process where there are factors, constructs, elements of pressure, consequences and follow-up of that failure. Our consideration regarding entrepreneurial failure as a process will make us understand that there is a possible link between entrepreneurial failure and the process of serial entrepreneurship. Our model also implies that there is a close association between entrepreneurial failure, the learning process and serial entrepreneurship in the entrepreneurial journey for nascent entrepreneurs. We view that the creation of tough entrepreneurs can only be achieved if nascent entrepreneurs can learn from failure, —and for this objective, they need to understand that failure is part of the process in the entrepreneurial journey. Failures need psychological preparation, rearrangement of business, management of social perception and overview regarding entrepreneurship asa responses to tackle those failures.

As this paper is a conceptual paper, there will be an opportunity to test this model in a broader empirical study which involves all stages in the entrepreneurial failure process. That will be a possible future research agenda regarding this topic that can be pursued by other scholars.

REFERENCES

Afraz, N., Hussain, S.T., & Khan, U., (2014). Barriers to the Growth of Small Firms in Pakistan: A Qualitative Assessment of Selected Light Engineering Industries, *The Lahore Journal of Economics*, 19:SE, pp. 135–176.

American Psychological Association (APA). (2010). The road to resilience: Resilience factors & strategies. Retrieved 2 September 2017, from http://www.apa.org/helpcenter/road- resilience.aspx

Artinger, S., & Powell, T., C., (2015). Entrepreneurial failure: Statistical and psychological explanations. *Strategic Management Journal* (DOI: 10.1002/smj.2378).

Bartlett, W., Popa, A., & Popovski, V., (2013). Business Culture, Social Networks and SME Development in the EU Neighborhood, WP5/18 SEARCH (Sharing Knowledge Assets Interregionally Cohesive neighborhoods) Working Paper, EU Seventh Framework Programme.

Cardon, M. S., Stevens, C. E., & Potter, D. R., (2011). Misfortunes or mistakes?: Cultural sense making of entrepreneurial failure. *Journal of Business Venturing*, 26, 79–92.

Chang, W., W.G.H. Liu, & Chiang, S., (2014). A study of the relationship between entrepreneurship courses and opportunity identification: An empirical survey, *Asia Pacific Management Review*, 19 (1), pp. 1–24.

Cordes, C., Richerson, P., McElreath, R., & Strimling, P., (2010). How Does Opportunistic Behaviour Influence Firm Size? An Evolutionary Approach to Organizational Behaviour, *Journal of Institutional Economics*, pp. 1–21.

Delmar, F. K., Hellerstedt, K., & Wennberg, K., (2006). The evolution of firms created by the science and technology labor force in Sweden 1990—2000. *Managing Complexity and Change in SME's: Frontiers in European Research*, pp. 69102.

De Vries, M. F. R. K., (1977). The entrepreneurial personality: A person at the crossroads. *Journal of Management Studies*, 14(1), pp. 34–57.

Dijkstra, S.G., Kemp, R.G.M., & Lutz, C. E., (2006). Do entry barriers, perceived by SMEs, affect real entry? Some evidence from the Netherlands, SCALE Project (Scientific Analysis of Entrepreneurship and SMEs), EIM Business and Policy Research.

Gratzer, K., (2001). The fear of failure: Reflections on business failure and entrepreneurial activity. In M. Henrekson, M. Larsson, & H. Sjoegren (Eds.), *Entrepreneurship in business and research*. Stockholm: Institute for Research in Economic History.

Hammer, M.H.M., (2012). How Business Management benefit from Entrepreneurship, in Proceedings of *10th Conference of Management, Enterprise and Benchmarking*, Budapest, Hungary, pp. 175–182.

Hammer, M.H.M. & Khelil, N., (2014). Exploring the Different Patterns of Entrepreneurial Exit; the Causes and Consequences, in *Proceedings of 59th ICSB World Conference on Entrepreneurship*, Dublin, June 10–15.

Herath, H. M. T. S., (2014). Conceptualizing the role of opportunity recognition in entrepreneurial career success. *International Journal of Scientific Research and Innovative Technology*, 1(3), 73–82.

Hills, G.E., & Shrader, R.C., (1999). Successful entrepreneurs' insights into opportunity recognition, *Frontiers of Entrepreneurship Research*, Babson College, Wellesley, MA.

Hofstede, G., (2017). National Culture of Indonesia. Retrieved from http://geert-hofstede.com/indonesia.html.

Jenkins, A. S., Wiklund, J., & Brundin, E., (2014). Individual responses to firm failure: Appraisals, grief, and the influence of prior failure experience. *Journal of Business Venturing*, 29, 17–33.

Khelil, N., (2012). What are we talking about when we talk about entrepreneurial failure?. *RENTXXVI—Research in Entrepreneurship and Small Business*, Lyon.

Khelil, N., (2016). The many faces of entrepreneurial failure: Insights from an empirical taxonomy. *Journal of Business Venturing*, 31, 72–94.

Lafontaine, F., & Shaw, K., (2014). Serial Entrepreneurship: Learning by Doing? *NBER Working Paper No. 20312*. Cambridge, MA. P.

Lutz, C. E., Kemp, R.G.M., & Dijkstra, S.G., (2010). Perceptions regarding strategic and structural entry barriers *Small Bus Economics*, 35:19–33.

Mangundjaya, W.L.H., (2010). Is There Cultural Change In The National Cultures Of Indonesia? Paper Presented at International Conference on Association of Cross Cultural Psychology (IACCP), Melbourne, Australia.

Mantere, S., Aula, P., Schildt, H., & Vaara, E., (2013). Narrative attributions of entrepreneurial failure. *Journal of Business Venturing*, 28, 459–473.

Martin A, & Orlando, M., (2004). Barriers to Network-Specific Innovation, Federal Reserve Bank of Kansas City. Research Working Paper, RWP 04–11.

Mellahi, K., & Wilkinson, A., (2004). Organizational failure: a critique of recent research and a proposed integrative framework, *International Journal of Management Reviews*, 5, 2141.

Rahman, H., (2016). Merantau—An Informal Entrepreneurial Learning Pattern in the Culture of Minangkabau Tribe in Indonesia, DeReMa Journal of Management, 11 (1), pp. 15–34.

Singh, S., Corner, P, & Pavlovich, K., (2015). Failed, not finished: A narrative approach to understanding venture failure stigmatization. *Journal of Business Venturing*, 30 (1):150–166.

Smida, A. & Khelil, N., (2010). Repenser l'echec entrepreneurial des petites entreprises emergentes. Proposition d'une ty- pologie s'appuyant sur une approche integrative. *Revue Internationale P.M.E.: economie et gestion de la petite et moyenne entreprise*, 23(2), 65–106.

Ucbasaran, D., Westhead, P., Wright, M., & Flores, M., (2010). The nature of entrepreneurial experience, business failure and comparative optimism. *Journal of Business Venturing*, 25, 541–555.

van Gelder, M., Thurik, R., & Bosma, N., (2007). Success and Risk Factors in the Pre-Startup Phase. *Small Business Economics*, 319–335.

Wadhwa, V., Aggarwal, R., Holly. H., & Salkever, A., (2009). Anatomy of an Entrepreneur: Family Background and Motivation, The Ewing Marion Kauffman Foundation Research Report, July 2009.

Wasdani, K.P. & Mathew, M., (2014). Potential for opportunity recognition: differentiating entrepreneurs, *International Journal of Entrepreneurship and Small Business*, 23(3), pp.336–362.

Wennberg, K. & DeTienne, D.R., (2012). What do we really mean when we talk about 'exit'? A critical review of research on entrepreneurial exit. *International Small Business Journal*, 32(1): pp. 4–16.

Wennberg, K., Wiklund, J., DeTienne, D. R., & Cardon, M. S, (2010). Reconceptualizing entrepreneurial exit. Divergent exit routes and their drivers. *Journal of Business Venturing*, 25(4): pp. 361–375.

Williamson, O.E., (1993). Opportunism and its Critics, *Managerial and Decision Economics*, 14, 97–107.

Williamson, O.E., (1999). Strategy Research; Governance and Competitive Perspectives, *Strategic Management Journal*, 20, 1087–1108.

Yamakawa, Y., & Cardon, M. S. (2015). Causal ascriptions and perceived learning from entrepreneurial failure. *Small Business Economics*, 44, 797–820.

Advances in Business, Management and Entrepreneurship – Hurriyati et al (eds)
© *2020 Taylor & Francis Group, London, ISBN 978-0-367-27176-3*

Application of freedom of contract principle in agreements in the Indonesian business environment

A. Amuninggar & A. Aris
Universitas Airlangga, Surabaya, Indonesia

ABSTRACT: In this modern era, business has become the main activity for many people, and many variety of business activities have been undertaken. Through business, many people have gained a lot of benefits and prosperity, although many of them also suffer from losses. Disputes can arise as a result of conducting business, and most of them are caused by the abuse of trusts or broken promises. Legal disputes can disrupt business activities of the perpetrators and a lot must be sacrificed, including the company's reputation. However, disputes can be minimalized through making agreements or contracts. The parties that make the agreement or contract will have the freedom to put whatever is deemed necessary to secure the their rights according to the content of the contract. Freedom possessed by the parties in making the agreement is based on the freedom of contract principle in legal agreements.

1 INTRODUCTION

Business is an activity that has been performed by almost all people in the world. The reason people conduct a wide range of businesses is in order to meet their needs. Humans are not solitary creatures like spiders, even though sometimes humans are also able to live alone for a relatively short period of time. However, humans have some needs that must be met, and they can fulfill their needs by contacting other human beings. Building relationships or interaction with others in the context of mutual cooperation is one of the ways to make ends meet. Interactions that are intended to meet the needs of the involving parties should be balanced. Relationships created by people can be of two different forms: first, social relationships, where each party is regulated by ethics; and second, legal relationships, regulated by law. When one party betray the others, they have to face legal consequences.

Each party should be in a balanced position in relationships based on mutual needs. However, the creation of a balanced relationship is very difficult to do, if not impossible. The reason is because there are different bargaining powers between parties. The presence of a powerful party who feels that he or she is more needed by the opposing party makes them having more authority to determine things. The authors of this paper feel that such cases should be eliminated, or regulated, in order to maintain justice between the parties involved.

Agreements and contracts can solve such problems, and these tools also can act as a risk management for the involving parties, because they contain clauses and conditions that are needed to regulate the relation between the parties. Each party can list all the things related with the relationship to help secure their rights in the event of a breach.

2 FRAME AGREEMENT AS A BUSINESS ACTIVITIES

Any business activity is prone to dispute due to the divergence of interests between parties. Definition of the term "business" itself is trade activity, industrial activity or financial activity related to the production or exchange of goods or services (Abdurrahman, 1991:150), by putting some money in certain businesses owned by entrepreneurs with the motive to earn profit (Friedman, 1987:66). A businessman's basic purpose to make profit also contributes to the occurance of dispute among them as they try to take control over another using their bargaining power in order to make profit.

The most powerful controller that can be utilized for business activities is an agreement or a contract. Every human being ought to have legal relationships with others at least once in his/her lifetime. This legal relationship is in fact a form of agreement or contract. The existence of an agreement will automatically bring up rights and obligations for the involved parties. Each party is bound to the other party to carry out his or her obligations.

An agreement is one of the sources of the origin of an engagement, and an agreement or contract is covered by the law. There are some known terms in agreements, including:

a. Understanding agreement according to Civil Code Article 1313:

"An agreement is an act by which one or more people bind themselves to one or more other people."

b. Understanding agreement according to Rutten:
"Agreement is a legal action that occurs in accordance with the formalities of the existing laws, whose statement will depend on the conformity of two or more people who are devoted to the emergence of the legal consequences for the sake of one party and at the expense of the other party or for the benefit of and at the expense of each party reciprocally."

c. Understanding agreement according to the Indigenous Treaty:
"Agreement whereby homeowners give permission to others to use his home as a place of residence with rental payments afterwards (or also upfront payment)."

However, many people do not realize that business contracts can be made by anyone; not only lawyers or notaries are able to draw up a contract. Because the content of the agreement is the will of all parties, it follows that in the process of making a business contract, the most important thing is to reach an agreement between parties. An agreement by the parties is possible in both written or oral form. Both forms have the same legal force. The involved parties have the freedom to make their own form of agreement in accordance with their needs. In addition, the content of the agreement is dependent on the will of each party, because they are the one who best understand what should be included in the clause of the contract. From the description, it can be concluded that the contract is actually the application of the precautionary principle.

3 CONTRACT IS A REALIZATION OF PRUDENTIAL PRINCIPLES IN BUSINESS

Before doing business transactions, the related parties should apply the precautionary principle, where the contents of an agreement is the will of the parties, upon which they have agreed. Thus, only contents that the related parties are able to make and understand are stated in the agreement. Each person also must be able to make a contract. However, many people have been mistaken by the belief that only lawyers and notaries can drarw up a contract. According to their view, not everyone understands the language of the law. In fact, the contract language actually is the language that is understood by the related parties. Thus, drawing up a contract is actually not that difficult to do for the related parties.

Difficulties that may arise when drawing up a contract is that each party should think of the potential harms that can come to them. The will or desire of each party is certainly different, so they have to negotiate with each other. During the process of negotiation, each party tends to try to maintain his will and make the other party agree to follow. Therefore, in order to succeed in the negotiation process, each party should have the ability to communicate well with their counterparts because the negotiation process is a process of mutually influencing one another.

The end result of the negotiation process is a contract, and the deal is included as the contract's clauses. After the parties mutually agree on all clause contents and sign it, they are mutually bound to each other. Thus, there will be legal consequences that accompany the event of a breach (broken promises) between the parties. That is because the relationship that has been created, is a legal relationship, which is affected by the law.

When drawing up a contract, each of the related parties have the freedom on how to arrange the contract, as there are no standard guidelines regarding the preparation of a contract. The Indonesian Civil Code does not describe in detail how to make and draft a contract. The reason is because the Indonesian Civil Code only serves as a complement and does not have the force of nature. The contract has a lot of meaning for business people, including:

1. Guidelines for the parties to conduct business
It should be written in the contract who the parties are who conduct the business transactions as well as the statement of rights and obligations and the conditions that should not be done by the parties. If we think of this like a locomotive, then the contract will serve as its rails.

2. Evidence for the parties
If a party breaches the agreement and causes a dispute, then the contract can be used as evidence for the purpose of self-defense.

The contents of the contract is a blend of the will of all the parties that are beyond the negotiation process. Therefore, the contents of a contract have a strong appeal for a law-maker. The will of the parties that have been embodied in the contract includes the intangible as well as the tangible rights and obligations between the parties.

4 FREEDOM OF CONTRACT PRINCIPLE IN AGREEMENTS IN THE INDONESIAN BUSINESS ENVIRONMENT

Contracts are like buildings that require some pillars to be able to stand firmly. To draw up a good contract or agreement, the principles of contract law should be applied properly. Contract law has some contract principles, including:

1. Consensual Principle
This principle states that a constract is valid and has a binding form only when tied to the agreement of each related party. In this principle, it is obvious that the agreement of each party plays an important role (Legal basis of Article 1320 Indonesia Civil Code).

2. Pacta Sunt Servanda

This principle states that the strength of binding agreements is as strong as the law. In addition, the contract that has been made cannot be terminated unilaterally (Legal basis Article 1338, paragraph 2 Indonesian Civil Code).

3. Freedom of Contract

Among others, this one is the most prominent between these principles. This principle gives the freedom and flexibility to the related parties in the process of making the agreement (Legal basis Article 1338, paragraph 1 Indonesian Civil Code).

4. Good Faith

Good faith should exist from the pre-contractual phase to the post-contractual phase (Legal basis Article 1338, paragraph 3 Indonesian Civil Code).

The principle of freedom of contract has a dominant position among the others, because this principle provides flexibility and freedom to the parties involved. The meaning of freedom of contract is that the parties have freedom according to the law to be given by the public. The freedom of making the agreement can be about anything, as long as it does not contradict the existing laws and regulations, decency and public order (Article 1338 of the Civil Code Jo 1337). Principle of freedom of contract is essential for the related individual to develop themselves in both private life and social life in society, so based on this thinking, the principle can be categorized as a part of human rights which should be respected.

Countries that have a common law system of freedom of contract should be familiar with the terms freedom of contract or laissez-faire, which is created by Jessel M.R. in the case of "Printing and Numerical Registering Co. Vs. Samson "Jessel in Haridjan Rush," the Law of Treaties of Indonesia and the Common Law," Pustaka Sinar Harapan, Jakarta, 1993, p-39, "... men of full age understanding shall have the utmost liberty of contracting, and that contracts are freely and roomates voluntarily entered into and shall be held by the courts onforce you are not lightly to interfere with this freedom of contract."

This principle is what makes the parties have the freedom to draw up a contract according to their needs. Sometimes they even do a lot of creations and unusual things, which is okay because according to the Indonesian Civil Code article 1338, paragraph 1, it is not clearly explained how to draft a contract, and the law only states that there should be an agreement between the related parties.

One of the creations of the parties in making the contract is the birth of the Standard Contract. Standard Contract is a contract drawn up by only one party, resulting in almost the whole set or even entire clauses being determined by only that particular party. The party setting the contents of the contract is usually the one who has more bergaining power than the other counterpart. Freedom is owned by one party whose bergaining power is higher, and the other parties just decided whether they agree to the contents of the contract or not. Therefore, the standard contract is often referred to as a take-it or leave-it contract. No agreement is created.

The fact that standard contract is only made by one party increases the potential for injustice between the related parties. It is very reasonable to conclude thusly, because the contents of the agreement would be in the favor of its maker, whereas the other party does not have the chance to negotiate their position. However, standard contract is often found to be very attractive to businesses because it can save both the time and the expense to draw up one. Standard contracts typically include contracts for mass-shaped form, where the contract maker only needs to change the dressing of the contract and does not need to create a new contract for different partners.

Contracts can be made in two forms, being oral and/or written. A written contract is also called the deed. Contracts in those two forms have the same legal force. There are two kinds of written contract, namely:

1. Deed under the hand

Written contract made without involving the state officials.

2. Authentic deed

Written contract made by the state officials and the notary or a Land Deed Official (PPAT)

In fact, the Convention on the International Sale of Goods 1980 also recognizes oral contracts. However, even if the parties are free to make contracts in any form, if the contract is in writing, it is relatively easy to prove its existence. This does not mean that an oral contract cannot be proven, but it is relatively more difficult because they have to gather more evidence as stated by the Indonesian Civil Code section 1866.

In addition, as disclosed by Erman Rajagukguk, a contract is considered to be a legal institution that can put the related parties even beneficial because:

1. They have the freedom to determine the content of the agreement according to their interests

2. Allow other parties to demonstrate their confidence in the market

3. Treat principle knowing each other so that the contract could run better.

This principle says that the contract made by the parties manages to tie them together in a strong force like the law, so that the cancellation is not easy.

4. The free set of dispute resolution options.

The parties related to the agreement may make a choice in the event of a dispute based on the contract, whether the disputed will be settled by way of peace or by some legal actions.

Freedom for the parties to draw up a contract, which is given by the principle of freedom of contract, is not absolute. There are some limitations to the application of the principle of freedom of contract that cannot be violated, such as:

1. Regulations
   The content of the contract is not allowed to violate the state regulation. If this is done, then the clause or the contract becomes void.
2. General order
3. Decency

After having seen the above, if you want to make arrangements to be present in it they are:

a. The Related Parties
   Parties who may make the arrangements are only legal subjects. Groups that can be categorized as legal subjects are people and legal entities.
b. Agreement
   Agreement in making a contract becomes a very important factor. In the absence of agreement between the parties, the contract will not be born.
c. Achievement
   Achievement is any obligation that must be met to fulfill the contract.
d. Use Either the Written or Oral Form
   The parties have the freedom to decide which form to use. Both forms have the same legal force What distinguishes both forms is the strength of the evidence alone. A written contract is relatively easy to be proved in the event of a dispute.
e. Possible Set Terms
   The terms of the contract must be made possible to be conducted by the parties.
f. Stated Objectives to be Achieved
   The purpose of a contract should actually be clear from its contents.

The principle of freedom of contract is a reflection of human rights, because its principles clearly give enormous flexibility to the parties in the making of the contract. There are some other things that make this a manifestation of the principle of human rights, namely:

1. Freedom to make or not to make the appointment.
2. Freedom to choose a party with whom he wants to make a deal
3. Freedom to determine or select the clause of the contract to be made.
4. Freedom to define the object of the contract.
5. Freedom to determine the form of the contract.
6. Freedom to accept or deviate Law Act provisions that are optional.

As for human rights, this principle also has some limitations as discussed earlier; one limitation inferred by Article 1338 paragraph (3) which states that a contract is only carried out in good faith. Under these conditions, it means there must be a balance between the parties by basing itself on good faith. If in a contract, the contract itself contains elements of fraud, the status of the contract will become void.

Another limitation to the principle of freedom of contract is the authority of the judge. The judge can be authorized to enter or examine the contents of a contract if there are some indications that its contents are contrary to certain values in society. Authority possessed by the judge is based on the interpretation of the law. Based on the interpretation of the law, the judge has the authority to pass the preceded judgment by actions to observe and examine the position of the parties and their freedom is cascrated.

5 CONCLUSION

Applicability of the principle of freedom of contract in Indonesia is reflected in the provisions of Article 1338 of the Civil Code, as well as the consensual principle that has been established. This principle does not give you unlimited freedom, and there are some limitations that should not be violated.

REFERENCES

Abdulkadir Muhammad, 1990 "Hukum Perikatan," PT. Citra Aditya bhakti, Bandung 1990.

Abdulkadir Muhammad, Hukum Perjanjian, Alumni, Bandung, 1980.

Dirdjosisworo,Soedjono. Kontrak Bisnis (menurut Sistem Civil Law, Common Law, dan Praktek Dagang Internasional). Bandung: Mandar Maju, 2003.

Friedman, W. Teori dan Filsafat Hukum [Legal Theory]. Diterjemahkan oleh Muhamad Arifin. Jakarta: Rajawali, 1990.

Fuady, Munir. Hukum Bisnis dalam Teori dan Praktek, Buku Ke Empat. Bandung: Citra Aditya Bakti, 1997.

Fuady Munir, Hukum Kontrak (Dari Sudut Pandang Hukum Bisnis), Cetakan Pertama, Bandung: PT Citra Aditya Bakti, 1999.

Munir Fuady, Hukum Kontrak (Dari Sudut Pandang Hukum Bisnis, Citra Aditya Bakti, Bandung, 2007.

Ibrahim, Johanes dan Lindawaty Sewu. Hukum Bisnis: Dalam Persepsi Manusia Modern. Bandung: Refika Aditama, 2004.

Jessel dalam Haridjan Rusli, "Hukum Perjanjian Indonesia dan Common Law," Pustaka Sinar Harapan, Jakarta, 1993.

Kartini Mulyadi, Gunawan Widjja, Perikatan yang Lahir dari Perjanjian, Raja Grafindo, Jakarta, 2003.

Mahmud Marzuki Peter, Kontrak Bisnis Internasional (Bahan Kuliah Magister Hukum Universitas Airlangga), Surabaya, 2001.

Rahman Hasanuddin, Contract Drafting, Bandung: Citra Aditya Bakti, 2003.

Rusli, Hardijan. Hukum Perjanjian Indonesia dan Common Law. Jakarta: Pustaka Sinar Harapan, 1996.

Setiawan, "Menurunnya Supremasi Azas Kebebasan Berkontrak." PPH Newsletter (April 2008): 1–10.

Salim. H, Perkembangan Hukum Kontrak Di Luar KUH Perdata, Raja Grafindo Persada, 2004, hal 22.

Shippey Karla C., J.D., Menyusun Kontrak Bisnis Internasional, Cetakan Pertama, Jakarta: PPM, 2001.

Shidarta, *Hukum* Perlindungan *Konsumen Indonesia*, Gramedi Widiasarana Indonesia, Jakarta, 2004.

Sjahdeini, Sutan Remy. Kebebasan Berkontrak dan Perlindungan yang Seimbang bagi Para Pihak dalam Perjanjian Kredit Bank di Indonesia. Jakarta: Institut Bankir Indonesia, 1993.

Subekti, R, Prof, S.H. dan Tjitrosudibio, R, 2001, *Kitab Undang Undang Hukum Perdata*, Cetakan ke-31, PT Pradnya Paramita, Jakarta.

Subekti, R, Prof, S.H., *Hukum Perjanjian*, Cetakan ke-VIII, PT Intermasa.

Subekti, *"Hukum Perjanjian,"* Intermasa, Jakarta, 1987.

Sudikno Mertokoesumo, *"Mengenal Hukum,"* Liberty, Yogyakarta, 1999.

Suharnoko, *Hukum Perjanjian*. Prenada Media, Jakarta, 2004.

Soebagjo, Felix S."Perkembangan Asas-asas Hukum Kontrak dalam Praktek Bisnis ."

Soedjono Dirdjosisworo, Misteri Dibalik Kontrak Bermasalah, Mandar Maju, Bandung, 2002.

Rajagukguk Erman, dalam Jurnal Magister Hukum, 1999.

Black, Henry Campbell, 1968: 394.

Gifis, Seteven H., 1984: 94.

Author Index

Ruhayati, S. 961
Rusmita, S.A. 442, 449
Ryandono, M.N.H. 343

Saefullah, K. 553
Sandroto, C.W. 627
Saputri, M.E. 44, 47
Saragih, R. 690
Sari, A.P. 727
Sari, M. 453
Sari, P.K. 8
Saribanon, E. 39
Sasikirono, N. 418
Sedarmayanti, S. 673
Semiarty, R. 592
Senen, S.H. 636, 679, 746
Septiarini, D.F. 301
Sern, L.C. 544
Setiadi, F.A. 957
Setianto, R.H. 281
Setiawan, A.R. 329
Setiawati, C.I. 881
Setiyono, W.P. 406
Setyawati, I. 338
Sharif, O.O. 843
Shofawati, A. 286
Siahaan, E. 617, 623
Siddiq, A.M. 632, 686
Sihombing, N.A.M. 881
Silvia, D. 160
Silvianita, A. 703
Sinulingga, S. 890
Siregar, A. M. 334
Slamet, A. 764
Soebandhi, S. 97
Soeharto, S.M. 366
Soeroto, W.M. 487
Soewarno, N. 457, 533
Sofiyah, S. 746
Solihat, A. 498
Sopacua, E. 549
Suastini, N.L.P. 321
Sudana, I.M. 494
Sudarsono, R. 553
Sudaryanto, E. 770
Sudika, I.G.M. 644
Sudirman, I. 873
Sudrajat, C.T. 32
Sugianti, I. 251

Sugianto, I. 274
Sugiarto, S. 389
Sugiharto, S. 890
Suharto, D.G. 847
Suhud, U. 125
Sukmana, R. 522
Sukmaningrum, P.S. 433
Sulastri, S. 172, 188, 952, 961
Sulistiawan, J. 971
Sulistiyo, H. 415
Sulistyowati, C. 965
Sultan, M.A. 25, 32, 39, 84,
     93, 209, 570
Sumani, S. 371
Sumaryadi, S. 150
Sumaryanto, T. 867
Sumawidjadja, R.N. 908
Sumiyati, S. 698, 754
Supriyadi, D. 256, 983
Surachim, A. 957
Surachman, E. 952
Surya, S. 323, 538
Suryana, S. 905, 908, 948
Sutarto, J. 978
Sutawijaya, A.H. 585
Sutikno, B. 13
Suwandhayani, A. 429
Suwatno, S. 199, 739, 775
Swidarto, S. 867
Syahidah, R.K. 226
Syahyono, S. 927

Tambunan, M.M.L. 102
Tanuatmodjo, H. 754
Tarmedi, E. 957
Tetep, T. 983
Tjahjadi, B. 457, 533
Triksina, E.R. 971
Trisnasih, F. D. 334

Usman, I. 71, 549
Utama, A.A.G.S. 52, 577
Utama, R.D.H. 938

Verinita, V. 183, 236
Vitus, H. 533

Wahyono, W. 978
Wahyuniardi, R. 873

Wahyuningtyas, R. 759
Waluyo, J. 121
Wardani, R.M. 588
Waspada, I. 385
Wenatri, L. 323, 538
Wibawa, B.M. 840
Wibawa, B.W. 673
Wibowo, L.A. 20, 28, 35,
     150, 574, 597, 859, 927
Wibowo, S.F. 125
Widianto, T. 64
Widiastuti, R. 724
Widiastuti, T. 314, 410, 414,
     487, 734
Widjaja, Y.R. 636
Widjajanta, B. 754
Widnyani, I.A.P.S. 663
Widodo, A. 68
Widodo, J. 764, 867, 978
Widyanti, R. 649
Widyastuti, A. 553
Wijaya, F. 245
William, W. 686
Winata, H. 437
Windijarto, W. 479
Wirata, G. 321, 663, 671
Wisudanto, W. 314
Wurjaningrum, F. 603

Yasmin, R.P. 719
Yola, F. 183
Yudiardi, D. 817
Yuliadi, H. 698
Yulianti, P. 654, 659, 666,
     727, 730
Yuliawati, A.K. 948
Yuniarsih, T. 353
Yuniarsih, Y. 739
Yunita, I. 396
Yunizar, Y. 553
Yuris, N.H. 353
Yusiana, R. 68
Yustian, O.R. 912
Yusup, I. 172, 952, 961
Yuvita, K. 494

Zakaria, H.M. 376
Zunairoh, B. 329